# BIOCHEMISTRY

# CONTRIBUTING AUTHORS

WAYNE M. BECKER

RAYMOND L. BLAKLEY

JAMES W. BODLEY

FREDERICK J. BOLLUM

RONALD BRESLOW

ANN BAKER BURGESS

RICHARD R. BURGESS

JAMES P. FERRIS

PERRY A. FREY

IRVING GEIS

MAX GOTTESMAN

SUSAN GOTTESMAN

LLOYD L. INGRAHAM

GARY R. JACOBSON

JULIUS MARMUR

RICHARD PALMITER

WILLIAM W. PARSON

R. C. PETERSON

MILTON H. SAIER, JR.

F. RAYMOND SALEMME

JOE SAMBROOK

JACK STROMINGER

H. EDWIN UMBARGER

DAVID A. USHER

DENNIS E. VANCE

GEOFFREY ZUBAY

# BIOCHEMISTRY

*Coordinating Author* GEOFFREY ZUBAY

COLUMBIA UNIVERSITY

ADDISON-WESLEY PUBLISHING COMPANY

READING, MASSACHUSETTS ▲▼ MENLO PARK, CALIFORNIA ▲▼ LONDON ▲▼ AMSTERDAM ▲▼ DON MILLS, ONTARIO ▲▼ SYDNEY

| | |
|---|---|
| Sponsoring Editor | *Bob Rogers* |
| Production Editor | *Marcia Mirski* |
| Copy Editor | *James K. Madru* |
| Text Designer | *Vanessa Piñeiro* |
| Illustrator | *Illustration Concepts, Michael Ockler* |
| Cover Designer and Illustrator | *Hannus Design Associates, Richard Hannus* |
| Art Coordinator | *Kristin Belanger* |
| Production Manager | *Karen M. Guardino* |
| Production Coordinator | *Peter Petraitis* |

*The text of this book was composed in Trump by York Graphic Services.*

Illustrations rendered and copyrighted by Irving Geis: Figures 1.1, 1.2, 1.3, 1.6, 1.7, 1.8, 1.9, 1.10, 1.11, 1.13, 1.14, 1.15, 1.16, 1.17, 1.18, 1.19, 1.20, 1.21, 1.23, 1.26, 1.27, 1.28, 1.29, 1.32, 1.33, 3.4, 3.7, 3.16, 3.17(a), 3.43, 3.44, 3.45, 3.46, 3.47, 3.54(a), 3.58(lower half), 3.59, 4.7, 4.8, 4.9, 4.10, 4.15, 4.21, 10.9, 12 opener, 18.16, 18.35(b).

**Library of Congress Cataloging in Publication Data**

Zubay, Geoffrey L.
  Biochemistry.

  Includes bibliographies and index.
  1. Biological chemistry.   I. Title.
QP514.2.Z83 1983        574.19′2        82-18502
ISBN 0-201-09091-0

ISBN 0-201-09091-0
ABCDEFGHIJ-DO-89876543

# PREFACE

The goal in preparing this textbook of biochemistry was to produce a uniformly authoritative, up-to-date book written in a style that would make it suitable for a wide range of students. In this preface I explain how this book was written and the various ways in which it can be used.

The author of a biochemistry textbook faces two problems that are virtually unique to the subject: its enormous scope and the rapid rate at which new information is generated. The task of preparing a satisfactory book on this subject has passed the point at which it can be done effectively by one person no matter how skilled or devoted that person may be. By some means, the expertise of many biochemists working in different areas must be pooled into a collective effort. This text represents the combined effort of twenty-six scholars who were invited to participate because of their knowledge of the subject and their dedication to teaching. Coordination of the writing has been assured by several devices. Prior to writing, each author received a detailed description of the overall organization of the book. During the writing there has been an ongoing communication between individual authors and myself and many reviews of individual chapters and parts of the text by expert biochemists and teachers. The authors themselves have been a marvelously cooperative group, alert to criticisms and willing to make modifications, even where they amounted to concessions, in the interest of the overall book.

The thirty-two chapters of the textbook have been organized into five parts. Part I deals with protein structure and function, including the coenzymes. Part II is concerned primarily with the production of biochemically useful energy; it also includes selected aspects of the structure and metabolism of sugars and carbohydrates. Part III contains a description of the structure and metabolism of lipids and lipoidal substances, as well as the structure and function of lipoprotein membranes. Part IV treats the basic aspects of nucleic acid structure and metabolism. Part V includes a variety of special topics that represent some of the major focal points of interest in modern biochemistry.

It is envisaged that this textbook will be used in a variety of ways, depending upon the level of the student, the emphasis on the subject, and the particular goals of the instructor. What follows is my personal view of how the text might be used in different situations. Three broad areas of usage are considered: the comprehensive one-year undergraduate course, the undergraduate survey course, and graduate and professional school courses.

Parts I through IV include the core material for the comprehensive undergraduate course. This is not to say that all of these chapters need be assigned or that the order of reading must be the order given in the text. The text contains more material than can be covered in a comprehensive one-year course, thereby providing the instructor with a great amount of flexibility in making selections. I would suggest that all such courses include Part I in its entirety and Chapters 8, 9, 10, and 11 of Part II in the prescribed order. Chapter 7 of Part II can be considered optional, since it presents a more detailed treatment of thermodynamics than some instructors may wish to present to their students. Chapter 12 of Part II can also be considered optional, since it deals with structural and metabolic aspects of carbohydrates that are not essential to understanding subsequent material. The basic understanding of protein behavior and bioenergetics conveyed in Parts I and II qualifies the student to proceed directly to either Part III or Part IV. Both parts are important, but they may be attacked in either order. Within each of these parts there is a great deal of latitude concerning which chapters or parts of chapters are assigned. This would depend very much on the individual goals of the course or the instructor. The same is even more true of Part V. For example, if emphasis is being put on lipids and membranes, one might wish to include all of Part III and the related chapters on neurotransmission, hormones, and vision in Part V. Alternatively, if there is an emphasis on nucleic-acid–related metabolism, the chapters on $\lambda$ bacteriophage and animal viruses in Part V could be included.

Clearly this text contains much more material than would be needed in an undergraduate survey biochemistry course. Nevertheless, by judicious selection of chapters, the thought-provoking explanations and other advantages offered by this text can be comfortably enjoyed by those seeking a less intensive course. The text has been designed so that a selective group of chapters from Parts I through IV can be assigned to provide an overview of biochemistry. Starting with Part I, Chapters 1, 4, and 5 would give an overview of structure and function. Most of Chapters 8 through 10 would be useful in such a course. Chapter 11, concerned with photosynthesis, could be considered optional. In Part III, one could jump immediately to Chapters 16 and 17, which deal with membrane structure and function. In Part IV one would definitely want to cover major portions of Chapters 18, 20, 21, and 24 to obtain an overview of nucleic acid structure and function; but the remainder need not be covered. In making these suggestions, I have considered the extent to which this selection would make it more difficult to understand certain chapters. No special problems of understanding would be created by the combination I have given.

The comprehensive treatment of core subjects in Parts I through IV, together with the presentation of special topics in Part V, gives this book the necessary depth for most graduate-level courses. To enhance and extend the coverage for some courses, a supplementary text has been prepared that contains selected readings from various scientific journals relating to each chapter and critical lists of references to the original literature. This book, *Selected Readings in Biochemistry*, is edited by Professors Julius Marmur, Sam Seifter, and Sasha England of the Department of Biochemistry in the Albert Einstein School of Medicine.

To aid all students in understanding the subject matter, problems are included at the end of each chapter. The problems have been contributed by the authors and myself. They have been edited for relevance and clarity by a group of graduate students in the Department of Biological Sciences at Columbia. These students, Cindy Hemenway, Richard Jove, John Noble, John D. Oberdick, and Robert Rooney, have examined each problem, worked through the solutions, and published a solutions guide to accompany the main text, which includes summaries of each chapter, a restatement of the problems, and detailed solutions for all the problems.

In addition to the authors of the main text and the two associated supplementary volumes mentioned above, we have received tremendous assistance from a wide body of teacher-scientists in text preparation.

Reviewers have been involved every step of the way, from the first draft to the final product. They were recruited by the authors and myself on an individual basis for the purpose of examining segments of chapters, whole chapters, or major parts of the book. We asked them to comment freely on the thoroughness, accuracy, and readability of the text. Their job was done gallantly and effectively, and their criticisms were needed. Those involved in the formal review process included H. Guy Williams-Ashman, University of Chicago; Jonathan Beckwith, Harvard Medical School; Robert M. Bell, Duke University Medical Center; Sherman Beychok, Columbia University; Konrad E. Bloch, Harvard University; David Brindley, University of Nottingham; Michael Cashel, National Institutes of Health; Roderick K. Clayton, Cornell University; Thomas Ebrey, University of Illinois; Robert H. Fillingame, University of Wisconsin; P. A. George Fortes, University of California–San Diego; Perry A. Frey, University of Wisconsin; Govindjee, University of Illinois; Jonathan Greer, Columbia University; Gary R. Jacobson, Boston University; W. Terry Jenkins, Indiana University–Bloomington; Jeremy R. Knowles, Harvard University; Marilyn S. Kozak, University of Pittsburgh; Robert Lefkowitz, Duke University Medical Center; Richard Malkin, University of California–Berkeley; James L. Manley, Columbia University; Julius Marmur, Albert Einstein College of Medicine; Christopher Miller, Brandeis University; Leslie E. Orgel, Salk Institute; Mary J. Osborn, University of Connecticut Health Center; Robert E. Parks, Jr., Brown University; William W. Parson, University of Washington; Efraim Racker, Cornell University; Charles C. Richardson, Harvard Medical School; Jane Richardson, Duke University Medical Center; William Scovell, Bowling Green State University; William L. Smith,

Michigan State University; Charles C. Sweeley, Michigan State University; H. Edwin Umbarger, Purdue University; Arthur Weissbach, Roche Institute of Molecular Biology; Herbert Weissbach, Roche Institute of Molecular Biology; and Juan Yguerabide, University of California–San Diego.

In many cases we found it necessary to consult individual experts on specific topics. This second group of advisors provided useful suggestions at many stages during the development of the text. The following scientists were contributors in this respect: Sidney Bernhard, University of Oregon; Martin Gellert, National Institutes of Health; Bernhard Horecker, Roche Institute; Roger Kornberg, Stanford University; Steve Lippard, Columbia University; William Lipscomb, Harvard University; Sanford Moore, Rockefeller University; Fred Richards, Yale University; Michael Rossmann, Purdue University; Tom Steitz, Yale University; Alex Tzagaloff, Columbia University; and David Ward, Yale University.

From the beginning of manuscript preparation to the completion of the text, we the authors have received tremendous assistance from a host of people within Addison-Wesley Publishing Company. During the early stages of author recruiting Laura Rich Finney, Acquisitions Editor, was most helpful. Throughout most of the book's development, Bob Rogers, Senior Editor, was a constant source of advice and encouragement. Other principals for whose aid we are most grateful include: Karen Guardino, Production Manager; Kris Belanger, Art Coordinator; and Vanessa Piñeiro, Designer.

The size and complexity of this book required the recruitment of several specialists. Unqualified praise is due to Marcia Mirski, our Production Editor. The expertise and energies she committed to this book greatly shortened the production time without sacrificing quality. She was aided by James Madru, who did a flawless job of copy editing.

The molecular art presented in the text was greatly enhanced by many drawings from Irving Geis and the computer graphic art from Gary Quigley and Will Gilbert of the Massachusetts Institute of Technology. Jane Richardson generously contributed a large number of her unique ribbon drawings of protein structure to Chapter 3. The large body of diagrammatic art seen throughout the text was the product of Michael Ockler of Illustration Concepts.

I would like to give my personal thanks to Victor Helu, Beth Jacober, Linda Sproviero, and Bongsoon Zubay for the help they gave me in manuscript preparation. Dennis Vance would like to acknowledge the expert assistance he received from Jean E. Vance in editing Chapters 13, 14, and 15.

*New York*
*January, 1983*                                                                      G.Z.

# LIST OF
# CONTRIBUTORS

**WAYNE M. BECKER**
CHAPTERS 8, 9, 10
Department of Botany, University of Wisconsin, Madison, Wisconsin 53706

**RAYMOND L. BLAKLEY**
CHAPTER 19
Department of Biochemistry, St. Jude Children's Research Hospital, Memphis, Tennessee 38101

**JAMES W. BODLEY**
CHAPTERS 24, 25
Department of Biochemistry, University of Minnesota, Minneapolis, Minnesota 55455

**FREDERICK J. BOLLUM**
CHAPTER 20
Chairman, Department of Biochemistry, Uniformed Service University of Health Sciences, Bethesda, Maryland 20014

**RONALD BRESLOW**
CHAPTERS 4, 5
Department of Chemistry, Columbia University, New York, New York 10027

**ANN BAKER BURGESS**
CHAPTER 21
Associate Chairperson, Biology Core Curriculum, University of Wisconsin, Madison, Wisconsin 53706

**RICHARD R. BURGESS**

CHAPTER 21

Department of Oncology, McArdle Laboratory For Cancer Research, University of Wisconsin, Madison, Wisconsin 53706

**JAMES P. FERRIS**

CHAPTER 32

Department of Chemistry, Rensselaer Polytechnic Institute, Troy, New York 12181

**PERRY A. FREY**

CHAPTER 6

Department of Biochemistry, Institute for Enzyme Research, University of Wisconsin, Madison, Wisconsin 53706

**IRVING GEIS**

CHAPTER 1

Associate in Department of Biophysics and Theoretical Biology, University of Chicago, Chicago, Illinois 60637

**MAX GOTTESMAN**

CHAPTER 27

Head, Laboratory of Molecular Biology, National Cancer Institute, NIH, Bethesda, Maryland 20014

**SUSAN GOTTESMAN**

CHAPTER 27

Laboratory of Molecular Biology, National Cancer Institute, NIH, Bethesda, Maryland 20014

**LLOYD L. INGRAHAM**

CHAPTER 7

Department of Biochemistry and Biophysics, University of California at Davis, Davis, California 95616

**GARY R. JACOBSON**

CHAPTERS 16, 17, 30

Department of Biology, Boston University, Boston, Massachusetts 02215

**JULIUS MARMUR**

CHAPTER 18

Department of Biochemistry, Albert Einstein College of Medicine, Bronx, New York 10461

**RICHARD PALMITER**

CHAPTER 29

Department of Biochemistry, University of Washington, Seattle, Washington 98195

**WILLIAM W. PARSON**

CHAPTERS 11, 31

Department of Biochemistry, University of Washington, Seattle, Washington 98195

**R. C. PETERSON**

CHAPTER 20

Department of Biochemistry, Uniformed Service University of Health Sciences, Bethesda, Maryland 20014

**MILTON H. SAIER, JR.**
CHAPTERS 16, 17, 30
Department of Biology, University of California at San Diego, LaJolla, California 92093

**F. RAYMOND SALEMME**
CHAPTERS 2, 3
Department of Biochemistry, University of Arizona, Tucson, Arizona 85724

**JOE SAMBROOK**
CHAPTER 28
Cold Spring Harbor Laboratory, Cold Spring Harbor, New York 11724

**JACK STROMINGER**
CHAPTER 12
Department of Biochemistry and Molecular Biology, Harvard University, Cambridge, Massachusetts 02138

**H. EDWIN UMBARGER**
CHAPTERS 22, 23
Department of Biological Sciences, Purdue University, West Lafayette, Indiana 47907

**DAVID A. USHER**
CHAPTER 32
Department of Chemistry, Cornell University, Ithaca, New York 14853

**DENNIS E. VANCE**
CHAPTERS 13, 14, 15
Head, Department of Biochemistry, University of British Columbia, Vancouver, British Columbia, Canada V6T 1W5

**GEOFFREY ZUBAY**
CHAPTERS 2, 3, 4, 5, 12, 18, 26
BioSciences Department, Columbia University, New York, New York 10027

# CONTENTS IN BRIEF

# CONTENTS

# PART II

## CARBOHYDRATE METABOLISM AND THE GENERATION OF CHEMICAL ENERGY   243

# PART III

## LIPIDS AND MEMBRANES   469

PART I

# PROTEIN STRUCTURE AND FUNCTION

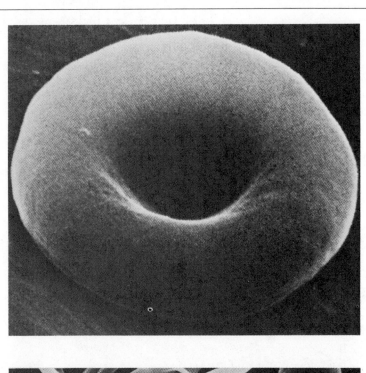

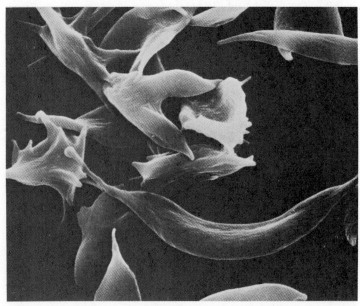

# 1

# PROTEINS:
# AN OVERVIEW

In the middle of the nineteenth century, the Dutch chemist Gerardus Mulder extracted a substance common to animal tissues and the juices of plants which he believed to be "without doubt the most important of all substances of the organic kingdom, and without it life on our planet would probably not exist." At the suggestion of Berzelius, the famous Swedish chemist, Mulder named this substance _protein_ (from the Greek _proteios_, meaning "of first importance"). Mulder ascribed to the substance he called protein a specific chemical formula ($C_{40}H_{62}N_{10}O_{12}$). He was wrong, of course, about the chemistry of proteins, but he was right about the existence of a type of molecular substance indispensable to living organisms. The term protein endures.

Proteins are the most abundant of cellular components and include enzymes, antibodies, peptide hormones, transport molecules, and even components for the skeleton of the cell itself. They are structurally and functionally the most diverse and dynamic of molecules and play key roles in nearly every biological process. Proteins are the workhorses of living systems. They are informational macromolecules, the ultimate heirs of the genetic information encoded in the sequence of nucleotide bases within the chromosomes. Essentially, proteins are complex macromolecules with exquisite specificity; each is a specialized player in the orchestrated activity of the cell. Together they tear down and build up molecules, extract energy, repel invaders, act as delivery systems, and even synthesize the genetic apparatus itself.

As late as 1930, there was considerable doubt that proteins possessed a distinct molecular structure. They were often thought of only as colloidal aggregations. In 1934, when Bernal and Crawfoot began the study of pro-

_A single amino acid change in the hemoglobin protein leads to a change in the shape of the red cell._ (Top) _Normal red cell._ (Bottom) _Red cell from a patient with sickle-cell anemia._

teins by x-ray analysis, the first structural concepts imagined regular geometric arrays, such as hexagonal rings. Early studies by Perutz (1949) suggested that hemoglobin might consist of a hexagonal lattice of parallel rods. The actual structures determined by x-ray crystallography in the 1960s turned out to be more interesting, with an architecture all their own.

Myoglobin was the first protein molecule to be understood in atomic detail thanks to the efforts of John Kendrew and his associates at Cambridge University. Five years later, the structure of the first enzyme was elucidated by D. C. Phillips and coworkers at the Royal Institution in London. Twenty years after the work with myoglobin, the specific architecture of more than a hundred protein molecules had been discovered. It has been estimated that there may be 100,000 different kinds of protein molecules in the human body. It might have turned out that the structure of each of the molecules discovered would be completely different, presenting a hopelessly chaotic version of molecular anatomy. Fortunately, this is not the case; the first two structures that were determined—myoglobin and hemoglobin—were seen immediately to have a strong family resemblance. This trend has continued, and most of the known structures are now studied as groups rather than as individual molecules.

## BUILDING THE POLYMER

*Every protein molecule can be considered as a polymer of amino acids.* There are 20 amino acids, each with a common backbone combined with one of 20 different side chains (also called R groups). Figure 1-1a shows the

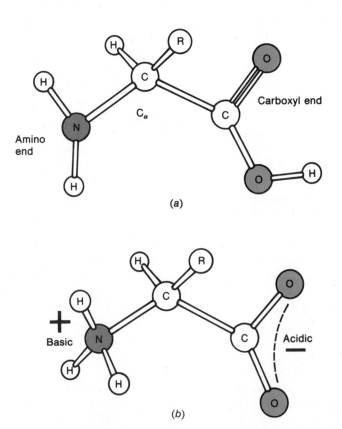

**Figure 1-1**
Amino acid anatomy. (a) Uncharged amino acid. (b) Doubly charged zwitterion.

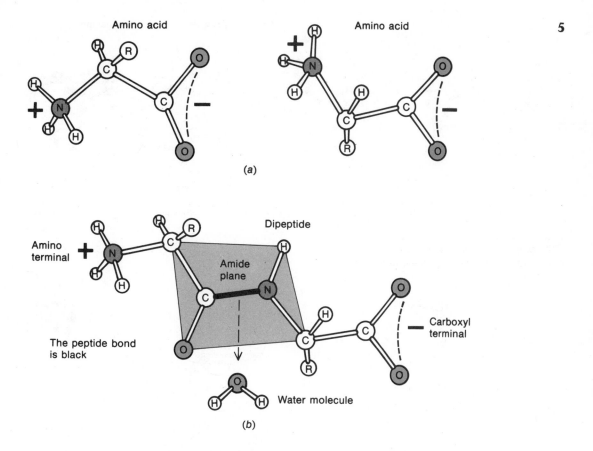

Amino acid

Amino acid

*(a)*

Amino
terminal

Dipeptide

Amide
plane

The peptide bond
is black

Carboxyl
terminal

Water molecule

*(b)*

structure of a single amino acid. At the center is a tetrahedral carbon atom called the $\alpha$ carbon ($C_\alpha$). It is covalently bonded on one side to an amino group ($NH_2$) and on the other side to a carboxyl group (COOH). A third bond is always hydrogen, and the fourth bond is to a variable side chain (R). In neutral solution (pH 7), the carboxyl group loses a proton and the amino group gains one. Thus an amino acid in solution is a neutral but also a doubly charged species called a zwitterion (Figure 1-1*b*).

Amino acids are joined by a peptide bond between the carboxyl end of one amino acid and the amino end of another. This bond is formed by the loss of a water molecule. Figure 1-2 shows two amino acids coming together to form a dipeptide. The peptide linkage (—CO—NH—) has partial double-bond character that results from resonance and keeps it planar. Any number of amino acids can be joined in this fashion to form a polypeptide chain. Short chains up to a length of about 20 amino acids are called *peptides*. A small protein molecule may consist of a chain of 50 to 100 amino acids; a large protein, 300 amino acids. One of the largest single polypeptide chains is that of the muscle protein myosin, which consists of approximately 1750 amino acids. Figure 1-3 shows a section of a protein chain as a linear array with $\alpha$ carbons and planar amides as repeating units of the main chain. Different side chains are attached to each $\alpha$ carbon. The variable side chains help direct the folding and, ultimately, are responsible for the functional diversity of protein molecules. In Figure 1-3, ball-and-stick

**Figure 1-2**
Formation of a dipeptide from two amino acids. (*a*) Two amino acids. (*b*) A peptide bond (CO—NH) links amino acids by joining the carboxyl terminal of one with the amino terminal of another. A water molecule is lost in the reaction.

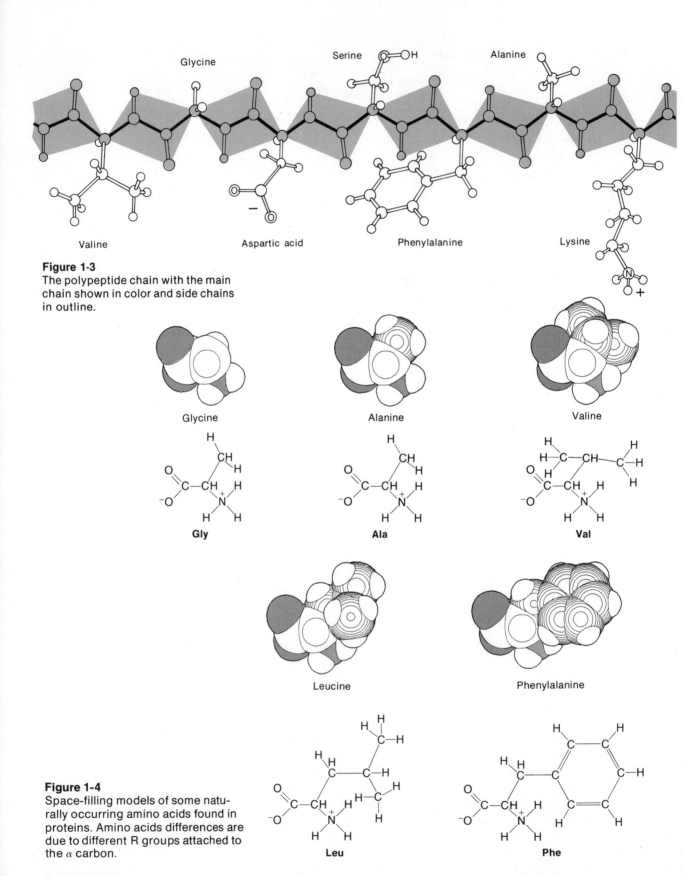

**Figure 1-3**
The polypeptide chain with the main chain shown in color and side chains in outline.

**Figure 1-4**
Space-filling models of some naturally occurring amino acids found in proteins. Amino acids differences are due to different R groups attached to the α carbon.

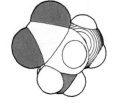
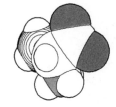

L-Alanine          D-Alanine

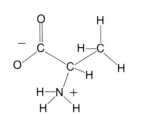

**Figure 1-5**
L-Alanine and its mirror image
D-alanine.

models represent the side chains, but the space-filling models of Figure 1-4 present a more realistic picture. Note the difference in size of the five side chains in Figure 1-4, varying from glycine, the smallest, to the bulky aromatic ring of phenylalanine.

The molecular weights of amino acids vary from 75 for glycine to 204 for tryptophan. The molecular weight of the relatively small protein ribonuclease, with 124 amino acids, is 13,700. From this it may be calculated that the molecular weight of the average amino acid residue in ribonuclease (RNAse) is 110. If one adds on 18 for the water molecule lost in polymerization, then the average molecular weight of an amino acid in RNAse is 128, or about the same as leucine shown in Figure 1-4.

All the 20 amino acids commonly found in proteins except glycine have an asymmetric $\alpha$ carbon atom connected to four different substituents; isoleucine and threonine each have an additional asymmetric center. The asymmetric center about the $\alpha$ carbon gives the amino acid a chiral or handed appearance. Figure 1-5 shows the two possible isomers of alanine called L- and D-alanine. Only chiral L-amino acids with the arrangement of substituents about the $\alpha$ carbon seen for L-alanine are found in proteins. Some D-amino acids are found in bacterial cell walls and certain antibiotics.

## FOLDING THE POLYMER

So far we have considered the polypeptide chain as a linear array. Actually, the chain is folded in three dimensions. The amide planes are free to rotate about single bonds to the connecting $\alpha$ carbon (Figure 1-6a). This twisting of amide planes about connecting $\alpha$ carbons gives the main-chain backbone of a protein its three-dimensional conformation. Figure 1-6b shows a helical conformation made only from planar amides and their connecting $\alpha$ carbons. Even in this highly schematized form, the sense of a folded chain is clearly depicted. A complete protein molecule can be seen only after all the

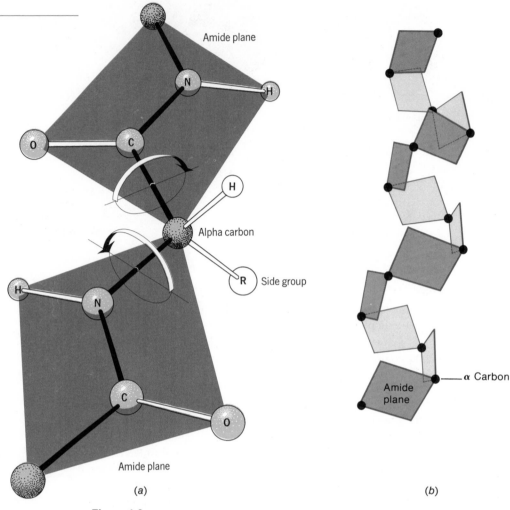

**Figure 1-6**
Polypeptides may be thought of as amides free to rotate about two single
bonds. (*a*) Two amide planes joined by an α carbon. Arrows indicate rotatable
single bonds. (*b*) Twisting amide planes can make a helix in three dimensions.

variable side chains are added at each α carbon. Not only can there be a
different side chain at each α carbon, but each can assume a variety of
conformations. Thus it might seem that the number of possible protein
chain conformations would be virtually infinite. In fact, the protein chain
seeks the conformation of lowest energy, and this makes certain folding
patterns virtually forbidden and others highly favored.

Figure 1-7 shows one of nature's favorites, the α *helix*, discovered by
Pauling and Corey. This structure contains 3.6 amino acid residues per
helical turn. In the α helix, the carbonyl oxygens are in a favorable position
to make hydrogen bonds with the hydrogen atom from a nitrogen three
residues away. *Hydrogen bonds are much weaker than covalent bonds, but
they are a major directive force in determining folding patterns in pro-*

*teins*. The α helix was the first folding pattern predicted for proteins, and fortuitously, the first protein (myoglobin) to be deciphered proved to be made almost entirely of segments of α helix.

For extended polypeptide chains, Pauling proposed another structure in 1951 that he called the *antiparallel β-pleated sheet*, in which polypeptide chains running in opposite directions make hydrogen bonds in the manner shown in Figure 1-8. Since the chains are not completely extended, the resulting sheet appears pleated. While only the α helix was found in the first observed protein structures, myoglobin and hemoglobin, the β-pleated sheet was found in lysozyme, the first enzyme structure to be elucidated by x-ray crystallography. It is becoming increasingly apparent, however, that the β sheet is a central organizing feature of a large number of proteins.

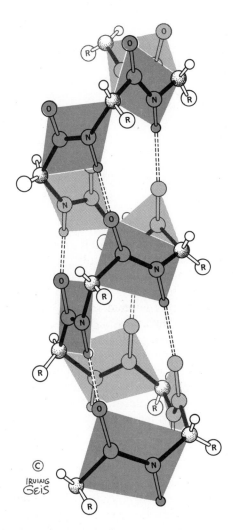

**Figure 1-7**
The α helix with 3.6 residues per turn. Hydrogen bonds are shown as dashed lines.

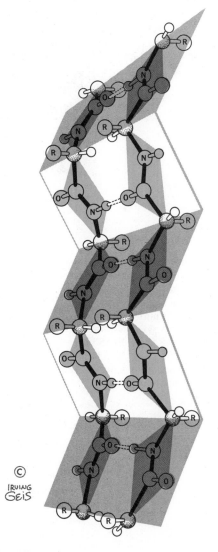

**Figure 1-8**
The antiparallel β strands. Hydrogen bonds are shown as dashed lines.

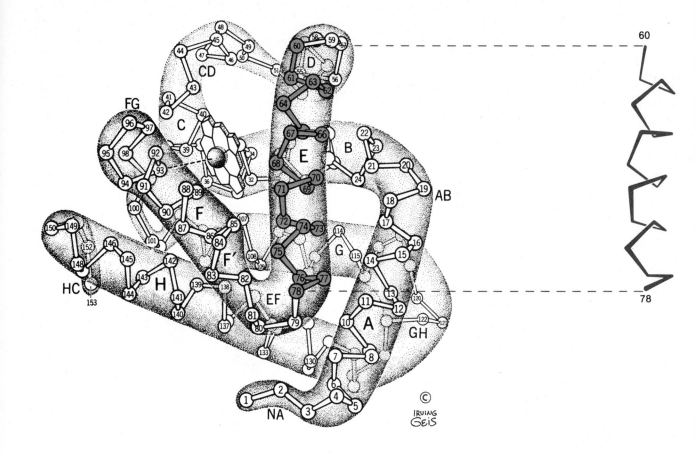

**Figure 1-9**
The tertiary structure of myoglobin. The eight helical segments are labeled by single letters (A through H) and the interhelical segments by double letters. The number of α-carbon positions begins at the N-terminal (NA) end and finishes with 153 at the carboxyl terminal (HC). An α helical portion of the molecule is abstracted at the right of the diagram.

Figure 1-9 shows the protein chain of myoglobin, the molecule that stores oxygen in muscle. This scale drawing shows only the positions of the α carbons joined by straight lines. The numbering is from the amino terminal (1) to the carboxyl terminal (residue 153). Myoglobin is 85 percent α helix. There are eight helical segments lettered from A to H, with the interhelical regions identified by double letters (CD and EF, for example). Myoglobin's function is to provide a pocket for the heme group with its central iron atom, where oxygen is bound. The octahedrally coordinated iron atom is linked to a histidine side chain of myoglobin at position 93. Four ligands of the iron are nitrogen atoms in the heme, and the sixth coordination position binds an oxygen molecule.

Is this unique folding found in purified protein crystals the same conformation that occurs in solution? Protein crystals are at least 30 to 40 percent solvent and sometimes 70 to 80 percent solvent, and the crystal-packing contacts are relatively weak compared with the contacts that hold individual molecules together. For these reasons, it seems likely that major structural features observed for proteins in crystals will be preserved in solution. Several lines of evidence indicate that this is the case. Similar spectroscopic patterns are seen in both crystals and solutions of myoglobin. Optical studies used to estimate helical content agree with the results of x-ray crystallography. Furthermore, even in the crystal state, myoglobin can still bind oxygen. The most convincing evidence of all comes from related molecules of different species that show great similarities of conformation. Myoglobin molecules from sperm whales and seals show similar folding patterns. As will be discussed later, myoglobin and its oxygen-binding cousin hemoglobin show striking similarities from widely separated species.

The fact that a protein molecule has a specific conformation does not mean it is absolutely rigid. It is more like a piece of soft sculpture than unyielding stone. As a matter of fact, specific motions within and between macromolecules are often the key to their function.

## SEQUENCE DICTATES FOLDING

*The specific three-dimensional folding of a protein is determined by the sequence of amino acids in the polypeptide chain.* The elucidation of the first three-dimensional structures confirmed earlier beliefs that *hydrophobic residues inhabit the interior of protein molecules and hydrophilic residues are seen on the surface in contact with water.*

The specific chemical characteristics of each of the 20 amino acids are shown in Figure 1-10 as they might be located in a protein molecule. *External side chains* are those which are normally charged (see Table 1-1) and/or polar, i.e., negatively charged aspartic and glutamic acids (at pH 7) and positively charged lysine and arginine (also at pH 7). Histidine can play a dual role, being either neutral or positively charged, depending on the exact pH. Asparagine and glutamine, although neutral, are so polar that they are classified as external. All these side chains are hydrophilic and are normally found on the surface in contact with water. *Internal side chains* include the uncharged hydrocarbons: phenylalanine, leucine, isoleucine, valine, and methionine (which acts like a hydrocarbon). All these side chains are generally found buried inside a protein. *Ambivalent side chains*

**Table 1-1**
The p$K$ Values for the Ionizing Groups of Some Amino Acids, Including All Commonly Occurring Amino Acids with Ionizable Side Chains

| Amino Acid | p$K_1$<br>$\alpha$-COOH | p$K_2$<br>$\alpha$-NH$_3{}^+$ | p$K_3$<br>R group |
|---|---|---|---|
| Glycine | 2.3 | 9.6 | — |
| Serine | 2.2 | 9.1 | — |
| Glutamine | 2.2 | 9.1 | — |
| Aspartate | 2.1 | 9.8 | 3.9 |
| Glutamate | 2.2 | 9.7 | 4.2 |
| Histidine | 1.8 | 9.2 | 6.0 |
| Cysteine | 1.7 | 10.8 | 8.3 |
| Tyrosine | 2.2 | 9.1 | 10.1 |
| Lysine | 2.2 | 8.9 | 10.5 |
| Arginine | 2.2 | 9.0 | 12.5 |

# EXTERNAL

## ACIDIC

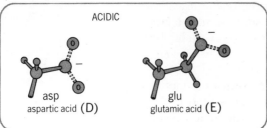

asp
aspartic acid (D)

glu
glutamic acid (E)

## BASIC

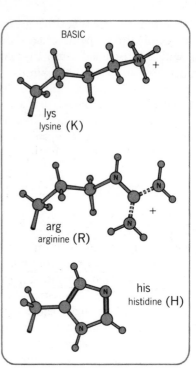

lys
lysine (K)

arg
arginine (R)

his
histidine (H)

## NEUTRAL

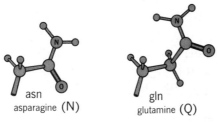

asn
asparagine (N)

gln
glutamine (Q)

# AMBIVALENT

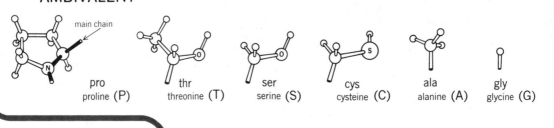

main chain

pro
proline (P)

thr
threonine (T)

ser
serine (S)

cys
cysteine (C)

ala
alanine (A)

gly
glycine (G)

# INTERNAL

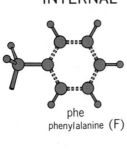

tyr
tyrosine (Y)

trp
tryptophan (W)

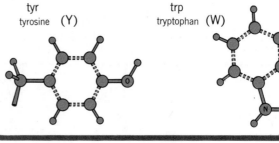

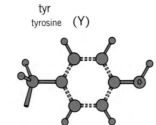

phe
phenylalanine (F)

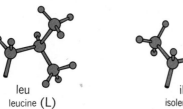

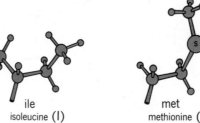

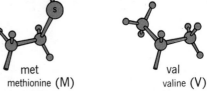

leu
leucine (L)

ile
isoleucine (I)

met
methionine (M)

val
valine (V)

include the small glycine (a single hydrogen atom) and alanine, serine and threonine (with attached hydroxyls), and cysteine (with its sulfhydryl). Proline has a hydrocarbon side chain, but its conformational properties put it at corners and therefore often outside.

Results of x-ray crystallography show these classifications by polarity and location to be valid in general for soluble globular proteins. The structures of myoglobin and hemoglobin, lysozyme, and cytochrome *c* all have buried hydrophobic side chains with hydrophilic side chains on the surface. Figure 1-11 shows the positions of all 104 side chains for horse heart cytochrome *c*. This is a protein with a heme group like myoglobin, but with an entirely different function. It is one of a chain of molecules that transports electrons in the mitochondria. Hydrophobic side chains (colored) pack inside the molecule, especially against the left side of the heme ring, and hydrophilic side chains (grey) are distributed over the surface of the molecule. This is a clear example of one way in which sequence dictates folding.

Other side chains have pronounced effects on three-dimensional conformation, particularly *proline* and the sulfur-containing *cysteine*. The side chain of proline contains a portion of the main chain and thus tends to change the direction of the main chain. Proline is often used to produce a bend in the protein chain, and many of the $\alpha$ helices in myoglobin and hemoglobin begin with a proline residue. The side chain —SH of cysteine can make a covalent —S—S— linkage with a similar residue from another protein chain (Figure 1-12). After the protein chain has reached its optimal low-energy conformation, the disulfide bonds can increase its stability. The enzyme ribonuclease contains four such *disulfide bridges*. If the —S—S— linkages are broken and the protein chain is made to unfold in the presence of a denaturing agent, such as urea, would it refold when the denaturing chemicals were removed? Christian Anfinsen and coworkers answered this question in the affirmative in the early 1960s with a classic set of experiments.

We have seen that sequence determines folding, but, in fact, it does more than that. It determines a unique folding pattern. The importance of the folding pattern can be appreciated through a consideration of the protein's function. Enzymes, for example, are molecular machines that operate with great precision on other molecules called *substrates*. Chymotrypsin is one of a class of pancreatic digestive enzymes that cuts other protein chains. The substrate is a polypeptide chain that is held on the surface of the enzyme so that a peptide bond can be cleaved. It is necessary that the substrate mesh with the enzyme in an exact lock-and-key fashion. In chymotrypsin there is a *specificity pocket* that fits an aromatic ring side chain of the substrate. Immediately adjoining the specificity pocket is an *active site* that assists in cutting a peptide bond near the bound aromatic ring.

◀ **Figure 1-10**
The 20 amino acid side chains classified by their probable position in the protein molecule. Three-letter and one-letter codes are given for each. The forms shown here are the most prevalent at pH 7. Note that histidine can play a dual role—neutral (as shown here) or positively charged.

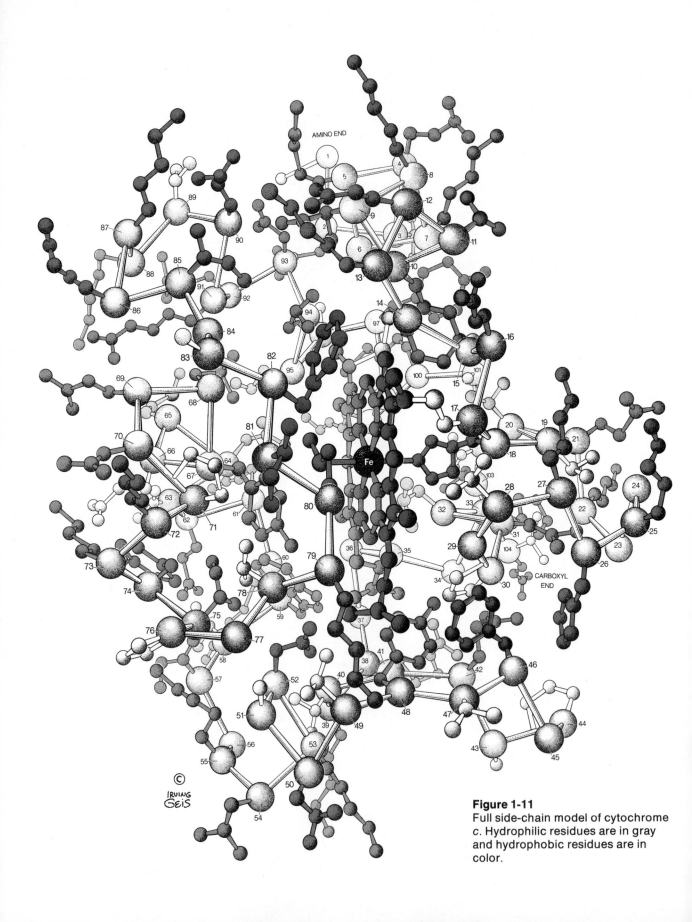

AMINO END

CARBOXYL END

Fe

© IRVING GEIS

**Figure 1-11**
Full side-chain model of cytochrome
*c*. Hydrophilic residues are in gray
and hydrophobic residues are in
color.

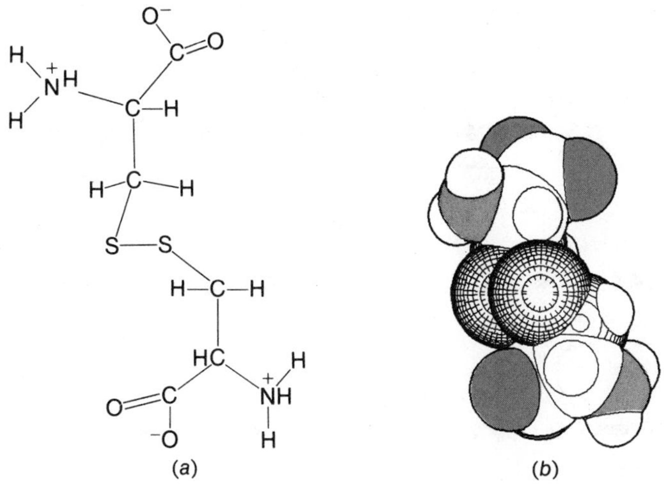

(a)                              (b)

Three side chains of the enzyme His 57, Asp 102, and Ser 195 comprise this active site. They are shown in color in the chymotrypsin sequence diagram in Figure 1-13a. It is obvious that they are far apart in the sequence, but close together in the folded chain (Figure 1-13b). The linear sequence has been translated into three-dimensional information, an active site with stereochemical specificity, and a pocket that fits only a certain type of side chain.

**Figure 1-12**
Disulfide bridges formed between two cysteines. (a) Ball-and-stick and (b) space-filling models of the disulfide bond.

## CLASSIFICATION OF PROTEINS

Because of the great structural and functional diversity of proteins, it is difficult to capture the important features or the whole range of them within any one simple classification scheme. Earlier in this century, attempts were made to characterize proteins by their physicochemical properties, particularly their solubilities: "globular" proteins are generally water-soluble, whereas "fibrous" proteins are not. As knowledge of protein structure and function has become more sophisticated, many other more specialized ways of grouping proteins have come into use, each of which has been helpful for at least some subset of examples. In this section proteins are discussed according to their function.

### Enzymes

The requirements of the cell for growth, reproduction, synthesis and breakdown of metabolites, and the generation and utilization of energy are met by myriad chemical reactions, nearly all of which are catalyzed by protein enzymes. Frequently rate enhancements of a million-fold or more are observed when enzymes are used to catalyze a reaction. Enzymes display a remarkable specificity for the reactions they catalyze, and therefore only one or very few structurally similar compounds will serve as substrates (or reactants) for any given enzyme. This high specificity, which extends to particular stereoisomeric forms, demands that the cell have thousands of different enzymes to perform its manifold functions, and well over 2000

AMINO ACID SEQUENCE OF CHYMOTRYPSIN

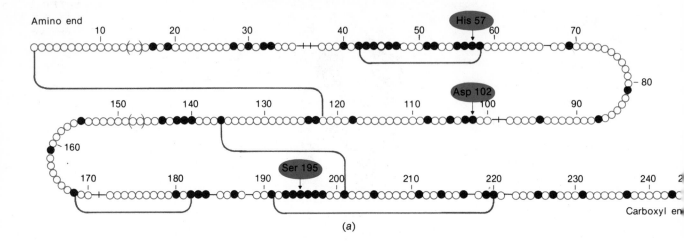

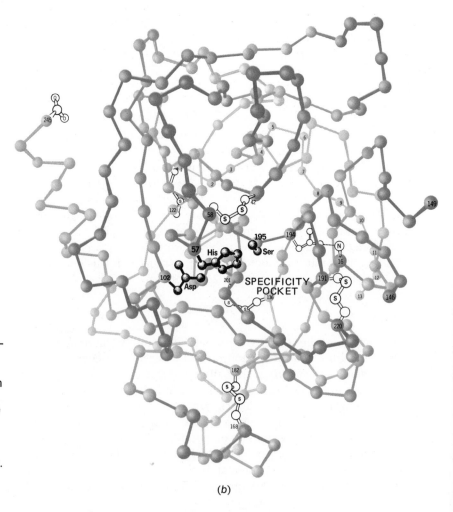

**Figure 1-13**
(a) Schematic diagram of the chymotrypsin structure. Three residues at the active site of chymotrypsin, His 57, Asp 102, and Ser 195, far apart in the sequence, are close together in the folded molecule. Circles indicate amino acid residues, with solid circles representing homologies with trypsin and elastase. (b) The three-dimensional folding of chymotrypsin. The three residues of the active site are shown in black.

enzymes have been isolated and characterized. By virtue of its army of enzymes the living cell is able to carry out a host of chemical reactions from the simple hydrolysis of sucrose to the complex biosynthesis of a chromosome, all under the very mild conditions of temperature, pressure, and pH that are compatible with life itself. This role of specifying the pattern of chemical transformations in the cell is one of the most important functions of the proteins, and much of biochemistry today is concerned with how enzymes function as catalysts and how their functions are controlled.

Lysozyme is an enzyme that cleaves a crucial bond in a complex sugar—a component of bacterial cell walls. Figure 1-14a illustrates the stereochemical specificity of an enzyme-substrate complex. The side view of a space-filling model clearly shows how the substrate fits into a cleft in the enzyme in lock-and-key fashion. Figure 1-14b shows a front view of the enzyme and substrate with framework models. The substrate shown is a hexasaccharide of alternating sugar rings of N-acetylglucosamine (GlcNAc) and N-acetylmuramic acid (MurNAc). They are held to the surface of the enzyme by hydrogen bonds so that a crucial glycosidic bond between the sugar rings can be cleaved.

There exists in the blood a series of inactive enzymes that can be activated by vascular damage to generate, ultimately, a clot at the site of the wound. The enzymes are almost all proteinases of highly refined specificity, each one of which can activate the next inactive enzyme of the series by cleaving it in a specific way. Since an enzyme can turn over many substrates, each step in the series can be at a higher concentration, so that the effects are multiplied into a cascade. The first step is triggered (by means of one of several mechanisms) by the vascular damage. The final enzyme in the cascade is thrombin, a serine protease similar to chymotrypsin, but highly specific for the soluble protein fibrinogen, whose cleavage

**Figure 1-14**
The structure of lysozyme. Enzyme alone (*left*); enzyme-substrate complex (*right*).
(*Continued*)

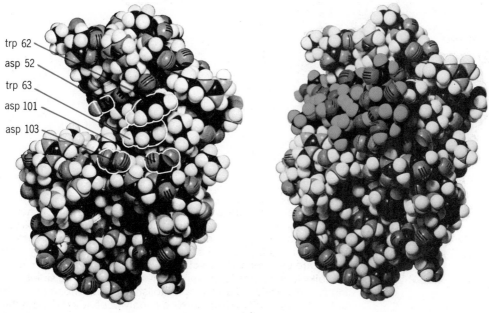

trp 62
asp 52
trp 63
asp 101
asp 103

(*a*)

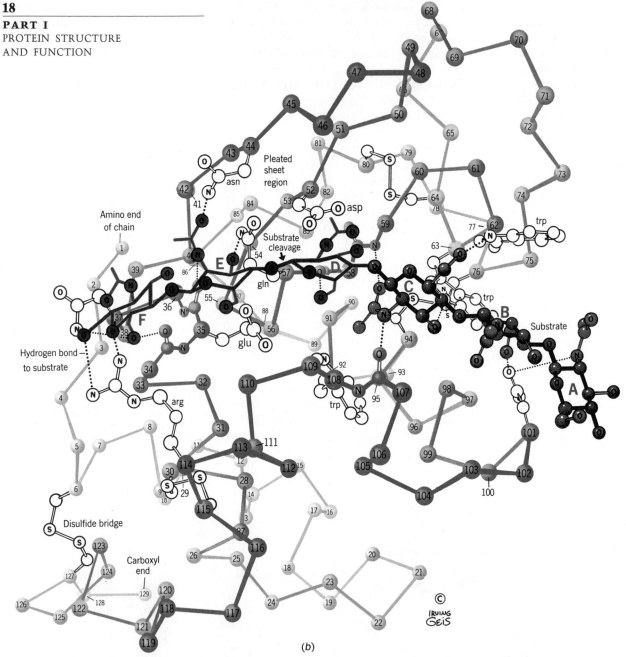

(b)

**Figure 1-14** (*Continued*)
(*b*) Enzyme-substrate complex. The crevice that forms the active site runs
horizontally across the molecule. The hexasaccharide substrate is shown in a
darker color. Rings A, B, and C to the right come from the observed trimer
binding. Rings D, E, and F are inferred from model building. The side chains,
which are believed to interact with the substrate, are shown in line. Ile 98 is
so bulky that it helps to prevent MurNAc, with its large side group, from binding at ring position C and thus establishes the arrangement on the molecule
of the alternating GlcNAc-MurNAc copolymer and points out the locus of
cleavage in the active site. (Coordinates courtesy Dr. D. C. Phillips, Oxford.)

product is fibrin. Fibrin has the ability to cross-link with other fibrin molecules to form the dense matrix of the actual blood clot which helps to stem the flow of blood from a wound. Fibrin is a highly insoluble protein, but it is eventually dissolved after further cleavage during the healing process. *The enzyme cascade of the blood coagulation system is an example of the more general phenomenon of control by limited and specific proteolysis at one or a few sites on a protein.*

## Transport Proteins

*Hemoglobin* is the best-known of the protein molecular carriers. Its chief function is to pick up oxygen in the lungs, where it is plentiful, and deliver it to tissues. Figure 1-15 shows the central structural feature of hemoglobin (and myoglobin as well). This is a water-free pocket for the heme, with its central iron atom located where oxygen is bound. The structure of this hydrophobic pocket is dictated by the sequence of amino acids in the hemoglobin genes. Hydrophobic residues direct the precise folding and positioning of the $\alpha$ helices and also direct that the lining of the heme pocket be hydrophobic. The result of this specific folding is to create an environment where the iron can bind oxygen reversibly without itself being oxidized.

The circulatory system in vertebrates provides the thoroughfare for metabolic "commerce," and it is not surprising that many of the transport proteins in addition to hemoglobin are found in blood. *Serum albumin* binds numerous substances, such as free fatty acids, which are transported between *adipose tissue* and other parts of the body. Iron is carried in the blood by a protein called *transferrin*, and *plasma lipoproteins* are really protein-encapsulated assemblies of *cholesterol*, *triglycerides*, *steroids*, and *phospholipids* that can deliver their lipid components to various organs and tissues.

**Figure 1-15**
The heme pocket. The helices of myoglobin (and hemoglobin) form a hydrophobic pocket for the heme and provide an environment where the iron atom can be oxygenated when it reversibly binds oxygen.

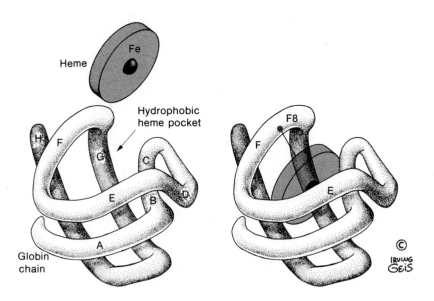

Another extremely important group of transport proteins deals with moving things in and out of cells through the membrane, e.g., ion-transport systems and systems that transport specific sugars or amino acids into cells.

### Storage Proteins

Storage proteins combine with specific substances for storage purposes. For example, iron is stored in the spleen by a protein called _ferritin_. Other storage proteins constitute reservoirs of amino acids that can be released as nutrients during growth and development; examples are _ovalbumin_ of egg white, _casein_ of milk, and _zein_ in wheat seedlings.

### Proteins of Contractile Systems

Skeletal muscle constitutes a large portion of the total mass of vertebrates, and the two major components of this contractile system are the proteins _myosin_ and _actin_ (see Chapter 3). Muscle contraction is accomplished by the energy-dependent sliding motion of these parallel filamentous proteins relative to one another. Myosin molecules constitute a thick filament of helical double strands that provide the drive stroke to move thin filaments, of which actin is the chief component.

Motion is characteristic of a number of subcellular processes, such as mitosis and the ciliary or flagellary propulsion of sperm and certain bacteria. Most of the movable or contractile elements of these systems are also multicomponent assemblies of protein filaments organized in flexible parallel arrays.

### Protein Toxins

Numerous proteins and peptides display toxic properties. Plants, for example, often produce proteins that are toxic in small amounts to insect predators. Many animal venoms contain postsynaptic or presynaptic neurotoxins that are relatively small proteins highly cross-linked with disulfide bonds. Bacterial toxins include diphtheria toxin and the toxin associated with bacterial food poisoning from the anaerobe _Clostridium botulinum_.

### Antibodies

One of the major defense systems of vertebrates against invading foreign substances, such as viruses, bacteria, and cells from other organisms, is the immune system which produces proteins called _antibodies_ that are released in response to the introduction of these foreign substances (usually proteins themselves) called _antigens_. Each antibody is highly specific for the particular antigen in question and combines with it to form an _antibody-antigen complex_ in which the invading substance is rendered inactive and later removed from the organism. Although the antibodies comprise a family of different molecules which, taken together, provide a broad spectrum of functional specificity, the antibody molecules are all variations of a common structural theme. Individual differences are restricted to a region of the molecule called the _antigen combining site_. Figure 1-16 gives a simplified overview of the "gross anatomy" of immunoglobulins. The

**Figure 1-16**
Immunoglobulin molecules (antibodies).

## STRUCTURE OF AN ANTIBODY

Antigen binding site

Light chain

Antigen binding site

S S

S S

-S S-
-S S-

Hinge region

Heavy chain

Black lines connecting the chains are disulfide bridges

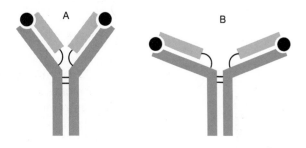

The antibody molecule is flexible and can bind antigens that are spaced differently (A and B)

A

B

## ANTIBODY WITH ANTIGEN

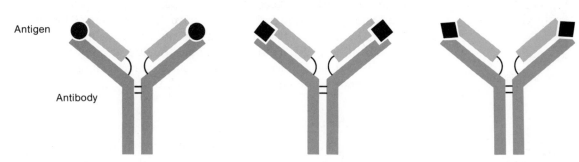

Antigen

Antibody

Antibody binding sites are different and are complementary to the bound antigen

## TYPES OF IMMUNOGLOBULINS

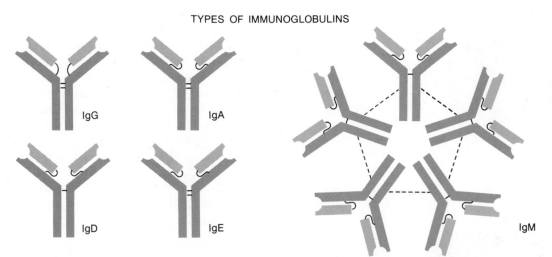

IgG

IgA

IgD

IgE

IgM

IgG molecules are internal serum proteins; IgA is found in internal secretions such as tears.

IgE is involved in allergies and IgM (a pentamer) responds most quickly to antigen.

IgD appears before IgM as a development marker that is not found in the mature B lymphocyte.

**22**

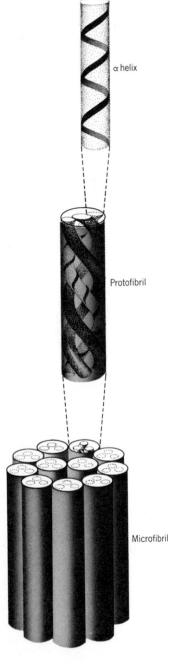

α helix

Protofibril

Microfibril

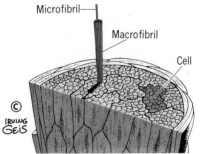

Microfibril

Macrofibril

Cell

© IRVING GEIS

antibody molecule is composed of two light and two heavy chains (*light* and *heavy* refer to relative size or molecular weight) shaped in the form of a Y. The antigen binding sites are at the ends of the Y. A flexible-hinge region allows the molecule to selectively bind antigens that may be spaced differently. Each pair of antigen binding sites is specific for a particular antigen. There may be as many as a million different types of antibodies, each with different binding sites. How this diversity may be generated by the transfer of information from gene to protein is described later.

## Hormones

*Hormones* are substances made in particular cells that function metabolically by interaction with a target cell or organ elsewhere in the system. Many hormones are proteins. Recent studies have provided important new insights regarding protein hormone biosynthesis and function. Some hormones, such as glucagon and insulin, are called *peptide hormones* and are correlated in function with diabetes mellitus and other metabolic diseases. Others, such as gastrin, growth hormone, and thyroid-stimulating hormone, are proteins with molecular weights ranging from 2000 to 30,000. Proopiomelanocortin has been shown to contain within its primary structure the amino acid sequences of corticotrophin hormone, β-lipotropin, endorphin, α and β melano-stimulating hormones, γ-lipotropin, and enkephalin, all well-characterized hormones. Implicit in the concept of a hormone is that of a hormone receptor at the target cell, and as these receptor substances are currently understood, they too are proteins.

## Structural Proteins

Most of the functional classes described earlier are composed of globular proteins, and all are characterized by a recognitional aspect (i.e., their function in association with other molecules), be they substrates, antigens, hormones, toxin receptor sites, storage or transport ligands, or other proteins in a contractile superassembly. In a sense, recognitional phenomena characterize all proteins. Nevertheless, one can distinguish a class of proteins that assumes primarily structural roles in living systems. Structural proteins are analogous in many ways to building materials. Both α-keratin, the major protein component of hair, and collagen, the major extracellular protein in connective tissue, are composed of polypeptide chains arranged to form long, twisted cables. The assembly of hair keratin begins with the α helix as a basic unit. Helices are twisted into protofibrils, which are bundled into microfibrils. These, in turn, are packaged into the macrofibrils of hair cells. The complete assembly of a single hair is diagramed in Figure 1-17. The modular unit of collagen is the threefold superhelix of tropocollagen (Figure 1-18). Each of the three strands is in itself an extended helix. Long tropocollagen molecules are assembled in parallel arrays to form the collagen fibrils of connective tissue.

**Figure 1-17**
The assembly of hair α-keratin from one α helix to protofibril, to microfibril, and finally, to a single hair.

**Table 1-2**
Molecular Weight and Subunit Composition of Selected Proteins

| Protein | Molecular Weight | Number of Subunits |
|---|---|---|
| Glucagon | 3,300 | 1 |
| Insulin | 11,466 | 2 |
| Cytochrome *c* | 13,000 | 1 |
| Ribonuclease A (pancreas) | 13,700 | 1 |
| Lysozyme (egg white) | 13,900 | 1 |
| Myoglobin | 16,900 | 1 |
| Chymotrypsin | 21,600 | 1 |
| Carbonic anhydrase | 30,000 | 1 |
| Rhodanese | 33,000 | 1 |
| Peroxidase (horseradish) | 40,000 | 1 |
| Hemoglobin | 64,500 | 4 |
| Concanavalin A | 102,000 | 4 |
| Hexokinase (yeast) | 102,000 | 2 |
| Lactate dehydrogenase | 140,000 | 4 |
| Bacteriochlorophyll protein | 150,000 | 3 |
| Ceruloplasmin | 151,000 | 8 |
| Glycogen phosphorylase | 194,000 | 2 |
| Pyruvate dehydrogenase (*E. coli*) | 260,000 | 4 |
| Aspartate carbamoyltransferase | 310,000 | 12 |
| Phosphofructokinase (muscle) | 340,000 | 4 |
| Ferritin | 440,000 | 24 |
| Glutamine synthase | 600,000 | 12 |
| Satellite tobacco necrosis virus | 1,300,000 | 60 |
| Tobacco mosaic virus | 40,000,000 | 2,130 |

The principal difference between the compact globular proteins described earlier and these fibrous structural proteins is that the latter are generally built from repeating motifs which enable them to grow into large structural assemblies. Silk fibroin is just such a large assembly, but with a different motif from collagen and α-keratin. Instead of an α helix, silk uses the extended β-pleated sheet of Figure 1-8.

## HIERARCHIES OF STRUCTURE

Proteins differ enormously in size and the number of subunits they contain. Proteins with defined numbers of subunits vary in molecular weight from 11,466 for insulin, with two subunits, to Tobacco mosaic virus, which has 2130 subunits and a molecular weight of 40,000,000 (Table 1-2).

Many proteins with a single, simple task to perform consist of a single polypeptide chain, or monomer. Examples include myoglobin, which binds oxygen in a reversible manner, and enzymes, such as lysozyme, which cuts

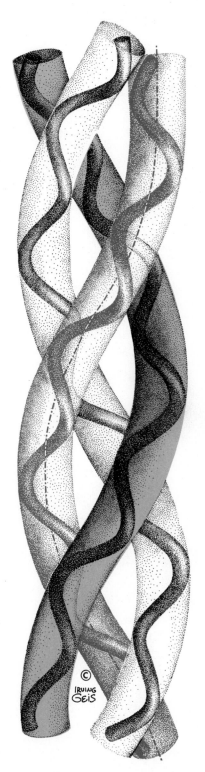

**Figure 1-18**
The triple helix of collagen.

a single specific glycosidic bond. Other molecules, such as hemoglobin, have more complex tasks; hemoglobin functions as a tetramer with four symmetrically associated subunits, each a monomer similar to myoglobin. Multisubunit proteins are called oligomers. A number of oligomeric proteins are listed in Table 1-2 together with their molecular weights and subunit content. The combination of protein subunits into larger aggregates has made possible some important refinements in biological function. Some enzymes, for example, contain in addition to a catalytic subunit, a regulatory subunit that can combine with metabolites in such a way as to facilitate or inhibit catalysis by the functional subunit. Such multisubunit proteins behave as single entities in which the individual chains are tightly associated.

At a higher level of organization supramolecular assemblies of proteins with different functions are often isolated as a multifunctional unit. One example of such a multifunctional aggregate is the ribosome, which is responsible for the synthesis of polypeptides using messenger RNA as a template for protein synthesis. Another such complex is the series of seven enzymes that together catalyze the biosynthesis of fatty acids. Presumably the tight packing of enzymes that catalyze a sequential series of reactions, in which the product of one reaction is the substrate for the next, provides a more efficient catalytic unit.

Finally, there are very large macromolecular assemblies that often serve a structural function. A striking example is the association of two similar protein subunits (protomers) $\alpha$- and $\beta$-tubulin to form microtubules that act as support elements for the cell, including the filaments of the mitotic spindle that maneuver the chromosomes when a cell divides. Yet other examples comprise the highly symmetrical protomers that make up the protein coats of viruses. Figure 1-19 presents a synopsis of the increasingly complex arrangement of structural elements from the amino acids of the primary sequence to the folding patterns of secondary and tertiary structures, the arrangement of subunits, and finally the arrangement of macromolecular assemblies.

## PROTEINS AS INFORMATIONAL MACROMOLECULES

*Proteins are informational macromolecules that are linear translations of a colinear sequence of building blocks in the genetic material specifying the protein sequence.* This genetic material, deoxyribonucleic acid (DNA), is the storehouse of information specifying all facets of the cell's existence. The process of information transfer from DNA to protein is illustrated in Figure 1-20. DNA and RNA are very similar structurally. Each is a linear polymer of four distinctive building blocks. DNA is composed of deoxyribonucleotides, which are named after the unique base in each: adenine (A), thymine (T), guanine (G), and cytosine (C). The messenger RNA (mRNA) is made of ribonucleotides with the same bases, except for thymine, which is replaced by uracil (U). mRNA is synthesized in the process of transcription as a strand that is complementary to the portion of DNA being transcribed. The generation of a complementary mRNA molecule involves the following specific pairing of bases in DNA and RNA, respec-

**Figure 1-19**
Hierarchies of structure.

A

PRIMARY STRUCTURE    (Amino acid sequence in the protein chain)

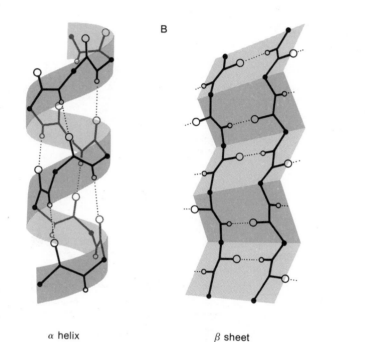

$\alpha$ helix                    $\beta$ sheet

SECONDARY STRUCTURE

Domains (dark color) in
an antibody molecule

LOCAL FOLDING

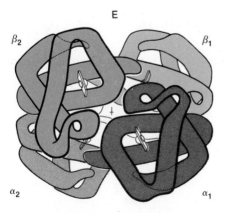

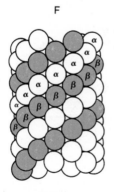

TERTIARY
STRUCTURE

One complete protein chain
($\beta$ chain of hemoglobin)

QUATERNARY STRUCTURE

The four separate chains
of hemoglobin assembled
into an oligomeric protein

MACROMOLECULAR
ASSEMBLY

$\alpha$ (white) and $\beta$ (color)
tubulin molecules in a
microtubule.

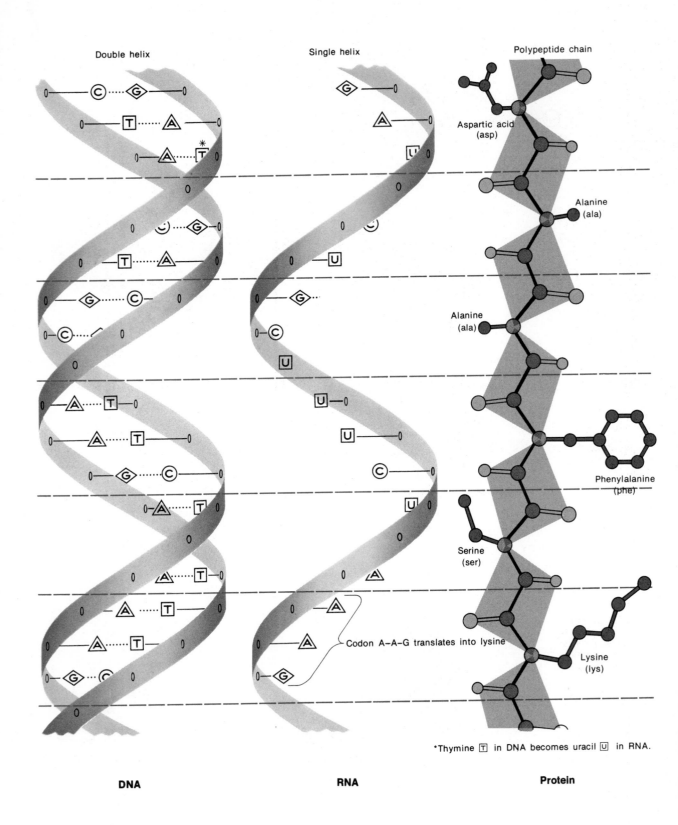

Double helix     Single helix     Polypeptide chain

Aspartic acid (asp)

Alanine (ala)

Alanine (ala)

Phenylalanine (phe)

Serine (ser)

Lysine (lys)

Codon A–A–G translates into lysine

*Thymine T in DNA becomes uracil U in RNA.

**DNA**        **RNA**        **Protein**

**Figure 1-20**
Transfer of information from DNA sequence to amino acid sequence.

tively: A with U, G with C, C with G, and T with A. Thus the mRNA is built to look like the DNA *gene* or *cistron* that codes for the protein product, but the mRNA is free to leave the nucleus and take its information to the cytoplasmic protein biosynthetic units called the ribosomes. It is here that the information coded in a linear nucleic acid sequence is translated into a linear sequence of amino acids.

The correspondence between a linear array of nucleotides and a linear sequence of amino acids involves a code relating the two. Each amino acid in a protein chain is represented by a *coding triplet* that comprises three successive nucleotides in DNA. For example, the triplet A–A–A in DNA is transcribed by base pairing the *codon* as U–U–U in mRNA and is translated as phenylalanine in the growing protein chain.

It is easy to see that from four different bases one can generate $4^3$, or 64, coding triplets or codons. Since only 20 amino acids are commonly found in proteins, it is clear that some amino acids must be coded for by more than one triplet, and this gives rise to the so-called *degeneracy* of the genetic code. This has some important implications in considering the relationship between information stored in nucleic acid and that in protein. Knowledge of the sequence of nucleotides in DNA or mRNA will establish precisely the sequence of amino acids in the protein. Because of the redundancy in coding triplets, however, the reverse is not true. Alterations in the structure of the genetic material arise naturally and can be induced by x-irradiation and certain chemical compounds. The DNA structure can therefore be changed by deletion or insertion of a nucleotide or by modification of a particular base. Since the linear array of nucleotides is *commaless*, these changes could have profound effects on the end product proteins:

Sequence 1   TAAGCTAGCCGTAAGCCTATCATGTCA . . .

Sequence 2   TAAGCTATGCCGTAAGCCTATCATGTCA . . .
⟍ ↑
Insertion

A single base insertion can change the reading (frame shift) and thus code for an entirely different set of amino acids in sequence 2 compared with sequence 1. Clearly, the insertion or deletion of a DNA deoxyribonucleotide unit would lead to a protein product that would be totally different and, most likely, nonfunctional. However, in rare instances such alterations in DNA structure provide the potential for creating new functional entities in the cell or for refining existing proteins.

In all prokaryotic cells the DNA and RNA nucleotide sequences are colinear with the amino acid sequence in the protein chain. It came as a surprise in 1977 to find that in eukaryotic cells, the nucleotide sequences in the messenger RNA frequently have large gaps when compared with the DNA sequences in the chromosome. For example, each of the genes that codes for the subunits of hemoglobin (the $\alpha$ and $\beta$ chains) is separated by long stretches of nucleotides that are not found on the messenger RNA. The RNAs that are initially transcribed are considerably longer than is necessary to code for the actual protein chain. The gene for the $\beta$ chain of hemoglobin is nearly 1500 bases long, when in fact less than 500 bases are needed to code for one complete $\beta$ chain. The entire gene is transcribed into RNA and the unneeded sequences, called _introns_, are cut out. The remaining segments that code for the protein chain, called _exons_, are spliced together and translated into the familiar sequence of amino acids that makes the functioning $\beta$-chain subunit of hemoglobin. There are three exons that code for three separate domains in the $\beta$ subunit (Figure 1-21). These three domains correspond to functional portions of the molecule; domain 2, in particular, includes the oxygen-binding heme pocket and the interface between subunits.

Studies aimed at explaining the origin of diversity in antibody molecules have shown one role for RNA splicing as well as uncovering another type of splicing that occurs at the DNA level. Genes that encode antibodies are unusual in that they are segmented and distributed in segments over a broad stretch of chromosome in the germ-line tissue. During the course of differentiation, certain somatic cells develop the potential to make specific antibodies. Accompanying this process of differentiation, the antibody gene segments become rearranged and mutate so that any particular antibody-forming cell can only make antibodies of one specific type.

The most common antibodies are tetramers that contain two identical heavy chains ($M_r \approx 53,000$ to $70,000$) and two light chains ($M_r = 23,000$). Each heavy and each light chain has an N-terminal region of about 110 amino acids, the _variable_ (_V_) _region_, that exhibits extensive sequence diversity and a C-terminal portion, the _constant_ (_C_) _region_, that exhibits little diversity. The V regions account for the ability of different antibodies to react with different foreign antigens. _In germ-line DNA, the V and C regions are encoded by gene segments that are distant from one another._ During development, a V gene is selected and associated with its C gene, generating a functional immunoglobulin gene and determining the antigen specificity of the lymphocyte and its progeny (Figure 1-22, step 1). This happens for both heavy- and light-chain genes. Light-chain V-gene segments are encoded by two distinct gene segments, the V gene proper and the joining (_J_) gene segment, which encodes the last 13 amino acids. The intervening sequence between VJ and C that is retained during V/J joining is transcribed (Figure 1-22, step 2) and removed from the pre-mRNA by splicing (Figure 1-22, step 3).

Figure 1-22 illustrates the changes that take place at the DNA, RNA, and protein levels in the production of a light chain ($V_L$) in the mouse. The situation for the $V_L$ chain is less complicated than for most antibody proteins, because the germ-line tissue only contains one gene for the $V_L$ chain. Nevertheless, the discrete base changes (mutations) that occur in this gene as the germ-line cells multiply and differentiate leads to a large family of lymphocytes that has the potential for producing antibodies with different specificities. Posttranslational processing that occurs in antibody polypeptide chains is also indicated in Figure 1-22 (step 5) and is discussed below.

Exon 1

Exon 2

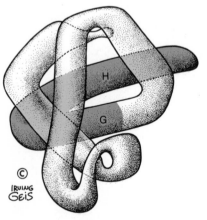

IRVING GEIS

Exon 3

**Figure 1-21**
Exon-encoded domains in hemoglobin. Only a single $\beta$ subunit is shown. The three domains are highlighted in color in three separate frames.

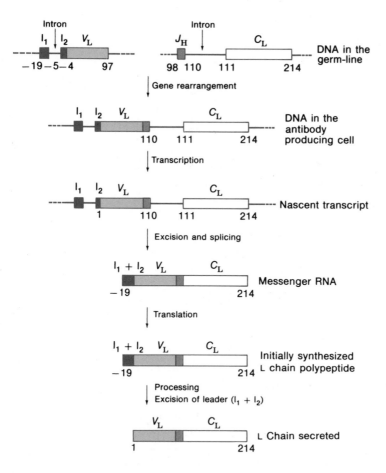

**Figure 1-22**
Origin of the antibody light chain (mouse L chain). Steps include rearrangement of the germ-line DNA. Transcription of the antibody gene, splicing of the nascent transcript to give messenger RNA, translation of the mRNA, and finally processing of the polypeptide chain.

## FINAL ADJUSTMENTS: POSTTRANSLATIONAL MODIFICATIONS

Posttranslational modification of proteins has become an area of intense study in recent years, and it would appear that *few proteins are in a finished state as they roll off the ribosomal "assembly line."* The newly synthesized protein may undergo a variety of changes in its life as an active biological entity. Chymotrypsin, for example, is manufactured in the pancreas as the inactive *zymogen* chymotrypsinogen. From the pancreas it is exported to the small intestine, where it acts as a digestive enzyme in concert with other proteases by cutting up ingested proteins into the constituent amino acids. The manner in which chymotrypsinogen is converted into an active enzyme is diagramed in Figure 1-23. Trypsin, which begins the activation of chymotrypsin, must itself first be activated. Activated trypsin cleaves a single bond between residues 15 and 16 in chymotrypsinogen. This induces a change in the specificity pocket where the substrate is bound. Other cleavages follow that result from autodigestion. Thus we see that highly active cutting enzymes must be exported with their "swords" sheathed; otherwise they might wreak internal havoc. Many secretory proteins have been shown to be synthesized with a so-called signal peptide extension, by which proteins are labeled "made for export." This peptide segment located at the amino-terminal end of the

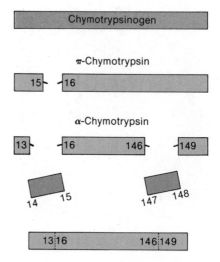

**Figure 1-23**
Posttranslational modification of chymotrypsinogen yielding chymotrypsin. Inactive chymotrypsinogen is shown in gray. Π- and α-chymotrypsin are active and shown in color. Trypsin converts chymotrypsinogen into Π-chymotrypsin. Π-Chymotrypsin converts other Π-chymotrypsin molecules into α-chymotrypsin.

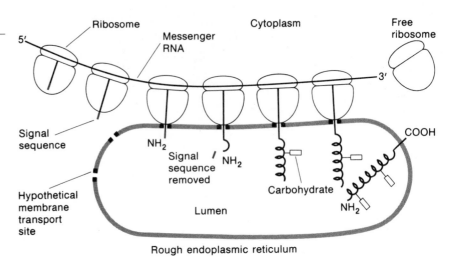

**Figure 1-24**
Posttranslational modification of a protein destined for export. The amino-terminal signal sequence (shown in color) is deleted during transport.

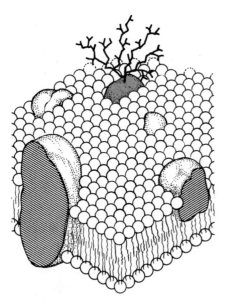

**Figure 1-25**
Posttranslational modification of a membrane-associated protein. Certain membrane-associated proteins (in color) serve as anchors for the construction of simple or branched-chain carbohydrate construction. (From S. J. Singer and G. L. Nicolson, *Science* 175: 723, 1972.)

polypeptide is functional in transferring the nascent polypeptide through membrane structures and into the secretory granules, after which it is removed proteolytically (Figure 1-22, step 5, and Figure 1-24).

The more that is learned about the details of protein structure, the more extensive becomes the catalog of posttranslational modifications. These modifications include the covalent attachment of carbohydrate, lipid, nucleic acid, phosphate, sulfate, carboxyl, methyl, acetyl, and hydroxyl functions. Proteins so modified are endowed with new functional capabilities in terms of binding and catalysis, regulation, physical properties, and so on. Figure 1-25 shows the addition of a carbohydrate tree to a membrane-associated protein. These carbohydrate moieties serve specific recognition functions on the cell surface.

## FAMILIES OF PROTEINS

As we have already seen, the structures of myoglobin and hemoglobin show a strong family resemblance. Sequences of their amino acids also show many homologies. The sequences of sperm whale myoglobin with the α- and β-hemoglobin chains of humans and horses show striking similarities. As more sequences have accumulated from other species, the number of similarities has been found to correlate with the relatedness of the species. Thus the β-chain sequence is identical for humans and chimpanzees; humans and gorillas differ by only one residue; humans and horses by 17; and humans and frogs by 46 residues out of the complete 146 amino acids in the hemoglobin β chain. In spite of the differences, identical residues remain for all species of hemoglobin examined so far. Of all known α chains, 23 of 141 amino acids are identical; of all β chains, 20 of 146 are identical. If only mammals are compared, there are 50 identical residues in α chains and 51 in β chains.

These identities, called *invariants*, are plotted on a map of the hemo-globin structure in Figure 1-26. Here the invariant positions are shown by colored dots. In a sense, the positions of these dots is a diagram of the working machinery of hemoglobin, for they line the heme pockets where oxygen is found and the crucial $\alpha_1 \beta_2$ interface, which changes its orienta-tion when oxygen becomes bound (see Chapter 3). Electrostatic forces and hydrogen bonds stitch this interface together in the deoxy conformation when the molecule gives up its oxygen to the tissues. If there are changes, resulting from mutations, at any of these positions, then trouble is likely to

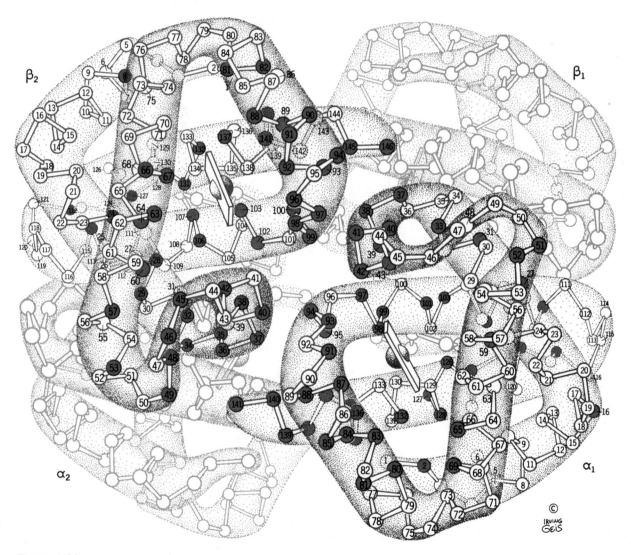

**Figure 1-26**
Invariant residues in the $\alpha$ and $\beta$ chains of mammalian hemoglobin. The col-ored dots indicating the positions of invariant residues line the heme pockets as well as the crucial $\alpha_1 \beta_2$ interface.

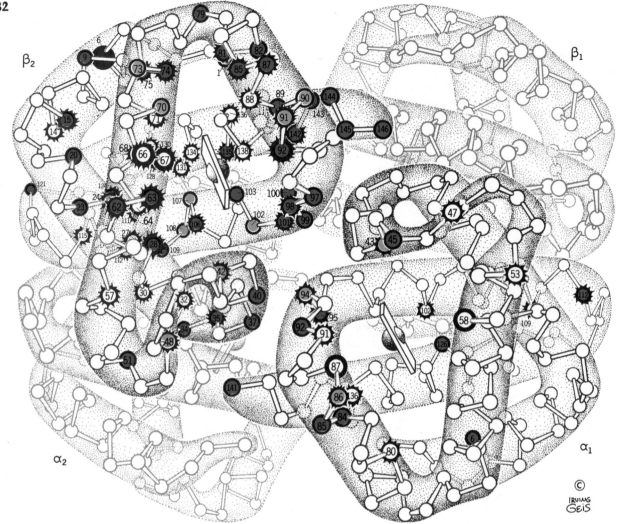

**Figure 1-27**
Positions of pathological mutations in hemoglobin. Comparison with Figure
1-26 shows that these mutations, in general, show the same pattern as the distribution of invariant positions. (Dark circles indicate positions of abnormal residues; solid black dot indicates the valine β6 mutation in sickle-cell anemia;
heavy circles indicate M (Met) hemoglobin; and jagged perimeter indicates unstable hemoglobin. Dark color indicates increased oxygen affinity; light color
indicates decreased oxygen affinity.)

develop. This is just what happens, as can be seen in Figure 1-27, which
shows the positions of pathological mutations in hemoglobin. Where there
are changes in the heme pockets or in the $\alpha_1\beta_2$ interface, hemoglobin
abnormalities occur; many of these are associated with serious diseases.

There are some exceptions to this rule. One striking exception occurs
at position 6 of the $\beta$ chain: a hydrophobic valine residue is substituted for
glutamic acid (a charged side chain) with disastrous results. The homozygous carrier of this mutation has the disease sickle cell anemia, while the

heterozygous carrier is virtually symptom free. The specific consequence of the $\beta$-6 mutation is to cause hemoglobin tetramers to aggregate when they are in the deoxy state. The aggregates form long fibers that stiffen the normally flexible red blood cell. When the disease becomes acute, the fibers distort the cells into "sickle" shapes. These distorted cells occlude capillaries, preventing the proper delivery of oxygen to the tissues. Figure 1-28 shows the aggregation of tetramers in a double strand, which is presumably the common module of the sickled fiber. Current theories hold that the

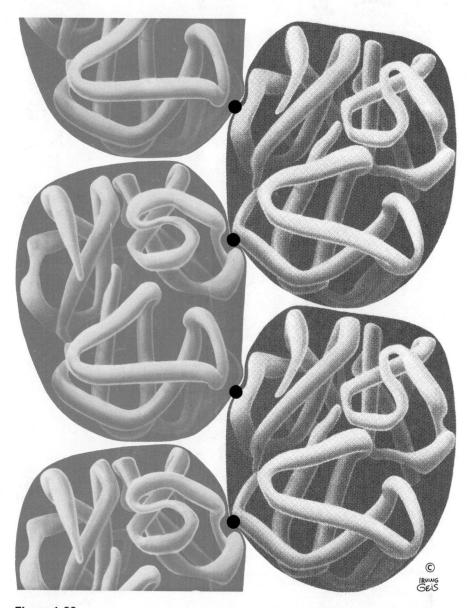

**Figure 1-28**
The "sickled" double strand of hemoglobins. The aggregation of hemoglobin molecules in sickle cell anemia. Colored dots indicate the positions of the mutant valine at the dot (position B-6). This mutant side chain interacts with a hydrophobic pocket in a molecule in an adjoining strand. Note that only one B-6 side chain from each molecule is involved in the polymerization.

**Figure 1-29**
Portion of the framework model of
the enzyme lysozyme used by D. C.
Phillips in 1966. Photo by Irving Geis.

fibers are composed of either seven or eight of these double strands. Thus the fiber is believed to be composed of either 14 or 16 strands. Sickle cell anemia is obviously a deleterious mutation. However, it has persisted because it is thought that in the heterozygous state it has provided protection against malaria.

The differences we see in hemoglobin (and other proteins) sequences among species are due to the accumulation of mutations that account for the differences among species. The longer ago they diverged from a common ancestor, the greater the number of mutations. Thus family trees can be made based on amino acid differences among species that share a particular common molecule. Such trees are especially convincing where a great many sequences are available, as in the case with cytochrome *c*. More than 80 sequences of cytochrome *c* have been compared from as many species. The resulting relationships compare very favorably with taxonomic relationships in living species and even with the fossil record from millions of years ago.

## VISUALIZING MOLECULAR STRUCTURES

The primary data of protein crystallography yield a three-dimensional electron-density map, which must be interpreted in terms of a complex three-dimensional model of all atom positions in the protein. In the past this has been done using actual wire models, whereas now it is generally done by computer graphics. Figure 1-29 shows a small portion of the original working model of the enzyme lysozyme. The atomic positions also can be represented by a space-filling model, as shown for lysozyme in Figure 1-14*a* and for ribonuclease in Figure 1-30. Each style has its uses: the space-filling model is excellent for displaying the volume occupied by molecular constituents and the shape of their outer surface, while the framework model shows connectivities and allows one to look inside.

The difficulty with complete models is that even for these relatively small proteins, it is almost impossible to comprehend such complexity. Therefore, another step of interpretation is made, in which only selected parts of the molecule are shown (such as only the polypeptide backbone) and particular features of interest may be highlighted. With a mind to aiding the student to "read" molecular structures, four different presentations are shown for ribonuclease, and each model is seen from the same perspective. Figure 1-31 shows a stereo pair of a space-filling model from exactly the same perspective. Seen with stereo glasses the three-dimensional illusion is striking. Figure 1-32 shows the polypeptide chain, with labeled N and O atoms, and with dotted lines for hydrogen bonds. In addition, all $\alpha$-carbon positions are numbered. Figure 1-33 shows an abstraction of the polypeptide chain in which the $\beta$ strands are characterized as flat arrows and the $\alpha$ helix as spiral ribbons. This simplified style has proved useful in classifying and comparing proteins according to their secondary- and tertiary-structure folding patterns.

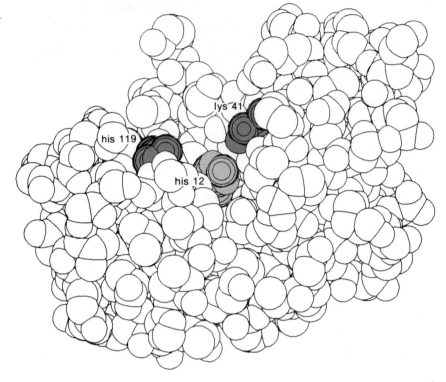

**Figure 1-30**
Space-filling model of ribonuclease S
with three critical side chains, His 119,
His 12, and Lys 41, shown in color.

The 3-D illusion can be seen without stereo glasses by the following
method. With your eyes about 10 inches from the page, stare at the
drawings below as if you were looking straight ahead at a far-away object. A double image will form and the central pair should drift together
and fuse. Then the illusion becomes apparent. Adjust the page, if necessary, so that a horizontal line is perfectly parallel with the eyes. As an
aid to seeing one image with each eye, use two cardboard tubes from
toilet-paper rolls. Close the right eye and focus the left eye on the left
image. Do the same for the other eye. Now with both eyes together, the
3-D illusion should appear.

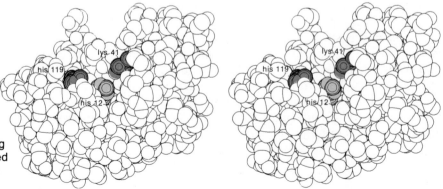

**Figure 1-31**
Stereo views of the same space-filling
model shown above. The three colored
side chains are part of the active site
crevice.

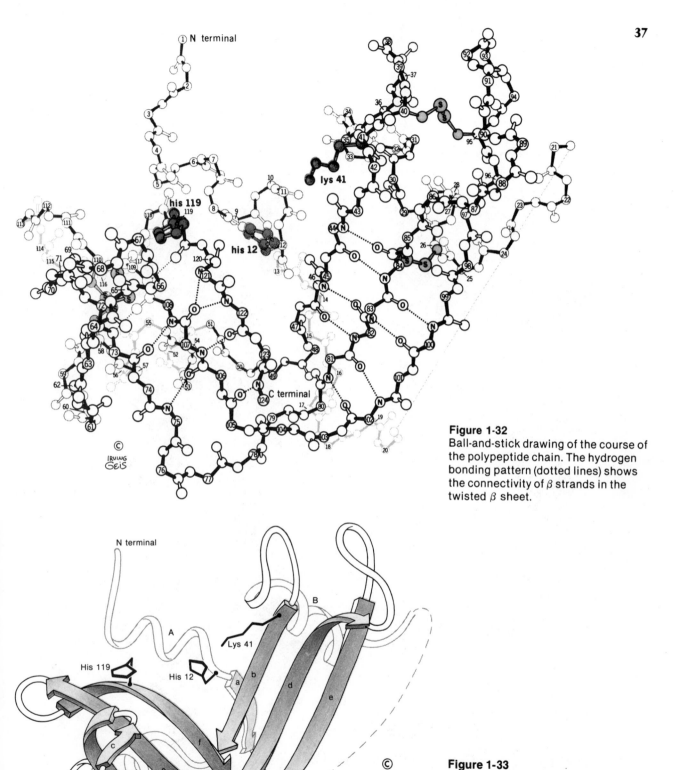

**Figure 1-32**
Ball-and-stick drawing of the course of the polypeptide chain. The hydrogen bonding pattern (dotted lines) shows the connectivity of β strands in the twisted β sheet.

**Figure 1-33**
Ribbon drawing of ribonuclease S. The three significant side chains are in solid color and a tint of color defines the β strands. Helices are designated sequentially by capital letters and β strands by lowercase letters.

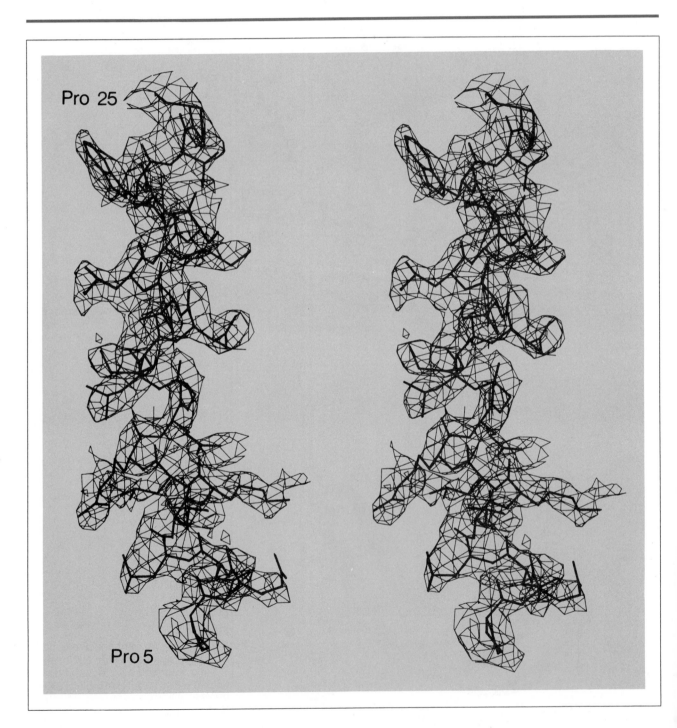

# 2

# ANALYTICAL TECHNIQUES
# OF PROTEIN STRUCTURE

The success of modern biochemistry has in large part depended on the ability of scientists to break living systems down into their constituent molecular components and investigate their properties individually. In this chapter, some approaches to the characterization of proteins will be described in a context that emphasizes the underlying physical and chemical principles involved. Virtually all the techniques described have as their objective the determination of some physical or chemical property of the protein that may be of ultimate use in understanding its function. The techniques fall into two categories. The first category includes techniques aimed at the general characterization of the protein's physical properties, such as its charge, molecular weight, and solubility properties. These characteristics of the protein are most useful in providing an initial identification and form the basis for a wide variety of strategies useful in their purification. These techniques include ultracentrifugation and various electrophoretic and chromatographic methods, all of which exploit differences in the physical properties of proteins as a means of uniquely identifying them. The second class of techniques described is aimed at a determination of a protein's detailed structural properties and so depends on first having an effective purification scheme. These techniques include chemical methods for determination of a protein's primary amino acid sequence, together with those for peptide synthesis, which allow the consequences of amino acid substitutions to be specifically examined. Finally, the technique of x-ray crystallography, which allows the examination of the native protein structure in its full, three-dimensional complexity, is discussed.

*Electron density for cytochrome c' α-helix viewed on computer graphics. The cage-like surface represents the electron density for an α-helix in cytochrome c' as viewed by computer graphics. The stick bonds represent the structure conformation as fitted to the observed electron density.*

# THE CHEMICAL STRUCTURE OF PROTEINS

## Determination of Amino Acid Composition

Each protein is uniquely characterized according to its amino acid composition and sequence. A protein's amino acid composition is defined simply as the number of each type of amino acid comprising the polypeptide chain. In order to determine a protein's amino acid composition it is necessary to (1) break down the polypeptide chain into its constituent amino acids, (2) separate the resulting free amino acids according to type, and (3) quantitatively estimate the relative quantities of each amino acid. Cleavage is usually achieved by boiling the protein in 6 N HCl, which causes hydrolysis of the peptide bonds and the consequent release of free amino acids (Figure 2-1). Although acid hydrolysis is the most frequently used means of breaking a protein into its constituent amino acids, it results in the partial destruction of the indole ring of tryptophan, which must consequently be estimated by an alternative method (e.g., spectroscopic absorption). In addition, acid hydrolysis of the amino acids glutamine and asparagine results in the loss of ammonia from their side-chain amide groups, with the consequent production of glutamic and aspartic acids (Figure 2-2). Consequently, composition estimates based on acid hydrolysates measure glutamine and glutamic acid combined as glutamic acid and asparagine and aspartic acid combined as aspartic acid.

Separation of amino acids for quantitative analytical purposes is usually achieved by ion-exchange chromatography. The general efficacy of chromatographic techniques is based on a difference in affinity that exists between the compounds to be separated and an immobile phase or resin. The resin might be composed of some otherwise chemically inert polymer

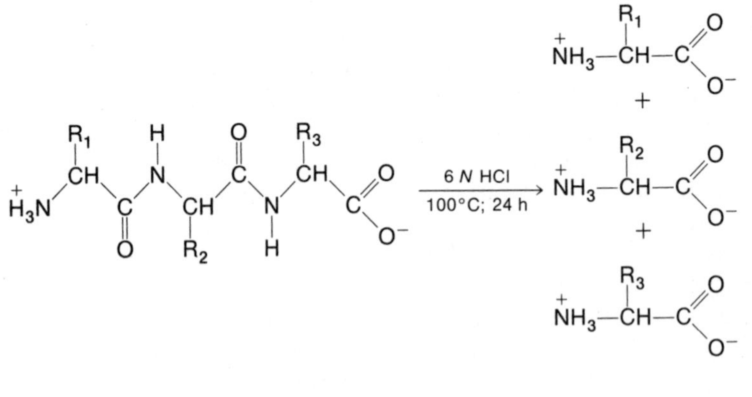

**Figure 2-1**
Acid hydrolysis of a protein or polypeptide to yield amino acids.

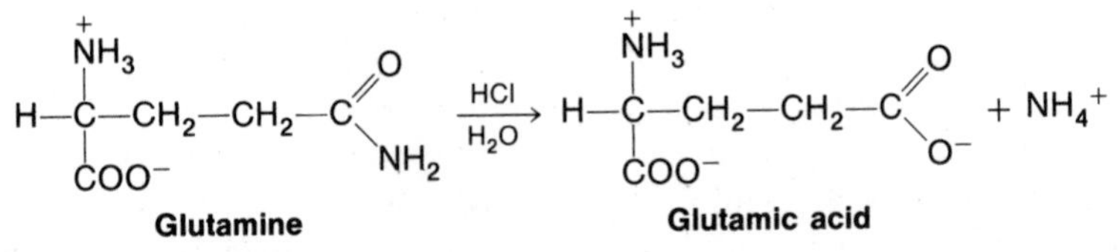

**Figure 2-2**
Acid hydrolysis of glutamine and asparagine result in formation of glutamic and aspartic acids.

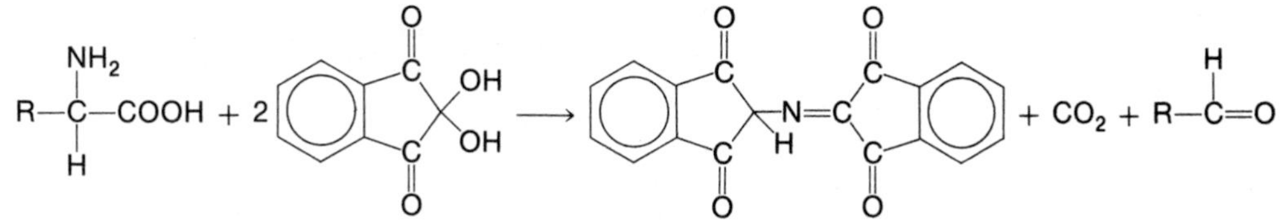

that has weakly basic side-chain constituents that are positively charged at pH 7. If we were to add some of this resin to a solution containing free aspartic acid and lysine, it is evident that the negatively charged aspartic acid would tend to bind to the resin, whereas the positively charged lysine would not. Consequently, if we were to pump a solution of these two amino acids over such a basically charged column of resin, the progress of the aspartic acid through the column would be retarded relative to the lysine owing to the greater affinity of the aspartic acid for the resin. Resins have been developed that have differential binding affinities for all the naturally occurring amino acids. Such resins are effective in separating an amino acid mixture into its components. It should be emphasized that the detailed nature of the forces responsible for the differential binding of amino acids to an ion-exchange resin are quite complicated and depend additionally on side-chain polarity, subtle differences in $\alpha$-amino and carboxyl group pKs, solvation effects, and other factors. Such separation techniques frequently exploit changes in the pH of the eluting solution buffer to enhance the separation properties of the column. For example, a column might initially be run with an eluting buffer at a pH that results in some amino acids being so strongly bound to the resin that they are essentially immobile. However, after the separation and elution of the less strongly bound amino acids, the eluting buffer pH can be appropriately shifted to lessen the charge difference between the resin and the strongly bound amino acids. Subsequently, these previously bound compounds can be eluted and separated according to the newly established pattern of binding affinities between the amino acids and the resin.

Quantitative detection of the separated amino acids is achieved by their reaction with ninhydrin to produce a colored reaction product. As shown in Figure 2-3, this reaction abstracts an amino group from each amino acid, so that the amount of colored product formed is proportional to the amount of amino acid initially present.

Quantitative estimates of amino acid composition are generally carried out on an _amino acid analyzer_, which is basically a device to automate the previously described operations. As illustrated in Figure 2-4, this device consists of an ion-exchange column through which the appropriate eluting buffer is pumped after introduction of the hydrolysate at the top of the column. As the separated amino acids emerge, they are mixed with ninhydrin solution and passed through a heated coil of tubing to allow the formation of the colored ninhydrin reaction product. The separated ninhydrin reaction products then pass through a cell that measures their optimal absorbance at 570 and 440 nm and plots the results on a strip-chart recorder. The absorbance is measured at two wavelengths because proline, which is substituted at its amino group, forms a different ninhydrin reaction product with a correspondingly different absorption maximum from

**Figure 2-3**
Reaction of ninhydrin with an amino acid yields a colored complex.

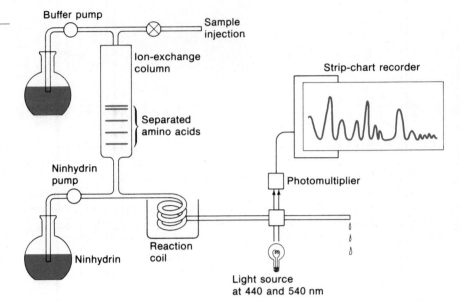

**Figure 2-4**
Schematic diagram of an amino acid analyzer.

**Table 2-1**
Amino Acid Content of Proteins (in percent)

| Constituent | Insulin (Bovine) | Ribonuclease (Bovine) | Cytochrome (Equine) |
|---|---|---|---|
| Alanine | 4.6 | 7.7 | 3.5 |
| Amide $NH_3$ | 1.7 | 2.1 | 1.1 |
| Arginine | 3.1 | 4.9 | 2.7 |
| Aspartic acid | 6.7 | 15.0 | 7.6 |
| Cysteine | 0 | 0 | 1.7 |
| Cystine | 12.2 | 7.0 | 0 |
| Glutamic acid | 17.9 | 12.4 | 13.0 |
| Glycine | 5.2 | 1.6 | 5.6 |
| Histidine | 5.4 | 4.2 | 3.4 |
| Isoleucine | 2.3 | 2.7 | 5.4 |
| Leucine | 13.5 | 2.0 | 5.6 |
| Lysine | 2.6 | 10.5 | 19.7 |
| Methionine | 0 | 4.0 | 2.1 |
| Phenylalanine | 8.6 | 3.5 | 4.5 |
| Proline | 2.1 | 3.9 | 3.3 |
| Serine | 5.3 | 11.4 | 0 |
| Threonine | 2.0 | 8.9 | 8.4 |
| Tryptophan | 0 | 0 | 1.5 |
| Tyrosine | 12.6 | 7.6 | 4.9 |
| Valine | 9.7 | 7.5 | 2.4 |

the remaining amino acids. Usually the analyzer is first standardized by running a sample containing known relative quantities of amino acids in order to account for any differences in their ninhydrin reaction properties. In this way it is possible to directly relate the amount of amino acid present to the amount of colored product formed, which can be measured from the area under the "peak" produced in the strip-chart recorder. Similarly, the hydrolysate of a protein of unknown composition can be run through the analyzer, and the relative peak areas can be used to estimate the ratios of the different amino acids present. Conversion of the relative ratios of amino acids into an estimate of composition requires some sort of additional information concerning the protein's molecular weight; e.g., an analysis giving relative ratios of Ala (1.0), Gly (0.5), and Lys (2.0) could correspond to compositions $Ala_2$-Gly-$Lys_4$ or any multiple thereof. Nevertheless, the required information is usually available, and in any case, an estimation of composition based on a minimum molecular weight of the protein is always possible. Results for three proteins are shown in Table 2-1.

## Determination of Primary Sequence

The most important properties of a protein are determined by the sequence of amino acids in the polypeptide chain. We know the sequences for hundreds of peptides and proteins, largely through the use of methods developed in F. Sanger's laboratory and first used to determine the sequence of the peptide hormone insulin in 1953. The determination of the order of amino acids fundamentally involves the sequential removal and identification of successive amino acid residues from one or the other free terminals of the polypeptide chain. However, in reality it is extremely difficult to get the required specific cleavage reaction of the desired products to go with 100 percent yield. This raises problems when sequencing long polypeptides, because the fraction of the total material of minimum polypeptide chain length becomes constantly smaller as the successive removal of terminal residues proceeds. Conversely, the amino acid released from the minimum length chain becomes increasingly contaminated with amino acids released from previously unreacted chains. This fundamental chemical limitation is what determines the strategy of amino acid sequence determination. It involves breaking the polypeptide chain down into sequences short enough for the chemistry to produce reliable results and then reassembling the results to obtain the overall sequence. The steps actually involved in protein sequencing (Figure 2-5) are (1) purification of the protein, (2) cleavage of all disulfide bonds, (3) determination of the amino- and carboxy-terminal amino acids, (4) specific cleavage of the polypeptide chain into small fragments in at least two different ways, (5) independent separation and sequence determination of peptides produced by the different cleavage methods, and (6) reassembly of the individual peptides with appropriate overlaps to determine the overall sequence.

Starting from the purified protein, the first step in primary structure determination requires disulfide bond cleavage. This can be achieved either by oxidation of the disulfide linkage with performic acid or by reduc-

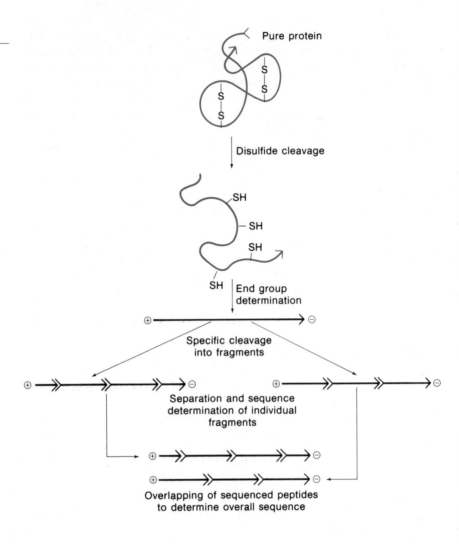

**Figure 2-5**
Steps involved in sequencing a protein.

tion of the bond followed by subsequent reaction to prevent re-formation of disulfide linkages (Figure 2-6).

The next step involves determination of the polypeptide chain end groups. The amino-terminal amino acid can be identified by reaction with fluorodinitrobenzene (FDNB) (Figure 2-7). Subsequent acid hydrolysis releases the FDNB-labeled amino-terminal amino acid, which is highly colored and can be identified by its characteristic migration behavior on thin-layer chromatography or paper electrophoresis.

Chemical methods for carboxyl end-group determination are considerably less satisfactory. Treatment of the peptide with anhydrous hydrazine at 100°C results in conversion of all the amino acid residues to amino acid hydrazides except for the carboxyl-terminal residue, which remains as the free amino acid and can be isolated and identified chromatographically. Alternatively, the polypeptide can be subjected to limited proteolysis with the enzyme carboxypeptidase. This results in release of the carboxy-terminal amino acid as the major free amino acid reaction product. The amino acid type can then be identified chromatographically.

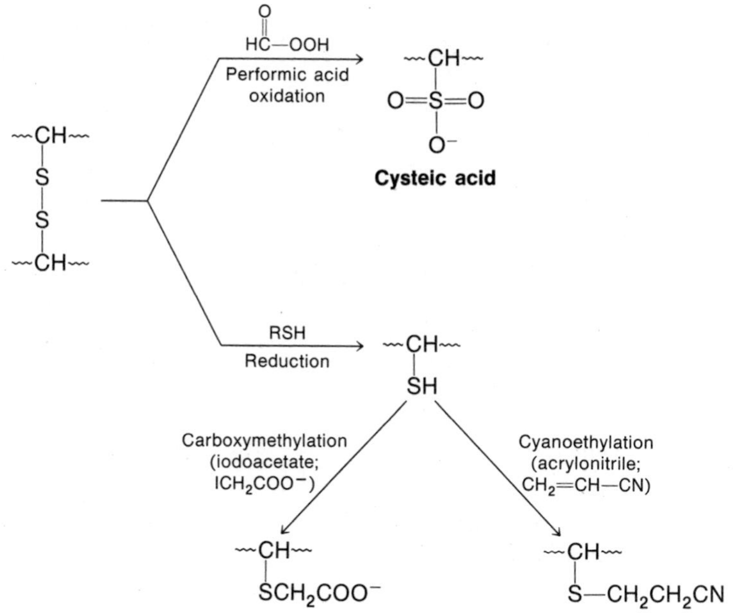

**Figure 2-6**
Disulfide cleavage reactions.

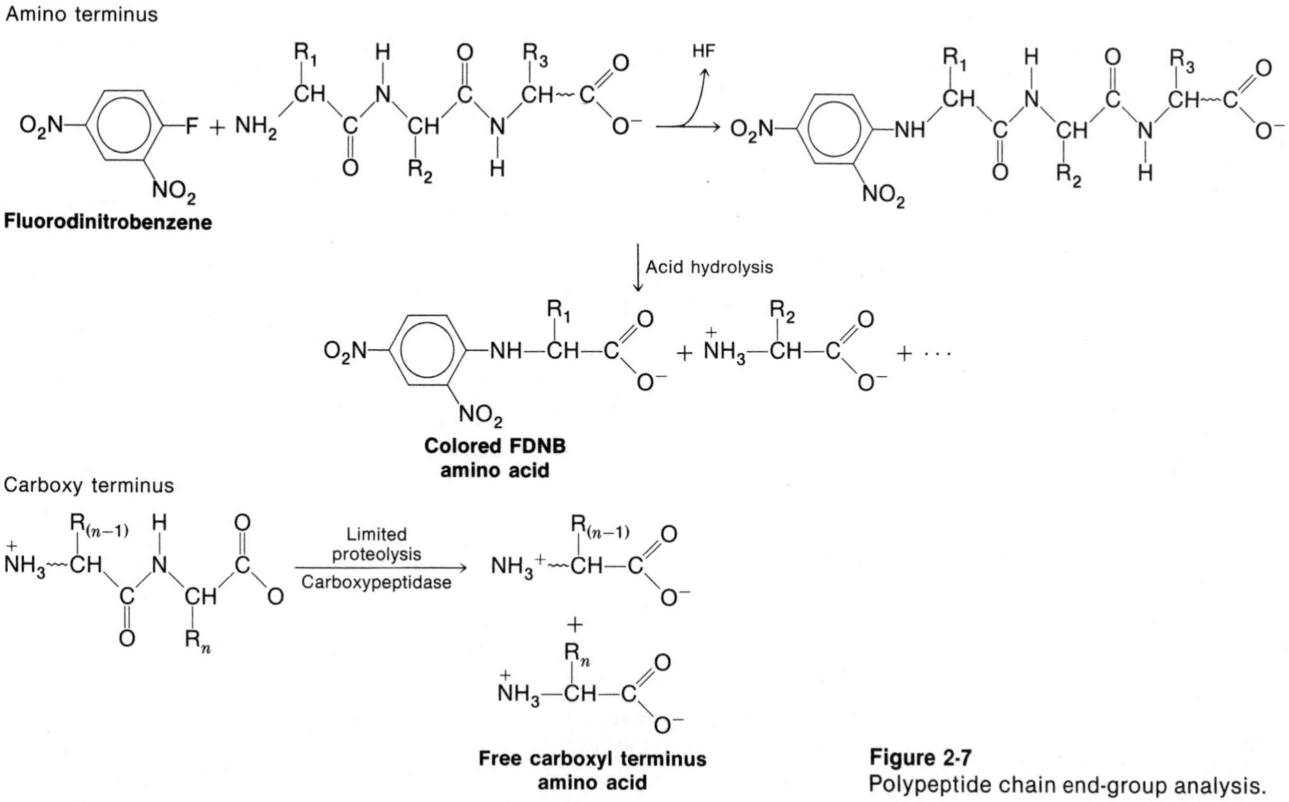

**Figure 2-7**
Polypeptide chain end-group analysis.

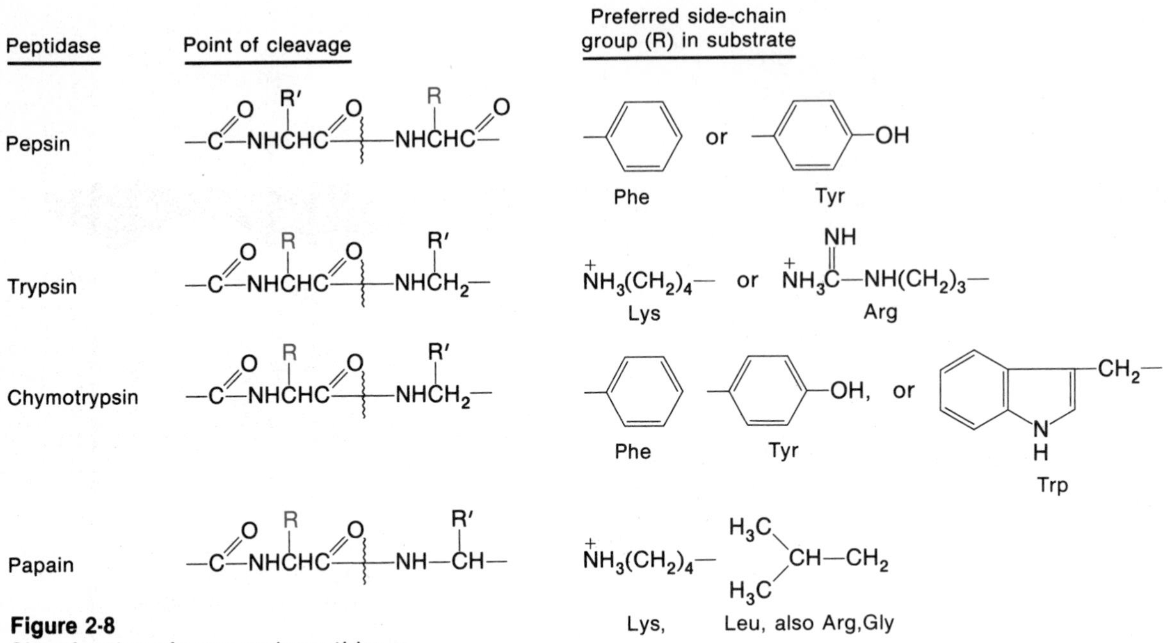

**Figure 2-8**
Site of action of some endopeptidases
used in sequence determination.

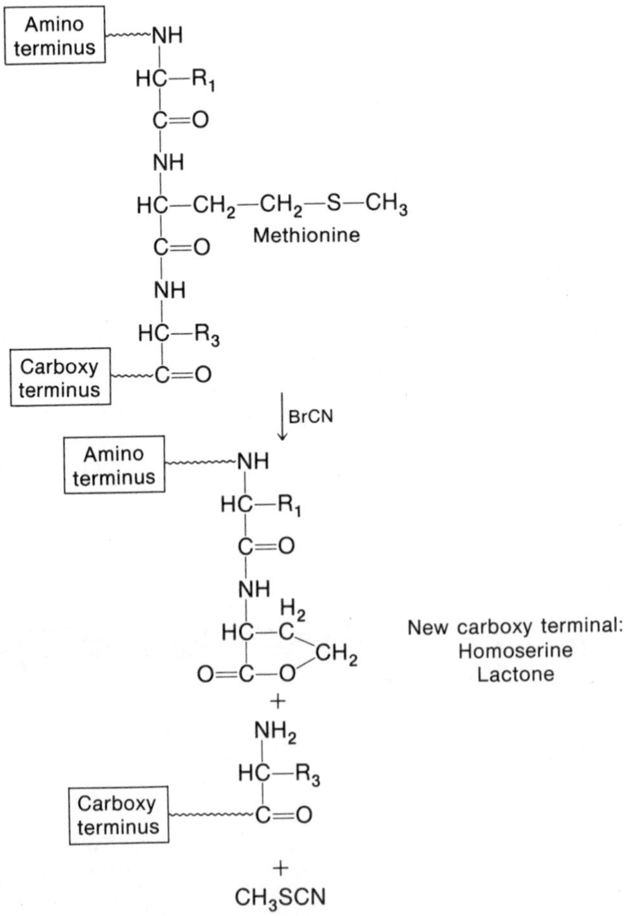

**Figure 2-9**
The cyanogen bromide reaction for
cleaving peptides at the methionine
residues.

The next step involves breaking the polypeptide chain down into shorter, well-defined fragments for subsequent sequence analysis. This can be achieved by the use of endopeptidases, which are enzymes that catalyze polypeptide-chain cleavage at specific sites in the protein. Figure 2-8, for example, shows the specificity of four enzymes commonly used for this purpose. A specific chemical method for polypeptide-chain cleavage involves reaction with cyanogen bromide. This reaction cleaves specifically at the methionine residues, with the accompanying conversion of free carboxy-terminal methionine to homoserine lactone (Figure 2-9). Although this methionine reaction product differs from the 20 naturally occurring amino acids, it is nevertheless readily identified by subsequent conversion to homoserine.

Peptides resulting from cleavage of the intact protein are generally separated by ion-exchange chromatographic methods. The isolated peptides may then be analyzed both to determine their amino acid composition and, by independent means, their sequence. Sequence determination involves the stepwise removal and identification of successive amino acids from the polypeptide amino terminal by means of the Edman degradation (Figure 2-10). This is carried out by reaction of the free amino terminal group with phenylisothiocyanate to form a peptidyl phenylthiocarbamoyl derivative. Gentle hydrolysis with hydrochloric acid releases the amino-terminal amino acid as the phenylthiohydantoin derivative. These in turn can be identified by their properties on thin-layer chromatography (Figure 2-11).

Devices called _sequenators_ are available that automate the Edman degradation procedure. The success of these automated procedures depends in

**Figure 2-10**
The Edman degradation method for $NH_2$-terminal residue determination. The residual peptide chain remains intact after removal of the $NH_2$-terminal residue.

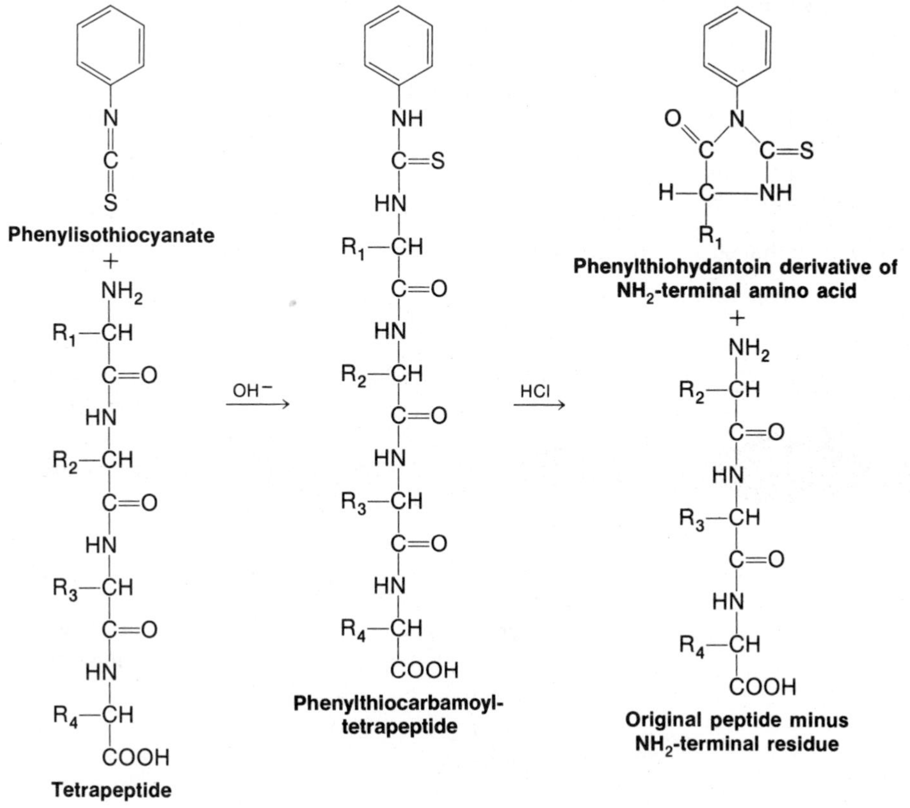

Phenylisothiocyanate + Tetrapeptide

$\xrightarrow{OH^-}$

Phenylthiocarbamoyl-tetrapeptide

$\xrightarrow{HCl}$

Phenylthiohydantoin derivative of $NH_2$-terminal amino acid + Original peptide minus $NH_2$-terminal residue

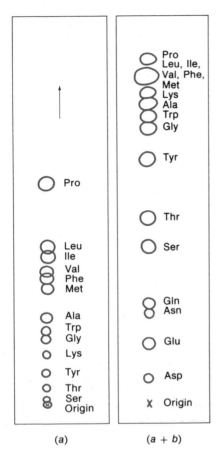

(a)

(a + b)

**Figure 2-11**
Chromatography of amino acid–phenylthiohydantoin derivatives in silica gel plates. (a) Separation is done in a 98 : 2 mixture of chloroform and ethanol. (b) This is followed by further separation using a 88 : 2 : 10 mixture of chloroform, ethanol, and methanol.

large part on the technical innovation of covalently linking the peptide to be sequenced through its carboxyl terminal group to glass beads. Attachment of the peptide being sequenced to this immobile phase facilitates the complete removal of potential contaminating reaction products during successive stages of the degradation reaction.

Having established the sequences of the individual peptides, it is next necessary to establish how they are connected together in the intact protein. It is this step which necessitates carrying out the sequence analysis on isolated peptides obtained by two different specific cleavage methods. This is sufficient to establish the overall sequence, since the two sets of sequences have _overlapping sequences_; i.e., the free amino and carboxy residues of peptides originally interconnected in the intact protein and liberated by one specific cleavage method will recur in the internal sequences of the peptides liberated by a second specific method.

Once having determined the protein's primary sequence, the location of disulfide bonds in the intact protein can be established by repeating a specific enzymatic cleavage on some protein whose disulfide bonds have not previously been cleaved. Separation of the resulting peptides will show the appearance of one new peptide and the disappearance of two other peptides when compared with the enzymatic digestion product of the material whose disulfide bonds have first been chemically cleaved. In fact, these difference techniques are generally useful in the detection of sites of mutations in protein molecules of previously known sequence, since a single substitution will generally affect the chromatographic properties of only a single peptide released during proteolytic digestion.

Finally, it should be mentioned that great progress has been made in recent years in procedures for sequencing the DNA encoding for proteins (Chapter 20). Knowing the coding triplets allows the amino acid sequence of the protein in question to be directly read off from the encoding DNA sequence. Nevertheless, such studies have produced the remarkable observation that some eukaryotic DNA sequences coding for proteins are not continuous, but instead contain untranslated _intervening_ DNA sequences. While these results have profound implications for protein evolution, they obviously confound the general applicability of DNA-sequencing methods for the purposes of protein primary-structure determination.

## CHEMICAL SYNTHESIS OF PEPTIDES

Investigation of the structure-function interrelationships in proteins and peptides has encouraged the development of techniques for the organo-chemical synthesis of peptides and proteins with a predetermined sequence. Chemical synthesis of peptides involves overcoming several problems that relate to preventing undesired groups from reacting. The amino and carboxyl groups that are not to be linked must be blocked, together with all reactive side chains. Some protecting groups for carboxyl and amino groups are shown in Figures 2-12 and 2-13, respectively. After coupling, one generally activates the carboxyl group; two methods for doing this are shown in Figure 2-14. It is of interest that carboxyl-group activation is also employed in biosynthesis. After peptide synthesis, the protecting groups must be removed by a mild method. The overall process comprising protection, activation, coupling, and unblocking is shown in Figure 2-15.

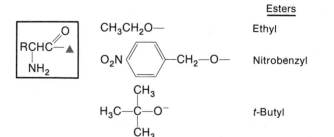

Esters

CH$_3$CH$_2$O—  Ethyl

O$_2$N ⟨benzene⟩ —CH$_2$—O—  Nitrobenzyl

H$_3$C—C(CH$_3$)(CH$_3$)—O$^-$  t-Butyl

**Figure 2-12**
Carboxyl protecting groups used in
peptide synthesis.

Carbobenzoxy  ⟨benzene⟩—CH$_2$OC—X ‖O

X = $\boxed{\text{RCHCOOH} \atop \text{NH—}}$

t-Butyloxycarbonyl  (CH$_3$)$_3$COC—X ‖O

Tosyl  CH$_3$—⟨benzene⟩—S(=O)(=O)—X

Trifluoroacetyl  F$_3$CC—X ‖O

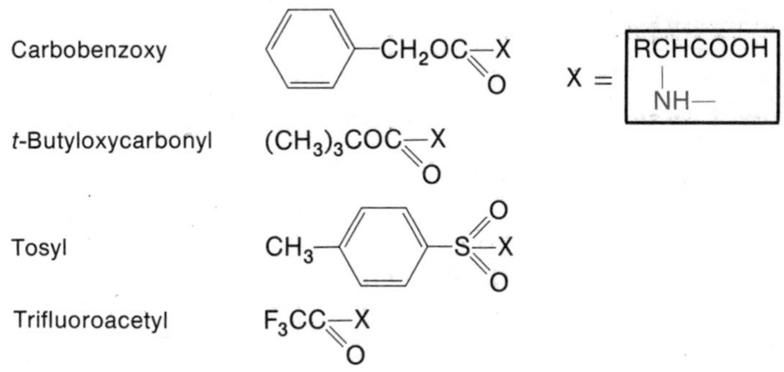

**Figure 2-13**
Amino protecting groups used in peptide synthesis.

Mixed
anhydride  RC(=O)—OH + Cl—C(=O)—OR' ⟶ RC(=O)—O—C(=O)—OR' + Cl$^-$

Carbodiimide  R'NHC(=O)NHR' + C$_6$H$_5$SO$_2$Cl ⟶ R'N=C=NR' + C$_6$H$_5$SO$_3$H

R'—N=C=N—R' + RCOOH ⟶ R'N=C—NHR' with RC(=O)—O

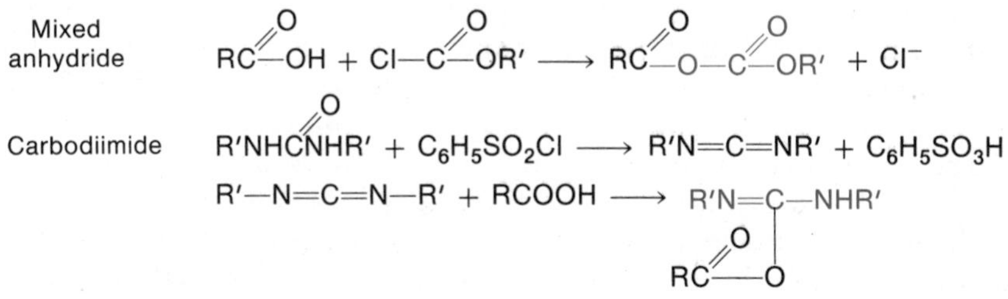

**Figure 2-14**
Different ways of activating the carboxyl group for peptide synthesis.

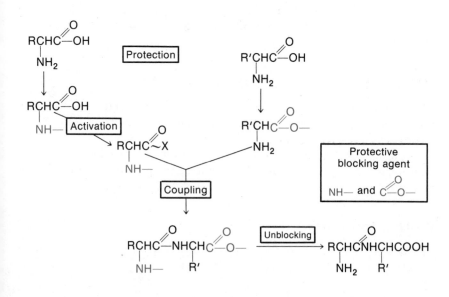

**Figure 2-15**
Schematic diagram illustrating the
chemical method for peptide
synthesis.

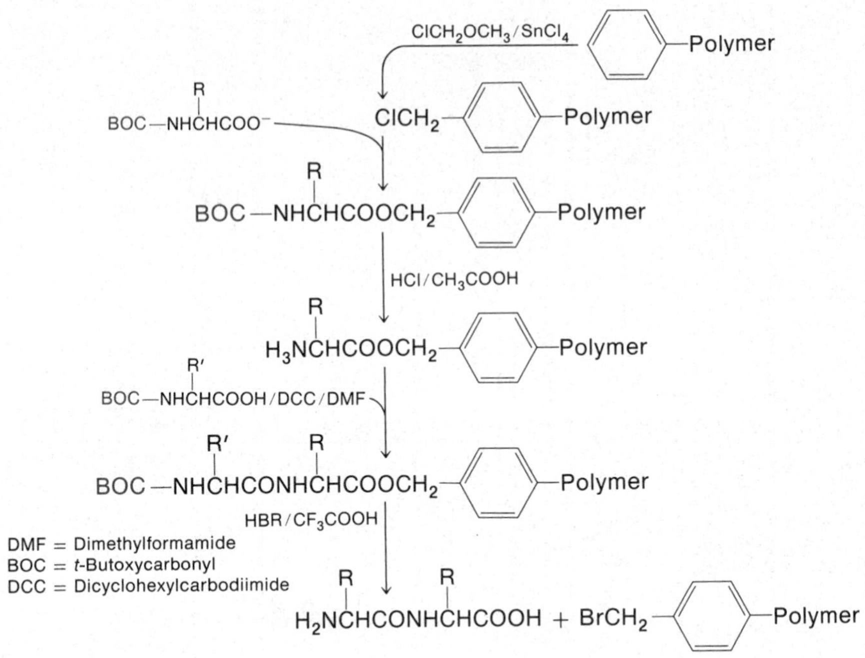

**Figure 2-16**
Merrifield procedure for solid-state dipeptide synthesis.

A variation of the usual methods of peptide synthesis involves attaching a protected ($t$-butoxycarbonyl group) amino acid to a solid polystyrene resin, removal of the amino protecting group, condensation with a second protected amino acid, and so on. In the last step, the finished peptide is cleaved from the resin. This method (outlined in Figure 2-16) has the advantage that cumbersome purification between steps, often resulting in serious losses, is replaced by mere washing of the insoluble resin. Since each reaction is essentially quantitative, very long peptides, and even proteins, can be made by this method. Indeed Li has synthesized the 39-amino acid peptide ACTH by this method, and Merrifield has synthesized bovine pancreatic ribonuclease. A number of variants of ribonuclease that contain one or more changes in amino acid sequence also have been made by this method.

## SOLUTION BEHAVIOR OF PROTEINS

The stabilization of a protein in its functionally active state reflects a delicate balance among a variety of different energetic contributions. These ultimately depend on interactions made both internally within the protein and between it and the surrounding solvent (see Chapter 3). Consequently, it might be anticipated that changing the protein's solvent or thermal environment could affect both its solubility and structure.

### Protein Solubility

Changes in protein solubility that do not destroy the molecule's inner structural integrity can occur in several ways.

**Isoelectric Precipitation.** Proteins typically have charged amino acid side chains on their surfaces that make energetically favorable polar interactions with the surrounding water. The total charge on the protein is just the sum of the charges contributed by the individual side chains. However, the actual charge on the weakly acidic and basic groups characterizing ionic amino acid side chains also depends on the solution pH. For each protein there will be a particular pH value where the sum of the positive side-chain charges exactly equals the sum of the negative charges, so that the net charge on the protein is zero. Since an uncharged molecule will not migrate in an imposed electric field, the pH value at which this occurs is termed the protein's _isoelectric point_ (abbreviated pI). As shown in Figure 2-17 for β-lactoglobulin, proteins tend to be least soluble at their isoelectric pH. This decrease in solubility reflects the fact that the individual protein molecules, which would all have similar charges at pHs away from their isoelectric points, cease to electrostatically repel each other at their isoelectric points.

**Salting In and Salting Out.** Proteins also show a variation in solubility depending on their solution ionic environments. These frequently complex effects may involve either specific interactions between charged side chains and solution ions or, particularly at high salt concentrations, reflect more comprehensive changes in the solvent properties. Figure 2-17 illustrates the effect of salt concentration on the solubility of β-lactoglobulin. Most globulins are sparingly soluble in pure water. The effect of salts such as sodium chloride on increasing the solubility of globulins is often referred to as _salting in_. The salting in effect is related to the nonspecific effect the salt has on increasing the _ionic strength_. The ionic strength $\mu$ is given by the expression

$$\mu = \frac{1}{2} \sum_i M_i Z_i^2$$

where $M$ is the molarity and $Z$ the charge of the ion. For a univalent salt at low concentration, $\mu$ is approximately equal to the molarity. Increasing the ionic strength decreases the "sphere of influence" of each charged site on the protein. The effective sphere of influence of a charge in solution is approximated by the Debye length $1/b$, where $b$ is calculated from the expression

$$b^2 = \frac{4\pi e^2}{\varepsilon kT} \sum_i M_i Z_i^2$$

where $e$ is the charge of the electron, $k$ is the Boltzmann constant, $T$ is the absolute temperature, and $\varepsilon$ is the dielectric constant of the medium (a measure of the charge-shielding effect of the medium).

The last term in this expression

$$\sum_i M_i Z_i^2$$

is twice the ionic strength $\mu$. Thus the Debye length $1/b \propto 1/\sqrt{\mu}$. In a 1 M aqueous solution of sodium chloride at 25°C, $1/b = 3.1$ Å. In a 0.01 M solution, $1/b = 31$ Å. As described earlier, many globular proteins precipi-

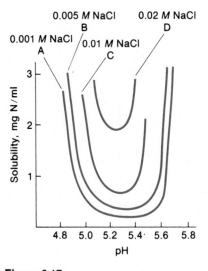

**Figure 2-17**
Solubility of β-lactoglobulin as a function of pH and ionic strength.

tate out at very low ionic strengths or in pure water. This happens because oppositely charged sites on different proteins are able to interact favorably, leading to an electrostatic complex. When this complex formation is extended between many protein molecules, it can lead to protein precipitation. Increasing the ionic strength tends to break up this complex because it decreases the Debye length.

Some salts, such as high concentrations of ammonium sulfate, have general effects on solvent structure that lead to decreased protein solubility and *salting out*. In this case, the protein molecules tend to associate with each other because protein-protein interactions become energetically more favorable than protein-solvent interactions. Proteins have characteristic salting out points, and these are used in protein separations in crude extracts.

### Protein Denaturation

Although protein denaturation also frequently results in loss of solubility, it represents the situation in which the protein's tertiary structure is destroyed. Protein denaturation can occur when there is a change in the solvent environment that is sufficiently large to upset the forces that keep the protein's structure intact.

**Thermal Denaturation.** As described in Chapter 1 (also see Chapters 3 and 7), stabilization of a protein's tertiary structure depends on a balance between the energy contributions arising from interatomic interactions (enthalpy) and its extent of disorder (entropy). Since native proteins are highly ordered structures, the entropic contributions to their stabilization are, to begin with, highly unfavorable. However, these are balanced by compensating enthalpic and entropic contributions arising from interactions both within the folded protein and with the surrounding water (see Chapters 3 and 7). When a protein is heated, this balance of forces is upset, resulting in the thermal unfolding of the protein's native structure. If the conditions are sufficiently drastic, the unfolded protein molecules will associate with each other randomly. This leads to irreversible thermal denaturation and insolubility. Specific proteins usually unfold over a characteristic narrow temperature range.

**Solvent Induced Denaturation.** Since the maintenance of a protein's structure depends in large part on interactions made with solvent, protein denaturation also can result from changes in solvent environment. This can happen in a variety of ways that correspond to disrupting one or more of the types of interactions stabilizing the protein's folded state. Extremes of solvent pH, for example, can upset the balance of a protein's charge interactions sufficiently to cause it to unfold. Similarly, changes in solvent polarity or hydrogen-bonding characteristics, caused by the addition of alcohols or strong hydrogen-bonding compounds such as urea, can cause proteins to unfold by providing alternative, more energetically favorable hydrogen-bonded interactions than those stabilizing the protein's native structure. Alternatively, detergents can be used to denature proteins. Typical detergent molecules are composed of long apolar, aliphatic hydrocarbon chains terminated by a strongly charged ionic end group. Molecules such as sodium dodecyl sulfate (SDS; see Figure 2-18) denature a protein by introduc-

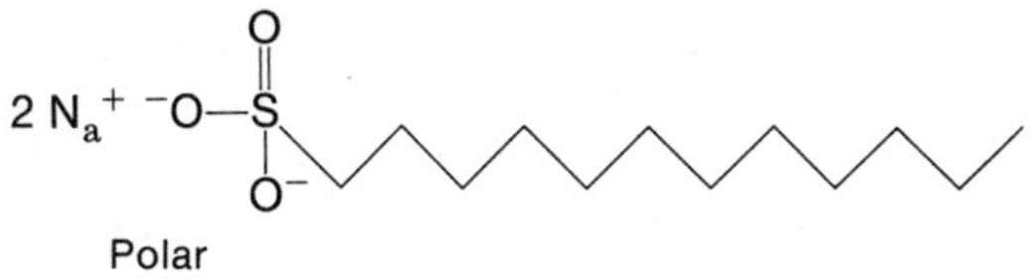

**Figure 2-18**
Sodium dodecyl sulfate.

ing their hydrophobic tails into the protein's interior, thereby disrupting the protein's structure.

Obviously, a protein's susceptibility to denaturation can cause problems during attempts to isolate and characterize new protein functions. However, denaturation frequently constitutes an initial step in sequence analysis and, as described below, forms the basis of a particularly useful technique for estimating protein molecular weights.

## PHYSICAL CHEMICAL SEPARATION AND CHARACTERIZATION TECHNIQUES

The success of any protein-isolation technique depends ultimately on exploiting the variations in physical or chemical properties among different protein molecules. Obviously, the same properties that make proteins different are also the ones that identify them uniquely. Frequently then, there is a direct correspondence between properties that uniquely identify proteins and allow them to be separated from each other. In fact, many protein-isolation schemes begin with steps that achieve initial purification on the basis of differences in protein solubility or resistance to denaturation, as described earlier. However, these techniques are usually relatively nonselective, so that alternative methods, usually combinations of different procedures, must be used to finally isolate and characterize a protein in a pure state.

### Methods of Separation and Characterization Based on Protein Charge

The surfaces of most globular proteins are populated by charged amino acid side chains. Although each ionizable side chain of a free amino acid has a well-defined p$K$ (Table 1-2), the specific environment of amino acid side chains on a protein's surface causes slight alterations in their p$K$ values. For example, a neutral histidine side chain that is partially buried at the protein's surface may have a decreased tendency to accept an additional proton, since the formation of the resulting positively charged species is less favorable in a partially apolar environment than in pure water. As a result, the p$K$ of the histidine can be lowered relative to its usual value in water. Similarly, interaction between ionic groups at the protein's surface can influence individual p$K$ values.

Owing to the sorts of environmental effects that perturb the p$K$s of protein amino acid side chains, even proteins that might on the basis of composition be expected to have identical formal charges are in fact slightly different in their charge properties. These differences form the basis for a variety of methods of protein isolation and characterization.

**Ion-Exchange Chromatography.** As described previously, amino acids can be separated according to their varying affinities for an ion-exchange resin immobilized in a column. Ion-exchange chromatography also can be used to separate proteins, but significant differences arise as a result of the large size of proteins. This means that different column procedures must be used. First of all, cross-linked resins are rarely used for protein separations because proteins are too large to penetrate the resin beads (however, see the discussion of high-performance liquid chromatography below). Instead, finely divided celluloses containing either positively or negatively charged groups are most commonly used to make such columns. Second, at a given salt concentration, protein binding tends to be an all-or-nothing phenomenon rather than an equilibrium phenomenon as with amino acids on resins. Consequently, the only way to achieve separations of proteins on charged cellulose columns is by changing the salt concentration. This is done in either a continuous (gradient elution) or a discontinuous (step elution) manner.

**Gel Electrophoresis.** An alternative technique used to separate proteins according to their differences in net charge is gel electrophoresis. The success of this method depends on the fact that proteins will migrate in an imposed electric field at rates proportional to their net charges (and inversely proportional to their effective size). Separations are typically carried out on either slabs or cylinders of an appropriate supporting medium, such as a gel composed of agarose or polyacrylamide. The percentage gel used is gauged according to the size of proteins being separated. Increasing the gel concentration lowers the mobility of the protein in a given electric field. For the finest separations, gradient gels are made with a continuous increase in gel percentage along the length of the slab. At the completion of a run, edge strips can be sliced from the gel and stained with a dye that reacts with proteins. This establishes the locations of the protein bands in the remainder of the gel, which can then be sliced apart so that individual proteins can be eluted (Figure 2-19). Gel electrophoresis is more frequently used as an analytical technique than as a preparative technique (e.g., see Figure 27-12).

A widely used analytical variation of polyacrylamide gel electrophoresis involves the use of a stacking gel. The basic idea behind this technique is to increase the resolution in the gel by creating a situation in which the protein mixture to be separated is first compressed into a very narrow start-

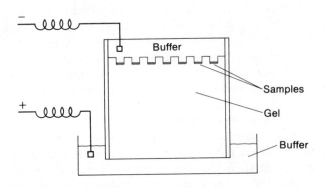

**Figure 2-19**
Apparatus for slab-gel electrophoresis.

ing zone. This is achieved by forming the gel in two layers. The upper, or "stacking," gel is generally less extensively polymerized than the lower, or "resolving," gel. More important, the pH of the solution in the stacking gel is adjusted so that the mobilities of proteins to be separated are higher in the stacking gel than in the resolving gel. As a consequence of this gel arrangement, a relatively dilute protein mixture introduced into the stacking gel becomes compressed into a narrow zone on entering the resolving gel. This happens because the proteins move rapidly and easily through the stacking gel, but accumulate at the interface with the resolving gel, owing to both its difference in pH and smaller pore size.

**Isoelectric Focusing.** Another method frequently useful in isolating or characterizing proteins on the basis of differences in their isoelectric points is isoelectric focusing. In this method, a solution of several charged organic molecules called _ampholytes_ is subjected to an electric field. After a period of time, the ampholyte molecules become continuously distributed in the electric field according to their relative charges. This, in turn, creates a continuous pH gradient from one end of the gel to the other. At this stage, a solution of proteins can be introduced into the gel. The proteins subsequently migrate in the electric field until they reach a point at which the pH resulting from the ampholyte gradient exactly equals their isoelectric points. Isoelectric focusing provides a way of both accurately determining a protein's isoelectric point and affecting separations among proteins whose isoelectric points may differ by as little as a few hundredths of a pH unit.

## Methods Based on Molecular Weight

Proteins also can be usefully separated and characterized on the basis of differences in their molecular weights.

**Gel-Exclusion Chromatography.** This widely used technique exploits the availability of both natural polysaccharide and synthetic polymers that can be formed into beads with varying pore sizes depending on the extent of interchain polymer cross-linking. For example, suppose you have some polysaccharide beads with average maximum pore sizes of 30 Å and you add these to a mixed solution of proteins of molecular weights 10,000 and 50,000. The final volume of the mixture equals the sum of the volume required to hydrate and fill the beads plus the remaining excess solution volume that fills the spaces between the beads. That is, the total solution volume $V_T = V_i + V_o$, where $V_i$ and $V_o$, respectively, represent the solution volumes entrained inside and outside the beads. The important point is that the lower-molecular-weight protein is sufficiently small that it can readily penetrate into the beads; as a result, it is uniformly distributed over the total volume $V_T$. The larger protein, by contrast, cannot penetrate the beads and so is concentrated in the solution volume $V_o$. By simply filtering out the beads, then, some separation could be achieved between the molecules based on their molecular weight. By repeating this process many times, a situation can eventually be arrived at where the concentration of the smaller protein in $V_o$ is vanishingly small. However, it is simpler and more efficient to construct a column of the beads and pass the protein solution through it in a continuous fashion. In this case, the smaller mole-

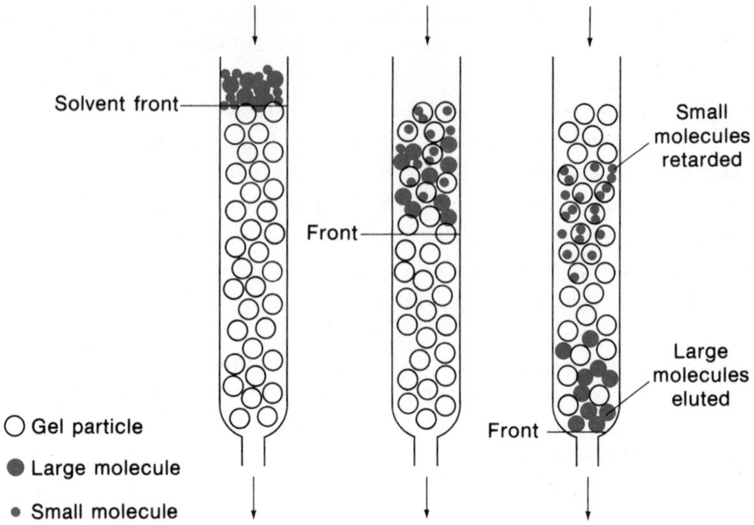

**Figure 2-20**
Sephadex column showing separation
of small and large molecules.

cule penetrates the entire column volume $V_T$, while the larger molecule, which is restricted to $V_o$, passes through the column at a more rapid rate, thus effecting a separation on the basis of molecular weight (Figure 2-20). Although this process has been described in terms of the molecules either penetrating or not penetrating the beads, it is clear that by careful selection of the bead pore size, we can create a situation in which the various molecules penetrate the beads to varying extents and, consequently, migrate through the column at varying rates. By observing the elution pattern of a mixture of proteins of known molecular weights, the behavior of the column can be standardized so that the molecular weight of an unknown protein can be approximately estimated (Figure 2-21).

**Osmotic Pressure.**  For proteins with molecular weights in the range of 10,000 to 100,000, osmotic-pressure techniques give a reasonable molecular-weight estimate. This method is rarely used, but it is of some theoretical interest (see Chapter 7). The general technique consists of putting the protein-containing solution on one side of a semipermeable membrane. The membrane is permeable to the water and any salts or buffer present, but not to the protein. After equilibration, a greater pressure will exist on the protein side of the membrane; this may be measured by the height of solution in a capillary tube (see Figure 2-22). This pressure, called the *osmotic pressure* and symbolized by $\pi$, is related to the concentration $C$ of protein (in grams per liter) by the van't Hoff law:

$$\lim_{C \to 0} \frac{\pi}{RTC} = \frac{1}{M_r}$$

where $R$ is the gas constant, $T$ is the absolute temperature, and $M_r$ is the molecular weight of the protein (see Chapter 7 for the derivation of this relationship). In practice, the osmotic pressure is measured over a range of concentrations; the values for $\pi/C$ are plotted against $C$, and the value of $\pi/C$ extrapolated to zero concentration is used to calculate the molecular weight.

**Ultracentrifugation.** Information concerning protein molecular weight also can be obtained by observing protein behavior in an intense centrifugal field. To get a qualitative understanding of how this method works, it is first necessary to recognize that protein molecules are generally slightly denser than water. However, the molecules in a protein solution seldom settle out in the earth's gravitational field (1g) since they are constantly being stirred up by collisions with surrounding solvent molecules. Nevertheless, protein molecules can be made to sediment if they are subjected to very high centrifugal force fields (~100,000g), such as can be attained by spinning protein solutions in the rotor of an ultracentrifuge (Figure 2-23).

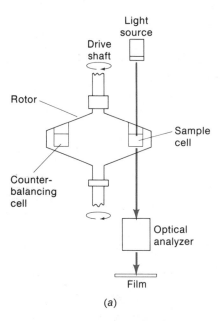

(a)

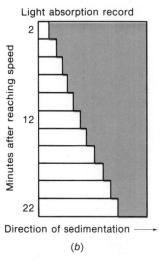

(b)

**Figure 2-21**
Plots of elution volume versus molecular weight for proteins on a Sephadex G-75 column. (After P. Andrews, Estimations of the molecular weights of proteins by Sephadex gel filtration, *Biochem. J.* 91:222, 1964.)

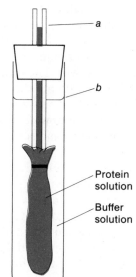

**Figure 2-22**
Apparatus for measuring osmotic pressure. Protein sample is placed inside a semipermeable membrane with buffer on both sides of the membrane. Osmotic pressure generated by the sample is proportional to the difference in height of the protein solution (a) and the buffer solution (b).

**Figure 2-23**
Apparatus for analytical ultracentrifugation. (a) The centrifuge rotor and method of making optical measurements. (b) The optical recordings as a function of time of centrifugation.

The protein molecules, which are denser than water, slowly migrate toward the bottom of the tube in the centrifuge. The rate of sedimentation may be recorded by optical methods that do not interfere with the operation of the centrifuge. From this rate, the *sedimentation constant*, or sedimentation coefficient, is evaluated. The sedimentation constant S (proportional to rate of sedimentation) is given by the equation

$$S = \frac{dx/dt}{\omega^2 x}$$

where $dx/dt$ is the rate at which the particle travels at distance x from the center of rotation, $\omega$ is the angular velocity of the rotor in radians per second, and $t$ is the time of centrifugation in seconds. The sedimentation constant is usually given in Svedberg units (S), where $1\,S = 10^{-13}\,s$. The molecular weight, in turn, can be calculated from the measured sedimentation constant, according to the formula

$$M_r = \frac{RTS}{D[1 - (\rho_S/\rho_P)]}$$

where $\rho_S$ is the density of the solvent, $\rho_P$ is the density of the protein, and $D$ is the diffusion constant. Obviously, the diffusion constant must be known before this relationship can be used to calculate the molecular weight. The diffusion constant is usually evaluated with the help of Fick's first law of diffusion:

$$\frac{dn}{dt} = -DA\left(\frac{dC}{dx}\right)_t$$

This equation states that the amount $dn$ of a substance crossing a given area $A$ in time $dt$ is proportional to the concentration gradient $dC/dx$ across that area. The diffusion constant $D$, which is related to molecule size and shape, is the proportionality constant. The diffusion constant can be measured by observing the spread of an initially sharp boundary between the protein solution and a solvent as the protein diffuses into the solvent layer. Once this has been evaluated, the information can be combined with the sedimentation data for purposes of molecular-weight calculation according to the preceding equation (see Table 2-2).

Sometimes the technique of equilibrium sedimentation is used to measure molecular weight. The ultracentrifuge is maintained at a constant speed until the distribution of the protein molecules becomes constant. At this time, the downward movement owing to sedimentation is exactly counterbalanced by the upward movement owing to diffusion. This is a most rigorous method for molecular-weight determination because it is an equilibrium method.

Frequently, sedimentation analysis is carried out in gradients, usually linear gradients of sucrose or glycerol. First, such a gradient is made in the centrifuge tube with the greatest density at the bottom of the tube. A small volume of the protein solution is layered on the top of the tube, and the tube is spun for the desired length of time. The protein will travel down the tube at a rate roughly proportional to its sedimentation constant. This arrangement has the advantage that the gradient stabilizes the solution so that the centrifuge may be stopped at any time without disturbing the distribution. The solution can be carefully removed from the tube by

**Table 2-2**
Physical Constants of Some Proteins

| Protein | Molecular Weight | Diffusion Constant ($D \times 10^7$) | Sedimentation Constant (S) | pI (Isoelectric) |
|---|---|---|---|---|
| Cytochrome $c$ (bovine heart) | 13,370 | 11.4 | 1.17 | 10.6 |
| Myoglobin (horse heart) | 16,900 | 11.3 | 2.04 | 7.0 |
| Chymotrypsinogen (bovine pancreas) | 23,240 | 9.5 | 2.54 | |
| β-Lactoglobulin (goat milk) | 37,100 | 7.5 | 2.9 | |
| Serum albumin (human) | 68,500 | 6.1 | 4.6 | 4.8 |
| Hemoglobin (human) | 64,500 | 6.9 | 4.5 | 6.9 |
| Catalase (horse liver) | 247,500 | 4.1 | 11.3 | 5.6 |
| Urease (jack bean) | 482,700 | 3.46 | 18.6 | 5.1 |
| Fibrinogen (human) | 339,700 | 1.98 | 7.6 | 5.5 |
| Myosin (cod) | 524,800 | 1.10 | 6.4 | |
| Tobacco mosaic virus | 40,590,000 | 0.46 | 198 | |

punching a hole in the bottom and collecting fractions in separate tubes. This technique is used for both sedimentation-constant estimation and preparative fractionation.

Frequently, only the approximate sedimentation constant is measured and the molecular weight is estimated using a protein "marker" of known sedimentation constant in the same tube. For a large number of protein molecules, the sedimentation constant is roughly proportional to the two-thirds power of the molecular weight, giving the relation

$$\frac{S(\text{of unknown})}{S(\text{of standard})} = \left[\frac{M_r(\text{of unknown})}{M_r(\text{of standard})}\right]^{2/3}$$

This is not a rigorous relationship, and the approximation is best for molecules that are spherical. Most globular proteins give a reasonable fit, and most fibrous proteins, which are highly asymmetrical in shape, give a poor fit.

**Sodium Dodecyl Sulfate (SDS) Gel Electrophoresis.** Another method widely used for protein molecular-weight determination is SDS gel electrophoresis. The same types of polyacrylamide gels as described earlier are used. In this method, the mixture of proteins to be characterized is first completely denatured by the addition of the detergent sodium dodecyl sulfate and mercaptoethanol and a brief heating step. Denaturation of the proteins by SDS is caused by the hydrophobic association of the apolar aliphatic tails of the SDS molecules with protein hydrophobic groups as explained earlier. Any cystine disulfide bonds are cleaved by a disulfide interchange reaction with mercaptoethanol. The resulting unfolded polypeptide chains consequently have relatively large numbers of SDS molecules bound to them. The success of the technique depends on the facts that (1) each bound SDS molecule contributes two negative charges to the denatured protein complex, so that the charge of the protein in its native state is effectively masked by the more numerous charged groups of the associated detergent molecules; and (2) the total number of detergent molecules bound is proportional to the polypeptide-chain length or, what is equivalent, the protein's molecular weight. As a result, the SDS-denatured protein molecules acquire net negative charges that are approximately proportional to their molecular weights and so will migrate in an imposed electric field at rates also approximately proportional to their molecular weights. Typically, the behavior of an SDS gel electrophoresis system is calibrated by concurrently running standards of known molecular weight and then comparing the migration behavior of the unknowns with the standards. Although SDS gel electrophoresis gives only an approximate measure of protein molecular weight, depending on experimental conditions, its experimental simplicity and high resolving power have led to its widespread use for the characterization of a wide variety of protein mixtures. In particular, it is applicable to the study of membrane-protein components, whose normal insolubility makes them difficult to handle in the absence of detergent.

## Methods of Separation Based on Specific Protein Interactions

**Affinity Chromatography.** In addition to the previously described techniques for the isolation and characterization of proteins, there are several more recently developed methods that exploit specific binding properties that may be characteristic of a given protein. Many enzymes, for example, reversibly bind organic cofactor molecules such as adenosine triphosphate or pyridine nucleotides in order to catalyze chemical reactions of their substrates. In many cases, the separation of such enzymes from contaminating proteins can be readily achieved by preparing a chromatographic column whose resin is first chemically reacted with a suitable cofactor derivative. On passage of the mixture over the resin having the cofactor molecule covalently bound to it, those molecules having specific binding affinities for the cofactor bond to the column, while the others pass through. The cofactor binding protein can subsequently be washed from the column by elution with a solution of soluble cofactor in a manner similar to that used to elute proteins from ion-exchange columns. The power of these affinity chromatography techniques lies in the endless variety of specific interactions that characterize the functional properties of protein molecules. For example, proteins with chemically attached sugar groups (glycoproteins) can frequently be separated on columns whose resin has first been covalently reacted with a polysaccharide binding protein such as concanavalin A. In this case, the bound glycoproteins can be washed from the column with an appropriate sugar solution.

**High-Performance Liquid Chromatography.** Thus far we have described three types of column chromatography for purification and characterization of proteins: gel-exclusion chromatography, ion-exchange chromatography, and affinity chromatography. High-performance liquid chromatography (HPLC) is not so much a new type of chromatography as a new way of looking at old chromatographic techniques. The same principles are involved, but the column materials used usually consist of more finely divided particles made of physically stronger materials that can withstand high pressures (5000 to 10,000 p.s.i.) without changing their structure. The column apparatus itself also must be designed to withstand high pressures. Finely divided column materials lead to slower flow rates, but this can be more than compensated for by applying high hydrostatic pressures. As a rule, much better separations are achieved in much shorter time with the proper applications of HPLC. The approach is producing a revolution in the techniques used in protein purification.

## Techniques that Measure the Absorption or Emission of Radiation

To understand the value of radiation techniques in studying the properties of proteins, it is essential to appreciate the nature of radiation and how it interacts with matter. These general points are examined in greater depth in Chapters 11 and 31.

Radiation is an electromagnetic field that oscillates sinusoidally in space and time. It interacts with matter in discrete packets called photons that contain a definite amount of energy. The relationship between the energy $E$ of a photon and the frequency $\nu$ of the oscillating field is given by the expression

$$E = h\nu$$

where $h$ is Planck's constant ($6.67 \times 10^{-27}$ erg $\cdot$ s). The frequency $\nu$ is the number of oscillations per second at a given point in space. It is related to the wavelength $\lambda$ of the oscillation, the distance in space between successive peaks in the amplitude of the field, and to the velocity $V$ at which the peaks move through space:

$$\nu = \frac{V}{\lambda}$$

Comparison of the two previous equations reveals that the energy of a photon is reciprocally related to the wavelength of the radiation.

Whereas radiation is never used as a tool for protein separations, a plethora of methods are available in which the interaction of radiation with proteins permits characterization of specific molecular properties. These may be classified into three main categories according to the measured property of the radiation that emerges from the exposed sample:

1. Fraction of incident radiation absorbed or scattered by the sample.
   a. Light microscopy
   b. X-ray diffraction
   c. Nuclear magnetic resonance spectrometry (NMR)
   d. Electron paramagnetic resonance spectrometry (EPR)
   e. Optical absorption spectroscopy
   f. Infrared absorption spectroscopy

2. Radiation emitted by protein at wavelengths other than those used for excitation is measured.
   a. Fluorescence
   b. Raman scattering

**Figure 2-24**
Ultraviolet absorption spectra of tryptophan, tyrosine, and phenylalanine at pH 8. (After J. S. Fruton and S. Simmonds, *General Biochemistry*, 2d Ed., Wiley, New York, 1958.)

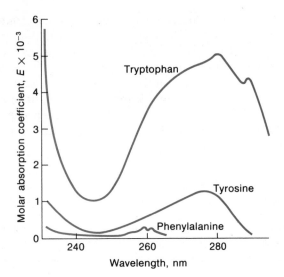

**3.** Effect of protein on the polarization of emerging radiation.
  a. Optical rotatory dispersion (ORD)
  b. Circular dichroism (CD)
  c. Fluorescence polarization

We will not explain all these methods and their applications to protein characterization (see Selected Readings). However, many applications of most of these methods are either described or alluded to in various chapters of this text. In this chapter we shall limit ourselves to a description of ultraviolet (UV) absorption spectroscopy and x-ray diffraction—two of the bread-and-butter techniques of modern biochemistry that illustrate entirely different uses of radiation.

**UV Absorption Spectroscopy.** The aromatic amino acids phenylalanine, tyrosine, and typtophan all possess absorption maxima in the near-ultraviolet range (see Figure 2-24). These absorption bands arise from the interaction of radiation with $\pi$ electrons in the aromatic rings. The near-ultraviolet absorption properties of proteins are determined solely by their content of aromatic amino acids. In dilute solutions, this absorption can be quantified and used as a measure of the concentration of the protein with the help of a conventional spectrophotometer. The general quantitative relationship that governs all absorption processes is called the Beer-Lambert law:

$$I = I_0 10^{-\varepsilon cd}$$

where $I_0$ is the intensity of the incident radiation, $I$ the intensity of the radiation transmitted through a cell of thickness $d$ (in centimeters) that contains a solution of concentration $c$ (expressed either in moles per liter or in grams per 100 ml), and $\varepsilon$ is the extinction coefficient, a characteristic of the substance being investigated. The spectrophotometer usually is capable of recording absorbance $A$ directly, which is related to $I$ and $I_0$ by the equation

$$A = \log_{10}(I_0/I)$$

Hence $A = \varepsilon cd$ and $A$ is a direct measure of concentration. It can be seen from Figure 2-24 that the $\varepsilon$ values are largest for tryptophan and smallest for phenylalanine. Since protein absorption maxima in the near-ultraviolet are determined by the content of the aromatic amino acids and their respective $\varepsilon$ values, most proteins have absorption maxima in the 280-nm region.

**Protein Tertiary-Structure Determination X-Ray Crystallography.**
Much of the fundamental understanding gained in recent years concerning the catalytic and functional properties of proteins has derived from a detailed knowledge of their tertiary structures. Given the enormous impact of this structural information on modern biochemistry, it is appropriate to present a qualitative description of the techniques involved in the determination of a protein's tertiary structure.

The interest here is in obtaining a three-dimensional image of a protein molecule in its native state at a sufficient level of detail to individually locate its constituent atoms. The way this is done can most easily be appreciated by considering the more familiar problem of how we obtain a magnified image of an object in a conventional light microscope. In a visible-light microscope, light from a point source is projected on the object we wish to examine. When the light waves hit the object, they are scattered from it so that each small part of the object essentially serves as a new source of light waves. The important point is that the light waves scattered from the object contain all the information about its structure. For this reason, the scattered waves can be collected and recombined by a lens to produce a magnified image of the object (Figure 2-25).

Given this picture, we might then ask what prevents us from simply putting a protein molecule in place of our object and viewing its magnified image? The basic problem here is one of resolution. The resolution, or extent of detail, that can be recovered from any imaging system depends on the wavelength of light incident upon the object. Specifically, the resolu-

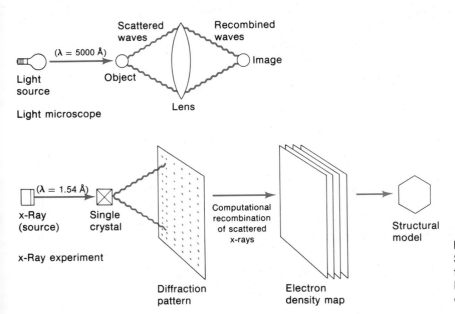

**Figure 2-25**
Schematic diagram of the procedures followed for image reconstruction in light microscopy (*above*) and x-ray crystallography (*below*).

tion obtainable equals $\lambda/2$, or one-half the wavelength of the incident light. Since $\lambda$ lies in the range of 4000 to 7000 Å for visible light, a visible-light microscope clearly does not have the resolving power to distinguish the atomic structural detail of molecules. What is needed is a form of incident radiation with a wavelength comparable to interatomic distances. X-rays emitted from excited metal atoms that have wavelengths in the range of one to a few angstroms would be most suitable. However, simply replacing a visible-light source with an x-ray source does not solve all the problems. For example, to get a three-dimensional view of a protein, it is clear that some provision must be made for looking at it from all possible angles, an obvious impossibility when dealing with a single molecule. Furthermore, when x-rays interact with proteins, only a small fraction of the rays are scattered. While most x-rays pass through the protein, a relatively large fraction interacts destructively with the protein, so that a single molecule would be destroyed before scattering enough x-rays to form a useful image. Both these problems are overcome by replacing a single protein molecule with an ordered three-dimensional array of molecules that scatters x-rays essentially as if they were one molecule. This three-dimensionally ordered array of protein molecules is a single crystal of protein molecules, so the general technique is called protein x-ray crystallography.

The problems do not end here, because although the protein crystal will readily scatter incident x-rays, there are no lens materials available that can recombine the scattered x-rays to produce an image. Instead, the best that can be done is to directly collect the scattered x-rays in the form of a diffraction pattern. Although recording the diffraction pattern results in loss of some important information, experimental techniques have been developed for recovering this lost information. This, in turn, eventually allows the scattered waves to be mathematically recombined in a computational analog of a lens. By collecting the diffraction pattern of the crystal in many orientations, it is then possible to reconstruct a three-dimensional image of the protein molecule.

Crystals suitable for protein x-ray studies may be grown by a variety of techniques that generally depend on solvent perturbation methods for rendering proteins insoluble in a structurally intact state. The trick is to create a situation in which the molecules associate with each other in a specific fashion to produce a three-dimensionally ordered array. A typical protein crystal useful for diffraction work is about 0.5 mm on a side and contains about $10^{12}$ protein molecules (an array $10^4$ molecules long along each crystal edge). It is important to recognize that protein crystals are from 20 to 70 percent solvent by volume. A crystalline protein consequently is in an environment that is not substantially different from free solution.

The x-ray irradiation usually employed for protein crystallographic studies is derived from the bombardment of a copper target with high-voltage ($\sim$50 kV) electrons, which produces a characteristic copper x-ray with $\lambda = 1.54$ Å. A schematic illustration of an x-ray diffraction pattern from a protein crystal is shown in Figure 2-26. There are several features about this pattern that bear explanation. First, it is apparent that the diffraction pattern consists of a regular lattice of spots of different intensities. *Diffraction from a crystal produces a series of spots owing to destructive interference of waves scattered from the repeating unit of the crystal.* For

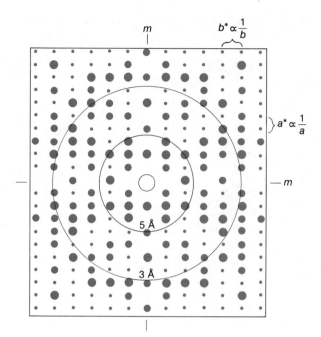

**Figure 2-26**
A schematic view of an x-ray diffraction pattern. The spacing of the spots is reciprocally related to the dimensions of the repeating unit cell of the crystal. The symmetry of the spots (e.g., the mirror planes in the sample shown) and the pattern of missing spots (alternating spots along the mirror axes) give information on how molecules are arranged in the unit cell. Information concerning the structure of the molecule is contained in the intensities of the spots. Spots closest to the center of the film arise from larger-scale or low-resolution structural features of the molecule, while those farther out correspond to progressively more detailed features.

the crystal whose diffraction pattern is shown, the repeating unit (or crystal unit cell) contains four symmetrically arranged protein molecules. This results in the appearance of corresponding symmetrical features in the spot-intensity pattern of the crystal's diffraction pattern. Further, _the lattice spacing of the diffraction spots is inversely proportional to the actual dimensions of the crystal's repeating unit or unit cell._ Consequently, both the crystal's unit-cell dimensions and general molecular packing arrangement can be derived from inspection of the crystal's diffraction pattern.

_Information concerning the detailed structural features of the protein is contained in the intensities of the diffraction spots._ A three-dimensional structural determination requires that all these spots be measured either by scanning x-ray films with a densitometer or by measuring the diffraction spots individually with a scintillation counter. In this connection, it is important to realize that all the atoms in the protein structure make individual contributions to the intensities of each diffraction spot. Conversely, all the intensity data have to be collected to reconstruct the protein's structure.

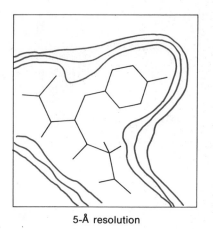

5-Å resolution

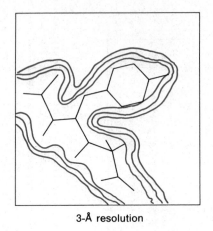

3-Å resolution

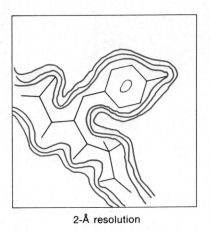

2-Å resolution

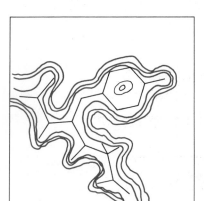

1.5-Å resolution

**Figure 2-27**
Views of crystallographic electron-density maps showing how the structural detail revealed depends on the resolution of the data used to compute the maps.

Initial studies of a protein's tertiary structure are generally carried out at low resolution. This corresponds to reconstructing the protein's structure using intensity data near the origin (center) of the diffraction pattern. Diffraction data near the origin reflect long-range structural features of the molecule, while those nearer the edge correspond to progressively more detailed features. Figure 2-27 shows some examples of electron-density maps calculated at different resolutions, and these illustrate how various levels of structural detail appear at different degrees of resolutions.

A powerful aspect of protein crystallography is that once the native structure is known, various cofactors or enzyme substrate analogs can be bound to the molecule in the crystal. By simply remeasuring the diffraction intensities, a new map can then be computed that allows direct and explicit examination of the structural interactions between the native protein and its substrate or cofactor molecules. The ability to analyze these interactions in detail has provided much of the foundation for our current understanding of a multitude of protein catalytic and functional properties.

## SELECTED READINGS

Ackers, G. K.  Analytical gel chromatography of protein. *Adv. Protein Chem.* 24:343, 1970.

Cantor, C. R., and Schimmel, P. R.  *Biophysical Chemistry*. San Francisco: Freeman, 1980; especially volume 2, entitled *Technique for Study of Biophysical Structure and Function*.

*Methods in Enzymology*. New York: Academic Press. A continuing series of over 50 volumes that discusses most methods at a level of detail suitable for laboratory application.

Regnier, F. E., and Gooding, K. M.  High-performance liquid chromatography of proteins. *Anal. Biochem.* 103:1, 1980.

Tristram, G. R., and Smith, R. H.  Amino acid composition of proteins. *Adv. Protein Chem.* 18:227, 1963.

## PROBLEMS

1. An organism of unknown origin produces a potent inhibitor of nerve conduction which you wish to sequence. Amino acid analysis shows the peptide's composition to be $Ala_5 Lys_1 Phe_1$. Reaction of the intact peptide with FDNB followed by acid hydrolysis liberates free FDNB alanine. Trypsin cleavage gives a tripeptide and a tetrapeptide, with compositions $Ala_3 Phe_1$ and $Lys_1, Ala_2$. Reaction of the intact peptide with chymotrypsin yields a hexapeptide plus free alanine. What is the inhibitor's sequence?

2. You have isolated an octapeptide from a rare fungus that prevents baldness, and you are interested in its commercial possibilities. Analysis gives the composition $Lys_2, Asp_1, Tyr_1, Phe_1, Gly_1, Ser_1, Ala_1$. Reaction of the intact peptide with FDNB followed by acid hydrolysis liberates FDNB-alanine. Cleavage with trypsin gives peptides with compositions $Lys_1, Ala_1, Ser_1$ and $Gly_1, Phe_1, Lys_1$, plus a dipeptide. Reaction with chymotrypsin liberates free aspartic acid, a tetrapeptide with composition $Lys_1, Ser_1, Phe_1, Ala_1$ and a tripeptide which liberates FDNB-glycine on reaction with FDNB followed by acid hydrolysis. What is the sequence?

3. A South American beetle produces a substance that extracts gold from seawater and you desire to know its structure in order to carry out its large-scale synthesis. Analysis gives a composition $Lys_1, Pro_1, Arg_1, Phe_1, Ala_1, Tyr_1, Ser_1$. Reaction with FDNB gives no products, unless the material is first reacted with chymotrypsin, in which case both DNP serine and DNP lysine are formed from peptides with compositions $Ala_1, Tyr_1, Ser_1$, and $Pro_1, Phe_1, Lys_1, Arg_1$. Reaction with trypsin also produces two peptides, with compositions Arg, Pro and Phe, Tyr, Lys, Ser, Ala. What is the structure of this peptide?

4. An enzyme exhibits a maximum in catalytic rate at pH 6.5. Amino acid analysis of the native protein shows that it contains two histidine residues. Some protein inactivated by chemical modification gives two products separable by chromatography. Both products, when analyzed, show the loss of one histidine. What conclusions do you reach concerning the active site of this enzyme?

5. You have isolated a potent toxin from a root found in a remote area of Arizona. In an initial SDS gel electrophoresis run, the protein runs midway between the marker proteins myoglobin and $\beta$-lactoglobulin. However, material treated first with mercaptoethanol and iodoacetate gives a single broad band on SDS gel electrophoresis which runs close to a cytochrome c marker. Further experiments show that reaction of the native material with FDNB followed by acid hydrolysis liberates DNP-glycine and DND-

tyrosine. What conclusions can you derive about the structure of this protein?

6. In carrying out some experiments on DNA-replication, you have isolated a protein complex that sediments in the ultracentrifuge similarly to a hemoglobin marker. However, when the same complex is run in the presence of 2 $M$ NaCl, other conditions being equivalent, the sedimentation behavior resembles that of a myoglobin marker. What conclusions do you reach concerning the properties of this complex?

7. A mutant form of alkaline phosphatase is run in an isoelectric focusing experiment and found to differ from the native material by an amount corresponding to one additional positive charge relative to the native material. Assuming the mutant's properties to reflect the consequences of a single amino acid substitution, what are the possibilities?

8. Osmotic pressure experiments on a protein in the presence of the membrane permeable cation $Ca^{2+}$ give plots of $\pi/C$ versus $C$ whose extrapolated intercepts on the $\pi/C$ axis differs by a fraction of 6. What do you conclude about this protein?

9. You are working in a laboratory that is investigating the functional properties of mutant forms of ribonuclease. An SDS gel run using previously purified ribonucleases as a marker shows an initial salt-precipitated fraction to be contaminated with two additional proteins. One contaminant has a $M_r$ of about 13,000 (similar to ribonuclease) and a pI 4 pH units more acidic than that of ribonuclease. The second contaminant is a large protein of $M_r$ 89,000. Describe an efficient purification protocol for your mutant ribonuclease.

10. You have a mixture of proteins with the following properties:

    a. $M_r = 12,000$, pI $= 10$
    b. $M_r = 62,000$, pI $= 4$
    c. $M_r = 28,000$, pI $= 7$
    d. $M_r = 9,000$, pI $= 5$.

    Other factors aside, what order of emergence would you expect from these proteins when run on (a) an anion exchange resin such as DEAE-cellulose with linear salt gradient elution and (b) a Sephadex G-50 gel exclusion column?

11. You are interested in rapidly effecting high purification of an enzyme that binds ATP from a crude extract containing several contaminating proteins. Since you are interested in large-scale purifications, it is worthwhile to consider some sophisticated strategies. Which would you consider?

# 3

# ANATOMY OF PROTEIN STRUCTURE

Before the first x-ray diffraction results were understood, it was imagined that protein structures were relatively simple geometric arrangements of polypeptide chains, such as geometric cages, repeating zigzags, or uniform arrays of parallel rods. Indeed the first structures determined for fibrous proteins were of this type, namely, the $\alpha$ helix and the $\beta$ sheet (Chapter 1). These structures were deduced by Pauling and Corey using information from various sources: (1) the structures of peptides (from small-molecule crystallography), which indicated a planarity of the peptide bond and gave accurate bond lengths and angles, (2) the importance of H bonds in determining the orientation of amino acids, peptides, and even water in simple crystals, (3) the interpretation of a few spacings in the diffraction patterns of certain fibrous proteins, and finally, (4) putting all this information together with molecular models to produce structures in reasonable agreement with all the available facts. This was a historic achievement. It is not surprising that such repeating structures with long-range order were the first protein structures to be understood. The demands on the available technology were minimal. Much more sophisticated technology was required to interpret the diffraction patterns of most proteins, which have less long-range repetition. Even Kendrew and Perutz, who led their research teams to a solution of the first structures to be determined, myoglobin and hemoglobin, were shocked by the gradual realization of the seemingly chaotic arrangements in such structures. Kendrew was once introduced at Harvard in the early 1960s as the man who proved that proteins were ugly, and Perutz, addressing his initial disappointment, said, "Could the search for ultimate truth really have revealed so hideous and visceral-looking an object?" Whether Kendrew and Perutz were truly as disappointed as they

*Electron micrograph showing a striated muscle sarcomere in longitudinal section. (Micrograph by H. E. Huxley.)*

appeared to be, or whether they were secretly delighted that their efforts of more than a quarter century had led them to structures that were so complex that no person could have predicted them is for science historians to determine. Suffice it to say that this crowning achievement of protein structure determination by x-ray diffraction has been repeated many times since with less ado. The enormous advances in protein chemistry and computer technology have systematized the necessary research and greatly decreased the amount of work and time required to determine a protein structure. Accurate structure determinations have been made for close to 200 different proteins. Although it now appears that detailed three-dimensional structures of proteins also may be determined either by a combination of electron diffraction and low-dose imaging in the electron microscope or by high-resolution two-dimensional nuclear magnetic resonance spectroscopy, essentially all the information summarized here comes from the results of x-ray crystallography. This wealth of data has been carefully examined and *patterns of structure are becoming apparent which, among other things, suggest that the final folding arrangements of proteins might some day be predictable from the amino acid sequences of the polypeptide chains.*

The first chapter of this book presented an overview of a wide variety of protein structures and functions. In this chapter a more rigorous and detailed consideration will be given to both the nature of the forces that cause proteins to fold and how these forces, in concert with what are basically geometric properties of the polypeptide chain, result in the highly organized structures that characterize functionally active proteins. This problem will be approached from the standpoint of an architect. That is, first the basic properties of the structural material of proteins will be examined to see what sorts of local structural arrangements are possible for polypeptide chains. Subsequently the way in which these locally organized structural units can be most efficiently assembled into progressively larger and more complicated arrangements will be described. Frequent excursions will be made to consider the relationships among a protein's sequential, functional, and structural properties.

## THE FOLDED STRUCTURE OF A PROTEIN IS DETERMINED BY THE PRIMARY STRUCTURE

The formation of a native or highly organized protein structural configuration reflects a delicate balance among a variety of interaction forces made both within the folded protein's interior and with surrounding solvent. As a consequence, perturbations of the protein's solvent environment can disrupt the protein's native tertiary conformation, resulting in loss of function and the production of a partially unfolded, or *denatured*, protein.

Proteins vary tremendously in their susceptibility to denaturation. For example, some secretory proteins, such as pancreatic ribonuclease and trypsin, can withstand exposure to strong mineral acid without irreversible loss of enzymatic activity, whereas most proteins would be irreversibly denatured by such conditions. Although the native and denatured states of proteins are frequently related in a reversible manner, the magnitude of the structural perturbation required for loss of function may vary appreciably. One could consider a spectrum of intermediates between two extreme forms—the native and so-called random-coil denatured state. Accordingly,

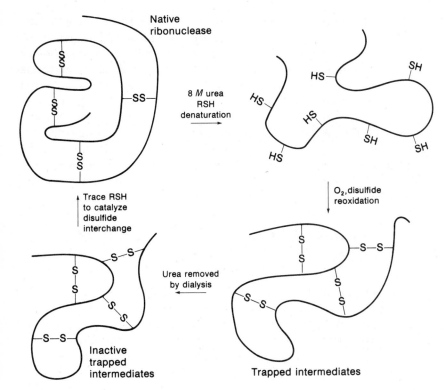

**Figure 3-1**
Schematic representation of an experiment to demonstrate that the information for folding into a biologically active conformation is contained in the protein's amino acid sequence.

the conditions giving rise to partial denaturation and consequent loss of function may be subtle, such as a small change in pH, temperature, ionic strength, or dielectric constant of the medium. Conversely, prolonged boiling or exposure to thiol-containing compounds and detergents such as sodium dodecyl sulfate or chaotropic reagents such as urea or guanidine hydrochloride may be required for complete denaturation.

It occurred to biochemists several years ago that the native conformation of proteins might be solely determined by the amino acid sequence. This was demonstrated to be the case by a series of classic experiments performed in the early 1960s by F. White and C. B. Anfinsen at the National Institutes of Health. They chose to study bovine pancreatic ribonuclease (see Figures 1-30 through 1-33), an enzyme containing 124 amino acid residues with four disulfide bridges. The experiment is described schematically in Figure 3-1. First, the enzyme was denatured in a solution containing 8 $M$ urea and $\beta$-mercaptoethanol, a thiol reagent that reduces —S—S to —SH + SH—, thus cleaving the covalent cross-links. These conditions have been used since as a general means of denaturing proteins by disrupting completely the conformation without giving rise to coagulation or precipitation. The reduced, denatured ribonuclease is biochemically inactive because its native structure has been destroyed. Next, the protein was allowed to slowly air oxidize, resulting in the formation of a variety of different intermediates with randomly distributed disulfide bonds. These trapped intermediates remained inactive after removal of the denaturant urea by dialysis through a semipermeable membrane. Native structures could be recovered by exposing these inactive intermediates to a trace amount of the reducing agent mercaptoethanol. This served to catalyze the rearrangement of the disulfide cross-links, resulting finally in the spontaneous generation of an essentially fully active product. Analysis by several methods showed that the renatured product was indistinguishable from native ribonuclease and that all the correct disulfide pairings had been reestablished. Thus *the information for folding to the native conformation is present in the amino acid sequence*, and of the many possible disulfide paired ribonuclease "isomers" that are possible, only the one correctly paired product was formed in major yield. The renaturation experiment can be carried out in a less cumbersome manner by removing the urea before air oxidation; in this case, the native structure is a major product.

## THE CHEMICAL PROPERTIES OF WATER ARE AN IMPORTANT COMPONENT IN DETERMINING PROTEIN CONFORMATION

Properties of peptide linkages and the amino acid side chains were discussed in Chapter 1. The peptide linkage was described as a rigidly planar structure. The amino acid side chains differ not only in size and shape, but also in the charge they carry, their general affinity for water (hydrophilicity), or their general abhorrence of water (hydrophobicity). The native conformation of proteins is a strong function of the interactions that

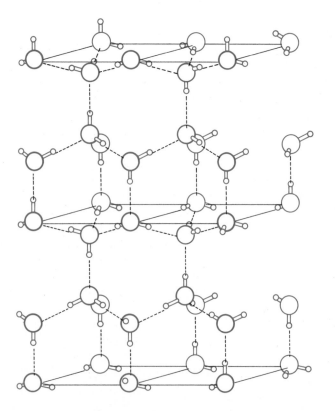

**Figure 3-2**
The arrangement of molecules in an ice crystal. The orientation of the water molecules, as represented in the drawing, is arbitrary; there is one proton along each oxygen-oxygen axis that is closer to one or the other of the two oxygen atoms.

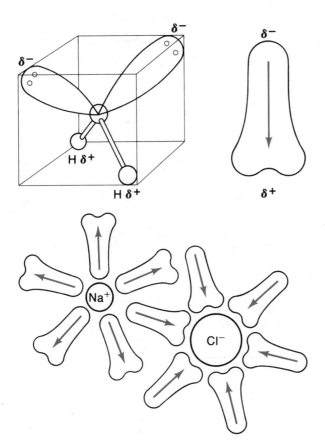

**Figure 3-3**
The water molecule is composed of two hydrogen atoms covalently bonded to an oxygen atom with tetrahedral ($sp^3$) electron orbital hybridization. As a result, two lobes of the oxygen $sp^3$ orbital contain pairs of oxygen electrons, thus giving rise to a dipole in the molecule as a whole. The presence of an electric dipole in the water molecule allows its easy solvation of charged ions because the water dipoles can orient to form energetically favorable electrostatic interactions with dissolved charged ions.

take place within and between polypeptide chains; it is also highly dependent on the interactions that take place with water since proteins exist in an aqueous environment. In order to appreciate the central role of water, it is essential to consider the properties of water and the types of interactions that water makes with polypeptides.

Ice is a highly organized structure in which individual molecules are held together by hydrogen bonds (H bonds) in a regular three-dimensional lattice (Figure 3-2). Water is also a highly H-bonded structure with a somewhat less regular and (judging by its greater density) more condensed structure. Judging by the fact that the heat of liquefaction of ice is 75 kcal/g, whereas the heat of vaporization of water is 540 kcal/g, most of the H bonds present in ice are also present in water. Hence water is a highly H-bonded structure, not too different from ice, but a structure in which the individual molecules have much more mobility and, consequently, a much higher entropy (entropy is a thermodynamic measure of the number of states a molecule can occupy without any change in enthalpy; see Chapter 7). Polypeptides also contain groupings that can form H bonds (see Chapter 1), and under certain conditions, polypeptides form H-bonds with water.

Individual water molecules have a significant dipole. This is due to the greater electronegativity of the oxygen atom over the hydrogen atoms (see Figure 3-3). The dipolar properties of water molecules are responsible for the dissolution of salts, such as sodium chloride in water (see Figure 3-3).

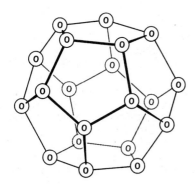

**Figure 3-4**
Clathrate structures are ordered
cages of water molecules around
hydrocarbon chains. The dodecahe-
dral cage (*above*) of water molecules
is a common building block in clath-
rates. To the right is a portion of the
cage structure of $(nC_4H_9)_3S^+F^- \cdot$
$23H_2O$. The trialkyl sulfur ion nests
within the hydrogen-bonded framework of water molecules. In the intact
framework, each oxygen is tetrahedrally coordinated to four others. One
such oxygen atom and its associated
hydrogens are shown by the arrow.

The kind of ion-dipole interactions that are made between water and simple ions such as $Na^+$ and $Cl^-$ are also an important factor in the interactions between the charged amino acid side chains of proteins and water.

Amino acid side chains containing charged groups, H-bond forming
groups, or dipoles also could be classified as hydrophilic. In the form of
small molecules, such groups tend to be highly soluble in water. By contrast, neutral organic side chains, which are hydrocarbon in character and
do not contain significant dipoles or the capacity for forming H bonds, have
little to gain by interacting with water. As a result, they are poorly soluble
in water. When such hydrophobic molecules are present in water, the
water forms a rigid so-called clathrate structure around them which is believed to be even more ordered than the structure of ice (see Figure 3-4 and
also the discussion in Chapter 7). This description of the properties of
water is an important prelude to a consideration of the forces involved in
determining protein conformation.

## THE FORCES THAT DETERMINE
## PROTEIN CONFORMATION

It has been seen that after the strenuous treatment leading to denaturation
or unfolding, a complex globular protein such as ribonuclease can be induced to renature or refold to its native conformation. This observation
leads to two very important conclusions about protein structure: *(1) the
conformation adopted by a protein is determined by its primary amino
acid sequence, and (2) the spontaneity of the process indicates that the*

*folding operation is energetically favorable*, i.e., $\Delta G$, the free energy change in going from the unfolded to the folded state, must be negative. Recall that $\Delta G$ is a composite of two very different terms that dictate the spontaneity of observing a transition between two states:

$$\Delta G = \Delta H - T \Delta S$$

where $\Delta H$ is the enthalpy, which is approximately equal to the difference in bond energies; and $\Delta S$ is the entropy, which is a measure of the order in the system. For a given transition, a negative $\Delta H$ indicates a system going from weaker to stronger bonded interactions. A positive $\Delta S$ indicates a system going from a more-ordered to a less-ordered state (these matters are discussed at greater length in Chapter 7).

In discussing polypeptide-chain folding it is often easy to forget the effects of water. The protein folding of concern is taking place in an aqueous environment, not in some other solvent or a vacuum. In evaluating the physicochemical forces involved in protein folding, account must be taken of what is happening to the water as well as to the polypeptide chain.

Other factors in aqueous solution, such as small anion and cation interactions, also can play a significant role in determining structural stability. Another factor that should be taken into account but is not in the presentation given here is the difference in the environment of an isolated protein in solution or in a protein crystal and its actual surroundings in vivo inside the cell. Except for membrane proteins, this potentially important consideration is not discussed because it is difficult to assess. However, this omission should be recognized as an embarrassment to the arguments that can be presented here. This factor is probably least important for proteins that function as nonaggregated molecules and most serious for proteins that function in aggregates, such as membrane proteins or enzymes that are believed to function in large multienzyme complexes. How such multimolecular aggregation affects the structures as we know them is a question for future study.

Covalent bond energies range in value from 30 to 230 kcal/mol between most atoms of biological interest (Table 3-1). Intermolecular bond energies between non-covalently-linked atoms range in value from 0.1 to 6 kcal/mol. Despite this, and except for the disulfide linkages formed between two cysteines, protein folding is mainly directed by the relatively weak intermolecular forces between noncovalently bonded atoms. Even in

**Table 3-1**
Bond Energies between Some Atoms of Biological Interest

| Energy Values for Single Bonds (kcal/mol) | | | | | |
|---|---|---|---|---|---|
| C—C | 82 | C—H | 99 | S—H | 81 |
| O—O | 34 | N—H | 94 | C—N | 70 |
| S—S | 51 | O—H | 110 | C—O | 84 |

| Energy Values for Multiple Bonds (kcal/mol) | | | | | |
|---|---|---|---|---|---|
| C=C | 147 | C=N | 147 | C=S | 108 |
| O=O | 96 | C=O | 164 | N≡N | 226 |

the case of the disulfide linkages, the "correct" linkages form because the appropriate cysteines are brought closely together by the weaker secondary forces. This was exemplified in the renaturation study of ribonuclease discussed earlier. Intermolecular forces of interest in protein structure may be classified into the following four categories:

**1.** Electrostatic forces. In this category we may consider charge-charge interactions, charge-dipole interactions, and dipole-dipole interactions (see Table 3-2). Each of these three types of interactions shows a different functional dependence on distance between the interacting groups that are summarized in Table 3-2. For example, the energy of interaction between two charges $U(R)$, $Q_1$ and $Q_2$, is proportional to the product of the charges and inversely proportional to the distance $R$ between them:

$$U(R) \propto \frac{Q_1 Q_2}{R}$$

In solution this interaction is reduced by the dielectric constant $\varepsilon$ of the surrounding medium:

$$U(R) \propto \frac{Q_1 Q_2}{\varepsilon R}$$

If the two charges in question are relatively buried within a protein, their interaction energy can be substantially increased because the dielectric constant in the regions inaccessible to water is much lower than the dielectric constant of water.

A number of amino acid side chains are charged at physiological pHs (lysine, arginine, aspartate, glutamate, and sometimes histidine). However, favorable charge-charge interactions between oppositely charged amino acid side chains are rarely as significant in determining protein folding as are the ion-dipole interactions between the charged groups and water. As mentioned earlier, individual water molecules have a dipolar character that leads to favorable electrostatic interactions with the charged groups of the polypeptide. From such considerations it seems likely that charged amino acid side chains would increase protein structural stability if they were present on the surface of a protein where they could interact with the water. As a general rule, this is the case.

**2.** van der Waals forces. The term *van der Waals forces* refers to two types of interactions, one attractive and one repulsive. The attractive forces are

**Table 3-2**
Range of Some Intermolecular Interactions Expressed as the Power of the Intermolecular Separation

| Range of Interaction | Type of Interaction |
|---|---|
| $1/R$ | Charge-charge |
| $1/R^2$ | Charge-dipole |
| $1/R^3$ | Dipole-dipole |
| $1/R^6$ | van der Waals (dipole-induced dipole) attractive forces |
| $1/R^{12}$ | van der Waals repulsive forces |

**Table 3-3**
Radii for Covalently Bonded and Nonbonded Atoms

| Element | Single Bond | Double Bond | Triple Bond |
|---|---|---|---|
| **Covalent bond radii (in Å)** | | | |
| Hydrogen | 0.30 | | |
| Carbon | 0.77 | 0.67 | 0.60 |
| Nitrogen | 0.70 | | |
| Oxygen | 0.66 | | |
| Phosphorous | 1.10 | | |
| Sulfur | 1.04 | | |
| **van der Waals radii (in Å)** | | | |
| Hydrogen | 1.2 | | |
| Carbon | 2.0 | | |
| Nitrogen | 1.5 | | |
| Oxygen | 1.4 | | |
| Phosphorous | 1.9 | | |
| Sulfur | 1.8 | | |

due to favorable interactions among the induced instantaneous dipole moments that arise from fluctuations in the electron charge densities of neighboring nonbonded atoms. Such forces tend to be small, yielding energies of 0.1 to 0.2 kcal/mol, but these can add up as the number of interactions between two molecules increases. Such forces explain why hexane is primarily a liquid rather than a gas at room temperature. In this example, we are considering the transition between liquid and noninteracting molecules in the gaseous state. In aqueous solution, van der Waals attractive forces between the hydrophobic groups are far less important in influencing the protein structure. In fact, the interaction of so-called hydrophobic molecules such as hexane with water has a favorable enthalpy over the interaction of hexane with itself (see Chapter 7). Other factors (entropic) therefore determine why liquid hexane and water are virtually immiscible (see below and Chapter 7).

Repulsive van der Waals forces arise when non-covalently-bonded atoms or molecules come too close together. An electron-electron repulsion arises when the charge clouds between two molecules begin to overlap. If two molecules are held together exclusively by van der Waals forces, their average separation will be governed by a balance between the van der Waals attractive and repulsive forces. The distance is known as the *van der Waals separation*. Some van der Waals radii for biologically important atoms are given in Table 3-3. The van der Waals separation between two nonbonded atoms is given by the sum of their respective van der Waals radii.

**3.** Hydrogen bond forces. The strength of the hydrogen bond is due to the partially unshielded nature of the single proton that makes up the hydrogen nucleus. An attractive interaction exists between the lone pair of electrons of either a nitrogen atom or an oxygen atom and the hydrogen atom in either an N—H or an O—H chemical bond. The attraction is usually directed along the lone-pair orbital axis of the H-bond acceptor group. Evidence for a significant H bond comes from the observation of a decreased distance between the donor and acceptor groups making up the H bond. Thus from the van der Waals radii given in Table 3-3, we can calculate that the distances between nonbonded H and O and H and N atoms are 2.6 and 2.7 Å, respectively. When an H bond is present, this distance is usually reduced to 1.8 to 1.9 Å. Some of the more important H-bond donors and acceptors are shown in Figure 3-5. As a rule the angle between the N or O acceptor and the N—H or O—H donor is close to 180°.

Polypeptides carry a number of H-bond donor and acceptor groups both in their backbone structure and in their side chains. Water also contains a hydroxyl donor group and an oxygen acceptor group for making H bonds (Figure 3-2). Formation of the maximum number of H bonds between a polypeptide chain and water would obviously require the complete unfolding of the polypeptide chain. However, it is not obvious that this would result in a net energy gain. This is so because water is a highly H-bonded structure, and for every H bond formed between water and protein, an H bond within the water structure itself must be broken. *The strategy followed by most proteins is to maximize the number of intramolecular H bonds between the backbone peptide groups, but to keep most of the po-*

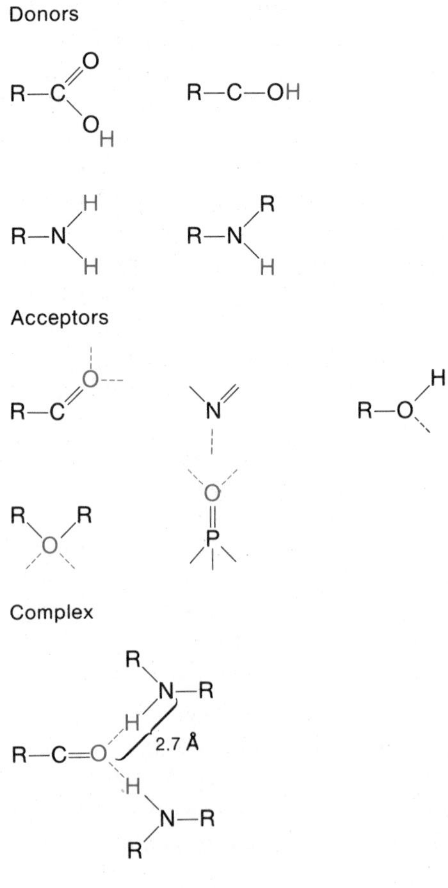

**Figure 3-5**
Major hydrogen-bond donor and acceptor groups.

*tential H-bond forming side chains on the protein surfaces for interaction with water.* It seems likely that such side-chain water interactions will occur both because of the H bonds and dipole-dipole interactions.

**4.** Hydrophobic forces. These are both the poorest understood and the hardest to appreciate type of interaction. The term is not a very good one either, since it is somewhat misleading. Whereas the previous forces discussed are due primarily to enthalpic factors, hydrophobic forces relate primarily to entropic factors. Furthermore, the entropic factors primarily concern the solvent, not the solute. It was stated earlier that hexane has a small but favorable enthalpy for solution in water, but that the entropic factor was sufficiently unfavorable to make hexane highly insoluble in water. What is this mysterious force? Owing to the weak enthalpic interactions between a hydrophobic molecule such as hexane and water, the water tends to withdraw to some extent in the region of the apolar hydrophobic molecule and form a relatively rigid hydrogen-bonded network with itself (for example, see the clathrate structures illustrated in Figure 3-4). This effectively restricts the number of possible orientations of the water molecules forming the water–hydrophobic group interface relative to the much larger number of possible alternatives that could be continuously sampled in bulk water. The energetically unfavorable character of these interactions stems basically from the fact that virtually all chemical systems tend spontaneously toward a state of maximum disorder (see Chapter 7). Solvent exposure of hydrophobic groups results in the introduction of some extent of ordering in the surrounding water and is energetically unfavorable because this unfavorable entropic effect is greater than the small, favorable enthalpic effect. This point can be difficult to grasp because one observes hydrophobic groups clustering together in aqueous solution and it is tempting to attribute this mainly to van der Waals attractive force between the hydrophobic groups. However, thermodynamic measurements do not support this explanation (see Chapter 7).

Proteins contain a number of amino acids with predominantly apolar side chains (e.g., alanine, valine, isoleucine, leucine, and phenylalanine). From what has been said, it might be anticipated that most of these side chains would be located internally in the native protein structure, and this appears to be the case.

In concluding this section on the intermolecular forces involved in folding one must refer back to the obvious fact stated at the outset that the energetically favorable state for a protein at physiological pHs and temperatures is the folded state. Because of experimental difficulties, the thermodynamic parameters $\Delta H$ and $\Delta S$ for the transition between unfolded to folded states have rarely been measured. However, in the limited number of cases where data are available (see Chapter 7), it appears that the overall entropy of folding is slightly negative (unfavorable) and the overall enthalpy is also slightly negative (favorable). Thus from thermodynamics the conclusion is reached that *folding is opposed by the entropy but favored by the enthalpy change and occurs because the latter factor outweighs the former.* More thermodynamic data on a number of proteins could alter this view as to which thermodynamic parameter dictates a negative free energy for folding in specific instances.

The main difficulty in making predictions from a detailed consideration of the different intermolecular forces involved is a quantitative one.

For example, it is known that the entropy effect owing to interaction between apolar side chains and water will encourage folding, but it is not known how big this factor is compared with the ordering effect in the polypeptide chain that should discourage folding. Again we know that intramolecular H bonds formed by the polypeptide chain constitute an enthalpic factor that encourages folding, but it is not known if the net gain here (which would have to take into account the breaking of H bonds between the polypeptide backbone and water) will be enough to overcome the negative overall entropy effect on folding.

Finally, it should be noted that *the native folded state of a protein reflects a delicate balance between opposing energetic contributions of enormously large magnitude*. Specifically, the total free energy of stabilization of a globular protein in the folded state relative to its unfolded form is typically equivalent to the free energy of formation of less than 10 hydrogen bonds or, alternatively, a fraction of a single covalent C—H bond. This delicate energy balance has important manifestations that relate to both the functional properties and stability of proteins in their native states.

## PROTEIN FOLDING AND THE HIERARCHY OF PROTEIN STRUCTURAL ORGANIZATION

In the preceding section the energetics of polypeptide-polypeptide interactions and polypeptide-water interactions leading to the stabilization of the native folded state of a protein were described. Although this treatment gave a general idea of why protein folding takes place, it obviously neglected the important question of how it takes place. To appreciate the magnitude of this problem, consider a hypothetical example involving the spontaneous folding of a polypeptide of 100 amino acid residues. In the native folded state of such a protein, each residue is spatially fixed relative to the others to produce a unique three-dimensional structure. For the sake of simplicity, the spatial orientation of each residue in the folded structure can be described in terms of three geometric parameters. (These parameters might, for example, correspond to the values of some rotational angles about single covalent bonds in the polypeptide chain.) In order to completely describe the folded protein, it is therefore necessary to uniquely specify 300 internal geometric parameters. Even if each of the parameters can assume only two possible values, the total number of potentially possible geometric arrangements of the structure is $2^{300} = 2 \times 10^{90}$. If folding the protein involved random sampling of all these possible arrangements at a rate corresponding to typically observed frequencies of rotational rearrangements about single covalent bonds (about $10^{-13}$ s), the estimated time required to fold the protein into the correct arrangement would exceed the age of the earth. In actuality, newly synthesized polypeptide chains typically fold in seconds. This means that *protein folding must in reality be a highly directed and cooperative process*. Although much remains to be learned about the details of this process, it appears probable that both its speed and facility reflect the existence of a sequential set of folding intermediates, each of which is successively more highly structurally organized (for example, see Figure 3-6).

In what follows it will be described how the forces that stabilize proteins in concert with related energetic and geometric factors give rise to

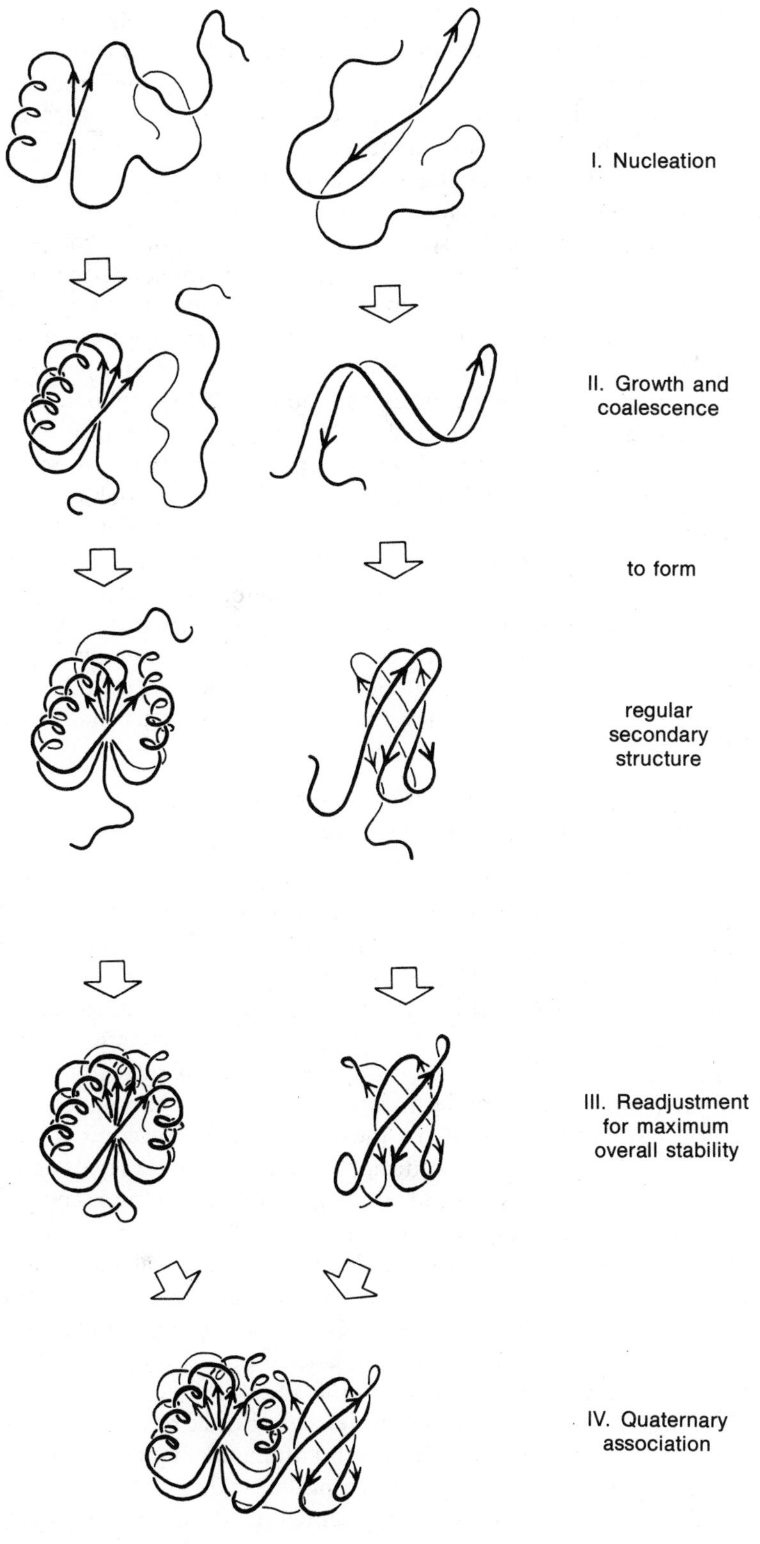

I. Nucleation

II. Growth and
coalescence

to form

regular
secondary
structure

III. Readjustment
for maximum
overall stability

IV. Quaternary
association

**Figure 3-6**
Possible successive steps in the protein-folding process as they might apply to
a typical example of each of the four major categories of structure (see text for
fuller explanation).

successively larger and more complex protein structural arrangements. To begin with, the geometric properties of the polypeptide chain will be examined; steric interactions restrict its accessible conformations and reflect features of the protein's amino acid sequence, or *primary structure*. Next, it will be shown how the requirements for hydrogen-bond preservation in the folded structure result in the cooperative formation of regular structural regions in proteins. This situation arises principally owing to the regularly repeating geometry of the hydrogen-bonding groups of the polypeptide backbone and leads to the formation of regular hydrogen-bonded *secondary structures*. Association between elements of secondary structure, in turn, results in the formation of *structural domains* whose properties are determined both by *chiral* properties of the polypeptide chain and *packing requirements* that effectively minimize the molecule's hydrophobic surface area. Further association of domains results in the formation of the protein's *tertiary structure*, or overall spatial arrangement of the polypeptide chain in three dimensions. Likewise, protein subunits can pack together to form *quaternary structures*, which can serve either a structural role or provide a structural basis for modification of the protein's functional properties. The final results of protein folding, therefore, reflect the participation of several interdependent effects that give rise to structures of increasing complexity. An understanding of protein organization begins with a description of the conformational properties of the polypeptide chain, which governs at the most basic level how proteins are structurally assembled.

## Steric Restrictions on Possible Conformations of the Polypeptide Chain

Figure 3-7 shows a ball-and-stick model of a short section of a polypeptide chain. Many geometric features of this structure are fixed owing to bonded interactions between adjacent atoms. The fixed geometric features include all the bond lengths and bond angles, which vary only slightly irrespective of the sort of protein structure in which they occur. Additionally, the backbone peptide bond has substantial double-bond character owing to electron delocalization over a $\pi$ orbital system involving the carbonyl oxygen, carbonyl carbon, and amide nitrogen atoms of the backbone peptide bond. As a result, all the atoms of the peptide bond, together with the connected $\alpha$-carbon atoms (conventionally labeled $C_\alpha$), lie in a common plane with the carbonyl oxygen and amide hydrogen in the trans configuration. Consequently, the only adjustable geometric features of the polypeptide-chain backbone involve rotations about the single covalent bonds that connect each residue's $C_\alpha$ to the adjacent planar peptide groups. Rotations about the $C_\alpha$—N bond are labeled with a Greek letter $\phi$ (phi), and rotations about the $C_\alpha$—carbonyl carbon bond are labeled $\psi$ (psi). In principle, both $\phi$ and $\psi$ can have any value between $-180$ and $+180$ degrees, so that all possible conformations of the polypeptide chain can be described in terms of their $\phi,\psi$ conformational angles, which automatically takes account of the fixed geometric features of the polypeptide backbone. As a result, any polypeptide conformation can be represented as a point on a plot of $\phi$ versus $\psi$, where $\phi$ and $\psi$ have values that range from $-180$ to $+180$ degrees. By convention, the conformation corresponding to $\phi = 0$ degrees, $\psi = 0$ degrees is one in which both peptide planes connected to a common $C_\alpha$

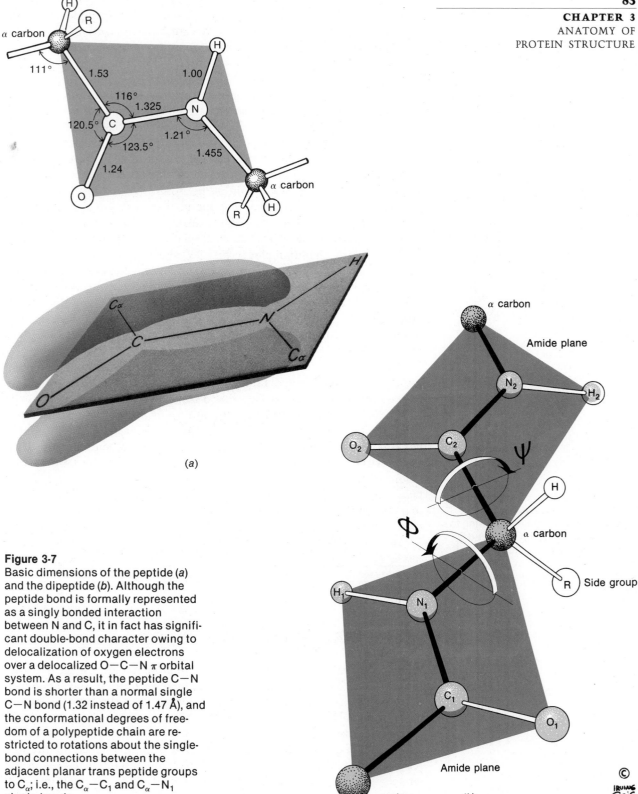

**Figure 3-7**
Basic dimensions of the peptide (a) and the dipeptide (b). Although the peptide bond is formally represented as a singly bonded interaction between N and C, it in fact has significant double-bond character owing to delocalization of oxygen electrons over a delocalized O—C—N π orbital system. As a result, the peptide C—N bond is shorter than a normal single C—N bond (1.32 instead of 1.47 Å), and the conformational degrees of freedom of a polypeptide chain are restricted to rotations about the single-bond connections between the adjacent planar trans peptide groups to $C_\alpha$; i.e., the $C_\alpha$—$C_1$ and $C_\alpha$—$N_1$ single bonds.

atom lie in the same plane, as shown in Figure 3-8. Positive variations in $\phi$ correspond to clockwise rotations of the preceding peptide about the $C_\alpha$—$N_1$ bond when viewed from $C_\alpha$ toward $N_1$ (see Figure 3-7b). Positive variations in $\psi$ correspond to clockwise rotations of the succeeding peptide about the $C_\alpha$—$C_2$ bond when viewed from $C_\alpha$ toward $C_2$ (see Figure 3-7b).

Experiments with models that approximate the polypeptide atoms as hard spheres having their appropriate van der Waals radii quickly reveal that many $\phi,\psi$ angular combinations are impossible owing to steric collisions between atoms along the backbone or between backbone atoms and the side chain R group. For example, it is clear that the $\phi = 0$ degrees, $\psi = 0$ degrees conformation shown in Figure 3-8 is impossible because it results in non-covalently-bonded interatomic contacts that are considerably less than the sum of the van der Waals radii of the atoms involved. If all possible $\phi,\psi$ combinations are investigated, it is found that owing to various sorts of unfavorable steric interactions, only a relatively restricted number of conformations are possible. The conformationally allowed regions of the $\phi,\psi$ in a so-called Ramachandran plot (Figure 3-9) shows explicitly how the accessible regions of $\phi,\psi$ space are limited by steric interactions among the polypeptide backbone and side-chain groups, assuming that the atomic groups behave as rigid spheres having appropriate van der Waals radii. In fact, the atoms in molecules do not behave as rigid spheres, so the dihedral angles found in real proteins span a slightly greater range of values than suggested by this plot (see Figure 3-10).

Figure 3-9 shows the distribution of some observed $\phi,\psi$ conformational values for proteins whose three-dimensional structures are known from crystallography (also shown in Figure 3-10). The great majority of these lie within the bounds defined by allowable steric interactions. The exceptional residues are usually glycines. Glycine frequently can assume conformations that are sterically hindered in other amino acids because its R group, a hydrogen atom, is considerably smaller than the $CH_2$ or $CH_3$ groups connected to $C_\alpha$ in all the remaining amino acids. This removes many of the steric interactions between the backbone atoms and R group that restrict the possible conformations of amino acids with R groups other than hydrogen.

In summary, it can be seen that *owing to the basic geometric properties of the polypeptide chain, its sterically allowed conformations are severely restricted as a result of the occurrence of unfavorable steric interactions between its various atomic groups.*

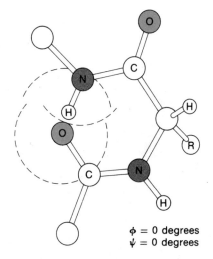

$\phi = 0$ degrees
$\psi = 0$ degrees

**Figure 3-8**
The conformation corresponding to $\phi = 0$ degrees, $\psi = 0$ degrees. This conformation is disallowed owing to steric overlap between the H and O atoms of adjacent peptide planes. Rotation of both $\phi$ and $\psi$ by 180 degrees gives the fully extended conformation seen in Figure 3-7b.

## Secondary Structures

As described earlier, a major driving force in protein folding is the necessity to minimize the extent of hydrophobic-group exposure to solvent. Nevertheless, there is a simultaneous necessity to preserve the favorable energy contributions of the hydrogen-bonded interactions formed between the polypeptide backbone and surrounding water in the protein's unfolded state. Since most proteins are typically large structures, meeting the first requirement will generally involve isolation of substantial regions of the polypeptide backbone from contact with solvent water. *Preservation of the energy balance in the system necessitates that the backbone polypeptide*

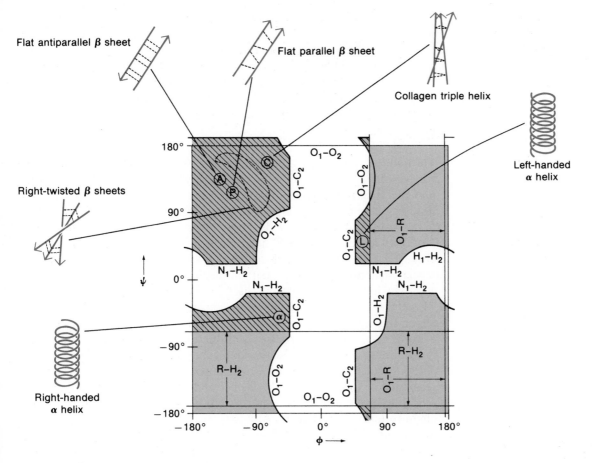

Flat antiparallel β sheet

Flat parallel β sheet

Collagen triple helix

Right-twisted β sheets

Left-handed α helix

Right-handed α helix

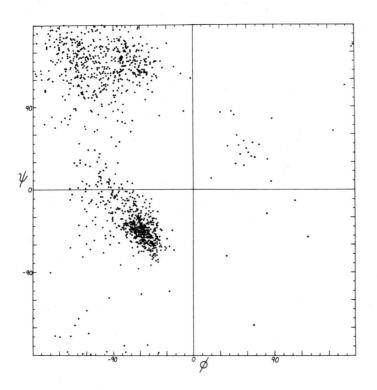

### Figure 3-9
"Derivation diagram" showing which atomic collisions (using a hard-sphere approximation) produce the restrictions on main-chain dihedral angles $\phi$ and $\psi$. The cross-hatched regions are allowed for all residues, and each boundary of a prohibited region is labeled with the atoms that collide in that conformation. Additional shaded regions are for glycine residues only. The numbering scheme for amide atoms used in the derivation diagram are given in Figure 3-7b. Each boundary of a prohibited region is labeled with the atoms which collide in that conformation.

### ◄ Figure 3-10
Plot of main-chain dihedral angles $\phi$ and $\psi$ experimentally determined for approximately 1000 nonglycine residues in eight proteins whose structures have been refined at high resolution (chosen to be representative of all categories of tertiary structure).

*groups form alternative hydrogen-bonded interactions between themselves in the protein's folded state.* This requirement has important consequences for protein folding because the backbone peptide carbonyl oxygen and peptide N—H bonding groups are regularly arranged along the polypeptide backbone. Extended arrangements having optimum hydrogen-bonded interactions among residues either within or between polypeptide chains reflect this periodic regularity and result in the formation of protein secondary structures.

**The α Helix.** One of the most commonly observed protein secondary structures is the α helix. In an α helix, the polypeptide backbone follows the path of a right-handed helical spring to form an arrangement in which each residue's carbonyl group forms a hydrogen bond with the amide NH group of the residue four amino acids farther along the polypeptide chain (Figure 3-11). All residues in an α helix have nearly identical conformations, averaging $\psi = -45$ to $-50$ degrees and $\phi = -60$ degrees, so they lead to a regular structure in which each 360 degrees of helical turn incorporates approximately 3.6 residues and rises 5.6 Å along the helix-axis direction. The advance per amino acid residue along the helix axis is 1.5 Å. Although alternative helical arrangements having both different hydrogen-bonding patterns and geometries are also conformationally possible, the α helix is by far the most commonly observed helical arrangement found in proteins. The particular stability of the α helix appears to be related not only to the formation of good hydrogen bonds between all the backbone carbonyl and NH groups, but also stems from the tight packing achieved in folding the chain to form the structure. Alternative arrangements, in contrast, either have inferior hydrogen bonds or are not as tightly packed as the α helix. Figures 3-9 and 3-10 show that there is a possible but small and shallow energy minimum at the left-handed α-helical ($L_\alpha$) position for nonglycine residues and that only 1 to 2 percent of the non-Gly residues are $L_\alpha$. However, for the symmetrical glycine, whose R group is the same as its $C_\alpha H$ and therefore has no hand at $C_\alpha$, left-handed conformations are exactly equivalent to right-handed ones, and in fact, about half the glycines have positive $\phi$ values. Extended $L_\alpha$ helices have not been observed, since they are difficult for any usual sequence of side chains, but isolated $L_\alpha$ residues are fairly common and are quite important in producing a greater diversity of backbone conformations. They occur frequently in some of the types of so-called β bends to be described below, as well as at other places where the backbone changes direction, such as at the C terminals of α helices. Most but not all of these $L_\alpha$ residues are glycines.

An important property stemming from the conformational regularity of the α helix, which applies to other secondary structures as well, is *cooperativity* in folding. For example, once a single turn of α helix has been formed, addition of successive residues becomes much more likely and faster because the first turn of the helix forms a template upon which to erect successive helical residues. Owing to steric restrictions, the torsional angle $\phi$ is approximately correct for each additional residue, so that each addition mainly involves sampling various conformations of $\psi$ until the residue is "captured" in the correct conformation by the formation of a hydrogen bond to a group that is already fixed in the helical conformation.

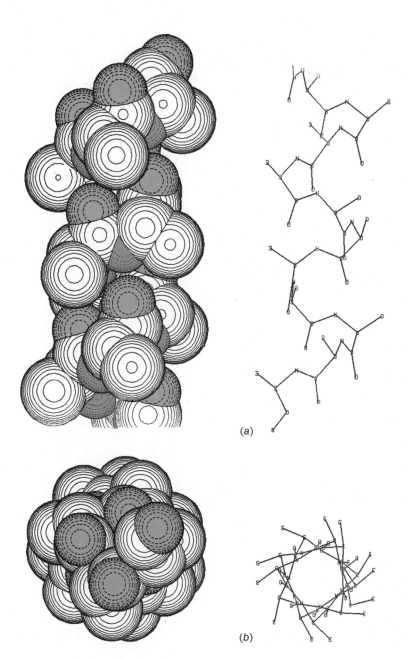

(a)

(b)

**Figure 3-11**
The Pauling-Corey α-helix structure.
(a) Side view. (b) Tube view. (Also
see Figure 1-7.) Hydrogens not
shown.

**β Sheets.** A second type of commonly occurring protein secondary structure is the β-pleated sheet (Figure 3-12a). β Sheets are formed when two or
more almost fully extended polypeptide chains are brought together side by
side, so that regular hydrogen bonds can form between the peptide backbone amide NH and carbonyl oxygen groups of adjacent chains. Notice
that since each backbone peptide group has its NH and carbonyl groups in
a trans orientation, it is possible to extend a β sheet into a multistranded
structure simply by adding successive chains to the sheet. β Sheets can
occur in two different arrangements. In the first of these, the chains are
arranged with the same N-to-C polypeptide sense to produce a <u>parallel</u> β

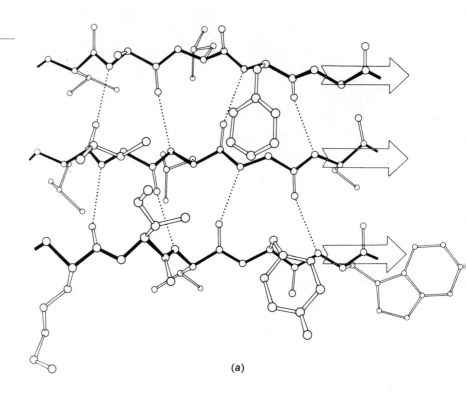

(a)

**Figure 3-12**

β-Sheet structures. (a) An example of a parallel β sheet from flavodoxin (residues 82–86, 49–53, and 2–6). In the pattern characteristic of a parallel β sheet, the hydrogen bonds are evenly spaced but slanted in alternate directions. Since both sides of the sheet are covered by other main-chain bonds (as is almost always true for parallel sheet), side groups that point in both directions are predominantly hydrophobic except at the ends of the strands. (b) An example of an antiparallel β sheet, from Cu, Zn superoxide dismutase (residues 93–98, 28–33, and 16–21). Arrows show the direction of the chain on each strand. Main-chain bonds are shown solid, and hydrogen bonds are dotted. In the pattern characteristic of an antiparallel β sheet, pairs of closely spaced hydrogen bonds alternate with widely spaced ones. The direction of view is from the solvent, so that side chains that point up are predominantly hydrophilic and those which point down are predominantly hydrophobic.

(b)

sheet. Alternatively, the chains can be aligned with opposite N-to-C sense to produce an *antiparallel* β sheet (Figure 3-12*b*). As illustrated, parallel and antiparallel β sheets are both composed of polypeptide chains that have conformations pointing alternate R groups to opposite sides of the sheet, but have their peptide planes nearly in the sheet plane to allow good interchain hydrogen bonding. Nevertheless, the chain conformation that produces the best interchain hydrogen bonding in parallel sheets is slightly less extended than that for the antiparallel arrangement. As a result, the parallel β sheet has both a shorter repeat period (6.5 Å per residue pair versus 7.0 Å for antiparallel) and more pronounced pleats than the antiparallel sheet.

**β Bends.**   Thus far the geometry of protein secondary structures that resemble long rods or flat sheets has been described. Obviously, folding a polypeptide chain to a compact globular form will necessitate some way to change the direction of the polypeptide chain, as might, for example, be required to connect adjacent ends of the polypeptide chains in an antiparallel β sheet. A commonly observed and particularly efficient way to do this is by formation of a tight loop in which a residue's carbonyl group forms a hydrogen bond with the amide NH group of the residue three positions farther along the polypeptide chain. The resulting so-called β bend reverses the direction of the polypeptide chain. Several conformational variations of the β bend have been observed which are a function of the bend amino acid sequence (Figure 3-13). In particular, it has been observed that the amino acids glycine and proline occur frequently in β bends. As noted earlier, glycine is conformationally more flexible than other amino acids and so can serve as a flexible hinge between regions of polypeptide chains whose steric interactions would otherwise keep them in more extended conformations. Proline, in contrast, is more conformationally restricted than other amino acids, since its cyclically bonded structure fixes its $\phi$ conformational degree of freedom. In a sense, then, part of the geometry that results in bend formation is preformed in proline-containing sequences. It appears probable that either situation might promote the formation of a

**Figure 3-13**
The two major types of tight turn of β bends (I and II). In type II (*bottom*), $R_3$ is generally glycine.

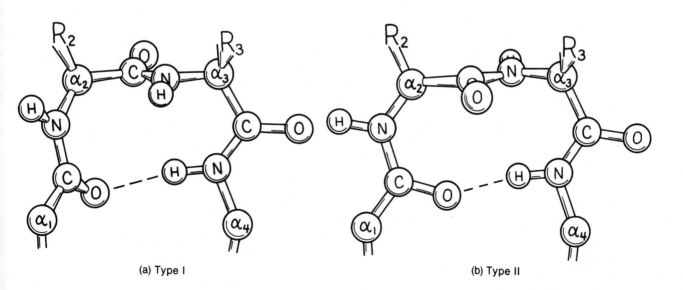

(a) Type I                              (b) Type II

bend during initial stages of protein folding and so cause structures such as antiparallel $\beta$ sheets to cooperatively assemble in a manner resembling the closure of a zipper.

The secondary structures described in this section are the most common structural elements found in most real proteins. Before discussing specific protein structures the roles played by the different amino acid residues in building these structural elements will be considered.

### Some Relationships between Amino Acid Sequence and Secondary Structure Formation

Examination of a large number of known three-dimensional protein structures has shown that the various amino acids have different secondary-structure-forming tendencies. As shown in Figure 3-14, glutamic acid, methionine, and alanine appear to be the strongest $\alpha$-helix formers and valine and isoleucine the most probable $\beta$-sheet formers, while proline, glycine, asparagine, and serine occur most frequently in $\beta$-bend conformations. This information is becoming of increasing value in its application to the prediction of secondary structural regions of proteins from their amino acid sequences. The observed frequencies of occurrence of each amino acid in a given conformation can be equated with the probabilities of that amino acid behaving similarly in a sequence whose actual tertiary structure is unknown. In order to predict the secondary structure from the sequence, it is consequently only necessary to sequentially plot the individual amino acid probabilities or, better, a local average over a few adjacent residues to account for the cooperative nature of secondary-structure formation. In such a scheme, sequences such as Gly–Pro–Ser and Ala–His–Ala–Glu–Ala give high joint probabilities for being, respectively, in $\beta$-bend and $\alpha$-helical conformations. However, comparisons of predicted versus directly observed polypeptide conformations give mixed results. This situation is a consequence of the facts both that several amino acids are somewhat ambiguous in their secondary-structure-forming tendencies and that

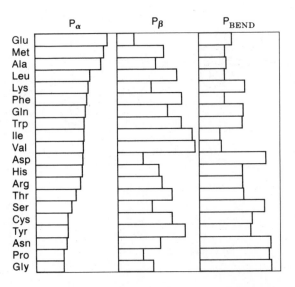

**Figure 3-14**
Relative probabilities that any given amino acid will occur in either $\alpha$-helical, $\beta$-sheet, or $\beta$-hairpin-bend secondary structural conformations.

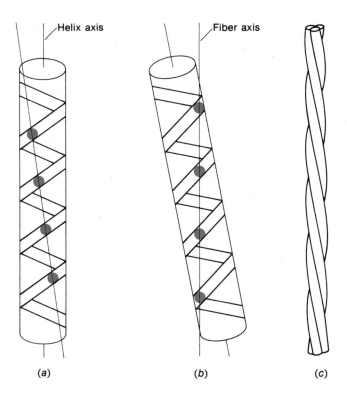

(a)  (b)  (c)

**Figure 3-15**
Coding of α helices in α-keratins. Residues on the same side of an α helix form rows that are tilted relative to the helix axis. Packing helices together in fibers is optimized when the individual helices wrap around each other so that rows of residues pack together along the fiber axis.

strong β-bend formers do occasionally turn up in the middle of α helices. Nevertheless, it is clear that the structural arrangement of a protein in its folded state must depend ultimately on its sequence. Although it is apparent that given sequences have general tendencies toward forming particular sorts of structures, the attainment of a particular folding arrangement must additionally involve effects that depend on details of both the short- and long-range interactions (e.g., charge pairing and close packing) that uniquely characterize and stabilize each protein's structure. At present it is difficult to evaluate these effects in the detail required to actually predict a protein's structure from its sequence.

## Secondary Structures and Fibrous Proteins

Fibrous proteins can be assembled in two ways: either extended or α-helical polypeptide chains can twist together or bundle into fibers or, alternatively, globular subunits can be helically arranged to make fibers (e.g., see actin, discussed below, or tubulin, discussed in Chapter 1). In this section the focus will be on the first type, which are predominantly based on regular secondary structures.

**Keratins.** One of the most diverse classes of fibrous proteins is the keratins, which are major constituents of such biological tissues as hair, scales, horns, hooves, wool, beaks, nails, and claws. The most prevalent forms of keratin are composed largely of polypeptide chains in the α-helical conformation (α-keratins), although alternative forms exist in bird feathers, which appear to be composed of stacked and folded β sheets (β-keratins). Figure 3-15 illustrates the arrangement of the α helices in typical α-keratins such as hair or wool. These materials are composed of extremely long α-helical polypeptide chains that are arranged side by side to form long cables. The α helices in α-keratins are spirally twisted so that the resulting cable has an overall left-handed twist. The formation of a cable with twisted α helices appears to be the result of optimization of packing among the amino acid side-chain residues between the helices. Side chain residues of an α helix are arranged in a spiral fashion so that residues falling on the same side of a helix generally do not lie along a line parallel to the helix axis (Figure 3-15a). Packing helices together is consequently optimized when the helices interact at an angle of about 18 degrees. Obviously, if the α helices making such a packing interaction are straight, they will eventually become separated. However, this packing interaction can be preserved if the helices are slightly twisted around each other, with the resultant formation of the left-twisted cable structure characteristic of the α-keratins (Figure 3-15c). The coiled character of the α helices in such fibers consequently represents a tradeoff between some local deformations that coil the α helix and the optimization of extended side-chain packing interactions in the cable as a whole.

The springiness of hair and wool fibers results from the tendency of the α-helical cables to untwist when stretched and spring back when the external force is removed. In many forms of α-keratin, the individual α helices or fibers are covalently linked by disulfide bonds formed between cysteine residues of adjacent polypeptide chains. The pattern of these covalent interactions serves to both influence and fix the extent of curliness in the hair fiber as a whole. Chemical reactions and mechanical processes involving reductive cleavage, reorganization, and reoxidation of these interhelix disulfide bonds form the basis of the "permanent wave."

**Silk.** A second variety of fibrous proteins are the silk fibroins, which are produced by insects. Silks are structurally composed of aligned and stacked antiparallel β sheets (Figure 3-16). Sequence analysis of silk proteins show them to be largely composed of glycine, serine, and alanine, where every other residue is glycine. Since the side-chain groups of a flat antiparallel β sheet point alternately upward and downward from the plane of the sheet, all the glycine residues are arranged on one surface of each sheet and all the substituted amino acids on the other. Two or more such sheets can consequently be intimately packed together to form an arrangement of stacked sheets in which two adjacent glycine-substituted or alanine-substituted sheet surfaces interlock with each other. Owing to both the extended conformations of the polypeptide chains in the β sheets and the interlocking of the side chains between sheets, silk is a mechanically rigid material that tends to resist stretching.

**Collagen.** An additional variety of structural protein is collagen, a particularly rigid and inextensible material that is a major constituent of tendons and many connective tissues. Analysis of collagen amino acid sequences shows them to be characterized by a repetitious tripeptide sequence, Gly-X-Pro or Gly–X–Hydroxyproline, where X can be any amino acid, and hydroxyproline is a hydroxylated derivative of proline. Owing to the repeating proline residue collagen polypeptide chains cannot adopt either an α-helical or a β-sheet conformation. Instead individual collagen polypeptide chains tend to assume a left-handed helical conformation, in which

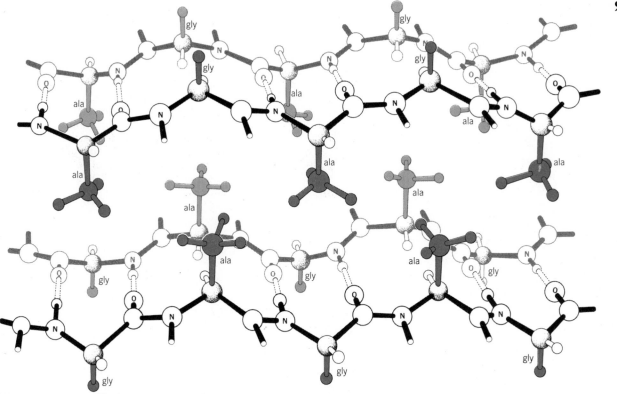

**Figure 3-16**
The three-dimensional architecture of silk. The side chains of one sheet nestle quite efficiently between those of neighboring sheets. The cut bonds extend to neighboring chains in the same sheet.

successive side-chain groups point toward the corners of an equilateral triangle when viewed down the polypeptide-chain axis (Figure 3-17). The glycine every three residues is strictly required because there is no room for any other amino acid inside the triple helix, where the glycine R groups are located. The collagen helix is already very extended, so it cannot easily stretch further like the $\alpha$ helix. Moreover, in contrast to the $\alpha$ helix, formation of a single-chain collagen helix is not accompanied by the formation of hydrogen bonds among residues within each polypeptide chain. Instead, three collagen chains associate in a three-stranded cable with hydrogen bonds between each chain and its neighbors. This produces a highly interlocked fibrous structure that is admirably suited to its biological role of providing rigid connections between muscles and bones, as well as structural reinforcement for skin and connective tissues.

Although there exist additional types of protein, as well as polysaccharide-based structural materials in living organisms, attention has been focused here on three arrangements whose structural properties are currently best understood. Two of these, the $\alpha$-keratins and the silks, incorporate polypeptide secondary structures that also commonly occur in globular proteins. Collagen, in contrast, is a protein that evolution has developed to play a more specialized role.

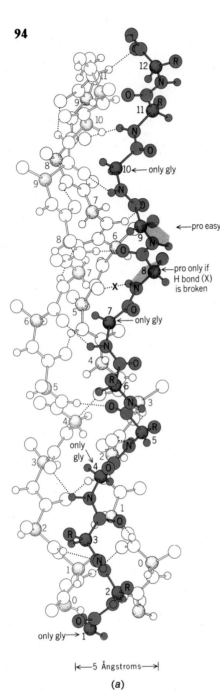

only gly → 1

|←—5 Ångstroms—→|

(a)

Figure 3-17
The basic coiled-coil structure of collagen. Three left-handed single-chain helices wrap around one another with a right-handed twist. (a) Ball-and-stick single-collagen chain. (b) View from top of helix axis. Note that glycines are all on the inside.

(b)

## STRUCTURAL ORGANIZATION IN GLOBULAR PROTEINS

Globular proteins, as their name implies, differ from fibrous proteins in that they generally have a more or less spherical shape. Nevertheless, *three-dimensional structural studies of globular proteins and enzymes show that they incorporate many of the secondary structural features that typify the fibrous proteins*. Figure 3-18, for example, illustrates the first enzyme whose three-dimensional structure was determined, the 129-residue protein lysozyme. This protein has local regions of ordered α-helical and antiparallel β-sheet secondary structures, but it has, in addition, several extended regions incorporating β bends and extended polypeptide chains with a less regular conformation.

The approximately 200 protein structures determined have revealed a rich variety of alternative structural arrangements. Each different polypeptide sequence is associated with the formation of a unique tertiary structure. Nevertheless, careful comparisons have shown that many proteins share some fundamental structural similarities. The recurrence of common structural features among proteins that otherwise show little similarity in sequence or function suggests that these recurring features have common physical origins. These similarities appear to reflect two different sorts of physical effects. The first of these is a chiral effect, or tendency for extended structural arrangements in proteins to be "handed." Such tendencies for structural regions to have a given handedness reflect the fact that their constituent polypeptide chains are composed of chiral L-amino acids. Chiral effects manifest themselves in both the manner in which regions of secondary structure are interconnected in globular proteins and in the geometric properties of globular-protein β sheets. The second effect of importance in tertiary structural organization relates to how secondary structural regions, such as α helices and β sheets, can most efficiently pack together so as to minimize the protein's solvent-accessible surface area. Both of these effects are considered in the following discussion.

## The Influence of Polypeptide-Chain Chiral Properties on Protein Structural Organization

In the discussion of polypeptide conformation it was found that the number of actual polypeptide conformations that could occur were restricted owing to unfavorable steric interactions. While this gives some indication of which conformations are possible, it is evident that the frequent occurrence of structures such as α helices represents situations in which an arrangement is not only allowed, but is particularly stable. As in the case of the α helix, the relative stability of a particular conformation is governed by the details of the interaction forces among the atoms comprising the polypeptide chain. Given the fact that proteins are composed primarily of chiral L-amino acids, it is consequently not surprising that the most stable conformations of extended polypeptide chains are not straight. Instead, detailed conformational energy calculations have shown that _extended polypeptide chains prefer to be slightly twisted in a right-handed sense when viewed down the polypeptide-chain axis._ Since the residues of straight, extended polypeptide chains alternate in position by 180 degrees (e.g., see Figure 3-12), _the cumulative effect of this tendency toward right twisting is to produce extended structures that are coiled in a right-handed direction_ (Figure 3-19a).

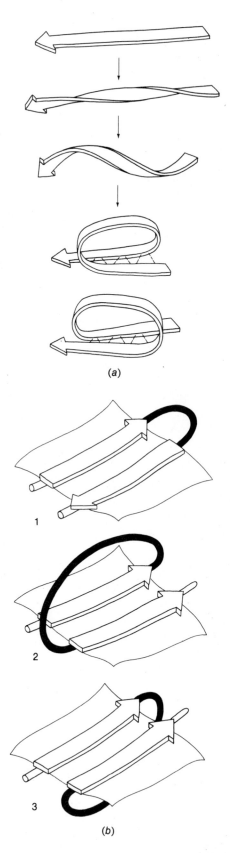

(a)

(b)

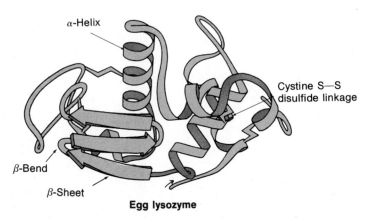

**Figure 3-18**
Lysozome; in these and succeeding figures the polypeptide backbone is represented as a ribbon to allow the polypeptide-chain course to be followed easily.

**Figure 3-19 ▶**
The natural tendency for the polypeptide chain to twist in a right-handed direction produces structures with an overall right-handed connectivity. (a) Illustrates how right-handed twisting leads to right-handed connectivity. On the right is shown an example of left-handed connectivity for comparison; (b) shows three ways of making connections between β strands: (1) a hairpin same-end connection, (2) a right-handed crossover connection, and (3) a left-handed crossover connection.

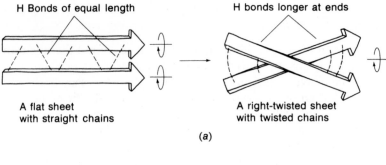

H Bonds of equal length

A flat sheet
with straight chains

H bonds longer at ends

A right-twisted sheet
with twisted chains

(a)

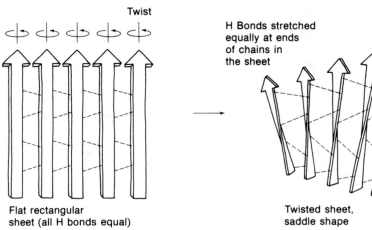

Twist

Flat rectangular
sheet (all H bonds equal)

H Bonds stretched
equally at ends
of chains in
the sheet

Twisted sheet,
saddle shape

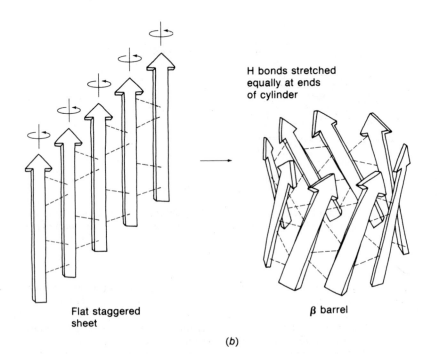

Flat staggered
sheet

H bonds stretched
equally at ends
of cylinder

β barrel

(b)

**Figure 3-20**
Different conformations adopted by arrays of parallel β sheets.

The effects of this tendency for extended-chain structures to form right coiled or twisted structures manifests itself in two related but different structural features that are common to virtually all known proteins. The first of these relates to the connectivity between the ends of parallel polypeptide strands that form β sheets in globular proteins. The connection from one β strand to the next can stay at the same end of the sheet in a simple hairpin turnaround only in antiparallel β sheets. In parallel β sheets a "crossover" connection to the other end of the sheet is required; this can be either right-handed or left-handed (Figure 3-19b). *Virtually all crossover connections observed in protein structures are right-handed, irrespective of whether or not the β strands they join are adjacent.* This invariant pattern most likely results from the energetically favored nature of the right-handed crossover.

A second consequence of the tendency of extended polypeptide chains to twist in order to minimize their conformational energy is reflected in the geometry of globular-protein parallel β sheets. In this case, it is observed that *the β sheets of globular proteins are always twisted in a right-handed sense when viewed along the polypeptide-chain direction* (Figure 3-20). This twisting behavior of β sheets is an important feature in protein structural architecture, since twisted β sheets frequently constitute the backbone of protein structures. Figure 3-21 shows schematic illustrations of the

**Figure 3-21**
A comparison of parallel β-sheet structures forming the backbone structures in different enzymes (or parts of enzymes): (a) β-barrel arrangement; (b) saddle shape.

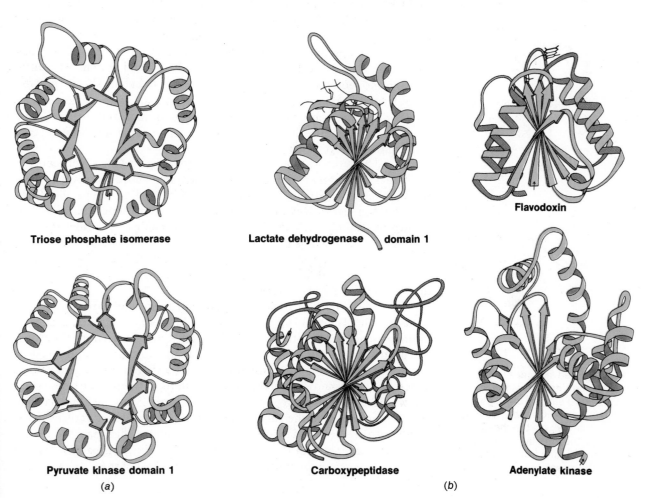

**Triose phosphate isomerase**  **Lactate dehydrogenase** domain 1  **Flavodoxin**

**Pyruvate kinase domain 1**  **Carboxypeptidase**  **Adenylate kinase**

(a)  (b)

polypeptide-chain folding of six proteins that incorporate twisted parallel or mixed $\beta$ sheets. These include the exoprotease carboxypeptidase A, the electron-transport protein flavodoxin, and the glycolytic enzyme triose phosphate isomerase. Although these proteins all have right-twisted $\beta$ sheets, it is clear that their overall geometries differ. That is, the $\beta$ sheets in carboxypeptidase and flavodoxin are smoothly twisted to form saddle-shaped surfaces, while the $\beta$ sheets in triose phosphate isomerase take the form of a cylinder or $\beta$ barrel.

Within each overall type of parallel $\beta$-sheet organization, the detailed hydrogen-bond pattern can be understood in terms of the forces acting within and between the polypeptide chains. In the case of the roughly rectangular sheets in carboxypeptidase and lactate dehydrogenase, the observed geometry reflects a competition between the tendency of the individual chains to twist and the preservation of the interchain hydrogen bonds. Basically, the interchain hydrogen bonds tend to stretch when the sheet is twisted and so resist introduction of twist into the sheet. The observed saddle-shaped geometry consequently reflects the uniform distribution of these conflicting forces throughout the sheet, as shown in Figure 3-20$b$. The sheet forming the barrel in triose phosphate isomerase differs from those just described in having an hourglass-shaped surface with cylindrical curvature. Twisted $\beta$ strands with a staggered hydrogen-bond pattern (see Figure 3-20$b$) automatically produce a cylindrical curvature, and conversely, twisted strands on a cylindrical surface necessitate a staggered hydrogen-bond pattern. Again, there is a compromise of twist versus H bonding between approximately straight chains that produce the hourglass shape with somewhat stretched H bonds at the top and bottom. The differences in the geometries of rectangular and staggered plane sheets therefore result from differences in how adjacent sheet strands are hydrogen bonded together. In either case, the operative forces are similar, and the final result reflects a compromise between chain twisting and preservation of good interchain hydrogen bonds.

Chiral preferences affect the connectivity as well as the sheet geometry in parallel $\beta$ proteins; according to the rule stated above crossover connections between the parallel $\beta$ strands are right-handed. In the parallel $\beta$ barrels, the connections cannot go down the center, which is only large enough to accommodate the hydrophobic side chains. As a rule, the polypeptide backbone winds in a simple right-handed spiral around the barrel, moving over by one $\beta$ strand at a time and packing helices or loops around the outside. Thus, although these structures tend to be large, their organization (see Figure 3-22$a$) is very simple.

The saddle-shaped parallel $\beta$ sheets, such as those in carboxypeptidase or flavodoxin (Figure 3-21), have a layer of helices and loops on each side. In order to accomplish this with right-handed crossover connections, the polypeptide chain must sometimes move along the sheet in one direction and sometimes in the other direction. The most common organizational pattern found in known protein structures starts in the middle of the sheet and winds toward one edge, with right-handed crossovers packing a layer of helices on one side of the sheet; then the polypeptide chain returns to the middle of the sheet and winds out to the opposite edge, packing helices against the other side of the sheet (see Figure 3-22$b$). This pattern is often known as a "nucleotide-binding domain," since most of these proteins bind

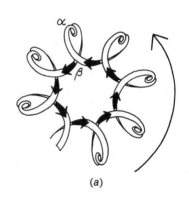

(a)

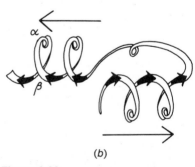

(b)

**Figure 3-22**
Highly simplified sketches (viewed from the C-terminal end of the $\beta$ strands) of (a) a singly wound parallel $\beta$ barrel; (b) a classic doubly wound $\beta$ sheet. Thin arrows next to the diagrams show the direction in which the chain is progressing from strand to strand in the sheet.

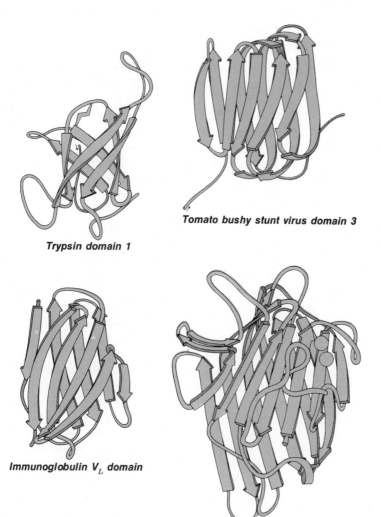

*Trypsin domain 1*

*Tomato bushy stunt virus domain 3*

*Immunoglobulin V$_L$ domain*

*Concanavalin A*

**Figure 3-23**
Examples of proteins containing anti-parallel β-sheet domains.

a mononucleotide or dinucleotide cofactor in the middle of the C-terminal end of the β sheet.

In antiparallel β proteins, chiral preferences also affect the twist of the strands around the barrel-shaped structures (as in Figure 3-23; also see pyruvate kinase, domain 2 in Figure 3-57). In addition, there is a strong chiral preference for the way the strands are connected in these antiparallel barrels. The second most common organizational pattern in the known protein structures is a barrel (or double sheet) of antiparallel β structure in which some of the connections between strands cross the top or bottom of the barrel (see examples in Figure 3-23) rather than going directly to their nearest neighbor strands in the sheet. If the barrel were opened up, laid out flat, and viewed from the solvent side, then the strand connections would show a pattern such as that in Figure 3-24. This is known as a "Greek key" topology because it resembles the border motif that is common on Greek vases. It is a handed pattern when it occurs on a surface with distinguishable sides (i.e., outside versus inside of a barrel); the swirl of the pattern is almost always counterclockwise as shown schematically in Figure 3-24.

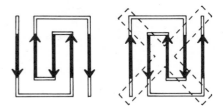

**Figure 3-24**
"Greek key" topology.

## Packing between Secondary Structures and Domain Formation

In the preceding section it was described how energetically preferred chiral properties of the polypeptide chain manifest themselves in the pattern of strand connectivity and the overall geometry of globular-protein $\beta$ sheets. An additional important factor in protein structural organization is the efficient packing together of secondary structural elements to form larger units. Figure 3-25 shows one of the most commonly observed arrangements, which is a special case of the right-handed crossover connections described in the previous section. This is composed of a pair of adjacent hydrogen-bonded parallel $\beta$ strands that are right connected with a stretch of $\alpha$ helix that packs tightly on the surface of the sheet. Generally, the $\beta$-sheet strands in a local $\beta\alpha\beta$ loop are part of a much larger $\beta$ sheet. Structures such as triose phosphate isomerase (Figure 3-21) can be viewed as a series of overlapping $\beta\alpha\beta$ structural units. This produces a final structural arrangement having an inner barrel composed of $\beta$-sheet strands which packs tightly with an outer barrel composed of $\alpha$ helices.

Although tertiary structures composed of $\beta\alpha\beta$ loops are perhaps the most commonly observed pattern in known protein structures, other arrangements are found in proteins having either predominantly antiparallel $\beta$- or predominantly $\alpha$-helical conformations.

The most common structural organization in antiparallel $\beta$ proteins has two layers of $\beta$ sheets that can range from a pair of essentially separate sheets whose relative orientation is determined by complementary packing between the saddle-shaped sheet surfaces to twisted barrels with continuous, staggered hydrogen bonding all the way around (see examples in Figure 3-23). Other antiparallel $\beta$ proteins have a single twisted $\beta$ sheet that is covered on only one side by a layer of helices and loops (see Figure 3-26).

*No protein is stable as a single-layer structure, since it requires at least two layers to bury the hydrophobic core. Antiparallel $\beta$ proteins are typically two-layer structures, since antiparallel sheets are apparently quite happy with one side exposed to solvent. Parallel sheets require at least one additional layer besides the sheet in order to make the crossover connections between the parallel $\beta$ strands. In contrast to antiparallel $\beta$ sheets, they apparently cannot tolerate solvent exposure on even one side and are always found as a structural "backbone" in protein interiors*, with other

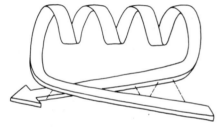

**Figure 3-25**
A $\beta\alpha\beta$ loop. This arrangement forms the basis of many of the more extended structural arrangements, such as those shown in Figure 3-21a.

**Figure 3-26**
Examples of antiparallel $\beta$ proteins that are covered on only one side by larger helices and loops.

**Streptomyces subtilisin inhibitor**

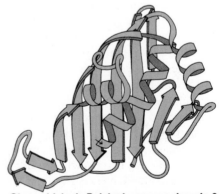

**Glyceraldehyde P dehydrogenase domain 2**

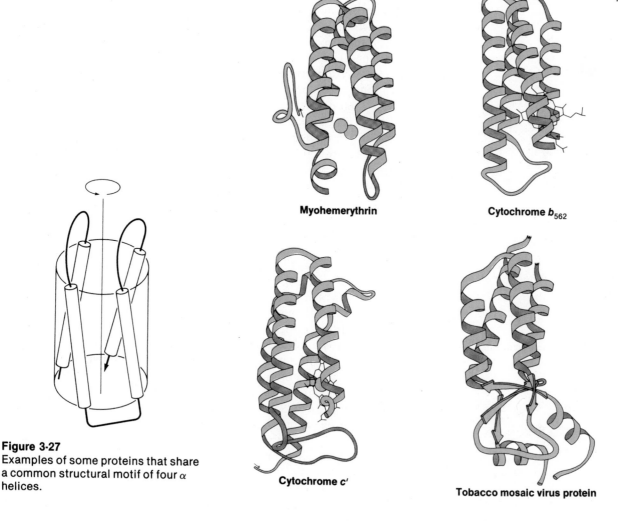

**Myohemerythrin**

**Cytochrome $b_{562}$**

**Cytochrome $c'$**

**Tobacco mosaic virus protein**

**Figure 3-27**
Examples of some proteins that share a common structural motif of four $\alpha$ helices.

layers of structure on both sides. Therefore, proteins with $\beta\alpha\beta$ loops generally either have three layers, as in the nucleotide-binding domains, or four layers, as in the parallel $\beta$ barrels (see Figure 3-21). Usually those outer layers are formed of $\alpha$ helices, which must pack against one another and also against the surface formed by the $\beta$-sheet side chains.

*The structural geometry of proteins exclusively incorporating regions of $\alpha$-helical secondary structure is also largely determined by requirements for efficient packing between the helices.* One particularly stable interhelical packing arrangement has already been encountered in the discussion of fibrous proteins. In the $\alpha$-keratins, adjacent helices pack together with an interaction angle between helices of about 18 degrees. This extended-helix interaction pattern forms the basis for a recurring protein structural motif seen in globular proteins, the 4-$\alpha$-helical bundle. In this arrangement, 4-$\alpha$ helices sequentially connected to their nearest neighbors pack together to form an array with a roughly square cross section. Since each helix interacts with its neighbors at about 18 degrees, the overall bundle has a left-handed twist (Figure 3-27). While this commonly observed

**Pancreatic trypsin inhibitor**

**Wheat germ agglutinin domain 2**

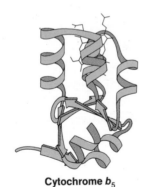

**Cytochrome $b_5$**

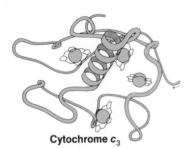

**Cytochrome $c_3$**

**Figure 3-28**
Examples of some small proteins or
domains where disulfide bonds or
prosthetic groups are a
dominant factor.

folding domain clearly represents a minimum accessible surface area arrangement for four sequentially connected $\alpha$ helices of approximately equal length, many $\alpha$-helical proteins that lack these features have more complex and irregular geometries (Figure 3-27). However, even in these cases it appears that the relative orientations of adjacently packed helices reflect geometric restrictions that accompany close packing between helices. Another commonly observed geometry for helix packing has an angle of about $-60$ degrees between adjacent helices; this arrangement is the usual type of helix contact found in the $\beta\alpha\beta$ proteins (Figure 3-21a).

In addition to the packing of elements of secondary structure, which is a dominant feature in most proteins, there are some cases, especially among the smallest structures, where the geometry and packing of disulfide bonds or prosthetic groups are a dominant factor. Figure 3-28 shows examples of this sort, in which the secondary structures are short and irregular and cannot assume their native structures if the disulfides are broken or the prosthetic groups are missing.

In summary, examination of a large number of protein tertiary structures has shown that they incorporate several different sorts of recurring structural arrangements. In general, these owe their origins to physical effects, some of which predispose the most stable conformations of the polypeptide chain, and some of which govern the formation of intimately packed tertiary structures.

The patterns of tertiary structure just described frequently constitute an entire protein. However, within a single subunit, contiguous portions of the polypeptide chain often fold into compact local units called *domains*, each of which might consist, for example, of a four-helix cluster or a barrel of antiparallel $\beta$ sheet. Sometimes the domains within a protein are very different from one another, such as within papain (Figure 3-29), but often they resemble each other very closely, such as in rhodanese (Figure 3-30). Similar pairs of domains are often related by an approximate twofold axis.

The separateness of two domains within a subunit varies all the way from independent globular domains joined only by a flexible length of

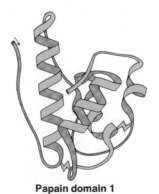

**Papain domain 1**

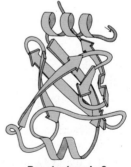

**Papain domain 2**

**Figure 3-29**
Papain, a protein in which the domains are very different from one another.

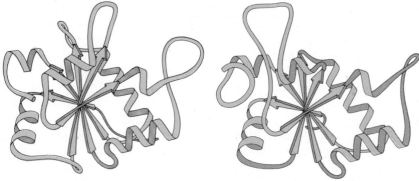

**Rhodanese domain 1**          **Rhodanese domain 2**

**Figure 3-30**
Rhodanese domains 1 and 2 as an example of a protein with two domains
that resemble each other extremely
closely.

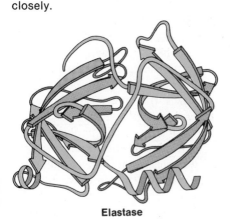

**Elastase**

**Figure 3-31**
Schematic backbone drawing of the
elastase molecule showing the similar
β-barrel structures of the two domains. The outside surfaces of the β
barrels are stippled.

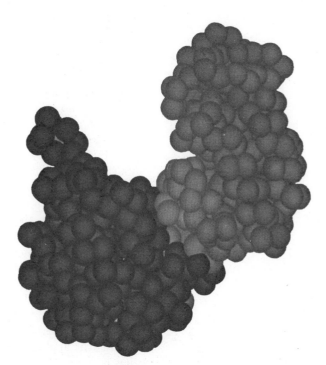

**Figure 3-32**
The "dumbbell" domain organization
of phosphoglycerate kinase with a relatively narrow neck between two well-
separated domains.

polypeptide chain to domains with tight and extensive contact and a
smooth globular surface for the outside of the entire subunit (as, for example, in Figure 3-31). An intermediate level of domain separateness is common in known protein structures, with an elongated overall subunit shape
and a definite neck or cleft between the domains, such as phosphoglycerate
kinase (Figure 3-32). Flexibly hinged domains are probably underrepres-
ented among known structures, since it is generally more difficult to crystallize such proteins. For tightly associated domains it may be difficult to
make a choice as to how many domains should be said to be present, but
the domain concept has proved so fruitful in explaining both the structure
and function of proteins that it is sure to survive in one form or another.

Domains as well as subunits can serve as modular bricks to aid in
efficient assembly. Undoubtedly the existence of separate domains is important in simplifying the protein folding process into separable, smaller

steps, especially for very large proteins. The most common domain size is between 100 and 200 residues, but apparently there is no strict upper limit on practicable folding size, since domains vary in size all the way from about 40 residues to over 400.

The other important function of domains is to provide motion. Completely flexible hinges would be impossible between subunits because they would simply fall apart, but they can exist between covalently linked domains. Limited flexibility between domains is often crucial to substrate binding, allosteric control (discussed below), or assembly of large structures. In hexokinase, the two domains within the individual's subunits hinge toward each other upon binding of glucose, enclosing it almost completely (see Figure 3-33). In this manner the substrate glucose can be bound in an environment which excludes water as a competing substrate.

## SPECIAL STRUCTURAL FEATURES OF MEMBRANE PROTEINS

In the discussion of the forces responsible for protein folding it was emphasized that the final structural arrangement reflected a situation that minimized the protein's water-accessible hydrophobic surface area, but situated charged amino acid side chains at the molecular surface. Nevertheless, it is clear from a variety of studies that highly structured proteins occur in the hydrophobic environment that characterizes the interior of biological membranes. Clearly such _membrane proteins are not being forced to assume structures that minimize exposure of their hydrophobic groups, since they are, in fact, in the highly hydrophobic environment of the membrane interior._ Consequently, it would appear that there is something basically different about how membrane proteins fold relative to most other proteins.

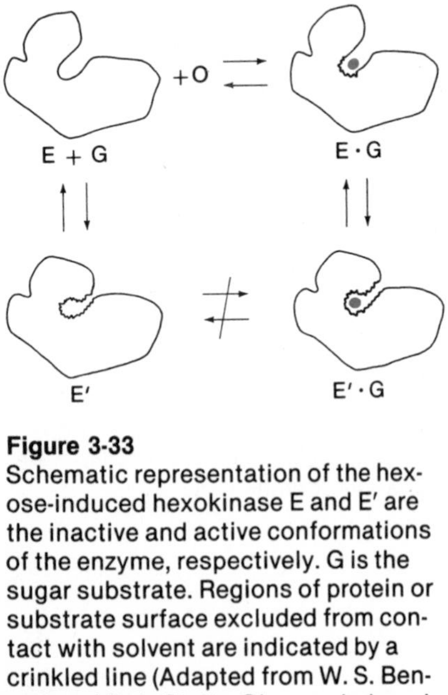

The most thoroughly studied membrane protein in this regard is bacteriorhodopsin. Bacteriorhodopsin gets its name from its incorporation of a retinal prosthetic group that is similar to the one found in the visual receptor protein rhodopsin (see Chapter 31). However, bacteriorhodopsin is found in the membranes of a bacteria that lives in salt marshes, where it functions as a light-driven transmembrane proton pump (see Chapter 17). Through the application of sophisticated electron microscope techniques, it has been possible to obtain a preliminary view of the three-dimensional structure of this protein as it exists in the membrane. These results suggest that the 247-residue polypeptide chain that comprises the molecule is organized as a bundle of seven α helices whose long axes are roughly perpendicular to the membrane surfaces (Figure 3-34). Estimates of the number of residues required to make seven α helices long enough to span the membrane (about 40 Å) suggest that virtually all the polypeptide chain is required to form the α helices, so that the interconnections between them must be quite short. This result has motivated model-building studies aimed at fitting the known amino acid sequence of the molecule to the observed pattern of packed α helices. Although the resolution of the current structure determination is not yet sufficiently detailed to allow structure examination at a level where individual residues can be distinguished, these studies have led to several important suggestions concerning the

**Figure 3-33**
Schematic representation of the hexose-induced hexokinase E and E′ are the inactive and active conformations of the enzyme, respectively. G is the sugar substrate. Regions of protein or substrate surface excluded from contact with solvent are indicated by a crinkled line (Adapted from W. S. Bennett and T. A. Steitz, Glucose-induced conformational change in yeast hexokinase, *Proc. Natl. Acad. Sci. USA* 75: 4848, 1978. Used with permission).

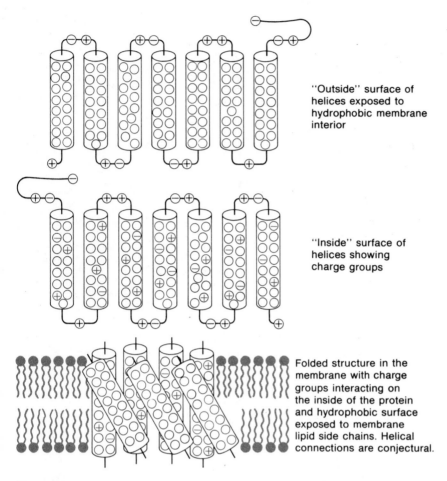

"Outside" surface of
helices exposed to
hydrophobic membrane
interior

"Inside" surface of
helices showing
charge groups

Folded structure in the
membrane with charge
groups interacting on
the inside of the protein
and hydrophobic surface
exposed to membrane
lipid side chains. Helical
connections are conjectural.

**Figure 3-34**
A model for the structure of bacteriorhodopsin.

structure of membrane proteins. First is the observation that the *regions of structure within the membrane appear to be primarily α-helical.* As outlined previously, α helices are secondary structures whose backbones are completely hydrogen bonded, except at their ends. Consequently, they might be readily inserted into membranes, since there are no unsatisfied hydrogen-bonded groups whose stabilization would require the formation of hydrogen-bonded interactions with water. Second, it has been shown that by fitting the known sequence to the observed pattern of helices it is possible to find an arrangement that *exposes only hydrophobic side chains to the hydrophobic environment of the membrane interior* (Figure 3-35). This orientation agrees with the results from neutron-diffraction experiments that show that the valines are located on the outer surface. This arrangement would necessitate that the charged amino acid side chains that do occur in the α helices interact in complementary fashion with each other and with the retinal group on the protein interior. Put simply, the

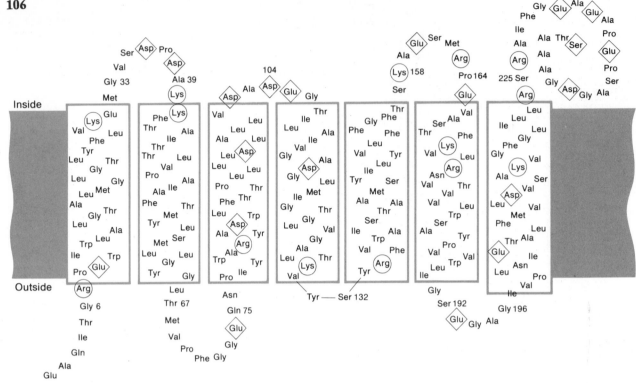

**Figure 3-35**
The amino acid sequence of bacterio-rhodopsin, arranged as unrolled α helices situated in the bacterial membrane. (Adapted From D. M. Engelman and G. Zaccai, Bacteriorhodopsin is an inside-out protein, *Proc. Natl. Acad. Sci. USA* 77: 5897, 1980).

membrane protein bacteriorhodopsin appears to be inside-out relative to globular proteins that exist in a polar water environment.

## QUATERNARY STRUCTURE

Although many globular proteins function as monomers, biological systems abound with examples of complex protein assemblies. This higher-order organization of several globular subunits to form a functional aggregate is referred to as the quaternary structure of the protein. Protein quaternary structures can be classified as being of two fundamentally different types. The first of these involves the assembly of different types of subunits. Examples range from dimeric molecules that contain different molecular subunits to complex assemblies such as ribosomes (which also contain ribonucleic acid as a structural component). The organization of these sorts of quaternary structures depends on the specific nature of each interaction made between the different molecular subunits and their neighbors. Each intermolecular interaction generally occurs only once within a given aggregate arrangement, so that the overall complex structure has a highly irregular geometry.

A second, more commonly observed pattern of quaternary structure is typified in molecular aggregates composed of multiple copies of one or more different kinds of subunits. Owing to the recurrence of specific structural interactions between the subunits, such aggregates typically form regular geometric arrangements. Recognizing that proteins are fundamentally asymmetrical objects (owing to their incorporation of chiral L-amino acids), it is clear that the simplest pattern of quaternary structure involves formation of a linear aggregate. As illustrated in Figure 3-36a, the formation of such an aggregate results from the repetition of one sort of specific structural interaction between adjacent subunits of the assembly. Such structures are extendable by the addition of successive subunits.

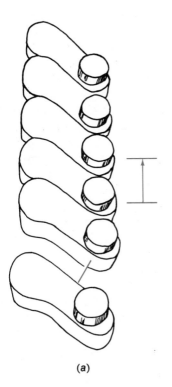

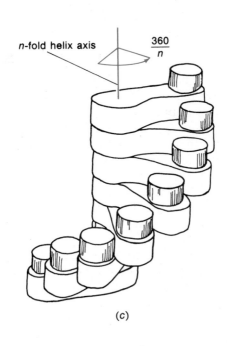

(c)

(a)

**Figure 3-36**
Linear and helical quaternary aggregates of protein molecules. (*a*) A linear arrangement of hypothetical protein subunits (illustrated as simplified right shoes). The interactions in such a linear arrangement are all identical, and the structure lends itself to the formation of an indefinitely long linear structure whose subunits are related by translation in one dimension. (*b*) A helical arrangement in which equivalently interacting subunits are related by unit translations along the helix axis followed by a 180-degree rotation about the helix axis. (*c*) A helical arrangement in which subunits are related by unit translation plus a rotation of 360 degrees/*n* to give an *n*-fold helix. (*d*) An *n*-fold multiple-start helix, an arrangement in which subunits form different equivalent interactions with their nearest neighbors. Filled and unfilled arrows indicate two types of identical contact points.

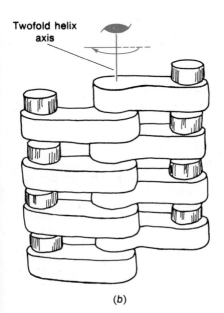

(b)

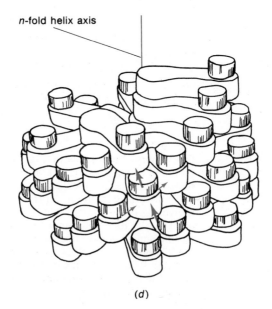

(d)

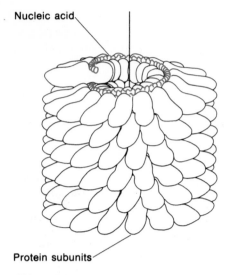

Nucleic acid

Protein subunits

**Figure 3-37**
Tobacco mosaic virus coat protein, an
example of a helical virus.

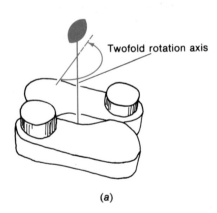

Twofold rotation axis

(a)

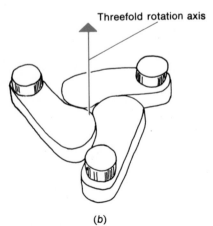

Threefold rotation axis

(b)

Somewhat more frequently observed than linear arrangements are helical arrangements of identical molecular subunits. As is the case for the amino acid residues in the $\alpha$ helix, the formation of helical quaternary structures reflects a situation in which the individual subunits make different, local interactions with their nearest neighbors (Figure 3-36b, c, and d). However, the pattern of nearest neighbor interactions is repeated for each subunit. Helical molecular aggregates are frequently associated with self-assembling molecular structures. Some outstanding examples of such aggregates are the rod-like and filamentous viruses, where the helical aggregation of the protein-coat subunits form a cylindrical container for the virus's nucleic acid (Figure 3-37). Since the entire coat is assembled from multiple copies of the same protein, it is evident that this represents a very efficient utilization of the information content of the virus nucleic acid. Helical quaternary structures are characterized by a repeating interaction that results in structures whose subunits are related both by some rise along and twist around a central axis. They are similar to linear arrangements, which have repeating rising interactions along the axis of the molecular chain, in that they are potentially indefinitely extendable. In fact, in helical viruses, the length of the coat-protein structure is not determined by the protein, but rather by the fixed length of the virus nucleic acid.

We also can imagine patterns of repeating molecular interactions that involve only twists between the subunits and so result in the formation of quaternary structures that are essentially like flat rings. Such cyclically repeating interactions typically give rise to symmetrical molecular dimers and trimers. Larger aggregates, in contrast, most frequently do not form flat-ring structures, since such structures would not have enough total contact surface to stabilize such an open, extended arrangement. Instead, they form arrangements that resemble geometric polyhedra. The formation of polyhedral aggregates reflects the fact that the molecular subunits can have more than one type of intermolecular interaction, so allowing multiple flat-ring arrangements to be assembled into larger, aggregated polyhedral assemblies. Figure 3-38 illustrates some of the types of structures observed. Such arrangements can be composed of identical subunits or of different types of subunits. One property that distinguishes the polyhedral and ring quaternary structures from linear and helical types is that they incorporate fixed numbers of subunit copies.

It is particularly manifest in the structures of helical and polyhedral viruses that quaternary structures play a central role in the self-assembly of very large biological structures from individual molecular subunits. The

**Figure 3-38**
Quaternary structures with rotational and polyhedral symmetry. (a) Arrangements of two molecules related by twofold rotational symmetry to form symmetrical dimers. (b) Symmetrical trimers. In these arrangements, the intersubunit contacts are all identical. (c) The most common arrangement for tetrameric molecules (as in hemoglobin), where each subunit makes three different interactions with its neighbors: a "side-by-side" interaction, a "toe-to-toe" interaction, and a "heel-to-heel" interaction. (d) A common arrangement for hexameric molecules. (e) Octameric molecules. (f) A cubic quaternary structure with 24 subunits as found in some iron-storage proteins. (g) An icosahedral quaternary structure with 60 subunits, 3 to each triangular face. This pattern is frequently seen in viruses.

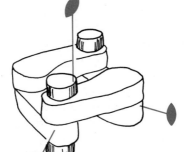

(c)

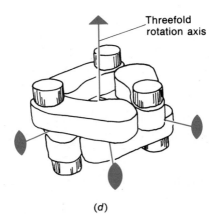

Threefold
rotation axis

(d)

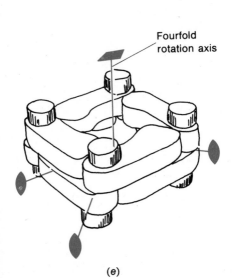

Fourfold
rotation axis

(e)

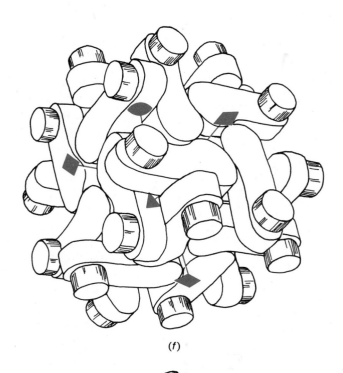

(f)

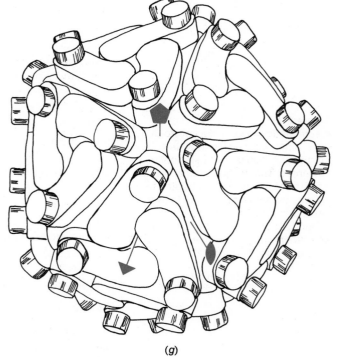

(g)

assembly of these symmetrical aggregates reflects the structural stabilization that occurs when all the subunits interact in geometrically similar ways, i.e., essentially like the atoms in a salt crystal. However, *one of the surprising results from x-ray crystallography concerns the frequency with which quaternary interactions are not symmetrical and equivalent, even between chemically identical subunits.* The simplest sort of nonequivalence occurs at some dimer contacts, where individual side chains close to the twofold axis (which in such cases is only approximate) are forced to take up different positions in order to avoid overlapping. Sometimes the departures from exact symmetry are purely local, in which case they probably have no functional consequences, but sometimes the nonequivalence extends to other parts of the subunit (e.g., in insulin and in malate dehydrogenase), where it can produce such effects as different binding constants for ligands. It is even easier, of course, for contacts between nonidentical subunits to be asymmetrical, such as, for instance, in hemoglobin, where the contact between the two $\beta$ chains is wider than that between the two $\alpha$ chains and produces the binding site for several important effector molecules (see below). An even more extreme case of asymmetrical association occurs in the dimer of yeast hexokinase, where in place of the pure 180-degree rotation of a twofold axis, the subunits are related by a rotation of 156 degrees plus a translation of 13.8 Å. Although this is basically a helical contact relationship, it cannot be extended past the dimer because a third subunit binding by the same rule would collide with the first one as in the model illustrated in Figure 3-39. In hexokinase the asymmetrical association creates a significant conformational difference between two initially equivalent subunits so that the two active sites in the dimer have quite different functional properties. Yet another type of nonequivalent association occurs in icosahedral viruses with only 60 symmetry-equivalent positions (see Figure 3-38g) but with more than 60 subunits. One way of reconciling this apparent contradiction is shown by the 180 subunits of tomato bushy stunt virus, which are placed so that 5 subunits are in contact around each fivefold axis, while 6 other subunits have a distinct but similar contact around each threefold axis. The versatility that permits such nonequivalent associations thus allows assembly of larger and more complex structures and may be even more common in biological structures too large to have been examined crystallographically.

The formation of a subunit aggregate can have extremely important functional results, as described below and in Chapter 5. In particular, it will be seen that the contact interactions provide a means of communication between the individual subunits. As a result, effects arising from the interaction of a ligand or substrate molecule with one subunit of an aggregate can influence the course of subsequent events in other subunits of the aggregate. Such interactions form the basis for cooperativity in biochemical systems and are of universal importance because they provide many of the control mechanisms for regulating biochemical processes.

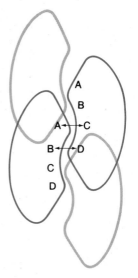

**Figure 3-39**
A schematic drawing of a heterologous dimer interaction in which infinite polymerization is sterically prevented. The dashed lines show that addition of a third subunit to bind to the free binding sites results in overlap of proteins. This arrangement of subunits is observed in the hexokinase dimer. (Adapted from a drawing obtained from T. A. Steitz).

## PROTEINS WHOSE FUNCTIONS RELATE TO MOVABLE PARTS

Thus far the static aspects of protein structure, i.e., the causes for folding and the characteristics of the folded state have been emphasized. A large number of proteins undergo reversible structural changes that are crucial to the functions they serve. The list of such proteins is long and varied; it includes ordinary enzymes, regulatory enzymes (Chapters 4 and 5), light-sensory proteins (Chapter 31), antibodies (Chapter 1), and active-transport membrane-bound proteins (Chapter 17). This subject will be introduced here by considering two systems that have been characterized at the molecular level: hemoglobin oxygen binding and the actinomyosin sliding filament of muscle.

## Hemoglobin: An Allosteric Protein

Hemoglobin is a four-subunit molecule with a molecular weight of 64,500. Each subunit is capable of binding a single molecule of oxygen. The main task of hemoglobin is to transport oxygen from the lungs to the capillaries of the tissues in order to meet the requirements of cellular metabolism. In muscle cells, a reserve oxygen store is provided by the myoglobin molecule, which has a similar structure to hemoglobin except that it exists as a monomer. While the components of myoglobin and hemoglobin are remarkably similar, their physiological responses are very different. On a weight basis, either molecule can bind about the same amount of oxygen at high oxygen tensions (pressures). At low oxygen tensions, however, hemoglobin gives up its oxygen much more readily. These differences are reflected in the oxygen-binding curves of the purified proteins in aqueous solution (Figure 3-40). The oxygen-binding curve for myoglobin is hyperbolic in shape, as

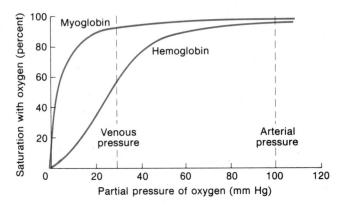

**Figure 3-40**
Equilibrium curves measure the affinity for oxygen of hemoglobin and of the simpler myoglobin molecule. Myoglobin, a protein of muscle, has just one heme group and one polypeptide chain and resembles a single subunit of hemoglobin. The vertical axis gives the amount of oxygen bound to one of these proteins, expressed as a percentage of the total amount that can be bound. The horizontal axis measures the partial pressure of oxygen in a mixture of gases with which the solution is allowed to reach equilibrium. For myoglobin, the equilibrium curve is hyperbolic. Myoglobin absorbs oxygen readily, but becomes saturated at a low pressure. The hemoglobin curve is sigmoid; initially hemoglobin is reluctant to take up oxygen, but its affinity increases with oxygen uptake. At arterial oxygen pressure, both molecules are nearly saturated, but at venous pressure, myoglobin would give up only about 10 percent of its oxygen, whereas hemoglobin releases roughly half. At any partial pressure, myoglobin has a higher affinity than hemoglobin, which allows oxygen to be transferred from blood to muscle.

would be expected for simple one-to-one association of myoglobin heme and oxygen:

$$Mb + O_2 \rightleftharpoons MbO_2$$

$$K_f = \frac{[MbO_2]}{[Mb][O_2]} = \text{equilibrium formation constant}$$

If $y$ is the fraction of myoglobin molecules saturated, and if the oxygen concentration is expressed in terms of the partial pressure of oxygen ($pO_2$), then

$$K_f = \frac{y}{(1 - y)pO_2} \quad \text{and} \quad y = \frac{KpO_2}{1 + KpO_2}$$

This is the equation of a hyperbola, as shown in Figure 3-40.

Hemoglobin behaves differently. Its sigmoidal binding curve can be fitted by an association-constant expression with a greater-than-first-power dependence on the oxygen concentration:

$$K_f = \frac{[HbO_2]}{[Hb][O_2]^n} \quad \text{and} \quad y = \frac{KpO_2{}^n}{1 + KpO_2{}^n}$$

The value of $n$ is generally quoted as 2.8, indicating that the binding of oxygen molecules to the four hemes in hemoglobin is not independent and that binding to any one heme is affected by the state of the other three. It seems likely that the first oxygen attaches itself with the lowest affinity, and each successive oxygen is bound with a higher affinity. The exact value of $n$ for hemoglobin is a function of the extent of oxygen binding as well as the presence of other factors discussed below. In general, a value of $n > 1$ indicates cooperative binding between small-molecule ligands (see Chapter 7), and a value of $n < 1$ indicates anticooperative binding. For myoglobin $n = 1$.

*The cooperative binding of oxygen by hemoglobin is ideally suited to its role as an oxygen transporter; in the lung, where the oxygen tension is relatively high (100 mm Hg), hemoglobin can become nearly saturated with oxygen, whereas in the tissues, where the oxygen tension is relatively low (~40 mm Hg), hemoglobin can release about half its oxygen* (see Figure 3-40). If myoglobin were used as the oxygen transporter, it can be seen that only about 10 percent of the oxygen would be released under similar conditions.

*The binding of a number of small molecules to hemoglobin influences the oxygen binding in a negative way.* Lowering the pH (i.e., increasing the $H^+$ concentration) decreases the affinity between hemoglobin and oxygen. For example, at pH 7.6 and 40 mm Hg oxygen tension, hemoglobin retains more than 80 percent of its oxygen; if the pH is lowered to 6.8, only 45 percent of the oxygen is retained. Likewise, 2,3-diphosphoglycerate (Figure 3-41) causes a decreased affinity between oxygen and hemoglobin (Figure 3-42). Neither $H^+$ nor diphosphoglycerate at the usually observed concentrations in vivo would interfere with the near-saturation uptake of oxygen that takes place in the lung tissue. However, at the lower oxygen tension that exists in oxygen-consuming tissues, increasing concentrations of $H^+$ or diphosphoglycerate cause a greater release of hemoglobin-bound oxygen. The net effect is that small shifts in pH (7.6 to 6.8) or low concentrations of

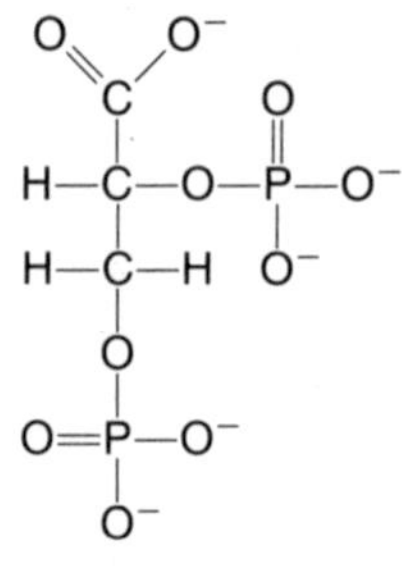

**Figure 3-41**
The structure of 2,3-diphosphoglycerate.

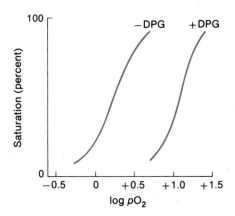

**Figure 3-42**
Oxygen-binding curve. $O_2$ concentrations in millimeters of mercury. Diphosphoglycerate, when present, is 5 m$M$, pH is 7. (Adapted from C. Bonaventura, J. Bonaventura, G. Amiconi, L. Tentori, M. Brunori, and E. Antonini, Hemoglobin Abruzzo. Consequences of altering the 2,3-diphosphoglycerate binding site, *J. Biol. Chem.* 250:6273, 1975.)

diphosphoglycerate (5 m$M$) make hemoglobin a more efficient transporter of oxygen. Both of these effects are likely to be physiologically significant. Proton concentrations tend to rise under the oxygen-demanding conditions of vigorous metabolic activity, and diphosphoglycerate found normally in red cells is present in higher concentrations in organisms living at high altitudes, where the oxygen supply is lower.

The cooperative binding of oxygen to hemoglobin as well as the inhibition of oxygen binding by protons or diphosphoglycerate can be understood in terms of the effects these small-molecule ligands have on the hemoglobin structure. Each of these ligands, i.e., oxygen, protons, and diphosphoglycerate, binds at different, nonoverlapping sites on the protein, meaning that the influence of one ligand binding on another must be transmitted through the protein polypeptide chains. Proteins that show this type of behavior are said to be _allosteric_, and the small-molecule ligands that influence their structure are referred to as _allosteric effectors_. Allostery is believed to be a mechanism used by a large number of proteins, including regulatory enzymes (Chapter 5), membrane transport proteins (Chapter 17), and gene regulatory proteins (Chapter 26).

Hemoglobin consists of two $\alpha$ subunits, each with 141 amino acids, and two $\beta$ subunits, each with 146 amino acids. These subunits are arranged with approximate 222 symmetry (see Figure 3-38c), as shown in Figure 3-43.

X-ray diffraction studies on fully oxygenated hemoglobin and deoxygenated hemoglobin have shown that the molecule is capable of existing in two states with significant differences in tertiary and quaternary structures. Further studies on partially oxygenated hemoglobin may reveal further intermediate structures between these two extremes. Until these can be characterized in structural terms, the two-state model serves as a useful conceptual framework for explaining the allosteric mechanism of the hemoglobin system.

The hemoglobin tetramer is composed of two identical halves (dimers) with the $\alpha_1\beta_1$ subunits in one dimer and the $\alpha_2\beta_2$ subunits in the other (Figure 3-44). The subunits within these dimers are tightly held together; the dimers themselves are capable of motion with respect to one another. _The interface between the movable dimers contains a network of salt bridges and hydrogen bonds when hemoglobin is in the deoxy conforma-_

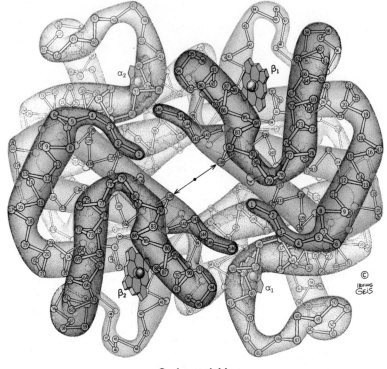

**Oxyhemoglobin**

**Figure 3-43**
Three-dimensional structure of oxy
and deoxy hemoglobin as determined by x-ray crystallography. This
is a view down the dyad axis with the
$\beta$ chains on top. In the oxy-deoxy
transformation (quaternary motion)
$\alpha_1\beta_1$ and $\alpha_2\beta_2$ dimers move as units
relative to each other. This allows
DPG to bind to the larger central cavity in the deoxy conformation. (See
Figure 3-46.)

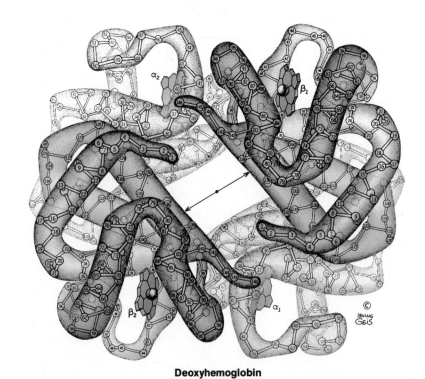

**Deoxyhemoglobin**

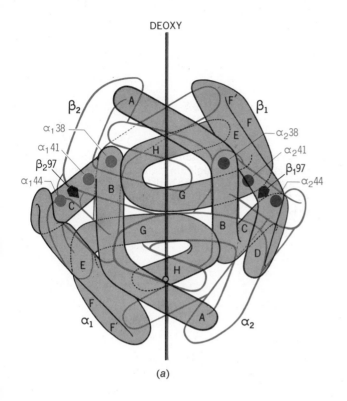

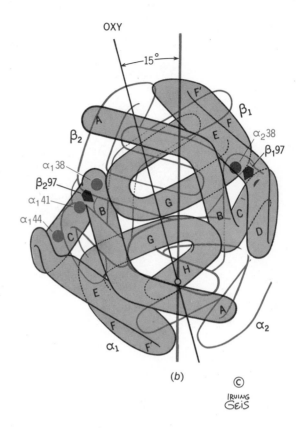

(a)

(b)

© IRVING GEIS

tion (Figure 3-45). *The quaternary transformation which takes place on binding of oxygen necessitates the breakage of these bonds.*

The effects of H⁺ and diphosphoglycerate on oxygen binding can be understood in terms of their stabilizing effect on the deoxy conformation. The decreased oxygen binding as the pH is lowered from 7.6 to 6.8 suggests the involvement of histidine side chains, because these are the only side-chain groups in proteins that can change their charge in this pH range. Certain histidines in the charged form make salt linkages that contribute to the stability of the deoxy form. As the pH is lowered, these histidines tend to become charged, which increases the stability of the deoxy form. This should discourage a structural transition to the oxy form and thereby lower the affinity of the protein for oxygen. Similarly, diphosphoglycerate binds most strongly to the deoxy form (Figure 3-46) and thereby discourages the transition to the oxy form, which lowers the affinity for oxygen.

The oxygen binding at the heme group itself initiates the tertiary- and quaternary-structure changes that are responsible for the cooperative effect seen on oxygen binding. The heme group contains an $Fe^{2+}$ ion located near the center of a porphyrin ring. This $Fe^{2+}$ makes four single bonds to the nitrogens in the heme ring, and a fifth bond to a histidine side chain of the F helix, F8 histidine. When oxygen is present it binds at the sixth coordination position of the iron on the other side of the heme. Movement of the

**Figure 3-44**
The deoxy-to-oxy shift upon binding oxygen in one hemoglobin molecule. The $\alpha_1\beta_1$ dimer moves as a unit relative to the $\alpha_2\beta_2$ dimer. The interface between the two dimers is crucial to the cooperativity effect in hemoglobin. (See Figure 3-45.)

116

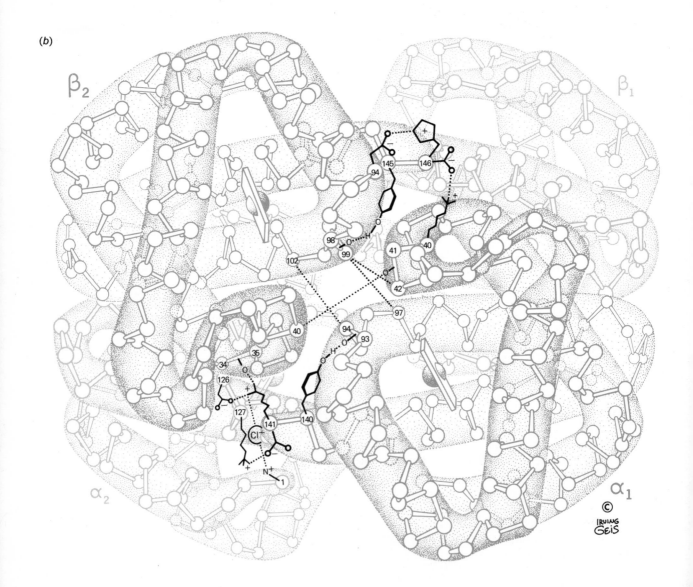

**Figure 3-45**
(a) The $\alpha_1\beta_2$ (and $\alpha_2\beta_1$) interface is shown schematically at left and in detail below. This is the regulatory zone of the hemoglobin molecule which contains crucial hydrogen bonds and salt bridges. (b) All the hydrogen bonds and salt bridges (dotted lines) shown here exist only in the deoxy state, with the exception of $\alpha_1 41 - \beta_2 40$ and $\alpha_1 94 - \beta_2 102$.

iron upon oxygen binding (or release) pulls the F8 histidine and the F helix to which it is covalently attached (Figure 3-48). *The tertiary-structure change in the F helix induces a strain in the rest of the protein that facilitates the conversion of the deoxy to the oxy structure. This favors the binding of further oxygen at other unoccupied sites on the tetramer.*

It seems likely that $Fe^{2+}$ was carefully chosen for its job in hemoglobin not only because it has a natural affinity for oxygen, but also because it changes its electronic structure in a highly significant way in so doing. $Fe^{2+}$ is normally paramagnetic as a result of four unpaired electrons in its outer $d$ electronic orbitals. As such, it is too large to sit precisely in the plane of a porphyrin, as studies with model compounds have shown. The $Fe^{2+}$ is also paramagnetic when it is pentacoordinated in deoxyhemoglobin, and as expected, it is displaced from the plane of the porphyrin by a few tenths of an angstrom unit. When $O_2$ becomes bound, the $Fe^{2+}$ becomes hexacoordinated and diamagnetic. This results in a major reorganization of its outer $d$ orbitals, which decreases the radius of the $Fe^{2+}$ so that it can move to an energetically more favorable position in the center of the porphyrin.

The structural arguments advanced here to explain oxygen binding by hemoglobin are supported by amino acid sequences of $\alpha$ and $\beta$ chains from a large number of vertebrate as well as invertebrate hemoglobins. As was shown in Chapter 1, data from 52 species of $\alpha$ chains and 49 species of $\beta$ chains reveal 49 invariant positions in the hemoglobin molecule. A glance back at Figure 1-26 will show that many of these invariant residues line the heme pockets and the $\alpha_1\beta_2$ interface. Furthermore, a comparison with

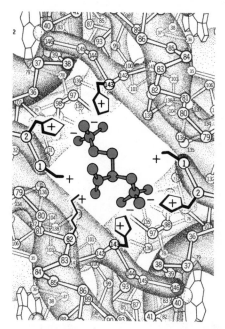

**Figure 3-46**
The binding of diphosphoglycerate in the central cavity of deoxyhemoglobin between $\beta$ chains. The surrounding positively charged residues are the amino terminal, His 2, Lys 82, and His 143.

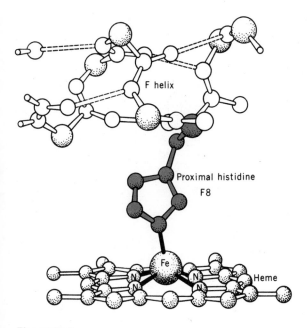

**Figure 3-47**
A close-up view of the iron-porphyrin complex with the F helix in deoxyhemoglobin. Note the out-of-plane iron and the side of approach of $O_2$.

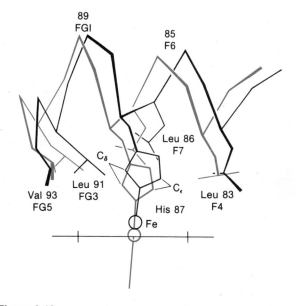

**Figure 3-48**
Oxy and deoxy superimposed in the region of the heme and F helix. Movement of His F8 is transmitted to FG5 valine, straining and breaking the hydrogen bond to the penultimate tyrosine. Only the $\beta$ chain is shown here.

Figure 1-27 shows that changes (mutations) in the region of this movable interface often cause altered oxygen affinity resulting in serious pathologies.

## Muscle: A Supermolecular Aggregate

Vertebrate muscle represents a remarkable example of a supermolecular aggregate capable of undergoing a reversible reorganization. Voluntary muscle tissue is arranged into fibers that are surrounded by an electrically excitable membrane called the _sarcolemma_. A fiber is composed of many _myofibrils_, which when viewed in the light microscope present a striated and banded appearance. As shown in Figure 3-49, each myofibril exhibits a longitudinally repeating structure called the _sarcomere_. This 23,000-Å-long repeating unit is characterized by the appearance of several distinct bands: the less optically dense band being referred to as the I band and the more dense one as the A band. Furthermore, a dense line appears in the center of the I band, called the Z line, and a dense narrow band somewhat similar in appearance also occurs in the center of the A band, called the M line. Adjacent to the M line are regions of the A band that appear less dense than the remainder and are termed the H zone.

Transverse sections of the sarcomere reveal that these band patterns result from the interdigitation of two sets of filaments (Figure 3-49). For example, when a sarcomere is sectioned in the I band, a somewhat disordered arrangement of thin filaments (about 70 Å in diameter) is seen. In contrast, when sectioned in the H zone, a hexagonal array of thick filaments (about 150 Å in diameter) is apparent. The substantive observation is that a transverse section in the dense region of the A band shows a

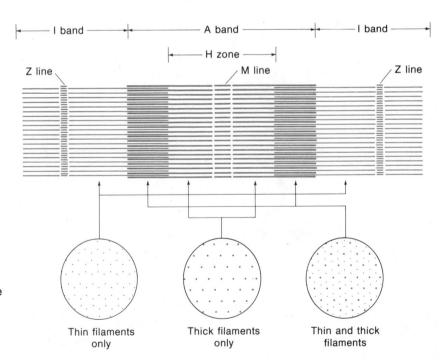

**Figure 3-49**
A schematic view of a striated muscle sarcomere showing the appearance of filamentous structures when cross-sectioned at the locations illustrated below.

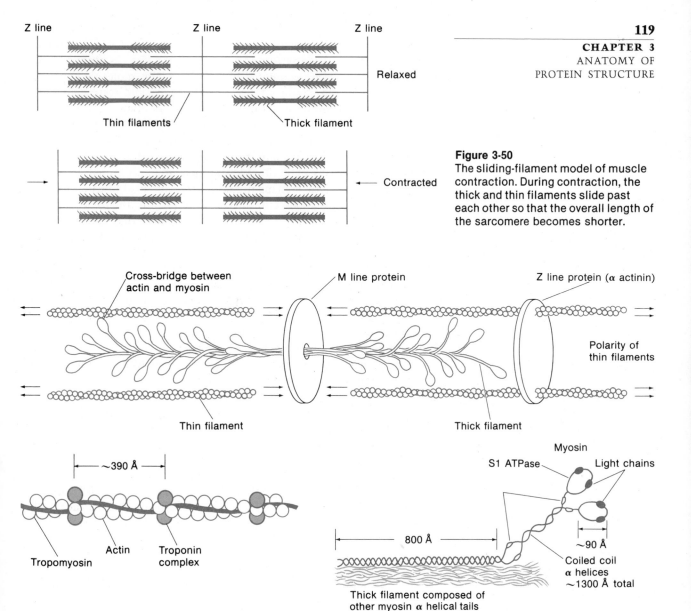

**Figure 3-50**
The sliding-filament model of muscle contraction. During contraction, the thick and thin filaments slide past each other so that the overall length of the sarcomere becomes shorter.

**Figure 3-51**
A molecular view of muscle structure.

regularly packed array of interdigitating thick and thin filaments. It was these observations that led H. E. Huxley to propose that the process of muscle contraction involved sliding the thick and thin filaments past each other, as shown in Figure 3-50.

Subsequent analyses have shown that the thin filaments are primarily composed of three proteins. The basic filamentous structure is a double-stranded helical arrangement of actin molecules with a molecular weight of 42,000 (Figure 3-51). Every turn of the double-stranded actin helix incorporates about 14 individual actin molecules and two molecules of a 400-Å-long filamentous molecule of *tropomyosin* that fits into the grooves of the actin helix. Two molecules of *troponin* ($M_r = 79,000$) bind to the actin filament at each helical repeat. These additional proteins, tropomyosin and troponin, play a regulatory role in muscle contraction, as described later.

Thick filaments are composed of myosin, a large molecule containing two identical heavy chains ($M_r = 200,000$ each) and four light chains ($M_r = 20,000$ each). The structural organization of myosin is illustrated in Figure 3-51. The molecule has two identical globular head regions that incorporate the light chains and a large fraction of the heavy chains. However, the tails of the heavy chains form very long $\alpha$ helices that are wrapped around each other to make double-stranded $\alpha$-helical coiled coils about 1350 Å long. Proteolytic enzyme digestion experiments indicate that about 800 Å of the $\alpha$-helical coiled tail is involved in forming the backbone of the thick filament, while the remainder forms an arm that can provide an extension for the globular head away from the body of the thick filament.

Given this marvelous piece of molecular architecture, the question remains as to how it actually works. The answer lies in the observation that actin cyclically binds the globular myosin headgroup to form cross-bridges in a reaction that depends on the myosin-catalyzed hydrolysis of adenosine triphosphate (ATP) (Figure 3-52). The cyclic binding of actin to myosin is driven by the energy-releasing hydrolysis of ATP catalyzed by the myosin headgroup in a manner that causes rearrangement of the actin-myosin cross-bridges between the thin and thick filaments. When muscle is completely relaxed, there is a minimum number of cross-bridges and the muscle is fully stretched; however, when the muscle is activated and under tension, it contracts and more cross-bridges are formed as the region of overlap between actin and myosin increases. At each stage of the contraction process it is essential to break the existing bridges with the help of ATP hydrolysis before new ones can be formed. It is important to realize that although ATP encourages more bridges to be formed, the ATP is required to break the bridges, not to form them. The breaking of bridges is required so that new and more numerous bridges can be formed. This means that cross-bridge formation is energetically favored. In this connection it is interesting to note that after death, when the ATP supply is exhausted, muscle enters a state of _rigor_ in which the muscle is fully contracted and the maximum number of bridges are formed.

The overall process of muscle contraction is initiated by depolarization of the sarcolemma membrane that surrounds the muscle fibers. This depolarization, which results from a nerve impulse communicated to the muscle, is accompanied by the release of $Ca^{2+}$ ions into the sarcoplasm, or cellular fluid, surrounding the muscle fibers. The $Ca^{2+}$ ions released bind to the troponin complex on the thin filaments, which, in turn, effects a structural change in the tropomyosin. In the absence of $Ca^{2+}$, tropomyosin prevents the binding of myosin to actin filaments. However, when $Ca^{2+}$ is present, a structural change takes place in tropomyosin that makes the myosin binding sites on actin accessible and so turns on the process of muscle contraction.

## FUNCTIONAL DIVERSIFICATION OF PROTEINS

The functional diversification of proteins has had its roots in a long evolutionary process. In addition to proteins which are structurally and functionally related, this has resulted in proteins with similar structures and

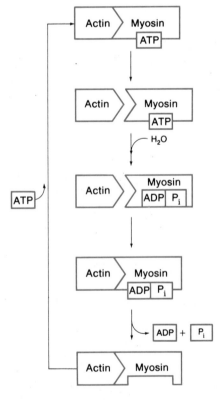

**Figure 3-52**
The sequence of events associated with cyclic ATP–dependent binding of actin to myosin.

dissimilar functions as well as proteins with dissimilar structures and similar functions.

## Proteins with Prosthetic Groups

The normal complement of amino acids that comprise proteins provide a variety of functional groups that can participate in enzymatic catalytic processes. However, the functional scope of proteins is considerably extended by their possession of favorable structures for the incorporation of metal ions or various larger compounds that have potentially useful chemical properties. In fact, several proteins incorporating metal ions or larger coenzyme or prosthetic groups have already been encountered, and additional examples will arise in subsequent descriptions of a variety of biochemical catalytic processes (especially see Chapters 5 and 6). There exist a relatively small number of available prosthetic groups that are either directly obtainable from the organisms' environment (e.g., metal ions) or require their own biosynthetic pathways to produce such complex compounds as flavins and hemes. However, *the functional properties of a given sort of prosthetic group can be greatly modified according to the type of interactions it makes with its surrounding polypeptide chain in a molecular complex.* The extent of structural and functional diversity that can arise when the same prosthetic group is wrapped up in different polypeptide chains is exemplified by some comparative properties of different heme-containing proteins.

Figure 3-53 shows the structure of protoporphyrin IX, one of the most commonly occurring of a number of porphyrin structural variants synthesized by living organisms. This is a macrocyclic ring system of four pyrrole rings with two vinyl and two propionic acid side chains attached to its periphery. Chelation of the dianionic form of protoporphyrin IX with an iron atom produces protoheme IX. Since the iron atom can potentially have different oxidation states, the net charge on the heme (neglecting the propionic acid side chains) reflects the difference between the porphyrin dianionic charge and the iron atom charge. Combination of the porphyrin dianion with $Fe^{2+}$ consequently produces a neutral species.

Three representative examples of heme-containing proteins of known structure are illustrated in Figure 3-54. Shown first is the structure of myoglobin. The ability of the heme iron in myoglobin to reversibly bind oxygen without itself becoming oxidized is a property conferred by interactions with residues of the surrounding polypeptide chain. On occasion, when the iron in myoglobin does become oxidized, a special enzyme system reduces it back to the $Fe^{2+}$ state.

The second heme protein shown in Figure 3-54 is mitochondrial cytochrome *c*. This molecule has 104 amino acid residues and contains regions of both α-helical and extended polypeptide-chain conformation. In contrast to myoglobin, the heme group in cytochrome *c* is covalently bound to the polypeptide backbone through thioether linkages formed by condensation of the protoheme IX vinyl groups with two cysteine side chains. In addition, the heme iron of cytochrome *c* forms two coordinate bonds with polypeptide side-chain groups, a histidine imidazole nitrogen and a methionine sulfur atom. Cytochrome *c* functions principally as an electron carrier in the mitochondrial electron-transport chain, during which time its heme iron is reversibly oxidized and reduced between the $Fe^{2+}$ and $Fe^{3+}$ oxida-

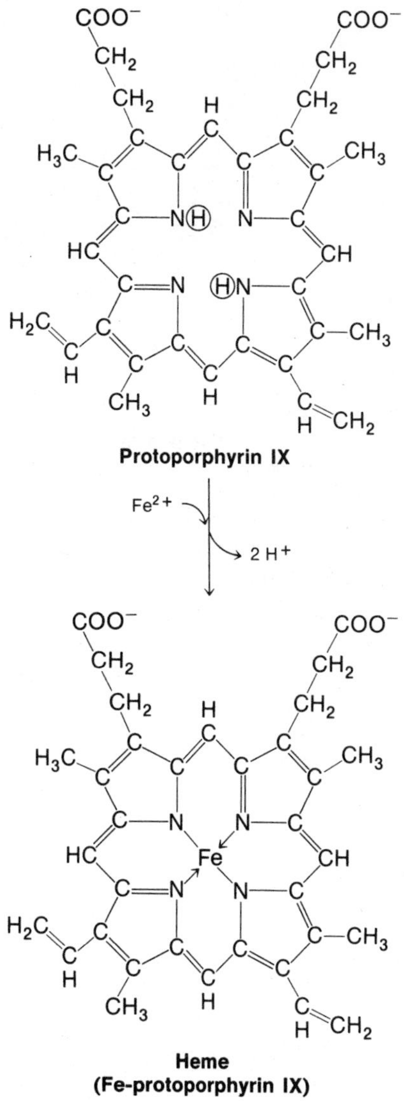

**Protoporphyrin IX**

**Heme
(Fe-protoporphyrin IX)**

**Figure 3-53**
Ligation of protoporphyrin IX by an iron atom results in the formation of the heme prosthetic group.

MYOGLOBIN

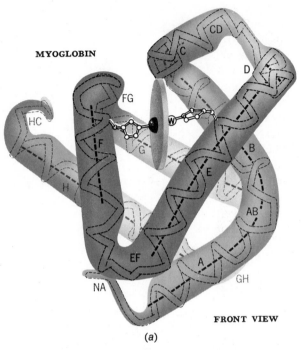

FRONT VIEW

*(a)*

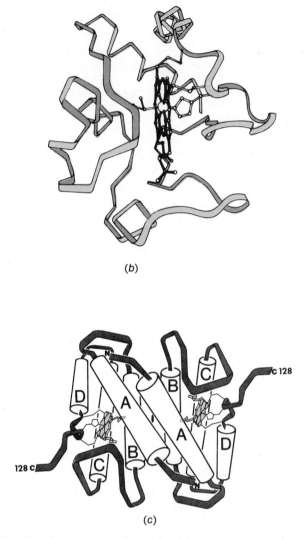

**Figure 3-54**
Some representative examples of
structurally and functionally different
heme proteins: (*a*) myoglobin, (*b*) mitochondrial cytochrome *c*, and (*c*) a dimeric cytochrome *c'* from photosynthetic bacteria.

tion states. In this molecule, the pattern of covalent interactions made
between the polypeptide chain and the heme group serves to both regulate
the iron atom's electromotive potential and orient the heme so that electron transfer can occur through the surface-exposed edge of the delocalized
porphyrin orbital system.

The third illustration in Figure 3-54 shows the structure of a dimeric
cytochrome *c'* from a photosynthetic bacterium. Each subunit is composed
primarily of four α helices. The subunit packing makes the hemes simultaneously accessible from the same side of the molecule. This gives the molecule the potential to simultaneously transfer two electrons (one from each
heme) to a physiological acceptor molecule.

The examples described here are but a few of the many functionally
different sorts of protein molecules that incorporate heme prosthetic
groups. Although all these molecules share a common prosthetic group, it is
in every case the specific pattern of interactions made between the heme
and the polypeptide chain that confers different functional properties on
the molecules as a whole.

An alternative way of viewing diversification of protein function involves an examination of how proteins change as a consequence of modification of their encoding DNA. In this context, the question of interest is how selective evolutionary pressures actually manifest themselves in the diversification of protein sequence, structure, and function.

Proteins serving similar or identical biological functions in different living organisms typically exhibit considerable similarity in both their amino acid sequences and tertiary structures. Such related families of molecules are generally assumed to result from processes of divergent evolution that involve the gradual and successive selective modification of an ancestral molecule owing to the fixation of mutations in the protein's encoding DNA. One of the most extensively studied protein families in this regard is the cytochrome *c* family. As indicated above, cytochrome *c* proteins function as electron carriers in the mitochondrial electron-transport chains of all multicellular organisms. Sequence comparisons of mitochondrial cytochrome *c* from organisms as diverse as humans and green plants reveal an extraordinary extent of sequence conservation. This suggests that the functional role of the molecule was highly refined by selective evolutionary pressures prior to the emergence of the first multicellular organisms. Some positions in the sequence are quite variable, whereas others are essentially invariant. Structural and chemical modification studies have shown that some of these invariant amino acid residues are associated with functionally important heme interactions, while others are important in governing the interactions of cytochrome *c* with its physiological oxidase and reductase.

The observation that the sequence and structure of mitochondrial cytochrome *c* appears to have been highly refined prior to the emergence of multicellular organisms suggests, of course, that its evolutionary precursors must exist in prokaryotic organisms. In fact, virtually all prokaryotic organisms that utilize oxidative or photosynthetic electron-transport chains to synthesize the high-energy intermediate adenosine triphosphate (ATP) incorporate molecules that are exceedingly similar to mitochondrial cytochrome *c*. However, as might be expected, cytochrome *c* proteins derived from prokaryotes exhibit considerably more sequence diversity than is typically seen among the proteins derived from higher organisms. In particular, the prokaryotic cytochrome *c* proteins frequently incorporate multiple amino acid insertions or deletions relative to mitochondrial cytochrome *c*. Tertiary-structure determination of several of these prokaryotic proteins has nevertheless revealed that these molecules are all variations on a basic structural theme (Figure 3-55). Further, these prokaryotic molecules all exhibit a strong conservation of amino acid residues that make functionally important interactions with the protein's heme group. The observed sequential and structural conservation of cytochrome *c* proteins consequently appear to represent slight variations of the basic structural theme to optimize the molecule's function in different organisms.

In addition to the cytochrome *c* proteins, several other families of proteins have been found to share both recognizable extents of amino acid sequence homology and tertiary-structural similarity. Again, the observed differences in sequence and structure among individual members reflect

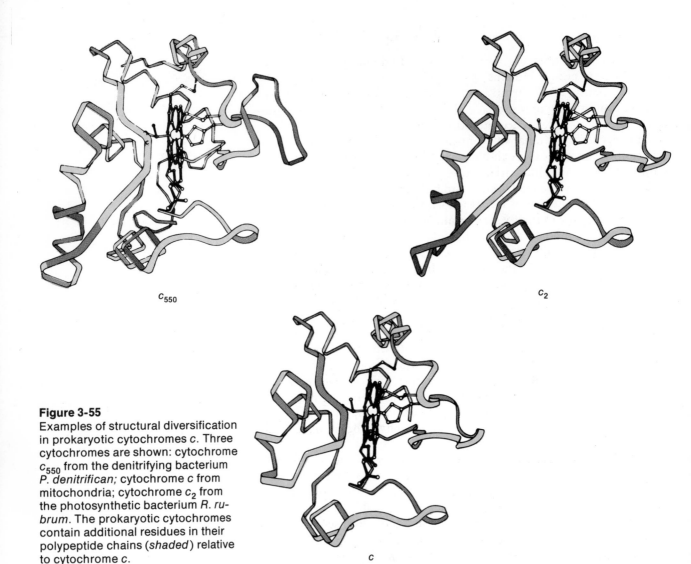

**Figure 3-55**
Examples of structural diversification in prokaryotic cytochromes $c$. Three cytochromes are shown: cytochrome $c_{550}$ from the denitrifying bacterium *P. denitrifican;* cytochrome $c$ from mitochondria; cytochrome $c_2$ from the photosynthetic bacterium *R. rubrum.* The prokaryotic cytochromes contain additional residues in their polypeptide chains (*shaded*) relative to cytochrome $c$.

$c_{550}$

$c_2$

$c$

evolutionary pressures that modified a basic structural arrangement in order to diversify the functional properties of the molecules. Examples of such structurally and functionally related families include the oxygen-binding globins (see Chapter 1), the dehydrogenases, and the serine protease enzyme families. Generally, related members within a given enzymatic family catalyze chemically similar reactions but exhibit varying specificities for structurally different substrate molecules. For example, while all serine proteases catalyze the hydrolytic cleavage of peptide bonds, different members of this molecule family cleave polypeptides at different locations determined by the nature of the amino acid side chains adjacent to the cleavage site (see Chapter 4).

Although divergent evolutionary processes generally appear to result in gradual changes in protein function, many situations arise where mutations occur that radically alter protein function. In fact, such mutations frequently result in the synthesis of functionally defective molecules and so constitute one cause of inheritable disease. Other examples exist where amino acid substitutions in related proteins result in the generation of new functions. The enzyme lysozyme, which binds and subsequently cleaves polysaccharide chains, and the protein α-lactalbumin, which transports sugars, show considerable similarity in both sequence and structure. In this case, it appears that relatively slight modifications of a common ancestral precursor has resulted in selection for molecules with quite different functions.

Although many functionally related families of proteins appear to have arisen from a process of divergent evolution from a common ancestor, there are other cases where proteins having extensive functional or structural similarities appear to have arisen independently. An outstanding example of a situation where two molecules are functionally similar but otherwise radically different in sequence and structure occurs in the serine proteases (Figure 3-56). As described earlier, many members of the serine protease family are closely related in sequence and structure. However, *Bacillus subtilis* produces a serine protease subtilisin which, while having an essentially identical arrangement of amino acid residues at the active site as the other serine proteases, otherwise differs from them completely in sequence and tertiary structure. This situation presumably manifests convergent evolution on a particular active-site arrangement required for the protein's catalytic function.

More frequently it is observed that proteins differing completely in sequence and function have quite similar tertiary structures. In these cases, the observed structural similarities most probably reflect selection of a particularly stable structural arrangement. Examples of such structurally related molecules include those with similarly twisted β sheets (Figure 3-21) and proteins organized as a bundle of four closely packed α helices (Figure 3-27).

## Molecular Evolution and Gene Splicing

The preceding descriptions of protein diversification had the common feature that they envisioned evolutionary processes to occur as a consequence of the continuing selection of individual point mutations in the protein's encoding DNA. However, many proteins exhibit structural characteristics that suggest that they have resulted from processes of gene splicing. In particular, a surprisingly large fraction of known protein structures incorporate multiple copies of structurally similar domains. In many cases it appears that these molecules have resulted from the splicing together of duplicate or multiple copies of a gene coding for a given structural domain, followed by the essentially independent fixation of mutations throughout the spliced genome. The eventual result of this is a protein composed of sequentially different but structurally similar repeating domains.

Additional evidence for the role of gene splicing in protein evolution comes from the observation that some large proteins are composed of several different structural domains, each of which may individually structurally resemble either parts or the entirety of other known proteins. A good example of this is the glycolytic enzyme pyruvate kinase. This large protein (Figure 3-57) is organized as three structural domains, two of which show convincing structural similarities to, respectively, the β barrel of triose

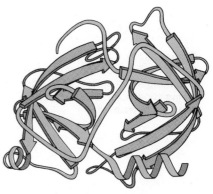

**Elastase**

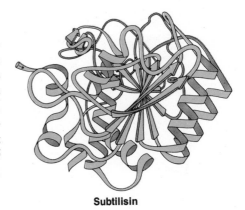

**Subtilisin**

**Figure 3-56**
Schematic illustration of the serine protease elastase and subtilisin. Their molecules differ totally in sequence and tertiary structure but have catalytic sites that are nearly identical. The configuration of the active site for elastase is described in Chapter 4.

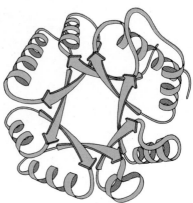

**Pyruvate kinase domain 1**

**Pyruvate kinase domain 2**

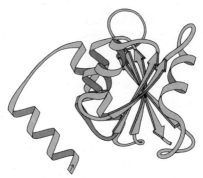

**Pyruvate kinase domain 3**

**Figure 3-57**
Pyruvate kinase domains 1, 2, and 3 as an example of a protein whose domains show no structural resemblance whatsoever.

phosphate isomerase (domain 1) and a twisted $\beta$-sheet domain common to many dehydrogenases (domain 3). The third domain of pyruvate kinase (domain 2) also has a convincing similarity to a common structural type: the Greek key antiparallel $\beta$ barrel.

## Protein Diversification in Action: Antibodies

In the preceding section, examples were given that illustrated that protein functional evolution involved both the fixation of individual point mutations and larger-scale rearrangements of a protein's encoding DNA. Further, it was apparent that the rate of fixation of mutations differed widely for different regions within a protein's sequences. For example, even in the case of "highly evolved" molecules such as cytochrome $c$, it was stated that some sequence positions showed much greater extents of interspecies variability than others. However, in all these cases it was implied that the fixation of new mutations was a relatively infrequent event, resulting in the gradual evolution of proteins such as cytochrome $c$. Nevertheless, one of the important biological defense mechanisms of higher organisms, the immune response, essentially depends on the very rapid generation of structurally novel molecules that can recognize and bind foreign substances that may be deleterious to the organism. The molecules responsible for the initial recognition and binding of foreign substances are the immunoglobins. These molecules are composed of two pairs of polypeptide chains of different length that are interconnected by covalent cysteine disulfide linkages (Figure 3-58). Sequence studies of various immunoglobins have shown that both the longer or heavy ($M_r \sim 50,000$) and shorter or light ($M_r = 25,000$) polypeptide chains contain repeating homologous sequences about 110 residues in length. Structural studies of the immunoglobin molecule show that the sequentially homologous regions fold individually into similar structural domains arranged as a bilayer of antiparallel $\beta$ sheets. The molecule in its entirety is consequently formed of 12 similar structural domains, of which 8 are formed by the two heavy chains and 4 by the two light chains.

**Immunoglobulin $V_L$ domain**

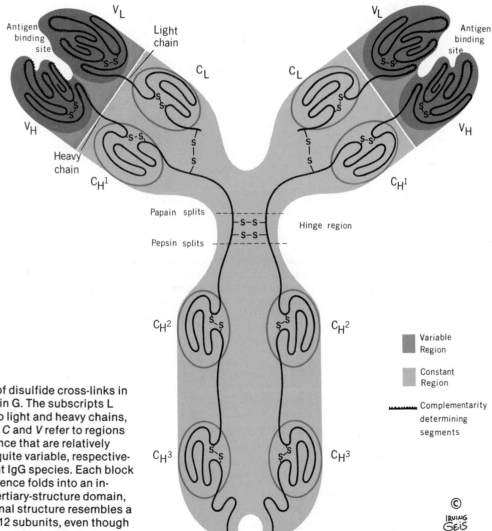

V_L

Antigen
binding
site

Light
chain

$C_L$

$C_L$

V_L

Antigen
binding
site

S-S

S-S

S-S

S-S

V_H

Heavy
chain

$C_H^1$

S-S

S-S

S-S

S-S

$C_H^1$

V_H

Papain splits

Pepsin splits

S-S

S-S

Hinge region

$C_H^2$

S-S

S-S

$C_H^2$

$C_H^3$

S-S

S-S

$C_H^3$

Variable
Region

Constant
Region

Complementarity
determining
segments

© IRVING GEIS

**Figure 3-58**
The pattern of disulfide cross-links in immunoglobin G. The subscripts L and H refer to light and heavy chains, respectively; *C* and *V* refer to regions of the sequence that are relatively constant or quite variable, respectively, in different IgG species. Each block of each sequence folds into an independent tertiary-structure domain, so that the final structure resembles a protein with 12 subunits, even though it is a single covalent entity.

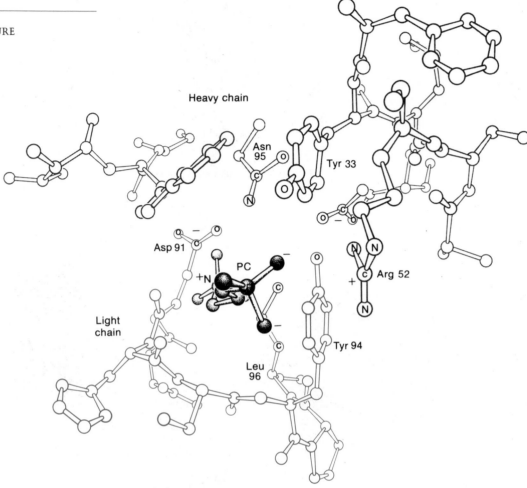

**Figure 3-59**
Frontal view of an antibody combining site. Heavy-chain complementary-determining regions occupy the upper portion. Light-chain residues contributing to the binding site are located mainly in the lower portion. The phosphorylcholine is centrally located. Phosphorylcholine (PC) is shown in black. (Adapted from an illustration contributed by David Davies.)

Immunoglobins expressing different specificities for binding foreign substances exhibit considerable sequence variability in the amino-terminal domains of both the heavy and light chains. It is these variable regions that form the binding sites between the immunoglobin molecule and foreign substances that trigger the immune response (Figure 3-59). The remarkable property of this system is that it can rapidly diversify the sequence of variable regions by mutation, gene splicing, and RNA splicing (also see Chapter 1). The net result is that the organism possesses the capacity to produce an enormous variety of antibodies from a quite limited amount of informational DNA originating in the germ-line tissue.

## SELECTED READINGS

Baldwin, J., and Chothia, C. Hemoglobin: The structural changes related to ligand binding and its allosteric mechanism *J. Mol. Biol.* 129:175, 1979.

Dickerson, R. E., and Geis, I. *Hemoglobin: Structure, Function, Evolution and Pathology.* Menlo Park, Calif.: Benjamin/Cummings, 1982.

Klotz, I. M., Langerman, N. R., and Darnell, D. W. Quaternary structure of proteins. *Ann. Rev. Biochem.* 39:25, 1970.

Pauling, L. *The Nature of the Chemical Bond,* 3d Ed. Ithaca, N.Y.: Cornell University Press, 1960.

Perutz, M. F. Regulation of oxygen affinity of hemoglobin: Influence of structure of the globin on the heme iron. *Ann. Rev. Biochem.* 48:327, 1979.

Richardson, J. S. The anatomy and taxonomy of protein structure. *Adv. Protein Chem.* 34:167, 1981.

Rossmann, M. G., and Argos, P. Protein folding. *Ann. Rev. Biochem.* 50:497, 1981.

## PROBLEMS

1. Which of the naturally occurring amino acid side chains are charged at pH 7, 2, and 13?

2. Which amino acid side chains can function as hydrogen bond donors, acceptors, or both?

3. What is the principal driving force for protein folding?

4. Outline the hierarchy of protein structural organization.

5. Many proteins are anchored to membranes by sections of polypeptide chain on their amino terminals. What would one expect the probable structure to be for a sequence Met–Ala–Leu–Phe–Leu–Leu–Met–Ala-Ala-Leu-Gly-Pro-Asn-Ala-Met-Leu-Phe-Leu-Leu-Ala-Ala-Met-and why would it tend to insert?

6. One of the proteins shown in Figure 3-56 illustrating the handedness of crossover connections in $\beta$ sheets has a left-handed connection. Which one?

7. You have isolated several multimeric proteins with the following subunit compositions: (a) $\alpha\beta\gamma$, (b) $\alpha_2\beta_2$, (c) $\alpha_3\beta_5$, (d) $\alpha_6\beta_6$, (e) $\alpha_{24}\beta_{24}$, and (f) $\alpha_{60}$. Which of these might you expect to have regular polyhedral quaternary structures?

8. Write a short essay outlining the relationship between protein structure, function, and evolution.

9. If you separated hemoglobin into its respective subunits, would you expect it to bind more or less oxygen at low oxygen tensions? Explain your answer. What effect would you expect DPG to have on the oxygen binding to monomers?

10. What is the cause of rigor after death?

11. Which is more stable, a right-handed or a left-handed $\alpha$-helix of polyglycine?

12. Polyhistidine is insoluble at pH 7.8 but soluble at pH 5.5. Suggest an explanation.

13. The right handed $\pi$ helix has 4.4 amino acid residues per turn with acceptable $\psi$ and $\phi$ values and good interchain hydrogen bonds. However, the $\pi$ helix has a 1 Å hole down the core which makes it considerably less stable than the $\alpha$ helix. Why?

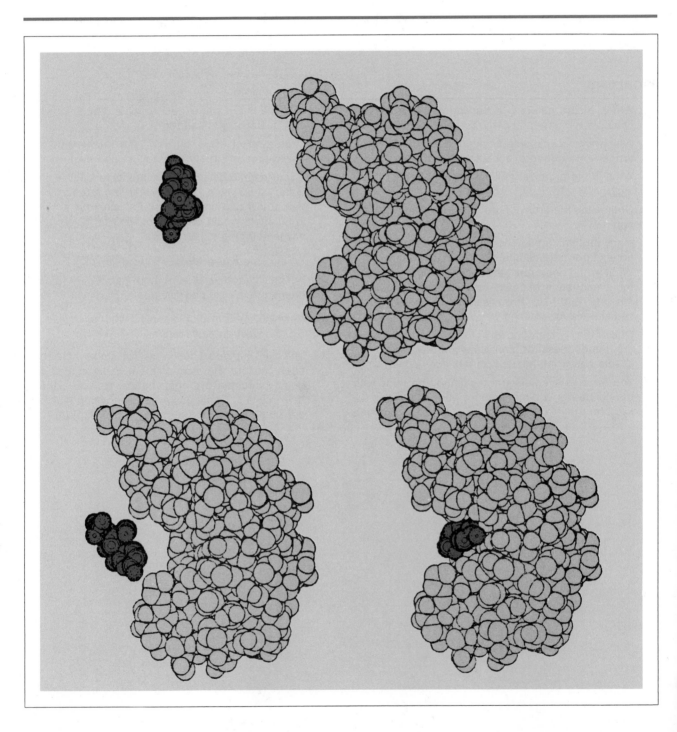

# 4

# ENZYME CATALYSIS I

Many chemical reactions take place spontaneously when the reactants are mixed together. The reaction of hydrogen ions with hydroxide ions to make water is an example. Mixing a strong acid with a strong base in aqueous solution will result in a very rapid neutralization reaction. The extent of the reaction is controlled solely by the amount of reactants added to the system. Other reactions, such as that between $H_2$ and $O_2$ to make water, will not occur spontaneously at ambient temperatures, but will take place with explosive violence if the temperature is raised. Both of these reactions are energetically favorable. In thermodynamic terms, they both occur with a large release of free energy (see Chapter 7). The latter reaction does not occur rapidly at ambient temperatures because of a "kinetic barrier."

In the living cell, the vast majority of chemical reactions that are of value to the organism have a kinetic barrier preventing their spontaneous occurrence. This kinetic barrier is overcome by the enzyme catalyst. The directive role of the enzyme catalyst is to make energetically favorable reactions take place at a rate that is conducive to sustaining and promoting growth.

In this chapter we will discuss general aspects of enzyme catalysis and reactions in which a single substrate molecule is being acted on. In the following chapter we will discuss reactions that involve more than one substrate molecule and regulation of enzyme activity.

## WHAT KINDS OF REACTIONS DO ENZYMES CATALYZE?

There are thousands of known enzymes and many more catalyzed reactions for which the enzymes are still to be discovered. No system of classification is perfect, but most of them have some didactic value. The currently

*Computer graphic display showing the substrate (3'-uridine-monophosphate) "docking" into the active site of the enzyme bovine ribonuclease A. (Courtesy William Gilbert of the Massachusetts Institute of Technology.)*

favored scheme of classification was adopted by the International Union of Biochemistry in 1961. Each enzyme is designated by four numbers separated by periods: the main class, the subclass, the sub-subclass, and the serial number. There are six main classes: oxidoreductases, transferases, hydrolases, lyases, isomerases, and ligases or synthases. Oxidoreductases catalyze oxidation-reduction reactions. Transferases catalyze the transfer of a molecular group from one molecule to another. Hydrolases catalyze cleavage by the introduction of water. Lyases catalyze reactions involving removal of a group from a double bond or addition of a group to a double bond. Isomerases catalyze reactions involving intramolecular rearrangements. Finally, ligases or synthases catalyze reactions joining together two molecules. A sampling of some of the divisions of main classes into further subdivisions with specific examples is given in Table 4-1. Enzymes are indicated by two names, a systematic name determined by the Enzyme Commission and a common name, which is used more frequently. This list is intended to give some sense of order to the seemingly chaotic array of enzyme-catalyzed reactions. All these reactions will be discussed at some point in various chapters of this text.

## SOME GENERAL ASPECTS OF CHEMICAL CATALYSIS

### The Rate of Chemical Reactions

A rate equation for a reaction expresses the rate of disappearance of reactant(s) or the rate of appearance of product(s). The simplest type of reaction we can imagine is that for an irreversible unimolecular process such as the decay of an unstable isotope. $^{32}P$ decays with a fixed rate constant:

$$\text{Rate of decay} = \frac{-d[^{32}P]}{dt} = k[^{32}P]$$

where $k$ is the rate constant, $[^{32}P]$ is the concentration of radioactive isotope, and $(-d[^{32}P])/dt$ is the rate of disappearance of $^{32}P$. This rate is proportional to the first power of the concentration of isotope, and the reaction is accordingly described as a first-order reaction. Reactions that are proportional to the square of the concentration of the reactant are referred to as second-order reactions. The reaction of nitrous oxide with hydrogen to form nitrogen and water is second order with respect to the nitrous oxide concentration and first order with respect to the hydrogen concentration.

The overall reaction, however, is described as third order:

$$2\,NO + 2\,H_2 \longrightarrow N_2 + 2\,H_2O$$

$$\frac{-d[NO]}{dt} = k[NO]^2[H_2]$$

Notice in this case that the rate of reaction is first order with respect to the hydrogen concentration, even though two $H_2$ molecules appear in the overall reaction leading from reactants to products. A detailed kinetic analysis of this reaction has shown that it takes place in several steps. Each step has a characteristic rate constant associated with it. The slowest step involves

**Table 4-1**
Classification of Enzymes

| Number | Systematic Name | Common Name | Reaction |
|--------|-----------------|-------------|----------|
| 1. | Oxidoreductases | | |
| 1.1 | Acting on CH—OH group of donors | | |
| 1.1.1 | With NAD or NADP as acceptor | | |
| 1.1.1.1 | Alcohol:NAD oxidoreductase | Alcohol dehydrogenase | Alcohol + NAD $\rightleftharpoons$ aldehyde or ketone + NADH |
| 1.1.3 | With $O_2$ as acceptor | | |
| 1.1.3.4 | $\beta$-D-glucose:oxygen oxidoreductase | Glucose oxidase | $\beta$-D-glucose + $O_2$ $\rightleftharpoons$ D-glucono-$\delta$-lactose + $H_2O_2$ |
| 1.2 | Acting on the aldehyde or keto-group of donors | | |
| 1.2.1 | With NAD or NADP as acceptor | | |
| 1.2.3 | With $O_2$ as acceptor | | |
| 1.2.3.2 | Xanthine:oxygen oxidoreductase | Xanthine oxidase | Xanthine + $H_2O$ + $O_2$ $\rightleftharpoons$ urate + $H_2O_2$ |
| 1.3 | Acting on the CH—CH group of donors | | |
| 1.3.1 | With NAD or NADP as acceptor | | |
| 1.2.1.1 | 4,5-Dihydrouracil:NAD oxidoreductase | Dihydrouracil dehydrogenase | 4,5-Dihydrouracil + $NAD^+$ $\rightleftharpoons$ uracil + NADH |
| 1.3.2 | With a cytochrome as an acceptor | | |
| 1.4 | Acting on the CH—$NH_2$ group of donors | | |
| 1.4.3 | With $O_2$ as acceptor | | |
| 1.4.3.2 | L-Amino acid:oxygen oxidoreductase (deaminating) | L-Amino acid oxidase | L-amino acid + $H_2O$ + $O_2$ $\rightleftharpoons$ 2-oxo-acid + $NH_3$ + $H_2O_2$ |
| 2. | Transferases | | |
| 2.1 | Transferring C-1 groups | | |
| 2.1.1 | Methyltransferases | | |
| 2.1.1.2 | S-Adenosylmethionine: guanidinoacetate N-methyltransferase | Guanidinoacetate methyltransferase | S-Adenosylmethionine + guanidinoacetate $\rightleftharpoons$ S-adenosylhomocysteine + creatine |
| 2.1.2 | Hydroxymethyltransferases and hydroxyformyltransferases | | |
| 2.1.2.1 | L-Serine:tetrahydrofolate 5,10-hydroxymethyltransferase | Serine hydroxymethyl-transferase | L-Serine + tetrahydrofolate $\rightleftharpoons$ glycine + 5,10-methylenetetra-hydrofolate |
| 2.1.3 | Carboxyltransferases and carbamoyltransferase | | |
| 2.2 | Transferring aldehydic or ketonic residues | | |
| 2.3 | Acyltransferases | | |

(Continued)

**Table 4-1** (*Continued*)

| Number | Systematic Name | Common Name | Reaction |
|--------|-----------------|-------------|----------|
| 2.4 | Glycosyltransferases | | |
| 2.6 | Transferring N-containing groups | | |
| 2.6.1 | Aminotransferases | | |
| 2.6.1.1 | L-Aspartate : 2-oxoglutarate | Aspartate aminotransferase | L-Aspartate + 2-oxoglutarate $\rightleftharpoons$ oxaloacetate + L-glutamate |
| 2.7 | Transferring P-containing groups | | |
| 2.8 | Transferring S-containing groups | | |
| 3. | Hydrolases | | |
| 3.1 | Cleaving ester linkage | | |
| 3.1.1 | Carboxylic ester hydrolases | | |
| 3.1.1.7 | Acetylcholine acetylhydrolase | Acetylcholinesterase | Acetylcholine + $H_2O$ $\rightleftharpoons$ choline + acetic acid |
| 3.1.3 | Phosphoric monoester hydrolases | | |
| 3.1.3.9 | D-Glucose-6-phosphate phosphohydrolase | Glucose-6-phosphatase | D-Glucose-6-phosphate + $H_2O$ $\rightleftharpoons$ D-glucose + $H_3PO_4$ |
| 3.1.4 | Phosphoric diester hydrolases | | |
| 3.1.4.1 | Orthophosphoric diester phosphohydrolase | Phosphodiesterase | A phosphoric diester + $H_2O$ $\rightleftharpoons$ a phosphoric monoester + an alcohol |
| 4. | Lyases | | |
| 4.1 | C—C lyases | | |
| 4.1.1 | Carboxy lyases | | |
| 4.1.1.1 | 2-Oxo-acid carboxy-lyase | Pyruvate decarboxylase | A 2-oxo-acid $\rightleftharpoons$ an aldehyde + $CO_2$ |
| 4.1.2 | Aldehyde lyase | | |
| 4.1.2.7 | Ketose-1-phosphate aldehydelyase | Aldolase | A ketose-1-phosphate $\rightleftharpoons$ dihydroxyacetone phosphate + an aldehyde |
| 4.2 | C—O lyases | | |
| 4.2.1 | Hydrolases | | |
| 4.3 | C—N lyases | | |
| 4.3.1 | Ammonia-lyases | | |
| 4.3.1.3 | L-Histidine ammonia-lyase | Histidine ammonia-lyase | L-Histidine $\rightleftharpoons$ urocanate + $NH_3$ |
| 5. | Isomerases | | |
| 5.1 | Racemases and epimerases | | |
| 5.1.3 | Acting on carbohydrates | | |
| 5.1.3.1 | D-Ribulose-5-phosphate 3-epimerase | Ribulose phosphate epimerase | D-Ribulose-5-phosphate $\rightleftharpoons$ D-xylulose-5-phosphate |
| 5.2 | Cis-trans isomerases | | |

(*Continued*)

**Table 4-1** (*Continued*)

| Number | Systematic Name | Common Name | Reaction |
|---|---|---|---|
| 5.3 | Intramolecular oxidoreductases | | |
| 5.3.1 | Interconverting aldoses and ketoses | | |
| 5.4 | Intramolecular transferases | | |
| 6. | Ligases | | |
| 6.1 | Forming C—O bonds | | |
| 6.1.1 | Amino acid–RNA ligases | | |
| 6.1.1.1 | L-Tyrosine:tRNA ligase (AMP) | Tyrosyl-tRNA synthase | ATP + L-tyrosine + tRNA $\rightleftharpoons$ AMP + pyrophosphate + L-tyrosyl-tRNA |
| 6.2 | Forming C—S bonds | | |
| 6.3 | Forming C—N bonds | | |
| 6.3.1 | Acid-ammonia ligases | | |
| 6.3.2 | Acid–amino acid ligases | | |
| 6.4 | Forming C—C bonds | | |
| 6.4.1 | Carboxylases | | |
| 6.4.1.2 | Acetyl-CoA:carbon dioxide ligase (ADP) | Acetyl-CoA carboxylase | ATP + acetyl-CoA + $CO_2$ + $H_2O$ $\rightleftharpoons$ ADP + orthophosphate + malonyl-CoA |

the reaction of a dimer of the NO molecule with one $H_2$ molecule. Other steps in the reaction take place much more rapidly. The various steps in the reaction are

$$2\,NO \underset{k_{-1}}{\overset{k_1}{\rightleftharpoons}} N_2O_2 \qquad \text{Rapid equilibrium}$$

$$N_2O_2 + H_2 \xrightarrow{k_2} N_2O + H_2O \qquad \text{Slow, rate-limiting step}$$

$$N_2O + H_2 \xrightarrow{k_3} N_2 + H_2O \qquad \text{Rapid step}$$

The rate constant that is actually measured if one only observes the rate of formation of products is that associated with step 2 above. This rate will be proportional to the first power of the hydrogen concentration and to the first power of the $N_2O_2$ concentration. The concentration of $N_2O_2$ should be proportional to the square of the concentration of NO according to the first step. Thus the predicted reaction rate should be third order, as is observed. This example illustrates two important rules about reaction rates: (1) *the reaction rate tells us more about the mechanism whereby a reaction proceeds than the overall stoichiometry, and (2) in a multistep process, the overall reaction is rate-limited by the slowest step*. Clearly, one must be able to measure intermediates in a multistep process if one is to decipher the mechanism. This is often a most challenging problem for the kineticist.

## Rates Are Determined by Effective Collisions between Reactants

In the reaction just described, the reactants involved in the rate-limiting step are both gases. Reaction requires that the gaseous molecules $N_2O_2$ and $H_2$ make an *effective collision* to become converted into the immediate products $N_2O$ and $H_2O$. Two factors must be scrutinized in determining why this is the rate-limiting step. First, how many collisions are these two molecules likely to make in a given time? And second, what percentage of these collisions is effective in leading to products? The number of collisions is a function of the concentration of the two reactants and the speed of the molecules. The probability that a given collision will be effective depends on the so-called *activation-energy barrier*. The activation energy $E_a$ is the difference between the free energy of the reactants and the highest free-energy state that the reactants will have to achieve before going to products. Since the activation energy is a free energy, both enthalpic and entropic terms are involved (see Chapter 7). This highest free-energy state is referred to as the *transition state*, or the *activated complex*. Kinetic theory states that the fraction of molecules $f$ that have enough kinetic energy to attain this state is a function of the temperature according to the following equation:

$$f = e^{-E_a/(RT)} \tag{1}$$

where $e$ is the base of the natural logarithm, $R$ is the molar gas constant, and $T$ is the absolute temperature. The rate constant for the reaction is related to the activation energy by the following equation:

$$k = Ae^{-E_a/(RT)} \qquad \text{or} \qquad \ln k = \ln A - \frac{E_a}{RT} \tag{2}$$

where $A$ is a constant related to the collision frequency. $A$ also depends on the temperature, because the number of collisions increases with temperature. However, this dependence is minimal and can be ignored in most treatments of interest to biochemists because they work over a narrow range of temperatures. According to Equation (2), the rate constant is a very sensitive function of the activation energy. It is clear that if one had a way of lowering the activation energy, the rate constant would be greatly increased. Likewise, raising the temperature would make the reaction go faster. In the reaction of hydrogen ions and hydroxide ions to make water, the $E_a$ is very small, so that essentially all collisions are effective. The reaction rate is diffusion-limited; that is, it is equal to the number of collisions made in a given time. In the case of the reaction of $H_2$ and $O_2$ to make water, the $E_a$ is substantial, and it is necessary to raise the temperature.

### The Function of a Catalyst

In industrial processes, reactions are usually accelerated by raising the temperature and adding a catalyst. The catalyst is a "third party" that lowers the activation energy of a reaction without itself being consumed (Figure 4-1). It forms an intermediate complex with one or more of the reactants. Biological systems are strongly dependent on catalysts to make reactions go faster because they must function at moderate temperatures. It is remarka-

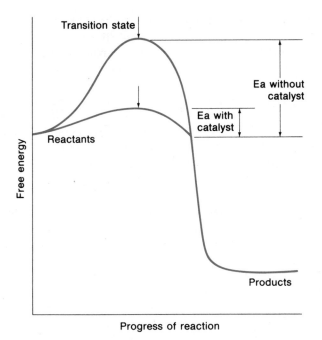

**Figure 4-1**
The energy of reactants and products and the activation energy. At equilibrium, the ratio of products to reactants is a function of the energy difference between reactants and products. The rate of reactants going to products is a function of the activation energy (see text). A catalyst does not change the energy of reactants or products; rather it lowers the activation energy.

ble that this concept of activation energy, borrowed from the kinetic molecular theory of gases, has been so useful to biochemists in accounting for the properties of enzyme catalysts.

## KINETICS OF ENZYME-CATALYZED REACTIONS

In the reaction of nitrous oxide with hydrogen to form nitrogen and water, we illustrated the necessity of kinetic analysis as an aid in determining the mechanism of the reaction. The same is true of catalyzed reactions. A kinetic analysis of an enzyme-catalyzed reaction is indispensable in determining the mechanism of the reaction. This is why we must devote some effort to this subject. We will first consider a simple model used to describe the rate of reaction as a function of the concentrations of enzyme and substrates (reactants).

### The Michaelis-Menten Treatment

At moderate concentrations of substrate and very low concentrations of enzyme, the reaction velocity (i.e., the rate of increase in product concentration) reaches a maximum independent of further increase in the substrate concentration. This observation suggests that at higher substrate concentrations, all the enzyme sites contain bound substrate; hence, further increases in substrate concentration do not lead to an increase in the reaction rate. Experiments demonstrating this saturation phenomenon were done by L. Michaelis and M. L. Menten in 1913. At a low, fixed enzyme concentration, the reaction velocity increases hyperbolically with increasing substrate concentration, reaching a concentration-independent maximal velocity at high substrate concentration. Such behavior is characteristic of enzyme-catalyzed reactions; it was unanticipated on the basis of the kinetics of catalyzed reactions studied in ordinary homogeneous solution under conditions of large molar excesses of substrate over catalyst. In homogeneous-solution reactions catalyzed by small molecules, the velocity usually increases linearly (without limit) as the substrate concentration is increased. *The hyperbolic approach to a constant maximal reaction velocity with increasing substrate concentration is a strong indication of the formation of an intermediate enzyme-substrate complex.* As we shall see subsequently, the exactness of the fit of the experimental velocity-concentration data to a hyperbolic equation is a measure of the independence of events at one enzyme site from the state of events at other sites. The hyperbolic substrate-concentration dependence is also a measure of the catalytic homogeneity of the enzyme sites. The dependence of enzyme-catalyzed reaction velocity on substrate concentration does not always strictly adhere to this hyperbolic form, especially with enzymes that play an important role in metabolic regulatory process (see the case of aspartate carbamoyltransferase in Chapter 5). However, virtually all enzymes exhibit the phenomenon of substrate saturation.

The hypothesis that the enzyme binds the substrate has led to a simple method for analyzing enzyme kinetics. Assume the enzyme and the substrate reversibly form an enzyme-substrate complex:

$$\text{Enzyme (E)} + \text{substrate (S)} \rightleftharpoons \text{enzyme-substrate (ES)} \qquad (3)$$

The formation of products P can proceed only through the enzyme-substrate complex, as indicated by the following reaction, so that the enzyme E is regenerated in the final step:

$$\text{ES} \longrightarrow \text{P} + \text{E} \qquad (4)$$

For simplicity, the arrow points only one way, as if the reaction goes to completion in this direction, which is essentially true if the reaction in one direction is strongly favored. The ES complex should not be confused with the activated complex or transition state described earlier. The activated complex is the highest energy state assumed by the reactants in making the transition to products. The enzyme-substrate complex ES represents the total concentration of substrate reversibly bound at the active site. After binding, the substrate is presumed to undergo a series of transformations, including the activated complex, before being converted to product (P). In the final step of a reaction, the enzyme-product complex (EP) dissociates to yield product (P) and regenerate the active enzyme (E). The main assumption of the Michaelis-Menten treatment is that the speed of the overall reaction is proportional to the concentration of enzyme-substrate complex. From Equations (3) and (4) it is obvious that as the concentration of enzyme is raised, more enzyme-substrate complex will be formed. This should increase the reaction rate, as shown in Figure 4-2. For similar reasons, increasing the substrate concentration at a fixed level of enzyme increases the reaction velocity. Maximum velocity is achieved when all the enzyme is in the form of the ES complex. The reaction approaches $V_{\text{max}}$ as [S] approaches infinity, as illustrated by Figure 4-3. The approximately hyperbolic curve shown in Figure 4-3 represents the situation in which all the substrate binding sites in the enzyme have an equal affinity for substrate.

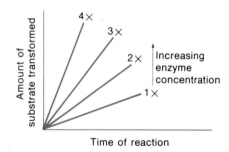

**Figure 4-2**
Variation of rates of substrate conversion at different levels of enzyme.

For purposes of kinetic analysis, it is convenient to examine that period of the enzymatic reaction known as the *steady state*. This occurs shortly after mixing enzyme and substrate and constitutes the time interval when the rate of reaction is a linear function of time. The concentration of the ES complex and other intermediates involved in the reaction is also nearly a constant during this time period. Usually substrate is present in much higher molar concentrations than enzyme, and since the initial period of the reaction is being examined, the free substrate concentration is approximately equal to the total substrate added to the reaction mixture. Although most reactions use more than one substrate in concert, we will first consider the case in which a single substrate is involved:

Enzyme (E) + substrate $\rightleftharpoons$ enzyme-substrate complex (ES) $\rightleftharpoons$
$$\text{enzyme + products (P)} \quad (5)$$

There are four rate constants associated with these reactions:

$$E + S \underset{k_{-1}}{\overset{k_1}{\rightleftharpoons}} ES \underset{k_{-2}}{\overset{k_2}{\rightleftharpoons}} P + E \quad (6)$$

If we confine ourselves to measuring the initial rate of substrate reaction, the rate constant $k_{-2}$ may be ignored, because not enough product will be present to make the reverse reaction proceed at a significant rate. For the rate of formation $V_f$ of ES, we may write

$$V_f = k_1[E][S] = k_1([E]_t - [ES])[S] \quad (7)$$

where $[E]_t$ represents the total enzyme concentration, i.e., the sum of the free and combined enzyme; $[ES]$ is the concentration of the enzyme-substrate complex; $([E]_t - [ES])$ represents the concentration of free or uncombined enzyme, since enzyme is assumed to be in either of two states, free or complexed with substrate; and $[S]$ represents the substrate concentration.

The rate of disappearance $V_d$ of ES is

$$V_d = k_{-1}[ES] + k_2[ES] \quad (8)$$

since ES can disappear to give the initial reactants ($k_{-1}$) or by the formation of products ($k_2$).

When the rates of formation and disappearance of ES are equal, i.e., during that phase of a reaction when the concentration of ES is virtually a constant, $V_f$ equals $V_d$. Equations (7) and (8) may be combined to describe this steady-state situation:

$$k_1([E]_t - [ES])[S] = k_{-1}[ES] + k_2[ES] \quad (9)$$

Rearrangement of Equation (9) gives

$$\frac{[S]([E]_t - [ES])}{(ES)} = \frac{k_{-1} + k_2}{k_1} = K_m \quad (10)$$

The constant $K_m$ is called the Michaelis constant, and it is a particularly useful parameter characteristic of the enzyme-substrate complex under conditions of the steady state. The same enzyme has characteristic Michaelis constants with different substrate molecules. By rearrangement

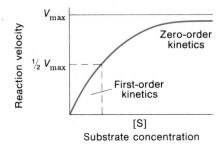

**Figure 4-3**
Reaction velocity as a function of substrate concentration. Note that as the concentration of substrate approaches zero, the system approaches first-order kinetics; i.e., the rate is proportional to the first power of the substrate concentration. As the substrate concentration approaches infinity, the system approaches zero-order kinetics; i.e., the rate is independent of substrate concentration.

of Equation (9), the steady-state concentration of the enzyme-substrate complex can be expressed in terms of E, S, and $K_m$:

$$[ES] = \frac{[E]_t[S]}{K_m + [S]} \qquad (11)$$

It is always best if the concentration of ES can be measured directly. However, this is frequently difficult to do, so it is helpful to derive another expression that relates $K_m$ to more readily measurable parameters $[E]_t$, $[S]$, and the reaction velocity.

If $v$ is the initial rate of formation of product, then from Equation (6) it can be seen that

$$v = k_2[ES] \qquad (12)$$

and since the maximum initial rate $V_{max}$ should be obtained when all the enzyme is in the form of the ES complex,

$$V_{max} = k_2[E]_t \qquad (13)$$

By solving Equations (12) and (13) for $[ES]$ and $[E]_t$, respectively, and inserting in Equation (11), we obtain

$$v = \frac{V_{max}[S]}{K_m + [S]} \quad \text{or} \quad K_m = [S]\frac{V_{max} - 1}{v} \quad \text{or} \quad v = \frac{V_{max}}{1 + K_m/[S]} \qquad (14)$$

Equation (14) shows three different ways of expressing the so-called Michaelis-Menten equation. From Equation (14) it can be seen that $K_m$ is equal to the concentration (expressed in moles per liter) of the substrate, which gives half the numerical maximum velocity. It should be emphasized that we are speaking of the velocity under steady-state conditions, i.e., when ES is approximately a constant. $K_m$ is a complex constant, since

$$K_m = \frac{k_{-1} + k_2}{k_1}$$

according to Equation (10). Three conditions are possible:

$$k_1 \gg k_2 \qquad \text{then} \qquad K_m = k_{-1}/k_1$$
$$k_2 \gg k_{-1} \qquad \text{then} \qquad K_m = k_2/k_1$$

or $k_{-1}$ and $k_2$ are of the same order of magnitude, then all three reaction constants determine the value of $K_m$. In practice, all three possibilities have been observed with different enzyme-substrate complexes.

## Applying the Michaelis-Menten Equation

For purposes of plotting experimental data, it is convenient to algebraically rearrange Equation (14). Data may be conveniently represented by the Lineweaver-Burk equation:

$$\frac{1}{v} = \frac{K_m}{V_{max}}\frac{1}{[S]} + \frac{1}{V_{max}} \qquad (15)$$

The data for $1/v$ and $1/[S]$ are plotted on a graph, as indicated in Figure 4-4, to give a straight line. The slope is $K_m/V_{max}$, the intercept on the ordinate is $1/V_{max}$, and the intercept on the abscissa is $-1/K_m$.

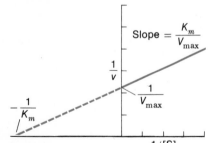

**Figure 4-4**
$1/v$ versus $1/[S]$ for an idealized enzyme-catalyzed reaction.

The value of $K_m$ depends on the enzyme, the substrate, and other conditions of the reaction, such as pH, temperature, etc. For most enzyme-substrate complexes, the value for $K_m$ lies between $10^{-1}$ and $10^{-6}$ $M$. As noted earlier when $k_1 \gg k_2$, then $K_m \cong k_{-1}/k_1$. In this situation $K_m = ([E][S])/[ES]$, which is the dissociation constant $K_s$ for the enzyme-substrate complex. The lower the $K_m$, the stronger the complex between enzyme and substrate. The kinetic constant $k_2$ is referred to as the turnover number and is sometimes symbolized by $k_{cat}$. From Equation (13), that is, $V_{max} = k_2[E]_t$, it can be seen that the turnover number of an enzyme is equal to the number of substrate molecules converted into product per second per mole of enzyme present when the enzyme is fully complexed with substrate. For most enzymes this number falls in the range of $10^1$ to $10^4$ $s^{-1}$. Some of the advantages and disadvantages of this steady-state approach to analyzing enzyme reactions are presented below.

## Additional Characteristic Features of Enzyme-Catalyzed Reactions

*Enzymes and small-molecule solution catalysts have many properties in common.* Enzymes are the most complex catalysts known, but they obey many of the same basic rules followed by small-molecule catalysts.

All chemical bonds are formed by electrons. The rearrangement or breakage of these bonds starts with the migration of electrons. In the most general terms, reactive groups can be said to function either as *electrophiles* or *nucleophiles*. The former are electron-deficient substances that attack electron-rich substances. The latter are electron-rich substances that attack electron-deficient substances. In enzymes, metal ions such as $Mg^{2+}$, $Mn^{2+}$, and $Fe^{3+}$ are examples of electrophilic groups, as are $-NH_3^+$ or $-COOH$ groups of proteins. Protein molecules contain a number of potentially nucleophilic groups in the amino acid side chains, particularly serine and tyrosine hydroxyl groups, aspartate or glutamate carboxylate groups, histidine imidazole groups, and cysteine sulfhydryl groups. In nucleophilic catalysis, the roles of catalyst and substrate are simply the reverse of that defined by electrophilic catalysis. Frequently the main job of the catalyst is to make a potentially reactive center more attractive for reaction, i.e., to accentuate the electrophilic or nucleophilic potential of the reactive center of the substrate.

In aqueous solution, proton or hydroxide ions are the most common catalysts for such purposes. Generalized acids or bases, defined as compounds capable of yielding proton or hydroxide ions themselves or through interaction with water, are also common forms of solution catalysts. The way in which such catalysts work is illustrated in Figure 4-5 for the hydrolysis of an ester linkage. As a result of the electronegativity of the oxygen atom in the ester C=O group, the oxygen of the carboxyl group has a fractional negative charge $\delta^-$ and the carbon has a fractional positive charge $\delta^+$. Hydrolysis of the ester can be accelerated by either acid or base catalysis. In acid catalysis, a proton or a generalized acid acting as an electrophile is attracted to the oxygen. This leads to an intermediate that accentuates the positive charge on the carbon atom, making it more attractive to a nucleophile, in this case water. The water is a poor nucleophile and would attack the carbon very slowly without this inducement. This is the key step in the catalysis. The remaining reactions leading to ester hydrolysis and regeneration of the catalyst occur rapidly and spontaneously. In hydroxide ion catalysis or general base catalysis of the ester, the carbon atom is attacked more directly by a stronger nucleophile, either $OH^-$ itself or a water molecule converted into a hydroxide ion by the presence of the generalized base. Again, after hydrolysis, the catalyst is regenerated. General acid-base catalyses are much more common in biochemical reactions because of the restricted range of pH over which such systems operate. In this pH range near neutrality, only low concentrations of free $H^+$ and $OH^-$ are present.

Thus far we have been discussing properties that enzyme catalysis shares in common with ordinary chemical catalysis. Let us now consider some of the unique features of enzyme catalysis that are shared by many reactions.

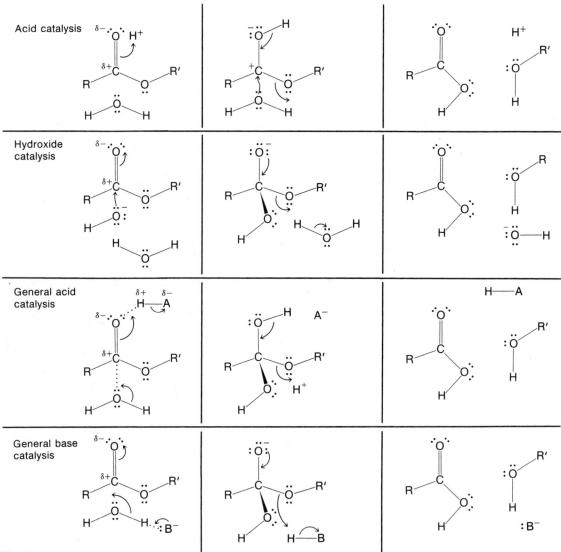

**Figure 4-5**
Different ways in which ester hydrolysis can be catalyzed.

**1.** Two classes of enzyme reactions can be distinguished: reactions involving only electron transfer, which have rates or turnover numbers on the order of $10^8\,s^{-1}$, and reactions involving transfer of both electrons and protons or other groups, which have rates on the order of $10^3\,s^{-1}$. Most reactions belong to the second class.

**2.** Enzymic catalysis is mediated by the functional groups found in amino acid side chains and *coenzymes*. Notably, the amino acid side chains of histidine, serine, cysteine, lysine, glutamate, and aspartate are frequently directly involved in the catalytic process. Coenzymes or metals work in conjunction with enzymes to provide a greater variety of functional groups than is available through amino acid side chains alone. They are discussed at length in Chapter 6.

**3.** Rates of enzyme-catalyzed reactions usually show narrow pH optima.

**4.** Enzyme molecules are very large in comparison with their substrates. The active site is usually somewhat larger than the substrate. This is because the active site partially surrounds the substrate in most cases. Also, a good deal of the enzyme structure is required to stabilize the conformation of the active site.

**5.** In addition to possessing the essential reactive groupings for carrying out the catalytic reaction, enzymes have other characteristics that facilitate such reactions and provide avenues for more complex reactions involving more than one substrate, which are beyond the scope of simpler catalysts. The complex folded nature of enzymes makes this possible. Four main advantages possessed by enzymes are (1) the presence of more than one catalytic grouping in the active center, so that *concerted catalysis* can occur; (2) the presence of a binding site, so that the substrate molecule can become bound near the active center in the proper *orientation* for reaction; (3) the presence of binding sites for more than one substrate molecule in cases where the reaction involves two or more substrate molecules; and (4) on occasion, substrate is bound to the enzyme in such a way that a bond strain is created in the substrate that favors the formation of the transition-state complex. These advantages of enzyme catalysis are best appreciated through the consideration of specific examples.

## THE TRYPSIN FAMILY OF ENZYMES

*Trypsin*, *chymotrypsin*, and *elastase* represent a group of closely related digestive enzymes whose role is to hydrolyze polypeptide chains. They are synthesized in the pancreas as inactive zymogens, or preenzymes, and then are secreted into the digestive tract and activated just before use. These three enzymes work as a team; each cleaves a protein chain at an internal peptide linkage next to a different type of amino acid side group (Figure 4-6). Trypsin cuts a chain just past the carbonyl group of a basic amino acid, either lysine or arginine. Chymotrypsin cuts a polypeptide chain next to an aromatic amino acid. Elastase is less discriminating in its choice of cleavage point, but it tends to cut preferentially adjacent to small, uncharged side chains. These enzymes have been intensively studied because they are small (approximately 250 amino acids), easily obtainable in quantity, and relatively stable. Chymotrypsin has been the subject of more chemical

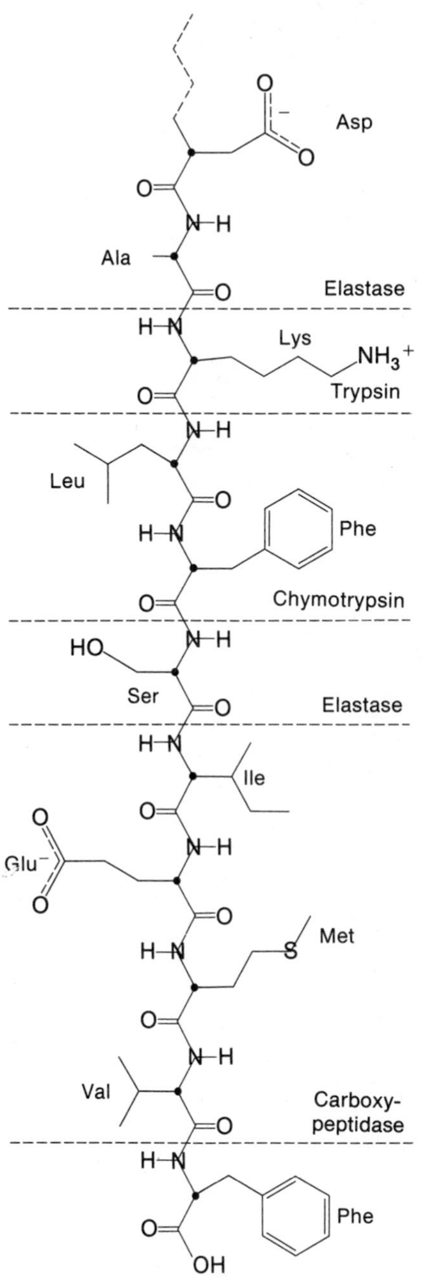

Protein-cleaving enzymes

**Figure 4-6**
Chymotrypsin, trypsin, and elastase each cut a polypeptide chain adjacent to a different type of amino acid side chain. Chymotrypsin prefers aromatic rings; trypsin favors positively charged groups; and elastase cuts best next to small, nonpolar side chains. Carboxypeptidase cleaves one amino acid at a time from the carboxy-terminal end of the chain.

studies than the others because it is less likely to digest itself in the test tube. Protein-digesting enzymes, being proteins themselves, have a tendency toward self-destruction. Chymotrypsin is reasonably safe from this because the large aromatic groups that it favors are usually buried inside the molecule, whereas the groups that trypsin or elastase prefer are more likely to be exposed on the enzyme surface.

The fact that the trypsin family operates together during digestion suggests that these enzymes might be related in some way. Their amino acid sequences show striking similarities, as illustrated in Figure 4-7. They have identical amino acids at 62 of the 257 positions. Although the preenzymes are different, all three active enzymes begin a polypeptide chain at the same place (residue 16). The four disulfide bridges that connect distant parts of the chain in elastase also are present in the other two enzymes, with chymotrypsin having one more disulfide bridge and trypsin having two. All three enzymes have a histidine at position 57, aspartic acid at 102, and serine at 195, which are the main groups involved in the catalytic mechanism.

X-ray crystal structure analyses have revealed that these three proteolytic enzymes are folded the same way in three dimensions. The backbone skeletons are shown for trypsin and chymotrypsin in Figure 4-8, with numbered balls for $\alpha$ carbons and connecting sticks for the —CO—NH— amide groups. The only amino acid side chains illustrated are His 57, Asp 102, and Ser 195 at the active site and the various disulfide bridges. Because the chains are folded in the same way, positions corresponding to disulfide bridges in one enzyme also are close together even in an enzyme that does not have that particular bridge. Another important feature is revealed by the three-dimensional structure that was apparent from the amino acid sequences alone. The chain is folded back on itself in such a way that the three catalytic side chains (57, 102, and 195) are brought close together at a depression on the surface of the molecule. This is the active site of the enzyme, where the substrate that will be cut during digestion is bound.

Very little $\alpha$ helix is present in these three proteins. A short helix is found near residues 169–170 at the bottom of the molecule, and the chain ends with a helix at residues 230–245 at the left rear. A more important structural feature is a silk-like twisted $\beta$ sheet of nearly parallel extended chains. One of these can be seen at the upper left of trypsin in the chains that contain residues 60, 89, 105, and 53. Another twisted sheet in the lower right of the molecule is harder to see in this view. These two cores of twisted $\beta$ sheets define the shape of the molecule, and the active site sits in a groove between them. The interiors of these enzymes are packed with hydrophobic residues, and charged side chains lie on the outside.

### The Chymotrypsin Catalytic Mechanism

Trypsin, chymotrypsin, and elastase not only share a common structure, they also share a common catalytic mechanism. The mechanism has been determined from chemical and x-ray studies. The polypeptide substrate binds to the molecule with one portion hydrogen-bonded to residues 215–219 in an antiparallel manner, like adjacent chains in silk (see Chapter 3). This helps to hold the substrate in place. At the bend in the substrate chain, the NH—CO bond that will be cut is brought close to His 57 and Ser 195.

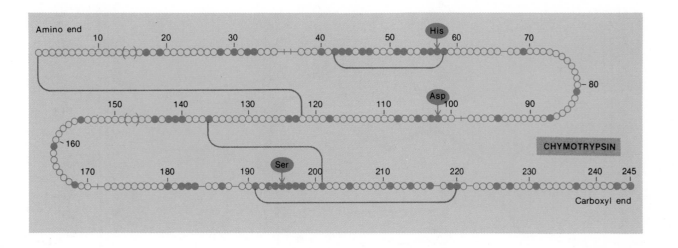

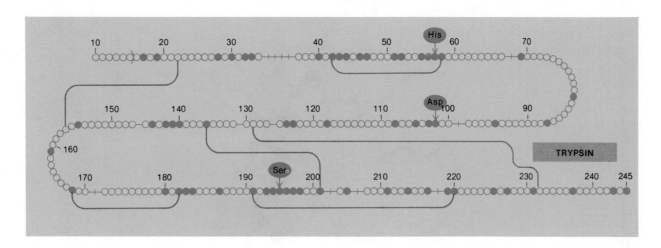

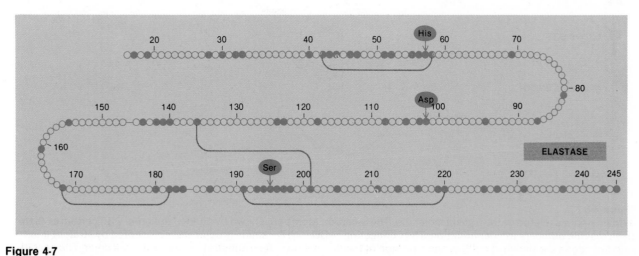

**Figure 4-7**
Schematic diagram of the amino acid sequence of the three protein-digesting enzymes. Each circle represents one amino acid. Amino acid positions that are identical in all three proteins are in color. Long connections between nonadjacent amino acids represent disulfide bonds. Locations of the three catalytically important side chains are marked.

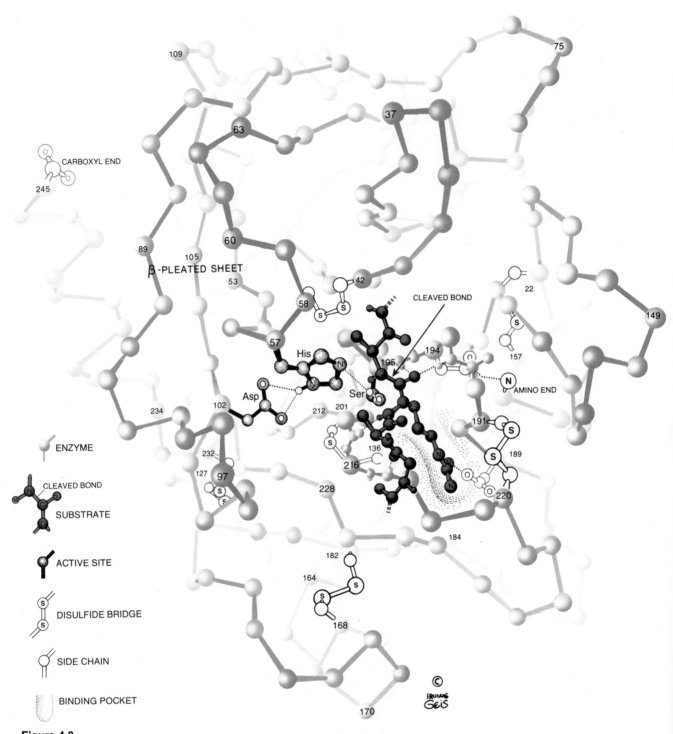

**Figure 4-8**

(*a*) Main-chain skeleton of the trypsin molecule. The $\alpha$-carbon atoms are shown by shaded spheres, with certain of them given residue numbers for identification. The connecting —CO—NH— amide groups are represented by straight lines. Disulfide bridges are shown in outline, and a portion of the polypeptide chain substrate appears in dark color. The specificity pocket is sketched in shading, with a lysine side chain from the substrate molecule inserted. The catalytically impor-

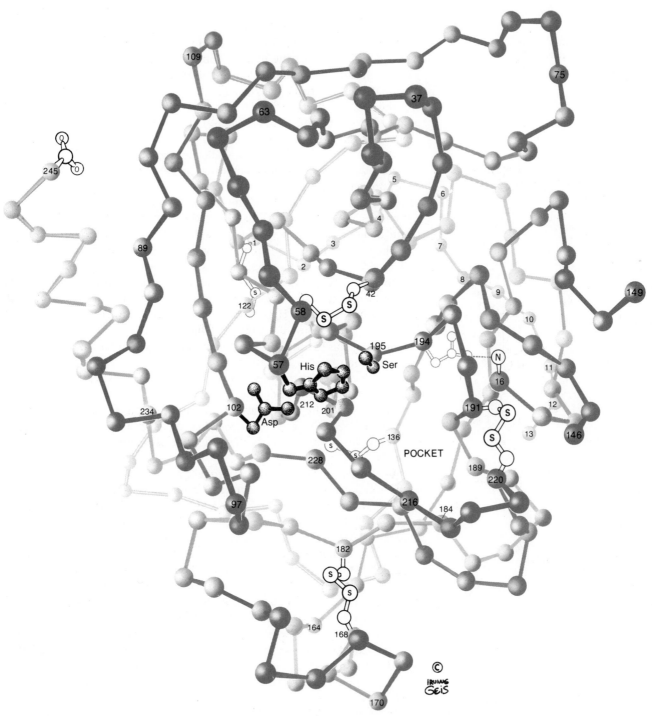

tant Asp, His, and Ser are poised for cleavage of the peptide bond marked by an arrow. (b) Folding of the main chain in chymotrypsin. All of the comments made about chain folding and pseudo disulfide bridges for elastase hold equally well for chymotrypsin. Notice the cut chain ends where residues 147–148 and 14–15 have been enzymatically removed in the process of activating chymotrypsin from its precursor, chymotrypsinogen. Activation of trypsinogen involves cleaving six residues from the amino terminal end of the chain.

Specificity pockets of trypsin,
chymotrypsin, and elastase

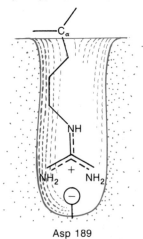

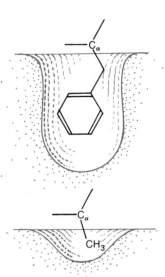

**Figure 4-9**
The size of each pocket and the nature
of the side chains lining it determine
what kind of amino acid chain will be
held best. This in turn determines at
which position along a substrate
chain cleavage will occur.

The side group of the substrate just prior to this bond is inserted into a pocket in the surface of the enzyme molecule. This pocket, bordered by residues 214–220, a disulfide bridge, and residues 191–195, can be seen in Figure 4-8. The rim of the pocket from residues 215 to 219 is the binding site for the substrate. The pocket provides an explanation for the different specificities of the three enzymes for the bonds that they will cut. In trypsin, the specificity pocket is deep and has a negative charge from aspartic acid 189 at the bottom (see Figure 4-9). The pocket is designed to accept a long, positively charged basic side chain: lysine or arginine. In chymotrypsin, the corresponding pocket is wider and completely lined with hydrophobic side chains, thereby providing an efficient receptacle for a large, bulky aromatic group. In elastase, the pocket is blocked by valine and threonine at the positions where the other two enzymes have only glycine, which has no side chain. This closes the pocket in elastase, so any side chain of appreciable size will be unable to bind to the enzyme surface.

The proposed mechanism of catalysis is shown in Figure 4-10. In steps 1 and 2, a polypeptide chain approaches and binds to the active site of the enzyme, with the proper type of side chain inserted into the specificity pocket. After substrate binding, the three catalytically important groups on the enzyme—Asp 102, His 57, and Ser 195—are connected by hydrogen bonds in what D. Blow has called a "charge-relay system." In step 3 of the mechanism, the histidine nitrogen acts first as a general base, pulling the serine proton toward itself, and then as a general acid, donating the proton to the lone electron pair on the nitrogen atom of the polypeptide bond to be cleaved. The aspartic acid group is pictured as helping the histidine to attract the proton by making an electrostatic linkage with the other proton on the imidazole ring. The importance of aspartate in this position is shown by the fact that almost all serine proteases have an aspartate in this position.

As the serine H—O bond is broken, a bond is formed between the serine oxygen and the carbonyl carbon on the polypeptide chain. This carbon becomes tetrahedrally bonded, and the effect of the negative charge on Asp 102 has in a sense been relayed to the carbonyl oxygen atom of the substrate. This negative oxygen is stabilized by hydrogen bonds to N—H groups on the enzyme backbone. The transition state of step 3, termed the tetrahedral intermediate because of the bonding around the carbonyl carbon atom, is short-lived and cannot be isolated, but there is chemical evidence for its presence. The enzyme passes quickly through the tetrahedral intermediate to step 4. As the polypeptide N accepts the proton from histidine, the N—C bond is weakened and finally broken. One-half of the polypeptide chain falls away as a free amine, R—$NH_2$. The other half remains bound covalently to the enzyme as an acylated intermediate. This acyl-enzyme complex is stable enough to be isolated and studied in special cases where further reaction is blocked.

The steps to deacylate the enzyme and restore it to its original state (steps 5 to 8) are like the first four steps run in reverse, with $H_2O$ playing the role of the missing half chain. A water molecule attacks the carbonyl carbon of the acyl group in step 5 and donates one proton to histidine 57 to form another tetrahedral intermediate (step 6). This intermediate breaks down when the proton is passed on from histidine to serine (step 7). The second half of the polypeptide chain falls away (step 8), aided by charge repulsion between the carboxyl group and the negative charge on aspartic acid 102. The enzyme is restored to its original state.

Chemical experiments have been done to find out how choosy chymotrypsin is among its substrates. It has been known for a long time that chymotrypsin hydrolyzes suitably substituted esters at comparable rates to peptide linkages. Many experiments using model ester substrates have been done to explore the question of substrate specificity. Small organic esters (see Figure 4-11 and Table 4-2) have the general structure $R_1CHR_2COOR_3$. Results with these model substrates are reported in terms of turnover number (also referred to as the catalytic rate constant) $k_{cat}$, where $k_{cat}$ is $V_{max}/[E]_t$, and where $[E]_t$ refers to the molar concentration of active sites (see Equation 13). Two main points are established by the results presented in Table 4-2. First, $k_{cat}$ is very sensitive to the side-group composition of the ester, and second, enantiomers react at very different rates, testifying to the stereospecificity of the reaction.

Although hydrolysis studies of model ester compounds have been useful in arriving at the forgoing conclusions, their use also initially led to a good deal of confusion when steady-state analyses were attempted. To understand this, we must recall that the mechanism of hydrolysis proposed earlier can be divided conceptually into three steps: (1) formation of the enzyme-substrate complex, (2) acylation, and (3) deacylation. A careful kinetic analysis has shown that for esters, deacylation is the rate-limiting step, whereas for peptides, acylation is the rate-limiting step. Consider the following set of reactions for hydrolysis of acetylphenylalanine methyl ester by chymotrypsin.

$$E + S \underset{k_{-1}}{\overset{k_1}{\rightleftharpoons}} ES; \qquad K_s = \frac{k_{-1}}{k_1} \tag{16}$$

$$ES \underset{k_2}{\overset{k_2}{\longrightarrow}} \overset{P_1}{\nearrow} EP_2 \tag{17}$$

$$EP_2 \overset{k_3}{\longrightarrow} E + P_2 \tag{18}$$

where $K_s$ is the dissociation constant of the enzyme-substrate complex, S is the concentration of acetylphenylalanine methyl ester, $P_1$ is methanol, $P_2$ is acetylphenylalanine, and ES is the reversibly formed enzyme-substrate complex.

The steady-state kinetic constants pertaining to Equations (16) to (18) are

$$k_{cat} = \frac{V_{max}}{[E]_t} = \frac{k_2 k_3}{k_2 + k_3} \tag{19}$$

$$K_{m(app)} = K_s \left( \frac{k_3}{k_2 + k_3} \right) \tag{20}$$

In the hydrolysis of esters, the acylation rate constant $k_2$ is much greater than the deacylation rate constant $k_3$. When this is the case, the preceding equation reduces to a more familiar expression:

$$k_{cat} = k_3 \tag{21}$$

$$K_{m(app)} = K_s(k_3/k_2) \tag{22}$$

Here $K_{m(app)}$ is the apparent Michaelis constant determined by the usual steady-state kinetic analysis and the Lineweaver-Burk plot described earlier.

In the hydrolysis of amides or peptides, the acylation rate constant $k_2$ is much smaller than the deacylation rate constant $k_3$, leading to

$$k_{cat} = k_2 \tag{23}$$

$$K_{m(app)} = K_s \tag{24}$$

It is clear from the preceding that the apparent Michaelis constant $K_{m(app)}$ for ester hydrolysis is going to be much smaller than the actual reversible dissociation constant $K_s$. Physically, the bound substrate is partly bound in a reversible fashion and partly bound as an intermediate by covalent linkage. By contrast, for amides, most of the bound substrate is bound in a reversible fashion, so that $K_{m(app)}$ is essentially $K_s$. The most important point to be learned from this comparison of ester and peptide hydrolysis is the danger of concluding that the $K_{m(app)}$ determined from the Lineweaver-Burk plot is in fact related to the $K_m$ in the Michaelis-Menten treatment. In either case, a linear Lineweaver-Burk plot may be obtained, but the $K_{m(app)}$ can have a different meaning depending on the rate-limiting

Substrate diffuses to enzyme.

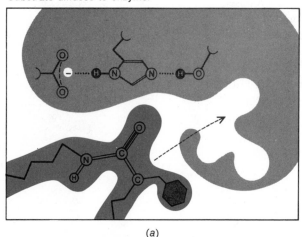

(a)

Substrate binds to enzyme.

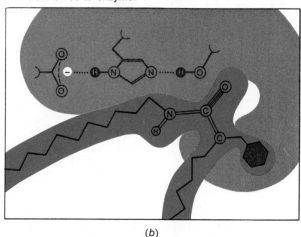

(b)

Water molecule approaches.

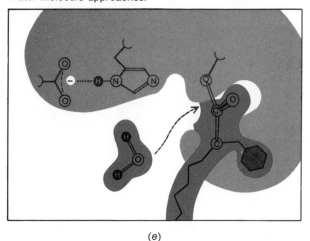

(e)

Water reacts with acyl enzyme.

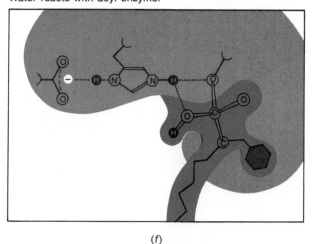

(f)

**Figure 4-10**
Proposed steps in chymotrypsin-catalyzed peptide hydrolysis.

step or steps in the reaction. The Michaelis-Menten treatment only strictly applies to situations like that represented by Equation (6), where the rate-limiting step involves the conversion of the substrate in one step into products that rapidly dissociate from the enzyme. This treatment works for peptide hydrolysis because $k_2$ is rate-limiting, but not for ester hydrolysis, where $k_3$ is rate-limiting. Further kinetic analyses of the chymotrypsin-catalyzed reaction are discussed below.

## Kinetic Measurements of Chymotrypsin Activity

Probing the mechanism of chymotrypsin action exemplifies some of the problems associated with measuring reaction rates. As indicated above, chymotrypsin hydrolyzes both peptides (or amides) and esters. Kinetic analysis of the individual steps in the reactions were needed to show that although the initial and final products of both reactions are homologous, the rate-limiting steps are quite different for the two types of substrates.

Substrate bond is cleaved.

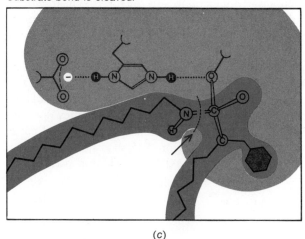

(c)

First product leaves.

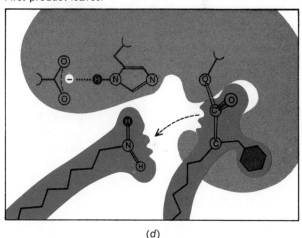

(d)

Acyl bond to enzyme is broken.

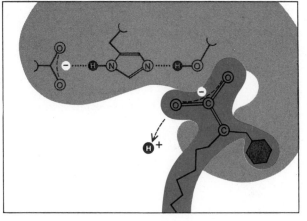

(g)

Second product leaves.

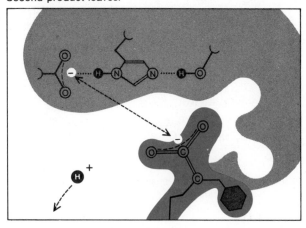

(h)

I

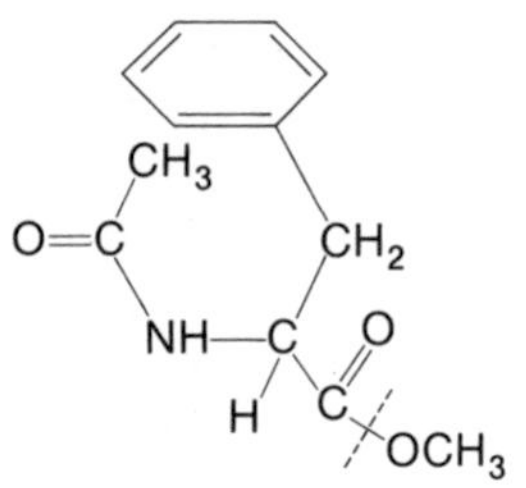

**Acetylphenylalanine methyl ester**

II

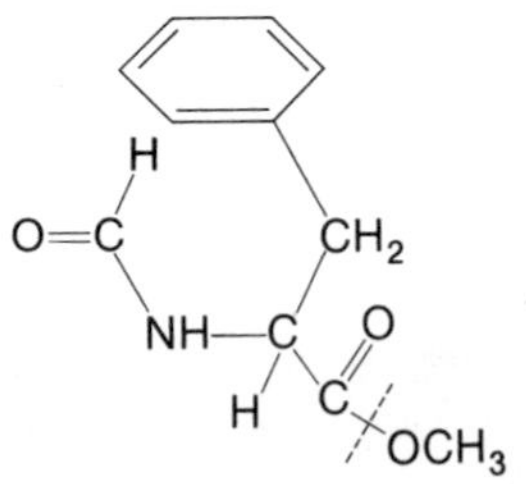

**Formylphenylalanine methyl ester**

III

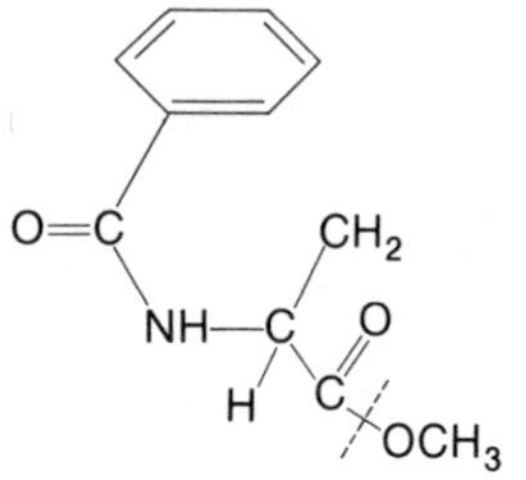

**Benzoylalanine methyl ester**

IV

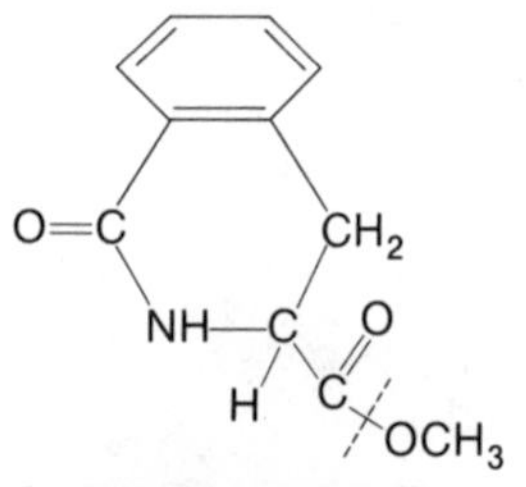

**3-Carbomethoxy-3,4-dihydro-isocarbostyril**

**Figure 4-11**
Structures of some substrates of chymotrypsin. Dashed lines indicate the point of enzymatic cleavage.

In the hydrolysis of the ester *p*-nitrophenyl acetate, the liberation of *p*-nitrophenol is biphasic, corresponding to a rapid liberation of *p*-nitrophenol in a concentration roughly equivalent to the concentration of enzyme and a slow, steady-state liberation of acetate resulting from the hydrolysis of the acyl-enzyme (Figure 4-12). This result shows that the acylation rate for ester hydrolysis is quite rapid, but that the enzyme very quickly becomes blocked for further reaction because of the slow rate of deacylation. In the steady-state situation that develops after about 2 minutes (see Figure 4-13), the reaction cannot proceed further until deacylation occurs. In the cyclic process, the deacylation reaction very quickly becomes the rate-limiting step. B. S. Hartley and B. A. Kilby believe that this is the case for most esters that can be hydrolyzed by chymotrypsin. Very little of the enzyme is available for reversible binding to substrate.

The key to analyzing this situation according to Equations (16) to (18) is to show that the alcohol is released in a fast step characterized by the rate constant $k_2$ and that the acid is released in a slower step. This information was obtained by measurements made before the system reached a steady state. This required the use of very rapid measuring techniques. Advantage was taken of the fact that absorbancy changes in the enzyme near 290 nm occur on binding of substrate. It is possible to witness four phases in the ester hydrolysis reaction. First, there is an initial rapid increase in absorbance that is complete in less than 2 ms. This phase of the reaction is considered to result from the reversible formation of enzyme-substrate complex. Second, a slower increase in absorbance leads to the formation of a steady-state intermediate (after about 20 s), which is considered to be $EP_2$ of Equation (17). Third is a period of time during which the absorbancy does not change and product is produced at a linear rate. The time interval corresponds to the steady-state phase of the reaction. Fourth, a slow decrease in absorbancy occurs when the substrate is exhausted. This phase may be identified with the decay of the steady-state intermediate to give free enzyme and product.

**Table 4-2**
Rates of Hydrolyses of the D and L Forms of Various Esters by Chymotrypsin

| Substrate | $k_{cat}$ (s$^{-1}$) |
|---|---|
| Acetylphenylalanine methyl ester | |
| L | 63 |
| D | Very low |
| Formyl-phenylalanine methyl ester | |
| L | Very high |
| D | 0.0034 |
| Benzoylalanine methyl ester | |
| L | 0.26 |
| D | 0.011 |
| 3-Carbomethoxy-3,4-dihydroisocarbostyril | |
| L | 0.12 |
| D | 22.7 |

**Table 4-3**
Rate and Equilibrium Constants Pertaining to Chymotrypsin-Catalyzed
Hydrolysis of Ester Substrates at pH 5.0 and 25°C

| Substrate | $k_2$ (s$^{-1}$) | $k_3$ (s$^{-1}$) | $K_s$ (mM) | $K_{m(app)}$ (mM) |
|---|---|---|---|---|
| N-Acetyl-L-Trp-ethylester* | 35 | 0.84 | 2.1 | 0.08 |
| N-Acetyl-L-Phe-ethylester† | 13 | 2.2 | 7.3 | 1.3 |

Note: $k_2$ and $K_s$ were measured directly by rapid kinetic methods; $k_{cat}$ and $K_{m(app)}$ were determined under conditions of steady state. For the reactions shown it can be seen that $k_3 \gg k_2$ and therefore $k_{cat} \approx k_3$.
*Data from K. G. Brandt, A. Himoe, and G. P. Hess.
†Data from A. Himoe, K. G. Brandt, R. J. DeSa, and G. P. Hess.

From this information, all the kinetic parameters for Equations (16) to (18) have been evaluated. Data for the two different esters shown in Table 4-3 indicate that the rate constant for the formation of the steady-state intermediate $k_2$ is larger than the rate constant $k_3$ for the decomposition of this intermediate. The values of $K_s$, the dissociation constant of the enzyme-substrate complex, obtained by rapid kinetic measurements, are much larger than the $K_{m(app)}$ value obtained from steady-state measurements.

The chymotrypsin-catalyzed hydrolysis of amide is also believed to follow the pathway described by Equations (16) to (18), but in this case, the rate of formation of the acyl-enzyme is believed to be rate-limiting. When the acylation rate constant $k_2$ is smaller than the deacylation rate constant $k_3$ for the pathway shown by Equations (16) to (18), the steady-state kinetic parameter $K_{m(app)}$ is equal to $K_s$, the enzyme-substrate dissociation constant. In this case, the value of $K_{m(app)}$ determined from steady-state measurements is similar to the value of $K_s$ determined from equilibrium measurements.

This single example illustrates both the merits and the pitfalls of using steady-state methods to evaluate kinetic parameters and the value of being able to make very rapid measurements during the initial stages of a reaction after mixing enzyme and substrate. Steady-state measurements are much easier to make, but more sophisticated kinetic measurements are frequently necessary to get an accurate description of the reaction.

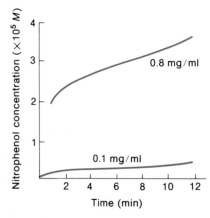

**Figure 4-12**
p-Nitrophenol formation as a function of time at different enzyme concentrations in the chymotrypsin-catalyzed hydrolysis of p-nitrophenyl acetate. (Data of Hartley and Kilby)

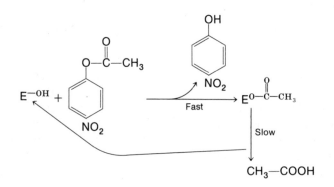

**Figure 4-13**
Suggested steps in the chymotrypsin-catalyzed hydrolysis of p-nitrophenyl acetate.

## Types of Groups Involved at the Active Site

In the trypsin family we have seen a case made for the involvement of three amino acids at the active site: serine, histidine, and aspartic acid. The variety of functional groups provided by protein side chains for reactive centers is really quite limited. The inert side chains of the amino acids glycine, alanine, phenylalanine, leucine, valine, isoleucine, and proline are not involved in chemical catalysis because they cannot function as Lewis acids or Lewis bases. Of the remaining amino acids, a number are of very similar type: aspartate and glutamate, asparagine and glutamine, and threonine and serine. Certain amino acid side-chain residues, most notably those of histidine, serine, cysteine, tyrosine, glutamate, aspartate, and lysine, are known to be individually involved in catalysis by specific enzymes. This seeming scarcity of functional groups at active sites is greatly augmented by metals, coenzymes, prosthetic groups, and other factors that work in conjunction with enzymes at the active site (see Chapter 6).

Pinpointing the active site of an enzyme is greatly aided by structural studies, where one determines the binding position of a substrate or a substrate analog by x-ray diffraction. It also depends on the use of various inhibitors that influence the enzyme activity. A potent inhibitor can interfere either with the binding of a substrate or with the functional groups in the active site, as described in the next section.

## ENZYME INHIBITION

Enzyme inhibition (and enzyme activation) is a phenomenon of great practical importance, as well as a useful probe for determining the active site on the enzyme and other factors that control enzyme activity. In this section we will consider some general aspects of enzyme inhibition, leaving more sophisticated aspects of enzyme activity regulation to Chapter 5.

The active site of an enzyme usually represents a small part of the whole enzyme molecule. Strong evidence for the localization of active centers comes from studies with specific inhibitors of enzymes. Inhibition of enzymes may be reversible or irreversible. Reversible inhibitors work by many different mechanisms, and these can sometimes be distinguished by kinetic analysis. Competitive and noncompetitive situations can be readily distinguished. In competitive inhibition, the binding of inhibitor and substrate to the enzyme is mutually exclusive. Most often, but not always, this results from an inhibitor that structurally resembles the substrate and binds at the active center. With some proteins, especially regulatory proteins (discussed in Chapter 5), competitive inhibition can result from the inhibitor binding at another location on the enzyme, thereby altering the configuration at the active site so that substrate cannot bind. Such an enzyme may be thought of as having two different configurations, one that permits binding of inhibitor and one that permits binding of substrate. Binding is mutually exclusive and therefore competitive, as in simpler situations where the substrate and inhibitor are competing for binding to the same site. In these more complex examples of competitive inhibition, the structure of the inhibitor usually does not resemble that of the substrate.

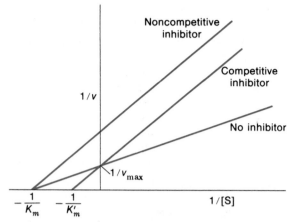

In this plot, $K'_m = K_m(1 + (I)/K_I)$. See Equation 15'.

**Figure 4-14**
Reciprocal plots of $v$ and [S] in the presence of a fixed amount of a competitive or a noncompetitive inhibitor.

According to the simple Michaelis-Menten treatment, the sequence of reactions leading from substrate to products is

$$E + S \rightleftharpoons ES \longrightarrow E + P \qquad (6')$$

The equation for the formation of an enzyme-inhibitor complex like an enzyme-substrate complex is

$$E + I \rightleftharpoons EI \qquad (25)$$

where I is the inhibitor and EI is the enzyme-inhibitor complex. When using a competitive inhibitor, the rate of the catalyzed reaction is dependent on the relative concentrations of substrate and inhibitor. The Lineweaver-Burk plot may be used to characterize competitive or noncompetitive inhibitors, as indicated in Figure 4-14. The derivation of the relevant equations runs parallel to the derivation of the original Michaelis-Menten equations.

The equation in which both competitive inhibitor and substrate binding must be considered should be compared with that in which only substrate is involved:

$$\frac{1}{v} = \frac{K_m}{V_{max}} \frac{1}{[S]} + \frac{1}{V_{max}} \qquad (15')$$

$$\frac{1}{v} = \frac{K_m}{V_{max}} \frac{1}{[S]} \left(1 + \frac{[I]}{K_I}\right) + \frac{1}{V_{max}} \qquad (26)$$

where I is the concentration of inhibitor and $K_I$ is the dissociation constant of the enzyme-inhibitor complex.

If a competitive inhibitor is present, high concentrations of substrate can be used to overcome the inhibition effect. A reciprocal plot of $1/v$ against $1/[S]$ in the presence of this type of inhibitor will result in two nonparallel straight lines that intercept on the $y$ axis at infinitely high substrate concentration, i.e., when $1/[S] = 0$, as shown in Figure 4-14. The apparent $K_m$ is altered to $K'_m$, where $K'_m = K_m\{1 + ([I])/K_i)\}$ according to

Equation (26). It should be noted from this relationship that $K'_m$ is only constant if the inhibitor concentration is a constant. The higher the inhibitor concentration, the larger will be the observed value for $K'_m$.

Inhibitors that are noncompetitive work in many different ways. Some noncompetitive inhibitors block substrate binding, but do not bind to the same site as the substrate (an example of this is given for aspartate carbamoyltransferase in Chapter 5). Other noncompetitive inhibitors (sometimes called uncompetitive) do not prevent substrate binding, but rather prevent substrate from reacting after it has bound by binding to the enzyme-substrate complex. In the simplest cases of noncompetitive inhibition, the $K_m$ is not altered. The Lineweaver-Burk plot for such a situation is shown in Figure 4-14. Both the slope and the intercept on the $y$ axis are increased. The hallmark of noncompetitive inhibition is that no amount of substrate can relieve the inhibition.

Irreversible inhibition can occur at the active site on the enzyme or elsewhere. It cannot be analyzed by the preceding methods; the site of binding of the irreversible inhibitor is best determined by direct chemical methods. Irreversible inhibitors often give valuable information about the active site on the enzyme. Thus certain amino acids in the enzyme can be reacted selectively with irreversible inhibitors to see if they are essential to enzyme activity. Some of these are listed in Table 4-4; others are discussed in the text.

### Reversible and Irreversible Inhibitors of Trypsin

As indicated earlier, competitive inhibitors frequently resemble part of the structure of the substrate. This should not be surprising, since they most frequently compete with the substrate for binding at the active site. The benzamidine ion is one such competitive inhibitor for trypsin. When protonated, the amidine end of the molecule has the same flat, delocalized electron structure found in protonated arginine side chains. To the specificity pocket of trypsin, a benzamidine ion looks like the outer half of an arginine side chain and is accepted and bound as such (Figure 4-15). A molecule inhibited in this way is useless for further catalysis. One might think that a benzamidine-inhibited trypsin thereafter would function as elastase, cutting chains with small side groups. Careful x-ray examination of the inhibited trypsin has shown, however, that the benzamidine ring is

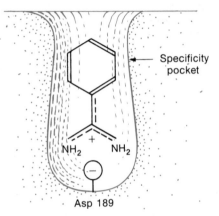

Specificity pocket

NH₂    NH₂

Asp 189

Benzamidine ion inhibitor

$- - - =$ Partial double bond

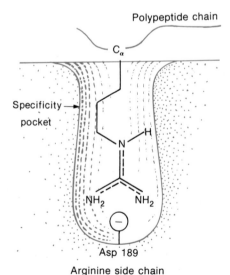

Polypeptide chain

Cα

Specificity pocket

H

N

NH₂    NH₂

Asp 189

Arginine side chain

**Figure 4-15**
The specificity pocket of trypsin can accommodate an arginine side chain of the substrate or a benzamidine ion inhibitor.

**Table 4-4**
Some Inhibitions of Enzymes that Form Covalent Linkages with Functional Groups on the Enzyme

| Inhibitor | Enzyme Group that Combines with Inhibitor |
|---|---|
| Cyanide | Fe, Cu, Zn, other transition metals |
| p-Mercuribenzoate | Sulfhydryl |
| Diisopropylphosphofluorophosphate | Serine hydroxyl |
| Iodoacetate | Sulfhydryl, imidazole, carboxyl, thiol ether |

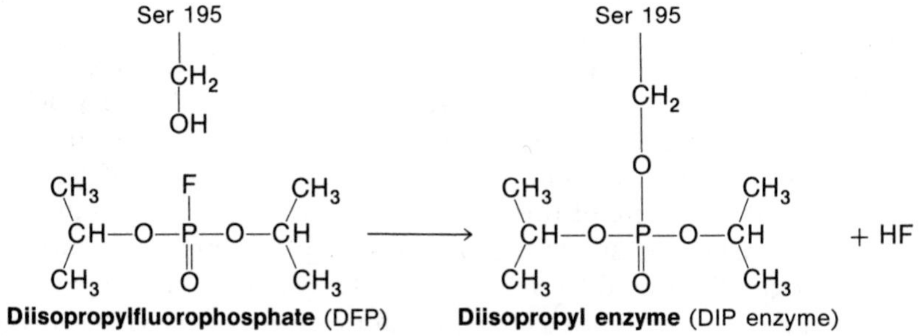

Ser 195
|
$CH_2$
|
OH

$CH_3$   F   $CH_3$
CH—O—P—O—CH
$CH_3$   O   $CH_3$

**Diisopropylfluorophosphate** (DFP)

Ser 195
|
$CH_2$
|
O
|
$CH_3$     $CH_3$
CH—O—P—O—CH
$CH_3$   O   $CH_3$

**Diisopropyl enzyme** (DIP enzyme)   + HF

**Figure 4-16**
DFP is an irreversible inhibitor with trypsin, chymotrypsin, and elastase. It forms a covalent bond with the Ser 195 side chain, thereby preventing subsequent catalysis. The DIP enzyme is permanently inhibited.

just a little too large and gets in the way of the $\alpha$ carbon of any potentially binding polypeptide chain, even if the side chain is only the —H of glycine.

Other irreversible inhibitors inactivate trypsin permanently by chemically modifying part of the active site. Diisopropylfluorophosphate (DFP) is an irreversible inhibitor of proteases like trypsin that contain an essential serine at the active site. It binds covalently and irreversibly to serine 195 to form diisopropyl-serine-195-trypsin (Figure 4-16). DFP also reacts with the same serine groups in chymotrypsin and elastase. The correlation between loss of enzyme activity and reaction with this serine group was instrumental in determining that the serine is important to enzyme activity.

The compound L-1-tosylamino-S-phenylethylchloromethylketone (TPCK) alkylates histidine 57 in chymotrypsin with concomitant loss of enzyme activity (Figure 4-17). Like DFP, the TPCK compound was a useful probe to pinpoint one of the groups involved at the active site. Many other reagents are available for reacting with other functional groups. Some of these will be considered later.

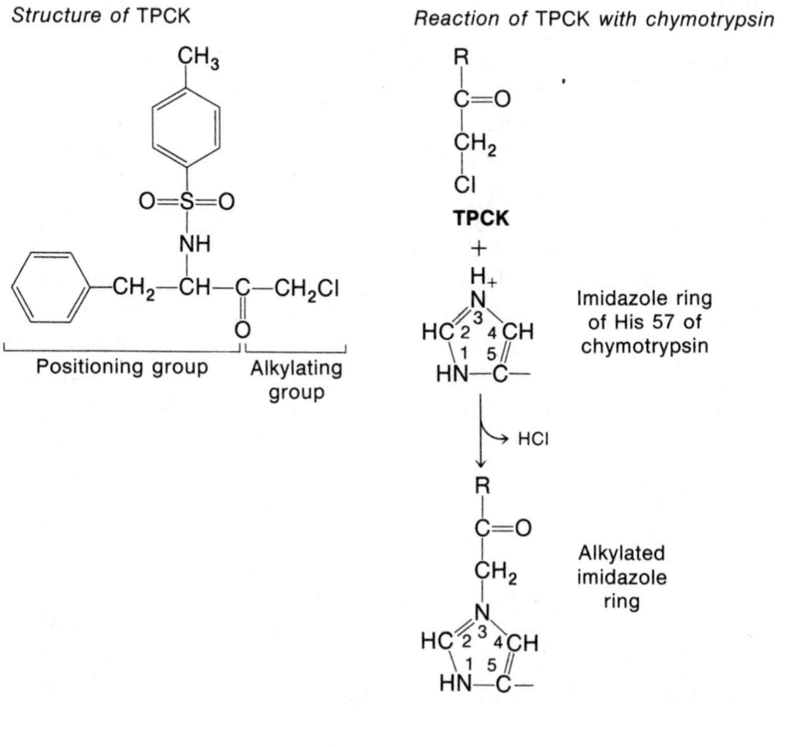

*Structure of TPCK*

$CH_3$

O=S=O
|
NH
|
—$CH_2$—CH—C—$CH_2Cl$
|
O

Positioning group    Alkylating group

*Reaction of TPCK with chymotrypsin*

R
|
C=O
|
$CH_2$
|
Cl

**TPCK**
+
H+
N
HC 2 ³ 4 CH
1 5
HN—C—

Imidazole ring of His 57 of chymotrypsin

→ HCl

R
|
C=O
|
$CH_2$
|
N
HC 2 ³ 4 CH
1 5
HN—C—

Alkylated imidazole ring

**Figure 4-17**
Structure of TPCK and reaction with His 57 in chymotrypsin.

**Effects of pH on Enzymatic Activity**

Most enzymes have a pH optimum of the general character indicated in Figure 4-18. The pH activity relationship for any particular enzyme depends on the p$K$ of the ionizing groups on the enzyme and the substrate that participates in the reaction. The optimum pH, unlike the $K_m$, is usually a characteristic more of the enzyme than of the particular substrate. Thus for the enzyme pepsin, the optimum is around 2; for catalase, around 7; and for arginase, around 10. Whereas the pH sensitivity of the enzyme is most often an indication of an ionizable group at the site of enzyme reaction, it can also be a measure of some change in the tertiary structure of the enzyme that affects the active site. For this reason, the sensitivity of enzyme activity to pH must be interpreted with caution. The pH effects on chymotrypsin activity suggest two such contrasting effects of pH.

Chymotrypsin-catalyzed reactions have bell-shaped pH profiles, such as the one shown in Figure 4-18 for the hydrolysis of a neutral substrate acetyl-L-tryptophan amide. Bell-shaped pH-rate profiles with neutral substrates have often been interpreted as evidence that two ionizing groups of the enzyme are important in the catalytic reaction. The midpoint of the left-hand side of the pH-rate profile indicates that the rate of the catalytic reaction increases as an ionizing group with an apparent p$K$ (p$K_{app}$) of about 7. The midpoint of the right side of the curve indicates that the rate of the reaction decreases as an amino acid residue with a p$K_{app}$ of about 8.5. The p$K$ around 7 could be identified with an imidazole group of histidine that would be unprotonated in the catalytically active form of the enzyme (Figure 4-10). No other amino acid side chains have titratable groups around this pH.

Other experiments reported earlier also implicated a histidine residue (His 57) in the active site (see Figure 4-17). As stated earlier, the activity for amide hydrolysis on the high pH side decreases with an approximate p$K_{app}$ of 8.5. The only amino acid side chain with a p$K$ in this region is cysteine. However, the amino-terminal group Ile 16 has a p$K$ with this value, and of course, in the neutral range this group would be charged. When the preenzyme chymotrypsinogen is activated by trypsin hydrolysis (see Figure 1-23) the Arg 15–Ile 16 linkage is cleaved to produce this amino-terminal linkage. Crystal structure analysis indicates that Asp 194 and His 40 form an ion pair in chymotrypsinogen, and in so doing, Asp 194 occupies part of the substrate binding site. Hydrolysis of the Arg 15–Ile 16 linkage creates a positively charged terminal amino group on Ile 16. This group displaces His 40 in the salt linkage with Asp 194. In so doing, the movement in the polypeptide chain establishes a substrate binding site. The importance of the Asp 194–Ile 16 linkage is underscored by the fact that chemical modification of either of these groups to prevent ionization results in a severe lowering of enzyme activity. Hence the pH inactivation effect on the basic side appears to be due to the removal of the positive charge from the amino-terminal isoleucine, thereby interfering with the substrate binding site. Further measurements indicate that when this happens, the substrate binding site returns to the conformation it had in the preenzyme.

This interpretation is supported by a kinetic analysis of the values of $k_{cat}$ and $K_m$ for the hydrolysis of N-acetyl-L-tryptophan amide as a function of pH. Between pH 8 and 10.5, $k_{cat}$ is almost constant, whereas $K_m$ increases severalfold (Figure 4-19). Thus the substantial decrease in enzyme reaction

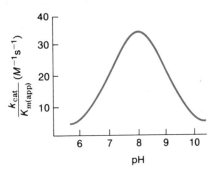

**Figure 4-18**
A pH-rate profile typical of reactions catalyzed by chymotrypsin. This curve was drawn from steady-state kinetic data on the α-chymotrypsin-catalyzed hydrolysis of N-acetyl-L-tryptophan amide at 25°C. Continuous measurements of released ammonia were made at each pH level over a range of substrate concentrations.

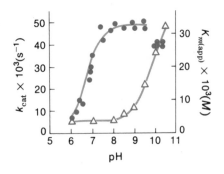

**Figure 4-19**
The pH dependence of the steady-state kinetic parameters of the α-chymotrypsin-catalyzed hydrolysis of N-acetyl-L-tryptophan amide at 25°C ($\bullet$ = $k_{cat}$ and $\triangle$ = $K_{m\,(app)}$).

rate in this pH range appears to be purely a function of decreased affinity of substrate and not related to a decreased effectiveness of the catalytic site.

It also can be shown that the decreased binding of substrate directly parallels a conformation change in the protein. This was done by measuring the circular dichroism of the protein and the substrate affinity over the same pH range. A sizable change in circular dichroism indicates an appreciable conformational change in the protein. The two changes do indeed parallel one another.

As stated at the outset of this section, pH effects on enzyme activity must be cautiously interpreted. They could be due to the effects of groups directly involved at the catalytic center or to the effects of other groups located elsewhere. In chymotrypsin, we see examples of both. One is reminded of the situation in hemoglobin (see Chapter 3), where the lowering of pH is believed to titrate an electrostatic linkage at a distant location from the oxygen binding site. The resulting alteration in tertiary and quaternary structures lowers the affinity for oxygen.

## CARBOXYPEPTIDASE A

In the preceding discussion we have described a group of closely related endopeptidases secreted by the pancreas, namely, trypsin, chymotrypsin, and elastase. These digestive enzymes are accompanied in the secretions by certain exopeptidases, i.e., carboxypeptidases A and B. Carboxypeptidase A (CPA) and carboxypeptidase B (CPB) finish the job started by the endopeptidases; they hydrolyze the oligopeptides one at a time from the C-terminal end of the polypeptide chain. CPA has a preference for aromatic residues and CPB functions best with basic residues, reminiscent of the chymotrypsin-trypsin complementarity. All these digestive enzymes are synthesized as inactive precursors and are converted to the active enzymes just before use. In the case of CPA, the harmless precursor is a complex assembly of three subunits that is split by trypsin through a complex series of reactions in which only one of the subunits eventually becomes trimmed down to a single uninterrupted chain with 307 residues. The amino acid sequence of CPA is known, and its three-dimensional structure, both with and without bound substrate, has been determined. In addition, a great deal of work has gone into studying the mechanism of hydrolysis, making CPA one of the better-understood enzymes. The enzyme differs in two major respects from the other enzymes we have been discussing; it is a zinc metalloenzyme and it undergoes a large conformational change upon binding of the substrate that serves the purpose of bringing together the components of the active site.

### Substrate Specificity

The peptide bond that is to be hydrolyzed must be adjacent to a terminal free carboxy group, as shown in Figure 4-20. As stated earlier, CPA preferentially hydrolyzes peptides when the terminal residue is hydrophobic; either aromatic or branched aliphatic groups make favorable substituents. The binding is also stereospecific, as that grouping must be in the L configuration. Integrity of the second peptide bond is also important for rapid hydrolysis. Thus dipeptides having a free amino group are hydrolyzed slowly,

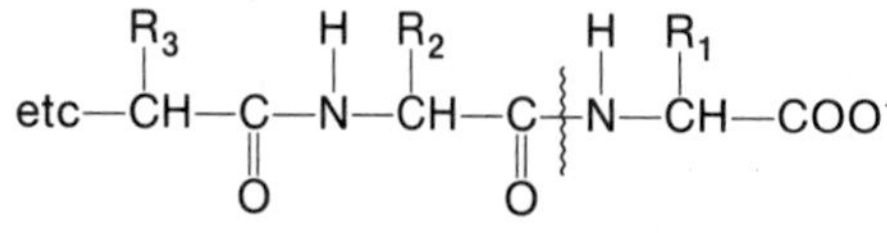

**Figure 4-20**
Substrate for carboxypeptidase, to be split at the wavy line.

but if this group is blocked by *N*-acylation, the hydrolysis is rapid. CPA also possesses esterase activity. As in the case of peptide hydrolysis, the L configuration of the C-terminal residue and an aromatic C-terminal side chain are important for rapid ester hydrolysis. Interesting differences between the mechanism of peptide hydrolysis and ester hydrolysis have been instrumental to an understanding of the reaction, and these differences are discussed below.

### Structure of the Enzyme-Substrate Complex

The three-dimensional structure of the enzyme has been determined in the free state and when complexed with the dipeptide glycyl-L-tryosine. As with all dipeptides, this is a poor substrate, but it is believed that it complexes with the enzyme in a way that closely resembles complexing with a good substrate. In Figure 4-21, a view of the structure with inhibitor in place is shown. Positions of Arg 135, Tyr 248, and Glu 270 before binding of substrate are shown. The inhibitor molecule shown is carbobenzoxy-Ala-Ala-Try, or

$$C_6H_5—CH_2—O—CO—[—NH—CH(CH_3)—CO—]_2$$
$$—NH—CH(CH_2—C_6H_4—OH)—COOH$$

The location of the left half of the inhibitor is predicted from model-building studies. The aromatic C-terminal side group of the substrate fits into a pocket in the interior of the molecule whose rim is composed of the chain of residues 245 to 251.

A good deal of movement in the protein occurs on binding substrate. Arg 145 moves 2 Å closer to interact with the substrate's terminal carboxyl group. Tyr 248 swings about 13 Å down to place its hydroxyl near the nitrogen of the bond to be split, and the neighboring carboxyl oxygen becomes complexed with the $Zn^{2+}$. Recent refinements in x-ray diffraction studies indicate certain minor errors in this model as drawn. The $Zn^{2+}$ is probably hexacoordinated rather than tetracoordinated as shown. It complexes with both oxygens of the Glu 72 side chain as well as with the amino nitrogen and the carboxyl of the dipeptide, as shown in Figure 4-22. In the absence of substrate, the $Zn^{2+}$ is pentacoordinated, two linkages to His 69 and His 196 as shown, two linkages to Glu 72, and one linkage to a water molecule that is expelled when substrate is present. Thus the $Zn^{2+}$ changes its coordination number from 5 in the native state to 6 in the complex with substrate. The overall effect of the binding of substrate to CPA is the conversion of the enzyme cavity from a water-filled to a hydrophobic region. At least four water molecules must be expelled when a substrate C-terminal side chain such as Tyr is inserted into the pocket, and one water molecule is displaced from the $Zn^{2+}$ when the carboxyl group of the substrate is bound. In addition, the charge of Arg 145 is neutralized by its electrostatic interaction with the terminal carboxylate ion of the substrate.

Finally, Tyr 248, in making its conformational change, closes the enzyme cavity. It seems likely that the displacement of water upon binding the substrate and the resultant conversion of the active center of the enzyme to a hydrophobic area provide a major part of the driving force for substrate binding. The presence of a dead-end pocket, in addition to a groove, provides an explanation for the observation that CPA is an exopeptidase and not an endopeptidase. Internal peptide units in a polypeptide

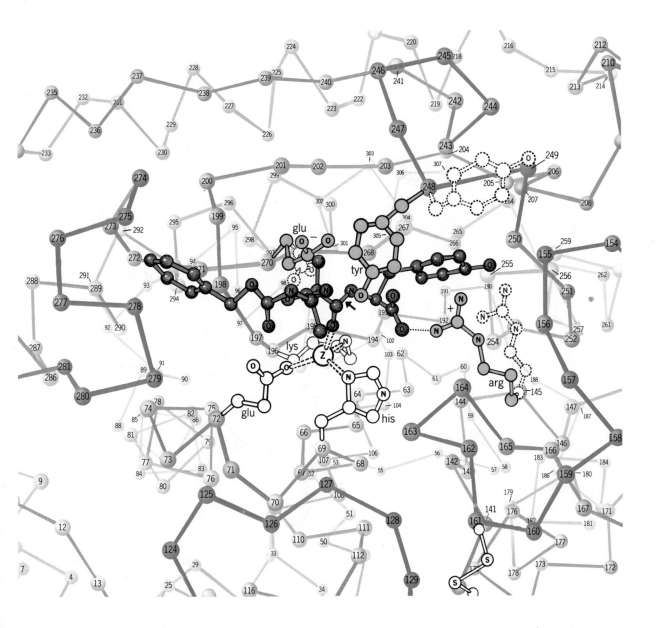

 *Inhibitor*

 *Zn-liganding side chains*

 *Moving side chains* without *inhibitor*

 *Same,* with *bound inhibitor*

**Figure 4-21**
The active site with substrate in place. Arg 145, Tyr 248, and Glu 270 are shown before substrate binding (*dotted line*) and after (*solid line*). Note the tetrahedral coordination of the zinc atom (Z) to His 69, Glu 72, Lys 196, and the carbonyl oxygen of the bond to be cleaved in the substrate.

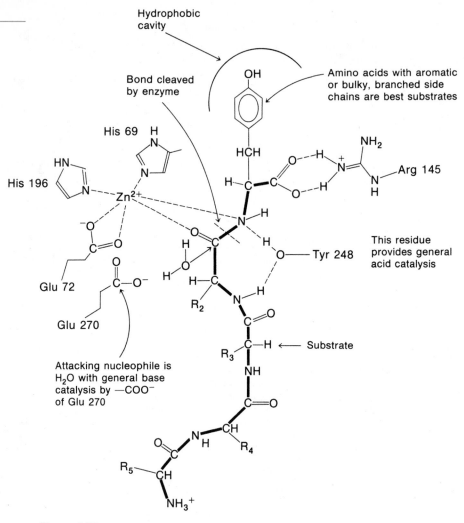

**Figure 4-22**
Expanded schematic view of the active site showing a hexacoordinated $Zn^{2+}$.

chain could never be effectively fitted into or around this binding site with the proper juxtaposition of the peptide bond to the active center. They also lack the negatively charged carboxylate for making the electrostatic linkage to Arg 145. Model-building studies show that it is difficult to fit the R group from a C-terminal D residue into the pocket, which explains the stereospecific preference of the enzyme for substrates with the L configuration.

## Mechanism of the Reaction

Schematic views of the bound substrate derived from structural studies and of other groups important at the active site are shown in Figure 4-22. The structure of the complex immediately suggests that the $Zn^{2+}$ plays a key role as an electrophilic catalyst, and this view has been supported by chemical studies. The favored mechanism proposed by R. Breslow and D. L. Wernick is shown in Figure 4-23. The peptide substrate binds so as to displace water from the $Zn^{2+}$. This makes the carbon of the carbonyl attractive for nucleophilic attack. Then a water molecule is delivered by the glutamate carboxylate (Glu 270). After delivery of the hydroxyl to the carbonyl and pickup of the first proton by Glu 270, there must be an additional proton transfer to permit cleavage, so both protons of $H_2O$ eventually are released. This second proton transfer is mediated through the hydroxyl group of Tyr 248, which functions as a general acid catalyst. The proton transfer from the tyrosine to the peptide N is encouraged by the complex formed between the $Zn^{2+}$ and the peptide N. The second proton originating from the water is then transferred to the tyrosine.

It was mentioned earlier that certain esters are also hydrolyzed by CPA. However, there is strong evidence that ester hydrolysis proceeds by a substantially different pathway. Thus selective acetylation or diazotization of Tyr 248 results in a total loss of peptidase activity but actually enhances esterase activity by twofold to threefold. Clearly, Tyr 248 is not used in ester hydrolysis. Another remarkable clue was the finding that replacing the $Zn^{2+}$ in the enzyme with other metals had different effects on esterase and peptidase activity. To explain this contrasting behavior, Breslow and Wernick suggested that in both reactions the substrate carbonyl and a water molecule are always aligned between Glu 270 and the $Zn^{2+}$, but that the sequence is different. Peptide substrates bind so as to displace water from $Zn^{2+}$. Then a water molecule is delivered by the glutamate carboxylate:

$$E-CO_2^- \quad H-O \quad C=O\text{---}Zn^{2+}-$$
$$\phantom{E-CO_2^- \quad H-} H \quad NH$$

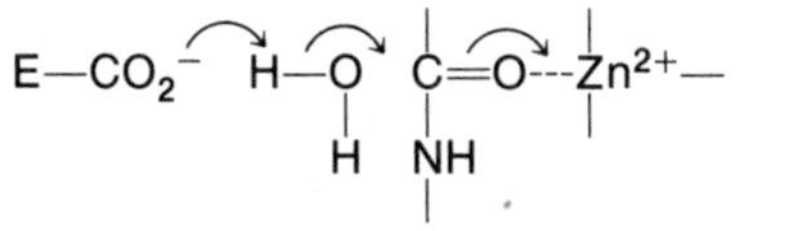

Ester substrates may bind without displacing the water from $Zn^{2+}$. This then puts them in position for a nucleophilic attack by the glutamate:

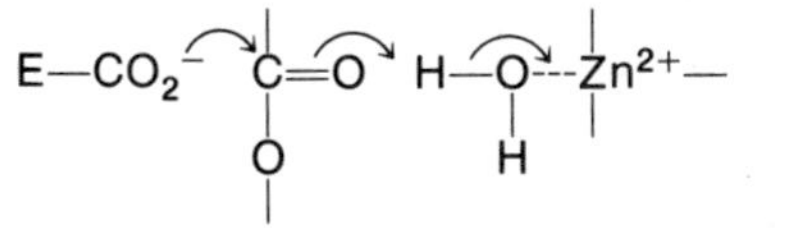

$$E-CO_2^- \quad C=O \quad H-O\text{---}Zn^{2+}-$$
$$\phantom{E-CO_2^- \quad C} O \phantom{ \quad H-} H$$

The ester is a weaker ligand and cannot displace the water molecule complexed with the $Zn^{2+}$. The reaction with ester should result in an interme-

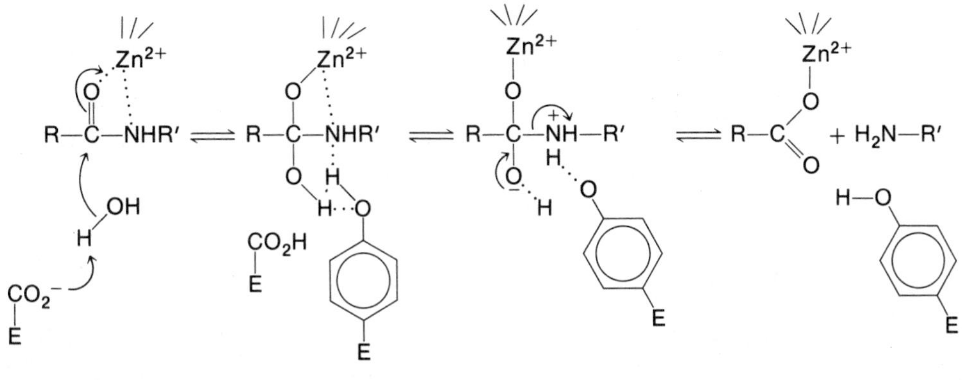

**Figure 4-23**
Proposed mechanism for carboxypeptidase A–catalyzed peptide hydrolysis.

diate enzyme-bound anhydride that is subsequently hydrolyzed by addition of water. Such an intermediate anhydride in ester hydrolysis has been detected by Makinen and Kaiser. This comparison of the ester and the peptide hydrolysis, as in the case of chymotrypsin, has been both interesting and informative in clarifying the mechanism of hydrolysis.

## PANCREATIC RNAse A

Ribonuclease is a hydrolytic enzyme cleaving RNA at the 3' P—O bond on the far side of pyrimidine nucleotides (Figure 4-24). The reaction is believed to occur in two discrete chemical steps with a cyclic 2',3'-phosphate intermediate.

S. Moore and W. H. Stein, who had earlier developed ion-exchange methods for the analysis of amino acids and peptides, determined the amino acid sequence of the 124-residue enzyme from the bovine pancreas (Figure 4-25). This was only the second protein and the first enzyme to be sequenced. For their efforts they were awarded the Nobel Prize in chemistry in 1972 together with C. B. Anfinsen.

Although RNAse A was available in large quantities and could be crystallized, the early chemical investigations of this enzyme preceded the technological developments in x-ray crystallography necessary for a three-dimensional structural analysis. A great deal was deduced about the catalytic mechanism from chemical and kinetic studies, and this was confirmed by future developments that led to a description of the three-dimensional structure (see Figures 1-31 to 1-34).

A discovery useful for correlating structure with function was the specific cleavage of ribonuclease between residues 20 and 21 by the bacterial proteolytic enzyme subtilisin. The resultant two peptides could be separated chromatographically in denaturing solvents. Separated peptides are catalytically inactive even after removal of the denaturant. On reincubation of the two peptide fragments in the absence of denaturant, a specific

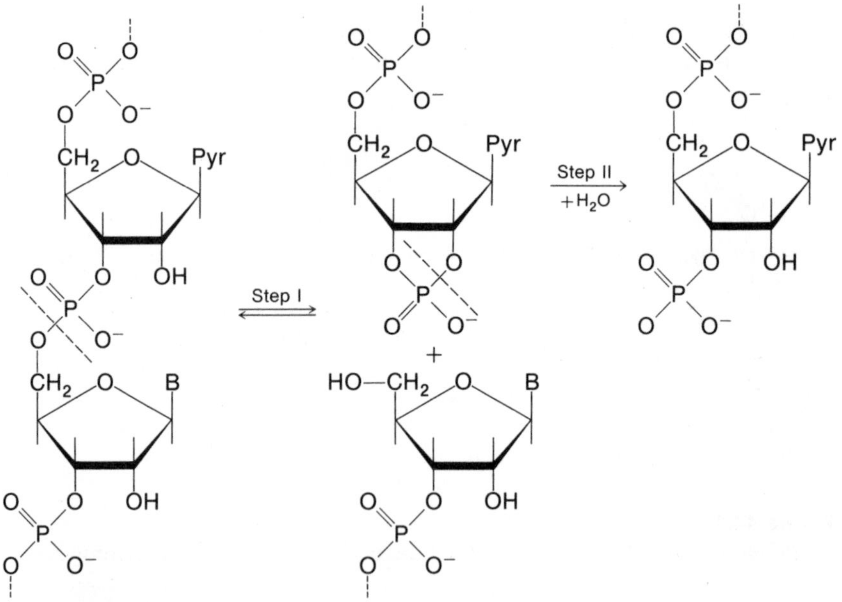

**Figure 4-24**
Ribonucleic acid chain indicating points of cleavage by pancreatic RNAse. Pyr refers to pyrimidine and B refers to purine or pyrimidine. The reaction is believed to be divided into two steps, as shown.

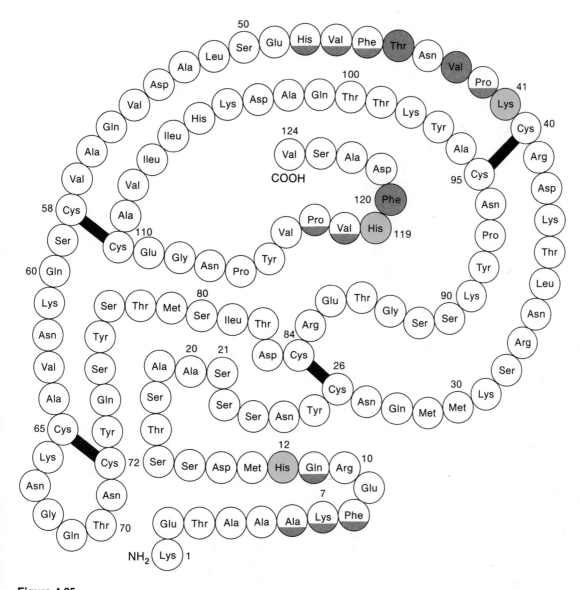

**Figure 4-25**
Amino acid sequence of bovine ribonuclease. Ribonuclease was the first protein to be completely synthesized from its amino acids. The final product is enzymatically indistinguishable from native ribonuclease. Four disulfides are indicated by gray bars. Three residues, His 12, His 119, and Lys 41, important
in the active site are indicated in color. Other groups involved in binding the substrate to the active site,
Thr 45, Val 43, and Phe 120, are shown in gray.

but noncovalent recombination takes place to yield a fully active ribonuclease called RNAse S. Aside from the important bearing of this result on
the existence of complementary intrapolypeptide chain interactions in enzyme proteins, it presents a method for specific chemical modifications of
amino acid residues on each of the two polypeptide chains independently
and thereby allows a study of the effect of such individual modifications on
the activity of the reconstituted active enzyme protein. Systematic chemical modification of the S peptide has permitted an estimation of the contribution of each residue to the peptide-peptide interaction and to the activity. From such studies it has been concluded that His 12 is required for

enzyme activity and that several other residues in the S peptide are important for reconstitution to form the RNAse S protein.

Chemical probes that react with specific side chains were most useful in predicting groups important in the catalytic action of the enzyme. In treatment of native RNAse with iodoacetate in equimolar quantities, carboxymethylated His 119 is the major product and carboxymethylated His 12 is a minor product (Figure 4-26). The other histidines in RNAse are much less reactive toward this reagent. Either of these derivatives has lost the majority of its catalytic activity, suggesting that both these groups are important in the active site. Support for this idea comes from the observation that the alkylation is inhibited by small molecules, such as cytidine-3'-phosphate, which is believed to bind to the active site. The Lys 41 residue was similarly implicated in the active site through the observation that the enzyme is inactivated under conditions in which the reagent fluorodinitrobenzene reacts selectively with the ε-amino group of this lysine (Figure 4-27).

An important indication as to the mechanism of action of ribonuclease was provided by detection of cyclic 2',3'-phosphates in RNAse digests. Simple nucleotide cyclic 2',3'-phosphates can serve as substrates of RNAse. It should be remembered that the first step in the digestion of RNA by the enzyme leads to the formation of a cyclic intermediate, as shown in step 1 of Figure 4-24; the second step results in the conversion of the cyclic 2'3'-phosphate to a cyclic 3'-phosphate derivative.

The pH dependence of the hydrolysis of these synthetic 2'3' cyclic nucleotide substrates as well as of RNA is shown in Figure 4-28. The bell-shaped profile is suggestive of a concerted acid-base catalyzed reaction (at least in step 2 of Figure 4-24), and the apparent $pK_a$'s are compatible with the involvement of two histidine residues, one functioning in the charged form as a general acid, the other functioning in the neutral form as a general base. The previously described evidence that histidine residues of the RNAse molecule are essential to catalysis strengthens this supposition. Direct determination of the $pK_a$'s of His 12 and His 119 has led to values of 5.8 and 6.2, respectively. This was done by observing the changes in nuclear magnetic resonance spectra of bands characteristic of these amino acids as a function of pH. Further discussion of the involvement of histidine in the catalytic process is presented below.

Despite the enormous number of chemical studies done in advance, the crystal-structure studies of RNAse and of RNAse complexed with specific inhibitors have been an indispensable aid in understanding how the enzyme works. There really is no substitute for this type of precise structural information, and it is clear that the increasing ease of determining structures by x-ray diffraction sets a trend for future investigations. The three-dimensional structure of the enzyme confirms the closeness of His 119, His 12, and Lys 41, which on the basis of chemical data were predicted to make up a critical part of the active center (see Figures 1-31 to

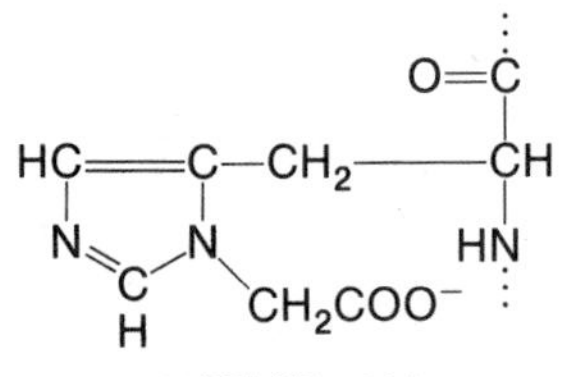

**1-CM His 119**

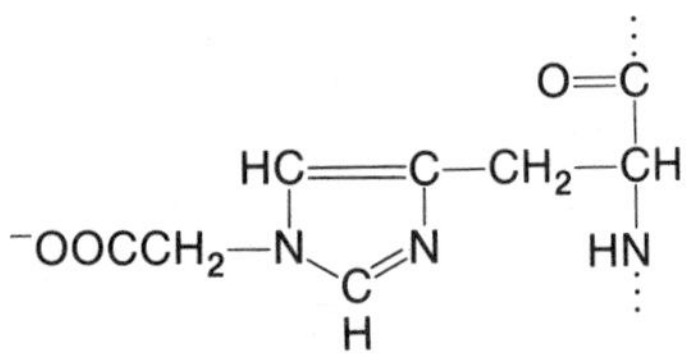

**3-CM His 12**

**Figure 4-26**
When RNAse is treated with equimolar amounts of iodoacetate (ICH₂COO⁻), two major products obtained are the carboxymethylated derivatives of His 119 and His 12.

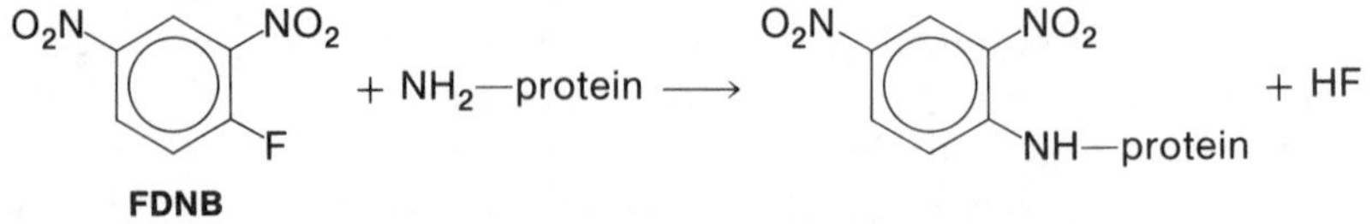

**Figure 4-27**
Reaction of fluorodinitrobenzene (FDNB) with the ε-amino group of lysine.

**FDNB**

1-34). These functional groups are distributed around a depression that occurs near the middle of the molecule. The structure of the cocrystal made from RNAse S and an inhibitor of RNAse (UpA with a methylene carbon atom replacing the oxygen between the phosphorus and the $5'$-$CH_2$ of the ribose of adenosine (A) and designated as $UpCH_2A$) shows the probable binding site of substrate (Figure 4-29). This inhibitor should resemble the substrate, but be nonhydrolyzable, because of the —O— replacement by —$CH_2$—. His 12 and His 119 are in close juxtaposition to the phosphate ester linkage that is hydrolyzed. The pyrimidine and purine rings fit into specific regions on the enzyme surface. Hydrogen bonds are formed between the pyrimidine ring and both the side chain —OH and the peptide —NH— of Thr 45. Phe 120 is located on one side of the pyrimidine ring and Val 43 on the other; together they form a groove for pyrimidine binding. F. Richards and colleagues found that when the pyrimidine is replaced by a purine, the compound still binds to the active site, but the distance

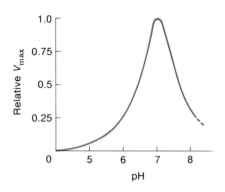

**Figure 4-28**
The dependence of $V_{max}$ on pH in the ribonuclease-catalyzed hydrolysis of nucleoside cyclic $2',3'$-phosphate.

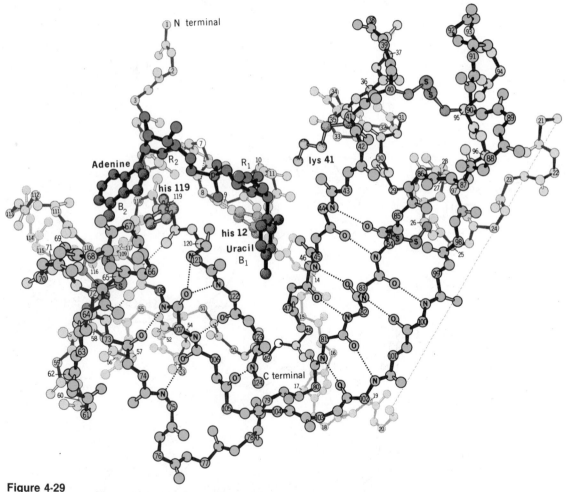

**Figure 4-29**
A schematic model of the binding of the dinucleotide phosphate $UpCH_2A$ (solid bonds) to RNAse S. His 119 is in one of four possible locations found on crystallographic analysis. In the native protein, the $\epsilon$-$NH_3^+$ of Lys 41 is down and projects toward the phosphate group but is not in contact with it. The —$CH_2$— attached to the phosphate of the inhibitor has perturbed the position of His 119. [From F. M. Richards and H. W. Wyckoff, Bovine Pancreatic Ribonuclease. In P. D. Boyer (Ed.), *The Enzymes*, Vol. 3, p. 647, Academic Press, New York, 1971. Used with permission.]

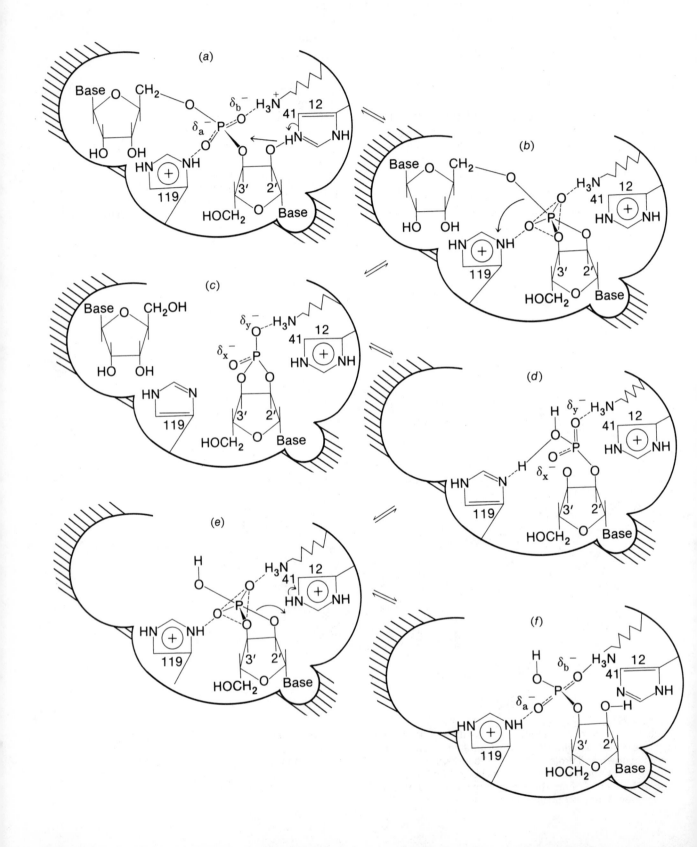

between the His 12 and the C-2' OH of the ribose is increased by about 1.5 Å. This probably explains why the enzyme is specific for a pyrimidine group, either C or U in this position.

It has been proposed that His 12 and His 119 act by general acid-base catalysis (Figure 4-30). In the first step, which leads to a cyclic phosphate, the His 12 could serve as a base (*a* to *b*), assisting in the removal of a proton from the ribose C-2' OH that couples to the phosphate to form the cyclic 2',3'-phosphate; His 119 could then act as an acid catalyst (*b* to *c*), protonating the leaving group to give a ribose C-5' OH. In the second step (opening of the cyclic phosphate ring), these roles could be reversed, with His 12 acting as an acid and His 119 as a base. It has been suggested that the function of the positively charged group on Lys 41 is to stabilize the transient pentacovalent phosphorus formed by an OH (from the C-2' OH in the cyclization step) or from water (in the ring-opening step). A schematic diagram illustrating this proposed mechanism is presented in Figure 4-30. It seems highly likely that this mechanism or something very close to it accurately depicts the mechanism of hydrolysis.

## LYSOZYME

Lysozyme like all of the previous enzymes discussed in this chapter is also a hydrolytic enzyme; in this case, the substrate is a polysaccharide chain. The bacteriolytic properties of hen eggwhite lysozyme were first reported by P. Laschtchenko in 1909, long before A. Fleming published his observations that similar bacteriolytic agents are widespread in biological tissues and secretions. It is still not clear if the function of the enzyme is to kill bacteria or to aid in their breakdown once they have been killed by other agents, or possibly both. The bacteriolytic action of lysozyme is accounted for by the ability of the enzyme to hydrolyze glycosidic linkages in bacterial cell walls. Some aspects of the structure of lysozyme have already been considered in Chapters 1 and 2 (see Figures 1-14 and 2-18). Other properties of lysozyme as an agent for cell-wall destruction are considered in Chapters 12 and 16. Here we will focus on the mechanism of action of the enzyme. Most of this work has been done on hen eggwhite lysozyme because it provides a bountiful source of the enzyme that is easy to purify.

Eggwhite lysozyme is a globular protein of molecular weight 14,600 with 129 amino acids in a single polypeptide chain and four disulfide linkages (Figure 4-31). Lysozyme attacks many bacteria by lysing the mucopolysaccharide structure of the cell wall (Chapter 12). Susceptible bacterial cell

◄ **Figure 4-30**
The substrate binds (*a*) at the active site with the purine and pyrimidine rings in tight binding subsites and Lys 41 $\epsilon$-NH$_3^+$ projecting toward the negatively charged phosphate along with His 12 and His 119. The concerted action of His 12, acting as a base to accept a proton from the ribose-2'-OH, and His 119, acting as an acid to form a hydrogen bond with a phosphate O atom as in (*b*), leads to a transition-state complex with a pentacoordinate phosphorus, represented by a trigonal bipyramid. The formation of the cyclic 2', 3'-phosphoribose intermediate (*c*) is accompanied by loss of a proton from His 119 and the uptake of a proton by His 12. Water then enters the site (*d*), donating a proton to His 119 and an —OH to the phosphate to form the trigonal pyramid structure in (*e*), which rearranges with the aid of the acidic His 12 to form the pyrimidine ribose-3'-phosphate product shown in (*f*). (From G. C. K. Roberts, E. A. Dennis, D. H. Meadows, J. S. Cohen, and O. Jardetsky, *Proc. Natl. Acad. Sci. USA* 62: 1151, 1969. Used with permission.)

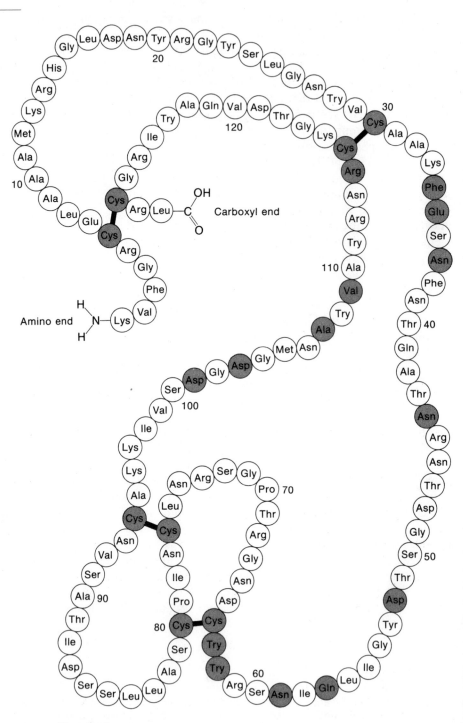

**Figure 4-31**
The primary sequence of amino acid residues in eggwhite lysozyme. Functional groups of the active site are indicated in
color. Some residues important in substrate binding are
shaded in gray.

walls are built from β-hydroxyl monomers called β-1,4-polymers. The monomers are arranged in a strict alternating sequence of N-acetylglucosamine (GlcNAc) and N-acetylmuramic acid (MurNAc) residues. These chains are cross-linked by short polypeptides that are attached to the —OR side chains of MurNAc by peptide bonds. Lysozyme cuts the polysaccharide chain on the far side of a linking oxygen atom that is attached to the C-4 of a GlcNAc residue. For an alternating GlcNAc–MurNAc–GlcNAc–MurNAc polymer, the cut is between the C-4 of MurNAc and the chain-linking oxygen, as shown in Figure 4-32.

The GlcNAc–GlcNAc–GlcNAc trimer was found to form a stable inhibited complex, and the structure of the enzyme was determined with and without the trimer present. Longer GlcNAc polymers were substrates, being cleaved with increasing rapidity up to the hexasaccharide. Although it has not been possible to observe the enzyme crystallographically with more than the trimer in place, it has been possible by model building to work out the probable mode of binding of the longer polysaccharides (see Figure 1-14b) and from this to suggest a catalytic mechanism. A crevice runs horizontally across the molecule, as shown in Figure 1-14. The fact deduced from x-ray diffraction that the tri-GlcNAc inhibitor binds in the top half of the crevice of the molecule, as shown in Figure 1-14, suggests that the crevice contains the active site. Most of the hydrophobic groups that are on the outside of the enzyme molecule appear as lining for the crevice; this leads to an energetically favorable interaction with the organic substrate molecule. There are small changes in the amino acid side-chain positions when the trimer inhibitor binds. Particularly noticeable is the movement of Trp 62 by about 0.75 Å toward the trimer B ring; the entire side of the crevice closes down slightly on the substrate. As stated earlier, it is possible to build three more sugar rings into the left half of the crevice to form a hypothetical complex for the hexasaccharide substrate. A schematic view of the hexamer and its interactions with the enzyme is shown in Figure 4-33 (also see the more precise representation presented in Figure 1-14b). Six subsites A through F on the enzyme bind the sugar residues. The nature and number of these contacts are indicated in Table 4-5. Alternate sites interact with the acetamide side chains (a) of the GlcNAc residues. These sites are unable to accommodate MurNAc residues because of their bulky lactyl side chains (P). This structure is consistent with the fact that β(1 → 4) linked hexoses prefer the almost fully extended state with adjacent hexose units flipped over 180 degrees relative to one another (e.g., see Figure 12-16). Site D cannot bind a sugar residue without distortion, and the glycosidic linkage that is cleaved binds between sites D and E, as shown by the arrow. The distortion forced on the residue binding to site D is believed to occur mostly in the sugar residue in such a way as to favor formation of the transition state of the complex. Therein is the crux of the mechanism. Binding to the D site is energetically unfavorable and most probably results in a distortion of the normal chair form of the sugar. Remember that the proposed distortion of the sugar has not been directly observed; it is based on model-building studies alone. This distortion is presumed to lower the activation energy for hydrolysis of the bound substrate. The other sugar residues bind favorably to their sites and provide enough energy to compensate for the unfavorable binding to the D site.

Conventional drawing of hexose sugar ring

Actual chain conformation of hexose sugar ring

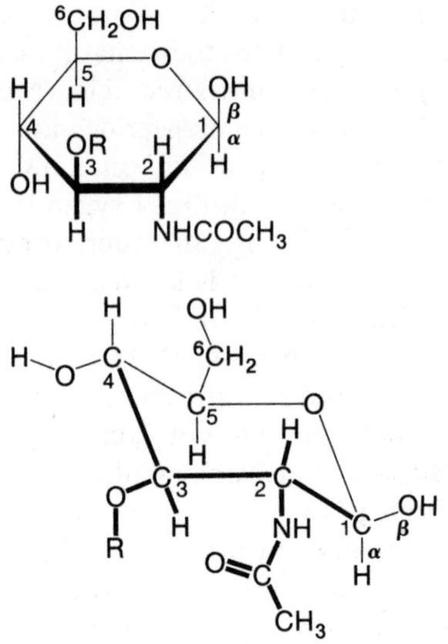

—OR = —OH in GlcNAc
—OR = $—OCH(CH_3)COOH$ in MurNAc

(*a*)

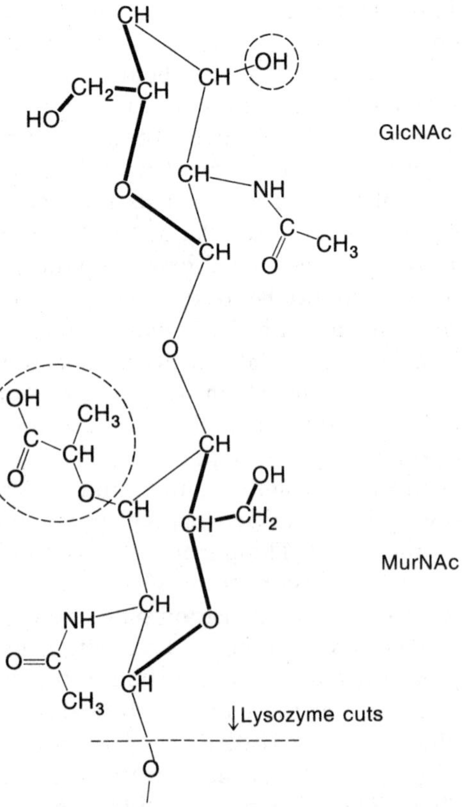


(*b*)

The substrate of lysozyme is an alternating copolymer of GlcNAc and MurNAc. The differences between GlcNAc and MurNAc are shown by dashed circles.

**Figure 4-32**
The conformations of *N*-acetylglucosamine amine (GlcNAc) and *N*-acetylmuramic acid (MurNAc) as monomers (*a*) and when linked (*b*). Position of cleavage by lysozyme is indicated in (*b*).

**Table 4-5**
The Main Interatomic Contacts between the Hexasaccharide
Substrate and the Lysozyme Molecule

| Site | Polar Contacts | Total Number of van der Waals Contacts $< 4 \text{A}$ |
|------|----------------|--------------------------------------|
| A | NH–Asp 101 | 7 |
| B | $O_6$–Asp 101 | 11 |
| C | $O_6$–Trp 62<br>$O_3$–Trp 63<br>NH–Co 107<br>CO–NH 59 | 30 |
| D | $O_6$–Co 57<br>$O_1$–Glu 35 | 35 |
| E | $O_3$–Gln 57<br>NH–CO 35<br>CO–Asn 44 | 45 |
| F | $O_6$–CO 34<br>$O_6$–Asn 37<br>$O_5$–Arg 114<br>$O_1$–Arg 114 | 13 |

(Data from J. A. Rupley, *Proc. Roy. Soc.* B167:416, 1967.)

In Table 4-6, the results of a study of the rates of hydrolysis of a homologous series of compounds containing from 3 to 6 residues of GlcNAc and the normal GlcNAc–MurNAc hexamer are indicated. It can be seen that the catalytic rate does not become significant until the pentamer. The lower oligomers bind to the enzyme, but they do not bind across the D site because this is energetically unfavorable. Therefore they are not properly oriented about the active site with a strained substrate favorably disposed for hydrolysis.

It can be seen (Table 4-6) that the hexamer hydrolysis occurs at the highest rate and exclusively between the fourth and fifth residues. This position in the enzyme-substrate complex must be located adjacent to the active site. Model building suggests that the susceptible $C_1$—$O_4$ bond between saccharide units D and E falls between Asp 52 and Glu 35. The favored proposed mechanism, illustrated in Figure 4-34, takes into account the closeness of these two acidic amino acid side chains, the strained configuration of the sugar, and the fact that the hydrolyzed tetramer product retains the same configuration after hydrolysis. Thus the leaving group ($-OR_1$) and the adding group ($-OH$) have to approach the sugar from the same side. After binding, the substrate undergoes a bond rearrangement to yield a carbonium ion at a rate enhanced by at least three factors:

1. A ring conformation distorted toward that of the transition state.

2. A glutamic acid 35 that acts as a general acid catalyst by donating a proton to the glycosidic oxygen.

3. A negatively charged asparate that stabilizes the positively charged carbonium ion.

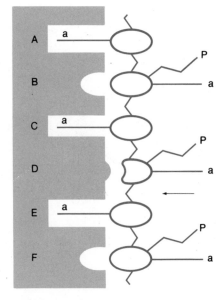

**Figure 4-33**
Schematic diagram showing the specificity of HEW lysozyme for hexasaccharide substrates. Six subsites (A–F) on the enzyme bind the sugar residues. Alternate sites interact with the acetamido side chains (a), and these sites are unable to accommodate MurNAc residues with their lactyl side chains (P). Site D cannot bind a sugar residue without distortion, and the glycosidic linkage that is cleaved binds between sites D and E, as shown by the arrow. (Adapted from drawing of T. Ionoto, L. N. Johnson, A. C. T. North, D. C. Phillips and J. A. Rupley, Vertebrate Enzymes. In P. D. Boyer (Ed.) *The Enzymes*, Vol. 3, p. 713, Academic Press, New York, 1971).

**Table 4-6**
Rates of Reaction and Cleavage Patterns Shown by Different Substrates

| Compound | Turnover Number $k_{cat}$ (s$^{-1}$) | Cleavage Pattern |
|---|---|---|
| (GlcNAc)$_3$ | $8.3 \times 10^{-6}$ | $X_1$—$X_2$—$X_3$ |
| (GlcNAc)$_4$ | $6.6 \times 10^{-5}$ | $X_1$—$X_2$—$X_3$—$X_4$ |
| (GlcNAc)$_5$ | 0.033 | $X_1$—$X_2$—$X_3$—$X_4$—$X_5$ |
| (GlcNAc)$_6$ | 0.25 | $X_1$—$X_2$—$X_3$—$X_4$—$X_5$—$X_6$ |
| (GlcNAc–MurNAc)$_3$ | 0.50 | $X_1$—$X_2$—$X_3$—$X_4$—$X_5$—$X_6$ |

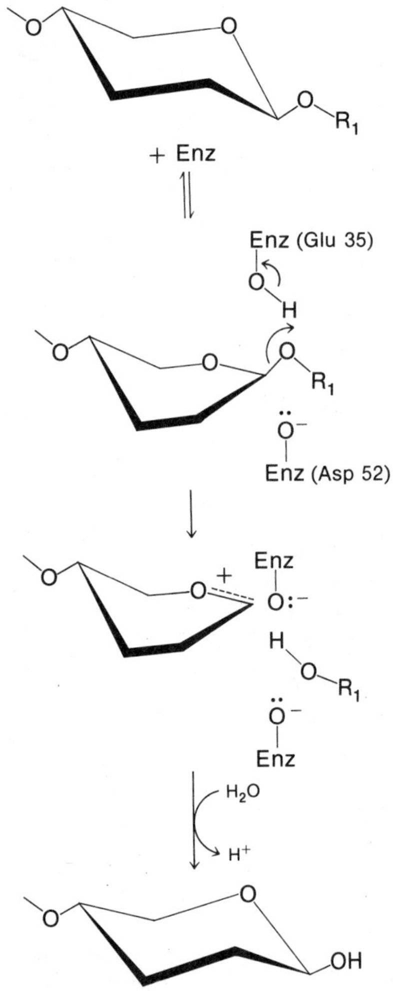

**Figure 4-34**
Proposed mechanism for lysozyme-catalyzed hydrolysis. Binding of the hexasaccharide to the enzyme distorts the reacting sugar (bound at the D site in Figure 4-33). This conformation is presumed to approach that of the positively charged intermediate. The final product produced by the addition of an —OH group from water restores the chair form in the product, as well as preserving the original stereochemistry in the glycosidic bond.

The resulting aglycone (HOR$_1$) diffuses away, and the process is completed by reaction with water or another acceptor.

None of the factors involved in lysozyme catalysis seems able alone to account for the observed rate enhancement, but concerted attack involving all three contributions mentioned earlier appears to be necessary. Despite this, lysozyme does not appear to be a very efficient enzyme as far as its turnover number ($k_{cat}$) is concerned. The $k_{cat}$ value of lysozyme (0.5 s$^{-1}$) is to be compared with that of chymotrypsin (100 s$^{-1}$) or lactate dehydrogenase (1000 s$^{-1}$), to be discussed in Chapter 5. If the main job of lysozyme is to rupture bacterial cell walls, it may not have to break many linkages to accomplish this. Thus the "physiological efficiency" of lysozyme may be much higher than its efficiency in terms of turnover number.

Two other aspects of lysozyme action are deserving of comment. First, the bond-strain mechanism used by lysozyme guarantees the safety of disaccharides, trisaccharides, and tetrasaccharides in its presence. It is not clear if this serves a useful purpose or merely results from an incidental property of the enzyme. Second, the pH optimum for lysozyme action is 5, and activity falls off rapidly on either side of this pH. The rapid fall off on the high side is consistent with Glu 35 functioning as a generalized acid catalyst. For this purpose, it must possess a proton. Various measurements indicate that the p$K$ of Glu 35 is near 6 in the free protein and is increased somewhat by saccharide binding. On the low side, the fall off of reaction rate could be due to protonation of Asp 52 as well as other carboxylates (e.g., Asp 101 and Asp 66) in the protein. Lysozyme is active in secretions that tend to have pHs on the acid side of neutrality. Thus if it is going to employ an amino acid for general acid catalysis, glutamic acid is a far better choice than histidine (which is used in RNAse), since the higher p$K_a$ of the latter amino acid would substantially raise the pH optimum.

## SELECTED READINGS

Blackburn, P., and Moore, S. Pancreatic Ribonuclease. In P. D. Boyer (Ed.), *The Enzymes,* Vol. 4. New York: Academic Press, In press.

Breslow, R., and Wernick, D. L. Unified picture of mechanisms of catalysis by carboxypeptidase. *Proc. Natl. Acad. Sci. USA* 74:1303, 1977.

Dixon, M., and Webb, E. C. *Enzymes,* 3d Ed. London: Longmans, 1979.

Jencks, W. P. *Catalysis in Chemistry and Enzymology.* New York: McGraw-Hill, 1969.

Imoto, T., Johnson, L. N., North, A. C. T., Phillips, D. C., and Rupley, J. A. Vertebrate Enzymes. In P. D. Boyer (Ed.), *The Enzymes,* Vol. 3. New York: Academic Press, 1971.

Reis, D. C., and Lipscomb, W. N. Binding of ligands to the active site of carboxypeptidase A. *Proc. Natl. Acad. Sci. USA* 78:5455, 1981.

Steitz, T. A., and Shulman, R. G. Crystallographic and NMR studies of the serine proteases. *Ann. Rev. Biophys. Bioeng.* 11:419, 1982.

## PROBLEMS

1. What is meant by the order of a reaction?

2. Define $K_m$, $K_s$ $k_{cat}$.

3. $K_m$ is all too frequently equated with $K_s$. In fact, in most reactions there is an appreciable disparity between the values for $K_m$ and $K_s$. For the reaction $A \rightarrow B$ define conditions under which $K_m = K_s$. Describe three conditions under which this is not true.

4. What is the difference between the activated complex and the enzyme substrate complex?

5. What is the steady-state approximation, and under what conditions is it valid?

6. In what ways are chymotrypsin, trypsin, and elastase similar as catalysts? In what ways do they differ? What factors in the enzyme structure are responsible for these differences?

7. For many enzymes, $V_{max}$ is dependent upon pH. At what pH would you expect the $V_{max}$ of RNAse to be optimal? Explain.

8. If histidines were substituted for Glu 35 and Asp 52 at the active site in lysozyme, would you expect the modified enzyme to be functional? If so, would the optimum pH be altered?

9. Assume that an enzyme catalyzed reaction follows Michaelis-Menten kinetics with a $K_m$ of $1 \times 10^{-6}$ $M$. If the initial reaction rate is 0.1 $\mu$mol/min at 0.1 $M$, what would it be at 0.01 $M$, $10^{-3}$ $M$, and $10^{-6}$ $M$?

10. If the $K_m$ of an enzyme for its substrate is $10^{-5}$ $M$ and the $K_I$ of the enzyme for a competitive inhibitor is $10^{-6}$ $M$, what concentration of inhibitor will be necessary to lower the rate of the reaction by a factor of 10 when the substrate concentration is 0.1 $M$? 0.01 $M$?

11. If the activation free energy, $E_a$, of an enzyme-catalyzed reaction is 10 kcal/mol, how much difference in the initial reaction velocity would you expect between 20°C and 30°C? Between 30°C and 40°C? Make the same calculation for an $E_a$ of 20 kcal/mol. If an enzyme lowers the $E_a$ from 20 kcal/mol to 2 kcal/mol, what is the rate enhancement factor achieved by using the enzyme?

12. Your mentor has just handed you a highly purified enzyme which catalyzes the reaction $A \rightarrow B$ and asked you to make the necessary measurements for a Lineweaver-Burk plot. Explain what you would do and what the results should tell you about the reaction.

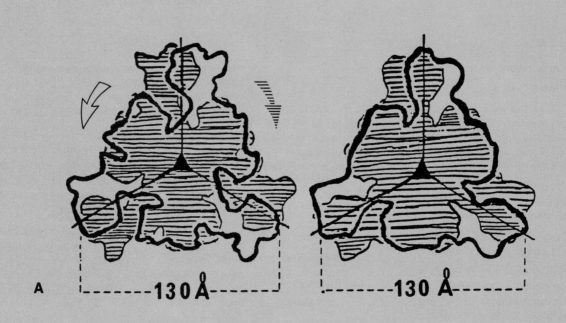

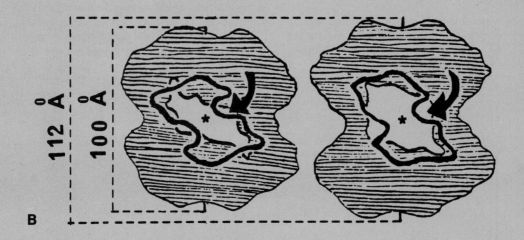

# 5

# ENZYME CATALYSIS II

In Chapter 4 the mechanisms for several hydrolytic enzymes were explored. These enzymes are all relatively simple, where the focus is on the interaction between a single substrate and the enzyme. Strictly speaking, even hydrolases use two substrates, water being the second, but because the concentration of water is so high and reactions have a negligible effect on its concentration, water is rarely considered as a second substrate. Furthermore, the reacting water is usually not bound to a specific site on the enzyme surface.

In this chapter we will discuss lactate dehydrogenase as an example of an enzyme that uses two substrates, aspartate carbamoyltransferase as an example of a two-substrate regulatory enzyme, and RNA polymerase as an example of a two-substrate regulatory enzyme that requires a template.

## MECHANISTIC CONSIDERATIONS FOR REACTIONS INVOLVING TWO SUBSTRATES

There are various models for representing enzyme-catalyzed pathways involving two substrates:

**1.** Random-order pathway:

$$S_1 + S_2 + E \rightleftharpoons ES_1 + S_2$$
$$\Updownarrow \qquad\qquad \Updownarrow$$
$$S_1 + ES_2 \rightleftharpoons ES_1S_2 \longrightarrow E + P$$

Changes in relative positions of the subunits of aspartate carbamoyltransferase on binding CTP (left) or PALA (right). (a) View down the molecular threefold axis. The lower $C_3$ and the three lower r subunits are represented by the shaded area. The upper $C_3$ and the three upper r subunits are represented by the heavily outlined area. The directions of $C_3$ rotation that accompany conversion from the inactive to the active state are indicated by the open arrow for the upper $C_3$ and the shaded arrow for the lower $C_3$. (b) View down the molecular twofold axis. Four C chains, those closest to the viewer, are represented by the shaded area. One $r_2$ is represented by the heavily outlined area. The open space is the central cavity. Access to the cavity is marked with an arrow. See the text for further discussion of this enzyme. (From J. E. Lander, J. P. Kitchell, R. B. Honzatka, H. M. Ke, K. W. Volz, A. J. Kalb, R. C. Lander, and W. N. Lipscomb, Proc. Natl. Acad. Sci. USA 79:3125, 1982.)

**2.** Ordered pathway:

$$\begin{array}{c}
S_2 \\
+ \\
E + S_1 \rightleftharpoons ES_1 \\
\big\Updownarrow \\
ES_1S_2 \longrightarrow E + P
\end{array}$$

**3.** Reactive intermediate pathway:

$$\begin{array}{c}
S_2 \\
+ \\
E + S_1 \rightleftharpoons ES_1 \longrightarrow EP_1 \rightleftharpoons EP_1S_2 \longrightarrow E + P_1 + P_2
\end{array}$$

**4.** Reactive intermediate pathway with the formation of an altered enzyme:

$$E + S_1 \rightleftharpoons ES_1 \rightleftharpoons E'P_1 \overset{P_1}{\rightleftharpoons} E' + S_2 \longrightarrow E'S_2 \overset{P_2}{\longrightarrow} E$$

In the random-order pathway (1), the order of binding of the substrates $S_1$ and $S_2$ is random and reaction occurs after the two substrates become bound to the enzyme. In the ordered pathway (2), one substrate, say $S_1$, must be bound before the second substrate, $S_2$, can become bound, after which reaction occurs as in pathway 1. In the reactive intermediate pathway (3), one substrate is bound and undergoes a reaction with the enzyme before the second substrate becomes bound, leading to additional reaction(s). In the reactive intermediate pathway that leads to an altered enzyme (4), we have another variation of the ordered pathway in which $S_1$ binds first and then reacts to form an altered enzyme $E'$ and a product $P_1$ that dissociates, making way for the altered enzyme to bind $S_2$ and convert it into product $P_2$. Finally, $P_2$ dissociates from the enzyme and the original enzyme is regenerated. This pattern of reaction is referred to as the "Ping-Pong mechanism" to emphasize the alternation of the enzyme between two states. Each of these four models is characterized by a different set of steady-state kinetic equations which gives rise to a characteristic Lineweaver-Burk plot [see Equation (15) in Chapter 4]. Frequently the situations are very complex, and additional information is almost always necessary to reach the correct interpretation of the steady-state kinetic data. Lactate dehydrogenase and aspartate carbamoyltransferase, both considered below, are examples of the simple ordered pathway (2). In this case, the characteristic kinetic equation for the initial reaction velocity is given by the general expression

$$\frac{1}{v} = \frac{K_{mS_1}}{V_{max}} \frac{1}{S_1} + \frac{K_{mS_2}}{V_{max}} \left(\frac{1}{S_2}\right)\left(1 + \frac{K_{eqS_1}}{S_1}\right) + \frac{1}{V_{max}}$$

where $K_{mS_1}$ and $K_{mS_2}$ are the Michaelis constants for $S_1$ and $S_2$, respectively, and $K_{eqS_1}$ is the dissociation constant for the $ES_1$ complex. It should be noted that the expression differs from the one for the single-substrate reaction by the presence of the middle term on the right side [compare with Equation (15) in Chapter 4]. The Ping-Pong mechanism is character-ized by the kinetic equation

$$\frac{1}{v} = \frac{K_{mS_1}}{V_{max}}\frac{1}{S_1} + \frac{K_{mS_2}}{V_{max}}\frac{1}{S_2} + \frac{1}{V_{max}}$$

Reactions involving pyridoxal, which frequently proceed through such an alternating cycle, are discussed in Chapter 6. Certain membrane-transport proteins are also believed to function by the Ping-Pong mechanism (see Chapter 17).

## LACTATE DEHYDROGENASE

Many enzymes use nicotinamide adenine dinucleotide (NAD) (see Figure 5-1) in oxidation reactions in which $NAD^+$ reacts with another substrate to accept a hydride ion (two electrons and one proton) in one or more steps. The resulting NADH (reduced form of the coenzyme) must be reoxidized so that it can be used again. In glycolysis (Chapter 8) in many organisms this is accomplished by L-lactate dehydrogenase (LDH), which catalyzes the reduction of pyruvate to lactate according to the reaction

$$\underset{\textbf{Lactate}}{\overset{\displaystyle CH_3}{\underset{\displaystyle COOH}{HCOH}}} + NAD^+ \rightleftharpoons \underset{\textbf{Pyruvate}}{\overset{\displaystyle CH_3}{\underset{\displaystyle COOH}{C{=}O}}} + NADH + H^+$$

For this reaction, the equilibrium at neutral pH is strongly favored to the left. In spite of this, the initial rate of the reaction can be studied in the rightward direction by addition of the appropriate substrates at elevated pHs. Physiologically it also appears that the enzyme can play useful roles in both directions (see below), although the conversion of pyruvate to lactate under anaerobic conditions is certainly the best-understood function. Lactate produced in such tissues functioning anaerobically is secreted and absorbed by other tissues functioning aerobically. LDH can then convert the lactate back to pyruvate for further utilization in the Krebs cycle (Chapter 9).

### Kinetic Studies on the NAD⁺-Lactate Reaction

The forward reaction for the oxidation of lactate by LDH and $NAD^+$ fits the general form of the steady-state kinetic expression for pathway 2 described earlier. Equilibrium binding experiments were used to demonstrate the order of binding. The coenzyme $NAD^+$ binds first; this by itself is believed to produce an alteration in the enzyme structure that creates a binding site for the lactate. The same is true for the reverse reaction; i.e., the NADH binds before pyruvate.

Optical properties of the protein and the coenzyme have been instrumental in making pre-steady-state kinetic measurements, which have been most important in determining the steps that take place after binding.

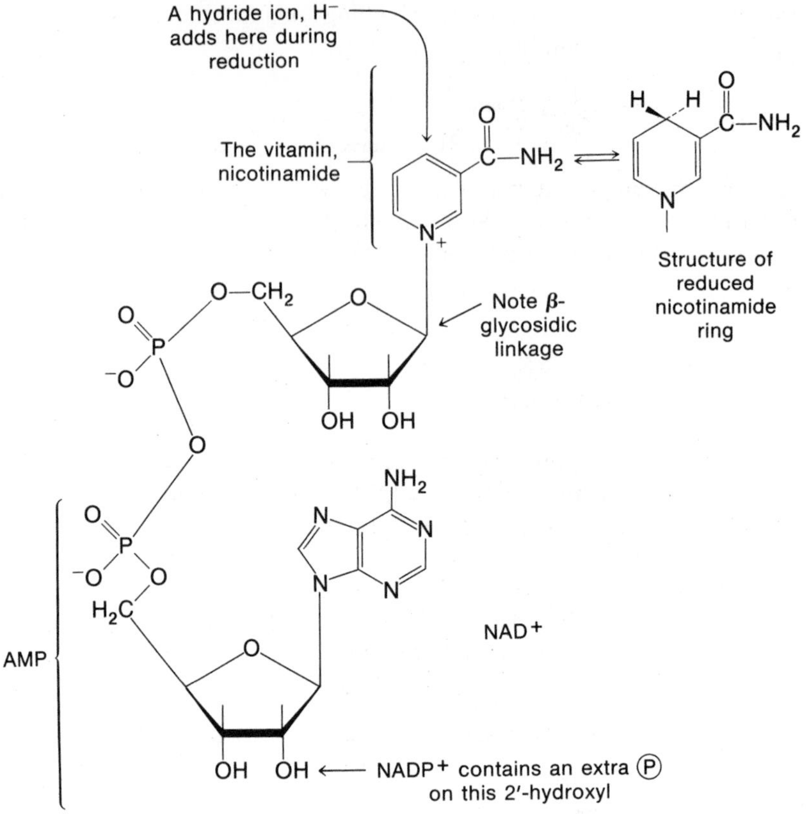

**Figure 5-1**
The hydrogen-carrying coenzymes
NAD$^+$ (DPN$^+$) and NADP$^+$ (TPN$^+$).
Note use of the abbreviations NAD$^+$
and NADP$^+$, even though the net
charge on the entire molecule
at pH 7 is negative.

NADH has a characteristic absorption band at 340 nm. This compound fluoresces weakly at 470 nm when excited with its absorption band, but this fluorescence is greatly increased on combination with the enzyme. The enzyme fluoresces at about 350 nm when excited at 270 to 305 nm. This fluorescence is due to the tryptophan residues in the enzyme. In LDH there is a substantial drop in this fluorescence on binding coenzyme that is probably due to the transfer of some of the trytophan residues to a more hydrophilic environment. Despite the fact that the exact cause of this drop in fluorescence is unclear, the change may be used to determine the protein state, i.e., whether or not it is bound to coenzyme and substrate. In addition to this, the release of H$^+$, a product of the reaction, has been measured optically by combination of H$^+$ with a dye molecule. All these optical changes have been measured as a function of time immediately after mixing the enzyme with saturating amounts of NAD$^+$ and lactate. Such measurements suggest at least 3 phases in the approach to the steady state. In the first phase, which takes less than 1 millisecond a small amount of NADH can be detected with no release of proton. In the second phase, which follows a first-order rate with respect to lactate concentration, NADH and H$^+$ are produced in equimolar amounts; only the proton is liberated. The third phase, which follows zero-order kinetics, begins after about 40 milliseconds. During this steady-state phase of the reaction, NADH and H$^+$ are liberated at the same rate. The proposed phases in the reaction after binding substrate and coenzyme are indicated by the following series of reactions:

| Phase I | Phase II | Phase III |
|---|---|---|

Here $\diagup$CHOH represents lactate and $\diagup$C=O represents pyruvate. One of the important conclusions from this experiment is that the rate-limiting step after reaching the steady state appears to be the release of the NADH product from the enzyme. Fitting the Lineweaver-Burk equation could never have yielded this information. Furthermore, the use of the Lineweaver-Burk plot does not yield a true $K_m$ for the reaction because the correct usage of this approach requires a system where the rate-limiting step is the conversion of reactions to products, not the release of product or products from the enzyme.

## Conformational Changes Take Place on Binding Coenzyme and Substrate

The strategy of all NAD-dependent dehydrogenases is to orient the coenzyme and the substrate on the enzyme surface so that the C-4 atom on the nicotinamide is pointed toward the reactive carbon of the substrate. Different dehydrogenases have remarkably similar protein domains for binding $NAD^+$ and dissimilar sites for binding the cosubstrate that are dependent on the structure of the substrate. The coenzyme binding domains are compared for lactate dehydrogenase and two other dehydrogenases in Figure 5-2.

In LDH binding of NADH is about 400 times stronger than the binding of $NAD^+$ and about 50 times stronger than that of NADPH (the same molecule with an additional phosphate; see Figure 5-1). Some enzymes can use either $NAD^+$ or $NADP^+$ as coenzymes, but for LDH the binding data show that there is a strong preference for the former. A conformational change takes place in the enzyme when the apoenzyme interacts with coenzyme NADH (or $NAD^+$) and a suitable substrate, usually pyruvate (or lactate). A minor movement of Asp 53 and its associated polypeptide chain occurs on binding the adenosine part of the coenzyme. There is a larger movement of the loop connecting $\beta$D and $\alpha$E, including the helix $\alpha$D (residues 98 to 120; see Figure 5-2), involving main-chain displacement of up to 11 Å. In ternary complexes this flexible loop drops down and encloses the coenzyme and substrate. In the apoenzyme it extends into the solvent. Arg 101 forms an ion pair with the pyrophosphate group of the coenzyme in the ternary complex (Figure 5-3). The formation of this ion pair may be the driving force for the conformational change of this loop and the subsequent rearrangements in the subunit. The guanidinium group of Arg 109 moves 14 Å and changes from being completely exposed to the solvent to

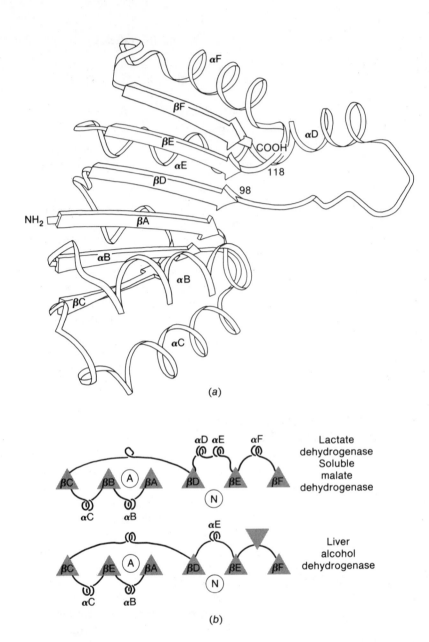

(a)

(b)

**Figure 5-2**
Nucleotide-binding protein domains and related tertiary structures in a series of proteins. (a) Schematic tertiary structure of part of lactate dehydrogenase. Arrows show strands of $\beta$ sheets drawn from the amino to the carboxyl end. (b) Cartoons of similar tertiary structure patterns in lactate dehydrogenase and other proteins. Triangles indicate $\beta$ sheets viewed with the N terminal closest to the observer. Coils indicate $\alpha$ helices. A and N indicate positions of the known binding sites for the adenine moiety of $NAD^+$ or ATP and for the nicotinamide moiety of $NAD^+$. (Adapted from M. G. Rossmann, In P. D. Boyer (Ed.), *The Enzymes*, 3d Ed., Vol. 9, p. 61, Academic Press, New York, 1975.)

having a close interaction with groups around the substrate site. The two connected helixes $\alpha$D and $\alpha$E are more angled to each other in the ternary complex. The movement of the N-terminal part of helix $\alpha$D is associated with a movement of the neighboring C-terminal part of helix $\alpha$H (located in the substrate-binding domain, not shown). Thus $\alpha$H pulls the top of the loop $\beta$K-$\beta$L along by about 3 Å. The helix $\alpha$2G and to some extent helices $\alpha$1G and $\alpha$3G move small but significant amounts in order to approach the coenzyme. A small but very important conformational change is related to the loop $\beta$G-$\beta$H. This piece of polypeptide chain stretches out more in the ternary complex than in the apoenzyme, which results in His 195 pointing deeper into the active site and into contact with the substrate. The overall picture of these complex conformational changes in the protein involves a movement toward the substrate and a contraction of the subunit on binding coenzyme and substrate.

In the native state, the enzyme exists as a tetramer containing four identical subunits. With all the conformational changes taking place in going from an apoenzyme to a ternary complex it is natural to suspect that there might be some cooperative (or anticooperative) effects between different binding sites by analogy with the examples of hemoglobin (Chapter 3) or aspartate carbamoyltransferase (discussed below). However, careful equilibrium binding studies of the interaction between enzyme and

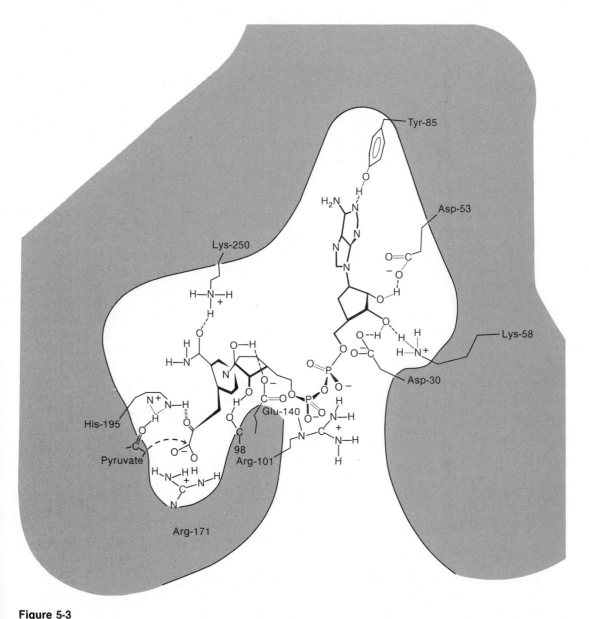

**Figure 5-3**
Diagrammatic representation of binding of pyruvate and NAD$^+$ to lactate dehydrogenase. [Adapted from J. J. Holbrook, A. Liljas, S. J. Steindel, and M. G. Rossmann, in P. D. Boyer (Ed.), *The Enzymes,* 3d Ed., Vol. 11, p. 240, Academic Press, New York, 1975.]

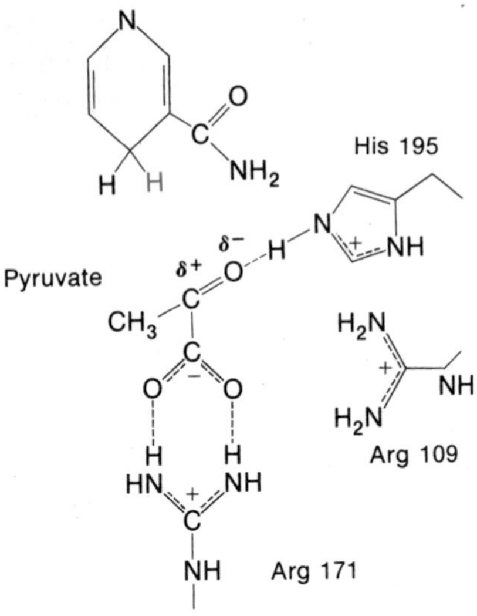

**Figure 5-4**
Expanded view of ternary complex at the active site.

NAD$^+$ or NADH have led to the conclusion that the intact enzyme contains four independent noninteracting binding sites. Existence of the enzyme as a tetramer must have other advantages. Sometimes dimers or tetramers are preferable simply because they are more stable and therefore less susceptible to denaturation or degradation.

## Hypothesized Mechanism of Enzyme Reaction

An expanded view of the active site in the ternary complex inferred from the x-ray structure studies is shown in Figure 5-4. The central carbonyl carbon atom of pyruvate is oriented so that it can accept a hydride ion from the C-4 of nicotinamide and the carbonyl O is hydrogen-bonded to His 195, which is protonated. A positive charge on the reactive carbon is favored by this interaction with histidine, which functions as a general acid catalyst; the partial neutralization of the carboxylate anion by Arg 171 and the closeness of Arg 109 also encourage stabilization of a positive charge on the reactive carbon. The reduction of pyruvate takes place with the overall result shown in Figure 5-5. Chiral aspects associated with NAD-dependent oxidation-reduction reactions are discussed in Chapter 6.

## Isoenzymes of Lactate Dehydrogenase and Their Function

Most vertebrates possess (at least) two genes for lactate dehydrogenase that make similar but nonidentical polypeptides called M and H. In embryonic tissue, both genes are equally active, resulting in equimolar amounts of the two gene products and a statistical array of tetramers ($M_4$, $M_3H_1$, $M_2H_2$, $H_3M_1$, and $H_4$ in the ratios of $1:4:6:4:1$). These so-called isoenzymes, or isozymes, can usually be detected by differing electrophoretic mobilities. As embryonic tissue multiplies and differentiates, the relative amounts of the M and H forms change. In pure heart tissue, which is considered aerobic, the $H_4$ tetramer predominates. In skeletal muscle, which functions anaerobically under stress, the $M_4$ isozyme predominates. It seems likely that the M and H forms were designed to serve different functions. A clue to these functions may be revealed by the inhibiting effect of pyruvate on the dehydrogenase. Pyruvate can form a covalent complex with NAD$^+$ at the active site of the enzyme according to the following reaction:

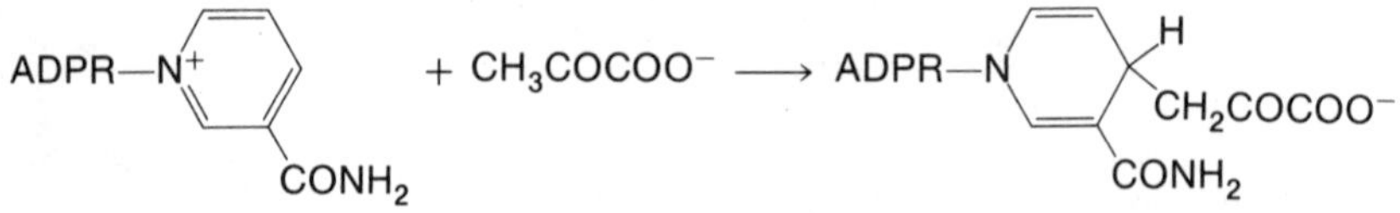

The $H_4$ tetramer shows a much greater inhibition by this compound than the $M_4$ tetramer. Active muscle tissue is anaerobic and produces a good deal of pyruvate. Inhibition of lactate dehydrogenase under anaerobic conditions would shrink the supply of NAD$^+$ and shut down the glycolytic pathway (see Chapter 8) with disastrous consequences. In fact, active muscle tissue has augmented levels of pyruvate but converts this readily to lactate. This is possibly because the predominant form of lactate dehydrogenase in muscle is $M_4$, which is only poorly inhibited by excess pyruvate. The function of LDH in aerobic tissue is less clear. Heart muscle is aerobic

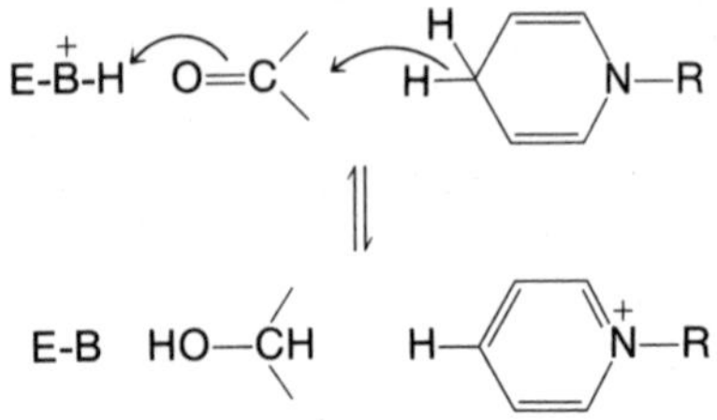

**Figure 5-5**
The protonated histidine (E—$\overset{+}{\text{B}}$—H) facilitates hydride transfer by acting as a generalized acid catalyst.

tissue, and consequently, most of its pyruvate is funneled into the Krebs cycle for greater energy production (see Chapter 9). The lactate dehydrogenase of heart muscle, $H_4$, might be inhibited by pyruvate to prevent waste of this potential high-energy carbon source or excessive buildup of pyruvate resulting from the conversion of incoming lactate to pyruvate. Possibly the $H_4$ enzyme is used in such tissues to convert absorbed lactate into pyruvate. This explanation for the two forms of the enzyme is speculative.

## REGULATION OF ENZYME ACTIVITY

One of the main ways that the metabolic activities of the cell or the organism are controlled is by regulating enzyme activity. In the case of the pancreatic zymogens discussed in Chapter 4, it was seen that this is done by preserving the enzyme in an inactive preenzyme form until it is needed. Lysozyme, also discussed in Chapter 4, may be pH-regulated in the sense that it has a low activity around physiological pH 7 and high activities around pH 5, which is common in extracellular secretions. The differing sensitivities of the $M_4$ and $H_4$ forms of lactate dehydrogenase to inhibition by pyruvate described earlier most likely serves an important regulatory function for this enzyme.

A most efficient situation for cellular metabolism would be to have the activity of each enzyme proportional to its need at any particular time. For example, if the cell needs histidine, the nine enzymes involved in histidine biosynthesis (in *E. coli*) should all be active (Chapter 22). If the cell has adequate histidine, the same enzymes should be inactive. When the cell needs energy, the mitochondrial enzymes should be producing ATP. If the cell accumulates a noxious metabolite in large quantity, it should be able to dispose of this by activating the appropriate degradative enzymes. In order to mobilize enzymes in this way, some form of reversible control of enzyme activity is needed.

*Small molecules frequently modulate enzyme activity*. Diffusible small molecules are more mobile than macromolecules and easier to synthesize and degrade than the latter. Because of this, small molecules that penetrate membranes within or between cells are most frequently the triggers that set various control processes in motion. In many cases, a long chain of events intercedes between the trigger and the modulation of the metabolic processes. The action of the small molecule also can be direct, as in cases where it binds to a protein, either covalently or noncovalently, thereby changing the protein structure in a way that increases or decreases its enzyme activity.

*Some enzymes are controlled by supply of substrate or other modulators in a special way*. In special cases, enzyme activity is modified by certain regulatory metabolites also termed *effectors*, *modifiers*, or *modulators*. The effector usually modifies the affinity of the enzyme for its substrate and/or other reaction components, but it also may operate by altering the rate at which the bound substrate reacts. The terms *positive* or *negative effector* indicate whether the compound has a stimulating or an inhibiting effect on enzyme activity. *The most common characteristic of regulatory enzymes is their ability to be activated or inhibited by effectors other than their substrates*. Often there is no structural similarity between the effector and the substrate. Thus, L-isoleucine is a specific inhibitor and L-valine a specific activator of L-threonine deaminase, the first enzyme involved specifically in the synthesis of isoleucine. It could be argued that these compounds are analogs of L-threonine because they share a carboxyl and an amino group in the same configuration. However, threonine deaminase is not inhibited by any amino acid other than L-isoleucine. Many other cases of strict specificity together with absence of a structural relationship between modulator and substrate are known. In the case of aspartate carbamoyltransferase (ACT), the first enzyme in the biosynthetic sequence for pyrimidines in *E. coli*, the enzyme is inhibited by cytidine triphosphate and activated by adenosine triphosphate (discussed below). Examination of the structural formulas of the substrates and effectors is enough to show that no similarity exists between them. Likewise, in the bacterium *Rhodopseudomonas spheroides*, hemin inhibits the synthesis of δ-aminolevulinic acid from glycine and succinyl-CoA. This reaction is the step at which porphyrin synthesis diverges from the common catabolic pathway (Chapter 23).

### Patterns of Regulation of Enzyme Activity

*Regulatory enzymes occupy key positions in metabolic pathways.* The structural elements needed to make a regulatory enzyme are used with great selectivity and imposed only on those enzymes with which it will result in the maximum economy of the cell's resources. An understanding of a particular multienzyme pathway usually suggests which enzymes might make attractive candidates for regulation. For example, many biosynthetic pathways involve a long chain of single chemical steps, each carried out by a discrete enzyme. Consider the nine-step transformation in which 5-phosphoribosyl-1-pyrophosphate is converted into histidine (Chapter 22). The first step in this series of reactions, which is catalyzed by phosphoribosylpyrophosphate-ATP phosphorylase, is inhibited by histidine. The other enzymes in the pathway are not. Inhibition of an enzyme by the product of a reaction is called *feedback inhibition*. It is obviously most economical for the first enzyme of the pathway to be inhibited once sufficient end product is present. Otherwise, appreciable energy would be wasted in making intermediate compounds that would accumulate without being utilized. Such intermediates in sufficient concentration often have a damaging effect on other essential functions of the cell. When an enzyme serves more than one pathway, a more complex pattern of feedback inhibition can occur. For example, β-aspartyl phosphokinase from *Rhodopseudomonas capsulatus* is affected only slightly by either threonine or lysine, both of which are synthesized by different pathways from β-aspartyl phosphate. However, when both amino acids are present, strong inhibition of the kinase occurs. This type of inhibition is called *concerted feedback inhibition*.

As a rule, in a pathway where the only function is to produce a single end product, the end product will inhibit the first enzyme of the pathway. In a pathway with a branch point where two end products result, both end products are required for inhibition of an enzyme before the branch. Concerted feedback inhibition ensures that the necessary enzymes will not be inhibited until an adequate supply of products from all branch points has been achieved.

These generalities about feedback inhibition are not to be taken as ironclad rules. The evolution of regulatory mechanisms has shown a great deal of divergence, so that schemes for inhibition (or activation) that obey the same general rules are quite different in different organisms. Patterns of inhibition and activation observed in amino acid biosynthetic pathways have been described in some detail in Chapter 22. Frequent examples of regulation of enzyme activity are presented at many points in this book.

## Aspartate Carbamoyltransferase: An Example of an Allosterically Regulated Enzyme

*A protein for which the binding of a ligand at one location influences its binding of another ligand at a nonoverlapping site is called an allosteric protein*. Aspartate carbamoyltransferase of *E. coli* is one of the better understood allosterically regulated enzymes, and we shall discuss it as an example of this class of regulatory enzymes. This enzyme catalyzes the first committed reaction in the biosynthetic pathway for pyrimidines (see Figure 5-6), which involves the formation of carbamoyl aspartate from carbamoyl phosphate and aspartate (the entire pathway is described in Chapter 19). The carbamoyl group appears to be transferred directly from the carbamoyl phosphate to L-aspartate without the formation of any intermediate products or a carbamoyl enzyme.

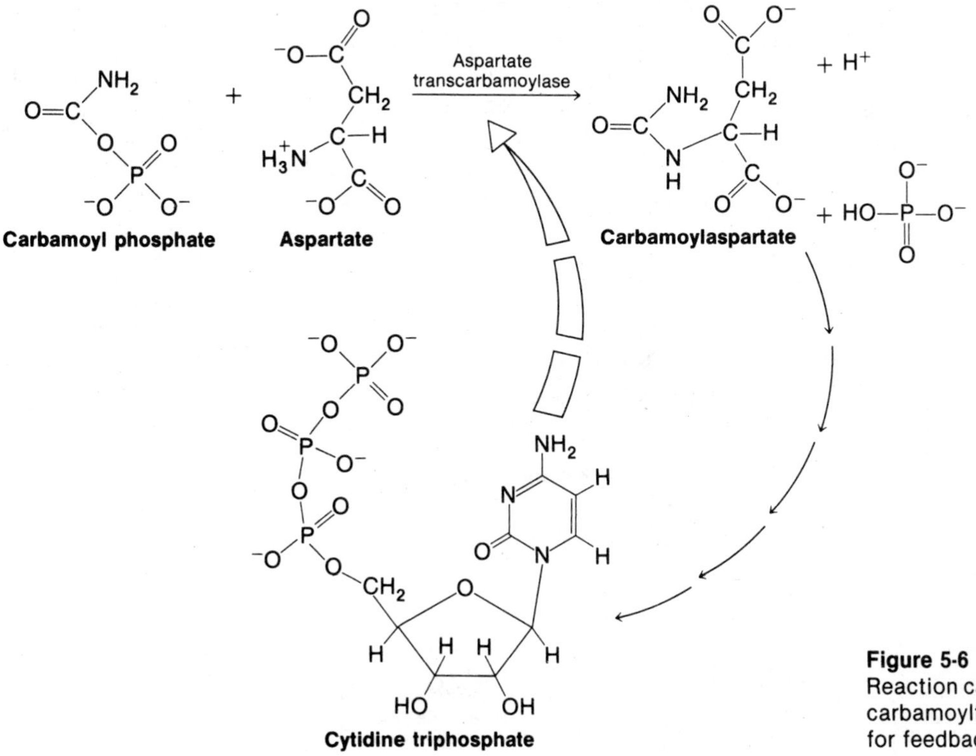

**Figure 5-6**
Reaction catalyzed by aspartate carbamoyltransferase and scheme for feedback inhibition in *E. coli*.

As with many other bisubstrate reactions, two pathways are possible for the order of binding of the two reactants to the enzyme. In fact, the upper pathway, shown in Figure 5-7, in which carbamoyl phosphate binds first, appears to predominate. Evidence favoring this order of binding comes from interaction studies of the two substrates in the presence of the transition-state, or bisubstrate, analog $N$-phosphonacetyl-L-aspartate (PALA), whose structure is shown in Figure 5-8. This compound is a strong competitive inhibitor for carbamoyl phosphate binding and a strong noncompetitive inhibitor for L-aspartate binding. The interpretation of the first result is that the inhibitor and carbamoyl phosphate compete for the same site on the enzyme. Thus a high concentration of one compound should be able to displace the other compound according to the law of mass action. However, although PALA strongly inhibits L-aspartate binding, the effect cannot be overcome by high levels of L-aspartate. If carbamoyl phosphate is required for aspartate binding, and if PALA prevents binding of the former, no amount of aspartate will displace PALA from the active site.

A schematic representation of the binding site and the partial mechanism of the reaction are shown in Figure 5-9. At the active site, the free enzyme contains positively charged sites for substrate binding, the general acid $BH^+$ and a steric constraint on the binding site for L-aspartate. Carbamoyl phosphate binds and interacts with $BH^+$ through its carbamoyl oxygen. The steric constraint on L-aspartate binding is relieved, so that L-aspartate can bind. Its $\alpha$-amino group is in a position to react with the bound and activated carbamoyl phosphate. The two substrates are pushed toward one another, aided by a conformational change of the enzyme. Transition from the hypothesized tetrahedral intermediate ($d$) to products could invoke additional base catalysis. Further information is required to verify this mechanism. For example, little is known about the actual functional groups at the active site, although the pH dependence of the binding of the substrate analog succinate implicates a positively charged histidine. The high-resolution x-ray crystallographic investigation of this enzyme that is in progress in W. Lipscomb's laboratory should be a great help in understanding the mechanism of the enzyme reaction as well as the allosteric regulatory mechanism discussed below.

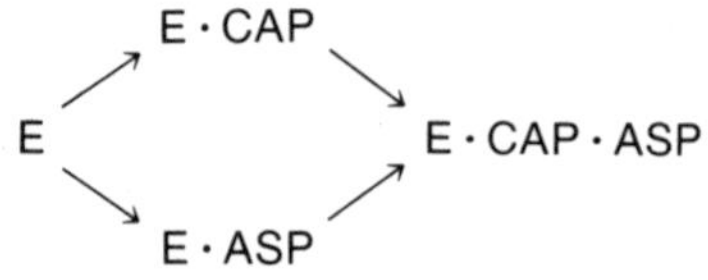

**Figure 5-7**
Possible pathways leading to an enzyme-carbamoyl-P-L-aspartate complex. A random mechanism would have significant contributions from either pathway, whereas an ordered mechanism would go by one route or the other. In fact, the ordered mechanism shown by the pathway above appears to predominate. CAP stands for carbamoyl phosphate and ASP stands for aspartate.

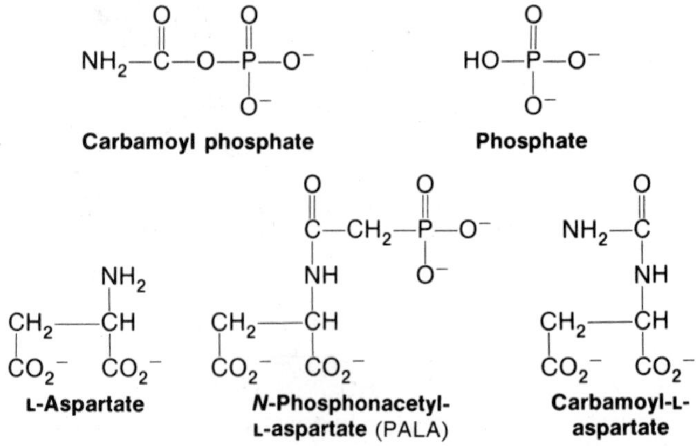

**Figure 5-8**
The structures of PALA and the substrates and products of the aspartate carbamoyltransferase reaction.

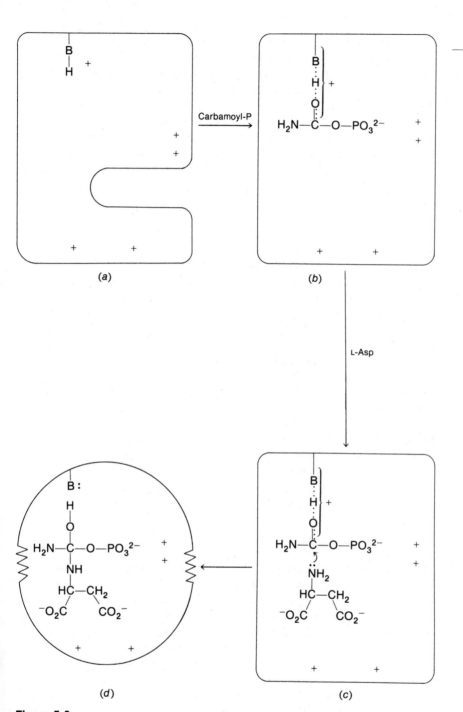

**Figure 5-9**
Schematic representation of the compression mechanism for ATCase.
(a) Unliganded enzyme with positively charged sites for substrate binding the
acid $BH^+$ and a steric constraint on the binding site for L-aspartate.
(b) Carbamoyl-P binds and interacts with $BH^+$ through its carbonyl oxygen.
The steric constraint on L-aspartate binding is relieved. (c) L-Aspartate binds.
Its $\alpha$-amino group is in a position to react with the bound and activated carbamoyl-P. (d) Compression of the two substrates together by a conformational change of the enzyme aids reaction. There is virtually no evidence for
the tetrahedral intermediate; it is shown only for purposes of illustration.

There is abundant evidence that aspartate carbamoyltransferase is allosterically regulated. The binding of aspartate to one substrate binding site influences the binding to other sites on the enzyme. Moreover, both ATP and CTP activate and inhibit this binding, respectively, by interaction at distant sites on the enzyme.

Some aspects of the theory of binding of small molecules or ligands to proteins are discussed in detail in Chapter 7 and applied to the case of hemoglobin oxygen binding in Chapter 3. If we have a protein with $n$ binding sites, all of which have an equal affinity for the small molecule or ligand in question, then the binding curve may be expressed by the following equation:

$$\nu = \frac{nK_f[B]}{1 + K_f[B]}$$

where $\nu$ is the average binding number, i.e., the ratio of the number of binding sites occupied to the total number of binding sites; $K_f$ is the affinity constant, or formation constant, for the binding to a single site, and [B] is the free molar concentration of the small molecule that binds [see Equation (58) in Chapter 7]. This is the equation for a hyperbola, as shown for the case of oxygen binding to myoglobin (see Figure 3-3), where $n = 1$. Even if $n$ is not equal to 1, as in a case such as lactate dehydrogenase, where $n$ is 4, a hyperbolic binding curve is obtained. As long as the binding sites are equivalent and binding at one site does not affect binding at another site on the same protein molecule, a hyperbolic binding curve is observed.

At the other extreme we can imagine a protein with more than one binding site, where binding at one site influences the binding at other sites on the same protein so strongly that either all the binding sites are occupied or none of them are occupied. This is the extreme case of positive cooperativity between binding sites, and the relevant equation [Equation (63) in Chapter 7] to describe the situation is

$$\nu = \frac{nK_f[B]^n}{1 + K_f[B]^n}$$

When $\nu$ is plotted against [B], a sigmoidal dependence on binding is observed (e.g., see the curve for oxygen binding to hemoglobin; Figure 3-3). Hemoglobin has four nearly identical binding sites for hemoglobin, so we might anticipate that the value of $n$ for oxygen binding would be 4. However, the experimentally determined value of $n$ in the exponent for the preceding equation is 2.8. This indicates some positive cooperativity between binding sites. It is not clear physically exactly how to translate this result into a molecular picture. Although we know the average amount of $O_2$ found per hemoglobin, we do not know the distribution of oxygen binding among the different hemoglobin molecules. This is a recurrent problem in analyzing binding curves that show some positive cooperativity. At the present time, we must content ourselves with a qualitative understanding of such situations: a hyperbolic binding curve indicates little or no cooperativity between binding sites and a sigmoidal curve indicates some positive cooperativity. The numerical value of $n$ gives us a qualitative idea of the extent of cooperativity. Values of $n$ of less than 1 indicate negative cooperativity between binding sites.

In the case of aspartate carbamoyltransferase, some positive cooperativity on aspartate binding is reflected by the pronounced sigmoidal character of the reaction curve (Figure 5-10). ATP eliminates the positive cooperativity and CTP augments it. Furthermore, ATP reverses the CTP effect, as would be expected if these two nucleotides compete for the same binding site on the enzyme. Both these nucleotide effects make good physiological sense. An abundance of CTP would decrease the need for further pyrimidine synthesis, whereas an abundance of ATP might be a good signal for favoring more pyrimidine synthesis, which, in turn, should favor more anabolic metabolism.

The inhibiting effect of CTP is proportional to the concentration of CTP. However, even the effects of elevated levels of CTP ($>0.5$ mM) are overcome by increasing the concentration of aspartate, showing that the inhibitory effect is a competitive one. It is clear from the results presented below that the CTP binding site in the enzyme is far removed from the active site. This shows that competitive inhibition need not result from binding of the inhibitor at a site coincident with the substrate binding site. Allosteric enzymes in different instances show either competitive or noncompetitive behavior toward inhibitors.

Direct evidence for the allosteric nature of aspartate carbamoyltransferase began with the observation that the regulatory behavior of the enzyme disappears when it is treated with organic mercurials such as *p*-hydroxymercuribenzoate (see Figure 5-10). After such treatment, ATP and CTP at discrete concentrations no longer influence the enzyme activity, and the binding of aspartate no longer shows any positive cooperativity. Other measurements show that exposure to mercurials results in the dissociation of the enzyme into two fragments. In the native enzyme, each molecule contains 6 R-protein subunits ($M_r = 17,000$) and 6 C-protein subunits ($M_r = 34,000$). Treatment of the enzyme with mercurials leads to a dissociation into two trimers of the C subunit and three dimers of the R subunit. These can be observed as separate peaks by sedimentation velocity analysis in the ultracentrifuge. The separated $C_3$ promoters show good enzyme activity, while the separated $R_2$ promoters show no enzymatic activity but good binding affinity for the usual allosteric effectors ATP and CTP. Sensitivity of enzyme activity to the allosteric effectors ATP or CTP requires reconstitution of the enzyme from the R- and C-containing portions in the presence of $Zn^{2+}$, which is a constituent of the native enzyme.

X-ray diffraction studies at a resolution of about 3 Å on the free enzyme and the enzyme complexed with CTP show that the C subunits are centrally located (Figure 5-11). A good deal of $\beta$-sheet structure and $\alpha$-helix structure is present in both subunits (Figure 5-12). In the regulatory chain there is a CTP-binding domain that interacts with an adjacent regulatory subunit and a zinc-binding domain that interacts with the catalytic subunit. In the catalytic chain a polar domain shows interactions between adjacent pairs of C chains that form each $C_3$ trimer, while an equatorial domain shows intramolecular $C_3$—$C_3$ interactions. The active site is at or near the interface between adjacent C chains within the trimers. It is highly likely that each active center involves amino acid residues from adjacent C chains. Although the existing degree of resolution of structural detail leaves many questions unanswered, it is clear that in the intact mole-

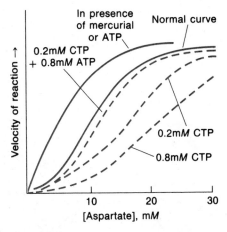

**Figure 5-10**
Effect of CTP, ATP, and mercurials on the affinity of aspartate carbamoyl-transferase for aspartate.

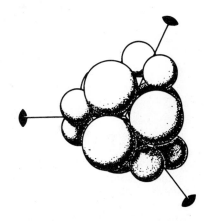

**Figure 5-11**
Subunit assembly to two $C_3$ and three $R_2$ subunits in aspartate carbamoyl-transferase. Each larger sphere represents a catalytic monomer C, and each smaller sphere represents a regulatory monomer R. The view is along the threefold axis, and the twofold axes are indicated by the usual symbol. This choice of enantiomer has been made on the basis of right-handed $\alpha$-helices in the high-resolution structures. (Adapted from H. L. Monaco, J. L. Crawford, and W. N. Lipscomb, Three-dimensional structures of aspartate carbamoyltransferase from *Escherichia coli* and of its complex with cytidine triphosphate, *Proc. Natl. Acad. Sci. USA* 75: 5277, 1978. Used with permission.)

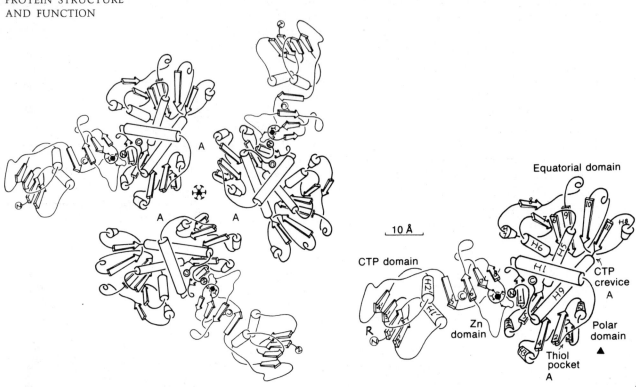

**Figure 5-12**
(*Left*) The "upper half" $C_3R_3$ of the molecule of aspartate carbamoyltransferase. Labeled RC unit. (*Right*) The CTP domain of the R chain is furthest to the left, while the Zn domain is adjacent and connected. The chain drawn slightly heavier is the catalytic monomer, in which the polar domain contains helices H1–H4 and the equatorial domain contains helices H6–H8. The active site is near the thiol pocket and the CTP (second site) crevice at a boundary between adjacent C monomers within a $C_3$ trimer. Helices and strands of $\beta$ sheet have been numbered in order from $NH_2$ terminal (N) of the C chain and similarly, with primes added, for the R chain. (Adapted from H. L. Monaco, J. L. Crawford and W. N. Lipscomb. Three-dimensional structures of aspartate carbamoyltransferase from *Escherichia coli* and of its complex with cytidine triphosphate, *Proc. Natl. Acad. Sci. USA* 75, 5277, 1978. Used with permission.)

cule the CTP binding sites are located a considerable distance from the active sites. Therefore, the CTP inhibitory effect as well as the stimulating effect of ATP must result from an alteration in the protein structure that is transmitted through the polypeptide chains to the active site (i.e., a conformational change).

## Different Models for How Allostery Influences Enzyme Activity

Aspartate carbamoyltransferase shows a positive cooperativity for the binding of aspartate, just as hemoglobin shows a positive cooperativity for the binding of oxygen. The distance between adjacent binding sites precludes direct interaction as an explanation for this stimulation effect. One is drawn to the conclusion that binding at one site directly alters the conformations of the protein subunit(s) that comprises that site and also influences the conformation of proteins of adjacent subunits. In hemoglobin this effect of ligand binding on protein conformation has been directly observed by the different structures of the deoxy and oxy forms of the protein. For aspartate carbamoyltransferase it is still a matter of conjecture, although it seems highly likely. Not resolved in either case is the question of the number of similar binding sites whose structures are affected. At one extreme we have the model that says that binding of a ligand at one site results in a conformational change of that site and an identical conformational change of all the other unoccupied binding sites on the same molecule. At the other extreme we have the model that says that ligand binding at one site changes the conformation of that site and facilitates a conformational change at other similar ligand binding sites that does not take place until additional ligands are bound. It is possible that the actual situation lies somewhere between these two extremes.

## E. COLI RNA POLYMERASE

All the enzymes discussed thus far in Chapters 4 and 5 have three-dimensional structures that are understood in some detail thanks to x-ray diffraction studies. As the technology for determining enzyme structure has advanced, enzyme kineticists have come to rely more and more heavily on this type of information in planning investigations and interpreting their results. Nevertheless, there are many important enzymes for which such structural investigations are in their infancy, and frequently biochemists are too impatient to wait for definitive structural information. The bacterial DNA-dependent RNA polymerase is an example of this sort. Whereas detailed structural information is lacking for this complex zinc-metalloenzyme, further research continues to reveal new and exciting aspects of its behavior.

RNA polymerase is a nucleotidyl transferase that catalyzes the following reaction:

$$-\text{pXpY} + 2\,\text{TP} \longrightarrow -\text{pXpYpZ} + \text{PP}_i$$

The normal substrates for this enzyme are the four commonly occurring nucleotide triphosphates: ATP, GTP, UTP, and CTP (Figure 5-13). The functionally reactive parts of triphosphates are the 3'-OH of the ribose and the triphosphate attached to the 5' of the ribose. Catalysis results in the polymerization of nucleotides into long chains with 3', 5' phosphodiester linkages between individual residues (Figure 5-14); for each linkage made, a pyrophosphate is cleaved from the added substrate. Chain growth occurs in the 5'→3' direction and the polymerase remains firmly attached to the growing chain from the point of initiation to the point of termination. A divalent cation, either $Mg^{2+}$ or $Mn^{2+}$, is required as a cofactor for phosphodiester bond formation.

Figure 5-15 illustrates a primitive model for the reaction mechanism and the role of the enzyme. The enzyme orients the nucleotide triphosphate and the growing chain in the proper juxtaposition so that reaction can take place. One group on the enzyme, possibly a neutral histidine, functions as a generalized base catalyst, making the 3'-OH group a better nucleophile. The metal cofactor complexes with the triphosphate region, thereby increasing the positive charge on the reactive phosphate and making it more attractive for nucleophilic attack. Most probably the $Zn^{2+}$ in the enzyme assists the free metal cofactor in this regard.

The reactive center on the enzyme must contain a firm binding site for the growing polymer, a binding site for the nucleotide triphosphate substrate, and a means of translocating the growing chain after addition so that

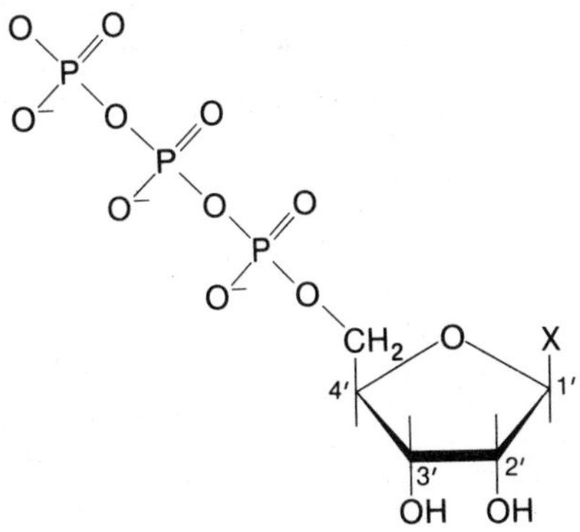

**Figure 5-13**
Generalized structure of a ribonucleotide triphosphate. See Chapter 18 for detailed structures of the four compounds.

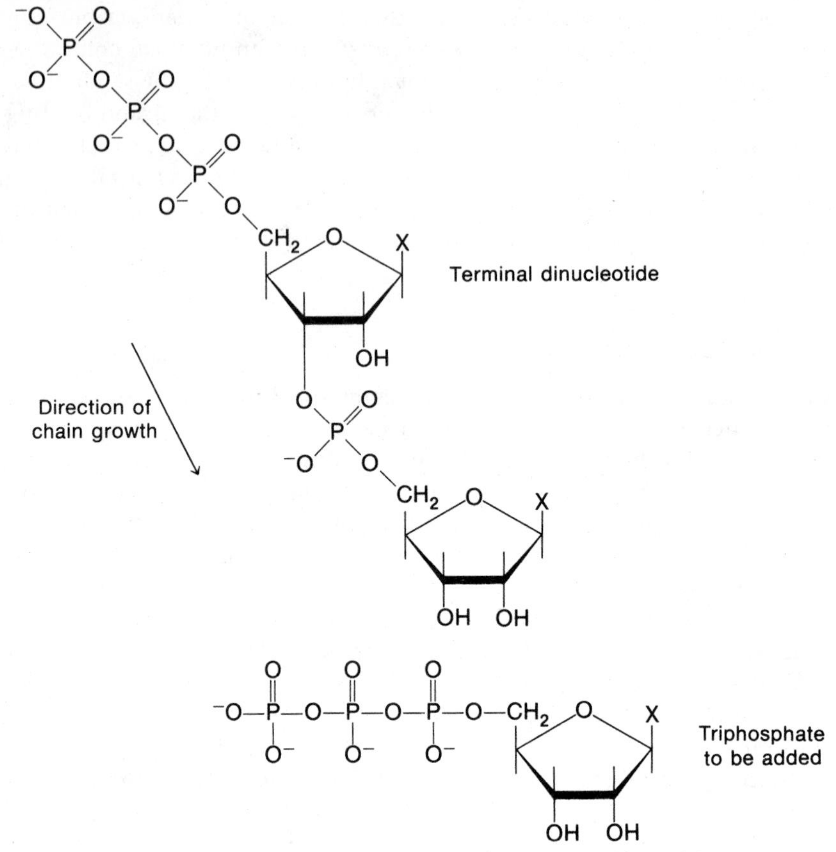

Terminal dinucleotide

Direction of
chain growth

Triphosphate
to be added

**Figure 5-14**
Growth of a polynucleotide chain.

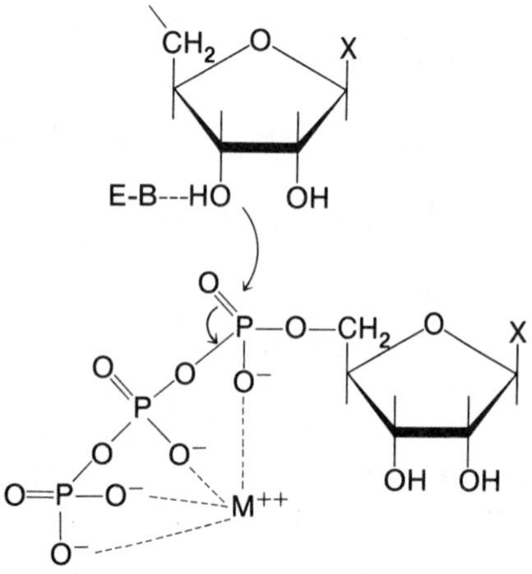

**Figure 5-15**
Hypothetical model of the mechanism for phosphodiester bond formation.

the reaction can be repeated indefinitely. The enzyme also must contain a binding site for a DNA template strand that precisely dictates the order of addition of bases to the growing strand (see Chapter 21) at a rate of about 30 residues per second. Other parts of the enzyme ensure that correct initiation and termination sites are recognized on the DNA template (see Chapters 21, 26, and 27). It is hardly surprising that this enzyme, which operates with an awesome efficiency in this complex process, should itself have a complex structure. The molecule contains five subunits and has a composite molecular weight of about 500,000 (see Chapter 21). One of these subunits, called σ, dissociates soon after initiation, which suggests that it only functions in initiation. In support of this it has been demonstrated that the enzyme lacking the σ subunit can polymerize subunits in an order dictated by the sequence of bases on the DNA template, but that the initiation points are recognized with a greatly decreased efficiency.

RNA polymerase is subject to a wide variety of regulatory effectors that influence its ability to initiate, elongate, and terminate the synthesis of polypeptide chains (Chapters 21, 26, and 27); subunit replacement, covalent modification, and a number of small molecule allosteric effectors play a role in regulating the behavior of the polymerase. It is hard to imagine a more complex or interesting enzyme.

## SELECTED READINGS

Evans, P. R., and Hudson, P. J.  Structure and control of phosphofructokinase from *Bacillus stearothermophilus*. A well-understood allosteric enzyme. *Nature* 279:500, 1979.

Holbrook, J. J., Liljas, A., Steindel, S. J., and Rossman, M. G.  Lactate Dehydrogenase. In P. D. Boyer (Ed.), *The Enzymes*, 3d Ed., Vol. 11. New York: Academic Press, 1975. Pp. 191–293.

Jacobson, G. R., and Stark, G. P.  Aspartate Transcarbamylases. In P. D. Boyer (Ed.), *The Enzymes*, 3d. Ed., Vol. 9. New York: Academic Press, 1973. Pp. 225–308.

Monaco, H. L., Crawford, J. L., and Lipscomb, W. N.  Three-dimensional structures of aspartate carbamoyltransferase from *Escherichia coli* and of its complex with cytidine triphosphate. *Proc. Natl. Acad. Sci. USA* 75:5276, 1978.

Monod, J., Changeux, J.-P., and Jacob, F.  Allosteric proteins and cellular control systems. *J. Mol. Biol.* 6:306, 1963.

See Selected Readings at the ends of Chapters 21, 26, and 27.

## PROBLEMS

1.  $$A \xrightarrow{1} B \xrightarrow{2} C \xrightarrow{3} D \xrightarrow{4} E \xrightarrow{5} F \xrightarrow{6} G$$
    with $D \xrightarrow{7} I \xrightarrow{8} J$ and $F \xrightarrow{9} H$

    Assume that the above flow diagram relates to an amino acid biosynthetic pathway, where J, G, H are all amino acid end products and A is their common precursor. The numbers over the arrows refer to enzymes that catalyze discrete steps. Suggest a plausible scheme in which certain enzymes are regulated by end-product inhibition. Indicate which enzymes these are and by what end products they are inhibited. If this were a degradative pathway rather than a synthetic pathway, would you expect to see the same pattern of regulation?

2.  Aspartate carbamoyltransferase is an allosteric enzyme in which the active site and the binding site for the allosteric effector are on different subunits. Is it possible to imagine an allosteric enzyme in which the two sites are on the same subunit? Explain.

3.  Some allosteric enzymes are inhibited competitively, and some are inhibited noncompetitively by the binding of an allosteric effector at a nonoverlapping site. Explain, in qualitative terms, the sets of conditions that could produce the two different results.

4.  Explain what is meant by positive and negative cooperativity.

5.  Mutants in hemoglobin which decrease the positive cooperativity of hemoglobin oxygen binding are commonly found at sites other than the oxygen binding sites. Explain. Is it likely that variants of regulatory enzymes will show similar behavior?

6.  Why should elevated pHs facilitate observation of the conversion of lactate to pyruvate by L-lactate dehydrogenase? You may wish to refer to the stoichiometric equation for this reaction in the text.

7.  Lactate dehydrogenase is an example of a bisubstrate enzyme that follows a simple ordered pathway for binding of substrates. Explain how this works. If you had an enzyme that uses two substrates and wanted to determine whether it followed a random-order pathway or an ordered pathway for binding of substrate, what measurements would you make?

8.  An enzyme inhibitor can act in a reversible or an irreversible manner. How would you distinguish between an irreversible inhibitor and a reversible noncompetitive inhibitor? Assume you have an abundant supply of pure enzyme to work with.

9.  Using equations given in this chapter, plot the average binding number, as a function of the free molar concentration of the small molecule which binds (B) for $n = 1$ and $n = 4$. Use a value for the formation constant, $K_f$, of $10^6$. Refer to relevant equations in the text. What is the physical meaning of these two curves?

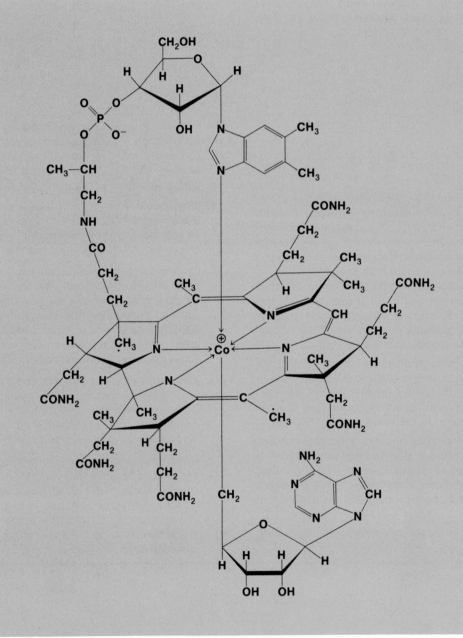

# 6

# STRUCTURE AND FUNCTION OF COENZYMES

Many metabolic reactions involve chemical changes that could not be brought about by the amino acid side-chain functional groups of enzymes acting as general acid-base or nucleophilic catalysts. The available amino acid side chains do not include the wide range of functional groups possessing the requisite chemical properties for catalyzing many of the known metabolic reactions. Enzymes, in catalyzing such reactions, do not operate outside the limits of chemical feasibility, however. Instead, they act in cooperation with other smaller molecules called _coenzymes_ that have the required chemical reactivities.

In a few cases, molecules that have come to be recognized as coenzymes do not exhibit specialized chemical properties unrepresented by the functional groups of proteins. For example, the catalytically functional groups in 4-phosphopantetheine coenzymes and $\alpha$-lipoic acid are sulfhydryl and disulfide groups also found in cysteine and cystine, respectively. In these cases, the structures of these coenzymes give them group translocation capabilities that are not shared by cysteine or cystine and that are essential aspects of catalysis in certain critically important metabolic reactions involving multienzyme complexes.

_A coenzyme may be defined as a molecule possessing physicochemical properties not found in the polypeptide chain of the enzyme that acts together with the enzyme to catalyze a biochemical reaction._* In this chapter we discuss the structures and functions of coenzymes, which will be seen to represent a wide diversity of chemical and structural properties.

---

*Tightly bound coenzymes are sometimes referred to as prosthetic groups.

_Vitamin $B_{12}$, the most intricate coenzyme known._

**Table 6-1**
Water-Soluble Vitamins and
Associated Nutritional Diseases

| Vitamin | Deficiency Disease |
|---|---|
| Thiamine (B$_1$) | Human beriberi |
| Riboflavin (B$_2$) | Rat growth factor |
| Pyridoxal (B$_6$) | Rat pellagra |
| Nicotinamide | Human pellagra |
| Pantothenic acid | Chick dermatitis |
| Biotin | Human dermatitis |
| Folic acid | Tropical macrocytic anemia |
| Vitamin B$_{12}$ | Pernicious anemia |

These do not exhaust the list of molecules that have been called coenzymes. Many, such as ATP, are now more properly classified as metabolites; at earlier stages in the development of biochemistry they appeared to be coenzymes because they activated enzymatic processes. As these processes became better understood, the true roles of such molecules became clear.

Coenzymes were originally discovered as _vitamins_ or growth factors in nutritional and medical studies. Most coenzymes are modified forms of vitamins, including thiamine pyrophosphate, pyridoxal phosphate, nicotinamide coenzymes, phosphopantetheine coenzymes, tetrahydrofolate, flavin coenzymes, and vitamin B$_{12}$ coenzymes. $\alpha$-Lipoic acid is a growth factor for certain bacteria, but it is not a vitamin because it is not a required constituent of the human diet. Vitamins were originally discovered in nutritional studies, in which they were purified from foodstuffs and shown to cure various disorders in animals maintained on deficient diets. Table 6-1 lists the water-soluble vitamins discussed in this chapter and the nutritional diseases they cure. Vitamin deficiencies in humans and animals most often result from dietary deficiencies. However, in certain cases, the deficiency may arise from a genetic fault so that the animal fails to absorb the vitamin from a normal diet. A case in point is vitamin B$_{12}$ deficiency, which usually results from a failure to absorb the vitamin. Intrinsic factor is an intestinal protein that binds the vitamin as a part of the absorption process. When intrinsic factor is absent or defective, the vitamin is not absorbed, leading to a deficiency.

This chapter emphasizes the _principles_ underlying the mechanisms of action of coenzymes and so does not delineate the detailed mechanisms of all the various reactions in which coenzymes play essential roles. The student who wishes to gain proficiency in the writing of reaction mechanisms, and thereby achieve an improved understanding, should work the exercises at the end of the chapter.

## THIAMINE PYROPHOSPHATE (TPP)

The structure of thiamine pyrophosphate is given in Figure 6-1. The vitamin, thiamine or vitamin B$_1$, lacks the pyrophosphoryl group. _Thiamine pyrophosphate is the essential coenzyme involved in the actions of enzymes that catalyze cleavages of the bonds indicated in color in Figure 6-1._ The bond scission in Figure 6-1b is representative of those in many $\alpha$-ketoacid decarboxylations, nearly all of which require the action of TPP. The phosphoketolase reaction involves both cleavages shown in Figure 6-1c, while the transketolase reactions (Chapter 8) involve the cleavage of the carbon-carbon bond but not the elimination of —OH. Acetoin and acetolactate arise by the formation of the carbon-carbon bond in Figure 6-1c.

The catalytic pathways in these reactions could be understood in chemical terms if the carbon-carbon bond cleavages in Figure 6-1b and c proceeded by heterolytic scission to produce acyl anionic species as reaction intermediates; these could react in subsequent steps with protons or other electrophiles or undergo elimination reactions to form products. However, such acyl anions are so high in energy that they cannot be produced at significant rates under physiological conditions. In fact the function of thiamine pyrophosphate is to promote the bond cleavages indicated in Figure 6-1b and c. This was first understood when it was shown by

R. Breslow, in one of the earliest applications of nuclear magnetic resonance to biochemical mechanisms, that the proton bonded to C-2 in the thiazolium ring is readily exchangeable with the protons of $H_2O$ and deuterons of $D_2O$ in the base-catalyzed reaction of Equation (1).

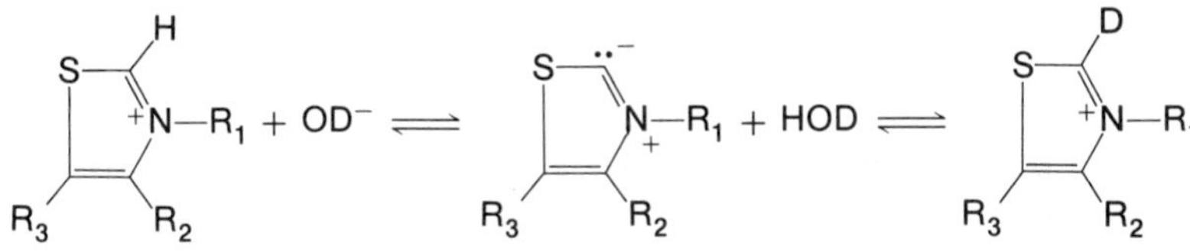

$$\text{(1)}$$

The ylid-like intermediate undergoes nucleophilic addition to the $\pi$ bond of polar carbonyl groups in substrates to produce intermediates such as those in Equation (2) that possess the chemical reactivities required for cleaving the bonds indicated in Figure 6-1b and c.

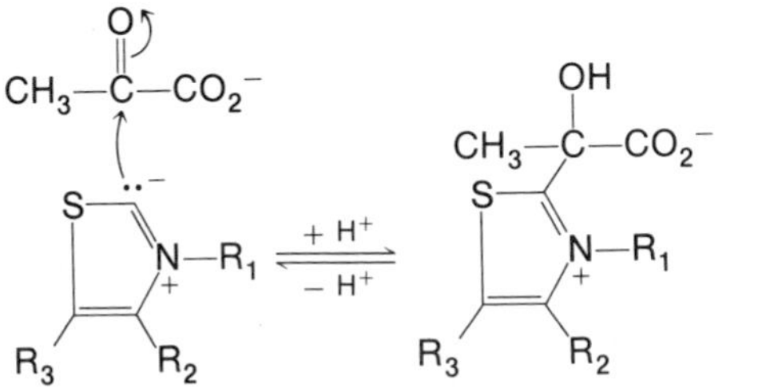

$$\text{(2)}$$

This is illustrated in Figure 6-2, which shows that the electron pair resulting from the carbon-carbon bond scission in Figure 6-1b, i.e., decarboxylation of pyruvate, is stabilized by the thiazolium ring. The immediate decarboxylation product (Figure 6-2a) is a resonance hybrid in which the electron pair is stabilized by delocalization into the thiazolium ring. The intermediate is common to many enzymatic reactions involving the decarboxylation of pyruvate. As shown in Figure 6-2d, it may be protonated to $\alpha$-hydroxyethyl thiamine pyrophosphate; it may react with other electrophiles, such as the carbonyl groups of acetaldehyde or pyruvate, to form the species in Figure 6-2b and c, respectively; or it may be oxidized to acetylthiamine pyrophosphate (Figure 6-2e), depending on the reaction specificity of the enzyme.

The $\alpha,\beta$-dihydroxy intermediate analogous to that in Figure 6-2a is shown in Figure 6-2f, and it results from carbon-carbon bond cleavage in reactions of thiamine pyrophosphate with ketose sugars, such as fructose-6-phosphate. This intermediate can undergo all the reactions cited earlier for the intermediate in Figure 6-2a, but in addition, it can eliminate the $\beta$-hydroxyl group by using the stabilized electron pair as the driving force to produce acetylthiamine pyrophosphate (Figure 6-2e), initially as its enol form. Acetylthiamine pyrophosphate can react with nucleophiles to produce, for example, acetate or acetyl phosphate, as illustrated in Figure 6-2, depending on the reaction specificity of the enzyme.

The essential aspects of the mechanisms of the best-known biological reactions involving the action of thiamine pyrophosphate are illustrated in Figure 6-2. The experimental support for these mechanisms is, in addition to Equations (1) and (2), that hydroxyethyl-TPP and $\alpha,\beta$-dihydroxyethyl-TPP have been synthesized and shown to be utilized as precursors of enzymatic products and of the central intermediates shown in Figure 6-2a and f.

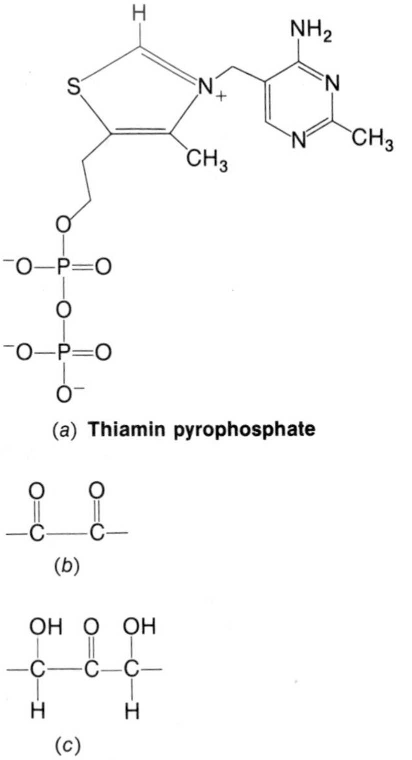

(a) **Thiamin pyrophosphate**

(b)

(c)

**Figure 6-1**
Structure of thiamine pyrophosphate and the bonds it cleaves or forms. Reactive part of the coenzyme and the bonds subject to cleavage in (b) and (c) are indicated in color.

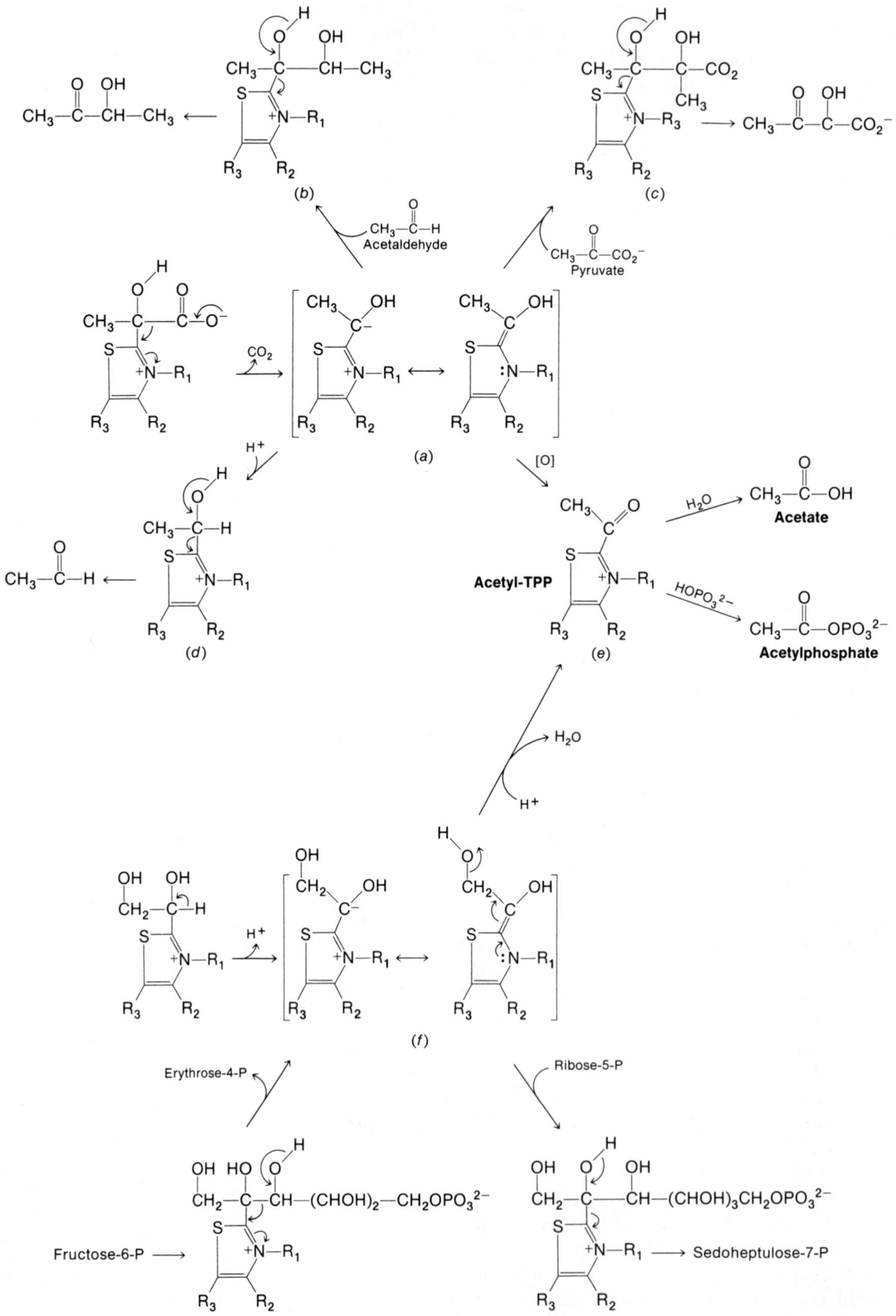

(b)

$CH_3-\overset{O}{\overset{\|}{C}}-\overset{OH}{\underset{|}{CH}}-CH_3$ ←

$CH_3-\overset{OH}{\underset{|}{\overset{|}{C}}}-\overset{OH}{\underset{|}{CH}}-CH_3$

$\overset{+}{N}-R_1$  S  $R_3$  $R_2$

$CH_3-\overset{O}{\overset{\|}{C}}-H$  Acetaldehyde

(c)

$CH_3-\overset{OH}{\underset{|}{\overset{|}{C}}}-\overset{OH}{\underset{|}{C}}-CO_2$

$\overset{|}{CH_3}$

$\overset{+}{N}-R_3$  S  $R_3$  $R_2$

→ $CH_3-\overset{O}{\overset{\|}{C}}-\overset{OH}{\underset{|}{C}}-CO_2^-$

$CH_3-\overset{O}{\overset{\|}{C}}-CO_2^-$  Pyruvate

$CH_3-\overset{H}{\underset{|}{\overset{OH}{C}}}-\overset{O}{\overset{\|}{C}}-O^-$

$\overset{+}{N}-R_1$  S  $R_3$  $R_2$

$\xrightarrow{CO_2}$

$\left[ \begin{array}{c} CH_3 \\ | \\ C^- \end{array} \begin{array}{c} OH \end{array} \right.$ $\overset{+}{N}-R_1$ ↔ $\begin{array}{c} CH_3 \\ \| \\ C \end{array} \begin{array}{c} OH \end{array}$ $:N-R_1$ $\left. \right]$

$R_3$  $R_2$  $R_3$  $R_2$

(a)

[O]

$H^+$

$CH_3-\overset{H}{\underset{|}{\overset{O}{C}}}-H$

$\overset{+}{N}-R_1$  S  $R_3$  $R_2$

$CH_3-\overset{O}{\overset{\|}{C}}-H$ ←

(d)

**Acetyl-TPP**

$CH_3-\overset{O}{\overset{\|}{C}}$

$\overset{+}{N}-R_1$  S  $R_3$  $R_2$

(e)

$\xrightarrow{H_2O}$ $CH_3-\overset{O}{\overset{\|}{C}}-OH$  **Acetate**

$\xrightarrow{HOPO_3^{2-}}$ $CH_3-\overset{O}{\overset{\|}{C}}-OPO_3^{2-}$  **Acetylphosphate**

→ $H_2O$

$H^+$

$\overset{OH}{\underset{|}{CH_2}}-\overset{OH}{\underset{|}{\overset{|}{C}}}-H$

$\overset{+}{N}-R_1$  S  $R_3$  $R_2$

$\xrightarrow{H^+}$ $\left[ \begin{array}{c} OH \\ | \\ CH_2 \end{array} \begin{array}{c} OH \\ | \\ C^- \end{array} \right.$ $\overset{+}{N}-R_1$ ↔ $\begin{array}{c} OH \\ | \\ CH_2 \end{array} \begin{array}{c} OH \\ | \\ C \end{array}$ $:N-R_1$ $\left. \right]$

$R_3$  $R_2$  $R_3$  $R_2$

(f)

Erythrose-4-P ↖

Ribose-5-P ↘

$\overset{OH}{\underset{|}{CH_2}}-\overset{HO}{\underset{|}{\overset{|}{C}}}-\overset{O-H}{\underset{|}{CH}}-(CHOH)_2-CH_2OPO_3^{2-}$

$\overset{+}{N}-R_1$  S  $R_3$  $R_2$

Fructose-6-P →

$\overset{OH}{\underset{|}{CH_2}}-\overset{O}{\underset{|}{\overset{H}{C}}}-\overset{OH}{\underset{|}{CH}}-(CHOH)_3CH_2OPO_3^{2-}$

$\overset{+}{N}-R_1$  S  $R_3$  $R_2$  → Sedoheptulose-7-P

The essence of the coenzymatic function of thiamine pyrophosphate is outlined in Equations (1) and (2) and Figures 6-1 and 6-2. Dissociation of the C-2 proton from thiamine pyrophosphate in Equation (1) generates the ylid-like zwitterion. This ion undergoes nucleophilic addition to the carbonyl group of a substrate in Equation (2) to produce the intermediate shown. Carbon-carbon bond cleavages in these intermediates produce the central intermediates in Figure 6-2a and f, in which the electron pairs resulting from the carbon-carbon bond cleavages are stabilized by delocalization into the electron-deficient thiazolium ring while being kept available for subsequent reactions. In its capability to function in this way, thiamine pyrophosphate is unique in the biosphere.

## PYRIDOXAL-5′-PHOSPHATE

Pyridoxal-5′-phosphate is the coenzyme form of vitamin $B_6$ and has the structure shown in Figure 6-3. Vitamin $B_6$ refers to any of a group of related compounds lacking the phosphoryl group, including pyridoxal, pyridoxamine, and pyridoxine (pyridoxol).

Pyridoxal-5′-phosphate is involved in a variety of reactions in the metabolism of $\alpha$-amino acids, including transaminations, $\alpha$-decarboxylations, racemizations, $\alpha,\beta$ eliminations, $\beta,\gamma$ eliminations, aldolizations, and the $\beta$ decarboxylation of aspartic acid. Equations (3) through (9) illustrate the variety of reactions in which pyridoxal-5′-phosphate exerts its essential coenzymatic functions.

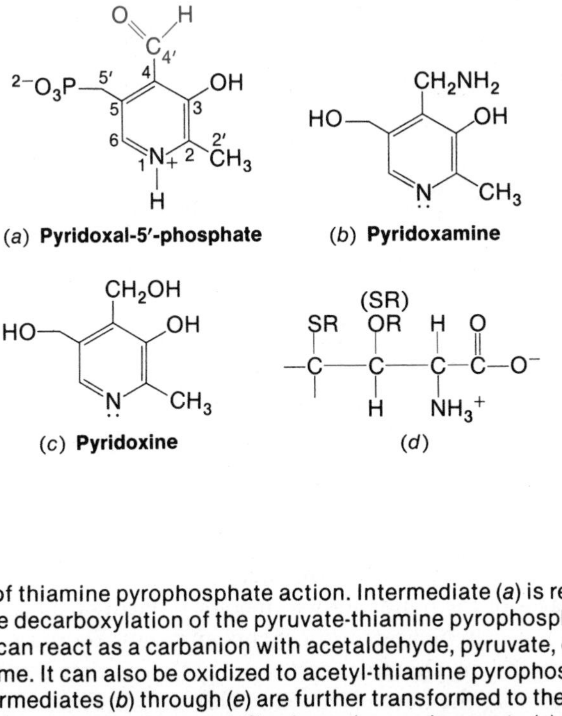

(a) **Pyridoxal-5′-phosphate**     (b) **Pyridoxamine**

(c) **Pyridoxine**                 (d)

**Figure 6-3**
Structures of vitamin $B_6$ derivatives and the bonds cleaved or formed by the action of pyridoxal phosphate (a). The reactive part of the coenzyme is shown in color in (a), and the bonds shown in color in (d) are the types of bonds in substrates that are subject to cleavage.

◀**Figure 6-2**
Mechanisms of thiamine pyrophosphate action. Intermediate (a) is represented as a resonance-stabilized species. It arises from the decarboxylation of the pyruvate-thiamine pyrophosphate addition compound shown at the left of (a) and in Equation 2. It can react as a carbanion with acetaldehyde, pyruvate, or $H^+$ to form (b), (c), or (d), depending on the specificity of the enzyme. It can also be oxidized to acetyl-thiamine pyrophosphate (e) by other enzymes, such as pyruvate oxidase. The intermediates (b) through (e) are further transformed to the products shown by the actions of specific enzymes. Intermediate (f) is the resonance-stabilized species analogous to (a) but derived from the cleavage of addition compounds formed between thiamine pyrophosphate and ketose sugars, such as fructose-6-P. Transketolases catalyze the reaction of (f) with aldose phosphates to form other ketose phosphates, such as sedoheptulose-7-P, as shown. Phosphoketolase catalyzes the dehydration of (f) to (e) and its further reaction with $P_i$ to acetyl-P. The addition and removal of protons indicated in the interconversion of intermediates and their conversions to products are catalyzed by general acids and bases in the enzymes that catalyze the reactions.

$$R_1 - \overset{\underset{\displaystyle |}{H}}{\underset{\underset{\displaystyle NH_3^+}{|}}{C}} - CO_2^- + R_2 - \overset{\overset{\displaystyle O}{\|}}{C} - CO_2^- \xrightleftharpoons{\text{Transaminase}}$$

$$R_1 - \overset{\overset{\displaystyle O}{\|}}{C} - CO_2^- + R_2 - \overset{\underset{\displaystyle NH_3^+}{|}}{\overset{\overset{\displaystyle H}{|}}{C}} - CO_2^- \quad (3)$$

$$R - \overset{\underset{\displaystyle NH_3^+}{|}}{\overset{\overset{\displaystyle H}{|}}{C}} - CO_2^- + H^+ \xrightleftharpoons{\text{Decarboxylase}} CO_2 + R - CH_2 - NH_3^+ \quad (4)$$

$$R - \overset{\underset{\displaystyle NH_3^+}{|}}{\overset{\overset{\displaystyle H}{|}}{C}} - CO_2^- \xrightleftharpoons{\text{Racemase}} R - \overset{\underset{\displaystyle H}{|}}{\overset{\overset{\displaystyle NH_3^+}{|}}{C}} - CO_2^- \quad (5)$$

$$R - \overset{\overset{\displaystyle OH}{|}}{CH} - \overset{\underset{\displaystyle NH_3^+}{|}}{\overset{\overset{\displaystyle H}{|}}{C}} - CO_2^- \xrightleftharpoons{\text{Dehydratase}} R - CH_2 - \overset{\overset{\displaystyle O}{\|}}{C} - CO_2^- + NH_4^+ \quad (6)$$

$$R - \overset{\overset{\displaystyle OH}{|}}{CH} - \overset{\underset{\displaystyle NH_3^+}{|}}{\overset{\overset{\displaystyle H}{|}}{C}} - CO_2^- \xrightleftharpoons{\text{Aldolase}} R - CHO + \overset{\underset{\displaystyle NH_3^+}{|}}{CH_2} - CO_2^- \quad (7)$$

$$H_2O + \overset{\overset{\displaystyle CH_2 - \overset{\overset{\displaystyle NH_3^+}{|}}{CH} - CO_2^-}{|}}{S} - CH_2 - CH_2 - \overset{\underset{\displaystyle NH_3^+}{|}}{CH} - CO_2^- \xrightleftharpoons{\text{Cystathionase}} \overset{\overset{\displaystyle SH}{|}}{CH_2} - \overset{\underset{\displaystyle NH_3^+}{|}}{CH} - CO_2^- +$$

$$CH_3CH_2 - \overset{\overset{\displaystyle O}{\|}}{C} - CO_2^- + NH_4^+ \quad (8)$$

$$^-O_2C - CH_2 - \overset{\underset{\displaystyle NH_3^+}{|}}{\overset{\overset{\displaystyle H}{|}}{C}} - CO_2^- + H^+ \xrightleftharpoons{\text{Aspartate-}\beta\text{-decarboxylase}}$$

$$CO_2 + CH_3 - \overset{\underset{\displaystyle NH_3^+}{|}}{\overset{\overset{\displaystyle H}{|}}{C}} - CO_2^- \quad (9)$$

Equations (3) through (9) show that pyridoxal-5′-phosphate is involved in the cleavages of the bonds shown in color in Figure 6-3d. This represents a versatility not matched by most other coenzymes. Yet these reactions can all be understood in chemical terms on the basis that pyridoxal-5′-phosphate can promote the heterolytic bond cleavages in Figure 6-3d by stabilizing the resulting electron pairs at the $\alpha$- or $\beta$-carbon atoms of $\alpha$-amino acids. This can be seen to be well within the chemical capabilities of pyridoxal-5′-phosphate once it is realized that the aldehyde group of the coenzyme can react with the $\alpha$ amino group of $\alpha$-amino acids to produce an aldimine (Figure 6-4a) or Schiff's base, which is internally stabilized by H bonding.

Loss of the α hydrogen as H⁺ produces a resonance-stabilized species (Figure 6-4b) in which the electron pair is stabilized by delocalization into the pyridinium system. This species may undergo further reactions at the α carbon to form products determined by the reaction specificity of the enzyme. If, for example, the enzyme is a racemase, the species resulting from the loss of the proton from the α carbon may accept a proton from the opposite side to produce ultimately the enantiomer of the amino acid.

When the substrate is substituted at the β carbon with a potential-leaving group, such as —OH, —SH, —$OPO_3^{2-}$ (Figure 6-3d), the corresponding α-carbanionic intermediate (Figure 6-4b) can eliminate this group, and this is an essential step in α,β eliminations, such as the serine dehydratase and threonine deaminase reactions. Upon hydrolysis, the elimination intermediate produces pyridoxal-5'-phosphate and the substrate-derived enamine, which spontaneously hydrolyzes to ammonia and an α-keto acid (Equation 6).

In transamination, the intermediate (Figure 6-4b) is protonated at the aldimine carbon-4' of pyridoxal-5'-phosphate, producing the intermediate in Figure 6-4c. Upon hydrolysis, this forms an α-keto acid and pyridoxamine-5'-phosphate. As shown in Figure 6-5, the microscopic reverse of this sequence with a second α-ketoacid accounts for the transamination reaction.

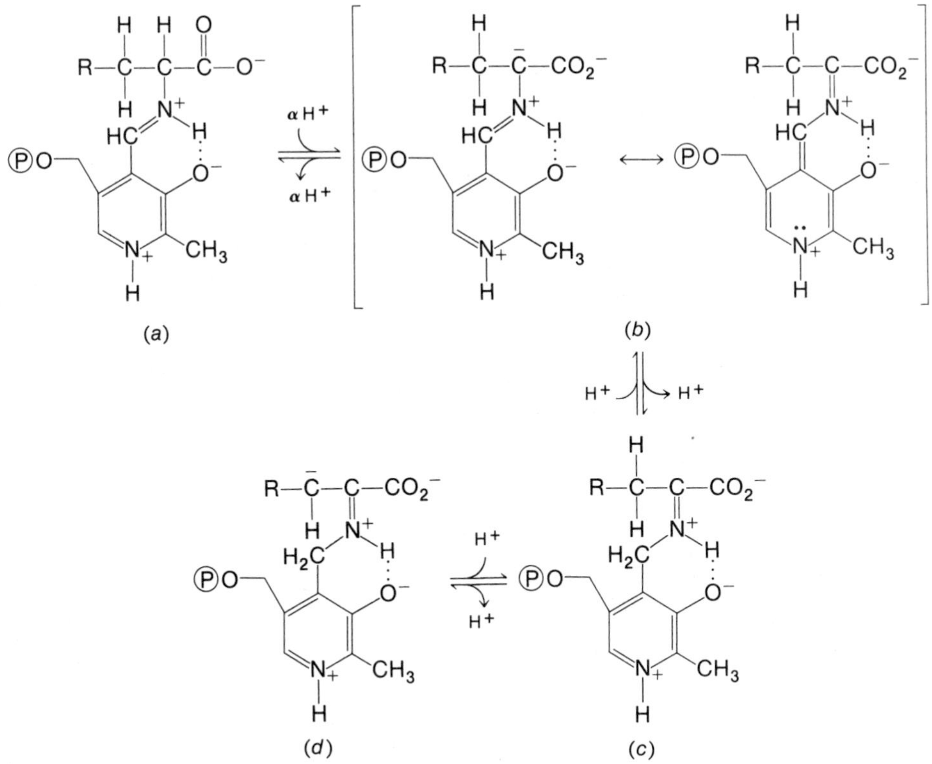

**Figure 6-4**
Structures of catalytic intermediates in pyridoxal phosphate-dependent reactions. The initial aldimine intermediate (a) is converted to the resonance-stabilized intermediate (b) by loss of a proton at the alpha carbon. Further enzyme-catalyzed proton transfers to intermediates (c) and (d) may occur, depending upon the specificity of a given enzyme. The enzymes use their general acids and bases to catalyze these proton transfers.

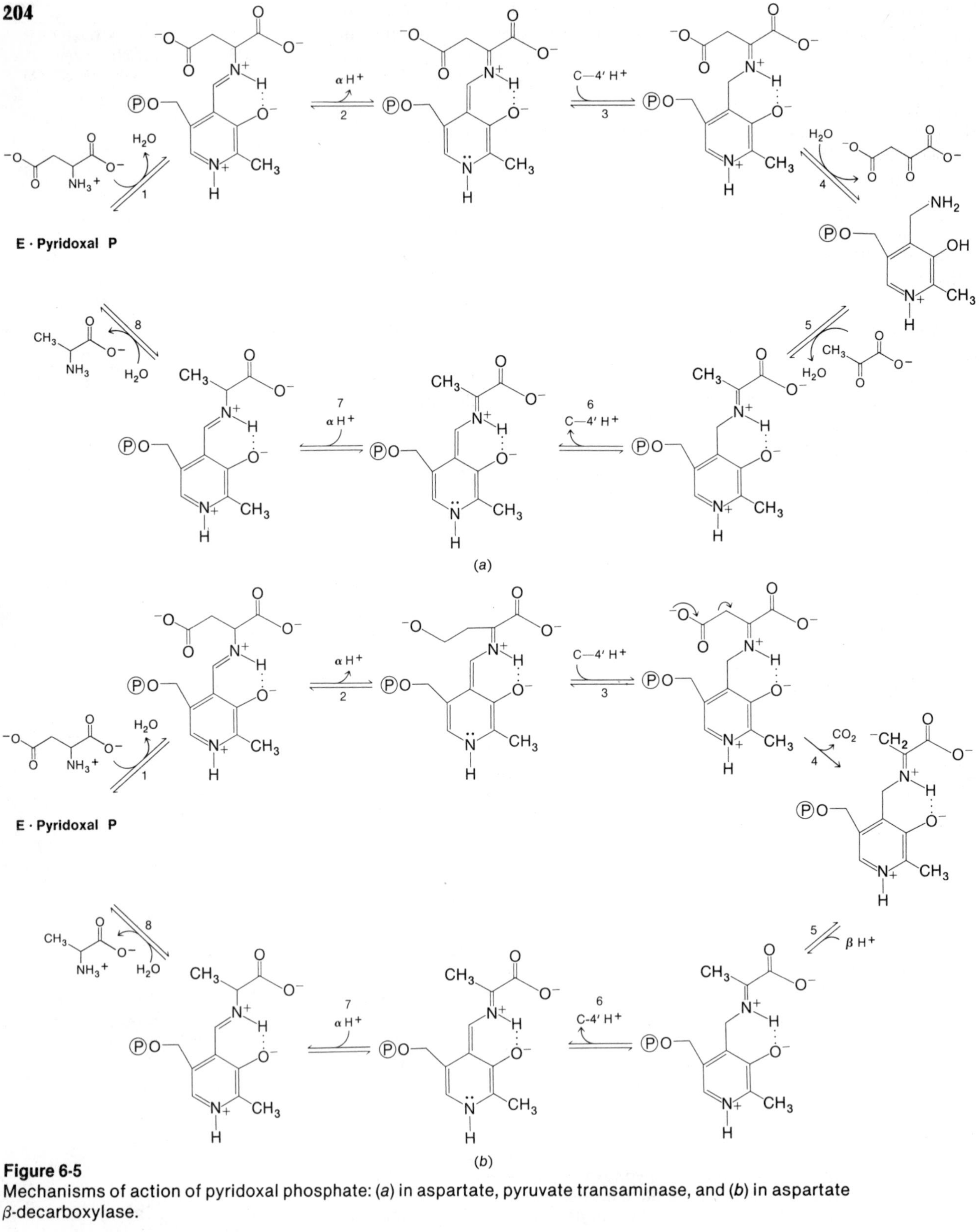

**Figure 6-5**
Mechanisms of action of pyridoxal phosphate: (*a*) in aspartate, pyruvate transaminase, and (*b*) in aspartate β-decarboxylase.

An intermediate analogous to that in Figure 6-4*b* but generated from glycine and so lacking the $\beta$ and $\gamma$ carbons can react as a carbanion with an aldehyde to produce ultimately a $\beta$-hydroxy-$\alpha$-amino acid. These reactions are catalyzed by aldolases such as threonine aldolase and serine trans-hydroxymethylase.

$\alpha$-Decarboxylases (Equation 4) generate intermediates analogous to that in Figure 6-4*b* by catalyzing the elimination of $CO_2$ instead of $H^+$ from the intermediate in Figure 6-4*a*. Protonation of the $\alpha$-carbanionic intermediates by protons from $H_2O$ followed by hydrolysis of the resulting imines produces the amines corresponding to the replacement of the carboxylate group in the substrate by a proton.

The stability of the resonance hybrid (Figure 6-4*b*) accounts for the catalytic action of pyridoxal-5'-phosphate in the reactions in Equations (3) through (7). The question of whether the carbanionic or the quinonoid form of the hybrid is the more important contributor to this structure is beyond the scope of this chapter. In certain cases, spectral data support the quinonoid structure as the main contributor.

$\beta$ Decarboxylation of aspartate and $\beta,\gamma$ elimination by cystathionase can be understood on the basis of the stabilization of amino acid $\beta$-carbanions by pyridoxal-5'-phosphate. Returning to the intermediate in Figure 6-4*c*, it can be seen that elimination of a $\beta$ proton produces a $\beta$-carbanion (Figure 6-4*d*) that is stabilized by resonance with the neighboring protonated imine. This carbanion can eliminate a good leaving group from the $\gamma$ carbon, exemplified by the $\gamma$-cystathionase-catalyzed elimination of cysteine from cystathionine in Equation (8). The $\beta$ decarboxylation of aspartate (Equation 9) proceeds by elimination of a $\beta$-carbanionic intermediate like that in Figure 6-4*d* from the ketimine analogous to the intermediate produced by loss of the $\alpha$ proton from the aldimine of aspartate with pyridoxal-5'-phosphate. This mechanism is outlined in greater detail in Figure 6-5.

*Most of the enzymatic reactions that involve pyridoxal-5'-phosphate are catalyzed by the coenzyme itself in the absence of any enzyme, albeit slowly, at high temperatures, and in the presence of metal ions.* The metal ions stabilize the imine complexes of the coenzyme with $\alpha$-amino acids by chelation of the $\alpha$-carboxylate, the imine nitrogen, and the phenolic oxygen, which is illustrated in Figure 6-6*a*. These reactions have been studied as models of the enzymatic reactions to gain insight into the mechanism of action of the coenzyme.

In view of the fact that the coenzyme alone catalyzes these reactions, it is pertinent to discuss the role of the enzymes in reactions involving the action of pyridoxal-5'-phosphate. The enzymes provide specificity and rate acceleration as in other enzymatic reactions. Substrate specificity is determined by binding specificity for $\alpha$-amino acids, and reaction specificity is determined in substantial part by the orientations of general acids and bases in the active sites of the enzymes. Comparisons of their mechanisms show that these reactions are often distinguished by the nature and orientation of the proton-transfer processes. These are catalyzed by general acids and bases present in the active site, the positioning and states of protonation of which can determine the reaction that is catalyzed, e.g., transamination, racemization, dealdolization, etc.

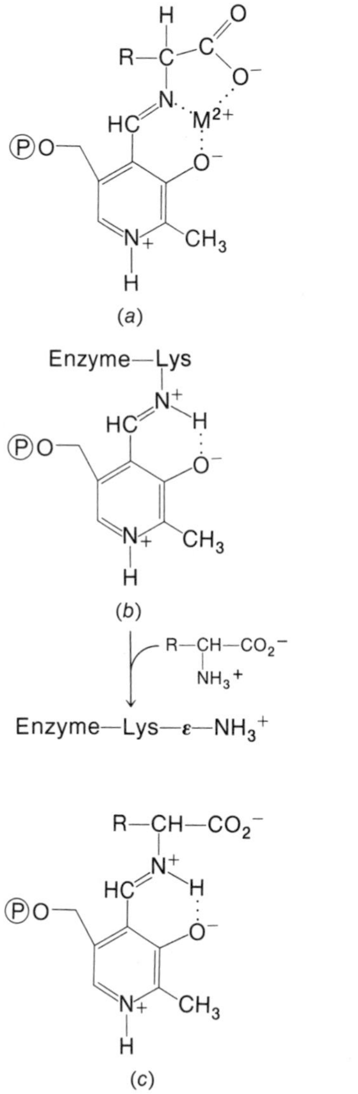

**Figure 6-6**
Structures of metal and enzyme complexes of pyridoxal phosphate.

The enzymes also provide rate acceleration by mechanisms other than general acid-base catalysis. Two are of special importance in pyridoxal-5'-phosphate–dependent reactions. The enzymes use their binding properties to stabilize the aldimine complexes of $\alpha$-amino acids and pyridoxal-5'-phosphate. This is accomplished without the involvement of metal ions. The enzymes also promote the formation of the aldimines by binding pyridoxal-5'-phosphate as a preformed imine between the $\varepsilon$ amino group of a lysine and the carboxaldehyde group of the coenzyme, illustrated in Figure 6-6b. Imine formation is known to involve nucleophilic addition of the amino group to the carbonyl group, as in Equation (10). A closely analogous reaction is Equation (11), between an imine and an amine.

$$R-NH_2 + \overset{O}{\underset{}{C}} \rightleftharpoons \underset{RNH}{-\overset{OH}{\underset{}{C}}-} \rightleftharpoons \underset{RN}{\overset{}{C}} + H_2O \qquad (10)$$

$$R_1-NH_2 + \overset{R_2NH^+}{\underset{}{C}} \rightleftharpoons \underset{R_1NH}{-\overset{R_2NH_2^+}{\underset{}{C}}-} \rightleftharpoons \underset{R_1NH^+}{\overset{}{C}} + R_2NH_2 \qquad (11)$$

The protonated imine in Equation (11) is more reactive in nucleophilic addition reactions than is the carbonyl group. Thus the enzymes catalyze "transimination" reactions analogous to Equation (11) in forming the amino acid–pyridoxal-5'-phosphate aldimine complexes.

*The most fundamental biochemical function of pyridoxal-5'-phosphate is the formation of aldimines with $\alpha$-amino acids that stabilize the development of carbanionic character at the $\alpha$ and $\beta$ carbons of $\alpha$-amino acids in intermediates* such as those in Figure 6-4b. Enzymes acting alone cannot stabilize these carbanions and so cannot, by themselves, catalyze reactions involving their formation as intermediates.

Pyridoxal-5'-phosphate is not quite unique in this function, however. There are a few enzymes that catalyze reactions similar to pyridoxal-5'-phosphate–dependent reactions but which do not involve this coenzyme. In each case, the enzyme contains a carbonyl or carbonyl-like functional group not normally found in proteins. For example, histidine decarboxylase from *Lactobacillus* and *S*-adenosylmethionine decarboxylase from *E. coli* and mammalian tissue contain $\alpha$-ketoacyl groups that act in place of pyridoxal phosphate. Histidine and phenylalanine ammonia lyases catalyze the reactions shown in Equations (12) and (13). These enzymes contain functional groups related to dehydroalanyl groups whose precise structures and functions remain to be determined.

$$\underset{N \quad N-H}{\quad} \underset{NH_3^+}{\overset{O}{\underset{}{\parallel}}}O^- \xrightarrow[\text{ammonia-lyase}]{\text{Histidine}} NH_4^+ + \underset{N \quad N-H}{\overset{O}{\underset{}{\parallel}}}O^- \qquad (12)$$

$$\underset{NH_3^+}{\overset{O}{\underset{}{\parallel}}}O^- \xrightarrow[\text{ammonia-lyase}]{\text{Phenylalanine}} NH_4^+ + \overset{O}{\underset{}{\parallel}}O^- \qquad (13)$$

# NICOTINAMIDE COENZYMES

Nicotinamide adenine dinucleotide ($NAD^+$), also known as diphospho-pyridine nucleotide ($DPN^+$), is one of the two coenzymatic forms of niacin (Figure 6-7). The other is nicotinamide adenine dinucleotide phosphate ($NADP^+$), also known as triphosphopyridine nucleotide ($TPN^+$), which differs from NAD by the presence of a phosphate group at C-2′ of the adenosyl moiety.

*The nicotinamide coenzymes are biological carriers of reducing equiv-alents, i.e., electrons, which in most of their reactions function as cosub-strates rather than as true coenzymes. The most common function of $NAD^+$ is to accept two electrons and a proton ($H^-$ equivalent) from a substrate undergoing metabolic oxidation to produce NADH, the reduced form of the coenzyme.* This then diffuses or is transported to the terminal-electron transfer sites of the cell and reoxidized by terminal-electron accep-tors, $O_2$ in aerobic organisms, with the concomitant formation of ATP (Chapter 10). Equations (14), (15), and (16) are typical reactions in which $NAD^+$ acts as such an acceptor.

$$NAD^+ + CH_3CH_2OH \underset{\text{dehydrogenase}}{\overset{\text{Alcohol}}{\rightleftharpoons}} CH_3-\overset{O}{\overset{\|}{C}}H + NADH + H^+ \quad (14)$$

$$^-O_2C(CH_2)_2\overset{NH_3^+}{\overset{|}{C}}HCO_2^- + NAD^+ + H_2O \overset{\text{Glutamate}}{\underset{\text{dehydrogenase}}{\longrightarrow}}$$

$$^-O_2C(CH_2)_2\overset{O}{\overset{\|}{C}}CO_2^- + NADH + NH_4^+ + H^+ \quad (15)$$

$$HPO_4^- + {}^-O_3POCH_2\overset{OH}{\overset{|}{C}}H-\overset{O}{\overset{\|}{C}}H + NAD^+ \overset{\text{Glyceraldehyde-3P}}{\underset{\text{dehydrogenase}}{\rightleftharpoons}}$$

$$^-O_3POCH_2\overset{OH}{\overset{|}{C}}H\overset{O}{\overset{\|}{C}}OPO_3^- + NADH + H^+ \quad (16)$$

The chemical mechanisms by which $NAD^+$ is reduced to NADH in Equations (14) through (16) are probably similar, as represented in general-ized forms in Equation (17).

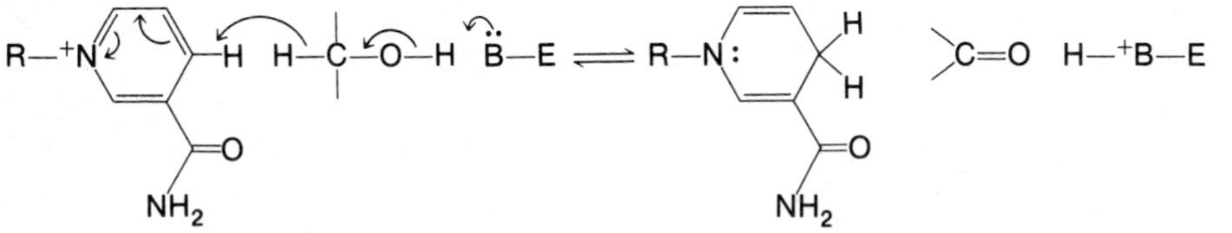

$$(17)$$

According to this formulation, the immediate oxidation product in Equa-tion (15) [$-NH_2$ replaces $-OH$ in Equation (17)] would be the imine of α-ketoglutarate, which would quickly undergo hydrolysis to α-ketoglutarate and ammonia in aqueous solution. The oxidation of an aldehyde group catalyzed by glyceraldehyde-3-P dehydrogenase (Equation 16) also can be understood on the basis of this formulation once it is realized that there is an essential $-SH$ group at the active site that is transiently acylated during the course of the reaction. The $-SH$ group reacts with the aldehyde group of glyceraldehyde-3-P according to Equation (18), forming a thiohemiacetal

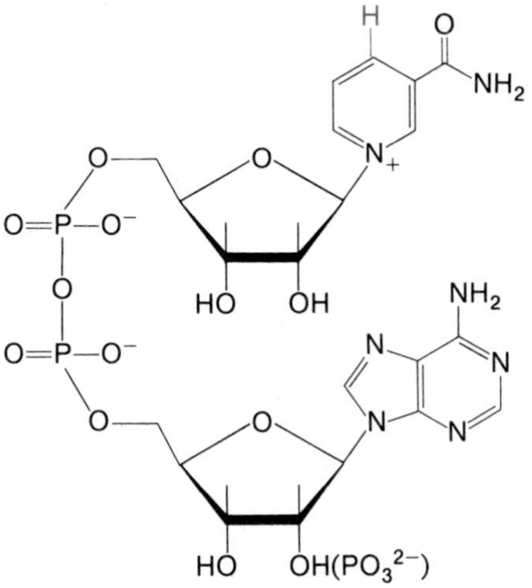

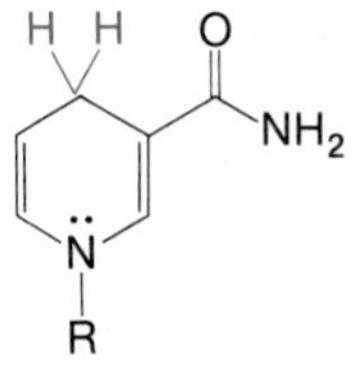

**Figure 6-7**
Structures of nicotinamide coen-zymes. The reactive centers of the coenzymes are shown in color.

that is the species oxidized. The resulting acyl-enzyme then reacts with phosphate to produce 1,3-diphosphoglycerate.

$$
E{-}SH + \;
\begin{matrix} O \\ \parallel \\ C{-}H \\ | \\ CHOH \\ | \\ {}^{2-}O_3POCH_2 \end{matrix}
\;\rightleftharpoons\;
\begin{matrix} OH \\ | \\ E{-}S{-}C{-}H \\ | \\ CHOH \\ | \\ {}^{2-}O_3POCH_2 \end{matrix}
\xrightleftharpoons[\;]{NAD^+ \;\; NADH}
\begin{matrix} O \\ \parallel \\ E{-}S{-}C \\ | \\ CHOH \\ | \\ {}^{2-}O_3POCH_2 \end{matrix}
\xrightarrow{HPO_4^{2-}}
E{-}SH + \;
\begin{matrix} O \\ \parallel \\ C{-}OPO_3^{2-} \\ | \\ CHOH \\ | \\ {}^{2-}O_3POCH_2 \end{matrix}
\quad (18)
$$

Equation (17) implies that the hydrogen atom and two electrons are transferred in a concerted process, i.e., as a hydride equivalent, with the quaternary nitrogen in the pyridinium ring serving as an electron sink. The hydrogen atom is certainly transferred directly, and this is discussed in the following section, but it is not yet known whether both electrons are transferred with it in a single step or in a two-step sequence as a hydrogen atom and an electron.

In a series of classic experiments, Westheimer, Vennesland, and co-workers demonstrated that hydrogen transfer between $NAD^+$ and substrates such as those in Equations (14) through (16) are direct hydrogen transfers and occur with stereospecificity. The experiments with alcohol dehydrogenase were the first to establish these points, and they were the earliest to define the remarkable stereospecificity of enzymatic action at "prochiral" centers involving chemically equivalent hydrogen atoms. Once the regiospecificity of hydrogen transfer to $NAD^+$ and the absolute configurations of the molecules were established, it became possible to formulate these processes as set forth in Equations (19) through (21) for alcohol dehydrogenase.

$$(19)$$

$$(20)$$

$$(21)$$

The tracing of deuterium through these transformations established quite clearly that enzymes are stereospecific in abstracting chemically equivalent hydrogens from prochiral centers and in transferring hydrogens specifically to one face of planar molecules, even in molecules as small as acetaldehyde.

This specificity is thought to be a natural consequence of the fact that enzymes are asymmetrical molecules that form highly stereoselective complexes with their substrates, even those that have planes of symmetry or are themselves planar molecules. This is illustrated schematically in Figure 6-8, which shows how specific binding interactions can lead to stereospecific hydrogen transfer between acetaldehyde and NADH.

*In addition to acting as a cellular electron carrier, NAD$^+$ also acts as a true coenzyme with certain enzymes.* Enzymes are sometimes confronted with the problem of catalyzing such reactions as epimerizations, aldolizations, and eliminations on substrates lacking the intrinsic chemical reactivities required for these reactions to occur at significant rates. Sometimes such reactivities can be introduced into the substrate by oxidizing an appropriate alcohol group to a carbonyl group, and the enzyme is then found to contain NAD$^+$ as a tightly bound coenzyme. NAD$^+$ functions

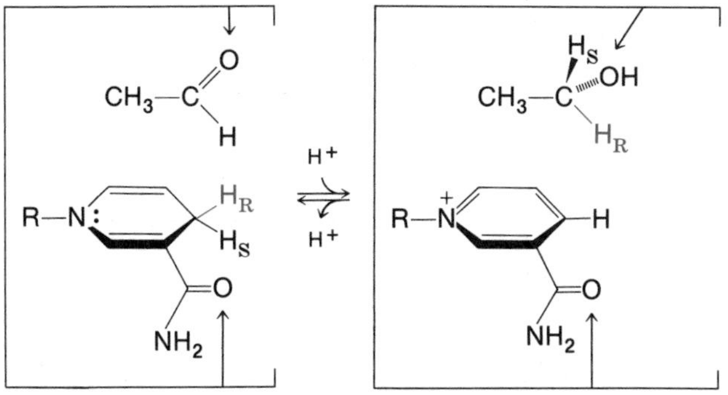

**Figure 6-8**
Stereospecificity of hydrogen transfer in nicotinamide coenzymes. The arrows represent hypothetical enzyme-binding interactions. The subscripted stereochemical symbols R and S are explained in the footnote.

---

*\*Chiral* and *prochiral* are derived from the Greek word χειρ, meaning "hand," and they refer to "handedness" in stereoisomeric or potentially stereoisomeric centers. The stereochemical symbols frequently used for designating two stereoisomers are *R* and *S*, which are assigned as follows. For a tetrahedral carbon (or other tetrahedral atom), the four different substituents are assigned relative priorities by the application of rules that generally accord higher priority to groups having the larger summation of atomic weights. The article by G. Popják listed in Selected Readings should be consulted for these priority rules. Once the priorities are assigned, the atom is viewed from the side opposite the lowest priority group, and the symbol *R*, for *rectus*, is assigned if the remaining groups appear in clockwise order from highest to lowest priority. The symbol *S*, for *sinister*, is assigned if they appear in counterclockwise order. For atoms whose substituent groups are $a < b < c < d$ in order of increasing priority, the following structures are those of the *R* and *S* isomers:

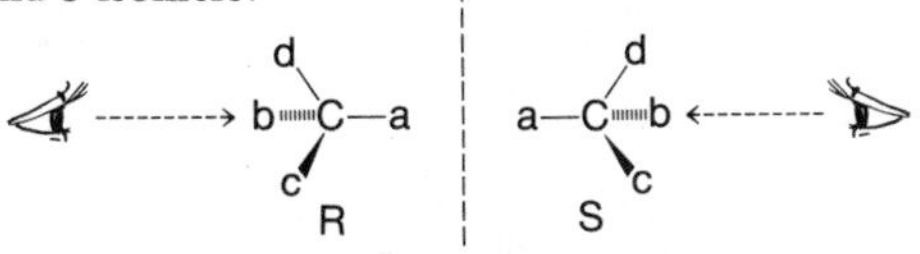

Atoms in which two of the groups are identical are called *prochiral*, since the elevation of one of the identical groups to a higher priority would lead to a chiral center. Carbon-1 in ethanol and nicotinamide C-4 in NADH are prochiral centers, since they each have two hydrogens and two other substituents. In Figure 6-8, these two hydrogens are subscripted *R* or *S* for the configurational designations that would result from according higher priority to one or the other. One means of *granting* such priority is by the use of isotopes of hydrogen. Thus deuterium has a higher priority than protium, so that the configuration of deuterioethanol in Equation (20) is *R*, while in Equation (21) it is *S*.

coenzymatically by transiently oxidizing the key alcohol group to the carbonyl level, producing an oxidatively activated intermediate whose further transformation is catalyzed by the enzyme. In the last step, the carbonyl group is reduced back to the hydroxyl group by the transiently formed NADH.

Typical examples are the enzymes UDPgalactose-4-epimerase and *S*-adenosylhomocysteinase, both of which contain tightly bound NAD$^+$ (Figure 6-9). The NAD$^+$ in *S*-adenosylhomocysteinase oxidizes carbon atom 3 of the ribose ring in the substrate. The resulting carbonyl group renders the

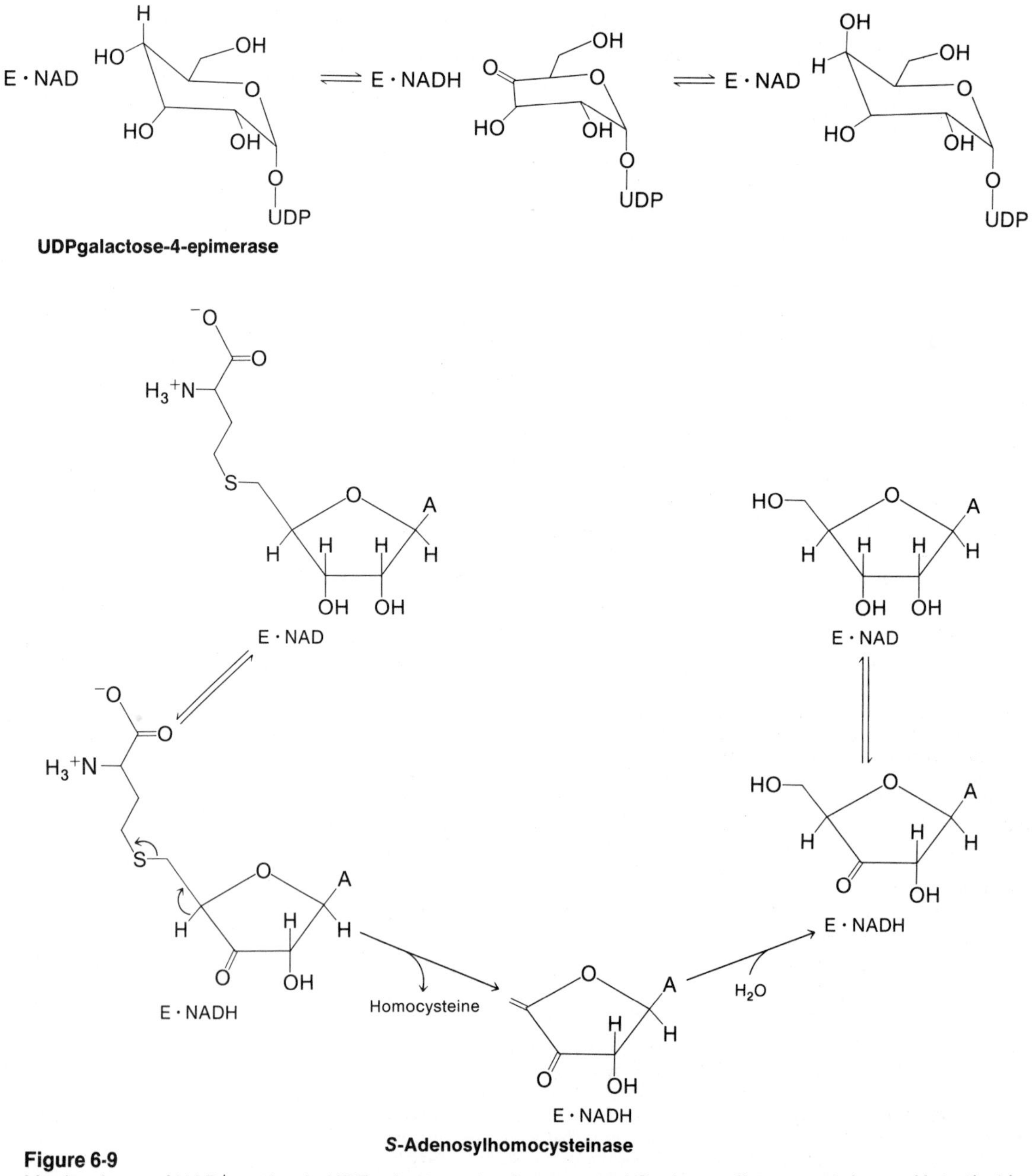

**Figure 6-9**
Mechanisms of NAD$^+$ action in UDPgalactose-4-epimerase and *S*-adenosylhomocysteinase. Note that in the latter reaction the catalytic pathway involves the intermediates analogous to (*a*), (*c*), and (*d*) in Figure 6-10.

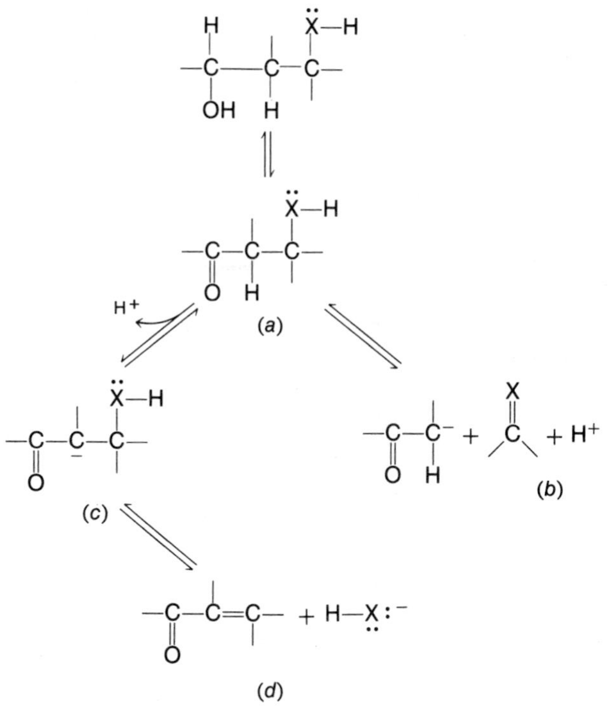

**Figure 6-10**
Bond cleavages and mechanisms
involving oxidative activation by
$NAD^+$. Bonds susceptible to cleavage
are shown in color.

C-4 proton enolizable and promotes the elimination of homocysteine. Addition of water and reduction of the ketone by NADH completes the process. The mechanisms of many of these reactions, generalized in Figure 6-10, are based on the stabilization of a carbanionic intermediate by the carbonyl group produced by oxidative activation.

In NADH, the two hydrogens bonded to nicotinamide carbon atom 4 are chemically equivalent, so that from a purely chemical standpoint either could be transferred to an aldehyde or ketone. It is their topographic inequivalence that leads to stereospecificity in enzymatic reactions; however, the enzymes do not all exhibit the same stereospecificity for catalyzing hydrogen transfer from this center or in forming this center from $NAD^+$. Those enzymes such as alcohol dehydrogenase which catalyze transfer of the pro-$R$ hydrogen in NADH (see Figure 6-8) are known as $R$-side-specific enzymes, and those transferring the other hydrogen, the pro-$S$ hydrogen, are known as $S$-side-specific enzymes. Table 6-2 lists the $RS$ specificities of some oxidoreductases.

## FLAVIN COENZYMES

### Structure

Flavin adenine dinucleotide (FAD) (Figure 6-11) and flavin mononucleotide (FMN) are the coenzymatically active forms of vitamin $B_2$, riboflavin. Riboflavin is the $N^{10}$-ribityl isoalloxazine portion of FAD, which is enzymatically converted into its coenzymatic forms first by phosphorylation of the ribityl C-5' hydroxy group to FMN and then adenylylation to FAD. FMN and FAD appear to be functionally equivalent coenzymes, and which is involved with a given enzyme appears to be a matter of enzymatic binding specificity.

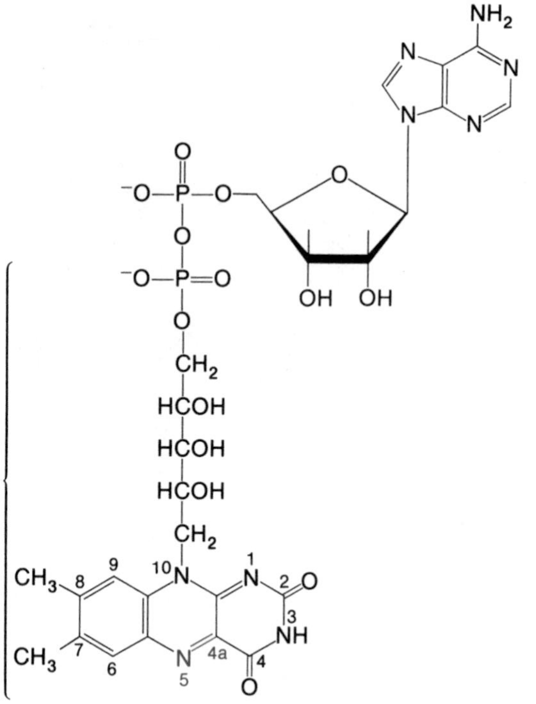

**Figure 6-11**
Structures of flavin coenzymes.

**Flavin adenine dinucleotide**
(FAD)

The catalytically functional portion of the coenzymes is the isoalloxazine ring, specifically N-5 and C-4a, which are thought to be the immediate loci of catalytic function, although the entire chromophoric system extending over N-5, C-4a, C-10a, N-1, and C-2 should be regarded as an indivisible catalytic entity, as are the nicotinamide, pyridinium, and thiazolium rings of NAD$^+$, pyridoxal-P, and thiamine pyrophosphate.

*Flavin-dependent enzymes are called flavoproteins and, when purified, normally contain their full complements of FAD or FMN as tightly bound coenzymes.* Flavoproteins are the bright yellow color of isoalloxazine chromophore in its oxidized form. In a few flavoproteins, the coenzyme is

**Table 6-2**
Nicotinamide Side-Specificities of Dehydrogenases

| Enzyme | Reaction | RS Specificity |
|---|---|---|
| Alcohol dehydrogenase | Ethanol + NAD$^+$ $\rightleftharpoons$ acetaldehyde + NADH + H$^+$ | $R$ |
| Lactate dehydrogenase | $S$-Lactate + NAD$^+$ $\rightleftharpoons$ pyruvate + NADH + H$^+$ | $R$ |
| Malate dehydrogenase | $S$-Malate + NAD$^+$ $\rightleftharpoons$ oxalacetate + NADH + H$^+$ | $R$ |
| Aldehyde dehydrogenase | Acetaldehyde + NAD$^+$ $\rightleftharpoons$ acetate + NADH + H$^+$ | $R$ |
| Triose-P dehydrogenase | R-Glyceraldehyde-3-P + NAD$^+$ + HPO$_4^{2-}$ $\rightleftharpoons$ 1,3-diphospho-R-glycerate + NADH + H$^+$ | $S$ |
| Glutamate dehydrogenase | $S$-Glutamate + NAD$^+$ $\rightleftharpoons$ $\alpha$-ketoglutarate + NADH + H$^+$ | $S$ |
| Glucose-6-P dehydrogenase | D-Glucose-6-P + NADP$^+$ $\rightleftharpoons$ 6-phosphogluconolactone + NADPH + H$^+$ | $S$ |

known to be covalently bonded to the protein by means of an enzymatic sulfhydryl or imidazole group at the C-8a methyl group and in at least one case at C-6. In most flavoproteins, the coenzymes are tightly but noncovalently bound, and many can be resolved into apoenzymes that can be reconstituted to holoenzymes by readdition of FAD or FMN.

Flavin coenzymes exist in three spectrally distinguishable oxidation states that account in part for their catalytic functions; the yellow oxidized form, the red or blue one-electron-reduced form, and the colorless two-electron-reduced form. Their structures are depicted in Figure 6-12. These and other less well-defined forms often have been detected spectrally as intermediates in flavoprotein catalysis.

*Flavins are very versatile redox coenzymes.* Flavoproteins are dehydrogenases, oxidases, and oxygenases that catalyze a variety of reactions on an equal variety of substrate types. Since these classes of enzymes do not consist exclusively of flavoproteins, it is difficult to define catalytic specificity for flavins. *Biological electron acceptors and donors in flavin-mediated reactions can be two-electron acceptors, such as $NAD^+$ or $NADP^+$, or a variety of one-electron acceptor systems, such as cytochromes ($Fe^{2+}/Fe^{3+}$) and quinones, and molecular oxygen is an electron acceptor for flavoprotein oxidases as well as the source of oxygen for oxygenases.* The only obviously common aspect of flavin-dependent reactions is that all in which flavins are direct catalytic participants are redox reactions.

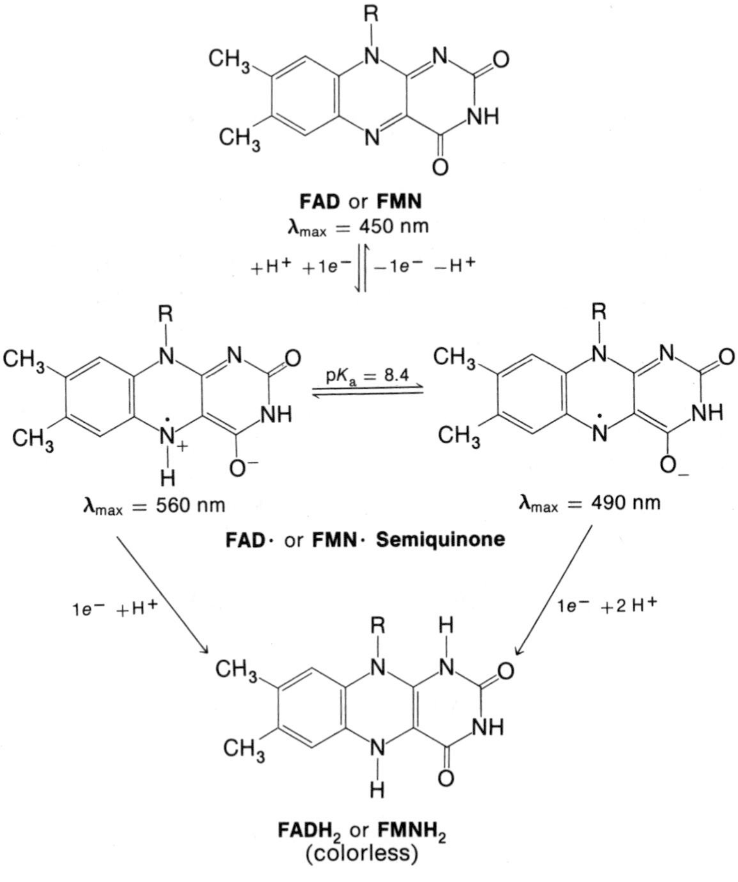

**Figure 6-12**
Oxidation states of flavin coenzymes. The one-electron reduced semiquinone is shown at two protonation levels with a $pK_a$ of 8.4 for the dissociation of the proton. The undissociated form is blue ($\lambda_{max} = 560$ nm) while the dissociated form is red ($\lambda_{max} = 490$ nm). In each of these forms the unpaired electron is delocalized between N–5 and C–4a. In the alternative resonance structures the unpaired electron is at C–4a.

Typical reactions catalyzed by flavoproteins are listed in Table 6-3, which groups flavoproteins into those that do not utilize molecular oxygen as a substrate and those that do. The significance of this is best appreciated when one realizes that $FADH_2$, a likely intermediate in many flavoprotein reactions, spontaneously reacts with $O_2$ to produce $H_2O_2$. In the case of the dehydrogenases, therefore, either $FADH_2$ is not an intermediate or it is somehow prevented from reacting with $O_2$. Among the dehydrogenases are several that utilize the two-electron acceptor substrates $NAD^+$ or $NADP^+$, and it is reasonable to suppose that the two-electron reduction of $NAD^+$ by an intermediate $E \cdot FADH_2$ might be involved. Also listed in Table 6-3 are other dehydrogenases for which the electron acceptors from $E \cdot FADH_2$ are not given. These enzymes are membrane-bound and transfer electrons directly to membrane-bound acceptors, mainly one-electron acceptors such as quinones and cytochromes ($Fe^{2+}/Fe^{3+}$). The stability of the flavin semiquinone, $FAD \cdot$ and $FMN \cdot$ in Figure 6-12, gives flavins the capability to interact with one-electron acceptors in electron-transport systems.

The other classes of flavoproteins in Table 6-3 interact with molecular oxygen either as the electron-acceptor substrate in redox reactions catalyzed by oxidases or as the substrate source of oxygen atoms for oxygenases. Molecular oxygen also serves as an electron acceptor and source of oxygen for metalloflavoproteins and dioxygenases, which are not listed in Table 6-3. These enzymes catalyze more complex reactions involving catalytic redox components such as metal ions and metal-sulfur clusters in addition to flavin coenzymes.

**Table 6-3**
Reactions Catalyzed by Flavoproteins

| Flavoprotein | Reaction |
| --- | --- |
| **Dehydrogenases** | |
| Glutathione reductase | $H^+ + GSSG + NADPH \rightleftharpoons 2\ GSH + NADP^+$ |
| Acyl-CoA dehydrogenases | $RCH_2CH_2COSCoA + NAD^+ \rightleftharpoons RCH{=}CHCOSCoA + NADH + H^+$ |
| Succinate dehydrogenase | $^-O_2CCH_2CH_2CO_2^- + E \cdot FAD \rightleftharpoons\ ^-O_2CCH{=}CHCO_2^- + E \cdot FADH_2$ |
| D-Lactate dehydrogenase | $CH_3{-}CHOH{-}CO_2^- + E \cdot FAD \rightleftharpoons CH_3{-}{-}CO{-}CO_2^- + E \cdot FADH_2$ |
| **Oxidases** | |
| Amino acid oxidases | $\overset{\overset{\displaystyle NH_3^+}{\vert}}{R{-}CH}{-}CO_2^- + O_2 \longrightarrow R{-}CO{-}CO_2^- + H_2O_2 + NH_4^+$ |
| Glucose oxidase | D-Glucose $+ O_2 \longrightarrow$ D-gluconolactone $+ H_2O_2$ |
| Monoamine oxidase | $R{-}CH_2NH_2 + O_2 + H_2O \longrightarrow R{-}CHO + H_2O_2 + NH_3$ |
| **Monooxygenases** | |
| Lactate oxidase | $CH_3{-}CHOH{-}CO_2^- + O_2 \longrightarrow CH_3{-}CO_2^- + CO_2 + H_2O$ |
| Salicylate hydroxylase | |
| Ketone monooxygenase | |

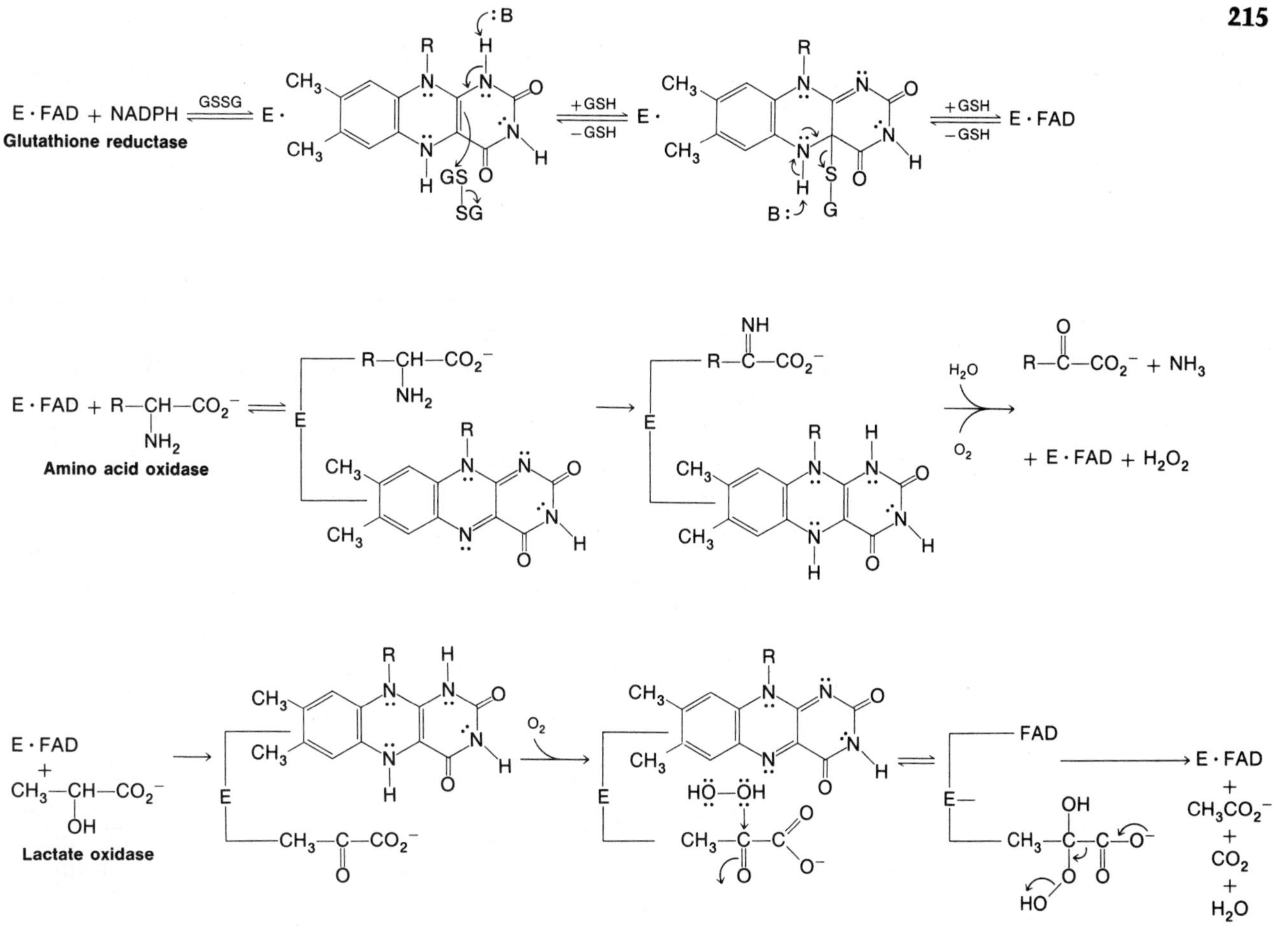

**Figure 6-13**
Mechanisms of flavin-dependent reactions. In the glutathione reductase reaction, the first steps, not shown, involve the reduction of FAD to $FADH_2$ by NADPH and the binding of glutathione. The mechanism by which oxidized glutathione is reduced by the $E\cdot FADH_2$ complex is shown. In the amino acid oxidase reaction, the first step shown is the binding of the substrate by the $E\cdot FAD$ complex. In the second step, the amino acid is oxidized to an imino acid and FAD is reduced to FADH. In the third step, the imino acid is released from the enzyme and hydrolyzed by water to an $\alpha$-ketoacid; and $FADH_2$ is oxidized by $O_2$ to FAD and $H_2O_2$. In the lactate oxidase reaction, the first step shown involves the binding and oxidation of lactate, producing a complex containing $FADH_2$ and pyruvate. In the second step $O_2$ oxidizes $FADH_2$ and is thereby reduced to $H_2O_2$. The hydrogen peroxide so produced then reacts with pyruvate at the active site, resulting in its oxidative decarboxylation to acetate and $CO_2$.

The mechanisms of action of flavin coenzymes are currently under active investigation. A recurrent theme appears to be the probable involvement of $FADH_2$ or $FMNH_2$ as transient intermediates in a variety of flavoprotein reactions. Figure 6-13 illustrates reasonable catalytic pathways for three of the enzymes listed in Table 6-3, one from each of the three classes, and shows the probable involvement of $E\cdot FADH_2$ in each case.

Several points should be understood about Figure 6-13. One is that the detailed mechanisms by which $E\cdot FADH_2$ arises need not be the same in all the reactions. Note, for example, that in the case of glutathione reductase, the mechanisms by which $E\cdot FAD$ is reduced to $E\cdot FADH_2$ by NADPH in the forward direction and by glutathione in the reverse direction are undoubtedly different. The mechanism shown is one recently proposed based on the nonenzymatic reaction of 3-methyl riboflavin with dithiothreitol. The reduction of FAD by NADPH might be expected to occur by concerted transfer of a C-4 proton and two electrons, a hydride equivalent, from NADH to N-5 of FAD. Transfer could not be detected directly using [4-$^3$H]NADPH because upon transfer to N-5 in FADH$_2$, the $^3$H would quickly exchange with protons in water. Such a direct transfer can be observed, however, using a 5-deaza analog of the coenzymes so that the transferred $^3$H in 5-deaza-FADH$_2$ is bonded to carbon and non-exchangeable.

In other cases, E $\cdot$ FADH$_2$ may result from a concerted two-electron reduction of FAD or by two successive one-electron reductions involving the free-radical semiquinone form of the coenzyme as an intermediate. The detailed mechanism by which oxygen reacts with FADH$_2$ to produce H$_2$O$_2$ and FAD is not known with certainty, although there is good reason based on nonenzymatic model reactions to expect C-4a flavin hydroperoxide addition compounds to be involved.

It is interesting to consider the relationship between the amino acid oxidases and lactate oxidases. They differ essentially in that H$_2$O$_2$ and the $\alpha$-keto acid dissociate quickly as products from the amino acid oxidases but not from lactate oxidase, which catalyzes further reaction of H$_2$O$_2$ with the enzyme-bound pyruvate to produce acetate and CO$_2$. Labeling experiments with $^{18}$O$_2$ are in accord with the proposed mechanism. However, they cannot distinguish it from closely related mechanisms involving flavin hydroperoxides acting in the capacity shown for H$_2$O$_2$ in Figure 6-13.

*The biochemical importance of flavin coenzymes appears to be their versatility in mediating a variety of redox processes, including electron transfer and the activation of molecular oxygen for oxygenation reactions.* The detailed mechanisms of oxygen activation are not well understood. *An especially important manifestation of their redox versatility is their ability to serve as the switch point from the two-electron transfer processes, which predominate in cytosolic carbon metabolism, to the one-electron transfer processes, which predominate in membrane-associated terminal electron-transfer pathways.* In mammalian cells, for example, the end products of the aerobic metabolism of glucose are CO$_2$ and NADH (Chapter 10). The terminal electron-transfer pathway is a membrane-bound system of cytochromes, nonheme iron proteins, and copper-heme proteins—all one-electron acceptors that transfer electrons ultimately to O$_2$ to produce H$_2$O and NAD with concomitant production of ATP from ADP and P$_i$. The interaction of NADH with this pathway is mediated by NADH dehydrogenase, a flavoprotein that couples the two-electron oxidation of NADH with the one-electron reductive processes of the membrane.

## PHOSPHOPANTETHEINE COENZYMES

4'-Phosphopantetheine coenzymes are the biochemically active forms of the vitamin pantothenic acid, which consists of the pantoate and $\beta$-alanyl portions. The two main classes are structurally illustrated in Figure 6-14, in which 4'-phosphopantetheine is shown as covalently linked to an adenylyl group in coenzyme A or to a functional group of a protein such as a serine hydroxyl group in acyl carrier protein (ACP). It is also found bonded to proteins that catalyze the activation and polymerization of amino acids to polypeptide antibiotics. Coenzyme A was discovered, purified, and structurally characterized by F. Lipmann and colleagues in work for which Lipmann was awarded the Nobel Prize in 1953.

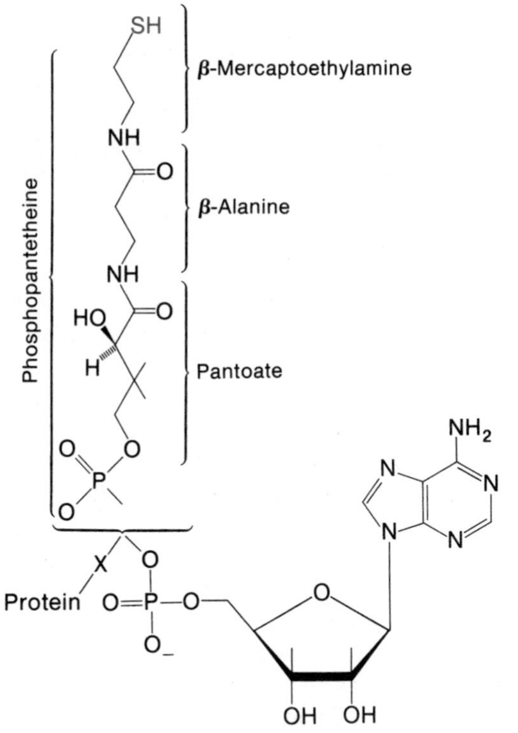

**Coenzyme A**

**Figure 6-14**
Structures of 4′-phosphopantetheine coenzymes. The reactive group is shown in color.

*The sulfhydryl group of the β-mercaptoethylamine (or cysteamine) moiety of phosphopantetheine coenzymes is the functional group that is directly involved in the enzymatic reactions for which they serve as coenzymes.* From the standpoint of the chemical mechanism of catalysis, it is the essential functional group, although it is now recognized that phosphopantetheine coenzymes have other functions as well. Many reactions in metabolism involve acyl-group transfer or enolization of carboxylic acids that exist as unactivated carboxylate anions at physiological pH. The predominant means by which these acids are activated for acyl transfer and enolization is esterification with the sulfhydryl group of pantetheine coenzymes.

The mechanistic importance of activation is exemplified by the condensation of two molecules of acetyl-coenzyme A to acetoacetyl-coenzyme A catalyzed by β-ketothiolase (Equation 22):

$$CH_3-\overset{O}{\overset{\|}{C}}-SCoA + CH_3-\overset{O}{\overset{\|}{C}}-CSoA \rightleftharpoons$$

$$CH_3-\overset{O}{\overset{\|}{C}}-CH_2-\overset{O}{\overset{\|}{C}}-SCoA + CoASH \quad (22)$$

The two important steps of the reaction depend on both acetyl groups being activated, one for enolization and the other for acyl-group transfer. In the first step, one of the molecules must be enolized by the intervention of a base to remove an α proton, forming the enolate in Equation (23).

$$B: \overset{\frown}{H}-\overset{\frown}{CH_2}-\overset{O}{\overset{\|}{C}}-SCoA \rightleftharpoons [B\cdots H\cdots\overset{\delta^+}{CH_2}\cdots\overset{\delta^-}{\overset{O}{\overset{\|}{C}}}-SCoA] \rightleftharpoons \overset{+}{B}-H + CH_2\cdots\overset{O}{\overset{-\|}{C}}-SCoA \quad (23)$$

The enolate is stabilized by delocalization of its negative charge between the $\alpha$ carbon and the acyl oxygen atom, making it thermodynamically accessible as an intermediate. Moreover, this developing charge is also stabilized in the transition state preceding the enolate, so it is also kinetically accessible; that is, it is rapidly formed. Consideration of the same enolization reaction with acetate anion reveals a starkly contrasting picture, for enolization of the acetate anion would result in the generation of a second negative charge in the enolate, an energetically and kinetically unfavorable process.

The second stage of the condensation is the reaction of the enolate anion with the acyl group of the second molecule of acetyl-CoA (Equation 24):

$$
\text{CH}_3\overset{O}{\overset{\|}{C}}-\text{SCoA} + \underset{\text{CH}_2-\overset{O}{\overset{\|}{C}}-\text{SCoA}}{} \rightleftharpoons \left[ \begin{array}{c} \text{CH}_3-\overset{O^-}{\overset{|}{C}}-\text{SCoA} \\ | \\ \text{CH}_2-\overset{O}{\overset{\|}{C}}-\text{SCoA} \end{array} \right] \rightleftharpoons \begin{array}{c} \text{CH}_3-\overset{O}{\overset{\|}{C}} \\ | \\ \text{CH}_2-\overset{O}{\overset{\|}{C}}-\text{SCoA} \end{array} + \text{CoASH} \quad (24)
$$

Nucleophilic addition to the neutral activated acyl group is a favored process, and coenzyme A is a good leaving group from the tetrahedral intermediate. Consideration of the occurrence of this process with the acetate anion, i.e., acetate reacting with an enolate anion, again provides a sharp contrast with the process of Equation (24), for it would entail the nucleophilic addition of an anion to an anionic center generating a dianionic transition state, an unfavorable process from both thermodynamic and kinetic standpoints. Moreover, the resulting intermediate would not have a very good leaving group other than the enolate anion itself, so the transition-state energy for acetoacetate formation would be high. Finally, the $K_{eq}$ for the condensation of 2 moles of acetate to 1 mole of acetoacetate is not favorable in aqueous media, whereas the condensation of 2 moles of acetyl-CoA to produce acetoacetyl-CoA and coenzyme A is thermodynamically spontaneous.

The maintenance of metabolic carboxylic acids involved in enolization and acyl-group transfer reactions as coenzyme A esters provides the ideal lift over the kinetic and thermodynamic barriers to these reactions. The foregoing discussion, in emphasizing the purely electrostatic energy barriers, does not address the question of whether there is an activation advantage in thiol esters relative to oxygen esters. Why thiol esters in preference to oxygen esters? Thiol esters are more readily enolized than oxygen esters. They are more "ketone-like" because of their electronic structures, in which the degree of resonance-electron delocalization from the sulfur atom to the acyl group resulting from overlapping of the occupied $p$ orbitals of sulfur with the acyl-$\pi$ bond is less than that of oxygen esters.

$$
\left[ \text{R}_1-\overset{O}{\overset{\|}{C}}-\overset{..}{\underset{..}{S}}-\text{R}_2 \longleftrightarrow \text{R}_1-\overset{O^-}{\overset{|}{C}}=\overset{+}{\underset{..}{S}}-\text{R}_2 \right] \quad \left[ \text{R}_1-\overset{O}{\overset{\|}{C}}-\overset{..}{\underset{..}{O}}-\text{R}_2 \longleftrightarrow \text{R}_1-\overset{O^-}{\overset{|}{C}}=\overset{+}{\underset{..}{O}}-\text{R} \right]
$$

Another statement of this is that the charge-separated resonance form is a smaller contributor to the electronic structure in thiol esters than in oxygen esters. The reasons for this are not fully understood, but it is thought that one factor may be the larger size of sulfur relative to carbon and oxygen. This may result in a poorer energy match for the overlapping orbitals in thiol esters relative to oxygen esters.

*While the pantetheine sulfhydryl group has the appropriate chemical properties for activating acyl groups, this is not unique to pantetheine coenzymes in the biosphere.* Both glutathione and cysteine, as well as cysteamine, would serve, so the chemistry does not itself explain the importance of these coenzymes. Coenzyme A has many binding determinants in its large structure, especially in the nucleotide moiety, so it may serve a specificity function in the binding of coenzyme A esters by enzymes. *It also may serve as a binding "handle" in cases in which the acyl group must have some mobility in the catalytic site,* i.e., if it must enolize at one subsite and then diffuse a short distance to undergo an addition reaction to a ketonic group of a second substrate.

One system in which pantetheine almost certainly performs such a transport role is the fatty acid synthase multienzyme complex from *E. coli* in which 4'-phosphopantetheine is a component of the acyl carrier protein (see also Chapter 13). The complex consists of the acyl carrier protein (ACP) surrounded by the other six enzymes (Figure 6-15). The process of fatty acid chain elongation is represented by Equations (25) through (32).

$$CH_3-\overset{O}{\overset{\|}{C}}-SCoA + ACP-SH \xrightleftharpoons{\text{ACP-acyltransferase}} CH_3-\overset{O}{\overset{\|}{C}}-SACP + CoA \tag{25}$$

$$\beta\text{-ketoacyl-ACP synthase} \overset{|}{\underset{SH}{}} + CH_3-\overset{O}{\overset{\|}{C}}-SACP \rightleftharpoons \beta\text{-ketoacyl-ACP synthase} \underset{O=\overset{|}{\overset{S}{C}}-CH_3}{} + ACP-SH \tag{26}$$

$$CH_3-\overset{O}{\overset{\|}{C}}-SCoA + HOCO_2^- + ATP \xrightarrow{\text{Acetyl-CoA carboxylase}} \overset{CO_2^-}{\underset{}{CH_2}}-\overset{O}{\overset{\|}{C}}-SCoA + ADP + P_i \tag{27}$$

$$\overset{CO_2^-}{\underset{}{CH_2}}-\overset{O}{\overset{\|}{C}}-SCoA + ACP-SH \xrightarrow{\text{ACP-acyltransferase}} \overset{CO_2^-}{\underset{}{CH_2}}-\overset{O}{\overset{\|}{C}}-SACP + CoA \tag{28}$$

$$\beta\text{-ketoacyl-ACP synthase} \underset{O=\overset{|}{\overset{S}{C}}-CH_3}{} + ACPS-\overset{}{\underset{O}{\overset{\|}{C}}}-CH_2-CO_2^- \xrightarrow[\text{transferase}]{\text{ACP-malonyl}}$$

$$\beta\text{-ketoacyl-ACP synthase} \overset{|}{\underset{SH}{}} + ACP-S-\overset{O}{\overset{\|}{C}}-CH_2-\overset{O}{\overset{\|}{C}}-CH_3 + CO_2 \tag{29}$$

$$ACP-S-\overset{O}{\overset{\|}{C}}-CH_2-\overset{O}{\overset{\|}{C}}-CH_3 + NADPH + H^+ \xrightarrow[\text{reductase}]{\beta\text{-Ketoacyl-ACP}} ACP-S-\overset{O}{\overset{\|}{C}}-CH_2-\overset{OH}{\underset{}{CH}}-CH_3 + NADP^+ \tag{30}$$

$$ACP-S-\overset{O}{\overset{\|}{C}}-CH_2-\overset{OH}{\underset{}{CH}}-CH_3 \xrightarrow{\text{Enoyl-ACP hydrase}} ACP-S-\overset{O}{\overset{\|}{C}}-CH=CH-CH_3 + H_2O \tag{31}$$

$$ACP-S-\overset{O}{\overset{\|}{C}}-CH=CH-CH_3 + NADPH + H^+ \xrightarrow[\text{reductase}]{\text{Enoyl-ACP}} ACP-S-\overset{O}{\overset{\|}{C}}-CH_2-CH_2-CH_3 + NADP^+ \tag{32}$$

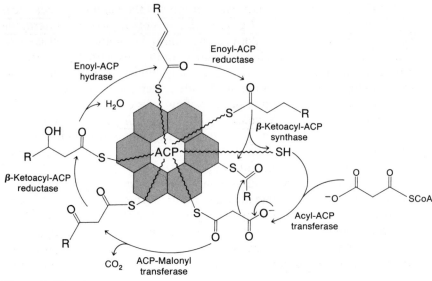

**Figure 6-15**
Interactions of 4'-phosphopantetheine in ACP with enzymes of *E. coli* fatty acid synthase. The central hexagon represents ACP and the shaded hexagons represent the other enzymes. The single 4'-phosphopantetheine in ACP is illustrated in its various orientations and acylated states interacting with the peripheral subunits.

This sequence involves the interaction of the 4'-phosphopantetheinyl moiety of ACP with each of the six enzymes in the complex, either as an acyl-group donor/acceptor or as a carrier of acyl groups to the active sites of the enzymes that catalyze the reduction of β-ketoacyl-ACP to fatty acyl-ACP. This involves the physical transport of acyl groups from one active site to another. As depicted in Figure 6-15, this role is played by the phosphopantetheine coenzyme. 4'-Phosphopantetheine is structurally suited to this role because it consists of a 2.0-nm (20-Å) chain of atoms with torsional freedom about at least nine single bonds. This enables it to sweep out a large volume of space, encompassing the active sites of all six enzymes.

## α-LIPOIC ACID

α-Lipoic acid is the internal disulfide of 6,8-dithioctanoic acid, whose structural formula is given in Figure 6-16. It is the coupler of electron and group transfers catalyzed by α-keto acid dehydrogenase multienzyme complexes. The pyruvate and α-ketoglutarate dehydrogenase complexes are centrally involved in the metabolism of carbohydrates by the glycolytic pathway (Chapter 8) and the tricarboxylic acid cycle (Chapter 9). They catalyze two of the three decarboxylation steps in the complete oxidation of glucose, and they produce NADH and activated acyl compounds from the oxidation of the resulting aldehydes:

$$\text{R}-\overset{\text{O}}{\underset{}{\text{C}}}-\text{CO}_2^- + \text{NAD}^+ + \text{CoASH} \longrightarrow \text{CO}_2 + \text{NADH} + \text{R}-\overset{\text{O}}{\underset{}{\text{C}}}-\text{SCoA} \qquad (33)$$

*The chemical aspect of the coenzymatic action of α-lipoic acid is to mediate the transfer of electrons and activated acyl groups resulting from*

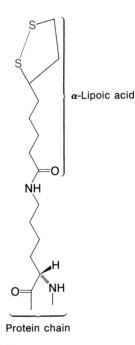

**Figure 6-16**
Structure of the α-lipoyl enzyme, showing the reactive disulfide in color.

*the decarboxylation and oxidation of α-keto acids within the complexes.
In this process, lipoic acid is itself transiently reduced to dihydrolipoic
acid, and this reduced form is the acceptor of the activated acyl groups.
This dual role of electron and acyl-group acceptor enables lipoic acid to
couple the two processes.*

The interactions of α-lipoic acid in the E. *coli* pyruvate dehydrogenase complex exemplify its coenzymatic functions. The complex consists of three proteins: a pyruvate dehydrogenase, which is thiamine pyrophosphate-dependent and designated $E_1$-TPP; dihydrolipoyl transacetylase, designated $E_2$-lipoyl-$S_2$, which contains α-lipoic acid covalently bonded through amide linkage with the ε amino group of a lysine residue (see Figure 6-17); and dihydrolipoyl dehydrogenase, a flavoprotein designated $E_3 \cdot$ FAD. A single particle of the complex consists of at least 24 chains of each of the first two enzymes and 12 of the flavoprotein. The core is composed of 24 subunits of $E_2$-lipoyl-$S_2$ arranged in cubic symmetry and containing 48 α-lipoyl groups, and associated with this core are 12 dimers of $E_1 \cdot$ TPP and 6 dimers of $E_3 \cdot$ FAD. Equations (34) through (38) represent the reaction sequence by which the complex catalyzes the reaction of pyruvate with $NAD^+$ and CoA.

$$H^+ + CH_3COCO_2^- + E_1 \cdot TPP \longrightarrow CO_2 + E_1 \cdot \text{hydroxyethyl-TPP} \quad (34)$$

$$E_1 \cdot \text{hydroxyethyl-TPP} + E_2\text{-lipoyl-}S_2 \longrightarrow$$
$$E_1 \cdot TPP + E_2\text{-lipoyl(SH)}-S-\overset{\overset{\displaystyle O}{\|}}{C}-CH_3 \quad (35)$$

$$E_2\text{-lipoyl(SH)}-S-\overset{\overset{\displaystyle O}{\|}}{C}-CH_3 + CoA \rightleftharpoons$$
$$E_2\text{-lipoyl(SH)}_2 + CH_3-\overset{\overset{\displaystyle O}{\|}}{C}-SCoA \quad (36)$$

$$E_2\text{-lipoyl(SH)}_2 + E_3 \cdot FAD \rightleftharpoons E_2\text{-lipoyl-}S_2 + \text{dihydro-}E_3 \cdot FAD \quad (37)$$

$$\text{Dihydro-}E_3 \cdot FAD + NAD^+ \rightleftharpoons E_3 \cdot FAD + NADH + H^+ \quad (38)$$

These equations show how α-lipoic acid couples the electron and group transfer processes involved. Note further, however, that lipoic acid must interact at active sites on all three enzymes. α-Lipoic acid bonded to $E_2$ is as shown in Figure 6-17, bonded through an amide linkage to a lysyl-ε-$NH_2$ group that places the reactive disulfide at the end of a flexible chain of atoms with rotational freedom about as many as 10 single bonds. When fully extended, this chain is 1.4 nm long, giving α-lipoic acid the potential capacity to sweep out a space having a spherical diameter of 2.8 nm. This distance turns out to be inadequate to account fully for the transport of electrons in this complex because the average distance between TPP on $E_1$ and FAD on $E_3$ has been estimated at between 4.5 and 6.0 nm by fluorescence energy-transfer measurements. The problem of long-distance interactions is overcome in the complex by the fact that each $E_2$ subunit contains two lipoyl groups that interact with each other and with α-lipoyl groups on other subunits according to Equation (39). This interaction facilitates the transport of electrons and acetyl groups through a network of α-lipoyl

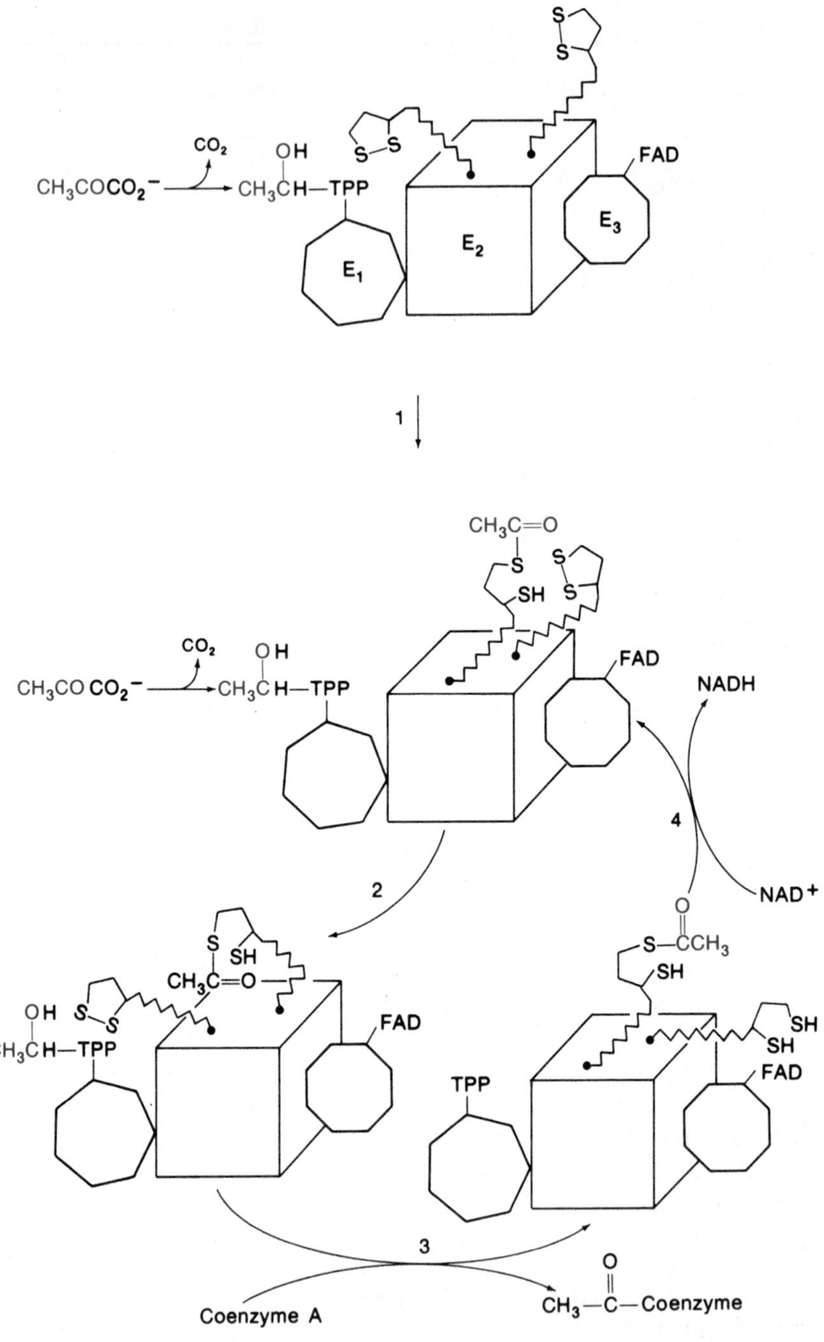

**Figure 6-17**
Interactions of $\alpha$-lipoyl groups in the pyruvate dehydrogenase complex. The
cubic structure represents the 24 subunits of $E_2$, dihydrolipoyl transacetylase,
which constitutes the core of the complex. Two of the 48 lipoyl groups in the
core are shown interacting with one of the 24 pyruvate dehydrogenases ($E_1 \cdot$TPP)
and one of the 12 dihydrolipoyl dehydrogenase ($E_3 \cdot$FAD) subunits. Note the
interaction of the lipoyl groups in relaying electrons over the long distance
between TPP and FAD.

groups encompassing the entire core of $E_2$. *By relaying electrons and acetyl groups through two or more α-lipoyl groups, the coenzyme is able to span the distances among sites.* The relay process is illustrated schematically in Figure 6-17, which shows how two or more S-acetyldihydrolipoyl groups can interact to transport electrons over the large distances separating the sites for TPP on pyruvate dehydrogenase and FAD on dihydrolipoyl dehydrogenase.

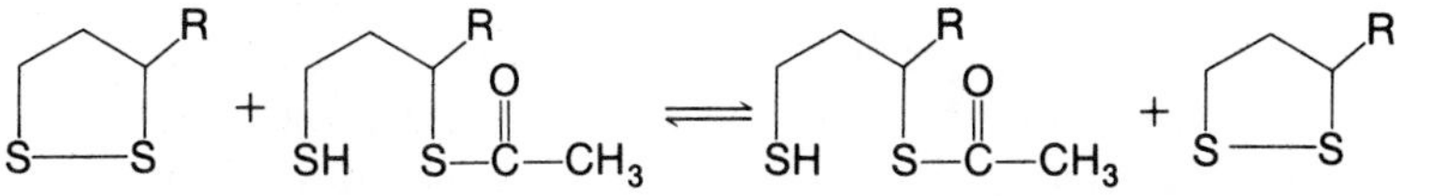

$$(39)$$

The coenzymatic capabilities of α-lipoyl groups result from a fusion of its chemical and physical properties, the ability to act simultaneously as both electron and acyl-group acceptor, the ability to span long distances to interact with sites separated by up to 2.8 nm, and the ability to act cooperatively with other α-lipoyl groups by disulfide interchange to relay electrons and acyl groups through distances that exceed its reach.

## BIOTIN

The biotin structure shown in Figure 6-18 is an imidazolone ring that is cis fused to a tetrahydrothiophene ring substituted at position 2 by valeric acid. In carboxylase enzymes, biotin is covalently bonded to the proteins by an amide linkage between its carboxyl group and a lysyl-ε-$NH_2$ group in the polypeptide chain. This places the imidazolone ring at the end of a long, flexible chain of atoms extending a maximum of about 1.4 nm from the α carbon of lysine.

Biotin is the essential coenzyme for carboxylation reactions involving bicarbonate as the carboxylating agent. Five such reactions have been described in which ATP-dependent carboxylation occurs at carbon atoms activated for enolization by ketonic or activated acyl groups, while in one other a nitrogen atom of urea is carboxylated. One other reaction, the transcarboxylase reaction of oxalacetate with propionyl-CoA to produce pyruvate and methylmalonyl-CoA, differs from the others in that it does not require ATP.

A general formulation of the ATP-dependent carboxylation of an α carbon by oxygen-18–enriched bicarbonate is given by Equation (40).

$$RCH_2-\overset{O}{\overset{\|}{C}}-SCoA + ATP + HC^{18}O_3^- \xrightarrow{\text{Biotinyl carboxylase}}$$

$$R-\underset{\underset{C^{18}O_2}{|}}{CH}-\overset{O}{\overset{\|}{C}}-SCoA + ADP + HP^{18}O_4^{2-} \quad (40)$$

The appearance of $^{18}O$ in inorganic phosphate verifies that the function of ATP in the reaction is essentially the "dehydration" of bicarbonate. This also explains the lack of an ATP requirement by the transcarboxylase, since in that case the carboxyl-group donor is a carboxylic acid in which the

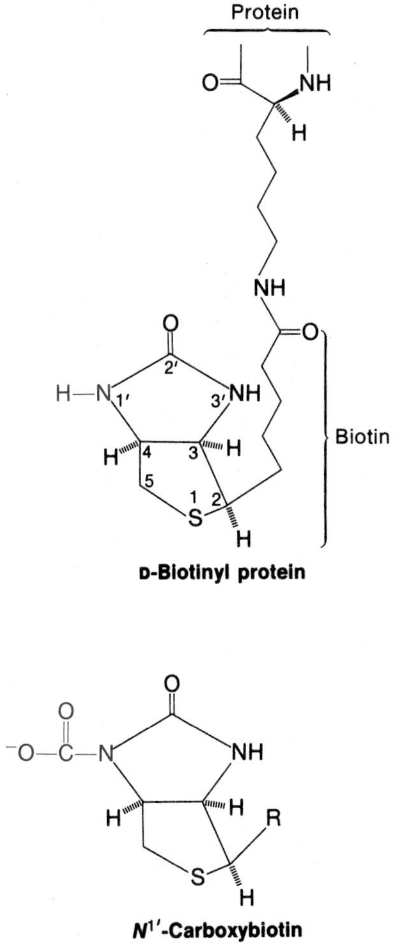

**Figure 6-18**
Structures of biotinyl enzyme and $N^1$-carboxybiotin. The reactive portions of the coenzyme and the active intermediate are shown in color.

carboxyl group is not hydrated as it is in bicarbonate. Thus the transfer of the carboxyl group from methylmalonyl-CoA to C-3 of pyruvate in Equation (41) is nearly isoenergetic and does not require ATP.

$$CH_3-\overset{\overset{\displaystyle CO_2^-}{|}}{CH}-\overset{\overset{\displaystyle O}{\|}}{C}-SCoA + CH_3-\overset{\overset{\displaystyle O}{\|}}{C}-CO_2^- \rightleftharpoons$$

$$CH_3-CH_2-\overset{\overset{\displaystyle O}{\|}}{C}-SCoA + {}^-O_2C-CH_2-\overset{\overset{\displaystyle O}{\|}}{C}-CO_2^- \quad (41)$$

Biotin-dependent carboxylation reactions proceed in two stages, the carboxylation of imidazolone-$N^1$ in biotin to form $N^1$-carboxybiotin followed by the carboxylation of the substrate by $N^1$-carboxybiotin. These processes can be separated and studied as individual steps defined by Equations (42) and (43) for acetyl-CoA carboxylase.

$$ATP + HOCO_2^- + \text{biotinyl-E} \rightleftharpoons ADP + P_i + N^1\text{-carboxybiotinyl-E} \quad (42)$$

$$N^1\text{-carboxybiotinyl-E} + CH_3-\overset{\overset{\displaystyle O}{\|}}{C}-SCoA \rightleftharpoons$$

$$\text{biotinyl-E} + {}^-O_2C-CH_2-\overset{\overset{\displaystyle O}{\|}}{C}-SCoA \quad (43)$$

The first step is the ATP-dependent carboxylation of biotin by bicarbonate. This is believed to involve the transient formation of carbonic-phosphoric anhydride, or "carboxyphosphate," as an active carboxylation intermediate according to Equation (44):

$$HO-\overset{\overset{\displaystyle O}{\|}}{C}-O^- \xrightarrow[ADP]{ATP} {}^-O-\overset{\overset{\displaystyle O}{\|}}{\underset{\underset{\displaystyle OH}{|}}{P}}-O-\overset{\overset{\displaystyle O}{\|}}{C}-O^- \xrightarrow[P_i]{\text{Biotinyl-E}} N^1\text{-carboxybiotinyl-E} \quad (44)$$

$N^1$-Carboxybiotinyl enzymes have been isolated using [$^{14}$C]bicarbonate as the carboxylation substrate, and the $^{14}$C has been shown to be bonded to $N^1$ of biotin. The [$^{14}$C]$N^1$-carboxybiotinyl enzymes have been shown to transfer their [$^{14}$C]carboxyl groups to the appropriate substrates, forming the corresponding [$^{14}$C]carboxylation products. The identification of $N^1$ as the carboxylation site in biotin was first achieved by F. Lynen and associates, who found that $\beta$-methylcrotonyl-CoA carboxylase would catalyze the ATP-dependent carboxylation of *free* biotin by [$^{14}$C]bicarbonate. The product of this reaction was methylated with diazomethane to stabilize the [$^{14}$C]carboxyl group, and the methylation product was identified as the methyl ester of $N^1$-[$^{14}$C]carboxybiotin by comparison with an authentic synthetic sample. [$^{14}$C]Carboxybiotinyl enzymes have subsequently been degraded to $N^1$ carboxylated products. Synthetic $N^1$-carboxybiotin derivatives have since been shown to transfer carboxyl groups to acceptor substrates in the presence of the appropriate enzymes.

On the evidence of the foregoing experiments, _the coenzymatic function of biotin appears to be to mediate the carboxylation of substrates by accepting the ATP-activated carboxyl group and transferring it to the carboxyl acceptor substrate._ There is good reason to believe that the enzymatic sites of ATP-dependent carboxylation of biotin are physically separated from the sites at which $N^1$-carboxybiotin transfers the carboxyl group to acceptor substrates, i.e., the transcarboxylase sites. In fact, in the case of the acetyl-CoA carboxylase from *E. coli*, these two sites reside on two different subunits, while the biotinyl group is bonded to a third, a small subunit designated biotin carboxyl carrier protein. Transcarboxylase is also a multisubunit protein, one subunit being a small biotinyl protein.

*Biotin appears to have just the right chemical and structural properties to mediate carboxylation. It readily accepts activated carboxyl groups at $N^1$ and maintains them in an acceptably stable yet reactive form for transfer to acceptor substrates. Since biotin is bonded to a lysyl group, the $N^1$-carboxyl group is at the end of a 1.6-nm chain with bond rotational freedom about nine single bonds, giving it the capability to transport activated carboxyl groups through space from the carboxyl activation sites to the carboxylation sites.*

## FOLATE COENZYMES

Tetrahydrofolic acid and its derivatives $N^5,N^{10}$-methylenetetrahydrofolate, $N^5,N^{10}$-methenyltetrahydrofolate, $N^{10}$-formyltetrahydrofolate, and $N^5$-methyltetrahydrofolate are the biologically active forms of folic acid, a four-electron-oxidized form of tetrahydrofolate. The structural formulas are given in Figure 6-19, which also shows how they arise from tetrahydrofolate. The structures are shown polyglutamylated on the carboxyl group of the *p*-aminobenzoyl group, these being the most active forms. The functions and lengths of the polyglutamyl chains are not known, although the triglutamates appear to be good substrates.

The tetrahydrofolates are the only coenzymes discussed in this chapter which are not known to function as tightly enzyme-bound coenzymes. They are specialized cosubstrates for a variety of enzymes involved in one-carbon metabolism. $N^{10}$-formyltetrahydrofolate and $N^5,N^{10}$-methenyltetrahydrofolate are formyl-group donor substrates for transformylases. In living cells, $N^{10}$-formyltetrahydrofolate is produced enzymatically from tetrahydrofolate and formate in an ATP-linked process in which formate is activated by phosphorylation to formyl phosphate; the formyl group of formyl phosphate is then transferred to $N^{10}$ of tetrahydrofolate. $N^{10}$-formyltetrahydrofolate is a formyl donor substrate for some enzymes and is interconvertible with $N^5,N^{10}$-methenyltetrahydrofolate by the action of cyclohydrolase. The methenyl derivative is also a formyl-donor substrate for other transformylases. $N^{10}$-formyltetrahydrofolate and $N^5,N^{10}$-methenyltetrahydrofolate also can be synthesized in nonenzymatic reactions of tetrahydrofolate with free formic acid.

$N^5,N^{10}$-Methylenetetrahydrofolate is a hydroxymethyl-group donor substrate for several enzymes and a methyl-group donor substrate for thymidylate synthase (see Figure 6-20). It arises in living cells from the reduction of $N^5,N^{10}$-methenyltetrahydrofolate by NADPH and also by the serine *trans*-hydroxymethylase-catalyzed reaction of serine with tetrahydrofolate. It also can be synthesized nonenzymatically by direct reaction of tetrahydrofolate with formaldehyde.

$N^5$-Methyltetrahydrofolate is the methyl-group donor substrate for methionine synthase, which catalyzes the transfer of the $N^5$-methyl group

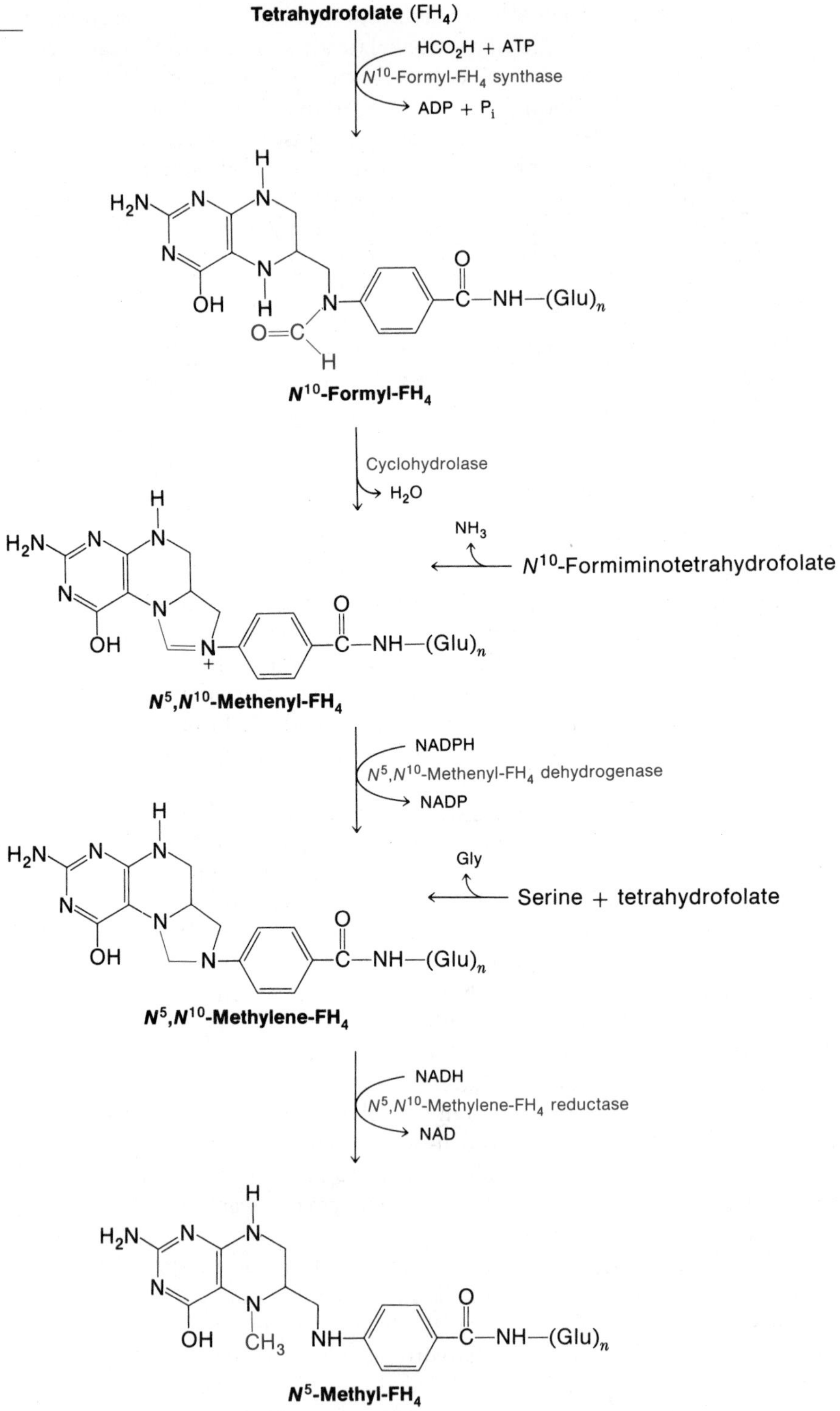

**Figure 6-19**
Structures and enzymatic interconversions of folate coenzymes. The reactive centers of the coenzymes are shown in color.

to the sulfhydryl group of homocysteine. This and selected reactions of the other folate derivatives are outlined in Figure 6-20, which emphasizes the important role tetrahydrofolate plays in nucleic acid biosynthesis by serving as the immediate source of one-carbon units in purine and pyrimidine biosynthesis.

Note that in the thymidylate synthase reaction, $N^5,N^{10}$-methylenetetrahydrofolate is the methyl-group donor to dUMP. The source of reducing equivalents to reduce the methylene group to the methyl level is tetrahydrofolate itself, so that the folate product of the reaction is dihydrofolate. The mechanism of the thymidylate synthase reaction is outlined in Figure 6-21.

The mechanism of action by thymidylate synthase exemplifies the reactions of $N^5,N^{10}$-methylenetetrahydrofolate. Figure 6-21 condenses the mechanism into four steps mainly for brevity but also because it is not known in precisely what sequence some of the steps occur. The reaction begins with the nucleophilic addition of an enzymatic sulfhydryl group to C-6 of the uracil ring, forming an enolate species that is potentially carbanionic at C-5. Meanwhile $N^5,N^{10}$-methylenetetrahydrofolate, upon protonation, isomerizes to a positively charged imine, which reacts with the enolate of uracil to produce a methylene-bridged intermediate. The bridged intermediate eliminates the enzymatic sulfhydryl group with concomitant loss of the C-5 proton, and either simultaneously or in a separate step, the bridging methylene group is reduced to a methyl group by the tetrahydropyrazine ring by a mechanism involving the conservation of the C-7 hydrogen, which ultimately appears as one of the three hydrogens in the methyl group.

The reduction of dihydrofolate produced by thymidylate synthase back to tetrahydrofolate by dihydrofolate reductase and NADH is an exceedingly important reaction because it maintains folate in the tetrahydro form and thereby facilitates the maintenance of all the tetrahydrofolates in

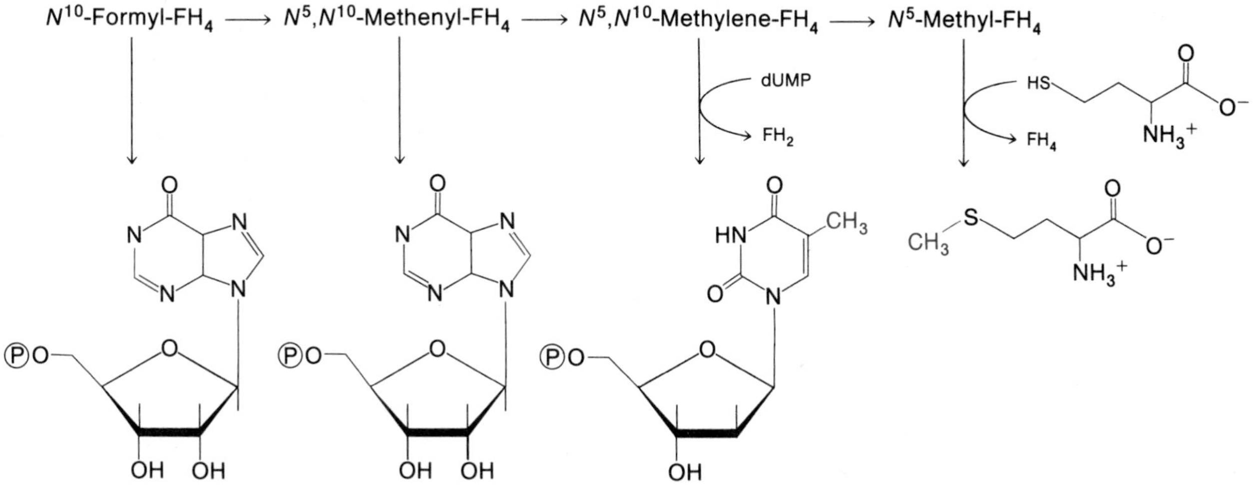

**Figure 6-20**
Involvement of folate coenzymes in one-carbon metabolism. Shown in color are the one-carbon units of the end products that originate with the reactive one-carbon units of the folate coenzymes.

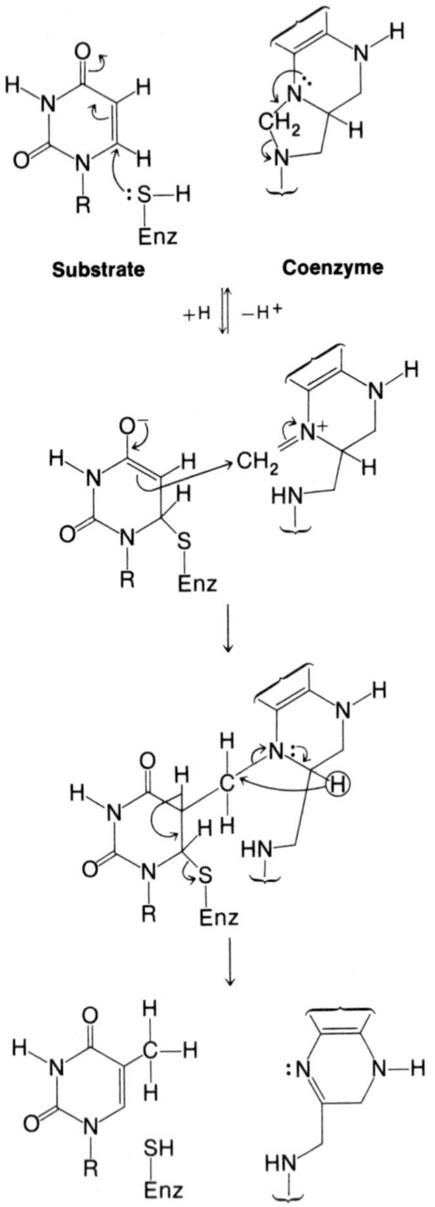

**Figure 6-21**
The mechanism of the thymidylate synthase reaction. The mechanism is discussed in the text. Note that in the reaction of 5-fluoro-UMP the decomposition of the last intermediate to the product is blocked by the presence of F in place of the H normally eliminated as a proton from uracil-C-5. 5-Fluoro-UMP inhibits thymidylate synthase by blocking the active site in this way.

Figure 6-19 at required levels. The importance of the tetrahydrofolates to cellular proliferation is highlighted by the fact that a potent inhibitor of dihydrofolate reductase, methotrexate (Figure 6-22), is one of the most effective and widely used drugs for cancer chemotherapy (see Chapter 19). 5-Fluorouracil (Figure 6-22) is also widely used in cancer chemotherapy, and it has been found that 5-fluoro-2'-dUMP is an exceedingly potent inhibitor of thymidylate synthase. 5-Fluoro-2'-dUMP reacts as if it were a substrate for thymidylate synthase through the first two steps of the mechanism outlined in Figure 6-21. However, the methylene-bridged intermediate cannot undergo the next step of the reaction, since it contains fluorine at C-5 instead of hydrogen and the next step involves the loss of the C-5 hydrogen as $H^+$. Fluorine *cannot* be eliminated as $F^+$. Thus the reaction is frozen at that step with the active site of the enzyme blocked.

Formaldehyde is a toxic substance that reacts spontaneously with amino groups of proteins and nucleic acids, hydroxymethylating them and forming methylene-bridged cross-links between them. Free formaldehyde, therefore, wreaks havoc in living cells and could not serve as a useful hydroxymethylating agent. In the form of $N^5,N^{10}$-methylenetetrahydrofolate, however, its chemical reactivity is attenuated but retained in a potentially available form where needed by specific enzymatic action. Formate, however, is quite unreactive under physiological conditions and must be activated to serve as an efficient formylating agent. As $N^{10}$-formyltetrahydrofolate and $N^5,N^{10}$-methenyltetrahydrofolate it is in a reactive state suitable for transfer to appropriate substrates. *The fundamental biochemical importance of tetrahydrofolate is to maintain formaldehyde and formate in chemically poised states, not so reactive as to pose toxic threats to the cell but available for essential processes by specific enzymatic action.*

## VITAMIN B$_{12}$ COENZYMES

The principal coenzymatic form of vitamin $B_{12}$ is 5'-deoxyadenosylcobalamin, whose structural formula is given in Figure 6-23. The structure involves a cobalt-carbon bond between the 5' carbon of the 5'-deoxyadenosyl moiety and the cobalt atom of cobalamin. Vitamin $B_{12}$ itself is cyanoco-

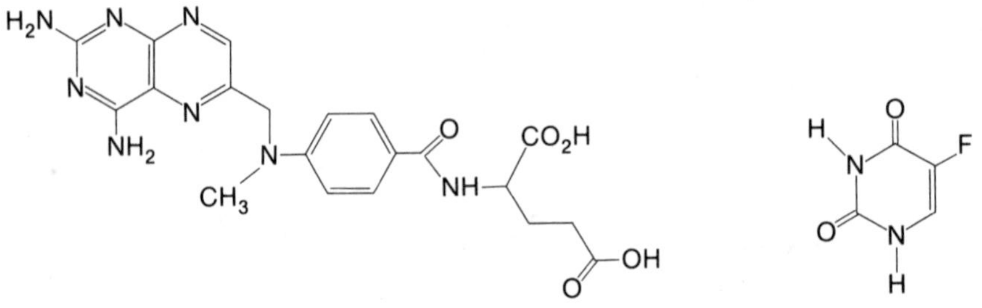

**Methotrexate**

**5-Fluorouracil**

**Figure 6-22**
Structures of methotrexate and 5-fluorouracil.

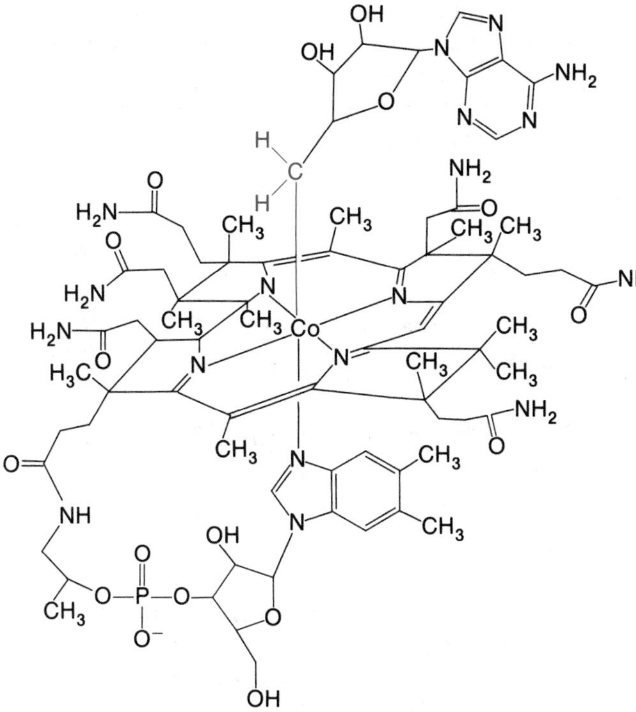

**Figure 6-23**
Structure of 5'-deoxyadenosylcobala-
min, vitamin $B_{12}$ coenzyme. The reactive groups are shown in color.

balamin in which the cyano group is bonded to cobalt in place of the
5'-deoxyadenosyl moiety. Other forms of the vitamin have water
(aquocobalamin) or the hydroxyl group (hydroxycobalamin) bonded to
cobalt.

The vitamin was discovered in liver as the antipernicious anemia factor in 1926, but the determination of its complete structure had to await its
purification, chemical characterization, and crystallization, which required
more than 20 years. Even then the determination of such a complex structure proved to be an elusive goal by conventional approaches of that day
and had to await the elegant x-ray crystallographic study of Lenhert and
Hodgkin in 1961, for which D. Hodgkin was awarded the Nobel Prize in
1964.

5'-Deoxyadenosylcobalamin is the first substance to be discovered to
contain a stable cobalt-carbon bond. The coenzyme was discovered by
H. A. Barker and coworkers as the activating factor for glutamate mutase in
*Clostridium*. The direct bonding between cobalt and the 5'-deoxyadenosyl
group was established by x-ray crystallography.

*All but one of the 5'-deoxyadenosylcobalamin–dependent enzymatic*
*reactions are rearrangements that follow the pattern of Equation (45),* in
which a hydrogen atom and another group (designated X) bonded to an
adjacent carbon atom exchange positions, with the group X migrating from
$C_\alpha$ to $C_\beta$.

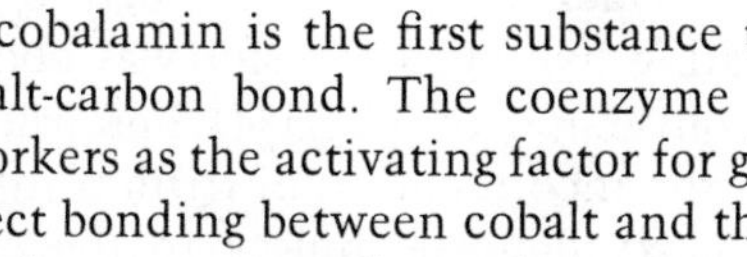

$$a \overset{b}{\underset{\underset{X}{|}\alpha}{-}}C \overset{c}{\underset{\underset{H}{|}\beta}{-}}C - d \rightleftharpoons a \overset{b}{\underset{\underset{H}{|}\alpha}{-}}C \overset{c}{\underset{\underset{X}{|}\beta}{-}}C - d \qquad (45)$$

Three specific examples of rearrangement reactions are given in Equations (46) through (48). It is interesting and significant that the migrating groups —OH, —COSCoA, and —CH(NH$_2$)CO$_2$ have little in common and that the hydrogen atoms migrating in the opposite direction are often chemically unreactive.

$$^-O_2C-CH_2CH_2-\underset{\underset{NH_3^+}{|}}{CH}-CO_2^- \;\underset{\text{mutase}}{\overset{\text{Glutamate}}{\rightleftharpoons}}\; {}^-O_2C-\underset{\underset{NH_3^+}{|}}{\overset{\overset{CH_3}{|}}{CH}}-CH-CO_2^- \quad (46)$$

$$^-O_2C-CH_2CH_2-\overset{O}{\overset{\|}{C}}-SCoA \;\underset{\text{mutase}}{\overset{\text{Methylmalonyl-CoA}}{\rightleftharpoons}}\; {}^-O_2C-\overset{\overset{CH_3}{|}}{CH}-\overset{O}{\overset{\|}{C}}-SCoA \quad (47)$$

$$CH_3\underset{\underset{OH}{|}}{CH}CH_2OH \;\xrightarrow{\text{Dioldehydrase}}\; CH_3CH_2\underset{\underset{OH}{|}}{CH}-OH \;\xrightarrow{\text{Dioldehydrase}}$$

$$CH_3CH_2CHO + H_2O \quad (48)$$

The hydrogen migrations in all the B$_{12}$ coenzyme-dependent rearrangements proceed without exchange with the protons of water; i.e., isotopic hydrogen in the substrates is conserved in the products even though, as discussed later, these migrations are not simply intramolecular 1,2 shifts. The hydrogen migrations in Equations (46) through (48) are stereospecific, as are the migrations of the —X groups. The migrations of the —CHNH$_2$CO$_2$H and —OH groups catalyzed by glutamate mutase and dioldehydrase occur with overall inversion of configuration at the terminals of migration, while that of the —COSCoA group catalyzed by methylmalonyl-CoA mutase proceeds with retention of configuration. The significance of these differences is not understood in mechanistic terms, but any general concept of the mechanisms of these rearrangements must be consistent with the stereochemistry.

The work of R. H. Abeles and coworkers studying the mechanism of action of dioldehydrase shed the first light on the role of 5′-deoxyadenosylcobalamin in the rearrangement reactions. Moreover, their findings have been confirmed in other rearrangement reactions.

The first direct evidence of the role played by 5′-deoxyadenosylcobalamin in mediating hydrogen transfer was obtained in experiments showing that the conversion of [1-$^3$H$_2$]1,2-propanediol to [1,2-$^3$H]propionaldehyde led to the incorporation of tritium into *both* C-5′ hydrogen positions of the 5′-deoxyadenosyl moiety in the coenzyme. It also was shown that [5′-$^3$H]5′-deoxyadenosylcobalamin, either isolated from the foregoing enzymatic experiments or prepared by chemical synthesis, could transfer all of its tritium into propionaldehyde when used as the coenzyme with unlabeled 1,2-propanediol and the dehydrase. These results accounted for the finding that hydrogen transfer catalyzed by this enzyme is not compulsorily intramolecular. It was subsequently shown that hydrogen transfer in a given turnover could be either intermolecular or intramolecular and that the rates of tritium transfer into the coenzyme from substrate and from the coenzyme into product account for the rate of appearance of tritium in the product.

The involvement of adenosyl-C-5′ in hydrogen transfer and the fact that hydrogen transfer can be either intermolecular or intramolecular in a given turnover strongly imply that the cobalt-carbon bond in 5′-deoxyadenosylcobalamin is transiently cleaved and that the C-5′ carbon transiently becomes a methyl group in the catalytic process.

Cobalamins lacking the 5′-deoxyadenosyl moiety, i.e., the vitamin itself or hydroxycobalamin, are known to exist in three oxidation states: the Co(III) state, known as $B_{12a}$ or $B_{12b}$ for hydroxycobalamin or aquocobalamin; the one-electron reduced form or Co(II) state, known as $B_{12r}$; and the two-electron reduced Co(I) state, known as $B_{12s}$. The three forms are distinguishable by their visible and ultraviolet absorption spectra ($B_{12a}$ is a red compound, $B_{12r}$ is yellow, and $B_{12s}$ is described as grey-green in color) and by the fact that $B_{12r}$ is paramagnetic, exhibiting a strong electron-spin resonance signal.

Spectroscopic data have implicated the Co(II) form of cobalamin as a catalytic intermediate in several 5′-deoxyadenosylcobalamin–dependent reactions, including the dioldehydrase reaction. The visible absorption spectrum of the enzyme-coenzyme complex is very similar to that of 5′-deoxyadenosylcobalamin itself, but upon addition of substrate, it quickly shifts to a spectrum characteristic of $B_{12r}$ and then reverts almost to its initial state upon complete conversion of substrate to product. In similar experiments monitored by electron-spin resonance spectroscopy, a signal corresponding to that of $B_{12r}$ is observed upon adding substrate to the enzyme-coenzyme complex, and signals corresponding to other free-radical species are also detected. The rates of appearance of these signals are on the catalytic time scale, indicating that the $B_{12r}$ and other free-radical species are catalytic intermediates.

The mechanisms of the 5′-deoxyadenosylcobalamin–dependent rearrangements are not fully understood, but from the preceding hydrogen-transfer and spectroscopic experiments, the outline of a mechanism is emerging. This is given in Figure 6-24, in which the substrate and product are shown as the generalized forms of Equation (45) and the structure of the coenzyme is abbreviated to emphasize the importance of the cobalt-carbon bond.

The one known 5′-deoxyadenosylcobalamin–dependent enzyme that does not catalyze a rearrangement is the ribonucleotide reductase from *Lactobacillus leichmanii*, which catalyzes the reduction of ribonucleoside triphosphates to deoxynucleoside triphosphates (Equation 49).

$$\text{(49)}$$

The reducing agent may be a vicinal dithiol, such as dihydrolipoic acid, or a vicinal dithiol protein, such as thioredoxin (see Chapter 19). This reaction is fundamentally different from the rearrangements, the one point of similarity being that both involve hydrogen transfer. The reductase catalyzes a net reduction, however, and so hydrogen transfer is compulsorily intermolecular. In [$^3$H]$H_2O$, the enzyme catalyzes the incorporation of $^3$H

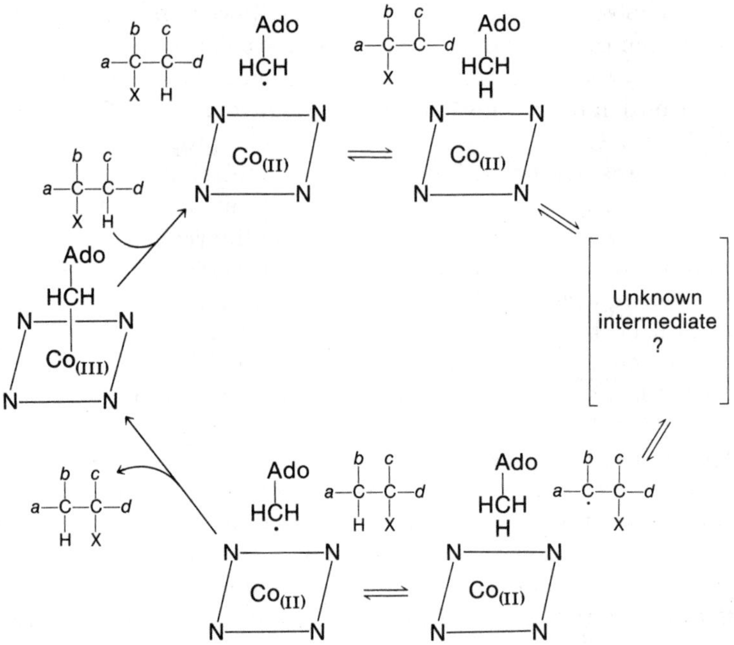

**Figure 6-24**
Hypothetical partial mechanism of vitamin $B_{12}$-dependent rearrangements. The structure of an intermediate illustrated by the bracketed question mark is presently unknown. The designations Co(III) and Co(II) refer to species that are spectrally and magnetically similar to $Co^{3+}$ and $Co^{2+}$, respectively. Co(III) is diamagnetic and red while Co(II) is paramagnetic (unpaired electron) and yellow. The metal does not undergo a change in electrostatic charge when the cobalt-carbon bond breaks homolytically (i.e., without charge separation), since one electron remains with the metal and the other with 5'-deoxyadenosine. One or more unknown intermediates symbolized by the brackets may be involved in the rearrangement.

into the C-5' position of the coenzyme in the presence of the vicinal dithiol, which is in exchange equilibrium with $[^3H]H_2O$ and probably the immediate source of $^3H$ for the coenzyme. 5'-Deoxyadenosine is also thought to be an intermediate in this reaction.

Another coenzymatic form of vitamin $B_{12}$ is methylcobalamin, which has a methyl group in place of the 5'-deoxyadenosyl moiety. Methylcobalamin arises as an intermediate in the methionine synthase reaction (Figure 6-20). This enzyme is a $B_{12}$ protein that must be maintained in its $B_{12s}$-reduced state to be active. The cobalamin mediates methyl transfer according to Equations (50) and (51), in which the nucleophilic Co(I) form of cobalamin accepts the methyl group from $N^5$-methyl-$FH_4$ and transfers it to homocysteine.

$$N^5\text{-Methyl-FH}_4 + B_{12s} \text{ enzyme} \longrightarrow FH_4 + \text{methyl-B}_{12} \text{ enzyme} \tag{50}$$

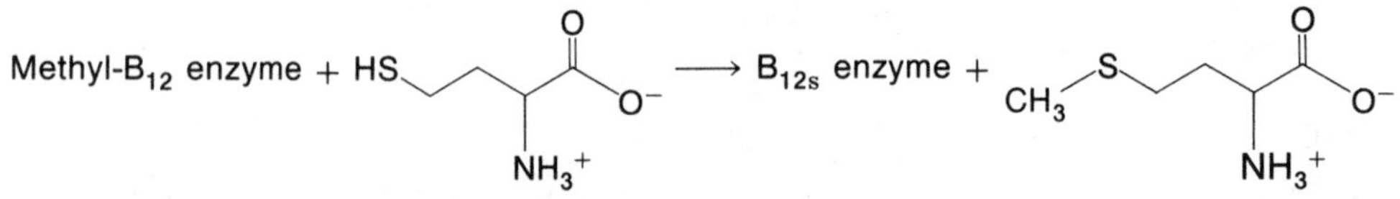

$$\tag{51}$$

Methylcobalamin is also thought to be an intermediate in the bacterial production of $CH_4$ from one- and two-carbon precursors.

## METAL IONS AS COENZYMES

Metal ions are involved in a wide variety of biochemical processes. An estimated one-third of all enzymes require a metal ion in one or more phases of the catalytic process. Metal ions control catalysis by binding the substrate directly to the active site or indirectly by maintaining the enzyme structure in a poised conformation to permit binding. Serving as essential structural components, metal ions participate in oxidative and hydrolytic reactions, in some cases undergoing reversible changes in oxidation state. Many metabolites, especially nucleotides, exist as metal complexes, such as Mg · ATP, and these complexes rather than the nucleotides themselves are the actual substrates for enzymatic reactions. Thus metal ions are capable of exerting their own catalytic effects by altering the chemical properties of the uncomplexed substrate. An important observation is that some enzyme-metal complexes have the potential for regulating levels of enzyme activity through specific control of rates of enzyme synthesis, degradation, or both of the protein moiety. This facet of enzyme-metal interaction is least understood, but it tends to underscore the importance of metal ions in enzyme expression and regulation.

## Metalloenzymes versus Metal-Activated Enzymes

Although many enzymes require metal ion cofactors, one can subdivide the list between what are called *metalloenzymes* and *metal-activated enzymes* on the basis of the strength of metal binding. Metalloenzymes generally engage stoichiometric amounts of the metal cofactor quite tightly and fail to show activity enhancement upon addition of the free metal ions. Metal-activated enzymes retain their metals in an equilibrium with binding groups on the surface. Often the metal ion is lost during purification and must be added back to restore the catalytic activity. The metal-activated enzymes, however, have a simple stoichiometric relationship among the enzyme site, the bound metal, and the substrate which is commonly 1:1:1. Various configurations of the active ternary complex are shown below. What is called a *substrate-bridge complex* (A) involves the binding of a substrate-metal complex to the enzyme. The *metal bridge* (B) positions the metal to engage both the substrate and the enzyme. This complex may include further contact of the substrate with the enzyme through a *nonmetal bond* (C). The *enzyme-bridge complex* (D) shows a contact of metal only with the enzyme. With the exception of the substrate bridge, these complexes also are seen with metalloenzymes. The substrate bridge, however, limits contact of the metal to only the substrate, which, of course, is not possible for metalloenzymes.

$$E\!-\!S\!-\!M \qquad E\!-\!M\!-\!S \qquad \text{or} \qquad E\!\underset{S}{\overset{M}{\diagdown\!|}} \qquad M\!-\!E\!-\!S$$

$$(a) \qquad\qquad (b) \qquad\qquad\qquad (c) \qquad\quad (d)$$

A survey of many enzymes reveals that the metals in the first transition series of elements of the periodic table, which includes Mn, Fe, Co, and Cu, comprise those found in most of the metalloenzymes. Zinc in its ionic form ($Zn^{2+}$) fits into the general class of metals in this series. The alkaline earth metals, such as Ca and Mg, are also very important cofactors in biological systems. However, rarely are these metals seen in metalloenzymes. Rather, they along with Na and K serve aptly as cofactors for metal-activated enzymes. There are exceptions, such as Ca bound tightly to the structure of α-amylase, and thermolysin. Some metalloenzymes have metal ion cofactors associated with the enzyme surface through a prosthetic group. The familiar heme complex associated with iron-containing enzymes and the less common corrin ring system for cobalt enzymes (methylmalonyl-CoA mutase, glycerol dehydratase, and ribonucleotide reductase) have already been mentioned. Prosthetic groups tend to minimize direct contact of the metal ion with sensitive binding groups on the protein, ensuring both a stronger coordination complex and, with redox metals, less opportunity for chemical modification of the protein structure.

### Iron-Containing Enzymes

Iron as a cofactor in catalysis is receiving increasing attention. The metal exists in two oxidation states: $Fe^{2+}$ and $Fe^{3+}$. Iron complexes are nearly all octahedral, and practically all are paramagnetic (caused by unpaired electrons in the $3d$ orbital). The most common form of iron in biological systems is heme. Heme groups ($Fe^{2+}$) and hematin ($Fe^{3+}$) most frequently involve a complex with protoporphyrin IX (Figure 6-25). They are the coenzymes (prosthetic groups) for a number of redox enzymes, including catalase, which catalyzes dismutation of hydrogen peroxide (Equation 52), and peroxidases, which catalyze the reduction of alkyl hydroperoxides by such reducing agents as phenols, hydroquinones, and dihydroascorbate [represented as $AH_2$ in Equation (53)].

$$2\,H_2O_2 \rightleftharpoons 2\,H_2O + O_2 \tag{52}$$

$$R\!-\!O\!-\!O\!-\!H + AH_2 \longrightarrow A + R\!-\!O\!-\!H + H_2O \tag{53}$$

Heme proteins exhibit characteristic visible absorption spectra as a result of protoporphyrin IX; their spectra differ depending on the identities of the lower axial ligand donated by the protein and the oxidation state of the iron as well as the identities of the upper axial ligands donated by the substrates. Spectral data show clearly that the heme coenzymes participate directly in catalysis; however, the mechanisms of action of hemes are not so well understood as those of other coenzymes.

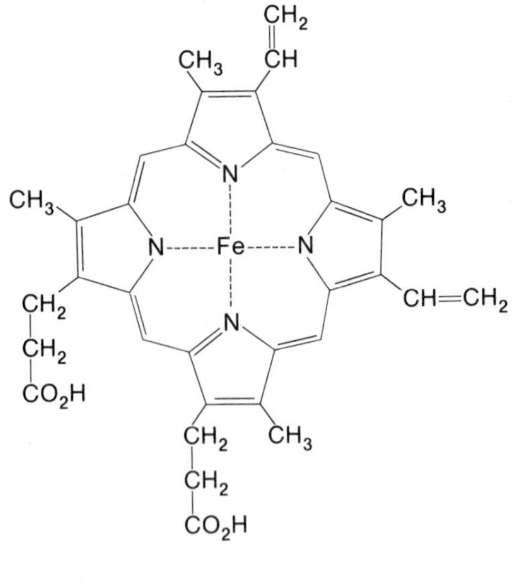

**Figure 6-25**
Structure of protoporphyrin IX.

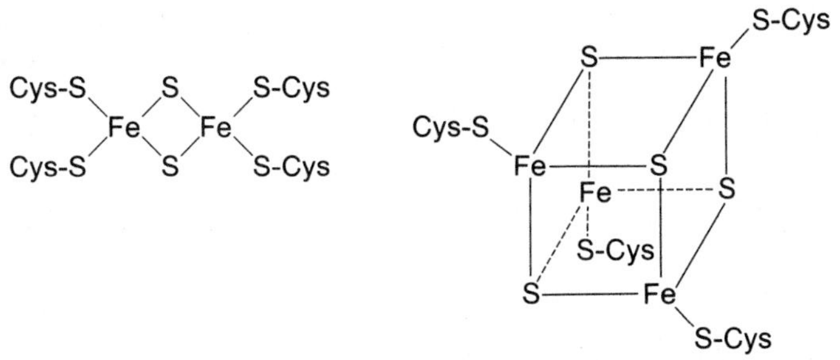

**Figure 6-26**
Structures of iron-sulfur clusters.

Many redox enzymes contain iron-sulfur clusters that mediate one-electron transfer reactions. The clusters consist of two or four irons and an equal number of inorganic sulfide ions clustered together with the iron, which is also liganded to cysteinyl-sulfhydryl groups of the protein (see Figure 6-26). The enzyme nitrogenase, which catalyzes the reduction of $N_2$ to $2 NH_3$, contains such clusters in which some of the iron has been replaced by molybdenum (Chapter 22). Electron-transferring proteins involved in one-electron transfer processes often contain iron-sulfur clusters. These proteins include the mitochondrial membrane enzymes NADH dehydrogenase and succinate dehydrogenase (Chapter 10), which are flavoproteins, and the small-molecular-weight proteins ferredoxin, rubredoxin, adrenodoxin, and putidaredoxin (Chapters 10, 11, and 15).

Heme coenzymes, iron-sulfur clusters, flavin coenzymes, and nicotinamide coenzymes cooperate in multienzyme systems to catalyze the chemically remarkable hydroxylations of hydrocarbons such as steroids (Chapter 15). In these hydroxylation systems, the heme proteins are known as cytochrome *P*-450, named for the wavelength corresponding to the most intense absorption band of the carbon monoxide-liganded heme, an inhibited form. The reactions catalyzed by these systems are represented in generalized form by Equation (54).

$$H^+ + R-CH_2-R' + O_2 + NADPH \longrightarrow R-\overset{\overset{\displaystyle OH}{|}}{C}H-R' + NADP^+ + H_2O \quad (54)$$

The enzymes involved usually include a cytochrome *P*-450, an iron-sulfur cluster–containing protein such as adrenodoxin or putidaredoxin, and a flavoprotein reductase. The detailed mechanisms by which these proteins cooperate in catalyzing hydroxylations are not understood. Cytochrome *P*-450 interacts directly with $O_2$ and receives electrons from the iron-sulfur protein, which is in turn reduced by NADPH by the action of the flavoprotein reductase. An oxygenating form of cytochrome *P*-450 is thereby produced, but its chemical nature and the mechanism by which it hydroxylates substrates are not known.

The cytochrome *P*-450 hydroxylases include those in adrenal cells which hydroxylate sterols and steroids in the production of steroid hormones, such as aldosterone and hydrocortisone (Chapter 15). The liver microsomal systems detoxify amines and polycyclic hydrocarbons by hydroxylating them in preparation for further transformations and eventual elimination. The hydroxylation of a secondary amine results in the dealkylation of the amine according to Equation (55). The immediate hydroxylation product, an $\alpha$-hydroxyamine, is an addition compound of a primary amine and an aldehyde, and it spontaneously dissociates.

$$H^+ + R{-}NH{-}CH_2{-}R' + O_2 + NADPH \longrightarrow$$

$$
\begin{array}{c}
OH \\
| \\
R{-}NH{-}CH{-}R' + NADP^+ + H_2O \quad (55) \\
\downarrow \\
R{-}NH_2 + R'{-}CHO
\end{array}
$$

Bacterial hydroxylation systems such as those found in *Pseudomonas* catalyze the hydroxylation of hydrocarbons, the first step in the oxidative degradation of hydrocarbons to produce energy for the organism. The hydroxylated hydrocarbons are further oxidized and degraded by more conventional oxidative processes involving nicotinamide and flavin coenzymes, the reduced forms of which are reoxidized by terminal electron acceptors in reactions coupled to the production of ATP.

**Zinc Metalloenzymes**

The activities of more than 80 enzymes require zinc. As a cofactor, zinc is the most versatile of metals. Existing entirely as $Zn^{2+}$ in ionic form, the metal, unlike copper, iron, or manganese, has no redox capabilities. Because zinc has a filled $3d$ shell (the two $4s$ electrons are lost upon ionization), zinc complexes are entirely without color and exhibit no special spectrophotometric properties. The most common coordination number for zinc is 4, and the metal prefers tetrahedral configurations. Frequently, zinc is found at the active site of enzymes serving as a bridge between the enzyme and substrate (see discussion of carboxypeptidase in Chapter 4). Both the binding groups on the substrate and those on the enzyme enter within the coordination sphere of the metal. This is illustrated in Figure 6-27, which depicts the mechanism of carbonic anhydrase, a zinc metalloenzyme. Note that three of the binding groups are on the enzyme, while the fourth is —OH. In the reaction, the metal is believed to convert coordinated $H_2O$ to the more potent nucleophile —OH, which attacks $CO_2$, resulting in $HCO_3^-$ formation.

A partial listing of well-characterized zinc metalloenzymes is given in Table 6-4. Carbonic anhydrase, the first to be discovered (Keilin and Mann, 1940), has a molecular weight of 30,000 (common to this enzyme from various sources) and contains one zinc bound to the single polypeptide chain. Alcohol dehydrogenase from yeast and from horse liver binds four zincs per molecule. The yeast enzyme, however, is a tetramer with a molecular weight of 140,000, while that of the horse liver, a dimer, is around 80,000. Alkaline phosphatase from *E. coli* shows zinc-dependent stability and catalytic activity. The enzyme's two subunits have a common genetic origin but develop heterogeneity after translation. The alkaline phosphatases from other bacteria, fungi, and higher animals also appear to be zinc metalloenzymes. The enzyme is not found in higher plants, nor is it present in erythrocytes and muscle cells. Aspartate carbamoyl transferase from *E. coli* binds six zincs per molecule (see Chapter 5). Both DNA polymerase and DNA-dependent RNA polymerase from *E. coli* are zinc enzymes. DNA

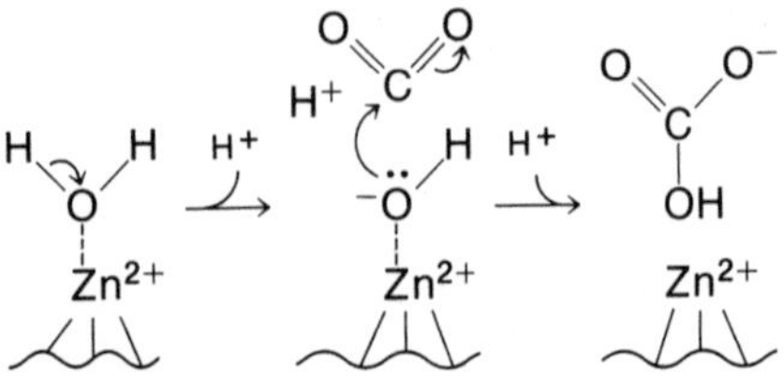

**Figure 6-27**
Proposed role of $Zn^{2+}$ in carbonic anhydrase. $Zn^{2+}$ forms a complex with water. This leads to proton displacement and a hydroxy ion. Nucleophilic attack of $CO_2$ by the hydroxy ion results in bicarbonate in formation.

**Table 6-4**
Zinc Metalloenzymes

| Enzyme | Source | | Zinc Atoms per Molecule |
|---|---|---|---|
| Alcohol dehydrogenase | Yeast | 150,000 | 4 |
| Alcohol dehydrogenase | Horse liver | 84,000 | 4 |
| Alkaline phosphatase | E. coli | 80,000 | 4 |
| Carbonic anhydrase | Erythrocytes | 30,000 | 1 |
| Carboxypeptidase A | Pancreas | 34,600 | 1 |
| Carboxypeptidase B | Pancreas | 34,300 | 1 |
| Glutamic dehydrogenase | Bovine liver | 1,000,000 | 2–6 |
| Leucine aminopeptidase | Porcine kidney | 300,000 | 4–6 |
| Neutral protease | B. subtilis | 44,700 | 1–2 |
| Thermolysin | B. thermoproteolyticus | 37,500 | 4 |

polymerase I contains 2 atoms of zinc. The DNA polymerase from rat liver is also a zinc enzyme. Zinc is also present in the $\beta$ subunit of DNA-dependent RNA polymerase from *Bacillus subtilus* and appears to be directly involved in phosphodiester bond formation. The reverse transcriptase induced by avian myeloblastosis virus has been shown to require zinc. The list is certainly not complete. For example, all bacterial neutral proteases characterized thus far have shown at least one zinc atom associated with the enzyme along with tightly bound calcium ions. Thermolysin is but one example of the enzymes in this group. Finally, recent studies have revealed a need for zinc in elongation factor I (EF-I) from rat liver (see Chapter 25). The highly purified factor contains 1 zinc per 54,000 daltons.

## Copper Metalloenzymes

Copper exists in two principal oxidation states: $Cu^+$ and $Cu^{2+}$. A third, $Cu^{3+}$, is postulated to be a fleeting intermediate. Depending on oxidation state, the metal shows either tetrahedral ($Cu^+$) or square planar ($Cu^{2+}$) geometry. The most common coordination number is 4. Cupric copper can form a distorted octahedral complex (four short and two long; what is referred to as the Jahn-Teller effect). Cupric copper is nearly always paramagnetic because of the unpaired electron in the $3d$ orbital. Exceptions arise when two $Cu^{2+}$ centers in the same enzyme are in close proximity and one center effectively cancels the "unpaired" electron effect of its neighbor.

The specificity of copper in catalysis is related to its multivalent oxidations states and the ease with which copper ions are reduced and reoxidized. Hence copper ions in combination easily accommodate electrons removed from the substrate and just as readily transfer them to a molecule of oxygen. Many copper enzymes are thus classified as hydroxylases and oxidases (Table 6-5), denoting their involvement with molecular oxygen in

**Table 6-5**
Copper Metalloenzymes

| Enzyme | Source | | Copper Atoms per Molecule |
|---|---|---|---|
| Ascorbic acid oxidase | Squash | 140,000 | 8 |
| Ceruloplasmin | Plasma | 134,000 | 6–7 |
| Cytochrome $c$ oxidase | Mitochondria | 340,000 | 1 |
| Diamine oxidase | Kidney | 185,000 | 2 |
| Dopamine-$\beta$-hydroxylase | Adrenal gland | 290,000 | 2 |
| D-Galactose oxidase | Fungi | 60,000 | 1 |
| Laccase | Lac tree | 120,000 | 6 |
| Lysyl oxidase | Chick aorta and cartilage | 60,000 | 1 |
| Superoxide dismutase | Liver | 34,000 | 2 |
| Uricase | Liver | 12,000 | 1 |

the catalytic process. Dopamine-$\beta$-hydroxylase, for example, uses $O_2$ to synthesize norepinephrine according to the following equation:

$$\text{L-Dopamine} + O_2 + \text{ascorbate} \longrightarrow$$
$$\text{norepinephrine} + H_2O + \text{dehydroascorbate} \quad (56)$$

The enzyme in the adrenal gland contains two copper atoms and in addition requires ascorbate for maximum activity. The amine oxidases are a broad category of copper enzymes (an exception is monoamine oxidase in mitochondria, which is a flavoprotein) that catalyze the oxidative deamination of primary and secondary amines to aldehydes according to Equation (57).

$$\text{R}-\text{CH}_2-\text{NH}_2 + O_2 + H_2O \longrightarrow \text{R}-\text{CHO} + NH_3 + H_2O_2 \quad (57)$$

Hydrogen peroxide and $NH_3$ are released in stoichiometric amounts. A connective-tissue amine oxidase known as lysyl oxidase catalyzes the posttranslational oxidation of proelastin and tropocollagen, converting peptidyl lysine residues to the corresponding aldehydes. These aldehyde groups so formed become the active centers for condensation reactions that lead to the formation of cross-linking groups that stabilize the extracellular matrixes of connective tissue. Other important copper enzymes include cytochrome $c$ oxidase, referred to as a terminal oxidase, in the electron-transport chain (Chapter 10). The enzyme is directly involved in generating the electrochemical potential, a proton gradient, across the inner membrane of the mitochondria. Superoxide dismutase, a copper-zinc enzyme in the cytosol of many cells, catalyzes dismutation of the oxygen radical *superoxide anion* according to Equation (58).

$$O_2^- + O_2^- + 2\,H^+ \longrightarrow H_2O_2 + \tfrac{1}{2}O_2 \quad (58)$$

The reaction prevents superoxide buildup, which would lead to the formation of potentially damaging radicals such as OH · and singlet oxygen. Ceruloplasmin (the name denotes both the celestial blue color and origin of this plasma enzyme) is a transporter of both copper and oxidase for important biogenic amines. The enzyme contains six or seven copper atoms bound to a single polypeptide chain, with a combined molecular weight of about 134,000 daltons. Both cupric and cuprous copper are present in the structure. Ceruloplasmin also has been called "ferroxidase," denoting its ability to catalyze oxidation of ferrous iron to the ferric form.

## Other Metalloenzymes and Metals

Zinc, copper, and iron constitute the metals found in most metalloenzymes. Pyruvate carboxylase from chick liver was the first manganese metalloenzyme to be found. The enzyme contains four subunits, four Mn ions, and four biotins. Interestingly, there are fewer manganometalloenzymes in nature than other metalloenzymes, zinc and copper, for example. One explanation is that because of the unfavorable distribution of bonding electrons, this transition metal lacks the ability to form tight complexes with ligands. Histidine ammonia-lyase requires Mn for activity. The reaction follows:

$$\text{(imidazole)}-CH_2-\overset{\overset{\displaystyle H}{|}}{\underset{\underset{\displaystyle NH_2}{|}}{C}}-COOH \longrightarrow \text{(imidazole)}-CH=CH-COOH + NH_3 \quad (59)$$

Mn may take part in the withdrawal of electrons at the imidazole ligand, facilitating elimination of ammonia and formation of the double bond. A related enzyme, phenylalanine ammonia-lyase, requires no metal, presumably because the substrate (phenylalanine) lacks a prominent side-chain ligand. Many metal-activated enzymes that use Mg have been shown to function with Mn as well. For example, both creatine kinase and pyruvate kinase use Mg · ATP as a substrate. Both enzymes will bind Mn in the presence of the nucleotide and effectively remove the free metal from solution. Arginase from rat liver, an enzyme in the urea cycle (Chapter 25), forms tight complexes with Mn. In arginase, the metal performs both stability and catalytic functions. Molybdenum is strongly complexed to the structure of xanthine oxidase. This enzyme, which also contains iron, is involved in the oxidation of purines (Chapter 19). Chromium is not found in the structure of enzymes. This metal, however, has been identified as a key component in a complex known as glucose tolerance factor. The complex appears to facilitate the uptake of blood glucose by extrahepatic tissue, a biological function very similar to the action of insulin. Nickel has no specific recognized biochemical function. The metal has been shown to replace Zn and Mg as a nonspecific activator of certain enzymes. Of greater significance, urease from jack bean reportedly contains two atoms of bound nickel. The precise function of the nickel, however, has not been determined.

## SELECTED READINGS

Bruice, T. C., and Benkovic, S. J. *Bioorganic Chemistry,* Vols. 1 and 2. Menlo Park, CA: Benjamin, 1966. A detailed discussion of the mechanisms of bioorganic reactions, including those involving coenzymes.

Jencks, W. P. *Catalysis in Chemistry and Enzymology.* New York: McGraw-Hill, 1969. A detailed analysis of mechanisms of enzymatic and nonenzymatic reactions, including those involving coenzymes.

Phipps, D. A. *Metals and Metabolism,* Oxford Chemistry Series. Oxford: Clarendon Press, 1976. Examines the importance of metal ions in metabolic processes.

Popjak, G. Stereospecificity of Enzymic Reactions. In P. D. Boyer (Ed.), *The Enzymes,* Vol. 2. New York: Academic Press, 1970. P. 115.

Walsh, C. T. *Enzymatic Reaction Mechanisms.* San Francisco: Freeman, 1977. Provides an up-to-date discussion of the mechanisms of enzymatic reactions. An in-depth treatment of coenzymes.

## PROBLEMS

1. The following reactions are catalyzed by pyridoxal-5'-phosphate–dependent enzymes. Write reaction mechanisms for them showing how pyridoxal-5'-phosphate is involved in catalyzing them.

   a. $CH_3-\overset{\overset{\displaystyle H}{|}}{\underset{\underset{\displaystyle NH_2}{|}}{C}}-CO_2H \rightleftharpoons CH_3-\overset{\overset{\displaystyle NH_2}{|}}{\underset{\underset{\displaystyle H}{|}}{C}}-CO_2H$

   b. $H_2N-(CH_2)_4-\overset{\overset{\displaystyle}{|}}{\underset{\underset{\displaystyle NH_2}{|}}{CH}}-CO_2H \longrightarrow$
   $CO_2 + H_2N-(CH_2)_5-NH_2$

   c. $HO_2C-CH_2-\overset{\overset{\displaystyle}{|}}{\underset{\underset{\displaystyle NH_2}{|}}{CH}}-CO_2H \longrightarrow$
   $CO_2 + CH_3-\overset{\overset{\displaystyle}{|}}{\underset{\underset{\displaystyle NH_2}{|}}{CH}}-CO_2H$

2. Thiamine pyrophosphate–dependent enzymes catalyze the following reactions. Write mechanisms for them showing the catalytic role of the coenzyme.

   a. $CH_3-\overset{\overset{\displaystyle O}{\|}}{C}-CO_2H \longrightarrow CO_2 + CH_3-\overset{\overset{\displaystyle O}{\|}}{C}-H$

   b. $\textcircled{P}OCH_2-(CHOH)_2-\overset{\overset{\displaystyle OH}{|}}{CH}-\overset{\overset{\displaystyle O}{\|}}{C}-CH_2OH$
   $+ HOPO_3^{2-} \longrightarrow \textcircled{P}O-CH_2-(CHOH)_2-CHO$
   $+ CH_3-\overset{\overset{\displaystyle O}{\|}}{C}-OPO_3^{2-} + H_2O$

3. What do biotin, lipoic acid, and phosphopantetheine coenzymes have in common?

4. How is NAD involved in the following enzymatic reactions? Write the mechanisms.

   a. $\longrightarrow CO_2 +$

   b. $\underset{\underset{\displaystyle NH_2}{|}}{CH_2}-CH_2-CH_2-\overset{\overset{\displaystyle}{|}}{\underset{\underset{\displaystyle NH_2}{|}}{CH}}-CO_2H \longrightarrow$ $-CO_2 + NH_3$

5. How is biotin involved in the following enzymatic reactions? Write the mechanisms.

   a. $CH_3-\overset{\overset{\displaystyle O}{\|}}{C}-SCoA + HCO_3^- + ATP \longrightarrow$
   $HO_2C-CH_2-\overset{\overset{\displaystyle O}{\|}}{C}-SCoA + ADP + HOPO_3^{2-}$

   b. $CH_3-CH_2-\overset{\overset{\displaystyle O}{\|}}{C}-SCoA + HO_2C-CH_2-\overset{\overset{\displaystyle O}{\|}}{C}-CO_2H \rightleftharpoons$
   $HO_2C-\overset{\overset{\displaystyle}{|}}{\underset{\underset{\displaystyle CH_3}{|}}{CH}}-\overset{\overset{\displaystyle O}{\|}}{C}-SCoA + CH_3-\overset{\overset{\displaystyle O}{\|}}{C}-CO_2H$

6. Explain the importance of phosphopantetheine coenzymes for activating carboxylic acids toward enolization.

7. How is flavin adenine dinucleotide involved in the following reaction? Show the mechanism.

   $\alpha\text{-D-Glucose} + O_2 \longrightarrow$
   $\alpha\text{-D-gluconolactone} + H_2O_2$

8. For each of the following enzymatic reactions, identify the coenzymes involved.

   a. $CH_3-\overset{\overset{\displaystyle}{|}}{\underset{\underset{\displaystyle NH_2}{|}}{CH}}-CH_2-\overset{\overset{\displaystyle}{|}}{\underset{\underset{\displaystyle NH_2}{|}}{CH}}-CH_2-CO_2H \rightleftharpoons$
   $\underset{\underset{\displaystyle NH_2}{|}}{CH_2}-CH_2-CH_2-\overset{\overset{\displaystyle}{|}}{\underset{\underset{\displaystyle NH_2}{|}}{CH}}-CH_2-CO_2H$

   b. $HO-CH_2-\overset{\overset{\displaystyle}{|}}{\underset{\underset{\displaystyle NH_2}{|}}{CH}}-CO_2H \longrightarrow CH_3-\overset{\overset{\displaystyle O}{\|}}{C}-CO_2H + NH_3$

c. 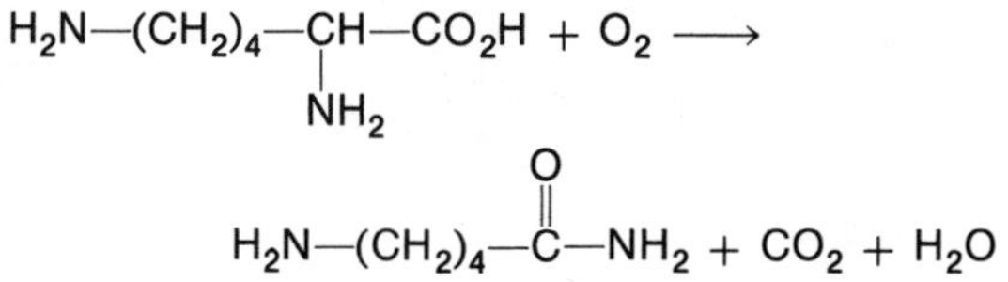 $H_2N-(CH_2)_4-\underset{\underset{NH_2}{|}}{CH}-CO_2H + O_2 \longrightarrow$

$H_2N-(CH_2)_4-\overset{\overset{O}{||}}{C}-NH_2 + CO_2 + H_2O$

d.

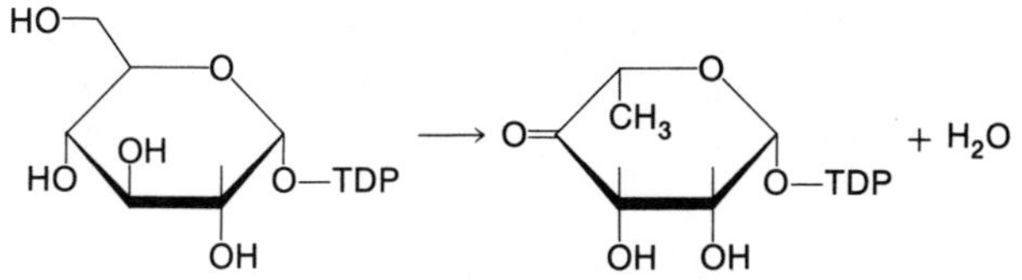

$\longrightarrow$

$+ H_2O$

e. 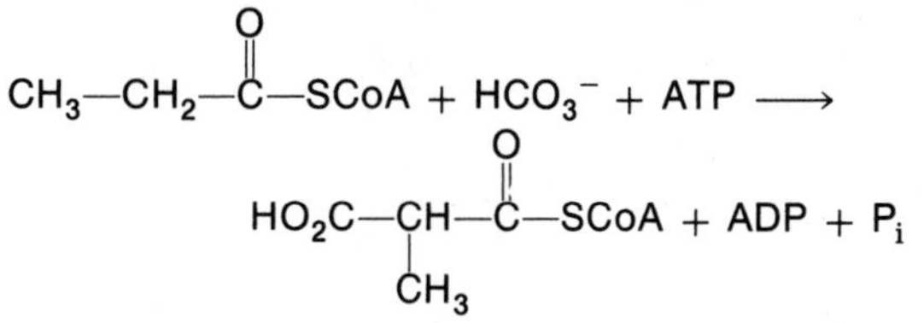 $CH_3-CH_2-\overset{\overset{O}{||}}{C}-SCoA + HCO_3^- + ATP \longrightarrow$

$HO_2C-\underset{\underset{CH_3}{|}}{CH}-\overset{\overset{O}{||}}{C}-SCoA + ADP + P_i$

f. 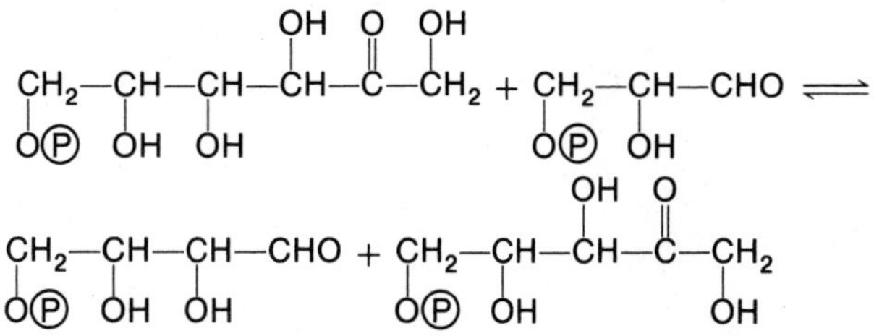 $\underset{\underset{O\circledP}{|}}{CH_2}-\underset{\underset{OH}{|}}{CH}-\underset{\underset{OH}{|}}{CH}-\underset{\overset{|}{OH}}{CH}-\overset{\overset{O}{||}}{C}-CH_2 + \underset{\underset{O\circledP}{|}}{CH_2}-\underset{\underset{OH}{|}}{CH}-CHO \rightleftharpoons$

$\underset{\underset{O\circledP}{|}}{CH_2}-\underset{\underset{OH}{|}}{CH}-\underset{\underset{OH}{|}}{CH}-CHO + \underset{\underset{O\circledP}{|}}{CH_2}-\underset{\underset{OH}{|}}{CH}-\underset{\overset{|}{OH}}{CH}-\overset{\overset{O}{||}}{C}-\underset{\underset{OH}{|}}{CH_2}$

# PART II

## CARBOHYDRATE METABOLISM AND THE GENERATION OF CHEMICAL ENERGY

- THERMODYNAMICS IN BIOCHEMISTRY
- ANAEROBIC PRODUCTION OF ATP
- AEROBIC PRODUCTION OF ATP: THE TCA CYCLE
- AEROBIC PRODUCTION OF ATP: ELECTRON TRANSPORT
- PHOTOSYNTHESIS
- STRUCTURE AND SYNTHESIS OF SUGARS

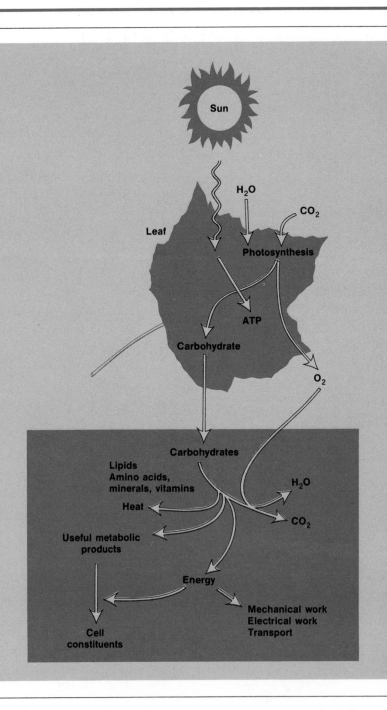

# 7
# THERMODYNAMICS IN BIOCHEMISTRY

Thermodynamics is a subject with an important message for all areas of physics and chemistry. In discussing the relationships between thermodynamics and biochemistry we must consider thermodynamics from two vantage points: the general nature of the laws of thermodynamics and the specific way in which these laws apply to biochemistry. In this chapter we present the laws of thermodynamics and the terms of thermodynamics as they affect biochemically important molecules in aqueous solution, leaving the details of such applications to other chapters.

*The primary usefulness of thermodynamics to the biochemist lies in the prediction of whether a process could or could not occur.* The word could is used because thermodynamics can only predict if a process could occur, not if it will occur. Whether or not a process will occur depends on the existence of a good chemical pathway for the process. Thermodynamics is probably even more important for the biochemist than it is for the chemist, because if a chemical reaction will not proceed because of an unfavorable free energy, the chemist can change the pressure or temperature or increase the concentration of reactants. An organism is under rigid constraints and must function at a fixed temperature and pressure and within a limited range of concentrations of reactants. Since organisms live on chemical energy, it is important to know which reactions are capable of producing energy. Thermodynamics is not an esoteric subject for an organism living on chemical energy—it is a matter of life or death.

A simple use of thermodynamics would be to predict what compounds could possibly be used as energy sources for an organism. It is well known that most oxidations by molecular oxygen produce energy. For example, wood and coal burn with a large output of energy. Similarly, organisms can

*Flow of energy in the biosphere.*

consume carbohydrates, fats, and proteins to produce energy by aerobic oxidations. Some organisms oxidize hydrocarbons, some oxidize reduced forms of sulfur or elemental sulfur, and others oxidize iron. No organisms oxidize molecular nitrogen for energy, and the explanation lies in the thermodynamics. The reaction simply does not produce energy. This exemplifies the importance of thermodynamics in controlling all life.

## CONCEPTS OF THERMODYNAMICS

### Thermodynamic Quantities

Thermodynamics was developed from a study of heat machines and Carnot cycles. It is usually discussed from this standpoint both because of historical significance and because of interest in the abstract beauty of thermodynamics that does not depend on molecular structure. However, we are interested in how organisms put molecules to work, and as a result, we are more interested in the molecular basis of thermodynamics.

Thermodynamic quantities are properties of the state of a substance and do not depend on how a substance was made or reached a certain state. The difference between the initial and the final state is emphasized; the pathway that was taken to get from the initial state to the final state is of no importance (Figure 7-1). Expressed in chemical terms, thermodynamics is usually not concerned with the mechanism of a reaction, but rather with the difference in energy between reactants and products.

The properties of a substance may be classified as *extensive* or *intensive*. Extensive properties depend on the amount of material. Many of the thermodynamic properties that we will discuss depend on the amount of material. Volume, weight, and energy are all properties that depend on the amount of material. Intensive properties that are intrinsic to the substance do not depend on the amount of material. Examples of these are density, pressure, temperature, and concentration.

### Internal Energy and the First Law of Thermodynamics

Of the most important thermodynamic quantities, *internal energy*, designated by $E$, is perhaps the easiest to understand in terms of molecular forces. Internal energy is an extensive property of molecules. The internal energy includes the *translational*, *rotational*, and *vibrational* energy of the molecule. It also includes the *electronic energy* of the molecule. Electronic energies involve electron-electron interactions, electron-nucleus interactions, and nucleus-nucleus interactions. Since internal energy is always given as a difference in energy between initial and final states upon reaction, all internal nuclear terms, most electron-nucleus terms, and many electron-electron terms on the same molecule are eliminated because these terms do not change during the course of a chemical reaction. Electronic energies are very much larger than the translational, rotational, and vibrational energies. Because of this, *the electronic terms contribute the major share to the internal energy of a chemical reaction, even though many electronic terms do not change on reaction.*

*Energy cannot be created or destroyed in chemical reactions;* it is conserved. This is the first law of thermodynamics. The total internal en-

The State Function Maize

State A

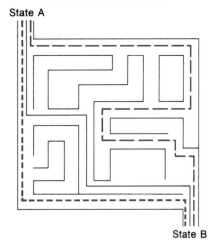

State B

**Figure 7-1**
Thermodynamics considers only the difference between initial and final states but gives us no information on the pathway.

ergy of the universe can be considered as divided into two functional areas, the system plus that of the environment. This energy remains constant even though energy may flow from the system to the environment or from the environment to the system. There are two means by which any specific system under consideration may gain or lose energy to the environment. The system may gain or lose heat to the surroundings or do work on or have work done on it by the surroundings. For a change in state, the first law of thermodynamics can be expressed by the equation

$$\Delta E = q - w \tag{1}$$

where $\Delta E$ is the change in internal energy of the system, $q$ is the heat flow, and $w$ is the work done. In this expression, the convention is adopted that heat absorbed by a system and work done by a system are positive quantities (Figure 7-2). For a small differential change,

$$dE = dq - dw \tag{2}$$

If the only kind of work done involves change in the volume of the system against a fixed external pressure (known as $P\,dV$ work), then

$$dE = dq - P\,dV \tag{3}$$

## Enthalpy

The _enthalpy_ of a compound, e.g., an ideal gas, is the internal energy of the compound plus the changes in pressure and volume of the whole assembly of molecules. Enthalpy takes into account the interactions between the molecules as well as the energy of the molecules themselves. There is interaction between the molecules as a result of collisions and this does not require any specific chemical interaction. The change in enthalpy in a reaction is designated by the symbol $\Delta H$, and the relationship between enthalpy and internal energy changes is given by the equation

$$\Delta H = \Delta E + \Delta PV \tag{4}$$

For small changes in enthalpy,

$$dH = d(E + PV) = dE + P\,dV + V\,dP \tag{5}$$
$$= dq - dw + P\,dV + V\,dP$$

For a system doing only $P\,dV$-type work, $dw = P\,dV$, and

$$dH = dq + V\,dP \tag{6}$$

From Equation (3) we can see that at constant volume,

$$dE = dq \quad \text{or} \quad \Delta E = q_V \tag{7}$$

For processes occurring at constant pressure, Equation (6) becomes

$$dH = dq \quad \text{or} \quad \Delta H = q_P \tag{8}$$

That is, the internal energy of a reaction is equal to the heat absorbed by the system at constant volume, whereas the enthalpy of a reaction is measured as the heat absorbed by the system at constant pressure. The latter is more relevant to most biochemical reactions. Nevertheless, for most biochemical reactions, there is little change in either pressure or volume and the difference between $\Delta H$ and $\Delta E$ can be regarded as insignificant.

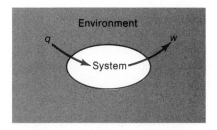

**Figure 7-2**
The system and the environment. Heat flow into the system is designated as a positive quantity. Work done on the environment is designated as a positive quantity.

A reaction system that absorbs heat causes the surroundings to cool, and the change in heat content or enthalpy of the system is positive. When heat is evolved from the system, the $\Delta H$ of the reaction is negative. Under these conditions, energy is lost from the system to the surroundings. Most reactions that proceed spontaneously evolve heat, but a number of them absorb heat. An example, well known to biochemists, of a heat-absorbing process that proceeds spontaneously is the dissolving of ammonium sulfate in water. The solution cools as the ammonium sulfate dissolves. The system is actually going uphill as far as internal energy, or enthalpy, is concerned, but this is more than compensated for by entropy changes, which are explained below.

*Since enthalpy is a state function, it is dependent only on the differences in heat content of the reactants and products and does not depend on the pathway or mechanism of the reaction. This allows us to determine the heat of a reaction by taking the difference between the enthalpy of formation of the products and the enthalpy of formation of the reactants.* Such information is usually recorded with respect to some arbitrarily chosen standard state for each molecule. The standard enthalpy of formation, $\Delta H_f^{\circ}$, is the enthalpy difference between compounds in their standard state (usually the pure compound) and the standard state of the elements. The standard states usually chosen for gases such as oxygen and nitrogen are $O_2$ and $N_2$, respectively. For liquids, the standard state is usually the pure liquid; for solids, the pure solid; and for solutes, the concentration of 1 $M$. All the standard states are defined as the form of the substance at 25°C and 1 atm pressure.

To illustrate the calculation of $\Delta H$ for a reaction, let us use enthalpies of formation to calculate the enthalpy of the reaction of gaseous ammonia with liquid acetic acid to form solid glycine and hydrogen:

$$NH_3(g) + CH_3COOH(l) \longrightarrow glycine(s) + H_2(g)$$

The heat of formation of solid glycine is $-125.46$ kcal/mol. This value is for the reaction of gaseous molecular nitrogen, oxygen, and hydrogen with graphite (carbon) to form glycine. The heat of formation of liquid acetic acid is $-117.20$ kcal/mol, and that for gaseous ammonia is $-10.95$ kcal/mol. Since gaseous hydrogen is the standard state of hydrogen, its heat of formation is 0. Therefore, the enthalpy of the reaction for the formation of glycine from ammonia and acetic acid is $-125.46 + 117.20 + 10.95 = +2.69$ kcal/mol. The reaction has a positive enthalpy and therefore would absorb heat if it were to occur.

Of the four types of energy in a molecule—translational, rotational, vibrational, and electronic—the latter, as indicated above, makes by far the largest contribution to the internal energy and thus to the enthalpy of a molecule. *Commonly it is possible to get a good value for the enthalpy of a reaction by considering only the electronic terms. Electronic terms include bond energies, resonance, steric hindrance, effects of solvation, and phase changes.* If there is no change in resonance, steric hindrance, or solvation during the course of a reaction and there are no problems of phase changes, then the differences in these terms disappear and bond energies may be used to determine the enthalpy of a reaction. Bond energies are positive and have larger positive values for more stable bonds. They

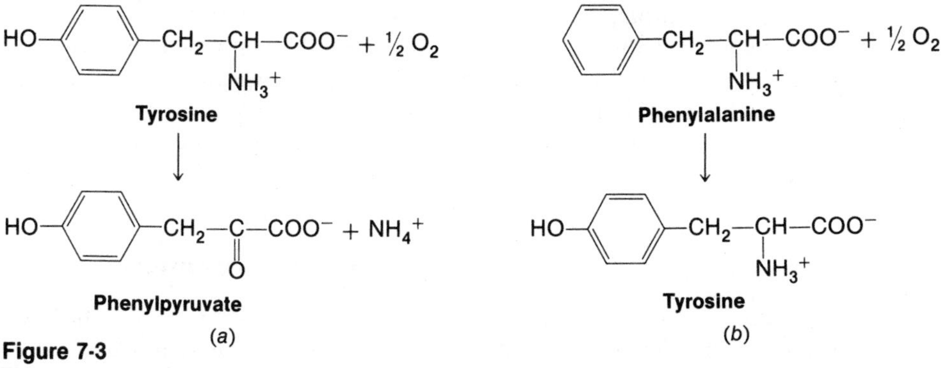

**Figure 7-3**
The oxidation of tyrosine to phenylpyruvate by molecular oxygen (a) and the oxidation of phenylalanine to tyrosine by molecular oxygen (b).

range in value from 34 to 110 kcal/mol for most single bonds. Bond energies for some bonds of biological interest are given in Chapter 3 (see Table 3-1). If the difference between the bond energies of products and reactants is positive, the enthalpy of the reaction is negative and the reaction produces heat. Bond energies are not referenced to standard states, as are heats of formation, but are referenced to the completely separate atoms.

As stated earlier, the use of bond energies to determine the enthalpy of a reaction is subject to error when there are changes in resonance or solvation during the course of the reaction. Let us consider a few cases where effects other than bond energies make a significant contribution to the enthalpy of a reaction.

If tyrosine is oxidatively deaminated to phenylpyruvate (Figure 7-3a), the resonance energy of the benzene ring is not altered during the reaction and bond energies may be used without a correction for the resonance to determine the enthalpy of the reaction. However, in aqueous solution, there would be a solvation correction between the zwitterion and the separated p-hydroxyphenylalanine anion and the ammonium ion. When the charges are separated, there is increased hydration. Hydration is an important factor in many biochemical reactions because most reactions occur in water, which is an excellent solvating agent.

In the oxidation of the phenylalanine by molecular oxygen to tyrosine (Figure 7-3b), there is little change in solvation upon reaction because both product and reactant are zwitterions and have about the same solvation (hydration) enthalpy. In this case, bond energies give a good estimate for the enthalpy of the reaction.

Bond energies could not be used without considerable correction for the estimation of the enthalpy of the oxidation of an alcohol by $NAD^+$. The enthalpy of the reduction of $NAD^+$ to NADH (Figure 7-4) could not be estimated from bond energies alone not only because of the destruction of the resonance of the pyridine ring, but also because the highly solvated cation is reduced to a neutral molecule. The reactants and products are not equally hydrated. The proton that is a product of the reaction is much more hydrated than the $NAD^+$ ion because of its small size. The hydration terms do not compensate for each other. In this example, resonance and hydration have opposing effects on the enthalpy of the reaction.

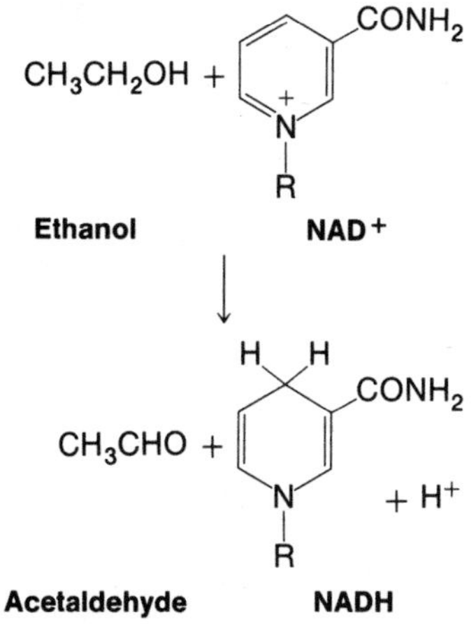

**Figure 7-4**
The reduction of $NAD^+$ by ethanol.

Whereas enthalpy changes are useful for calculating heat changes during the course of a reaction, they do not provide us with a reliable criterion for the spontaneity of a reaction. For example, LiCl and $(NH_4)_2SO_4$ are both readily soluble in water. The former reaction releases heat, whereas the latter absorbs heat. The latter reaction is driven by the entropy effect, which we will now discuss in some detail.

### Entropy and the Second Law of Thermodynamics

The second law of thermodynamics states that systems tend to proceed from a state of low probability (ordered) to a state of high probability (disordered). This effect is measured by a term called _entropy_ (denoted by the symbol $S$). A reaction that increases in entropy (positive $\Delta S$) tends to proceed over one that decreases in entropy (negative $\Delta S$). If $W$ is the number of ways of arranging a system without changing the internal energy, or enthalpy, we may define the absolute entropy per molecule in the system by the equation

$$S = k \ln W \qquad (9)$$

where $k$ is Boltzman's constant; $k \simeq 3.4 \times 10^{-24}$ cal/°C. For a mole of substance,

$$S = Nk \ln W = R \ln W \qquad (10)$$

where $N$ is the number of molecules in a mole ($6 \times 10^{23}$), and $R$ is the gas constant [$R \simeq 2$ cal/(°C · mol)]. Quantitative values for entropies are usually given in entropy units; 1 eu = 1 cal/(°C · mol).

If we have $N$ total molecules in $T$ energy states labeled 1 through $T$ and $n_i$ molecules in the $i$th energy state, we find that the total number of ways $W$ of arranging the system is

$$W = \frac{N!}{n_1! n_2! n_3! \cdots n_T!} \qquad (11)$$

For example, if we have three molecules in two energy levels with one molecule in the upper state and two molecules in the lower state, e.g.,

we find that $W = 3!/(2!1!) = 3$. The three ways of arranging the system correspond to placing any one of the three molecules in the upper state and the other two in the lower state. As the temperature increases, the molecules obtain more energies, so they occupy other levels that are higher in E. Entropy increases as $W$ increases. At absolute zero, when all molecules are in their lowest internal energy states, $W$ adopts its lowest value, i.e., $W = N!/N! = 1$, and the entropy is 0. $W$ can never be less than 1, and hence the absolute entropy of a compound can never be negative. Nevertheless, the entropy of formation of a compound or the entropy change in a reaction may be negative because this involves the difference in entropies between reactants and products.

A greater number of energy levels, particularly low-lying energy levels, give a higher entropy. This fact is evident from Equation (11). Translational and rotational energy levels are very close, vibrational energy levels are further apart, and electronic energy levels are usually very far apart. Because of this, translational levels contribute greatly to the entropy and electronic states contribute the least; usually only one electronic state (the lowest) is available and the entropy is zero as far as the electronic states are concerned.

The absolute entropy of a compound may be calculated from statistical mechanical methods if all the vibrational and rotational parameters are known. These include force constants, moments of inertia, and rotational barriers. Statistical mechanical calculations of entropy are only possible for relatively small molecules and certainly could not be used for the usual biological reactant. The entropy contributions to propane that have been calculated by means of statistical mechanics are listed in Table 7-1. These values show a large translational and rotational contribution to the entropy, a small vibrational contribution, and no electronic contribution. This qualitative distribution of entropy is typical for small molecules such as propane or for larger molecules found in biochemical systems. Thus _when we think of entropy we should associate it primarily with translation and rotation. This is very different from enthalpy, in which electronic terms are all important and the other terms are insignificant by comparison._

There are several molecular properties that are important in determining the entropy of the compound. If we consider some of these properties and how they affect the entropy, we will have a better qualitative appreciation of entropy.

From statistical mechanics we find that translational entropy, which is still the most important entropy term even in large molecules, is proportional to $\frac{3}{2}R \ln M$ (plus other terms), where $R$ is the gas constant (defined above) and $M$ is the molecular weight. Thus the entropy per mole increases with molecular weight. For example, the entropy of hydrocarbons increases by 5.8 eu for each $CH_2$ in the solid phase, 7.7 eu in the liquid phase, and 10.0 eu in the gas phase. The entropy of a liquid hydrocarbon is equal to $25.0 + 7.7n$, where $n$ is the number of carbon atoms. Hexyl bromide has the same number of atoms and bonds as hexane, but as a consequence of the higher molecular weight and correspondingly higher translational entropy, the entropy is 108.33 eu compared with 70.6 eu for normal hexane.

Structural features that make molecules more rigid reduce rotational and vibrational contributions to entropy. A double bond will reduce the entropy by about 3.5 eu, a triple bond by about 4.5 eu, a ring by about 14 eu, and a branch in a chain by about 3.0 eu. The reduction of entropy in forming a ring depends on the size of the ring, as demonstrated by the cyclization reactions in Table 7-2.

The physical state is also a very important factor in the entropy of a compound. A gas has much more translational and rotational freedom than does a liquid and a liquid has much more freedom than a solid. As a result, the relative magnitude of entropy decreases from gas to liquid to solid. The entropy increase on vaporization and melting may be determined from the heat of vaporization divided by the boiling point and the heat of fusion divided by the melting point, respectively. The entropy increase on vapori-

**Table 7-1**
Entropy of Propane at 231 K

|  | kcal/(°C·mol) |
|---|---|
| Translational | 36.04 |
| Rotational | 23.38 |
| Vibrational | 1.05 |
| Electronic | 0.0 |
| Total | 60.47 |

**Table 7-2**
Entropy Decrease in Ring
Formation Depends on Size of Ring

| | ΔS (eu) |
|---|---|
| △ | −6.9 |
| □ | −10.3 |
| ⬠ | −13.1 |
| ⬡ | −21.1 |

zation of water is 26 eu/mol and on melting ice is 5.3 eu/mol. It can be shown that for an isothermal reversible process, such as the melting of ice at 0°C (273 K), the entropy increase is given by

$$\Delta S = \frac{q_{\text{melting}}}{T} = \frac{\Delta H_{\text{melting}}}{T_{\text{melting}}} \tag{12}$$

where $T$ is the absolute temperature in degrees Kelvin.

*The entropy of solutions is markedly affected by mixing two solvents, by solvation of solutes, and by hydrogen bonding and other associations in the solvent or between solute and solvent.* The probability of two miscible liquids being mixed is greater than each staying separate in the same container. The entropy change on going from the pure liquid state to the mixed state is

$$\Delta S = n_a R \ln \frac{1}{X_a} + n_b R \ln \frac{1}{X_b} \tag{13}$$

$$= -n_a R \ln X_a - n_b R \ln X_b$$

when $n_a$ and $n_b$ are the number of moles of a and b that are mixed and $X_a$ and $X_b$ are the corresponding mole fractions. Since $X_a$ and $X_b$ are always less than 1, $\ln X_a$ and $\ln X_b$ will be negative. This means that the dilution of each component resulting from mixing will make a positive contribution to the entropy. This equation applies to the mixing of ideal solutions in which there is no interaction between the molecules. Any intermolecular interaction will result in a decrease in the entropy of mixing because it will restrict the translational and rotational freedom of the individual molecules.

Solvation, i.e., the interaction between solute and solvent molecules, makes an important negative contribution to the entropy of a compound in solution. When a solute is added to a solvent and solvation occurs, the entropy decreases because of the restricted movement of the solute as well as because of the restricted movement of the solvent in the vicinity of the solute. However, for simplicity in bookkeeping, all the decrease in entropy is usually ascribed to the solute (Figure 7-5). Small ions are more highly solvated or hydrated than large ions with the same charge, and anions are more solvated than cations in water. The entropy of a monatomic ion in water is reduced by $-270Z/r_e$ eu, where $Z$ is the absolute value of the charge on the ion and $r_e$ is the effective radius of the ion in angstroms. The effective radius is equal to the crystal radius plus 1 Å for anions or plus 2 Å for cations. This correction to the radius takes into account the fact that anions are solvated more than cations. Although this equation applies quantitatively only to monatomic ions, we see that *in general the entropy of solvation in water becomes a larger negative number with increasing charge and decreasing radius.* In addition to ions, molecules that have dipoles or hydrogen-bond donor or acceptor groups tend to interact strongly with water, and consequently, they restrict the mobility of the water. Similar decreases in entropy usually result from the hydration of such molecules.

*It is interesting to note that although enthalpy and entropy depend on entirely different terms, solvation affects both values because it contains*

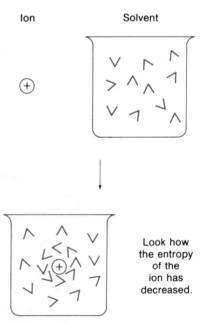

Ion          Solvent

Look how
the entropy
of the
ion has
decreased.

**Figure 7-5**
A bookkeeper's simple view of solvation.

*both electronic and translational terms. Enthalpies and entropies of solvation most frequently tend to negate each other. For charged species, the more negative the enthalpy of solvation (favorable), the more negative (unfavorable) the entropy of solvation.* In the solvation of ions in water, the enthalpy term usually overbalances the entropy term, so the ion dissolves.

*An entirely different type of solvation occurs when an apolar molecule is added to water. The result is a decrease in entropy, but for very different reasons. The water orients on the surface of the apolar molecule, forming a relatively rigid cage held together by hydrogen bonds* (see Figure 3-4). The explanations for this phenomenon are varied, but certainly an important factor is the low dielectric in the vicinity of the hydrophobic surface that should result in stronger hydrogen bonding within the water structure itself. This kind of solvation effect becomes a very important factor for the large nonpolar areas of proteins and other biological macromolecules (discussed below).

As a rule, when a molecule divides into two or more molecules, the entropy increases. It is easy to see how the translational entropy increases on partitioning a molecule by the following equation, which compares the relative translational entropy of molecules with molecular weights $M_1$ and $M_2$:

$$\tfrac{3}{2}R \ln M_1 + \tfrac{3}{2}R \ln M_2 = \tfrac{3}{2}R \ln (M_1 M_2) > \tfrac{3}{2}R \ln (M_1 + M_2)$$

Clearly then, when proteins, carbohydrates, or other macromolecules are degraded to smaller molecules during metabolism, there will be an increase in translational entropy.

From the preceding one might expect that the ionization of a proton from an acid, which produces more particles, would lead to an increase in entropy. However, this effect is more than counterbalanced by the hydration effects. The production of the charged anion and proton in water "freezes out" a large amount of water in the form of water of hydration, which results in a large decrease in entropy (Figure 7-6). Thus when a weak acid ionizes there is a decrease in the total number of free molecules instead of an increase as one might expect. The entropy of ionization of a typical weak acid in water is usually about $-22$ eu. The first and the second ionizations of an amino acid (Figure 7-7) show much less of a decrease in entropy than this because the zwitterion is highly hydrated before the proton is lost; this results in a less negative entropy of ionization than predicted. The entropy change for the first ionization of the glycine cation to a zwitterion is $-7.3$ eu, and the entropy change for the second ionization of the zwitterion to the anion is $-9.2$ eu.

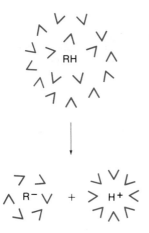

**Figure 7-6**
The number of free particles decreases upon ionization instead of increasing as one might imagine at first.

$$\Delta S° \text{ eu}$$

$$\underset{\substack{|\\ NH_3^+}}{CH_2}-COOH \longrightarrow \underset{\substack{|\\ NH_3^+}}{CH_2}-COO^- + H^+ \qquad -7.3$$

$$\underset{\substack{|\\ NH_3^+}}{CH_2}-COO^- \longrightarrow \underset{\substack{|\\ NH_2}}{CH_2}-COO^- + H^+ \qquad -9.2$$

**Figure 7-7**
The entropy of differences on the ionization of glycine.

The reduction of a pyruvate ion by NADH

$$H^+ + NADH + Pyr^- \longrightarrow lactate^- + NAD^+$$

occurs with an entropy increase of 16.5 eu. This is mainly due to the large size of the $NAD^+$ ion compared with the proton. Many water molecules are freed during the reaction, and the entropy increases. The hydration of the lactate ion is approximately equal to that of the pyruvate ion, so the hydration of these two ions tends to compensate in entropy. The neutralization of methylamine in aqueous solution

$$H^+ + CH_3NH_2 \longrightarrow CH_3NH_3^+$$

also takes place with an increase in entropy of 4.7 eu because the proton is much more highly hydrated than the methylammonium ion. This more than overcomes the loss of entropy resulting from the association of the proton and the methylamine.

The entropy changes on binding or absorption of molecules on the surface of a macromolecule are important to a biochemist because these are commonly occurring phenomena in the cell. Of particular interest is binding of a substrate or an inhibitor to an enzyme catalyst. When gas molecules are absorbed on a solid catalytic surface, there is a large decrease in entropy (Figure 7-8). The large translational entropy of the gas molecules disappears when they are held fixed on a surface, and the thermodynamics of the absorbed gas is much more like that of a solid. The negative change in entropy forms a barrier to the very common types of industrial use of solid catalysts acting on gases that must be overcome by a strongly negative enthalpy for the absorption of the gas by the catalyst or by increased pressure. Therefore, the binding energy of the solid catalyst for the gas must be strong enough to overcome the improbable situation of gas molecules sitting on the surface. By analogy, one might expect a highly negative entropy when a substrate is absorbed on an enzyme surface. However, these processes frequently have positive entropies, the reason being that *a large number of water molecules are displaced when a substrate is bound to the surface of an enzyme. This large release of absorbed water causes an increase in translational entropy.* Binding between pepsin and an ester substrate has an increase of 20.6 eu and urea combining with urease has an increase of 13.3 eu. *Because of the favorable entropy, the binding enthalpy of an enzyme for substrate need not be nearly so large as that of industrial*

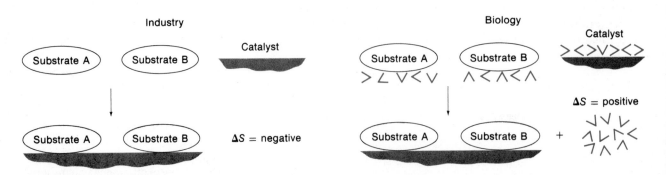

**Figure 7-8**
A comparison of the entropy binding to an industrial catalyst as compared with a biological catalyst.

**Table 7-3**                                                                                           **255**

Enthalpy and Entropy Changes on Forming Complexes between Cadmium Ions
and Methylamine or Ethylenediamine

| | Values for Association | | | |
|---|---|---|---|---|
| | $\Delta H$ | $\Delta G$ | $\Delta S$ | $T \Delta S$ |
| $Cd^{2+} + 2\,CH_3NH_2 \longrightarrow Cd(CH_3NH_2)_2^{2+}$ | $-7.03$ | $-6.56$ | $-1.58$ | $-.47$ |
| $Cd^{2+} + en^* \longrightarrow Cd(en)^{2+}$ | $-7.02$ | $-7.96$ | $+3.15$ | $+.94$ |
| $Cd^{2+} + 4\,CH_3NH_2 \longrightarrow Cd(CH_3NH_2)_4^{2+}$ | $-13.7$ | $-8.93$ | $-16.0$ | $-4.77$ |
| $Cd^{2+} + 2\,en \longrightarrow Cd(en)_2^{2+}$ | $-13.5$ | $-14.48$ | $+3.29$ | $+.98$ |

*en = ethylenediamine.
   Data from Spike and Parry, Thermodynamics of chelation I. The statistical factor in chelate ring
formation, *J. Am. Chem. Soc.* 75:2726, 1953.

*catalysts working on gases*. Industrial and biological catalysts may seem
superficially to be similar, but there are qualitative differences in the thermodynamics of each.

   Another most important entropy effect when an enzyme and substrate
combine could be called the *chelation effect*. This effect is best understood
by discussing a relatively simple case of metal chelation. Cadmium ion
tends to be quadrivalent, so if there are two amino groups in one molecule,
such as in ethylenediamine, the cadmium will combine with two molecules. If there is only one amino group in a molecule, such as in methylamine, the cadmium ion requires four molecules, as shown in Table 7-3.
Notice that the entropy is much more favorable for the combination with
ethylenediamine than with methylamine, resulting in stronger binding to
the ethylenediamine than to the methylamine. Water molecules are released from cadmium ions when the ligands are added, but the increase in
entropy from this factor is the same whether methylamine or ethylenediamine is added. The much less favorable entropy factor resulting from association of the methylamine results from the larger number of molecules
that combine with the cadmium ions when methylamine is the complexing agent and the correspondingly larger loss of translational entropy. If a
molecule has *n* points of attachment to another molecule (a metal or a
protein), then it will bind much more strongly than *n* molecules with one
point of attachment even though the enthalpy of each point of attachment
is the same. The chelation effect is a big factor in substrate binding to an
enzyme when there are several binding sites. Several weak bindings can
produce an overall tight binding because of the additive effect of many
small enthalpy contributions and the lack of a proportional decrease in
entropy. This effect is even greater in binding between two proteins or a
protein and a nucleic acid (e.g., see the case of *lac* repressor binding to
DNA in Chapter 26), because there are many interactions between such
large molecules.

## Free Energy: A Criterion for Spontaneity

As we have seen in the previous discussion, a system tends toward the
lowest enthalpy and the highest entropy. Neither of these terms alone
makes it possible to predict whether a reaction will proceed or not. It is

clear that a firmer criterion is needed to predict whether or not a reaction will proceed. J. W. Gibbs was the first to appreciate that in reactions occurring at equilibrium and constant temperature (where no net reaction is taking place), such as the phase change of ice going to water at 0°C, that $\Delta S = \Delta H / T$ (see Equation 12). The change in entropy is numerically equal to the change in enthalpy divided by the absolute temperature.

This equation can be transposed to

$$\Delta H - T \, \Delta S = 0 \qquad (14)$$

In search of a criterion for spontaneity, Gibbs proposed a new function called the *free energy*, defined by the equation

$$\Delta G = \Delta H - T \, \Delta S \qquad (15)$$

For a reaction occurring at equilibrium, such as the melting of ice at 0°C, the change in free energy would be zero. For the same reaction occurring in a nonequilibrium situation, such as an ice and water mixture at 10°C, the free energy change would be negative for melting. The melting would occur spontaneously. In the reverse reaction, the conversion of water to ice, there would be a positive free energy change and the reaction would not occur.

*Free energy is a valuable concept not only because it allows us to determine whether a reaction could proceed, but also because it allows us to calculate other useful parameters of the reaction, such as the equilibrium constant, which tells us the quantitative extent to which a reaction can proceed.* Free energy, like internal energy, enthalpy, and entropy is a state function and an extensive property of a system. Provided a good pathway exists and the free energy change is favorable (negative), the reaction will occur. If the free energy change is favorable and no pathway is available for the conversion, a catalyst may be added that provides an acceptable pathway. However, if the free energy change is unfavorable (positive), no catalyst can ever make the reaction proceed.

Free energy is the most useful of all thermodynamic terms, but it is the least easily understood because it is a combined function of both enthalpy and entropy. The preceding discussion has provided a qualitative understanding of enthalpic and entropic contributions—that changes in state of molecules in aqueous solution are likely to make to the free energy.

The free energy of formation of many compounds can be found in tabular form in books on thermodynamics. These are given as the free energy of the reaction of the elements in their standard states to form the desired compound in its standard state. The standard states are the same as those described in the discussion on enthalpy. By subtracting the sum of the free energy of formation of the reactants from the sum of the free energy of formation of the products it is possible to determine the free energy change of any reaction for which the free energy of formation of all reactants and products are known. Some standard free energies of formation are given in Table 7-4. From the values listed in the table we can calculate the standard free energy change for the reaction oxaloacetate$^{2-}$ + H$^+$($10^{-7}$ M) → CO$_2$(g) + pyruvate$^-$ as $-113.44 - 94.45 + 9.87 + 190.62 = -7.4$ kcal/mol. Thus the free energy change of this reaction for 1 M oxaloacetate anion, $10^{-7}$ M hydrogen ions (note: in biochemistry the standard state for hydrogen in solution is usually defined for a $10^{-7}$ M solution because this is close to the concentration in most

**Table 7-4**
Free Energies of Formation
of Some Compounds of
Biological Interest

| Substance | $\Delta G°_f$ (kcal/mol) |
|---|---|
| Lactate ions | −123.76 |
| Pyruvate ions | −113.44 |
| Succinate dianions | −164.97 |
| Glycerol (1 M) | −116.76 |
| Water | −56.69 |
| Acetate anions | −88.99 |
| Oxaloacetate dianions | −190.62 |
| Hydrogen ions ($10^{-7}$ M) | −9.87* |
| Carbon dioxide (gas) | −94.45 |
| Bicarbonate ions | −140.49 |

*This is the value for hydrogen ions at a concentration of $10^{-7}$ M. The free energy of formation at unit activity (1 M) is 0.

**Table 7-5**
Determination of the Standard Free Energy of ATP Hydrolysis
by Adding the Standard Free Energies of Two Other Reactions

| | $\Delta G°'$(kcal/mol) |
|---|---|
| Glucose + ATP $\rightleftharpoons$ Glucose-6-P + ADP | −4.7 |
| Glucose-6-P + $H_2O$ $\rightleftharpoons$ Glucose + P | −3.1 |
| ATP + $H_2O$ $\rightleftharpoons$ ADP + P | −7.8 |

systems of interest to biochemists), carbon dioxide at 1 atm, and 1 $M$
pyruvate anions is −7.4 kcal. From this it would appear that the reaction
will proceed at these concentrations. However, the concentrations are not
all expressed in a realistic manner. At pH values where the dianion of
oxaloacetate would be formed, the carbon dioxide would not be a gas but
would be bicarbonate. One must make some further calculations before the
free energy we have found is of any real use to us. One of these would be to
add the free energy of the reaction of carbon dioxide gas to give bicarbonate
anion:

$$CO_2 + H_2O \longrightarrow HCO_3^- + H^+$$

This would amount to a correction of $-140.49 - 9.87 - (-56.69) -$
$(-94.43) = 0.8$ kcal. The free energy change for the same reaction
to form 1 $M$ bicarbonate ions instead of carbon dioxide is $-7.4 + 0.8 =$
$-6.6$ kcal/mol.

*The standard free energy for a reaction may be determined by adding*
*or subtracting the standard free energies of two other reactions that will*
*combine to give the desired reaction* if the free energies of the two reactions are known. An example is the calculation of the free energy of hydrolysis of ATP at pH 7 from the free energies of formation of glucose-6-phos-
phate and the hydrolysis of glucose-6-phosphate at pH 7 as shown in
Table 7-5.

## Solution Thermodynamics and the Chemical Potential

In pure liquids or pure solids, free energy is the most valuable parameter for
predicting the behavior of a system. However, in multicomponent systems
in which the amount of any one component may vary independently, it is
convenient to introduce the concept of the partial molar free energy, which
is a free energy function relating to a single component. Most biochemical
reactions take place in multicomponent solutions, and so an understanding
of solution thermodynamics is a central concern in applying thermodynamics to biochemistry.

The change in free energy with the change in amount of any one
component, while everything else is held constant, is called the *partial*
*molar free energy*. Partial molar quantities are intensive quantities in contrast to the usual extensive thermodynamic quantities $E$, $H$, $G$ and $S$. At
constant temperature and pressure, all partial molar quantities $\bar{Z}_i$ are defined in terms of the extensive quantity $Z$ by the general equation

$$\left(\frac{\partial Z}{\partial n_i}\right)_{T, P, n_j \neq n_i} = \bar{Z}_i \tag{16}$$

The partial molar free energy where $n_i$ is the number of moles of the $i$th component, often called the *chemical potential*, is given by the special symbol $\mu_i$ and is defined by the equation

$$G_i = \mu_i = \left(\frac{\partial G}{\partial n_i}\right)_{P,\,T,\,n_j \neq n_i} \tag{17}$$

It follows from summing Equation (17) over all components that the total free energy of the solution is

$$G = \sum_{i=1}^{n} n_i \mu_i \tag{18}$$

where $n$ is the total number of components in the solution. The change in $G$ for an infinitesimal reversible change in the state of a system in which the $T$, $P$, or amounts of components can change is given by

$$dG_i = \left(\frac{\partial G_i}{\partial T}\right)_{P,\,n_i} dT + \left(\frac{\partial G_i}{\partial P}\right)_{T,\,n_i} dP + \left(\frac{\partial G_i}{\partial n_i}\right)_{T,\,P,\,n_j} dn_i \tag{19}$$

It can be shown that

$$\left(\frac{\partial G_i}{\partial T}\right)_{P,\,n_i} = -\bar{S}_i \quad \text{and} \quad \left(\frac{\partial G_i}{\partial P}\right)_{T,\,n_i} = \bar{V}_i \tag{20}$$

where $\bar{S}_i$ and $\bar{V}_i$ are the partial molar entropy and volume of the $i$th component.

Therefore

$$\partial G_i = -\bar{S}_i\,dT + \bar{V}_i\,dP + \mu_i\,dn_i \tag{21}$$

Summing over all components for a small change in free energy, we get

$$dG = -S\,dT + V\,dP + \sum_i \mu_i\,dn_i \tag{22}$$

At constant $T$ and $P$, this reduces to

$$dG = \sum_i \mu_i\,dn_i \tag{23}$$

However, if we differentiate Equation (18), we get another expression for $dG$, which is

$$dG = \sum_i \mu_i\,dn_i + \sum_i n_i\,d\mu_i \tag{24}$$

Comparison of Equations (23) and (24) shows that

$$\sum n_i\,d\mu_i = 0 \tag{25}$$

*This relationship, known as the Gibbs-Duhem equation, tells us that variations in the different chemical potentials are not independent of one another. A change in the chemical potential of one component will be reflected by an equal and opposite change in the chemical potentials of the other components in a system.*

The chemical potential measures the increment of free energy that accompanies an infinitesimal reversible change in the amount of one particular component in a system. This most useful term can be applied to many biochemical problems of phase equilibria and chemical equilibria, as will be seen in what follows.

# APPLICATION OF THERMODYNAMICS
# IN BIOCHEMISTRY

**259**

**CHAPTER 7**
THERMODYNAMICS
IN BIOCHEMISTRY

In this text thermodynamics is applied to three main areas of solution biochemistry: (1) phase equilibria where the phases are bounded by a semipermeable membrane, (2) chemical reactions and association and disassociation reactions, and (3) the energetics of biopolymer conformations.

## Phase Equilibria

If we have a solute in one phase, $\alpha$, in equilibrium with the same solute in another phase, $\beta$, such as the equilibrium of a compound between two immiscible solvents (e.g., water and $CHCl_3$), then

$$dG = \mu_i{}^\alpha \, dn_i{}^\alpha + \mu_i{}^\beta \, dn_i{}^\beta \tag{23'}$$

at constant $T$ and $P$. Since at equilibrium $dG$ is 0 and $dn_i{}^\alpha = -dn_i{}^\beta$ because the total number of moles of $n_i$ does not change,

$$0 = \mu_i{}^\alpha - \mu_i{}^\beta = dn_i{}^\alpha \tag{26}$$

Therefore

$$\mu_i{}^\alpha = \mu_i{}^\beta \tag{27}$$

*Thus the chemical potential of each component in one phase is equal to the chemical potential of the same component in a second phase at equilibrium.* This equation is used as a starting point for most problems involving equilibria across membranes.

## The Dependence of the Chemical Potential
## of the Solvent on Solute Concentration

From Equation (13) we know that the entropy of mixing components in an ideal solution where there is no reaction between components is given by

$$\Delta S_m = -R \sum_{i=1}^{n} n_i \ln X_i \tag{13'}$$

For an ideal solution where there is no interaction between components on mixing, $\Delta H_m = 0$. As a result,

$$\Delta G = \Delta H_m - T \, \Delta S_m = RT \, n_i \ln X_i \tag{28}$$

The thermodynamic treatment of solutions presented here will be restricted to ideal solutions, keeping in mind that deviation from ideality could result in significant corrections.

In general, the free energy of mixing is defined by the equation

$$\Delta G_m = G_{\text{solution}} - \sum_i G_{\text{pure components}} \tag{29}$$

The free energy of a particular component will equal the number of moles of that component times the free energy per mole [see Equation (18)]. If the standard free energy of the pure substance is referred to by the symbol $\mu_i^\circ$, then

$$\Delta G_m = \sum_i n_i \mu_i - \sum_i n_i \mu_i^\circ \tag{30}$$

or

$$\sum_i n_i(\mu_i - \mu_i^\circ) = RT \sum_i n_i \ln X_i$$

Since the components of a solution can be considered independently, it follows that

$$\mu_i - \mu_i^\circ = RT \ln X_i \qquad (31)$$

In discussing some properties of macromolecules, such as the osmotic pressure, the primary interest is in the chemical potential of the solvent water. For dilute solutions it is convenient to take the standard state of the solvent as pure solvent. If $X_1$ and $X_2$ are the mole fractions of solvent and solute, respectively, in a two-component system, then

$$X_1 = 1 - X_2$$

From Equation (31) the chemical potential of the solvent is given by

$$\mu_1 - \mu_1^\circ = RT \ln X_1 \qquad (32)$$

When $X_2$ is small, $\ln X_1 = \ln (1 - X_2) \cong -X_2$. Therefore

$$\mu_1 - \mu_1^\circ = -RTX_2 \qquad \text{or} \qquad \mu_1 = \mu_1^\circ - RTX_2 \qquad (33)$$

Thus chemical potential of solvent is reduced by the term $-RTX_2$ on adding solute. It can also be shown that

$$X_2 \cong \frac{C_2 V_1^\circ}{M_2} \qquad (34)$$

where $C_2$ is the concentration of solute in grams per liter, $M_2$ is the molecular weight of the solute, and $V_1$ is the molar volume of pure solvent. This allows us to rewrite Equation (33) as

$$\mu_1 = \mu_1^\circ - \frac{RTV_1^\circ C_2}{M_2} \qquad (35)$$

for a two-component system, where the concentration of component 2 is much lower than the concentration of component 1.

## Membrane Equilibria

Cells and organelles within cells are bounded by semipermeable membranes that permit free diffusion of some molecules but not others. The most important application of the thermodynamics of phase equilibria in biochemistry is to situations where the phases are defined by a membrane boundary.

**Membrane Permeable to Both Solvent and Solute.**  One of the simplest applications of chemical potential is to equilibrium across a membrane in which both solvent (water) and solute (say, glucose) can pass through the membrane. If we start out with a glucose solution on one side of the membrane and pure water on the other side, we know from experience that at equilibrium the concentration of glucose will be equal on both sides of the membrane. Initially there will be a net flow of water and glucose in opposite directions. This is so because the chemical potential of glucose is higher on one side of the membrane while the chemical potential of water is higher on the other side. The unrestricted flow of a component will always go from a region of high chemical potential to one of low chemical potential. Eventually the mole fraction of glucose and water will become equal on both sides of the membrane, resulting in equal chemical potentials for each component and no net flow of either component. This situation is so simple and intuitively obvious that you may be asking yourself why we went through such a sophisticated treatment to reach the result. The answer is that we did so because the treatment of this simple situation is formally equivalent to the treatment of more complex situations where the outcome is less than obvious.

**Membrane Permeable to Solvent Only.** Imagine a simple membrane situation again, but this time let the two components be water solvent and a solute such as protein that is present on one side of the membrane but too large to pass through the membrane. This is a common situation for biological membranes, which also may be studied in a cell-free system. As before, the solvent will flow from the side containing pure water to the side containing the protein. Given enough time, all the water will pass through the membrane to the side containing the protein. The volume of water will keep on increasing on the side containing the protein solution.

Now let us ask what would happen if the containers on either side of the membrane were closed and filled so that the volume could not change. In this case, at time zero water would start to flow from the side containing pure water to the side containing the protein solution. Since the containers are closed, this would lead to a rise in pressure on the solution side that would eventually stop the net flow of water. The system would reach equilibrium when the chemical potential of water becomes the same on both sides of the membrane. The lower mole fraction of water on the solution side would be compensated for by the higher pressure. The pressure difference attained at equilibrium is known as the _osmotic pressure_. Osmotic pressure is an important characteristic displayed by all living cells; it also may be used in a carefully designed apparatus to determine the molecular weights of macromolecules (see Chapter 2). A thermodynamic treatment of this situation is essential to understanding the quantitative relationship between the pressure and the molecular weight of the solute.

First of all, we know that the variation of the free energy of the solvent with pressure is given by Equation (20):

$$\left(\frac{\partial G_1}{\partial P}\right)_T = \bar{V}_1 \tag{20'}$$

where $\bar{V}_1$ is the partial molar volume of solvent. On one side of the membrane, where we have pure solvent,

$$\mu_1{}^\beta = \mu_1^\circ \tag{36}$$

On the other side of the membrane, where the chemical potential is reduced by a dilution factor but increased by a hydrostatic pressure, we find that

$$\mu_1{}^\alpha = \mu_1^\circ - RTX_2 + \int_{P_0}^{P_0 + \pi} \bar{V}_1 \, dP \tag{37}$$

where $P_0$ is the pressure at time zero and $\pi$ is the increase in pressure after equilibrium, called the osmotic pressure.

Water is not very compressible, so for low pressure and dilute solute, $\bar{V}_1 \cong \bar{V}_1^\circ$. Thus after integration this equation becomes

$$\mu_1{}^\alpha = \mu_1^\circ - RTX_2 + V_1^\circ\pi \qquad (38)$$

At equilibrium, $\mu_1{}^\beta = \mu_1{}^\alpha$, but since $\mu_1{}^\beta = \mu_1^\circ$, $\mu_1^\circ = \mu_1{}^\alpha$. Substituting in Equation (38), we get

$$\mu_1^\circ = \mu_1^\circ - RTX_2 + V_1^\circ\pi \quad \text{and} \quad \pi = \frac{RTX_2}{V_1^\circ} \qquad (39)$$

From Equation (34) we know that $X_2 \cong (C_2V_1^\circ)/M_2$, so that the osmotic pressure also may be expressed by the equation

$$\pi = RT\left(\frac{C_2}{M_2}\right) \qquad (40)$$

The osmotic pressure $\pi$ is the pressure difference required to equate the chemical potential of solvent on the two sides of the membrane. This equation shows that $\pi$ is inversely proportional to the molecular weight of the solute. Osmotic pressure is a simple way of measuring the molecular weight of a pure macromolecule (see Chapter 2), although it has been used less in recent years because of the availability of more efficient methods.

**Active Transport.**   Cells or organelles found in cells frequently have the capacity to concentrate small molecules uphill against a concentration gradient. From Equation (27) it is clear that transfer from a region of low concentration to a region of high concentration produces an increase in free energy. Since such processes do not occur spontaneously, free energy must be supplied from another source. For this reason the name _active transport_ has been used to refer to this type of reaction. A number of different mechanisms exist for active transport across membranes, and these are discussed later in this chapter and elsewhere in the book (see Chapters 10, 17, and 30).

## Chemical Equilibria in Solution

**The Relationship between Free Energy and the Equilibrium Constant.** Suppose we have a chemical reaction with the following stoichiometry:

$$a\text{A} + b\text{B} \rightleftharpoons c\text{C} + d\text{D}$$

where $a$, $b$, $c$, and $d$ refer to the moles of A, B, C, and D, respectively, at concentration $C_A$, $C_B$, $C_C$, and $C_D$. If the reaction occurs at constant $P$ and $T$, the free energy change $\Delta G$ for the reaction is

$$\Delta G = G_{\text{final state}} - G_{\text{initial state}} \qquad (41)$$

From Equation (18) for the total free energy of solution, we see that

$$\Delta G = d\mu_D + c\mu_C - a\mu_A - b\mu_B \qquad (42)$$

where $\mu_D$, $\mu_C$, $\mu_B$, and $\mu_A$ are the chemical potentials at the concentrations existing in the corresponding solutions. Recalling that $\mu_i = \mu_i^\circ + RT\ln C_i$

and substituting this equation for each component in Equation (42), we obtain

$$\Delta G = d\mu_D^\circ + c\mu_C^\circ - a\mu_A^\circ - b\mu_B^\circ + RT \ln \frac{C_D{}^d C_C{}^c}{C_A{}^a C_B{}^b} \tag{43}$$

The first four terms in this expression are simply the standard-state free energy change $\Delta G^\circ$ for the reaction, so that

$$\Delta G = \Delta G^\circ + RT \ln \frac{C_D{}^d C_C{}^c}{C_A{}^a C_B{}^b} \tag{44}$$

At equilibrium $\Delta G = 0$ and

$$\frac{C_D{}^d C_C{}^c}{C_A{}^a C_B{}^b} = K_{eq}$$

Therefore

$$\Delta G^\circ = -RT \ln K_{eq} \tag{45}$$

The equilibrium constant for a reaction may be used to estimate the free energy of a reaction. More frequently, free energies are used to estimate equilibrium constants. Indeed, Equation (45) is probably the most important equation for predicting the direction and extent of a chemical reaction. Because of the logarithmic relationship, the equilibrium constant has a very high dependence on the free energy. A reaction that proceeds to 99 percent completion is, for most practical purposes, a quantitative reaction. This requires an equilibrium constant of 100 but a free energy difference of only $-2.7$ kcal, which is little more than half the free energy for the formation of a hydrogen bond.

**Variables Affecting the Free Energy of a Reaction.** The determination of free energy values under conditions deviating from the standard state is of great importance because we seldom study reactions under standard conditions. The free energy under conditions other than the standard-state concentrations is written as $\Delta G$ with the superscript degree omitted. For the reaction of A and B to form C and D, with small letters representing the number of moles of each component, we return to Equation (44):

$$\Delta G = \Delta G^\circ + RT \ln \frac{C^c D^d}{A^a B^b} \tag{44'}$$

If all the concentrations are at the standard state, i.e., 1 $M$, then the second term on the right side of this equation becomes 0 and $\Delta G = \Delta G^\circ$. At concentrations other than 1 $M$, the value of $\Delta G$ may be calculated from this equation, from $\Delta G^\circ$, and measured or estimated values of the concentrations. This has been done below to obtain an estimate of the free energy of hydrolysis of ATP at concentrations closer to those existing in the cell.

The easiest way to remember the sign of this correction is to use simple logic. If, for example, the products are at a lower concentration than that of the standard state, the reaction will have a greater tendency to proceed and the free energy of the reaction will be more negative. The opposite applies to diluting the reactants. The free energy of dilution of hydrogen ions from 1 $M$ to $10^{-7}$ $M$ at 37°C is equal to $RT \ln 10^{-7} = 1.987 \times 31 \times 2.3 \times$

$(-7) = -9.87$ kcal. This is a large term. It actually produces more free energy than the hydrolysis of 1 $M$ adenosine triphosphate.

Free energy also depends on both the temperature and the pressure. The dependence of the free energy on temperature can be determined by the entropy of the reaction at constant pressure [see Equation (20)]. Conversely, Equation (20) can be useful for determining the entropy of a reaction by measuring the free energy at various temperatures.

Often the entropy of the reaction is not known, but the enthalpy is known, or one again wishes to determine the enthalpy from measurements of the free energy at various temperatures. The *Gibbs-Helmholtz equation* relates the temperature dependence of the free energy to the enthalpy of the reaction:

$$\left[ \frac{\delta(\Delta G/T)}{\delta(1/T)} \right]_P = \Delta H$$

A straight line is obtained over a range of temperatures for which the enthalpy is relatively constant if the free energy at each temperature divided by that temperature is plotted against the reciprocal of the absolute temperature.

The free energy also depends on the pressure because the enthalpy depends on the pressure, but this is usually of little consequence to biochemists except in the case of osmotic pressure, which was dealt with earlier.

**Free Energy and Electrical Work.** Since the free energy change determines the maximum work other than $P \, dV$ work that a reaction can produce, we can use it to calculate energy available to carry out electrical work, i.e., the ability of a system to transfer electrons. For this purpose we introduce a new term, called the *electromotive force* (or *electromotive potential*). The electromotive force of a molecule is a measure of the tendency of the molecule to release or to accept electrons. The standard free energy of a reaction involving a transfer of electrons is proportional to the negative of the number of electrons transferred times the electromotive force. The equation that expresses this is

$$\Delta G° = -nFE° \tag{46}$$

where $n$ is the number of electrons and $F$ is the proportionality constant between electron volts ($nE°$) and calories. The proportionality constant $F$, called the *Faraday constant*, is equal to 23,060 cal/eV. The symbol $E°$ is for the standard potential of the reaction. This is the potential observed in a cell where all reactants and products are in their standard states. The electromotive force is comparable to pressure in a water pipe. The electromotive force is independent of the number of electrons, just as the pressure is independent of the amount of water. However, the work the water can do depends on the amount of flow times the pressure, just as the free energy depends on the number of electrons times the electromotive force. This is the reason that we must know the number of electrons when free energy is converted to electromotive force or electromotive force is converted to free energy. The use of standard potentials to calculate standard free energies for oxidation-reduction reactions is described in Chapter 10.

Another useful expression is that between the equilibrium constant and the standard electromotive potential:

$$E° = \frac{RT}{nF} \ln K_{eq} = 2.3 \frac{RT}{nF} \ln K_{eq} \tag{47}$$

This relationship is easy to derive from previously described relationships between the equilibrium constant and the free energy (Equation 43) and the standard potential and the free energy (Equation 46). The logarithmic relationship means that a small change in potential will correspond to a large change in equilibrium constant. A reaction that goes to 99 percent completion ($K_{eq} = 100$) requires a driving potential of only 118 mV.

The electromotive potential at concentrations of reactants and products other than $1 M$ is given by the Nernst equation:

$$E = E° - \frac{RT}{nF} \ln \frac{C_D{}^d C_C{}^c}{C_A{}^a C_B{}^b} \tag{48}$$

for the reaction $aA + bB \rightleftharpoons cC + dD$.

Electric potentials are used to drive a number of reactions at membrane surfaces. The lipid membranes that exist at cell surfaces or organelle surfaces (e.g., in mitochondria and chloroplasts) do not permit free ion flow between the inside and the outside of membrane-bound structures. This creates a situation where it is possible to achieve stable transmembrane concentration gradients and electric potentials. In order to do this, selective pumps that translocate specific ions across membranes are used. The transmembrane potentials, in turn, can be used to provide energy for driving reactions that occur at the membrane surfaces. Numerous examples relating to the usefulness of transmembrane potentials are given in Chapters 10, 17, and 30. At constant $T$ and $P$, the general relationship for the difference in electrochemical potential $\Delta\bar{\mu}_j$ of an ion species $j$ across the membrane is given by the equation

$$\Delta\bar{\mu}_j = Z_j F \Delta\psi + RT \ln \frac{C_j^{outside}}{C_j^{inside}} \tag{49}$$

where $C_j^{outside}$ and $C_j^{inside}$ are the concentrations of ion species $j$ on either side of the membrane-bound structure, $Z_j$ is the valency (with sign), and $\Delta\psi$ is the transmembrane potential. If a proton is the ion species in question, this equation becomes

$$\Delta\bar{\mu}_{H^+} = F \Delta\psi - 2.3RT \Delta pH \tag{50}$$

since $Z = +1$ and $pH = -\log (H^+)$. These two equations have been used as the starting point for explaining the energetics of electrochemical potentials, how they are used in active transport (Chapters 17 and 30), and how they are believed to be coupled to endothermic reactions such as the phosphorylation of ADP to ATP at the membrane surface (see Chapters 10 and 17).

**Coupled Reactions.** Chemists commonly study isolated reactions. However, reactions in living systems are usually not isolated but are part of a long, complicated sequence. The crucial thermodynamics is not that of individual reactions but of the initial and final states of a related sequence.

An important consequence of this is that an individual reaction with an unfavorable equilibrium may proceed if the total sequence has a favorable equilibrium constant. Biochemical pathways frequently use this strategy, so that *if the cell needs a reaction to proceed that will not proceed in isolation* $(+\Delta G)$, *the reaction is coupled to another favorable reaction* $(-\Delta G)$.

The advantages of coupling can be appreciated by consideration of the equilibrium constants for each reaction of a coupled series. For example, let us consider the following sequence:

$$A \rightleftharpoons B \rightleftharpoons C$$

We shall assume that the equilibrium constant between A and B is $K_1 = 0.1$ and the equilibrium constant between B and C is $K_2 = 100.0$. Because of the low equilibrium constant, only small quantities of A would be consumed if the conversion of A to B were not followed or driven by the subsequent conversion of B to C.

At equilibrium,

$$K_1 = \frac{B}{A} = 0.1 \quad \text{and} \quad K_2 = \frac{C}{B} = 10.0$$

The equilibrium constant for the overall process is the product of the equilibrium constants for the individual reactions:

$$K_1 K_2 = \frac{C}{A} = 100$$

Thus when the reaction A to B is driven by a subsequent favorable reaction B to C, we find that most of A reacts. This type of coupling is a common occurrence in biochemical sequences. Consider, for example, the condensation of acetyl-CoA with oxaloacetate to form citric acid. The hydrolysis of the coenzyme A off the citric acid provides the driving force for the reaction (Figure 7-9). The first reaction in this sequence has a free energy change of only $-0.05$ kcal/mol. However, the second reaction in this sequence, involving the hydrolysis of acetyl-CoA at pH 7, has a free energy change of $-8.4$ kcal/mol, which causes the overall reaction to proceed.

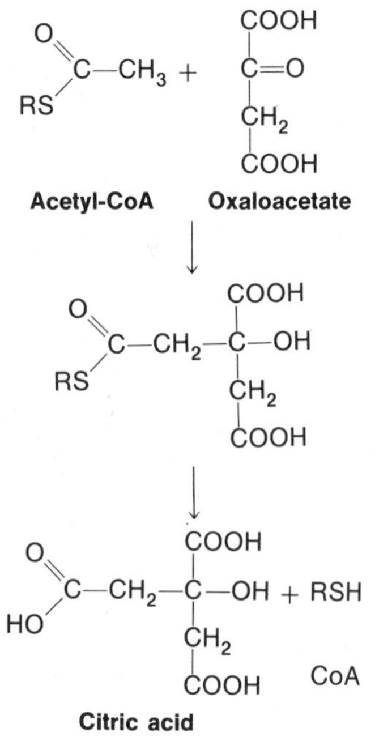

Another and perhaps more common type of coupling reaction is that in which the two reactions occur simultaneously. For example, assume the reaction $A \rightleftharpoons B$ has an equilibrium constant again of only 0.1 and the reaction of $C \rightleftharpoons D$ has an equilibrium constant of 100. If these two reactions occur simultaneously, the reaction of A to B will be driven by the C to D reaction:

$$A + C \overset{K_3}{\rightleftharpoons} B + D$$

$$K_1 = \frac{B}{A} \quad K_2 = \frac{D}{C} \quad K_3 = K_1 K_2 = 10 = \frac{[B][D]}{[A][C]}$$

An example of this type of reaction is the phosphorylation of glucose by ATP to produce ADP and glucose-6-phosphate. The reaction of glucose with phosphoric acid to form glucose-6-phosphate and water has an unfavorable equilibrium constant of $0.0062\ M^{-1}$. The hydrolysis of ATP has a quite favorable equilibrium constant equal to $9.2 \times 10^5$. When these two

**Figure 7-9**
The condensation of acetyl-CoA with oxaloacetic acid followed by the hydrolysis of the acetyl-CoA ester.

reactions are simultaneously coupled, the overall reaction is favorable with an equilibrium constant equal to $5.7 \times 10^3$:

$$\text{Glucose} + \text{ATP} \rightleftharpoons \text{G-6-P} + \text{ADP}$$

Notice from the two previous examples that when free energy is being calculated we add the free energies for the individual reactions, but that when the equilibrium constant is being calculated we multiply the equilibrium constants for the individual reactions.

**Binding Phenomena.** Biochemists are often faced with the problem of binding small molecules or ligands to proteins or nucleic acids. Proteins in particular commonly bind many ligands, including substrates, inhibitors, and activators. Often these ligands are bound to more than one site. Hemoglobin binds a total of four oxygen molecules (see Chapter 3), and serum albumin binds an extremely varied group of substances at a large number of available sites.

Binding studies require a measurement of the free ligand [B] in the presence of the macromolecule [A]. From this we may determine the *average binding number*, called $\nu$. The value of $\nu$ is equal to the ratio of the total number of B molecules bound to the total number of binding sites. If there is only one binding site per A molecule, the expression for $\nu$ is rather simple:

$$\nu = \frac{[AB]}{[A] + [AB]} \tag{51}$$

The value of $[A] + [AB]$ is known from the total amount of A added to the solution. The value of $[AB]$ is equal to the total amount of B added minus the experimentally observed concentration of free B after A is added:

$$\nu = \frac{\text{total B} - \text{free B}}{\text{total A}} \tag{52}$$

Let us first consider the simple binding of one ligand B to a protein molecule A with a formation constant $K_f$:

$$A + B \xrightleftharpoons{K_f} AB \qquad K_f = \frac{[AB]}{[A][B]} \tag{53}$$

From Equations (51) and (53) we can express $\nu$ in terms of $K_f$ or $K_d$, the *dissociation constant:*

$$\nu = \frac{K_f[B]}{1 + K_f[B]} \tag{54}$$

or since $K_f = 1/K_d$,

$$\nu = \frac{[B]}{[B] + K_d} \tag{55}$$

Taking the reciprocal of both sides,

$$\frac{1}{\nu} = 1 + K_d\left(\frac{1}{[B]}\right) \tag{56}$$

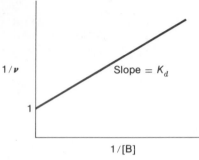

**Figure 7-10**
A binding plot to determine the dissociation constant $K_d$ for the simple
situation where there is one ligand
binding site. $\nu$ is the average binding
number and [B] is the concentration
of ligand.

By plotting the experimentally determined value of $1/\nu$ against $1/[B]$, one
obtains a straight line with a slope equal to the dissociation constant $K_d$.
The intercept on the ordinate should be 1 (Figure 7-10).

When there is more than one site on A for binding B, the equations
become more complex and in general the plots are not linear. If we consider
the average binding number for $n$ sites on a molecule, the total average
binding number is the sum of the binding number for each of these sites.

$$\nu = \sum_{i=1}^{n} \frac{K_{fi}[B]}{1 + K_{fi}[B]} \tag{57}$$

There are two situations in which the data for multiple binding may be
treated rather simply. First, when all binding sites bind B with the same
energy, all terms are equal and the solution to the sum is simply $n$ times
each term:

$$\nu = \frac{nK_f[B]}{1 + K_f[B]} \tag{58}$$

Again, since $K_f = 1/K_d$,

$$\nu = \frac{n[B]}{[B] + K_d} \tag{59}$$

and taking the reciprocal of both sides,

$$\frac{1}{\nu} = \frac{1}{n} + \frac{1}{n}K_d\left(\frac{1}{[B]}\right) \tag{60}$$

If $1/\nu$ is plotted against $1/[B]$, the slope is equal to $K_d/n$ and the intercept
(when $1/[B]$ approaches 0) is $1/n$. Alternatively, the data may be plotted as
$\nu$ versus $\nu/[B]$ (Figure 7-11), in which case the slope is $-K_d$ and the intercept at $\nu/[B] = 0$ is $n$, since

$$\nu = n - \frac{\nu K_d}{[B]} \tag{61}$$

Second, in one other situation, multiple binding can be treated rather
simply. If the sites interact so strongly that only the fully saturated product
$AB_n$ is formed, the data may be treated in the following way. The formation
of only one major product means that the binding of the ligand molecule to
the macromolecule greatly enhances the further binding of additional ligands such that at equilibrium, only three species exist in significant concentrations: A, B, and $AB_n$:

$$A + nB \rightleftharpoons AB_n$$

$$\nu = \frac{n[AB_n]}{[AB_n] + A} \tag{62}$$

$$\nu = \frac{nK_f[B]^n}{1 + K_f[B]^n} \tag{63}$$

$$\frac{1}{\nu} = \frac{1}{n} + \frac{1}{nK_f[B]^n} \tag{64}$$

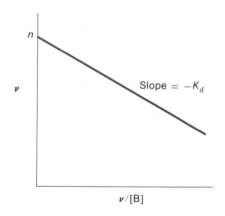

**Figure 7-11**
A binding plot to determine the dissociation constant $K_d$ and the number of
binding sites $n$ for the situation where
there are $n$ ligand binding sites per
macromolecule with identical
binding affinities.

Under these conditions, a plot of $1/v$ against $1/[B]$ is not linear. However, since

$$\frac{v}{n - v} = K_f[B]^n \tag{65}$$

and

$$\log \frac{v}{n - v} = \log K_f + n \log [B] \tag{66}$$

a plot of $\log v/(n - v)$ versus $\log [B]$ is linear. The intercept of this plot at $\log [B] = 0$ is $\log K_f$ and the slope is $n$. This is called a *Hill plot* (Figure 7-12). It usually requires a guess at the value of $n$. By trial and error, the value of $n$ is varied to give a linear plot. Often nonintegral values of $n$ are reported that describe the data but lack precise meaning. For example, in the case of hemoglobin binding of oxygen, a value of 2.8 is reported for $n$, and this indicates some cooperativity in the binding (see Chapter 3).

If none of the preceding simplifying conditions apply, it is still possible to determine all the equilibrium constants. More advanced texts should be consulted for the use of these methods.

**Use of Equilibrium Constants to Calculate Equilibrium Concentrations for a Reaction.** The equilibrium constant is most useful for calculating the concentrations of reactants and products for a reaction at equilibrium. Assume the initial concentrations before reaction are known and we wish to determine the equilibrium concentrations. For example, in the reaction $A + B \rightleftharpoons C$, we start with a concentration of A equal to $A_0$ and a concentration of B equal to $B_0$. Let $X$ be the equilibrium concentration of C. According to the stoichiometric equation for the reaction, when 1 mol C is formed, 1 mol A and 1 mol B is consumed. We can write $A_0 - X$ for the equilibrium concentration of A and $B_0 - X$ for the equilibrium concentration of B. Then the equilibrium expression for the reaction becomes

$$K = \frac{[C]}{[A][B]} = \frac{[X]}{([A_0] - [X])([B_0] - [X])} \tag{67}$$

and

$$KA_0B_0 - X(KA_0 + KB_0 + 1) + KX^2 = 0 \tag{68}$$

This expression is a quadratic in $X$ and is inconvenient to solve. If the reaction does not proceed very far and $X$ is small compared with $A_0$ and $B_0$, the term $KX^2$ can be neglected and we obtain the expression

$$KA_0B_0 - X(KA_0 + KB_0 + 1) = 0 \tag{69}$$

or

$$X = \frac{KA_0B_0}{A_0K + B_0K + 1} \tag{70}$$

$X$ is a function of the equilibrium constants and the initial concentrations of A and B, and it is easy to compute from this equation. If the solution showed that $KX^2$ was not negligible, one would have to go back and use a more complex expression. This type of approximation is very common and often quite accurate if the equilibrium constant is either large or small.

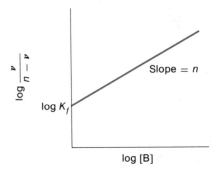

**Figure 7-12**
A Hill plot

One should always consider first what reasonable approximations can be made before solving quadratic equations derived from equilibrium expressions.

The most common equilibria that biochemists encounter are those of acids and bases. The dissociation of an acid may be written as

$$HA \rightleftharpoons H^+ + A^-$$

So the equilibrium is written as

$$K = \frac{[H^+][A^-]}{[HA]} \qquad (71)$$

The equilibrium constant written in this manner is called the _acidity constant_ and is represented by the symbol $K_a$.

The dissociation of water, which could be considered a weak acid, can be represented by the following equilibrium:

$$H_2O \rightleftharpoons H^+ + OH^-$$

for which the equilibrium expression would be

$$K = \frac{[H^+][OH^-]}{[H_2O]} \qquad (72)$$

Because the concentration of water is high and never varies significantly for chemical reactions in aqueous solutions, the denominator in this equation is effectively constant and for simplicity is given a value of unity. The constant $K_w$ for the dissociation of water is redefined by the expression

$$K_w = [H^+][OH^-] = 10^{-14} \text{ (mol/liter)}^2 \qquad (73)$$

at 25°C.

Strong acids dissociate completely into anions and protons. In solutions of strong acids, the concentration of hydrogen ion $[H^+]$ is equal to the total concentration $C_{HA}$ of the acid HA that is added to the solution. Thus the pH of the solution of a strong acid is simply $-\log C_{HA}$. The pH of the solution of a weak acid is not quite so simple because the acid is incompletely dissociated. The dissociation constant of a weak acid may be written in terms of the species present as in Equation (71). By taking the negative logarithm on both sides of this equation and transposing we obtain the _Henderson-Hasselbach equation_:

$$pH = pK_a + \log\left[\frac{A^-}{HA}\right] = pK_a + \log\left[\frac{base}{acid}\right] \qquad (74)$$

When the concentration of anion or base is equal to the concentration of undissociated acid (i.e., _when the acid is half neutralized), the Henderson-Hasselbach equation shows that the pH of the solution is equal to the pK of the acid._ A typical pH dependence curve for the titration of a weak acid by a strong base is shown in Figure 7-13. The point where the concentration of the salt and acid are equal is when the acid is exactly half neutralized. Notice at this point on the curve of Figure 7-13 that the pH is almost independent of added base (or acid). _Biochemical reactions are commonly highly dependent on the pH of the solution. Because of this, it is frequently advantageous to study reactions in such buffered solutions._

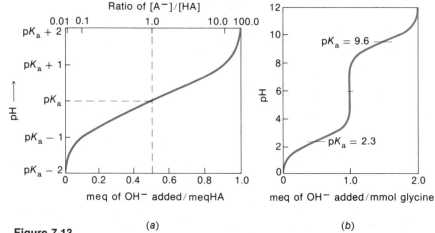

**Figure 7-13**
The dependence of pH on the equivalents of base added to a typical weak acid
(a) and to a dibasic acid glycine (b). The amino acid has two $pK_a$'s, one corresponding to the titration of the $\alpha$-COOH group ($pK_a = 2.35$) and the $\alpha$-NH$_3^+$
group ($pK_a = 9.6$) respectively.

**Equilibrium and Thermodynamics Associated with ATP Hydrolysis.** By
far the most important compound for energy storage in the cell is adenosine triphosphate (ATP). ATP serves as a common medium of exchange of
energy between energy-producing and energy-consuming systems (see
Chapter 8). Energy-producing systems make ATP and energy-requiring systems consume ATP in coupled reactions. In humans about 5 lb of ATP is
consumed every day to satisfy normal energy needs. Adenosine triphosphate can hydrolyze by two different reactions (Figure 7-14). The $\alpha$, $\beta$
linkage can be hydrolyzed to form adenosine monophosphate (AMP) and
pyrophosphate ion or the $\beta$, $\gamma$ linkage can be hydrolyzed to form adenosine
diphosphate and phosphate ion. The free energy of hydrolysis for these
reactions at pH 7 is about equal. In the first reaction the pyrophosphate
initially produced can be subsequently hydrolyzed to ensure the irreversibility of the reaction. The first reaction is used primarily in situations
where any reversibility would be intolerable, such as in nucleic acid synthesis. The second reaction is used in most processes where a limited
amount of reversibility is tolerable. It releases about the same amount of
energy as the first reaction and has the advantage that less energy is required to recharge the ADP product than would be required to recharge AMP.

The free energy values for these hydrolyses and a number of other
organophosphate compounds are given in Table 7-6. It can be seen that
considerable free energy is released when the phosphate group(s) is hydrolyzed from either ATP or ADP but not from AMP. For this reason the $\alpha$, $\beta$
and the $\beta$, $\gamma$ linkages (see Figure 7-14) are often referred to as high-energy
linkages. In fact, the reasons for the relatively large negative free energy for
the first two hydrolyses must be discussed from two standpoints: the factors leading to the relative instability of the reactant, ATP, and the factors
leading to the relative stability of the products. Comparison of these reac-

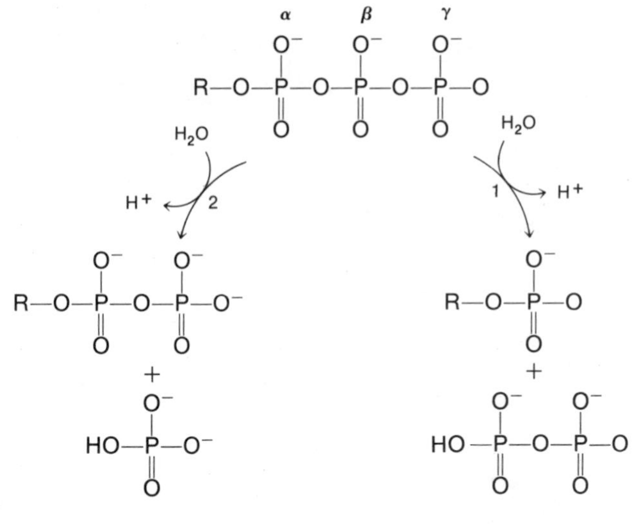

**Figure 7-14**
Alternate routes of ATP hydrolysis. The adenosine group is represented by R. The charge groups shown are the main ones that exist at physiological pH and ionic strength. The phosphate groups are often referred to as $\alpha_1$, $\beta$, and $\gamma$ as shown.

tions in Table 7-6 suggests two things. The reacting species that yield the higher free energies on hydrolysis are more highly negatively charged and produce a proton on hydrolysis. Resonance also stabilizes the products more than the reactants, which increases the negative free energy change on hydrolysis. The relative contributions of these different factors to the free energy of hydrolysis is still unclear.

The main event taking place when ATP hydrolyzes to ADP and phosphate at neutrality is described by the following equation:

$$ATP^{4-} + H_2O \rightleftharpoons ADP^{3-} + HPO_4^{2-} + H^+$$

The equilibrium constant $K$ for this reaction is equal to 0.63:

$$K = 0.63 = \frac{[ADP^{3-}][HPO_4^{2-}][H^+]}{[ATP^{4-}]} \tag{75}$$

**Table 7-6**
Approximate Free Energies of Hydrolysis for ATP and Some Other High-Energy Phosphate Compounds of Biological Interest

| Principal Reacting Species | Hydrolysis Products | $\Delta G^{\circ\prime}$ (kcal/mol) |
|---|---|---|
| $ATP^{4-}$ | $ADP^{3-} + HPO_4^{2-} + H^+$ | $-8.2$ |
| $ATP^{4-}$ | $AMP^{2+} + H_2P_2O_7^{3-} + H^+$ | $-8.4$ |
| $ADP^{3-}$ | $AMP^{2-} + HPO_4^{2-} + H^+$ | $-8.6$ |
| $AMP^{2-}$ | Adenosine $+ HPO_4^{2-}$ | $-2.2$ |
| $HP_2O_7^{3-}$ | $2 HPO_4^{2-} + H^+$ | $-7.9$ |
| Acetyl phosphate | Acetate$^- + HPO_4^{2-} + H^+$ | $-11.3$ |
| Phosphoenolpyruvate | Pyruvate$^- + HPO_4^{2-}$ | $-14.7$ |
| Phosphocreatine | Creatine$^+ + HPO_4^{2-}$ | $-10.2$ |
| Phosphoarginine | Arginine$^+ + HPO_4^{2-}$ | $-9.1$ |

Biochemists commonly study reactions at or close to neutrality. Instead of continually correcting from one normal acid or base to neutrality, the values of free energy and the equilibrium constant are commonly reported for solutions at neutrality, i.e., pH 7. These values are designated by a prime and written as $\Delta G^{\circ\prime}$, $\Delta G'$, and $K'$. For the equilibrium constant $K'$ we redefine the standard state of proton as $10^{-7}$ $M$; it should be noted that for this reaction, $K' = K/[H^+]$. One must be very careful in evaluating the free energies $\Delta G^{\circ}$ or $\Delta G^{\circ\prime}$ to use the equilibrium constants $K$ or $K'$, respectively, since these are very different quantities for reactions such as ATP hydrolysis that involve the uptake or release of protons. Further complications in evaluating the equilibrium constant for ATP hydrolysis arise because $ATP^{4-}$, $ADP^{3-}$, and $HPO_4^{2-}$ are not the only species present at pH 7. Each of these phosphate-containing moieties is partially protonated. The most useful equilibrum constant at pH 7 would be for the reaction that takes the concentration of all species of ATP, ADP, and P into account. This equilibrium constant is designated by the symbol $K^{\dagger}$ and represented by the equation

$$K^{\dagger} = \frac{[\text{ADP, all forms}][\text{P, all forms}]}{[\text{ATP, all forms}]} \tag{76}$$

The values of $K^{\dagger}$ may be calculated from $K$ by the following expression, where $f$ATP, $f$ADP, and $f$P are the ratios of $ATP^{4-}$ to the total ATP, of $ADP^{3-}$ to the total ADP, and of $HPO_4^{2-}$ to the total inorganic phosphate, respectively:

$$K^{\dagger} = \frac{K}{[H^+]} \frac{f\text{ATP}}{f\text{ADP} \cdot f\text{P}} \tag{77}$$

To evaluate the factors $f$ATP, etc. requires a knowledge of the ionization constants of ATP and other species:

$$[\text{ATP}] \simeq [\text{ATP}^{4-}] + [\text{ATP}^{3-}] + [\text{ATP}^{2-}] \tag{78}$$

$$[\text{ATP}] \simeq [\text{ATP}^{4-}]\left[1 + \frac{(H^+)}{(K_1\text{ATP})} + \frac{(H^+)_2}{(K_1\text{ATP})(K_2\text{ATP})}\right] \tag{79}$$

In this expression, $K_1$ATP and $K_2$ATP are the dissociation constants for $HATP^{3-}$ and $H_2ATP^{2-}$, respectively. The values of these dissociation constants are given in Table 7-7. The factor $f$ATP for any given pH can be determined from these dissociation constants and the pH:

$$f\text{ATP} = \frac{1}{1 + \dfrac{(H^+)}{(K_1\text{ATP})} + \dfrac{(H^+)^2}{(K_1\text{ATP})(K_2\text{ATP})}} \tag{80}$$

Similar expressions may be derived for $f$ADP and $f$P. Actually, at pH 7, the differences between $K$ and $K^{\dagger}$ only produce a difference in free energy of hydrolysis of ATP of about 2 percent. At lower pHs, the difference is greater, whereas at higher pH values, the discrepancy is less.

The preceding discussion has only considered the association of protons with the phosphates involved in ATP hydrolysis. Almost all reactions involving ATP hydrolysis are carried out in the presence of magnesium ions or other divalent cations that have significant associations with both ATP and ADP. Similar terms to those used for protons may be added to

**Table 7-7**
Acid Dissociation Constants at 25°C and 0.2 Ionic Strength

| | |
|---|---|
| $HATP^{3-} \rightleftharpoons H^+ + ATP^{4-}$ | $1.12 \times 10^{-7}$ |
| $H_2ATP^{2-} \rightleftharpoons H^+ + HATP^{3-}$ | $8.71 \times 10^{-5}$ |
| $HADP^{2-} \rightleftharpoons H^+ + ADP^{3-}$ | $1.32 \times 10^{-7}$ |
| $H_2ADP^- \rightleftharpoons H^+ + HADP^{2-}$ | $1.18 \times 10^{-4}$ |
| $H_2PO_4^- \rightleftharpoons H^+ + HPO_4^{2-}$ | $1.66 \times 10^{-7}$ |

$f$ATP, $f$ADP, and $f$P for the magnesium chelate species and used in the equation for $K^\dagger$ to determine the $K^\dagger$ at any pH in the presence of magnesium or any other ion. For example, the total concentration of ATP in terms of $ATP^{4-}$ in the presence of $Mg^{2+}$ ions has two extra terms:

$$[ATP] = [ATP^{4-}]\left[1 + \frac{Mg^{2+}}{(K MgATP)} + \frac{(H^+)}{(K_1 ATP)} + \frac{Mg^{2+}}{(K MgHATP)} + \frac{(H^+)^2}{(K_1 ATP)(K_2 ATP)}\right] \quad (81)$$

The constants $K$MgATP and $K$MgHATP are the equilibrium constants for the dissociation of magnesium ions from MgATP and from MgHATP, respectively:

$$MgATP^{2-} \longrightarrow Mg^{2+} + ATP^{4-}$$

$$MgHATP^- \rightleftharpoons Mg^{2+} + HATP^{3-}$$

The effect of magnesium ions on the free energy of hydrolysis of ATP is complex, since it depends on the pH and the ionic strength. When all the relevant calculations are made, it is found that magnesium ions cause slightly less negative values for the free energy of hydrolysis of ATP to ADP and phosphate. A value of $-8.4$ kcal/mol for the hydrolysis of ATP to ADP at pH 7, 0.2 ionic strength, 25°C, and in the absence of magnesium becomes $-7.7$ kcal/mol in the presence of 0.001 $M$ magnesium ions and decreases further to $-7.5$ kcal/mol in the presence of 0.01 $M$ magnesium ions.

Concentrations of the various reactants and products in the ATP hydrolysis can have a much larger effect on the computed free energy of hydrolysis. Assuming that all phosphate-containing molecules (ATP, ADP, and P) are present at the same concentration (this is a gross assumption), the free energies of hydrolysis at concentrations other than the standard state may be calculated from Equation (44) (see Table 7-8). The value of $-8.4$ kcal/mol at 1 $M$ becomes $-12.5$ kcal/mol at 1 m$M$. In fact, the concentrations of the ATP, ADP, and P in most cells are probably somewhere between 2 and 10 m$M$. Also shown in Table 7-8 are the variations in the free energy of hydrolysis over a range from pH 6 to 8.

### Energetics of Biopolymer Conformation

Whereas biopolymer formation is a highly endothermic reaction requiring the input of vast sums of energy, once formed, most biopolymers will adopt their natural three-dimensional conformations spontaneously. The primary structure is designed so that the transition from the extended poly-meric chain to the final tertiary conformation is mildly exothermic. A thermodynamic analysis of this transition is both instructive and has pre-dictive value. The conformations adopted by biopolymers (proteins, nu-cleic acids, polysaccharides, and lipids) depends on three general types of energy contributions. These are (1) steric effects, (2) noncovalent bonds, and (3) entropy effects. The importance of these factors in determining protein conformation or folding has been discussed in Chapter 3. Owing mainly to x-ray diffraction studies, more is known about the conformation of proteins than about the conformations of other types of biopolymers. At the same time, proteins are more varied in the types of structures they form because of the diversity of side-chain composition and sequences. Because of this, more has been learned about the principles controlling biopolymer structure from proteins than from all other biopolymers collectively.

Each type of biopolymer has a characteristic repeating backbone link-age: peptide bonds in proteins, phosphodiester linkages in nucleic acids, glycosidic linkages in polysaccharides, and carbon-carbon bonds in satu-rated fatty acids. Each biopolymer has different types of chemical groupings that are attached to the repeating backbone. Each side chain must be con-sidered individually for the effect it will have on determining the native conformation adopted by the various biopolymers; the underlying thermo-dynamic principles controlling the outcome are the same. Let us first re-view these principles as they apply to proteins (see Chapters 1 and 3) and follow this by a consideration of the conformations adopted by lipids (dis-cussed in Chapters 13 and 16), carbohydrates (discussed in Chapter 12), and nucleic acids (discussed in Chapter 18).

The polypeptide chain has certain dimensions that are fixed by cova-lent bond distances and bond angles (see Figure 3-7). The peptide grouping itself is planar because of resonance (see Figure 3-7). Those linkages in the

**Table 7-8**
Free Energy of Hydrolysis for ATP as a Function
of pH and Concentration

| pH | $-\Delta G^{\dagger}$ | | | |
|---|---|---|---|---|
| | **1 M** | **0.1 M** | **0.01 M** | **0.001 M** |
| 6.0 | 7.89 | | | |
| 6.2 | 7.94 | | | |
| 6.4 | 8.01 | | | |
| 6.6 | 8.12 | | | |
| 6.8 | 8.26 | | | |
| 7.0 | 8.40* | 9.78 | 11.15 | 12.52 |
| 7.2 | 8.59 | | | |
| 7.4 | 8.81 | | | |
| 7.6 | 9.06 | | | |
| 7.8 | 9.32 | | | |
| 8.0 | 9.56 | | | |

*Calculated from R. W. Guynn and R. L. Veech, The equilibrium constants of the adenosine triphosphate hydrolysis and the adenosine triphosphate citrate lyase reactions, *J. Biol. Chem.* 248:6966, 1973.

backbone which are single-bond in character can rotate freely as long as steric repulsion effects that would result from different groupings coming too close together do not interfere with these rotations. The distance at which a steric repulsion effect becomes serious is just below the distance computed from the van der Waals radii of the respective nonbonded atoms (see Table 3-3). Conceivably this distance of closest approach could be accommodated by changing some of the covalent bond distances or covalent bond angles. However, little flexibility is provided here because deviating from the favored bond distances or bond angles for the covalently bonded atoms requires large energies. These initial steric considerations place rather strict limitations on the kinds of structures that are possible. Subject to such limitations, the weak intermolecular interaction forces we must consider are those which form between various groups of the polypeptide chain, within the water, and between the water and the polypeptide. These weak forces include hydrogen and electrostatic bonds of various types, i.e., ion-ion, ion-dipole, and dipole-dipole interactions. Arguments are raised below and in Chapter 3 that van der Waals attractive forces are not a significant factor in determining conformation in aqueous solution, so that to a good approximation they can be ignored. Two types of entropy effects are highly significant: one concerns the polypeptide chain; the other concerns the water. An ordered polypeptide chain has a lower entropy than a disordered polypeptide chain, making the transition of a polypeptide chain to an ordered state energetically unfavorable as far as this entropy effect on the polypeptide chain is concerned. Hydrogen and electrostatic bonds formed between water and polar groups on the polypeptide chains have a favorable enthalpy of formation in many cases, but such interactions invariably restrict the mobility of the water and therefore have an unfavorable entropy for formation. The enthalpy term is usually more important than the entropy term, leading to an overall free energy that is favorable for hydration. The hydration of apolar groups has been discussed earlier in this chapter. Water in the immediate vicinity of an apolar amino acid side chain is believed to form a rigid ice-like structure that lowers its entropy. In such cases the small favorable negative $\Delta H$ owing to the formation of the rigid water structure is more than counterbalanced by the unfavorable entropy, so that such hydrated structures are energetically unfavored. This conclusion is substantiated by the measurement of thermodynamic parameters associated with the transfer of the apolar molecules, such as methane and ethane, from an apolar solvent to water (see Table 7-9). The enthalpy for these transfers ($\sim -2.6$ kcal/mol) is weakly favorable, but the entropies are unfavorable (about $-19$ eu). In such cases, the entropies dominate the situation, leading to unfavorable free energies for these transfers. It has

**Table 7-9**
Thermodynamic Parameters for Transferring Methane and Ethane from Benzene to Water at 25°C

| Transferred Molecule | $\Delta H$ (kcal/mol) | $\Delta S$ [cal/(°C · mol)] | $\Delta G$ (kcal/mol) |
|---|---|---|---|
| $CH_4$ | $-2.8$ | $-18$ | $+2.6$ |
| $C_2H_6$ | $-2.2$ | $-20$ | $+3.8$ |

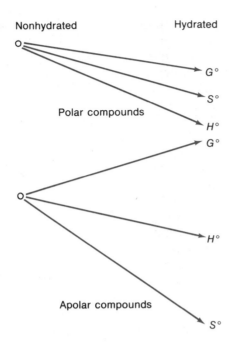

Nonhydrated          Hydrated

$G°$

$S°$

Polar compounds

$H°$

$G°$

$H°$

Apolar compounds

$S°$

**Figure 7-15**
A comparison of the thermodynamics of hydration of polar and apolar molecules.

been calculated that the same general situation holds for the hypothetical transfer of the side chains of apolar amino acids. The thermodynamic situation for apolar and polar compounds in making the transition from a nonhydrated to a hydrated state is summarized in Figure 7-15. As discussed earlier, the signs of both the enthalpy change and the entropy change are the same as those for the solvation of polar and apolar groups (all negative), but the difference lies in the fact that for the apolar molecule, the entropy term is more important than the enthalpy term and the overall free energy is positive. Because of the large entropy decrease and not because of the enthalpy, apolar molecules have a low solubility in water. When two apolar groups of a biopolymer approach each other in an aqueous solution, the opposite occurs. The enthalpy of the process is positive (unfavorable), but the release of water molecules hydrogen bonded around the apolar group increases the entropy enough to make the overall interaction between the apolar groups favorable. This interaction between apolar groups, called _hydrophobic bonding_, is an entropy effect.

As stated in Chapter 3, the greatest difficulty in evaluating the relative contributions of different factors to the final enthalpy and entropy and free energy is a quantitative one. We know which type of interactions will lead to an increase in entropy on folding and which will lead to a decrease, but it is not possible to predict the net effect with sufficient precision to know if the overall $\Delta S$ will be positive or negative. The same is true for the $\Delta H$. Within the biopolymers, added complications result when considering proteins because of the widely differing character of the various amino acid side chains.

Enthalpies and entropies of denaturation for whole proteins are difficult if not impossible to measure under physiological conditions of temperature and ionic strength because the equilibrium usually lies too far in the direction of the folded structure. There is not enough of the unfolded structure or the various intermediates between the unfolded and folded structures to allow a determination of the equilibrium constant. In the few cases that have been studied, the conditions are far from normal, so the results must be cautiously interpreted. For example, the denaturation of chymotrypsin at pH 3 has an enthalpy increase of +14.4 kcal and an entropy increase of 438 eu. Similarly, the enthalpy increase for trypsin denaturation at pH 2 is 68 kcal and the entropy increase is 213 eu. These terms are indicative of the large structural organization lost on denaturation. Part of the loss of organization is counterbalanced by an increase in solvation in the denatured state. However, denaturations commonly have large positive values for both enthalpy and entropy. Conversely, folding under these conditions should have large negative values for both enthalpy and entropy. The overall free energy is favorable for folding because the favorable enthalpy factor dominates the unfavorable entropy factor. A detailed analysis of the location of individual side chains in the three-dimensional native protein structures shows that they are located where they would be expected to contribute maximum stability to the native structure. Thus the apolar groups are buried inside the protein structure and usually are closely associated with other apolar groups, not water. The polar groups are mostly located on the surface, where they can and do associate with water. When polar groups, particularly the peptide backbones, are not associated with water, they are invariably strongly associated with each other, frequently leading to extensive repeating structures, such as the $\alpha$ helix or the $\beta$ sheet.

We turn now to a discussion of the other three main classes of biopolymers. Phospholipids are the structural components of lipid bilayers, which are a major component of biological membranes. They contain long, saturated carbon chains with charged groups at one end. Purified phospholipids form bilayers (micelles) spontaneously in aqueous solution (see Chapters 13 and 16). The structures that are formed are exactly what would be expected from the principles discussed earlier. The hydrophobic apolar carbon chains are located on the inside of the structures away from the water and in a nearly closely packed arrangement. As with hydrophobic side chains in proteins, the driving force here is not the van der Waals attractive forces but rather the energetically favorable situation that results from avoiding water contacts. The straight-chain hydrocarbon structures formed are regular and reasonably closely packed. Anything less than a closely packed arrangement would be energetically unfavorable in solution because it would require additional free energy to maintain such gaps. On the exterior side of the micelle, the charged groups of the phospholipids are located where they can make energetically favorable contacts with water.

Polysaccharides vary greatly in the types of monomeric building blocks, and it would not be fruitful to dwell on a detailed discussion of these here. In fact, little detailed information is available concerning the precise structures of these molecules. Steric hindrance between bulky sugar groups greatly limits the number of accessible conformations. This is particularly apparent in comparing the structures formed by the main structural polysaccharide, cellulose, and the main energy-storage polysaccharide, glycogen. These contain $\beta(1 \rightarrow 4)$- and $\alpha(1 \rightarrow 4)$-glycosidic linkages, respectively (see Figure 12-16). In Chapter 12 it is pointed out that the stereochemistry of the former is suited to making stiff straight-chain structures, whereas the latter is suited to making coiled or helical structures.

Next to proteins, more is known about the conformations of nucleic acids than about those of any other types of biopolymers. A variety of conformations adopted by nucleic acids is discussed in Chapter 18. The dominating factors that determine nucleic acid conformation are the limitations imposed by the stereochemistry of the polynucleotide chains, the high negative charge resulting from the regularly repeating phosphate groups, and the ambivalent nature of the purine and pyrimidine bases. Polar character is displayed by the H-bonding groups and dipoles within the planes of the bases; apolar character is imparted by the aromatic aspect of the purine and pyrimidine ring structures. The charge on the phosphates thermodynamically favors placing these groups on the outside of the structure. The negative charge density resulting from the phosphates is so great a repulsive force that ordered nucleic acid molecules tend to denature spontaneously in salt-free water. Addition of some salt or basic proteins either screens or neutralizes this charge sufficiently that the native three-dimensional structure is stable. In the DNA double-helix structure, the bases are regularly hydrogen bonded in the interior of the molecule and the planes of the bases are stacked with a mean separation of 3.4 Å along the helix axis; this is the expected van der Waals separation for planar organic ring compounds. Three factors account for the stable associations observed between the A-T and G-C bases in the interior of the double-helix structure. There are three H-bonds between a G-C base pair and two between an A-T base pair. Stacked base pairs in a typical polymeric structure are also stabilized by dipole-dipole interactions and hydrophobic forces. Thermal denaturation measurements (see Chapter 18) show that there is a proportionality between the denaturation temperature and the $G+C$ content for a given DNA. This would seem to follow from the fact that a G-C base pair has three H bonds whereas the A-T base pair only has two H bonds. Closer investigation of nearest neighbor interactions shows that the stability of an individual base pair is influenced by the immediately adjacent base pairs in a specific way. This has been interpreted as being the result of the directional character of the dipoles in the different base pairs. As with proteins, the net enthalpy favors the folded conformation and the net entropy is unfavorable for folding. Fortunately for us, the enthalpy term is larger.

## SELECTED READINGS

Alberty, R. A., and Daniels, F. *Physical Chemistry*, 5th Ed. New York: Wiley, 1975. A well used classic textbook.

Cantor, C. R., and Schimmel, P. R. *Biophysical Chemistry*. San Francisco: Freeman, 1980. A three-volume, up-to-date treatment covering a broad range of topics in an authoritative manner.

Dickerson, R. E. *Molecular Thermodynamics*. New York: Benjamin, 1969. One of the best of the "newer" books on thermodynamics.

Ingraham, L. L., and Pardee, A. B. Free Energy and Entropy in Metabolism. In D. M. Greenberg (Ed.), *Metabolic Pathways*, Vol. 1. New York: Academic Press, 1967. An excellent review of thermodynamics stressing empirical relationships useful to biochemists.

Latimer, W. M. *Oxidation Potentials*. Englewood Cliffs, N.J.: Prentice-Hall, 1952. Although this is an old book, it does contain much useful information on oxidation-reduction potentials and free energies, enthalpies, and entropies of various compounds.

Tanford, C. *The Hydrophobic Effect*, 2d Ed. New York: Wiley, 1980. A quite useful book but on a more specialized topic.

Van Holde, K. E. *Physical Biochemistry*. Englewood Cliffs, N.J.: Prentice-Hall, 1971. A lot of information presented quite precisely and mathematically.

## PROBLEMS

1. In some respects thermodynamic considerations are more important to a biochemist than to a chemist. Why is this so?

2. Name three extensive and three intensive properties which relate to thermodynamic quantities. What is the basic difference between the two types of properties?

3. As a rule, what are the most significant parameters to be considered in evaluating the enthalpy and the entropy of a biochemical reaction?

4. What is meant by a state function? Why is enthalpy a state function? In what way is it useful to know that you are dealing with a state function?

5. For most biochemical reactions, internal energy and enthalpy may be equated. Why?

6. In thermodynamics, it is important to distinguish between the total system and the system being studied. Why?

7. Why do enthalpies and entropies of solvation tend to negate one another?

8. Why does the entropy of a weak acid decrease on ionization?

9. Proteins that serve as gene repressors frequently have two identical binding sites which interact with complementary sites on the DNA. If a repressor protein is cut in half without damaging either of its DNA binding sites, how will this affect the binding to DNA? How will the binding be affected if one of the two binding sites on the DNA is eliminated by changing the nucleotide sequence? Discuss the enthalpy, entropy, and free-energy effects.

10. What causes osmotic pressure? Estimate the osmotic pressure for a 1% solution of a protein with a molecular weight of $2 \times 10^4$.

11. Cite some of the factors that make ATP an ideal source of biochemical energy.

12. Table 7-8 gives the free energy for hydrolysis of ATP at pH 7.0 for standard conditions (1 $M$) and for more dilute solutions. Explain how the latter values are obtained.

13. What is the physical meaning of the chemical potential?

14. Tom and Mary both ate a complete basal diet containing all nutrients, but they decided to gain weight by eating an extra 3000 K calories a day. Tom chose to get his 3000 K calories in glycine and Mary took her 3000 K calories in glucose. Recall that nutritional calories are measured by heats of combustion to form nitrogen instead of urea as is done in the body and that the body gains weight on free energy, not enthalpy. Therefore, the caloric values of their diets are not equivalent. Who gained the most weight?

Data:

| | | |
|---|---|---|
| Glucose | $\Delta H_{combustion}$ = | $-675.4$ Kcal/mol |
| | $\Delta G_{combustion}$ = | $-686.5$ Kcal/mol |
| Glycine | $\Delta H_{combustion}$ = | $-232.6$ Kcal/mol |
| | $\Delta G_{combustion}$ = | $-241.0$ Kcal/mol |
| Urea | $\Delta G_{combustion}$ = | $-158.9$ Kcal/mol |

15. If we ignore solvent effects, how would this affect the enthalpy or entropy of unfolding of a polypeptide chain?

16. Why do most proteins denature on heating?

17. Why are apolar compounds sparingly soluble in water? How does this influence protein structure? Lipid structure?

18. Calculate the free-energy change involved in transporting 1 mol of $H^+$ from a region with a concentration of $10^{-7}$ $M$ to a region with a concentration of 1 $M$.

19. Explain why a closed system cannot be in a steady state. Remember that a steady state refers to a nonequilibrium situation that prevails because of a balance between reactions that supply and remove substances.

20. In Chapter 9, the final reaction in the trichloroacetic acid cycle (TCA-8) involves regeneration of oxaloacetate from malate. This reaction has a $\Delta G^{\circ\prime}$ of $+7.1$ kcal/mol. Suggest reasons why this reaction goes in the forward direction in the cell.

21. Hemoglobin has four essentially identical sites for binding $O_2$. Explain how you would determine whether the binding of $O_2$ at one site influenced the binding at another site. What are the relevant equations?

22. The oxidation of one mole of palmitate to $CO_2$ and water forms 129 moles of ATP from ADP and inorganic phosphate at pH 7. The free energy ($\Delta G^{\circ\prime}$) is $-2,340$ kcal/mol for the oxidation of palmitate by molecular oxygen.
    a. What is the free energy of the overall process to form ATP?
    b. What is the efficiency of the storage of energy as ATP?

23. Could an aerobic organism obtain energy by the oxidation of ammonia at 0.01 atm to form atmospheric nitrogen? If so, how much? Data for $\Delta G^{\circ}$ of formation:

| | |
|---|---|
| $NH_3(g)$ | $-3.94$ kcal/mol |
| $H_2O(l)$ | $-55.69$ kcal/mol |

24. Pyridine binds to an enzyme with a free energy of

−16 kcal/mol and benzaldehyde binds to the same enzyme with a free energy of −20 kcal/mol. The substrate for this enzyme is compound A.

A

Estimate the range of the free energy of binding for this substrate to the enzyme. Assume no interaction between the pyridine and benzaldehyde on the enzyme.

25. Calculate the pH of the following solution of acetic acid ($pK = 5$).
   a. 0.01 $M$ acetic acid.
   b. $10^{-4}$ $M$ acetic acid.

26. Calculate a dissociation constant for magnesium ion from $ATP^{-4}$ from the following data. Give an estimate of the enthalpy of the reaction.

| (ATP) $\times 10^5$ | Total (Mg) $\times 10^5$ | Free (Mg) $\times 10^5$ |
|---|---|---|
| 75 | 20 | 2 |
| 41 | 20 | 4 |
| 28.6 | 20 | 6 |
| 21.4 | 20 | 8 |
| 16.4 | 20 | 10 |

27. How much sodium hydroxide should be added to 0.1 $M$ acetic acid ($pK = 5.0$) to make a buffer of pH = 4.0?

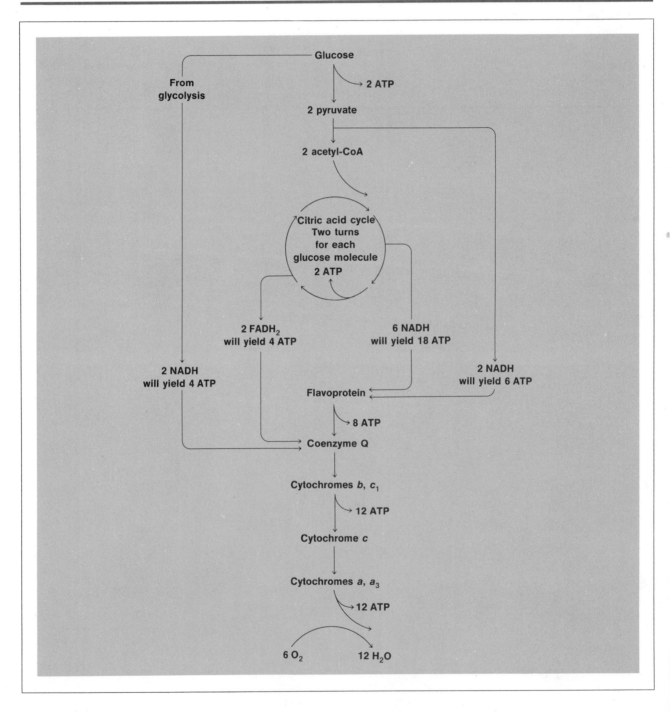

# 8

# ANAEROBIC PRODUCTION OF ATP

One of the most fundamental properties of living organisms is their requirement for energy. As we have discussed in the preceding chapter, a living cell is a highly improbable entity, maintaining as it does a myriad of chemical reactions at a steady-state position far removed from equilibrium. To do so requires the uptake from the environment of copious quantities of energy. In some organisms, the _phototrophs_, this need for energy is met by absorption of quanta of solar radiation, while in others, the _chemotrophs_, energy is obtained by oxidation of pre-formed organic molecules. It is to this latter mode of meeting energy needs that the present discussion addresses itself, so that our topic becomes _chemotrophic energy metabolism_.

## CHEMOTROPHIC ENERGY METABOLISM

### The Need for Energy

If there is a single most characteristic property of life, it is certainly the need and ability to effect change. To produce the changes necessary to sustain and propagate life requires the availability of energy.

Phrasing energy needs in this way, we can ask what sorts of changes living systems must effect or what kinds of work they must do. The answer comes in terms of three major categories of cellular activities that involve change. These will be discussed in more detail elsewhere in the text, but they can be intimated here briefly. In terms of changes effected, living systems require energy for the following purposes:

1. _Mechanical work: Changes in location or position_. Mechanical work involves a change in the location or orientation of an organism, a cell,

_Catabolism and the production of ATP._

or a part thereof. As examples, consider the contracting muscles of a jogger, the motion of a flagellated protozoan in a pond, the migration of chromosomes toward the opposite poles of the mitotic spindle, or the movement of a ribosome along a strand of messenger RNA.

2. *Osmotic and electrical work: Changes in concentration.* Concentration work always involves the movement of chemical compounds or ions against a prevailing concentration gradient, thereby establishing the localized concentrations of specific molecules and ions upon which most essential life processes depend. Concentration work is sometimes also referred to, although less satisfactorily, as osmotic work. Examples include the uptake of glucose from the blood by the cells of your body, the pumping of sodium out of a marine microorganism, and the movement of nitrate from the soil into the cells of a plant root. When the species being concentrated is a charged ion, the resulting concentration gradient is also a potential gradient, so that electrical work becomes a subset of concentration work for charged species. Although the most dramatic example of this is the large potential developed by the electric eel, electrical work is a common phenomenon and is in fact the mechanism of membrane excitation and of conduction of nerve impulses.

3. *Synthetic work: Changes in bonds.* Synthetic work is that involved in the formation of the energy-rich chemical bonds necessary to fabricate and assemble the complex organic molecules of which cells are composed. These are in general molecules of greater complexity and energy content than the simple molecules available to organisms from their environment, so that energy must clearly be expended in their synthesis. Although synthetic work is most obvious during periods of growth of an organism, it also occurs in nongrowing mature organisms, in which existing structures must be continuously repaired and replaced. The continuous expenditure of energy to elaborate and maintain ordered structures out of less-ordered raw materials is, in fact, one of the most characteristic properties of living systems.

## Sources of Energy

*As already noted, organisms fall into two categories in terms of meeting energy needs (Figure 8-1). The phototrophs (literally, "light feeders") depend for their energy supplies directly on the radiant energy of the sun.* These marvelous green forms of life use solar energy to form energy-rich, oxidizable organic (or, in the case of some microorganisms, inorganic) compounds, with chemical bonds that are responsible for storage, transferral, and release of energy. *The second category of organisms includes those of us who meet our energy needs by oxidizing the energy-rich compounds we obtain, directly or indirectly, from the phototrophs.* We are chemotrophs (literally, "chemical feeders"), since we require the intake of oxidizable chemical compounds. Most commonly, these oxidizable compounds are organic, and the organisms that utilize them are *chemoorganotrophs*. Some microorganisms oxidize inorganic compounds instead and are called *chemolithotrophs* (from the Greek, *lithos* = stone). Both the writer and the reader of these lines, together with all other animals and most microorganisms, are chemotrophs; we are, in a sense, parasites living off the energy that has been packaged for us by the phototrophs into the

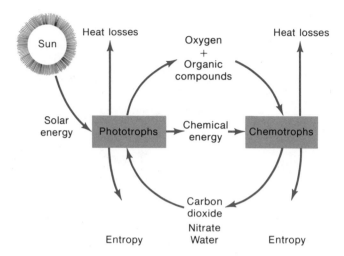

**Figure 8-1**
Flow of energy and mass in the
biosphere.

chemical bonds of the food molecules we consume. A world composed
only of chemotrophs would last only as long as preexisting food supplies
held out. Then all of us would die for lack of energy on a planet that is
flooded daily with solar energy, but in a form that we can use for nothing
more productive than a suntan.

In light of these two quite different answers to a common need for
energy, any discussion of energy metabolism must, of necessity, deal with
both the phototrophic and the chemotrophic answers. So phototrophic energy metabolism will be dealt with in Chapter 11, and in this and the
subsequent two chapters the focus will be on chemotrophic energy metabolism.

## Metabolic Pathways

In Part I, you were introduced to enzyme-catalyzed reactions, and to the
concept of metabolic pathways as a sequence of consecutive enzyme-catalyzed reactions functioning in concert to accomplish a specific chemical
task. In biochemical terms, life might well be defined as a network of integrated and regulated metabolic pathways, each contributing to the sum of
all the activities that every living system must carry out.

Metabolic pathways are of two types. Those concerned with synthetic
work are termed _anabolic_ pathways (from the Greek _ana_ = up), while
those involved in degradative processes are called _catabolic_ pathways (from
the Greek _kata_ = down). Anabolic pathways usually involve an increase
in atomic order (i.e., a decrease in entropy), are often reductive in nature,
and are almost always energy-requiring (_endergonic_). Examples are the synthetic reactions leading to the nucleic acids (Chapters 20 and 21) and the
proteins (Chapters 24 and 25). Catabolic pathways, however, are usually
energy-liberating (_exergonic_) because they involve the oxidative release of
chemical bond energies and a decrease in atomic order (increase in entropy). Catabolic and anabolic pathways sometimes share a common partial sequence, which functions in one direction for synthesis and in the
opposite direction for degradation. However, _the catabolic route is never
exactly the reverse of the anabolic pathway, since both need to be exer-_

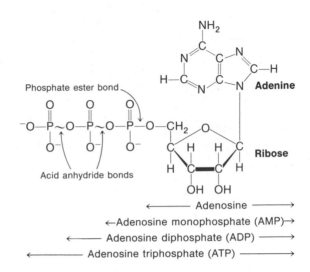

**Figure 8-2**
The structure of adenosine and its
phosphorylated forms.

*gonic in their respective directions.* In general, this is accomplished by several steps that are highly exergonic in one direction and then are circumvented by an alternative reaction or reaction sequence in the opposite direction. Such steps are also often sites of regulation, since they represent reactions at which control can be exerted in one direction without directly affecting the other direction. For example, the pathways for glucose synthesis (*gluconeogenesis*) and glucose degradation (*glycolysis*) share many reactions in common, but they have several steps that are unique to the synthetic or degradative direction and are accordingly steps that ensure thermodynamic irreversibility and sites at which regulation can be effected (see Chapter 12).

To a good first approximation, cells carry out anabolic reactions to accomplish the growth and repair characteristic of all living systems, and they engage in catabolic reactions to obtain the energy needed to drive the anabolic reactions. *Crucial to the thermodynamic success of every living system, then, is the efficient coupling of energy-yielding to energy-requiring processes.* This coupling depends on a common intermediate—a common chemical entity that serves as the "currency" of the cellular energy economy.

## Adenosine Triphosphate (ATP): The Currency of the Realm

Clearly, if energy is obtained in one form (be it from the sun or from a chemical bond) and is to be used in another form (to move a muscle, form a bond, or transport a molecule), there must be some way of coupling the energy-yielding and the energy-requiring processes—some way of temporarily storing the energy derived from an exergonic reaction for use in driving what would otherwise be an endergonic reaction. This becomes especially clear when you realize that in the absence of some such coupling mechanism, the energy of exergonic reactions would be released as heat. This would have some limited utility in maintaining the body temperature of warm-blooded animals or in melting overlying snow to speed emergence of plants in spring, but it would otherwise represent a waste of energy. Heat, in other words, is not a generally useful form of energy to an isothermal system. *The purpose of coupling exergonic reactions with endergonic reactions is to use the energy of the former to drive the latter without losing all the energy as heat.*

Essential for this purpose is some common intermediate—some common energy currency. In biochemistry, this common intermediate is usually either a reduced coenzyme or a high-energy phosphate bond. Reduced coenzymes will be discussed later; here the emphasis is on high-energy phosphate bonds in general and on the specific high-energy phosphorylated compound called *adenosine triphosphate* (ATP) in particular. As shown in Figure 8-2, ATP consists of an aromatic base called *adenine,* a five-carbon sugar called *ribose,* and a series of three phosphate groups linked to each other by phosphoanhydride bonds and to the ribose by an ester bond. The compound formed by linking adenine and ribose is called *adenosine,* which may occur in the unphosphorylated form or with one, two, or three phosphates attached to carbon 5 of the ribose, forming, respectively, adenosine monophosphate (AMP), diphosphate (ADP), and triphosphate (ATP). In most cells, the total concentration of all three species (i.e., AMP + ADP + ATP) is usually in the range 2 to 10 m$M$, although the proportions will obviously vary considerably with the energy status of the cell.

Crucial to an understanding of the key role that ATP plays as an intermediate in energy metabolism is an appreciation of the unstable, energy-rich nature of the anhydride bonds that link the phosphate groups. The so-called energy-rich bonds in ATP are indicated by the wavy lines in Figure 8-2. The explanation of why hydrolytic cleavage of these bonds releases so much potential energy has been given in Chapter 7. In this chapter we shall use $-7.3$ kcal/mol for the standard free energy of hydrolysis ($\Delta G^{\circ\prime}$) of ATP to ADP and $P_i$ (see Figure 8-3), recognizing that the actual free-energy change in the cell may be considerably greater than this (see Chapter 7).

The term *energy-rich bond* needs qualification from a quantitative standpoint, since it takes in a variety of compounds with quite a range of free-energy changes of hydrolysis. Figure 8-4 summarizes this information for several phosphorylated intermediates common to energy metabolism.

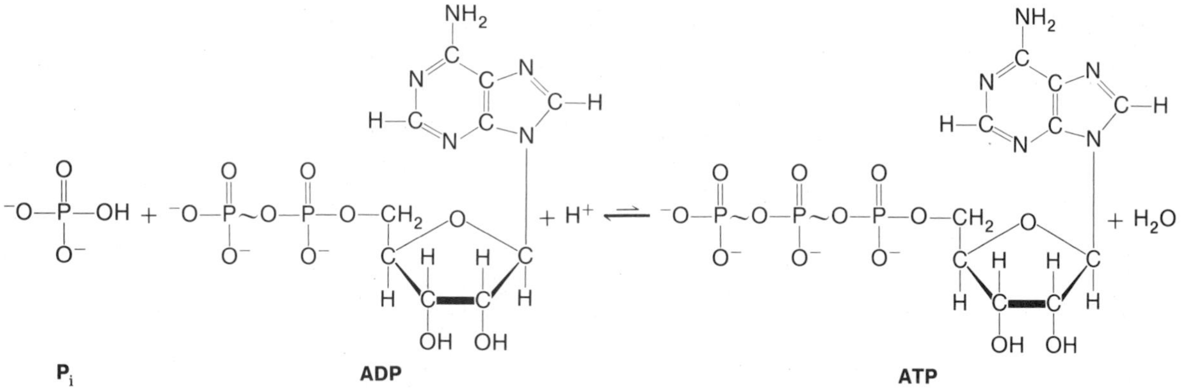

**Figure 8-3**
Formation (*to the right*) and hydrolysis (*to the left*) of ATP.

In general, the relative positions of two compounds on the scale dictates the ease with which phosphate can be transferred from one to the other. Unless concentration conditions result in $\Delta G$ values that actually invert the rankings shown in Figure 8-4, it is generally true that any compound in the figure can be exergonically phosphorylated at the expense of any compound that lies above it (i.e., has a more negative $\Delta G°'$ value), but not by a compound that lies below it. Thus ADP can be phosphorylated to ATP by transfer of phosphate from glycerate-1,3-diphosphate, but not from glucose-6-phosphate. Similarly, ATP can be used to phosphorylate glycerol, but not to generate phosphoenolpyruvate from pyruvate.

Although phosphorylated compounds in cells were originally classified as either low-energy or high-energy compounds based on their $\Delta G°'$ values, it should be clear from Figure 8-4 that such a distinction is at best quite arbitrary and perhaps even misleading, since phosphorylated compounds actually display a broad range of $\Delta G°'$ values, with no sharp dividing line. More important than any arbitrary classification is the crucial intermediate position that ATP occupies on the scale. _The whole role of the ATP-ADP system is to serve as an acceptor and donor of energy-rich phosphate groups. Essential to that function is an intermediate position between the higher-energy compounds from which ATP accepts phosphate and the lower-energy compounds to which it donates phosphate._

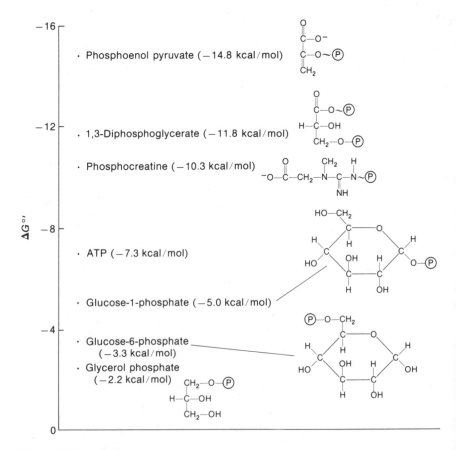

**Figure 8-4**
Standard free energies of hydrolysis for some common phosphorylated compounds.

ATP is the immediate source of energy for most energy-requiring reactions in most cells. Moreover, since energy is needed continuously by all living cells, it follows that every cell must make provision for the continual generation of ATP and for careful adjustment of ATP formation to actual energy needs. The importance of continuous ATP synthesis and the immediacy with which ATP utilization is linked to its formation can be seen in the very rapid rate of ATP turnover in cells. In an actively respiring bacterial cell, the half-time of ATP turnover (i.e., the time for one-half of all the ATP molecules present in the cell to be used in driving endergonic reactions and then regenerated in exergonic reactions) is often only a few seconds. Even in a far more sluggish cell, half the ATP molecules can turn over in a matter of minutes. Obviously, then, cells would live for only a few seconds or minutes if suddenly deprived of further ATP-generating ability. The turnover is awesome, and the balance is fine.

## Biological Oxidation

The essence of chemotrophic energy metabolism is the harnessing of exergonic catabolic reactions to drive the various energy-requiring reactions in the cell, with the ATP-ADP system as the crucial intermediate. The next question then becomes: What sorts of catabolic reactions are typically involved? Or, more personally, what are the catabolic processes by which the cells of your own body make use of the food you eat to meet your energy needs? The answer is that, by and large, _chemotrophs meet their energy needs by releasing free energy from the bonds of the food molecules available to them and the release of such bond energies almost invariably involves oxidation_. That food molecules such as sugars, fats, and proteins are sources of energy that can be liberated upon oxidation is readily apparent when you consider the ease with which a marshmallow (sugar), a drop of vegetable oil (fat), or a steak (protein) burns over an open fire. So to say that such molecules have energy stored in their bonds is another way of saying that these are oxidizable organic compounds and that their oxidation is a highly exergonic process that can be harnessed to do useful work.

To say that something is "oxidizable" means that it is possible to remove one or more electrons from it. For example, the ferrous ion ($Fe^{2+}$) is an oxidizable species because it can lose an electron to become a ferric ion ($Fe^{3+}$):

$$Fe^{2+} \xrightleftharpoons[\text{Reduction}]{\text{Oxidation}} Fe^{3+} + e^-$$

In organic and biological chemistry, _oxidation_ is defined in exactly the same way; it is the removal of electrons. However, for organic molecules, oxidation frequently involves the removal of hydrogen ions (protons) in addition to electrons, such that, in effect, it is the equivalent of a hydrogen atom (one electron plus one proton) that is removed. This is illustrated by the oxidation of an alcohol such as ethanol to the corresponding aldehyde:

$$CH_3-CH_2-OH \xrightleftharpoons[\text{Reduction}]{\text{Oxidation}} CH_3-\overset{\displaystyle H}{\underset{}{C}}=O + 2\,e^- + 2\,H^+$$

**Alcohol**        **Aldehyde**

Thus, for biochemical reactions involving organic molecules, oxidation can be defined in a secondary way as the removal of hydrogen atoms. Or, in other words, *biological oxidation reactions are frequently also dehydrogenation reactions*. Similarly, *reduction* is defined as the addition of electrons, but in biological reactions, this is frequently accompanied by the addition of protons, so that the overall effect is a *hydrogenation* reaction, as shown by the reverse direction in the preceding reaction. Also illustrated in the same equation is the further general feature that *biological oxidation-reduction reactions almost always involve two-electron transfers*.

Obviously, oxidation and reduction reactions must always take place concomitantly; any time an oxidation event occurs, a concomitant reduction also must take place, since the electrons removed from one molecule must be added to another molecule. *In chemotrophic energy metabolism, the ultimate acceptor of electrons is frequently (although by no means always) oxygen*. Rarely, however, are electrons passed directly to oxygen from an oxidizable molecule. Rather, the immediate electron acceptor in most biological oxidations (and indeed the immediate electron donor in most biological reductions) is one of several *coenzymes*—specialized molecules that function specifically as carriers of electrons (and protons, since such coenzymes become reduced by being reversibly hydrogenated). As already discussed in Chapter 6, the most common coenzymes involved in energy metabolism are nicotinamide adenine dinucleotide ($NAD^+$), nicotinamide adenine dinucleotide phosphate ($NADP^+$), and flavin adenine dinucleotide (FAD). Both $NAD^+$ and $NADP^+$ serve as electron acceptors by adding two electrons and one proton to the ring of the nicotinamide portion of the molecule, thereby generating the reduced form, NADH or NADPH, with concomitant release of a proton into the medium. FAD, however, accepts two electrons and both of the protons to generate $FADH_2$:

$$NAD^+ + 2\,[H] \rightleftharpoons NADH + H^+$$

$$NADP^+ + 2\,[H] \rightleftharpoons NADPH + H^+$$

$$FAD + 2\,[H] \rightleftharpoons FADH_2$$

In the case of $NAD^+$ and $NADP^+$, reduction involves a carbon atom as the electron acceptor (see Figure 6-7), while for FAD, the electrons are added to nitrogen atoms (see Figure 6-12). This distinction will turn out to be important in understanding why $NAD^+$ is used in some oxidation reactions and FAD in others. Another useful general distinction can be made between $NAD^+$, which is commonly used in catabolic pathways (and therefore figures heavily in the present discussion), and $NADP^+$, which is generally preferred in anabolic sequences. Also worthy of mention in passing is that both the nicotinamide of $NAD^+$ and $NADP^+$ and the flavin of FAD are derivatives of compounds (nicotinamide and riboflavin, respectively) recognized as B vitamins, which must be present in the diets of humans and other vertebrates. It is precisely their involvement in energy metabolism as parts of indispensable coenzymes that makes these substances essential in the diet of any organism that cannot manufacture its own supply.

## Biological Oxidation of Glucose

The significance of oxidation for our present discussion is that it is the means whereby chemotrophs obtain the energy they need to effect desired changes. A great variety of organic compounds can be used as substrates for biological oxidation, as evidenced by a brief review of the many different things you likely ate within the last 24 hours. There are, then, a great many oxidative processes by which chemotrophs can obtain energy. Some microorganisms can use reduced inorganic compounds (such as reduced forms of iron, nitrogen, or sulfur) as energy sources. These organisms play important roles in the inorganic economy of the biosphere and obviously utilize rather specialized oxidative reactions. Most chemotrophs, however, depend for energy on organic food molecules, with carbohydrates, fats, and proteins as the three major categories. To simplify our discussion initially and provide a unifying metabolic theme, we will consider the biological oxidation of the six-carbon sugar glucose ($C_6H_{12}O_6$), since this is quantitatively the single most important substrate for energy metabolism in many (although not all) chemotrophs. In many vertebrates, including humans, glucose is the single most important sugar in the blood, and in plants, glucose makes up one-half the transport disaccharide sucrose. Furthermore, once the catabolic fate of glucose is understood, it becomes a much easier matter to consider the catabolism of alternative substrates, such as fats or amino acids, since these turn out to be oxidatively degraded by pathways that almost invariably feed into the mainstream pathway for glucose catabolism. Far from looking at the fate of a single esoteric sugar, then, we are in fact considering a metabolic sequence that is at the very heart of chemotrophic energy metabolism.

As discussed in Chapter 12, glucose is a six-carbon aldosugar with the open-chain and hemiacetal-ring forms shown in Figure 8-5. We will usually use the open-chain representation for present purposes because of its pedagogic appeal, but you should keep in mind that relatively little glucose actually exists in this form in solution because the ring forms are thermo-

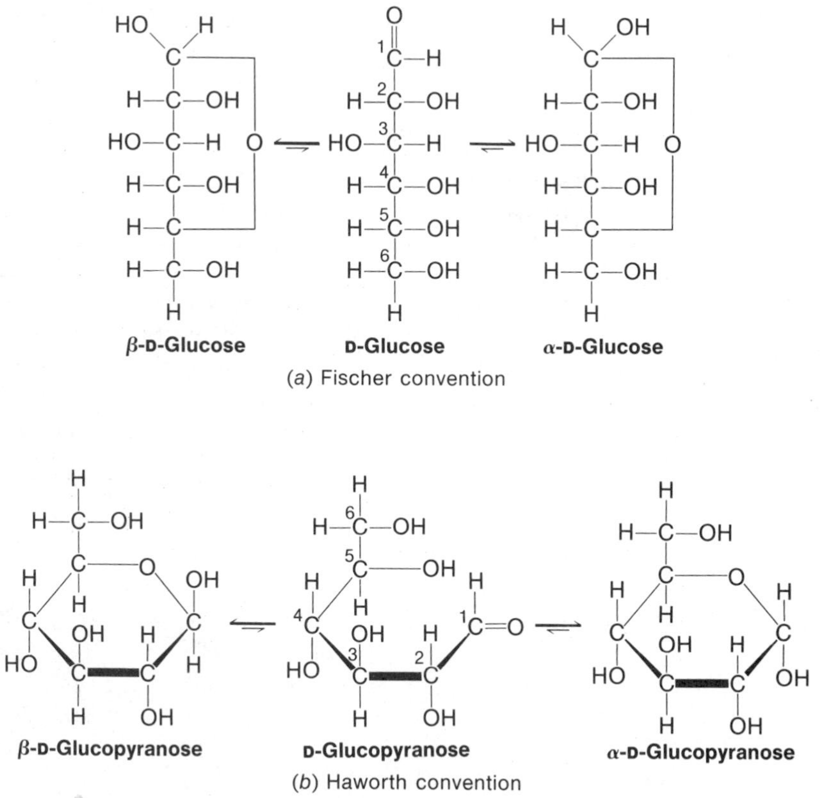

(a) Fischer convention

β-D-Glucose    D-Glucose    α-D-Glucose

β-D-Glucopyranose    D-Glucopyranose    α-D-Glucopyranose

(b) Haworth convention

**Figure 8-5**
Structure of glucose as represented by the older Fischer convention (a) and by the more accurate Haworth pyranose structures (b). The anomeric α and β forms predominate over the straight-chain configuration.

dynamically favored. Note that the glucose molecule contains four asymmetric carbon atoms (numbers 2, 3, 4, and 5). There are therefore $2^4 = 16$ stereoisomers of the molecule, but of these 16 possibilities, glucose is the single most widespread sugar in the entire biological world. Most commonly, the glucose that organisms use as substrate for energy metabolism comes from (1) direct uptake or ingestion of glucose or dissacharides such as sucrose or maltose that can be broken down to yield glucose, (2) the breakdown of ingested or stored polysaccharides (principally starch and glycogen), (3) the conversion of other foodstuffs into glucose, and (4) photosynthetic sugar synthesis.

That the oxidation of glucose is a highly exergonic process can be seen from the $\Delta G^{\circ\prime}$ value for its complete combustion to carbon dioxide and water:

$$C_6H_{12}O_6 + 6\,O_2 \longrightarrow 6\,CO_2 + 6\,H_2O \qquad \Delta G^{\circ\prime} = -686\,\text{kcal/mol}$$

Since $\Delta G^{\circ\prime}$ is a thermodynamic parameter, it will be unaffected by route and will therefore have the same value whether the oxidation is by rapid combustion (such as a burning marshmallow over a campfire) or by biological oxidation, with maximal trapping of energy as ATP. For biological systems, however, route is of paramount importance, since uncontrolled combustion would be incompatible with life, while _controlled, stepwise oxidation mediated by a series of enzyme-catalyzed reactions represents an isothermal process that can be coupled with ATP generation_ and subjected to careful regulation. We need, therefore, to examine the pathways for biological oxidation of glucose in terms of the specific reactions involved, the means whereby the energy released by oxidation is conserved as high-energy phosphate bonds of ATP, and the regulatory mechanisms operative to ensure that oxidation rates are carefully adjusted to meet actual ATP needs.

## Respiration and Fermentation

The most important factor in determining the amount of energy that can be obtained from the oxidation of glucose is the nature of the electron acceptor. Access to the full 686 kcal/mol indicated as the $\Delta G^{\circ\prime}$ value associated with the complete oxidation of glucose to $CO_2$ presumes the availability of oxygen as an electron acceptor and therefore requires aerobic conditions. Aerobic energy metabolism involves several multistep processes which collectively are termed _respiration_. Under anaerobic conditions, oxygen is not available as an electron acceptor, and the electrons are passed instead to some organic molecule, usually one generated by the actual catabolic process itself. All such anaerobic processes are called _fermentations_, and they are usually further identified in terms of the principal end product (i.e., the reduced form of the organic electron acceptor). Thus the usual pathway for anaerobic glucose catabolism in animal cells is termed _lactate fermentation_ because the end product in such cells is lactate, which is obtained by reduction of pyruvate. Similarly, in yeast the process is termed _alcoholic fermentation_ because the end product is ethanol. _In such fermentative processes, there is no net oxidation. Rather, internal oxidoreduction occurs such that the oxidation of one part of the starting glucose molecule involves the reduction of another part in a way that releases free energy and thereby allows ATP generation._

*Organisms with an absolute requirement for oxygen are called strict or obligate aerobes.* Most higher animals are obviously in this category. *Strict* or *obligate anaerobes*, however, cannot tolerate the presence of oxygen. Not surprisingly, such organisms occupy environments from which oxygen is generally excluded, such as deep waters, deep soils, or deep wounds. Most are bacteria, including the soil *Clostridia* responsible for gangrene and food poisoning and those involved in denitrification and methane production. *Facultative organisms are those which can exist anaerobically,* extracting energy from glucose (or other substrate) by fermentative processes, *but which also can function in the presence of oxygen, in which case they carry out the full respiratory sequence.* Many bacteria and fungi are facultative. It is also important to note that some cells or tissues of otherwise aerobic organisms can function anaerobically if conditions require it. Thus your muscle cells, although obviously part of a strictly aerobic organism, can operate under temporarily anaerobic conditions, as for example when oxygen demand exceeds supply during periods of strenuous exertion. The red blood cells of your circulatory system also function in what is de facto an anaerobic mode, although in this case for the quite different reason that they lose their mitochondria during normal differentiation and are therefore devoid of the capacity to transfer electrons to oxygen.

In this chapter we will consider the anaerobic generation of ATP, with lactic acid and alcoholic fermentation as the main processes of interest. The following chapter will then be given over to aerobic energy metabolism. It seems appropriate to consider the anaerobic pathways first, since even under aerobic conditions, the basic fermentation pathways are still operative, but serving as just the first phase of respiration rather than representing the full catabolic capability of the cell. Thus to understand fermentation is not only to understand anaerobic energy metabolism, but also to have in hand the first phase of aerobic respiration.

## ATP PRODUCTION BY GLYCOLYSIS

Whether aerobic or anaerobic, the process of glucose catabolism involves a 10-step reaction sequence that is without doubt the single most ubiquitous pathway in all energy metabolism. Glycolysis can be characterized as a nearly universal process, since it occurs in almost every living cell. It is, as already noted, anaerobic, in that the pyruvate produced by the pathway can be reduced to lactate (or to ethanol with evolution of $CO_2$), allowing the overall process to occur in the absence of oxygen. It is generally regarded as a primitive process, both in the sense that it occurs in the cytosol rather than being compartmentalized into a specific organelle within the eukaryotic cell and also because it is thought to have arisen early in biological history, before the advent of eukaryotic organelles.

The glycolytic pathway was the first major metabolic sequence to be elucidated. Most of the decisive work was done in the 1930s by the German biochemists G. Embden, O. Meyerhof, and O. Warburg, two of whom gave the sequence its alternative name, the *Embden-Meyerhof pathway.*

The glycolytic pathway appears in detail in Figure 8-6 and is shown in the context of overall chemotrophic energy metabolism in Figure 9-1. The essence of the process is suggested by the name, since *glycolysis* comes from the Greek roots *glykos,* meaning "sweet," and *lysis,* meaning "loosing." Literally, then, *glycolysis* is the loosing, or splitting, of something sweet, which is, of course, the starting sugar. From Figure 8-6 it is clear that the actual splitting occurs at step 4 (labeled Gly-4 both in the figure and in the following text), since it is at this point that a six-carbon sugar is cleaved to yield two three-carbon compounds, one of which, glyceraldehyde-3-phosphate, is the only oxidizable molecule in the whole pathway. Subsequent to the cleavage of step 4, two successive ATP-generating steps occur, one at Gly-7 and the other at Gly-10. These obviously represent the energy payoff of the process, since these are the only ATP-yielding reactions of the pathway under anaerobic conditions. We will therefore consider the overall pathway in three segments, consisting, respectively, of phosphorylation and cleavage (Gly-1 through Gly-5), the first ATP-generating sequence (Gly-6 and Gly-7), and the second ATP-generating events (Gly-8 through Gly-10).

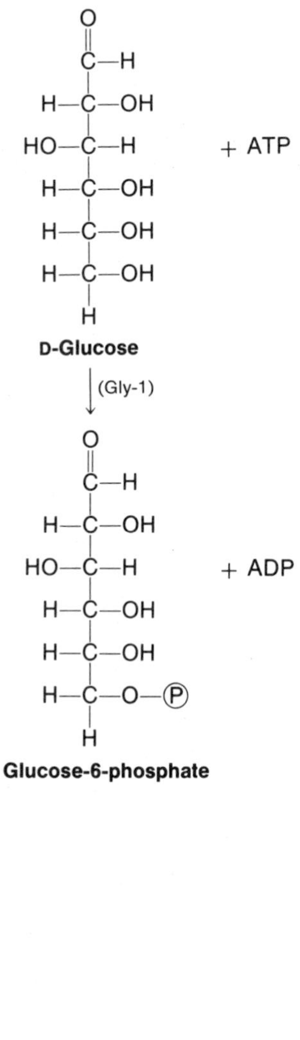

### The First Phase of Glycolysis

The purpose of the first several steps of the glycolytic sequence is to prepare a six-carbon sugar for its subsequent splitting into two interconvertible three-carbon molecules, one of which is subject to further metabolism. We can therefore begin our analysis of the pathway by noting that fructose-1,6-diphosphate, the molecule that actually undergoes lysis, differs from glucose principally in the presence of two phosphate groups, one on each of its terminal carbon atoms (carbons 1 and 6). Thus the first three steps in the sequence can be understood in terms of getting the starting hexose into a doubly phosphorylated form.

**Gly-1.** The first reaction in the sequence involves phosphorylation of glucose on carbon atom 6. Chemically, this can be accomplished easily, since that carbon atom already possesses a free hydroxyl group that can be readily phosphorylated to form a phosphoester. Thermodynamically, however, the direct addition of inorganic phosphate to glucose would be unfavorable ($\Delta G^{\circ}{}' = +3.3$; see Figure 8-4). Instead, the addition of phosphate at carbon 6 is achieved by transfer of a phosphate group from ATP.

The transfer of phosphate from the anhydride linkage of ATP ($\Delta G^{\circ}{}' = -7.3$ kcal/mol) to the phosphoester bond of glucose-6-phosphate ($\Delta G^{\circ}{}' = -3.3$ kcal/mol) is thermodynamically feasible under most conditions. The free-energy loss ($\Delta G^{\circ}{}' = -4.0$ kcal/mol) serves as the driving force for the reaction and renders it essentially irreversible. In fact, the equilibrium for the reaction lies so far to the right ($K = 660$ at 25°C) that virtually all the glucose that enters a cell is almost immediately converted to the phosphorylated form. In addition to activating the glucose for subsequent cleavage, this also ensures that the sugar molecule is effectively trapped within the cell once it enters, because the phosphate group is highly polar and prevents the sugar molecule from passing across the plasma membrane again. In fact, a quick look at the glycolytic pathway reveals that all the intermediates between glucose (which usually enters the cell initially by crossing the plasma membrane) and pyruvate (which in aerobic eukaryotic cells must cross the mitochondrial membrane for further catabolism) are present in phosphorylated form.

Reaction Gly-1 can be catalyzed by either of two separate enzymes, hexokinase or glucokinase. These differ in specificity and in their relative affinities for glucose, but both require $Mg^{2+}$. Hexokinase is, as the name

D-Glucose

(Gly-1)

Glucose-6-phosphate

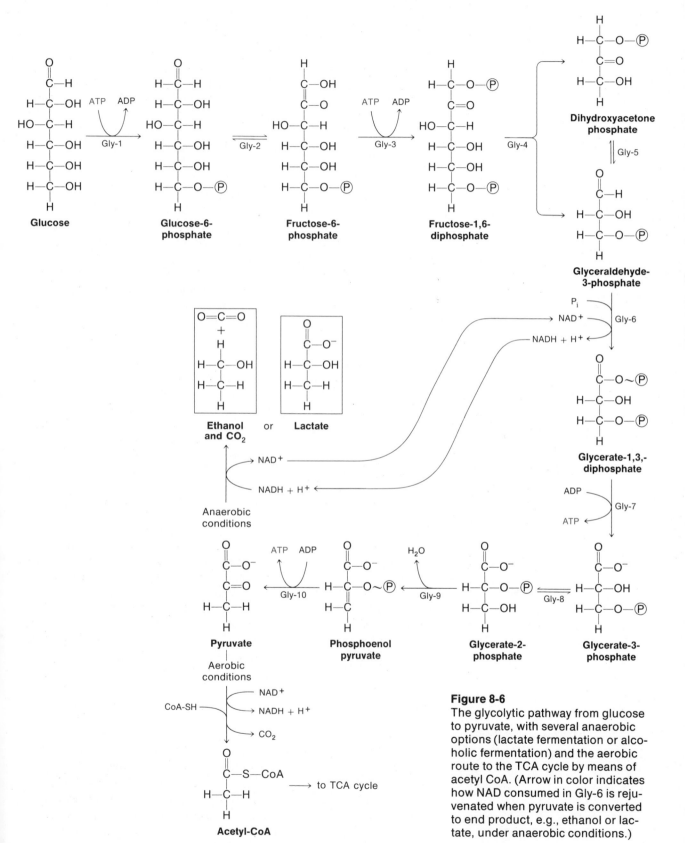

**Figure 8-6**

The glycolytic pathway from glucose to pyruvate, with several anaerobic options (lactate fermentation or alcoholic fermentation) and the aerobic route to the TCA cycle by means of acetyl CoA. (Arrow in color indicates how NAD consumed in Gly-6 is rejuvenated when pyruvate is converted to end product, e.g., ethanol or lactate, under anaerobic conditions.)

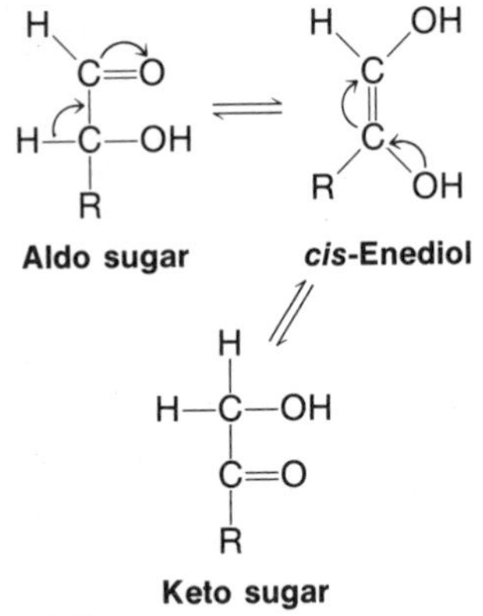

**Aldo sugar**     *cis*-Enediol

**Keto sugar**

**Figure 8-7**
Mechanism of interconversion of an aldo and a keto sugar.

implies, an enzyme capable of phosphorylating a variety of hexoses, including not only glucose, but also fructose, mannose, and galactose. Mammalian hexokinase exists in several isozymic forms that differ in their affinities for glucose. Hexokinase is a regulatory enzyme in the sense that it is allosterically inhibited by its products, glucose-6-phosphate and ADP. Glucokinase, however, is specific for D-glucose and is not inhibited by glucose-6-phosphate. It has a much higher $K_m$ for glucose than does hexokinase (5 to 10 m$M$ versus 0.1 m$M$) and consequently plays an important metabolic role only under conditions of high glucose concentration. Glucokinase is found only in the liver, a tissue that serves as a "reservoir" of glucose by trapping it and storing it as glycogen.

**Gly-2.** With the phosphate on one end of the glucose molecule in place, attention now focuses on the other end of the molecule. The carbonyl group (or, in the ring structure, the hemiacetal) present on carbon 1 is not as readily phosphorylated as a hydroxyl group, so the next reaction involves an isomerization, in effect the moving of the carbonyl group from carbon 1 to carbon 2, thereby converting the molecule into the ketosugar fructose with a free hydroxyl group on carbon 1.

**Glucose-6-phosphate**     **Fructose-6-phosphate**

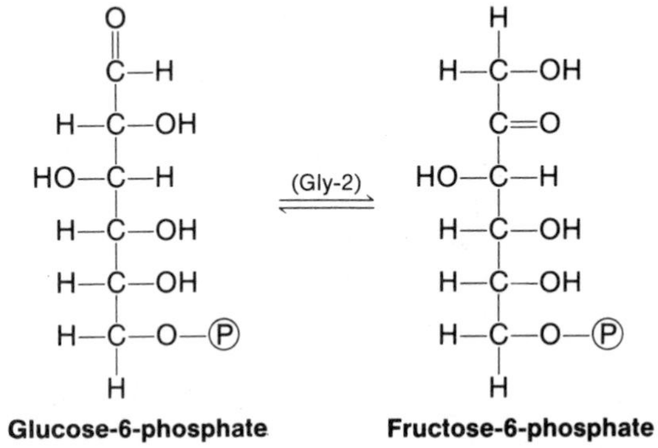

In addition to preparing the molecule for phosphorylation on carbon 1, this isomerization also ensures the presence of a carbonyl group on carbon 2, which turns out to be necessary for the $\beta$ cleavage by which two three-carbon fragments will eventually be generated. The reaction is catalyzed by phosphoglucoisomerase. The mechanism involves an enzyme-bound enediol intermediate that can be readily converted to either the corresponding aldo or keto sugars, as shown in Figure 8-7. The isomerization is accompanied by a small change in free energy ($\Delta G°' = +0.4$ kcal/mol) and is therefore freely reversible.

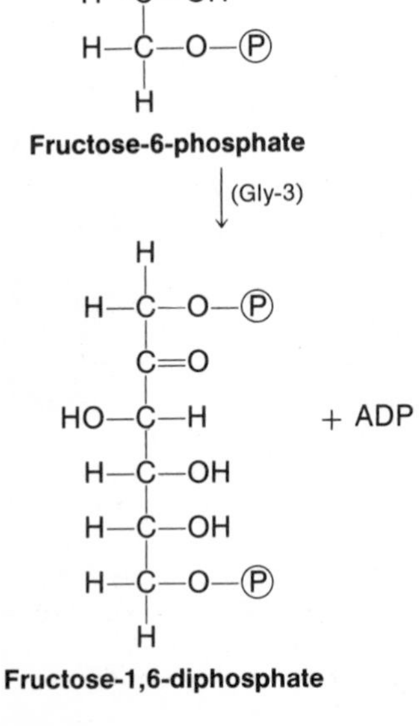

**Gly-3.** A primary hydroxyl group on carbon 1 is now available to be phosphorylated, yielding the doubly phosphorylated sugar fructose-1,6-diphosphate. As in Gly-1, ATP serves as the source of the phosphate group and in the process also provides the energy that drives the reaction.

Again, the difference in free energy of hydrolysis between the anhydride bond of ATP and the phosphoester bond on carbon 1 of fructose-1,6-diphosphate renders the reaction exergonic ($\Delta G°' = -3.4$ kcal/mol) and irreversible in the glycolytic direction in vivo. This illustrates well the general principle that any reaction that constitutes a major metabolic commitment is usually sufficiently exergonic to ensure irreversibility. *Phospho-*

**Fructose-6-phosphate**

(Gly-3)

**Fructose-1,6-diphosphate**

*fructokinase, the enzyme responsible for this reaction, requires $Mg^{2+}$ and is a large, allosterically regulated enzyme* (see Chapter 5 for a definition of allosteric enzyme). *This is the single most important control point of the glycolytic sequence.* Although ATP is itself a substrate for the enzyme, phosphofructokinase is subject to allosteric inhibition by ATP, consistent with the overall role of the glycolytic pathway in ATP generation. *The inhibitory effect of ATP is lessened by ADP and especially by AMP. Citrate, however, accentuates ATP inhibition, thereby linking regulation of the glycolytic pathway with the status of the tricarboxylic acid (TCA) cycle into which it flows under aerobic conditions.*

**Gly-4.** The actual cleavage of sugar from which glycolysis derives its name, fructose-1,6-diphosphate, is split reversibly by aldolase to yield two three-carbon sugars that turn out to be isomers of each other. Note that carbons 1, 2, and 3 of the hexose give rise, respectively, to carbons 3, 2, and 1 of dihydroxyacetone phosphate, while carbons 4, 5, and 6 of the hexose become carbons 1, 2, and 3 of glyceraldehyde-3-phosphate.

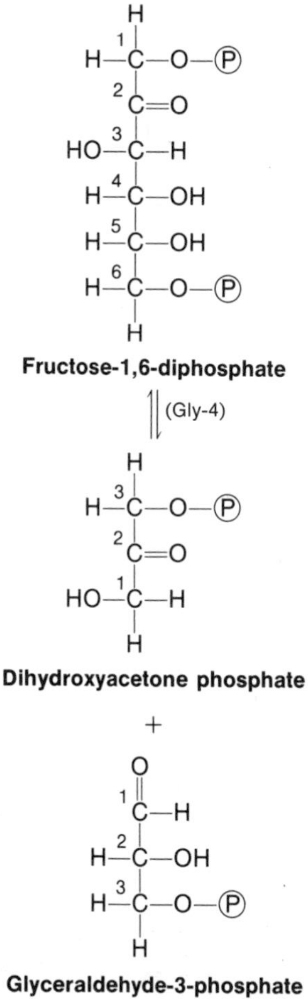

The reaction is thermodynamically very unfavorable as written ($\Delta G^{\circ\prime} = +5.7$ kcal/mol), but it is driven in the direction of triose formation by the exergonic nature of both fructose-1,6-diphosphate formation (Gly-3) and subsequent oxidation of the glyceraldehyde-3-phosphate (Gly-6), which has a very large negative $\Delta G^{\circ\prime}$. Two quite different classes of aldolases are known. Class I aldolases are characteristic of animal tissues and are tetrameric proteins. Muscle, liver, and brain tissues each contain distinctive aldolase isozymes. As shown in Figure 8-8, the reaction mechanism for class I aldolases involves formation of a Schiff's base between the amino group of a lysine residue at the active site of the enzyme (position 227 in a chain of 361 amino acids for rabbit muscle aldolase) and the carbonyl group of the substrate. Class II aldolases occur in yeast and bacteria and are dimeric proteins. They require a divalent cation ($Zn^{2+}$ is probably the natural activator) and do not involve Schiff's base formation in their mechanism of action.

**Gly-5.** The two trioses formed in Gly-4 bear the same isomeric relationship to each other as do glucose-6-phosphate and fructose-6-phosphate. It is therefore not surprising that dihydroxyacetone phosphate and glyceraldehyde-3-phosphate are readily interconvertible by an isomerization reaction analogous to that of Gly-2.

This interconversion is catalyzed by the enzyme triose phosphate isomerase. As with the isomerase of Gly-2, the mechanism here involves an enediol intermediate (see Figure 8-7). The equilibrium of Gly-5 actually lies to the left ($\Delta G^{\circ\prime} = +1.8$ kcal/mol), but the reaction is drawn to the right by the continual depletion of glyceraldehyde-3-phosphate in Gly-6 and Gly-7.

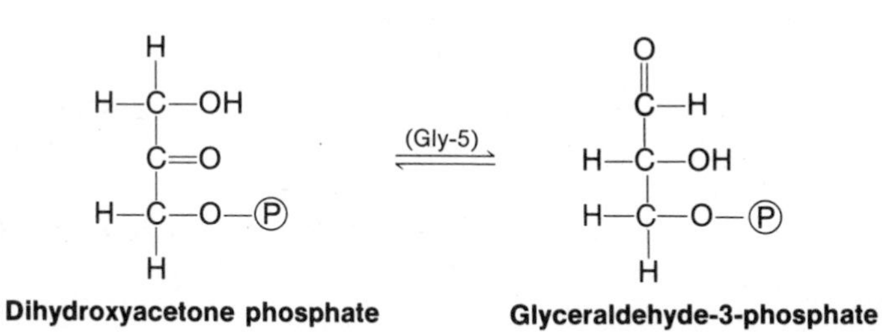

**Fructose-1,6-diphosphate**

(Gly-4)

**Dihydroxyacetone phosphate**

+

**Glyceraldehyde-3-phosphate**

**Dihydroxyacetone phosphate**          **Glyceraldehyde-3-phosphate**

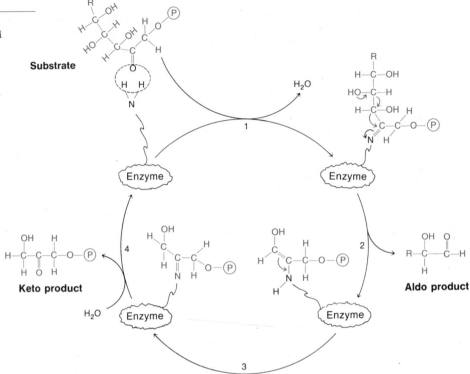

**Figure 8-8**
Mechanism of action of class I
aldolases.

## The Second Phase of Glycolysis

Thus far, 5 of the 10 steps of glycolysis have been accounted for, and the
original glucose molecule has been doubly phosphorylated and cleaved into
two interconvertible molecules, each of which can therefore be subjected
to the same further catabolic fate. Note also that the net ATP yield so far is
$-2$, since two ATPs have been consumed per glucose up to this point. Now,
however, the ATP debt is about to be turned into a profit in the two energy-
yielding phases of glycolysis. In the first instance (Gly-6 and Gly-7), ATP
production is linked to an oxidative event, while in the second case (Gly-8
through Gly-10), it is the highly unstable nature of the enol form of pyruvate that serves as the driving force behind ATP generation.

**Gly-6.** The oxidation of glyceraldehyde-3-phosphate to the corresponding
acid is highly exergonic—sufficiently so, in fact, to drive both the reduction
of the coenzyme NAD$^+$ (Gly-6) and the phosphorylation of ADP (Gly-7).
Historically, this was the first example of a reaction mechanism by which
the energy liberated upon oxidation of an organic molecule could be conserved as ATP. Crucial to this conservation of energy is the coupling in
Gly-6 of the actual oxidation event to the formation of a high-energy
phosphoanhydride bond on carbon 1, which then allows generation of ATP
by subsequent transfer of the phosphate to ADP. In overall formulation, the
oxidative event appears as shown in Equation (Gly-6).

The enzyme responsible for this reaction is glyceraldehyde-3-phosphate
dehydrogenase. The reaction is inhibited both by alkylating agents and by
heavy metals, consistent with the involvement of a sulfhydryl group at the

active site, as shown in the reaction mechanism of Figure 8-9. The reaction is initiated (Figure 8-9, step 1) by addition of the sulfhydryl group of the enzyme to the carbonyl group of the substrate to form the corresponding thiohemiacetal. Oxidation then occurs (step 2) as two electrons and one proton (in effect, a hydride ion, $H^-$) are transferred to an enzyme-bound molecule of $NAD^+$, with concurrent release of a proton ($H^+$). This results in a high-energy thioester bond between the enzyme sulfhydryl group and the newly generated carboxyl group of the substrate. The reduced coenzyme remains attached to the enzyme surface, but becomes reoxidized by passing the hydride ion to a molecule of free $NAD^+$ (step 3). The glycerate molecule is then cleaved from the enzyme surface by phosphorolysis (cleavage of the thioester bond by inorganic phosphate) (step 4). This conserves the energy of the thioester bond as a high-energy phosphoanhydride bond on carbon 1 of the glycerate and leaves the enzyme free to begin another catalytic cycle. The net result of the cycle of events depicted in Figure 8-9 is the coupling of a highly exergonic reaction, the oxidation of glyceraldehyde-3-phosphate ($\Delta G^{\circ\prime} = -10.3\,\text{kcal/mol}$), to what would otherwise be a very endergonic reaction, the formation of a phosphoanhydride bond ($\Delta G^{\circ\prime} = +11.8\,\text{kcal/mol}$). The overall reaction is slightly endergonic under standard conditions ($\Delta G^{\circ\prime} = +1.5\,\text{kcal/mol}$), but it is readily "drawn" in the desired direction by the exergonicity of the next reaction in the sequence.

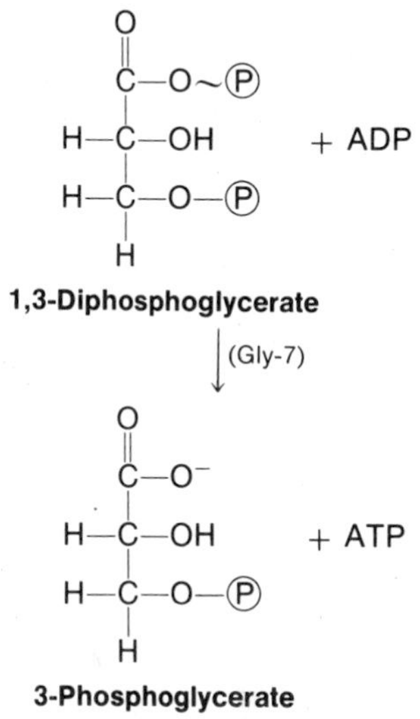

**Gly-7.** Attention in this reaction focuses on the phosphate group on carbon 1 of 1,3-diphosphoglycerate because it is linked to the carboxyl carbon by a high-energy anhydride bond. The bond is sufficiently energy-rich ($\Delta G^{\circ\prime} = -11.8\,\text{kcal/mol}$) that the phosphate group can be readily transferred to ADP to generate ATP ($\Delta G^{\circ\prime} = +7.3\,\text{kcal/mol}$).

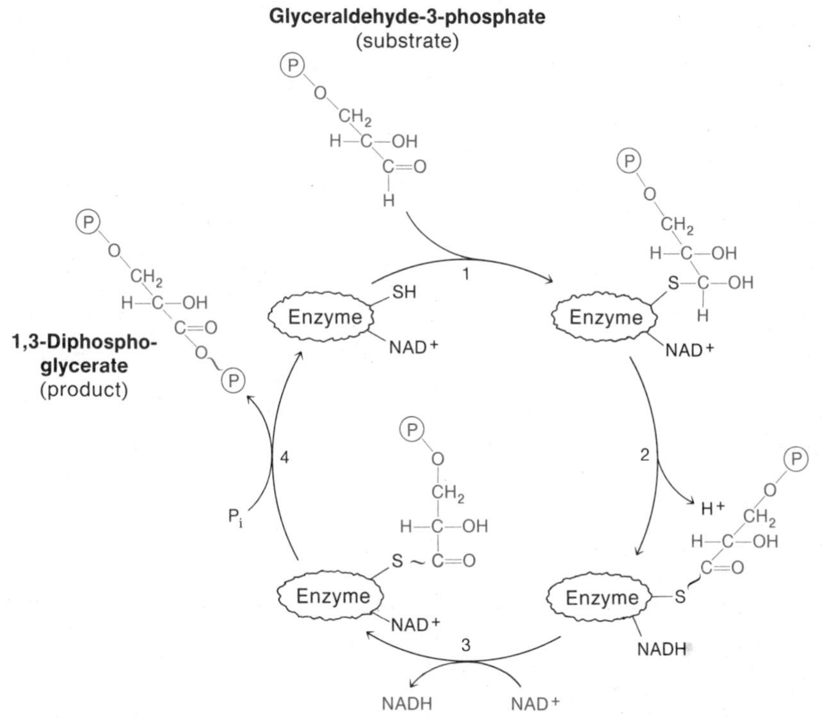

**1,3-Diphosphoglycerate**

(Gly-7)

**3-Phosphoglycerate**

Glyceraldehyde-3-phosphate
(substrate)

1,3-Diphospho-
glycerate
(product)

**Figure 8-9**
Mechanism of action of glyceraldehyde-3-phosphate dehydrogenase.

This reaction, catalyzed by phosphoglycerokinase, is sufficiently exergonic under cellular conditions to ensure that the preceding two reactions are drawn in the forward direction, thereby ensuring the continued oxidation of glyceraldehyde-3-phosphate and dihydroxyacetone phosphate.

*Whenever ATP generation is coupled directly with a specific reaction in a metabolic pathway, as is clearly the case here in glycolysis, it is referred to as substrate-level phosphorylation, to distinguish it from oxidative phosphorylation, which is driven by the transport of electrons* from reduced coenzymes (such as NADH) to oxygen, as will be discussed in Chapter 10. Additional examples of substrate-level phosphorylation will be encountered in Gly-10 (see below) and TCA-5 (see Chapter 9). The example represented by Gly-6 and Gly-7 serves as a particularly beautiful prototype of coupled reactions in bioenergetics, illustrating the simple yet elegant way in which an otherwise heat-liberating reaction has been harnessed to conserve the energy as ATP.

You should note that both the trioses formed on cleavage of the hexose in Gly-4 yield an ATP as they pass through Gly-7, such that the ATP yield from this stage of glycolysis is two ATPs per starting glucose molecule. The two ATPs initially invested in the phosphorylation reactions of Gly-1 and Gly-3 are therefore recovered at this point, and the net ATP yield through this stage is zero. Thus far, 7 of the 10 reactions of glycolysis have been used to convert one molecule of glucose into two molecules of 3-phosphoglycerate, but we have as yet nothing to show for it in terms of net ATP generation. That comes only as we consider the last phase of the pathway.

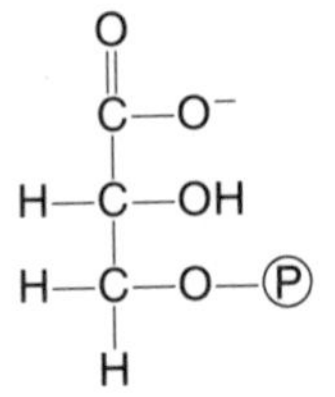

## The Third Phase of Glycolysis

Generation of another molecule of ATP at the expense of 3-phosphoglycerate depends on activation of the phosphate group on carbon 3. The ester bond that links this phosphate group to the glycerate is relatively low in energy ($\Delta G^{\circ\prime} = -3.3 \text{ kcal/mol}$), but in the last phase of the glycolytic pathway, this ester bond is converted to a high-energy phosphoenol bond by a rearrangement of energy in what is in effect an intramolecular oxidation.

**Gly-8.** The first step in this second ATP-generating sequence of glycolysis involves the movement of the phosphate group from its original position on carbon 3 to a new location on carbon 2. The reaction is catalyzed by the enzyme phosphoglyceromutase, which requires $Mg^{2+}$ and involves 2,3-diphosphoglycerate as an intermediate. The equilibrium lies slightly to the left ($\Delta G^{\circ\prime} = +1.1 \text{ kcal/mol}$), but the reaction is readily drawn to the right by the highly exergonic nature of the final reaction in the sequence.

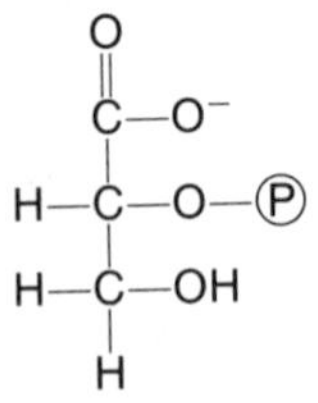

**Gly-9.** A double bond is now created by the removal of water from 2-phosphoglycerate, thereby generating phosphoenolpyruvate. The dehydration will be readily recognized by the chemist as a standard $\alpha,\beta$ elimination. Enolase, the enzyme responsible for this reaction, requires a divalent cation ($Mg^{2+}$ or $Mn^{2+}$) for activity and is strongly inhibited by fluoride.

The $\Delta G^{\circ\prime}$ for this reaction is only $+0.4 \text{ kcal/mol}$, but with the generation of the double bond, the phosphate group on carbon 3 takes on what might be defined as a distinguishing characteristic of a high-energy phos-

**3-Phosphoglycerate**

(Gly-8)

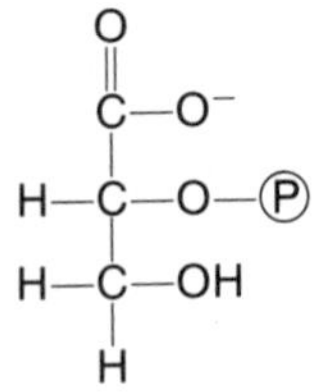

**2-Phosphoglycerate**

**2-Phosphoglycerate**

(Gly-9)

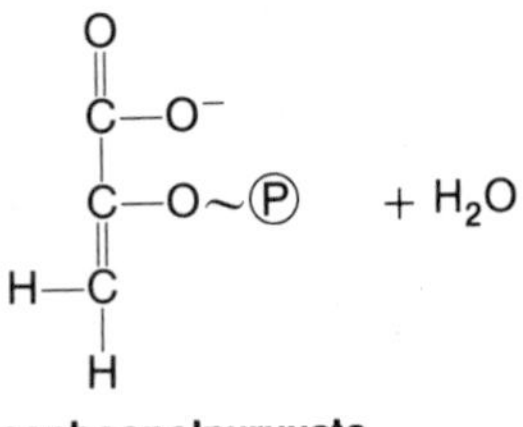

**Phosphoenolpyruvate**

phate bond: a phosphate group adjacent to a double bond. In fact, phospho-enolpyruvate has one of the highest-energy phosphate bonds known in biological systems, with a $\Delta G^{\circ\prime}$ for hydrolysis of $-14.8$ kcal/mol (see Figure 8-4).

To understand how a low-energy phosphate ester bond can become a high-energy phosphoenol bond just by removal of water requires a closer look at the reaction and a recognition that what has happened is in essence an intramolecular oxidation-reduction reaction. This is easier to see when 2-phosphoglycerate is compared not with the phosphoenol form of pyruvate, which is the immediate product of the reaction, but with the keto form of pyruvate, which represents the end product of the glycolytic pathway. Both carbons 2 and 3 of 2-phosphoglycerate are at the same oxidation level, with a single bond to oxygen in each case. Not so for pyruvate, however, which has carbon 2 at the carbonyl level (two bonds to oxygen) and carbon 3 at the hydrocarbon level (no bonds to oxygen). Conversion of 2-phosphoglycerate to pyruvate therefore involves an internal oxidation-reduction event, with carbon 2 oxidized from the hydroxyl to the carbonyl level while carbon 3 is reduced from the hydroxyl to the hydrocarbon level. It is the energy of this internal oxidative event that becomes concentrated in the phosphate bond of phosphoenolpyruvate and makes this such a high-energy bond.

To understand more fully why phosphoenolpyruvate is so high in energy, we need to recognize that the pyruvate molecule can exist in either the enol or the keto form, but that the equilibrium greatly favors the latter (Figure 8-10). Since the equilibrium lies far to the right, the keto form of pyruvate is by far the more stable, while the enol form is a highly unstable, thermodynamically unlikely configuration for the molecule. Thus any compound in which pyruvate is chemically trapped in the enol form will be highly unstable because of the strong tendency of the enol form to tautomerize (undergo transition) to the keto form. That is precisely what happens in Gly-9: When water is removed from 2-phosphoglycerate, the product is pyruvate, but it is "locked" in the enol form by the presence of a phosphate group on carbon 2 which prevents transition to the keto form.

**Gly-10.** To say that the phosphate bond of phosphoenolpyruvate has a highly negative $\Delta G^{\circ\prime}$ of hydrolysis is to make the next step in the sequence entirely reasonable, involving as it does the transfer of that high-energy phosphate to ADP, thereby generating another molecule of ATP.

The difference in energy between hydrolysis of phosphoenolpyruvate ($\Delta G^{\circ\prime} = -14.8$ kcal/mol) and generation of ATP ($\Delta G^{\circ\prime} = +7.3$ kcal/mol) means that the preceding reaction is highly exergonic ($\Delta G^{\circ\prime} = -7.5$ kcal/mol) and therefore essentially irreversible in the direction of pyruvate formation. Pyruvate kinase, the enzyme responsible for this reaction, has a requirement both for $K^+$ and for a divalent cation ($Mg^{2+}$ or $Mn^{2+}$). It is a regulatory enzyme, subject to allosteric inhibition by high ATP concentration and to activation by fructose-1,6-diphosphate and/or by phosphoenolpyruvate.

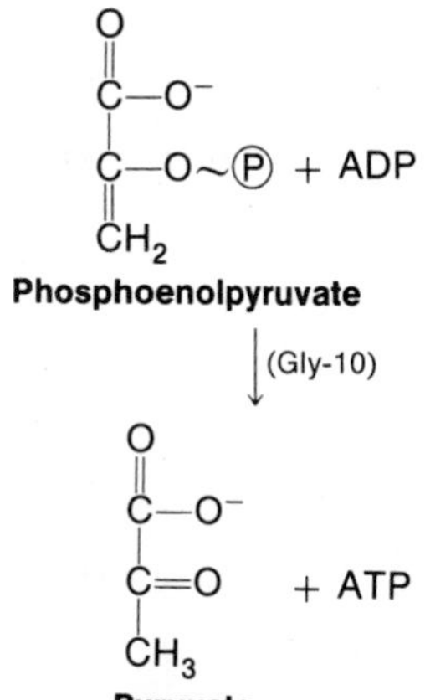

As for Gly-7 previously, the ATP yield for Gly-10 is two ATPs per glucose, since both the triose molecules generated at Gly-4 can be metabolized by the same pathway. Since the ATP initially invested in Gly-1 and Gly-3 was recouped in the first phosphorylation event (Gly-7), the two ATPs

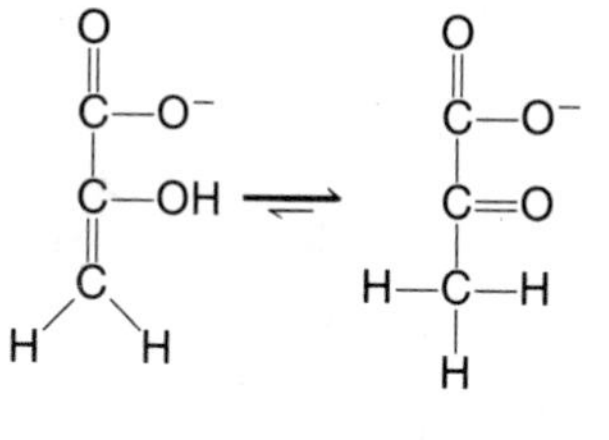

**Figure 8-10**
Tautomerization of the enol and keto forms of pyruvate.

formed here in Gly-10 represent the net yield of the glycolytic pathway. We can therefore write a summary equation for the glycolytic sequence from glucose to pyruvate as follows:

$$C_6H_{12}O_6 + 2\,NAD^+ + 2\,ADP + 2\,P_i \xrightarrow{\text{Gly-1 through Gly-10}} 2\,C_3H_4O_3 + 2\,NADH + H^+ + 2\,ATP + 2\,H_2O$$

## FERMENTATION: THE ANAEROBIC OPTION

The glycolytic sequence is one of the most universal metabolic pathways known, since virtually all cells possess the capability of converting glucose to pyruvate with the extraction of energy in the form of two ATPs per glucose. What happens thereafter, however, depends critically on the availability of oxygen, since pyruvate is the branching point at which aerobic catabolism takes a different direction than that possible under anaerobic conditions.

### Pyruvate as a Branching Point

Pyruvate occupies a key position at the crossroads of several metabolic options. *The fate of the pyruvate depends crucially on the conditions under which metabolism is taking place as well as on the particular organism involved.* The most critical factor is oxygen availability. In the presence of oxygen as an electron acceptor, pyruvate is channeled in the direction of aerobic metabolism, and the glycolytic pathway becomes just the first of three major segments of the overall process of respiratory metabolism. As we shall see in the next chapter, this results in the complete oxidation of pyruvate to $CO_2$, with the generation of much higher ATP yields than can be obtained by glycolysis alone.

An important feature of glycolysis, however, is that it also can be carried out under anaerobic conditions, as is always the case for strict anaerobes and as can be the case for facultative cells if they are functioning in the absence of oxygen. Under these conditions, further oxidation of pyruvate is not possible, and no further ATP can be generated. Instead, *the anaerobic cell must content itself with the two ATPs of glycolysis and uses pyruvate only as an acceptor molecule for the hydrogen atoms that must be removed continuously from NADH in order to generate the NAD$^+$ necessary for continued oxidation of glyceraldehyde-3-phosphate.*

### Lactate Fermentation

A common mechanism for anaerobic reoxidation of NADH at the expense of pyruvate involves the reduction of the carbonyl group of the pyruvate to form lactate.

Lactate formation is thermodynamically favored ($\Delta G^{\circ\prime} = -6.0$ kcal/mol), so substantial lactate might be expected to accumulate in the presence of pyruvate, even under aerobic conditions. The reaction is catalyzed by lactate dehydrogenase, which in higher animals exists in five different isozymic forms, representing all possible tetrameric combinations of the A and B (or, in the alternative terminology of some investigators, the M and H) subunits: $A_4$, $A_3B$, $A_2B_2$, $AB_3$, and $B_4$. All five isozymes catalyze the

O
||
C—O$^-$
|
C=O  + NADH + H$^+$
|
CH$_3$
**Pyruvate**

↓

O
||
C—O$^-$
|
H—C—OH + NAD$^+$
|
CH$_3$
**L-Lactate**

same reaction, but each has its own characteristic $K_m$ values for the substrates (either pyruvate and NADH or lactate and $NAD^+$). The $A_4$ and $A_3B$ forms of the enzyme have low $K_m$ values (i.e., high affinities) for pyruvate and predominate in skeletal muscle and other tissues that are highly dependent on glycolysis for energy. Conversely, the $AB_3$ and $B_4$ forms have higher $K_m$ values (i.e., lower affinities) for pyruvate and are characteristic of purely aerobic tissues, such as heart muscle, where the lower affinity for pyruvate ensures that pyruvate is not diverted into lactate (as would be thermodynamically favored) but is instead maximally available for the enzyme responsible for channeling it in the aerobic direction (pyruvate dehydrogenase; see Chapter 9). The A and B subunits of lactate dehydrogenase are known to be coded for by different genes, such that the relative abundance of subunits, and hence the predominant isozymes of the enzyme present in a given tissue, is genetically controlled.

Since the glycolytic sequence generates NADH and pyruvate on an equimolar basis, both the NADH molecules generated per glucose can be reoxidized at the expense of the two pyruvate molecules, and the overall equation for the metabolism of glucose to lactate under anaerobic conditions becomes

$$C_6H_{12}O_6 + 2\,ADP + 2\,P_i \longrightarrow 2\,C_3H_6O_3 + 2\,ATP + 2\,H_2O$$

*This process of anaerobic glycolysis terminating in lactate is called lactate fermentation.* It is the major energy-yielding pathway in many bacterial and animal cells operating under anaerobic or relatively anaerobic conditions.

Lactic acid fermentation is very important economically, since bacteria capable of the process are responsible for the production of cheeses, yogurts, and other foods obtained by fermentation of the lactose of milk. Another familiar example of lactate fermentation involves skeletal muscles during periods of particularly strenuous exertion. Whenever the muscle cells use oxygen at a faster rate than it can be supplied by the circulatory system, the cells begin functioning anaerobically, reducing the pyruvate to lactate instead of oxidizing it further, as would occur if oxygen supplies were adequate. (Recall that skeletal muscle has the isozymes of lactate dehydrogenase with the greatest affinity for pyruvate, thereby facilitating the anaerobic reduction of pyruvate to lactate.) Lactate thus accumulates in the cell, diffuses out into the bloodstream, and eventually finds its way to the liver, where it is reoxidized to pyruvate and converted back to glucose by the gluconeogenic pathway operative there.

## Alcoholic Fermentation

An alternative fermentation process that also involves pyruvate and NADH but with different end products is alcoholic fermentation. In this case, the pyruvate is first decarboxylated to acetaldehyde, and the latter then serves as the electron acceptor, giving rise to ethanol, the alcohol from which the process derives its name.

As indicated, the reductive decarboxylation of pyruvate involves a two-step sequence catalyzed by two separate enzymes. In the first step, pyruvate is irreversibly decarboxylated to acetaldehyde and $CO_2$ by the enzyme pyruvate decarboxylase. The reaction mechanism involves the coenzyme thiamine pyrophosphate (TPP), which is bound to the surface of the decarboxylase enzyme. TPP is a form of thiamine (vitamin $B_1$), one of the water-soluble vitamins required in the diet of most vertebrates, including humans. The structure of TPP is shown in Figure 8-11. TPP is involved in many decarboxylation reactions, and in each case the mechanism is similar to that shown for pyruvate decarboxylation in Figure 8-12. Note that only the thiazolium ($N═C—S$) portion of the thiazole ring of TPP is actually depicted. The reaction is initiated by attack on the carbonyl carbon of the substrate by the carbanion in position 2 of the thiazole ring. $CO_2$ is then eliminated, leaving a hydroxyethyl group bound to the TPP. This then leaves as acetaldehyde, with the TPP ready to accept another pyruvate molecule.

The acetaldehyde formed in this manner is then reduced to ethanol by the enzyme alcohol dehydrogenase. Alcoholic fermentation thus leads to the formation of ethanol and $CO_2$ in equimolar quantities. The overall equation for the fermentation of glucose by this route is as follows:

$$C_6H_{12}O_6 + 2\,ADP + 2\,P_i \longrightarrow 2\,C_2H_5OH + 2\,CO_2 + 2\,ATP + 2\,H_2O$$

Alcoholic fermentation is of great economic significance, because it is the key process on which both the baking and the brewing industries depend. For the baker, the $CO_2$ is the desired end product. The yeast cells added to bread dough function anaerobically, generating both $CO_2$ and ethanol. The $CO_2$ becomes entrapped within the proteinaceous mass of dough, causing it to rise, and the alcohol is driven off harmlessly during the subsequent baking process. For the brewer, the interest focuses on the ethanol produced during microbial fermentation of fruit juices, and the $CO_2$ does nothing more than blow the cork out of the keg. (Amateur wine makers should note the importance of keeping the cork in, since the microorganisms responsible for anaerobic production of alcohol are likely to make acetic acid by oxidation of the acetaldehyde if oxygen is available, resulting in a keg of vinegar instead of wine!)

## Other Fermentation Options

Although lactate and alcoholic fermentations are the most common and are of the greatest economic significance, they by no means exhaust the microbial options with respect to fermentation. In addition to homolactic bacteria, which produce only lactate from pyruvate, for example, hetero-

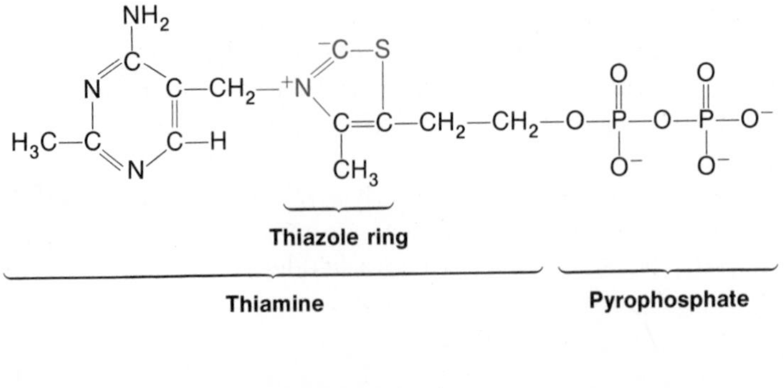

**Figure 8-11**
Structure of the coenzyme thiamine pyrophosphate. The three atoms shown in color represent the "active" part of the coenzyme and are the only part of the molecule shown in Figure 8-12.

Thiazole ring

Thiamine          Pyrophosphate

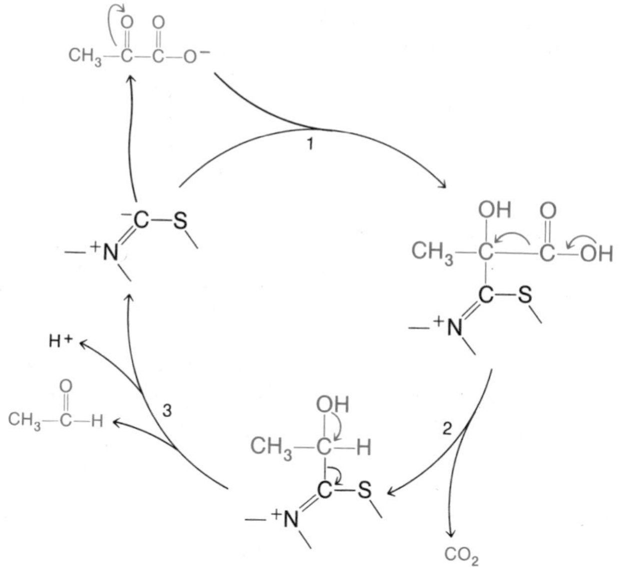

**Figure 8-12**
Mechanism of action of pyruvate decarboxylase. The N=C—S structure
represents the active portion of the
thiamine pyrophosphate coenzyme,
as shown in color in Figure 8–11.

lactic bacteria are also known, and they produce ethanol as well. Propionic
bacteria convert pyruvate reductively to propionate, an important reaction
in the production of Swiss cheese. Many bacteria that are responsible for
food spoilage do so by butylene glycol fermentation, which has as its end
product 2,3-butylene glycol. Other fermentation processes yield butyrate
(the cause of rancid butter), acetone, or isopropanol. All are, however, just
metabolic variations on the common theme of reoxidizing NADH by transfer of electrons to an organic acceptor, which, in essence, is the definition
of fermentation.

## The Energetics of Fermentation

*Fermentation is necessarily a low-energy process, since no net oxidation
occurs.* In fact, the complete aerobic oxidation of lactate has a $\Delta G^{\circ\prime}$ of
about $-319.5$ kcal/mol, so the two lactate molecules that result from lactic
acid fermentation of glucose account for 639 of the 686 kcal of free energy
present per mole of glucose. This means that about 93 percent (639/686 $\times$
100%) of the original free-energy content of glucose is still present in the
product molecules and that *anaerobic fermentation is therefore able to tap
only about 7 percent (47 kcal/mol) of the free energy potentially available from glucose.* As a corollary, an anaerobic cell must therefore consume
much larger quantities of glucose (or other fuel) per unit of time to accomplish the same amount of cellular work as an aerobic cell. This serves to
underscore the tremendous advantage of aerobic metabolism and assigns
considerable significance to the process of respiration, whereby all the remaining energy of the glucose can be tapped upon complete oxidation to
$CO_2$ and $H_2O$, as discussed in the next chapter.

Worth noting, however, is how efficiently the anaerobic cell handles
the 47 kcal/mol of energy available to it. The energy is, of course, partially
conserved in the generation of two ATPs per glucose. Based on standard
free-energy changes (which are admittedly somewhat misleading), the two

ATPs, with a $\Delta G°'$ of $-7.3$ kcal/mol, represent an efficiency of energy conservation of about 31 percent ($14.6/47 \times 100\%$). This may be a low estimate for fermentation, because actual $\Delta G'$ values for ATP hydrolysis under cellular conditions are probably substantially more negative than $-7.3$ kcal/mol (see Chapter 7). If, for example, we assume a $\Delta G'$ value of about $-12$ kcal/mol for ATP hydrolysis, then the efficiency of energy conservation can easily exceed 50 percent.

## ALTERNATIVE SUBSTRATES FOR GLYCOLYSIS

Thus far we have assumed glucose to be the starting point for glycolysis (and therefore, by implication, for all cellular energy metabolism, both aerobic and anaerobic). While it is true that glucose represents a quantitatively very significant substrate for energy metabolism in a variety of organisms and tissues, it is clearly not the only such substrate and is in fact not very important at all in some tissues and under some circumstances. So we need to ask what the major alternatives to glucose are, and how they are handled by the cell. A principle that quickly emerges from such a consideration is that regardless of the chemical nature of the alternative substrates, they are converted as quickly as possible into a form that can be channeled into the mainstream glycolytic pathway with which we are already familiar (Figure 8-13). To illustrate this, we will consider three classes of alternative substrates: other simple sugars, storage carbohydrates, and glycerol.

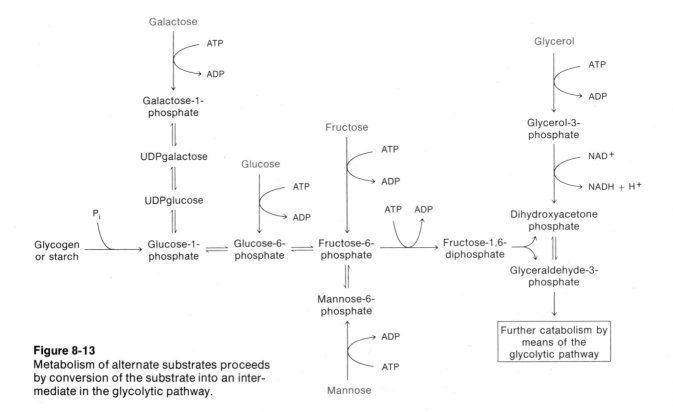

**Figure 8-13**
Metabolism of alternate substrates proceeds by conversion of the substrate into an intermediate in the glycolytic pathway.

## Catabolism of Other Sugars

*A variety of sugars other than glucose is available to cells,* either by ingestion or upon degradation of storage carbohydrates. Most of these are either hexoses or pentoses, with the former predominating. Ordinary table sugar (sucrose), for example, consists of the hexoses glucose and fructose, and milk sugar (lactose) contains glucose and galactose. Organisms consuming these disaccharides must cope not only with the glucose, but with the fructose or the galactose as well. Other sugars with which cells must frequently contend include the hexose mannose and the pentoses ribose, ribulose, and xylulose. *In general, each of these has a specific reaction sequence that brings it as quickly as possible into the glycolytic sequence,* as shown in Figure 8-13.

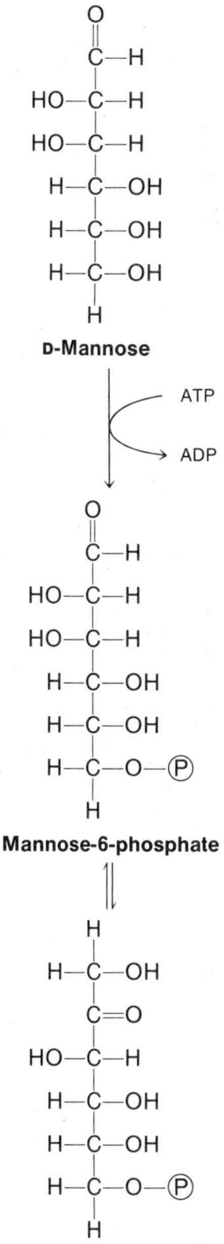

D-Mannose

Mannose-6-phosphate

Fructose-6-phosphate

**Mannose.** Mannose illustrates this principle well. Like glucose, it is first phosphorylated by hexokinase, the relatively nonspecific nature of which has already been noted. The resulting mannose-6-phosphate is then converted by phosphomannoisomerase into fructose-6-phosphate, which is already an intermediate in glycolysis.

**Fructose.** Fructose is most commonly obtained by ingestion and hydrolysis of sucrose. Its entry into the glycolytic pathway is direct indeed, since like glucose and mannose, it also can be phosphorylated at position 6 by hexokinase.

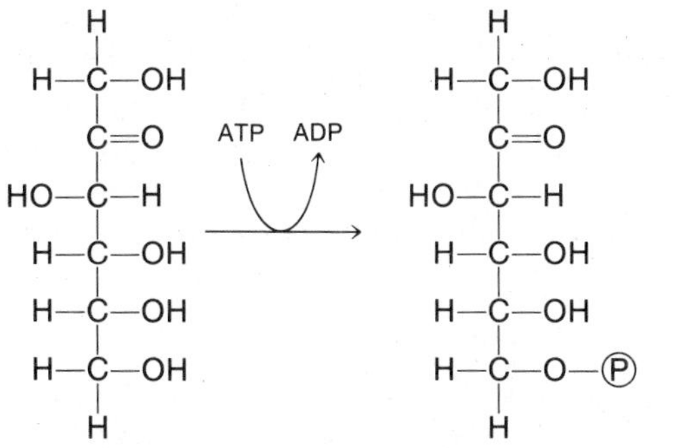

D-Fructose    Fructose-6-phosphate

An alternative pathway exists in mammalian liver, where phosphorylation occurs on carbon 1 instead, catalyzed by the enzyme fructokinase. The fructose-1-phosphate that results can then undergo cleavage by a specific aldolase to dihydroxyacetone phosphate and free glyceraldehyde. The latter compound is phosphorylated to glyceraldehyde-3-phosphate by glyceraldehyde kinase, so that the overall process from hexose to two triose phosphates requires two ATPs, just as for glucose in the initial steps of glycolysis, but with the phosphorylation events occurring in this case in a different sequence:

$$\text{D-Fructose} + \text{ATP} \longrightarrow \text{fructose-1-phosphate} + \text{ADP}$$
$$\text{Fructose-1-phosphate} \rightleftharpoons \text{glyceraldehyde} + \text{dihydroxyacetone phosphate}$$
$$\text{Glyceraldehyde} + \text{ATP} \longrightarrow \text{glyceraldehyde-3-phosphate} + \text{ADP}$$
$$\text{D-Fructose} + 2\ \text{ATP} \longrightarrow \begin{matrix}\text{glyceraldehyde}\\ \text{-3-phosphate}\end{matrix} + \begin{matrix}\text{dihydroxyacetone}\\ \text{phosphate}\end{matrix} + 2\ \text{ADP}$$

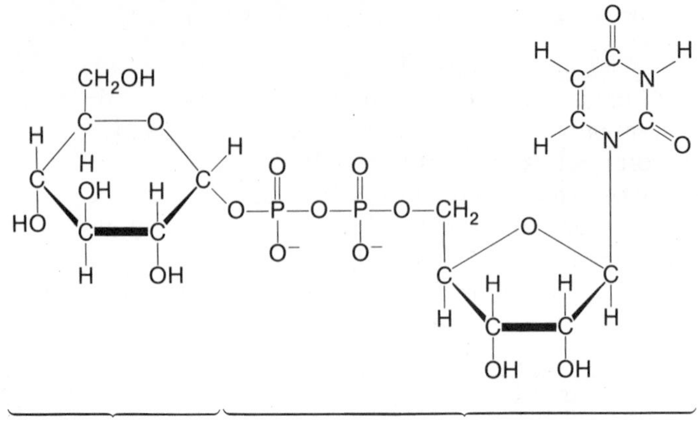

**Figure 8-14**
Structure of UDPglucose.

**Glucose**      **Uridine diphosphate**

**Galactose.** The most common dietary source of galactose is the disaccharide lactose. Galactose is metabolized by phosphorylation and conversion of glucose, but the reaction sequence is somewhat complicated because the conversion to glucose (an *epimerization* reaction on carbon 4) occurs while the sugar is attached to the carrier uridine diphosphate (UDP, a compound related in structure to ADP; see Figure 8-14). The reaction begins with phosphorylation of galactose on carbon 1, catalyzed by galactokinase, a liver enzyme. The galactose is then exchanged for the glucose of UDPglucose in a reaction catalyzed by phosphogalactose uridyl transferase. This results in UDP-bound galactose, which can undergo epimerization on carbon 4 to form UDPglucose, mediated by the enzyme UDPglucose epimerase. The glucose-1-phosphate that results from the transferase reaction is in the meantime converted by the enzyme phosphoglucomutase into glucose-6-phosphate, at which point entry is gained into the glycolytic pathway:

$$\text{D-Galactose} + \text{ATP} \longrightarrow \text{galactose-1-phosphate} + \text{ADP}$$

$$\text{Galactose-1-phosphate} + \text{UDPglucose} \rightleftharpoons \text{glucose-1-phosphate} + \text{UDPgalactose}$$

$$\text{UDPgalactose} \rightleftharpoons \text{UDPglucose}$$

$$\underline{\text{Glucose-1-phosphate} \rightleftharpoons \text{glucose-6-phosphate}}$$

$$\text{D-Galactose} + \text{ATP} \longrightarrow \text{glucose-6-phosphate} + \text{ADP}$$

The congenital disease *galactosemia* is due to a genetic absence of the transferase enzyme. The result in infants ingesting significant quantities of milk is an accumulation of high levels of galactose in the blood (designated by the *-emia* ending). This gives rise to mental disorders, cataracts of the eye, and other characteristic symptoms. Treatment of galactosemic infants involves the obvious remedy of eliminating milk and other galactose sources from the diet during childhood.

**Pentoses.** The reaction mechanisms whereby phosphorylated pentoses and hexoses can be interconverted are the topic of a later section in this chapter. Suffice it here to note that the phosphogluconate pathway discussed below is a means whereby the monophosphorylated forms of the pentoses D-xylulose, D-ribose, and D-ribulose (as well as the tetrose D-ery-throse and the heptose D-sedoheptulose) can be rearranged and, if need be, converted to fructose-6-phosphate for catabolism.

## Catabolism of Storage Polysaccharides

Although glucose is the immediate substrate for both fermentation and respiratory metabolism in many cells and tissues, it is not present in the cell to any large extent as the free monosaccharide. Wherever it is desirable for the cell to accumulate and store glucose for future catabolic use, the glucose is polymerized to form polysaccharides. This reduces the osmotic pressure of the stored sugar, since a polysaccharide consisting of 1000 glucose units will exert an osmotic pressure only 1/1000th that which would result if the glucose units were all present as separate molecules. In addition, the polymerized form of glucose turns out to contain more free energy than the free form, so that catabolism of a polysaccharide actually yields one additional ATP per glucose unit compared with the free sugar, as we will see shortly.

The two major storage polysaccharides are starch in plant cells and glycogen in animal cells (see Chapter 12). Both are $\alpha(1\rightarrow4)$ polymers with $\alpha(1\rightarrow6)$ linkages at branch points, as shown in Figure 8-15. The two polysaccharides differ primarily in their chain lengths and branching patterns. Glycogen is highly branched, with an $\alpha(1\rightarrow6)$ linkage occurring every 8 to 10 glucose units along the backbone, giving rise in each case to short side chains of about 8 to 12 glucose units each. Starch occurs both as un-

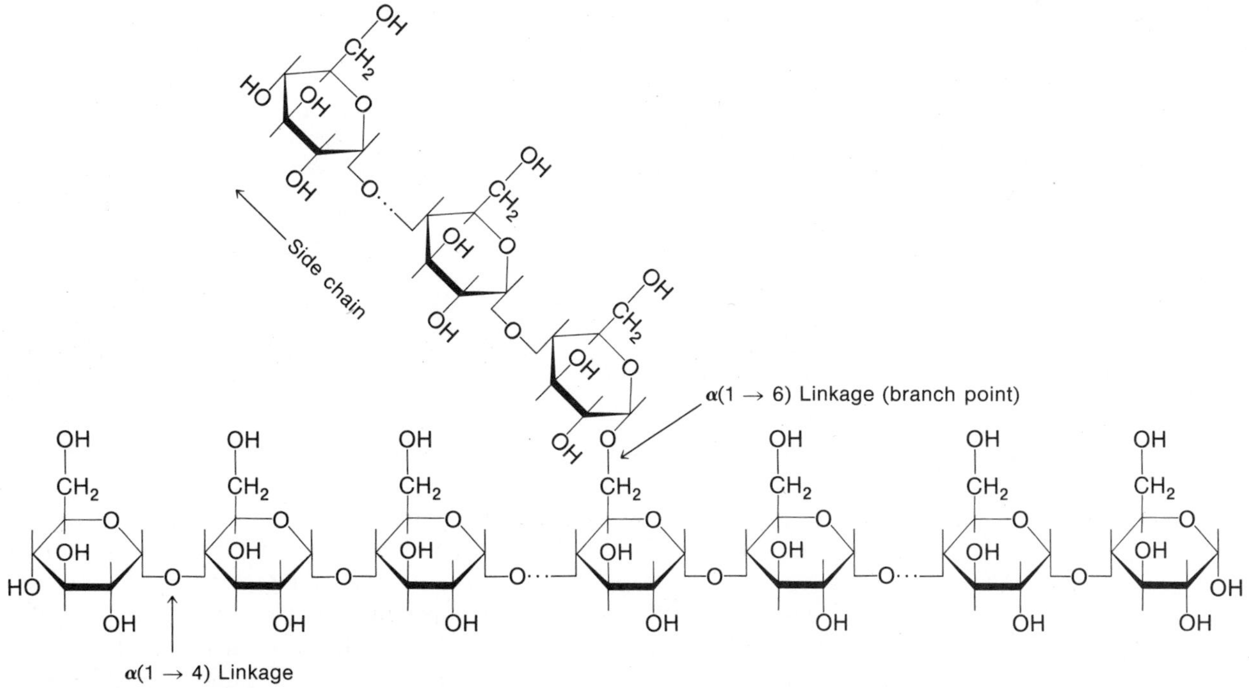

**Figure 8-15**
Structure of the storage polysaccharides glycogen and starch.

branched *amylose* and as branched *amylopectin*. Like glycogen, amylopectin has $\alpha(1\rightarrow6)$ branches, but these occur less frequently along the molecule (once every 12 to 25 glucose residues) and give rise to longer side chains (lengths of 20 to 25 glucose units are common). Starch deposits are usually about 10 to 30 percent amylose and 70 to 90 percent amylopectin. Amylose chains coil into helices and range in molecular weight from a few thousand to a few hundred thousand, while that of amylopectin can be up to a million. Both forms of starch react with iodine, which changes optical properties when it enters the intrahelical space of the starch molecule; amylose gives a characteristic blue color, while amylopectin produces a reddish-violet hue.

Storage of starch in plant cells always occurs as granules in the stroma of plastids, either in the chloroplasts, which are the sites of carbon fixation and reduction in photosynthetic tissue, or in the amyloplasts, which are plastids specialized for starch accumulation and found with special prominence in storage tissues such as potato tubers. Glycogen is also stored in granules, but these are located in the cytoplasm of animal cells rather than in an organelle. Glycogen storage is most pronounced in the liver, where the glycogen is used as a source of glucose to maintain blood sugar levels, and in muscle tissue, where it serves as a fuel source to generate the ATP required for muscle contraction.

The mobilization of storage polysaccharides is depicted in Figure 8-13. Glycogen and starch are both split into their monosaccharide components by stepwise phosphorolytic cleavage of successive glucose units in reactions catalyzed by the enzymes glycogen phosphorylase and starch phosphorylase, respectively. Both are examples of a class of enzymes called $\alpha(1\rightarrow4)$-glucan phosphorylases. With glucose represented in its ring form, a single such cleavage step can be shown as follows:

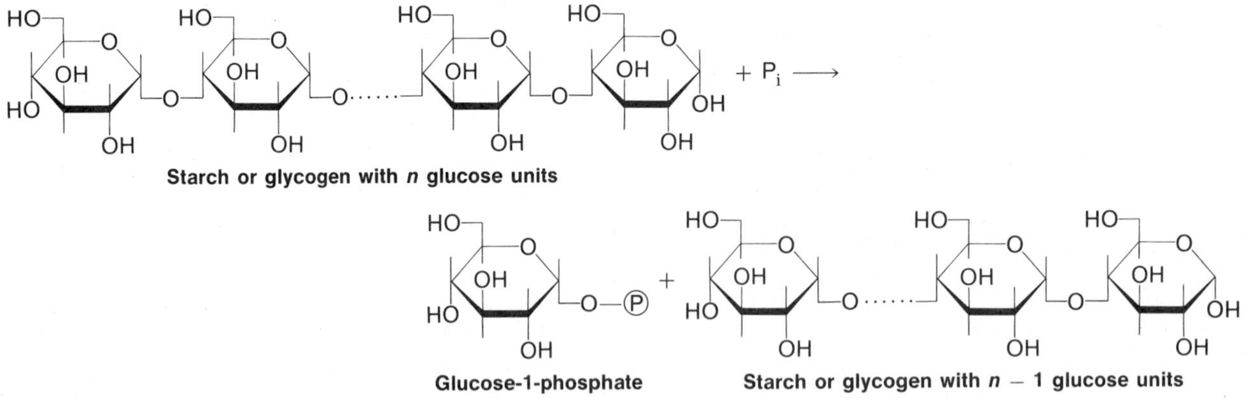

**Starch or glycogen with *n* glucose units**

**Glucose-1-phosphate**        **Starch or glycogen with *n* − 1 glucose units**

This reaction is carried out repeatedly, each time removing a glucose residue from the end with the free hydroxyl group on carbon 4, thereby generating a polysaccharide chain with one less glucose. The $\alpha(1\rightarrow6)$ branch points of glycogen or amylopectin cannot be attacked by phosphorylase, so the end product of exhaustive digestion of either glycogen or amylopectin by the respective phosphorylase is a *limit dextrin* which then requires the action of an $\alpha(1\rightarrow6)$glucosidase to hydrolyze the $1\rightarrow6$ linkages at the branch points and thereby to allow further digestion of the resulting

$\alpha(1\rightarrow4)$ stretches. The net result is the regulated provision of glucose-1-phosphate from the storage polysaccharide.

The glucose-1-phosphate liberated by phosphorylase activity can be converted by the phosphoglucomutase reaction encountered earlier (Figure 8-13) into glucose-6-phosphate, which can then enter the glycolytic pathway for further catabolism.

Note that glucose stored in polymerized form enters the glycolytic pathway as glucose-6-phosphate without the input of ATP that would be required for initial phosphorylation of the free sugar. Consequently, the overall energy yield for glucose is greater by one ATP when it is catabolized from the polysaccharide level than it is with the free sugar as the starting point.

## Catabolism of Glycerol

Glycerol is derived from hydrolysis of triglycerides, a process that will be encountered in more detail in the following chapter. As shown in Figure 8-13, free glycerol is prepared for entry into the glycolytic sequence by phosphorylation, catalyzed by glycerol kinase. The resulting glycerol-3-phosphate (which also can be formed by hydrolysis of phosphoglycerides) is then oxidized by glycerophosphate dehydrogenase to dihydroxyacetone phosphate, an intermediate in glycolysis that can be converted to glyceraldehyde-3-phosphate for further oxidation.

## REGULATION OF GLYCOLYSIS

Since the purpose of the glycolytic pathway under both anaerobic and aerobic conditions is to generate ATP, it is crucial that the functioning of the pathway be continuously adjusted to meet cellular ATP needs. This is accomplished in two major ways: (1) regulation of the extent to which glycogen or starch is mobilized to release glucose for catabolism and (2) control of the rate at which the glucose is converted to pyruvate.

## Control of Glycogen Mobilization

The mobilization of carbohydrate reserves involves the phosphorolytic cleavage of glycogen or starch by phosphorylase. Glycogen phosphorylase is an important regulatory enzyme capable of existing in two forms: phosphorylase $a$ (the more active form) and phosphorylase $b$ (the less active form). Phosphorylase $a$ from skeletal muscle consists of two (possibly four) identical subunits, each with a phosphorylated serine residue (at position 14) that is important for enzyme activity. Conversion of phosphorylase $a$ to the $b$ form of the enzyme involves hydrolysis of the phosphate group of each subunit by the enzyme phosphorylase $a$ phosphatase. Conversion of phosphorylase $b$ back to the active form requires the input of ATP, catalyzed by phosphorylase $b$ kinase.

Phosphorylase kinase, in turn, also exists in an active and an inactive form, with conversion from the inactive to the active form mediated by calcium or by the enzyme protein kinase. Activation of this latter enzyme is effected by _3',5'-cyclic AMP_ (cAMP), the "second messenger" of mam-

Glycerol

Glycerol-3-phosphate

Dihydroxyacetone phosphate

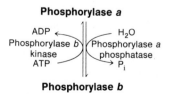

**Phosphorylase a**

**Phosphorylase b**

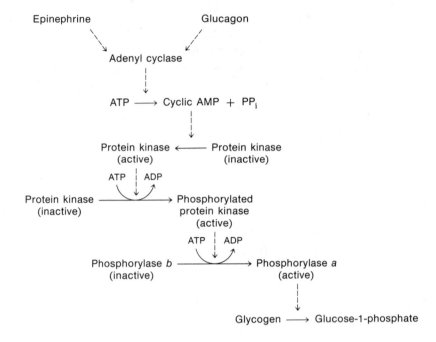

**Figure 8-16**
Cascade regulation of glycogen phosphorylase activity in skeletal muscle.

malian hormonal control systems (see Chapter 29). Cyclic AMP levels are increased by the hormones *epinephrine* and *glucagon*. These hormones serve rather different purposes, since glucagon affects glycogenolysis only in the liver, whereas epinephrine will act on both liver and muscle. In both cases, however, the hormone stimulates the adenyl cyclase responsible for synthesis of cAMP from ATP and thereby activates glycogen breakdown indirectly. The sequence of events from hormonal stimulation to enhanced glycogen degradation is shown in Figure 8-16. Glucagon or epinephrine stimulates adenyl cyclase, thereby augmenting the level of cAMP within the cell. The cAMP is an allosteric effector that activates protein kinase; the protein kinase, in turn, activates phosphorylase kinase. As a consequence, inactive phosphorylase *b* is converted by phosphorylation to the active *a* form, and the breakdown of glycogen to glucose-1-phosphate is thereby promoted. In addition to being activated by phosphorylation, phosphorylase *a* is inhibited by both glucose and nucleosides that bind near the catalytic site. AMP is an allosteric activator of phosphorylase *b* that raises the level of activity of the enzyme in the absence of the hormonally induced *b* to *a* conversion, thus providing an alternative mechanism for stimulating glycogen breakdown. It is not uncommon for regulatory proteins such as glycogen phosphorylase to be responsive to more than one type of control factor. This serves the function of making the enzyme sensitive to different metabolic needs of the organism.

The same hormones that stimulate glycogen breakdown in the liver also inhibit its synthesis. Thus the synthesis and breakdown of glycogen are coordinately regulated, preventing the waste that would result if both processes were to occur simultaneously. Synthesis of glycogen occurs by a different biochemical pathway consistent with the principle stated early in this chapter that the catabolic route is never exactly the reverse of the anabolic pathway, since both need to be exergonic in their respective directions. As will be seen in Chapter 12, the synthesis of glycogen results from the stepwise polymerization of "activated glucose" (UDPglucose) subunits and is catalyzed by glycogen synthase. The same cAMP-dependent protein kinase that phosphorylates glycogen phosphorylase to an active form phosphorylates the glycogen synthase. However, in the latter case, the phosphorylation has the opposite effect on the enzyme; it converts the active form of the synthase into an inactive form.

## Control of the Glycolytic Pathway

*The major pacemaker (rate-limiting) enzyme of the glycolytic sequence is phosphofructokinase.* This enzyme is allosterically activated by ADP and AMP and is inhibited by ATP, giving the enzyme great sensitivity to the ATP-ADP status of the cell. The allosteric inhibition by ATP of an enzyme for which ATP is itself a required substrate (for phosphorylation of fructose-6-phosphate) creates an apparent contradiction of effects, since increasing levels of substrate should increase the rate of an enzyme-catalyzed reaction, while increasing levels of an allosteric inhibitor should render the enzyme less active. This is possible for phosphofructokinase because the active (or catalytic) and effector (allosteric) sites of the enzyme differ in their affinities for ATP. The active site has a high affinity (i.e., a low $K_m$) for ATP, while the affinity of the effector site for ATP is lower. Thus, *at low ATP levels, binding occurs at the catalytic site, but not at the allosteric site, so the enzyme remains in the active form and functions. At high ATP levels, however, binding is promoted at the effector site, converting the enzyme to its inactive form* and thereby serving as a throttle for the whole glycolytic sequence.

*In addition to its sensitivity to ATP, ADP, and AMP, phosphofructokinase is also allosterically inhibited by citrate, NADH, and fatty acids. The sensitivity to citrate and NADH provides a crucial regulatory link between glycolysis and the TCA cycle, of which citrate is an intermediate and NADH is in a sense a product* (see Chapter 9). *This ensures that the two stages of respiration are carefully regulated with respect to each other.* When the level of citrate rises as a result of excessive glycolytic activity, the citrate binds to its effector site on phosphofructokinase, thereby converting the enzyme reversibly to an inactive form and ensuring that the glycolytic pathway is adjusted downward toward actual cellular needs. Similarly, when fatty acids are available for use in ATP generation by the β-oxidation pathway (see Chapter 9), their inhibitory effect on phosphofructokinase activity will cause the glycolytic pathway to be shut down, thereby facilitating the use of fatty acids rather than carbohydrate as an energy source.

Further allosteric regulation of the glycolytic pathway occurs at the step catalyzed by glyceraldehyde-3-phosphate dehydrogenase, which is subject to activation by $NAD^+$. This confers upon the pathway further sensitivity to the coenzyme status of the cell. If $NAD^+$ levels are high, glycolysis is stimulated to generate more reduced coenzyme, which can serve as a source of ATP production in the presence of oxygen (see Chapter 10). If $NAD^+$ levels are low, glycolysis is shut down until sufficient coenzyme is reoxidized again.

Another important regulatory enzyme in the glycolytic sequence is pyruvate kinase. This enzyme is inactivated by high ATP concentrations in all tissues. The isozyme present in the liver is subject in addition to allosteric activation by fructose-1,6-diphosphate, thereby providing for coordinated control of the several segments of the glycolytic pathway.

The overall effect of the multiple regulatory sites of the glycolytic pathway is to render the sequence sensitive to cellular levels of ATP and of reduced coenzyme, as well as to the next stage of aerobic energy metabolism (the TCA cycle) and to the availability of alternate substrates (such as fatty acids). As a result, *the activity of the pathway is subject to continuous adjustment, so that the production of pyruvate and ATP is carefully paced to meet the energy needs of the cell.*

## THE PHOSPHOGLUCONATE PATHWAY

An alternative to the glycolytic pathway (and therefore to the whole process of respiratory metabolism of which it is a part under aerobic conditions) as a means of oxidizing glucose is the *phosphogluconate pathway*

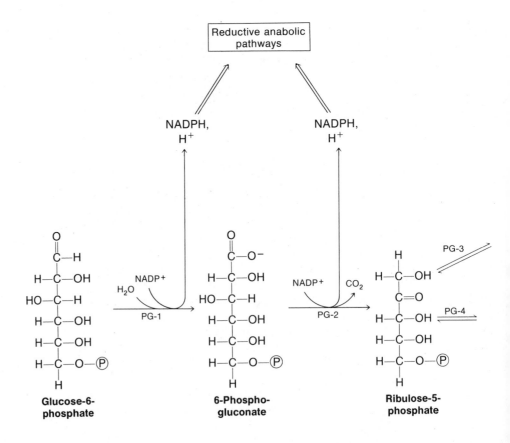

**Figure 8-17**
The phosphogluconate pathway.

shown in Figure 8-17. This pathway is also called the _hexose monophosphate shunt_ or the _pentose phosphate pathway. The multiplicity of names derives from the variety of functions which the pathway can serve under different conditions._ In fact, we are dealing here not so much with a single unidirectional pathway as with a series of interrelated reaction sequences that can be channeled toward several different ends. Most straightforwardly, this pathway represents an alternative mode of glucose oxidation, coupled with formation of the reduced coenzyme NADPH. The immediate product of glucose oxidation by this route is phosphogluconate, thereby accounting for the name of the pathway. The NADPH generated in the formation and further oxidation of phosphogluconate is widely used by cells to drive reductive anabolic pathways, notably the synthesis of fatty

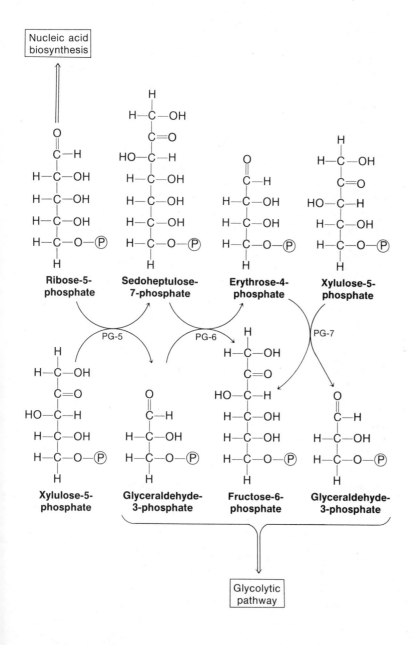

acids and steroids. Accordingly, this pathway is of greatest significance in tissues that are actively involved in fat and steroid synthesis, such as the mammary gland, the liver, the adrenal cortex, and adipose tissue. More generally, however, *the phosphogluconate pathway is the principal source of reducing power for biosynthetic reactions in most cells.* The pathway is also important in providing for the interconversion of hexoses and pentoses, thereby explaining the alternative names it bears. It therefore plays a key role in the synthesis of pentoses (used for nucleic acid synthesis; see Chapter 19), the catabolism of pentoses (by conversion to hexoses and trioses for entry into the glycolytic pathway), and the regeneration of the phosphorylated pentose ribulose-1,5-diphosphate required for carbon fixation in photosynthesis.

### Glucose Oxidation and NADPH Generation

A major function of the phosphogluconate pathway is to oxidize glucose in a manner that provides for generation of the reduced coenzyme NADPH. The pathway is initiated by oxidation of glucose-6-phosphate on carbon 1 to form 6-phosphogluconate.

Although shown as a single step, this reaction is actually carried out on the ring form of glucose, and the immediate product of oxidation is 6-phosphoglucono-$\delta$-lactone (see Figure 19-1). This molecule is unstable and undergoes spontaneous hydrolysis to the free acid, although a lactonase enzyme is known that speeds the hydrolysis. The 6-phosphogluconate is then oxidatively decarboxylated to ribulose-5-phosphate, thereby opening the gateway to pentose generation. The NADPH generated in these reactions is used to drive a variety of reductive processes, with the synthesis of fatty acids a significant example. All the enzymes of the phosphogluconate pathway occur in the cytoplasm, which is also the intracellular locale of fatty acid synthesis.

Glucose-6-phosphate $+ NADP^+ + H_2O$ $\xrightarrow{(PG-1)}$ 6-Phosphogluconate $+ NADPH + H^+$

6-Phosphogluconate $+ NADP^+ \xrightarrow{(PG-2)} CO_2 +$ Ribulose-5-phosphate $+ NADPH + H^+$

### Interconversion of Pentoses

With ribulose-5-phosphate available from PG-2, several related pentose phosphates can be generated readily. Ribose-5-phosphate is the corresponding aldosugar and can be formed from ribulose-5-phosphate by the enzyme phosphopentose isomerase. The reaction is entirely analogous to those with which we are already familiar for the interconversion of keto and aldo forms of hexoses (Gly-2) and trioses (Gly-5). Ribose-5-phosphate is an especially important pentose because of its use in the synthesis of nucleic acids.

```
        H                              O
        |                              ‖
    H—C—OH                          C—H
        |                              |
        C=O              (PG-3)    H—C—OH
        |              ⇌               |
    H—C—OH                          H—C—OH
        |                              |
    H—C—OH                          H—C—OH
        |                              |
    H—C—O—P                         H—C—O—P
        |                              |
        H                              H
  Ribulose-5-phosphate          Ribose-5-phosphate
```

Xylulose-5-phosphate is another phosphorylated pentose that can be formed from ribulose-5-phosphate, in this case by epimerization on carbon 3. The reaction is catalyzed by phosphopentose epimerase using a mechanism reminiscent of the interconversion of galactose and glucose (see Figure 8-13), although in that case the epimerization occurs on the UDP-bound sugar.

```
        H                              H
        |                              |
    H—C—OH                          H—C—OH
        |                              |
        C=O              (PG-4)        C=O
        |              ⇌               |
    H—C—OH                        HO—C—H
        |                              |
    H—C—OH                          H—C—OH
        |                              |
    H—C—O—P                         H—C—O—P
        |                              |
        H                              H
  Ribulose-5-phosphate          Xylulose-5-phosphate
```

As a result of PG-3 and PG-4, ribulose, ribose, and xylulose all exist as an equilibrium mixture of the monophosphates.

## Further Interconversions of Sugar Phosphates

These pentose phosphates serve as the starting point for a series of further interconversions capable of generating three-, four-, five-, six-, and seven-carbon sugar phosphates, provided only that the enzymes transketolase and transaldolase are present. As shown in Figure 8-18, transketolase is capable of moving two-carbon fragments (as the glycolaldehyde group) between sugars, and transaldolase is responsible for shifting three-carbon fragments (as the dihydroxyacetone group).

**Transketolase.** Transketolase can move a glycolaldehyde group from any of a number of 2-keto sugar phosphates onto a variety of aldosugar phosphate acceptors, as shown in Figure 8-18. The enzyme uses thiamine pyrophosphate (TPP) as its coenzyme. The TPP is tightly bound to the enzyme surface and carries the glycolaldehyde group as the dihydroxyethyl derivative in a way analogous to the mechanism already described for pyruvate decarboxylase (see Figure 8-12).

## Xylulose-5-phosphate

```
        H                      O
        |                      ‖
   H—C—OH                      C—H
        |                      |
       C=O                H—C—OH
        |                      |
  HO—C—H          +      H—C—OH
        |                      |
   H—C—OH                 H—C—OH
        |                      |
   H—C—O—Ⓟ                H—C—O—Ⓟ
        |                      |
        H                      H
```

**Xylulose-5-phosphate**     **Ribose-5-phosphate**

‖(PG-5)

```
                               H
                               |
                          H—C—OH
                               |
                              C=O
                               |
                         HO—C—H
        O                      |
        ‖                 H—C—OH
        C—H                    |
        |                 H—C—OH
   H—C—OH                      |
        |                 H—C—OH
   H—C—O—Ⓟ                     |
        |                 H—C—O—Ⓟ
        H                      |
                               H
```

**Glyceraldehyde-3-**        **Sedoheptulose-7-**
**phosphate**                **phosphate**

```
        H
        |
   H—C—OH
        |
       C=O
        |
  HO—C—H                       O
        |                      ‖
   H—C—OH          +           C—H
        |                      |
   H—C—OH                 H—C—OH
        |                      |
   H—C—OH                 H—C—O—Ⓟ
        |                      |
   H—C—O—Ⓟ                     H
        |
        H
```

**Sedoheptulose-7-**         **Glyceraldehyde-3-**
**phosphate**                **phosphate**

‖(PG-6)

```
                               H
                               |
        O                 H—C—OH
        ‖                      |
        C—H                   C=O
        |                      |
   H—C—OH          +     HO—C—H
        |                      |
   H—C—OH                 H—C—OH
        |                      |
   H—C—O—Ⓟ                H—C—OH
        |                      |
        H                 H—C—O—Ⓟ
                               |
                               H
```

**Erythrose-4-phosphate**    **Fructose-6-phosphate**

A specific example of the general transketolase reaction shown in Figure 8-18 involves the transfer of two carbons from xylulose-5-phosphate to ribose-5-phosphate, thereby generating the seven-carbon sugar sedoheptulose-7-phosphate and the three-carbon compound glyceraldehyde-3-phosphate.

Since this reaction interconverts pentose and triose phosphates, with the latter (glyceraldehyde-3-phosphate) an intermediate in glycolysis, this can serve either as a means of converting pentose to triose for catabolic entry into the glycolytic pathway or as a way of generating pentoses from glycolytic (or, as it turns out, from photosynthetic) intermediates in a synthetic mode. Another example of a transketolase reaction is that designated in Figure 8-17 as PG-7, in which xylulose-5-phosphate donates a two-carbon fragment to the four-carbon sugar erythrose-4-phosphate to generate glyceraldehyde-3-phosphate and fructose-6-phosphate.

**Transaldolase.** The sedoheptulose-7-phosphate and glyceraldehyde-3-phosphate generated in the transketolase reaction just shown can serve in turn as substrates for the transaldolase reaction. This enzyme transfers a three-carbon dihydroxyacetone group between a variety of sugar phosphates, as shown in Figure 8-18. For the specific seven- and three-carbon substrates available from PG-5, the three-carbon transfer takes place as shown in reaction PG-6.

## Summary of the Phosphogluconate Pathway

Given the enzymatic capability to oxidize glucose by means of phosphogluconate, to interconvert pentoses, and to move two-carbon and three-carbon fragments between sugar phosphates, the cell can accomplish an amazing variety of related but often distinctly different purposes. For example, if the need is for reducing power and ribose-5-phosphate for nucleotide synthesis, PG-1, PG-2, and PG-3 can be used, and the overall equation becomes

Glucose-6-phosphate $+ 2\,NADP^+ \longrightarrow$
$$\text{ribose-5-phosphate} + CO_2 + 2\,NADPH + 2\,H^+$$

If, however, the cell faces a need to catabolize pentoses, PG-3 through PG-7 can be combined to accomplish the desired end. (Note that although the example here is for catabolism of ribose, other pentoses can be handled similarly because of the facile interconversions possible by means of PG-3 and PG-4.)

$$2\text{ Ribose-5-phosphate} \rightleftharpoons 2\text{ ribulose-5-phosphate} \quad \text{(PG-3)}$$

$$2\text{ Ribulose-5-phosphate} \rightleftharpoons 2\text{ xylulose-5-phosphate} \quad \text{(PG-4)}$$

Xylulose-5-phosphate + ribose-5-phosphate $\rightleftharpoons$
sedoheptulose-7-phosphate + glyceraldehyde-3-phosphate  (PG-5)

Sedoheptulose-7-phosphate + glyceraldehyde-3-phosphate $\rightleftharpoons$
erythrose-4-phosphate + fructose-6-phosphate  (PG-6)

Erythrose-4-phosphate + xylulose-5-phosphate $\rightleftharpoons$
glyceraldehyde-3-phosphate + fructose-6-phosphate  (PG-7)

---

3 Ribose-5-phosphate $\rightleftharpoons$
2 fructose-6-phosphate + glyceraldehyde-3-phosphate

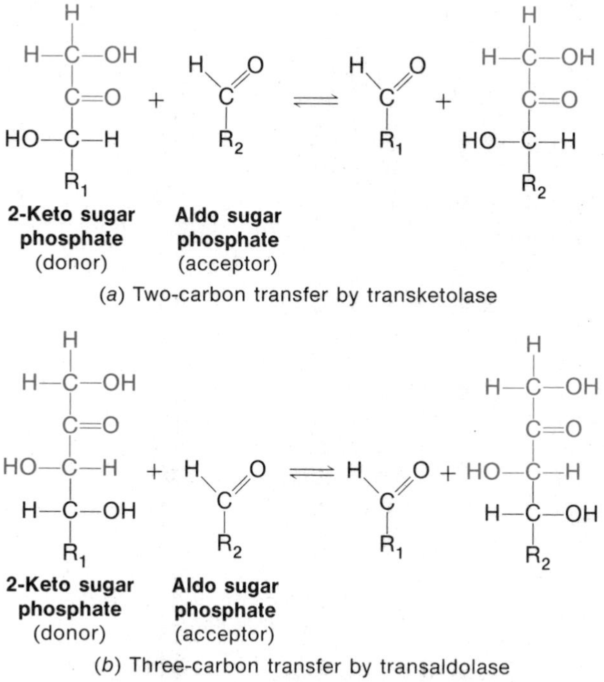

(a) Two-carbon transfer by transketolase

(b) Three-carbon transfer by transaldolase

**Figure 8-18**
Reactions catalyzed by (a) transketolase and (b) transaldolase.

Since each of the preceding reactions is readily reversible, the sequence could equally well be used to generate ribose-5-phosphate (or another pentose phosphate) from glycolytic intermediates. Similar interconversions are involved in the Calvin cycle for photosynthetic $CO_2$ fixation (see Chapter 11) to account for the continual regeneration of the doubly phosphorylated pentose ribulose-1,5-diphosphate that serves as the acceptor of $CO_2$.

By summing the reactions for the oxidative generation of ribose-5-phosphate from glucose-6-phosphate and the reactions for the production of fructose-6-phosphate and glyceraldehyde-3-phosphate from 3 ribose-5-phosphate, we can write an overall equation for the oxidation of glucose-6-phosphate by means of the phosphogluconate pathway, but with eventual return of the carbon to the glycolytic sequence. This requires only that the first reaction be multiplied by 3 and added to the second reaction; the result is summarized as follows:

$$3 \text{ Glucose-6-phosphate} + 6 \text{ NADP}^+ \longrightarrow 3 \text{ CO}_2 + 2 \text{ fructose-6-phosphate}$$

$$+ \text{ glyceraldehyde-3-phosphate} + 6 \text{ NADPH} + 6 \text{ H}^+$$

It is even possible to devise a sequence in which glucose-6-phosphate is oxidized completely to $CO_2$. This requires oxidation of six molecules of glucose-6-phosphate by means of phosphogluconate to six molecules of ribulose-5-phosphate (PG-1 and PG-2 carried out six times), followed by conversion of the six resulting pentose molecules back to five hexose molecules (by judicious combination of PG-3 through PG-7 and partial reversal of the glycolytic sequence from glyceraldehyde-3-phosphate to glucose). The summary equation for such an oxidative process would be

$$\text{Glucose-6-phosphate} + 12 \text{ NADP}^+ \longrightarrow$$

$$6 \text{ CO}_2 + 12 \text{ NADPH} + 12\text{H}^+ + \text{P}_i$$

However, both the preceding reactions presume the existence of some concomitant cytoplasmic process that requires reducing power and thereby provides for the continual regeneration of $NADP^+$. The extent to which these alternative means of glucose oxidation actually occur in cells appears to be highly dependent on the need for NADPH to drive anabolic processes, especially the synthesis of fats and steroids. In the absence of such need, the glycolytic pathway remains the mechanism of choice for the catabolism of glucose under both anaerobic and aerobic conditions.

## SELECTED READINGS

Atkinson, D. E. *Cellular Energy Metabolism and its Regulation.* New York: Academic Press, 1977. A solid introduction to the intricacies of energy metabolism by an author who is very much a part of the scene.

Broda, E. *The Evolution of Bioenergetic Processes.* New York: Pergamon Press, 1975. Energy metabolism from an evolutionary perspective.

Dagley, S., and Nicolson, D. E. *Metabolic Pathways.* New York: Wiley, 1970. An older reference, but still useful as a compendium of pathways.

Fletterick, R. J., and Madsen, N. B. The structure and related functions of phosphorylase *a. Ann. Rev. Biochem.* 49:31, 1980. An up-to-date treatment of one of the more current research topics from this chapter.

Hofmann, E. The significance of phosphofructokinase to the regulation of carbohydrate metabolism. *Rev. Physiol. Biochem. Pharmacol.* 75:1, 1976. Recommended for its treatment of a clearly important regulatory enzyme of the glycolytic pathway.

Kalcker, H. M. (Ed.). *Biological Phosphorylations.* Englewood Cliffs, N.J.: Prentice-Hall, 1969. A highly useful collection of original publications that are now recognized as classics.

Ramaiah, A. Regulation of glycolysis in skeletal muscle. *Life Sci.* 19:455, 1976. A detailed look at the control of the glycolytic pathway.

## PROBLEMS

1. *Lactate fermentation.* An anaerobic bacterial culture is known to be accumulating lactate as fermentation progresses. Decide which of the following statements about the culture are true (T) and which are false (F).
   a. The culture is almost certainly growing on glucose, since few other compounds can be fermented by bacteria.
   b. The product(s) of fermentation are no more highly oxidized than the starting compounds, since no external electron acceptor is available.
   c. The culture cannot be producing any $CO_2$.
   d. If the culture is opened and air is bubbled through continuously, the lactate level in the culture medium will continue to increase.
   e. Addition of fluoride ions will result in a rapid increase in the glycerate-2-phosphate : phosphoenol pyruvate ratio in the bacterial cells.

2. *Eat, drink, and be merry.* In the absence of oxygen, chemotrophic cells give rise to fermentation products that can be associated with bread, wine, and tired muscles. Explain each of these associations.

3. *Alcoholic fermentation.* A cell-free extract of yeast known to contain all the enzymes required for alcoholic fermentation is incubated anaerobically in 100 ml of a medium initially known to contain 200 m$M$ glucose, 20 m$M$ ADP, 40 m$M$ ATP, 2 m$M$ NADH, 2 m$M$ NAD$^+$, and 20 m$M$ $P_i$.
   a. Assuming that the ethanol is removed continuously from the incubation medium as soon as it is formed, what is the maximum amount of ethanol (in millimoles) that can be formed? Explain your answer.
   b. Which of the following changes would be most likely to allow the production of the most additional ethanol, once the culture had reached the maximum of part (*a*)? Explain your answer.
      i. Doubling the glucose concentration of the medium.
      ii. Addition of 20 m$M$ glyceraldehyde-3-phosphate.
      iii. Addition of 20 m$M$ pyruvate.
      iv. Addition of ATPase, an enzyme known to degrade ATP to ADP and $P_i$.

c. What is the maximum amount of ethanol that can be formed after making the change indicated in part (b)?

4. *More alcoholic fermentation.* If a bacterial culture is carrying out alcoholic fermentation of $^{14}$C-labeled glucose, in which position(s) of the glucose molecule would the radioactive $^{14}$C atoms have to be located in order to ensure that the $CO_2$ produced during the fermentation process is labeled with $^{14}$C?

5. *Glyceraldehyde oxidation.* If a mutant form of glyceraldehyde-3-phosphate dehydrogenase were discovered that cleaved the oxidized product from the enzyme surface by hydrolysis rather than by phosphorolysis, what effect, if any, would this have on an obligately anaerobic organism that had the mutant form of the enzyme?

6. *Sucrose hydrolysis.* The disaccharide sucrose can be cleaved by either of two alternative means:

$$\text{Sucrose} + H_2O \xrightarrow[\text{Invertase}]{} \text{glucose} + \text{fructose}$$

$$\text{Sucrose} + P_i \xrightarrow[\text{Sucrose phosphorylase}]{}$$

$$\text{glucose-1-phosphate} + \text{fructose}$$

a. If the $\Delta G^{\circ\prime}$ value for the invertase reaction is $-7.0$ kcal/mol, what would you estimate the $\Delta G^{\circ\prime}$ value for the sucrose phosphorylase reaction to be?

b. What advantage would there be to a cell to carry out intracellular sucrose cleavage by the sucrose phosphorylase reaction rather than by the invertase route?

7. *Erythrose synthesis.* An organism that possesses all the enzymes of the glycolytic and phosphogluconate pathways has remarkable metabolic flexibility. Consider, for example, the need that bacteria and plants have for erythrose-4-phosphate in the synthesis of aromatic amino acids (see Chapter 22).

a. Devise a metabolic sequence that will accomplish net synthesis of erythrose-4-phosphate from glucose without the accumulation of any other carbon-containing compounds (except for coenzymes, if needful).

b. Write an overall equation for the pathway.

c. What assumptions have you made in devising the pathway?

8. *Sedoheptulose catabolism.* Suppose that a culture of facultative bacteria is maintained under anaerobic conditions with the seven-carbon sugar sedoheptulose as its sole energy source. Assume that the organism possesses all the enzymes of the glycolytic and phosphogluconate pathways, as well as a heptokinase capable of phosphorylating sedoheptulose at the expense of ATP.

a. Devise a pathway for the fermentation of sedoheptulose to lactate.

b. Write an overall equation for the pathway.

c. How does the ATP yield of sedoheptulose fermentation per carbon atom compare with that for glucose fermentation?

9. *Glucose-6-phosphate oxidation.* The metabolic flexibility of the phosphogluconate pathway is illustrated by the ability to carry out net oxidation of glucose-6-phosphate to $CO_2$. Devise a metabolic sequence that will accomplish the complete oxidation of glucose-6-phosphate to $CO_2$ by this route.

10. *Arsenate toxicity.* Arsenate ($AsO_4^{3-}$) is chemically similar to phosphate and can replace it in most, if not all, phosphorolytic reactions. However, arsenate esters are far less stable than phosphate esters and undergo spontaneous hydrolysis. For example, glyceraldehyde-3-phosphate dehydrogenase can use arsenate instead of phosphate to cleave the newly oxidized molecule from the enzyme surface (arsenolysis instead of phosphorolysis). The product, glycerate-1-arseno-3-phosphate, can then undergo nonenzymatic hydrolysis into glycerate-3-phosphate and free arsenate.

a. In what sense might arsenate be called an uncoupler of substrate-level phosphorylation?

b. Why is arsenate such a toxic substance for an organism that depends critically on glycolysis to meet its energy needs?

c. Can you think of other reactions that are likely to be uncoupled by arsenate in the same way that the glyceraldehyde-3-phosphate dehydrogenase reaction is?

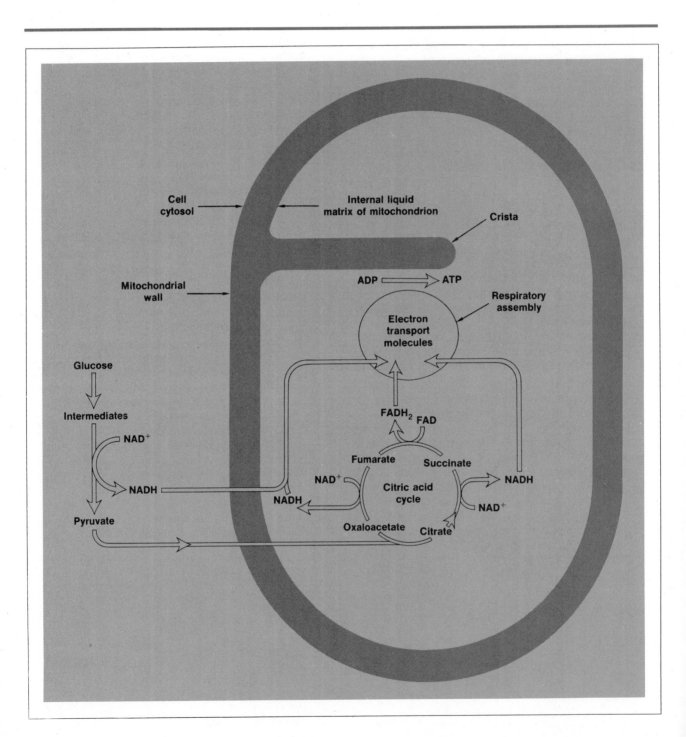

# 9

# AEROBIC PRODUCTION OF ATP: THE TCA CYCLE

In the previous chapter we saw that only a limited portion of the total free-energy content of glucose (or any other oxidizable substrate) can be released under anaerobic conditions. This is so because of the limited extent to which the oxidation of an organic substrate can occur in the absence of oxygen. Essentially, catabolism under anaerobic conditions means that every oxidative event in which electrons are removed from an organic compound must be accompanied by a reductive event in which electrons are returned to another organic compound, often closely related to the first compound. Since the difference in free energy in such coupled oxidation-reduction reactions is small, relatively little energy can be obtained from the starting substrate under these conditions. Thus a cell carrying out fermentation of glucose to lactate essentially transfers electrons from glyceraldehyde-3-phosphate by means of NADH to pyruvate and has access to only about 7 percent of the total free-energy content of the glucose molecule. The cell must accordingly content itself with the generation of only two ATP molecules per molecule of glucose fermented. Most of the energy of the glucose molecule remains untapped in the lactate molecule, which is chemically almost as complex as glucose.

Given access to oxygen, however, the cell can do much more with the oxidizable organic molecules available to it, and the energy yield increases dramatically. With oxygen available as the electron acceptor, the carbon atoms of glucose (or another substrate) can be oxidized fully to $CO_2$, and all the electrons that are removed during the multiple oxidation events are transferred ultimately to oxygen. In the process, all the free energy of the organic substrate is released, and the ATP yield per glucose is almost 20 times greater than that possible under anaerobic conditions. Therein lies the advantage of the aerobic way of life.

*Role of the mitochondrion in the complete oxidation of glucose.*

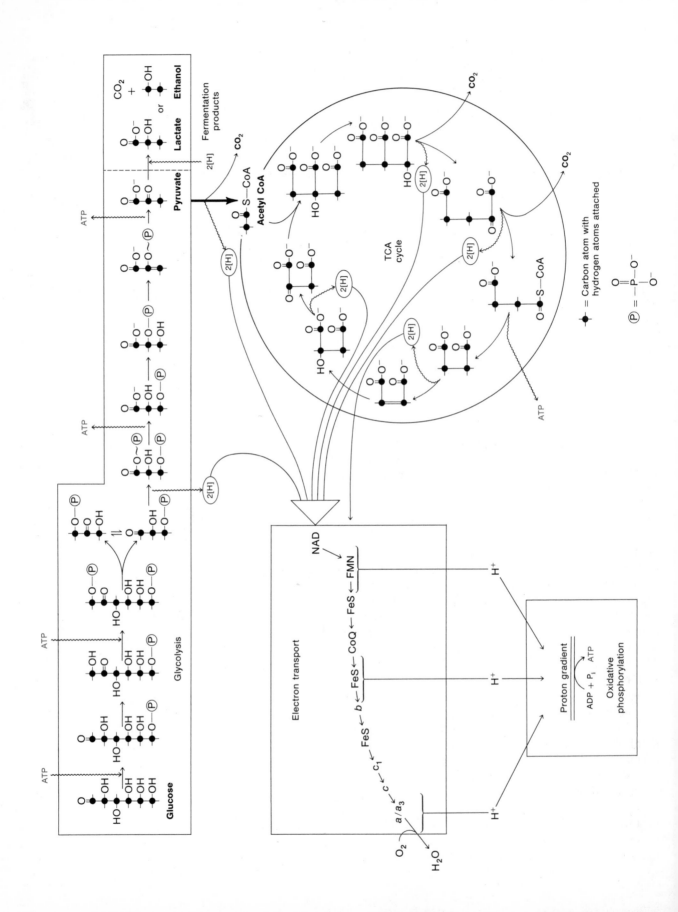

It is not surprising, then, that aerobic processes capable of extracting further energy from pyruvate have come to play so prominent a role in chemotrophic energy metabolism. The overall process is _aerobic respiratory metabolism_, and its distinguishing characteristics are (1) the use of oxygen as the ultimate electron acceptor, (2) the complete oxidation of organic substrates to $CO_2$ and water, and (3) the conservation of much of the free energy as ATP.

The oxygen required for aerobic metabolism actually serves as the terminal electron acceptor only, providing for the continuous reoxidation of reduced coenzyme molecules (most prominent of which are NADH and $FADH_2$). It is these coenzyme molecules (in their oxidized form) that actually carry out the stepwise oxidation of organic intermediates derived from pyruvate. _Respiratory metabolism can therefore be thought of in terms of two separate but intimately linked processes: the actual oxidative metabolism, in which electrons (hydrogen atoms, really) are removed from organic substrates and transferred to coenzyme carriers, and the concomitant reoxidation of the reduced coenzymes by transfer of electrons to oxygen, accompanied indirectly by the generation of ATP._

Under aerobic conditions, the glycolytic pathway becomes the initial phase of glucose catabolism. As shown in Figure 9-1, the other three components of respiratory metabolism are the _tricarboxylic acid (TCA) cycle,_ which is responsible for further oxidation of pyruvate, the _electron-transport chain,_ which is required for the reoxidation of coenzyme molecules at the expense of molecular oxygen, and the _oxidative phosphorylation_ of ADP to ATP, which is driven by a "proton gradient" generated in the process of electron transport. Overall, this leads to the potential formation of 36 molecules of ATP per molecule of glucose in the typical eukaryotic cell. (In prokaryotes, the generally accepted value is 38 ATPs per glucose, for reasons that will be discussed in Chapter 10.)

In this chapter, discussion will focus on the TCA cycle and its central role in the aerobic catabolism not only of carbohydrates, but of fatty acids and amino acids as well. The following chapter will then explain how the free energy present in the reduced coenzymes that are generated by glycolysis and the TCA cycle is conserved as ATP during the companion processes of electron transport and oxidative phosphorylation.

## THE TRICARBOXYLIC ACID CYCLE

Given the availability of oxygen, glucose catabolism does not stop with pyruvate, nor is the pyruvate reduced to lactate or ethanol, as would occur anaerobically. Instead, the carbon atoms of pyruvate are oxidized fully to $CO_2$ in a cyclic process that is at the very heart of energy metabolism in almost all aerobic organisms. A key intermediate in this cyclic series of reactions is citrate, which has three carboxylic acid groups. For this reason, the cyclic pathway is usually called the _tricarboxylic acid_ (TCA) _cycle_ or,

◄**Figure 9-1**
The components of respiratory metabolism include glycolysis, the TCA cycle, the electron-transport chain, and the oxidative phosphorylation of ADP to ATP. Glycolysis converts glucose to pyruvate, the TCA cycle oxidizes the pyruvate (by means of acetyl-coenzyme A) fully to $CO_2$ by transferring electrons stepwise to coenzymes, the electron transport chain reoxidizes the coenzymes at the expense of molecular oxygen, and the energy of coenzyme oxidation is conserved in the form of a transmembrane proton gradient that is then used to generate ATP.

sometimes (originally, in fact), the *citric acid* (or citrate) *cycle.* Yet another name for it is the *Krebs cycle*, in honor of Sir Hans Krebs, the German-born scientist who received a Nobel Prize in 1953 for his efforts in the elucidation of this process. The careful observations and brilliant reasoning that led Krebs to postulate the TCA cycle in 1937 place his work among the classic investigations of biochemistry and become all the more impressive with the realization that this was the last major metabolic pathway worked out without use of the radioactive tracers to which so much present experimentation makes routine recourse.

The TCA cycle is shown in Figure 9-2. In outline form, the cycle begins with acetyl-coenzyme A, which is obtained either by oxidative decarboxylation of pyruvate available from glycolysis or by oxidative cleavage of fatty acids, as described later in this chapter. Regardless of source, acetyl-CoA transfers its acetate to a four-carbon acceptor (oxaloacetate), thereby generating citrate, the six-carbon compound for which the cycle was originally named. In a cyclic series of reactions, the citrate is subjected to two successive decarboxylations and several oxidative events, leaving a four-carbon compound from which the starting oxaloacetate is eventually regenerated. Each turn of the cycle involves the entry of two carbons as the acetate from acetyl-CoA and the release of two carbons as $CO_2$, with provision for regeneration of the oxaloacetate with which the cycle was initiated. As a result, the cycle is balanced with respect to carbon flow and functions without net consumption or buildup of oxaloacetate (or any other intermediate), unless side reactions occur that either feed carbon into the cycle or drain it off into alternative pathways. The TCA cycle can therefore be thought of as functioning in a catalytic manner, since a small amount of oxaloacetate (or any other intermediate in the cycle) can be used to carry out the oxidation of as much acetyl-CoA as may be desired.

### The Oxidative Decarboxylation of Pyruvate

Although the acetyl-CoA with which the cycle is initiated also may be derived catabolically from fatty acids or amino acids, the usual source in most cells is the pyruvate available from the glycolytic breakdown of carbohydrate. The gap from the glycolytic pathway to the TCA cycle is bridged by the oxidative decarboxylation of pyruvate (with which glycolysis ends) to yield acetate in the form of acetyl-CoA (with which the TCA cycle commences). This is accomplished by a complicated sequence of events catalyzed by a cluster of enzyme activities called the *pyruvate dehydrogenase complex*. In eukaryotic cells, this enzyme complex is located in the mitochondria, as are all the other reactions of aerobic energy metabolism beyond pyruvate. Since the glycolytic pathway occurs in the cytoplasm, it is as pyruvate that the carbon derived from glucose (or other carbohydrate substrates) enters the mitochondria.

The overall reaction catalyzed by the pyruvate dehydrogenase complex can be written as follows:

$$\underset{\text{Pyruvate}}{\underset{\displaystyle |}{\overset{\displaystyle O}{\underset{\displaystyle CH_3}{\overset{\displaystyle \|}{\underset{\displaystyle |}{\overset{\displaystyle C-O^-}{\underset{\displaystyle C=O}{}}}}}}} + \underset{\text{Coenzyme A}}{CoA-SH} + NAD^+ \longrightarrow \underset{\text{Acetyl-CoA}}{CoA-S-\overset{\displaystyle O}{\overset{\displaystyle \|}{C}}-CH_3} + CO_2 + NADH + H^+$$

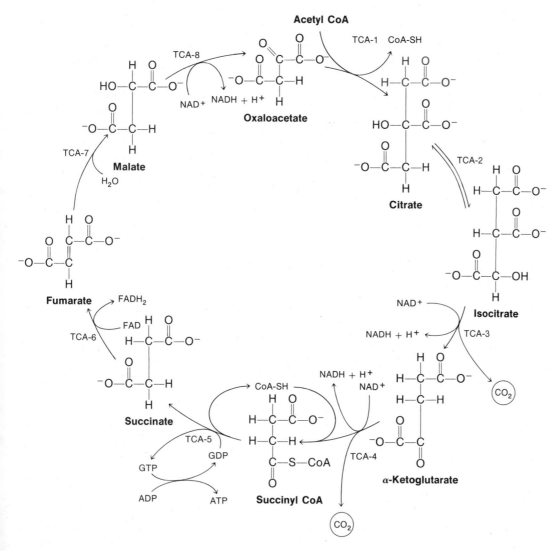

**Figure 9-2**
The tricarboxylic acid (TCA) cycle. Acetyl-CoA from the glycolytic pathway or from $\beta$ oxidation of fatty acids enters and two fully oxidized carbon atoms leave (as $CO_2$) per turn of the cycle. ATP is generated at one point in the cycle and coenzyme molecules are reduced.

The reaction is highly exergonic ($\Delta G^{\circ\prime} = -8.0$ kcal/mol) and is essentially irreversible. The actual mechanism involves three different enzyme activities and five different coenzymes, all of which are carefully ordered into the pyruvate dehydrogenase complex. The sequence of reactions is illustrated schematically in Figure 9-3. The initial step in the sequence (reaction 1 of Figure 9-3) is identical to that of pyruvate decarboxylase (see the section Alcoholic Fermentation in Chapter 8): the active carbon on the thiazole ring of the thiamine pyrophosphate (TPP) cofactor bound to the pyruvate dehydrogenase (PDH) enzyme reacts with the carbonyl group of pyruvate to yield $CO_2$ and a hydroxyethyl group bound to the thiazole ring (see Figure 8-12 for greater detail). Instead of simple hydrolysis to yield free acetaldehyde, however, the hydroxyethyl group is in this case transferred to and dehydrogenated by the lipoic acid cofactor associated with the enzyme dihydrolipoyl transacetylase (DLT), which is also an integral part of the same enzyme complex. In the process, the hydroxyethyl group from pyruvate is oxidized to an acetyl group, and the lipoic acid group on the enzyme is reduced from the disulfide to the dithiol form (reaction 2 of Figure 9-3). The acetyl group is then transferred to the thiol group of coenzyme A (a straightforward transfer from one sulfur to another) and leaves the enzyme surface as the product acetyl-coenzyme A (reaction 3). The dithiol form of the lipoic acid is reoxidized to the disulfide form in reaction 4 by transfer of electrons (and protons) to the coenzyme FAD, which is tightly bound to dihydrolipoyl dehydrogenase (DLDH), the third enzyme in the complex. Reoxidation of the $FADH_2$ then occurs by electron transfer to $NAD^+$, generating the NADH, which, along with acetyl-CoA and $CO_2$, are the end products of the overall reaction sequence. The flow of electrons

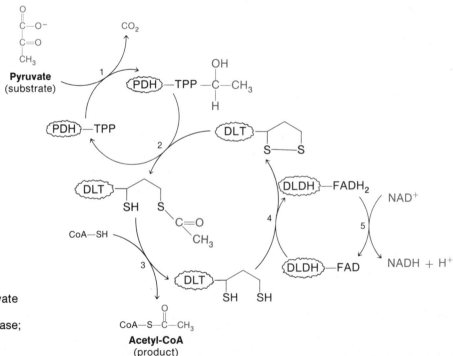

**Figure 9-3**
Mechanism of action of the pyruvate dehydrogenase (PDH) complex. (DLT = dihydrolipoyl transacetylase; DLDH = dihydrolipoyl dehydrogenase.)

in this complex oxidation sequence is therefore from pyruvate to lipoic acid to FAD to NAD$^+$.

The pyruvate dehydrogenase complex of mammalian heart and kidney mitochondria has a molecular weight of over 7 million and consists of a core of 60 DLT molecules with which are associated about 20 molecules of PDH and a lesser number of DLDH molecules, as well as several molecules of the regulatory enzymes pyruvate dehydrogenase kinase and pyruvate dehydrogenase phosphatase. These latter two enzymes comprise a reversible phosphorylation/dephosphorylation control system that renders the complex sensitive to inhibition by ATP. In the presence of high ATP concentrations, the kinase catalyzes an ATP-dependent phosphorylation of one of the pyruvate dehydrogenase subunits, rendering the dehydrogenase inactive. Dephosphorylation of the inactive phosphoenzyme is readily accomplished by the phosphatase, so the regulatory effect of ATP is readily reversible. At high ATP levels, the complex is rendered inoperative by phosphorylation, thereby shutting off the supply of acetyl-CoA to the TCA cycle. As ATP levels drop, the complex is activated by phosphatase-mediated dephosphorylation, thus allowing resumption of acetyl-CoA generation from pyruvate.

The critical feature of the overall pyruvate dehydrogenase reaction is that the free energy liberated as a result of the oxidation of pyruvate is conserved as the high-energy thioester linkage of acetyl-CoA ($\Delta G^{\circ\prime} = -7.5$ kcal/mol). This energy is not used to generate ATP (although that would be energetically feasible), but rather to energize or activate the acetate molecule for entry into the TCA cycle.

*The oxidative decarboxylation of pyruvate is the aerobic gateway to the TCA cycle* and occurs twice per glucose molecule (i.e., once per triose). By doubling the coefficients in this reaction and adding it to an earlier expression for the glycolytic pathway, we arrive at a summary equation for the oxidation of glucose to acetyl-CoA:

Glucose $+ 4\,NAD^+ + 2\,ADP + 2\,P_i + 2\,CoA{-}SH \longrightarrow$
$\qquad\qquad 2\,$acetyl-CoA $+ 2\,CO_2 + 4\,NADH + 4\,H^+ + 2\,ATP$

The NADH will have to wait until Chapter 10 for its reoxidation, but the acetyl-CoA is ready for entry into the TCA cycle.

### The First Phase of the TCA Cycle: Entry of Acetate

The TCA cycle is essentially a disassembly-line-in-the-round (Figure 9-2). *Per turn of the cycle, two carbon atoms enter as acetate, two carbon atoms leave in the fully oxidized form of $CO_2$, and oxidation occurs at four separate dehydrogenation events.* Our consideration of the cycle begins with the entry of acetate as acetyl-CoA.

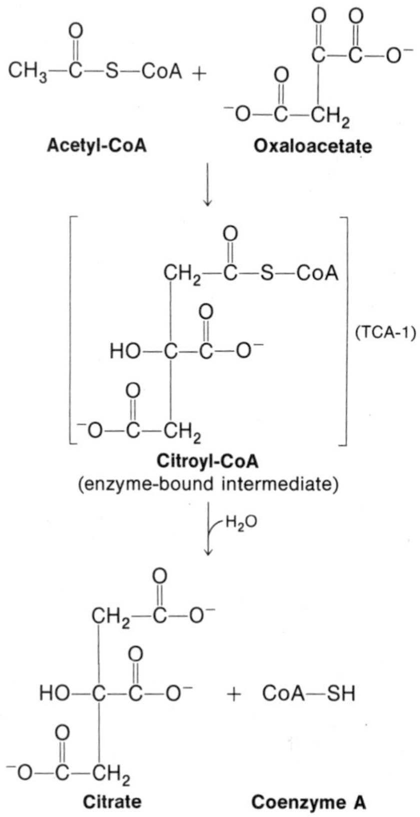

**TCA-1.** As its CoA thioester, the incoming acetate possesses sufficient energy to add across the carbonyl group of oxaloacetate. The reaction is an aldol-type condensation, in which a hydrogen ion is first removed from the methyl carbon of the acetyl group, followed by nucleophilic attack of the resulting anion on the carbonyl bond of oxaloacetate. This generates citroyl-CoA as an enzyme-bound intermediate. Hydrolysis of the high-energy thioester bond of citroyl-CoA liberates free citrate and ensures that the reaction will proceed irreversibly to the right.

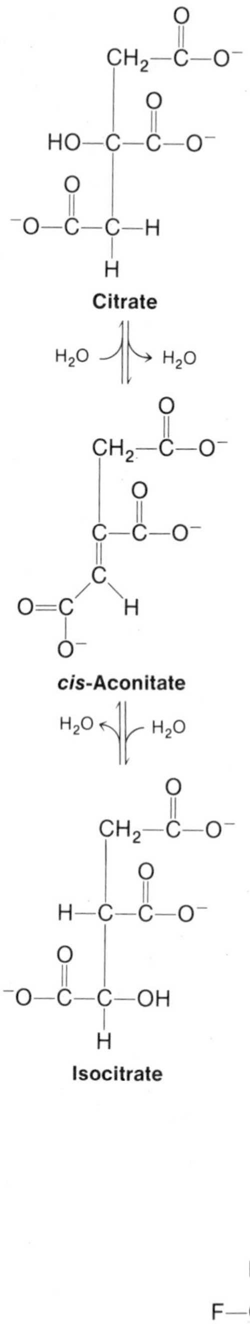

Citrate

$H_2O$ → $H_2O$

(TCA-2)

cis-Aconitate

$H_2O$ → $H_2O$

Isocitrate

Fluoroacetate    Fluorocitrate

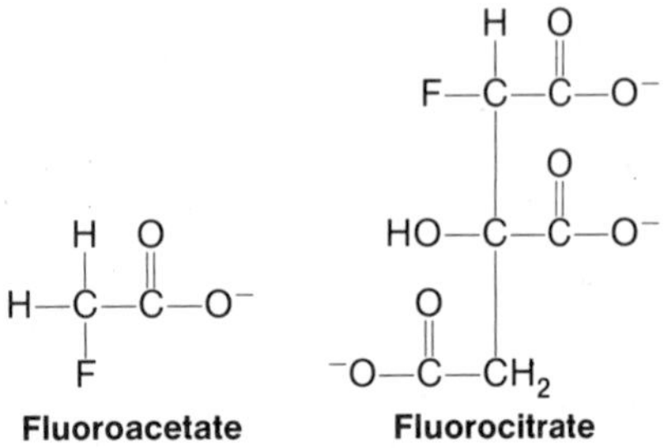

**Figure 9-4**
Structures of fluoroacetate and its lethal metabolite, fluorocitrate, a potent inhibitor of aconitase.

This reaction is catalyzed by the enzyme citrate synthase (originally called condensing enzyme) and is strongly exergonic ($\Delta G^{\circ\prime} = -7.7$ kcal/mol). As might be expected of an enzyme that initiates a pathway, citrate synthase is a regulatory enzyme, with its rate affected as usual by the levels of acetyl-CoA and oxaloacetate as substrates, but also by NADH and succinyl-CoA, the latter generated by TCA-5 (see below).

### The Second Phase of the TCA Cycle: The Oxidative Decarboxylation Steps

Having seen how the number of carbon atoms of the TCA cycle intermediates increases from four (oxaloacetate) to six (citrate) upon entry of acetate (as acetyl-CoA) in phase 1, we come now to a series of reactions in which the number of carbon atoms is reduced from six back to four by two successive decarboxylations, each of which is also an oxidative event. The starting compound, citrate, is first converted to isocitrate, a more readily oxidizable molecule. In two oxidative decarboxylation steps, isocitrate (six carbons) is converted by way of α-ketoglutarate (five carbons) to succinate (four carbons). The latter is actually generated as the coenzyme A derivative, with enough energy conserved in its thioester bond to drive the synthesis of ATP as succinyl-CoA is hydrolyzed to succinate. This sequence of events requires four separate steps, identified here as TCA-2, TCA-3, TCA-4, and TCA-5.

**TCA-2.** Citrate cannot itself be readily oxidized because it is a tertiary alcohol with the most obvious target of oxidation, the hydroxyl group, in a position that precludes oxidation. The conversion of citrate to isocitrate is therefore best understood as a means of converting a nonoxidizable tertiary alcohol into an oxidizable secondary alcohol. This involves the movement of the hydroxyl group from its internal location to an adjacent carbon by a mechanism that requires successive dehydration and rehydration reactions with cis-aconitate as the unsaturated intermediate.

Both of these reactions are catalyzed by the enzyme _aconitase_ (also called _aconitate hydratase_). Aconitase contains a ferrous ion, which is apparently important in binding of the substrate to the enzyme surface. Like every step in the TCA cycle, the aconitase reaction is completely stereospecific: in both directions, the $H^+$ and $OH^-$ of water are added and removed trans with respect to each other.

The thermodynamics of the reaction are such that an equilibrium mixture consists of 90 percent citrate, 4 percent cis-aconitate, and 6 percent isocitrate. Clearly, it is the exergonicity of both citrate formation (TCA-1) and isocitrate oxidation (TCA-3) that drives the aconitase reaction in the desired direction.

A specific and potent inhibitor of aconitase is fluorocitrate, which forms in the cell upon condensation of fluoroacetate with oxaloacetate, catalyzed by the enzyme citrate synthase (Figure 9-4). Fluoroacetate occurs naturally in the leaves of a variety of poisonous plants and is used as a rat poison. It is one of the most deadly simple molecules known, with an $LD_{50}$ (the dose required to kill half the animals to which it is fed) of 0.2 mg/kg of body weight in rats. This is an order of magnitude lower than the $LD_{50}$ for the potent nerve poison diisopropylphosphofluoridate. Fluoroacetate is itself apparently nontoxic to cells until it is converted to fluorocitrate, which then binds to and blocks the active site of aconitase and thereby poisons aerobic energy metabolism.

**TCA-3.** Once it has been formed by the aconitase reaction, isocitrate becomes the substrate for the first of the two oxidative decarboxylation events of the TCA cycle. Attention in this reaction focuses on the hydroxyl group, now in a secondary rather than a tertiary position and therefore readily oxidizable to the carbonyl level. The reaction is catalyzed by the enzyme *isocitrate dehydrogenase*. The immediate product of oxidation is the keto acid oxalosuccinate, but this compound is rapidly decarboxylated on the enzyme surface to yield the corresponding five-carbon compound α-ketoglutarate. The reaction is thermodynamically favorable in the oxidative direction, with a $\Delta G^{\circ\prime}$ of $-5.0$ kcal/mol.

Isocitrate dehydrogenase exists in several isozymic forms, which are specific for either $NAD^+$ or $NADP^+$. The NAD-specific isozyme is found exclusively in the mitochondria of eukaryotic cells and is presumed to be the more important form in TCA cycle activity.

The isocitrate dehydrogenase reaction is an excellent example of an important biochemical sequence—the oxidation of a β-hydroxy acid to a β-keto acid followed by decarboxylation. Note that with respect to the "internal" carboxyl carbon (i.e., the one that leaves as $CO_2$), the hydroxyl group of isocitrate is on a β carbon. Upon oxidation, this becomes a β-carbonyl group, which in general facilitates cleavage of an adjacent carbon-carbon bond. Such β cleavage is one of the most common biological answers to the problem of breaking (and in the reverse direction, of forming) carbon-carbon bonds. It might even be argued that the TCA cycle, with its provision of β cleavage here at step 3, is nature's solution to the problem of oxidizing and degrading the two-carbon acetyl group. The covalent bond that links the two carbons of acetate cannot be broken by any common cleavage mechanism; β cleavage is certainly not possible within a two-carbon molecule. Once condensed onto another compound, however, the acetyl group becomes susceptible to β cleavage. In fact, the specific carbon-carbon bond that is broken here in the isocitrate dehydrogenase reaction is that which linked the two carbons of the acetyl group that entered the cycle on the previous turn.

Isocitrate dehydrogenase is an allosterically regulated enzyme. It is activated by ADP, which lowers the $K_m$ of the enzyme for isocitrate by at least an order of magnitude. The enzyme is also subject to allosteric inhibition by NADH and/or ATP, depending on the tissue.

**TCA-4.** The second of the two oxidative decarboxylation reactions of the TCA cycle is that catalyzed by *α-ketoglutarate dehydrogenase*. The reaction sequence is entirely analogous to that of pyruvate dehydrogenase (Figure 9-3), complete with conservation of some of the energy of the oxidation as the high-energy thioester bond of the coenzyme A derivative of the product, which in this case is succinyl-CoA.

**Isocitrate**

$NAD^+$
$(NADP^+)$

$NADH, H^+$
$(NADPH, H^+)$

(TCA-3)

**Oxalosuccinate**
(enzyme-bound intermediate)

$CO_2$

**α-Ketoglutarate**

**α-Ketoglutarate** $+ NAD^+ + CoA—SH$ $\xrightarrow{\text{(TCA-4)}}$ **Succinyl-CoA** $+ NADH + H^+ + CO_2$

The oxidation of α-ketoglutarate is sufficiently energy-yielding to drive both reduction of the coenzyme NAD$^+$ and formation of the thioester bond of succinyl-CoA and still have enough free energy "left over" to ensure that the reaction is highly exergonic in the oxidative direction, with a $\Delta G°'$ of $-8.0\,$kcal/mol.

The α-ketoglutarate dehydrogenase enzyme, like pyruvate dehydrogenase, occurs as a large multienzyme complex of three enzyme activities and their associated cofactors (TPP, etc.) organized together into a functional complex. Thus the mechanism shown in Figure 9-3 for the pyruvate dehydrogenase complex is entirely applicable to α-ketoglutarate oxidation also.

**TCA-5.** For both pyruvate and α-ketoglutarate oxidation, the product is released from the enzyme surface as the CoA thioester. In the case of pyruvate oxidation, the energy of the thioester bond is used to activate the acetyl group for entry into the TCA cycle by condensation onto oxaloacetate. Unlike acetyl-CoA, however, succinyl-CoA is not destined to be condensed onto another molecule, and the free energy is instead conserved by coupling hydrolysis of the thioester to the synthesis of ATP, either directly, as in bacteria and higher plants, or indirectly by means of GTP, as in mammals:

**Succinyl-CoA** + (ADP) GDP + P$_i$ $\xrightarrow{\text{(TCA-5)}}$

**Succinate** + (ATP) GTP + CoA—SH

This reaction has a $\Delta G°'$ of $-0.7\,$kcal/mol and is therefore readily reversible. The enzyme responsible for this activity is *succinyl-CoA synthase* (also called *succinyl thiokinase*). Figure 9-5 illustrates the mechanism whereby hydrolysis of the thioester bond is coupled to GTP (or ATP) synthesis. In mammals, succinyl-CoA synthase activity results in the synthesis of GTP rather than ATP directly. Transfer of the terminal high-energy phosphate group from GTP to ADP requires the enzyme *nucleoside diphosphokinase* (not an integral part of the TCA cycle). By the combined activities of succinyl-CoA synthase and nucleoside diphosphokinase, the net result of succinyl hydrolysis is the generation of one molecule of ATP, as summarized in TCA-5a:

$$\text{Succinyl-CoA} + \text{GDP} + \text{P}_i \longrightarrow \text{succinate} + \text{CoA} + \text{GTP}$$
$$\text{GTP} + \text{ADP} \longrightarrow \text{GDP} + \text{ATP}$$
$$\overline{\text{Succinyl-CoA} + \text{ADP} + \text{P}_i \xrightarrow{\text{(TCA-5a)}} \text{succinate} + \text{CoA} + \text{ATP}}$$

This is therefore a further example of substrate-level phosphorylation comparable with the ATP-generating steps of the glycolytic sequence (Gly-7 and Gly-10 in Chapter 8), in the sense that ATP synthesis occurs as an integral part of a catabolic pathway rather than as a result of coenzyme reoxidation (oxidative phosphorylation; see Chapter 10).

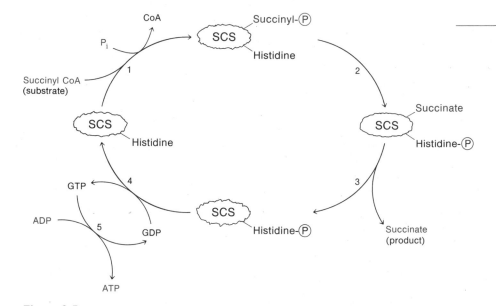

**Figure 9-5**
Mechanism of action of succinyl-CoA synthase (SCS). Succinyl phosphate, a high-energy mixed anhydride, is formed from succinyl-CoA and inorganic phosphate (step 1), and the high-energy phosphate is then transferred to a histidine residue of the enzyme to form 3-phosphohistidine (step 2). The free succinate is liberated (step 3), and the phosphohistidine residue donates its phosphate to GDP, forming GTP (step 4). Nucleoside diphosphokinase is the enzyme responsible for subsequent phosphorylation of ADP at the expense of GTP (step 5).

## The Third Phase of the TCA Cycle:
## Regeneration of Oxaloacetate

In the first five steps of the TCA cycle, the entry of two carbons as acetyl-CoA has been balanced by the release of two carbons as $CO_2$; two molecules of reduced coenzyme have been generated by the accompanying oxidative events, and one molecule of ATP has been formed by substrate-level phosphorylation. In the process, the four-carbon molecule oxaloacetate was converted to the six-carbon compound citrate, and this in turn eventually became succinate, back at the four-carbon level. Now all that remains is to convert succinate to oxaloacetate, thereby regenerating the acceptor molecule with which the cycle was initiated.

As can be seen from their structures (Figure 9-2), succinate and oxaloacetate are identical except for one carbon atom (C-2), which in oxaloacetate carries a carbonyl group and in succinate is at the methylene ($—CH_2—$) level. Conversion of succinate to oxaloacetate therefore requires two oxidative events, one to bring carbon atom 2 up to the hydroxyl level and the other to raise it to the carbonyl level. This is accomplished by a characteristic three-reaction oxidative sequence (TCA-6, TCA-7, and TCA-8), which will be encountered again later in this chapter in our consideration of fatty acid oxidation.

**TCA-6.** The sequence begins with an oxidation reaction in which a hydrogen atom is removed from each of the two internal carbon atoms of succinate, generating the unsaturated compound fumarate.

Of the four oxidative steps in the TCA cycle, this reaction, catalyzed by *succinate dehydrogenase,* is unique in that both electrons come from adjacent carbon atoms rather than one from a carbon atom and the other from an adjacent oxygen. Oxidation of a carbon-carbon bond is in general not sufficiently energetic to allow transfer of electrons to $NAD^+$ ($\Delta G°'$ of reoxidation: $-52.6$ kcal/mol), so the acceptor is flavin adenine dinucleotide (FAD), a related but lower-energy coenzyme ($\Delta G°'$ of reoxidation: $-43.4$ kcal/mol). The "cost" of this substitution will become apparent in Chapter 10, where the ATP yield upon coenzyme reoxidation will turn out to be 3 for NADH and 2 for $FADH_2$.

Unlike most of the enzymes of the TCA cycle, which are generally considered to be located in the matrix of the mitochondria in eukaryotic cells, succinate dehydrogenase is an integral membrane protein, firmly embedded in the inner mitochondrial membrane. The enzyme contains a covalently linked flavin that has been shown to be a modified form of FAD, 8-histidyl-FAD.

It is important to note that the FAD coenzyme bound to succinate dehydrogenase can be reoxidized *only* by the electron-transport chain that you will meet in detail in the next chapter (see Figure 10-2). This step therefore represents a direct physical link between the TCA cycle described here and the electron-transport chain to be discussed in the following chapter. Unless oxygen is available and the electron-transport chain is functioning, the coenzyme of succinate dehydrogenase will accumulate in the reduced form ($FADH_2$) and the TCA cycle will cease operating because of an inability to accept further electrons from succinate molecules.

Succinate dehydrogenase is stereospecific; the two hydrogen atoms that are removed come from trans positions, and the resulting unsaturated compound is always the trans isomer fumarate and never the cis form maleate (see Figure 9-6 for structures). Maleate is, in fact, not recognized by the enzyme. Malonate, a structural analog of succinate (Figure 9-6), is a competitive inhibitor of the enzyme. Krebs put this fact to important use in 1940 when he showed that concentrations of malonate adequate to inhibit succinate oxidation (and thereby causing succinate accumulation) effectively blocked tissue respiration. This provided powerful confirmatory proof of the importance of the TCA cycle in cellular respiration, in accord with the role of the TCA cycle he had postulated 3 years earlier.

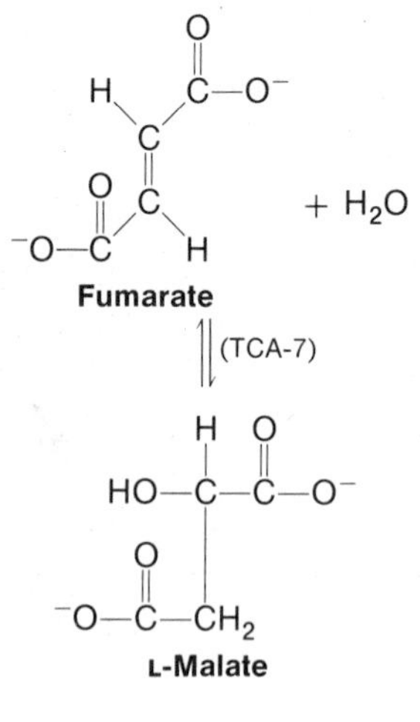

**TCA-7.** In the next step in the sequence, fumarate is hydrated to form L-malate. The enzyme responsible for this hydration is *fumarase* (or fumarate hydratase). The reaction is freely reversible, with $\Delta \overline{G°'} \simeq 0$. Again, the enzyme is strictly stereospecific, with the $H^+$ and $OH^-$ added trans across the double bond in the formation of L-malate.

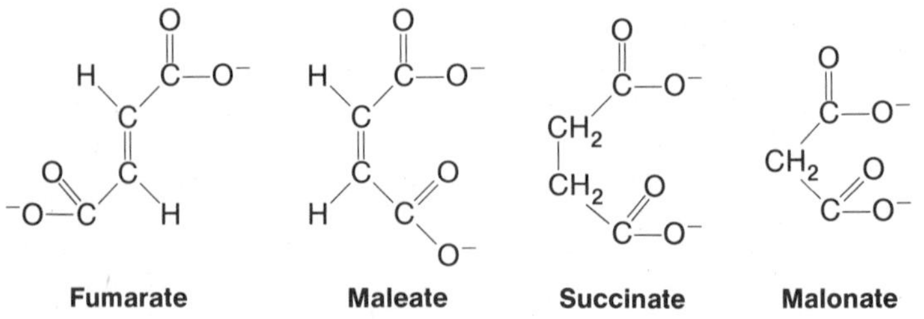

**Figure 9-6**
Structure of several compounds relevant to the succinate dehydrogenase reaction.

**TCA-8.** Finally, oxaloacetate is regenerated from malate by oxidation of the hydroxyl group to the carbonyl level, with $NAD^+$ again serving as the electron acceptor. This reaction, catalyzed by _malate dehydrogenase_, is thermodynamically much more favorable in the reverse direction, since the $\Delta G°'$ for malate oxidation is $+7.1$ kcal/mol. Despite the highly unfavorable $\Delta G°'$ value, this reaction is drawn in the direction of malate oxidation by the continued consumption of oxaloacetate by reaction TCA-1, a highly exergonic step.

## Summary of the TCA Cycle

With the regeneration of oxaloacetate, the compound destined to have the next acetyl-CoA added across its carbonyl group, one full round of the cycle is complete. We can summarize what has been accomplished by noting the following characteristics of the cycle:

1. Acetate enters the cycle as acetyl-CoA by condensation onto a four-carbon acceptor molecule.

2. Decarboxylation occurs at two steps in the cycle so that the input of a two-carbon acetate unit is balanced by the loss of two carbons as $CO_2$ (although the two carbons released in any given cycle do not actually come from the acetate added during that cycle).

3. Oxidation occurs at four steps, with $NAD^+$ serving as the electron acceptor in three cases and FAD being used in one case.

4. ATP is generated (by means of GTP in the case of mammals) at one point.

5. The cycle is completed upon regeneration of the original acceptor.

By summing the eight component reactions of the TCA cycle, we arrive at the following overall summary reaction:

$$\text{Acetyl-CoA} + 2\,H_2O + 3\,NAD^+ + FAD + ADP + P_i \longrightarrow$$
$$2\,CO_2 + 3\,NADH + 3\,H^+ + FADH_2 + CoA + ATP$$

Since the cycle must, in effect, turn twice to metabolize both the acetyl-CoA molecules derived from one glucose, the overall reaction on a per-glucose basis is obtained by multiplying all the coefficients in the preceding reaction by 2. If we do so and then add to it the overall expression for all steps through pyruvate oxidation (Figure 9-2 or 9-3), the overall equation for the entire sequence from glucose through the TCA cycle becomes

$$\text{Glucose} + 6\,H_2O + 10\,NAD^+ + 2\,FAD + 4\,ADP + 4\,P_i \longrightarrow$$
$$6\,CO_2 + 10\,NADH + 10\,H^+ + 2\,FADH_2 + 4\,ATP$$

Looking at this summary reaction, one is struck by the lack of evidence for the substantially greater ATP yield that is supposed to be characteristic of aerobic metabolism. As it stands, the equation indicates only a modest enhancement in ATP yield over glycolysis (an extra two ATPs per glucose), which seems hardly to justify the metabolic complexity of the TCA cycle. Where, we need to ask, is all the energy?

And the answer, of course, is that it is stored in the reduced coenzyme molecules on the right-hand side of the reaction, since these are high-energy compounds in their own right, capable of liberating substantial

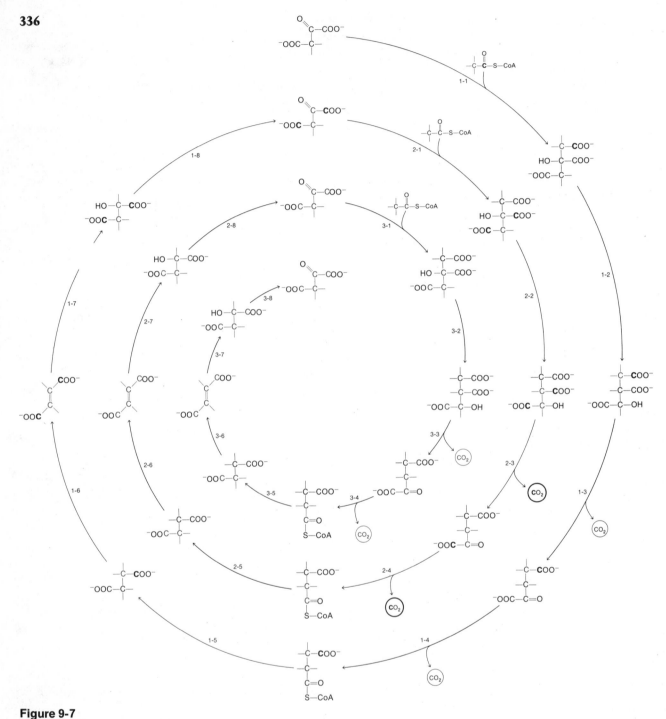

**Figure 9-7**
The pathway and fate of carbon introduced into the TCA cycle as the methyl (color) or carboxyl (boldface) carbon of acetyl-CoA. Note that the $CO_2$ released in steps 3 and 4 of round 2 are both derived from the carboxyl carbon, while the release of the methyl carbon does not occur until round 3.

amounts of free energy upon reoxidation. For the release of that energy we must wait for the final phases of chemotrophic energy metabolism in Chapter 10, since it is only as the electrons are transferred stepwise from the high-energy coenzyme carriers to molecular oxygen that the coupled generation of ATP occurs. It is this final series of oxidations that accounts for the vast majority of the ATPs produced during the complete oxidation of glucose.

Before leaving the TCA cycle, however, we need to look in some detail at several additional points, including flow of carbon through the cycle and regulation of TCA cycle activity.

## The Path of Carbon through the TCA Cycle

Although elucidated without the aid of radioactive tracer molecules, the TCA cycle has since been abundantly confirmed by the use of acetate labeled with $^{14}C$ in either the methyl carbon or the carboxyl carbon. Labeled acetate is provided to cells or tissues carrying out respiratory metabolism, and various TCA cycle intermediates are then isolated and degraded to determine in which position(s) the labeled carbon is found. Figure 9-7 traces flow of the methyl carbon (in color) and the carboxyl carbon (in boldface) of an incoming acetyl group through three successive turns of the cycle, indicating the steps at which the labeled carbon is subsequently released as $CO_2$.

The points of greatest interest with respect to understanding carbon flow arise at TCA-2 (aconitase) and TCA-6 (isocitrate dehydrogenase), since it is at these two steps that the substrate is symmetrical. Because of this symmetry, enzymatic discrimination between the two ends of the molecule was not expected, but has in fact turned out to be the case for aconitase. The substrate for the aconitase reaction is citrate. Because of the symmetry of the citrate molecule, it was initially expected that aconitase would move the hydroxyl group in either direction with equal probability, such that carbon atoms from the newly added acetate group would be just as likely to be on the "bottom" (with the hydroxyl group) as on the "top" (without the hydroxyl group) of the isocitrate molecule as it appears in Figure 9-7. In fact, however, isotopic tracer experiments have clearly established that aconitase always moves the hydroxyl group of citrate away from the newly added acetate carbons, so that the oxidative decarboxylations of the succeeding two reactions involve carbon atoms derived from oxaloacetate rather than from acetyl-CoA. This means that the $CO_2$ released in TCA-4 always comes from the oxaloacetate of the preceding TCA-1 step and never from the incoming acetyl-CoA. This finding led initially to serious doubts as to whether citrate (or any other symmetrical molecule) could be an intermediate in this part of the cycle. A possible way out of the dilemma was suggested by Ogston, who proposed three-point attachment of the citrate to the aconitase surface as a way in which a symmetrical molecule could be recognized and acted upon asymmetrically. It is now recognized that citrate belongs to a class of molecules which are themselves not asymmetrical, but have the potential to react in an asymmetrical manner. Such molecules are called _prochiral_ to indicate that they have a potential "handedness" to them (_chiral_ comes from the Greek word for hand). The prochirality of citrate comes about because the two halves of the molecule have a mirror-image relationship to each other that makes them stereochemically distinct. Aconitase is apparently an enzyme that can distinguish the two different halves of the prochiral citrate molecule, rather like other enzymes can distinguish between two separate molecules that are enantiomorphs (mirror images) of each other. The ability of the enzyme to differentiate between the "upper" and "lower" halves of citrate therefore resides in the prochiral nature of the citrate molecule itself and does not require (nor, for that matter, does it exclude) a three-point attachment of substrate to enzyme.

As a consequence of the addition of acetyl-CoA to the end of the citrate molecule that is not subject to oxidation in the immediately succeeding steps, both the carbon atoms introduced in a given turn of the cycle remain in organic form through the remainder of that cycle. As shown in Figure 9-7, they become the $\gamma$ and $\delta$ carbons of $\alpha$-ketoglutarate and the distal two carbons of succinyl-CoA.

The free succinate formed by TCA-5 is also a symmetrical molecule, but in this case, the succinate dehydrogenase enzyme that acts upon it shows no ability to distinguish between the two ends of the molecule. As a result, the two internal carbons of succinate and fumarate are equivalent, as are also the two carboxyl carbons. Radioactively labeled carbon atoms are therefore randomized at the four-carbon level in tracer experiments. This is indicated in Figure 9-7 by a randomization of color between the two internal carbons (color) and between the two carboxyl carbons (boldface) beyond succinate.

Both carboxyl carbons of oxaloacetate are released as $CO_2$ in the next turn of the cycle, thereby accounting for the carbon derived from the carboxyl group of the original acetyl-CoA. Thus any $^{14}C$ introduced by means of the carboxyl carbon of acetyl-CoA is always eliminated in the two decarboxylation steps in the succeeding turn of the cycle. $^{14}C$ originally derived from the methyl carbon of acetyl-CoA persists through two turns of the cycle and is randomized between the two internal carbons of succinate in the first cycle and between all four carbons of succinate on the second turn. This means that 25 percent of the original label will be eliminated at each of the two decarboxylation steps on the third turn of the cycle. Thereafter, one-half the remaining label will be released on each successive turn, always with an equal contribution from steps 3 and 4. Presuming only that no intermediates are drawn off or added to the cycle, the $CO_2$ eliminated at either of the decarboxylation steps of the TCA cycle has, in theory, a 50 percent probability of coming from the carboxyl group of the acetyl-CoA added in the previous turn of the cycle, a 25 percent chance of representing the methyl carbon of the acetyl-CoA of two cycles ago, a $12\frac{1}{2}$ percent chance of coming from the methyl carbon of three cycles ago, a $6\frac{1}{4}$ percent chance of being the methyl group from four cycles earlier, and so forth. Pursuing carbon atoms around the cycle in this manner may seem a somewhat esoteric exercise, but it does represent an excellent way to acquire a real understanding of the intricacies of TCA cycle functioning.

### Regulation of TCA Cycle Activity

Like all metabolic pathways, the TCA cycle must be carefully regulated to ensure that its level of activity corresponds closely to cellular needs. Basically, *the cycle serves two functions: (1) furnishing reducing equivalents (as NADH and to a lesser extent as FADH₂) to the electron-transport chain, and (2) by means of side reactions, providing substrates for biosynthetic reactions*. Both these functions are reflected in the regulation to which the cycle is subject.

In its primary role as a means of oxidizing acetyl groups to $CO_2$ and water, the TCA cycle is sensitive both to the availability of its substrate, acetyl-CoA, and to the accumulated levels of its principal end product, NADH (and to a lesser extent, ATP). *The key energy-linked regulatory factor determining TCA cycle flux is the $[NAD^+]/[NADH]$ ratio*, both because $NAD^+$ is required as an electron acceptor for several dehydrogenase reactions within the cycle and because several enzymes are allosterically sensitive to $NAD^+$ and/or NADH. The TCA cycle is also sensitive to the $[ATP]/[ADP]$ ratio, but this appears to be less important for regulatory purposes than the $[NAD^+]/[NADH]$ ratio, mainly because the mitochondrial $[ATP]/[ADP]$ ratio actually changes relatively little over the physiologically relevant range of cellular $[ATP]/[ADP]$ ratios. Other regulatory parameters to which the TCA cycle is sensitive include the ratios of acetyl CoA to free coenzyme A, succinyl CoA to free coenzyme A, acetyl CoA to succinyl CoA, and citrate to oxaloacetate. The major known sites of regulation are shown in Figure 9-8.

*TCA cycle activity is regulated most critically at the reactions catalyzed by citrate synthase, isocitrate dehydrogenase, and α-ketoglutarate dehydrogenase*. Sensitivity to acetyl-CoA concentration resides primarily with citrate synthase, the enzyme for which it serves as substrate. Citrate synthase activity is inhibited by physiological concentrations of NADH, NADPH, and succinyl-CoA, all of which have the effect of increasing its apparent $K_m$ for acetyl-CoA. (However, the purported inhibition of citrate synthase by ATP that is noted in most texts is no longer considered valid, since Mg—ATP, the predominant form of ATP in the mitochondria, is not inhibitory.) Citrate synthase activity is also controlled by the available pool of its other substrate, oxaloacetate, the concentration of which is frequently in the regulatory $K_m$ range.

Isocitrate dehydrogenase is another enzyme that is strongly affected by the $[NAD^+]/[NADH]$ ratio, owing mainly to product inhibition by NADH. Isocitrate dehydrogenase is also allosterically activated by ADP in a $Mg^{2+}$-dependent manner. The relative importance of the $[NAD^+]/[NADH]$ and $[ATP]/[ADP]$ ratios in the regulation of this enzyme appears to vary with the tissue in which it is studied.

The $[NAD^+]/[NADH]$ ratio also exerts a regulatory influence on α-ketoglutarate dehydrogenase, which is subject to product inhibition by both NADH and succinyl-CoA. The purified enzyme is also inhibited by a high $[ATP]/[ADP]$ ratio, but again the physiological significance of this is uncertain.

The overall availability of acetyl-CoA is regulated primarily by pyruvate dehydrogenase. As already noted, this enzyme complex is deactivated by ATP-dependent phosphorylation and is reactivated by dephosphorylation as the ATP level falls again. The reaction is also inhibited by high ratios of $[acetyl\text{-}CoA]/[CoA]$ and of $[NADH]/[NAD^+]$. Both acetyl-CoA and NADH affect the pyruvate dehydrogenase enzyme by modulating the activity of either the kinase or the phosphatase enzyme responsible for phosphorylation or dephosphorylation, respectively. Acetyl-CoA is in turn

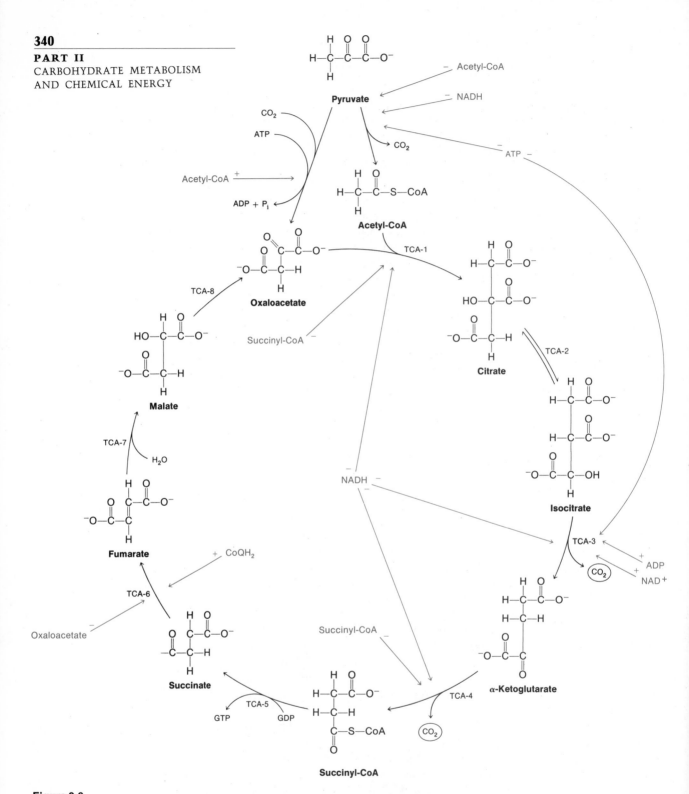

**Figure 9-8**
Major regulatory sites of the TCA cycle. Inhibitory effects are designated with a  −; activators are
indicated with a  +  and appear in color.

a positive effector of pyruvate carboxylase, thereby activating the replenishment of the oxaloacetate needed for continued TCA cycle activity.

Intracyclic coordination is achieved by the inhibitory effects of oxaloacetate on succinate dehydrogenase and of succinyl-CoA on citrate synthase, the latter effect owing to competition between succinyl-CoA and acetyl-CoA for the active site of the enzyme. Feedback control from the TCA cycle to the glycolytic pathway is provided by the inhibitory effect of citrate on phosphofructokinase, as already noted in Chapter 8.

## THE AMPHIBOLIC NATURE OF THE TCA CYCLE

The sole purpose of the TCA cycle when it is functioning as such is the oxidation of acetate to $CO_2$, with concomitant conservation of the energy of oxidation as reduced coenzymes and eventually as ATP. Strictly speaking, then, the TCA cycle has but a single substrate, and that is acetyl-CoA. Because of the cyclic nature of the pathway and the consequent regeneration of oxaloacetate, neither oxaloacetate nor any of its precursors can be said to "enter" the cycle in any net sense. It is, in other words, not possible for the TCA cycle to accomplish the net intake and catabolism of any substrate other than acetyl-CoA.

*In most cells, however, there is considerable flux of four-, five-, and six-carbon intermediates into and out of the cycle*, which occurs in addition to (rather than as part of) the primary catabolic function of the cycle. *Such side reactions serve two main purposes: (1) to provide for the synthesis of compounds derived from any of several intermediates of the cycle, and (2) to replenish and augment the supply of intermediates in the cycle, as needed*. Because the TCA cycle can function both in a catabolic mode and as a source of precursors for anabolic pathways, it is often called an *amphibolic* pathway (from the Greek, *amphi* = "both"). We will examine several biosynthetic pathways that begin with intermediates in the TCA cycle and then go on to look at ways in which the supply of intermediates in the cycle is replenished and maintained. Most of these possibilities are indicated in Figure 9-9.

### Anabolic Pathways Originating with
### TCA Cycle Intermediates

A variety of reactions and pathways can be cited to illustrate the amphibolic role of the TCA cycle. For example, two of the intermediates of the TCA cycle, oxaloacetate and α-ketoglutarate, are the α-keto equivalents of the amino acids aspartate and glutamate, respectively, and can be converted to these amino acids by transamination reactions. Transamination will be discussed in more detail in Chapter 23; for the present, it is adequate to note that transamination involves the conversion of an α-keto compound into its α-amino equivalent, at the expense of another α-amino acid, which becomes in the process the corresponding α-keto compound. A common (although by no means the only) source of the amino group is the amino acid alanine, which is thereby converted to pyruvate, itself an inter-

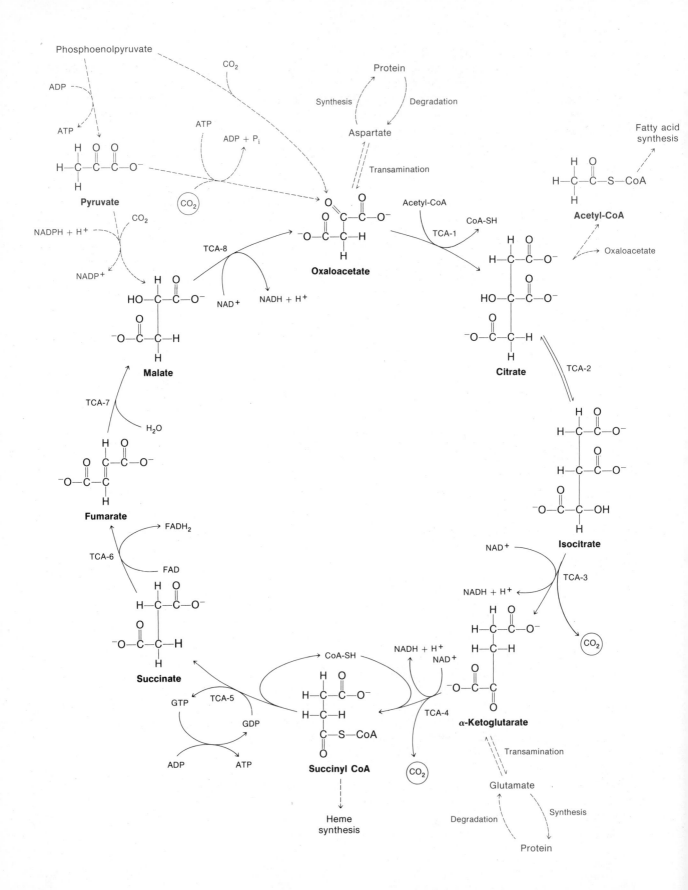

mediate in aerobic catabolism. The transamination reactions responsible for glutamate and aspartate synthesis are as follows:

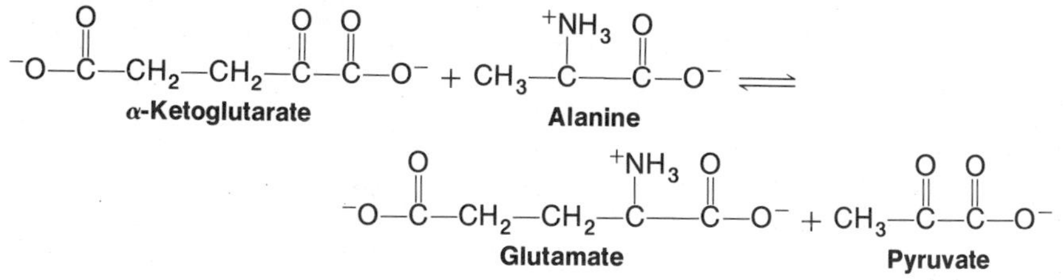

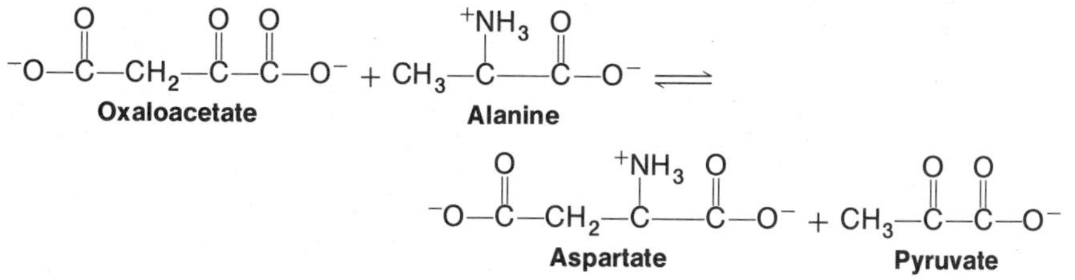

$$\underset{\alpha\text{-Ketoglutarate}}{{}^-O-\overset{O}{\overset{\|}{C}}-CH_2-CH_2-\overset{O}{\overset{\|}{C}}-\overset{O}{\overset{\|}{C}}-O^-} + \underset{\text{Alanine}}{CH_3-\overset{{}^+NH_3}{\underset{|}{C}}-\overset{O}{\overset{\|}{C}}-O^-} \rightleftharpoons$$

$$\underset{\text{Glutamate}}{{}^-O-\overset{O}{\overset{\|}{C}}-CH_2-CH_2-\overset{{}^+NH_3}{\underset{|}{C}}-\overset{O}{\overset{\|}{C}}-O^-} + \underset{\text{Pyruvate}}{CH_3-\overset{O}{\overset{\|}{C}}-\overset{O}{\overset{\|}{C}}-O^-}$$

$$\underset{\text{Oxaloacetate}}{{}^-O-\overset{O}{\overset{\|}{C}}-CH_2-\overset{O}{\overset{\|}{C}}-\overset{O}{\overset{\|}{C}}-O^-} + \underset{\text{Alanine}}{CH_3-\overset{{}^+NH_3}{\underset{|}{C}}-\overset{O}{\overset{\|}{C}}-O^-} \rightleftharpoons$$

$$\underset{\text{Aspartate}}{{}^-O-\overset{O}{\overset{\|}{C}}-CH_2-\overset{{}^+NH_3}{\underset{|}{C}}-\overset{O}{\overset{\|}{C}}-O^-} + \underset{\text{Pyruvate}}{CH_3-\overset{O}{\overset{\|}{C}}-\overset{O}{\overset{\|}{C}}-O^-}$$

In general, the transaminase (or aminotransferase) enzymes that catalyze such reactions are substrate-specific, so two different enzymes would be required to carry out the preceding reactions. The glutamate and aspartate formed in this way are 2 of the 20 amino acids found in proteins and are therefore required by the cell for protein synthesis. Since α-ketoglutarate and oxaloacetate are TCA intermediates found predominantly in the mitochondria and the bulk of cellular protein synthesis takes place in the cytoplasm, the glutamate and aspartate must be translocated to the cytoplasm before they can be used in protein synthesis.

Other amphibolic precursors in the TCA cycle include succinyl-CoA, which serves as a starting point for the biosynthesis of heme (see Chapter 23), and citrate, which is transported out of the mitochondria and is used as a source of acetyl-CoA for cytoplasmic synthesis of fatty acids (see Chapter 13). In each of these cases, intermediates of the TCA cycle are drawn off for biosynthetic purposes. This in turn dictates the need for mechanisms whereby the cycle can be continuously replenished. Indeed, the very occurrence of biosynthetic reactions originating with TCA cycle intermediates implies concurrent replenishment of such intermediates in order to preserve the primary acetate-oxidizing function of the cycle.

## Replenishment of TCA Cycle Intermediates

Regeneration of intermediates in the TCA cycle is necessary both to replace compounds that are drawn off in side reactions and to augment the levels of intermediates when necessary to meet cellular needs for elevated TCA cycle activity. Reactions or pathways that accomplish this replenishment have been termed *anaplerotic*, from a Greek root that means "filling

◀ **Figure 9-9**
Amphibolic nature of the TCA cycle and anaplerotic reactions responsible for replenishment of intermediates. Solid arrows depict the functional TCA cycle; dashed arrows indicate side reactions that deplete and/or replenish intermediates.

up." The most important anaplerotic reactions are those in which oxaloace-tate is formed from either pyruvate or phosphoenolpyruvate (PEP).

In animals, oxaloacetate is formed from pyruvate by action of the mito-chondrial enzyme pyruvate carboxylase, with ATP as the energy source:

$$\underset{\textbf{Pyruvate}}{\overset{\displaystyle O\;\;O}{\underset{CH_3}{\|\;\;\|}{C-C-O^-}}} + HCO_3^- + ATP + H_2O \rightleftharpoons \underset{\textbf{Oxaloacetate}}{C-C-O^- + ADP + P_i}$$

The bicarbonate ion required for carboxylation is initially bound cova-lently to the enzyme surface by means of a biotin prosthetic group present at the active site of each of the four subunits of the multimeric enzyme. ATP is required to couple the bicarbonate to the biotin, forming carboxy-biotin. The pyruvate then binds to the active site, and the carboxyl group of carboxybiotin is transferred to it.

Acetyl-CoA is required as an allosteric activator of pyruvate carboxyl-ase. In the absence of acetyl-CoA, the enzyme is almost completely inactive and, in fact, went undetected for many years for this reason. As soon as acetyl-CoA becomes available, however, the enzyme is activated and main-tains oxaloacetate concentrations at a level high enough to allow the TCA cycle to function effectively and, if necessary, to provide for replacement of intermediates drained off for anabolic purposes.

In plants and bacteria, oxaloacetate also can be generated from phospho-enolpyruvate:

$$\underset{\textbf{Phosphoenolpyruvate}}{P-O-\underset{CH_2}{C}-C-O^-} + HCO_3^- \rightleftharpoons \underset{\textbf{Oxaloacetate}}{C-C-O^- + P_i}$$

The reaction is catalyzed by PEP carboxylase, an enzyme not found in animal tissues or in fungi. Unlike the pyruvate carboxylase of animal tis-sue, PEP carboxylase does not utilize the vitamin biotin as a prosthetic group. ATP is not required either, presumably because phosphoenolpyru-vate is itself a high-energy compound. In green plants, PEP carboxylase is important not only in replenishing the TCA cycle, but also as the initial $CO_2$-fixing enzyme in the Hatch-Slack pathway of $C_4$ plants (see Chap-ter 11).

A related reaction that also serves an anaplerotic function is that cata-lyzed by malic enzyme, known more officially (but less commonly) as _malate dehydrogenase (decarboxylating; NADP)_, which accomplishes both the carboxylation and the reduction of pyruvate, thereby generating malate instead of oxaloacetate:

$$\underset{\textbf{Pyruvate}}{\overset{\displaystyle O\;\;O}{\underset{CH_3}{\|\;\;\|}{C-C-O^-}}} + HCO_3^- + NADPH + H^+ \rightleftharpoons \underset{\textbf{L-Malate}}{HO-CH-C-O^- + NADP^+}$$

Further sources of TCA cycle intermediates include the transamination reactions already discussed but functioning for anaplerotic purposes in the reverse direction to yield oxaloacetate and/or α-ketoglutarate at the expense of the analogous amino acids. In germinating seedlings of fat-storing plant species and in many microorganisms that have access to acetate or can generate it metabolically, TCA cycle intermediates also can be replenished by means of the glyoxylate cycle to be discussed later in this chapter. Contrary to the impression given in some texts, however, this is not the main function of the glyoxylate cycle in such organisms.

By means of a variety of side reactions, then, the TCA cycle acquires a metabolic versatility beyond its primary catabolic function. It is important to reiterate, however, that acetyl-CoA oxidation is the only function that actually involves the cycle as such. In all the amphibolic roles of the cycle, one intermediate is removed for synthetic purposes, another is replenished by an anaplerotic mechanism, and only that portion of the cycle which links the two intermediates actually operates. For purposes of the biosynthetic pathway, in fact, it is in a sense mere coincidence that several of the enzymes in the pathway happen also to have a role in the TCA cycle as well.

## ALTERNATIVE SOURCES OF ACETYL-COENZYME A FOR THE TCA CYCLE

An important point to underscore with respect to the TCA cycle is its central role in all aerobic energy metabolism. Thus far, glucose has been presumed to be the primary substrate for respiration. True, a variety of alternative carbohydrate substrates were considered in Chapter 8 (see Figure 8-13), but most summary equations that are written for chemotrophic energy metabolism (e.g., see Figure 9-1) seem to assume glucose as the starting compound. In a sense, the assumption is entirely reasonable, since glucose is quantitatively the single most important source of energy for the chemotrophic world. However, it is also important to note the role of other substrates in cellular energy metabolism and the centrality of the TCA cycle in the catabolism of such alternative fuel molecules, especially fats and proteins. _The TCA cycle is, in fact, a major unifying feature of aerobic metabolism, around which nearly all catabolic pathways are ordered._

The involvement of the TCA cycle in the catabolism of fats and proteins becomes clear when it is recognized that the starting point for the cycle itself is acetyl-CoA, and that the oxidative decarboxylation of pyruvate is only one possible source of acetyl-CoA. Other prominent sources include the stepwise oxidative degradation of fatty acids and the catabolism of amino acids. Far from being a minor route for the catabolism of a single sugar, then, the TCA cycle represents the very heart and mainspring of aerobic energy metabolism.

### Fat as a Source of Energy

Fat serves a variety of secondary biological functions, such as cushioning, contouring, and insulating, but its primary role in most cases is that of energy storage. Fat is in fact the principal long-term storage strategy of most organisms. Storage polysaccharides such as starch and glycogen represent localized, readily mobilizable energy reserves, but for long-term energy reserves, fat is almost always the storage form of choice in both animals and plants. In animals, fat reserves are laid down as *adipose tissue* at various locations throughout the body in amounts that vary greatly, depending on the nutritional status of the organism (Chapter 13). (All too many of us, in fact, attest with our waistlines to the propensity of the body to accumulate fat reserves indiscriminately under conditions of dietary excess.) Fat reserves are especially important in hibernating animals or migrating birds, for under these conditions they become the main, if not only, source of energy. Among plant species, fat is the most common single means by which energy and carbon are stored in the seed as the food reserves necessary to support the eventual germination and early growth of the seedling. Examples of species in which fat storage is especially prominent include oil-bearing seed plants such as soybean, peanut, sunflower, and corn (maize). For both plants and animals, fats are well-suited for this storage function because they allow the packaging of a maximum of calories in a minimum of volume and weight, a feature of obvious significance for both animal motility and seed dispersal. Fat is a compact storage form both because of its high caloric value (about 9 kcal/g versus about 4 kcal/g for carbohydrates) and because the hydrophobicity (aversion to water) of the neutral fats allow their deposition without accompanying water, which would otherwise add weight and volume. The high caloric value of fat is due to its more reduced state (see Problem 10 at the end of the chapter).

Most fat is stored as deposits of triglycerides, which are neutral triesters of glycerol and fatty acids, as shown in Figure 9-10. Fatty acids, in turn, are long-chain hydrocarbons, but with a carboxylic acid group on one end. Most fatty acids have an even number of carbon atoms, usually in the range 12 to 20, with 16- and 18-carbon fatty acids especially prominent in higher plants and animals. Since the three fatty acids present in a triglyceride need not be identical, considerable chemical heterogeneity is possible for triglycerides. Further variety can arise as a result of the presence of one or more double bonds in the unsaturated or polyunsaturated fatty acids that are so characteristic of vegetable oils (an *oil* is just a neutral fat that is liquid at room temperature, usually a result of such unsaturation).

Physically speaking, a fatty acid is a strategic combination of a highly reduced hydrocarbon chain (which yields a maximum amount of energy per unit mass upon eventual oxidation) and a polar carboxyl group to render at least one end of the molecule soluble in water. The structure, chemistry, and metabolism of fatty acids and the fats of which they are a part will be dealt with in detail in Chapter 13. Here it will be adequate simply to note the basic features of triglyceride structure and of fatty acid oxidation, since that will underscore the importance of fat as an alternative source of acetyl-CoA for the TCA cycle.

Triglycerides may be obtained either from the diet directly or by mobilization of fat reserves in adipose tissue. In either case, the catabolism of triglycerides begins with their hydrolysis into glycerol and free fatty acids. This hydrolysis is catalyzed by lipases and proceeds through the diglyceride and monoglyceride forms as the fatty acids are removed stepwise. The glycerol is channeled efficiently into the glycolytic pathway by oxidative conversion to dihydroxyacetone phosphate.

Fatty acids also illustrate well the general rule of conversion to an intermediate of the mainstream respiratory pathway, since they are degraded by a stepwise process involving successive removal of two-carbon units that enter the TCA cycle as acetyl-CoA. The sequential process of fatty acid

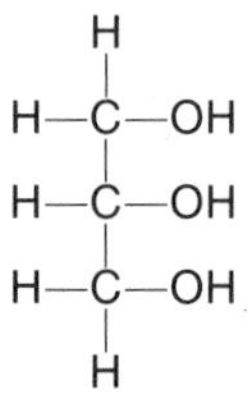

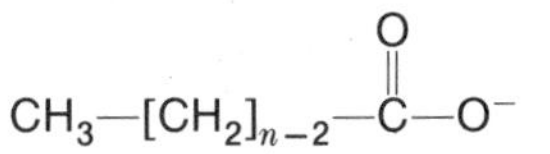

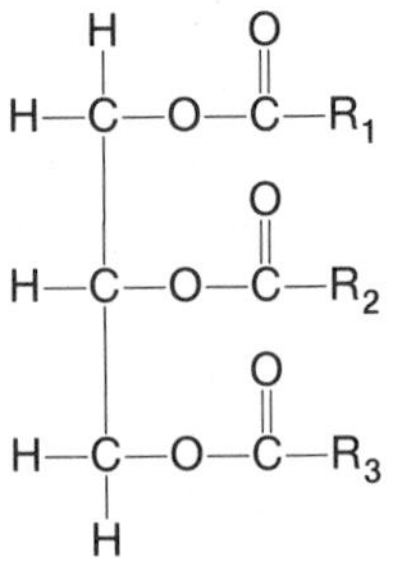

**Figure 9-10**
The structures of glycerol, a fatty acid, and a triglyceride. The common fatty acids range in length from 12 to 20 carbon atoms and may have one or more sites of unsaturation. The three fatty acids of a given triglyceride need not be identical; $R_1$, $R_2$, and $R_3$ may differ both in length of the carbon chain and in occurrence of carbon-carbon double bonds.

catabolism to acetyl-CoA is called $\beta$ oxidation because the initial oxidative event occurs on the carbon atom in the $\beta$ position to (i.e., the second carbon from) the carboxylic acid group. In vertebrates, $\beta$ oxidation is a potentially important source of energy for virtually all tissues with the important exception of the brain, which requires a continuous supply of blood glucose as its source of energy.

## $\beta$ Oxidation of Fatty Acids

The process of $\beta$ oxidation involves several sequential cycles of oxidative attack on the long-chain fatty acid, beginning from the carboxyl end, as shown in Figure 9-11. Each cycle begins with oxidation of the $\beta$ carbon, results in the release of two carbons as acetyl-CoA, and leaves the fatty acid shortened by two carbons, but ready for another round of reactions. Leaving the details of the reactions to Chapter 13, we can look briefly at the overall process to see how the acetyl-CoA arises.

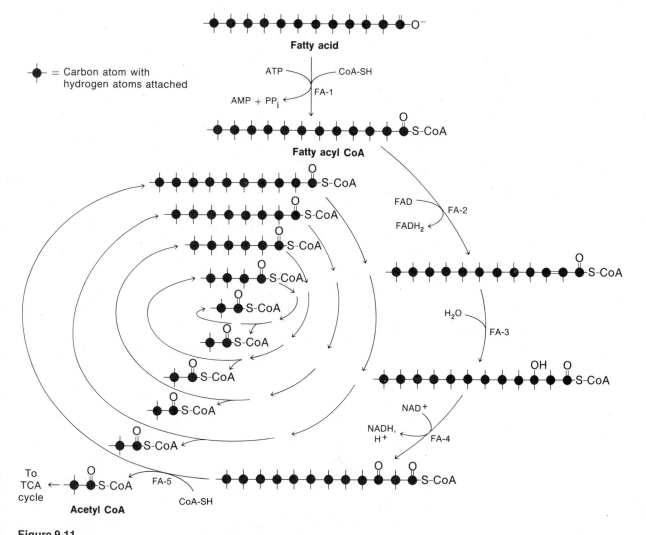

**Figure 9-11**

The process of $\beta$ oxidation, illustrating the spiral or cyclic nature of the stepwise disassembly of a fatty acid into acetyl-CoA units. Reaction numbers and chemical details are shown for the first cycle only. The particular fatty acid shown is lauric acid or laurate, with 12 carbon atoms, so five cycles of $\beta$ oxidation are required, yielding 6 molecules of acetyl-CoA.

**FA-1.** The $\beta$ oxidation of a fatty acid begins with an activation step in which the energy of ATP is used to form the fatty acyl-CoA derivative, with a high-energy thioester bond:

$$CH_3-(CH_2)_x-CH_2-CH_2-\overset{\overset{\displaystyle O}{\|}}{C}-O^- + CoA-SH + ATP + H^+ \xrightarrow{\text{(FA-1)}}$$

**Fatty acid**

$$CH_3-(CH_2)_x-CH_2-CH_2-\overset{\overset{\displaystyle O}{\|}}{C}-S-CoA + AMP + PP_i$$

**Fatty acyl-CoA**

Note that the subscript x in this and each of the following reactions represents four less than the total number of carbon atoms in the starting fatty acid (or in subsequent cycles, two less than the x of the previous cycle). The three carbon atoms on the carboxyl end of the molecule are shown explicitly because both the $\alpha$ and $\beta$ carbons are involved in the oxidative events that follow.

This activation step occurs only once per fatty acid molecule, as the initial reaction in the whole catabolic sequence. It is catalyzed by acyl-CoA synthase (alternate name: fatty acid thiokinase). Several such enzymes are known, each specific for fatty acids of different chain lengths. The driving force for this reaction is the pyrophosphate cleavage of ATP to AMP and inorganic pyrophosphate ($PP_i$). Since pyrophosphate can itself be further hydrolyzed (by pyrophosphatase) into two phosphate groups, both anhydride bonds of ATP are effectively available to drive synthesis of the thioester bond:

$$PP_i + H_2O \longrightarrow 2\,P_i$$

**FA-2.** Fatty acyl-CoA is oxidized by a series of four reactions, three of which (FA-2, FA-3, and FA-4) parallel exactly the sequence whereby succinate is converted into oxaloacetate in the TCA cycle (see TCA-6, TCA-7, and TCA-8). Just as that series begins with an FAD-mediated $\alpha,\beta$ dehydrogenation of succinate, so here the fatty acyl-CoA is dehydrogenated to the corresponding $\alpha,\beta$-unsaturated acyl-CoA, with enzyme-bound FAD as the electron acceptor:

$$CH_3-(CH_2)_x-CH_2-CH_2-\overset{\overset{\displaystyle O}{\|}}{C}-S-CoA + FAD \xrightarrow{\text{(FA-2)}}$$

**Fatty acyl-CoA**

$$CH_3-(CH_2)_x-CH=CH-\overset{\overset{\displaystyle O}{\|}}{C}-S-CoA + FADH_2$$

**$\Delta^2$-trans-Enoyl-CoA**

The product is named with the use of a $\Delta$ to indicate the position of the double bond and is always the trans stereoisomer. The enzyme that catalyzes this reaction is fatty acyl-CoA dehydrogenase, and again several forms are known, each with maximal activity for fatty acids of a different chain length.

Like the succinate dehydrogenase reaction of the TCA cycle, this step of fatty acid oxidation represents a direct link with electron transport, since the enzyme-bound FAD can be reoxidized only by passing its electrons to coenzyme Q of the transport chain (see Figure 10-2). Fatty acid oxidation can therefore take place only if the electron-transport chain is also functioning.

**FA-3.** As with fumarate in the TCA cycle, the unsaturated compound generated by the preceding reaction is hydrated to form the corresponding hydroxylated derivative, in this case $\beta$-hydroxy fatty acyl-CoA:

$$CH_3-(CH_2)_x-CH=CH-\overset{\overset{\displaystyle O}{\|}}{C}-S-CoA + H_2O \xrightarrow{\text{(FA-3)}}$$
$$\Delta^2\text{-}\textit{trans}\text{-}\textbf{Enoyl-CoA}$$

$$CH_3-(CH_2)_x-\overset{\overset{\displaystyle OH}{|}}{CH}-CH_2-\overset{\overset{\displaystyle O}{\|}}{C}-S-CoA$$
$$\boldsymbol{\beta}\text{-}\textbf{Hydroxyacyl-CoA}$$

The enzyme is called enoyl-CoA hydratase, and the product is always the L stereoisomer, since the water is added stereospecifically across the double bond.

**FA-4.** The hydroxyl group formed in the previous reaction now becomes the target of oxidative attack, forming the $\beta$-ketoacyl-CoA, with $NAD^+$ as the electron acceptor:

$$CH_3-(CH_2)_x-\overset{\overset{\displaystyle OH}{|}}{CH}-CH_2-\overset{\overset{\displaystyle O}{\|}}{C}-S-CoA + NAD^+ \xrightarrow{\text{(FA-4)}}$$
$$\boldsymbol{\beta}\text{-}\textbf{Hydroxyacyl-CoA}$$

$$CH_3-(CH_2)_x-\overset{\overset{\displaystyle O}{\|}}{C}-CH_2-\overset{\overset{\displaystyle O}{\|}}{C}-S-CoA + NADH + H^+$$
$$\boldsymbol{\beta}\text{-}\textbf{Ketoacyl-CoA}$$

The enzyme responsible for this reaction is $\beta$-hydroxyacyl-CoA dehydrogenase; it is specific for the L stereoisomer of hydroxyacyl-CoA.

**FA-5.** The $\beta$-keto acid generated in this way is not very stable. In the final step of the cycle, it is cleaved with the uptake of another molecule of coenzyme A into acetyl-CoA and an acyl-CoA compound with two fewer carbon atoms than the original molecule:

$$CH_3-(CH_2)_x-\overset{\overset{\displaystyle O}{\|}}{C}-CH_2-\overset{\overset{\displaystyle O}{\|}}{C}-S-CoA + CoA-SH \xrightarrow{\text{(FA-5)}}$$
$$\boldsymbol{\beta}\text{-}\textbf{Ketoacyl-CoA}$$

$$CH_3-(CH_2)_x-\overset{\overset{\displaystyle O}{\|}}{C}-S-CoA + CH_3-\overset{\overset{\displaystyle O}{\|}}{C}-S-CoA$$
$$\textbf{Fatty acyl-CoA} \qquad\qquad \textbf{Acetyl-CoA}$$

The enzyme involved here is called _thiolase_ (more formally, _acetyl-CoA acetyltransferase_). Again, there is a family of enzymes with differing chain-length specificities. The acetyl-CoA produced in this step can enter the TCA cycle directly, since both $\beta$ oxidation and the TCA cycle are located in the mitochondron of the eukaryotic cell. (For an exception to this generalization, see the case of $\beta$ oxidation in the glyoxysomes of fat-storing seeds, to be discussed later in this chapter.)

**Subsequent cycles.** Notice that the newly formed acyl-CoA compound in the preceding reaction is identical to the activated starting compound that served as the substrate for FA-2, except that it is two carbons shorter. Clearly, this new acyl-CoA molecule can again undergo the series of reactions represented by FA-2 through FA-5, thereby removing another two-carbon unit as acetyl-CoA. This establishes the cyclic or repetitive nature of the process, which continues down the fatty acid backbone, removing two carbon atoms at a time. The complete $\beta$ oxidation of a fatty acid with $n$ carbon atoms requires $n/2 - 1$ such cycles of oxidation and gives rise eventually to $n/2$ molecules of acetyl-CoA, each of which can be catabolized by means of the TCA cycle to $CO_2$ and water. (The process actually gets somewhat more complicated when the molecule gets down to the four-carbon level, but, in essence, we can assume that the process runs to completion, with all carbons of the fatty acid chain eventually appearing as acetyl-CoA.) The repetitive nature of $\beta$ oxidation is illustrated in Figure 9-11 for a fatty acid with 12 carbon atoms.

## Proteins as a Source of Energy

Proteins are not usually considered primarily as an energy source, since they clearly function more fundamentally in their roles as enzymes, hormones, antibodies, transport proteins, structural proteins, and components of contractile systems in the cell. However, proteins (or more specifically, the amino acids of which they are composed) also can be used as a source of energy, especially in animals under dietary conditions where intake of proteins and/or amino acids exceeds cellular needs for protein synthesis.

Amino acids that are catabolized for energy come from three different sources: dietary proteins, storage proteins, and metabolic turnover of endogenous proteins. Catabolism of dietary proteins and amino acids is obviously a characteristic of higher animals, while the use of storage protein for energy is best illustrated during the germination of protein-storing seeds such as beans or peas. In addition, all cells undergo metabolic turnover of most proteins and protein-containing structures, and the amino acids to which they are degraded can either be recycled into proteins or degraded oxidatively to yield energy.

Protein catabolism begins with hydrolysis of the covalent _peptide bonds_ that link successive amino acid residues together in a polypeptide chain. This process is termed _proteolysis_, and the enzymes responsible for this action are called _proteases_. For ingested proteins, proteolysis occurs in the gastrointestinal tract and depends on proteases secreted by the stomach, pancreas, and small intestine. For endogenous protein, proteolytic digestion occurs within the cell.

The products of proteolytic digestion are free amino acids and small peptides. Further digestion of peptides then depends on _peptidases_, which are characteristic of the intestinal mucosa. Peptidases act on their substrate either by hydrolyzing internal peptide bonds (_endopeptidases_) or by removing successive amino acids from the end of the peptide (_exopeptidases_). Exopeptidases are either _aminopeptidases_ or _carboxypeptidases_, depending on the end of the peptide from which digestion proceeds.

## Catabolism of Amino Acids

Free amino acids, whether obtained by digestion of dietary protein in the gastrointestinal tract or by intracellular proteolysis, become the substrates for catabolic pathways in which the energy of their oxidative degradation is conserved as ATP.

The pathways by which amino acids are catabolized illustrate well the general principle that alternative substrates are converted to intermediates of mainstream catabolism in as few steps as possible. In the case of amino acids, however, this principle is complicated by the presence in proteins of 20 different kinds of amino acids, most of which require separate degradative pathways. At first glance, this diversity of amino acids seems a needless extravagance that violates the conservatism of biological systems. Moreover, if proteins were used primarily for energy, it would in fact be needlessly extravagant to construct them of 20 different kinds of monomers, each with its own catabolic pathway. However, proteins serve other, much more fundamental functions in cells, and it is not surprising that the performance of so many different functions requires many proteins varying in size, shape, and charge. This great diversity of function and structure is possible precisely because of the chemical diversity of the amino acids from which proteins are constructed. In a sense, then, the complexity of the metabolic pathways necessary for amino acid catabolism is the price a cell must pay for the structural and functional diversity it requires of its proteins.

It is beyond the intended scope of the present discussion to examine the detailed pathways necessary for the oxidation of 20 different amino acids, since that falls within the province of Chapter 23. Here it will suffice to note several general principles and consider a few illustrative examples. In spite of their number and diversity, all the pathways for amino acid degradation lead eventually to a few key intermediates in the TCA cycle, notably acetyl-CoA, oxaloacetate, fumarate, $\alpha$-ketoglutarate, and succinyl-CoA.

**Removal of the Amino Group.** The catabolic metabolism of amino acids always involves the removal of the amino group somewhere in the sequence, usually near the beginning. This removal can be accomplished either by the process of _transamination_ discussed earlier or by direct _oxidative deamination_. Transamination involves the transfer of an amino group from an amino acid to an $\alpha$-keto acid acceptor, such that the _deamination_ of the amino acid is accompanied by the _amination_ of the $\alpha$-keto acid. This is a convenient, much-used mechanism for shifting amino groups between carbon skeletons and is used by the cell not only in degradative pathways, but also in synthetic routes. Transamination does not effect the net removal of nitrogen from the pool of organic molecules, as clearly must occur if net catabolism of amino acids is to occur. It does, however, allow for the collection of amino groups on a common carbon skeleton, that of the five-carbon amino acid glutamate, from which the amino group is oxidatively liberated as free ammonia by the enzyme glutamate dehydrogenase.

Glutamate is the main amino acid that has a specific deaminating enzyme. Thus the amino groups collected from the various amino acids by transamination reactions ultimately appear on glutamate and are liberated

**Glutamate**

Glutamate
dehydrogenase

**α-Ketoglutarate**

as ammonia (or ammonium ions, $NH_4^+$) by the glutamate dehydrogenase reaction. The $\alpha$-ketoglutarate formed in the process either can be used as an amino acceptor in further transamination reactions or can function in the TCA cycle, of which it is an integral component. The ammonia liberated by oxidative deamination either is excreted as such (as is the case for most microorganisms and many aquatic animals) or is first "packaged" into the organic molecule urea (or in some species, uric acid) for eventual excretion in the urine (as occurs in most terrestrial vertebrates). Ammonia is highly soluble in water, but is toxic to cells. Urea, however, is also very soluble, but nontoxic.

**Further Catabolism of Carbon Skeletons.** Of the 20 amino acids, 3 give rise to TCA cycle intermediates or precursors directly upon transamination or deamination. These are alanine, aspartate, and glutamate, which give rise, respectively, to pyruvate, oxaloacetate, and $\alpha$-ketoglutarate. All the other amino acids require considerably more complicated pathways, often with many intermediates. In fact, some of the catabolic pathways for amino acids appear at first glance to be almost unnecessarily long and complicated for what is accomplished. Often, however, the intermediates in such pathways have other functions in the cell, such as starting points for biosynthetic routes to other essential compounds. A number of these catabolic pathways are discussed in detail in Chapter 23. As you encounter them, be careful to note how many of them yield end products that eventually feed into the TCA cycle, because that will serve to underscore yet again the centrality of the TCA cycle for all cellular energy metabolism, regardless of the starting substrate.

## THE GLYOXYLATE CYCLE: AN ALTERNATIVE FATE FOR ACETYL-COENZYME A

Thus far we have seen that acetyl-CoA can arise from a variety of catabolic processes, but we have considered only a single fate for the acetyl-CoA. Whether arising as a result of glycolysis, $\beta$ oxidation, or catabolism of specific amino acids, acetyl-CoA has been presumed to become a substrate for further catabolism by the TCA cycle, with $CO_2$ and reduced coenzymes as the end products. However, acetyl-CoA also can be used for biosynthetic purposes. The most common anabolic route that starts with acetyl-CoA is fatty acid synthesis, a process that in general formulation (although not in specific details) is the reverse of $\beta$ oxidation. In fatty acid synthesis, acetyl-CoA molecules (frequently derived from sugar catabolism) are successively condensed together and reduced (using the reducing power of NADPH, often generated by the phosphogluconate pathway) to yield long-chain fatty acids. This will be taken up in Chapter 13. Nothing further need be said here, except perhaps to note that it is the ready conversion of dietary carbohydrate by means of acetyl-CoA to storage triglycerides that accounts for the weight-watcher's concern for the starch and sugar content of foods.

In certain species of plants and microorganisms, an alternative fate for acetyl-CoA exists that is relevant to this unit because it draws on some of the reactions of the TCA cycle—but in a way that is anabolic rather than catabolic. The result is a pathway called the *glyoxylate cycle*, named for glyoxylate, a two-carbon intermediate in the cycle. The glyoxylate cycle makes possible the synthesis of four-carbon (and, as we shall see shortly, six-carbon) compounds from the two-carbon level of acetyl-CoA.

# The Glyoxylate Cycle as an Anabolic Pathway

Usually, condensation of acetyl-coenzyme A with oxaloacetate to form citrate is a signal that the metabolic fate of the acetyl carbons is sealed—the inevitable result by means of the TCA cycle is their oxidation and eventual release as $CO_2$. However, the glyoxylate cycle, shown in Figure 9-12, represents an alternative pathway that also begins with citrate formation, but results in anabolism to the four-carbon level rather than catabolism to the one-carbon level. Comparison of the glyoxylate cycle (Figure 9-12) with the TCA cycle (Figure 9-2) reveals that two of the five reactions of the glyoxylate cycle (labeled Glx-3 and Glx-4) are unique to this pathway, while the other three (Glx-1, Glx-2, and Glx-5) are also part of the TCA cycle. Specifically, _the glyoxylate cycle effectively bypasses the two steps of the TCA cycle in which $CO_2$ is released._ Furthermore, two molecules of acetyl-CoA are taken in per turn of the cycle rather than just one, as in the TCA cycle.

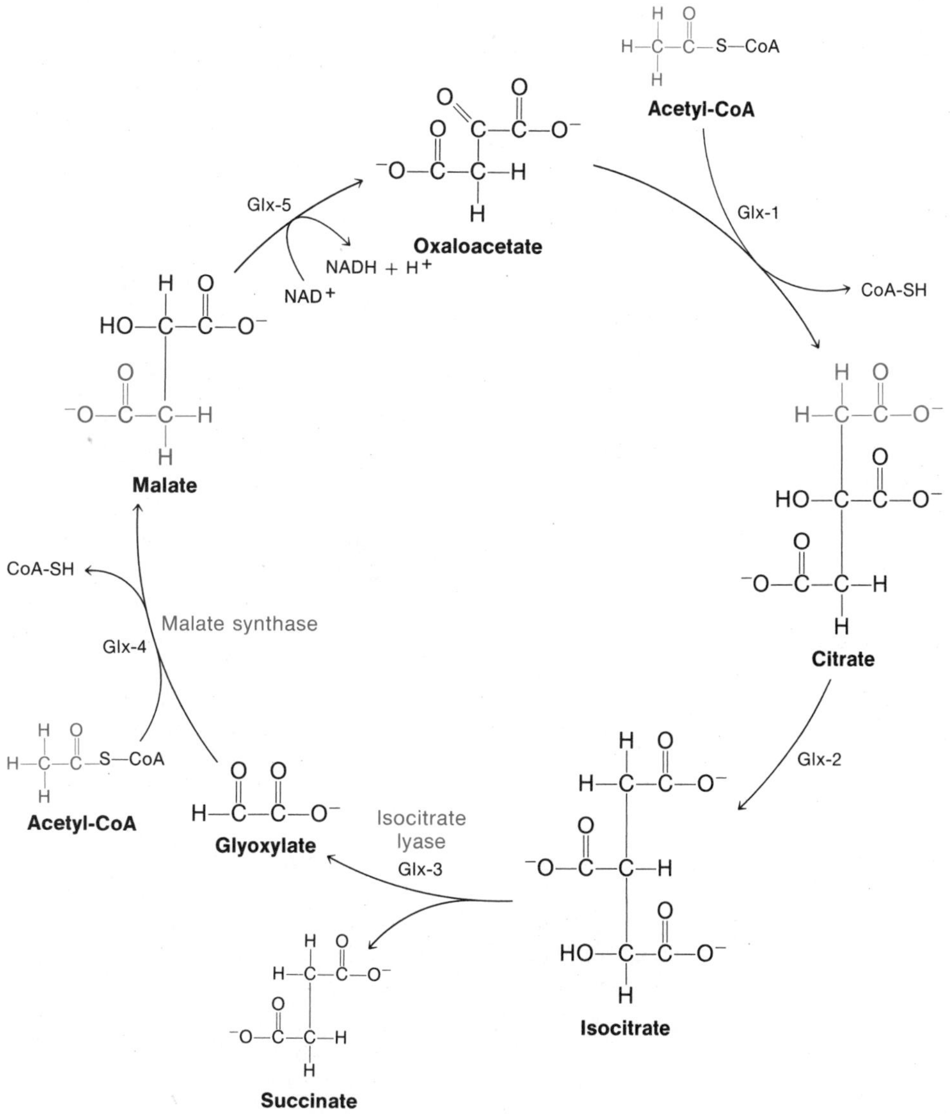

**Figure 9-12**
The glyoxylate cycle. The two enzymes unique to the glyoxylate cycle are indicated in color next to the reactions they catalyze.

The net result is the conversion of two molecules of two carbons each (i.e., the acetate of acetyl-CoA) into one four-carbon compound, succinate.

*The glyoxylate cycle is an indispensable metabolic capability for those species of bacteria, protozoa, fungi, and algae that grow on a two-carbon substrate such as acetate or ethanol.* It is also an essential reaction sequence for seedlings of fat-storing plant species that must effect net synthesis of sugars and other cellular components from the acetyl-CoA produced by $\beta$ oxidation of storage triglycerides. In such plant seedlings, and all other eukaryotic organisms that possess this capability, *the enzymes of the glyoxylate cycle (and those of related metabolic pathways to be discussed below) are compartmentalized together in specialized organelles called glyoxysomes.*

**Formation of Acetyl-CoA.** Species capable of growth on two-carbon substrates need in addition to the glyoxylate cycle some preparatory sequence for converting the substrate into acetyl-CoA. If the substrate is acetate, activation requires only formation of the CoA derivative, catalyzed by acetate thiokinase, with ATP hydrolysis as the driving force:

$$\underset{\textbf{Acetate}}{CH_3-\overset{\overset{\textstyle O}{\|}}{C}-O^-} + CoA-SH + ATP \longrightarrow \underset{\textbf{Acetyl-CoA}}{CH_3-\overset{\overset{\textstyle O}{\|}}{C}-S-CoA} + AMP + PP_i$$

If the substrate is ethanol, it must first be oxidized in two steps to the level of acetate:

$$\underset{\textbf{Ethanol}}{CH_3-CH_2-OH} \xrightarrow{\;NAD^+\;\;NADH,\,H^+\;} \underset{\textbf{Acetaldehyde}}{CH_3-\overset{\overset{\textstyle O}{\|}}{C}-H} \xrightarrow{\;H_2O,\,NAD^+\;\;NADH,\,H^+\;} \underset{\textbf{Acetate}}{CH_3-\overset{\overset{\textstyle O}{\|}}{C}-O^-}$$

Other two-carbon substrates are also possible; each requires specific processing to convert it to the acetyl-CoA with which the glyoxylate cycle itself begins.

**Glx-1 and Glx-2.** Both the initial formation of citrate from oxaloacetate and acetyl-CoA (Glx-1) and its subsequent conversion by means of aconitate to isocitrate (Glx-2) are already familiar from the TCA cycle (see TCA-1 and TCA-2). The only difference is that in eukaryotic cells the citrate synthase and aconitase enzymes that carry out these reactions in the glyoxysomes are organelle-specific isozymes, differing in physical and enzymatic properties from the enzymes responsible for the same reactions that occur in the mitochondria as part of the TCA cycle.

**Glx-3 and Glx-4.** The two reactions unique to the glyoxylate cycle are those responsible for generation and subsequent utilization of glyoxylate, the two-carbon compound from which the cycle (and the organelle) derives its name. In Glx-3, isocitrate is split into two molecules rather than being oxidatively decarboxylated, as would occur in the TCA cycle. The products are glyoxylate and succinate, with two and four carbons, respectively. The reaction is catalyzed by *isocitrate lyase,* an enzyme found only in those microbial and plant species which are able to carry out net growth (and hence synthesis of higher-carbon compounds) from the two-carbon level.

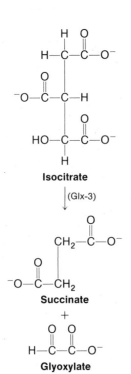

As implied by the preceding formulation, the succinate arises from the "upper" four carbon atoms of isocitrate, while the glyoxylate corresponds to the "lower" two carbons. The succinate represents the immediate product of the glyoxylate cycle and becomes in turn the starting point for synthesis of other compounds needful to the cell or organism. The glyoxylate, however, becomes the acceptor for the acetate group from the second acetyl-CoA molecule that enters the cycle in Glx-4:

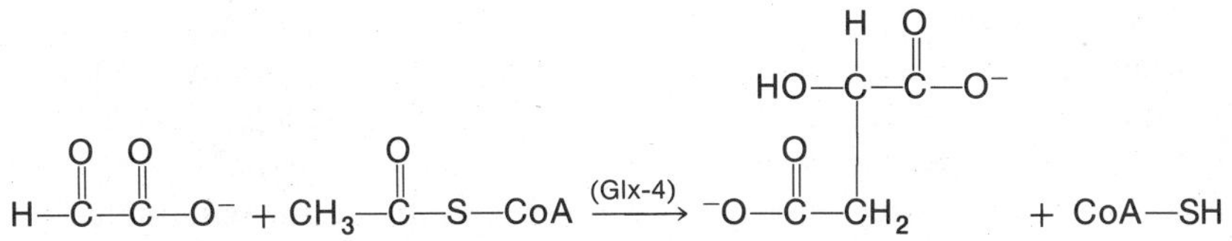

The enzyme responsible for this reaction is _malate synthase._ Like its companion enzyme isocitrate lyase, malate synthase is found only in species capable of net growth from the two-carbon level. The mechanism of action is thought to be analogous to that of citrate synthase, involving nucleophilic addition of the anionic form of acetyl-CoA to the carbonyl carbon of the acceptor, which in this case is glyoxylate.

**Glx-5.** The glyoxylate cycle is completed by the oxidative conversion of malate to oxaloacetate, a reaction catalyzed here, as in the TCA cycle, by malate dehydrogenase, with $NAD^+$ as the electron acceptor. In fat-storing plant species, the isozyme of malate dehydrogenase that is involved in the glyoxylate cycle is specific for the glyoxysomes in which the process is localized.

## Summary of the Glyoxylate Cycle

The glyoxylate cycle itself can be summarized by the following overall reaction:

$$2 \text{ Acetyl-CoA} + NAD^+ \longrightarrow \text{succinate} + NADH + H^+ + 2 \text{ CoA—SH}$$

By adding to the preceding expression the reaction for acetyl-CoA generation by acetate activation, we arrive at a general expression for the synthesis of succinate from acetate in organisms capable of this process:

$$2 \text{ Acetate} + 2 \text{ ATP} + 2 \text{ H}_2\text{O} + NAD^+ \longrightarrow$$
$$\text{succinate} + NADH + H^+ + 2 \text{ ADP} + 2 \text{ P}_i$$

## Gluconeogenesis from Succinate

Thus the glyoxylate cycle is the mechanism by which biosynthesis of more complex molecules from the two-carbon level is initiated, and succinate is the immediate product of the cycle. It therefore becomes relevant to examine the pathways by which other compounds can be synthesized from succinate. Since succinate is also an intermediate in the TCA cycle, we are in effect dealing with an aspect of the amphibolic nature of the TCA cycle. This means that any of the side reactions shown in Figure 9-9 that divert

TCA cycle intermediates in synthetic directions are relevant as pathways for carbon flow from succinate. In this sense, the glyoxylate cycle may properly be described as an anaplerotic reaction sequence, since it replenishes TCA cycle intermediates. However, to suggest, as some textbooks do, that this is the main or only role of the glyoxylate cycle is to misrepresent the importance and purpose of the cycle. *Of greatest significance to an organism that depends on a two-carbon substrate for all its carbon needs is the gluconeogenic (sugar-synthesizing) route from succinate to the hexose level,* since access to the six-carbon level essentially guarantees access to all other biosynthetic pathways in the cell. Gluconeogenesis in organisms that do not possess the enzymes specific to the glyoxylate cycle usually proceeds from molecules with at least three carbon atoms, such as pyruvate or lactate. The strategy of gluconeogenesis is considered in greater detail in Chapter 12.

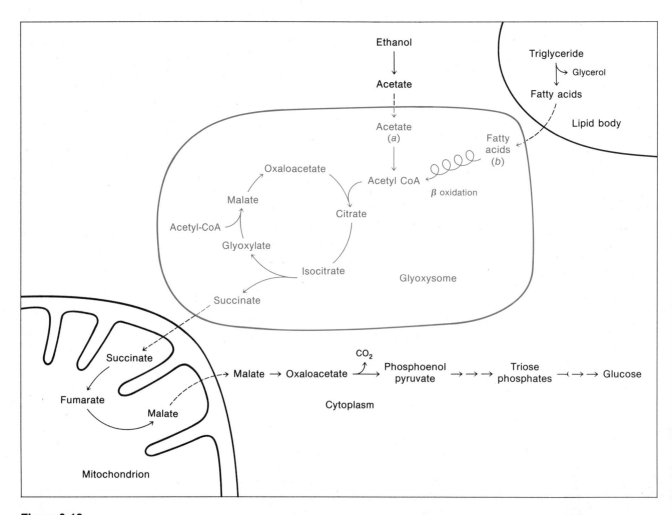

**Figure 9-13**
Role of the glyoxylate cycle in gluconeogenesis from (*a*) two-carbon compounds or (*b*) fatty acids. The pathway involves reaction sequences that in eukaryotic cells are localized within lipid bodies, glyoxysomes, mitochondria, and the cytoplasm.

Synthesis of hexoses from succinate proceeds by means of phospho-enolpyruvate by the pathway shown in Figure 9-13. (The figure also indicates the localization in a eukaryotic cell of the several component parts of this overall sequence.) In outline form, the succinate arising from glyoxylate cycle activity is converted by means of fumarate and malate to oxaloacetate by TCA cycle enzymes located in the mitochondria. The oxaloacetate is then decarboxylated to phosphoenolpyruvate by reversal of the PEP carboxylase reaction discussed earlier in an anaplerotic context. From phosphoenolpyruvate, the route to glucose is direct, involving in essence a reversal of the glycolytic pathway. Since the NADH generated by malate oxidation in the initial conversion of succinate to oxaloacetate is consumed as phosphoglycerate is reduced to glyceraldehyde-3-phosphate during the gluconeogenic reversal of the glycolytic sequence, the only net coenzyme involvement is that of two FAD reduced per glucose. In summary form, the synthesis of glucose from succinate can be represented as follows:

$$2 \text{ Succinate} + 2 \text{ FAD} + 2 \text{ ATP} + 4 \text{ H}_2\text{O} \longrightarrow$$
$$\text{glucose} + 2 \text{ CO}_2 + 2 \text{ FADH}_2 + 2 \text{ ADP} + 2 \text{ P}_i$$

By multiplying the previous reaction by 2 and adding it to this reaction, we arrive at the following overall expression for the synthetic route from acetate to glucose:

$$4 \text{ Acetate} + 2 \text{ NAD}^+ + 2 \text{ FAD} + 2 \text{ ATP} + 4 \text{ H}_2\text{O} \longrightarrow$$
$$\text{glucose} + 2 \text{ CO}_2 + 2 \text{ NADH} + 2 \text{ H}^+ + 2 \text{ FADH}_2 + 2 \text{ ADP} + 2 \text{ P}_i$$

## Gluconeogenesis from Fatty Acids

The preceding expression is adequate to summarize the overall process of gluconeogenesis from acetate (or other two-carbon substrates) in lower organisms capable of using such substrates to meet all their energy and carbon needs. Plant species characterized by the presence of storage triglycerides in the seed also must effect gluconeogenesis from acetyl-CoA during early seedling growth. In this case, however, the acetyl-CoA is derived not from two-carbon substrates available to the organism exogenously, but from concomitant $\beta$ oxidation of fatty acids that are in turn derived from storage triglycerides deposited as lipid bodies in the fat-storing cells of the endosperm or cotyledons during the latter stages of seed maturation. Thus the full metabolic process which such fat-storing seeds carry out during early postgerminative growth (i.e., before becoming green and photosynthetically competent) involves $\beta$ oxidation of fatty acids to acetyl-CoA and subsequent synthesis of hexoses (and all other compounds needed for early seedling growth), as shown in Figure 9-13.

The overall process is gluconeogenic conversion of fat to carbohydrate, but it involves both a catabolic component ($\beta$ oxidation) and an anabolic sequence (synthesis of sugar from acetyl-CoA). Appropriately enough, all the enzymes of $\beta$ oxidation are packaged together with those of the glyoxylate cycle in the glyoxysomes present in the fat-storing tissue of such plant species. The whole process from fatty acid to succinate therefore occurs within the glyoxysome.

The fat-storing plants are unique among higher organisms in their ability to carry out net conversion of fat to carbohydrate. The conversion of carbohydrate to fat is nearly universal. However, the reverse process of net synthesis of carbohydrate from fat is denied all higher plants and animals except for the fat-storing plants. Moreover, even in these species, the requisite glyoxylate cycle enzymes and $\beta$ oxidation activities are only transiently present during the relatively short postgerminative period during which the seedling is dependent on triglyceride reserves to meet all or most of its carbon needs. As soon as the seedling emerges from the soil and becomes photosynthetically competent, glyoxylate cycle activity and $\beta$ oxidation are lost, and they do not appear again until the next generation of seeds germinates.

## SELECTED READINGS

Atkinson, D. E. *Cellular Energy Metabolism and Its Regulation.* New York: Academic Press, 1977. A solid, detailed discussion of energy metabolism by an author who is very much a part of the scene.

Broda, E. *The Evolution of Bioenergetic Processes.* New York: Pergamon Press, 1975. An excellent discussion of cellular energetics from an evolutionary perspective.

Krebs, H. A. The history of the tricarboxylic acid cycle. *Perspect. Biol. Med.* 14:154, 1970. An engaging historical account by the man who masterminded much of it.

Mehlman, M., and Hanson, R. W. *Energy Metabolism and the Regulation of Metabolic Processes in Mitochondria.* New York: Academic Press, 1972. The title speaks for itself—a good reference.

Williamson, J. R., and Cooper, R. V. Regulation of the citric acid cycle in mammalian systems. *FEBS Lett.* 117(Suppl.):K73, 1980. A well-rounded, current review of TCA cycle regulation as a contemporary research theme from a symposium dedicated to Hans Krebs.

## PROBLEMS

1. *True, false, or eeny meeny.* Mark each of the following statements with one of the following: T (for true), F (for false), or E (for eeny meeny, if you cannot tell whether the statement is true or false on the basis of the information given):
   a. All five cofactors or prosthetic groups of the pyruvate dehydrogenase complex are directly involved in electron transport.
   b. Thermodynamically, acetyl-CoA ought to be capable of driving the phosphorylation of ADP (or GDP) just as succinyl-CoA does.
   c. The methyl carbon of each acetyl-CoA entering the TCA cycle comes from carbon atom 3 of pyruvate.
   d. Even if aconitase were incapable of discriminating between the two ends of the citrate molecule, the $CO_2$ evolved at step TCA-3 would still inevitably come from the oxaloacetate rather than the acetyl-CoA substrate of the preceding TCA-1 reaction.
   e. Activation of pyruvate carboxylase means that one or more intermediates of the TCA cycle are being drawn off for biosynthetic purposes.
   f. Malate cannot be converted to fumarate in the mitochondria of an animal cell because the TCA cycle operates only unidirectionally.

2. *TCA cycle activity.* Assume that you have a solution containing the pyruvate dehydrogenase complex and all the enzymes of the TCA cycle but none of the metabolic intermediates.
   a. If you add 3 m$M$ each of pyruvate, coenzyme A, NAD, FAD, ADP, GDP, and $P_i$, what will happen? How much $CO_2$ will be evolved?
   b. If in addition to the reagents specified in part (*a*) you add 3 m$M$ each of citrate, isocitrate, $\alpha$-ketoglutarate, succinate, fumarate, and malate, how much $CO_2$ will be evolved?
   c. If in addition to the reagents specified in part (*a*) you could add only *one* of the intermediates specified in part (*b*), which one would you choose? Why? Now how much $CO_2$ will be evolved?
   d. If in addition to the reagents specified in part (*a*) you add pyruvate carboxylase, what effect would you expect that to have on $CO_2$ evolution?

3. *Label chasing.* Suppose you feed glucose[1-$^{14}$C] to an aerobic culture of bacteria and then extract and separate intermediates of the glycolytic pathway and the TCA cycle. Which carbon atom(s) would you expect to find labeled *first* with $^{14}$C in each of the following molecules?

a. Glyceraldehyde-3-phosphate
b. Phosphoenolpyruvate
c. $\alpha$-Ketoglutarate
d. Oxaloacetate

4. *More label chasing*. In which carbon atom(s) of glucose fed to an aerobic bacterial culture would $^{14}C$ have to be located to ensure that labeled $CO_2$ was released *first* at the step catalyzed by
   a. Pyruvate dehydrogenase
   b. Isocitrate dehydrogenase
   c. $\alpha$-Ketoglutarate dehydrogenase

5. *Glutamate synthesis*. Glutamate synthesis is an example of a pathway that exploits the amphibolic nature of the TCA cycle.
   a. Devise a pathway for the synthesis of glutamate using pyruvate and alanine as the starting materials. Identify the enzymes you will need and write the complete reaction for each step.
   b. Write a balanced summary equation for the synthesis of glutamate from pyruvate and alanine.

6. *Malonate inhibition*. In his landmark experiments which led to the elucidation of the TCA cycle, Sir Hans Krebs observed that addition of malonate to suspensions of pigeon flight muscle inhibited pyruvate utilization and resulted in the accumulation of succinate upon oxidative catabolism of a variety of added substrates.
   a. Why do you suppose Krebs used minced flight muscle preparations in these studies?
   b. Why does malonate inhibit pyruvate utilization by such preparations?
   c. What would Krebs have been able to conclude when he found that succinate accumulated in malonate-treated muscle following addition of citrate, isocitrate, or $\alpha$-ketoglutarate?
   d. Why was the accumulation of succinate in malonate-treated muscle even more significant when the added substrate was fumarate, malate, or oxaloacetate instead of those mentioned in part (*c*)?
   e. Krebs also found that inhibition of pyruvate utilization by malonate could be overcome by adding oxaloacetate, malate, or fumarate along with the pyruvate. How would you explain these results?

7. *Clever, smart, or wise?* The normal function of the acetyl-CoA formed in aerobic cells is to add acetate to the carbonyl carbon of oxaloacetate to form citrate. Recently, Dr. Melvin Clever announced to his biochemistry class that he was working with a strain of bacterial cells that had no detectable citrate synthase activity, but possessed an enzyme capable of catalyzing the following reaction under aerobic conditions:

   Acetyl-CoA + ADP + $P_i$ $\longrightarrow$
   acetate + CoA + ATP

Dr. Clever said that the free acetate formed in this way is excreted from the cell into the medium, and he claimed that bacterial cells that possess this enzyme can derive more energy (i.e., make more ATP) from glucose under aerobic conditions by this route than by conventional respiratory metabolism. Sam Smart, a student in his class, said he thought Dr. Clever's claim was incorrect for aerobic conditions, but that the preceding reaction would certainly increase ATP yield under anaerobic conditions by allowing the cell to oxidize glucose to acetate instead of to lactate. Wilma Wise, however, insisted that the preceding reaction would not improve ATP yields under either condition.
   a. Is the preceding reaction likely to be thermodynamically feasible as written? How do you know?
   b. Do you agree with Sam, Wilma, or Dr. Clever? Explain.

8. *Hydroxypyruvate metabolism*. A principle of cellular energy metabolism is that alternative substrates are catabolized by bringing them as quickly as possible into the mainstream glycolytic/TCA cycle pathway. Hydroxypyruvate is such an alternative substrate. It is a three-carbon compound with the following structure:

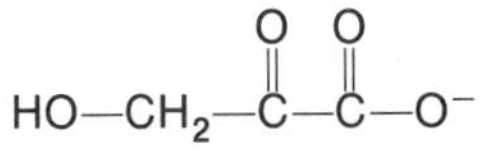

$$HO-CH_2-\overset{\overset{O}{\|}}{C}-\overset{\overset{O}{\|}}{C}-O^-$$
**Hydroxypyruvate**

Despite the similarity between hydroxypyruvate and pyruvate, the only cellular route by which the former can be converted into the latter is a five-step pathway involving the four intermediates shown below. (These intermediates are labeled A, B, C, and D, although not necessarily in the order in which they appear in the pathway.) The pathway begins with an NADH-mediated reaction and requires the presence of both ADP and ATP, although with no net consumption or generation of either.

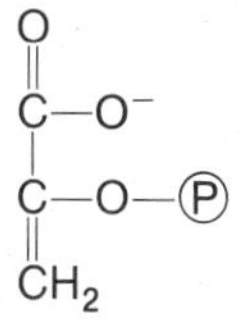

A

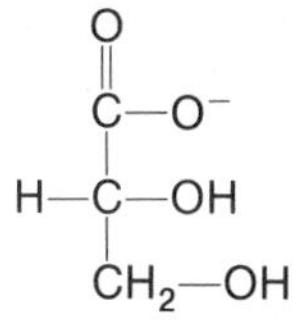

B

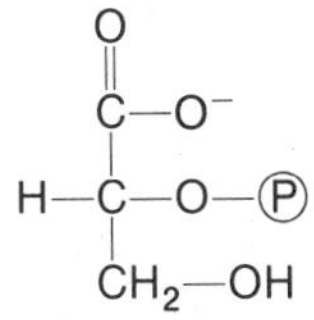

C

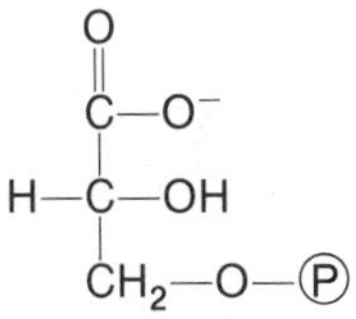

D

a. Order the intermediates A, B, C, and D in the proper sequence to account for the stepwise conversion of hydroxypyruvate into pyruvate, and write a balanced equation for each of the five reactions.

b. Write an overall equation for the conversion of hydroxypyruvate to pyruvate.

**9.** *Palmitate oxidation.* Palmitate is a 16-carbon straight-chain fatty acid ($C_{16}H_{32}O_2$) that can be oxidized completely to $CO_2$ and $H_2O$ by a combination of $\beta$ oxidation and the TCA cycle.

a. Calculate the net number of ATP molecules generated by the complete catabolism of palmitate to the two-carbon (acetyl-CoA) level, assuming (as will become clear in Chapter 10) that the ATP yield upon coenzyme oxidation is 3 for NADH and 2 for $FADH_2$.

b. Next, calculate the number of ATP molecules generated by the further oxidation of the eight resulting acetyl-CoA molecules to $CO_2$ and $H_2O$, making the same assumptions as in part (a).

c. Finally, calculate the total number of ATP molecules generated by the complete oxidation of one molecule of palmitate to $CO_2$ and $H_2O$, and write a balanced overall equation for this process.

**10.** *Comparative energy content of carbohydrates and fats.* Higher plants and animals store energy reserves preferentially as fat rather than as carbohydrate because fat has a higher energy content per unit weight. The calculations specified here are designed to quantitate this difference.

a. Consider first the utilization of glucose ($C_6H_{12}O_6$; $M_r = 180$) as an energy source and calculate the moles of ATP generated during the complete oxidative metabolism of 1 g glucose. (You will need an equation from the next chapter for this purpose.) Assuming that the $\Delta G$ value for the hydrolysis of ATP under physiological conditions is $-12$ kcal/mol, how much free energy is conserved as ATP upon aerobic catabolism of 1 g glucose?

b. Now repeat the calculations of part (a), but for the complete oxidation of 1 g palmitate ($C_{16}H_{32}O_2$; $M_r = 256$). [See part (c) of Problem 9.]

c. How much more efficient on a per-gram basis is fat as a form of energy storage compared with carbohydrate, assuming the values for glucose and palmitate to be representative of carbohydrates and fats in general? Why do organisms as diverse as castor beans and humans prefer fat as a means of storing energy reserves?

d. If a friend laments to you that she is "getting fat" and is already 5 kg overweight, you can console her that she would be even more overweight if she were "getting carbohydrate" instead. Why is this so? How much overweight would she in fact

be if the human body stored energy reserves as carbohydrate instead of fat?

e. Bearing in mind that respiratory metabolism is essentially an oxidative process, how might you explain the difference in energy content of carbohydrate and fat on a per-gram basis?

**11.** *Itsafarcin inhibition.* Itsafarcin (IT) is a potent metabolic poison prepared by aqueous extraction of biochemistry texts. Because they suspected IT to be an inhibitor of one of the enzymes of respiratory metabolism, Drs. G. Whiz and Y. Bother grew cultures of an aerobic bacterial strain on a glucose-containing medium and showed that addition of IT to the medium resulted in complete cessation of oxygen consumption (i.e., coenzyme reoxidation; see Chapter 10) by the bacterial cells (see the following table, line 1). The same effect of IT was seen with cells grown on pyruvate (line 2). When intermediates of the TCA cycle were added along with either the glucose or the pyruvate in pairwise combination, Whiz and Bother found that for some substrate pairs, oxygen consumption was stopped by IT (lines 3 and 6), while for other pairs, IT had no effect at all on oxygen consumption (lines 4 and 5). In cases where IT had no inhibitory effect, it was noted that an $\alpha$-keto acid accumulated in the culture medium as oxygen consumption proceeded.

|  | Oxygen Consumption | |
| Substrate(s) in Culture Medium | Without IT | With IT |
| --- | --- | --- |
| **1.** Glucose only | Yes | No |
| **2.** Pyruvate only | Yes | No |
| **3.** Glucose and citrate | Yes | No |
| **4.** Glucose and succinate | Yes | Yes |
| **5.** Pyruvate and malate | Yes | Yes |
| **6.** Pyruvate and $\alpha$-ketoglutarate | Yes | No |

a. From the data on lines 1 and 2 only, Whiz and Bother concluded that the enzyme affected by itsafarcin is not part of the glycolytic pathway. Do you agree with their conclusion? Explain.

b. Assuming that itsafarcin exerts its effect by inhibiting a single reaction of the TCA cycle, which reaction is it? Explain.

c. Assuming that the ATP yield upon coenzyme oxidation is 3 for NADH and 2 for $FADH_2$, what is the ATP yield (in molecules of ATP per molecule of glucose consumed) when the bacterial culture is grown aerobically on glucose plus succinate in the presence of itsafarcin?

d. Would you expect itsafarcin to kill the cells, presuming that its only effect is that elucidated in part (b)?

**12.** *Isoleucine catabolism.* Isoleucine is one of the 20 different kinds of amino acids used by cells to synthesize proteins. If necessary, however, isoleucine also can be catabolized to meet cellular energy needs (as in the case of starvation, for example). The catabolic pathway (shown below) begins with a transamination step (Reaction 1) that converts the amino acid into the corresponding α-keto acid. Four further steps (Reactions 2, 3, 4, and 5) are then necessary to convert the α-keto acid into α-methyl-acetoacetyl-CoA.

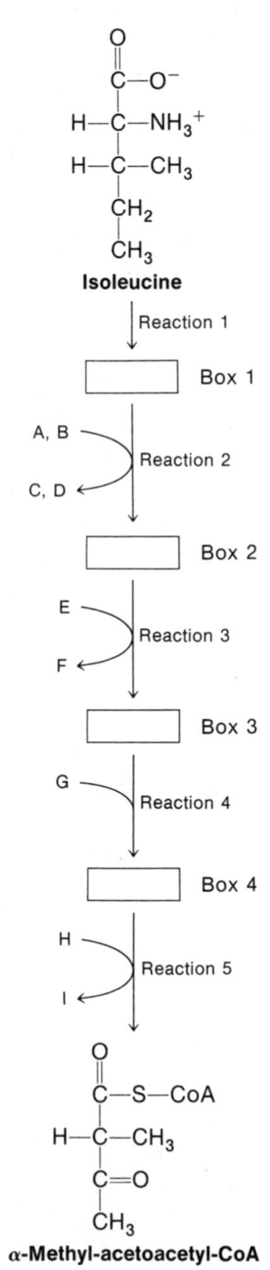

**a.** The four missing intermediates in the pathway are shown below, although not in any special order. Order the intermediates in the pathway by putting the correct letter (P, Q, R, or S) in boxes 1, 2, 3, and 4 (one letter per box).

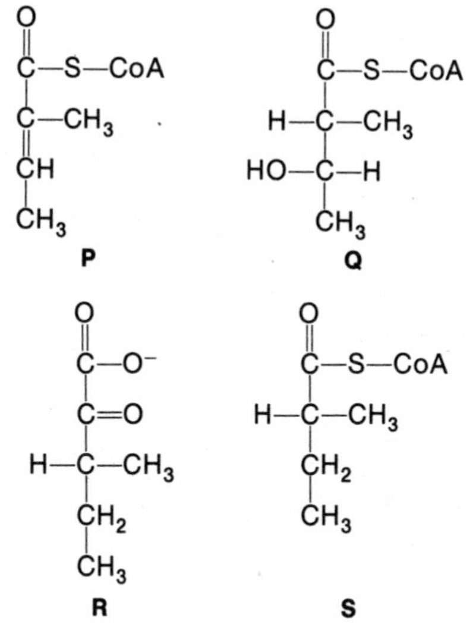

P

    O
    ‖
    C—S—CoA
    |
    C—CH₃
    ‖
    CH
    |
    CH₃

Q

    O
    ‖
    C—S—CoA
    |
H—C—CH₃
    |
HO—C—H
    |
    CH₃

R

    O
    ‖
    C—O⁻
    |
    C=O
    |
H—C—CH₃
    |
    CH₂
    |
    CH₃

S

    O
    ‖
    C—S—CoA
    |
H—C—CH₃
    |
    CH₂
    |
    CH₃

**b.** Several coenzymes and other substances are involved in the reaction sequence, as indicated by the letters A through I in the pathway. Identify the substance to which each letter corresponds.

**c.** The α-methyl-acetoacetyl-CoA formed by this sequence of reactions is further converted (by an additional three-step sequence not shown here) to one molecule each of acetyl-CoA and succinyl-CoA. In what sense are these "appropriate" end products of amino acid catabolism?

Isoleucine

    O
    ‖
    C—O⁻
    |
H—C—NH₃⁺
    |
H—C—CH₃
    |
    CH₂
    |
    CH₃

↓ Reaction 1

[  ] Box 1

A, B ⟩ Reaction 2
C, D ←

[  ] Box 2

E ⟩ Reaction 3
F ←

[  ] Box 3

G ⟩ Reaction 4

[  ] Box 4

H ⟩ Reaction 5
I ←

α-Methyl-acetoacetyl-CoA

    O
    ‖
    C—S—CoA
    |
H—C—CH₃
    |
    C=O
    |
    CH₃

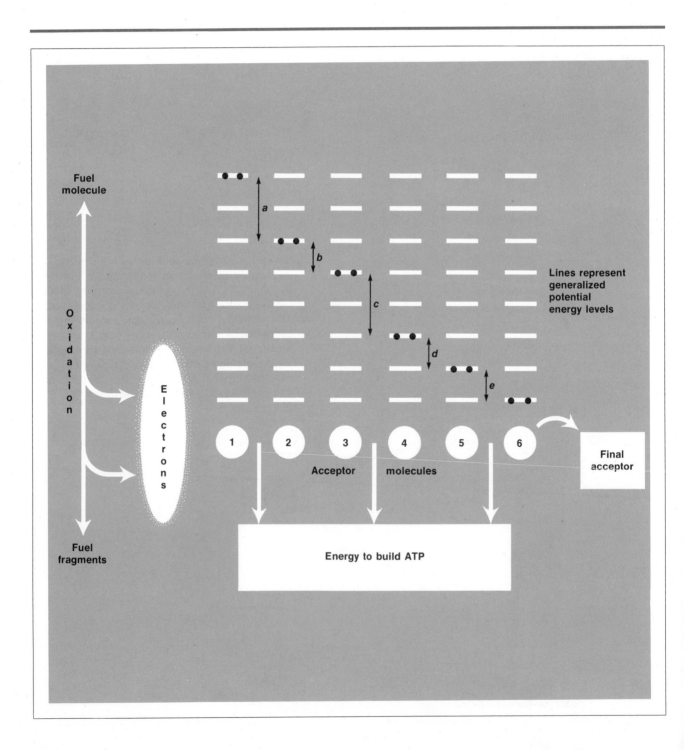

# 10

# AEROBIC PRODUCTION OF ATP:
# ELECTRON TRANSPORT

Chemotrophic energy metabolism has thus far been characterized by the stepwise oxidation of organic fuel molecules, always accompanied by the reduction of a coenzyme (either $NAD^+$ or FAD) that serves as acceptor both of electrons and of hydrogen ions. Glucose catabolism, for example, involves six different oxidative events, one during glycolysis, one at the level of pyruvate oxidation, and the remaining four as components of the TCA cycle. In the process, all six carbon atoms of the glucose molecule are oxidized completely to $CO_2$, and 12 pairs of electrons and hydrogen atoms are removed successively and transferred to $NAD^+$ (in 10 of the 12 cases) or to FAD (in the remaining 2 cases). Similarly, oxidation of fatty acids is accompanied by the reduction of one molecule each of $NAD^+$ and FAD per cycle of oxidation (i.e., per molecule of acetyl-CoA released) and of additional coenzyme molecules as the resulting acetyl-CoA is oxidized further in the TCA cycle. Amino acids also are catabolized by oxidative pathways that use these same coenzymes as electron acceptors.

Clearly, the continued availability of oxidized coenzyme molecules for use as electron acceptors depends on the concomitant reoxidation of reduced coenzymes, so that the reduction of coenzymes during substrate oxidation is accompanied and balanced by an ongoing process whereby the oxidized forms of the coenzymes are being regenerated. It is in this process of coenzyme reoxidation that the oxygen requirement for aerobic energy metabolism finally becomes apparent, because the terminal electron acceptor for coenzyme oxidation is molecular oxygen. Since the transfer of electrons from NADH or $FADH_2$ to oxygen is highly exergonic, the process of coenzyme oxidation also turns out to account for most of the ATP yield of respiratory metabolism (32 out of 36 ATPs per glucose in the eukaryotic cell).

*Electron flow and ATP synthesis.*

Our task in this chapter is twofold: to understand how coenzymes are reoxidized by transfer of electrons to molecular oxygen and to consider the mechanism whereby the energy of such transport is coupled with ATP generation. The first of these processes is called *electron transport* and involves the stepwise transfer of electrons from reduced coenzyme by means of a chain of reversibly oxidizable electron carriers to molecular oxygen. The accompanying process of ATP generation is called *oxidative phosphorylation* because it involves phosphorylation of ADP to form ATP, accompanied by the reoxidation of reduced coenzymes as electrons are transferred down the chain to oxygen. For convenience of understanding, we will consider these processes as separate entities, but we have to keep in mind that they obviously occur in the cell not as isolated events, but as integral parts of respiratory metabolism. It may, for purposes of discussion, be convenient to consider component phases separately as we are doing here and have done in the preceding two chapters, but it would be a mistake to think of them as occurring in isolation or in sequential order, since all four phases—glycolysis (or for fatty acids, $\beta$ oxidation), the TCA cycle, electron transport, and oxidative phosphorylation—must occur in a continuous, highly integrated manner to accomplish respiratory metabolism and thereby meet the energy needs of the cell.

## ELECTRON TRANSPORT

Since electron transport involves the reoxidation of coenzymes at the expense of molecular oxygen, the process can be summarized by the following overall reactions:

$$NADH + H^+ + \tfrac{1}{2} O_2 \longrightarrow NAD^+ + H_2O \qquad \Delta G^{\circ\prime} = -52.6 \, \text{kcal/mol}$$

$$FADH_2 + \tfrac{1}{2} O_2 \longrightarrow FAD + H_2O \qquad \Delta G^{\circ\prime} = -43.4 \, \text{kcal/mol}$$

As these summary reactions indicate, electron transport is responsible not only for the reoxidation of coenzymes and the consumption of oxygen, but also for the generation of the water, which we recognize along with $CO_2$ as one of the two expected end products of aerobic energy metabolism. The $CO_2$ is generated in the TCA cycle, but the water arises only upon reduction of oxygen in the process of electron transport.

The most important aspect of the preceding reactions, however, is the large amount of free energy released upon oxidation of $FADH_2$ and NADH. Both these coenzymes are energy-rich molecules because each contains a pair of electrons with a high transfer potential. It is in the stepwise transfer of electrons from these reduced coenzyme molecules to molecular oxygen that the free energy is released that powers the synthesis of most of the ATP formed during respiratory metabolism.

As indicated by the $\Delta G^{\circ\prime}$ values, the oxidation of a coenzyme is a highly exergonic process. In fact, most of the free energy of the glucose molecule is still present in the reduced coenzymes that are generated during glycolysis and the TCA cycle. Complete oxidation of a glucose molecule results in the formation of 10 NADH and 2 $FADH_2$, with a standard free-energy content of $10(-52.6) + 2(-43.4) = -613 \, \text{kcal/mol}$. Since the total change in free energy accompanying glucose oxidation is about $-686 \, \text{kcal/mol}$, it is clear that *about 90 percent of the free energy origi-*

*nally present in the oxidizable bonds of the glucose molecule is conserved in the reduced coenzyme molecules and is tapped by the cell* only during the final two phases of respiratory metabolism.

Clearly, the free-energy changes that would accompany the direct oxidation of NADH and $FADH_2$ by oxygen are much larger than those usually encountered in biological reactions and are, in fact, sufficient to drive the synthesis of at least several ATPs. Hence it should not be surprising to find that electrons are not passed directly from reduced coenzymes to oxygen. Rather, the transfer is accomplished stepwise, by means of a series of reversibly oxidizable electron acceptors. In this way, *the total free-energy difference between reduced coenzymes and oxygen is parcelled out among a series of intermediates and is released in increments to maximize the opportunity for ATP generation. Specifically, the energy of electron transport appears to be used to build up and maintain an electrochemical potential across a membrane* (also see Chapter 17). In the case of prokaryotes, the cytoplasmic membrane is involved; for eukaryotic cells, it is the inner membrane of the mitochondria. The electrochemical potential is related to a *proton motive force* (pmf) that is thought to drive the actual synthesis of ATP (see Chapters 7 and 17 for a fuller discussion of pmf). Especially important steps in the electron-transport process are therefore those in which electron transfer between two adjacent intermediates in the transport chain is coupled with the uptake or release of $H^+$, since these are the steps that can be harnessed to achieve the vectorial transport of $H^+$, upon which maintenance of the transmembrane proton motive force (and hence the ability to generate ATP) depends.

## Reduction Potentials

To understand the order of intermediates in the chain of acceptors responsible for the transport of electrons from reduced coenzymes to oxygen requires an appreciation for the standard reduction potential that is used as a measure of the electron-transfer potential of a given oxidizable compound (see Chapter 7). The *standard reduction potential*, designated as $E_0'$, is a measure of the electromotive force generated at 25°C and pH 7.0 by a sample half-cell with respect to a reference half-cell (Figure 10-1). To make such a measurement, the sample half-cell is filled with a solution that is 1 $M$ with respect to both the oxidized and the reduced forms of the substance whose reduction potential is to be determined. The electrode immersed in the sample half-cell is then connected by means of a potentiometer to an electrode immersed in a reference half-cell that contains 1 $M$ $H^+$ in equilibrium with $H_2$ at a pressure of 1 atm. By convention, the $H^+/H_2$ couple is considered the reference standard in such electrochemical measurements and is assigned an $E_0$ value of 0.0 V. All other reducible/ oxidizable (redox) couples therefore have standard reduction potentials that are either positive or negative with respect to the $H^+/H_2$ standard, depending on the direction of electron flow. If electron flow is *toward* the sample half-cell, the redox couple is assigned a *positive* $E_0'$ value, since reduction is occurring in the half-cell at the expense of the $H^+/H_2$ couple. If electrons flow *away* from the sample half-cell, the couple has a *negative* reduction potential, since reduction does *not* occur in the sample half-cell. The numerical value of the reduction potential is the observed voltage

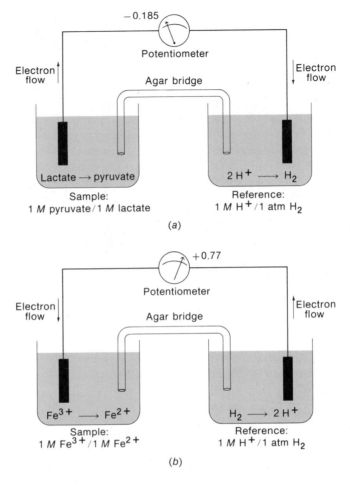

**Figure 10-1**
Sample and reference half-cells for the determination of the standard reduction potential of (a) the pyruvate/lactate ($E'_0 = -0.185$ V) and (b) the $Fe^{3+}/Fe^{2+}$ couple ($E'_0 = +0.77$ V). In both cases, the sample half-cell contains both the oxidized and the reduced species at an initial concentration of 1 $M$, and the reference half-cell contains 1 $M$ $H^+$ and $H_2$ at a pressure of 1 atm. Measurements are made at standard conditions of pH (7.0) and temperature (25°C).

recorded by the potentiometer at the beginning of the experiment, while oxidants and reductants are still present at a concentration of 1 $M$ (or, for gases, at a pressure of 1 atm).

As an example, the sample half-cell of Figure 10-1a contains 1 $M$ each of lactate and pyruvate. With both electrodes in place and electrical continuity between the two half-cells ensured by an agar bridge, electrons will begin to flow in response to the electromotive force generated between the two half-cells. Because pyruvate is less readily reduced than $H^+$, the reactions in the two cells will be as follows:

Sample half-cell:  Lactate $\longrightarrow$ pyruvate $+ 2\,H^+ + 2\,e^-$

Reference half-cell:  $2\,H^+ + 2\,e^- \longrightarrow H_2$

Since electrons are being released in the sample cell and consumed in the reference cell, electron flow will be *away* from the sample cell, as shown in Figure 10-1a. The potentiometer reading at the outset of the experiment will be 0.185 V on the negative side, so the $E'_0$ value for the lactate/pyruvate couple is $-0.185$ V.

In the case of Figure 10-1b, the sample half-cell contains 1 $M$ each of ferrous and ferric ions. Because $Fe^{3+}$ is more readily reduced than $H^+$, the direction of electron flow is *toward* the sample half-cell, and the reactions in the two cells are as follows:

Sample half-cell: $Fe^{3+} + e^- \longrightarrow Fe^{2+}$

Reference half-cell: $H_2 \longrightarrow 2\,H^+ + 2\,e^-$

The potentiometer reading in this case will be 0.77 V on the positive side, so the $E_0'$ value for the $Fe^{2+}/Fe^{3+}$ couple is $+0.77$ V.

The meaning of the standard reduction potential should now be clear. If the value is negative, the oxidized form of the couple has a lower affinity for electrons than does $H^+$ (or equivalently, the reduced form of the couple has a greater potential for electron transfer than does $H_2$). Conversely, a positive reduction potential means that the oxidized form of the couple has a greater electron affinity than does $H^+$ (or that the reduced form has a lesser potential for donating electrons than $H_2$). Put more simply, *a strong reducing agent (i.e., a good electron donor) has a negative reduction potential, while a strong oxidizing agent (i.e., a good electron acceptor) has a positive reduction potential.*

The standard reduction potentials (at 25°C, 1 M concentration, 1 atm pressure, and pH 7.0) for a number of biologically important redox couples are presented in Table 10-1. It is important to note that these are reduction potentials, and therefore they apply specifically to the half-reaction *written in the direction of reduction*, as shown in the table. For a half-reaction written in the reverse direction (i.e., in the direction of oxidation) instead, the sign would be reversed and it would be called an *oxidation potential* instead. Because the redox couples in Table 10-1 are arranged with the most negative reduction potentials (i.e., the strongest reducing agents) at the top, simple inspection of the relative positions of two couples in the table allows prediction of the direction of the reaction between the two couples under standard conditions, since a given redox couple will tend to reduce any redox couple that is below it in the table (i.e., that has a more positive reduction potential) under standard conditions. Thus NADH will spontaneously reduce oxaloacetate or pyruvate, but not acetate or α-ketoglutarate.

The relative positions of redox couples in the table are therefore indicators of the thermodynamic spontaneity of an oxidation-reduction reaction between them. The standard free-energy change ($\Delta G^{\circ\prime}$) for an oxidation-reduction reaction is in fact linearly related to the difference between the reduction potentials of the two half-reactions (see Chapter 7). The equation relating $\Delta G^{\circ\prime}$ to the difference in reduction potentials between the two redox couples is as follows:

$$\Delta G^{\circ\prime} = -nF\Delta E_0' = -nF(E_{0,\text{ acceptor}}' - E_{0,\text{ donor}}')$$

where $n$ = number of electrons transferred in the half-reactions
$F$ = Faraday constant (23,062 cal mol$^{-1}$ volt$^{-1}$)
$\Delta E_0'$ = difference in reduction potentials between the two redox couples ($E_{0,\text{ acceptor}}' - E_{0,\text{ donor}}'$)

As an example, consider the NADH-mediated reduction of pyruvate to lactate as it occurs during anaerobic fermentation:

Pyruvate + NADH + $H^+$ $\longrightarrow$ lactate + $NAD^+$

The two half-reactions, both written in the direction of reduction, are as follows:

$$\text{Pyruvate} + 2\,H^+ + 2\,e^- \longrightarrow \text{lactate} \qquad E_0' = -0.185\ \text{V}$$

$$NAD^+ + 2\,H^+ + 2\,e^- \longrightarrow NADH + H^+ \qquad E_0' = -0.32\ \text{V}$$

In this case, NADH is the electron donor and pyruvate is the electron acceptor, so $\Delta E_0'$ is calculated as follows:

$$\Delta E_0' = E_{0,\ \text{acceptor}}' - E_{0,\ \text{donor}}' = -0.185 - (-0.32) = +0.135\ \text{V}$$

The standard free-energy change for the reaction is then:

$$\Delta G^{\circ\prime} = -nF\Delta E_0' = -2(23{,}062)(0.135) = -6{,}227\ \text{cal/mol}$$
$$= -6.2\ \text{kcal/mol}$$

Note that $\Delta G^{\circ\prime}$ and $\Delta E_0'$ have opposite signs. A reaction that is spontaneous under standard conditions is characterized by a *negative* $\Delta G^{\circ\prime}$, but a *positive* $\Delta E_0'$.

**Table 10-1**
Standard Reduction Potentials for Redox Couples of Biological Relevance

| Half-Cell Equation | $E_0'{}^*$ | $n\dagger$ |
|---|---|---|
| Acetate + $CO_2$ + $2\,H^+$ + $2\,e^-$ $\longrightarrow$ pyruvate + $H_2O$ | $-0.70$ V | 2 |
| Succinate + $CO_2$ + $2\,H^+$ + $2\,e^-$ $\longrightarrow$ $\alpha$-ketoglutarate + $H_2O$ | $-0.67$ | 2 |
| Acetate + $2\,H^+$ + $2\,e^-$ $\longrightarrow$ acetaldehyde + $H_2O$ | $-0.58$ | 2 |
| 3-Phosphoglycerate + $2\,H^+$ + $2\,e^-$ $\longrightarrow$ glyceraldehyde-3-P + $H_2O$ | $-0.55$ | 2 |
| $\alpha$-Ketoglutarate + $CO_2$ + $2\,H^+$ + $2\,e^-$ $\longrightarrow$ isocitrate | $-0.38$ | 2 |
| $NAD^+$ + $2\,H^+$ + $2\,e^-$ $\longrightarrow$ $NADH$ + $H^+$ | $-0.32$ | 2 |
| 1,3-Diphosphoglycerate + $2\,H^+$ + $2\,e^-$ $\longrightarrow$ glyceraldehyde-3-P + $P_i$ | $-0.29$ | 2 |
| Acetaldehyde + $2\,H^+$ + $2\,e^-$ $\longrightarrow$ ethanol | $-0.197$ | 2 |
| Pyruvate + $2\,H^+$ + $2\,e^-$ $\longrightarrow$ lactate | $-0.185$ | 2 |
| FAD + $2\,H^+$ + $2\,e^-$ $\longrightarrow$ $FADH_2$ | $-0.18\ddagger$ | 2 |
| Oxaloacetate + $2\,H^+$ + $2\,e^-$ $\longrightarrow$ malate | $-0.166$ | 2 |
| Fumarate + $2\,H^+$ + $2\,e^-$ $\longrightarrow$ succinate | $-0.031$ | 2 |
| **Standard hydrogen half-cell $E_0 = 0.00$§** | | |
| Cytochrome $b$ ($Fe^{3+}$) + $e^-$ $\longrightarrow$ cytochrome $b$ ($Fe^{2+}$) | $+0.06$ V | 1 |
| Ubiquinone + $2\,H^+$ + $2\,e^-$ $\longrightarrow$ ubiquinal | $+0.10$ | 2 |
| Cytochrome $c$ ($Fe^{3+}$) + $e^-$ $\longrightarrow$ cytochrome $c$ ($Fe^{2+}$) | $+0.22$ | 1 |
| Cytochrome $a$ ($Fe^{3+}$) + $e^-$ $\longrightarrow$ cytochrome $a$ ($Fe^{2+}$) | $+0.29$ | 1 |
| $Fe^{3+}$ + $e^-$ $\longrightarrow$ $Fe^{2+}$ | $+0.77$ | 1 |
| $\frac{1}{2}\,O_2$ + $2\,H^+$ + $2\,e^-$ $\longrightarrow$ $H_2O$ | $+0.816$ | 2 |

*$E_0'$ values are determined at pH 7.0 and 25°C relative to the standard hydrogen half-cell. For a two-electron reaction, a difference in $E_0'$ values of 0.10 V corresponds to a $\Delta G^{\circ\prime}$ of $-4.6$ kcal/mol.

†$n$ represents the number of electrons involved in the half-cell reaction.

‡The value for FAD/$FADH_2$ presumes the free coenzyme; when bound to a flavoprotein, the coenzyme has an $E_0'$ value in the range from 0.0 to $+0.3$ V, depending on the specific protein.

§The standard hydrogen half-cell requires that $[H^+] = 1.0\ M$ and therefore specifies pH 0.0. The reduction potential is therefore $E_0$ and not $E_0'$. At pH 7.0, the value for the $2\,H^+/H_2$ couple is $-0.421$ V.

The standard reduction potentials in Table 10-1 apply only to oxidation-reduction reactions proceeding under standard conditions of temperature, pH, and concentration. Under nonstandard conditions, the *Nernst equation* is used to calculate the actual reduction potential, $E'$:

$$E' = E_0' + \frac{2.303\, RT}{nF} \log \frac{\text{(electron acceptor)}}{\text{(electron donor)}}$$

At 25°C, the term $2.303\, RT/nF$ has the value 0.03 when $n = 2$. Thus, for a two-electron transfer, the $E'$ value is calculated from the $E_0'$ and the prevailing concentration of the electron-accepting and the electron-donating species by the simple expression

$$E' = E_0' + 0.03 \log \frac{\text{(electron acceptor)}}{\text{(electron donor)}}$$

## The Electron-Transport Chain

Having seen how reduction potentials are calculated and what they mean, we are now in a position to appreciate Figure 10-2, in which the major components of the electron-transport chain that link NAD ($E_0' = -0.32$ V) with $O_2$ ($E_0' = +0.816$ V) are ordered according to their reduction potentials. These carriers include flavin mononucleotide (FMN), coenzyme Q (CoQ), and the cytochromes $b$, $c_1$, $c$, $a$, and $a_3$, as well as several iron-sulfur proteins that are not shown in the figure. We will look at the chemical nature of these intermediates shortly; for the present, just note that the position of each intermediate in the sequence is determined by its reduction potential, such that *the transport chain is really a series of electron carriers arranged in order of increasing electron affinity.*

As it turns out, *the order of intermediates shown in Figure 10-2 expresses more than just the thermodynamic relationship between intermediates. It also reflects accurately their physical proximity to each other in the cell.* All these intermediates are membrane components. In the case of eukaryotic cells, they are localized within the inner mitochondrial membrane. For prokaryotes, they are found in the cytoplasmic membrane. In

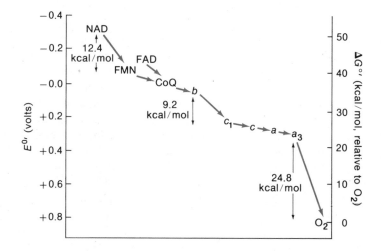

**Figure 10-2**
The energetics of electron transport. The major intermediates in the transport of electrons from NADH ( − 0.32 V) to oxygen ( + 0.82 V) are shown at positions appropriate to their energy level, as measured by the standard reduction potential. The right-hand axis provides a measure of the free-energy release upon oxidation by $O_2$. At three points along the chain, the electron transfer between successive carriers is sufficiently exergonic to drive the transport of protons upon which ATP generation depends, according to the chemiosmotic model.

either case, *the electron carriers are thought to be organized within the membrane into discrete clusters or complexes called respiratory assemblies*. Within each such respiratory assembly, the electron carriers appear to be physically positioned in such a way that electron flow always occurs as shown in Figure 10-2. Thus electrons from NADH are always passed first to FMN and then down the rest of the chain in the indicated order. Direct transfer from NADH to other carriers in the sequence simply does not occur, almost certainly because the physical ordering of the intermediates in the membrane is that dictated by their reduction potentials, and *transfer of electrons is possible only between contiguous carriers*.

Figure 10-2 also illustrates another important feature of the electron-transport chain that will become important to our understanding of its function in energy conservation and ATP synthesis. At several points along the chain, the transfer of electrons from one carrier to another is accompanied by a change in free energy that is large enough to drive the phosphorylation of ADP to form ATP. Three such sites can be readily identified: one from NAD to FMN, one from cytochrome $b$ to cytochrome $c_1$, and one from cytochrome $a_3$ to oxygen. This agrees well with the observation that three ATP molecules are formed as each pair of electrons flows exergonically from NADH to oxygen. Not at all apparent in Figure 10-2, however, is the mechanism whereby electron transport is coupled to ATP formation. According to the theory that is at present most widely accepted, this coupling depends on a directional pumping of protons from one side of the membrane to the other as electrons flow down the transport chain localized within the membrane. The proton motive force thereby established across the membrane is thought to be the immediate driving force responsible for the synthesis of ATP that must eventually accompany electron transport. This, too, we will have opportunity to look at in more detail shortly. For now, it will suffice that we note that electron transport within the membrane leads to a difference in electric charge and proton concentration across the membrane, and it is the resulting proton motive force that serves as the immediate driving force for ATP generation.

## The Electron Carriers of the Transport Chain

To understand the functioning of the electron-transport chain in more detail requires a familiarity with the carriers that comprise the chain. These include flavoproteins, iron-sulfur proteins, cytochromes, and a quinone. Except for the quinone, all are proteins with specific prosthetic groups capable of being reversibly oxidized and reduced. Because the whole process of electron transport in which these intermediates participate is localized in membranes, it is not surprising that these proteins and other carriers are hydrophobic molecules. This has made them difficult to isolate, solubilize, and study. As a result, considerable uncertainty still exists concerning details of the carriers and of the electron-transport process in which they participate. This is in fact an area of substantial current research interest. Our present understanding of electron transport and the carriers involved is likely to be correct in its overall formulation, but is almost certain to require considerable modification and clarification as research in this area continues.

**Flavoproteins.** Several flavin-linked enzymes participate in energy metabolism. The prosthetic group is either flavin mononucleotide (FMN) or flavin adenine dinucleotide (FAD). In either case, it is the isoalloxazine ring of the riboflavin that actually accepts the electrons (see Figure 6-12). In most cases, the flavin nucleotide is tightly (although not always covalently) bound to the protein and does not leave the enzyme surface after the electron-transfer reaction. For this reason, flavins are sometimes referred to as prosthetic groups rather than as coenzymes. Examples of flavoproteins that have already been encountered in aerobic energy metabolism include succinate dehydrogenase (TCA-6 of Chapter 9), the dihydrolipoyl dehydrogenase activity associated with the oxidation of both pyruvate and $\alpha$-ketoglutarate (see Figure 9-3), and the fatty acyl-CoA dehydrogenase involved in $\beta$ oxidation of fatty acids (FA-2 of Chapter 9).

The specific flavoprotein involved directly in the electron-transport chain is NADH dehydrogenase (also called NADH-Q reductase), a multimeric enzyme with at least 16 subunits. Unlike the preceding dehydrogenases, this enzyme uses FMN rather than FAD as its prosthetic group. NADH dehydrogenase contains a second type of prosthetic group, an iron-sulfur center, and is therefore also an iron-sulfur protein, as discussed below. Electrons are initially accepted by NADH dehydrogenase on its FMN group and then are transferred by means of several iron-sulfur complexes to coenzyme Q.

**Iron-Sulfur Proteins.** These are a family of proteins containing iron and sulfur atoms complexed with four cysteine residues of the protein. They are also called _nonheme iron proteins_ to distinguish them from the better-known heme proteins that also contain iron, but as part of a heme prosthetic group (as in the case of the cytochromes to be discussed shortly). The iron and sulfur atoms of the nonheme iron proteins are usually present in equimolar amounts ($Fe_2S_2$ and $Fe_4S_4$ are the most common), but some iron proteins contain a single iron atom tetrahedrally coordinated to the sulfhydryl groups of four cysteine residues. The iron atoms in the iron-sulfur centers of these proteins can be in either the oxidized ($Fe^{3+}$) or the reduced ($Fe^{2+}$) state and are the actual electron acceptors and donors when an iron-sulfur protein is functioning as an electron carrier.

The best-known examples of iron-sulfur proteins are the ferredoxins found in bacteria and chloroplasts, but it is now clear that the mitochondrial electron-transport chain also contains iron-sulfur proteins. NADH dehydrogenase, for example, is an iron-sulfur protein possessing both $Fe_2S_2$ and $Fe_4S_4$ centers (see Figure 6-26). Iron-sulfur proteins are also involved in the transfer of electrons from succinate to coenzyme Q and from coenzyme Q to cytochrome $c$.

**Cytochromes.** Better known than the iron-sulfur proteins, cytochromes also contain iron, but always as part of the porphyrin prosthetic group commonly called _heme_ (see Figure 10-3). The cytochromes are unique to aerobic cells and play a prominent role in the transfer of electrons from coenzyme Q to oxygen. Some cytochromes occur in the endoplasmic reticulum, where they are involved in a variety of hydroxylation reactions, but interest here focuses on those of the respiratory chain.

Cytochromes were initially classified by D. Keilin into three groups designated $a$, $b$, and $c$, based on the relative positions of their characteristic $\alpha$, $\beta$, and $\gamma$ (or Soret) absorption bands in the visible range (Figure 10-4). Current practice, however, is to designate new cytochromes by specifying the actual wavelength of the $\alpha$ band. The mitochondrial electron-transport chain of higher animals contains at least five different cytochromes, designated by the original Keilin nomenclature as cytochromes $b$, $c$, $c_1$, $a_1$, and $a_3$. Cytochromes $b$, $c$, and $c_1$ all contain iron-protoporphyrin IX, the same prosthetic group found in hemoglobin and myoglobin (Figure 10-3). Cytochromes $a_1$ and $a_3$, however, contain a modified prosthetic group called heme A, differing from the heme of the other cytochromes by the presence

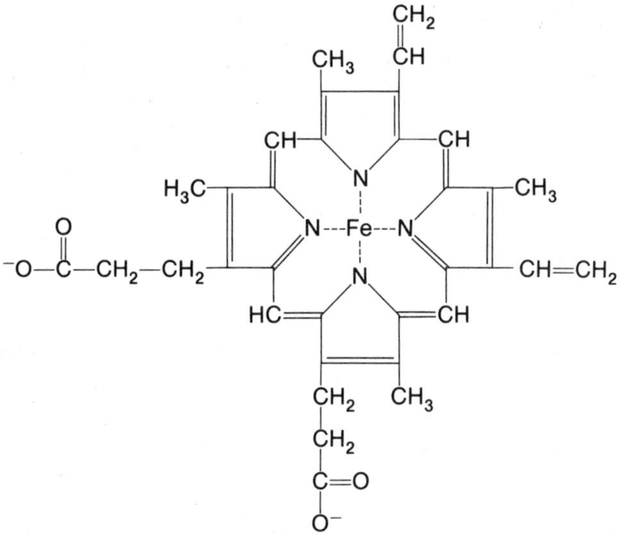

**Heme**
**(iron-protoporphyrin IX)**

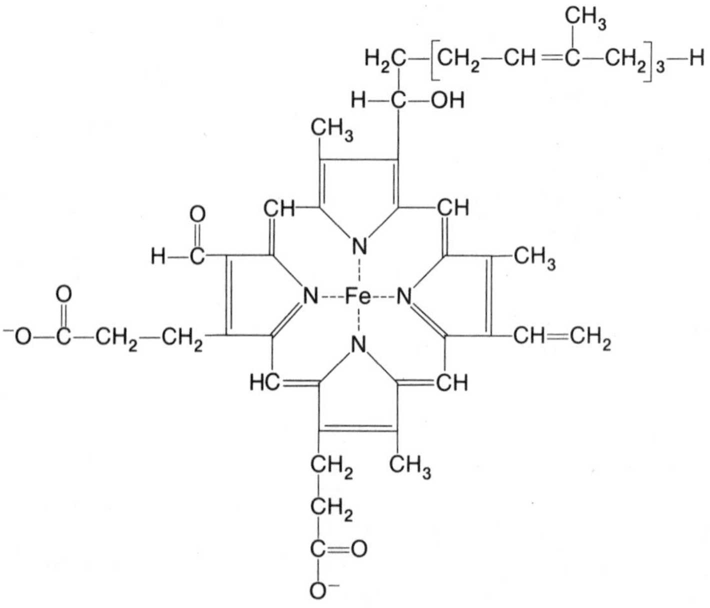

**Heme A**

**Figure 10-3**
Structures of heme (*upper*) as found in cytochromes $b$ and $c$ and of heme A (*lower*) as found in cytochromes $a$ and $a_3$. The heme of cytochromes $c$ and $c_1$ (but not of cytochromes $b$, $a$, or $a_3$) is covalently attached to the protein by thioester bonds between the sulfhydryl groups of two cysteine residues in the protein and the vinyl groups of the heme.

of a formyl group instead of a methyl group at position 8 and a long hydrophobic chain instead of a vinyl group at position 2 (Figure 10-3).

Cytochromes $b$, $c_1$, $a_1$, and $a_3$ are integral membrane proteins, the latter two occurring together at the end of the transport chain as a complex called cytochrome $c$ oxidase. The hydrophobic nature of these cytochromes and their tight association with the inner mitochondrial membrane make them difficult proteins to solubilize and study. Cytochrome $c$, however, is soluble in water and is a peripheral membrane protein. This property allows cytochrome $c$ to function as a carrier of electrons between two segments of the electron-transport chain. Because cytochrome $c$ is a small, soluble protein ($M_r = 13,000$), its amino acid sequence has been determined for a wide variety of species. It is therefore an important protein in studies that attempt to elucidate evolutionary relationships between species based on amino acid sequences of common proteins.

Like the iron of the iron-sulfur complex, that of the heme prosthetic group is also reversibly oxidizable and serves as the actual electron acceptor for the cytochrome. Both cytochromes and iron-sulfur proteins are therefore one-electron carriers.

**Coenzyme Q (Ubiquinone).** Coenzyme Q, also called _ubiquinone_ because of its ubiquitous occurrence, is the only nonprotein component of the electron-transport chain. It was first characterized in 1958, and it was shown to be a benzoquinone with an isoprenoid side chain (see Figure 10-5). Several ubiquinones are known that differ only in the number of isoprenoid units contained on the side chain. Chain length is specified by subscripting the number of isoprenoid units to the Q, which is the official abbreviation for coenzyme Q. Thus $Q_{10}$ stands for a coenzyme Q molecule with 10 isoprenoid units in its side chain. $Q_{10}$ is in fact the most common form of the carrier in mammalian mitochondria.

Coenzyme Q can be reversibly reduced to the hydroquinone, as shown in Figure 10-5. This is the basis for its function in electron transport. As Figure 10-2 indicates, coenzyme Q occupies a central position in the electron-transport chain, since it is the collection point for electrons from both NADH and the FADH$_2$-linked dehydrogenases. Coenzyme Q is in fact the only electron carrier in the whole transport chain that is not tightly bound or covalently linked to a protein. This allows coenzyme Q to play a strategic role as a mobile carrier of electrons between flavoproteins and cytochromes.

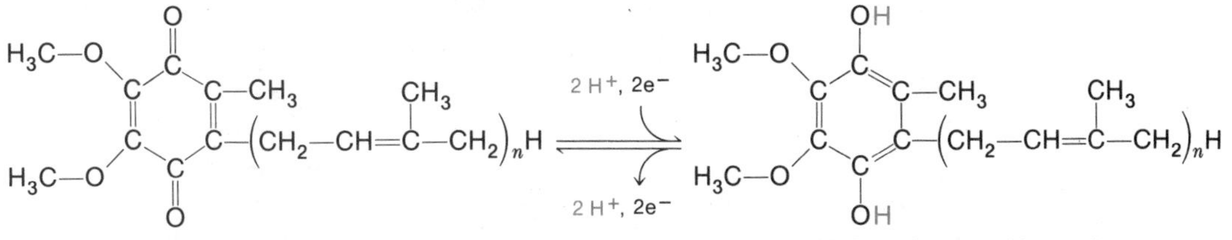

| | $\gamma$ | $\beta$ | $\alpha$ |
|---|---|---|---|
| Cytochrome $a$: | 439 | — | 600 |
| Cytochrome $b$: | 429 | 532 | 563 |
| Cytochrome $c$: | 415 | 521 | 550 |
| Cytochrome $c_1$: | 418 | 524 | 554 |

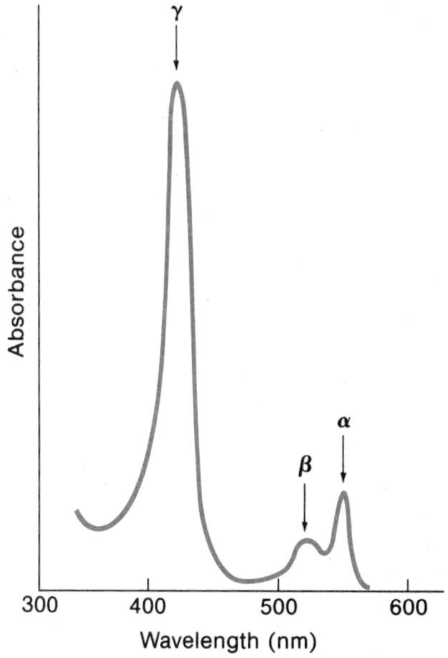

**Figure 10-4**
The absorption spectrum for cytochrome $c$. Indicated in the table above the spectrum are the wavelengths of the $\alpha$, $\beta$, and $\gamma$ (Soret) peaks for cytochromes $a$, $b$, and $c_1$ also.

**Coenzyme Q** (ubiquinone)　　　　　　**Hydroquinone** (ubiquinol)

**Figure 10-5**
Structure and oxidation-reduction of coenzyme Q. The side chain can have a variable number of isoprenoid units. $Q_{10}$, the most common form of the coenzyme in mammalian tissue, has $n = 10$. For bacteria, $Q_6$, with $n = 6$, is the predominant form. Reduction of coenzyme Q (ubiquinone) to the corresponding hydroquinone (ubiquinol) is a two-electron process, with a semiquinone as the postulated one-electron intermediate.

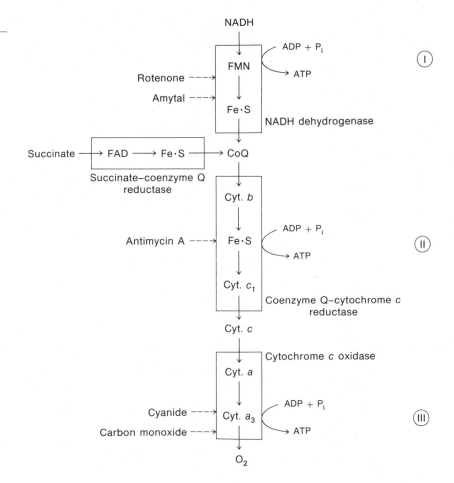

**Figure 10-6**
Ordering of intermediates in the electron-transport chain. The four major enzyme complexes are indicated in boxes, with the strategic roles of coenzyme Q and cytochrome c thereby emphasized. Sites of inhibition are indicated by dotted lines. Roman numerals identify the three sites of coupling with ATP synthesis.

## Determining the Sequence of the Carriers

Ordering the foregoing components of the electron-transport chain into their correct sequence has been the subject of a great deal of research effort over the past half-century or more. Figure 10-6 depicts our current understanding of carrier sequence and therefore represents the fruit of such work. Essentially, *four different approaches have proven useful in determining the sequence of intermediates in functional mitochondria: determination of reduction potentials, measurement of difference spectra, use of inhibitors, and isolation of submitochondrial membrane fragments that retain an ability to carry out some portion of the overall transport process.* Each of these will be discussed briefly before we look at the detailed sequence of events that has been elucidated by these techniques.

**Reduction Potentials.** As mentioned earlier, we ought to expect the sequence of electron carriers to be such that their reduction potentials increase steadily from NADH ($E_0' = -0.32$ V) to $O_2$ ($E_0' = +0.816$ V), since electron transfer must be thermodynamically favored at every step. This is complicated somewhat by the drastic change in environment that an electron carrier undergoes when it is extracted from its usual location within the membrane and solubilized for experimental purposes, since the reduction potential can be very substantially influenced by the environment in which the molecule finds itself. Nonetheless, there appears to be good general agreement between experimentally determined reduction potentials and the order of intermediates within the chain, as can be seen by comparing Figures 10-2 and 10-6.

**Difference Spectra.** An important advance in determining carrier sequence came with the development by B. Chance of difference spectrophotometry. This is a clever technique that allows assessment of the redox status of specific mitochondrial electron carriers within the intact organelle, provided only that a specific carrier undergoes a change in absorbance at some characteristic wavelength when it is converted from the oxidized to the reduced form, or vice versa. Standard spectrophotometry is inadequate for measuring the absorption spectra of electron carriers in intact mitochondria because the preparation is turbid and will absorb and scatter far too much light extraneously. However, by determining the absorbance of a mitochondrial suspension against that of a control suspension in which all the carriers have been deliberately converted to their oxidized form, it is possible to "see" spectrophotometrically only genuine absorbance differences between the two preparations, with all the background absorption and scattering common to both preparations eliminated. By varying wavelength, a _difference spectrum_ can be obtained that is a plot versus wavelength of the difference between the spectra of the reduced and the oxidized forms of the various carriers, each with one or more peaks at specific wavelengths (Figure 10-7). When such mitochondria are allowed to

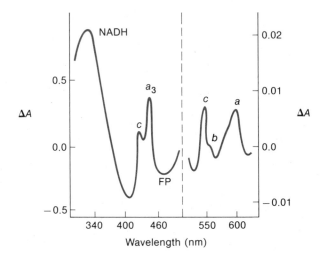

**Figure 10-7**
A difference spectrum of the electron carriers in isolated mitochondria. Plotted as a function of wavelength is the difference in absorbance ($\Delta A$) between a mitochondrial suspension maintained in the reduced state (excess substrate, no oxygen) and a control suspension kept in a fully oxidized state (fully saturated with oxygen). Most of the carriers, including the cytochromes $a$, $a_3$, $b$, and $c$, absorb more strongly in the reduced form and appear as peaks. Flavoprotein (FP) appears as a trough because it absorbs less strongly in the reduced form. Note the difference in scale for $\Delta A$ on the left and right ordinates which refer to the carriers on the left and right sides of the plot, respectively (separated by a dashed line).

oxidize substrate aerobically, the electron carriers on the substrate end of the chain will be largely in the reduced form, while those at the oxygen end will be mainly in the oxidized form. As an alternative approach, difference spectra can be monitored after oxygen is admitted to previously anaerobic suspensions of mitochondria. In this case, all carriers will start out fully reduced, and their sequence in the chain can be deduced from the order in which they become reoxidized. Further information can be obtained with this technique by the use of artificial electron donors that introduce electrons at specific points along the chain.

**Transport Inhibitors.** A great deal of information on carrier sequence has been obtained by the use of inhibitors that block specific transfer steps in the chain. The actual site of inhibition is usually determined by difference spectra made of inhibited versus uninhibited mitochondrial preparations. *Once the site of action is known for a specific inhibitor, it becomes possible to deduce whether a given carrier is "downstream" or "upstream" from that site of inhibition depending on whether the carrier accumulates in the oxidized or reduced form in the presence of the inhibitor.*

The structures of several such inhibitors are shown in Figure 10-8, and the site of inhibition for each is indicated in Figure 10-6. Most transport inhibitors fall into one of three categories depending on where the site of

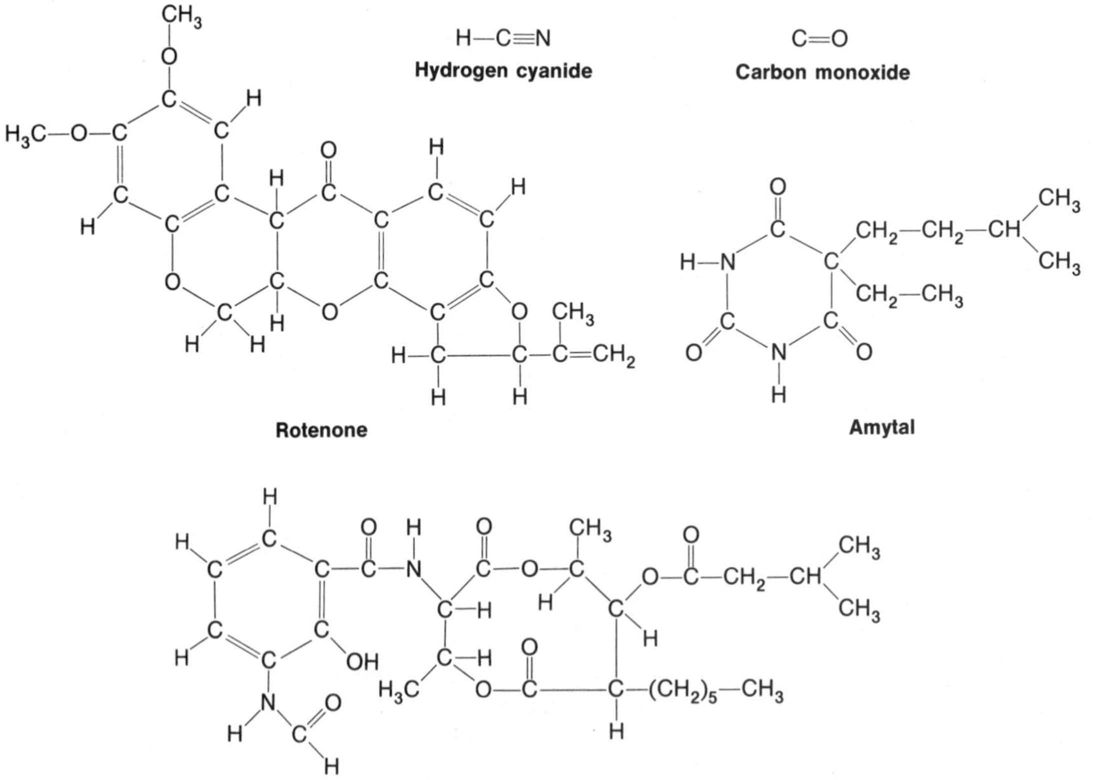

**Figure 10-8**
The structures of some inhibitors of electron transport. Rotenone and amytal both block electron transport between NADH and coenzyme Q (site I), antimycin A blocks between coenzyme Q and cytochrome *c* (site II), and HCN and CO block at the final step of electron transport from cytochrome oxidase to oxygen (site III).

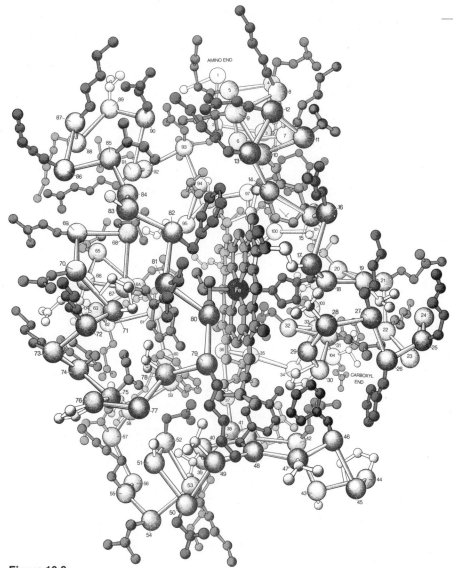

**Figure 10-9**
The three-dimensional structure of cytochrome *c* from horse heart. The planar
heme group is located in the center of the molecule, with its iron atom bonded to
the sulfur atom of the methionine residue at position 80 and to the nitrogen atom
of the histidine residue at position 18. The polypeptide chain is wrapped around
the heme, giving the heme a very hydrophobic environment and a correspondingly more positive reduction potential than it would have in an aqueous medium.

inhibition is located along the chain. *Rotenone* is an example of an inhibitor that acts near the coenzyme (reduced) end of the chain. It is a plant
toxin that is extremely potent as a fish poison and an insecticide. *Antimycin A*, an antibiotic isolated from *Streptomyces griseus*, blocks in the
middle of the chain, while both hydrogen cyanide and carbon monoxide
act at the oxygen end of the chain between the final cytochrome carrier
and oxygen. It is in fact blockage of the vital respiratory link to oxygen that
makes cyanide compounds such a potent poison for aerobes. Carbon mon-

oxide, however, is toxic because of its affinity for the oxygen binding sites of hemoglobin. It therefore kills by preventing delivery of oxygen to tissues rather than by poisoning the respiratory chain.

**Submitochondrial Particles.** Yet another approach in ordering transport intermediates involves disruption of mitochondrial membranes and isolation of membrane fragments that are still able to carry out some limited portion of the overall transport process. Disruption can be achieved by homogenization, sonication, or detergent treatment. _Digitonin_ is a detergent of choice for this purpose, because it solubilizes the outer mitochondrial membrane, while the fragmental inner membrane still retains electron-transport activity. Each of the four major complexes shown in Figure 10-6 has been isolated from mitochondria after extraction of submitochondrial particles with detergents.

## Organization and Function of the Electron-Transport Chain

From studies of reduction potentials, difference spectra, inhibitor effects, and submitochondrial particles, the various intermediates of the electron-transport chain have been ordered into a sequence on which there is good general agreement, although the location of some of the intermediates, especially the iron-sulfur centers, is still somewhat tentative. _As shown in Figure 10-6, the transfer of electrons from reduced coenzymes to molecular oxygen can be considered to occur in three successive membrane-bound transfer sequences,_ with NADH, coenzyme Q, and cytochrome _c_ (Figure 10-9) involved as important intermediates in the process. Thus electrons flow from NADH to coenzyme Q, from coenzyme Q to cytochrome _c_, and from cytochrome _c_ to oxygen. The enzyme complexes responsible for these three segments of the transport chain are NADH dehydrogenase, coenzyme Q–cytochrome _c_ reductase, and cytochrome _c_ oxidase, respectively. In each case, electron transport is accompanied by proton uptake and/or release, so that _the passage of electrons through the complex is coupled obligatorily to the pumping of protons across the membrane in which the complex is embedded._ Each of these complexes therefore represents an energy-conserving site along the chain, since it is at these three sites that the free energy of exergonic electron transfer is coupled with the pumping of protons and therefore with the electrochemical energy that is stored as a proton motive force at the membrane surface.

In addition to the three complexes needed for electron transport from NADH to oxygen, another enzyme complex, succinate-coenzyme Q reductase, is involved in transferring to coenzyme Q the electrons derived from succinate. (Similar but separate complexes are required for transfer to coenzyme Q of electrons from other FAD-linked dehydrogenases, such as those involved in the oxidation of fatty acids or glycerol phosphate.) Each of these portions of the transport chain will be considered in turn.

**From NADH to Coenzyme Q: The NADH Dehydrogenase Complex.** Electrons from the reduced coenzyme NADH are passed to coenzyme Q by the NADH dehydrogenase complex. As noted earlier, NADH dehydrogenase is both a flavoprotein (with FMN as the prosthetic group) and an iron-sulfur protein (with both $Fe_2S_2$ and $Fe_4S_4$ centers). The NADH dehydrogenase transport sequence can therefore be represented as follows:

$$\text{NADH} \searrow \text{H}^+ \nearrow \text{FMN} \rightleftarrows \text{2 Fe}^{2+} \searrow \text{2 H}^+ \nearrow \text{CoQ}$$
$$\text{Fe} \cdot \text{S}$$
$$\text{NAD}^+ \nearrow \searrow \text{FMNH}_2 \rightleftarrows \text{2 Fe}^{3+} \nearrow \searrow \text{CoQH}_2$$
$$\text{2 H}^+$$

Both the FMN prosthetic group and the iron-sulfur centers of the NADH dehydrogenase complex participate in the electron-transport process. The role of the iron-sulfur centers in mediating electron flow from $FMNH_2$ to coenzyme Q is well-documented. In addition, one of the iron-sulfur centers has been implicated in the initial transfer of electrons from NADH to FMN, but this is not included in the simplified mechanism as shown.

Important in the preceding scheme is the alternation in electron carriers between those (such as FMN) which accept protons as well as electrons and those (such as the FeS centers) which need only electrons. This means that electron flow from NADH to coenzyme Q is coupled with the uptake and release of protons. NADH dehydrogenase is an integral membrane protein, and its orientation in the membrane may be such that transfers requiring protons draw them from one side of the membrane, while those which release protons do so to the other side. It has not been shown directly that the protons used in the alternate reduction and oxidation of these carriers are actually the protons pumped from one side of the membrane to the other. However, the NADH dehydrogenase complex clearly couples proton pumping with electron flow, and the direct coupling implied by this model has been suggested as a likely mechanism. According to this model, passage of a pair of electrons from NADH to coenzyme Q would result in the pumping of two protons from one side of the membrane to the other. The actual stoichiometry of proton pumping per site is still a very controversial issue, however.

**From $FADH_2$ to Coenzyme Q: The Succinate–Coenzyme Q Reductase Complex.** In addition to accepting electrons from NADH by means of the flavoprotein NADH dehydrogenase, coenzyme Q also serves as the central collection point for electrons from the prosthetic groups of a variety of dehydrogenases that use covalently bound FAD rather than free $NAD^+$ as the electron acceptor. A key example is succinate dehydrogenase, the only FAD-linked dehydrogenase of the TCA cycle (see TCA-6 in Chapter 9). Succinate dehydrogenase, unlike the other enzymes of the TCA cycle, is an integral membrane protein. It is in fact intimately associated in the membrane with an iron-sulfur protein. Together, the two proteins make up the succinate–coenzyme Q reductase complex that is responsible for transport of electrons from succinate to coenzyme Q. The preceding scheme illustrates

$$\text{Succinate} \searrow \nearrow \text{FAD} \rightleftarrows \text{2 Fe}^{2+} \searrow \text{2 H}^+ \nearrow \text{CoQ}$$
$$\text{Fe} \cdot \text{S}$$
$$\text{Fumarate} \nearrow \searrow \text{FADH}_2 \rightleftarrows \text{2 Fe}^{3+} \nearrow \searrow \text{CoQH}_2$$
$$\text{2 H}^+$$

trates some of the present uncertainties in our understanding, because it implies the possibility of proton translocation in this segment of the electron-transport chain, but such a coupling site does not exist. Other FAD-linked dehydrogenases are also integral membrane proteins, and they transfer their electrons to coenzyme Q in an analogous manner. These include fatty acyl-CoA dehydrogenase (FA-4 in Chapter 9) and glycerol phosphate dehydrogenase (discussed below).

**From Coenzyme Q to Cytochrome *c*: The Coenzyme Q–Cytochrome *c* Reductase Complex.**  Whether collected from free NADH or enzyme-bound $FADH_2$, the electrons of reduced coenzyme Q are eventually passed to cytochrome *c* by means of a membrane-bound complex that contains two cytochromes and an iron-sulfur protein. The complex is called coenzyme Q–cytochrome *c* reductase, and the transfer can be represented as follows:

$$
\begin{array}{ccccccc}
CoQH_2 & 2\,Fe^{3+} & 2\,Fe^{2+} & 2\,Fe^{3+} & 2\,Fe^{2+} \\
 & Cyt.\ b & Fe\cdot S & Cyt.\ c_1 & Cyt.\ c \\
CoQ & 2\,Fe^{2+} & 2\,Fe^{3+} & 2\,Fe^{2+} & 2\,Fe^{3+} \\
2\,H^+
\end{array}
$$

Two molecules of cytochrome *b* are involved in the oxidation of reduced coenzyme Q, since the former carries a single electron, while the latter carries two. The postulated intermediate in the stepwise oxidation of $CoQH_2$ is $CoQH\cdot$, a free-radical semiquinone. Proton release appears to occur at a single site in this sequence, since all the transfers beyond cytochrome *b* involve electrons only. Again, however, the scheme illustrates well how incomplete our understanding of energy coupling is, because no net translocation of protons seems possible in this segment of the electron-transport sequence, yet this complex is known to be a coupling site. This has led some investigators to postulate an additional intermediate X, as shown in Figure 10-13, but there is little or no direct evidence to suggest that X actually exists.

Unlike the other cytochromes of the respiratory chain, cytochrome *c* is a peripheral rather than an integral membrane protein. Because it can be easily solubilized, cytochrome *c* has been purified and crystallized and its structure is known in great detail. Shown in Figure 10-9 is the structure of cytochrome *c* from horse heart. It consists of a single polypeptide with 104 amino acids and a heme group. The molecule is approximately spherical, with the heme sequestered in the hydrophobic interior of the molecule and the polypeptide chain more or less wrapped around it. The iron atom of the heme is bonded to a nitrogen atom of the histidine residue at position 18 of the polypeptide and to the sulfur atom of the methionine residue at position 80. The recognition and binding of cytochrome *c* by the reductase and oxidase complexes of the transport chain appears to depend crucially on clusters of positively charged residues, especially of lysine, on the surface of the enzyme. One of these clusters is thought to interact with the reductase and another with the oxidase. Several of the lysine and arginine residues of these clusters are among the 26 invariant residues of the protein for which sequencing studies with more than 80 eukaryotic species have shown the amino acid at that position to be constant across the phylogenetic spectrum.

**From Cytochrome *c* to Oxygen: The Cytochrome *c* Oxidase Complex.** The last segment of the electron-transport sequence carries out the oxidation of reduced cytochrome *c* at the expense of molecular oxygen. The complex is called cytochrome *c* oxidase and consists of cytochromes *a* and $a_3$. In addition to its heme prosthetic group, cytochrome $a_3$ contains a copper atom that also participates in electron transfer, probably between the heme of cytochrome $a_3$ and molecular oxygen. The sequence can therefore be represented as follows:

$$
\begin{array}{ccccc}
2\ Fe^{3+} & 2\ Fe^{2+} & 2\ Fe^{3+} & 2\ Cu^{+} & 2\ H^{+}\ \ \tfrac{1}{2}\ O_2 \\
Cyt.\ c & Cyt.\ a & Cyt.\ a_3 & Cyt.\ a_3 & \\
2\ Fe^{2+} & 2\ Fe^{3+} & 2\ Fe^{2+} & 2\ Cu^{2+} & H_2O
\end{array}
$$

Of all the electron-transferring intermediates involved in respiratory metabolism, only the cytochrome $a_3$ component of cytochrome *c* oxidase is a _terminal oxidase_ capable of direct transfer of electrons to molecular oxygen. Almost every electron extracted from any oxidizable organic molecule anywhere in the aerobic cell must eventually pass through cytochrome $a_3$, for this is the only link between respiratory metabolism and the oxygen that makes it all possible.

## Summary of Electron Transport

Taken together, the preceding components comprise the complete electron-transport chain that provides for an orderly, stepwise transfer of electrons from reduced coenzymes to molecular oxygen. In the process, _the total free-energy difference between reduced coenzymes and oxygen is partitioned over a number of individual electron transfers, several of which are coupled with the pumping of protons and thereby with the establishment and maintenance of a proton motive force that is thought to be responsible for ATP generation_. The process of electron transport therefore accounts for the following aspects of aerobic energy metabolism:

1. Reoxidation of reduced coenzymes, thereby providing for the continuous regeneration of the pool of acceptor molecules required for oxidative reactions within the cell.

2. Consumption of oxygen, obviously a key feature of aerobic energy metabolism.

3. Formation of water, one of the expected end products of cellular respiration.

4. Stepwise release of the free energy of coenzyme oxidation.

5. Coupled unidirectional pumping of protons across the membrane in which the transport intermediates are embedded, such that an electrochemical potential for protons is established and maintained by the transport process.

# OXIDATIVE PHOSPHORYLATION

We are now ready to examine the mechanism whereby the exergonic transport of electrons from coenzymes to oxygen is coupled with the generation of ATP, as we know it must be if we are eventually to account for the high ATP yield of aerobic energy metabolism. *Since these ATPs are formed by phosphorylated reactions coupled with oxidative electron transport, the process is called oxidative phosphorylation.* It is thereby distinguished from substrate-level phosphorylation, which occurs as an integral part of a specific reaction in a metabolic pathway. (The generation of ATP at Gly-7 and Gly-10 of the glycolytic pathway and at TCA-5 during the TCA cycle are examples of substrate-level phosphorylation.) A great deal of controversy has arisen from research on oxidative phosphorylation over the years, centered primarily on the mechanism responsible for the coupling of electron transport with ATP generation. Alternative models still have their ardent proponents, but it is now generally agreed that the crucial link between electron transport and ATP formation is the proton motive force established by the transmembrane proton movement that accompanies electron transport. We will return to this model shortly, but first we need some appreciation for the historical background out of which the concept arose.

## P/O Ratios and Sites of ATP Synthesis

That oxygen consumption by aerobic cells was linked somehow to the generation of ATP from ADP and inorganic phosphate has been known for a long time. Since the work of S. Ochoa in the early 1940s, it has been possible to express this relationship in terms of the *P/O ratio*, which quantitates the number of moles of ATP generated per atom of oxygen utilized in respiration. Since two electrons are required per atom of oxygen reduced, *the P/O ratio corresponds to the number of molecules of ATP generated as a pair of electrons passes through the chain from reduced coenzyme (or another electron donor) to oxygen.* Few areas of research have been fraught with as much uncertainty and strong emotion over the years as this one, mainly because of the experimental difficulties encountered in making these measurements properly. However, it is generally agreed that the oxidation of NADH by electron transport to oxygen occurs with a P/O ratio of 3 (which, incidentally, was the original value announced by Ochoa back in 1941 for the NAD-linked oxidation of pyruvate to acetyl-CoA). This led to the concept of three discrete sites of ATP generation en route from NADH to oxygen (see Figure 10-6). These have been localized to specific energy-conserving regions along the transport chain by a variety of ingenious experiments in which electrons were allowed to flow through only a portion of the chain by judicious use of artificial electron acceptors and/or donors in the presence or absence of specific transport inhibitors.

The first of the three sites, site I, was localized to the segment of the chain between NADH and coenzyme Q when it was found that the P/O ratio for succinate oxidation (which funnels electrons into coenzyme Q by means of $FADH_2$) was only 2. Confirmation of this was provided by using ferricyanide as an artificial electron acceptor in the presence of antimycin

A, an inhibitor known to block transport between cytochromes $b$ and $c_1$ (see Figure 10-6). The result was a P/O ratio of 1 for NADH oxidation. Resolution of sites II and III became possible when A. Lehninger showed that ascorbate could be used to introduce electrons into the chain at cytochrome $c$ with a resulting P/O ratio of 1. This localized site II between coenzyme Q and cytochrome $c$ and placed site III between cytochrome $c$ and oxygen. This localization of site II was further strengthened by the finding that the passage of electrons from succinate to cytochrome $c$ resulted in the generation of one ATP per pair of electrons.

As a result of these and similar findings, it has been possible to associate one ATP-generating event with each of what are now recognized as the three main enzyme complexes of electron transport: NADH dehydrogenase (site I), cytochrome $c$ reductase (site II), and cytochrome $c$ oxidase (site III). From Figure 10-2 it is clear that each of these three segments of the transport chain is characterized by a sufficiently large free-energy change to drive the ATP generation attributable to it. Confirmation of these complexes as ATP-generating sites has been provided by the reconstitution experiments of E. Racker and colleagues. Each complex has, in turn, been incorporated with the mitochondrial ATP-synthesizing complex into synthetic phospholipid vesicles. Upon addition of the appropriate oxidizable substrate, such vesicles are capable of generating one ATP per pair of electrons that passes through the complex.

## Coupling of ATP Synthesis with Electron Transport

Under normal physiological conditions, electron transport is tightly coupled with phosphorylation. *Not only is ATP generation critically dependent on electron flow, but the tightness of coupling also means that electron flow occurs only when ATP can be synthesized.* This is an important regulatory mechanism, since it means that the [ATP]/[ADP] ratio in the cell plays an important role in regulating the activity of the electron-transport chain (and, by means of accumulation or depletion of reduced coenzymes, that of the entire respiratory apparatus). The requirements for oxidative phosphorylation are NADH (or another electron source), $O_2$, ADP, and $P_i$. Of these, the most important for actually regulating the rate of oxidative phosphorylation is ADP, since the other three are rarely present in rate-limiting concentrations. This regulation of the rate of oxidative phosphorylation by ADP is called *respiratory control*. Its physiological significance should be clear. When ATP is being consumed rapidly to perform cellular work, the level of ATP drops, the level of ADP increases, and the availability of ADP means that phosphorylation is favored both thermodynamically and kinetically. ATP synthesis is therefore stimulated, electrons are transported, coenzymes are recycled, substrates are oxidized, and oxygen is consumed. However, when ATP accumulates in the cell, the concentration of ADP drops reciprocally, electron transport is slowed or halted, coenzymes build up in the reduced form, oxidized coenzymes are no longer available as electron acceptors, and the whole process of respiratory metabolism is effectively throttled. *Oxidative phosphorylation is therefore coupled carefully with cellular ATP needs, such that electron flow from organic fuel molecules to oxygen is adjusted to the energy needs of the cell by ADP-mediated respiratory control.*

## Respiratory States and Conformational Changes

By measuring the rate of oxygen consumption by isolated mitochondria, B. Chance and G. R. Williams were able to define five states of respiratory function, as shown in Figure 10-10. In the absence of both oxidizable substrate and ADP, mitochondria display state 1 respiration, with a low rate of oxygen consumption attributable to oxidation of endogenous substrates. Addition of ADP leads to state 2 respiration, an initial but short-lived stimulation of endogenous substrate oxidation. If both ADP and an oxidizable substrate are added, state 3 respiration is attained, characterized by a rapid rate of oxygen consumption that continues until the ADP is consumed. At that point, oxygen uptake again slows down and the mitochondria are said to be in state 4. State 5 occurs under the special condition of oxygen depletion, in which case no respiration is possible regardless of what other components of the system are present.

Of these five states, only states 3 and 4 are commonly mentioned in the current literature. The dependence of state 3 respiration on the presence of ADP is further evidence of respiratory control by ADP. Respiratory control can, in fact, be quantitated by determining the ratio of the rate of oxygen consumption in the presence of ADP (state 3) to that observed in its absence (state 4). For intact mitochondria with tight coupling of phosphorylation with electron transport, this ratio can be as high as 10 or more. In damaged mitochondria, it can be as low as 1, indicating that the rate of electron transport is virtually independent of ATP generation (i.e., little or no coupling).

Ultrastructural studies have since led to the suggestion that each of these several respiratory states is characterized by a distinctive morphologic appearance. For example, state 4 mitochondria display an orthodox configuration such as is seen in most electron micrographs of fixed tissue. State 3 mitochondria, however, are characterized by a condensed configuration. The transition from orthodox to condensed configurations involves a decrease in the internal (matrix) volume of the organelle and a corresponding increase in the space between the inner and outer membranes. Such reversible changes in ultrastructural appearance have been invoked by some investigators as support for the hypothesis (to be discussed shortly) that such conformational changes are an integral part of the mechanism whereby ATP generation is coupled with electron transport.

## Uncouplers and Inhibitors of Oxidative Phosphorylation

Studies of oxidative phosphorylation have been greatly aided by the availability of a variety of agents that affect the process in several different ways. Two different types of compounds are known that are classified either as uncouplers or as inhibitors of oxidative phosphorylation.

**Uncouplers.** Uncouplers of oxidative phosphorylation function, as the name suggests, by dissociating the electron-transport process from the generation of ATP with which it is normally tightly coupled. An uncoupling agent therefore frees electron transport from its normal throttle and allows it to proceed at an uncontrolled pace, unaccompanied by ATP production

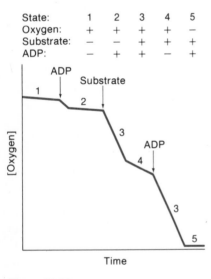

| State: | 1 | 2 | 3 | 4 | 5 |
|---|---|---|---|---|---|
| Oxygen: | + | + | + | + | − |
| Substrate: | − | − | + | + | + |
| ADP: | | − | + | + | − | + |

**Figure 10-10**
The five states of mitochondrial respiration. The plot is of oxygen consumption of a mitochondrial suspension. Addition of ADP and substrate is indicated by the arrows. Inorganic phosphate is present in excess throughout.

(See Figure 10-11*a*). The result is an excessive consumption of oxygen, an unimpeded utilization of substrate, and dissipation of energy as heat. The best-known such uncoupler is 2,4-dinitrophenol (DNP), but other acidic aromatic compounds also are effective (see Figure 10-12 for the structures of several). The uncoupling effect of these compounds is specific for the respiratory chain; they are without effect on substrate-level phosphorylation. This has made them very useful reagents for studying oxidative phosphorylation.

Although DNP and related compounds are of experimental interest only (and can, in fact, be fatal if consumed), physiological uncoupling of oxidative phosphorylation is known and is useful as a means of generating heat. This is important in such diverse species as hibernating animals, cold-adapted mammals, and plants characterized by emergence in early spring when they must literally melt their way through the snow. An especially dramatic example is the spadix, or floral spike, of the skunk cabbage *Symplocarpus foetidus*, which can maintain a temperature of 10 to 25°C above ambient. In animals, this process of thermogenesis is characteristic of brown adipose tissue that is rich in mitochondria.

**Inhibitors.**  Agents in this class prevent both oxygen consumption and ATP generation, but they do so without any direct inhibitory effect on the electron carriers of the transport chain and can thereby be differentiated from inhibitors of electron transport, such as rotenone, antimycin A, or cyanide. Instead, they appear to interfere directly with the process of ATP synthesis. *Oligomycin* is an example of this group of inhibitors. It is possible to distinguish between such inhibitors of oxidative phosphorylation and those which block electron transport directly by observing the effect of adding an uncoupling agent such as DNP along with the inhibitor (See Figure 10-11*b*). Inhibition of oxygen consumption will be relieved by an uncoupling agent in the former case, but not in the latter.

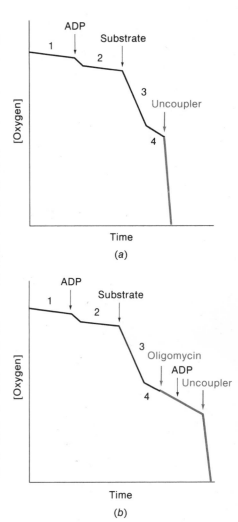

**Figure 10-11**
Different states of mitochondrial respiration. The plot is of oxygen consumption of a mitochondrial suspension. Addition of ADP and substrate is indicated by the arrows. Inorganic phosphate is present in excess throughout.

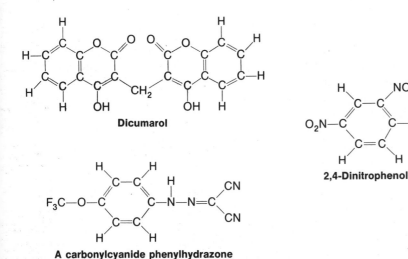

**Dicumarol**

**2,4-Dinitrophenol**

**A carbonylcyanide phenylhydrazone**

**Figure 10-12**
Structures of some commonly used uncoupling agents.

## The Mechanism of Coupling of Oxidative Phosphorylation

Although the obligatory nature of the coupling of ATP synthesis with electron transport is well-established, the actual mechanism whereby the transfer of electrons from one intermediate in the chain to the next can drive the phosphorylation of ADP has been until quite recently a source of considerable uncertainty and controversy. Much of this was due to the lack of definitive proof for any of the three alternative models that had been proposed. Each hypothesis appeared to be the best explanation for some of the available data while remaining in apparent conflict with other findings. Each had (and to some extent still has) its passionate devotees and its skeptics. We will examine each briefly for their historical interest and then concentrate on the mechanism that appears most likely to be the correct explanation.

**Chemical Coupling.** According to this hypothesis, first proposed in 1953 by E. C. Slater, oxidative phosphorylation is thought to occur in a manner similar to the phosphorylation reactions that occur during glycolysis and the TCA cycle, involving the formation of a high-energy phosphorylated intermediate followed by transfer of the activated phosphate to ADP. An especially attractive prototype is provided by the oxidation of glyceraldehyde-3-phosphate to 3-phosphoglycerate (Gly-6 and Gly-7 in Chapter 8), which proceeds by means of a high-energy phosphorylated intermediate, 1,3-diphosphoglycerate. Chemical coupling would thus allow the free energy of electron transfer down the chain to be conserved as a high-energy bond that could then drive the phosphorylation of ADP to ATP.

A model sequence for such a process might look as follows, using as an example the transfer of electrons from NADH to the FMN of NADH dehydrogenase at the first step in the transport chain:

$$NADH + H^+ + FMN + X \longrightarrow NAD^+{\sim}X + FMNH_2$$
$$NAD^+{\sim}X + P_i \longrightarrow NAD^+ + X{\sim}P$$
$$\underline{X{\sim}P + ADP \longrightarrow X + ATP}$$
$$NADH + H^+ + FMN + ADP + P_i \longrightarrow NAD^+ + FMNH_2 + ATP$$

Compound X in this scheme represents a hypothetical high-energy coupling intermediate, and the structures $NAD^+{\sim}X$ and $X{\sim}P$ denote high-energy phosphorylated intermediates. Similar reaction sequences would presumably occur at the other two steps in the chain where coupling to a phosphorylation event is indicated. Numerous variations on this theme were advanced, but they all involved one or more such hypothetical high-energy intermediates, and they all therefore shared in the uncertainty generated by the inability to isolate such intermediates despite years of intense effort. Many investigators now feel that such intermediates have never been isolated because they do not exist. Despite the tremendous stimulus to research in this area that the chemical-coupling hypothesis model provided over the years, it is now generally accepted that the coupling of oxidative phosphorylation with electron transport probably does not involve high-energy phosphate bonds as intermediates. Slater himself has acknowledged that the chemical-coupling hypothesis is "almost certainly incorrect."

**Conformational Coupling.** A more recent model attributable to P. Boyer envisions macromolecular components within the membrane that have two different three-dimensional configurations, one representing a high-energy form. Electron transport would then result in a conversion of the low-energy configuration to the high-energy form, while the reverse process would drive the phosphorylation of ADP by an enzyme included in the conformational unit. The reversible conformational changes between the orthodox (state 3) and condensed (state 4) configurations mentioned earlier are considered by some investigators to be consistent with such a model. These characteristic morphologic changes were in fact the main impetus that led to the formulation of the model. At best, however, the supportive evidence has been circumstantial, and most serious proponents of the conformational hypothesis suggest that energy is stored in the conformation of specific proteins and not in a grossly different ultrastructural configuration of the mitochondrion itself.

**Chemiosmotic Coupling.** An attractive alternative to the preceding models, the chemiosmotic hypothesis, was first advanced in 1961 by P. Mitchell, a British biochemist. Mitchell's model, which has gained substantial support in recent years, proposes that *electron transport is accompanied by a net translocation of protons from the inside to the outside of the inner mitochondrial membrane. The resulting electrochemical potential then provides the driving force for ATP synthesis*. To explain how a gradient of protons can be established and maintained across the membrane, the theory assumes that *the carriers involved in the hydrogen-transferring steps of electron transport are positioned asymmetrically in the membrane in such a way that protons are always taken up from the inside of the mitochondrion and released to the outside* (Figure 10-13). *By coupling the transfer of electrons with a directional pumping of protons in this way, some of the free energy released in the transport process is stored as an electrochemical potential that can be used in turn to drive ATP synthesis, provided only that the ATP-generating enzyme system is itself also asymmetrically organized in the membrane*. The chemiosmotic model therefore dispenses with the need for a high-energy chemical or conformational intermediate and depends instead on an electrochemical potential as the link between electron transport and ATP formation. This radical departure from bioenergetic orthodoxy earned the Mitchell hypothesis fierce resistance when it was initially proposed, but it has since been largely vindicated as supportive findings have accumulated. The chemiosmotic hypothesis is now supported by a considerable weight of evidence and appears to be the most likely explanation of coupling as well as an important influence on the way we think about energy conservation in biological systems. We will therefore examine the model in further detail, looking specifically at the predictions it makes and the extent to which they have been borne out by experimental findings.

## THE CHEMIOSMOTIC MODEL OF COUPLING

Features crucial to the chemiosmotic hypothesis include the presence of a proton motive force across the membrane of a closed compartment, the vectorial transport of protons across the membrane, which is essential to establishing this gradient, and the ability of the resulting electrochemical

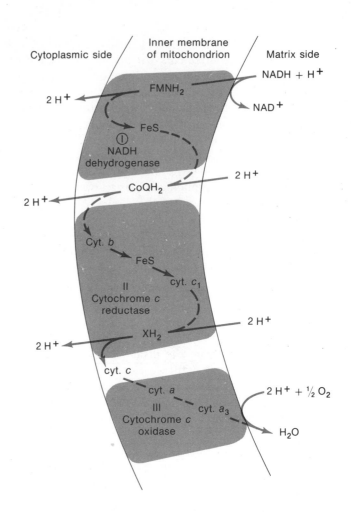

**Figure 10-13**
Arrangement of the electron carriers of the transport chain within the inner mitochondrial membrane to provide for the vectorial pumping of protons from inside to outside as electrons flow from reduced coenzyme to oxygen. Note the crucial role played by coenzyme Q and cytochrome *c* as intermediates between the three complexes responsible for electron transport. The intermediate labeled X in the cytochrome *c* reductase complex has been postulated to account for the proton translocation attributed to this complex, but its identity is not known and its existence is questioned.

potential to drive the synthesis of ATP. The evidence supporting the model can be summarized conveniently by looking in turn at each of these features of the model and at the experimental evidence that bears on it.

## The Transmembrane Proton Motive Force

The existence of an electrochemical potential for protons across any membrane in which electron transport is occurring is clearly a key feature of the chemiosmotic hypothesis, since such a potential represents the means of conserving the free energy of electron transport in this model. The total electrochemical potential resulting from a proton motive force increases with the difference in proton concentration across the membrane and is proportional to the *proton motive force* (pmf). *The pmf is due in part to the pH gradient across the membrane and in part to the membrane potential attributable to the pumping of positive charge by the electron-transfer system* (see Chapter 7, Equation 50)

$$\Delta p = \Delta \psi - 2.3RT\, \Delta pH/F$$

where $\Delta p$ is the pmf, $\Delta \psi$ is the difference in electrical potential across the membrane, and $F$ is the Faraday constant.

Initially, it proved difficult to detect and measure the differences in pH and membrane potential predicted by the model. This led to considerable disenchantment with the Mitchell hypothesis, especially among biochemists who, in Racker's words, "wanted chemical intermediates and not an elusive proton motive force." It is now clear that electron-transport activity is capable of generating a pH gradient of about 1.4 units across the inner mitochondrial membrane. The accompanying membrane potential $\psi$ is about 0.14 V, and _the total electrochemical potential represented by these two components of the gradient is in the right range to drive the generation of one ATP per electron pair at each of the three energy-conserving sites of the transport chain_. In addition, it has been established that a pmf is also generated across the cytoplasmic membrane of bacteria as a result of electron transport within the membrane and across the thylakoid membrane of chloroplasts during photophosphorylation. Interestingly, the direction of the pH gradient is reversed in the case of the chloroplast, with the pH within the thylakoid disk being lower than that of the surrounding cytoplasm rather than higher, as in the case of the mitochondrion.

Unique to the Mitchell hypothesis is the integral part that the membrane plays in the coupling mechanism. To maintain the desired pmf, such membranes would have to be impermeable to protons (to prevent leakage back across the membrane and consequent dissipation of the gradient), and they would also have to define some sort of enclosed compartment from which (or into which) protons can be pumped.

Both these predictions have been borne out in the several systems that are known to involve pmfs. In the case of the mitochondrion, the inner membrane is in fact impermeable to protons and can therefore be expected to maintain a stable gradient. This permeability barrier can, however, be breached by uncoupling agents such as dinitrophenol (DNP) that are known to allow protons to cross the otherwise impermeable mitochondrial membrane. This results in dissipation of the pmf. The consequence is a continued pumping of protons as electron transport proceeds, but without the expected buildup of a pmf and therefore without the expected ATP production. This both explains the uncoupling action of agents such as DNP that render the membrane proton-permeable and provides further support for the existence and importance of a pmf in oxidative phosphorylation.

The requirement for intactness of the membrane-bounded compartment is best illustrated by the submitochondrial particles that can be prepared by disruption of the mitochondrial inner membrane. Oxidative phosphorylation can be demonstrated only when such membrane fragments seal to form continuous, closed vesicles, thereby creating two compartments—the inside and the outside of the vesicle—which permits the establishment of a pmf across the membrane. Interestingly, such vesicles seal with the membrane oriented "inside out." As a result, the surface of the membrane that normally faces the matrix of the mitochondrion (the M surface) now faces out, while the surface that normally faces outward toward the cytoplasm (the C surface) now faces in. As might be predicted, the direction of the gradient across the membrane of such submitochondrial vesicles is just the reverse of that found in intact mitochondria—the pH is lower inside the vesicle.

## The Asymmetry of Carriers and the Vectorial Pumping of Protons

A further feature of the Mitchell hypothesis directly related to the pH gradient is the vectorial, or directional, transport of protons that is essential for the establishment and maintenance of the pmf. Since the electron carriers of the respiratory chain serve in this model as an active transport system to pump protons across the membrane, it is essential that the transport be vectorial rather than nondirectional, as would be the case for reactions in solution. This in turn requires that for each of the three energy-conserving sites of the transport chain, the carriers must be oriented asymmetrically within the membrane, and they must physically span the membrane if actual pumping of protons is to occur. A variety of techniques has been employed to examine the position of specific carriers within the inner mitochondrial membrane. These include binding of specific antibodies, binding of lectins (proteins with a high affinity for specific sugar residues of glycosylated proteins), cleavage by proteolytic enzymes, and labeling by reagents that cannot penetrate into the membrane. In each case, the basic intent is the same: to determine which carriers are at the surface of the membrane and therefore available for binding, cleavage, or labeling. By using both intact mitochondria (with the inner membrane right-side out) and vesicles reconstituted from sonicated inner-membrane fragments (with the membrane inside out), both the matrix (M) and cytoplasmic (C) surfaces of the membrane are amenable to study. From such investigations it has become clear that each of the energy-conserving sites spans the membrane and is asymmetrically ordered within the membrane. For example, the cytochrome $c$ oxidase complex binds cytochrome $c$ only on the C surface, binds oxygen only on the M surface, and pumps protons in one direction only. These and similar results for each of the other sites have made it possible to construct a detailed model of the inner mitochondrial membrane, specifying the spacial arrangement of the carriers within the membrane, as shown in Figure 10-13.

The actual mechanism for proton pumping is still a matter of considerable uncertainty and probably will remain so until the related controversy concerning the number of protons transported per electron pair is resolved.

## The Role of the Proton Motive Force in ATP Synthesis

The final (and in a sense the most crucial) feature of the Mitchell hypothesis concerns the role of the pmf in the synthesis of ATP. For if this is in fact to serve as the sole link between electron transport and phosphorylation of ADP, it is essential to demonstrate not only that the generation of a pmf is an integral part of electron transport, but also that the pmf is both necessary and sufficient to account for ATP generation.

A crucial experimental test of this part of the model came when investigators demonstrated that a pH gradient imposed on isolated mitochondria or chloroplasts was adequate to cause ATP synthesis in the absence of electron transport. This was first shown in 1966 by A. Jagendorf for chloroplasts. Recall that in the chloroplast, the pH gradient is just the reverse of that in the mitochondrion, with the proton concentration higher inside the thylakoid disk than outside of the membrane. Jagendorf induced a transient pH gradient across the thylakoid membrane by first incubating chloroplasts at pH 4 for several hours to equilibrate the contents of the thylakoid disk at this pH and then suddenly transferring the chloroplasts to a buffer of pH 8. As a result, the stroma on the outside of the thylakoid disks was substantially more alkaline than the inside of the disk. Upon addition of ADP and inorganic phosphate, a burst of ATP synthesis was observed, and the pH gradient disappeared. Similar results have since been obtained for mitochondria, and it is now clear that in both systems, a pmf is generated by electron transport (whether oxidative or photosynthetic) and is used to drive ATP generation. The only difference between the two organelles lies in the direction of the gradient. _In the mitochondrion, ATP synthesis is accompanied by the inward movement of protons across the inner membrane, whereas in the chloroplast, ATP formation is driven by an outward flow of protons across the thylakoid membrane._

Further confirmation of the pmf as the immediate driving force in ATP synthesis came from studies with bacteriorhodopsin, a protein in the purple membrane fraction of certain halophilic ("salt-loving") bacteria. Such bacteria synthesize ATP by oxidative phosphorylation under aerobic conditions, but they switch to a photosynthetic mode when illuminated in the absence of oxygen. Under the latter conditions, the role of the bacteriorhodopsin is to pump protons from the inside to the outside of the cell. The pmf generated in this way can be used to drive ATP synthesis in the complete absence of respiratory-chain function. The same ATP-generating complex is used for both oxidative phosphorylation and photophosphorylation, and the only common feature of the two processes is the generation of a pmf. It therefore seems clear from this work that the pmf is the most likely link between electron transport and ATP formation, with no common intermediate, whether chemical or conformational, being necessary.

### The Mitochondrial ATP-Generating System: $F_1$ and $F_0$

To be driven by the pmf, the ATP-forming enzyme system of both the mitochondrion and the chloroplast (as well as that associated with the cytoplasmic membrane of the bacterial cell) must be oriented vectorially in the membrane, just as the electron carriers are. This is now well-established and can, in fact, even be visualized in the electron microscope. When intact mitochondria are examined under the right conditions, spherical knob-like projections (sometimes irreverently called "lollipops") with a diameter of about 85 Å can be observed on the cristae, supported on stalks that are about 50 Å long and 30 Å wide (Figure 10-14a). When submitochondrial particles prepared from the inner membrane are examined, such knobs also can be seen, but now they appear on the outer surface, in keeping with the inside-out orientation of the membrane in such particles (Figure 10-14b). Similar structure can be seen on the inner surface of the cytoplasmic membrane of bacterial cells.

Interest in these structures was heightened when E. Racker found that they could be dislodged from the membrane of submitochondrial particles by mechanical agitation, leaving stripped membrane that could still carry out electron transport but could no longer synthesize ATP (Figure 10-15).

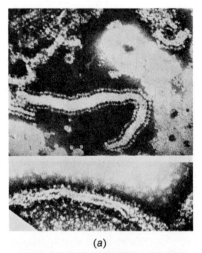

_(a)_

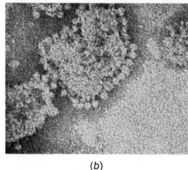

_(b)_

**Figure 10-14**
Electron micrographs showing the $F_1$ spheres that are responsible for mitochondrial ATP synthesis. (a) The $F_1$ spheres line the surface of the crista bordering the mitochondrial matrix. The micrograph is of a negatively stained bee mitochondrion. (Reproduced with permission from Chance and Parsons, Cytochrome function in relation to inner membrane structure of mitochondria, _Science_ 142: 1176, 1963.) (b) The $F_1$ spheres associated with submitochondrial particles prepared from rat liver mitochondria by sonication. The ATP synthase activity of the particles is crucially dependent on retention of the $F_1$ spheres during the isolation procedure. (Micrograph courtesy of E. Racker.)

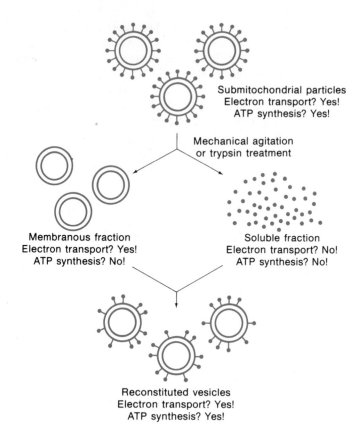

Submitochondrial particles
Electron transport? Yes!
ATP synthesis? Yes!

Mechanical agitation
or trypsin treatment

Membranous fraction
Electron transport? Yes!
ATP synthesis? No!

Soluble fraction
Electron transport? No!
ATP synthesis? No!

Reconstituted vesicles
Electron transport? Yes!
ATP synthesis? Yes!

**Figure 10-15**
Dissociation and reassociation of submitochondrial particles, with accompanying loss and restoration of ATP-synthesizing activity. Submitochondrial particles can be dissociated by mechanical agitation or trypsin treatment into a membranous fraction devoid of ATP-synthesizing capacity and a soluble fraction with synthase activity. Mixing of the two reconstitutes the particles and restores ATP synthase activity.

The capacity for ATP generation was restored upon addition of the knobs back to the stripped particles, suggesting strongly that the spherical projections are an integral part of the ATP synthase complex of the membrane. These spheres are now called _coupling factor 1_, or $F_1$. (Alternatively, they are also referred to as _mitochondrial ATPase_, because of their ATP-cleaving activity when isolated in solubilized form, but this term has potential for confusion, because the ATPase activity observed in vitro is the exact opposite of the presumed physiological role of the $F_1$ spheres.) The $F_1$ structure consists of five different polypeptides, with an aggregate mass of about 360,000 daltons.

$F_1$ is only part of the ATP-synthesizing system of the mitochondrion. The other major component is called $F_0$ and is embedded in the inner mitochondrial membrane (Figure 10-16). $F_0$ serves as the channel through which protons flow when the pmf across the membrane is being used to drive ATP synthesis by $F_1$. The stalk that supports $F_1$ and links it physically to the $F_0$ component is itself composed of several proteins that appear to regulate proton flow and thereby to control the rate of ATP generation.

Two critical questions that remain as yet unanswered with respect to pmf-driven ATP synthesis concern the number of protons that must pass through the $F_0$ channel to cause one phosphorylation event and the mechanism whereby proton passage actually drives ATP formation. Estimates of the $[H^+]/[ATP]$ ratio range from 2 to 4. Which turns out to be correct obviously bears both on the number of protons that must be transported per electron pair at each of the three energy-conserving sites of the transport chain and also on the mechanism of ATP synthesis.

With respect to mechanism, it is still a matter of conjecture how the exergonic ("downhill") movement of protons through the $F_0$ channel drives the otherwise endergonic phosphorylation of ADP to ATP by $F_1$. An early model envisioned the condensation of ADP and $P_i$ by removal not of a water molecule, but of one $H^+$ ion and one $OH^-$ ion, the former drawn inward by the alkaline state of the mitochondrial matrix and the latter drawn outward by the acidity on the outside of the membrane. The driving force for the condensation would then presumably derive from the attraction of the inner and outer milieus for the $H^+$ and $OH^-$ ions, respectively. More recent proposals involve a direct interaction of the proton flux at the active site of $F_1$ (Mitchell, 1974), conformational coupling transmitted through the enzyme complex (Boyer, 1977), and a model in which the function of the pmf of the gradient is to displace $Mg^{2+}$ from the ATP-synthesizing enzyme of $F_1$, thereby allowing a cycle of $Mg^{2+}$-proton interaction that is essential to ATP generation (Racker, 1980).

Discussing these alternatives in a review article in 1980, Racker concluded with some comments on the importance of testable hypotheses that are relevant not only to oxidative phosphorylation and bioenergetics, but also to every aspect of biochemistry as an experimental discipline:

> *What we can learn from a study of the history of bioenergetics is that we should keep an open mind and be susceptible to novel ideas regardless of how foreign they are to current concepts. . . . As experimentalists, we should be willing to tentatively accept a hypothesis, even a wrong hypothesis, if it allows us to design significant experiments. When a hypothesis has thus aided us, it is very tempting to cling to it. An old Indian story tells of a man who escaped from a lion by jumping on a boat that carried him across a river. He was so grateful to the boat that he carried it on his shoulder the rest of his life. We can learn two important lessons from the history of bioenergetics: (1) let us not miss the boat because of its unfamiliar form; and (2) let us get rid of it once we are safely across the river and it turns out to be of no further use.*

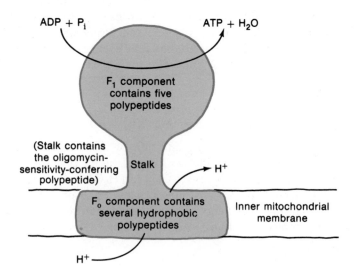

**Figure 10-16**
Structure of the $F_1/F_0$ ATP synthase system of the mitochondrion. $F_1$ is a spherical projection found especially on the cristae and supported by a stalk that links it physically and functionally with the $F_0$ complex embedded in membrane of the crista. Proton flux through $F_0$ is thought to provide the driving force for phosphorylation of ADP by $F_1$.

# SUMMARY OF RESPIRATORY METABOLISM

The processes of electron transport and oxidative phosphorylation can be summarized by placing them in their proper perspective in the overall respiratory metabolism of the aerobic cell. As the glycolytic pathway and the TCA cycle (or other catabolic pathways such as $\beta$ oxidation or amino acid degradation) proceed in the cell, coenzyme molecules are continuously being reduced. These coenzymes, notably free NADH and enzyme-bound $FADH_2$, represent a storage form of much of the energy of substrate oxidation—energy that can be tapped and released as the coenzymes are themselves reoxidized by molecular oxygen by means of the electron-transport system. In prokaryotic cells, substrate oxidation takes place in the cytoplasm, and electron transport occurs within the cytoplasmic membrane. In eukaryotic cells, most catabolic metabolism except for the glycolytic pathway is localized within the matrix of the mitochondrion, and electron transport takes place within the inner mitochondrial membrane. In either case, electron transport from NADH to oxygen involves passage of electrons through three successive complexes of carriers. Each such complex represents a site of energy conservation, because transport of a pair of electrons through the complex is coupled with the directional pumping of protons (at least two per electron pair in each of the three complexes) outward across the membrane. The amount of free energy released by electron transfer through the complex is in each case more than adequate to pump the required number of protons actively against the prevailing electrochemical gradient.

The resulting pmf most likely serves as the driving force for ATP synthesis. Under most conditions, a steady-state pmf will be maintained across the membrane, with the transfer of electrons from coenzymes to oxygen carefully and continuously adjusted so that the resultant outward pumping of protons will balance the inward flux of protons necessary to synthesize ATP at the desired rate.

Assuming that the inward movement of two protons is adequate to account for the generation of one ATP molecule (a reasonable although not universally accepted estimate), then the two protons pumped outward per electron pair at each of the three energy-conserving sites between NADH and oxygen account nicely for the observed P/O ratio of three for NADH oxidation. For $FADH_2$-linked oxidations, electrons enter the chain at coenzyme Q and therefore bypass the first ATP-generating site, resulting in a P/O ratio of 2 for each FAD-mediated oxidative event. Thus the ATP yield per coenzyme is three for NADH and two for $FADH_2$.

## The ATP Yield of Respiratory Metabolism

We are now ready to return to the question of the total ATP yield per molecule of glucose under aerobic conditions. Recall that the complete oxidation of one molecule of glucose to $CO_2$ and $H_2O$ results in the generation of four molecules of ATP by substrate-level phosphorylation, with most of the remaining free energy of glucose oxidation stored in the form of 12 coenzyme molecules—10 of NADH and 2 of $FADH_2$. Since the ATP yield is three for NADH and two for $FADH_2$, the reoxidation of the 12 coenzyme molecules can be represented as follows:

$$10\,NADH + 10\,H^+ + 5\,O_2 + 30\,ADP + 30\,P_i \xrightarrow[\text{at least 60 protons}]{\text{Mediated by pumping of}}$$

$$10\,NAD^+ + 10\,H_2O + 30\,ATP + 30\,H_2O$$

$$2\,FADH_2 + O_2 + 4\,ADP + 4\,P_i \xrightarrow[\text{at least 8 protons}]{\text{Mediated by pumping of}}$$

$$2\,FAD + 2\,H_2O + 4\,ATP + 4\,H_2O$$

Addition of the preceding two reactions to the summary equation for glycolysis and the TCA cycle leads to the following overall reaction for the complete aerobic respiration of glucose in a prokaryotic cell:

$$C_6H_{12}O_6 + 6\,O_2 + 38\,ADP + 38\,P_i \longrightarrow 6\,CO_2 + 38\,ATP + 44\,H_2O$$

For eukaryotic cells, the ATP yield of glucose oxidation is reduced from 38 to 36 because the two molecules of NADH that are generated in the cytoplasm require the equivalent of one ATP each to effect transport of their electrons into the mitochondria, as discussed later in this chapter.

### The Efficiency of Respiratory Metabolism

To determine the efficiency of aerobic respiration, recall that the complete oxidation of glucose to $CO_2$ and $H_2O$ occurs with a standard free-energy change of 686 kcal/mol. Since the actual free-energy change under cellular conditions is about the same as this (i.e., $\Delta G' \cong \Delta G^{\circ\prime}$), we can use this value as the basis for our calculations. Assuming a $\Delta G$ value of $-12$ kcal/mol for the hydrolysis of ATP under cellular conditions, the 38 ATPs produced per glucose correspond to about 456 kcal of energy conserved per mole of glucose oxidized. The efficiency of the process is therefore about 67 percent, far in excess of that obtainable with the most efficient of artificial machines.

## THE LOGISTICS OF RESPIRATORY METABOLISM: MITOCHONDRIAL STRUCTURE AND FUNCTION

An understanding of respiratory metabolism as it occurs in eukaryotes requires an appreciation for the structure and function of the mitochondrion, since much of aerobic energy metabolism is localized within this organelle. Of special interest to us will be the overall organization of the mitochondrion, the localization within the organelle of the various components of the respiratory process, and the mechanisms whereby reactants (especially pyruvate and electrons from cytoplasmically generated NADH) get in and products (especially ATP) get out.

### Discovery and Early Studies of Mitochondria

Mitochondria have been known and studied by biologists for more than a century. Their interest to biochemists is more recent, dating from the 1940s, when the development of differential centrifugation made organelle isolation possible. As early as 1850, Kölliker described the presence of ordered arrays of particles in muscle cells. Upon isolation, these particles were shown to swell when placed in water, suggesting the presence of a limiting membrane with osmotic activity. A variety of names were given to such particles in the early work, but the term _mitochondrion_ (meaning "thread-like granule") has gradually eclipsed other names since it was first introduced in 1898 and is now the universally recognized term.

Early evidence suggesting a role for the organelle in oxidative events included the finding by L. Michaelis in 1900 that mitochondria stained with the dye Janus green B lose their color as oxygen consumption proceeds, as well as O. Warburg's discovery in 1913 that cellular oxygen consumption could be associated with particles obtained by filtration of homogenates. Most of our present understanding of the role of the mitochondrion in energy metabolism dates from the development of the technique of differential centrifugation. This was pioneered by A. Claude and led in 1948 to the isolation of intact mitochondria by G. H. Hogeboom, W. C. Schneider, and G. E. Palade, who were the first to use sucrose in their homogenization and fractionation media. Shortly thereafter, E. Kennedy and A. Lehninger showed that all the reactions of the TCA cycle, electron transport, and oxidative phosphorylation can be carried out by isolated mitochondria.

## Occurrence and Size

Mitochondria are present in virtually all aerobic cells of eukaryotes. They are found in both chemotrophic and phototrophic cells, serving in the latter as a reminder that photosynthetic organisms are perfectly capable of respiration, an option they in fact depend on critically to meet energy needs during periods of darkness. The number of mitochondria per cell is highly variable, ranging from a few per cell in some protists, fungi, and algae up to several hundred or even a few thousand per cell in higher plants and animals. Human liver cells, for example, contain about 500 to 1000 mitochondria each.

Except for the nucleus, the mitochondrion is the largest organelle in most animal cells. (In plant cells, chloroplasts are also generally larger than mitochondria.) Typically, a mitochondrion has a diameter of about 0.5 to 1.0 $\mu$m and a length of several microns (more rarely, up to 10 $\mu$m). The organelle is therefore similar in size to an entire bacterial cell, to which it bears an evolutionary relationship according to one theory on the origins of eukaryotic organelles.

## Structure and Function

A characteristic feature of mitochondrial structure is the presence of two membranes (see Figure 10-17). This was first revealed by the pioneering electron microscopic observations of G. E. Palade and F. S. Sjöstrand. The outer membrane is smooth, has no folds, and is readily permeable to molecules with masses up to about 5000 daltons, which obviously includes all metabolites of relevance to the organelle. The outer membrane is therefore not generally regarded as a permeability barrier of any significance. Few enzyme activities appear to be associated with the outer membrane. By contrast, the inner membrane is highly folded, presents a permeability barrier to many solutes, and is the location of much of electron transport and oxidative phosphorylation. The inner membrane also can be distinguished from the outer membrane by its high content of the unusual lipid cardiolipin, its low ratio of phospholipid to protein (0.27 instead of 0.82), and its higher density (1.2 versus 1.1 $g/cm^3$).

The distinctive infoldings of the inner membrane are called *cristae*. They serve to increase the surface area of the inner membrane greatly, thereby allowing it to accommodate many more respiratory assemblies. The interior of the mitochondrion is filled with a semifluid *matrix* that is more like a gel than a true solution. Also within the matrix are the DNA molecules and the ribosomes that confer genetic semiautonomy upon the organelle. In most mammals, the mitochondrial genome consists of a circular DNA molecule of about 11 million daltons. This represents a coding capacity of about a dozen proteins and agrees well with evidence showing that isolated mitochondria incorporate labeled amino acids into about 10 major resolvable polypeptides. These include hydrophobic subunits of $F_0$, cytochrome oxidase, and the cytochrome $b$ complex. These polypeptides account for about 20 percent of the total protein of the inner membrane. Most of the rest of the membrane proteins and all the matrix enzymes are coded for by the nuclear genome and are synthesized by cytoplasmic ribosomes.

The mitochondrion functions as the chemotrophic powerhouse of the eukaryotic cell. *All of respiratory metabolism beyond pyruvate occurs within the mitochondrion, and all but two of the ATPs produced by the complete catabolism of glucose are generated within this organelle.* This role is often reflected both in the localization of mitochondria within the cell and in the extent to which the inner membrane is infolded into cristae.

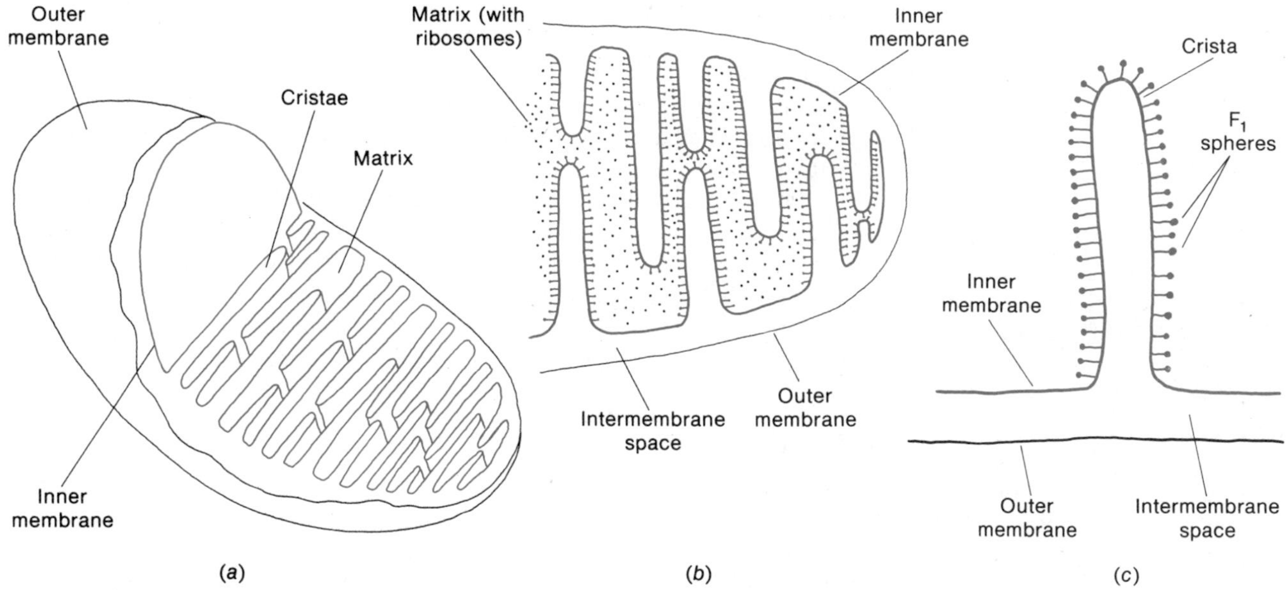

**Figure 10-17**
Structure of the mitochondrion (*a*) in a cut-away view and (*b*) in cross section. Shown in (*c*) is an enlargement of a single crista showing the $F_1$ (ATP synthase) spheres that line each crista on the matrix side.

**Figure 10-18**
Electron micrograph of frog skeletal muscle showing parallel arrays of myofibrils along with the sacroplasmic reticulum (SR) and the transverse tubule (TT)
system. Note especially the intimate relationship between the mitochondria (M)
responsible for ATP production and the fibrils that require ATP for contraction.
(From C. Franzini-Armstrong, Studies of the triad. I. Structure of the junction in
frog twitch fibers, *J. Cell Biol.* 47: 488, 1970; reprinted with permission.)

Frequently, mitochondria appear to cluster within the cell in the regions of
greatest metabolic activity (i.e., where the ATP need is the greatest). The
contractile fibrils of muscle tissue shown in Figure 10-18 represent an especially good example of such localization. The mitochondria in such cells
are organized in rows along the myofibrils, presumably to minimize the
distance over which ATP molecules must diffuse en route from the site of
generation in the mitochondrion to their site of utilization in the contracting muscle fibril. A similar ordering of mitochondria is seen in sperm tails,
flagella, and cilia and at the base of kidney tubule cells, where exchange
with the blood is most rapid.

The prominence of cristae within the mitochondrion frequently reflects the relative metabolic activity of the cell or tissue in which the organelle is located. Cells of the heart, kidney, and muscle (especially flight
muscle) have high respiratory activity and correspondingly large numbers
of prominent cristae. Liver cells and plant cells in general are characterized
by a lower rate of respiratory activity and by relatively fewer cristae.

### Localization of Function within the Mitochondrion

Enzyme and electron-transfer activities have been localized within the
mitochondrion by disruption of the organelle and/or by partial solubilization of one or both of its membranes. Enzymes or other activities that are
readily solubilized upon disruption of the mitochondria are assumed to be
either in the matrix or only loosely associated with the membrane. *All the
mitochondrial enzymes involved in the TCA cycle, β oxidation, and
amino acid catabolism are regarded as soluble matrix enzymes except for
the FAD-linked dehydrogenases, which are membrane-bound.*

As already noted, *the electron carriers of the respiratory chain are integral components of the inner membrane. They are arranged within the membrane in discrete functional units called respiratory assemblies,* each occupying an area of membrane several hundred angstrom units square. A single mitochondrion is thought to have 5000 to 20,000 such respiratory assemblies on its inner membrane and cristae. Also *embedded in the membrane of the cristae is the $F_0$ component of the ATP synthase complex.* As already discussed, the exergonic flux of protons through $F_0$ provides the driving force for the phosphorylation of ADP to ATP by the *protruding $F_1$ spheres that line the cristae.*

The organization of the carriers within the membrane is such that the binding sites for NADH and oxygen are localized on the inner side of the membrane, as are the membrane-associated dehydrogenases that use covalently bound FAD as the electron acceptor. This is the same side of the membrane on which most reduced coenzyme molecules and substrates for flavin-linked dehydrogenases are generated during energy metabolism, except for the NADH produced by the glycolytic pathway in the cytoplasm of eukaryotic cells.

## Entry of Pyruvate and Other Substrates into the Mitochondrion

In general, *the outer membrane of the mitochondrion is freely permeable to most solutes of low molecular weight, but the inner membrane is not.* However, *a variety of organic molecules can be transported across the inner membrane because of the presence in the membrane of specific transport systems* (also see Chapter 17). Pyruvate, citrate, isocitrate, malate, fumarate, succinate, aspartate, glutamate, $\alpha$-ketoglutarate, and fatty acids are in this class, as are inorganic phosphate, ADP, and ATP.

Frequently, the carrier systems are such that the inward transport of one solute is coupled with the concomitant outward movement of another. Thus the dicarboxylate carrier allows for the reciprocal exchange of malate, succinate, and fumarate with each other or of any of them with inorganic phosphate. Similarly, the tricarboxylate carrier provides for the exchange of citrate and isocitrate with each other or with a dicarboxylic acid.

For other small organic molecules, transport is coupled with and driven by the movement of protons into the mitochondrion. Pyruvate, for example, is produced in the cytoplasm by glycolysis, but is consumed in the mitochondrion by the TCA cycle. The inward transport of pyruvate is in symport with (i.e., is coupled with the simultaneous inward transport of) a proton, so that pyruvate transport depends on both the existence and size of the pH gradient across the mitochondrial membrane.

## Entry of Electrons into the Mitochondrion

In contrast to the small organic molecules discussed earlier, *neither $NAD^+$ nor NADH can cross the inner mitochondrial membrane because the membrane does not have specific carriers for these coenzymes. As a result, the intramitochondrial and extramitochondrial pools of this coenzyme represent physically separate compartments. (The same is true for $NADP^+$, NADPH, coenzyme A, and acyl-coenzyme A*, to which the inner membrane is similarly impermeable because of the lack of specific car-

H
|
H—C—OH
|
C=O          + NADH + H$^+$
|
H—C—O—(P)
|
H

**Dihydroxyacetone
phosphate**

In cytoplasm ⇅

H
|
H—C—OH
|
H—C—OH      + NAD$^+$
|
H—C—O—(P)
|
H

**Glycerol-3-phosphate**

H
|
H—C—OH
|
H—C—OH
|
H—C—O—(P)   +    FAD
|              (enzyme-bound)
H

**Glycerol-3-phosphate**

Within inner
membrane ↓

H
|
H—C—OH
|
C=O          +    FADH$_2$
|              (enzyme-bound)
H—C—O—(P)
|
H

**Dihydroxyacetone
phosphate**

riers.) All NADH generated in the cytoplasm must therefore pass electrons inward to the electron-transport chain in the inner membrane without physical movement of the coenzyme across or into the membrane. Several *shuttles* exist to accomplish this purpose. In each case the principle is the same: cytoplasmic NADH transfers its electrons to an organic molecule, which then carries the electrons into the mitochondrion (or at least to the inner membrane). There it is reoxidized, giving up the electrons to a coenzyme molecule within the organelle.

The most common such carrier system is the dihydroxyacetone phosphate/glycerol-3-phosphate shuttle. The electrons of cytoplasmically generated NADH are transferred to dihydroxyacetone phosphate, which is thereby reduced to glycerol-3-phosphate by the cytoplasmic enzyme glycerol phosphate dehydrogenase.

The glycerol-3-phosphate then diffuses through the outer mitochondrial membrane and is reoxidized by a FAD-linked glycerol phosphate dehydrogenase located on the external side of the inner membrane.

As with NADH dehydrogenase and succinate dehydrogenase, so also with glycerol phosphate dehydrogenase: the electrons are passed to coenzyme Q by a reductase complex located within the inner membrane. The dihydroxyacetone phosphate formed in this way is free to return to the cytoplasm and serve as the electron acceptor for the reoxidation of another molecule of NADH. This sets up a shuttle for the entry of electrons from extramitochondrial NADH into the electron transport of the inner membrane, as shown in Figure 10-19.

This inward flux of electrons is exergonic, driven by the difference in the free energy of the NADH generated in the cytoplasm and that of the FADH$_2$ of the membrane-bound enzyme. The energy difference is in fact sufficiently large to ensure that the inward transport of electrons by this shuttle is irreversible and therefore unidirectional. Since electrons entering in this way are transferred directly from FADH$_2$ of the membrane-bound dehydrogenase to coenzyme Q, they bypass the first energy-conserving site of the transport chain and yield only two molecules of ATP per electron pair. The ATP yield of NAD-linked oxidations is therefore three for mitochondrial dehydrogenases, but only two for cytoplasmic dehydrogenases, with the difference representing the energy required to drive the shuttle that brings electrons into the mitochondrion. Since the glycolytic pathway generates two extramitochondrial NADH molecules per glucose, the ATP yield for the aerobic catabolism of glucose is only 36 for eukaryotic cells instead of 38, as for prokaryotes.

An ATP yield of three instead of two is possible for extramitochondrial NADH, but only if the molecules of the shuttle system actually gain entry into the cytoplasmic matrix and if NADH-linked dehydrogenases function on both sides of the membrane. Such is clearly not the case for the glycerol phosphate/dihydroxyacetone phosphate shuttle, since dihydroxyacetone phosphate is reoxidized by a flavin-linked enzyme on the outer surface of the inner membrane and never actually gains entry into the interior of the mitochondrion. An alternative is the malate/aspartate shuttle, which is found especially in the liver and heart. In this shuttle (Figure 10-20), electrons from extramitochondrial NADH are used to reduce oxaloacetate to malate, catalyzed by an isozyme of malate dehydrogenase specific to the

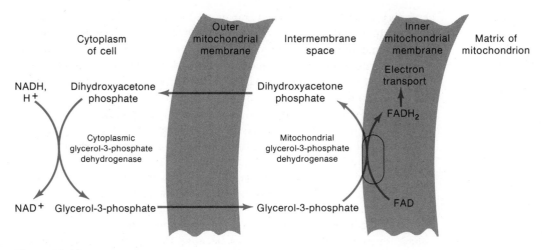

**Figure 10-19**

The glycerol-3-phosphate–dihydroxyacetone phosphate shuttle for the transport of electrons from cytoplasmic NADH into the electron-carrier chain of the inner mitochondrial membrane. Electrons are moved inward as glycerol-3-phosphate, which can be oxidized to dihydroxyacetone phosphate by an FAD-linked dehydrogenase on the outer surface of the inner membrane. The outer mitochondrial membrane is not a permeability barrier and does not require carrier proteins to allow passage of small organic molecules:

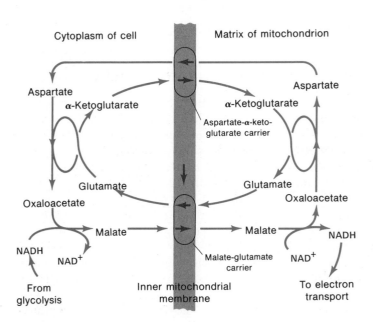

**Figure 10-20**

The malate-aspartate shuttle for the transport of electrons from NADH across the inner membrane of the mitochondrion. Electrons are moved inward as malate, which can be oxidized to oxaloacetate to generate NADH intramitochondrially. Transamination of the resulting oxaloacetate to aspartate is necessary because the membrane is not permeable to oxaloacetate. The shuttle is shown here for inward movement of electrons, but is readily reversible to accomplish outward electron flow.

cytoplasm. The malate is then transported across the inner membrane by one of the specific carrier systems mentioned earlier. Once inside the organelle, malate is reoxidized to oxaloacetate by the mitochondrial isozyme of malate dehydrogenase. Since both the cytoplasmic and mitochondrial isozymes of malate dehydrogenase are NAD-linked, the net result is a shuttle of electrons from NADH in the cytoplasm to NADH in the matrix. The ATP yield is therefore three, as it is for any other molecule of NADH within the mitochondrion. Since the same coenzyme is used on both sides of the membrane, the only driving force is the relative concentration of NADH in the cytoplasm and mitochondrion. The shuttle is therefore bidirectional and can be used to move reducing power either into or out of the mitochondrion, depending on the needs of the moment.

The relatively greater complexity of the malate/aspartate shuttle (Figure 10-20) compared with the glycerol phosphate/dihydroxyacetone phosphate shuttle (Figure 10-19) arises because the inner membrane is impermeable to oxaloacetate. Oxaloacetate is therefore transaminated to aspartate, which then crosses the membrane in the opposite direction to malate and is again transaminated on the other side to regenerate the oxaloacetate required as the actual electron acceptor. Malate and aspartate therefore comprise the primary shuttle, but a concomitant and stoichiometric transmembrane flux of glutamate and $\alpha$-ketoglutarate is required to provide for the transamination of oxaloacetate and aspartate on both sides of the membrane. Transport of each of these species is possible because of the presence in the membrane of the specific carrier systems for these solutes, as noted earlier.

## Transport of ATP Out of the Mitochondrion

All the ATP generation that accompanies electron transport occurs within the mitochondrion, since the $F_1$ spheres responsible for ATP synthase activity are located on the matrix side of the cristae. Most of this ATP must be transported out of the organelle, since most cellular ATP-requiring processes are extramitochondrial. *ATP flux out of the mitochondrion is mediated by a specific transport system, the ATP-ADP carrier, which couples the outward translocation of an ATP to the inward movement of an ADP, such that the total concentration of ADP + ATP within the mitochondrion remains constant.* This is an electrogenic carrier because it transports ATP$(-4)$ in one direction and ADP$(-3)$ in the other. It is therefore driven by the membrane potential that exists across the mitochondrial membrane. Specific inhibitors of this carrier which have proven useful in studies of mitochondrial ATP transport include the fungal antibiotic *bongkrekic acid* and the plant toxin *atractyloside*. The structures of both appear in Figure 10-21.

Inward transport of phosphate takes place by means of the *phosphate carrier*, which exchanges phosphate for hydroxide ions and is driven by the pH gradient across the membrane. Continued ATP production within the mitochondrion is therefore accompanied by an exchange across the inner membrane of ATP outward for ADP inward and by an equimolar inward movement of phosphate balanced by an efflux of hydroxide.

We have already examined in Chapters 8 and 9 the regulatory mechanisms that are operative to adjust the level of glycolysis and TCA cycle activity to cellular energy needs. Crucial to such regulation are the relative concentrations of ATP and ADP and of NADH and $NAD^+$ within the cell and within the mitochondrion because of the important role these compounds play as allosteric regulators of key enzymes in both glycolysis and the TCA cycle, as well as in a variety of other metabolic pathways. Earlier in this chapter, the obligatory coupling of electron transport with ATP generation and the role of ADP availability in respiratory control were noted as means of adjusting oxygen consumption to cellular energy needs. Here, attention focuses on several ways of expressing the energy status of the cell and on the integration of the cytoplasmic and mitochondrial components of energy metabolism.

## The Energy Charge

Because of the importance of ATP, ADP, and AMP as allosteric regulators of a number of reactions, D. E. Atkinson has proposed that the energy status of the cell be expressed by what he refers to as _energy charge_, calculated as

$$\text{Energy charge} = \frac{\frac{1}{2}\,[\text{ADP}] + [\text{ATP}]}{[\text{AMP}] + [\text{ADP}] + [\text{ATP}]}$$

The energy charge can range in value from 0 (all adenylate nucleotides present as AMP) to 1.0 (all present as ATP). Most cells maintain a steady-

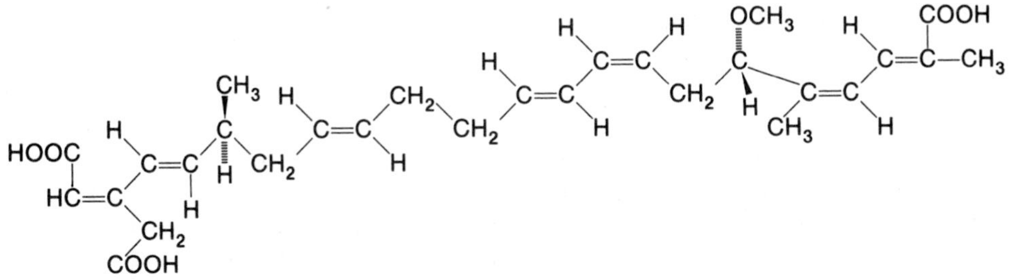

**Bongkrekic acid**

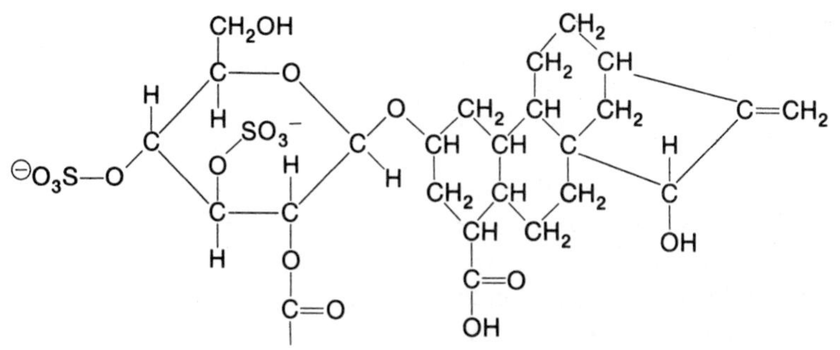

**Atractyloside**

**Figure 10-21**
Structures of two inhibitors of electron transport-driven ATP synthesis. Atractylic acid (atractyloside), isolated from a species of thistle, binds to the form of the exchanger which has a high affinity for ADP. Bongkrekic acid, synthesized by a fungus found in decaying coconut meal (_bongkrek_ in Indonesian), binds to the conformation of the adenine nucleotide carrier which prefers ATP. These inhibitors have been useful in investigating the structure and transport mechanism of the ATP/ADP exchanger.

state energy charge in the fairly narrow range from 0.8 to 0.95. Atkinson went on to demonstrate that ATP-producing and ATP-consuming pathways and reactions within the cell respond to the energy-charge status of the cell in the characteristic ways illustrated in Figure 10-22. At high values of energy charge, ATP-generating processes are inhibited, whereas ATP-utilizing processes are stimulated. When the energy charge is low, the opposite effects are seen. As the figure illustrates, the two response curves intersect at an energy charge of about 0.85. This is the steady-state level at which ATP consumption is just balanced by ATP generation. The steepness of the response curves in this region reflects the strong resistance the system offers to any deviation from the steady state. In other words, *the energy status of the cell is buffered to remain within a fairly narrow range*. Energy-charge values are in fact remarkably similar for a variety of cell types under a variety of conditions and are therefore less useful as an indicator of the energy status of the cell than might be initially inferred from Figure 10-22.

### The Phosphorylation Potential

An alternative measure of the cellular energy status is the phosphorylation potential, expressed as

$$\text{Phosphorylation potential} = \frac{[\text{ATP}]}{[\text{ADP}][\text{P}_i]}$$

Unlike the energy charge, the phosphorylation potential also takes into consideration the inorganic phosphate concentration, clearly an important substrate in ATP synthesis. The phosphorylation potential varies over a wider range than the energy charge and is therefore the more sensitive indicator of the cellular energy status.

### The Pasteur Effect

The Pasteur effect, first observed by L. Pasteur during his studies of yeast fermentation more than a century ago, is an especially good example of the integration of two distinct but related processes by means of the energy status of the cell. When a facultative cell is fermenting glucose anaerobically, it consumes substrate at a much greater rate than it would under aerobic conditions, because of the much lower ATP yield for glycolysis alone compared with complete respiratory catabolism (2 versus 36 ATPs per glucose for the eukaryotic cell). When such a cell is suddenly exposed to oxygen, *the onset of respiration leads to a rapid and dramatic reduction in glucose consumption and an equally rapid cessation of lactate accumulation, attesting to the tight integration of glycolysis and respiration*. It is this inhibition of glucose consumption and lactate accumulation by oxygen availability that has come to be known as the *Pasteur effect*. Since the phenomenon was first reported by Pasteur in 1861, a great deal of effort has been invested in attempts to understand how the onset of oxygen consumption at the terminus of the electron-transport chain in the mitochondrion could affect glucose consumption by the glycolytic pathway in the cytoplasm. A major part of the explanation seems to lie in the effect of oxygen consumption on the energy status of the cell and in the allosteric sensitiv-

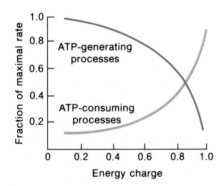

**Figure 10-22**
Effect of the energy charge of a cell on the reaction rates of ATP-generating and ATP-consuming processes. The steady state at which ATP production and utilization are balanced usually occurs at an energy charge of about 0.85.

ity of phosphofructokinase to ATP and ADP. Recall from Chapter 8 that phosphofructokinase is the key regulatory enzyme in the glycolytic sequence, sensitive both to activation by ADP and inhibition by ATP. When previously anaerobic cells are exposed to oxygen, the onset of oxidative phosphorylation results in a rapid elevation in the [ATP]/[ADP] ratio of the cell, owing to the high affinity of the mitochondrial ATP synthase complex for ADP. Since phosphofructokinase is sensitive to both ATP and ADP, it is severely inhibited by the higher [ATP]/[ADP] ratio characteristic of aerobic conditions, and the rate of glycolysis is sharply reduced.

A further partial explanation of the Pasteur effect may lie in the sensitivity of phosphofructokinase to allosteric inhibition by citrate, since mitochondrial citrate levels usually rise when the TCA cycle is functioning actively and citrate can move out into the cytoplasm by means of the tricarboxylate carrier in the inner mitochondrial membrane. Other mechanisms may sometimes also contribute to the overall Pasteur effect, but allosteric inhibition of phosphofructokinase appears to be the major explanation in most cases. It is, in any event, an excellent example of the regulation and interplay between two related but distinct metabolic capabilities that operate in separate compartments of the eukaryotic cell and yet are carefully integrated to ensure that cellular energy needs are met, whether in the presence or absence of oxygen.

## SELECTED READINGS

Boyer, P. O., Chance, B., Ernster, L., Mitchell, P., Racker, E., and Slater, E. C. Oxidative phosphorylation and photophosphorylation. *Ann. Rev. Biochem.* 46:955, 1977. An unparalleled collection of individual review articles published back-to-back in a single annual review volume.

Cross, R. L. The mechanism and regulation of ATP synthesis by $F_1$-ATPases. *Ann. Rev. Biochem.* 50:681, 1981. An up-to-date evaluation of what the author regards as "the best working model" for the mechanism of ATP synthesis by $F_1$.

Fillingame, R. The proton-translocating pumps of oxidative phosphorylation. *Ann. Rev. Biochem.* 49:1079, 1980. An excellent state-of-the-art review.

Hinkle, P. C., and McCarty, R. E. How cells make ATP. *Sci. Am.* 238(3):104, 1978. A very readable account of ATP generation, available as *Sci. Am.* offprint 1383; a good place to begin further reading.

Mitchell, P., Keilin's respiratory chain concept and its chemiosmotic consequences. *Science* 206:1148, 1979. The chemiosmotic model reviewed as a Nobel lecture by the laureate responsible for it.

Racker, E. *A New Look at Mechanisms in Bioenergetics.* New York: Academic Press, 1976. An indepth update by one of the primary movers in the field.

Stoeckenius, W. The purple membrane of salt-loving bacteria. *Sci. Am.* 234(6):38, 1976. A fascinating account of the role that this halobacterial membrane has played in elucidating the coupling of electron transport with ATP synthesis; available as *Sci. Am.* offprint 1245.

Wikström, M., Krab, K., and Saraste, M. Proton-translocating cytochrome complexes. *Ann. Rev. Biochem.* 50:623, 1981. A good indepth look at the role of membranous cytochrome complexes in the generation, transmission, and utilization of $H^+$ currents.

## PROBLEMS

1. *Energy storage.* Energy can be transferred and stored in cells in a variety of forms, of which the most immediately available is ATP. In order of decreasing immediacy of availability, several important energy and storage forms commonly used by cells are ATP, NADH, acetyl-CoA, pyruvate, and glucose.

   a. Calculate the number of ATP equivalents represented by each of these molecules, assuming complete oxidation under aerobic conditions by a eukaryotic cell (example: glucose = 36 ATPs).

   b. Assuming the $\Delta G^{\circ\prime}$ for the hydrolysis of ATP to be $-10$ kcal/mol under physiological conditions, calculate the weight in grams of each compound that would be required to obtain 1 kcal of usable (free) energy. Molecular weights for ATP, NADH, and acetyl-CoA are 507, 662, and 809, respectively.

c. Can you suggest at least two reasons why glucose is the form in which energy is transported in your body rather than, for example, ATP or NADH?

2. *Reduction potentials*. Assume that two half-cells are set up as in Figure 10-1, but that one (half-cell A) initially contains 1 $M$ each of 3-phosphoglycerate and glyceraldehyde-3-phosphate, while the other (half-cell B) initially contains 1 $M$ each of $NAD^+$ and NADH. Answer each of the following questions as true (T) or false (F).

a. The reaction occurring in half-cell A is an oxidation.

b. The concentration of NADH in half-cell B will decrease with time.

c. Electrons will flow *from* half-cell A *to* half-cell B.

d. The $\Delta E_0'$ for the overall reaction (half-cell A plus half-cell B) will be $+0.23$ V.

e. The $\Delta G^{\circ\prime}$ value for $NAD^+$-mediated oxidation of glyceraldehyde-3-phosphate can be calculated from the preceding $\Delta E_0'$ value.

f. If the initial concentrations of both compounds in each half-cell are reduced from 1 $M$ to 1 m$M$, the $\Delta G$ values for the overall reaction (half-cell A plus half-cell B) will be $\Delta G^\circ \times 10^{-3}$.

3. *More reduction potentials*. For each of the four oxidative reactions in the TCA cycle (TCA-3, TCA-4, TCA-6, and TCA-8), predict whether the reaction will be exergonic or endergonic in the oxidative direction under standard conditions.

4. *Reference potential*. By convention, the standard reduction potential $E_0$ of the $2H^+/H_2$ redox couple is used as the reference potential in determining reducing potentials. It has the value of 0.0 V under standard conditions. If the $H^+$ concentration is adjusted from the standard condition of 1.0 $M$ to a pH of 7.0, the reduction potential of the couple becomes $-0.42$ V. Show why this is so.

5. *Electron transport*. Shown below is the postulated electron-transport chain for *Itsa faka*, an extinct fungus. Assuming the chemiosmotic model for coupling ATP synthesis with electron transport, what is wrong with the scheme (i.e., why is *Itsa faka* extinct)?

MADH₂ ⟋ FUN ⟋ CoGH₂ ⟋ CoP ⟋ RoBH₂ ⟋ ½ O₂
MAD ⟍ FUNH₂ ⟍ CoG ⟍ CoPH₂ ⟍ RoB ⟍ H₂O

6. *Rotenone poisoning*. Rotenone is an extremely potent insecticide and fish poison. At the molecular level, its mode of action is to block electron transport from the FMN of NADH dehydrogenase to coenzyme Q.

a. Why do fish and insects die after digesting rotenone?

b. Would you expect the use of rotenone as an insecticide to be a potential hazard to other forms of animal life as well (people, for example)? Explain.

c. Would you expect the use of rotenone as a fish poison to be a potential hazard to aquatic plants that might be exposed to the compound? Explain.

7. *Pasteur effect*. In his studies of alcoholic fermentation by yeast, Louis Pasteur noted that the sudden addition of oxygen ($O_2$) to a previously anaerobic culture of fermenting grape juice resulted in a dramatic decrease in the rate of glucose consumption. This Pasteur effect can be counteracted by the addition of 2,4-dinitrophenol, an uncoupler of oxidative phosphorylation.

a. Why would the yeast cells consume less glucose in the presence of oxygen? Can you estimate how much less glucose they would use?

b. Why would 2,4-dinitrophenol counteract or prevent the Pasteur effect?

8. *Reduction potentials for cytochromes*. Cytochromes $a$, $b$, and $c$ all use the iron atom as the reducible species in their heme prosthetic group, and they all have very similar hemes. Their reduction potentials, however, are substantially less positive than that of the inorganic $Fe^{3+}/Fe^{2+}$ couple, and they differ from each other by more than 0.2 V. How can these observations be explained?

9. *Electron-transport chain*. A novel aerobic bacteria has just been discovered that possesses five previously unknown electron carriers, designated as $m$, $n$, $o$, $p$, and $q$.

a. When the transport chain is isolated, provided with NADH as the electron donor, and then treated with one of several different respiratory inhibitors, spectrophotometric assay shows each carrier to be either in the fully reduced ($+$) or fully oxidized ($-$) form, as indicated in the following table:

| | Oxidation State of Carriers | | | | |
| Inhibitor | *m* | *n* | *o* | *p* | *q* |
| --- | --- | --- | --- | --- | --- |
| Antimycin | + | + | + | − | + |
| Cyanide | + | + | + | + | + |
| Rotenone | − | − | + | − | − |
| Barbiturate | + | − | + | − | − |

Indicate the order of the carriers in the transport chain and the direction of electron flow, as well as the point at which each inhibitor acts.

b. When succinate is substituted for NADH as the electron donor, the following data were obtained using the same inhibitors:

| Inhibitor | Oxidation State of Carriers | | | | |
|---|---|---|---|---|---|
| | $m$ | $n$ | $o$ | $p$ | $q$ |
| Antimycin | + | + | − | − | + |
| Cyanide | + | + | − | + | + |
| Rotenone | − | − | − | − | − |
| Barbiturate | − | − | − | − | − |

What more can you now say about electron transport in this organism?

10. *Dinitrophenol as an uncoupler.* Oxidative phosphorylation is normally tightly coupled with electron transport in the sense that no electrons will flow if ATP synthesis cannot occur concomitantly (owing, for example, to low levels of either ADP or $P_i$). Uncoupling agents dissociate ATP synthesis from electron transport, allowing the latter to occur in the absence of the former. One such uncoupling agent is 2,4-dinitrophenol (DNP), which is highly toxic to humans, causing a marked increase in metabolism and temperature, profuse sweating, collapse, and death. For a brief period in the 1940s, however, sublethal doses of DNP were actually prescribed as a means of weight reduction in humans.

a. Why would an uncoupling agent such as DNP be expected to cause an increase in metabolism, as evidenced by consumption of oxygen or catabolism of foodstuffs?

b. Based on what you know about allosteric regulation of the glycolytic pathway and the TCA cycle, why would an uncoupling agent such as DNP be likely to have greater, more far-reaching effects on respiratory metabolism than might be predicted by the simple lack of control of the rate of electron transport?

c. Why would consumption of DNP lead to an increase in temperature and to profuse sweating?

d. DNP has been shown to carry protons across biological membranes. How might this observation be used to explain its uncoupling effect?

e. Why would DNP have been considered as a drug for weight-reduction purposes? Can you guess why it was abandoned as a reducing aid?

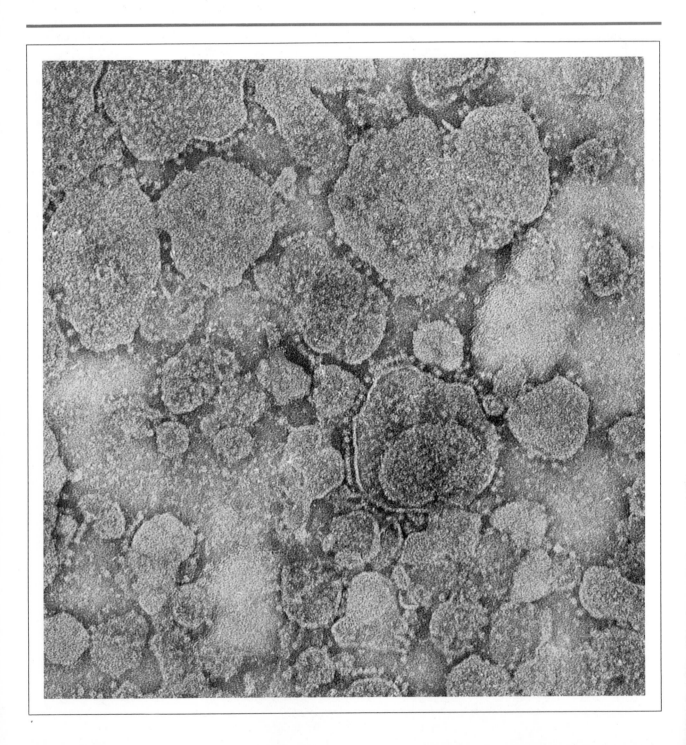

# 11

# PHOTOSYNTHESIS

Animals live by degrading complex molecules that are provided by other organisms. Life on earth obviously could not continue indefinitely in this manner, unless there were some independent mechanism for the creation of complex molecules from simple ones. Such a mechanism requires the input of energy, which the sun provides through the process of _photosynthesis_. About $10^{11}$ tons of carbon are converted annually into organic compounds as a result of photosynthesis. In spite of the enormous magnitude of this conversion, the total amount of fixed carbon on the earth appears to be decreasing as a result of consumption. As our reserves of energy and food diminish, it becomes increasingly important that we understand how photosynthesis works and how our activities affect it. The question is also an intriguing one intellectually: How can light be made to do chemistry?

When most of us think of photosynthesis, we think of green trees or perhaps of fields of grain. Actually, on the order of one-third of the carbon fixation that occurs on earth takes place in the oceans and is carried out by microorganisms. In addition to plants and algae, many species of bacteria are capable of photosynthesis. Photosynthetic bacteria have been extremely useful for the study of photosynthesis, because their photosynthetic apparatus is simpler than that of plants. Unlike plants, photosynthetic bacteria do not evolve $O_2$, but the photochemical reactions they use for capturing the energy of light are similar to those which occur in plants.

## IN PHOTOSYNTHESIS ELECTROMAGNETIC ENERGY IS CONVERTED INTO CHEMICAL ENERGY

_Photosynthetic bacteria and plants use the electromagnetic energy of sunlight to drive a chemical reaction in the direction away from equilibrium._

_Electron micrograph of vesicles prepared from chloroplast thylakoid membranes. (Magnification 150,000×.) Knobs protruding from the membrane are visible against the dark background around the edges of the vesicles. The knobs are part of the ATPase that synthesizes ATP when the chloroplasts are eliminated. Light-driven electron-transfer reactions in the membrane result in the movement of protons across the membrane into the thylakoid space. The flow of protons back out by way of the ATPase drives the formation of ATP (Courtesy Dr. Richard McCarty.)_

An overall equation of the process that occurs in plants is

$$6\,CO_2 + 6\,H_2O + light \longrightarrow C_6H_{12}O_6 + 6\,O_2$$

Or more generally

$$CO_2 + H_2O + light \longrightarrow (CH_2O) + O_2$$

where $(CH_2O)$ represents part of a carbohydrate molecule. These equations describe what is called the fixation of carbon into carbohydrate. If one leaves out the light, the equilibrium for this process lies vanishingly far to the left: $\Delta G^{\circ\prime}$ for the synthesis of glucose from $CO_2$ and $H_2O$ is about $+686$ kcal/mol. The equilibrium constant at 27°C is $10^{-496}$. However, the energy of red light (700 nm) is 41 kilocalorie per einstein or per mole of photons. This means that, in principle, it would be possible to drive the biosynthesis of glucose from $CO_2$ and $H_2O$ if cells were able to capture all the energy of about 17 einsteins of red light per mole of hexose or about 3 einsteins per mole of $(CH_2O)$ units. In practice, more light than this is needed in order to make the process work.

### A Physical Definition of Light

*Light is an electromagnetic field that oscillates sinusoidally in space and time. It interacts with matter in packets, or quanta, called photons, each of which contains a definite amount of energy.* A mole of photons is called an *einstein*. The relationship between the energy $\varepsilon$ of a photon and the frequency $\nu$ of the oscillating field is given by the expression

$$\varepsilon = h\nu$$

where $h$ is Planck's constant ($6.63 \times 10^{-27}$ erg $\cdot$ s or $4.12 \times 10^{-15}$ eV $\cdot$ s). The energy per einstein is

$$E = N\varepsilon = Nh\nu$$

where $N$ is Avogadro's number ($6.02 \times 10^{23}$ photons per einstein). The frequency $\nu$ is the number of oscillations per second at a given point in space. It is related to the wavelength $\lambda$ of the oscillation (the distance in space between successive peaks in the amplitude of the field) and to the velocity $v$ at which the peaks move through space:

$$\nu = v/\lambda$$

The velocity $v$ is $3 \times 10^{10}$ cm s$^{-1}$ in a vacuum. In a dense medium such as water, the velocity is less, depending inversely on the refractive index. Blue light has a wavelength in the region of 450 nm ($\nu \cong 6.7 \times 10^{14}$ s$^{-1}$) and an energy of about 2.8 eV ($2.7 \times 10^{12}$ ergs/einstein, or 64 kcal/einstein). At the other end of the visible region of the spectrum, red light has a wavelength of about 700 nm.

Radiation of wavelengths much below 400 nm or above 750 nm is invisible to the human eye, and some authors prefer not to call this light. However, such radiation can be important biologically. Many photosynthetic bacteria are adapted to use radiation in the 800- to 900-nm region, and some species do well with even longer wavelengths.

The oscillating electric and magnetic fields of light are perpendicular to each other and are both in a plane perpendicular to the direction of the light ray. A light beam is said to be *polarized* if the orientations of the fields in this plane are fixed. In an unpolarized beam, there are equally strong fields with all different orientations in the plane.

## Photosynthesis Takes Place in Membranes

In green plants, the reactions of photosynthesis take place in specialized subcellular organelles, the *chloroplasts*. Figure 11-1 is an electron micrograph of a chloroplast from a lettuce leaf. Chloroplasts have two outer membranes and an extensive internal membrane called the *thylakoid membrane*. In electron micrographs of thin-sectioned chloroplasts, the thylakoid membrane gives the appearance of a large number of flattened vesicles. Actually, it probably is a single membrane that is highly folded. The membrane is topologically closed, so that it has a definite inside and outside. In places, the folded membrane is tightly stacked into structures called *grana*. The chlorophyll found in chloroplasts is bound to hydrophobic proteins in the thylakoid membrane, and it is here that the photochemical reactions of photosynthesis occur. The thylakoid membrane also contains a collection of electron carriers and an ATPase similar to the proton-linked ATPase of mitochondria (see Chapters 10 and 17). The enzymes responsible for the actual fixation of $CO_2$ and the synthesis of carbohydrates are soluble proteins and reside in the *stroma* that surrounds the thylakoid membrane.

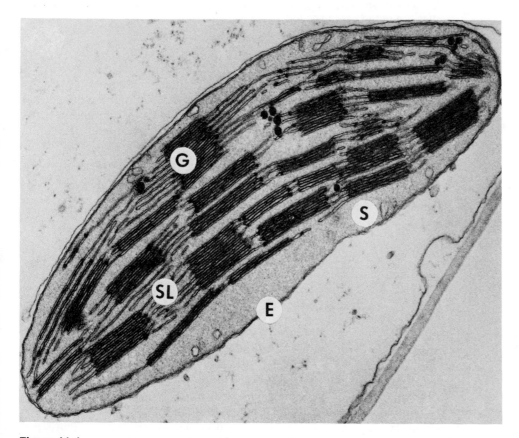

**Figure 11-1**
Cross-sectional view of a chloroplast in a partially disrupted lettuce leaf. The organelle is shaped like a flattened sausage with a width of about 10 $\mu$m and a thickness of about 3 $\mu$m. It is surrounded by an outer membrane or envelope (E). The stroma (S) contains DNA, ribosomes, and the soluble enzymes of $CO_2$ fixation. Extending throughout the stroma is the *thylakoid* membrane system, which is differentiated into stacked membranes or *grana* (G) and unstacked *stroma lamellae* (SL). (Courtesy Dr. Charles Arntzen.)

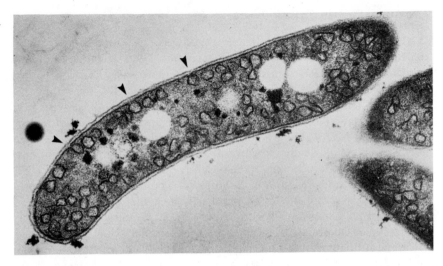

**Figure 11-2**
Cross-sectional view of a cell of *Rhodospirillum rubrum*, a purple photosynthetic bacterium. A double-membrane system surrounds the cell. The inner membrane is extensively invaginated into tubules (*arrow*). These look circular when they are cut in cross section. (Courtesy Dr. Gerald Peters.)

The photochemical reactions of photosynthetic bacteria take place in the membrane that encloses the cell. The membrane is highly invaginated (Figure 11-2). If the cells are broken by sonic oscillation, membrane fragments pinch off and seal to form small vesicles called *chromatophores*. These are useful for study, because they contain all the membrane-bound components of the photosynthetic apparatus but are free of many of the cellular components that are not directly involved in photosynthesis.

## THE PHOTOCHEMISTRY OF CHLOROPHYLLS

### Photosynthesis Depends on the Photochemical Reactivity of Chlorophyll

*With the exception of certain halophilic bacteria (Chapter 31), all known photosynthetic organisms take advantage of the photochemical reactivity of one or another type of chlorophyll.* Figure 11-3 shows the structures of several of the principal types of chlorophyll. The chlorophylls are basically tetrapyrroles. They resemble hemes, the prosthetic groups of the cytochromes and hemoglobin, and are derived biosynthetically from protoporphyrin IX. However, they differ from hemes in four major respects. First, chlorophylls contain an additional ring (ring V) with carbonyl and carbomethoxy substituents. Second, the central metal atom is magnesium rather than iron. Third, in the case of chlorophyll *a* and chlorophyll *b*, one of the pyrrole rings (ring IV) is reduced by the addition of two hydrogens. In bacteriochlorophyll *a*, two of the rings are reduced (rings II and IV). Finally, the propionyl side chain of ring IV is esterified with a polyisoprenoid alcohol. Chlorophyll *a* and *b* contain the alcohol *phytol*; bacteriochlorophyll *a* has either phytol or geranylgeraniol, depending on the species of bacteria.

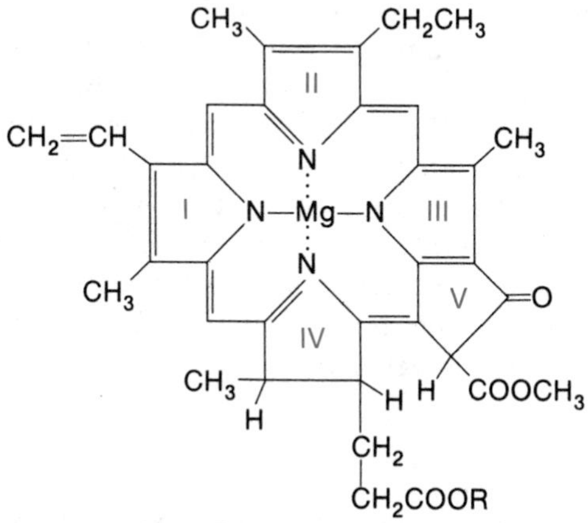

**Chlorophyll a**

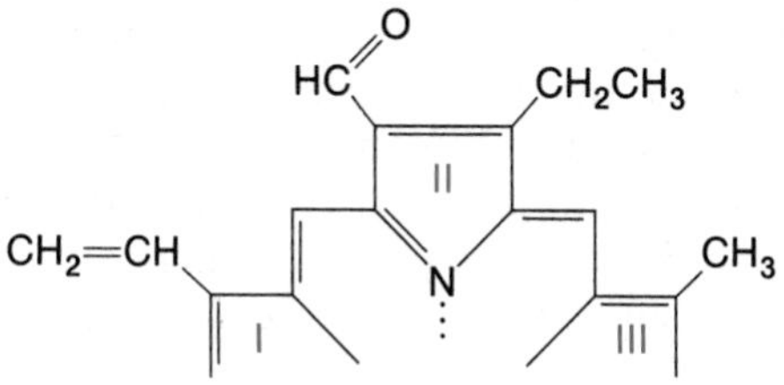

**Bacteriochlorophyll a**

**Chlorophyll b**

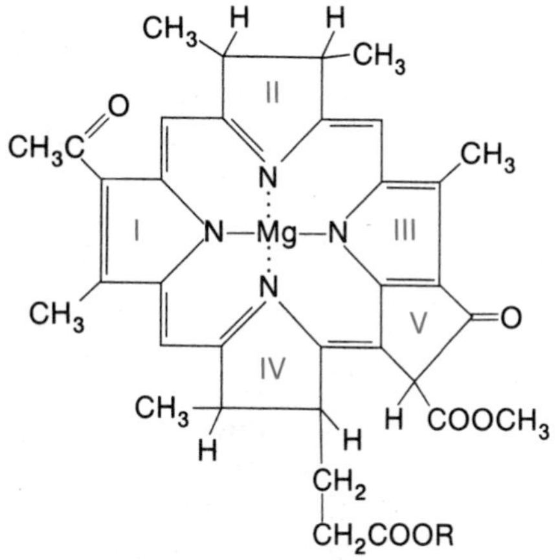

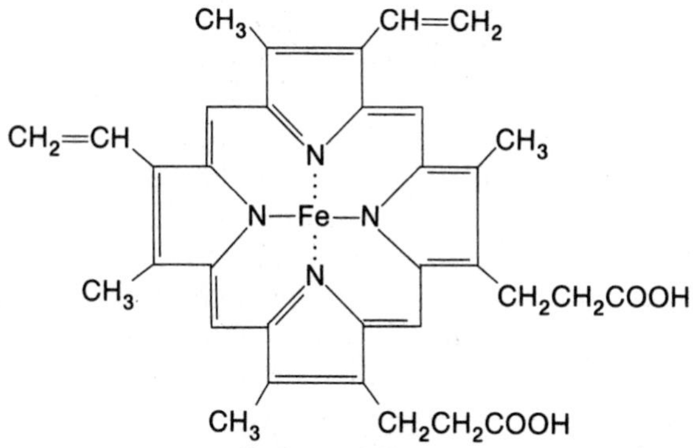

**Protoporphyrin IX**

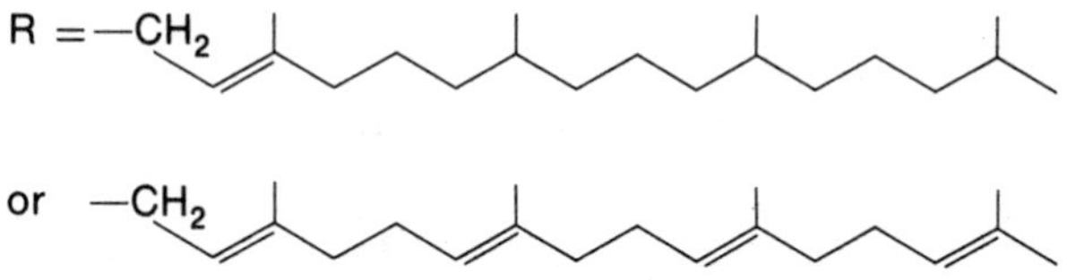

$R = -CH_2$ ... Phytyl side chain

or $-CH_2$ ... Geranylgeranyl side chain

**Figure 11-3**
Structures of chlorophyll *a*, chlorophyll *b*, bacteriochlorophyll *a*, and protoporphyrin IX (the prosthetic group of hemoglobin, myoglobin, and the *c*-type cytochromes). The numbering system used for the five rings is shown. Chlorophyll *b* is the same as chlorophyll *a*, except for the substituents on ring II.

The reduction of one of the pyrrole rings in chlorophyll *a* or *b* decreases the size of the conjugated aromatic ring system and, more important, makes it decidedly asymmetrical. This changes the optical absorption spectrum of the molecule dramatically. Whereas the long-wavelength absorption bands of the heme (the α bands) are relatively weak, chlorophyll *a* has an intense absorption band at 685 nm (Figure 11-4). Chlorophyll *b* has a similar band at 650 nm. Bacteriochlorophyll *a*, in which the asymmetry of the conjugated system is even more pronounced, has an extremely strong absorption band at 780 nm (Figure 11-4). The chlorophylls thus absorb light very well, particularly at relatively long wavelengths.

## How Light Interacts with Molecules

The distribution of electrons in atoms or molecules is described quantum mechanically by mathematical functions called *wave functions*, which are solutions to the Schrödinger wave equation. The probability of finding an electron at a given place at any particular time is proportional to the square of the wave function at that place and time. The detailed form of the Schrödinger equation that one must solve to find the wave functions for any system of interest is determined by how the potential energy of the system varies with space and time. In an isolated molecule, the potential energy of an electron depends mainly on the coulombic interactions of the electron with the nuclei and other electrons and is more or less independent of time. For such a system, the Schrödinger equation has a set of different solutions that correspond to different electronic orbitals. Each orbital is associated with a particular value of the total energy of the system. In other words, *the energy of the molecule is quantized. An electron placed in any one of the orbitals will remain there indefinitely, as long as the potential energy does not change.*

In the presence of light, the oscillating electric field makes the potential energy strongly time-dependent, so that the original solutions to the Schrödinger equation are no longer completely satisfactory. The electric field distorts the wave functions, and the probability of finding an electron at a given place changes with time. *Light thus can cause an electron to move from one orbital to another.* A more sophisticated analysis shows that two requirements must be met in order for this to occur. First, *the difference between the energies of the two orbitals must be the same as the photon's energy hν.* This is why molecules absorb light of some wave-

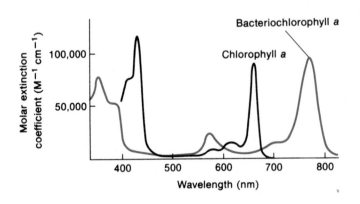

**Figure 11-4**
Absorption spectra of chlorophyll *a*
and bacteriochlorophyll *a* in ether.
(After R. K. Clayton, *Photosynthesis:
Physical Mechanisms and Chemical
Patterns*, Cambridge University
Press, Cambridge, 1980.)

lengths and not of others. The second requirement has to do with the shapes of the two orbitals, the disposition of the orbitals in space, and the orientation of the oscillating electric field. These must be such that the distortions caused by the electric field increase the extent to which the two orbitals resemble each other. This is why some absorption bands are stronger than others. It also accounts for the fact that a molecule with a definite orientation in space generally will absorb only light that is polarized in a particular way with respect to the molecular axes. In the case of bacteriochlorophyll, the electric field of 780-nm light must be polarized in the plane of the molecule and approximately along the axis connecting rings I and III, that is, parallel to the long axis of the $\pi$-electron system.

The ability of a molecule to absorb light of a given wavelength is characterized by the molecule's _molar extinction coefficient_ $\varepsilon$, which has the dimensions $M^{-1}\,cm^{-1}$. If a light beam of intensity $I_0$ is incident on a sample of concentration $c$ (in moles per liter) and thickness $d$ (in centimeters), the intensity $I$ of the light that is transmitted by the sample is given by Beer's law:

$$I = I_0 10^{-\varepsilon cd} \qquad \varepsilon cd = \text{O.D.}$$

The dimensionless product $\varepsilon cd$ is called the _absorbance_, or optical density (O.D.), of the sample.

_When chlorophyll (or any other molecule) absorbs a photon, an electron is excited from one molecular orbital to another orbital that has a higher energy._ The excited molecule can decay back to the ground state by releasing energy in several different ways. One possibility is simply to emit a photon; this is _fluorescence_. The wavelength of the fluorescence is generally slightly longer than that of the original absorption, because nuclear relaxations decrease the energy of the excited molecule somewhat before the molecule fluoresces. The extra energy is given off to the environment as heat. If the molecule has several absorption bands, as the chlorophylls do, the wavelength of the fluorescence is usually slightly longer than that of the longest-wavelength absorption band. The molecule relaxes to the lowest, or "first," excited singlet electronic state before the emission occurs. In some cases, the excited molecule can decay all the way to the ground state by radiationless processes, so that the excitation energy is converted entirely into heat. A third decay mechanism is to transfer the energy to a neighboring molecule by the process of _resonance energy transfer_. This phenomenon is extremely important in photosynthesis, and we shall return to it shortly. A fourth possibility is for the excited molecule to transfer an electron to a neighboring molecule.

## Light Causes an Electron-Transfer Reaction

_Electron transfer can be a favorable path for the decay of an excited molecule, because an electron in the upper orbital is bound less tightly than one in the lower, ground-state orbital._ The midpoint redox potential $E_m$ for removal of an electron from the excited molecule is more negative than the $E_m$ for the molecule in the ground state, by approximately $h\nu/e$. Here $h\nu$ again is the excitation energy, which can be expressed in electron volts, and $e$ is the charge of an electron. In the case of chlorophyll $a$, the $E_m$ for oxidation in the ground state is approximately $+0.5$ V, and $h\nu$ for the long-

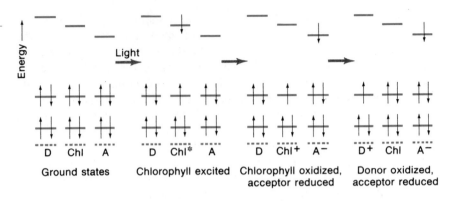

**Figure 11-5**
The photochemical process that initiates photosynthesis is an electron-transfer reaction. (Chl = chlorophyll molecule or complex; A = initial electron acceptor; D = electron donor.) Horizontal bars in this diagram represent molecular orbitals for electrons. Each orbital can hold two electrons with antiparallel spins (arrows pointing upward and downward). Only the top two filled orbitals are indicated in the figure. The absorption of light by chlorophyll excites an electron to an orbital with a higher energy. (The choice of an upward or downward arrow is arbitrary, but there is no change of spin during the excitation or during the initial electron-transfer reactions. As long as this condition holds, so that the spins of the two unpaired electrons continue to be antiparallel, the chlorophyll or Chl $^+$ A $^-$ complex is said to be in a "singlet" state.) The excitation increases the partial molecular free energy of the chlorophyll by approximately $h\nu$, making electron transfer to A thermodynamically favorable.

wavelength absorption band is about 1.5 eV. The $E_m$ of the excited molecule is therefore approximately $-1.0$ V. This means that in the excited state, chlorophyll $a$ is a very strong reductant. From Chapter 10 recall that $NAD^+/NADH$ has an $E_m$ of $-0.32$ V. This is the basic principle underlying the photochemistry of photosynthesis. *Excitation causes a molecule of chlorophyll or bacteriochlorophyll (or a complex of several such molecules) to release an electron. The chlorophyll is oxidized, and some other molecule (A) becomes reduced* (Figure 11-5). The oxidized chlorophyll complex (Chl$^+$) is a relatively strong oxidant and can extract an electron from a third molecule (D).

The idea that light drives the formation of oxidants and reductants was first advanced by C. B. van Niel in the 1920s. It was known at that time that photosynthetic bacteria thrive only if they are provided with a reduced substrate, such as an organic acid. Some species grow well on a reduced inorganic material, such as $H_2S$. Van Niel noticed that although the bacteria do not evolve $O_2$, the reactions they carry out have a formal similarity to the process that occurs in plants. If one uses $H_2B$ to represent a reduced substrate and B for an oxidized product, the reaction that occurs in bacteria can be written

$$CO_2 + 2\,H_2B \xrightarrow{h\nu} (CH_2O) + H_2O + 2\,B$$

The equation presented earlier for $CO_2$ fixation in plants can be put in a similar form by replacing $H_2B$ and B with $H_2O$ and O:

$$CO_2 + 2\,H_2O \xrightarrow{h\nu} (CH_2O) + H_2O + O_2$$

To van Niel, this suggested that the essence of photosynthesis in both plants and bacteria is the photochemical separation of oxidizing and reducing power. The substance that is reduced in the photochemical reaction could be used to reduce $CO_2$ to carbohydrates by enzyme-catalyzed reactions that do not require light. The material that is oxidized photochemically could be discharged by the oxidation of $H_2O$ to $O_2$ in plants or by the oxidation of some other material in bacteria (Figure 11-6).

Support for these ideas came from experiments by R. Hill in 1939. Hill discovered that isolated chloroplasts would evolve $O_2$ if they were illuminated in the presence of an added nonphysiological electron acceptor such as ferricyanide [$Fe(CN)_6^{3-}$] or benzoquinone. The electron acceptor became reduced in the process. Since there was no fixation of $CO_2$ under these conditions, it was clear that the photochemical reactions of photosynthesis can be separated from the reactions that involve $CO_2$. Similarly, H. Gaffron found that if green algae are illuminated under certain conditions, they will transfer electrons from $H_2$ to $CO_2$ without evolving $O_2$.

## Photooxidation of Chlorophyll Creates a Cationic Free Radical

When chlorophyll or bacteriochlorophyll is oxidized, the product is a cationic free radical. The situation differs subtly from the oxidation of a cytochrome. When a cytochrome is oxidized, the electron that is removed comes from the iron, which changes from ferrous ($Fe^{2+}$) to ferric ($Fe^{3+}$). In chlorophyll, the electron is removed not from the magnesium, but rather from the aromatic $\pi$-electron system. The positive charge and the spin of the unpaired electron that remains behind in the radical are delocalized extensively over the $\pi$-electron system. This can be shown by studying the electron spin resonance (ESR) and the electron nuclear double-resonance (ENDOR) spectra of the radical.

The photooxidation of chlorophyll in photosynthetic systems can be detected by measuring changes in the optical absorption spectrum of the molecule. The oxidation causes the loss of the strong absorption band at long wavelengths. The first measurements of this sort were made in the 1950s by both B. Kok and L. Duysens. Duysens found that the illumination of photosynthetic bacteria caused a small absorbance decrease at 870 nm. He suggested that the bleaching reflected the photooxidation of a bacteriochlorophyll complex, which he called P870. P stood for "pigment," and 870 for the wavelength at which the unoxidized complex had its main absorption band. Kok made similar observations in chloroplasts. Here the reactive chlorophyll complex absorbed at 700 nm, and Kok called it P700. A second reactive complex in chloroplasts, P680, was discovered subsequently by H. Witt and his colleagues. P700 and P680 are parts of two separate photosystems, called *photosystem I* and *photosystem II*.

## The Reactive Chlorophyll Is in Complexes Called Reaction Centers

The chlorophyll or bacteriochlorophyll that undergoes photooxidation is bound to a protein in a complex called the *reaction center*. Reaction centers have been purified from both bacteria and plants by disrupting chro-

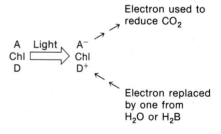

**Figure 11-6**
(*a*) Van Niel proposed that the reductant generated in the photochemical reaction (A$^-$) is used to reduce $CO_2$ to carbohydrate. He suggested that the oxidant (D$^+$) oxidizes $H_2O$ to $O_2$ in plants and oxidizes some other material ($H_2B$) in bacteria. Only the initial charge separation requires light. (Chl = Chlorophyll.)

matophores or chloroplasts with detergents and then applying conventional techniques of protein purification. This was first achieved with photosynthetic bacteria, particularly by R. Clayton, and bacterial reaction centers are still much better understood than are those of plants. _Purified bacterial reaction centers are capable of carrying out the initial photochemical separation of electric charges_, but they are free of most of the secondary electron carriers and the antenna pigments that are discussed later.

Reaction centers isolated from the purple bacteria _Rhodopseudomonas sphaeroides_ or _Rhodospirillum rubrum_ contain three hydrophobic polypeptides with molecular weights of approximately 21,000, 24,000, and 28,000. Bound noncovalently to the protein are four molecules of bacteriochlorophyll and two molecules of bacteriopheophytin. Bacteriopheophytin is identical to bacteriochlorophyll except that two H atoms replace the Mg in the center. The reaction center also contains two molecules of ubiquinone and one nonheme iron. In some other species, a menaquinone (a naphthoquinone, or vitamin K) replaces one of the ubiquinones.

The most highly purified reaction-center preparations from plant photosystem I contain several different polypeptides and approximately 40 molecules of chlorophyll _a_, of which only a small part (P700) is photochemically active. Recent photosystem II preparations also contain about 40 molecules of chlorophyll _a_ per P680, along with several polypeptides.

As with bacteriochlorophyll in vitro, the photooxidation of P870 generates a cationic radical (P870$^+$) in which the unpaired electron is delocalized over the macrocyclic ring system. In fact, the ESR and ENDOR spectra of the P870$^+$ radical indicate that the unpaired electron is delocalized over two of the four bacteriochlorophyll molecules of the reaction center. These two evidently form a special dimer. Exactly how the two molecules of bacteriochlorophyll in P870 are arranged with respect to each other is not yet clear. P700 of chloroplast photosystem I has been thought to contain a similar dimer of chlorophyll _a_ molecules, but recent evidence suggests that it consists of only a single chlorophyll. P680 most likely also consists of a single chlorophyll molecule.

Table 11-1 collects some of the main features of the reaction center of purple photosynthetic bacteria and the reaction centers of chloroplast photosystems I and II. The reactive chlorophyll of photosystem II, P680, differs from P700 and P870 in that it forms a much stronger oxidant. The initial electron acceptors in all the reaction centers appear to be chlorophylls or pheophytins. Pheophytins _a_ and _b_ have the same relationship to chlorophylls _a_ and _b_ that bacteriopheophytin has to bacteriochlorophyll: two H atoms replace the Mg. In general, pheophytins are somewhat easier to reduce than the corresponding chlorophylls. The secondary electron acceptors are quinones in bacteria and in photosystem II and iron-sulfur proteins in photosystem I. All these components are discussed in more detail later.

## An Antenna System Transfers Energy to the Reaction Centers

The chlorophyll or bacteriochlorophyll molecules of P700, P680, or P870 make up only a small fraction of the total pigment in photosynthetic membranes. Chloroplasts contain on the order of 400 chlorophyll molecules per P700 and P680, and chromatophores of wild-type strains of _Rhodopseudo-_

**Table 11-1**
Reaction Centers of Bacteria and of the Two Photosystems
of Chloroplasts

| | Purple Bacteria | Chloroplasts | |
| --- | --- | --- | --- |
| | | System I | System II |
| Reactive chlorophyll | P870 (BChl)$_2$ | P700 (Chl) | P680 (Chl) |
| $E_m$ for $P^+ + e^- \rightarrow P$ (in ground state) | $\sim +0.45$ V | $\sim +0.5$ V | $> +0.82$ V |
| Initial electron acceptors | BChl and BPh | Chl | Phe |
| Secondary electron acceptors | Quinones | Fe-S proteins | Quinones |
| Secondary electron donors | $c$-Type cytochromes | Plastocyanin (Cu protein) | Mn protein and $H_2O$ |

Abbreviations: BChl = bacteriochlorophyll; BPh = bacteriopheophytin; Chl = chlorophyll $a$; Phe = pheophytin $a$.

*monas sphaeroides* have about 100 bacteriochlorophylls per P870. Most of the chlorophyll or bacteriochlorophyll is not photochemically active. Instead, it plays the role of a large antenna. _Energy absorbed from light by the antenna molecules migrates rapidly from molecule to molecule by resonance energy transfer until it is trapped by the photochemical reaction that occurs in the reaction centers._

Measurements of the lifetime of fluorescence from the antenna system indicate that the energy is trapped in the reaction centers within about $10^{-10}$ s after light is absorbed in chromatophores and within about $5 \times 10^{-10}$ s in chloroplasts. When isolated bacterial reaction centers are excited directly with light, the oxidation of P870 occurs within about 1 ps ($10^{-12}$ s).

The distinction between the antenna system and the reaction centers grew out of experiments by R. Emerson and W. Arnold in the 1930s. Emerson and Arnold measured the amount of $O_2$ evolution that occurred when they excited suspensions of a green algae repeatedly with short flashes of light. In order to obtain maximum $O_2$ per flash, they found that they had to allow a period of about 20 ms of darkness between successive flashes. The amount of $O_2$ evolved per flash decreased if the flashes were spaced too closely together. Each flash evidently generated a product that had to be consumed before the photosynthetic apparatus was ready to work again. If a second flash arrived too soon, its energy was wasted, mainly as heat and fluorescence. We know now that the _chlorophyll complexes that undergo oxidation in the reaction centers have to be returned to their reduced states before they can react again. The electron acceptors that take electrons from the chlorophyll complexes also have to be reoxidized._

Emerson and Arnold measured the amount of $O_2$ that was evolved per flash when the timing of the flashes was optimal. The amount depended on the strength of the flash, but it reached a plateau when the flashes were made sufficiently strong. The surprising observation was that even with flashes of saturating intensity, the amount of $O_2$ was very small relative to the chlorophyll content of the algae. The cells contained approximately 2500 molecules of chlorophyll for each molecule of $O_2$ they evolved. This was extremely puzzling at the time, because it was generally accepted that chlorophyll was intimately involved in the photochemistry of $CO_2$ fixation and $O_2$ evolution. Now it is clear that _most of the chlorophyll is part of the antenna system_. When the flash intensity is high, the amount of $O_2$ evolution that can occur on each flash is limited by the concentration of reaction centers in the sample, and this is much smaller than the total concentration of chlorophyll. The discrepancy between the number of chlorophylls per P700 or P680 (400) and the number of chlorophylls per $O_2$ (2500) will make more sense after we have discussed the fact that chloroplasts have to absorb about 8 photons for each molecule of $O_2$ that is evolved.

The structures of the antenna systems are still not well understood. The pigments generally are bound noncovalently to hydrophobic proteins with molecular weights of 10,000 to 40,000, and each protein carries only a small number of pigments. These small complexes aggregate into larger arrays that allow the excitation energy to hop rapidly from complex to complex. Antenna complexes isolated from wild-type _Rhodopseudomonas sphaeroides_ have three bacteriochlorophyll molecules. A complex that contains three molecules of chlorophyll _a_, three of chlorophyll _b_, and one carotenoid has been isolated from chloroplasts. The green photosynthetic bacterium _Prosthecochloris aestuarii_ contains a complex with seven bacteriochlorophyll _a_ molecules that is unusual because it is water-soluble. This complex has been crystallized and its structure determined by x-ray diffraction.

In addition to bacteriochlorophyll or chlorophyll, antenna systems contain a variety of other pigments, such as carotenoids, which transfer energy to the chlorophylls. These _accessory pigments fill in the absorption spectrum in regions where the chlorophylls do not absorb well_. The importance of the accessory pigments becomes clear when one realizes that after sunlight has passed through a meter or so of seawater, it is green in color. Chlorophyll _a_ absorbs red or blue light well, but not green (Figure 11-4). The transfer of energy from the accessory pigments to chlorophyll is downhill thermodynamically because the first excited singlet state of chlorophyll lies at a comparatively low energy.

Carotenoids are linear polyenes (see Figure 31-3). _In addition to transferring excitation energy to the chlorophylls, they play an important role in protecting the cell against damage by $O_2$ at high light intensities_. If the antenna system is flooded with light too rapidly for the reaction centers to keep pace, the antenna chlorophylls discharge most of the extra energy as fluorescence and heat. However, the excited molecules also have an opportunity for "intersystem crossing" into an excited _triplet state_, in which the spins of the two unpaired electrons are parallel. To visualize this, invert one of the unpaired arrows in the diagram for excited chlorophyll in Figure 11-5. Excited triplet states are relatively long-lived. They cannot decay to the ground (singlet) state unless the electronic spin changes again, and this does not happen readily. One way they decay is by reacting with molecular $O_2$, which has a triplet ground state. The reaction returns the chlorophyll to its ground state and promotes the $O_2$ to its excited singlet state. Singlet $O_2$ is extremely toxic: it reacts irreversibly with a variety of groups in proteins, nucleic acids, and lipids. Carotenoids can intervene to prevent these destructive processes by quenching the excited triplet chlorophylls before they have a chance to react with $O_2$, as well as by quenching singlet $O_2$ itself. Both types of quenching involve elevation of the carotenoid to its excited triplet state. This lies below singlet $O_2$ in energy, and it decays harmlessly to the ground state.

## THE BACTERIAL ELECTRON-TRANSPORT SYSTEM

### The Electron Released by P870 Goes to Bacteriochlorophyll, to Bacteriopheophytin, and Then to Quinones

What is the acceptor that removes an electron from the reactive bacteriochlorophyll or chlorophyll complex in the reaction center? And how does the oxidized complex return to its reduced state? The answers to these questions are clearer for bacteria than they are for plants, partly because most bacteria appear to contain only a single type of reaction center, and partly because the bacterial reaction centers can be isolated in a simpler form. Recent studies with picosecond and nanosecond spectroscopic techniques indicate that the initial electron acceptor in bacterial reaction centers (or at least the earliest yet detected) is one of the other bacteriochlorophylls. Recall that the reaction centers contain two bacteriochlorophylls and two bacteriopheophytins in addition to the pair of bacteriochlorophylls that make up P870. When isolated reaction centers are excited with a subpicosecond flash, a transient state that appears to be a $P870^+ BChl^-$ charge-transfer state is formed within about $10^{-12}$ s (Figure 11-7). Here $BChl^-$ represents an anionic free radical of bacteriochlorophyll. The formation of the $P870^+ BChl^-$ charge-transfer state can be detected spectrophotometrically by the bleaching of absorption bands due to P870 and the bacteriochlorophyll and the formation of a new absorption band that probably is due to $BChl^-$.

The initial charge-transfer state appears to decay in about $4 \times 10^{-12}$ s by the transfer of an electron from the $BChl^-$ to one of the bacteriopheophytins. The formation of the anionic radical of the bacteriopheophytin ($BPh^-$) is reflected by the bleaching of absorption bands due to the bacteriopheophytin and the reappearance of bands due to the bacteriochlorophyll. The $P870^+ BPh^-$ state appears to be slightly below the $P870^+ BChl^-$ state in free energy, which is in turn slightly below the lowest excited singlet state of P870 (Figure 11-7). The $P870^+ BPh^-$ state lasts about $2 \times 10^{-10}$ s, decaying by the movement of an electron from $BPh^-$ to one of the quinones ($Q_A$).

The components involved in the reactions to this point must all be held close together in the reaction center with very little freedom of motion. One indication of this is that the electron-transfer reactions, in addition to being phenomenally fast, are almost independent of temperature. Some of the reactions even increase in speed with decreasing temperature. Further, the reactions are amazingly efficient. This is generally expressed in terms of the _quantum yield_ of $P870^+ Q_A^-$, which is the number of moles of $P870^+ Q_A^-$ formed per einstein of light absorbed. The measured quantum yield of $P870^+ Q_A^-$ in isolated reaction centers is $1.02 \pm 0.04$. _Essentially every time the reaction center is excited, an electron moves from P870 to $Q_A$._ The probability of competing processes such as fluorescence or wasteful back-reactions is virtually nil.

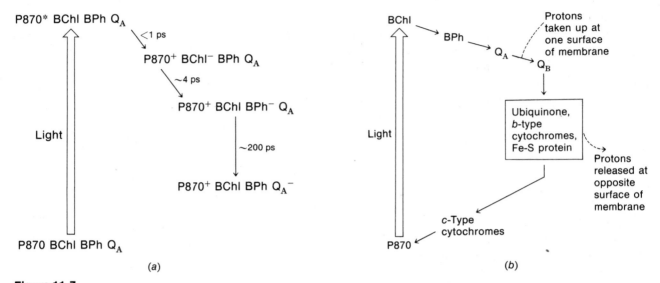

**Figure 11-7**

(a) Initial electron-transfer sequence in reaction centers isolated from *Rhodopseudomonas sphaeroides*. ($P_{870}^*$ = excited state of $P_{870}$; BChl = bacteriochlorophyll; BPh = bacteriopheophytin; $Q_A$ = ubiquinone.) In this scheme, the different states of the photosynthetic apparatus are positioned vertically according to their energies, with the states that have the highest energies at the top. The actual energy spacing is still subject to considerable uncertainty. (After V. A. Shuvalov and W. W. Parson, Energies and kinetics of radical pairs involving bacteriochlorophyll and bacterio-pheophytin in bacterial reaction centers, *Proc. Natl. Acad. Sci. USA* 78:957, 1981). Green photosynthetic bacteria contain a different series of electron acceptors, which are similar to those of plant photosystem I. (b) In purple photosynthetic bacteria, electrons return to $P_{870}$ via of a chain of electron carriers that includes a second ubiquinone ($Q_B$), additional quinones, and cytochromes. In this scheme, the solid arrows represent paths of electron flow. The electron carriers are positioned vertically according to their $E_m$ values, with the strongest reductants (most negative $E_m$ values) at the top. Electron flow downward is thermodynamically spontaneous. When $Q_B$ is reduced by two electrons, two protons are taken up at the inner surface of the membrane that bounds the cell (*dashed arrow*). The protons are translocated across the membrane and released at the opposite surface, generating an electrochemical potential gradient for protons across the membrane. The gradient can be used to drive the synthesis of ATP.

### An Electron-Transfer Chain Returns Electrons to P870 and Moves Protons Across the Membrane

From $Q_A^-$, an electron moves to a second quinone ($Q_B$). This step takes about $10^{-4}$ s, considerably longer than the earlier steps, and it becomes even slower with decreasing temperature. The nonheme iron atom of the reaction center may play a role in electron transfer between the two quinones, but the evidence on this point is inconclusive. (Treatments that remove the iron block the reaction.) $Q_B$ normally remains in the semiquinone form ($Q_B^-$) until the reaction center is excited a second time. While these reactions are occurring on the reducing side of the reaction center, a *c*-type cytochrome replaces the electron that was removed from P870, preparing the reaction center to operate again. Electron transfer between the cytochrome and $P870^+$ takes $10^{-6}$ to $10^{-3}$ s, depending on the bacterial species. In some species, the reaction centers contain two different types of *c*-cytochromes, either of which can react with $P870^+$. Several cytochrome molecules of each type frequently are present.

When the reaction center is excited a second time, a second electron is pumped from P870 through the bacteriochlorophyll, bacteriopheophytin, and $Q_A$ to $Q_B$. This places $Q_B$ in the fully (two-electron) reduced form, $Q_B^{2-}$. The uptake of two protons transforms the reduced quinone to the dihydroquinone, $Q_B H_2$. In intact bacteria, the protons that are taken up in this step come from the aqueous space inside the cell.

The electrons that are transferred to $Q_B$ return to P870 by a chain of additional electron carriers. The details of this chain have not yet been fully worked out. In some species, electrons appear to move from $Q_B H_2$ to a *b*-type cytochrome, from there to a third molecule of ubiquinone, to an iron-sulfur protein, and then to one of the *c*-type cytochromes that communicate with P870. There may be branches in this path and parallel routes through which electrons move in pairs. Chromatophores contain additional *b*-type cytochromes and a large pool of ubiquinones, whose roles in the cycle are still unclear.

The most important features of the bacterial electron-transfer chain are first that it is cyclic and second that the protons taken up by $Q_B H_2$ from the interior of the cell are released on the outside of the cell when $Q_B H_2$ or one of the later carriers is reoxidized. The cyclic flow of electrons therefore drives the movement of protons across the membrane, generating a transmembrane electrochemical potential gradient. The inside of the cell becomes alkaline relative to the outside, and it also becomes negatively charged. Proton flow back into the cell is therefore downhill thermodynamically. The movement of protons into the cell is conducted by an ATPase similar to the $H^+$-conducting ATPase of the mitochondrial inner membrane (see Chapters 10 and 17) and is linked to the formation of ATP. The ATP, or the electrochemical potential gradient itself, can be used to drive electrons from various reductants to NADP for the reduction of $CO_2$ to carbohydrates.

# THE CHLOROPLAST ELECTRON-TRANSPORT SYSTEMS

## Chloroplasts Have Two Photosystems Linked in Series

*The photosynthetic electron-transport chains of plants differ from those of bacteria in being predominantly linear rather than cyclic.* The reduced components generated in the reaction center are used more directly for the reduction of other materials, such as NADP. The oxidized components return to the reduced state by extracting electrons from water. However, some cyclic electron flow probably occurs in plants also.

We mentioned earlier that *plants contain two types of reaction centers, P700 and P680. Together with their antennas and their initial electron acceptors and donors, the two photosystems are called photosystem I and photosystem II, respectively, or more simply system I and system II.* Much evidence indicates that systems I and II are connected in series, as shown in Figure 11-8. Excitation of P700 (system I) generates a strong reductant, which transfers electrons to NADP by way of several secondary electron carriers. Excitation of P680 (system II) generates a strong oxidant, which oxidizes $H_2O$ to $O_2$. The reductant formed in system II injects electrons into a chain of carriers that connects the two photosystems. This scheme, which is frequently called the *Z scheme*, was first suggested by R. Hill and F. Bendall in 1960.

Figure 11-9 gives a more detailed picture of the electron carriers that are believed to participate in the Z scheme. The chain of carriers between the

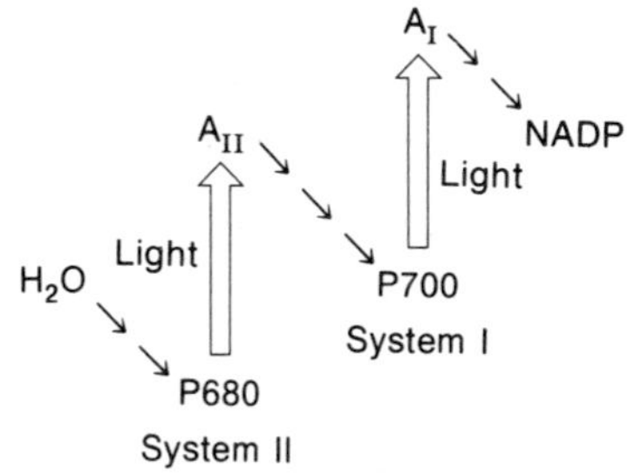

**Figure 11-8**
The Z scheme for the photosynthetic apparatus of plants. Two photochemical reactions are required to drive electrons from $H_2O$ to NADP. In this scheme, the arrows represent paths of electron flow. The electron carriers are positioned vertically according to their $E_m$ values, with the strongest reductants (most negative $E_m$ values) at the top. Electron flow downward is thermodynamically spontaneous.

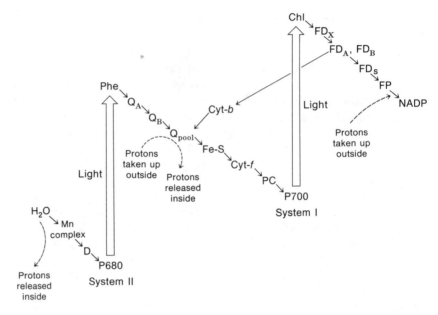

**Figure 11-9**
A more detailed version of the Z scheme. Most of the electron carriers shown are discussed in the text. The component called $D_1$ has not been identified. [Phe = pheophytin $a$; $Q_A$ and $Q_B$ = two molecules of plastoquinone; $Q_{pool}$ = a large pool of plastoquinones; Fe-S = an iron-sulfur protein (sometimes called M); cyt-$b$ = cytochrome $b_{563}$; cyt-$f$ = cytochrome $f$; PC = plastocyanin; Chl = chlorophyll $a$; $FD_X$, $FD_A$, and $FD_B$ = iron-sulfur proteins (bound ferredoxins); $FD_S$ = soluble ferredoxin; FP = flavoprotein (ferredoxin-NADP oxidoreductase).] $FD_X$ is sometimes called X or $A_2$; D is sometimes called Z. Electron transport is coupled with proton translocation across the thylakoid membrane (*dashed arrows*).

two systems includes several quinones (plastoquinones), a $c$-type cytochrome (cytochrome $f$), a copper protein (plastocyanin), and probably also an iron-sulfur protein. Additional details on these components, as well as on the electron carriers between P700 and NADP and between $H_2O$ and P680, will be filled in later, after we discuss some of the evidence supporting the Z scheme.

Some of the earliest observations that led to the Z scheme came from measurements of how the quantum yield of $O_2$ evolution in algae depends on the excitation wavelength. The *quantum yield*, again, is the molar ratio of $O_2$ evolved to light absorbed. In green algae, the quantum yield is relatively independent of wavelength between 400 and 675 nm, but falls off drastically in the far-red region near 700 nm (Figure 11-10). This is odd, because chlorophyll $a$ absorbs 700-nm light quite well (Figure 11-4). Why would light absorbed by chlorophyll $a$ be used less efficiently than light absorbed by carotenoids or other accessory pigments? A clue came from R. Emerson's finding in 1956 that 700-nm light could be used much more efficiently if it were superimposed on a background of weak green light (Figure 11-10). J. Myers and C. S. French then discovered that the green light actually did not have to be presented simultaneously with the red light. Illumination with green light improved the utilization of far-red light even if the green light was turned off several seconds before the red light was turned on.

These observations can be explained by the Z scheme, if the absorption spectra of the antennas of systems I and II are somewhat different. Since the two systems must operate in series, light will be used most efficiently when the flux of electrons through system II is equal to that through system I. If light of some wavelengths excites one of the systems more frequently than it does the other, some of the light will be wasted.

The effectiveness of light of different wavelengths at exciting the two photosystems can be explored by blocking one of the systems. For example, herbicides such as 3-(3,4-dichlorophenyl)-1,1-dimethylurea (DCMU) block electron flow between two quinones at a site close to system II. System I continues to function well in the presence of DCMU if one adds a reductant that can donate electrons to one of the carriers in the chain connecting the two systems. By varying the excitation wavelength, one can show that system I absorbs and uses virtually all wavelengths of light up to about 740 nm. Similar studies of system II show that it is not excited well by far-red light. The pool of electron carriers between the two systems thus will be drained of electrons during illumination with far-red light. P700 will remain oxidized and will be unable to respond to the light. System II does absorb green light well, so illumination with green light can replenish the pool. Electrons remain in the pool when the green light is turned off, explaining why far-red light can be used effectively for a time even after a period of darkness.

Subsequent experiments by L. Duysens provided strong support for the idea that _system I withdraws electrons from carriers situated between the two systems, while system II feeds electrons to these components._ Duysens and colleagues measured the redox state of several of the electron carriers directly. The easiest component to measure is cytochrome _f_, because its absorption spectrum changes markedly when the cytochrome undergoes oxidation. (Refer to Figure 11-9 to see the position of cytochrome _f_ in the Z scheme.) Illumination of algae with far-red light causes essentially all the cytochrome _f_ in the cells to become oxidized (Figure 11-11). When a supplementary green light is turned on, some of the cytochrome returns

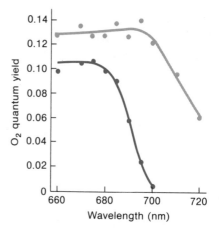

**Figure 11-10**
The quantum yield of $O_2$ evolution from suspensions of green algae (_Chlorella pyrenoidosa_) as a function of the excitation wavelength. Measurements were made without any supplementary light (_lower curve_) and with supplementary yellow-green light (_upper curve_). Without the supplementary light, the quantum yield falls off precipitously at wavelengths above 680 nm. The supplementary light greatly increases the quantum yield obtained with light in the 680- to 710-nm region. The quantum yield was calculated as moles of $O_2$ evolved per einstein of light incident on the sample. The $O_2$ evolution caused by the supplementary light alone was subtracted. All the incident light was absorbed at wavelengths below 700 nm, but at longer wavelengths, some was transmitted. This explains why the apparent quantum yield decreases at wavelengths above 700 nm, even in the presence of the green light. (After R. Emerson et al., Some factors influencing the long-wavelength limit of photosynthesis, _Proc. Natl. Acad. Sci. USA_ 43:133, 1957).

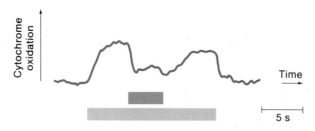

**Figure 11-11**
Cytochrome _f_ oxidation and reduction are measured by absorbance changes at 422 nm when a suspension of red algae (_Porphyridium cruentum_) is illuminated. Cytochrome oxidation gives an upward deflection of the trace (an absorbance decrease). Illumination with red (680-nm) light causes cytochrome oxidation. Green (562-nm) light superimposed on top of the red light causes a reduction. In the absence of red light, when the cytochrome is largely reduced, green light can cause an oxidation. The vertical bar represents an absorbance change of 0.001. (After L. N. M. Duysens and J. Amesz, Function and identification of two photochemical systems in photosynthesis, _Biochim. Biophys. Acta_ 64:243, 1962.)

quickly to the reduced state. This makes sense if green light can drive system II. During the illumination, individual cytochrome molecules will cycle repeatedly through the oxidized and reduced states, so that the cytochrome population as a whole will be in a steady state of partial oxidation and partial reduction. When the green light is turned off, the cytochrome returns to the fully oxidized state. Green light actually can cause either a net oxidation or a net reduction of cytochrome $f$, depending on the initial conditions, so it must be able to excite either system II or system I. Far-red light, however, can only cause oxidation of the cytochrome.

### $O_2$ Evolution Requires the Accumulation of Four Oxidizing Equivalents in System II

The oxidation of $H_2O$ to $O_2$ requires the removal of four electrons for each $O_2$ produced. Because each electron must traverse the photochemical reactions of both system I and system II, one would expect that at least 8 photons would have to be absorbed for each $O_2$ that is released. Currently accepted measurements of the quantum requirement for $O_2$ production are indeed on the order of 8 to 12. (The *quantum requirement* is the reciprocal of the quantum yield.) The quantum requirement will exceed 8 if some of the electrons pumped through the photosystems are not removed to NADP, but instead cycle back into the electron-transport chain between the two photosystems. Such cycling evidently does occur.

If the photosystem II reaction center can transfer only one electron at a time, how does the photosynthetic apparatus assemble the four oxidizing equivalents that are needed for the oxidation of $H_2O$ to $O_2$? One possibility would be that several different system II reaction centers cooperate, but this seems not to happen. Instead, each reaction center progresses through a series of oxidation states, advancing to the next state each time it absorbs a photon. $O_2$ evolution occurs only when the center has accumulated four oxidizing equivalents. This conclusion comes principally from measurements of the amount of $O_2$ that is evolved on each flash when algae or chloroplasts are excited with a series of short flashes after a period of darkness. P. Joliot found that essentially no $O_2$ is released on the first or second flash (Figure 11-12). On the third flash, however, there is a burst of $O_2$. After this, the amount of $O_2$ released on each flash oscillates, going through a maximum every fourth flash.

**Figure 11-12**
Little or no $O_2$ is evolved on the first two flashes, after isolated chloroplasts have been kept in the dark for a time. $O_2$ evolution peaks on the third flash and on every fourth flash thereafter. The oscillations in $O_2$ yield are damped, and after many flashes, the yields converge to the level indicated by the dashed horizontal line. (After B. Forbush et al., Cooperation of charges in photosynthesis $O_2$ evolution-II. Damping of flash yield, oscillation, deactivation, *Photochem. Photobiol.* 14:307, 1971.)

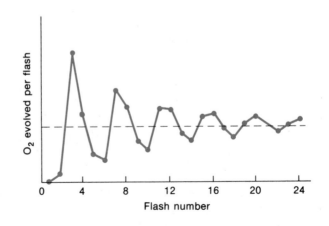

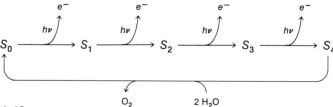

**Figure 11-13**
The $S$ states of the $O_2$-evolution apparatus. One electron ($e^-$) is removed photochemically in each of the transitions $S_0 \rightarrow S_1$, $S_1 \rightarrow S_2$, $S_2 \rightarrow S_3$, and $S_3 \rightarrow S_4$. $S_4$ decays spontaneously to $S_0$ by oxidizing 2 $H_2O$ to $O_2$. The protons are not all released in this step, but come off one or two at a time in several earlier steps.

B. Kok pointed out that this pattern can be explained if the reaction center cycles through five different oxidation states, $S_0$ through $S_4$, as shown in Figure 11-13. When the system reaches state $S_4$, $O_2$ is given off, and the reaction center returns to state $S_0$. The fact that the first burst of $O_2$ comes on the third flash instead of the fourth can be explained if, during the dark period before the flashes, the algae relax mainly into $S_1$ rather than $S_0$. By adjusting the ambient redox potential, one can in fact shift the equilibrium between $S_0$ and $S_1$ in either direction. The gradual damping out of the oscillations is due to reaction centers that get out of phase, either because they miss being excited on one of the flashes or because they are excited twice and advance two steps.

What are the species that actually undergo oxidation and reduction when the reaction center progresses from one of the oxidation states to the next? The answer to this is not yet known. The oxidant that strips electrons from $H_2O$ is certainly not P680$^+$ itself or the electron donor that reacts most directly with P680$^+$. The initial donor, which is called $D$ in Figure 11-9, can be detected by an ESR signal, but its identity is not certain either. There is evidence, however, that the final oxidant involves a complex of several atoms of manganese. Chloroplasts contain approximately four equivalents of manganese per P680 that appear to play a role in $O_2$ evolution, in addition to manganese that is bound to other proteins. Extracting one or more of the four special manganese atoms abolishes $O_2$ evolution. Manganese atoms are seen to undergo changes in oxidation state when chloroplasts are excited with short flashes of light, but whether these atoms are part of the $O_2$-evolving apparatus is still uncertain.

The $E_m$ for the reaction P680$^+$ + $e^- \rightarrow$ P680 must be substantially above +0.8 V, because the secondary oxidants that P680$^+$ generates are powerful enough to oxidize water. The reaction 4 $H^+$ + 4 $e^-$ + $O_2 \rightarrow$ 2 $H_2O$ has an $E_m$ of +0.82 V at pH 7. How P680$^+$ can be such a strong oxidant is not yet clear. One possibility is that the radical cation is formed in an extremely hydrophobic environment or in a region that is already positively charged.

## System II Reduces Pheophytin and Then Quinones

The initial electron acceptors on the reducing side of photosystem II are very similar to those of the bacterial reaction center. The earliest recognized electron acceptor appears to be a molecule of pheophytin $a$. The second and third acceptors are probably both molecules of plastoquinone. Plastoquinone is similar to ubiquinone, with variations in the substituents

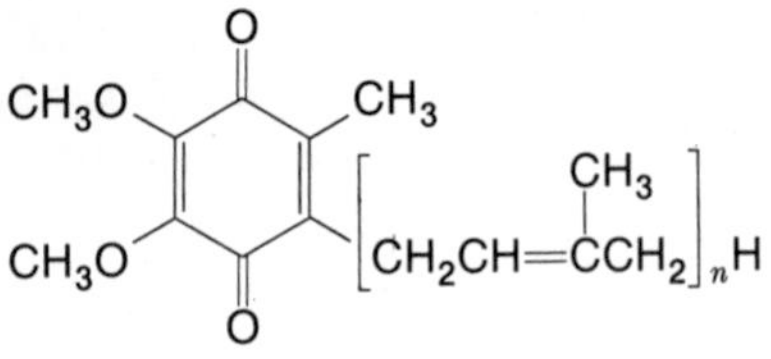

**Ubiquinone**

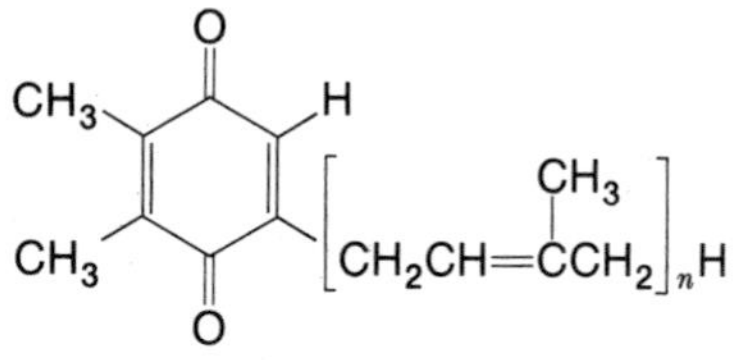

**Plastoquinone**

**Figure 11-14**
Structures of ubiquinone and plastoquinone. The number $n$ of prenyl units in the side chain varies from species to species, but it is generally 7 to 10 for ubiquinone and 9 for plastoquinone.

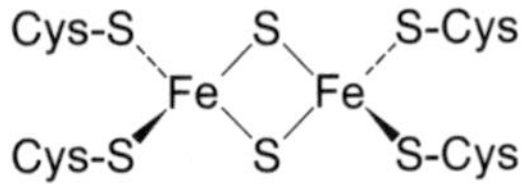

(a)

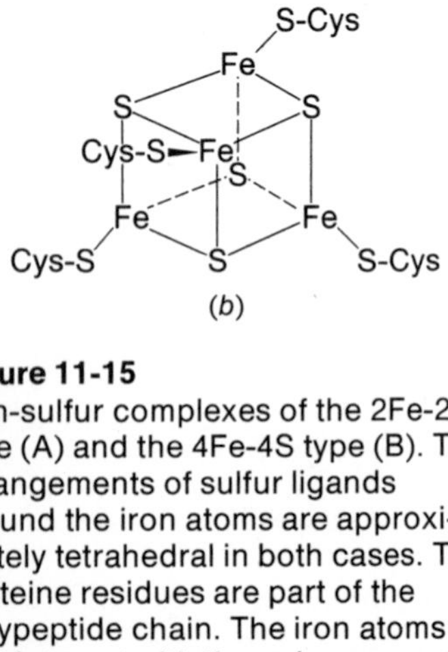

(b)

**Figure 11-15**
Iron-sulfur complexes of the 2Fe-2S type (A) and the 4Fe-4S type (B). The arrangements of sulfur ligands around the iron atoms are approximately tetrahedral in both cases. The cysteine residues are part of the polypeptide chain. The iron atoms that interact with the quinones on the reducing side of photosystem II (FeS in Figure 11-9) or in photosynthetic bacteria are in a different type of structure that has not yet been solved.

on the quinone ring (Figure 11-14). As in bacteria, an iron atom is near the two quinones and may play a role in the movement of electrons between them. Electrons are transferred one at a time from the first plastoquinone to the second. When the second quinone becomes doubly reduced, it transfers a pair of electrons to a large pool of plastoquinones. The pool contains 5 to 10 molecules of plastoquinone per reaction center and seems to be fed by many different reaction centers. From the plastoquinone pool, electrons move to cytochrome $f$, probably by way of an iron-sulfur protein. Cytochrome $f$ reduces the copper protein plastocyanin, which in turn supplies an electron to P700. In some species of algae, a second $c$-type cytochrome replaces plastocyanin.

*The flow of electrons from photosystem I to photosystem II is linked to the transport of protons across the thylakoid membrane.* As in bacteria, the quinones appear to play a major role in the proton movement. Protons evidently are taken up at the outer surface of the thylakoid membrane when plastoquinone is reduced by photosystem II and are released at the inner surface when the plastoquinone is reoxidized. The stoichiometry of proton translocation is still in dispute, but it is probably 1 $H^+$ per electron. Protons also are released inside when $H_2O$ is oxidized to $O_2$ and are bound outside when NADP is reduced to NADPH. Proton movement back out through the thylakoid membrane is conducted by an ATPase, and this movement drives the formation of ATP (see the chapter-opening photograph).

### System I Reduces Chlorophyll and Then Iron-Sulfur Proteins

On the reducing side of photosystem I, the earliest recognized acceptor appears to be a molecule of chlorophyll $a$. In this respect, the reaction center of system I resembles those of system II and photosynthetic bacteria, which use pheophytin $a$, bacteriochlorophyll, and bacteriopheophytin as initial electron acceptors (Table 11-1). From this point on, however, system I is different. The next set of electron carriers consists of *iron-sulfur proteins* instead of quinones.

The first of the iron-sulfur proteins to be characterized was a soluble protein called *ferredoxin*. Several different proteins now go under this name. The soluble ferredoxin obtained from chloroplasts contains two iron atoms and two atoms of inorganic sulfide. These are bound to the sulfur atoms of four cysteine residues, as shown in Figure 11-15$a$. Photosystem I also contains three additional iron-sulfur proteins that are frequently called "bound" ferredoxins because they cannot be solubilized readily. These are designated $FD_x$, $FD_A$, and $FD_B$ in Figure 11-9. $FD_A$ and $FD_B$ each probably contains four iron atoms and four inorganic sulfides held by four cysteines in the cubic structure shown in Figure 11-15$b$. Structures of this sort are found in some of the soluble ferredoxins isolated from other sources. Soluble ferredoxins obtained from some bacteria contain two of the 4 Fe-4 S clusters. $FD_x$ also is likely to contain a 4 Fe-4 S cluster, but it is possible that it has only two irons. If one treats a ferredoxin with mild acid, the iron-sulfur complex decomposes and the inorganic sulfides are released as $H_2S$. The smell of the $H_2S$ is a simple but sensitive assay for the protein.

Ferredoxins undergo one-electron oxidation-reduction reactions. The $E_m$ values for these transitions range from about $-0.4$ V for the soluble ferredoxin of chloroplasts to as low as $-0.7$ V for $FD_x$. (Iron-sulfur proteins with much more positive $E_m$ values also are known; the iron-sulfur protein that appears to operate between the two photosystems has an $E_m$ near +0.3 V.) $FD_A$ and $FD_B$ have $E_m$ values of approximately $-0.54$ and $-0.59$ V. Reduction of a ferredoxin can be detected by the ESR signal of the reduced species and by a small decrease in absorption near 430 nm. $FD_x$ appears to be reduced by the chlorophyll *a* molecule that reacts with P700. It then transfers an electron to ferredoxins A and B, which in turn reduce the soluble ferredoxin ($FD_S$ in Figure 11-9). From the soluble ferredoxin, electrons move to a flavoprotein (*ferredoxin-NADP oxidoreductase*) and then to NADP.

Whether ferredoxins A and B operate in series or in parallel is not yet clear. It is possible that one of them serves mainly to transfer electrons to a component other than the soluble ferredoxin. When the availability of NADP is limiting, electrons ejected by the reaction center of photosystem I can be passed to a *b*-type cytochrome (cytochrome *b*563) and then returned to the plastoquinone pool between the two photosystems (Figure 11-9). This cyclic electron flow is coupled with proton translocation across the thylakoid membrane and probably occurs to some extent even when NADP is not limiting. Its role presumably is to bolster the supply of ATP to drive the fixation of $CO_2$ or other energy-requiring processes. As we shall see shortly, the conversion of $CO_2$ into fructose-6-phosphate requires two equivalents of NADPH and three equivalents of ATP for each $CO_2$ that is fixed. During continuous illumination, the NADPH and approximately two of the three equivalents of ATP are probably obtained from noncyclic electron transport, and the third equivalent of ATP is obtained from cyclic electron transport.

## CARBON FIXATION

### The Reductive Pentose Cycle

*The ATP and NADPH generated by the photosynthetic electron-transfer reactions are used to drive the fixation of $CO_2$.* Reactions in which $CO_2$ is incorporated into carbohydrates were discovered in the early 1950s by M. Calvin and coworkers in a series of studies that were among the first to use radioactive tracers. A key experiment was to expose algae to a brief period of illumination in the presence of $^{14}CO_2$ and then to disrupt the cells quickly and search for organic molecules that had become labeled with $^{14}C$. The material that turned out to be labeled most rapidly was 3-phosphoglyceric acid, and almost all its $^{14}C$ was found to be in the carboxyl group. To identify the precursor of the 3-phosphoglycerate, Calvin's group looked for changes in the steady-state concentrations of other compounds when they turned a continuous light on or off or when they suddenly raised or lowered the concentration of $CO_2$. These experiments showed that 3-phosphoglycerate is formed from ribulose-1,5-diphosphate. A molecule of $CO_2$ is incorporated in the process, so that one molecule of the five-carbon sugar ribulose-1,5-diphosphate gives rise to two molecules of the three-carbon 3-phosphoglyceric acid (Figure 11-16). The reaction is catalyzed by *ribulose diphosphate carboxylase*, which is found in the stromal region of chloroplasts.

Further studies by Calvin's group revealed that 3-phosphoglycerate can be converted back to ribulose-1,5-diphosphate under the influence of addi-

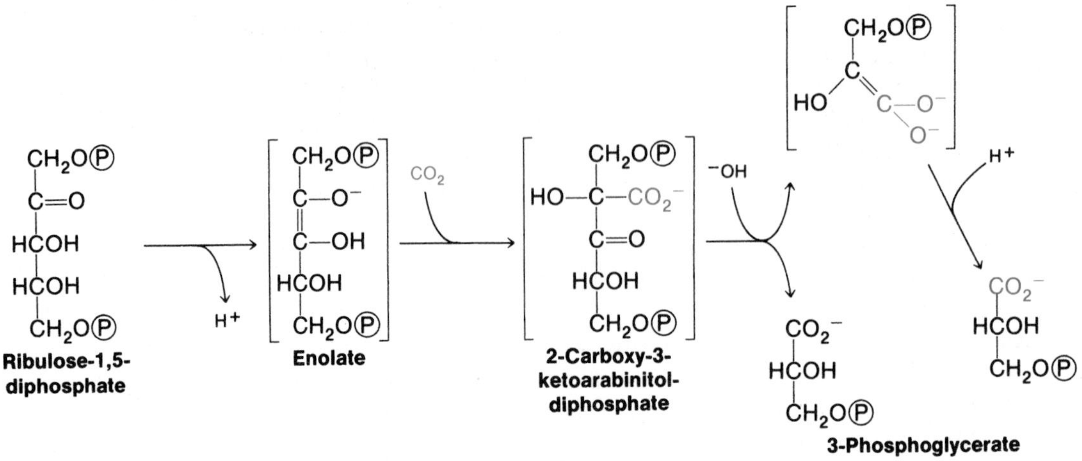

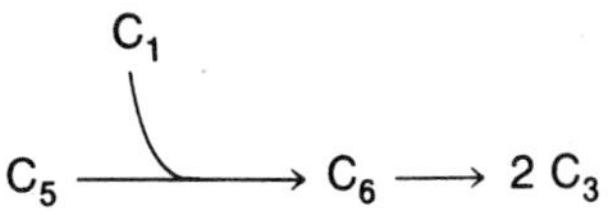

**Figure 11-16**
The ribulose diphosphate carboxylase reaction probably involves 2-carboxy-3-ketoarabinitol-1,5-diphosphate as an enzyme-bound intermediate. This could form by the addition of $CO_2$ to the enolate of ribulose-1,5-diphosphate. The substrate is known to be $CO_2$ rather than bicarbonate.

tional enzymes in the stroma. The reactions of this _reductive pentose cycle_, or _Calvin cycle_, are shown in Figure 11-17. The 3-phosphoglyceric acid is first phosphorylated to 1,3-diphosphoglycerate at the expense of ATP and then reduced to glyceraldehyde-3-phosphate by NADPH. (Note that the nucleotide specificity in the reductive step differs from that of the cytoplasmic enzyme glyceraldehyde-3-phosphate dehydrogenase, which uses NAD and NADH.) Glyceraldehyde-3-phosphate and its isomerization product dihydroxyacetone phosphate can combine to form fructose-1,6-diphosphate under the influence of aldolase. Fructose-6-phosphate, formed by hydrolysis of the fructose-1,6-diphosphate, combines with another molecule of glyceraldehyde-3-phosphate, generating xylulose-5-phosphate and erythrose-4-phosphate. This reaction involves thiamine pyrophosphate and is similar to the steps catalyzed by transketolase in the phosphogluconate pathway (Chapter 10). Erythrose-4-phosphate combines with dihydroxyacetone phosphate to form sedoheptulose-1,7-diphosphate in a second reaction catalyzed by aldolase. After hydrolysis to sedoheptulose-7-phosphate, the seven-carbon sugar reacts with glyceraldehyde-3-phosphate in another transketolase reaction, forming ribulose-5-phosphate and a second molecule of xylulose-5-phosphate. Xylulose-5-phosphate can be isomerized to ribulose-5-phosphate. Finally, ribulose-5-phosphate is phosphorylated to ribulose-1,5-diphosphate at the expense of additional ATP, completing the cycle.

**Figure 11-17** ▶
The reductive pentose cycle, or Calvin cycle. The number of arrows drawn at each step in the diagram indicates the number of molecules proceeding through that step for every three molecules of $CO_2$ that enter the cycle. The entry of three molecules of $CO_2$ results in the net formation of one molecule of glyceraldehyde-3-phosphate (_box on right_).

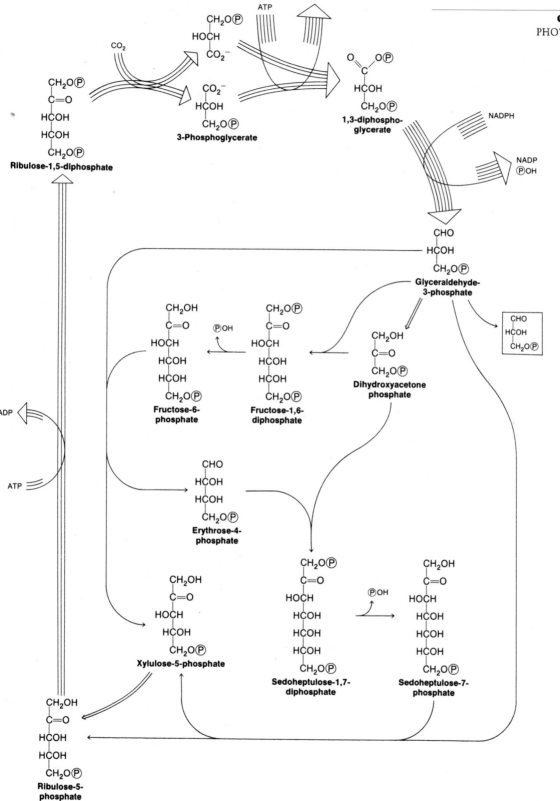

In Figure 11-17, note that three molecules of ribulose-1,5-diphosphate are regenerated for every three that are carboxylated. In the process, three molecules of $CO_2$ are taken up, and there is a gain of one molecule of glyceraldehyde-3-phosphate. The expenses are nine molecules of ATP converted to ADP and six molecules of NADPH oxidized to NADP, or three molecules of ATP and two of NADPH consumed per $CO_2$ fixed. The triose phosphate that is saved is exported from the chloroplast to the cytoplasm or converted to hexoses for storage in the chloroplast as starch.

## Photorespiration and the C-4 Cycle

Ribulose diphosphate carboxylase frequently accounts for more than half the soluble protein in a leaf. It is surely the world's most abundant enzyme, and probably the most abundant protein. The enzyme found in higher plants and in some species of bacteria contains eight 70,000-dalton protomers, each consisting of one large subunit with a molecular weight of 56,000 and one smaller subunit with a molecular weight of 14,000. The larger subunit of the plant enzyme is synthesized within the chloroplast under the direction of chloroplast DNA and ribosomes. The smaller subunit is synthesized on cytoplasmic ribosomes under the direction of nuclear DNA and has to be brought into the chloroplast for assembly of the intact enzyme. The large subunit is catalytically active in the absence of the smaller one, and what role the smaller subunit plays is unknown.

In addition to serving as a substrate, $CO_2$ activates ribulose diphosphate carboxylase by binding to the ε-amino group of a lysyl residue in the large subunit to form a carbamate (lysine—NH—$CO_2^-$). This process requires $Mg^{2+}$, which probably binds to the carboxyl group of the carbamate. The $Mg^{2+}$ in turn probably forms part of the binding site for a second molecule of $CO_2$, which acts as the substrate in the carboxylase reaction.

Studies of ribulose diphosphate carboxylase have taken on additional complexity with the discovery that the enzyme catalyzes a competing reaction in which $O_2$ takes the place of $CO_2$ as a substrate. The products of the _oxygenase_ reaction are 3-phosphoglycerate and a two-carbon acid, phosphoglycolate (Figure 11-18). Phosphoglycolate is oxidized to $CO_2$ by $O_2$ in a series of additional reactions involving enzymes in the cytoplasm, mitochondria, and another organelle, the peroxisome. This _photorespiration_ appears to constitute a severe drain on chloroplast metabolism. In some plants, on the order of one-third of the $CO_2$ fixed is released again by

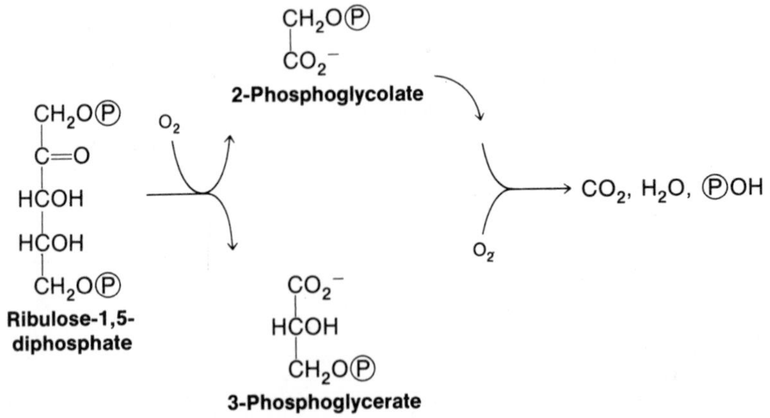

**Figure 11-18**
Photorespiration results from the oxygenation reaction catalyzed by ribulose-diphosphate carboxylase/oxygenase. Phosphoglycolate generated in the reaction is oxidized further to $CO_2$ by other enzymes. The oxygenation reaction, like carboxylation, does not require light directly. It occurs mainly during illumination, however, because the formation of the substrate, ribulose-1, 5-diphosphate requires ATP (see Figure 11-17).

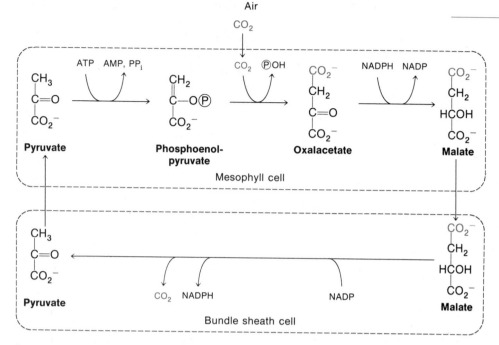

**Figure 11-19**
The C-4 pathway involves the cooperation of two types of cells: mesophyll cells (*top*) and bundle sheath cells (*bottom*).

photorespiration. If photorespiration has any significant benefit to the plant, it is not obvious.

The outcome of the competition between $CO_2$ and $O_2$ in the ribulose diphosphate carboxylase/oxygenase reactions depends on the concentrations of the two gases. The $K_m$ of the activated enzyme for the substrate molecule of $CO_2$ is about 20 $\mu M$ and that for $O_2$ is about 200 $\mu M$, so $CO_2$ is the preferred substrate. In illuminated chloroplasts, however, the concentration of $O_2$ is elevated as a result of the photosynthetic splitting of $H_2O$, while the concentration of $CO_2$ is kept low by the formation of carbohydrates. The concentrations of the two gases are frequently on the order of the $K_m$ values, making $O_2$ a serious competitor.

If photorespiration is only a liability, one might think that plants would have evolved a ribulose diphosphate carboxylase that had minimal activity as an oxygenase, but there is no evidence that evolution has moved in this direction. The oxygenase activity of the enzyme isolated from photosynthetic bacteria is similar to that of the enzyme from higher plants. *Some species of plants have, however, evolved an alternative strategem for favoring the carboxylase reaction over the oxygenase: they increase the concentration of $CO_2$ in the region of the enzyme*. These plants, which include corn, sugar cane, and numerous tropical species, have a layer of specialized mesophyll cells at the outer surface of the leaf. The mesophyll cells take up $CO_2$ from the air by the carboxylation of phosphoenolpyruvate to obtain oxalacetate (Figure 11-19). Oxalacetate is reduced to malate, which then moves from the mesophyll cells to the bundle sheath cells that surround the vascular structures in the interior of the leaf. Here the malate is decarboxylated to pyruvate in an oxidative reaction that reduces NADP to NADPH. The pyruvate returns to the mesophyll cell, where it is phosphorylated to phosphoenolpyruvate. The phosphorylation is driven by the splitting of ATP to AMP and pyrophosphate and the subsequent hydrolysis of the pyrophosphate to inorganic phosphate.

The result of the cycle, shown in Figure 11-19, is the delivery of $CO_2$ and reducing power (NADPH) to the bundle sheath cell at the cost of two high-energy phosphate bonds. The bundle sheath cell fixes the $CO_2$ by the ribulose diphosphate carboxylase reaction and the reductive pentose cycle. Plants that have this mechanism for concentrating $CO_2$ can grow considerably more rapidly than species that do not, particularly in hot, sunny climates. The auxiliary cycle is called the _C-4 cycle_ because it involves the four-carbon acids malate and oxalacetate. In some species, the oxalacetate is transaminated to aspartate rather than being reduced to malate, and it is aspartate that moves to the bundle sheath cells. There is currently interest in the possibility of increasing agricultural productivity by using genetic-engineering techniques to introduce the C-4 pathway into plant species that lack it.

## SELECTED READINGS

Clayton, R. K. _Photosynthesis: Physical Mechanisms and Chemical Patterns_. Cambridge: Cambridge University Press, 1980. An excellent general treatment of photosynthesis.

Duysens, L. N. M., and Amesz, J. Function and identification of two functional systems in photosynthesis. _Biochim. Biophys. Acta_ 64:243, 1962. Evidence supporting the Z scheme is obtained by measuring oxidation and reduction of cytochrome _f_ in algae illuminated with light of various colors.

Emerson, R. The quantum yield of photosynthesis. _Ann. Rev. Plant Physiol._ 9:1, 1958. The quantum yield of $O_2$ declines in far-red light, but it can be elevated by a weak background light of shorter wavelengths.

Forbush, B., Kok, B., and McGloin, M. Cooperation of charges in photosynthetic $O_2$ evolution: II. Damping of flash yield, oscillation, deactivation. _Photochem. Photobiol._ 14:307, 1971. Oscillations in $O_2$ yield during a train of flashes provide evidence for the four S states.

Govindjee (Ed.). _Photosynthesis: Energy Conversion by Plants and Bacteria_. New York: Academic Press,

1982. A collection of review articles discussing current work on many aspects of photosynthesis.

Lorimer, G. H. The carboxylation and oxygenation of ribulose-1,5-bisphosphate: The primary events in photosynthesis and photorespiration. _Ann. Rev. Plant Physiol._ 32:349, 1981. Mechanisms and biochemical significance of the carboxylase and oxygenase reactions.

Shuvalov, V. A., and Parson, W. W. Energies and kinetics of radical pairs involving bacteriochlorophyll and bacteriopheophytin in bacterial reaction centers. _Proc. Natl. Acad. Sci. USA_ 78:957, 1981. The roles of bacteriochlorophyll, bacteriopheophytin and $Q_A$ as early electron acceptors are explored by exciting reaction centers with short flashes under conditions that block electron transfer to the quinone.

Shuvalov, V. A., Dolan, E., and Ke, B. Spectral and kinetic evidence for two early electron acceptors in photosystem I. _Proc. Natl. Acad. Sci. USA_ 76:770, 1979. Fast absorbance changes are measured to investigate the roles of chlorophyll and ferredoxins as electron acceptors.

## PROBLEMS

1. The lowest-energy absorption band of P870 is at 870 nm.
   a. Calculate the energy of an einstein of light with this wavelength.
   b. Estimate the effective midpoint potential $E_m$ of P870 in its first excited singlet state if the $E_m$ for oxidation in the ground state is $+0.45$.

2. The traces below show measurements of optical absorbance changes at 870 and 550 nm when a suspension of chromatophores was excited with a short flash. Downward deflections of the traces represent absorbance decreases. Explain the observations. (Absorption spectra of a _c_-type cytochrome in its reduced and oxidized forms are shown in Chapter 10.)

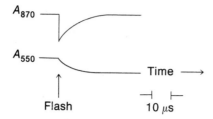

$A_{870}$

$A_{550}$

Time →

Flash    ⊢  ⊢
         10 μs

3. Ubiquinone has an absorption band at 275 nm. This band bleaches when the quinone is reduced to either semiquinone or dihydroquinone. The anionic semiquinone has an absorption band at 450 nm, but neither the quinone nor the dihydroquinone has a band at this wavelength. If a suspension of purified bacterial reaction centers is supplemented with extra ubiquinone and reduced cytochrome $c$ and is then excited with a series of short flashes, the absorbance at 275 nm decreases on the odd-numbered flashes as shown in the first curve below. The absorbance at 450 nm increases on the odd-numbered flashes, but returns to its original level on the even-numbered flashes, as shown in the second curve.
   a. Explain the patterns of absorbance changes at the two wavelengths.
   b. Why is it necessary to have reduced cytochrome $c$ present in order to see these effects?

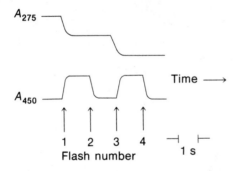

$A_{275}$

Time →

$A_{450}$

↑  ↑  ↑  ↑
1  2  3  4    ⊢  ⊢
              1 s
Flash number

4. A suspension of algae is excited with a train of 1000 short flashes spaced 20 ms apart.
   a. Draw a graph showing how the total amount of $O_2$ that the algae evolve during the flash train will depend on the intensity of the flashes. Use meaningful units for the scales on the abscissa and ordinate of the graph.
   b. How could you use the data from such a graph to calculate the average quantum yield of $O_2$ evolution?
   c. How could you use the data to calculate the effective size of the antenna system?
   d. Why is it necessary to use a large number of flashes for these determinations?

5. You add a nonphysiological electron donor to a suspension of chloroplasts. When you illuminate the chloroplasts, the donor becomes oxidized. How could you determine whether this process involves both photosystem I and photosystem II, only photosystem I, or only photosystem II? (In principle, the donor could transfer electrons either to some component on the $O_2$ side of photosystem II or to a component between the two photosystems.)

6. The photoreduction of the early electron carrier $FD_x$ in photosystem I can be detected by the optical absorption spectrum of the reduced species. The reduced carrier is not ordinarily seen in chloroplasts that are illuminated under physiological conditions, but it can be seen when chloroplasts are illuminated at very low temperatures after the ambient redox potential has been poised at a strongly negative level.
   a. Why might it be necessary to lower the temperature and the redox potential?
   b. How negative would you predict the redox potential would have to be in order to make the photoreduction of $FD_x$ readily observable?
   c. Does the fact that the photoreduction of $FD_x$ can be seen under these conditions prove that $FD_x$ participates in the Z scheme under physiological conditions? Why or why not?

7. When chloroplasts are illuminated with weak continuous light, the intensity of fluorescence from the antenna chlorophylls is relatively low. (About 2 percent of the photons absorbed are reemitted as fluorescence.) The intensity of the fluorescence from the antenna chlorophyll of photosystem II increases several-fold if the ambient redox potential is lowered below about 0 mV. Explain these observations.

8. When chloroplasts are illuminated with a strong continuous light, the intensity of fluorescence from the chlorophyll of the photosystem II antenna increases with time, as shown in the graph below (curve $A$).
   a. Explain the increase in fluorescence.
   b. In the presence of 3-(3,4-dichlorophenyl)-1,1-dimethylurea (DCMU), the fluorescence increases more abruptly (curve $B$). Explain why there is a lag in the increase in the absence of DCMU (curve $A$) but not in the presence of DCMU (curve $B$).

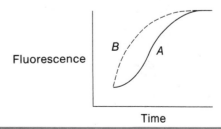

Fluorescence

B   A

Time

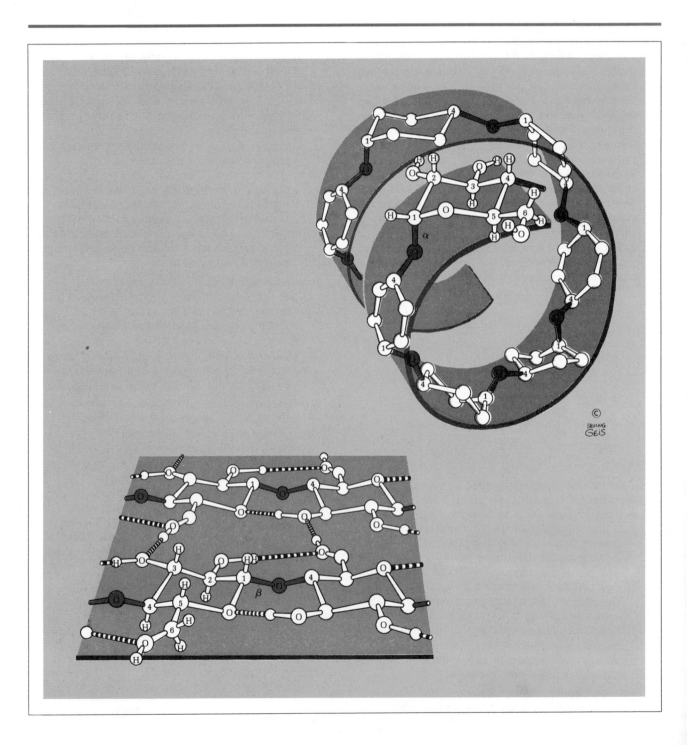

# 12

# STRUCTURE AND SYNTHESIS OF SUGARS

Carbohydrates make up the bulk of organic substance on the earth. As we have seen in the preceding four chapters, they are an important source of carbon compounds and chemical energy either for immediate use in such compounds as glucose or fructose or for storage in the form of polysaccharides such as starch or glycogen. Many carbohydrates, such as cellulose and chitin, serve a structural function. Structural polysaccharides are frequently found in combination with proteins (glycoproteins or mucoproteins) or lipids (lipopolysaccharides). For classification purposes, carbohydrates may be divided into monosaccharides, oligosaccharides (i.e., disaccharides and trisaccharides, etc.), and high-molecular-weight polysaccharides. In this chapter we will discuss some of the basic structural characteristics of naturally occurring carbohydrates as well as their biosynthesis.

## MONOSACCHARIDES AND RELATED COMPOUNDS

Monosaccharides are either polyhydroxy aldehydes, aldoses, or polyhydroxy ketones, i.e., ketoses. They may be thought of as derived from polyalcohols (polyols) by oxidation of one carbinol group to a carbonyl group. For example, the simple three-carbon triol gycerol can be converted either to glyceraldehyde or dihydroxyacetone by the loss of two hydrogens (see Figure 12-1).

Since the middle carbon atom of glyceraldehyde is connected to four different substituents, it is a chiral center. This single chiral center leads to two possible forms of glyceraldehyde known as the D and L forms. D-glyceraldehyde is illustrated such that in the so-called Fischer projection formula, the OH group attached to the central carbon atom points to the

*D-Glucose molecules can be α(1→4)-linked or β(1→4)-linked structures in polyglucose. The former polymers tend to adopt helical conformations (above) and the latter (below) tend to adopt extended chain conformations (also see Figure 12-16 and the accompanying discussion).*

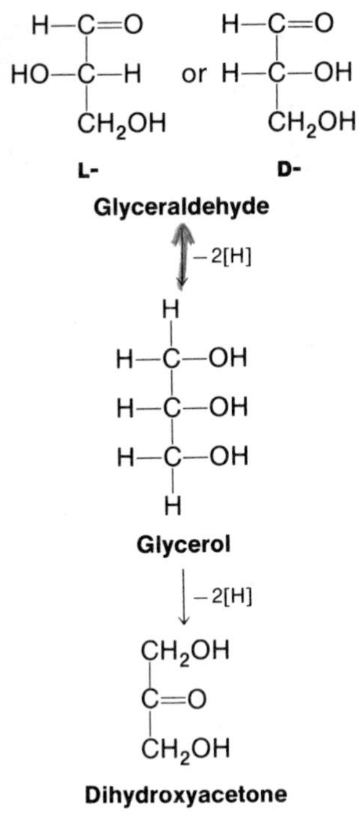

H—C=O     H—C=O
HO—C—H   or   H—C—OH
CH₂OH       CH₂OH

**L-**         **D-**

**Glyceraldehyde**

↕ −2[H]

H
H—C—OH
H—C—OH
H—C—OH
H

**Glycerol**

↓ −2[H]

CH₂OH
C=O
CH₂OH

**Dihydroxyacetone**

**Figure 12-1**
Loss of two hydrogens by glycerol leads to the formation of glyceraldehyde or dihydroxyacetone depending on whether the two hydrogens are lost from the end or middle position, respectively.

right. If the central carbon were in the plane of the paper with tetrahedrally arranged substituents, the H and OH would project above the plane of the paper and the other two substituents would project below the plane of the paper. A molecule such as glyceraldehyde having one center of asymmetry (chiral center) is optically active, and the two forms of the molecule can be described as the dextrorotatory or the levorotatory forms according to the way in which they rotate plane-polarized light. This is testable by making a solution of an optically active compound and measuring the rotation of plane-polarized light through it. The apparatus used for this purpose is called a _polarimeter_. A mixture containing equal amounts of the two forms is optically inactive and is referred to as a _racemic mixture_. The symbol $d$ or $(+)$ refers to dextrorotatory rotation and $l$ or $(-)$ refers to levorotatory rotation. The specific rotation $[\alpha]$ is defined as $[\alpha] = (100 \times A)/(c \times l)$, where $A$ is the observed rotation in degrees, $c$ is the concentration of the optically active substances in grams per 100 ml of solution, and $l$ is the path length in decimeters of the solution through which the rotation is observed.

The convention for the nomenclature of sugars has been devised based on configurational rather than optical properties. The actual sign of rotation may still be indicated by the italic letters $d$ and $l$, but the configuration is shown by the prefixed symbols in small roman capitals D and L. For the common sugars, the rule relates to that center of asymmetry most remote from the aldehydic end of the molecule. The conventions as applied to the glyceraldehydes are such that D-glyceraldehyde is the $d$ form and L-glyceraldehyde is the $l$ form. Other sugars terminating in these configurations are said to belong to the D or L configurational series. We may visualize the $C_4$, $C_5$, and $C_6$ sugars as arising from the trioses through the stepwise condensation of formaldehyde to either glyceraldehyde or dihydroxyacetone. The actual synthesis of the sugars is carried out by other means.

## SUGARS CAN FORM INTRAMOLECULAR HEMIACETALS

Aldehydes can add hydroxyl compounds to the carboxyl group. If a molecule of water is added, the product is an aldehyde hydrate, as shown in Figure 12-2. If a molecule of alcohol is added, the product is a hemiacetal. Addition of a second alcohol produces an acetal. Sugars can often form _intramolecular hemiacetals_, and they do so readily in water in cases where the resulting compound has a five- or six-membered ring.

That sugars can exist in more than one configurational state as a result of hemiacetal formation first became apparent from observations with D-glucose that the rotation of a freshly dissolved sample slowly changes. This is due to the existence of two different forms that are convertible in solution; they are referred to as α-D-glucose, where $[\alpha] = 112$ degrees, and β-D-glucose, where $[\alpha] = 19$ degrees. A freshly prepared solution of either

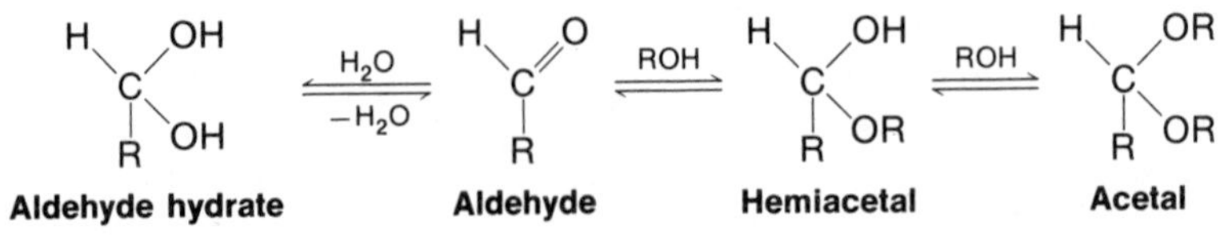

**Figure 12-2**
Aldehydes can add H₂O to form hydrates or alcohols to form hemiacetals and acetals.

**Aldehyde hydrate**     **Aldehyde**     **Hemiacetal**     **Acetal**

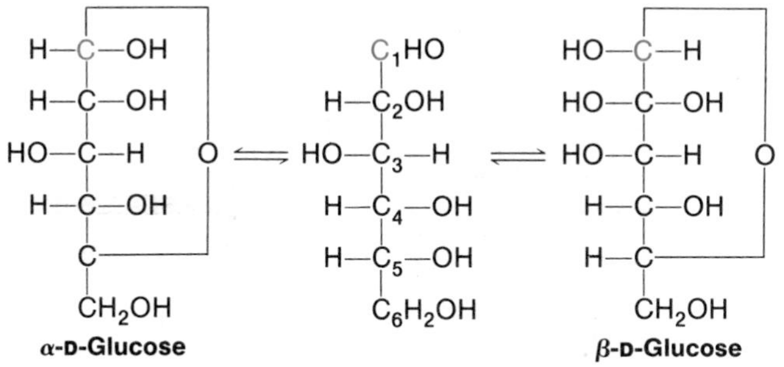

$\alpha$-D-Glucose            $\beta$-D-Glucose

**Figure 12-3**
Different forms of glucose that result from dissolving glucose in water. The "anomeric" carbon is shown in color.

compound will eventually approach an intermediate value. The exact intermediate value depends on the equilibrium between the two forms. Formulas for the two hemiacetals are indicated in Figure 12-3.

The $\alpha$ designation is used to indicate that the aldehyde or C-1 hydroxyl group is on the same side of the structure as the ring oxygen, and the $\beta$ designation indicates the reverse situation. The convention for numbering hexoses is indicated in the central structure of Figure 12-3. When the sugar is dissolved in water, the hemiacetal is in equilibrium with the straight-chain hydrated form. From the straight-chain form either hemiacetal can be formed. The conversion of one stereoisomer to another in solution is referred to as _mutarotation_. The two different forms are referred to as _anomers_, and the _anomeric carbon_ is that carbon which contains the reactive carbonyl. An equilibrium situation is reached without added catalyst in a few hours at room temperature. The open chain usually does not represent more than a small fraction of the total.

Hemiacetals with five-membered rings are called _furanoses_, and hemiacetals with six-membered rings are called _pyranoses_. In cases where either five- or six-membered rings are possible, the five-membered ring usually predominates, but there are notable exceptions, such as glucose. Studies with three-dimensional modular models suggest that there is usually less crowding or steric repulsion in the five-membered ring. The furanoses or pyranoses are more realistically represented by pentagons or hexagons by the Haworth convention, as shown in Figure 12-4. The lower edge of the ring (heavy line) is to be thought of as projecting out, and the other edge is to be thought of as projecting behind the plane of the paper. Haworth structures are easy to draw and unambiguous in depicting configurations, but they do not show correctly the spacial relationship of groups attached to rings.

Whereas the aromatic compound benzene forms a perfect hexagon, the normal valence angle of the saturated carbon of 109 degrees prevents a stable planar arrangement for cyclohexane or the related pyranose molecule. The two most likely conformations are the so-called chair and boat forms. Usually the chair form is considerably more stable than the boat form. The 12 substituent atomic groups of the ring carbons fall into two classes, those which are approximately perpendicular to the plane of the ring, i.e., axial, and those which are parallel to the plane of the ring, i.e., equatorial (see Figure 12-5). As a rule, a substituent is at a lower energy state in the equatorial location because of less chance of steric hindrance with other substituents. This becomes more important with larger sub-

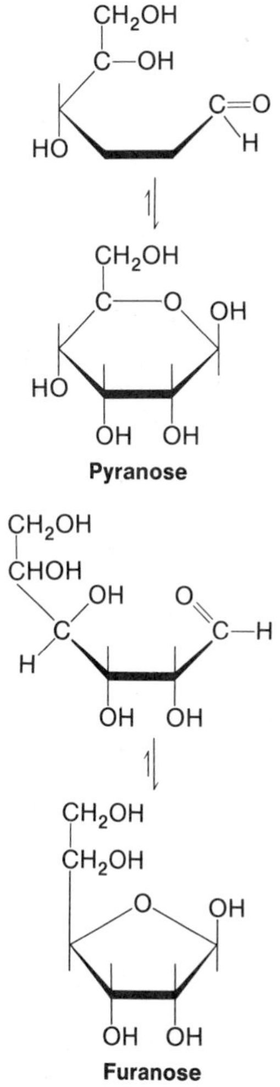

**Pyranose**

**Furanose**

**Figure 12-4**
Formation of a pyranose and a furanose from open-chain forms—an intramolecular reaction.

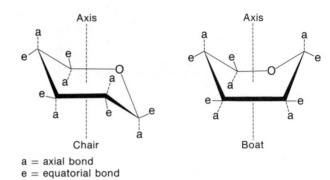

**Figure 12-5**
Chair and boat forms for a generalized pyranose ring structure.

a = axial bond
e = equatorial bond

stituents. The two anomers for the favored chair-form configuration of D-glucose are shown in Figure 12-6. In sugar chemistry, conformational formulation of aldohexoses is important in interpreting reactivity of hydroxyl groups and other sizable substituents. To find the most stable configuration for a particular aldohexose, one examines, by trial and error, that chair form in which the maximum number of substituents larger than hydrogen are in the equatorial position.

## FAMILIES OF SUGARS ARE STRUCTURALLY RELATED

There are about 20 naturally occurring monosaccharides. The aldose D-glyceraldehyde is an intermediate in the degradation of carbohydrates that is in equilibrium with dihydroxyacetone phosphate (Figure 12-1) by means of an enzyme. The tetroses, pentoses, and hexoses related to D-glyceraldehyde are shown in Figure 12-7. A similar series exists for L-glyceraldehyde. The ketoses (e.g., fructose) are similarly related to dihydroxyacetone.

## SUGARS ARE LINKED BY GLYCOSIDIC BONDS IN DISACCHARIDES AND POLYSACCHARIDES

Warming glucose in methanol and acid produces a mixture of two new substances, $\alpha$- and $\beta$-methylglucosides (Figure 12-8). In general, derivatives of glucose are referred to as _glucosides_, derivatives of galactose as _galactosides_, and so on. The bond between the sugar and the alcohol is

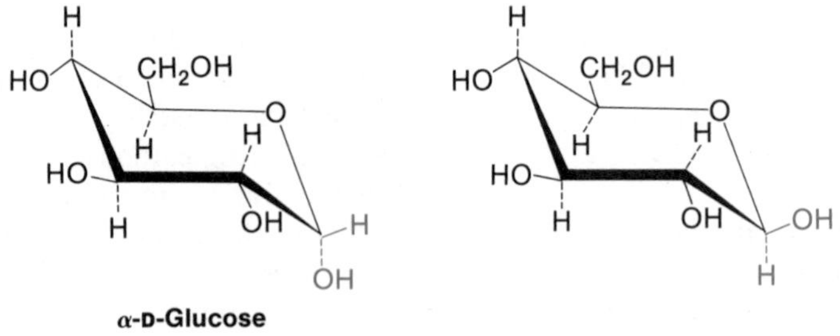

**Figure 12-6**
Chair configuration for the two anomers of D-glucose.

**α-D-Glucose**

referred to as the _glycosidic bond_, and the compound by the generic name _glycoside_. The formation of glycoside from sugar and methanol by acid catalysis is identical to the formation of acetal from aldehyde and alcohol (see Figure 12-2). While the two forms of glucose in solution are in equilibrium through mutarotation, the corresponding glucosides are locked into one configuration. This is understandable, because mutarotation requires that, in the intermediate, the anomeric carbon adopt a carbonyl structure. A glycoside can be formed with aliphatic alcohols, phenols, and carboxylic acids. The sugarless moiety of a glycoside is sometimes called the _aglycon_.

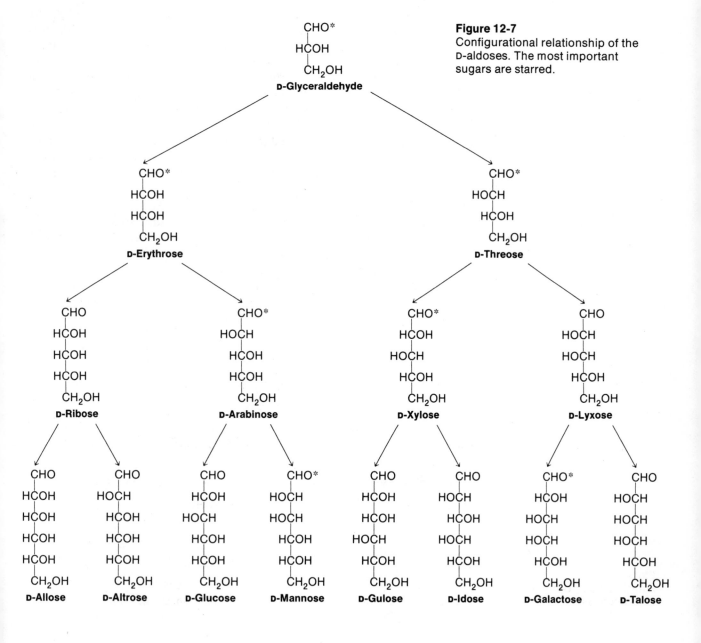

**Figure 12-7**
Configurational relationship of the D-aldoses. The most important sugars are starred.

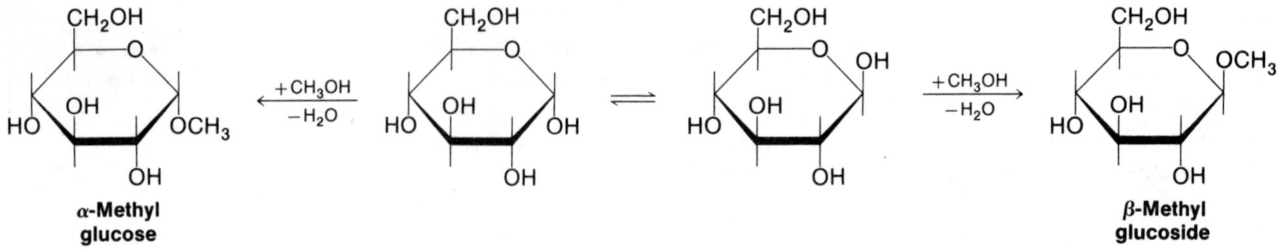

**Figure 12-8**
Formation of methyl glucosides.

## Disaccharides

Sugars are glycosidically linked in disaccharides (Figure 12-9) and polysaccharides. For instance, the disaccharide maltose contains a glucosidic bond between the C-1 of one glucose molecule and the C-4 of another glucose molecule. The compound is said to have an $\alpha(1{\rightarrow}4)$ glucosidic linkage because the C-1 carbon of the glucosidic bond is in the $\alpha$ configuration. Maltose possesses one potentially free aldehyde group, which makes it a reducing sugar. Because it can undergo mutarotation, the configuration about the hemiacetal hydroxyl group has not been specified in Figure 12-9. Maltose is a degradation product of starch. The disaccharide cellobiose is identical with maltose except for having a $\beta(1{\rightarrow}4)$ glucosidic linkage. Cellobiose is a degradation product of cellulose. Lactose is a disaccharide found exclusively in the milk of mammals. Lactose contains a $\beta(1{\rightarrow}4)$ glycosidic linkage between galactose and glucose. Sucrose is found in abundance in sugar beets and sugar cane. On acid hydrolysis it yields 1 mol of both D-glucose and D-fructose. Sucrose is a nonreducing sugar. Since the reducing groups are found on the C-1 and C-2 carbon atoms, respectively, of glucose and fructose, the glycosidic bond of the disaccharide must be between these two carbon atoms. For some reason, fructose possesses a furanose ring structure in the disaccharide, even though the pyranose ring is energetically favorable in the free ketohexose.

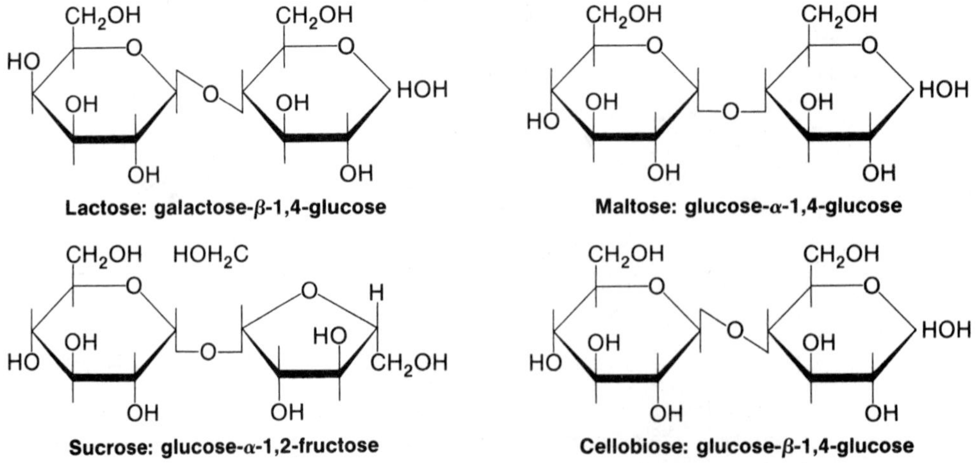

**Figure 12-9**
Four commonly occurring disaccharides.

Lactose: galactose-β-1,4-glucose

Maltose: glucose-α-1,4-glucose

Sucrose: glucose-α-1,2-fructose

Cellobiose: glucose-β-1,4-glucose

# Polysaccharides

Most carbohydrates in nature exist as high-molecular-weight polymers called polysaccharides. *Polysaccharides exist for two quite distinct purposes: some serve as a means for storage of chemical energy (see Chapter 8) and others serve a structural function.* It is much more convenient to store potential energy in the high-polymer form; more energy can be stored in a compact area, and the high molecular weight of a polymer minimizes osmotic pressure effects and diffusion. Most frequently, D-glucose is the building block of polysaccharides, although D-mannose, D- and L-galactose, D-xylose, L-arabinose, as well as D-glucuronic, D-galacturonic and D-mannuronic acids, D-glucosamine, D-galactosamine, and neuraminic acid also are used. Polymers that use one type of building block are called *homopolymers*, and those which use more than one are called *heteropolymers*.

**Structurally Important Polymers.** Cellulose is found in most plants in the cell wall. Cotton is 90 percent cellulose. On hydrolysis, cellulose yields the monomer glucose and some of the dimer cellobiose. When a sugar is reacted with dimethylsulfate and alkali, all the free hydroxyl groups are converted to methyl ethers. The stability of the O-methyl ether renders these derivations very useful in structural analysis. Thus fully methylated cellulose gives 2,3,6-tri-O-methylglucose on hydrolysis. Since the 1 and 4 positions do not react with dimethylsulfate in the polymer form, they are probably involved in the glycosidic links holding the polymer together. In this way it was deduced that the glycosidic linkages in cellulose are of the 1,4 type. Other measurements show that the 1,4 linkages are $\beta$. The repeating unit of cellulose is indicated in Figure 12-10.

Cellulose is insoluble in water. The polymeric fibers have a molecular weight of 50,000 or greater. The chain molecules of cellulose occur in parallel bundles of about 2000 chains, a bundle having a diameter of 100 to 250 Å. Each bundle of 2000 comprises a single microfibril. Many microfibrils arranged in parallel comprise a macrofibril, which can be seen under the light microscope. Figure 12-11 shows the inner secondary walls of the plant *Valonia*; the fibrils in the secondary wall are almost pure cellulose.

In the woody material of plants, cellulose is held together by the complex polymer *lignin*, which appears to consist of a repeating unit derived from coniferyl alcohol (Figure 12-12). *Chitin*, found in the shells of crustaceans and insects, is a linear $\beta(1\rightarrow4)$ polymer of N-acetyl-D-glucosamine (Figure 12-13).

Complexes between proteins and polysaccharides are termed *mucoproteins* or *glycoproteins*. Acid mucopolysaccharides are found in connective tissue, bacterial cell walls, eyeball synovial fluid, and the circu-

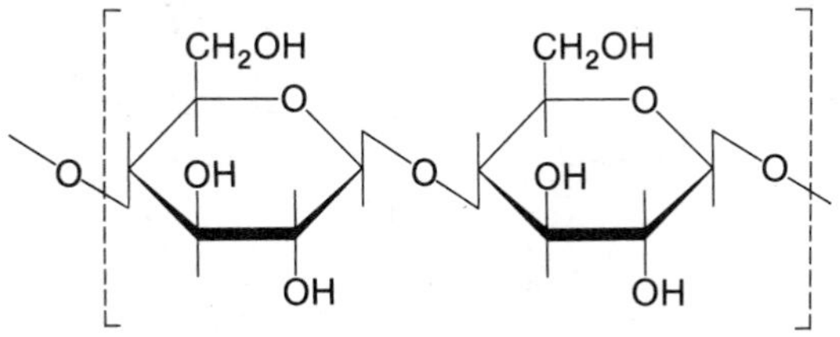

**Figure 12-10**
Structure of the repeating unit of cellulose. As can be seen, the monomeric residues are connected by $\beta$ (1→4) linkages.

**Figure 12-11**
Fibril arrangements in the cell wall of
*Valonia* (12,000 ×): (*a*) dispersed in the
primary wall; (*b*) parallel in the second-
ary wall. (From A. Frey-Wyssling and
K. Mühlethaler, *Ultrastructural Plant
Cytology*, Amsterdam, Elsevier, 1965.
P. 298. Reprinted with permission.)

latory system in humans. One of their functions that takes advantage of
their large size is to give a high viscosity to the surrounding medium. They
are usually found in combinations with proteins. Hyaluronic acid is one of
the better-known members of this group. It contains D-glucuronic acid and
N-acetyl-D-glucosamine in alternating 1,4 and 1,3 linkages, as shown in Figure 12-14. Chondroitin, chondroitin sulfate A and C, and keratosulfate are
other acid mucopolysaccharides. The anticoagulant heparin is also an acid
mucopolysaccharide.

The sugar derivatives muramic and neuraminic acids are important
building blocks in the structural polysaccharides found in the cell walls of
bacteria and the cell coats of higher animal cells, respectively. Both are
nine-carbon sugar acid derivatives that may be thought of as consisting of
a six-carbon amino sugar linked to a three-carbon acid. The amino groups
are usually acetylated, as shown in Figure 12-15.

Most bacteria share, as a common structural feature in their cell wall, a
large sac-like molecule that is a complex polysaccharide peptide called a
*peptidoglycan* or *murein*. The repeating unit in the polysaccharide chain is
the muropeptide. It consists of a disaccharide of N-acetyl-D-glucosamine in
$\beta(1{\to}4)$ linkage with N-acetylmuramic acid. To the carboxyl group of the
lactic acid substituent of muramic acid is attached a tetrapeptide side

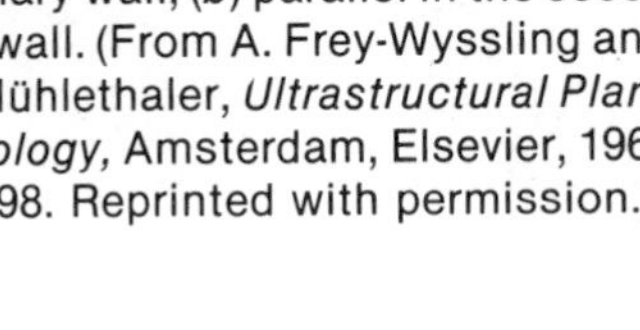

**Figure 12-12**
Coniferyl alcohol.

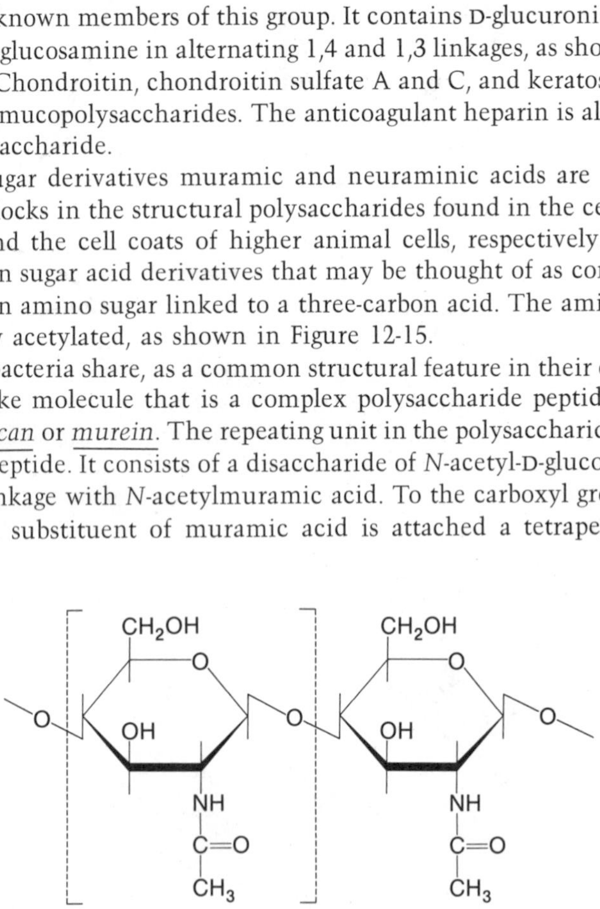

**Figure 12-13**
Structure of the repeating
unit of chitin.

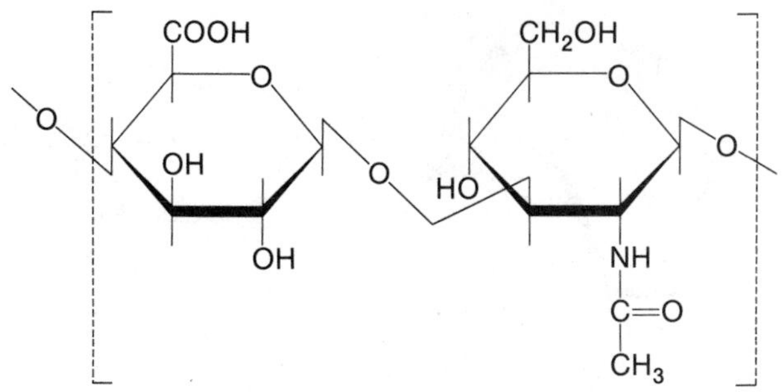

D-Glucuronic acid    N-Acetyl-D-glucosamine

**Figure 12-14**
Structure of the repeating unit of hyaluronic acid.

chain containing L-alanine, D-alanine, D-isoglutamine, and either DL-*meso*-diaminopimelic acid or its decarboxylation product L-lysine. The long, parallel polysaccharide chains are cross-linked by their peptide side chains. The structure and synthesis of peptidoglycan is a fascinating subject taken up in detail later.

**Energy-Storage Polysaccharides.** Starch is the main energy-storage polysaccharide of plants. Like cellulose, it is an $\alpha(1\rightarrow4)$ glucosidic polymer. Unlike cellulose, the units are joined in an $\alpha(1\rightarrow4)$ glycoside rather than a $\beta(1\rightarrow4)$ glycoside. Also in contrast to cellulose, starches often contain branching. Native starches are separable into amylose, a long, unbranched chain, and amylopectin, which has been shown by methylation studies to contain an average of one $\alpha(1\rightarrow6)$ branch point for every 30 glucose residues. The primary structures of glycogen and starch were discussed in Chapter 8 (see Figure 8-16). Glycogen is the main energy-storage polysaccharide in animals. It is found in significant amounts in liver and muscle and is particularly abundant in mollusks. Glycogen is a branched-chain polysaccharide that resembles amylopectin and contains 8 to 12 glucose residues per nonreducing end group.

Dextrans are branched-chain polysaccharides found in yeast and bacteria. The distance between branch points is characteristic of the strain and species. Dextrans are most useful in the laboratory for fractionation of macromolecules on columns, a technique that is explained in Chapter 2.

**Secondary Structures of Glycogen and Cellulose Dictate Their Roles.** It is a remarkable fact that the main energy-storage polysaccharides and the main structural polysaccharides found in nature both have a primary structure of 1,4-linked polyglucose. Why should two such closely related compounds be used in totally different roles? A closer look at the stereochemistry of the $\alpha$ and $\beta$ glycosidic linkage for polyglucose suggests why this is so. Recall that D-glucose exists in the chair form of a pyranose ring (Figure 12-6). The ring has a rigid character to it. We can think of it as a structural building block in a polysaccharide chain just as we think of the rigid peptide planar grouping as a structural building block in a polypeptide chain. It is also possible to specify two torsion angles $\phi$ and $\psi$ for rotation about the glycosidic ether linkage. This is used extensively to discuss polypeptide configuration in proteins where a great deal of information is available (see Chapter 2). It has limited value in discussing polysaccharide structures

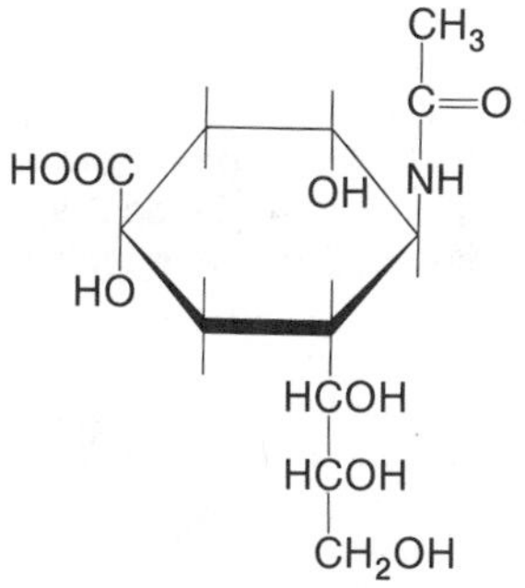

*N*-Acetylneuramic acid (sialic acid)

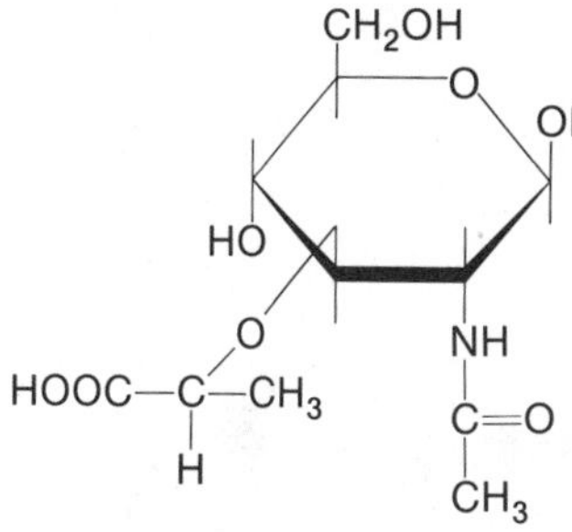

*N*-Acetylmuramic acid (O-lactyl-GlcNAc)

**Figure 12-15**
*N*-Acetylneuraminic acid and *N*-acetylmuramic acid are important building blocks of the structural polysaccharides found in animal and bacterial cell coats, respectively.

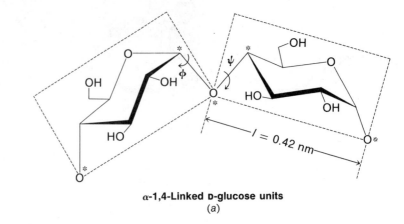

**α-1,4-Linked D-glucose units**
*(a)*

**β-1,4-Linked D-glucose units**
*(b)*

**Figure 12-16**
Energetically favored conformations
of α(1→4)-linked D-glucose (*a*) and
β(1→4) linked D-glucose units (*b*).

because much less information is available. Even though a detailed under-standing is lacking, it is clear that only *the β(1→4)-linked polyglucose has the capacity for forming straight chains (see Figure 12-16). This is accomplished by flipping each glucose unit over 180 degrees relative to the previous one. Such a structure should result in an almost fully extended polysaccharide chain, which is a known property of cellulose.* Evidently this conformation is energetically favored. *By contrast, α(1→4)-linked units in a polyglucose results in a natural turning of the chain (see Figure 12-16). Consistent with this is the observation that amylose adopts a coiled helical configuration.* It seems highly likely that the extended-chain form of polyglucose has been exploited in nature for structural purposes, leaving by default the coiled form as an energy-storage macromolecule. Correlated with this is the omnipresence of degrading enzymes for glyco-gen and starch and the very limited phylogenetic distribution of compara-ble enzymes for cellulose. Cellulose is degraded in the gastrointestinal tract of herbivores, such as the cow, or in insects, such as termites, by a proto-zoan which synthesizes the enzyme cellulose.

## GLUCONEOGENESIS

The central role of glucose in the synthesis of other sugars has been empha-sized in the four preceding chapters. Glucose synthesis from $CO_2$ and $H_2O$, as it occurs in phototrophs, has already been considered in Chapter 11. Chemotrophs also carry out the synthesis of glucose from simpler starting compounds (gluconeogenesis), but the process is more restrictive in that the starting materials are not $CO_2$ and $H_2O$. *Chemotrophs that have the two essential enzymes of the glyoxylate cycle can make glucose from $C_2$ carbon sources (see Chapter 10). Most chemotrophs do not possess these enzymes and require compounds with at least three carbon atoms to carry out the net synthesis of glucose.* The ability to synthesize glucose is important to mammals on a daily basis, since certain tissues, particularly the brain and red blood cells, are almost solely dependent on glucose as an energy source. In the normal adult human under fasting conditions, about 80 percent of the glucose consumed is utilized by the brain. The glycogen reservoir in the liver contains no more than a half-day supply for the brain alone. In periods of dietary glucose deprivation, the organism must be able to make glucose from sources other than this limited glycogen reservoir.

The synthesis of glucose from a three-carbon source such as pyruvate or lactate proceeds by a route that is essentially the reverse of glycolysis, consuming ATP and NADH instead of producing them. The lactate for gluconeogenesis in the liver is frequently supplied by way of the circulatory system as a secretory product from skeletal muscle which produces a great deal of lactate when it is functioning anaerobically (see Chapter 5). There are, however, several distinctions between the glycolytic and the gluconeogenic directions that are important from both the thermodynamic and the regulatory points of view and illustrate general principles of biosynthetic pathways as well (see Figure 12-17). Recall from Chapter 8 that the glycolytic sequence is designed to accomplish the oxidative degradation of glucose and includes several steps that are so highly exergonic in the direction of degradation that they render the whole pathway essentially irreversible. That is, of course, a sound and very general design feature for a metabolic pathway; virtually every pathway in the cell is so contrived that at one or more steps it includes a reaction characterized by such a large release of free energy that there is essentially no chance for reversibility. This guarantees the unidirectionality of the entire pathway, since it is a dictate of thermodynamics that no pathway can proceed in a given direction unless every component reaction in that direction is exergonic under prevailing conditions.

To reverse most metabolic pathways requires that such irreversible steps be bypassed. In the case of glycolysis, these steps can be readily pinpointed: the ATP-driven phosphorylation of both glucose and fructose-6-phosphate and the conversion of phosphoenolpyruvate to pyruvate (Gly-1, Gly-3, and Gly-10 in Figure 8-7 are highly exergonic and irreversible in vivo). The two phosphorylation reactions (Gly-1 and Gly-3, catalyzed in the glycolytic direction by enzymes called *kinases*) are irreversible because they are coupled to ATP hydrolysis and would therefore have to drive the synthesis of ATP in the reverse direction. These steps can be rendered thermodynamically favorable in the gluconeogenic direction by using not kinases, but *phosphatases*, enzymes that remove the phosphate group as inorganic phosphate. (This, you may recall, is also the way that the same reactions are rendered thermodynamically possible in the Calvin cycle. See Chapter 11.) Thus the sequence between glucose and fructose-1,6-diphosphate can be exergonic in either direction, depending on whether the phosphorylating (kinase) or phosphate-cleaving (phosphatase) enzymes are active, as shown on the next page.

In the glycolytic direction:

$$\text{Glucose} \xrightarrow[\text{Kinase}]{\text{ATP} \quad \text{ADP}} \text{glucose-6-P} \xrightarrow[\text{Isomerase}]{} \text{fructose-6-P} \xrightarrow[\text{Kinase}]{\text{ATP} \quad \text{ADP}} \text{fructose-1,6-diP}$$

Net equation:

$$\text{Glucose} + 2\,\text{ATP} \longrightarrow \text{fructose-1,6-diP} + 2\,\text{ADP} \qquad \Delta G° = -7.3\,\text{kcal/mol}$$

In the gluconeogenic direction:

$$\text{Fructose-1,6-diP} \xrightarrow[\text{Phosphatase}]{\text{P}_i} \text{fructose-6-P} \xrightarrow[\text{Isomerase}]{} \text{glucose-6-P} \xrightarrow[\text{Phosphatase}]{\text{P}_i} \text{glucose}$$

Net equation:

$$\text{Fructose-1,6-diP} \longrightarrow \text{glucose} + 2\,\text{P}_i \qquad \Delta G° = -7.3\,\text{kcal/mol}$$

The other essentially irreversible reaction in the glycolytic pathway is the step from phosphoenolpyruvate to pyruvate; even though already harnessed to the generation of ATP, it is still sufficiently exergonic to be virtually unidirectional. To bypass this step, the gluconeogenic pathway uses a two-step sequence, with both steps coupled to the hydrolysis of a high-energy phosphate bond (one from ATP, the other from GTP). Involved is a carboxylation/decarboxylation sequence with the four-carbon compound oxaloacetate (of TCA-cycle fame) as the intermediate. Thus a reaction that generates one high-energy phosphate bond in the catabolic direction is driven in the anabolic direction by a sequence that uses two high-energy phosphate bonds.

In the glycolytic direction:

$$\text{Phosphoenolpyruvate} \xrightarrow[\text{Pyuvate kinase}]{\text{ADP} \quad \text{ATP}} \text{pyruvate}$$

Net equation:

$$\text{Phosphoenolpyruvate} + \text{ADP} \longrightarrow \text{pyruvate} + \text{ATP} \qquad \Delta G° = -7.5\,\text{kcal/mol}$$

In the gluconeogenic direction:

$$\text{Pyruvate} \xrightarrow[\text{Carboxylase}]{\text{CO}_2 \;\text{ATP}\;\; \text{ADP} + \text{P}_i} \text{oxaloacetate} \xrightarrow[\text{Carboxykinase}]{\text{GTP}\;\text{GDP}\;\text{CO}_2} \text{phosphoenolpyruvate}$$

Net equation:

$$\text{Pyruvate} + \text{ATP} + \text{GTP} \longrightarrow \text{phosphoenolpyruvate} + \text{ADP} + \text{GDP} + \text{P}_i \qquad \Delta G° = +0.2\,\text{kcal/mol}$$

Note, then, that the synthesis of glucose from pyruvate occurs by a route that is essentially the reverse of glycolysis (and can, as a result, use 7 of 10 already existing enzymes), but those reactions which would otherwise ensure that the sequence is highly exergonic in the catabolic direction are bypassed by alternative reactions made exergonic in the anabolic direction by the input of energy from ATP and GTP.

By using similar but not identical pathways for the opposite processes of glycolysis and gluconeogenesis, the cell enjoys several advantages: (1) since most of the pathway is common to both directions, reversal of direction requires a minimum of new enzymes and is therefore genetically conservative; (2) yet at the same time the reaction sequence can be rendered essentially irreversible in a given direction simply by specifying which set of enzymes is available or active for the key bypass reactions; such that (3) the direction of the sequence can be controlled within a cell

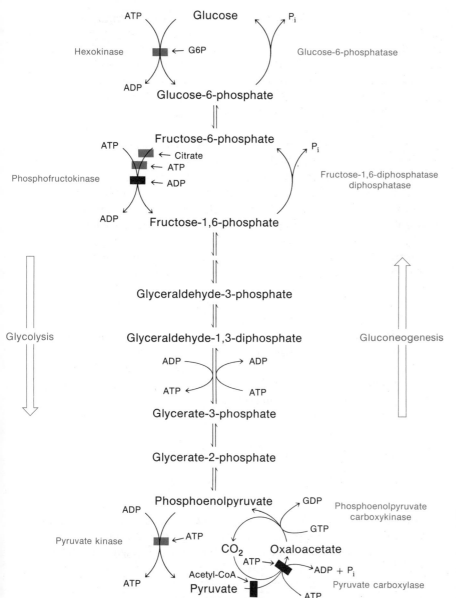

**Figure 12-17**
The glycolytic and gluconeogenic pathways. Enzymes that are unique to either pathway are indicated in color. Points of allosteric regulation by small-molecule inhibitors (■) or activators (■) are indicated.

or subcellular location by regulating the presence or activities of these key enzymes. The actual direction in which the sequence proceeds is then dependent on the set of enzymes present or active under specific circumstances.

Of the four enzymes required uniquely for gluconeogenesis (glucose-6-phosphatase, fructose-1,6-diphosphate diphosphatase, phosphoenolpyruvate carboxykinase, and pyruvate carboxylase; See Figure 12-17), only pyruvate carboxylase is mitochondrial. This enzyme was discovered by Merton Utter in 1959 as a result of investigations aimed at explaining the intramitochondrial conversion of pyruvate to oxaloacetate. Oxaloacetate itself cannot pass through the mitochondrial membrane (Chapter 10). To overcome this problem, the oxaloacetate is converted to malate, which is transportable. Once in the cytoplasm, the malate is reoxidized to oxaloacetate and the remaining reactions leading to glucose follow.

Gluconeogenesis in the liver is somehow strongly stimulated by the same hormones that stimulate glycogen breakdown, namely, glucagon and epinephrine. However, as long as the glycogen supply is adequate to meet the needs of the organism, these hormones do not have much effect on stimulating gluconeogenesis. Evidently other signals are required to supplement these hormonal signals. Indeed it is known that cortisol is needed to induce synthesis of specific enzymes required for gluconeogenesis, e.g., PEP carboxykinase. Not surprisingly, the steps at which biosynthetic and degradative pathways differ for thermodynamic reasons are also the points at which the pathways are usually regulated allosterically. This makes good sense, of course, since it allows control to be exerted in a specific direction. In fact, it is mainly in the context of Figure 12-17 that phosphofructokinase, the enzyme that phosphorylates fructose-6-phosphate to fructose-1,6-diphosphate (Gly-3 in Chapter 8), is seen to be a particularly logical choice for allosteric regulation of glycolysis. The phosphofructokinase reaction is one of the three steps unique to the glycolytic direction, such that regulation of its activity affects the flow along the pathway in the catabolic direction only. (As shown in Figure 12-17, the other enzymes unique to the glycolytic direction are also allosteric control points, although perhaps of lesser overall significance than phosphofructokinase.)

By the same reasoning, the most logical candidates for allosteric regulation in the gluconeogenic direction ought to be the enzymes unique to that direction. Figure 12-17 indicates that the prediction is a good one: regulation of glucose synthesis is in fact exerted at the phosphate-cleaving step that bypasses the phosphofructokinase reaction and at the pyruvate-carboxylating step used at the other bypass.

## SYNTHESIS OF GLYCOSIDIC BONDS

### Formation of Activated Monomeric Building Blocks

Glycoside synthesis occurs in all living cells, since these compounds are universally utilized as a form of energy storage, as constituents of cell membranes, and in fungi and plants in the formation of cell walls. The hydrolysis of a typical glycoside, such as maltose, proceeds with a free-energy change $\Delta G'$ of $-4$ kcal/mol. Thus *the formation of a glycosidic bond requires energy, and this energy is provided through the formation of a high-energy intermediate, the nucleoside diphosphate sugar.* Most oligosaccharides and polysaccharides use this type of intermediate. The general features common to all the nucleoside diphosphate sugars are indicated in Figure 12-18. Depending on which polysaccharide is being made, different purine and pyrimidine bases and sugars are used in the donor molecule.

The first of the nucleoside diphosphate sugars discovered was uridine diphosphate glucose (UDPG). For this discovery L. Leloir was awarded the Nobel Prize in 1970. UDPG is synthesized from glucose-1-phosphate and uridine triphosphate (UTP) in a reaction catalyzed by UDPglucose pyrophosphorylase:

$$\text{UTP} + \text{glucose-1-phosphate} \longrightarrow \text{UDPG} + \text{PP}_i$$

Nucleoside diphosphate sugars are not only the most common building blocks used in oligosaccharide and polysaccharide synthesis; they also are

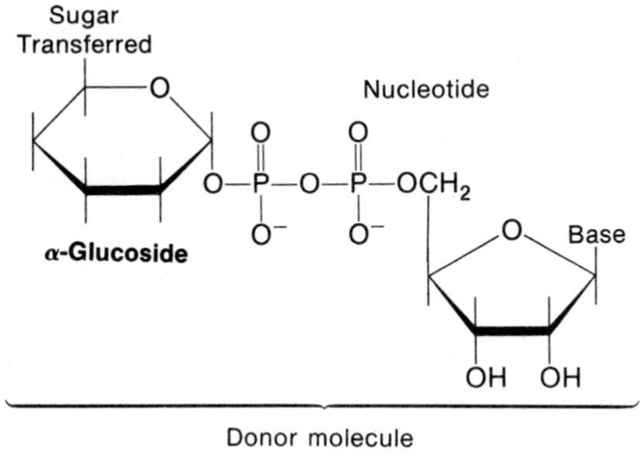

Donor molecule + acceptor molecule $\xrightarrow{\text{transglycosylase}}$ glycoside

Specific for: (1) Acceptor, (2) sugar transferred, and (3) nucleotide carrier

**Figure 12-18**
Generalized structure of activated
building block used in most types of
polysaccharide synthesis.

frequently used in the interconversion of sugars. Conversion of galactose to glucose occurs while they are both activated as nucleotides (Chapter 8). Other interconversions mediated through structurally similar complexes are discussed later in this chapter.

### Disaccharide Synthesis

The synthesis of some of the better-known disaccharides proceeds through a nucleoside diphosphate sugar intermediate. Sucrose is a dissacharide containing glucose and fructose with an $\alpha(1\rightarrow4)$ linkage from the C-1 of glucose to the C-4 of fructose. Fructose-6-phosphate is produced in one enzymatic step from glucose-6-phosphate. The latter is also converted by means of glucose-1-phosphate to uridine diphosphate glucose (UDPglucose), which then reacts with fructose-6-phosphate to give UDP and sucrose-6'-phosphate. The phosphate is removed by a single enzymatically catalyzed hydrolysis to yield sucrose and inorganic phosphate. In some plants, sucrose is formed simply by the reaction of UDPglucose with fructose.

The disaccharide lactose is formed in the mammary gland from D-glucose and UDPgalactose by the action of two proteins which together constitute the lactose synthase system. The first protein, protein A, otherwise known as galactosyltransferase (which is found not only in the mammary gland, but also in the liver and small intestine), catalyzes the reaction:

UDPgalactose + N-acetyl-D-glucosamine $\longrightarrow$ UDP + N-acetyllactosamine

The second protein, protein B, is known as the $\alpha$-lactalbumin of milk. It has no catalytic activity of its own. This protein alters the specificity of protein A so that it utilizes D-glucose instead of N-acetyl-D-glucosamine as the galactose acceptor, causing the protein A to make lactose instead of N-acetyllactosamine:

UDPgalactose + D-glucose $\longrightarrow$ UDP + lactose

This is a fascinating case of the alteration of the specificity of a catalytic subunit by an auxiliary protein subunit.

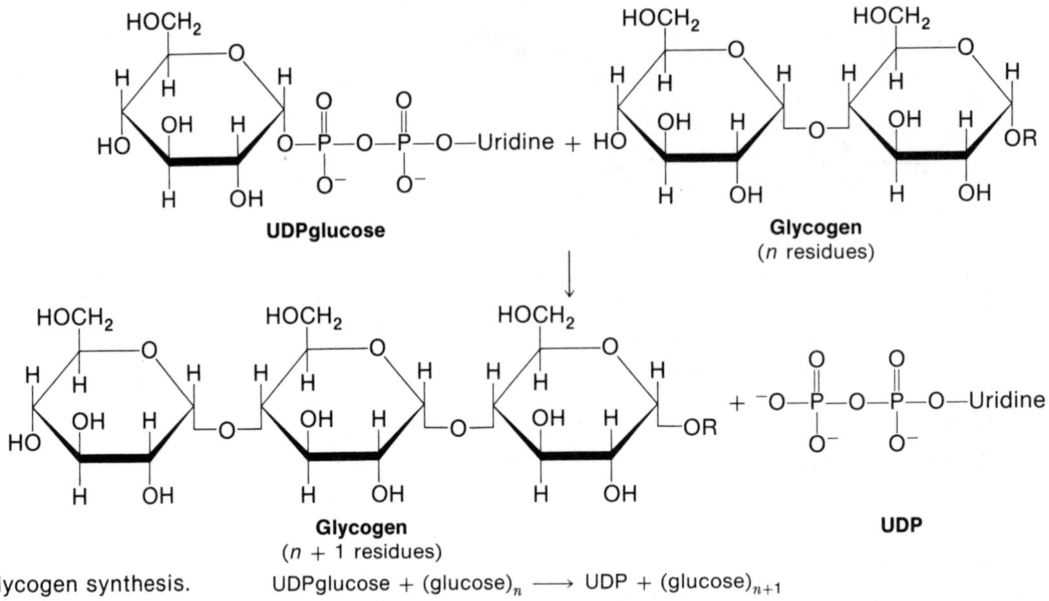

**Figure 12-19**
Elongation step in glycogen synthesis.

UDPglucose + (glucose)$_n$ ⟶ UDP + (glucose)$_{n+1}$

### Synthesis of Energy-Storage Polysaccharides

Polysaccharides are the most common form of energy storage in all cells. Some examples are listed in Table 12-1. Glycogen is formed under the influence of _glycogen synthase_ by reaction of UDPglucose with a polymer of indefinite length, as indicated in Figure 12-19. The $\Delta G°'$ of this reaction is about $-3.2$ kcal/mol. The enzyme glycogen synthase requires as a _primer_ an $\alpha(1{\rightarrow}4)$ polyglucose chain having at least four glucose residues, to which it adds successive glucosyl groups. It works best with long-chain glucose polymers as primers. In animal tissues, ADPglucose is about 50 percent as active as UDPglucose as a donor in this reaction; in lower forms, ADPglucose is more reactive than UDPglucose. In addition to $\alpha(1{\rightarrow}4)$ bonds, glycogen contains $\alpha(1{\rightarrow}6)$ bonds. The latter bonds are made by the branching enzyme amylo(1,4-1,6)-_trans_-glycosylase. This enzyme transfers

**Table 12-1**
Some Storage Polysaccharides

| Source | Polysaccharide | Monosaccharide Component(s) | Glycosyl Donor | Polymer Structure |
|---|---|---|---|---|
| Primarily muscle and liver cells of animals | Glycogen | Glucose | UDPglucose | $\alpha(1 \rightarrow 4)$ with $\alpha(1 \rightarrow 6)$ branch points |
| Bacterial glycogen | Glycogen | Glucose | ADPglucose | $\alpha(1 \rightarrow 4)$ with $\alpha(1 \rightarrow 6)$ branch points |
| Green algae | Amylose | Glucose | ADPglucose | Linear $\alpha(1 \rightarrow 4)$ |
| Leaves, stem, roots, and seeds of higher plants | Amylopectin | Glucose | ADPglucose | Linear $\alpha(1 \rightarrow 4)$ with $\alpha(1 \rightarrow 6)$ branch points |
| Some bacteria | Dextran | Glucose | Sucrose | Linear $\alpha(1 \rightarrow 6)$ with $\alpha(1 \rightarrow 2)$, $\alpha(1 \rightarrow 3)$ or $\alpha(1 \rightarrow 4)$ branch points |

a terminal oligosaccharide fragment of six or seven glucosyl residues from the end of the main glycogen chain to the 6-hydroxyl group of a glucose residue somewhere in a glycogen chain. This reaction produces a branched-chain polymer from a straight-chain polymer, as shown in Figure 12-20. As might be expected, the free-energy change in this reaction is very small, since very similar chemical linkages are involved.

In plant tissues, starch synthesis occurs by an analogous pathway that is catalyzed by amylose synthase; ADPglucose is the preferred glucose donor.

## Glycogen Breakdown

Since glycogen is just a convenient way of storing sugar until it is required for energy purposes, its breakdown is as important as its synthesis. This proceeds by a different route involving the action of inorganic phosphate and the enzyme *glycogen phosphorylase* on the polymer (see Chapter 8). One of the most interesting aspects of glycogen metabolism has to do with the intricate mechanism that controls synthesis and breakdown, which relates to the organism's energy requirements. This is discussed in Chapter 8.

**Figure 12-20**
Schematic diagram showing the action of the "branching enzyme" in glycogen formation.

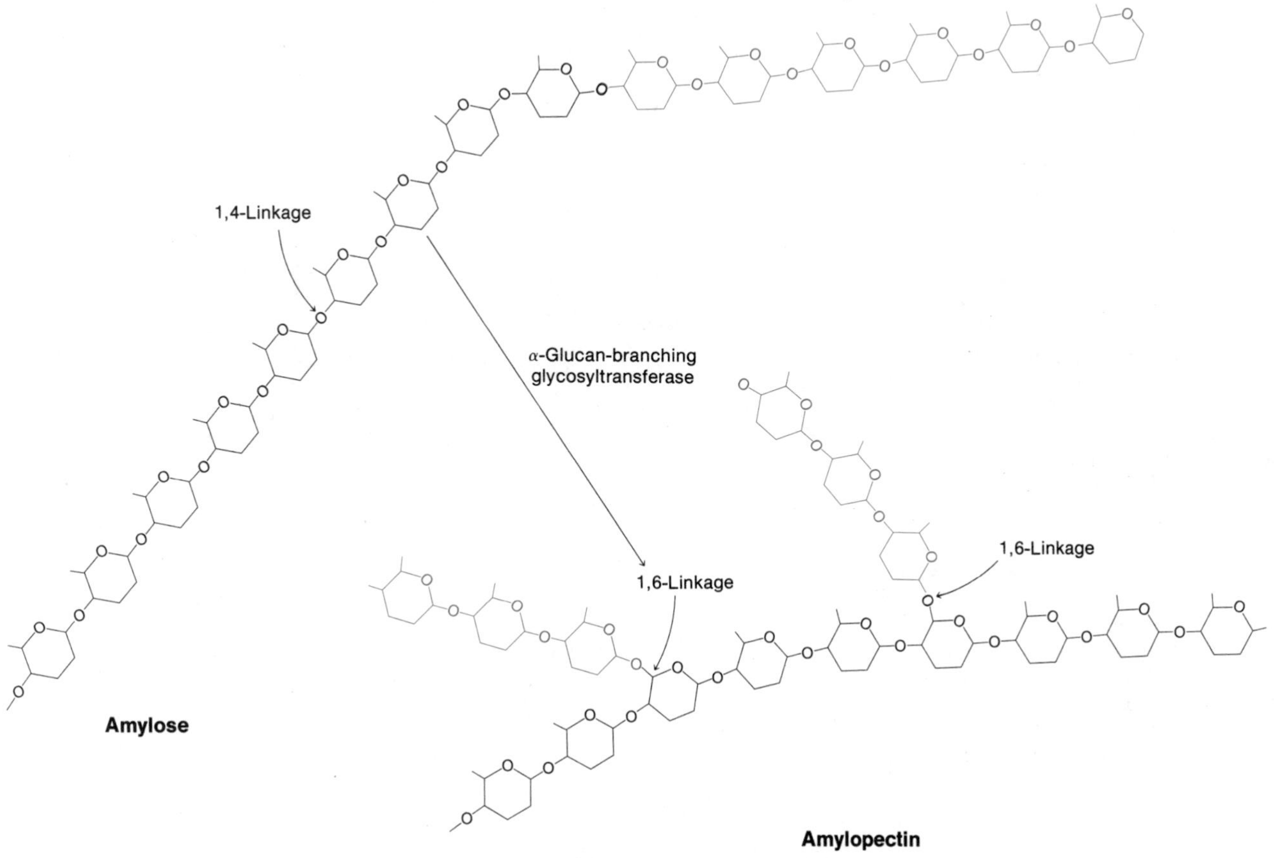

1,4-Linkage

α-Glucan-branching glycosyltransferase

1,6-Linkage

1,6-Linkage

**Amylose**

**Amylopectin**

## Synthesis of Structural Polysaccharides

In plant cellulose, a straight-chain homopolymer of glucose with a $\beta(1\rightarrow4)$ linkage is formed by the same general mechanism using nucleoside diphosphate sugars. In addition, chitin in insects, which is a $\beta(1\rightarrow4)$ homopolymer of N-acetylglucosamine, is formed in a similar reaction from UDP-N-acetylglucosamine. The animal polysaccharide hyaluronic acid presents a slight variation. It consists of an alternating copolymer of D-glucuronic acid and N-acetylglucosamine formed by successive alternating reactions of UDPglucuronic acid and UDP-N-acetylglucosamine with the end of the chain. Two enzymes, one specific for UDPglucuronic acid and one specific for UDP-N-acetylglucosamine, are used in the polymerization stage.

In most polysaccharide syntheses, nucleoside diphosphate sugars are used as activated substrates. A striking exception is seen in the case of certain bacteria that synthesize dextran, a predominantly $\alpha(1\rightarrow6)$ polymer of glucose. The substrate for dextran synthesis is sucrose, and the energy of the glycosidic bond between glucose and fructose in this disaccharide drives the reaction:

$$n\text{-Sucrose} \xrightarrow{\text{Dextran sucrase}} \text{dextran} + n\text{-fructose}$$

Dextrans formed by bacteria growing on the surface of teeth are an important component of dental plaque. Hence the precaution of dentists not to eat too much candy is a sensible one.

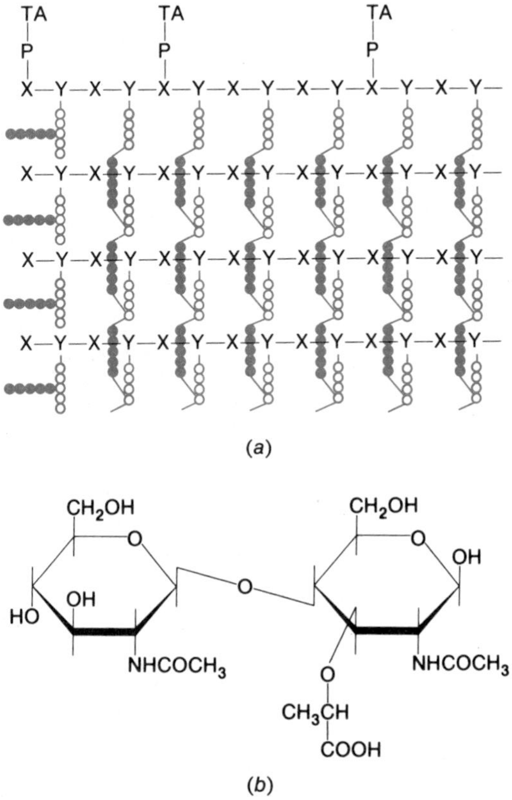

**Figure 12-21**
Structure of the peptidoglycan of the cell wall of *Staphylococcus aureus*. (a) In this representation, X (acetylglucosamine) and Y (acetylmuramic acid) are the two sugars in the peptidoglycan. Open circles represent the four amino acids of the tetrapeptide L-alanyl-D-γ-glutamyl-L-lysyl-D-alanine. Closed circles are pentaglycine bridges that interconnect peptidoglycan strands. The nascent peptidoglycan units bearing open pentaglycine chains are shown at the left of each strand. TA-P is the teichoic acid antigen of the organism, which is attached to the polysaccharide through a phosphodiester linkage. (b) The structure of X (N-acetylglucosamine) and Y (N-acetylmuramic acid) which are linked by $\beta(1\rightarrow4)$ linkages and alternate in the glycan strand.

# BACTERIAL CELL-WALL SYNTHESIS

Most bacteria have a cell wall around an inner membrane. A notable exception includes *Mycoplasma*, which have no cell wall, and this makes them very sensitive to osmotic pressure changes and highly deformable. *Gram-negative bacteria* have both an inner membrane and an outer membrane on the other side of the cell wall (see Figure 16-4). The cell wall in Gram-negative bacteria consists of *peptidoglycan* and associated proteins. *Gram-positive bacteria* have no outer membrane, and their cell walls tend to be much thicker, consisting of peptidoglycan and polyol phosphate polymers called *teichoic acids* (Figure 12-21, and see also Figure 16-6a). *The biosynthesis of peptidoglycan poses some virtually unique problems in biopolymer synthesis because it involves the construction of an ordered two-dimensional network in which part of the synthesis must take place outside the cell.* It will be considered in some detail.

Investigations on bacterial cell-wall synthesis were actually initiated in the hope of discovering the mode of action of the "miracle drug" penicillin. These investigations ultimately led to the uncovering of the structure of bacterial cell walls and their synthesis. However, the detailed mechanism of action of penicillin and other antibiotics is still not completely understood (see below). It is in some ways fortunate that penicillin inhibits the terminal reactions in cell-wall synthesis, since otherwise the picture of the myriad of enzymatic reactions that are involved might never have been developed.

## Structure of Peptidoglycan

Peptidoglycan is a polymeric structure of a glycan composed of amino sugars in one dimension and cross-linked through branched polypeptides in the other (Figure 12-21). The amino sugars are N-acetylglucosamine and N-acetylmuramic acid (which is the 3-O-D-lactic acid ether of N-acetylglucosamine). These amino sugars strictly alternate in the polymer, forming the glycan strands. The carboxyl group of the lactic acid moiety of the acetylmuramic acid is substituted by a tetrapeptide which in the bacterium *Staphylococcus aureus* has the sequence L-alanyl-D-$\gamma$-isoglutaminyl-L-lysyl-D-alanine. All the muramic acids are substituted in this way to form peptidoglycan strands. Some close variant of this structure is characteristic of all bacterial species. The peptidoglycan strands are further linked to each other by means of an interpeptide bridge. In *S. aureus*, this bridge is a pentaglycine chain that extends from the terminal carboxyl group of the D-alanine residue of one tetrapeptide to the $\varepsilon$-NH$_2$ group of the third amino acid, L-lysine, in another tetrapeptide. The third dimension is probably built up by bridges extending in different planes. This gigantic macromolecule has the mechanical stability required for the cell wall.

## Biosynthesis of Peptidoglycan

The stages of synthesis of peptidoglycan were worked out over a period of almost 20 years. *The biosynthesis can be conveniently broken down into three stages that occur at three different places in the cell:* (1) synthesis of

UDP-*N*-acetylmuramylpentapeptide, (2) polymerization of *N*-acetylglucosamine and *N*-acetylmuramylpentapeptide to form the linear peptidoglycan strands, and (3) cross-linking of the peptidoglycan strands.

**Synthesis of UDP-*N*-acetylmuramylpentapeptide.** *The first stage (Figure 12-22) involves the synthesis of UDP-N-acetylmuramylpentapeptide,* the major compound that accumulates in penicillin-treated cells. First, the condensation of *N*-acetylglucosamine-1-phosphate with UTP leads to the formation of UDP-*N*-acetylglucosamine (Figure 12-23). A specific transferase catalyzes a reaction with phosphoenolpyruvate to give the 3-enolpyruvylether of UDP-*N*-acetylglucosamine. The pyruvyl group is then reduced to lactyl by an NADPH-linked reductase, thus forming the 3-O-D-lactylether

**Figure 12-22**
The first stage of cell-wall synthesis: formation of UDP-*N*-acetylmuramyl-pentapeptide (structure shown). Points of inhibition by the antibiotics D-cycloserine and phosphonomycin are indicated.

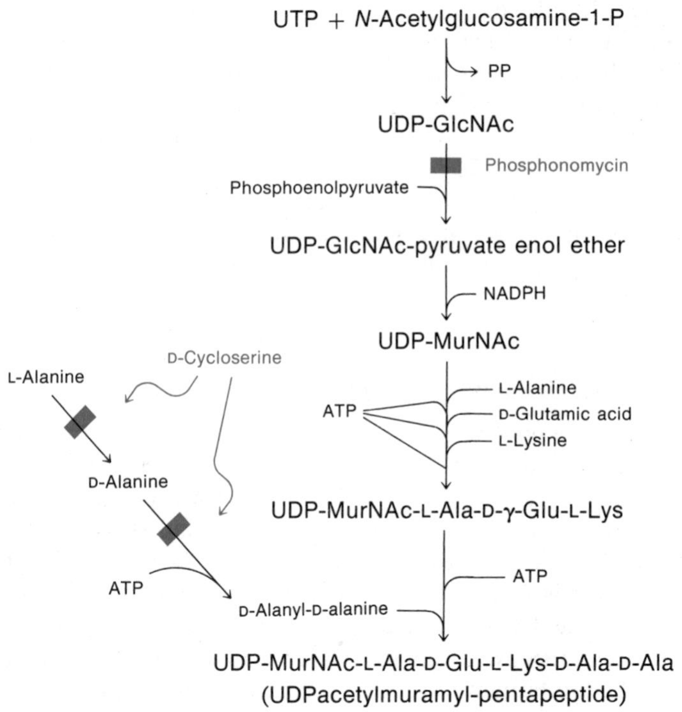
UTP + N-Acetylglucosamine-1-P
PP
UDP-GlcNAc
Phosphonomycin
Phosphoenolpyruvate
UDP-GlcNAc-pyruvate enol ether
NADPH
UDP-MurNAc
D-Cycloserine
L-Alanine
ATP
L-Alanine
D-Glutamic acid
L-Lysine
D-Alanine
UDP-MurNAc-L-Ala-D-γ-Glu-L-Lys
ATP
ATP
D-Alanyl-D-alanine
UDP-MurNAc-L-Ala-D-Glu-L-Lys-D-Ala-D-Ala
(UDPacetylmuramyl-pentapeptide)

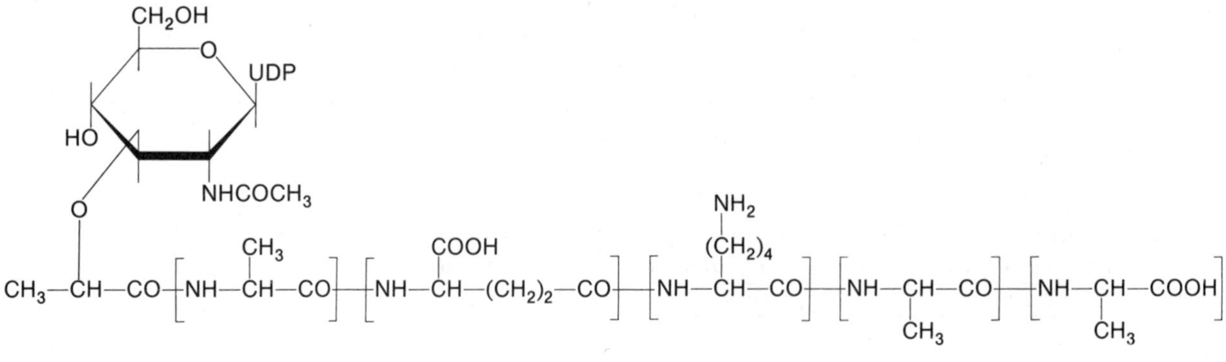
CH2OH
O
UDP
HO
NHCOCH3
O
CH3—CH—CO
NH—CH—CO
CH3
NH—CH—(CH2)2—CO
COOH
NH—CH—CO
NH2
(CH2)4
NH—CH—CO
CH3
NH—CH—COOH
CH3

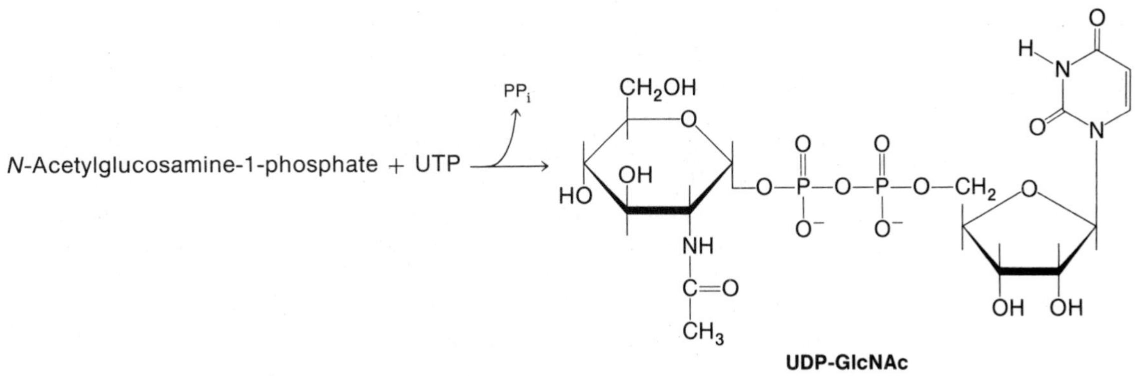

N-Acetylglucosamine-1-phosphate + UTP $\longrightarrow$

**UDP-GlcNAc**

**Figure 12-23**
Synthesis of UDP-GlcNAC (first step in Figure 12-22).

of N-acetylglucosamine; this compound is known as UDP-N-acetyl-muramic acid. Conversion of this compound to its pentapeptide form occurs by the sequential addition of the requisite amino acids. Each step requires ATP and a specific enzyme that ensures the addition of amino acids in the proper sequence. The addition of L-alanine occurs first, followed by D-glutamic acid (later amidated to D-isoglutamine), L-lysine (attached by its α amino group to the γ carboxyl group of the glutamic acid), and finally the dipeptide D-alanyl-D-alanine as a unit. The latter dipeptide is formed by two enzymatic reactions: conversion of L-alanine to D-alanine by a racemase, followed by the linking of the two alanine residues in an ATP-requiring reaction to form D-alanyl-D-alanine. These reactions, all of which occur in the soluble compartment of the bacterial cell, have been fairly well characterized.

**Formation of Linear Peptidoglycan Strands.** The second stage of peptidoglycan synthesis occurs in the membrane. It is the polymerization of N-acetylglucosamine and N-acetylmuramylpentapeptide to form the linear peptidoglycan strands. This reaction turned out to be far more complicated than was at first apparent. Initially it was believed that the polysaccharide polymers were formed by simple and direct transglycosylations involving UDPacetylmuramylpentapeptide and UDP-N-acetylglucosamine. However, when the stoichiometry of this reaction was examined in detail using a variety of radioactive precursors, the products turned out to be somewhat unexpected. UDP was derived from UDP-N-acetylglucosamine, but the products formed from UDPacetylmuramylpentapeptide were UMP and $P_i$. This result was inconsistent with a direct transglycosylation reaction. When the substrates and products were separated from one another by paper chromatography, there was always a trace of radioactivity present at the solvent front. For some time this radioactive "contaminant" was ignored, but it proved to be the key to the mechanism of the reaction. The "contaminant" was a lipid intermediate in the reaction sequence (Figure 12-24). First, UDPacetylmuramylpentapeptide reacts with a phospholipid in the membrane, transferring phosphoacetylmuramylpentapeptide, a reaction that also leads to the formation of UMP. N-Acetylglucosamine is then added to the lipid intermediate by means of a typical transglycosylation from UDP-N-acetylglucosamine; UDP is released. The disaccharide-pentapeptide unit is then transferred from the lipid intermediate to the peptidoglycan, and lipid pyrophosphate is generated. Elimination of one $P_i$ regenerates the phospholipid, which then can react once again with UDPacetylmuramylpentapeptide and participate in another cycle, resulting in the addition of a new unit to the growing peptidoglycan strand. The antibiotic bacitracin is a specific inhibitor of the dephosphorylation of the pyrophosphate form of the lipid.

The next problem was the identification of the specific phospholipid of the membrane involved in the peptidoglycan-synthesizing reaction. It took several years to separate the phospholipid in a pure form, since it represented only about 0.1 percent of the total phospholipid of the membrane. It took a much shorter time to identify it from its mass spectrum, an illustra-

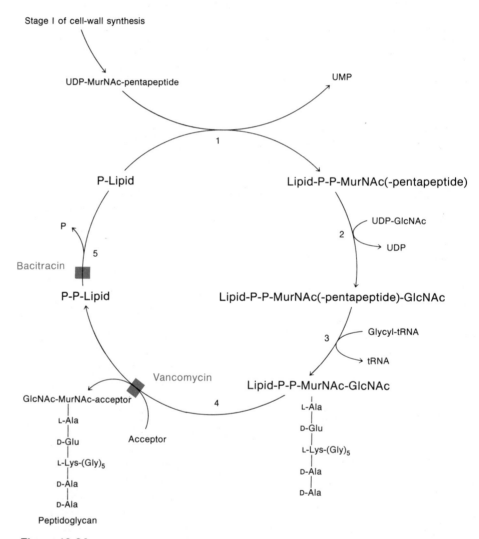

**Figure 12-24**
The second stage of cell-wall synthesis in *S. aureus*. An ATP-requiring amidation of glutamic acid that occurs between steps 2 and 3 has been omitted. Step 3 is the sequential addition of five glycine residues to add the elements of the interpeptide bridge to the lipid intermediate. Points of action of the antibiotic inhibitors bacitracin and vancomycin are indicated (see steps 4 and 5).

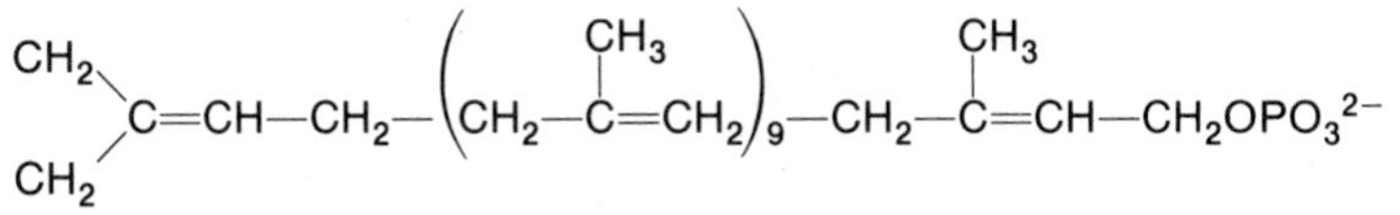

$$CH_2 \diagdown \atop CH_2 \diagup C{=}CH{-}CH_2{-}\left(CH_2{-}\underset{\underset{CH_3}{|}}{C}{=}CH_2\right)_9{-}CH_2{-}\underset{\underset{CH_3}{|}}{C}{=}CH{-}CH_2OPO_3{}^{2-}$$

**Figure 12-25**
Structure of undecaprenol phosphate ($C_{55}$-isoprenol phosphate, bactoprenol phosphate). (See Chapter 15 for the mode of synthesis of undecaprenol phosphate.)

tion of the power of the tools of modern biochemistry. The compound was $C_{55}$-isoprenyl alcohol, consisting of 11 isoprene units ending in an alcoholic function, to which the pyrophosphate and disaccharide-pentapeptide are attached (Figure 12-25). The same alcohol also was identified as the lipid component of an intermediate in bacterial lipopolysaccharide synthesis (discussed below).

Since the identification of this isoprene alcohol, several other syntheses of polysaccharides have been shown to occur through the intermediate action of isoprene alcohols. For example, the synthesis of mannan in *Micrococcus lysodeikticus* occurs by direct transfer of the sugar from $C_{55}$-isoprenyl phosphate, although this reaction sequence is of a different type from the one involved in cell-wall synthesis. Alcohols with 18 to 20 isoprene units, known as *dolichols*, have been known for some time to be present in mammalian liver. The phosphate esters of these alcohols are involved in the synthesis of some types of glycoproteins in liver and other cells (for example, surface glycoproteins such as histocompatibility antigens, in human lymphocytes) in a reaction cycle similar to the cycle of bacterial peptidoglycan synthesis. The isoprene alcohols seem to have a rather general function in the synthesis of a variety of substances in bacteria, plants, and animals. They probably serve a membrane-transport function. The syntheses involved are concerned with the formation outside the cell of products that are synthesized from intermediates inside the cell. The function of the isoprene alcohol appears to be that of a carrier for the activated sugar fragments from the inside to the outside of the cell; i.e., it is the lipid-soluble substrate in the membrane for the water-soluble nucleoside diphosphates.

**Cross-Linking of the Peptidoglycan Strands.** All but one of the series of reactions of cell-wall synthesis have been considered to this point. None of these was found to be penicillin-sensitive. Only one more reaction remains—the final cross-linking of the peptidoglycan chains to form the gigantic macromolecule. Suffice it to say that this reaction is indeed the one sensitive to penicillins and the structurally related cephalosporins.

*The cross-linking of peptidoglycan strands takes place outside the cell membrane at the site of the preexisting wall.* Since there is no ATP or other obvious energy source available at this site, a mechanism independent of any external energy source has apparently been evolved for this reaction. The reaction is a transpeptidation in which the terminal amino end of an open cross-bridge attacks the terminal peptide bond in an adjacent strand to form a cross-link. The terminal alanine residue from the strand that becomes cross-linked is thus eliminated (Figure 12-26). It is this reaction that is sensitive to penicillins and cephalosporins (Figure 12-27).

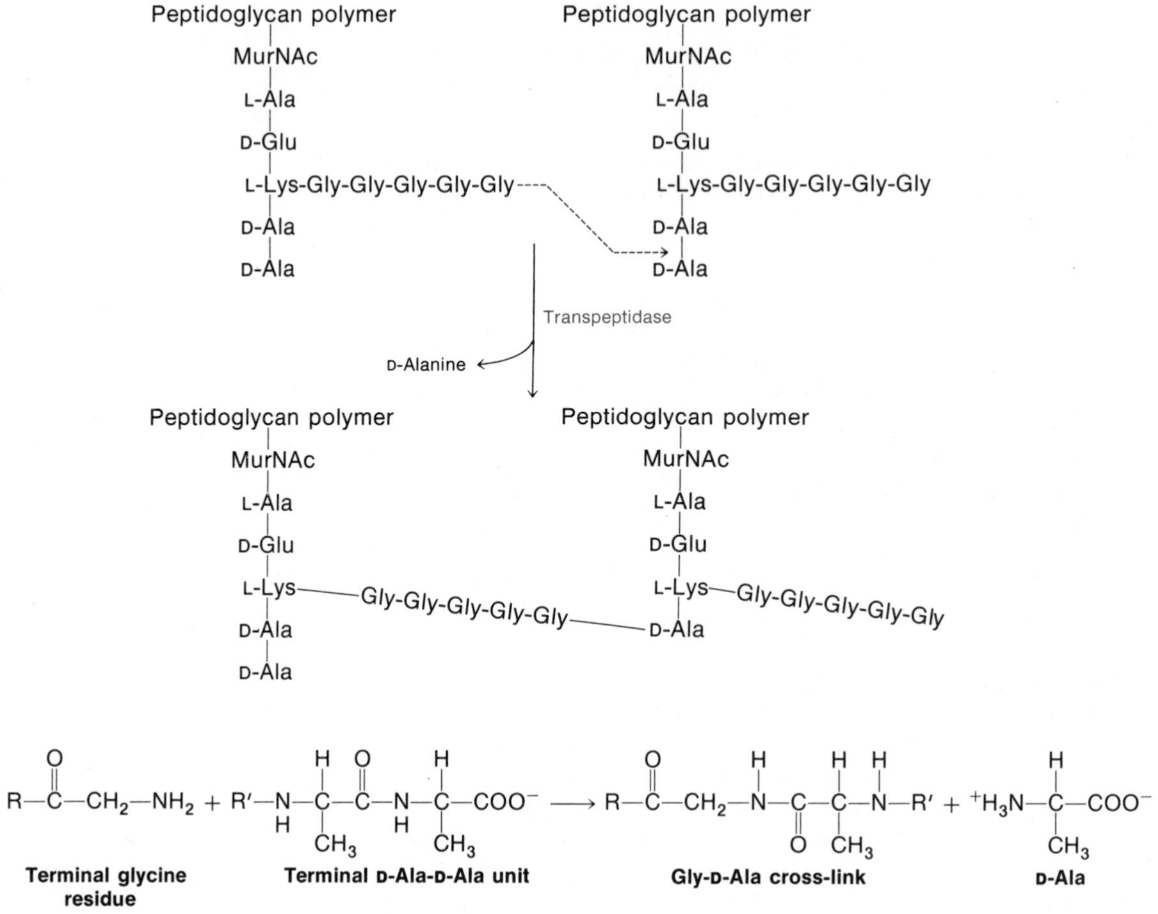

**Figure 12-26**
The third stage of cell-wall synthesis in *S. aureus*: (*a*) cross-linking of peptidoglycan polymers by trans-peptidation with the elimination of D-alanine; (*b*) the amino group of the pentaglycine bridge attacks the peptide bond between D-Ala residues to form a cross-link.

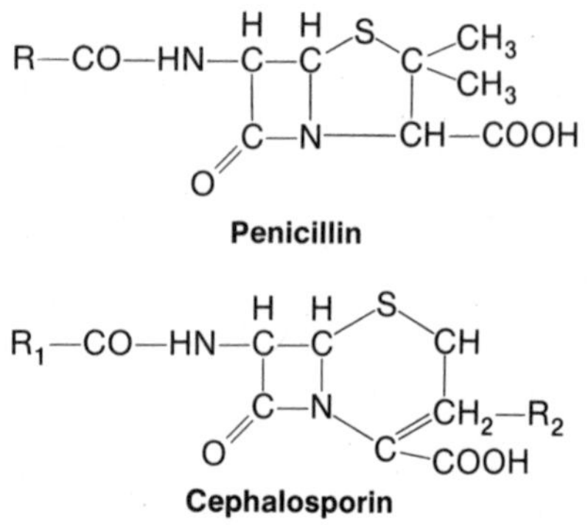

**Figure 12-27**
The chemical structures of penicillin and cephalosporin.

**Mode of Action of Penicillin.** The penicillins and cephalosporins have been unequaled for their usefulness in combatting bacterial diseases and infections. During the 50 years since Fleming brought penicillin to the attention of microbiologists, many biochemists and pharmacologists have been interested in the mechanism by which this potent antibiotic kills bacteria. The details of this process are still being actively investigated.

It has been proposed that penicillin inhibits the reaction by acting as a structural analog of the terminal D-alanyl-D-alanine residue of the peptido-glycan strand. The similarity between the conformation of the penicillin molecule and one of the conformations of the dipeptide D-alanyl-D-alanine is shown in Figure 12-28. It was suggested, therefore, by analogy with simi-lar reactions that the transpeptidase (TPase) would first react with the substrate to form an acyl enzyme intermediate, with the elimination of D-alanine, and that this active intermediate would then react with another strand to form the cross-link and regenerate the enzyme. If the penicillin were an analog of alanylalanine, it should fit the substrate-binding site with the highly reactive —CO—N— bond in the β-lactam ring in the same position as the bond involved in the transpeptidation. It might therefore acylate the enzyme, forming a penicilloyl enzyme, and thereby inactivate it (Figure 12-29). In the past few years, many of the details of this hypothesis have been verified experimentally. Penicilloyl is the piece of the antibiotic found in inhibited enzymes. An acyl group derived from the substrate is used to form an acyl enzyme intermediate. Most important, the antibiotic-derived penicilloyl moiety and the substrate-derived acyl moiety are substituted on the same site in the penicillin-sensitive enzymes from a variety of genera of bacteria.

Initial studies on the site of penicillin action were hampered by the inability to find a cell-free enzyme system that would catalyze the transpeptidation. The synthesis of peptidoglycan by membranes of E. coli was first demonstrated in vitro in 1968. The cross-linking TPase, but not the sugar-polymerizing activity, was inhibited by penicillin. At this time, a second, related, penicillin-sensitive activity was discovered in these membranes. This enzyme specifically hydrolyzed the terminal D-alanine from cell-wall precursor or from nascent peptidoglycan and was called D-alanine carboxy-peptidase (CPase). TPase and CPase were assumed to be different enzymes because CPase was 150-fold more sensitive to penicillin G than TPase in E. coli membranes. TPase was presumed to be the penicillin "killing site" because the concentrations of various penicillins and cephalosporins required to inhibit the growth of E. coli more closely matched the concentrations required to inhibit TPase than those required to inhibit CPase.

Both peptidoglycan-polymerizing activity and TPase activity were lost upon solubilization of membranes with nonionic detergents. In contrast,

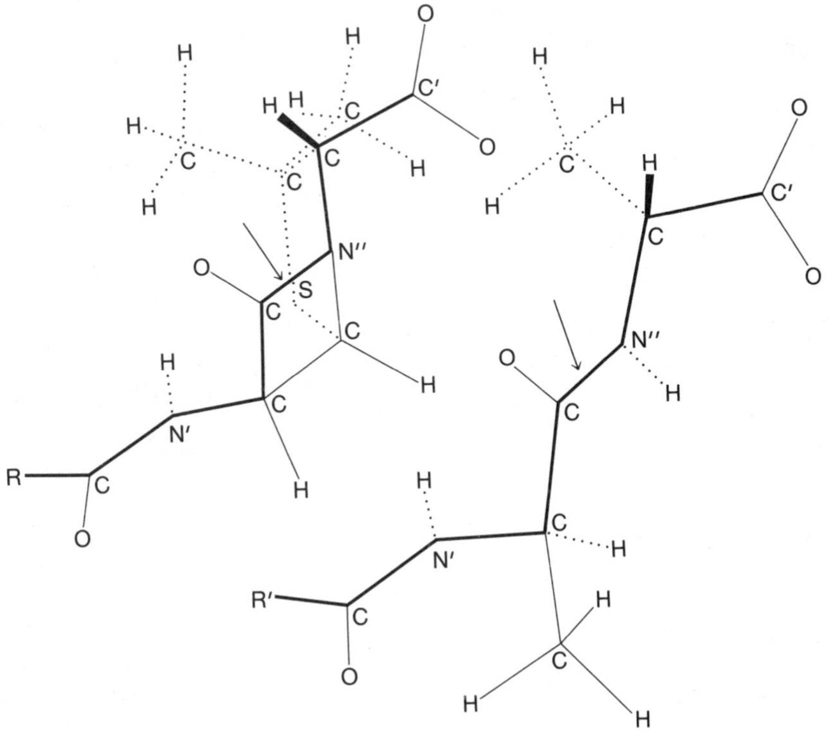

**Figure 12-28**
Dreiding stereomodels of penicillin (*left*) and of the D-alanyl-D-alanine end of the peptidoglycan strand (*right*). Arrows indicate the position of the CO-N bond in the β-lactam ring of penicillin and of the CO-N bond in D-alanyl-D-alanine at the end of the peptidoglycan strand.

**Figure 12-29**
Proposed mechanism of cell-wall transpeptidase and carboxypeptidase action and their inhibition by penicillin in *S. aureus:* (a) the end of the peptide side chain of a glycan strand; (b) the end of the pentaglycine substituent from an adjacent strand.

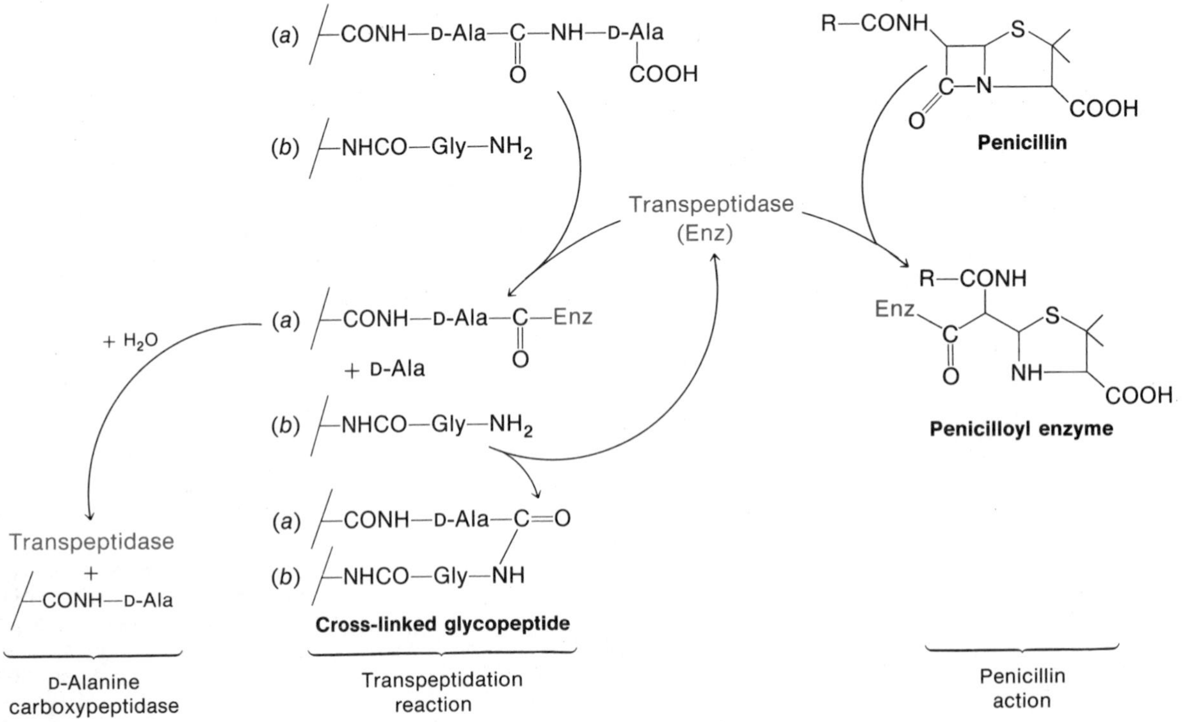

the CPase activity is stable to detergent solubilization, and penicillin-sensitive CPases have been purified from several species of bacteria.

The finding of more than one penicillin-sensitive enzyme activity in bacterial membranes foreshadowed additional complications. In 1972 it was shown that membranes of several bacterial species each contained several distinct proteins that bind penicillin G covalently. Thus began the last and current phase of penicillin research, during which many penicillin-binding proteins (PBPs) have been purified and studied. The affinity for a specific penicillin or cephalosporin varies greatly among the PBPs. For example, PBPs 1 through 4 of *Bacillus stearothermophilus* are 27,000-fold more sensitive to cephalothin than PBP 5, the CPase. Thus different $\beta$-lactam antibiotics kill a particular bacterium most likely by binding to different subsets of the PBPs; i.e., there are multiple killing sites for $\beta$-lactam antibiotics. The function of this large number of PBPs (five or more in different bacteria) is best understood in *E. coli*, where genetic experiments have led to the idea that among the seven PBPs found, PBP 1b is involved in elongation, PBP 2 in initiation of septation, and PBP 3 in completion of septation. Systems for demonstrating in vitro reactions of these PBPs have only recently been developed. In both *E. coli* and bacilli, some of these PBPs appear to be bifunctional enzymes that catalyze both penicillin-insensitive polymerization of peptidoglycan (transglycosylation) and penicillin-sensitive cross-linking (transpeptidation). The functions of many of the PBPs are still unknown, however.

Since CPase is a major PBP that can be purified easily in milligram amounts by penicillin-affinity chromatography, and since it catalyzes a penicillin-sensitive reaction that utilizes substrates related to the cell wall, it has so far been the model system of choice for studying the molecular details of the mode of action of penicillin. Penicillin binds covalently to CPase with complete loss of activity accompanying formation of a stoichiometric penicilloyl-enzyme complex. Although penicillin binding was originally thought to be irreversible, it was later discovered that the penicilloyl moiety is enzymatically released—either rapidly in the presence of hydroxylamine or slowly in the absence of nucleophiles (with a concomitant reactivation of CPase).

In general it has been difficult to obtain true penicillin-resistant mutants as opposed to mutants that produce penicillinase. Penicillin resistance is gained stepwise, a little at a time, as each of the PBPs mutates in turn. Using a series of five selections, a mutant of *B. subtillis* was obtained that was 180-fold more resistant to cloxacillin than the parent. This mutant was missing PBP 1 and had an altered PBP 2a. However, the resistance was clearly dependent on the bulky synthetic substituent of cloxacillin because the mutant retained wild-type sensitivity to other penicillins. Perhaps penicillin is such a close analog of the cell-wall substrate that any alternation of an active site that would lead to resistance would also render the enzyme unable to catalyze the intended essential reaction.

Is inhibition of cell-wall biosynthesis by penicillin sufficient to explain cell lysis? Penicillin-induced lysis probably results from the disruption of a balance between the cell-wall biosynthetic and autolytic enzymes. A mutant *Pneumococcus* that lacks an autolysin specific for its cell wall is not lysed by penicillin. Other experiments suggest that penicillin actually stimulates autolytic activity. However, several other inhibitors of cell-wall synthesis (D-cycloserine, phosphonomycin, and vancomycin) also cause stimulation of cell-wall autolysins. Therefore, this stimulation is probably a secondary effect of the action of penicillin.

In summary, *the mechanism by which β-lactams kill bacteria has turned out to be enormously complex. Penicillin and other β-lactam antibiotics acylate several penicillin-binding proteins in the bacterial membrane. These PBPs are thought to be penicillin-sensitive enzymes involved in cell-wall biosynthesis, including the glycanpolymerase as well as the transpeptidase and D-alanine carboxypeptidase activities.* These enzymes have functions at precise places and stages of cell-wall synthesis and in fact are determinants of cellular morphology. Many aspects of their functions remain to be elucidated.

A number of other antibiotics (D-cycloserine, phosphonomycin, bacitracin, and vancomycin) block other stages in cell-wall synthesis. Their points of action are indicated in Figures 12-19 and 12-21. In addition to their biological and medical importance these antibiotics have been extremely useful in elucidating the biosynthetic pathway.


## O-ANTIGEN SYNTHESIS

The outer membrane of Gram-negative bacteria is composed of phospholipid and various proteins and lipopolysaccharides (Figure 12-30). The lipopolysaccharides contain repeating oligosaccharide units that are attached to a basal core polysaccharide, which in turn is attached to a com-

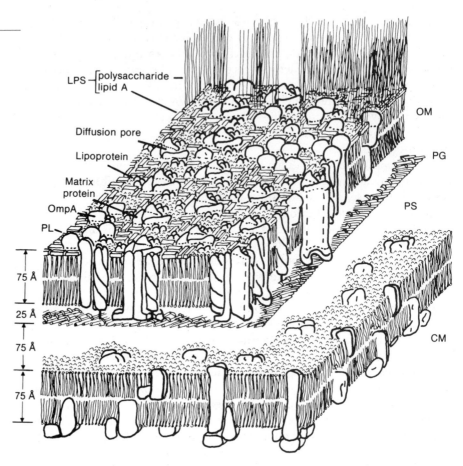

**Figure 12-30**
Diagram of a Gram-negative cell envelope. Components are listed on the right. The trimers of matrix protein of the outer membrane are associated with lipoprotein and with LPS (of variable polysaccharide length), and lipoprotein is covalently bound to peptidoglycan. Diagram also illustrates some general properties of membranes. Phospholipid molecules are illustrated with a *circle* for the polar groups and a *line* for each fatty acid acyl moiety. (Courtesy M. Inouye.)

plex lipid conglomerate known as lipid A (see Figure 16-6b). The repeating oligosaccharide units attached to the core polysaccharide protrude as fibers of indefinite length from the outer membrane surface. Because of their structure and location, these fibers are highly antigenic and are known as O antigens. O antigens vary widely in composition in different bacterial strains, thus providing many dozens of specific O-antigen determinants. A wide variety of cells, including mammalian cells, have glycan-containing substances at their external surface. In higher organisms they are involved in some manner in a variety of immune phenomena, including recognition of self and rejection of foreign cells, as occurs after transfusion of unmatched red blood cells or transplantation of organs. The complexity of such surface antigens is staggering. Over 100 immunologically distinct forms of these substances have been detected on human cells.

Infection of a vertebrate host by a particular strain of bacteria leads to an immune response to that organism. Antisera against different O anti-

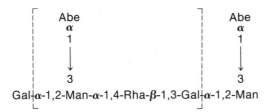

**Figure 12-31**
Repeat unit in O antigen of
*S. typhimurium*. (Abe = abequose;
Man = mannose; Rha = rhamnose;
Gal = galactose.)

gens are intentionally produced in this way. Such antisera are extremely important medically in the classification of bacterial pathogens. Since the immune system has evolved largely to protect vertebrates against bacterial pathogens, it seems very likely that antisera are tailor made to react specifically with O antigens. Conversely, it seems highly likely that the large variety of O antigens found even within a single species of bacteria represents a coevolutionary event designed to allow the bacteria to escape detection by the immune system.

The origin of O-antigen variability is a fascinating subject in itself. The genes that control O-antigen variability appear to be unusually active in switching from the production of one predominant O-antigen type to another. The rate of switching far exceeds normal mutation frequencies, suggesting a recombination mechanism of the type involved in producing different antibodies (Chapters 1 and 2) or a transposition-type process akin to that observed in the control of mating type in yeast (Chapter 26). The types of changes seen within a given bacteria include the addition or deletion of a single hexose in the O-antigen repeat unit, the replacement of one sugar by another, or the wholesale replacement of one O antigen by another. The length of different O-specific chains using the same repeating unit also varies considerably (up to 40 repeat units per chain), but this latter type of variability is probably a function of a rather haphazard mode of synthesis that does not lead to a precise control of length.

*Each of the four sugars in the O-antigen repeat unit of S. typhimurium (Figure 12-31) is formed from a substrate of the glycolytic pathway through a series of conversions into an activated nucleoside diphosphate sugar,* as shown in Figure 12-32.

Determination of the mechanism of synthesis of the tetrameric repeat unit was accomplished by a combination of genetic and biochemical investigations. A mutant of *S. typhimurium* that was incapable of making GDPmannose was isolated. Using membrane-containing extracts of this

**Figure 12-32**
Biosynthesis of the precursors of a specific wall polysaccharide (O antigen) of group B *Salmonella*.

UDPglucose ⟶ UDPgalactose

Glucose-1-phosphate ⟶ TDPglucose ⟶ TDP-4-keto-6-deoxyglucose ⟶ ⟶ TDPrhamnose

Glucose-6-phosphate    CDPglucose ⟶ CDP-keto-6-deoxyglucose ⟶ ⟶ CDPabequose

Fructose-6-phosphate ⟶ Mannose-6-phosphate ⟶ Mannose-1-phosphate ⟶ ⟶ GDPmannose

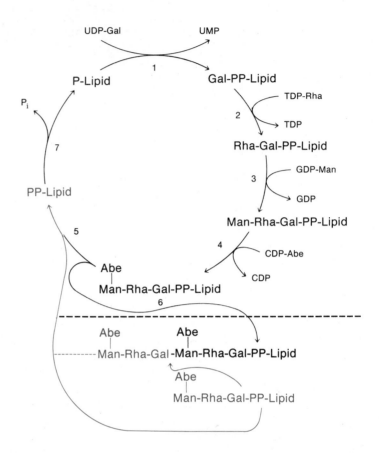

**Figure 12-33**
Steps in the synthesis of O antigen in *S. typhimurium*. The upper portion refers to reactions taking place on the inner side of the cytoplasmic membrane. The lower portion below the dashed line refers to reactions taking place within or outside of the cytoplasmic membrane. In step 6, the components being added to the growing polymer are indicated in color.

mutant, it was possible to demonstrate incorporation of [14]C-labeled mannose from GDPmannose [[14]C] into the lipopolysaccharide. This incorporation was stimulated by the presence of TDPrhamnose and UDPgalactose, suggesting that the latter two sugars are incorporated first during synthesis of the repeat unit. From this and other observations, the scheme for the synthesis of the activated repeat unit was determined (steps 1–4 in Figure 12-33). The activated monomers are condensed in stepwise fashion (UDP-Gal, TDP-Rha, GDP-Man, and CDP-Abe) onto a carrier $C_{55}$-isoprene alcohol phosphate, the same compound as is involved in peptidoglycan synthesis.

The activated tetrasaccharide is next transferred from the inner face of the cytoplasmic membrane to the outer face. Since both the monomer repeat unit and the growing polymer are activated at the galactose end, the growth step in polymer synthesis could conceivably occur either by transfer of the monomer repeat unit to the growing polymer or by transfer of the polymer to the activated monomer. In O-antigen synthesis, the latter is the case. Hence the energy for the condensation reaction comes from the activated polymer. Insofar as we know, this type of "tail-to-head" polymerization is unique to O-antigen synthesis in the carbohydrates, although it is also encountered in both lipid and protein synthesis. It seems likely that in the case of O-antigen synthesis, this mode of chain growth is advantageous because it keeps the chemical reaction involved in chain elongation close to the membrane surface, where the polymerizing enzyme probably functions best.

## SELECTED READINGS

Hanson, R. W., and Mehlman, M. A. (Eds.) *Gluconeogenesis, Its Regulation in Mammalian Species*. New York: Wiley, 1976.

Hubbard, S. C., and Ivatt, R. J. Synthesis and processing of asparagine-linked oligosaccharides. *Ann. Rev. Biochem.* 50:555, 1981. A recent review of glycoprotein synthesis in eukaryotes.

Kochetkov, N. K., and Shibaev, V. N. Glycosyl esters of nucleoside pyrophosphates. *Adv. Carbohydr. Chem. Biochem.* 28:307, 1973. A concise review of the chemistry and biochemistry of nucleoside pyrophosphate sugars and derivatives.

Tipper, D. J., and Wright, A. The Structure and Biosynthesis of Bacterial Cell Walls. In J. R. Sokatch and L. N. Ornston (Eds.). *The Bacteria,* Vol VII, *Mechanisms of Adaptation*. New York: Academic Press, 1979. Pp. 291–426.

Van Schaftingen, C., and Hers, H.-G. Synthesis of a stimulation of phosphofructokinase, most likely fructose 2,6-bisphosphate from phosphoric acid and fructose 6-phosphoric acid. *Biochem. Biophys. Res. Commun.* 96:1524, 1980.

## PROBLEMS

1. Define *gluconeogenesis*.

2. Why was it necessary for cells to evolve alternate pathways for synthesizing glucose rather than merely reversing glycolysis?

3. What is the relative requirement of energy for synthesizing glucose from lactate compared to the yield of energy in degrading glucose to lactate?

4. List some biologically important hexoses derived from glucose and their functions. What is the predominant nucleoside triphosphate utilized in their synthesis?

5. What are the key steps in the conversion of lactate to glucose?

6. Penicillin only kills growing cells. Explain.

7. Penicillin has been helpful in elucidating cell-wall synthesis but not in obtaining mutants in enzymes involved in cell-wall synthesis. Explain.

8. Name two antibiotics other than penicillin that inhibit cell-wall synthesis and indicate what type of intermediates you might expect to accumulate in their presence.

9. Cell-wall and O-antigen synthesis are both unusual in that part of the synthesis is carried on outside the cell. What complications does this introduce in making polymeric linkages in the two cases, and how are these complications overcome?

PART **III**

# LIPIDS AND MEMBRANES

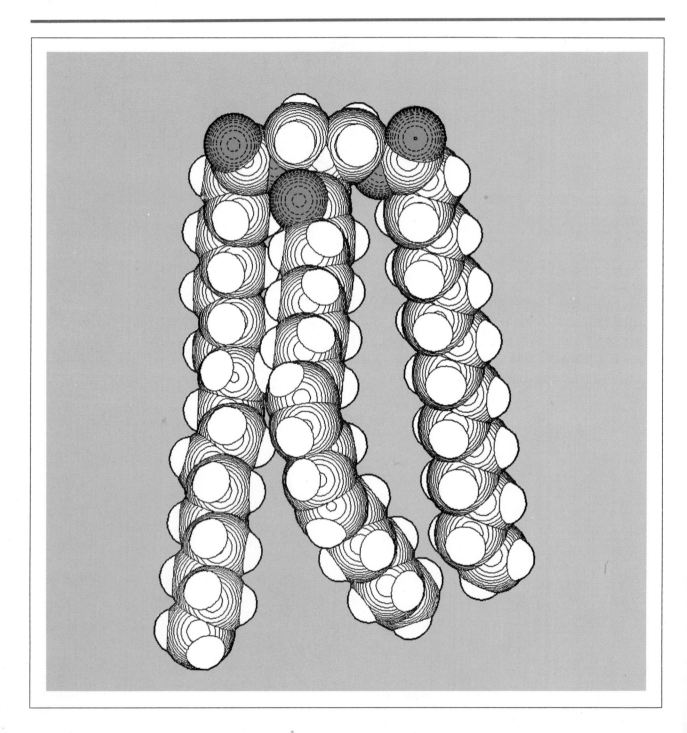

# 13

# METABOLISM OF FATTY ACIDS AND TRIACYLGLYCEROLS

_Lipids_ are broadly defined as biological molecules that are soluble in organic solvents. Although the lipids encompass a large and diverse group of compounds, they have only four major biological functions: (1) in all cells, the major structural elements of the membranes are composed of lipids; (2) certain lipids (the _triacylglycerols_) serve as efficient reserves for the storage of energy; (3) many of the vitamins and hormones found in animals are lipids or derivatives of lipids, and (4) the bile acids help to solubilize the other lipid classes during digestion.

This and the following two chapters present an introduction to the biochemistry of the lipids. The subject has been somewhat artificially divided. The advantage of such a division is to package the subject into a smaller and more easily digestible format. Obviously, this division does not occur in cells and organisms. The metabolism of fatty acids, phospholipids, triacylglycerols, and steroids impinge on one another. Similarly, the metabolism of carbohydrates has striking effects on lipid metabolism. With this caveat in mind, let us begin our discussion of fatty acids.

## FATTY ACIDS

### Structure of Fatty Acids

Compounds with the structural formula $CH_3(CH_2)_n COOH$ that contain no carbon–carbon double bonds are known as _saturated fatty acids_. The two most abundant saturated fatty acids are _palmitic_ and _stearic_ acids

_Space-filling model of triacylglycerol, a most efficient molecule for storing chemical energy._

## Table 13-1
Fatty Acids

| Common Name | Systematic Name | Structure | Abbreviation |
|---|---|---|---|
| **Saturated fatty acids** | | | |
| Myristic acid | n-Tetradecanoic acid | $CH_3(CH_2)_{12}COOH$ | 14:0 |
| Palmitic acid | n-Hexadecanoic acid | $CH_3(CH_2)_{12}CH_2CH_2COOH$ | 16:0 |
| Stearic acid | n-Octadecanoic acid | $CH_3(CH_2)_{12}CH_2CH_2CH_2CH_2COOH$ | 18:0 |
| Arachidic acid | n-Eicosanoic acid | $CH_3(CH_2)_{12}CH_2CH_2CH_2CH_2CH_2CH_2COOH$ | 20:0 |
| Behenic acid | n-Docosanoic acid | $CH_3(CH_2)_{12}CH_2CH_2CH_2CH_2CH_2CH_2CH_2CH_2COOH$ | 22:0 |
| Lignoceric acid | n-Tetracosanoic acid | $CH_3(CH_2)_{12}CH_2CH_2CH_2CH_2CH_2CH_2CH_2CH_2CH_2CH_2COOH$ | 24:0 |
| Cerotic acid | n-Hexacosanoic acid | $CH_3(CH_2)_{12}CH_2CH_2CH_2CH_2CH_2CH_2CH_2CH_2CH_2CH_2CH_2CH_2COOH$ | 26:0 |
| **Unsaturated fatty acids** | | | |
| Palmitoleic acid | cis-9-Hexadecenoic acid | $CH_3(CH_2)_5\overset{\text{H}}{C}{=}\overset{\text{H}}{C}(CH_2)_7COOH$ | 16:1$^{\Delta 9}$ |
| Oleic acid | cis-9-Octadecenoic acid | $CH_3(CH_2)_7\overset{\text{H}}{C}{=}\overset{\text{H}}{C}(CH_2)_7COOH$ | 18:1$^{\Delta 9}$ |
| Vaccenic acid | cis-11-Octadecenoic acid | $CH_3(CH_2)_5\overset{\text{H}}{C}{=}\overset{\text{H}}{C}(CH_2)_9COOH$ | 18:1$^{\Delta 11}$ |
| Linoleic acid | cis,cis-9,12-Octadecadienoic acid | $CH_3(CH_2)_4C{=}C{-}CH_2{-}C{=}C(CH_2)_7COOH$ (all cis, H on double-bond carbons) | 18:2$^{\Delta 9,12}$ |
| α-Linolenic acid | All-cis-9,12,15-Octadecatrienoic acid | $CH_3CH_2C{=}C{-}CH_2{-}C{=}C{-}CH_2{-}C{=}C(CH_2)_7COOH$ | 18:3$^{\Delta 9,12,15}$ |
| Arachidonic acid | All-cis-5,8,11,14-Eicosatetraenoic acid | $CH_3(CH_2)_4C{=}C{-}CH_2{-}C{=}C{-}CH_2{-}C{=}C{-}CH_2{-}C{=}C(CH_2)_3COOH$ | 20:4$^{\Delta 5,8,11,14}$ |
| | All-cis-4,7,10,13,16,19-Docosahexaenoic acid | $CH_3CH_2C{=}C{-}CH_2{-}C{=}C{-}CH_2{-}C{=}C{-}CH_2{-}C{=}C{-}CH_2{-}C{=}C{-}CH_2{-}C{=}C{-}(CH_2)_2COOH$ | 22:6$^{\Delta 4,7,10,13,16,19}$ |
| **Some unusual fatty acids** | | | |
| | 2,4,6,8-Tetramethyl decanoic acid | $CH_3CH_2{\left(\overset{CH_3}{CH}{-}CH_2\right)}_3\overset{CH_3}{CH}{-}COOH$ | |
| Lactobacillic acid | | $CH_3(CH_2)_5\overset{\displaystyle CH_2}{\overbrace{CH{-}CH}}(CH_2)_9COOH$ | |
| An α-mycolic acid | | $CH_3(CH_2)_{17}\overset{\displaystyle CH_2}{\overbrace{CH{-}CH}}(CH_2)_{10}{-}\overset{\displaystyle CH_2}{\overbrace{CH{-}CH}}(CH_2)_{17}{-}\overset{OH}{CH}{-}\underset{\underset{CH_3}{(CH_2)_{23}}}{CH}{-}COOH$ | |

(Table 13-1). Some other saturated fatty acids present in smaller quantities in mammalian tissues also are shown in Table 13-1. The *sphingolipids*, which will be discussed in Chapter 14, contain longer-chain fatty acids ($n = 20$–24), as well as palmitic and stearic acids. In some tissues, short-chain fatty acids also are found, such as decanoic acid (10:0) in milk.

Fatty acids with double bonds in the aliphatic chain are called unsaturated fatty acids. *Monounsaturated* fatty acids have one double bond, and *polyunsaturated fatty acids* contain more than one double bond. The double bonds in virtually all naturally occurring fatty acids are cis. Fatty acids are often abbreviated as shown in Table 13-1. The number to the left of the colon indicates the number of carbon atoms of the fatty acid and the number to the right indicates the number of double bonds. The numbering begins from the carboxyl group. The position of the double bond is shown by a superscript $\Delta$ followed by the number of carbons between the double bond and the carboxyl group. The double bonds in polyunsaturated fatty acids are always separated by one methylene group. Mammalian tissues contain all the unsaturated fatty acids listed in Table 13-1 with the exception of vaccenic acid, which is present in *Escherichia coli* and other bacteria. *E. coli* does not contain polyunsaturated fatty acids. Oleic acid is the most common monounsaturated fatty acid in mammals. Two unsaturated fatty acids, *linoleic* and *linolenic* acids, are not synthesized by mammals and are therefore important dietary requirements. Like vitamins, these two fatty acids are required for growth and good health. Hence they are called *essential fatty acids*. Plants are able to synthesize linoleic and linolenic acids and are the original source of these fatty acids in our diet.

Sometimes unsaturated fatty acids are numbered from the terminal methyl group. In this instance, the numbering is preceded by a lowercase omega ($\omega$). Thus linoleic acid might be called $\Delta$9,12-octadecadienoic acid or $\omega$-6,9-octadecadienoic acid.

In addition to the commonly occurring fatty acids, many structural variations have evolved. There are well over 100 other fatty acids found in various creatures and organisms, often associated with specialized functions. For instance, branched-chain fatty acids are found in many different tissues. The uropygial gland of the duck produces such a fatty acid (2,4,6,8-tetramethyldecanoic acid). The duck uses the fatty acids secreted by this gland to preen its feathers and thereby ensure that water continues to "run off its back." In the bacterial genus *Bacillus*, the monoenoic fatty acids are replaced by branched-chain fatty acids in which a methyl group is adjacent to either the terminal methyl group (the *iso* series) or the terminal ethyl group (the *ante-iso* series). Another example is fatty acids with a *cyclopropane ring* in the alkyl chain found in many bacteria. The bacterium that causes the disease tuberculosis, *Mycobacterium tuberculosis*, produces a family of complex fatty acids known as *mycolic acids*, which contain cyclopropane rings. One class of these is the $\alpha$-mycolic acids and an example is given in Table 13-1. Many structurally related $\alpha$-mycolic acids are found in the mycobacteria and other related bacteria (nocardiae and corynebacteria). These mycolic acids appear to have a structural function in the outer part of the bacterial cell wall. There is much evidence to suggest that a major drug used in the treatment of tuberculosis (Isoniazid) functions by the inhibition of an early reaction of $\alpha$-mycolic acid biosynthesis.

$$CH_3-CH-\overset{\displaystyle CH_3}{|}$$

*Iso* **series**

$$CH_3CH_2-CH-\overset{\displaystyle CH_3}{|}$$

*Ante-iso* **series**

*Fatty acids are usually found as components of complex lipids,* and hence, only a very small percent exists as unesterified (free) fatty acids. Nevertheless, it is worth noting that the $pK_a$ for dissociation of the acid proton is around 4.7. Therefore, at pH 7.0, the fatty acid exists primarily in the dissociated form ($RCOO^-$):

$$CH_3(CH_2)_nCOOH \rightleftharpoons CH_3(CH_2)_nCOO^- + H^+$$

Because it exists as a salt at neutral pH, it is not easily extracted from an aqueous medium by organic solvents such as hexane. However, if the pH is lowered by the addition of HCl or another strong acid, the fatty acid becomes protonated and is easily extracted by organic solvents.

Another property of fatty acids that should be noted is the physical form of the fatty acids at room temperature. If $n$ is equal to 8 or less, the fatty acid is a liquid, whereas if $n$ is equal to 10 or more, the fatty acid is a solid. If a fatty acid has a double bond, it has a lower melting point than the saturated fatty acid with the same number of carbons. Unsaturated fatty acids are more condensed in length than the corresponding saturated fatty acids (Figure 13-1). The double bonds prevent the tight packing within membranes that occurs with saturated fatty acids (see Chapter 16).

When fatty acids were first discovered and chemists were involved in structural determinations, a major problem was separation of the various fatty acids into pure compounds. Consider the difficulties that were encountered in the separation of stearic acid from palmitic acid by the tech-

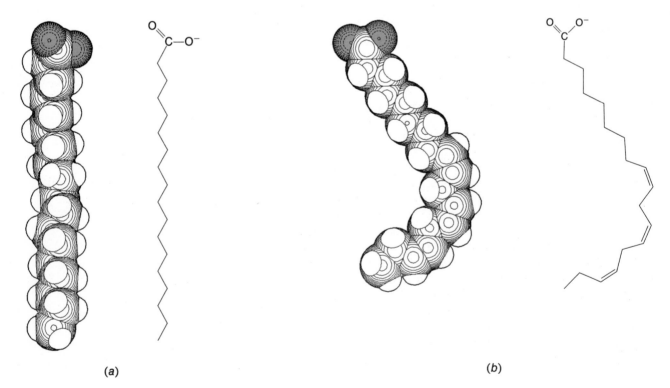

(a)                                                                  (b)

**Figure 13-1**
Space-filling model of stearic and linolenic acids.

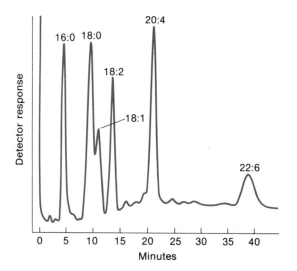

**Figure 13-2**
Gas chromatogram of methyl esters of fatty acids derived from total lipids of rat liver. The analysis was done on a 6-ft column packed with 15% ethylene glycol succinate on 80/100 Gas Chrom P. The temperature was 150°C for 15 min, programmed to 180°C at 6°C per min and held at 180° until 22:6 had eluted.

niques available in the early 1900s (organic extractions and crystallization). However, with the advent of gas-liquid chromatography in the 1950s, it became relatively easy to analyze and purify fatty acids from complex mixtures. Today fatty acids are commonly analyzed by gas chromatography of the methyl esters. These are formed by esterification of the fatty acids with methanol [R represents hydrogen (H) or any group to which the fatty acid is esterified]:

$$\underset{\text{HCl or BF}_3}{CH_3(CH_2)_n \overset{\overset{\displaystyle O}{\|}}{C}\text{—OR} + CH_3OH \longrightarrow CH_3(CH_2)_n \overset{\overset{\displaystyle O}{\|}}{C}OCH_3 + ROH}$$

An analysis of the fatty acids from rat liver is shown in Figure 13-2.

## Triacylglycerols (Triglycerides)

Fatty acids are major components of the *triacylglycerols* (Table 13-2 and Figure 13-3) and most of the complex lipids present in membranes. *It is in the triacylglycerols that fatty acids are stored as an energy reserve*. Triacylglycerols are the major uncharged glycerol derivatives found in animals. The monoacylglycerols and diacylglycerols are metabolites of triacylglycerols (and of phospholipids, as discussed in Chapter 14) and are normally present in cells in very small quantities.

Because the substituents esterified to the first and third carbons are usually different, the second carbon of the glycerol derivative is asymmetric. In naming and numbering these glycerol derivatives, a special convention has been adopted: The prefix *sn-* (for *stereospecifically numbered*) immediately precedes "glycerol" and differentiates the naming of the compound from other approaches, such as the R*S* system described in Chapter 6. The glycerol derivative is drawn in a Fischer projection with the secondary hydroxyl to the *left* of the central carbon, and the carbons are numbered 1, 2, and 3 from the top to the bottom. The prefix *rac-* (for *racemo*) precedes the name if the compound is an equal mixture of antipodes. If the configuration is unknown or not specified, x- precedes the name.

**Table 13-2**
Neutral Glycerides

| Common Name | Systematic Name | Structure |
|---|---|---|
| Triglyceride | 1,2,3-Triacyl-*sn*-glycerol | 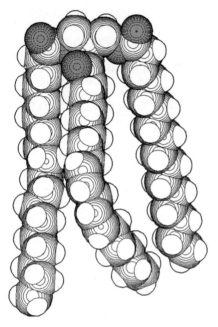 |
| Diglyceride | 1,2-Diacyl-*sn*-glycerol | |
| Monoglyceride | 1-Monoacyl-*sn*-glycerol | |

Although triacylglycerols are found in the liver and intestine, they are primarily found in adipose tissue (fat), which functions as a storage depot for this lipid. The specialized cell in this tissue is called the *adipocyte*. The cytoplasm of the cell is full of lipid vacuoles that are almost exclusively triacylglycerols (Figure 13-4) and serve as an energy reserve for mammals. At times when the diet or glycogen reserves are insufficient to supply the body's need for energy, the fuel stored as fatty acyl components of the triacylglycerols is mobilized and transported to other tissues in the body. A second important function of adipose tissue is insulation of the body from cold. This function is most obvious in such cold-water mammals as the arctic whales (Beluga whales), which have vast stores of fat (blubber).

## MOBILIZATION AND TRANSFER OF FATTY ACIDS FROM ADIPOSE TISSUE

### Mobilization of Fatty Acids from Triacylglycerols

*When the energy supply from diet becomes limited, the animal responds to this deficiency with a hormonal signal* that is transmitted to the adipose tissue by the release of *epinephrine*, *glucagon*, and other hormones. The hormones bind to the plasma membrane of the adipocyte and stimulate the synthesis of *cyclic AMP* (cAMP), as previously discussed for the mobilization of glycogen (Chapter 8). As shown in Figure 13-5, this process involves the activation by cAMP of a protein kinase that phosphorylates and activates hormone-sensitive *triacylglycerol lipase*. The latter enzyme hydrolyzes the triacylglycerol to diacylglycerol with release of a fatty acid from carbon 1 or 3 of the glycerol backbone. This reaction is thought to be the rate-limiting step in the complete hydrolysis of the triacylglycerols. The diacylglycerols and monoacylglycerols are rapidly hydrolyzed to yield fatty

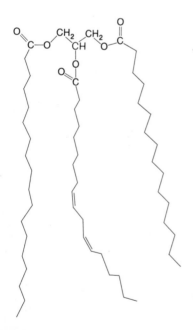

**Figure 13-3**
Space-filling model of triacylglycerol.

acids and glycerol. At this juncture it is not clear whether the diacylglycerol lipase is a separate enzyme or the same enzyme as triacylglycerol lipase. However, monoacylglycerol lipase is a separate enzyme, at least in chicken liver.

The unesterified fatty acids move through the plasma membranes of the adipocytes and endothelial cells of the blood capillaries and are bound by _albumin_ in plasma. The mechanism for transfer of these fatty acids from inside the adipocytes to the plasma compartment is thought to involve passive diffusion. Hence the rate of transfer depends on the concentrations of fatty acids both in the adipocytes and in the plasma. Albumin carries the fatty acids to other tissues in the body. The glycerol also can be released into the plasma and be taken up by the liver for glucose production.

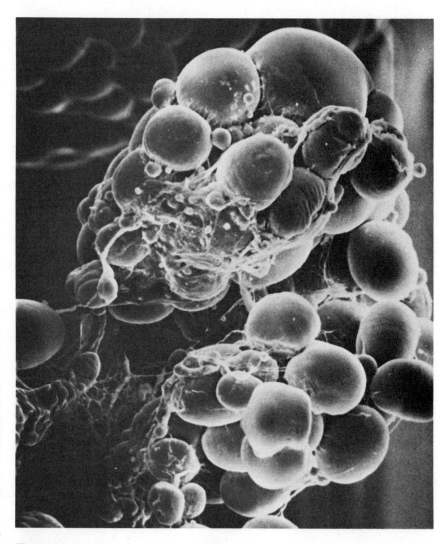

**Figure 13-4**
Scanning electron micrograph of white adipocytes from rat adipose tissue (500×). (Courtesy Dr. A. Angel and Dr. M. J. Hollenberg of the University of Toronto.)

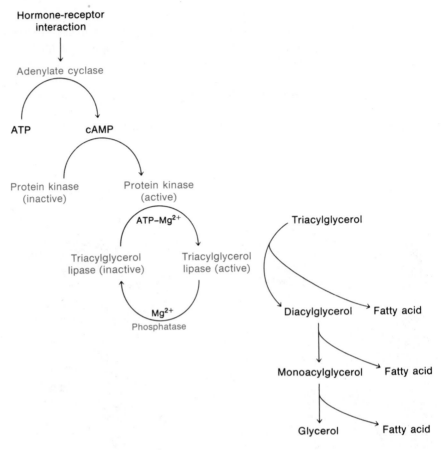

**Figure 13-5**
Activation of triacylglycerol lipase in adipose tissue.

## Albumin: A Principal Carrier of Fatty Acids in the Plasma

Albumin is a quantitatively significant protein in humans because it constitutes about 50 percent (4 g/dl) of the plasma proteins. The protein has a molecular weight of 66,200 and is a single polypeptide chain that is linked together by 17 disulfide bridges. Each albumin molecule has the capacity to bind 10 molecules of fatty acid, although normally only 0.5 to 2 mol of fatty acid is bound. There appear to be two or three primary binding sites. Longer-chain fatty acids bind more readily to albumin than do shorter-chain fatty acids (e.g., stearate binds more readily than palmitate). Mono-unsaturated fatty acids bind with higher affinity to albumin than saturated fatty acids (e.g., oleate binds more readily than stearate). However, linoleate binds less readily than stearate. The time (half-life) it takes one-half the fatty acids bound to albumin in plasma to be taken up by various tissues is 1 to 2 min.

Although albumin has long been considered an essential protein of plasma, a few people with depressed levels of albumin (4.6 to 24 mg/dl compared with 4 g/dl) have been described (*analbuminemia*). Curiously, these people are virtually asymptomatic and their problem is usually diagnosed as a result of routine blood analyses. Clearly, large quantities of albumin are not essential for life. The transport function of albumin in people suffering from analbuminemia is most likely assumed by the lipoproteins.

*Albumin carries the fatty acids to energy-deficient tissues, where the fatty acids move from the plasma compartment into the tissues by a process of diffusion.* The amount of fatty acid removed by a tissue depends on the relative concentrations both in the plasma and in the cells of the tissues. Cardiac and red muscle utilize fatty acids as the major oxidative source of adenosine triphosphate (ATP) and therefore remove large amounts from the circulation.

## Acyl-CoA Synthases: The Enzymes that Activate Fatty Acids

The fatty acids that are taken into the cells are activated in the cytosol by reaction with *coenzyme A* (CoA) and ATP to yield fatty acyl-CoA in a reaction catalyzed by *acyl-CoA synthase* (also known as *thiokinase*):

$$RCOO^- + ATP + CoA \xrightleftharpoons{Mg^{2+}} RCO-CoA + PP_i + AMP$$

In addition to the cytosolic enzyme, there are several synthases that activate fatty acids in the mitochondria and at least one thiokinase associated with microsomes. As is the case with other reactions in which $PP_i$ is a product, this is rapidly hydrolyzed to $2 P_i$, and therefore, the formation of the fatty acyl-CoA derivative is highly favored. Hence two high-energy bonds are hydrolyzed for the synthesis of one acyl-CoA. The concentration of $PP_i$ in the cells (0.01 m$M$ in rat liver) is very low; thus the pyrophosphorolysis of the acyl-CoA by the synthase is prevented.

## Transport of Fatty Acids into the Mitochondria

Fatty acids cannot be utilized for the energy requirements of cells until they have been transported into the mitochondria, the major sites of $\beta$ oxidation. *Fatty acyl-CoAs, which cannot cross the inner membranes of mitochondria, are converted to their acyl carnitine derivatives, which can cross this membrane:*

$$RCO-CoA + (CH_3)_3\overset{+}{N}-CH_2\underset{\underset{\text{OH}}{|}}{C}HCH_2COO^- \rightleftharpoons$$

**Carnitine**

$$(CH_3)_3\overset{+}{N}-CH_2\underset{\underset{\underset{RC=O}{|}}{\overset{|}{O}}}{C}HCH_2COO^- + CoA$$

**Acyl carnitine**

This reaction is catalyzed by *carnitine acyltransferase I*. There are at least three acyltransferases associated with mitochondria: one specific for short-chain fatty acids (*carnitine acetyltransferase*) and two specific for the longer-chain fatty acids (*carnitine acyltransferases I and II*). There also is evidence for an acyltransferase with intermediate-chain-length specificity.

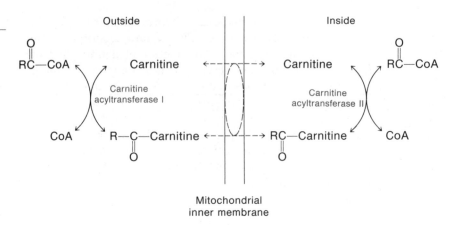

**Figure 13-6**
Transport of acyl derivatives across the mitochondrial membrane.

There is a protein carrier in the inner mitochondrial membrane that can transport carnitine, acetyl-carnitine, and short- and long-chain acyl carnitine derivatives across the membrane. The transfer of fatty acyl carnitine into the mitochondria appears to involve an exchange with free carnitine, as illustrated in Figure 13-6. Once inside the mitochondria, the reaction is reversed by _carnitine acyltransferase II_ to yield a fatty acyl-CoA inside the mitochondria. Thus there are at least two distinct pools of acyl-CoA in the cell, one in the cytosol and the other in the mitochondria.

## OXIDATION OF FATTY ACIDS

### Historical Background

Our current understanding of fatty acid oxidation did not start to develop in detail until the early 1950s. However, in 1905, F. Knoop reported a series of experiments that indicated that fatty acids were oxidized by removal of two carbons at a time. He fed rabbits fatty acids in which the methyl group had been replaced with a phenyl group. If the altered fatty acid contained an even number of carbons [for example, $C_6H_5$—$CH_2(CH_2)_2COOH$], the primary metabolite was phenylacetic acid ($C_6H_5$—$CH_2COOH$), which was excreted as its glycine conjugate, phenylaceturic acid ($C_6H_5$—$CH_2CO$—$NHCH_2COOH$). When he fed the rabbits a phenyl derivative of a fatty acid with an odd number of carbons [for example, $C_6H_5$—$CH_2(CH_2)_3COOH$], he found benzoic acid ($C_6H_5COOH$), which was excreted in the urine as its glycine conjugate, hippuric acid ($C_6H_5CO$—$NH$—$CH_2COOH$). Knoop postulated that _fatty acids were oxidized at the β carbon (hence β oxidation) and degraded to acetic acid and a fatty acid with two fewer carbons:_

$$RCH_2CH_2COOH \longrightarrow R\overset{O}{\overset{\|}{C}}CH_2COOH \longrightarrow R\overset{O}{\overset{\|}{C}}OH + CH_3COOH$$

The next major experimental step was the demonstration in 1944 by Luis LeLoir that fatty acids could be oxidized in a cell-free system. This was followed by Albert Lehninger's demonstration that the process of fatty acid oxidation occurred in liver mitochondria and, apparently, involved an "active acetate." Experiments by Fritz Lipmann proved that CoA was involved in the formation of "active acetate":

$$\text{Acetate} + \text{ATP} + \text{CoA} \longrightarrow \text{"active acetate"}$$

Subsequently, in 1951, Feodor Lynen, working in Munich with yeast, demonstrated that "active acetate" was acetyl-CoA. At this stage, several laboratories conceived the idea that CoA may play a role in the activation of fatty acids for $\beta$ oxidation, and by 1954, the basic outline of $\beta$ oxidation as we know it today was developed.

## Overall Scheme for $\beta$ Oxidation, the Principal Route for Catabolism of Fatty Acids

The outline for $\beta$ oxidation of a saturated fatty acid is shown in Figure 13-7. In the first reaction, the acyl-CoA is dehydrogenated to yield the $\alpha$–$\beta$(or 2–3)-*trans*-enoyl-CoA and $FADH_2$. This enoyl-CoA is subsequently hydrated stereospecifically to yield the 3-L-hydroxyacyl-CoA. The hydroxyl group is oxidized by $NAD^+$ and a dehydrogenase to yield $\beta$-ketoacyl-CoA and NADH. The final step in the sequence involves a thiolytic cleavage to form acetyl-CoA and an acyl-CoA that is two carbons shorter than the initial substrate for $\beta$ oxidation (Figure 13-7). This acyl-CoA can undergo another round of $\beta$ oxidation to yield $FADH_2$, NADH, acetyl-CoA, and acyl-CoA. The enzymatic steps are repeated until in the last sequence of reactions, butyryl-CoA ($CH_3CH_2CH_2CO$—CoA) is degraded to two acetyl-CoAs.

The equations for the complete oxidation of palmitoyl-CoA are shown in Table 13-3. In Equation (1), the oxidation of palmitoyl-CoA by the enzymes of $\beta$ oxidation is shown. Each of the products of Equation (1) is further oxidized by the respiratory chain (Equations 2 and 3) or by the tricarboxylic acid cycle and the respiratory chain (Equation 4). When these reactions (Equations 1 to 4) are added together, the result is Equation (5). Hence 1 molecule of palmitoyl-CoA can be oxidized to yield 16 $CO_2$ + 131 ATP + 146 $H_2O$ and CoA. If the starting material were palmitic acid, its complete oxidation would yield 129 ATP + 16 $CO_2$ + 145 $H_2O$. The formation of the CoA derivative of the fatty acid requires the hydrolysis of two high-energy bonds, and one molecule of $H_2O$ is consumed in the hydrolysis of the $PP_i$ produced by the thiokinase reaction. Interestingly, oxidation of fatty acids can be used as a major source of $H_2O$, e.g., in the killer whale (which is actually a dolphin). This animal lives in the sea, but does not drink the seawater. Instead, the whale obtains a significant amount of its water from dietary fatty acids.

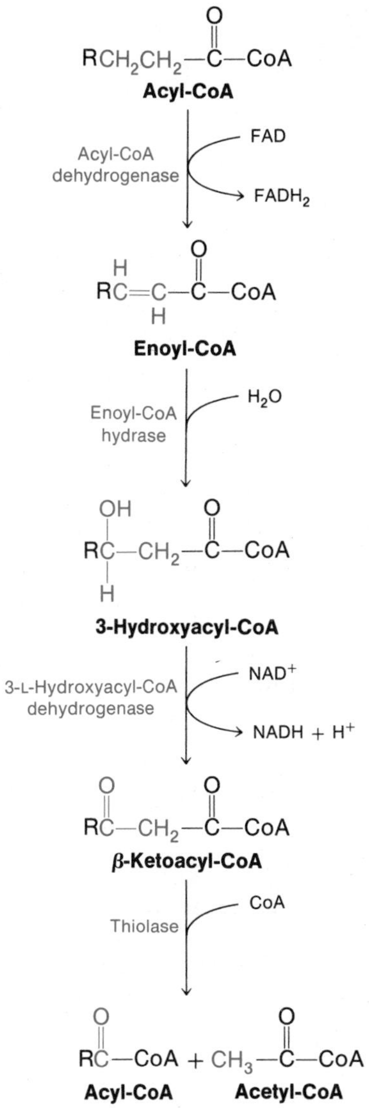

**Figure 13-7**
Scheme for $\beta$ oxidation of fatty acids.

**Table 13-3**
Equations for the Complete Oxidation of Palmitoyl-CoA to $CO_2$ and $H_2O$

$$CH_3(CH_2)_{14}CO—CoA + 7\,FAD + 7\,H_2O + 7\,NAD^+ + 7\,CoA \longrightarrow$$
$$8\,CH_3COSCoA + 7\,FADH_2 + 7\,NADH + 7\,H^+ \quad (1)$$

$$7\,FADH_2 + 14\,P_i + 14\,ADP + 3\tfrac{1}{2}\,O_2 \longrightarrow 7\,FAD + 21\,H_2O + 14\,ATP$$

$$7\,NADH + 7\,H^+ + 21\,P_i + 21\,ADP + 3\tfrac{1}{2}\,O_2 \longrightarrow$$
$$7\,NAD^+ + 28\,H_2O + 21\,ATP \quad (3)$$

$$8\,\text{Acetyl-CoA} + 16\,O_2 + 96\,P_i + 96\,ADP \longrightarrow$$
$$8\,CoA + 104\,H_2O + 16\,CO_2 + 96\,ATP \quad (4)$$

$$CH_3(CH_2)_{14}CO—CoA + 131\,P_i + 131\,ADP + 23\,O_2 \longrightarrow$$
$$131\,ATP + 16\,CO_2 + 146\,H_2O + CoA \quad (5)$$

The yield of ATP from the oxidation of palmitic acid can be compared with that from glucose. Both are major sources of energy in our body. Since palmitate has 16 carbons and glucose has 6 carbons, the comparison should be made between 1 palmitate and $2\frac{2}{3}$ glucose molecules. As learned in Chapter 10, 36 ATPs are produced from the complete oxidation of glucose to $CO_2$ and $H_2O$. The yield from $2\frac{2}{3}$ glucose molecules is therefore 96 ATPs. Thus oxidation of palmitate yields an additional 33 ATPs. Therefore, palmitate is a more efficient molecule than glucose for storage of energy. Lipid is also less hydrated than carbohydrate and therefore takes up less space. These two factors are probably why fat is the major storage form of energy. The chemical reason for the difference in oxidative energy yield between glucose and palmitate is that palmitate is almost completely in the reduced state, whereas glucose is partially oxidized with six oxygens in the molecule.

## Enzymology of $\beta$ Oxidation

Three different flavoproteins have been found in mitochondria that catalyze the initial dehydrogenation of acyl-CoA. The enzymes show specificity for either short-, medium-, or long-chain acyl-CoAs and specifically yield the $trans$-$\alpha,\beta$-enoyl-CoA. The molecular weight of all three enzymes is in the range of 140,000 to 200,000. A second dehydrogenase, called the electron-transferring flavoprotein (ETF), oxidizes the acyl-CoA dehydrogenases so that they can participate in another round of $\beta$ oxidation.

Only a single enoyl hydrase with very broad specificity for the acyl group has been identified. The enzyme has a molecular weight of 210,000. Similarly, a single $\beta$-hydroxyacyl-CoA dehydrogenase appears to be involved in the oxidation of the L-hydroxy group. NAD$^+$ is the specific electron acceptor for this reaction. The last reaction in the fatty acid oxidation cycle is catalyzed by $\beta$-ketoacyl-CoA thiolase. A second thiolase has been identified that is specific for acetoacetyl-CoA.

In recent years, the enzymes of $\beta$ oxidation also have been found in peroxisomes. At the present time, the role of these organelles in the metabolism of fatty acids has not been well-defined.

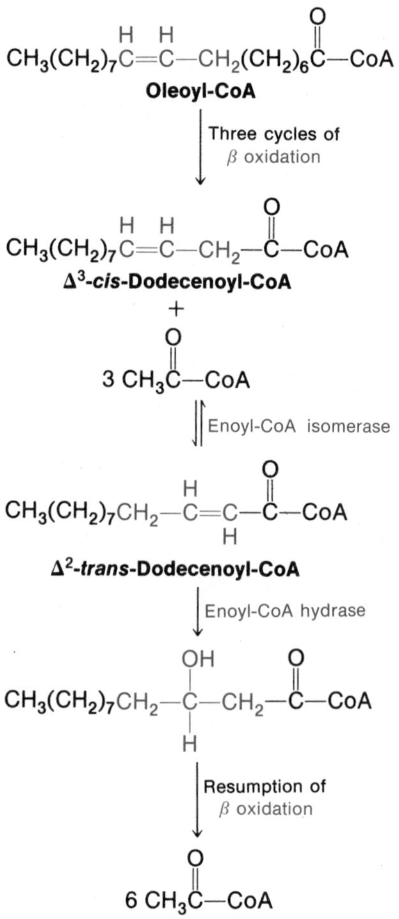

## Oxidation of Unsaturated Fatty Acids

Unsaturated fatty acids are similarly degraded by $\beta$ oxidation. As seen in Figure 13-8, oleic acid can be degraded in the same manner as stearic acid through the first three cycles of $\beta$ oxidation. The resulting $cis$-3-dodecenoic acid ($12:1^{\Delta 3}$), however, is not a substrate for the acyl-CoA dehydrogenase. This step is bypassed by an isomerization of the double bond by _enoyl-CoA isomerase_ to the $trans$-2-dodecenoyl-CoA, which is a normal substrate for enoyl-CoA hydratase, and the normal route for $\beta$ oxidation resumes.

Polyunsaturated fatty acids also are degraded by $\beta$ oxidation, but the process is aided by two additional enzymes, _enoyl-CoA isomerase_ and _3-hydroxyacyl-CoA epimerase_ (Figure 13-9). The degradation of linoleoyl-CoA begins, as with oleoyl-CoA, with three rounds of $\beta$ oxidation and results in a $cis$-3-unsaturated fatty acid that is not a substrate for the acyl-CoA dehydrogenase. Isomerization of the double bond to the $trans$-2 posi-

**Figure 13-8**
The $\beta$ oxidation of oleoyl-CoA.

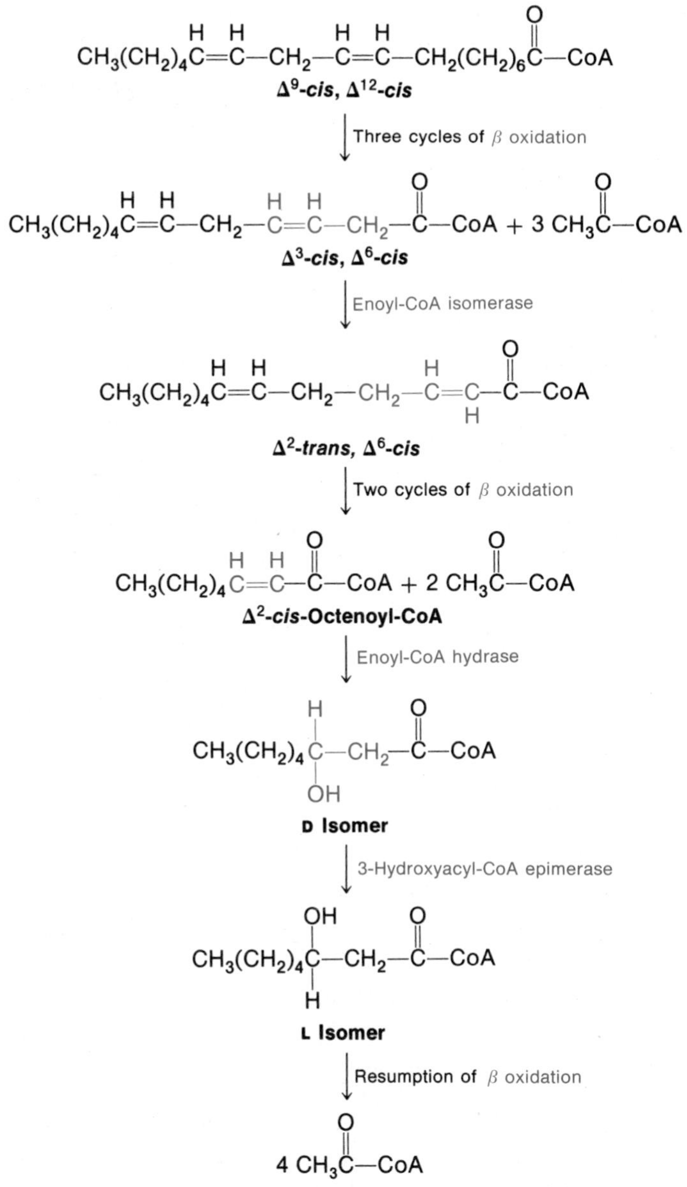

$$CH_3(CH_2)_4\overset{H\ H}{C=C}-CH_2-\overset{H\ H}{C=C}-CH_2(CH_2)_6\overset{O}{\overset{\|}{C}}-CoA$$

$\Delta^9$-*cis*, $\Delta^{12}$-*cis*

Three cycles of $\beta$ oxidation

$$CH_3(CH_2)_4\overset{H\ H}{C=C}-CH_2-\overset{H\ H}{C=C}-CH_2-\overset{O}{\overset{\|}{C}}-CoA + 3\ CH_3\overset{O}{\overset{\|}{C}}-CoA$$

$\Delta^3$-*cis*, $\Delta^6$-*cis*

Enoyl-CoA isomerase

$$CH_3(CH_2)_4\overset{H\ H}{C=C}-CH_2-CH_2-\overset{H}{\underset{H}{C=C}}-\overset{O}{\overset{\|}{C}}-CoA$$

$\Delta^2$-*trans*, $\Delta^6$-*cis*

Two cycles of $\beta$ oxidation

$$CH_3(CH_2)_4\overset{H\ H}{C=C}-\overset{O}{\overset{\|}{C}}-CoA + 2\ CH_3\overset{O}{\overset{\|}{C}}-CoA$$

$\Delta^2$-*cis*-Octenoyl-CoA

Enoyl-CoA hydrase

$$CH_3(CH_2)_4\underset{OH}{\overset{H}{C}}-CH_2-\overset{O}{\overset{\|}{C}}-CoA$$

**D Isomer**

3-Hydroxyacyl-CoA epimerase

$$CH_3(CH_2)_4\overset{OH}{\underset{H}{C}}-CH_2-\overset{O}{\overset{\|}{C}}-CoA$$

**L Isomer**

Resumption of $\beta$ oxidation

$$4\ CH_3\overset{O}{\overset{\|}{C}}-CoA$$

**Figure 13-9**
Pathway for the $\beta$ oxidation of
linoleoyl-CoA.

tion by enoyl-CoA isomerase allows for the resumption of two cycles of $\beta$ oxidation. The product, *cis*-2-octenoyl-CoA, has the double bond in the correct position of the chain, but is cis rather than trans. The enoyl-CoA hydrase will hydrate this fatty acid. However, the product is in the D configuration and will not be further degraded by the 3-hydroxyacyl-CoA dehydrogenase, which is specific for the L configuration. This problem is resolved by the presence of 3-hydroxyacyl-CoA epimerase, which catalyzes the inversion of configuration at carbon 3. Subsequently, $\beta$ oxidation can continue and completely degrade the rest of the acyl chain to 4 acetyl-CoAs.

## Oxidation of Fatty Acids with an Odd Number of Carbons

The amounts of fatty acids with an uneven number of carbons are very low in many mammalian tissues. However, in ruminant mammals, the oxidation of odd-chain fatty acids can account for as much as 25 percent of their energy requirements. Consequently, straight-chain fatty acids with 17 carbons will be oxidized by the normal $\beta$-oxidation sequence and give rise to 7 acetyl-CoAs and 1 propionyl-CoA:

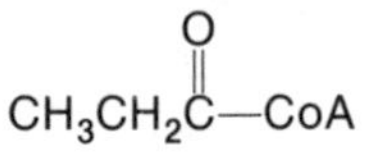

$$CH_3CH_2\overset{\overset{\displaystyle O}{\|}}{C}\text{—CoA}$$

This three-carbon acyl-CoA also is a product of degradation of the amino acids valine and isoleucine (see Chapter 23). The propionyl-CoA is converted to succinyl-CoA by three enzymatic steps, as indicated in Figure 13-10. The initial carboxylation is catalyzed by *propionyl-CoA carboxylase*, which utilizes biotin as a cofactor. In the second reaction, D-methylmalonyl-CoA is converted to its optical isomer, L-methylmalonyl-CoA by *methylmalonyl-CoA racemase*. The last step in the sequence involves an unusual migration of the carbonyl-CoA group to the methyl group in an exchange for hydrogen. The product, succinyl-CoA, can be metabolized in the tricarboxylic acid cycle.

*Methylmalonyl-CoA mutase* is a mammalian enzyme that requires cobalamin (see Chapter 6) for activity. The enzyme has been purified from sheep liver and human placenta. The human enzyme has a molecular weight of 145,000. The absence of this enzymatic activity in children with *congenital methylmalonicaciduria* results in death during childhood.

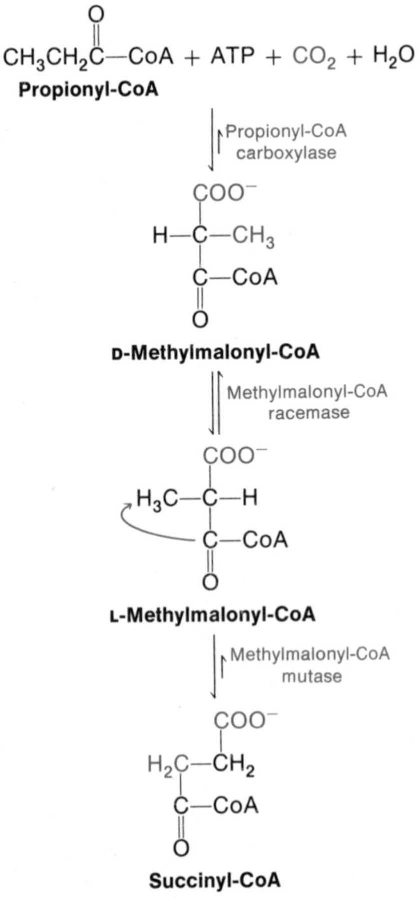

### α Oxidation and Refsum's Disease

Although $\beta$ oxidation is quantitatively the most significant pathway for catabolism of fatty acids, $\alpha$ oxidation of some fatty acids is essential to our well-being. In a normal diet, small amounts of *phytol*, a component of chlorophyll, are ingested. As shown in Figure 13-11, this long-chain alcohol is oxidized to *phytanic acid*, which is a more important dietary component present in ruminant fat and dairy products. The estimated daily intake of phytanic acid is somewhere between 50 and 100 mg. Because of the methyl substitution on carbon 3, phytanic acid is not a substrate for acyl-CoA dehydrogenase, the first enzyme in $\beta$ oxidation. This step is circumvented by another mitochondrial enzyme that hydroxylates the $\alpha$ carbon of phytanic acid. The hydroxy intermediate is decarboxylated to yield *pristanic acid* and $CO_2$ (Figure 13-11). Pristanic acid is unsubstituted at carbon 3 and can be oxidized by acyl-CoA dehydrogenase and the normal enzymes of $\beta$ oxidation to produce propionyl-CoA and an acyl-CoA. The latter can be degraded by four cycles of $\beta$ oxidation, which yield, alternately, acetyl-CoA and propionyl-CoA. The final sequence of reactions produces acetyl-CoA and isobutyryl-CoA. The latter acyl-CoA can be converted into succinyl-CoA and subsequently metabolized by means of the tricarboxylic acid cycle.

Our current understanding of how humans metabolize phytol and phytanic acid came largely as a result of the studies of Daniel Steinberg and co-workers. The impetus for their experiments resulted from studies

$$CH_3CH_2\overset{\overset{\displaystyle O}{\|}}{C}\text{—CoA} + ATP + CO_2 + H_2O$$

**Propionyl-CoA**

Propionyl-CoA carboxylase

$$\begin{array}{c} COO^- \\ | \\ H\text{—}C\text{—}CH_3 \\ | \\ C\text{—CoA} \\ \| \\ O \end{array}$$

**D-Methylmalonyl-CoA**

Methylmalonyl-CoA racemase

$$\begin{array}{c} COO^- \\ | \\ H_3C\text{—}C\text{—}H \\ | \\ C\text{—CoA} \\ \| \\ O \end{array}$$

**L-Methylmalonyl-CoA**

Methylmalonyl-CoA mutase

$$\begin{array}{c} COO^- \\ | \\ H_2C\text{—}CH_2 \\ | \\ C\text{—CoA} \\ \| \\ O \end{array}$$

**Succinyl-CoA**

**Figure 13-10**
Conversion of propionyl-CoA to succinyl-CoA.

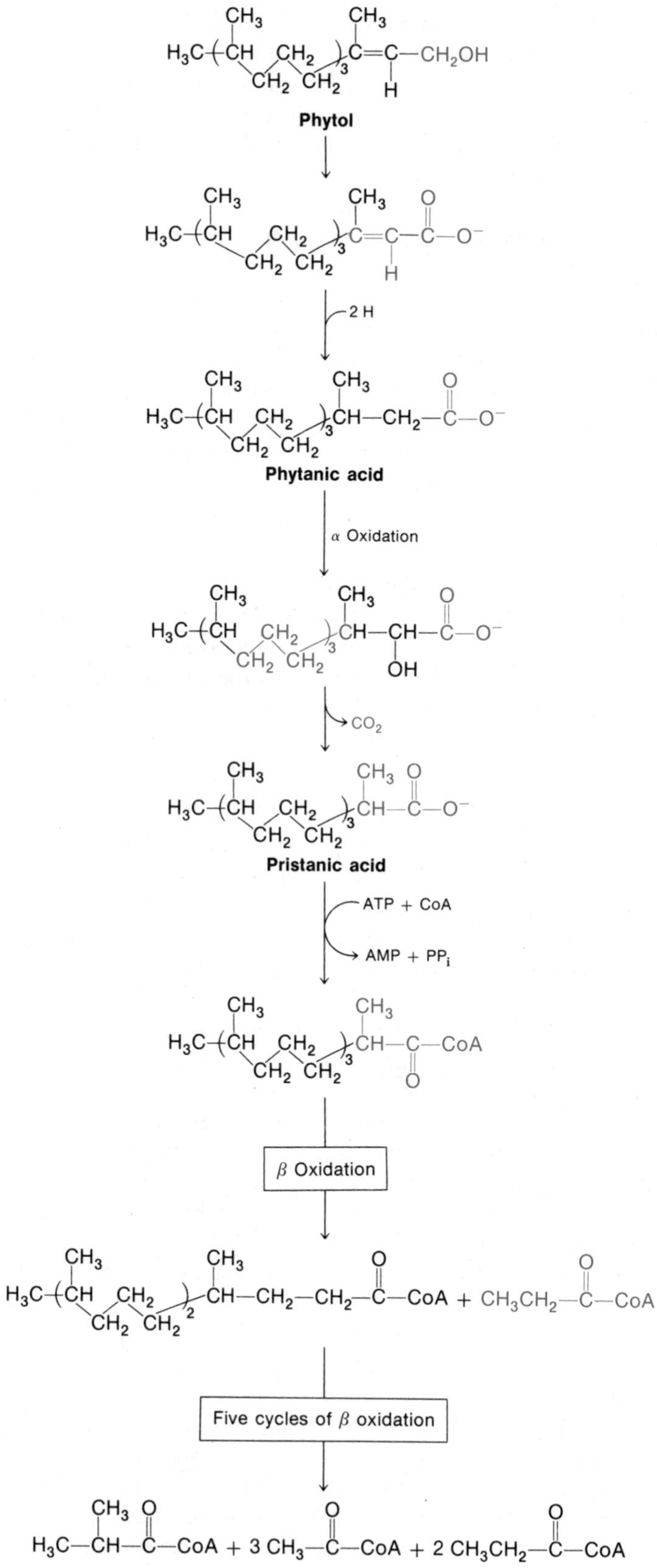

**Figure 13-11**
Scheme for oxidation of phytol.

on *Refsum's disease*, an inherited and extremely rare disorder characterized by numerous neurologic malfunctions—tremors, unsteady walking, constricted visual field, and poor night vision. The symptoms are probably due to an accumulation of phytanic acid throughout the nervous system. In these patients, 5 to 30 percent of the plasma fatty acids (20–100 mg/100 ml) and approximately 50 percent of the liver fatty acids was phytanic acid. In contrast, the normal amount of phytanic acid in plasma is below 1 percent (0.3 mg/100 ml). The disease is now known as *phytanic acid storage syndrome*. A series of biochemical studies has demonstrated that this fatty acid accumulates because of a deficiency in α oxidation of phytanic acid to pristanic acid. The metabolic defect is most likely in the α hydroxylation of phytanic acid, since people with the disorder are able to oxidize phytol to phytanic acid and pristanic acid to $CO_2$ and $H_2O$. Once these patients are identified, the symptoms of the disease can be diminished by a strict dietary regimen in which foods that contain phytanic acid are restricted.

### ω Oxidation

A minor pathway for the oxidation of fatty acids has been observed in rat liver microsomes. This involves oxidation of the terminal methyl (ω carbon) or adjacent methylene carbon of fatty acids by NADPH and molecular oxygen (Figure 13-12). This pathway is probably not quantitatively significant for the oxidation of long-chain fatty acids. However, ω oxidation may be important for the metabolism of short-chain fatty acids ($C_6$ to $C_{10}$).

### Formation and Utilization of Ketone Bodies

Once fatty acids are degraded in the mitochondria, the acetyl-CoA can undergo a number of metabolic fates. Of central importance, as we have learned in Chapter 9, is utilization of acetyl-CoA by the tricarboxylic acid cycle. An alternate fate is the synthesis of *ketone bodies*, which takes place only in the mitochondria, as depicted in Figure 13-13. In the first reaction catalyzed by *acetoacetyl-CoA thiolase*, two acetyl-CoAs condense to form *acetoacetyl-CoA*. A third molecule of acetyl-CoA reacts with acetoacetyl-CoA to yield *β-hydroxy-β-methyl-glutaryl-CoA* (HMG-CoA) in a reaction catalyzed by *HMG-CoA synthase*. As we shall see in Chapter 15, the same two reactions are the first steps in cholesterol biosynthesis; however, these reactions take place only in the cytosol. In the formation of ketone bodies, the next reaction is catalyzed by *HMG-CoA lyase* and yields *acetoacetate* and acetyl-CoA. The acetoacetate can be reduced to *β-hydroxybutyrate* by the mitochondrial enzyme, D-*β-hydroxybutyrate dehydrogenase*. Although the acetoacetate also can be decarboxylated to form acetone, this is normally of minor importance.

Ketone body synthesis is primarily a liver function, since HMG-CoA synthase is present in large quantities only in this tissue. Acetoacetate and β-hydroxybutyrate diffuse into blood, where they are carried to other tissues and converted into acetyl-CoA, as described in Figure 13-14. The reactions catalyzed by *β-hydroxybutyrate dehydrogenase* and *thiolase* are common to both the synthesis and degradation of the ketone bodies. However, the second enzyme in the sequence (Figure 13-14), *β-oxoacid-CoA-transferase*, is present in all tissues but liver. Hence the ketone bodies are largely made in the liver and are metabolized to $CO_2$ and energy in nonhepatic tissues.

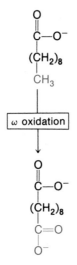

**Figure 13-12**
ω Oxidation of decanoic acid.

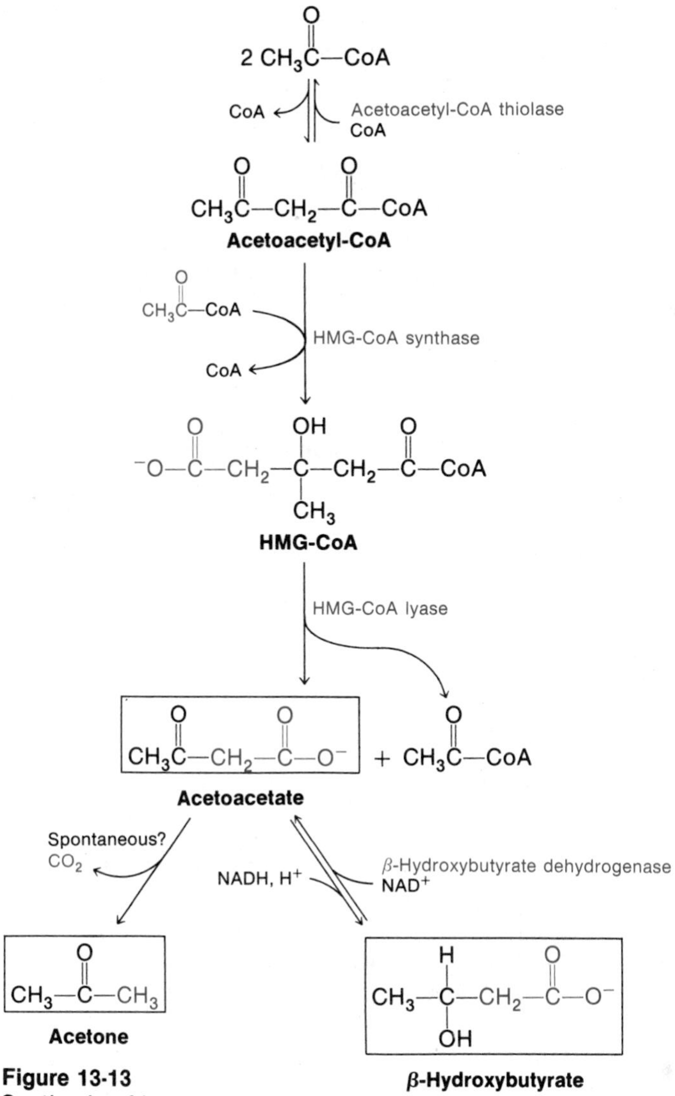

$2\ CH_3\overset{O}{\underset{}{C}}{-}CoA$

Acetoacetyl-CoA thiolase
CoA ← CoA

$CH_3\overset{O}{C}{-}CH_2{-}\overset{O}{C}{-}CoA$
**Acetoacetyl-CoA**

$CH_3\overset{O}{C}{-}CoA$

HMG-CoA synthase
CoA ←

$^-O{-}\overset{O}{C}{-}CH_2{-}\overset{OH}{\underset{CH_3}{C}}{-}CH_2{-}\overset{O}{C}{-}CoA$
**HMG-CoA**

HMG-CoA lyase

$CH_3\overset{O}{C}{-}CH_2{-}\overset{O}{C}{-}O^-$ + $CH_3\overset{O}{C}{-}CoA$
**Acetoacetate**

Spontaneous?
$CO_2$ ←

NADH, H⁺

β-Hydroxybutyrate dehydrogenase
NAD⁺

$CH_3\overset{O}{C}{-}CH_3$
**Acetone**

$CH_3\overset{H}{\underset{OH}{C}}{-}CH_2{-}\overset{O}{C}{-}O^-$
**β-Hydroxybutyrate**

**Figure 13-13**
Synthesis of ketone bodies.

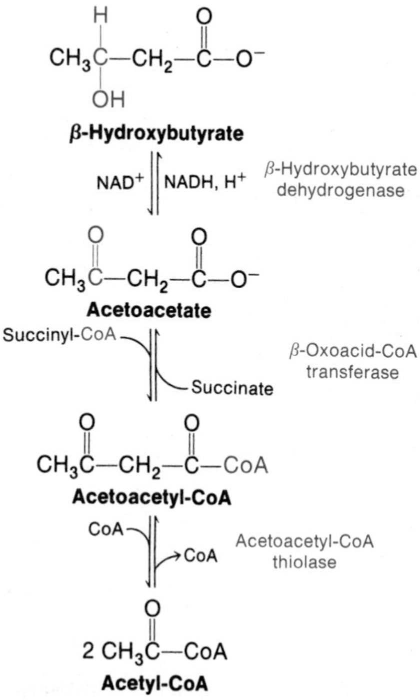

$CH_3\overset{H}{\underset{OH}{C}}{-}CH_2{-}\overset{O}{C}{-}O^-$
**β-Hydroxybutyrate**

NAD⁺ ‖ NADH, H⁺    β-Hydroxybutyrate dehydrogenase

$CH_3\overset{O}{C}{-}CH_2{-}\overset{O}{C}{-}O^-$
**Acetoacetate**

Succinyl-CoA
Succinate    β-Oxoacid-CoA transferase

$CH_3\overset{O}{C}{-}CH_2{-}\overset{O}{C}{-}CoA$
**Acetoacetyl-CoA**

CoA
CoA    Acetoacetyl-CoA thiolase

$2\ CH_3\overset{O}{C}{-}CoA$
**Acetyl-CoA**

**Figure 13-14**
Metabolism of ketone bodies.

## BIOSYNTHESIS OF FATTY ACIDS

### Historical Developments

*Once it was established that most fatty acids contained an even number of carbon atoms from $C_4$ to $C_{20}$, it was postulated by Rapier in 1907 that they were produced by condensation of a highly reactive two-carbon compound.* After the introduction of isotopes in the late 1930s and early 1940s as a fundamental tool for the biochemist, experiments were performed that clearly implicated an acetate derivative as the two-carbon compound. This was demonstrated by David Rittenberg and Konrad Bloch in 1944 and 1945 by a series of experiments in which they fed mice acetate labeled with deuterium and $^{13}C$ ($C^2H_3\ ^{13}COOH$) and found both isotopes incorporated into fatty acids. After Lynen discovered that "active acetate" was *acetyl-CoA*, a central role for this compound in fatty acid biosynthesis was demonstrated. Precisely how acetyl-CoA was converted into fatty acids eluded workers until the late 1950s, when the involvement of _malonyl-CoA_ was discovered. Subsequently, progress was rapid and the scheme for fatty acid biosynthesis as we know it today was elucidated.

### Reactions of Saturated Fatty Acid Biosynthesis

All cells, from _E. coli_ to liver, appear to synthesize fatty acids by the same chemical reactions. However, there are fascinating variations in the structure and organization of the enzymes that catalyze the reactions. Best understood is the pathway for the biosynthesis of palmitic acid as its _acyl carrier protein(ACP)-thioester_ in _E. coli_, as shown in Figure 13-15. This

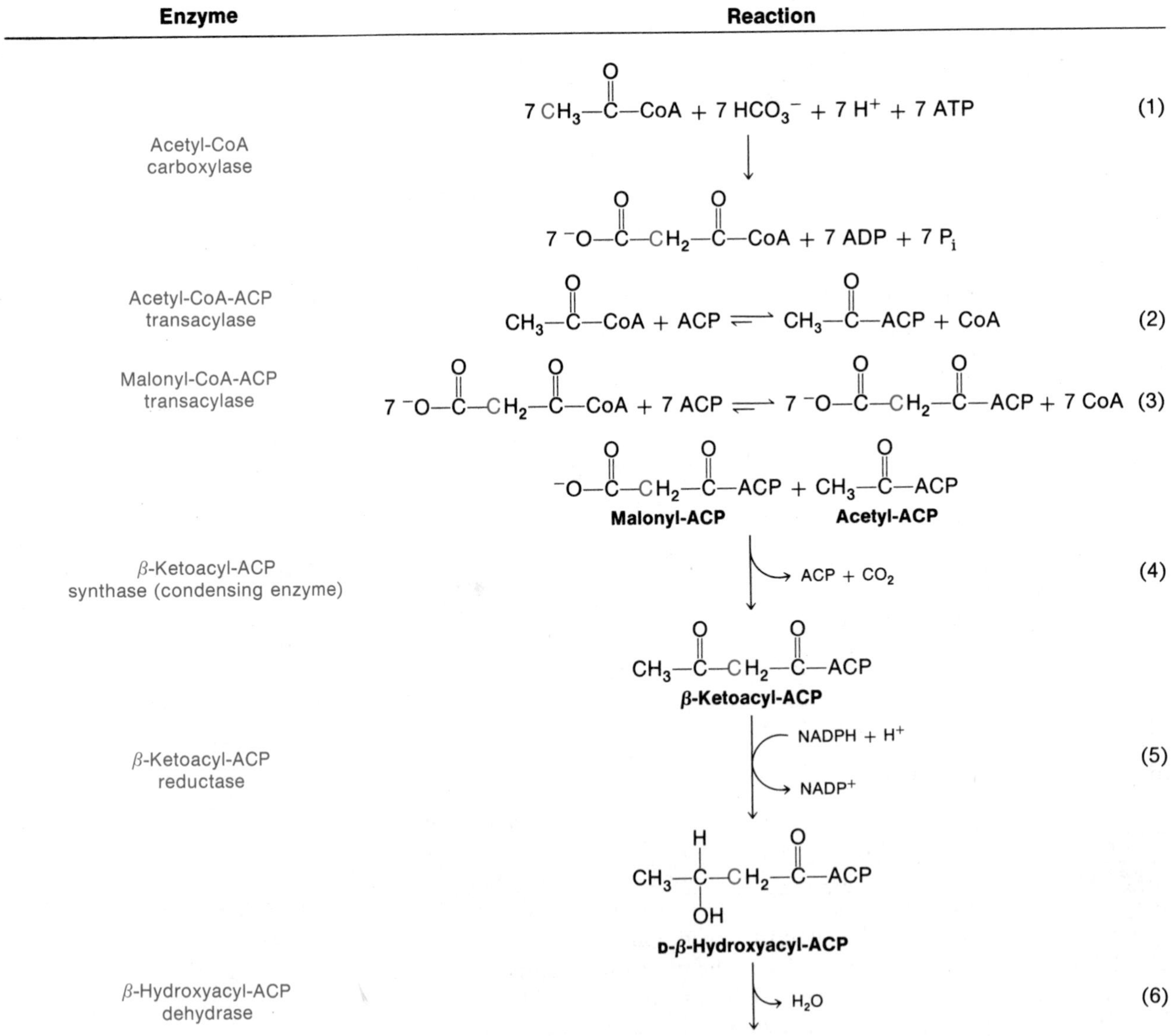

| Enzyme | Reaction |
|---|---|

$$7\ CH_3-\overset{O}{\underset{\parallel}{C}}-CoA + 7\ HCO_3^- + 7\ H^+ + 7\ ATP \quad (1)$$

Acetyl-CoA carboxylase

$$7\ ^-O-\overset{O}{\underset{\parallel}{C}}-CH_2-\overset{O}{\underset{\parallel}{C}}-CoA + 7\ ADP + 7\ P_i$$

Acetyl-CoA-ACP transacylase

$$CH_3-\overset{O}{\underset{\parallel}{C}}-CoA + ACP \rightleftharpoons CH_3-\overset{O}{\underset{\parallel}{C}}-ACP + CoA \quad (2)$$

Malonyl-CoA-ACP transacylase

$$7\ ^-O-\overset{O}{\underset{\parallel}{C}}-CH_2-\overset{O}{\underset{\parallel}{C}}-CoA + 7\ ACP \rightleftharpoons 7\ ^-O-\overset{O}{\underset{\parallel}{C}}-CH_2-\overset{O}{\underset{\parallel}{C}}-ACP + 7\ CoA \quad (3)$$

$$^-O-\overset{O}{\underset{\parallel}{C}}-CH_2-\overset{O}{\underset{\parallel}{C}}-ACP + CH_3-\overset{O}{\underset{\parallel}{C}}-ACP$$

**Malonyl-ACP**   **Acetyl-ACP**

β-Ketoacyl-ACP synthase (condensing enzyme)

$$\searrow ACP + CO_2 \quad (4)$$

$$CH_3-\overset{O}{\underset{\parallel}{C}}-CH_2-\overset{O}{\underset{\parallel}{C}}-ACP$$

**β-Ketoacyl-ACP**

β-Ketoacyl-ACP reductase

$$\begin{array}{c} NADPH + H^+ \\ \searrow \\ NADP^+ \end{array} \quad (5)$$

$$CH_3-\overset{H}{\underset{\underset{OH}{|}}{C}}-CH_2-\overset{O}{\underset{\parallel}{C}}-ACP$$

**D-β-Hydroxyacyl-ACP**

β-Hydroxyacyl-ACP dehydrase

$$\searrow H_2O \quad (6)$$

**Figure 13-15**
Scheme for biosynthesis of palmitoyl-ACP in _E. coli_.

protein has a molecular weight of 8847 in *E. coli* and functions as the carrier of the acyl residue during fatty acid biosynthesis. There are approximately $15 \times 10^6$ molecules of *acyl carrier protein* per cell, the most abundant protein found in *E. coli*. ACP, like CoA, contains one *phosphopantetheine* per molecule of protein (Figure 13-16). It is the sulfhydryl group of the phosphopantetheine to which acetyl and acyl groups are esterified.

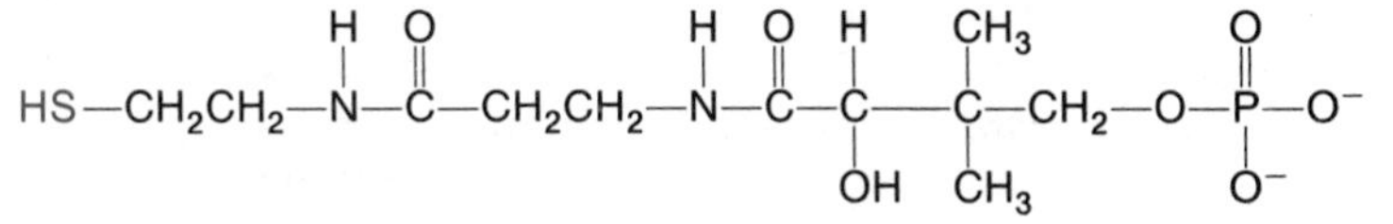

$$HS-CH_2CH_2-\overset{H}{\underset{}{N}}-\overset{O}{\underset{}{C}}-CH_2CH_2-\overset{H}{\underset{}{N}}-\overset{O}{\underset{}{C}}-\overset{H}{\underset{OH}{C}}-\overset{CH_3}{\underset{CH_3}{C}}-CH_2-O-\overset{O}{\underset{O^-}{P}}-O^-$$

**Figure 13-16**
Structure of phosphopantetheine.

| Enzyme | Reaction |
|---|---|

$$CH_3-\overset{H}{\underset{H}{C}}=C-\overset{O}{\underset{}{C}}-ACP$$

**$\alpha,\beta$-*trans*-Enoyl-ACP**

Enoyl-ACP reductase

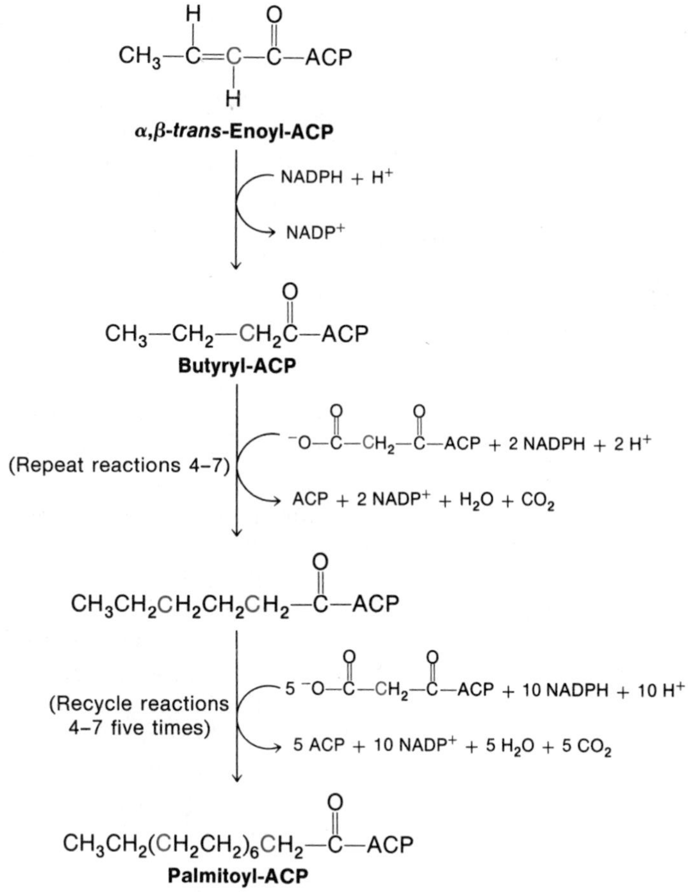

NADPH + H⁺

NADP⁺

(7)

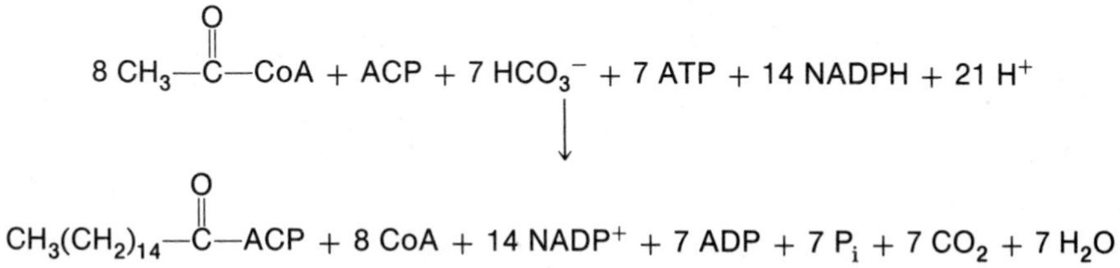

$$CH_3-CH_2-CH_2\overset{O}{\underset{}{C}}-ACP$$
**Butyryl-ACP**

(Repeat reactions 4–7)

$$^-O-\overset{O}{\underset{}{C}}-CH_2-\overset{O}{\underset{}{C}}-ACP + 2\,NADPH + 2\,H^+$$

$$ACP + 2\,NADP^+ + H_2O + CO_2$$

$$CH_3CH_2CH_2CH_2CH_2-\overset{O}{\underset{}{C}}-ACP$$

(Recycle reactions 4–7 five times)

$$5\,^-O-\overset{O}{\underset{}{C}}-CH_2-\overset{O}{\underset{}{C}}-ACP + 10\,NADPH + 10\,H^+$$

$$5\,ACP + 10\,NADP^+ + 5\,H_2O + 5\,CO_2$$

$$CH_3CH_2(CH_2CH_2)_6CH_2-\overset{O}{\underset{}{C}}-ACP$$
**Palmitoyl-ACP**

$$8\,CH_3-\overset{O}{\underset{}{C}}-CoA + ACP + 7\,HCO_3^- + 7\,ATP + 14\,NADPH + 21\,H^+$$

(8)

$$CH_3(CH_2)_{14}-\overset{O}{\underset{}{C}}-ACP + 8\,CoA + 14\,NADP^+ + 7\,ADP + 7\,P_i + 7\,CO_2 + 7\,H_2O$$

The first reaction in the biosynthetic sequence is the carboxylation of acetyl-CoA to malonyl-CoA (Reaction 1, Figure 13-15). Subsequently, acetyl-CoA and malonyl-CoA are transacylated to yield the corresponding ACP-thioesters (Reactions 2 and 3) that condense in an essentially irreversible reaction to form $\beta$-ketoacyl-ACP (Reaction 4). This $\beta$-ketoacyl-ACP is reduced with NADPH to D-$\beta$-hydroxyacyl-ACP (Reaction 5), which is stereospecifically dehydrated to yield the $\alpha,\beta$-trans-enoyl-ACP (Reaction 6). The L-$\beta$-hydroxyacyl-ACP is not a substrate for this dehydration. The double bond is reduced with NADPH (Reaction 7), and the resulting butyryl-ACP serves as a substrate for another condensation with malonyl-ACP (Reaction 4.) This reaction sequence (Reactions 4–7) continues to recycle until palmitoyl-ACP is produced after a total of 43 reactions have taken place. The palmitoyl-ACP is either transacylated to the CoA derivative or used directly in phospholipid biosynthesis, as discussed in Chapter 14.

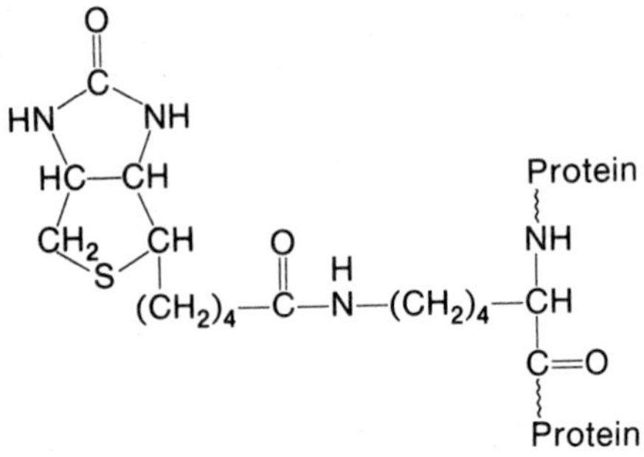

## Acetyl-CoA Carboxylase

The initial reaction of fatty acid biosynthesis in all cells is catalyzed by *acetyl-CoA carboxylase* (Reaction 1, Figure 13-15). The activity of this enzyme appears to play an important role in the control of fatty acid biosynthesis in mammals, yeast, and probably most organisms and tissues. The enzyme contains the vitamin biotin covalently linked by means of the $\varepsilon$ amino group of a lysine in the protein (Figure 13-17). The carboxyl group is initially transferred to the biotin moiety by an ATP-requiring step. In a second reaction, the carboxyl is transferred to the methyl carbon of acetyl-CoA.

Of the many acetyl-CoA carboxylases known, the enzyme from *E. coli* is perhaps best understood. The enzyme consists of three protein components: *biotin carboxyl carrier protein* (BCCP) ($M_r = 22,500$), *biotin carboxylase* ($M_r = 98,000$; two subunits of 49,000), and *carboxyltransferase* ($M_r = 130,000$). The carboxyltransferase component has an $A_2B_2$ structure, and the molecular weights of the two types of subunits are 35,000 and 30,000. The reaction sequence (Figure 13-18) involves an initial carboxylation of BCCP, catalyzed by biotin carboxylase. Subsequently, the carboxyltransferase transfers the $CO_2$ from BCCP to acetyl-CoA. The probable mechanism for the carboxyltransferase reaction is shown in Figure 13-19.

Acetyl-CoA carboxylase is found in the cytosol of animal livers and has been purified from rat and chicken liver. The rat liver enzyme has a molecular weight of 460,000 and is composed of two subunits ($M_r = 230,000$). There is one biotin per subunit. It is possible that the three functional parts of acetyl-CoA carboxylase occur as a single multifunctional polypeptide in rat liver. Although this form of the enzyme ( *protomer*) is essentially inac-

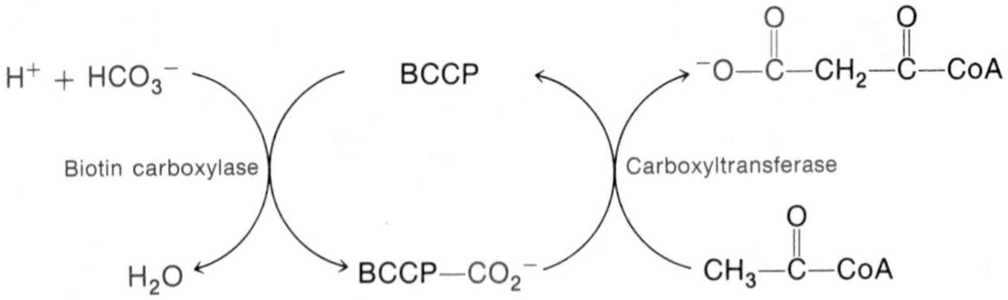

**Figure 13-17**
Structure of biotin linked to BCCP.

**Figure 13-18**
Reactions catalyzed by acetyl-CoA carboxylase from *E. coli*.

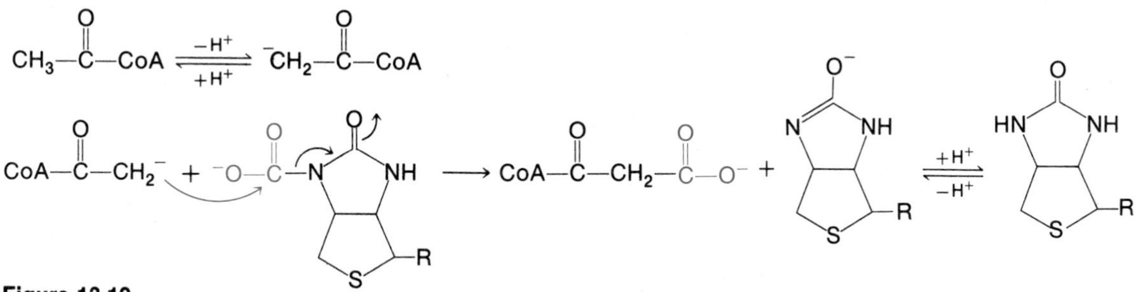

**Figure 13-19**
Carbanion mechanism for the carboxylation of acetyl-CoA.

tive, incubation with citrate results in polymerization to an active form with a molecular weight of between 4 and 8 million (Figure 13-20). The enzyme is deactivated and depolymerized when incubated with malonyl-CoA or palmitoyl-CoA. The significance of these modulators on the rate of fatty acid biosynthesis will be discussed in the section entitled Control of Fatty Acid Metabolism. Citrate will not cause polymerization-depolymerization of the *E. coli* enzyme.

### The Fatty Acid Synthase of *E. coli*

Nature has developed a splendid diversity in the organization of the enzymes that catalyze Reactions 2 to 7 of Figure 13-15. Although composed of at least six enzymatic activities, the enzyme is usually referred to in the singular as *fatty acid synthase*. The activity of fatty acid synthase is assayed by the incorporation of [2-$^{14}$C]malonyl-CoA or [$^3$H]acetyl-CoA into fatty acid, as shown in Figure 13-21. This scheme illustrates a common principle often utilized in the assay of lipid biosynthetic enzymes. The radioactive substrate is a water-soluble molecule that can easily be separated from the lipid product by extraction of the reaction mixture with an organic solvent, such as petroleum ether. When the fatty acid synthase is highly purified, it also can be assayed by a spectrophotometric method in which the oxidation of NADPH is followed. As NADP$^+$ is formed, there is a decrease in the absorbance at 340 nm.

The fatty acid synthase from *E. coli* exists as a group of enzymatic activities that can be separated from one another by conventional methods of purification. Table 13-4 summarizes some of the main properties of the *E. coli* enzymes.

The ACP from *E. coli* has been purified to homogeneity, and the amino acid sequence has been determined (Figure 13-22). The phosphopantetheine is linked to Ser-36. ACP and various analogs have been chemically synthesized by the solid-phase method developed by Bruce Merrifield. This has facilitated studies on the structural requirement of the ACP for biological activity. The three C-terminal amino acids could be removed with no loss of biological activity. Although removal of 16 amino acids from the C-terminal resulted in a decrease in activity, the polypeptide still functioned at a low rate. In contrast, as the N-terminal amino acids were removed from ACP, there was a progressive loss of activity. The protein without six N-terminal amino acids was inactive. The details of how the acyl-ACP interacts with the fatty acid biosynthetic enzymes are unknown.

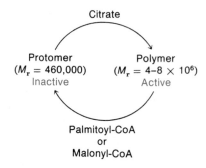

**Figure 13-20**
Activation of acetyl-CoA carboxylase from rat liver.

**Step 1.** Prepare incubation mixture, which contains:
0.3 m$M$ NADPH
0.1 $M$ phosphate buffer, pH 6.8
5 $\mu M$ mercaptoethanol
3 $\mu M$ EDTA
50 $\mu M$ malonyl-CoA
12 $\mu M$ [$^3$H]Acetyl-CoA (specific radioactivity = 2.0 $\times$ 10$^6$ dpm/$\mu$mol).
Distilled H$_2$O, so final volume will be 1 ml.

**Step 2.** Equilibrate mixture for 5 min at 37°C in a shaking water bath.

**Step 3.** Add enzyme (e.g., 2 mg protein), mix thoroughly, and incubate at 37°C for 5 min.

**Step 4.** Stop the reaction by the addition of 0.1 ml 18% perchloric acid.

**Step 5.** Add 1 ml ethanol and 2 ml petroleum ether, mix thoroughly, and allow the phases to separate.

**Step 6.** Transfer the ether layer to a liquid scintillation vial. Extract the aqueous phase two more times with petroleum ether.

**Step 7.** Evaporate the combined petroleum ether extracts, add liquid scintillation fluid, and determine the radioactivity by liquid scintillation spectrophotometry.

**Step 8.** Calculate the specific activity of the enzyme. The specific activity is calculated by dividing the dpm incorporated into the fatty acid by the specific radioactivity of the acetyl-CoA, the time of the incubation, and the milligrams of protein added to the assay.

Example:

$$\text{Specific activity} = \frac{\text{dpm in fatty acid}}{\text{specific radioactivity of acetyl-CoA} \cdot \text{min} \cdot \text{mg protein}}$$

$$= \frac{50,000 \text{ dpm}}{2 \times 10^6 \text{ dpm}/\mu\text{mol} \cdot 5 \text{ min} \cdot 2 \text{ mg protein}}$$

$$= 2.5 \times 10^{-3} \ \mu\text{mol fatty acid formed/min} \cdot \text{mg protein}$$

**Figure 13-21**
An assay for fatty acid synthase from rat liver.

$$\overset{+}{NH_3}\text{—}\overset{1}{\text{Ser}}\text{–Thr–Ile–Glu–Glu–}\overset{6}{\text{Arg}}\text{–Val–Lys–Lys–}\overset{10}{\text{Ile}}\text{–Ile–Gly–Glu–}$$

$$\overset{20}{\text{Gln–Leu–Gly–Val–Lys–Gln–Glu–Glu–Val–Thr–Asp–Asn–Ala–Ser–}}$$

$$\overset{30}{\text{Phe}}\text{–Val–Glu–Asp–Leu–Gly–Ala–}\overset{34}{\text{Asp}}\text{–}\overset{\text{(P)-Pantetheine}}{\underset{|}{\text{Ser}}}\text{–Leu–Asp–Thr–Val–Glu–}$$

$$\overset{44}{\text{Leu}}\text{–Val–Met–Ala–Leu–Glu–}\overset{50}{\text{Glu}}\text{–Glu–Phe–Asp–Thr–Glu–Ile–}\overset{55}{\text{Pro}}\text{–}$$

$$\overset{60}{\text{Asp}}\text{–Glu–Glu–Ala–Glu–Lys–Ile–Thr–Thr–Val–Gln–Ala–Ala–Ile–}$$

$$\overset{70}{\text{Asp}}\text{–Tyr–Ile–Asn–Gly–His–}\overset{77}{\text{Gln}}\text{–Ala—COO}^-$$

**Figure 13-22**
The amino acid sequence of ACP from *E. coli*.

**Table 13-4**
The Enzymes of Fatty Acid Synthase from *E. coli*

| Enzyme | $M_r$ | Subunits | Specificity | Miscellaneous |
|---|---|---|---|---|
| Acetyl-CoA-ACP transacylase | – | – | Acetyl-CoA 100% <br> Butyryl-CoA 10% <br> Hexanoyl-CoA 4.5% | Acetyl-CoA + enz. $\rightleftharpoons$ acetyl-enz. + CoA-SH <br> Acetyl-enz. + ACP $\rightleftharpoons$ acetyl-$S$-ACP + enz. |
| Malonyl-CoA-ACP transacylase | 36,700 | None | Acetyl-CoA is a competitive inhibitor; $K_i = 115\,\mu M$ | Malonyl-CoA + enz. $\rightleftharpoons$ malonyl-enz. + CoA-SH <br> Malonyl-enz. + ACP-SH $\rightleftharpoons$ malonyl-ACP + enz. <br> Malonate is esterified to serine on the enzyme |
| $\beta$-Ketoacyl-ACP synthase I | 80,000 | $2 \times 40,000$, apparently identical | Active with $C_2$–$C_{14}$ ACP, but inactive with $C_{16}$ ACP; inactive with CoA derivatives | Acetyl-$S$-ACP + enz.-SH $\rightleftharpoons$ acetyl-$S$-enz. + ACP-SH <br> Acetyl-$S$-enz. + malonyl-$S$-ACP $\rightleftharpoons$ acetoacetyl-$S$-ACP + $CO_2$ + enz.-SH |
| $\beta$-Ketoacyl-ACP synthase II | 88,000 | $2 \times 44,000$ | Active with $16:1^{\Delta 9}$-ACP as substrate | |
| $\beta$-Ketoacyl-ACP reductase | – | – | Specific for NADPH; active with $C_4$–$C_{16}$ $\beta$-ketoacyl-ACP | The product is the D configuration |
| $\beta$-Hydroxyacyl-ACP dehydrase | – | – | Specific for D isomer; inactive with L-$\beta$-hydroxyacyl-ACP; active with $C_4$–$C_{16}$ derivatives; lowest activity with $C_{10}$ substrate; inactive with CoA derivatives | The product is trans |
| Enoyl-ACP reductase | – | – | There may be two enzymes, one specific for NADH and one for NADPH | |

Of the enzymes of fatty acid synthase, only *malonyl-CoA: ACP transacylase* and *β-ketoacyl-ACP synthases I and II* have been purified to homogeneity. The other four enzymes have been partially purified and some properties have been determined. None of the enzymes has been sequenced nor have the three-dimensional structures been determined. Moreover, it is not clear whether the enzymes exist separately or as an aggregate in the intact organism. Clearly, we have only a primitive understanding of how these enzymes synthesize fatty acids.

### The Fatty Acid Synthase of *Mycobacterium smegmatis*

Most bacteria have a fatty acid synthase that resembles the *E. coli* enzyme. However, Konrad Bloch and co-workers discovered that the phylogenetically more advanced bacteria such as mycobacteria have high-molecular-weight fatty acid synthases with multifunctional polypeptides. Bloch's studies on the enzyme from *Mycobacterium smegmatis* have revealed an enzyme with many unusual features. The enzyme has a molecular weight of $2.0 \times 10^6$ and is composed of six to eight copies of apparently identical subunits with a molecular weight of 290,000. The major products of the synthase are $C_{16}$-, $C_{18}$-, $C_{22}$-, and $C_{24}$-CoAs. At a concentration of acetyl-CoA (50 $\mu M$) for which other fatty acid synthases are fully active, the mycobacterial enzyme has very low activity. Another unusual feature is that the synthase is markedly stimulated by the addition of _methylated polysaccharides_, either a polymer with units of 3-O-methylmannose ($M_r = 2100$) or a polymer that contains 6-O-methylglucose as the major unique sugar ($M_r = 4100$). These polysaccharides, which can be isolated from the mycobacteria and other closely related bacteria, have hydrophobic sites that bind equimolar amounts of long-chain acyl-CoA. Without the polysaccharides, the highly hydrophobic acyl-CoA products ($C_{22}$- or $C_{24}$-CoA) do not readily diffuse from the enzyme and, thereby, feedback inhibit fatty acid synthesis. Thus the rate-limiting step for the synthesis of fatty acids in this organism is the diffusion of the very long chain fatty acyl-CoAs from the active site of the enzyme. The hydrophobic polysaccharides apparently "catalyze" this diffusion and relieve feedback inhibition of the fatty acid synthase by $C_{22}$- and $C_{24}$-CoAs.

### Eukaryotic Fatty Acid Synthases

Similar to the mycobacterial enzyme, the fatty acid synthase of most types of eukaryotic cells exists as a multienzyme complex. (Plant cells have an ACP-dependent synthase that is reminiscent of the _E. coli_ enzyme.) One of the best-characterized is the enzyme from yeast. In 1961, Lynen published a classic article that described many of the important features of the enzyme crystallized from yeast. This enzyme has a very high molecular weight ($M_r = 2.3 \times 10^6$) and catalyzes the same reactions (Reactions 2–7) as described for the enzymes from _E. coli_ (Figure 13-15).

Elucidation of the subunit structure of the yeast enzyme presented many difficulties. Lynen and others viewed the synthase as an aggregate of seven distinct enzymes, each present in several copies. The complex was thought to be held together by strong noncovalent interactions. Even though repeated attempts to dissociate the various polypeptides failed, the "aggregate" view of the structure remained popular until the mid-1970s. Eventually, Eckhart Schweizer and colleagues demonstrated through genetic and structural studies that the fatty acid synthase was expressed from two genetically unlinked loci that code for the two different _multifunctional polypeptides_. Thus the subunit structure of the yeast enzyme is now thought to be $A_6B_6$. Subunit A is a multifunctional polypeptide with a molecular weight of 185,000 that contains the ACP region and the $\beta$-ketoacyl synthase and $\beta$-ketoacyl reductase activities. Subunit B is also a multifunctional polypeptide with a molecular weight of 175,000 that contains the remaining activities of the synthase. Among these activities is an enzyme that transfers palmitate from ACP to CoA to give the major product, palmitoyl-CoA. Moreover, there is evidence that the malonyl transacylation and palmitoyl transacylation are catalyzed by the same enzyme in the giant polypeptide B. Thus, with an $A_6B_6$ structure, the yeast enzyme has the theoretical capacity to synthesize six fatty acids at one time. However, the number of fatty acid chains that are actually synthesized at one time, under optimal conditions, has not been determined.

A possible structural arrangement of the yeast fatty acid synthase is shown in Figure 13-23. The actual arrangement of the enzymatic activities and the ACP on the two peptides has not been established. It is also unclear how the acyl residue on ACP can serve as a substrate for the various enzymatic reactions. An understanding of the structure of the fatty acid synthase remains a difficult, but fascinating problem.

The fatty acid synthases from rat liver, pigeon liver, and many other tissues are found in the cytosol and have been purified. The liver enzyme has a molecular weight of around 500,000, with two apparently identical polypeptides ($M_r = 240,000$). Each giant peptide contains one molecule of phosphopantetheine. It is thought that each multifunctional polypeptide contains the ACP region, all the activities for fatty acid synthesis, and a palmitoyl esterase that hydrolyzes the palmitoyl residue from the enzyme. Recent evidence suggests that in the presence of CoA, the palmitoyl moiety is transferred to CoA. The three-dimensional arrangement of the enzymatic activities in these multifunctional polypeptides is a problem of current interest.

The apparently rapid evolution of the multienzyme complexes in bacteria and the retention in most eukaryotes suggest that this arrangement of the fatty acid synthase may have conferred a selective advantage. Clearly, the multifunctional polypeptides avoid the accumulation of intermediates and provide equivalent stoichiometry for each of the component enzyme activities.

## Biosynthesis of Monounsaturated Fatty Acids

Two chemically distinct mechanisms exist for the introduction of the cis double bond into monounsaturated fatty acids: the *anaerobic pathway*, as typified in *E. coli*, and the *aerobic pathway*, found in eukaryotes.

The anaerobic pathway has been studied most intensely in the laboratory of Konrad Bloch at Harvard University. As the name "anaerobic" implies, the double bond of the fatty acid is inserted in the absence of oxygen. Biosynthesis of these fatty acids follows the pathway described previously for saturated fatty acids in *E. coli* until the intermediate *β-hydroxydecanoyl-ACP* is reached (Figure 13-24). At this point, there is an apparent competition between *β-hydroxyacyl-ACP dehydrase*, which forms an α,β-trans double bond, and *β-hydroxydecanoylthioester dehydrase*, which forms a β,γ-cis double bond. This β,γ unsaturated acyl-ACP is subsequently elongated by the fatty acid synthase to yield palmitoleoyl-ACP ($16:1^{\Delta 9}$). The conversion of this compound to the major unsaturated fatty acid of *E. coli*, *cis*-vaccenic acid ($18:1^{\Delta 11}$) appears to involve β-ketoacyl-ACP synthase II, which shows in vitro a preference for palmitoleoyl-ACP as a substrate (see Table 13-4). The subsequent conversion of β-keto-*cis*-vaccenyl ACP to *cis*-vaccenic acid is catalyzed by the usual enzymes of fatty acid biosynthesis.

β-Hydroxydecanoylthioester dehydrase has been purified. It has a molecular weight of 36,000 with two apparently identical subunits. The enzyme is highly specific for the dehydration of $C_{10}$-β-hydroxyacyl-ACP and has no activity with the $C_8$ or $C_{12}$ homologs.

In contrast to the anaerobic pathway, the aerobic pathway in eukaryotic cells introduces double bonds *after* the saturated fatty acid has been synthesized. In rat liver and other eukaryotic cells, an enzyme complex

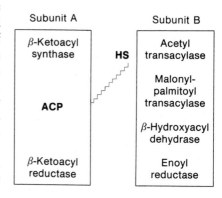

**Figure 13-23**
Possible structural arrangement of A-B subunit of yeast fatty acid synthase.

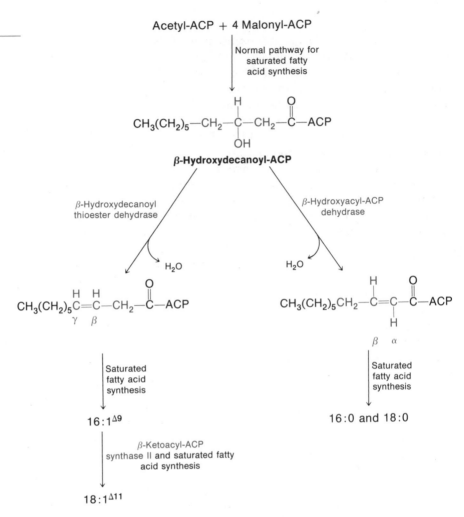

**Figure 13-24**
Anaerobic pathway for biosynthesis of monounsaturated fatty acids.

associated with the endoplasmic reticulum desaturates stearoyl-CoA to oleoyl-CoA. This reaction requires NADH and $O_2$ and results in the remarkable formation of a double bond in the middle of an acyl chain with no activating groups nearby. Although many elegant experiments have been performed, the chemical mechanism for desaturation of long-chain acyl-CoAs remains unclear.

Desaturation requires the cooperative action of two enzymes and cytochrome $b_5$. A scheme for this set of reactions is shown in Figure 13-25. _Cytochrome $b_5$ reductase_ ($M_r = 43,000$) is a flavoprotein that transfers electrons from NADH by means of flavin (F) to _cytochrome $b_5$_, a heme-containing protein ($M_r = 16,700$) in which $Fe^{3+}$ is reduced to $Fe^{2+}$. Both cytochrome $b_5$ and the reductase are _amphipathic proteins_ with a hydrophobic peptide tail that anchors the protein into the membrane and a hydrophilic portion that is outside the membrane surface (see Chapter 16 for details about membrane-bound proteins). _Stearoyl-CoA desaturase_ utilizes two electrons from cytochrome $b_5$ coupled with an atom of oxygen to form a cis double bond in the $\Delta^9$ position of stearoyl-CoA. The desaturase

$(M_r = 53,000)$ has 62 percent nonpolar amino acids, which is probably the main reason it is tightly embedded in the membrane. There is also one atom of non-heme iron per molecule of enzyme.

## Biosynthesis of Polyunsaturated Fatty Acids

*E. coli* does not have polyunsaturated fatty acids, whereas eukaryotes produce a large variety of polyunsaturated fatty acids. Animals cannot desaturate beyond the $\Delta^9$ position of an acyl chain, whereas plants have the enzymes to desaturate at positions $\Delta^{12}$ and $\Delta^{15}$. Thus animals have a dietary requirement for linoleic and linolenic acids. However, enzyme complexes occur in the endoplasmic reticulum of animal cells that desaturate at $\Delta^5$ if there is a double bond at the $\Delta^8$ position or $\Delta^6$ if there is a double bond at the $\Delta^9$ position. These enzymes are different from $\Delta^9$-desaturase, but they also appear to utilize cytochrome $b_5$ reductase and cytochrome $b_5$.

The major polyunsaturates of animals are either derived from diet or from desaturation and elongation of $18:2^{\Delta 9,12}$ or $18:3^{\Delta 9,12,15}$. A scheme for the synthesis of arachidonic acid ($20:4^{\Delta 5,8,11,14}$) from linoleic acid is shown in Figure 13-26, which illustrates the principle by which polyunsaturated fatty acids are made in animals. The elongation step is catalyzed by a series of membrane-bound enzymes that are present in the endoplasmic reticulum. These enzymes use malonyl-CoA as the donor for the $C_2$ unit, and the chemical mechanism seems to be similar to that described earlier for fatty acid synthesis (Figure 13-15). The liver enzymes also will elongate other polyunsaturated fatty acyl-CoAs. In addition, microsomal enzymes will elongate $C_{16}$- and $C_{18}$-acyl-CoAs to produce the $C_{22}$- and $C_{24}$-CoAs characteristic of sphingolipids (see Chapter 14). The latter elongation enzymes are most active in brain tissue during the synthesis of myelin.

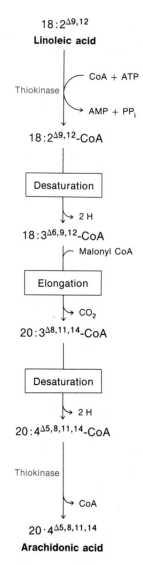

**Figure 13-26**
Synthesis by mammalian tissues of arachidonic acid from linoleic acid.

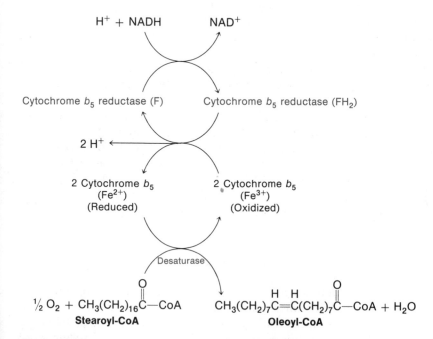

**Figure 13-25**
Formation of oleoyl-CoA in eukaryotes. (In some instances, NADPH has been found to be the electron donor for desaturation.)

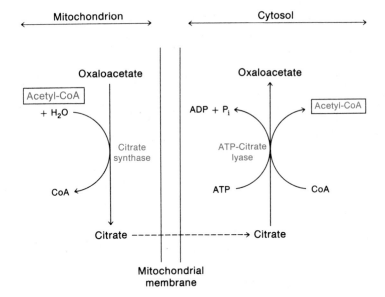

**Figure 13-27**
Transfer of acetyl-CoA from a mitochondrion to the cytosol.

### Metabolism Related to Fatty Acid Synthesis

*During periods of excess food supply, an animal will store energy as fat.*
Thus excess carbohydrate will be converted to pyruvate, which will in turn
be degraded to acetyl-CoA by pyruvate dehydrogenase in the mitochondria.
(Acetyl-CoA is also produced from the degradation of certain amino acids.)
Acetyl-CoA is not readily transported into the cytosol, where the enzymes
of fatty acid synthesis occur. However, in the mitochondria, *citrate synthase* will convert acetyl-CoA and oxaloacetate into citrate, which will
cross the mitochondrial membranes. In the cytosol, *ATP-citrate lyase* catalyzes the formation of acetyl-CoA and oxaloacetate from citrate (Figure
13-27). Thus the acetyl-CoA generated in the mitochondria can be used for
fatty acid synthesis in the cytosol. The oxaloacetate produced in the
cytosol cannot cross the mitochondrial membranes, but it can be returned
to the mitochondria after conversion to malate (Figure 13-28). Alternatively, the malate can be oxidatively decarboxylated to pyruvate by *malic
enzyme* in the cytosol (Figure 13-28). The pyruvate can be transported back
to the mitochondria and converted to acetyl-CoA by pyruvate dehydrogenase or to oxaloacetate by pyruvate carboxylase. The decarboxylation of
malate by malic enzyme appears to play an important role in fatty acid
synthesis by generation of NADPH, which is an essential reducing agent for
fatty acid synthesis. Other reactions, notably the one catalyzed by *glucose-
6-P dehydrogenase*, also appear to be important sources of NADPH for fatty
acid synthesis.

### CONTROL OF FATTY ACID METABOLISM

As the student will now appreciate, the rates of metabolism are controlled
by a variety of different mechanisms, and fatty acid metabolism is no exception. The major mechanisms for control of any metabolic pathway are
given in Table 13-5. In this section we will discuss these possible control
mechanisms as applied to fatty acid synthesis in animals and, where appro-

**Table 13-5**
General Mechanisms for Control of a Metabolic Pathway

1. Change in the concentration of active enzyme:
   Affected by enzyme synthesis and degradation.
   Affected by covalent modifications.
2. Change in the concentration and availability of substrates and products.
3. Change in the supply of covalent and noncovalent cofactors.
4. Change in activators and inhibitors: active site and allosteric.

priate, comment on their physiological importance. Fatty acid synthesis in *E. coli* seems to be linked closely with phospholipid synthesis; thus control of both these biosynthetic pathways will be discussed together in Chapter 14.

A first consideration in the control of a metabolic pathway is: which enzyme catalyzes the rate-limiting reaction? The slowest reaction in a metabolic sequence is the most appropriate point for control, because this "bottleneck" determines the overall flux of the pathway. There would be little point in altering the rate of a reaction if it were in any case faster than the slowest step in the pathway. For example, *there is evidence that acetyl-CoA carboxylase catalyzes the rate-limiting reaction for fatty acid synthesis in a number of tissues and cells* (e.g., liver). *Thus acetyl-CoA carboxylase would be the most suitable enzyme for control of fatty acid synthesis.*

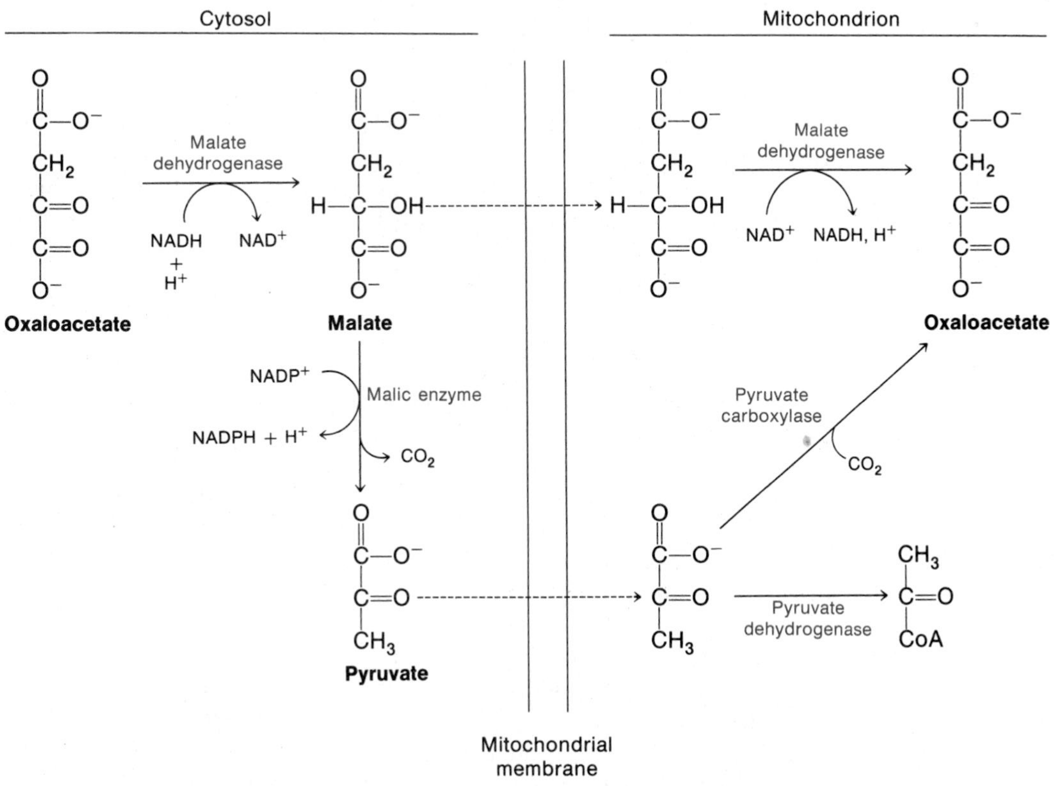

**Figure 13-28**
Oxaloacetate metabolism in the cytosol and mitochondrion.

Nevertheless, there is at least one exception, because acetyl-CoA carboxylase does not catalyze the rate-limiting reaction for fatty acid synthesis in *M. smegmatis*.

## Concentration of Acetyl-CoA Carboxylase and Fatty Acid Synthase

*One major mechanism for control of fatty acid synthesis is a change in the concentration of active enzyme(s) available for catalysis.* Genetic factors, stage of development, hormones, and energy supply are important elements that dictate the amount of acetyl-CoA carboxylase, fatty acid synthase, malic enzyme, and other enzymes related to fatty acid synthesis. For example, the concentrations of fatty acid synthase and acetyl-CoA carboxylase in rat liver are reduced fourfold to fivefold after fasting. When the rat is allowed to eat again, the concentrations of fatty acid synthase and acetyl-CoA carboxylase rise dramatically. If the rat is fed a fat-free diet, the concentration of fatty acid synthase is 14-fold higher than in a rat maintained on a normal rat chow. *Current evidence strongly suggests that the levels of these enzymes are governed by the rate of enzyme synthesis, not degradation.* Thus a decreased level of acetyl-CoA carboxylase occurs because less enzyme is made, while the enzyme is degraded at the normal rate. Similarly, higher concentrations of enzyme are due to an increased rate of synthesis. It appears that the change in the rate of enzyme synthesis is due to a fluctuation in the supply of mRNA, which in turn is governed by the rate of transcription of DNA. A question of current interest is how this transcription of mRNA is controlled.

The alteration in enzyme levels is an *adaptive*, or long-term, change, since the response occurs over a period of hours or days. A faster response can be mediated by activation or inactivation of an enzyme. There is evidence that the activity of acetyl-CoA carboxylase in rat liver is regulated by a covalent phosphorylation-dephosphorylation mechanism. The less active form of the enzyme is phosphorylated. *Insulin*, which stimulates fatty acid synthesis, causes dephosphorylation of acetyl-CoA carboxylase and thereby activation of the enzyme. *Glucagon* causes the reverse effect by promoting the synthesis of cyclic AMP, which activates a protein kinase that phosphorylates and inactivates acetyl-CoA carboxylase.

## Substrates and Products

A second general mechanism for control of a metabolic pathway is a change in the concentration and availability of substrates and products. Clearly, a cell will not synthesize fatty acids without an adequate supply of cytosolic acetyl-CoA and malonyl-CoA, nor will $\beta$ oxidation proceed without sufficient acyl-CoA in the mitochondria. Regulation of the supply of acetyl-CoA would not be a particularly good candidate for control of the rate of fatty acid synthesis, since it is a central metabolite involved in many different reactions (for example, synthesis of isoprene compounds and sterols; see Chapter 15). *Considerable evidence suggests that the rate of fatty acid synthesis and degradation is determined by the supply of malonyl-CoA.* Thus the rate of fatty acid synthesis increases as the concentration of malonyl-CoA increases in the cytosol. At the same time, malonyl-CoA inhibits carnitine acyltransferase I and thereby reduces the transfer of acyl-CoA into the mitochondria for $\beta$ oxidation.

The major product of fatty acid synthesis, palmitoyl-CoA, is a potent inhibitor of many enzymes, including acetyl-CoA carboxylase, fatty acid synthase, citrate synthase, and glucose-6-P dehydrogenase. It appears that the carboxylase is the most sensitive to palmitoyl-CoA and is inhibited by dissociation into its inactive protomer. Inhibition of this activity will cause the decreased synthesis of malonyl-CoA, which will in turn limit fatty acid synthesis. This is not the case in *M. smegmatis*, where the fatty acid synthase is more sensitive to palmitoyl-CoA than the carboxylase. As discussed earlier, a unique regulatory mechanism has evolved for control of fatty acid synthesis in this organism.

## Supply of Cofactors

A third mechanism for control of fatty acid metabolism is a change in the rate of supply of such cofactors as biotin, NADPH, pantetheine, and carnitine. If an animal becomes deficient in biotin, clearly this will affect the activity of biotin-dependent carboxylases. Thus a biotin-deficient animal may not synthesize fatty acids under conditions that normally favor this process because of a deficiency of *holo*-acetyl-CoA carboxylase. Similarly, a limited supply of pantetheine in the diet would adversely affect the supply of CoA and *holo*-fatty acid synthase and would eventually have dramatic effects on fatty acid metabolism. However, these possibilities for control probably operate only in extreme circumstances.

The level of carnitine also can be important in the control of fatty acid oxidation. In addition, if sufficient adenosine diphosphate (ADP) were not present, the oxidation of NADH and $FADH_2$ would not proceed, and therefore, the rate of $\beta$ oxidation would be decreased.

## Activators and Inhibitors

A fourth category for control of metabolism is alteration in the concentration and availability of activators and inhibitors. As already mentioned, citrate plays a key role as an allosteric activator of acetyl-CoA carboxylase (Figure 13-20), as well as in the transport of acetate across mitochondrial membranes. *There is much evidence, particularly in chicken liver, for a central role for citrate in modulation of the rate of fatty acid synthesis.* The cytoplasmic level of citrate is decreased by glucagon, which causes a rapid inhibition of fatty acid synthesis in chicken liver cells that are maintained in culture. The effect of glucagon is mediated by cAMP, which somehow causes a decrease in cytoplasmic citrate. It is postulated that the lower citrate levels result in a depolymerization and, as a result, inactivation of the acetyl-CoA carboxylase.

Methylmalonyl-CoA can inhibit fatty acid synthesis as an active-site competitor with malonyl-CoA. In vitamin $B_{12}$ deficiency, methylmalonyl-CoA accumulates because of a decreased activity of methylmalonyl-CoA mutase (Figure 13-10). However, the $B_{12}$-deficient rat, at least, compensates for possible inhibition of fatty acid synthesis by methylmalonyl-CoA with an increase in the concentration of acetyl-CoA carboxylase.

## Conclusion

The rate of fatty acid synthesis and degradation is governed by a complex set of interactions among DNA, RNA, proteins, carbohydrates, and lipids. Although desirable, it would be an extremely complex task to set forth a

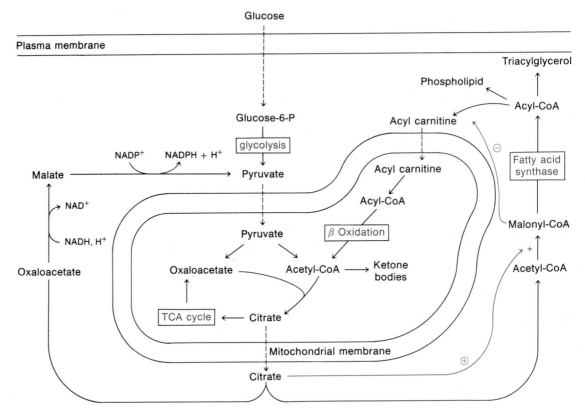

**Figure 13-29**
Conversion of glucose to lipid in liver cells.

unifying and comprehensive scheme for control of fatty acid metabolism. However, a scheme for the major metabolic events in the conversion of glucose to fatty acid is outlined in Figure 13-29. By glycolysis, plus the action of three mitochondrial enzymes, glucose is converted into citrate. The citrate migrates into the cytosol, where it is metabolized to acetyl-CoA, which is carboxylated to form malonyl-CoA. These two latter substrates are used for fatty acid synthesis. The fatty acyl-CoAs are used for triglyceride and phospholipid synthesis and are diverted from β oxidation by inhibition of carnitine acyltransferase I by malonyl-CoA. In glucose deprivation, the levels of citrate and malonyl-CoA in the cytosol would fall and the rate of fatty acid synthesis would decrease. At the same time, fatty acids would be mobilized from triglyceride stores and would be transported into the mitochondria for β oxidation.

## SELECTED READINGS

Bloch, K., and Vance, D. E. Control mechanisms in the synthesis of saturated fatty acids. *Ann. Rev. Biochem.* 46:263, 1977. An article in which control mechanisms are emphasized.

Deuel, H. J. *The Lipids: Biochemistry,* Vol 3. New York: Interscience, 1957. This is a comprehensive and classical treatise on the biochemistry of lipids until the mid-1950s.

Jeffcoat, R. The biosynthesis of unsaturated fatty acids and its control in mammalian liver. *Essays Biochem.* 15:1, 1979. A very good review on unsaturated fatty acid synthesis.

McGarry, J. D., and Foster, D. W. Regulation of hepatic fatty acid oxidation and ketone body production. *Ann. Rev. Biochem.* 49:395, 1980. A recent review in which the role of malonyl-CoA is discussed.

Najjar, V. A. *Fat Metabolism.* Baltimore, Md.: The Johns Hopkins Press, 1954. A good summary of the early work on fatty acid metabolism.

Stanbury, J. B., Wyngaarden, J. B., and Fredrickson, D. S. *The Metabolic Basis of Inherited Disease.* New York: McGraw-Hill, 1978. This excellent book contains chapters on Refsum's disease and deficiencies of circulating enzymes and plasma proteins.

## PROBLEMS

1. If a person eats only meat, eggs, and dairy products, will he or she eventually develop essential fatty acid deficiency?

2. What would be the net yield of ATP from the complete oxidation of oleic acid?

3. Why do patients with phytanic acid storage syndrome not excrete the excess quantities of phytanic acid in the urine?

4. Draw a scheme that illustrates how mammals would synthesize $22:6^{\Delta 4,7,10,13,16,19}$ from linolenic acid.

5. There was a student whose only vice in life was eating odd-chain fatty acids. The student was discovered to be deficient in vitamin $B_{12}$. After consulting a physician, the student was told to stop this unusual eating habit until the vitamin deficiency was improved. What is the biochemical reasoning for the doctor's advice? What are the consequences of this deficiency on fatty acid metabolism?

6. For an in vitro synthesis of fatty acids with purified fatty acid synthase, the acetyl-CoA was supplied as the tritium derivative:

$$C^{3}H_3\overset{O}{\underset{\|}{C}}-CoA$$

   The other reactants, including malonyl-CoA, were not radioactive.
   a. Where would the tritium label be found in the product palmitic acid?
   b. If malonyl-CoA were supplied as the only labeled compound in which both hydrogens on carbon 2 were replaced by tritium,

$$^{-}O\overset{O}{\underset{\|}{C}}-C^{3}H_2-\overset{O}{\underset{\|}{C}}-CoA$$

   how many tritium atoms would be incorporated per molecule of palmitate?

c. If [3-$^{14}$C]malonyl-CoA, i.e.,

$$^{-}O-\overset{O}{\underset{\|}{{}^{14}C}}-CH_2\overset{O}{\underset{\|}{C}}-CoA$$

   were used in the reaction, which carbons in palmitate would be labeled?

7. Explain why a variation in the concentration of oxaloacetic acid might affect the rate of fatty acid synthesis.

8. Is oxygen required for the synthesis of unsaturated fatty acids in *E. coli*? Explain your answer.

9. How many molecules of ATP, NADPH, and acetyl-CoA would be required for the synthesis of a molecule of palmitic acid from acetyl-CoA:
   a. By an *E. coli* cell?
   b. By a liver cell?

10. How many molecules of ATP, NADPH, and acetyl-CoA would be required for a liver cell to synthesize a molecule of:
    a. Oleic acid?
    b. Linoleic acid?

11. There is a rare disease called "thin-person syndrome." (It is so rare we are still looking for the first case.) The problem with these people is that they are unable to make fatty acids. A new drug called "gainabit" was tested in animal models and caused an induction of fatty acid synthase. Treatment of people with "thin-person syndrome" with this new drug resulted in a fivefold higher level of fatty acid synthase than normal levels. However, the drug had no effect on fatty acid synthesis in these patients. Offer an explanation for why this drug was ineffective. What other enzyme should have been studied prior to the human drug trial?

12. Obesity is a general disorder in metabolism that results from a variety of causes, for example, overeating. An inherited disorder that results in obesity might be discovered some day in which there is a deficiency of acyl-carnitine transferase I. Explain why a defect in this enzyme would result in obesity.

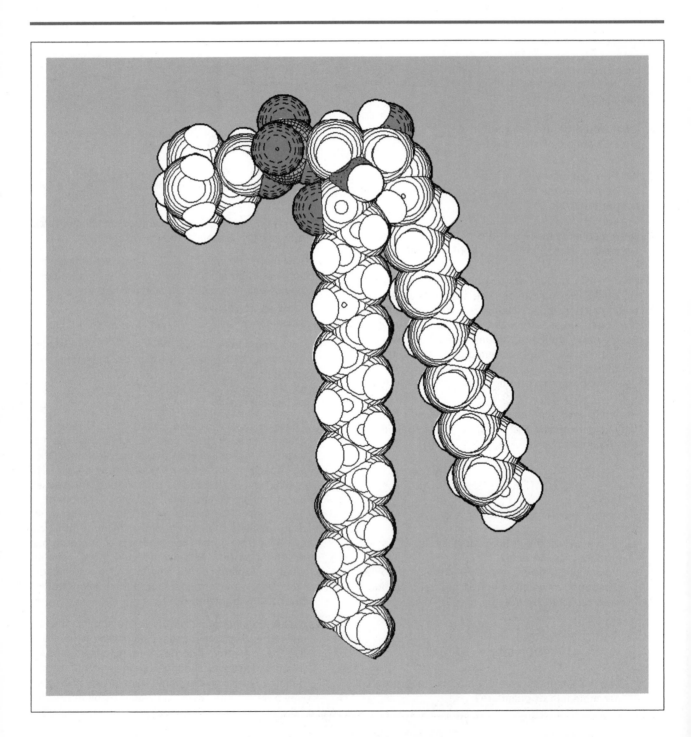

# 14

# METABOLISM OF GLYCEROLIPIDS, SPHINGOLIPIDS, AND PROSTAGLANDINS

The last chapter was primarily concerned with the metabolism of fatty acids as it relates to the storage and release of energy. Fatty acids are also a major component of *phospholipids*, which are the major structural lipids of all biological membranes in eukaryotic and prokaryotic cells. In eukaryotic cells, fatty acids are also an important component of another class of membrane lipids, the *sphingolipids*. This chapter will provide an introduction to the biochemistry of these complex lipids. In addition to a functional role in membranes, some $C_{20}$ unsaturated fatty acids (particularly arachidonic acid) are also precursors to a group of substances (*prostaglandins*) with very potent biological activities. This chapter will also provide an introduction to the biochemistry of prostaglandins and related substances.

## GLYCEROLIPIDS

Glycerolipids are those lipids which contain glycerol as a backbone. The major glycerolipids include triacylglycerol (the structure of which was discussed in Chapter 13) and phospholipids. The importance of phospholipids to living organisms is underscored by the nearly complete lack of genetic defects in the metabolism of these lipids in humans. Presumably, any such defects are lethal at any early stage of development and, therefore, are never observed.

### Structures of Phospholipids

By definition, phospholipids are those lipids which contain phosphate. The *phosphoglycerides* are quantitatively the most important structural group in this lipid class; hence the first part of this chapter will be restricted to the

*Computer graphics display of a space-filling model of sphingomyelin, an important structural component of many membranes.*

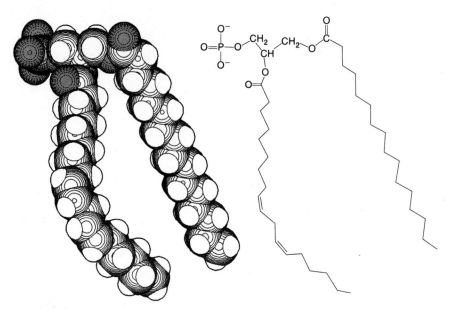

**Figure 14-1**
The structure of phosphatidic acid, a phosphoglyceride.

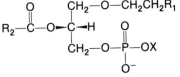

**Phospholipid with an alkyl ether**

**Phospholipid with an alkenyl ether**

**Figure 14-2**
Phospholipids with alkyl or alkenyl ether substituents.

phosphoglycerides. _Sphingomyelin_ is also an important phospholipid, and it will be discussed with the _sphingolipids_.

All the phosphoglycerides have a _glycerol-3-phosphate_ backbone, as shown in Figure 14-1, and the _sn_ nomenclature discussed in Chapter 13 is also used for the systematic naming of the various phospholipids. The hydroxyls on carbons 1 and 2 are usually acylated with fatty acids. In most phospholipids, the fatty acid substituent at carbon 1 is saturated and at carbon 2 is unsaturated. In some instances, the substituent on carbon 1 is an _alkyl ether_ or an _alkenyl ether,_ as shown in Figure 14-2. The alkenyl ether phospholipids are also known as _plasmalogens_.

_The phosphoglycerides are classified according to the substituent (X) on the phosphate group_ (Table 14-1). If X is a hydrogen, the compound is called _3-sn-phosphatidic acid_. If X—OH is choline, the lipid is called _3-sn-phosphatidylcholine_ (Figure 14-3), which is often referred to by its original name, _lecithin_, and is the most abundant phospholipid in animal tissues. The name "lecithin" is derived from the Greek name for egg yolk, the source from which the lipid was first isolated. In addition to its role in membrane structure, phosphatidylcholine is an important structural component of the plasma lipoproteins and bile. The other major phospholipids are listed in Table 14-1.

Although phospholipids are essential components of all biological membranes, there is great variation in the composition of lipids that make up the membranes around and within different cells. This is illustrated in Table 14-2. The lipid composition of red blood cell membranes is typical of the plasma membranes of most animal cells. The major lipids are phosphatidylcholine, phosphatidylethanolamine, sphingomyelin, and cholesterol. (Cholesterol is the major topic of Chapter 15.) The membrane of the myelin sheath is characterized by a relatively high percentage of glyco-sphingolipids, which are discussed later in this chapter. Many animal vi-ruses, e.g., Semliki Forest virus (Table 14-2), are surrounded by a mem-brane. The viral membrane is usually derived from the plasma membrane of the host cell and, hence, has a similar lipid composition. In contrast to eukaryotic membranes, the membranes of *Escherichia coli* have a simple lipid composition and lack phosphatidylcholine, phosphatidylinositol, cho-lesterol, and sphingolipids. The major phospholipid of *E. coli* is phos-phatidylethanolamine.

**Table 14-1**
Major Classes of Phospholipids

| Name of X—OH | Formula of X | Name of Phospholipid |
| --- | --- | --- |
| Water | —H | Phosphatidic acid |
| Choline | $-CH_2CH_2\overset{+}{N}(CH_3)_3$ | Phosphatidylcholine (lecithin) |
| Ethanolamine | $-CH_2CH_2\overset{+}{N}H_3$ | Phosphatidylethanolamine |
| Serine | $-CH_2-\underset{COO^-}{\overset{\overset{+}{N}H_3}{CH}}$ | Phosphatidylserine |
| Glycerol | $-CH_2CH(OH)CH_2OH$ | Phosphatidylglycerol |
| Phosphatidyl-glycerol | | Diphosphatidylglycerol (cardiolipin) |
| *myo*-Inositol | | Phosphatidylinositol |

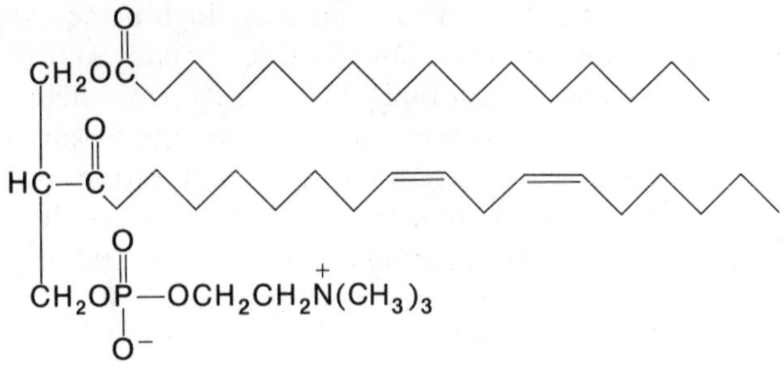

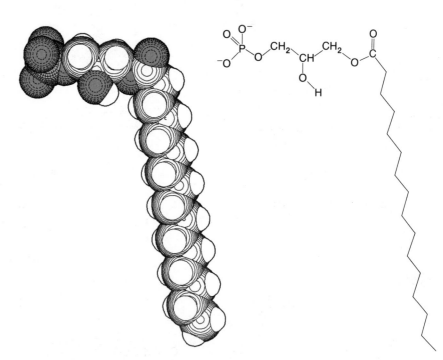

**Figure 14-3**
Structure of phosphatidylcholine.

**Figure 14-4**
Basic structure for lysophospholipids.

**Table 14-2**
Lipid Composition of Selected Membranes

| Lipid | Human Red Blood Cells (percent) | Human Myelin (percent) | Semliki Forest Virus (percent) | E. coli (percent) |
|---|---|---|---|---|
| Phosphatidylcholine | 19 | 13 | 17 | 0 |
| Phosphatidylethanolamine | 16 | 15 | 12 | 75–85 |
| Phosphatidylserine | 10 | 1 | 7 | Trace |
| Phosphatidylinositol | 1 | 0 | 1 | 0 |
| Phosphatidic acid | 1 | 0 | 0 | Trace |
| Phosphatidylglycerol | 0 | 0 | 0 | 10–20 |
| Cardiolipin (diphosphatidylglycerol) | 0 | 0 | 0 | 5–15 |
| Sphingomyelin | 15 | 5 | 10 | 0 |
| Glycosphingolipids | 8 | 20 | 4 | 0 |
| Cholesterol | 24 | 38 | 49 | 0 |
| Other | 6 | 2 | 0 | 0 |

Another important group of phospholipids is the _lysophospholipids_, in which one of the acyl substituents (usually from position 2) is missing, as shown in Figure 14-4. If the acyl substituent at carbon 1 were removed, the acyl group from position 2 would spontaneously migrate to position 1. One can differentiate between deacylation at position 1 or position 2 by analysis of the fatty acids derived from the lysophospholipid. If the fatty acids on the lysophospholipid are mostly saturated, it is likely that the fatty acids from position 2 of the phospholipid are cleaved. However, if the fatty acids on the lysophospholipid are mostly unsaturated, the fatty acids from position 1 are probably cleaved and migration of the fatty acids from position 2 has occurred. The lysophospholipids are named by simply adding the prefix _lyso-_ to the name of the original phospholipid (e.g., _lysophosphatidylcholine_). The lysophospholipids account for only 1 or 2 percent of the total phospholipids in animal cells.

The phospholipids are _amphipathic_ (dual sympathy) molecules, because they have both polar and nonpolar portions. The _polar headgroups_, as they are called, prefer an aqueous environment (hydrophilic), whereas the nonpolar (hydrophobic) acyl substituents do not. It is this amphipathic property that causes most phospholipids to form spontaneously a _bilayer_ when suspended in an aqueous environment, as shown in Figure 14-5. Hence the lipid bilayer is thought to be responsible for the major structural organization of most membranes in which proteins are embedded. This is discussed in more detail in Chapter 16. It is interesting that one quantitatively important phospholipid, phosphatidylethanolamine, spontaneously adopts at room temperature what is called the _hexagonal II phase_ (Figure 14-5). Whether or not this structural arrangement of lipids is found in biological membranes is presently under active investigation.

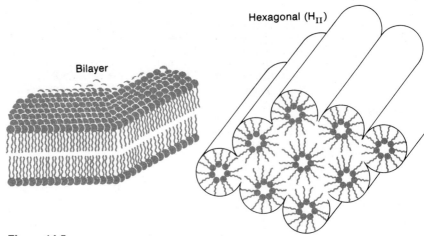

**Figure 14-5**
Structure of phospholipid bilayer and hexagonal II phase.

Although the phospholipids are often referred to in the singular (e.g., phosphatidylcholine), we are actually describing a complex mixture of phospholipids with the same headgroup (choline) but with a variety of different fatty acid substituents. Thus human red blood cells have 21 different molecular species of phosphatidylcholine that differ in fatty acid substituents at either position 1 or position 2, or both. This is true in a similar fashion for the other red blood cell phospholipids. Therefore, in the membranes of red blood cells, there are only six major classes of phospholipids, but over 100 different molecular species. What initially appears to be a simple phospholipid composition for the red blood cell is actually a very complex mixture. The advantage of such complexity is not readily apparent. However, it is clear that the fatty acid composition of the phospholipids is very important for the normal operation of membrane-associated functions such as active transport. This will be discussed in more detail in Chapter 17.

Lipids are extracted from cells or tissues with organic solvents (for example, $CHCl_3$). Phospholipids are resolved from uncharged lipids (neutral lipids) by adsorption column chromatography (usually silicic acid) or by adsorption thin-layer chromatography. The various classes of phospholipids are usually separated on a small scale by thin-layer chromatography, as shown in Figure 14-6. Large-scale preparations usually involve column chromatography or high-pressure liquid chromatography.

## Phospholipid Synthesis in *E. coli*

*E. coli* contains only three major classes of phospholipids: phosphatidylethanolamine (75–85%), phosphatidylglycerol (10–20%), and cardiolipin (5–15%). All three phospholipids share the same biosynthetic pathway up to the formation of CDPdiacylglycerol, where the pathways branch (Figure 14-7).

The enzymes are located on the inner membrane of the cell, except phosphatidylserine synthase, which is mostly found associated with the ribosomes. Because of their association with membranes, the enzymes have been difficult to study. Consequently, our knowledge of them has lagged behind that of other important enzymes, such as those involved in glycolysis. An additional problem involved in the study of most of the reactions is that at least one of the substrates and one of the products is not readily soluble in an aqueous phase. Despite these difficulties, several of the enzymes have been obtained in pure form.

*Glycerol-3-P acyltransferase* preferentially utilizes saturated fatty acyl derivatives such as palmitoyl-CoA or palmitoyl-ACP for the initial acylation of glycerol-3-P. The purified enzyme has a molecular weight of 92,000. The pure protein has no activity unless reconstituted with phospholipid and is most active with phosphatidylglycerol. The second enzyme, *1-acylglycerol-3-P acyltransferase*, prefers acyl-CoAs with a double bond. Apparently, this specificity is at least partially responsible for the distribution of saturated and unsaturated fatty acids in phospholipids described earlier in this chapter. The next enzyme in the sequence, *phosphatidate cytidylyltransferase*, will utilize either cytidine triphosphate *(CTP)* (Fig-

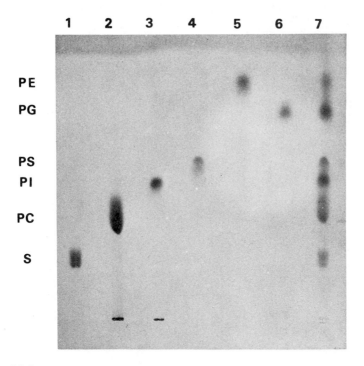

**Figure 14-6**
Thin-layer chromatography of the major phospholipids. Lanes 1 to 6 contain 0.1 mg of each lipid. Lane 7 is a mixture of the six phospholipids. PE = phosphatidylethanolamine; PG = phosphatidylglycerol; PS = phosphatidylserine; PI = phosphatidylinositol; PC = phosphatidylcholine; S = sphingomyelin. Each compound was spotted on a silica gel G60 thin-layer plate that had been activated at 100°C 1 h before the analysis. The plate was developed in a solvent that contained $CHCl_3:CH_3OH:CH_3COOH:H_2O$ (50:25:8:4 by volume). The compounds were visualized after spraying the plate with dilute sulfuric acid and heating in the oven.

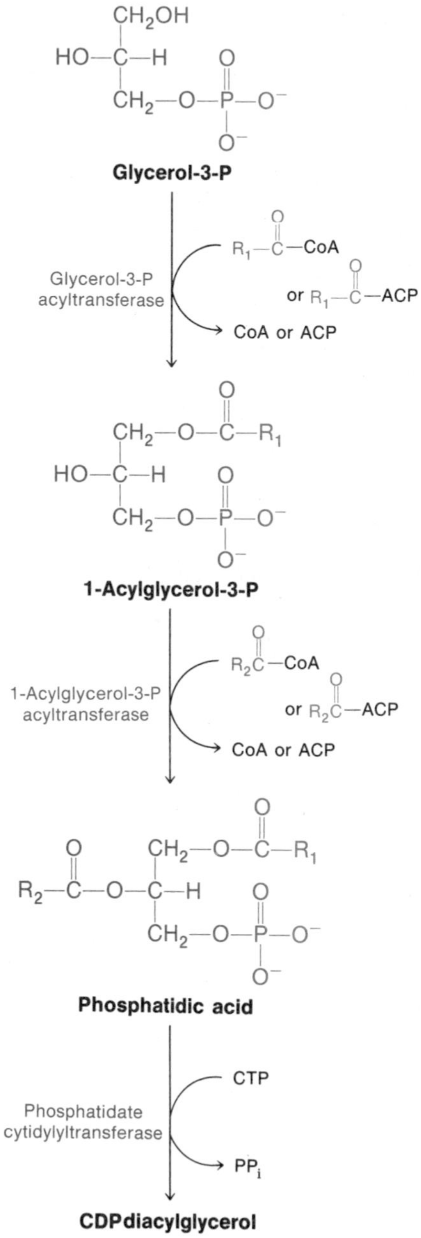

**Figure 14-7**
Scheme for the biosynthesis of phospholipids in *E. coli*. (*Continued on next page.*)

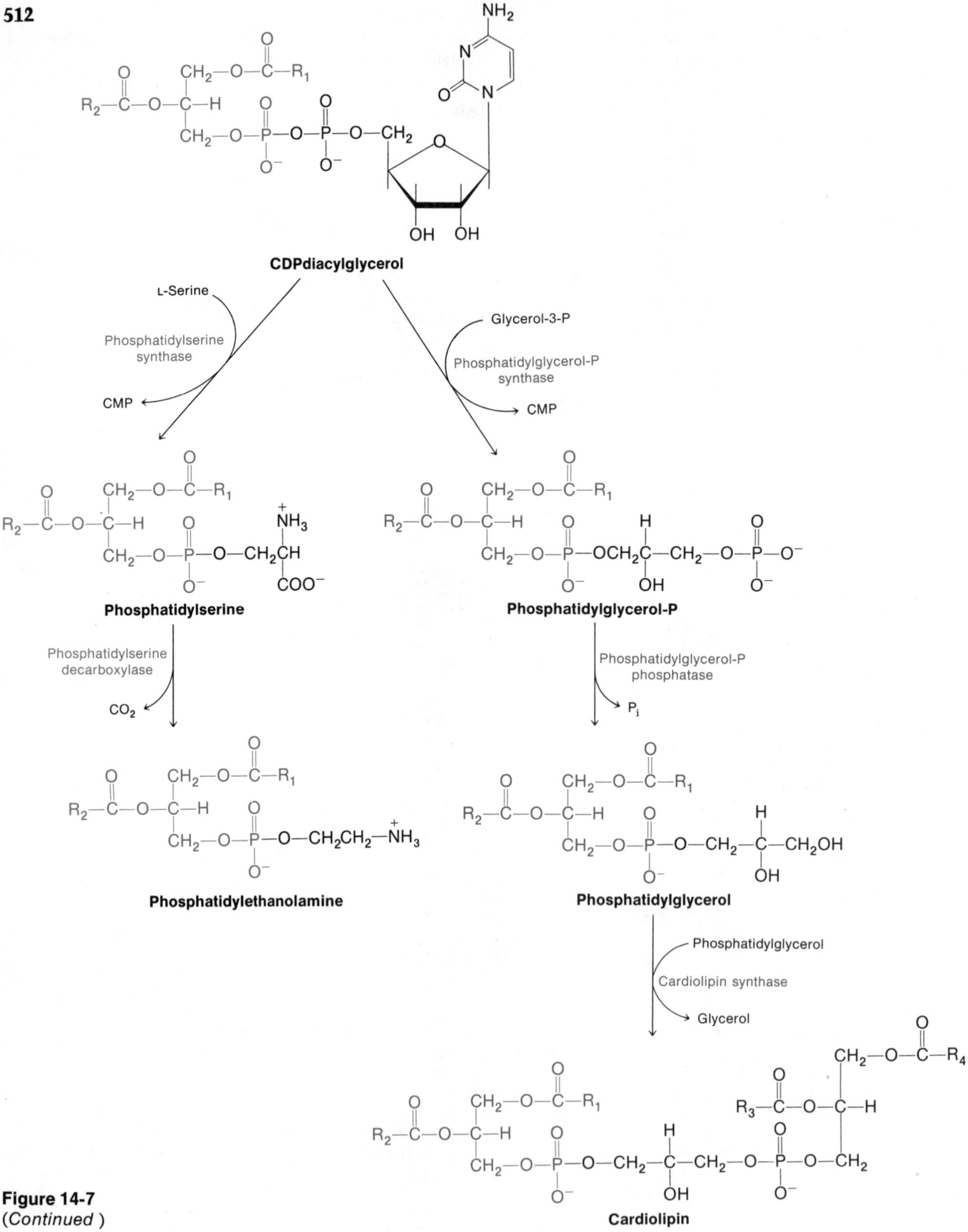

**Figure 14-7**
(*Continued*)

ure 14-8) or deoxy CTP (_dCTP_) (Figure 14-9) as a substrate, and both lipo-nucleotides are present in _E. coli_. A functional difference between _dCDPdiacylglycerol_ and _CDPdiacylglycerol_ has not been demonstrated.

One fate for CDPdiacylglycerol is a reaction with serine to yield phos-phatidylserine. _Phosphatidylserine synthase_ has been extracted from ribo-somes and purified to homogeneity. The molecular weight of the native enzyme is unknown, although the subunits have a molecular weight of 54,000. It has been estimated that an _E. coli_ cell contains only 800 subunits of this enzyme. This points to another problem in the study of these phos-pholipid biosynthetic enzymes—the relatively small quantity of enzyme per cell. Recently, gene cloning techniques have been developed for the over-production of this enzyme, making purification a simpler proposition. In the final reaction of this sequence, phosphatidylserine is subsequently de-carboxylated by _phosphatidylserine decarboxylase_ (subunit $M_r = 36,000$) to yield phosphatidylethanolamine.

Alternatively, the CDPdiacylglycerol can be converted to phosphatidyl-glycerol-P by _phosphatidylglycerol phosphate synthase_. This enzyme shows high specificity for the glycerol-3-P but will utilize in vitro ADPdiacylglycerol or UDPdiacylglycerol. The product, phosphatidylglyc-erol-P, is dephosphorylated by _phosphatidylglycerol phosphate phospha-tase_. Subsequently, cardiolipin can be made by the action of _cardiolipin synthase_.

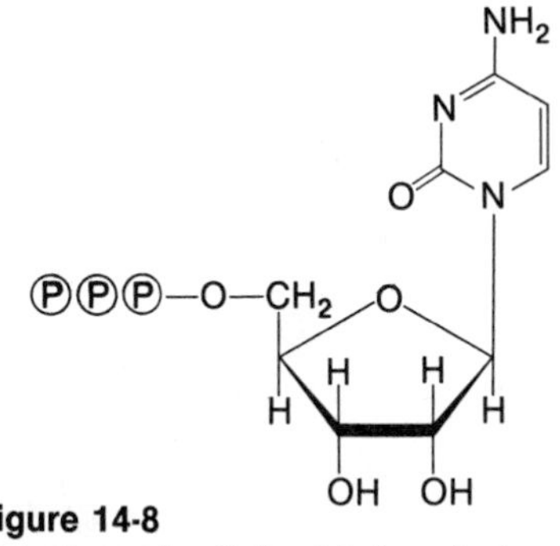

**Figure 14-8**
Structure of cytidine triphosphate (CTP).

## Glycerolipid Synthesis in Eukaryotes

In eukaryotes there are six major phospholipid classes, and biosynthesis involves many more enzymatic steps than found in _E. coli_. Just as in bacte-ria, phosphatidic acid is a central biosynthetic intermediate and CTP is the nucleotide required for phospholipid biosynthesis. In addition, many eu-karyotic (but not prokaryotic) cells synthesize triacylglycerol from phos-phatidic acid.

The pathway for the synthesis and catabolism of phosphatidic acid is shown in Figure 14-10. One major biosynthetic route is similar to that observed in _E. coli_, where glycerol-3-phosphate is acylated by the successive action of _glycerol-3-phosphate acyltransferase_ and _1-acylglycerol-3-phos-phate acyltransferase_. Alternatively, dihydroxyacetone phosphate can serve as a precursor of phosphatidic acid. In this pathway, an initial acyla-tion precedes the reduction of the ketone to yield _1-acylglycerol-3-phos-phate_. A third route for phosphatidic acid synthesis is by means of the phosphorylation of diacylglycerol. Once phosphatidic acid is made, it is rapidly converted to _diacylglycerol_ or _CDPdiacylglycerol_, which are me-tabolized as described in the following discussion.

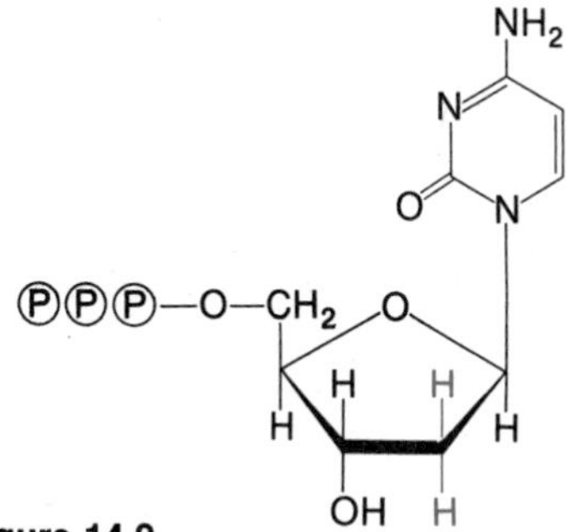

**Figure 14-9**
Structure of deoxycytidine triphos-phate (dCTP).

## Synthesis of Triacylglycerols

Triacylglycerol is a major product made from diacylglycerol in liver, intes-tine, and adipose tissues. The important biosynthetic pathway in liver and adipose tissues involves the acylation of diacylglycerol by _diacylglycerol acyltransferase_ (Figure 14-11). This enzyme is tightly bound to the cyto-solic side of the endoplasmic reticulum.

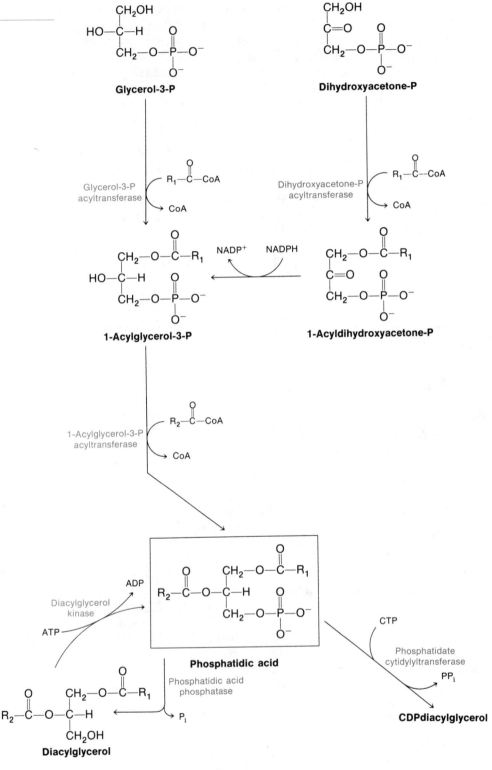

**Figure 14-10**
Metabolism of phosphatidic acid in eukaryotes.

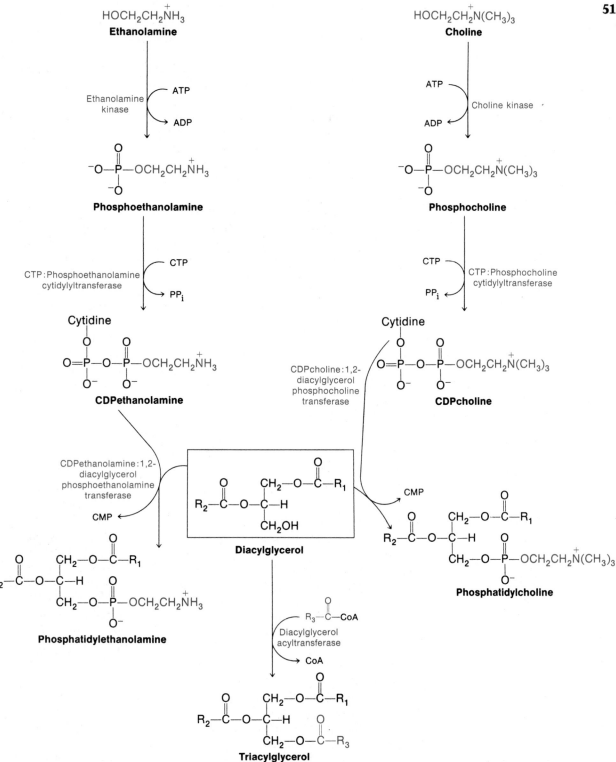

**Figure 14-11**
Metabolism of diacylglycerol in eukaryotes.

Several mechanisms appear to operate for the control of triacylglycerol synthesis. Of major importance is the supply of the substrates diacylglycerol and fatty acyl-CoA. In situations of energy excess (e.g., after a meal), the supply of fatty acids would be increased as a result of both diet and biosynthesis from carbohydrate. Increased insulin levels after a meal stimulate the synthesis of malonyl-CoA, which (as noted in Chapter 13) promotes fatty acid synthesis and inhibits fatty acid oxidation. Thus acyl-CoA would be available for both the synthesis of phosphatidic acid (Figure 14-10) and the acylation of diacylglycerol. The conversion of phosphatidic acid to diacylglycerol by *phosphatidic acid phosphatase* also appears to be regulated. The activity of this enzyme increases and decreases with the rate of synthesis of triacylglycerol under a number of different metabolic situations. In addition, recent studies indicate that the activity of this phosphatase may be regulated by a class of steroid hormones (*glucocorticoids*).

A different biosynthetic pathway for triacylglycerol synthesis is found in the intestine of *nonruminant* animals. Dietary triacylglycerol is degraded in the intestinal lumen to 2-monoacylglycerol and fatty acids by *pancreatic lipase*. These products enter the epithelial cells of the small intestine, where triacylglycerols are resynthesized by direct acylation of monoacylglycerol and diacylglycerol, as shown in Figure 14-12. This pathway accounts for 80 percent of the triacylglycerol made in this tissue. The remainder is made from glycerol phosphate and dihydroxyacetone phosphate, as described previously. In *ruminant* animals, the bacteria in the rumen degrade the monoacylglycerol to glycerol and fatty acid. Hence the pathway for acylation of monoacylglycerol and diacylglycerol is not important in these animals.

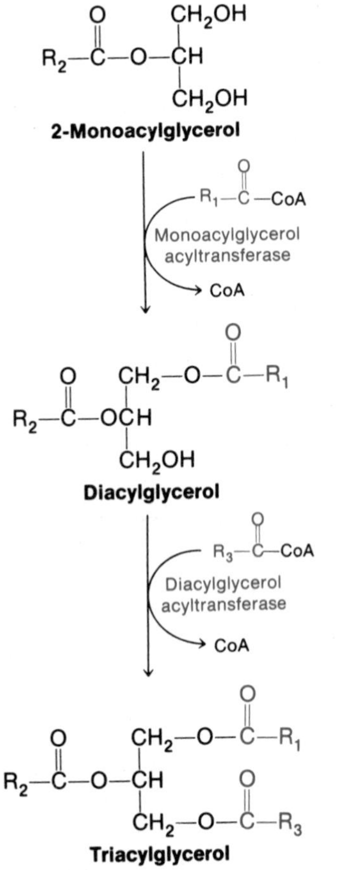

## Biosynthesis of Phospholipids in Eukaryotes

Phosphatidylcholine and phosphatidylethanolamine, which are quantitatively the most important phospholipids in eukaryotic cells, also are derived from diacylglycerol, as shown in Figure 14-11. The biosynthesis of phosphatidylcholine begins with the transport of choline into the cell. *Choline* is an essential ingredient in the human diet and cannot be made by animals except indirectly, as noted below. Once inside the cell, the choline is immediately phosphorylated to *phosphocholine* by the action of *choline kinase*, a cytosolic enzyme. In the liver, the choline may be transported alternatively into the mitochondria and oxidized to *betaine* (Figure 14-13), which can serve as a methyl donor in the biosynthesis of methionine. The other product of the *betaine-homocysteine methyltransferase* reaction, *dimethylglycine*, can be further metabolized first to *sarcosine* by *dimethylglycine dehydrogenase* and then to *glycine* by *sarcosine dehydrogenase* (Figure 14-13). Both these reactions can generate $5,10—CH_2—H_4—$ PteGlu, which enters the one-carbon pool as discussed in Chapter 6. It appears that the phosphorylation of choline is favored over oxidation when choline is in short supply.

After the formation of phosphocholine, the next reaction in the biosynthesis of phosphatidylcholine is catalyzed by the cytosolic enzyme *CTP:phosphocholine cytidylyltransferase* (Figure 14-11). The product, *CDPcholine* (Figure 14-14), immediately reacts with diacylglycerol in a

**Figure 14-12**
Biosynthesis of triacylglycerol in intestinal epithelial cells.

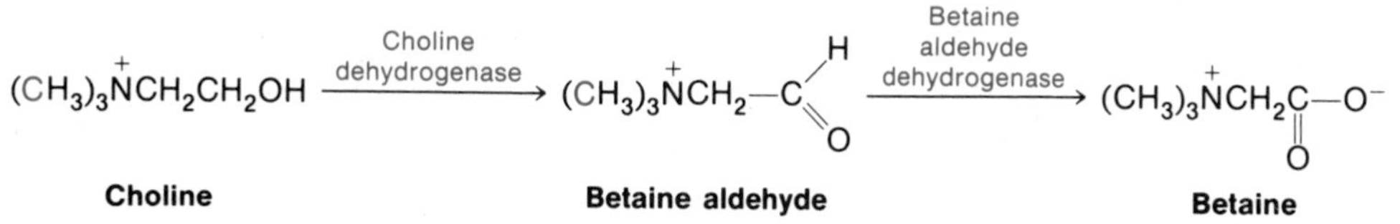

$$(CH_3)_3\overset{+}{N}CH_2CH_2OH \xrightarrow[\text{dehydrogenase}]{\text{Choline}} (CH_3)_3\overset{+}{N}CH_2-\overset{\displaystyle H}{\underset{\displaystyle O}{C}} \xrightarrow[\text{dehydrogenase}]{\begin{array}{c}\text{Betaine}\\ \text{aldehyde}\end{array}} (CH_3)_3\overset{+}{N}CH_2\overset{\displaystyle }{\underset{\displaystyle O}{C}}-O^-$$

**Choline**        **Betaine aldehyde**        **Betaine**

reaction catalyzed by *CDPcholine:1,2-diacylglycerol phosphocholine transferase*, an enzyme localized on the endoplasmic reticulum.

The biosynthesis of phosphatidylethanolamine (Figures 14-11 and 14-15) proceeds from ethanolamine in a parallel series of reactions catalyzed by *ethanolamine kinase* (cytosol), *CTP:phosphoethanolamine cytidylyltransferase* (cytosol), and *CDPethanolamine:1,2-diacylglycerol phosphoethanolamine transferase* (endoplasmic reticulum). Although the reactions are similar to those involved in phosphatidylcholine biosynthesis, the enzymes are different.

What regulates the flux of diacylglycerol to phosphatidylcholine, phosphatidylethanolamine, or triacylglycerol is not well understood. Apparently, at least in liver, the requirements for the synthesis of phosphatidylcholine and phosphatidylethanolamine are met before much triacylglycerol is made.

In liver, yeast, and the bacteria *Pseudomonas*, there is a pathway for the conversion of phosphatidylethanolamine to phosphatidylcholine in which methyl groups are transferred from *S-adenosylmethionine* (*AdoMet*) to phosphatidylethanolamine in three consecutive reactions (Figure 14-16). *Phosphatidylethanolamine-N-methyltransferase* is located on the endoplasmic reticulum. In liver it is not known whether the N-methylations are catalyzed by one or more separate enzymatic activities. By the methylation of phosphatidylethanolamine, liver can produce phosphatidylcholine and, indirectly, choline.

Lung tissue manufactures and secretes a specialized species of phosphatidylcholine called *dipalmitoylphosphatidylcholine*, in which palmitic acid is the fatty acyl substituent on both the *sn*-1 and *sn*-2 positions of the glycerol backbone. This species is the major component (approximately 50–60%) of *lung surfactant*, a substance that maintains surface tension in the lung alveoli. The secretion of lung surfactant is defective in

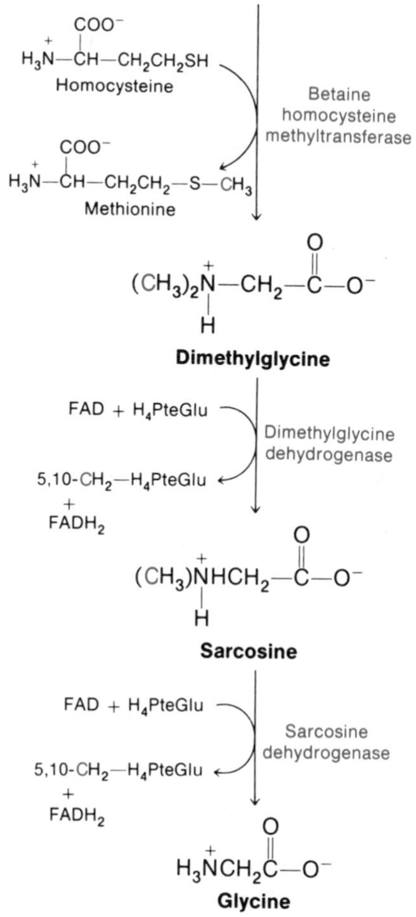

**Figure 14-13**
Metabolism of betaine in liver mitochondria.

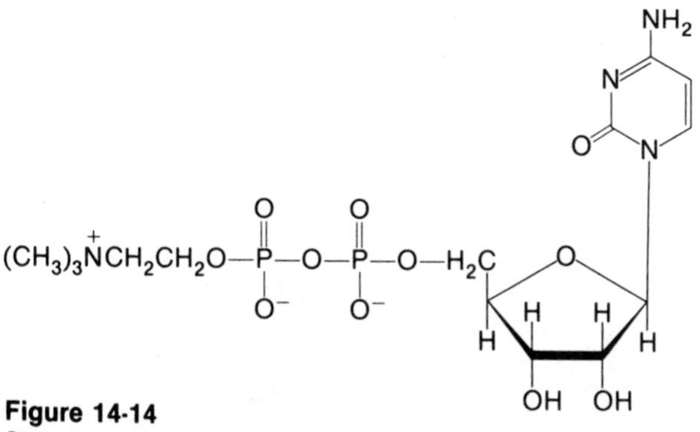

**Figure 14-14**
Structure of cytidine diphosphocholine (CDPcholine).

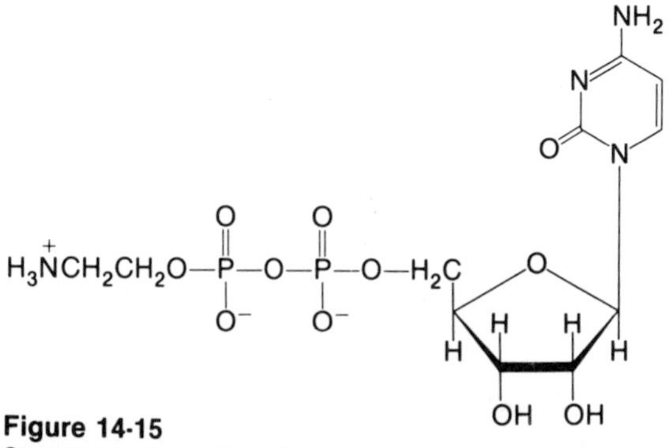

**Figure 14-15**
Structure of cytidine diphosphoethanolamine.

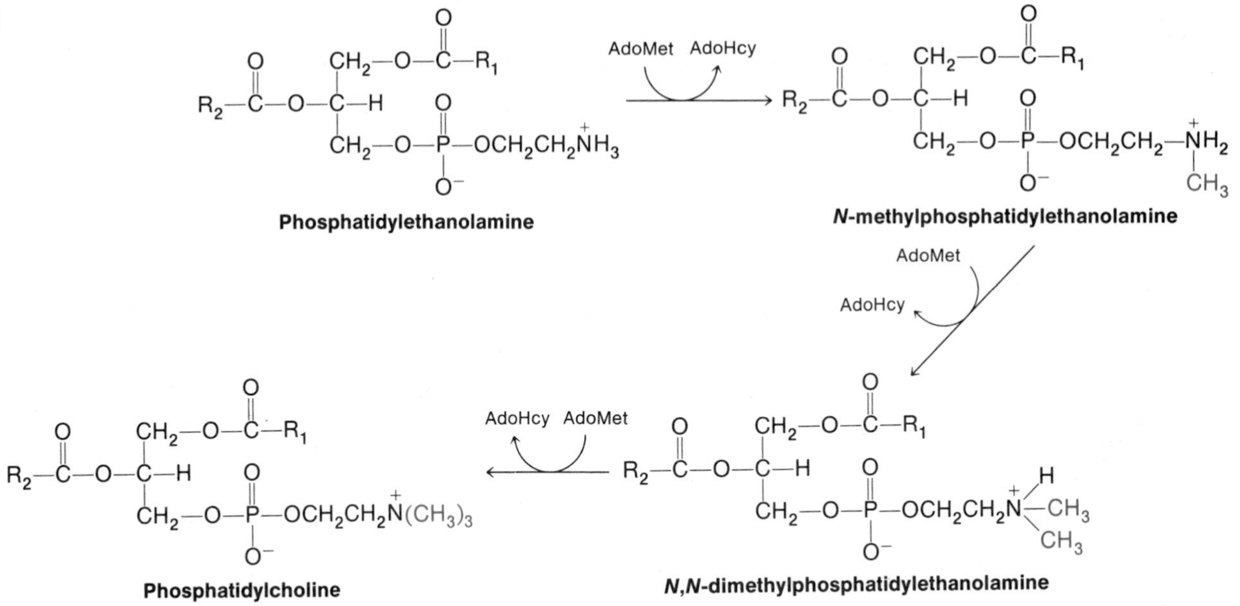

**Figure 14-16**
Conversion of phosphatidylethanolamine to phosphatidylcholine by phosphatidylethanolamine-*N*-methyltransferase. (AdoMet is a standard abbreviation for *S*-adenosyl-L-methionine and AdoHcy is a standard abbreviation for *S*-adenosyl-L-homocysteine.)

newborns with *respiratory distress syndrome*. The probable pathway for the synthesis of dipalmitoylphosphatidylcholine is illustrated in Figure 14-17. The starting species of phosphatidylcholine is made by the CDPcholine pathway (Figure 14-11). The fatty acid at the *sn*-2 position is hydrolyzed by *phospholipase A₂*, and the lysophosphatidylcholine is reacylated with palmitoyl-CoA. This allows alteration of the properties of the phospholipid

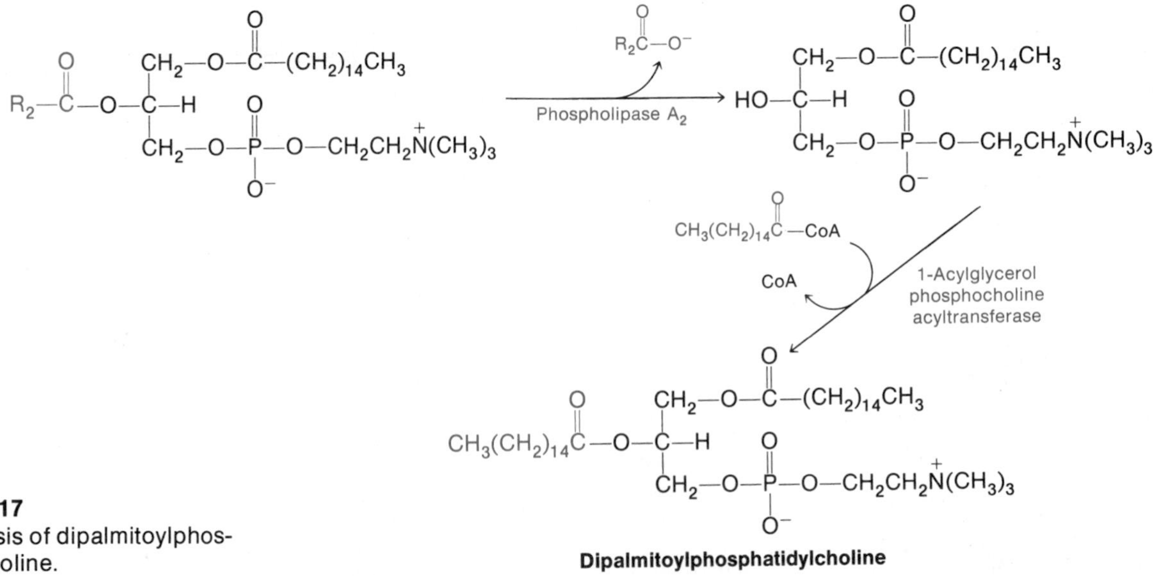

**Figure 14-17**
Biosynthesis of dipalmitoylphosphatidylcholine.

**Dipalmitoylphosphatidylcholine**

without resynthesis of the entire molecule. The synthesis of dipalmitoyl-phosphatidylcholine in lung tissue is a specific example of the modulation of the fatty acid composition of a phospholipid. The deacylation-reacylation reaction also occurs in other tissues and provides an important route for the introduction of polyunsaturated fatty acids into the *sn*-2 position of phospholipids.

The remaining four major classes of phospholipids in eukaryotes are synthesized from CDPdiacylglycerol as shown in Figure 14-18. Phos-

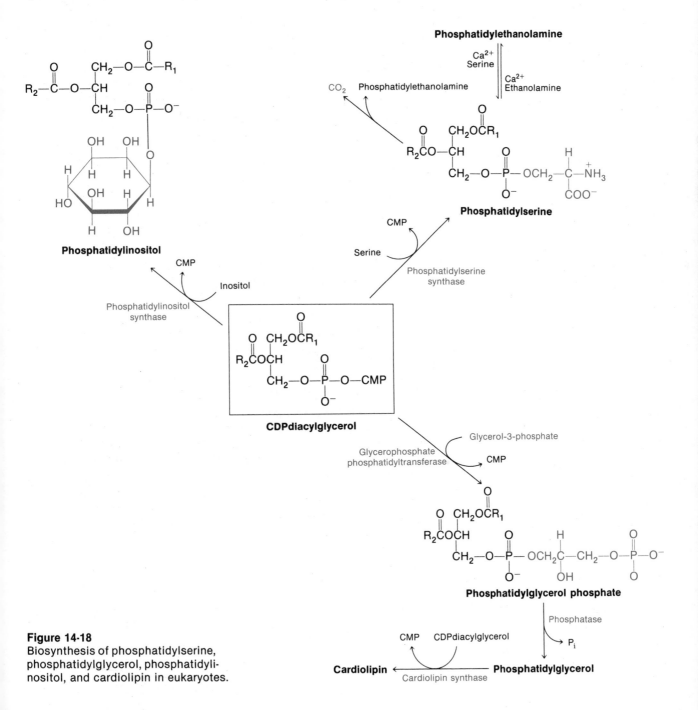

**Figure 14-18**
Biosynthesis of phosphatidylserine, phosphatidylglycerol, phosphatidylinositol, and cardiolipin in eukaryotes.

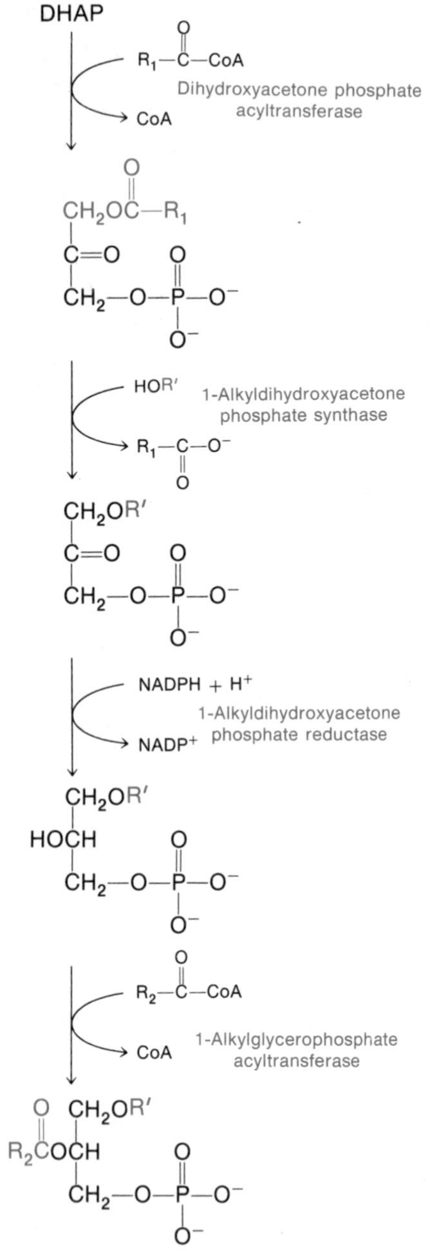

DHAP

**Figure 14-19**
Biosynthesis of the alkyl ether species of phosphatidic acid.

phatidylserine is made either by a reaction between CDPdiacylglycerol and serine catalyzed by *phosphatidylserine synthase* or from phosphatidylethanolamine by a base-exchange reaction with serine. The phosphatidylserine synthase activity in rat liver is only 2 percent of the base exchange activity. Hence the biosynthesis of phosphatidylserine by means of CDPdiacylglycerol is not likely to be a major pathway. The microsomal enzyme that catalyzes the base-exchange reaction requires $Ca^{2+}$. Reversal of this $Ca^{2+}$-mediated reaction is a second pathway for the biosynthesis of phosphatidylethanolamine. Decarboxylation of phosphatidylserine, which appears to be limited to the mitochondria, is a third route to phosphatidylethanolamine. Phosphatidylglycerol synthesis in eukaryotes (Figure 14-18) is similar to that in *E. coli* (Figure 14-7). However, the formation of cardiolipin in animals seems to involve the reaction of phosphatidylglycerol with CDPdiacylglycerol rather than the reaction of two phosphatidylglycerol molecules, as in *E. coli* (Figure 14-7). The synthesis of phosphatidylinositol (Figure 14-18) involves the transfer of phosphatidic acid from CMP to L-*myo*-inositol, a hexahydroxy cyclohexane that is biosynthesized from D-glucose-6-P. Many animal cells make small amounts of two additional phosphorylated derivatives of phosphatidylinositol (phosphatidylinositol-4-P and phosphatidylinositol-4,5-di-P) by phosphorylation with ATP. These *di-* and *tri-phosphoinositides* turn over very rapidly in nerve cells and have intrigued biochemists for a long time. However, the function of these phospholipids has still not been clearly established.

### Biosynthesis of Alkyl and Alkenyl Ethers

The biosynthetic pathway for the alkyl ether species of phosphatidic acid in eukaryotes is summarized in Figure 14-19. The initial step is the acylation of dihydroxyacetone phosphate, a reaction already discussed in connection with phosphatidic acid biosynthesis (Figure 14-10). Subsequently, an exchange reaction replaces the 1-acyl group with an alkyl group derived from an alcohol (Figure 14-19). The long-chain alcohol used in this reaction is formed by reduction of an acyl-CoA with 2 mol of reduced pyridine nucleotide (NADPH or NADH). The subsequent reduction of the ketone and acylation of the 2-hydroxyl group are the same as previously noted (Figure 14-10). Once the 1-alkyl ether derivative of phosphatidic acid is formed, it is used for the synthesis of other phospholipids, as already discussed. In many tissues, 1-alkyl-2-acylphosphatidylethanolamine can be desaturated by a microsomal enzyme to yield the corresponding *plasmalogen* derivative (Figure 14-20). The desaturation reaction is catalyzed by *1-alkyl-2-acylglycerophosphoethanolamine desaturase*, a microsomal enzyme. This enzyme requires $O_2$, NADH, and cytochrome $b_5$, the same cofactors required for the desaturation of stearoyl-CoA. However, stearoyl-

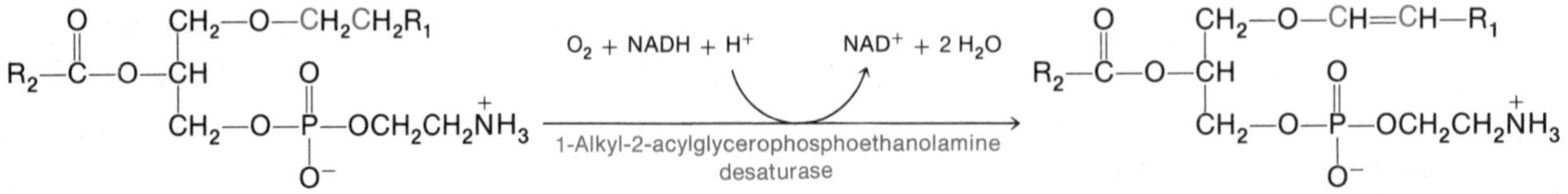

**Figure 14-20**
Formation of plasmalogens.

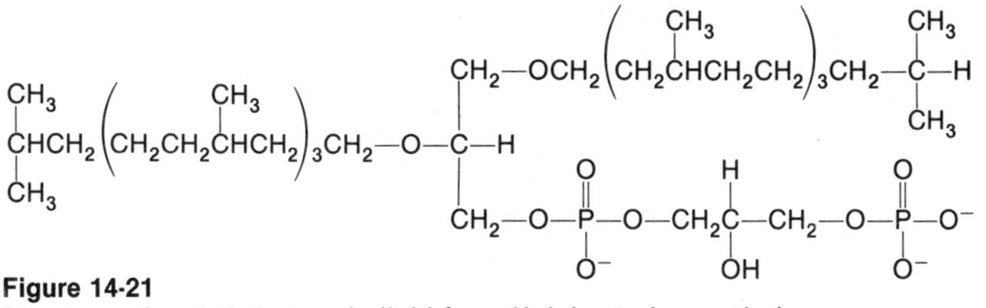

$$CH_3 \quad \Big( \quad CH_3 \quad \Big) \quad CH_2—OCH_2\Big(CH_2CHCH_2CH_2\Big)_3CH_2—C—H$$

$$CHCH_2\Big(CH_2CH_2CHCH_2\Big)_3CH_2—O—C—H$$

**Figure 14-21**
Structure of a dialkyl phospholipid from *Halobacterium cutirubrum.*

CoA desaturase appears to be different from the alkyl-acylglycerophospho-ethanolamine desaturase. Although the plasmalogens are minor constituents in many tissues, nearly 50 percent of the phospholipids in heart tissue contain the alkenyl ether at position *sn*-1. The functional significance of the ether phospholipids remains an enigma.

The ether phospholipids also are found in microorganisms, notably protozoa and *Halobacterium cutirubrum*. This bacterium is halophilic (it requires 4 $M$ NaCl for growth) and has the unusual phosphatidylglycerol-P shown in Figure 14-21. The alkyl ether bonds are stable to alkaline hydrolysis and relatively stable to acid hydrolysis when compared with esters (alkenyl ethers are more labile than alkyl ethers to acid hydrolysis). Perhaps the ability of *H. cutirubrum* to survive in such high concentrations of salt is related to the stability of these phospholipids.

An unusual species of phosphatidylcholine that possesses extremely high biological activity has been recently discovered independently by Fred Snyder's and Donald Hanahan's laboratories. This compound, *1-alkyl-2-acetylglycerophosphocholine* (Figure 14-22), has a striking ability to decrease blood pressure in hypertensive rats and, at concentrations of $10^{-10}$ $M$, will cause blood platelets to aggregate. The biosynthesis of this lipid from 1-alkyl-2-lysophosphatidylcholine is catalyzed by *acetyl-CoA:1-alkyl-2-lysoglycerophosphocholine transferase* (Figure 14-23). Significant activity of this enzyme has been found in microsomes from rat spleen, kidney, and other tissues. An enzyme that cleaves the acetyl group (*acetylhydrolase*) has been found in cytosol from kidney and other tissues. The discovery of this new derivative of phosphatidylcholine opens a new and unexpected chapter in phospholipid metabolism. Knowledge of the structure and metabolism of this compound should facilitate investigations into its biological function which could be extremely important.

## Control of Phospholipid Biosynthesis in Eukaryotes

Our knowledge about the control of the synthesis of phospholipids is in a rudimentary stage for both eukaryotes and prokaryotes. In eukaryotes, *the rate of phosphatidylcholine biosynthesis can be regulated by the activity*

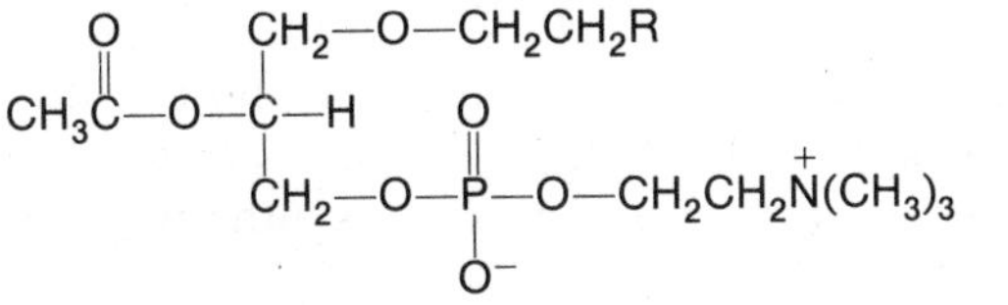

**Figure 14-22**
Structure of 1-alkyl-2-acetylglycero-phosphocholine.

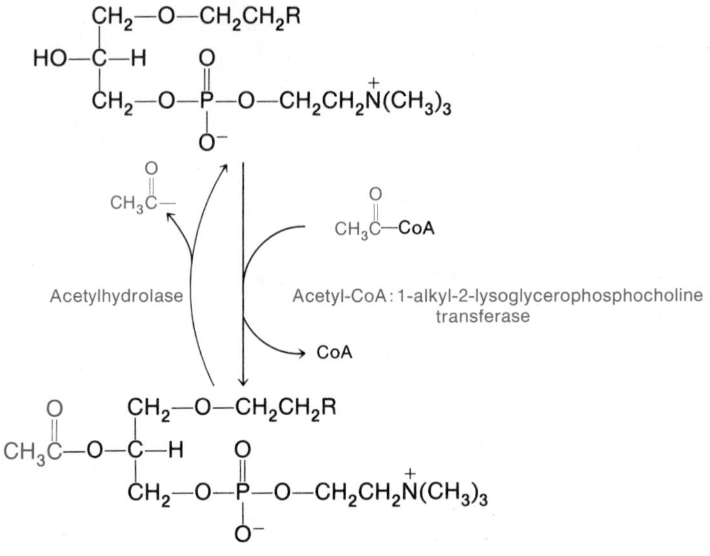

**Figure 14-23**
Biosynthesis of 1-alkyl-2-acetylglycer-ophosphocholine.

of CTP:phosphocholine cytidylyltransferase (Figure 14-11). This enzyme isolated from rat liver has a molecular weight of 200,000 and when purified is nearly inactive. However, the enzyme is markedly stimulated by lyso-phosphatidylethanolamine, phosphatidylserine, phosphatidylglycerol, and phosphatidylinositol, but inhibited by lysophosphatidylcholine and, to a lesser extent, phosphatidylcholine and phosphatidylethanolamine. It is not clear whether or not any of these lipid modulators plays a functional role in the regulation of phosphatidylcholine biosynthesis in vivo. In several different systems, the supply of substrates (CTP or phosphocholine) for the phosphocholine cytidylyltransferase appears to regulate the rate of phosphatidylcholine biosynthesis. Similarly, for phosphatidylethanolamine biosynthesis in liver cells, evidence suggests that the activity of the CTP:phosphoethanolamine cytidylyltransferase can be rate-limiting. No activators or inhibitors of this enzyme have been described. It is likely that the supply of diacylglycerol also can influence the rate of phosphatidylcholine and phosphatidylethanolamine biosynthesis.

It is noteworthy that major fluctuations in the amounts of phospholipid biosynthetic enzymes are not usually detected. This is in contrast with some of the enzymes of fatty acid and cholesterol biosynthesis, the amounts of which fluctuate with different metabolic states. Only recently have there been reports of the involvement of protein kinases in the alteration of the activity of phospholipid biosynthetic enzymes. There is still much to be learned about the various mechanisms for control of phospholipid biosynthesis.

## Control of Fatty Acid and Phospholipid Biosynthesis in E. coli

The lack of information on the control of phospholipid metabolism in eukaryotes can be partially excused, since even in the relatively simple organism E. coli, control of phospholipid synthesis is not well-explained.

*The control of phospholipid and fatty acid synthesis appears to be coordinated in E. coli.* This is demonstrated by a mutant of *E. coli* that is defective in the conversion of dihydroxyacetone-P into glycerol-3-P. Such a mutation in a eukaryotic cell would probably not be detected, because dihydroxyacetone-P can be acylated and converted into phospholipid independently of glycerol-3-P. However, the *E. coli* mutant can survive when grown on glycerol, because glycerol can be phosphorylated by *glycerol kinase*. Withdrawal of glycerol from the growth medium prevents the synthesis of glycerol-3-P and, therefore, phospholipids. In addition, fatty acid synthesis is inhibited in the glycerol-starved mutants. Apparently, *E. coli* has an unidentified mechanism for coupling the biosynthesis of fatty acids and phospholipids. This makes good sense, because almost all the fatty acids in *E. coli* are used for phospholipid synthesis. Another example of the coordination of fatty acid and phospholipid synthesis can be observed with certain strains of *E. coli* that require a particular amino acid for growth. In response to a deficiency of this amino acid, fatty acid and phospholipid synthesis, as well as protein and RNA synthesis, are inhibited. It has been postulated that the inhibition of all these processes is due to the accumulation of a nucleotide known as *ppGpp* (guanosine-5′-3′-diphosphate-diphosphate).

Not only is our knowledge of the regulation of phospholipid synthesis limited, but the rate-limiting reactions for fatty acid and phospholipid synthesis in *E. coli* also have not been identified clearly. In the case of fatty acid synthesis, the carboxylation of acetyl-CoA is a likely candidate.

The control of the length of the fatty acids found in *E. coli* is most likely due to the specificity of *β-ketoacyl-ACP synthase*, which under most conditions has very low activity with palmitoyl-ACP. As a result, this product would not be elongated and would thus be available for phospholipid synthesis.

Another unsolved problem is how the unsaturation of fatty acids is controlled in *E. coli*. The ratio of monounsaturated compared with saturated fatty acids in this cell is inversely related to temperature: the higher the temperature, the lower the content of unsaturated fatty acid. Once a fatty acid is made, its incorporation into phospholipids is regulated by the activity of the two acyltransferases (Figure 14-7). At lower temperatures, these enzymes show preference for unsaturated acyl-CoAs. It is likely that control of the synthesis of fatty acids and their incorporation into phospholipids in *E. coli* are both important factors for the maintenance of constant fluidity (or viscosity) of the bacterial membrane (see Chapter 16 for more on membrane fluidity).

## Phospholipases

The enzymes that degrade phospholipids are called *phospholipases*. They are classified according to the bond cleaved in a phospholipid, as shown in Figure 14-24. Thus phospholipases $A_1$ and $A_2$ selectively remove fatty acids from the *sn*-1 and *sn*-2 positions, respectively. Phospholipase C cleaves between glycerol and the phosphate; phospholipase D hydrolyzes the X moiety from the phospholipid. The lysophospholipids are degraded by phospholipases $L_1$ and $L_2$, also known as *lysophospholipases*. Phospholipases are found in all types of cells and in various locations within eukaryotic cells.

Some of these enzymes show specificity for different polar head groups; others are nonspecific. The enzymes from pancreatic tissue and snake venom have been studied extensively because of the relatively high concentrations of phospholipases in these sources.

Phospholipase $A_2$ from porcine pancreatic tissue has been sequenced, and the structure of its zymogen precursor is shown in Figure 14-25. The zymogen, which has very low activity, can be converted into its active form by the removal of a heptapeptide from the N terminal by _trypsin_. The phospholipase shows great stability, is not denatured by 8 $M$ urea, and retains 85 percent of its activity in a solution of 2% sodium dodecylsulfate. Such properties are attractive attributes for an enzyme that is involved in phospholipid degradation in the intestine.

### Synthesis and Assembly of Phospholipids in Membranes

The final reactions for the biosynthesis of phosphatidylcholine, phosphatidylethanolamine, phosphatidylserine, and phosphatidylinositol occur on the outer surface of the endoplasmic reticulum in rat liver (Figure

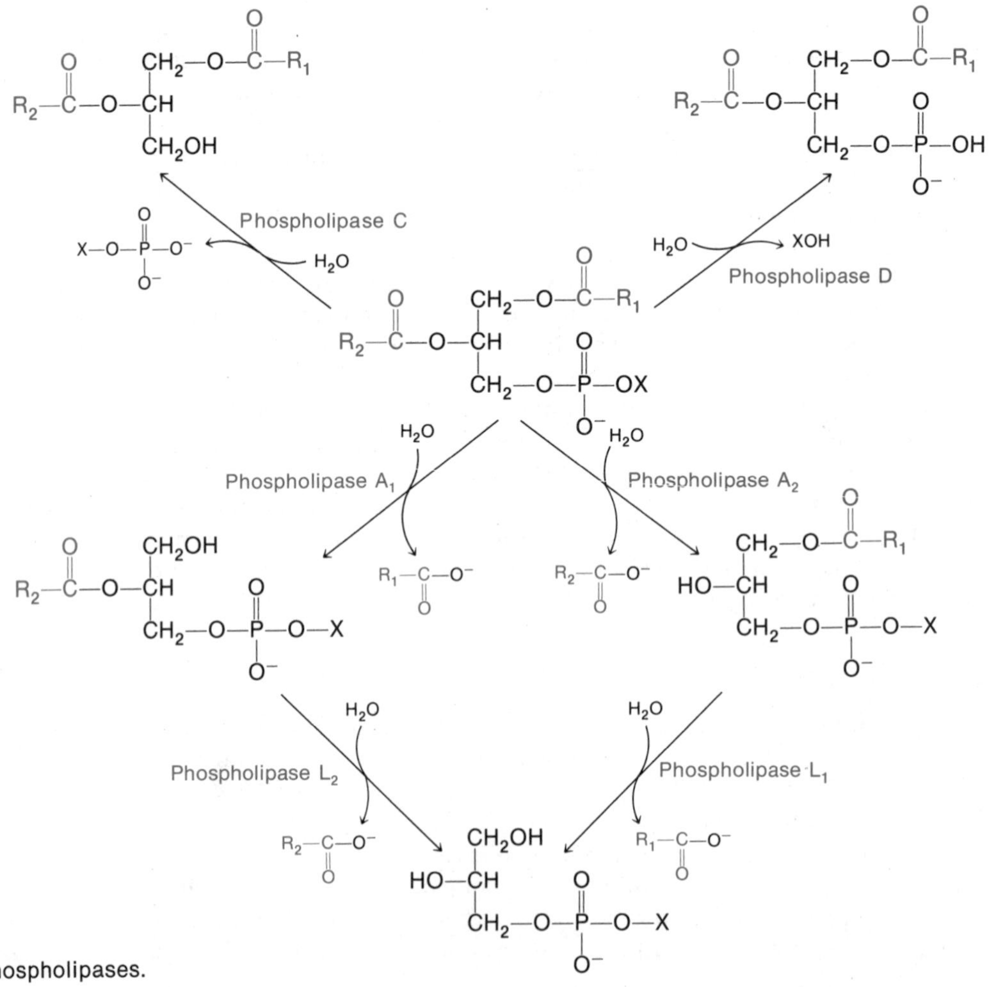

**Figure 14-24**
Reactions catalyzed by phospholipases.

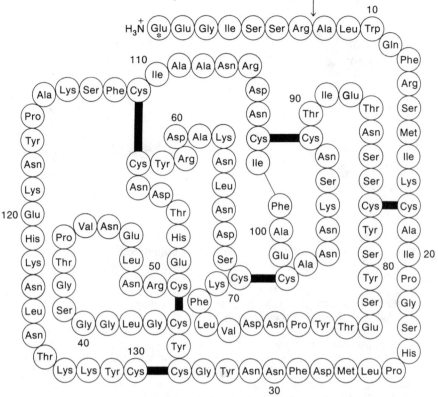

**Figure 14-25**
Primary structure of porcine pancreatic prophospholipase A₂.

14-26). (Phosphatidylglycerol is synthesized in and remains largely in the mitochondria.) The other organelles of the cell (e.g., nucleus, cytoplasmic membrane) contain these phospholipids in various proportions. How does the cell sort and transport these phospholipids from the site of synthesis, and what regulates this process? One possibility is that lipid vesicles with specific proteins bud from areas of the endoplasmic reticulum. The vesicle would move to and fuse with another organelle in the cell. It is possible that a particular protein might direct the vesicle to a specific organelle. A second possible mechanism for lipid transfer would be by phospholipid exchange proteins.

## Phospholipid Exchange Proteins

*Phospholipid exchange proteins are a class of proteins found in the cytosol of eukaryotic cells that catalyzes the exchange of phospholipid species between two membranes*, as shown in Figure 14-27. In this example, a phosphatidylcholine molecule on the endoplasmic reticulum exchanges with a molecule of phosphatidylcholine on the exchange protein. The protein moves to a mitochondrion, where another exchange occurs. Thus a phosphatidylcholine molecule can be moved from the endoplasmic reticulum to a mitochondrion. Some exchange proteins have been isolated that

* (Stands for pyroglutamic acid.)

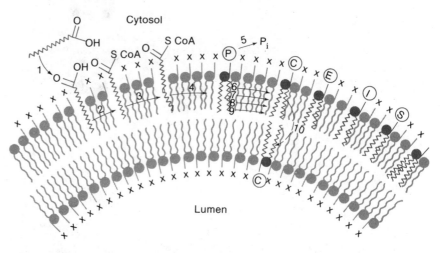

**Figure 14-26**
Glycerolipid synthesis on the endoplasmic reticulum from rat liver. Fatty acids are inserted into the cytoplasmic surface of the endoplasmic reticulum (1) and activated to form acyl-CoA thioesters (2). The acyl chains may be elongated and/or desaturated (3). Glycerol-P undergoes acyl-CoA dependent esterification to form phosphatidic acid (4). The action of phosphatidic acid phosphatase (5) forms diacylglycerols that are converted to phosphatidyl-choline and phosphatidylethanolamine by acquisition of phosphocholine and phosphoethanolamine polar headgroups (6). Phosphatidyl serine synthesis occurs by base exchange (7). Triacylglycerol synthesis occurs by esterification of diacyglycerol (8). CDP-diacylglycerol is an intermediate in the synthesis of phosphatidylinositol (9). Once formed, the glycerolipids may move to the lumenal surface of the endoplasmic reticulum (10). (C = choline; E = ethanolamine; I = inositol; S = serine; X = polar headgroup C, E, I, or S; P = PO$_4$) (From R.M. Bell, L.M. Ballas and R.A. Coleman, Lipid topogenesis, *J. Lipid Res.* 22:391, 1981.)

show nonspecificity for the polar headgroup; other proteins are highly specific. The phosphatidylcholine exchange protein from beef liver has been purified and partially sequenced. The protein is present in very small amounts in the cells. The protein has a molecular weight of 28,000 and is highly specific for phosphatidylcholine. There appears to be one hydrophobic binding site for phosphatidylcholine per molecule of protein. The amino acid sequence at this site is

$$R_1-N-Val-Phe-Met-Tyr-Tyr-Phe-\overset{O}{\overset{\|}{C}}-R_2$$

The functional importance of these proteins to the cell has not been established. Since most phospholipids are synthesized on the endoplasmic

$$ER-PC + protein-PC \longrightarrow ER-PC + protein-PC \qquad (1)$$

$$Protein-PC + mitochondrion-PC \longrightarrow protein-PC + mitochondrion-PC \quad (2)$$

$$ER-PC + mitochondrion-PC \xrightarrow{Protein-PC} ER-PC + mitochondrion-PC \qquad (1)+(2)$$

**Figure 14-27**
An example of phospholipid exchange between two membranes. (ER = endoplasmic reticulum; PC = phosphatidylcholine.)

reticulum, a mechanism is required for transfer of these lipids to other parts of the cell, such as the nucleus, cytoplasmic membrane, etc. The exchange proteins are candidates for this job. However, for the membranes to grow, a net transfer of lipid is required, and this has been difficult to demonstrate with these proteins. At a minimum, these proteins could function in renewal of the phospholipids in various membranes by exchange, since the phospholipids in eukaryotic cells do turn over.

## SPHINGOLIPIDS

### Structures

*The common structural feature of sphingolipids is a long-chain, hydroxylated secondary amine.* There are three major <u>long-chain bases</u> (as they are called) (Table 14-3) that contain 18 carbons and a number of other bases that differ in chain length, number of double bonds, or branching of the alkyl chain. *Sphingosine* (*4-sphingenine*) is quantitatively the most important long-chain base (usually 90 percent or more) in animal cells, whereas *phytosphingosine* (4-*hydroxysphinganine*) is characteristically found in plant tissues. Most bacteria, including *E. coli*, do not contain sphingolipids, whereas yeast cells do.

The sphingolipid bases occur as components of more complex lipids. The bases themselves are toxic to cells and, therefore, are present only in trace quantities. The sphingolipid base is acylated on the amine with a fatty acid to give *ceramide* (Figure 14-28), which is common to all the sphingolipids. The fatty acid substituents are mainly $C_{16}$, $C_{18}$, $C_{22}$, or $C_{24}$, saturated or monounsaturated. In many tissues, the acyl group also may contain an $\alpha$-hydroxy residue.

Ceramide occurs as part of more complex lipids that have on the primary hydroxyl group either phosphocholine to give *sphingomyelin* (Figure 14-29) or carbohydrate to give the class called *glycosphingolipids*. The carbohydrates most often associated with the glycosphingolipids are glu-

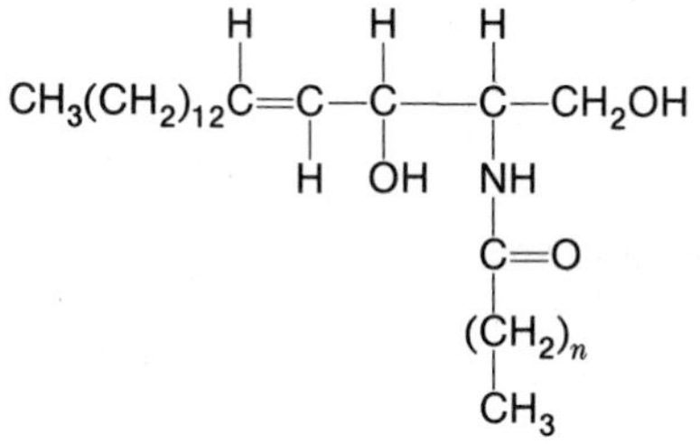

**Figure 14-28**
Structure of ceramide with sphingosine as the long-chain base.

**Table 14-3**
Three Important Sphingolipid Bases

| Structure | Systematic Name | Common Name |
|---|---|---|
| 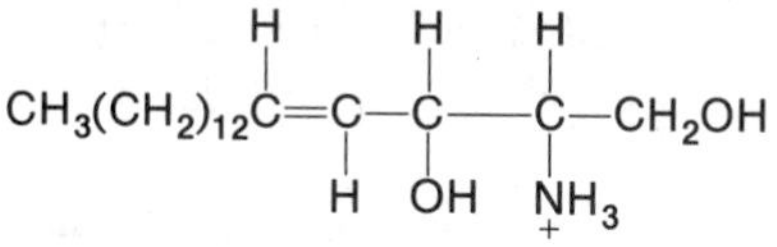 | 4-Sphingenine | Sphingosine |
| 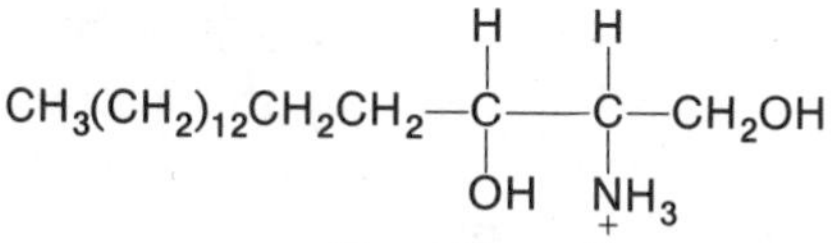 | Sphinganine | Dihydrosphingosine |
| 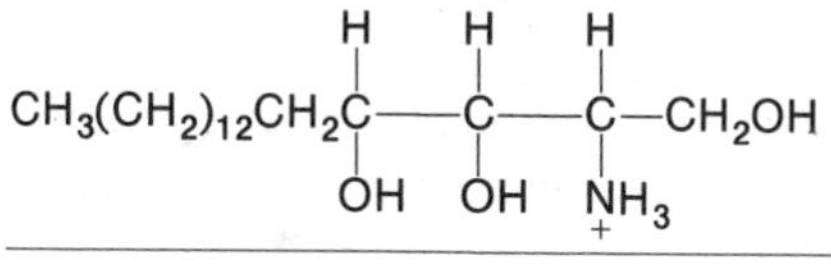 | 4-Hydroxy-sphinganine | Phytosphingosine |

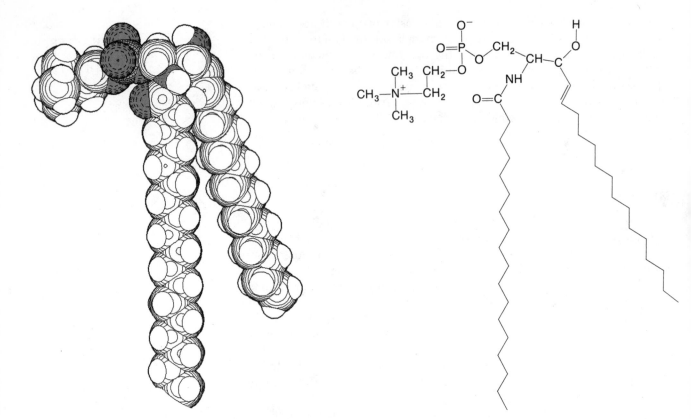

**Figure 14-29**
Structure of sphingomyelin.

cose, galactose, _N-acetylglucosamine_, and _N-acetylgalactosamine_ (Figure 14-30). There is a subdivision of the glycosphingolipids called _gangliosides_, and these also contain one or more molecules of _N-acetylneuraminic acid_ (_sialic acid_) (Figure 14-31) in addition to other carbohydrates. As the name implies, the gangliosides were first isolated from nerve tissue; subsequently, however, they were found in most other animal tissues. Some of the major glycosphingolipids are listed in Table 14-4. The structures of two important glycosphingolipids, globoside and $GM_2$, are shown in Figures 14-32 and 14-33. As with the phospholipids, each class of sphingolipids is a mixture of many molecular species with different fatty acid substituents and, to a lesser extent, different long-chain bases. In addition, at least 50 separate classes of glycosphingolipids have been identified that differ in the structure of the oligosaccharide. Consequently, this chapter provides only a modest introduction to these lipids.

Sphingolipids are important components of the myelin sheath, a multi-layered membranous structure that protects and insulates nerve fibers. The lipids in human myelin contain 5 percent sphingomyelin (the original source of this lipid, as the name implies) and 15 percent _galactosyl-ceramide_ (_galactocerebroside_) (Figure 14-34). In addition, a sulfate deriv-ative, _3'-sulfate-galactosylceramide_, makes up 5 percent of the myelin lipid. The sphingolipids are found in plasma as components of _lipoproteins_, primarily _low-density lipoproteins_, which are discussed in Chapter 15.

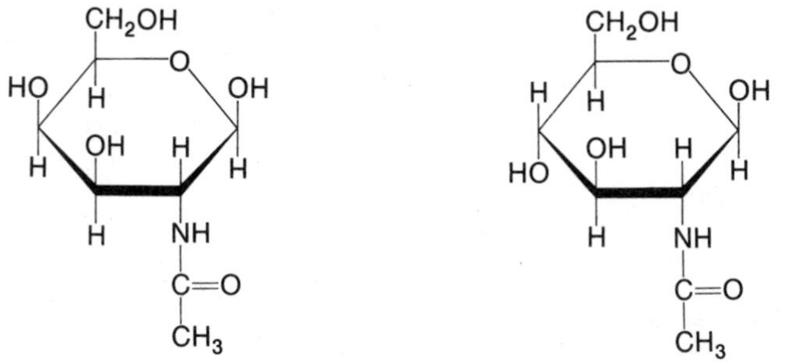

**β-N-Acetylgalactosamine**    **β-N-Acetylglucosamine**

**Figure 14-30**
Structure of β-N-acetylgalactosamine and β-N-acetylglucosamine.

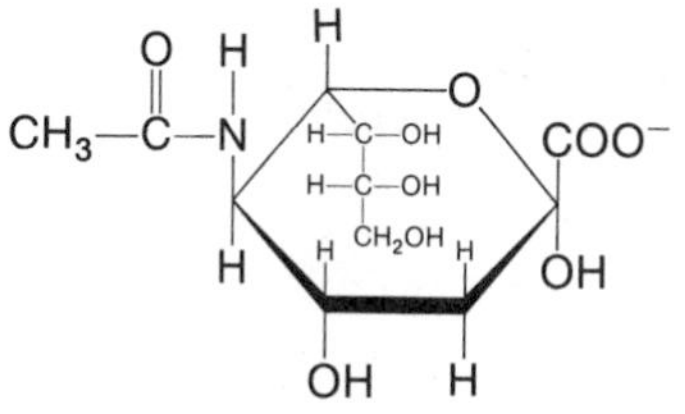

## Biosynthesis of Sphingolipids

The biosynthesis of sphinganine (Figure 14-35) occurs on the endoplasmic reticulum and involves a condensation of *palmitoyl-CoA* with *serine*, catalyzed by *3-ketosphinganine synthase* and a subsequent reduction of the 3-ketone by *3-ketosphinganine reductase*. The 3-ketosphinganine synthase has pyridoxal phosphate as an essential cofactor. Although the biosynthesis of sphingosine also involves serine, the origin of the trans double bond appears to differ in yeast and animals. In yeast, palmitoyl-CoA can be desaturated to yield *trans*-2-hexadecenoyl-CoA, which condenses with serine to yield sphingosine. Once the long-chain base is formed, it is quickly acylated to form ceramide (Figure 14-36). In animals it appears that the trans double bond of sphingosine is introduced after ceramide has been synthesized.

**Figure 14-31**
Structure of α-N-acetylneuraminic acid.

**Table 14-4**
Some Major Glycosphingolipids

| Structure | Common Name | Abbreviated Name |
|---|---|---|
| Galβ1-1ceramide | Galactosylceramide | — |
| Glcβ1-1ceramide | Glucosylceramide | — |
| Galβ1-4Glcβ1-1ceramide | Lactosylceramide | — |
| Galα1-4Galβ1-4Glcβ1-1ceramide | Trihexosylceramide | — |
| GalNAcβ1-3Galα1-4Galβ1-4Glcβ1-1ceramide | Globoside | — |
| NeuAcα2-3Galβ1-4Glcβ1-1ceramide | Hematoside | GM3 |
| GalNAcβ1-4Galβ1-4Glcβ1-1ceramide<br>3<br>↑<br>α2<br>NeuAc | Tay Sachs ganglioside | GM2 |
| Galβ1-3GalNAcβ1-4Galβ1-4Glcβ1-1ceramide<br>3<br>↑<br>α2<br>NeuAc | — | GM1 |

Note: Gal = galactose; Glc = glucose; GalNAc = N-acetylgalactosamine; NeuAc = N-acetylneuraminic acid.

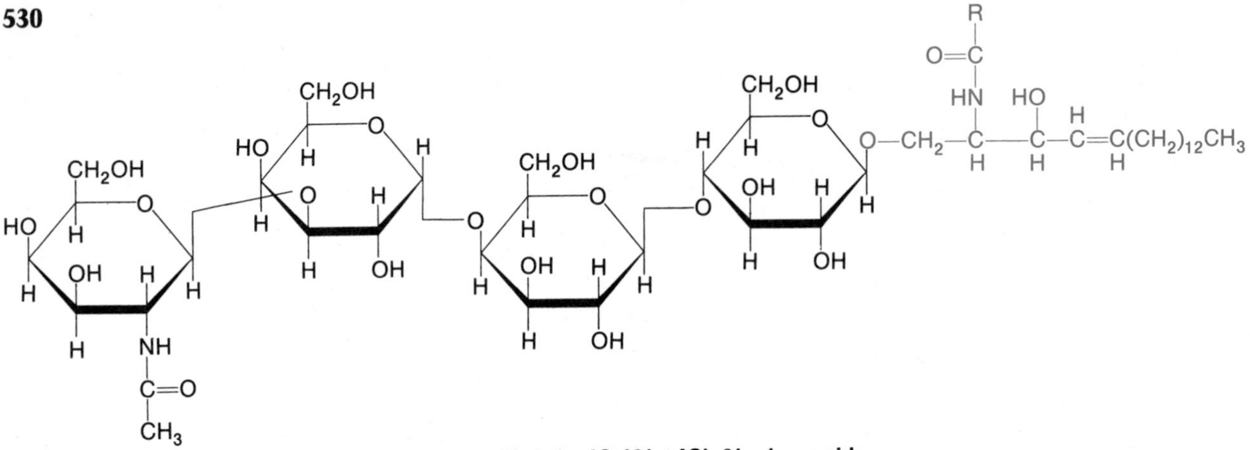

**GalNacβ1→3Galα1→4Galβ1→4Glcβ1→1ceramide**

**Figure 14-32**
Structure of globoside.

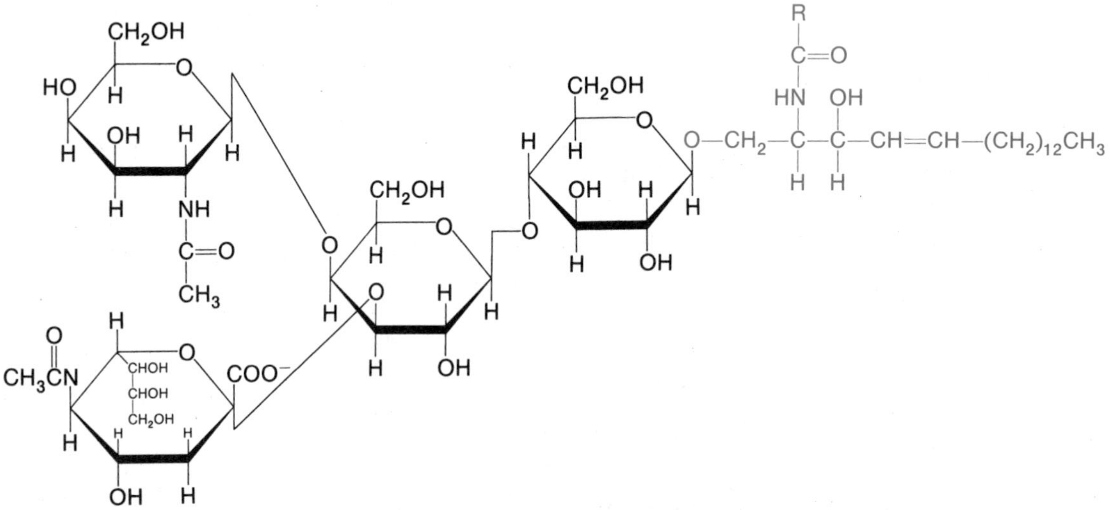

**GalNAcβ1→4Galβ1→4Glcβ1→1ceramide**

$$\begin{pmatrix} 3 \\ \uparrow \\ \alpha 2 \end{pmatrix}$$

**NeuAc**

**Figure 14-33**
Structure of Tay-Sachs ganglioside (GM₂).

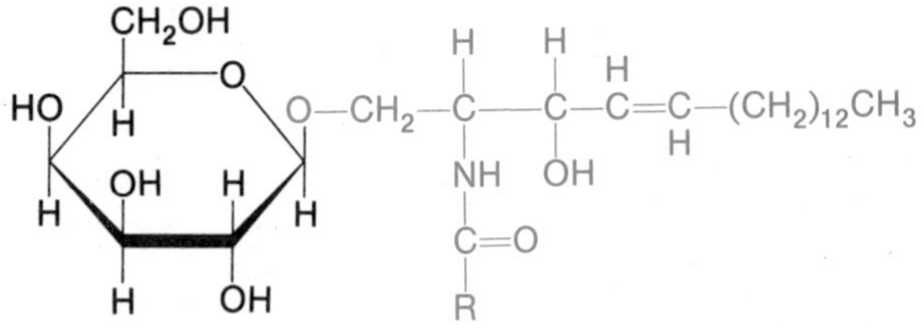

**Figure 14-34**
Structure of galactosylceramide.

Two different pathways have been suggested for the biosynthesis of sphingomyelin (Figure 14-37). Recent evidence from several laboratories indicates that the transfer of phosphocholine from phosphatidylcholine to ceramide is the correct pathway.

The biosynthesis of some of the glycosphingolipids is shown in Figure 14-38. The stereochemistry of the products is not shown. The principle governing the synthesis of these lipids is that the carbohydrates are added to the acceptor lipid by transfer from a sugar nucleotide. Thus *UDPglucose*, which is the glucosyl donor for the synthesis of glycogen (Chapter 12), is also the donor of glucose for the synthesis of the glycosphingolipids. The enzymes involved in these reactions are called *glycosyltransferases* and are thought to be specific for each reaction. There has been much debate about the subcellular localization of these enzymes. However, recent results are consistent with the *Golgi apparatus* being the major site for the synthesis of glycosphingolipids.

Unfortunately, none of the enzymes involved in the biosynthesis of the sphingolipids has been purified. The major difficulties are that the enzymes are present in small quantities, are membrane-bound, and use amphipathic substrates.

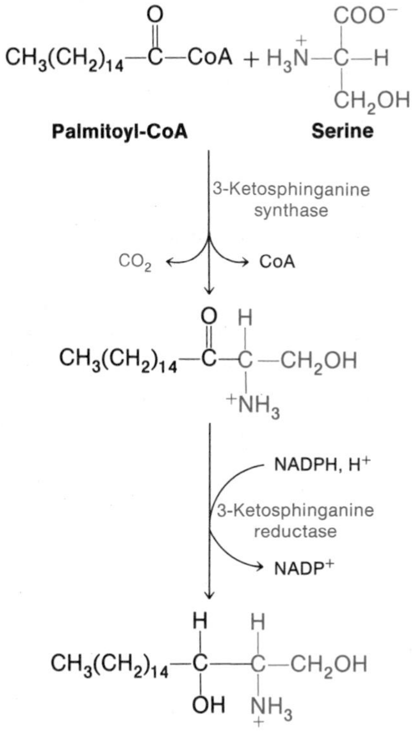

## Catabolism of Sphingolipids and Related Metabolic Diseases

The degradation of long-chain sphingosine bases begins with a phosphorylation by *sphinganine kinase* to yield *sphinganine-1-phosphate*. This is subsequently cleaved by *sphinganine phosphate lyase* to form palmitaldehyde and phosphoethanolamine (Figure 14-39). The palmitaldehyde can either be reduced to the $C_{16}$-alcohol or oxidized to palmitic acid. The phosphoethanolamine appears to enter the major metabolic pool of this compound and is therefore converted to phosphatidylethanolamine.

Investigations on the structure and biochemistry of sphingolipids have been stimulated by the occurrence of a number of genetic diseases, the *sphingolipidoses*, each of which results from the accumulation of one of these lipids. In all but a few cases, these diseases are caused by a deficiency in the activity of one of the enzymes involved in the catabolism of the sphingolipids. The degradation of glycosphingolipids proceeds in a stepwise fashion, as shown in Figure 14-40. These catabolic enzymes are localized in the lysosomes. Hence the sphingolipidoses are part of a larger group of metabolic diseases that result from impaired function of lysosomal enzymes.

The assay of these *glycosyl hydrolases* in vitro is complicated because, in aqueous systems, the glycolipids will aggregate to form micelles. As a result, these glycolipids are not readily degraded by the glycosyl hydrolases.

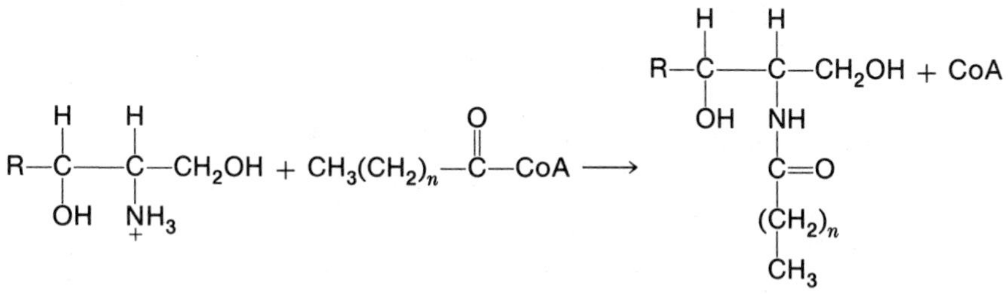

**Figure 14-35**
Biosynthesis of sphinganine.

**Figure 14-36**
Biosynthesis of ceramide.

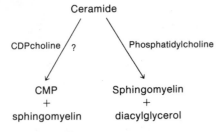

**Figure 14-37**
Two pathways for sphingomyelin biosynthesis. The pathway on the right is well established. The pathway on the left is still questionable.

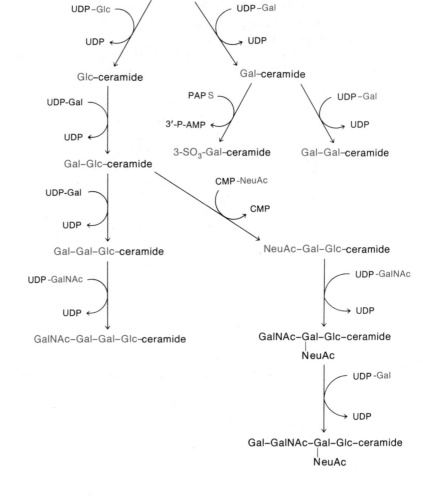

**Figure 14-38**
Outline of biosynthesis of some glycosphingolipids. (Glc = glucose, Gal = galactose, GalNAc = N-acetylgalactosamine, NeuAc = N-acetylneuraminic acid, PAPS = 3'-phosphoadenosine, 5'-phosphosulfate.)

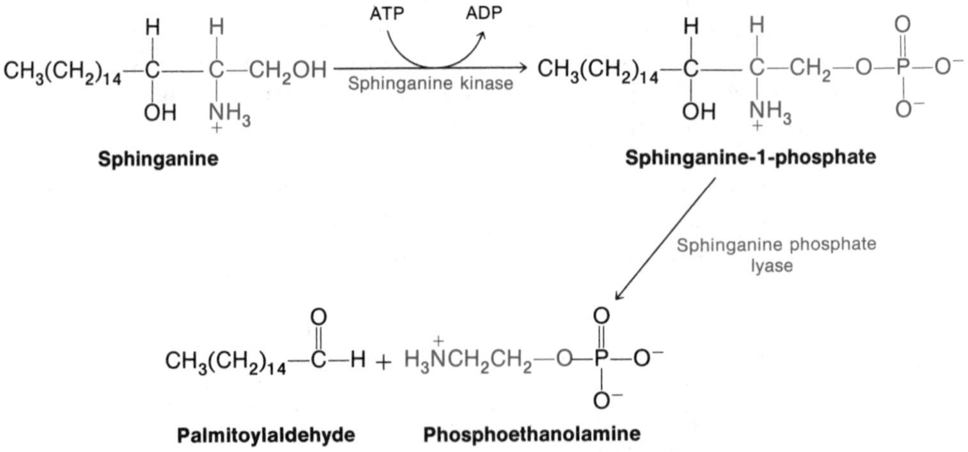

$CH_3(CH_2)_{14}$—

**Sphinganine**

ATP    ADP

Sphinganine kinase

$CH_3(CH_2)_{14}$—

**Sphinganine-1-phosphate**

Sphinganine phosphate lyase

$CH_3(CH_2)_{14}$—C—H  +  $H_3\overset{+}{N}CH_2CH_2$—O—P—O$^-$

**Palmitoylaldehyde**    **Phosphoethanolamine**

**Figure 14-39**
Catabolism of sphinganine.

This problem is alleviated by the addition of detergents (such as _bile acids_; see Chapter 15), which make the lipids more soluble in aqueous environments. Lysosomes do not contain bile acids; yet these organelles are the sites for degradation of the glycosphingolipids. In the past few years it has been discovered that in place of detergents, certain activator proteins occur in the soluble portion of the lysosomes. The lipid binds to one of these proteins, and as a result, degradation by the hydrolase is greatly facilitated. Separate activator proteins (with low $M_r \sim 25,000$) have been identified which aid the hydrolysis of gangliosides $GM_1$ and $GM_2$, glucosylceramide, and 3'-sulfate galactosylceramide.

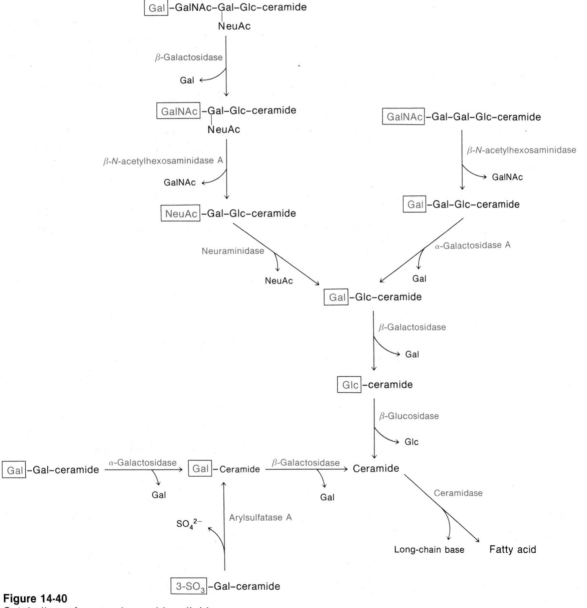

**Figure 14-40**
Catabolism of some glycosphingolipids.

The major diseases in which a defect in the catabolism of a sphingo-lipid has been detected are shown in Table 14-5. Many of these diseases have variations that are not included in the table. A complete discussion of these diseases is clearly beyond the scope of this book, and the student is referred to *The Metabolic Basis of Inherited Disease* (see Selected Readings). A brief summary of the biochemistry of two of these disorders follows.

*Fabry's disease* was first described in 1898 by J. Fabry in Germany and independently by W. Anderson in England. Some characteristic symptoms are skin rash, pain in the extremities, and renal impairment which leads to hypertension. Patients lead a reasonably normal life until approximately their fourth decade, when the kidneys usually fail. Little progress was made in understanding the cause of this disease until Charles Sweeley and Bernard Klionsky described, in 1963, the structure of *Gal-Gal-Glc-ceramide* as the major lipid that accumulates in the Fabry kidneys. Subsequently, in 1967, Roscoe Brady demonstrated that the enzymatic defect was a deficiency of the enzyme that degrades trihexosylceramide, and this was later shown by J. A. Kint to be *α-galactosidase A* (Figure 14-40). It is now possible to diagnose this *X-chromosome-linked* disease with biochemical tests based on these discoveries. For example, in prenatal diagnosis, cells are obtained from the amniotic fluid by *amniocentesis* and cultured; then the activity of α-galactosidase A is determined. Unfortunately, there is presently no effective treatment for this disease. Preliminary studies have suggested the feasibility of infusing purified α-galactosidase A into the Fabry patients (*enzyme replacement therapy*). Such studies have had limited success because of the short life time of the infused enzyme in the blood. Fortunately, Fabry's disease is rare, with only a few hundred cases reported throughout the world.

**Table 14-5**
Inherited Diseases of Sphingolipid Catabolism

| Disease | Enzyme Activity that Is Deficient | Reaction |
|---|---|---|
| **1.** Ceramidase deficiency: Farber's lipogranulomatosis | Ceramidase | Ceramide $\longrightarrow$ fatty acid + long-chain base |
| **2.** Sphingomyelin lipidosis: Niemann-Pick disease | Sphingomyelinase | Sphingomyelin $\longrightarrow$ ceramide + phosphocholine |
| **3.** Glucosylceramide lipidosis: Gaucher's disease | Glucocerebroside β-glucosidase | Glc-ceramide $\longrightarrow$ Glc + ceramide |
| **4.** Galactosylceramide lipidosis: globoid cell leukodystrophy | Galactocerebroside β-galactosidase | Gal-ceramide $\longrightarrow$ Gal + ceramide |
| **5.** Sulfatide lipidosis: metachromatic leukodystrophy | Arylsulfatase A | $3'\text{-}SO_3{}^-$-Gal-ceramide $\longrightarrow$ Gal-ceramide + $SO_4{}^{2-}$ |
| **6.** Fabry's disease | α-Galactosidase A | Gal-Gal-Glc-ceramide $\longrightarrow$ Gal + Gal-Glc-ceramide |
| **7.** $GM_1$ gangliosidosis | $GM_1$-β-galactosidase | $GM_1 \longrightarrow$ Gal + $GM_2$ |
| **8.** Tay-Sachs disease ($GM_2$ gangliosidosis) | Hexosaminidase A | $GM_2 \longrightarrow GM_3$ + GalNAc |
| **9.** Sandhoff's disease | Hexosaminidases A + B | $GM_2 \longrightarrow GM_3$ + GalNAc |

The disease described by Warren Tay in 1881 and now known as _Tay-Sachs disease_ is quantitatively more important. It is estimated that 30 to 50 children with Tay-Sachs disease are conceived each year in the United States, largely by Jewish parents. It is a devastating disease, because the children usually do not survive beyond the age of 3. As is clear from Table 14-5, in Tay-Sachs disease there is an accumulation of ganglioside $GM_2$ (see Table 14-4), especially in the brain, as a result of a deficiency of hexosaminidase A. There is no useful treatment for the disease. Moreover, enzyme replacement is not considered a likely remedy because infused enzyme would not penetrate the blood-brain barrier. The best hope for preventing Tay-Sachs disease is prenatal detection followed by therapeutic abortion.

## Specialized Function of Glycosphingolipids

It is clear that a major function of these lipids in some membranes is structural. Galactosylceramide represents 15 percent of the myelin membrane and globoside approximately 7 percent of the human red blood cell membrane. Moreover, in all instances examined, the glycolipids are oriented asymmetrically in the bilayer of the plasma membrane, facing the outside of the cell.

Many intriguing observations indicate that the glycosphingolipids are more than just structural lipids and have specific cellular functions that are presently not understood. A number of complex glycosphingolipids have been identified as blood-group-active substances. It has also been demonstrated that ganglioside $GM_1$ can act as a receptor for cholera toxin. (The binding of this toxin provides no obvious advantage to the cell.) In human blood cells, lymphocytes have neutral glycosphingolipids similar to those presented in Table 14-4. In contrast, neutrophils do not contain glycosphingolipids with _N_-acetylgalactosamine. Rather, these blood cells have glycosphingolipids with _N_-acetylglucosamine (for example, Gal$\beta$1-4GlcNAc$\beta$1-3Gal$\beta$1-4Glc$\beta$1-1ceramide). The reason for these differences in glycosphingolipid composition is not apparent.

Viral transformation (conversion of a normal cell to a cell with an indefinite ability to multiply, which will usually cause tumors) vastly alters the glycolipid composition of a cell. For example, transformation by SV40 virus of a number of different cell lines derived from mice caused a loss of all the major gangliosides in the cell with increases in the amount of simpler glycolipids. What role does this change in glycolipid composition play in the transformation of the cell? The answer is not known.

Finally, many workers believe that the relatively high concentrations of gangliosides in neuron cells suggest a function for these lipids in nerve transmission. Clearly, there is still much to be learned about the function of these complex lipids in various cells and tissues.

## PROSTAGLANDINS AND RELATED COMPOUNDS

Up to this point the focus has been on the metabolism of lipids as related to the energy and structural requirements of the cell. We are now going to switch gears and explore a class of lipids that are structurally very different and biologically active at minute concentrations. The prostaglandins have been known for decades, yet research on the physiological role of these and related compounds remains an area of tremendous activity in the 1980s.

## Historical Developments

Prostaglandins were first extracted from human semen in the 1930s by Ulf Von Euler in Sweden. He found that when components in this fluid were injected into animals, the uterus contracted and blood pressure was lowered. He believed the potent compounds originated from the _prostate_, hence the name _prostaglandin_. The prostaglandins in semen were later shown to be from the _seminal gland_. It is now known that prostaglandins are present in most tissues of both male and female animals.

Von Euler was able to establish that the compounds were hydroxy fatty acids. However, the techniques available at that time did not permit complete purification and structural analysis. With the advent of _gas chromatography_ and _mass spectrometry_, the first structures were determined in the early 1960s by Sune Bergstrom and Jon Sjövall, also from Sweden. At that time, two compounds were isolated. One was more soluble in ether and was referred to as _prostaglandin E_. The other was more soluble in phosphate buffer (_fosfat_ in Swedish) and was named _prostaglandin F_. There followed a virtual explosion of activity in prostaglandin research in the mid-1960s. Bengt Samuelsson joined with Bergstrom and his colleagues, and together they elucidated the structures of other prostaglandins. With the help of D. A. van Dorp in the Netherlands, the biosynthetic and metabolic pathways of these compounds were deduced.

Another discovery that should be noted was by John Vane in Britain. In 1971 he reported that _aspirin_ blocks the synthesis of prostaglandins. Aspirin had been used since the nineteenth century without anyone understanding how it worked. Many of its pharmacologic actions can now be explained by its inhibition of the synthesis of prostaglandins and related compounds. In the mid-1970s, the discovery of new compounds (_thromboxanes_, _prostacyclin_, and _leukotrienes_) continued to make prostaglandin research one of the most rapidly developing in biochemistry. Because this field is still moving so rapidly, the interested student will have to refer to the recent literature to find the latest developments.

## Structures

Prostaglandins are cyclopentanoic acids that are biosynthetically derived from $C_{20}$ polyunsaturated fatty acids. The compounds are named from a hypothetical compound called _prostanoic acid_ (Figure 14-41). The various classes of prostaglandins differ from each other in the structure of the substituted cyclopentane ring. There are at least nine different cyclopen-

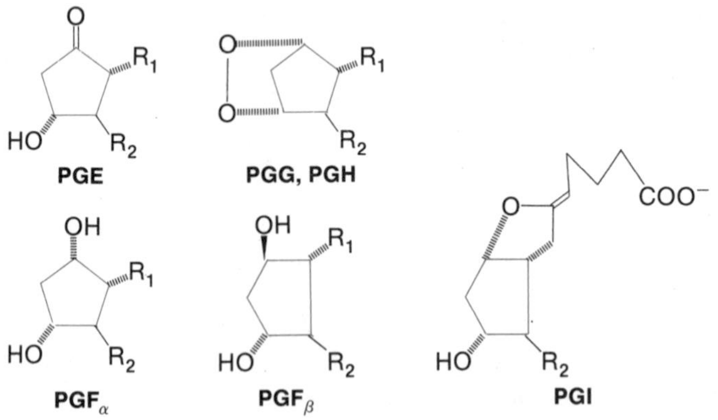

**Figure 14-41**
Prostanoic acid, a hypothetical compound.

**Figure 14-42**
Structures of cyclopentane rings in prostaglandins.

tane ring structures, five of which are shown in Figure 14-42. The names of the prostaglandins are abbreviated to PG, followed by a letter of the alphabet, e.g., E, F, G, H, or I. One rarely sees the systematic chemical names of these compounds in the literature.

The $R_1$ and $R_2$ groups attached to the cyclopentane ring differ in the various types of prostaglandins. All prostaglandins contain an L-hydroxyl substituent at C-15, except for PGG, which contains an L-peroxide (—O—OH) at C-15. The prostaglandins are named according to the number of carbon-carbon double bonds in groups $R_1$ and $R_2$, as shown in Table 14-6. The subscript numerals 1, 2, or 3 refer to the number of these double bonds in the molecule. Once the structures in Figures 14-41 and 14-42 and Table 14-6 are understood, you should be able to draw the structure of any of these compounds, e.g., PGE$_2$, as shown in Figure 14-43.

_Thromboxanes_ (TX) are compounds closely related to prostaglandins. They were first isolated from _thrombocytes_ (blood platelets), hence the name. They differ from prostaglandins in the ring structure, which is a _cyclic ether_ (oxane ring). The structures of the rings for TXA and TXB are shown in Figure 14-44. The naming of thromboxanes follows the same rules as for prostaglandins for localization of the double bonds in the $R_1$ and $R_2$ groups (Table 14-6). They also have the hydroxyl substitution at C-15. The structure of TXA$_2$ is shown in Figure 14-45.

There is a third class of related compounds, the _leukotrienes_, which are substituted derivatives of arachidonic acid. The structures of these compounds will be presented in the section entitled Biosynthesis.

## Analysis

Several major problems have made research on prostaglandins difficult. One is the instability of many of the compounds. For example, the half-life ($t_{1/2}$) for PGG$_2$ in H$_2$O at 37°C is 5 min. TXA$_2$ is more unstable ($t_{1/2}$ = 30–40 s) and decomposes to TXB$_2$.

A second problem is quantitative analysis. The normal in vivo concentrations are difficult to estimate, since the amounts of prostaglandins in a tissue vary according to the procedure used for preparation of the tissue. For example, less than 10 ng/g of tissue of PGE$_2$ could be detected in rat tissues homogenized in ethanol, whereas 200 to 300 ng/g of tissue of PGE$_2$ was obtained after homogenization in 0.9% saline at 0°C for 5 min. These changes in prostaglandins are due to the rapid synthesis of prostaglandins in recently excised tissue. A third difficulty results from the low concentrations of prostaglandins found in tissues. Accurate measurements can be made by gas chromatography with an electron capture detector, by gas

**Table 14-6**
Double Bonds in Prostaglandins

| Classification | Location of Double Bonds |
| --- | --- |
| PG$_1$ | _trans_-$\Delta^{13}$ |
| PG$_2$ | _trans_-$\Delta^{13}$, _cis_-$\Delta^5$ |
| PG$_3$ | _trans_-$\Delta^{13}$, _cis_-$\Delta^5$, _cis_-$\Delta^{17}$ |

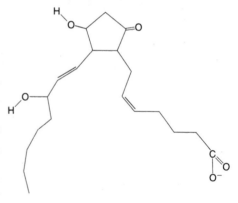

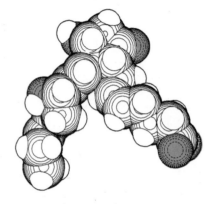

**Figure 14-43**
Structure of PGE$_2$.

**Thromboxane A (TXA)**   **Thromboxane B (TXB)**

**Figure 14-44**
Ring structure for the thromboxanes.

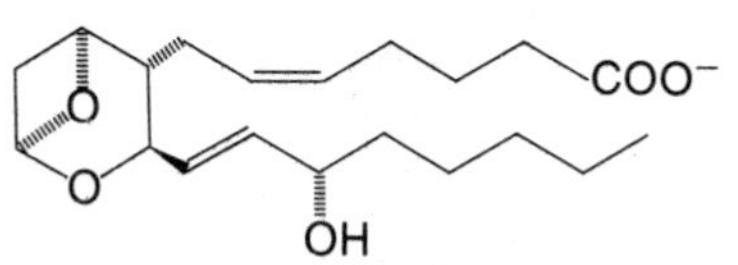

**Figure 14-45**
Structure of TXA$_2$.

chromatography–mass spectrometry, or by radioimmunoassay. With these procedures it has been possible to determine the concentration of $PGF_{2\alpha}$ in human serum to be 2 pg/ml (1 pg = $10^{-12}$ g).

### Biosynthesis

Prostaglandins and thromboxanes arise from $C_{20}$ polyunsaturated fatty acids and contain two fewer double bonds than the parent fatty acids (Figure 14-46). Because _arachidonic_ acid is the major $C_{20}$ polyunsaturated fatty acid in most mammals, it is not surprising that the $PG_2$ and $TX_2$ series are the predominant classes of these compounds. The fatty acid precursors are derived from phospholipids, possibly by the release of the fatty acid by a specific phospholipase $A_2$. Alternatively, the phospholipid may be degraded by a phosphatidylinositol-specific phospholipase C to yield diacylglycerol, which is subsequently catabolized by diacylglycerol lipase to yield arachidonic acid. At the present time, both possible routes should be considered.

The initial step for the synthesis of all prostaglandins and thromboxanes is the oxidation and cyclization of arachidonic acid to yield $PGG_2$ and then $PGH_2$ in reactions catalyzed by _prostaglandin endoperoxide synthase_, a microsomal enzyme (Figure 14-47). This enzyme, which has two catalytic activities, has been purified from bovine and sheep vesicular glands. The formation of $PGG_2$ is catalyzed by the _fatty acid cyclooxygenase_ component, and the subsequent formation of $PGH_2$ is catalyzed by a _peroxidase_ component. The synthase has a molecular weight of approximately 130,000 and contains 2 mol of hemin per mole of synthase. Tryptophan or one of several other aromatic compounds is required for the peroxidation step, but the nature of the electron donor in vivo is not known. The enzyme is inactivated during catalysis by an undefined mechanism. _Aspirin_ inhibits the formation of $PGG_2$ by acetylation of a serine residue in prostaglandin endoperoxide synthase.

$PGH_2$ is converted into the various prostaglandins and thromboxanes as shown in Figure 14-48. As indicated, each step is catalyzed by a separate enzyme. There are different amounts of these enzymes in various tissues. Thus blood platelets have _$TXA_2$ synthase (prostaglandin endoperoxide: thromboxane A isomerase)_, which makes $TXA_2$, whereas arterial walls have _$PGI_2$ synthase (prostaglandin endoperoxide I isomerase)_, which makes $PGI_2$. The tissue localization of the enzyme relates to the roles of the prostaglandins; for example, $TXA_2$ causes platelet aggregation, whereas $PGI_2$ inhibits this process.

In addition to the formation of prostaglandins and thromboxanes, arachidonic acid can be metabolized to several other compounds. Human platelets have the capacity to form 12-L-hydroxy-5,8,10,14-eicosatetraenoic acid (_12-HETE_) and 12-L-hydroxy-5,8,10-heptadecatrienoic acid (_12-HHT_), as shown in Figure 14-49. The synthesis of 12-HETE is not blocked by aspirin. The metabolic function of these compounds is presently unknown.

For many years, scientists have struggled to understand the biochemical processes involved in asthma. Possibly involved in this disease is a compound called _slow-reacting substance_, an extremely potent muscle contractant that can severely constrict the small airways of the lung. Slow-reacting substance has now been identified as _leukotriene C_ (Figure 14-50), another metabolite of arachidonic acid. The leukotrienes have been

$$C_{20:3}^{\Delta8,11,14} \longrightarrow PG_1 \text{ or } TX_1$$

$$C_{20:4}^{\Delta5,8,11,14} \longrightarrow PG_2 \text{ or } TX_2$$

$$C_{20:5}^{\Delta5,8,11,14,17} \longrightarrow PG_3 \text{ or } TX_3$$

**Figure 14-46**
Fatty acid precursors of prostaglandins.

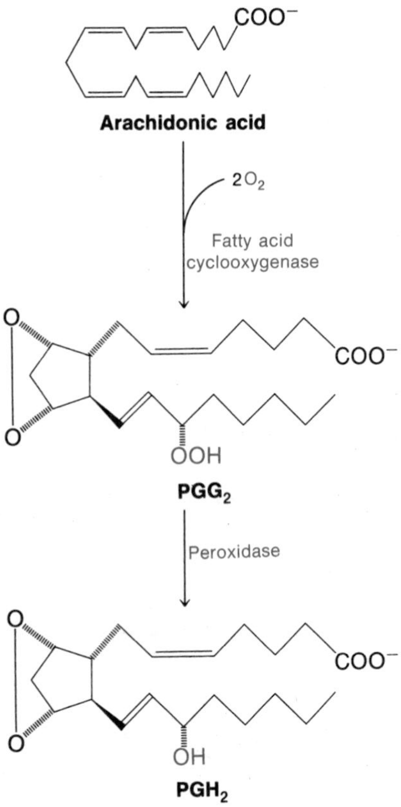

**Figure 14-47**
Reaction catalyzed by prostaglandin endoperoxide synthase.

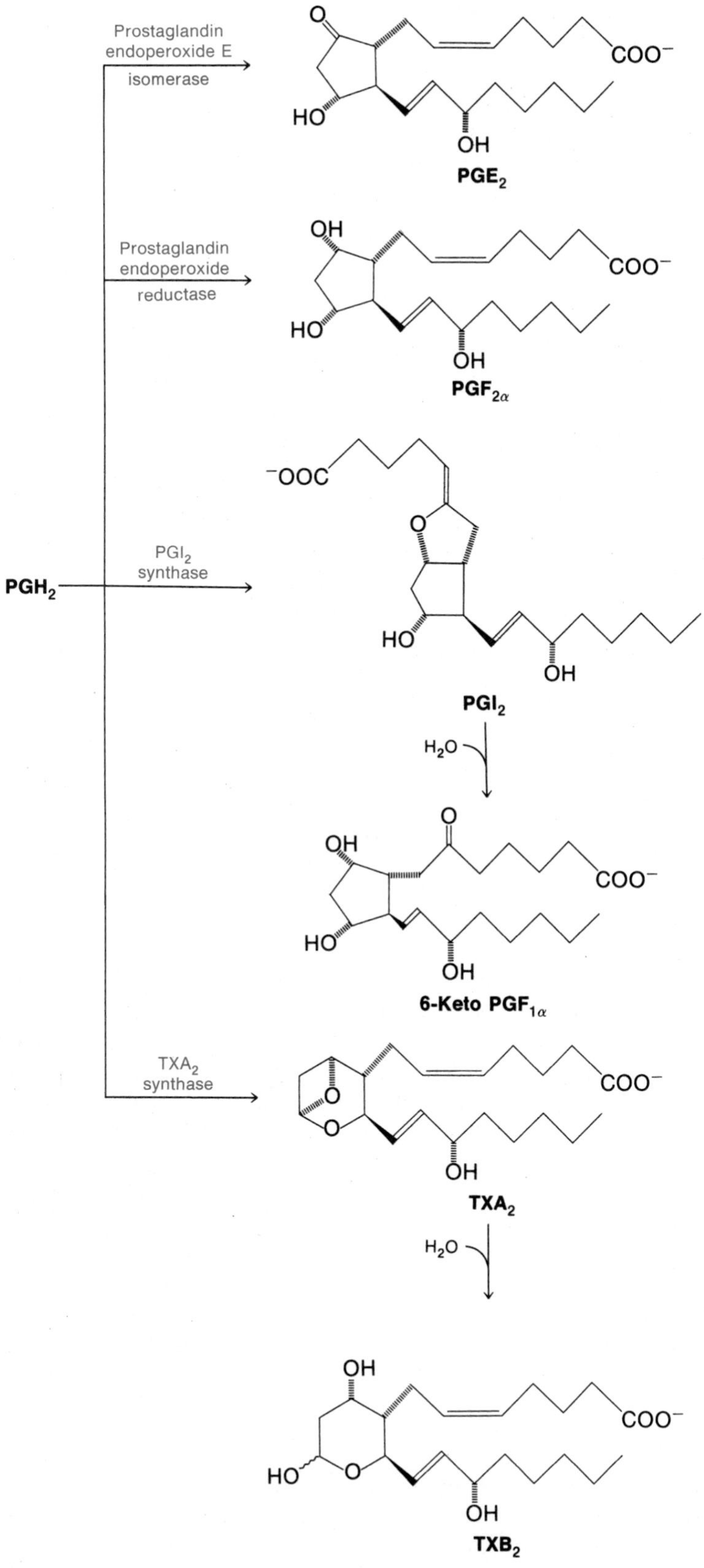

**Figure 14-48**
Formation of prostaglandins and
thromboxanes.

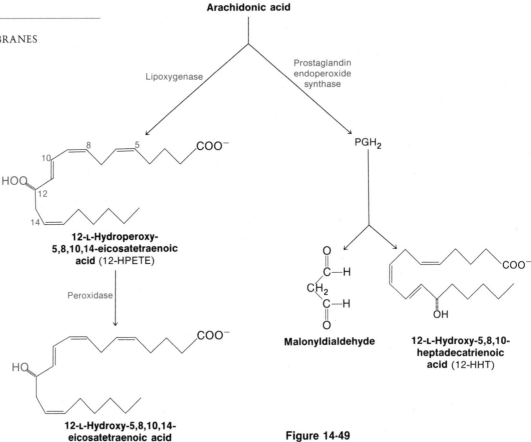

Figure 14-49
Additional metabolites of arachidonic acid.

found in polymorphonuclear leukocytes and mast cell tumor cells. The structures and biosynthesis of the leukotrienes are shown in Figure 14-50. In the initial reaction, catalyzed by a *lipoxygenase*, molecular oxygen is added to C-5 of arachidonic acid with an isomerization of the double bond to the $\Delta^6$ position. A subsequent dehydration results in the formation of the 5,6-epoxide, with isomerization of the adjacent double bonds to give *trans*-$\Delta^7$ and *trans*-$\Delta^9$ double bonds in conjugation with the *cis*-$\Delta^{11}$ double bond (leukotriene A). *Leukotriene C* is formed by addition of glutathione to the C-6 position of the epoxide and formation of a hydroxyl at C-5. Transfer of glutamic acid from leukotriene C catalyzed by $\gamma$-glutamyltranspeptidase results in the formation of *leukotriene D*. A dipeptidase present in many tissues can convert leukotriene D to leukotriene E.

### Catabolism

The prostaglandins have short half-lives in vivo and are rapidly metabolized to physiologically inactive compounds. The PGE and PGF compounds do not normally survive a single pass through the circulatory system. They are rapidly catabolized to 15-keto-13,14-dihydro derivatives by lung (Figure

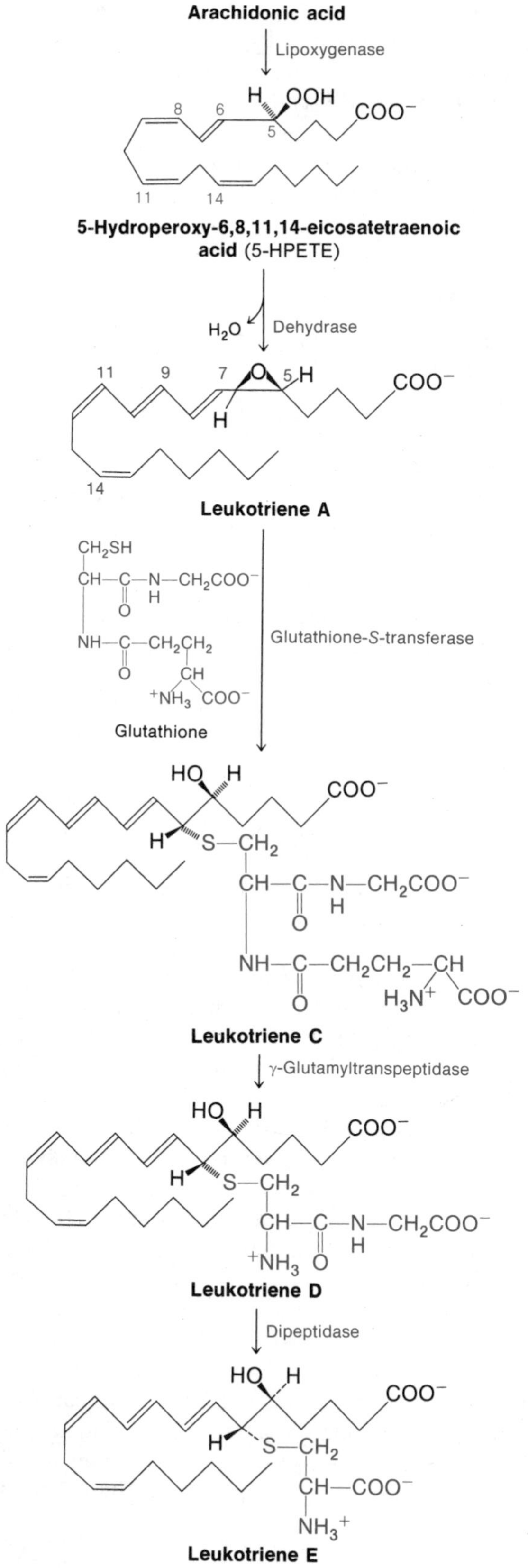

**Figure 14-50**
Biosynthesis of the leukotrienes.

14-51). The number of final metabolites is huge. For example, in the human female, at least 15 metabolites of $PGF_{2\alpha}$ have been identified. It would serve no useful purpose to elaborate on all these metabolic transformations. However, as one example, the pathway for the metabolism of $PGF_{2\alpha}$ to $5\alpha,7\alpha$-dihydroxy-11-ketotetranor prostane-1,16-dioic acid is shown in Figure 14-51. The latter compound is the major human urinary metabolite of $PGF_{2\alpha}$ that has been identified.

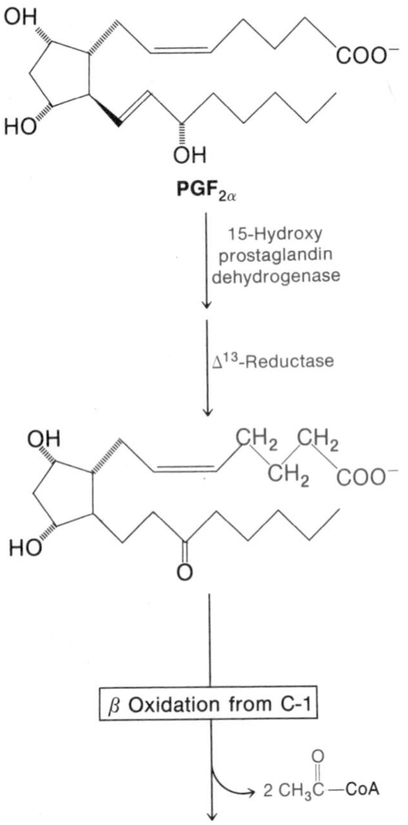

## Prostaglandin Function

Prostaglandins and related compounds have a wide range of biological effects, some of which have already been mentioned. _They are generally considered to be hormones that exert their effect locally without altering functions throughout the body._ Unfortunately, the mechanism by which these compounds exert their effects is not clear. Hence a comprehensive scheme that depicts the mode of action of these compounds is not yet available. However, a few general points can be made.

First, it appears that many of the effects of these compounds are mediated by _cAMP_. Specific binding of PGE to a variety of cells correlates with an activation of adenylate cyclase and the accumulation of cAMP. In contrast, PGF has no such effect. Moreover, different cell types respond differently to PGE. In human adipocytes, $PGE_1$ causes a 15-fold increase in cAMP. Fibroblasts isolated from human adipose showed a 95-fold increase in cAMP after exposure to $PGE_1$. In platelets, $PGI_2$ rather than $PGE_2$ appears to mediate the increase in cAMP.

$PGE_1$ relaxes, whereas $PGF_{2\alpha}$ contracts venous smooth muscle. $PGF_{2\alpha}$ has been shown to increase the levels of _cGMP_ in several tissues and cells. Whether or not the antagonistic effects of PGE and PGF are mediated by cAMP and cGMP, respectively, remains to be established.

Although the mechanisms by which prostaglandins exert their effects are not fully established, there is evidence for specific, high-affinity receptors for these compounds. $PGF_{2\alpha}$ appears to inhibit progesterone secretion and regression of the corpus luteum. Analysis of bovine corpus luteum has shown that a specific receptor for $PGF_{2\alpha}$ is localized in the plasma membrane. However, the biochemical events subsequent to the binding of $PGF_{2\alpha}$ to the receptor remain to be delineated.

The possible function of prostaglandins in blood clotting is currently generating much research and speculation. $PGI_2$ lowers blood pressure, relaxes coronary arteries, and inhibits platelet aggregation. $TXA_2$ has the opposite effects. $PGI_2$ is made in the endothelial lining of blood vessel walls. It appears to inhibit platelet aggregation by binding to a receptor on the plasma membrane, and this subsequently results in an increase of cAMP in platelets. $TXA_2$ appears to suppress the increase in cAMP caused by $PGI_2$. It has been speculated that the synthesis of $PGI_2$ may prevent platelets from binding to arterial walls. In damaged areas of arteries, the synthesis of $PGI_2$ may be decreased and the presence of $TXA_2$ would cause the platelets to aggregate, leading to the formation of a thrombus (blood clot). The physiological relevance of these observations has been questioned, and further studies are needed to provide a clear picture of the function of these two compounds in blood clotting.

**Figure 14-51**
Major metabolic pathway of $PGF_{2\alpha}$ in humans.

## SELECTED READINGS

Bell, R. M., and Coleman, R. A. Enzymes of glycerolipid synthesis in eucaryotes. *Ann. Rev. Biochem.* 49:459, 1980. A recent review on the synthesis and assembly of eukaryotic glycerolipids.

Martonosi, A. *The Enzymes of Biological Membranes,* Vol. 2: *Biosynthesis of Cell Components.* New York: Plenum, 1976. This book contains review chapters on the enzymes of phospholipid metabolism and ether-lipid metabolism.

Raetz, C. R. H. Enzymology, genetics and regulation of membrane phospholipid synthesis in *Escherichia coli. Microbiol. Rev.* 42:614, 1978. A thorough discussion of phospholipid metabolism in *E. coli.*

Rossiter, R. J. Metabolism of Phosphatides. In *Metabolic Pathways,* 3d Ed., Vol. 2. New York: Academic Press, 1968. A review of phospholipid metabolism as understood in the 1960s.

Samuelsson, B., Goldyne, M., Granstrom, E., Hamberg, M., Hammarstrom, S., and Malmsten, C. Prostaglandins and thromboxanes. *Ann. Rev. Biochem.* 47:997, 1978. A review on the developments in this field during the mid-1970s.

Stanbury, J. B., Wyngaarden, J. B., and Fredrickson, D. S. *The Metabolic Basis of Inherited Disease.* New York: McGraw-Hill, 1978. This book contains 15 chapters on disorders characterized by abnormal lipid metabolism.

## PROBLEMS

1. Draw the structure of the major class of phospholipids found in *E. coli.*

2. If (1-$^{14}$C)sphingosine were injected into a rat, would you expect to find any radioactivity in phosphatidylcholine?

3. Why would an enzymatic defect in the biosynthesis of CDPethanolamine most likely be lethal to a liver cell? Would such a defect be lethal to *E. coli*?

4. Arachidonic acid is the major precursor of prostaglandins and thromboxanes. If a person were unable to absorb arachidonic acid from the diet, but could absorb linoleic acid, would that person still be able to make $PGE_2$?

5. After the discovery of $TXA_2$, it was postulated that regular treatment of people with aspirin might prevent platelet aggregation and thus help prevent thromboses from forming. Subsequently, $PGI_2$ and its biosynthetic pathway was discovered. In light of this new knowledge, would continuation of aspirin treatment for $TXA_2$-mediated thrombosis still make sense?

6. A person was admitted to the emergency room of a hospital with extremely low levels of fatty acids in plasma. The attending physician was going to attempt to increase fatty acid levels by injecting the person with either $PGE_1$ or $PGF_1$. Which prostaglandin would he choose and how might it help?

7. If you discovered a disease in which a lipid with the structure gal($\beta$1-4)glc($\beta$1-1) ceramide accumulated abnormally because of an enzymatic defect, what enzymatic reaction would you expect to be defective?

8. A graduate student in biochemistry was asked to devise an experiment to prove that the synthesis of phosphatidylcholine in vitro occurs on the outer surface of microsomes. Can you help?

9. Until recently the biosynthesis of sphingomyelin was thought to occur by the transfer of phosphocholine from CDPcholine to ceramide. Phosphatidylcholine is now thought to be the immediate donor of the phosphocholine moiety. Suggest an experiment that would help prove that phosphatidylcholine is the immediate precursor of sphingomyelin.

10. An investigator was presented with a liver biopsy from a patient with Gaucher's disease. Much to his surprise when he assayed the glucocerebroside $\beta$-glucosidase, he found normal enzyme activity. As always, his assay included a bile acid to aid in the solubilization of the substrate, glucocerebroside. Suggest an explanation for the results observed.

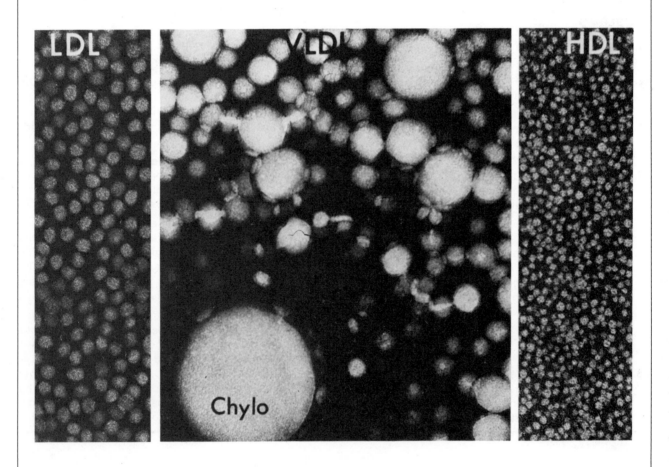

# 15

# METABOLISM OF STEROIDS AND LIPOPROTEINS

_Cholesterol_ is the most prominent member of the steroid family and is infamous among the public because of the relationship between an elevated level of serum cholesterol and an increased risk of cardiovascular disease. Approximately 1100 mg of sterol is excreted each day from a normal human adult. In a typical American diet, this is replaced by 250 mg in the diet and 850 mg by means of biosynthesis. Cholesterol is also the precursor for other important steroids in animals: _bile acids_ and _steroid hormones_. In addition, cholesterol is an important structural component of some eukaryotic membranes, but is generally absent from bacterial membranes. Hence the major theme of this chapter will be the biosynthesis, transport, and catabolism of cholesterol in animals. The last section of this chapter will discuss briefly the structure and metabolism of _terpenes_, organic-soluble vitamins, and _acetogenins_.

## CHOLESTEROL AND RELATED STEROLS

### Structure of Cholesterol

The name "cholesterol" is derived from the Greek words _chole_, for bile, and _stereos_, for solid. Highly purified cholesterol is a white powder at room temperature. Cholesterol is a member of the class of lipids called _steroids_, which are derivatives of the tetracyclic hydrocarbon _perhydrocyclopentanophenanthrene_ (Figure 15-1). The four rings are identified by the first four letters of the alphabet and the carbons are numbered in the sequence shown in Figure 15-2. In addition to the basic ring structure, cholesterol contains a hydroxyl group at C-3, an aliphatic chain at C-17, methyl groups at C-10 and C-13, and a $\Delta^5$ double bond.

**Phenanthrene**

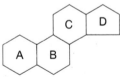

**Perhydrocyclopentanophenanthrene**

**Figure 15-1**
Structures of phenanthrene and perhydrocyclopentanophenanthrene.

_Electron micrographs of low-density lipoproteins (LDL), very-low-density lipoproteins (VLDL), high-density lipoproteins (HDL), and chylomicrons (Chylo). These particles are the main vehicle for transport of lipids in the plasma. (Photograph provided by Dr. Robert Hamilton, University of California, San Francisco.)_

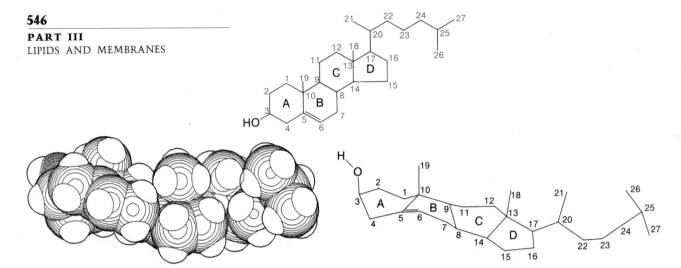

**Figure 15-2**
Structure of cholesterol.

The cyclohexane rings of the steroids can adopt either the chair or the boat conformation. The chair conformation is more stable and is the preferred conformation of steroids. The conformations of *cholestanol* and *coprostanol*, the two saturated derivatives of cholesterol, are shown in Figure 15-3. The A and B rings can be joined in a trans configuration, as in cholestanol, or in a cis configuration, as in coprostanol. It is obvious that the spatial orientation of the A and B rings of these two stereoisomers is very different.

Although it is important that we recognize the three-dimensional structures of steroids, such structures are too cumbersome for most uses in biochemistry. Hence a configurational convention for steroids has been

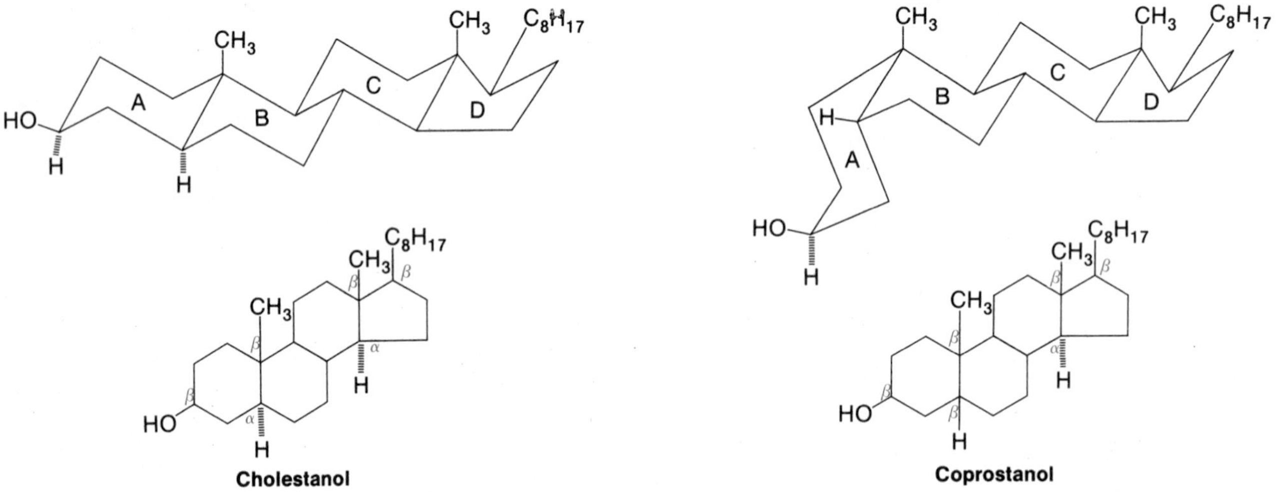

**Cholestanol**

**Coprostanol**

**Figure 15-3**
Conformational and conventional structures of cholestanol and coprostanol.

adopted in which structural formulas are more easily drawn and recognized. The substituents of the steroid rings are related to the $CH_3$ group at position 10, which by definition projects above the plane of the rings. This methyl, which is said to be a $\beta$ substituent, is indicated in structural formulas by a solid line (—). Similarly, other groups that are above the plane of the rings are referred to as $\beta$. Those substituents below the plane of the rings are called $\alpha$ and are indicated in structural formulas by a dashed line (---). Examples of $\alpha$ and $\beta$ substituents are shown in Figure 15-3. In *cholestanol*, the hydroxyl at C-3, the methyl groups at C-10 and C-13, and the aliphatic side chain at C-17 are all $\beta$-oriented substituents.

## Biosynthesis of Cholesterol

**Historical Developments.** Early in the 1930s, the structure of cholesterol was determined, thus bringing to an end a brilliant chapter in structural chemistry. However, the solution of one problem often leads to the formulation of a new one. The biosynthesis of cholesterol presented an exceptionally difficult question, because it was not clear how such a complex structure could be assembled from small molecules. Work on the biosynthesis of cholesterol began in earnest after Rudolf Schoenheimer and David Rittenberg, at Columbia University, developed isotopic tracer techniques for the analysis of biochemical pathways.

In 1941, Rittenberg and Konrad Bloch were able to show that deuterium-labeled acetate ($C^2H_3COO^-$) was a major precursor of cholesterol in rats and mice. Subsequently, in collaboration with Edward Tatum and others, Bloch proved that the sterol of *Neurospora crassa* (ergosterol) was entirely derived from acetate. In 1949, James Bonner and Barbarin Arreguin postulated that three acetates could combine to form a single five-carbon unit called *isoprene*:

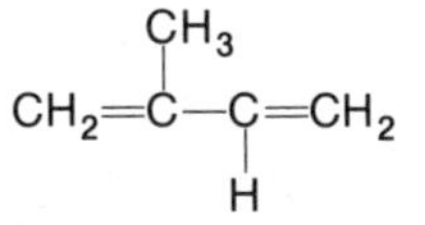

$$CH_2{=}\underset{}{C}{-}\underset{H}{\overset{CH_3}{C}}{=}CH_2$$

This proposal agreed with an earlier prediction of Sir Robert Robinson that cholesterol was a cyclization product of squalene, a 30-carbon polymer of isoprene units. Thus Bloch postulated a scheme for the biosynthesis of cholesterol (Figure 15-4). In 1952, Bloch and Robert Langdon demonstrated that squalene could readily be converted into cholesterol.

Two major problems remained. What was the structure of the isoprenoid intermediate, and how did squalene cyclize to form cholesterol? In 1953, R. B. Woodward and Bloch postulated a cyclization scheme for squalene (Figure 15-5) that was later shown to be correct. In 1956, the unknown isoprenoid precursor was identified as *mevalonic acid* by Karl Folkers and others at Merck, Sharpe and Dohme Laboratories. Interestingly, mevalonic acid (mevalonate) also was found to be a necessary growth factor for certain strains of *Lactobacillus*, bacteria that do not synthesize sterols. The discovery of mevalonate provided the missing link in the basic outline of cholesterol biosynthesis. Since that time, the sequence and the stereochemical course for the biosynthesis of cholesterol have been defined.

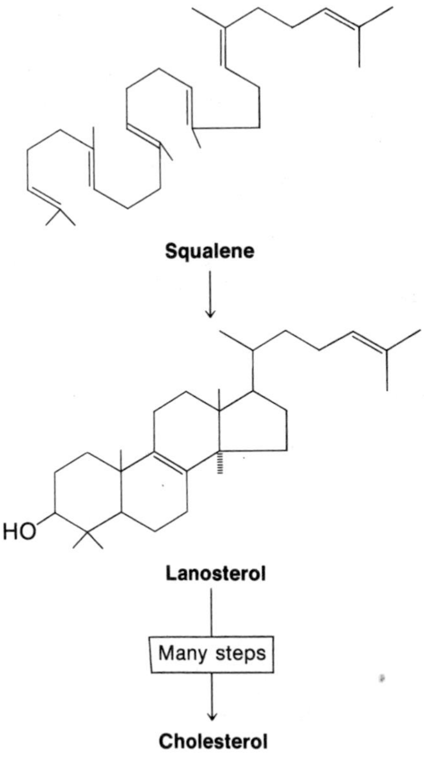

Acetate

↓

Isoprenoid intermediate (5 carbons)

↓

Squalene (30 carbons)

↓

Cyclization product (30 carbons)

↓

Cholesterol (27 carbons)

**Figure 15-4**
The basic outline for cholesterol biosynthesis (in 1952).

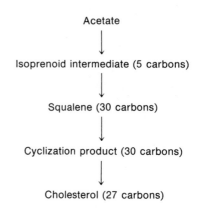

**Squalene**

↓

**Lanosterol**

| Many steps |

↓

**Cholesterol**

**Figure 15-5**
Cyclization of squalene as proposed by Woodward and Bloch in 1953.

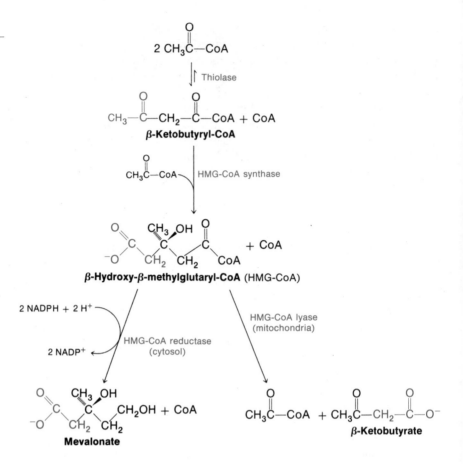

**Figure 15-6**
Formation of mevalonate.

**Formation of Mevalonate, a Key Intermediate.** The outline for the biosynthesis of mevalonate is shown in Figure 15-6. The sequence begins with a condensation in the cytosol of two molecules of acetyl-CoA catalyzed by *thiolase*. In the next step, *β-hydroxy-β-methylglutaryl-CoA (HMG-CoA) synthase* catalyzes the condensation of a third acetyl-CoA with β-ketobutyryl-CoA to yield *HMG-CoA*. This product is reduced to mevalonate by *HMG-CoA reductase*. The activity of this reductase is primarily responsible for control of the rate of cholesterol biosynthesis.

*HMG-CoA is an important intermediate for the biosynthesis of both cholesterol and ketone bodies* (see Chapter 13). The biosynthesis of cholesterol is catalyzed by enzymes in the cytosol and enzymes bound to the endoplasmic reticulum, whereas ketogenesis is restricted to the mitochondrial matrix. Thus thiolase and HMG-CoA synthase are found in both mitochondria and cytosol of rat liver. In contrast, *HMG-CoA lyase*, which cleaves HMG-CoA to ketone bodies (see Chapter 13), is located only in mitochondria, whereas HMG-CoA reductase is bound to the endoplasmic reticulum.

HMG-CoA synthase, purified from chicken liver mitochondria, has a molecular weight of 105,000 and is composed of two subunits. Four isozymes of HMG-CoA synthase have been identified in chicken liver cytosol, all of which are different from the mitochondrial enzyme. The molecular weights of the cytosolic enzymes are approximately 100,000, and each enzyme is composed of two subunits. Unlike avian liver, rat liver has only a single cytosolic species of HMG-CoA synthase.

**HMG-CoA Reductase Activity and the Regulation of Cholesterol Biosynthesis.** The thiolase and HMG-CoA synthase exhibit some regulatory properties in rat liver (cholesterol feeding causes a decrease in these enzyme activities in cytosol, but not in mitochondria). However, *major regulation of cholesterol biosynthesis appears to be centered on the HMG-CoA reductase reaction*. Dietary cholesterol causes a reduction in the synthesis of cholesterol in rat liver, and this coincides with a decrease in HMG-CoA reductase activity. Precisely how this effect is mediated is not presently known.

HMG-CoA reductase from rat liver has a molecular weight of 104,000 and is composed of two subunits ($M_r = 52,000$). The activity of the enzyme in rat liver displays a *circadian (diurnal) rhythm*. During a 24-h period, the activity fluctuates by fourfold to fivefold, with a peak of activity around midnight. The rhythm appears to be independent of food intake, and the diurnal variation continues when the animal is left in complete darkness or light. Although the reason for the change in activity during this cycle is not known, current evidence suggests that it is due to a fluctuation in the amount of active species of enzyme.

In the late 1970s, a phosphorylation-dephosphorylation scheme for the activation and inactivation of the reductase was discovered, as outlined in Figure 15-7. *HMG-CoA reductase kinase kinase* catalyzes the phosphorylation and activation of *HMG-CoA reductase kinase*. This kinase phosphorylates HMG-CoA reductase, which results in an inactivation of the reductase. Both kinases are cAMP-independent. These effects are reversed by the action of *HMG-CoA reductase kinase phosphatase* and *HMG-CoA reductase phosphatase*. It is thought that these kinases and phosphatases mediate short-term changes in the activity of HMG-CoA reductase.

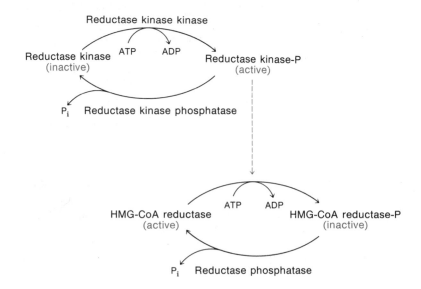

**Figure 15-7**
Scheme for modulation of the activity of HMG-CoA reductase.

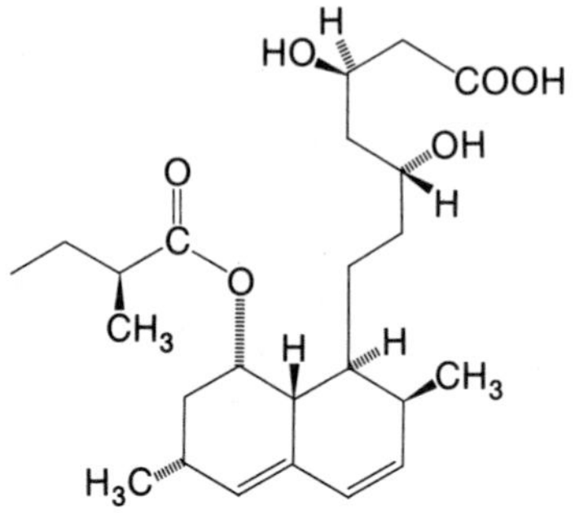

Very elegant control of cholesterol biosynthesis by plasma low-density lipoproteins has been demonstrated in many nonhepatic tissues. This will be discussed in the context of lipoprotein metabolism later in this chapter.

It has long been recognized that a correlation exists between a high level of serum cholesterol and cardiovascular disease. Most serum cholesterol is made in the liver; thus a drug that would specifically reduce cholesterol biosynthesis has been sought. It would be logical for such a drug to inactivate HMG-CoA reductase, since this enzyme catalyzes the key regulatory step in this pathway. A number of fungal metabolites have been isolated that are competitive inhibitors of HMG-CoA reductase. One of the most active compounds is *mevinolinic acid* (Figure 15-8), which competes favorably ($K_i = 0.6$ n$M$) with HMG-CoA for the reductase. Small doses of this drug (8 mg/kg of body weight) will lower the levels of plasma cholesterol in dogs by 30 percent. Although there is still much testing to be done, this or a similar drug may markedly reduce the incidence of cardiovascular disease in the 1980s.

**Figure 15-8**
Structure of mevinolinic acid.

**Conversion of Mevalonate to Lanosterol.** *Mevalonate is converted to squalene, which is cyclized to form lanosterol.* The first step in this sequence of reactions is the synthesis of the five-carbon isoprenoid intermediates *isopentenyl pyrophosphate* and *dimethylallyl pyrophosphate* from mevalonate, as shown in Figure 15-9. The action of *mevalonate kinase* and *phosphomevalonate kinase* produces *5-pyrophosphomevalonate*. The third enzyme (Figure 15-9), *pyrophosphomevalonate decarboxylase*, catalyzes a decarboxylation and elimination of the 3-hydroxyl group. The decarboxylase probably acts by an initial phosphorylation of the 3-hydroxyl group with ATP followed by the trans elimination of the carboxyl and phosphate to give *isopentenyl pyrophosphate*. This intermediate can be enzymatically isomerized to *3,3-dimethylallyl pyrophosphate*. These two

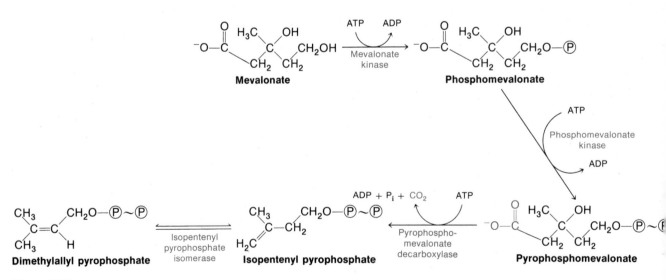

**Figure 15-9**
Conversion of mevalonate to isopentenyl pyrophosphate and dimethylallyl pyrophosphate.

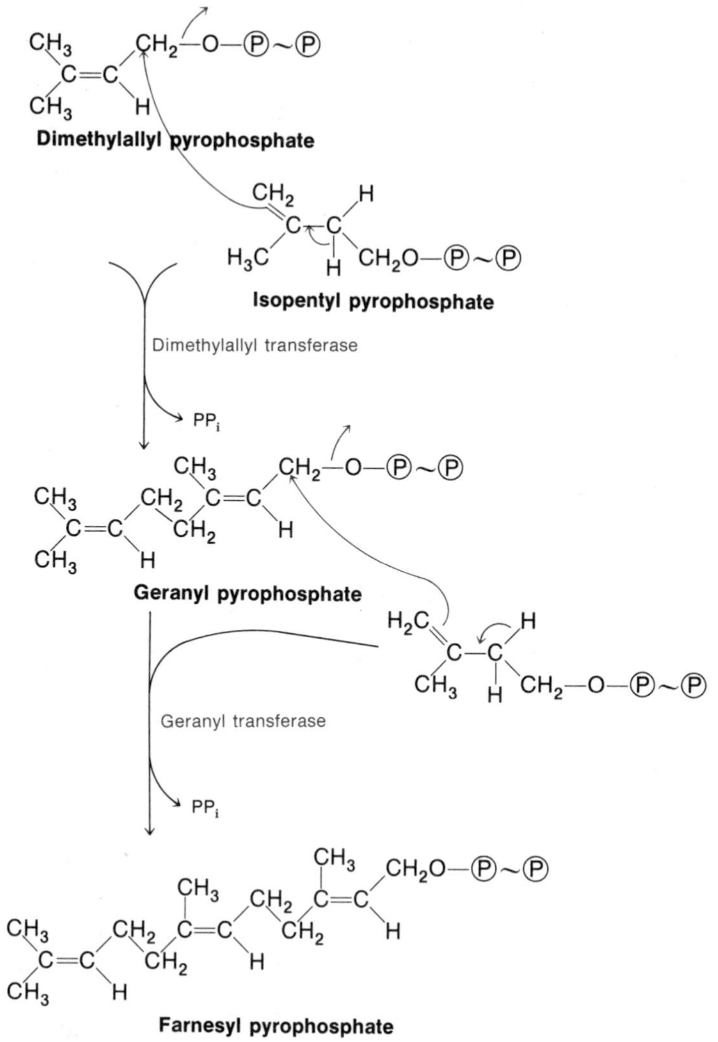

**Figure 15-10**
Biosynthesis of farnesyl
pyrophosphate.

$C_5$ isoprenoid pyrophosphates react as shown in Figure 15-10 to produce the $C_{10}$ intermediate *geranyl pyrophosphate*. Subsequently, a $C_{15}$ intermediate, *farnesyl pyrophosphate*, is formed by the reaction of another molecule of isopentenyl pyrophosphate with geranyl pyrophosphate. Two molecules of farnesyl pyrophosphate react to form *presqualene pyrophosphate*, which rearranges with the elimination of $PP_i$ to yield the $C_{30}$ intermediate *squalene* (Figure 15-11).

The enzymes that convert mevalonate to farnesyl pyrophosphate are probably cytosolic, whereas *farnesyl transferase* (squalene synthase) is tightly associated with the microsomal membrane. Even though such membrane-bound enzymes are difficult to solubilize, this enzyme has been isolated from yeast. In a polymeric form and in the presence of NADPH, farnesyl transferase will convert two farnesyl pyrophosphates to squalene. When dissociated into its protomeric form, the farnesyl transferase catalyzes only the formation of presqualene pyrophosphate.

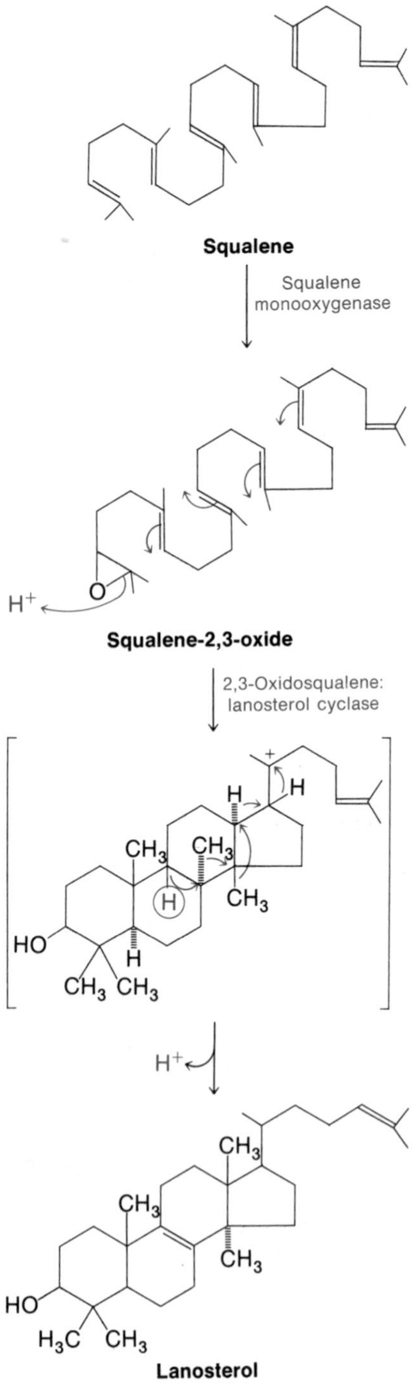

**Squalene**

Squalene
monooxygenase

**Squalene-2,3-oxide**

2,3-Oxidosqualene:
lanosterol cyclase

H⁺

**Lanosterol**

**Figure 15-12**
The transformation of squalene
into lanosterol.

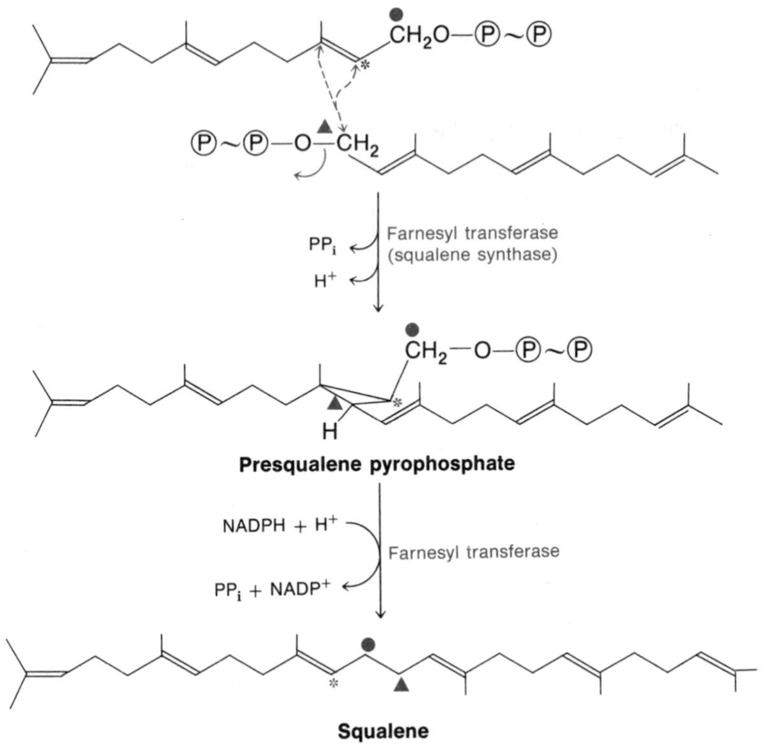

**Figure 15-11**
Formation of squalene from farnesyl pyrophosphate.

The two remaining reactions in the biosynthesis of lanosterol are shown in Figure 15-12. In the first of these reactions, *squalene-2,3-oxide* is formed from squalene by *squalene monooxygenase*, a microsomal enzyme, in a reaction that requires $O_2$, NADPH, FAD, phospholipid, and a cytosolic protein. As can be seen in Figure 15-11, squalene is a symmetrical molecule; hence the formation of squalene oxide can be initiated from either end of the molecule. The oxide is converted into lanosterol by a microsomal enzyme, *2,3-oxidosqualene lanosterol cyclase*. The reaction can be formulated as proceeding by means of a protonated intermediate, as shown in Figure 15-12, that undergoes a series of remarkable *trans-1,2* shifts of methyl groups and hydride ions to produce lanosterol.

**Conversion of Lanosterol to Cholesterol.** *The last sequence of reactions in the biosynthesis of cholesterol involves approximately 20 enzymatic steps starting with lanosterol.* In mammals (e.g., the rat), the major route is shown in Figure 15-13 and involves a series of double-bond reductions and demethylations. The exact position in the scheme for the reduction of the $\Delta^{24}$ double bond is not established. Otherwise, the sequence of reactions involves the oxidation and removal of the 14α-methyl group followed by the oxidation and removal of the two methyl groups at position 4 in the sterol. The final reaction is a reduction of the $\Delta^7$ double bond in *7-dehydrocholesterol*. An alternative metabolic route for lanosterol (Figure 15-13) initially involves three demethylations to give _zymosterol_ and then isomerization of the $\Delta^8$ double bond to the $\Delta^5$ position to produce _desmosterol_. The final reaction in this pathway is the reduction of the $\Delta^{24}$ double bond. The enzymes involved in the transformation of lanosterol to cholesterol are all located on the microsomes, and none of them has been purified. In addition to these enzymes, two cytosolic proteins have been found that stimulate several of the microsomal reactions that convert squalene to cholesterol. How these soluble proteins actually function in cholesterol biosynthesis is a problem of current interest.

**Stereochemistry of the Conversion of Mevalonate to Squalene.** As we have seen, the biosynthesis of cholesterol from acetyl-CoA is accomplished by means of a complicated route that involves more than 30 different enzymes, numerous cofactors, and at least two cytosolic proteins. Although this certainly represents enough complexity for most people, George Popjak and John Cornforth, in Britain, recognized that it was possible to define the conversion of mevalonate to squalene in a more precise and elegant man-

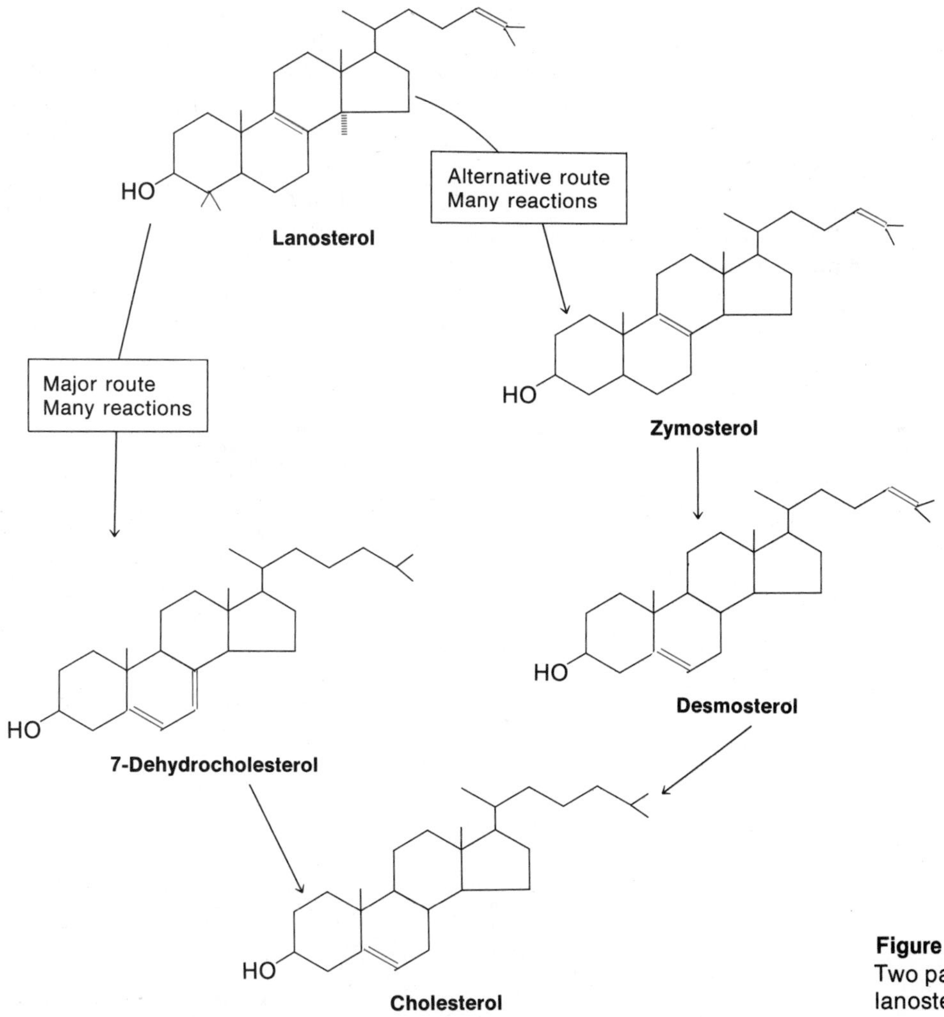

**Figure 15-13**
Two pathways for the conversion of lanosterol to cholesterol.

ner. These two scientists observed in the 1960s that there were 14 "stereo-chemical ambiguities" in the conversion of pyrophosphomevalonate to squalene. In other words, there were $2^{14}$, or 16,384, theoretically possible stereochemical pathways by which mevalonate could be transformed into squalene. This in itself was a remarkable observation. However, these two men and their collaborators subsequently were able to define precisely which one of these 16,384 possible stereochemical pathways actually occurred.

It is beyond the scope of this introductory text for us to examine each of the 14 stereochemical ambiguities and their resolution. Two examples should demonstrate the principles involved. The reaction catalyzed by _pyrophosphomevalonate decarboxylase_ could involve either a cis or trans elimination of the carboxyl and hydroxyl groups to produce isopentenyl-pyrophosphate (Figures 15-9 and 15-14). Cornforth et al. solved this stereochemical ambiguity by the synthesis of a stereospecifically deuterium-labeled pyrophosphomevalonate (Figure 15-14) that was incubated with the decarboxylase. The product of the reaction was isolated, and after several chemical transformations, Cornforth and co-workers were able to distinguish which of the two possible isomers of isopentenylpyrophosphate was formed. The product was solely the result of a trans elimination (Figure 15-14). Thus the first stereochemical ambiguity was resolved.

Another stereochemical problem was to determine which of the two hydrogens from C-2 of isopentenyl pyrophosphate was lost in its isomerization to dimethylallyl pyrophosphate (Figure 15-15). Isopentenyl pyrophosphate was chemically labeled with deuterium on the 2-R or 2-S position (the use of the RS terminology is explained in Figure 6-8). Incubation of this substrate with the isomerase and subsequent characterization of the product demonstrated that the proR hydrogen ($H_R$) was specifically removed during the isomerization reaction. Further information on the 14 stereochemical ambiguities and their resolution can be found in the book by Ronald Bentley (see Selected Readings).

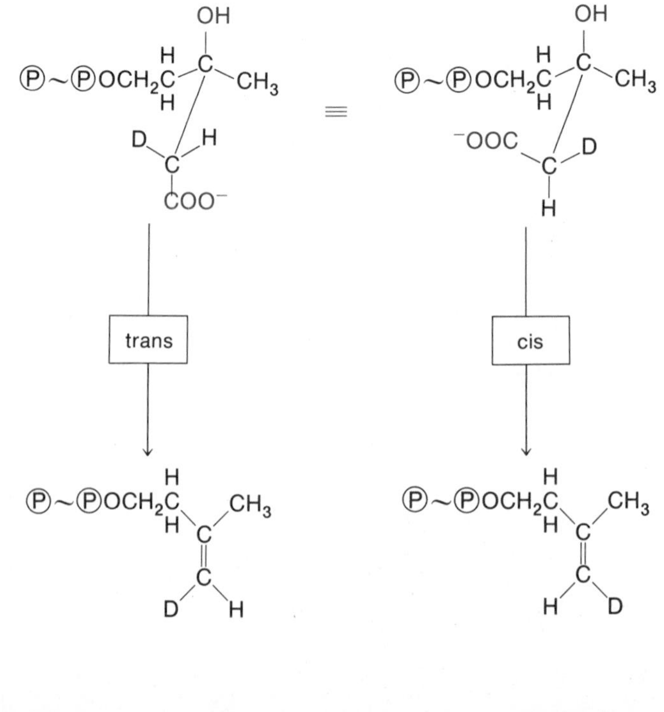

**Figure 15-14**
Does the phosphomevalonate decarboxylase reaction proceed by a cis or a trans elimination?

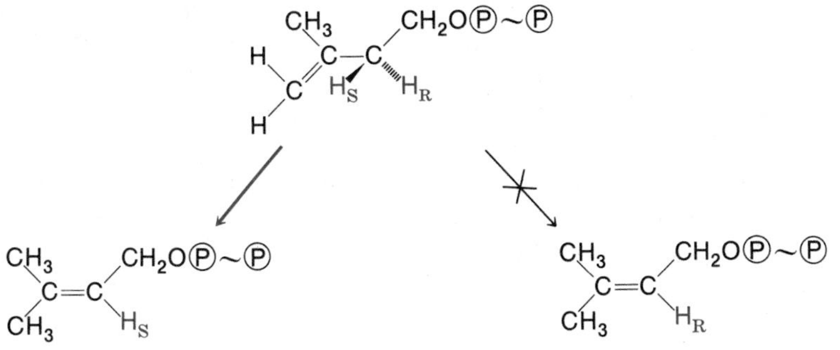

**Figure 15-15**
Which of the two hydrogens is lost in the isomerization of isopentenyl pyrophosphate?

## LIPOPROTEIN METABOLISM

We noted in Chapter 13 that unesterified fatty acids are carried in plasma by albumin. The plasma also transports more complex lipids among the various tissues as components of *lipoproteins* (particles composed of lipids and proteins). This section is concerned with the structure and metabolism of these lipoproteins.

### Structure and Classification of Lipoproteins

*The amounts and types of lipids found in human plasma fluctuate according to the dietary habits and metabolic states of the individual.* The normal ranges for the lipid levels in plasma are shown in Table 15-1. In plasma these lipids are associated with proteins in the form of lipoproteins, which are classified into four major types on the basis of their density (Table 15-2). The lipoproteins of lowest density, the *chylomicrons*, are the largest in size and contain the most lipid and the smallest percentage of protein. At the other extreme are the *high-density lipoproteins* (HDL), which are the smallest particles and contain the highest percentage by

**Table 15-1**
Normal Concentrations of the Major Lipid Classes in Plasma

| Lipid | Concentration (g/l) |
|---|---|
| Total lipid | 3.6–6.8 |
| Cholesterol and cholesterol ester | 1.3–2.6 |
| Triacylglycerol | 0.8–2.4 |
| Phospholipid | 1.5–2.5 |

**Table 15-2**
Composition and Density of Human Lipoproteins

| | Chylomicron | VLDL | LDL | HDL |
|---|---|---|---|---|
| Density (g/ml)* | <0.95 | 0.95–1.006 | 1.019–1.063 | 1.063–1.210 |
| Diameter (nm) | 500–1000 | 30–70 | 20–25 | 10–15 |
| Components (% dry weight) | | | | |
| Protein | 1–2 | 10 | 25 | 33 |
| Triacylglycerol | 83 | 50 | 10 | 8 |
| Cholesterol and cholesterol esters | 8 | 22 | 46 | 30 |
| Phospholipids | 7 | 18 | 22 | 29 |
| Apoprotein composition | A-I, A-II B C-I, C-II, C-III | B C-I, C-II, C-III E | B | A-I, A-II C-I, C-II, C-III D E |
| Classification by electrophoresis | Omega | Pre-beta | Beta | Alpha |

*A lipoprotein of intermediate density (IDL, 1.006–1.019) is also found in human plasma but is not a major lipoprotein class.

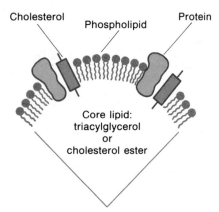

Cholesterol   Phospholipid   Protein

Core lipid:
triacylglycerol
or
cholesterol ester

**Figure 15-16**
Generalized structure of human
lipoproteins.

weight of protein and lowest percentage of lipid. Between these two classes, in both size and composition, are the _low-density lipoproteins_ (LDL) and the _very-low-density lipoproteins_ (VLDL).

The structures of the various lipoproteins appear to be similar (Figure 15-16). Each of the lipoprotein classes contains a neutral lipid core composed of triacylglycerol and/or cholesterol ester. Around this core is a layer of protein, phospholipid, and cholesterol oriented with the polar portions exposed to the surface of the lipoprotein.

There are at least seven apoproteins (Table 15-2) associated with the lipoproteins, as well as several enzymes and a _cholesterol ester transfer protein_. The structure and function of these apoproteins has been intensely studied in the past decade, and some of the properties of these apoproteins are summarized in Table 15-3. Most of the apoproteins have been sequenced and contain regions that are rich in hydrophobic amino acids, which facilitate binding of phospholipid.

## Synthesis and Secretion of Lipoproteins

Of the various lipid components of the lipoproteins, only the biosynthesis of _cholesterol esters_ has not been mentioned. Cholesterol ester is the storage form of cholesterol in cells. These esters are synthesized from choles-

**Table 15-3**
Properties of the Apoproteins of the Major Human Lipoprotein Classes

| Apoprotein | $M_r$ | Plasma Concentration (mg/100 ml) | Miscellaneous |
|---|---|---|---|
| A-I | 28,300 | 130 | Major protein in HDL (64%); contains 245 amino acids and no carbohydrate; activates LCAT* |
| A-II | 17,400 | 40 | Mainly in HDL (20% of dry mass); two identical chains with 77 amino acids each, joined by a disulfide at residue 6 |
| B | | 80 | Major protein in LDL; very difficult to solubilize in detergents |
| C-I | 7,000 | 6 | Contains 57 amino acids |
| C-II | 10,000 | 3 | Contains 80–85 amino acids; activates lipoprotein lipase |
| C-III | 9,300 | 12 | Contains 79 amino acids and Gal, GalNAc, and NeuNAc |
| D | 35,000 | 10 | Also known as cholesterol ester transfer protein; associated with HDL and LCAT |
| E | 33,000 | 5 | Also known as arginine-rich lipoprotein |

*LCAT = Lecithin-cholesterol acyltransferase.

terol and acyl-CoA by *acyl-CoA : cholesterol acyltransferase* (ACAT) (Figure 15-17), which is located on the cytosolic surface of liver microsomes. The synthesis of the apoproteins takes place on ribosomes that are bound to the endoplasmic reticulum. As mentioned previously, the biosynthesis of cholesterol, triacylglycerols, and phospholipids also occurs on the endoplasmic reticulum.

How the various components of the lipoproteins are assembled and secreted into the plasma is not known. Current ideas for this process suggest the transfer of the components from the endoplasmic reticulum to the Golgi apparatus, where secretory (membrane encapsulated) vesicles are formed. These vesicles would subsequently fuse with the plasma membrane and release their lipoprotein contents into plasma.

The plasma lipoproteins appear to be made mainly in the liver and intestine. In the rat, approximately 80 percent of the plasma apoproteins originate from the liver; the rest come from the intestine. Most of the components of chylomicrons, including apoprotein A, apoprotein B, phospholipid, cholesterol, cholesterol ester, and triacylglycerols, are products of the intestine. The chylomicrons are secreted into lymphatic capillaries, which eventually enter the blood stream at the large subclavian veins and therefore bypass the liver. The liver appears to be the major source of VLDL and HDL, which include apo A-I, A-II, B, C-I, C-II, C-III, and E and the lipid components of these lipoproteins. Low-density lipoprotein is produced from VLDL, as will be discussed below.

## Catabolism of Chylomicrons and Very-Low-Density Lipoproteins (VLDL)

*Chylomicrons serve as the mode of transport of triacylglycerol and cholesterol ester from the intestine to other tissues in the body.* Very-low-density lipoprotein functions in a similar manner for the transport of lipid from the liver to other tissues. These two types of triacylglycerol-rich particles are initially degraded by the action of *lipoprotein lipase,* an extracellular enzyme that is most active within the capillaries of adipose tissue, cardiac and skeletal muscle, and lactating mammary gland. Lipoprotein lipase catalyzes the hydrolysis of triacylglycerols to yield fatty acids and 2-monoacylglycerols. The enzyme is specifically activated by apoprotein C-II, which is associated with chylomicrons and VLDL (Table 15-2). As a result, this lipase could supply the heart and adipose tissue with fatty acids, derived from these lipoproteins in plasma. In both the heart and adipose tissue, the fatty acids produced by lipoprotein lipase can be used for energy or stored as a component of triacylglycerols. Alternatively, the fatty acids could be bound by albumin and transported to other tissues.

As the lipoproteins are depleted of triacylglycerol, the particles become smaller. Some of the surface molecules (apoproteins, phospholipids) are transferred to HDL. In the rat, "remnants" that result from chylomicrons and VLDL catabolism are taken up by the liver. In humans, the uptake of remnant VLDL seems to be of minor importance. Instead, most of the VLDL appears to be gradually degraded, so that LDL particles are produced. The half-life for clearance of chylomicrons and remnants from plasma of man is 4 to 5 min. The clearance value for VLDL is 1 to 3 h.

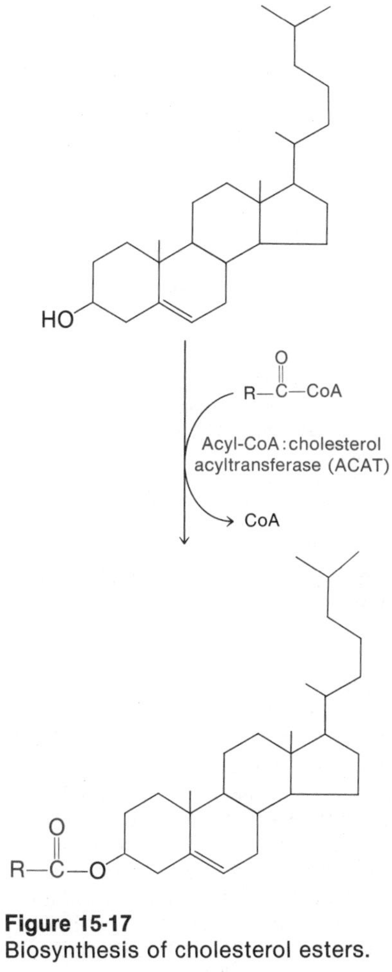

**Figure 15-17**
Biosynthesis of cholesterol esters.

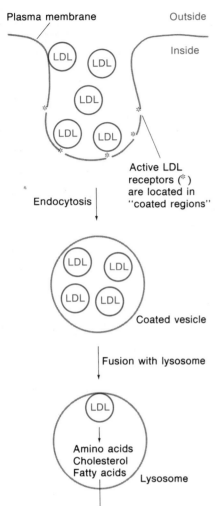

Plasma membrane

Outside

Inside

Active LDL receptors (*) are located in "coated regions"

Endocytosis

Coated vesicle

Fusion with lysosome

Amino acids
Cholesterol
Fatty acids

Lysosome

Cholesterol or a derivative:
(a) Decreases HMG-CoA reductase activity
(b) Stimulates ACAT activity
(c) Decreases synthesis of LDL receptors

**Figure 15-18**
Receptor-mediated uptake of LDL by human skin fibroblasts.

**Figure 15-19**
Electron micrograph of LDL particles (made electron-dense with covalently bound ferritin) bound to coated regions of a human skin fibroblast (97,000×). (From R. G. W. Anderson, M. S. Brown, and J. L. Goldstein, Role of the coated endocytic vesicle in the uptake of receptor-bound low density lipoprotein in human fibroblasts, *Cell* 10:351, 1977. Reprinted with permission.)

## Catabolism of Low-Density Lipoproteins (LDL)

Each day approximately 45 percent of the plasma pool of *low-density lipoprotein is removed from human plasma by both the liver and extrahepatic tissues* (particularly the adrenals and adipose tissue). The mechanism for the uptake of LDL by liver is a problem of current interest in many laboratories. The mechanism for the uptake of LDL in extrahepatic tissues has been extensively described by Michael Brown and Joseph Goldstein. They studied the uptake of LDL by human skin fibroblasts grown in cultures in Petri dishes. LDL particles bind to the cell surface by specific receptors that congregate in areas of the plasma membrane called *"coated regions"* (Figures 15-18 and 15-19). These areas of plasma membranes engulf the LDL particles (a process called *endocytosis*) to form "coated vesicles," which are somehow directed toward and fuse with lysosomes. The LDL particles are degraded within the lysosomes by the action of proteases and *lysosomal acid lipases* (lipid degradative enzymes). The cholesterol, or a derivative of cholesterol, diffuses from the lysosomes, suppresses the activity of *HMG-CoA reductase*, and stimulates the activity of *acyl-CoA : cholesterol acyltransferase* (ACAT). ACAT catalyzes the synthesis of cholesterol esters (Figure 15-17), which are then stored within the cell. The cholesterol (or its derivative) also suppresses the synthesis of the LDL receptors and thereby limits the uptake of LDL.

The *LDL receptor* is a glycoprotein that shows specificity for LDL and VLDL. The number of binding sites per cell can vary between 15,000 and 70,000, depending on the cell's requirement for cholesterol. Once an LDL molecule is bound to the fibroblast receptor, it is rapidly internalized by endocytosis ($t_{1/2} = 3$ min).

## Familial Hypercholesterolemia: A Serious Disease Resulting from Cholesterol Deposits

Brown and Goldstein were able to deduce the pathway for the uptake and catabolism of LDL largely as a result of their studies on the inherited disease *familial hypercholesterolemia*. Patients with the *homozygous* (two defective genes) form of this disease have grossly elevated levels of plasma cholesterol (650 to 1000 mg/100 ml) (see Table 15-1 for normal values), which is largely carried by elevated concentrations of LDL. Cholesterol is

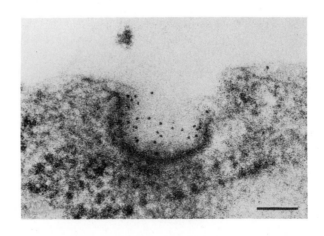

deposited in skin (xanthomas) in various areas of the body. Of greater consequence is the deposit of cholesterol in arteries, which results in _atherosclerosis_, a disease of the arteries that is the underlying cause for most cardiovascular diseases. In fact, patients with homozygous familial hypercholesterolemia have symptoms of heart disease by the early teens and usually die from cardiovascular disease before the age of 20. The _heterozygotes_ (one normal and one defective gene) manifest similar but less severe symptoms. Their plasma cholesterol is in the range of 250 to 550 mg/100 ml of plasma, and heterozygotes generally do not have a heart attack before the age of 40.

Three different biochemical mutations have been shown to cause familial hypercholesterolemia. In one group, the fibroblasts from these patients show no detectable LDL receptor activity. In a second group, the LDL receptor activity is 1 to 10 percent of normal. In one case, the binding of LDL to the cell surface was normal, but endocytosis did not occur. The frequency of the heterozygous form of familial hypercholesterolemia has been estimated at 1 in 500, and the homozygote frequency could be expected to be 1 in 1 million.

A related disorder, _Wolman's disease_, has provided further evidence for the receptor-mediated pathway of LDL uptake (Figure 15-18). Wolman's disease is a very rare inborn error of metabolism (approximately 20 cases diagnosed since 1956) characterized by the accumulation of cholesterol esters and triacylglycerols in various tissues. The disease can be diagnosed within several weeks of birth and is fatal, usually within 6 months. The disease is caused by a complete lack of a _lysosomal acid lipase_, which is responsible for the normal catabolism of cholesterol esters and triacylglycerols in lysosomes. _Cholesterol ester storage disease_ is a related disorder caused by a substantial reduction in the activity of lysosomal acid lipase (1 to 20 percent of normal). The symptoms are far less severe, and patients have survived to the age of 40.

## Catabolism of High-Density Lipoproteins (HDL)

The catabolism of high-density lipoproteins is a complex process that is currently under investigation. The half-life of HDL in human plasma (5 to 6 days) is much longer than the other lipoproteins. When HDL is secreted into plasma from liver, it has a discoid shape and is almost devoid of cholesterol ester. These _"nascent" HDL_ particles are converted into spherical particles by the accumulation of cholesterol ester in the interior. The cholesterol ester is derived from cholesterol and phosphatidylcholine on the surface of the HDL particle in a reaction catalyzed by _lecithin:cholesterol acyltransferase_ (LCAT) (Figure 15-20). LCAT is a glycoprotein (24 percent carbohydrate by weight) with a molecular weight of 59,000. This enzyme is associated with HDL in plasma and is activated by apoprotein A-I (Table 15-3). Associated with the LCAT-HDL complex is _apoprotein D_, also known as _cholesterol ester transfer protein_. It is currently thought that after cholesterol esters are formed in HDL, apoprotein D catalyzes the transfer of cholesterol esters from HDL to VLDL or LDL. In the steady state, cholesterol esters that are synthesized by LCAT would be transferred to these other lipoproteins and catabolized as noted earlier. The

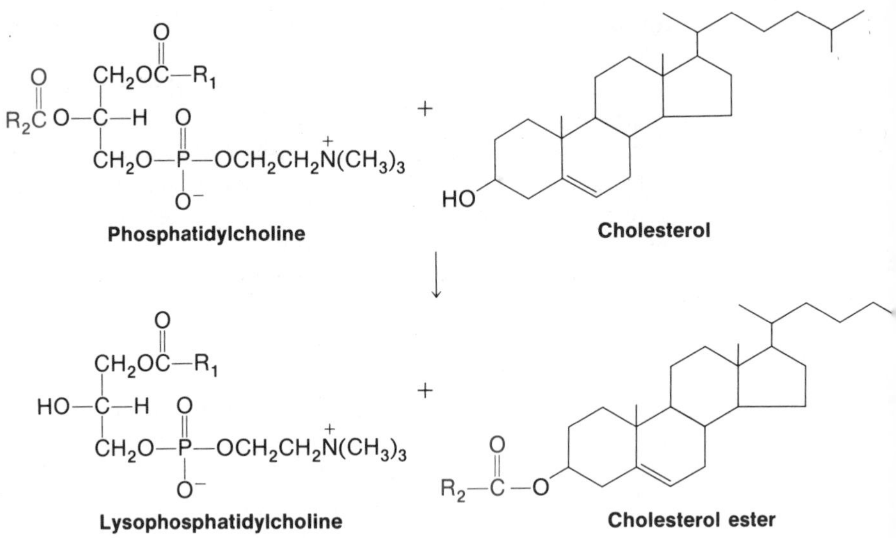

Phosphatidylcholine

Cholesterol

Lysophosphatidylcholine

Cholesterol ester

**Figure 15-20**
Reaction catalyzed by lecithin:
cholesterol acyltransferase (LCAT).

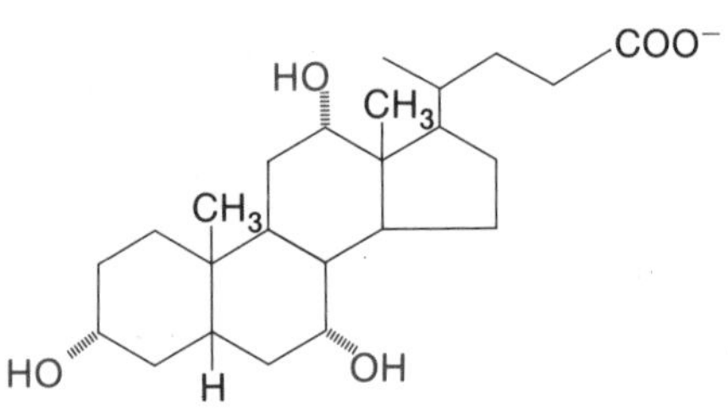

Cholic acid

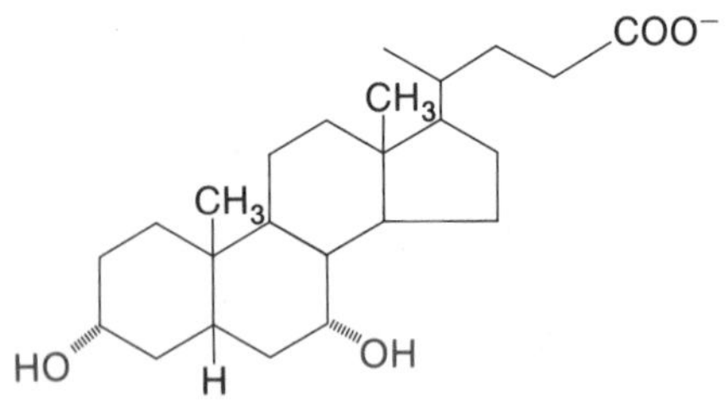

Chenodeoxycholic acid

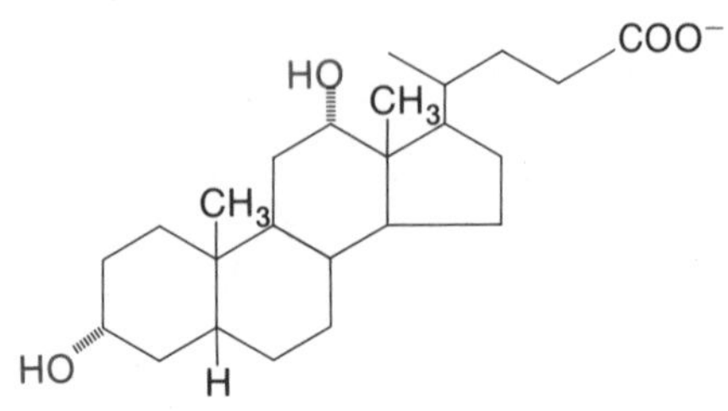

Deoxycholic acid

**Figure 15-21**
Structures of three bile acids.

HDL particles themselves turn over, but where they are degraded is not firmly established.

Although elevated levels of cholesterol and LDL in human plasma are linked with an increased incidence of cardiovascular disease, recent data have shown that an increase in concentration of HDL in plasma is correlated with a lowered risk of coronary artery disease. Why does an elevated HDL in plasma appear to protect against cardiovascular disease, whereas an elevated LDL seems to cause this disease? The answer to this question is not known. An explanation currently favored is that HDL functions to return cholesterol to the liver, where it is metabolized and excreted. The net effect would be a decrease of plasma cholesterol available for deposit in arteries.

## BILE ACIDS

*The bile acids, the major degradation products of cholesterol, are dihydroxylated and trihydroxylated steroids with 24 carbons.* All hydroxyl groups of the bile acids are α in orientation, and the two methyl groups are β (Figure 15-21); thus these molecules have a polar and a nonpolar face. Another distinguishing feature is the C-24 carboxylic acid group. In addition, the major bile acids have the A and B rings joined in a cis configuration (see Figures 15-3 and 15-21). The important bile acids in humans are *cholic acid* and *chenodeoxycholic acid* (Figure 15-21). *Deoxycholic acid* is found in bile of other mammals and is also an important reagent used in the laboratory for solubilization of membrane proteins and enzymes. Cholate is also an important solubilization reagent. The bile acids are present in bile mostly as conjugates of *taurine* or *glycine* (Figure 15-22), and these derivatives are referred to as *bile salts*. The structures for *taurocholate* and *glycocholate* are given in Figure 15-22.

*Bile* consists of a mixture of organic and inorganic compounds. As shown in Table 15-4, the bile salts and phosphatidylcholine are quantitatively the most important constituents of bile. The bile, made in the liver, is stored in the <u>gall bladder</u> and passes along the <u>common bile duct</u> into the duodenum when food is present. *The bile salts act as detergents in the small intestine and thereby aid the solubilization of lipids, which are more easily degraded by intestinal lipases.* The ability of the bile salts to solubilize lipids can be attributed to the distinctly amphipathic nature of these compounds. The polar groups, which favor an aqueous environment, all lie on the $\alpha$ side of the steroid nucleus. The nonpolar side interacts with hydrophobic lipids. As a result of these properties and interactions, the nonpolar dietary lipids are suspended in a uniform dispersion in the aqueous medium.

The digested lipids are absorbed primarily in the upper part of the intestine, whereas most of the bile salts are resorbed in the lower intestine and returned to liver by means of the portal blood. The bile salts are resecreted into bile and recirculate through the intestine. This cycle is called the <u>*enterohepatic circulation*</u> of bile salts. In humans, approximately 20 to 30 g of bile salts is secreted by the liver, but only 0.5 g appears in the feces each day. Approximately half the cholesterol in bile is excreted in the feces either as cholesterol or, after its conversion by intestinal bacteria, as *coprostanol*. This represents a major mechanism for elimination of cholesterol from the body.

**Table 15-4**
Organic Components in
Human Bile

| Compound | Concentration (g/l) |
|---|---|
| Bile salts | 1.2–18.0 |
| Phosphatidylcholine | 1.4–8.0 |
| Cholesterol | 1.0–3.2 |
| Bilirubin | 0.1–0.7 |
| Protein | 0.3–3.0 |

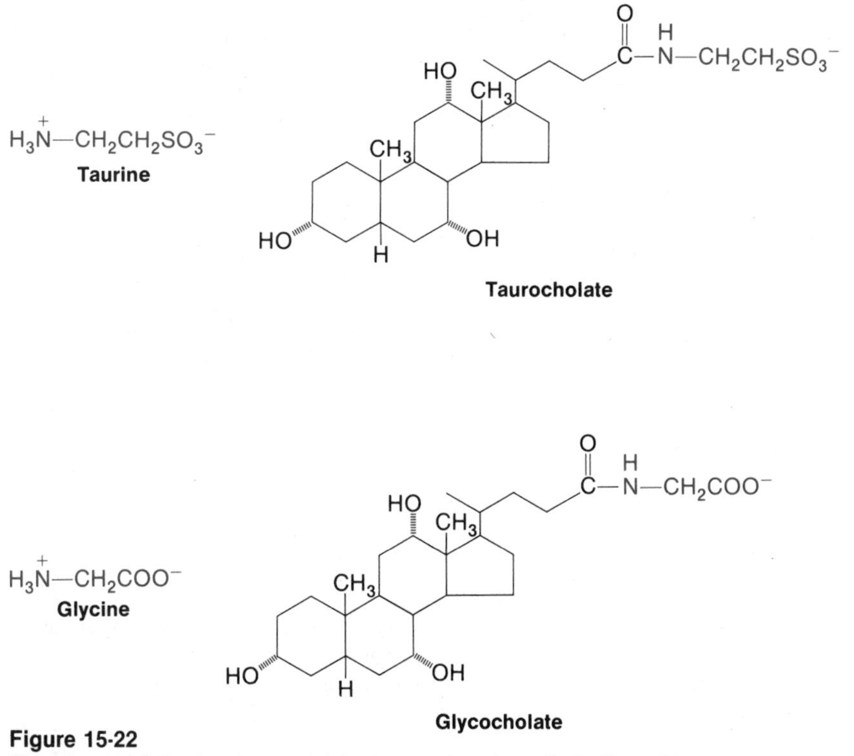

**Figure 15-22**
Structures of the taurine and glycine conjugates of cholic acid.

**Table 15-5**
Major Steroid Hormones

| Steroid | Function |
|---------|----------|
| **Progesterone** | Precursor of other steroids; prepares uterus for implantation of an egg; prevents ovulation during pregnancy |
| **Aldosterone** (a mineralocorticoid) | Increases retention of sodium ions by the renal tubules |
| **Cortisol** (a glucocorticoid) | Promotes gluconeogenesis; suppresses inflammatory reactions |
| **Testosterone** (an androgen) | Promotes male sexual development; promotes and maintains male sex characteristics |
| **Estradiol** (an estrogen) | Responsible for sexual development in the female; promotes and maintains female sex characteristics |

*The steroid hormones are biosynthesized from cholesterol* (see Figure 29-5). The adrenal cortex and the gonads are the major organs involved in the biosynthesis of these steroids. In the pregnant female, the placenta also manufactures steroid hormones. The structures and functions of five major hormones are given in Table 15-5. The biosynthesis and the action of steroid hormones is covered in Chapter 29.

## MAMMALIAN CHOLESTEROL METABOLISM

The metabolism of cholesterol in mammals is extremely complex. A summary sketch (Figure 15-23) may help draw the major metabolic interrelationships together. Cholesterol is taken in through the diet or biosynthesized largely in the liver. From the intestine it is secreted into the plasma mainly as a component of chylomicrons. These are quickly degraded by lipoprotein lipase and the remnants are removed by the liver. Apoproteins and lipid components of the chylomicrons and remnants appear to exchange with HDL. Cholesterol made in the liver can be secreted as a component of HDL and VLDL, can be stored as cholesterol esters, used as a structural component of membranes or converted to bile salts. In plasma VLDL is degraded to IDL and LDL by the action of lipoprotein lipase and through exchange reactions with HDL. The LDL serves as a major carrier of cholesterol to extrahepatic cells which includes the adrenals and gonads. LDL and HDL are also returned to the liver. The steroids made in the adrenals and gonads are delivered to various target tissues and promote a wide range of metabolic effects. The steroid hormones are eventually excreted as glycosyl conjugates in the urine. The bile acids made in the liver are delivered to the upper intestine, where they aid in the solubilization of dietary lipid. Most of the bile acids are resorbed in the lower intestine and returned to the liver by the portal vein.

## LIPID-SOLUBLE VITAMINS, TERPENES, AND ACETOGENINS

The emphasis of the three chapters on lipid metabolism has been on the primary lipids used for membrane structures and for storage and generation of energy. There are many other diverse and important lipids and related compounds that will not be mentioned. This section describes some properties of lipid-soluble vitamins, terpenes, and acetogenins.

### Lipid-Soluble Vitamins

*Vitamin $D_3$ (cholecalciferol)* is not truly a vitamin, because it can be made in the skin from 7-dehydrocholesterol (Figure 15-13) in the presence of ultraviolet light (Figure 29-8). Vitamin $D_3$ is formed by the cleavage of ring B of 7-dehydrocholesterol. Vitamin $D_3$ made in skin or absorbed from the small intestine is transported to the liver and hydroxylated at C-25 by a microsomal mixed-function oxidase. *25-Hydroxyvitamin $D_3$* appears to be biologically inactive until it is hydroxylated at C-1 by a mixed-function oxidase in kidney mitochondria. The *1α,25-dihydroxyvitamin $D_3$* is delivered to target tissues for the regulation of calcium and phosphate metabo-

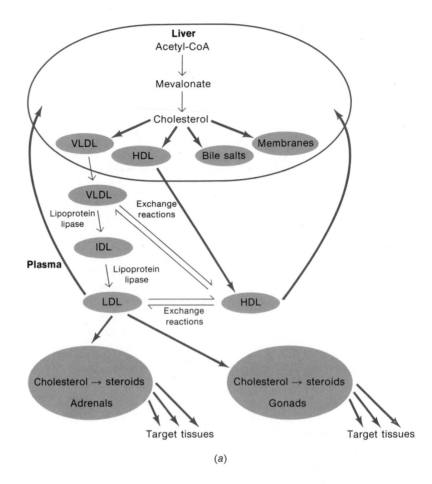

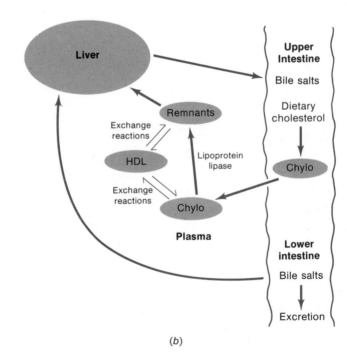

**Figure 15-23**
Fate of cholesterol. (*a*) Cholesterol
biosynthesized in the liver; (*b*) dietary
cholesterol.

lism. The mode of action of 1,25-dihydroxyvitamin $D_3$ is analogous to that of steroid hormones, as discussed in Chapter 29.

Vitamin K was discovered by Henrik Dam in Denmark in the 1920s as a fat-soluble factor important in blood coagulation (K is for koagulation). The structures of vitamins $K_1$ and $K_2$ (Figure 15-24) were elucidated by Edward Doisey. Vitamin $K_1$ is found in plants; vitamin $K_2$, in animals and bacteria. How this vitamin functions in blood coagulation eluded scientists until 1974, when a requirement for vitamin K was shown for the formation of γ-carboxyglutamic acid (Figure 15-25) in certain proteins. γ-Carboxyglutamic acid specifically binds calcium, which is important for blood coagulation. Such modified glutamic acid residues appear to be important in many other processes involving calcium-transport and calcium-regulated metabolic sequences.

Vitamin E (α-tocopherol) (Figure 15-26) was recognized in 1926 as an organic-soluble compound that prevented sterility in rats. The function of this vitamin still has not been clearly established. A favorite theory is that it is an antioxidant that prevents peroxidation of polyunsaturated fatty acids. Tocopherol certainly prevents peroxidation in vitro, and it can be replaced by other antioxidants. However, other undescribed functions still seem likely, since other antioxidants will not relieve all the symptoms of vitamin E deficiency.

Vitamin A (trans-retinol) is an isoprenoid alcohol (Figure 15-27) that has been known for many decades to be important for vision and for animal growth and reproduction. The vitamin is either biosynthesized from β-carotene (Figure 15-27), a polyisoprenoid compound, or is absorbed in the diet. Vitamin A is stored in liver predominantly as an ester of palmitic acid. The form of the vitamin active in the visual process is 11-cis-retinal

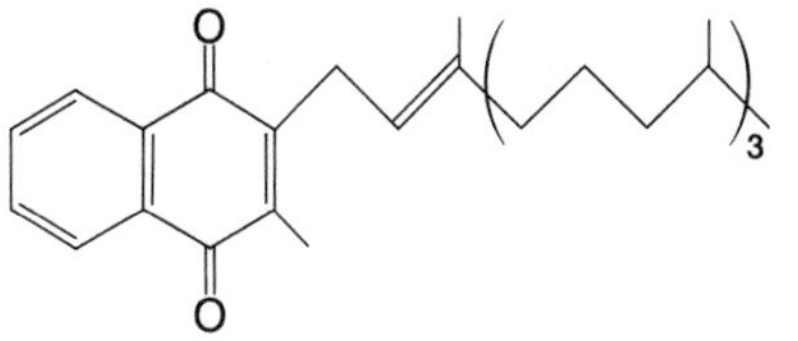

**Vitamin $K_1$**
(phylloquinone)

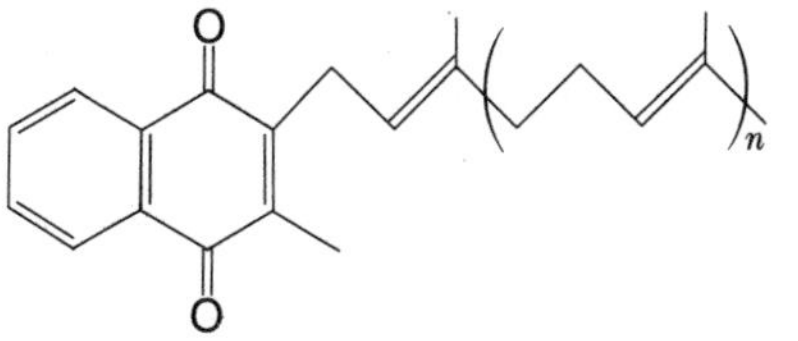

**Vitamin $K_2$**
(menaquinone series)

**Figure 15-24**
Structures of vitamins $K_1$ and $K_2$.

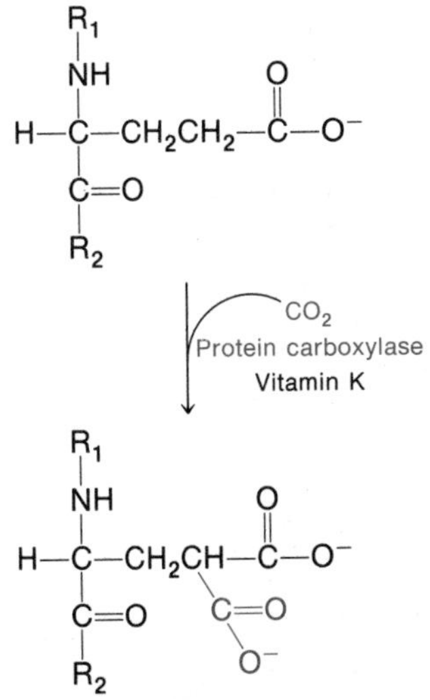

γ-**Carboxyglutamic acid
in a protein**

**Figure 15-25**
Vitamin K–dependent carboxylation of a glutamic acid residue in a protein.

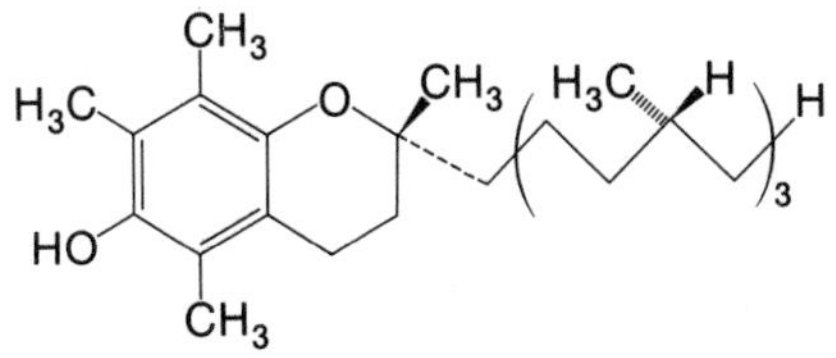

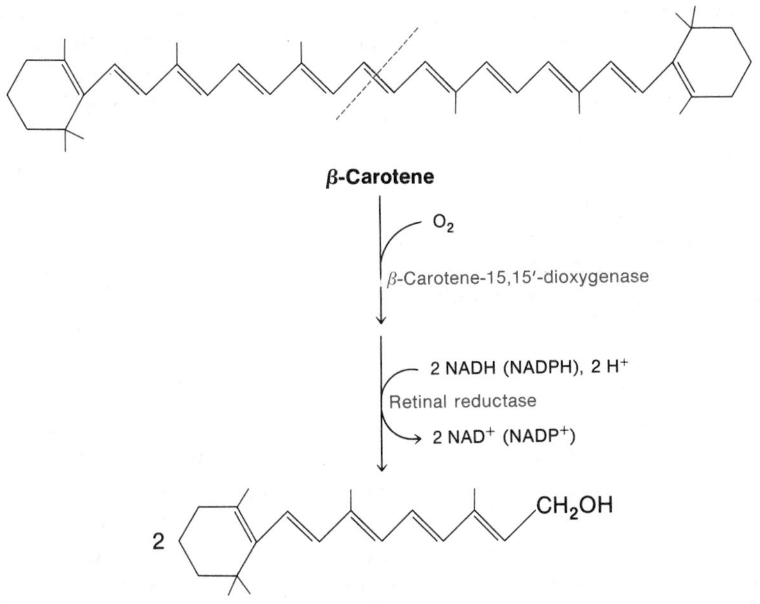

**β-Carotene**

$O_2$

β-Carotene-15,15′-dioxygenase

2 NADH (NADPH), 2 $H^+$

Retinal reductase

2 $NAD^+$ ($NADP^+$)

$CH_2OH$

**Vitamin A**
(trans-retinol)

**Figure 15-27**
Biosynthesis of vitamin A from β-carotene.

**Figure 15-26**
Structure of vitamin E (α-tocopherol).

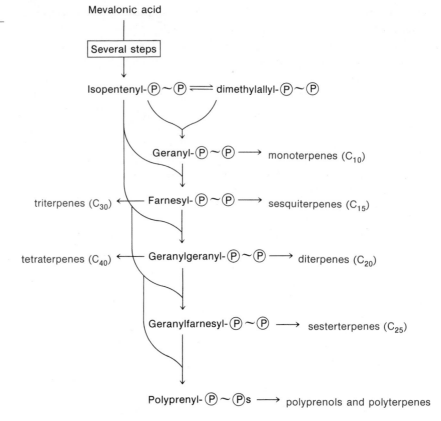

**Figure 15-28**
General scheme for the biosynthesis
of terpenes.

(Figure 31-3), which combines with the protein *opsin* to form *rhodopsin*.
11-*cis*-Retinal is produced from *trans*-retinol by means of 11-*cis*-retinol, as
shown in Figure 31-3. In the visual process, there is photochemical isomerization of 11-*cis*-retinal in rhodopsin to *trans*-retinal (see Chapter 31). The
trans isomer of retinal can be converted back into the cis isomer by *retinal
isomerase*. Alternatively, the *trans*-retinal can be reduced to *trans*-retinol
by *retinal reductase* in the retina. It is believed that isomerization of *trans*-
retinol to 11-*cis*-retinol occurs in the liver by *retinol isomerase*. The product, 11-*cis*-retinol, is returned to the retina and oxidized to 11-*cis*-retinal,
which is ready to undergo another reaction in the visual cycle. Retinol is
carried in plasma by *retinol-binding protein*.

### Terpenes

Terpenes form an extraordinarily diverse group of compounds that are biosynthesized from isoprene precursors. In the section on cholesterol (considered to be a terpene) biosynthesis, we have already devoted considerable
attention to the isoprene precursors (isopentenyl-PP and dimethylallyl-PP)
derived from mevalonate. These same precursors are involved in the biosynthesis of the terpenes, as indicated in Figure 15-28. The terpenes, which
have a wide range of functions, include primary metabolites and secondary
metabolites. Examples of primary metabolites are steroids, certain hormones (juvenile hormone in insects), and precursors to hormones ($\beta$-caro-
tene). The major diversity found among terpenes occurs in *secondary me-*

tabolites (compounds that serve no obvious function in the life of the organism that produces them), which are made by plants and microorganisms.

Monoterpenes ($C_{10}$) are found in all higher plants. One example is limonene (Figure 15-29), which is largely responsible for the characteristic odor of lemons. A bicyclic monoterpene, pinene (Figure 15-29), is the major constituent of turpentine, the solvent obtained from pine resins. Sesquiterpenes ($C_{15}$) also are widely distributed in the plant kingdom; sev-

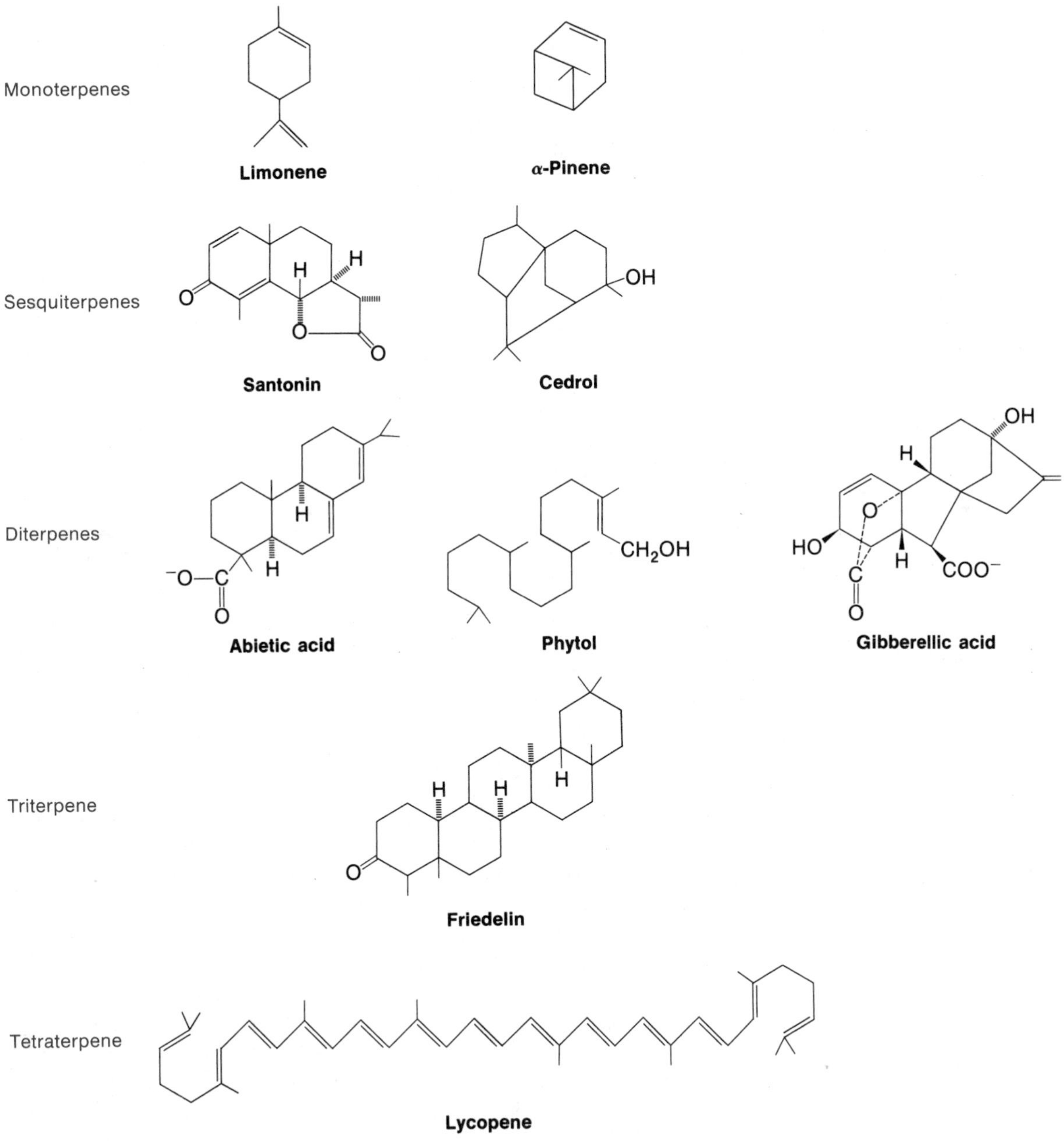

Monoterpenes

**Limonene**

**α-Pinene**

Sesquiterpenes

**Santonin**

**Cedrol**

Diterpenes

**Abietic acid**

**Phytol**

**Gibberellic acid**

Triterpene

**Friedelin**

Tetraterpene

**Lycopene**

**Figure 15-29**
Structures of selected terpenes.

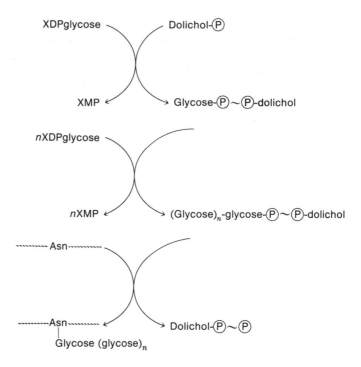

**Figure 15-30**
Generalized scheme for the glycosylation of proteins.

eral thousand individual sesquiterpenes have been identified. Two examples are shown in Figure 15-29. *Santonin* has been used in India for the treatment of intestinal worms, and *cedrol* is a constituent of cedarwood oil. The *diterpenes* are $C_{20}$ terpenes. *Abietic acid* (Figure 15-29) from pine resin is one of the most prevalent terpenes in this group. Another diterpene, *phytol*, is a component of the chlorophyll molecule. Also in this group are the *gibberellins*, which are important plant growth hormones. One of these hormones, *gibberellic acid*, is shown in Figure 15-29. *Triterpenes* are $C_{30}$ terpenes, of which *lanosterol* is an important example that was discussed earlier. There are a large number of triterpenes in plants, most of which are pentacyclic. *Friedelin* (Figure 15-29) is one example of a triterpene extracted from cork. Of the *tetraterpenes* ($C_{40}$), the *carotenes* predominate. *β-Carotene* (Figure 15-27), from carrots, has been mentioned in connection with vitamin A biosynthesis. *Lycopene* (Figure 15-29), a pigment from tomatoes, is similar to β-carotene, but lacks the cyclohexene rings.

Among the various *polyprenols* (polyisoprenoid alcohols), the *dolichols* are currently under intensive investigation. Dolichols are long-chain polyprenols that consist of 16 to 22 isoprene units. They function in the form of *dolichyl phosphate* as carriers of carbohydrates in the biosynthesis of *glycoproteins* in animals. Bacteria use a similar lipid carrier for carbohydrates, *undecaprenol*, in which the number of isoprene units is 11 (see Chapter 12). A generalized scheme for the biosynthesis of these *glycosylphosphodolichols* and the transfer of the oligosaccharide to a protein are shown in Figure 15-30. *XDPglycose* represents one of the sugar nucleotides (e.g., GDPmannose, UDPgalactose), and the number of carbohydrates linked to the lipid can reach 20. The oligosaccharide, once it is synthesized on the lipid, is transferred to an asparagine residue in a protein.

One sesquiterpene that has attracted much attention in the past two decades is *juvenile hormone III* and two closely related juvenile hormones, I and II (Figure 15-31). These hormones maintain insects in the larval form (see Chapter 29). During normal metamorphosis, the blood (hemolymph) levels of juvenile hormone fall and the insect matures. Consequently, this hormone inhibits the transformation of these larva into the pupa and adult forms of the insect. The hormone appears again in the adult female and plays a role in maturation of the reproductive system.

## Acetogenins (Polyketides)

*Acetogenins are complex molecules derived from condensation of three or more units of acetyl-CoA. The acetogenins are responsible for many of the brilliant colors that abound in nature.* They account for most colors of flowers, autumn leaves, rhubarb, sea urchins, lichens, molds, and fungi. The other two major groups of pigments that are not acetogenins are the tetrapyrroles (e.g., heme and chlorophyll) and the carotenes. The *polyphenols* account for more than three-quarters of the greater than 1,000 acetogenins identified. Most of these phenols are biosynthesized from acetyl-CoA, malonyl-CoA, and shikimic acid–derived acyl-CoAs (see Chapter 22 for shikimic acid). The biosynthetic enzymes appear to occur as aggregates, as in the case of fatty acid synthase. However, unlike fatty acid synthesis, the reduction reactions are relatively few, and hence the compounds retain many oxygenated functional groups.

The largest subgroup of acetogenins is the *flavonoids*, which are plant, flower, and fruit pigments with the skeleton of *flavone* (Figure 15-32). One example is *naringenin* (Figure 15-32). Another important group of acetogenins is the *quinones*. Compounds in this category are coenzyme Q (Chapter 6) and vitamin K, which is a *napthoquinone*. Another napthoquinone is the yellow dye *lapachol* (Figure 15-32). The mold metab-

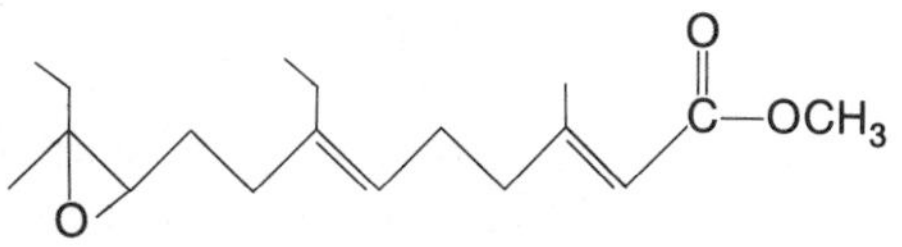

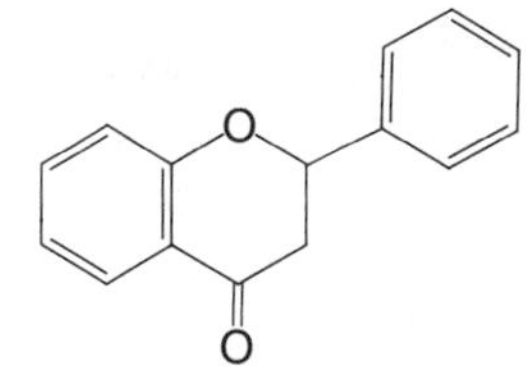

**Flavone**

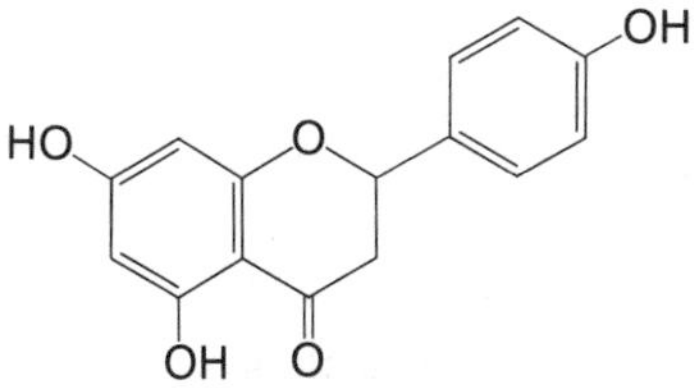

**Naringenin**

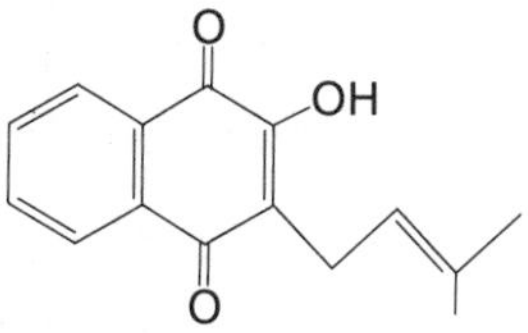

**Lapachol**

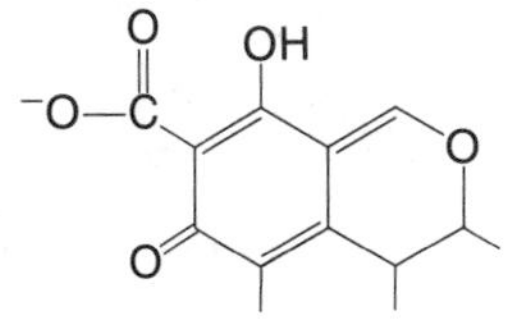

**Citrinin**

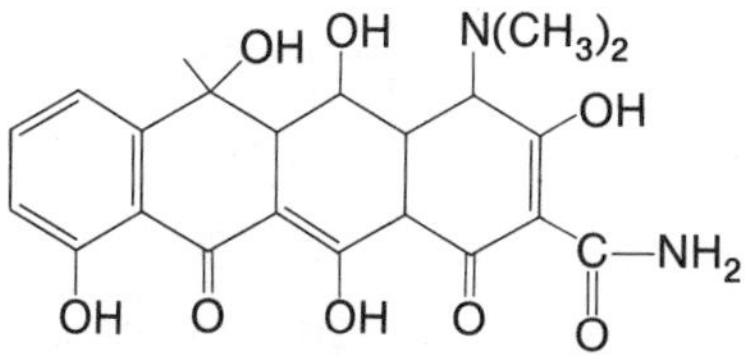

**Terramycin**

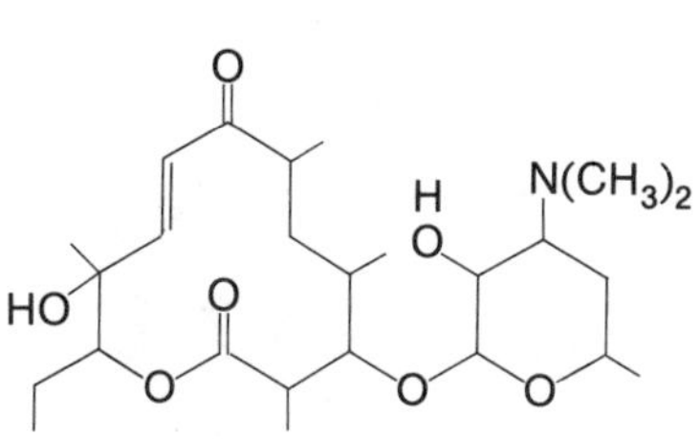

**Methymycin**

**Figure 15-32**
Structures of selected acetogenins.

**Juvenile hormone I**

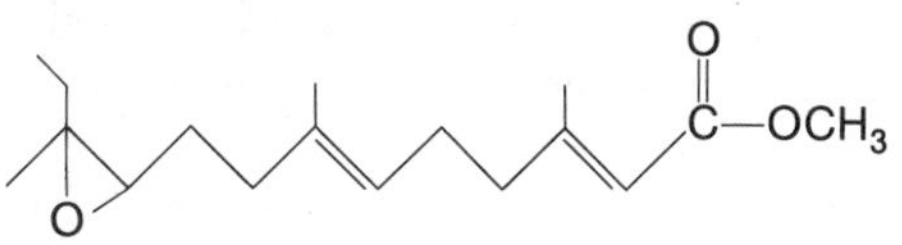

**Juvenile hormone II**

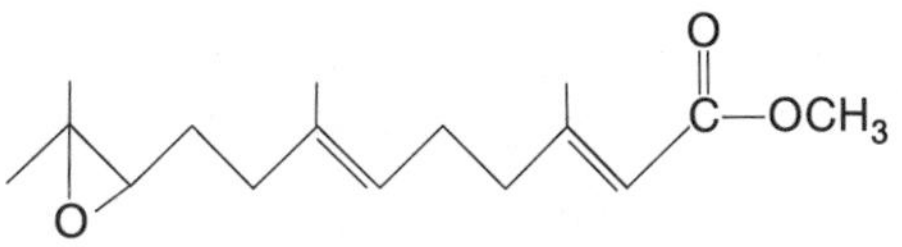

**Juvenile hormone III**

**Figure 15-31**
Structures of juvenile hormones.

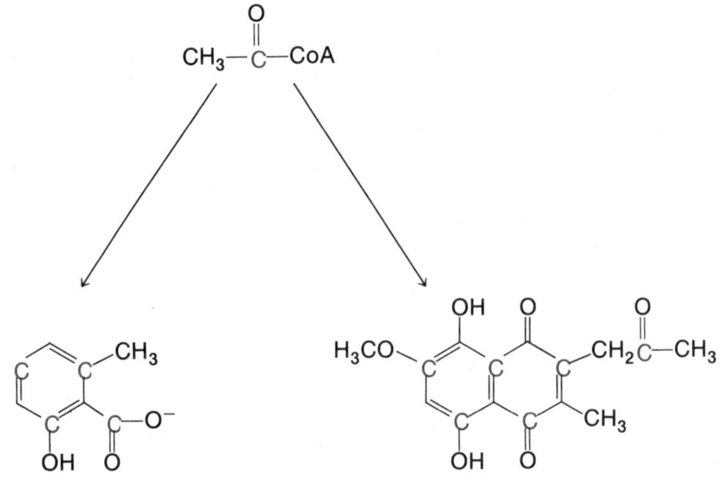

**Figure 15-33**
Labeling of two acetogenins by [1-$^{14}$C]acetate.

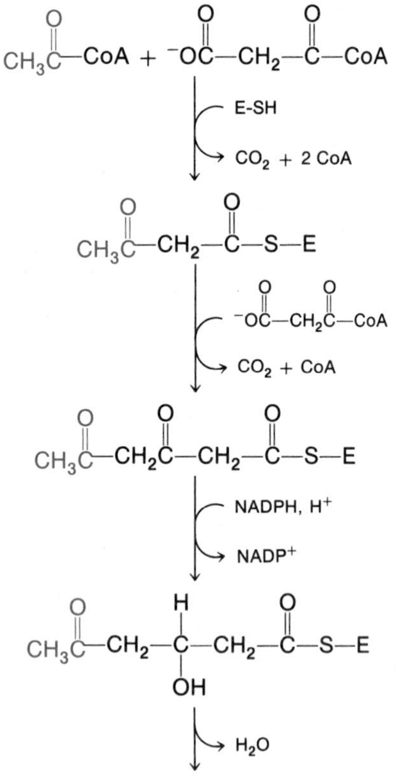

6-Methylsalicylate

Javanicin

olite _citrinin_ is related to this class of acetogenins. The _tetracyclines_, which are important drugs for the treatment of bacterial infection, belong to another group of acetogenins. _Terramycin_ (Figure 15-32) was the first compound in this group for which the structure was determined. Another group of acetogenins of great importance in medicine is the _macrolide_ antibiotics, one example of which is _methymycin_.

The biosynthesis of acetogenins has been studied mostly with tracer experiments in fungi. For example, when [1-$^{14}$C]acetate was added to the growth medium, the labeling patterns in _6-methylsalicylate_ and _javanicin_ (Figure 15-33) were subsequently deduced by chemical degradations. The enzymes involved in acetogenin biosynthesis have not been well-described. An exception is _6-methylsalicylate synthase_, which has been purified from the fungus _Penicillium patulum_. The enzyme is similar to fatty acid synthase from yeast, because it is a multienzyme complex that requires acetyl-CoA, malonyl-CoA, and NADPH and contains phosphopantetheine. However, the molecular weight of 6-methylsalicylate synthase ($1.3 \times 10^6$) is about half the molecular weight of fatty acid synthase ($2.3 \times 10^6$) from yeast. The reaction sequence (Figure 15-34) takes place on the enzyme without the release of intermediates. Only one of the ketone groups is reduced. The polyketone nature of the intermediates gives rise to the alternative name for this group of compounds, _polyketides_.

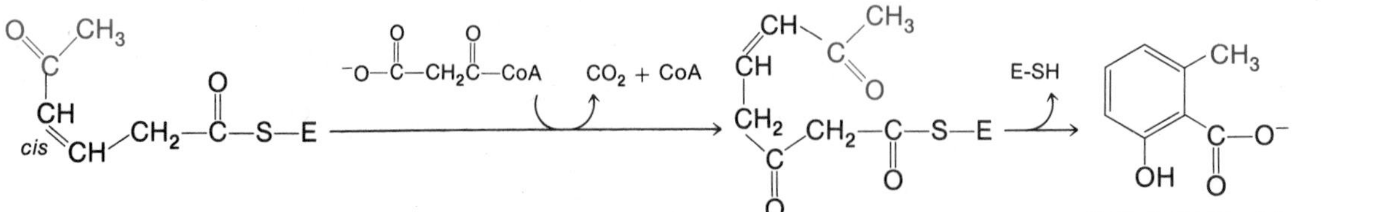

**Figure 15-34**
Reaction sequence catalyzed by 6-methylsalicylate synthase.

# SELECTED READINGS

Bentley, R. *Molecular Asymmetry in Biology,* Vol. 2. New York: Academic Press, 1970. This book contains a very lucid explanation of the stereochemistry of cholesterol biosynthesis.

Brown, M. S., Kovanen, P. T., and Goldstein, J. L. Regulation of plasma cholesterol by lipoprotein receptors. *Science* 212:628, 1981. A review which provides a good summary of recent developments in the control of cholesterol metabolism.

Danielsson, H., and Sjovall, J. Bile acid metabolism. *Ann. Rev. Biochem.* 44:233, 1975. This review provides a good summary of the developments in this area in the early 1970s.

Gower, D. B. *Steroid Hormones.* London: Croom Helm, 1979. A very good summary on the metabolism and function of steroid hormones.

Nes, W. R., and McKean, M. L. *Biochemistry of Steroids and Other Isopentenoids.* Baltimore: University Park Press, 1977. This is a complete and authoritative text on the structure and metabolism of steroids.

Stumpf, P. K. *The Biochemistry of Plants,* Vol. 4: *Lipids: Structure and Function.* New York: Academic Press, 1980. This is a general summary on the biochemistry of plant lipids.

# PROBLEMS

1. Which enzyme catalyzes the rate-limiting reaction in cholesterol biosynthesis? Draw the structures of the substrates and products for this reaction.

2. During routine investigations, the plasma from a family of rats was found to have very low concentrations of cholesterol. When the microsomal HMG-CoA reductase from liver was assayed, there was extremely low activity. When the cytosol from normal rats was added to the microsomes which had low HMG-CoA reductase activity, the enzyme activity was gradually restored to normal values. What enzyme activity(ies) might be deficient in these unusual rats?

3. Why are children who live in Mexico less likely to develop rickets from vitamin $D_3$ deficiency than children who live in northern Canada?

4. What is the hereditary defect in Wolman's disease? Would you expect HMG-CoA reductase activity to be high or low in skin fibroblasts cultured from patients with this disorder? Would the number of LDL receptors be high or low in these fibroblasts?

5. A deficiency of apoprotein C-II results in the disease hyperlipoproteinemia type I, in which there is a massive increase in the concentration of plasma triacylglycerol. Provide an explanation for this clinical finding.

6. There was a person with diabetes who even when in diabetic shock showed no signs of ketone bodies in plasma. Which enzymes in ketone body synthesis might be deficient? If cholesterol synthesis were normal, would this be a clue as to which of these enzymes might be deficient?

7. There is an inherited disease in which lecithin:cholesterol acyltransferase (LCAT) is deficient. What effect would you expect this deficiency to have on the composition of HDL and other lipoproteins in plasma?

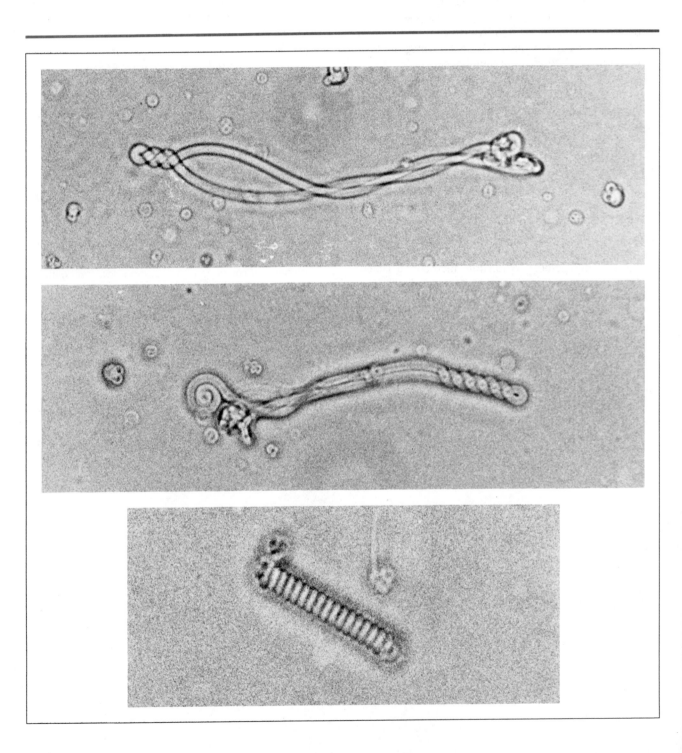

# 16

# BIOLOGICAL MEMBRANES: STRUCTURE AND ASSEMBLY

All cells are characterized by a cytoplasmic membrane often referred to as a *plasma membrane*, and this membrane encapsulates the cytoplasm and creates internal compartments in which essential functions are carried out. In addition to its role as a physical barrier that maintains the integrity of the cell, the plasma membrane provides functions necessary for the survival of the cell, including exclusion of harmful substances, acquisition of nutrients and energy sources, disposal of unusable and toxic materials, reproduction, locomotion, and interaction with components in the environment. All these functions require coordination both for short-range activities, such as sensation, and for long-range processes, such as growth and differentiation. The deceptively simple appearance of a typical plasma membrane in the electron microscope (Figure 16-1) reveals few clues as to how these processes might be carried out, and it has been largely left to biochemists and biophysicists to investigate the molecular architecture of this complex—and vital—macromolecular structure.

Most *eukaryotic* (nucleated) cells also contain numerous intracellular *organelles* of widely differing structure and function, each of which is specialized in its function: digestion (lysosomes), respiration (mitochondria), photosynthesis (chloroplasts), secretion (endoplasmic reticulum and Golgi apparatus), or nucleic acid biosynthesis (nucleus). In turn, each of these organelles is bounded by its own specialized membrane system that has evolved to participate in these specialized functions (Figure 16-2). In contrast, *prokaryotic* cells (bacteria) typically have all these functions integrated into the plasma membrane and lack specialized intracellular organelles. It is clear, however, that these differences in cell structure do not indicate divergent fundamental biochemical mechanisms, but merely the

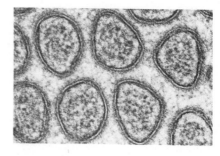

**Figure 16-1**
Cross section of microvilli of cat intestinal epithelial cells, showing the trilaminar structure of the cytoplasmic membranes (165,000×). (Courtesy Dr. S. Ito.)

*Mixtures of cardiolipin and dimyristoyl phosphatidylcholine form helical liposomes in the presence of Ca²⁺. (Magnification about 1200×.) (Courtesy Robert M. Weis of Stanford University.)*

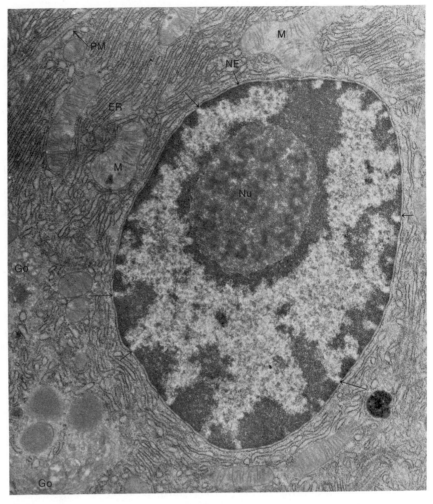

**Figure 16-2**
Electron micrograph of a cell from the rat pancreas, showing several different intracellular organelles (PM = plasma membrane; NE = nuclear envelope; N = nucleus; M = mitochondrion; ER = endoplasmic reticulum; Go = Golgi apparatus; arrows show pore complexes in the nuclear envelope; 24,000×). (From S. L. Wolfe, *Biology of the Cell*, 2d Ed., Wadsworth, Belmont, Calif., 1981. Reprinted with permission.)

presence or absence of compartments specifically designed to fulfill essential functions. In the eukaryotic cell, each process is performed in a spatially isolated domain, whereas these processes operate largely within a single compartment in the prokaryotic cell. In either case, the membrane biochemist's task is a formidable one: to deduce how individual processes take place within a water-insoluble matrix.

In this chapter and the next, we shall examine the structures, assembly, and functions of biological membranes. Since most of our current understanding of membrane structure and function stems from research done only over the last two decades, this field continues to be one of the most rapidly expanding in biochemistry. Nevertheless, *a number of fundamen-*

*tal principles have emerged that appear to apply to most membrane, systems that have been studied.* For this reason, we shall examine aspects of membrane structure and function in systems as seemingly divergent as bacteria and mammalian mitochondria. We will find that in both, the biosynthesis of ATP is a membrane-associated process that occurs by a universal mechanism. Similarly, membranes of *Escherichia coli* contain a protein that behaves in artificial membrane systems much like nerve cell membrane channels, which are, in part, responsible for propagation of the nerve impulse (Chapters 17 and 30). Thus many mechanisms responsible for complex membrane phenomena will undoubtedly be used repeatedly throughout the living kingdom. We shall therefore treat the structure and functions of the *E. coli* cell envelope on an equal basis with those of the human erythrocyte membrane.

## CONSTITUENTS OF BIOLOGICAL MEMBRANES

Typically, a biological membrane contains lipid, protein, and carbohydrate in ratios varying with the source of the membrane (Table 16-1). Nearly always, the carbohydrate is covalently associated with protein (*glycoproteins*) or with lipid (*glycolipids* and *lipopolysaccharides*), so that the membrane can be thought of as a lipid-protein matrix in which specific functions are carried out by proteins, while the permeability barrier and the structural integrity of the membrane are provided by membrane lipids. As we shall see, current evidence is in favor of a model of membrane structure that agrees with this overall interpretation of the roles of protein and lipid in biological membranes.

### Membrane Isolation from Eukaryotic Cells

*In order to study the structure of biological membranes, it is first necessary to isolate them in a more or less intact form* from the cell. In eukaryotic cells, this problem is complicated by the existence of several different membrane systems in addition to the plasma membrane, each surrounding a specific organelle. Separation of membrane fractions initially requires the

**Table 16-1**
Chemical Compositions of Some Cell Membranes

| Membrane | Protein (%) | Lipid (%) | Carbohydrate (%) |
|---|---|---|---|
| Myelin | 18 | 79 | 3 |
| Human erythrocyte plasma membrane | 49 | 43 | 8 |
| Amoeba plasma membrane | 54 | 42 | 4 |
| *Mycoplasma* cell membrane | 58 | 37 | 1.5 |
| *Halobacterium* purple membrane | 75 | 25 | 0 |

Adapted from G. Guidotti, Membrane proteins. *Ann. Rev. Biochem.* 41:731, 1972.

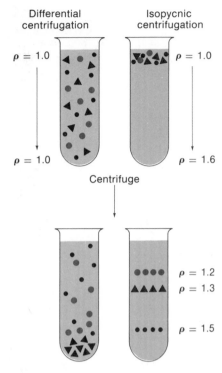

Differential centrifugation

$\rho = 1.0$

$\rho = 1.0$

Isopycnic centrifugation

$\rho = 1.0$

$\rho = 1.6$

Centrifuge

$\rho = 1.2$

$\rho = 1.3$

$\rho = 1.5$

**Figure 16-3**
Comparison of differential centrifugation (*left*), which separates on the basis of size, and isopycnic centrifugation (*right*), which separates on the basis of density.

disruption of the plasma membrane under conditions that leave subcellular organelles intact. One common procedure involves <u>mild homogenization</u> in a slightly hypotonic solution, while another method, <u>nitrogen cavitation</u>, involves forcing nitrogen gas into the cells under pressure and then rapidly releasing the pressure to "explode" the cell membrane.

Organelles can be isolated from disrupted cells by <u>differential centrifugation</u>, which separates them on the basis of size (Figure 16-3). Ruptured plasma membrane fragments can be purified from the same mixture by equilibrium-density-gradient (<u>isopycnic</u>) centrifugation because of their low density (high lipid content) relative to intact organelles (Table 16-2, column 4). This technique relies on centrifuging the sample into a preformed gradient of a solute, such as sucrose. When equilibrium is reached, each type of membrane or organelle will be found in the region of the gradient corresponding to its own density (Figure 16-3). More recently, gradients of synthetic sucrose polymers (Ficoll) or colloidal silica particles (Percoll) have been used in these separations because of their inertness, ability to form more stable gradients, and impermeability to biological membranes. The properties of isolated rat liver organelles are summarized in Table 16-2. The entries in column 2 provide some idea of the relative proportions of these organelles in the mammalian liver. For example, mitochondria represent 25 percent of the total cell protein, while lysosomes comprise less than 10 percent. Interestingly, the plasma membrane, which completely surrounds the cell, represents only 2 percent of the total protein. Although soluble proteins comprise only 30 percent of the total, the isolated organelles also contain soluble protein, so that somewhat less than 50 percent of the total cell protein is probably membrane-associated.

In column 3 of Table 16-2, the relative sizes of the organelles are indicated. It should be noted that the sedimentation behavior of an organelle in

**Table 16-2**
Properties of Rat Liver Organelles

| Organelle | Percent of Cell Protein | Diameter ($\mu$m) | Equilibrium Density in Sucrose (g/ml) | Organelle-Specific Enzyme Marker |
|---|---|---|---|---|
| Liver cell | 100 | 20 | 1.20 | — |
| Nuclei | 15 | 5–10 | 1.32 | DNA polymerase |
| Golgi apparatus | 2 | 2 | 1.10 | Glycosyl transferases |
| Mitochondria | 25 | 1 | 1.20 | Monoamine oxidase (outer membrane); cytochrome $c$ (inner membrane) |
| Lysosomes | 2 | 0.5 | 1.20 | Acid phosphatase |
| Endoplasmic reticular vesicles | 20 | 0.1 | 1.15 | Cytochrome $b_5$ reductase and cytochrome $b_5$; glucose-6-phosphatase |
| Cytoplasmic membrane | 2 | — | 1.15 | Na$^+$-K$^+$ ATPase; viral receptors |
| Soluble protein | 30 | <0.01 | — | — |

From M. H. Saier, Jr. and C. D. Stiles, *Molecular Dynamics in Biological Membranes*, Heidelberg Science Library, Vol. 22, Springer Verlag, New York, 1975. Reprinted with permission.

**Table 16-3**
Protein and Lipid Content of Organellar Membranes

| Membrane | Approximate Protein/Lipid Ratio (wt/wt) | Approximate Cholesterol/ Other Lipids (Molar Ratio) |
|---|---|---|
| Golgi apparatus | 0.7 | 0.08 |
| Liver plasma membrane | 1.0 | 0.40 |
| Endoplasmic reticulum | 1.0 | 0.06 |
| Mitochondrial outer membrane | 1.0 | 0.05 |
| Mitochondrial inner membrane | 3.0 | 0.03 |
| Nuclear membrane | 3.0 | 0.11 |
| Lysosomal membrane | 3.0 | 0.16 |

Adapted from M. H. Saier, Jr., and C. D. Stiles, *Molecular Dynamics in Biological Membranes*, Heidelberg Science Library, Vol. 22, Springer Verlag, New York, 1975.

a sucrose density gradient (column 4) does *not* correlate with size, but rather with its density, which is determined by its chemical composition (Figure 16-3). Nucleic acid ($\rho \sim 1.7$) is more dense than protein ($\rho \sim 1.25$), and protein is more dense than lipid ($\rho = 0.9$–$1.1$). These facts account for the relatively high density of nuclei and the low density of the Golgi apparatus, which has a relatively high lipid content. Since each organelle has a specific function, it also must possess a unique complement of enzymes. This prediction has been amply verified by the subcellular localization of numerous enzymes (Table 16-2, column 5), and these specific associations have greatly facilitated the assay and isolation of organelles from eukaryotic cells.

Once the subcellular organelles have been separated, their membranes can be isolated. For those organelles enclosed by a single membrane, treatment in hypotonic buffer (*osmotic shock*) followed by centrifugal separation of the membrane ghosts from the intraorganellar soluble proteins allows one to study membrane composition. Nuclei and mitochondria, however, possess two membranes, and these must be separated before their chemical and physical properties can be studied. In these cases, selective solubilization of the outer membrane can be obtained by treatment with appropriate detergents allowing purification of intact inner membranes. Procedures such as these have allowed detailed analyses of the lipid and protein contents of organellar membranes (Table 16-3) and have provided experimental systems in which to study the structures and functions of each different membrane system.

## Constituents of Bacterial Cell Envelopes

In contrast to animal cells, most prokaryotic cells are surrounded by a rather complex and rigid *cell wall*, which allows bacteria to live in a hypotonic environment without bursting and confers upon these cells their characteristic shape (rod, sphere, or spiral). In 1884, Christian Gram dis

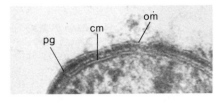

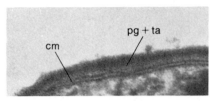

**Figure 16-4**
Electron micrographs of sections through the surface layers of a Gram-positive (a) and a Gram-negative (b) bacterium (cm = cytoplasmic membrane; om = outer membrane; Pg = peptidoglycan; ta = teichoic acid). Note the thick cell wall in (a) compared with the distinct inner and outer trilaminar membranes separated by a thin peptidoglycan layer in (b). (Courtesy J. Stolz; 150,000×.)

covered that bacteria could be divided into those which retained a crystal violet–iodine dye complex after washing with alcohol (*Gram positive*) and those which did not (*Gram negative*). Even today, the Gram stain reaction is a useful tool in classifying bacteria, and this difference in staining has been found to correlate with a fundamental difference in cell wall structure between Gram-positive and Gram-negative cells (Figure 16-4). Gram-positive cells are surrounded by a cytoplasmic membrane and a thick cell wall consisting of a sugar–amino acid heteropolymer, or *peptidoglycan* (Figure 16-5) and polyol phosphate polymers called *teichoic acids* (Figure 16-6a). Gram-negative bacteria have a much thinner cell wall consisting entirely of peptidoglycan and associated proteins, and this cell wall is surrounded by a second, outer membrane comprised of lipid, *lipopolysaccharide* (Figure 16-6b), and protein. The biosynthesis of the peptidoglycan and the outer membrane lipopolysaccharide are discussed in Chapter 12. The space between the inner and outer membranes, or *periplasmic space*, also contains proteins that have a variety of functions (see the following discussion).

In order to examine the compositions and functions of the various cell layers of Gram-negative bacteria, it is necessary to first separate these layers. This has been accomplished by treatment of the cells with *lysozyme* (which hydrolyzes peptidoglycan) and EDTA (which destabilizes the outer membrane) in isoosmotic sucrose solutions (Figure 16-7). *Periplasmic proteins* are released by this first step and can be separated by sedimenting the resulting *spheroplasts*, which have lost any nonspherical shape characteristic of the original cell because their peptidoglycan cell wall has been digested. Subsequent treatment of spheroplasts with high-frequency sound (*sonication*) ruptures both outer and inner membranes which quickly reseal into smaller spherical, closed *vesicles* (Figure 16-7). Because of their higher carbohydrate content, vesicles derived from the outer membrane have a higher density than those derived from the inner membrane and thus can be separated from them by isopycnic centrifugation in a sucrose density gradient. By these techniques, electron transport chains, ATP synthesizing enzymes, many transport proteins, and other enzymes have been

-GlcNAc–MurNAc–GlcNAc–MurNAc–GlcNAc–MurNAc-

L-Ala  -GlcNAc–MurNAc–GlcNAc–MurNAc-

D-Glu            L-Ala

DAP            D-Glu

D-Ala—C—NH—DAP

D-Ala

**Figure 16-5**
Two-dimensional representation of a bacterial peptidoglycan network. Chains of repeating *N*-acetylglucosamine (GlcNAc) and *N*-acetylmuramic acid (MurNAc) residues are linked by means of amide bonds between D-alanine and diaminopimelic acid (DAP) residues of tetrapeptides attached to the MurNAc units. Some tetrapeptides are not so linked (*vertical chains*).

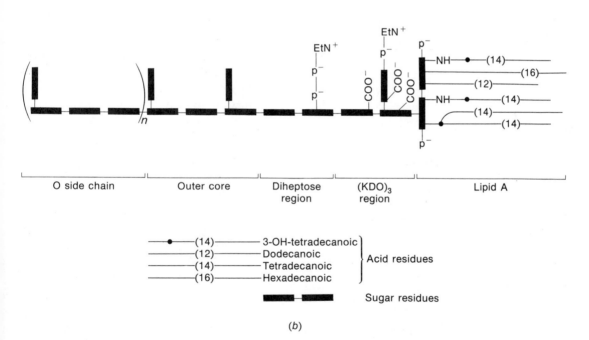

**Figure 16-6**
Structures of some bacterial cell envelope constituents. (*a*) Some teichoic acids of Gram-positive bacteria: (*i*) *Lactobacillus casei* (R = D-alanine); (*ii*) *Actinomyces antibioticus* (R = D-alanine); (*iii*) *Staphylococcus lactis* (R = D-alanine); (*iv*) *Bacillus subtilis* (R = glucose); (*i*)-(*iii*) are composed of repeating glycerol units, while (*iv*) is a ribitol teichoic acid to which D-alanine may be attached at either position 3 or 4 of the pentitol. (Adapted from R. Y. Stanier, E. A. Adelberg, and J. L. Ingraham, *The Microbial World,* 4th Ed., Prentice-Hall, Englewood Cliffs, N.J., 1976. Used with permission.) (*b*) Schematic illustration of the structure of lipopolysaccharide in the outer membrane of *Salmonella typhimurium.* (EtN = ethanolamine; KDO = 2-keto-3-deoxyoctonic acid.) (Adapted from H. Nikaido, Biosynthesis and assembly of lipopolysaccharide, in L. Leive (Ed.), *Bacterial Membranes and Walls,* Dekker, New York, 1973.)

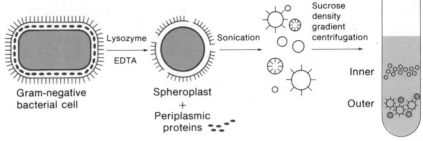

**Figure 16-7**
Separation of periplasmic and inner and outer membranes of a Gram-negative bacterium.

localized to the cytoplasmic membrane of Gram-negative bacteria, while the outer membrane has been shown to harbor receptors for bacteriophage and bacteriocins, certain other transport proteins, and various phospholipases. The periplasmic space has been shown to contain hydrolytic enzymes as well as nutrient-binding proteins involved in transmembrane transport and chemotaxis (Chapter 17). These observations on protein localization in Gram-negative bacteria are summarized in Table 16-4. The organization of proteins in Gram-positive bacteria is usually much simpler because these cells are surrounded by a single membrane. Thus soluble and cytoplasmic membrane proteins carry out functions similar to those of Gram-negative bacteria, while macromolecular hydrolases are exported into the extracellular medium where they function to scavenge nutrients from the environment.

**Table 16-4**
Subcellular Location of Proteins in Gram-Negative Bacteria

| **Extracellular** | **Inner Cytoplasmic Membrane** |
|---|---|
| Proteases | Electron-transfer chain |
| Lipases | Proton-translocating ATPase |
| Carbohydrases | Some transport proteins |
| Nucleases | Lipid and cell envelope biosynthetic enzymes |
| **Outer Lipopolysaccharide Membrane** | **Cytoplasm** |
| Receptor proteins for viruses and bacteriocidal agents | Enzymes that catalyze the synthesis and degradation of soluble substrates |
| Phospholipase A and lysophospholipase | DNA |
| | RNA |
| **Periplasm** | Ribosomes |
| Solute-binding proteins involved in transmembrane transport and chemotaxis | Enzymes involved in DNA replication, transcription, translation, etc. |
| Phosphatases and esterases | |

From M. H. Saier, Jr., and C. D. Stiles, *Molecular Dynamics in Biological Membranes*, Heidelberg Science Library, Vol. 22, Springer Verlag, New York, 1975. Reprinted with permission.

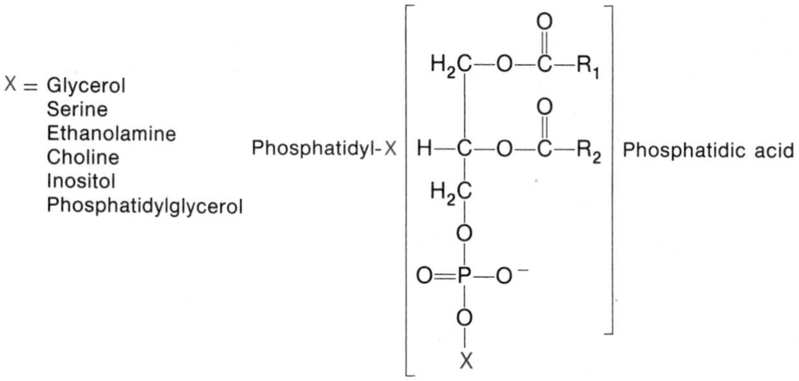

X = Glycerol
Serine
Ethanolamine
Choline
Inositol
Phosphatidylglycerol

Phosphatidyl-X

Phosphatidic acid

**Figure 16-8**
The structures of phospholipids. The head groups, X, are attached through their hydroxyl groups by means of phosphate ester linkages to phosphatidic acid. $R_1$ and $R_2$ are hydrocarbon chains of the esterified fatty acids.

## Membrane Lipids

Once isolated by any of the procedures just described, membrane fractions can be analyzed biochemically for their content of lipids and proteins. Lipids are usually extracted from proteins by treatment of membranes with organic solvents, such as chloroform and methanol. Numerous studies have shown that the most common lipids of biological membranes are the _phospholipids_, which can be subdivided into the _phosphoglycerides_ and _sphingomyelin_. Although the structures and biosynthetic pathways of these compounds were considered in Chapters 13 through 15, we shall review some of their properties, especially those which are most important in our consideration of membrane structure and function.

All phosphoglycerides are synthesized from a core molecule of L-glycerol-3-phosphate, to which two fatty acids have been esterified at the C-1 and C-2 hydroxyl groups of the glycerol moiety (1,2-difattyacyl-L-glycerol-3-phosphate or L-_phosphatidic acid_) (Figure 16-8). Phosphatidic acid itself is only a very minor component of most biological membranes and is esterified on the phosphoryl group by a variety of compounds to make up the majority of naturally occurring phosphoglycerides. These compounds all contain at least one hydroxyl group and include glycerol, serine, ethanolamine, choline, and inositol. The intact phospholipids resulting from esterification of phosphatidic acid by these compounds are called _phosphatidylglycerol, phosphatidylserine, phosphatidylethanolamine, phosphatidylcholine (or lecithin), and phosphatidylinositol_, respectively (Figure 16-8). Alternatively, a molecule of phosphatidylglycerol can be esterified through the C-1 hydroxyl group of the glycerol moiety to the phosphoryl group of another phosphatidic acid molecule, resulting in _diphosphatidylglycerol_ or _cardiolipin_ (Figure 16-8), another common phosphoglyceride of biological membranes. Finally, nerve and muscle cell membranes contain significant amounts of _plasmalogens_—phosphoglycerides in which the C-1 fatty acid ester is replaced by a long-chain, $\alpha$-$\beta$ unsaturated alkene group in ether linkage to the glycerol backbone (Figure 16-9).

The second group of phosphorus-containing lipids, which is especially important in membranes of higher animals, particularly in nervous tissue, consists of the sphingomyelins. These compounds are built from an aminoalcohol, _sphingosine_, to which is attached a fatty acid by means of an amide linkage to form a _ceramide_. Esterification of a molecule of

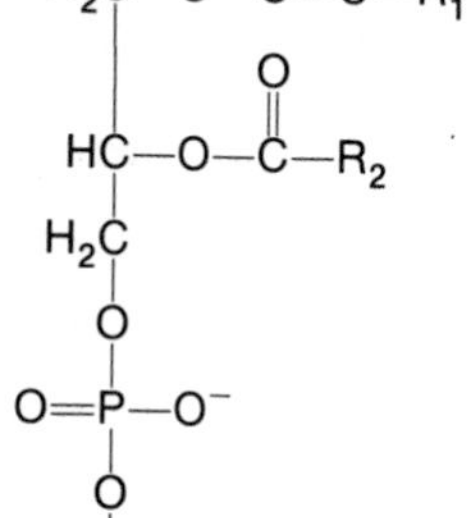

**Figure 16-9**
A plasmalogen. Note the $\alpha$-$\beta$ unsaturated alkene group in ether linkage to the C-1 hydroxyl of the glycerol moiety.

phosphorylethanolamine or phosphorylcholine to the 1-hydroxyl group of a ceramide completes the structure of a sphingomyelin (Figure 16-10).

Because a variety of fatty acids (Chapter 14) may be found in any of the phosphoglycerides or sphingomyelin, one usually isolates a mixture of compounds when purifying a particular phospholipid class from a biological membrane. Thus, in theory at least, a sample of purified lecithin could consist of a mixture of 1-palmitoyl-2-stearoyl-phosphatidylcholine and 1-oleoyl-2-linoleoyl-phosphatidylcholine (Figure 16-11). In practice, however, one usually finds that phosphoglycerides contain one saturated fatty acid at position 1 and an unsaturated fatty acid or cyclopropane fatty acid (in bacteria) in position 2 of the glycerol moiety (Figure 16-12). Thus a common constituent of bovine lecithin is found to be 1-palmitoyl-2-oleoylphosphatidylcholine, while a typical bacterial phosphatidylethanolamine might have the structure shown in Figure 16-13.

An examination of the fatty acid compositions that have been determined for various phospholipids from biological sources indicates the variety of compounds that a given class of phospholipids can encompass because of variable fatty acid compositions (see Chapter 14 and Table 14-2). Nearly all fatty acids found in phospholipids from biological sources have an _even_ number (14–24) of carbon atoms (except for cyclopropane fatty acids) (Figure 16-13). An exception to this rule is found in some marine organisms, which contain significant amounts of fatty acids with odd numbers of carbons. The unsaturated fatty acids of phospholipids may have one, two, three, or even four double bonds. In most bacteria, however, unsaturated fatty acids containing more than one double bond are lacking. Finally, it should be noted that nearly all unsaturated fatty acids found in naturally occurring phospholipids are of the cis configuration. The importance of this fact will become evident when we discuss the physical properties of membranes.

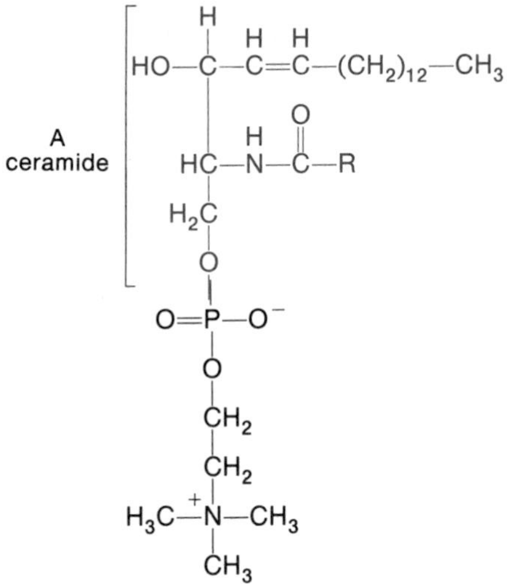

**Figure 16-10**
A sphingomyelin. The sphingosine moiety is shown in color.

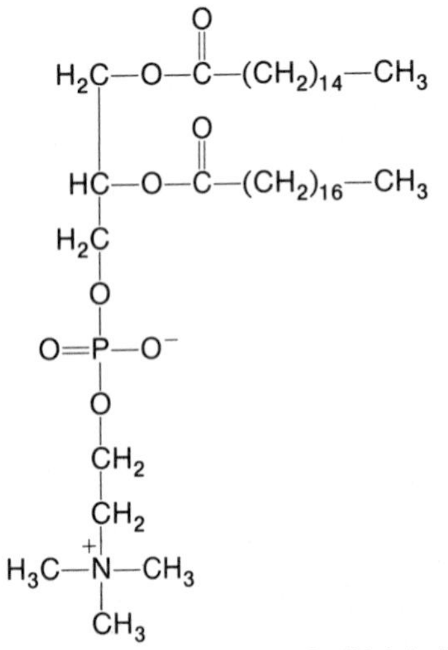

1-Palmitoyl-2-stearoyl-phosphatidylcholine

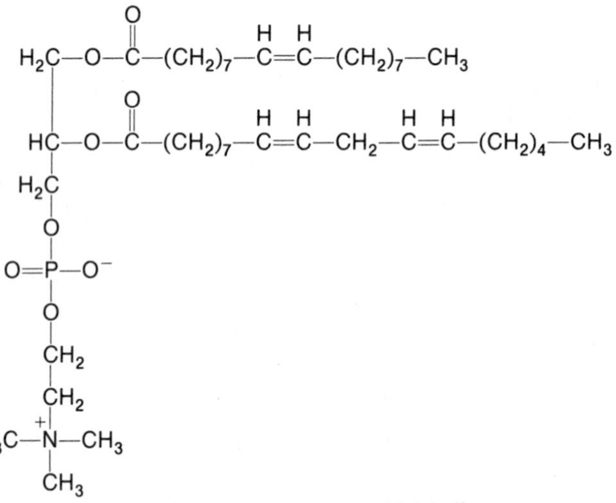

1-Oleoyl-2-linoleoyl-phosphatidylcholine

**Figure 16-11**

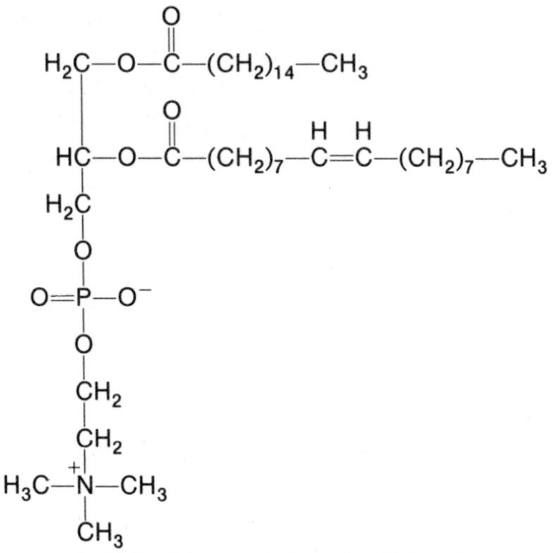

$$H_2C-O-\overset{\displaystyle O}{\overset{\|}{C}}-(CH_2)_{14}-CH_3$$

$$HC-O-\overset{\displaystyle O}{\overset{\|}{C}}-(CH_2)_7-\overset{H}{C}=\overset{H}{C}-(CH_2)_7-CH_3$$

$$H_2C$$
$$O$$
$$O=P-O^-$$
$$O$$
$$CH_2$$
$$CH_2$$
$$H_3C-\overset{+}{N}-CH_3$$
$$CH_3$$

**1-Palmitoyl-2-oleoyl-phosphatidylcholine**

**Figure 16-12**
A common constituent of bovine lecithin.

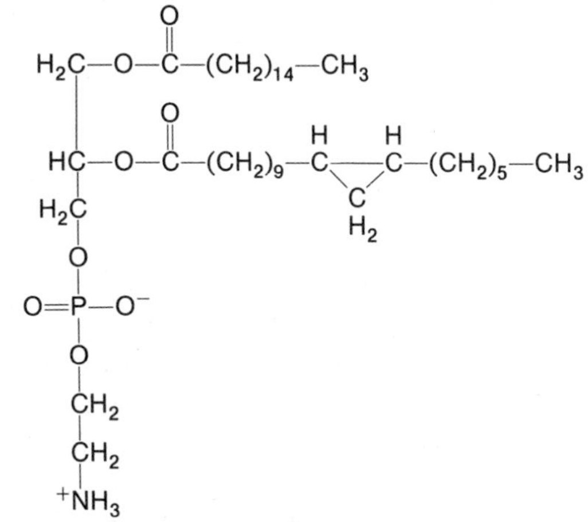

$$H_2C-O-\overset{\displaystyle O}{\overset{\|}{C}}-(CH_2)_{14}-CH_3$$

$$HC-O-\overset{\displaystyle O}{\overset{\|}{C}}-(CH_2)_9-\overset{H}{\underset{\displaystyle \overset{C}{H_2}}{C}}\overset{H}{-C}-(CH_2)_5-CH_3$$

$$H_2C$$
$$O$$
$$O=P-O^-$$
$$O$$
$$CH_2$$
$$CH_2$$
$$^+NH_3$$

**1-Palmitoyl-2-lactobacilloyl-phosphatidylethanolamine**

**Figure 16-13**
A bacterial phospholipid containing the cyclopropane fatty acid *lactobacillic acid* at the C-2 position.

A second class of lipids that are found in biological membranes is comprised of the *glycosphingolipids*. These compounds are important components of nerve and muscle cell membranes, as is sphingomyelin, and are also derived from a ceramide backbone (Figure 16-10). Instead of containing phosphorylcholine or phosphorylethanolamine, however, these glycolipids are characterized by one or more sugar residues bound in a β-glycosidic linkage to the 1-hydroxyl group of the ceramide (see Table 14-4). If a single sugar residue, either glucose or galactose, is bound, the molecule is termed a *cerebroside* (Figure 16-14). Galactocerebrosides are abundant in the brain and nervous system and may be esterified at position 3 of the galactose residue to yield a *sulfatide* (Figure 16-14). More complex glycosphingolipids include the *gangliosides*, which contain three or more sugar residues attached to the ceramide backbone, including at least one residue of a *sialic acid* (Figure 16-15). Gangliosides are found in membranes of nerve endings, especially in the brain, and may be important in transmission of nerve impulses (Chapter 30).

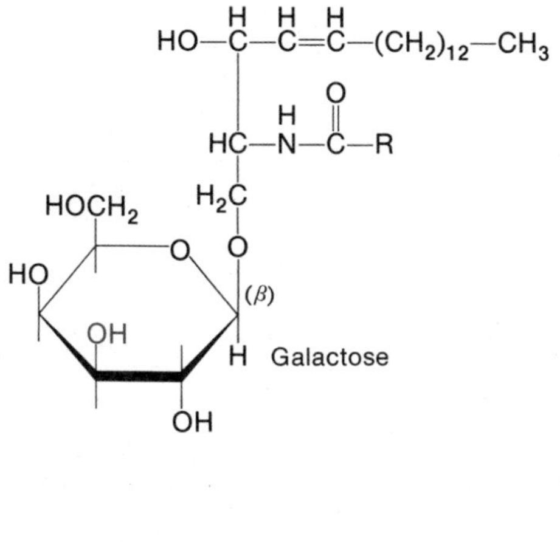

$$HO-\overset{H}{C}-\overset{H}{C}=\overset{H}{C}-(CH_2)_{12}-CH_3$$

$$\overset{H}{HC}-N-\overset{\displaystyle O}{\overset{\|}{C}}-R$$

$$H_2C$$

HOCH₂ ... HO ... OH ... (β) ... H Galactose ... OH

**Figure 16-14**
A galactocerebroside. The sulfate ester of the hydroxyl group at position 3 (*shown in color*) is termed a sulfatide.

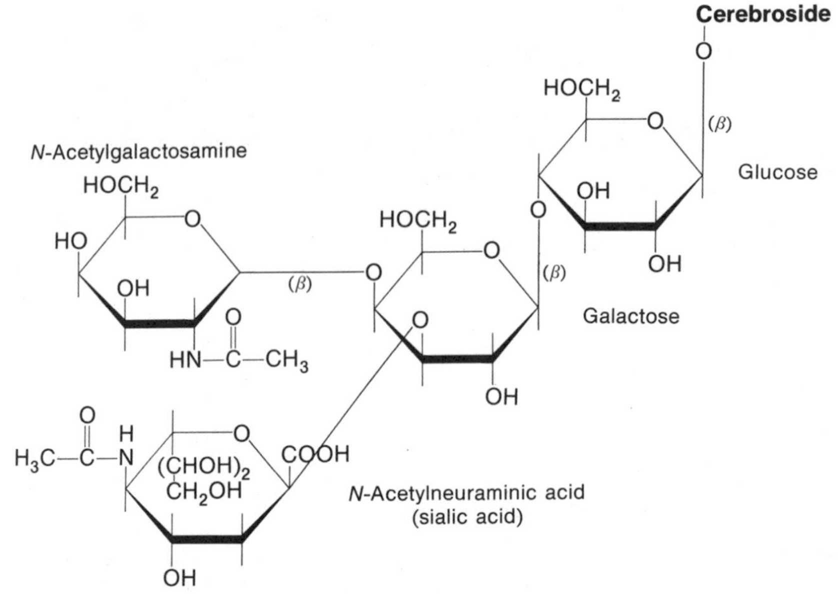
Cerebroside
HOCH2
(β)
Glucose
N-Acetylgalactosamine
HOCH2
HO
OH
HOCH2
OH
OH
(β)
(β)
Galactose
HN—C—CH3
OH
H3C—C—N
(CHOH)2
CH2OH
COOH
N-Acetylneuraminic acid
(sialic acid)
OH

**Figure 16-15**
A ganglioside.

*Lipopolysaccharide,* discussed earlier as a component of the outer membrane of Gram-negative bacteria, appears to be limited to this class of organisms. Although it generally has a higher content of carbohydrate than lipid by weight, it also is often classified as a membrane lipid because its fatty acyl chains constitute an integral part of the structure of the Gram-negative outer membrane. Among other functions, lipopolysaccharide confers upon a bacterium its characteristic antigenicity because the terminal carbohydrate chains (O antigens) extend to the surface of the cell (Figure 16-6). Lipopolysaccharide is also the major component of *endotoxins,* which are released upon lysis of Gram-negative bacteria. Endotoxins are potent pyrogens (fever producers) and contribute to the pathogenicity of a number of these organisms.

The final class of lipids found in biological membranes that we will consider are the *steroids* and their biosynthetic precursors, the *isoprenoids* or *terpenes.* The most common membrane steroid is *cholesterol* (Figure 16-16), which is found in animal cell membranes and a few microorganisms. In addition, steroids are common components of some of the lipoproteins of blood plasma (Chapter 15). Some of the more important terpenes found in biological membranes include *11-cis-retinal,* a derivative of vitamin A found in the vertebrate retina (Chapter 31); *phytol,* a compo-

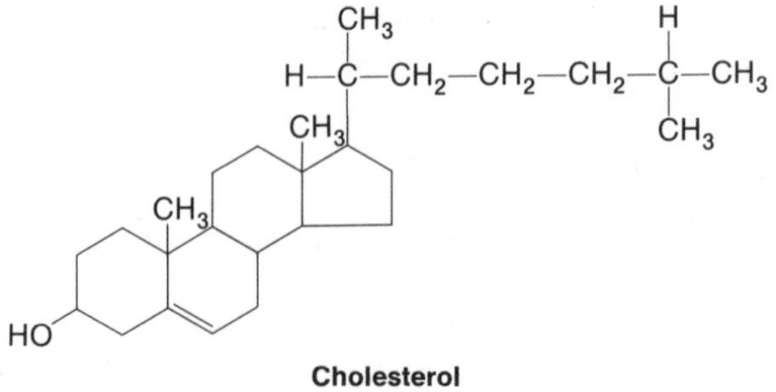
CH3
H
H—C—CH2—CH2—CH2—C—CH3
CH3
CH3
CH3
HO

**Cholesterol**

**Figure 16-16**

nent of the chlorophylls (Chapter 11); *undecaprenyl alcohol* or *bactoprenol* (Figure 16-17), an important coenzyme in cell wall biosynthesis in bacteria; and the *ubiquinones* or *coenzymes Q*, which function as hydrogen carriers in the inner mitochondrial membrane (Chapter 10).

Table 16-5 compares the lipid compositions of membranes from a number of biological sources. A number of generalizations can be made from the data in this table. First, phosphatidylcholine is the major phospholipid found in membranes of animal cells, while phosphatidylethanolamine predominates in bacteria. Second, in addition to cholesterol, both sphingomyelin and glycolipids (except for lipopolysaccharides) are usually *absent* from prokaryotic membranes. Most of the membranes represented in the table, however, have a characteristic trilaminar appearance in the electron microscope (Figure 16-1) and yet have widely varying lipid compositions. Thus either membrane lipids have little to do with determining membrane structure, or they all have certain properties in common which allow them

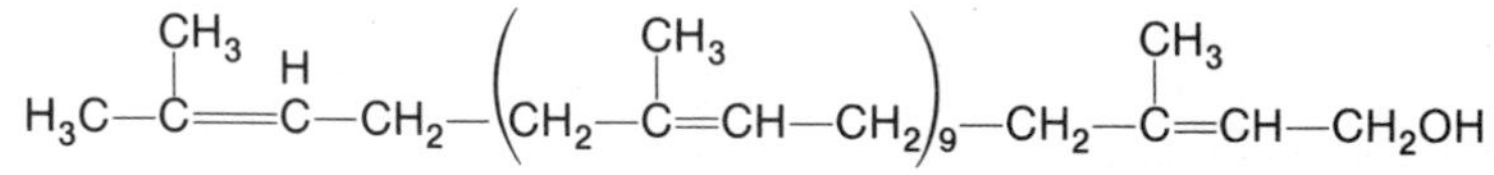

**Undecaprenyl alcohol** (bactoprenol)

**Figure 16-17**

**Table 16-5**
Lipid Compositions of Membrane Preparations

| Source | Choles-terol | PC | SM | PE | PI | PS | PG | DPG | PA | Glyco-lipids |
|---|---|---|---|---|---|---|---|---|---|---|
| Rat liver | | | | | | | | | | |
| Cytoplasmic membrane | 30.0 | 18 | 14.0 | 11 | 4.0 | 9.0 | — | — | 1 | — |
| Endoplasmic reticulum (rough) | 6.0 | 55 | 3.0 | 16 | 8.0 | 3.0 | — | — | — | — |
| Endoplasmic reticulum (smooth) | 10.0 | 55 | 12.0 | 21 | 6.7 | — | — | 1.9 | — | — |
| Mitochondria (inner) | 3.0 | 45 | 2.5 | 25 | 6.0 | 1.0 | 2.0 | 18.0 | 0.7 | — |
| Mitochondria (outer) | 5.0 | 50 | 5.0 | 23 | 13.0 | 2.0 | 2.5 | 3.5 | 1.3 | — |
| Nuclear membrane | 10.0 | 55 | 3.0 | 20 | 7.0 | 3.0 | — | — | 1.0 | — |
| Golgi | 7.5 | 40 | 10.0 | 15 | 6.0 | 3.5 | — | — | — | — |
| Lysosomes | 14.0 | 25 | 24.0 | 13 | 7.0 | — | — | 5.0 | — | — |
| Rat brain | | | | | | | | | | |
| Myelin | 22.0 | 11 | 6.0 | 14 | — | 7.0 | — | — | — | 21 |
| Synaptosome | 20.0 | 24 | 3.5 | 20 | 2.0 | 8.0 | — | — | 1.0 | — |
| Rat erythrocyte | 24.0 | 31 | 8.5 | 15 | 2.2 | 7.0 | — | — | 0.1 | 3 |
| Rat rod (outer segment) | 3.0 | 41 | — | 37 | 2.0 | 13.0 | — | — | — | — |
| *E. coli* cytoplasmic membrane | 0 | 0 | — | 80 | — | — | 15.0 | 5.0 | — | — |
| *Bacillus megaterium* cytoplasmic membrane | 0 | 0 | — | 69 | — | — | 30.0 | 1.0 | — | Trace |

*PC = phosphatidylcholine; SM = sphingomyelin; PE = phosphatidylethanolamine; PI = phosphatidylinositol; PS = phosphatidylserine; PG = phosphatidylglycerol; DPG = diphosphatidylglycerol (cardiolipin); PA = phosphatidic acid.

Adapted from M. K. Jain and R. C. Wagner, *Introduction to Biological Membranes*, Wiley, New York, 1980; the data for *B. megaterium* are from J. E. Rothman and E. P. Kennedy, Asymmetrical distribution of phospholipids in the membrane of *Bacillus megaterium*. *J. Mol. Biol.* 110:603, 1977.

to contribute in a similar manner to the characteristic architecture of biological membranes. As we shall see in the following sections, the latter explanation is undoubtedly the correct one.

### Membrane Lipids Form Ordered Structures Spontaneously

One characteristic common to all the membrane lipids we have discussed is the presence in the same molecule of both hydrophilic and hydrophobic groups. Such molecules are said to be *amphipathic*. Phospholipids consist of long hydrophobic *"tails,"* the fatty acyl chains, and a compact *polar "headgroup"* consisting of the phosphate group and the esterified alcohol. At neutral pH, the polar headgroup may have a net negative charge (phosphatidic acid, phosphatidylglycerol, diphosphatidylglycerol, phosphatidylinositol, phosphatidylserine), positive charge (*O-lysylphosphatidylglycerol*), or no net charge (phosphatidylcholine, phosphatidylethanolamine). Sphingomyelin, although differing considerably in chemical structure from the phosphoglycerides, nevertheless appears very similar to the other phospholipids in space-filling models (see Chapter 14). Similarly, glycolipids and bacterial lipopolysaccharide have the hydrocarbon tails and polar headgroups (in this case, sugar residues) as do the phospholipids. Even cholesterol has hydrophobic and hydrophilic *"sides,"* although differing altogether in chemical structure from the fatty acid–containing lipids.

Because of their amphipathic nature, membrane lipids exhibit only a limited solubility in aqueous solution. Thus, when phospholipid is added to water, very few of the lipid molecules exist freely in solution as monomers because of the large hydrophobic surface area of the molecule. Instead, a "film" of phospholipid tends to form on the water-air interface, and physical studies have shown that this film is a *monolayer* of phospholipid arranged such that the polar headgroups are in contact with water, while the hydrocarbon tails extend up into the air phase (Figure 16-18). When more phospholipid is added to the solution, saturating the air-water interface, other assemblages of phospholipids are formed, including *micelles* and *bilayers* (see Figure 16-18). Both these structures maximize hydrophobic and van der Waals interactions between the fatty acyl chains, effectively excluding water from their vicinity, and allow the polar headgroups to interact with water molecules. Monolayers, micelles, and bilayers are the favored forms of phospholipid in aqueous solution because their formation results in an *increase* in entropy, resulting from the fact that water molecules need not order themselves around the hydrophobic hydrocarbon tails of the phospholipid monomer.

As shown in Figure 16-18, phospholipid bilayers in aqueous solution are actually spherical "bubbles" or vesicles with water inside and out. This structure is favored over a planar bilayer because exposed hydrocarbon tails, which would occur at the periphery of a planar sheet of phospholipid, are not present. In vesicular structures, no hydrophobic groups need to be exposed to water molecules. In addition, most naturally occurring phospholipids prefer to form vesicular bilayers instead of micelles in water solution because more efficient packing of the molecules can take place in the bilayer vesicle. In contrast, *lysophospholipids*, which lack one fatty acyl chain, free fatty acids, and detergents (see below) form micelles more readily than bilayers because of their geometry, which includes a smaller hydrophobic surface area relative to the phospholipids.

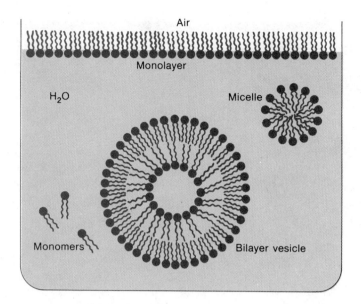

**Figure 16-18**
Structures formed by phospholipids in aqueous solution. Each molecule is depicted schematically as a polar headgroup ( ● ) attached to two fatty acyl hydrocarbon chains ( ).

The implications of these properties of phospholipids and glycolipids with respect to biological membrane structure should be obvious:

1. In aqueous solution, these molecules *spontaneously* form bilayer structures; i.e., the $\Delta G$ for this process is negative owing to an increase in entropy, $\Delta S$ (recall that $\Delta G = \Delta H - T\,\Delta S$).

2. A cell is essentially a plasma membrane–encapsulated vesicle, although it is usually not a perfect sphere owing to other structural contributions to its shape.

3. The arrangement of phospholipid polar headgroups in bilayers suggests an explanation for the trilaminar appearance of biological membranes in the electron microscope (Figure 16-1). Indeed, electron micrographs of purified phospholipid dispersed in water have this same appearance (Figure 16-19).

4. Phospholipid bilayers are relatively impermeable to most hydrophilic substances because of their hydrophobic interiors. This is also a property of biological membranes, unless a *transport system* recognizes the hydrophilic molecule (Chapter 17).

We are left with the conclusion that *biological membranes must themselves consist at least partially of lipid bilayers*, and that lipids must have a very important role in determining membrane structure. As we shall see shortly, both these conclusions are overwhelmingly confirmed by physical and chemical studies of biological membrane structure.

## Proteins of Biological Membranes

**Isolation and General Properties.** Although some characteristics of biological membranes can be explained by the properties of membrane lipids in aqueous solution, it is clear that others, especially specific functions such as transport and enzymatic activities, must rely on membrane-associated proteins. Therefore, in any model of membrane structure, we also must consider the properties of proteins found in biological membranes. Two general

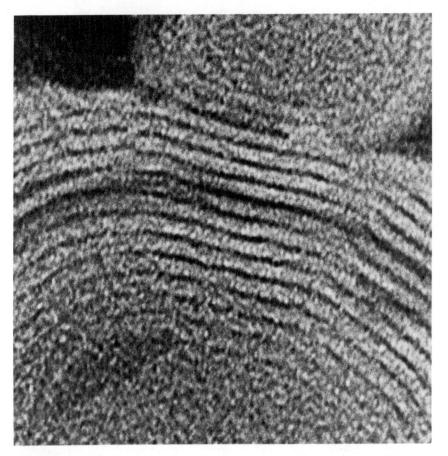

**Figure 16-19**
Multilayered vesicles formed from sonically dispersed lecithin (containing 10% dicetylphosphate) in 2% potassium phosphotungstate. Note the trilaminar appearance of each layer. (From A. D. Bangham, M. M. Standish, and J. C. Watkins, Diffusion of univalent ions across the lamellae of swollen phospholipids, *J. Mol. Biol.* 13: 238, 1965. Reprinted with permission.)

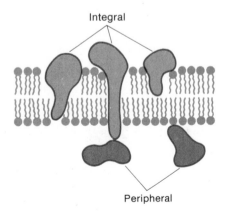

**Figure 16-20**
Schematic illustration of integral and peripheral membrane proteins.

types of such proteins have been found: those which are dissociable from the isolated membrane by agents that tend to disrupt ionic or hydrogen bonds (high salt, EDTA, which chelates $Ca^{2+}$ and $Mg^{2+}$ ions, or urea); and those which can be released from the membrane only by disruption of the hydrophobic interactions of membrane lipids by organic solvents or detergents. The former proteins are referred to as _peripheral_ membrane proteins, while the latter appear to be deeply embedded in the membrane and are called _integral_ membrane proteins. Peripheral proteins probably are bound to the membrane as a result of specific interactions with exposed, hydrophilic portions of integral membrane proteins. In contrast, significant hydrophobic interactions with membrane lipids and proteins appear to best explain the interaction properties of integral membrane proteins. Figure 16-20 illustrates these differences, assuming that the fundamental structure of the biological membrane is, indeed, a lipid bilayer.

In order to understand the physical and chemical properties of integral membrane proteins, it is necessary to purify them. First, the proteins must be dissociated from the membrane matrix. As mentioned earlier, integral membrane proteins often can be brought into solution using organic solvents or _detergents_. Detergents are naturally occurring or synthetic amphipathic molecules that disrupt membranes by intercalation into the membrane matrix and solubilization of the component lipids and proteins. Examples of detergents that are natural products include _lysolecithin_ (Figure 16-21) and _sodium deoxycholate_, a steroid bile salt (Figure 16-22). Both are _ionic_ detergents. Synthetic detergents include _Triton X-100_ (Figure 16-23) and _octylglucoside_ (Figure 16-24), two _nonionic_ detergents, and the ionic compounds _cetyl trimethylammonium bromide_ and _sodium dodecylsulfate_ (SDS) (Figure 16-25), which is also an extremely effective protein denaturant.

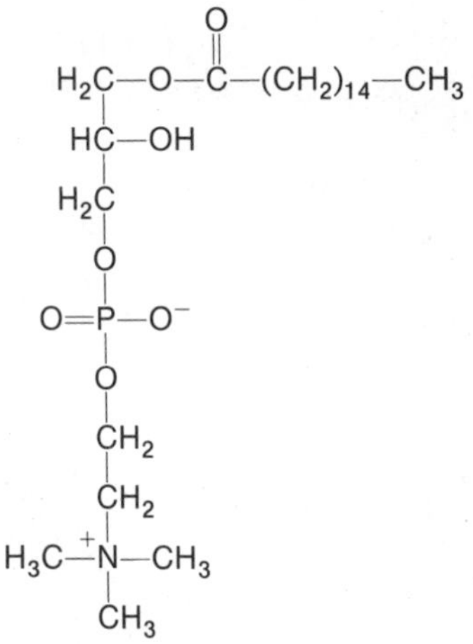

**A lysolecithin**

**Figure 16-21**

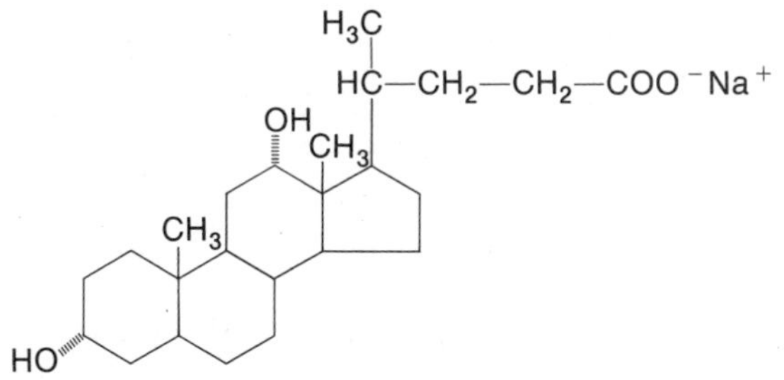

**Figure 16-22**        **Sodium deoxycholate**

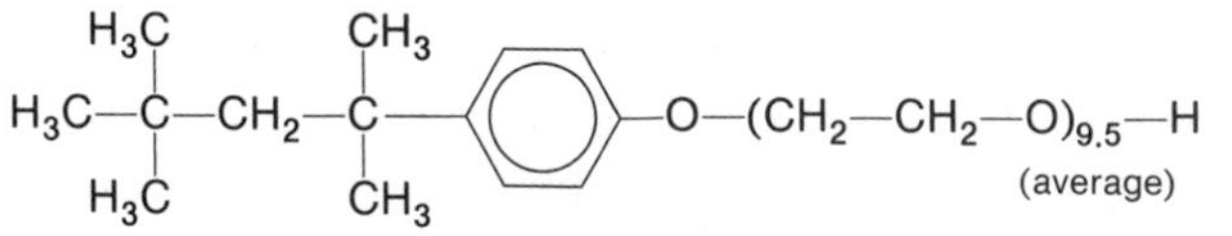

**Triton X-100**

**Figure 16-23**      [polyoxyethylene(9.5)_p-t_-octylphenol]

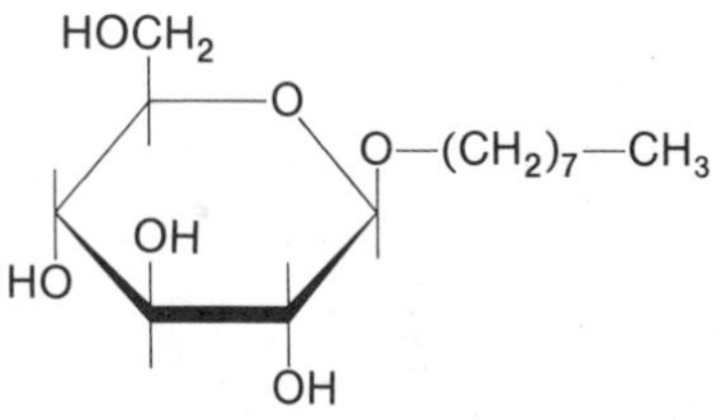

**Octylglucoside**
(octyl-_β_-D-glucopyranoside)

**Figure 16-24**

Many detergents, including the nonionic ones and lysolecithin, dissolve membranes by forming detergent-lipid and detergent-lipid-protein _mixed micelles_ (Figure 16-26). This is a result of the fact that these detergents prefer to form micelles rather than bilayers because of their particular geometries, as discussed earlier. Thus an excess of these compounds, when added to a membrane suspension, will tend to shift the equilibrium of _all_ amphipathic molecules in the mixture from bilayer to mixed micelle (Figure 16-26). Each detergent has a characteristic _critical micellar concentration_ (CMC), above which it exists in aqueous solution almost entirely in a micellar form, and a characteristic _hydrophilic-lipophilic balance_ (HLB), which is defined as the ratio of the molecular weight of the hydrophilic portion of the molecule to that of the hydrophobic portion. Both these properties appear to be important in determining the effectiveness of a particular detergent in dissolving membranes, although different membrane systems often respond differently to the same detergent. Some of the properties of detergents commonly used in membrane biochemistry are listed in Table 16-6. Because ionic detergents are more likely to alter the conformation of hydrophilic portions of membrane proteins, which are often responsible for catalytic activity, they are more likely to destroy bio-

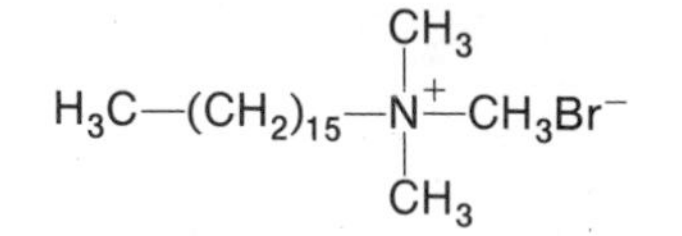

**Cetyltrimethylammonium bromide**

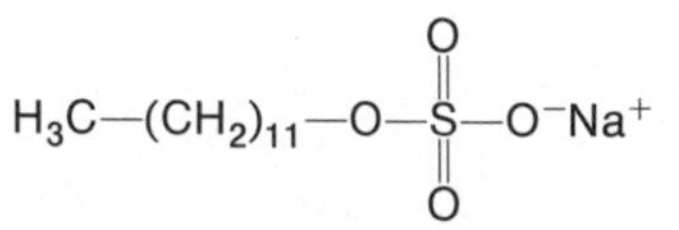

**Sodium dodecylsulfate** (SDS)

**Figure 16-25**

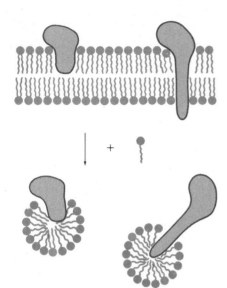

**Figure 16-26**
Detergent solubilization of biological membranes yields detergent-lipid-protein mixed micelles.

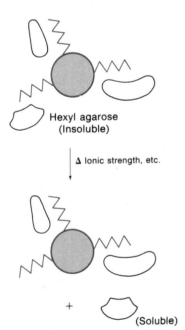

Hexyl agarose
(Insoluble)

Δ Ionic strength, etc.

+

(Soluble)

**Figure 16-27**
Hydrophobic interaction chromatography. Proteins bound by hydrophobic interactions to insoluble alkyl agarose beads can be differentially removed by changes in ionic strength, hydrophobicity, pH, or temperature.

**Table 16-6**
Properties of Some Detergents

| Detergent | Monomer $M_r$ | CMC (mM)* | Micellar $M_r$ (average)* |
|---|---|---|---|
| Nonionic | | | |
| Triton X-100 | 625 | 0.24 | 90,000 |
| Octyl-$\beta$-D-glucoside | 292 | 25 | — |
| Ionic | | | |
| Deoxycholate (anion) | 392 | 4 | 800 |
| Lysolecithin (egg; mixture) | 500–600 | 0.02–0.2 | 95,000 |
| SDS | 288 | 8 | 17,000 |

*It should be noted that both CMC and micellar size are dependent on a number of factors, including ionic strength and pH. These numbers should therefore be used only for general comparisons, since each was determined under slightly different conditions.

logical function. This is in agreement with the observation that only ionic detergents bind to and denature soluble proteins. Little interaction is usually observed between typical hydrophilic proteins and nonionic detergents.

Once dissociated from the membrane with detergent, specific membrane proteins can be isolated by a variety of separation techniques provided an assay is available for the protein of interest. Gel filtration and other procedures that separate proteins on the basis of size and shape (Chapter 3) are often <u>not</u> particularly useful in this regard because inclusion of integral membrane proteins in detergent-lipid micelles, which can have molecular weights of $10^5$ or greater, tends to mask individual size differences. Only with detergents possessing a high CMC, such as deoxycholate and octylglucoside, are separations based on size generally possible. Ion-exchange chromatography (Chapter 3) can be useful in separating solubilized membrane proteins if the detergent used is nonionic, such as Triton X-100. These difficulties have led to the search for techniques that might be especially suited to membrane protein isolation.

One such tool found to be particularly useful in this regard is <u>hydrophobic interaction chromatography</u>. This technique separates proteins on the basis of their relative hydrophobicities and uses insoluble supports, such as agarose or polyacrylamide, to which hydrophobic alkyl or aryl groups have been covalently attached. Liquid chromatography columns of such <u>hydrophobic resins</u> have been used in the purification of a number of integral membrane proteins. Proteins bound to the resin by means of hydrophobic interactions can then often be eluted sequentially, depending on the strength of this interaction, by gradual changes in hydrophobicity, ionic strength, pH, or temperature of the eluting buffer (Figure 16-27). <u>Affinity chromatography</u>, a technique that is very useful in purifying soluble enzymes (Chapter 3), also has been applied successfully to a number of integral membrane proteins, such as transport proteins (Chapter 17), which interact with specific metabolites. Ammonium sulfate and isoelectric precipitation, isoelectric focusing, and preparative gel electrophoresis also have been used in the purification of integral membrane proteins in the presence of suitable solubilizing agents (see Chapter 3).

The molecular weights of membrane proteins and the complexity of a specific membrane can be estimated by the technique of polyacrylamide gel electrophoresis after complete solubilization of the membrane in the ionic detergent SDS (Chapter 3). With this procedure it has been possible to show, for example, that myelin, a membrane surrounding some nerve cell axons (Chapter 30), and mammalian muscle sarcoplasmic reticular membrane both contain three major proteins. The most abundant sarcoplasmic reticular protein has been identified as the calcium translocating ATPase (Chapter 17), whereas a single protein, rhodopsin (Chapter 31), is the major species of the rod outer segment membrane of the mammalian retina. Only about a dozen proteins are distinguishable by this procedure in human erythrocyte membranes (Figure 16-28). By contrast, most bacterial cytoplasmic membranes are far more complex, containing well over 100 different protein species with none predominating. The large number of bacterial membrane proteins was to be predicted from the multiplicity of cytoplasmic membrane–related functions in prokaryotes (Table 16-4). In many but not all cases, it appears that cellular differentiation and specialization tend to render a membrane structurally less complex.

In Table 16-7, some physical and chemical properties of a number of integral membrane proteins that have been purified to apparent homogeneity are compared. No single property appears to be unique to these proteins compared with soluble enzymes, although a few, such as bacteriorhodopsin and the lactose permease from *E. coli*, are highly enriched in hydrophobic amino acids. This property may reflect the proportion of the polypeptide chain embedded in the hydrophobic portion of the membrane (see the following discussion). One property that does, however, seem to be com-

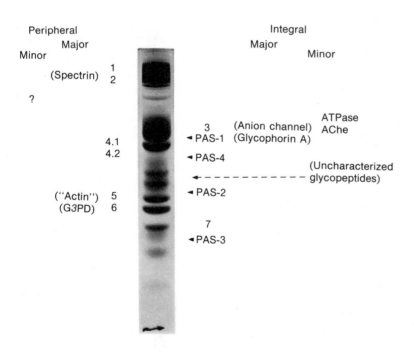

Polypeptides of the human erythrocyte membrane

**Figure 16-28**
Proteins of the human erythrocyte membrane resolved by polyacrylamide gel electrophoresis in the presence of SDS. Protein bands were stained with the dye Coomassie brilliant blue. Stained bands are numbered 1 to 7 (after G. Fairbanks, T. L. Steck, and D. F. H. Wallach, Electrophoretic analysis of the major polypeptides of the human erythrocyte membrane, *Biochemistry* 10: 2606, 1971). PAS-1 to PAS-4 are sialoglycoproteins, which stain heavily with a reagent that detects carbohydrate. On this electropherogram, the anion channel (band 3) and glycophorin A (PAS-1) are not completely resolved. (From V. T. Marchesi, H. Furthmayr, and M. Tomita, The red cell membrane, *Ann. Rev. Biochem.* 45: 667, 1976. Reprinted with permission.)

mon to many isolated integral membrane proteins is not reflected by the data in Table 16-7. Techniques that remove detergent and residual phospholipid from these proteins often lead to _inactivation_ of any enzymatic or other biological activity they possess, and this inactivation is often accompanied by aggregation or precipitation of the protein from solution. From this fact we can conclude that _integral membrane proteins usually require a hydrophobic environment for maintenance of their biologically active structures._

**Arrangement of Integral Proteins within the Membrane.** In order to understand how proteins contribute to the structure of biological membranes, it is also necessary to examine their _topographies in situ,_ i.e., their three-dimensional structures with respect to the membrane system in which they are found. This has been done for a number of proteins that are major components of particular biological membrane systems. Techniques for examining membrane protein topography include the use of proteases and other membrane-impermeable modifying reagents coupled with amino acid sequence analysis to determine which portions of the polypeptide chain are exposed on the outside or inside of the membrane and which are buried within the lipid portion of the membrane structure. In a few cases in which a membrane contains a single type of protein, it has even been possible to use electron microscopic techniques to examine the disposition of the polypeptide with respect to the membrane.

**Table 16-7**
Properties of Some Purified Membrane Proteins

| Protein | Source | Monomer $M_r$ | Subunit Structure | Percent Hydrophobic Amino Acids* | Covalent Carbohydrate |
|---|---|---|---|---|---|
| Cytochrome $b_5$ | Liver endoplasmic reticulum | 16,000 | Dimer (0.4% deoxycholate) | 40 | — |
| Cytochrome $b_5$ reductase | Liver endoplasmic reticulum | 43,000 | Monomer (0.4% deoxycholate) | 48 | — |
| Anion transport (band 3) protein | Human erythrocytes | 95,000 | Dimer | 48 | + |
| Glycophorin | Human erythrocytes | 31,000 | Dimer | 38 | + + |
| Bacteriorhodopsin | _Halobacterium halobium_ | 27,000 | Trimer | 57 | — |
| Lactose permease | _E. coli_ plasma membrane | 46,000[†] | ? | 59[†] | ? |
| Mannitol permease | _E. coli_ plasma membrane | 60,000 | ? | 46 | ? |
| Porin | _E. coli_ outer membrane | 36,000 | Trimer | 34 | — |

*Mole percent of Pro, Ala, ½ Cys, Val, Met, Ile, Leu, Phe, and Trp.

[†]Deduced from the DNA sequence of the _lacY_ gene. (D. E. Büchel, B. Gronenborn, and B. Müller-Hill, Sequence of the lactose permease gene. _Nature_ 283:541, 1980.)

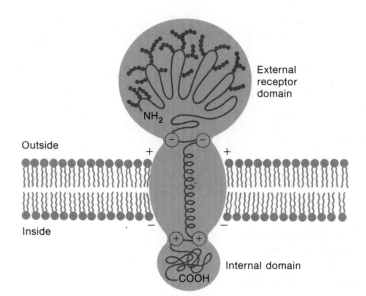

External receptor domain

NH₂

Outside

Inside

Internal domain

COOH

**Figure 16-29**
Topography of glycophorin in the mammalian erythrocyte membrane. Carbohydrate residues (●●●●) are all in the N-terminal domain on the outside of the cell and are attached mainly to hydroxyl groups of serine and threonine residues of the protein. (Adapted from M. H. Saier, Jr., and C. D. Stiles, *Molecular Dynamics in Biological Membranes,* Heidelberg Science Library, Vol. 22, Springer-Verlag, New York, 1975. Used with permission.)

These kinds of experiments have shown that *two basic types* of integral membrane proteins exist: those with a large proportion of their mass extending beyond the hydrophobic interior of the membrane into the aqueous medium, and those in which most of the polypeptide is embedded in the membrane and is thus inaccessible to hydrophilic labeling reagents. Examples of the former include two major proteins found in the red blood cell membrane: a major sialoglycoprotein, *glycophorin*, and the so-called *band 3*, or *anion-transport*, protein (Figure 16-28).

Glycophorin has been isolated from the red blood cell membrane and has a molecular weight of about 31,000. It is 40 percent protein and 60 percent carbohydrate on a weight basis and bears the ABO- and MN-blood group antigenic specificities of the cell as well as serving as the receptor for influenza virus. The probable structure for this protein is shown in Figure 16-29. It consists of a single polypeptide chain to which short carbohydrate chains are covalently attached to amino acid residues comprising the N-terminal region of the protein. This portion of the molecule is polar by virtue of the presence of sugar and hydrophilic amino acid residues. The carboxyl end of the polypeptide is also rich in polar amino acids, but the central region of the protein contains an extremely hydrophobic stretch of about 20 amino acid residues. Labeling experiments have shown that the carbohydrate-rich N terminus is localized on the external surface of the red blood cell and that this region of the protein comprises over 80 percent of the mass of glycophorin. Because the carboxyl terminus of the protein has been shown to be exposed to the cytoplasm of the red blood cell, *each molecule of glycophorin must span the membrane*.

Other studies have shown that the hydrophobic portion of glycophorin has a high affinity for phospholipids and cholesterol, the two principal lipid constituents of the red blood cell membrane. Moreover, this portion of the molecule forms a very stable α helix in a hydrophobic environment . The length of this helix is about 40 Å, slightly less than the known width of biological membranes. These observations provide strong experimental evidence for the structural model proposed in Figure 16-29.

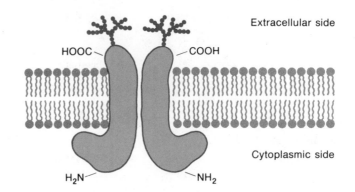

**Figure 16-30**
Disposition of the anion-channel (band 3) protein in the human erythrocyte membrane. (Adapted from G. Guidotti, The structure of intrinsic membrane proteins, *J. Supramol. Struct.* 7:489, 1977.)

A second well-characterized integral membrane protein is the anion-channel, or band 3, protein (Figure 16-28) of the human erythrocyte membrane. Labeling studies using intact cells and unsealed membrane ghosts have shown that this 95,000-dalton glycoprotein also spans the membrane with the carbohydrate residues on the outside of the cell (Figure 16-30). In contrast to glycophorin, however, nearly half the mass of the polypeptide is exposed on the *inside* of the cell, including the N terminus of this protein, and the polypeptide exists as a dimer within the red blood cell membrane. The C-terminal third of the molecule, including most of the amino acids to which sugar residues are attached, is at least partially available to labels from the outside of the cell, while a 17,000-dalton segment has been shown to completely penetrate the membrane. Both these domains are relatively rich in hydrophobic amino acid residues compared with the completely water-soluble cytoplasmic domain (Table 16-8). Recent evidence suggests that within these segments of the molecule, *the polypeptide chain of the anion channel traverses the membrane at least three times.*

In a few unusual instances it has been possible to isolate a biological membrane that contains only a single kind of protein. The so-called *purple membrane* from the halophile (salt lover) *Halobacterium halobium* consists of lipid and only one protein, *bacteriorhodopsin*. This protein functions as a proton pump in response to light and is extremely important to this organism for ATP synthesis under anaerobic conditions (Chapter 17).

**Table 16-8**
Domains of the Anion Channel (Band 3) Protein

| Domain | $M_r$ | Isolation | Location | Carbohydrate | Percent Hydrophobic Amino Acids* |
|--------|-------|-----------|----------|--------------|-------------------------------|
| C-terminal | 38,000 | Chymotryptic digestion from outside the cell | Exterior and intramembrane | + + | 52 |
| Transmembrane | 17,000 | Chymotryptic digestion from both sides of the membrane | Intramembrane | + | 50 |
| N-terminal | 41,000 | Trypsin digestion on the cytoplasmic side of the membrane | Cytoplasm | − | 44 |

*Defined as in Table 16-7.

Adapted from T. L. Steck, The band 3 protein of the human red cell membrane: A review. *J. Supramol. Struct.* 8:311, 1978.

Because it forms a highly ordered hexagonal lattice within the plane of the purple membrane, it has been possible to deduce the structure of bacteriorhodopsin from electron micrographs and diffraction patterns to a resolution of 7 Å. One molecule of this protein spans the membrane seven times, and each segment traversing the membrane has been shown to be an $\alpha$ helix (Figure 16-31). Most of the mass of bacteriorhodopsin is embedded within the lipid matrix of the membrane, and this protein has an unusually high content of hydrophobic amino acids (Table 16-7).

Finally, some integral membrane proteins appear not to span the membrane at all. Thus current evidence on the topography of the electron-transport chain in the inner mitochondrial membrane indicates that some integral proteins are localized at the inner face and some at the outer face, while still others appear to penetrate the entire structure. This arrangement is consistent with the currently accepted model of electron transport and proton pumping during respiration (Chapter 10). Similarly, liver endoplasmic reticular membranes contain an electron-transport chain that functions in the desaturation of fatty acids. Electrons pass sequentially from NADH to *cytochrome $b_5$ reductase*, then to *cytochrome $b_5$*, and finally, to the desaturase enzyme. The catalytic sites of all these proteins are on the cytoplasmic side of the endoplasmic reticulum. Recent experiments on the structure of cytochrome $b_5$ ($M_r = 16,000$) have revealed that it is <u>anchored</u> in the membrane by a 5000-dalton segment of the polypeptide that contains a large proportion of hydrophobic amino acid residues. This anchor apparently does <u>not</u> span the membrane, but loops into and out of the membrane such that both N-terminal and C-terminal ends are on the cytoplasmic surface of the endoplasmic reticulum.

Thus the topographies of proteins found in biological membranes are considerably more variable compared with those of membrane lipids. The polypeptide may or may not span the membrane; considerable portions of the protein may project outside or inside the cell, and the polypeptide may have a relatively hydrophobic amino acid composition; and one protein may have its N terminus outside the cell and its C terminus projecting into the cytoplasm, while the reverse may be true for a different integral polypeptide. The only generalizations that appear valid are <u>(1) *a significant, if small, proportion of the polypeptide must interact strongly with membrane lipids through hydrophobic interactions, (2) $\alpha$ helices are common intramembrane structures of polypeptides, and (3) the carbohydrate portions of plasma membrane glycoproteins are always found on the outside of the cell.*</u>

## THE STRUCTURE OF BIOLOGICAL MEMBRANES

As we have seen, the most probable arrangement of lipids in biological membranes is a bilayer, based on structures formed by pure lipids in aqueous solution. The idea of a lipid bilayer was first proposed by the Dutch investigators E. Gorter and F. Grendel in 1925. By measuring the total surface area of a red blood cell and the area occupied by a monolayer of lipids isolated from the same cell, they came to the conclusion that the membrane must, in fact, be made up of a lipid bilayer. Although their calculations contained a number of systematic errors (they assumed the wrong shape for the red blood cell and failed to extract all the membrane lipid),

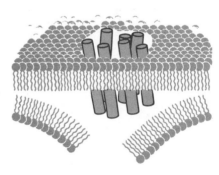

**Figure 16-31**
Diagrammatic representation of a single molecule of bacteriorhodopsin in the purple membrane of *Halobacterium halobium*. The seven $\alpha$-helical segments run perpendicular to the plane of the membrane, and all but the very top and bottom of the molecule is in contact with lipid. Polypeptide connections between the $\alpha$ helices are not shown. (Adapted from R. Henderson, The purple membrane from *Halobacterium halobium*, *Ann. Rev. Biophys. Bioeng.* 6:87, 1977.)

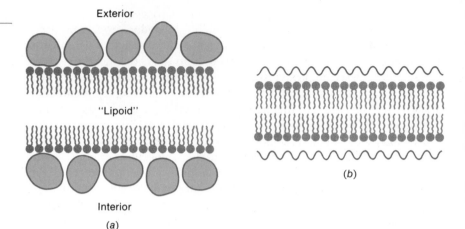

**Figure 16-32**
Early models of membrane structure. (*a*) The Davson-Danielli model included a phospholipid bilayer separated by "lipoid" material, and globular proteins were hypothesized to be present only at the two surfaces of the lipid domain. (Adapted from J. F. Danielli and H. Davson, A contribution to the theory of permeability of thin films, *J. Cell Physiol.* 5: 495, 1935. (*b*) The "unit membrane" model of J. D. Robertson correctly depicted the arrangement of lipids as a single bilayer, but also excluded proteins and other nonlipid molecules from the hydrophobic interior of the membrane. (Adapted from J. D. Robertson, The ultrastructure of cell membranes and their derivatives, *Biochem. Soc. Symp.* 16:3, 1959.)

they nevertheless were the first to suggest that a biological membrane could not be a simple monolayer of lipid molecules. Subsequent models of membrane structure by J. Danielli and H. Davson in 1935 and by J. D. Robertson in 1959 took this fact into account, as shown in Figure 16-32. However, none of these investigators had the information we now possess on the topographies of membrane lipids and proteins.

### The Fluid Mosaic Model

The currently accepted conception of biological membrane structure, proposed by J. S. Singer and G. L. Nicolson in 1972, is shown in Figure 16-33. This *fluid mosaic model* suggests that the essential structural repeating unit is the phospholipid molecule in a bilayer arrangement with a thickness of about 50 Å. Integral membrane proteins are "dissolved" in the bilayer in a seemingly random fashion. Some proteins (such as certain mitochondrial cytochromes) are localized at one or the other of the two surfaces of the lipid bilayer; other proteins (such as glycophorin and the anion channel) may pass from one side of the membrane to the other; and still others (such as bacteriorhodopsin) may be largely embedded in the hydrophobic matrix. Although most of the membrane phospholipids are in the bilayer array, some may be specifically associated with integral membrane proteins and essential for their biological activities. Also part of the fluid mosaic model is the idea that the entire structure is <u>dynamic</u> rather than static, with most components capable of relatively rapid lateral diffusion and of rotational motion about an axis perpendicular to the plane of the bilayer.

Rotation of lipids and proteins through the plane of the bilayer, however, is proposed to be a rare event. Thus Figure 16-33 depicts a hypothetical biological membrane at one point in space and time.

The fluid mosaic model incorporates a number of well-known features of biological membranes. The bilayer nature of membrane lipids has been well established by physical techniques. For example, the hydrocarbon "tails" of phospholipids and glycolipids have a lower electron density than the polar headgroups. Experiments that measure the diffraction by electrons of x-rays beamed at low incident angles on biological membranes have shown a "trough" of electron density at the center of the membrane bounded by two peaks of diffraction at the two peripheries (Figure 16-34). The dimensions of these electron-rich and electron-poor areas are similar to those of the bilayer structures that pure phospholipids form in aqueous solutions. Other physicochemical measurements are fully consistent with the view that most, if not all, biological membranes contain a lipid bilayer as an essential structural feature (see below).

It is also well-established that the lipid bilayer acts as a *permeability barrier* for hydrophilic solutes, such that only those water-soluble molecules recognized by specific integral transport proteins can easily permeate the membrane (Chapter 17). In turn, these membrane proteins only function efficiently in a hydrophobic environment provided by their interactions with the nonpolar portions of membrane lipids in the interior of the bilayer. Unidirectional membrane processes, such as proton pumping out of the mitochondrion (Chapter 10), suggest an *asymmetric* organization of proton carriers across the inner mitochondrial membrane. The fact that proteins such as glycophorin and the anion channel *are always oriented in a single direction* with respect to the inside and outside of the cell shows that such asymmetry can exist. Such an arrangement can be maintained only if rotation of integral membrane proteins around an axis parallel to the lipid bilayer (flip-flopping) does not occur (see the following discussion).

In contrast, the abilities of membrane proteins and lipids to diffuse laterally, parallel to the plane of the bilayer, have been amply demonstrated by a variety of techniques. One of the more straightforward of these is the *fluorescence photobleach recovery technique*. In this method, fluorescently labeled membrane components are bleached (their fluorescence is destroyed) in a small area of the membrane by a short, intense flash of

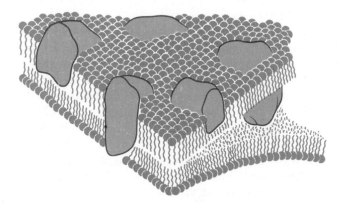

**Figure 16-33**
The fluid mosaic membrane as envisioned by Singer and Nicolson. (Adapted from S. J. Singer and G. L. Nicolson, The fluid mosaic model of the structure of cell membranes, *Science* 175: 720, 1972.)

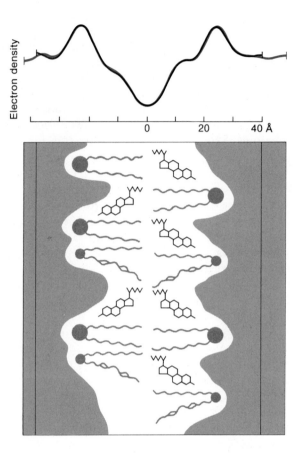

**Figure 16-34**
(*Top*) Relative electron densities of rabbit optic (*black line*) and sciatic (*color*) nerve myelin membranes as a function of distance from the center, measured by x-ray diffraction. The density profile is due mainly to lipids because of the high lipid/protein ratio in myelin membranes (Table 16-1). (*Bottom*) Structural interpretation of the density profile using the approximate known ratios of lipids in mammalian nerve myelins (6 cholesterols, 5 glycerolipids, and 4 sphingolipids). (Adapted from D. L. D. Caspar and D. A. Kirschner, Myelin membrane at 10 Å resolution, *Nature* 231:46, 1971.)

light. By monitoring this region as a function of time using a fluorescence microscope, one observes a reappearance of fluorescence (recovery), which is due to unbleached molecules diffusing into the treated area. If a specific membrane component is so labeled, its diffusion coefficient in the plane of the membrane can be calculated from the time course of fluorescent recovery. By using this technique, lateral movements of both proteins and lipids have been demonstrated in a number of biological membranes. For example, rhodopsin, a light-receptor protein of vertebrate retinal membranes (Chapter 31), has been shown to be highly mobile within the lipid bilayer by a photobleach recovery method utilizing its intrinsic absorbance, while other proteins vary enormously in their intramembrane diffusion coefficients (see the following discussion).

In an elegant experiment also designed to test protein mobility in biological membranes, L. Frye and M. Edidin fused a mouse cell and a human cell with the aid of a virus, called Sendai virus, that facilitates such fusion. Membrane proteins of the mouse cell were labeled with specific antibodies that fluoresced green, while similar labeling of the human cells was with red fluorescent antibody. If membrane proteins failed to diffuse, then the *heterokaryon* resulting from the fusion of these two cells should remain "half red" and "half green," even after long periods of incubation at physiological temperatures (37°C). In this experiment, however, significant intermixing of red and green labels was observed in the light microscope in less than 30 min, and complete *mosaics* of human and mouse proteins were seen in all heterokaryons within 1 h. These observations, therefore, also provided direct evidence for the lateral mobility of integral membrane proteins and indirect evidence for the movement of the lipid components of the bilayer itself. It should be noted, however, that the rates of diffusion in this experiment were somewhat less than expected if the protein molecules were as highly diffusible in the bilayer as, for example, rhodopsin. The reasons for this may be the facts that some integral membrane proteins appear to self-associate while others interact with structural elements of the cytoplasm (see the next section).

Evidence for the diffusion of lipid molecules within the plane of the membrane also has been obtained by a variety of other spectroscopic techniques. One such method, _electron spin resonance_ (ESR) spectroscopy, measures the energy absorbed by an unpaired electron of a free radical as a function of the magnitude of an externally applied magnetic field. Since biological membranes do not normally contain such paramagnetic groups, it is necessary to introduce into the system being examined a molecule containing a stable free radical. _Spin labels_ containing _nitroxide_ groups are often used for this purpose and may be either hydrophobic themselves (Figure 16-35a) or attached to a normal lipid component of the membrane, such as phosphatidylcholine, as shown in Figure 16-35b. Energy absorption by the unpaired electron in nitroxides is influenced by the nitrogen nucleus, which splits the spectrum into three peaks, and environmental factors, such as viscosity and polarity of the surrounding medium, which affect the shape and relative positions of these peaks (Figure 16-36). ESR spectroscopy has therefore been useful in examining the mobility of components of biological membranes.

Egg lecithin-cholesterol

(4:1)

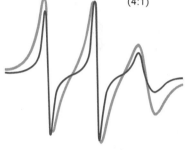

Sarcoplasmic reticulum

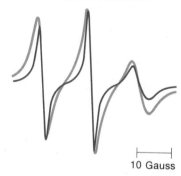

$\vdash\!\!-\!\!-\!\!\dashv$
10 Gauss

**Figure 16-36**
ESR spectra of the compound shown in Figure 16-35b in egg lecithin-cholesterol (4:1) and in muscle sarcoplasmic reticular membranes (lighter color = 5% analog, darker color = 0.25% analog). (Adapted from C. J. Scandella, P. Devaux, and H. M. McConnell, Rapid lateral diffusion of phospholipids in sarcoplasmic reticulum, _Proc. Natl. Acad. Sci. USA_ 69:2056, 1972.)

**2,2,6,6-Tetramethylpiperidine-1-oxyl** (TEMPO)

(a)

**Spin-label analog of phosphatidylcholine**

**Figure 16-35**
(b)

Using the spin label TEMPO (Figure 16-35a), which is spontaneously incorporated into some biological membranes, W. Hubbell and H. McConnell showed in 1968 that the hydrophobic regions of nerve and muscle cell membranes have a low-viscosity, fluid-like nature. In parallel experiments, they also observed that vesicles of purified soybean phospholipid affected the ESR spectrum of TEMPO in a like manner. Similarly, nitroxide derivatives of phospholipids, such as that shown in Figure 16-35b, have been used to show that the diffusion rates of these paramagnetic probes in the plane of typical biological membranes are on the order of *several micrometers per second* at 37°C. Thus it is possible for a phospholipid molecule to travel from one side of an average-sized animal cell to the other in a few minutes, while the same process could take less than a second in a typical bacterium.

The types of experiments just described provide unequivocal evidence for the lateral motion of proteins and lipids in biological membranes, a major feature of the fluid mosaic model. Likewise, other physical measurements, such as [$^{13}$C]nuclear magnetic resonance spectroscopy, have shown that membrane lipids and proteins are free to rotate about an axis perpendicular to the plane of the bilayer. However, the specific orientations maintained by integral membrane proteins with respect to the bilayer suggest that rotation of these molecules *through* the plane of the bilayer (*flip-flop* or *transverse motion*) does not occur. Since most integral membrane proteins have at least some hydrophilic surface area exposed at one or both membrane faces (see above), a transverse rotation would be an energetically unfavorable event, requiring transient interaction of these polar groups with the hydrophobic interior of the bilayer. Indeed, physical and chemical measurements have failed to detect such motions of proteins in biological membranes. Comparable measurements for phospholipids have revealed that flip-flop of these molecules can occur, but that half-times for this process can be as long as days at physiological temperatures. These rates are many orders of magnitude smaller than those of lateral diffusion and again can be explained by high activation energies for these processes. Hence *the asymmetric topography of biological membranes, which is essential to many membrane functions* (see below), *is maintained by the amphipathic nature of the membrane's lipid and protein constituents.*

## Factors Affecting the Physical Properties of Membranes

As emphasized earlier, biological membranes are fluid, dynamic structures, and these properties are undoubtedly important for a number of processes carried out on and within them (Chapter 17). It is therefore important that we examine those factors which influence or change these properties if we are to understand relationships of structure to function in membrane biochemistry.

Diffusion rates in lipid bilayers are a function both of temperature and of the composition of the membrane being examined. Bilayers consisting of a single type of phospholipid typically show an abrupt change in physical properties over a characteristic and narrow temperature range ($T_m$). These temperature-dependent phase transitions are due to an organizational change in the fatty acyl side chains and can be detected by a variety of physical techniques including x-ray diffraction, ESR spectroscopy, and *dif-*

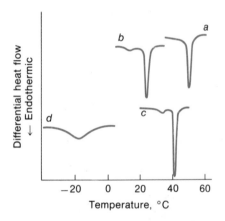

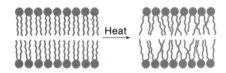

**Figure 16-37**
(*Top*) Differential scanning calorimetry of various phospholipids dispersed in water: (*a*) dipalmitoyl phosphatidylethanolamine; (*b*) dimyristoyl lecithin; (*c*) dipalmitoyl lecithin; (*d*) egg lecithin (plus ethylene glycol to prevent freezing). (Adapted from D. L. Melchior and J. M. Steim, Thermotropic transitions in biomembranes, *Ann. Rev. Biophys. Bioeng.* 5:205, 1976. Used with permission.) (*Bottom*) Molecular interpretation of the heat-absorbing reaction during the phase transition.

*ferential scanning calorimetry*. The last procedure measures energy (heat) absorption as a function of temperature and shows that over the temperature range of the phase transition, a relatively large amount of heat is absorbed per degree of temperature change compared with temperature ranges well above or below the transition (Figure 16-37). It is thought that the membranes pass from a state in which the fatty acyl side chains are highly ordered (*gel phase*) to one in which they are far more mobile (*liquid crystalline phase*). This process is illustrated in Figure 16-37. It is accompanied by increased rotational motion about the carbon–carbon bonds of the hydrocarbon chains of the phospholipids, allowing them to assume more random, disordered conformations.

In contrast to pure phospholipid bilayers, membranes isolated from cells usually undergo such phase transitions over a much broader ($\geq 10°C$) temperature range. In some instances, distinct transitions cannot even be distinguished, because of the heterogeneity of lipids found in most biological membranes and because integral membrane proteins may decrease the mobility of lipids in their immediate vicinity. In general, lipids bearing *short* or *unsaturated* fatty acyl chains undergo phase transitions at *lower temperatures* than those containing long-chain saturated fatty acids. This is because short hydrocarbon chains have a smaller surface area with which to undergo hydrophobic interactions that stabilize the gel state and because cis unsaturation introduces a "kink" in the fatty acyl chain that also leads to more disorder in the bilayer (Figure 16-38 and Table 16-9). Therefore, both the length of fatty acyl groups present and the proportion of unsaturated fatty acids (as well as the position of the double bond along the hydrocarbon chain) will affect the fluidity (viscosity) of a biological membrane at a given temperature. Since different lipids will "melt" at different temperatures, *broad phase transitions are a general characteristic of cellular membranes*.

It is apparent from the data in Table 16-9 that the midtransition temperature of a pure phospholipid suspension also depends on the nature of the polar headgroup. Thus the dipalmitoyl esters of phosphatidic acid and phosphatidylethanolamine "melt" about 20°C higher than the same derivatives of phosphatidylcholine and phosphatidylglycerol. *Divalent cations*, such as $Ca^{2+}$ and $Mg^{2+}$, also affect membrane fluidity, presumably by forming ionic bonds with neighboring phosphoryl headgroups, tending to "tie" phospholipid molecules together and decrease their mobility. This is undoubtedly one of the reasons that divalent cations are well-known stabilizers of biological membranes and their removal often facilitates lysis of cells as well as dissociation of peripheral membrane proteins.

Finally, cholesterol is a well-known modulator of membrane structure in eukaryotes and in one type of bacteria, the *Mycoplasmas*. Cholesterol intercalates among the fatty acyl chains, with its polar hydroxyl group interacting with the polar headgroups of membrane lipids. At low concentrations of cholesterol in phospholipid bilayers, separate *domains*, or patches, of cholesterol plus phospholipid and pure phospholipid appear to exist, with a resulting broadening of the phase transition profile compared with phospholipid alone. Its effects on various physical parameters of membranes depends on its proportion to other lipid components, as well as on the membrane system examined. However, a general property of membranes containing a high concentration of cholesterol appears to be an

**Figure 16-38**
Introduction of a cis double bond into a fatty acyl chain results in an inflexible kink in the phospholipid tail (*right*) compared with a phospholipid containing only saturated fatty acids (*left*).

**Table 16-9**
Midtransition Temperatures for Aqueous Suspensions of Phospholipids

| Phospholipid* | $T_m$ (°C) |
|---|---|
| Di-14:0 PC | 24 |
| Di-16:0 PC | 41 |
| Di-18:0 PC | 58 |
| Di-22:0 PC | 75 |
| Di-18:1 PC | −22 |
| 1-18:0, 2-18:1 PC | 3 |
| Di-14:0 PE | 51 |
| Di-16:0 PE | 63 |
| Di-14:0 PG | 23 |
| Di-16:0 PG | 41 |
| Di-16:0 PA | 67 |

*Phospholipid abbreviations are as in Table 16-5; additionally, Di-14:0, for example, refers to dimyristoyl (14 carbons, 0 double bonds).

Adapted from M. K. Jain and R. C. Wagner, *Introduction to Biological Membranes*, Wiley, New York, 1980.

*inhibition* of processes dependent on a fluid environment. For example, the permeability of vesicles made from purified egg phosphatidylcholine to both water and glucose decreases when greater than 20 mol % cholesterol is incorporated into the vesicles. Intercalated cholesterol apparently restricts the freedom of motion of the phospholipid hydrocarbon side chains, thereby decreasing the mobilities of membrane constituents.

Most cells maintain a lipid composition that allows for relatively rapid lateral diffusion of many membrane components at the growth temperature. Membrane-associated processes, such as the vectorial reactions catalyzed by some transmembrane transport systems (Chapter 17), and endo- and exocytosis rely on a semifluid environment for their operation. Consequently organisms have evolved intricate mechanisms to maintain this environment under a variety of conditions. One of the most remarkable of these is the ability of a variety of plant, animal, and bacterial cells to increase the proportion of membrane unsaturated fatty acids in response to a decrease in temperature. This ensures proper functioning of the membrane at the lower temperature. In bacteria such as *E. coli* and *Bacillus megaterium*, this modulation appears to be the result of temperature effects on the activities and/or on the induction of synthesis of enzymes involved in the biosynthesis of phospholipids containing unsaturated fatty acyl chains. In plants and yeast, increased solubility of $O_2$ at low temperatures apparently increases the proportion of unsaturated fatty acids because $O_2$ is a substrate of the desaturase that leads to their biosynthesis. Other factors that affect unsaturated fatty acid biosynthesis, and thus membrane fluidity, in various systems include light (in plants), nutrition, developmental stage, and aging.

Despite the large body of evidence in favor of the lateral mobility of many membrane constituents at physiological temperatures, it would be an oversimplification to view a biological membrane only as a random "sea" of lipids with proteins floating aimlessly about in them. In recent years it has become clear that nearly all eukaryotic cells contain within their cytoplasm a *cytoskeleton* made up of *microtubules* and *microfilaments* consisting primarily of the proteins *tubulin* and *actin*, respectively. Microfilaments, which are structurally similar to actin filaments of muscle cells, have been shown to form bundles just beneath the plasma membrane of many cells (Figure 16-39). These are believed to have an important role in such processes as locomotion and phagocytosis, which involve local or general changes in the shape of the cell surface and thus in the plasma membrane. In a few cases, direct association of microfilaments with the plasma membrane has been demonstrated. For example, *fibronectin*, a peripheral, cell-surface glycoprotein of many animal cells, is believed to have a role in cell–cell and cell–substratum adhesion. A *transmembrane* association of fibronectin with cytoskeletal microfilaments has been deduced from immunofluorescent microscopy (Figure 16-40), presumably by means of one or more integral membrane proteins. As a consequence, lateral diffusion of fibronectin has been shown to be at least *5,000 times slower* than freely diffusible membrane lipids and proteins. It is very likely that other peripheral and integral membrane proteins also interact with the cytoskeleton, and these interactions may similarly reduce their mobilities, thereby affecting processes that must be localized at specific points on the cytoplasmic membrane.

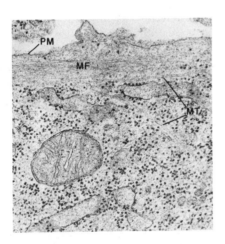

**Figure 16-39**
Membrane-associated cytoskeletal components of cultured mouse cells. (PM = plasma membrane; MF = microfilaments; MT = microtubules; 54,000 ×). (From G. L. Nicolson, Transmembrane control of the receptors on normal and tumor cells. I. Cytoplasmic influence over cell surface components, *Biochim. Biophys. Acta* 457: 57, 1976. Reprinted with permission.)

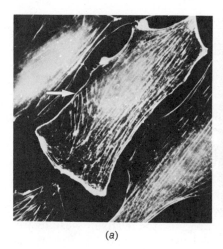
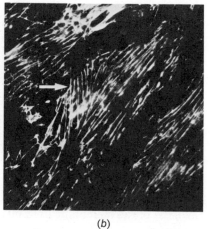

(a)                            (b)

**Figure 16-40**
Colinearity of actin and fibronectin fibrils in cultured hamster fibroblast cells
as seen by immunofluorescence in the light microscope. (*a*) The fluorescence
is due to actin antibodies (raised in rabbits) bound to intracellular microfilaments which have been stained with fluorescent goat antirabbit IgG. (*b*) The
fluorescence is that of fluorescent antifibronectin antibodies bound to extracellular fibronectin fibrils. Note that good correspondence is seen between
the arrangement of actin and fibronectin filaments, strongly suggesting a
transmembrane association between the two proteins. [From R. O. Hynes and
A. T. Destree, Relationships between fibronectin (LETS protein) and actin,
*Cell* 15:875, 1978.]

It also should be recognized that membrane lipids may not always be
freely diffusible in the plane of a biological membrane, even above the
phase transition temperature. Many purified integral membrane proteins
have been shown to retain bound phospholipid molecules after solubilization.
For example, _cytochrome oxidase_, isolated from beef heart mitochondria,
spontaneously forms vesicular structures in aqueous suspension owing to
associated lipid. When a nitroxide derivative of stearic acid is added to
such a suspension, ESR spectroscopy reveals two mobility classes of the spin
label—one highly diffusible and one less mobile. These observations have
been interpreted as evidence for an associated boundary of lipid molecules
surrounding the enzyme (Figure 16-41). These "lipids of solvation" are believed to be essential for maintaining the structural and functional integrity of integral membrane proteins such as cytochrome oxidase. It should
be emphasized, however, that the situation depicted in Figure 16-41 is probably not a static one. Rather, it is likely that protein-associated lipids are
more or less freely exchangable with the bulk of the lipid molecules in the
fluid bilayer array. Furthermore, "patches" of membrane lipids differing in
overall composition from the membrane as a whole also have been shown
to exist in some membrane systems. An example was given earlier: at low
concentration, cholesterol induces phospholipid-cholesterol "patches" in
an otherwise pure phospholipid model membrane system. Within such
areas, phospholipid molecules would be expected to be less mobile than the
bulk of the phospholipid in the membrane. Indeed, this interpretation may
provide the explanation for the broad phase transition profiles of such
membrane systems. _Thus temperature, ionic environments, and composi-_

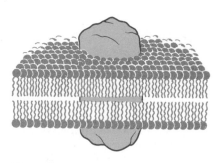

**Figure 16-41**
Schematic representation of cyto-
chrome oxidase in the inner mito-
chondrial membrane surrounded by
an associated boundary of lipid.
(Adapted from P. C. Jost, O. H.
Griffith, R. A. Capaldi, and G.
Vanderkooi, Evidence for boundary
lipids in membranes, *Proc. Natl.
Acad. Sci. USA* 70:480, 1973.)

*tion can affect the general physical state of a biological membrane, while local mobilities of membrane components can be influenced by protein–protein, lipid–protein, and lipid–lipid interactions.*

## The Asymmetry of Biological Membranes

Chemical probes of membrane protein structure have provided ample evidence for the unidirectional, asymmetric orientation of proteins with respect to the lipid bilayer, as described earlier. This asymmetry also has been more directly observed by the technique of *freeze-fracture electron microscopy*. In this procedure, whole cells or membranes are rapidly frozen, and the specimen is then struck with a sharp knife or microtome. Very often, the fracture plane actually passes between the outer and inner monolayers (*leaflets*) of the membrane lipid bilayer because the relatively weak hydrophobic interactions between the fatty acyl chains of the two leaflets offer a "path of least resistance" (Figure 16-42a). The inner surfaces of the two leaflets can then be viewed after heavy-metal shadowing in the electron microscope. This technique is especially useful in examining membrane morphology, because it avoids the potentially destructive fixing, embedding, and staining steps of more conventional sample-preparation procedures. Many biological membranes, when split in half by freeze-fracture, can be seen to have quite different morphologies of the inner and outer

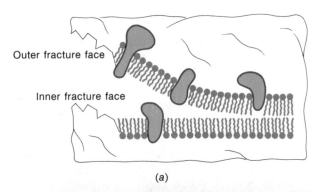

Outer fracture face

Inner fracture face

*(a)*

**Figure 16-42**
Freeze-fracture electron microscopy. (*a*) When struck with a sharp knife, membranes embedded in ice usually fracture between the monolayer leaflets of the lipid bilayer. (*b*) Freeze-fracture electron micrograph of the plasma membrane of *Streptococcus faecalis* showing a large number of protrusions (presumably proteins) on the outer fracture face and the relative lack of such particles on the inner fracture face (*inset*). (From H. C. Tsien and M. L. Higgins, Effect of temperature on the distribution of membrane particles in *Streptococcus faecalis* as seen by the freeze-fracture technique, *J. Bacteriol.* 118:725, 1974. Reprinted with permission.)

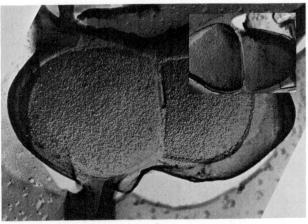

*(b)*

**Table 16-10**
Lipid Asymmetry in Biological Membranes

| Membrane | Preferential Outside | Preferential Inside | Equal |
|---|---|---|---|
| Various erythrocytes | PC, Sph,* glycolipids, cholesterol | PE, PS | — |
| Rabbit sarcoplasmic reticulum | PE | PS | PC, lyso PC |
| Mouse LM cell plasma membrane | Sph | PE | PC |
| E. coli outer membrane | — | PE | — |
| Bacillus megaterium | — | PE | — |
| Micrococcus lysodeikticus | PG | PI | DPG |
| PC/PE artificial vesicles | PC | PE | — |

*Sphingomyelin; other abbreviations are as in Table 16-5.

Adapted from J. A. F. Op den Kamp, Lipid asymmetry in membranes. *Ann. Rev. Biochem.* 48:47, 1979.

leaflet surfaces owing to preferential adherence of some membrane proteins to one of the two monolayers (Figures 16-42b). This presumably reflects the asymmetric and unidirectional disposition of these polypeptides across the lipid bilayer.

Proteins are not the only membrane constituents to show asymmetric orientations. Recent evidence favors an *unequal distribution of certain lipids between the inner and outer leaflets of the bilayer* in many biological membranes. For example, in 1972, M. S. Bretscher showed that most of the phosphatidylethanolamine and all the phosphatidylserine of human erythrocyte membranes are *inaccessible to chemical modification from outside* the cell. Both phosphoglycerides could be modified, however, in *erythrocyte ghosts*, in which both inner and outer leaflets were exposed to the chemical reagent. These results suggested that lipid asymmetry also could exist in a biological membrane. Subsequent experiments using enzymes that degrade phospholipids (*phospholipases* and *sphingomyelinase*) showed that phosphatidylcholine and sphingomyelin appear to be preferentially found in the *outer leaflet* of human red blood cell membranes (Table 16-10). Thus, although total membrane lipid is equally distributed between outer and inner monolayers in the erythrocyte membrane, the *lipid composition of each leaflet appears to be different*.

Lipid asymmetry has now been found in membranes from a variety of biological sources, and some examples are given in Table 16-10. A favorite probe for examining the asymmetry of phospholipids containing a primary amino group (phosphatidylethanolamine and phosphatidylserine) has been *2,4,6-trinitrobenzene sulfonic acid* (TNBS) (Figure 16-43). Because of its polarity, TNBS does not penetrate most biological membranes rapidly at low temperature, and the fact that TNBS reacts rapidly with primary amino groups of lipids in intact cells or sealed vesicles reveals that these molecules are in the outer leaflet of the bilayer, as shown in Figure 16-43. Separation and quantitation of both modified and unmodified phosphatidylserine and

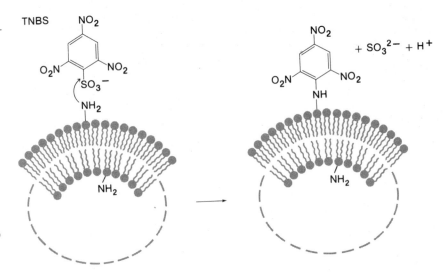

**Figure 16-43**
TNBS reacts rapidly at low temperatures only with free primary amino groups exposed on the outside of a biological membrane. It can therefore be used to determine the distribution of lipids containing free amino groups between outer and inner leaflets of the bilayer.

phosphatidylethanolamine after reaction of cells with TNBS can then give an estimation of their relative proportions in each half of the membrane. The distribution of other membrane lipids can be determined enzymatically, as described earlier, or by a variety of physical techniques including NMR and ESR spectroscopy and x-ray diffraction analysis. Such experiments strongly suggest that the *asymmetric partitioning of lipids between the bilayer leaflets may be a general property of biological membranes.*

Given the asymmetric arrangements of proteins in membranes along with the nearly unidirectional nature of many membrane-associated processes (Chapter 17), it is perhaps not surprising that certain lipids might be found predominantly on one side or the other of the bilayer. This could be a consequence of the fact that some integral membrane proteins may associate with specific lipids. Thus the orientation of the protein could influence the orientation of bound lipid with respect to the bilayer. Furthermore, the solutions that contact each monolayer are usually quite different (e.g., extracellular milieu versus cytoplasm), so that lipid asymmetry also could be induced by compositional and/or biological activity differences between the inside and outside environments. It is therefore reasonably clear from energetic arguments, how membrane asymmetry can be maintained, and from functional requirements, why it exists. An equally intriguing question, however, is: *How are asymmetric membrane structures formed in the first place?* To answer this, we must consider how biological membranes are synthesized and assembled by cells.

## BIOSYNTHESIS AND ASSEMBLY OF BIOLOGICAL MEMBRANES

In terms of its overall composition, the plasma membrane is certainly one of the most complex cellular structures. For example, a typical bacterial cytoplasmic membrane contains over 100 different proteins and several classes of phospholipids, each of which may be subdivided into species of different fatty acid compositions. Furthermore, as we have just seen, these molecules are not arranged symmetrically in the membrane, but rather are found in specific orientations with respect to the inside and outside of the cell. Obviously, then, the process of membrane assembly, or *biogenesis*, must have programmed within it not only methods for ensuring insertion of proteins and lipids destined to become membrane components, but also mechanisms for attaining the proper orientations of these molecules in the bilayer.

A second problem related to membrane biogenesis concerns the mechanism of biosynthesis of macromolecules destined to be *exported* through the membrane of a cell or organelle. For example, how do components of the outer membrane, the peptidoglycan and periplasmic proteins, reach their ultimate destinations in Gram-negative bacteria, and how do hydrophilic, water-soluble proteins excreted by secretory organelles, such as the endoplasmic reticulum of eukaryotic cells, traverse the hydrophobic lipid bilayer? These events would seem to have very unfavorable activation energies, and cells must have evolved mechanisms to circumvent this problem.

In this section, we shall examine what is known about the biosynthesis and assembly of biological membranes in a number of systems, especially as they relate to the structural features we have already pointed out. Our discussion will be confined mainly to the cytoplasmic and outer membranes of Gram-negative bacteria and to the endoplasmic reticulum in eukaryotic cells (the precursor of the plasma membrane), because most information currently available is about these systems. It is likely, however, that mechanisms of biogenesis of other membrane systems will turn out to be related to those found in the better-characterized systems.

## Topography and Coordination of Membrane Assembly

All growing cells need to synthesize new plasma membrane components as they enlarge and eventually divide into daughter cells. Because the membrane must maintain its permeability barrier function throughout this process, it is clear that newly synthesized membrane proteins and lipids must be inserted into the preexisting bilayer without disrupting its structural continuity. Thus membrane biosynthesis involves *expansion* as well as self-assembly. Further, the biosynthetic machinery must be so organized as to ensure efficient insertion of membrane lipids and the proper orientations of integral membrane proteins with respect to the bilayer.

Not surprisingly, most of the biosynthetic enzymes for membrane lipids have been found themselves to be integral membrane proteins in both prokaryotic and eukaryotic cells. In many bacteria, nearly all the membrane lipid is phospholipid (Table 16-5), and fatty acylation of glycerol-3-phosphate and all subsequent steps in phospholipid biosynthesis (Chapter 14) occur on the cytoplasmic membrane. In eukaryotic cells, cytoplasmic membrane assembly is believed to occur by expansion of the endoplasmic reticulum followed by the migration of vesicles derived from this organelle to the cytoplasmic membrane, with which they fuse (see the following discussion). Supporting this conclusion is the observation that lipid biosynthetic enzymes have been shown to be tightly associated with the endoplasmic reticular membrane in animal cells. In both eukaryotic and prokaryotic systems, the active sites of these enzymes have been shown to be on the *cytoplasmic face* of the membrane. This result was to be expected, since the precursors of membrane lipids, many of which are hydrophilic, are synthesized in the cytoplasm. Thus lipid is apparently inserted into biological membranes passively, as a direct result of the site of its biosynthesis.

An interesting question is posed by this topographic organization of membrane lipid biosynthesis. *Because flip-flop of preexisting membrane lipids is generally a slow process* (see the earlier discussion), *how do newly synthesized molecules become inserted into the outer leaflet of the bilayer?* This question was explored by J. Rothman and E. Kennedy in the Gram-positive bacterium *Bacillus megaterium*. Using the membrane-impermeable probe TNBS, they demonstrated that 33 percent of the total membrane phosphatidylethanolamine was in the outer leaflet of the cytoplasmic membrane, while the rest was in the inner leaflet. By labeling *newly synthesized* phosphatidylethanolamine with a short pulse of [$^{32}$P]phosphate and then immediately treating with TNBS, they showed that none of the radioactive phospholipid was initially found in the outer leaflet. After 30 min at 24°C, however, about one-third of the [$^{32}$P]phosphatidylethanolamine was labeled with TNBS, the same ratio as for the preexisting, unlabeled molecules. Their experiments lead to two important conclusions about membrane lipid biosynthesis in *B. megaterium*: (1) *newly synthesized phosphatidylethanolamine is incorporated initially only in the inner leaflet, and* (2) *its appearance in the outer leaflet occurs nearly* $10^5$ *times faster than known flip-flop rates for phospholipids in model bilayers*. Recent experiments with endoplasmic reticular membranes suggest that the same conclusions may hold true in eukaryotic cells. This leads to the possibility that transverse rotation of lipids through the bilayer may be catalyzed by one or more proteins in biological membranes. If such proteins exist (there is as yet no convincing evidence for them), then their activities could play a role in establishment of asymmetric lipid distributions between the two monolayers.

In contrast to membrane phospholipids, *synthesis and insertion of integral membrane proteins appear to take place by at least two pathways* (see Chapter 25). Some are synthesized by polyribosomes in the cytoplasm, after which they are incorporated into the membrane (*posttranslational insertion*). Other membrane proteins are synthesized by polyribosomes that are bound to the membrane, and these polypeptides appear to insert as they are being translated (*cotranslational insertion*). Membrane-bound polysomes have been isolated from bacteria and from endoplasmic reticulum and have been shown to synthesize membrane and secreted proteins almost exclusively. The converse is not true, however, because isolated free (cytoplasmic) polysomes direct the synthesis of both cytoplasmic and membrane proteins. *Whether a membrane protein is synthesized on membrane-associated or free polyribosomes seems to be a function solely of the amino acid sequence of the protein itself*, as we will discuss later in this section.

Because both membrane lipids and some membrane proteins are synthesized at the membrane itself, and because normal membrane biogenesis during cell growth involves insertion of both, it is reasonable to ask whether *coordinate* synthesis of lipid and protein is necessary for membrane expansion. It is clear that the *protein composition* in cell membranes can change considerably depending on age and environmental factors. For example, within 5 days after birth, levels of the enzyme glucose-6-phosphatase in rat liver endoplasmic reticular membranes increase about 30-fold. Over the same time period, NADH:cytochrome $c$ reductase and cytochrome $b_5$ are also induced, but the increase in amount of these two electron carriers is neither coordinate with nor equivalent to that of glucose-6-phosphatase. In bacteria, rates of synthesis of many membrane proteins can be altered by changing the medium in which the cells are grown. For example, specific transport proteins for many sugars and amino acids (Chapter 17) appear in the membrane only if induced by the appropriate substrate. Thus *the rates of synthesis and insertion of integral membrane proteins are not generally coordinate with one another.*

To investigate whether ongoing membrane lipid synthesis is necessary for synthesis and insertion of membrane proteins, mutants of *E. coli* defective in phospholipid biosynthesis under some conditions have been especially useful. These mutants are dependent on exogenous glycerol in the medium for the synthesis of phosphoglycerides, and when glycerol is removed, net phospholipid synthesis ceases immediately. Upon glycerol starvation, these cells continue to grow, however, for about one generation, during which time the proportion of protein to lipid in both inner and outer membranes *increases about 50 percent* as a result of continued protein synthesis. This causes an increase in buoyant density of both membrane fractions, as shown in Figure 16-44. These results demonstrate that *coordinate synthesis of membrane lipid is not necessary for synthesis and insertion of membrane protein*, at least in bacterial cells. Conversely, if total protein synthesis is blocked by the antibiotic *chloramphenicol* in the presence of glycerol, phospholipids continue to be synthesized and inserted into the cytoplasmic membrane without the concomitant incorporation of protein. Hence *membrane lipid synthesis can occur independently of the biosynthesis of membrane protein.*

These observations on the assembly of biological membranes seem to rule out a mechanism in which nonmembranous intracellular protein-lipid complexes are first formed, followed by their insertion into the growing membrane. They are consistent with the fact that biosynthesis of many membrane components takes place on the membrane itself. Nevertheless, *regulation of the differential rates* of membrane protein and lipid synthesis must occur to achieve relatively stable ratios of these molecules in the bilayer during cell growth. Thus *E. coli* phospholipid biosynthetic mutants that have been starved for glycerol and therefore have abnormally protein-rich membranes show a higher rate of phospholipid synthesis upon readdition of glycerol to the medium than unstarved cells. In this case, one or more of the membrane-bound phospholipid biosynthetic enzyme activities is presumably controlled by the protein-lipid ratio in the membrane.

Finally, it should be emphasized that the assembly and maintenance of membrane structures in cells is a *dynamic* process. Not only are components synthesized and inserted into a growing membrane, but they also are being continuously *degraded*, albeit at a slower rate. This *turnover* process varies with each individual type of molecule. In general, phospholipids have a *shorter half-life* in the membrane (higher turnover) than membrane proteins, which themselves vary enormously in "life expectancy" depending on the specific protein examined. This constant turnover allows cells to rapidly adjust membrane composition in response to changes in the environment (temperature, nutrition, and so forth).

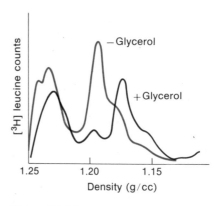

**Figure 16-44**
Isopycnic centrifugation of membranes from a phospholipid biosynthetic mutant of *E. coli*. In the absence of glycerol, both inner (right-hand peak of [³H]leucine radioactivity) and outer (left-hand peak) membranes are *more dense* than the corresponding fractions from cells supplemented with glycerol. This shows that both membranes become richer in protein after phospholipid synthesis is stopped in this mutant and thus that membrane proteins can be synthesized and inserted under these conditions. (Adapted from T. M. McIntyre and R. M. Bell, Mutants of *Escherichia coli* defective in membrane phospholipid synthesis. Effect of cessation of net phospholipid synthesis on cytoplasmic and outer membranes, *J. Biol. Chem.* 250:9053, 1975.)

## The Problem of Membrane Protein Biogenesis

Phospholipids are relatively small molecules, with molecular weights that are usually less than 1000. It is not difficult to imagine their spontaneous incorporation into the bilayer after their synthesis by membrane-bound enzymes. Most integral membrane proteins, however, have molecular weights that range between $10^4$ and $10^5$ and can contain a reasonably high proportion of hydrophilic amino acids (Table 16-7). Furthermore, their insertion into the membrane also must result in an asymmetric, unidirectional orientation with respect to the bilayer. It is not surprising, therefore, that the problem of membrane protein biogenesis is a good deal more complex than that of many membrane lipids. Only recently have enough clues been obtained to the mechanism of this process to enable the formulation of the models we will consider. The furious pace of research in this area, however, ensures a more complete understanding of membrane protein biosynthesis in the very near future.

Related to this problem is the mechanism of biosynthesis and assembly of macromolecules and supramolecular structures whose ultimate location is *exterior to the plasma membrane*. Included in this category are the outer membrane, cell wall, and periplasmic proteins of Gram-negative bacteria as well as proteins secreted from some eukaryotic cells, including hormones, hydrolytic enzymes, and antibodies. In both cases, the question is similar: *How do large, often hydrophilic molecules partially or completely traverse the hydrophobic barrier of the membrane bilayer from their sites of synthesis within the cell?* In the case of the bacterial peptidoglycan envelope, it seems likely that a $C_{55}$ isoprenoid lipid carrier, bactoprenol (Figure 16-17), is involved in carrying the peptidyl-disaccharide "building block" across the plasma membrane to be added to the growing cell wall. Similarly, the terminal carbohydrate portions of the lipopolysaccharide of the Gram-negative outer membrane (O antigens) are attached to bactoprenol before being added to the lipid A core polysaccharide unit on the outside of the cytoplasmic membrane (Chapter 12). A compound related in structure to bactoprenol, _dolichol_, similarly participates in the transfer of carbohydrate residues to membrane glycoproteins in eukaryotic cells.

Although these types of mechanisms can explain bacterial cell wall biosynthesis and glycoprotein carbohydrate translocation in eukaryotic cells, they are not applicable to the problem of integral membrane protein insertion or of the secretion of some proteins by both types of cells. In the remainder of this chapter, we consider current models for membrane and secreted-protein biogenesis and some of the evidence in favor of these mechanisms.

### The Signal Hypothesis

One of the first important clues to the way in which some proteins might be inserted into or through a membrane was obtained by C. Milstein and G. Brownlee and their co-workers in 1972. They were studying mouse myeloma tumor cells, which synthesize and secrete large amounts of a particular immunoglobulin light chain. Synthesis of this and other secreted proteins occurs on endoplasmic reticular (ER) membranes covered with bound ribosomes (_rough ER_); the polypeptide is extruded through the membrane into the lumen of the organelle and is thought to be eventually secreted from the cell, after processing in the Golgi apparatus, by fusion of vesicles derived from the Golgi with the cytoplasmic membrane (Figure 16-45). These workers found that polyribosomes removed from the rough ER of these myeloma cells directed the synthesis of light-chain polypeptide that was _larger by about 1500 daltons_ than the protein synthesized by intact rough ER. This extension was found to be at the N terminus of the protein and was proposed to be cleaved off during transit through the rough ER membrane. It was further suggested that this N-terminal extension might act as a _"signal"_ for the insertion of the growing polypeptide chain into the membrane, and when it was no longer necessary, it was cleaved off (_processed_) by a protease, presumably associated with the membrane.

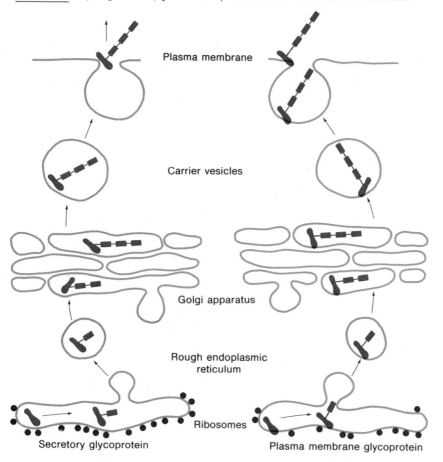

**Figure 16-45**
Secreted and plasma membrane proteins of animal cells are synthesized by ribosomes bound to the endoplasmic reticulum. These proteins are thought to be transferred sequentially to the Golgi apparatus and then to the plasma membrane by carrier vesicles. Fusion of these vesicles with the cell membrane deposits secreted proteins outside the cell (_left_) or results in integral glycoproteins being inserted into the plasma membrane with their carbohydrate residues outside the cell (_right_). Glycosylation reactions take place both in the lumen of the endoplasmic reticulum and in the Golgi apparatus. (Adapted from C. P. Leblond and G. Bennett, Role of the Golgi apparatus in terminal glycosylation, In _International Cell Biology_, Rockefeller University Press, New York, 1977.)

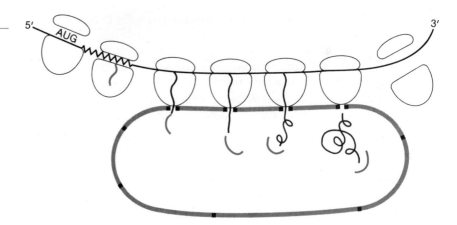

**Figure 16-46**
The signal hypothesis as proposed by Blobel and Dobberstein. Polyribosomes are bound to the membrane by means of an N-terminal signal sequence ( ⌣ ) on the secreted protein that is hypothesized to interact with a transmembrane pore. In the example shown, this sequence is cleaved by a membrane-bound protease before translation is complete. (Adapted from G. Blobel and B. Dobberstein, Transfer of proteins across membranes: I. Presence of proteolytically processed and unprocessed nascent immunoglobulin light chains on membrane-bound ribosomes of murine myeloma, *J. Cell. Biol.* 67:835, 1975.)

In 1975, these observations were confirmed and extended by G. Blobel and B. Dobberstein, who proposed a mechanism for the export of proteins through the rough ER membrane, which they termed the *signal hypothesis*. The essential elements of this proposal (Figure 16-46) are as follows:

1. mRNAs, which code for proteins to be exported through the ER membrane, contain a unique sequence of codons immediately following the initiation codon called *signal codons*.

2. Translation of this mRNA is initiated by "free" (non-membrane-bound) ribosomes, which first synthesize a *signal sequence* of amino acids corresponding to the signal codons.

3. Emergence of this signal sequence from the ribosome triggers its attachment to the membrane, as well as the formation of a *transmembrane pore* around the signal sequence.

4. As translation continues, the nascent polypeptide chain is extruded through the pore, and the signal sequence is cleaved off before chain completion.

5. Termination of translation results in complete deposition of the protein into the rough ER lumen, dissolution of the pore, and dissociation of the ribosomal subunits from the mRNA, and thus the membrane.

Only proteins to be exported are hypothesized to contain a signal sequence, and it is this sequence that signals the attachment of the nascent chain, and thus the ribosome, to the membrane.

The signal hypothesis has since been extended to account for the insertion of a number of integral membrane proteins by assuming, in such cases, that the pore ceases to function and/or the ribosome dissociates from the membrane _before_ completion of the polypeptide. An example of this is the spike protein (G protein) of _vesicular stomatitis virus_ (VSV). VSV infects some animal cells, and after transcription, translation, and replication of its RNA genome, it leaves the host cell by a budding off of the ribonuclear protein complex through the cytoplasmic membrane. The membrane contains large amounts of G protein, whose synthesis was directed by the viral genome (Figure 16-47). This membrane now becomes part of the virus particle with its G protein forming spike-like protrusions on the outside. G protein has been shown to span the membrane and is synthesized on the rough ER of cells infected with VSV. Extensive studies have shown that G protein is synthesized as a precursor with a signal sequence and that insertion of the protein into the membrane and removal of the signal sequence are cotranslational. G protein is a glycoprotein, and carbohydrate residues are attached to it in the rough ER lumen _before translation is complete_, demonstrating that the _nascent polypeptide spans the membrane_ (Figure 16-48).

By analogy with secreted proteins, VSV G protein is sequentially transferred from the rough ER through the Golgi apparatus and finally is inserted into the cytoplasmic membrane by fusion of vesicles, with the G protein carbohydrate on the _inside_, with the cytoplasmic membrane. This results in G protein oriented in the cytoplasmic membrane with its carbohydrate _outside_ the cell. Other membrane glycoproteins of eukaryotic cells are presumed to be synthesized and inserted in a similar manner (Figure 16-45).

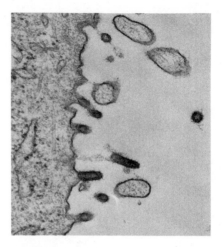

**Figure 16-47**
Electron micrograph of a cell infected by vesicular stomatitis virus (VSV). The small, dark oblong objects are virus particles emerging outside the cell (a few of them are sectioned end on and appear spherical). Budding off results in the virus particle being surrounded by the host cell membrane. G protein is encoded in the viral genome and is the only surface protein found in this membrane. It is responsible for spike-like protrusions on the surface of the virus particle. Because the VSV membrane contains such an abundance of a single protein, it is a convenient system in which to study the assembly of an integral protein into a biological membrane. (H. F. Lodish and J. E. Rothman, The assembly of cell membranes, _Sci. Am._ 240:48, 1979.) (Micrograph courtesy Dr. D. M. Knite.)

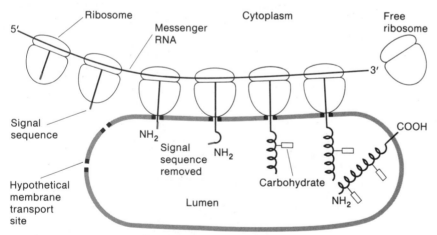

Rough endoplasmic reticulum

**Figure 16-48**
The signal mechanism applied to an integral membrane protein, the G (spike) protein of vesicular stomatitis virus. Carbohydrate is added to the protein in the lumen of the rough ER before translation is complete, showing that the nascent polypeptide spans the membrane. In contrast to secreted proteins, the completed polypeptide remains lodged in the membrane and eventually is transported to the plasma membrane by way of the Golgi apparatus (see Figure 16-45). (Adapted from H. F. Lodish and J. E. Rothman, The assembly of cell membranes, _Sci. Am._ 240:48, 1979.)

A considerable amount of evidence has accumulated in favor of at least certain elements of the signal hypothesis in both eukaryotes and prokaryotes. A number of mitochondrial and chloroplast membrane proteins, as well as proteins of the outer membrane of E. coli, have been shown to be synthesized as precursors that are then cleaved into "mature" proteins during or after translation. Several periplasmic proteins of E. coli, which must be secreted across the inner membrane, are also translated with signal sequences by membrane-bound ribosomes, and these sequences are rich in hydrophobic amino acids, which might be expected to "lead" the nascent polypeptide along with the ribosome to the membrane. It also has been shown directly using membrane-impermeable labeling reagents that nascent polypeptide chains of exported proteins of E. coli are exposed on the outside of the cell while they are still attached to ribosomes on the inside. Therefore, in this system as well, protein extrusion through the membrane is cotranslational, and the growing polypeptide chain spans the membrane before synthesis is complete.

Finally, elegant genetic studies by J. Beckwith, T. Silhavy, and coworkers, also using E. coli cells, have demonstrated the importance of the signal sequence in the membrane insertion and export of some proteins. In these experiments, most of the gene for the soluble enzyme β-galactosidase (lacZ) was fused to portions of each of several genes coding for proteins involved in the transport of the disaccharide maltose. These maltose-specific proteins, products of the genes malF, malE, and lamB, are normally located in the inner membrane, periplasmic space, and outer membrane, respectively (Chapter 17). Cells containing these hybrid genes synthesize hybrid proteins containing portions of the membrane or exported protein at the N terminus and a covalently linked, functional β-galactosidase molecule that is easy to assay spectrophotometrically. With these cells, the localization of the three hybrid proteins could easily be determined by cell fractionation and β-galactosidase activity measurements. As long as a sufficient amount of the N terminus of the maltose transport proteins was

**Figure 16-49**
The use of gene fusion to study membrane protein localization and assembly in bacteria. Genetic techniques have been developed for the fusion of specific genes in E. coli. In this example, part of the gene for an inner membrane protein involved in maltose transport (malF) is fused to most of the β-galactosidase gene (lacZ). β-Galactosidase, normally a soluble protein, becomes membrane-bound in this case, showing that the information for membrane localization of the malF product is carried by the amino acid sequence of this portion of the malF protein itself. Fusions of various parts of the malF gene to lacZ can therefore be used to determine which portions of the protein are important for insertion simply by determining the location of β-galactosidase, which is easy to assay.

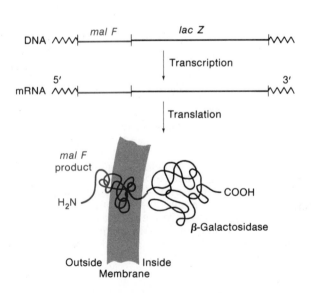

**Table 16-11**
Localization of Hybrid Proteins Produced by Various Maltose
Gene-*lacZ* Fusion Strains

| Fusion | Approximate Amount of Maltose Gene DNA in Hybrid Gene | Cellular Localization of Hybrid Protein ($\beta$-Galactosidase Activity) |
|---|---|---|
| *malF* (inner membrane protein)-*lacZ* | 1/11 | Cytoplasm (92%) |
| | 9/11 | Inner membrane (80%) |
| *malE* (periplasmic protein)-*lacZ* | 2/13 | Cytoplasm (88%) |
| | 4/13 | Inner membrane (65%) |
| *lamB* (outer membrane protein)-*lacZ* | 1/9 | Cytoplasm (90%) |
| | 6/9 | Outer membrane (80%) |

Data from T. J. Silhavy, P. J. Bassford, Jr., and J. R. Beckwith, A genetic approach to the study of protein localization in *Escherichia coli*. In M. Inouye (Ed.), *Bacterial Outer Membranes: Biogenesis and Functions*, Wiley, New York, 1979.

present, $\beta$-galactosidase activity was found in the inner membrane for the first two fusion products and in the outer membrane for the last, even though $\beta$-galactosidase is normally a soluble protein (Figure 16-49 and Table 16-11). Thus, in accordance with the signal hypothesis, the signal sequence was able to direct membrane or exported proteins to the membrane, even though a large hydrophilic molecule was attached to it. Furthermore, mutations were isolated in which the *lamB-lacZ* hybrid protein was entirely cytoplasmic. These were mapped at the extreme N terminus of the *lamB* portion, showing that defects in the signal sequence can lead to defective insertion of membrane proteins.

## Alternatives to the Signal Mechanism

Despite the attractiveness of the signal hypothesis in explaining protein secretion especially and integral membrane protein insertion in some instances, this model does not adequately account for observations on the modes of synthesis and insertion of several integral membrane proteins that have been studied. These proteins include some mitochondrial and chloroplast membrane proteins, cytochrome $b_5$, and NADH:cytochrome $b_5$ reductase of endoplasmic reticular membranes as well as several cytoplasmic membrane proteins of *E. coli* and other bacteria. The problems that arise are as follows:

1. Some integral membrane proteins are synthesized on soluble, non-membrane-bound ribosomes and are inserted into the membrane posttranslationally.

2. Quite a number of integral membrane proteins are apparently synthesized without a signal sequence.

3. Genetic studies in bacteria show that in certain instances the signal sequence is necessary but not sufficient for correct protein insertion or secretion.

4. The topographies of some proteins in the membrane are not easy to explain by the vectorial "threading" mechanism inherent in the original signal hypothesis.

These observations have led W. Wickner to propose an alternative to the signal hypothesis for membrane protein assembly. This _membrane trigger hypothesis_ suggests that some membrane proteins that are synthesized on soluble polysomes can assume two conformations, one more stable in aqueous solutions and the other induced (triggered) by contact with the hydrophobic environment of the membrane. The soluble precursor is proposed to diffuse to the membrane and undergo a conformational change as it self-inserts into the bilayer. This induced conformation and association of the protein with the membrane are thought to be stabilized by proteolytic cleavage of the N-terminal extension of amino acids (Figure 16-50b). This hypothesis differs from the signal mechanism in that _membrane protein synthesis need not occur on membrane-bound ribosomes_ (although it may in some cases) and that membrane protein insertion is a _self-assembly process_ rather than one that is catalyzed by membrane-bound translation machinery and transmembrane pores.

Several membrane proteins have been studied that do not conform to the signal mechanism of assembly. For example, the lactose transport protein of E. coli (_lactose permease_; Chapter 17) is an integral cytoplasmic membrane protein that appears not to be synthesized in precursor form and therefore lacks a signal, or "leader," sequence. Similar observations have been made for several other bacterial membrane proteins, as well as for some proteins found in mitochondrial, chloroplast, and endoplasmic reticular membranes. Furthermore, a major lipoprotein found in the E. coli outer membrane, which is synthesized with a removable leader sequence, has been found in its correct location even when a mutation was introduced in the signal sequence that blocked removal of the N-terminal extension. These results show that _neither a signal sequence nor precursor processing is necessary for the proper insertion of some membrane proteins._

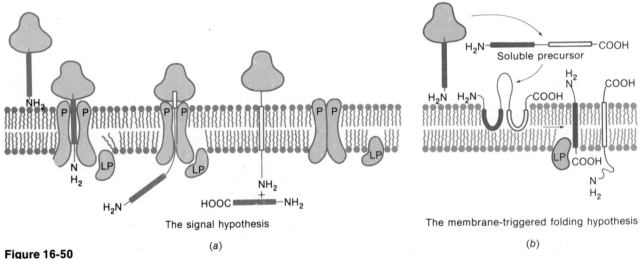

The signal hypothesis

(a)

The membrane-triggered folding hypothesis

(b)

**Figure 16-50**
The signal hypothesis (a) compared with the membrane-triggered hypothesis (b) for the assembly of proteins into membranes. In (b), the protein is synthesized in a water-soluble form of unknown conformation and then is post-translationally inserted into the membrane by means of a conformational change "triggered" by contact with the membrane. The solid thick portion of the protein is the hydrophobic signal or leader sequence, while the unshaded thick region is a hydrophobic, membrane-spanning sequence. (P = pore protein; LP = leader peptidase). (Adapted from W. Wickner, Assembly of proteins into membranes, _Science_ 210:861, 1980.)

Genetic studies of some bacterial membrane and periplasmic proteins have led to the conclusion that even in the case of those polypeptides synthesized with an N-terminal extension, *the leader sequence is not by itself sufficient to ensure proper localization*. When the N-terminal half (including the signal sequence) of the periplasmic protein necessary for maltose transport in *E. coli* was fused to β-galactosidase, the hybrid protein was found in the inner membrane, not in the periplasm (Table 16-11). More recently, mutations have been isolated in β-lactamase (penicillinase) of the Gram-negative bacterium *Salmonella typhimurium*, resulting in the loss of about 10 percent of the C-terminal amino acid residues of this protein. β-Lactamase is synthesized as a precursor and is normally found in the periplasmic space in a processed form. The chain-termination mutant proteins, however, were found in the cytoplasm and were therefore not secreted across the inner membrane. Hence *an intact C terminus is necessary for β-lactamase export*.

Finally, quite a few integral membrane proteins are oriented topographically in a manner that is difficult to reconcile with the signal hypothesis as originally proposed. We have already considered two prominent examples, namely, the erythrocyte anion-transport protein (band 3) and bacteriorhodopsin. The former protein spans the membrane in an orientation opposite to that predicted by the signal hypothesis because its C-terminal amino acid residues are on the outside of the cell, while its N terminus is exposed in the cytoplasm (Figure 16-30). In the case of bacteriorhodopsin, the polypeptide traverses the membrane seven times (Figure 16-31), which is very difficult to explain with the single-pore extrusion mechanism of the original signal hypothesis. Alternative mechanisms, such as the membrane trigger hypothesis or a revised signal mechanism, may better account for the dispositions of these proteins in the membrane.

In conclusion, it seems likely that the complex problems of membrane protein assembly and the secretion of extracellular polypeptides may be solved by the cell in more than one way. Alternative or revised explanations to the ones we have considered will unquestionably be forthcoming in the near future. The rapid pace of research in this area reflects the central role that these problems play in our understanding of biological membrane structure and function.

## SELECTED READINGS

Fleischer, S., and Packer, L., (Eds.). *Methods in Enzymology*, Vols. 31 and 32. New York: Academic Press, 1974. Volume 31 of this invaluable series focuses on subcellular fractionation and membrane isolation techniques, including isopycnic and differential centrifugation. Volume 32 deals with the composition and characterization of various membranes and membrane components, including articles on model membrane systems.

Guidotti, G. The structure of intrinsic membrane proteins. *J. Supramol. Struct.* 7:489, 1977. Review of integral membrane protein structure with emphasis on the erythrocyte anion transport protein (Band 3).

Helenius, A., and Simons, K. Solubilization of membranes by detergents. *Biochim. Biophys. Acta* 415:29, 1975. Lengthy review of the properties of detergents and their use for membrane solubilization.

Jain, M. K., and Wagner, R. C. *Introduction to Biological Membranes*. New York: Wiley, 1980. Chapters 1–8. Includes an up-to-date survey of the electron microscopy of biological membranes, isolation and characterization of membrane components, properties of lipid bilayers and membrane biogenesis.

Lodish, H. F., and Rothman, J. E. The assembly of cell membranes. *Sci. Am.* 240:48, 1979. A clearly written account of some of the problems of membrane assembly with a focus on the synthesis and membrane insertion of G protein of vesicular stomatitis virus.

Rothman, J. E., and Lenard, J. Membrane asymmetry. *Science* 195:743, 1977. This review discusses evidence for membrane asymmetry with an emphasis on lipids; also presented is an extension of the signal hypothesis to explain how some integral membrane proteins could be inserted into the bilayer.

Singer, S. J. The molecular organization of membranes. *Ann. Rev. Biochem.* 43:805, 1974. Excellent review of membrane structure through 1973.

## PROBLEMS

1. Considerable evidence, much of it circumstantial, suggests that mitochondria may have evolved from endosymbiotic aerobic bacteria within ancestral eukaryotic cells (and likewise, that chloroplasts evolved from a photosynthetic prokaryote).

   a. From what you know about membrane structure, defend this hypothesis.

   b. Assuming that this hypothesis is correct, explain the differences that are observed in envelope structure between mitochondria and present-day bacteria.

2. Predict the effects of the following on the phase transition temperature ($T_m$) and phospholipid mobility in pure dipalmitoylphosphatidylcholine vesicles ($T_m = 41°C$).

   a. Raising the temperature from 30° to 50°C.

   b. Introducing dipalmitoleoylphosphatidylcholine into the vesicles.

   c. Adding dimyristoylphosphatidylcholine to the vesicles.

   d. Introducing a high concentration of cholesterol into the vesicles.

   e. Incorporating integral membrane proteins into the vesicles.

3. a. Speculate on the basis for the differences in $T_m$ between the dipalmitoyl esters of phosphatidic acid, phosphatidylethanolamine, and phosphatidylcholine (Table 16-9).

   b. Why do you think that phosphatidylethanolamine partitions preferentially into the inner leaflet in mixed PC/PE artificial vesicles (Table 16-10)?

4. Lidocaine and phenethylalcohol are both local anesthetics that have been shown to interact with nerve cell and other membranes. Although the precise mechanisms of their anesthetic action are not known, it seems reasonable to conclude that they involve changes in membrane structure and physical properties. Both these compounds have been shown to inhibit processing (cleavage) of signal sequences and in some cases also to affect final localization of certain membrane and periplasmic proteins of *E. coli*. From the chemical structures of these compounds given below, postulate reasonable mechanisms for their inhibition of membrane protein processing and assembly.

**Lidocaine**     **Phenethylalcohol**

5. Triton X-100 ($M_r = 625$; CMC = 0.24 mM) and sodium deoxycholate ($M_r = 414$; CMC = 4 mM) are both often used to solubilize crude membrane fractions before purification of constituent proteins is attempted.

   a. Which of these detergents would be most easily removed by dialysis (diffusion through a membrane permeable to molecules with molecular weights less than about 10,000 daltons) from a 0.1% (weight/volume) aqueous solution?

   b. In which of these detergents would ion-exchange chromatography probably be more successful for protein purification? How might you exchange the less favorable for the preferred detergent if this were necessary?

   c. Gel filtration chromatography would be likely to give a closer approximation to the molecular weight of an integral membrane protein in a 0.1% solution of which of these detergents? Why?

6. In 1925, Gorter and Grendel measured the total area occupied by a monolayer of red blood cell lipids using a device known as a Langmuir trough. In one experiment, they found that a monolayer of lipids was achieved in an area of 0.89 m$^2$ starting with lipids isolated from $4.74 \times 10^9$ human red blood cells. The surface area they measured for one cell was 99.4 $\mu$m$^2$. Using this information, show how they concluded that the membrane covering a human red blood cell was two lipid molecules thick.

7. Suppose you treated a hypothetical cell that was grown at 30°C with TNBS at 3°C and found that about 75 percent of the total membrane phosphatidylethanolamine (PE) could be labeled. At 15°C, however, you could label nearly all the PE in several hours, but 75 percent reacted with a half-time of 2 min, while the other 25 percent had a half-time of reaction of 30 min.

a. Assuming that no new phospholipid could be synthesized at either 3°C or 15°C, how would you interpret the preceding results?

b. Suppose that at 30°C in growing cells, 75 percent of the newly synthesized PE flip-flops with a half-time of 5 min from the inside monolayer to the outside one, while the rate of the reverse process is negligible. What would the distribution of PE synthesized at "time 0" be 5 min later under these conditions?

c. Can you think of any reason why 25 percent of newly synthesized PE does not flip-flop at a high rate?

d. Give an explanation other than specific catalysis by a protein for the high rates of transverse diffusion seen in some systems for newly synthesized membrane phospholipids.

8. When about half the gene for the periplasmic maltose binding protein (*malE*) is fused to most of the gene for β-galactosidase (*lacZ*), the hybrid protein product is found "stuck" in the plasma membrane of *E. coli* (i.e., never is exported into the periplasm). In this same strain, the product of the *lamB* gene, an outer membrane protein, is found in a precursor form in the cytoplasm (Silhavy et al., 1979; see legend to Table 16-11). Explain these results in light of the signal hypothesis.

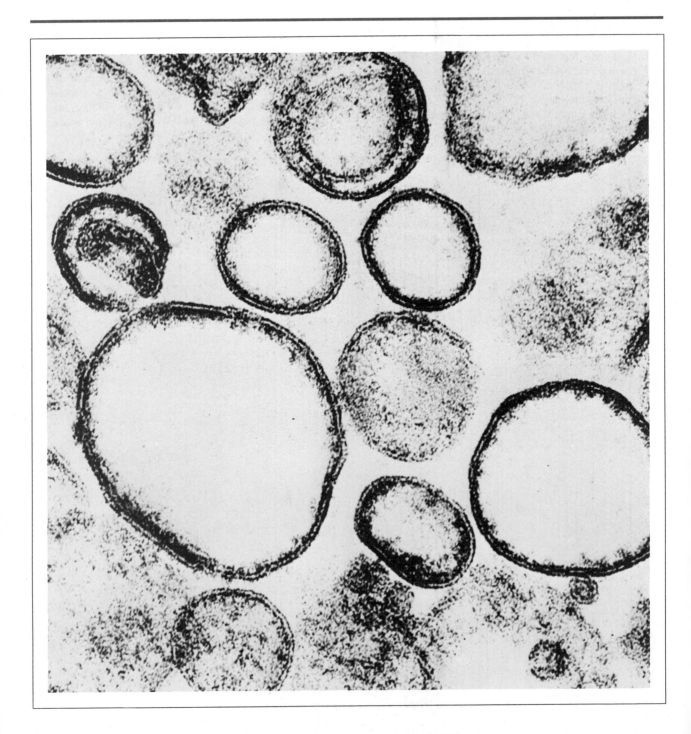

# 17

# BIOLOGICAL MEMBRANES: TRANSPORT

In the preceding chapter we considered the current state of knowledge about the structure and assembly of biological membranes. Two of the most important conclusions to be drawn are that membranes are both _dynamic_ and _asymmetric_ structures. As we shall see, these properties allow _transmembrane_ processes to be carried out, often, but not always, in a unidirectional manner. The most prominent of these processes are the transport of nutrients and inorganic electrolytes into the cell and the export of toxic substances and waste materials out of the cytoplasm. In both instances, the problem is usually the same: how do small, hydrophilic metabolites traverse the hydrophobic and generally impermeable lipid bilayer at rates sufficient to satisfy the maintenance and growth needs of the cell? Often this problem is solved by integral membrane proteins, which, by analogy to enzymes, specifically interact with their substrates, but, unlike enzymes, have as their primary purpose the movement of a molecular species across a membrane rather than the catalysis of a chemical reaction.

Only recently have sufficiently general enough techniques been developed for the purification of integral membrane proteins, not the least of which is the use of detergents to allow their study and separation in a soluble form (Chapter 16). Accordingly, our understanding of the molecular mechanisms of transport proteins, or _permeases_, has lagged considerably behind that of soluble enzymes. Nevertheless, in several instances, preparations of sufficient purity have been obtained to conduct physical, chemical, and mechanistic studies that should eventually reveal the molecular mechanisms of the "catalysis" of transport.

In this chapter, we shall consider the theory, energetics, and mechanisms of transmembrane transport in cells. We shall further explore a num-

_Electron micrograph of membrane vesicles obtained from E. coli. Preparations of this sort have been most useful as a model system for studying membrane transport processes. (From H. R. Kaback.)_

ber of biological processes that rely on transport, directly or indirectly, for their proper functioning (Chapters 29 to 31). Examples of these seemingly diverse phenomena include nerve impulse propagation, cellular behavior and differentiation, and sensory reception and transduction. Indeed it should become apparent that nearly all cellular and organismal functions are dependent on biological transport and its efficient functioning. As in Chapter 16, examples from both prokaryotic and eukaryotic worlds will show us that many fundamental principles concerning the biochemistry of transport and related functions are probably valid throughout the living kingdom.

## THE THEORY AND THERMODYNAMICS OF BIOLOGICAL TRANSPORT

Before we consider specific examples of transport across biological membranes, it is important that we discuss the thermodynamics of this process, as well as specific ways in which such a transfer can occur. With this foundation, it then becomes possible to assign certain predicted characteristics to each kind of transport mechanism. Experiments with a given system can therefore be designed based on these predictions that will give useful information regarding the events responsible for the transmembrane movement of a solute.

### Simple Diffusion versus Carrier-Mediated Transport

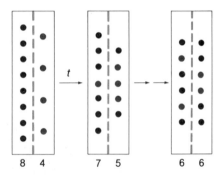

**Figure 17-1**
Illustration of simple diffusion in a two-compartment system divided by a semipermeable membrane. After time $t$, one-fourth of the molecules present originally on one side of the membrane have diffused by random thermal motion to the other side. After many such intervals, equilibrium is approached. The molecules in this illustration are identical, but different colors are used to indicate their initial locations. In simple diffusion, the *number* of molecules diffusing from one side to the other in any given time interval is directly proportional to their concentration on the first side. (Adapted from K. D. Neame and T. G. Richards, *Elementary Kinetics of Membrane Carrier Transport*, Blackwell Scientific Publications, Oxford, 1972.)

In principle, a solute could move across a barrier such as a membrane by either of two fundamentally different mechanisms: *simple* (or *passive*) *diffusion* or *carrier-mediated transport*. In the first of these, the molecule in question must be somewhat soluble in the hydrophobic phase of the bilayer, while in the second mechanism, the solute's journey across the membrane is facilitated by another molecule, in most cases a protein. For these reasons, simple diffusion is usually of importance in biological transport systems only for molecules having a largely hydrophobic character or in membrane systems possessing nonspecific aqueous pores, as will be discussed in a later section. An exception to this rule, however, is water, which traverses many biological membranes by simple diffusion. Most other hydrophilic molecules, which comprise the majority of species transported into and out of cells, are usually recognized by transport systems that include one or more protein carriers.

In order to illustrate the characteristics of simple diffusion, let us consider two dilute solutions of a solute $So$, one twice as concentrated as the other, which are separated by a barrier that is permeable to $So$, as in Figure 17-1. As a consequence of random thermal motion, a certain proportion of molecules originally present on one side of the barrier (one-quarter in Figure 17-1) will move to the other side in a specified length of time $t$. This *flux* is bidirectional, so that after time $t$, the ratio of the concentration of $So$ on the left side of the barrier to that on the right will have been reduced from 2:1 to 7:5. After many such time intervals, this ratio will approach 1:1, or equilibrium. The number of molecules of $So$ that move from left to right in time interval $t$ is therefore directly proportional to the initial concentration of $So$ on the left side:

$$v \propto [So] \qquad (1)$$

while a similar relationship holds true for movement in the opposite direction:

$$v' \propto [So'] \qquad (2)$$

where $v$ and $v'$ refer to rates of transfer from left to right and right to left, respectively, and $[So]$ and $[So']$ are the initial concentrations of the solute on the left and right sides of the barrier. The net rate of transfer $V$ from the more concentrated to the more dilute solution is therefore the difference between these two component rates:

$$V = v - v' \qquad \text{or} \qquad V \propto [So] - [So'] \qquad (3)$$

Hence, at equilibrium, $V = 0$.

If the barrier in Figure 17-1 is, for example, a biological membrane of finite thickness $l$, we also must take this parameter into account in the expression for overall transfer rate:

$$V \propto \frac{[So] - [So']}{l} \qquad (4)$$

Finally, the rate of transfer of $So$ will depend on such factors as temperature, viscosity of the solvent, and the solubilities of $So$ in the solvent and in the membrane. In a given membrane system, these factors often can be held relatively constant, so that we can express relationship (4) as an equation:

$$V = \frac{D([So] - [So'])}{l} \qquad (5)$$

where $D$ is a constant, the _diffusion coefficient_, which has the dimensions of area per unit of time (e.g., $cm^2/s$). Equation (5) is essentially _Fick's law_, and it shows that the net rate of diffusion is directly proportional to the concentration gradient of $So$ across the membrane barrier.

It should be apparent from Equation (5) that in a system in which $So$ traverses the membrane only by simple diffusion, a plot of $V$ versus $[So] - [So']$ should give a straight line that passes through the origin with a slope of $D/l$ (Figure 17-2a). In practice, it is often possible to measure $V$ as a function of the concentration gradient of $So$ across, for example, the plasma membrane of a cell by methods to be considered later in this chapter. Provided that a wide range of solute concentration gradients is studied, _a linear plot of $V$ versus $[So] - [So']$ is evidence that $So$ is taken up by simple diffusion_.

The uptake kinetics shown in Figure 17-2a for simple diffusion are, in fact, seen only in a minority of cases for transport of molecules across cytoplasmic membranes of cells. More often, the curve approximates that of a _rectangular hyperbola_, as illustrated in Figure 17-2b. In this case, the transport rate approaches a limiting value at high $[So] - [So']$ that is termed $V_{max}$ by analogy to steady-state enzyme kinetics (Chapter 4). This _saturation_ phenomenon implies that uptake of $So$ is dependent on a dis-

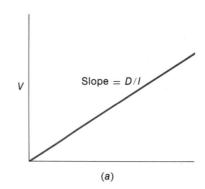

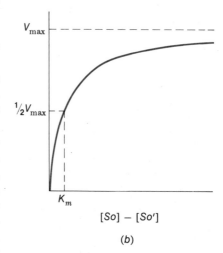

**Figure 17-2**
Initial rate kinetics of simple diffusion (a) and carrier-mediated transport (b). Plots of velocity versus the difference in substrate concentrations on the two sides of the membrane give a straight line with a slope of $D/l$ in (a) and a rectangular hyperbola approaching a maximum velocity ($V_{max}$) at high $[So] - [So']$ in (b).

crete number of *carriers* in the membrane, all of which become filled at high concentration gradients of So. This situation is formally analogous to saturation of an enzyme by its substrate, and the rate equation for this process, *carrier-mediated transport*, is essentially the same as the Michaelis-Menten equation derived in Chapter 4:

$$V = \frac{V_{max}[So]}{K_m + [So]} \tag{6}$$

In the derivation of Equation (6), [So'] is assumed to be negligible (just as is [P] in enzyme kinetics) and is taken to be the initial concentration of So inside the cell or vesicle being studied. Experimentally, this situation is usually easy to attain, so that [So] equals the concentration initially present outside, [So'] = 0, and V is an *initial rate of transport* of So into the system enclosed by the membrane. As in enzyme kinetics, $K_m$ is the solute concentration at which $V = \frac{1}{2}V_{max}$ (Figure 17-2b). It should be noted that the value of $K_m$, although often assumed to be a measure of the affinity of a carrier for its "substrate," is really a ratio of rate constants and is not equivalent to the dissociation constant of the solute from its carrier.

From the foregoing discussion it is clear that in many cases, by studying the rate of transport, V, as a function of [So], simple diffusion can be distinguished from carrier-mediated transport. Often, however, the data so obtained are plotted in a double-reciprocal fashion (Lineweaver-Burk plot; Chapter 4). Plots of $1/V$ against $1/[So]$ yield straight lines in both instances, but the line for simple diffusion passes through the origin, while that for carrier-mediated transport intersects the ordinate at $1/V_{max}$ and the abscissa at $-1/K_m$ (Figure 17-3a). This, of course, reflects the fact that carrier systems are saturable ($V_{max}$ is finite), while simple diffusion systems are not ($V_{max} = \infty$; $1/V_{max} = 0$).

Unfortunately for biochemists interested in studying a single biological transport system, double-reciprocal plots for the uptake of a specific solute into a cell, or vesicles derived from cells, are sometimes nonlinear. This usually indicates that *more than one process is involved in the transport of this substance into the cell*. These different processes may include both simple diffusion and carrier mediation, or both may be carrier-mediated (Figure 17-3b). In either case, however, it is first necessary to find experimental conditions under which only the system of interest is functional before conclusions concerning this transport mechanism can be drawn. This may be accomplished by working at solute levels recognized only by one carrier, by the use of mutants lacking one system, by using specific inhibitors, or even by purification and reconstitution of the carrier of interest (see the following discussion).

Finally, carrier-mediated transport can in many cases be distinguished from simple diffusion by its stereospecificity. For example, in a membrane system that is permeable to glucose only by simple diffusion, both D- and L-glucose are found to traverse the membrane at equal rates. In contrast, many biological membranes contain carriers specific for glucose that recognize only the D-stereoisomer. In these cells, the initial transport rate for D-glucose is much higher than that for the L-isomer and shows saturation kinetics. In fact, the rate of transport of L-glucose is often used in such

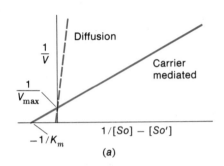

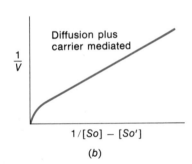

**Figure 17-3**
Double-reciprocal (Lineweaver-Burk) plots of carrier-mediated transport and simple diffusion. For diffusion, the line passes through the origin, while for carrier-mediated processes the x intercept is equal to $-1/K_m$ and the y intercept is equal to $1/V_{max}$, as shown in (a). In (b), the transport kinetics for a cell transporting a given solute by *both* simple diffusion and a carrier-mediated system are illustrated. The line shown is a combination of the processes plotted in (a) and passes through the origin. In the case of two carrier-mediated systems, a similar line would be obtained that would intersect the ordinate above the origin. (Adapted from K. D. Neome and T. G. Richards, *Elementary Kinetics of Membrane Carrier Transport*, Blackwell Scientific Publications, Oxford, 1972. Used with permission.)

instances to correct the total rate observed for diffusion in order to obtain the "true" carrier-mediated rate for the D-isomer. *Stereospecificity in biological transport, as in enzyme-catalyzed reactions, is a consequence of the highly selective configuration of the "active site" of the transport protein for its "substrate."*

## The Energetics of Transmembrane Transport

In the preceding section we saw that a solute will tend to diffuse "down" its concentration gradient (from a region of higher to a region of lower concentration) spontaneously, until equilibrium is reached. Such a process requires no outside input of energy into the system and thus occurs with a negative change in free energy $\Delta G$. This is a consequence of the fact that the equilibrium situation ($[So] = [So']$ in Figure 17-1) is more "random" or "disordered" than the starting situation in which ($[So] > [So']$) and thus has relatively more entropy $S$. Because $\Delta G = \Delta H - T\,\Delta S$, an increase in entropy results in a negative free energy change (Chapter 7). The magnitude of $\Delta G$ can be calculated from the expression

$$\Delta G = 2.3RT \log \frac{[So']}{[So]} \tag{7}$$

for the situation illustrated in Figure 17-1 (see Chapter 7). *Net transport from left to right will only occur spontaneously in this system if $[So']/[So] < 1$, while at equilibrium, $\Delta G = 0$ ($[So'] = [So]$).*

In biological systems, both simple diffusion and *facilitated diffusion* are transport mechanisms that result in net transport of a solute only in the direction of negative $\Delta G$, i.e., usually from higher to lower concentrations. In contrast to simple diffusion, facilitated diffusion is a carrier-mediated process and hence is saturable (Figure 17-4). Both it and simple diffusion generally result in *equilibration* of a solute across a membrane, unless the solubility properties of the solute are significantly different on both sides of the membrane or the solute has a net charge. *If the solute is charged, both types of diffusion can lead to an equilibrium situation in which $[So] \neq [So']$ if there is an independently maintained electrical potential, $\Delta\psi$, across the membrane* (see Equation 49, Chapter 7):

$$\Delta G = 2.3RT \log \frac{[So']}{[So]} + ZF\,\Delta\psi \tag{8}$$

In this equation, $Z$ is the net charge on $So$, and $F$ is the Faraday constant. Thus, at equilibrium ($\Delta G = 0$), unequal concentrations of $So$ can exist on either side of a membrane as long as the term $ZF\,\Delta\psi$ is not equal to zero (Figure 17-5). The relationship between free energy and membrane potential is very similar to the relationship between free energy and electromotive potential discussed in Chapter 7 and again in Chapter 10.

Both simple and facilitated diffusion are examples of *energy-independent* biological transport mechanisms. Although they occur in fundamentally different ways (one is carrier-mediated and one is not), neither requires the input of energy, nor can either lead to concentration of the solute on one side of the membrane relative to the other in the absence of a charged solute and an electrical potential across the membrane. Most

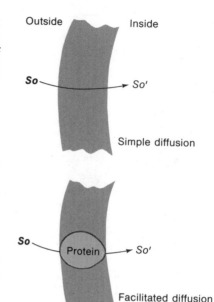

**Figure 17-4**
Schematic illustration of diffusion processes in biological transport. In both simple and facilitated diffusion, the concentrations of the solute on both sides of the membrane are the same at equilibrium, and no metabolic energy is expended.

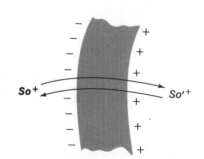

**Figure 17-5**
At equilibrium, a *charged* solute, $So^+$, may be unequally distributed across a biological membrane if an independently maintained potential exists across it. This situation can occur even if $So^+$ is transported by simple or facilitated diffusion. Bold type indicates higher concentration.

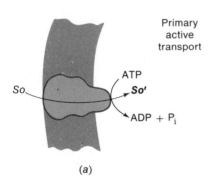

Primary active transport

(a)

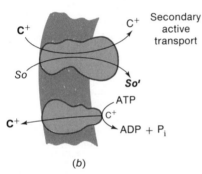

Secondary active transport

(b)

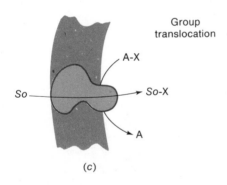

Group translocation

(c)

**Figure 17-6**
Schematic illustrations of primary active transport (a), secondary active transport (b), and group translocation (c). In (a), the energy source can be ATP, as shown, or electron transport or light can be used. In (b), active accumulation of So is energized by cotransport of an ion that flows down its concentration gradient. In (c), So is accumulated as a derivative, So-X, which is formed by a chemical reaction catalyzed by the carrier during transport through the membrane.

cells, however, can maintain relatively high intracellular concentrations of some metabolites and ions relative to the extracellular environment. For example, many cells have an intracellular level of $K^+$ that is at least 30 times the concentration of this ion in the surrounding medium. It therefore follows that if carriers for accumulated molecules, such as $K^+$, exist in the cell membrane, they must be able to transport their substrates in the direction of a positive $\Delta G$, i.e., _up_ a concentration gradient of the solute. In order to accomplish this thermodynamically unfavorable movement, _such processes must be coupled to events that have negative $\Delta G$ values_, i.e., that can yield energy to "push" the solute "uphill." _Concentrative accumulation of a solute inside a cell, organelle, or vesicle with the consequent expenditure of metabolic energy is termed active transport_. In a related process, _group translocation_, energy for the transmembrane accumulation of a solute is provided by chemical modification of the molecule as it passes through the membrane.

## Active Transport and Group Translocation

The use of metabolic energy to drive active transport implies that the transport system itself must somehow be able to harness the energy in a particular form and use it to do work, i.e., to transport a solute against its concentration gradient. The molecular mechanisms by which these interconversions take place are still, in most cases, obscure, although a few clues are emerging that we will consider later in this chapter. Because systems that carry out active transport are invariably carrier-mediated (almost by definition), it seems likely that in most cases integral membrane permeases themselves will directly participate in this energy interconversion process. A major direction of research in membrane-transport biochemistry in recent years has therefore been to study the energetics of active transport and, whenever possible, to isolate and characterize the proteins that comprise a particular system.

Two basic types of active transport have been recognized. In _primary active transport_, energy is provided directly by the hydrolysis of ATP, by electrons flowing down an electron transport chain, or by light. Examples of these include the $Na^+$-$K^+$ ATPase of animal cells, which pumps $K^+$ into the cell and $Na^+$ out of the cell at the expense of ATP; the respiratory electron-transport chain, which pumps $H^+$ out of the mitochondrion (Chapter 10), and the light-driven $H^+$ pump of _Halobacterium_, bacteriorhodopsin. In _secondary active transport_, ion gradients across the membrane, themselves created by active transport systems, are used to drive the concentrative uptake of other ions or metabolites. Because an ion gradient is a form of potential energy, a process involving its collapse releases this energy, which can be used to perform work. In mitochondria and bacteria, the potential energy stored in gradients of $H^+$ is used during respiratory metabolism to synthesize ATP (Chapter 10). As we shall see, gradients of $H^+$ as well as other ions also are an important source of energy for active transport in both prokaryotic and eukaryotic cells.

Related to primary active transport is the process of _group translocation_, a term coined by P. Mitchell over 20 years ago. While molecules transported by active systems are translocated across the membrane without chemical changes occurring, in group translocation, the transported solute is chemically modified on its journey through the membrane. The best-documented example of group translocation occurs in many bacteria, which convert sugars taken up from the surrounding medium into sugar monophosphates simultaneously with their transport through the membrane. Energy is expended in the process, since the phosphoryl donor is the high-energy phosphate glycolytic intermediate, phosphoenolpyruvate. In this case, the permease for the sugar is also an enzyme that catalyzes both the transport and the phosphorylation of its substrate. Figure 17-6 illustrates schematically the differences between primary and secondary active transport and group translocation, while Table 17-1 summarizes modes of energy coupling for selected transport systems of both animal and bacterial cells.

**Table 17-1**
Energy-Coupling Modes of Well-Studied Transport Systems

| Solute | Polarity of Pump | Energy Coupling | Organism/Tissue |
|---|---|---|---|
| $Na^+$-$K^+$ | Out-in | Primary active (ATP) ($Na^+$-$K^+$ ATPase) | Animal cells |
| $Ca^{2+}$ | Out | Primary active (ATP) ($Ca^{2+}$ ATPase) | Sarcoplasmic reticulum |
| $H^+$ | Out | Primary active (ATP) ($H^+$ ATPase) | Plasma membrane of stomach epithelial cells |
| $H^+$ | Out | Primary active (light) (bacteriorhodopsin) | *Halobacterium halobium* |
| $H^+$ | Out | Primary active (electron transport) | Mitochondria, bacteria |
| $H^+$ | Out | Primary active (ATP) ($F_1/F_0$ ATPase) | Bacteria (anaerobic) |
| Various sugars and amino acids | In | Primary active (chemical energy) (binding protein systems) | Gram-negative bacteria |
| Lactose | In | Secondary active ($H^+$ cotransport) | *E. coli* and some other bacteria |
| Glucose | None | Facilitated diffusion | Most animal cells |
| Glucose | In | Secondary active ($Na^+$ cotransport) | Certain animal cells |
| Glucose | In | Group translocation (phosphotransferase system) | *E. coli* and some other bacteria |
| $HCO_3^-$-$Cl^-$ | None | Facilitated diffusion (anion channel) | Erythrocytes |
| ATP-ADP | Out-in | Facilitated diffusion, but sensitive to membrane potential (ATP/ADP exchanger) | Mitochondria |

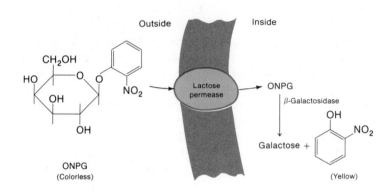

**Figure 17-7**
o-Nitrophenyl-β-galactoside (ONPG) is a convenient substrate for measuring transport by the lactose permease in *E. coli.* Upon entering the cell, it is cleaved by β-galactosidase, yielding galactose and o-nitrophenol, which can be quantitated spectrophotometrically.

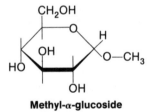

**Methyl-α-glucoside**

**2-Deoxyglucose**

**Figure 17-8**
Methyl-α-glucoside and 2-deoxyglucose are two analogs that are useful in the study of glucose transport in · many cells. Usually they are not metabolized beyond the hexose-phosphate stage, which avoids the complications of subsequent reactions that might affect the measurement of transport rates.

## ENERGY COUPLING MECHANISMS IN BIOLOGICAL TRANSPORT

It is the ultimate goal of the biochemist interested in transmembrane transport mechanisms to understand at the molecular level the physicochemical properties of a transport protein that allow it to carry out its function. In this section, we shall briefly describe experimental systems commonly used to study biological transport and show how these systems can be used to investigate energy-coupling mechanisms at the molecular level. In so doing, we shall place emphasis on some of the best-characterized biological transport systems. With this as a background, we will then venture one step further in the next section and attempt to explain how the structures and properties of integral membrane permeases permit them to carry out the myriad of transport processes so vital to the cell.

### Experimental Approaches to the Study of Membrane Transport

In order to measure the rate of transport of a solute into a cell, some method must be available for easily identifying it and distinguishing molecules that have indeed crossed the membrane. Most commonly, radioactively labeled solutes are employed for this purpose, and time-dependent uptake of radioactivity into a cell is evidence that the compound is being transported through the cell membrane. With many animal cells grown in culture, it is often sufficient to separate intracellular label from that remaining in the medium by quickly washing the cells, which adhere to the vessel in which they are grown, with buffer or medium that is free of the solute being tested. With bacteria and small, free-living eukaryotes such as yeast and protozoa, separation of cells from the medium for this purpose can be accomplished quickly by centrifugation or filtration.

In a few special instances, rapid and simple spectrophotometric assays have been developed to measure transport rates into cells. The most prominent example of this has been the use of *o-nitrophenyl-β-galactoside* (ONPG) to study lactose transport in *E. coli.* ONPG is transported into the cell by means of the *lactose permease* (*lacY* gene product) and is subsequently cleaved by β-galactosidase inside the cell to galactose and o-nitrophenol (Figure 17-7), which is yellow and can be quantitated spectrophotometrically. As long as transport is the rate-limiting step in this process, the appearance of o-nitrophenol in cell suspensions can be used to measure the transport rate of ONPG by the lactose carrier. The convenience of this assay has allowed detailed studies of the mechanism of lactose transport in *E. coli.*

One of the problems associated with measuring transport rates, especially in whole cells, is the fact that subsequent *metabolism* of the transport substrate may affect the rate observed. In order to uncouple transport from metabolism, *nonmetabolizable analogs* of some solutes have been used. Examples of compounds that have been employed to study the transport of glucose, a common carbon and energy source for many cells, are *methyl-α-glucoside* and *2-deoxyglucose* (Figure 17-8). Both compounds cannot be metabolized beyond the hexose-phosphate stage by most cells. Another technique that has been useful in this regard, especially in bacteria, is the isolation of mutants that are able to transport a given solute, but which lack one or more enzymes for the further metabolism of the compound. By these procedures, it is often possible to determine many characteristics of the transport system itself, without any complications owing to other reactions that take place after the solute enters the cell.

Finally, in recent years, vesicles derived from the membrane system of interest have become favorite tools of the biochemist interested in membrane transport phenomena. Cytoplasmic membrane vesicles often can be isolated from eukaryotic cells after mild homogenization or $N_2$ cavitation (Chapter 16), which leave many of the intracellular organelles intact. In bacteria, the use of membrane vesicles was pioneered in the 1960s by H. R. Kaback, who developed a method, similar to that shown in Figure 16-7, for preparing membrane vesicles from *E. coli*. Most of these vesicles (Figure 17-9) appear to be "right side out"; i.e., the orientations of integral membrane proteins appear to be the same as those in the intact cell membrane. Kaback and others have shown that many substances transported by *E. coli* also are taken up by these membrane vesicles as long as the required energy sources are provided. These vesicles have the advantage of lacking most cytoplasmic enzymes, so that metabolism of the transport substrate is severely limited. Furthermore, they are especially suited for the study of energy-coupling mechanisms in active transport because the investigator can easily introduce various energy sources into the vesicle preparations and test their effects on the transport system of interest. Such experiments are sometimes difficult to interpret with whole cells, again because of the complications of multiple metabolic events occurring in the cytoplasm.

## Facilitated-Diffusion Transport Systems

Transport systems that appear to operate only by facilitated diffusion have been identified both in bacterial and animal cells. Glycerol, which can act as the sole carbon and energy source for many bacterial cells, has been shown to be transported in several bacteria by nonconcentrative systems. Saturation studies as well as genetic analyses using *E. coli* by E. Lin and his collaborators have provided evidence for a "facilitator protein" which allows glycerol permeation in this organism. The facilitator increases the rate at which glycerol crosses the phospholipid bilayer but cannot accumulate it within the cell.

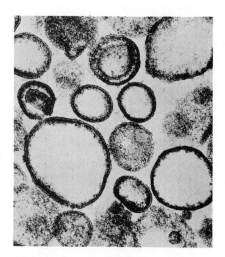

**Figure 17-9**
Electron micrograph of membrane vesicles obtained from *E. coli* spheroplasts subjected to osmotic lysis. These closed, spherical structures have been shown to transport a variety of compounds normally accumulated by whole cells as long as any required energy source is provided. They also have the advantage of lacking cytoplasmic components that could affect the transport rate ($77,000 \times$). (From H. R. Kaback, Transport studies in bacterial membrane vesicles, *Science* 186:882, 1974. Reprinted with permission.)

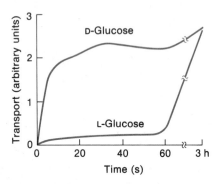

**Figure 17-10**
Facilitated diffusion carried out by the glucose carrier from human erythrocytes. The solubilized carrier was introduced into phospholipid vesicles, and transport was measured with D-glucose and L-glucose as substrates. The initial transport rate was much greater for D-glucose, but L-glucose, which accumulates in these vesicles by simple diffusion, eventually reaches the same interior concentration as D-glucose. (Adapted from M. Kasahara and P. C. Hinkle, Reconstitution and purification of the D-glucose transporter from human erythrocytes, *J. Biol. Chem.* 252: 7384, 1977. Used with permission.)

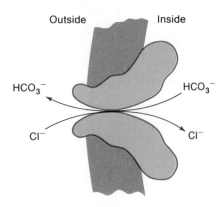

**Figure 17-11**
Exchange diffusion of anions carried out by the human erythrocyte anion-channel (band 3) protein. This exchange is important in lung capillaries for the removal of $CO_2$ generated by cellular metabolism from the red cell cytoplasm.

Recent experimental work on the glycerol facilitator has shown that this permease exhibits characteristics of a fairly nonspecific pore. The transport system allows a variety of straight-chain polyols (erythritols, pentitols, and hexitols, in addition to glycerol) to enter the bacterial cell. The analogous sugars, present in a ring form, cannot be transported, presumably owing to steric hindrance. Other small, neutral, or zwitterionic molecules including urea, glycine, and D,L-glyceraldehyde are transported by the system. Moreover, glycerol and xylitol do not effectively compete with each other as transport substrates, and these compounds do not exhibit saturation kinetics at physiological concentrations. Transport is relatively insensitive to temperature, which is suggestive of a diffusion process. Taken together, these observations suggest that the glycerol facilitator is a nonspecific aqueous channel with an inner diameter of about 0.4 nm. No other nonspecific pore in the inner cytoplasmic membrane of *E. coli* is known.

In animal cells, glucose, a common energy source, appears to be transported by a variety of mechanisms depending on cell type. Both active (see the following discussion) and passive glucose transport systems have been characterized, sometimes both in the same type of cell. In the human erythrocyte, glucose is apparently transported only by facilitated diffusion. The carrier for this sugar in red blood cells has been isolated in Triton X-100 by P. Hinkle and co-workers and reintroduced into phospholipid vesicles. The reconstituted carrier was shown to rapidly equilibrate glucose across the bilayer, but no evidence was found for its ability to carry out active transport, as shown in Figure 17-10.

A third example of facilitated diffusion in biological membranes is the process carried out by the anion-transport (band 3) protein of erythrocytes, whose structure we considered in Chapter 16. The anion transporter allows $CO_2$ accumulated in the erythrocyte from respiring tissues to rapidly diffuse out of the red blood cell in the lung capillaries in the form of the bicarbonate anion $HCO_3^-$. This process has been shown to be accompanied by influx of $Cl^-$ into the erythrocyte. Much work has demonstrated that *the anion transporter catalyzes the one-for-one exchange of $HCO_3^-$ for $Cl^-$*. These two molecules always flow in a fashion determined by the sums of their concentration gradients, and thus this protein catalyzes a type of facilitated diffusion. Transport of $HCO_3^-$ out of the cell is *obligatorily coupled* to the influx of a monovalent anion such as $Cl^-$, as shown in Figure 17-11. This process is called *exchange diffusion*, which is an example of a transport process termed *antiport*. *Antiport involves the coupled transport of two different molecules which move in opposite directions across a membrane*. "Antiporters" may carry out facilitated diffusion, as does the anion-transport protein, or they can catalyze active transport, as discussed below.

### Primary Active Transport: The Na⁺-K⁺ ATPase

One of the first biological transport systems to be studied in detail was that responsible for maintaining the normally high concentrations of $K^+$ and low levels of $Na^+$ inside animal cells. In order for this situation to occur, both cations must be transported essentially unidirectionally against their concentration gradients. A breakthrough in explaining how this could be

$$-O\overset{\overset{O}{\|}}{\underset{\underset{O^-}{|}}{P}}-O\overset{\overset{O}{\|}}{\underset{\underset{O^-}{|}}{P}}-O^-$$

8 Fe
4 H
50
1 S

8 H₂O's

8 Fe        8 Fe
18 H        18 H
12 O        12 O
1 S          1 S

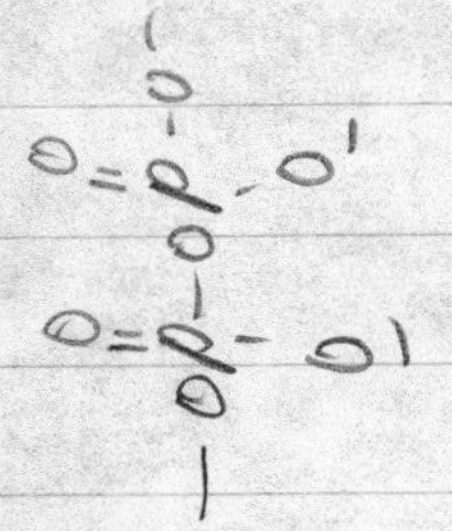

8 Fe
18 H
12 O
1 S

June 19, 1986

Dear Reg:

All went well in D.C. The ATPase grant is hanging in - Pete rattles his career plans. Khalil stays in any case he can't move before Sept. 1. So Nothing new except he was very pleased we'll your help and is ready to publish. If you got some labeled Refield(5) I can try to separate the components here. Any thing new from there?

The Bieber manuscript is enclosed. It turns out to be too long for FEBS Letters, it appears. But there are several points no matter where we publish.

accomplished came in the 1950s when J. C. Skou identified an enzyme from crab nerves that *hydrolyzes ATP to ADP only in the presence of both* $Na^+$ *and* $K^+$. This *sodium-potassium ATPase* has subsequently been shown to be a *transmembrane protein* that pumps $Na^+$ out of the cell and $K^+$ into the cell at the expense of the energy contained in the $\beta$-$\gamma$-phosphate anhydride bond of ATP. This coupling of chemical energy to concentrative transport is an example of primary active transport, as well as of an active antiport system. The same protein catalyzes both the ion translocations and the hydrolysis of ATP, as shown in Figure 17-12. By a mechanism we will explore later in this chapter, the energy released by ATP hydrolysis is "translated" by the $Na^+$-$K^+$ ATPase into potential energy in the form of $Na^+$ and $K^+$ chemical gradients across the membrane. These gradients may then be used to drive secondary active transport processes, as we shall discuss shortly.

An extremely important tool in the study of transport mechanisms has been the use of *specific inhibitors*. In the case of the $Na^+$-$K^+$ ATPase, *ouabain* (Figure 17-13), a compound isolated from certain plants, specifically binds to the outer surface of the enzyme and blocks both ion translocation and ATP hydrolysis. This observation independently confirms the obligatory coupling of these three processes as catalyzed by a single protein. Furthermore, it has been demonstrated that $K^+$ has a much higher affinity for the outside surface of $Na^+$-$K^+$ ATPase, while $Na^+$ binds much more tightly to the cytoplasmic domain. Finally, the site of ATP hydrolysis has been localized to the inner surface of the cytoplasmic membrane. These observations support the schematic illustration shown in Figure 17-12 and demonstrate that the $Na^+$-$K^+$ ATPase must be *asymmetrically oriented* in the cytoplasmic membrane.

Because ouabain binds on the outside of the $Na^+$-$K^+$ ATPase and yet inhibits ATPase activity on the cytoplasmic surface, information must be conducted through the membrane by means of *conformational changes* in the enzyme. Studies with purified $Na^+$-$K^+$ ATPase have shown that the protein is *covalently phosphorylated* during its catalytic cycle, the site of phosphorylation being an aspartic acid residue. Phosphoryl $Na^+$-$K^+$ ATPase is formed in the presence of $Na^+$ and ATP (plus $Mg^{2+}$), while the dephosphorylation step is $K^+$-dependent. Ouabain binds most tightly to the enzyme-phosphate intermediate and prevents dephosphorylation in the presence of $K^+$.

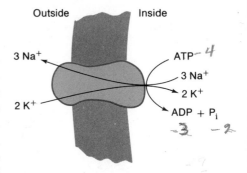

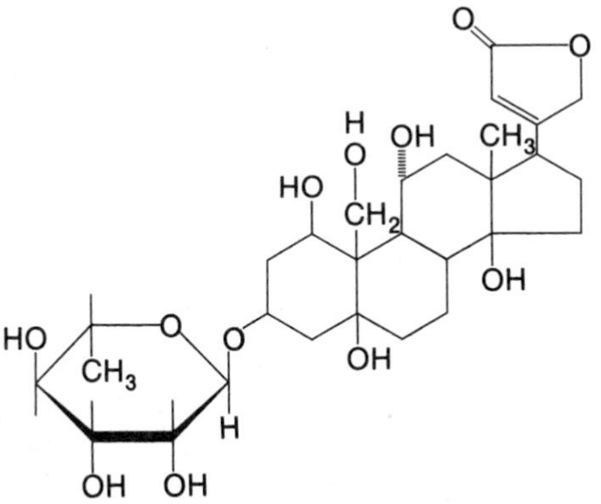

**Figure 17-12**
Primary active transport of $Na^+$ and $K^+$ by the sodium-potassium ATPase found in many animal cells. This enzyme maintains the normally high $K^+$ and low $Na^+$ concentrations within the cell at the expense of chemical energy stored in ATP. Each cycle of the enzyme results in the extrusion of three molecules of $Na^+$ out of the cell, the transport of two molecules of $K^+$ into the cell, and the hydrolysis of one molecule of ATP. The subunit structure and stoichiometry of this enzyme have not yet been unambiguously established, although a large catalytic subunit ($M_r = 95,000$) and a smaller glycoprotein ($M_r = 50,000$) are integral parts of this membrane-spanning protein.

**Figure 17-13**
The structure of ouabain, a specific inhibitor of the $Na^+$-$K^+$ ATPase. Ouabain is a member of a class of compounds called *cardiotonic steroids*. *Digitalis*, a cardiotonic steroid which inhibits the $Na^+$-$K^+$ ATPase indirectly, causes an increase in intracellular $Ca^{2+}$, which stimulates contraction in muscle cells. This explains the clinical use of digitalis in treating such conditions as congestive heart failure.

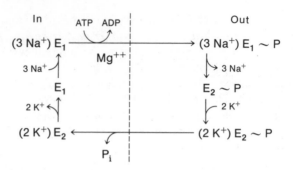

**Figure 17-14**
Postulated mechanism for the functioning of the $Na^+,K^+$ ATPase. The dashed line separates events in which cation association or dissociation occurs on the inside and outside surfaces of the membrane, while the arrows relating $E_1$ and $E_2$ denote conformational changes in the enzyme, but are not meant to imply its physical translocation across the membrane. The enzyme is hypothesized to have at least 2 conformations, $E_1$ and $E_2$, both of which can exist in a phosphorylated state. Only $E_1$ possesses high affinity for $Na^+$, while only $E_2$ possesses high affinity for $K^+$. Phosphorylation of $E_1$ drives the cyclical process in a unidirectional fashion, causing $Na^+$ to be translocated from the inside to the outside. Conversion of $E_1$-P to $E_2$-P decreases the affinity of the enzyme for $Na^+$ and increases its affinity for $K^+$. Dephosphorylation of $E_2$ drives translocation of $K^+$ from the outside to the inside. Conversion of $E_2$ to $E_1$ decreases the affinity of the enzyme for $K^+$ and enhances its affinity for $Na^+$. Release of $K^+$ from $E_2$ is greatly accelerated by ATP binding. Translocation of $Na^+$ out of the cell and $K^+$ into the cell is the net consequence of this series of conformational changes in the protein. Ouabain inhibits by binding to $E_2$-P, preventing dephosphorylation in the presence of $K^+$.

These observations, and others, have led to the formulation of the model depicted in Figure 17-14 for the functioning of the $Na^+$-$K^+$ ATPase. Transport of $Na^+$ out of the cell and of $K^+$ into the cytoplasm is thought to be mediated by conformational changes in the enzyme triggered by phosphorylation, dephosphorylation, and the binding of ligands at both surfaces. While Figure 17-14 illustrates a reasonable mechanism for $Na^+$-$K^+$ ATPase, it says little about how conformational changes in the protein can lead to unidirectional transport of ions through the membrane. Although this question has yet to be satisfactorily answered in any system, a number of models have emerged for the molecular mechanisms of such translocation events. These will be considered later in this chapter, after we describe a number of other well-characterized transport systems.

It is interesting to note that $K^+$ is transported and accumulated by a $K^+$-dependent, $Na^+$-_independent_ ATPase in _E. coli_. This multicomponent enzyme system is known to be phosphorylated by ATP in a process that is coupled to $K^+$ permeation. To what extent this bacterial transport system resembles the $Na^+$-$K^+$ ATPase from animal tissues remains to be determined.

### Other Ion-Translocating ATPases

Muscle cells contain a specialized endoplasmic reticular membrane system called the _sarcoplasmic reticulum_. This organelle acts as a repository for $Ca^{2+}$ in resting muscle, and electric stimulation of muscle cells causes rapid release of this ion from the sarcoplasmic reticulum into the sarcoplasm, where it triggers contraction of the myofibrils. Reuptake of $Ca^{2+}$ into the reticular lumen is catalyzed by a *calcium ATPase*, which catalyzes the following reaction:

$$2\ Ca^{2+}(\text{out}) + ATP(\text{out}) \xrightarrow{Mg^{2+}} 2\ Ca^{2+}(\text{in}) + ADP(\text{out}) + P_i(\text{out})$$

In this reaction sequence, (out) refers to the outside of the sarcoplasmic reticulum (sarcoplasm), and (in) corresponds to the lumen of the system.

The $Ca^{2+}$ ATPase has been extensively characterized, and like the $Na^+$-$K^+$ enzyme, it spans the membrane in an asymmetric fashion. In the presence of $Ca^{2+}$ and ATP, formation of a covalent phosphoryl enzyme also can be demonstrated in this system. Thus many of the aspects of $Ca^{2+}$ transport in the sarcoplasmic reticulum parallel those of $Na^+$ and $K^+$ translocation in the cytoplasmic membranes of animal cells. Again, conformational changes in the protein triggered by bound ligands and the state of phosphorylation of the enzyme are hypothesized to be involved in the translocation event, as shown in Figure 17-15.

A third ion-translocating ATPase is the $H^+$-translocating enzyme found in the cytoplasmic membrane of epithelial cells lining the stomach. This enzyme is responsible for acidification of the lumen and appears to be structurally related to the $Na^+$-$K^+$ and $Ca^{2+}$ ATPases. For example, each of the three enzymes has a large, catalytic polypeptide chain with a molecular weight of about 100,000 that is transiently phosphorylated during transport. $H^+$-translocating ATPases of similar structure have also been identified in fungi and in yeast cells. All these enzymes, however, appear to be quite distinct in structure from the $H^+$-translocating ATPases found in bacteria and eukaryotic organelles (see the following discussion).

A fourth well-characterized ion-translocating ATPase was discussed in Chapter 10. The $F_1/F_0$ *ATPases* (or *ATP synthases*) of mitochondria, chloroplasts, and bacteria appear to have a similar function in all three systems.

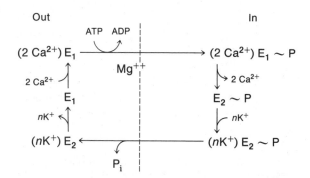

**Figure 17-15**
Proposed mechanism for the $Ca^{2+}$ ATPase of sarcoplasmic reticulum. Phosphorylation of the enzyme by ATP and the binding of $Ca^{2+}$ and of ATP plus $Mg^{2+}$ are all believed to play roles in causing the enzyme to cycle in an essentially irreversible fashion from its conformation with the cation binding site on the inside of the membrane to that with its cation binding site exposed to the outside, and back again. All interconversions are believed to occur by mechanisms that are analogous to those described in Figure 17-14 for the $Na^+,K^+$ ATPase. The reactions in this sequence have been shown to be reversible, but are driven in the direction shown by the hydrolysis of ATP. An obligatory involvement of $K^+$ has yet to be established. The dashed line and arrows relating $E_1$ and $E_2$ have the same significance as in Figure 17-14.

Proton gradients formed as a result of electron transport (see the following discussion) are used to synthesize ATP from ADP and $P_i$ as a consequence of $H^+$ flow _down_ its electrochemical gradient through the $F_1/F_0$ ATPase (Figure 17-16_a_).

It is also possible, however, for the $F_1/F_0$ ATPase to work in the reverse direction, i.e., to hydrolyze ATP and translocate protons _against_ a concentration gradient (Figure 17-16_b_). This fact has been established in some bacteria under conditions where respiratory electron transport does not occur (e.g., anaerobiosis or the absence of a functional electron-transport chain as in some strict anaerobes). Thus the $F_1/F_0$ ATPase also can be classified as a protein complex carrying out primary active transport by pumping $H^+$ against a concentration gradient using the chemical energy stored in ATP. Unlike $Na^+$-$K^+$, $Ca^{2+}$, and $H^+$ ATPases of animal cell plasma membranes, however, there is no evidence that $F_1/F_0$ ATPases are covalently phosphorylated during proton translocation. As we shall see in a later section, proton gradients generated by the $F_1/F_0$ ATPase can be used to provide energy for the uptake of nutrients into bacteria under anaerobic conditions.

The structure of the mitochondrial $F_1/F_0$ ATPase was described in Chapter 10, and the chloroplast and bacterial enzymes appear to have very similar structures. The $F_1$ portion, which possesses the ATPase activity, consists of five different subunits (named $\alpha$ through $\varepsilon$), each of which has a similar molecular weight from each of these three sources, as summarized in Table 17-2. The $F_0$ portion, which presumably spans the membrane, has been shown to be the proton channel. It consists of two polypeptides with molecular weights of about 13,000 and 5,000 in a subunit ratio of 3:6. While the larger subunits are believed to function in binding $F_1$ to $F_0$, the small polypeptides, which contain bound lipid, function as the proton channel. Removal of $F_1$ from membranes results in an increased permeability of the bilayer to protons, suggesting that $F_1$ acts as a "cap" that couples proton translocation to ATP synthesis or degradation. The $F_0$ portion, which is rich in hydrophobic amino acids, has been shown to cause proton conductivity when incorporated into artificial membrane vesicles.

Elegant reconstitution studies by Y. Kagawa and co-workers have led to a model of how the various protein subunits of the $F_1/F_0$ ATPase may be arranged on the membrane. As a result of information gained by purifying each subunit from the thermophilic bacterium PS3 and recombining them in various ways, the structure depicted in Figure 17-16_c_ was proposed. In

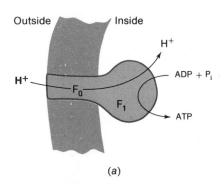

Outside    Inside

$H^+$

$H^+$

$F_0$

$F_1$

ADP + $P_i$

ATP

(a)

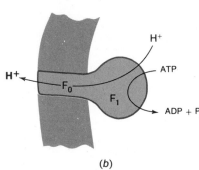

$H^+$

$H^+$

$F_0$

$F_1$

ATP

ADP + $P_i$

(b)

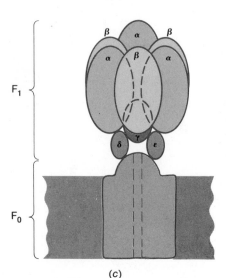

$F_1$

$\beta$    $\alpha$    $\beta$

$\alpha$    $\beta$    $\alpha$

$\delta$    $\gamma$    $\varepsilon$

$F_0$

(c)

**Figure 17-16**
The $F_1/F_0$ ATP synthase or ATPase of bacterial, mitochondrial, and chloroplast membranes. (_a_) A proton gradient generated by electron transport can be used for ATP synthesis by means of chemiosmotic coupling. (_b_) Alternatively, in bacteria growing anaerobically, ATP hydrolysis can be used to _generate_ a proton gradient in the reverse reaction. This gradient can then be used to energize secondary active transport systems as well as for other processes in bacteria that use chemiosmotic energy such as flagellar motility. (_c_) Reconstitution studies of the $F_1/F_0$ ATPase from the thermophilic bacterium PS3 have led to a model for the arrangement of the subunits in the $F_1$ portion of this enzyme. (Adapted from M. Yoshida, H. Okamoto, N. Sone, H. Hirata, and Y. Kagawa, Reconstitution of thermostable ATPase capable of energy coupling from its purified subunits, _Proc. Natl. Acad. Sci. USA_ 74:936, 1977.)

**Table 17-2**
Subunit Compositions of $F_1$ ATPases

| Source | Molecular Weight ($\times 10^{-3}$) | Subunit Sizes ($\times 10^{-3}$)* | | | | |
|---|---|---|---|---|---|---|
| | | $\alpha$ | $\beta$ | $\gamma$ | $\delta$ | $\varepsilon$ |
| E. coli | 380 | 58 | 52 | 31 | 18 | 16 |
| Thermophilic bacterium (PS3) | 380 | 56 | 53 | 32 | 15 | 11 |
| Beef heart mitochondrion | 360 | 54 | 50 | 33 | 17 | 11 |
| Chloroplast | 325 | 59 | 56 | 37 | 17 | 13 |

*The stoichiometry of subunits in the $F_1$ ATPase is still controversial; it may be $\alpha_3\beta_3\gamma\delta\varepsilon$ for the bacterial enzymes.

Adapted from D. B. Wilson and J. B. Smith, Bacterial transport proteins, in B. P. Rosen (Ed.), *Bacterial Transport*, Dekker, New York, 1978.

this model, the complex of $\alpha$ and $\beta$ subunits ($\alpha_3\beta_3$), which possesses ATPase activity by itself, is bound to the $F_0$ portion through subunits $\delta$ and $\varepsilon$. Subunit $\gamma$ is believed to act as the "cap" or _gate_ through which protons flow. This conclusion was reached through experiments that showed that membranes containing $F_0$ and the $\delta$ plus $\varepsilon$ subunits of $F_1$ were permeable to protons, but addition of the $\gamma$ subunit blocked proton conductance. Because the subunit structures and functions of $F_1/F_0$ ATPases from a number of sources are very similar, it seems reasonable to expect that the functions and topographies of the individual protein subunits will be related as well.

## Primary Systems Driven by Electron Transport and Light

Membrane-bound electron-transport chains have as a primary function the extrusion of protons out of mitochondria and bacteria and into chloroplasts of plants and algae (Chapter 10). These gradients can then be used for ATP biosynthesis, as described earlier, or for secondary active transport processes linked to ion translocation (see the following discussion). The exact mechanisms of proton translocation during electron transport remain a matter of controversy (see Chapter 10), and both unidirectional protein channels and coenzyme Q may be involved. In any case, energy derived from the oxidation-reduction reactions that take place within the membrane is more or less directly transformed into a chemical gradient of protons. For this reason, electron transport is generally thought of as a primary active transport event.

Light energy can likewise be used to drive active transport in a number of systems. For example, in photosynthesis, light energy absorbed by chlorophyll molecules is converted into a proton gradient at least partly by means of the electron-transport reactions that subsequently take place. In halophilic bacteria, such as *Halobacterium halobium*, a remarkable _light-driven proton pump_, bacteriorhodopsin, has evolved in part to allow this organism to synthesize ATP under anaerobic conditions. The structure of bacteriorhodopsin, which we considered in Chapter 16, in some manner allows it to _actively transport protons from inside the cell to the outside in response to light absorption_.

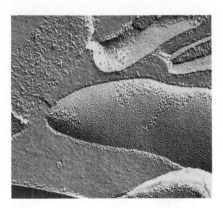

Freeze-fracture electron micrograph of the membrane of *H. Halobium*. Fine-grain regions are patches of purple membrane containing bacteriorhodopsin. (Magnification 48,000×.) (Courtesy Walter Stoeckenius.)

In 1971, W. Stoeckenius and co-workers showed that if $O_2$ was limited in cultures of *Halobacterium*, a _purple membrane_ appeared on the surface of the cell in addition to its normally red, carotenoid-containing membrane. The purple membrane is 75 percent protein by weight, and bacteriorhodopsin is the sole protein component. The purple color ($\lambda_{max} = 570$ nm) is due to the covalent association of one molecule of _retinal_, a derivative of vitamin A, to a lysine residue of bacteriorhodopsin by means of a Schiff base linkage (Figure 17-17*a*). It is the retinal moiety that absorbs photons and is responsible, at least partly, for the conversion of this form of energy into a proton gradient that can be used for ATP synthesis.

Although the photochemical reactions of bacteriorhodopsin are extremely complex, it is believed that _light absorption by retinal leads to deprotonation of the Schiff base_ to a form that absorbs light maximally at 412 nm. Reassociation of a proton with the retinal moiety then takes place slowly, and the bacteriorhodopsin protein cycles back to the 570-nm form through several intermediate conformations that can be detected spectrophotometrically (Figure 17-17*b*). It is highly probable that this protonation-deprotonation cycle involving multiple protein conformational states is a key to the mechanism of proton translocation and thus energy transformation carried out by bacteriorhodopsin. A molecular model for how this process may occur will be discussed later in this chapter.

## Bacterial Binding Protein Transport Systems

The last primary active transport mechanisms we shall consider are the so-called _binding protein transport systems_ of the Gram-negative bacteria. As discussed in Chapter 16, proteins that bind specific sugars, amino acids, and inorganic ions have been shown to be localized to the periplasmic spaces of these organisms. Some of the compounds for which such binding proteins have been demonstrated are listed in Table 17-3. Cold osmotic shock of cells or the transformation of the culture to spheroplasts releases these proteins, and a concomitant decrease in the ability of these cells to transport binding protein-linked substrates is observed. Thus these periplasmic proteins apparently are important either in concentrating the substrate at the cell surface or in interacting with the transmembrane permeases (or both). In all cases that have been studied, however, the binding

*(a)*

*(b)*

**Figure 17-17**
Bacteriorhodopsin contains a covalently bound molecule of retinal, a derivative of vitamin A, attached by means of a Schiff base linkage to a lysine residue of the protein. Both the all-*trans* isomer (*a*) and the 13-*cis* form can be isolated from the protein. Retinal is also the light-gathering chromophore of the vertebrate retina where it exists as the 11-*cis* isomer attached by means of a Schiff base to the protein opsin (Chapter 31). Light absorption by the retinal moiety of bacteriorhodopsin results in a transient conversion of the form absorbing maximally at 570 nm ($BR_{570}$) to one having a maximum absorption at 412 nm ($BR_{412}$). This transition is accompanied by a deprotonation of the Schiff base, as shown in (*b*), in a process believed to be involved in proton translocation by bacteriorhodopsin.

**Table 17-3**
Binding Protein Transport Systems in Gram-Negative Bacteria

| Organism | Substrate | Molecular Weight of Binding Protein $(\times 10^{-3})$ |
|---|---|---|
| *E. coli* | Phosphate | 41 |
| | Leucine/isoleucine | 36 |
| | Glutamine | 25 |
| | Lysine/arginine/ornithine | 27 |
| | Arabinose | 38 |
| | Ribose | 30 |
| | Maltose | 37 |
| | Galactose/glucose | 35 |
| *Salmonella typhimurium* | Sulfate | 31 |
| | Histidine | 26 |
| | Ribose | 31 |
| *Salmonella enteritidis* | Galactose/glucose | 35 |

Adapted from D. B. Wilson and J. B. Smith, Bacterial transport proteins, in B. P. Rosen (Ed.), *Bacterial Transport*, Dekker, New York, 1978.

proteins are not themselves responsible for the transmembrane transport of the molecules they recognize. Rather, specific integral membrane proteins exist for this purpose (Figure 17-18).

The integral membrane proteins that, in part, comprise the binding protein-dependent systems have been characterized in several instances. One such system is the maltose transport system in *E. coli*. It is known that four distinct proteins are required for the translocation of maltose across the cytoplasmic membrane. These are illustrated in Figure 17-18. In addition to the maltose-binding protein in the periplasm (the E protein), two integral membrane proteins (F and G) probably span the membrane. While the E protein associates with the transmembrane transport components on the external surface, the K protein is localized to the inner surface, probably owing to a specific association with the G protein (Figure 17-18). The K protein functions not only as an essential transport component, but also in the regulation of transport activity.

Energy for the active accumulation of binding protein substrates is apparently provided by ATP or a high-energy metabolic derivative of it. The evidence for this is that arsenate, which greatly reduces the intracellular ATP pool, is generally a strong inhibitor of binding protein-dependent transport. Furthermore, compounds such as fructose, which increase the ATP concentration in the cell by means of substrate-level phosphorylation, are stimulatory to the uptake of some binding protein substrates. However, an electron donor such as D-lactate is unable to stimulate these systems in *E. coli* cells that lack a functional $F_1/F_0$ ATPase. Since D-lactate is metabolized mainly by oxidative phosphorylation in *E. coli*, the proton gradient resulting from its metabolism is unable to form ATP in such mutants and thus cannot promote uptake of the binding protein substrate.

**Figure 17-18**
Schematic illustration of solute transport by the binding protein systems of Gram-negative bacteria using the maltose system as an example. Maltose passes through a pore in the outer membrane (the *lamB* gene product, which is also the receptor for bacteriophage λ), interacts with its periplasmic binding protein (E) and is then transported by inner membrane permeases (F and G) in a process dependent on a high-energy phosphate compound. The K protein is a peripheral membrane protein that is necessary for maltose transport, but its exact role in this process is unknown (see text). The localizations and interactions of these proteins have been determined by biochemical and genetic analyses (see also Chapter 16). The *malE* binding protein has been shown to interact both with the *lamB* protein pore in the outer membrane and with the maltose permease complex in the cytoplasmic membrane. It may act to facilitate and provide stereospecificity to the transport processes across both membranes.

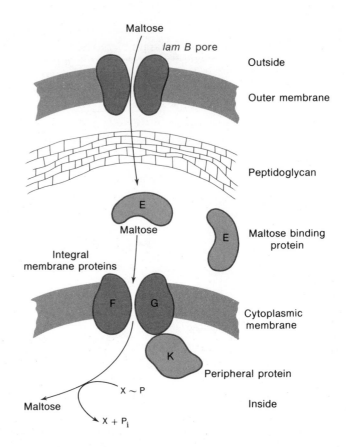

The evidence just discussed appears to rule out any mechanism in which these binding protein systems are energized solely by a secondary mechanism, such as a proton gradient (see the following discussion). At least one exception appears to exist to this rule, however. The transport system for dicarboxylic acids such as succinate in *E. coli* has been shown to have a periplasmic binding protein and yet to probably be energized by a proton gradient. The method by which ATP or another high-energy phosphate compound donates its energy to support active transport in the majority of binding protein systems remains obscure. This is mainly due to the unavailability of purified integral membrane permeases from any of these systems. Such purified preparations will be necessary to determine, for example, if transient phosphorylation of the transport protein is a key step in the translocation event, as it is for the $Na^+$-$K^+$ and $Ca^{2+}$ ATPases.

## Secondary Active Transport: Ion and Electrical Gradients

In Chapter 10 we learned that a gradient of protons across a biological membrane is a form of potential energy that can be used to drive an endergonic process, the synthesis of ATP from ADP and $P_i$. Expulsion of protons out of mitochondria or bacterial cells during electron transport results in both a difference in $H^+$ concentration [ΔpH] and in electric charge (ΔΨ; interior negative) across the membrane. Recall that both these components contribute to the _proton motive force_ (pmf or $\Delta p$; see Chapter 12), which can be expressed as

$$\Delta p = \Delta \Psi - \frac{2.3RT}{F} \Delta pH \qquad (9)$$

The pmf, or $\Delta p$, is related to the free energy change experienced by protons tending to drive them across the cellular or organellar membrane (see Equation 50, Chapter 4). Since lipid bilayers are relatively impermeable to protons, such a flow must be mediated by membrane proteins. In the example of the $F_1/F_0$ ATPase, the exergonic flow of $H^+$ through the enzyme, driven by $\Delta p$, is used to drive the endergonic synthesis of ATP. It also should be apparent that _$\Delta p$ could be used to drive active transport if the inward flow of $H^+$ were coupled to the uptake of a particular solute._ Examples of such proton _cotransport_ or _symport_ mechanisms have now been established in mitochondria, bacteria, and lower eukaryotic cells. Because the primary active transport of $H^+$ is responsible for $\Delta p$, active transport systems dependent on $\Delta p$ for energization are said to carry out _secondary active transport_ by means of _chemiosmotic coupling_.

Undoubtedly the most well-characterized $H^+$ symport system is the lactose transport system of _E. coli_. This system has been extensively studied by P. Mitchell, T. H. Wilson, and H. R. Kaback, among many others. Evidence that the active accumulation of this sugar by means of the _lacY_ gene product involves the cotransport of $H^+$ can be summarized as follows:

1. Agents such as 2,4-dinitrophenol (Figure 10-13) which collapse proton gradients across membranes inhibit lactose transport in whole cells and membrane vesicles.

2. Uptake of lactose, or the nonmetabolizable analog _thiomethyl-$\beta$-D-galactopyranoside_ (TMG; Figure 17-19), is stimulated by reducing the extracellular pH of energy-depleted cells.

3. A 1:1 ratio of entry of TMG and $H^+$ has been demonstrated in energy-depleted cells.

4. Lactose transport in _E. coli_ membrane vesicles, which cannot synthesize ATP in the absence of a source of ADP, is greatly stimulated by compounds such as D-lactate, which are oxidized and donate electrons to the electron-transport chain in this system.

5. A mutation in the _lacY_ gene has been isolated in which lactose uptake is not tightly coupled to $H^+$ entry (i.e., is relatively insensitive to $\Delta p$). This mutant protein is severely impaired in the active transport of lactose, but has an increased ability to facilitate its diffusion across the membrane compared with the normal permease.

6. The lactose permease has been purified and the transport function of the pure protein has been reconstituted in a vesicular phospholipid membrane. The active accumulation of lactose in response to an artificially imposed $\Delta p$ has been demonstrated.

The preceding results establish that the active accumulation of lactose in _E. coli_ is obligatorily coupled to the simultaneous entry of $H^+$ flowing down its concentration and/or electrical gradients (Figure 17-20a). If this is

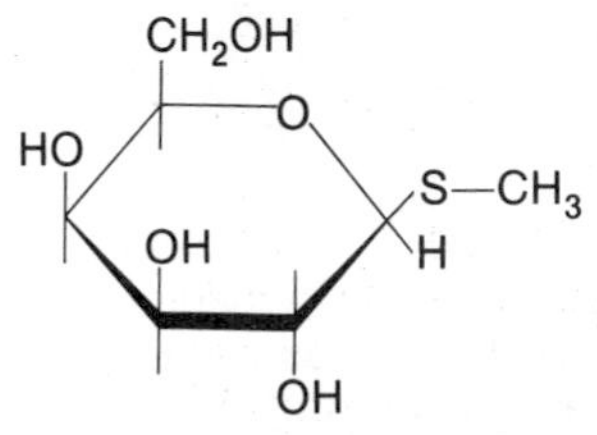

**Figure 17-19**
Thiomethyl-$\beta$-D-galactopyranoside (TMG), a nonmetabolizable substrate of bacterial lactose permeases. TMG has been useful in elucidating the $H^+$ symport mechanism of these proteins in the active transport of lactose.

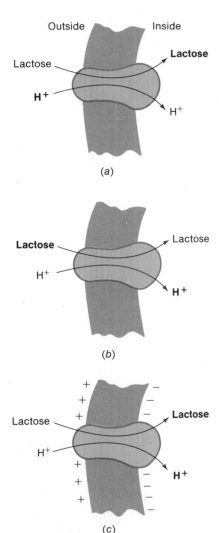

Outside     Inside

**(a)**

**(b)**

**(c)**

**Figure 17-20**
Transport events carried out by the
bacterial lactose permeases. (a) In
response to a proton gradient, lactose is actively accumulated in the
cell. (b) In the presence of a lactose
gradient, protons can be concentrated in the cytoplasm. (c) In the
presence of a $\Delta\psi$, *both* lactose and
protons can be actively accumulated.
The physiological situation in respiring cells corresponds to (a), in which
a $\Delta\psi$ is usually also present.

true, then *a gradient of lactose (or TMG) should produce a $\Delta p$ in cells in which this value is initially zero*. Indeed, it has been demonstrated that addition of TMG to the outside of energy-depleted cells causes acidifica-tion of the cytoplasm as a result of $H^+$-TMG symport, as mentioned earlier (Figure 17-20b).

Transport systems, such as the *E. coli* lactose permease, in which elec-trical charges are carried across the membrane without simultaneous com-pensation of the electrical potential so generated (for example, by the movement of other charged species across the membrane) are termed *electrogenic. Electrogenic transport systems are always affected by the electric potential $\Delta\Psi$ across a biological membrane, even if the trans-ported ion is not $H^+$*. In the lactose transport system, a $\Delta\Psi$ (interior nega-tive) is sufficient to drive active accumulation of this sugar in the absence of $\Delta pH$ because $\Delta p$ has both $\Delta\Psi$ and $\Delta pH$ components (Equation 9), as shown in Figure 17-20c. In animal cells and in some bacteria such as *Halo-bacterium* that live in high concentrations of $Na^+$, $Na^+$-solute symport systems are common means for the accumulation of salts, carbohydrates, and amino acids. These systems are also electrogenic and thus are sensitive to both $\Delta\Psi$ and $\Delta[Na^+]$, but not to $\Delta pH$. The $Na^+$-$K^+$ ATPase in animal cells ensures that $Na^+$ flowing in through these $Na^+$-solute "symporters" is rapidly pumped back out to maintain a high $[Na^+]out/[Na^+]in$ ratio (Fig-ure 17-21).

These results establish the essential mechanistic features of cation-solute symport. Thus in the case of the lactose permease, a single protein alone catalyzes the tightly coupled, unidirectional transport of sugar and a proton. The stoichiometry of this process is 1:1. While these facts appear well-established, further work will be required to ascertain the detailed mechanistic features of the translocation process. For example, it is not yet known if the protein functions by a carrier-type or a channel-type mecha-nism, or by a process that incorporates mechanistic features of both mod-els. These problems will be considered in greater detail later in this chapter. A list of some of the better-characterized secondary active transport sys-tems is given in Table 17-4.

### The Mitochondrial ATP/ADP Exchanger

Because ATP is synthesized within the mitochondria of eukaryotic cells, while metabolic processes that utilize ATP are largely confined to the cyto-plasm, a mechanism must exist for transport of this molecule across the inner mitochondrial membrane. The protein responsible for this function has been shown to be an *ATP/ADP exchanger* that carries out exchange of intramitochondrial ATP for ADP formed from metabolic reactions in the cytoplasm. This antiport is electrogenic because, at pH 7, ATP molecules average about one more negative charge than do ADP molecules (Figure 17-22).

Despite the fact that the [ATP]/[ADP] ratio is usually higher in the cytoplasm than inside mitochondria in actively respiring cells, the ATP/ADP exchanger still preferentially expels ATP from the organelle with the concomitant inward movement of ADP. This can be explained by the presence of a membrane potential $\Delta\Psi$ (inside negative), resulting, in

**Table 17-4**
Some Secondary Active Transport Systems

| Substrate | Cotransported Ion | Organism/Tissue |
|---|---|---|
| Neutral amino acids | $Na^+$ | Eukaryotic cells |
| Glucose | $Na^+$ | Some animal cells (intestine, kidney, choroid plexus) |
| Lactose | $H^+$ | E. coli and some other bacteria |
| Dicarboxylic acids | $H^+$ | E. coli |
| Proline | $H^+$ | E. coli |
| Glutamate | $Na^+$ | E. coli and H. halobium |
| Melibiose | $Na^+$ | E. coli and S. typhimurium |
| Alanine | $H^+$ | Thermophilic bacterium PS3 |

part, from the $\Delta p$ formed across the mitochondrial membrane during respiration. This potential favors outward transport of ATP and inward transport of ADP because this process results in the net translocation of one negative charge from inside to outside (Figure 17-22).

The ATP/ADP exchanger is an example of a transport system that does not fit neatly into either the category of facilitated diffusion or of active transport as we have defined them. On the one hand, the transport can be thought of as active because concentrative accumulation can occur and metabolic energy is sacrificed ($\Delta\Psi$ is partially collapsed as a result of ATP/ADP exchange). On the other hand, the ATP/ADP exchanger essentially corresponds to our definition of facilitated diffusion, in which the solute is charged, and an independent potential is maintained across the membrane (Equation 7). This situation reflects more the somewhat arbitrary definitions we have made by convention rather than any real confusion as to the overall transport reaction carried out by this protein.

The ATP/ADP exchanger has recently been purified by M. Klingenberg and his associates and has been shown to be a dimer of identical 30,000-dalton polypeptides in detergent solution and possibly in the membrane as well. Its dimeric structure may help explain the molecular mechanism by which the exchange reaction takes place, as we shall discuss in a later section.

### Group Translocation: The Bacterial PEP:Sugar Phosphotransferase System

The final mechanism of energy coupling in transport processes that we shall consider is group translocation. In this mechanism, chemical modification of the transported solute is part of the translocation step itself (Figure 17-6). The most famous example of group translocation is the _phosphoenolpyruvate(PEP)-dependent sugar phosphotransferase system_ (PTS) in bacteria, which was discovered by S. Roseman and co-workers in 1964. The PTS both transports and phosphorylates sugars as they pass through the cytoplasmic membrane, with PEP as the phosphoryl donor. It consists of

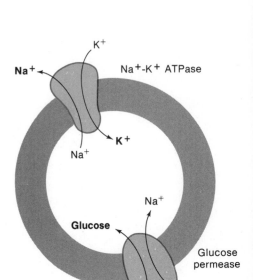

**Figure 17-21**
$Na^+$ symport is a common means of actively transporting sugars and amino acids in animal cells. For example, glucose is actively accumulated by this mechanism in intestinal and renal epithelial cells. The $Na^+$-$K^+$ ATPase maintains the $Na^+$ gradient essential to these secondary active transport systems.

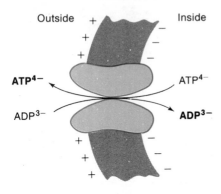

**Figure 17-22**
The ATP/ADP exchanger of the mitochondrial inner membrane. This dimeric protein carries out the exchange of intramitochondrial ATP for ADP formed in the cytoplasm by metabolic reactions. A membrane potential (interior negative) favors this exchange because of its electrogenic nature at pH values near neutrality.

several soluble and membrane-bound proteins in *E. coli* that catalyze the following set of reactions:

$$\text{PEP} + \text{Enzyme I} \rightleftharpoons \text{Enzyme I} \sim \text{P} + \text{pyruvate} \quad (10)$$

$$\text{Enzyme I} \sim \text{P} + \text{HPr} \rightleftharpoons \text{HPr} \sim \text{P} + \text{Enzyme I} \quad (11)$$

$$\text{HPr} \sim \text{P} + \text{sugar}^a_{\text{(out)}} \xrightarrow[\text{(Enzyme III}^a)]{\text{Enzyme II}^a} \text{HPr} + \text{sugar}^a_{\text{(in)}}\text{-P} \quad (12)$$

In this reaction scheme, Enzyme I and HPr are soluble, cytoplasmic proteins that participate in phosphoryl transfer reactions common to all sugars transported by the PTS in *E. coli*. In contrast, Enzyme II is an integral membrane protein that is usually specific for only one sugar (as indicated by the superscript) and acts as the permease as well as the phosphoryl transfer enzyme. Enzyme III, which is required for the transport of some but not all sugars, also is sugar-specific and may or may not be membrane-bound. The spacial arrangement of the reactions that occur during PTS sugar transport in enteric bacteria is depicted in Figure 17-23. The PTS allows for unidirectional sugar transport because the sugar-phosphate product is very impermeable to the lipid bilayer and is "trapped" once it is transported into the cell. The cost to the cell is one ATP equivalent (in the form of PEP) for each molecule of sugar accumulated by means of the PTS. The importance of this fact will be discussed in the next section.

The phosphotransferase systems in *E. coli* and *Salmonella typhimurium* have been among the most extensively studied. All the general pro-

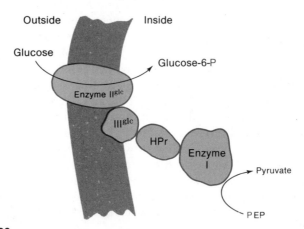

**Figure 17-23**
Schematic representation of the reactions carried out by the bacterial sugar phosphotransferase system (PTS). A phosphoryl moiety from PEP is sequentially transferred to Enzyme I, HPr, and Enzyme III$^{\text{glc}}$ before its ultimate transfer to glucose by an integral membrane protein specific for this sugar (Enzyme II$^{\text{glc}}$). Although the general PTS enzymes, Enzyme I, and HPr, are soluble proteins, there is some evidence that they are associated peripherally with the membrane as shown. In *E. coli* there are probably seven different Enzymes II, each specific for the sugars recognized by the PTS in this organism: glucose, mannose, fructose, *N*-acetyglucosamine, mannitol, glucitol, and galactitol. While HPr has been shown to be a monomer ($M_r = 9,500$) and Enzyme I a dimer of 70,000-dalton subunits, the subunit structures of the Enzymes II are as yet unknown. Recently, however, the mannitol-specific Enzyme II (or mannitol permease) has been shown to consist of a single kind of polypeptide chain ($M_r = 60,000$) (see Table 16-8).

teins have been purified and shown to be transiently phosphorylated, as shown in Reactions (10) through (12). The phosphate is attached to a histidine residue in both Enzyme I and HPr, and it is also probable that the Enzymes II and III are similarly transiently modified before the phosphate is transferred to the incoming sugar. Such enzyme phosphorylation reactions may provide the driving force for transport by means of conformational changes in the Enzymes II, as it apparently does for the ion-translocating ATPases. The recent purification of the Enzyme II specific for mannitol from *E. coli* membranes should help to answer some of these questions about the mechanism of PTS-mediated transport.

Group translocation mechanisms also have been proposed for the uptake of fatty acids and purine and pyrimidine bases in bacteria. The evidence for these is much weaker than that for the PTS. In the *E. coli* PTS, the one permease that has been purified, the mannitol-specific Enzyme II, has been shown to possess coupled sugar-phosphorylating and transport activities in a reconstituted phospholipid-Enzyme II vesicular proteoliposome system. This establishes the tight and obligatory coupling of transport and phosphorylation by a single membrane protein. In order to demonstrate true group translocation in other systems, similar properties (i.e., tightly coupled translocation and enzymatic activities) must be observed for the permeases involved.

## Energy Interconversion and Active Transport in Bacteria

In Chapter 16, the structural similarities between mitochondria and bacteria were pointed out. In this chapter we have seen that both mitochondria and many bacteria possess similar $F_1/F_0$ ATPases and electron-transport chains in their membranes. In energy generation, however, *facultatively anaerobic bacteria* (able to live either aerobically or anaerobically) such as *E. coli* are considerably more versatile than mitochondria. They can generate energy either by aerobic respiration, as do mitochondria, or by anaerobic glycolysis, which does not take place in the eukaryotic organelle. This situation can best be illustrated by considering the pathways of energy interconversion that exist in many facultative anaerobes, as illustrated in Figure 17-24.

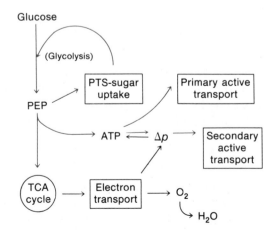

**Figure 17-24**
Energy interconversion pathways in anaerobic and facultatively anaerobic bacteria growing on glucose as the sole carbon and energy source. Note that ATP can be used to create a proton motive force by means of the $F_1/F_0$ ATPase under anaerobic conditions. Energy flow that occurs in the absence of $O_2$ is indicated by the colored arrows, while all these reactions except formation of a $\Delta p$ by ATP hydrolysis occur in the presence of oxygen.

During aerobic respiration, a $\Delta p$ generated by electron transport is used to drive ATP synthesis by means of the $F_1/F_0$ ATPase and to energize $H^+$ symport and other electrogenic transport systems. Under anaerobic conditions, bacteria obtain energy in the form of ATP by substrate-level phosphorylation during glycolytic fermentation of sugars. *In order for $\Delta p$-linked processes to occur in the absence of $O_2$, protons must be expelled from the cell by reversal of the $F_1/F_0$ ATP synthase reaction.* This has been shown to occur in *E. coli*, as well as in anaerobic bacteria, such as *Streptococcus lactis*, which completely lacks a respiratory electron-transport chain and has as its only source of $\Delta p$ the $F_1/F_0$ ATPase. The energy-conversion reactions that take place under anaerobic conditions in these bacteria are indicated by the colored arrows in Figure 17-24.

Because respiration of a sugar such as glucose results in many more ATP equivalents than its fermentation, *E. coli* cells grow much more slowly in the absence of $O_2$. *In order to grow anaerobically, bacteria must make judicious use of the ATP formed.* An illustration of this is the fact that the *sugar-phosphotransferase transport system (PTS) is almost exclusively found in bacteria that can live anaerobically.* It is generally absent in strictly aerobic bacteria. The probable reason for this can be demonstrated by considering the energetics of these situations. Bacteria growing anaerobically on glucose use one ATP equivalent (as PEP) to both transport and phosphorylate this sugar by means of the PTS. Bacteria lacking a PTS must use one ATP equivalent to actively transport the sugar and a second ATP in the hexokinase reaction to phosphorylate it in preparation for glycolysis. Thus the PTS is likely to have evolved as an efficient means of conserving energy during anaerobic growth of bacteria on a fermentable sugar.

Finally, it is reasonable to ask what dictates the direction of operation of the $F_1/F_0$ ATPase. The answer appears to be that *there is a threshold value of $\Delta p$ below which the enzyme does not work in the direction of ATP synthesis but only works in the direction of ATP hydrolysis.* This threshold value has been shown to be about 200 mV (negative inside) for *Streptococcus lactis* (Figure 17-25). Recent measurements by E. Kashket and her co-workers of $\Delta p$ values in several actively growing facultative anaerobes are consistent with this threshold value. Under aerobic conditions, values of 200 mV or greater were measured, while $\Delta p$ dropped considerably below this value when these bacteria were grown in the absence of $O_2$. Thus the pathways of energy flow and transport are finely regulated in these organisms to allow adaptation to a variety of growth conditions.

## Regulation of Permease Function

In virtually all cells that have been studied, transport processes have been found to be subject to regulation. In animal cells, hormones can influence both the activities and the rates of synthesis of the permease proteins. Numerous examples of transport regulation have been documented in both prokaryotic and eukaryotic cells, and in a few cases, the mechanisms have been established.

One of the best-studied examples of carbohydrate transport regulation in *E. coli* involves the inhibition of lactose uptake by extracellular glucose. While lactose enters the cell by proton symport, glucose is transported by the phosphotransferase-catalyzed group translocation process. It is now

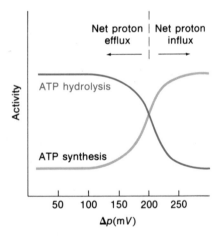

**Figure 17-25**
Illustration of the functioning of the bacterial $F_1/F_0$ ATPase as a function of $\Delta p$. The enzyme has a "threshold" of around 200 mV, above which net proton influx and net ATP synthesis occur. Below this value, the enzyme hydrolyzes ATP to pump protons out of the cell. The latter situation generally occurs under anaerobic conditions. This is a highly schematic diagram, since the threshold $\Delta p$ can vary depending on the organism and the cell growth conditions.

clear that glucose influences the phosphorylation state of a _cytoplasmic regulatory protein_, RPr. When RPr is fully phosphorylated, the lactose permease is not inhibited and exhibits maximal activity. When the regulatory protein is dephosphorylated, however, RPr binds to an allosteric regulatory site on the cytoplasmic surface of the lactose permease, thereby inducing a less active conformation of the protein that transports lactose at low rates. Extracellular glucose, during transport into the cell, is thought to promote the dephosphorylation of RPr, thereby inhibiting lactose uptake. The allosteric regulatory protein, RPr, has recently been shown by direct biochemical means to be the glucose Enzyme III of the PTS. This example illustrates that in bacteria, as well as in animal cells, complex regulatory phenomena have evolved to allow integration of the multiple facets of the biochemical machinery.

## MOLECULAR MECHANISMS OF BIOLOGICAL TRANSPORT

In the last section we examined the wide variety of energy-coupling mechanisms used by cells to accumulate and expel nutrients and ions. In order to fully understand transport at the molecular level, however, it is necessary to determine how the structures of transport permeases allow the often unidirectional translocation of hydrophilic molecules through the hydrophobic lipid bilayer. It is reasonable to assume that such proteins will contain specific binding sites for their transport substrates by analogy to enzymes, but that unlike most soluble enzymes, permeases also must be able to carry out a _vectorial_ process: "picking up" a substrate on one side of a membrane and depositing it on the other. It will become clear in our discussion that in no instance do we yet know the molecular details of how this process is accomplished. However, as more transport proteins are purified and their structures determined, clues are emerging that will undoubtedly allow this question to be answered for at least a few systems in the near future. In this section we shall examine these clues as well as models for the function of transport permeases.

### Conceptual Models: Mobile Carrier versus Pore

Two conceptual models have been put forth to explain the transport of a substance through a biological membrane. In the _mobile-carrier_ mechanism, the transporter or a solute-binding moiety of the permease is assumed to physically change its orientation in the membrane during translocation, either by shuttling back and forth across the bilayer or by rotation through the plane of the membrane (Figure 17-26a). On the other hand, the _pore_ model assumes that the protein is more or less fixed in its intramembrane orientation and forms a hydrophilic pore that is stereospecific for its transport substrate. This pore is usually thought of as being _gated_ in the sense that it opens only transiently in response to proper solute recognition (Figure 17-26b).

As model systems for studying biological transport, the _ion-translocating antibiotics_ have provided evidence for both of the mechanisms illustrated in Figure 17-26. _Valinomycin_, an antibiotic isolated from a species of the bacterium _Streptomyces_, is a cyclic depsipeptide containing the

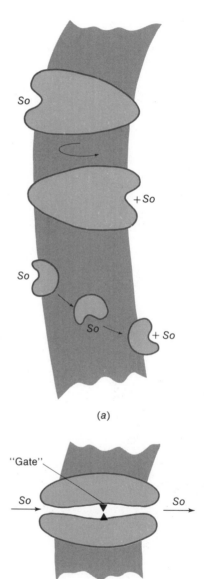

(a)

(b)

**Figure 17-26**
In theory, a solute could be transported across a membrane by a mobile carrier that either rotates through or traverses the plane of the membrane (a), or by a more or less fixed channel or pore that may be controlled by a gate (b). Examples of both have been found among the ion-translocating antibiotics.

sequence D-valine, L-lactate, L-valine, and D-hydroxyisovalerate repeated three times (Figure 17-27a). It is extremely selective for binding $K^+$ and makes both biological membranes and artificial bilayers permeable to this cation. This can be explained by the fact that the valinomycin-$K^+$ complex effectively shields the hydrophilic groups of valinomycin on the interior of the molecule, while the hydrophobic side chains are exposed to solvent, as shown in Figure 17-27b. For this reason, the valinomycin-$K^+$ complex is soluble in lipid bilayers, while free $K^+$ is highly water-soluble.

A second well-studied antibiotic is _gramicidin A_, also isolated from a bacterium (_Bacillus brevis_). It is a 15 amino acid linear polypeptide that is a cation-specific _ionophore_ with much less specificity than valinomycin for any particular cation (Figure 17-28). The mechanisms of ion translocation carried out by both valinomycin and gramicidin A have been extensively studied. In one experiment, the conductance of an artificial phospholipid bilayer ($T_m = 41°C$) to $K^+$ was measured as a function of temperature in the presence and absence of these antibiotics. The results showed that $K^+$ conductance was high in this model system throughout the temperature range studied with gramicidin A. In contrast, the $K^+$ permeability of the bilayer in the presence of valinomycin was low (below 41°C) but sharply increased when the temperature was increased above this value. Hence ion translocation by valinomycin apparently depends on the physical state of the phospholipid bilayer, while the permeability induced by gramicidin A is insensitive to this parameter. This and many other experi-

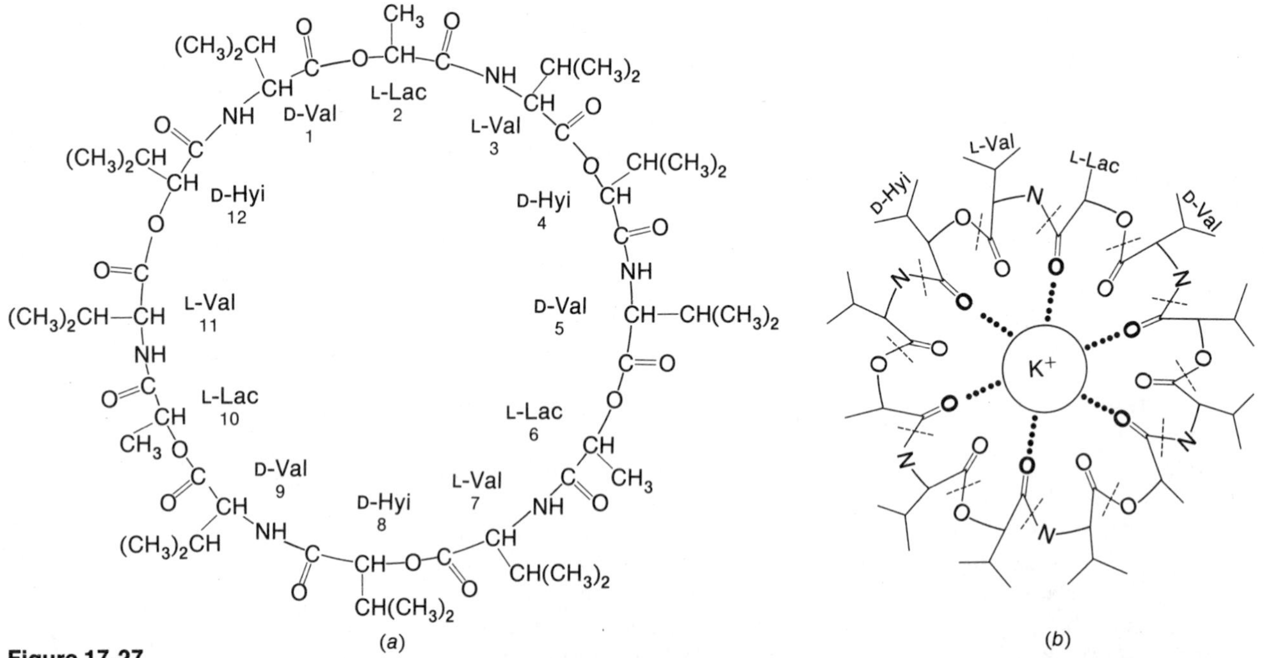

**Figure 17-27**
Valinomycin, an ionophore specific for $K^+$. (_a_) Its chemical structure is a cyclic peptide containing the sequence D-valine (D-Val), L-lactate (L-Lac), L-valine (L-Val), and D-hydroxyisovalerate (D-Hyi) repeated three times. (_b_) Its complex with $K^+$ is illustrated, showing that the hydrophilic groups complexing the ion are "buried" inside the "donut-shaped" molecule, while the hydrophobic side chains are exposed around the perimeter of the molecule. This makes the $K^+$-valinomycin complex highly soluble in lipid bilayers. (The structure in _b_ is adapted from B. C. Pressman, Biological applications of ionophores, _Ann. Rev. Biochem._ 45:501, 1976.)

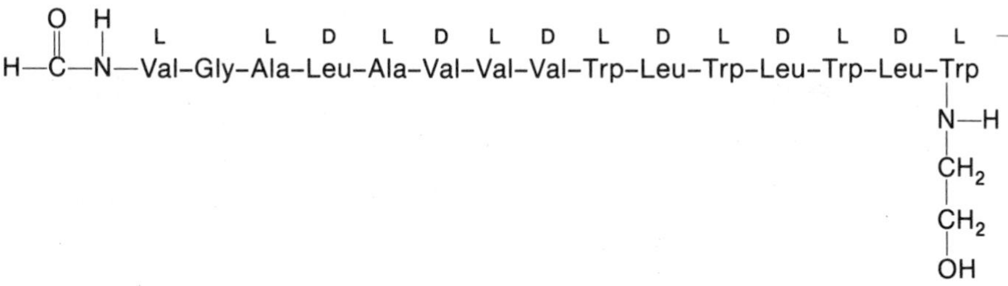
O H
L L D L D L D L D L D L D L
H—C—N—Val-Gly-Ala-Leu-Ala-Val-Val-Val-Trp-Leu-Trp-Leu-Trp-Leu-Trp
N—H
CH2
CH2
OH

**Figure 17-28**
The structure of gramicidin A, a linear antibiotic that facilitates the transmembrane transport of many cations.

ments, including kinetic and physicochemical measurements, have led to the conclusion that *valinomycin is a mobile carrier that can diffuse rapidly through the bilayer only above the $T_m$, while gramicidin A forms a static pore through the bilayer and does not require an environment of low viscosity for its function.*

The mechanisms of ion translocation by valinomycin and gramicidin A as they are believed to occur are illustrated in Figures 17-29 and 17-30. A single valinomycin-$K^+$ complex traverses the bilayer by simple diffusion. In contrast, physical measurements show that gramicidin A forms a transmembrane pore by means of head-to-head dimerization of helical monomers, each of which has a hydrophilic aqueous channel through the axis of the helix. Thus, at least in these model systems, both mobile-carrier and pore mechanisms of solute translocation can be readily demonstrated.

A vast array of both natural and synthetic ionophores of varying specificities has now been studied. In Chapter 10, proton ionophores such as 2,4-dinitrophenol were shown to be uncouplers of oxidative phosphoryla-

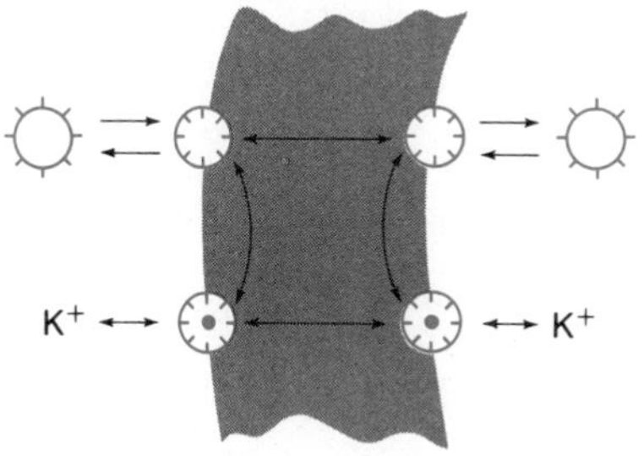
K+
K+

**Figure 17-29**
Mechanism of ion translocation by valinomycin. The carbonyl oxygens chelating $K^+$ are shown as "spokes" on the circular antibiotic molecule. In the aqueous phase, these oxygens are more freely available to the solvent; *i.e.*, valinomycin undergoes a conformational change upon going from a lipid to an aqueous environment, as illustrated in the top of the figure. The lipid-soluble form is the one that complexes $K^+$ and traverses the membrane by diffusion, as shown in the lower part of the figure. (Adapted from Y. A. Ovchinnikov, Physico-chemical basis of ion transport through biological membranes: Ionophores and ion channels, *Eur. J. Biochem.* 94:321, 1979.)

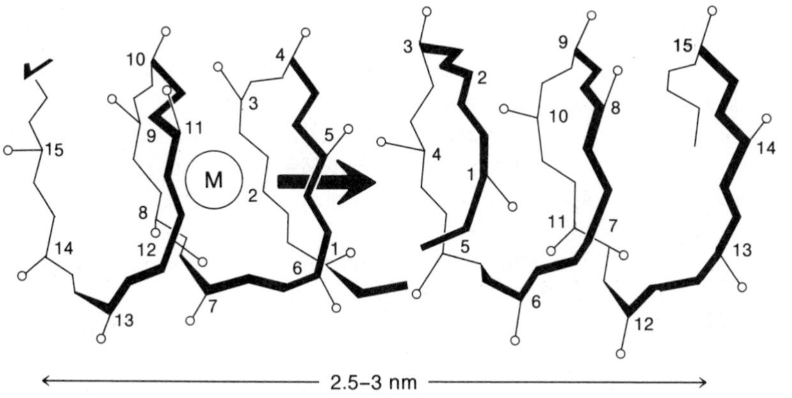
M
2.5–3 nm

**Figure 17-30**
Head-to-head dimer channel formed by gramicidin A in lipid bilayers. Each monomer is a so-called $\pi_{L,D}$ helix formed by intramolecular hydrogen bonds between amino acid residues of the gramicidin molecule (numbered 1–15 from the L-valyl terminus). The length of the dimer, about 3 nm, is consistent with the length of the hydrophobic region of a typical phospholipid bilayer. (Adapted from Y. A. Ovchinnikov, Physico-chemical basis of ion transport through biological membranes: Ionophores and ion channels, *Eur. J. Biochem.* 94:321, 1979. Used with permission.)

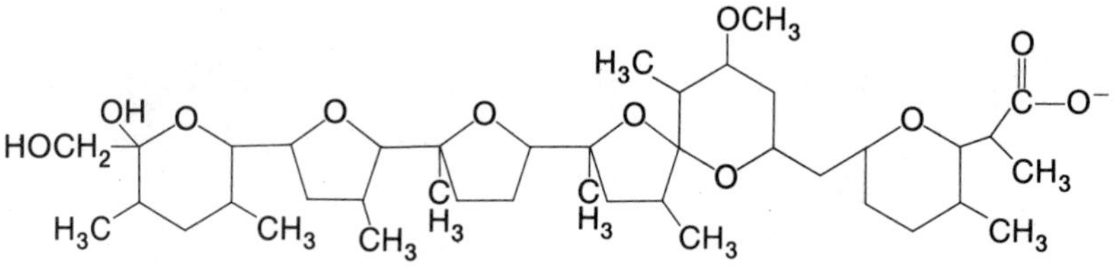

**Figure 17-31**
The structure of nigericin, a polycyclic ether carboxylic acid that exchanges $H^+$ for $K^+$ across membranes.

tion. Because of its size and lipid solubility, 2,4-dinitrophenol is undoubtedly a mobile $H^+$ carrier. Another ionophore that also transports its substrates by diffusion is *nigericin*, a polycyclic ether carboxylic acid (Figure 17-31). Nigericin has the interesting property of catalyzing the one-for-one exchange of $H^+$ and $K^+$ across biological and artificial membranes. The reason for this is indicated in Figure 17-32. The nigericin-$K^+$ complex is circular, much like that formed with valinomycin, and the negatively charged carboxylate interacts with the cation in this state. This complex is freely soluble in the bilayer, so that $K^+$ is conducted to the other side. In order to return to the first side, this carboxylate anion must be neutralized. A nigericin-$H^+$ complex is one form in which it can diffuse back. The net result is the exchange of $K^+$ for $H^+$, and transport by nigericin, in contrast to that by valinomycin or gramicidin, is by necessity electroneutral.

### Transport Proteins: Mobile Carriers or Pores?

Studies on the topographies of integral membrane proteins that we considered in the last chapter suggest a more or less static and asymmetric disposition of these polypeptides across biological membranes. Furthermore, any large movements, such as rotations through the membrane plane (Figure 17-26a), are unfavorable, especially for charged membrane proteins with a considerable proportion of their mass in contact with the aqueous phase. In fact, considerable evidence leads to the conclusion that several classes of proteins that carry out transport do so by means of transmembrane pores or channels. These structures may or may not be static. Nonspecific transport, where solute recognition is minimal, is presumably mediated by relatively static pore structures. By contrast, stereospecific solute permeation may involve conformational changes in the permease that transiently "open" the channel only in response to binding of the specific solute recognized by the transport system.

*In all cases for which detailed structure-function information is available, the evidence is in favor of a pore-type mechanism for biological transport proteins.* In several instances, presumptive transmembrane pores formed by an integral membrane protein have been visualized by electron microscopy. The outer membrane of *E. coli* has been shown to be *permeable to hydrophilic molecules with molecular weights below about 600.* This permeability is conferred by a class of proteins called *porins*, which are integral constituents of the outer membrane. Porin molecules are arranged in a hexagonal lattice of trimers in the outer membrane, and optical filtration and computer processing of many negatively stained preparations viewed in the electron microscope give images that have been interpreted as representing transmembrane pores (Figure 17-33). The diameter of these pores is about 1 nm, which is consistent with the molecular-weight exclu-

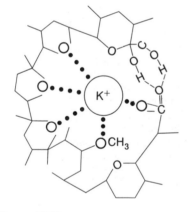

**Figure 17-32**
Circular complex formed between nigericin and $K^+$. Note that the negatively charged carboxylate oxygen is involved in chelating the cation. In order for nigericin, which has deposited $K^+$ on one side of the membrane, to return to the other side, this carboxylate must be protonated (neutral), resulting in a 1:1 exchange of $H^+$ for $K^+$. (Adapted from B.C. Pressman, Biological application of ionophores, *Ann. Rev. Biochem.* 45: 501, 1976. Used with permission.)

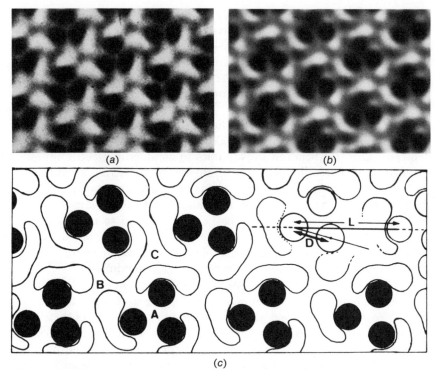

(a)    (b)

(c)

**Figure 17-33**
Arrangement of porin "pores" in the outer membrane of *E. coli*. Computer processing of many electron micrographs gives the image shown in (*a*), while a technique involving optical diffraction and filtering of micrographs gives the image shown in (*b*). A schematic interpretation of these images, shown in (*c*), suggests "triplet indentations" that are penetrated by the negative stain (*dark circles*) and which are the presumptive pores. A, B, and C are centers of local threefold symmetry and L, the lattice constant, is 7.7 nm. The center-to-center spacing of the pores is 3 nm, and each is partially surrounded by a kidney-shaped region that excludes the negative stain and is probably part of a single molecule of porin. The view is perpendicular to the plane of the outer membrane. (From A. C. Steven, B. ten Heggeles, R. Müller, J. Kistler, and J. P. Rosenbusch, Ultrastructure of a periodic protein layer in the outer membrane of *Escherichia coli*, *J. Cell Biol.* 72:292, 1977. Reprinted with permission.)

sion limit given earlier. *When incorporated into artificial phospholipid vesicles, porin molecules allow rapid diffusion of most small hydrophilic molecules through the bilayer, and thus solute recognition by this protein is minimal.* Channels similar to those formed by bacterial porins are also present in the outer membranes of chloroplasts and mitochondria. *Thus, for many hydrophilic molecules, specific transport systems reside only in the inner membranes of Gram-negative bacteria, chloroplasts, and mitochondria.*

Nonspecific transmembrane pores also have been demonstrated in the plasma membranes of eukaryotic cells. Well-studied examples of these pores are the so-called *gap junctions*, which connect neighboring cells of similar types in tissues such as liver, intestine, kidney, brain, and cardiac muscle. In the region of the gap junction, the plasma membranes of two different cells are closely apposed and appear to have channels connecting

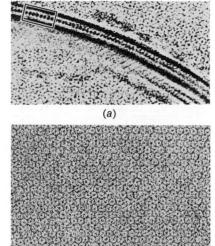

(a)

(b)

**Figure 17-34**
Isolated gap-junction sheets viewed in the electron microscope (*a*) parallel and (*b*) perpendicular to the planes of the two apposed membranes. The view in (*a*), part of which is outlined within a rectangle, arose from a curling up of a sheet on its edge and shows links (gaps or pores) penetrated by negative stain bridging the space between the two cell membranes. The perpendicular view in (*b*) shows annuli (rings), each composed of a protein hexamer surrounding a pore that has been filled with negative stain. The specimens were stained with uranyl acetate (300,000 ×). (From P.N.T. Unwin and G. Zampighi, Structure of the junction between communicating cells, *Nature* 283: 545, 1980. Reprinted with permission.)

them (Figure 17-34a). In the surface view of a gap junction, the region appears as a hexagonal lattice of protein hexamers, each of which seems to form a hydrophilic pore (Figure 17-34b). Ultrastructural studies have produced detailed models of the structure of the gap junction, one of which is shown in Figure 17-35. *The protein hexamer of each membrane is thought to span the bilayer, and apposition of two such hexamers forms an aqueous channel between the cells.*

Gap junction pores have been shown to be permeable to hydrophilic molecules with masses as large as 1000 to 2000 daltons, depending on the cell type. They are believed to play roles in conducting electric impulses through particular cell types (nerve, muscle, or epithelia), as well as in providing avenues for the free flow of small metabolites between cells. The permeability of gap junction pores is regulated by the cytoplasmic concentration of $Ca^{2+}$; low concentrations ($<10^{-7} M$) lead to open channels, while higher concentrations tend to close the channels in a graded manner. This sensitivity to $[Ca^{2+}]$ may play a role in the regulation of intercellular communication between cells connected by gap junctions.

Although porins and gap junctions are relatively *nonspecific* aqueous channels in biological membranes, it is reasonable to imagine that *solute-specific* pores can be formed by permease proteins as a result of their binding specificities. It further seems reasonable that such pores might be formed easily *between* subunits of oligomeric transmembrane proteins. Based on these considerations, a number of workers have suggested similar models for how specific permeases recognize and translocate their substrates. Aqueous channels present between transmembrane subunits may respond to solute binding by undergoing *conformational changes that alter the relative positions of the subunits to each other and thereby open the channels* (Figure 17-36). This conformational change might be triggered solely by the substrate itself, in which case facilitated diffusion would be the consequence. Alternatively, metabolic energy might expedite this process, in which case active transport could result. Although conformational changes resulting in solute translocation are most easily envisioned between subunits of an oligomeric protein, there is also evidence that at least some permeases may accomplish this process in monomeric form.

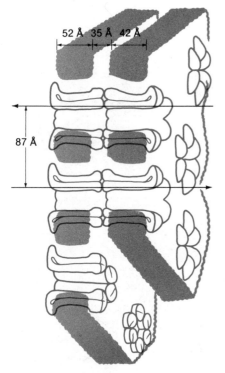

**Figure 17-35**
Molecular model of gap-junction pores as inferred from chemical, electron microscopic, and x-ray diffraction studies. A cross-section, nearly perpendicular to the two membrane planes, is shown. The arrows, representing aqueous channels, are drawn through cross sections of two pores, formed by apposition of two hexamers (or *connexons*), each spanning its own lipid bilayer membrane. The end-on views of each connexon at the right and bottom of the illustration show that they protude beyond the lipid bilayer on both sides of the plasma membrane. (Adapted from L. Makowski, D. L. D. Caspar, W. C. Phillips, and D. A. Goodenough, *J. Cell Biol.* 74:629, 1977.)

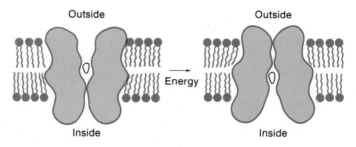

**Figure 17-36**
Generalized molecular model for transport involving a conformational rearrangement of subunits of an oligomeric carrier protein. In active transport, energy (such as that from the hydrolysis of ATP) could be used to effect this conformational change as shown. Alternatively, binding of the transport substrate could be enough to trigger this rearrangement in systems carrying out facilitated diffusion. (Adapted from S. J. Singer, The molecular organization of membranes, *Ann. Rev. Biochem.* 43:805, 1974. )

Pore-type transport mechanisms appear well established for nonspecific transport systems (porins and gap junctions) as well as for certain stereospecific transport systems (such as those in bacteria which utilize periplasmic binding proteins). However, other permeases, such as those which catalyze solute-cation cotransport, may function by mechanisms which incorporate features of the carrier model. Thus, while the *E. coli* glycerol channel, bacterial porins and periplasmic binding protein dependent transport systems are relatively insensitive to phospholipid temperature transitions, solute-cation symport systems are generally more sensitive to these physical changes in the membrane. Moreover, detailed kinetic analyses of lactose transport in *E. coli* are most easily interpreted in terms of a carrier-type mechanism. Finally, it should be noted that in addition to naturally occurring ionophore antibiotics, the coenzymes Q and plastoquinones are examples of nonprotein carriers that probably operate in a mobile fashion. These molecules play a role in $H^+$ translocation during electron transport in respiration and photosynthesis, respectively. They are believed to be able to diffuse freely in the hydrophobic phase of the phospholipid bilayer, although the exact mechanism by which they function in unidirectional $H^+$ transport is still unclear.

## Molecular Transport Mechanisms in Biological Membranes

*Several lines of evidence have been used to lend support to pore-type transport mechanisms for specific solutes and to the importance of protein conformational changes in the overall process.* E. Racker and P. Hinkle obtained support for a channel-type mechanism of $H^+$ translocation by bacteriorhodopsin in 1974. They incorporated the purified protein into vesicles of dimyristoylphosphatidylcholine ($T_m = 24°C$). In these vesicles, bacteriorhodopsin molecules were shown to have the *opposite* orientation when compared with *Halobacterium* cells, so that $H^+$ was accumulated *inside* the vesicles in response to light. By measuring $H^+$ uptake as a function of temperature, they showed that *the $H^+$ pumping ability of bacteriorhodopsin was insensitive to the phase of the bilayer* (gel or liquid crystalline), as shown in Figure 17-37. A mobile-carrier mechanism in which the protein undergoes large changes in orientation in the membrane thus appeared to be ruled out by this experiment.

It should be pointed out, however, that this type of experiment does not always lead to unequivocal results concerning transport mechanism. For example, many transport proteins do undergo abrupt changes in their activation energies for transmembrane transport at a temperature that does not correspond to the $T_m$ of the membrane system in which they are being studied. Nevertheless, these and other experiments strongly suggest that bacteriorhodopsin contains a proton channel, with the protonated Schiff base, formed between retinal and the protein, acting as a gate. One attractive mechanism is that the Schiff base may change its physical position in the membrane in response to light absorption as a consequence of a conformational change in the protein. This might "deliver" the proton to hydrogen-bonded groups near the outer surface of the membrane. Loss of the proton might then be accomplished by a return of the chromophore to its original position, where it would pick up a proton from hydrogen-bonded

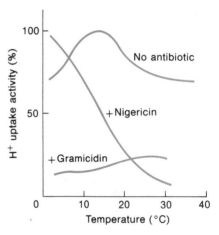

**Figure 17-37**
The effect of temperature on $H^+$ translocation by bacteriorhodopsin inserted into vesicles of dimyristoyl-phosphatidylcholine ($T_m = 24°C$). The top curve shows that proton transport is relatively insensitive to temperature, suggesting a pore-type mechanism. Nigericin, a mobile carrier that would collapse any proton gradient formed, does so only at higher temperatures, demonstrating that these membranes do, indeed, undergo a phase transition in the range studied. Gramicidin, a pore-forming antibiotic, inhibits proton translocation at all temperatures as expected. The curves with antibiotic present are plotted as percentages of the activity with no antibiotic, while the upper curve is in arbitrary units. (Adapted from E. Racker and P. C. Hinkle, Effect of temperature on the function of a proton pump, *J. Memb. Biol.* 17:181, 1974.)

groups nearer the inner membrane surface. This light-driven "proton shuttle" is a plausible mechanistic model for the functioning of bacteriorhodopsin, but it has not yet been proven experimentally.

In primary active transport systems driven by the hydrolysis of ATP, ample evidence has accumulated that the translocation steps take place by means of a gated-channel mechanism. The structure of the $F_1/F_0$ ATPase (Figure 17-16c), with its large, hydrophilic $F_1$ portion projecting into the cytoplasm, makes any mobile-carrier mechanism extremely unlikely for the translocation of protons by this protein. Similarly, experiments with both the $Na^+$-$K^+$ and $Ca^{2+}$ ATPases have shown that under certain conditions, _antibody molecules that are bound to these proteins at the membrane surface do not interfere with the functions of these proteins_. Again, because of the large amount of activation energy that would be required to translocate hydrophilic antibodies across the bilayer, pore mechanisms can be postulated for these ATPases as well. This is consistent with their asymmetric orientations in the membrane, as was discussed earlier.

The translocation mechanisms of both the $Na^+$-$K^+$ and $Ca^{2+}$ ATPases have been studied by substrate-binding experiments, as well as by physical and kinetic measurements. In both instances, _the translocation of ions is believed to occur by means of a conformational change in the protein that moves the ion-binding site from one surface of the membrane to the other._ This is similar to the model presented in Figure 17-36, in which elements of the classical channel and carrier models can be recognized. Both the $Na^+$-$K^+$ and $Ca^{2+}$ ATPases have been shown to be oligomeric, so a slight shift in the orientations of subunits with respect to each other could be the conformational change associated with translocation. Alternatively, the channel could be formed within a single protein subunit that spans the membrane and translocates ions as a result of a conformational change. A mechanism consistent with these possibilities was first proposed by O. Jardetzky in 1966 for the functioning of the $Na^+$-$K^+$ ATPase and is shown in Figure 17-38. A conformational change in the _phosphorylated_ $Na^+$-enzyme complex is thought to expel $Na^+$ from the cell, and interaction of $K^+$ with this form of the enzyme on the outside of the cell causes dephosphorylation and translocation of $K^+$ into the cytoplasm. Essential to this model is that _the affinity of the enzyme is higher for $Na^+$ than for $K^+$ on the inside, while the opposite is true when the binding site is exposed to the outside._ This ensures unidirectional transport of both ions.

Although Jardetzky's model was based primarily on theoretical considerations, it remains consistent with most of the experimental evidence obtained on this system. The energy for active transport is thus obtained by phosphorylation of the enzyme by ATP. Phosphorylation is stimulated by $Na^+$ and subsequently drives the conformational change that is responsible for translocation against a concentration gradient. The return cycle is triggered by dephosphorylation, which is stimulated by $K^+$ and allows the enzyme to return to its original state. Such a mechanism is analogous to a _Ping-Pong_ mechanism in an enzyme-catalyzed reaction (see Chapter 5), except that unidirectional transport rather than chemical transformation of the substrates is involved. A similar series of events has been proposed for $Ca^{2+}$ translocation by the $Ca^{2+}$ ATPase, and it is likely that other primary transport systems that use ATP also may operate by this type of conformational coupling mechanism.

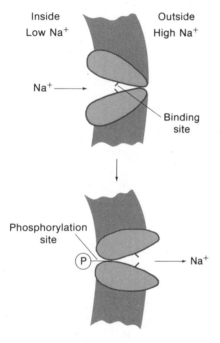

**Figure 17-38**
Model proposed by O. Jardetzky for the functioning of the $Na^+$-$K^+$ ATPase. The binding of $Na^+$ to a high-affinity site on the inside triggers phosphorylation by ATP, which in turn causes a conformational change exposing the cation binding site to the outside. The latter conformation has a low affinity for $Na^+$, but a high affinity for $K^+$. The binding of $K^+$ to this form reverses the conformational change and is accompanied by dephosphorylation of the enzyme and pumping of $K^+$ into the cytoplasm (not shown). (Adapted from O. Jardetzky, Simple allosteric model for membrane pumps, _Nature_ 211:969, 1966.)

As a final example of a molecular model for membrane transport, let us consider the ATP/ADP exchange protein of mitochondria. As mentioned earlier, this protein is a dimer of identical subunits. Because the protein loses its binding affinity for transport substrates if it is dissociated into monomers, a dimer may be necessary for carrying out the exchange process. Biochemical and kinetic studies on the ATP/ADP exchanger support a two-state gated-pore mechanism for this protein. The evidence for this suggestion can be summarized as follows:

1. There appears to be only one nucleotide binding site per dimeric ATP/ADP exchanger.

2. The nucleotide binding site has a higher affinity for ADP on the outer surface of the membrane than on the inner surface, while the opposite is true for ATP.

3. This site was shown never to be on both sides of the membrane simultaneously.

4. The respiratory poison _atractylic acid_ binds to the form preferring ADP, while the antibiotic _bongkrekic acid_ (see Figure 12-22) binds to the form that shows specificity toward ATP.

These results strongly support the transport model presented in Figure 17-39. A single nucleotide binding site per dimer is involved in transport by this protein. The configuration of this site, and thus its binding specificity, depends on the side of the membrane that it faces, and these two states are interconvertible through protein conformational changes that also result in translocation of any nucleotide bound to the protein. A conformational change can be triggered by binding of either ADP at the outside or ATP at the inside of the membrane surface. Once transported through the membrane, the nucleotide is now in a binding site that has little affinity for it, and the complex dissociates. Accordingly, the two forms of the adenine nucleotide exchanger appear to have quite different conformations. Antibodies prepared against the atractylic acid-protein complex do not cross-react with the form that binds bongkrekic acid, and vice versa. These two forms are, however, interconvertible by adding the appropriate inhibitor or adenine nucleotide. This conformational change is possibly the cause of structural alterations seen in the inner mitochondrial membrane when exposed to adenine nucleotides.

All the transport permeases we have discussed in this section have been purified. They are among the transport systems which are best understood at the molecular level. Secondary active transport and group translocation systems also should be amenable to detailed mechanistic studies now that some of these transport proteins are available in purified form. Such studies will be aided by reconstitution of isolated transport proteins into artificial membrane systems, a powerful investigative tool developed in recent years by biochemists interested in transport. This technique will be discussed in the following section.

## Reconstitution of Purified Transport Proteins

In Chapter 16, the importance of phospholipids in maintaining the structural and functional integrity of integral membrane proteins was stressed. This can be best illustrated by considering the lipid requirements of those

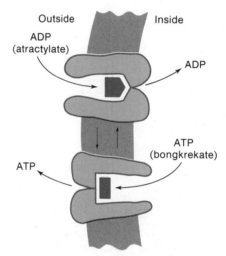

**Figure 17-39**
Molecular model for the functioning of the ATP/ADP exchanger of the mitochondrial inner membrane. The dimeric protein appears to have only one binding site for adenine nucleotides, perhaps involving amino acid residues from both subunits. When facing the outer surface of the membrane, this site has a high affinity for ADP (_pentagon_), and a low affinity for ATP (_rectangle_). The opposite specificities are observed when the site faces the lumen of the organelle. These states are interconvertible, even in the purified carrier, demonstrating that they arise from different conformations of the same protein. This model is analogous to those shown in Figures 17-36 and 17-38 except that energy is apparently not required for the conformational interconversion. (Adapted from M. Klingenberg, Membrane protein oligomeric structure and transport function, _Nature_ 290:449, 1981.)

transport permeases which still exhibit enzymatic activity after dissociation from the membrane, such as the ion-translocating ATPases. For example, the Na⁺-K⁺ ATPase isolated from rabbit kidney loses ATPase activity if residual bound phospholipid molecules are removed from the purified protein. About 90 mol of phospholipid per mole of enzyme is required for maximal activity. Similarly, a molar ratio of 30:1 (phospholipid to protein) is necessary for optimal functioning of purified $Ca^{2+}$ ATPase from sarcoplasmic reticulum. By replacing bound phospholipid in these purified preparations with lipids of known structure, it has further been possible to define which of these promote optimal activity of ATPase function in the isolated state, and presumably in the membrane as well.

These types of studies have set the stage for a fairly recent development in experimental membrane transport biochemistry, the reconstitution of isolated permeases into artificial membrane systems. By the introduction of such purified proteins into pure phospholipid bilayers, it has become possible to study the functional properties of a single transport system without interference by other membrane proteins or associated metabolic processes. Accordingly, _reconstitution of membrane transport systems offers one of the more powerful tools available to membrane biochemists_ who are interested purely in the mechanism of the transport event itself. The results of several of these types of experiments have been referred to elsewhere in this chapter. We shall briefly consider the types of information that reconstitution has contributed to our understanding of transport mechanisms.

E. Racker and co-workers have been successful in incorporating many of the enzymes involved in oxidative phosphorylation and electron transport, as well as other transport proteins, into artificial phospholipid vesicles. A number of techniques, including sonication of purified membrane proteins with phospholipid and dilution or dialysis of protein-detergent complexes in solutions containing phospholipid, have led to the preparation of _proteoliposomes_ (protein-phospholipid vesicles) capable of transporting the appropriate solutes (Figure 17-40). In a particularly important experiment, Racker and W. Stoeckenius reconstituted both purified bacteriorhodopsin and the mitochondrial $F_1/F_0$ ATPase into proteoliposomes. The orientations of both these molecules were shown to be opposite to those in the bacterial and mitochondrial membranes, respectively. _In the presence of light, these reconstituted vesicles catalyzed the synthesis of ATP from ADP and $P_i$, and this reaction required both proteins to be present simultaneously_ (Figure 17-41). Thus, in one reconstitution experiment, both the role of bacteriorhodopsin as a proton pump and the chemiosmotic hypothesis of Mitchell were confirmed (see Chapter 10).

A second useful reconstitution method in membrane biochemistry is the introduction of purified permeases into _planar phospholipid bilayers_. An apparatus for accomplishing this is schematically illustrated in Figure 17-42. Typically, a solution of phospholipid in an organic solvent is painted over an aperture in a thin wall separating two compartments of the apparatus that contain aqueous solutions. After several minutes, a bilayer of the phospholipid spontaneously forms between the two halves of the apparatus. Membrane proteins often can be introduced into the bilayer as protein-detergent or protein-lipid complexes added to either aqueous compartment. Alternatively, the phospholipids can be mixed with the protein of interest before the bilayer is formed. The potential advantages of this system over proteoliposomes include:

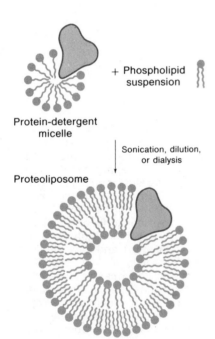

+ Phospholipid suspension

Protein-detergent micelle

Sonication, dilution, or dialysis

Proteoliposome

**Figure 17-40**
Schematic illustration of the reconstitution of a purified transport protein (in a detergent micelle) into a phospholipid vesicle to form a proteoliposome. Such proteoliposomes are very useful in studying transport mechanisms in a well-defined system.

1. Ability to easily control and determine the composition of the solutions on both sides of the reconstituted membrane.

2. Ease of applying voltages across the reconstituted bilayer in the study of transport proteins sensitive to the membrane potential.

3. Possibility of measuring transport continuously by monitoring current flow if the transported species has a net charge.

4. The ability, in some cases, to control the orientation of the reconstituted protein in the bilayer.

By the use of such planar bilayer systems, it has been possible to study the functional properties of a number of solute pumps and carriers. For example, planar membranes into which bacteriorhodopsin is incorporated in predominantly one orientation induce a potential between the two aqueous compartments in response to light energy applied to the aperture supporting the membrane. This potential has been shown to be due to the net pumping of protons into one of the compartments of an apparatus similar to that shown in Figure 17-42. Recently, a purified *E. coli* porin preparation (Figure 17-33) was incorporated into planar bilayers. _When an electric potential was externally applied across these membranes, an ionic current was measured._ This current was not constant, but _fluctuated in a stepwise manner_ about the steady state level (Figure 17-43). This has been interpreted as the opening and closing of single porin channels in the reconstituted membranes. This is supported by the fact that similar results are obtained when pore-forming antibiotics such as gramicidin A are introduced into a planar bilayer system. In fact, such incremental changes in the permeability of a membrane for a particular solute are often taken as evidence for the transport protein acting as a gated pore.

Several additional purified transport proteins have been incorporated into proteoliposomes and/or planar bilayers in functionally active states. These techniques therefore hold the promise of solving many unanswered questions about the mechanisms of biological transport. It should be pointed out, however, that reconstituted membranes are indeed artificial

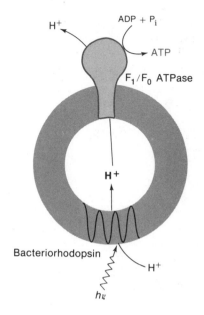

**Figure 17-41**
Reconstitution experiment of Racker and Stoeckenius that demonstrated that a proton gradient formed by illumination of bacteriorhodopsin in a proteoliposome could drive the synthesis of ATP by the mitochondrial $F_1/F_0$ ATPase incorporated into the same membrane. ATP was synthesized in the extravesicular space only if both proteins were included in the reconstitution and only in the presence of light. Both proteins in the reconstituted vesicles assumed orientations opposite to those they normally have in their native membranes. This was one of several experiments which were instrumental in proving one of the essential concepts of Mitchell's chemiosmotic hypothesis of oxidative phosphorylation.

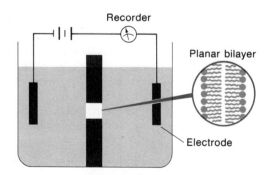

**Figure 17-42**
Cross section of an apparatus used to form planar lipid bilayers. The partition separating the two aqueous compartments has a circular aperture over which the bilayer film is formed. Proteins can be introduced into such bilayers either during or after their formation. This apparatus has the advantages of convenience in controlling and measuring the compositions of the solutions on both sides of the membrane, and of the ability of the investigator to measure electrical potentials and current flux through the reconstituted membrane.

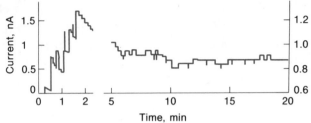

**Figure 17-43**
Current measurements on planar phospholipid bilayers containing *E. coli* porin, using an apparatus similar to that depicted in Figure 17-42. Porin channels are initially closed, but are induced to open by application of a potential across the membrane (240 mV at time zero in this experiment). After about 5 min, the current fluctuates in a stepwise manner about a steady-state value. (Adapted from H. Schindler and J. P. Rosenbusch, Matrix protein from *Escherichia coli* outer membranes forms voltage-controlled channels in lipid bilayers, *Proc. Natl. Acad. Sci. USA* 75:3751, 1978.)

systems, so that what one measures in a particular experiment may not occur under "real" conditions in a biological membrane. Reconstitution, therefore, has the potential to tell a great deal about transport per se, but it must be combined with observations on unfractionated membranes and even whole cells in order for one to ascertain the role of a particular transport system in the overall physiology of an organism.

## CONCLUDING REMARKS

In this chapter and the preceding one we have examined the structure and assembly of biological membranes and the mechanisms by which solutes are carried across these hydrophobic barriers. It should be obvious that certain areas, especially membrane assembly and transport mechanisms, are still rapidly evolving subjects of investigation in biochemistry. In the near future, therefore, we should obtain answers to many of the questions that have been raised in our treatment of these subjects, and we should be able to pose new questions that we are not yet knowledgeable enough to ask.

## SELECTED READINGS

Bronner, F., and Kleinzeller, A. (Eds.). *Current Topics in Membranes and Transport.* New York: Academic Press. Continuing series reviewing some of the current problems in biological transport.

Dills, S. S., Apperson, A., Schmidt, M. R., and Saier, M. H., Jr. Carbohydrate transport in bacteria. *Microbiol. Rev.* 44:385, 1980. Review of carbohydrate transport mechanisms and regulation in bacteria.

Ghosh, B. K. (Ed.) *Organization of Prokaryotic Cell Membranes,* Vols. I and II. Boca Raton, Florida: CRC Press, 1981. Reviews in these volumes deal with transport of small molecules across the inner and outer membranes of *E. coli* as well as the structures of these membranes.

Hobbs, A. S., and Albers, R. W. The structure of proteins involved in active membrane transport. *Ann. Rev. Biophys. Bioeng.* 9:259, 1980. Includes recent information on the $Na^+$-$K^+$ and $Ca^{2+}$ ATPases, the $F_1/F_0$ ATPase, and bacteriorhodopsin.

Oxender, D., and Fox, C. F. (Eds.). *Progress in Clinical and Biological Research,* Vol. 22: *Molecular Aspects of Membrane Transport.* New York: Alan Liss, 1977. Assemblage of papers on many transport systems that are being studied.

Rosen, B. P. (Ed.). *Bacterial Transport.* New York: Dekker, 1978. Comprehensive collection of articles reviewing many aspects of bacterial transport, including experimental systems, energetics, and discussions of specific classes of transport systems.

## PROBLEMS

1. a. Using Fick's law, show that the diffusion coefficient $D$ has the dimensions of area per unit time.
   b. The diameter of a porin channel is about $10^{-9}$ m and its length is $4 \times 10^{-9}$ m. In planar bilayers, glucose traverses this channel at the rate of about 50 molecules per channel per second at room temperature when the concentration of glucose is $3 \times 10^{-6}$ M on one side of the membrane. Calculate the diffusion coefficient for glucose through porin channels under these conditions.

2. The Nernst equation relates the electric potential $\Delta\Psi$ resulting from an unequal distribution of a charged solute $So$ across a membrane permeable to $So$ to the ratio of the concentration of $So$ on one side to that on the other:

$$m\,\Delta\Psi = \frac{-2.3RT}{F} \log \frac{[So]_1}{[So]_2}$$

where $m$ is the charge on the solute, $2.3RT/F$ has a value of about 60 mV at 37°C, and $[So]_1$ and $[So]_2$ refer to the concentrations of $So$ on either side of the membrane. Consider a planar phospholipid bilayer separating two compartments of equal volume, one of which (side 1) contains 50 mM KCl and 50 mM NaCl, while the other (side 2) contains 100 mM KCl.
   a. If the membrane is made permeable *only* to $K^+$, e.g., by addition of valinomycin, what will be the magnitude of $\Delta\Psi$?
   b. If the membrane is made permeable to $H^+$ and $K^+$, in which direction will $H^+$ initially flow?
   c. If the membrane could be made to be selectively permeable to both $K^+$ *and* $Cl^-$, what would be the value of $\Delta\Psi$ and the ion concentrations on both sides of the membrane at equilibrium? (*Hint:* Initially, $K^+$ would diffuse down its concentration gradient accompanied by an equivalent amount of $Cl^-$. Equilibrium would be established when the potentials due to $K^+$ and $Cl^-$ were equal to each other and to the overall membrane potential.)

3. Membrane vesicles of *E. coli* that possess the lactose permease are preloaded with KCl and are suspended in an equal concentration of NaCl. It is observed that these vesicles actively, although transiently, accumulate lactose if valinomycin is added to the vesicle suspension. No such active uptake is observed if KCl replaces NaCl in the suspending medium. Explain these results in light of what you know about the mechanism of lactose transport and the properties of valinomycin.

4. Intracellular vacuoles in the yeast *Saccharomyces cerevisiae* are membrane-bounded organelles that are known to concentrate within them a variety of basic amino acids, including arginine (net charge $= +1$). Vesicles prepared from these vacuoles lack an electron-transport chain, and arginine uptake into them is dependent on extravesicular ATP. A membrane potential $\Delta\Psi$ has no effect on ATP-dependent arginine uptake in the absence of a proton gradient, while proton ionophores and dicyclohexylcarbodiimide (a known inhibitor of the $F_1/F_0$ ATPase) greatly inhibit accumulation of arginine by this system. Upon addition of ATP in the absence of arginine, the intravesicular pH of these vesicles drops. Describe a mechanism for the energization of arginine transport in this system taking into account all these observations.

5. In *E. coli*, lactose is taken up by means of proton symport, maltose by means of a binding protein system, melibiose by means of $Na^+$ symport, and glucose by means of a phosphotransferase system (PTS). Although this bacterium normally does not transport sucrose, suppose you have isolated a strain that does. How would you determine if one of the four mechanisms just listed is responsible for sucrose transport in this mutant strain?

6. In some instances, the *efflux* of a radioactively labeled transport substrate out of preloaded cells or vesicles is transiently *stimulated* by addition of the same nonradioactive transport substrate to the *outside*. This phenomenon is known as *trans-stimulation* and occurs with transport systems that are reversible (i.e., can operate in either direction). Can you think of an explanation for trans-stimulation in view of what is known about the molecular mechanisms of transmembrane transport?

7. Outline a molecular mechanism by which, and the conditions under which, an $H^+$ symport system, such as the *E. coli* lactose permease, might operate to actively accumulate a metabolite such as lactose.

8. Predict the effects of the following on the initial rate of glucose transport into vesicles derived from animal cells that accumulate this sugar by means of $Na^+$ symport. Assume that initially $\Delta\Psi = 0$, $\Delta pH = 0$ (pH $= 7$), and the outside medium contains 0.2 M $Na^+$, while the vesicle interior contains an equivalent amount of $K^+$.
   a. Valinomycin
   b. Gramicidin A
   c. Nigericin
   d. Preparing the membrane vesicles at pH 5 (in 0.2 M KCl), resuspending them at pH 7 (in 0.2 M NaCl), and adding 2,4-dinitrophenol.

# PART IV

## NUCLEIC ACIDS AND PROTEIN METABOLISM

- STRUCTURE OF NUCLEIC ACIDS AND NUCLEOPROTEINS
- NUCLEOTIDE METABOLISM
- DNA METABOLISM
- RNA METABOLISM
- BIOSYNTHESIS OF AMINO ACIDS
- UTILIZATION OF AMINO ACIDS
- THE GENETIC CODE
- MECHANISMS OF PROTEIN SYNTHESIS
- GENE EXPRESSION IN MICROORGANISMS

# 18

# STRUCTURE OF NUCLEIC ACIDS AND NUCLEOPROTEINS

Nucleic acids are long-chain polymers composed of nucleotides. The sequence of nucleotides is the repository of all genetic information carried by chromosomes. Despite this, not all nucleic acid is informational, nor is all the informational nucleic acid found in the chromosome. Examples of non-informational nucleic acid include ribosomal RNA and centromeric DNA, whose functions are primarily structural. Examples of informational nucleic acids not found in chromosomes include nucleic acids of mitochondria, chloroplasts, plasmids, and viruses. Most of the chapters in Part IV and Chapters 27 and 28 in Part V are devoted to explaining the ways in which nucleic acids are replicated and transmit their genetic information for use in the cell. In this chapter the focus is on the basic structural properties of nucleic acids in the free-solution state and as they exist in protein complexes in cells.

## NUCLEOTIDES, THE BUILDING BLOCKS OF NUCLEIC ACIDS

There are two chemically different types of nucleic acids, deoxyribonucleic acid (DNA) and ribonucleic acid (RNA). Both DNA and RNA contain four different nucleotides. Each _nucleotide_ contains a nitrogenous base known as a _purine_ or a _pyrimidine_; a sugar, _ribose_ in RNA; _deoxyribose_ in DNA; and a phosphoryl group. The nucleotide may be converted to a _nucleoside_ by removal of the phosphate. The primary structure of the four commonly occurring deoxyribonucleotides found in DNA are shown in Figure 18-1.

_Electron micrograph of a human chromosome in late-prophase. (Magnification 21,000×.) The chromosome consists of two identical chromatids united at their centromeres. The chromatin consists primarily of a complex of DNA and histone (see text). It is still a mystery as to what forces cause the condensation of the nucleohistone into this highly condensed form. (Micrograph obtained from Gunter F. Bahr, M.D.)_

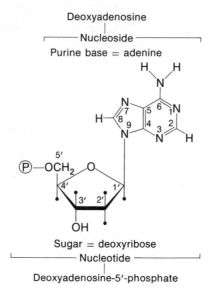

**Figure 18-1**
Structure of four deoxyribonucleotides found in DNA. Note the numbering system for the carbon and nitrogen atoms of the purine and pyrimidine bases. The carbon atoms of the sugars are usually given prime designations. Note that the nitrogen bases are cis relative to the C-5′ and trans relative to the C-3′-OH.

**Table 18-1**
Ionization Constants of the Ribonucleotides (Presented as p$K$ Values)

|  | Base | Secondary Phosphate | Primary Phosphate |
|---|---|---|---|
| Adenosine-5′-phosphate (5′-AMP)* | 3.8 | 6.1 | 0.9 |
| Uridine-5′-phosphate (5′-UMP) | 9.5 | 6.4 | 1.0 |
| Cytidine-5′-phosphate (5′-CMP) | 4.5 | 6.3 | 0.8 |
| Guanine-5′-phosphate (5′-GMP) | 2.4, 9.4 | 6.1 | 0.7 |

*5′-AMP (or 5′-rAMP) refers to the ribonucleotide. The comparable deoxyribonucleotide (deoxynucleotide) is indicated by the symbol 5′-dAMP.

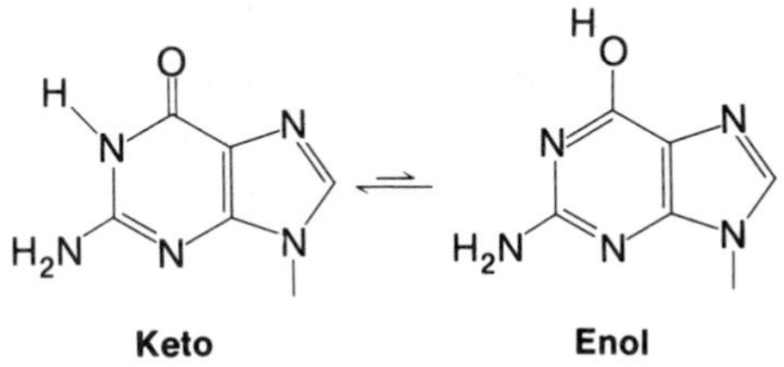

**Figure 18-2**
Keto-enol isomerization of guanine.

Compositional variability in the nucleotide is accounted for solely by the purine or pyrimidine base attached to the C-1′ position of the sugar (atoms in the sugar are commonly given a prime designation). These bases are either the purines, _adenine_ and _guanine_, or the pyrimidines, _thymine_ and _cytosine_. The same bases are found in RNA, except that thymine is replaced by _uracil_, which has an H— group instead of a $CH_3$— group on the C-5 position of the pyrimidine. The sugar in ribonucleotides is also different in having an additional HO group on the C-2′ which is cis with respect to the C-3′-OH.

All the commonly occurring nucleosides and nucleotides are capable of existing in two tautomeric forms. For example, guanosine (G) can undergo the keto-enol shift shown in Figure 18-2. The keto form is strongly favored, so much so that it is difficult to detect even trace amounts of the enol form. Similarly, the keto forms of thymidine (T) or uridine (U) are strongly preferred. Adenosine (A) and cytidine (C) can isomerize to imino forms (not shown), but once again the amino forms (shown) are strongly preferred. Even though the unusual tautomers are present in very small amounts, it is conceivable that they are contributors to the mutation process.

Some nucleotides undergo protonation in acid and some undergo deprotonation in base; the relevant p$K$s are indicated in Table 18-1. At neutrality there is no charge on any of the bases. Three of the bases undergo protonation as the pH is lowered (A, C, and G). X-ray diffraction and spectroscopic analysis (nuclear magnetic resonance and infrared spectroscopy) have been used to show that adenosine protonates on the N-1 position of the purine rather than on the amino group (see Figure 18-3). The charged form is stabilized by the resonance hybrids shown. On cytidylic

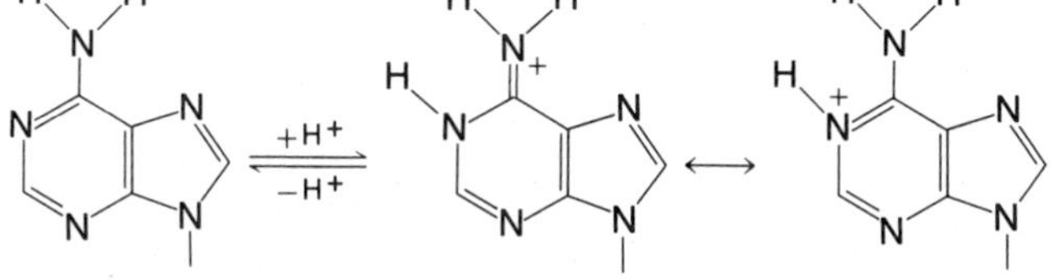

**Figure 18-3**
Uncharged and protonated forms of adenosine. The charged base resonates between the two structures shown on the right.

**Figure 18-4**
Uncharged and protonated forms of guanosine.

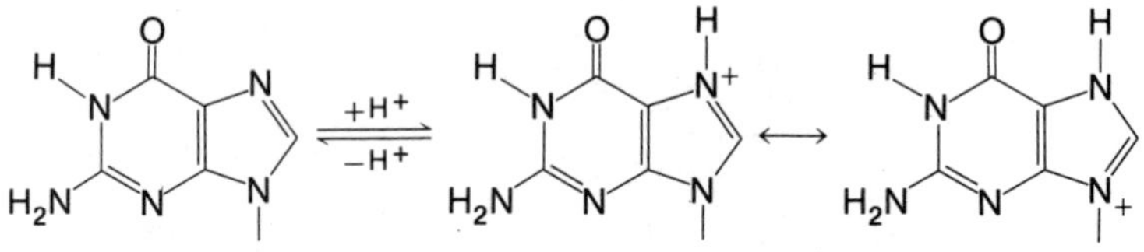

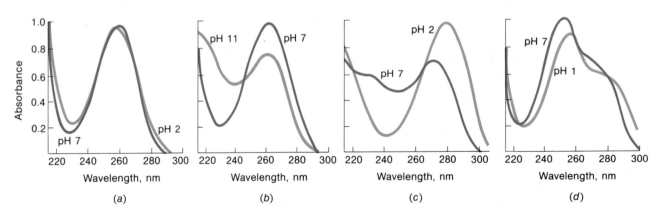

acid the proton adds to the comparable N-3 ring nitrogen. In guanosine the proton adds to the N-7 rather than the —NH₂ side chain (Figure 18-4), again indicating the unusually low basicity of the amino groups on the nucleotides compared with primary aliphatic amines. On the basic side of neutrality both 5'-UMP and 5'-GMP release a proton, probably from the imino nitrogens at positions 3 and 1, respectively. As might be expected, the ionization constants for the primary and secondary dissociations of the phosphate group do not differ appreciably for the various nucleotides.

Owing to the large number of conjugated double bonds in their nitrogen bases, all nucleotides show absorption maxima in the near ultraviolet range (see Figure 18-5). The spectrum is pH-dependent because protonation or deprotonation changes the electronic distribution in the base rings. The ultraviolet absorption of the nucleotides has been useful in many ways for the study of mononucleotides and polynucleotides. [One generally measures absorbancy ($A$) or optical density (O.D.) directly in a 1-cm² cell filled with a dilute solution in a spectrophotometer. The O.D. is equivalent to the $\log(I_0/I)$, where $I_0$ is the intensity of incident monochromatic light falling on the cell and $I$ is the intensity of the transmitted light. Further, $\log(I_0/I) = Ecd$, where $E$ is the molar extinction coefficient, or molar absorbancy; $c$ is the concentration of the absorbing species in moles per liter; and $d$ is the thickness of the light-absorbing sample, usually 1 cm.]

All the common 5'-nucleotides also occur in cells as 5'-diphosphates and 5'-triphosphates (Figure 18-6). The nucleotide diphosphates (NDPs) and the nucleotide triphosphates (NTPs) dissociate three and four protons,

**Figure 18-5**
Curves showing the molar absorbancies ($\times 10^{-3}$) of the 5'-ribonucleotides at several pHs: (*a*) 5'-AMP, (*b*) 5'-UMP, (*c*) 5'-CMP, and (*d*) 5'-GMP.

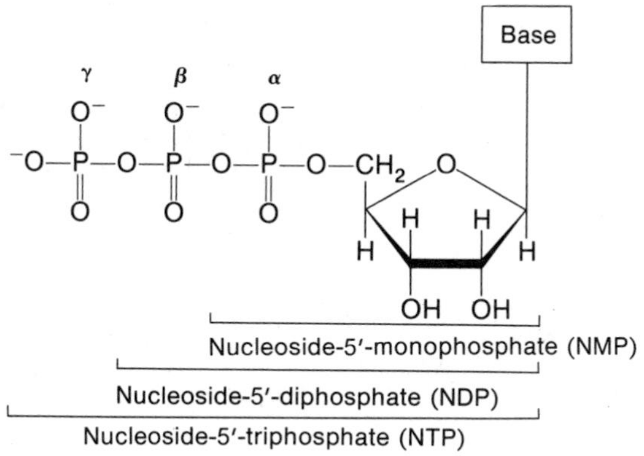

Nucleoside-5′-monophosphate (NMP)

Nucleoside-5′-diphosphate (NDP)

Nucleoside-5′-triphosphate (NTP)

**Figure 18-6**
The general structure of a nucleotide monophosphate, diphosphate, and triphosphate.

respectively, from their phosphate groups. The first has a p$K$ of 0.9; the second, third, and fourth have p$K$s that range from 6.1 to 6.7. The NDPs and NTPs can form complexes with $Mg^{2+}$ or $Ca^{2+}$ and probably exist as such in the cell. The NDPs and NTPs have a number of important functions in the cell; they serve as energy-carrying enzyme cofactors (for example, see Chapters 8 through 11) and the immediate precursors of polymeric nucleic acids (see Chapters 20 and 21).

## STRUCTURAL PROPERTIES OF DNA

If one takes a 5′-mononucleotide and joins it by a phosphoester linkage to the 3′-OH group of a second mononucleotide, a dinucleotide is formed. Repetition of this process will lead to the formation of polydeoxyribonucleotides or polyribonucleotides; the primary structure of a polydeoxyribonucleotide chain is shown in Figure 18-7. The directional sense of the sugar to phosphate linkages constituting the backbone of DNA and RNA should be noted.

Most DNA in the nucleus is found in the form of chromosomes. Much smaller amounts of DNA are present in cellular organelles, such as mitochondria and chloroplasts. Our knowledge of DNA has been largely obtained from studies on nuclear DNA, DNA isolated from viruses or plasmids, or synthetically produced deoxyribopolynucleotides. DNAs from different sources differ widely in amounts per cell. This is especially interesting, since one would conjecture that the amount of genetic information in a cell is roughly proportional to the amount of DNA it contains. The DNA content per cell for a number of cell types is indicated in Figure 18-8. The spread is impressive; thus mammalian cells contain about three orders of magnitude more DNA than bacterial cells. Interestingly enough, some amphibians, fishes, and algae contain considerably more DNA than mammals. Bacterial viruses such as the T-type bacteriophages (or phages) that infect *E. coli* contain about an order of magnitude less DNA than the bacterial chromosome. The DNA of the smallest viruses is about a tenth the size of the T phage DNA or only enough to accommodate about three to four genes. This is consistent with the fact that viruses do not contain sufficient genetic information for independent growth. Plasmids exist as circular DNA duplexes found in the nucleus or the cytoplasm.

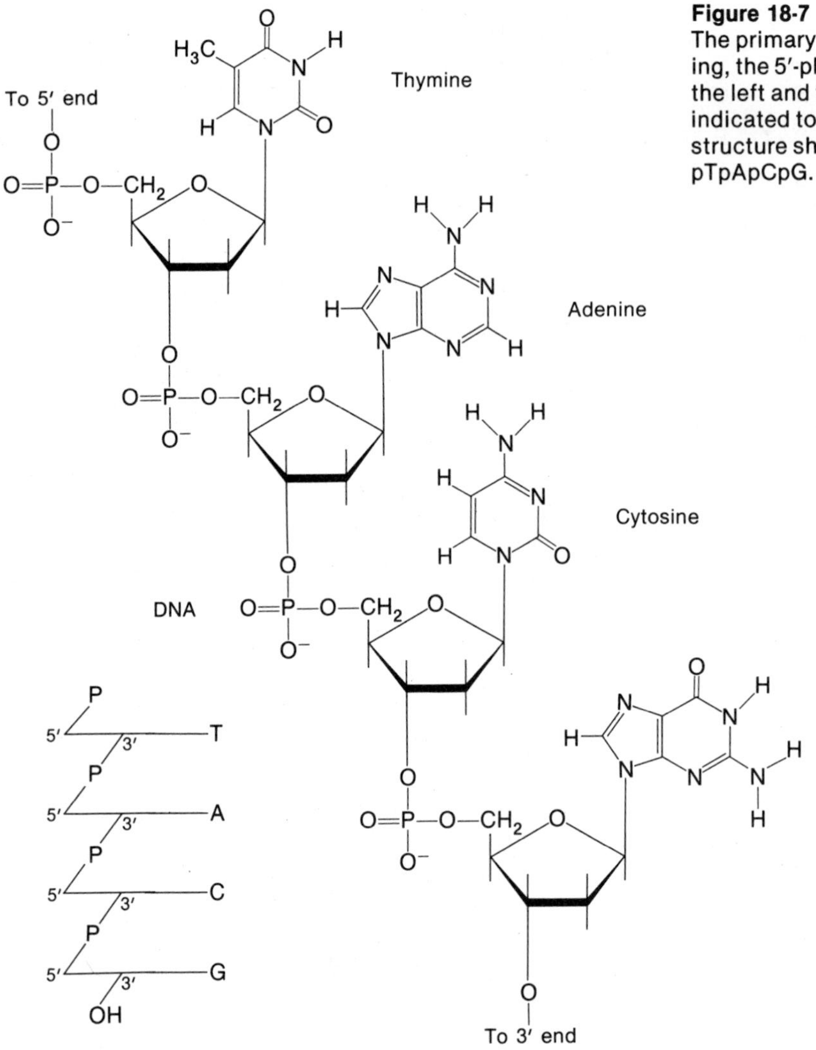

## Figure 18-7

The primary structure of DNA. In writing, the 5'-phosphate is indicated to the left and the 3'-phosphate is indicated to the right. The illustrated structure should be written as pTpApCpG.

To 5' end

Thymine

Adenine

Cytosine

DNA

Guanine

To 3' end

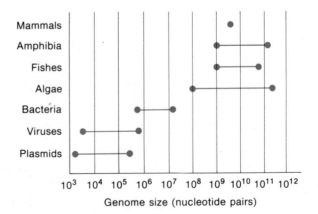

## Figure 18-8

Genome size in different cells, viruses, and bacterial plasmids.

Mammals
Amphibia
Fishes
Algae
Bacteria
Viruses
Plasmids

$10^3$  $10^4$  $10^5$  $10^6$  $10^7$  $10^8$  $10^9$  $10^{10}$  $10^{11}$  $10^{12}$

Genome size (nucleotide pairs)

A variety of procedures exist for degrading high-molecular-weight DNAs to demonstrate the overall structure. These procedures are either physical, chemical, or enzymatic (see Chapter 20). Mononucleotides resulting from exhaustive degradation of DNA can be separated on resins by ion-exchange chromatography and the relative proportions of the bases determined by the relative amounts of each base found. Detection and identification is made spectrophotometrically, taking advantage of the high absorption of nucleotides in the ultraviolet region and the characteristic changes in absorption in acidic and basic buffer. DNAs from different sources differ widely in base composition, as indicated in Table 18-2. In spite of this variation, examination of the data shows that for most DNAs, the amount of A is very nearly equal to the amount of T and the amount of G is very nearly equal to the amount of G + 5-MC (5-methylcytosine). 5-MC usually results from postreplicative modification of certain C residues (see Chapter 20). These two equalities were the first indication that regular complexes occur between these bases in most DNA structures. J. D. Watson realized in model-building studies that H-bonded base-paired structures could be formed between A and T and G and C that have the same overall dimensions (see Figure 18-9). Such complexes were consistent with Watson and Crick's proposal, made in 1953, that _DNA exists as a regular two-chain structure with H-bonds formed between opposing bases on the two chains_. Detailed model construction showed that this hydrogen bonding is possible only when the directional senses of the two interacting chains are opposite or antiparallel (Figure 18-10). Two chains with antipar-

**Table 18-2**
Base Composition of DNAs from Different Sources

| | (A) Adenine | (G) Guanine | (C) Cytosine | (5-MC) 5-Methyl cytosine | (T) Thymine | $\dfrac{A + T}{G + C + 5\text{-MC}}$ |
|---|---|---|---|---|---|---|
| Human | 30.4 | 19.6 | 19.9 | 0.7 | 30.1 | 1.53 |
| Sheep | 29.3 | 21.1 | 20.9 | 1.0 | 28.7 | 1.38 |
| Ox | 29.0 | 21.2 | 21.2 | 1.3 | 28.7 | 1.36 |
| Rat | 28.6 | 21.4 | 20.4 | 1.1 | 28.4 | 1.33 |
| Hen | 28.0 | 22.0 | 21.6 | | 28.4 | 1.29 |
| Turtle | 28.7 | 22.0 | 21.3 | | 27.9 | 1.31 |
| Trout | 29.7 | 22.2 | 20.5 | | 27.5 | 1.34 |
| Salmon | 28.9 | 22.4 | 21.6 | | 27.1 | 1.27 |
| Locust | 29.3 | 20.5 | 20.7 | 0.2 | 29.3 | 1.41 |
| Sea urchin | 28.4 | 19.5 | 19.3 | | 32.8 | 1.58 |
| Carrot | 26.7 | 23.1 | 17.3 | 5.9 | 26.9 | 1.16 |
| Clover | 29.9 | 21.0 | 15.6 | 4.8 | 28.6 | 1.41 |
| _Neurospora crassa_ | 23.0 | 27.1 | 26.6 | | 23.3 | 0.86 |
| _Escherichia coli_ | 24.7 | 26.0 | 25.7 | | 23.6 | 0.93 |
| T4 bacteriophage | 32.3 | 17.6 | | 16.7* | 33.4 | 1.91 |

*In T bacteriophage all of the 5-MC exists in the 5-hydroxymethyl form 5-HMC.

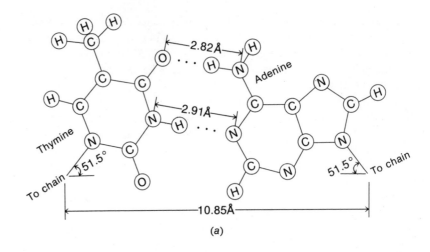

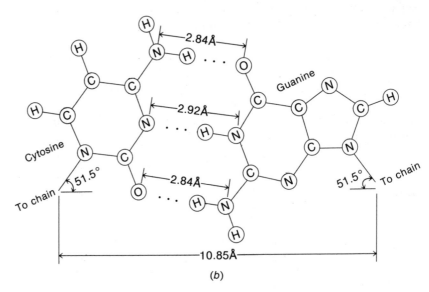

**Figure 18-9**
Dimensions and hydrogen bonding
of (a) thymine to adenine and (b) cy-
tosine to guanine. (Adapted from
M. H. F. Wilkins and S. Arnott,
*J. Mol. Biol.* 11:391, 1965.)

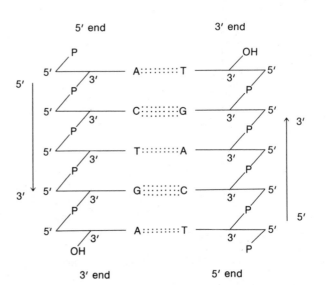

**Figure 18-10**
Segment of DNA duplex illustrating
antiparallel orientation of the complementary chains.

allel orientation can adopt different types of conformations: (1) One extreme possibility would be a stepladder-like structure (Figures 18-10 or 18-11a) in which the chains lie straight in a fully extended conformation with a distance of 6.8 Å between residues in the direction of the long axis. In this event, adjacent base pairs are also 6.8 Å apart, leaving an appreciable gap between them that would presumably have to be filled by water. This would not be an energetically favorable structure, however, since the planes of the bases are hydrophobic in character and would prefer to be in contact with other like hydrophobic surfaces. (2) The stepladder structure can be converted into a helix structure by a simple right-handed twist. When this is done, the distance between base pairs decreases until they are stacked about 3.4 Å apart, which allows for close contact between adjacent base pairs. Such a structure repeats itself after about 10 residues, or once every 34 Å along the helix axis, and this distance is referred to as the _pitch length_. Watson and Crick were the first to appreciate the significance of strong 3.4-Å and 34-Å spacings and the central cross-like pattern that reflects a helix structure in the x-ray diffraction pattern of DNA (Figure 18-12). They interpreted this as arising from a hydrogen-bonded antiparallel double-helix structure, as shown in Figure 18-11b.

**Figure 18-11**
Different possible conformations of base-paired DNA: (*a*) stepladder; (*b*) helix.

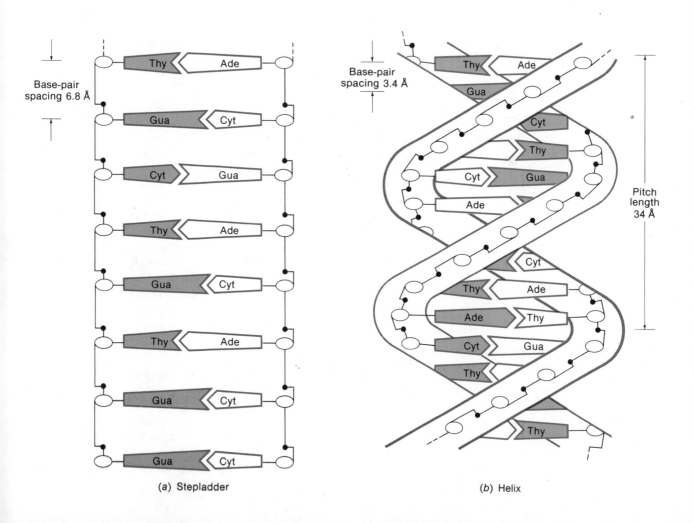

(a) Stepladder        (b) Helix

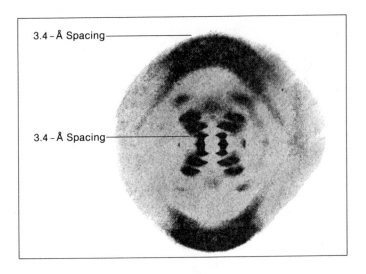

3.4 – Å Spacing

3.4 – Å Spacing

**Figure 18-12**
X-ray fiber diagram of the B structure
of DNA. (Courtesy M. H. F. Wilkins.)

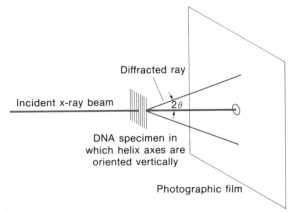

Diffracted ray

Incident x-ray beam

$2\theta$

DNA specimen in
which helix axes are
oriented vertically

Photographic film

**Figure 18-13**
Experimental arrangement for obtain-
ing x-ray diffraction pattern shown in
Figure 18-12.

The x-ray pattern of DNA illustrated in Figure 18-12 is obtained by
holding a stretched fiber containing many DNA molecules in a vertical
direction and exposing it to a collimated monochromatic beam of x-rays
(see Figure 18-13). Only a small percentage of the x-ray beam is diffracted.
Most of the beam travels through the specimen with no change in direction. A photographic film is held in back of the specimen; a hole in the
center of the film allows the incident undiffracted beam to pass through.
Coherent diffraction only occurs in certain directions, specified by Bragg's
law: $2d \sin \theta = n\lambda$, where $d$ is the distance between identical repeating
structural elements, $\theta$ is the angle between the incident beam and the
regularly spaced diffracting planes, $\lambda$ is the wavelength of x-rays used, and $n$
is the order of diffraction, which may equal any integer but is usually
strongest for $n = 1$. For small $\theta$, $\sin \theta \approx \theta$ and $d \approx 1/\theta$, so that a spot far out
on the photographic film is indicative of a repeating element of small di-
mension.

Fiber-diffraction analysis, as discussed earlier, where one merely looks
for allowed spacings is a crude type of crystallographic analysis. In more
refined work, one compares the intensities of allowed diffractions with a
proposed molecular structure by procedures described in Chapter 2 for
proteins. A rigorous analysis of this type, as explained in Chapter 2, requires crystals with three-dimensional order that do not exist in samples of
stretched DNA fibers. Finally, in 1980, R. E. Dickerson and his coworkers
were able to form suitable crystals from the self-complementary dodecamer

sequence d(CpGpCpGpApApTpTpCpGpCpG). A rigorous analysis of the diffraction pattern from such crystals confirmed the essential correctness of the proposal made by Watson, Crick and Wilkins over a quarter of a century earlier.

## Conformational Variants of the Double-Helix Structure

The same base-pairing arrangement is found in all naturally occurring double-helix structures. However, *the inherent flexibility in the pyranose ring and the degrees of freedom generated by six rotatable single bonds per residue, five in the sugar phosphate backbone and one in the C-1'-N-glycosidic linkage (see Figure 18-14) has led to considerable variation in the conformations adopted by double-helix structures.* Four puckered conformations for the sugar, with small differences in stability, are shown in Figure 18-15. In the double-helix structure described earlier with 10 base pairs per turn and the planes of the bases nearly perpendicular to the helix axis, the C-2'-endo conformation is observed. This is believed to be the

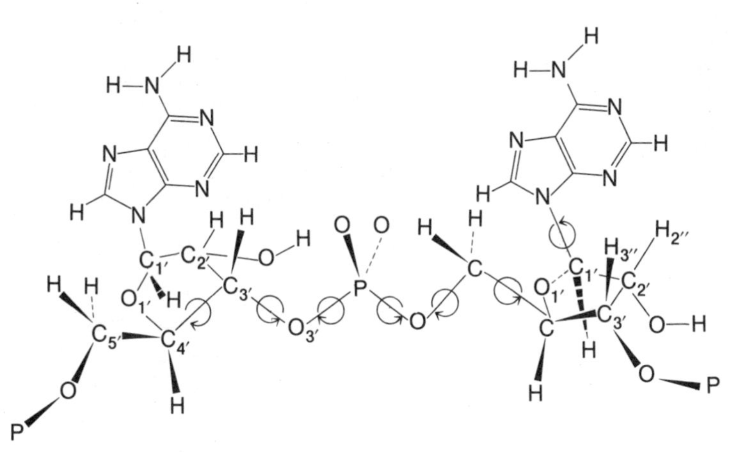

**Figure 18-14**
A segment of polynucleotide chain with rotatable bonds indicated by curved arrows. (Adapted from W. K. Olson and P. J. Flory, *Biopolymers* 11:1, 1972.)

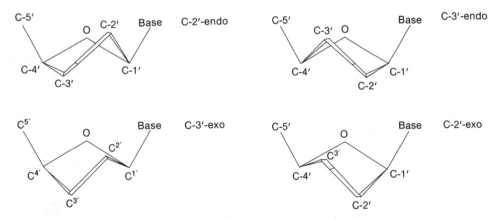

**Figure 18-15**
Four puckered conformations of the pyranose rings of ribose and deoxyribose.

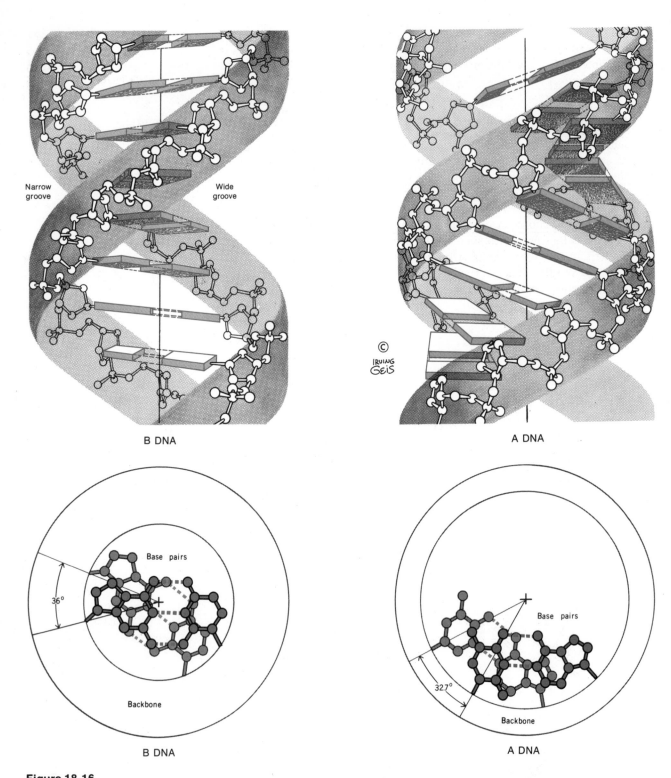

Narrow
groove

Wide
groove

B DNA

A DNA

Base pairs

36°

Backbone

B DNA

Base pairs

32.7°

Backbone

A DNA

**Figure 18-16**
The A and B forms of DNA. (*Top*) A view perpendicular to the helix axis. (*Bottom*) A view of two adjacent base pairs looking down the helix axis. A single pitch length of DNA is shown for each duplex. Note that the B DNA has a large groove and a narrow groove. Different parts of the base pairs are exposed in the large and narrow grooves (see Figure 18-9).

major structure of DNA in solution. When some of the water is removed from the hydrated DNA fibers, the double helices push closer together and the structure changes to the A form, which has about 11 bases per turn and base pairs that are tilted about 20 degrees with respect to the helix axis (Figure 18-16). In the A form, the pyranose rings have changed their pucker to the C-3'-endo conformation. The pyranose rings in RNA have a stronger preference for the C-3'-endo conformation, with the result that RNA double helices adopt a structure similar to the DNA A form even at high degrees of hydration.

The most strikingly different structure that has been observed for a DNA double helix with Watson-Crick base pairing is referred to as the Z form. This structure was first detected by A. Rich and his coworkers for the deoxyoligonucleotide d(CpGpCpGpCpG), which crystallizes into an antiparallel double helix with a left-handed rather than a right-handed twist (see Figure 18-17). The Z form is a considerably slimmer helix than the B form, containing 12 base pairs per turn rather than 10. In the Z form, the planes of the base pairs are rotated approximately 180 degrees with respect to the helix axis from their orientation in the B form (Figure 18-18). The flipping of the base pairs involves different conformational changes in the G and C residues in the alternating GC structure observed by Rich. In the case of G residues, the base is rotated by 180 degrees about the glycosidic bond, resulting in the so-called syn conformation for the nucleotide. Model-building studies indicate that the anti conformation has less steric crowding than the syn conformation but that it is easier for a purine nucleotide to adopt the syn conformation than a pyrimidine. The entire cytidine residue rotates in going from the B to the Z form, so that the cytidine remains in the anti conformation (Figure 18-19). The different conformations observed for the G and C residues in the Z-form DNA of alternating G–C DNA gives rise to a zigzag course for the sugar phosphate backbone (Figure 18-17), justifying *Z DNA* as an appropriate descriptive designation for this structure. Because of the difficulty in getting the pyrimidine residues to adopt the syn conformation, it may be that Z DNA can form only when the primary structure has an alternating purine-pyrimidine sequence. Nevertheless, recent observations suggest that left-handed helical DNA may exist in specific regions of naturally occurring DNA and that this unusual DNA conformation may play an important role in regulating gene expression (see below and Chapter 26).

A number of the structural parameters associated with the A, B, and Z helices are summarized in Table 18-3. In addition to these helices, many other conformations have been observed for DNA; all, however, preserve the Watson-Crick hydrogen bonding. Synthetic polynucleotide complexes have been made that simulate the normal duplex structure, for example, poly(A) and poly(U) in the ratios of 1:1 base pair to form a duplex resembling the A form of DNA. In the ratios 1:2 they form a triple helix with one A chain and two U chains that display additional kinds of H bonding. Two chains of A make a double helix, and four chains of G can form a quadruple helix. A great variety of structures and hydrogen-bonding patterns is possible with polymers that have special sequences, but it is not clear that any of these unusual helix structures are of any biological significance. Unusual pairings are also observed in the structure for transfer RNA (described later).

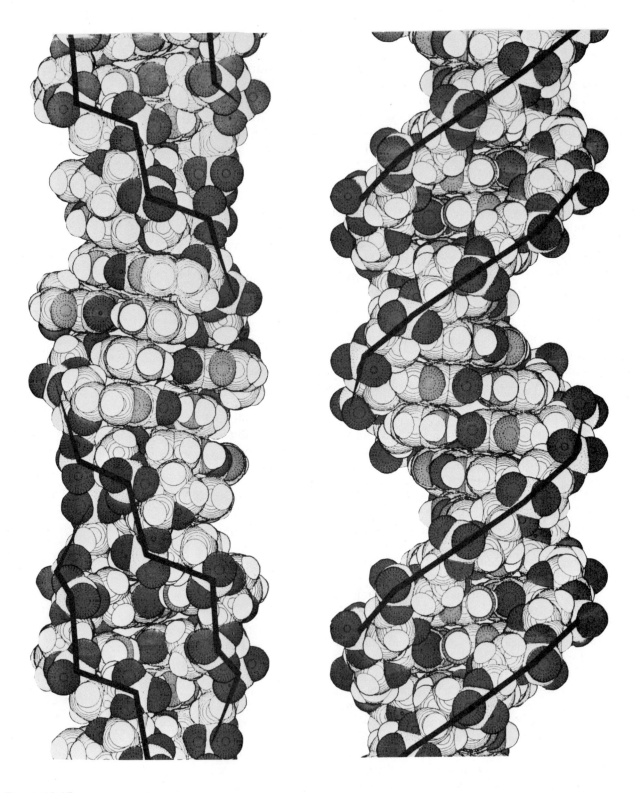

**Figure 18-17**
Van der Waals side views of Z DNA (*left*) and B DNA (*right*). The irregularity of the Z DNA backbone is illustrated by the heavy lines that go from phosphate to phosphate residues along the chain. In contrast, B DNA has a smooth line that connects the phosphate groups and two grooves, neither one of which extends into the helix axis of the molecule.

## Figure 18-18

A diagram illustrating the change in topological relationship if a four base-pair segment of B DNA were converted into Z DNA. This conversion could be accomplished by rotation of the bases relative to those in B DNA. This rotation is shown diagrammatically by shading one surface of the bases. All the dark-shaded areas are at the bottom in B DNA. In the segment of Z DNA, however, four of them are turned upwards. The turning is indicated by the curved arrows. Rotation of the guanine residues about the glycosidic bond produces deoxyguanosine in the syn conformation, while for dCyt residues, both cytosine and deoxyribose are rotated. The altered position of the Z DNA segment is drawn to indicate that these bases will not be stacking directly on the base pairs in the B DNA segment.

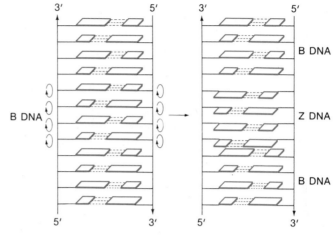

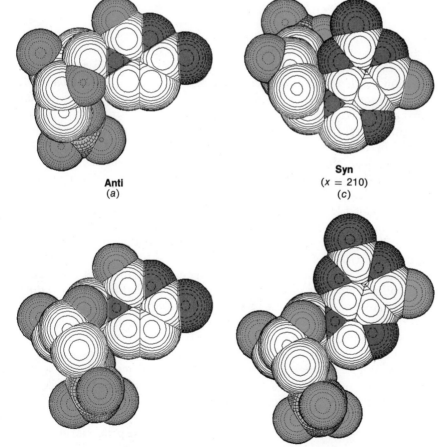

**Anti**
(*a*)

**Syn**
($x = 210$)
(*c*)

**Anti**
(*b*)

**Anti**
($x = 0$)
(*d*)

## Figure 18-19

The conformations of the G and C bases in the Z and B forms of DNA. Deoxycytidilic acid in Z (*a*) and B (*b*) DNA. Deoxyguanylic acid in Z (*c*) and B (*d*) DNA.

**Table 18-3**
Some Structural Parameters of A, B, and Z DNA

|  | A DNA | B DNA | Z DNA |
|---|---|---|---|
| Helix sense | Right-handed | Right-handed | Left-handed |
| Residues per turn | 11 | 10 | 12 (6 dimers) |
| Rise per residue | 2.55 Å | 3.4 Å | 3.7 Å |
| Helix pitch | 28 Å | 34 Å | 45 Å |
| Base pair tilt | 20 degrees | 6 degrees | 7 degrees |
| Rotation per residue | 33 degrees | 36 degrees | −60 degrees (per dimer) |
| **Glycosidic conformation** | | | |
| Deoxycytidine | *Anti* | *Anti* | *Anti* |
| Deoxyguanosine | *Anti* | *Anti* | *Syn* |
| **Sugar pucker** | | | |
| Deoxycytidine | *C-3′-endo* | *C-2′-endo* | *C-2′-endo* |
| Deoxyguanosine | *C-3′-endo* | *C-2′-endo* | *C-3′-endo* |

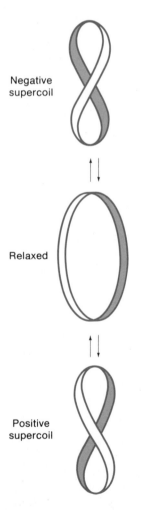

Negative
supercoil

Relaxed

Positive
supercoil

**Figure 18-20**
Topology of a negative and a positive
supercoil.

## Supercoiled DNA

The finding of different conformations of DNA underscores the inherent flexibility built into the double-helix structure. All the conformations discussed thus far involve regular linear duplexes. Energetically favorable interactions with other molecules, particularly proteins, can induce additional conformations that do not result in major changes in either base pairing or stacking. Several conformations are believed to play important roles in different situations. Bends are known to be important structures formed by chromosomes (discussed later in this chapter). Cruciforms in which a single chain folds back on itself into a hairpin-like duplex are important as intermediates in DNA and RNA synthesis (see Chapters 20, 21, 26, and 27).

Supercoiled DNA is a common type of tertiary structure observed in DNAs that are topologically constrained by being covalently closed and circular or by being complexed to proteins so that the ends of the DNA cannot rotate freely. DNA can form right-handed (negatively supercoiled) or left-handed (positively supercoiled) supercoils (Figure 18-20). *Negative supercoiling imparts a torsional stress to the DNA that favors unwinding, whereas positive supercoiling favors tighter winding of the double helix.* Supercoiling imparts a more compact structure to a circular duplex, which makes it sediment more rapidly in a centrifuge. Thus either a positively or a negatively supercoiled DNA will sediment more rapidly than a relaxed circular duplex. J. Vinograd and coworkers demonstrated this by adding ethidium bromide (Figure 18-21) to a small circular phage PM2 DNA that has about 40 negative supercoiled turns when it is isolated from cells. Addition of ethidium bromide to duplex DNA leads to a type of binding known as *intercalation*. The ethidium binds between two base pairs. Normally the base pairs are nearly closely packed and this would not be possible. In order to accommodate the ethidium, the duplex must unwind by about −27 degrees per base pair. This unwinding reduces the negative supercoiling in naturally occurring circular duplexes such as PM2 DNA. As more ethidium is added, enough unwinding takes place to eliminate all the supercoiling. At this point the DNA has its slowest sedimentation rate (Figure 18-22). Addition of further ethidium causes the DNA to adopt a positively supercoiled form, which again increases the sedimentation rate.

Supercoiling of a circular duplex DNA is quantitatively considered in terms of the *linking number* ($L$), an integer which specifies the number of times two strands are intertwined. *The linking number cannot change as long as no covalent linkages are broken*. If a molecule of DNA is projected onto a two-dimensional surface, the linking number is defined as the excess of right-handed over left-handed crossings of one strand over the other. DNA in solution adopts a conformation close to the B form, with 10 base pairs per turn (most experts say this number should be 10.4 instead of 10). A closed circular duplex with this twist is presumed to be under no torsional strain and is said to be relaxed. Because B DNA is a right-handed helix, the linking number is normally positive. The linking number of relaxed DNA, $L°$, will be distributed over a narrow range of integral values centered around 1/10 per base pair. DNA species with a mean linking number smaller than this are termed *negatively supercoiled*, or *underwound*; DNA with a larger linking number is termed *positively supercoiled*, or *overwound*. The deviation of the linking number from its relaxed value, $\Delta L = (L - L°)$, can be partitioned between *twist* (altered pitch of the helix) and *writhe* (supercoiling):

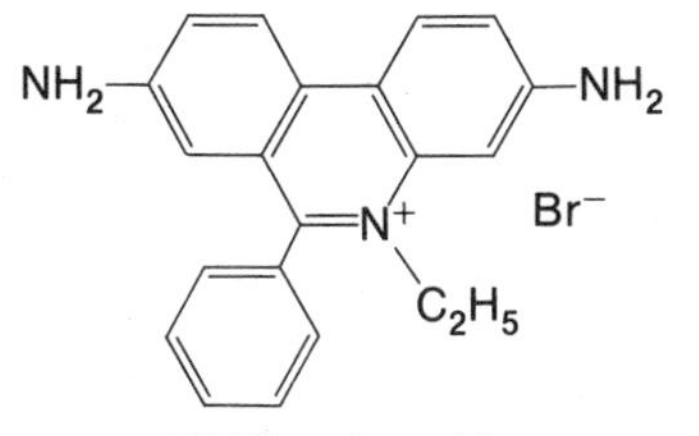

$$\Delta L = \text{twist}(T) + \text{writhe}(W)$$

At the present time it is not known precisely how $L$ will partition between twist and writhe for helices under torsional stress. This can best be explained by considering the hypothetical situation illustrated in Figure 18-23. A 360 base-pair structure in the circular relaxed form ($L = 36$, $T = 36$, and $W = 0$) is indicated to the left. Exposure to the bacterial enzyme DNA gyrase would introduce negative supercoils into such a structure in an ATP-dependent reaction (see Chapter 20). If four negative supertwists are introduced, $L$ will be reduced to 32. Barring other changes, $T$ remains fixed and $W$ becomes $-4$. The torsional strain introduced by the negative supertwists will tend to reduce $T$, i.e., to cause a partial unwinding of the duplex. At one extreme, this could lead to the unwinding of four helical turns, in which case the supercoiling would disappear entirely ($L = 32$, $T = 32$, and $W = 0$). In fact, the actual situation would probably lead to a reduction in the negative value of $W$ and a concomitant reduction in $T$ that would be spread over the entire duplex without any localized total unwinding as pictured. Note that the linkage number $L$ only changes when covalent bonds are broken, such as in the case of gyrase treatment.

A convenient way of expressing the degree of supercoiling that is independent of molecular length is given by the specific linking difference $\Delta L/L°$. For DNA with a molecular weight of less than $10^7$, agarose gel electrophoresis is a most effective method for assessing the extent of supercoiling. In a suitable range of $\Delta L/L°$, DNA isomers differing by 1 in linking number form separate bands in the gel (Figure 18-24). The more highly supercoiled molecule will migrate more rapidly through the gel as a result of its more compact structure. For the same reason supertwisted DNA molecules sediment faster than relaxed circular molecules.

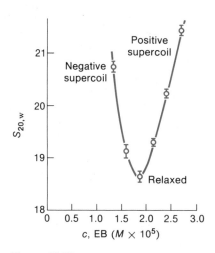

**Ethidium bromide**

**Figure 18-21**
The structure of ethidium bromide.

**Figure 18-22**
The sedimentation coefficient of closed circular phage PM2 DNA in 2.85 *M* CsCl containing varying amounts of ethidium bromide (EB). (Adapted from J. Vinograd, *Nature [New Biol.]* 229:10, 1971.)

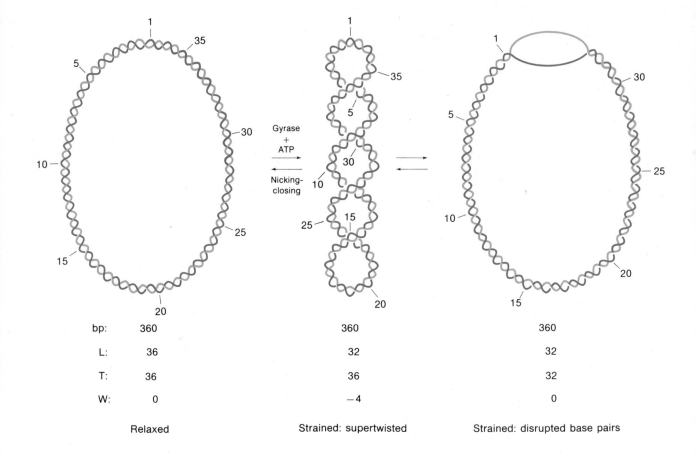

| | | | |
|---|---|---|---|
| bp: | 360 | 360 | 360 |
| L: | 36 | 32 | 32 |
| T: | 36 | 36 | 32 |
| W: | 0 | −4 | 0 |

| Relaxed | Strained: supertwisted | Strained: disrupted base pairs |
|---|---|---|

**Figure 18-23**
A circular duplex molecule in different topological states. (Adapted from a diagram supplied by M. Gellert; see text for explanation.)

The excess free energy of a highly supercoiled DNA is appreciable. It has been found to be closely proportional to the square of the linking difference. Per unit length of DNA, this can be expressed as

$$\Delta G/L° = k(\Delta L/L°)^2$$

where $k$ is a constant independent of length. For DNA with $\Delta L/L° = -0.06$, a typical value for natural DNA species, such as the PM2 DNA discussed earlier, an increase of one unit in $L$ (toward the relaxed state) is favored by about 9 kcal/mol. Correspondingly, the unwinding of one turn of duplex DNA (e.g., by a bound protein) also would be favored by 9 kcal/mol relative to the unwinding of a nonsupercoiled DNA. Clearly, supercoiling will greatly influence the equilibrium state of such reactions. Initiation of DNA synthesis for certain chromosomes (see Chapter 20) and initiation of RNA transcription for certain genes (see Chapter 26) has been found to be strongly dependent on negative supercoiling. In both types of initiation processes, unwinding of at least one turn of the double helix has been postulated.

Enzymes called _topoisomerases_ that relax positively or negatively supercoiled DNA as well as topoisomerases that generate negatively supercoiled DNA are discussed in Chapter 20. Topoisomerases that catalyze negative supercoiling only occur in bacteria. Consistent with this it is found that all duplex DNA that is topologically constrained, such as circular DNA, is negatively supercoiled in bacteria. Circular DNA (such as SV40 viral DNA; see Chapter 28) isolated from eukaryotic cells is frequently found to be negatively supercoiled. In such instances, the DNA is not supercoiled in cells. Rather the supercoiling results from removal of the histone proteins that normally are bound to the DNA. The structure of DNA-histone complexes is discussed below.

## Denaturation and Renaturation

Disruption of the secondary structure of DNA is referred to as _denaturation_. The reassembly of two separated polynucleotide strands in perfect register is referred to as _renaturation_ or hybridization if the DNA strands originate from different sources or if a DNA strand base pairs with a complementary RNA strand. Renaturation was first discovered using transforming DNA as a biological assay. The basis of the assay was that denatured _Diplococcus pneumoniae_ DNA is inactive in DNA-mediated transformation, whereas native DNA is active in transferring genetic traits such as streptomycin resistance from a donor to a recipient strain. When transforming DNA was heated and rapidly cooled it was biologically inactive; however when denatured single-stranded DNA was slowly cooled a large fraction of the initial transforming activity was recovered.

A decrease in the UV absorbance of purines and pyrimidines (hypochromic shift) occurs when the bases stack in a duplex structure. This gross change in absorbancy can be used to measure the transition between denatured and native DNA. The transitions between native and denatured DNA may be most conveniently measured spectrophotometrically in a thermostated cell, which allows one to determine the absorbancy in the ultraviolet region of a solution of DNA at any designated temperature. When a solution of native DNA is heated, the absorption at the ultraviolet maximum (~260 nm) remains constant until an elevated temperature is reached (in the region of 80°C, the exact point depending on the base composition of the DNA), at which point the absorbance increases sharply over a narrow temperature range by about 40 percent. This is followed by a plateau region (see curve I in Figure 18-25). The rise in absorbance coincides with the separation of the strands; after a rise of about 40 percent, the strands of linear DNA molecules are completely separated. The $T_m$ of the DNA is the temperature at the midpoint of the absorbancy increase. If the solution is rapidly cooled, the absorbance decreases, but only by about three-fourths of the total original increase, and the decrease occurs over a much broader

**Figure 18-24**
Gel electrophoretic separation of circular DNA according to its linking number. DNA is visualized by ethidium bromide staining and ultraviolet photography. Covalently closed circular DNA can be converted from a relaxed form to a highly supercoiled form by the enzyme DNA gyrase, which is inhibited by the drug novobiocin (Chapter 20). Channel (a) intracellularly supercoiled ColE1 DNA; (b) relaxed ColE1 DNA; (c–h) relaxed ColE1 DNA incubated with DNA gyrase from a normal strain and varying concentrations of novobiocin; (i–n) relaxed ColE1 DNA incubated with DNA gyrase purified from a novobiocin-resistant strain. Novobiocin concentrations were (c) and (i), none; (d) and (j), 0.3 mg/ml; (c) and (k), 1 mg/ml; (f) and (l), 1.3 mg/ml; (g) and (m), 10 mg/ml; and (h) and (n), 30 mg/ml. (Adapted from M. Gellert, M. W. O'Dea, T. Itoh, and J. Tomizawa, Novobiocin and courmermycin inhibit DNA supercoiling catalyzed by DNA gyrase, _Proc. Natl. Acad. Sci. USA_ 73:4474, 1976.)

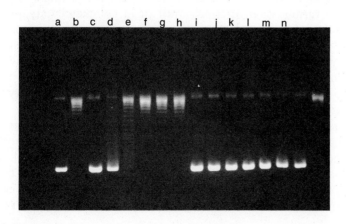

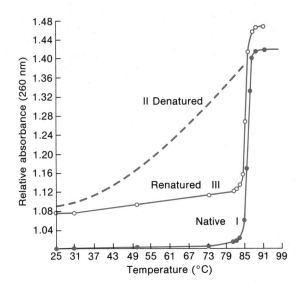

**Figure 18-25**
Effect of temperature on the relative absorbance of native, renatured and denatured DNA. (Adapted from P. Doty, in D. J. Bell and J. K. Grant (Eds.), *The Structure and Biosynthesis of Macromolecules,* Biochemical Society Symposium No. 21, Cambridge Univ. Press, New York, 1962, p. 8.)

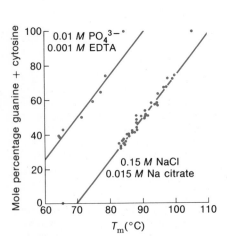

**Figure 18-26**
Dependence of the temperature midpoint $T_m$ of DNA on the content of guanine and cytosine. (Adapted from P. Doty, in D. J. Bell and J. K. Grant (Eds.), *The Structure and Biosynthesis of Macromolecules,* Biochemical Society Symposium No. 21, Cambridge Univ. Press, New York, 1962.)

range of temperatures, as shown in Figure 18-25. On subsequent heating and recooling, the absorbance follows the original cooling curve, indicating that denaturation has resulted in an irreversible change. If the temperature of the denatured DNA solution is maintained at 25°C below the $T_m$ for a long period of time, there is a gradual decrease in the absorbance of the solution. A further heating cycle of the material shows in a partial or almost complete return (if the complementary strands are matched in length) to the original denaturation profile (see curve II in Figure 18-25), indicative of a restoration of the native structure. The explanation for what has happened is as follows. Heating of native DNA does not break many hydrogen bonds until the sharp rise in absorbancy begins. The breakage of the secondary bonds is abrupt and parallels the rise in absorbancy. The sharpness of the disruption of the regularly hydrogen-bonded cooperatively base paired native structure may be likened to the "melting" of a pure organic compound. Subsequent fast cooling leads to re-formation of hydrogen bonds, but in an irregular manner, as indicated by the broad melting curve of denatured DNA (see curve II in Figure 18-25). The optimum "annealing temperature" is ideal for the disruption of an irregular H-bonded structure, but below the temperature at which the regular double helix is unstable. Prolonged exposure at the annealing temperature allows the bases to explore various configurations until small complementary regions or nuclei are formed between otherwise separated strands. After nuclei are formed, "zippering up" of the molecule occurs rapidly to generate a duplex helix that can take place in a concentration-independent manner. Covalently closed circular DNA molecules or DNA with cross links renature readily even when heated and fast cooled since the complementary strands do not separate from one another.

There are many interesting aspects to denaturation and renaturation. The midpoint of denaturation ($T_m$) is precisely correlated with the average base composition of the DNA; the higher the GC/AT base ratio, the higher the $T_m$ (see Figure 18-26). This seems reasonable since the GC base pair contains three H bonds instead of the two contained by the AT base pair (see Figure 18-9), and DNA with a greater GC content would be expected to result in greater stability. Base stacking is also believed to contribute to the stability of the duplex structure. In general, the interaction energy gained by stacking between adjacent GC base pairs is greater than that gained by interaction between AT base pairs (see Chapter 7). The denaturation temperature is also a function of the ionic strength. This is due to the repulsive electrostatic interaction between the negatively charged phosphate groups. Salt shields charge interaction (see Chapter 7), which results in a stabilization of the duplex structure. DNA in 0.15 $M$ NaCl denatures at a $T_m$ about 20°C higher than DNA in 0.01 $M$ phosphate (see Figure 18-26). In pure water with no salt present DNA denatures at room temperature. Extremes of pH also may be used to disrupt the double-helix structure. Thus, pHs above 11.5 or below 2.3 lead to extensive deprotonization or protonization of the bases (p$K$s listed in Table 18-1) and consequent breakage of the H-bonded structure. Alkali is an excellent DNA denaturant since it does not degrade the separated strands. Exposure of RNA to alkali degrades it to mononucleotides. A variety of microligands and macroligands lower the denaturation temperature by forming complexes with the purines and the pyrimidines (see Chapter 20).

Any small molecule that can form H bonds with a nucleic acid base or interfere with more generalized stacking interactions is a potential denaturant. The denaturant effectiveness of microligands can be evaluated by placing a solution of DNA in a UV spectrophotometer thermostated at about 15°C below the thermal denaturation temperature of the DNA and monitoring absorbancy at 260 nm as a function of denaturant concentration. In general, high concentrations of microligand are required to denature duplex DNA in this way (see Table 18-4). Experimental procedures for separation of RNA mixtures or denatured DNA on sucrose gradients or by electrophoresis often specify addition of reagents such as urea or formamide to prevent aggregation. Appropriate concentrations of denaturants are also frequently used in DNA-DNA annealing studies or in DNA-RNA hybridization studies to permit the use of lower temperatures.

In order to be an effective DNA denaturant, small ligands need to be present at high concentration to exert a competitive advantage against the bonds that stabilize the double helix. The limited solubility of macrosolutes serves to restrict the potential effectiveness of most potential macroligands. If, however, a macroligand has other chemical features that enhance binding to DNA, such as a high proportion of basic amino acids or a site-specific affinity, then a high local concentration of interacting, competing functional groups can be achieved. Many proteins that interact with DNA but do not denature it produce compact structures in chromosome condensation (see below) and DNA packaging in viruses. Local denaturation is known to be induced by RNA polymerase when it binds to DNA prior to transcription (Chapter 21) and by other proteins that are involved in DNA replication (Chapter 21). Proteins that encourage denaturation of duplex DNA bind preferentially to single-stranded DNA. Gene 32 of bacteriophage T4 encodes a DNA binding protein essential in bacteriophage T4 DNA replication. This protein induces the local unwinding of the DNA duplex that is thought to facilitate DNA replication (Chapter 20). The action of gene 32 protein can be demonstrated in vitro by observing denaturation of DNA in a thermostated spectrophotometer, as described for microligands. In this case, low salt concentrations are used to reduce the melting temperature of the DNA, so that protein denaturation can be

**Table 18-4**
DNA Denaturing Reagents

| Compound | $M$* |
| --- | --- |
| Methanol | 3.5 |
| Ethanol | 1.2 |
| Formamide | 1.9 |
| Dimethyl formamide | 0.6 |
| Urea | 1.0 |
| Urethane | 0.50 |
| Cyanoguanidine | 0.21 |

*Molar concentration required to produce 50 percent denaturation of bacteriophage T7 DNA at 73°C, ionic strength 0.043.

avoided. Addition of appropriate amounts of gene 32 protein leads to complete denaturation of individual DNA molecules. This is due to the fact that gene 32 protein exhibits cooperative binding; binding of one gene 32 protein molecule to DNA facilitates binding of a second gene 32 molecule at a neighboring site and continued binding at adjacent sites. A large number of DNA-binding proteins similar to the gene 32 protein have been found in both prokaryotic and eukaryotic cells. Gene 5 protein, encoded by the related filamentous phages M13 and fd of *E. coli*, is another helix-destabilizing protein known to be involved in DNA replication. Tyrosine and phenylalanine residues in gene 5 protein are thought to be essential in this interaction with DNA, suggesting that interaction with the aromatic purines and pyrimidines is an important component in the binding. Single-strand binding protein (SSB) is essential for replication of *E. coli* chromosome or bacteriophage $\phi$X174 in vitro (see Chapter 20).

Not all DNAs occur naturally in the double-helix form. For instance, DNA of the *E. coli* bacteriophage $\phi$X174 and fd each exist as a single circular polynucleotide strand. The ultraviolet adsorption heat curve for the DNA of these phages is similar to that of denatured DNA. Broad melting curves also are characteristic of most RNAs, which rarely have regions of regular base pairing that extend for more than 10 or 20 residues.

The extent of denaturation or renaturation as a function of time may be measured spectrophotometrically or in other ways that distinguish between single- and double-stranded DNA. One of these involves digestion of the nucleic acid with a nuclease that hydrolyzes single-stranded but not double-stranded DNA. For this purpose, the $S_1$ nuclease derived from the mold *Aspergillus* is commonly used. The amount of DNA resistant to digestion by this nuclease gives an accurate measure of the amount of duplex structure. A more commonly used method involves binding to a column made from a calcium phosphate gel known as hydroxyapatite. For reasons that are still not completely understood, duplex DNA binds to hydroxyapatite under salt conditions where single-stranded DNA does not. This method must be used with caution as a quantitative measure of renaturation, because a DNA molecule that is partially duplex and partially single-stranded also will bind to hydroxyapatite. Hydroxyapatite chromatography is useful also as a preparative technique for separating rapidly reassociating and slowly reassociating fragments (discussed below).

### Renaturation Rate and Sequence Heterogeneity

The renaturation rate of DNA has been used extensively to study DNA sequence heterogeneity. For a given concentration of DNA, the more homogeneous a DNA sample, the more rapidly it will renature. Thus denatured T4 bacteriophage DNA renatures much more rapidly than does

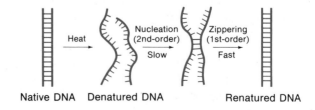

**Figure 18-27**
Steps in denaturation and renaturation of a DNA duplex.

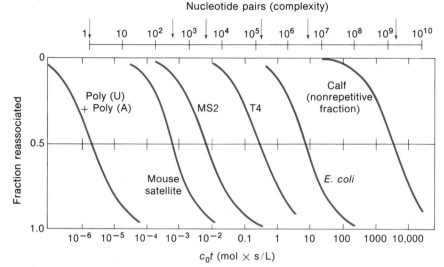

**Figure 18-28**
Reassociation of double-stranded nucleic acids from various sources. The genome size is indicated by the arrows near the upper nomographic scale. Over a factor of $10^9$, this value is proportional to the cot required for half reaction. The DNA was sheared and the other nucleic acids are reported to have approximately the same fragment size (about 400 nucleotides, single-stranded). Correction has been made to give the rate that would be observed at 0.18 $M$ sodium ion concentration. No correction for temperature has been applied since it was approximately optimum in all cases. (Adapted from R. J. Britten and D. E. Kohn, *Science* 161:529, 1968.)

*E. coli* DNA, and *E. coli* DNA renatures much more rapidly than mammalian DNA. The reason for this is explained below.

The steps involved in denaturation and renaturation are indicated in Figure 18-27. *Renaturation may be thought of as a two-step process involving first nucleation and second zippering. The first step is believed to be rate-limiting, and since it involves interaction between two molecules, it should occur at a rate proportional to the square of the DNA concentration.*

If $c$ is the concentration of single-stranded DNA at time $t$, then the second-order rate equation for a loss of single-stranded DNA is

$$\frac{-dc}{dt} = k_2 c^2$$

where $k$ is the second-order rate constant. Starting with a concentration $c_0$ of completely denatured DNA, the amount of single-stranded DNA left at time $t$ after starting renaturation is given by

$$\frac{c}{c_0} = \frac{1}{1 + k_2 c_0 t}$$

where $k_2$ is the bimolecular rate constant.

At half renaturation, $c/c_0 = 0.5$ and $t = t_{1/2}$, from which it follows that

$$c_0 t_{1/2} = \frac{1}{k_2}$$

As a rule $c/c_0$ is plotted as a function of $c_0 t$ (referred to as a *cot curve*) on a semilogarithmic plot, as shown in Figure 18-28. Data are shown for DNAs and RNAs with varying *nucleotide pair complexity N, which is defined as the number of base pairs in a nonrepeating helical structure*. It can be seen that $cot_{1/2}$ is proportional to $N$ for these samples. This family of curves has been used as a calibration curve for more complex situations discussed below. For a given nucleotide complexity, the renaturation rate is also a function of the length $L$, i.e., the actual number of bases per single-stranded nucleic acid molecule present in a renaturation mixture. It can be shown that

$$k_2 \propto \frac{L^{0.5}}{N}$$

and therefore $cot_{1/2} \propto N/L^{0.5}$. In comparing DNAs of different initial length, the effect of length is usually eliminated by shearing the DNA of a sample to a more or less uniform length, so that the $cot_{1/2}$ may be used directly as a measure of $N$ when comparing different samples.

Simple monophasic cot curves are obtained when most prokaryotic DNAs are examined (Figure 18-28). In such instances, $N$ is directly proportional to the size of the genome. However, when the total DNA from a complex eukaryotic organism such as human is analyzed, the curves are more complex. A precise interpretation of the cot curve for human DNA, shown in Figure 18-29, is not possible, but the following interpretation has been made. About 2 percent of the DNA renatures very rapidly. This is due most likely to foldback DNA and could result from single-stranded DNA that renatures by forming a hairpin duplex structure. It will include those DNA species which have *inverted repeating sequences called palindromes*. The next class of DNA sequences to renature (labeled fast) accounts for about 5 percent of the total and has a $cot_{1/2}$ value of about $10^{-2}$. The low amount of DNA together with the rapid reassociation kinetics suggest that about 5 percent of the nuclear DNA is present in many copies per genome (*highly repetitive* DNA). In the same figure, a zone labeled *intermediate* can be seen that accounts for an additional 20 percent of the DNA (*middle repetitive* DNA). About 70 percent of the DNA has reassociation kinetics in the range expected for a single-copy or unique DNA, i.e., DNA whose sequence is only present as a single copy per haploid genome. This type of renaturation kinetic analysis has been used for the gross characterization of DNA in many cell types.

Bacterial DNA contains very little repetitive DNA (0.3 percent). This is mainly accounted for by the eight genes for ribosomal RNA that have nearly identical sequences. In eukaryotes, single-copy DNA accounts for 40 to 70 percent of the DNA, most of the remainder being somewhat arbitrarily divided between middle repetitive ($<10^4$ copies per genome) and highly repetitive ($>5 \times 10^4$ copies per genome). Further analyses of eukaryotic gene structure by a variety of other techniques have shown that most structural genes that encode proteins belong to the unique class. The middle repetitive class encodes transfer RNA genes and ribosomal RNA genes as well as the structural genes for histones, the main chromosomal proteins. Highly repetitive DNA sequences frequently appear as spacer DNA between structural genes.

Some highly repetitive sequences that occur in tandem, referred to as *satellite sequences*, are separable from the main nuclear DNA by cesium chloride density-gradient centrifugation. The same principle used to detect satellite DNA was used in the Meselson-Stahl experiment to separate DNAs according to their density difference (see Figure 20-1). In this instance, the density difference results from the significantly different base composition of the satellite DNA from the bulk nuclear DNA. Mouse satellite DNA, which comprises about 10 percent of the nuclear DNA, has been purified and characterized. Treatment of high-molecular-weight mouse satellite DNA with the restriction enzyme Eco R2, which breaks the DNA at a

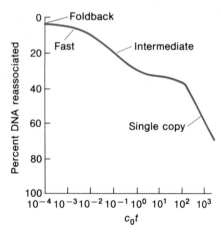

**Figure 18-29**
The cot curves for total human nuclear DNA. (Adapted from unpublished data of A. R. Mitchell.)

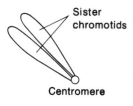

Sister chromotids

Centromere

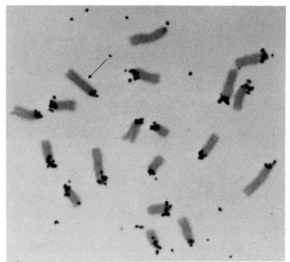

specific pentanucleotide sequence, produces a low-molecular-weight homogeneous fragment that contains about 250 base pairs. This shows that this satellite DNA is composed of a tandem repeating sequence of this length.

The location of the mouse satellite DNA in chromosomes has been determined by a technique known as in situ hybridization. First labeled RNA is transcribed from the purified satellite DNA by in vitro synthesis using radioactive substrates. Next a chromosome preparation is fixed on a glass slide and the DNA is denatured by brief alkali treatment; the preparation is annealed with radioactive RNA (radioactive DNA representing the same sequences would also be effective in such an experiment). The chromosome preparation is finally autoradiographed to determine what part of the chromosome is radioactive. The result (Figure 18-30) clearly demonstrates that only the DNA in the immediate vicinity of the centromere becomes labeled, and therefore, only this region of the chromosomes contains satellite sequences. Similar experiments have shown that some satellite DNAs from a variety of plants and animals behave in a similar fashion; i.e., they bind in the vicinity of the centromere. Many satellite DNA do not code for any proteins, which is not surprising in view of its short repetitious sequence. It seems likely that the function of the main satellite DNAs is structural and has something to do with centromere formation or function.

**Figure 18-30**
Autoradiograph of a mouse metaphase cell hybridized with radioactive cRNA copied in vitro from mouse satellite DNA. In condensed chromosomes, each can be seen to have a very narrow region where the spindle fibers attach, the centromere. These are instrumental in separating the chromosomes during mitosis. In different chromosomes, the centromere may be located at different sites along the chromosome. Mice are unusual in that their centromeres are attached to one end of the chromosome. They are described as telomeric for this region (schematic diagram to the left). (Data obtained from M. L. Pardue and J. G. Gall.)

## NUCLEOPROTEIN COMPLEXES

All nucleic acids form complexes with proteins which are essential to their function. Viruses are frequently encapsulated in their mature form by a protein shell that serves a protective function and often facilitates the infectious process (e.g., see Chapters 27 and 28). In some viruses, specific proteins covalently linked to the nucleic acids play a crucial role in viral nucleic acid replication (e.g., see adenovirus in Chapter 28). Ribosomal RNA forms an intimate complex with about 50 different proteins; the final structure of the ribosome is determined by the complex interaction between the RNA and the proteins (see Chapter 24). Eukaryotic chromosomes are composed of DNA, basic histone proteins (discussed below), and some other nonhistone proteins about which little is known. The primary function of the histone is probably to stabilize the DNA in a compact form without interfering unduly with DNA or RNA synthesis. They may play a more positive role in DNA or RNA synthesis but it is too early to say. Regulatory proteins form sequence-specific complexes that either facilitate or inhibit DNA or RNA expression (for examples, see below and Chapters 26, 27, and 28). Single-stranded binding proteins facilitate DNA unwinding by a preferential binding to single-stranded DNA (see Chapter 20).

Little is known about the precise groups that are responsible for the formation of stable complexes between nucleic acids and proteins, but a few generalizations can be made. Many nucleic acid-binding proteins have an excess of basic amino acid residues, and the complexes they make are dissociated by high salt. In such cases, salt linkages formed between positively charged basic amino acid side chains and the nucleic acid phosphate groups are a major stabilizing factor. The core in a duplex structure presents certain groupings that could interact to form either H bonds or hydrophobic bonds with the appropriate amino acid side chains on proteins. Single-stranded nucleic acids obviously present more possibilities for interaction. For example, aromatic amino acid side chain interactions with purines or pyrimidines could be an important component in the affinity between DNA and single-stranded binding proteins.

## Interaction between DNA and Regulatory Proteins

The precise modes of protein binding to double-stranded DNA have not yet been visualized by cocrystallization of the proteins with fragments of appropriate DNA, but some features of the interaction have been inferred from recent x-ray crystallographic studies (Figure 18-31) of the *cro repressor* of bacteriophage λ (Chapter 27) and the *catabolite gene activator protein* (CAP) of *E. coli* (Chapter 26).

The *cro* repressor is a small protein containing only 66 amino acid residues. In a 2.8-Å resolution x-ray study, the polypeptide chain is seen to be organized in a three-stranded segment of antiparallel β-sheet structure, together with three α helices and a C-terminal tail. The *cro* repressor is known from methylation studies (the use of chemical methylation to determine binding site is discussed in Chapters 20, 21, and 26) to bind to a 17 base-pair length of DNA in such a way as to prevent purine alkylation in the large groove (such as guanine N-7), but to permit purine alkylation in the narrow groove (such as adenine N-3). The suggestion is that the *cro* helix extending from amino acid residue 27 to residue 36 binds to the large

**Figure 18-31**
Schematic drawings comparing the backbone conformations in the presumed DNA-binding domain of CAP (*left*) and cro repressor (*right*). In each case, dimers of the protein are depicted as seen along their respective twofold symmetry axes with the presumed DNA-binding α helices (*F* in CAP and $\alpha_3$ in cro) toward the viewer. The difference in the tilt of these α helices is apparent. The tilt shown by cro protein would enable the $\alpha_3$ cylinders to fit into adjacent large grooves of a right-handed DNA duplex; the tilt shown by the *F* α helices in CAP would be most adaptable to binding in the adjacent large grooves of a comparable left-handed DNA duplex. (Illustration generously donated by T. A. Steitz, D. H. Ohlendorf, B. B. McKay, W. F. Anderson, and B. W. Matthews.)

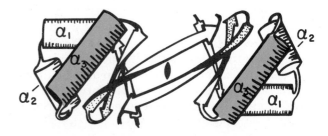

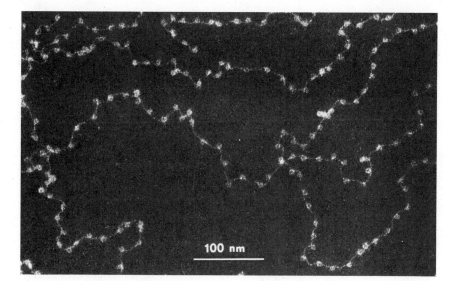

**Figure 18-32**
Swollen fibers of chromatin from the nucleus of the chicken red blood cell. The electron micrograph is enlarged about 325,000 × and negatively stained with uranyl acetate. (Micrograph generously donated by A. I. Olins and D. E. Olins.)

groove of B DNA (see Figure 18-16 for large and narrow grooves of B DNA). The *cro* molecule exists as a dimer and the symmetry-related α helix could bind to the next part of the major groove 3.4 nm along the DNA molecule. The CAP protein has an analogous structure: a dimeric form with two protruding α helices. However, these helices are oriented so that they would need to bind to a left-handed DNA duplex with a 3.4-nm pitch. Whether this prediction is accurate or not will have to be checked by other experiments that explore the DNA-protein structure directly. Additional properties of DNA-regulatory protein complexes are discussed in Chapters 26 and 27.

## Nucleohistone and Chromatin

DNA in eukaryotic chromosomes exists in a highly compacted form known as *nucleohistone*, a complex of DNA with an approximately equal weight of five identifiable proteins known collectively as *histones*. Since the same kind of compaction occurs with duplex DNA of almost any sequence, it is presumed that a repeating aspect of the structure is being recognized by the protein.

Electron microscopic and x-ray diffraction studies on compacted chromatin have led to the proposal that chromatin exists as a coiled-coil structure. This information has been scant, and some of the x-ray spacings observed still require explanation. In the 1950s, M. H. F. Wilkins, a codiscoverer of the DNA structure, once said when looking at some of the electron micrographs of chromatin, "It looks like scrambled eggs. How do you explain the structure of scrambled eggs?" Fortunately, the situation proved to be not quite so complex, but it is fair to say that further understanding of the molecular structure of chromatin has tested the ingenuity of interested investigators to the limit, necessitating the development of new approaches. A breakthrough came when D. E. Olins and A. L. Olins observed that chromatin viewed after sudden swelling in water showed a beaded structure (Figure 18-32). The beads, called *nucleosomes*, contain most of the histone; they are about 10 nm in diameter, and the spacing between the beads is about 14 nm. Brief enzymatic digestion of chromatin with micrococcal nuclease or some other endonucleases fragments this structure. The DNA from this partial digestion gives rise to a banded pattern on agarose gel electrophoresis that indicates nucleoprotein structures containing 200 base pairs of DNA or multiples thereof (400, 600, 800 bp, etc.). The products of partial micrococcal-nuclease digestion were fractionated by ultracentrifugation. Examination of the fractions in the electron microscope shows a direct correlation between the size of the DNA estimated on gels and the number of nucleosomes. Thus the most rapidly moving DNA band seen on gels was derived from a structure containing one nucleosome, and the second fastest migrating species contains nucleosome dimers, etc. Evidently, the brief treatment with endonuclease preferentially cleaves DNA in the internucleosomal region, where the DNA is least likely to be protected from enzyme attack. More exhaustive nuclease treatment gives rise to a single band on gels that contain a single nucleosome and 140 base pairs of DNA. The effect of the more exhaustive nuclease treatment is to degrade all the DNA that is not in direct contact with the nucleosome. Combined electrophoresis and electron microscopy results have led to a model for the beaded structure. In this model the nucleosome core particles contain clumps of histone complexed to 140 base pairs of DNA duplex, with about 60 base pairs of duplex serving as linkers between the core particles.

The histones present in chromatin have been extensively characterized, and their sequences are known. How they are classified and their molecular weights are listed in Table 18-5. The lysine-rich histone H1 is not present in the nucleosome core particle, as evidenced by its release on extensive nuclease treatment. Consistent with this is the finding that H1 is the only histone that readily exchanges between free and chromatin-bound histone. The other eight histones, two each of the four histones, form a tightly complexed core particle that is conserved when chromosomes duplicate.

Many fine points about the structure of chromatin remain to be determined. However, it is clear that we have come a long way from Wilkins' "scrambled eggs." An illustration of the probable DNA coiled-coil structure of the nucleosome core particle is presented in Figure 18-33. There are 140 base pairs of DNA in a nuclease-resistant _nucleosome core_ that make about

**Table 18-5**
Characteristics of Histones

| Name | Ratio of Lysine to Arginine | $M_r$ | Copies per Nucleosome |
|---|---|---|---|
| Histone H1 | 20 | 21,000 | 1 (not in bead) |
| Histone H2a | 1.2 | 14,500 | 2 (in bead) |
| Histone H2b | 2.5 | 13,700 | 2 (in bead) |
| Histone H3 | 0.7 | 15,300 | 2 (in bead) |
| Histone H4 | 0.8 | 11,300 | 2 (in bead) |

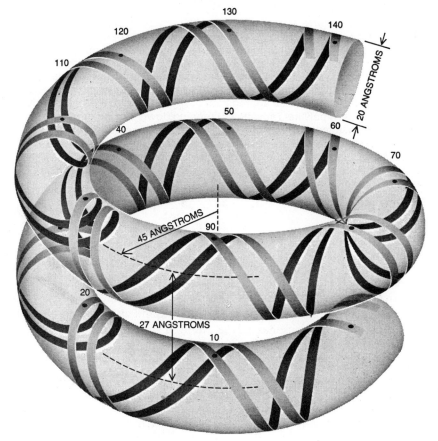

**Figure 18-33**
Path of DNA that can account for the bipartite structure of the nucleosome core is a super helix with an external diameter of 110 Å and a pitch of 27 Å; the turns of the 20-Å-wide DNA helix are nearly in contact. There are about 80 nucleotide pairs of DNA per turn; the nucleosome core, an enzymatically reduced form of the nucleosome consisting of some 140 nucleotide pairs, has about one and three-quarter turns wrapped on it. The histone octamer complex containing two each of histones H2a, H2b, H3, and H4 is packed on the inside of the DNA coiled-coil structure. (From R. D. Kornberg and A. Klug, *The Nucleosome, Sci. Am.*, February 1981, p. 59. Used with permission.)

1.75 superhelical turns about a histone octamer. An additional 60 base pairs of spacer DNA connect the core particles. Electron microscopic observations indicate that the amount of linker DNA is actually different in different species and in different tissues of the same species, varying from about 20 to 95 base pairs.

Nuclear magnetic resonance spectroscopy indicates that nucleosomal DNA has a secondary structure quite similar to B DNA. The DNA is wound around the histone complex. Treatment with pancreatic DNAse I (instead of with micrococcal nuclease) produces single-stranded fragments which are multiples of about 10 nucleotides. The current interpretation of this observation is that the digestion by DNAse I is confined to the exposed side of the DNA in the nucleosome core (see Figure 18-33), the approach of the enzyme to the other side being hindered by the presence of the histones.

Salt bridges between positively charged basic amino acid side chains of histones and the DNA phosphates play an important role in stabilizing the DNA-histone complex. It was proposed over 20 years ago that the main interaction of histone with DNA involves an α-helix component of the protein fitting into the large groove. This aspect of the structure is remarkably similar to the proposed binding between *cro* protein and DNA dis-

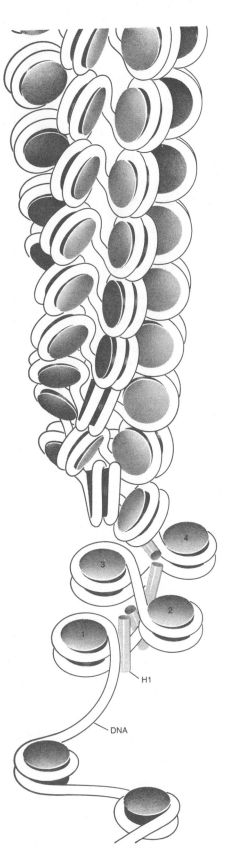

**Figure 18-34**
Helical superstructures might be formed with increasing salt concentration (*bottom to top*) as is suggested here. The zigzag pattern of nucleosome (1,2,3,4) closes up, eventually to form a solenoid, a helix with about six nucleosomes per turn. (The helix is probably more irregular than it is in this drawing.) Cross-linking data indicate that HI molecules on adjacent nucleosomes make contact. Extrapolation from the zigzag form to the solenoid suggests (but does not prove) that the aggregation of HI at higher ionic strengths gives rise to a helical HI polymer (not shown) running down the center of the solenoid. In the absence of HI (*bottom*) no ordered structures are formed. The details of HI associations are not known at this time; the drawing is meant to indicate only that HI molecules contact one another and linker DNA. (From R. D. Kornberg and A. Klug, *The Nucleosome, Sci. Am.*, February 1981, p. 64. Used with permission.)

cussed earlier; further structural information on cocrystals of protein and DNA are needed to evaluate both proposals.

Higher-order structures beyond that of the nucleosome itself have been inferred from light, x-ray, and neutron scattering and visualized by electron microscopy. The nucleosome can be arranged in a zigzag fashion to produce a *fibril* that is 10 nm wide in which the nucleosomes are packed edge to edge rather than face to face. When the ionic strength is raised, the fibrils reversibly condense into an irregular supercoiled *fiber* about 30 nm in diameter. The nucleosome particles in this fiber are thought to have their cylindrical axes approximately perpendicular to the long axis of the fiber with six to seven nucleosomes per turn (Figure 18-34).

## STRUCTURAL PROPERTIES OF RNA

The differences in primary structure between DNA and RNA were discussed earlier. RNA, like DNA, is a polymer of four different nucleotides. In this sense it has the same potential for carrying genetic information as DNA. Nevertheless, cellular genomes are invariably composed of DNA, and it has been argued that this is due to the greater chemical stability of a chain lacking the C-2′-OH group on the pyranose. Despite this, the genomes of a number of viruses in both prokaryotes and eukaryotes are composed of RNA. In *E. coli*, all known RNA phages exist as single strands in the mature form. Double stranded phages have been isolated from other bacteria such as *Pseudomonas*. In plants and animals, RNA viruses of the single-strand or duplex types have been found (e.g., see Chapter 28).

Single-stranded viruses are all believed to form partial duplex intermediates during replication. DNA-RNA hybrid duplexes are also well known; they are intermediates in retrovirus DNA synthesis (see Chapters 20 and 28) and they are presumed to be intermediates (containing about 10 base pairs) in all reactions involving DNA-directed RNA synthesis. Furthermore, DNA-RNA hybrids are made extensively in the laboratory for analytical purposes (see Chapters 20 and 21). Watson-Crick base pairing is universal in all natural duplex structures involving RNA; a conformation of the A

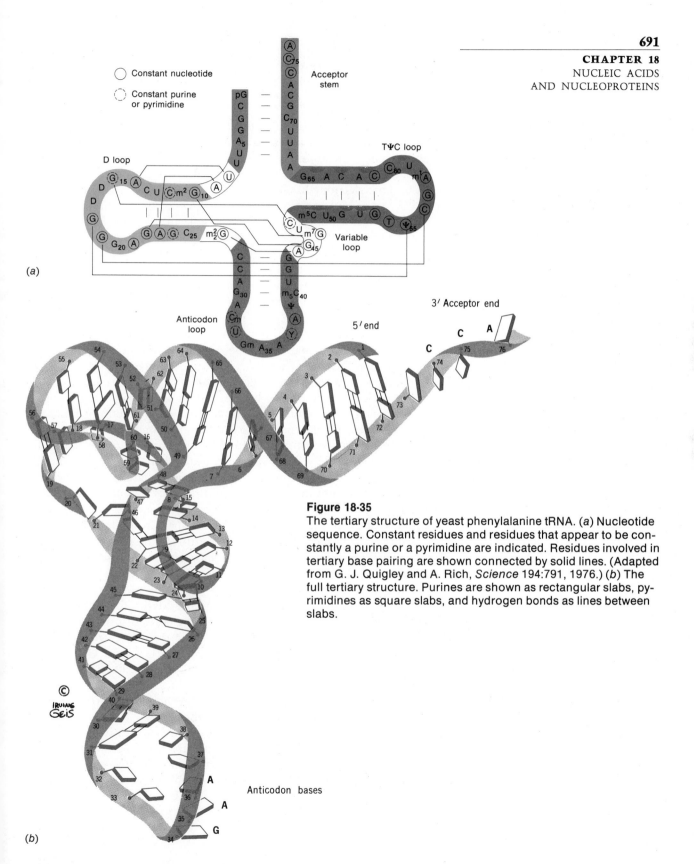

**Figure 18-35**
The tertiary structure of yeast phenylalanine tRNA. (*a*) Nucleotide sequence. Constant residues and residues that appear to be constantly a purine or a pyrimidine are indicated. Residues involved in tertiary base pairing are shown connected by solid lines. (Adapted from G. J. Quigley and A. Rich, *Science* 194:791, 1976.) (*b*) The full tertiary structure. Purines are shown as rectangular slabs, pyrimidines as square slabs, and hydrogen bonds as lines between slabs.

type is favored, as stated earlier, because of the preference of ribose for the C-3′-endo conformation (Figure 18-15).

A wide range of nongenetic RNA exists in the cell that functions in messenger RNA processing and protein synthesis (see Chapters 21, 24, and 25). Many aspects of RNA function are taken up in subsequent chapters. In this section, the structural properties of transfer RNA (tRNA) will be discussed. Transfer RNA carries activated amino acids to the messenger RNA template, where the amino acids are ordered and linked to form polypeptide chains. The complete three-dimensional structure of the yeast transfer RNA for phenylalanine has been determined by x-ray crystallography. The primary, secondary, and tertiary structures of phenylalanine tRNA, tRNA$^{Phe}$, are shown in Figure 18-35. This molecule contains 76 nucleotides in a single chain with four loops. Twelve of the bases are abnormal, resulting from posttranscriptional modification (see Chapter 21 for a discussion of these).

Most of the bases are involved in a complex secondary and tertiary structure. The four loops in the molecule (Figure 18-35) are referred to as the T$\psi$C loop, the anticodon loop, the D loop, and the variable loop. The variable loop is so named because it contains a different number of bases in different tRNAs. The remainder of the molecule (with the exception of the D loop in some instances) contains the same number of residues in all tRNAs, even though there is considerable sequence variability. The tRNA$^{Phe}$ molecule contains 20 base pairs that are H bonded in Watson-Crick fashion. It contains an additional 40 or so H bonds, most of which are not of the Watson-Crick type. These additional H bonds and the accompanying base stacking stabilize the tRNA in the complex folded structure shown in Figure 18-35b. Some of the unusual H-bonded interactions involved in stabilizing the complex tertiary structure are shown in Figure 18-36. These structures involve two to four residues from different regions of the RNA. Not only are the H-bonding patterns unusual, but in some cases phosphate and the C-2′-OH group participate in the H bonding. It is believed that the tertiary structures of most tRNAs are very similar except for the variable loop even though tRNAs show considerable sequence variability.

The importance of base stacking in stabilizing regular nucleic acid structures was stressed earlier. Despite the irregularity of the H bonding found in tRNA, base stacking, assessed by a close scrutiny of the three-dimensional structure, is a prominent feature of the tRNA structure. Only four of the 76 bases in the molecule (D$^{16}$, D$^{17}$, U$^{47}$, and G$^{20}$) do not participate in stacking. The D$^{16}$ and D$^{17}$ dihydrouracil residues cannot stack under any conditions, because these residues no longer have a planar configuration. It is believed that stacking and H bonding contribute about equally to the conformational stability of the tRNA.

The detailed structural investigations on tRNA greatly broaden our expectations in considering the less well understood structures formed by other nuclear and cytoplasmic RNAs. This single example illustrates the fact that nucleic acids with a properly adjusted primary sequence can adopt complex secondary and tertiary structures. Little is known in detail about the tertiary structures of most other RNAs.

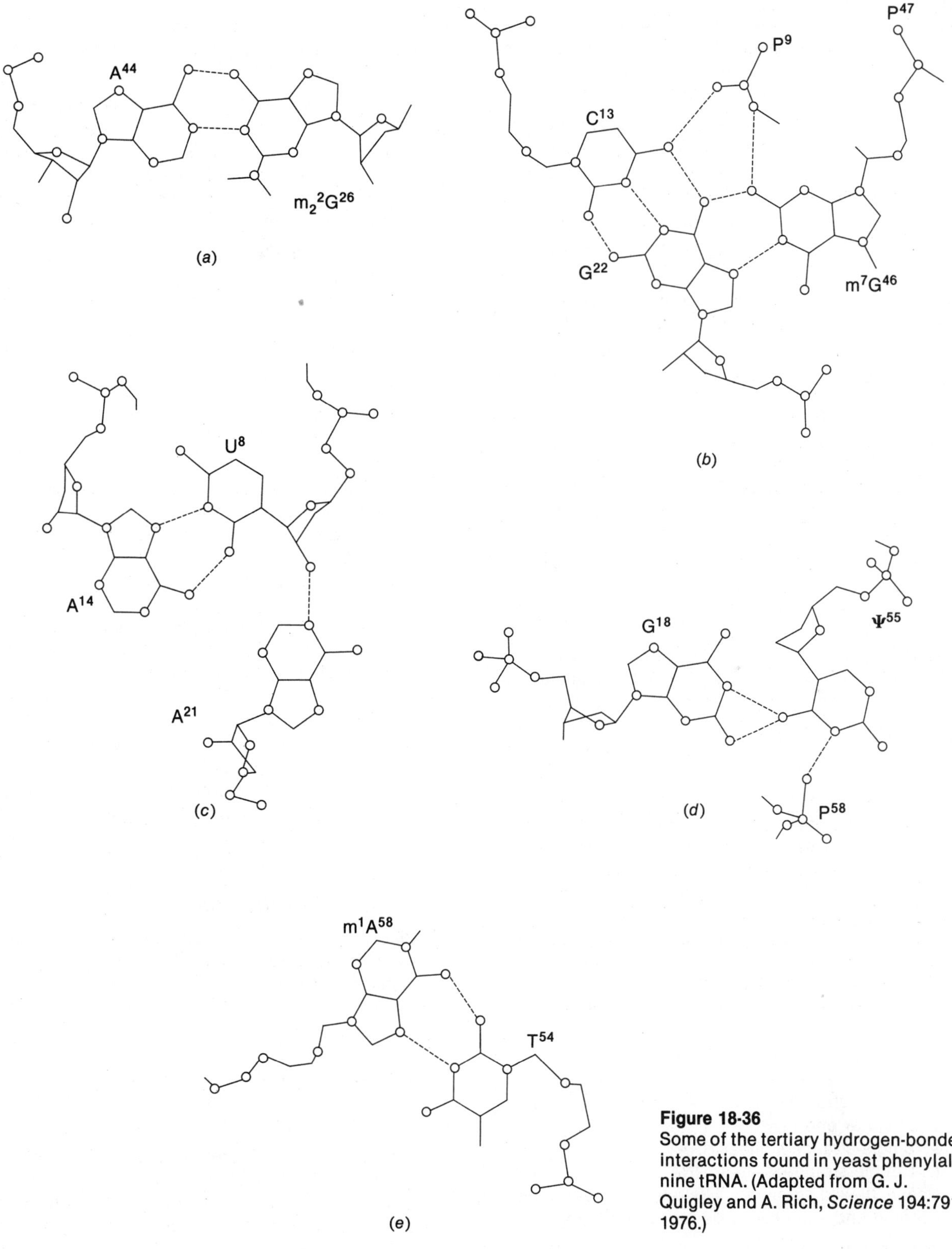

(a)

(b)

(c)

(d)

(e)

**Figure 18-36**
Some of the tertiary hydrogen-bonded interactions found in yeast phenylalanine tRNA. (Adapted from G. J. Quigley and A. Rich, *Science* 194:791, 1976.)

## SELECTED READINGS

Cantor, C. R., and Schimmel, P. R. *Biophysical Chemistry,* Part I: *The Conformation of Biological Macromolecules.* San Francisco: Freeman, 1980.

Kornberg, R. D., and Klug, A. The nucleosome. *Sci. Am.* 244:52, 1981. An excellent review of nucleohistone structure.

*Movable Genetic Elements. Cold Spring Harbor Symposium on Quantitative Biology XLV.* New York: Cold Spring Harbor, 1981. Includes many timely papers on nucleic acid structure.

Steitz, T. A., Ohlendorf, D. H., McKay, D. B., Anderson, W. F., and Mathews, B. W. Structural similarity in the DNA binding domains of catabolite gene activa-tor and cro repressor proteins. *Proc. Natl. Acad. Sci. USA* 79:3097, 1982.

Stephenson, E. C., Erba, H. P., and Gall, J. G. Histone gene clusters of the newt *Notophthalmus* are separated by long tracts of satellite DNA. *Cell* 24:639, 1981. An example of a recent paper that uses several techniques to determine the arrangement of repetitive sequence in eukaryotic DNA.

Wang, A. H. J., Quigley, G. J., Kolpak, F. J., Crawford, J. L., vanBloom, J. H., van der Marel, G., and Rich, R. Molecular structure of a left-handed double helical DNA fragment at atomic resolution. *Nature* 282:680, 1979.

## PROBLEMS

1. Why does linear duplex DNA separate into its constituent strands when the DNA is put into pure water?

2. Which of the following lowers the $T_m$ of duplex DNA? Explain their mode of action.

   SSB, histone, NaCl, formamide, and alkali

3. Linear duplex DNA can bind more ethidium bromide than closed covalent circular DNA of the same molecular weight. Suggest an explanation for this. How might advantage be taken of this to separate the two types of DNA on a CsCl gradient? Hint: The ethidium cation has a density lower than water.

4. You are given a mixture of the following nucleic acid species, all in the same container, and all approximately of the same molecular weight. How would you separate the species from one another (assuming that your separation scheme is a very efficient one, with a logical basis) and specifically identify each species. Assume that each nucleic acid species is derived from a phage or virus and is homogeneous in size and composition.
   a. Circular phage, φX174 DNA (isolated from the phage)
   b. Double-stranded, linear phage DNA
   c. Supercoiled DNA
   d. Double-stranded, linear DNA with hairpin loops at either end
   e. Single-stranded, linear RNA
   f. Double-stranded, linear RNA
   Use a flow diagram to illustrate your separation scheme and state the basis for the fractionation.

5. What is the structure of the following nucleic acid (structure A) based on the clues given below?

   a. It sediments as a single species in a neutral sucrose density gradient.
   b. In an alkaline sucrose gradient, one-half (structure B) of the mass of (A) sediments down the tube, the other half remains at the miniscus.
   c. The buoyant density of (A) is much higher than that of duplex DNA of the same G + C content.
   d. Whereas structure (A) gives a sharp thermal transition profile, (B)'s profile is broad.
   e. When (A) is heated and fast cooled it gives rise to (B) and (C) whose buoyant densities are quite different, that of (C) being heavier than (B). (B) and (C) when added together will renature when exposed to a temperature 20°C below the $T_m$ of (A). What does each clue signify? Diagram the structure of (A), (B), and (C). Hints: Alkali degrades RNA but not DNA. Single-stranded DNA is more dense than duplex DNA, and RNA is more dense than DNA.

6. Size is a critical parameter when separating DNAs by sucrose gradient sedimentation but not when separating DNAs by CsCl gradient sedimentation. Explain. Hint: CsCl gradient sedimentation is an equilibrium technique but sucrose gradient sedimentation is not. Some aspects of sedimentation are explained in Chapter 2 and the use of CsCl density gradient sedimentation for the separation of DNAs is explained in Chapter 20.

7. DNA-DNA duplexes are less stable than DNA-RNA duplexes. How might advantage be taken of this fact in the preferential formation of DNA-RNA hybrids in a mixture containing complementary single strands of DNA and RNA complementary to one of the DNA strands?

8. Renaturation of randomly sheared denatured DNA can be measured by hypochromic shift in the ultraviolet absorbance, $S_1$ nuclease resistance, or retention on hydroxyapatite. Which method is likely to give an overestimate for the extent of renaturation?

9. Give the relative times for 50 percent renaturation of the following pairs of denatured DNAs at the same concentrations.

   a. T4 DNA and *E. coli* DNA both sheared to an average single-strand length of 400 nucleotides.

   b. Unsheared T4 DNA and T4 DNA sheared as above.

10. When SV40 viral DNA is isolated as a nucleosomal complex from infected cells, the duplex DNA is in the relaxed circular conformation. If the histone is removed, the DNA becomes highly negatively supercoiled. Explain this observation. What does this observation suggest about the packaging of DNA into nucleosomal complexes?

695

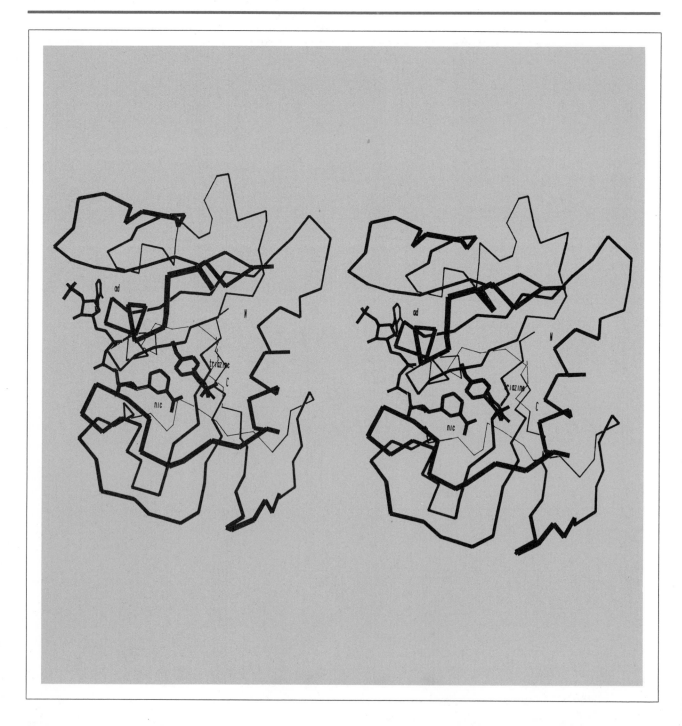

# 19

# NUCLEOTIDE METABOLISM

This chapter deals with the biosynthesis of ribonucleotides and deoxyribonucleotides, their role in metabolic processes, and the pathways for their degradation. The biosynthesis of nucleotides is a vital process for all cells, since cells are unable to take up nucleotides from the surrounding medium even though nucleotides are indispensable precursors for the synthesis of both RNA and DNA. Without RNA synthesis, protein synthesis is halted, and unless DNA can be synthesized, the cell cannot divide.

It is not surprising that inhibitors of nucleotide biosynthesis are very toxic to cells, and as described in this chapter, advantage has been taken of this cytotoxicity in the use of some of these inhibitors in the treatment of cancer, while others are effective in the treatment of diseases resulting from infections by bacteria or protozoa.

Nucleotides play important roles in all major aspects of metabolism. ATP, an adenine nucleotide, is the major substance used by all organisms for the transfer of chemical energy from energy-yielding reactions to energy-requiring reactions such as biosynthesis. Other nucleotides are activated intermediates in the synthesis of carbohydrates, lipids, proteins, and nucleic acids. Adenine nucleotides are components of many major coenzymes, such as $NAD^+$, $NADP^+$, FAD, and CoA. The critical role played by nucleotides as regulators of metabolism in both prokaryotic and eukaryotic organisms also will be described briefly in this chapter.

Finally, pathways for the metabolic degradation of nucleotides are also very important to the organism, as demonstrated by the fact that several genetic defects causing blocks in these pathways have serious consequences for the health of the organism.

*Stereo diagram of dihydrofolate reductase backbone with NADPH and a triazene inhibitor bound in the catalytic site. (Courtesy Dr. David Mathews.)*

# SYNTHESIS OF RIBOSE AND OTHER PENTOSES

The central role of glucose-6-phosphate in carbohydrate metabolism and energy generation has been emphasized in Chapters 8, 10, and 12. The oxidative decarboxylation of glucose-6-phosphate with the formation of ribose phosphate and other pentoses is another important way in which this intermediate is used in animals and a wide variety of microorganisms. The series of reactions involved are known by various names: the pentose phosphate pathway, the hexose monophosphate pathway, the pentose shunt, or the phosphogluconate pathway. Note that in higher plants and other photosynthetic organisms, the Calvin cycle of reactions (Chapter 11), by which pentoses and $CO_2$ are converted to hexoses, represents a kind of modified pentose phosphate pathway.

The pentose phosphate pathway serves several important roles, which have been discussed in Chapter 8. Of particular concern here is its role in the synthesis of pentoses, which are essential for nucleotide synthesis and (in many organisms) for histidine biosynthesis.

The reactions of the first part of the phosphogluconate pathway, from glucose-6-phosphate to ribose-5-phosphate, are shown in Figure 19-1. In the first step, which is freely reversible, the hydroxyl group at C-1 of the pyranose ring of glucose-6-phosphate is oxidized by a dehydrogenase utilizing $NADP^+$. This produces a carbonyl group at C-1, and since it is attached to the ring oxygen, the structure is an intramolecular ester or lactone: 6-phospho-D-gluconolactone. This δ-lactone is very unstable and hydrolyzes spontaneously, but the process is accelerated by a specific lactonase. The overall equilibrium for these first two steps lies far in the direction of 6-phosphogluconate formation. In the third step, the 3-OH group of 6-phosphogluconate is oxidized to a carbonyl with simultaneous loss of C-1 as $CO_2$. The dehydrogenase requires $NADP^+$ and $Mn^{2+}$, and the reaction closely resembles the β-hydroxyacid oxidative decarboxylations catalyzed by malic enzyme and isocitrate dehydrogenase (Chapter 10). The carbonyl group becomes C-2 of the pentose produced, ribulose-5-phosphate, the name implying that it is the ketose analog of ribose-5-phosphate. For the three reactions, the overall equilibrium is far to the right. Finally ribose phosphate isomerase reversibly transforms ribulose-5-phosphate to ribose-5-phosphate (Figure 19-2). This product is the source of pentose for nucleotide and nucleic acid synthesis.

The second part of the pathway involves an intriguing set of rearrangements of the sugar phosphates, the result of which is the recovery of any excess pentose phosphates by conversion to hexose phosphates (Figure 19-2). This part of the pathway is used under conditions where the major function being served by the pathway is the provision of NADPH for use in the reductive steps of fatty acid and steroid biosynthesis, so that flux through the pathway far exceeds the requirement for pentose phosphate. Three enzymes are involved in this latter part of the pathway: ribulose-5-phosphate-3-epimerase, transketolase (which contains tightly bound thiamine diphosphate and $Mg^{2+}$), and transaldolase. In the transketolase reaction, a two-carbon unit, the glycoaldehyde group ($CH_2OH—CO—$), is transferred from the 2-keto sugar phosphate to the thiazolium ring of thiamine diphosphate to form the α,β-dihydroxyethyl derivative of the latter. This is analogous to the formation of the α-hydroxyethyl derivative of thiamine diphosphate during the action of pyruvate dehydrogenase de-

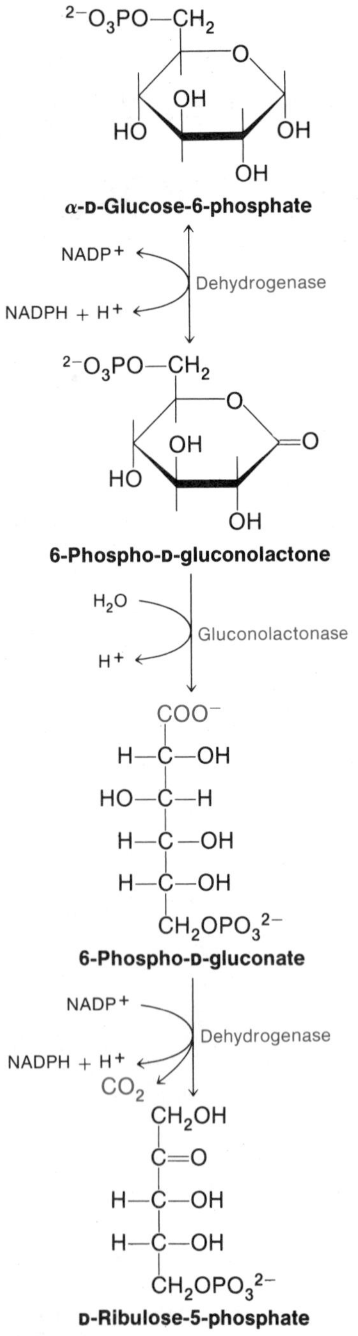

**Figure 19-1**
Stage 1 of the phosphogluconate pathway.

scribed earlier (Figure 9-3). The glycolaldehyde group is then transferred to the anomeric carbon of either ribose-5-phosphate or erythrose-4-phosphate. In the aldolase reaction, the dihydroxyacetone group (corresponding to carbons 1, 2, and 3 of sedoheptulose-7-phosphate) is transferred to glyceraldehyde phosphate to yield fructose-6-phosphate. The remainder of the sedoheptulose-7-phosphate molecule becomes erythrose-4-phosphate.

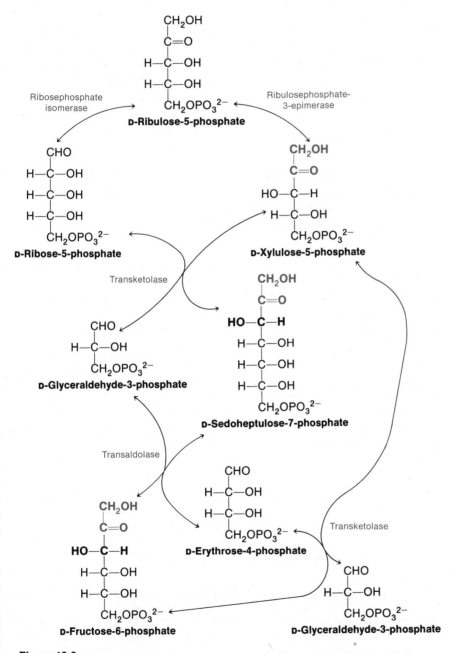

**Figure 19-2**
Stage 2 of the phosphogluconate pathway. The groups with color background are those transferred in the ketolase reactions. The groups in bold type are transferred in the aldose reaction.

The overall equation can be written

3 Ribulose-5-phosphate $\rightleftharpoons$

2 fructose-6-phosphate + glyceraldehyde-3-phosphate

Since two molecules of glyceraldehyde-3-phosphate can be converted to fructose-6-phosphate by the gluconeogenesis pathway (Chapter 12), six molecules of ribulose phosphate can yield five molecules of fructose-6-phosphate (and hence of glucose-6-phosphate) and one molecule of inorganic phosphate.

It should be noted that all the reactions of this second part of the pathway are reversible. It follows that ribulose-5-phosphate can be synthesized from the glycolytic intermediates fructose-6-phosphate and glyceraldehyde-3-phosphate, according to the reverse of the preceding equation, in any cell equipped with transaldolase, transketolase, ribose phosphate isomerase, and ribulose-5-phosphate-3-epimerase. This nonoxidative pathway for ribose phosphate formation from hexose phosphate is an alternate route to the phosphogluconate pathway, which is oxidative. The relative contributions of the oxidative and nonoxidative pathways in animal tissues is uncertain, but according to some estimates, 50 to 70 percent of the pentose is derived by means of this nonoxidative pathway. A further function of the nonoxidative pathway in microorganisms is the provision of erythrose-4-phosphate for aromatic amino acid biosynthesis and sedoheptulose-7-phosphate for cell wall biosynthesis.

### Synthesis of 5-Phospho-α-D-Ribosyl-1-Pyrophosphate (PRPP)

Formation of phosphoribosyl pyrophosphate from ribose-5-phosphate is an essential preliminary step for de novo biosynthesis of purine and pyrimidine nucleotides, for the conversion of purine bases to their nucleotides, and (in microorganisms) for the synthesis of histidine and tryptophan. The formation of PRPP from ribose-5-phosphate and ATP is catalyzed by ribose-5-phosphate pyrophosphokinase (Figure 19-3). This is an unusual kinase because the pyrophosphoryl group is transferred rather than the phosphoryl group. As might be anticipated for an enzyme with a key position in several biosynthetic pathways, its activity is regulated by a number of metabolites and agents. Inorganic phosphate is an activator, and the $Mg^{2+}$ ion is both cofactor and activator. Inhibitors include both ADP and 2,3-diphosphoglycerate, which are competitive with respect to ribose-5-phosphate, and the noncompetitive inhibitors AMP and GDP. At any specific time, the concentration of these agents as well as that of the substrates will determine the activity of the kinase. In particular, the nucleotides will curtail synthesis of PRPP when the energy stores of the cell are low.

## SYNTHESIS OF PURINE RIBONUCLEOTIDES

In early work, the ultimate precursors of the purine ring were established by administering isotopically labeled compounds to pigeons and determining the incorporation of labeled atoms into the purine ring of uric acid. Birds were used in these experiments because they excrete waste nitrogen largely as uric acid, a purine derivative that is easily isolated in pure form.

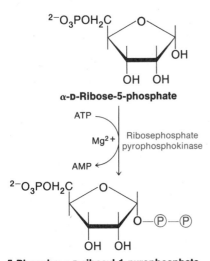

**α-D-Ribose-5-phosphate**

**Figure 19-3**
Synthesis of phosphoribosyl pyrophosphate (PRPP).

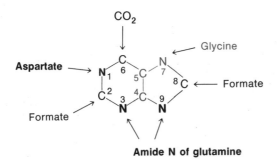

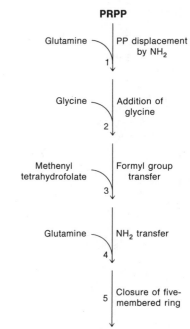

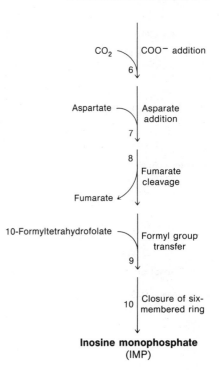

## Figure 19-4

Precursors of the purine ring of uric acid in pigeons as determined by isotope labeling experiments. The indicated precursor substances were administered one at a time to pigeons. Each precursor was labeled with isotopic nitrogen or carbon. Excreted uric acid was purified and degraded chemically. The isotope content of the various degradation products indicated that precursors contributed the specific atoms indicated.

Chemical degradation of the uric acid revealed the origins of the atoms as depicted in Figure 19-4. A flow scheme showing the successive incorporation of atoms from these precursors is shown in Figure 19-5.

## Synthesis of Inosine Monophosphate (IMP)

The pathway from PRPP to the first complete purine nucleotide, inosine monophosphate, involves 10 steps and is shown in Figure 19-6. Contrary to expectations that the purine would be built up first followed by addition of ribose-5-phosphate, the starting point is PRPP, to which the imidazole ring is added, followed by building up of the six-membered ring. The first step, in which phosphoribosylamine is formed, is catalyzed by glutamine phosphoribosyl pyrophosphate amidotransferase, an enzyme containing nonheme iron. The reaction involves inversion of configuration at C-1 of the ribose and leads to the $\beta$ configuration characteristic of naturally occurring nucleotides. This step involves commitment of PRPP to the purine biosynthetic pathway and, as might be expected, is subject to important feedback inhibitory effects by purine nucleotides, as discussed more fully below.

In step 2, an amide bond is formed by the synthase between the carboxyl group of glycine and the amino group of phosphoribosylamine, with ATP supplying energy and being hydrolyzed to ADP and inorganic phosphate. After these two steps have introduced atoms 4, 5, 7, and 9 of the purine ring, the remaining atoms are introduced one by one (steps 3, 4, 6, 7, and 9). Carbon 8 is introduced as a formyl group that is transferred from 5,10-methenyl-tetrahydrofolate (5,10-methenyl-tetrahydropteroylglutamate; 5,10-methenyl-$H_4$PteGlu). The role of tetrahydrofolate derivatives in accepting a one carbon unit from donors such as serine, glycine, or formate and transferring it to a suitable acceptor in biosynthetic reactions is discussed in Chapters 6 and 23. The five-membered (imidazole) ring is now ready for closure, but before this occurs in step 5, N-3 of the purine ring is introduced by transfer of another amide group from glutamine to phosphoribosyl formylglycinamide. ATP provides energy for the amido group transfer, being itself hydrolyzed to ADP and phosphate.

## Figure 19-5

Summary of incorporation of precursors into the purine ring of IMP. Formate as well as other indirect donors of one carbon, such as serine, donate their atoms by means of the methenyl and formyl groups of folate derivatives in steps 3 and 9.

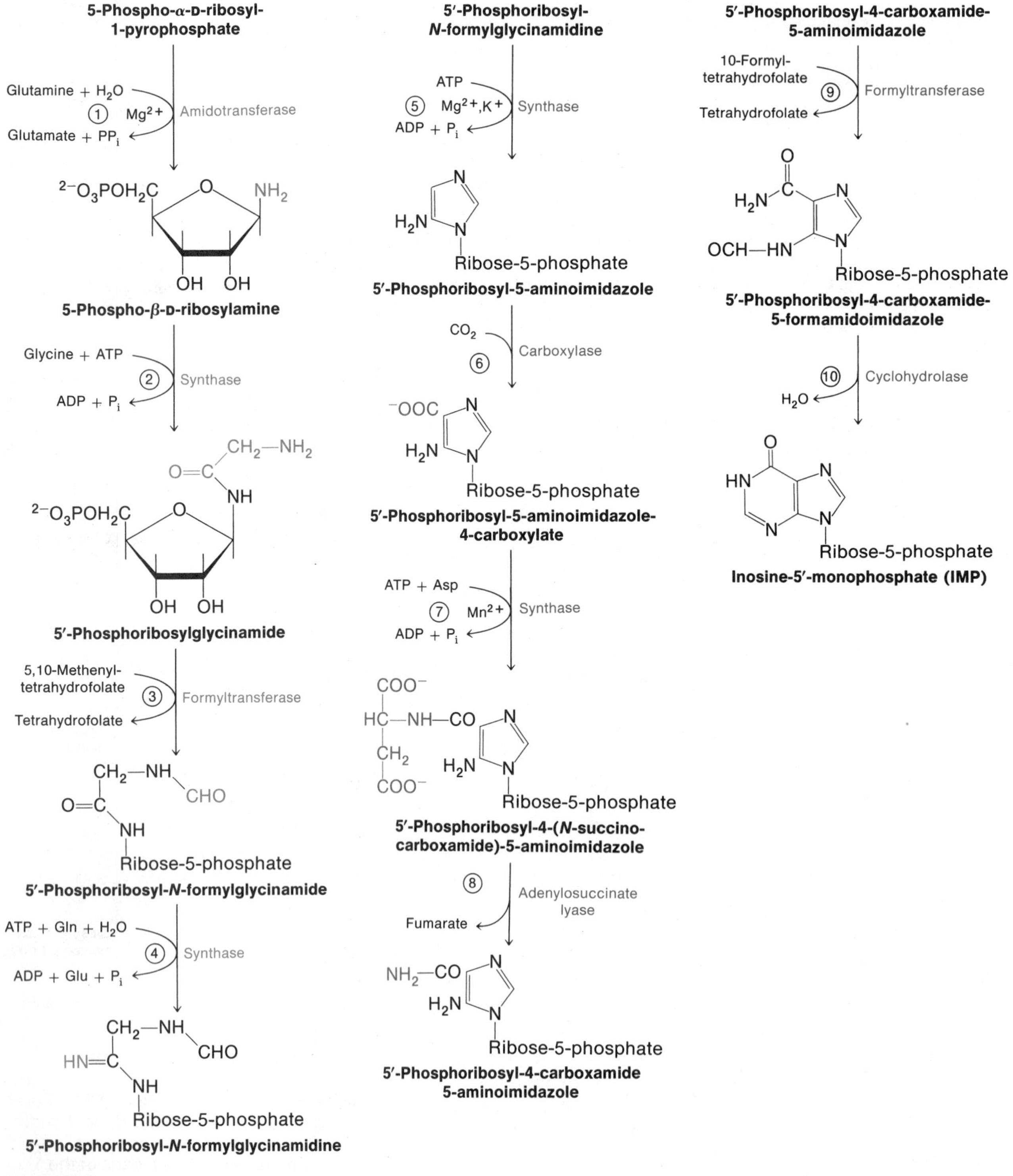

702

**5-Phospho-α-D-ribosyl-1-pyrophosphate**

Glutamine + H₂O
① Mg²⁺ | Amidotransferase
Glutamate + PPᵢ

$^{2-}O_3POH_2C$ — O — NH₂
OH  OH

**5-Phospho-β-D-ribosylamine**

Glycine + ATP
② Synthase
ADP + Pᵢ

CH₂—NH₂
O=C
NH
$^{2-}O_3POH_2C$ — O
OH  OH

**5'-Phosphoribosylglycinamide**

5,10-Methenyl-tetrahydrofolate
③ Formyltransferase
Tetrahydrofolate

CH₂—NH
            CHO
O=C
NH
Ribose-5-phosphate

**5'-Phosphoribosyl-N-formylglycinamide**

ATP + Gln + H₂O
④ Synthase
ADP + Glu + Pᵢ

CH₂—NH
            CHO
HN=C
NH
Ribose-5-phosphate

**5'-Phosphoribosyl-N-formylglycinamidine**

---

**5'-Phosphoribosyl-N-formylglycinamidine**

ATP
⑤ Mg²⁺, K⁺ | Synthase
ADP + Pᵢ

N
H₂N      N
Ribose-5-phosphate

**5'-Phosphoribosyl-5-aminoimidazole**

CO₂
⑥ Carboxylase

⁻OOC      N
H₂N      N
Ribose-5-phosphate

**5'-Phosphoribosyl-5-aminoimidazole-4-carboxylate**

ATP + Asp
⑦ Mn²⁺ | Synthase
ADP + Pᵢ

COO⁻
HC—NH—CO      N
CH₂
COO⁻    H₂N      N
Ribose-5-phosphate

**5'-Phosphoribosyl-4-(N-succino-carboxamide)-5-aminoimidazole**

⑧ Adenylosuccinate lyase
Fumarate

NH₂—CO      N
H₂N      N
Ribose-5-phosphate

**5'-Phosphoribosyl-4-carboxamide 5-aminoimidazole**

---

**5'-Phosphoribosyl-4-carboxamide-5-aminoimidazole**

10-Formyl-tetrahydrofolate
⑨ Formyltransferase
Tetrahydrofolate

O
H₂N—C      N
OCH—HN      N
Ribose-5-phosphate

**5'-Phosphoribosyl-4-carboxamide-5-formamidoimidazole**

⑩ Cyclohydrolase
H₂O

O
HN      N
N      N
Ribose-5-phosphate

**Inosine-5'-monophosphate (IMP)**

---

**Figure 19-6**
Biosynthetic pathway to inosine monophosphate. Color indicates the group or atom introduced at each step.

After closure of the imidazole ring in an essentially irreversible cyclization requiring the presence of $Mg^{2+}$ and $K^+$, C-6 of the purine ring is introduced by addition of bicarbonate in the presence of a specific carboxylase. This carboxylation is unusual in that it does not seem to involve biotin and is not coupled with any energy-yielding process such as ATP hydrolysis. As might be expected, the equilibrium is unfavorable for the formation of the carboxylate. In vivo the reaction proceeds at physiological concentrations of bicarbonate because of coupling with subsequent steps that are thermodynamically favorable. (Enzyme reactions are coupled if the product of one is the substrate for the next.) Next, in steps 7 and 8, N-7 of the purine ring is contributed by aspartate. Aspartate forms an amide with the 4-carboxyl group, and the succinocarboxamide so formed is then cleaved with release of fumarate. Energy for carboxamide formation is provided by ATP hydrolysis to ADP and phosphate. These reactions resemble the conversion of citrulline to arginine in the urea cycle (Chapter 23) and the conversion of IMP to AMP (see Figure 19-8). It is worth noting that fumarate released in all these synthetic pathways can be converted to oxaloacetate by fumarase and malate dehydrogenase, and in the process, $NAD^+$ is reduced to NADH (by malate dehydrogenase). Reoxidation of the NADH by the electron-transport chain then generates three ATP from ADP and $P_i$. This supplies some of the energy required for purine synthesis, and the oxaloacetate formed can be used to replenish the supply of aspartate by transamination with glutamate (Figure 19-7).

The final atom of the purine ring is provided in step 9 by donation of a formyl group from 10-formyltetrahydrofolate (10-formyltetrahydropteroyl-glutamate or 10-formyl-$H_4$PteGlu) to the 5-amino group of the almost completed ribonucleotide. In the final step, ring closure is effected by elimination of water to form IMP (inosine monophosphate), the first product with a complete purine ring. Although this final ring closure does not require energy from ATP, closure of the imidazole ring does, and the synthesis of IMP from ribose-5-phosphate requires a total of six high-energy phosphate groups from ATP (assuming hydrolysis of pyrophosphate released during phosphoribosylamine synthesis; step 1 of Figure 19-6).

## Synthesis of Ribonucleotides of Adenine and Guanine

IMP does not accumulate in the cell, but is converted to AMP, GMP, and the corresponding diphosphates and triphosphates. The two steps of the pathway from IMP to AMP (Figure 19-8) are typical reactions by which the amino group from aspartate is introduced into a product. The 6-hydroxyl group of IMP (tautomeric with the 6-keto group) is first displaced by the amino of aspartate to give adenylosuccinate, and the latter is then cleaved nonhydrolytically by adenylosuccinate lyase to yield fumarate and AMP. In the condensation of aspartate with IMP, cleavage of GTP to GDP and phosphate provides energy to drive the reaction.

Conversion of IMP to GMP also proceeds by a two-step pathway: first, dehydrogenation of IMP to xanthosine-5'-phosphate (XMP), and second, transfer of an amide group from glutamine to C-2 of the xanthine ring to yield GMP. The second reaction also involves the cleavage of ATP to AMP and inorganic pyrophosphate. The latter is in turn hydrolyzed to inorganic phosphate by the ubiquitous inorganic pyrophosphatase in a reaction with

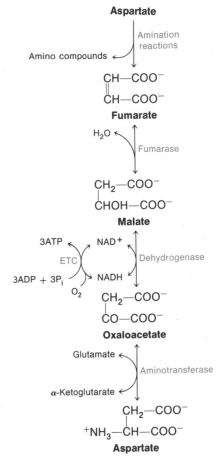

**Figure 19-7**
Utilization of fumarate formed in reactions in which aspartate acts as donor of an amino group to generate ATP and regenerate aspartate (ETC = electron-transport chain).

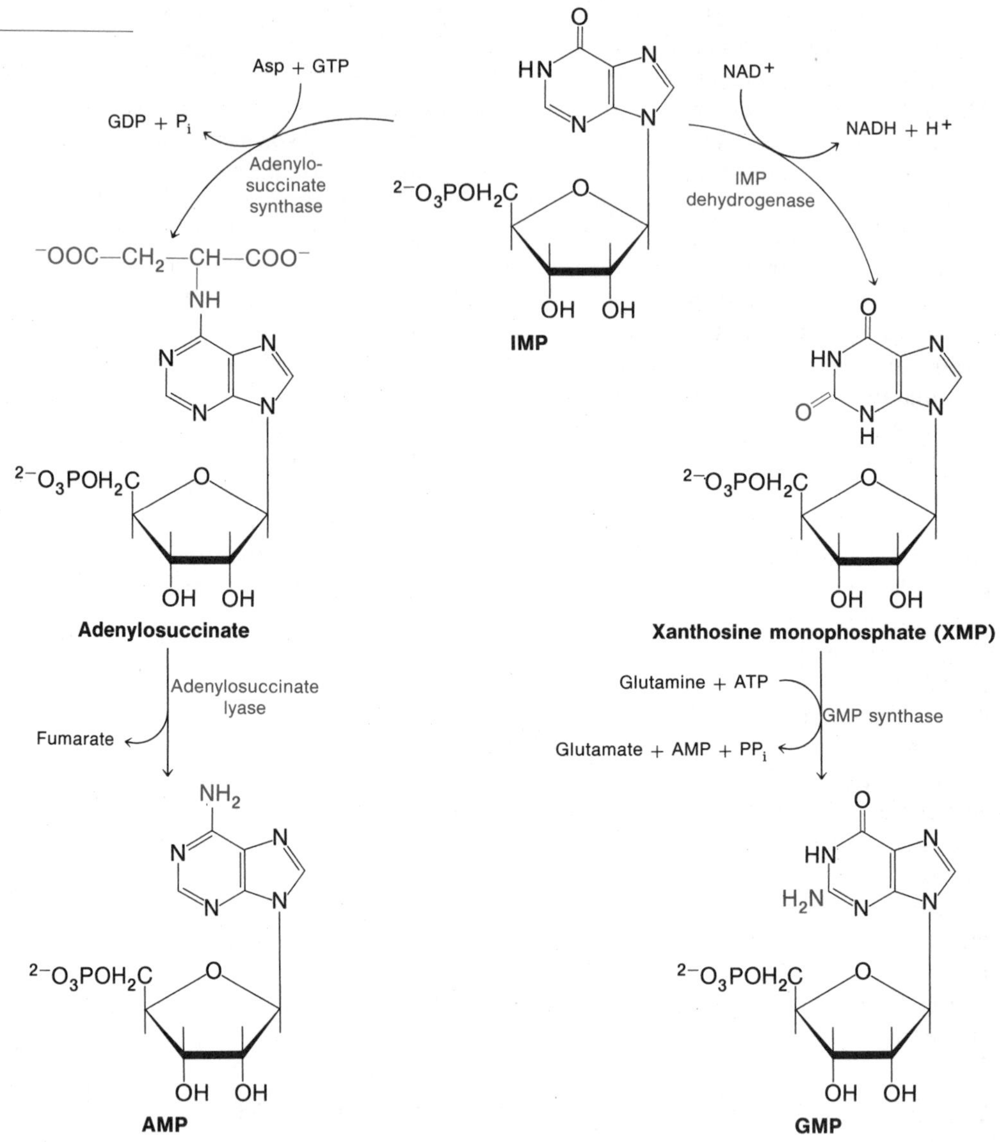

**Figure 19-8**
Conversion of IMP to AMP and GMP.

a very favorable equilibrium. This hydrolysis is coupled with the GMP synthase reaction because pyrophosphate is a product of the latter and a substrate of the pyrophosphatase. The net result is that the release of two high-energy phosphate groups is used to drive the GMP synthase reaction to completion.

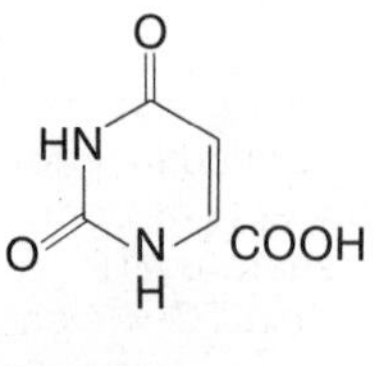

## SYNTHESIS OF PYRIMIDINE RIBONUCLEOTIDES

The biosynthetic pathway to pyrimidine nucleotides is simpler than that for purine nucleotides, reflecting the simpler structure of the base. *In contrast to the biosynthetic pathway for purine nucleotides, in the pyrimidine pathway, the pyrimidine ring is first constructed before ribose-5-phos-*

**Figure 19-9**
Orotic acid.

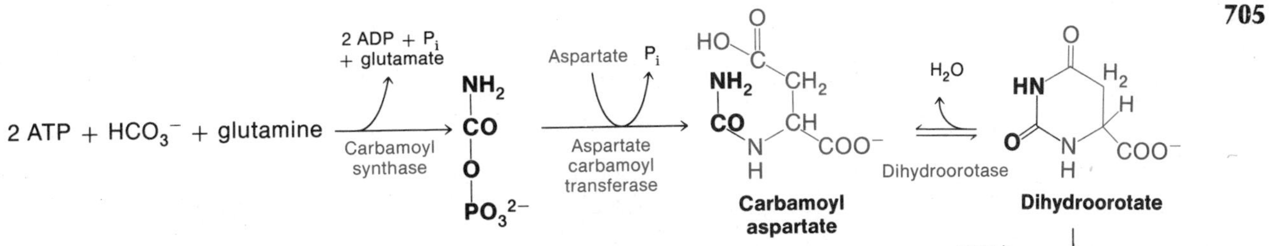

$$2\ ATP + HCO_3^- + glutamine \xrightarrow[\text{Carbamoyl synthase}]{2\ ADP + P_i + glutamate} \underset{PO_3^{2-}}{\overset{NH_2}{\underset{|}{\overset{|}{CO}}}\,\overset{}{\underset{}{O}}} \xrightarrow[\text{Aspartate carbamoyl transferase}]{Aspartate \quad P_i} $$

**Carbamoyl aspartate**

**Dihydroorotate**

Dihydroorotase $\quad H_2O$

Dihydroorotate dehydrogenase $\quad NAD^+ \rightarrow NADH + H^+$

**Orotate**

Orotate phosphoribosyl transferase $\quad$ Phosphoribosyl pyrophosphate $\rightarrow PP_i$

**OMP**

OMP decarboxylase $\quad CO_2$

**UMP**

*phate is incorporated into the nucleotide*. The first pyrimidine mononucleotide to be synthesized is orotidine-5′-monophosphate (OMP), and from this compound, pathways lead to nucleotides of uracil, cytosine, and thymine. OMP thus occupies a central role in pyrimidine nucleotide biosynthesis somewhat analogous to the position of IMP in purine nucleotide biosynthesis. Like IMP, OMP is found only in low concentrations in cells and is not a constituent of RNA.

Early clues to the nature of the pyrimidine pathway were provided by the observation that orotic acid (6-carboxyuracil; Figure 19-9) can satisfy the growth requirement of mutants of *Neurospora* unable to make pyrimidines and that isotopically labeled orotate is an immediate precursor of pyrimidines in *Neurospora* and a number of bacteria.

## Synthesis of Uridine Monophosphate

The pathway shown in Figure 19-10 starts with the synthesis of carbamoyl phosphate catalyzed by carbamoyl phosphate synthase (glutamine). This enzyme is present in microorganisms and in the cytosol of all eukaryotic cells capable of forming pyrimidine nucleotides. Eukaryotes also have a carbamoyl phosphate synthase (ammonia), a distinct enzyme that uses ammonia as a substrate instead of glutamine. It is associated with citrulline formation on the pathway for arginine biosynthesis, and in mammals, this is a mitochondrial enzyme present predominantly in the liver, where it catalyzes a step in the urea cycle (Chapter 23).

Carbamoyl phosphate synthase does not contain biotin and is not activated by it. The product of this enzyme, carbamoyl phosphate, next reacts with aspartate to form carbamoylaspartate. The reaction is catalyzed by aspartate carbamoyl transferase (aspartate transcarbamoylase or aspartate transcarbamylase), and the equilibrium greatly favors carbamoylaspartate synthesis.

In the third step, the pyrimidine ring is closed by dihydroorotase with the formation of L-dihydroorotate. The equilibrium in this reaction is pH-dependent, becoming more favorable at low pH. At pH 6, the ratio of L-dihydroorotate to carbamoylaspartate at equilibrium is 1.5. Presumably at physiological pH, the equilibrium is favorable enough for the reaction to proceed, especially when coupled with the subsequent reactions in the pathway.

Dihydroorotate is oxidized to orotate by dihydroorotate dehydrogenase. This flavoprotein in some organisms contains FMN and in others both FMN and FAD. It also contains nonheme iron and sulfur. In eukaryotes it is a lipoprotein associated with the inner membrane of the mitochondria. In the final two steps of the pathway, orotate phosphoribosyltransferase yields orotidine-5′-phosphate (OMP), and a specific decarboxylase then produces UMP.

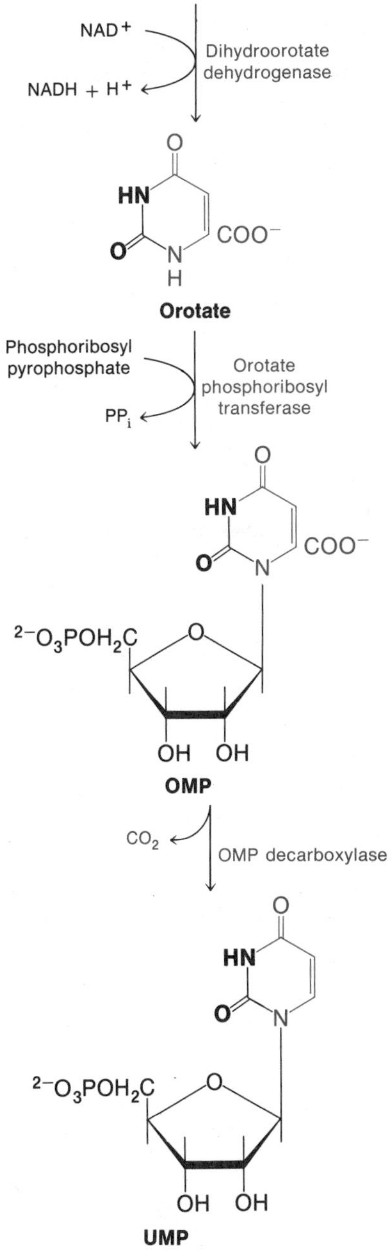

**Figure 19-10**
Biosynthesis of UMP. Color indicates the parts of the intermediates derived from aspartate. Bold type indicates atoms derived from carbamoyl phosphate.

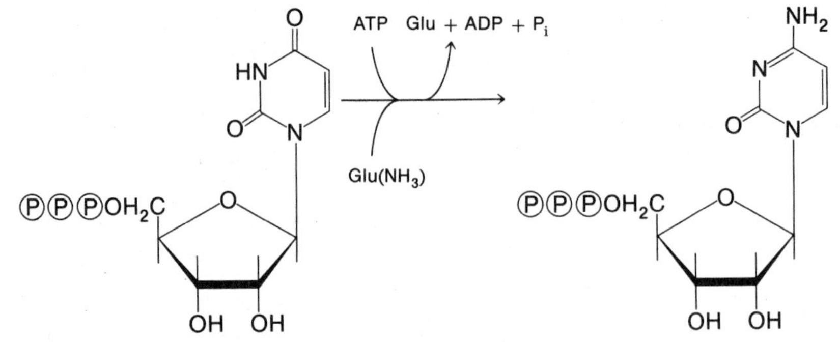

**Figure 19-11**
UTP amination by CTP synthase.
ⓅⓅⓅ represents the triphosphate
group.

A genetic disease in children characterized by abnormal growth, megaloblastic anemia, and the excretion of large amounts of orotate is associated with low activities of orotidine phosphate decarboxylase and (usually) orotate phosphoribosyltransferase. When these children are fed a pyrimidine nucleoside, usually uridine, the anemia improves and the excretion of orotate diminishes. The improvement is belived to be due to phosphorylation of administered uridine to UMP, which is then converted to other pyrimidine nucleotides, permitting nucleic acid and protein synthesis to be resumed. In addition, the increased intracellular concentrations of pyrimidine nucleotides would inhibit carbamoyl phosphate synthase (see below).

### Synthesis of Cytidine Triphosphate

After phosphorylation of UMP to UTP, CTP is formed by reaction of UTP with glutamine in a reaction driven by the concomitant hydrolysis of ATP to ADP and inorganic phosphate (Figure 19-11). Both the mammalian and bacterial cytidine triphosphate synthases can use ammonia as a donor in place of glutamine, but this reaction is of no physiological significance because the $K_m$ for ammonia is very high and the reaction rate low. With glutamine as the amino donor, GTP is an allosteric activator for CTP synthase from *E. coli* and probably for the mammalian enzyme as well (Figure 19-12). However, there is no stimulation of CTP synthesis from ammonia, a result interpreted to mean that the allosteric effect of GTP is specifically on the release of ammonia from glutamine. This ammonia is then channeled into reaction with UTP in the enzyme active site.

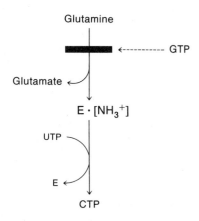

**Figure 19-12**
Allosteric activation of CTP synthase
by GTP.

### BIOSYNTHESIS OF DEOXYRIBONUCLEOTIDES

2'-Deoxyribonucleotides are not built up from 2-deoxyribose-5-phosphate by analogous biosynthetic pathways to those used for ribonucleotides, as was once supposed. Instead, tracer studies with isotopically labeled precursors have shown that *in both mammalian tissues and in microorganisms, deoxyribonucleotides are formed from corresponding ribonucleotides* by replacement of the 2'-OH group with hydrogen.

There are two types of ribonucleotide reductase that catalyze this reduction of the ribose ring. The type that is widely distributed in nature and present in mammalian cells contains nonheme iron, is composed of more than one type of polypeptide chain, and is specific for the reduction of diphosphates (ADP, GDP, CDP, and UDP). The second type appears to be restricted to certain microorganisms, including species of the bacteria *Clos-*

*tridium, Lactobacillus,* and *Rhizobium,* a number of species of algae, such as *Euglena* and *Phormidium,* and the fungus *Pithomyces chartarum.* Ribonucleotide reductase of this type requires adenosylcobalamin (coenzyme $B_{12}$) as obligatory coenzyme, does not contain nonheme iron, uses either nucleoside diphosphates or triphosphates depending on the source of the enzyme, and consists of only one type of polypeptide chain (although this may form oligomers in some cases).

The most completely studied ribonucleotide reductase of the first type is that from *E. coli.* It consists of two distinct proteins, B1 and B2, each consisting of two identical or very similar polypeptide chains and both apparently contributing to the catalytic site. The best-studied example of the second type is that from *Lactobacillus leichmannii,* which is a monomeric protein and reduces ribonucleoside triphosphates. Both types of reductase are very unusual in that they appear to catalyze reactions involving radical mechanisms. The *E. coli* reductase B2 subunit contains a stable free organic radical, and the unpaired electron appears to be localized on the benzene ring of a tyrosine side chain. In the case of the *L. leichmannii* enzyme, a radical pair is generated by homolytic cleavage of the C—Co bond of the adenosylcobalamin coenzyme. How the tyrosyl radical plus iron or the deoxyadenosyl radical plus cobalt participates in the reduction of the ribose ring is still obscure.

For both types of reductase, the physiological reducing substrate is a small-molecular-weight (13,000 daltons) electron-transport protein, thioredoxin. Thioredoxin has two half-cystine residues separated in the polypeptide chain by two other residues. The oxidized form of thioredoxin, with a disulfide bridge between the half-cystines, is reduced by NADPH in the presence of a flavoprotein, thioredoxin reductase. The reduced form of thioredoxin, with two cysteine residues present, is the reducing substrate for ribonucleotide reduction. The flow of electrons from NADPH to ribose is shown in Figure 19-13.

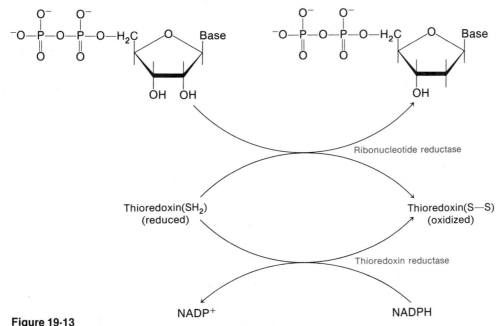

**Figure 19-13**
Ribonucleotide reductase and thioredoxin system.

An alternative electron-transport system for ribonucleotide reduction has been discovered in *E. coli*. In this case, the ultimate source of electrons is again NADPH, but they are passed to glutathione (catalyzed by glutathione reductase); reduced glutathione, in turn, reduces a small protein called glutaredoxin; and reduced glutaredoxin acts as the reducing substrate in ribonucleotide reduction. The distribution of glutaredoxin in nature and the relative importance of the glutaredoxin and thioredoxin systems in *E. coli* remain to be determined.

## THYMIDYLATE BIOSYNTHESIS

Since DNA contains thymine (5-methyluracil) as a major base instead of uracil, the synthesis of thymidine monophosphate (dTMP or thymidylate) is therefore essential in order to provide dTTP (thymidine triphosphate), which is required for DNA replication together with dATP, dGTP, and dCTP.

### Synthesis of dUMP

*Thymidylate is synthesized from dUMP, and there are several pathways by which the latter may be formed in cells.* A possible pathway is the deamination of dCMP according to a reaction catalyzed by deoxycytidylate deaminase. This enzyme is widely distributed in animal tissues, but several lines of evidence suggest that it is not the major source of dUMP.

$$dCMP + H_2O \longrightarrow dUMP + NH_3$$

Deamination of 5-methyldeoxycytidylate and 5-hydroxymethyldeoxycytidylate is also catalyzed by this enzyme.

A route to dUMP that is probably of greater importance is the reduction of UDP to dUDP, followed by phosphorylation of dUDP to dUTP (or direct reduction of UTP to dUTP in some microorganisms). The dUTP is then hydrolyzed to dUMP. This circuitous route to dUMP is dictated by two considerations. First, the ribonucleotide reductase in most cells only acts on ribonucleoside diphosphates, probably because this permits better regulation of its activity. Second, cells contain a powerful deoxyuridine triphosphate diphosphohydrolase (dUTPase) for another purpose. It prevents the incorporation of dUTP into DNA by keeping intracellular levels of dUTP low by means of the reaction

$$dUTP + H_2O \longrightarrow dUMP + PP_i$$

### Thymidylate Synthase

Methylation of dUMP to give thymidylate is catalyzed by thymidylate synthase and utilizes 5,10-methylenetetrahydrofolate as the source of the methyl group. This reaction is unique in the metabolism of folate derivatives in that the folate derivative acts both as a donor of the one-carbon group and also as its reductant, using the reduced pteridine ring as the source of reducing potential. Consequently, in this reaction, unlike any other in folate metabolism, dihydrofolate is a product (Figure 19-14). Since folate derivatives are present in cells at very low concentrations, continued synthesis of thymidylate requires regeneration of 5,10-methylenetetrahydrofolate from dihydrofolate. As shown in Figure 19-14, this occurs in two steps catalyzed by the enzymes dihydrofolate reductase, an NADP$^+$-linked dehydrogenase, and serine hydroxymethyltransferase. Inhibition of these steps indirectly interrupts thymidylate synthesis.

## FORMATION OF NUCLEOSIDE MONOPHOSPHATES FROM BASES (SALVAGE PATHWAYS)

*In addition to the pathways for synthesis of nucleotides* de novo *described in the preceding sections of this chapter, enzymes are widely distributed in both mammalian tissues and microorganisms that will catalyze the synthesis of mononucleotides from purine and pyrimidine bases.* Phosphoribosyltransferases catalyze reactions of the following type:

$$\text{Base} + \text{PRPP} \rightleftharpoons \text{ribonucleoside-5'-phosphate} + \text{PP}_i$$

The equilibrium of this reaction is in favor of nucleotide synthesis, and since the inorganic pyrophosphate released is rapidly hydrolyzed by inorganic pyrophosphatase, the coupling of these reactions makes the synthesis of nucleotide irreversible.

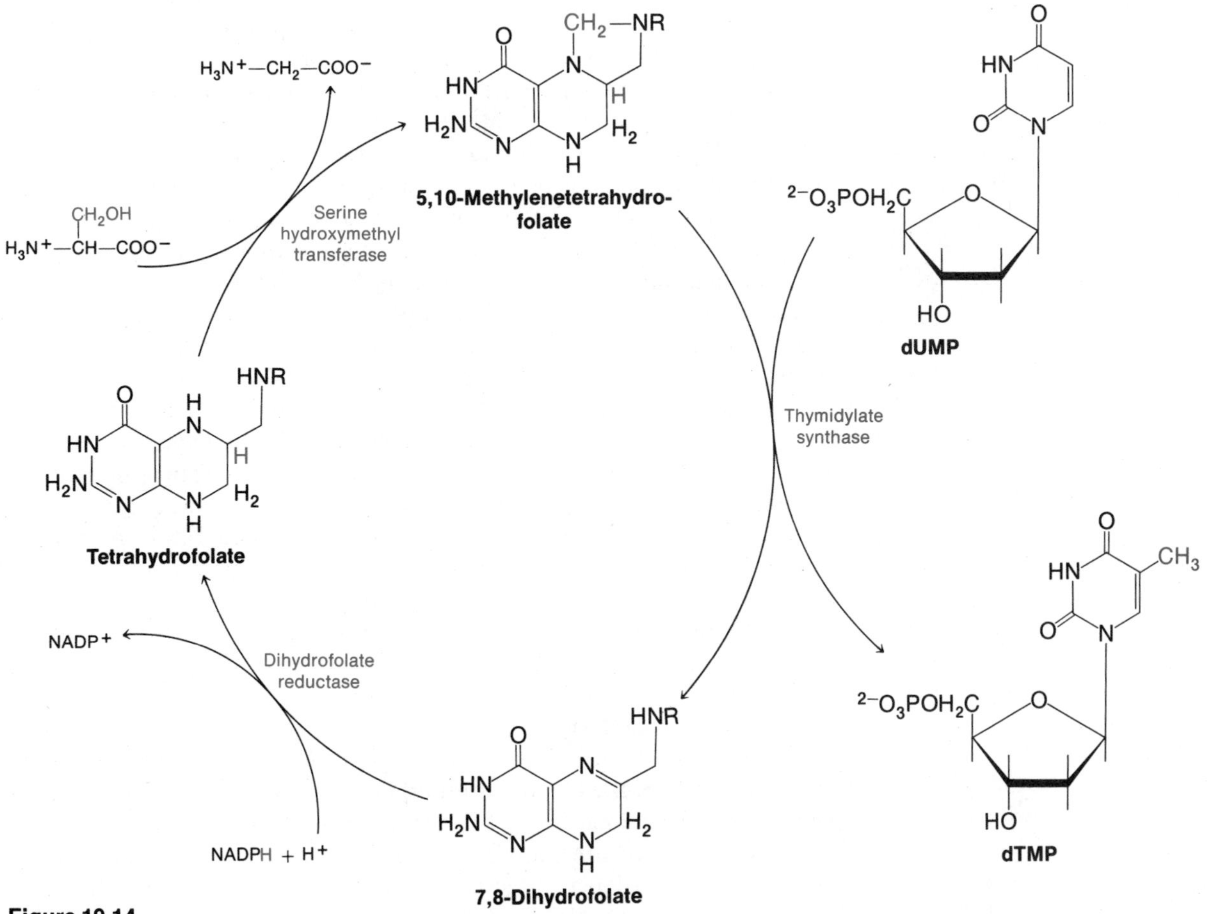

**Figure 19-14**
Thymidylate biosynthesis. R = *p*-aminobenzoyl-L-glutamate. Colored background indicates atoms that are precursors of the methyl group of thymidylate.

## Purine Phosphoribosyltransferases

In mammals, two phosphoribosyltransferases that convert purine bases to nucleotides are present in many organs. The first, adenine phosphoribosyltransferase, catalyzes the formation of AMP from adenine, but it also will accept 4-aminoimidazole-5-carboxamide as an alternate substrate. The second, hypoxanthine-guanine phosphoribosyltransferase, catalyzes the conversion of hypoxanthine to IMP and guanine to GMP and is probably the enzyme that converts xanthine to XMP. Some microorganisms contain a separate xanthine phosphoribosyltransferase in addition to the adenine and hypoxanthine-guanine phosphoribosyltransferases.

The exact role of these enzymes in metabolism remains controversial. Originally they were considered to function solely for the recovery of bases released during the degradation of nucleic acids. A role in such salvage of bases seems necessary, since turnover of some types of mRNA is rapid, and maturation of red cells is accompanied by considerable nucleic acid loss. To test this hypothesis, purine excretion (as uric acid) in subjects who lack hypoxanthine-guanine phosphoribosyltransferase has been compared with purine excretion in normal subjects. In addition, purine metabolism in cultures of cells from both groups has been studied. The results indicate that most of the purines formed by degradation of nucleic acids (Chapters 20 and 21) are utilized for the resynthesis of nucleotides and nucleic acids. Measurements also have been made by the isotope-dilution technique of the amounts of xanthine and hypoxanthine formed daily by human subjects who lack the enzyme xanthine oxidase (discussed below) and consequently excrete purines as xanthine. Comparison of the results with the amounts of xanthine and hypoxanthine that these individuals excreted indicated that about 90 percent of the purines formed were reutilized. This high recovery is presumably possible because of the restricted distribution in higher animal tissues of catabolic enzymes acting on free purines.

Since hypoxanthine and xanthine appear to be the major purine bases produced and released by cells with little or no adenine release (discussed below), adenine phosphoribosyltransferase does not seem to have a significant role in salvage of nucleotides. This enzyme perhaps serves mainly to utilize small amounts of adenine that are produced during intestinal digestion of nucleic acids or in the metabolism of 5'-deoxy-5'-methylthioadenosine, a byproduct of polyamine synthesis. Hypoxanthine-guanine phosphoribosyltransferase probably also functions to convert to nucleotides the small amounts of hypoxanthine and guanine released into the circulation by intestinal degradation of dietary nucleic acids.

Besides this salvage role, however, hypoxanthine-guanine phosphoribosyltransferase is probably important also for the transfer of purines from liver to extrahepatic tissues. Purine biosynthesis de novo is especially active in the liver, and there is evidence to suggest that extrahepatic cells that have a low capacity for the synthesis of purines de novo, such as erythrocytes and bone marrow cells, depend on uptake of hypoxanthine and xanthine from the blood to fulfill their needs for purine nucleotides. It seems likely that blood levels of xanthine and hypoxanthine, which are normally about 0.04 m$M$, are maintained by release of these bases from the liver. Some evidence suggests that the bases released by the liver are largely taken up by red blood cells and converted to purine nucleotides that are later released to tissues. If this is the case, the factors regulating their release and breakdown are unknown. The uptake of purine bases by extra-hepatic tissues, as well as by bacteria, appears to be closely linked to the activity of the purine phosphoribosyltransferases. At least in some cases, this is so because the transferases are membrane proteins.

The importance of the purine phosphoribosyltransferases in human metabolism of purine nucleotides was dramatically demonstrated by the discovery that the neurologic disorder of children called the _Lesch-Nyhan syndrome_ is due to a congenital lack of hypoxanthine-guanine phospho-ribosyltransferase. The disorder is characterized by aggressive behavior, mental retardation, spastic cerebral palsy, and self-mutilation. Purine me-tabolism is profoundly disturbed, with greatly increased de novo biosynthe-sis of purines (200 times normal), overproduction of uric acid (6 times normal), and elevated blood levels of uric acid. Severe gout is caused by the latter in some individuals. The increased rate of purine biosynthesis is probably due to several factors (Figure 19-15). Decreased levels of nucleo-tide production by phosphoribosyltransferases lift feedback inhibitory control of PRPP amidotransferase, and decreased utilization of PRPP by phosphoribosyltransferase makes a higher intracellular concentration of PRPP available for PRPP amidotransferase. At the same time, decreased nucleotide pools also would increase the PRPP level by deregulating the activity of ribose-5-phosphate pyrophosphokinase.

Although purine nucleosides are intermediates in the catabolism of nucleotides and nucleic acids in higher animals and humans, these nucleo-sides do not accumulate and are normally present in blood and tissues only

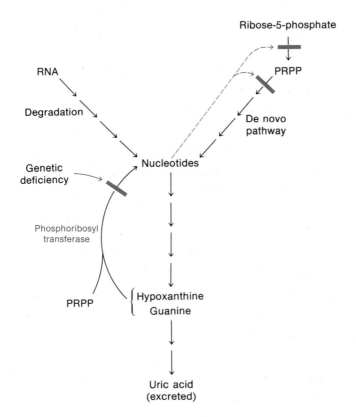

**Figure 19-15**
Mechanism of overproduction of pu-rine nucleotides in congenital defi-ciency of hypoxanthine-guanine phos-phoribosyltransferase.

in trace amounts. Nevertheless, cells of many vertebrate tissues contain kinases capable of converting purine nucleosides to nucleotides. Typical of these is adenosine kinase, which catalyzes the reaction:

$$\text{Adenosine} + \text{ATP} \longrightarrow \text{AMP} + \text{ADP}$$

These kinases have not been studied extensively and their specificity and role in various tissues and organs is uncertain. Their physiological function may be to prevent the accumulation of such nucleosides.

## Conversion of Pyrimidines to Mononucleotides

Many bacteria take up pyrimidine bases efficiently for nucleotide synthesis, and a phosphoribosyltransferase for uracil has been identified in a few species of bacteria. By contrast, pyrimidines are utilized poorly by mammalian cells but orotate phosphoribosyltransferase appears to accept uracil as a substrate. An alternative route for conversion of uracil to UMP both in bacteria and higher animals is by successive reactions catalyzed by uridine phorphorylase and uridine kinase, respectively.

$$\text{Uracil} + \text{ribose-1-phosphate} \rightleftharpoons \text{uridine} + \text{P}_i$$

$$\text{Uridine} + \text{ATP} \xrightarrow{\text{Mg}^{2+}} \text{UMP} + \text{ADP}$$

Mammalian uridine phosphorylase also will accept deoxyuridine as a substrate, and thymidine is slowly attacked, but other nucleosides are not substrates. Uridine kinase activity has been demonstrated in a variety of bacteria and animal cells, including tumors, and is especially high in cells of high growth rate. Cytidine is the only other physiological nucleoside to act as a substrate for uridine kinase, but a variety of nucleoside triphosphates can act as phosphoryl donors.

Thymine is similarly converted to dTMP according to the following reactions:

$$\text{Thymine} + \text{deoxyribose-1-phosphate} \rightleftharpoons \text{thymidine} + \text{P}_i$$

$$\text{Thymidine} + \text{ATP} \longrightarrow \text{dTMP} + \text{ADP}$$

Thymidine phosphorylase has been purified from bacteria and mammalian tissues. Thymidine is the preferred substrate, but deoxyuridine and 5-substituted deoxyuridines are also cleaved. Deoxythymidine kinase is widely distributed. Its activity in cells increases dramatically during rapid growth and DNA synthesis. Infection of cells with any of several viruses also induces thymidine kinase activity, together with certain other enzymes concerned with deoxyribonucleotide synthesis. The viral thymidine kinase is quite different from the host enzyme and, in some cases, also has deoxycytidine kinase and even thymidylate kinase activity. Significantly, the viral enzymes are subject to none of the allosteric controls of the host enzymes.

A deoxycytidine kinase has been purified from certain bacteria, from thymus, and from tumor cells. It catalyzes the reaction:

$$\text{Deoxycytidine} + \text{ATP} \longrightarrow \text{dCMP} + \text{ADP}$$

Frequently several isoenzymes are present in cells and differ in Michaelis constants and substrate specificity. Thus mitochondria contain an isoenzyme different from the cytosolic kinase. Deoxyadenosine and deoxyguanosine are also substrates for deoxycytidine kinase, but the purine deoxyribonucleosides have much higher $K_m$ values than deoxycytidine for the cytosol enzyme.

## CONVERSION OF NUCLEOSIDE MONOPHOSPHATES TO DIPHOSPHATES AND TRIPHOSPHATES

The products of the biosynthetic pathways discussed in the preceding sections are, in most cases, mononucleotides. *A series of kinases (phosphotransferases) is present in cells that converts these mononucleotides to their metabolically active diphosphate and triphosphate forms.*

### Nucleoside Monophosphate Kinases

Bacteria and other microorganisms, as well as animal cells, contain a variety of kinases that catalyze reactions of the general type

$$(d)NMP + ATP \rightleftharpoons (d)NDP + ADP$$

Four types of nucleoside monophosphate kinases are known. These catalyze, respectively, the phosphorylation of (1) GMP and dGMP, (2) AMP and dAMP, (3) dCMP, CMP, and UMP, and (4) dTMP. In the case of the third kinase, there is conflicting evidence concerning whether a single enzyme catalyzes phosphorylation of all three substrates. Perhaps in some cells a single enzyme is responsible, whereas in other cells UMP kinase and CMP kinase are distinct enzymes. T-even bacteriophages induce a kinase that catalyzes phosphorylation of 5-hydroxymethylCMP, dGMP, and dTMP, and the kinase induced by the T5 bacteriophage has activity toward dAMP, dTMP, dGMP, and dCMP.

The second of the kinases mentioned, which uses AMP as substrate, is referred to as *adenylate kinase.* Its activity is high in tissues where the turnover of energy from adenine nucleotides is great, e.g., in liver and muscle, and its activity is also high in mitochondria. In these tissues its function is to make more energy available as ATP bond energy. When ADP is formed from ATP in energy-consuming reactions, more ATP can be formed from ADP according to the reaction

$$2\,ADP \rightleftharpoons AMP + ATP$$

Under conditions where energy-generating reactions convert intracellular ADP to ATP, AMP will be phosphorylated by running the preceding reaction from right to left. Adenylate kinase is therefore important in biological systems for maintaining equilibrium among adenine nucleotides as they are depleted or formed by energy transfers.

### Nucleoside Diphosphate Kinases

Enzymes of this type are ubiquitous in nature, having been found in many tissues of animals, plants, and microorganisms. They catalyze reactions of the following general type:

$$N_1TP + N_2DP \rightleftharpoons N_1DP + N_2TP$$

where $N_1$ and $N_2$ are purine or pyrimidine ribonucleosides or deoxyribonucleosides. The activity of NDP kinases is relatively high, usually 10- to

100-fold greater than the activity of the monophosphate kinases. As a result, intracellular concentrations of triphosphates are normally much higher than those of diphosphates, which in turn are often higher than those of monophosphates. Unlike the monophosphate kinases which are substrate-specific, NDP kinases from all sources are active with a wide range of nucleoside diphosphates and triphosphates. They require a divalent cation for activity, and although many metal ions can satisfy this requirement, $Mg^{2+}$ is the physiological cofactor.

Nucleoside diphosphate kinases from many sources have been shown to function by means of the formation of a phosphoryl-enzyme intermediate as shown.

$$E + N_1TP \rightleftharpoons E{\sim}P + N_1DP$$

$$E{\sim}P + N_2DP \rightleftharpoons E + N_2DTP$$

In several cases it has been shown that the phosphoryl group is attached to N-1 of a histidine side chain.

## INHIBITORS OF NUCLEOTIDE SYNTHESIS

*There are several distinct types of inhibitors of nucleotide biosynthesis, each type acting at different points in the pathways to purine or pyrimidine nucleotides.* Inhibitors of all these types are very toxic to cells, especially rapidly growing cells such as tumors or bacteria, because interruption of the supply of any class of nucleotides seriously limits the cell's capacity to synthesize the nucleic acids necessary for protein synthesis and replication. In some cases, this toxicity permits the clinical use of the inhibitors in the chemotherapy of cancer or in the treatment of bacterial infections.

Reactions involving glutamine as a substrate are inhibited by the glutamine analogs azaserine and 6-diazo-5-oxo-L-2-aminohexanoic acid (Figure 19-16). Inhibition is irreversible, with formation of a covalent bond between the inhibitor and an amino acid side chain at the catalytic site. The specific reactions inhibited are those catalyzed by glutamine PRPP amidotransferase and phosphoribosylglycinamidine synthase in the de novo purine pathway and carbamoyl phosphate synthase and CTP synthase in the pyrimidine pathway.

Another group of inhibitors prevents nucleotide biosynthesis indirectly by depleting the level of intracellular tetrahydrofolate derivatives, including methenyltetrahydrofolate and 10-formyltetrahydrofolate (required for purine synthesis) and methylenetetrahydrofolate (required for thymidylate synthesis). As seen from Figure 19-16, sulfonamides are structural analogs of p-aminobenzoic acid, and they competitively inhibit the bacterial biosynthesis of folic acid at a step at which p-aminobenzoic acid is incorporated into folic acid. Sulfonamides are widely used in medicine because they inhibit growth of many bacteria. When cultures of susceptible bacteria are treated with sulfonamides, they accumulate 4-carboxamide-5-aminoimidazole in the medium, because of a lack of 10-formyltetrahydrofolate for the penultimate step in the pathway to IMP (Figure 19-6). Methotrexate, trimethoprim, and a number of related compounds inhibit the reduction of dihydrofolate to tetrahydrofolate catalyzed by dihydrofolate reductase. These inhibitors are structural analogs of folic acid (Figure 19-16) and bind at the catalytic site of dihydrofolate reductase, an enzyme catalyzing one of the steps in the cycle of reactions involved in thymidylate synthesis (Figure 19-14). These inhibitors therefore prevent synthesis of thymidilate and purine nucleotides in replicating cells. Methotrexate is used as an anticancer drug, and trimethoprim, which specifically inhibits bacterial dihydrofolate reductase, is used in combination with sulfonamides for treating bacterial infections.

Other inhibitors that interfere with thymidylate synthesis are 5-fluorouracil and 5-fluorodeoxyuridine, both of which have their major inhibitory effects after being converted in the cell to 5-fluoro-2'-deoxyuridine-5'-monophosphate (Figure 19-16). The latter acts as an analog of dUMP and binds very tightly to thymidylate synthase, forming a covalent complex with the enzyme that is unable to undergo the normal catalytic reaction. Both these inhibitors are used in cancer chemotherapy.

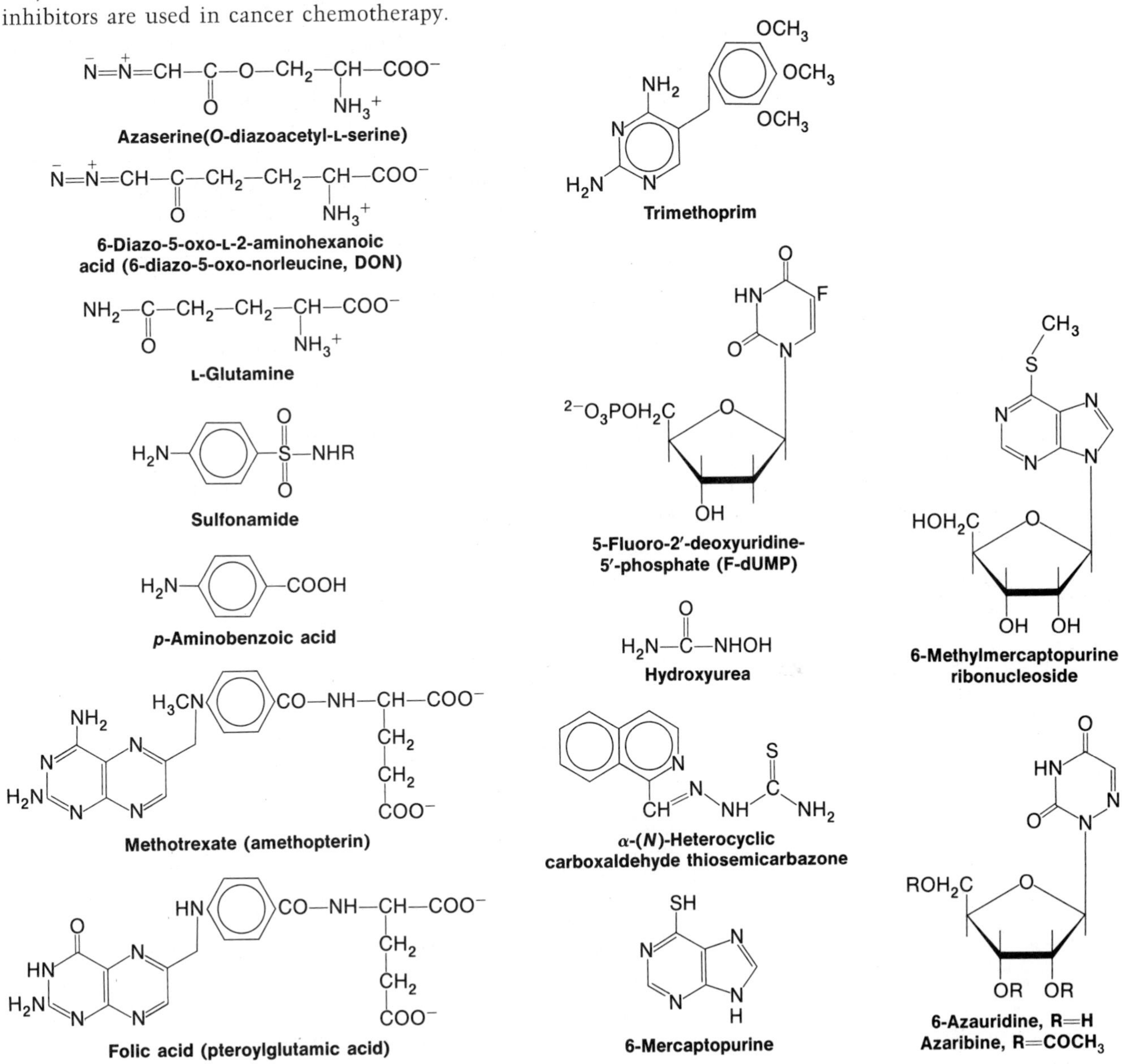

**Figure 19-16**
Inhibitors of nucleotide metabolism, together with physiological metabolites (glutamine, *p*-aminobenzoic acid, folic acid) that some of the inhibitors resemble.

Hydroxyurea and $\alpha$-($N$)-heterocyclic carboxaldehyde thiosemicarba-zones (Figure 19-16) interfere with the synthesis of deoxyribonucleotides by inhibiting ribonucleotide reductase of mammalian cells, an enzyme that is crucial and probably rate-limiting in the biosynthesis of DNA. These inhibitors, which probably act by chelating the iron present in the reduc-tase, are in clinical use as anticancer agents.

6-Mercaptopurine, 6-methylmercaptopurine ribonucleoside (Figure 19-16), and related thiopurines are potent inhibitors of nucleotide biosyn-thesis, but they are inactive until they are converted to the corresponding ribonucleoside 5'-phosphates. 6-Mercaptopurine is converted by the action of hypoxanthine-guanine phosphoribosyltransferase to the nucleotide 6-thioinosine-5'-monophosphate (T-IMP). 6-Methylmercaptopurine ribo-nucleoside is probably phosphorylated by ATP through the action of adeno-sine kinase, another salvage enzyme of purine metabolism. The monophos-phate analogs inhibit several enzymes of purine biosynthesis, and it is uncertain which effect is primarily responsible for the toxicity. T-IMP blocks conversion of IMP to adenylosuccinate and to XMP, key reactions in the formation of AMP and GMP. T-IMP and methylmercaptopurine ribo-nucleotide are also capable of "pseudo-feedback inhibition" of glutamine PRPP amidotransferase, the first committed step in the purine nucleotide pathway. This enzyme is highly responsive to intracellular concentrations of both normal ribonucleoside 5'-monophosphates and analogs. 6-Mercap-topurine is used clinically in the treatment of leukemia.

6-Azauridine (Figure 19-16), after conversion to the 5'-phosphate by uridine kinase, blocks the pyrimidine de novo pathway of synthesis by specifically inhibiting OMP decarboxylase, the last enzyme in the pathway to UMP. 2',3',5'-Triacetyl-6-azauridine (azaribine), which is hydrolyzed to 6-azauridine in the body but causes less toxicity than the latter, is used to treat several disease states. Patients treated with azaribine excrete orotate and orotidine in the urine.

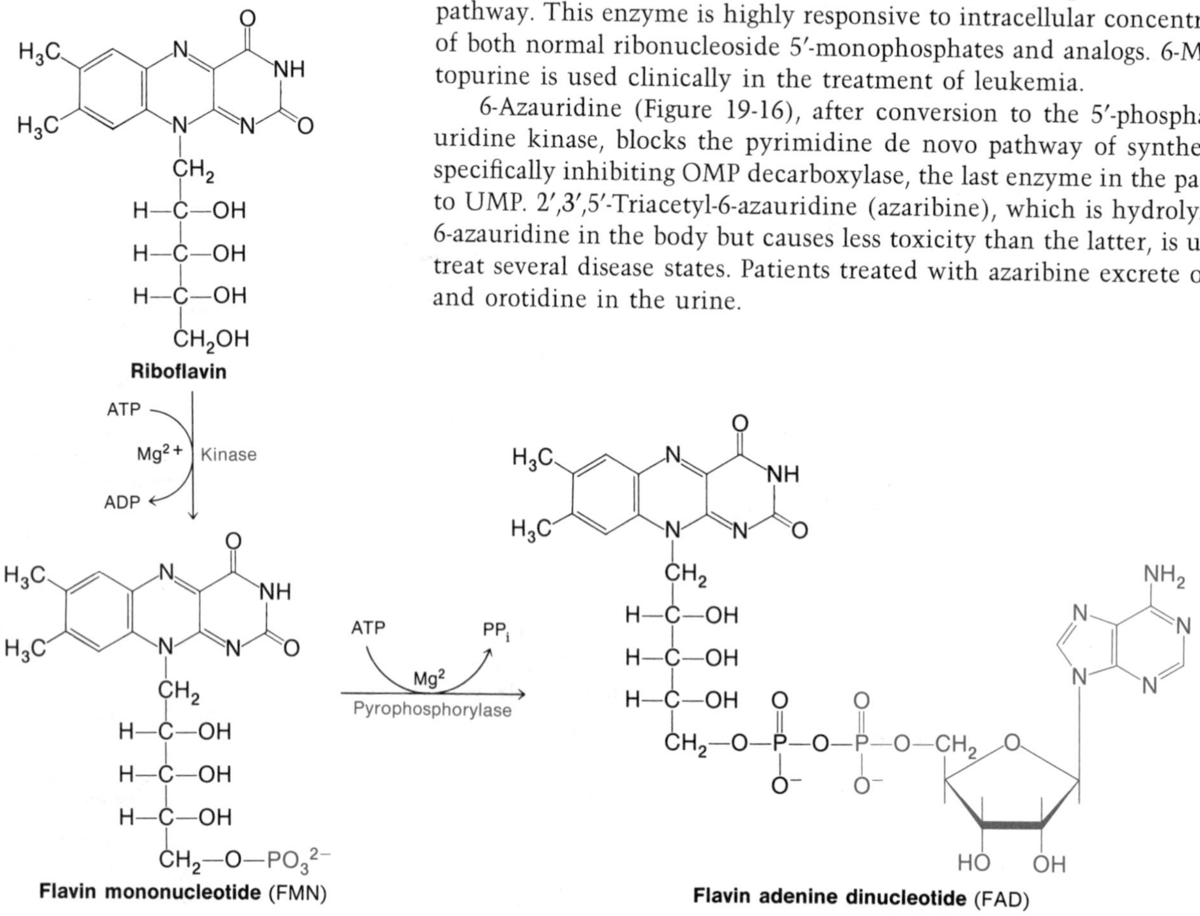

**Figure 19-17**
Biosynthesis of NMN and FAD from riboflavin.

# BIOSYNTHESIS OF NUCLEOTIDE COENZYMES

Many of the nucleotides considered thus far play important roles in metabolism and are discussed in other chapters. However, nucleotide coenzymes such as flavin nucleotides, $NAD^+$, $NADP^+$, and coenzyme A are also extremely important in metabolism. In the following, the pathway for the completion of each nucleotide is discussed.

The synthesis of riboflavin, i.e., 7,8-dimethyl-10-(1'-D-ribityl)isoalloxazine, by microorganisms such as the fungus *Eremothecium* and mutants of the yeast *Saccharomyces* has been shown to start from GTP. Riboflavin is an essential dietary constituent for mammals and is converted in the body to the mononucleotide or dinucleotide forms that function as the prosthetic groups of many enzymes. Riboflavin is converted to riboflavin-5'-phosphate, more commonly called *flavin mononucleotide* (FMN), by flavokinase (ATP:riboflavin phosphotransferase), as shown in Figure 19-17. The enzyme has been purified from yeast, plants, and liver. It is also present in a variety of other animal tissues (kidney, brain, spleen, and heart).

The other nucleotide form of riboflavin, flavin adenine dinucleotide (FAD), is formed from FMN in a reversible reaction catalyzed by flavin nucleotide pyrophosphorylase (Figure 19-17). This enzyme is also widely distributed in nature and has been observed in plants, yeast, lactobacilli, and many animal tissues.

The nicotinamide moiety of the coenzymes nicotinamide adenine dinucleotide ($NAD^+$) and nicotinamide adenine dinucleotide phosphate ($NADP^+$) is synthesized by several routes. In liver and other animal tissues, tryptophan degradation forms, among other products, quinolinic acid (Chapter 23), which is converted to nicotinate mononucleotide (deamidonicotinamide mononucleotide, deamido-NMN) by quinolinate phosphoribosyltransferase (Figure 19-18). Nicotinate is apparently synthesized by other pathways in plants and microorganisms. In the cytosol of cells of many mammalian tissues and in yeast and other microorganisms there is present a nicotinate phosphoribosyltransferase that also forms deamido-NMN (Figure 19-18). A very similar phosphoribosyltransferase present in the cytosol of all animal tissues investigated acts on nicotinamide. These transferases are responsible for utilization of nicotinate and nicotinamide in the diet. The role of ATP in these reactions is unclear. Some transferases do not require it, for others it seems to be an allosteric regulator, and in yet other cases ATP seems to be hydrolyzed to yield ADP and $P_i$ in equimolar amounts with deamido-NMN formation.

The mononucleotides so formed are converted to the corresponding dinucleotides by NMN adenylyltransferase (Figure 19-18). In mammalian cells it appears to be a single enzyme that catalyzes both reactions, but an adenylyl transferase acting only on NMN has been isolated from some bacteria (*Lactobacillus fructosus*). A cytoplasmic $NAD^+$ synthase present in yeast, liver, and other tissues transfers the amino group from glutamine at the expense of ATP hydrolysis (Figure 19-18).

A cytoplasmic kinase present in liver, mammary gland, and brain is responsible for the formation of $NADP^+$ from $NAD^+$:

$$NAD^+ + ATP \xrightarrow{Mg^{2+}} NADP^+ + ADP$$

NADH is not a substrate and inhibits competitively with respect to $NAD^+$.

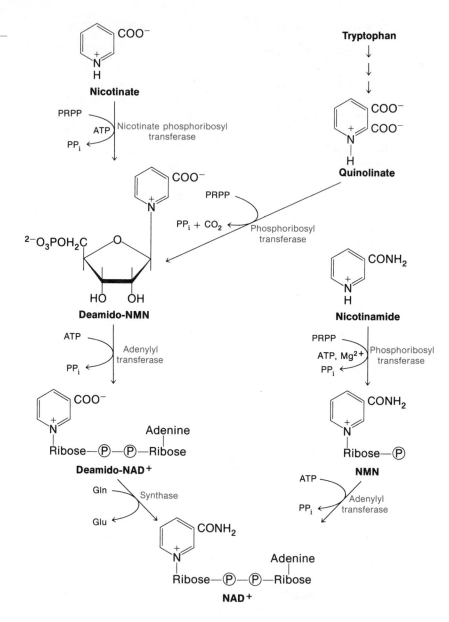

**Figure 19-18**
Biosynthesis of NAD$^+$ from nicotinate and nicotinamide.

Coenzyme A is synthesized in the mammalian liver from pantothenic acid (pantoyl-$\beta$-alanine), which is required in the mammalian diet. The five steps in the synthesis are shown in Figure 19-19. In the last step, a specific kinase transfers a phosphoryl group to the 3'-hydroxyl of the adenylate portion of the molecule.

## SYNTHESIS OF SPECIAL NUCLEOTIDES WITH A REGULATORY ROLE IN METABOLISM

In addition to their role in energy transfer and as coenzymes, it has been discovered in recent years that *a number of special nucleotides and oligonucleotides play a vital role in the regulation of many aspects of metabolism in all types of cells.* In this section, the synthesis of some of these nucleotides is briefly described.

# Adenosine 3′,5′-Cyclic Monophosphate (Cyclic AMP or cAMP)

Adenosine 3′,5′-cyclic monophosphate is a regulatory molecule found in most cells, where it controls diverse metabolic processes in both prokaryotic and eukaryotic organisms. Its formation from ATP is catalyzed by the enzyme adenylate cyclase (also called adenyl cyclase and, most accurately, adenylyl cyclase). The reaction is shown in Figure 19-20. In animal cells this enzyme is bound to the cytoplasmic membrane and can be stimulated by one or more of a large number of hormones, as discussed in other sec-

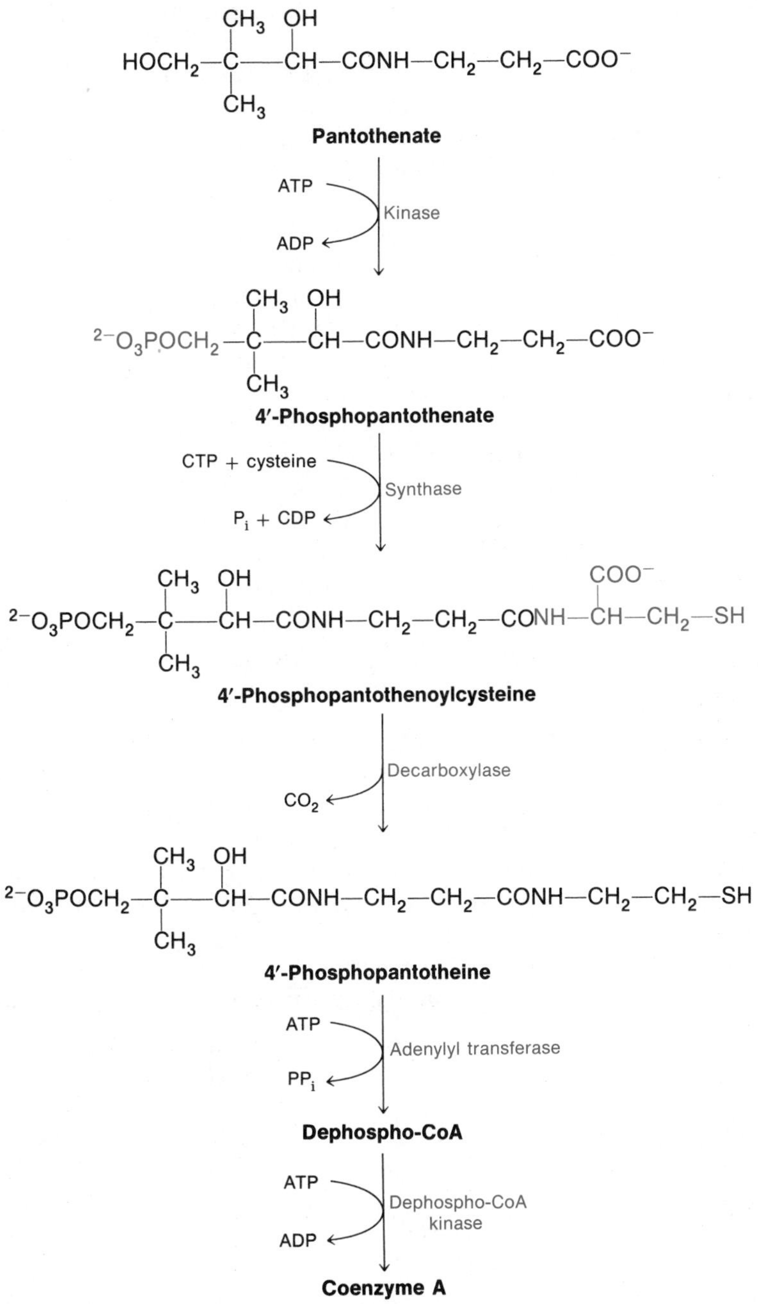

**Figure 19-19**
Biosynthesis of coenzyme A from pantothenate. Color indicates the groups introduced at the synthase step.

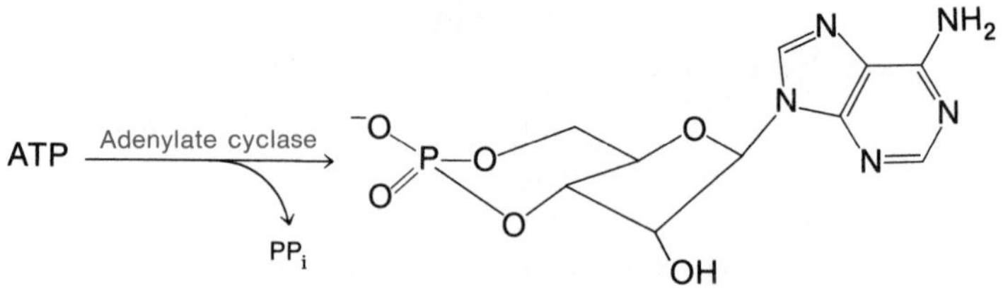

**Figure 19-20**
Synthesis of cyclic AMP.

tions. Cyclic AMP is sometimes called an intracellular "second messenger" because it carries the "message" brought by the "first messenger" (the hormone) to metabolic processes within the cell. The effects of cyclic AMP on metabolism in eukaryotes are primarily mediated through activation of protein kinases (see Chapter 8).

In the cytoplasmic membrane, adenyl cyclase is associated with two other proteins that are required for its activation. The first of these is a guanine nucleotide-binding protein G/F, which, when GTP is complexed with it, stimulates the cyclase. This activation is terminated upon hydrolysis of the bound GTP. Regeneration of the G/F-GTP complex is catalyzed by a complex formed by the binding of the hormone messenger to the second auxiliary protein, which is a hormone receptor. The hormone-receptor complex accelerates dissociation of GDP from G/F, thus allowing more GTP to bind. All three proteins are intrinsic constituents of the membrane and require proper integration in the membrane for hormonal regulation of the cyclase to take place.

Cyclic AMP is present in many bacterial species and related microorganisms, but apparently not in all. Much, and in some species all, of the adenyl cyclase activity is found in the bacterial cytoplasm, and the soluble enzyme has been purified to homogeneity and crystallized from some species. A major regulatory role for cAMP in some bacteria is participation in catabolite repression, which is discussed in Chapter 26.

## Guanosine 3′,5′-Cyclic Monophosphate (Cyclic GMP or cGMP)

Although cyclic GMP was first isolated from urine in 1963 and is known to be widely distributed in nature, the physiological roles of this nucleotide remain unknown. Guanylate cyclase, which catalyzes the reaction shown in Figure 19-21, is found in essentially all animal tissues and in most phyla. In most tissues, activity is present both in the cytoplasm and bound to membranes, but much of the latter activity is unexpressed until preparations are treated with detergents. In some tissues (liver and platelets), the activity is predominantly soluble, whereas in others (intestinal mucosa and fibroblast cultures), it is predominantly particulate. Although hormones

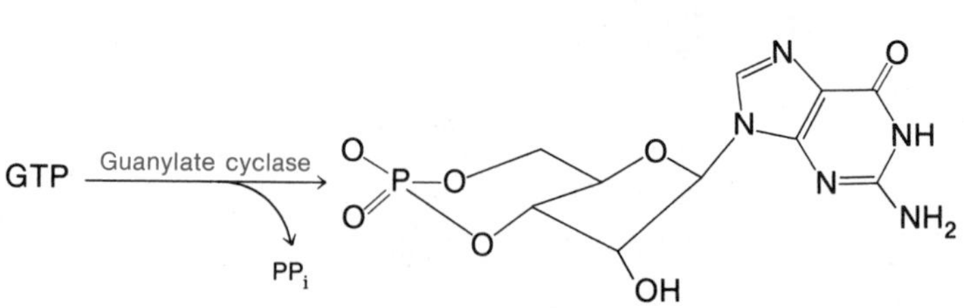

**Figure 19-21**
Synthesis of cyclic GMP.

alter the levels of cyclic GMP in animal cells, the mechanism and significance of this are unknown, since in cell-free preparations it has been impossible to obtain reproducible hormone effects on guanylate cyclase that could be considered physiologically significant.

## Guanosine 5'-Diphosphate 3'-Diphosphate (ppGpp) and Guanosine 5'-Triphosphate 3'-Diphosphate (pppGpp)

These nucleotides are the effectors in what is called the *stringent response* of *E. coli*. This is a response of the bacteria to amino acid starvation, and the primary result is curtailment of the synthesis of stable RNA. Secondarily, there is a reduction in the rate of synthesis of many metabolites, and there are other metabolic effects, such as an increase in the rate of protein turnover.

The enzyme catalyzing the synthesis of pppGpp and ppGpp has been called the *stringent factor* and is removed from ribosomes by washing with 0.5 $M$ NH$_4$Cl. It has been purified to near homogeneity and catalyzes the synthesis of pppGpp from ATP and GTP and ppGpp from ATP and GDP

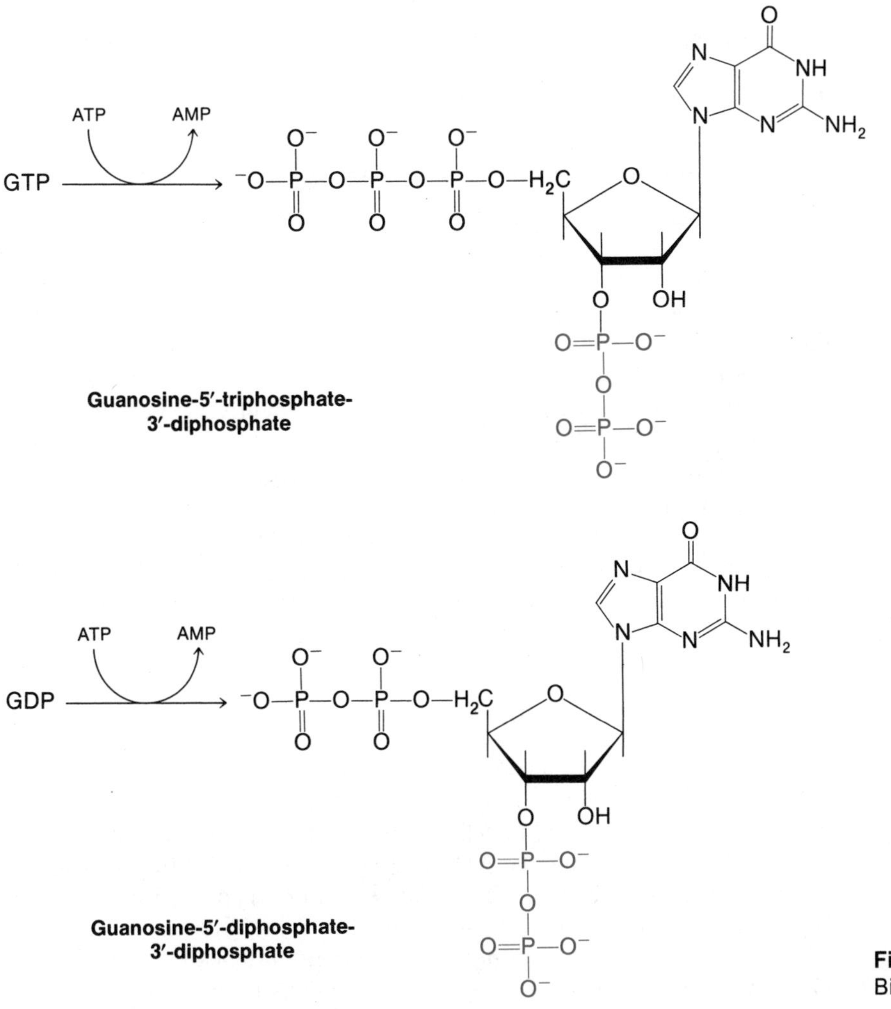

Guanosine-5'-triphosphate-3'-diphosphate

Guanosine-5'-diphosphate-3'-diphosphate

**Figure 19-22**
Biosynthesis of pppGpp and ppGpp.

(Figure 19-22). The reaction requires the presence of ribosomes that have mRNA bound to them and which have uncharged tRNA on the A site. Alternatively, below 30°C, a ribosome-independent reaction occurs in vitro that is stimulated by certain ribosomal proteins or other agents.

ppGpp inhibits the synthesis of ribosomal RNA but, in addition, exerts an inhibitory effect on a number of metabolic steps in nucleotide biosynthesis [IMP dehydrogenase and adenylosuccinate synthase (Figure 19-8) and membrane-bound nucleoside phosphorylases], lipid biosynthesis (acetyl-CoA carboxylase), and protein biosynthesis (formation of the initiation complex and elongation).

## Diadenosine 5′,5″-$P^1$,$P^4$-Tetraphosphate (A5′pppp5′A or Ap$_4$A)

This nucleotide is formed in a variety of mammalian cells by reaction of ATP with the amino acid adenylate formed on the active site of aminoacyl-tRNA synthases. Aminoacyl-tRNA synthesis occurs according to the following reactions:

$$\text{Amino acid} + \text{ATP} \rightleftharpoons \text{aminoacyl adenylate} + \text{PP}_i$$

$$\text{Aminoacyl adenylate} + \text{tRNA} \rightleftharpoons \text{aminoacyl-tRNA} + \text{AMP}$$

The aminoacyl adenylate formed in the first reaction can react with ATP as follows:

$$\text{Aminoacyl adenylate} + \text{ATP} \rightleftharpoons \text{amino acid} + \text{Ap}_4\text{A}$$

Concentrations of Ap$_4$A detected in mammalian cells range from 0.8 to 1.2 $\mu M$ in certain tumors to 0.03 to 0.06 $\mu M$ in normal mouse or rat liver. A variety of nucleoside 5′-diphosphates and 5′-triphosphates can substitute for ATP in the preceding reaction, as shown in Figure 19-23, but no other products have been shown to accumulate in mammalian cells. Levels of Ap$_4$A in cells are inversely related to their doubling time. Arresting the growth of mammalian cells in culture by serum deprivation or amino acid starvation causes Ap$_4$A levels to drop 30- to 50-fold, and inhibitors of protein synthesis cause a 100-fold drop. Levels of Ap$_4$A are in part controlled by its rapid enzymatic hydrolysis.

These observations suggest that Ap$_4$A may act in mammalian cells as a positive growth signal. In this respect it is of interest that DNA polymerase $\alpha$ from HeLa tumor cells has a highly specific site for noncovalent binding of Ap$_4$A and another site for binding tryptophanyl-tRNA.

## pppA2′p5′A2′p5′A ("Two-Five A")

This nucleotide (Figure 19-24) is produced as part of the mechanism by which interferon enables cells to resist attack by RNA viruses. Interferon induces in mammalian cells a synthase that, after activation by double-stranded RNA, converts ATP to this nucleotide with the unusual 2′-5′phosphodiester bonds. The related dimer (pppA2′p5′A), tetramer [ppp(A2′p)$_3$A], and higher oligomers are also formed in smaller amounts. Oligomers higher than the dimer activate an endonuclease that hydrolyzes phosphodiester bonds in mRNA and thus blocks translation.

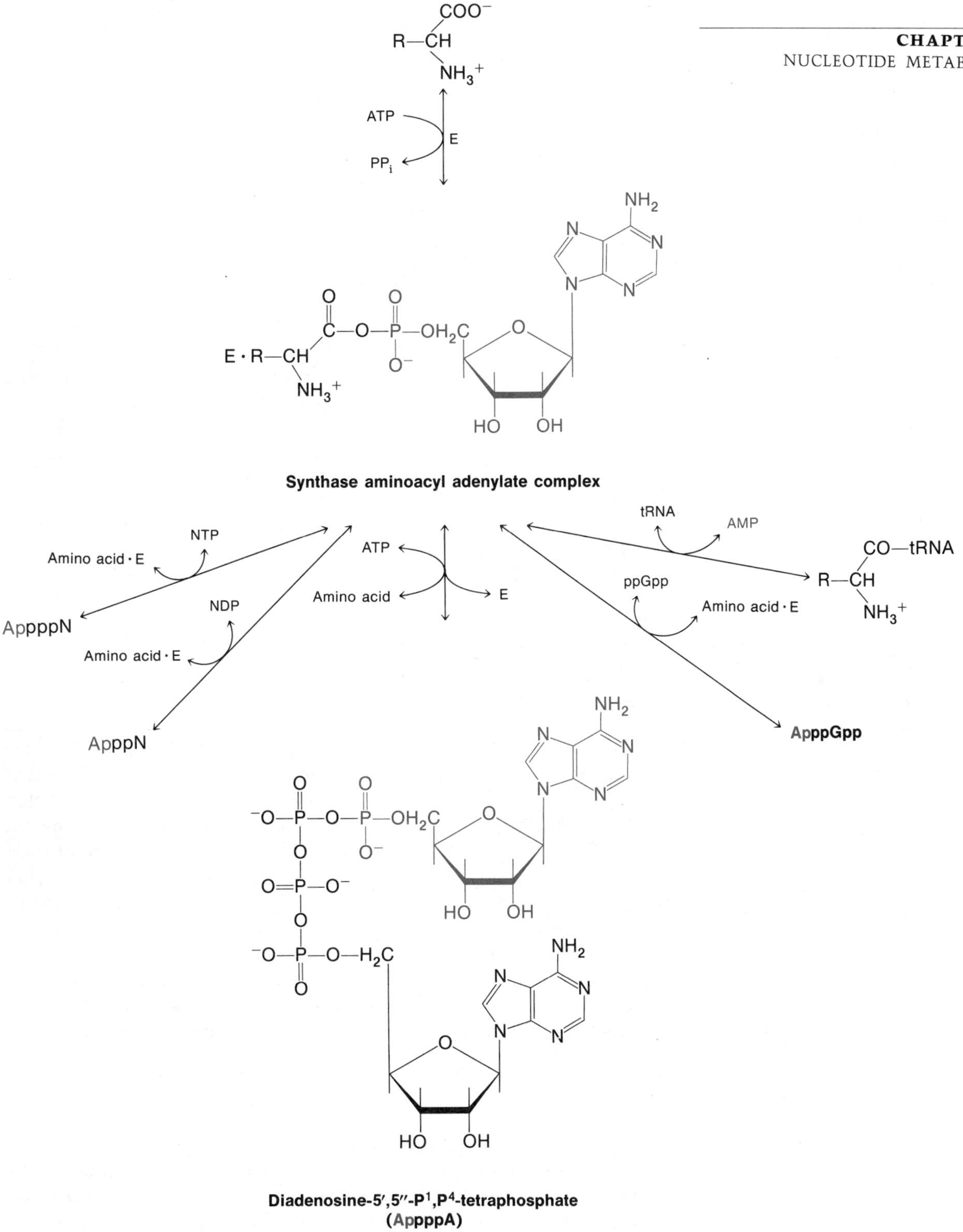

**Synthase aminoacyl adenylate complex**

**Diadenosine-5',5''-$P^1$,$P^4$-tetraphosphate**
**(ApppppA)**

**Figure 19-23**
Reaction of aminoacyl adenylate bound to aminoacyl-tRNA synthase with tRNA or (under conditions of limiting protein synthesis) with a variety of pyrophosphate-containing nucleotides.

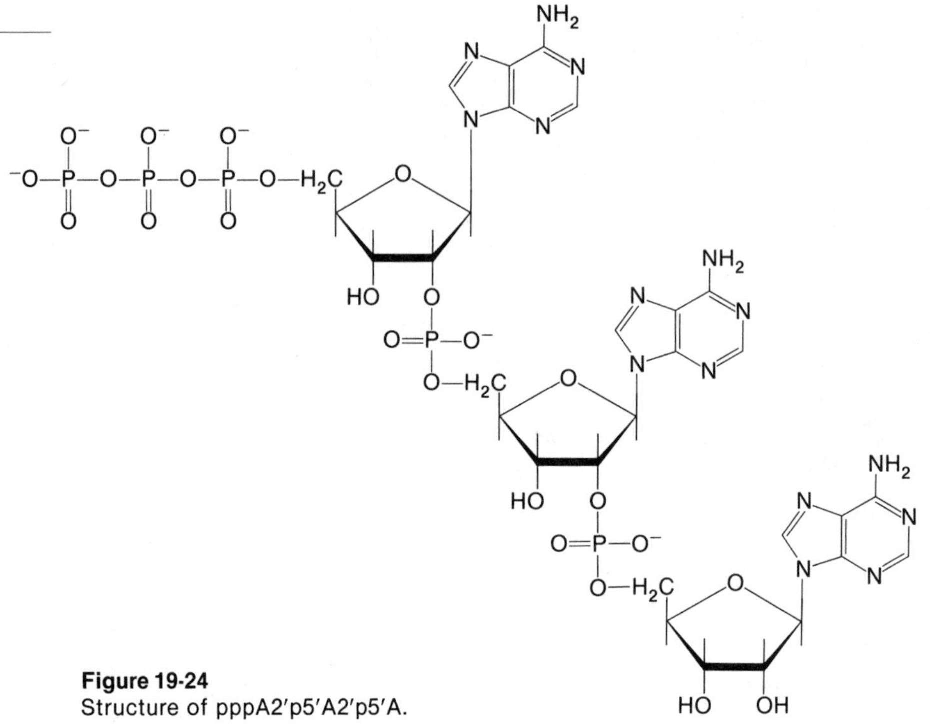

**Figure 19-24**
Structure of pppA2′p5′A2′p5′A.

## CATABOLISM OF NUCLEOTIDES

### Catabolism of Ingested Nucleic Acids

Dietary nucleic acids are unaffected by gastric enzymes, but in the small intestine, ribonuclease and deoxyribonuclease I secreted in the pancreatic juice hydrolyze nucleic acids mainly to oligonucleotides. The oligonucleotides are further hydrolyzed by phosphodiesterases, also secreted by the pancreas, to yield 5′- and 3′-mononucleotides. The major portion of these nucleotides is hydrolyzed to nucleosides by various group-specific nucleotidases or by a variety of nonspecific phosphatases, and the nucleosides may be absorbed intact by the intestinal mucosa or they may undergo phosphorolysis by nucleoside phosphorylases and by nucleosidases to free bases. Little is known about these enzymes and their specificity. The small intestinal mucosa are rich in nucleoside phosphorylase, and most of the remaining nucleoside is probably hydrolyzed to bases in this tissue.

$$\text{Nucleoside} + P_i \rightleftharpoons \text{base} + \text{ribose-1-phosphate}$$

$$\text{Nucleoside} + H_2O \longrightarrow \text{base} + \text{ribose}$$

Experiments with labeled nucleic acid indicate that both purines and pyrimidines of ingested nucleic acids are used only to a small extent for synthesis of tissue nucleic acids, and in the case of purines, most of the bases were shown to be catabolized. This is consistent with the presence in intestinal mucosa of a high level of xanthine oxidase, a catabolic enzyme (see the following discussion).

# Catabolism of Purines

After conversion of purine nucleotides to the corresponding nucleosides by 5′-nucleotidases and by phosphatases, inosine and guanosine are readily cleaved by the widely distributed purine nucleoside phosphorylase. The corresponding deoxynucleosides yield deoxyribose-1-phosphate and base with the phosphorylase from most sources. Adenosine and deoxyadenosine are not attacked by the phosphorylase of mammalian tissue, but much AMP is converted to IMP by an aminohydrolase (deaminase), which is very active in muscle and other tissues (Figure 19-25). It has recently been

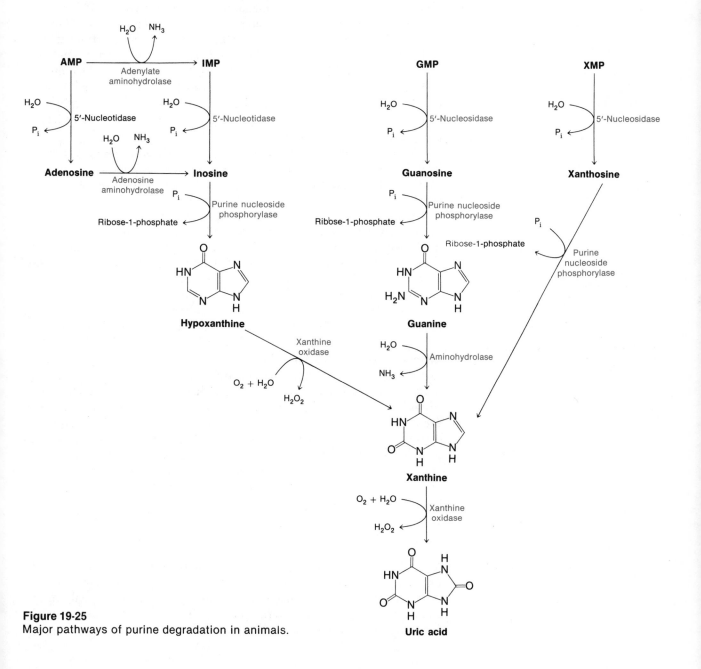

**Figure 19-25**
Major pathways of purine degradation in animals.

discovered that inherited deficiency of purine nucleoside phosphorylase is associated with a deficiency in the cellular type of immunity, but not in humoral immunity.

An adenosine aminohydrolase (deaminase) is also present in many mammalian tissues. This enzyme is of interest because hereditary deficiency of the enzyme is linked to a severe (usually fatal) immunodeficiency marked by a serious deficiency in lymphocytes and consequent inability to combat infections. Inosine formed by either route is then phosphorolyzed to yield hypoxanthine.

Although we have previously seen that much of the hypoxanthine and guanine produced in the mammalian body is converted to IMP and GMP by a phosphoribosyltransferase, about 10 percent is catabolized. Xanthine oxidase, an enzyme present in large amounts in liver and intestinal mucosa and in traces in other tissues, oxidizes hypoxanthine to xanthine and xanthine to uric acid (Figure 19-25). Xanthine oxidase contains FAD, molybdenum, iron, and acid-labile sulfur in the ratio $1:1:4:4$, and in addition to forming hydrogen peroxide, it is also a strong producer of the superoxide anion $\cdot O_2$, a very reactive species. The enzyme oxidizes a wide variety of purines, aldehydes, and pteridines.

Guanine aminohydrolase (guanine deaminase or guanase), present in liver, brain, and other mammalian tissues, provides another pathway to xanthine, this time from guanine. Subsequent oxidation of xanthine to uric acid then occurs.

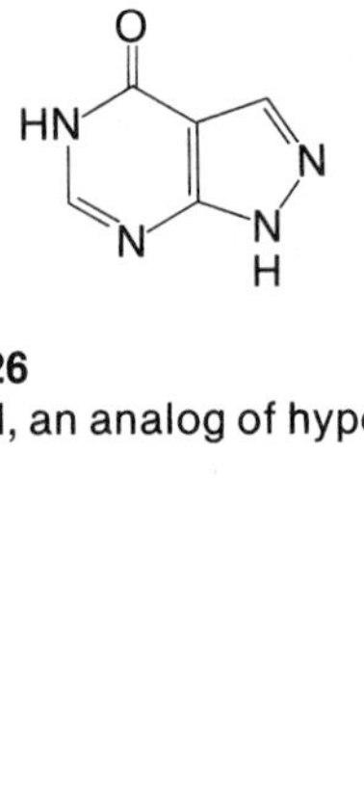

A relatively common ($\sim$3 per 1000 persons) derangement of purine metabolism known as *gout* is associated with elevated plasma levels of monosodium urate. This results in painful deposits of monosodium urate in the cartilage of joints, especially of the big toe. Uric acid deposits also may occur as calculi in the kidney, with resultant renal damage. The genetics are complex and incompletely understood. Individuals suffering from the high plasma urate levels of gout may be treated with the xanthine oxidase inhibitor allopurinol (Figure 19-26), an isomer of hypoxanthine. This produces a gradual decrease in urate levels in blood and urine and increased excretion of xanthine.

Mammals other than primates further oxidize urate by a liver enzyme urate oxidase, a copper protein. The product, allantoin, is excreted. Humans and other primates, as well as birds, lack urate oxidase and hence excrete uric acid as the final product of purine catabolism. In many animals other than mammals, allantoin is metabolized further to other products that are excreted: allantoic acid (some teleost fish), urea (most fishes, amphibia, some mollusks), and ammonia (some marine invertebrates, crustaceans, etc.). This pathway of further purine breakdown is shown in Figure 19-27.

**Figure 19-26**
Allopurinol, an analog of hypoxanthine.

## Pyrimidine Catabolism

A number of deaminases present in many cells are able to deaminate cytosine or its nucleosides or nucleotides to the corresponding uracil derivatives. Cytosine aminohydrolase (deaminase) appears to occur only in microorganisms (yeast and bacteria), but cytidine aminohydrolase is widely distributed in bacteria, plants, and mammalian tissues. A distinct deoxy-

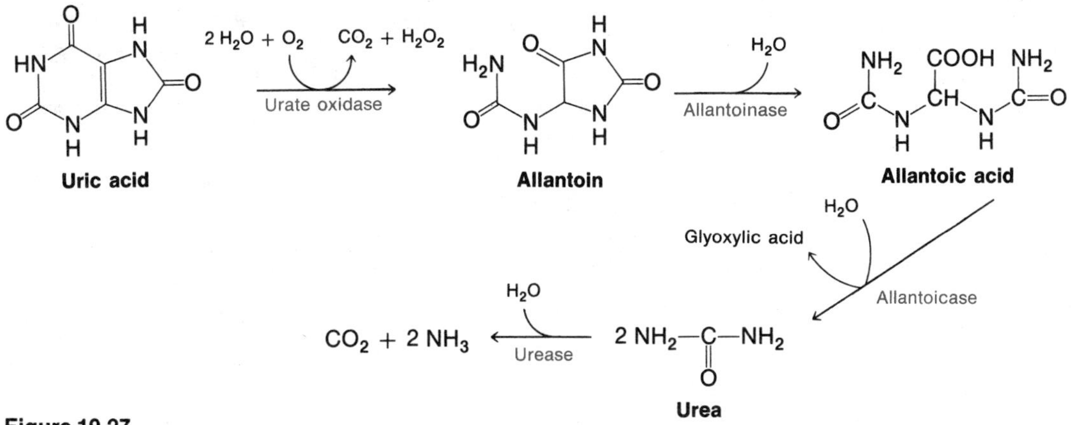

**Figure 19-27**
Degradation of uric acid to excretory products.

cytidine aminohydrolase is present in various mammalian tissues and tumors, in plants, and in bacteria. A deoxycytidylate aminohydrolase that is similarly distributed produces dUMP, which is susceptible to attack by 5′-nucleotidase to give deoxyuridine. Although the physiological function of these aminohydrolases is not completely understood, the uridine and deoxyuridine formed can be further degraded by uridine phosphorylase (see above) to uracil, so that these reactions provide a pathway for converting nucleotides of uracil and cytosine to uracil and ribose-1-phosphate or deoxyribose-1-phosphate (Figure 19-28). Similarly, thymine nucleosides and nucleotides can be converted by 5′-nucleotidase and phosphorylase to thymine.

Enzymes present in mammalian liver are capable of the catabolism of both uracil and thymine. The first reduces uracil and thymine to the corresponding 5,6-dihydro derivatives. This hepatic enzyme uses NADPH as the reductant, whereas a similar bacterial enzyme is specific for NADH. Similar enzymes are apparently present in yeast and plants. Hydropyrimidine hydrase then opens the reduced pyrimidine ring, and finally the carbamoyl group is hydrolyzed off from the product to yield $\beta$-alanine or $\beta$-aminoisobutyric acid, respectively, from uracil and thymine (Figure 19-28).

## REGULATION OF NUCLEOTIDE METABOLISM

Among the reaction pathways described earlier are many possibilities for "futile cycles" in which nucleotides built up in the biosynthetic pathways are broken down in catabolic pathways to products closely related to the starting materials. As an example, AMP synthesized from IMP by adenylosuccinate synthase and adenylosuccinate lyase (Figure 19-8) may be hydrolyzed back to IMP by adenylate aminohydrolase (Figure 19-25). The net result is the conversion of aspartate to fumarate and ammonia and the hydrolysis of GTP to GDP and $P_i$. To avoid such futile cycles, both biosynthetic and catabolic processes are under tight regulatory controls. The efficiency of these controls is demonstrated by the increased activity of many enzymes involved in nucleotide biosynthesis when cells are proliferating.

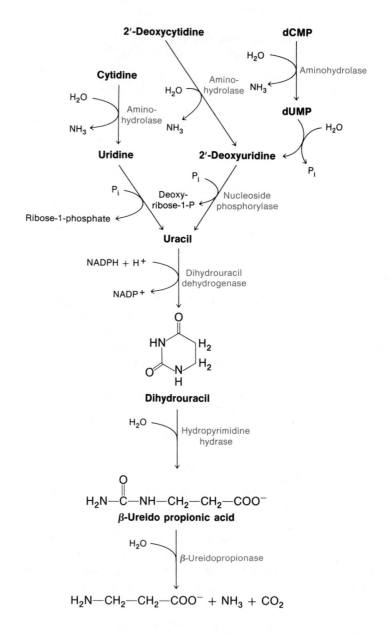

**Figure 19-28**
Degradation of pyrimidine bases.

Evidently, regulatory mechanisms increase nucleotide biosynthesis as intracellular nucleotides are removed for the synthesis of RNA and DNA. Further indications of the importance of these regulatory mechanisms are found in the drastic consequences that attend impairment of the control machinery, as in the Lesch-Nyhan syndrome (discussed earlier) or intervention with drugs such as 6-mercaptopurine (discussed earlier).

Although much remains to be discovered about the details of the regulation of nucleotide metabolism, a number of important control points are rather well understood and will be discussed together with their known effects on intracellular nucleotide pools.

# Regulation of Purine Biosynthesis

Many lines of evidence indicate that *the first committed step in de novo purine nucleotide biosynthesis, glutamine PRPP amidotransferase, is rate-limiting for the entire sequence*. Consequently, regulation of this enzyme is probably the most important factor in control of purine synthesis de novo (Figure 19-29). The enzyme is inhibited by purine-5′-nucleotides, but the nucleotides that are most inhibitory vary with the source of the enzyme. Enzyme from many sources, including pigeon liver, is effectively inhibited by AMP, GMP, and IMP. Some diphosphates and triphosphates also are effective as inhibitors of the enzyme from some species. Inhibition constants ($K_i$) are usually in the range $10^{-3}$ to $10^{-5}$ M. The maximum effect of this end-product inhibition is produced by certain combinations of nucleotides (e.g., AMP and GMP) in optimum concentrations and ratios, apparently indicating two kinds of inhibitor binding sites. This is an example of a *concerted* feedback inhibition.

The rate of the amidotransferase reaction is also governed by intracellular concentrations of the substrates L-glutamine and PRPP. Competing metabolic reactions or drugs that alter the supply of these substrates also affect the rate of IMP synthesis.

*The second important level of regulation of purine nucleotide synthesis is in the branch pathways from IMP to AMP and to GMP* (Figure 19-8). The first of the two reactions leading from AMP to IMP is the irreversible synthesis of adenylosuccinate. This requires GTP as a source of energy and is inhibited by AMP. Of the two reactions required to convert IMP to GMP, the first is irreversible and is inhibited by GMP and the second requires ATP as a source of energy. Thus there are two types of regulation of this level of purine nucleotide synthesis: a "forward" control, by which increased GTP accelerates AMP synthesis and increased ATP accelerates GMP synthesis, and feedback inhibition, by which AMP and GMP each regulate their own synthesis. Excess AMP also may be converted to IMP by adenylate aminohydrolase and thus can serve as a source of GMP. Adenylate aminohydrolase is activated by ATP and inhibited by GTP, which may serve to control this potential conversion of adenine nucleotides to guanine nucleotides. Finally, when the energy reserves of the cell are low, feedback inhibition of ribose-5-phosphate pyrophosphokinase by ADP and GDP restricts the synthesis of PRPP.

## Regulation of Pyrimidine Biosynthesis

In bacteria, the first committed step in pyrimidine nucleotide biosynthesis is the formation of carbamoyl aspartate from carbamoyl phosphate and aspartate. In *E. coli*, the enzyme catalyzing this step, aspartate carbamoyl transferase, is powerfully inhibited by CTP, a major mechanism for this being a decrease in the affinity of the enzyme for aspartate. ATP has the opposite effect, activating the enzyme by increasing affinity for aspartate. Concentrations of ATP and CTP in *E. coli* are high enough for these nucleotides to influence the intracellular activity of aspartate carbamoyl transferase. However, this is not a regulatory enzyme in all bacterial species and is not involved in regulation of pyrimidine nucleotide synthesis in animal cells.

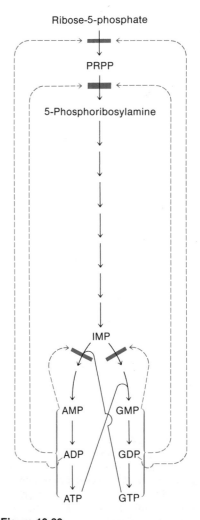

**Figure 19-29**
Regulation of purine biosynthesis. Red bars show points of feedback inhibition.

In eukaryotes, carbamoyl phosphate synthase is inhibited by pyrimidine nucleotides and stimulated by purine nucleotides and appears to be the most important site of feedback inhibition of pyrimidine nucleotide biosynthesis in mammalian tissues. However, it has been suggested that under some conditions, orotate phosphoribosyltransferase may be a regulatory site as well.

### Regulation of Ribonucleotide Reduction

Regulation of the reduction of ribonucleotides to deoxyribonucleotides has been studied with reductases from relatively few species. The enzymes from *E. coli* and from Novikoff rat liver tumor have a complex pattern of inhibition and activation (Figure 19-30). In both cases, dATP inhibits the reduction of all substrates, dTTP inhibits the reduction of CDP and UDP but activates the reduction of ADP and GDP, and ATP activates the reduction of UDP and CDP. Since there is evidence that ribonucleotide reductase may be the rate-limiting step in deoxyribonucleotide synthesis in at least some animal cells, these allosteric effects may be important in controlling deoxyribonucleotide synthesis.

The adenosylcobalamin-requiring ribonucleotide reductases from lactobacilli and certain other microorganisms have a different pattern of allosteric effects, the principal one being specific activation effects. For example, in the case of the *L. leichmannii* enzyme, dGTP activates ATP reduction, dATP activates CTP reduction, and dCTP activates UTP reduction. It is suggested that these effects serve to adjust the relative rates of reduction of the various substrates to more equal values. In addition, the synthesis of the *L. leichmannii* enzyme is repressed by the presence in the growth medium of an excess of vitamin $B_{12}$ (cyanocobalamin) or of a deoxyribonucleoside such as thymidine. The repressor for enzyme synthesis is probably dTTP or a closely related nucleotide, which accumulates in

**Figure 19-30**
Proposed scheme for the physiological regulation of deoxyribonucleotide synthesis in *E. coli* and mammalian cells. Black bars and colored bars indicate points of activation and inhibition, respectively. (From L. Thelander and P. Reichard, Reduction of ribonucleotides, *Ann. Rev. Biochem.* 48:133, 1979.)

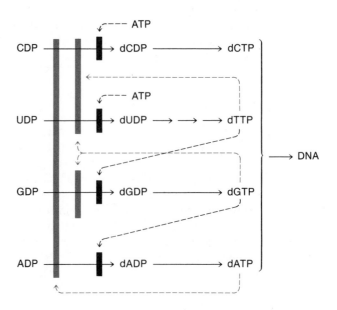

the cell when rapid ribonucleotide reduction occurs as a result of an ample cobalamin supply or when deoxynucleoside is supplied. Further deoxyribonucleotide synthesis is then slowed by the decreased rate of reductase synthesis.

## Channeling in Nucleotide Biosynthetic Pathways

In addition to the regulatory controls described in the preceding sections, which are mainly allosteric feedback mechanisms, evidence is accumulating that nucleotide biosynthetic pathways are closely controlled through the phenomenon of *channeling*. This involves an assembly of enzymes catalyzing successive steps in the biosynthetic pathway so that metabolic intermediates pass directly from one enzyme to the next. Thus the *metabolites are channeled along the metabolic pathway, with restricted opportunity for their diffusion into the medium or entry into the general metabolic pool in the cell.*

There are two major ways in which channeling of nucleotide precursors is achieved. The first involves multifunctional enzymes, in which several catalytic activities occur on a single polypeptide chain. The second involves noncovalent association of many pathway enzymes in complexes that may sometimes be concentrated in a particular intracellular location. In addition, channeling may be assisted by complete or partial confinement of metabolites within a specific intracellular compartment, e.g., within the nucleus.

Examples of multifunctional enzymes are provided by the pyrimidine biosynthetic pathway. Although in most prokaryotes, six structural genes code for the six enzymes involved in the de novo synthesis of UMP, in eukaryotes, the number of genes is reduced because of the production of multifunctional proteins. In *Neurospora*, a single protein has both carbamoyl phosphate synthase activity and aspartate carbamoyl transferase activity, but in mammalian cells, a single protein not only has both these activities, but also has dihydroorotase activity. The latter protein is an oligomer (probably a trimer) of a 200,000-dalton polypeptide, and there is evidence to indicate that the multifunctional polypeptide is a single gene product. This multifunctional enzyme channels carbamoyl phosphate and carbamoyl aspartate, provided dihydroorotate is rapidly removed, which is the case in the normal cell.

In mammalian cells, the last two steps of the pathway to UMP (Figure 19-10) also appear to be catalyzed by a multifunctional protein that is the product of a single gene. This protein therefore has both orotidine phosphoribosyltransferase and OMP decarboxylase activities. Although added OMP is accepted as substrate for the decarboxylase, OMP formed from orotate and PRPP is not released, but is preferentially utilized at the decarboxylase site for UMP formation.

In the purine pathway, a trifunctional enzyme catalyzes three reactions concerned with generation of the formyl donors for steps 3 and 9 (Figure 19-6). The enzymatic reactions catalyzed by this protein are shown in Figure 19-31. Since 10-formyltetrahydrofolate inhibits the cyclohydrolase, this may serve to regulate the relative amounts of the two formyl donors that are available. Channeling is further enhanced by noncovalent association

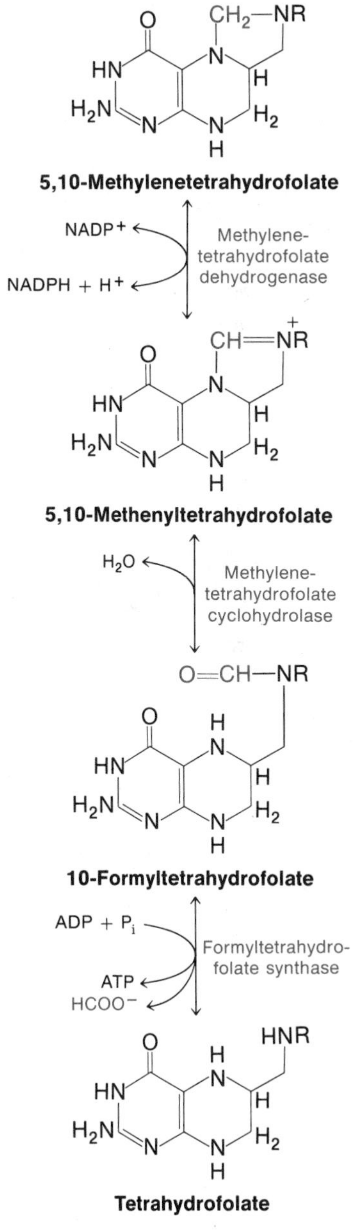

**Figure 19-31**
Formation of formyl group donors for purine biosynthesis (R = *p*-aminobenzoyl-L-glutamate).

of other enzymes of the pathway with the dehydrogenase-cyclohydrolase-synthase enzyme. Evidence has been obtained that this loose complex contains serine hydroxymethyltransferase (which generates methylenetetrahydrofolate), the transformylases catalyzing steps 3 and 9 of the purine pathway and probably all the other enzymes of the pathway. These enzyme activities largely remain associated through certain mild purification procedures. This association of pathway enzymes greatly increases channeling and permits the efficient generation and use of the hydrolytically and oxidatively unstable formyl donors.

Research with fibroblast cells in culture suggests that during the DNA-synthesizing phase of the cell cycle (S phase), a complex of at least six enzymes associated with DNA synthesis is present in the cell nucleus. Possibly its assembly signals the initiation of the S phase. The complex apparently involves DNA polymerase and many enzymes involved in the synthesis of deoxyribonucleoside triphosphates. These include thymidine kinase, dCMP kinase, nucleoside diphosphate kinase, thymidylate synthase, and dihydrofolate reductase. When the cells are quiescent, or in the $G_1$ phase, these enzymes are largely present in the cytoplasm and are no longer associated in a complex. Channeling by the nuclear complex is indicated by the observation that ribonucleoside diphosphates are incorporated more efficiently into DNA than deoxyribonucleoside triphosphates. This observation also suggests that ribonucleotide reductase is part of the complex.

### Intracellular Nucleotide Pools

Methods are now available for analysis of nucleotides in cells and tissues, and many reports have appeared on concentrations in various organisms under a variety of conditions. Concentrations are frequently expressed in terms of picomoles per $10^6$ cells or picomoles per microgram of DNA, since it is easier to express analyses on this basis than as intracellular molar concentration. However, some estimates in molar terms are available, and an example is given in Table 19-1. Deoxyribonucleotide concentrations in nondividing cells such as erythrocytes, unstimulated lymphocytes, or cultured cells blocked in $G_1$ are low but significant. dTTP is frequently present at the highest concentration (although not in Table 19-1), and dGTP is usually lowest in hamster or mouse cells, whereas dCTP is as low as dGTP, or lower, in human cell lines. _As cells enter the S phase there is an increase in the amount of each deoxyribonucleoside triphosphate in the cell_, the increase being greatest for dCTP and dGTP. _There is also a marked change in the relative distribution of nucleotides between the nucleus and the cytoplasm_ (Table 19-1), with a much greater increase in concentrations in the nucleus. This is particularly marked in the case of dTTP, which moves almost entirely into the nucleus by the end of the S phase.

In microorganisms, the levels of deoxyribonucleotides vary from undetectable to 200 picomoles per $10^6$ cells (compared with 3 to 30 picomoles of dATP, dCTP, and dGTP per $10^6$ eukaryotic cells). As in eukaryotic cells, dTTP is usually present at higher concentrations than the other deoxyribonucleoside triphosphates.

The arrest of cell growth by many agents is associated with depletion of one or more of the deoxyribonucleotide pools. Thus thymidine at millimolar concentrations arrests cell growth and decreases the concentration of dCTP dramatically, whereas the concentrations of dATP, dGTP, and especially dTTP increase. This effect is considered to be mediated by the allosteric inhibition of ribonucleotide reductase, referred to earlier. Hydroxyurea, another agent that arrests cell growth by blocking DNA synthesis, depletes the pools of dATP and dGTP, and it is the effect on the latter, also brought about by ribonucleotide reductase inhibition, that is probably critical.

Ribonucleotides are present in cells in much higher concentrations than deoxyribonucleotides, the general range of the concentration of the triphosphates in mammalian cells being 600 to 4000 picomoles per $10^6$ cells. This is equivalent to about 0.2 to 10 m$M$. The relative concentrations of the four triphosphates vary with conditions and the cell type, but ATP is normally highest in concentration. Levels of ribonucleotides are also perturbed by inhibitors. For example, they are increased by inhibitors of ribonucleotide reductase or decreased by inhibitors of the biosynthetic pathways.

Evidence has been obtained suggesting that equilibration of ribonucleotides in the nuclei of mammalian cells with those in the cytoplasm occurs at different rates in different cell types. In most cultured cell lines it is very rapid, but in others (Novikoff hepatoma) it is quite slow, and in still others (HeLa) equilibration occurs at an intermediate rate. This has been determined by examining the extent to which the incorporation into RNA of intracellular radiolabeled uridine is affected by putting the cells in a medium containing a high concentration of unlabeled uridine. The rapid influx of unlabeled uridine into the cytoplasm can decrease incorporation of labeled uridine in the nucleus only as cytoplasmic and nuclear nucleotides equilibrate.

## Effect of T-Even Phages on Nucleotide Metabolism

Infection of *E. coli* by T4 phage results in the induction of nearly 30 proteins. Many of these are enzymes that ensure an abundant supply of deoxynucleotides above those produced by the host. As a result, deoxynu-

**Table 19-1**
Nuclear and Cytoplasmic Concentrations of Deoxyribonucleoside Triphosphates in Chinese Hamster Ovary Cells in the $G_1$ and S (DNA-Synthesizing) Phases of the Cell Cycle

| | In $G_1$ Phase | | | In S Phase | | |
|---|---|---|---|---|---|---|
| | Nuclear Concentration ($\mu M$) | Cytoplasmic Concentration ($\mu M$) | Total Amount (pmol/ $\mu$g DNA) | Nuclear Concentration ($\mu M$) | Cytoplasmic Concentration ($\mu M$) | Total Amount (pmol/ $\mu$g DNA) |
| dATP | 6 | 12 | 3.2 | 40 | 19 | 3.4 |
| dCTP | 14 | 13 | 4.0 | 140 | 55 | 10.4 |
| dGTP | 1 | 0.7 | 0.16 | 10 | 2.5 | 0.76 |
| dTTP | 38 | 7 | 3.2 | 100 | <1 | 3.1 |

Source: From B. Bjursell and L. Skoog, Control of nucleotide pools in mammalian cells, *Antibiotics Chemother.* 28:78, 1980.

cleotide concentrations are increased many times. Phage-coded enzymes and related proteins include thioredoxin, ribonucleotide reductase, dihydrofolate reductase, dCMP deaminase, thymidylate synthase, and deoxyribonucleotide kinase. However, in some instances, the phage relies completely on host enzymes, which are normally present at high levels. Examples of such enzymes are adenylate kinase and nucleoside diphosphate kinase. Completely novel enzymes coded by the phage are endonucleases II and IV, which supply nucleotides directly by degrading host DNA.

The phage DNA contains no cytosine; instead hydroxymethylcytosine is incorporated. To accomplish this the phage induces enzymes that hydrolyze dCTP and dCDP, synthesize 5-hydroxymethyl dCMP from dCMP and methylenetetrahydrofolate, and phosphorylate hydroxymethyl dCMP. All these enzymes help to ensure rapid and specific synthesis of phage DNA while preventing synthesis of host DNA.

## SELECTED READINGS

Anderson, E. P. Nucleoside and Nucleotide Kinases. In P. D. Boyer (Ed.), *The Enzymes,* 3rd Ed., Vol. 9. New York: Academic Press, 1973. Pp. 46–96.

Becker, M. A., Raivio, K. D., and Seegmiller, J. E. Synthesis of phosphoribosylpyrophosphate in mammalian cells. *Adv. Enzymol.* 49:281, 1979.

Jones, M. E. Pyrimidine nucleotide biosynthesis in animals: Genes, enzymes and regulation of UMP biosynthesis. *Ann. Rev. Biochem.* 49:253, 1980.

Koshland, D. E., Jr., and Levitzke, A. CTP Synthetase and Related Enzymes. In P. D. Boyer (Ed.), *The Enzymes,* 3rd Ed., Vol. 10. New York: Academic Press, 1974. Pp. 539–559.

Parks, R. E., Jr., and Agarwal, R. P. Nucleoside Diphosphokinases. In P. D. Boyer (Ed.), *The Enzymes,* 3rd Ed., Vol. 8. New York: Academic Press, 1973. Pp. 307–333.

Thelander, L., and Reichard, P. Reduction of ribonucleotides. *Ann. Rev. Biochem.* 48:133, 1979.

## PROBLEMS

1. [3-$^{14}$C]Glucose is used for pentose phosphate synthesis by a cell extract in the presence of a concentration of fluoride sufficient to inhibit enolase. What positions of the pentose phosphates formed will be radiolabeled?
   a. If synthesis is only by the oxidative pathway, i.e., by means of 6-phosphogluconate.
   b. If synthesis is entirely nonoxidative, i.e., by means of glycolytic intermediates.

2. An analog of 2′-deoxyadenosine is extremely cytotoxic to human cells in culture.
   a. Suggest a possible mechanism involving nucleotide metabolism that might be responsible for the cytotoxicity.
   b. How would you obtain data to support the proposed mechanism?
   c. Deoxycoformycin, an inhibitor of adenosine deaminase, increases the cytotoxicity of deoxyadenosine but not of the analog. Explain.

3. Compare the number of high-energy phosphate bonds required for the synthesis of GTP and the synthesis of CTP. Assume that PRPP and folate one-carbon derivatives are available.

4. Allopurinol administered together with 6-mercaptopurine under certain conditions enhances the anticancer effectiveness of the latter. How can the known site of action of allopurinol explain this?

5. Which atoms of nucleotide bases isolated from a hydrolysate of DNA would be labeled by the following precursors?
   a. [3-$^{14}$C]Serine
   b. [$^{15}$N]Serine
   c. [$^{15}$N]Aspartic acid
   d. [2-$^{14}$C]Glucose
   e. [$^{14}$C]CO$_2$

6. A pig liver multienzyme protein preparation was incubated with 22 $\mu M$ 5,10-methylenetetrahydrofolate and

36 $\mu M$ NADP$^+$. After 3 min, the amount of 5,10-methenyltetrahydrofolate formed was 1.0 nmol and the amount of 10-formyltetrahydrofolate was 2.5 nmol. Over this interval, the formation of both compounds proceeds at a constant rate. What does this indicate about the kinetic behavior of this multienzyme protein? What result would be expected in the absence of this phenomenon?

7. A rat was injected subcutaneously with 13.5 $\mu$mol of cytidine labeled in all carbons with $^{14}$C. The specific radioactivity of the base moiety of this cytidine was 144 cpm/$\mu$mol, while the pentose moiety had 126 cpm/$\mu$mol. Twenty hours after the injection the animal was killed, and DNA and RNA from the tissues were converted to nucleosides. Specific activity of base (B) and pentose (P) moieties of nucleosides were as follows: cytidine: B 45, P 40; deoxycytidine: B 30, P 24. What conclusion can be drawn about deoxyribonucleotide synthesis in the rat? Do the results exclude the possibility of a pathway not involving ribose compounds?

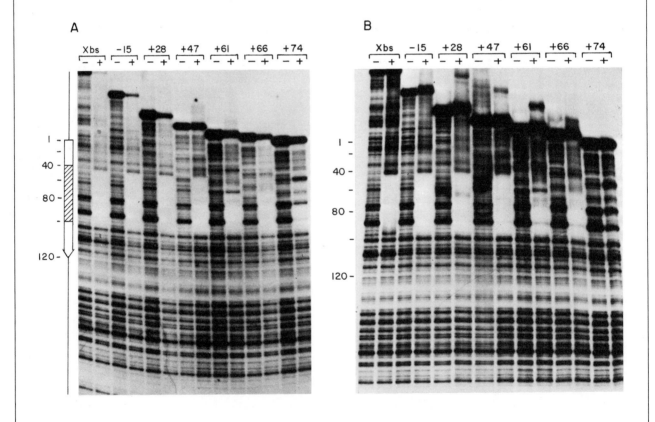

# DNA METABOLISM

Prior to 1944, chromosomes were generally thought to consist of linear arrangements of specific proteins held in place by a nonspecific DNA scaffolding. Interest in DNA was aroused by the experiments of Avery and coworkers that demonstrated that genetic information is not carried by protein but by nucleic acid. They isolated DNA from one strain of pneumococcus bacteria, exposed a second strain of pneumococcus bacteria to the DNA, and thereby transformed a small fraction of cells of the second strain into the genotype of the DNA donor strain. Related experiments by Hershey and Chase on bacteriophage T2 and by Gierer and Schramm and Frankel-Conrat and Singer on tobacco mosaic virus demonstrated that the genetic information of these viruses is carried by their nucleic acid. The flurry of excitement initiated by these observations has led to an ever-increasing interest in nucleic acids, making them one of the main topics in both biochemistry and genetics.

In biochemistry this interest is currently focused in two main channels, the first of which is to understand the normal metabolism of DNA. This includes all those biochemical reactions which result in the formation, maintenance, and degradation of DNA in the living cell. The second main channel was initiated around 1970 when Paul Berg and others proved the feasibility of manipulating genes. By chemical and biochemical procedures it is now possible to make new arrangements of genes that can be reinjected into cells or studied in vitro. In a sense, nucleic acid biochemists have become geneticists. However, unlike traditional geneticists seeking new gene arrangements made naturally, the biochemists have learned to make their own.

*DNAse I protection experiment on 5 S DNA of* Xenopus. *The diagram on the left indicates the region on the gel that corresponds to the 5 S RNA gene. Arrow points in the direction of transcription. Cross-hatched area indicates the region that binds transcription factor protein. Column labeled Xbs refers to an intact gene containing 160 bp of the 5′ flanking sequence, the 5 S RNA gene (120 bp), and the 3′ flanking sequence (138 bp). The various deleted 5 S DNAs are preceded by 74 pb of the plasmid pBR322 sequence. Numbers in other columns refer to portions of the 5 S gene that have been deleted. All samples are subject to partial digestion with DNAse I and electrophoresed and autoradiographed. In + columns, transcription factor protein was added before DNAse I treatment. In B, half as much transcription factor protein was used. (Courtesy Donald D. Brown of the Carnegie Institute of Washington.)*

## DNA METABOLISM

### DNA Synthesis

**The Universality of Semiconservation Replication.** The complementary duplex nature of the DNA structure and the necessity for precise doubling of DNA prior to cell division led to the suggestion that *DNA synthesis in vivo occurs by the unwinding of the double helix, absorption of complementary mononucleotides to each polynucleotide strain, and polymerization.* In this mode of semiconservative replication, *newly duplicated DNA should consist of one old strand and one newly synthesized strand.* Evidence that this is the case for bacterial DNA was provided by the elegant experiments of M. Meselson and F. Stahl. They conceived of the idea of synthesizing DNA of one density by growing cells on medium containing only $N^{15}$ in the form of ammonium chloride and then transferring the cells to medium of normal $N^{14}$ and allowing the DNA to duplicate for one or more generations. DNA of different densities can be separated in the ultracentrifuge by the technique of *density-gradient centrifugation.* A small amount of DNA in a concentrated solution of cesium chloride is centrifuged until equilibrium is closely approached. The opposing processes of sedimentation and diffusion produce a stable concentration gradient of the cesium chloride. The concentration and pressure gradients result in a continuous increase of density along the direction of centrifugal force. Macromolecules of DNA present in this density gradient are driven by the centrifugal field into the region where the solution density is equal to their own buoyant density. This concentrating tendency is opposed by diffusion, with the result that at equilibrium a single species of DNA is distributed over a band whose width is inversely related to the molecular weight of that species. Bands for pure $N^{15}$ and pure $N^{14}$ DNA are shown in Figure 20-1 (frames 1 and 2, respectively), as well as for DNA from the same

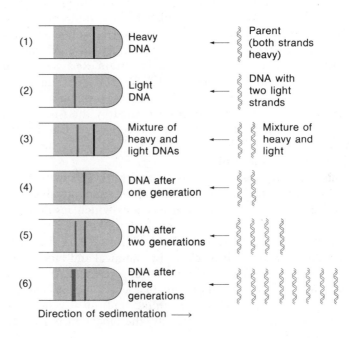

**Figure 20-1**
The Meselson-Stahl experiment illustrating semiconservative replication in *E. coli.*

culture at one, two, and three generations after transferring to an $N^{14}$ medium (frames 4, 5, and 6). Bands of varying intensity appear at points corresponding to pure $N^{15}$ DNA, and $N^{15}N^{14}$ DNA. After precisely one generation time in $N^{14}$, the only band visible is that corresponding to $N^{15}$-$N^{14}$ hybrid DNA. These data argue strongly in favor of the semiconservative mode of DNA replication.

Similar experiments have been done on mammalian cells grown in tissue culture using bromouracil as a density label. Bromouracil, when incorporated by tissue culture cells in the presence of methotrexate (see Chapter 19), can replace most of the thymidine, leading to a substantially higher density of the resulting DNA. The results on eukaryotic DNA parallel those on *E. coli*. Both these experiments suffer from one big disadvantage. During isolation, chromosomal DNA is susceptible to shear, so the largest pieces of DNA that can be conveniently isolated without fragmentation are in the size range of $10^7$ to $10^8$ daltons. This is far below the size of the average chromosome, meaning that it is impossible to know if the DNA in a whole chromosome is replicated semiconservatively over its entire length. J. H. Taylor, P. Woods, and P. Hughes devised a method to overcome this problem. Chromosomal DNA was labeled by the incorporation of tritiated deoxynucleosides, and the distribution of the radioactive label was visualized in autoradiographs of chromosomes during the subsequent replication of cells in the absence of tritiated nucleoside (Figure 20-2). Bean seedlings were grown in medium containing [$^3$H]thymidine for a period sufficient to allow some of the cells in the seedlings to undergo one round of DNA duplication and cell division. After this, the seedlings were transferred to a fresh solution containing colchicine without the $^3$H label. Colchicine is a plant alkaloid that inhibits normal mitosis by interacting with microtubule proteins, keeping the chromosomes from moving apart. This prevents the separation of daughter cells, but does not inhibit DNA synthesis or the ability of chromosomes to replicate. After colchicine treatment, chromosomes are found in the metaphase state with the sister chromatids paired. Cells that were examined shortly after the transfer from the [$^3$H]-thymidine medium were stopped at the first mitotic event, and all the chromosomes appeared to be labeled uniformly. When the cells were allowed to duplicate their chromosomes in the unlabeled medium, only one of the two chromatids in each chromosome pair was labeled, demonstrating semiconservative replication for entire chromosomes. Similar results have been obtained for other eukaryotic cell types.

**DNA Duplication in Bacteria.** The overall simplicity of the DNA replication process is not attended by a simple enzymology. The evolutionary advantages of rapid and accurate replication coupled with the topological problems of unwinding long helical molecules have resulted in one of the most complex biochemical processes known. Conceptually, replication can be divided into three stages: *initiation*, *elongation*, and *termination*. Each stage may be subdivided further into several different events that involve many different proteins.

DNA replication in prokaryotes, particularly in *E. coli*, has been most extensively studied. Here it is possible to exploit both genetic and biochem-

**Figure 20-2**
Autoradiographs of *Vicia faba* chromosomes labeled with [³H] thymidine. The labeled thymidine becomes incorporated into the chromosomal DNA. A suitably labeled preparation is flattened and subject to film exposure. Small dots indicate radioactive disintegration in the exposed film: (*a*) shows the first metaphase after replication in the presence of [³H] thymidine; (*b*) shows the second metaphase after an additional replication in nonradioactive medium; (*c*) shows a diagrammatic interpretation of the results shown in (*a*) and (*b*). Radioactive single strands of DNA are shown in color. Radioactive chromatids at metaphase are indicated in color. Colchicine has been used to inhibit spindle fiber formation and thus the anaphase separation of sister chromatids. Under these "C-metaphase" conditions, separation of sister chromatids is delayed. In (*a*) both sister chromatids are labeled uniformly. In (*b*) the sister chromatids are not labeled uniformly. The large chromosome at the top has one labeled chromatid and one virtually unlabeled. The homolog to its right has two exchanges (a labeled segment moved into the lower chromatid). The two small chromosomes to the lower left of it are lying one on top of the other. Both have one sister chromatid exchange. The small chromosome to the upper left has one sister chromatid exchange and the one at the lower left is lightly labeled but probably has two exchanges. (Autoradiographs generously donated by J. H. Taylor.)

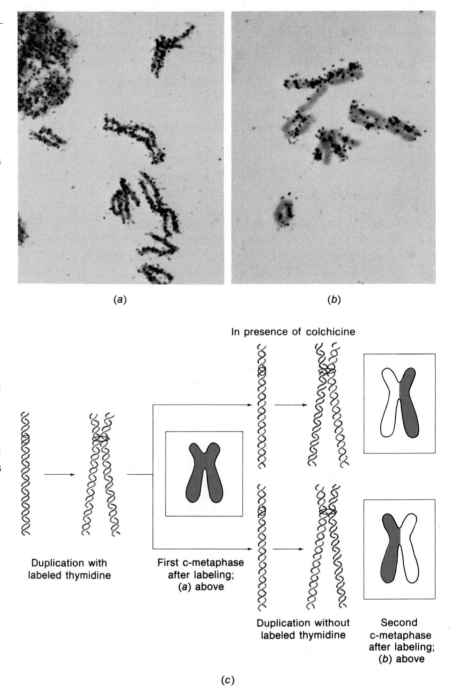

(*a*)          (*b*)

In presence of colchicine

Duplication with labeled thymidine

First c-metaphase after labeling; (*a*) above

Duplication without labeled thymidine

Second c-metaphase after labeling; (*b*) above

(*c*)

ical techniques to attack the problem. Replication of viruses and plasmids that infect *E. coli* as well as the cellular chromosome itself have been examined. Whereas the semiconservative mode of replication seems to be universal, the details of the replication process of each system examined show interesting differences.

DNA replication in *E. coli* begins at a unique locus on the circular *E. coli* chromosome known as *oriC* (Figure 20-3). In different systems, the initiation of DNA synthesis at the origin of replication leads to one or two replication forks (Figure 20-4) depending on whether the synthesis is unidirectional or bidirectional. If a single fork is created, replication will proceed unidirectionally away from the origin of replication. If two forks are created, replication will proceed bidirectionally. For a circular DNA, such as the *E. coli* chromosome, the bidirectional model predicts that the relative frequency of genes near the origin during early stages of replication will be higher than the relative frequency of those distant from the origin. This was shown to be the case in *E. coli*. Additional cytologic evidence for bidirectional replication in *E. coli* and other bacteria was obtained. When the bacteria are grown for a very short time in the presence of radioactive thymidine and the replicating chromosomes are examined by autoradiography, both forks visible in the replicating structures are intensely labeled. This indicates that both forks are active during replication. Continuous synthesis on both strands of the replication fork would require synthesis in the 5′→3′ direction on one strand and in the 3′→5′ direction on the other strand because of the antiparallel nature of the DNA duplex. All known polymerases work in the 5′→3′ direction (see below). Attempts to isolate the "missing" 3′→5′ polymerase failed. Finally it was appreciated that *replication is discontinuous on at least one of the branches at the replication fork*. Careful electron microscopic examination of replication forks in bacteriophages showed that transient gaps may appear on one of the DNA strands near the replication fork. Observations such as this led to the concept of *leading strand* and *lagging strand synthesis*. Synthesis of the leading strand (5′→3′) could occur continuously in the direction of unwinding at the replication fork. Synthesis of the lagging strand (5′→3′) could occur in discontinuous spurts in the opposite direction. This concept was supported by the finding that at least half the newly synthesized DNA was first made in small pieces that later become incorporated into large pieces of DNA (Figure 20-5). Nascent replication fragments were first obtained in the laboratory of R. Okazaki by labeling of *E. coli* with tritiated thymidine

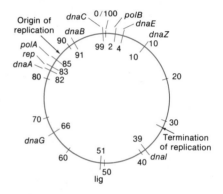

**Figure 20-3**
The circular *E. coli* chromosome. The origin and approximate region of termination of replication are indicated. Locations of some genes involved in DNA replication are also indicated.

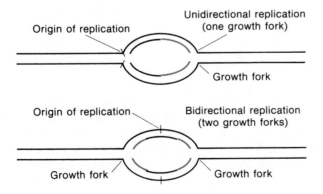

**Figure 20-4**
Schematic diagrams of two different modes of DNA synthesis at the replication fork for a circular chromosome: unidirectional and bidirectional. Newly synthesized DNA is shown in color.

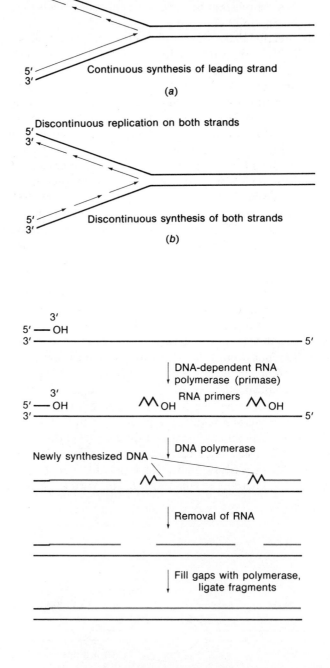

**Figure 20-5**
Models for synthesis at the replication fork. (*a*) Discontinuous synthesis of lagging strand. (*b*) Discontinuous synthesis of lagging and leading strands.

**Figure 20-6**
Proposed mechanism for the synthesis and knitting together of Okazaki fragments.

for a very short time (2 to 30 s), followed by rapid isolation of the radioactively labeled DNA. After longer periods of time (1 to 2 min), most of the labeled *Okazaki fragments* were found in large continuous pieces of DNA. Sedimentation analysis on alkaline sucrose gradients (the alkali denatures the DNA into single strands) provided an estimated length of 1000 to 2000 bases for bacterial Okazaki fragments. The conclusion was reached that the replicating polymerase must be operating in an on-and-off (dispersive) manner, synthesizing short fragments on the lagging strand as new points for the initiation are presented by the progressive unwinding of the double helix at the replication fork. A closer *examination of the Okazaki fragments showed that short stretches of ribonucleotides are covalently linked at the 5' ends*. This indeed was a happy day, because it was known that DNA polymerases have an absolute requirement for primer at least in vitro (see below). It seemed likely that a primer RNA was made either by RNA polymerase or by another enzyme. A detailed model for discontinuous synthesis could now be proposed (Figure 20-6). First primers are made on the single-strand region of the template; then DNA is synthesized. Finally, the RNA is removed from the fragments and the gaps are filled and ligated. An understanding of the detailed biochemistry of this process has come largely from analysis of simpler viral replicating systems (see below).

**Analysis of Proteins Involved in DNA Replication.** Table 20-1 contains a list of some proteins implicated in DNA replication by mutant studies and their alleged function. How does one go about analyzing such a complex situation? Historically, there are three general methods that have been used for the identification and characterization of proteins involved in DNA replication: purification, reconstitution, and mutation. Insofar as possible,

**Table 20-1**
A Partial List of Protein Factors Involved in DNA Replication in *E. coli*

| Protein | Locus | Size | Function |
|---|---|---|---|
| A | *dnaA* | — | Unknown initiation function |
| B | *dnaB* | 48,000 subunits | Interacts with primase |
| C | *dnaC* | 25,000 | Interacts with dnaB protein |
| RNA polymerase | *rpo* | 460,000 | Initiation at *oriC* |
| G (primase) | *dnaG* | 65,000 | Synthesizes primers |
| | *dnaZ* | 52,000 | Subunit of polymerase III holoenzyme |
| DNA polymerase III | *dnaE* (*polC*) | 140,000 | Subunit of polymerase III holoenzyme |
| DNA polymerase I | *polA* | 109,000 | Gap filling and repair synthesis |
| DNA ligase | *lig* | 74,000 | Joins fragments |
| | *dnaX* | 32,000 | Subunit of polymerase III holoenzyme |
| | *dnaN* | 37,000 | Subunit of polymerase III holoenzyme |
| Topoisomerase II | | | |
| A subunit | *gyrA* (*nal*) | 105,000 | Component sensitive to nalidixic acid; breakage and rejoining activity |
| B subunit | *gyrB* (*cou*) | 95,000 | Component sensitive to coumermycin and novobiocin; hydrolyzes ATP |
| Helicase | *rep* | 67,000 | Unwinds DNA at replication forks; hydrolyzes ATP |
| Single-stranded DNA-binding protein | *SSB* | 18,500 | Stabilizes single-stranded DNA |

all three methods are used together. The first method involves isolation of proteins with enzymatic activities that are logically related to the replication process, such as DNA polymerases and ligases. This is the classical biochemical approach and can be applied to any biological system. After isolation and characterization, several approaches may be used to demonstrate that the purified enzyme is active in the replication process in vivo. Sometimes this can be done by using inhibitors that act on both the purified protein and the cellular process. The concentration of inhibitor required to inhibit DNA replication in vivo should be approximately the same as that required to inhibit the purified enzyme. In prokaryotes, mutations have been very useful for confirming the functions of isolated proteins in the replication process. The induction of a new enzyme activity associated with a biological process, such as virus infection or cell proliferation, also provides useful evidence.

A second method uses reconstitution to aid in identifying proteins necessary to restore activity to an in vitro replication system. Crude extracts containing the replication system are fractionated and the DNA replication process is then reconstituted with various combinations of the purified or partially purified proteins. Components of the replication system are recognized on the basis of their ability to reconstitute overall activity in vitro. This procedure can be applied to any organism, even when genetic mutants are not available.

A third method uses genetic mutation as the primary tool. This approach is most effective with organisms having a well-characterized genetic map and requires the isolation of conditionally lethal mutants, i.e., mutants that grow normally under one set of conditions but not under another set of conditions. Most commonly, temperature-sensitive (ts) DNA replication mutants are used. Such mutants have been isolated in *E. coli*, and they grow normally at a low permissive temperature (~33°C) but poorly or not at all at a nonpermissive temperature (~41°C). Preliminary analysis of the temperature-sensitive step provides clues to the stage affected. For example, the length of time that is required for DNA synthesis to stop after cells are shifted from permissive to nonpermissive temperatures can indicate whether the mutation occurs in a protein involved in initiation or elongation. If the mutation is in the gene for a protein required for elongation, DNA synthesis will stop almost immediately at the nonpermissive temperature. If the mutation is in a protein required only for initiation of replication, DNA synthesis will continue for some time and stop when the round of replication in progress is completed. In vitro complementation assays are then used to aid in purifying the corresponding proteins from wild-type cells. Extracts from cells with the temperature-dependent defect will not synthesize DNA at the elevated temperature, but activity can be restored by adding the corresponding protein from wild-type cells. The complementation effect can be used as an assay for the purification of particular replication proteins. To prove that the correct protein has been purified from wild-type cells, the protein from the temperature-sensitive mutant also must be purified and shown to be abnormal, frequently exhibiting unusual thermolability.

**DNA Polymerase I: The Kornberg Enzyme.** The Watson-Crick proposal of a complementary duplex structure for DNA stimulated a search for a DNA

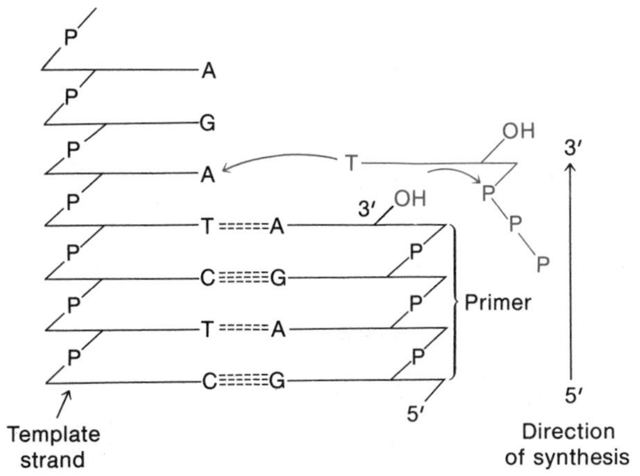

**Figure 20-7**
Template and priming strands of DNA in cell-free synthesis.

synthase with certain implied properties. The enzyme should require an intact DNA chain to serve as a template for the absorption of complementary bases, and the de novo synthesized DNA should be a complement of one of the template DNA chains. A. Kornberg and coworkers isolated a DNA-synthesizing enzyme from cells of *E. coli* that satisfied these requirements and named it *DNA polymerase*; it is now known as *DNA polymerase I*, or the *Kornberg enzyme*. Similar enzymes of viral origin have been isolated from some virus-infected bacterial cells. In the pure state, the Kornberg enzyme requires a DNA template, the four commonly occurring deoxynucleotide triphosphates, and $Mg^{2+}$ ions for making DNA. The enzyme catalyzes the addition of mononucleotides to the 3' end of a growing chain (see Figure 20-7). The reaction probably occurs as a nucleophilic attack by the 3'-hydroxyl group of the terminal mononucleotide residue at the growing end of the chain on the α-phosphorus atom of the entering nucleoside-5'-triphosphate, causing displacement of its pyrophosphate group and formation of the internucleotide linkage (Figure 20-8). If no 3' end is available, such as when a closed circular single-stranded DNA template is used, the enzyme must start from mononucleotides. In this case, there is a long lag period in the initiation of synthesis. The bases added to the growing chain are determined by the sequence of bases in the DNA template. Complementary bases are added so that the single-stranded DNA template gradually becomes converted to double helix in which one chain has resulted from cell-free synthesis. Ordinarily the cell-free synthesis proceeds beyond the double-helix stage at a considerably reduced rate. Physicochemical studies have shown that the enzyme has a complex surface with specific attachment sites for template chain, growing primer chain, and monomer triphosphate. The enzyme is highly selective, since it connects only a base that is properly paired to the template strand. As the primer chain becomes lengthened by synthesis, the enzyme must move along the template one base at a time. That the newly synthesized DNA is a faithful complementary replica of the template strand has been shown in a number of ways: by gross base-composition analysis, by melting-curve profile (see Chapter 18 for explanation of nucleic acid melting curves) of the hybrid formed between the template strand and the newly synthesized primer strand, and finally, through a series of manipulations, by the de novo synthesis of genetically active DNA.

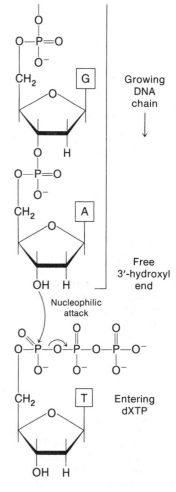

**Figure 20-8**
Mechanism of chain extension by DNA polymerase.

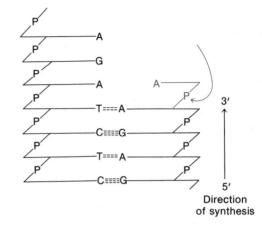

**Figure 20-9**
DNA polymerase I "proofreading." The residue incorporated is mispaired and removed by 3'-exonuclease proofreading activity of enzyme.

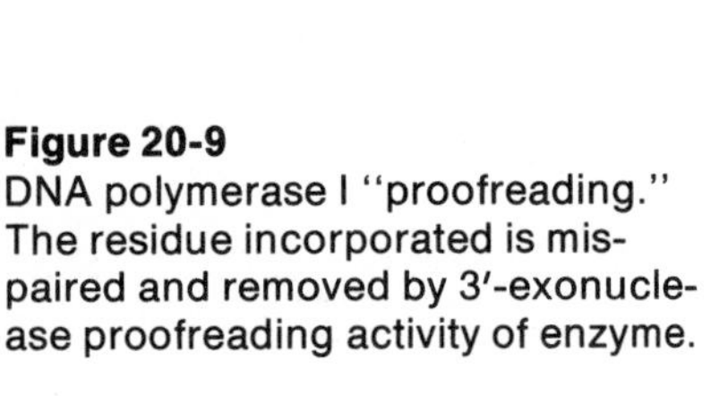

**Figure 20-10**
Nick translation. DNA polymerase I is the only enzyme that is able to promote replication at a nick unaided by other proteins. This entails unwinding of the duplex beyond the nick and progressive strand displacement of the 5' chain. The 5'→3' exonuclease activity appears to operate simultaneously. The net result is that nick translation results in a movement of the nick site down the chain without any change in the mass of the DNA. Nick translation has been a useful way for preparing highly radioactive DNA. The desired template is randomly nicked with pancreatic DNAse I and incubated with radioactively labeled triphosphates. Since the nick sites are randomly disposed, the final DNA is labeled over the entire genome.

In addition to the characteristic template-directed dNTP polymerization, DNA polymerase I has an associated 3'→5' exonuclease activity. The associated exonuclease can remove a 5'-nucleotide from the 3' end of a double-stranded template molecule and effect an apparent reversal of the polymerization process. The catalytic site for the 3'→5' exonuclease activity is thought to be separate from but close to the polymerization center on the enzyme molecule. For chain propagation to occur beyond the 3'-OH initiation point, the polymerization rate must exceed the depolymerization rate. Although the ratio of polymerization to depolymerization rates varies with different prokaryotic DNA polymerases (extremely high depolymerization rates are found in some bacteriophage-induced polymerases and even higher rates are found in so-called antimutator strains), polymerization is much faster than depolymerization if the correct base-paired nucleotide is inserted in the growing chain. If a mismatched base is accidentally inserted, the opposite is true and the mismatched base is usually removed. *The combined polymerization-depolymerization reaction has been viewed as a "proofreading" error-reducing mechanism in DNA synthesis* (Figure 20-9).

*E. coli* DNA polymerase I also has a second associated exonuclease activity catalyzing 5'→3' degradation of DNA. Whereas the 3'→5' nuclease activity of the enzyme is much more effective on unpaired or mispaired bases, the 5'→3' nuclease cleaves only at base-paired regions. The ability of DNA polymerase I to degrade DNA in either direction as well as to carry out a polymerization reaction would suggest a possible function in removing and restoring damaged bases as part of a DNA repair system (discussed below).

The ability of the enzyme to excise nucleotides at a nick and add new nucleotides to the 3'-OH at the gap so formed is used extensively for producing highly radioactive DNA molecules for use in hybridization experiments (Figure 20-10). This process has been called _nick translation_ because the position of the original nick is moved along the DNA chain. The 5'→3' exonuclease function can be removed from polymerase I by subtilisin cleavage, which produces a small fragment (30,000 daltons) with 5'→3' exonuclease activity and a large fragment (70,000 daltons) exhibiting the polymerization and 3'→5' exonuclease activities.

**Establishing the Physiological Roles of DNA Polymerases I and III.** Final proof that an enzyme or protein functions in the same capacity in vivo requires that mutants be isolated that affect specific proteins. Our current understanding of the physiological role of DNA polymerase I exemplifies the importance of correlating the effects of mutants with the biochemical behavior.

From the time of its initial discovery it took about 20 years to reach our current understanding of the physiological role of the DNA polymerase I enzyme. Kornberg and coworkers discovered DNA polymerase I in 1956. The enzyme was highly purified from cell-free extracts of *E. coli;* it had all the polymerization properties initially anticipated for an enzyme responsible for chromosome replication, as explained earlier. Fifteen years after Kornberg discovered DNA polymerase I, J. Cairns and P. De Lucia laboriously scanned several thousand strains of heavily mutagenized *E. coli* and found some that contained almost no DNA polymerase I polymerizing activity (1 to 2 percent of normal). Such mutants grew normally under optimal conditions, but were particularly sensitive to ultraviolet radiation. These properties strongly suggested that DNA polymerase I was not the chromosome-replicating polymerase, but might be involved in repairing chromosome damage resulting, for example, from ultraviolet radiation. The discovery of the polymerase mutant had a profound effect on thought and experimental design in studies on DNA biosynthesis. An important general principle of biochemical research was underscored by this unexpected finding. A function should not be assigned to an enzyme on the basis of its presence and its in vitro properties alone. Only genetic mutants make meaningful in vivo correlates possible. In a cell-free extract from an *E. coli* strain that did not contain a masking amount of DNA polymerase I polymerizing activity it was subsequently possible to detect two additional DNA polymerizing enzymes. These were named DNA polymerases II and III (see Table 20-2). The behavior of conditional lethal mutants of DNA polymerase III has led to the conclusion that DNA polymerase III (the largest subunit of which is encoded by *dnaE*) is the main replicating enzyme.

For a few years following the observations of De Lucia and Cairns it was assumed that DNA polymerase I was not important in replication but only

**Table 20-2**
Properties of Polymerases I, II, and III of *E. coli*

|  | Pol I | Pol II | Pol III |
| --- | --- | --- | --- |
| Molecules per cell | 400 | — | 15 |
| Turnover number* | 600 | 30 | 9000 |
| Structural gene† | *polA* | *polB* | *polC* |
| Conditional lethal mutant | + | — | + |
| 5'→3' Polymerizing activity | + | + | + |
| 3'→5' Exonuclease activity | + | + | + |
| 5'→3' Exonuclease activity | + | — | + |

*Nucleotides polymerized/min/molecule of enzyme at 37°C.
†Only the structural gene for the largest protein subunit in the enzyme is recorded.

in repair. However, further genetic and biochemical studies provided convincing evidence that DNA polymerase I is a multifunctional enzyme possessing, in addition to 5′→3′ polymerizing activity, the 3′→5′ and 5′→3′ exonuclease activities described earlier (see Table 20-2). The original mutant isolated by Cairns and De Lucia was inactivated only in the polymerizing function. Subsequently, mutants in DNA polymerase I that affect the 5′→3′ exonuclease function were found that are conditionally lethal and do not permit elongation of DNA synthesis under nonpermissive conditions; this demonstrated that this activity of the DNA polymerase I enzyme is indispensable for chromosome replication.

DNA polymerase III, which functions as the main replicating enzyme, is believed to be composed of at least eight different protein subunits. The complete enzyme, called the _holoenzyme_, is assayed by its requirement in the conversion of primed, single-stranded circular phage $\phi$X174 to the duplex form (discussed below). The holoenzyme consists of a tripolypeptide DNA polymerase III core ($\alpha$, $\varepsilon$, and $\theta$ subunits that are 140,000, 25,000, and 10,000 daltons, respectively) and the following separable subunits: $\beta$, 37,000 daltons; $\gamma$, 52,000 daltons; $\delta$, 32,000 daltons; $\tau$, 83,000 daltons; and $\zeta$. Genetic loci for some of these polypeptides have been identified (see Table 20-1). Mutants affecting the remaining subunits as well as the genes encoding them are unknown, and therefore their physiological significance has not been definitely established.

**The Joining Enzyme: Polynucleotide Ligase.** As mentioned earlier, discontinuous DNA synthesis necessitates the existence of an enzyme for joining the newly synthesized segments together (see Figure 20-6). Such an enzyme has been found in a variety of cell types and is generically referred to as _polynucleotide ligase_. DNA ligases cannot execute a monomer polymerization reaction like the DNA polymerases. _The net reaction with DNA ligase involves the formation of a single phosphodiester bond between long runs of discontinuous chains held in the proper juxtaposition by a_

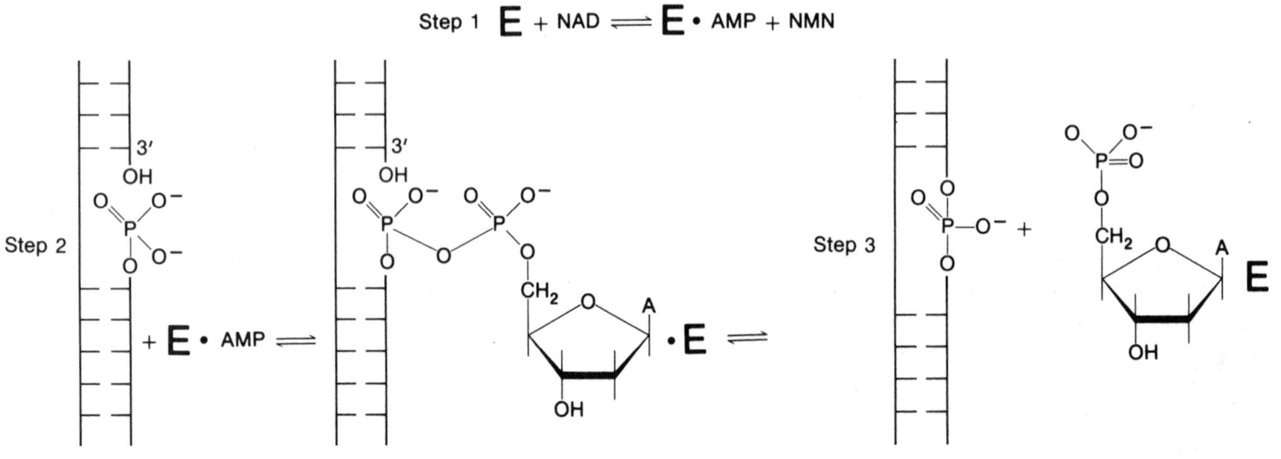

**Figure 20-11**
Steps in the DNA ligase-catalyzed sealing of a DNA nick. The bacterial ligase uses NAD $^+$ to make an enzyme-AMP intermediate. Mammalian DNA ligases and bacteriophage T4 ligase use ATP for the same purpose.

Step 1  $E + NAD \rightleftharpoons E \cdot AMP + NMN$

template chain (Figure 20-11). This joining reaction is vital to DNA replication. Thus it is not surprising to find that DNA ligases are ubiquitous in living cells. They differ according to the source of activation energy used for the phosphodiester bond formation. The *E. coli* bacterial ligase (encoded by the *lig* gene; Table 20-1) uses NAD$^+$ as the source of energy and transfers AMP to a lysine residue of the ligase, forming a phosphoramidate bridge. The 3',5'-phosphate bond is finally closed, releasing AMP (step 3). The T4 bacteriophage and mammalian DNA-associated ligases use ATP in an entirely analogous manner, releasing PP$_i$ in the first step. DNA ligases are necessary for converting the discontinuous daughter chains formed as a result of multiple initiation during DNA replication into continuous double helices. They also are used to re-form continuous chains in the ultimate step in DNA repair and recombination. In vitro, DNA ligases have been used as important reagents for the closure of annealed sequences in the combined chemical and enzymatic synthesis of tRNA genes, as well as to join recombinant DNA fragments to vector DNA for genetic-engineering studies (discussed below).

**The Topoisomerases.** Supercoiled helixes are frequently found to be present in closed circular (or otherwise constrained segments of) DNA molecules. Supercoiling of the helical strands is induced by applying torsional energy to the DNA helix, resulting in more or less than 10 base pairs per helical turn. The torsional tension is partially relieved by "supertwisting" of the molecule in the opposite sense (see Chapter 18).

The availability of supercoiled DNA and advances in the technology of DNA analysis by agarose gel electrophoresis (Figure 18-25) has led to the discovery of several kinds of enzymes that can either reduce or increase the winding number (linkage number) of a supercoiled helix. These enzymes, called *topoisomerases*, were first identified as activities capable of relaxing negatively supercoiled DNA in *E. coli* and in mouse embryo cells. Topoisomerases are thought to be essential for DNA metabolism and they are widely distributed in living organisms. The enzymes can break and rejoin DNA repeatedly without any added cofactor to supply the energy for rejoining, leading to the conclusion that the reaction intermediate conserves the energy of the DNA phosphodiester bond. The hypothesized mechanism involves the covalent-bond formation between the protein and the broken end of the DNA, followed by resealing of the DNA and dissociation of the enzyme after the linking number has been changed by one or more units. A 3'-phosphotyrosine linkage has been detected as an intermediate. *Topoisomerases are classified as type I or type II according to whether they change the linking number in steps of one or steps of two, respectively. Type I enzymes produce transient single-strand breaks in the double helix, whereas type II enzymes produce transient double-strand breaks* (Figure 20-12). Type I topoisomerase of *E. coli*, also known as ω protein, is the best understood enzyme of this class; its molecular weight is 110,000 and the native protein exists as a monomer in solution. The enzyme shows a preference for highly negatively supercoiled DNA and it is inactive on positively supercoiled DNA. This enzyme can also catalyze the *catenation* of double-stranded circular DNA, or catenanes can be separated

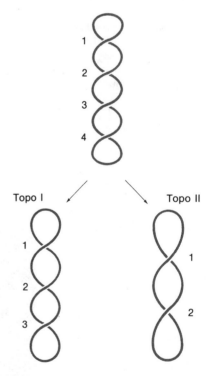

**Figure 20-12**
Type I and type II topoisomerases relax negatively supercoiled DNA in steps of one and steps of two, respectively.

**Figure 20-13**
Catenation by topoisomerases. (*a*)
Two circular DNAs can be catenated
by type I topoisomerase only if one of
the DNAs is nicked. This is not necessary when using a type II topoisomerase. (*b*) Electron micrographs
of catenated DNA before (*i*) and after
(*ii*) incubation with DNA gyrase. The
catenane contains one large, circular
λ DNA and one small, circular pBP66
plasmid DNA. (Adapted from M. Gellert, L. M. Fisher, H. Ohmori, M. H.
O'Dea, and K. Mizuuchi, DNA gyrase:
Site-specific interactions and transient double-strand breakage of
DNA, *Cold Spring Harbor Symp.
Quant. Biol.* 45:391, 1981.)

into simple circles provided at least one circle contains a single-stranded break (Figure 20-13). Type II topoisomerases also can carry out the catenation reaction; in this case, no single-strand breaks are required. Type I topoisomerases that have been isolated from eukaryotic cells differ in two important respects from the *E. coli* ω protein; they do not require $Mg^{2+}$ ions and they can relax positively as well as negatively supercoiled DNA.

*DNA gyrase, a type II topoisomerase found only in prokaryotes, differs from other topoisomerases in being able to catalyze the conversion of relaxed duplex DNA into a high-energy negatively superhelical form* (Figure 20-14). This is an ATP-dependent reaction. In the absence of ATP, gyrase can still catalyze relaxation of superhelical DNA or the catenation reactions described earlier (Figure 20-14). The gyrase enzyme contains two different subunits encoded by the *gyrA* and the *gyrB* genes (Table 20-1). It normally exists as a tetramer with two subunits of each type.

**Proteins Involved in Prokaryotic DNA Replication.** Our understanding of prokaryotic DNA replication has proceeded most rapidly through investigations on viruses because of their relative simplicity. Bacteriophages that infect *E. coli* vary considerably in size and structure. Each type replicates its chromosome in a unique manner and relies on the host enzymatic machinery in a way that suits its needs. In general there is an inverse relationship between the size of the virus chromosome and the degree of dependency on the host enzymes. Large viruses such as bacteriophage T4 encode several genome-replicating functions that have parallel properties to closely related host proteins but for subtle reasons are preferred by the virus. The bacteriophage φX174 has a small, closed single-stranded circular DNA chromosome (5386 bases) that is dependent on many host functions for replication (see Table 20-3). φX174 DNA has the added fascination of passing from a single-stranded to a double-stranded form and then back to a single-stranded form during replication. The replication process is particularly well under-

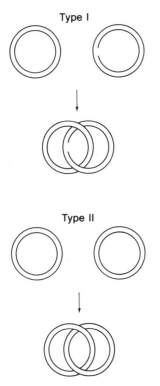

Type I

Type II

(*a*)

(*i*)

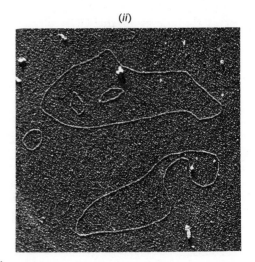

(*ii*)

(*b*)

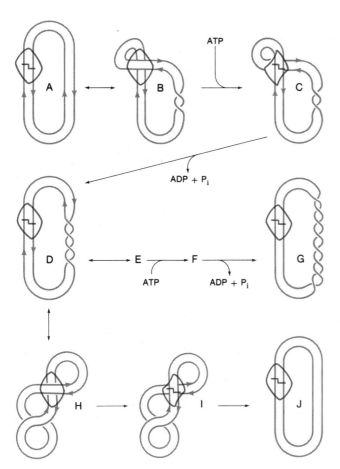

**Figure 20-14**
A model for DNA gyrase-induced DNA supercoiling by means of transient double-strand breaks. The enzyme binds preferentially to certain sites on DNA and induces a left-handed (positive superhelical) wrapping of a local DNA region (B). ATP binding then leads to transport of the upper double helix through the lower via a transient double-strand break (C) with an accompanying conformational change in the enzyme. This reaction decreases the linking number of the DNA by two. Subsequent hydrolysis of ATP and release of the transported DNA segment prepares the system for another cycle of supercoiling (D through G). During relaxation of negatively supercoiled DNA (*bottom series of drawings*), the superhelical coiling causes a loop of DNA to fold over the enzyme with the opposite (right) handedness to that used in the supercoiling reaction (H). Transport through a transient double-strand break causes an increase of linking number (relaxation) by two units (I and J). (Diagram courtesy M. Gellert.)

stood, and it is believed that the findings not only reveal the replication process of the virus but reflect on how the host proteins involved are used in host chromosome replication.

When ϕX174 infects *E. coli* it loses its protein coat, so that only the circular viral DNA enters the cytoplasm. During the first phase of the replication process, the single-stranded viral DNA must be converted to a double-stranded form. This double-stranded form serves two purposes: it is an intermediate in replication, and it is a template for transcription of viral mRNA. Since no viral proteins can be made until after the viral DNA is converted to the double-stranded form, it follows that the single- to double-strand transition must be carried out by preexisting host proteins. This has been verified by showing that the double-stranded form of the virus is formed in the presence of the antibiotic chloramphenicol (CAM), which inhibits protein synthesis. Further steps in viral DNA replication cannot occur in the presence of this antibiotic.

Kornberg and coworkers have reconstructed a cell-free system from ϕX174 DNA and purified proteins that can carry out many of the steps believed to be involved in ϕX174 DNA replication in vivo. Before DNA synthesis can be initiated on the single-stranded circular DNA an RNA

**Table 20-3**
Replication Proteins of *E. coli* Used by Phage φX174

| Polypeptide | Mass (kdal) | Function | Molecules per cell |
|---|---|---|---|
| SSB | 74 | Single-strand binding | 300 |
| Protein i | 80 | Prepriming | 150 |
| Protein n | 25 | Prepriming | |
| Protein n' | 75 | Site recognition, ATPase | 80 |
| Protein n'' | 11 | Prepriming | |
| dnaC | 29 | Prepriming | |
| dnaB | 250–300 | Mobile promoter, ATPase | 20 |
| Primase | 60 | Primer formation | 100 |
| Pol III holoenzyme α | 140 | | |
| β | 37 | | |
| γ | 52 | | |
| δ | 32 | Synthesis | 20 |
| ε | 25 | | |
| θ | 10 | | |
| Pol I | 109 | | 300 |
| Ligase | 74 | Ligation | 300 |
| Gyrase | 400 | | |
| nalA(A) | 210 | Supertwisting | |
| cou(B) | 190 | | |
| rep | 65 | Helicase | 50 |

Adapted from A. Kornberg, *DNA Replication*, Freeman, San Francisco, 1980.

primer must be synthesized. This is synthesized by the enzyme primase. Two types of complexes have been made in which primase is active. The first complex is formed by adding primase to a preformed complex of DNA and dnaB protein. General priming occurs in this system at random sites on the DNA. In the second complex, which is sequence-specific, several additional proteins are required before dnaB protein or primase can bind (Figure 20-15). These include n, n', n'', i, dnaC, and single-stranded binding (SSB) proteins. When the DNA is uniformly coated with SSB, general priming is prevented. This is so because the dnaB protein cannot displace SSB from the DNA to form a complex. Under these conditions n' makes an initial complex at a specific site on the DNA (around base 2300 in Figure 20-15). This causes the displacement of some SSB protein. Other proteins join this complex: n, n'', i, and dnaC. Finally, dnaB becomes a part of the complex, resulting in the displacement of more SSB. When primase is added, a short oligoribonucleotide is made. DNA polymerase III holoenzyme uses this primer as an initiation point for DNA synthesis. At the same time, *the multiprotein complex known as the primosome moves along the template DNA strand in the opposite direction to DNA chain propagation, creating new sites for primer synthesis.* DNA synthesis also initiates from the new primer-containing sites. This movement of the primosome is an energy-dependent reaction requiring ATP. Although primase initiates oligoribonucleotide synthesis at several locations on the template, the necessary primosome can be assembled only at a unique site around base 2300. The primosome complex is very stable and can survive many rounds of replication. The Okazaki fragments formed by primase and DNA polymerase III require DNA polymerase I and ligase for completion. The exonuclease activity of DNA polymerase I closes the gaps as it removes the RNA present on the 5′ ends of the fragments. Ligase knits the block copolymers of DNA together. The closed covalent duplex formed is referred to as the _replicative form_ (RF). The steps leading from single strand to RF are summarized in Figure 20-16. If gyrase is present, this molecule will become negatively supercoiled.

The RF is used for viral transcription that is mediated exclusively by host RNA polymerase. Transcription and translation of viral proteins are essential for subsequent steps in the viral DNA replication process.

After SSDNA has been converted to the replicative form duplex (SSDNA→RF), the parental RF is duplicated to produce multiple copies of RF (RF→RF). In this reaction, the viral and complementary strands are replicated by distinct mechanisms. _Synthesis of the viral plus strand begins with the viral gene A protein–induced cleavage of the viral plus strand in the RF at position_ 4305–4306. The bifunctional A protein cleaves the plus strand at this point and becomes covalently attached to the 5′ end of the interrupted strand (see Figure 20-17). More copies of the replicative duplex form are initiated from this point using the 3′-OH end of the plus strand as a primer and the so-called rolling circle mode of replication. Host functions required during this phase of replication include the host rep protein, called _helicase_, DNA polymerase III, and single-stranded binding protein. The function of helicase is to facilitate unwinding of the duplex. It is not absolutely required for DNA synthesis using a duplex template, but it greatly accelerates the process. The phage-encoded A protein plays a crucial role in maintaining the integrity of the virus plus strand during synthesis

**Figure 20-15**
Scheme for assembly of the primosome at the origin of complementary strand replication. (Adapted from K. Arai, R. Low, J. Kobori, J. Shlomai, and A. Kornberg, _J. Biol. Chem._ 256: 5280, 1981.)

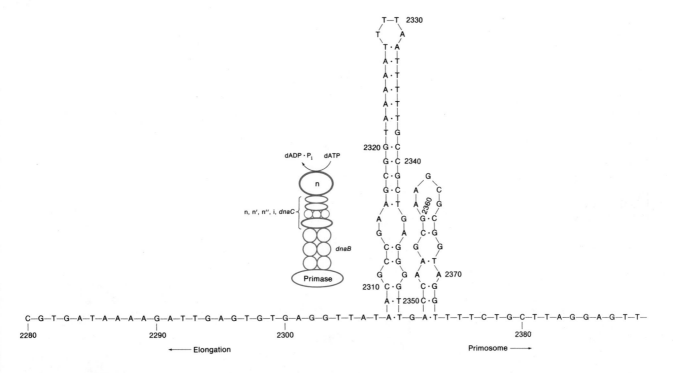

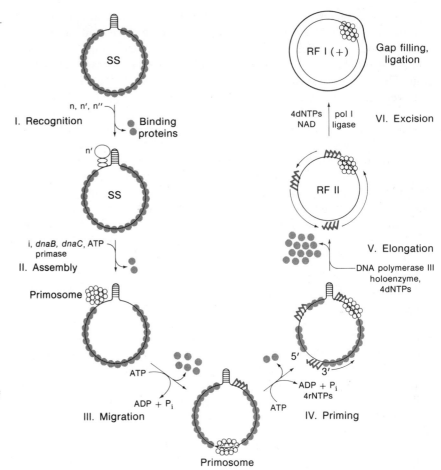

**Figure 20-16**
Scheme for assembly and migration of the primosome and the stepwise displacement of SSB in the SS→RF reaction. (Adapted from K. Arai, R. Low, J. Kobori, J. Shlomai, and A. Kornberg, *J. Biol. Chem.* 256: 5280, 1981.)

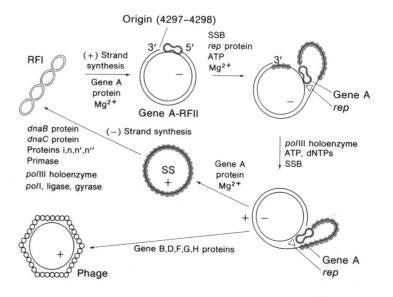

**Figure 20-17**
Scheme for φX174 RF replication in two stages. Continuous replication initiated by gene A protein cleavage generates viral (+) circles, and discontinuous replication of the viral circles by the SS→RF system produces RF. In the presence of phage-encoded maturation and capsid proteins, viral circles are encapsulated rather than replicated. (Adapted from A. Kornberg, *DNA Replication,* W. H. Freeman and Co., San Francisco, 1980, p. 510.)

and ensures closure at the proper point without releasing itself from the replicative-form template. The displaced plus-strand circles are converted to duplex forms, as in the initial stages of infection.

The description of further steps in the replication process has resulted primarily from an analysis of the events occurring in vivo. After about 60 copies of replicative form have been made (20 minutes after infection), the infected cell switches to making single-stranded circles exclusively. Presumably some regulatory device is involved to ensure that further single-stranded circles do not get converted into duplex forms at this stage. Indeed, these plus-strand circles get packaged into the phage heads as they are being synthesized in preparation for making mature viruses. This concludes the replication cycle for $\phi$X174 DNA.

Other DNA phages whose replication cycles have been intensively studied include T7, T4, and $\lambda$. The $\lambda$ replication cycle is discussed in Chapter 27.

Replication of the *E. coli* chromosome is much more difficult to study partly because of its size. As stated earlier, it seems likely that host proteins involved in $\phi$X174 replication also will be involved in similar ways in *E. coli* replication. *The backward movement of the primosome complex suggests a mechanism for discontinuous DNA synthesis on the lagging strand.* At the replication fork, the primosome could migrate processively on the lagging strand as the template unwinds (see Figure 20-18), stopping periodically to allow primase to act. All the elongation steps so clearly elucidated in $\phi$X174 minus-strand synthesis could account for lagging strand synthesis at the *E. coli* replication fork.

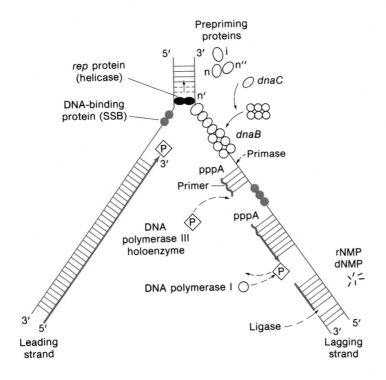

**Figure 20-18**
Role of the primosome at the bacterial replication fork. The primosome, assembled at or near the DNA replication origin, migrates processively on the lagging strand coupled with movement of the replication fork. (Adapted from K. Arai, R. L. Low, and A. Kornberg, Movement and site selection for priming by the primosome in phage $\phi$X174 DNA replication, *Proc. Natl. Acad. Sci. USA* 78: 711, 1981.)

This leaves two aspects of *E. coli* chromosome replication to be accounted for. Termination is yet to be studied. To explore the mechanism of initiation, Kornberg and coworkers developed a cell-free system for bidirectional replication using small circular DNA molecules (hybrid plasmids) that contain the origin of the *E. coli* chromosome (*oriC*) and crude extracts of *E. coli*. These ongoing studies are beginning to give a picture of how bacterial chromosome synthesis is initiated. The use of the plasmid is crucial in such studies because it greatly facilitates detection of initiation on a much smaller and tractable piece of DNA. The plasmid contains about 245 base pairs from the *oriC* region of the *E. coli* chromosome. Initiation of the replication process depends on RNA polymerase and DNA gyrase, as judged by specific antibiotic inhibitions by rifampicin and nalidixic acid. Rifampicin is known to specifically inhibit RNA polymerase (see Chapter 21), and nalidixic acid is known to specifically inhibit DNA gyrase (see below). The reaction also requires replication proteins dnaB, single-stranded binding protein, and dnaA protein. Further properties of this crude cell-free system are under intensive investigation.

**DNA Replication in Eukaryotic Cells.** Genetics has not been much help thus far in finding proteins with known functions in eukaryotic DNA replication. Isolation of proteins having activities logically related to DNA replication and the isolation of replicative intermediates has been the major approach to understanding DNA replication in eukaryotes. Different forms of DNA polymerase, DNA ligases that require ATP, topoisomerases, single-stranded DNA-binding proteins, unwinding enzymes, and exonucleases have been isolated from eukaryotic cells.

Multiple species of eukaryotic DNA polymerases have been described. These polymerases have been named $\alpha$, $\beta$, and $\gamma$ in order of discovery (much like polymerase I, II, and III in *E. coli*) and to ensure distinction from prokaryotic enzymes. The eukaryotic DNA polymerases share with prokaryotic enzymes the features of template-directed polymerization of dNTPs and the inability to initiate new chains, but there are many important differences. One outstanding difference is that DNA polymerases $\alpha$, $\beta$, and $\gamma$ do not contain any of the associated exonuclease activities present in the bacterial enzymes. DNA polymerases $\beta$ and $\gamma$ are present in eukaryotic cells in resting and differentiated states. The levels of $\beta$ and $\gamma$ enzymes do not change markedly in proliferating cells. By contrast DNA polymerase $\alpha$ varies greatly with cell type and the physiological state of the cell. Whereas all three polymerases require template and primer, the $\alpha$ enzyme is the only one that can use an oligoribonucleotide primer. There are several lines of evidence that indicate that *DNA polymerase $\alpha$ is involved in the replication of cellular DNA and some mammalian viruses*, such as SV40 and polyoma (also see Chapter 28):

1. The level of DNA polymerase $\alpha$ activity is correlated with cellular proliferation and is absent in certain terminally differentiated cells, such as neurons.

2. The inhibition of cellular DNA replication by several compounds correlates with the action of these inhibitors on DNA polymerase $\alpha$ (see below).

**3.** DNA polymerase $\alpha$ copurifies with replicating chromosomes of the SV40 animal virus (see Chapter 28).

The function of the $\beta$ enzyme is less clear, although its localization in the nucleus suggests that it may be involved in repair or replication processes. *DNA polymerase $\gamma$ is usually localized in mitochondria, and it is believed to be involved exclusively in mitochondrial DNA replication. In adenovirus DNA replication (see Chapter 28), both the $\alpha$ and the $\gamma$ enzymes have been implicated.*

Although general features of DNA replication in eukaryotes are thought to be similar to those of prokaryotes, there are some important differences. The chromosomes of higher eukaryotic organisms are quite large, at least a thousand times larger than their bacterial counterparts. In order to replicate these large DNA molecules in a reasonable length of time, multiple origins of replication are used. The simultaneous synthesis of DNA in several replication units has been demonstrated by incorporating radioactive nucleotides into DNA for a short period and then observing the distribution of radioactive DNA by autoradiography (Figure 20-19*a*). Multiple regions of DNA synthesis can be observed, indicating that many replication units are involved simultaneously in DNA synthesis. Replication proceeds bidirectionally from these origins. Termination of replication occurs when the replication forks from two adjacent replication units meet (Figure 20-19*b*). DNA on at least the lagging strand of a fork is made discontinuously. The Okazaki fragments are much shorter than those found in prokaryotes, averaging about 100 to 200 nucleotides in length. Synthesis of these DNA fragments is probably initiated by RNA, which is found covalently attached to the 5' end of newly synthesized fragments.

DNA in eukaryotic chromosomes is associated with histones in the form of nucleosomes and chromatin (see Chapter 18). These nucleoprotein complexes serve to condense exceedingly long DNA molecules into much shorter structures, but during replication the nucleosomes must be disassembled so the DNA strands can be separated. The disassembly of the DNA-histone complex presumably occurs directly in front of the replication fork. Before the newly replicated DNA is reassembled into nucleosomes, the RNA primers are removed by an unknown enzyme, the gaps are filled in by DNA polymerase, and the replication fragments are linked together by DNA ligase. The slower rate of migration of replication forks in eukaryotes (with nucleosomes) than in prokaryotes (without nucleosomes) and the similarity in size of the replication fragments and the length of DNA that is associated with a single nucleosome ($\sim$200 base pairs) suggest that nucleosome disassembly may be a rate-limiting step in the migration of the replication fork in chromatin.

Mitochondria in animal cells contain a circular DNA with a molecular weight of $10^7$ that is replicated by a *displacement loop, or D-loop, mechanism* (Figure 20-20). There are two origins of replication for mitochondrial DNA, one for each strand. The origin of replication for one strand is at a different location from the origin for the other strand. DNA replication is initiated at the first site and proceeds unidirectionally, but with synthesis of only the leading strand and displacement of what would be the lagging strand, without synthesis on the displaced strand. Replication continues in

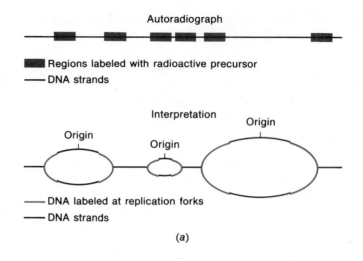

(a)

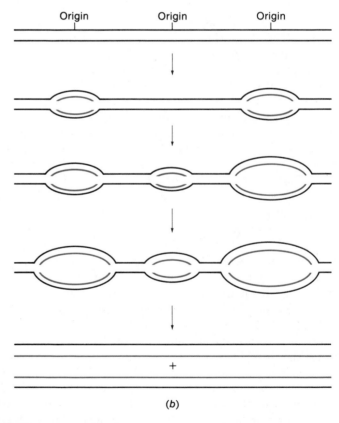

(b)

**Figure 20-19**
Multiple-origin model for eukaryotic chromosomal DNA replication. (a) Pulse
labeling of a eukaryotic chromosome during replication and its interpretation.
(b) Overall replication scheme for a eukaryotic chromosome.

this manner until the origin of replication on the displaced strand is passed, and DNA synthesis then begins there. DNA replication on the second strand proceeds in the opposite direction, while being continuously displaced on the other side of the origin of replication. Mitochondrial DNA may be synthesized as a continuous polynucleotide rather than as Okazaki fragments.

Oncogenic RNA viruses contain an unusual DNA polymerase called _reverse transcriptase_ (Chapter 28). This enzyme carries out template-directed polymerization of dNTPs and has many general features of the cellular DNA polymerase reaction. The preferred template for reverse transcriptase is RNA, but only dNTPs are polymerized. Heteropolymeric or homopolymeric RNAs with appropriate initiation sites will serve as a primer-template and an RNA-DNA hybrid is the intermediate product. A second round of DNA synthesis on the hybrid produces a double-stranded DNA. In this manner, viral RNA-encoded sequences can be transcribed into a DNA sequence that is subsequently integrated into the host chromosome (see Chapter 28 for a detailed description of reverse transcriptase action). Reverse transcriptase is used extensively for conversion of RNA species into DNA sequences before insertion into plasmid vectors for replication in _E. coli_ for cloning, sequencing studies, and hybridization probes (discussed below).

The only deoxynucleoside triphosphate polymerizing enzyme that does not require a template for polymerization is called _terminal deoxy-nucleotidyl transferase_. This enzyme will extend the 3'-OH of an oligodeoxyribonucleotide or polydeoxyribonucleotide with any dNTP, singly or in combinations. The biological function of this enzyme has not been established; but its occurrence in primitive lymphocyte populations from primary lymphoid organs (thymus and marrow) suggests a special role in differentiating lymphocytes. In acute lymphoblastic leukemia, the population of cells containing terminal transferase is greatly expanded. Measurement of this enzyme in blood and marrow cells is frequently used in differential diagnosis of leukemia. In the laboratory, the enzyme is used extensively for producing homopolymer tails on restriction fragment DNA and the plasmid vector to facilitate formation of joined molecules prior to insertion into _E. coli_ for recombinant DNA studies (discussed below).

**Control of DNA Replication.**   Once DNA synthesis has begun, it continues to completion. Only extreme conditions, such as substrate starvation, will interfere with elongation. All indications point to the initiation site as being the primary point for regulating DNA replication, even though very little is known about the factors involved.

In prokaryotes such as _E. coli_, the rate of reinitiation is a sensitive function of the cell division time. Under conditions of rapid growth, the chromosome replication time (40 min at 37°C) actually exceeds the cell division time (as low as 20 min). In such situations, initiation of replication for future cell divisions occurs before the division of the current generation. Each of the daughter cells then receives a chromosome that is already in the process of replication. The generation time can be so short that multiple rounds of replication are occurring simultaneously on the same chromosome structure with the production of multiple replication forks.

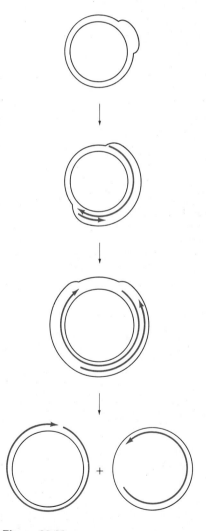

**Figure 20-20**
Replication scheme found for mammalian mitochondrial DNA synthesis.

In eukaryotes, the replication of DNA occurs only during one phase of the cell cycle in actively dividing cells (see Figure 20-21). Following mitosis (M) and cell division, there is a gap in time ($G_1$ phase) before DNA replication begins (S phase). Upon completion of DNA synthesis, there is another pause ($G_2$ phase) before mitosis can occur. Actively dividing cells grown under the same conditions spend about the same time in each phase of the complete cell cycle. Therefore, the initiation of replication is not a random event in the life of a cell and presumably occurs in response to some intracellular signal. It is known that RNA and protein synthesis must occur before DNA synthesis can begin, but the exact nature of the signals for initiation of DNA replication is still a mystery. Furthermore, in eukaryotes, protein synthesis is required during S phase. Chromosomal proteins, especially histones, are usually synthesized at the same time as the DNA.

In cultured mammalian cells there are approximately 20,000 origins of replication for each haploid equivalent of DNA in the genome. If replication were to proceed evenly throughout the S phase, all the replication units would be active simultaneously. In fact, this is not the case. Considerable variation occurs with different types of cells and culture conditions. During the S phase of the cell cycle, all the DNA in the genome is replicated once, and (with the exception of certain special cases noted below) each section of DNA is replicated only once. Some replication units complete DNA synthesis before other units initiate, indicating the existence of a mechanism that distinguishes between the origins of replication on DNA that has already been replicated during the current cell cycle.

Some specialized cells require the products of specific genes in very large amounts. In order to achieve a high rate of RNA synthesis, these genes are selectively amplified so that many more copies exist at this stage of development than at other times. For example, in the South African frog *Xenopus laevis* the genes for the ribosomal RNAs are amplified during the maturation of oocytes. At this stage there are about 4000 times more ribosomal RNA genes than normally occur in the genome. A mechanism that normally ensures a single replication cycle must be circumvented to allow for specific *gene amplification*.

**Postreplicative Modification of DNA.** *Whereas the sequence of bases in DNA is restricted by the requirement for Watson-Crick base pairing, the purines and pyrimidines occasionally are modified.* The observed modification is a function of the cell type in which the DNA is replicated, as well as the sequence immediately surrounding the modified base.

The DNA of bacteriophages T2, T4, and T6 all contain 5-hydroxymethylcytosine (HMC) instead of cytosine (see Chapter 19). The HMC DNA of the T-even bacteriophages is normally present in a glycosylated form. Specific phage-induced α- and β-glucosyl transferases carry out glycosylation on DNA using UDPglucose as substrate and produce a specific pattern of glucosylation. Glucosylation appears to protect the infecting phage DNA from restriction endonucleases of the host bacteria (discussed below).

Most animal cell, bacterial, and viral DNAs contain small amounts of the methylated bases 5-methylcytosine (5-MC) and 6-methyladenine. Enzymes that carry out the methylation reactions are called *DNA methylases*.

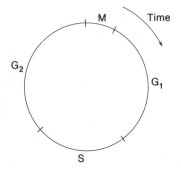

**Figure 20-21**
The cell cycle for a eukaryotic cell.

They use *S*-adenosylmethionine as the source of active methyl groups. Many of the bacterial DNA methylases are highly sequence-specific. If a methylase from one source, such as *Micrococcus luteus*, is used to methylate *M. luteus* DNA, no reaction occurs. If the same *M. luteus* methylase is used to methylate *E. coli* DNA, then some methylation does occur. The interpretation of this result is that the *M. luteus* enzyme has already methylated *M. luteus* DNA *in vivo* at certain specific sites, so that these sites are not available for in vitro reaction. *E. coli* DNA, presumably already methylated by its cognate methylase at sites it recognizes in vivo, can be further methylated in vitro at sites recognized by the *M. luteus* methylase. The specificity of the methylation reaction seems quite clear in such systems, and the rationale for specific methylation may be related to a protective function. Host-modification enzymes protect the DNA from related host-restriction enzymes. *Restriction-modification systems come in matched pairs*, and it seems likely that their primary function is to protect the cell against foreign invading DNA. Different strains of bacteria possess restriction-modification systems with different specificities. This can be illustrated by considering the efficiency of λ bacteriophage infection on two different strains of *E. coli*, say, strain B and strain K12: λ grown on strain B infects B cells with a high efficiency but K12 cells with a very low efficiency. The reason for this is that DNA replicated in strain B is modified at specific sites and thereby protected against the B restriction endonuclease, which otherwise would attack and degrade the DNA at those sites. When the λ DNA enters the K12 cell, most of it is destroyed before the DNA can be appropriately methylated. The few phages that do manage to survive and replicate will now multiply efficiently on K12 cells but will have lost the ability to grow efficiently on the B strain for parallel reasons.

In vertebrates, 5-MC is the main modified base. Most 5-MC occurs in the dinucleotide 5'-CpG, and in mammals and birds approximately 50 to 70 percent of all such dinucleotides are modified. The distribution of 5-MC within specific genes can be probed with the use of bacterial restriction endonucleases, such as HpaII, which cleaves recognition sequences containing 5'-CCGG, but not if they are methylated to 5'-C^mCGG. It has been shown that *vertebrate systems have a passive methylation maintenance system* by injecting methylated or nonmethylated DNA into mammalian systems. If the injected DNA is methylated, the pattern of methylation will be maintained through many DNA duplications. If the DNA is nonmethylated, this structure also will be maintained. *This situation contrasts with host-modification systems found in bacteria that are much more specific and methylate appropriate sequences in nonmethylated DNA.* Methylation patterns observed in mammalian DNA are tissue-specific, and some evidence has accumulated that there is an inverse relationship between the extent of methylation of a gene and its activity in transcription.

**Inhibitors of DNA Replication.** The replication of DNA can be inhibited at many different steps in the process. Inhibitors can be divided into three general categories:

1. *Inhibitors that interact with the DNA*. Compounds such as acridine, actinomycin D, and ethidium bind between the stacked bases of the DNA duplex. Hydrophobic planar portions of these molecules become

**Table 20-4**
DNA Replication Inhibitors that Act on DNA Polymerase

| Inhibitor | Polymerase Affected | Action |
|---|---|---|
| Aphidicolin | Mammalian DNA polymerase $\alpha$ | Competitive with dCTP |
| Arylhydrazinopyrimidines | B. subtilis DNA polymerase III | Competitive with dGTP or dATP |
| Arabinosylcytosine*; arabinosyladenine* | DNA polymerases; several other enzymes in nucleic acid metabolism | Chain terminator |
| 2',3'-Dideoxythymidine triphosphates† | Mammalian DNA polymerases ($\gamma > \beta \gg \alpha$); prokaryotic DNA polymerases | Competitive with dTTP (or other dNTP); chain terminator |
| N-Ethylmaleimide | Mammalian DNA polymerase ($\alpha > \gamma \gg \beta$) | Reacts with sulfhydryl groups |

*Probably converted to triphosphate in cells.
†Only works in vitro because triphosphates cannot pass through the cell membrane.

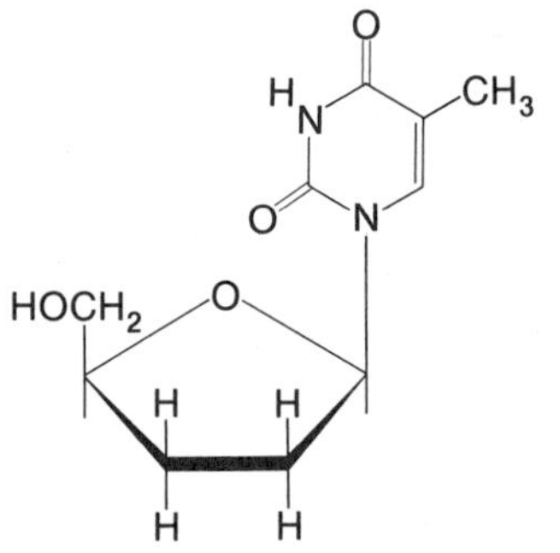

**2',3'-Dideoxythymine**

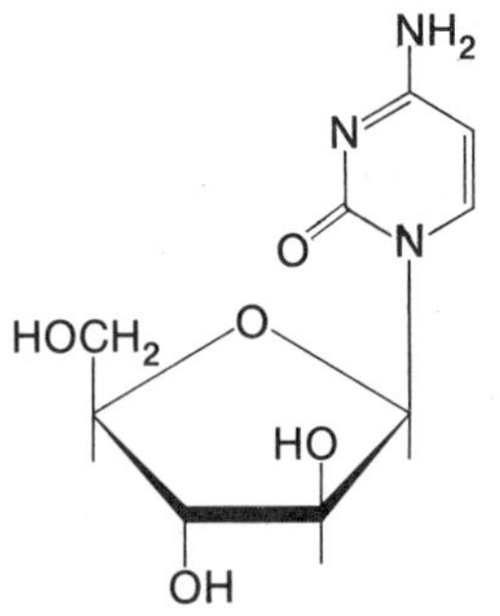

**β-Arabinofuranosylcytosine**

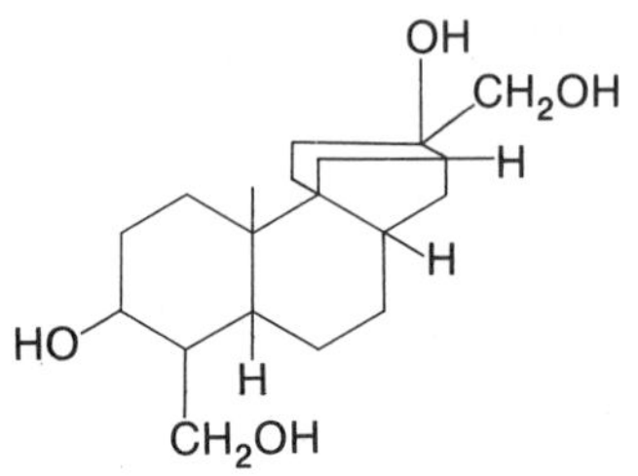

**Aphidicolin**

**Figure 20-22**
Selective inhibitors of eukaryotic DNA polymerases.

inserted between the hydrophobic faces of the adjacent base pairs in the DNA. This intercalation disrupts the normal structure of the DNA, necessitating some unwinding of the DNA helix (see Chapter 18). Netropsin and a related antibiotic distamycin A bind tightly through hydrogen bonding and van der Waals forces to the exterior of the DNA duplex along the minor groove to regions rich in A and T residues.

The binding of these compounds inhibits the use of the DNA as a template for both replication and transcription by a simple blocking reaction. At low concentrations, acridine is a mutagen that produces additions or deletions of single bases during the replication process. This effect of acridine to cause addition or deletion of bases is believed to be intimately associated with the way in which acridine binds to the DNA.

Bleomycin and the antibiotic protein neocarzinostatin introduce breaks in the DNA. The reaction of bleomycin is mediated by a metal chelate that generates a reduced form of oxygen near susceptible sites on DNA. The breaks caused by neocarzinostatin and bleomycin cannot be sealed by DNA ligase nor act as substrates for DNA polymerase, so they are not the result of simple phosphodiester hydrolysis. Recent work has shown that bleomycin splits the C-3'–C-4' bond in deoxyribose. These breaks disrupt the continuity of the DNA template and result in nonproductive binding of DNA polymerase.

2. *Inhibitors that affect the synthesis of deoxyribonucleotides.* The synthesis of new DNA requires deoxynucleoside triphosphates, and anything that affects the production of these precursors can have an effect on DNA replication. Such inhibitors have been discussed in Chapter 19.

3. *Inhibitors of enzymes involved in DNA synthesis.* Several compounds that act directly on the DNA polymerases show some selectivity toward the individual polymerases (Table 20-4). The differences in sensitivity to some of these compounds have been most useful in assigning the functions of DNA polymerases in vivo, especially in cases where mutants are not available.

The structures of several compounds that show selective inhibition of the eukaryotic polymerases are shown in Figure 20-22. The 2'-3'-dideoxy derivative of thymidine lacks the 3'-hydroxyl group of the sugar moiety.

When a dideoxynucleotide is incorporated into a growing chain, polymerization is terminated (see below for use in DNA sequencing). DNA polymerase $\gamma$ is much more sensitive to ddTTP than DNA polymerase $\alpha$. The sensitivity is directly related to the extent to which the polymerase recognizes the dideoxytriphosphate as a substrate for incorporation into a growing DNA chain. ddTTP also inhibits adenovirus and mitochondrial DNA replication at the same concentration that it inhibits DNA polymerase $\gamma$ in vitro, but it has no effect on SV40 virus or cellular DNA replication. This has been taken as evidence that DNA polymerase $\gamma$ participates in mitochondrial DNA replication. The arabinose analogs of CTP and ATP have a hydroxyl group in the trans configuration at the 2'-position of the sugar. This altered configuration affects the ability of the arabinosyl derivatives to act as nucleotide acceptors when they are incorporated into a growing DNA chain. DNA polymerase $\alpha$ is more sensitive than DNA polymerase $\beta$ to inhibition by $\beta$-arabinosyl CTP and $\alpha$-arabinosyl ATP. The arabinosyl nucleoside analogs, most likely as triphosphate derivatives, also specifically inhibit cellular DNA replication.

$N$-Ethylmaleimide (NEM) reacts generally with sulfhydryl groups of proteins, so it is not a specific DNA polymerase inhibitor. It is useful for comparing enzymes. SV40 viral and cellular DNA synthesis in vitro is as sensitive to NEM as DNA polymerase $\alpha$, DNA polymerase $\gamma$ is less sensitive, and polymerase $\beta$ is completely resistant. Aphidicolin is a highly specific inhibitor of DNA polymerase $\alpha$. This compound blocks chain elongation in SV40 viral and cellular DNA replication.

The action of the arylhydrazinopyrimidines on *Bacillus subtilis* DNA polymerase III has been studied by enzymatic, genetic, and physical methods (Figure 20-23). These compounds have been used to isolate mutants of *B. subtilis* polymerase III and to establish the role of this enzyme in DNA replication. The inhibitory action of the arylhydrazinopyrimidines is believed to be due to the formation of ternary 1:1:1 complexes with DNA polymerase III and the DNA template primer. The presence of primer is required for formation of this ternary complex. Inhibitor binding to the template is a base-specific interaction, with hydroxyphenylhydrazino uracil [OHPhe(NH)$_2$Ura] hydrogen bonding to C residues and the isocytosine [OHPhe(NH)$_2$Iso] derivative hydrogen bonding to T residues.

Inhibitors affecting DNA metabolism in bacteria have been very useful in two ways. On wild-type cells or cell extracts, specific inhibitors have been helpful in pinpointing particular functions. Mutant strains resistant to an inhibitor frequently lead to the characterization of a specific protein involved in DNA metabolism. For example, specific inhibitors of DNA gyrase have been of great value in characterizing the genes encoding the enzyme and in characterizing its action (see Table 20-1). The supercoiling reaction in *E. coli* is inhibited by the antibiotics novobiocin and coumermycin, which inhibit competitively with ATP. The $K_I$ for novobiocin is $10^{-8}\ M$, while the $K_m$ for ATP is $3 \times 10^{-4}\ M$. Supercoiling activity is also inhibited by oxolinic and nalidixic acids. The relaxation activity is inhibited only by the latter two antibiotics. Mutant studies indicate that novobiocin and coumermycin react primarily with the gyrA protein and that oxolinic acid and nalidixic acid react primarily with the gyrB protein. Thus resistant mutants to these drugs have been found to result in changes at the corresponding genetic loci.

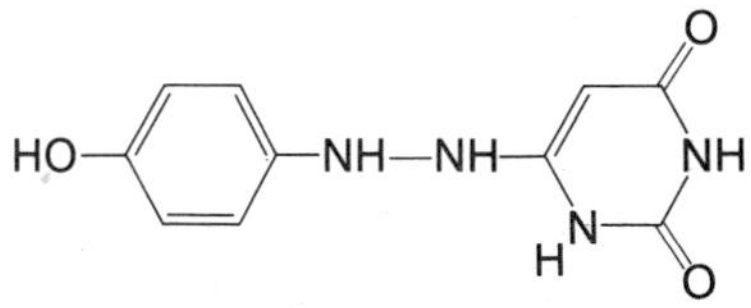

**6-($p$-Hydroxyphenylhydrazino)-uracil [OHPh(NH)$_2$Ura]**

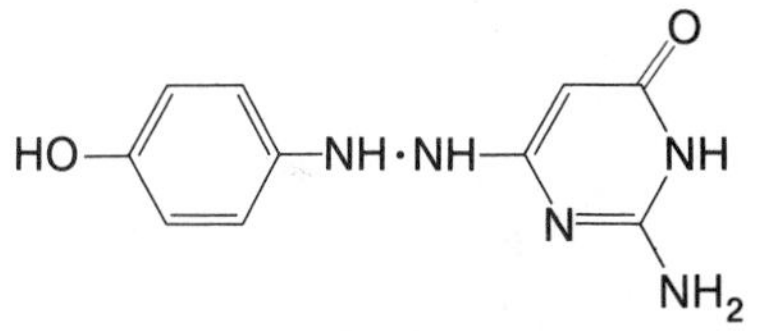

**6-($p$-Hydroxyphenylhydrazino)-isocytosine [OHPh(NH)$_2$Iso]**

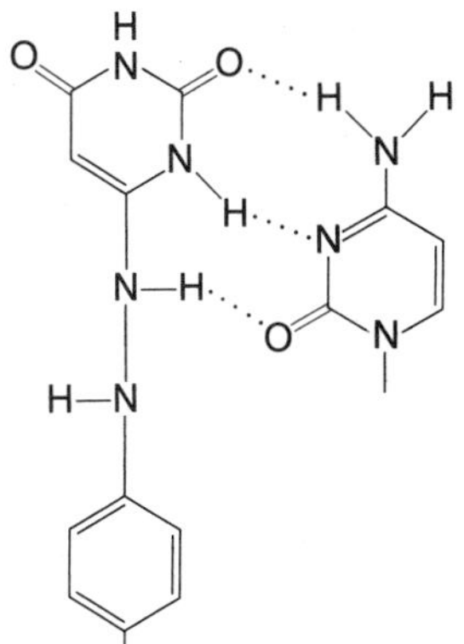

Hydrogen bonding between OHPh(NH)$_2$Ura and cytidine

**Figure 20-23**
Arylhydrazinopyrimidines: inhibitors of *Bacillus subtilis* DNA polymerase III.

**Breaking Phosphodiester Bonds.** Certain enzymes involved in DNA bio-synthesis also have a capacity for cleaving phosphodiester linkages. The bacterial polymerases I and III both have exonucleolytic activity, removing nucleotides one at a time from either end of a polymer. All topoisomerases have endonucleolytic activity, making transient breaks which they invaria-bly mend. There are also a large number of nucleases that are exclusively degradative enzymes. It is presumed that these enzymes serve functions useful to the cell; many of them also have been useful in DNA manipula-tion (discussed in the second half of this chapter).

Some nucleases attack phosphodiester bonds in a polynucleotide chain from one end (or the other); others carry out a random attack on internal bonds. Hydrolysis of an internal linkage is classified as endonucleolytic and is carried out by enzymes called _endonucleases_. The site of phosphodiester bond cleavage may be on the 3'-phosphate side, leading to the formation of 5'-phosphates, or on the 5'-phosphate side, leading to the formation of 3'-phosphates. The attack from the ends of chains is an exonucleolytic proc-ess, and the enzymes are grouped as _exonucleases_. Exonucleases may be further distinguished according to whether the direction of cleavage is from the 3' end or the 5' end or in some cases from both ends of the polynucleotide chain. The initial products of endonucleases are mixtures of oligonucleotides of varying chain lengths. The products of exonucleases are usually either 3'- or 5'-phosphomononucleotides.

A phosphodiesterase isolated from rattlesnake venom (_Crotalus adamanteus_) is a 3'→5' exonuclease (3'-exonuclease) that acts on single polydeoxynucleotide or polyribonucleotide chains to produce only 5'-nucleotides. The enzyme has an alkaline pH optimum and requires no added divalent metal ions for activity. The opposite specificity is exhibited by a phosphodiesterase isolated from bovine spleen. Spleen phosphodiester-ase is a 5'→3' exonuclease (5'-exonuclease) that acts on DNA or RNA substrates to produce only 3'-nucleotides. Most rapid action is observed if the 5'-terminal phosphate of the polynucleotide is first removed by a monophosphatase. _E. coli_ exonuclease I exhibits 3'→5' exonuclease activ-ity, much like venom phosphodiesterase. In contrast to the venom enzyme, exonuclease I requires $Mg^{2+}$, will not split the terminal 5'-dinucleotide of a DNA chain, and will hydrolyze glycosylated DNAs (encountered in the T-even series of _E. coli_ bacteriophages). _E. coli_ exonuclease III and bacterio-phage T4 exonuclease are only capable of degrading double-stranded DNA. Both are 5'-nucleotide formers, but T4 exonuclease attacks at the 5' end of a polynucleotide, whereas exonuclease III starts at the 3' end. If the DNA chain has a 3'-phosphate, exonuclease III is able to hydrolyze that phospho-monoester bond before beginning 3'→5' degradation to 5'-nucleotides. These well characterized exonucleases are listed in Table 20-5.

Higher degrees of structural complexity frequently affect the specificity of nuclease cleavage. Some DNA endonucleases hydrolyze double-stranded structures most rapidly (Table 20-6); others prefer single strands.

Deoxyribonuclease I, a digestive enzyme secreted by the pancreas, acts initially on double-stranded DNA. Initially, only one chain of the double helix is cleaved, leaving 5'-P and 3'-OH terminals at the point of cleavage. Continued digestion eventually leads to mixtures of di-, tri-, tetra-, and some higher oligonucleotides. Deoxyribonuclease II, a lysosomal enzyme

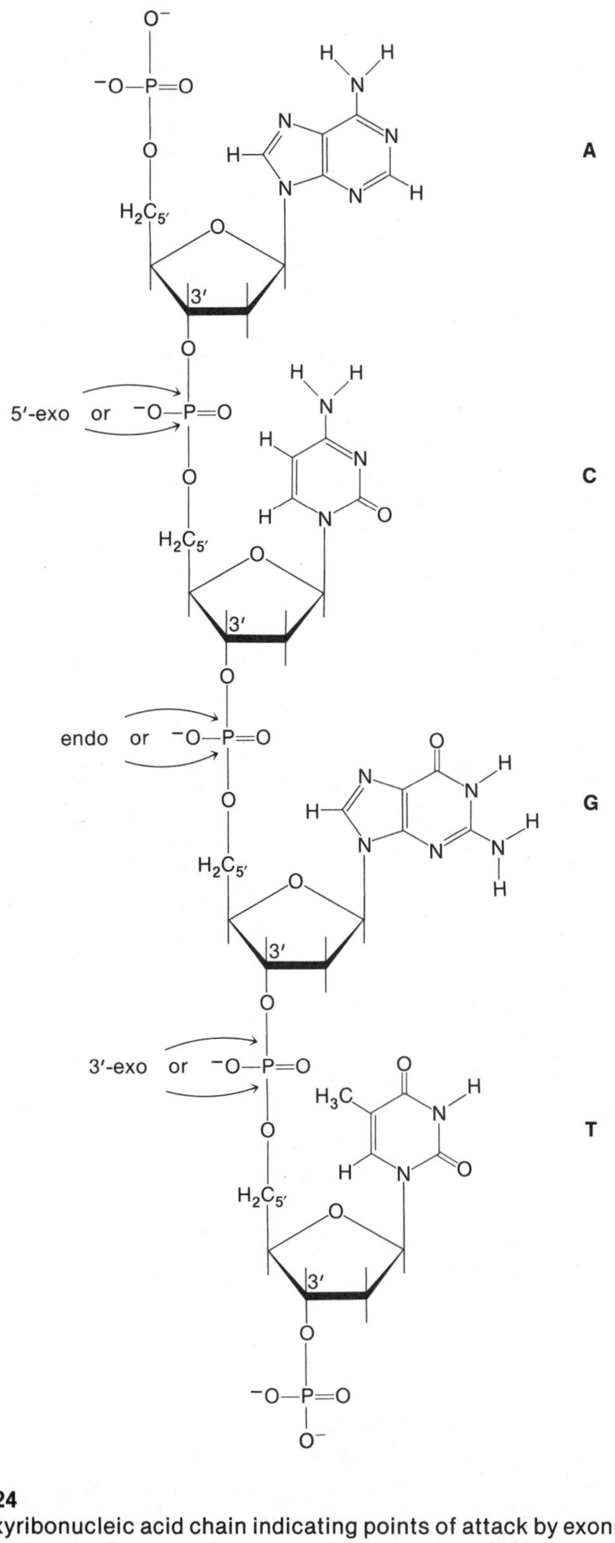

**Figure 20-24**
A polydeoxyribonucleic acid chain indicating points of attack by exonucleases and endonucleases. Enzymes that attack the 5′ end are referred to as 5′-exonucleases and those which attack the 3′ end are referred to as 3′-exonucleases. Depending on the P–O linkage attacked, these produce 3′- or 5′-phosphomononucleotides.

**Table 20-5**
Exonucleases

| Name | Source | Substrate | Product | Direction | Divalent Cation Requirement |
|------|--------|-----------|---------|-----------|------------------------------|
| Venom phosphodiesterase | Snake venom | Oligo or poly | 5′-dXMP | 3′→5′ | − |
| Spleen phosphodiesterase | Beef spleen | Oligo or poly | 3′-dXMP | 5′→3′ | − |
| *E. coli* Exo I | Bacteria | Oligo or Poly | 5′-dXMP | 3′→5′ | + |
| *E. coli* Exo III* | Bacteria | Double-stranded DNA | 5′-dXMP | 3′→5′ | + |
| λ-Exonuclease | λ-Infected *E. coli* | Double-stranded DNA | 5′-dXMP | 5′→3′ | + |

*Has an associated 3′-phosphatase activity.

**Table 20-6**
Endonucleases

| Name | Source | Substrate | Product | Sequence Specificity | Divalent Cation Requirement |
|------|--------|-----------|---------|----------------------|------------------------------|
| Pancreatic DNAse I | Beef pancreas | Double- (or single-) stranded DNA | 5′-P-ended oligos | AT rich | + |
| Spleen DNAse II | Beef spleen | Double-stranded DNA | 3′-P-ended oligos | None described | − |
| S1 endo | *Aspergillus oryzae* | Single-stranded DNA | 3′-P-ended oligos | None described | − |
| EcoR1 | *E. coli* | Double-stranded DNA | p-AATTCXXXX HO-GXXXX | ↓ p-XGAATTCXX HO-XCTTAAGXX ↑ | + |
| Pst I | *Providencia stuartii* | Double-stranded DNA | p-GXXX HO-ACGTCXXX | ↓ p-XCTGCAGXX HO-XGACGTCXX ↑ | + |
| Hae II | *Haemophilus egyptius* | Double-stranded DNA | p-CCXXXXXX HO-GGXXXXXX | ↓ p-XGGCCXXXX HO-XCCGGXXXX ↑ | + |
| AP-endonuclease | Bacteria and animal cells | Apurinic or apyrimidinic DNA | Chain cleavage | Free sugar in DNA chain | − |

isolated from the spleen, splits both chains of DNA simultaneously. S1 endonuclease is a single-strand–specific enzyme that attacks both DNA and RNA. It is often used to remove single-stranded (nonhybridized) regions after solution-hybridization experiments (also see Chapters 18 and 21).

The highest degree of structure specificity is demonstrated by site-specific endonucleases (Table 20-6) known as *restriction enzymes*. These enzymes (already mentioned in Chapter 18 and in the section entitled

Postreplicative Modification of DNA) _recognize unique sequences in dou-_
_ble-stranded DNA molecules and cleave only in or near these sites._ A
sequence of at least four bases is generally required for recognition. Phosphodiester bonds on both sides of the helix are cut by the enzymes, forming
5′-phosphate ends. The products are double-stranded DNA molecules with
specific cleavages, usually zero to four bases apart on each strand of the
double helix. These molecules dissociate into large fragments owing to instability of the short regions of overlap. Simple DNAs produce a unique
fragmentation pattern with each restriction endonuclease, the number of
fragments being determined by the number of specific sites in the molecule. The restriction patterns are produced by electrophoresis of the digest
on agarose gels followed by staining with ethidium bromide (Figure 20-25).

Restriction enzymes are frequently sensitive to the presence or absence
of methylated bases in the susceptible "site" (discussed earlier in the section entitled Postreplicative Modification of DNA). For example, if the first
A residue is methylated in the EcoR1 site, it is not cleaved by this restriction enzyme. The EcoR1 cleavage produces a 5′-phosphate "overhang" with
a 3′-OH four bases back on the opposite chain (see Table 20-6). A restriction
enzyme from _Providencia stuartii_, Pst I, in addition to having a different
specific cleavage site, produces molecules that have a 3′-OH "overhang"
with a 5′-phosphate four bases back from the cleavage site. An enzyme from
_Haemophilus egyptius_, Hae III, recognizes the sequence -GGCC- and
cleaves it in the center, generating blunt-ended fragments. The restriction
endonucleases show their greatest use to biochemistry in facilitating the
preparation of large, well-defined fragments of DNA (see below). Over 200
enzymes with different specificities have been isolated.

Whereas the great interest in restriction enzymes stems from their ability to cleave DNA at specific points, it should be mentioned that there are a
number of bacteria that harbor restriction enzymes that cleave DNA at
nonspecific sequences. In such cases, the restriction enzymes are sequence-
specific with respect to where they bind to the DNA but not with respect to
the actual cleavage point.

AP endonucleases (apurinic or apyrimidinic) exhibit specific cleavage
at DNA sites lacking a base. Sites of this kind are produced when DNA
glycohydrolases (described below) remove damaged or incorrect bases from
DNA. These enzymes are widely distributed and apparently participate in
certain DNA repair processes.

**Breaking Glycosidic Bonds.** The hydrolysis of the _N_-glycosidic bonds of
purine and pyrimidine mononucleosides is a well-established enzymatic
reaction in the degradation pathway for nucleic acids. Enzymes that degrade nucleosides do not act on intact nucleic acids, but other special nucleic acid–base glycohydrolases that can cleave _N_-glycosidic bonds in DNA
have been described. They appear to act only on DNA with abnormal bases
or abnormal base pairing and are thought to be part of the cellular mechanism for repair of DNA damage.

Uracil in DNA resulting from misincorporation of dUTP or deamination of cytosine can be removed by a uracil-DNA glycohydrolase. Hypoxanthine resulting from dITP incorporation or deamination of adenine can be
removed by a hypoxanthine-DNA glycohydrolase. Separate enzymes for
removing alkylated bases also have been discovered. All these enzymes

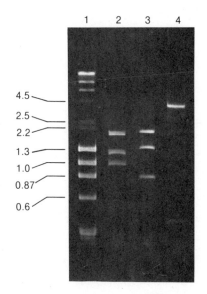

**Figure 20-25**
Restriction endonuclease digests of
yeast mitochondrial DNA (strain DS40
of _Saccharomyces cerevisiae_).
Samples of DS40 mtDNA were digested with _Hpa_II (Lane 2), _Hin_tI (Lane
3), and _Hind_III (Lane 4) and separated
by electrophoresis on 1% agarose gel.
A mixture of a Hae III digest of φX174
RF DNA and of a _Hind_III digest of λ
DNA was applied to Lane I as molecular weight standards. The sizes of the
φX174 RF and λ fragments are reported in kilobase pairs in the left mar-
gin. The restriction fragments were
visualized with ethidium bromide.
(Adapted from B. E. Thalenfeld and A.
Tzagoloff, Assembly of the mitochondrial membrane system, _J. Biol. Chem._
255: 6174, 1980.)

participate in the repair process by removing inappropriate bases in a process known as *base excision repair*. The apurinic or apyrimidinic site in DNA is then cleaved by an AP-endonuclease (Table 20-6) at the nearest phosphodiester bond, often followed by further exonuclease degradation. The gap produced can then be repaired by the sequential action of DNA polymerase and DNA ligase.

### DNA Repair

Abnormalities in DNA in the form of a mispaired base or biochemical alteration of the preformed structure would obviously cause high mutation rates or lethal effects if there was not some mechanism for correcting them. One mechanism, the proposed "proofreading" function of DNA polymerases I and III contributed by their 3'-exonuclease activities, has already been described as a mechanism of correcting some errors made during synthesis. For correcting damage to preformed DNA, more elaborate systems exist. The field of DNA repair has received increasing attention with the recognition that most types of reparable damage in DNA are both mutagenic and carcinogenic.

Damage to DNA is caused by a variety of agents, including UV light, ionizing radiation, and reactive chemicals. Damage can be in the form of a missing, incorrect, or modified base or an alteration in the structural integrity of the DNA strands by breaks, cross-links, or base dimers.

Some chemicals that modify the bases in nucleic acids are very specific for one base. Hydroxylamine ($NH_2OH$) forms a specific adduct with cytosine that can form a base pair with adenine. Nitrous acid, however, is a general reagent used to deaminate bases, converting adenine to hypoxanthine, guanine to xanthine, and cytosine to uracil. Other compounds that cause DNA damage alkylate the bases or form other covalent adducts. Nitrosamine derivatives usually must be converted to more reactive species by biological oxidation before becoming active as base-alkylating reagents. The formation of alkyl adducts may alter the pairing characteristics of the bases. *N*-methyl-*N*-nitrosoguanidine is a particularly potent nitrosamine that has found widespread use in the laboratory for generating mutations in bacteria. It acts with greater efficiency at the growing point during DNA synthesis.

Aromatic polycyclic hydrocarbons, such as 2-acetamidofluorene, benzo(a)pyrene, and the mycotoxin, aflatoxin B, form adducts with nucleic acids. Most of these compounds are inactive by themselves but can be converted to highly reactive derivatives within organisms. The reactive derivatives are nonselective and can modify most nucleic acid bases. The bulky groups that are introduced by these adducts apparently act by preventing base pairing rather than by causing errors of base pairing, such as those seen with the simpler alkylating agents.

*Repair of damaged bases consists of recognition of the damage, removal of the altered portion of the DNA, and replacement with the correct structure* (Figure 20-26). This process is called *excision repair* because the damaged region of DNA is excised before the correct sequence is restored. Base defects caused by the deamination of cytosine to uracil and adenine to hypoxanthine are removed by DNA glycohydrolases (described

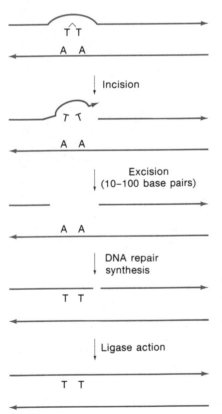

**Figure 20-26**
Scheme for the excision repair mechanism of an abnormal base.

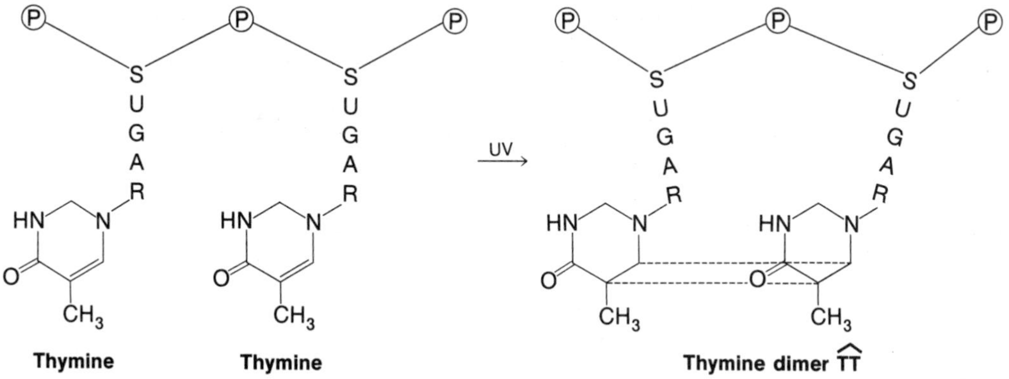

Thymine            Thymine                                      Thymine dimer T̂T

earlier), leaving the free deoxyribose phosphate in the chain. AP-endonuclease recognizes this lack of a base in the DNA strand and breaks the phosphodiester bond. Exonucleases then remove the part of the DNA that includes the modified or missing base, and polymerases fill in the gap.

In situations where the modified bases cannot be removed by specific DNA glycohydrolases, the structural modification is still recognized and removed. A modification inducing a structural irregularity might be recognized by a single-strand–specific endonuclease that would then introduce cuts in the phosphodiester backbone of the DNA and, with the aid of exonucleases, remove a segment of DNA that includes the modified base. The gap is then filled by polymerases and the DNA is joined by ligase.

Adjacent pyrimidine bases in a DNA strand can form dimers upon the absorption of ultraviolet light (Figure 20-27). These dimers can be removed by the general excision repair mechanism as described earlier or repaired directly by enzymatic photoreactivation (Figure 20-28). The photoreactivation enzyme binds to the DNA containing the pyrimidine dimer and uses visible light to cleave the dimer without breaking any phosphodiester bonds (see Chapter 31).

Simple single-strand breaks in the DNA can be repaired by ligation, or by excision repair if the ends are not substrates for ligase. Double-strand breaks would be much more difficult to repair by ligase, since the two ends would not be expected to remain in close proximity to each other. It is not clear what happens in such situations.

The well-defined genetic system of *E. coli* has promoted an understanding of the enzymatic steps involved in excision repair, particularly in response to UV-induced damage. The genes involved in DNA excision repair fall into three categories: those which are exclusively involved in repair processes, those which are involved in both repair and recombination processes, and those which are also involved in DNA replication (Table 20-7). The *uvrA* and *uvrB* gene products are absolutely required for making the initial <u>incision</u> at the site of DNA damage (Figure 20-26). The *uvrC* gene product greatly accelerates the process. Genes involved in <u>excision</u> and resynthesis are more difficult to determine for two reasons: (1) there are many genes that could provide the necessary functions, and (2) many of these genes are required for other life-sustaining processes. Indeed no single mutation has been shown to completely eliminate excision and repair in

**Figure 20-27**
Structure of thymine dimer formed by adjacent thymines when DNA is UV-irradiated.

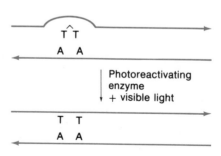

**Figure 20-28**
Thymine dimers may be removed from DNA by enzymatic photoreactivation.

**Table 20-7**
Some *E. coli* Genes that Affect Responses to DNA Damage

| Gene | Map Location | Function |
|---|---|---|
| **uvr Genes** | | |
| *uvrA* | 91 ⎫ | Gene products work together to make initial incision at or near the site of DNA damage. |
| *uvrB* | 17 ⎬ | |
| *uvrC* | 42 ⎭ | |
| **rec Genes** | | |
| *recA* | 58 | Structural gene for recA protein |
| *recB* | 60 ⎫ | Encodes two subunits of exonuclease V. |
| *recC* | 60 ⎭ | |
| **Other genes** | | |
| *LexA* | 90 | Controlling gene for *recA* and other SOS functions |
| *polA* | 85 | Structural gene for DNA polymerase I |
| *polC* | 4 | Structural gene for DNA polymerase III (same as *dnaE*) |
| *lig* | 51 | Structural gene for DNA ligase |

vivo once the initial incision has been made, and no single mutation affecting excision or resynthesis confers the degree of UV sensitivity characteristic of *uvr* strains. The *lig* gene is probably indispensable in making the final linkage in the repair process, but it is also required for normal replication. Therefore, serious defects in the *lig* gene would be expected to have lethal effects and not be detectable as lesions in the repair process. Both polymerases I and III have associated 5′-exonuclease activities and hence are attractive candidates for coupled excision-repair enzymes. Polymerase I is a more likely possibility because it can bind in vitro at nick sites generated by dimer-specific endonuclease. Furthermore, there are many more copies of polymerase I per cell than polymerase III (400 versus 15). Consistent with this, it was noted earlier that mutants of polymerase I that affect the polymerase function are much more sensitive to UV damage.

A very interesting aspect of the repair system is the regulation of expression of the *uvr* genes and other genes that are specifically involved in repair. Such genes are regulated by a combination of the _lexA_ and the _recA_ genes. Damage to DNA activates a protease function of the recA protein that hydrolyzes the _lexA_ gene product. _lexA_ gene product represses the expression of the *uvr* genes and a number of other genes associated with DNA repair. The net result is that biochemical insults that lead to DNA damage stimulate a much higher level of expression of those proteins associated with DNA repair. This regulatory mechanism is described more fully in Chapter 27.

In humans the inability to repair DNA damage is associated with several rare genetic syndromes. The best known is *xeroderma pigmentosum,* and there are several variations of the condition. People with this disease are unable to repair the damage caused by exposure to ultraviolet light and some, but not all, chemicals. Individuals with different syndromes associated with inactive DNA repair mechanisms show different sensitivities to damaging agents. This suggests that there are several enzyme systems for the repair of DNA in humans. Hypersensitivity to DNA-damaging agents may be caused by a defect in a single enzyme in any one of the different pathways used for DNA repair. The multiplicity of defects in repair in different patients is confirmed by complementation studies carried out with cells grown in culture. Cell-fusion experiments with various cultured cell lines indicate that there are about six to eight complementation groups (complementation analysis is discussed in Chapter 27), indicative of at least an equal number of loci that encode enzymes involved in repair of DNA damage in human cells.

## DNA Recombination

Recombination involves the rearrangement of DNA on the same or on different chromosomes. Three basically different types of recombination have been recognized: (1) site-specific recombination, such as is involved in λ integration (see Chapter 27) or immunoglobin gene translocation (see Chapter 1); (2) general recombination, which involves DNA breakage, homologous pairing, and repair; and (3) so-called illegitimate recombination, such as is involved in the translocation of transposons, the translocation of control loci for mating in yeast (see Chapter 26), or the integration of retrovirus DNA into host chromosomes (see Chapter 28). Each recombination system has different enzymes associated with it. Since recombination involves breakage and repair, it would not be surprising to find that some of the same proteins are involved in repair, replication, and recombination. This appears to be the case for the *recA* gene-related protein in *E. coli*, which is involved in both general recombination and repair.

## MANIPULATING DNA

There have been two major turning points in DNA research. The first came suddenly and unexpectedly when it was discovered that DNA was the informational portion of the gene. The second has come from the gradual realization that DNA can be manipulated. It is possible to take genes from their normal surroundings, analyze their sequence, alter the sequence, and reinsert the gene into an organism either to measure the effect of the alteration or to produce larger quantities of a desired gene or gene product. The potential usefulness of the new technology is almost overwhelming. It is now possible to produce vast quantities of precious human hormones by implanting the gene in microorganisms that can be grown on an industrial scale. Projects are in progress to cure certain molecular diseases by simple DNA transfection and to increase the tyrosine content of corn by suitable gene transplantation. Many of the basic skills required for DNA manipulation are discussed below. More advanced reviews should be consulted for fuller descriptions.

## Chemical Synthesis

Most of the chemical methods for making synthetic DNA originated in the laboratory of H. G. Khorana. Khorana's original goal was to chemically synthesize an entire gene and then to see if it would function properly when reinserted into the organism. He wisely chose a small gene for this purpose, the transfer RNA for alanine. His success in the total synthesis of a gene was a historic achievement and led others to refine the methods and apply them in other areas. The greatest value of the synthetic approach is not in the area of fabricating entire genes, but rather in the synthesis of small gene fragments which may then be inserted as parts of naturally occurring genes or hybrid genes by other methods.

Formation of a phosphodiester bond between two nucleotides requires activation of a monoester (5'-nucleotide) and reaction of the activated molecule with an alcoholic group (3'-OH) on another nucleotide. Chemical activation of phosphate groups produces species that will react with several of the functional groups present on the acceptor nucleotide, so protecting groups must be used to reduce undesired reactions. Several activating reagents have been investigated, and one of the most useful that has been found is triisopropylbenzenesulfonyl chloride (TPS). Amino groups can be protected by acetylation, anisoylation, or isobutyrylation. Hydroxyl groups in the 3' position are usually protected by acetylation. Monomethoxytrityl residues are used to protect the 5'-OH. Dinucleotides can be produced by the reaction shown in Figure 20-29, proceeding from blocked 5'-deoxynucleotides. Repetition of the condensation using blocked dinucleotides lead to tri-, tetra-, and eventually higher nucleotides of defined chemical sequence.

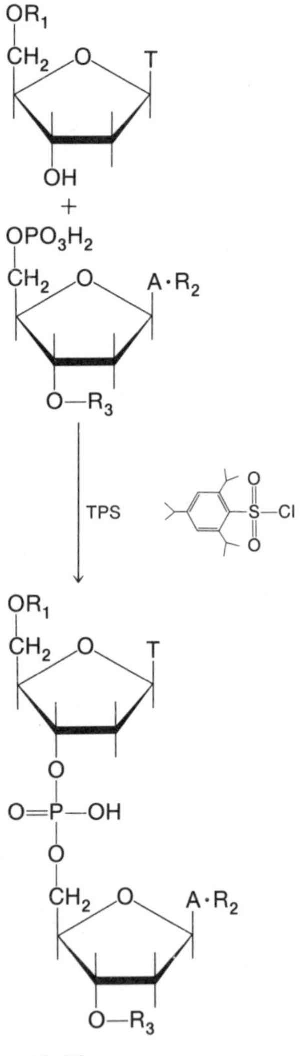

Another approach to the synthesis of oligodeoxynucleotides of defined sequence employing triesters proceeds from 3'-deoxynucleotides. The reactions are carried out by condensation of blocked diesters in the presence of triisopropylbenzenesulfonyl tetrazolide (TPSTe) to form products that are fully blocked triesters. The use of aromatic triesters for the intermediate products reduces side reactions at the phosphate, improves solubility in organic solvents, and enhances possibilities for rapid separation. Before use in biological systems, the triester oligodeoxynucleotides are completely deblocked to diesters. The synthetic reaction employing the triester method is shown in Figure 20-30. As with the phosphodiester method, it is possible to use the procedure repetitiously to produce oligonucleotides of the desired sequence and size. Drawing on the technology developed by Merriweather for the solid-phase synthesis of polypeptides (see Chapter 2), it is possible to couple the 5'-OH group of the initial nucleotide to a resin support and proceed with the addition of nucleotides. All mixing and washing steps are thereby greatly simplified, and the procedure has become amenable to automation; it is possible to produce oligonucleotides of the desired sequence without difficulty.

Having carried out the pioneering work on the chemical synthesis of defined oligodeoxynucleotide sequences, Khorana and coworkers attacked the problem of the combined chemical and enzymatic synthesis for alanine transfer RNA (tRNA$^{Ala}$). To accomplish this, 15 oligodeoxynucleotide sequences, each containing 6 to 20 nucleotides of strategically designed sequences comprising the complete tRNA gene, were chemically synthesized.

**Figure 20-29**
Chemical synthesis of a dinucleotide. The synthesis of TpA dinucleotide is sketched. Protecting groups are represented by $R_1$, $R_2$, and $R_3$. The 5'-phosphate on the suitably protected A residue is reacted with TPS and condensed with the T residue to make the dinucleotide. The process can be repeated by removal of the $R_3$ protecting group to make a trinucleotide, and so on.

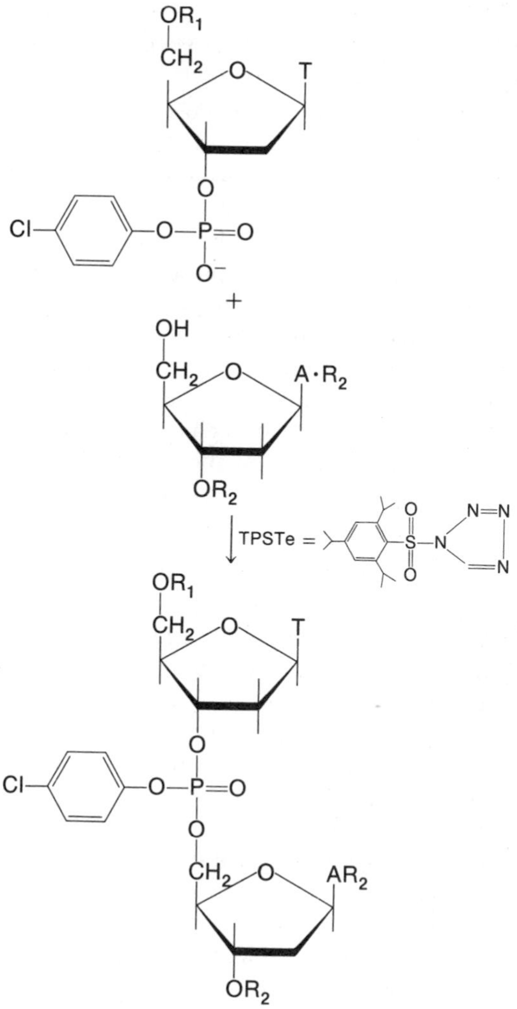

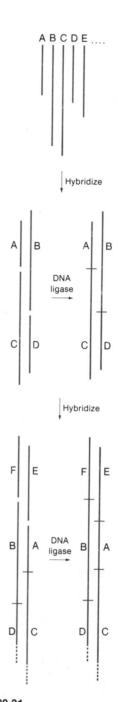

**Figure 20-30**
Triester method for chemical synthesis of a dinucleotide.

These complementary oligonucleotides had overlaps of four to five nucleotides to allow annealing to adjacent sequences. Appropriate double-stranded molecules were then formed by sequential hybridization (Figure 20-31) and covalently coupled by DNA ligase. Eventually, the complete double-stranded molecule was prepared and inserted into biological systems.

## Sequencing DNA

Eventually, one would hope to understand the meaning of DNA sequence in molecular terms. The sequence -TAC- has no intrinsic meaning in written languages, but the biochemist who understands how this may lead to an -AUG- sequence in an mRNA that may be an initiation point for translation into protein already has whetted his or her appetite for more and more complex sequence information.

**Figure 20-31**
Schematic diagram showing how the tRNA gene for alanine tRNA was synthesized. First, small oligonucleotides representing different fragments of the gene were synthesized. Then these were annealed and covalently linked with the help of DNA ligase.

Sequencing a trinucleotide illustrates a simple case that provides an introduction to sequencing. Ultraviolet irradiation is known to produce thymine dimers (T̂T) in DNA. In order to determine whether the thymine residues were from the same chain or opposite chains of the double helix, some structure information was required. DNA isolated from *E. coli* grown in $H_3{}^{32}PO_4$-labeled (p̊) medium was irradiated with 280-nm ultraviolet light and then exhaustively digested with DNAse I and venom diesterase. When the digest was separated on DEAE-cellulose sheets at pH 8.5 in 0.25 $M$ $NH_4HCO_3$, 75 percent of the $^{32}P$ was present in 5'-nucleotides and about 25 percent was present in more slowly migrating material, mostly expected to be trinucleotide. These oligonucleotides represent a protected sequence. The undigested material was eluted from the first DEAE chromatogram and rerun in two dimensions, the first at pH 8.5 in $NH_4HCO_3$ and the second in 0.2 $M$ ammonium formate at pH 3.5. Autoradiography of the two-dimensional DEAE chromatogram showed four separate spots (Figure 20-32). These were expected to be the trinucleotides p̊Ap̊Tp̊T, p̊Cp̊Tp̊T, p̊Tp̊Tp̊T, and p̊Gp̊Tp̊T, if they came from the same chain, or p̊Ap̊T̂Tp̊, p̊Tp̊T̂Tp̊, etc., if they came from separate chains. The four radioactive spots were eluted and irradiated at 240 nm, a wavelength known to reverse the thymine dimer. If the compounds came from separate chains, the compounds p̊Np̊T and p̊T would be expected, whereas continuous chains would produce true trinucleotides of the form p̊Np̊Tp̊T. Rechromatography of the 240-nm–irradiated UV products produced no p̊T. The compounds formed from 240-nm reversal were then found to be susceptible to degradation with venom phosphodiesterase, producing p̊N and p̊T in the ratio 1:2. Alkaline phosphatase (an enzyme commonly used to remove terminal 5'- or 3'-phosphates) digestion of 240-nm treated material followed by venom phosphodiesterase degradation and chromatography of the products produced only p̊T and p̊$_i$ on the autoradiogram. Thus the structure of the original trinucleotide products was pNpTp̂T, and the thymine residues were nearest neighbors on the same chain.

Sequencing intermediate-length oligodeoxynucleotides requires additional technology. An undecanucleotide representing the DNA sequence for residues −5 to +6 of the *E. coli* tyrosine suppressor tRNA was synthesized using the triester procedure described earlier. The negative number refers to bases immediately preceding the gene and the positive number to the tRNA gene itself. In all synthetic work, proof of structure is required. The sequence was expected to be GpGpApApGpCpGpGpGpGpC—OH. After chemical removal of all blocking groups, the free 5'-OH was labeled by phosphorylation with γ-$^{32}P$-ATP (p̊ppA) using the enzyme polynucleotide kinase to transfer the γ-phosphate group of the labeled ATP to the 5'-OH terminal of the polynucleotide chain. All the fragments produced by partial venom diesterase digestion could then be detected by autoradiography of two-dimensional chromatograms of the digestion products separated

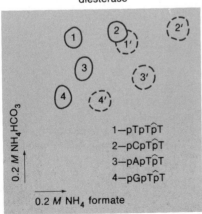

Irradiated native DNA plus venom
diesterase

0.2 $M$ $NH_4HCO_3$

0.2 $M$ $NH_4$ formate

1—pTpTp̂T
2—pCpTp̂T
3—pApTp̂T
4—pGpTp̂T

**Figure 20-32**
Two-dimensional chromatogram of the enzyme-resistant sequences found in *E. coli* DNA irradiated in the native state by 280-nm light. Identification of known products is given in the figure.

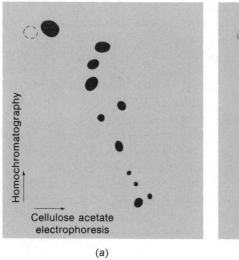

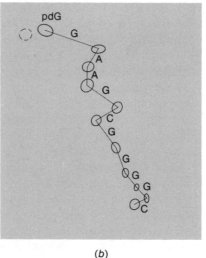

(a)                              (b)

**Figure 20-33**
Sequencing an undecanucleotide. (*a*) A two-dimensional fingerprint of a partial snake venom phosphodiesterase digest of the undecanucleotide dGGAAGCGGGGC. (*b*) An artist's conception of the two-dimensional fingerprint shown in (*a*). The dashed circle in (*a*) indicates the position of the dye marker xylene cyanol.

first by electrophoresis on cellulose acetate at pH 3.5 (which separates by charge) and then by homochromatography on DEAE-cellulose (which separates by size). Partial degradation by venom diesterase should produce a related family of products with decreasing chain lengths:

$$5'\text{-}\overset{*}{p}GpGpApApGpCpGpGpGpGpC\text{-}3'$$
$$5'\text{-}\overset{*}{p}GpGpApApGpCpGpGpGpG \qquad —pC$$
$$5'\text{-}\overset{*}{p}GpGpApApGpCpGpGpG \qquad —pG$$
$$\vdots$$

Removal of each terminal nucleotide reduces the chain length by one nucleotide, which increases mobility of the oligonucleotide in one dimension and results in a specific change in direction of movement in the other dimension. The result is a series of spots showing a staggered path on the chromatogram (Figure 20-33). Movement to the left denotes C removal. A removal results in no deviation and G or T removal results in rightward movement. The sequence is read directly from the autoradiogram by noting the stepwise changes in mobility.

*The most effective DNA sequencing methods depend on display of a continuous set of DNA fragments that differs in length by only one nucleotide and identifying the chemical nature of the end nucleotide by simple inspection of agarose gel autoradiographs of four separate reaction mixtures that somehow identify each of the four bases.* The method developed in F. Sanger's laboratory using chain-terminating dideoxynucleotide triphosphates to produce a continuous series of fragments in DNA polymerase reactions is straightforward and easy to understand.

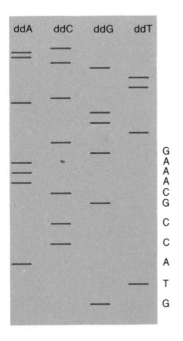

**Figure 20-34**
Dideoxy sequencing method applied to a DNA molecule. The autoradiogram is shown after gel electrophoresis of DNAs prepared in the presence of different dideoxynucleotides. (Redrawn from actual autoradiogram.)

Either DNA or RNA may be sequenced by the Sanger method. When RNA is being sequenced, reverse transcriptase is used to make a DNA copy of an RNA template. When DNA is being sequenced, DNA polymerase I is used to make a copy of a single-stranded DNA template. By choice of an appropriate primer, the region of the nucleic acid that is copied can be predetermined. Reaction mixtures are set up to contain four radioactive deoxyribonucleotide triphosphates and a single dideoxynucleoside triphosphate. The products of four separate reaction mixtures, each containing a different dideoxynucleoside triphosphate, are analyzed. By random termination, reaction 1, using ddATP, contains all A terminations; reaction 2, using ddCTP, contains all C terminations; and so on. After incubation, the reaction products are separated by electrophoresis on polyacrylamide gels. Gels containing 8 to 12 percent acrylamide are used for separating fragments 10 to 300 nucleotides long. Urea is added during gel electrophoresis to prevent aggregation of the fragments and to ensure separation of the oligonucleotides strictly according to size. The fragments are detected on the gel by autoradiography.

The sequence is read directly from the autoradiograph (Figure 20-34), starting with the fastest-moving fragment and moving up the gel. If the first band is in reaction 3, it is a G residue; the next highest band appearing from reaction 4 would be T; and so on. Up to 200 residues can be read from a single gel.

Another method for sequencing long stretches of DNA has been developed by A. Maxam and W. Gilbert. *The Maxam-Gilbert sequencing procedure merits description not only as an alternative procedure for sequencing, but also because it provides a technique for determining where proteins bind to the DNA.* For the initial sequencing of the DNA, either single- or double-stranded molecules may be used. First, a specific segment of DNA is labeled at one end. DNA labeled with [32]P at the 5' end by polynucleotide kinase, as described earlier, is used for most sequence analysis. If double-stranded DNA is used both 5' ends may be labeled. Such a molecule cannot be sequenced directly. The complementary strands of a doubly labeled DNA would have to be separated first. This can be done in two ways: (1) denaturation followed by gel electrophoretic fractionation of the individual chains, or (2) treatment of the DNA with a restriction enzyme that cleaves it into two segments that may be separated. Next the DNA is treated with a specific chemical reagent that specifically reacts with one of the four bases. This reaction is carried out for a limited period of time, so that on the average only a single residue in the polynucleotide reacts with the reagent. The modified base introduces a linkage amenable to cleavage by subsequent chemical treatment. To a first approximation, all bases of a given type in a chain are equally susceptible to modification. The net result is a family of products labeled at the 5' end with [32]P and terminating at the point of cleavage (see Figure 20-35). These may be separated and characterized as when using the Sanger method.

Four chemical reactions are used that cleave DNA preferentially at guanines (G > A), at adenines (A > G), at cytosines and thymines equally (C + T), and at cytosine alone (C). When the products of the four reactions are resolved according to size by gel electrophoresis, the DNA sequence can be read from the pattern of radioactive bands on an autoradiogram (Figure 20-36).

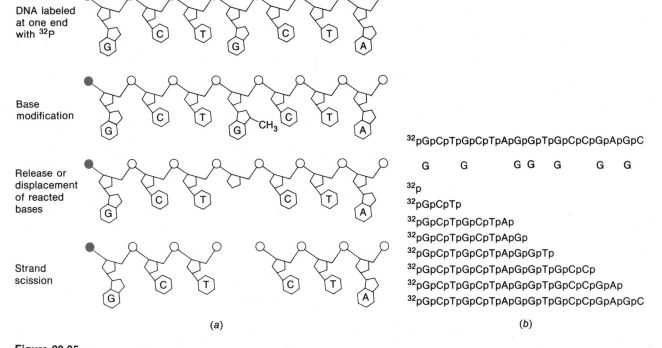

DNA labeled at one end with $^{32}$P

Base modification

Release or displacement of reacted bases

Strand scission

$^{32}$pGpCpTpGpCpTpApGpGpTpGpCpCpGpApGpC

G   G     G G   G    G   G

$^{32}$p
$^{32}$pGpCpTp
$^{32}$pGpCpTpGpCpTpAp
$^{32}$pGpCpTpGpCpTpApGp
$^{32}$pGpCpTpGpCpTpApGpGpTp
$^{32}$pGpCpTpGpCpTpApGpGpTpGpCpCp
$^{32}$pGpCpTpGpCpTpApGpGpTpGpCpCpGpAp
$^{32}$pGpCpTpGpCpTpApGpGpTpGpCpCpGpApGpC

(a)              (b)

**Figure 20-35**
Sequencing end-labeled DNA by limited, base-specific chemical cleavage. (a) On the left, three consecutive reactions cleave one DNA molecule at one guanine, and when these reactions cleave at a guanine in all such molecules, they generate the nested set of end-labeled fragments listed in (b) on the right.

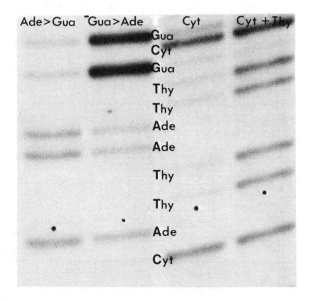

**Figure 20-36**
Autoradiogram of a sequencing gel according to Maxam and Gilbert. Only a portion of a gel is shown. The smallest labeled oligonucleotide moves the fastest and is shown at the bottom of the gel.

For the purine-specific reaction, an aliquot of the DNA is treated with dimethylsulfate, which methylates the guanines in DNA at the N-7 position and the adenines at the N-3 position. The glycosidic bond of a methylated purine breaks on heating at neutral pH, leaving the sugar free. Alkali at 90°C will cleave the sugar from the neighboring phosphate groups. When the resulting end-labeled fragments are resolved on a gel, the radioautogram contains a pattern of dark and light bands (Figure 20-36, G > A lane). An adenine-enhanced cleavage can be obtained by treating the methylated DNA with acid, which releases adenine preferentially (Figure 20-36, A > G lane).

Other aliquots are reacted with hydrazine, which cleaves cytosine and thymine. After a partial reaction in aqueous hydrazine, the phosphate backbone of the DNA is cleaved with 0.5 $M$ piperidine. The final gel pattern contains bands of similar intensity owing to the cleavages at cytosines and thymines (Figure 20-36, C + T lane). However, if 2 $M$ NaCl is included in the hydrazine reaction, the reaction of thymines is suppressed. Then the piperidine breakage produces bands only from cytosines (Figure 20-36, C lane).

If a protein is bound to the DNA, it will interfere with the base-modification reactions. Certain proteins, such as RNA polymerase (Chapter 21), *lac* repressor (Chapter 26), and T antigen (Chapter 28), bind firmly to specific sites on the DNA. When the Maxam-Gilbert treatment is applied to DNA fragments in such specific protein complexes, the bases in close contact with the protein will be apparent by a missing part of the autoradiogram leaving a so-called footprint on the pattern (see the chapter-opening photograph). Although this method says nothing about the linkages involved in a specific protein-DNA complex, it is the most precise method available for determining the location of protein binding.

The best sequencing methods (described earlier) are limited by their capacity to resolve polynucleotides. Fragments longer than about 200 resi-

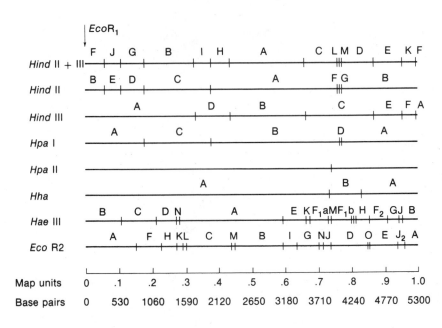

**Figure 20-37**
Cleavage map of the SV40 genome. The zero point of the map is the EcoR1 site. For clarity, the circular genome is shown opened at the R1 site, and the cleavage sites (and resulting fragments) for each restriction enzyme are indicated on a separate line.

dues are impossible to sequence directly. As in peptide-sequence analysis (see Chapter 2), well-defined fragments of this size or smaller must be obtained that have partially overlapping information. To obtain such fragments, restriction enzymes have been most useful. The first step in sequence analysis using the Maxam-Gilbert method on a virus the size of SV40 ($\approx$5300 base pairs) is to obtain a "restriction map" of the virus genome. The virus DNA is digested singly with a large number of restriction enzymes that cleave at different sequences on the DNA. The resulting fragments are resolved according to size by agarose gel electrophoresis. The order of fragments in the chromosome is determined by analysis of partial digestion products, by successive cleavage with multiple restriction endonucleases, by hybridization of individual fragments (or radioactive transcripts prepared from them with RNA polymerase) to reference fragments previously ordered, by pulse labeling of synthetic DNA, and by end labeling of DNA prior to cleavage or by other mapping techniques. A restriction map of SV40 DNA obtained in this way is shown in Figure 20-37. Fragments obtained by single- or multiple-enzyme digestion may be isolated directly from an agarose gel after electrophoresis. Such fragments may be sequenced directly or amplified first by the cloning procedures described below.

## Making Recombinant DNA

*Covalently joined DNA with sequences arising from different regions of the same organism or from more than one source (be it synthetic or natural) is defined as recombinant DNA.* DNA from viruses or plasmids that replicate extrachromosomally in *E. coli* are commonly used as vectors for making recombinant DNA, since they can be manipulated in vitro and reintroduced into the bacteria. The source of the DNA inserted into the vector can be derived from *E. coli* or other bacteria, plants, or animals. Small circular duplex plasmid DNAs from *E. coli*, especially modified derivatives of these plasmids, modified λ-phage DNA, and single-stranded M13-phage DNA have all been very useful as vectors in recombinant DNA work. All these vectors replicate autonomously in *E. coli* after transfection, yielding many copies per cell of the recombinant DNA.

In a simple procedure for cloning, vector DNA (e.g., an autonomously replicating plasmid) and insert DNA are cut with a restriction enzyme and then the pieces are annealed and joined together by the action of DNA ligase. The recombinant molecules are then placed into *E. coli*, where they replicate. When λ vectors are used, a population of cells is infected with the viruses and virus replication proceeds spontaneously. When plasmid vectors are used, a population of cells is bathed in the plasmid DNA containing the inserted DNA. Since only a small number of cells become transfected by this procedure, some means must be available for detecting those cells which carry the desired hybrid plasmid. For this purpose, special selection procedures have been devised. A particularly useful plasmid vector called pBR322, itself a hybrid plasmid, has been constructed to serve as a generally useful vector (Figure 20-38). This plasmid contains two antibiotic resistance genes: $Ap^r$, which confers resistance to penicillin, and $Tc^r$, which confers resistance to tetracycline. BamH1 restriction fragments of foreign DNA may be inserted into the unique BamH1 site on pBR322. This is done by digesting pBR322 with BamH1, mixing with the restriction frag-

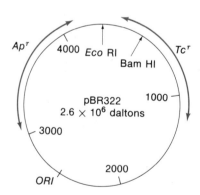

**Figure 20-38**
Map of the pBR322 plasmid. Indicated are the origin of replication, the genes $Ap^r$ and $Tc^r$, which confer resistance to the antibiotics penicillin and tetracycline, respectively, and the unique cleavage sites for the EcoR1 and BamH1 restriction enzymes.

ments at low temperatures to permit annealing to take place between the two DNAs, and finally ligating the annealed fragments with DNA ligase. The product will contain a mixture of original pBR322 and some pBR322 with the inserted foreign DNA. When this mixture is used in transfection, most cells will not be transfected, some will be transfected with pBR322, and some with the desired hybrid plasmid. The three cell types may be readily distinguished by their drug-resistance properties. Normal cells will be killed by penicillin. Transfected cells with the DNA inserted in the plasmid will be penicillin-resistant but tetracycline-sensitive, since the insert has disrupted the $Tc^r$ gene. Cells containing the desired plasmids are distinguished from those containing pBR322 by a technique called replica plating. On semisolid agarose plates containing growth medium, a large population of treated bacteria are spread and allowed to grow for several hours or overnight. A seemingly homogeneous lawn of cells grows on the surface of the gel. Actually, the lawn results from the growth of many microcolonies to the point of confluency. Following this, a piece of velvet is lightly pressed against the surface of the plate and this impression is transferred to other agarose plates containing growth medium with penicillin or penicillin plus tetracycline. Only the transfected cells will produce colonies on the plates containing antibiotics, and because of their small number, each of these will give rise to visibly detectable clones. Those clones present on the penicillin-containing plates which are missing on the penicillin plus tetracycline plates most likely contain the desired hybrid plasmids (see Figure 20-39). These clones are usually picked off the penicillin plates and retested to eliminate any uncertainty about the original drug testing. Once this is established, the plasmid-containing cells are grown in liquid culture. After a moderate density of growth has been achieved, the plasmid DNA is selectively amplified by adding chloramphenicol and continuing incubation of the liquid culture under otherwise normal growing conditions. Chloramphenicol inhibits protein synthesis. This prevents normal cellular metabolism, including cellular DNA replication, but permits the replication of the plasmid DNA. Plasmid DNA replication continues for several hours, until each cell contains 1000 to 2000 copies of the small circular plasmid DNA. This DNA is readily separable from the host DNA and can be characterized by its rate of migration on gel electrophoresis and other more specific tests to determine if it contains the inserted DNA sequence. If desired, the inserted sequence may be removed from the plasmid vector by digestion with BamH1, the restriction enzyme used in the initial construction of the hybrid plasmid.

**Figure 20-39**
Replica plating technique for detecting hybrid plasmid-containing cells.

Master plate,
normal media

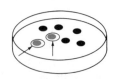

Normal media plus
penicillin

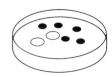

Normal media plus
penicillin plus
tetracycline

## Making a DNA Library

*A DNA library may be defined as a large collection of independently isolated random DNA fragments from a single organism, each associated with a plasmid vector.* The source may be the DNA from germ-line tissue or somatic cells, depending on the ultimate uses of the library.

A library of *Drosophila* DNA fragments was first produced by D. Hogness and coworkers by sonic fragmentation of the DNA. The 3'-OH ends of the fragments were exposed by digestion with λ-exonuclease, and a homopolymer tail of 100 to 200 dAMP residues was then added to the fragments by incubation with dATP and terminal deoxynucleotidyl transferase (Figure 20-40). A vector DNA was prepared by isolating pSC101 plasmid from *E. coli*, cleaving with EcoR1, and digesting away protruding 5'-phosphate terminals with λ-exonuclease (Table 20-5). The vector was then tailed with dTMP residues using terminal deoxynucleotidyl transferase (described earlier).

*Drosophila* library fragments tailed with dAMP were then mixed with pSC101 tailed with dTMP, and recombinant molecules with pSC101 containing *Drosophila* library inserts were formed by annealing (Figure 20-41) and sealed with polymerase and ligase. Obviously, a random mixture was formed. When the mixture containing some pSC101-*Drosophila* fragment chimeric molecules was introduced into *E. coli*, only those bacteria containing plasmid DNA could replicate in the presence of tetracycline because the plasmid contains the $Tc^r$ gene. The bacteria that grew all contained plasmids with *Drosophila* library inserts.

*To find bacteria that contain plasmids with specific pieces of Drosophila DNA inserted, a specific probe is needed.* One could, for example, label *Drosophila* embryos with $^{32}$P and isolate a highly radioactive 18 S ribosomal RNA for this purpose. *E. coli* cultures containing *Drosophila* library plasmids are then plated on nutrient agar and individual colonies are allowed to grow up on a petri plate. A nitrocellulose sheet is laid on the plate to absorb the bacterial colonies. The sheet is removed and the colonies are lysed with alkali, neutralized, and deproteinized by digestion with protease, leaving a denatured DNA replica on the sheet. The DNA replica sheet is then hybridized to the 18 S $^{32}$P rRNA probe. After hybridization and washing, the nitrocellulose sheet is autoradiographed. Only those colonies containing pSC101 plasmid with 18 S ribosomal DNA inserts will hybridize, producing radioactive images (Figure 20-42). These colonies are picked from the original agar plate and grown in mass culture. The *E. coli* selected in this manner will contain only those plasmids with 18 S ribosomal DNA gene sequences inserted. The plasmid DNA can be isolated, and vector DNA can be removed by restriction enzyme cleavage and fractionation. In this way, a large quantity of pure ribosomal DNA could be produced.

If one wishes to screen for colonies that produce a particular protein, then different detection procedures could be used. The colonies could be lysed and protein adsorbed onto a suitable matrix. This protein replica matrix could then be tested for a specific protein by application of monospecific antibodies and testing for antigen-antibody reactions on the matrix. If the cloned gene is expressed to produce a functional protein, then

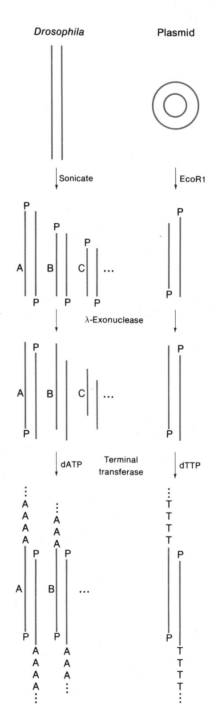

**Figure 20-40**
Preparation of tailed insert DNA and vector DNA, the first step in making a DNA library.

**Figure 20-41**
Hybridization of insert DNA to vector DNA, the second step in making a DNA library.

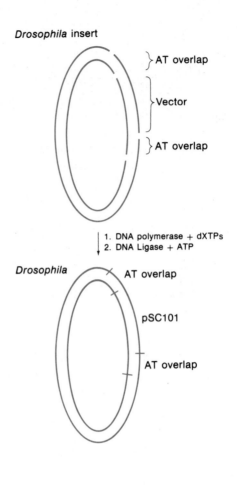

*Drosophila* insert

AT overlap

Vector

AT overlap

1. DNA polymerase + dXTPs
2. DNA Ligase + ATP

*Drosophila*    AT overlap

pSC101

AT overlap

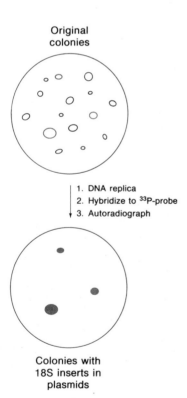

Original colonies

1. DNA replica
2. Hybridize to $^{33}$P-probe
3. Autoradiograph

Colonies with 18S inserts in plasmids

**Figure 20-42**
Selecting colonies containing 16 S inserts by colony hybridization, the third step in making a DNA library.

detection of the insert can be accomplished by complementation of a mutational lesion in the host organism harboring the chimeric plasmid.

Libraries similar to the one constructed for *Drosophila* have been made for a number of organisms, including *E. coli*, yeast, and human DNA. The vast majority of hybrid plasmid DNAs in these large libraries are uncharacterized. The wealth of information available in these libraries is incalculable. Given appropriate selection procedures, the desired hybrid plasmid may be isolated from these libraries and the DNAs characterized according to sequence and biologic behavior after injection into the organism. It is possible to inject DNA into bacteria, yeast, animal cells growing in tissue culture, or germ-line cells of multidifferentiated organisms such as mice. Frequently, the injected DNA "survives" by integrating into a host chromosome or by replicating independently.

## Directed Mutagenesis

In the preceding discussion, where the plasmid vector pBR322 was introduced as a cloning vector, it was shown that inserting foreign DNA at the

BamH1 restriction site of the plasmid inactivates the $Tc^r$ gene for tetracycline resistance. This is an example of _directed mutagenesis,_ the intentional alteration of a gene at a specific location by conscious design. Discrete alterations can be made in a variety of ways on any DNA in vitro, and the effect of such alterations can be subsequently tested in vivo. The biochemist has the edge over the geneticist when the intention is to determine the effects of a change at a specific gene locus because it is possible to carry out a reaction in vitro to obtain a specific change. To obtain the same change, a geneticist might have to screen $10^5$ or more mutants. Most of the techniques used rely upon the specificity of a restriction enzyme at some stage. As the number of specific restriction enzymes recognizing different sequences increases, this factor becomes less of a limitation on where mutations can be made.

In D. Nathan's laboratory it has been demonstrated that specific basepair changes can be made at a single point in the SV40 virus chromosome (Figure 20-43). The mature viral chromosome is treated with HpaII restriction enzyme in the presence of ethidium bromide. Ordinarily, restriction enzymes cut both DNA strands, but in the presence of ethidium bromide there is an increased probability of getting a single nick at a CCGG site recognized by the HpaII enzyme, as shown.

The nicked DNA is subject to limited treatment with DNA polymerase I in the presence of dTTP. The gapped molecules are treated with the single-strand specific mutagen sodium bisulfite, which deaminates cytosine to uracil, and the gaps are then filled in by incubating the DNA with DNA polymerase I in the presence of all four triphosphates. The final result will be the replacement of a GC base pair by an AU base pair, which should result in an AT base pair after replication in vivo. To enrich the mutagenized population of molecules, the same restriction enzyme used to nick the DNA at the beginning of the procedure is used to linearize those molecules whose restriction sites escaped mutagenesis. The modified DNA can then be used to infect cells and the resulting viruses reisolated and tested to see if their DNA is resistant to HpaII. Several mutants are possible depending on which DNA chain is cleaved by the restriction enzyme and which C residue(s) are deaminated by the bisulfite treatment. Other differences in the behavior of the mutant virus also can be studied.

This example of directed mutagenesis was picked to illustrate the approach. A variety of different procedures has been developed for making other types of base replacements as well as deletions and insertions of segments of DNA.

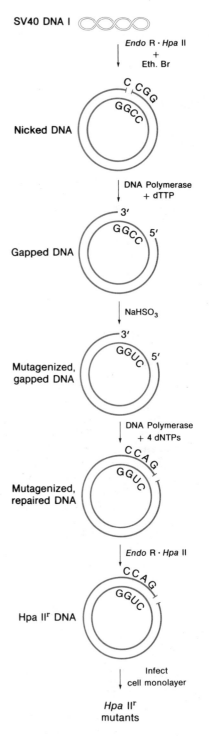

**Figure 20-43**
Steps used to generate SV40 mutants with base substitutions at a preselected site in viral DNA. HpaII$^r$ indicates resistance to the HpaII restriction endonuclease. (Adapted from D. Shortle and D. Nathans, _Proc. Natl. Acad. Sci. USA_ 75:2170, 1978.)

## SELECTED READINGS

Bollum, F. J. Mammalian DNA Polymerases in Progress. In W. E. Cohn (Ed.), *Nucleic Acid Research and Molecular Biology,* Vol. 15. New York: Academic Press, 1975. Pp. 109–144. A review of research on eukaryotic DNA polymerases.

Fuller, R. S., Kagumi, J. M., and Kornberg, A. Enzymatic replication of the origin of the *Escherichia coli* chromosome. *Proc. Natl. Acad. Sci. USA* 78:7370, 1981.

Grossman, L., and Moldave, K. (Eds.). *Methods in Enzymology,* Vol. 65, Part I: *Nucleic Acids.* New York: Academic Press, 1980. Discusses the use of restriction endonucleases and RNA and DNA sequencing techniques.

Hanawalt, P. C., Cooper, P. K., Ganesan, A. K., and Smith, C. A. DNA repair in bacteria and mammalian cells. *Ann. Rev. Biochem.* 48:783, 1979. Discussion of excision repair processes.

Kornberg, A. *DNA Replication.* San Francisco: Freeman, 1980. An enlarged version of "DNA Synthesis" treating prokaryotic DNA replication in outstanding detail.

Low, R. L., Arai, K.-I., and Kornberg, A. Conservation of the primosome in successive stages of $\phi$X174 DNA replication. *Proc. Natl. Acad. Sci. USA* 78:1436, 1981.

Maniatis, T. Recombinant DNA Procedures in the Study of Eukaryotic Genes. In L. Goldstein and D. M. Prescott (Eds.), *Cell Biology,* Vol 3. New York: Academic Press, 1980. Pp. 563–608.

Nathans D., and Smith, H. O. Restriction endonucleases in the analysis and restructuring of DNA molecules. *Ann. Rev. Biochem.* 44:273, 1975.

Weissbach, A. Eukaryotic DNA polymerases. *Ann. Rev. Biochem.* 46:25, 1977.

Wigler, M. H. The inheritance of methylation patterns in vertebrates. *Cell* 24:285, 1981.

Wu, R. *Methods in Enzymology,* Vol. 68. New York: Academic Press, 1979. State of the art treatment of the concepts of DNA sequencing and recombinant DNA technology.

For several articles on recombinant DNA, see the issue of *Science* for September 19, 1980, which is specifically devoted to articles on this subject.

## PROBLEMS

1. You have just isolated a novel recombinant clone, purified the desired insert from the vector as a 10.0 kilobase pair (kbp) linear duplex DNA, and wish to map the recognition sequences for restriction endonucleases A and B. The DNA was cleaved with these enzymes and the digestion products were fractionated according to size by gel electrophoresis. The following fragment sizes were observed:
   a. Digestion with A alone gave two fragments, 3.0 and 7.0 kbp
   b. Digestion with B alone gave three fragments, 0.5, 1.0, and 8.5 kbp
   c. Digestion with A plus B gave four fragments, 0.5, 1.0, 2.0, and 6.5 kbp.
   Diagram a restriction map for the insert, showing the relative positions of the cleavage sites with respect to one another.

2. Describe an enzymatic test for distinguishing between (a) single-stranded and double-stranded DNA; (b) single-stranded linear and single-stranded circular DNA; (c) duplex linear and duplex circular DNA.

3. λ-Phage DNA is linear and double-stranded with overlapping complementary ends ("sticky ends") as described in Chapter 27. DNA ligase converts linear λ DNA into closed circular DNA. Pretreatment of linear λ DNA with either bacterial alkaline phosphatase or spleen phosphodiesterase prevents circularization by DNA ligase. Explain.

4. A small circular duplex viral DNA with about 5000 base pairs contains a 72 base-pair segment tandemly duplicated. This 72 base-pair segment contains a cleavage site for the EcoRI restriction enzyme which is not present in the rest of the virus. In order to explore the function of the 72 base-pair repeat, it would be desirable to remove one of the 72 base-pair segments. Describe how you would construct such a modified viral chromosome and the enzymes that would be required.

5. Certain mutagens are more likely to damage DNA during the process of replication. Give an example of such a mutagen and explain its mode of action. Draw structures.

6. Diagram in detail the steps involved in the synthesis of the $\phi$X174 phage minus strand (SSDNA→RF), and briefly describe the function of each protein required at every step. What aspects of the replica-

tion of the *E. coli* chromosome are these steps likely to reflect and why?

7. T$_7$-bacteriophage DNA is a linear duplex which replicates bidirectionally starting at a point near the end of the duplex. There are special problems associated with the synthesis of DNA at the 3' ends of the template. What are these and how might they be resolved? You may wish to refer to Kornberg's book on DNA replication mentioned in the Selected Readings.

8. Describe the steps for sequencing a duplex DNA with the following structure

      5'—ATGCATTGAATTCTC—3'

      3'—TACGTAACTTAAGAG—5'

by the Maxam-Gilbert procedure.

9. Describe a protocol for constructing and isolating a hybrid plasmid containing the gene for *E. coli* DNA polymerase I.

10. Describe the preparation of pBR322 plasmid DNA with five negative supercoils starting from a preparation of relaxed DNA.

11. What is the role of DNA methylases in bacteria? Why does most methylation occur near the replication fork?

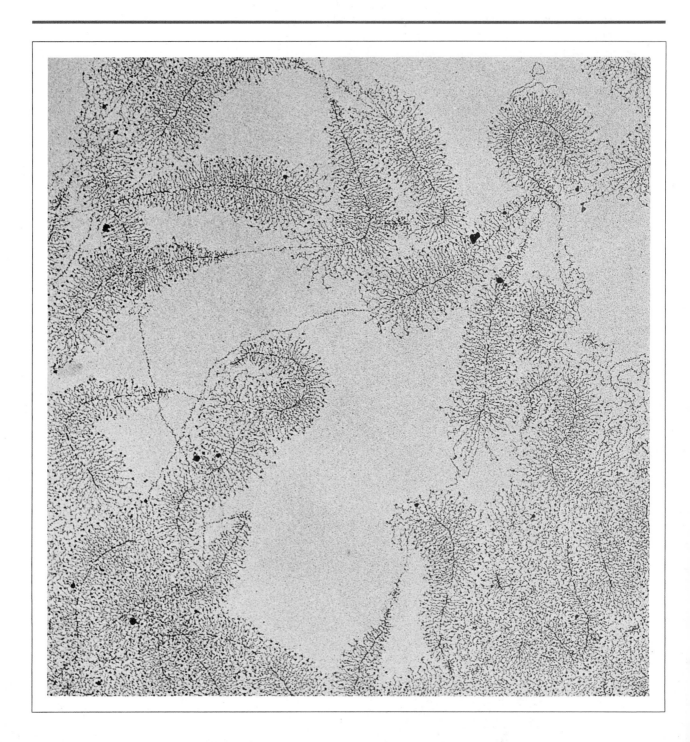

# 21

# RNA METABOLISM

In the early 1950s, Gamov suggested that DNA might serve as a template for protein synthesis. However, biochemical investigations showed that in eukaryotes the processes of RNA and protein synthesis are neatly compartmentalized in the nucleus and cytoplasm, respectively. Because of this, it seemed unlikely that DNA could serve directly as a template for protein synthesis. Base composition studies by E. Chargaff, which had provided a vital piece of information in the deduction of DNA structure and the suggestion of a mechanism for DNA replication, were less informative regarding both the structure and synthesis of RNA. Total cellular RNA did not reflect the base composition of the DNA. However, Volkin and Astrachan, in a monumental experiment, were able to show that when a T2 bacteriophage infects an *E. coli* cell, the newly synthesized RNA (which is exclusively transcribed from the phage DNA template) did reflect the base composition of the DNA. Revealing experiments such as this and an understanding of the DNA duplex structure led to the hypothesis that DNA, by unwinding, could serve as a template for its own synthesis and for the synthesis of RNA. Crick elegantly rephrased these ideas into the "*central dogma*," which states that genetic information flows from DNA to RNA and thence to protein:

$$\text{DNA} \longrightarrow \text{RNA} \longrightarrow \text{protein}$$

This dogma has provided the theoretical framework that experimentalists used to design ways of demonstrating the existence of a transcribing enzyme.

The first enzyme discovered that could catalyze polynucleotide synthesis was polynucleotide phosphorylase. This enzyme, isolated by S. Ochoa

*Electron micrograph of nucleolar DNA showing a cluster of ribosomal RNA genes and growing RNA chains. [From O. L. Miller, Jr., and B. R. Beatty, Portrait of a Gene, J. Cell. Physiol. 74 (Supple. 1): 225, 1969]*

and M. Grunberg-Manago in 1955, could make long chains of $5' \rightarrow 3'$ linked polyribonucleotides starting from nucleotide diphosphates. However, no template dependence could be found for this synthesis, and the sequence was uncontrollable except in a crude way by adjusting the relative concentrations of different nucleotides in the starting materials.

S. Weiss was the first investigator to obtain evidence for a true transcribing activity in cell-free extracts of rat liver (1959). His experimental design was influenced by the theoretical framework described earlier and by Kornberg's discovery that in vitro DNA synthesis required nucleoside triphosphates for substrates. With crude liver extracts, Weiss was able to demonstrate a capacity for RNA synthesis that was severely inhibited by DNAse. This was the beginning of systematic investigations of the biochemistry of RNA metabolism. A continuous expansion of research effort in related areas over the past quarter century has led us in many directions and has provided us with a wealth of understanding about the transcription process and other aspects of RNA metabolism.

Various aspects of nucleotide metabolism (Chapter 19) and RNA structure (Chapter 18) have already been considered. This chapter will focus on the basic aspects of RNA metabolism. In subsequent chapters, consideration will be given to various related topics: (1) regulation of transcription in bacteria and yeast (Chapter 26), (2) transcription in λ bacteriophage (Chapter 27), (3) transcription in animal viruses (Chapter 28), and (4) finally, regulation of transcription by hormones in multicellular organisms (Chapter 29).

## DIFFERENT CLASSES OF RNA

There are three major types of cellular RNA involved in protein synthesis: messenger RNA (mRNA), ribosomal (rRNA), and transfer RNA (tRNA) (see Chapters 24 and 25). In addition, there are other RNAs, some of which function as primers for DNA synthesis, as component parts of ribonucleases, and perhaps in processing of eukaryotic mRNA precursors. The properties of the RNAs found in E. coli are summarized in Table 21-1.

**Table 21-1**
Types of RNA in E. coli

| Type | Function | Number of Different Kinds | Number of Nucleotides | Percent of Synthesis | Percent of Total RNA in Cell | Stability |
|------|----------|---------------------------|----------------------|---------------------|------------------------------|-----------|
| mRNA | Messenger | Thousands | 500–6000 | 40–50 | 3 | Unstable ($T_{1/2}$ = 1 to 3 min) |
| rRNA | Structure and function of ribosomes | 3 { 23 S, 16 S, 5 S } | 2800, 1540, 120 | 50 | 90 | Stable |
| tRNA | Adapter | 50–60 | 75–90 | 3 | 7 | Stable |
| RNA primers | DNA replication | ? | <50 | <1 | <1 | Unstable |
| RNA component of RNAse P | ? | 1 or 2 | 250–350 | <1 | <1 | ? |

## Messenger RNA

Messenger RNA is transcribed from DNA by RNA polymerase, and its sequence contains the information for the sequence of amino acids in the protein product. Specific mRNAs are synthesized by the cell in response to the conditions under which it finds itself, and the cell can thereby control the kinds and amounts of proteins it produces. Prokaryotic cells have unstable mRNAs that turn over rapidly, with an average half-life of 1 to 3 min, whereas eukaryotic cells, which do not have to be able to respond as rapidly to changing conditions, often have more stable mRNAs. The mRNA fraction is heterogeneous in size, ranging from 500 to 6000 nucleotides in *E. coli*. This reflects not only the heterogeneity in size of the proteins of the cell, but also the fact that some prokaryotic mRNAs contain the information to encode more than one protein. Such mRNAs are referred to as *polycistronic mRNAs*, each cistron containing the information for the synthesis of a single polypeptide chain. In prokaryotes, ribosomes begin binding to mRNA and synthesizing protein even before the entire mRNA molecule has been synthesized (see Figure 24-1). This rapid utilization of nascent transcript minimizes the opportunity for processing of the transcript. In contrast, in eukaryotes, where the processes of transcription and translation are sharply divided, the nascent transcript undergoes an elaborate regimen of processing, usually at both ends and frequently internally, before it is transported to the cytoplasm for use as mRNA.

## Ribosomal RNA

Ribosomal RNA is also transcribed from DNA and is a structural and functional component of ribosomes, the cellular organelles responsible for protein synthesis. Ribosomes in prokaryotes are referred to as 70 S ribosomes, a measure of their rate of sedimentation in a centrifuge and hence their size (S refers to the Svedberg constant, which is defined in Chapter 3). A 70 S ribosome consists of two subunits, the 50 S subunit and the 30 S subunit, each of which is made up of RNA and protein. The 50 S subunit contains 23 S and 5 S rRNAs and 33 different ribosomal proteins. The 30 S subunit contains 16 S rRNA and 21 different ribosomal proteins (see Chapter 24 for further information on ribosomes.) Eukaryotic ribosomes are similar in structure, although they are somewhat larger (80 S) and contain larger RNAs (25 to 28 S, 18 S, 5.5 to 5.8 S, and an additional 5 S). Mitochondria and chloroplasts have ribosomes and rRNAs that are distinctly different from those present in the cytoplasm and resemble those of prokaryotes.

## Transfer RNA

The third major type of RNA is tRNA. It too is transcribed from a DNA template. tRNAs contain both a site for the attachment of an amino acid and a site, the *anticodon*, that recognizes the corresponding three-base codon on the mRNA (see Chapters 18 and 24). There are about 50 different kinds of tRNAs in a bacterial cell. Each amino acid is enzymatically attached to the 3′ end of one or more tRNAs by a specific aminoacyl-tRNA synthase that recognizes both the amino acid and the tRNA. This two-step process is discussed in Chapters 24 and 25.

A great deal is known about the structure of tRNA because the entire nucleotide sequences of a large number of tRNAs have been determined and the three-dimensional structures of some of them have been obtained by x-ray crystallography (Chapter 18). Four hydrogen-bonded hairpin stems are present, forming a structure that looks somewhat like a cloverleaf. In the three-dimensional structure of yeast phenylalanine tRNA, the loops at the ends of the stems are folded giving an L-shaped structure. The amino acid accepting site is at one end of the L and the anticodon is at the other (Figure 21-1).

Up to 5 percent of the ribonucleotide bases in tRNA differ from the usual four. tRNA is initially synthesized from the four normal bases, which are then enzymatically modified by a large variety of enzymes. Some of the more than 30 such unusual bases formed are shown in Figure 21-2. Their function is not understood, but it may involve interacting with the mRNA, the ribosomes, or the aminoacyl-tRNA synthases.

## Other RNAs

DNA synthesis requires a primer, an oligonucleotide that is base-paired with the template strand and provides a 3'-hydroxyl group upon which to start polymerizing a new DNA strand (see Chapter 20). The cell uses small RNAs as primers for DNA replication. In some cases these primers are synthesized by RNA polymerase (the main transcribing enzyme), but most primers in *E. coli* are made by a special enzyme, called *DNA primase*, coded by the *dnaG* gene (see below).

One of the processing ribonucleases in *E. coli*, RNAse P, contains a small RNA complexed to the protein. This RNA is about 350 nucleotides long and is essential for enzyme activity. It is not clear whether it plays a strictly structural role or contributes some specificity in helping the nuclease recognize its processing site.

Eukaryotic cells contain in their nuclei a unique collection of small RNAs, called *small nuclear RNAs* (snRNAs), which are complexed with certain proteins to form ribonucleoprotein particles. These RNAs have been named U1, U2, U3, U4, U5, and U6 RNA and range in size from about 110 to 220 bases. One, U3, is found in the nucleolus, the site of rRNA synthesis. All are very abundant, and there may be as many as 1 million copies per nucleus. Their base sequences are highly conserved between organisms, and all seem to contain unusual trimethylguanosine structures at their 5' ends. The role of these RNAs is not known, but one suggestion is that they are involved in processing RNA. This is discussed below. It seems likely that additional RNAs will be found that play regulatory or structural roles in the cell.

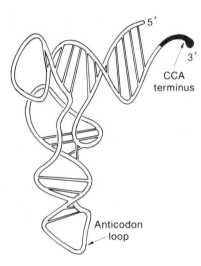

**Figure 21-1**
Schematic diagram of the three-dimensional structure of yeast phenylalanine tRNA. (Redrawn from S. H. Kim, J. L. Sussman, F. L. Suddath, G. L. Quigley, A. McPherson, A. H. J. Wang, N. C. Seeman, and A. Rich, The general structure of transfer RNA molecules. *Proc. Natl. Acad. Sci. USA* 71: 4970, 1974. Used with permission.)

## DNA-DEPENDENT SYNTHESIS OF RNA

### Overall Reaction

The enzymatic activity that Weiss discovered in rat liver extracts is capable of joining ribonucleoside triphosphates by 3'-5' phosphodiester bonds under the direction of a DNA template, releasing pyrophosphate. This en-

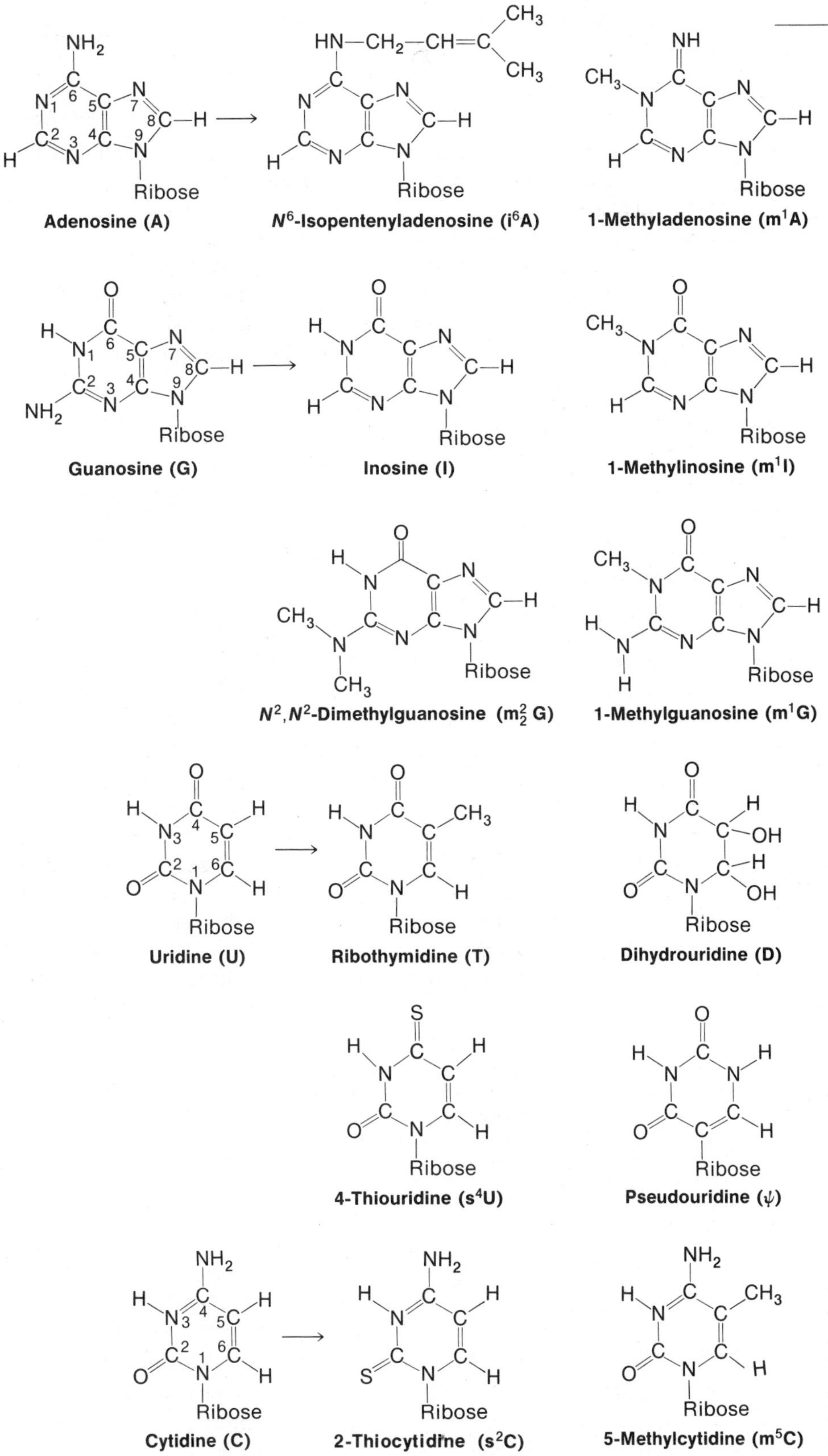

**Figure 21-2**
The structures of some modified bases found in tRNA. The parent ribonucleosides are shown with the numbering of the atoms in the purine and pyrimidine rings.

zyme is now referred to as DNA-dependent RNA polymerase and carries out the following reaction:

$$NTP + (NMP)_n \xrightarrow[\text{DNA}]{Mg^{2+}} (NMP)_{n+1} + PP_i$$

The DNA template strand determines which base will be added on to the growing RNA molecule by base pairing similar to that used to direct the semiconservative replication of DNA. For example, a cytosine in the template strand of DNA means that a complementary guanine will be incorporated into the corresponding position in the RNA. Synthesis proceeds in a 5′→3′ direction with each new nucleotide being added on to the 3′-OH end of the growing RNA chain. This reaction will be described in detail later in this chapter. Figure 21-3 summarizes a method for assaying this enzyme.

### RNA Polymerase

In this section we will focus on the process of transcription in the bacterium *E. coli*, since it is in this organism that it is best understood. This process is quite similar in other prokaryotes.

**Subunit Structure and Function.** Methods that allowed the purification of substantial amounts of RNA polymerase from *E. coli* have made it possible to study the structure of this enzyme. The active enzyme molecule is a complex of four different polypeptide chains with a total molecular weight of about one-half million. The subunits of the enzyme can be separated by electrophoresis on polyacrylamide gels, as is illustrated in Figure 21-4 (this technique was explained in Chapter 3). These five different polypeptide chains, termed $\beta'$, $\beta$, $\sigma$, and $\alpha$, have molecular weights of 155,000, 151,000, 70,000, and 36,500, respectively. An additional subunit called $\omega$ ($M_r = 11,000$) usually copurifies with the rest of the enzyme, but no function for $\omega$ has been established. Until a function has been established, we do not know if the presence of $\omega$ is accidental or reflects some function yet to be discovered. Some of the properties of these subunits are summarized in Table 21-2. A complex with the subunit structure $\alpha_2\beta\beta'\sigma$ can carry out the functions necessary for correct and efficient synthesis of RNA and is

**Table 21-2**
*E. coli* RNA Polymerase Subunits and Regulatory Factors

| Subunit or Protein | Gene Name | Map Position (min) | Polypeptide MW (daltons) | No. in Enzyme | Function | Properties |
|---|---|---|---|---|---|---|
| $\beta'$ (beta′) | *rpoC* | 89.5 | 155,000 | 1 | DNA binding? | Basic |
| $\beta$ (beta) | *rpoB* | 89.5 | 151,000 | 1 | Active site | Acidic |
| $\sigma$ (sigma) | *rpoD* | 66.5 | 70,000 | 1 | Promoter recognition, initiation | Acidic |
| $\alpha$ (alpha) | *rpoA* | 72 | 36,500 | 2 | ? | Acidic |
| $\rho$ (rho) | *rho* | 84.5 | 46,000 | 6 | Termination | Basic |
| CAP | *crp* | 73 | 23,000 | 2 | Activation | Basic |
| nusA | *nusA* | 65 | 69,000 | 1 | Termination | Acidic |

referred to as _holoenzyme_. Holoenzyme can be reversibly separated into two components by chromatography on a phosphocellulose column:

$$\alpha_2\beta\beta'\sigma \;\rightleftharpoons\; \alpha_2\beta\beta' \;+\; \sigma$$

**Holoenzyme**      **Core polymerase**    **Sigma**

One component, called _core polymerase_ ($\alpha_2\beta\beta'$), _retains the capability to synthesize RNA, but it is defective in the ability to bind and initiate transcription at the appropriate initiation sites_ (promoters) on the DNA. The other component, called _sigma ($\sigma$), has no RNA synthetic activity, but when added back to core polymerase, it re-forms holoenzyme with its_

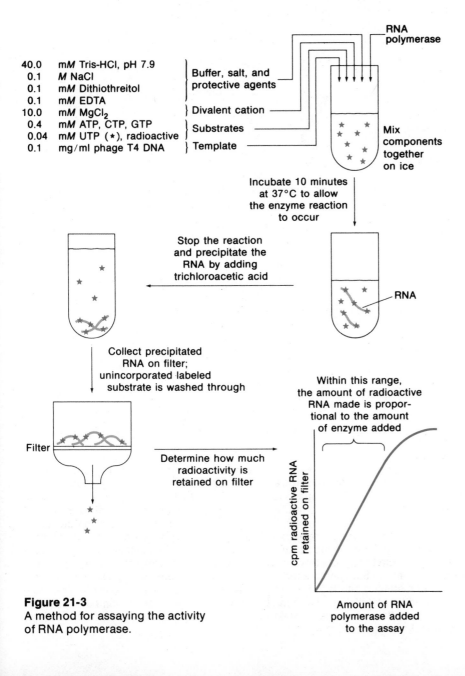

**Figure 21-3**
A method for assaying the activity
of RNA polymerase.

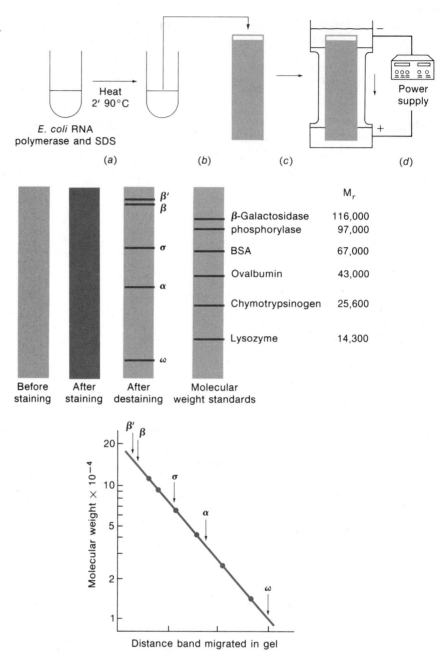

**Figure 21-4**
Analysis of the subunits of RNA polymerase by polyacrylamide gel electrophoresis.
(a) The protein is denatured by treatment with the detergent sodium dodecyl sulfate
(SDS). This causes the subunits to bind the negatively charged SDS and dissociate
from each other. (b) The denatured enzyme is layered on top of a gel of polymerized acrylamide in a cylindrical tube. (c) A voltage is applied across the gel, causing
the polypeptides, which are negatively charged owing to the bound SDS, to move
through the gel toward the positively charged anode. The smaller the protein, the
faster it is able to migrate through the gel. (d) After electrophoresis, the gel is removed from the tube and placed in dye that stains proteins. Excess dye is washed
away and stained bands, containing as little as 1 $\mu$g protein, are visible. The molecular weights of the stained subunit bands can be estimated by comparing their positions with those of proteins of known molecular weights electrophoresed under
identical conditions.

ability to bind tightly and selectively at promoters and initiate RNA chains efficiently. Because of its role in binding and initiation, sigma is often referred to as an initiation factor.

The precise functions of the various subunits of core are not known. $\beta'$ is a basic (positively charged) polypeptide thought to be involved in DNA binding. The $\beta$ subunit is the site of binding of several inhibitors of transcription and is thought to contain most or all of the active sites for phosphodiester bond formation. The $\alpha$ subunit is necessary in order to reconstitute active enzyme from separated subunits.

**Genetics of the Subunits.** Since RNA polymerase is an essential cellular enzyme, most mutations in the polymerase genes are lethal. Those which can be studied are the ones that affect the enzyme activity only under certain conditions, e.g., at higher than normal temperatures. Mutants that can grow normally at one temperature but not at another are called temperature-sensitive mutants. Another class of mutants is comprised of those which render the enzyme resistant to an inhibitor of RNA synthesis (such as the antibiotics rifampicin and streptolydigin), which acts by binding to the enzyme (see below).

Strains of E. coli have now been obtained with mutations in the $\alpha$, $\beta$, $\beta'$, and $\sigma$ subunits of RNA polymerase. These mutants have allowed the genes for these subunits to be mapped. The circular map of the E. coli genome is shown in Figure 21-5, along with the position of the genes known to be involved in RNA synthesis, modification, processing, and degradation. It is evident that these genes are scattered throughout the genome. It is surprising that the subunits of RNA polymerase are not all clustered together in the same transcription unit (operon). This is probably due to the fact that only $\beta$ and $\beta'$ are required in equimolar amounts. While $\beta$ and $\beta'$ are in the same operon at 89.5 min, this operon also contains two riboso-

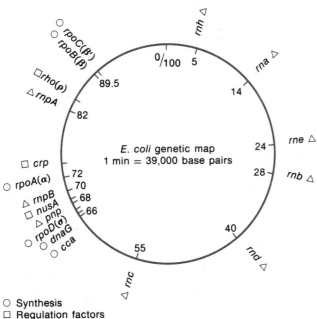

O Synthesis
□ Regulation factors
△ Processing and degradation

**Figure 21-5**
Genetic map of E. coli showing the location of the genes involved in RNA metabolism. The map is divided into 100 minutes. (From B. J. Bachmann and K. B. Low, Linkage map of E. coli K12, edition 6, Microbiol. Rev. 44:1, 1980.)

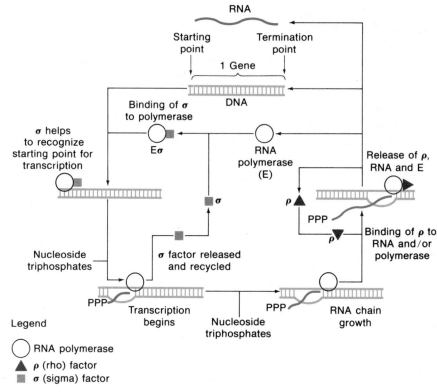

**Figure 21-6**
Schematic of the overall transcription cycle. (Adapted from J. D. Watson, *Molecular Biology of the Gene,* 3rd Ed., Benjamin, Menlo Park, Calif., 1976. Used with permission.)

mal proteins (see Chapter 26). Alpha is found far from $\beta$ and $\beta'$ at 72 min on the map and also is in an operon containing several ribosomal proteins. The gene for the sigma subunit is found at 66.5 min in an operon containing the DNA primase gene and a ribosomal protein.

## Steps in Transcription

The overall transcription cycle involving binding, initiation, elongation, and termination is shown in Figure 21-6 and is discussed in detail below.

**Transcription Unit and Promoter Signals.** In addition to the sequences that code for proteins, there are precise sequences along the DNA that signal RNA polymerase start and stop sites. *The promoter region contains the information that tells the RNA polymerase where to bind, how tightly to bind, and how frequently to initiate an RNA chain.* It also often contains sites at which additional regulatory proteins bind and influence binding and initiation by RNA polymerase. The terminator region contains DNA sequences that cause the RNA polymerase to stop transcribing. Then either spontaneously or with the aid of a termination factor, the RNA polymerase and RNA are released from the template. The transcribed region, including these signals, is called a *transcription unit,* or *operon.* It may include the structural genes for several proteins whose syntheses are controlled coordinately. An example of a transcription unit in *E. coli* is shown in Figure 21-7.

A large number of transcription units has been studied and the promoter and terminator regions sequenced. Common features have been identified in the sequence of over 50 different promoters, and in the case of several promoters, further information about DNA bases important for binding has been obtained. Promoter features are summarized in Figure 21-7.

RNA polymerase binding at the promoter protects about 60 base pairs of DNA from digestion with DNAse, from −40 (40 base pairs before the RNA chain starting site) to +20 (20 base pairs after the starting site). *Two sites within this 60 base pair region, one centered at −35 and one centered at −10, contain specific recognition sequences that are common to all bacterial promoters.* Mutations in these two sites affect the ability of RNA polymerase to bind and initiate. The DNA base sequence of each promoter differs somewhat from the "average" promoter sequence. This is to be expected because promoters differ tremendously in their "strength" or frequency of RNA initiation. Some promoters are very "weak," e.g., the promoter for the *lac* repressor produces an mRNA only once in 20 to 40 min. This is estimated as follows: Each *E. coli* contains about 12 molecules of *lac* repressor. Since each repressor is composed of four identical subunits, this represents about 48 molecules of repressor polypeptide. If each mRNA is translated by about 50 ribosomes before it decays, then only one mRNA is needed per generation of 20 to 40 min. However, the promoters for rRNA genes must be utilized once every second or two in order to produce over

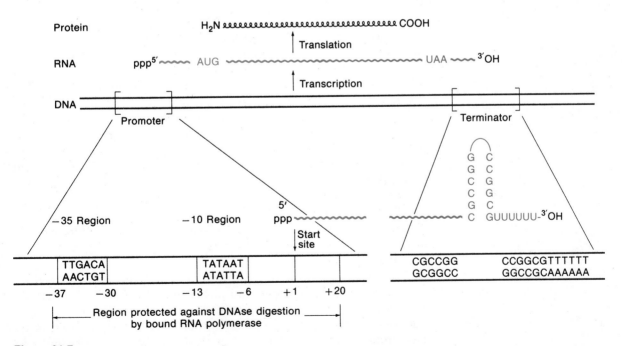

**Figure 21-7**
Important features of a typical transcription unit. DNA is shown with promoter and terminator regions expanded below. RNA is transcribed starting in the promoter region at +1 and ending after the stem and loop of the terminator. The protein resulting from translation of this RNA is shown above with its N and C termini indicated.

RNA polymerase–DNA binding interactions

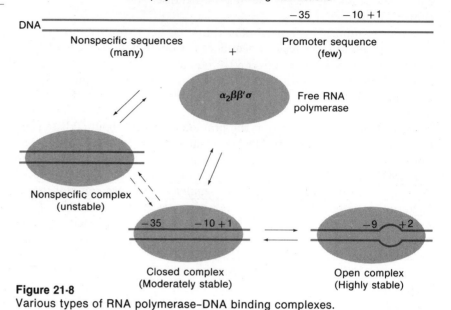

**Figure 21-8**
Various types of RNA polymerase–DNA binding complexes.

10,000 rRNAs per generation from the seven copies of the rRNA gene. This 2000-fold difference in promoter strength is encoded in the base sequence of the promoter.

**Binding at Promoters.** In the cell, RNA polymerase transcribes only selected regions of the DNA and makes RNA complementary to only one of the DNA strands in any particular region. This is possible because holo-enzyme is able to recognize and form a stable complex with DNA at specific promoter regions. RNA polymerase is able to form unstable nonspecific complexes at any place on the template, mainly by an interaction with the DNA phosphates, but it either rapidly dissociates and rebinds or slides along the DNA until it reaches a promoter region. It then forms a moderately stable complex with the promoter DNA, most likely interacting stereospecifically with particular nucleotides in the "−35 region" of the promoter. These complexes can form at 0°C, where the DNA remains double helical or unmelted, and are referred to as _closed promoter complexes_ (Figure 21-8). The next step involves a temperature-dependent melting of about 10 base pairs of DNA from position −10 to +1 and a conformational change in the polymerase. This very stable complex is termed the _open promoter complex_.

The existence of consensus sequences in both the −35 and −10 regions of the promoter carries with it the implication that these regions are important in forming the open promoter complex with RNA polymerase. One possibility is that both regions are complexed with polymerase in the open promoter complex. Another possibility is that one region serves for initial recognition of the polymerase, but that in the final complex the polymerase binds only to the other region. Early experiments aimed at determining the binding site were misleading. The complex was made and treated extensively with DNAse, and the so-called protected fragment was reisolated. This recovered fragment contained the −10 region and a substantial amount of DNA on either side of it, but it was lacking the −35 region. The fragment was incompetent in forming a strong complex with RNA polymerase. These two results led to the early suggestion that the −35 region was the site of initial recognition ("entrybox"), from which the polymerase moved along to make a strong complex finally in the −10 region ("Pribnow box"). This idea is no longer considered correct because of the more definitive experiments reported below.

The different approaches used to establish the regions of contact (or close approach) between polymerase and DNA more precisely were developed in W. Gilbert's laboratory. One strategy used was to make the polymerase-DNA complex and then determine which regions showed altered susceptibility to specific chemical probes. Another strategy was to carry out a partial reaction with a specific chemical probe and see which sites interfered with polymerase complex formation. For these experiments, two chemical probes have been most useful: (1) dimethylsulfate, which methylates the N-7 position of guanine in the large groove and the N-3 position of adenine in the minor groove (the methylated purine will depurinate on heating; alkali then can cause a β-elimination reaction and create a series of breaks in the DNA chain), and (2) ethylnitrosourea,

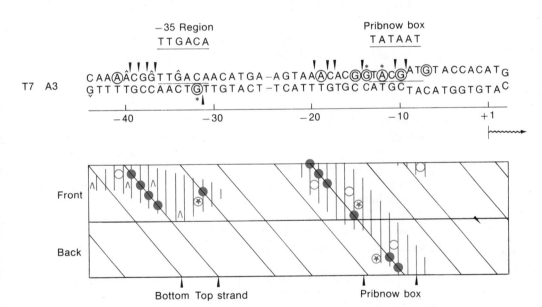

**Figure 21-9**
(*Upper*) Sequence of the T7 A3 promoter, showing the stronger contacts to the RNA polymerase. ❙s indicate phosphate contacts; ○ and ∧ indicate purines that the polymerase protects from methylation or whose susceptibility to this methylation is enhanced, respectively; ∗s indicate methylated purines that interfere with polymerase binding. The most probable bases for the Pribnow box and the −35 region are shown above the corresponding regions in the A3 promoter sequence; +1 represents the start of transcription. The minimal region unwound by the polymerase is represented by a separation of the strands. (*Lower*) Planar representation of the cylindrical projection of the DNA molecule [10.5 base pairs per turn], with contacts to the polymerase marked. •s represent phosphate contacts, ⋆s represent methylated purines that interfere with polymerase binding, and other symbols are as in *Upper*. Contact regions, strands, front view, back view, and initiation site are indicated. Regions likely to interact with polymerase are shaded with vertical lines. (From U. Siebenlist and W. Gilbert, Contacts between *E. coli* RNA polymerase and an early promoter of phage T7, *Proc. Natl. Acad. Sci. USA* 77:122, 1980.)

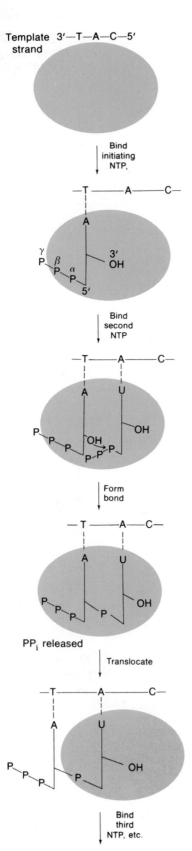

Template 3'—T—A—C—5'
strand

Bind
initiating
NTP,

Bind
second
NTP

Form
bond

PPᵢ released

Translocate

Bind
third
NTP, etc.

which preferentially ethylates the phosphates in DNA. The resulting phosphotriesters serve as cleavage sites when the DNA is exposed to alkali.

The results of these experiments are depicted in Figure 21-9. Domains of interaction are focused on the −35 region and the −10 region. _This result provides strong support for the conclusion that polymerase binds simultaneously to both of these regions._

Additional evidence emerged from the experiments with dimethylsulfate that _the polymerase produces an unwinding of the double helix in the −9 to +2 region._ Dimethylsulfate, which reacts with the N-1 of adenine and the N-3 of cytosine only in single-stranded DNA, reacted with this region in the presence of polymerase. In the double helix these nitrogens are normally unavailable for reaction.

Another type of experiment was used to establish the thymine contacts with polymerase. For this purpose, thymine was replaced in the DNA by 5-bromouracil. Ultraviolet light normally cleaves DNA at the bromouracils. A nearby bound protein can either protect the DNA from such cleavage or become cross-linked to the DNA at position 5' of the bromouracil. This approach was particularly valuable in suggesting which of the polymerase subunits are in direct contact with the DNA. Preliminary indications for the _lac_UV5 promoter show that the sigma subunit cross-links at position −3 of the sense strand and that the $\beta$ subunit cross-links at position +3 of the sense strand.

The chemical-probe approach used to establish points of close contact between polymerase and DNA also has been exploited to obtain useful information on regulatory protein interaction with DNA (see the discussion on CAP and _lac_ repressor in Chapter 26).

**Initiation at Promoters.** Once the polymerase binds to the promoter and strand separation occurs, initiation usually proceeds rapidly (1 to 2 s). The first, or initiating, nucleoside-5'-triphosphate, which is usually ATP or GTP, binds to the enzyme. The binding is directed by the complementary base in the DNA template strand at position +1, the start site. A second nucleoside-5'-triphosphate (NTP) binds, and initiation occurs upon formation of the first phosphodiester bond by a nucleophilic attack of the 3'-hydroxyl group of the initiating NTP on the $\alpha$ phosphorus atom of the second NTP. Inorganic pyrophosphate derived from the second NTP is a product of the reaction. This process is illustrated in Figure 21-10.

The initiation process can be followed during an in vitro transcription reaction in several ways. NTP radioactively labeled with ³²P in the $\beta$ or $\gamma$ phosphate positions can be used in transcription reactions. Since the initiating NTP retains its 5' triphosphate end, the beginning of the RNA chain, or the 5' triphosphate end, will be exclusively labeled. Another way is to add ³²P-labeled pyrophosphate (PPᵢ) to an otherwise unlabeled reaction and measure initiation by a reaction called _pyrophosphate exchange_. The labeled pyrophosphate participates in the reverse of the phosphodiester-bond

**Figure 21-10**
Details of phosphodiester bond formation. The circles represent NTP binding sites on the RNA polymerase. The $\alpha$, $\beta$ and $\gamma$ phosphates are indicated on the initiating NTP, which in this case is ATP.

formation reaction, allowing the 3′ terminal nucleotide to be removed and the label to be incorporated into its $\beta$ and $\gamma$ positions. Pyrophosphate exchange also can occur during chain elongation:

$$\text{pppApU} + {}^{32}\overset{**}{\text{PP}}_{\text{i}} \underset{\text{Bond formation}}{\overset{\text{Pyrophosphorolysis}}{\rightleftharpoons}} \text{pppA} + \overset{**}{\text{pppU}}$$

A third method for measuring initiation depends on the finding that in the absence of the nucleotide needed for the third position on the RNA chain, the dinucleotide formed, pppXpY, dissociates from the active site and the initiation process must start again. This is called *abortive initiation*. For example, if only ATP and [$\alpha$-$^{32}$P]UTP are added to a reaction mixture where an RNA chain that begins with the sequences pppApUpCp is being synthesized, the reaction produces the dinucleotide pppA$^{32}$pU. The production of this dinucleotide can be quantified to measure initiation of this chain. Some promoters in vitro undergo considerable abortive initiation, even in the presence of all four nucleoside triphosphates, releasing oligonucleotides of two to eight residues several times before finally succeeding in producing a long RNA chain. Therefore, *promoter strength should be equated not strictly with the frequency of initiation at a promoter, but rather with the frequency at which long RNA chains are produced.*

**Elongation of RNA.**  After initiation has occurred, chain elongation proceeds by the successive binding of the nucleoside triphosphate complementary to the next base in the template strand, bond formation with pyrophosphate release, and translocation of the polymerase one base farther along the template strand. Transcription proceeds in the 5′→3′ direction, antiparallel to the 3′→5′ strand of the template DNA. Once elongation has produced an RNA chain about 10 bases long, the $\sigma$ subunit dissociates from the holoenzyme to yield core polymerase, which continues the elongation reaction until a terminator signal is reached. The released $\sigma$ is available to bind to a free core polymerase and re-form holoenzyme capable of binding at a promoter and initiating a new RNA chain (see Figure 21-6).

In vivo mRNA chains grow at a rate of about 45 nucleotides per second, nicely matched with the rate of translation of 15 amino acids per second. rRNA appears to be synthesized about twice as fast as mRNA.

As the polymerase travels along the DNA, it must continually cause an opening or strand separation of the DNA so that a single DNA template strand is available at the active site of the enzyme. For the transcription reaction to be energetically feasible, one base pair must re-form behind the active site for every base pair opened in front of it. It is likely that a short transient RNA-DNA hybrid duplex forms between the newly synthesized RNA and the 10-base-long unpaired region of the DNA and helps hold the RNA to the elongating complex.

**Termination of Transcription.**  Termination of transcription involves stopping the elongation process at a region on the DNA template that signals termination and releases the RNA product and RNA polymerase. *The vast majority of terminators that have been studied are similar in that they code for a double-stranded RNA stem and loop structure just preceding the 3′ end of the transcript* (see Figure 21-7). It is thought that such a structure causes the RNA polymerase to pause or stop elongating. Two main types of terminators have been distinguished. The first is capable of termination with no accessory factors and contains about six uridine residues following the stem and loop. Apparently, a particularly weak interaction between the 3′ terminal oligo(U) and the oligo(dA) region in the template DNA strand allows the release of the RNA from the transcription complex. The second type of terminator lacks the oligo(U) region and requires *termination factor rho* for RNA chain release. Rho, discovered by J. Roberts in 1969, is a hexamer of 46,000-dalton subunits. Rho does not bind tightly to DNA or to RNA polymerase. It does bind tightly to RNA, especially C-rich RNA, and in its presence it hydrolyzes ribonucleoside triphosphates to nucleoside diphosphates. A current model is that rho binds to sites on RNA and, by hydrolyzing nucleoside triphosphates, either moves along the RNA or winds the RNA around it until it reaches an RNA polymerase stopped at a terminator, where it causes release.

Another termination factor, nusA, has been identified recently. This acidic protein with a molecular weight of 65,000 is able to bind to core polymerase and aid in the release of RNA and RNA polymerase at some terminators. Much more on rho and nusA is presented in Chapter 27.

**Other Factors Regulating Transcription.**   As more transcription units are studied in detail, it becomes apparent that the basic processes just described can be modulated in a great number of ways, both by including additional signals in the DNA sequence and by supplying additional regulatory factors. These regulatory mechanisms are discussed in more detail in Chapters 26 and 27.

Many transcription units have terminator signals called *attenuators* preceding the structural gene or between two structural genes. Attenuators sometimes cause termination and sometimes allow read-through to the structural part of the gene. The efficiency of termination at these sites is variable and can be regulated in response to changes in the growth conditions of the cell. Attenuators are discussed in Chapter 26.

Protein factors can act to inhibit transcription (negative control) or to stimulate it (positive control). The binding of a repressor molecule at a specific site (operator) overlapping the promoter region prevents transcription by sterically interfering with RNA polymerase binding. This is an example of negative control. An example of positive control is the stimulation of transcription caused by the binding of the catabolite activator protein (CAP) in the presence of cyclic AMP (cAMP) to sites just adjacent to the promoters of a number of genes coding for enzymes involved in the catabolism of a variety of sugars. In in vitro transcription reactions, these promoters bind RNA polymerase holoenzyme only weakly and are greatly activated by the addition of CAP and cAMP.

Although most transcription in *E. coli* appears to utilize $\sigma$ as an initiation factor, there may be other initiation factors that bind to core polymerase to form holo-like enzymes that recognize different promoters. This has been shown to be the case for the bacterium *Bacillus subtilis*. Several different sigma-like subunits are present in growing cells, and additional ones appear during sporulation to allow expression of sporulation-specific genes. In addition, the infection of *B. subtilis* with certain bacteriophages, such as SP01 or SP82, results in virus-coded sigma-like factors that bind to core polymerase in place of $\sigma$ and direct RNA polymerase to viral promoters.

## RNA Polymerase

Unlike prokaryotes, in which all major types of RNA are synthesized by one RNA polymerase holoenzyme, eukaryotic cells have become more specialized in their transcription capabilities. They contain at least four different DNA-dependent RNA polymerases, three of which are nuclear, each responsible for synthesizing a different class of RNA. The basic mechanism of RNA synthesis for all these enzymes is very similar to that of prokaryotic RNA polymerase.

**Nuclear RNA Polymerases.** Nuclear extracts can be fractionated by chromatography on DEAE-cellulose to give three peaks of RNA polymerase activity (the use of DEAE was explained in Chapter 2). These three peaks correspond to three different RNA polymerases (I, II, and III), which differ in relative amount, cellular location, type of RNA synthesized, subunit structure, response to salt and divalent cation concentration, and sensitivity to the mushroom-derived toxin $\alpha$-amanitin (see below). These enzymes and their properties are summarized in Table 21-3. *RNA polymerase I is located in the nucleolus and synthesizes a large precursor that is later processed to form rRNA*. It is resistant to inhibition by $\alpha$-amanitin at concentrations even greater than 1000 $\mu$g/ml. *RNA polymerase II is located in the nucleoplasm and synthesizes large precursor RNAs (sometimes called heterogeneous nuclear RNA or hnRNA) that are processed to form cytoplasmic mRNAs*. It is also responsible for synthesis of most viral RNA in virus-infected cells. It is very sensitive to $\alpha$-amanitin, being inhibited 50 percent by about 0.05 $\mu$g/ml. *RNA polymerase III is also located in the nucleoplasm and synthesizes small RNAs such as 5 S RNA and the precursors to tRNAs*. This enzyme is somewhat resistant to $\alpha$-amanitin, requiring about 5 $\mu$g/ml to reach 50 percent inhibition.

All three of these enzymes have been purified extensively and have complex subunit structures. All contain two subunits larger than 120,000 daltons and 6 to 10 smaller subunits. Although each has some unique subunits, some subunits are common to two or three. At present, almost nothing is known about the functions of the individual subunits. The subunit structure for RNA polymerase II is quite similar for enzymes purified from a variety of eukaryotes, including humans, calf, mouse, wheat, cauliflower, acanthamoeba, yeast, and slime molds.

Although purified RNA polymerases have not been shown to be capable of selective transcription of DNA in vitro, crude enzyme preparations

**Table 21-3**
Comparison of Eukaryotic DNA-Dependent RNA Polymerases

| Type | Location | RNAs Synthesized | Sensitivity to $\alpha$-Amanitin |
|------|----------|------------------|----------------------------------|
| RNA polymerase I | Nucleolus | Pre-rRNA | Resistant |
| RNA polymerase II | Nucleoplasm | hnRNA, mRNA | Sensitive |
| RNA polymerase III | Nucleoplasm | Pre-tRNA, 5 S RNA | Sensitive to very high levels |
| Mitochondrial | Mitochondria | Mitochondrial | Resistant |
| Chloroplast | Chloroplasts | Chloroplast | Resistant |

of RNA polymerases I, II, and III recently have been shown to be capable of selective transcription of defined DNA templates, initiating at sites known in some cases to be utilized in vivo. Fractionation of these extracts is beginning to reveal additional proteins that are necessary for in vitro selectivity. It is possible that the purified enzymes are more like core polymerases than holoenzymes and that additional components are needed to obtain selective binding and initiation in vitro.

The recent availability of a large number of cloned eukaryotic genes has made possible the comparison of DNA sequences preceding the genes that may act as promoter-like signals for RNA polymerase II. One feature that stands out is a common sequence, TATAAATA, often called a "TATA box," found about 25 to 30 bases before the transcription start site. In contrast to this, the region necessary for selective transcription of 5 S RNA by RNA polymerase III is located in a region 40 to 80 bases after the start site, well into the 5 S RNA transcript. One of the additional proteins needed for selective 5 S RNA transcription in vitro binds to this site and somehow directs RNA polymerase III to bind and start in the correct place. The binding site for this additional regulatory protein was determined by the DNA "footprinting" technique (see the chapter-opening photograph of Chapter 20.) The promoter regions for RNA polymerases II and III are shown in Figure 21-11.

**Organelle RNA Polymerases.**  In addition to nuclear RNA polymerases, eukaryotic cells also contain mitochondrial and, in plants, chloroplast RNA polymerases that are responsible for transcription of the mitochondrial and chloroplast DNAs. These enzymes are present in very small amounts and have proved quite difficult to purify. Despite these difficulties, L. Bogorad and coworkers have recently succeeded in isolating a complex polymerase from maize chloroplasts that faithfully initiates transcription from certain chloroplast genes in vitro. Mitochondrial RNA polymerase purified from several organisms appears to consist of a single polypeptide subunit of between 45,000 and 65,000 daltons.

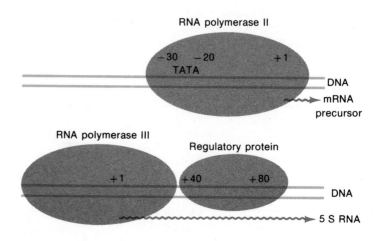

**Figure 21-11**
Promoter regions for eukaryotic RNA polymerases II and III. The polymerase is depicted as a large oval. The wavy line indicates the RNA starting at position + 1.

Although the basic mechanism by which RNA is synthesized is quite similar in prokaryotes and eukaryotes, there are several important differences. The DNA in eukaryotic cells is complexed with proteins, primarily small basic proteins called _histones_, to form chromatin (see Chapter 18). The chromatin, in turn, is condensed to form chromosomes. Only a small fraction of the chromosome is actively being transcribed. In transcriptionally active regions, the DNA must be partially exposed and be more accessible to RNA polymerase. The DNA in these regions is found to be more sensitive to cleavage by mild treatment with bovine pancreatic DNAse I and appears to contain bound RNA polymerase, additional nonhistone proteins, and perhaps modified histones. In addition, DNA in active regions is undermethylated when compared with DNA from regions not active in transcription. Although prokaryotic DNA is not complexed with histones, it may be coated with small basic proteins, although this is not firmly established yet. However, as far as we know, most parts of the bacterial DNA are accessible to RNA polymerase binding and transcription.

As stated earlier, transcription and translation occur simultaneously in prokaryotes. RNA polymerase moves along a gene generating an mRNA chain, and as soon as a bit of mRNA is synthesized, ribosomes bind to it and start translating it. The situation is very different in eukaryotes, where transcription and translation occur in separate compartments of the cell. The nucleus, where the chromosomes are located, is the site of DNA-dependent RNA synthesis. Large RNA precursors to mRNA are synthesized in the nucleus, become complexed with proteins to form ribonucleoprotein particles (RNPs), and then are modified, processed to form the smaller mRNAs, and transported across the nuclear membrane to the cytoplasm, where they bind to the ribosomes and are translated into protein. The processing steps that eukaryotic mRNAs undergo are described below.

## OTHER RNA SYNTHESIS

Although most RNA in cells is synthesized by cellular DNA-dependent RNA polymerases, there are several other enzymes that are capable of forming phosphodiester bonds and of synthesizing additional RNA in cells or cells infected with viruses. Some properties of these enzymes are summarized in Table 21-4.

### DNA Primase

RNA primers used for the initiation of DNA synthesis during replication are made by DNA primase, the 65,000-dalton product of the _dnaG_ gene of _E. coli_ (also see Chapter 20). It usually binds to DNA in association with another protein, the product of the _dnaB_ gene, although it can initiate primer synthesis by itself at certain hairpin structures in single-stranded DNA. The primase can use either NTPs or dNTPs as substrates in vitro and synthesizes primers 10 to 50 nucleotides long that are complementary to the DNA template. Synthesis is in the $5' \rightarrow 3'$ direction. As was mentioned earlier, RNA polymerase is also capable of synthesizing primers. The use of primase or RNA polymerase to make primer is controlled by the DNA. Either one or the other is used exclusively in specific instances.

**Table 21-4**
RNA Synthesizing Enzymes

| | Template | Primer | Molecular Weight of Subunit(s) | Gene Name | Substrate | Inhibition by Rifampicin |
|---|---|---|---|---|---|---|
| **Template-dependent** | | | | | | |
| Enzymes from bacteria | | | | | | |
| Holoenzyme (*E. coli*) | DNA | — | 155,000<br>151,000<br>70,000<br>36,500 | *rpoC*<br>*rpoB*<br>*rpoD*<br>*rpoA* | 4 NTPs | Yes |
| DNA primase | DNA | — | 65,000 | *dnaG* | 4 NTP,<br>4 dNTP | No |
| Enzymes from phage or phage-infected bacteria | | | | | | |
| T7 RNA polymerase | T7 DNA | — | 110,000 | T7 *gene1* | 4 NTPs | No |
| N4 RNA polymerase | N4 DNA | ? | 350,000 | Viral | 4 NTPs | No |
| Qβ replicase | Qβ RNA | — | 65,000<br>55,000<br>43,000<br>35,000 | *rps*A<br>Viral<br>*tuf*<br>*tsf* | 4 NTPs | No |
| **Template-independent** | | | | | | |
| CCA enzyme | — | 3' end tRNA | 45,000 | *cca* | CTP<br>ATP | No |
| Poly(A) polymerase (eukaryotic) | — | 3' end mRNA | | | ATP | No |
| Polynucleotide phosphorylase | — | 3' end RNA | 86,000<br>48,000 | *pnp* | 4 NDPs | No |

## Virus-Induced RNA Polymerases

There seem to be three types of DNA viruses, each having a different strategy to accomplish transcription of viral DNA. The first type utilizes the host RNA polymerase, in some cases modifying it or synthesizing new promoter-specificity factors to direct it to read the viral promoters. Examples of such viruses are bacteriophage φX174, λ, and T4 (of *E. coli*), and SP01 and SP82 (of *B. subtilis*), and animal viruses SV40 and adenovirus.

The second type utilizes the host RNA polymerase to transcribe some early expressed viral genes. One of these "early" genes codes for a new RNA polymerase that transcribes exclusively the remaining "late" viral genes. *E. coli* bacteriophages T7 and T3 are the best-known examples of this type. T7 RNA polymerase is a single polypeptide with a molecular weight of 110,000. It is not inhibited by the antibiotics rifampicin and streptolydigin, which inhibit host RNA polymerase. It recognizes specifically the T7 late promoters, all of which contain a nearly identical 18 to 22 nucleotide sequence just before the 5' triphosphate terminal GTP start site. T7 RNA polymerase is also capable of termination at specific points on the template. Thus it seems to be able to carry out the basic polymerization reaction and specific initiation and termination with a much simpler subunit structure than the host holoenzyme. However, with simplicity it loses versatility. Unlike the host polymerase, it is not able to recognize a wide variety of related but nonidentical promoter sequences, nor is it able to be regulated by positive and negative control factors that act at sites of initiation and termination.

A third type of virus, exemplified by bacteriophage N4, carries a viral RNA polymerase within its virion. This enzyme enters the cell along with the viral DNA and transcribes some early viral genes. Some of these genes code for specificity factors that direct the host RNA polymerase to transcribe late genes. Vaccinia virus also contains a virion encapsulated RNA polymerase that shares some of the properties of the N4 enzyme.

## RNA-Dependent RNA Polymerases of RNA Viruses

The RNA genomes of single-stranded RNA bacterial viruses (such as Q$\beta$, MS2, R17, and f2) are themselves mRNAs. Bacteriophage Q$\beta$ codes for a polypeptide of about 55,000 daltons that combines with three host proteins to form an RNA-dependent RNA polymerase (replicase). The three host proteins are ribosomal protein S1 and two elongation factors for protein synthesis, EF-Tu and EF-Ts (see Table 21-4). The Q$\beta$ replicase can use only the Q$\beta$ RNA (+ strand) as a template. It first makes a complementary RNA transcript (− strand) and ultimately uses it to make more viral RNA + strands. Like the DNA-dependent RNA polymerases, the replicase utilizes ribonucleoside-5'-triphosphates and transcribes in a 5'→3' direction. The phage RNA must first act as an mRNA to direct the synthesis of a component of the replicase, since uninfected cells do not have an RNA-dependent RNA polymerase or replicase.

RNA tumor viruses that infect animal cells exhibit a different replication strategy (see Chapter 28). They carry in their virion an enzyme that can use the RNA viral genome as a template to synthesize a DNA copy. Since this process is the reverse of transcription, the enzyme is called _reverse transcriptase_. Once a double-stranded DNA copy is made, it is integrated into the host genome and additional virus RNA is synthesized by the host RNA polymerase II in a normal DNA-dependent fashion.

## 3'-End Addition Enzymes

Three enzymes are known that add ribonucleotides posttranscriptionally to the 3'-hydroxyl end of RNA. None of them are DNA-dependent. One adds the terminal CCA sequence that all tRNAs have at their 3' ends. The 3' terminal A is the base to which the amino acid is covalently attached by the aminoacyl-tRNA synthase. The 3'-CCA is relatively unstable and is continually being added when needed by this enzyme, called the CCA enzyme, or tRNA nucleotidyltransferase.

In eukaryotes, 100 to 200 adenosine residues are added to the 3' end of most mRNAs by a poly(A) polymerase. This addition occurs in the nucleus before the mRNA is processed and transported to the cytoplasm, as described below. The third enzyme capable of adding nucleotides posttranscriptionally to the end of RNA is polynucleotide phosphorylase which is described in the next section.

## Polynucleotide Phosphorylase

Polynucleotide phosphorylase was the first enzyme found that could synthesize long polynucleotide chains in vitro. Two forms of the enzyme have been isolated, a form with three identical 86,000-dalton subunits and a form with these subunits and two additional 48,000-dalton subunits. Unlike all the enzymes we have discussed, polynucleotide phosphorylase utilizes ribonucleoside-5'-diphosphates instead of triphosphates as substrates for RNA synthesis. It catalyzes the reaction

$$NDP + (NMP)_n \xrightleftharpoons{Mg^{2+}} (NMP)_{n+1} + P_i$$

Polynucleotide phosphorylase does not require a template and randomly incorporates bases into RNA depending on the relative concentration of the four NDPs in the reaction medium. The enzyme takes advantage of a primer if one is available to provide a 3'-hydroxyl end and synthesizes RNA in the 5'→3' direction. The reaction is readily reversible, and it is not known whether this enzyme plays a role primarily in degradation or in synthesis of RNA in the cell. It may be involved in the synthesis of the poly(A) recently found on the RNA of E. coli. E. coli cells that have been treated with toluene become permeable to labeled ADP and ATP and have been used to study poly(A) synthesis. Mutants in polynucleotide phosphorylase are defective in this poly(A) synthesis.

### RNA Ligase Induced by Bacteriophage T4

Perhaps the most unusual enzyme capable of synthesizing RNA is the bacteriophage T4 RNA ligase. It can link together, in an ATP-dependent reaction, a 3'-OH terminus on an "acceptor" to a 5'-PO$_4$ terminus on a "donor," as shown below:

$$ATP + \ldots XpY\text{-}3'\text{-}OH + PO_4\text{-}5'Zp \ldots \longrightarrow$$

**Acceptor**      **Donor**

$$\ldots XpYpZp \ldots + ADP + PP_i$$

**Ligated product**

A template strand is not required to align the ends of the reactants. The donor and acceptor end can be on the same RNA molecule, in which case a circular RNA product is produced. The enzyme has been used to end-label the 3'-OH end of RNA molecules by the addition of the short donor [5'-$^{32}$P]Cp.

This enzyme is now widely used to synthesize defined sequences of RNA. The ligase also can accept deoxyribose polymers as donors. It is not known whether this enzyme carries out the ligation reaction in vivo after T4 infection of E. coli.

## POSTTRANSCRIPTIONAL MODIFICATION AND PROCESSING OF RNA

Almost all the major types of RNA synthesized by cellular DNA-dependent RNA polymerases undergo some modification and processing before they can carry out their functions. This is summarized in Table 21-5.

Transfer RNAs are processed from larger precursors in prokaryotic and eukaryotic cells. This processing involves two types of enzymes that can cleave phosphodiester bonds in RNA. Endoribonucleases cleave at internal sites in the RNA, resulting in two smaller RNAs. Exonucleases sequentially remove single nucleotides from one end of the RNA. Some of the processing ribonucleases of *E. coli* are shown in Table 21-6. As an example, the processing of the *E. coli* tyrosine $tRNA_1$ is diagrammed in Figure 21-12. The initial transcript has, in addition to the 85 nucleotides of the final product, 41 nucleotide residues at the 5′ end and 225 residues at the 3′ end. The initial transcript is able to fold up to form the typical cloverleaf tRNA structure. Processing begins by a specific endonuclease, called RNAse F, cleaving the precursor at a site three nucleotides beyond what will be the 3′ end of the mature tRNA. Another endonuclease, RNAse P, then cleaves the remaining RNA to produce the mature 5′ end. At the 3′ end, exonuclease RNAse D sequentially removes the additional nucleotides and stops, leaving the 3′ terminal CCA sequence. Individual bases on the tRNA molecule are then modified by a variety of enzymes, including methylases, deaminases, thiolases, pseudouridylating enzymes, and transglycosylases. Some of the modified bases are presented in Figure 21-2. In addition, some eukaryotic tRNAs are processed to remove internal RNA sequences, as will be discussed later.

**Table 21-5**
Summary of RNA Modification and Processing

| RNA | Precursor | Modification | Processing | Products |
|---|---|---|---|---|
| mRNA | | | | |
|   Prokaryotic | None | None? Polyadenylation? | In some cases specific cleavage by endoribonucleases | mRNAs |
|   Eukaryotic | hnRNA | Capping, methylation, polyadenylation | In most cases splicing out introns | mRNAs |
| rRNA | Pre-rRNA | Methylation | Specific cleavage | 16 S, 23 S, 5 S, spacer tRNA (18 S, 28 S in eukaryotes) |
| tRNA | Pre-tRNA | Many modified bases | Specific cleavage by endonucleases, trimming by exonucleases, CCA addition, removal of intervening sequences in eukaryotes | Mature tRNAs |

**Table 21-6**
Representative Enzymes Involved in Processing and Degradation

| Enzyme | For *E. coli* Genes | | Type | Product | Specificity |
|---|---|---|---|---|---|
| | Gene Name | Map Position | | | |
| **Processing** | | | | | |
| RNAse III | *rnc* | 55′ | endo | 3′-OH, 5′-PO$_4$ | Specific, long, double-stranded RNA |
| RNAse D | *rnd* | 40′ | 3′→5′ exo | 5′-NMPs | Nonspecific, but stops at CCA |
| RNAse E | *rne* | 24′ | endo | | Specific |
| RNAse F | *rnf* | ? | endo | | Specifically cuts 3′ to tRNA-like structures |
| RNAse P | { *rnpA* <br> { *rnpB* | 82′ <br> 70′ | endo | 3′-OH, 5′-PO$_4$ | Specifically cuts 5′ to tRNA-like structures |
| RNAse M16 | | | endo(s) | | Specific, cuts pre-16 S to 16 S RNA |
| RNAse M23 | | | endo(s) | | Specific, cuts pre-23 S to 23 S RNA |
| RNAse M5 | | | endo | | Specific, cuts pre-5 S to 5 S RNA |
| **Degradation** | | | | | |
| RNAse I | *rna* | 14′ | endo | 3′-PO$_4$ oligos | Nonspecific |
| RNAse II | *rnb* | 28′ | 3′→5′ exo | 5′-NMP | Nonspecific |
| Polynucleoside phosphorylase | *pnp* | 68′ | 3′→5′ exo | 5′-NDP | Nonspecific |
| RNAse H | *rnh* | 5′ | endo | | Nonspecific, digests RNA out of RNA-DNA duplex |
| Bovine pancreatic RNAse A | | | endo | Py-3′-PO$_4$ | Specific, cuts 3′ to pyrimidines |
| Aspergillus RNAse T1 | | | endo | G-3′-PO$_4$ | Specific, cuts 3′ to guanine |
| Aspergillus S1 nuclease | | | endo | | Nonspecific, cuts single-stranded RNA or DNA |
| Bovine spleen phosphodiesterase | | | 5′→3′ exo | 3′-NMP | Nonspecific |
| Snake venom phosphodiesterase | | | 3′→5′ exo | 5′-NMP | Nonspecific |

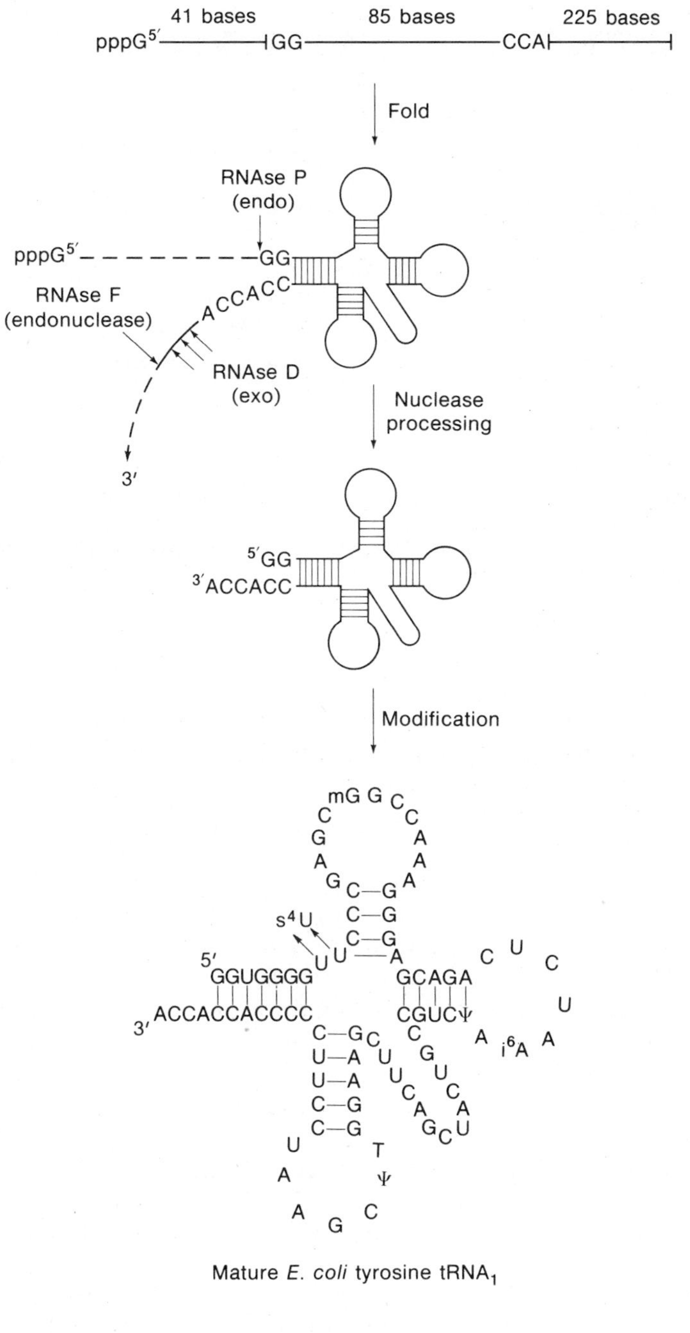

**Figure 21-12**
Processing and modification of *E. coli* tyrosine tRNA$_1$. Abbreviations are as follows: ribothymidine (T), pseudouridine ($\psi$), isopentyl adenosine (i$^6$A), methylguanosine (mG), and thiouridine (s$^4$U).

## Processing of Ribosomal RNA

Both eukaryotic and prokaryotic cells synthesize large precursors to rRNA that are processed to produce the mature rRNAs. The most detailed studies of the numerous enzymatic steps involved in this process have been carried out in *E. coli*. This processing scheme is summarized in Figure 21-13.

The initial transcript is over 5500 nucleotides long and includes, reading from the 5′ end of the RNA: the 16 S rRNA, a spacer region that includes one or two tRNAs, the 23 S rRNA, the 5 S rRNA, and in some cases

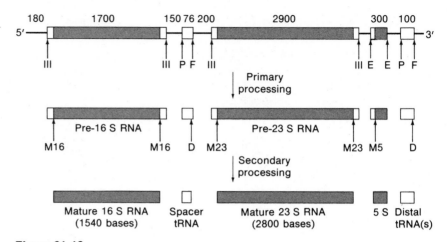

**Figure 21-13**
Processing of *E. coli* ribosomal RNA. The processing steps are discussed in the text. The various nucleases are indicated as in Table 21-6. (Adapted from D. Apirion and P. Gegenheimer, Processing of bacterial RNA, *FEBS Lett.* 125:1, 1981. Used with permission.)

one or two more tRNAs. Extra bases are found preceding and following each of these RNAs. Primary processing events include the endonucleolytic action by RNAse III to produce pre-16 S and pre-23 S rRNAs and then the action of specific ribonucleases to produce the tRNAs and pre-5 S rRNA. Secondary processing by endonucleases M16, M23, and M5 result in mature 16 S, 23 S, and 5 S RNAs, respectively. These three endonuclease activities, as well as that of RNAse F, have not been completely purified and characterized, and each may represent more than one enzyme. Extra bases on the 3′ end of the tRNAs are removed by exonuclease RNAse D. This processing scheme has been deduced by observing the accumulation of intermediates in mutant strains defective in one or more of the nucleases and by cleaving the intermediates in vitro with purified or partially purified nucleases.

## Modification and Processing of Messenger RNA

In prokaryotes, many mRNAs function in translation with no prior modification or processing. There is evidence, however, that some are processed by specific endonucleolytic cleavage, often cutting polycistronic messengers into smaller units (e.g., see the *β* operon in Chapter 26). In eukaryotes, a much more complex process occurs to produce functional mature mRNA. The discoveries of these modification and processing steps during the last decade have provided some surprising results and provocative new concepts.

**Capping the 5′ End.** In 1975, A. Shakin and coworkers found that most viral and cellular mRNAs contain an unusual methylated nucleotide at the 5′ terminus. This entire methylated terminal oligonucleotide is called a *cap structure* and is shown in Figure 21-14. This cap structure is formed in the nucleus by a series of enzymatic reactions outlined in Figure 21-15. It frequently takes place before the nascent transcript is completed. First, a triphosphatase converts a 5′-triphosphate to a 5′-diphosphate terminus. Second, a guanylyltransferase adds a GMP to the 5′ end to form an unusual 5′-5′ triphosphate bond, and this terminal guanosine is methylated in the N-7 position by a guanine-7-methyltransferase. Then the first nucleotide in the initial transcript is methylated in the 2′-0 position of the ribose to form what is called a *cap I structure*. Methyl groups are derived from the methyl donor *S*-adenosylmethionine. Some mRNAs, after transport to the cytoplasm, become methylated in the 2′-0 position of the second nucleotide of the initial transcript to form a *cap II structure*. The cap structure facilitates binding of ribosomes prior to initiation of translation of eukaryotic mRNAs; it may also function to stabilize the mRNA, since uncapped messengers have considerably reduced half-lives. rRNA and tRNA are not capped. Most small nuclear RNAs (snRNAs) have a different cap structure containing a trimethylguanosine.

**Polyadenylation of the 3′ End.** Most eukaryotic mRNAs found associated with ribosomes contain 50 to 150 adenine nucleotides on their 3′ ends. This poly(A) is not coded by the DNA template, but is added to the mRNA before it leaves the nucleus. In several cases studied, the process occurs in at least two steps. First, the RNA is cleaved about 12 nucleotides past an AAUAAA sequence near its 3′ end, and then 200 to 250 adenylate residues are added by a poly(A) polymerase. After the mature mRNA is transported to the cytoplasm, the poly(A) is shortened somewhat as the mRNA ages. Poly(A) addition does not appear to be essential for transport or translation of all mRNAs, since some eukaryotic mRNAs, including histone mRNAs, do not contain poly(A).

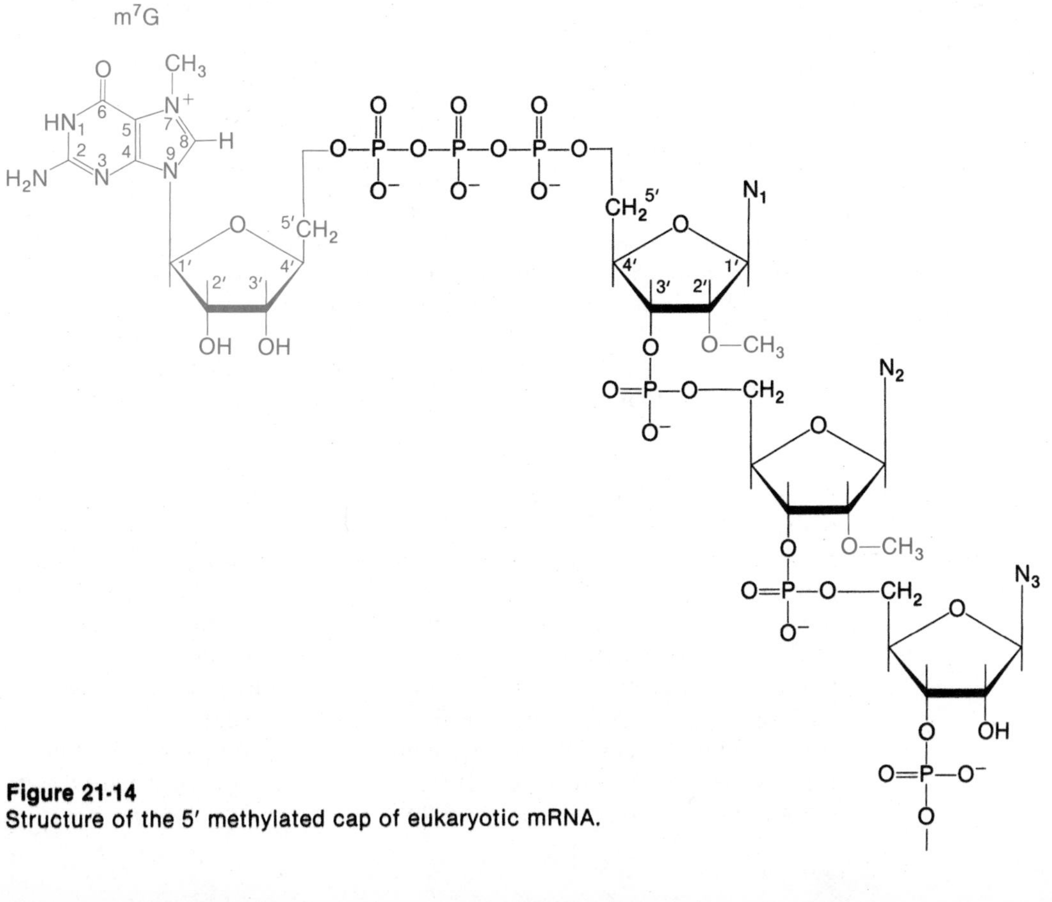

**Figure 21-14**
Structure of the 5′ methylated cap of eukaryotic mRNA.

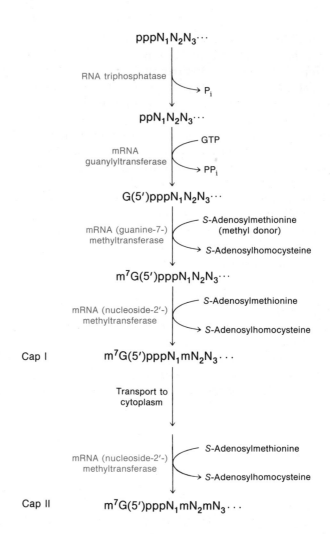

**Figure 21-15**
Proposed reaction sequence for cap formation in HeLa cells. The enzyme catalyzing each reaction is shown on the left. (Adapted from S. Venkatesan and B. Moss, Donor and acceptor specificities of HeLa cell mRNA guanylyl transferase, *J. Biol. Chem.* 255:2835, 1980. Used with permission.)

Recently it has been found that RNA isolated from *E. coli* by a new rapid procedure also contains poly(A) sequences. This poly(A) was not observed before because it was degraded or lost during the isolation methods that had been used. One current hypothesis is that this poly(A) is added to the mRNA by polynucleotide phosphorylase. Mutants defective in polynucleotide phosphorylase result in mRNA that degrades more rapidly, suggesting that the function of the poly(A) may be to prevent mRNA degradation by $3' \rightarrow 5'$ exonucleases.

**Removal of Intervening Sequences.** It had long been known that some of the mRNA precursors in the nucleus (hnRNAs) are much larger than the mRNAs found in the cytoplasm associated with ribosomes. Therefore, it came as no surprise to learn that processing occurs to remove parts of the precursor RNAs. What was surprising was the finding made simultaneously by T. Broker's laboratory and P. Sharp's laboratory in 1977 that the parts that are removed are not at the ends of the molecules but are interspersed with the coding regions. This was first observed for adenovirus transcripts (see Chapter 28). One of the early demonstrations of this was the finding that mouse β-globin precursor mRNA did not form a perfect hybrid with DNA complementary to mature mRNA. When the hybrid was observed with an electron microscope, a loop in the DNA of the heteroduplex appeared, which suggested that the precursor contained internal sequences not present in the mature mRNA (see Figure 21-16). These noncoding intervening sequences (also called _introns_) are interspersed with the coding sequences (_exons_) in a large number of eukaryotic mRNAs that have been studied. As an example, the primary transcript of the chicken ovalbumin gene (Figure 21-17) is 7700 bases long and has 7 introns and 8 exons. The intervening sequences, or introns, are removed and the exons are spliced together, giving a final product that is only 1872 nucleotides long. This includes 1158 nucleotides that code for the 386 amino acids of ovalbumin and untranslated regions at the 5' and 3' ends.

The function of intervening sequences is not yet understood. Some genes (e.g., histones) do not contain them, yet function well. Experimentally removing a particular intervening sequence from SV40 virus DNA before using it to infect cells resulted in the formation of unstable mRNA that did not transport from nucleus to cytoplasm. However, similar experiments with other mRNAs have not caused abnormal mRNA function. Frequently, splice points are correlated with "domains" that define structural regions of the protein that are often seen in different proteins. For example, the exons of hemoglobin contain three structural domains of different types (see Chapter 1), whereas the heavy-chain immunoglobulin exons encode domains that are quite similar in structure (see Chapters 1 and 2).

The mechanism by which splicing occurs is being investigated. Introns vary in base sequence and length but have some common features, particularly at their ends. The 5' end of one of the small nuclear RNAs (snRNAs), U1, contains sequences complementary to the nucleotide sequences found to be common at both ends of many introns, leading to the suggestion that U1 RNA may line up the adjacent ends of the introns with the center part looped out in position for splicing to take place (see Figure 21-18). The snRNAs are found as ribonucleoprotein particles (snRNPs) sedimenting at about 10 S. Some persons with the autoimmune disease lupus erythematosus carry antibodies which can precipitate the snRNPs. The same antibodies inhibit the splicing reaction in a permeablized nuclei system, lending further support to the idea that the snRNA U1 is involved in

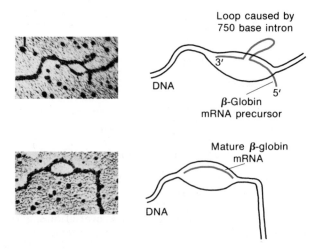

**Figure 21-16**
Electron micrograph showing mouse β-globin precursor mRNA hybridized with DNA complementary to mature mRNA _(upper photo)._ A control experiment _(lower photo)_ shows that mature mRNA forms a perfect hybrid with DNA complementary to mature mRNA as expected. Schematics indicating the RNA and DNA are shown on the right. A second small intron is present in the precursor mRNA near the 5' end and is the reason that the 5' end of the RNA is not hybridized to the DNA in the upper photo. (From A. Kinniburgh, J. Mertz, and J. Ross, The precursor of mouse β-globin contains two intervening sequences, _Cell,_ 14: 681, 1978. Used with permission.)

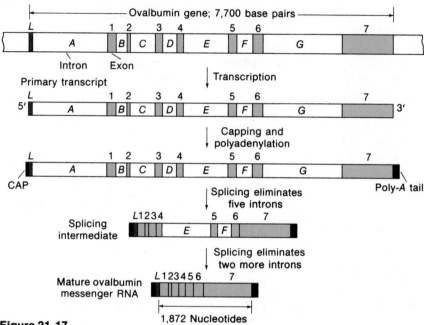

**Figure 21-17**
Maturation of ovalbumin mRNA. First, the entire ovalbumin gene is transcribed into a precursor RNA, the primary transcript. The transcript is capped
at the 5'end and the poly(A) tail is added at the 3' end. Then the transcripts of
the introns are excised and the adjacent exon transcripts are ligated in a series of splicing steps; an intermediate, from which five of the seven intron
transcripts have been eliminated, is illustrated. These steps are accomplished
in the cell nucleus. After splicing, mature messenger is transferred to cytoplasm. L indicates the 5' leader region, which is part of the mature RNA that is
not translated. (From P. Chambon, Split genes, *Sci. Am.* 244:60, 1980. Used
with permission.)

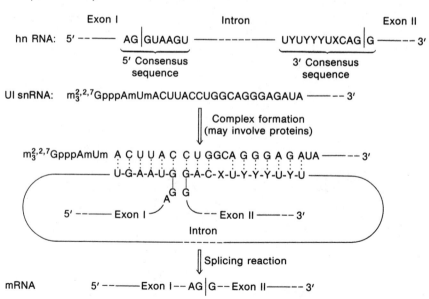

**Figure 21-18**
Proposed role of U1 RNA in splicing of introns from eukaryotic mRNA. A typical
hnRNA with two exons flanking an intron is shown. The 5' and 3' "consensus"
sequences shown have been found to be similar in over 30 introns. The 5' end of
U1 RNA is shown complexed with the exon-intron junctions.

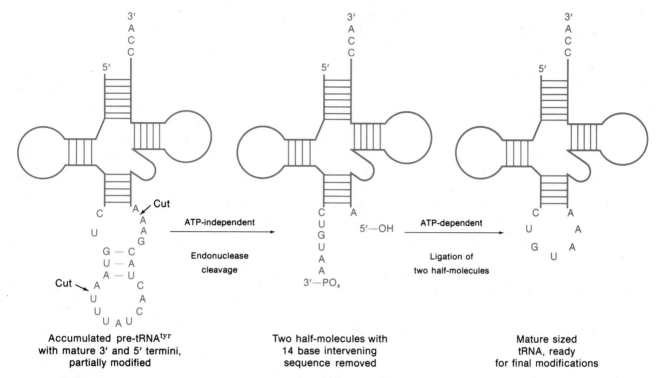

**Figure 21-19**
Processing of yeast tyrosine tRNA to remove intervening sequence.

the splicing reaction. Whatever the mechanism, splicing must be very exact. Removal of just one base too few (or too many) would shift the reading frame and result in a nonfunctional protein.

Abnormal mRNA splicing appears to be the cause of the human disease $\beta^+$-thalassemia. The inefficient splicing of $\beta$-globin mRNA precursors in affected individuals seems to be due to mutations in the intron and leads to very low levels of mature mRNA and thus to a $\beta$-globin deficiency.

Removal of intervening sequences in eukaryotes is not restricted to mRNA processing. It also occurs in the processing of rRNA* and some tRNAs. A temperature-sensitive mutant of yeast has been isolated that accumulates certain tRNA precursors at the nonpermissive temperature. One of these is a tyrosine tRNA precursor that has a 14-base intervening sequence that can be removed by cell extracts in vitro. The reaction occurs in two steps and is shown in Figure 21-19. First, a 14-base sequence is removed by an ATP-independent endonucleolytic cleavage to produce two half tRNA molecules. These cleavages produce 3'-$PO_4$ and 5'-OH termini, unlike the 3'-OH and 5'-$PO_4$ termini produced by most RNA processing enzymes such RNAse III or RNAse P. The two termini are then ligated in a second reaction that requires ATP. Some yeast tRNAs have intervening sequences of up to 60 bases long, but they seem to be processed by the same endonuclease, which appears to recognize a structure in the precursor rather than a particular sequence. The function of these intervening sequences is not known, nor is it known whether this splicing reaction is at all similar to the one used for mRNA.

---

*T. Cech has made the startling discovery that certain rRNAs in ciliated protozoa can splice themselves without any protein enzyme.

## Viroids May Originate from Introns and Interfere with Splicing

The initial products of mRNA splicing usually include an mRNA with the intron removed and in some cases a covalently closed circular single-stranded intron. Viroids, which cause infectious disease in higher plants, have such a structure. They are the smallest known infectious agents ($M_r \simeq 1.2 \times 10^5$), and they do not appear to code for any protein. The nucleotide sequence of the potato spindle tuber viroid (PSTV) is known, and comparison of this sequence with that of the 5' end of the U1 snRNA, which is believed to be involved in splicing, has revealed a region of striking similarity (Figure 21-20). This finding has led to the suggestion that viroids may have originated as "escaped introns" and that their pathologic effects may be due to interference with normal splicing. The mechanism whereby viroid RNA replicates is still unclear.

## DEGRADATION OF RNA BY RIBONUCLEASES

Although a large number of nonspecific exonucleases and endonucleases have been identified in many organisms, the role of most of them is not understood. Some are extracellular or secreted enzymes that presumably function in breaking down RNA to recycle the purine and pyrimidine bases (see Chapter 19). Intracellular ribonucleases also may be involved in recycling, as well as in some aspects of processing, as described earlier. Some of the better-known enzymes are listed in Table 21-6. *E. coli* RNAse I is an endonuclease that is located in the periplasmic space between the cell membrane and the cell wall. RNAse II is an intracellular 3' exonuclease that rapidly degrades RNA fragments. Mutations defective in these enzymes were very important in the development of in vitro translation systems, since extracts from cells defective in these enzymes allowed mRNA to be translated without being rapidly degraded.

RNAse H has the unusual property of degrading the RNA strand from a RNA-DNA hybrid molecule and may be involved in the removal of RNA primers during DNA replication. Some endonucleases have marked preferences for cleavage after certain bases. Pancreatic RNAse A cleaves at the 3' side of pyrimidines, while RNAse T1 cleaves only after guanosine residues.

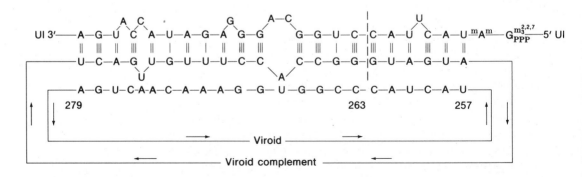

**Figure 21-20**
Possible base-pairing interactions between the PSTV complement and the 5' end of U1 RNA (---hypothetical splice junction). (From T. O. Diener, Are viroids escaped introns?, *Proc. Natl. Acad. Sci. USA* 78:5014, 1981. Used with permission.)

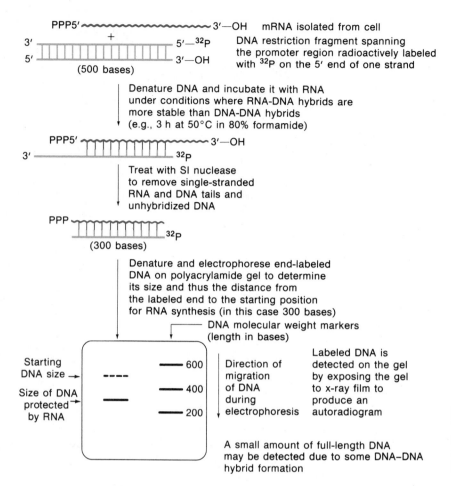

PPP5' ~~~~~~~~~~~~~~~~~~~3'—OH     mRNA isolated from cell

3' |||||||||||||||| 5'—³²P     DNA restriction fragment spanning
5' |||||||||||||||| 3'—OH     the promoter region radioactively labeled
(500 bases)     with ³²P on the 5' end of one strand

Denature DNA and incubate it with RNA
under conditions where RNA-DNA hybrids are
more stable than DNA-DNA hybrids
(e.g., 3 h at 50°C in 80% formamide)

PPP5' ~~~~~~~~~~~~~~~3'—OH
3' ————————————————³²P

Treat with SI nuclease
to remove single-stranded
RNA and DNA tails and
unhybridized DNA

PPP ~~~~~~~~~~³²P
(300 bases)

Denature and electrophorese end-labeled
DNA on polyacrylamide gel to determine
its size and thus the distance from
the labeled end to the starting position
for RNA synthesis (in this case 300 bases)

DNA molecular weight markers
(length in bases)

Starting
DNA size →      — 600     Direction of     Labeled DNA is
                         migration        detected on the gel
Size of DNA              of DNA           by exposing the gel
protected →     — 400     during           to x-ray film to
by RNA                   electrophoresis   produce an
                — 200                      autoradiogram

A small amount of full-length DNA
may be detected due to some DNA–DNA
hybrid formation

**Figure 21-21**
The use of S1 nuclease to locate the
5' end of in vivo mRNA, a technique
called *S₁ mapping,* first used by
P. Sharp and A. Berk.

Both leave 3'-phosphate termini. Because of their specificity, these enzymes
have proven very useful in the analysis of RNA sequences. S1 nuclease
digests single-stranded DNA or RNA. An example of the use of S1 nuclease
to locate the 5' end of in vivo mRNA is diagramed in Figure 21-21.

Finally, there are exonucleases that digest in the $3' \rightarrow 5'$ direction and
others that digest $5' \rightarrow 3'$. It is likely that degradation of mRNA proceeds by
numerous endonucleolytic cleavages and then exonuclease digestion of the
resulting fragments. All known intracellular exonucleases in *E. coli* produce 5'-NMPs, which can readily be recycled for use in RNA synthesis (see
Chapter 19 for a discussion of the salvage pathways that permit recycling.)

## INHIBITORS OF RNA METABOLISM

A large variety of inhibitors of RNA synthesis have been identified. These
inhibitors have proved useful in elucidating transcription mechanisms,
and some have allowed the isolation of mutant strains with enzymes that
are resistant to their inhibition. The inhibitors fall into several classes, as
described below. The structures of representative compounds are shown in
Figure 21-22.

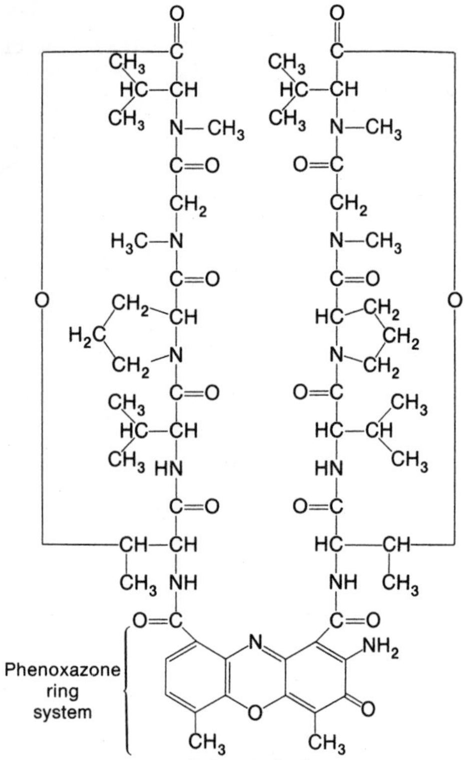

**Figure 21-22**
Inhibitors of RNA metabolism.

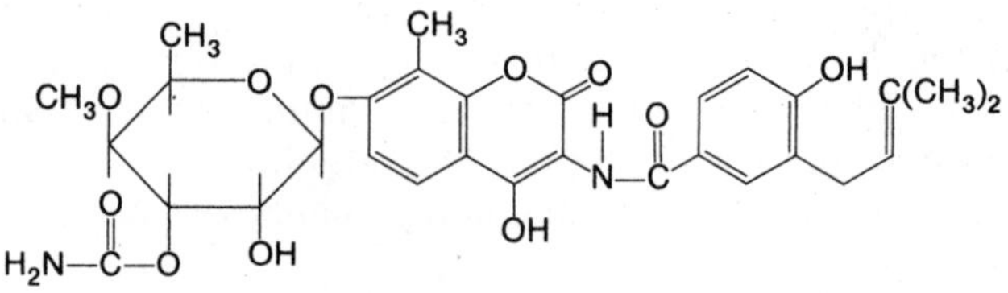

Actinomycin D

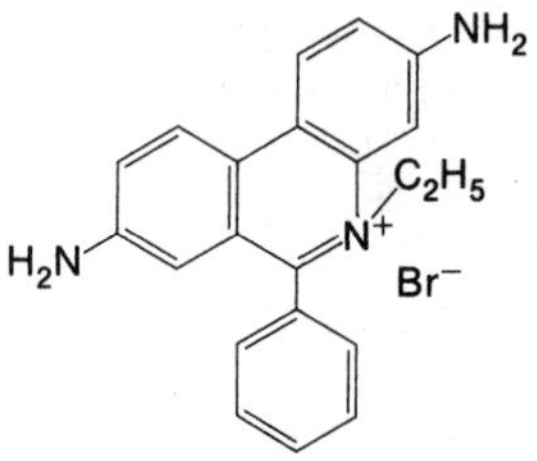

Phenoxazone ring system

Ethidium bromide

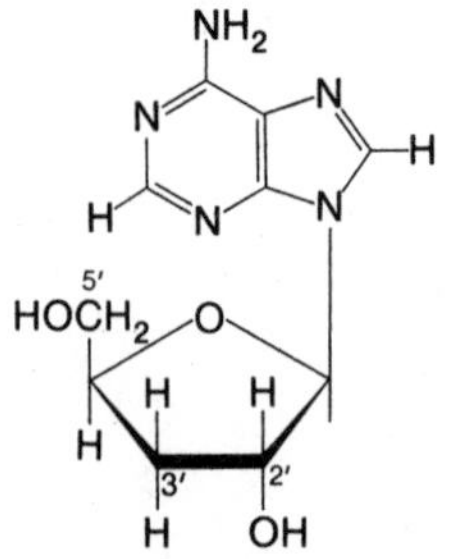

Cordycepin

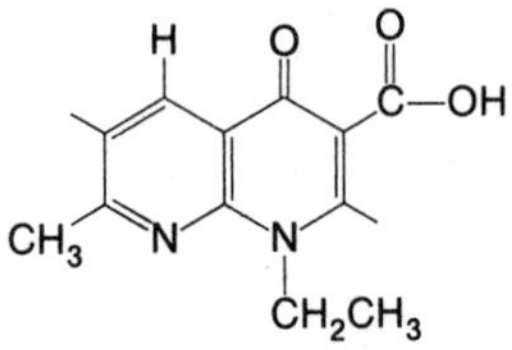

Naladixic acid

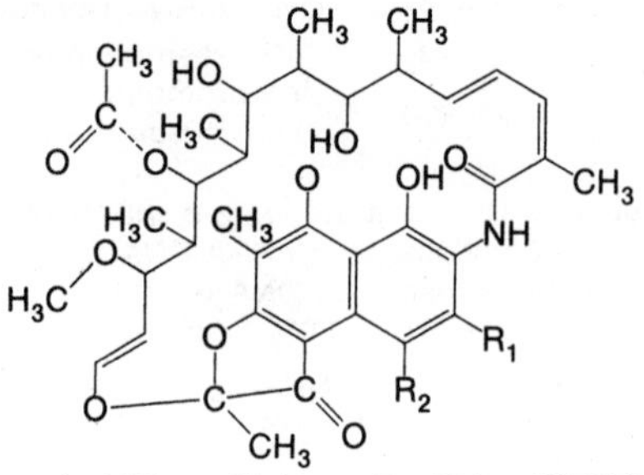

Rifamycin B($R_1$ = H; $R_2$ = O—CH$_2$—COOH)

Rifampicin ($R_1$ = CH=N$^+$N—CH$_3$; $R_2$ = OH)

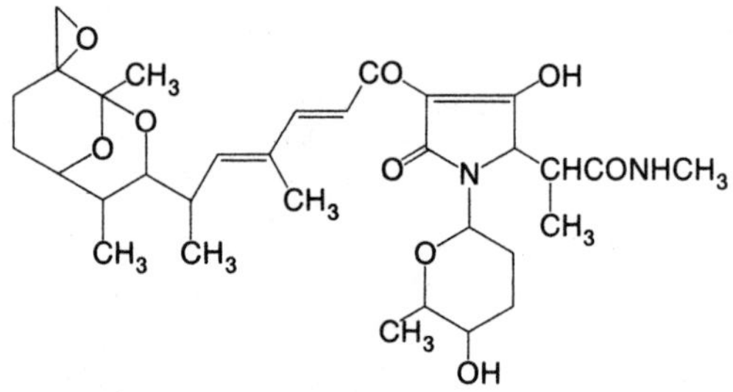

Streptolydigin

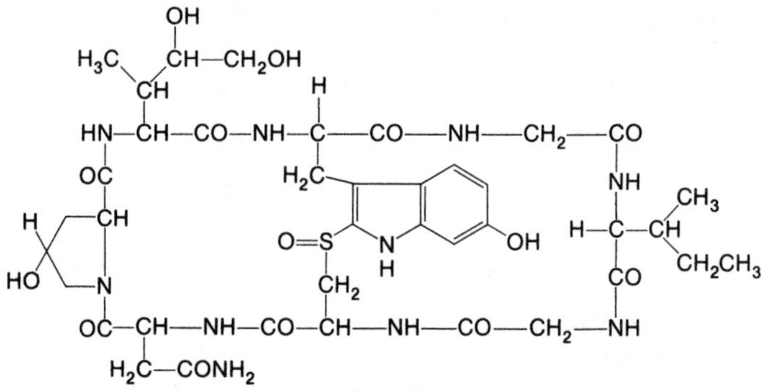

α-Amanitin

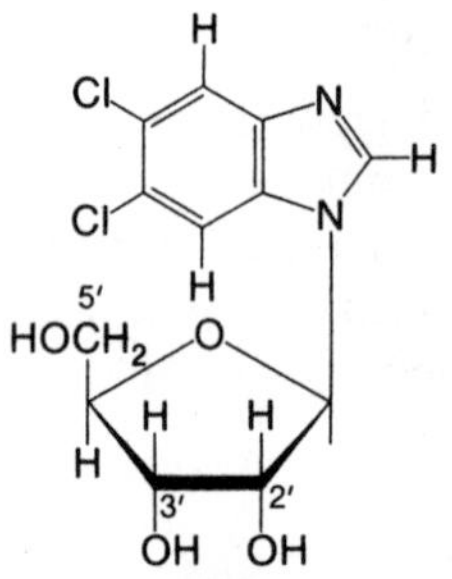

DRB

Novobiocin

## Inhibitors that Act by Binding Noncovalently to DNA and Blocking Template Function

The best-known example of this type of inhibitor is actinomycin D, an antibiotic produced by *Streptomyces antibioticus*. The inhibition of RNA synthesis is caused by the insertion (intercalation) of its phenoxazone ring between two G–C base pairs, with the side chains projecting into the minor groove of the double helix, hydrogen bonded to G residues. RNA polymerase binding to DNA containing actinomycin D is only slightly impaired, but RNA chain elongation in both eukaryotes and prokaryotes is blocked. Ethidium bromide also intercalates into DNA and at low concentrations preferentially binds to negatively supercoiled DNA (see Chapter 18). It has been used to selectively inhibit transcription in mitochondria, which contain supercoiled DNA.

## Inhibitors that Bind to and Inhibit the Action of DNA Gyrase

DNA gyrase is a protein in *E. coli* that is needed to maintain the proper degree of supercoiling in DNA. Transcription of some genes seems to be stimulated on supercoiled templates, and transcription of these genes is selectively inhibited by compounds that interfere with the action of DNA gyrase. Examples of compounds that inhibit DNA gyrase are naladixic acid, which binds to one of the two gyrase subunits, and coumermycin and novobiocin, which bind to the other (see Chapter 20).

## Inhibitors that Bind to RNA Polymerase

Rifampicin is a synthetic derivative of a naturally occurring antibiotic, rifamycin, that inhibits bacterial DNA-dependent RNA polymerase but not T7 RNA polymerase or eukaryotic RNA polymerases. It binds tightly to the $\beta$ subunit. While it does not prevent promoter binding or formation of the first phosphodiester bond, it effectively prevents synthesis of longer RNA chains. It does not inhibit elongation when added after initiation has occurred. Another antibiotic, streptolydigin, also binds to the $\beta$ subunit, but it is able to prevent all bond formation, whether involved in initiation or elongation.

The most useful inhibitor of eukaryotic transcription has been $\alpha$-amanitin, the major toxic substance in the poisonous mushroom *Amanita phalloides*. This toxin preferentially binds to and inhibits RNA polymerase II. At high concentrations it also can inhibit RNA polymerase III, but not RNA polymerase I or bacterial, mitochondrial, or chloroplast RNA polymerase.

Two inhibitors, cordycepin and DRB, in their 5'-triphosphorylated forms are substrate analogs and can be incorporated into growing RNA chains by most RNA polymerases. Cordycepin is a 3'-deoxyadenosine. It causes chain termination, since it does not contain the 3'-hydroxyl group necessary for the formation of the next phosphodiester bond. DRB is a 5,6-dichloro-1-$\beta$-D-ribofuranosylbenzimidazole and also may inhibit transcription by causing premature termination of RNA chains. RNA synthesized in the presence of DRB is much shorter than RNA synthesized in its absence.

## SELECTED READINGS

Apirion, D., and Gegenheimer, P.  Processing of bacterial RNA. *FEBS Lett.* 125:1, 1981. A summary of RNA processing in bacteria.

Chamberlin, M.  Bacterial RNA Polymerase and Bacteriophage RNA Polymerase. In P. D. Boyer (Ed.), *The Enzymes,* 3rd. Ed., Vol. 15B. New York: Academic Press, 1982. Pp. 61–108. The most recent review of prokaryotic and bacteriophage RNA polymerases.

Chambon, P.  Split genes. *Sci. Am.* 244:60, 1981. A description of the processing of eukaryotic mRNA, in particular the mRNA for chicken ovalbumin.

Lerner, M., Boyle, J., Mount, S., Wolin, S., and Steitz, J.  Are snRNPs involved in splicing? *Nature* 283:220, 1980.

Lewin, B.  Alternatives for splicing: Recognizing the ends of introns. *Cell* 22:324, 1980.

Lewis, M. K., and Burgess, R. R.  Eukaryotic RNA Polymerases. In P. D. Boyer (Ed.), *The Enzymes,* 3rd Ed., Vol. 15B. New York: Academic Press, 1982. Pp. 109–153. The most recent review of eukaryotic RNA polymerases.

Losick, R., and Pero, J.  Cascades of sigma factors. *Cell* 25:582, 1980. A mini review of the multiple sigmas found in *B. subtilis* during sporulation and after bacteriophage infection.

Siebenlist, U., Simpson, R., and Gilbert, W.  *E. coli* RNA polymerase interacts homologously with two different promoters. *Cell,* 20:269, 1980. A summary of the structure of promoters.

## PROBLEMS

1. One strand of a DNA molecule is completely transcribed into a messenger RNA by RNA polymerase. The base composition of the DNA template strand is G = 24.1%, C = 18.5%, A = 24.6%, T = 32.8%. The base composition of the newly synthesized RNA molecule is:
   a. G = 24.1%, C = 18.5%, A = 24.6%, U = 32.8%.
   b. G = 24.6%, C = 24.1%, A = 18.5%, U = 32.8%.
   c. G = 18.5%, C = 24.1%, A = 32.8%, U = 24.6%.
   d. G = 32.8%, C = 24.6%, A = 18.5%, U = 24.1%.
   e. None of these.

2. a. Calculate the moles of UTP incorporated in the RNA polymerase assay depicted in Figure 21-3, given a reaction volume of 1 ml, a specific activity of the radioactive UTP of 100 Ci/mol (1 Ci equals $2.2 \times 10^{12}$ disintegrations per minute), and 220,000 dpm of radioactivity incorporated into RNA in a 10-min reaction.
   b. Assuming that the RNA is 40% U, calculate the moles of total nucleotide incorporated into RNA.
   c. Why is the concentration of UTP lower than that of the other three NTPs?
   d. The molecular weight of RNA polymerase is 450,000 daltons. If 1 $\mu$g of polymerase was added to the preceding reaction and each polymerase molecule initiated at once and elongated steadily during the course of the reaction, how long is the average RNA chain and what is the average chain growth rate?

3. Cordycepin-5′-triphosphate is 3′-deoxyATP, a ribonucleoside triphosphate analog, and can be bound as a substrate by RNA polymerase. When it is added to an in vitro transcription reaction, it stops RNA synthesis and is incorporated into the RNA chain. This fact was used to argue that the direction of transcription is 5′→3′. Explain why this is a reasonable argument.

4. Explain why (Table 21-1) 40 to 50 percent of the RNA being synthesized in bacteria at a given time is mRNA but only about 3 percent of the total RNA in the cell is mRNA.

5. It is possible to determine which subunit of RNA polymerase is altered by a mutation affecting the activity of the holoenzyme. RNA polymerase purified from an *E. coli* strain resistant to antibiotic X is called X-resistant enzyme and its subunit structure is written $\alpha_2{}^R\beta^R\beta'^R\sigma^R$. Likewise, enzyme from a strain sensitive to X has the subunit structure $\alpha_2{}^S\beta^S\beta'^S\sigma^S$. The subunits can be separated from

| Subunit Combination | Sensitivity or Resistance of the Resulting Enzyme |
|---|---|
| $\alpha^S\beta^S\beta'^R\sigma^S$ | R |
| $\alpha^S\beta^R\beta'^S\sigma^S$ | S |
| $\alpha^R\beta^S\beta'^S\sigma^S$ | S |
| $\alpha^S\beta^S\beta'^S\sigma^R$ | S |
| $\alpha^S\beta^S\beta'^R\sigma^R$ | R |
| $\alpha^S\beta^R\beta'^R\sigma^S$ | R |
| $\alpha^R\beta^R\beta'^S\sigma^S$ | S |

each other and mixed together in various combinations to reconstitute active enzyme. The reconstituted enzymes are tested for resistance to antibiotic X. Using the results above, determine which subunit in the resistant enzyme is mutated and thus is responsible for the resistance.

6. Figure 21-4 shows a method for determining the molecular weight of a polypeptide by electrophoresis on SDS-containing polyacrylamide gels and comparison with proteins of known molecular weights. RNA polymerase II purified from wheat germ analyzed in this way produces 12 bands that migrate the following distances (in cm) from the top of the gel: 1–1.0, 2–2.6, 3–7.1, 4–8.5, 5–8.8, 6–9.4, 7–9.6, 8–9.8, 9–10.0, 10–10.2, 11–10.4, and 12–10.9. The molecular-weight marker proteins in Figure 21-4 electrophoresed on a parallel gel migrated as follows: $\beta$-galactosidase: 3.3; phosphorylase: 3.9; bovine serum albumin: 5.3; ovalbumin: 6.8; chymotrypsinogen: 8.7; and lysozyme: 10.8. Calculate the molecular weight of the 12 wheat germ RNA polymerase II subunits. Assuming one of each subunit is present in the enzyme, what is the molecular weight of the enzyme?

7. What is the maximum rate of initiation at a promoter assuming that the diameter of RNA polymerase is about 200 Å and the rate of RNA chain growth in vivo is 45 nucleotides/s?

8. A short RNA chain synthesized in vitro has the sequence 5′-AUGUACCGAAGUGGUUU-3′-OH.
   a. Draw the phosphate groups and star those which would be radioactive when the transcription is carried out in the presence of $[\gamma\text{-}^{32}P]$ATP.
   b. Do the same for $[\alpha\text{-}^{32}P]$UTP.

c. Show where RNAse T1 and pancreatic RNAse A would cleave this RNA.
d. Show the oligonucleotides produced by RNAse T1 digestion of the $[\alpha\text{-}^{32}P]$UTP-labeled RNA and indicate which are labeled.

9. The following hypothetical RNA is an $\alpha$-amanitin-sensitive primary transcript made in an eukaryotic nucleus:

5′-pppAUUAUGCCGAUAAGGUAAGUA($N_{50}$)
     AUCUCCCUGCAGGGCGUAACCAAUAAA
          CGACGACGACGUCCC. . .---3′OH

Indicate the final processed mRNA found in the cytoplasm on polysomes capable of directing the synthesis of a 5 amino acid peptide and point out important features.

10. Given two oligonucleotides, 5′$PO_4$-GUUC-3′-$PO_4$ and 5′OH-ACCG-3′-OH, describe how to synthesize 5′OH-ACCGGUUCAAAA---3′-OH. You may use enzymes described in this chapter and an enzyme called bacterial alkaline phosphatase that is capable of removing a 5′-$PO_4$ or 3′-$PO_4$ to produce a nucleotide 5′-OH or 3′-OH.

11. A particular eukaryotic DNA virus was found to code for two mRNA transcripts, one shorter than the other, from the same region of its genome. Analysis of the translation products revealed that the two polypeptides shared the same amino acid sequence at their amino termini, but differed at their carboxy termini. Surprisingly, the longer of the two polypeptides was coded for by the shorter mRNA. Suggest an explanation.

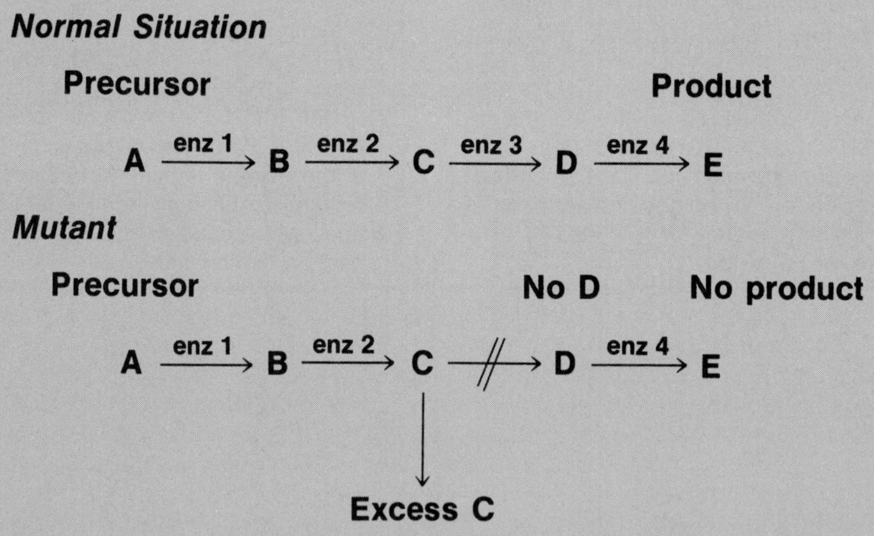

**Normal Situation**

Precursor                                                          Product

$$A \xrightarrow{\text{enz 1}} B \xrightarrow{\text{enz 2}} C \xrightarrow{\text{enz 3}} D \xrightarrow{\text{enz 4}} E$$

*Mutant*

Precursor                                          No D        No product

$$A \xrightarrow{\text{enz 1}} B \xrightarrow{\text{enz 2}} C \xrightarrow{\;\;/\!/\;\;} D \xrightarrow{\text{enz 4}} E$$

$$\downarrow$$

**Excess C**

# 22

# BIOSYNTHESIS OF AMINO ACIDS

Living cells that use a single carbon source, whether it be carbon dioxide, an organic acid, or a simple sugar such as glucose, must necessarily synthesize all the 20 different amino acids for their proteins. The pathways to these amino acids do not arise uniquely from the carbon source, but rather arise as branching pathways from a few key intermediates in the cental metabolic routes that are common to all cells regardless of the carbon source. These central metabolic routes are the *glycolytic pathway*, the *hexose monophosphate shunt* and the associated *pentose phosphate cycle*, and the *tricarboxylic acid cycle*, all of which have been discussed in Chapters 8 and 9.

Amino acid biosynthesis has been most actively studied in microorganisms, where genetic and biochemical techniques have been jointly used to obtain an understanding of the pathways. In applying mutant methodology, a large number of mutants were obtained that required a particular amino acid for growth. In one approach, the mutants were sorted out into different complementation groups by genetic analysis (complementation group analysis is discussed in Chapter 27 for λ bacteriophage mutants). Each complementation group is associated with a single gene-encoded step in the biochemical pathway; therefore, the number of complementation groups usually indicated the number of steps in the pathway. Alternatively, or in parallel, the mutants could be analyzed nutritionally by their response to suspected intermediates in the pathway or by the accumulation of intermediates in their culture fluids, which often fed other mutants requiring the same amino acid. Finally, each mutant could be analyzed enzymatically for its capacity to convert presumptive substrates to the amino acid or to other intermediates in the pathway. In this way, a large number of

*Immediate consequences of a point mutant on a biosynthetic pathway. The chain of reactions leading to the end product in the pathway is broken, and frequently excessive amounts of intermediates are produced.*

amino acid pathways have been elucidated in the enteric bacteria and other microorganisms. Extensions of this information to investigations using other organisms have been most fruitful.

In this chapter, the results of such labors will be discussed; both the pathways and the regulation of the pathways will be considered. Additionally, the biosynthesis of some amino acids that are not used in protein synthesis is described.

## THE BIOSYNTHETIC FAMILIES OF AMINO ACIDS

Figure 22-1 gives an abbreviated view of the way in which the amino acid biosynthetic pathways arise from a few metabolites that are intermediates in the central metabolic routes. From these few metabolites, all 20 amino acids are formed, so that the amino acids can be classified into families based on the points at which divergence of the pathway occurs. Thus the entire glutamate family arises initially from the removal of *α-ketoglutarate* from the tricarboxylic acid cycle. In all organisms, this family includes L-glutamate, L-glutamine, L-proline, and L-arginine. In one of the few departures from the unity of amino acid biosynthesis, α-ketoglutarate also provides the starting point for L-lysine biosynthesis in the fungi and *Euglena* and gives rise to four of the six carbons of lysine. Another intermediate of the tricarboxylic acid cycle that is drained in amino acid biosynthesis is *oxaloacetate*, which serves as the starting point in the formation of L-aspartate and other members of the aspartate family, e.g., L-asparagine, L-lysine (in bacteria and all plants except the fungi), L-methionine, L-threonine, and L-isoleucine.

It should be pointed out that it is exactly the magnitude of this kind of drainage from the tricarboxylic acid cycle that dictates the extent to which the anaplerotic pathways must function to replenish the oxaloacetate lost from the cycle (see Chapter 8).

A glycolytic intermediate, *3-phosphoglycerate*, is the intermediate from which L-serine and its derivative amino acids, L-cysteine and glycine, are formed. *Pyruvate* gives rise to the carbon skeletons of L-alanine and L-valine, to four of the six carbons of L-leucine, and to the two-carbon fragment condensed with the four carbons from threonine that give rise to the skeleton of isoleucine. Pyruvate also provides, on the average, two and one-half of the six carbons of lysine in those forms in which aspartate is converted to lysine. Thus both isoleucine and lysine can be considered members of both the pyruvate and aspartate families.

The three aromatic amino acids are derived from key intermediates in both the pentose cycle and the glycolytic pathway, *erythrose-4-phosphate* and *phosphoenolpyruvate* (PEP), respectively. In addition, L-tryptophan requires *phosphoribosylpyrophosphate* (PRPP), which is derived from the hexose monophosphate pathway. Tryptophan biosynthesis also requires serine, but in the process, a triose phosphate molecule is returned to the glycolytic pathway, so that the serine carbons can be effectively replaced.

Finally, by a pathway unrelated to those of the other amino acids, L-histidine biosynthesis involves the transfer of an N—C group from ATP to the ribosyl moiety of PRPP. The histidine pathway could thus be considered as a branch from the purine nucleotide pathway.

As will be described in this chapter, most of the carbon flow from the central metabolic routes is an irreversible flow. However, it also will be shown that in nearly all cases the flow is an orderly one that provides to the cell just those amounts of amino acids needed for cellular synthesis. The remarkable fact is that this orderly flow is achieved by relatively simple, yet almost faultless regulatory mechanisms.

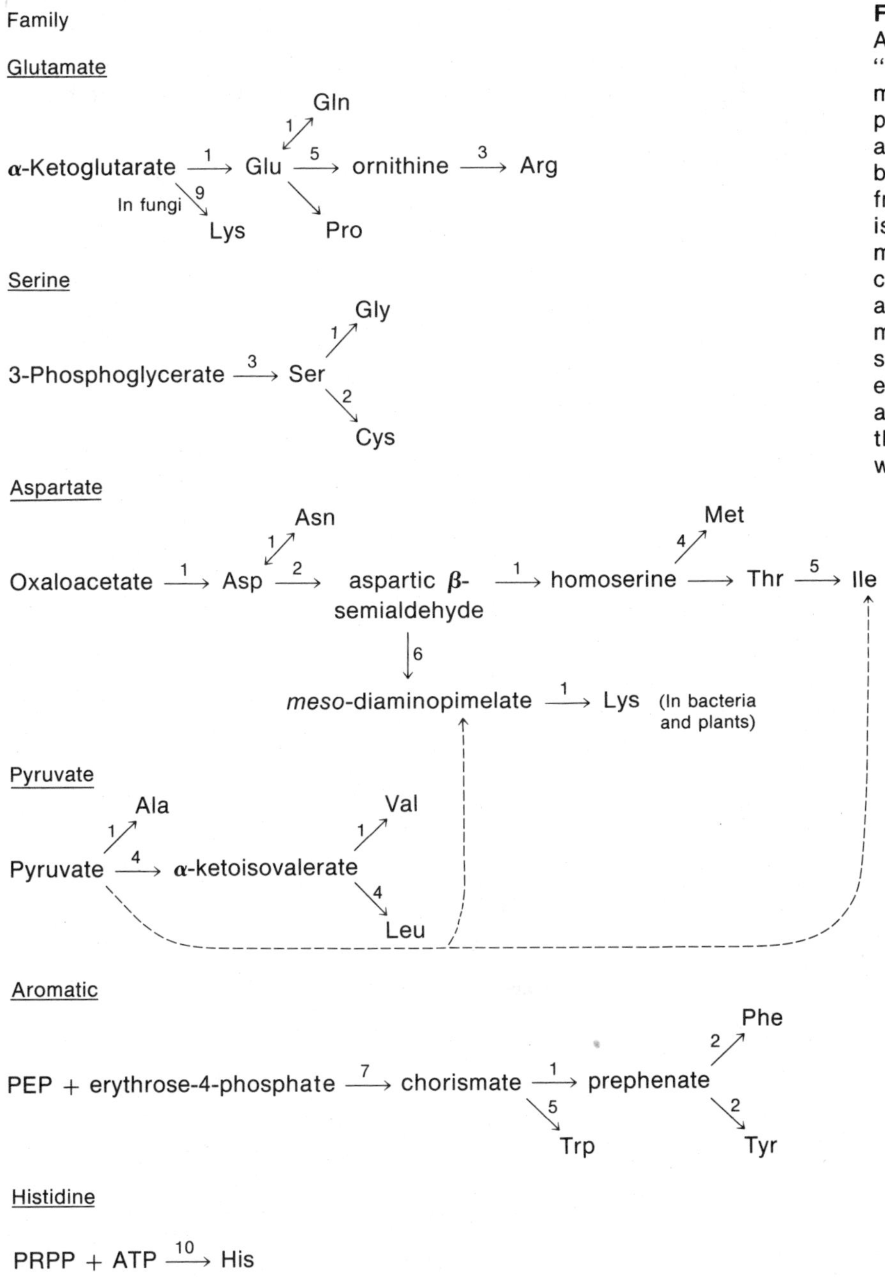

Family

Glutamate

Serine

Aspartate

Pyruvate

Aromatic

Histidine

$$PRPP + ATP \xrightarrow{10} His$$

**Figure 22-1**
Amino acids are ordered into "families" according to the central metabolites that serve as starting points for their synthesis. Solid arrows indicate one or more discrete biochemical steps. Dashed arrows from pyruvate to both lysine and isoleucine indicate that pyruvate makes a contribution of some of the carbon atoms of each of these amino acids. Only amino acids and key intermediates are specified. The number of steps between key intermediates or end points are indicated over the arrows. Notice that lysine is unique in that two completely different pathways exist for its biosynthesis.

# THE CONVERSION OF AMMONIA TO AMINO GROUPS

A common element in all amino acids is the $\alpha$-amino group. _Directly or indirectly, these amino groups are all derived from ammonia by way of the amino group of L-glutamate._

## The Direct Amination of $\alpha$-Ketoglutarate

The simplest route to glutamate (and therefore, amino group formation) is that exhibited by many bacteria when grown in a medium containing an ammonium salt as the sole nitrogen source. The reaction is a reductive amination reaction catalyzed by glutamate dehydrogenase (Figure 22-2). In organisms such as _E. coli_, the enzyme is specific for NADPH as the hydrogen donor, as might be expected of a biosynthetic reaction involving a reductive step.

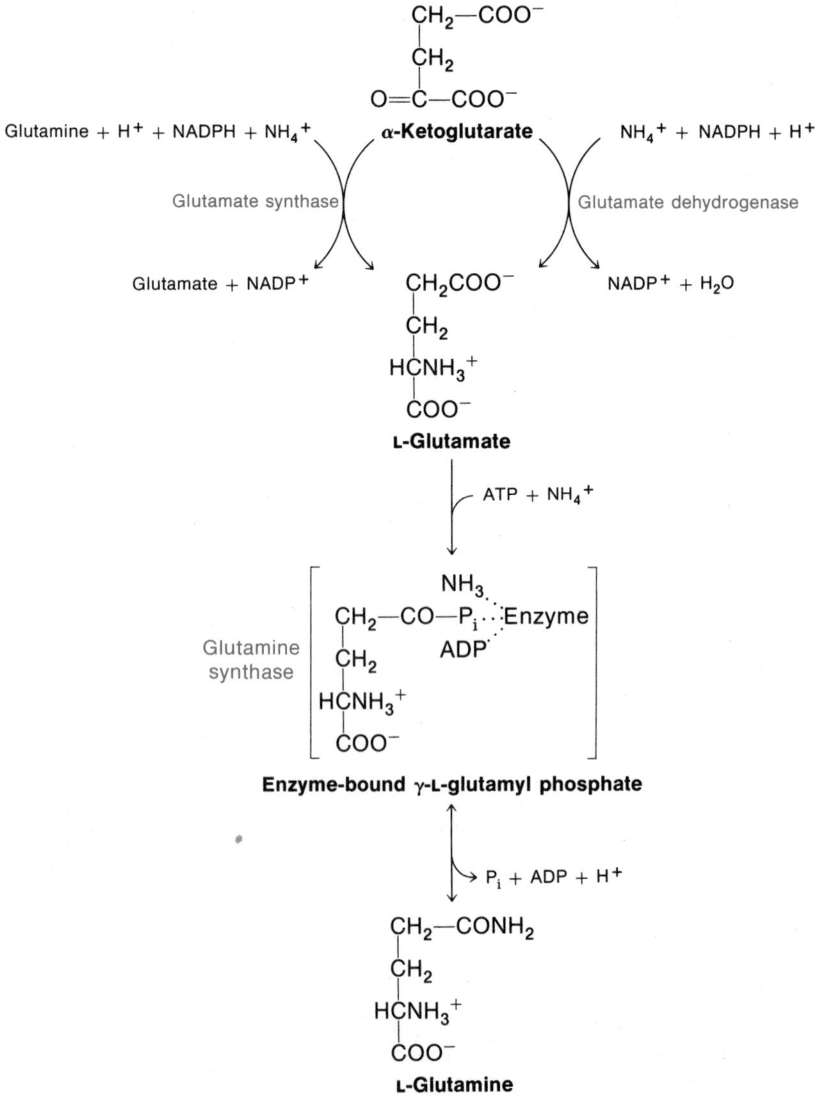

**Figure 22-2**
The conversion of ammonia into the $\alpha$-amino group of glutamate and into the amide group of glutamine. Note that the direct amination of $\alpha$-ketoglutarate by $NH_4^+$ occurs only under conditions of a high $NH_4^+$ concentration, as discussed in the text.

Some organisms have both an NAD/NADH- and an NADP/NADPH-dependent glutamate dehydrogenase. In such cases, the NAD-dependent enzyme is considered to play a catabolic role that converts glutamate back to $\alpha$-ketoglutarate. Both enzymes, however, catalyze reversible reactions, and assignment to anabolic or catabolic functions is largely inferential. However, in some organisms (e.g., certain water molds), the presumed catabolic enzyme is inhibited by ATP, CTP, and fructose diphosphate, indicators of a high-energy charge, while being stimulated by AMP.

Studies on the glutamate dehydrogenases from green plants revealed that the plant enzymes can use either NADH or NADPH as the hydrogen donor in the amination reaction. However, for reasons given below, it is unlikely that the glutamate dehydrogenase of plants plays a significant role in glutamate biosynthesis. It is more likely that in green plants, as in many bacteria and fungi, and in probably all bacteria when they are not being grown under $NH_4^+$ excess, the formation of the amino group of glutamate occurs not from $NH_4^+$, but from the amide group of glutamine (see the following discussion). In such cases, the primary conversion from $NH_4^+$ to organic nitrogen is catalyzed by glutamine synthase.

### The Formation of L-Glutamine

The reaction catalyzed by *glutamine synthase* is one that involves activation of the $\gamma$-carboxyl group of glutamate to yield a $\gamma$-glutamylenzyme complex and the cleavage of ATP to ADP and $P_i$ (Figure 22-2). In a second step, $\gamma$-glutamyl transfer to $NH_4^+$ occurs.

*Glutamine synthase is a key enzyme in the flow of ammonia nitrogen to organic compounds, and its activity is subject to elaborate controls that sense the cell's need for nitrogen-containing compounds.*

The enzyme from *E. coli* has been studied extensively, and much is known of its structure and its regulatory behavior. The studies of Stadtman and Ginsberg and their colleagues have shown that glutamine synthase of *E. coli* is composed of 12 identical 50,000-dalton subunits arranged symmetrically in two hexameric rings. *The activity of the enzyme is regulated in a complex pattern of feedback inhibition in which a partial inhibition is effected by each of eight different nitrogenous compounds:* carbamylphosphate, glucosamine-6-phosphate, tryptophan, alanine, glycine, histidine, cytidine triphosphate, and AMP. All these compounds except glycine and alanine receive the amide nitrogen of glutamine directly during their biosynthesis and are thus end products of reaction sequences leading from glutamine. Glycine and alanine, while not directly end products, could be looked upon as "indicators" of the sufficiency of the nitrogen supply of the cell.

Perhaps even more important in the regulation of *E. coli* glutamine synthase activity is the reversible ATP-dependent adenylylation of a specific tyrosyl residue on each subunit. As the enzyme becomes progressively more adenylylated (up to the fully adenylylated form of 12 AMP groups per enzyme molecule), the enzyme becomes progressively less active in an assay of glutamine biosynthetic activity in which $Mg^{2+}$ is the divalent cation. At the same time, the enzyme becomes progressively more active in an assay in which $Mn^{2+}$ is the divalent cation. In the presence of an excess of $Mg^{2+}$ ions over $Mn^{2+}$ ions, as would be found in the intact bacterial cell,

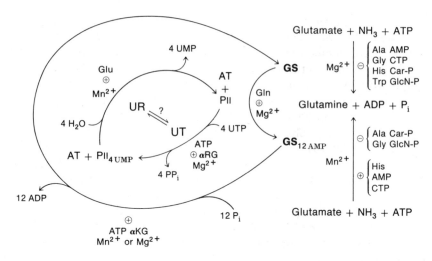

**Figure 22-3**

The regulation of glutamine synthase of *E. coli*. The $Mg^{2+}$-dependent ("active" form) glutamine synthase (GS) is converted to the adenylylated $Mn^{2+}$-dependent ("inactive" form) of glutamine synthase (GS · 12 AMP) by adenylyltransferase (AT) in the presence of the regulatory protein PII. Upon uridylylation of PII by uridylyltransferase (UT), AT catalyzes the phosphorolytic removal of the adenylyl groups from GS · 12 AMP to yield GS. The uridylylated form of PII (PII · 4 UMP) is hydrolytically deuridylated by the uridylyl-removing enzyme (UR). The stimulation of deadenylylation of GS · 12 AMP and of uridylylation of PII by $\alpha$-ketoglutarate ($\alpha$KG) and the stimulation of adenylylation of GS and deuridylylation of PII · 4 UMP by glutamine (Gln) provides a mechanism whereby the state of GS in the cell is controlled by the $\alpha$KG:Gln ratio. UR and UT may be the same protein catalyzing one reaction or the other depending on the $\alpha$KG:Gln ratio. The two forms of GS respond differentially to the eight small molecule effectors.

the $Mn^{2+}$-dependent activity of the adenylylated enzyme would be only about one-twentieth the $Mg^{2+}$-dependent activity of the deadenylylated enzyme. *Thus the adenylylation process serves physiologically, in effect, to inactivate the enzyme.* The two forms of the enzyme differ in many properties, including some differences in the degree to which the various end products affect the activity. However, relative to the effect of adenylylation itself on activity, these differences may be of minor importance.

*The adenylylation reaction and its reversal by a phosphorolytic deadenylylation are quite responsive to the nitrogen supply in the cell.* Both reactions are catalyzed by a protein (glutamine synthase adenylyltransferase) that contains *separate and noninteracting sites for adenylylation and deadenylylation. Which site is active is dependent* both on the $\alpha$-ketoglutarate : glutamine ratio and *on the state of a second, tetrameric protein, PII. In its native state, PII and a low $\alpha$-ketoglutarate : glutamine ratio (nitrogen excess) stimulate adenylylation.* The PII protein undergoes a UTP-dependent uridylylation reaction. *In its uridylylated state, PII and a high $\alpha$-ketoglutarate:glutamine ratio (nitrogen limitation) stimulate deadenylylation.* Both the conversion of PII to *uridylylated PII and its hydrolytic reversal are catalyzed by a protein that in the presence of $\alpha$-ketoglutarate is a uridylyltransferase and in the presence of glutamine is a uridylyl-removing enzyme.* This bicyclic cascade regulating glutamine synthase activity of *E. coli* is illustrated in Figure 22-3.

The control of glutamine synthase by this pattern of adenylylation and deadenylylation is not widespread but may be limited to some of the Gram-negative bacteria. There may, however, be different modifications of other glutamine synthases that are physiologically analogous. For example, at least some of the glutamine synthases of eukaryotic cells are octameric structures that respond to regulation by dissociation to a tetrameric form under conditions of restricted $NH_4^+$ supply.

### The Amination of $\alpha$-Ketoglutarate by the Amide Group of L-Glutamine

As implied earlier, the direct reductive amination of $\alpha$-ketoglutarate with ammonia is probably an exception in nature; only certain forms of life form glutamate this way, and they do so only when they are grown in the presence of a high concentration of $NH_4^+$. The more prevalent reaction is the amination of $\alpha$-ketoglutarate by glutamate synthase, a reductive reaction in which L-glutamine is converted back to L-glutamate (Figure 22-2). Although this mode of amino group formation requires a high-energy phosphate for the formation of the amide group, there is considerable advantage, since amide group formation can proceed at a much lower concentration of $NH_4^+$ than can amination by glutamate dehydrogenase. For example, the $K_m$ for $NH_4^+$ of the *E. coli* glutamate dehydrogenase is 1.1 m$M$, while that of the *E. coli* glutamine synthase is only 0.2 m$M$. (Furthermore, the equilibrium of the direct amination reaction strongly favors deamination.)

The hydrogen donor varies for the various glutamate synthases. For bacteria, it is NADPH, as it is for nearly all biosynthetic reductive reactions. For at least several fungi, however, the hydrogen donor is NADH. In plants, two kinds of the enzyme have been found: one is specific for reduced ferredoxin; the other functions in vitro with either NADH or NADPH. Since $NH_4^+$ is not often present in high concentration in soils (owing to microbial oxidation of $NH_4^+$ and regulatory mechanisms that control both $NO_3^-$ reduction and nitrogen fixation), it is likely that the formation of amino groups in plants invariably occurs in a low-$NH_4^+$ environment and that the amination of $\alpha$-ketoglutarate occurs at the expense of the amide group of glutamine.

## THE FORMATION OF AMMONIA FROM OTHER FORMS OF INORGANIC NITROGEN

*Although $NH_4^+$ is the form in which nitrogen is incorporated into organic materials, it is less often available to plants or bacteria for biosynthesis than other forms of nitrogen.* When present for any length of time in nature, $NH_4^+$ will be assimilated rapidly, but it also will be readily oxidized by nitrifying bacteria such as *Nitrosomonas* and *Nitrobacter* to nitrite and nitrate. Reduction of nitrate by plants or by bacteria seldom yields $NH_4^+$ in excess (for an exception, see below). Nitrogen fixation, by which nitrogen of the atmosphere is reduced to $NH_4^+$ by either free-living or symbiotic bacteria, is also regulated, so that $NH_4^+$ does not accumulate in high concentration. *Thus most plants must assimilate $NH_4^+$ from low pools generated by their own nitrate and nitrite reductases or from the reduction of nitrogen by microorganisms.*

## Nitrogen Fixation

*The biological fixation of nitrogen by both free-living and symbiotic nitro-gen-fixing bacteria involves an enzyme complex called nitrogenase.* Nitrogenases consist of two proteins. One is an Fe protein called *component II*, and the other is a Mo-Fe protein called *component I*. The Fe protein from *Clostridium pasteurianum* is a dimer of two identical subunits of 29,000 daltons surrounding an iron-sulfur center that contains four iron atoms and four acid-labile sulfur atoms. The Mo-Fe protein is considerably more complex and has not been completely characterized from any of the nitrogen-fixing organisms in which it has been studied. That from *C. pasteurianum* is a 220,000-dalton tetramer with two molybdenum atoms, about 30 iron atoms, and almost as many acid-labile sulfur atoms. The four subunits are of two kinds, two of 50,000 and two of 60,000 daltons. There are probably four iron-sulfur centers, each containing four iron atoms and four labile sulfur atoms, and two extractable iron-molybdenum-sulfur "cofactors" that contain eight iron atoms, six labile sulfur atoms, and one molybdenum atom.

*Nitrogen fixation is a metabolically costly process, with 12 high-energy phosphate bonds and 6 low-potential reducing equivalents being required for each mole of $N_2$ reduced.* In one study, it was found that *C. pasteurianum*, a common free-living nitrogen fixer, consumed 70 percent more carbon source when it grew under nitrogen-fixing conditions than when it was grown in excess $NH_4^+$.

For nitrogen fixation to occur, the Fe protein is reduced by one or both of two low-potential reductants, ferredoxin or flavodoxin, generated during carbohydrate fermentation in *C. pasteurianum* or in the presence of a high [NADPH]/[NADP] ratio. Electrons are transferred from the Fe protein to the Mo-Fe protein. This transfer is obligatorily coupled to the hydrolysis of four ATPs to four ADPs for each electron pair. The reduced Mo-Fe protein then reduces nitrogen to $NH_4^+$.

The study of nitrogenase has been greatly facilitated by the fact that nitrogenase also can reduce acetylene to ethylene, thus allowing convenient gas chromatographic techniques to be used to follow the reaction. Other reactions that can be catalyzed by nitrogenase are $CN^-$ to $CH_4$ and $NH_4^+$, $N_2O$ to $N_2$ and $H_2O$, and $H^+$ to $H_2$. It is this last reaction, the conversion of $H^+$ to $H_2$, that can make nitrogen fixation even more costly. Thus, when the ATP supply is low or the reduction of the Fe protein is retarded, the six-electron reduction of the Mo-Fe protein is sluggish and the two-electron reduction of protons would be favored. Furthermore, in the presence of $H_2$, one of the presumed intermediates, a diimine intermediate, NH=NH, is converted to $N_2$ and $2H_2$:

$$NH{=}NH + H : H \longrightarrow N_2 + 2H \cdot H$$

Such a futile cycle apparently always occurs to some extent, so that for each dinitrogen reduced, a pair of protons is also reduced.

*Nitrogenases appear to be highly conserved proteins. One of their characteristics is the extreme lability of the enzyme from all sources in the presence of oxygen.* For strict anaerobes, this feature is not a special problem, since the organism can grow only anaerobically. For facultative anaerobes, such as *Klebsiella pneumoniae*, nitrogen fixation occurs only under anaerobic conditions. Some nitrogen fixers, however, are strict aerobes. One genus, *Azotobacter*, protects its nitrogenase by virtue of a very active electron-transport system that removes oxygen at a rapid rate. Another genus, *Rhizobium*, which fixes nitrogen symbiotically in the root nodules of legumes, depends on the leghemoglobin of the plant to absorb $O_2$ and maintain a low oxygen tension.

## Nitrate and Nitrite Reduction

*Nitrate and nitrite reduction play two physiological roles. One, exhibited primarily by bacteria, is a dissimilatory one in which nitrate and nitrite serve as terminal electron acceptors.* In other words, the reduction serves to oxidize reducing equivalents, e.g., NADH, generated during oxidation of substrates. *The other, exhibited by bacteria, fungi, and plants, is an assimilatory role in which nitrite is formed from nitrate and is reduced to $NH_4^+$* at a rate no greater than that required for synthesis of nitrogenous compounds during growth.

In general, the assimilatory nitrate and nitrite reductases are soluble enzymes that utilize reduced pyridine nucleotides or reduced ferredoxin. In contrast, the dissimilatory nitrate reductases are membrane-bound terminal electron acceptors that are tightly linked to cytochrome $b_1$ pigments. Such complexes allow one or more sites of energy conservation (ATP generation) coupled with the electron transport.

The $NH_4^+$ produced by fermentative bacteria that utilize nitrite as an oxidant or produced during the decomposition of organic materials can be utilized by plants and bacteria for cell material. However, under suitably aerobic conditions, $NH_4^+$ is rapidly converted again to nitrite and nitrate by the nitrifying bacteria, such as *Nitrosomonas europaea* and *Nitrobacter agilis*, which are chemolithotrophs that gain energy from the oxidation of $NH_4^+$. It is because of the ubiquity of such organisms that the ability to assimilate nitrate and nitrite is so important to green plants and, ultimately, to animals. The so-called nitrogen cycle (Figure 22-4) involves the passage of nitrogen between different microorganisms and the atmosphere.

**Figure 22-4**
The nitrogen cycle. The flow of nitrogen in the biological world. (1) The proteins of animals and plants are cleaved by many microorganisms to free amino acids ($-3$ oxidation state) from which ammonia ($-3$ oxidation state) is released by deamination. Urea, the main nitrogen excretion product of animals, is hydrolyzed to $NH_3$ and $CO_2$. (2) *Nitrosomonas* soil bacteria obtain their energy by oxidizing $NH_3$ to nitrite, $NO_2^-$ ($+3$ oxidation state). (3) *Nitrobacter* obtain their energy by oxidizing nitrite, $NO_2^-$, to nitrate, $NO_3^-$ ($+5$ oxidation state). (4) Plants and many microorganisms reduce nitrate for incorporation into amino acids, completing the cycle. (5) Other microorganisms reduce nitrate partly to $NH_3$ and partly to $N_2$ (0 oxidation state), which is lost to the atmosphere. (6) The atmospheric nitrogen can be recaptured and converted after reduction into organic substances by a limited number of nitrogen-fixing bacteria and algae.

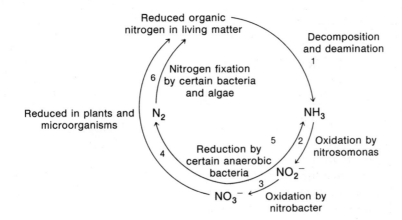

# THE BIOSYNTHESIS OF OTHER AMINO ACIDS OF THE GLUTAMATE FAMILY

Whereas most of the amino nitrogen distributed from glutamate to other amino acids occurs by means of transamination, the carbons as well as the nitrogens of glutamate are used in proline and arginine biosynthesis.

## Proline Biosynthesis

The pathway to L-proline is shown in Figure 22-5. The conversion of glutamate to proline involves the reduction of the γ-carboxyl group to yield glutamic-γ-semialdehyde, which spontaneously cyclizes to yield a five-membered ring compound, $\Delta^1$-pyrroline-5-carboxylate. A second reduction then yields proline.

In animals, 4-hydroxy-L-proline is an important constituent of some proteins, principally _collagen_. Hydroxyproline is not biosynthesized as such, but results from a posttranslational modification of proline residues in a precursor protein, such as protocollagen. The enzyme _proline hydroxylase_ is a mixed-function oxidase that requires $Fe^{2+}$ and a reducing agent such as ascorbate. In addition, the reaction requires α-ketoglutarate, which is oxidatively decarboxylated to succinate stoichiometrically with the hydroxylation of prolyl residues.

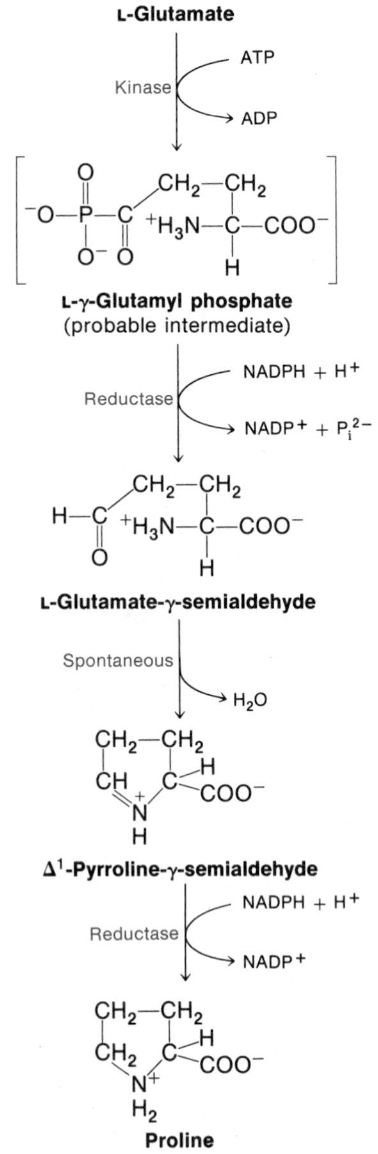

## Arginine Biosynthesis

Another pathway that involves an activation and reduction of the γ-carboxyl of glutamate is that leading to L-arginine by way of L-ornithine (Figure 22-6). For ornithine biosynthesis, however, the α-amino group is protected by acetylation prior to carboxyl activation and reduction. Ring closure is thus prevented, and glutamyl residues destined for arginine biosynthesis are effectively sequestered from those destined for proline biosynthesis.

Ornithine is converted to arginine by means of a series of reactions that will be encountered again in the following chapter, where urea formation is considered.

## L-Lysine Biosynthesis in Fungi

_A unique situation exists for lysine biosynthesis in that two totally different pathways have evolved in different organisms._ In bacteria and green plants, L-lysine is formed from the condensation of pyruvate with asparticβ-semialdehyde. In fungi (and _Euglena_), however, lysine is formed by a distinctly different pathway, involving α-ketoglutarate as the intermediate drained from the common metabolic routes. _The initial steps in the pathway result in the lengthening of the carbon chain of α-ketoglutarate to α-ketoadipate_ (Figure 22-7). These reactions follow a general mechanism for the lengthening of an α-keto acid carbon chain and is also encountered in the pathway to leucine. It is like the conversion of oxaloacetate to α-ketoglutarate in the tricarboxylic acid cycle, except that the α-carboxyl of α-ketoglutarate arises from the β-carboxyl of oxaloacetate rather than from the carboxyl of the transferred acetyl group. The α-ketoadipate so formed is aminated by transamination with glutamate.

**Figure 22-5**
The biosynthesis of proline.

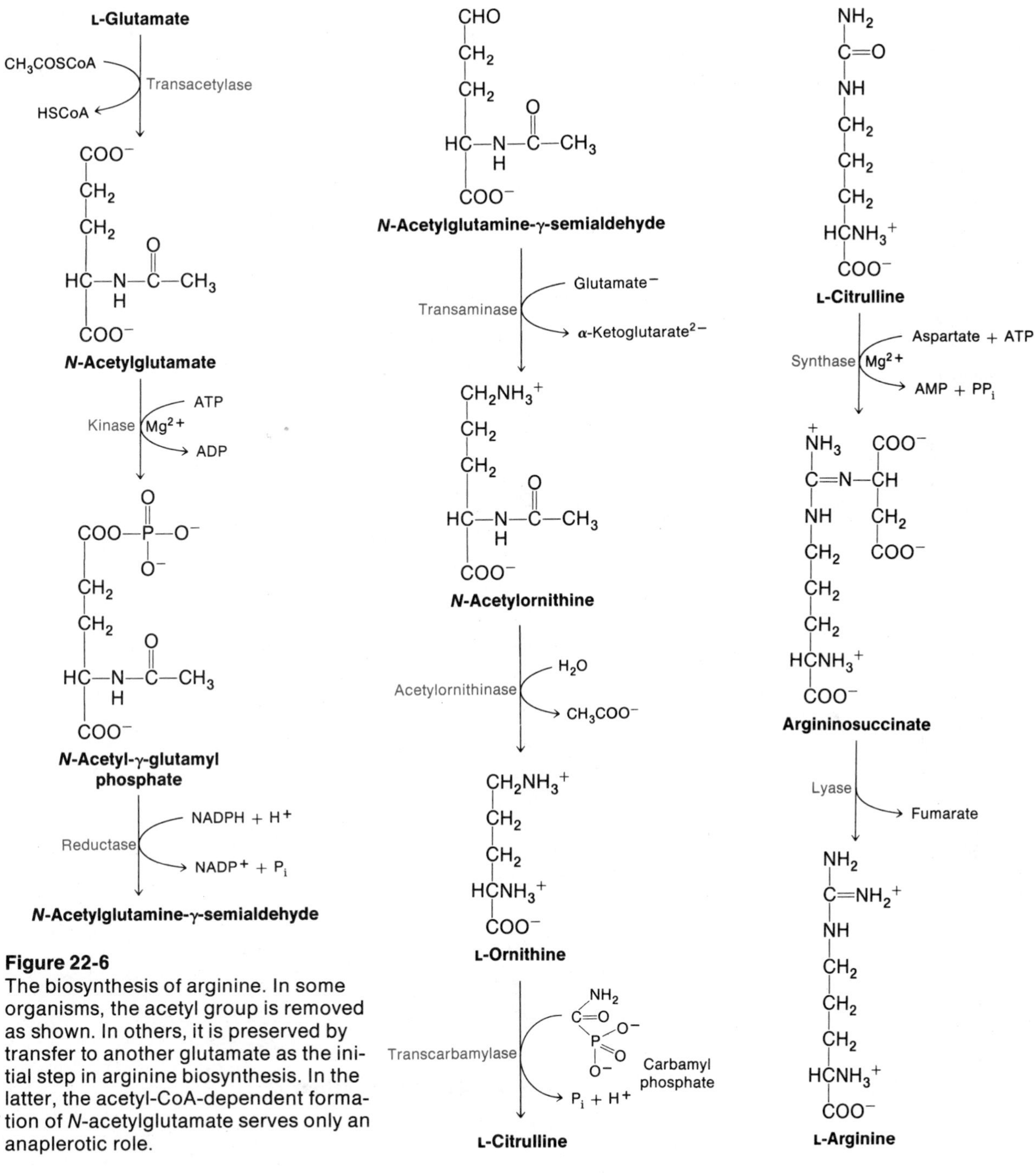

**Figure 22-6**
The biosynthesis of arginine. In some organisms, the acetyl group is removed as shown. In others, it is preserved by transfer to another glutamate as the initial step in arginine biosynthesis. In the latter, the acetyl-CoA-dependent formation of N-acetylglutamate serves only an anaplerotic role.

The next transformation is a reduction of the δ-carboxyl group. Activation of the carboxyl by ATP is necessary, but it is different from that encountered in proline and ornithine biosynthesis. The activation is thought to be achieved by formation of a mixed anhydride with AMP, δ-adenylyl-α-aminoadipate. Reduction with NADPH by δ-adenylyl-α-aminoadipate reductase is followed by release of the adenylyl group. Whether the adenylyl group is released concomitantly with reduction, as shown in Fig-

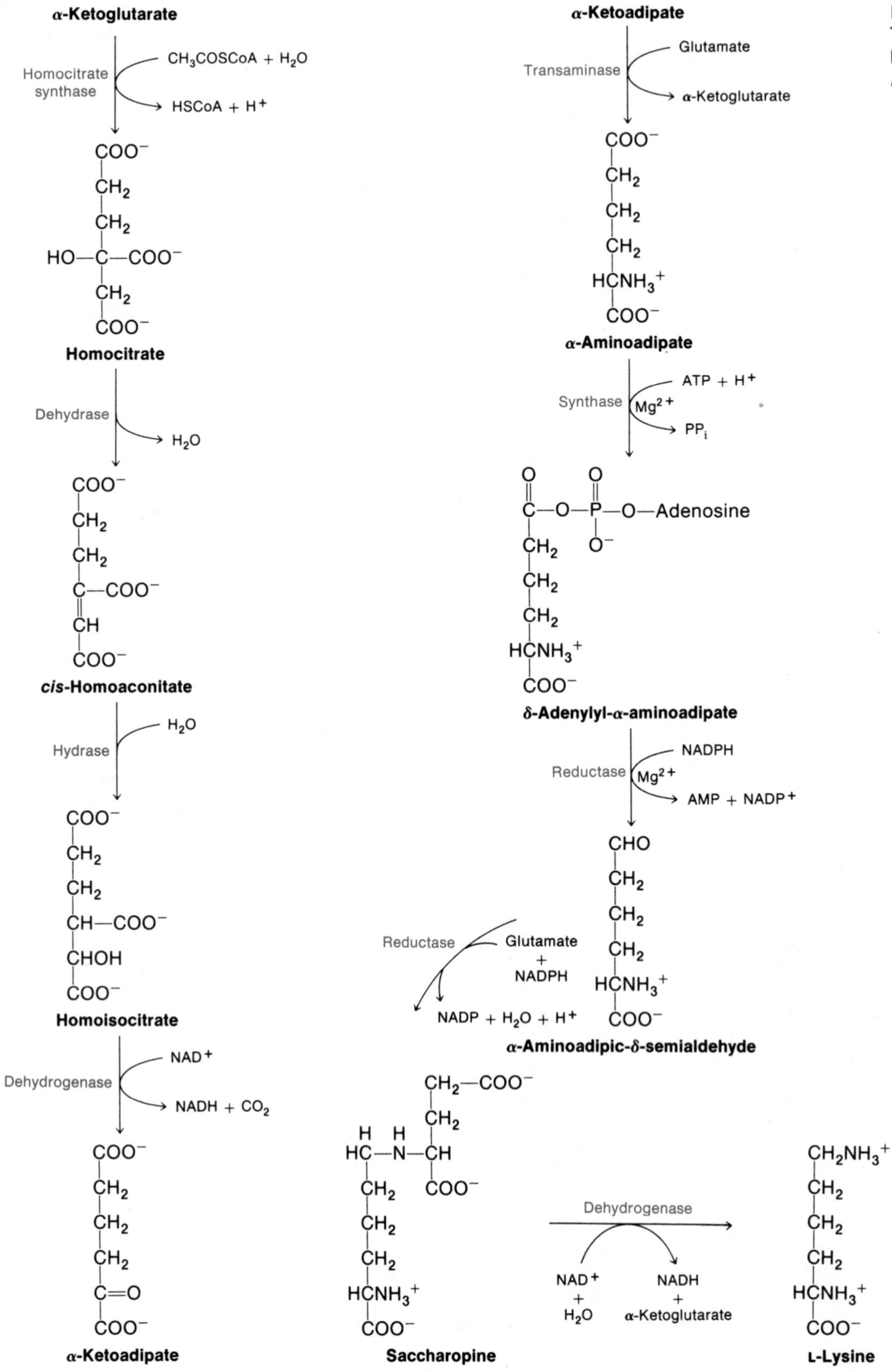

**Figure 22-7**
The biosynthesis of lysine in fungi and *Euglena*.

ure 22-7, or is cleaved from another postulated intermediate, δ-adenylyl-α-aminoadipate-δ-semialdehyde, by another enzyme is still unclear. Once formed, the semialdehyde group is aminated by the amino group of glutamate in a unique NADPH-specific reductive condensation reaction in which saccharopine is formed. Saccharopine is then oxidatively cleaved in an NAD-specific reaction to yield lysine and α-ketoglutarate. The two reactions are, in effect, a transamination, but quite different from the pyridoxal phosphate-dependent transaminases encountered in other pathways. The enzyme forming saccharopine has been termed α-aminoadipic-δ-semialdehyde-glutamate reductase; that cleaving saccharopine has been termed saccharopine dehydrogenase.

# THE BIOSYNTHESIS OF AMINO ACIDS OF THE SERINE FAMILY AND THE FIXATION OF SULFUR

The diversion of *3-phosphoglycerate* from the glycolytic pathway into the serine biosynthetic pathway is important, not only for the formation of L-*serine, L-cysteine, and glycine needed for incorporation into protein, but also for the many other functions these amino acids serve. For example, the conversion of serine to glycine serves to generate one-carbon units* that can be used in purine, thymine, and methionine biosynthesis as well as to replenish the methyl group transferred from methionine in many methylation reactions. The *carbons of glycine contribute to purine and heme-containing compounds,* to glutathione (as do those of cysteine), and in animals, to certain detoxification products. Oxidation of glycine provides an additional source of one-carbon units from the α carbon. *In many plants and microorganisms, sulfur in the form of sulfide is incorporated first into cysteine and then later is transferred to methionine and other sulfur-containing compounds.* In so doing, the carbons of cysteine are returned to the glycolytic pathway in the form of pyruvate. *Serine* itself *is incorporated directly into phospholipids* and into tryptophan, which, therefore, also might be considered a member of the serine family of amino acids. However, it will be more convenient to consider tryptophan biosynthesis along with the formation of the other aromatic amino acids.

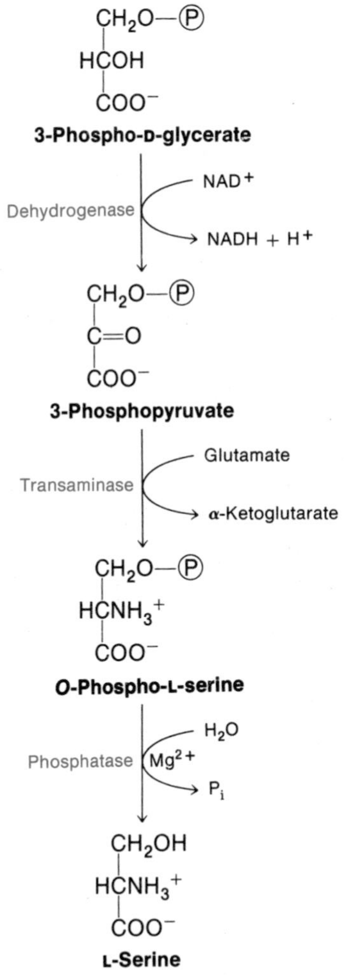

## Serine Biosynthesis

The initial reaction in L-serine biosynthesis is the oxidation of 3-phosphoglycerate by an NAD-linked dehydrogenase (Figure 22-8). Even though the reaction is freely reversible, the enzyme activity is regulated by the end product of the biosynthetic sequence, serine. The product of the reaction, O-phosphohydroxypyruvate, is converted to O-phosphoserine by a specific phosphoserine-glutamate transaminase. Removal of the phosphate group by a specific phosphoserine phosphatase yields serine. Phosphoserine itself is found as a protein constituent. However, this phosphorylation of serine occurs posttranslationally by the action of specific protein kinases. Such phosphorylation reactions serve an important function in regulating the activity of some enzymes.

The phosphorylated pathway described here appears to be the sole pathway to serine biosynthesis in green plants, as well as in microorgan-

**Figure 22-8**
The biosynthesis of serine. The end product serine inhibits the first enzyme in the pathway, 3-phosphoglycerate dehydrogenase.

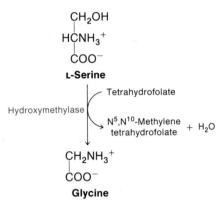

**Figure 22-9**
The biosynthesis of glycine.

isms. A series of reversible nonphosphorylated reactions and a route from glycine to serine that can be demonstrated in many microorganisms as well as mammals are probably catabolic reactions and will be considered in the next chapter.

## Glycine Biosynthesis

*The pathway from serine to glycine consists of a single but complex step catalyzed by serine hydroxymethylase* (Figure 22-9). The reaction is a pyridoxal phosphate-dependent aldol cleavage to yield glycine and an "active" formaldehyde that is transferred to a tetrahydrofolate cofactor. *This reaction is an important one in supplying one-carbon units for other biosynthetic reactions.*

## Cysteine Biosynthesis

The biosynthesis of L-cysteine consists essentially of a sulfhydryl transfer to an activated form of serine. The form of the sulfhydryl group actually used in cells is still unclear. For most plants and microorganisms, the sulfur source is sulfate, which must be reduced to the level of sulfide. This reduction is complex and poorly understood and will not be considered in this section. For some organisms, it appears that the initial sulfhydryl transfer is made not to a serine moiety, but to an activated form of homoserine, a methionine precursor. Transfer of the sulfur to cysteine would then occur by a transsulfuration pathway.

**Direct Sulfhydrylation Pathway.** The direct sulfhydrylation pathway to L-cysteine has been most thoroughly studied in *E. coli* and the related *S. typhimurium* and in *Chlorella*. The initial step, which is inhibited by cysteine, is a transfer of the acetyl group of acetyl-CoA to serine to yield O-acetylserine (Figure 22-10). The reaction is catalyzed by serine transacetylase. The formation of cysteine itself is catalyzed by O-acetylserine sulfhydrylase, a reaction in which O-acetylserine serves as a β-alanyl donor. In vitro, the enzyme will react with $H_2S$ as the β-alanyl acceptor. However, the experiments of Tsang and Schiff with *E. coli*, yeast, and *Chlorella* suggest that unless $H_2S$ or sulfite are supplied in the medium, a more likely sulfhydryl donor is a carrier-bound sulfide such as thioredoxin-S-S$^-$ (see Chapter 19) in *E. coli*. As will be discussed later, the carrier-bound sulfide is thought to be formed from a carrier-bound thiosulfonate by the enzyme traditionally identified as a sulfite reductase. Both possibilities are represented in Figure 22-10.

**The Transsulfuration Pathway.** The direct sulfhydrylation of O-acetylserine to yield cysteine is probably the most common mechanism for sulfide incorporation. In some forms it is possible that the major, if not the sole, mechanism for sulfur incorporation is by means of homocysteine synthase, an enzyme considered in the section L-Methionine Biosynthesis. Under such conditions, cysteine formation occurs by transsulfuration, with the intermediate formation of L,L-cystathionine (Figure 22-10). Cystathionine biosynthesis from serine and homocysteine is presumably catalyzed as a simple condensation by cystathionine-β-synthase. The

**Figure 22-10**
The biosynthesis of cysteine by direct sulfhydrylation and by a transsulfuration route in which the sulfur is derived from homocysteine. The direct sulfhydrylation pathway (*a*) is indicated as occurring either with $H_2S$ or a carrier-bound form of sulfide as the source of sulfur. The trans-sulfuration pathway (*b*) passes through homocysteine. Two routes leading to homocysteine are shown. The first starts with L-methionine and proceeds through reactions described in Figure 22-16. The second route involves the homocysteine synthase–catalyzed conversion of *O*-acetyl-L-homoserine as shown.

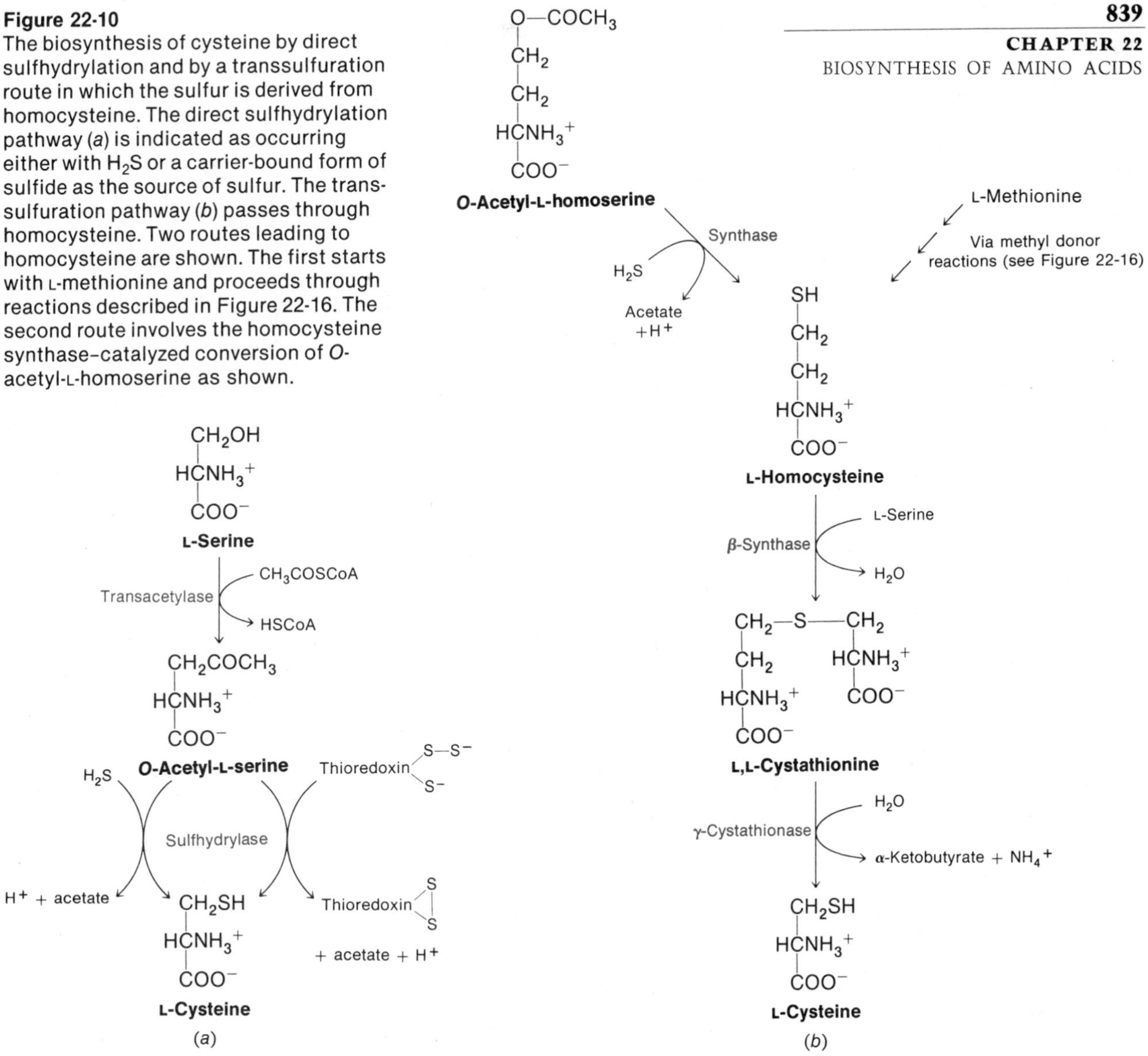

L-Serine

L-Cysteine
(*a*)

O-Acetyl-L-homoserine

L-Homocysteine

L,L-Cystathionine

L-Cysteine
(*b*)

cleavage of cystathionine to yield cysteine, $\alpha$-ketobutyrate, and $NH_4^+$ is catalyzed by $\gamma$-cystathionase, a pyridoxal phosphate-containing enzyme. This transsulfuration pathway from methionine will be considered again in the next chapter as one route for methionine catabolism.

**The Reduction of Sulfate.** The eight-electron reduction of $SO_4^{2-}$ to $H_2S$ is only poorly understood. Initially, an activation of sulfate (Figure 22-11) is necessary. The first step, catalyzed by adenylylsulfate pyrophosphorylase, yields an "active" form of sulfate, adenosine-5'-phosphosulfate (APS). Further activation is achieved by phosphorylation of the 3'-OH by APS kinase to yield 3-phosphoadenosine-5'-phosphosulfate (PAPS) with ATP as the

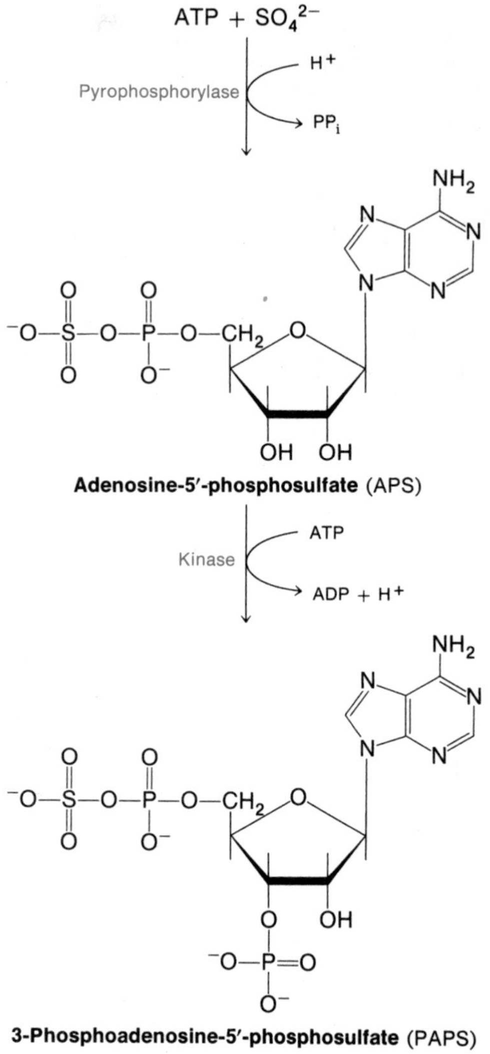

$$ATP + SO_4^{2-}$$

Pyrophosphorylase

H$^+$

PP$_i$

**Adenosine-5′-phosphosulfate** (APS)

Kinase

ATP

ADP + H$^+$

**3-Phosphoadenosine-5′-phosphosulfate** (PAPS)

**Figure 22-11**
Formation of 3′-phosphoadenosine-
5′-phosphosulfate, an activated intermediate in sulfate reduction.

donor. From this point on, the precise mechanism of sulfate reduction is
unclear. It is, however, being actively investigated.

Control of the conversion of sulfate to sulfide may occur in *E. coli* by
an inhibition by cysteine of the active transport of sulfate into the cell.

## THE BIOSYNTHESIS OF AMINO ACIDS OF THE ASPARTATE FAMILY

The formation of the aspartate family of amino acids and the conversion of
aspartate to the pyrimidine nucleotides account for an even more significant drain of carbon from the tricarboxylic acid cycle in organisms such as
*E. coli* than does that of the α-ketoglutarate family. In addition, the nitrogen of aspartate is used in both the formation of inosinate and its conversion to adenylate and in the conversion of citrulline to arginine.

## Aspartate Biosynthesis

L-Aspartate is formed from oxaloacetate in a transamination reaction with glutamate as the amino donor (Figure 22-12). In *E. coli*, the major protein exhibiting aspartate-glutamate transaminase activity is transaminase A, an enzyme that also exhibits activity with the aromatic amino acids but not with leucine. A minor aspartate-glutamate transaminase activity is exhibited by a "tyrosine-repressible" protein in *E. coli* that also can react with the aromatic amino acids and leucine (see the following discussion).

## Asparagine Biosynthesis

In most organisms it is likely that the formation of L-asparagine occurs by an ATP-dependent transfer of the amide group of glutamine to the $\beta$-carboxyl of aspartate by asparagine synthase (Figure 22-12). Some organisms have a second kind of asparagine synthase in which the nitrogen donor is specifically $NH_4^+$. In both reactions, the ATP is cleaved to AMP and $PP_i$. Thus the basic mechanism of amidation is probably the same in both reactions and different from that catalyzed by glutamine synthase. It probably involves $\beta$-aspartyladenylate as an enzyme-bound intermediate. The two asparagine synthases would differ in that one would have a high affinity for $NH_4^+$ and the other would involve amide-group cleavage to yield an enzyme-bound $NH_3$ comparable to many other glutamine-dependent amido-transferases.

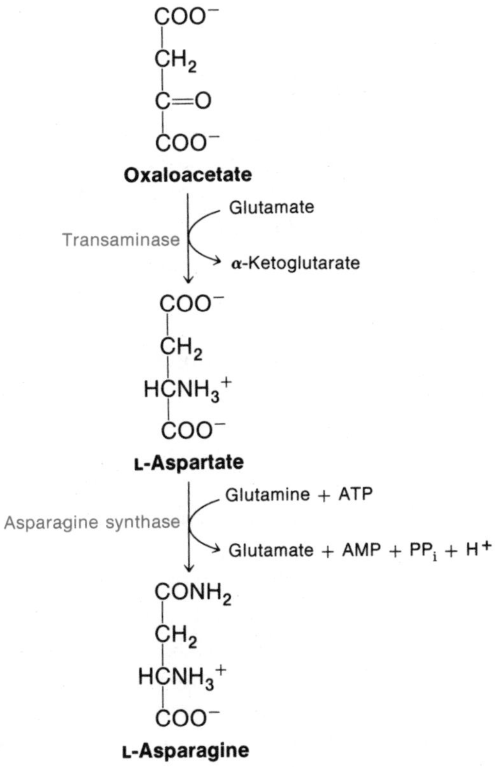

**Figure 22-12**
The biosynthesis of aspartate and asparagine. The more prevalent transamidation reaction is shown. In some organisms, an ammonia-dependent asparagine synthase has been found. That enzyme also catalyzes a pyrophosphate cleavage of ATP.

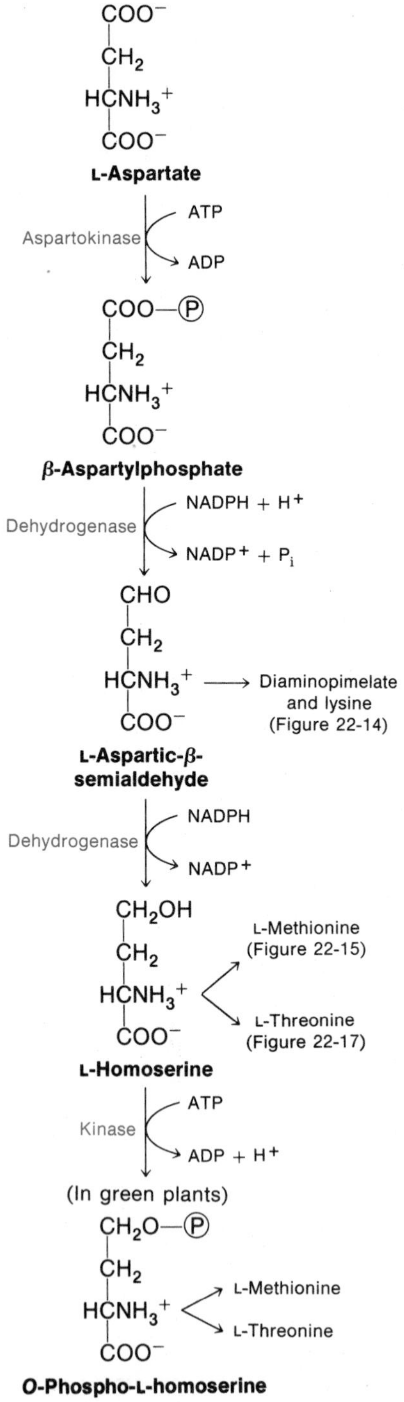

COO$^-$
|
CH$_2$
|
HCNH$_3^+$
|
COO$^-$

**L-Aspartate**

Aspartokinase ⟨ ATP → ADP

COO—Ⓟ
|
CH$_2$
|
HCNH$_3^+$
|
COO$^-$

**β-Aspartylphosphate**

Dehydrogenase ⟨ NADPH + H$^+$ → NADP$^+$ + P$_i$

CHO
|
CH$_2$
|
HCNH$_3^+$ ⟶ Diaminopimelate and lysine (Figure 22-14)
|
COO$^-$

**L-Aspartic-β-semialdehyde**

Dehydrogenase ⟨ NADPH → NADP$^+$

CH$_2$OH
|
CH$_2$
|
HCNH$_3^+$ ⟨ L-Methionine (Figure 22-15) / L-Threonine (Figure 22-17)
|
COO$^-$

**L-Homoserine**

Kinase ⟨ ATP → ADP + H$^+$

(In green plants)

CH$_2$O—Ⓟ
|
CH$_2$
|
HCNH$_3^+$ ⟨ L-Methionine / L-Threonine
|
COO$^-$

**O-Phospho-L-homoserine**

**Figure 22-13**
The common aspartate family pathway. In some green plants, O-phosphohomoserine rather than homoserine itself is the point at which methionine and threonine biosynthesis diverge.

## The Common Aspartate Family Pathway

In the conversion of L-aspartate to L-lysine, L-methionine, and L-threonine, a reduction of the β-carboxyl group is necessary. As in the case of the reduction of the glutamate γ-carboxyl group or in the conversion of 3-phosphoglycerate to glyceraldehyde-3-phosphate, an activation by ATP is needed (Figure 22-13). The reaction is catalyzed by aspartokinase.

The NADPH-dependent reduction of β-aspartyl phosphate by aspartic-β-semialdehyde dehydrogenase has been demonstrated in yeast. However, owing to the lability of β-aspartyl phosphate, the reaction is usually studied in the reverse of the biosynthetic direction. The reverse reaction is a phosphate-dependent reduction of NADP in the presence of aspartic-β-semialdehyde.

*Aspartic-β-semialdehyde* is a branch-point compound from which *lysine* biosynthesis in plants and bacteria proceeds (Figure 22-1). The common pathway is longer for *methionine* and *threonine* biosynthesis. In most forms, *homoserine* is the branch-point compound, although in some green plants still another step, the formation of homoserine phosphate, is required before the routes to methionine and threonine diverge. Homoserine itself is formed by the reduction of aspartic-β-semialdehyde by homoserine dehydrogenase. Whereas some homoserine dehydrogenases use either NADPH or NADH in vitro, others appear to be specific for one or the other. It would be anticipated that NADPH would more likely serve as the hydrogen donor in bacteria and in those eukaryotic cells in which the enzyme is not sequestered from the mitochondrial electron-transport system.

The intricacies of the control of carbon flow over the common aspartate family pathway will be considered after the specific branches have been described.

## The Pathway to L-Lysine in Bacteria and Plants

As mentioned earlier, the plant and bacterial pathway to L-lysine branches from the common aspartate family pathway with the appearance of aspartic-β-semialdehyde (as shown in Figure 22-14). The first specific step is a condensation of aspartic-β-semialdehyde and pyruvate with the elimination of water to yield a cyclic compound, 2,3-dihydrodipicolinate. (Thus lysine could be looked upon as a member of both the aspartate family and the pyruvate family of amino acids.) The enzyme dihydrodipicolinate synthase is inhibited by lysine. 2,3-Dihydrodipicolinate is itself a branch-point compound in the spore-forming bacilli and clostridia, since it is oxidized to *dipicolinate*, an essential constituent of bacterial spores. In at least some bacilli, the condensing enzyme is desensitized to lysine inhibition at the onset of sporulation.

2,3-Dihydrodipicolinate is reduced by an NADPH-requiring reductase to $\Delta^1$-piperideine-2,6-dicarboxylate (2,3,4,5-tetrahydrodipicolinate). This cyclic compound is then hydrolytically opened and maintained in a straight-chain form by succinylation with succinyl-CoA as the donor in some organisms, such as *E. coli*, to yield *N*-succinyl-ε-keto-L-α-aminopimelate. In other organisms, an acetyl group serves the same function. The ε-keto group is aminated by a specific transaminase to yield *N*-α-succinyl-L,L-α-ε-diaminopimelate. Cleavage by succinyl-diaminopimelate succinylase yields L,L-α-ε-diaminopimelate.

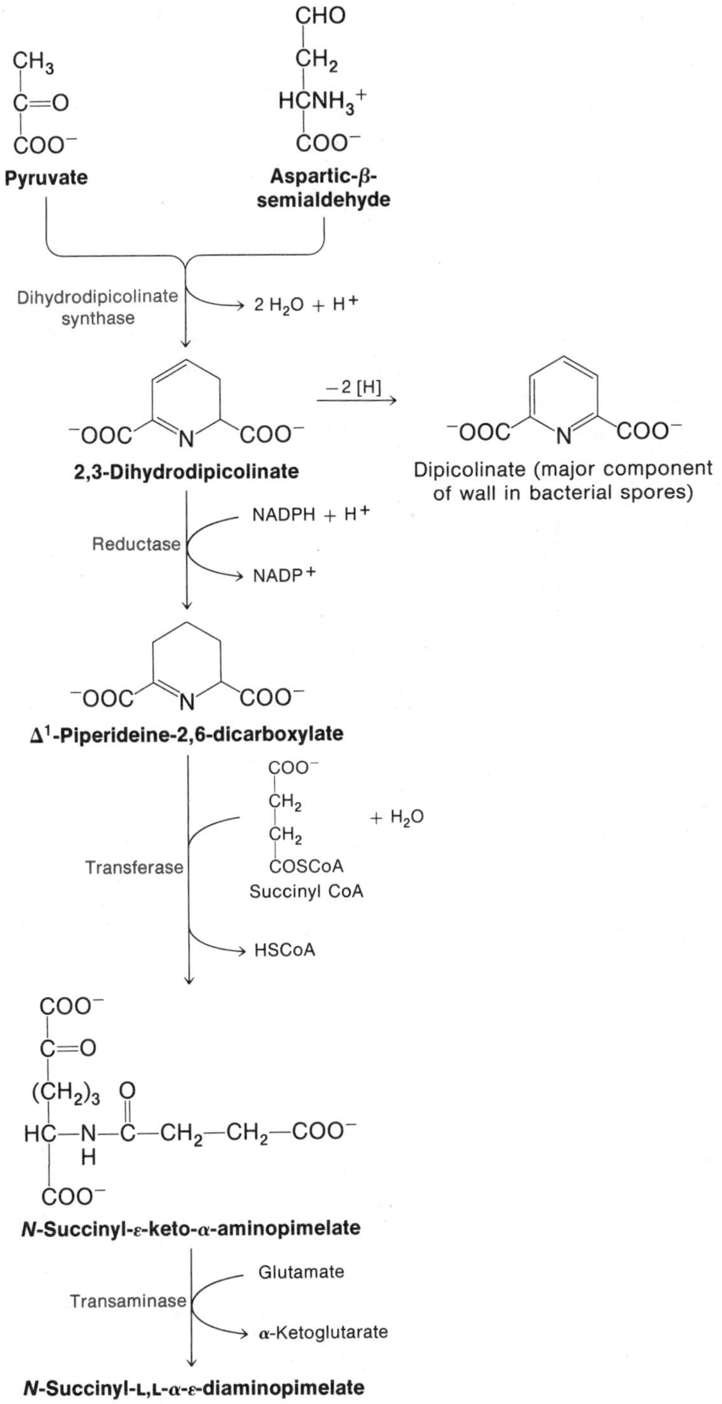

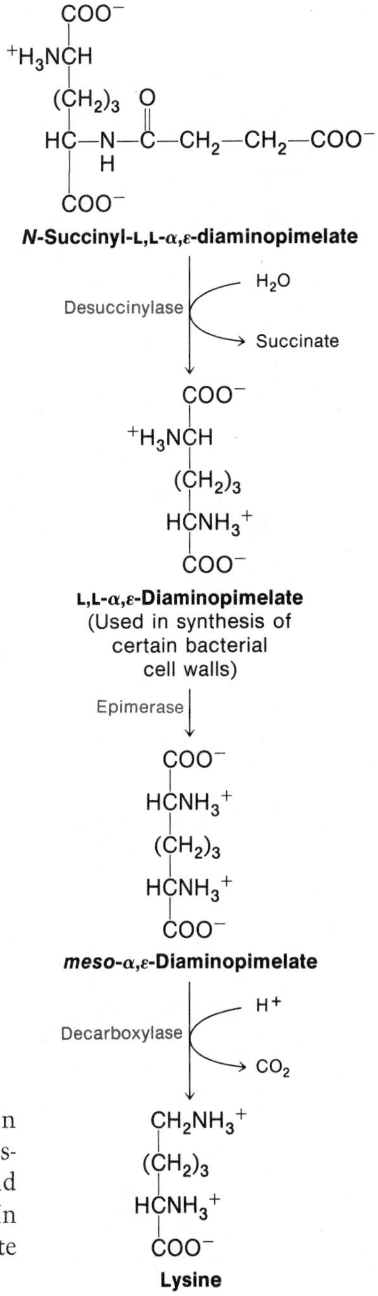

**Figure 22-14**
The biosynthesis of lysine in bacteria and in higher plants.

In bacteria, <u>diaminopimelate</u> is also a branch-point compound, since in many bacteria it is incorporated into the <u>peptidoglycan</u>, a highly cross-linked polymer of several amino acids (including some D-amino acids) and two amino sugars that form the rigid sacculus of the bacterial cell wall. In some bacteria, lysine itself is in the cell wall, instead of diaminopimelate (see Chapter 12).

The *meso* form of diaminopimelate is the substrate for diaminopimelate decarboxylase, the enzyme that forms lysine. In some bacteria, the decarboxylase is inhibited by lysine. Since lysine is formed by way of an L,L intermediate, it would be expected that one-half the carboxyl carbon of lysine is derived from the carboxyl carbon of pyruvate and one-half from the α-carboxyl of aspartate.

## Methionine Biosynthesis

The conversion of L-homoserine to L-methionine occurs in more than one way. One route, used in some bacteria, consists of a succinylation of the hydroxyl group of homoserine, a condensation with cysteine to yield cystathionine, cleavage to homocysteine, and methylation to yield methionine (see Figure 22-15). There are several variations to be found in the acylation reactions.

Although methionine is itself an end product of the pathway and is incorporated into protein, another important biosynthetic intermediate is *S-adenosylmethionine* (SAM). It is this intermediate that serves as a methyl donor in many reactions and as a precursor of the propylamine groups in spermidine and spermine, and in some organisms, it participates in the control of the pathway leading from homoserine to methionine (see

**Figure 22-15**
The biosynthesis of methionine. The acyl group employed to activate homoserine varies among different organisms. When a phosphoryl group is used, the intermediate, *O*-phosphohomoserine, is a branch-point compound that is converted to either threonine or methionine (see Figures 22-13 and 22-17.) The direct sulfhydrylation of activated homoserine by either $H_2S$ or carrier-bound sulfide found in some forms is indicated by a broken line. Finally, the methylation of homocysteine is shown to occur by either a cobalamin-dependent or a cobalamin-independent route.

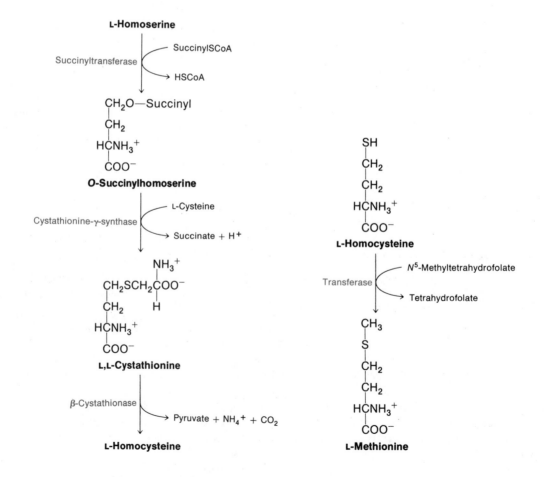

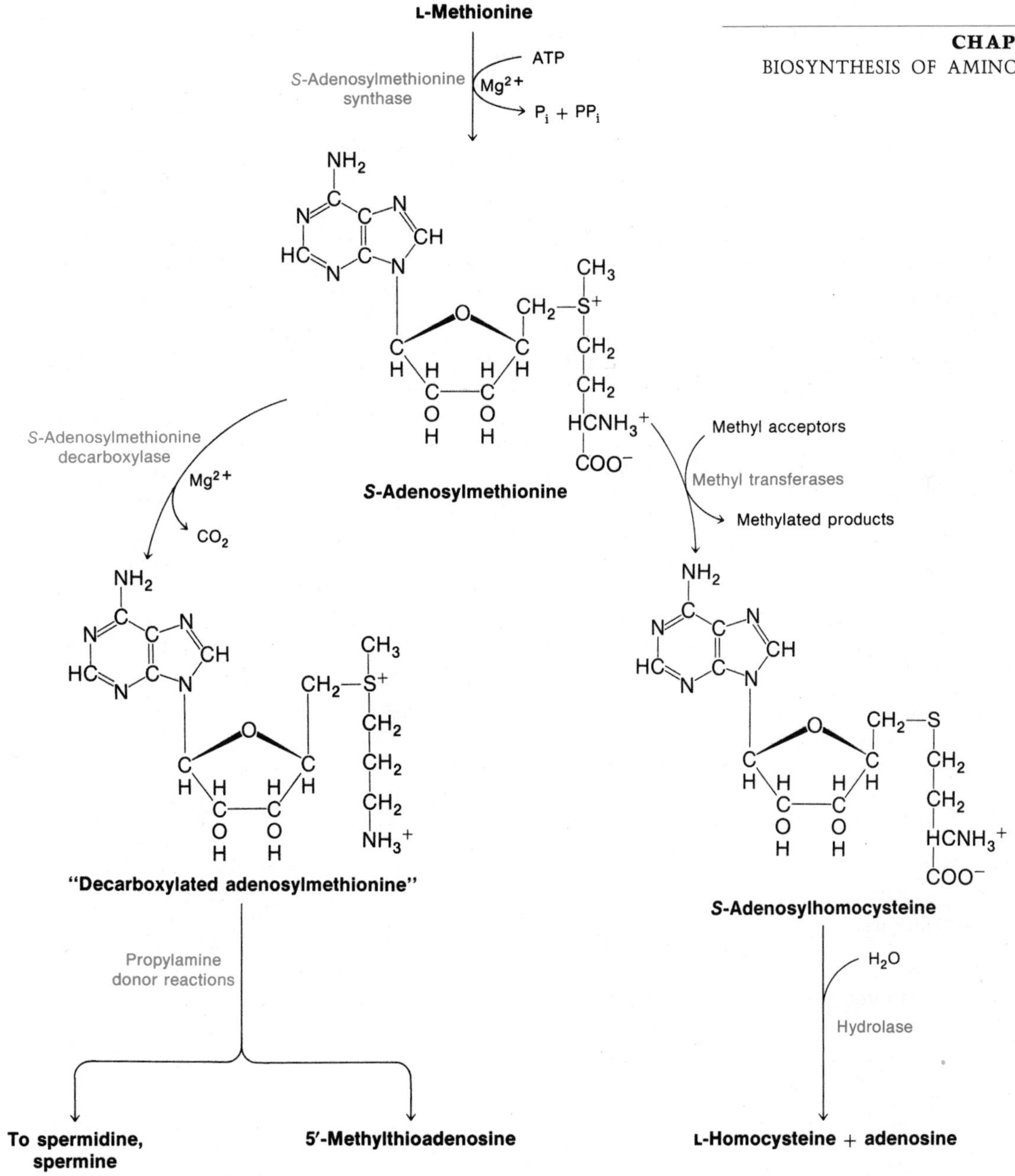

**Figure 22-16**
The formation of "active" methionine, S-adenosylmethionine (SAM), the donor of methyl and propylamine groups in biosynthetic reactions.

below). Thus adenosylation of methionine by SAM synthase serves to activate methionine for either methyl group or, with a decarboxylase, propylamine transfers. As Figure 22-16 shows, the activation is unusual in that all three phosphate groups of ATP are cleaved in the adenosylation reaction. In the methyl donor reaction, S-adenosylhomocysteine is liberated by one of several methyltransferases and then cleaved to homocysteine and adenosine by S-adenosylhomocysteine hydrolase. In the reactions in which a propylamine group is transferred to putrescine or to spermidine, methylthioadenosine is liberated.

**L-Homoserine**

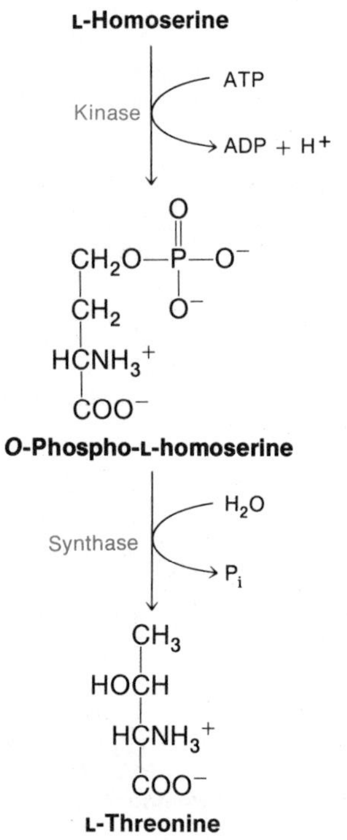

**Figure 22-17**
The biosynthesis of threonine.

## Threonine Biosynthesis

For the formation of L-threonine from L-homoserine, the —OH group must be transferred from the $\gamma$ to the $\beta$ carbon, as shown in Figure 22-17. This is done indirectly by phosphorylation of the hydroxyl group of homoserine by homoserine kinase. For most organisms that form threonine, the kinase reaction is the first specific step in the pathway, and the reaction is inhibited by threonine.

As indicated in Figure 22-15, in some green plants, O-phosphohomoserine is a branch-point compound, being a common intermediate for both methionine and threonine. *In some cases at least, the one-step route to threonine is regulated not by the end product, i.e., threonine, but by the "need" for the competing reaction* (transsulfuration to cystathionine), for which O-phosphohomoserine is also a substrate. Thus threonine synthase is inhibited by L-cysteine (the cosubstrate for transsulfuration) and stimulated by S-adenosylmethionine (a product derived ultimately by means of transsulfuration). *This general pattern might be sought in other systems when the more "typical" pattern of end product inhibition is not found.*

**The Formation of Glycine and Serine from Aspartate.** In *C. pasteurianum*, the carbon of aspartate is found to be heavily incorporated into both glycine and serine, indicating the possibility that these amino acids were members of the aspartate family in this organism. Examination of extracts of the organism failed to reveal any phosphoserine phosphatase, an essential enzyme for the conversion of 3-phosphoglycerate to L-serine described earlier. The extracts did contain an L-threonine aldolase, an enzyme that cleaves threonine to acetaldehyde and glycine. The enzyme will be discussed further in the next chapter, where it will be considered along with other enzymes of threonine catabolism. *It is interesting, however, that an enzyme that serves a catabolic function in some organisms has been mobilized to serve a biosynthetic role in C. pasteurianum.*

The incorporation of aspartate carbon into serine would be readily explained by the freely reversible serine hydroxymethylase, which in most organisms forms glycine from serine, but which, when glycine is supplied, catalyzes the reverse reaction.

## Isoleucine Biosynthesis

L-*Isoleucine is usually considered a member of the aspartate family*, since in most bacteria and plants four of its six carbons are derived from aspartate and only two from pyruvate. However, *since four of the five enzymes in isoleucine biosynthesis are also involved in valine biosynthesis, isoleucine biosynthesis is more appropriately considered with the pyruvate family of amino acids.*

## Regulation of Carbon Flow in the Common Aspartate Family Pathway

The unidirectional flow of carbon into the common *aspartate family pathway must be regulated in a way that provides ample amounts of the branch-point compounds aspartic-$\beta$-semialdehyde* (except in fungi) *and homoserine* (or O-phosphohomoserine in most green plants) *but does not* *lead to oversynthesis of either.* Since in most forms there appear to be negligible pools of either branch-point compound (perhaps only those amounts bound to product or substrate sites of the respective enzymes), feedback control cannot be exerted by these intermediates. Rather, in one way or another, *control must be exerted by the end products* (lysine, methionine, and threonine), which do accumulate in measurable pools.

Although an essentially common pathway has evolved for the formation of the aspartate family precursors, the patterns of feedback control have been found to vary considerably. They are of two general kinds, with considerable specific variation from the two basic patterns. In one, there are single aspartokinases, but the control of the enzyme is by *multivalent* or *synergistic inhibition*. The other basic pattern is one of *multiple aspartokinases*, with each being controlled differently.

The most extensively studied pattern is that of the enteric bacteria, as exemplified by *E. coli*. In *E. coli*, there are three different proteins with aspartokinase activity. One enzyme is inhibited by threonine. The enzyme is part of a protein that also exhibits homoserine dehydrogenase activity, which is also sensitive to inhibition by threonine. The enzyme is called aspartokinase I-homoserine dehydrogenase I. Another distinctly different protein, aspartokinase II-homoserine dehydrogenase II, is not inhibited by any of the multiple end products of the aspartate family pathway, but its synthesis is repressed by methionine (as are other methionine biosynthetic enzymes). The synthesis of aspartokinase I-homoserine dehydrogenase I is repressed by a combination of two end products, threonine and isoleucine (multivalent repression). The molecular mechanism of this repression will be considered later in the chapter.

The third aspartokinase in *E. coli* is inhibited by lysine. The inhibition is enhanced synergistically by phenylalanine, leucine, or, to a lesser extent, methionine. Any physiological advantage that results from this synergistic inhibition has not been explained. Unlike the other two aspartokinases, no other activity is associated with the protein. Its synthesis is also repressed by lysine.

It should be emphasized that the intermediates of the common aspartate family pathway are not being channeled into the branches leading to lysine, methionine, and threonine, but rather, there is probably a common pool of intermediates from which materials needed for the three specific pathways are drawn. In some strains of *E. coli*, lysine-sensitive aspartokinase III is the predominant aspartokinase, and there is very little of the methionine-repressible aspartokinase II-homoserine dehydrogenase II in an amino acid-free medium. Under conditions of strong inhibition and repression (i.e., repressed synthesis of the enzyme) of aspartokinase I-homoserine dehydrogenase I and of aspartokinase III, a starvation for methionine would be prevented by a derepression (i.e., increased synthesis of the enzyme) of the aspartokinase II-homoserine dehydrogenase II.

*In those organisms in which there are single aspartokinases, it is common to find that they are multivalently inhibited by lysine and threonine.* Usually lysine or threonine alone are weakly inhibitory, so that the pattern is actually a strongly synergistic one. In such organisms there is only a single homoserine dehydrogenase, and it is usually inhibited by threonine alone. In some cases, the addition of lysine and threonine is strongly inhibitory to growth, with the growth inhibition being reversed by methionine. It may be that aspartokinase II-homoserine dehydrogenase II of *E. coli* provided one means of avoiding this complication. In other organisms the inhibition is less severe, and the proteins are relatively insensitive to the feedback inhibitors after growth in the presence of the inhibitory amino acids. The physical basis of the desensitization is unknown.

The precise patterns of regulation of synthesis of amino acids of the aspartate family vary widely, and in few microorganisms or plants have the studies been as thorough as they have in *E. coli* and related organisms, which appear to have rather a unique pattern of regulation.

## THE BIOSYNTHESIS OF AMINO ACIDS OF THE PYRUVATE FAMILY

The pyruvate family of amino acids consists of L-alanine, L-valine, and L-leucine. In addition, pyruvate contributes two carbons to isoleucine and, on the average, two and one-half carbons to lysine. As mentioned earlier, the biosynthesis of isoleucine, a member of the aspartate family, is formed by a pathway that parallels that of valine. For this reason, the biosynthesis of isoleucine will be considered here. There are two additional ways that have been evolved to form isoleucine, and these will not be considered in this section.

### Alanine Biosynthesis

The formation of L-alanine occurs by a transamination reaction with glutamate as the amino donor and pyruvate as the acceptor [glutamate-alanine transaminase (Figure 22-18)]. Alanine may be formed by several different transaminases in some bacteria, such as *E. coli*, since no single-step mutants have ever been isolated that are unable to form alanine. There is no feedback control over alanine formation, and in many forms large intracellular pools of alanine are found unless the nitrogen supply is restricted. However, since the transaminases catalyze completely reversible reactions, this accumulation of alanine does not effect a drain on the supply of pyruvate.

### Isoleucine and Valine Biosynthesis

In all forms, *the four enzymes required for valine biosynthesis are required for the last four, parallel steps in isoleucine biosynthesis* (Figure 22-19). The first step in valine biosynthesis is a condensation between pyruvate and "active" acetaldehyde (probably hydroxyethyl thiamine pyrophosphate) to yield α-acetolactate. The enzyme acetohydroxyacid synthase requires thiamine pyrophosphate. The enzyme usually has a requirement for FAD, which, in contrast to most flavoproteins, is rather loosely bound to the protein. The same enzyme transfers the acetaldehyde group to α-ketobutyrate, yielding α-aceto-α-hydroxybutyrate, the isoleucine precursor. Since α-ketobutyrate, unlike pyruvate, is not one of the key intermediates in any of the central metabolic routes, it is necessary that a specific pathway to ketobutyrate be present.

For nearly all plants, fungi, and bacteria, the normal route to α-ketobutyrate is that from aspartate by way of threonine, which in turn is deaminated to α-ketobutyrate. The enzyme threonine deaminase contains pyridoxal phosphate and functions as a dehydrase, presumably liberating

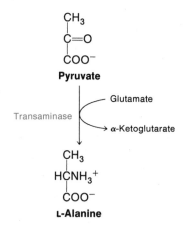

**Figure 22-18**
The biosynthesis of alanine.

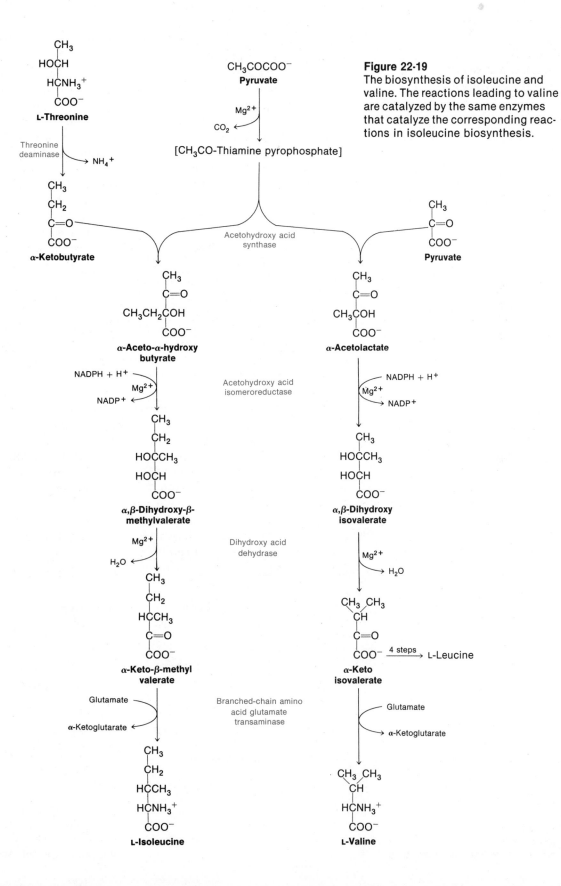

**Figure 22-19**
The biosynthesis of isoleucine and valine. The reactions leading to valine are catalyzed by the same enzymes that catalyze the corresponding reactions in isoleucine biosynthesis.

$\alpha$-aminocrotonate, which upon rearrangement to $\alpha$-iminobutyrate spontaneously yields $\alpha$-ketobutyrate and ammonia.

Conversion of the acetohydroxy acids to the $\beta$-dihydroxy acid precursors of valine and isoleucine is a complex reaction catalyzed by acetohydroxy acid isomeroreductase. Since the enzyme will reduce the rearranged, unreduced isomer of $\alpha$-acetolactate ($\alpha$-keto-$\beta$-hydroxyisovalerate) and will, under certain circumstances, convert it back to acetohydroxybutyrate, it is assumed that isomerization precedes reduction. However, any intermediate in the overall reaction is presumably enzyme-bound. The $\alpha,\beta$-dihydroxy acids are both converted to the $\alpha$-keto acid precursors of valine and isoleucine by a dihydroxy acid dehydrase. Finally, the two amino acids are formed in transamination reactions in which glutamate is the amino donor (branched-chain amino acid-glutamate transaminase).

*The pathways to isoleucine and valine illustrate well the way studies with nutritionally deficient mutants (auxotrophs), isotope incorporation experiments, and enzymatic analysis have been used to decipher the biosynthetic pathways to the amino acids.*

*Early studies with mutants of both Neurospora and E. coli revealed that certain mutants, presumably altered in but a single gene, required not one, but two amino acids,* isoleucine and valine, an apparent contradiction to the one gene, one enzyme concept. Examination of the mutants genetically and nutritionally revealed several classes of these doubly auxotrophic mutants. One class was found to accumulate material in the culture fluids that fed mutants of several other classes. Analysis revealed that the active material consisted of the $\alpha$-keto acid precursors of valine and isoleucine. One class of the mutants that responded to the keto acids accumulated material that fed other isoleucine and valine auxotrophs. This accumulated material was identified as $\alpha,\beta$-dihydroxyisovalerate and $\alpha,\beta$-dihydroxy-$\beta$-methylvalerate. Both these findings were followed by the demonstration of a lack of the branched-chain amino acid-glutamate transaminase in the class accumulating the $\alpha$-keto acids and the lack of what is now called dihydroxy acid dehydrase in the dihydroxy acid accumulators. *This loss of a single enzyme that catalyzes the corresponding step in both pathways thus accounted for the unexpected double auxotrophy.*

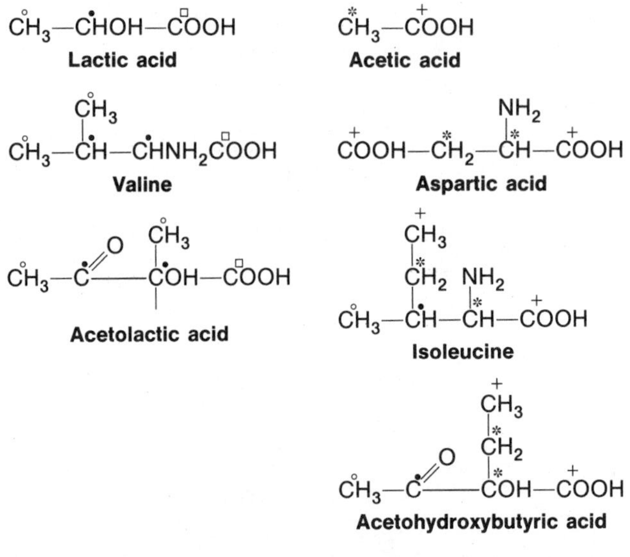

**Figure 22-20**
Incorporation of lactate and acetate carbon into valine and isoleucine.

The valine and isoleucine auxotrophs that responded to the dihydroxy acids did not feed any other class except one that appeared to be blocked only in isoleucine biosynthesis and which responded as well to $\alpha$-ketobutyrate as to $\alpha$-aminobutyrate. These singly blocked auxotrophs were later shown to lack threonine deaminase, the enzyme required only for isoleucine biosynthesis.

While these nutritional analyses were in progress, isotopic studies with *Neurospora* and yeast indicated that pyruvate carbons were being incorporated into valine and that exogenous nonradioactive aspartate, homoserine, threonine, and $\alpha$-ketobutyrate competed effectively with radioactive glucose for incorporation into isoleucine. The paradox was that when valine and isoleucine were degraded carbon by carbon after growth with either labeled lactate (metabolically equivalent to pyruvate) or labeled acetate, there was no sequence of three carbons in valine (see Figure 22-20) that could have arisen from pyruvate directly, nor was there any four-carbon sequence in isoleucine that was labeled with acetate carbon, as aspartate and threonine were labeled. It was proposed by Strassman and Weinhouse that a condensation of a two-carbon fragment derived from pyruvate with either pyruvate (to yield acetolactate) or $\alpha$-ketobutyrate (to yield acetohydroxybutyrate) followed by an intramolecular migration could account for the isotopic distribution in valine and isoleucine. This postulated rearrangement was shortly followed by the demonstration of acetolactate accumulating in the culture fluids of mutants unable to reduce and rearrange the molecule and by the demonstration of the condensing enzyme that formed both acetohydroxy acids and the isomeroreductase that led to the formation of the dihydroxy acids.

The control of metabolite flow over the pathways to valine and isoleucine is subject to one complication. In many organisms, acetohydroxy acid synthase is inhibited by valine. The complete inhibition of this enzyme not only would prevent oversynthesis of valine, but also would prevent isoleucine biosynthesis, for which acetohydroxy acid synthase is the second enzyme in the pathway. It is perhaps for this reason that many organisms also contain a valine-insensitive acetohydroxy acid synthase. Such an enzyme allows isoleucine synthesis to occur in the presence of valine, but also potentially allows an oversynthesis of valine under some conditions. Control of the synthesis of isoleucine itself is achieved by an inhibition of threonine deaminase, the first enzyme in the pathway to isoleucine.

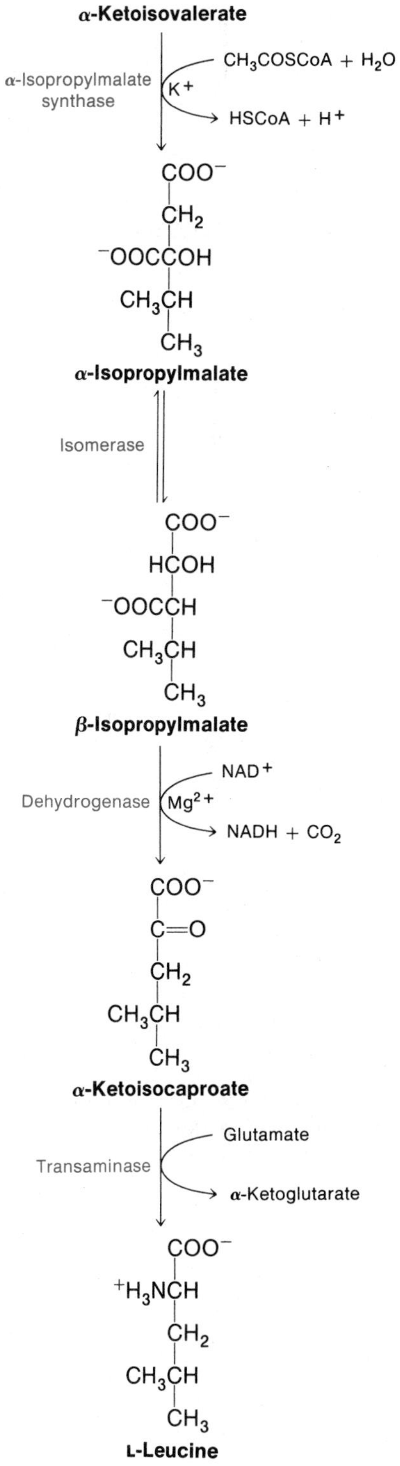

## Leucine Biosynthesis

The biosynthesis of L-leucine involves the lengthening of the carbon chain of $\alpha$-ketoisovalerate by one carbon to yield $\alpha$-ketoisocaproate (Figure 22-21). This lengthening is comparable to that by which $\alpha$-ketoglutarate is converted to $\alpha$-ketoadipate in the pathway to lysine in fungi (Figure 22-7).

## THE BIOSYNTHESIS OF THE AROMATIC FAMILY OF AMINO ACIDS

The aromatic amino acids *phenylalanine, tyrosine, and tryptophan* are all formed by means of a common pathway to build the benzene ring. This pathway, often referred to as the *shikimate pathway*, is also important for

**Figure 22-21**
The biosynthesis of leucine.

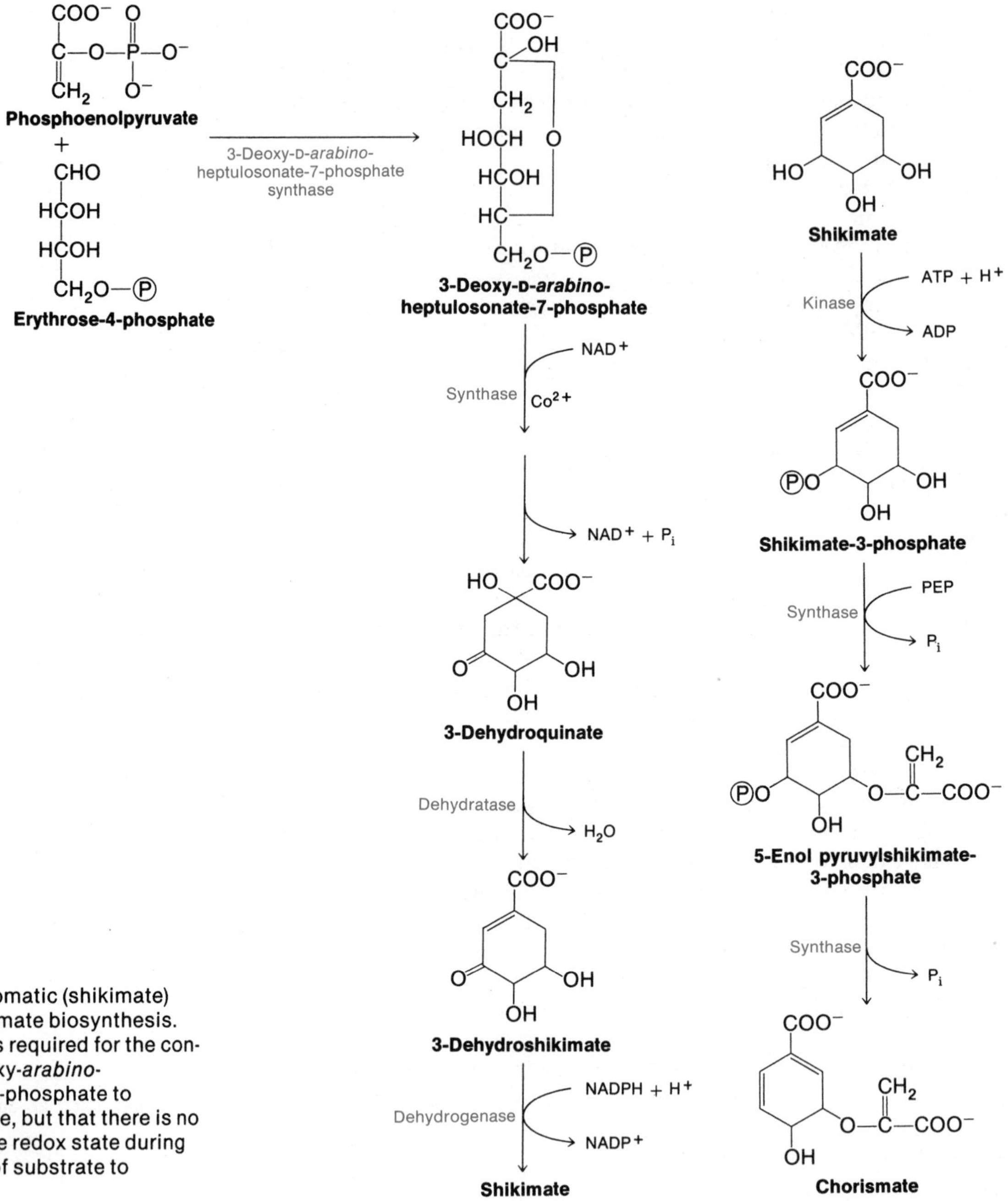

**Figure 22-22**
The common aromatic (shikimate) pathway: chorismate biosynthesis. Note that NAD is required for the conversion of 3-deoxy-*arabino*-heptulosonate-7-phosphate to 3-dehydroquinate, but that there is no net change in the redox state during the conversion of substrate to product.

the formation of the aromatic nuclei in or the aromatic precursor of *vitamins E and K, folic acid, ubiquinone, and plastoquinone* and certain metal chelators, such as *enterochelin*. The branch-point compound for all these diverse products is *chorismate*, which has a prearomatic cyclohexadiene nucleus.

# The Biosynthesis of Chorismate:
# The Common Aromatic Pathway

The overall route of chorismate synthesis beginning with the condensation of two intermediates from the central metabolic routes, phosphoenolpyruvate and erythrose-4-phosphate, is illustrated in Figure 22-22. Chorismate is the final product of the common pathways to aromatic compounds, including phenylalanine, tyrosine, and tryptophan.

The control of metabolite flow over the common pathway will be considered later, together with control over the branches to the aromatic amino acids.

## The Biosynthesis of Phenylalanine and Tyrosine

In some organisms, the pathways from chorismate to L-phenylalanine and L-tyrosine shown in Figure 22-23 are truly separate pathways, even though the first step from chorismate is the same for the two pathways. For example, in *E. coli* and related organisms, one protein, chorismate mutase P-prephenate dehydratase, converts chorismate to phenylpyruvate, with the intermediate, prephenate, being enzyme-bound. A second protein, the

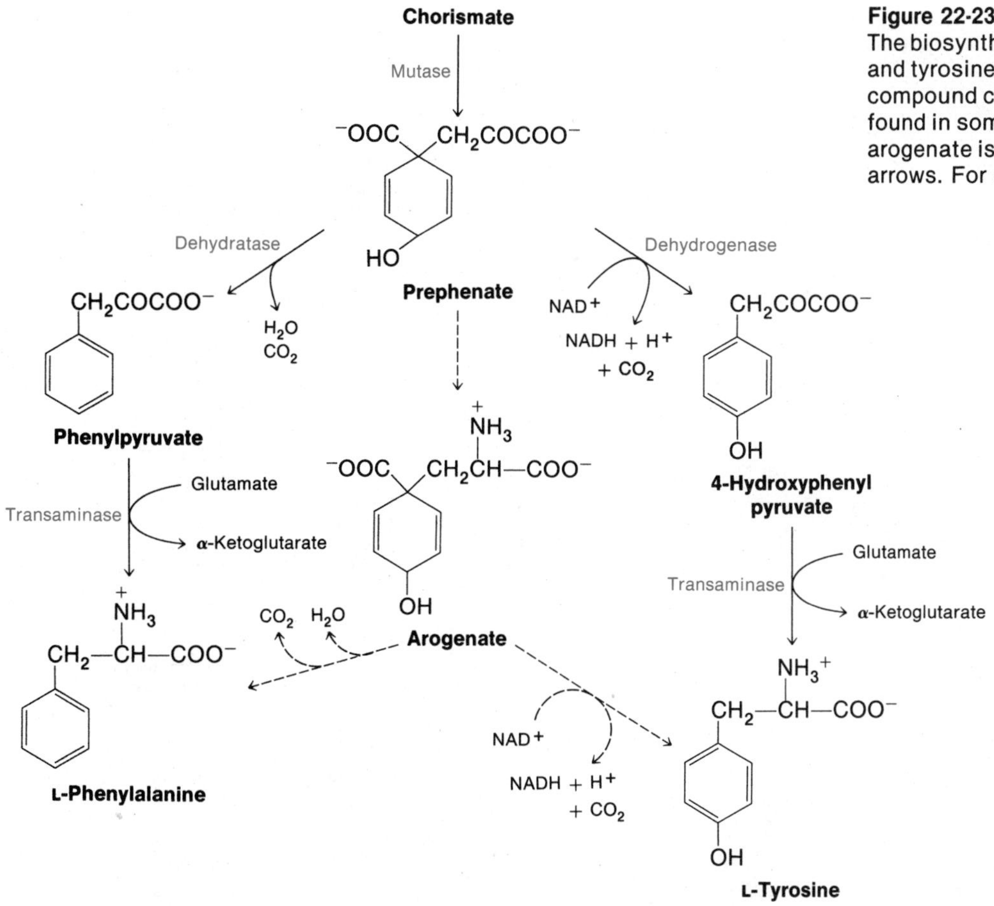

**Figure 22-23**
The biosynthesis of phenylalanine and tyrosine from the branch-point compound chorismate. The pathway found in some organisms by means of arogenate is indicated by broken arrows. For details, see text.

NAD-dependent chorismate mutase T-prephenate dehydrogenase, converts chorismate to 4-hydroxyphenylpyruvate, again with the formation of prephenate occurring as an enzyme-bound intermediate. Both enzymes utilize prephenate, the phenylalanine biosynthetic enzyme yielding phenylpyruvate and the tyrosine biosynthetic enzyme yielding 4-hydroxyphenylpyruvate. The final aromatization step, the removal of water from prephenate in phenylalanine biosynthesis or the removal of hydrogen in tyrosine biosynthesis, is accompanied by the loss of the ring carboxyl as $CO_2$.

### Tryptophan Biosynthesis

The pathway leading from chorismate to L-tryptophan (Figure 22-24) is the most thoroughly studied of any biosynthetic pathway. In *E. coli*, the details of the enzymatic steps, the correlation between DNA sequence and the

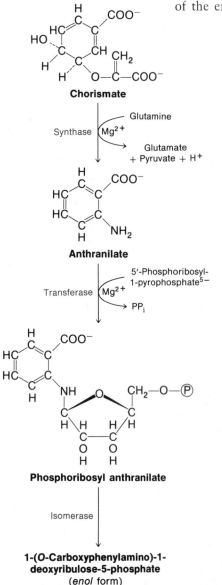

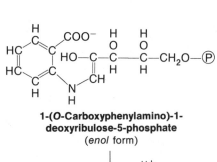

**Figure 22-24**
The biosynthesis of tryptophan from the branch-point compound, chorismate in *E. coli*.

protein products, and the factors controlling the transcription of the structural genes far exceed those of any other set of related genes. Although this had not been the work of any one group, the extensive gene-enzyme analysis of Yanofsky and coworkers has laid the foundation for others to explore details of some of the enzymatic steps by physical and kinetic approaches. Comparative studies in other bacteria and in fungi have revealed a variation upon the themes found in *E. coli*, particularly with respect to the distribution of the sequence of enzyme activities, which are identical in all forms, on one protein or another. In addition, these studies have also revealed differences in the way the genes are arranged in the DNA and in the way expression of those genes is controlled.

The first specific step in tryptophan biosynthesis is the glutamine-dependent conversion of chorismate to the simple aromatic compound *anthranilate*. Like most other glutamine-dependent reactions, the reaction can also occur with ammonia as the source of the amino group. However, the reaction is slower, and high concentrations of ammonia are required. Thus far, all the anthranilate synthases examined except that in *Euglena gracilis* have the glutamine amidotransferase activity (component II) and the chorismate to anthranilate activity (component I) on separate proteins. Component II by itself exhibits glutaminase activity under appropriate conditions, suggesting that its role is to transfer the amide group of glutamine to the $NH_3$ binding site of a component I.

Anthranilate is transferred to a ribose phosphate chain in a phosphoribosyl pyrophosphate–dependent reaction catalyzed by anthranilate phosphoribosyltransferase. *Phosphoribosylanthranilate* undergoes an *Amadori rearrangement*, in which the ribosyl moiety becomes a ribulosyl moiety. When this activity is exhibited as a separate protein, it is called phosphoribosylanthranilate isomerase. The product, 1-(O-carboxyphenyl-amino)-1-deoxyribulose-5′-phosphate, is cyclized to *indoleglycerol phosphate* by the removal of water and loss of the ring carboxyl by indoleglycerol phosphate synthase. The final step in tryptophan biosynthesis is a replacement reaction catalyzed by tryptophan synthase in which glyceraldehyde-3-phosphate is removed from indoleglycerol phosphate and the enzyme-bound indole is condensed with serine.

The distribution of these five enzyme activities on proteins is subject to considerable variation (Table 22-1). For example, in *E. coli*, indoleglycerol phosphate synthase catalyzes both the isomerization of phosphoribosylanthranilate and the cyclization step. Of particular interest is the occurrence on a single protein for nonconsecutive reactions in some cases. If in such cases the proteins were separate from each other in the cell, this arrangement, for example, in *Neurospora*, would necessitate the product of one reaction leaving the product site of one enzyme to be acted upon by another enzyme and then returning to the substrate site of a third enzyme on the same protein that exhibited the first enzyme activity. *The persistence of this arrangement during evolution makes attractive the idea that all the tryptophan biosynthetic enzymes exist in the cell as a single multienzyme (and multiprotein) complex.* However, if so, the complex must be quite labile, since individual gene products are so readily separated.

The tryptophan pathway provides another example in which nutritional studies with bacteria and biochemically defective mutants of *Neurospora* and *E. coli*, isotope incorporation studies, and finally, enzymatic anal-

**Table 22-1**
Distribution of Tryptophan Biosynthetic Enzyme Activities on Different Proteins in Bacteria and Fungi

| Organism | Anthranilate Synthase | | Phosphoribosyl-anthranilate Transferase | Phosphoribosyl-anthranilate Isomerase | Indole Glycerol-P Synthase | Tryptophan Synthase | |
|---|---|---|---|---|---|---|---|
| | CoI | CoII | | | | $\alpha$ | $\beta$ |
| *E. coli, S. typhimurium* | ⌐‒‒‒‒‒⌐ | | | | | | ⌐‒‒‒‒‒⌐ |
| *Seratia marcescens* | ⌐‒‒‒‒‒⌐ | | | | | | ⌐‒‒‒‒‒⌐ |
| *Pseudomonas putida* | ⌐‒‒‒‒‒⌐ | | | | | | ⌐‒‒‒‒‒⌐ |
| *Acinetobacter calcoaceticus* | ⌐‒‒‒‒‒⌐ | | | | | | ⌐‒‒‒‒‒⌐ |
| *B. subtilis* | ⌐‒‒‒‒‒⌐ | | | | | | ⌐‒‒‒‒‒⌐ |
| *N. crassa* | ⌐‒‒‒‒⌐⌐ | | | | | | |
| *S. cerevisiae* | ⌐‒‒‒‒⌐⌐ | | | | | | |

Note: ⌐____⌐ = covalent linkage; ⌐‒‒‒‒⌐ = obligatory association required for full activity.

yses have been exploited to reveal the steps in a biosynthetic pathway. For example, early studies with tryptophan-requiring organisms found in nature revealed that some could use indole (a compound known to be formed by the microbial degradation of tryptophan; see Chapter 23), while others could use anthranilate. Later, after Beadle and Tatum introduced the approach of studying metabolism with mutants of the bread mold *Neurospora*, tryptophan-requiring mutants of this organism were found that could use indole or either anthranilate or indole. Still later, similar mutants of *E. coli* were found, and mutants of both organisms were described that accumulated one or the other of these compounds. Clearly, *those mutants that grew on anthranilate or indole were blocked in some step before these compounds, and those that accumulated them were blocked in the step after them.*

Incorporation studies with isotopes showed that when anthranilate was converted to tryptophan, the carboxyl group of anthranilate was lost as carbon dioxide, but the nitrogen was retained. Because the enzymes in the tryptophan biosynthetic pathway have only a limited specificity, it was possible to substitute 4-methylanthranilate in *E. coli* extracts that could convert anthranilate to indole. This "nonisotope" label was conserved during the conversion to yield 6-methyl indole:

4-Methylanthranilate $\xrightarrow{C_2}$ 6-Methyl indole $+ CO_2$

It was thus clear that some two-carbon unit replaced the carboxyl carbon of anthranilate. Further studies with such *E. coli* extracts indicated that phosphoribosyl pyrophosphate was a good cosubstrate for the formation of indole from anthranilate. Fractionation of these extracts, as well as examination of mutants blocked between anthranilate and indole, revealed that an intermediate in this conversion was indole-3-glycerol phosphate. Extracts from one group of such mutants could not form indole-3-glycerol phosphate, while the other group could not convert it to indole and glyceraldehyde-3-phosphate. The latter group was found to accumulate the dephosphorylated derivative in culture fluids.

The two intermediates in the conversion of anthranilate to indole-3-glycerol phosphate, phosphoribosylanthranilate and 1-(*O*-carboxyphenyl-amino)-1-deoxyribulose-5-phosphate, were originally postulated to account for the involvement of phosphoribosyl pyrophosphate in indole-3-glycerol phosphate formation. Support for the postulate was obtained when the dephosphorylated derivative of the second of these intermediates was found in the culture fluids of certain bacterial tryptophan-requiring mutants. The corresponding derivative of the first intermediate has not been found, probably because of its instability. Indeed this compound, when formed in extracts, is rapidly broken down to anthranilate and ribose-5-phosphate.

For several years, indole, which was accumulated by some mutants and used to satisfy the tryptophan requirement by others, was considered as an intermediate in tryptophan biosynthesis. Such a role for indole would have been of interest, since it appeared to be an exception to a "rule of nature" that biosynthetic intermediates had to bear a charge. It was found that extracts of cells that utilized indole did indeed catalyze the condensation of indole with serine, and extracts of cells that accumulated indole catalyzed the cleavage of indole-3-glycerol phosphate to indole and glyceraldehyde-3-phosphate. Furthermore, *E. coli* mutants of these two classes clearly were affected in separate genes, i.e., *trpA* and *trpB*, respectively. However, the two products of these genes catalyzed their corresponding reactions faster when they were associated in a complex of the form $\alpha\beta_2\alpha$. The complex itself catalyzed the overall reaction

Indole-3-glycerol phosphate + serine $\longrightarrow$
$$\text{tryptophan} + \text{glyceraldehyde-3-phosphate}$$

faster than either of the separate reactions. The same was found with extracts of *Neurospora* in which the two partial reactions were catalyzed by the same protein and with which no evidence for indole as a free intermediate could be found. Thus it became clear that indole, while historically important in deciphering the pathway to tryptophan, occurs only as a bound intermediate, as implied in Figure 22-24.

### Regulation of Carbon Flow over the Common Aromatic Pathway and over the Specific Routes to Phenylalanine, Tyrosine, and Tryptophan

*Metabolite flow to tryptophan is controlled by inhibition of anthranilate synthase by tryptophan. Regulation of metabolite flow in phenylalanine and tyrosine biosynthesis varies from organism to organism, owing to the variety of enzyme patterns in the conversion of chorismate to the two amino acids.* In *E. coli* and related organisms, phenylalanine inhibits both activities of chorismate mutase P-prephenate dehydratase, while tyrosine inhibits only the mutase activity of chorismate mutase T-prephenate dehydrogenase.

There are two general patterns of control over the common aromatic pathway. One is that in *E. coli* and related organisms. The pattern is similar to that of the common aspartate family pathway of the same organism in that there are three isozymic deoxy-*arabino*-heptulosonate-7-phosphate synthases. Each is inhibited by one of the three aromatic amino acids. (There is, in addition, a tryptophan-specific repression of the tryptophan-sensitive enzyme, a tyrosine-specific repression of the tyrosine-sensitive enzyme, and a tryptophan plus phenylalanine-specific multivalent repression of the phenylalanine-sensitive enzyme.) As in the synthesis of the intermediates in the aspartate family common pathway, the three enzymes contribute to a common pool of deoxy-*arabino*-heptulosonate-7-phosphate that is drawn upon for all the compounds formed from chorismate. Indeed, in some strains, the phenylalanine-sensitive enzyme is predominant, while in others, the tyrosine-sensitive enzyme is predominant.

Another pattern is found in *B. subtilis*. The single deoxy-*arabino*-heptulosonate-7-phosphate synthase is carried on the same protein that exhibits chorismate mutase activity. The protein is complexed with another protein that exhibits shikimate kinase activity. Both the deoxy-*arabino*-heptulosonate-7-phosphate synthase activity and the shikimate kinase activity are inhibited by chorismate and prephenate, which may inhibit by virtue of binding to the substrate and product sites of the chorismate mutase. The chorismate mutase activity is inhibited by prephenate. Prephenate dehydratase is inhibited by phenylalanine, whereas prephenate dehydrogenase is inhibited by tyrosine. This overall pattern can be looked upon as an example of "sequential" feedback inhibition.

## THE BIOSYNTHESIS OF HISTIDINE

The pathway to histidine in all plants and bacteria involves the transfer of N-1 and C-2 of the adenine moiety of ATP to the ribose phosphate moiety of PRPP, as shown in Figure 22-25. This transfer is initiated by a condensation reaction between the two parent compounds. After a series of additional reactions involving ring opening, isomerization, and an amido-group transfer, the residue from the ATP molecule is released as 5-aminoimidazole-4-carboxamide ribotide. The latter, an intermediate in the purine nucleotide biosynthetic pathway (see Chapter 19), is then "recycled" to replace the ATP used in the condensation reaction. In this cyclic process, the 6-amino group of the adenine ring which had been derived from aspartate becomes the amide group of the 5-aminoimidazole-4-carboxamide ribotide. During purine biosynthesis, this amide group is also derived from aspartate and becomes N-1 of the purine ring. Thus there is no way to distinguish an adenosine derivative that has been formed directly by the de novo pathway from one that has been recycled from the histidine pathway.

Another interesting feature of the histidine pathway is the conversion of imidazoleglycerol phosphate to a carbonyl derivative that can undergo

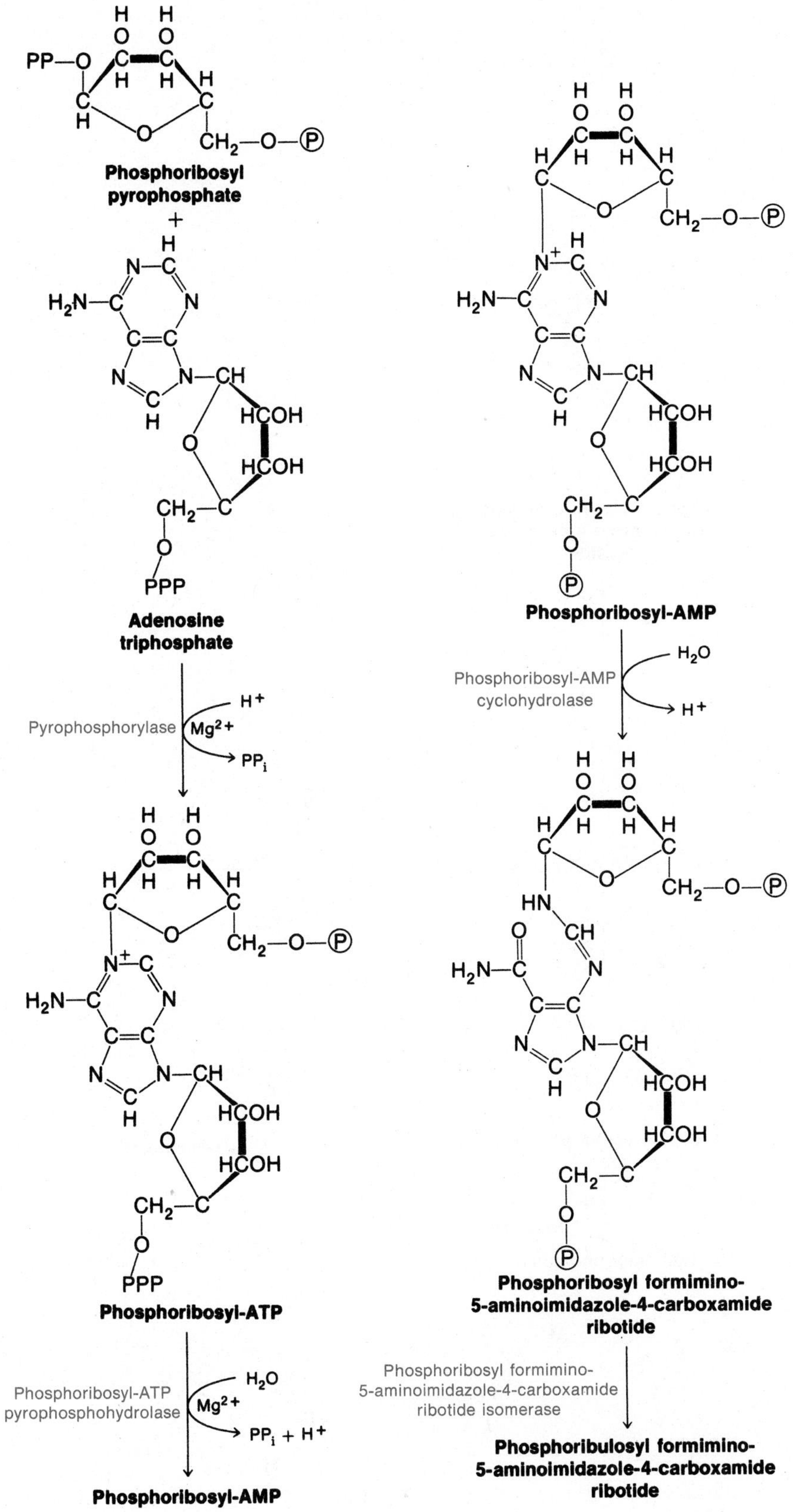

**Figure 22-25**
The biosynthesis of histidine. Note that the 5-aminoimidazole-4-carboxamide ribotide formed during the course of histidine biosynthesis is an intermediate in purine nucleotide biosynthesis, so that it can be readily regenerated to an ATP, thus replenishing that consumed in the first step in the histidine biosynthetic pathway. (*Continued*)

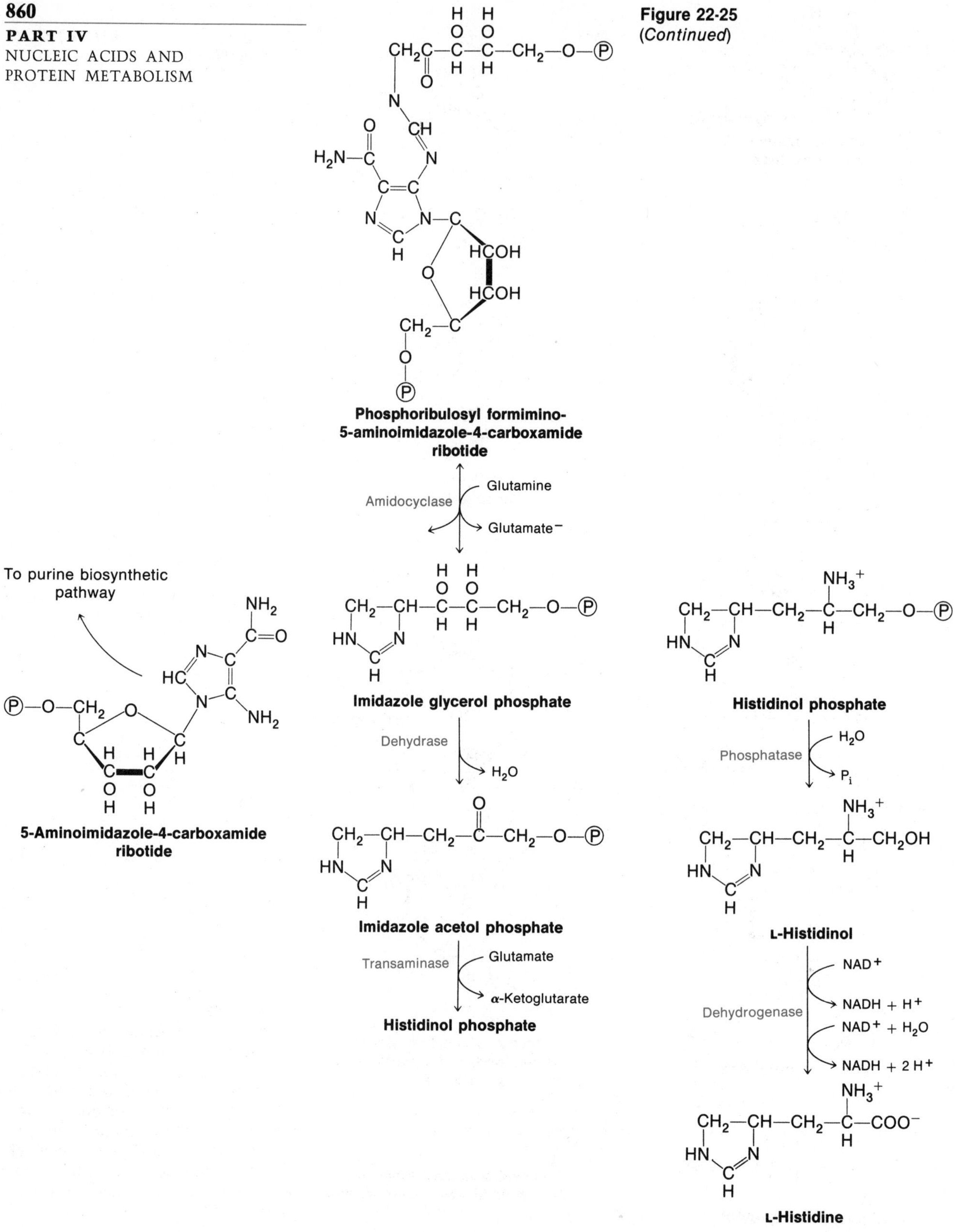

**Figure 22-25**
(*Continued*)

Phosphoribulosyl formimino-
5-aminoimidazole-4-carboxamide
ribotide

To purine biosynthetic
pathway

5-Aminoimidazole-4-carboxamide
ribotide

Amidocyclase — Glutamine → Glutamate⁻

Imidazole glycerol phosphate

Dehydrase → H₂O

Imidazole acetol phosphate

Transaminase — Glutamate → α-Ketoglutarate

**Histidinol phosphate**

Histidinol phosphate

Phosphatase — H₂O → Pᵢ

L-Histidinol

Dehydrogenase — NAD⁺ → NADH + H⁺, NAD⁺ + H₂O → NADH + 2 H⁺

L-Histidine

transamination. Following the incorporation of the α-amino group, which provides a positively charged molecule, the phosphate group is removed, thus illustrating the importance of charged groups on biosynthetic intermediates.

Histidine biosynthesis has been studied primarily in a few bacteria, *Neurospora*, and yeast. In several bacteria, the phosphatase and the dehydratase activities are carried on a single bifunctional protein. In yeast there is a single protein that exhibits the pyrophosphatase, cyclohydrolase, and dehydrogenase activities (the second, third, and tenth steps). These findings, along with the fact that significant "pools" of the intermediates are not found, raise the question of whether all the enzymes of histidine biosynthesis might not be arranged in a single complex, as has been suggested for the enzymes involved in tryptophan biosynthesis.

## THE REGULATION OF AMINO ACID BIOSYNTHESIS

The extent to which amino acid biosynthesis is controlled and the mechanisms that have evolved to effect the controls vary widely in the various pathways. They also vary widely for the same pathway in different plant and microbial systems. This variation in patterns of regulation results probably because the selection of control mechanisms is superimposed on and is secondary to the selection of the capacity to form a given amino acid. Thus the emergence of a variant route to an amino acid may make a preexisting control mechanism inefficient or even inoperative, and the emergence of a variant control mechanism may soon follow.

It would therefore not be possible in a short space to describe or even to catalog all the ways in which the biosynthesis of amino acids is controlled. In only the Enterobacteriaceae and specifically in *E. coli* and the closely related *S. typhimurium* has the regulation of all 20 amino acids been studied sufficiently to describe some details of the regulation. However, there have been enough studies with other bacteria, fungi, and plants to show that some of the *E. coli*-type control systems are far from being universally distributed. It is clear that no matter how well studied some regulatory interaction has been in *E. coli* or how elegant and potentially universal it seems to be, one should not conclude a priori that the same pattern is to be found in the pathway to any other amino acid in *E. coli* or in the same pathway in any other organism. It is against this background that the examples selected in this section should be viewed.

*Regulation of amino acid biosynthesis may occur at both the level of regulation of enzyme activity or metabolite flow over a pathway and at the level of regulation of enzyme amount.* The former will be considered first.

### Patterns of Control of Enzyme Activity

The production of an amino acid can be quenched by blocking any step within the pathway; the flow of metabolites into the pathway can be achieved only by blocking the first (usually irreversible) step that is specific for that amino acid. If an enzyme within the sequence were the inhibited one, an intermediate would accumulate and would probably be excreted unless the intermediate itself was an inhibitor of the first step in the pathway. Some of the general patterns that have been found to block metabolite flow into amino acid biosynthetic pathways are illustrated in Figure 22-26.

**Endproduct Inhibition.** *The simplest kind of feedback loop in the living cell is the direct inhibition of the initial enzyme in an amino acid biosynthetic sequence by the amino acid itself* (Figure 22-26). The mechanism by which such a feedback loop is consummated, like many other interactions between proteins and small molecules, may not at all be simple on the molecular level. That such interactions are significant has been demonstrated in several cases by the isolation of microbial mutants in which the loss of sensitivity of an initial enzyme to its respective end product is accompanied by an overproduction and excretion of that end product.

Those reactions known to be sensitive to end product inhibition in at least one organism have been so noted in the text. Alanine, aspartate, and glutamate are three amino acids for which no form of end product inhibition is known. These amino acids, however, are in an essential equilibrium by means of reversible reactions with their corresponding $\alpha$-keto acids, which are key intermediates in the central metabolic routes. In most bacteria, the pools of these three amino acids are fairly high, at least under conditions of excess nitrogen supply, and actually are "leaked" to the external medium in small amounts.

Glycine is another amino acid formed by a single reversible enzyme. Glycine pools are usually small, but the enzyme is not product-inhibited. It may be controlled in part by the availability of the acceptor for one-carbon units, tetrahydrofolate.

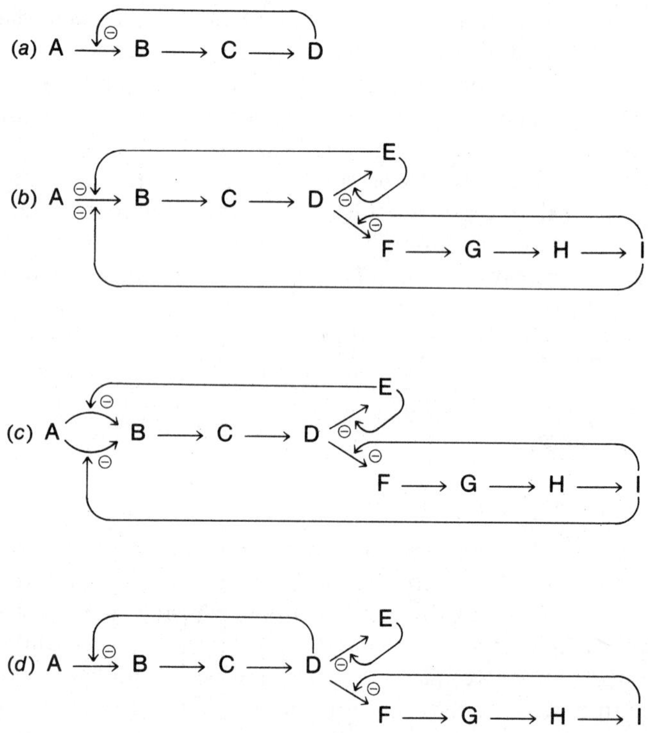

**Figure 22-26**
Patterns of end product inhibition in amino acid biosynthetic pathways.
(*a*) Simple end product inhibition: End product D inhibits the first enzyme in the pathway. (*b*) Concerted end product inhibition of a common pathway: End products E and I act in concerted fashion to inhibit first enzyme in a common pathway, and each inhibits the first enzyme in the branch pathway leading specifically to it. (A special case, not shown, would be synergistic inhibition in which E and I singly are only weak inhibitors, but when both present, a strong synergistic inhibition occurs.) (*c*) Multiple enzymes specifically controlled by end products of different branch pathways: Two enzymes convert A to B. One enzyme is inhibited by E, the end product of one branch pathway. The other enzyme is inhibited by I, the end product of the other branch pathway. The two branch pathways are controlled by simple end product inhibition. (*d*) Sequential end product inhibition: The ultimate end products E and I inhibit only the first enzyme in their own branch pathways. Intermediate D inhibits the first enzyme in the common pathway.

Metabolite flow into the biosynthetic pathways of all the remaining 16 amino acids is controlled by one means or another. The specific controls for these 16 amino acids, as found in *E. coli*, have been cited earlier. In addition, some of the examples in which there are different controls in other forms also have been cited.

Most of the departures from the patterns exhibited by *E. coli* are related to the different ways the common aspartate family and common aromatic pathways have evolved. It is of interest that in some forms, a means has evolved for sequential feedback inhibition to occur. The basic requirement for such a pattern of regulation to be effective is that the early enzyme be inhibited when the inhibitory intermediate is no longer being utilized. However, pool levels of biosynthetic intermediates are usually very small, so that fluctuations in pool size alone might not be sufficient to provide the range of control required. In *B. subtilis* this control may have been achieved by the single 3-deoxy-D-*arabino*-heptulosonate-7-phosphate synthase being part of the same protein that exhibits chorismate mutase activity and by both the substrate and product of the mutase being inhibitors of the synthase. Whether this is the general mechanism by which intermediates exert feedback inhibition has not been widely examined.

The multiple-enzyme pattern found in the *E. coli* common aspartate pathway might be considered costly, and in some organisms, a different pattern of regulation has been selected: the concerted (or highly synergistic) inhibition by lysine plus threonine. In some bacteria this pattern results in a methionine deficiency in a medium containing threonine plus lysine. That such organisms have survived in nature and that no organisms containing a threonine-plus-lysine–inhibited aspartokinase along with a methionine-repressed aspartokinase have been found may be evidence that the threonine plus lysine combination does not often occur in the absence of methionine. In one organism that has a threonine-plus-lysine–sensitive aspartokinase, *B. subtilis*, the need for diaminopimelate as a cell-wall constituent may have provided a more important selective pressure, since a second aspartokinase inhibited by diaminopimelate is found. The pattern found in some plants of two aspartokinases, one inhibited by threonine, the second by a synergistic effect of lysine plus *S*-adenosylmethionine, represents still another mechanism to achieve control in this highly branched pathway.

**Activation of Enzyme Activity.** *Just as end product inhibition* (or in the few cases, inhibition by biosynthetic intermediates) *has been invoked as an important mechanism for the control of amino acid biosynthesis, so has the less frequently encountered activation of enzyme activity by precursors or by intermediates in other pathways that depend on the function of a converging pathway for their further utilization* (Figure 22-27). The most striking example of the latter encountered in this chapter has been found in the regulation of those carbamoyl phosphate synthases that function for both pyrimidine and arginine biosynthesis. The inhibitor is UMP. Alone, this interaction could lead to a quenching of arginine biosynthesis. The inhibition, however, is reversed by ornithine. This is very important, because ornithine is the intermediate that would accumulate if (1) arginine were no longer in high enough concentration to block the first enzyme in the pathway and (2) there were no carbamoyl phosphate availa-

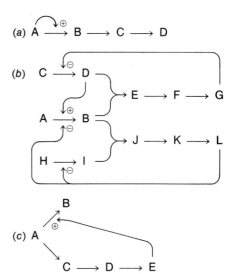

**Figure 22-27**
Patterns of enzyme activation in amino acid biosynthetic pathways. (*a*) Substrate activation: Substrate is also an activator of its enzyme, i.e., enzyme exhibits cooperative substrate binding. (*b*) Compensatory reversal of end product inhibition: The intermediate D, the formation of which is regulated by end product G, activates the formation of B, thus balancing the inhibitory effect of L, which inhibits formation of B, one of its precursors. (*c*) Stimulation of one branch pathway by the end product of another branch pathway: E, one of the two products derived from branch-point compound A, stimulates the conversion of A to B, the other product derived from A.

ble. The antagonism between ornithine (the positive effector) and UMP (the negative effector) is thus a *compensatory* mechanism controlling metabolite flow.

The activity of cystathionine-γ-synthase of *N. crassa*, an essential enzyme for methionine formation, is stimulated by 5-methyltetrahydrofolate. 5-Methyltetrahydrofolate might accumulate if the flow from cystathionine to homocysteine were retarded, and thus it might provide an appropriate "signal" for a methionine deficiency. In green plants, *S*-adenosylmethionine, the ultimate end product in one biosynthetic branch from phosphohomoserine, stimulates threonine synthase, which constitutes another (competing) branch leading from phosphohomoserine.

Other examples of enzyme activation by compounds other than the substrate itself are less readily explained with respect to any physiological or selective advantage. For example, the very efficient inhibition of *E. coli* threonine deaminase by isoleucine is antagonized by valine. The first enzyme in leucine biosynthesis, α-isopropylmalate synthase, is also activated by valine. The *B. subtilis* prephenate dehydratase is inhibited by tryptophan as well as phenylalanine. Tyrosine acts as a positive effector to antagonize the inhibition by tryptophan.

As has been mentioned elsewhere (Chapter 4), many (but certainly not all) end product-sensitive enzymes exhibit nonhyperbolic (sigmoid) substrate saturation curves. In some cases, the sigmoid curves (cooperative binding) are dependent on the presence of some end product inhibitor. Where cooperative binding does occur, the substrate itself can be looked upon as an activator.

**Modification of Enzymes as a Regulatory Mechanism.** Modification of enzymes through the covalent linkage of a phosphate group, for example, often plays a regulatory role in catabolic systems in animal tissue. The general mechanism has been less frequently encountered among amino acid biosynthetic enzymes. A striking exception is the *glutamine synthase* of *E. coli* and related organisms, which was described earlier (see Figure 22-3). *In this case, the modification of the protein is probably more important in the control of biosynthetic function than is the noncovalent binding of any of the several inhibitors.*

Two other examples of modification of protein that may have regulatory significance have been found in yeast. Whether a covalent modification is involved has not been established. The two examples are the initial enzymes in leucine (Figure 22-21) and lysine (Figure 22-7) biosynthesis, each of which involves the addition of an acetyl group to an α-keto acid. In yeast, both are mitochondrial enzymes and, upon incubation with $Zn^{2+}$ and coenzyme A, a product of the reaction, are slowly and reversibly inactivated. Reactivation requires incubation with ATP.

**Regulation of Enzyme Activity by Protein-Protein Interaction.** Both activation and inhibition of enzyme activity have been found to accompany binding to a second protein. A striking example is the inhibition of yeast ornithine transcarbamylase upon binding of arginase, an enzyme induced upon addition of arginine to the growth medium. Thus arginase, which converts arginine to ornithine, prevents the futile cycling of ornithine back to arginine.

There also are examples of stimulation of activity of enzymes upon binding to other proteins. The best examples might be those in which one component of a multienzyme complex exhibits a low activity alone but an enhanced activity in the complex, such as the stimulation of the ammonia-dependent anthranilate synthase activity of *E. coli* component I by component II. Many other examples of such an activation have been found, and it may be that, within the cell, many less tightly bound protein-protein associations can occur that do, in fact, greatly influence rates of enzyme-catalyzed reactions.

**General Controls on Biosynthetic Activity.** The control mechanisms just cited can all be considered specific controls that are exerted in response to the specific need for the end product of the pathway. There are other factors affecting the flow of metabolites over a pathway that are quite general in nature but might become the predominating factors. Their physiological importance, however, may still be considered somewhat hypothetical. These factors would include energy charge, reducing potential, and pH. In some cases, the first two effects might be considered special cases of substrate availability. These factors also could affect flow of metabolites through a pathway but not necessarily flow into a pathway, with the resulting accumulation of an intermediate. To what extent these potential regulatory parameters have been harnessed in achieving a control mechanism for any particular system remains to be demonstrated.

## Patterns of Control over Enzyme Amount

Another important aspect of the regulation of amino acid biosynthesis is the regulation of the amount of any of the biosynthetic enzymes. *The amount of an enzyme at any one time is a summation of the amount of the enzyme formed less the amount broken down.* The regulation of the amount formed is achieved at the level of gene expression. For only a few systems has the analysis been carried far enough to tell how this regulation occurs.

**End Product Repression of Enzyme Synthesis.** For many amino acid biosynthetic enzymes in bacteria it has become quite clear that *the level of a biosynthetic enzyme is regulated in response to the presence or absence of the end product in the medium.* Such amino acid biosynthetic enzymes are said to be *end product repressible*. The term *end product repressible* does not carry any connotation regarding the mechanism by which the repression occurs. Indeed, analysis may reveal activation of a negative control element or blocking of a positive control element. Repression may involve preventing the initiation of transcription, or it may involve the extent to which an initiated transcription process continues into the structural gene (attenuation). Although no such systems have been described, there is no a priori reason why, in some cases, specific repression of an amino acid biosynthetic enzyme might not involve regulation of the initiation of message translation, a pattern that functions in expression of a number of ribosomal proteins in *E. coli*.

The Jacob-Monod model proposed in 1961 postulated that repression of amino acid biosynthetic enzymes could be achieved by the specific binding of a repressor to a site adjacent to a structural gene (or gene cluster), thereby inhibiting the initiation of transcription. Another pattern of transcriptional control has been encountered in amino acid biosynthetic pathways of *E. coli* and related organisms. The hallmark or indicator that the mechanism may be functioning in a given pathway is the observation that repression of the enzyme or enzymes is dependent not only on the presence of the amino acid in excess, but also on its ready transferral to its cognate acceptor tRNA. These two important modes of regulating transcription will be discussed in detail in Chapter 26.

**Substrate Induction of Biosynthetic Enzymes.** *For several bacterial and fungal enzymes involved in amino acid biosynthesis, there is evidence that* rather than an end product repression of enzyme synthesis, *there is a substrate induction.* Just as end product repression might be achieved by any of several mechanisms, so could substrate induction. In the cysteine biosynthetic pathway, the *cysB* product provides at least positive control. The enzymes are induced by *O*-acetylserine (substrate) as well as being repressed by cysteine (or perhaps by sulfide derived from cysteine). One of the enzymes of the isoleucine and valine biosynthetic pathways in the enteric bacteria, acetohydroxy acid isomeroreductase, is induced by either of its two substrates, and tryptophan synthase of *P. putida* is induced by its substrate, indoleglycerol phosphate. In both cases, regulatory proteins serve as positive control elements.

In *N. crassa*, the second and third enzymes of the leucine biosynthetic pathway appear to be induced by the product of the first, α-isopropylmalate. Like control of gene expression in other eukaryotic systems, the mode of this substrate induction remains a complete mystery.

For several amino acids in *E. coli*, there appear to be no end product-specific controls over enzyme formation. However, the enzymes involved in the biosynthesis of these amino acids are still subject to regulation. Indeed, these enzymes as well as those referred to earlier are probably all subject to what might be termed general metabolic controls. In terms of the magnitude of the effects, the general metabolic controls may be of even greater importance than the end product-specific controls.

**General Metabolic Control of Enzyme Synthesis.** Only a few of the enzymes that do not appear to be controlled by specific mechanisms reflecting the biosynthetic need have been studied in much detail, even in *E. coli*. *The synthesis of all amino acid biosynthetic enzymes of E. coli appears to be strongly reduced when the cells are grown in a rich medium* in which the growth rate is very fast. This reduction has been termed *metabolic depression* of enzyme synthesis to differentiate it from repression which appears to be specific for the pathway irrespective of precise mechanism.

Metabolic depression of enzyme synthesis can be more pronounced than specific repression for a pathway for which both are demonstrable. It should be remembered that at fast growth rates, ribosomal RNA, ribosomal proteins, nucleic acid polymerases, and the soluble factors required for macromolecular synthesis all constitute a greater proportion of total cell protein than they do in cells growing at slow rates. Thus metabolic depression might, in part, be due to mere competition for RNA polymerases or ribosomes. What conditions might affect the selection of "housekeeping" genes over genes for biosynthesis of small molecules? One factor might be ppGpp, which is increased in amount during shiftdown of carbon source or during amino acid starvation. One model of promoter selection postulated that high ppGpp stimulates transcription of biosynthetic or catabolic operons and impedes transcription of the "housekeeping" genes. The faster the growth rate, presumably the lower is the ppGpp concentration. However, the "rich" medium effect cannot be explained solely by ppGpp-directed selection of promoters.

For only a few of the "nonspecifically" controlled biosynthetic systems has metabolic control been well-examined. In one of these, the serine biosynthetic pathway, carbon flow is very effectively controlled by end product inhibition, although there is no end product-specific repression by serine or by any combination of any of the compounds derived from serine. There is, instead, a strong correlation of level of serine biosynthetic enzymes with growth rate.

Metabolic control of enzyme formation is still an area in which there have been few studies, and it is still too early to state whether the general control mechanism is one of a single basic mechanism or whether a variety of physiological interactions have been mobilized during evolution to achieve metabolic control.

**Regulation of Degradation of Amino Acid Biosynthetic Enzymes.** While relatively few biosynthetic pathways have been analyzed sufficiently well even in *E. coli* to describe the pattern of regulation of enzyme synthesis, in even fewer pathways have the factors regulating enzyme breakdown been analyzed. In only a few cases has it been shown that loss in activity of a given enzyme is accompanied by disappearance of the protein itself. Nevertheless, the protection of a given enzyme against proteolysis by a specific mechanism related to its function in the cell could be an important regulatory process. One example that might be cited is tryptophan synthase of yeast, which, like many yeast enzymes, is markedly susceptible to proteolytic breakdown. However, in the cell, the proteases are confined to a vacuole and are controlled by protease inhibitors. Furthermore, the proteases are kept from tryptophan synthase much more effectively in minimal-grown cells than in cells grown in a rich medium.

This subject is, again, a topic in its infancy. The extent of the process of protein turnover as a regulatory mechanism still remains somewhat hypothetical. It is interesting that in organisms such as *E. coli*, in which a certain amount of turnover occurs with many (but not all) proteins, it increases during periods of carbon or nitrogen starvation and may be mediated by ppGpp. ppGpp may act by activating preexisting proteases rather than by stimulating protease synthesis.

## NONPROTEIN AMINO ACIDS

Not all amino acids in plants, bacteria, or animals are found as protein constituents. Some are found in peptide linkages in compounds that are important as cell-wall or capsular structures in bacteria or as antibiotic substances produced by bacteria and fungi. Others are found as free amino acids in seeds and other plant products. Among the amino acids that occur in these forms are not only the 20 normal amino acids, but also some amino acids never found in proteins. These _nonprotein amino acids_, numbering into the hundreds, _include precursors of normal amino acids_, such

**Table 22-2**
Some D-Amino Acids Found in Peptide Antibiotics

| Antibiotic | D-Amino Acids Present | Produced by |
|---|---|---|
| Actinomycin $C_1$ (D) | D-Valine | *Streptomyces parralus and others* |
| Bacitracin A | D-Asparagine, D-glutamate, D-ornithine, D-phenylalanine | *Bacillus subtilis* |
| Circulin A | D-Leucine | *Bacillus circulans* |
| Fungisporin | D-Phenylalanine, D-valine | *Penicillium* species |
| Gramicidin S | D-Phenylalanine | *Bacillus brevis* |
| Malformin $A_1$, C | D-Cysteine, D-leucine | *Aspergillus niger* |
| Mycobacillin | D-Aspartate, D-glutamate | *Bacillus subtilis* |
| Polymixin $B_1$ | D-Phenylalanine | *Bacillus polymyxa* |
| Tyrocidine A, B | D-Phenylalanine | *Bacillus brevis* |
| Valinomycin | D-Valine | *Streptomyces fulrissimus* |

as homoserine and diaminopimelate, *intermediates in catabolic pathways*, such as pipecolic acid, D-enantiomers of "normal" amino acids, *and amino acid analogs*, such as azetidine-2-carboxylic acid and canavanine, *that might be formed by unique pathways or by modification of normal amino acid biosynthetic pathways*.

## D-Amino Acids in Microbial Products

Certain D-amino acids along with some L-enantiomers are commonly found both in microbial cell walls (see Chapter 12) and in many peptide antibiotics. For example, the peptidoglycans of bacteria contain both D-alanine and D-glutamate. The latter is present in a γ-glutamyl linkage. In some forms, the α-carboxyl of the D-glutamyl residue is either amidated or in peptide linkage with glycine. D-Lysine or D-ornithine are found in the glycopeptide of some Gram-positive organisms. The capsule of the anthrax bacillus is composed of a nearly pure homopolymer of D-glutamate in γ-linkage. Other bacilli also produce γ-linked polyglutamates, some of which form separate D-glutamate and L-glutamate chains, while others form a copolymer of D- and L-glutamate. A wide variety of D-amino acids have been found in antibiotics, and some of these are listed in Table 22-2.

## The Formation of D-Amino Acids

At least some of the D-amino acids are formed by racemases as free intermediates and are subsequently incorporated into peptide bonds. An example is D-alanine found in the bacterial cell-wall peptidoglycan. L-Alanine is converted to D-alanine by a racemase that contains pyridoxal phosphate as a cofactor. The racemization is followed by the formation of a D-alanyl-

D-alanine dipeptide, which is accompanied by the conversion of ATP to ADP. The dipeptide is subsequently incorporated into the glycopeptide.

*In all cases of formation of D-amino acid–containing peptides studied thus far, the L form of the amino acid is the substrate for the incorporating enzyme.* In contrast, the free D-amino acid is ordinarily a poor substrate for the incorporation reaction. Whether the racemization occurs on the enzyme or inversion occurs afterwards remains to be determined in most cases. In the case of the D-valyl residue formed in penicillin, a tripeptide derivative containing L-valine is an intermediate, and conversion is thought to occur by way of an $\alpha$-$\beta$-dehydro form of the valyl residue.

## Naturally Occurring Amino Acid Analogs

Among the hundreds of nonprotein amino acids found in nature are many that might be considered naturally occurring amino acid analogs and many that are toxic and antagonistic to the usual 20 amino acids found in proteins. Some are found as components of antibiotics, but others have been identified as antibiotic substances themselves. The frequent occurrence of toxic amino acids in the seeds of plants suggests that they might play a role in the protection of the seeds from insects or other predators.

Some typical examples of these naturally occurring analogs are given in Table 22-3, along with their sources and the antagonistic L-amino acid, where such an antagonism is known. It should be pointed out, however, that not all the "analogs" are toxic. For example, pipecolic acid, the next higher homolog of proline and an intermediate in lysine degradation, does not interfere in any demonstrable way with proline metabolism.

**Table 22-3**
Some Naturally Occurring Nonprotein Amino Acids

| Compound | Occurrence | Remarks |
|---|---|---|
| **Branched-chain and cyclopropane amino acids** | | |
| $CH_3-CH_2-CH(CH_3)-CH_2-CHNH_2-COOH$<br>2-Amino-4-methylcaproic acid (homoisoleucine) | California buckeye | Leucine antagonizes toxicity |
| $(CH_3)_2-NCH_2-CHNH_2-COOH$<br>2-Amino-3-dimethylaminopropionic acid (azaleucine) | *Streptomyces neocaliberis* | Leucine antagonizes toxicity |
| $\underset{\textstyle CH_2=C-CHCHNH_2-COOH}{\overset{\textstyle CH_2}{\diagup\diagdown}}$<br>2-(Methylenecyclopropyl)glycine | Lychee seeds | Leucine antagonizes toxicity |
| **Sulfur-containing amino acids** | | |
| $CH_3-SCH_2-CHNH_2-COOH$<br>S-Methylcysteine | Broad bean<br>(*Phaseolus vulgaris*) | — |
| $CH_3-CH=CH-S-CH_2-CHNH_2-COOH$<br>S-(Prop-1-enyl)cysteine | Garlic | — |

*(Continued)*

**Table 22-3** (*Continued*)

| Compound | Occurrence | Remarks |
|---|---|---|
| **Aromatic and heterocyclic amino acids** | | |
| $CH_2$—$CHNH_2$—COOH (with benzene ring and COOH) 3-(3-Carboxyphenyl)alanine | Iris | — |
| $CH_2$—$CHNH_2$—COOH (pyridone structure with N, OH, O) β-N-(3-Hydroxy-4-pyridone)alanine (mimosine) | Mimosa tree | Tyrosine; toxic to nonruminants |
| Pipecolic acid (piperidine ring with CHCOOH, NH) | Widely distributed in plants | Probably not toxic; an intermediate in lysine catabolism |
| **Acidic amino acids** | | |
| HOOC—$CH(CH_3)$=$CH_2$=$CHNH_2$=COOH 4-Methylglutamic acid | Sweet pea | — |
| **Basic amino acids** | | |
| $H_2N$—C(=NH)—NH—O—$CH_2$—$CH_2$—$CHNH_2$—COOH Canavanine | Jack bean and other legumes | Arginine antagonizes toxicity |
| $CH_2NH_2$—$CH_2$—$CHNH_2$—COOH 2,3-Diaminopropionic acid | Seeds of acacia and mimosa | As the oxalyl derivative, acts as a neurotoxin |

# SELECTED READINGS

Cohen, G. N. *Biosynthesis of Small Molecules.* New York: Harper and Row, 1967. A small book reviewing some of the early literature and describing the way isotope incorporation and mutant and enzymatic analyses were used to unravel metabolic pathways.

Fowden, L., Lea, P. J., and Bell, E. A. The nonprotein amino acids of plants. *Adv. Enzymol.* 50:117, 1979. A discussion of the occurrence and biosynthesis of naturally occurring amino acid analogs in plants.

Katz, E., and Demain, A. L. The peptide antibiotics of *Bacillus:* Chemistry, biogenesis and possible functions. *Bacteriol. Rev.* 41:449, 1977. A description of several peptide antibiotics showing the distribution of D-amino acids in these compounds.

Meister, A., (Ed.). *Biochemistry of the Amino Acids,* Vols. 1 and 2. New York: Academic Press, 1965. The two-volume classic provides a thorough discussion of amino acid literature, occurrence, properties, and metabolism of amino acids up to that time.

Mifflin, B. J., and Lea, P. J. Amino acid metabolism. *Ann. Rev. Plant Physiol.* 28:299, 1977. A review of amino acid biosynthesis in the higher plants with emphasis on primary amino group formation.

Umbarger, H. E. Amino acid biosynthesis and its regulation. *Ann. Rev. Biochem.* 47:533, 1978. A review of amino acid biosynthesis in bacteria, fungi, and plants and a consideration of factors controlling metabolite flow and enzyme amount in these pathways.

Umbarger, H. E., and Davis, B. D. Pathways of Amino Acid Biosynthesis. In I. C. Gunsalus and R. Y. Stanier (Eds.), *The Bacteria*, Vol. III: *Biosynthesis*. New York: Academic Press, 1962. Pp. 167–251. Survey of the early studies, primarily with bacteria and fungi, that revealed most of the steps in amino acid biosynthesis.

## PROBLEMS

1. Carboxyl-labeled acetate (experiment A) and carboxyl-labeled pyruvate (experiment B) were provided as a 10-min pulse to an *E. coli* culture grown on glucose as the sole carbon and energy source. The protein fraction of the cells harvested at the end of the pulse was hydrolyzed by hydrochloric acid. Seventeen amino acids were separated (Which ones were missing? Why?) by column chromatography. The emerging samples were split for retention of the intact amino acid and for automated, quantitative estimation by the ninhydrin method. The intact samples were then used for determination of total radioactivity and for radioactivity in the $\alpha$-carboxyl carbons (again by a ninhydrin reaction). What would you predict for the specific activity (high, moderate, or low) for the two kinds of sample for each of the 17 amino acids in the two experiments?

2. Assume that you have two organisms that grow only anaerobically on glucose as a carbon source. Both can form methionine, but organism A employs a direct sulfhydrylation route to homocysteine and forms cysteine by transsulfuration, while organism B forms cysteine by direct sulfhydrylation and homocysteine by transsulfuration. Assume further the following for both organisms: (a) the organisms form lactate as the fermentation product; (b) NADPH needed for biosyntheses is formed by the hexose monophosphate shunt (and pentose phosphate cycle); (c) acetyl-CoA can be formed for biosynthetic needs by an anaerobic pyruvate cleavage reaction producing formate in the process (pyruvate-formate lyase reaction); (d) the *O*-acylation reactions use acetyl-SCoA; (e) the organisms also contain malate dehydrogenase and fumarate reductase, which permits reoxidation of NADH if pyruvate residues are not available; and (f) $SO_4^{2-}$ is the sulfur source. Calculate the moles of glucose required to form 1 mol of methionine in the two organisms by means of the pathways shown in Figures 22-10 and 22-11. [In addition to the 1 mol of methionine, you must calculate the amount of lactate, succinate, acetate, and $CO_2$ that had accumulated when all pyridine nucleotides were reoxidized. Remember, too, to regenerate any glutamate consumed in amino donor reactions and any *N*-5,10-methylene tetrahydrofolate consumed in methyl-group transfer

(from formate rather than from serine; see Chapter 6).]

3. Recent examination of some lunar samples revealed the presence of bacteria-like organisms that grew only on the surface of lunar soil samples under reduced atmospheric pressures and in the presence of visible light. The organisms contained lipid, carbohydrate, protein, and nucleic acids remarkably similar to prokaryotic cells found on earth. However, the proteins contained 21 amino acids but no isoleucine. The two unusual amino acids were the L-enantiomers of

$$CH_3CH_2\underset{\underset{NH_2}{|}}{C}HCOOH \quad \text{and} \quad CH_3CH_2CH_2\underset{\underset{NH_2}{|}}{C}HCOOH$$

On the basis of known amino acid biosynthetic pathways, would you expect these amino acids to belong to the pyruvate, the aspartate, or the glutamate family? By what route might they be formed? (Enlist only enzymes shown to exist in terrestrial plants or bacteria, postulating only "reasonable" modifications in specificity of the "parent" enzymes.)

   Indicate the labeling expected in the new compounds if you were able to use 1-$^{14}$C-labeled aspartate, glutamate, pyruvate, and acetate as tracers in a growth experiment.

4. At the beginning of the chapter a diagram was provided showing the flow of carbon into the various amino acids from key intermediates in the central metabolic routes. You should prepare an analogous diagram showing the flow of nitrogen from ammonia into all 20 amino acids. (Assume the organism is one that contains an NADPH-dependent glutamate dehydrogenase and is growing in a high ammonia environment. Remember, too, that some nitrogen is discarded during amino acid biosynthesis.)

5. The accumulation of biosynthetic intermediates, or of metabolites derived from those intermediates (as in the case of indole-3-glycerol accumulation by certain tryptophan auxotrophs), proved to be a valuable tool in the analysis of biosynthetic pathways. It was found, however, that these accumulations occurred only after the required amino acid had been consumed and growth had stopped. How do you account for this fact?

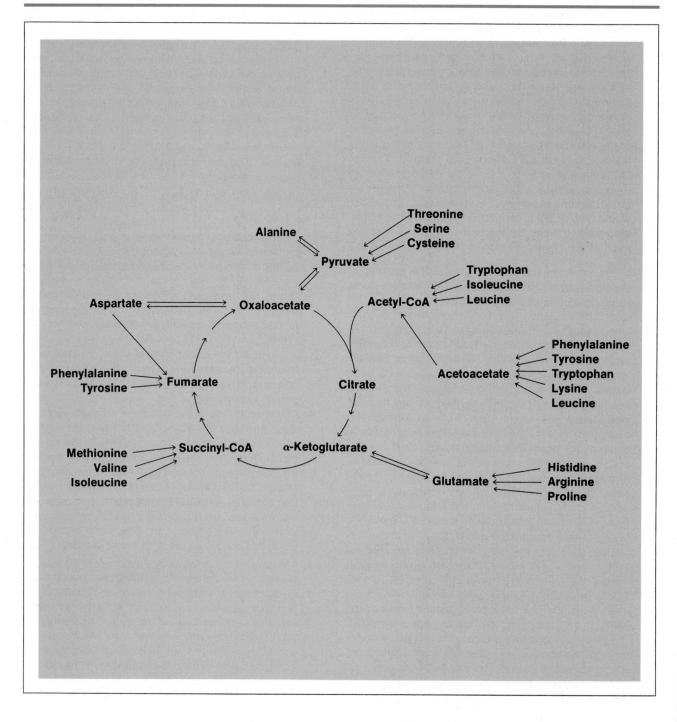

# 23

# UTILIZATION OF AMINO ACIDS

Clearly, the major fate of amino acids formed from carbon dioxide or other simple carbon sources by microorganisms and plants is their incorporation into protein. The activation of amino acids prior to incorporation into protein, their transfer to tRNA adapter molecules, and the mechanism of their polymerization into the primary structures specified by the genetic information in the various structural genes are all described in Chapters 24 and 25.

In Chapter 6, reference was made to the several vitamins and coenzymes that require an amino acid as a component of the coenzyme (e.g., folic acid) or as precursors in their biosynthesis (e.g., nicotinamide and pantothenate). In Chapter 12, the formation of bacterial cell walls was described. There are several other important biosynthetic processes initiated with preformed amino acids that are considered in this chapter on the utilization of amino acids. These include processes leading to the formation of the porphyrin nucleus found in many oxygen and electron-carrying proteins, of biologically active amines, of glutathione, of peptide antibiotics, and of several other important metabolites. Finally, this chapter describes the pathways by which the individual amino acids are degraded either as a nitrogen source or as a carbon and energy source. Indeed, amino acids constitute important carbon and energy sources for some cells, either because these cells cannot utilize fatty acids or carbohydrates or because these sources of energy are not available.

*Breakdown products resulting from amino acid catabolism funnel into the Krebs TCA cycle. Catabolism is a surprisingly orderly process, often leading to intermediates for new biosynthesis.*

## THE CONVERSION OF AMINO ACIDS TO OTHER AMINO ACIDS AND TO OTHER METABOLITES

### "Essential" versus "Nonessential" Amino Acids

The classical differentiation between "essential" and "nonessential" amino acids by H. Rose and colleagues was made on the basis of weight gain of growing white rats fed diets containing 19 of the 20 amino acids found in proteins. An analogous identification of amino acids essential and nonessential for the human was derived from a limited number of human studies based not on weight gain or loss, but on short-term maintenance of a positive nitrogen balance. A negative nitrogen balance (total nitrogen excretion exceeding total nitrogen intake) during the experimental period of a single amino acid deficiency indicated that tissue protein was being degraded during that period and used to supply the missing amino acid for those "high priority" proteins that must be continually resynthesized from dietary sources or at the expense of other body proteins. Table 23-1 lists the amino acids found on the basis of such simple experiments to be essential for the rat and for humans. For comparison, the amino acids found to be essential and nonessential for a strain (L) of mouse fibroblasts grown in cell culture are also included in Table 23-1.

These experimental results cannot be taken as evidence that rats have all the enzymes described in the preceding chapter needed for conversion of the several central metabolites to amino acids listed in column 2 of Table 23-1 or that the human has the enzymes needed to form those in column 4. Rather, they indicate that given 19 other amino acids, the particular amino acid can be formed either by the same _de novo pathway_ used by plants or by a _"salvage" pathway_ at the expense of another amino acid.

**Table 23-1**
The Essential Amino Acids

| For Weight Gain in Protein-Starved Adult Rats | | For Positive Nitrogen Balance in Adult Humans | | For Mouse L Cells in Culture | |
|---|---|---|---|---|---|
| **Essential** | **Nonessential** | **Essential** | **Nonessential** | **Essential*** | **Nonessential** |
| Histidine | Alanine | Isoleucine | Alanine | Arginine | Alanine |
| Isoleucine | Arginine† | Leucine | Arginine | Cysteine | Asparagine |
| Leucine | Asparagine | Lysine | Asparagine | Glutamine | Aspartate |
| Lysine | Aspartate | Methionine | Aspartate | Histidine | Glutamate |
| Methionine | Cysteine | Phenylalanine | Cysteine | Isoleucine | Glycine |
| Phenylalanine | Glutamate | Threonine | Glutamate | Leucine | Proline |
| Threonine | Glutamine | Tryptophan | Glutamine | Lysine | Serine |
| Tryptophan | Glycine | Valine | Glycine | Methionine | |
| Valine | Proline | | Histidine‡ | Phenylalanine | |
| | Serine | | Proline | Threonine | |
| | Tyrosine | | Serine | Tryptophan | |
| | | | Tyrosine | Tyrosine | |
| | | | | Valine | |

*The medium also contained 0.25 to 1 percent dialyzed horse serum.
† Arginine is required in the diet of young rats.
‡ Histidine is required in infant humans.

**Table 23-2**
"Salvage" Pathways Allowing the Formation of Certain Nonessential
Amino Acids from Essential Amino Acids

| Amino Acid Formed | Formed From | Enzymes Required |
|---|---|---|
| Arginine | Proline | Proline oxidase |
| | | Ornithine-glutamate transaminase |
| | | Ornithine transcarbamoylase |
| | | Argininosuccinate synthase |
| | | Argininosuccinate lyase |
| Cysteine | Methionine | S-adenosylmethionine synthase |
| | | A methyltransferase |
| | | S-adenosylhomocysteinase |
| | | Cystathionine-β-synthase |
| | | Cystathionine-γ-lyase |
| Tyrosine | Phenylalanine | Phenylalanine-4-monooxygenase |

It is true that many animals, including humans, do contain the enzymes of several of the "short" biosynthetic pathways—those to glutamate, glutamine, proline, aspartate, asparagine, serine, and glycine. The enzymes for proline biosynthesis, a kinase and two reductases, are present in animal tissues. In addition, animals contain a proline oxidase that yields $\Delta^1$-pyrroline-5-carboxylic acid which, because of the equilibrium with glutamic-γ-semialdehyde and a transaminase, allows for a nonacetylated route to ornithine. Ornithine, in turn, can be converted to arginine by means of the enzymes of liver that are essential for urea formation. Thus, although the pathway by which bacteria and plants form arginine is not present in animal tissue, arginine can be formed but, in some organisms (e.g., rats), not at a rate sufficient for growth. Two other amino acids normally present in the diet also can be formed from other amino acids by unidirectional routes. For example, cysteine can be formed by transsulfuration from serine and homocysteine (formed from dietary methionine following removal of its methyl group by a transmethylation step), and tyrosine can be formed by hydroxylation of phenylalanine by a pteridine-dependent monooxygenase. These routes and the corresponding enzymes are listed in Table 23-2.

## Porphyrin Biosynthesis

The early isotope tracer experiments of D. Shemin and colleagues at Columbia University permitted the elucidation of the formation of the immediate precursor of the porphyrins needed for the cytochromes and for hemoglobin. These studies indicated that the glycine methylene carbon and nitrogen were incorporated along with both carbons of acetate. Subsequent enzymic studies in both bacteria and animals revealed a condensation reaction between succinyl-CoA and glycine to yield *δ-aminolevulinate* and

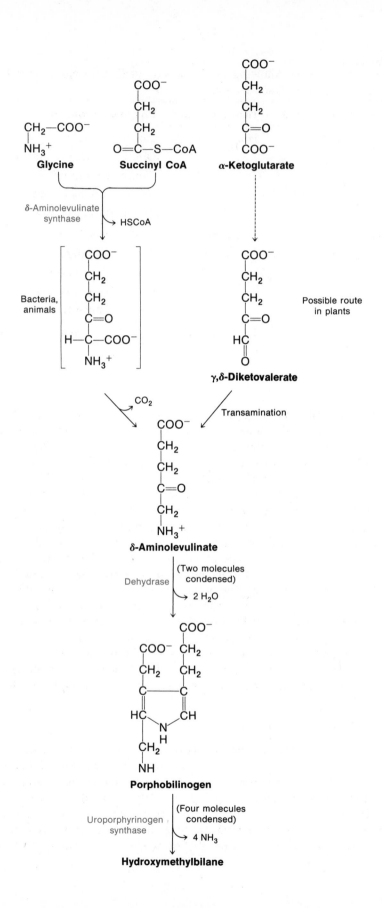

**Figure 23-1**
Tetrapyrrole biosynthesis. The sequence by which the four porphobilinogen residues are converted to uroporphyrinogen III is the sequential head-to-tail condensation of the four residues by uroporphyrinogen I synthase (porphobilinogen deaminase) to yield the unrearranged hydroxymethylbilane. This rather unstable intermediate is rearranged and cyclized by uroporphyrinogen III cosynthase to yield uroporphyrinogen III. Hydroxymethylbilane in the absence of cosynthase will spontaneously cyclize to yield the unrearranged urobilinogen I, which is not an intermediate in the pathway (hence the name uroporphyrinogen I synthase for the deaminase).

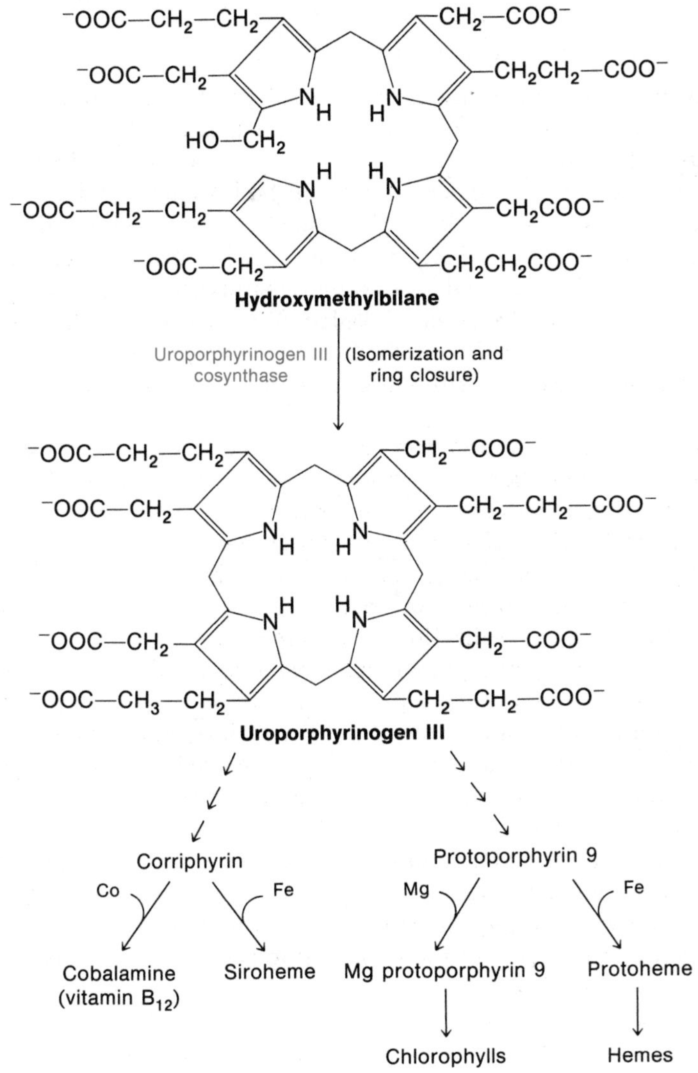

**Hydroxymethylbilane**

Uroporphyrinogen III cosynthase | (Isomerization and ring closure)

**Uroporphyrinogen III**

Corriphyrin      Protoporphyrin 9

Co      Fe      Mg      Fe

Cobalamine      Siroheme      Mg protoporphyrin 9      Protoheme
(vitamin $B_{12}$)

Chlorophylls      Hemes

$CO_2$ (presumably by way of an enzyme-bound $\beta$-keto acid, $\alpha$-amino-$\beta$-ketoadipate) (Figure 23-1).

$\delta$-Aminolevulinate is also the precursor to porphobilinogen in plants and in the blue green algae (cyanobacteria), but it is not formed by a condensation of glycine and succinyl-CoA. Rather, it is probably formed by the reduction of the $\alpha$-carboxyl group of $\alpha$-ketoglutarate to yield $\gamma$-$\delta$-diketovalerate followed by an $\omega$ amination to yield $\delta$-aminolevulinate.

The pyrrole monomer porphobilinogen arises from the condensation of two molecules of $\delta$-aminolevulinate with the loss of two water molecules. The reaction is catalyzed by $\delta$-aminolevulinate dehydrase. The condensation of four porphobilinogen molecules to yield the branch-point compound in tetrapyrrole synthesis, uroporphyrinogen III, is a complex reaction requiring two enzymes, uroporphyrinogen I synthase, which catalyzes a head-to-tail condensation of four porphobilinogen molecules, and uroporphyrinogen III cosynthase, which inverts one of the units and closes the ring.

## The Conversion of Amino Acids to Biologically Active Amines

Among the more important metabolites derived from amino acids are those derived directly or indirectly by decarboxylation. *Amino acid decarboxylases* often use *pyridoxal phosphate* as the coenzyme. The amines or their derivatives play a variety of roles in both bacteria and animals.

One of the simplest amines, ethanolamine, is usually formed by decarboxylation of the serine moiety of phosphatidylserine (Chapter 14). Upon the transfer of three methyl groups, phosphatidylcholine is formed. Choline can be liberated from the phosphatide to give rise to the important neurotransmitter *acetylcholine*, or it can be recycled into phospholipids. Ethanolamine is also a constituent of certain lipopolysaccharides of Gram-negative bacteria.

Two polyamines found complexed with DNA are spermidine and spermine. The formation of *spermidine* and *spermine* is dependent on the decarboxylation of both ornithine (or arginine) and *S*-adenosylmethionine. Ornithine is decarboxylated to *putrescine* in both animals and bacteria. A second route to putrescine is found in some bacteria and involves the decarboxylation of arginine. Both pathways will be considered in the section dealing with the utilization of arginine.

Several physiologically important metabolites are derived indirectly by amino acid decarboxylation. Among these are *adrenaline* (epinephrine) and *serotonin*, which are formed by the decarboxylation of hydroxylated derivatives of tyrosine and tryptophan, respectively. Their biosynthetic pathways are shown in Figure 23-2. The direct decarboxylation products of the aromatic amino acids phenylethylamine, tyramine, and tryptamine are, like adrenaline and serotonin, neurologically active compounds that serve as vasoconstrictors (i.e., constrictors of blood vessels) but with much less activity than either adrenaline or serotonin. Histamine, the direct decarboxylation product of the basic amino acid histidine, acts as a vasodilator.

## Glutathionine and Its Physiological Roles

The tripeptide γ-glutamylcysteinylglycine, or glutathione, is found in nearly all cells and plays a variety of roles. The tripeptide is formed in two steps catalyzed by ATP-requiring reactions. The first step is the condensation of glutamate with cysteine:

$$\text{Glutamate} + \text{cysteine} + \text{ATP} \xrightarrow{\substack{\text{γ-Glutamylcysteine} \\ \text{synthase}}} \text{γ-glutamylcysteine} + \text{ADP} + P_i$$

The second step is the condensation of the dipeptide with glycine:

$$\text{γ-Glutamylcysteine} + \text{glycine} + \text{ATP} \xrightarrow{\substack{\text{Glutathione} \\ \text{synthase}}} \text{glutathione} + \text{ADP} + P_i$$

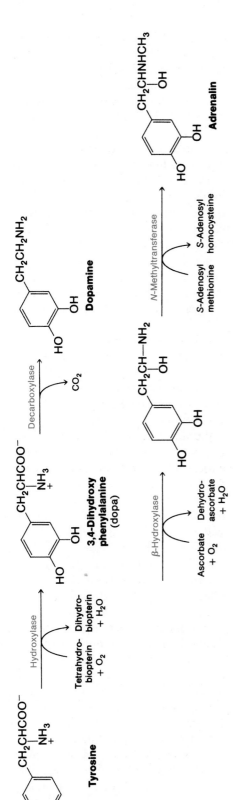

**Figure 23-2**
The conversion of tyrosine and tryptophan to adrenalin and serotonin.

Glutathione has often been considered important in maintaining the sulfhydryl groups of proteins in the cell in a reduced state, presumably by a nonenzymatic reaction. In contrast, an enzyme, protein-disulfide reductase, does catalyze sulfhydryl-disulfide interchanges between glutathione and proteins. The enzyme is important in insulin breakdown and is probably important in the reassortment of disulfide bonds during polypeptide chain folding.

Mutants of *E. coli* have been isolated that are essentially devoid of glutathione owing to the loss of one or the other of the two synthases. Such cells are viable and have normal growth rates, but they are more sensitive to sulfhydryl reagents, such as mercurials.

Glutathione also plays a role as a reduced carrier for the reduction of glutaredoxin, which, like thioredoxin, is a hydrogen donor for nucleotide reductase (see Chapter 19) and for the reduction of the carrier-bound sulfate reduction pathway of *E. coli* (see Chapter 22). Glutathione is also important in maintaining the iron of hemoglobin in the ferrous state. In all these roles, the reduction of oxidized glutathione by the NADPH-dependent glutathione reductase provides for the regeneration of reduced glutathione.

Glutathione also serves as a coenzyme in several reactions, including the two-step conversions of methylglyoxal to lactate by the two enzymes, glyoxylase I and glyoxylase II, and of formaldehyde to formate by formaldehyde dehydrogenase and *S*-formylglutathione hydrolase. In each case, an *S*-acylester of glutathione is formed as a free intermediate.

A role for glutathione that is independent of its reducing property is one as a γ-glutamyl donor in *the γ-glutamyl cycle*. The reactions of the γ-glutamyl cycle are shown in Figure 23-3. Of particular significance is the fact that cells exhibiting the activities of the cycle contain substantial amounts of γ-glutamyl transpeptidase activity on the outer surface of their cell membranes. This enzyme is thought to transfer the γ-glutamyl group of extracellular glutathione to an extracellular amino acid. The γ-glutamyl-amino acid is transported into the cell, where, as a substrate for γ-glutamyl cyclotransferase, the amino acid and 5-oxoproline are released. 5-Oxoprolinase is converted to glutamate in an ATP-dependent cleavage catalyzed by 5-oxoprolinase. The glutathione is then regenerated from glutamate and the cysteine and glycine that were released by cysteinylglycine dipeptidase.

The γ-glutamyl cycle enzymes are found in those tissues for which the transport of glutathione into cells is an important function. While glutathione is exported by most cells, it is efficiently transported only into cells that contain the membrane-bound γ-glutamyl transpeptidase. Thus the *γ-glutamyl transpeptidase appears to facilitate salvage of glutathione secreted by some tissues into the blood stream as well as to permit an energy-driven transport system for amino acids*. Either oxidized or reduced glutathione is a substrate for the transpeptidase. The enzyme also catalyzes the hydrolysis of glutathione to glutamate and cysteinylglycine and glutamine to glutamate and ammonia. Many amino acids can serve as acceptors for the γ-glutamyl group, including γ-glutamylamino acids (to yield γ-glutamyl-γ-glutamylamino acids) and even glutathione itself. *Such a transport across cell membranes is probably especially important for cysteine and methionine, as well as for glutathione.*

That this cycle does play a role in transport is shown by the fact that incorporating glutamyl transpeptidase isolated from kidney into erythro-

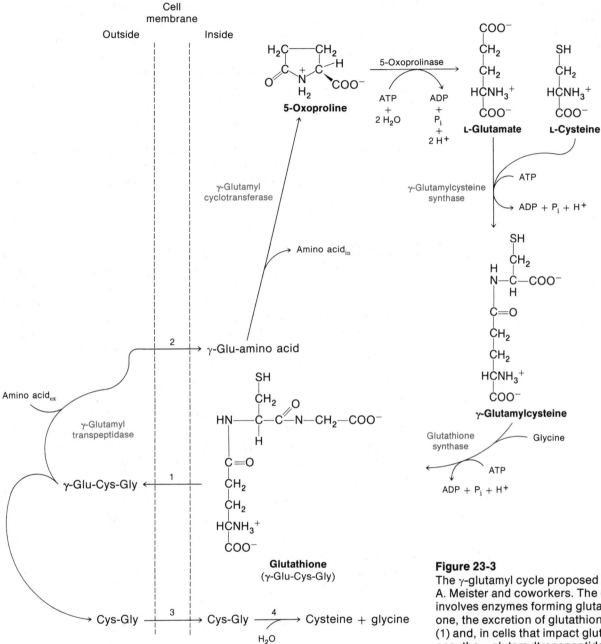

**Figure 23-3**
The γ-glutamyl cycle proposed by A. Meister and coworkers. The cycle involves enzymes forming glutathione, the excretion of glutathione (1) and, in cells that impact glutathione, the γ-glutamyltranspeptidase-dependent transport (2) of another amino acid (or peptide), the cleavage of the intracellular γ-glutamyl amino acid, and the ATP-dependent conversion of 5-oxoproline to glutamate. The cysteinylglycine (Cys-Gly) formed in the reaction is thought to be transported by an uncharacterized transport system (3) and cleaved by an intracellular protease (4) or cleaved by a membrane-bound protease and the free amino acids transported.

cyte membranes stimulates the transport of glutamate and alanine supplied along with glutathione to such preparations. Furthermore, patients with marked deficiencies in γ-glutamyl transpeptidase activities excrete glutathione, and because high concentrations of glutathione inhibit transport of γ-glutamylcysteine, the latter, formed in small amounts by residual levels of transpeptidase, is also excreted.

There is another class of enzymatic conversions involving glutathione that occurs in the cytosol of the liver. The enzymes are called glutathione S-transferases. They are similar enzymes, few in number, that exhibit a broad range of activities owing to their ability to bind many hydrophobic substances, such as bilirubin, steroids, and polycyclic aromatic hydrocarbons. Nearby this hydrophobic binding site is a second site specific for glutathione. Whether the hydrophobic ligand is also a substrate depends on whether it has an electrophilic atom that can undergo nucleophilic attack by glutathione resulting in conjugation with glutathione and release of the electrophilic atom:

$$RX + GSH \longrightarrow RSG + HX$$

The glutathione conjugation product may be further metabolized or excreted as such. The glutathione S-transferases thus serve to solubilize, detoxify, and initiate the catabolism of a wide variety of hydrophobic substances.

## Gramicidin Biosynthesis

The antibiotic gramicidin produced by the bacterium *Bacillus brevis* is a cyclic decapeptide composed of a repeated sequence of five amino acids $(-D\text{-Phe-L-Pro-L-Val-L-Orn-L-Leu-})_2$. Whereas in glutathionine biosynthesis the order of amino acids is determined by a specific enzyme for each peptide linkage made, in gramicidin the ordering is mainly a function of the attachment points of amino acids on the enzyme surface. Thus one of the two enzymes involved in gramicidin appears to be serving a dual role of enzyme and template. The synthesis of more complex polypeptides requires the intricate biochemical machinery used in protein synthesis. Gramicidin synthesis is divided into five phases.

**1.** Activation, thioesterification, and racemization of L-phenylalanine. The light enzyme of the gramicidin-forming system activates L-phenylalanine as the aminoacyl adenylate and transfers the phenylalanyl residue to a thiol group on the enzyme. Racemization of L-phenylalanine occurs at this thioester stage:

$$\text{ESH} + \text{ATP} + \text{L-Phe} \longrightarrow (\text{L-Phe} \sim \text{AMP})\text{ESH} + \text{PP}_i \longrightarrow$$
$$\text{ES} \sim \text{L-Phe} + \text{AMP} \longrightarrow \text{ES} \sim \text{D-Phe}$$

**2.** Activation and thioesterification of L-proline, L-valine, L-ornithine, and L-leucine. The heavy enzyme of the gramicidin-forming system activates the other four amino acids found in gramicidin S and transfers each to a specific thiol-containing site on the protein:

$$\frac{\text{E}}{\underset{H\ H\ H\ H}{S\ S\ S\ S}} + \text{Pro} + \text{Val} + \text{Orn} + \text{Leu} + 4\,\text{ATP} \longrightarrow$$

$$\frac{\text{E}}{\underset{\substack{P\ V\ O\ L \\ r\ a\ r\ e \\ o\ l\ n\ u}}{S\ S\ S\ S}} + 4\,\text{AMP} + 4\,\text{PP}_i$$

**3.** Transfer of D-phenylalanine to the heavy enzyme and initiation of peptide formation. The heavy enzyme contains a covalently linked 2-nm-long pantotheine arm that is thought to serve as a carrier of the growing peptide chain. It is to the —SH group of this pantotheine arm on the heavy enzyme that the D-phenylalanyl group is probably transferred from the light enzyme (transthiolation). The first peptide bond would then be formed by a transpeptidation reaction that liberates the pantotheine thiol group.

**4. Elongation.** The liberated pantotheine arm is now free to undergo transthiolation with the newly formed phenylalanylprolyl residue and to move to the valyl-thiol site to repeat the transpeptidation step. This step is repeated at the ornithinyl and leucyl sites.

**5. Cyclization.** After the pentapeptide is formed, it is cyclized with an identical peptide in a head-to-tail fashion. One possible mechanism, implied in Figure 23-4, involves the transfer of the first pentapeptide to a thiol "waiting" site, and when a second pentapeptide has been completed, cyclization occurs by two additional transpeptidation reactions (Figure 23-4e). Another possibility is that the cyclization occurs by an *inter*molecular transpeptidation involving two heavy enzymes, each containing one completed pentapeptidyl residue at the terminal thiol site (not shown).

## The Formation of Creatine

In skeletal muscle of vertebrates and, to a lesser extent, in other tissues, *creatine* (α-methylguanidoacetate) is an important reservoir of high-enery phosphate groups. In many invertebrates, arginine plays a similar role (see Chapter 8). Creatine is synthesized by means of a pair of reactions in which the amidino group of arginine is transferred to glycine to yield ornithine and guanidinoacetate, and a methyl group is transferred to the α nitrogen. These reactions are shown in Figure 23-5. The reversible phosphorylation of the terminal amino group results in the high-energy storage compound phosphocreatine. Creatine is lost from the metabolic pool by the spontaneous cyclization of either phosphocreatine or creatine itself to yield creatinine, which is excreted in the urine.

## Other Important Metabolites Derived from Amino Acids

The pathways to the purine and pyrimidine nucleotides described in Chapter 19 account for significant amounts of the glycine (for purine nucleotides) and aspartate (for both kinds of nucleotides) consumed for additional biosynthetic purposes. Several of the pathways to vitamins and

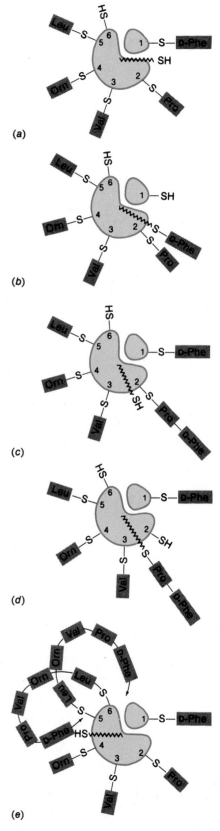

(a)

(b)

(c)

(d)

(e)

**Figure 23-4**
The formation of gramicidin S on a protein template. (*a*) The activated amino acids are held in thioester linkage on the light and heavy enzymes. The phenylalanyl residue has undergone racemization and is being transferred to the pantotheine arm on the heavy enzyme. (*b*) The first peptide bond is about to be formed by transfer of the phenylalanyl group from the pantotheine group to the prolyl residue. (*c*) The phenylalanylprolyl residue is about to be transferred to the free thiol group of the pantotheine arm. The light enzyme has accepted another phenylalanyl residue that has already undergone racemization. (*d*) The phenylalanylprolyl residue is about to be transferred to the valyl residue. (*e*) The first pentapeptidyl group after being made was transferred to the waiting site 6. The second pentapeptidyl group has just been completed and is about to be condensed with the first to yield the decapeptide gramicidin S. The pantotheine arm is now free to repeat the process.

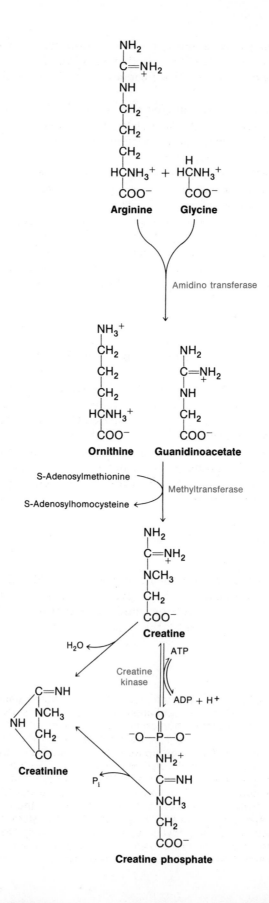

**Figure 23-5**
The formation of creatine phosphate
and creatinine.

coenzymes also rely on the availability of amino acids as direct precursors. Among these routes are those to pyridoxine (probably from serine), nicotinamide (from aspartate in bacteria and green plants; from tryptophan in fungi and, to a usually insufficient extent, in mammals), folic acid (from glutamate), and pantothenate (from valine or its precursor α-ketoisovalerate and from aspartate by way of β-alanine by a specific decarboxylase).

## DEGRADATION OF AMINO ACIDS

### Amino Acids as a Source of Carbon and Energy

*Just as amino acids are biosynthesized from a few of the intermediates in the central metabolic routes* (see Figure 22-1), *their complete degradation to carbon dioxide and water involves their conversion to intermediates in these same central pathways.* All amino acids can thus be degraded to carbon dioxide and water by appropriate enzyme systems. In every case, the pathways involve the formation, directly or indirectly, of a dicarboxylic acid intermediate of the tricarboxylic acid cycle, of pyruvate, or of acetyl-CoA.

Acetyl-CoA so formed can, of course, be oxidized to carbon dioxide by means of the tricarboxylic acid cycle or, when cycle function is restricted, can be converted to acetoacetate and lipid. Amino acids metabolized to acetoacetate and acetate are considered *ketogenic*. In contrast, α-ketoglutarate or the four-carbon dicarboxylic acids derived from amino acid breakdown can stimulate tricarboxylic acid function, since they play a catalytic role in the cycle. For their further metabolism they must leave the cycle by one of two routes (see Chapter 8). By one route, the conversion of oxaloacetate to phosphoenolpyruvate results in gluconeogenesis when carbohydrate utilization is restricted. For this reason, such amino acids are considered *glycogenic*. By the other route, pyruvate is formed and, after conversion of the latter to acetyl-CoA, can be oxidized completely to carbon dioxide and water.

*When amino acids are consumed as a carbon or energy source by either animal tissues or microorganisms, a large amount of nitrogen as ammonia is liberated.* Because ammonia is toxic, most terrestrial animals must convert it to urea or uric acid and dispose of it in that form. Microorganisms and many aquatic and marine animals, however, usually excrete nitrogen not needed for biosynthetic functions as free ammonia. Microorganisms using an amino acid as a carbon source utilize some of this liberated ammonia for biosynthesis of other amino acids and other nitrogenous compounds but excrete the rest into the medium. While some microorganisms are unable to utilize carbohydrates or readily available organic acids as a source of carbon and energy and must use amino acids, most microorganisms use carbohydrate carbon sources more readily and are even prevented from amino acid degradation when a readily utilizable carbohydrate is available. This repression of amino acid catabolic enzymes in microorganisms will be considered later in this chapter.

### Amino Acids as a Source of Nitrogen

Under the rather special condition in which there is no inorganic nitrogen source, some microorganisms can utilize amino acids as a nitrogen source while utilizing carbohydrates as the primary carbon source. Since some amino acid catabolic pathways are repressed in the presence of ample carbon sources (catabolite repression), there exists a block in the utilization of these amino acids that cannot be surmounted in some microorganisms. In other organisms, however, there is a "nitrogen limitation" signal that overcomes this catabolic block and allows the liberation of nitrogen from these amino acids even under conditions of an excess of readily utilizable carbon sources. When an amino acid is degraded as a source of nitrogen in the presence of a readily utilizable carbon source, the carbon skeleton of the amino acid is, of course, also utilized and contributes to the carbon metabolism of the cell.

### Some General Reactions in Amino Acid Degradation

With the exception of perhaps lysine, arginine, and threonine, the α-amino groups of all the amino acids found in proteins can be removed by transamination in one organism or another. Most *transaminases* have as one of the two different amino acid substrates *glutamate, aspartate, or alanine*, all of which can be deaminated by another means. Thus transamination can lead to a net conversion of α-amino nitrogen to ammonia. Nearly all transaminases examined contain pyridoxal-5′-phosphate as a prosthetic group. The general mechanism by which pyridoxal-5′-phosphate functions in transamination as a prosthetic group and other amino acid reactions has been considered in Chapter 6.

Many amino acids that can undergo transamination also can be deaminated by oxidative reactions either by a flavoprotein or by an NAD-linked enzyme of rather broad specificity, such as L-amino acid oxidase (FAD-linked) or L-amino acid dehydrogenase (NAD-linked). Some amino acids capable of undergoing transamination are substrates for more specific *oxidative deaminases*. Some of them will be considered later.

The broadly specific flavin-linked D-amino acid oxidases of animals as well as of microorganisms are of special interest, since D-amino acids are so seldom encountered by animals except as breakdown products of bacterial cell walls. Thus the physiological importance of the widely distributed and highly active D-amino acid oxidases remains obscure. The mammalian enzymes also exhibit activity against glycine.

*Whether transamination or direct deamination is more important as an initial step in amino acid breakdown probably depends on the organism or tissue under investigation*. However, where two mechanisms are available for one amino acid in a given cell type, it may well be that both mechanisms are employed.

Another general class of amino acid degradative enzymes would include the *decarboxylases*. Most of these also contain pyridoxal phosphate as a cofactor. The decarboxylases, however, do exhibit a more narrow specificity than do the L-amino acid dehydrogenases or oxidases. Except for some decarboxylases of some bacteria, the decarboxylases *do not appear to be important as degradative reactions but are*, as considered earlier, *involved in the generation of biologically important amines*. Certain bacteria, however, exhibit formation of amino acid decarboxylases only after the pH of the environment is low. It has been postulated that the production of amines in an acid environment serves as a selective homeostatic mechanism.

# SOME SPECIFIC ROUTES FOR AMINO ACID DEGRADATION

Space does not permit coverage of all aspects of amino acid degradation. This section is highly selective and considers only certain pathways in a limited amount of detail.

## Glutamate as an Amino-Group Carrier

The very important role for glutamate as an amino-group donor was considered in the previous chapter on amino acid biosynthesis. Under conditions of amino acid breakdown, glutamate also can be important in the removal of amino groups from amino acids serving as nitrogen or carbon sources. This role is important in cells that lack a specific deaminase for the amino acid being metabolized or when the activity of the specific deaminase is not sufficient. However, for glutamate to play a role in the net conversion of amino groups to ammonia, a mechanism for glutamate deamination is needed to regenerate the α-ketoglutarate. When glutamate can be readily deaminated, transamination can play a key role in the indirect deamination of many other amino acids (Figure 23-6).

## Methionine

The most important aspects of methionine breakdown have already been mentioned in Chapter 22 and involve the roles of methionine as a donor of methyl groups, as a propylamine donor in spermidine formation, and, in animals with low levels of cysteine in their diets, as a donor of organic sulfur. While methionine readily undergoes transamination, this reaction is probably less important than the other reactions it undergoes.

**Figure 23-6**
Routes for the oxidative deamination of an amino acid by transamination. (I) Route involving glutamate dehydrogenase. (II) Route involving aspartase. In either case, the net reaction is L-amino acid + $H_2O$ + $NAD^+$ ⟶ α-keto acid + $NADH$ + $H^+$ + $NH_3$.

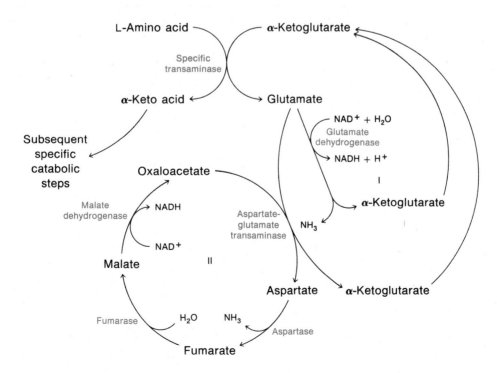

**The Role of Methionine in Methyl-Group Transfer.** Although during the biosynthesis of methionine its methyl group was derived from N-5 methyl-tetrahydrofolate, the transfer from methionine requires its activation by ATP to yield the methyl donor *S-adenosylmethionine*. Many enzymatic reactions have been observed in which *S*-adenosylmethionine serves as a methyl donor.

**Other Metabolic Routes from S-Adenosylmethionine.** Another route of *S*-adenosylmethionine metabolism is that in which *S*-adenosylmethionine serves as a donor of propylamine residues in spermidine and spermine biosynthesis. The 5'-methylthioadenosine formed in this transfer is further metabolized in the animal liver to yield methionine by a salvage pathway that converts the original methylthio group and four of the five ribose carbons to methionine. This role of *S*-adenosylmethionine will be considered further in the section on arginine utilization. Another pathway initiated from *S*-adenosylmethionine has been found to be an important one in plants. This is one leading to *ethylene*. Ethylene is important in stimulating the ripening and aging process in certain plants. It arises by cleavage of 5'-methylthioadenosine from *S*-adenosylmethionine to yield a four-carbon residue that cyclizes to 1-carboxy-1-amino cyclopropane. By an as yet undetermined route, this cyclic compound is cleaved to ethylene, $CO_2$, formate, and ammonia.

### Arginine

**Conversion to Ornithine.** *Arginine is converted to ornithine by two different routes. One provides a mechanism for urea formation and is important in the nitrogen metabolism in many animal species. The other route is one that provides a source of energy for many microorganisms.* The latter is called the *arginine dihydrolase pathway*. The first enzyme, arginine deiminase, converts arginine to citrulline, with the liberation of $NH_3$ (Figure 23-7). Citrulline can be cleaved by a degradative ornithine transcarbamoylase to yield ornithine and carbamoyl phosphate. The carbamoyl phosphate so formed serves as a high-energy phosphate donor for ATP formation in a reaction catalyzed by carbamate kinase. In some animal tissues it appears that the same reaction can be catalyzed by an acetate kinase.

The alternative formation of ornithine by the cleavage of urea from arginine is catalyzed by *arginase*. One role of this reaction is to serve as a means for the formation of urea as an end product of nitrogen metabolism and involves the cyclic resynthesis of arginine. This so-called urea cycle will be considered in detail in the next section. The arginase reaction is also the first in a series of reactions by which some microorganisms mobilize arginine as a nitrogen source.

**Urea Breakdown.** Urea is most simply broken down to $CO_2$ and $NH_3$ by the enzyme urease, which was the very first enzyme to be obtained in crystalline form. The enzyme is found in plants and many bacteria and in some fungi. In yeast, however, an entirely different mechanism for urea breakdown is found. The key enzyme in this pathway is a biotin-linked urea carboxylase, in which ATP is converted to ADP. The product, allophanate, is hydrolyzed by a second enzyme activity, allophanate hydrolase, which cleaves allophanate to two molecules of $CO_2$ and $NH_3$. In yeast, this

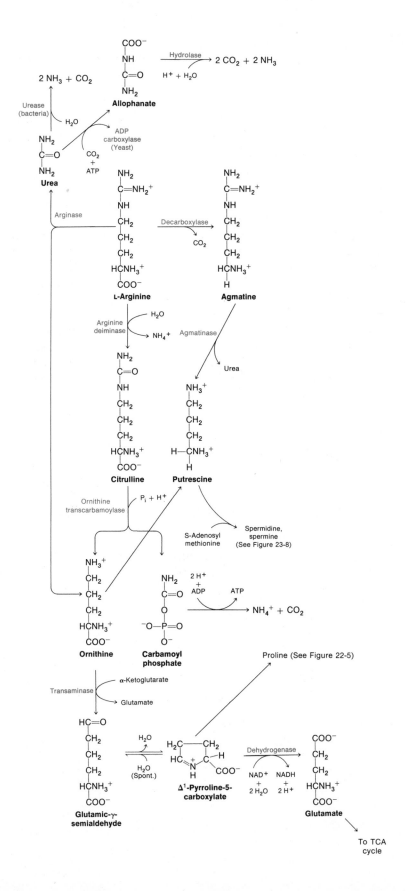

**Figure 23-7**
The breakdown of arginine and urea.

second activity is performed by the same protein that exhibits urea carboxylase activity. Whether the two activities are always part of the same protein or not is not clear. Both routes of urea breakdown are illustrated in Figure 23-7.

**The Degradation of Ornithine.** Ornithine, whether formed by the arginase or arginine dihydrolase pathway, is broken down in most organisms by a transaminase to yield glutamate-γ-semialdehyde or its cyclized derivative $\Delta^1$-pyrroline-5-carboxylate. This compound can be further oxidized to glutamate by $\Delta^1$-pyrroline-5-carboxylate dehydrogenase, an $NAD^+$-linked enzyme, or it can be reduced to proline by the normal proline biosynthetic enzyme $\Delta^1$-pyrroline-5-carboxylate reductase in an NADPH-requiring reaction (see Figure 22-5). This route to proline from ornithine accounts for the interconvertibility of ornithine and proline seen in many cells and tissues.

Other routes of ornithine degradation have been encountered in microorganisms, particularly among the anaerobes, but they will not be discussed here.

**Amines Produced from Arginine.** As mentioned in an earlier section, putrescine, the decarboxylated product of ornithine, is important as an intermediate in spermidine and spermine formation. The decarboxylation is catalyzed by ornithine decarboxylase, and in most bacteria this route is the

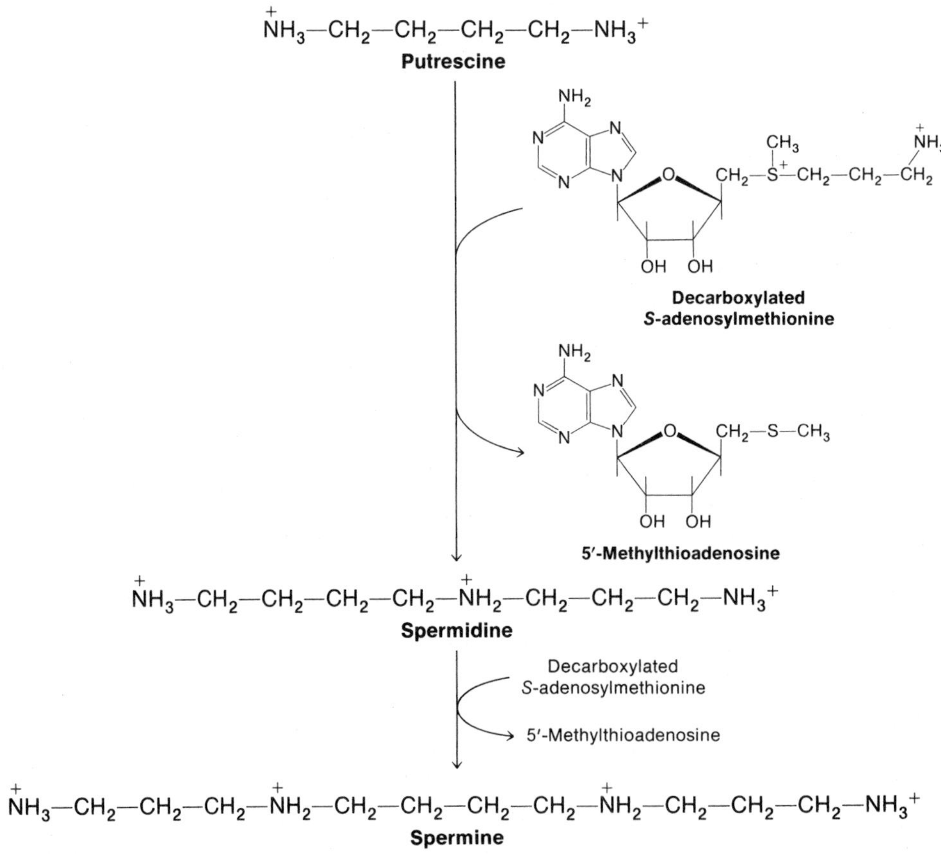

**Figure 23-8**
The utilization of putrescine and *S*-adenosylmethionine for the formation of spermidine and spermine.

sole or primary route of putrescine formation (see Figure 23-7). Another route is found in some organisms and is initiated by the conversion of arginine to agmatine by the action of arginine decarboxylase (Figure 23-7). Agmatine is cleaved to putrescine and urea by a urease-like enzyme, agmatinase.

**The Synthesis of Spermidine and Spermine.** The synthesis of the important polyamines spermidine and spermine requires not only the formation of putrescine, but the generation of a propylamine group. The propylamine donor is a product derived from S-adenosylmethionine by a specific decarboxylase (Figure 22-6). Spermidine is formed in a reaction catalyzed by aminopropyltransferase (see Figure 23-8). The same enzyme transfers an aminopropyl group to spermidine to yield spermine.

## Aromatic Amino Acid Catabolism

**Degradation of Tyrosine.** The oxidation of tyrosine (and phenylalanine; see below) by the liver proceeds by way of acetoacetate and a dicarboxylic acid. Thus tyrosine and phenylalanine are considered both _ketogenic_ and _glycogenic_. The role of tyrosine as a precursor of adrenaline and other physiologically active amines was cited in an earlier section of this chapter (The Conversion of Amino Acids to Biologically Active Amines).

**The Conversion of Tyrosine to Fumarate and Acetoacetate.** The first step in tyrosine catabolism is its conversion to 4-hydroxyphenylpyruvate by a tyrosine-glutamate transaminase of rather broad specificity (Figure 23-9). The next step is catalyzed by 4-hydroxyphenylpyruvate dioxygenase, a copper-containing enzyme that is stimulated by ascorbate. The product, homogentisate, results from oxidation of the aromatic ring and an oxidative decarboxylation and migration of the side chain. The aromatic ring is further oxidized and cleaved by homogentisate-1,2-dioxygenase to 4-maleylacetoacetate. The enzyme requires ferrous iron and is also stimulated by ascorbate. An isomerase, maleylacetoacetate isomerase, which requires glutathione as a cofactor, yields the trans compound 4-fumarylacetoacetate, which is hydrolytically cleaved to fumarate and acetoacetate by fumarylacetoacetate hydrolase.

**Melanin Formation.** Another route of tyrosine metabolism is that leading to melanin, which results from a two-stage attack on tyrosine by tyrosinase (Figure 23-10), yielding first dihydroxyphenylalanine (dopa). The latter is oxidized as a cosubstrate by tyrosinase to yield the 3,4-quinone. Thus catalytic amounts of dopa are required to sustain the reaction. The quinone form is unstable and undergoes a series of spontaneous reactions that result in the formation of indole-5,6-quinone, which is polymerized to melanin. Another enzyme forming dopa is tyrosine hydroxylase, which is important in dopamine (brain) and epinephrine (adrenal medulla) formation.

**Degradation of Phenylalanine.** Phenylalanine is broken down normally by way of tyrosine through the action of phenylalanine-4-monooxygenase, as indicated in Figure 23-11. The enzyme requires tetrahydrobiopterin as a cosubstrate, which is kept in the reduced form by NADPH, the ultimate

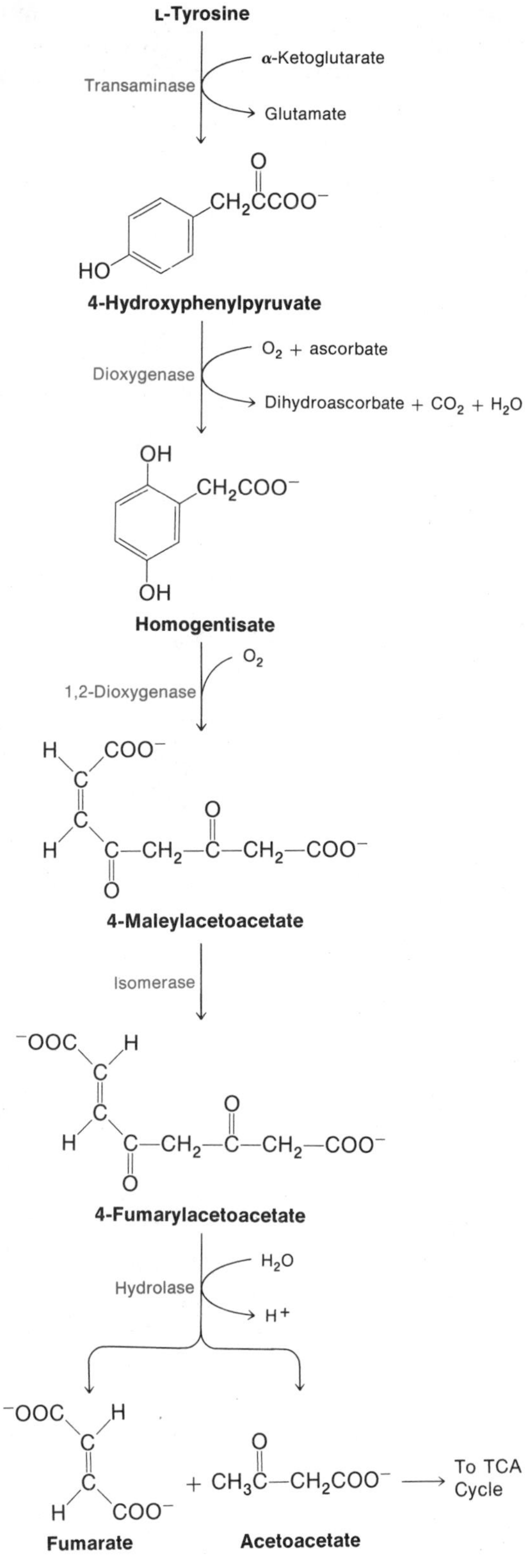

**Figure 23-9**
The conversion of tyrosine to
fumarate and acetoacetate.

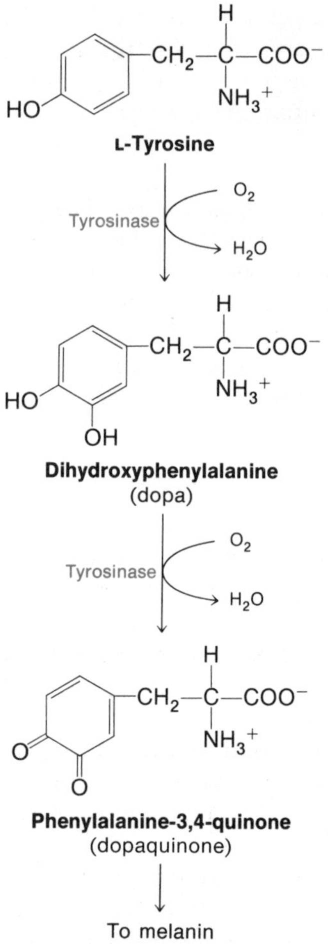

**Figure 23-10**
Melanin formation from tyrosine.

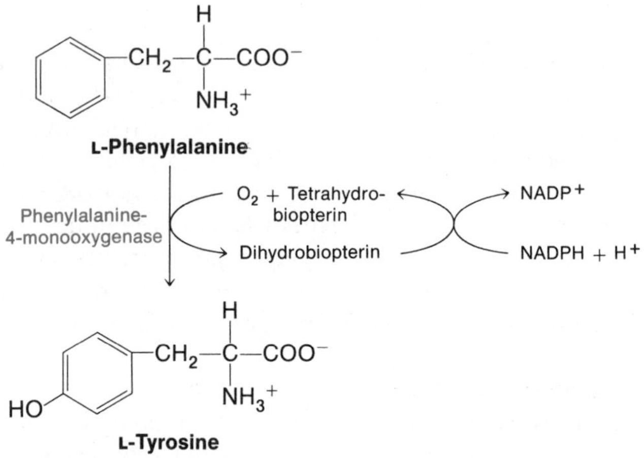

**Figure 23-11**
Degradation of phenylalanine.

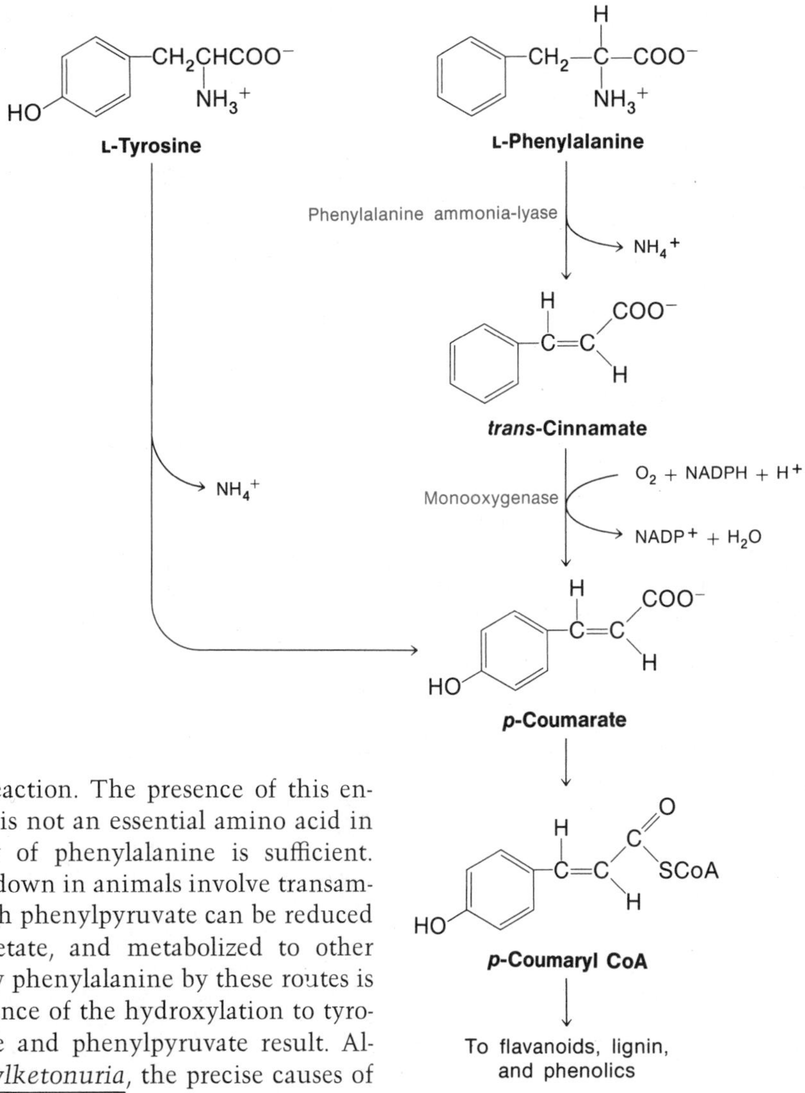

**Figure 23-12**
Conversion of phenylalanine and tyrosine to *p*-coumaryl-CoA, the precursor to flavanoids, lignin, and other compounds in plants.

hydrogen donor in the hydroxylation reaction. The presence of this enzyme accounts for the fact that tyrosine is not an essential amino acid in mammals, provided the dietary supply of phenylalanine is sufficient. Minor pathways for phenylalanine breakdown in animals involve transamination to yield phenylpyruvate. Although phenylpyruvate can be reduced to phenyllactate, oxidized to phenylacetate, and metabolized to other phenyl derivatives, the disposal of dietary phenylalanine by these routes is insufficient, so that in the inherited absence of the hydroxylation to tyrosine, high blood levels of phenylalanine and phenylpyruvate result. Although the condition is known as _phenylketonuria_, the precise causes of the mental retardation accompanying phenylketonuria are unknown, owing in part to the fact that there are several metabolic effects of the high levels of phenylalanine metabolites.

Heritable disorders in phenylalanine and tyrosine metabolism are among the most studied of the inborn metabolic errors in humans. These and other inborn errors will be considered in a later section of this chapter (Inborn Errors in Catabolism of Amino Acids in the Human).

Another important route for phenylalanine utilization, and probably for tyrosine breakdown as well, is found in plants in which the formation of flavonoids, lignin, and other derivatives of phenolic compounds plays an important role by means of the intermediate formation of *p*-coumarate coenzyme (Figure 23-12). A key reaction in the conversion of phenylalanine to *p*-coumarate-CoA is phenylalanine ammonia-lyase. The enzyme,

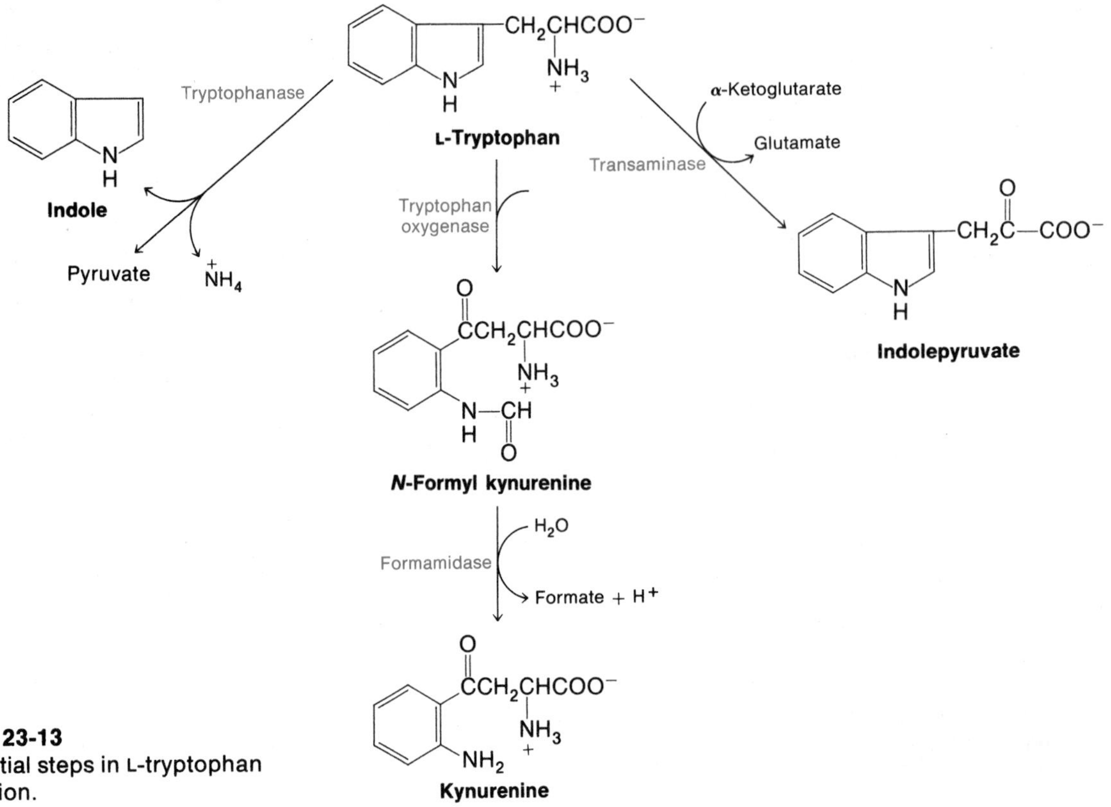

**Figure 23-13**
The initial steps in L-tryptophan utilization.

which catalyzes the removal of the hydrogen from the $\beta$ carbon that is cis to the amino group to yield *trans*-cinnamate, contains a dehydroalanine residue at the N-terminal end of the peptide chain. The amino group of the dehydroalanine is thought to be in an imine linkage with some other group on the protein that provides the electron sink required for elimination of the amino group from the $\alpha$ carbon. The *trans*-cinnamate is, in turn, oxidized to 4-hydroxycinnamate (*p*-coumarate) by *trans*-cinnamate-4-monooxygenase. The enzyme requires FAD and NADPH, which is oxidized in the hydroxylation process. *p*-Coumarate can also be formed directly from tyrosine through the action of phenylalanine ammonia-lyase on tyrosine. *p*-Coumarate is then converted to its coenzyme A derivative by *p*-coumaryl-CoA synthase. It is the coenzyme A derivative that is the branch point for a variety of biosynthetic routes found in plants.

**Degradation of Tryptophan.** The major pathways for tryptophan catabolism in the mammalian liver and for many microorganisms proceed by way of *kynurenine*. Kynurenine itself can be metabolized in liver by way of $\alpha$-ketoadipate, which is also an intermediate in lysine degradation. In certain microorganisms, kynurenine is metabolized by catechol, which, in turn, is metabolized by two different pathways. In other organisms, kynurenine is converted to kynurenate and further metabolized by a pathway yielding $\alpha$-ketoglutarate and aspartate. An interesting variant of the kynurenine pathway allows the synthesis of nicotinamide.

In many bacteria, tryptophan is initially cleaved by tryptophanase, which yields the readily utilized pyruvate and ammonium ions and indole, which would usually be accumulated in the medium. Some organisms utilize tryptophan as a source of nitrogen by means of an inducible aromatic amino acid–glutamate transaminase. The indolepyruvate so formed is excreted by the cells and is spontaneously polymerized to a brick-red pigment.

**The Route to Kynurenine.** The first step in the breakdown of tryptophan by many organisms is catalyzed by tryptophan oxygenase, which yields N-formylkynurenine (Figure 23-13). The enzyme is a dioxygenase and cleaves the indole ring by incorporating an oxygen atom on both C-2 and C-3 of the indole ring. Kynurenine itself is formed by the liberation of formate by kynurenine formamidase.

**The Conversion of Kynurenine to α-Ketoadipate in the Liver.** Kynurenine is converted to 3-hydroxykynurenine by the NADPH-dependent kynurenine-3-monooxygenase (Figure 23-14). Kynureninase, a pyridoxal phosphate enzyme, catalyzes a hydrolytic cleavage of the alanine side chain to yield 3-hydroxyanthranilate. The aromatic ring is cleaved to 2-amino-3-carboxymuconate-6-semialdehyde (ACS) by 3-hydroxyanthranilate oxygenase. Again, this enzyme is a dioxygenase, and oxygen atoms are incorporated on both the carbons at the site of ring cleavage. Ferrous ions are required by the enzyme. ACS can be spontaneously cyclized to quinolinate with the liberation of a molecule of $H_2O$. Quinolinate is an intermediate in the biosynthesis of nicotinamide, a synthesis that many animals have a

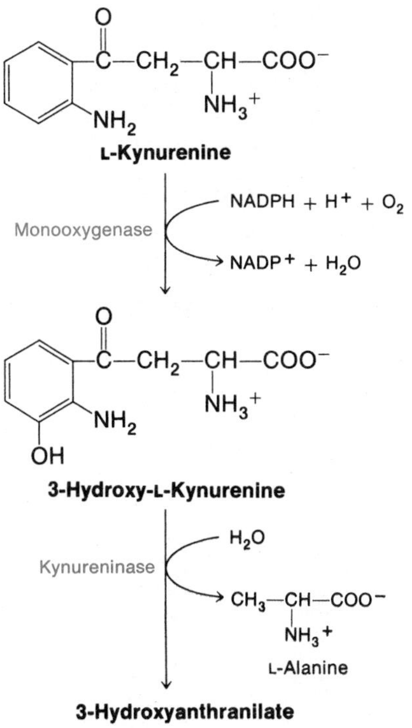

**Figure 23-14**
The conversion of kynurenine to α-ketoadipate in the liver. (*Continued*)

**Figure 23-14**
(*Continued*)

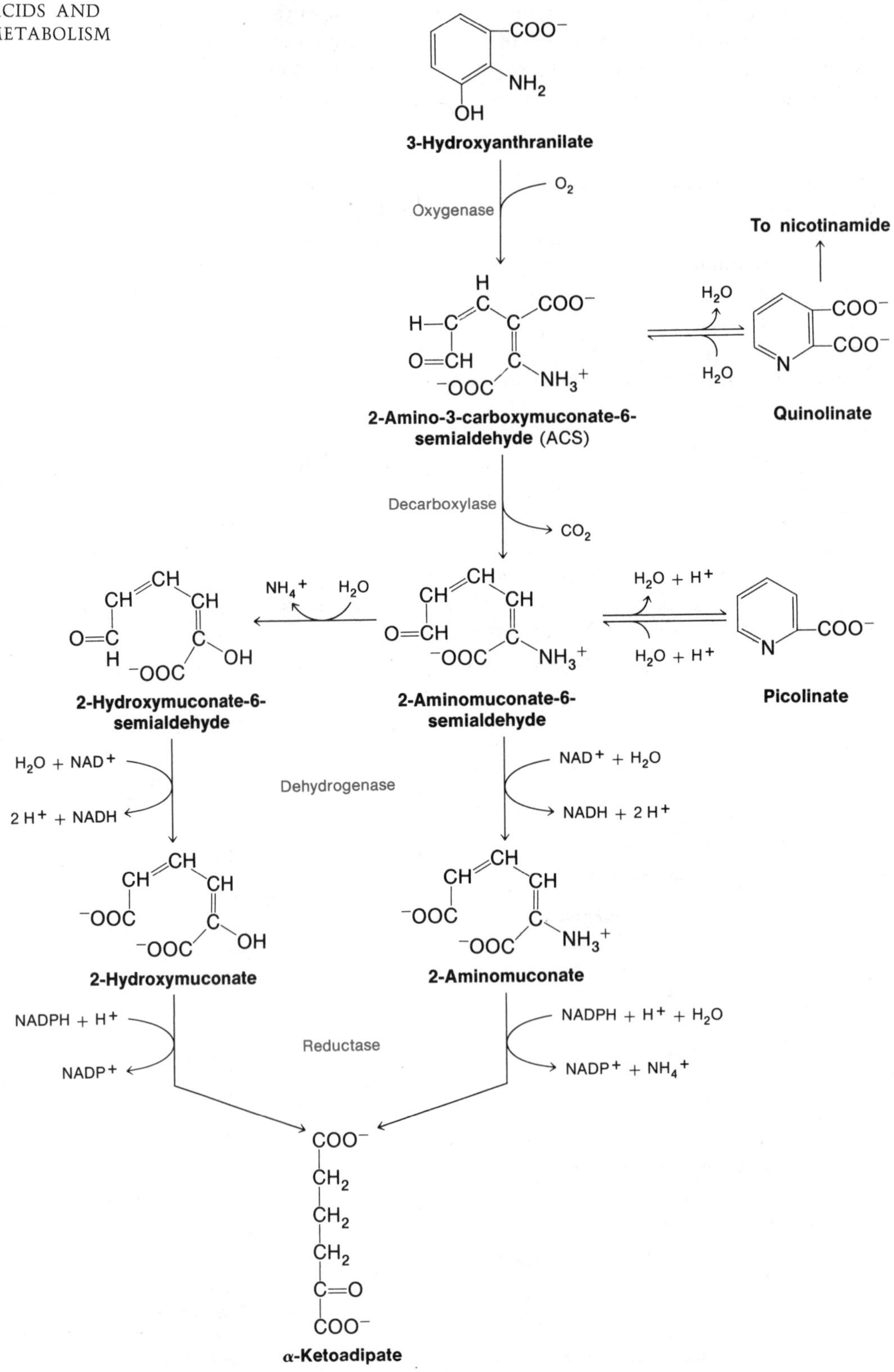

**3-Hydroxyanthranilate**

Oxygenase   $O_2$

**2-Amino-3-carboxymuconate-6-
semialdehyde** (ACS)

$H_2O$

$H_2O$

**To nicotinamide**

**Quinolinate**

Decarboxylase   $CO_2$

$NH_4^+$   $H_2O$

$H_2O + H^+$

$H_2O + H^+$

**2-Hydroxymuconate-6-
semialdehyde**

**2-Aminomuconate-6-
semialdehyde**

**Picolinate**

$H_2O + NAD^+$

$2 H^+ + NADH$

Dehydrogenase

$NAD^+ + H_2O$

$NADH + 2 H^+$

**2-Hydroxymuconate**

**2-Aminomuconate**

$NADPH + H^+$

$NADP^+$

Reductase

$NADPH + H^+ + H_2O$

$NADP^+ + NH_4^+$

**α-Ketoadipate**

limited capacity to perform. The ACS is decarboxylated by a specific decarboxylase to yield 2-aminomuconate-6-semialdehyde. 2-Aminomuconate-6-semialdehyde can undergo cyclization to picolinate, as indicated in Figure 23-14. Alternatively, 2-aminomuconate-6-semialdehyde can be oxidized by an NAD-dependent aminomuconate semialdehyde dehydrogenase. The same enzyme also attacks 2-hydroxymuconate-6-semialdehyde, a compound that could arise spontaneously by the hydrolytic removal of ammonia. 2-Amino(or 2-hydroxy)muconate is reduced to $\alpha$-ketoadipate by an NAD(P)H-dependent reductase.

**The Conversion of $\alpha$-Ketoadipate to Acetyl-CoA.** The further catabolism of $\alpha$-ketoadipate results in the liberation of two molecules of $CO_2$ and two of acetyl-CoA (Figure 23-15). The first step is a coenzyme A-dependent oxidative decarboxylation by an enzyme probably identical to $\alpha$-ketoglutarate dehydrogenase. The product, glutaryl-CoA, is oxidized by a flavin-linked dehydrogenase to a common intermediate in the $\beta$ oxidation of fatty acids, crotonyl-CoA. (The $\beta$-keto derivative glutaconyl coenzyme is probably an enzyme-bound intermediate in this reaction.) Two molecules of acetyl coenzyme are finally formed by the action of the fatty acid-oxidizing enzymes (see Chapter 13). $\alpha$-Ketoadipate is also formed in the liver by the breakdown of lysine by means of a pathway essentially the reverse of the fungal pathway for lysine biosynthesis (i.e., saccharopine is an intermediate). Thus the degradative pathways for lysine and tryptophan converge at the level of $\alpha$-ketoadipate.

**The Bacterial Catabolism of Kynurenine.** Kynurenine is formed from tryptophan by many soil pseudomonads by a route such as that found in liver. There are three major pathways by which kynurenine is further metabolized by microorganisms. In some forms, kynurenine undergoes transamination with $\alpha$-ketoglutarate as the preferred acceptor in a reaction catalyzed by a specific transaminase. In bacteria capable of metabolizing other aromatic compounds, the formation of catechol or a catechol derivative is a step frequently observed. Thus the steps in catechol utilization are not to be considered unique steps in tryptophan degradation; they also occur when other aromatic compounds, such as benzoate, serve as sources of carbon and energy for those organisms that can use them.

## THE FATE OF NITROGEN DERIVED FROM AMINO ACID BREAKDOWN

In the specific pathways of amino acid breakdown, the carbons of the amino acids are converted either to carbon dioxide or to intermediates in the central metabolic routes that can, in turn, be assimilated or oxidized to carbon dioxide. The nitrogen in each case is converted to ammonia either directly or indirectly (e.g., by means of a transamination to yield a readily deaminated product such as glutamate). In microorganisms using a single amino acid as a nitrogen source, the ammonia so liberated is assimilated and used to form other nitrogen-containing cell components. When the amino acid is a carbon source, much more ammonia is liberated than is needed for biosynthesis, and it is disposed of by excretion to the medium.

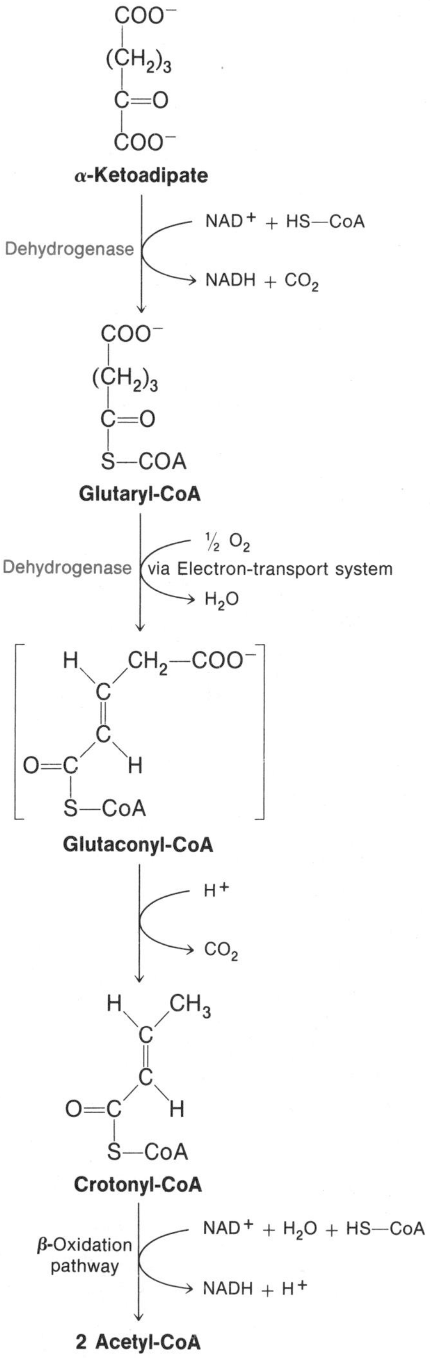

**Figure 23-15**
The conversion of $\alpha$-ketoadipate to acetyl-CoA.

This simple disposal mechanism is adequate *for free-living microorganisms,* since the *ammonia is carried away in the surrounding medium or escapes into the atmosphere.*

In some of the simpler aquatic and marine animal forms, such as protozoa, nematodes, and even bony fishes, aquatic amphibia, and amphibian larvae, ammonia is also the major nitrogenous end product. Such animals are called *ammonotelic.* However, *urea formation occurs in many animals in which removal of $NH_3$ by simple diffusion would be difficult.* Thus, in terrestrial snails and amphibia, as well as in other animals with environments in which water is limited, urea is the principal end product. Urea formation also serves to aid in maintaining osmotic balance with seawater in the cartilagenous fishes. In such animals, most of the urea secreted by the kidney glomerulus is readsorbed by the tubules. Indeed, the amount of nitrogen excreted by the kidneys of fishes is small compared with that excreted by the gills, and in most, ammonia is the major form of excreted nitrogen.

Another form of "detoxified" ammonia that is used in nitrogen excretion is uric acid (uricotelism). Uric acid is the predominant nitrogen excretory product in birds and terrestrial reptiles (turtles excrete urea, whereas alligators excrete ammonia unless dehydrated, when they, too, excrete uric acid). Since uric acid formed as a product of amino acid catabolism involves the de novo pathway of purine biosynthesis, its formation from $NH_3$ liberated upon amino acid catabolism will not be described here. In mammals, uric acid is exclusively an intermediate in purine catabolism, and in most mammals (primates excluded), it is further converted by uricase to alantoin.

### The Formation of Urea in the Animal Liver

The formation of urea in the liver involves the conversion of ornithine to arginine, as described in Chapter 22. Urea itself is formed from arginine by the action of arginase, which regenerates ornithine. The overall cyclic pathway was deduced by Krebs and Henseleit in 1932 with liver slices. The details, shown in Figure 23-16, have been developed through the efforts of many workers, particularly P. P. Cohen, S. Grisolia, and S. Ratner. The steps involved in the conversion of ornithine to arginine are identical to those found in forms that produce arginine and ornithine de novo from the carbon chain of glutamate (Figure 22-6). The formation of carbamoyl phosphate, however, is different. Rather than using glutamine, as does the enzyme found in plants and bacteria and that used by the liver for pyrimidine biosynthesis, the enzyme uses free ammonia. Another unique feature of the enzyme is that *N*-acetylglutamate is required for the function of the enzyme. The $NH_3$-dependent *N*-acetylglutamate-requiring enzyme is also found in the livers of ureotelic amphibia.

### Urea Formation in Marine Elasmobranchs and Invertebrates

As mentioned earlier, urea plays an important osmoregulatory role in the blood of marine elasmobranchs. The overall pathway in these organisms is again that postulated by Krebs and Henseleit. However, the enzyme catalyzing carbamoyl phosphate formation is different from either of those found

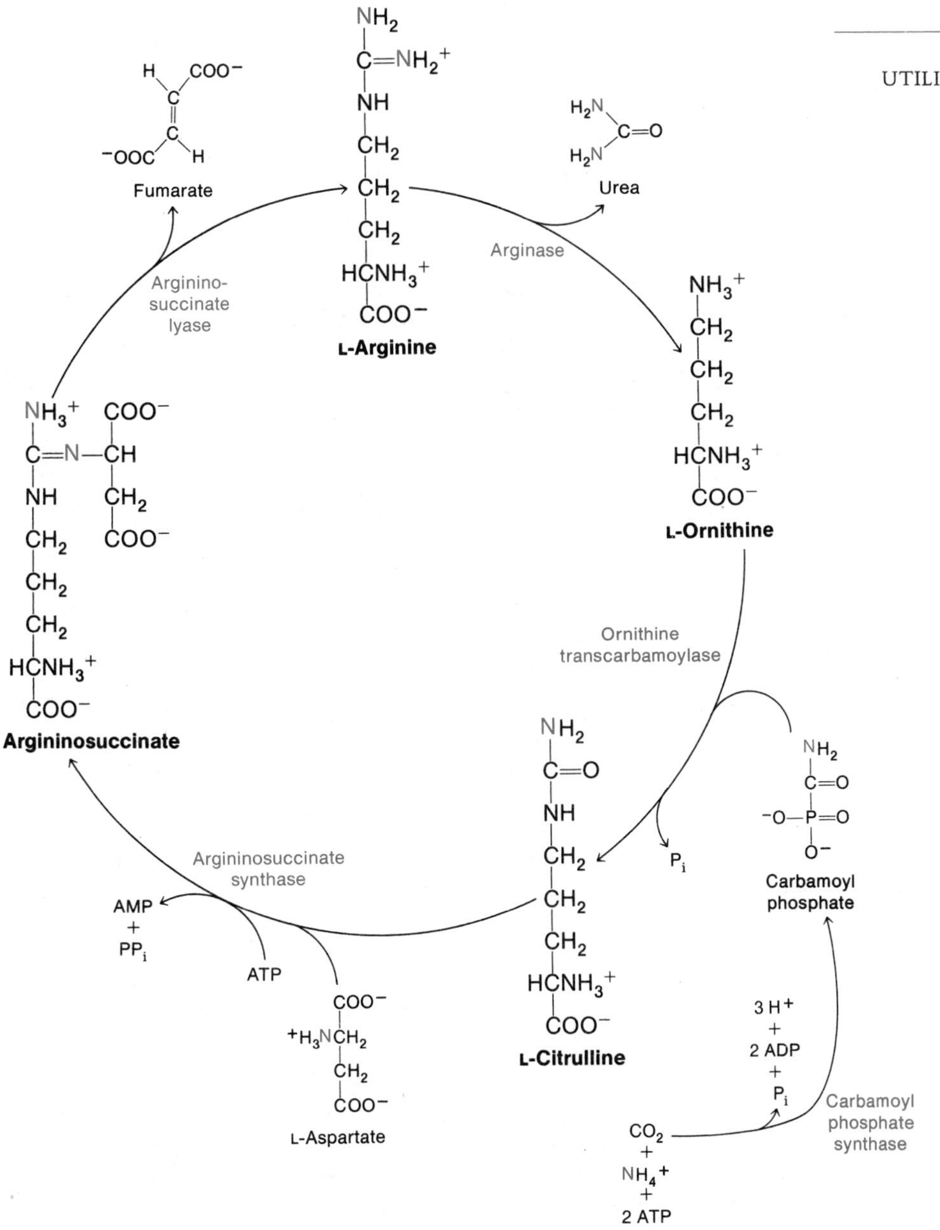

**Figure 23-16**
The urea cycle. A mechanism for removing unwanted nitrogen. Source of nitrogens involved in urea formation are drawn in color.

in the livers of other vertebrates. Rather than ammonia being used as substrate, the amide group of glutamine is used. However, like the enzymes found in the urea-forming systems in the livers of other vertebrates, the elasmobranch enzyme requires *N*-acetylglutamate as a cofactor. It is of interest that glutamine-utilizing *N*-acetylglutamate-stimulated carbamoyl phosphate synthases have been found in some invertebrates, such as terrestrial snails and earthworms, which shift from ammonotelism to ureotelism under conditions of restricted water supply.

The presence of a glutamine-dependent urea-forming system in elasmobranchs could be the result of retaining a system evolved in some invertebrates or it could be an adaptation related to the special role urea plays as an osmoregulator in these forms. It is interesting that the livers of urea-retaining elasmobranchs exhibited much higher glutamine synthase activities than do those of the non-urea-retaining teleosts, an observation that appears to be related to the important role of glutamine in urea synthesis by these animals.

## INBORN ERRORS IN CATABOLISM OF AMINO ACIDS IN THE HUMAN

It is interesting that the concept of a gene specifying the formation of a specific enzyme, which was exploited so successfully in analyzing biosynthetic pathways in bacteria and fungi, was introduced by Garrod in 1902 as a result of analyzing the occurrence of homogentisic acid excretion (alkaptonuria) in some of his patients and their families. He recognized the mendelian nature of the condition in families of several alkaptonuric patients and postulated that a genetically controlled enzymatic deficiency underlay this metabolic error.

Since then, *many metabolic diseases have been described that are due to the inability of the affected individuals to dispose of certain dietary components*. The diseases may be difficult to treat in the cases of errors in amino acid catabolism, since the culprit amino acid is one of the normal constituents of protein and is required for growth and development as well as for replacement of those body proteins that undergo rapid turnover. Because the affected fetus is usually carried by a mother heterozygous for the deficiency (carrying one normal and one defective gene) and whose own metabolism is essentially normal, the development of the fetus is essentially normal. Thus management of these diseases is possible, but it is dependent on a diagnosis soon after birth and the use of a low-protein diet carefully selected to supply enough of the culprit amino acid for protein formation but not enough to allow high plasma levels of it or of the offending metabolites. Supplements of nonoffending amino acids prepared by synthesis or by fermentation processes could be employed to compensate in part for the low-protein diet. Indeed, in Japan, the chemical industries have made such preparations available. Most Western countries have lagged behind in such developments.

Some of the diseases of amino acid catabolism are listed in Table 23-3. *Just as mutant methodology with fungi and bacteria helped to elucidate the pathways of biosynthesis, these naturally occurring mutants allow us to demonstrate conclusively the obligatory nature of some of the steps in amino acid breakdown*.

## REGULATION OF AMINO ACID BREAKDOWN

As pointed out earlier, some bacteria and fungi can use certain amino acids as a sole source of carbon and energy. In so doing, there is considerable nitrogen liberated, so that the amino acid used as a carbon source provides an excess of nitrogen, which is then excreted into the medium usually as ammonia. There are some microorganisms that can only utilize amino acids as a carbon source and are unable to attack carbohydrates. However, most forms will utilize a carbohydrate or an organic acid as a carbon and energy source in preference to an amino acid. In such forms, the induced formation of enzymes metabolizing the amino acid is prevented as long as the "preferred" carbon and energy source is present. Such *an antagonism of the induction of enzymes catabolizing one compound by a preferred or*

**Table 23-3**
Some Inborn Errors of Amino Acid Metabolism in Humans

| Amino Acid Catabolic Pathway Involved | Condition | Distinctive Clinical Manifestation | Enzymatic Block or Deficiency |
|---|---|---|---|
| Arginine and the urea cycle | Argininemia and hyperammonemia | Mental retardation | Arginase |
| | Hyperammonemia | Neonatal death, lethargy, convulsions | Carbamoyl phosphate synthase |
| | Ornithinemia | Mental retardation | Ornithine decarboxylase |
| Glycine | Hyperglycinemia | Severe mental retardation | Glycine-cleavage system |
| Histidine | Histidinemia | Speech defects, mental retardation in some; in others, none | Histidase |
| Isoleucine, leucine, and valine | Branched-chain ketoaciduria ("maple syrup disease") | Neonatal vomiting, convulsions, and death; mental retardation in survivors | Branched-chain keto acid dehydrogenase complex |
| Isoleucine, methionine, threonine, and valine | Methylmalonic acidemia | Similar to preceding except that methylmalonate accumulates | Methylmalonyl-CoA mutase (some patients respond to vitamin $B_{12}$ therapy) |
| Leucine | Isovaleric acidemia | Neonatal vomiting, acidosis, lethargy, and coma; survivors mentally retarded | Isovaleryl-CoA dehydrogenase |
| Lysine | Hyperlysinemia | Mental retardation and some noncentral nervous system abnormalities | Lysine-ketoglutarate reductase |
| Methionine | Homocystinuria | Mental retardation common; several eye diseases and thromboembolism common; osteoporosis and faulty bone structures | Cystathionine-$\beta$-synthase |
| Phenylalanine | Phenylketonuria and hyperphenyl-alanineuria | Vomiting is an early neonatal symptom, but mental retardation and other neurologic disorders develop in the absence of dietary treatment | Phenylalanine hydroxylase |
| Proline | Hyperprolinemia, type I | Probably not etiologically associated with any disease; proline excreted | Proline oxidase |
| Tyrosine | Alkaptonuria | Homogentisic acid in urine darkens on standing; in adult years, pigment deposits cause darkening of skin, cartilage; arthritis develops | Homogentisic acid oxidase |
| | Albinism | The most common type, oculocutaneous albinism, results in white hair, pink skin, and an extreme photophobia owing to lack of pigment in the eye | Tyrosinase of the melanocyte is absent |
| Glutathione and the γ-glutamyl cycle | 5-Oxoprolinuria | Metabolic acidosis; γ-glutamylcysteine accumulates because of a lack of glutathione, the feedback inhibitor of γ-glutamylcysteine synthase; the 5-oxoproline arises from activity of the cyclotransferase | Glutathione synthase |

*more readily used carbon source is an example of catabolite repression and has been described elsewhere* (Chapters 8 and 26).

Microorganisms also can occur in environments in which a good carbon source is available, but the only source of nitrogen is an amino acid. For some microorganisms, this nitrogen source would be unavailable, since its breakdown is prevented by catabolite repression. A fairly common finding, however, is that catabolite repression can be bypassed by an induced formation of the particular catabolic pathway as a result of a control signal that is conditioned through nitrogen starvation in the cell.

In animal cells, the breakdown of amino acids is also subject to regulation. The amino acid catabolic enzymes are much more often subject to a hormonal control than to a specific induction by substrate. Some catabolic enzymes attacking amino acids appear to be developmentally controlled and are formed only in certain tissues or at certain times during development. The developmentally programmed appearance of certain catabolic enzymes could, in fact, be mediated by hormonal signals that are themselves developmentally programmed.

### Carbon and Energy Control over the Induced Formation of Amino Acid Catabolic Enzymes in Microorganisms

A bacterium such as *E. coli*, growing in a medium containing a good carbon source such as glucose and a readily used nitrogen source such as ammonium ions, would exhibit, at most, only low activities of the amino acid catabolic enzymes that have been described in this chapter. Further, addition of tryptophan to the medium would not induce the enzyme that degrades it, i.e., tryptophanase. However, if glucose is consumed, or if the carbon source is a poor one, such as succinate, tryptophanase is induced. Thus tryptophan can serve as the sole source of nitrogen when succinate is the carbon source, but not when glucose is the carbon source. Like the catabolite repression of enzymes degrading certain carbohydrates that are used less well than glucose (discussed in Chapter 26), the carbon-source effect is achieved by preventing transcription of the structural gene. In addition, like other catabolite-repressible enzymes, the transcription of the structural gene for tryptophanase (*tna*) is dependent on cAMP, which is maintained at a low level when an ample carbon and energy source is available. In addition, a protein that binds cAMP is also required and serves as a positive control element that binds to a specific site near the promoter it activates.

A unique carbon-source effect is found in the regulation of asparaginase II and biodegradative threonine deaminase of *E. coli*, in which not only is the enzyme repressed by glucose, but its formation appears to be dependent on anaerobiosis and the presence of an amino acid mixture. Neither are specifically induced by their substrates. Table 23-4 lists some amino acid degradative enzymes that are subject to catabolite repression in bacteria and fungi.

### Nitrogen-Supply Control over the Induced Formation of Amino Acid Catabolic Enzymes in Microorganisms

While many amino acid degrading enzymes are catabolite-repressible and completely dependent on cAMP for their formation, there are others which, instead, can be induced only if the nitrogen supply is limited. The

**Table 23-4**
The Regulation of Some Typical Amino Acid Degradative Pathways in Bacteria and Fungi

| Amino Acid | Organism | Inducer | Key Enzymes in Degradation |
|---|---|---|---|
| **Enzymes induced when carbon and energy limit growth** | | | |
| Alanine | *E. coli* | D- or L-alanine | Alanine racemase, D-alanine dehydrogenase |
| Arginine | *S. cerevisiae* | Arginine | Arginase, ornithine transaminase |
| Asparagine | *E. coli* | None; stimulated by anaerobiosis | Asparaginase |
| Glutamate | Fungi | Glutamate | NAD-linked glutamate dehydrogenase |
| Histidine | *K. aerogenes* | Urocanate | Histidase, urocanase |
| Proline | *E. coli* | Proline | Proline oxidase |
| D-Serine | *E. coli* | D-Serine | D-Serine deaminase |
| Threonine | *E. coli* (growing anaerobically) | Amino acid mixture (including threonine) | Threonine deaminase |
| Tryptophan | *E. coli* | Tryptophan | Tryptophanase |
| **Enzymes induced when nitrogen limits growth** | | | |
| Arginine | *S. cerevisiae* | Arginine | Arginase, ornithine transaminase |
| Asparagine | *K. aerogenes* | None | Asparaginase |
| Histidine | *K. aerogenes* | Urocanate | Histidase, urocanase |
| Proline | *E. coli* | Proline | Proline oxidase |
| **Enzymes induced independently of carbon or nitrogen supply** | | | |
| Glycine | *E. coli* | Glycine | Glycine cleavage system |
| Serine | *E. coli* | Leucine | S-Serine deaminase |
| Threonine | *E. coli* | Leucine | Threonine dehydrogenase |

exact mechanism of nitrogen control over the enzyme formation is still being examined. However, it is clear that the capacity of the cell to form glutamine and amino groups is reflected in the nitrogen-control signal.

As discussed in Chapter 22, when nitrogen is in excess, the predominant mode of amino-group formation in many bacteria is by means of the reductive amination of $\alpha$-ketoglutarate by free $NH_3$. Under these conditions, less glutamine is required, glutamine synthase is repressed, and the protein is in its less active state. When nitrogen is limited, $NH_3$ is more efficiently utilized by the energy-dependent glutamine synthase reaction, and the amide group of glutamine is used in the reductive amination of $\alpha$-ketoglutarate catalyzed by glutamate synthase. Under these conditions, glutamine synthase is derepressed, and the protein is in its more active form. At the present time, it is not known whether the *"nitrogen limitation" signal is more directly related to the state of glutamine synthase, to one of the proteins involved in the cascade modifying the glutamine synthase protein, or to some other regulatory element that responds to the $\alpha$-ketoglutarate:glutamine ratio and which is also important in the control of glutamine synthase formation.* Amino acid catabolic enzymes subject to nitrogen control are also listed in Table 23-4. It should be noted that some are subject to both carbon control and nitrogen control. Such enzymes are derepressed by either nitrogen or carbon starvation. Thus, for these enzymes, catabolite repression is counteracted by nitrogen limitation, nitrogen repression is counteracted by carbon limitation, and the amino acids involved can serve as sole carbon or nitrogen sources when they are so supplied.

Not included in the second part of Table 23-4 are many other enzymes that are related to nitrogen metabolism and which are also dependent on a nitrogen-deficiency signal for their formation. Examples of these would be the nitrogen-fixation enzymes, nitrate and nitrite reductases, purine-degrading enzymes, and several transport systems in various microorganisms.

## Amino Acid Catabolic Enzymes Subject to neither Carbon nor Nitrogen Repression

In contrast to the enzymes that are dependent on a nitrogen or carbon limitation for induction, some amino acid degradative enzymes are not influenced by either the nitrogen supply or the carbon and energy supply. Among these in some bacteria are L-serine deaminase and threonine dehydrogenase. Surprisingly, both these enzymes are induced by growth in leucine. The glycine cleavage enzyme in *E. coli* is induced by glycine even in the presence of ample carbon and nitrogen supplies. The induction of this enzyme allows exogenous glycine to serve as a source of one-carbon units and, in conjunction with serine hydroxymethylase, as a source of serine in a glucose-ammonia salts medium. Although the enzyme is present under these conditions, it is probably kept from functioning beyond the need for supplying one-carbon units by the availability of free tetrahydrofolate as the obligatory acceptor of one-carbon units.

## Regulation of Amino Acid Catabolic Enzymes in Animals

Very few amino acid degradative enzymes have been studied in animals to ascertain the factors involved in their formation. The *presence of different degradative enzymes in different tissues clearly indicates a developmentally controlled program of gene expression*. A few enzymes degrading amino acids have been measured at various stages during the development

**Table 23-5**
Regulation of Some Typical Amino Acid Degradative Enzymes in Animal Cells and Tissue

| Amino Acid | Enzyme | Cell or Tissue | Factors Affecting Formation |
|---|---|---|---|
| Arginine and its precursor, ornithine | Ornithine-glutamate transaminase | Liver | High-protein diet; glucagon or cyclic AMP stimulate formation in vivo and in cell culture; corticosteroids and glucose repress in vivo |
| | Urea cycle enzymes | Liver | High-protein diet stimulates |
| Serine | Serine deaminase | Liver | High-protein diet stimulates |
| Tryptophan | Tryptophan oxygenase | Liver | Glucocorticoids stimulate formation |
| Tyrosine | Tyrosine-glutamate transaminase | Liver | Glucocorticoids stimulate formation |
| Threonine | Threonine deaminase | Liver | High-protein diet stimulates formation |

of animals from late fetal periods to the adult stage. Thus tryptophan oxygenase is very low in the newborn rat, but it increases after about 12 days concomitantly with a rise in adrenal activity. That at least part of the developmentally controlled formation of the enzyme is mediated by adrenal activity is indicated by the fact that glucocorticoids induce the de novo formation of the enzyme in young rats and stimulate its formation in adults. Liver serine deaminase also increases around this time, but it exhibits a transient increase in activity at birth. Ornithine transcarbamoylase and the other urea-forming enzymes appear shortly after birth and, like serine and threonine deaminases and ornithine transaminase, are further induced by high-protein diets.

Table 23-5 lists several amino acid degrading enzymes that have been shown to be induced or repressed in animal cells and tissues, and some of the factors affecting their formation.

## SELECTED READINGS

Barker, H. A. Amino acid degradation by anaerobic bacteria. *Ann. Rev. Biochem.* 50:23, 1981. A review of an important group of fermentation pathways of amino acid breakdown that occur in nature and could not be covered in this chapter.

Meister, A. *Biochemistry of the Amino Acids,* Vol. 2, 2d Ed. New York: Academic Press, 1965. Pp. 593–1084. A very complete survey of amino acid catabolic pathways as they were known up to that time.

Mazelis, M. Amino Acid Catabolism. In B. J. Mifflin (Ed.), *The Biochemistry of Plants,* Vol. 5. New York: Academic Press, 1980. Pp. 541–567. A survey of some of the amino acid catabolic pathways that have been found in plants.

Battersby, A. R., Fookes, C. J. R., Matcham, G. W. J., and McDonald, E. Biosynthesis of the pigments of life: Formation of the macrocycle. *Nature* 285:17, 1980. These two papers discuss the steps in tetrapyrrole biosynthesis and the pathways diverting this nucleus to chlorophylls, hemes, cytochromes, and other macrocyclic pigments.

Griffith, O., Bridges, R. J., and Meister, A. Formation of γ-glutamylcyst(e)ine *in vivo* is catalyzed by γ-glutamyltranspeptidase. *Proc. Natl. Acad. Sci. USA* 78:2778, 1981. This and earlier references cited therein provide a discussion of the physiological role of the γ-glutamyl cycle.

Wellner, D., and Meister, A. A survey of inborn errors of metabolism and transport in man. *Ann. Rev. Biochem.* 50:911, 1981. This review documents the importance of the pathways that break down amino acids in humans.

## PROBLEMS

1. Isoleucine, valine, threonine, and methionine are all "glycogenic" amino acids. From studies of human inborn errors (see Table 23-3), what can you suggest might be the common pathway to glucose-6-P into which the metabolism of these four amino acids converge?

2. How do you account for the fact that some patients who accumulate propionate and others who accumulate β-methylcrotonate can be treated successfully by biotin therapy? Why would vitamin $B_{12}$ therapy remedy some patients' methylmalonic acidemia?

3. On the basis of the several inborn metabolic errors affecting leucine catabolism, propose a pathway for leucine breakdown.

4. α-Ketoadipate is formed as an intermediate in the catabolism of lysine by the liver. Based on the way ketoadipate is converted to lysine in fungi (see Chapter 22), propose a pathway for its formation from lysine in liver. Consider carefully the choice of pyridine nucleotide. What is the basis for your choice? (Note: There is an NAD-linked α-aminoadipate-δ-semialdehyde dehydrogenase in liver.)

5. What advantage can you propose for the way arginine is formed from ornithine in the mammalian liver over the way it is formed from ornithine in *E. coli?* (See Chapter 22.)

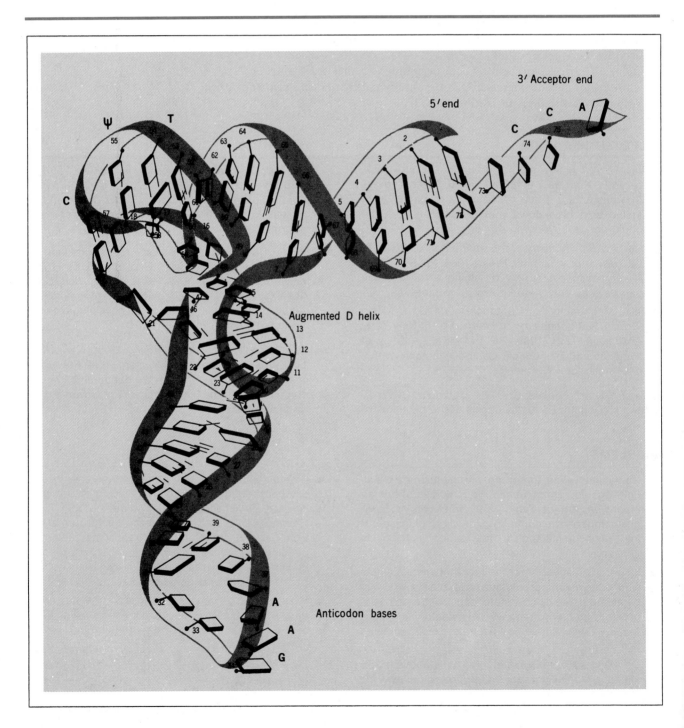

# 24

# THE GENETIC CODE

The end result of protein synthesis is the assembly, packaging, and distribution of the various catalytic and structural proteins essential to the life processes. Hundreds of macromolecules must function together to yield the thousands of different proteins in bacteria and tens of thousands of different proteins in higher organisms.

As analyzed in this and the following chapter, protein synthesis begins with transcripts of specific genes in the form of mRNA and ends with the assembly of free amino acids into active gene products. This overall reaction, which involves the rewriting of base sequences in nucleic acids as amino acid sequences in proteins, is termed _translation_. The catalogue of base sequences that specifies individual amino acids in proteins is known as the _genetic code_.

The assembly of amino acids into proteins occurs in all living systems on small particulate structures known as _ribosomes_. These structures are quite complex, some containing nearly 100 protein and RNA molecules; considerable effort is currently being expended to define their three-dimensional structure. Ribosomes serve to read the base sequence in mRNA, and in the course of this process they bind to and move along mRNA strands. Most commonly, a single molecule of mRNA is read simultaneously by a number of ribosomes, each engaged in the synthesis of a single polypeptide; the resulting structure resembles a string of beads and is known as a _polysome_ (Figure 24-1). Polysomes producing proteins that are destined for intracellular or extracellular transport are in turn bound to membranes. In higher organisms, these latter structures are known as _rough endoplasmic reticulum_ (Figure 24-2).

_The tertiary structure of yeast phenylalanine tRNA. Transfer RNAs are the key to translating a polynucleotide sequence into a polypeptide sequence._

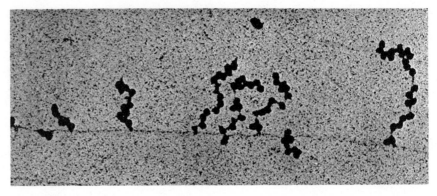

**Figure 24-1**
Electron micrograph of *E. coli* polysomes. Ribosomes are the dark structures connected by the faintly visible mRNA strand. A DNA strand connecting the polysomes from which the mRNA is being transcribed is visible as the diagonal line. (From O. L. Miller et al., Visualization of bacterial genes in action. *Science* 169: 392, 1970. Used with permission.)

The assembly of proteins on ribosomes in response to specific base sequences in mRNA does not involve free amino acids. Rather, these amino acids are first attached to small RNAs that act as adaptors in reading the genetic code of the message. The attachment of amino acids to these adaptors, or *transfer RNAs*, prior to their incorporation into protein is an important step in the translation process.

The ability to synthesize proteins is a property of all living systems. While the overall reaction is similar in all life forms, specific differences exist. The mechanism of protein synthesis is best understood as it occurs in bacterial, or *prokaryotic*, systems, and the ensuing discussion will focus on the process as it has been defined in the bacterium *Escherichia coli*. In higher organisms (*eukaryotes*), protein synthesis occurs in two different cellular locations, in the cytoplasm and in organelles known as mitochondria and chloroplasts. The distinguishing features of protein synthesis in these three systems, where known, will be pointed out throughout the text.

This chapter, then, focuses on three key elements of translation: the nature of the genetic code and how it was deduced, the structure and enzymology of transfer RNA, and the structure of the ribosome.

## THE GENETIC CODE

The demonstration by Sanger in 1953 that proteins are unique linear arrays of amino acids and the suggestion by Watson and Crick in the same year that DNA, the genetic material, is similarly a linear array of bases gradually led to the realization that genes are expressed by the linear translation of the base sequence of nucleic acids into the amino acid sequence of proteins. The *genetic code* is simply the sequence relationship between bases in genes (or mRNAs) and the amino acids in the proteins they encode.

How the chemical properties of nucleic acid bases are actually translated into amino acids in proteins was defined in a general way prior to the definition of the genetic code itself. A number of possibilities were considered in which amino acids fitted into a complementary pocket generated by a sequence of bases, but in 1957 Crick suggested that translation was actually accomplished by first attaching amino acids to "adaptor" molecules, small nucleic acids, which themselves paired with a template. Shortly thereafter, this _adaptor hypothesis_ was confirmed by the discovery of _transfer RNA (tRNA)_.

_Messenger RNA (mRNA)_ as the vehicle of genetic information in translation was not formally recognized until the proposal of its existence by Jacob and Monod in 1961. In the same year, Nirenberg and Matthaei made the startling discovery that _polyuridylic acid [poly(U)] codes for the synthesis of polyphenylalanine._

## Coding by Random-Sequence Synthetic Polynucleotides

The discovery that synthetic polynucleotides can act as messengers in protein synthesis was both serendipitous and crucial to the early elucidation of the code. This observation had its foundation in the earlier discovery by Grunberg-Manago and Ochoa of the enzyme polynucleotide phosphorylase, which catalyzes the conversion of nucleotide diphosphates into polyribonucleotides according to the following equation:

$$nNDP \rightleftharpoons RNA_n + nP_i$$

The reaction is reversible, and it was originally thought that this enzyme was responsible for RNA synthesis in vivo, but it is now believed to phosphorolytically degrade RNA in the cell. Nonetheless, it can be used in vitro to synthesize polynucleotides whose sequence is random (synthesis is independent of a template) and whose composition is determined by the available nucleotides. Polynucleotide phosphorylase has never been found in eukaryotes.

Nirenberg and Matthaei were attempting to test the messenger hypothesis by obtaining cell-free protein synthesis dependent on RNA. For this purpose, they chose to use _E. coli_ extracts and viral RNA (TMV) because both were relatively easy to obtain and the latter presumably functioned as a message during the replication cycle of the virus, albeit in plants, not bacteria. In order to improve the meaning of their experimental results, Nirenberg and Matthaei felt they should compare the behavior of message in their system with that of an RNA that would not act as a message. For this control purpose, they chose a synthetic RNA produced by polynucleotide phosphorylase that contained only uridine residues [poly(U)]. Very quickly they discovered that poly(U) stimulates the synthesis of polyphenylalanine and set in motion a series of experiments that culminated in the description of the genetic code in 1966.

_Because of the ease of preparing polynucleotides of different composition with polynucleotide phosphorylase, the relationship between base composition and amino acid incorporation into proteins could be quickly established._ RNA containing only adenine residues [poly(A)] was found to specify the incorporation of lysine and not phenylalanine. Similarly, poly(C) was found to code for proline. Incorporation experiments using templates containing two or more bases confirmed the earlier suspicion based on genetic observations that the code is a triplet.

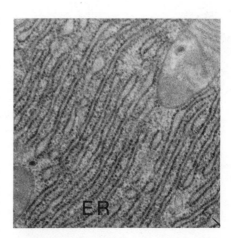

**Figure 24-2**
Electron micrograph of mammalian rough endoplasmic reticulum

**Table 24-1**
Frequency of Triplets in a Poly(AC) (5:1) Random Copolymer

| Composition | Number | Probability | Relative Frequency |
|---|---|---|---|
| 3A | 1 | 0.578 | 1.0 |
| 2A, 1C | 3 | $3 \times 0.116$ | $3 \times 0.20$ |
| 1A, 2C | 3 | $3 \times 0.023$ | $3 \times 0.04$ |
| 3C | 1 | 0.005 | 0.01 |

Tables 24-1 and 24-2 show examples of the types of experimental results that were crucial to this interpretation. Using a template that contains 5 parts A to 1 part C and assuming that these bases are randomly distributed along the RNA (this was shown to be true), it is possible to calculate the probability of encountering any particular triplet of bases. Such theoretical calculations can then be compared with the amount of incorporation observed with different labeled amino acids. This comparison depends on the assumption, which also was shown to be true, that the ribosome and its attendant translational machinery randomly encounter and translate these triplets.

The probability of encountering any particular triplet (its frequency) is simply the product of the probabilities of occurrence of its substituent bases. In this case, the probability of occurrence of adenine in any position is 5/6 and cytosine is 1/6. The probability of any triplet being ACA (or CAA or AAC), for example, is therefore $5/6 \times 1/6 \times 5/6 = 0.116$.

Comparison of the theoretical and experimental results was facilitated by comparing triplet frequencies relative to that of AAA and amino acid incorporation relative to that of lysine. The results of this comparison were consistent with the view that the amino acids Thr, Asn, and Gln are each encoded by one of the three 2A, 1C triplets (see Table 24-2). It is harder to decide the coding specificity of less abundant triplets, but numerous experiments of this general type provided an overall picture of the genetic code. The most serious limitation of these experiments was that because the

**Table 24-2**
Amino Acid Incorporation with Poly(AC) (5:1) as a Template

| Radioactive Amino Acid | CPM | | Observed Incorporation | Theoretical Incorporation |
|---|---|---|---|---|
| | (−) Template | (+) Template | | |
| Lysine | 60 | 4615 | 100.0 | 100 |
| Threonine | 44 | 1250 | 26.5 | 24 |
| Asparagine | 47 | 1146 | 24.2 | 20 |
| Glutamine | 39 | 1117 | 23.7 | 20 |
| Proline | 14 | 342 | 7.2 | 4.8 |
| Histidine | 282 | 576 | 6.5 | 4 |

*Source:* From J. F. Speyer et al., Synthetic polynucleotides and the amino acid code, *Cold Spring Harbor Symp. Quant. Biol.* 28:559, 1963. The theoretical incorporation shown in the last column was a correct interpretation of the genetic code as it was subsequently defined.

templates were random, they could give no sequence information. In other words, one could not decide whether Thr was specified by AAC, ACA, or CAA.

## Coding by Nucleotide Triplets of Defined Sequence

Since the polymerization of amino acids on aminoacyl-tRNAs occurs on the ribosome-message complex, the aminoacyl-tRNAs must bind to this complex prior to polymerization. Nirenberg's laboratory showed that this binding occurs in a message-specific way [i.e., Phe-tRNA binds to ribosomes with poly(U), but not with poly(A), for example] and that this binding does not require the complex mixture of components necessary for protein synthesis. Beyond ribosomes, template, and suitable buffer, the only requirement is for relatively high magnesium ion concentrations. Under these conditions, binding occurs without polymerization.

At that time, it became possible to chemically synthesize short oligonucleotides of defined sequence. Nirenberg and coworkers reasoned that while these polymers would not be long enough to support the polymerization reaction, they might be sufficient to support the binding reaction. They found, in fact, that no binding occurred with mono(U) or di(U), but that tri(U) was nearly as effective in promoting the binding of Phe-tRNA to ribosomes as were higher oligomers of U.

In order to test the binding specificity of triplets of different sequence, all that remained was to develop a rapid means of measuring binding. Because sucrose density gradient analysis, which readily separates ribosome complexes from unbound smaller molecules, is too time-consuming, they sought to retain ribosomes on Millipore filters. These filters are made of nitrocellulose and have pores that are sufficiently small to retain bacteria. These filters also retain ribosomes and their bound aminoacyl-tRNA, but not unbound aminoacyl-tRNA. We now know that nitrocellulose membranes do not retain ribosomes by virtue of filtration (their pore sizes are much too large), but rather, the ribosome is simply adsorbed by the filter as a result of the general affinity between protein and nitrocellulose. Not only did this technique allow the rapid assignment of amino acids to particular triplet code words (Table 24-3), but it also laid the groundwork for numerous other binding assays. It was subsequently found that many, but not all, proteins bind to nitrocellulose membranes and that their complexes are also retained. This phenomenon is now widely exploited in biochemical analysis.

**Table 24-3**
Aminoacyl-tRNA Binding to Ribosomes as Measured by the Millipore Assay

| Aminoacyl-tRNA | Triplet | | | |
| --- | --- | --- | --- | --- |
| | **None** | **UUU** | **AAA** | **CCC** |
| phe-tRNA | 0.34 pmol | 1.56 pmol | 0.20 pmol | 0.30 pmol |
| lys-tRNA | 0.80 | 0.56 | 6.13 | 0.60 |
| pro-tRNA | 0.24 | 0.20 | 0.18 | 0.73 |

*Source:* From M. Nirenberg and P. Leder, RNA codewords and protein synthesis, *Science* 145:1399, 1964.

With the aid of defined-sequence nucleotide triplets and the Millipore filter binding assay, the general sequence relationships of the genetic code were quickly established. However, there were two limitations to these results: some triplets were very inefficient in promoting aminoacyl-tRNA binding to the ribosome, while others appeared to promote nonspecific binding. Most troubling was the possibility of misinterpretation owing to the fact that not all the specificity of protein synthesis was involved in the assay. Nonetheless, these results gave a general picture of the sequence relations of the code, but they could not provide a complete or definite picture.

## Coding by Repeating-Sequence Synthetic Polynucleotides

The key set of experiments that completed the definition of the genetic code was performed by H. G. Khorana and coworkers and resulted in Khorana sharing the Nobel Prize with Nirenberg and Holly, the latter for having sequenced the first RNA molecule. Khorana's experiments were the culmination of years of painstaking work on the chemical synthesis of oligonucleotides of defined sequence and represented a biochemical *tour de force*.

These experiments were conducted in several discrete stages. In the first stage, chemical methods were used to assemble relatively short deoxyribonucleotide oligomers of defined and repeating sequence in relatively small amounts (see Chapter 20). In the second stage, the short, single-stranded oligomers were combined as double-standard structures, and these, in turn, were amplified in both length and quantity by DNA polymerase. In the third stage, the individual strands of the amplified double-stranded DNA were selectively transcribed by RNA polymerase. And in the final stage, the resulting RNA strands were translated in an in vitro polymerizing system similar to that developed originally by Nirenberg, and the nature of the translation products was defined. In a relatively short period of time, Khorana and coworkers prepared and translated a number of repeating di-, tri-, and tetraribopolynucleotides. In general, these were found to stimulate the incorporation of two, three, and four different amino acids, respectively.

A careful analysis of the coding specificity of the polynucleotides and the resulting polypeptides unambiguously defined the properties of the genetic code as well as the outlines of its mechanism of expression. The nature of these results is illustrated by one particularly clever experiment that characterized the translation products of a perfectly alternating copolymer of U and C. This template stimulates the incorporation only of the amino acids serine and leucine, and these are incorporated in equimolar amounts. Peptide bonds involving the amino group of serine are unusually sensitive to acid hydrolysis through $\beta$ elimination, and Khorana exploited this property to characterize the translation products. Upon partial acid hydrolysis, the in vitro translation products of poly(UC) were quantitatively converted to the dipeptide Ser-Leu. Hence, the original peptide was a perfectly alternating copolymer of the two amino acids (Figure 24-3). This result, in combination with earlier work, demonstrated unambiguously that translation involves the sequential reading of adjacent triplets in messenger RNA.

5′···TCTCTCTCTCTC···3′

3′···AGAGAGAGAGAG···5′

| RNA polymerase
| plus UTP and CTP

5′···UCUCUCUCUCUC···3′

| Translation with
| mixed radioactive
| amino acids

···Ser·Leu·Ser·Leu·Ser···

| Acid hydrolysis

Ser·Leu

**Figure 24-3**
Demonstration that poly(UC) encodes a Ser-Leu repeating peptide. (From S. Nishimura, D. S. Jones, and H. G. Khorana, Studies on polynucleotides XLVIII. The in vitro synthesis of a copolypeptide containing two amino acids in alternating sequence dependent on a DNA-like polymer containing two nucleotides in alternating sequence, *J. Mol. Biol.* 13:302, 1965.)

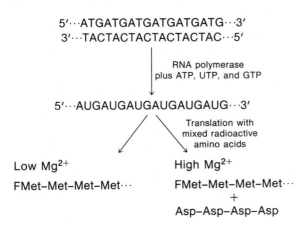

5'···ATGATGATGATGATGATG···3'
3'···TACTACTACTACTACTAC···5'

RNA polymerase
plus ATP, UTP, and GTP

5'···AUGAUGAUGAUGAUGAUG···3'

Translation with
mixed radioactive
amino acids

Low Mg²⁺

FMet–Met–Met–Met···

High Mg²⁺

FMet–Met–Met–Met···
+
Asp–Asp–Asp–Asp

**Figure 24-4**
Translation of poly(AUG). (From H. P. Ghosh, D. Soll, and H. G. Khorana, Studies on polynucleotides LXVII. Initiation of protein synthesis in vitro as studied by using ribopolynucleotides with repeating nucleotide sequences as messengers, *J. Mol. Biol.* 25:275, 1967.)

## The Identification of Start-and-Stop Triplets

The coding properties of the repeating triplet AUG illustrate a number of important features of the mechanism of protein synthesis. When the translation of poly(AUG) was tested by Khorana and coworkers (Figure 24-4), it was found that rather than coding for the incorporation of three amino acids, it specified the incorporation of only the amino acids methionine and aspartic acid. Moreover, methionine was much more effectively incorporated with this template than was aspartic acid, especially at low magnesium ion concentrations. Experiments with the AUG triplet and repeating copolymers of it revealed that it plays a unique role in the initiation of protein synthesis, but that this role can be overcome by the simple presence of higher magnesium ion concentrations. In retrospect, it is clear that Nirenberg's successful translation of synthetic polynucleotides depended on this suppression of natural initiation mechanisms by high magnesium ion concentrations and his fortuitous choice of this high magnesium ion concentration in his initial experimental systems. Perhaps most significant was the observation that poly(Met), particularly when formed at low magnesium ion concentrations, was always blocked at its amino terminus with a formyl group.

The reason that poly(AUG) does not specify the incorporation of a third amino acid is that the triplet UGA is one of three signals used to stop protein synthesis. In wild-type *E. coli* there are no aminoacyl-tRNAs with codons complimentary to these triplets.

## The Code Word Assignments

The culmination of all this work, the statement of the relationship between base sequences in mRNAs and amino acid sequences in proteins, is seen in Table 24-4. For reasons that will become clear later, *the coding relationship between the 64 triplets and 20 amino acids can be neatly summarized by grouping codons with similar first and second bases into a grid*. The four major horizontal rows are composed of codons with the same first base. The four major vertical columns are composed of codons with the same second base. The individual boxes generated by the intersection of the rows and columns are *code word families*, which differ only in their 3'-terminal

**Table 24-4**
The Genetic Code

| 5'-OH Terminal Base | Middle Base | | | | 3'-OH Terminal Base |
|---|---|---|---|---|---|
| | **U** | **C** | **A** | **G** | |
| U | Phe | Ser | Tyr | Cys | U |
| | Phe | Ser | Tyr | Cys | C |
| | Leu | Ser | Term | Term | A |
| | Leu | Ser | Term | Trp | G |
| C | Leu | Pro | His | Arg | U |
| | Leu | Pro | His | Arg | C |
| | Leu | Pro | Gln | Arg | A |
| | Leu | Pro | Gln | Arg | G |
| A | Ile | Thr | Asn | Ser | U |
| | Ile | Thr | Asn | Ser | C |
| | Ile | Thr | Lys | Arg | A |
| | Met* | Thr | Lys | Arg | G |
| G | Val | Ala | Asp | Gly | U |
| | Val | Ala | Asp | Gly | C |
| | Val | Ala | Glu | Gly | A |
| | Val* | Ala | Glu | Gly | G |

*Sometimes used as initiator codons.

base. For example, the code words UCU, UCC, UCA, and UCG comprise a family encoding leucine.

Despite the existence of code word families for most amino acids, different organisms show a strong bias for the use of certain code words within any given family. The precise reasons for codon preference are not known, but it is clear that codons are translated with markedly different efficiencies in different organisms.

## ATTACHMENT OF AMINO ACIDS TO ADAPTORS: THE STRUCTURE AND ENZYMOLOGY OF AMINOACYL-tRNA

Aminoacyl-tRNA occupies a pivotal position in the sequence of events involved in the translation of the genetic code. The enormous amount of work by Robert Holly and coworkers that went into sequencing the first tRNA was certainly justifiable on the basis of its central role in protein synthesis. It was originally believed that knowledge of the primary sequences of a small number of tRNAs would be a sufficient basis to define their role in gene expression. To date, well over 100 tRNAs or their genes have been sequenced, and such sequencing has become a rather commonplace accomplishment because of advances in the technology. Certainly a great deal of understanding has resulted from this work, particularly concerning the interactions between codons and anticodons, but at least as many important questions about the function of tRNA remain unanswered.

Transfer RNA was originally known as soluble RNA or sRNA. This designation arose from the fact that it is the most abundant RNA that remains in the soluble phase of cell extracts after the removal of particulate material by high-speed centrifugation and because all tRNAs have very similar chemical and physical properties. In *E. coli*, tRNA accounts for approximately 10 percent of the total cellular RNA, and upon sedimentation analysis, it migrates as a fairly homogeneous component with an average sedimentation coefficient value of about 4 S. In general, this same picture is observed in extracts of almost all cells.

This similarity in tRNA structure extends to all levels of analysis. Their molecular weights average about 25,000, and all of them fall within the range of 22,000 to 28,000. This molecular-weight range corresponds to between 73 and 93 nucleotides. In addition, all tRNAs are single-stranded in the sense that they give rise to only one polynucleotide chain upon complete denaturation. The very earliest studies concerned with the sequencing of tRNA, however, indicated that they contain a great deal of regular internal structure (see Chapters 18 and 21).

Many ribonucleases show a marked preference for the hydrolysis of single-stranded RNA molecules. The treatment of native tRNA with these nucleases under mild conditions invariably produces very specific initial cleavage patterns and frequently first cuts the molecule into nearly equal-sized pieces that can be separated following denaturation. This fact was not only important in the initial sequencing experiments, since it reduced tRNA to two half-size molecules that were individually easier to sequence, but it also defined the structure that must exist in solution. We now know that this initial cleavage frequently occurs in the anticodon region, that this region is generally single-stranded, and that it resides approximately in the middle of the polynucleotide chain.

## Primary and Secondary Structure

An alanine-specific tRNA (tRNA$^{Ala}$) from yeast was the first sequenced, and it contains 74 bases. Its primary structure immediately suggested to Holly and coworkers why certain regions of the molecule were particularly resistant to the action of nucleases. They recognized that the molecule could fold back on itself to form intramolecular base pairs and double-helical regions. On the basis of this first sequence, they considered a number of possible intramolecular base-pairing configurations and correctly recognized that maximum pairing and hence the most stable structure would be generated by a cloverleaf structure involving double-stranded stems and single-stranded loops. This structure (Figure 24-5), with minor variants, is now known to be a universal feature of the secondary structure of all tRNA molecules.

tRNAs specific for all amino acids and derived from all biological sources except possibly organelles like mitochondria are sufficiently similar in secondary structure that a single figure can describe the generalities of their structure. All these molecules fold back on themselves to form a cloverleaf with either four or five double-stranded stems and either three or four single-stranded loops. In all cases, they fold in such a way as to bring the 5' and 3' termini together in what is known as *the acceptor stem*. Also in every case the amino acid is covalently attached to tRNA at the 3' end of the primary sequence, and the structure of the acceptor stem is highly conserved. Nearly all acceptor stems are composed of seven regular Watson-Crick base pairs between regions at the 3' and 5' ends of tRNA.

The most striking conservation is the absolutely invariant sequence of three bases at the 3' terminus. In all tRNAs, this sequence is CCA and the amino acid is attached to the ribose of the terminal adenosine residue. Although found in all tRNAs, the reason for this sequence conservation at the 3' terminus is not entirely clear. The picture is further complicated by the fact that these three bases occur in the genes of some but not all tRNAs and by the occurrence of a ubiquitous enzyme, tRNA terminal transferase, that can readily remove and add this sequence to tRNAs that possess it or add it to tRNAs in which the sequence is not present in the original gene. The participation of all tRNAs in protein synthesis requires this sequence, and it is reasonable to assume that it constitutes part of the common structure recognized by the ribosome and its attendant factors.

The unpaired regions of tRNA, loops I to IV, are also designated according to their unique structural features. Loop I varies in size from 7 to 11 unpaired bases and frequently contains dihydrouracil and is thus designated as the *dihydro-U loop*. Loop II contains the sequence of three bases, the anticodon, which base pairs with messenger RNA and ultimately determines the functional specificity of the aminoacyl-tRNA. This loop is known as the *anticodon loop*. Loop III, the *variable loop*, may contain as few as 3 or as many as 21 bases and is thus the major site of size variability in tRNA. Loop IV frequently but not always contains pseudo-uridine in one

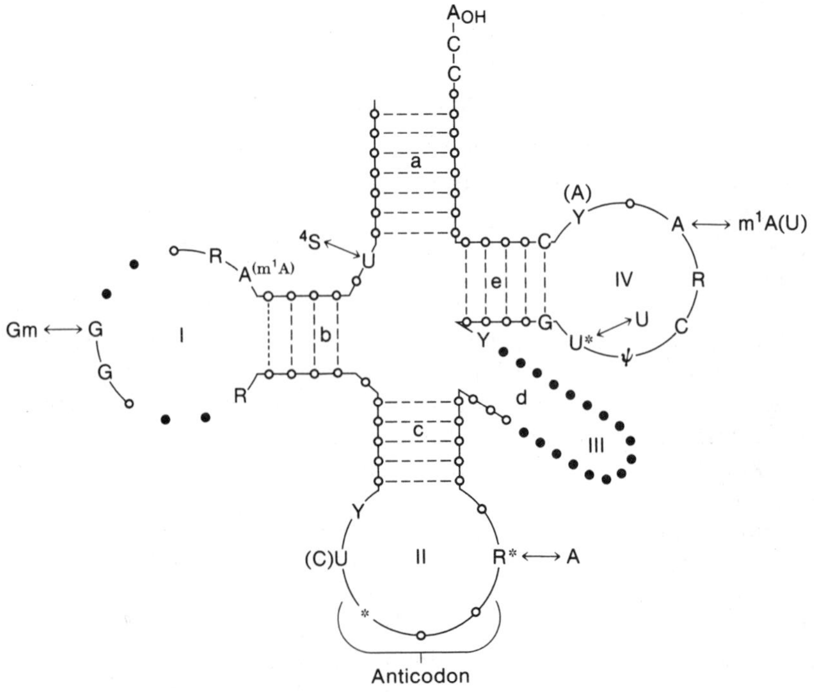

**Figure 24-5**
The primary and secondary structure of tRNA. The solid line is the phosphodiester backbone; base-paired nucleotides are indicated by dashed lines and nucleotides by circles or letters. $A_{OH}$ is the 3' end. Nucleosides common to all structures are indicated as Y, pyrimidine; R is purine; R* is modified purine; U* is modified uridine; $\psi$ is pseudo-uridine; $m^1A$ is 1-methyladenosine; * is a frequently modified purine or pyrimidine; and $^4S$ is 4-thiouridine. The ↔ means both nucleotides are frequently found; ( ) means the nucleotide in parentheses is the single exception to the common base indicated. The dotted line in stem *b* indicates that a base pair is found only in some tRNAs. The solid circles indicate nucleosides that may or may not be present. (From H. Weissbach and S. Pestka, *Molecular Mechanisms of Protein Biosynthesis*, Academic Press, New York, 1977. Used with permission.)

position, and it is thus known as the *pseudo-U loop*. Beyond the anticodon loop, the functional significance of the structural features common to the other loops has not been established.

## Tertiary Structure

Despite intensive investigations of its hydrodynamic properties, detailed knowledge of the three-dimensional conformation of tRNA was not forthcoming until the solution of the first crystal structure in 1974 (see Chapter 18). To date, only a small number of tRNA structures have been solved by x-ray crystallography. This has been due in part to difficulty in obtaining suitably large and stable crystals.

Although studies of tRNA tertiary structure are far from complete, it is reasonable to assume that the picture that has emerged to date will be common to most if not all tRNAs. There are two reasons for this assumption. First is the high degree of conservation in primary and secondary structure. The second stems from the common functional role of tRNA and the fact that while they must be discernible by the various aminoacyl-tRNA synthases, they also must have sufficient structural similarity in order to interact with the same sites on the ribosome and protein-synthesis factors.

The first tRNA to be crystallized, an *E. coli* tRNA specific for phenylalanine, is still the best understood (see Figure 18-36). In this structure, the elements formed by the cloverleaf secondary structure fold back on themselves to form a rather thick L-shaped rod-like structure. The anticodon is at one end of the L, while the CCA acceptor is at the other end. The molecule is about 75 Å in its longest dimension and about 25 Å thick.

The L-shaped structure of tRNA in solution is stabilized not only by the base-paired helical regions seen in the two-dimensional cloverleaf structure, but also by the stacking and pairing of bases that otherwise appear single-stranded in the cloverleaf structure (see Chapter 18). Some of these tertiary interactions appear to be of the Watson-Crick type, but other equally important interactions appear to involve nonstandard base pairing. This is, of course, to be expected, because there are numerous ways in which bases can pair and stack when the constraints of the double helix are removed. Close inspection of this structure reveals important pairing interactions between purine bases and between pyrimidine bases, as well as the pairing together of three bases.

It is reasonable to expect that the specific nature of these tertiary interactions will vary substantially from one tRNA to the next. It is likely that these subtle structural features are in fact responsible for the enzymatic discrimination among tRNAs, but at the moment, the rules that govern this discrimination are not clear. It seems equally possible that major structural differences among tRNAs will not be encountered. Hence the general structural characteristics of tRNA[Phe] are likely to be seen in the crystal structures of most if not all tRNAs. Moreover, there is a high degree of correlation of the solution properties of tRNA and the structure seen in crystals. It is probable, therefore, that tRNAs function within the cell in configurations similar to that deduced from the x-ray crystallography of tRNA[Phe].

## The Occurrence of Unusual Bases in tRNA

*One of the distinguishing characteristics of tRNA, apparent from the earliest studies, is the frequent occurrence of nonstandard or unusual bases* (see Figure 21-2). These unusual bases, of which more than 50 different examples have been found, arise by frequently complex posttranscriptional enzymatic modifications of the four standard bases originally incorporated by RNA polymerase (see Figure 21-5).

As noted earlier, unusual modifications of the same type frequently occur in the same position in different tRNAs. Dihydro-U and pseudouridine in loops I and IV, respectively, are examples. In addition, unusual bases occur in the anticodon or in the adjacent anticodon loop. As will be noted later, the occurrence of unusual bases in the anticodon loop in general has important ramifications with respect to the nature of the interactions between codons and anticodons on the ribosome.

Unusual bases occur more frequently, but not exclusively, in the unpaired regions. It is reasonable to assume, therefore, that these modifications play a role in both the formation of tertiary structure and in its recognition by various specific enzymes. To test these possibilities, numerous experiments have been performed that involve tRNAs lacking specific posttranscriptional modifications. Unfortunately, with the exception of modifications in the anticodon loop, the results of these experiments have been disappointing. In most cases, very little or very subtle perturbations in either structure or function have been seen to result. Recently, the gene for tyrosine suppressor tRNA in the sequence G–T–T–C corresponding to the universal G–T–$\psi$–C sequence of tRNAs has been altered to G–A–T–C by directed mutagenesis. Cells transformed by hybrid plasmids containing the mutant are phenotypically su⁻, showing that this single base change has inactivated the function of this tRNA. It is not clear whether the adverse effect in the mutant has something to do with the synthesis of the RNA or its functioning after it has been made.

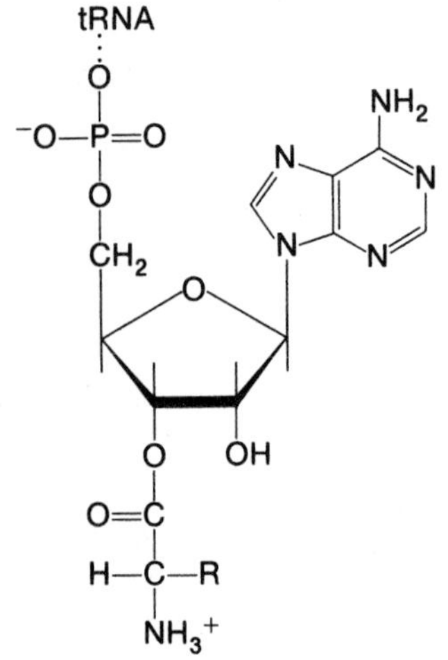

## The Mechanism of Attachment of Amino Acids to tRNA

*The incorporation of free amino acids into proteins requires their intermediate covalent attachment to tRNAs.* This covalent attachment is catalyzed by a class of enzymes known as aminoacyl-tRNA synthases. Most cells contain 20 different aminoacyl-tRNA synthases, one for each amino acid precursor of proteins. These synthases vary widely in size, and, of course, they have different specificities, but they all catalyze the same reaction by means of a generally similar mechanism.

The attachment of amino acids to tRNA involves the formation of an ester linkage between the carboxyl group of the amino acid and a hydroxyl group on the 3′ terminal ribose of tRNA (Figure 24-6). In all cases, this energetically unfavorable reaction is driven by the hydrolysis of ATP to AMP and pyrophosphate. In almost every case, the interaction among the synthase, the amino acid, ATP, and tRNA can and does proceed in the two separate steps:

$$AA + ATP \rightleftharpoons AA\text{-}AMP + PP_i \tag{1}$$

$$AA\text{-}AMP + tRNA \rightleftharpoons AA\text{-}tRNA + AMP \tag{2}$$

$$\text{Sum:} \quad AA + tRNA + ATP \rightleftharpoons AA\text{-}tRNA + AMP + PP_i \tag{3}$$

**Figure 24-6**
Covalent structure of the 3′ end of aminoacyl-tRNA. Here the amino acid is shown attached in ester linkage to the 3′-terminal hydroxyl of tRNA.

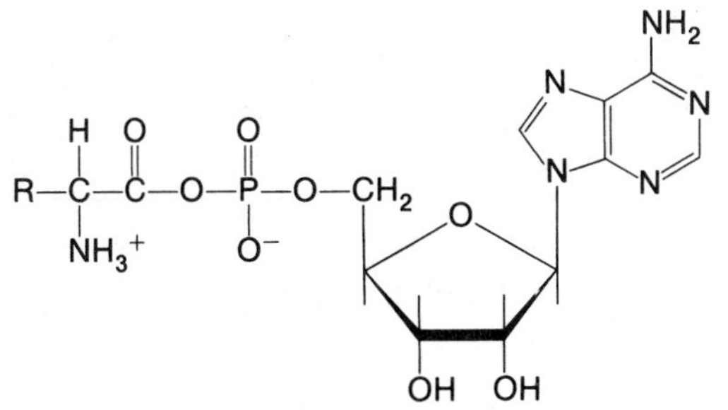

**Figure 24-7**
The structure of the aminoacyl-adenylate intermediate in aminoacyl-tRNA synthesis.

In the first step of the reaction, the synthase binds its cognate amino acid and ATP, forms an anhydride linkage between the amino acid carboxyl group and the nucleotide $\alpha$-phosphate (Figure 24-7), and liberates pyrophosphate. In most cases, this reaction is readily reversible and the aminoacyl-adenylate or "activated amino acid," remains tightly bound to the enzyme. In some cases, amino acid activation is much faster than the overall attachment of amino acids to tRNA. As a result of this, the ready reversibility of the reaction, and the fact that activation can proceed in the absence of tRNA, synthases readily catalyze amino acid–dependent ATP-$PP_i$ exchange reactions. The second reaction is also reversible, and it can take place independent of the first reaction. The equilibrium constants for both reactions are close to unity. Thus the free energy of hydrolysis of the ester linkage between the amino acid and tRNA is comparable with the free energy of hydrolysis of the pyrophosphate linkage in ATP. This ester linkage is therefore a high-energy bond, and as will be seen later, it provides the driving energy for the formation of the actual peptide bond on the ribosome.

The sum of reactions shows that one equivalent of ATP is cleaved per equivalent of aminoacyl-tRNA formed. Because the equilibrium constant for this overall reaction is also close to 1.0, it is not particularly favored. In the cell, the overall driving force for the reaction is the subsequent hydrolysis of pyrophosphate to inorganic phosphate. Thus the formation of aminoacyl-tRNA ultimately requires the hydrolysis of two high-energy bonds.

## The Anticodon Determines the Incorporation Specificity of Aminoacyl-tRNA

In 1962, a classic experiment was performed that demonstrated that once an amino acid is attached to tRNA, it no longer plays a role in the specificity of its subsequent incorporation into proteins. The experiment itself was very simple. First cysteinyl tRNA$^{Cys}$ was formed by incubating cysteine, ATP, and tRNA$^{Cys}$ with the cysteine-specific aminoacyl-tRNA synthase. Next cysteinyl tRNA$^{Cys}$ was reacted with hydrogen in the presence of Raney nickel, which reduced the cysteine sulfhydryl to hydrogen sulfide and replaced it with a proton. This chemical treatment converted cysteine to alanine without removing it from tRNA and thus yielded alanyl tRNA$^{Cys}$. Finally, it was determined whether the amino acid, now alanine, was incorporated in response to codons for cysteine or alanine. The results clearly showed the incorporation of alanine in response to cysteine codons and not in response to alanine codons. As a result of this type of experiment we know that the specificity of interaction between messenger RNA and aminoacyl-tRNA on the ribosome is limited to the pairing of the bases in the codon and anticodon; once it is incorporated into an aminoacyl-tRNA, the structural features of the amino acid no longer play a role in the specificity of its incorporation into protein. Thus mistakes by the aminoacyl-tRNA synthases that result in the incorrect attachment of amino acids to tRNA, by themselves, will lead to production of erroneous proteins. Both enzymatic specificity and error-correcting functions of the synthases guard against this possibility.

## Synthases Recognize Subtle Structural Differences in Amino Acids and tRNAs

*In general, cells contain a single aminoacyl-tRNA synthase but multiple tRNAs for each amino acid.* Since incorrect attachment of amino acids to tRNA inevitably leads to the production of faulty proteins, it is crucially important that this reaction proceed with high fidelity. This fidelity is achieved in two ways, both of which depend on the recognition by synthases of subtle structural differences between the various amino acids and the various tRNAs. Despite extensive structural studies, we are not yet in a position to define the molecular basis of this recognition. Nonetheless, it is clear that the enzymatic specificity of the aminoacyl-tRNA synthases is governed by the same rules of active-site geometry that apply to all enzymes.

Not only do the synthases employ this specificity in the attachment of the amino acid to tRNA, but they also are capable of recognizing and hydrolyzing inappropriate reaction products. These proofreading hydrolysis reactions, which are not a simple reversal of synthesis, are best understood in the case of valine and isoleucine synthases. Discrimination between these amino acids is the most demanding because they differ by only a single methylene group. If isoleucyl-tRNA synthase either erroneously forms or encounters valyl-tRNA$^{Ile}$ in free solution, it rapidly hydrolyzes it in the absence of ATP:

$$\text{Valyl-tRNA}^{Ile} + H_2O \longrightarrow \text{Val} + \text{tRNA}^{Ile}$$

Thus not only does the isoleucine tRNA synthase prefer isoleucine over valine in the synthetic reaction, but it also preferentially hydrolyzes incorrectly acylated tRNAs in an entirely separate reaction. This two-step discrimination is common to many aminoacyl-tRNA synthases and overall produces an error frequency of less than 1 in 10,000.

## Synthases Can Attach Amino Acids to Either or Both of the Terminal 2'- and 3'-Hydroxyls

An unambiguous demonstration of the site of initial attachment of amino acids to tRNA was surprisingly difficult. This difficulty arose from the fact that under physiological conditions, aminoacyl-tRNA esters are in rapid equilibrium between the 2'- and 3'-hydroxyls. The question then becomes to which hydroxyl group is the amino acid originally attached by the enzyme? Investigation of this question has revealed a complex picture. Some synthases attach the amino acid to the 3'-hydroxyl group, other synthases attach the amino acid to the 2'-hydroxyl group, and yet a third group of synthases can attach the amino acid to either the 2'- or 3'-hydroxyl group.

The clearest picture of this relationship came through the use of tRNA nucleotidyl transferase to synthesize tRNA analogs lacking either the 2'- or 3'-hydroxyl group. Clearly, if such a tRNA functions in the synthase reaction, the amino acid must have been attached to the single hydroxyl group that is present. Some synthases can esterify 2'-deoxy-tRNA but not 3'-deoxy-tRNA. Others show the opposite specificity, while some synthases will accept either 2'- or 3'-deoxy-tRNA. Of course, some synthases will accept neither 2'- nor 3'-deoxy-tRNA, and thus other approaches to their positional specificity must be employed. The overall picture that emerges from these studies is that synthases do not employ a universal mechanism in their action. There is also no basis for determining *a priori* which reaction mechanism a particular synthase will employ or why synthases are subject in general to such mechanistic complexity.

## Most Amino Acids Have Multiple tRNAs and Multiple Codons

The four common bases in messenger RNA can be arranged in $4^3$, or 64, different combinations of three bases. One of the questions that arose very early in the study of the genetic code was: How does this number 64 relate to the 20 different amino acids and the number of different tRNAs?

All the possible 64 base triplets are used in protein synthesis, and each has a specific informational meaning. As noted before, only three triplets (UAA, UGA, and UAG) do not have complementary tRNAs, and these three perform a special role in the termination of protein synthesis. Two other triplets (AUG and GUG) play a dual role in translation. In combination with other sequences, these triplets are used to initiate protein synthesis, but they also are used to specify amino acids within protein primary sequences. Thus 61 of the 64 possible triplets have one or more tRNAs with complementary anticodons. In this sense, the genetic code is redundant; there are more code words than amino acids, and therefore, at least some amino acids must be specified by more than one code word.

The distribution of the 61 triplets among the 20 amino acids is shown in Table 24-5. The most striking feature of this distribution and the specific code word assignment seen in Table 24-4 is that the 17 amino acids that have four or fewer codons always use the same two bases in the first and second positions of their code words. The three additional amino acids that have six code words use only two sequences in the first two bases. Thus the redundancy of the genetic code is absolutely minimized in the first two positions of the code and maximized in the third position. However, it should be noted that even in the third position there is a pattern; in cases of double redundancy, the third position is occupied by either a pyrimidine in both cases or a purine in both cases. Because of these reasons the coding library can be neatly arranged in the format shown in Table 24-4. This pattern of degeneracy also has important ramifications regarding the mechanism of interaction between codons and anticodons.

**Table 24-5**
Most Amino Acids Are Specified by Multiple Code Words

| Codons/AA | Number of AAs | Number of Codons | Minimum Number of Anticodons |
|---|---|---|---|
| 1 | 2 | 2 | 3* |
| 2 | 9 | 18 | 9 |
| 3 | 1 | 3 | 1 |
| 4 | 5 | 20 | 10 |
| 6 | 3 | 18 | 9 |
| | 20 | 61 | 32 |

*Methionine requires a separate initiator tRNA that also recognizes the AUG sequence. The fact that this initiator tRNA also can recognize the codon GUG has been overlooked for the purpose of this table. The minimum number of codons required for codon recognition is determined by applying the "wobble" rules outlined in Table 24-6.

### The 3' Terminal Codon Base Can "Wobble"

The 3' terminal redundancy of the genetic code and its mechanistic basis were first recognized by Francis Crick in 1965. The essence of his proposal was that *codons in mRNA and anticodons in tRNA interact in an antiparallel manner on the ribosome in such a way as to require normal Watson-Crick pairing in the first two positions of the codon, but to allow other pairs in the 3' terminal position.* This formation of nonstandard base pairs demands different geometry between the paired bases (Figure 24-8), and Crick's proposal was appropriately labeled the "wobble hypothesis."

A typical cell contains between 50 and 60 physically different tRNA molecules, but this number is not identical to the 61 code words of the genetic code and is only coincidently similar. At the time of Crick's proposal, it was recognized that some individual tRNA molecules can pair specifically with two or three different code words. Other tRNAs can interact with only a single codon. Crick examined the geometric relationship between bases paired in different ways, and on that basis and the pattern of redundancy known at that time, he correctly predicted the physical constraints that must apply on the ribosome in the pairing on the 3' terminal base. The physical basis of this pairing is still unknown, but the rules that Crick proposed have satisfactorily accounted for most coding redundancies subsequently observed (Table 24-6). His hypothesis stated that a C or A in the 5' position of an anticodon can only pair with a G or U, respectively, in the 3' position of a codon (codons and anticodons pair in an antiparallel manner), but U, G, or I in this position would permit pairing with two or three codons differing in their 3' terminal base. It is important to note in this regard that the very first tRNA sequenced contained an I base (inosine, derived by the posttranscriptional deamination of A) in its 5' terminal anticodon position. All tRNAs that pair with three code words (none have been found in bacteria or eukaryotic cytoplasm that pair with four code words) have subsequently been found to contain I in this position. Thus despite our lack of precise information regarding the forces that govern codon and anticodon pairing, they undoubtedly conform to Crick's empirical relationships embodied in the "wobble hypothesis."

**Table 24-6**
The "Wobble" Rules of
Codon-Anticodon Pairing

| 5' Base of Anticodon | 3' Base of Codon |
|---|---|
| C | G |
| A | U |
| U | A or G |
| G | C or U |
| I | U, C, or A |

## In Higher Organisms, Different Cells Frequently Contain Different tRNAs

The actual number of tRNAs with different anticodons in a given cell is usually larger than the minimum number of 32 required to translate the code but smaller than the total number of physically distinct tRNAs. Thus tRNAs are redundant in two ways. Different anticodons can pair with one codon, within the constraints of the "wobble hypothesis," and tRNAs with different sequences also have the same anticodons. *In every case, multiple tRNAs that correspond to a single amino acid are all recognized by a*

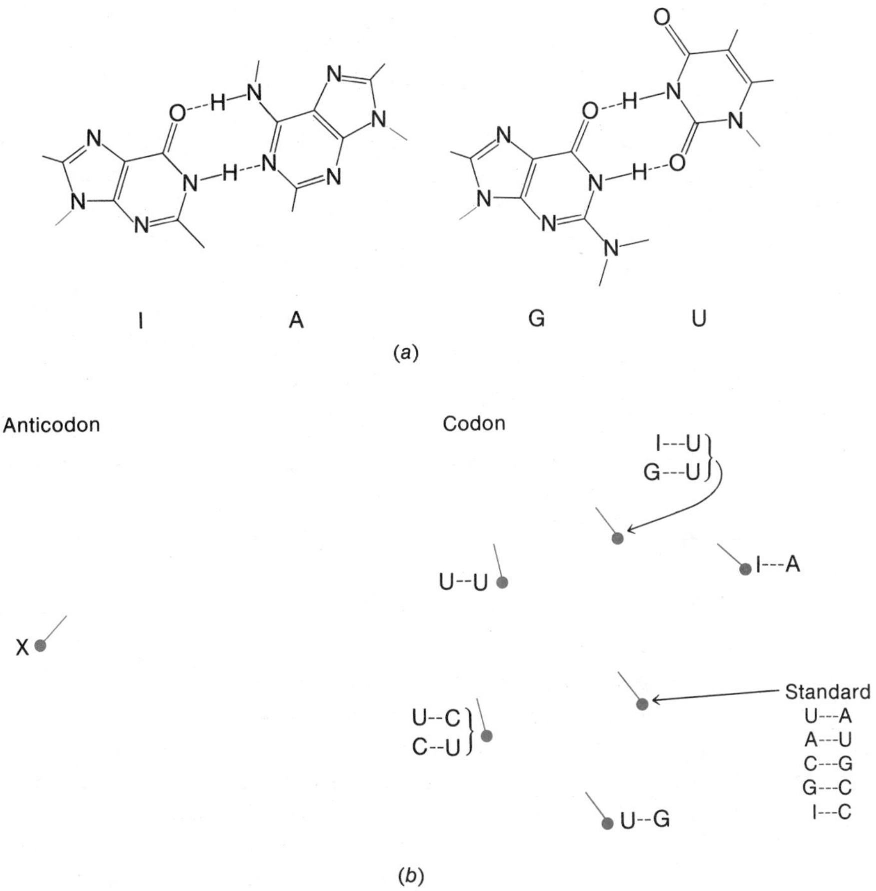

(*a*)

(*b*)

**Figure 24-8**
Dimensions of "wobble" base pairs. (*a*) Two "wobble" pairs: inosine-adenine and guanine-uracil. (*b*) Overall dimensions of standard Watson-Crick H-bonded base pairs and a variety of abnormal base pairs. Only the position of the glycosidic bonds are indicated. The point X represents the position of the C1′ atom of the glycosidic bond in the anticodon. All glycosidic bonds are indicated in color. The other label points show where the C1′ atom and glycosidic bond would fall in various base pairs. The "wobble" code uses the four codon positions to the right in the diagram but not the three close positions (U--U, U--C or C--U).

*single aminoacyl-tRNA synthase.* In higher organisms, the pattern of physically different tRNAs frequently changes with the developmental state and is usually different in different organs. It has been suggested that this changing pattern of tRNAs is in some way related to a changing pattern of gene expression.

## A Special tRNA Is Used to Initiate Protein Synthesis

Because of the common CCA terminus of all aminoacyl-tRNAs and the specificity of pancreatic ribonuclease, digestion of aminoacyl-tRNA labeled with radioactive amino acids gives rise to one radiolabeled aminoacyl-adenosine for each amino acid. With bacterial tRNA, the amino acid methionine is an exception to this rule. The discovery that bacterial tRNAs labeled with radioactive methionine yield two spots following pancreatic ribonuclease digestion provided an important insight into the mechanism of initiation of protein synthesis.

Despite the fact that only one codon corresponds to methionine, all protein synthesizing systems, with the exception of those in mitochondria (see below), contain two tRNAs that can accept methionine in response to a single synthase. One of these tRNAs is employed to read internal methionine codons, while the other plays a unique and indispensable role in the initiation of every protein in all biological systems. The reason that bacterial methionine labeled tRNA gives two spots following pancreatic ribonuclease digestion is that *bacteria contain an enzyme which formylates the terminal amino group of the initiator methionine tRNA* (Figure 24-9). In all respects, the initiator tRNA in the cytoplasm of higher organisms is similar except that it is not formylated; the cytoplasm lacks such a formylating enzyme.

$$\text{tRNA}_f{}^{\text{Met}} \longrightarrow \text{methionyl tRNA}_f{}^{\text{Met}} \longrightarrow \text{FMet-tRNA}_f{}^{\text{Met}}$$

$$\text{tRNA}^{\text{Met}} \xrightarrow{\text{Synthase}} \text{methionyl tRNA}^{\text{Met}} \xrightarrow{\text{Transformylase}} \text{no reaction}$$

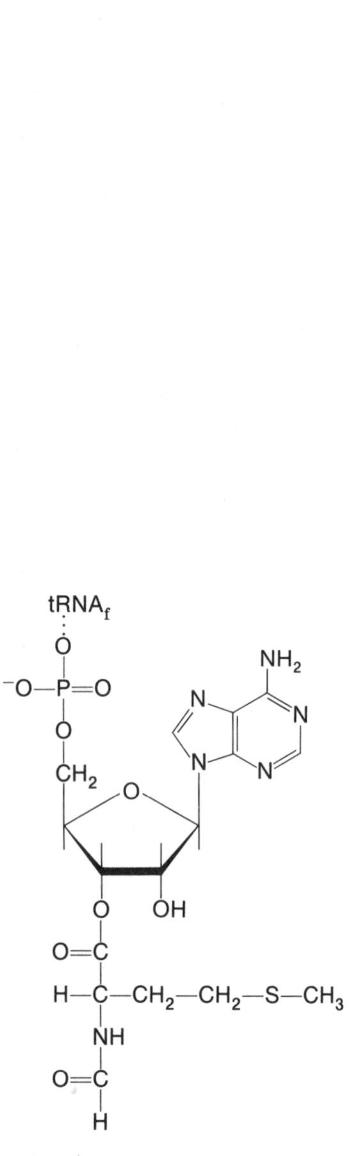

Formylation of the methionine amino group of the initiator tRNA, of course, absolutely prevents it from being incorporated into internal peptide bonds of proteins. In this way, the initiator tRNA in bacterial systems is completely prevented from participating in the translation of internal AUG sequences. Eukaryotes appear to have lost this requirement for the blocking of the initiator amino terminus. This may be due to the fact that initiation of protein synthesis in cytoplasmic systems is more complex than in bacterial systems and thus no longer requires the additional guarantee of a blocked amino terminus on its initiator tRNA. Interestingly, the cytoplasmic initiator tRNA from eukaryotes, while not normally formylated, is nonetheless a good substrate for the bacterial formylating enzyme. The methionine tRNA that codes for internal methionines in proteins is designated tRNA$^{\text{Met}}$ to distinguish it from the initiator tRNA, which is designated tRNA$_f{}^{\text{Met}}$.

The occurrence, structure, and function of these two closely related molecules, tRNA$^{\text{Met}}$ and tRNA$_f{}^{\text{Met}}$, presents an interesting problem in enzyme recognition of nucleic acids. On the other hand, both these tRNAs must be capable of interacting with the AUG codon and both are recog-

**Figure 24-9**
The 3'-terminal structure of bacterial initiator tRNA.

nized by the same methionyl-tRNA synthase, but tRNA$^{Met}$ is not formylated and does not participate in the initiation of protein synthesis. Although the primary sequences of several initiator tRNAs and the three-dimensional structure of one is known, it still is not clear how this tRNA can be distinguished from all others. The initiator appears to have the same general L-shaped structure common to all tRNAs. Clearly though, it does not appear identical to the enzymes that interact with tRNAs.

## The Genetic Code Is Only Quasi-Universal

The very earliest experiments concerning the genetic code, after some uncertainty, led to the conclusion that the genetic code was universal. The essential conclusion was that while the pattern of tRNAs and their anticodons might differ between organisms, each of the 61 triplets specified the same amino acid in all living systems. Thus, beyond special sequences that are involved in initiating synthesis, the messenger RNA for rabbit hemoglobin will elicit the production of globin in a bacterial extract. In this sense, the code was thought to have a universal meaning in all genetic systems. Recently, this conclusion has been shown not to apply completely to the genetic systems of the mitochondria and chloroplasts of eukaryotes.

Mitochondria and chloroplasts contain a genetic system that is semi-independent of that contained in the nucleus of eukaryotic cells. Exactly why this is true is not clear. Thus, for example, mitochondrial DNA codes for a limited number of intramitochondrial polypeptides, plus ribosomal RNA and mitochondrial tRNAs. The ribosomes, protein-synthesis factors, tRNA, and aminoacyl-tRNA synthases contained within the mitochondrion are different from those contained within the cytoplasm. Surprisingly, the genes for the aminoacyl-tRNA synthases and the majority of the ribosomal proteins and factors are contained within the cellular nucleus, and their mRNAs are translated in the cytoplasm, after which the proteins are transported into the mitochondria.

The first clue that something was peculiar about mitochondrial genetics derived from the realization that some mitochondria simply do not contain enough genetic material to encode enough tRNAs to permit genetic translation following the "universal" rules previously described. Recent detailed analysis of both mitochondrial tRNAs and their genes have shown, in fact, that they contain fewer than the 32 tRNAs minimally required for the translation of all 61 codons following Crick's "wobble" rules. One possible solution to this conundrum is that mitochondrial genes do not contain sequences for all 61 codons. Another possibility is that the rules that govern the interaction between codons and anticodons on mitochondrial ribosomes, i.e., "wobble," are different from those governing other protein-synthesizing systems. Recent and still incomplete information indicates that the latter possibility is true.

Eight amino acids are specified in the genetic code by codon families with the same two bases in the first two positions (Tables 24-4 and 24-5). The "wobble" rules require that at least two different tRNAs are needed to translate such families. Mitochondria appear to have developed a means of reading an entire codon family with a single tRNA and in this way have reduced the minimum number of code words from 32 to 24. In all cases, these tRNAs contain a modified U in the anticodon "wobble" position.

These tRNAs on mitochondrial ribosomes appear to follow a different set of "wobble" rules that allow all four bases to pair with the unmodified U in the "wobble" position of the anticodon. Six of the mitochondrial tRNAs with a U in the "wobble" position show normal behavior; these are all chemically modified (see Table 24-7).

In some mitochondria, the genetic code itself appears to have different meanings, but at the moment the generality of this picture is not clear. For example, yeast mitochondria appear to read the codon family CUN (where N is any base) as threonine rather than as leucine and UGA as tryptophan rather than as termination, while *N. crassa* mitochondria appear to read the same codon family as leucine.

## STRUCTURE AND ASSEMBLY OF THE SITE OF PROTEIN SYNTHESIS: THE RIBOSOME

The structural and functional definition of the ribosome represents a special challenge. On the basis of size considerations, it is between 10 and 100 times larger and more complex than a typical protein, but in contrast to still larger structures, such as the mitochondrion, for example, there is every reason to believe that it has a unique and regular structure comparable with that seen in proteins. Most challenging is the fact that the ribosome plays a key metabolic role in gene expression in all living systems. There are, of course, structural differences between ribosomes from different biological sources, but it appears that all ribosomes operate in a generally similar way. The most structural information is available concerning the ribosome from *E. coli*, and the ensuing discussion will focus on the ribosome from this organism.

**Table 24-7**
The Genetic Code of Yeast Mitochondria

| UUU<br>UUC | Phe | AAG | UCU<br>UCC | Ser | AGU | UAU<br>UAC | Tyr | AUG | UGU<br>UGC | Cys | ACG |
|---|---|---|---|---|---|---|---|---|---|---|---|
| UUA<br>UUG | Leu | AAU* | UCA<br>UCG | | | UAA<br>UAG | Term | | UGA<br>UGG | Trp | ACU* |
| CUU<br>CUC<br>CUA<br>CUG | Thr | GAU | CCU<br>CCC<br>CCA<br>CCG | Pro | GGU | CAU<br>CAC | His | GUG | CGU<br>CGC | Arg | GCA |
| | | | | | | CAA<br>CAG | Gln | GUU* | CGA<br>CGG | | |
| AUU<br>AUC | Ile | UAG | ACU<br>ACC | Thr | UGU | AAU<br>AAC | Asn | UUG | AGU<br>AGC | Ser | UCG |
| AUA<br>AUG | Met | UAC | ACA<br>ACG | | | AAA<br>AAG | Lys | UUU* | AGA<br>AGG | Arg | UCU* |
| GUU<br>GUC | Val | CAU | GCU<br>GCC | Ala | CGU | GAU<br>GAC | Asp | CUG | GGU<br>GGC | Gly | CCU |
| GUA<br>GUG | | | GCA<br>GCG | | | GAA<br>GAG | Glu | CUU* | GGA<br>GGG | | |

*Source:* Adapted from S. G. Bonitz et al., Codon recognition rules in yeast mitochondria, *Proc. Natl. Acad. Sci. USA* 77:3167, 1980. Codons and anticodons of the yeast mitochondrial code. The codons (5'→3') are at the left and the anticodons (3'→5') are at the right in each box (* designates U in the 5' position of the anticodon that carries the —CH$_2$NH$_2$CH$_2$COOH grouping on the 5 position of the pyrimidine).

It was originally thought that the ribosome might be structurally quite simple. This erroneous expectation was based on the overall similarity in size and chemical composition between the ribosome and small RNA viruses. These viruses usually contain only a small number of coat proteins that are arranged in a regular and repeating pattern around an RNA core. Despite original expectations to the contrary, no such regular structural pattern exists within the ribosome. Rather, the 50 or more proteins within the ribosome are not only themselves structurally unique, but each appears also to occupy unique structural and functional domains within the particle.

Despite its structural complexity, the ribosome has only three essential functions. These are the selection of appropriate regions on messenger RNA to begin translation, the correct pairing of codons and anticodons, and the actual formation of peptide bonds. The initiation events of protein synthesis are quite complex. These involve the selection of the appropriate start region within a messenger RNA and thus offer a potential locus for the regulation of the specificity of protein synthesis. The reactions of protein chain elongation, codon-anticodon matching, and peptide bond formation are conceptually much simpler. Overall, the ribosome might appear needlessly complex, but at least some of this complexity undoubtedly derives from the necessity of maintaining fidelity in protein synthesis.

Despite many years of intensive investigation, our knowledge of ribosome structure is embarrassingly incomplete. This embarrassment derives primarily from two factors. First, the ribosome is too large to be definable by the chemical and hydrodynamic approaches useful in the analysis of proteins, but it is slightly too small for detailed analysis by electron microscopy. Second, all attempts to crystallize the ribosome have failed, and hence the powerful techniques of x-ray crystallography have not yet been brought to bear on ribosome structure.

### Size and Composition of Ribosomes

The ribosomes in bacteria have a particle weight between 2.5 and 3 million daltons and are somewhat smaller than ribosomes found in the cytoplasm of eukaryotes, which have particle weights of approximately 4 million daltons. It is customary to designate ribosomes according to their sedimentation coefficients. Accordingly, bacterial ribosomes are designated as 70 S, and those from the cytoplasm of higher organisms are designated as 80 S. The ribosomes from bacteria are about one-third protein and two-thirds RNA, while those from the cytoplasm of eukaryotes contain about equal amounts of RNA and protein. A portion of the ribosomes from higher organisms, particularly, can be isolated in association with intracellular membranes. Except for this association and counterions that neutralize change, ribosomes contain only protein and RNA.

The ribosomes from intracellular organelles present a more confusing picture. The ribosomes from organelles of primitive eukaryotes resemble bacterial ribosomes in both composition and size. However, ribosomes from the mitochondria of higher eukaryotes are similar in size to those from bacteria, but they have substantially smaller (55 S) sedimentation coefficients. These reduced sedimentation coefficients derive from the fact that these ribosomes have lower buoyant densities because they contain a much higher percentage of protein.

## All Ribosomes Are Composed of Two Subunits

It was recognized very early that divalent cations, especially magnesium, play a very important role in ribosome structure. Complete removal of divalent ions invariably leads to major and usually irreversible alterations in the ribosome. A modest reduction in divalent ion concentration leads to the reversible dissociation of all ribosomes into two subunits of unequal size. In bacteria, these subunits have sedimentation coefficients of 30 S and 50 S, while in the cytoplasm of eukaryotes, the subunits have sedimentation coefficients of approximately 40 S and 60 S.

This division into subunits, one approximately twice the mass of the other, results from a universal feature of the mechanism of protein synthesis. A variety of sophisticated density labeling experiments, modified after those conducted by Meselson and Stahl in the analysis of DNA replication (see Chapter 20), have shown that in bacteria, the separation and rejoining of the subunits is an integral part of the selection of start signals in mRNA.

## The Ribosomal RNAs

The *E. coli* ribosome contains three ribosomal RNAs. A 16 S RNA of about 1540 bases is contained within the small ribosomal subunit, and the larger ribosomal subunit contains both a 23 S RNA of about 2900 bases and a 5 S RNA of 125 bases. The ribosomes in eukaryotic cytoplasm contain four ribosomal RNAs, and three of these correspond to those found in bacterial ribosomes. The smallest, 5 S RNA, is comparable in size and structure with the corresponding bacterial RNA, while the two larger RNAs, with sedimentation coefficients of about 18 S and 28 S, are larger than their bacterial counterparts. The fourth RNA has a sedimentation coefficient of 5.8 S.

The sequencing of all three *E. coli* ribosomal RNAs is essentially complete. The 5 S RNA in fact was the very first RNA molecule sequenced with the aid of radioactive tracer techniques. Despite this large amount of work, the role of ribosomal RNA in the structure and function of the ribosome remains largely obscure. Before the existence of messenger RNA was fully recognized, it was thought that ribosomal RNA might play a direct informational role in protein synthesis. The current view, with some specific exceptions, is that ribosomal RNA plays a largely structural role in which it provides a scaffold for the correct assembly and positioning of the ribosomal proteins.

A major exception to this largely structural role was discovered in relation to the mode of action of colicin E3, which is a bacteriocidal protein produced by certain strains of *E. coli* that contain the E3 plasmid. This protein enters the cytoplasm of susceptible bacteria, where it acts as a specific nuclease, cleaving 16 S RNA so that a single 49-base piece from the 3' end of the molecule is released from the ribosome. Colicin-treated ribosomes are active in all aspects of protein synthesis except those involved in specific initiation. Subsequent sequencing of bacterial mRNAs have shown them to contain an untranslated leader sequence on the 5' side of the AUG initiation sequence, a portion of which is complementary to part of the 16 S RNA that is released by colicin. This sequence in mRNA is known as the Shine-Dalgarno sequence (see Chapter 25), and it is protected from nuclease digestion upon formation of the initiation complex with the ribosome. Clearly an AUG sequence itself is an insufficient basis for specific initiation in prokaryotes. Pairing between mRNA leader sequences and 16 S RNA is believed to provide additional alignment information, but in addition, the degree of this pairing is thought to provide a means of selecting among messenger RNAs with different translational efficiencies. As will be seen later, this base-pairing mechanism does not operate during initiation in eukaryotic systems.

## The Ribosomal Proteins

The definition of the number and properties of the proteins contained within a particular ribosome has been difficult. This difficulty arises from the fact that these proteins have generally similar properties, they tend to be insoluble when separated from the ribosome, and some can be lost in purification of the ribosome. It has become customary to designate the ribosomal proteins with a shorthand convention that employs an L to designate those proteins from the large ribosomal subunit and an S to designate those from the smaller ribosomal subunit. These letter designations are followed by a number that designates the protein mobility in a two-dimensional electrophoresis system with isoelectric focusing in the first dimension and molecular sieving in the second (these techniques are explained in Chapter 3). Thus the slowest moving protein from the 30 S subunit is designated S1 (Figure 24-10).

With respect to protein composition, there appears to be only one type of ribosome within a given *E. coli* cell. This ribosome contains approximately 900,000 daltons of protein divided among 56 molecules. Most of these proteins have been either partly or entirely sequenced, and 53 of them appear to be unique. One protein partitions between subunits upon their dissociation and hence occurs partially in the small subunit (S20) and partially in the large subunit (L26). Another protein is present in four copies per large subunit and its amino terminus is partly acetylated so that it gives rise to two electrophoretic spots (L7/L12). Beyond these two exceptions, the small subunit contains one equivalent of 20 different proteins (S1 to S19 and S21) and the large subunit contains one equivalent each of 31 different proteins (L1 to L6, L8 to L11, L13 to L25, and L27 to L34).

The picture with respect to the protein composition of eukaryotic cytoplasmic ribosomes is not nearly so clear. In general, these ribosomes appear to contain more proteins, between 70 and 90, than bacterial ribosomes, and these proteins appear to have greater molecular weights on the average than those found in bacteria. Whether or not ribosomes in the cytoplasm of eukaryotes are heterogeneous with respect to their protein composition remains to be determined.

## The Assignment of Function to the Ribosomal Components

The catalytic center responsible for the formation of the peptide bond during protein synthesis clearly resides entirely within the large ribosomal subunit. Similarly, the specific binding of messenger RNA during the initi-

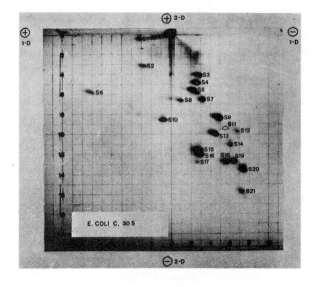

**Figure 24-10**
2-D gel electropherograms of the proteins from the small (*top*) and large (*bottom*) *E. coli* ribosomal subunits. (From E. Kaldschmidt and H. G. Wittmann, Ribosomal proteins, 12, *Proc. Natl. Acad. Sci. USA* 67:1276, 1970.)

ation of protein synthesis occurs entirely on the small ribosomal subunit. Beyond these two generalities, the assignment of specific functions to the various ribosomal components has been difficult. The biggest impediment to this functional assignment is the fact that very little function is retained when the ribosomal subunits are separated into their individual protein and RNA components. For example, the peptide-bond-forming activity of the large subunit, peptidyl transferase, is lost upon the removal of only a small number of proteins, and none of these individual proteins is capable of catalyzing this reaction by itself. Thus the ribosome is viewed as a highly cooperative structure in which its individual components participate together in its functions.

A number of approaches have been employed to assign function to the individual ribosomal components. Among these are the inhibition of function by protein-specific antibodies, the mapping of mutations to resistance or dependence on antibiotic inhibitors of ribosomal functions, the reaction with ribosomes of active-site-directed reagents, and the functional activities that remain upon partial disassembly of the ribosome. Using various combinations of these approaches, it has been concluded, for example, that the four copies of ribosomal protein L7/L12 play a conspicuous role in the interaction between the ribosome and the various nonribosomal factors. A number of these approaches also suggest that protein L16 is either responsible for the formation of the peptide bond itself or at least is near the location where this reaction takes place. A number of ribosomal mutations have been found (ribosomal ambiguity mutants, or ram) that result in increased errors in translational specificity. These mutations are mapped to protein S4. The antibiotic streptomycin itself also causes errors in translation upon binding to the small subunit, and resistance to the effects of this antibiotic result in alterations in either protein S4 or S12. Thus on the basis of these observations it is reasonable to conclude that proteins S4 and S12 play at least an indirect role in maintaining the proper alignment of codons and anticodons during translation.

## Three-Dimensional Arrangement of Ribosomal Components

The assignment of spatial location within the ribosome of its various components requires at the outset a clear three-dimensional picture of the overall structure of the particle. Because of the limitations cited earlier, an entirely satisfying three-dimensional picture has not yet emerged. On the basis of its hydrodynamic properties, the 70 S E. coli ribosome appears roughly spherical, with dimensions on the order of 150 to 200 Å in diameter. The individual ribosomal subunits, particularly when viewed by electron microscopy, appear to be substantially more irregular, and it has been difficult to distinguish those structural features which are intrinsic to the subunits from those which are introduced in the preparation of samples for microscopy (Figure 24-11).

One particularly striking feature of ribosomal architecture is that essentially all the 53 immunologically distinct ribosomal proteins have antigenic determinants accessible on the surface of the individual subunits. This fact has been exploited in the application of immunoelectron microscopy. In this technique, two ribosomal subunits are cross-linked by a single bivalent immunoglobulin that reacts with surface antigens on a single protein. The subunits cross-linked by this single immunoglobulin are then viewed in the electron microscope, thus providing a means of mapping the relative location on the surface of the ribosome of these antigenic determinants. Unfortunately, unambiguous spatial assignments require a clear three-dimensional picture of the overall structure of the subunits.

Another approach to the definition of the three-dimensional arrangements of proteins within the ribosome has involved establishing "neighbor relationships" among these proteins. Two approaches have been employed.

In one of these, the ribosomal proteins in the intact ribosome are chemically cross-linked to each other and then the cross-linked pairs are identified. In the other approach, fluorescent groups are attached to the individual proteins, and energy transfer between these groups is used to measure the distance between the proteins in the ribosome. Together these two techniques have provided a broad picture of the relative location within the ribosome of its various proteins.

One of the major limitations to all these studies is the uncertainty regarding the structure of the individual proteins within the ribosome. Clearly, at least some of these proteins are not simple spheres. In fact, some of the neighbor relationships already established demand that some proteins are decidedly nonspherical. A detailed molecular understanding of the mechanism of protein synthesis requires a clear three-dimensional picture of the ribosome and the components within it. Such a picture of even the *E. coli* ribosome is some distance in the future.

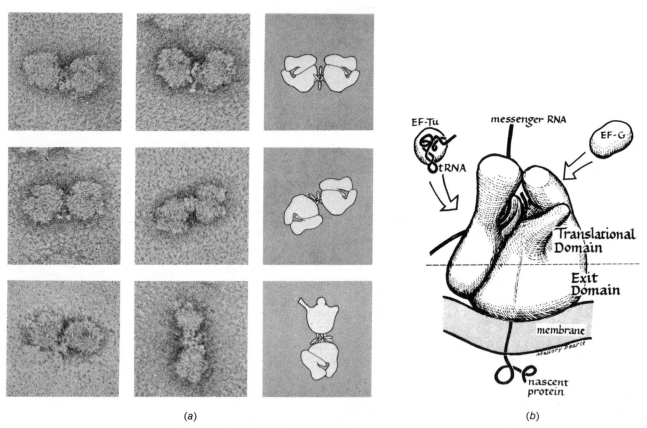

(a)

(b)

**Figure 24-11**
A model of the *E. coli* ribosome and the electron micrographs upon which it is based. (a) Electron micrographs of ribosomes linked by antibodies against nascent chains of β-galactosidase. Large subunits are linked by IgG antibody to monomeric ribosomes. (b) Diagrammatic representation of the exit and translational domains of the ribosome and their orientations with respect to the membrane binding site. The binding sites of mRNA and elongation factors EF-Tu and EF-G are those previously inferred from the locations of ribosomal proteins. The nascent protein is shown as an unfolded, extended chain during its passage through the ribosome. (From C. Bernabeu and J. H. Lake, *Proc. Natl. Acad. Sci. USA* 79:3111, 1982.)

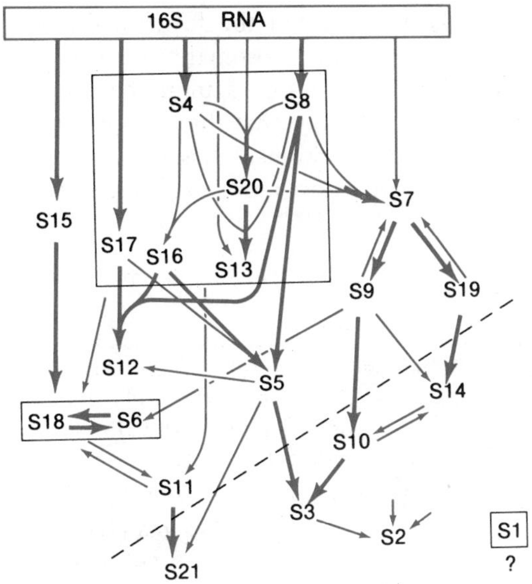

**Figure 24-12**
The assembly map of the *E. coli* 30 S subunit. (From M. Nomura and W. A. Held, Reconstitution of Ribosomes: Studies of Ribosome Structure, Function, and Assembly. M. Nomura, A. Tissieres, and P. Lengyel (Eds.), *Ribosomes,* Cold Spring Harbor Press, New York, 1974. Used with permission.)

## Ribosome Assembly

Experiments conducted in the mid-1960s largely by Masayasu Nomura and associates concerning the assembly of ribosomes have had a significant impact not only on our view of the ribosome, but also on our view of the general features of macromolecular assembly. Essentially stated, Nomura and coworkers showed that given the proper solvent environment, the information necessary for the assembly of the individual ribosomal components is entirely contained within their primary sequences. Thus *it is possible after separating the ribosome into its 59 individual macromolecular components to recombine these under conditions in which they will spontaneously reassemble into a fully functional particle.*

With various combinations of individual ribosomal components, Nomura was able to analyze their assembly and define the sequence of events that it involves (Figure 24-12). Assembly of the small ribosomal subunit begins with the independent binding of several ribosomal proteins to 16 S RNA. Additional proteins then add to this structure, ultimately giving rise to an assembly intermediate that requires a temperature-dependent structural rearrangement prior to the addition of the remaining proteins. A number of experiments suggest that this same assembly pathway occurs during the biosynthesis of the ribosome within the cell.

By conducting partial-assembly experiments in which one or more ribosomal protein was omitted, Nomura was able to construct an "assembly map." This map defines not only the approximate sequence of events in assembly, but also those proteins whose binding requires the prior assembly of other proteins. It is reasonable to conclude that these obligatory relationships result from the interactions among proteins within the ribosome and that proteins which depend on each other for assembly are in fact close to one another within the ribosome. In general, there is good agreement between neighbor relationships inferred from the assembly map and those established by the means described earlier.

Ribosome assembly also allowed the functional analysis of artificial particles deficient in one or more ribosomal protein. Analysis of the functional activities of such partial ribosomes, more than any other experimental approach, has enforced the view that the ribosome is a highly cooperative structure. With some exceptions, ribosomes deficient in individual proteins show impaired activity in more than one function.

## SELECTED READINGS

Barrell, B. G., et al. Sequence and organization of the human mitochondrial genome. *Nature*. 48:457, 1981. The sequence and coding properties of the 16,569 base mitochondrial genome.

Kudo, I., Leineweber, M., and Rajbhandary, U. L. Site-specific mutagenesis on cloned DNAs: Generation of a mutant of *E. coli* tyrosine suppressor tRNA in which the sequence G–T–T–C corresponding to the universal G–T–ψ–C sequence of tRNAs is changed to G–A–T–C. *Proc. Natl. Acad. Sci. USA*. 78:4753, 1981. Some functional consequences of a mutational change in tRNA.

Lake, J. A. The ribosome. *Sci. Am.* 245:84, 1981. A graphic picture of the structure and function of the bacterial ribosome.

Schimmel, P. R., Soll, D., and Abelson, J. N. *Transfer RNA: Structure, Properties and Recognition*. New York: Cold Spring Harbor Press, 1979. A collection of current experiments.

Soll, D., Abelson, J. N., and Schimmel, P. R. *Transfer RNA: Biological Aspects*. New York: Cold Spring Harbor Press, 1980. A collection of current experiments.

*The Genetic Code, Cold Spring Harbor Symposium on Quantitative Biology*, Vol. 31. New York: Cold Spring Harbor Press, 1965. A description of the original experiments that defined the genetic code.

## PROBLEMS

1. Assume that you have a random copolymer of A and U containing equimolar amounts of the two bases. When the copolymer is used as an mRNA, which amino acids would be incorporated and in what ratios?

2. Bromouracil is a base replacement mutagen while acridine causes single base additions or deletions. Mutations caused by acridine frequently can be compensated for by secondary mutations several nucleotides distant from the first mutation, while those caused by bromouracil usually require changes within the same codon. The difference in the requirement for recovery from mutations induced by the two agents derives from a fundamental feature of the genetic code. Explain.

3. The effect of single point mutations on the amino acid sequence of a protein can provide precise identification of the codon used to specify a particular residue. Assuming a single base change for each step, deduce the wild-type codon in each of the following instances.
   a. Gln ⟶ Arg ⟶ Trp
   b. Glu ⟶ Lys ⟶ Ile

c.
$$\text{Ser} \swarrow \overset{\text{Leu}}{\underset{\downarrow}{\text{Val}}} \searrow \text{Met}$$

d.
$$\text{Ile} \swarrow \overset{\text{Thr}}{\underset{\downarrow}{\text{Pro}}} \searrow \text{Lys}$$

4. Assuming that translation begins at the first codon, deduce the amino acid sequence of the polypeptide encoded by the following mRNA template:

   AUGGUCGAAAUUCGGGACACCCAUU
   UGAAGAAACAGAUAGCUUUCUAGUAA

   Although this polypeptide does not contain serine, a mutation involving a single base substitution alters the composition such that it contains a single serine residue. How else might the composition of the mutant protein be expected to differ from the wild type?

5. In Table 24-5, it is stated that 32 anticodons are the minimum required to decode all 61 codons. Explain.

6. Bearing in mind the rules known to govern the specificity of codon-anticodon interaction and the need for fidelity in translation, predict the anticodons used to specify the following amino acids:
   a. serine

b. tryptophan

c. isoleucine

Indicate the corresponding codons in each case.

**7.** In what way does the established pattern of codon assignments relate to the chemical nature of the amino acids encoded? How might this relationship be explained?

**8.** Is it necessary for an aminoacyl tRNA synthase to recognize the anticodon part of the tRNA? Explain.

**9.** The overall process of translation is characterized by an impressively low error frequency. Why is such a low error frequency necessary? Which crucial steps in the translation scheme determine this fidelity? For at least one step, indicate how this has been shown to be achieved.

**10.** Transfer RNA molecules are rather large considering the fact that the anticodon is a trinucleotide. Suggest probable reasons why this is the case.

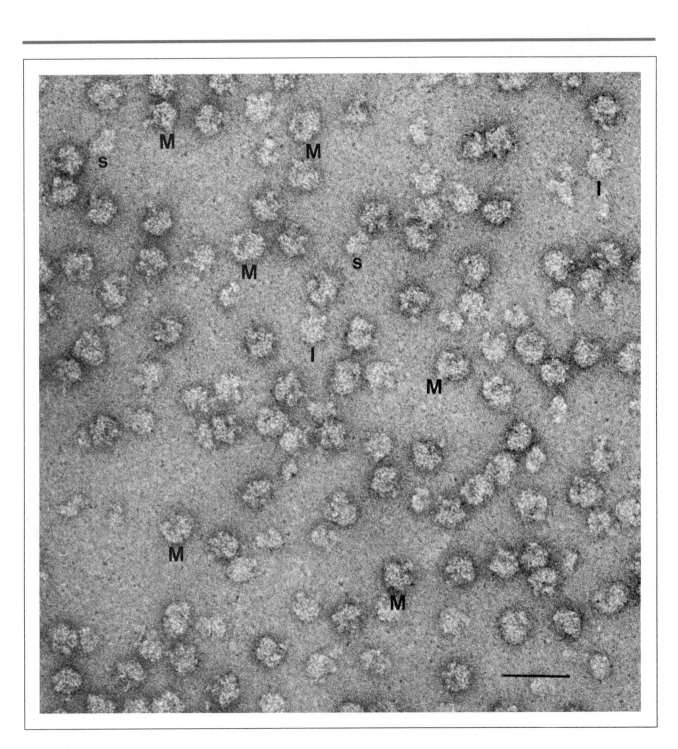

# 25

# MECHANISMS OF PROTEIN SYNTHESIS

A major challenge in biochemistry is to define the chemical and physical basis of the process by which genetic information is translated into active proteins. Chapter 24 outlined the nature of the genetic code and described the properties of two major participants in its translation, aminoacyl-tRNA and the ribosome. This chapter presents the current view of how these two components interact with other participants to achieve the synthesis of proteins.

The assembly of amino acids into the polypeptide chains is only part of the process of protein synthesis. During and following the assembly steps, the primary sequences must fold into their active configurations, and frequently they also undergo a variety of covalent alterations collectively termed posttranslational modifications. These modifications include the alteration of amino acid side chains as well as the cleavage of specific peptide linkages.

Finally, specific mechanisms exist for the degradation of proteins. In eukaryotic organisms in particular, protein metabolism is a dynamic process in which individual protein levels are determined by a balance between synthetic and degradative rates. In addition, separate degradative mechanisms exist in order to remove incorrectly assembled or damaged proteins from the cell.

Thus protein metabolism involves not only the complex reactions that lead to the polymerization of amino acids, but also conformational and covalent alterations to yield active proteins as well as specific degradative reactions.

*Electron micrograph of a field of 70S, 50S, and 30S ribosomes and ribosomal subunits negatively contrasted using the single-layer carbon technique. A few small and large subunits are indicated by the letters s and l, respectively. Some monomeric 70S ribosomes are indicated by the letter M. Adapted with permission from James A. Lake, J. Mol. Biol. 105:131, 1976. (The scale bar is 500Å.)*

# FROM GENES TO ACTIVE GENE PRODUCTS: THE REACTIONS

## Dissection of the Protein Synthesizing System

The ultimate objective of investigators concerned with the mechanism of protein synthesis is to define the process as it occurs in living cells. There are several obstacles to this objective. In the first place, nearly all mechanistic investigations require the formation of cell-free extracts. Upon breaking the cell membrane, even under mild conditions, the gross amount of protein synthesis drops drastically. Measured in terms of the rate of chain growth there is only about a 50 percent decrease in rate, and one is saved by the use of radioactively labeled substrates in many cases which allow one to distinguish the de novo synthesized material from what was present beforehand. Nevertheless, the results from cell-free experiments must be cautiously interpreted. The biochemist is accustomed to thinking of processes such as protein synthesis as involving only the interaction of soluble molecules. There is clearly a level of organization of these molecules within the cell that is both important to the overall rate of the process and which is destroyed by opening the cell. Not only do mechanistic investigations require the rupture of cells, but they frequently also require the purification of individual components. This purification separates one enzyme from another and also removes the substrates with which the enzymes interact. There is considerable evidence, for example, that in some cell extracts aminoacyl-tRNA synthases exist in a large complex. On the basis of such observations, it has been suggested that these enzymes may normally function as part of a multiprotein complex within the cell. In addition, the synthases probably always exist within an environment that is saturating with respect to their substrates. It is hard to imagine what changes they might undergo when these substrates are removed. In the final analysis, then, one is left in the uncomfortable position of having to project from observations on purified components about the nature of the complex and highly interacting systems within cells.

Approaches to the molecular dissection of the components essential to protein synthesis have divided along several natural lines. Most of this work had its inception in observations made in the early 1950s in the laboratory of Paul Zamecnik, who first succeeded in achieving cell-free protein synthesis in extracts of rat liver. Zamecnik and coworkers were able to show a requirement for ATP, GTP, ribosomes, and soluble components. With the discovery of tRNA and aminoacyl-tRNA synthases, the synthesis of aminoacyl-tRNA could be and was investigated independent of the overall process of protein synthesis. Similarly, Nirenberg's discovery of poly(U) translation (Chapter 24) opened the door to investigations of protein chain elongation, while the discovery of the initiator fMET-tRNA$^{fMet}$ and the initiating codon AUG allowed the independent investigation of the events in protein synthesis initiation.

The way in which the nature of the reactions involved in protein chain elongation was unraveled by Fritz Lipmann and coworkers in the middle 1960s provides an especially good example of the techniques and approaches involved in the dissection of complex metabolic processes. Lipmann and coworkers, building on earlier observations, began with Phe-tRNA, poly(U), the ribosome, GTP, and the soluble components of a cell extract. These components together were essential to the formation of polyphenylalanine, and the questions were: Through what sequence of steps does this reaction proceed and what participants are involved?

In order to approach answers to these questions, the soluble proteins of *E. coli*, which were essential for polyphenylalanine synthesis, were subjected to chromatography on an ion-exchange resin. Very little activity in polyphenylalanine synthesis was recovered following this chromatography, and closer investigation revealed that what small activity was observed resulted from the chromatographic overlap of two different components, both of which were essential to polypeptide synthesis. Once these two components were identified, they could be individually purified, and the activity of one could be assessed in the presence of a saturating amount of the other. The essential components obtained in this way were called factors because they were only later identified as proteins and enzymes. Even after these proteins were well-characterized, their designation as factors persisted, so that now all proteins that function with the ribosome in translation are designated as factors.

The first factor to elute from the ion-exchange resin was subsequently discovered to promote the binding of aminoacyl-tRNA to the ribosome in the presence of GTP. It became known as *elongation factor T* (EF-T) for this transfer function (Figure 25-1). Further purification later showed that EF-T could be separated into two proteins, *EF-Ts* and *EF-Tu* (the s and u designate the relative stability and instability of the proteins, respectively). The second protein to elute from the ion-exchange resin was subsequently discovered to hydrolyze GTP in the presence of the ribosome, and it was named *EF-G* for this GTPase activity.

These three bacterial proteins, in addition to GTP, are the only requirements beyond the ribosome, aminoacyl-tRNA, and messenger RNA for protein chain elongation. Considerable effort has since been expended to elucidate their roles in protein synthesis, and the conclusions from these experiments will be described later.

## The Rate of Protein Synthesis

The overall rate of protein synthesis in intact cells is very rapid considering the complexity of the process. The speed of the reaction is illustrated by the fact that within 1 to 2 minutes after exposure of a growing *E. coli* culture to isopropylthiogalactoside (IPTG), a gratuitous inducer of the *lac* operon, the

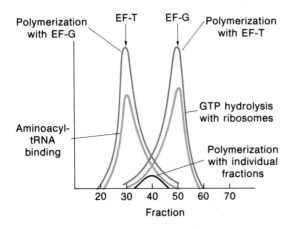

**Figure 25-1**
Chromatographic separation of the bacterial elongation factors.

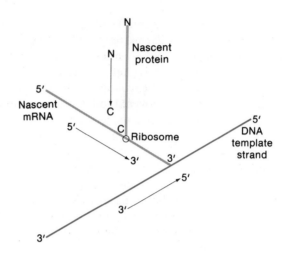

**Figure 25-2**
Polarity of transcription and
translation in bacteria.

enzyme β-galactosidase is first detectable within the cell (see Chapter 26). The enzyme is composed of four identical subunits, each with about 1200 amino acid residues. In this period of time, the inducer must enter the cell and remove a repressor from the promoter region of the gene in order to allow RNA polymerase to begin transcribing the structural gene for the enzyme. Then, while transcription is proceeding, the translational apparatus must initiate protein synthesis at the appropriate start signal on the nascent mRNA and begin and complete assembly of the polypeptide chain. During and after the release of the completed polypeptide from the ribosome the polypeptide must assume its appropriate configuration and assemble with other subunits to yield the functionally active tetramer. Simultaneously, aminoacyl-tRNA and the four ribonucleoside triphosphates must be regenerated. In accord with this rapid sequence of events, direct measurements of the rate of protein chain growth in rapidly growing *E. coli* cells show that amino acids can be added to a growing peptide chain as fast as 20 residues per second. In vitro rates under optimal conditions are about half this.

### The Polarity of Translation

All biopolymers have polarity in the sense that one end of the molecule is different from the other, and all are assembled in a stepwise manner, one monomeric unit at a time (Figure 25-2). In defining the biological mechanism of assembly of such a polymer, one of the most fundamental questions is its polarity of synthesis: To which end are the stepwise additions made? In addition, template-directed assembly reactions, such as translation, also must be defined in relation to the polarity of the template: Which end of the template is read first?

The polarity of amino acid assembly was first investigated by H. Dintzis in intact reticulocytes that produce hemoglobin almost exclusively. After a brief exposure to radioactive amino acids, hemoglobin molecules that had completed their synthesis were isolated from reticulocytes, and the protein was analyzed for the distribution of radioactivity in the polypeptide chain. Radioactive amino acid was found preferentially at the

C terminal of the protein. This means that the C-terminal residues are the last added, and therefore, synthesis must start at the N terminal in a sequential process.

The polarity of message translation could not be as readily tested in vivo. The in vitro translation of block copolymers such as 5′...UUUAAA...3′ was used in early experiments. Translation of this template gave rise to peptide products enriched with phenylalanine at their N terminal and lysine at their C terminal. Thus the message is read with 5′→3′ polarity, and it follows that in bacteria, nascent mRNA molecules can be translated while they are still being transcribed at their 3′ termini on DNA by RNA polymerase.

## Nascent Peptides Are Always Attached to tRNA

Throughout the translation of laboratory-synthesized messages, the nascent or growing peptide chain remains tightly bound to the ribosome message complex. This finding not only suggested that natural mRNA must contain a specific sequence not present in such artificial messages that signal the release of completed proteins, but also raised questions concerning the nature of the attachment of nascent proteins to the ribosome.

Nascent peptides can be released from ribosomes by treatment with relatively high salt concentrations. When ribosomes are treated in this way, the released peptides are invariably attached by their C termini to a tRNA molecule. Conditions that cleave tRNA or the ester bond that joins it to the nascent peptide also lead to the release of the peptide. Therefore, growing peptides covalently attached to tRNAs are bound to the ribosome message complex by noncovalent forces that involve primarily the tRNA.

## The Ribosome Must Bind Two tRNAs

Since nascent peptides are sequentially elongated at their C termini and are always covalently attached to a tRNA, the ribosome must be capable of simultaneously interacting with a minimum of two tRNAs: one that bears the peptide and one that bears the incoming amino acid that will serve to elongate the peptide (Figure 25-3). At the other extreme, the ribosome could function to align a large number of aminoacyl-tRNAs on the message in anticipation of peptide bond formation. In this latter case, one might visualize the ribosome as a cylinder with multiple sites of tRNA binding that rolls along the message forming peptide bonds at each tRNA binding site in turn.

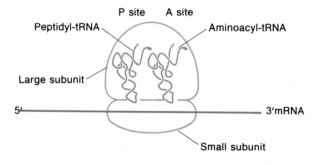

**Figure 25-3**
The two ribosomal sites for tRNA binding. Prior to peptide-bond formation, the peptidyl-tRNA is bound to the P site and the aminoacyl-tRNA is bound to the A site.

The essential distinction between these two possibilities is the number of tRNAs that can bind to the ribosome and the functional properties of these binding sites. From measurements that have been made, it appears that the ribosome can bind only two tRNAs simultaneously. In addition, a variety of functional tests, including the use of various antibiotics to be described later, are in agreement with the view that _the ribosome has only two sites for tRNA binding_.

This two-site model of ribosome structure is at the heart of the current view of its role in protein synthesis. In this model, _the peptidyl-tRNA is held prior to peptide formation in one site, termed the P site, and the aminoacyl-tRNA is brought into the second site, termed the A site_. After peptide-bond formation, this model requires that the peptidyl-tRNA, now elongated by one amino acid residue, return to the P site so that a new aminoacyl-tRNA can enter the A site.

These two tRNA binding sites on the ribosome are the key to understanding the mechanism of protein synthesis, for they are the points at which the genetic code is read. Therefore, they are to a large extent the raison d'être of the complex ribosome.

## THE INITIATION OF PROTEIN SYNTHESIS

The mechanism of initiation of protein synthesis is the most intrinsically interesting of the subreactions of protein synthesis because it is usually the rate-limiting step in translation and because it offers a potential site for regulating which messages will be expressed. In addition, since codons are translated as a continuous sequence of triplets, it is vital that initiation occur at the correct nucleotide, otherwise an entirely erroneous gene product could result. In light of the importance of the initiation reaction, it is not surprising that it is both complex and still not completely understood.

### All Proteins Begin with Methionine

In the early 1960s, a rather surprising observation was made that ultimately provided an important clue into the mechanism of protein synthesis initiation. Approximately 50 percent of all _E. coli_ proteins are amino terminated with methionine, while this amino acid accounts for only about 4 percent of the internal amino acids of these proteins. At about the same time, formylmethionyl-tRNA was discovered in bacterial extracts, and shortly thereafter, formylmethionine was recognized to be uniquely incorporated into the N-terminal positions of proteins. Over the several years that followed these discoveries, evidence gradually accumulated to suggest that almost all proteins are initiated with methionine: formylmethionine in bacteria and subcellular organelles and methionine itself in the cytoplasm of eukaryotes.

As noted in Chapter 24, the code word AUG in mRNA specifies internal methionines as well as the initiating methionine at the beginning of proteins. Occasionally, GUG is also read as an initiating methionine, but in internal positions it is always read as valine. One of the major demands of protein synthesis is to select the appropriate AUG (or infrequently, GUG) initiation code word.

Few mature proteins in _E. coli_ have formyl-blocked N termini, and about half have amino terminals other than methionine. Clearly, therefore, most *E. coli* proteins are posttranslationally modified to remove either or both the formyl or methionyl residues. A variety of enzymes have been described that can catalyze one or more of these reactions, sometimes while the substrate is still bound to the ribosome. Presumably, the three-dimensional structure of the protein substrate itself determines the site of N-terminal processing.

## Initiation Begins with Termination

The initiation of synthesis in all protein-synthesizing systems requires the individual participation of the separate ribosomal subunits. Both subunits are required to function together during protein chain synthesis, but they appear to separate individually from the message-ribosome complex at the end of each round of translation. Once separated, their rejoining is prevented by the binding of an "antiassociation factor" to the smaller ribosomal subunit. In bacteria, this factor (an initiation factor) is called IF-3, while in eukaryotes it is known as eIF-3 (the prefix e designates its eukaryotic origin).

In most cells, a small pool of ribosomal subunits held apart by the initiation factor stands ready to initiate protein synthesis on messenger RNA as it becomes available. This mechanism of initiation is presumably the reason ribosomes occur universally as two subunits.

## The General Nature of the Initiation Reactions

As noted in Chapter 24, the code word AUG in mRNA specifies internal methionines as well as the initiating methionine at the beginning of proteins. The initiation factors function with the ribosome to discriminate the correct starting point in messages. Specific differences, to be pointed out later, distinguish the initiation reactions in eukaryotes and prokaryotes. However, in all protein-synthesizing systems, *three general steps are required to initiate synthesis. The first two steps in initiation involve the binding of the small ribosomal subunit to the initiator tRNA and the mRNA at an appropriate initiator codon. In the third step, the large ribosomal subunit joins this complex and the initiation factors dissociate from it* (Figure 25-4).

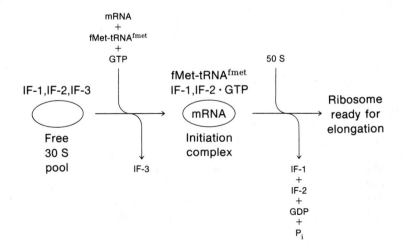

**Figure 25-4**
The initiation events in *E. coli.*

Following the addition of the large ribosomal subunit and the dissociation of the initiation factors from the assembly described earlier, the ribosome is poised to begin the elongation reactions. Stated another way, the initiator tRNA is bound in the P site and the penultimate N-terminal codon is prepared for translation in the A site. These relationships can be assessed in several different ways. As will be described later, the ability of a ribosome-bound tRNA to react with the antibiotic *puromycin* is one of the major criteria of tRNA binding to the P site. A-site bound tRNA cannot react with puromycin. At the conclusion of the initiation reactions, the initiator tRNA readily reacts with puromycin. Even more fundamental is the fact that the ribosome-message complex resulting from the initiation reactions readily binds the aminoacyl-tRNA specified by the penultimate N-terminal codon of the message. This binding, without further addition, is immediately followed by the formation of the first peptide bond.

## Base Pairing Helps to Align the Initiation Codon on Bacterial Ribosomes

How do bacterial ribosomes and factors select the appropriate initiating code word for translation? Several years ago, sequence analysis of 16 S RNA and mRNAs suggested an answer. It was noticed by Shine and Dalgarno that 16 S RNA contains a seven-base pyrimidine-rich sequence near its 3' terminus. Additionally, on the basis of the limited mRNA sequence information available at that time, Shine and Dalgarno observed that bacterial messages contain a complementary purine-rich region (which has subsequently become known as the Shine-Dalgarno sequence) centered approximately 10 bases toward the 5' side of the AUG sequence. On this basis they proposed that base pairing between these sequences could serve to align the initiating AUG for decoding. The degree of pairing would then determine the avidity of the ribosome for the message and hence the efficiency with which it is translated.

Subsequent sequence analysis of mRNAs has lent further support to the functional importance of the Shine-Dalgarno sequence. Some representative examples are shown in Table 25-1. The sequences are somewhat variable in both their length and position relative to the AUG codon and appear to involve G · U as well as conventional base pairs. Most important, it has been observed that short RNA sequences complementary to the pyrimidine-rich region of 16 S RNA specifically inhibit protein-synthesis initiation. Thus specific pairing with the Shine-Dalgarno sequence is important in the discrimination by bacterial ribosomes of message start signals. This pairing probably also affects the translational efficiency of messages, those messages whose Shine-Dalgarno sequences form the best match with 16 S RNA being read most often.

## Prokaryotic Initiation Factors

The recognition that formylmethionine is the first amino acid incorporated into every bacterial protein led to a search for protein factors that interact with fMet-tRNA$^{fMet}$ and promote its binding to the ribosome. In *E. coli* there are three of these protein factors. Work on these proteins was

**Table 25-1**
The Sequence of the 3′ End of *E. coli* 16 S RNA and Some
Shine-Dalgarno Sequences at the 5′ End of Bacterial mRNAs

|  | The pyrimidine-rich complement to the Shine-Dalgarno sequence |
|---|---|
| 16 S-RNA | 3′···HOAUUCCUCCACUA···5′ |
| *lacZ* | 5′···ACAC<u>AGGA</u>AACAGCUAAUG···3′ |
| *trpA* | 5′···ACG<u>AGGGG</u>AAAUCUGAUG···3′ |
| RNA polymerase β | 5′···GAGCU<u>GAGGA</u>ACCCUAUG···3′ |
| r-Protein L10 | 5′···CC<u>AGGAGC</u>AAAGCUAAUG···3′ |

The purine-rich Shine-Dalgarno sequence

The initiation codon

complicated by the fact that they are found associated with the ribosome and do not occur free to any appreciable extent in the soluble portion of the cell. It is known that these factors are bound to the small pool of 30 S subunits that is waiting to initiate the events of protein synthesis.

In order to study the functional properties of the initiation factors it is necessary to remove them from the 30 S subunit. This can be accomplished by washing the ribosomes with moderately high concentrations of monovalent salts, such as ammonium chloride. Unfortunately, this washing procedure also tends to release the loosely bound ribosomal proteins. As a practical matter it is sometimes difficult to distinguish a functionally essential nonstructural factor from a protein that is a structural component of the ribosome. By definition, the structural proteins of the ribosome are those which remain associated with the particle throughout its cycle of reactions in protein synthesis. By contrast, in order for a protein to qualify as a factor, it must be shown to cycle on and off the ribosome during the events of protein synthesis.

In *E. coli*, one of these factors is IF-3, which, as we have seen, serves to hold the subunits apart after termination. The two other factors, IF-1 and IF-2, function to promote the binding of fMet-tRNA$^{fMet}$ and mRNA to the 30 S subunit. This overall binding reaction requires the participation of GTP, which, upon addition of the 50 S subunit, is hydrolyzed to GDP and $P_i$ (Figure 25-4).

The exact temporal sequence of events involved in the initiation reactions and the mechanistic roles played by the individual factors is still not entirely clear. When removed from the 30 S subunit, IF-2 is able to bind both fMet-tRNA$^{fMet}$ and GTP. This factor is also essential to the hydrolysis of GTP, which occurs when the 50 S subunit is added to the initiation complex. Very likely, therefore, IF-2 is directly involved in these binding reactions and in the hydrolysis of GTP. However, IF-1 stimulates the rates of these reactions, but the mechanistic basis of its action is unclear.

## Eukaryotic Ribosomes Initiate at the First AUG Sequence

Bacterial messages are frequently polycistronic (encode more than one protein), and frequently the ribosome recognizes internal initiation sites as well as the 5′ proximal initiation site. Eukaryotic messages, in contrast, appear to specify only single proteins; usually only the 5′ proximal AUG is recognized as a translation start site. A fundamental mechanistic distinction between initiation in prokaryotes and eukaryotes was suggested by the failure of an extensive search to find complementarity between eukaryotic rRNA and messages comparable to that seen in prokaryotes.

The clearest demonstration of the basic distinction between the initiation process in prokaryotes and eukaryotes implied earlier is provided by an experiment performed by M. Kozak. She showed that circular RNA molecules readily serve to initiate protein synthesis with bacterial ribosomes but are ignored by eukaryotic ribosomes until a single nick is introduced into the template. Building on this basic observation, Kozak has obtained evidence that the small eukaryotic ribosomal subunit binds at or near the 5′ end of mRNA and moves along it until it encounters the first AUG sequence (in rare cases the second or even the third AUG from the 5′ end is preferentially recognized).

### Eukaryotic Initiation Factors

Eukaryotic cells contain a more complex spectrum of initiation factors. These factors are also found primarily associated with the small ribosomal subunit and are even more difficult to distinguish from ribosomal proteins than their bacterial counterparts. As many as 10 separable proteins have been suggested to function as initiation factors in eukaryotic systems. Several of these contain multiple subunits, and it is possible that some of the factors that can be separated normally function within the cell as protein complexes.

In eukaryotes, protein synthesis is initiated on the small ribosomal subunit by the binding of the ternary complex composed of an initiation factor protein, GTP, and methionyl-tRNA (eIF-2, GTP, and Met-tRNA$^{Met}$). This binding appears, in contrast to prokaryotes, to be a prerequisite to the interaction of the small subunit with mRNA (Figure 25-5).

**Figure 25-5**
The initiation events in eukaryotes.

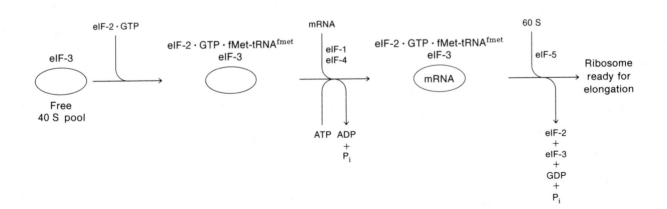

The binding of mRNA and the selection of the appropriate initiation site are the most interesting steps in eukaryotic initiation, and as noted before, the steps that differ to the greatest extent from bacterial initiation. Binding of the small subunit at or near the 5' end of mRNA occurs first, and this binding is promoted by several initiation factors. Many, but not all, eukaryotic mRNAs are terminated at their 5' end by the cap structure (see Chapter 21 for a description of the cap structure), and one of the proteins identified as an initiation factor binds to this portion of mRNA and assists in ribosome binding. It is not clear why some mRNAs are uncapped or how initiation can occur on these mRNAs with the participation of the cap binding protein.

In all cases, initiation in eukaryotes requires the participation of ATP and its hydrolysis to ADP and $P_i$. This nucleotide is not required for ribosomal events in prokaryotes. If ATP is replaced by a nonhydrolyzable analog, the small subunit can interact with message, but this interaction is weak and does not lead to initiation site selection. One attractive hypothesis is that ATP hydrolysis provides the motive force to propel the ribosome from the site of its initial binding at the message end to the first initiation codon. This model suggests that the ribosome is locked in place at the initiation sequence by the codon-anticodon pairing with the initiator tRNA.

Once the small subunit is properly positioned at the initiation site, the large ribosomal subunit joins the complex in a factor-assisted reaction, and GTP, originally bound in complex with eIF-2, is hydrolyzed to GDP and $P_i$. At this point, both eIF-3 and eIF-2 dissociate and the ribosome is primed to begin elongation.

Because of the plethora of factors involved in protein-synthesis initiation in eukaryotes and our limited knowledge of the structure of the eukaryotic ribosome, the mechanistic details of the reactions seen in Figure 25-5 will remain unknown for some time.

## THE ELONGATION STEPS OF PROTEIN SYNTHESIS

The ribosomal complex that results from the initiation steps is poised to accept the N-terminal penultimate aminoacyl-tRNA and form the first peptide bond of the protein. After peptide-bond formation, the ribosome moves along the message to the next adjacent codon triplet, a new aminoacyl-tRNA is bound, and the second peptide bond is formed. This series of three sequential reactions, known as the elongation cycle, is repeated with each adjacent codon and is outlined in Figure 25-6. The nonribosomal proteins that participate in these reactions, designated EF (elongation factors), are not functionally interchangeable between prokaryotes and eukaryotes. Nonetheless, the nature of elongation-cycle reactions in all protein-synthesizing systems appears mechanistically very similar.

### The Binding of Aminoacyl-tRNA

Although message-specific binding of aminoacyl-tRNA to the ribosome can occur in the absence of soluble factors, this reaction is neither sufficiently rapid nor specific to satisfy the demands of cellular protein synthesis. One

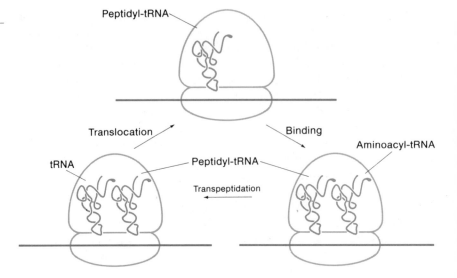

**Figure 25-6**
The elongation cycle.

of the factors, EF-Tu, identified by Lipmann as essential for poly(U) translation, catalyzes this reaction.

In *E. coli*, the aminoacyl-tRNA binding factor is composed of a single polypeptide chain ($M_r = 43,000$), and it promotes the binding to the ribosome of all aminoacyl-tRNAs except the initiator. The importance of this reaction to overall cellular economy is illustrated by the fact that EF-Tu is the single most abundant protein (approximately 10 percent of total protein) in rapidly growing *E. coli*.

The overall reaction scheme by which aminoacyl-tRNA binding is accomplished in prokaryotes is shown in Figure 25-7. In the first step of this cycle, EF-Tu bound to GTP interacts with aminoacyl-tRNA to form a ternary complex. The EF-Tu $\cdot$ GTP $\cdot$ aminoacyl-tRNA complex then interacts with a ribosome containing a complementary codon at its A site. Three things result from this interaction: the aminoacyl-tRNA binds to the ribosome; GTP is hydrolyzed; and $P_i$ and a complex of Tu $\cdot$ GDP dissociate from the ribosome. EF-Tu $\cdot$ GDP is unable to interact with aminoacyl-tRNA. Following the binding reaction, a complex series of reactions serves the seemingly simple purpose of converting EF-Tu $\cdot$ GDP back to EF-Tu $\cdot$ GTP. This exchange reaction involves a separate protein, EF-Ts, whose only known function is to promote the exchange of nucleotides bound to EF-Tu by the reaction sequence shown in Figure 25-7.

EF-Tu, in contrast to the aminoacyl-tRNA synthases, must be able to recognize and interact with all aminoacyl-tRNAs except the initiator tRNA. At present there are only hints as to how this recognition is accomplished. The primary interaction between EF-Tu and aminoacyl-tRNA in the ternary complex appears to involve the acceptor end of the molecule. The tRNA must be charged with an amino acid, and it, in turn, must have a free amino terminus (this terminus is, of course, blocked by an *N*-formyl group in the initiation tRNA of bacteria). In addition, only the acceptor end of the tRNA in the ternary complex is protected from RNAse attack.

It is difficult to understand how EF-Tu, by interacting with the acceptor end of tRNA, can facilitate the correct pairing of the anticodon at the other end of the molecule. Clearly, GTP hydrolysis has a major role in this process. If the anticodon of the tRNA in the ternary complex is completely noncomplementary to the A-site-bound codon, the complex will not interact with the ribosome. If the anticodon of the complex is a close, but not quite correct, match, transient binding will take place and GTP hydrolysis will occur, but the mismatched tRNA will be rejected from the ribosome. Aminoacyl-tRNA binding will persist after GTP hydrolysis only if the codon and anticodon are correctly paired. This "proofreading" mechanism at the expense of GTP hydrolysis catalyzed by EF-Tu enhances the accuracy of amino acid placement (fidelity) in protein synthesis.

In eukaryotic cytoplasm, a similar series of reactions takes place, but only a single multisubunit protein, EF-1, is involved. This one protein seems to combine the functions of the prokaryotic factors EF-Tu and EF-Ts. Although its mechanism of action is similar to that of the bacterial factors, it will not function in conjunction with bacterial ribosomes.

## Peptide-Bond Formation: Transpeptidation

Curiously, the actual formation of the peptide bond is the only subreaction of protein synthesis that does not require the participation of either a nonribosomal protein or GTP. This reaction, given appropriately bound substrates, is catalyzed by a component contained entirely within the large ribosomal subunit. The reaction does not require the participation of a nucleotide, and its energetic requirements are entirely satisfied by the cleavage of the high-energy bond through which the amino acid is attached to tRNA.

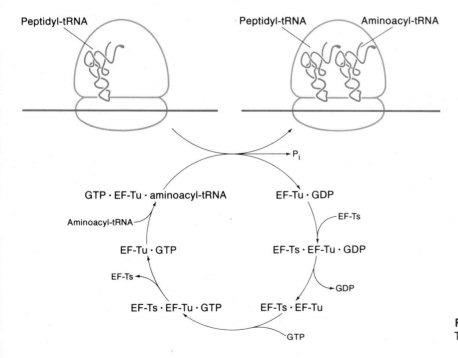

**Figure 25-7**
The EF-Tu·EF-Ts cycle.

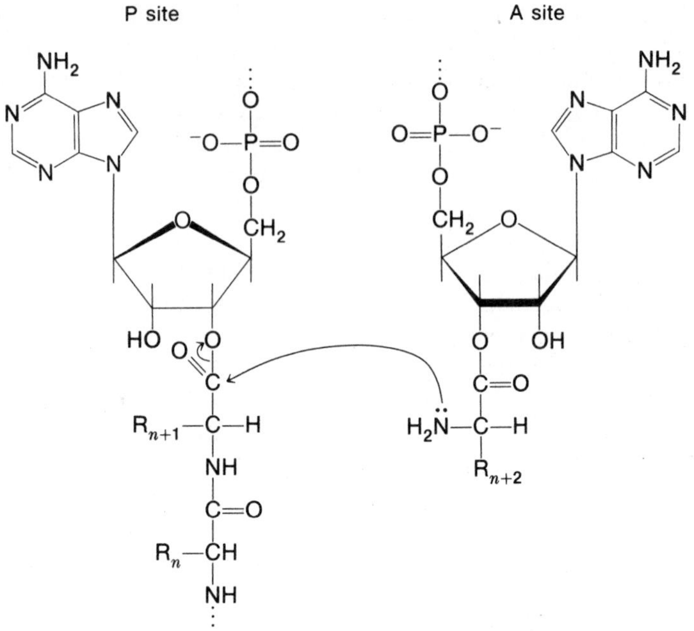

**Figure 25-8**
The peptidyl transferase reaction.

Transpeptidation is a nucleophilic displacement reaction. The free amino group of the newly bound aminoacyl-tRNA attacks the carbonyl group of the adjacently bound peptidyl-tRNA and replaces the ester bond with a peptide bond (Figure 25-8). The net result of this reaction is the transfer of the growing peptide chain from one tRNA to the next simultaneous with its single-step elongation.

As noted in Chapter 24, aminoacyl-tRNAs exist as a mixture of 2′ and 3′ esters regardless of their mechanism of synthesis. Rather surprisingly, the peptidyl transferases, at least on the *E. coli* ribosome, appear to utilize only 3′ esters. A similar specificity is observed on the part of the aminoacyl-tRNA binding factor EF-Tu. Thus, although aminoacyl-tRNA synthases can attach amino acids to either or both of the 2′ or 3′ hydroxyls of tRNA, only the 3′ esters are substrates for protein synthesis.

It is reasonable to expect that the transpeptidation reaction itself is catalyzed by one or more proteins of the large ribosomal subunit. As noted before, all efforts to conclusively identify this protein or proteins have failed. This failure presumably results from the fact that *the peptidyl transferase catalytic center, at least in the binding of substrates, is shared among several proteins.*

## The Puromycin Reaction

The antibiotic puromycin has played a key role in defining the mechanism of ribosomal reactions. It was observed in the early 1960s that *puromycin bears a structural resemblance to aminoacyl-tRNA and that in inhibiting protein synthesis it becomes covalently incorporated into peptide chains* (Figure 25-9). The nascent proteins with puromycin attached to their C-terminal residue are released from the ribosome. Puromycin competes with aminoacyl-tRNA as a substrate for peptidyl transferase.

This substrate role of puromycin can be clearly seen in its reaction with ribosome-bound fMet-tRNA$^{fMet}$. Incubation of puromycin with ribosomes bearing bound fMet-tRNA$^{fMet}$ from the initiation reaction yields fMet-puromycin without further additions. Thus, fMet-tRNA$^{fMet}$ is bound by initiation factors so that it can participate in peptide-bond formation, and this reaction is catalyzed entirely by the ribosome. If the ribosome, after initiation, is reacted with an appropriate aminoacyl-tRNA, a dipeptide is formed, and this dipeptide cannot react with puromycin. Thus puromycin can react with peptidyl-tRNA when it is bound in the P site, but not when it is bound in the A site.

A variety of experimental findings has served to localize peptidyl transferase to the large ribosomal subunit. Normally, of course, both subunits are required for this reaction, because of the contribution the small subunit makes to the binding of peptidyl-tRNA. However, it has been empirically observed that peptidyl-puromycin is formed when the large subunit alone is incubated with peptidyl-tRNA and puromycin in the presence of organic solvents. Apparently the organic solvent alters the structure of the subunit so that sufficient substrate binding occurs to allow slow catalysis.

Exactly why puromycin should be such an effective competitor with aminoacyl-tRNA for the ribosomal A site is not clear. Thus far, chemical modifications of puromycin have either altered its activity only slightly or significantly reduced it. Although the A site must have broad specificity so that it can accept any amino acid, puromycin must combine a number of structural features that uniquely facilitate its interaction in the absence of the remainder of the usual tRNA structure.

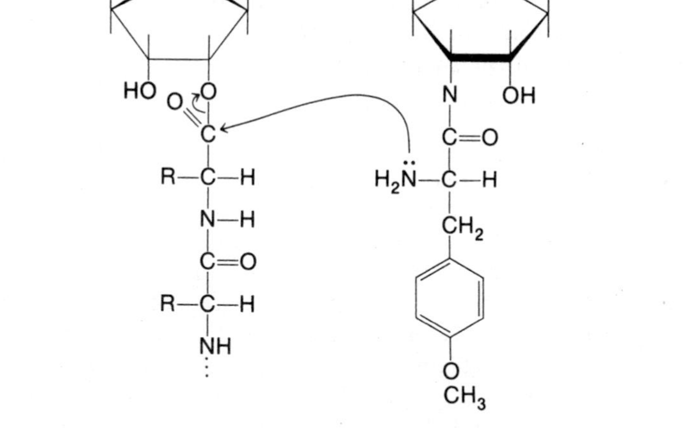

**Figure 25-9**
The puromycin reaction. Puromycin is mistaken for an aminoacyl-tRNA.

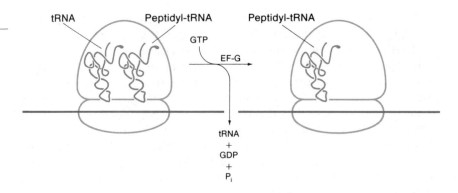

**Figure 25-10**
The translocation reaction.

## Movement of the Ribosome along mRNA: Translocation

The stepwise sequential reading of the genetic code requires the ribosome to move along the message. Each step of this movement must be precisely three bases, and during this process, the codon-anticodon interaction between peptidyl-tRNA and mRNA must be preserved.

At the conclusion of the transpeptidation reaction, the ribosomal P site is occupied by a deacylated tRNA, and the peptidyl-tRNA occupies the puromycin-unreactive A site. The translocation reaction can be viewed as a reordering of this binding so as to expel the vacant tRNA, reposition the peptidyl-tRNA in the P site, and place the next adjacent codon in the A site for its subsequent decoding. These collective reactions require the hydrolysis of GTP to GDP and $P_i$ and the participation of EF-G in bacteria and EF-2 in eukaryotes (Figure 25-10).

Translocation is one of the most mechanically complex reactions of protein synthesis, and the description of its physical basis represents a special challenge. Our present view of the process is based largely on observations of the hydrolysis of GTP catalyzed by EF-G and the ribosome, a reaction first observed by Lipmann and coworkers. Unfortunately, since this reaction proceeds in the absence of protein synthesis, it offers little insight into the nature of the mechanical events involved in translocation. Since mRNA interacts primarily with the small ribosomal subunit, it was rather surprising to discover that the uncoupled hydrolysis of GTP occurs perfectly well on the large subunit alone. There must be some means of transmitting the consequences of GTP hydrolysis on the large subunit to the site of mRNA binding on the small subunit.

## THE TERMINATION OF PROTEIN SYNTHESIS

Genetic point mutations, as far as their direct effect on protein structure is concerned, can be divided into three categories: missense, frame-shift, and nonsense mutations. The first two types of mutations can be understood in relation to what has already been said about the genetic code. *Missense mutations* are due to base-pair changes in the DNA that result in the replacement of one coding base by another. Such replacements, particularly if they do not involve a "wobble" base, usually result in an amino acid replacement in a particular protein structure. _Frame-shift mutations_ are due to the addition or deletion of base pairs in the DNA so that the downstream coding sequences are translated in a different reading frame. _Nonsense mutations_ produce shortened proteins as a result of the occurrence within a coding sequence of a codon that is normally read as a signal to terminate protein synthesis.

## The Mechanism of Protein-Synthesis Termination: The Release Reaction

The translation of poly(U) comes to a halt with polyphenylalanine still attached to tRNA and bound to the ribosome. The translation of natural messages, however, yields protein products that are both detached from tRNA and released from the ribosome. This overall reaction, the release reaction, comes about through the special reading of termination codons, a process that requires the participation of release factors (Figure 25-11).

The normal termination codons (UAA, UAG, and UGA) are the same ones that are responsible for nonsense mutations, and the only 3 of 64 codons which do not specify amino acids. Since there are no tRNAs with anticodons that pair with these sequences in normal cells, how are the termination codons read? Fractionation of cell extracts, using methods similar to those used to separate the elongation factors, revealed the presence of release factors (RFs), i.e., proteins that promote the release of nascent proteins from the ribosome.

If the detection of release factors had been limited to the overall reaction just described, the study of the functional properties of these proteins would have been slow indeed. Fortunately, a much simpler manifestation of the release reaction was discovered independently by T. Caskey and M. Capecchi and their coworkers. They observed that when the complex of

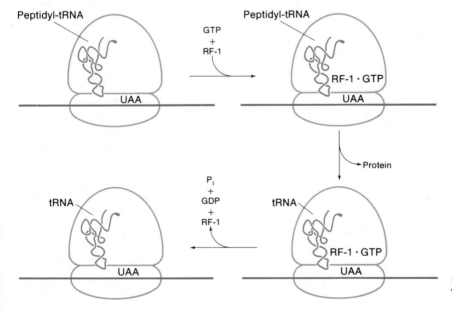

**Figure 25-11**
The release reaction.

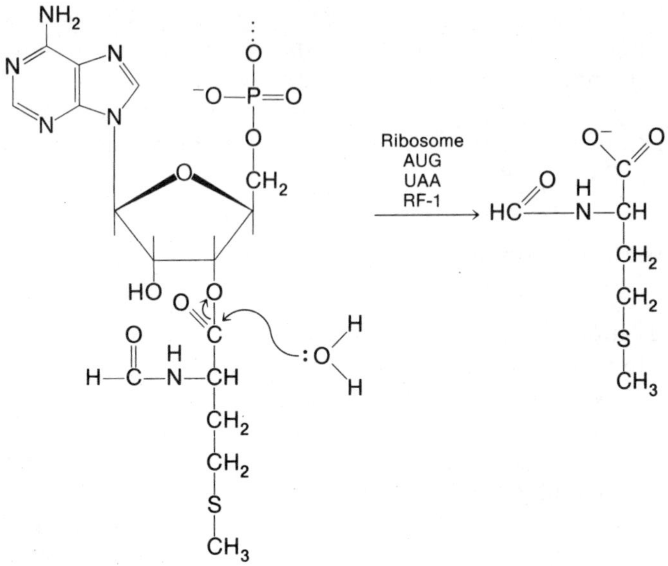

**Figure 25-12**
A simple *in vitro* assay for release factors.

fMet-tRNA$^{fMet}$, an AUG triplet, and the ribosome is incubated with a termination codon, formylmethionine is released (Figure 25-12). Two features of this reaction indicate that it provides a model for the normal release reaction. First, it requires a termination triplet; no other codon will promote the reaction. Second, the factors identified and obtained through the use of this assay promote the normal release reaction.

Through the use of the simplified release assay, Caskey and Capecchi isolated three release factors from *E. coli*. RF-1 responds specifically to the triplets UAA and UAG, while RF-2 responds specifically to UAA and UGA. RF-3 does not have release activity by itself, but stimulates the reaction catalyzed by either RF-1 or RF-2.

The release factors themselves recognize and bind weakly to termination codons in the absence of ribosomes. If the termination codons are bound to the decoding site on the ribosome, then the release factors, in turn, bind. This codon-directed binding of either RF-1 or RF-2 causes some as yet unknown change in peptidyl transferase so that it transfers the nascent peptide to water rather than to an amino group of aminoacyl-tRNA. Said another way, when a release factor occupies the ribosomal A site, peptidyl-tRNA is hydrolyzed. As expected, antibiotics that inhibit peptide-bond formation, e.g., sparsomycin, also inhibit the release reaction.

In eukaryotic cytoplasm, only a single release factor has been observed (it is designated simply RF), and it recognizes all three termination codons. In addition, this protein requires the presence of GTP, which, as in other factor reactions, is hydrolyzed to GDP and $P_i$. The function served by GTP hydrolysis in this reaction may be related to the recycling of the termination factor. The bacterial release factors show only a slight stimulation by GTP and then only in the presence of RF-3. Despite this, it is likely that GTP hydrolysis is an integral part of the release reaction in all systems in vivo.

*When nonsense mutations occur within essential genes, the result is usually lethal because the shortened protein is inactive. The lethal consequences of these mutations, however, can be overcome by nonsense suppressor genes* (see also Chapter 27). For many years, the nature of nonsense suppressor genes was a mystery. First of all, they occur entirely outside the genes that contain the nonsense mutation they suppress; i.e., they act intergenically, or between genes. In addition, three different types of nonsense suppressors were observed. An individual nonsense mutation could be suppressed only by a limited number of suppressor genes.

We now know that suppressor genes most frequently but not always encode tRNAs with altered properties that allow them to read nonsense or termination codons. *Many of these suppressor tRNAs have altered anticodons that permit them to pair with one of the termination codons.* One suppressor gene, for example, codes for a tRNA$^{Tyr}$ in which the anticodon is CUA rather than GUA. As a result, this suppressor tRNA can insert tyrosine in response to the UAG termination codon.

If nonsense codons are normally employed to release completed proteins from the ribosome, how can a cell tolerate the presence of a tRNA that reads one of these codons as a signal to incorporate an amino acid? While there is not yet a completely satisfying answer to this question, several observations bear on it. For one thing, suppressor-bearing strains tend to grow less well than wild-type bacteria, and the suppressors themselves never completely eliminate the abortive translation caused by release factors for nonsense codons. Suppression of UGA and UAG nonsense mutations can be most efficient (up to 50 percent), while UAA nonsense suppression usually does not exceed 1 to 5 percent. In addition, sequencing of genes has revealed that UAA is most commonly used as a termination codon, and termination codons in general occur frequently in pairs at the end of genes. Thus, while one might be suppressed, the second might serve to ensure release of the completed protein at normal termination sites.

## THE ROLE OF GTP IN RIBOSOMAL REACTIONS

GTP and its hydrolysis to GDP and P$_i$ play a conspicuous role in ribosomal reactions. As noted before, *each major step in protein synthesis, except peptide-bond formation itself, involves the hydrolysis of GTP to GDP.* For these reasons, a great deal of effort has been focused on explaining the enzymology of GTP in ribosomal reactions. A general picture of the role played by GTP has begun to emerge. As far as the chemistry of hydrolysis is concerned, none of these reactions has been found to involve a covalent intermediate. GTP hydrolysis is not used to phosphorylate a ribosomal protein, for example. What alteration, then, is brought about by GTP hydrolysis? Some answers to this question have come from experiments with the nonhydrolyzable analog of GTP, GMPPCP (Figure 25-13). This analog can satisfy the GTP requirements in several but not all ribosomal reactions. It will, for example, replace GTP in the binding of aminoacyl-tRNA catalyzed by EF-Tu. Most significantly, however, after tRNA binding, the EF-Tu · GMPPCP complex does not dissociate from the ribosome. From this

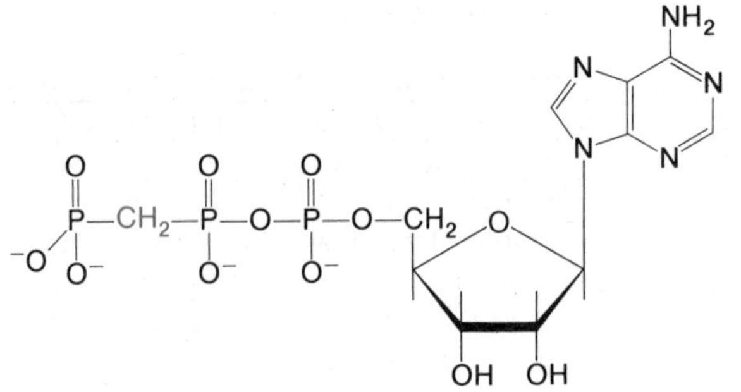

**Figure 25-13**
The structure of guanylyl methylene
diphosphonate (GMPPcP).

observation the view has developed that the purpose of GTP hydrolysis is to change the conformation of the factor so that it will dissociate from the ribosome. Another way of stating this in the case of EF-Tu is that when GTP is bound to the protein, it has high affinity for both the ribosome and aminoacyl-tRNA. When GDP is bound to the protein, its conformation is such that it will bind neither the ribosome nor aminoacyl-tRNA. Thus *GTP hydrolysis in general appears to change the conformation of the protein-synthesis factors so that they will cycle on and off the ribosome.*

Why do factors exist separate from the ribosome? Why must they cycle on and off, and why are their functions not performed by ribosomal proteins? Answers to these questions appear to lie in a certain economy of structure that results from the cycling of proteins on and off the ribosome. As the factors come and go from the ribosome, they each appear to occupy the same or a nearby site. For example, no two factors, at least those which employ GTP, appear to be able to interact simultaneously with the ribosome. In addition, efficient interaction with each of the GTP-requiring factors specifically requires the full complement of four copies of the ribosomal protein L7/12. Thus *the ribosome seems to have essentially a single site through which the various factors cycle.*

## INHIBITORS OF PROTEIN SYNTHESIS

The study of the mode of action of various inhibitors of protein synthesis has played a conspicuous role in the definition of the mechanistic basis of the individual reactions involved in the overall process of translation. A few of these inhibitors, such as the nonhydrolyzable analog of GTP have been intentionally synthesized for this purpose, but most of these compounds are of natural origin (Figure 25-14).

Because of the central role of translation in overall metabolism and the complexity of the process, it is not surprising that many antibiotics function by inhibiting translation. In addition, because bacterial ribosomes are structurally different from eukaryotic cytoplasmic ribosomes, these bacterial inhibitors frequently do not operate against eukaryotic protein synthesis. Finally, several toxic proteins have been discovered that inhibit protein synthesis, and these generally act against eukaryotic systems and are inactive against bacterial protein-synthetic components.

## The Tetracyclines

The tetracyclines are a family of chemically related antibiotics. These antibiotics inhibit the factor-dependent binding of aminoacyl-tRNA to the ribosome. Current evidence points to the fact that *tetracyclines interact primarily with the small ribosomal subunit so as to prevent the anticodon binding of aminoacyl-tRNA*. Two features of the action of these antibiotics have, however, limited a definitive assessment of their mode of action. First, they bind to many macromolecules in a nonspecific way, so it has been difficult to correlate function with binding. Second, it has been very difficult to obtain ribosomes bearing tetracycline-resistant mutations. Most bacterial resistance to this drug results from altered membrane permeability or antibiotic-inactivating enzymes. It is interesting to note that *eukaryotic ribosomes themselves are also sensitive to this drug, but its lack of permeability to eukaryotic membranes prevents in vivo inhibition of eukaryotic protein synthesis*.

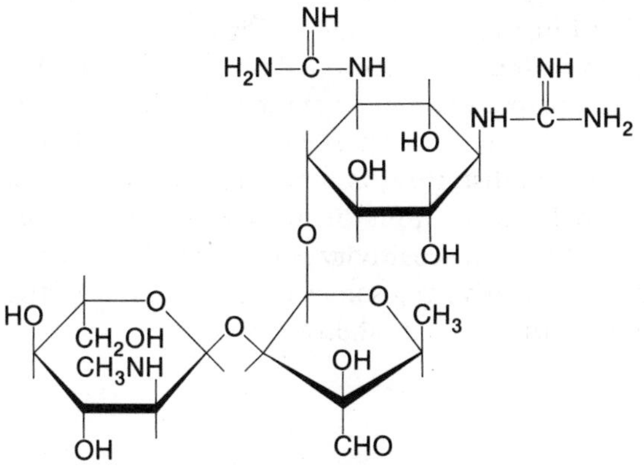
**Streptomycin**

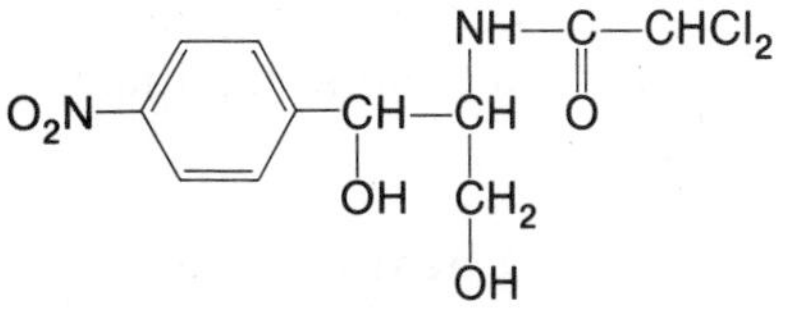
**Chloramphenicol**

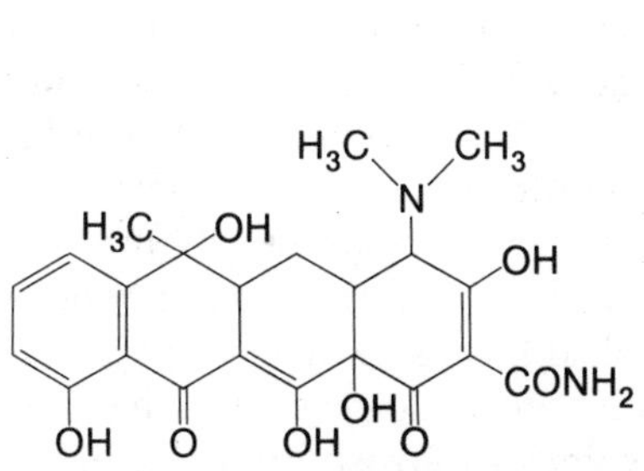
**Tetracycline**

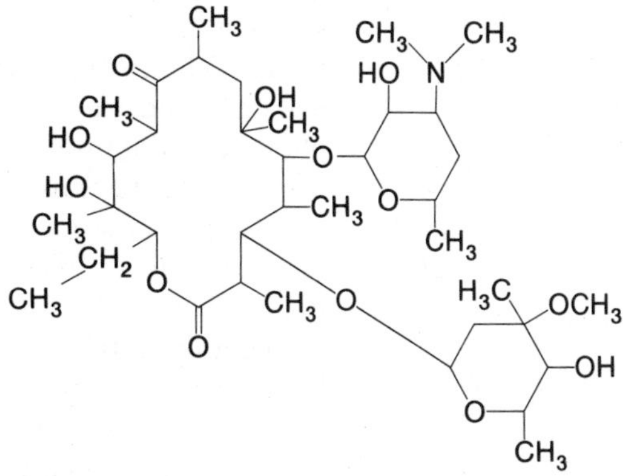
**Erythromycin**

**Figure 25-14**
The structures of some antibiotic inhibitors of protein synthesis.

## The Aminoglycosides

From the point of view of both structure and function, there is a wide range of antibiotic inhibitors of protein synthesis that are known as aminoglycosides. Of these, *streptomycin* is the best known and most extensively investigated in relation to its effect on protein synthesis.

Streptomycin is bacteriocidal in that it causes cell death, and although it affects a variety of cellular functions, its primary lethal action appears to be on protein synthesis at the level of the ribosome. Thus resistance to the effects of streptomycin can result from altered ribosomes. Ribosomes from streptomycin-sensitive but not streptomycin-resistant cells have a high affinity for the antibiotic, and binding occurs to the small ribosomal subunit. When bound to the ribosome, streptomycin produces a variety of functional alterations, none of which are fully understood. One of the first alterations to be recognized was the loss of translational fidelity of the ribosome. When bound to the small ribosomal subunit, streptomycin, in some unknown way, appears to distort the interaction between codons and anticodons so as to allow the incorporation of incorrect amino acids. Indeed, as noted in Chapter 24, mutations to streptomycin resistance frequently prevent antibiotic binding to the ribosome and involve alterations in proteins that are involved in maintaining ribosomal fidelity. The synthesis of erroneous proteins in the presence of streptomycin, however, is clearly not a sufficient explanation of its bacteriocidal action. Rather it appears that streptomycin binding distorts ribosome structure in such a way as to cause the ribosome to dissociate from the message and prevent its normal reinitiation. Thus streptomycin sensitivity is dominant over streptomycin resistance, because the sensitive ribosomes in the presence of the antibiotic reinitiates protein synthesis in an abortive manner.

## Chloramphenicol

Chloramphenicol inhibits the growth of a wide range of both Gram-positive and Gram-negative bacteria, and it was the first so-called broad-spectrum antibiotic to be used clinically. Although eukaryotic cells are generally resistant to the effects of the drug, its clinical utility has been severely curtailed as a result of toxic side effects commonly associated with its use. The toxic effects are at least partly due to the fact that mitochondrial ribosomes are sensitive to the antibiotic. The relative structural simplicity of chloramphenicol and the fact that only one of its four diastereoisomers is active in inhibiting protein synthesis might suggest that its mode of action by now should be well understood. This is not the case. At inhibitory levels, one molecule of the antibiotic binds to the 50 $S$ ribosomal subunit. A variety of tests indicate that chloramphenicol specifically inhibits the peptidyl transferase reaction. Because of these circumstances, it would seem that chloramphenicol provides the ideal means of dissecting the nature and location of this reaction. Unfortunately, however, this has not proved to be the case, and this failure typifies the frustration associated with the dissection of ribosomal reactions. For example, it has not been possible to show unambiguously that substrate binding is altered by chloramphenicol. Puromycin inhibits the binding of chloramphenicol, and under some conditions, chloramphenicol also inhibits the binding of aminoacyl-tRNA to the ribosome. In the same vein, the chemical attachment of chloramphenicol to proteins in the vicinity of its binding has led to conflicting results. At best, *these studies lead to the view that one molecule of chloramphenicol interacts with a variety of proteins in the vicinity of the peptidyl transferase, and this binding produces multiple effects on their functions.*

## Diphtheria Toxin

Diphtheria is among the best understood of infectious diseases. The pathogenesis of the disease results entirely from the elaboration by *Corynebacterium diphtheria* of a single exotoxin that kills cells by inhibiting protein synthesis. The pathogenic consequences of the disease are entirely prevented by immunization with toxoid, an inactivated form of the purified toxin. Curiously, the structural gene for the protein is carried by a phage, $\beta$, that must infect the bacterium to induce toxin production.

The toxin, as elaborated into the circulation of the infected host, is a single polypeptide chain ($M_r = 63,000$) with two intramolecular disulfide bonds. Its toxic action requires that a portion of the molecule, the A fragment ($M_r = 21,000$), enter the cytoplasm, where it acts catalytically and specifically to inactivate an essential component of protein synthesis, EF-2.

The entry of the A fragment into the cytoplasm (a single molecule within the cell is sufficient to kill it) proceeds by a poorly understood mechanism. Essentially stated, the intact toxin interacts with an unidentified receptor on the surface of sensitive cells and is both proteolytically cleaved and disulfide reduced to yield the A and a larger B fragment ($M_r = 42,000$). The B fragment then facilitates the penetration of the A fragment through the cell membrane. This penetration appears to involve formation by the B fragment of a pore in the membrane through which the A fragment passes.

Within the cell, the A fragment acts as a very specific protein-modifying enzyme. It catalyzes the ADP-ribosylation and consequent inactivation of EF-2 by the following reaction:

$$EF\text{-}2 + NAD^+ \rightleftharpoons ADP\text{-}ribosyl\text{-}EF\text{-}2 + nicotinamide + H^+$$

The reaction as shown is reversible when conducted in vitro, but under the intracellular conditions of pH and nicotinamide concentration, it is irreversible. Thus diphtheria toxin kills by irreversibly destroying the ability of EF-2 to participate in the translocation step of protein chain elongation.

The enzymatic specificity of diphtheria toxin deserves special comment. The catalytic A fragment in vitro will ADP-ribosylate EF-2 in the cytoplasm of apparently all eukaryotic cells, whether sensitive to toxin in vivo or not, but it will not modify any other protein, including the bacterial counterpart of EF-2. This narrow enzymatic specificity of the toxin has called attention to an unusual structure in EF-2 at its site of modification by the toxin. ADP-ribose is linked by toxin by means of an unusual amino acid in EF-2, *diphthamide*, which has the structure seen in Figure 25-15. This novel amino acid occurs at the site of toxin modification in all eukaryotic EF-2s. While the unique occurrence of diphthamide in EF-2 at least partially explains the specificity of the toxin, it raises questions about the function of the residue in the translocation step. Diphthamide is derived by the posttranslational modification of histidine. Rather interestingly,

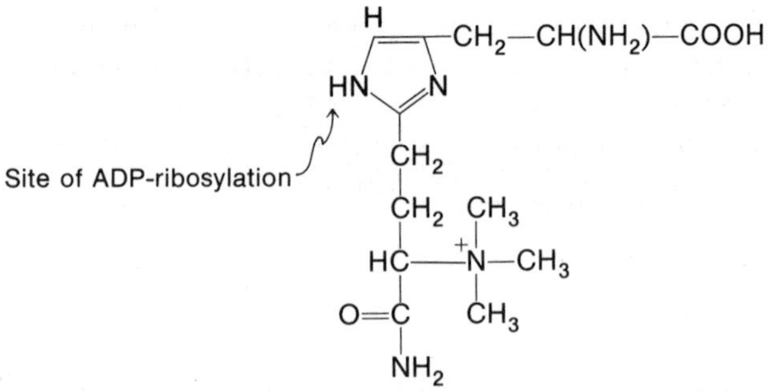

**Figure 25-15**
The structure of diphthamide.

some mutants of cultured animal cells selected for resistance to the toxin appear to lack the enzymes that carry out the posttranslational modification of EF-2 necessary for toxin action, but these cells seem perfectly capable of protein synthesis. Thus the raison d'être of diphthamide, as well as the biological origin of the toxin that modifies it, remains a mystery.

## POSTTRANSLATIONAL MODIFICATION OF PROTEINS

Only 20 amino acids (plus formylmethionine in prokaryotic systems) are directly specified by the genetic code. Yet analysis of proteins has revealed that they contain well over 100 different amino acids, all structural variants on the original 20. In addition, proteins infrequently function in their final form with the full complement of amino acids assembled during their translation. Most often their primary translation products are proteolytically cleaved in producing their final three-dimensional structure, in their transport, or in the activation of proenzymes or prohormones. This collection of structural alterations of proteins is considered together as posttranslational modifications.

### Three-Dimensional Structure

The question of how polypeptide chains arrive at their final three-dimensional structure within proteins has received a great deal of attention (see Chapter 3). Two observations bear directly on this question. It would appear that some enzymes can assume enzymatically active conformations even before they have been completely assembled and released from the ribosome. Additionally, many proteins, once assembled, can be fully denatured, and when the denaturant is removed, the proteins, albeit sometimes slowly and incompletely, will spontaneously reassume their native configuration.

From observations of this type, a picture emerges in which protein structure begins to develop as the polypeptide chain grows on the ribosome. The configuration of the polymer is presumed to result entirely from the developing sequence of amino acids, modified by whatever posttranslational alterations may have taken place. Thus far, all attempts to discover additional information input that might contribute to the three-dimensional structure of proteins beyond primary sequence have failed. Therefore, at present it is reasonable to picture protein structure developing on the ribosome like a metal filing peeling off a lathe. This is a difficult model to test experimentally, however. Clearly, some proteins, when denatured, would resume their native state like a spring. Other more complex structures might not so readily re-form because their original shape was determined as they were spun from the ribosome. Moreover, just as a lathe tool contributes to the shape of a metal filing, it would not be surprising if the ribosome played a role in passively shaping or limiting unwinding protein conformations.

## Covalent Modification of Amino Acids in Proteins

A partial list of covalently modified amino acids known to occur in proteins is shown in Table 25-2. A number of simple modifications are observed on different amino acids, particularly as they occur at the termini of proteins. Acetylated and amidated amino acids are examples. In addition, a number of interesting single amino acid transfer reactions have been observed at the N and C termini of proteins. These one-step reactions occur by nonribosomal means and result in amino acids that are joined at

**Table 25-2**
Some Amino Acid Derivatives Found During
Posttranslational Modification

| Parent Amino Acid | Derivative |
|---|---|
| Ala | $N$-acetylalanine, $N$-methylalanine |
| Arg | $N^{\omega}$-methylarginine, ADP-ribosyl-arginine, citrulline, ornithine |
| Asn | Aspartic acid, $N$-acetylglucosaminylarginine |
| Asp | Aspartic acid $\alpha$-amide, $N$-acetylaspartic acid |
| Cys | Cystine, $S$-galactosylcysteine |
| Glu | Glutamic acid $\alpha$-amide, $\gamma$-methylglutamic acid $\gamma$-carboxyglutamic acid |
| Gly | Glycinamide, $N$-formylglycine |
| His | $\pi$-Methylhistidine, diphthamide |
| Lys | $N^{\Sigma}$-trimethyllysine, $N^{\Sigma}$-phosphopyridoxyllysine, desmosine-$\delta$-hydroxylysine |
| Met | Methioninamide, $N$-acetylmethionine |
| Phe | Phenylalanine amide, $\beta$-hydroxyphenylalanine |
| Pro | 4-hydroxyproline, $O^4$-arabinosylhydroxyproline |
| Ser | Pyruvate, $O^{\beta}$-phosphonoserine, $O^{\beta}$-mannosylserine |
| Thr | $\alpha$-ketobutyrate, $O^{\beta}$-mannosylthreonine |
| Tyr | Tyrosine-$O^4$-sulfate, 3,5-diiodotyrosine, $O$-adenosyltyrosine |

*Source:* From F. Wold, Posttranslational modifications of proteins. *Ann. Rev. Biochem.* 50:783, 1981. Used with permission.

the ends of proteins by peptide bonds not specified by the message. In general, the functional significance of these terminal modifications is obscure, but it likely reflects some regulatory modification. In this context, it is interesting to consider the *E. coli* ribosomal protein L7/12. These two proteins are the product of the same gene and differ only by an N-terminal acetyl group. The fraction of total L7/12 that is acetylated varies with bacterial growth rate, but thus far no functional alteration has been found to result from this modification.

Most readily detected are modifications that result in altered protein functions or are essential to protein function. Regulation of protein function has been observed to result from methylation, phosphorylation, nucleotidylylation, and ADP-ribosylation. In these cases, usually quite specific enzymes modify regulatory targets in key enzymes. Moreover, in this category are the coenzymes that are essential to enzyme function and are covalently attached by amino acid side chains.

Carbohydrates of a wide variety become attached to proteins in many different ways. In some cases, the carbohydrate addition is essential to the participation of the protein in membrane or cell-wall structure. In higher organisms, the carbohydrate plays a key role in the circulation of blood proteins. These proteins may be targeted to certain tissues or cleared from the circulation by the liver on the basis of their carbohydrate content.

Some of the most exotic protein alterations are employed to form cross-links between proteins, particularly in the connective-tissue protein collagen. These cross-links are important to the mechanical stability of connective tissue. In this context it should be noted that cystine formed by the posttranslational cross-linking of two cysteine residues probably serves a similar stabilizing role, particularly in extracellular enzymes. The disulfide cross-linking of the two chains of insulin is a clear example (see Figure 29-10).

In general, the techniques of protein chemistry are tuned to the analysis of standard amino acids, and it is relatively easy to overlook unstable or infrequent derivatives. It is noteworthy that two of the most interesting modifications of proteins, γ-carboxyglutamic acid in coagulation proteins and diphthamide in EF-2, were discovered as a result of the study of the biological inhibitors dicumerol and diphtheria toxin, respectively. Dicumerol inhibits the formation of γ-carboxyglutamic acid in the vitamin K-dependent blood proteins, and diphtheria toxin inactivates EF-2 by modifying its single diphthamide residue, as described earlier. It is possible that the actual list of functionally important amino acid derivatives is much longer.

### Proteolytic Processing of Proteins

Proteolytic activation of zymogens or proproteins to functional proteins has been recognized for many years. First seen in the activation of proteolytic enzymes (e.g., trypsinogen → trypsin; prothrombin → thrombin), a similar phenomenon has since been observed with a variety of other proteins and hormones (e.g., proalbumin → albumin, proinsulin → insulin). The proteolytic cleavage or cleavages involved in proprotein → protein conversion can take a variety of forms, most of which lead to the removal of a peptide fragment or fragments. The propeptide of proproteins may be either terminal or internal to the protein chain. Both internal and terminal peptides are removed from prothrombin; an N-terminal 6-residue peptide is removed from proalbumin, and a 29-residue peptide (the C peptide) is removed from the interior of proinsulin. Many of the N-terminal proprotein cleavages occur on the C-terminal side of Arg–Arg sequences and appear to involve a common enzyme system.

The in vitro translation of messages for proteins destined for extracellular transport (e.g., proalbumin, proinsulin) has revealed yet another class of posttranslational proteolytic cleavages. Translation of these messages in protease-deficient in vitro translation systems (e.g., reticulocyte and wheat germ extracts) invariably yields translation products larger than the proproteins observed in vivo. These in vitro translation products, known as preproteins or preproproteins, in general, begin with Met and contain 16 to 30 additional amino acids at their amino terminals.

These presequences, or "signal" sequences, as they have been designated by G. Blobel, are rich in hydrophobic amino acids, especially leucine (see Chapter 17). The ribosomes that produce proteins containing signal sequences are generally attached to intracellular membranes, and the sequences are not normally seen on completed proteins because they are removed cotranslationally, i.e., while the nascent protein is still being assembled on the ribosome. When wheat germ or reticulocyte translation systems, which accumulate preproteins, are supplemented with endoplasmic membrane vesicles, typically from dog pancreas, the signal sequences are removed and the translation products are found within the vesicles.

Observations of this type led Blobel to propose the "signal hypothesis" (see also Chapter 17). According to this hypothesis, the presequence serves to guide the polysome producing the protein that is destined for extracellular transport to the endoplasmic reticulum, where the peptide inserts into the membrane because of its hydrophobic character. In some as yet undefined way, the nascent protein threads through the membrane as it is being assembled and the signal peptide is removed by a protease on the distal side of the membrane, where the completed protein is ultimately sequestered.

Thus, for example, insulin mRNA is translated in wheat germ extracts as a single polypeptide chain, preproinsulin, that contains the 84 residues of proinsulin plus a 23-residue signal peptide. Within the islets of Langerhans, the N-terminal sequence is removed cotranslationally in targeting proinsulin to the Golgi apparatus, and the two disulfides joining the ends of the molecule are formed. Following this, the C-peptide region of proinsulin is removed to yield the circulating form of insulin with 51 residues in two disulfide-linked peptides.

## INTRACELLULAR PROTEIN DEGRADATION

The proteins that comprise living cells are much more labile than the cells themselves and are subject to continual renewal; they are continually being degraded back to amino acids and replaced by new synthesis. For example, the average protein in rat liver has a half-life of approximately 1 day, and in brain or muscle, 3 or 6 days, while individual enzymes in these cells may turn over with half lives as short as 1 to 2 h. In bacteria, regulatory proteins have been identified that are completely hydrolyzed within minutes after their synthesis. At first glance this *continuous degradation of cell proteins appears to be a highly wasteful process*, although it obviously must provide the organism with a clear selective advantage. *In fact, this process is of major importance in the regulation of enzyme levels, in protecting the organism against the accumulation of abnormal proteins, in the control of tissue mass, and in the organism's ability to adapt to poor nutritional conditions.*

The first clear evidence for this dynamic state of cell constituents came in the early 1940s with the introduction of isotopic amino acids by Schoenheimer and coworkers, but biologists and biochemists were slow to recognize the fundamental importance of this phenomenon. In the 1950s, findings by Schimke, Segal, Knox, and others demonstrated that changes in the rates of degradation of a protein can be an important factor determining its cellular content. Although this development stimulated much research, it is only since the late 1970s that clear information has emerged concerning the selectivity, intracellular pathways, and mechanisms regulating this process.

### Protein Breakdown and the Control of Enzyme Levels

Whenever it has been investigated, the degradation of specific proteins in cells has been found to obey first-order kinetics like the decay of radioactive nuclei (see Chapter 4). Therefore, the rate of degradation is defined normally by the half-life of the protein, the time during which 50 percent of it is degraded. This time is independent of when the protein was synthesized; thus, cell proteins do not undergo an aging process, such as occurs for erythrocytes or human beings, which show an average life span and then a rapid rate of death. *Within a cell, the half-lives of individual cytoplasmic proteins are not uniform* (Table 25-3). For example, in rat liver, the half-lives of specific enzymes are known to range between 11 min for ornithine decarboxylase to 19 days for isozyme 5 of lactic dehydrogenase. In addition, different proteins within the same organelle can also differ in their rates of degradation; in heart mitochondria, for example, the enzymes δ-aminolevulinate synthase and ornithine transcarbamylase are degraded much more rapidly than the cytochromes. Thus, all mitochondrial proteins are not assembled and degraded as a unit. In addition, rates of degradation of specific proteins and the overall rate of protein degradation in a cell or tissue can vary under different physiological conditions, e.g., with starvation or in response to hormones.

*The level of any protein is determined by the balance between its rate of synthesis and degradation and, therefore, inherent differences in degradative rates of different proteins can have important implications for the regulation of enzyme levels.* A rapid rate of degradation ensures that the concentration of an enzyme falls very quickly when its synthesis is reduced. On the other hand, the concentration of a short-lived protein rises to a new steady-state level especially rapidly when its rate of synthesis is enhanced.

Such an analysis predicts that key rate-limiting enzymes that regulate the flow of substrates through biochemical pathways might have evolved especially short half-lives. The rates of degradation of large numbers of liver enzymes have now been measured in rat liver. By surveying these values, it is clear that *the enzymes degraded most rapidly in liver all catalyze reac-*

**Table 25-3**
Half-lives of Some Rat Liver Proteins

|  | Enzyme | Half-life (h) |
|---|---|---|
| Rapidly degraded | 1. Ornithine decarboxylase | 0.2 |
|  | 2. δ-Amino levulinate synthase | 1.1 |
|  | 3. RNA polymerase I | 1.3 |
|  | 4. Tyrosine aminotransferase | 2.0 |
|  | 5. Tryptophan oxygenase | 2.5 |
|  | 6. Deoxythymidine kinase | 2.6 |
|  | 7. β-Hydroxy-β-methylglutaryl coenzyme A reductase | 3.0 |
|  | 8. Phosphoenol pyruvate carboxykinase | 5.0 |
| Slowly degraded | 1. Arginase | 96 |
|  | 2. Aldolase | 118 |
|  | 3. Cytochrome $b_5$ | 122 |
|  | 4. Glyceraldehyde-3-phosphate dehydrogenase | 130 |
|  | 5. Cytochrome B | 130 |
|  | 6. Lactic dehydrogenase (isoenzyme 5) | 144 |
|  | 7. Cytochrome C | 150 |
|  | 8. β-glucoronidase | 240 |

*tions that are particularly important metabolic control points* (see Table 25-3). By contrast, the liver proteins with especially long half-lives are rarely, if ever, the sites of metabolic control. Included in the group of rapidly degraded proteins are the rate-limiting enzymes for polyamine synthesis, heme synthesis, RNA synthesis, amino acid breakdown, and cholesterol production. For example, the rate-limiting enzyme in cholesterol biosynthesis is hydroxyl-methyl-glutaryl-CoA reductase. Its level varies dramatically with diet and circadian periodicity, and it has a half-life of only 2 h in rat liver. This pattern suggests that rapid degradative rates for certain enzymes have evolved in order that their intracellular levels can adapt quickly to environmental changes.

*Alterations in the half-lives of specific enzymes have also been documented under various physiological conditions.* In a number of instances, decreased breakdown of an enzyme occurs in the presence of a cofactor or substrate. By itself, reduced degradation can lead to an increase in the level of an enzyme, even in the absence of any change in its synthesis. For example, the half-life of the amino acid-degrading enzyme, tryptophan oxygenase, is dramatically prolonged in the presence of its substrate tryptophan or its cofactor heme. In extracts, tryptophan or tryptophan analogs stabilize this enzyme against denaturation, and this effect may account for the stabilization against proteolysis in vivo. Many such examples have been found where substrates, cofactors, or even drugs can stabilize a protein against intracellular degradation. Physiologically, this effect ensures that levels of these enzymes will be higher when the substrates are present in large amounts.

A related phenomenon has been demonstrated for glutamine synthase in which the end-product of the enzymatic reaction, glutamine, increases the enzyme's degradation rate and thus reduces its level. In this way, glutamine retards its own rate of production. Although the generality of this type of control is still unclear, it would appear to be a very useful mechanism for ensuring that enzyme levels are appropriate for metabolic needs.

There is now extensive evidence that these large differences in the stability of proteins are determined largely by their different conformations, although the precise structural features that lead to a short or long half-life are uncertain. Various structural properties of protein have been found to correlate with intracellular stability; for example, intracellular half-lives correlate roughly with relative sensitivities of proteins to digestion by a wide variety of proteases in vitro and to their thermal stabilities. In addition, soluble proteins with large molecular weights tend to be degraded more rapidly than smaller ones, while acidic cellular proteins tend to be degraded more rapidly than more basic ones. Unfortunately, the biochemical basis for these interesting correlations remains unclear. Nevertheless, such findings, as well as the regulatory effects of substrates or products, do emphasize that degradative rates of proteins, like their enzymatic and allosteric properties, must also be determined largely by amino acid sequences and must also have been determined by evolutionary selection.

## The Selective Degradation of Abnormal Proteins

One very important function of intracellular protein breakdown in animal and bacterial cells is to protect the organism against the intracellular accumulation of polypeptides whose conformations are highly abnormal. This process thus serves as a sort of cellular "sanitation system" that helps prevent the accumulation of partially denatured, potentially harmful polypeptides. This protective function seems particularly important in complex organisms, such as humans, whose cells divide only very slowly or not at all, and thus cannot dilute out such abnormal polypeptides simply by cell division.

_Both bacterial and animal cells rapidly hydrolyze proteins with highly abnormal structures_ as may arise from nonsense or some missense mutations, from errors in RNA or protein synthesis, or by intracellular denaturation. Such mutant proteins are synthesized at the same rates as normal gene products but fail to accumulate to the usual extent because of their rapid hydrolysis. For example, in E. coli, incomplete chains of $\beta$-galactosidase lacking the normal carboxyl terminus are degraded with half-lives as short as a few minutes, even though the normal polypeptide is almost completely stable in these cells. Incomplete short-lived proteins may result from nonsense mutations, from biosynthetic errors, or as a consequence of incorporation of the antibiotic puromycin (which causes the premature termination of polypeptides and their release from the ribosome). Certain drugs (e.g., streptomycin) or mutations that affect the fidelity of the protein synthetic apparatus also promote the production of complete proteins that have highly abnormal conformations. Such error-containing polypeptides tend to be degraded rapidly, as are many new polypeptides generated by the techniques of genetic engineering.

Many studies in this area have followed the fate of proteins that have incorporated synthetic analogs of the natural amino acids. The incorporation of such analogs interferes with the normal tertiary folding of proteins and the resulting polypeptides are rapidly hydrolyzed in vivo. For example, hemoglobin is normally one of the most stable proteins in the organism and lasts the life span of the red cell (110 days in humans). However, globin containing the valine analog, aminochlorobutyrate, has a half-life as short as 10 or 12 min. Thus these aberrant molecules are degraded up to 1000 times faster than the normal hemoglobin within the same cell whose degradation is unaltered by such treatments. The rapid degradation of such abnormal proteins is also an important factor in many human diseases, such as certain hemoglobinopathies. Several human hemoglobin variants (e.g., the "unstable" hemoglobins) as well as the free $\alpha$-chains that are produced in excess in $\beta$-thalassemia are rapidly degraded within reticulocytes. The selective hydrolysis of such abnormal hemoglobins during maturation of the reticulocyte appears highly advantageous to the organism, since it should minimize the deleterious consequences of accumulating large amounts of partially denatured molecules.

It is still uncertain to what extent such "aberrant" polypeptides may be found in normal cells. Possibly abnormal proteins arise continuously through the spontaneous denaturation or chemical modification of cell enzymes. In fact, there is now strong evidence that oxidation of hemoglobin to met-hemoglobin or free-radical damage to the proteins from superoxides or peroxides can lead to rapid intracellular degradation. Thus, although cells contain many mechanisms to maintain protein conformations intact (e.g., met hemoglobin reductase, superoxide dismutase), cells also have mechanisms to quickly remove damaged proteins if these protective mechanisms fail. In fact, the rate-limiting step in the degradation of most cell proteins may be their irreversible unfolding and thus their assuming an "abnormal" conformation.

The process by which such mutant or damaged polypeptides, or short-lived normal enzymes, are recognized and selectively eliminated appears to be quite similar in animal and bacterial cells. Furthermore, isolated liver mitochondria have recently been shown also to contain such a degradative system for rapidly hydrolyzing incomplete or otherwise abnormal mitochondrial proteins all the way to amino acids. Within the mitochondrial space, as in the cytoplasm, proteins are present in what must be a highly reactive environment, where protein denaturation may be a frequent event. Thus in all cells and even in all membrane-enclosed organelles, the continued elimination of such damaged molecules may be essential for cell viability.

## ATP-Dependent Pathway for Protein Breakdown

Our relative lack of detailed knowledge about the pathway of intracellular proteolysis reflects the general inability of investigators, until recently, to prepare cell-free extracts that mimic the in vivo process. It had long been assumed that in animal cells, the lysosome is the only, or at least the primary, site for intracellular protein degradation. This membrane-enclosed organelle contains a number of proteinases, and thus has the capacity to hydrolyze proteins rapidly and completely. However, certain cells (e.g., *E. coli* or red cells) lack such degradative organelles and yet have the capacity for degrading abnormal proteins.

Recently, cell free systems have been established by A. Goldberg and colleagues from reticulocytes, *E. coli*, and mitochondria that can selectively degrade abnormal proteins in a fashion similar to intact cells. Such extracts do not degrade hemoglobins or β-galactosidase but are capable of digesting completely analog-containing or mutant polypeptides. This proteolytic pathway is found in the soluble fraction of the cytoplasm, and it does not appear to be enclosed within any sort of membrane. Unlike the lysosomal proteases which are only active at acid pH, this degradative system is optimal at pH 7.8. These preparations can degrade proteins completely to amino acids. Thus, in mammalian cytoplasm there exist at least two major systems for protein breakdown: (1) the lysosomal system, and (2) the soluble alkaline system (Figure 25-16) (in addition to the pathway in mitochondria for degrading organelle proteins).

The most interesting biochemical feature of these new degradative systems is that they require ATP. Extensive studies in intact animal and bacterial cells have shown that inhibitors of glycolysis and oxidative phosphorylation block the breakdown of abnormal proteins, as well as abnormal enzymes. When such inhibitors are removed, these cells regain this capacity for protein breakdown concomitant with the restoration of intracellular ATP levels. A similar ATP-requirement has also been demonstrated for the protein breakdown occurring in mitochondria. An energy requirement for protein catabolism, however, would not be anticipated on thermodynamic grounds, since the hydrolysis of peptide bonds is an exergonic reaction. This energy requirement thus became an important clue in the discovery of the responsible degradative systems and the demonstration that novel biochemical mechanisms must be operating in intracellular proteolysis.

In these cell-free preparations, the breakdown of abnormal proteins is completely dependent upon the supply of ATP. By following the disappearance of specific mutant polypeptides in vivo, it became clear that ATP is required for the initial cleavage reactions of the substrates. In *E. coli* extracts, the ATP appears to be essential for the function of a novel endoprotease, called protease La. A very similar ATP-dependent protease has also

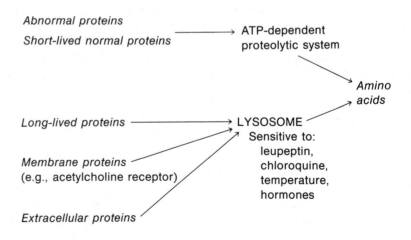

**Figure 25-16**
Pathways for protein degradation in mammalian cells.

been found in mammalian mitochondria, where it also appears to catalyze rate-limiting steps in protein breakdown. Compared to the digestive proteases from the pancreas and the lysosome, or the proteases secreted by microorganisms, these ATP-dependent proteases are unusually large. They have a molecular weight of 450,000 to 500,000, and the *E. coli* enzyme is comprised of four identical subunits.

*Of particular enzymological interest has been the demonstration that protease La hydrolyzes ATP and proteins by some kind of coupled mechanism.* Inhibitors of ATP hydrolysis, such as vanadate, block protein breakdown, and this requirement can account for the energy requirement for proteolysis in vivo. Some stoichiometric relationship exists between the extent of ATP cleavage and proteins, and a limited number of ATP molecules appears necessary for the breakdown of proteins to acid-soluble peptides. Furthermore, the cleavage of ATP rises severalfold when this enzyme digests protein substrates. Proteins (e.g., hemoglobin) that are not hydrolyzed do not activate ATP breakdown, while abnormal globins do. It appears likely that this protein-activated ATP cleavage triggers the subsequent proteolytic step. Thus activation of this ATPase by protein substrates may be a mechanism to ensure that unless an appropriate substrate is present, the protease does not assume an active conformation in vivo and does not degrade desirable cytoplasmic proteins.

In mammalian cells, the role of ATP in the soluble pathway for protein breakdown is less clear than in bacteria. These cells contain a large molecular weight ($M_r = 450,000$) alkaline protease that is activated severalfold directly by ATP and that seems to carry out the initial cleavage reactions of proteins in a similar fashion to protease La in *E. coli*. It also generates large peptides which are subsequently hydrolyzed by other cytoplasmic enzymes that are independent of ATP.

Hershko, Rose, and coworkers have demonstrated in reticulocyte extracts an additional process that seems to be involved in the soluble proteolytic pathway, and that also requires ATP. These workers have reported that the soluble proteolytic pathway requires a small, heat stable, nonenzymatic polypeptide. This polypeptide was discovered earlier as a constituent of many cells and had been named _ubiquitin_. In the presence of ATP, the carboxyl end of ubiquitin was shown to undergo an activation that releases pyrophosphate from ATP in reticulocyte extracts. This process is formally similar to the activation of amino acids prior to their charging to tRNA. The charged ubiquitin is then linked by the carboxyl group to the amino group of lysine in various proteins that are substrates for proteolysis. Ubiquitin ligation may serve an essential function to mark specific polypeptides for selective degradation. It is noteworthy that the ligation of ubiquitin to proteins has not been found in bacteria or mitochondria which are nevertheless capable of selectively degrading abnormal proteins. Thus, this interesting process may play a special role in eukaryotic cells in the digestion of specific polypeptides.

*Studies with various protease inhibitors indicate that multiple enzymes are involved in the soluble degradative process,* including both endoproteases with serine residues in their active sites, various metalloproteases, as well as various peptidases. These enzymes seem to act sequentially; first, the ATP-dependent endoproteases cleave protein substrates to large polypeptides, then cytoplasmic proteases independent of ATP generate smaller oligopeptides which are then rapidly hydrolyzed to free amino acids by various soluble cellular hexopeptidases. Surprisingly little is known about the other enzymes that comprise this pathway even in *E. coli*, which is surely the best characterized cell at the biochemical level. Only quite recently, *E. coli* was found to contain seven endoproteases, in addition to Protease La, and their precise function in the degradation of normal or abnormal proteins is not clear. Mammalian reticulocytes contain three endoproteases and a variety of peptidases that appear to be active in the removal of abnormal proteins. The final steps of this process by which oligopeptides are hydrolyzed to amino peptides has been shown in microorganisms to utilize the same enzymes that bacteria use when they grow on extracellular peptides as a nitrogen or carbon source.

It should also be emphasized that this ATP-dependent pathway is probably also responsible for the degradation of various short-lived normal enzymes, as well as abnormal polypeptides. For example, during the maturation of reticulocytes into adult erythrocytes, many enzymes, mitochondria, and ribosomes are lost. This programmed degradative process is catalyzed by the ATP-dependent pathway and not by the lysosomal mechanism. In *E. coli*, mutations affecting protease La (*lon* mutations) alter a number of important cellular processes, including the production of polysaccharides, recovery from UV-damage, and cell division, apparently because these various processes all involve some short-lived proteins degraded by protease La. In cultured mammalian cells, the degradation of various enzymes, like the degradation of abnormal proteins, requires continuous production of ATP. Furthermore, inhibitors of lysosomal function do not affect their degradative rates, although they do reduce other types of protein breakdown within the same cells (see the following discussion).

### Protein Breakdown within the Lysosome

The classic studies of De Duve and coworkers in the 1960s established that *mammalian cells contain a degradative organelle, the lysosome, in which proteases and other acid hydrolases are concentrated*. This membrane-enclosed organelle contains a number of proteases active at acidic pH, including cathepsin D (which resembles pepsin), a number of thiol proteases (such as cathepsin B, cathepsin H and L, which resemble papain in mechanism of action) and certain exoproteases (e.g., cathepsin A). This organelle thus has the capacity of hydrolyzing proteins rapidly to peptides and amino acids. There is very strong microscopic evidence implicating this organelle in the degradation of exogenous proteins taken up into cells by phagocytosis, e.g., bacteria engulfed by macrophages, or by pinocytosis, such as the receptor-mediated uptake of circulating polypeptides, including lipoproteins, transferrin, insulin, or other hormones. These endocytosed proteins are then incorporated into lysosomes, where they undergo rapid digestion. Various studies have also implicated this organelle in the breakdown of surface proteins, such as hormone receptors.

In addition, there is now strong evidence that the lysosome plays an important role in degradation of many intracellular proteins, especially under various catabolic conditions. For example, in the liver of rats during starvation or even in an overnight fast, the average rate of protein breakdown rises, probably to provide amino acids for gluconeogenesis or direct oxidation. In such livers, the lysosomes are particularly large and contain cytoplasmic or organelle debris which seem to be in the course of degradation. Such structures, called "autophagic vacuoles," appear to be the site of the accelerated protein breakdown. Mortimore and coworkers have shown that the number of autophagic vacuoles in electron micrographs of cells correlates closely with the overall rate of proteolysis. In addition, when isolated in this form, lysosomes contain large pools of amino acids and appear to be actively digesting proteins contained within them.

More direct evidence for a role of the lysosome in intracellular protein degradation has come from the use of inhibitors of lysosomal acidification or lysosomal proteases. The various lysosomal hydrolases have very low pH optima, and this organelle generates an acidic milieu within it apparently by an ATP-dependent proton pump. Weak bases, such as chloroquine or ammonia, that freely permeate membranes, accumulate within the acidic environment of the lysosome and raise intralysosomal pH. Treatment of mammalian cells with these agents has been found to reduce overall protein breakdown. In addition, certain highly specific protease inhibitors are known that can enter mammalian cells and selectively inhibit proteases within the lysosome; for example, the antibiotics, leupeptin and chymostatin, are both transition state analogs that can inhibit cathepsin B and L in vivo. Cells treated with these agents show a reduction in the overall rate of protein breakdown similar to that seen with the weak bases.

Use of such reagents has proven very useful in determining the role of the lysosomal apparatus in different proteolytic processes. These inhibitors of lysosomal proteolysis do not affect the rapid degradation of short-lived enzymes or the rapid degradation of abnormal proteins. Thus, in mammalian cells the two major degradative systems appear to have distinct functions. These inhibitors are most effective in reducing overall protein breakdown in catabolic states, where overall proteolysis is enhanced, e.g., in tissues of starving animals, in denervated muscles, or in cultured cells deprived of serum or insulin. Under such conditions, the lysosome seems to be the major site of protein breakdown. Exactly how parts of the cytoplasm are surrounded by a membrane and then incorporated into the lysosome are important unanswered questions. There is no evidence that this process is selective, unlike the ATP-dependent degradative process responsible for the rapid hydrolysis of short-lived enzymes or abnormal proteins. It is also noteworthy that protein breakdown by this pathway seems to be enhanced in various pathological states. Of particular interest is the possibility that inhibitors may be used therapeutically in many human diseases where protein breakdown occurs at an excessive rate, such as muscular dystrophy, in denervated muscle, or in patients with burns.

The quantitative importance of the lysosomal system in overall proteolysis depends on the nutritional and endocrine status. In rapidly growing cultured fibroblasts, inhibitors of lysosomal function have little or no effect on overall proteolysis. By contrast, in nongrowing cultures, protein breakdown rises and this increase is sensitive to inhibitors of lysosomal proteases. It is thus clear that the rate of proteolysis in the lysosome is subject to precise regulation. Increased numbers of autophagic vacuoles and more rapid proteolysis occur when virtually all mammalian cells are perfused in the absence of an adequate supply of amino acids. Many hormones also affect overall rates of protein catabolism in tissues. For example, in liver and skeletal muscle, insulin is probably the most important factor promoting protein accumulation in mammalian tissues. Insulin not only stimulates protein synthesis but also inhibits protein breakdown, and the latter effect involves an inhibition of intralysosomal proteolysis apparently through a reduced rate of autophagic vacuole formation. In addition, thyroid hormones, which stimulate protein breakdown in liver and muscle enhance synthesis of various lysosomal enzymes concomitantly with the acceleration of protein breakdown. Other physiological factors, including the level of use or disuse of a muscle or pathological processes involving prostaglandins, can promote overall protein breakdown in tissues apparently through this poorly understood (but very important) lysosomal pathway.

### Regulation of Overall Protein Breakdown in Microorganisms

In bacteria, as in mammalian cells, there also exist mechanisms that regulate precisely the overall rate of protein breakdown, even though these cells lack lysosomes. During exponential growth of bacteria, most cell proteins are very stable. However, starvation of bacteria for required amino acids, for nitrogen, or other essential nutrients prevents further growth and quickly leads to a two- to fourfold stimulation of overall protein degradation. This response to the lack of a full complement of amino acids is triggered by the associated decrease in any aminoacyl tRNA, the immediate precursors for proteins. This conclusion is based on use of temperature-sensitive mutants or inhibitors that prevent the charging reaction. When a cell lacks any specific leucyl tRNA, further protein synthesis cannot occur, but the accelerated degradation of cell protein can provide the amino acids required for further synthesis. Thus, one important role of protein breakdown in nutritionally poor environments is to provide precursors for further protein synthesis. Accordingly, mutants with defects in protein breakdown are incapable of synthesizing new proteins under starvation conditions.

When a bacterium lacks a complete set of charged tRNA, many growth-related processes are inhibited, including the synthesis of ribosomal RNA, ribosomal proteins, phospholipids, nucleotides, and many enzymes. The inhibition of these processes and the simultaneous acceleration of protein breakdown all are signaled by the accumulation of the unusual nucleotide, guanosine tetraphosphate. This compound (ppGpp) is synthesized in large amounts by the stalled ribosomes (see Chapter 26), and it serves as an intracellular messenger of the news that insufficient amino acids are available for growth. A rise or fall in ppGpp levels alters within minutes the rates of intracellular proteolysis by unknown mechanisms. This nucleotide also builds up dramatically, but by a different mechanism, when bacteria lacks sufficient ATP for growth, as occurs during starvation for a carbon source. Under these conditions, breakdown of cell proteins can provide metabolizable substrates in the form of amino acids.

A dramatic acceleration of proteolysis also occurs in many microorganisms under poor nutritional conditions when they undergo the process of differentiation called _sporulation_. In this process, the organism changes dramatically and assumes a more heat-resistant form in which metabolism is minimal. Sporulation involves extensive degradation of cell proteins, and an essential step in this process of sporulation is the induction of new proteases. The amino acids generated are then used for production of spore-specific proteins. Similarly, when environmental conditions improve and are appropriate for bacterial growth, the spores undergo germination in which specific new endoproteases are synthesized to digest the spore to provide amino acids for synthesis of proteins characteristic of vegetative cells. These bacterial examples are analogous to many other biological phenomena in which breakdown of certain proteins is linked to the synthesis of new cell enzymes (e.g., the degradation of egg albumin as the egg develops). Such phenomena emphasize an important additional function of intracellular protein degradation. For cells to undergo adaptations or dramatic alterations in shape or composition, it is essential to degrade preexistent cellular structures, both to remove them and to obtain the essential building blocks for new proteins.

## REGULATION OF TRANSLATION

In bacteria, the primary regulatory control over gene expression is at the level of transcription. Although translational control appears to be a major factor in coordinating ribosomal protein synthesis, there are few other precedents for translational regulation in bacteria (see Chapter 26). However, in eukaryotes, several well-characterized regulatory systems have been shown to modulate the overall rate of protein synthesis at the level of translation. Some of the best-understood examples of translational regulation are described below.

### Ribosomal Protein Synthesis Is Coordinated by Feedback Inhibition

The synthesis of the proteins that participate in protein synthesis is closely coordinated in bacteria. With the exception of EF-Tu (which is present in large amounts and for which there are two genes) and L7/L12 (which is present in four copies per ribosome), the elongation factors and ribosomal proteins are all produced in very nearly equimolar amounts. Moreover, their rates of synthesis rise and fall together when the number of ribosomes varies in response to changing growth conditions.

In light of this close coordination of synthesis, it was rather surprising to find that the genes for the r-proteins (and elongation factors) are spread throughout the E. coli chromosome and are organized into at least 10 different operons. Moreover, sequencing of the promoters for these operons failed to reveal obvious homologies. This and other evidence suggested that coordinated synthesis of ribosomal components must be achieved (at least to some extent) at the level of translation.

The way this coordination is achieved has recently been demonstrated in a series of elegant studies by M. Nomura and colleagues. These investigators have found that each operon thus far examined encodes at least one protein that suppresses the translation of all or part of the message for that operon. They propose, therefore, a feedback system in which a single ribosomal protein, when overproduced, acts as a translational repressor for its entire operon. How is this repression achieved? Nomura and coworkers have observed that each of these repressor proteins is itself an "initial binding protein" in the in vitro assembly of the ribosome that exhibits strong and specific binding to rRNA. They visualize a situation in which the repressor protein binds preferentially to ribosomal assembly intermediates when they are available, but when the supply of the repressor exceeds assembly demand, the ribosomal protein binds instead to its message. In line with this suggestion, Nomura and coworkers have recently observed striking sequence homologies between mRNA and r-protein binding sites in rRNA.

In principle, this type of feedback inhibition could regulate the translation of any protein that interacts strongly and specifically with nucleic acids. The message for the protein in question would simply contain a region of primary or secondary structure similar to that with which the protein normally binds. If this region in mRNA had a somewhat lower affinity for the protein than its normal substrate, it would be occupied only when the protein was overproduced. There is evidence that the synthesis of several phage proteins may be regulated in this way.

### Regulation by "Magic Spot" Nucleotides

No discussion of translational regulation would be complete without comment on the "magic spot" nucleotides that serve to link the rate of protein synthesis with the rates of a variety of other cellular processes in bacteria, including production of the protein-synthetic apparatus itself.

The two magic spot nucleotides (see Figure 19-22) are the 3' pyrophosphate derivatives of GDP (MSI = 5'ppG3'pp) and GTP (MSII = 5'pppG3'pp). They achieved this whimsical designation because they were observed to appear as if by magic in chromatograms of extracts of *E. coli* cells upon starving for an essential amino acid. Wild-type *E. coli* are said to be "stringent" (rel$^+$) because they cease producing rRNA and certain specific messages when they are deprived of an amino acid essential to protein synthesis. These cells show elevated levels of the magic spot nucleotides. Mutant bacteria are said to be "relaxed" (rel$^-$) because they continue to produce rRNA even when starved for an essential amino acid. These cells do not accumulate the magic spot nucleotides. These nucleotides are thus mediators of the "stringent" response, and circumstantial evidence at least suggests that they regulate numerous other metabolic processes as well (e.g., see Chapter 13).

The product of the *rel* gene, the stringency factor, catalyzes the synthesis of the magic spot nucleotides according to the following reaction:

$$\text{GDP (or GTP)} + \text{ATP} \longrightarrow \text{ppGpp (or pppGpp)} + \text{AMP}$$

While the factor alone will catalyze this reaction slowly, maximum activity requires the ribosome, mRNA, and deacylated tRNA. Thus maximum magic spot synthesis, both in vivo and in vitro, is brought about by any circumstance, such as amino acid starvation, that elevates the level of deacylated tRNA. The mechanism of magic spot synthesis as well as the way in which magic spot regulates gene expression, is further discussed in Chapter 26. The *rel* system thus provides a mechanism that links the overall rate of protein synthesis to the overall rate of production of the protein-synthetic apparatus, while feedback regulation by r-proteins ensures balanced production of the individual components of translation.

## Regulation of Hemoglobin Synthesis by Hemin

Much of our knowledge about the mechanism of protein synthesis in eukaryotes has come from the study of hemoglobin in reticulocytes, an ideal system because this one protein accounts for more than 90 percent of total protein synthesis. In this cell, globin synthesis is dependent on the presence of hemin, the prosthetic group of hemoglobin; in its absence, protein synthesis initiation is inhibited and polysomes disaggregate. The deficit in initiation has been traced to inactivation of eIF-2; the eIF-2 · GTP · Met-tRNA$^{Met}$ ternary complex is not present in hemin-deficient extracts, and biosynthetic activity in vitro is restored by adding the protein derived from normal cell extracts. In hemin-deficient extracts, the inactivation of eIF-2 is associated with the phosphorylation of one of the protein's subunits. Phosphorylated eIF-2 is unable to form a ternary complex and is thus inactive in initiation. The phosphorylation and inactivation of eIF-2 in response to a deficit in hemin appears to proceed in the two steps outlined in Figure 25-17. In the absence of hemin, a cAMP-dependent protein kinase, $R_2 C_2$, is activated, and this causes the phosphorylation and activation of a specific eIF-2 kinase, or hemin-controlled repressor. This activated kinase then phosphorylates and inactivates eIF-2. Presumably, phosphoprotein phosphatases reverse these reactions so that activity is reestablished upon the addition of hemin.

Control by phosphorylation of factors does not appear to operate in bacteria, but translational inhibitors with eIF-2 kinase activity have been observed in a variety of eukaryotic tissues. Moreover, it is possible that eIF-2 phosphorylation is a widely employed method of regulating translation in higher organisms.

### Interferon Affects Translation

Interferons are glycoproteins produced by animal cells in response to viral infection. Interferon is released by the cells that produce it and acts to induce an antiviral state in other cells. At least part of this antiviral state is associated with alterations in protein synthesis (see Chapter 29). The complex path by which the interferons are thought to affect protein synthesis is discussed in Chapter 29.

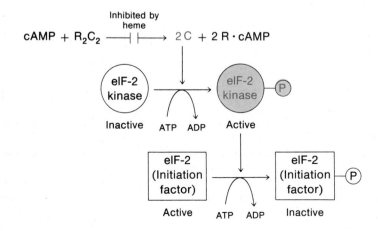

**Figure 25-17**
The regulation of eIF-2 function in reticulocytes by hemin.

## Not All Eukaryotic Messages Are Bound to Ribosomes

In growing *E. coli* it is believed that all existing mRNAs are being translated at any given time. The ribosome density on any given message may vary depending on the strength of its translation initiation signal, but there is probably not a significant pool of translationally inert mRNA.

There are a number of tantalizing suggestions that quite a different situation might exist in eukaryotes that could result from selective control of translation. In the cytoplasm of many eukaryotic tissues, a significant fraction of mRNA can be isolated without attached ribosomes. This mRNA is associated in complexes known as *messenger ribonucleoprotein particles* (mRNP) and can be readily translated if first deproteinized. It has been suggested that message-specific proteins or small RNA molecules (translational control RNAs) may regulate what messages are translated from this pool. Unfortunately, naked mRNA tends to bind a variety of molecules, and it can be difficult to distinguish specific from nonspecific complexes.

One of the clearest cases for translational regulation of preexisting messages is seen in the unfertilized oocyte. The unfertilized sea urchin egg, for example, contains a great deal of inert mRNA that is expressed upon fertilization. Whether this translational control results from protein or RNA binding or the activation of initiation factors, as seen for hemin and interferon, regulation remains to be seen.

Thus, while it seems likely that both proteins and regulatory RNAs exert control on protein synthesis by interacting with mRNA, there is a great deal to be learned about how this regulation is achieved.

## SELECTED READINGS

Caskey, C. T. Peptide chain termination. *Trends Biochem. Sci.* 5:234, 1980. A contemporary summary of the mechanism of protein chain termination.

Chambliss, G., Craven G., Davies, J., Davis, K., Kahan, L., and Nomura, M., (Eds.) *Ribosomes: Structure, Function and Genetics*. Baltimore: University Park Press, 1980. A collection of papers which provides a comprehensive and current picture of the ribosome.

Clark, B. The elongation step of protein synthesis. *Trends Biochem. Sci.* 5:207, 1980. A contemporary summary of the mechanisms of the elongation reactions.

Gale, E. F., Cundliffe, E., Reynolds, P. E., Richmond, M. H., and Waring, M. H. *The Molecular Basis of Antibiotic Action,* 2nd Ed. New York: Wiley, 1981. Pp. 402–549. A description of the ways in which antibiotics and toxins inhibit protein synthesis.

Hunt, T. The initiation of protein synthesis. *Trends Biochem. Sci.* 5:178, 1980. A contemporary summary of the mechanisms of the initiation reactions.

Weissbach, H., and Pestka, S. *Molecular Mechanisms of Protein Biosynthesis*. New York: Academic Press, 1977. A comprehensive view of protein synthesis.

Wold, F. In vivo chemical modification of proteins. *Ann. Rev. Biochem.* 50:783, 1981. Posttranslational modification of proteins.

## PROBLEMS

1. The following Abstract appeared in a paper by K. Itakura et al. (*Science* 198:1056, 1977).

*Abstract*. A gene for somatostatin, a mammalian peptide (14 amino acid residues) hormone, was synthesized by chemical methods. This gene was fused to the *Escherichia coli* β-galactosidase gene on the plasmid pBR322. Transformation of *E. coli* with the chimeric plasmid DNA led to the synthesis of a polypeptide, including the sequence of amino acids corresponding to somatostatin. In vitro, active somatostatin was specifically cleaved from the large chimeric protein by treatment with cyanogen bromide. This represents the first synthesis of a functional polypeptide product from a gene of chemically synthesized origin.

The synthetic gene in question had the following sequence in one chain.

5' AATTCATGGCTGGTTGTAAGAACTTCTT
   TTGGAAGACTTTCACTTCGTGTTGATAG '3

Given this information, deduce the C and N terminal amino acid residues of active somatostatin.

2. The construction of a chimeric plasmid DNA in which synthetic somatostatin sequences are fused in a continuous reading frame to a portion of the *E. coli* β-galactosidase gene offers particular advantages for the efficient expression of the synthetic gene in *E. coli*. Explain, confining yourself to a consideration of translation.

3. Of the following, which component in the presence of GTP is most likely to provide at least partial protection of Phe-tRNA$_{Phe}$ against nuclease digestion? Explain.
   a. EF-G          d. IF-2
   b. EF-Tu         e. RF-1
   c. EF-Ts

4. Protein factors play an important role in translation, cycling on and off the ribosome during initiation, elongation, and termination. Suggest a possible advantage afforded the system by the use of such factors.

5. The antibiotic viomycin, an inhibitor of protein synthesis, specifically affects translocation. Upon addition of viomycin to endogenous bacterial polysomes engaged in polypeptide synthesis *in vitro*, chain elongation is rapidly curtailed. What effect would you expect to result from addition of puromycin to viomycin-treated polysomes?

6. Describe all the reactions in protein synthesis in which GTP is involved. In which of these reactions is cleavage of GTP required?

7. As a rule, amber-specific suppressors are more effective than ochre suppressors. Suggest possible reasons for this.

8. Consider a monocistronic bacterial mRNA that contains 936 bases in its coding sequence, including one initiation and two termination codons.
   a. How many amino acids would the primary translation product of this mRNA contain? Assuming an average amino acid residue weight in protein of 110, what would be the molecular weight of this translation product?
   b. Calculate the number of molecules of ATP that would be consumed in the synthesis of a single such transcript, neglecting noncoding regions.
   c. Calculate the number of molecules of ATP that would be consumed in the synthesis of a single molecule of the protein product.
   d. Assuming an incorporation rate of 50 ms/amino acid, how long would it take for the production of one molecule of this translation product?
   e. Assuming that the rate of translation is not limited by initiation and that one mRNA can accommodate 20 ribosomes, how long would it take to produce 10 molecules of the protein with one molecule of mRNA and 10 ribosomes?

9. Approximately how much energy is required to synthesize a single peptide bond in protein synthesis? How does this compare with the free energy of formation of the peptide linkage which is about 5 kcal/mol? Explain qualitatively in thermodynamic terms why there is such a huge discrepancy.

10. The amino acid composition of a functional protein may in many instances differ from that prescribed by its mRNA. Explain.

11. In a hypothetical test of expression, it is discovered that *E. coli* transformed with the construction described in question (1) yield far less of the somatostatin product than expected. Suggest a reason for the discrepancy.

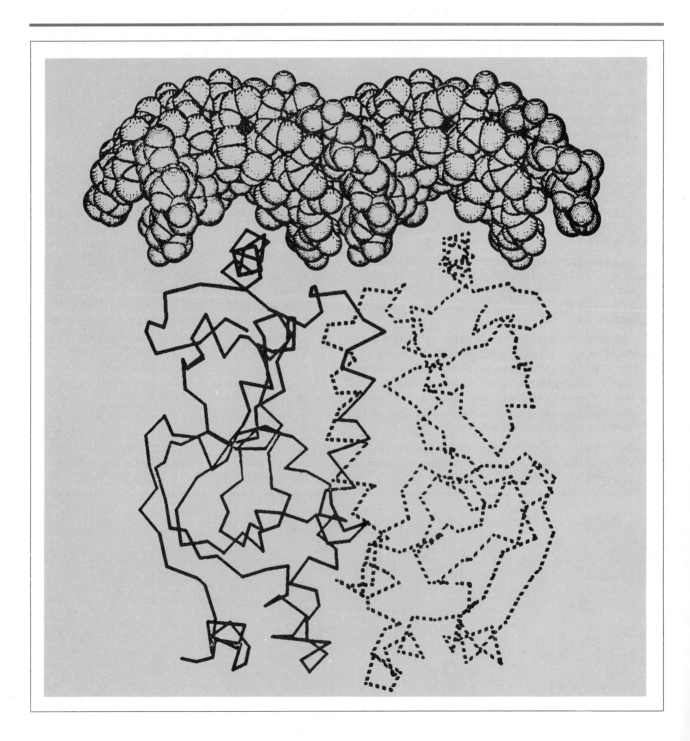

# 26

# GENE EXPRESSION IN MICROORGANISMS

Organisms that exist as single cells comprise the majority of living things in terms of both mass and numbers; they are called _protists_ to distinguish them from more complex multicellular organisms, in which individual cells undergo differentiation. The most basic division within protists is that between prokaryotes and eukaryotes. In the prokaryotes, each cell packages all its DNA into one chromosome, whereas in eukaryotes, several chromosomes are contained in a centrally located membrane-bounded nucleus. The nuclear membrane imposes a barrier that separates the activities of nucleic acid and protein synthesis into the nuclear and cytoplasmic compartments of the cell, respectively. In prokaryotes, these two activities are usually strongly coupled. That is, translation lags only a short distance behind the growing point on the mRNA. This difference in division of activities accentuates different regulatory modes for controlling gene expression. In prokaryotes, gene expression is usually regulated at the transcription level, whereas in eukaryotes, controls at the levels of messenger processing and translation are more often of major importance.

Despite these differences, the similarities in control mechanisms used by prokaryotes and eukaryotes are impressive. This stems from the fact that the macromolecules playing the main roles in the regulation process are the same in both cases. The physicochemical principles that govern the interactions between DNA, RNA, and proteins are much more basic than the division between prokaryotes and eukaryotes. In Chapter 21 it was shown how the DNA sequence signals starts and stops for transcription. Similarly, nucleotide sequences in DNA or RNA are arranged to signal the binding of regulatory proteins at specific locations. It also was shown how small molecules can alter the structure of an enzyme, thereby changing its ability to

_Computer graphics display of a segment of left-handed helical DNA and the gene regulatory protein CAP. The two protruding helical segments of CAP could bind in adjacent large grooves of the DNA. Donald McKay and Tom Steitz believe that cocrystals of DNA and CAP will be needed to evaluate their model. (Courtesy Tom Steitz of Yale University.)_

interact with substrate (in particular, see Chapter 5). Small-molecule effectors also can alter the structure of a regulatory protein, either enhancing or diminishing its potential to interact with a control site on the target nucleic acid.

Within the countless species that make up the protists, each has found the ecological niche where it can grow most favorably. Nevertheless, all cells must be able to adapt to a wide range of growth conditions because they are directly exposed to an external environment that can vary from extremely favorable for growth to one that is hostile for survival. Rapid adaptation to a changing environment is facilitated by a regulation of the cell's metabolic activities. Almost all metabolic activities are regulated; most are regulated in more than one way. In this chapter we shall focus on those metabolic activities concerned with the regulation of a gene's potential for expression.

## *E. COLI*, A PROKARYOTIC PROTIST

*E. coli* is a member of the Enterobacteriaceae family. It is a Gram-negative, non-spore-forming rod about 2 $\mu$m long and 0.5 $\mu$m in diameter. The single chromosome is centrally located and surrounded by a cytoplasm rich in ribosomes. Enclosing the cytoplasm is a complex cell-well–membrane structure. The semipermeable inner cytoplasmic membrane composed of lipoprotein directly encompasses the cytoplasm. A mucopolysaccharide cell wall overlays the inner cytoplasmic membrane and confers mechanical strength, so that the cell can withstand extreme osmotic pressures. Finally, an outer membrane surrounds the cell wall. The space between the inner and outer membranes is called the _periplasm_ and contains a number of enzymes and proteins not found in significant quantities in the main body of the cytoplasm. Flagellae, which confer motility, are distributed over the cell surface. *E. coli* is referred to as the colon bacillus because it is the predominant facultative species in the large bowel. This is the natural habitat in which it flourishes. In the laboratory, conditions have been found for growing *E. coli* on simple, defined media with doubling times as short as 20 minutes. The intensity of genetic and biochemical investigations that have been carried out on this ideal laboratory organism make it the best understood of living cells.

The *E. coli* chromosome contains about $2 \times 10^9$ daltons ($3 \times 10^6$ bps) of DNA, sufficient to code for approximately 2000 genes. Over half these genes have been accurately mapped on the *E. coli* chromosome. This complex system is regulated so that under conditions of active growth, _only about 5 percent of the genome is highly active in transcription at any given time_. The remainder of the genome is either silent or transcribing at a very low rate. When growth conditions change, some active genes are turned off and other inactive genes are turned on. The cell always retains its totipotency, so that within a short time (seconds to minutes in most cases), and given appropriate circumstances, any gene can be fully turned on. The maximal activity for transcription varies from gene to gene. For example, a fully expressing rRNA gene makes one copy per second, a fully turned on $\beta$-galactosidase gene makes about one copy per minute, and a fully turned on biotin synthase gene makes about one copy per 10 minutes. In the maximally repressed state, all these genes express less than one transcript per 10 minutes. *The level of transcription for any particular gene usually results from a complex series of control elements organized into a hierarchy* that coordinates all the metabolic activities of the cell. For example, when the rRNA genes are highly active, so are the genes for ribosomal proteins, and the latter are regulated in such a way that stoichiometric amounts of most ribosomal proteins are produced. When glucose is abundant, most genes involved in processing more complex carbon sources are turned off in a process called *catabolite repression*. If the glucose supply is depleted and lactose is present, then the genes involved in lactose catabolism are expressed. In *E. coli*, the production of most RNAs and proteins is regulated exclusively at the transcription level, although there are notable exceptions (see the following discussion of ribosomal proteins). *Rapid response to changing conditions is ensured partly by a short mRNA lifetime—on the order of 1 to 3 minutes for most mRNAs*. Some mRNAs have appreciably longer lifetimes (10 minutes or longer) and the consequent potential for much higher levels of protein synthesis per mRNA molecule. These atypical mRNAs, at least in some instances, also may be subject to translational control. Examples of all these situations will be considered below. Finally, the fine-level control for any particular enzyme system is subject to regulation by activators or inhibitors directly modulating the enzyme activities. This topic has been treated elsewhere in this text (especially in Chapter 4), and here, only instances where enzymes are directly involved in transcription or translation will be discussed.

## Basic Mechanics of the Initiation Process

The most commonly known way of regulating gene expression in bacteria involves controlling the rate of initiation of transcription. The basic mechanics of the transcription process were described in considerable detail in Chapter 21. For purposes of this discussion, it is sufficient to remember that *prior to initiation of transcription, the RNA polymerase holoenzyme becomes attached to a 35- to 45-nucleotide segment of the DNA called the promoter*. The affinity between DNA and polymerase is controlled by a sequence of bases in the DNA. Two main areas of contact have been recognized, one centered in the −35 region of the promoter with a favored sequence of TTGACA and one centered in the −10 region with the favored sequence TATAAT. These two hexanucleotide regions will be referred to as the polymerase binding site 1 (PBS1) and the polymerase binding site 2 (PBS2), respectively. Future research may show that other regions are also important in promoter–RNA polymerase interaction. From the time it first makes contact with the promoter to the time that it has achieved the proper orientation for initiation, the promoter-polymerase complex may go through several metastable states. *The final conformation of this complex adopted immediately before initiation is referred to as the rapid-start complex or open-promoter complex*. In this state, the polymerase is in contact with PBS1 and PBS2, and about 11 base pairs from the −9 to the +2 positions at the origin of transcription are unpaired (see Figure 26-1). Once the rapid-start complex has been formed, initiation of RNA synthesis is rapid, taking only a fraction of a second in the presence of ribonucleotide triphosphates. Since the spontaneous dissociation rate of the rapid-start complex is usually much longer than this, most rapid-start

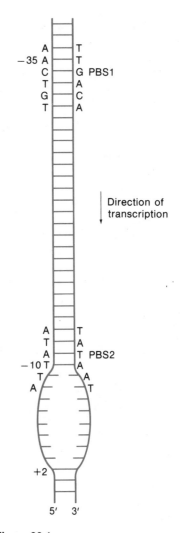

**Figure 26-1**
Schematic diagram of DNA conformation in the rapid-start complex. Two regions, PBS1 and PBS2, most important in polymerase binding, are lettered with most favored sequences. Unpaired region from −9 to +2 is so indicated. Transcription starts at the +1 base pair.

complexes once formed will initiate after they have been formed. This means that for most promoters, *the rate of formation of the rapid-start complex rather than the equilibrium constant of the complex is the critical parameter that determines the activity of a promoter*. The point of initiation of transcription is located eight or nine bases downstream from the center of the PBS2 site. The vast majority of RNAs initiate with a purine base A or G, and the product after dinucleotide synthesis preserves a triphosphate group on the 5′ end of the growing RNA according to the following reaction:

$$pppX + pppY \longrightarrow pppXpY + pp_i$$

## Regulating Initiation of Transcription

*The rate of initiation of transcription can be regulated in several ways, all of which influence the rate of formation of the rapid-start complex.* The primary sequence of nucleotides in the promoter region is the first factor that should be considered. The closer this sequence is to the favored sequence (or consensus sequence) described earlier, the greater will be the affinity of the polymerase for the promoter. For some promoters, negative supercoiling of the DNA serves as an appreciable stimulus to transcription. This probably results from the fact that negative supercoiling facilitates unwinding and unpairing of the double helix (see Chapter 18), such as is observed in the −9 to +2 region of the promoter in the rapid-start complex. The rate of initiation of transcription also can be altered by changes in the RNA polymerase structure. This structure can be altered by subunit replacement, subunit covalent modification, or small-molecule-induced allosteric transition. During sporulation of *Bacillus subtilus*, it is believed that the σ subunit is replaced, producing a change in the types of promoters recognized by the polymerase. In bacteriophage T4 infection, the α subunits of the polymerase become ribose-adenylated, lowering the affinity of polymerase for bacterial promoters and raising the affinity for phage promoters. Although not finally proven, it is believed that binding of guanosine tetraphosphate (ppGpp) to RNA polymerase changes its structure so that it has a greatly lowered affinity for rRNA, tRNA, and ribosomal protein promoters and at the same time a somewhat greater affinity for some other promoters. Finally, the rate of initiation of RNA synthesis can be controlled by auxiliary regulatory proteins that affect the rate of formation of the rapid-start complex in either a positive or a negative way; such regulatory proteins are known as *activators* or *repressors*, respectively. The *lac* repressor inhibits polymerase binding to the *lac* operon promoter. The CAP *apoactivator* stimulates polymerase binding to the same promoter. Strictly speaking, CAP should be referred to as an apoactivator rather than an activator because it only promotes transcription when complexed to the small-molecule *coactivator* 3′,5′-cyclic AMP (cAMP). These examples are discussed in considerable detail in the following discussion.

## The *lac* Operon and Some Other Genes Involved in Catabolism

The story of the *lac* operon reveals one of the clearest and best-understood pictures of a gene regulatory mechanism. The historical presentation given here is intended to give a sense of the close interplay between genetics and biochemistry and of the importance that various techniques have had to progress in this field. Significant events in elucidation of the concept and role of the operon have spanned a period of almost 40 years. The principal reason for the relatively slow rate of progress was the unavailability of the genetic and biochemical skills necessary to solve the problem. Improvement in technical skills and advances in our understanding of the *lac* operon have gone hand in hand. Knowledge of the genetic processes of conjugation and transduction, coupled with elucidation of the basic mechanisms of DNA, RNA, and protein synthesis, has been essential to progress. More recently, cell-free synthesis techniques, use of restriction enzymes for isolation and cloning of small discrete segments of DNA, and methods for determining nucleotide sequences have played important roles in advancing our knowledge of the *lac* operon.

The chronological list of major events in *lac* operon studies is presented in Table 26-1. This chronology can be separated into two major phases: first, the period culminating in the proposal of the operon model in 1961; second, the period from 1961 to the present, in which the concept of the operon has guided research. To a considerable extent, studies on the *lac* operon have served as a model for those studying other genetic regulatory mechanisms in protists and higher forms.

The *lac* DNA is a region of the *E. coli* chromosome with a molecular weight of about $4 \times 10^6$, constituting about 0.2 percent of the chromosome. The DNA is separated into two functional portions: the controlling elements of the operon, and the structural genes, which code for the three proteins specified by the *lac* operon, i.e., $\beta$-galactosidase, lactose permease, and thiogalactoside transacetylase (see Figure 26-2). $\beta$-Galactosidase hydrolyzes $\beta$-galactosides to produce monosaccharides. Thus it cleaves the disaccharide lactose to its component monosaccharides, glucose and galactose. The permease protein is associated with the $\beta$-galactoside active transport system. When present, induced cells can concentrate galactosides 100-fold over their concentration in the external medium (see Chapter 17 for further discussion of the permease system). The *a* gene codes for thiogalactoside transacetylase. Unlike the situation with the other structural genes of the operon, defective *a* gene mutants do not affect the ability of the bacteria to grow on lactose. Transacetylase is known to catalyze the transfer of an acetyl group from acetyl-CoA to a thiogalactoside to form an acetylthiogalactoside. Strangely enough, the physiological value of this enzyme is not clear despite all the work that has been done on the *lac* operon.

The elements controlling the operon consist of a promoter locus *p*, an operator locus *o* and a regulator gene *i*. These will be introduced here and discussed in greater detail later. The operator is the site on the chromosome where the *lac* repressor binds. The promoter contains a site for RNA polymerase binding and an adjacent site for binding the activator protein CAP. The regulator gene *i* is located near the operon and encodes the *lac* repressor protein.

**Enzyme Induction.** Ordinary *E. coli* cells grown in the absence of a galactoside contain an average of 0.5 to 5.0 molecules of $\beta$-galactosidase per cell, whereas bacteria grown in the presence of an excess of a suitable inducer of the *lac* operon contain 1000 to 10,000 molecules per cell. Radioactive amino acid has been used to show that the increase in enzyme activity observed on induction results from de novo protein synthesis. On addition

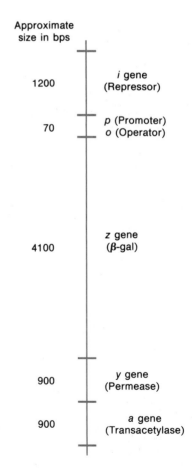

Approximate
size in bps

| | |
|---|---|
| 1200 | *i* gene (Repressor) |
| 70 | *p* (Promoter) *o* (Operator) |
| 4100 | *z* gene ($\beta$-gal) |
| 900 | *y* gene (Permease) |
| 900 | *a* gene (Transacetylase) |

**Figure 26-2**
Different elements of the *lac* operon, including the *i* gene repressor (bp refers to base pairs).

**Table 26-1**
Chronology of Major Advances in Our Understanding of the *lac* Operon

| | |
|---|---|
| 1946 | First regulatory mutants characterized. |
| 1947 | Diauxic growth observed. Cells given glucose and lactose as the main carbon sources will use up all the glucose before beginning to metabolize the lactose. |
| 1955 | Induction involves de novo synthesis of enzyme. |
| 1956 | Constitutive mutations expressed as the ability to synthesize large amounts of enzyme in the absence of inducer. Permease detected. |
| 1959 | Operator proposed as site where repressor binds to stop the flow of information from gene to protein. Regulator gene proposed. |
| 1960 | Operator constitutive mutants isolated. |
| 1961 | Operon model of control proposed. |
| 1963 | *lac*-Specific mRNA characterized by hybridization to DNA. |
| 1964 | The promoter is required for genetic expression of the operon and is probably the initiation point for transcription. |
| 1965 | Studies on temperature-sensitive repressor show that repressor is composed of subunits that combine directly with inducer. |
| 1965 | Catabolite repression is correlated with cAMP depletion. |
| 1966 | Incorporation of the *lac* operon into the transducing virus $\phi$80. |
| 1966 | Sequential transcription of the genes of the lactose operon observed. |
| 1966 | Binding of inducer used as assay to aid in repressor isolation. |
| 1967 | Repression and induction demonstrated in a DNA-directed cell-free system. |
| 1967 | Specific binding observed between repressor and operator containing DNA. |
| 1968 | Reversal of catabolite repression by adding cAMP. |
| 1968 | Promoter located on the opposite side of the operator from the $\beta$-galactosidase structural gene. |
| 1969 | Demonstration that cAMP is required for normal expression of the *lac* operon. |
| 1969 | Catabolite-sensitive site of the *lac* operon is part of the promoter. |
| 1969 | Catabolite gene activator protein (CAP) purified and characterized. |
| 1971 | CAP binds to DNA in the presence of cAMP. |
| 1975 | Nucleotide sequence of *lac* promoter-operator region determined. |
| 1975 | CAP binds preferentially to a specific region of the *lac* promoter. |
| 1976 | Precise location of repressor binding to DNA determined by chemical protection experiments. |
| 1977 | Precise location of CAP binding to DNA determined by chemical protection experiments. |
| 1981 | CAP crystallized and x-ray diffraction studies yield gross features of structure. |

of excess β-galactoside or inducer, enzyme activity increases at a rate proportional to the increase in total protein within the culture. Enzyme formation reaches its maximum rate within 3 minutes of addition of inducer at 37°C. After removal of inducer, enzyme synthesis ceases in about the same amount of time (see Figure 26-3). A large number of compounds have been tested for their capacity to induce β-galactosidase. All inducers contain an intact unsubstituted galactosidic residue. Many compounds that are not substrates for β-galactosidase, such as thiogalactosides, are good inducers. Lactose, the natural substrate of the operon, is not the inducer in vivo. Rather _allolactose_, formed as an intermediate in lactose metabolism with the help of the very limited amount of β-galactosidase present in uninduced cells, is believed to be the natural inducer (Figure 26-4). In fact, no correlation exists between affinity for β-galactosidase and capacity to induce. The three proteins of the _lac_ operon are coordinately induced; i.e., they are induced to the same extent by the same inducer. These results suggest that the receptor molecule for inducer is distinct from the structural components of the operon and that there is one site where inducer acts.

**Discovery of the Repressor Gene.** Two distinct types of mutants have been observed in the lactose system. One class of mutants includes structural gene mutations: (1) β-galactosidase mutations $(z^+ \rightarrow z^-)$, expressed as the loss of the capacity to synthesize active β-galactosidase; (2) permease mutations $(y^+ \rightarrow y^-)$, expressed as the loss of the capacity to concentrate lactose; and (3) transacetylase mutations $(a^+ \rightarrow a^-)$, expressed as the loss of the capacity to form thiogalactoside transacetylase. The other class of mutations involves controlling elements of the operon, such as $i$ gene _constitutive_ mutants $(i^+ \rightarrow i^-)$, expressed as the capacity to synthesize large amounts of β-galactosidase in the absence of inducer. _Structural gene mutants usually affect only the enzyme in whose gene the alteration occurs._ (This is strictly true only when synthetic inducer is used. As explained earlier, a small amount of β-galactosidase is required to convert lactose to allolactose, the natural inducer.) _In contrast, constitutive mutations in the i gene invariably affect the amounts of β-galactosidase, permease, and transacetylase synthesized, but not their structures._

Some of the most informative early studies were performed with partial diploids (merodiploids) that contained the relevant genes on both

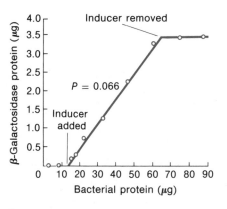

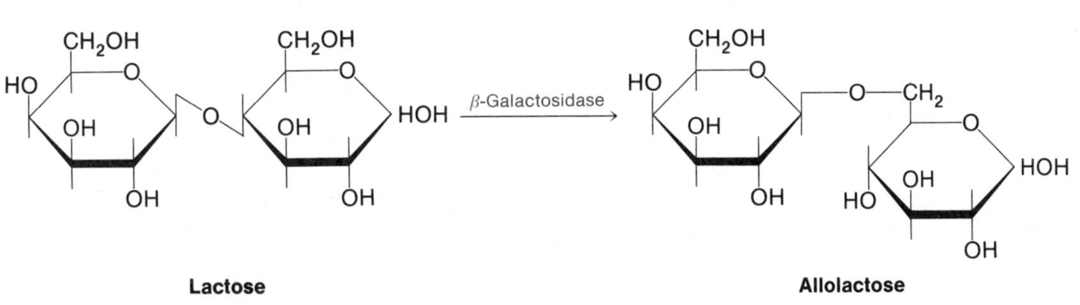

**Figure 26-3**
Kinetics of induced enzyme synthesis. Differential plot expressing accumulation of β-galactosidase as a function of increase in mass of cells in a growing culture of _E. coli_. Since abscissa and ordinates are expressed in the same units (micrograms of protein), the slope of the straight line gives galactosidase as the fraction (_P_) of total protein synthesized in the presence of inducer. (After Melvin Cohn, 1957.)

**Figure 26-4**
Conversion of lactose to allolactose, the natural inducer of the _lac_ operon.

Lactose

β-Galactosidase

Allolactose

cellular chromosome and F-factor plasmid. Merodiploids of the type $z^+y^-a^-/Fz^-y^+a^+$ or $z^-y^+a^+/Fz^+y^-a^-$ are wild-type; that is, they behave normally; they metabolize lactose and form normal amounts of both $\beta$-galactosidase and transacetylase. This complete complementation between structural gene mutants indicates that they belong to independent genes or cistrons. The most significant feature of $i^-$ mutations is that they simultaneously affect all three gene-product proteins, each independently determined by its different structural genes, $z$, $y$, or $a$. Most $i^-$ mutants synthesize more $\beta$-galactosidase to acetylase than induced wild-type cells, but the ratio of $\beta$-galactosidase to acetylase is the same in constitutive cells as in the induced wild-type, strongly suggesting that the mechanism controlled by the $i$ gene is related to inducer interaction (see Table 26-2).

The study of merodiploids of the types $i^+z^-/Fi^-z^+$ or $i^-a^+/Fi^+a^-$ demonstrated that the $i^+$ inducible allele is _dominant_ to the constitutive allele and that it is active on the same chromosome (cis) or on a different chromosome (trans) with respect to both $a^+$ and $z^+$ (see Table 26-2). _The fact that it acts in the trans position shows that i gene mutations belong to an independent cistron, governing the expression of z, y, and a through production of a diffusible cytoplasmic component. The dominance of the inducible to the constitutive allele suggests that the former corresponds to the active form of the i gene._

Further understanding of $i$ gene function came from study of a rare-type mutant gene designated $i^s$. This mutant has lost its capacity to synthesize all structural gene products. In merodiploids of the constitution $i^s/i^+$, $i^s$ is dominant; that is, the merodiploids cannot synthesize either galactosidase or transacetylase even when inducer is present (see Table 26-2). The most reasonable explanation for the $i^s$ mutant is that it is an allele of $i$ in which the structure of the repressor is changed so that it can no longer be antagonized by the inducer.

**Table 26-2**
Synthesis of $\beta$-Galactosidase and Galactoside Transacetylase by Haploids and Partial Diploids of E. coli Regulator Mutants

| Strain No. | Genotype | Galactosidase | | Galactoside-transacetylase | |
|---|---|---|---|---|---|
| | | Noninduced | Induced | Noninduced | Induced |
| 1 | $i^+z^+y^+$ | $<0 \cdot 1$ | 100 | $<1$ | 100 |
| 2 | $i_6^-z^+y^+$ | 100 | 100 | 90 | 90 |
| 3 | $i_3^-z^+y^+$ | 140 | 130 | 130 | 120 |
| 4 | $i^+z_1^-y^+/Fi_3^-z^+y^+$ | $<1$ | 240 | 1 | 270 |
| 5 | $i_3^-z_1^-y^+/Fi^+z^+y_U^-$ | $<1$ | 280 | $<1$ | 120 |
| 6 | $i_3z_1^-y^+/Fi^-z^+y^+$ | 195 | 190 | 200 | 180 |
| 7 | $\Delta_{izy}/Fi^-z^+y^+$ | 130 | 150 | 150 | 170 |
| 8 | $i^sz^+y^+$ | $<0 \cdot 1$ | $<1$ | $<1$ | $<1$ |
| 9 | $i^sz^+y^+/Fi^+z^+y^+$ | $<0 \cdot 1$ | 2 | $<1$ | 3 |

Bacteria are grown in glycerol as carbon source and induced, when stated, by isopropylthiogalactoside (IPTG), $10^{-4}\,M$. Values are given as a percentage of those observed with induced wild-type. $\Delta_{izy}$ refers to a deletion of the whole lac region. It would be noted that organisms carrying the wild allele of one of the structural genes ($z$ or $y$) on the F factor form more of the corresponding enzyme than the haploid. This is presumably due to the fact that several copies of the Flac plasmid are present per cell. In $i^+/i^-$ heterozygotes, values observed with uninduced cells are sometimes higher than in the haploid control. This is due to the presence of a significant fraction of $i^-/i^-$ homozygous recombinants in the population. (Data obtained from J. Monod.)

**Discovery of Operator Mutants.** In the *lac* system, rare dominant constitutive mutants ($o^c$s) were isolated by selecting for constitutivity in cells diploid for the *lac* region, including the *i* gene, thus virtually eliminating the much more frequently occurring recessive ($i^-$) constitutive mutants (since two copies of *i* gene are present in such cells, both *i* genes would have to mutate simultaneously to give a constitutive phenotype. If the probability of an $i^+ \rightarrow i^-$ mutation is $10^{-6}$, the probability of two such simultaneous events in the same cell would be $10^{-12}$). By recombination, the $o^c$ mutants were mapped in the *lac* region, between the *i* and *z* loci, generating the gene order shown in Figure 26-2. Genetic and biochemical evidence exists showing that the *o* locus is adjacent to but distinct from the *z* gene. Thus $o^c$ mutations affect the quantity of $\beta$-galactosidase synthesized, but not its structure.

In merodiploids of the type $o^c/o^+$, galactosidase and transacetylase are constitutively synthesized, showing that the $o^c$ mutation is dominant. In merodiploids of the type $o^c z^+/o^+ z^-$ and $o^c z^-/o^+ z^+$, $o^c$ is dominant in the former but recessive in the latter for galactosidase expression. Thus the $o^c$ mutation is only dominant in the cis position. From this evidence, Jacob and Monod inferred that the $o^+ \rightarrow o^c$ mutations correspond to a modification of the specific repressor-accepting structure of the operator. This identifies the operator locus, i.e., the genetic segment responsible for the structure of the operator, but not necessarily the operator itself.

**The Operon Hypothesis.** The behavior of the various mutations discussed earlier led Jacob and Monod to propose a model for the regulation of protein synthesis. The genetic elements of this model consist of a structural gene or genes, a regulator gene, and an operator locus. The structural gene produces a messenger RNA molecule that serves as the template for protein synthesis. The regulator gene produces a repressor that can interact with the operator locus. The operator is always adjacent to the structural genes it controls (Figure 26-5). The operator and its associated structural genes are referred to as the *operon*. Combination of the repressor molecule with the operator prevents the structural gene from synthesizing messenger RNA. In induction, the *inducer* (or *antirepressor*, a more descriptive term) is thought to combine with the repressor to prevent its interaction with the operator.

This operon hypothesis provided a tremendous stimulus for investigations directed toward understanding not only the *lac* system, but other genetic regulatory systems as well.

**Isolation and Properties of the *i* Gene Repressor.** According to the operon hypothesis, antirepressor is supposed to combine with repressor. Taking advantage of this fact, W. Gilbert and B. Muller-Hill used [14]C-labeled isopropylthiogalactopyranoside (IPTG), one of the strongest known antirepressors, to monitor repressor purification from a crude cell extract. Unlike the natural inducer allolactose, IPTG not only binds strongly to repressor, but it is completely stable in *E. coli* cells or crude extracts. A crude cell-free extract was fractionated by standard protein purification procedures, and the fraction containing the repressor was found that binds IPTG. Binding of any particular extract was measured by equilibrium dialysis. In this procedure, the extract was placed inside a semipermeable dialysis bag that was nonpermeable to protein but permeable to IPTG. This was allowed

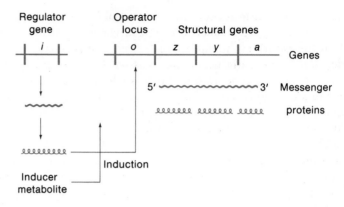

**Figure 26-5**
Schematic model illustrating the operon hypothesis. This is modified from the original proposed by Jacob and Monod, who thought *i* gene repressor was an RNA rather than a protein. The *i* gene encodes a repressor that binds tightly to the operator locus, thereby preventing transcription of the mRNA from the *z*, *y*, and *a* genes. Inducer metabolite combines with repressor, which changes the structure of the repressor so that it can no longer bind to the operator locus. Inducer also can remove repressor already complexed with the *o* locus.

to equilibrate with the external solution by gentle agitation in the presence of a buffer containing radioactive inducer. At the end of the experiment, a certain amount of free inducer ($I_f$) should be detectable outside the bag. The inducer inside the bag should exceed that outside by the amount that is bound to the protein ($I_b$). The strength of binding of repressor to IPTG could also be determined by equilibrium dialysis. When the ratio $I_b : I_f$ is plotted versus $I_b$, the slope should equal the negative of the formation constant for binding $K_f$, where $K_f$ is defined by the equation

$$K_f = \frac{(\text{repressor} + \text{IPTG})}{(\text{repressor})(\text{IPTG})}$$

This is a modified Scatchard plot. For a standard Scatchard plot see Chapter 7.

The formation constant for this complex using wild-type repressor is approximately $10^6 \, M^{-1}$, or $10^6$ moles$^{-1}$ liters; the $K_f$ for allolactose binding to repressor is about 100 times higher. Variants of wild-type repressor purified to the same degree also were examined. The mutant repressor $i^t$, which was isolated from strains that are easier to induce, had about double this affinity. The repressor encoded by $i^s$, isolated from noninducible strains, showed no affinity for IPTG, as predicted (see above).

Other physicochemical properties of repressor have been investigated. The repressor normally exists as a tetrameric protein composed of identical subunits, each with a molecular weight of 39,500. Pure repressor contains four IPTG binding sites and probably two operator binding sites, as discussed below.

The binding of purified repressor to DNA has been demonstrated in a number of ways. This was first shown by mixing $^{35}$S-labeled repressor with $\lambda$*lac* DNA (i.e., $\lambda$ DNA containing the *lac* operon) and centrifuging the mixture through a glycerol gradient for a limited time. In the absence of

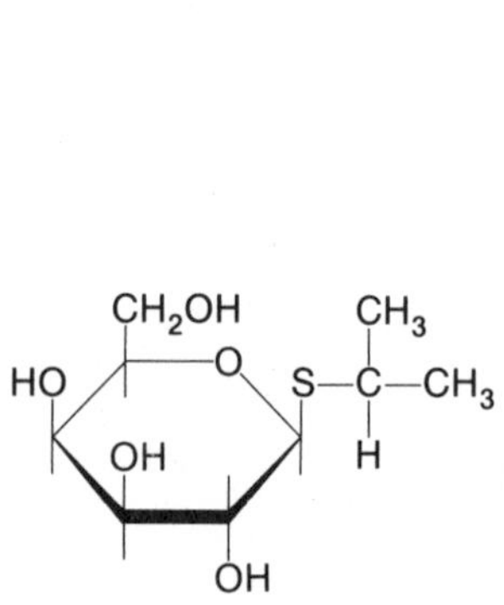

**Figure 26-6**
The structure of the gratuitous inducer isopropyl-$\beta$-D-thiogalactoside (IPTG). This inducer can induce the *lac* operon without itself being metabolized. Therefore it is called a *gratuitous inducer.*

binding, the repressor, detected by radioactivity, sediments much more slowly than the DNA. Under conditions favorable for binding of repressor to DNA, the repressor comigrates with the DNA. Because the DNA is so much larger than the repressor, the position of the DNA in the sedimentation pattern is only slightly affected by the binding of repressor. Proof that this activity represents specific binding of repressor to the *lac* operator was provided by several experiments: (1) the binding was eliminated in the presence of $1.2 \times 10^{-4}\,M$ IPTG; (2) $10^{-3}\,M$ *o*-nitrophenyl-$\beta$-D-fucoside (DNPF), which binds repressor but does not induce, has no effect on the binding of repressor to DNA; (3) the binding of repressor was greatly reduced if DNA containing an $o^c$ mutant was used; and (4) binding was not observed for λDNA not carrying the *lac* operon.

A faster and more versatile method known as the membrane-filter binding technique was also used to demonstrate the binding of repressor to DNA (this technique is also discussed in Chapter 21). This technique is based on the fact that most proteins bind to nitrocellulose filter membranes, but DNA does not. However, DNA complexed to protein does bind to the membrane filter. With this technique, radioactively labeled DNA is used in conjunction with unlabeled purified repressor. A solution containing the labeled DNA with or without added repressor is passed through the membrane filter with the help of mild suction. Only when repressor is present is the label retained by the filter. Other variables were tested, and the same conclusions were reached as when the centrifugation technique was used.

Binding of repressor is very sensitive to ionic strength; in $0.01\,M$ $MgCl_2$, the repressor-operator formation constant is about $3 \times 10^{11}\,M^{-1}$; in $0.01\,M$ $MgCl_2$ plus $0.15\,M$ KCl, it is reduced by about two orders of magnitude. The ionic-strength sensitivity of the complex probably results from the electrostatic interaction between the negatively charged phosphates of the DNA and the positively charged basic amino acid side chains on the protein. Binding of repressor to single-stranded DNA is poor and does not show significant site specificity. This observation plus the fact that binding is very rapid and complete at low temperature argues against major structural changes in the DNA as a result of repressor binding. Thus it seems likely that repressor binds to the duplex DNA without significant disruption of the double-helix structure. This does not rule out minor conformational changes, such as a slight bending or change in pitch of the double helix to obtain an optimal fit between the interacting sites on the DNA and the repressor.

Sequence analysis of the *lac* operon DNA in the region of the initiation site for transcription reveals a region of about 36 bases from about $-8$ to $+28$ (see Figure 26-1 for an explanation of the numbering system) that shows extensive symmetry (see Figure 26-7). Twenty-eight of the 36 bases in this region are arranged so that if the double helix is rotated 180 degrees, the same sequence is observed. This type of symmetry in DNA is referred to as *dyad symmetry*. Since the *lac* repressor is composed of an even number of identical subunits (four) and most proteins so composed also contain similar rotational axes of symmetry (see Chapter 3), it seems likely that this region may be the location for repressor binding. If this is the case, the repressor binding site may be thought of as two symmetrically disposed binding sites on the DNA surface that simultaneously bind to complemen-

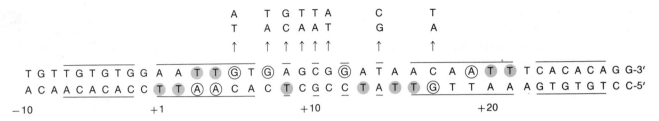

**Figure 26-7**

Presumptive *lac* repressor-binding region. Regions showing dyad symmetry are underlined and overlined. Arrows indicate point mutations leading to the constitutive phenotype ($o^c$). Circled bases are those groups which are strongly protected against reaction with dimethylsulfoxide when *lac* repressor is bound. Shaded circles indicate those groups which become cross-linked to repressor in the presence of ultraviolet light when thymine in the DNA is replaced by 5-bromouracil.

tary binding sites on the protein. This type of binding has all the thermodynamic advantages of chelate binding, which was discussed in Chapter 7 (see Table 7-3). Several pieces of evidence add support to the notion that this region includes the repressor binding site. First, all eight $o^c$ mutations thus far located involve base replacements near the center of this region. Second, the reactivity of certain purines in the DNA toward dimethylsulfate is changed by repressor binding. Dimethylsulfate reacts with the N-7 position on guanine and the N-3 position on adenine in the double helix (see Chapter 20). The reactivity of several bases (see Figure 26-7) in this region is substantially decreased by repressor binding. The method for determining the position in the sequence of reacted groups is described in Chapter 20. Finally, if all the thymines in the DNA are replaced by bromouracil, it can be shown that a number of the bromouracil bases become cross-linked to the repressor in the presence of ultraviolet light. Taking into account the normal twist of the double helix, *all the groups shown by chemical methods to be in the vicinity of the repressor are situated on one side of the double helix. This strongly supports the notion that repressor binds on one side of the double helix.*

The RNA polymerase and repressor binding sites have considerable overlap (see Figure 26-8), so that the binding of one protein would preclude the binding of the other. Whereas there are other DNA sequences that the repressor might bind to prevent initiation, this may be the most common

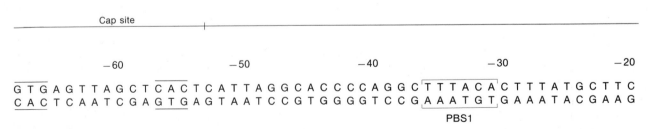

**Figure 26-8**

DNA sequence in the *lac* promoter region. Bases are numbered +1 for the first base transcribed and −1 for the base before that. CAP- and repressor-binding sites are indicated. Regions within these two sites showing dyad symmetry are underlined and overlined. Two regions, PBS1 and PBS2, where polymerase binds most strongly in a number of promoters,

situation for most repressors affecting the expression of adjacent structural genes. It does create one obvious complication, which is that the nucleotide sequences that interact specifically with RNA polymerase and repressor must overlap to some extent.

Numerous quantitative relationships have been established in cell-free systems that would be difficult or impossible to establish in whole cells. Thus it has been shown by varying repressor concentration that one repressor molecule is sufficient for inhibiting gene expression. It also has been shown that in the presence of excess repressor, *lac* gene expression is directly proportional to the square of the IPTG concentration. The latter fact suggests that derepression requires two antirepressor molecules, as represented by the following equation:

$$RO + 2\,I \rightleftharpoons RI_2 + O$$

Here $RO$ refers to the concentration of repressor and operator complex, $I$ to the free antirepressor, and $RI_2$ to the repressor-antirepressor complex. It has been proposed that the tetrameric repressor molecule contains two operator binding sites and four inducer binding sites and that each operator binding site is sensitive to two of the inducer binding sites so that the tetrameric molecule can be visualized as two functionally separate inducible DNA binding sites. Most arrangements for tetramers would lead to more than one operator binding site from symmetry considerations alone (see Chapter 3). Thus tetramers can be arranged with tetrahedral symmetry or other symmetries that give three orthogonally related dyad axes. In either case, more than one site for binding to operator would be generated by symmetry unless the binding site goes through the center of the molecule, which seems most unlikely. A complete understanding of the structure of repressor and its interaction with DNA will have to await a detailed crystallographic analysis of the structure. More on the symmetry of binding sites is presented below.

**Isolation and Properties of the Catabolite Gene Activator Protein CAP.** When *E. coli* cells are grown in the presence of glucose and lactose as the sole carbon sources, the cells will first utilize glucose and afterwards utilize the lactose. This phenomenon is referred to as *diauxic growth*. Careful

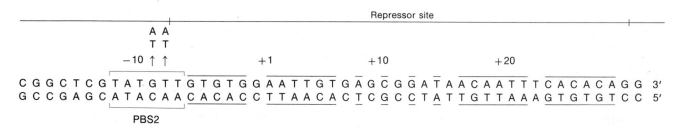

are bracketed in color. It should be noted that PBS1 has 5 out of 6 of the idealized sequence, whereas PBS2 has 4 out of 6. Arrows indicate base replacements in the mutant UV5 promoter that lead to high-level expression in the absence of cAMP-CAP.

analysis of the cells has shown that during the glucose phase of growth, the *lac* operon is poorly expressed and induction of the operon must occur before the lactose can be utilized. Repression of *lac* operon expression by glucose or its closely related derivatives, such as glucose-6-phosphate or fructose, is known by the general term catabolite repression (or glucose repression). Enzymes of the glycolytic pathway that directly utilize glucose are always present in the cell. The extra enzymes of the *lac* operon are not needed as long as this more directly utilizable carbon source is available. Many genes in *E. coli* associated with catabolism are subject to this type of repression for the same reason. A turning point in our understanding of catabolite repression was provided by the finding that the intracellular cAMP level in *E. coli* is drastically lowered in the presence of glucose, from about $10^{-4}$ to $10^{-7}$ $M$. Subsequently it was shown that large quantities of cAMP (Figure 26-9) in the growth medium could partially reverse the glucose catabolite repression effect on the *lac* operon. The stimulatory effect of cAMP on genes subject to catabolite repression was demonstrated more directly in a cell-free system. The system used contained $\lambda lac$ DNA, a cell-free extract of *E. coli*, and the substrates and other low-molecular-weight components necessary for transcription of the DNA and translation of the resulting messenger RNA (mRNA). In this system it was shown that $10^{-3}$ $M$ cAMP stimulated $\beta$-galactosidase synthesis as much as 30-fold. Subsequently, mutants of *E. coli* that were permanently catabolite-repressed were isolated by J. Beckwith and colleagues. The phenotype of such mutants was that they were incapable of expressing a number of genes that required çAMP for activation. These mutants fell into two categories, those which could be phenotypically corrected by growing in the presence of $5 \times 10^{-3}$ $M$ cAMP and those which could not. The first class of mutants was defective in the synthesis of cAMP, and the latter class of mutants was defective in the protein(s) with which cAMP interacts to bring about stimulation of $\beta$-galactosidase synthesis as well as the synthesis of other proteins whose synthesis is subject to catabolite repression. Further work showed that these two types of mutants originate in the genes *cya* and *crp*, respectively. When the cell-free system was made from extracts of *crp⁻* cells, it produced a low level of $\beta$-galactosidase that was not stimulated by adding cAMP. The defect could be corrected by adding soluble protein from *crp⁺* cells. Making use of this in vitro complementation assay, it was possible to fractionate the *crp⁺* cells and isolate a single protein termed *CAP* that was responsible for the activity. CAP is a dimer composed of identical subunits with a molecular weight of 22,000. It is a basic protein with an isoelectric point pI of 9.2. This high isoelectric point is due to the relatively low concentration of acidic amino acids; the CAP content of the basic amino acids lysine and arginine is not unusually high. At moderate salt concentrations, CAP interacts with cAMP with an intrinsic formation constant $K_f$ of $0.6 \times 10^5$ $M^{-1}$, where $K_f$ is defined by the equation

$$K_f = \frac{[CAP^* + cAMP]}{[CAP^*][cAMP]}$$

Here [CAP*] is the concentration of ligand-binding sites. Since CAP is a dimer, CAP* concentration is probably equal to double the CAP concentration. At very high salt concentrations, the cAMP affinity for CAP is decreased about five fold, attesting to an electrostatic component to the bind-

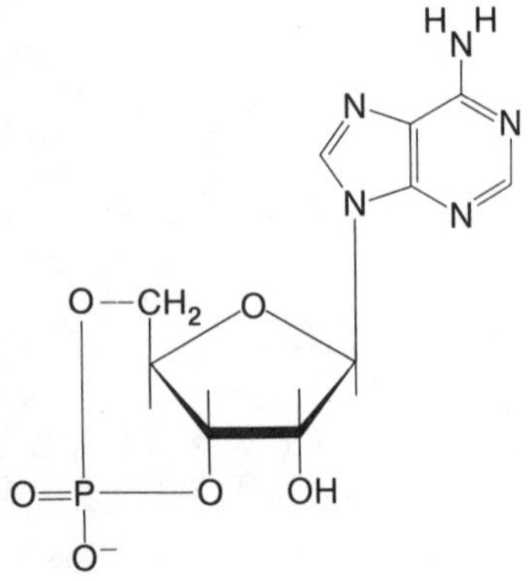

**cAMP**

**Figure 26-9**
The structure of 3′,5′-cyclic AMP (cAMP).

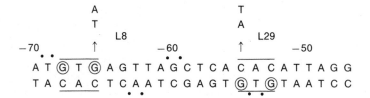

**Figure 26-10**
Presumptive CAP-binding region in the *lac* promoter. Regions showing dyad
symmetry that are believed to interact strongly with CAP are overlined and
underlined. Circled bases are those which are strongly protected from reaction
with dimethylsulfate when CAP is bound. Point mutations L8 and L29, which
produce a promoter that is not stimulated in transcription by cAMP, are indicated. Black dots indicate those phosphate positions which if ethylated by
ethylnitrosourea block CAP binding. If the area of interest exists as a normal
DNA double helix when binding CAP, then all the groups implicated in CAP binding would appear on one side of the DNA.

ing. Physical and chemical studies have been carried out that indicate that
structural alterations in CAP result from cAMP binding. CAP binds to
DNA, and this binding is greatly stimulated in the presence of cAMP. The
structural alterations produced in CAP by cAMP binding apparently alter
its conformation so that it can form a strong complex with DNA. CAP
binds preferentially to the *lac* promoter region, although the selectivity for
preferential binding to the specific site is substantially less than is the case
for *lac* repressor.

It is known from a series of genetic deletions that the site necessary for
CAP stimulation of the *lac* operon is in the $-50$ to $-80$ base-pair region.
Inspection of the sequence in this region reveals a 14-base segment between
$-55$ and $-68$ that shows dyad symmetry for 12 of the 14 base pairs (Figure
26-10). It is noteworthy that the *gal* and the *ara* operons (discussed below),
which are also stimulated by the cAMP-CAP complex, have similar regions
of dyad symmetry. Comparison of the three presumptive binding sites for
CAP indicates that only the outer three bases are identical in composition.
On this basis, it seems likely that the critical sequence of a CAP binding
site is

GTG ——————— CAC

CAC ——————— GTG

In *lac* and *gal* there are 10 base pairs within this critical sequence, whereas
in *ara* there are 8. As in the case with *lac* repressor, chemical-probe experiments have been carried out to characterize the region binding to CAP. In
these experiments, a small DNA segment containing the CAP binding site
was used. Dimethylsulfate (DMS), which reacts with the adenine N-3 and
the guanine N-7 positions, and ethylnitrosourea (EtNu), which reacts with
the phosphates, have been used. Experiments with the DMS probe show
that CAP binding protects the two outer guanines on each side of the
presumptive binding site from methylation. Conversely, if the DNA is first
premethylated, binding of CAP is strongly inhibited. The pattern of protection demonstrates symmetry and interaction in two adjacent, large grooves
of the DNA. Experiments with EtNu both confirm the importance of the
symmetry and show that, like *lac* repressor, CAP interacts with only one
side of the DNA helix, over a 14-base region covered by the symmetry axis.

In the absence of bound repressor, the full sequence of reactions involved in initiation of transcription is summarized by the following set of equations. First, *the coactivator cAMP combines with the apoactivator CAP, which then binds in the −60 region of the promoter. This complex stimulates the binding of RNA polymerase to an adjacent site on the promoter.* Although the mechanism whereby CAP stimulates binding of RNA polymerase is uncertain, it seems likely that it may result from an affinity between CAP and polymerase.

$$cAMP + CAP \rightleftharpoons cAMP - CAP$$

$$cAMP - CAP + DNA \rightleftharpoons cAMP - CAP - DNA$$

$$cAMP - CAP - DNA + polymerase \rightleftharpoons$$
$$cAMP - CAP - DNA - polymerase$$

The transcription rate of the *lac* operon in the presence of cAMP-CAP is enhanced 20- to 50-fold, and it may be assumed that this is directly due to the increased affinity of RNA polymerase holoenzyme for the promoter in the presence of CAP and cAMP. Although CAP and polymerase do not bind to one another when free in solution, the affinity required to enhance polymerase binding to the DNA promoter by the observed amount should be much less. Formation constants for different polymerase promoter constants have been detected in the range of $10^8$ to $10^{11} M^{-1}$. The *lac* promoter sequence is far from ideal for RNA polymerase binding, and so its affinity constant would probably fall on the lower side of this range, say, $10^8 M^{-1}$. From the Gibbs free-energy equation, $\Delta G = -1364 \log K$ at 25°C, or a $\Delta G$ of about −11 kcal (see Chapter 7). If the $\Delta G$ for CAP and polymerase interaction were −2.7 kcal, this would enhance the polymerase binding by 100-fold. Such an affinity by itself would never lead to a significant complex formation between free CAP and polymerase in solution at attainable concentrations. A $\Delta G$ of this magnitude would not require a highly specific interaction between the two proteins on the DNA surface. It is easy to visualize two proteins bound at nearby points on the DNA duplex adjusting to each other's structures by a small "wobble" that would lead to a favorable interaction of this magnitude. These calculations are only gross estimates, and further investigations will be required to confirm this highly plausible model. Cooperative interaction of proteins binding to the DNA surface has been observed in several other instances: the binding of histones to DNA (Chapter 18), the binding of T antigen to SV40 DNA (Chapter 28), and the binding of $\lambda C_I$ repressor to DNA (Chapter 27). A similar model has been proposed for multiple binding of $C_I$ repressor to the $\lambda$ operator (see Chapter 27).

The overall strategy in creating a promoter sequence responsive to cAMP-CAP activation could be summarized as follows. The nucleotide sequence in the RNA polymerase binding site is adjusted so that polymerase by itself produces a low level of transcription. An adjacent site for binding CAP is created so that when CAP is binding, the additional affinity contributed by favorable contacts between the CAP and the polymerase convert this into a high-level promoter. A mutant (UV5) containing two base replacements in the *lac* promoter produces high-level transcription in the absence of cAMP-CAP. These base replacements in the PBS2 site (see Figure 26-8) bring the promoter sequence closer to the consensus sequence and therefore should lead to a stronger polymerase binding site in the absence of the cAMP-CAP complex, as is in fact observed.

Recently the structure of CAP has been determined by x-ray diffraction. It has been proposed that CAP would bind best to a left-handed DNA duplex (see Chapter 18). Such a possibility raises new questions about how CAP works which have yet to be considered.

**Symmetry Considerations in the Binding of Activator or Repressor to DNA.** As discussed earlier, chemical probes such as dimethylsulfate and ethylnitrosourea have been most useful in localizing the sites of RNA polymerase and regulator protein binding on the DNA surface. However, such observations only lead to speculation about the actual chemical groups involved in stabilizing various DNA-protein interactions. This type of understanding may have to await more precise atomic localizations of the various protein-DNA complexes by x-ray diffraction. The most important conclusions to arise from sequence studies and the related chemical probe studies thus far are twofold: (1) all proteins (repressors, activators, and RNA polymerase) bind predominantly to one side of the double helix, and (2) there is a strong symmetry component to the binding of DNA sequences by regulator proteins, but not to the binding of polymerase. Theoretical considerations suggest why symmetry plays a major role in regulator protein binding.

CAP is composed of two identical subunits, whereas *lac* repressor is composed of four identical subunits. _All repressors that have been analyzed thus far are composed of an even number of identical subunits._ Structures of this type generally are highly symmetrical and as a rule have one or more dyad axes of symmetry (see Chapter 3 and Figure 3-36a and c). The sugar phosphate backbone of the DNA double helix also has a dyad axis of symmetry, since it is composed of two polydeoxynucleotide chains arranged in antiparallel fashion (see Chapter 18). It is known from chemical protection experiments that both CAP and *lac* repressor interact at their binding sites in approximately side-by-side fashion with the DNA with little, if any, disruption of the DNA structure. Because of the dyad symmetry in the protein, the binding site on the protein might be expected to show dyad symmetry. If the nucleotides on the DNA involved in regulator protein binding were arranged on a perfect dyad axis, it would permit binding to both elements of the symmetrically disposed binding site on the protein. That this may be the case is supported by the pronounced dyad symmetry displayed by the nucleotide sequences in the activator and re-

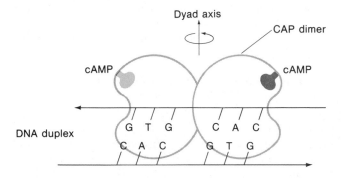

**Figure 26-11**
Schematic diagram of the cAMP-CAP-DNA complex. CAP is a dimer composed of identical subunits. The binding site on the DNA is arranged so that rotation of either the DNA or the CAP dimer by 180 degrees on the dyad axis should produce an identical structure.

pressor binding sites. The advantages in terms of strength of binding are enormous. The effect on the formation constant for the protein-DNA complex can be estimated with the help of the Gibbs free-energy equation (see above). If the $\Delta G$ for one binding site were, say, $-6.8$ kcal, then the $\Delta G$ for the two binding sites would be more than double this, or greater than $-13.6$ kcal. The enthalpy term is doubled. The entropy term is much more favorable for a dimeric repressor than it would be for two monomeric repressors for the same reason that ethylene diamine binds $Cd^{2+}$ ions more favorably than ethylamine (see Chapter 7). The corresponding formation constants would be $10^5$ and greater than $10^{10}$, respectively.

Another advantage resulting from symmetry in the regulator protein relates to the availability of symmetrically disposed binding sites for the small-molecule allosteric effector. For the sake of example, the situation for *lac* derepression by IPTG discussed earlier is reconsidered. The intrinsic formation constant for the repressor-inducer complex $K_f(RI)$ is $10^6\ M^{-1}$. The formation constant for the repressor-operator complex $K_f(RO)$ has never been measured for wild-type repressor under normal physiological conditions, but it is probably around $10^{11}\ M^{-1}$. If only one antirepressor molecule is involved in derepression, the equilibrium between the repressed and depressed states would be represented by the equation

$$RO + I \rightleftharpoons RI + O$$

Here $RO$ refers to the concentration of repressor-operator complex, $RI$ to repressor-antirepressor complex, $O$ to free operator, and $I$ to free antirepressor. The equilibrium constant for this reaction would be

$$K_{eq} = \frac{K_f(RI)}{K_f(RO)} = \frac{10^6}{10^{11}} = 10^{-5}$$

However, it has already been stated that the observed data suggest that the main equilibrium-describing derepression involves two antirepressor molecules, as indicated by the following equation:

$$RO + 2I \rightleftharpoons RI_2 + O$$

The equilibrium constant for this reaction would be

$$K_{eq} = \frac{K_f(RI)^2}{K_f(RO)} = \frac{10^{12}}{10^{11}} = 10\ M^{-1}$$

Physically, this means that the availability of two binding sites (instead of one) for antirepressor shifts the equilibrium constant in favor of the derepressed state at moderate concentrations of antirepressor. With the natural inducer allolactose, which has a $K_f(RI)$ of about $10^8$, much less inducer would be required for derepression.

**Influence of the Small-Molecule Effector on the Regulatory Protein.**
Most proteins, whether they serve a structural or an enzymatic function, have a unique conformation that is determined mainly by their primary amino acid sequence and their natural surroundings (Chapter 3). A special class of proteins, called *allosteric proteins*, contain functional and regulatory sites that are usually located at nonoverlapping regions on the protein surface. The regulatory site is designed to interact with a small-molecule

effector or allosteric effector. Binding of the allosteric effector produces a structural alteration in another part of the protein that contains the functional site. In the simplest examples, e.g., hemoglobin, which contains four sites for binding oxygen, the regulatory and the functional sites are the same (Chapter 3). The binding of oxygen at one site on the molecule facilitates the binding of oxygen at another site. In more complex situations, such as aspartate carbamoyltransferase (Chapter 5), the first enzyme in the biosynthetic pathway for CTP synthesis, CTP binds at a regulatory site on the enzyme that leads to an alteration in the structure of the enzymatic site, rendering it inactive. The mechanism by which these small molecules induce structural alterations was discussed in Chapters 3 and 5. Regulatory proteins, such as the *lac* repressor or CAP, are allosteric proteins. Their regulatory sites interact with allolactose and cAMP, respectively. Their functional sites interact with DNA. It is possible that there is some cooperativity between the binding of small-molecule effectors to the regulatory proteins. However, currently available data are not detailed enough to show if this is the case, so we have no way of evaluating it. Qualitatively, the effect of cooperativity would be to make the dependence of binding on inducer concentration greater than unity (see discussion on hemoglobin oxygen binding in Chapter 3).

The *lac* repressor is capable of adopting two conformations, one which is most stable in the absence of allolactose and binds strongly to the *lac* operator, and one, most stable when it is complexed with allolactose, which has a low affinity for DNA. A similar situation may be visualized for CAP, except that in this case the role of the allosteric effector is reversed. That is, the binding of cAMP converts the protein from a structure with a low affinity to one with a high affinity for a specific site on the DNA.

Little is known about the detailed conformation of proteins that regulate gene expression, but it seems highly likely that the same general principles guide their action as in the case of allosteric proteins such as hemoglobin and aspartate carbamoyltransferase. In this rapidly moving field, major advances in our understanding can be anticipated in the near future.

**Other Factors Affecting Expression of the *lac* Operon.** Several other factors have a differential effect on *lac* operon expression. Most of these have been studied only in vitro, so their importance in regulating *lac* expression in vivo remains to be explored. Guanosine tetraphosphate, or ppGpp (this compound is also known as magic spot), stimulates $\beta$-galactosidase mRNA synthesis by twofold to threefold in vitro. The concentration of this pleiotropic effector is inversely proportional to the gross rate of cellular RNA synthesis. Its role in inhibiting rRNA, tRNA, and ribosomal protein synthesis is well-established, as described below.

Supercoiling of the DNA (Chapter 18) strongly affects *lac* operon expression. Studies indicating this involves the use of linear $\lambda lac$ DNA in a coupled transcription-translation system. When the antibiotic novobiocin is present during the in vitro synthesis, only 20 percent of the normal level of $\beta$-galactosidase is obtained. Since novobiocin inhibits DNA gyrase (see Chapter 20), the enzyme that causes negative supercoiling of covalently closed circular DNA, the inference from these results is that supercoiled DNA is about 5 times more effective in *lac* operon expression than DNA lacking supercoils. Studies of other bacterial and plasmid genes indicate that supercoiling affects expression in a differential manner; some genes are insensitive to supercoiling, whereas sensitive genes show as much as a 10-fold increase in expression when the DNA goes from the fully relaxed to the highly supercoiled state. Negatively supercoiled DNA should require less energy for localized unpairing (see Chapter 18), possibly accounting for the cases where a stimulatory effect is observed.

With the use of a partially purified coupled transcription-translation system, it has been shown that $\beta$-galactosidase synthesis is strongly dependent on the presence of a so-called L factor. Many other genes do not share this requirement, although a full description of the effect is still lacking. L factor is known to be a product of the *nusA* gene of *E. coli*, which is required for the λN-gene–induced antitermination activity seen during normal λ infection (see Chapter 27). It is not clear what these two properties of L factor have in common. Recent evidence suggests that L factor forms a complex with RNA polymerase during the elongation phase of RNA synthesis, after the σ factor has been released (Chapter 27).

**Behavior of Some Other Genes Important in Catabolism.** The cAMP-CAP activator is an essential component for regulating a wide variety of genes involved in catabolism. It usually functions in series with a gene-specific regulator so that any particular gene subject to cAMP-CAP activation will only be fully expressed if both control switches are in the "on" position. The galactose operon encoding enzymes for galactose catabolism are regulated by cAMP-CAP as well as by a specific repressor that is antagonized by galactose. The *gal* operon appears to have a second promoter close by that is not subject to cAMP-CAP control; the function of this second promoter is not understood at the present time.

The arabinose operon also uses the pleiotropic cAMP-CAP activator and a specific regulator, but here the similarity with *lac* and *ara* ends. The specific regulatory protein, encoded by the *araC* gene, regulates its own synthesis by functioning as a repressor. The sugar L-arabinose serves as both antirepressor and coactivator. When it complexes with the araC protein, the complex dissociates from the repressor binding site and becomes converted into an activator complex that binds at a different site adjacent to the polymerase binding site for the operon. The operon contains three structural genes, *araB*, *araA*, and *araD*, whose gene products can bring about the stepwise conversion of L-arabinose into D-xylulose-5-phosphate. The regulatory locus, containing 167 base pairs, is bounded by the *araB* gene on one side and by the *araC* gene on the other side (see Figure 26-12). The repressor form of araC protein binds at a site that overlaps the cAMP-CAP binding site(s). This inhibits transcription of the *araC* gene to the left and of the *araBAD* genes to the right. L-Arabinose complexes with araC protein, releasing it from the repressor binding site and converting it into an activator, the L-arabinose-araC protein complex. This complex can bind at the *araI* site, which stimulates polymerase binding to an adjacent site for rightward transcription. Apparently the binding of the L-arabinose-araC complex to the *araI* site is greatly stimulated in the presence of cAMP-CAP, which complexes at a more distant location on the DNA. Thus optimal rightward transcription involves first the removal of araC repressor, which permits the binding of cAMP-CAP to an overlapping site. The binding of cAMP-CAP stimulates the binding of the L-arabinose-araC complex to an adjacent site, which in turn stimulates the nearby binding of RNA polymerase.

Optimal leftward transcription requires the removal of araC from the repressor binding site followed by binding of the cAMP-CAP complex and RNA polymerase. The behavior of the cAMP-CAP-DNA complex is unusual in two respects. For rightward transcription, the complex directly stimulates the binding of the specific activator L-arabinose-araC rather than RNA polymerase. For leftward transcription, it binds at a site that overlaps the polymerase binding site. If the binding sites were on the same side of the DNA, the cAMP-CAP activator would function as a repressor rather than as an activator. In this instance, it is presumed that the binding sites for polymerase and cAMP-CAP are not on the same sides of the DNA, so that both may bind simultaneously with favorable contacts between the two proteins. The studies on *lac* and *ara* have led to the impression that activators stimulate expression by an energetically favorable interaction between the protein surface of the activator and the protein surface of another activator or the polymerase bound to an adjacent site on the DNA. This protein-protein interaction need not be highly specific, since only a few kilocalories of free energy would be required to greatly enhance the DNA protein binding, as indicated earlier. Transcription of the *araBAD* operon, like *lac*, is stimulated by ppGpp. The *ara* operon control region is very complex, and much further work will have to be done before it is fully understood.

## Regulation of Amino Acid Biosynthesis

Amino acid synthesis in *E. coli* is regulated at the level of transcription and through the modulation of certain enzyme activities located at branch points in the various biosynthetic pathways. Some of the patterns of control have been discussed in a comprehensive fashion in Chapter 22. Here the detailed molecular aspects of transcription control of the messenger RNA involved in tryptophan biosynthesis will be considered.

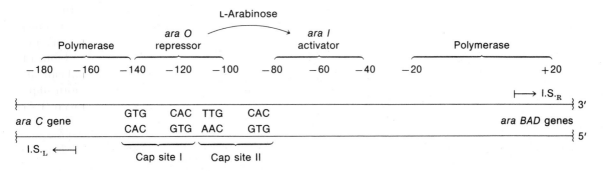

**Figure 26-12**
Schematic diagram of the arabinose operon control region. Initiation sites for leftward (I.S.$_L$) and rightward transcription (I.S.$_R$) are indicated. These transcripts produce the mRNA for the *araC*-encoded regulatory protein and the *araBAD*-encoded enzyme, respectively. The two binding sites for polymerase are indicated. In the absence of L-arabinose the araC Protein binds at the *araO* locus. This overlaps the two CAP sites and the polymerase-binding site for leftward transcription. In the presence of L-arabinose the araC protein is structurally altered so that it binds at the *araI* site only. This permits leftward transcription and the binding of CAP at the CAP site I. CAP site II is missing one base from the idealized sequence. Location of the binding sites for CAP have yet to be confirmed by direct measurements. It is hypothesized that the CAP site II can be occupied if CAP site I is occupied despite its imperfect sequence because of cooperative interaction of the proteins binding to the two locations. Only the crucial base sequences for presumptive CAP-binding sites are indicated.

Of all the genes concerned with anabolic processes, the genes involved in tryptophan biosynthesis are probably the best understood. This is mostly a result of the efforts of C. Yanofsky and colleagues, who have used a wide variety of genetic and biochemical techniques to probe the complexities of this system.

Wild-type *E. coli* has the capacity to synthesize all 20 amino acids from simpler substrates, but for many of them, it does so only when they are not available in adequate amounts from the external growth medium. For example, the production of enzymes for tryptophan synthesis is sharply reduced when the external tryptophan supply is high. A lowering of the available L-tryptophan selectively stimulates synthesis of the messenger RNA and the five polypeptide chains associated with the biosynthesis of tryptophan from chorismic acid (see Figure 26-13). *The five contiguous structural genes are transcribed as a single polycistronic messenger RNA* approximately seven kilobases in length. Initiation of transcription is regulated by the interaction of the tryptophan aporepressor, the protein product of the *trpR* gene, with its target site on the DNA, the *trp* operator. Binding of L-tryptophan to the aporepressor causes a structural alteration essential for strong specific binding to the *trpO* locus (see Figure 26-14). *Like lac repressor, the trp repressor binds at a site that overlaps the RNA polymerase binding site,* and the region where the repressor binds contains a number of bases arranged with dyad symmetry (see Figure 26-15). It seems likely that the bases on the dyad-symmetry axis are strongly involved in the binding site for repressor and probably reflect similar symmetry in the repressor itself. The repressor is composed of four identical subunits, each with a molecular weight of 12,400. The most significant difference between the action of the trp and lac repressors relates to the function of the small-molecule effector. In the case of *lac*, the effector molecule allolactose acts as an antirepressor, causing release of repressor from the operator; in the case of *trp*, the effector molecule L-tryptophan acts as a *corepressor*, stimulating the binding of repressor to the operator. It should be obvious that the difference in action of these small-molecule effectors, whose concentrations dic-

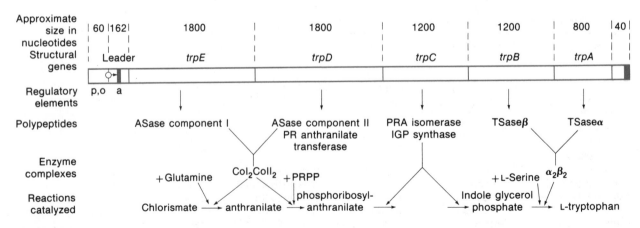

**Figure 26-13**
The tryptophan operon indicating the size of the different genes, the polypeptide chains, the resulting enzyme complexes, and the reactions catalyzed by the enzyme complexes. Control elements at the left represent the promoter-operator (*po*) region and the attenuator (*a*) region.

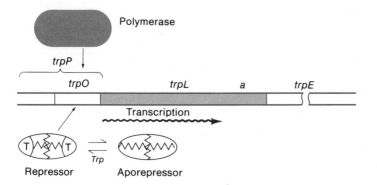

tate the level of operon activity, is well-suited to the different metabolic needs of the cell satisfied by the two operons.

From this point on, the close parallel between the regulation of the *trp* operon and the *lac* operon ends. The *trp* operon has no positive control system such as cAMP-CAP, but it does use another means of regulating transcription. This is a provisional stop signal called an *attenuator* located 141 bases downstream from the initiation site for transcription. *Only a fraction of the messenger RNAs (0.1 to 0.9 depending on metabolic conditions) that are initiated transcribe through this provisional stop signal to the end of the operon.* The existence of this stop signal was first suspected when it was discovered that a genetic deletion of some of the bases between the initiation site for transcription and the translation start of the first structural gene (*trpE*) raised the level of expression of the operon 8- to 10-fold. This was true even in strains with a defective repressor gene (*trpR⁻* strains). How could this be, and what could be the mechanism of action? An important clue was provided by sequence analysis of the 162 bases in the *trp leader region,* that is, the region between the initiation site for transcription and the initiation site for translation of the first structural gene (see Figure 26-16). This leader region contains a potential initiation codon (bases 27 to 29), two tandem *trp* codons (bases 54 to 59), and a terminator codon (bases 69 to 71). A so-called *leader peptide* (see Figure 26-16) of 14 amino acids would result from translation of this region. Al-

**Figure 26-14**
Schematic diagram of the repressor control of *trp* operon expression. The *trp* promoter (*p*) and *trp* operator (*o*) regions overlap. The *trp* aporepressor is encoded by a distantly located *trpR* gene. L-Tryptophan binding converts the aporepressor to the repressor that binds at the operator locus. The formation constant for the L-tryptophan aporepressor complex is $0.54 \times 10^5\ M^{-1}$. This complex prevents the formation of the polymerase-promoter complex and transcription of the operon that begins in the leader region (*trpL*).

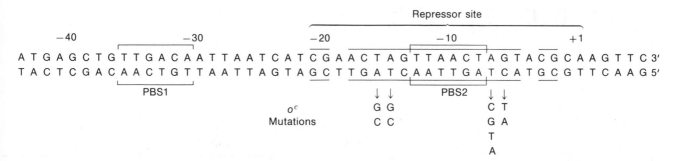

**Figure 26-15**
The promoter-operator region of the tryptophan operon. Two regions, PBS1 and PBS2, where polymerase binds most strongly in a number of promoters are bracketed. Regions within the repressor-binding site showing dyadic symmetry are underlined and overlined. Single base changes that lead to operator constitutive (*oᶜs*) are indicated.

**Table 26-3**

Attenuation of the *trp* Operon Is Affected by *trpT*, *trpX*, and *trpS*
Mutations in *trpR$^-$* Strains

| Strain (*trpR* Derivative) | Enzyme Levels (Normalized) | |
|---|---|---|
| | **Wild type (trpa$^+$)** | **(trp$\Delta$LD102 (trpa$^-$)** |
| Wild type | 1.0 | 7.1 |
| *trpT$_{ts}$* | 7.1 | — |
| *trpX* | 4.5 | 6.7 |
| *trpS9969* | 2.7 | 7.5 |

All strains were *trpR$^-$* (derepressed) and grown in the presence of excess tryptophan;
*trp$\Delta$LD102* deletes the attenuator. Enzyme levels were measured on anthranilate synthase
for the *trpT$_{ts}^-$* strain, and tryptophan synthase for the *trpX$^-$*, and *trpS9969* strains.
*trpT$_{ts}^-$* is a temperature-sensitive lesion in the tRNA$^{trp}$ gene. *trpX$^-$* leads to a defect in
the modification of the adenine adjacent to the anticodon sequence of the tRNA.
*trpS9969* is a mutant in the tryptophenyl-tRNA synthase gene; the resulting enzyme has
one one-hundredth the affinity for tryptophan shown by the wild-type enzyme, resulting
in much lower levels of charged tRNA$^{trp}$ than normal.

though this peptide has never been detected in whole cells, there are numerous reasons for believing that it is synthesized and that its translation
up to or through the *trp* codons regulates attenuation. First of all, selective
starvation of cells for tryptophan relieves attenuation and permits most
RNA polymerase molecules to read through the leader region. The only
other amino acid that relieves attenuation of the *trp* operon when it is
lacking is arginine; arginine starvation is about 80 percent as effective as
tryptophan starvation. It should be noticed that an *arg* codon is located
adjacent to the two *trp* codons in the leader region. Most telling of all is the
finding that a leader mutation resulting in the replacement of the AUG
start codon by AUA, which would eliminate translation of the leader peptide, also prevents transcription beyond the attenuator.

Other experiments indicate that the level of charging of tRNA$^{trp}$ is a
crucial factor in the attenuation response. This has been examined in vivo
by comparing the *trp* operon enzyme levels in *trpR$^-$* strains that are otherwise normal with strains that are defective in some respect in charged
tRNA$^{trp}$ (see Table 26-3). This work shows that structural defects in
tRNA$^{trp}$ or in the charging enzyme elevates expression, probably by permitting polymerase to transcribe through the attenuator.

**Figure 26-16**
The 5'-terminal region of tryptophan
operon mRNA.

Leader polypeptide (hypothetical)

+27              +54         +69

MET - LYS - ALA - ILE - PHE - VAL - LEU - LYS - GLY - TRP - TRP - ARG - THR - SER

pppAAGUU ~ AGGGUAUCGACA AUG AAA GCA AUU UUC GUA CUG AAA GGU UGG UGG CGC ACU UCC UGAAACGGGC ~

+163       trpE protei

+141

MET-GLN-THR-GLN-LYS-PR

UUCACCAUGCGUAAAGCAAUCAG ~ CCUAAUGAGCGGGCUUUUUUUUGAACAAAAUUAGAGAAUAACA AUG CAA ACA CAA AAA CC

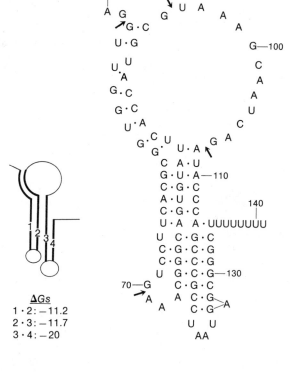

**Figure 26-17**
Proposed secondary structures in the leader RNA. Four regions can base pair to form three stem and loop structures. The arrows mark the RNAse T1 cleavage sites. The estimated free energies of formation of the various stem and loop structures are indicated.

The preceding results strongly support the hypothesis that *transcription read-through requires partial translation of the leader sequence*. However, only if the translation pauses or stops in the region where the *trp* or *arg* codons occur, is read-through favored. A careful examination of the secondary-structure possibilities in the attenuator region suggests why this is so. The leader region between bases 50 and 141 possesses the potential for forming a variety of base-paired conformations. Figure 26-17 illustrates the most likely secondary structures that form in terminated *trp* leader RNA based on analysis of regions of the transcript that show resistance to RNAse T1 digestion under mild conditions and the base-pairing rules established by studies of defined oligonucleotides. Four regions of base pairing capable of forming three stem and loop structures have been proposed. Region 1, which includes the tandem *trp* codons and the leader peptide translation stop codon (residues 54 to 68), can base pair with region 2 (residues 76 to 91). Region 2 (residues 74 to 85) also should be capable of base pairing with region 3 (residues 108 to 119). Stem and loop 2·3 has not been observed in vitro, presumably because stem and loop 3·4 and stem and loop 1·2 form preferentially. Region 3 (residues 114 to 121) can base pair with region 4 (residues 126 to 134). The existence of this stem and loop is inferred from the relative resistance of the GC-rich region from residue 107 to the 3′ end of the transcript to RNAse T1 digestion. The stem and loop structure formed between regions 3 and 4 followed by a sequence of U residues is a common structure used in transcription termination (see Chapter 21). Hence it is expected that conditions under which this structure is preserved would favor transcription termination. In support of this, a number of single-base replacement mutations have been isolated that lead to

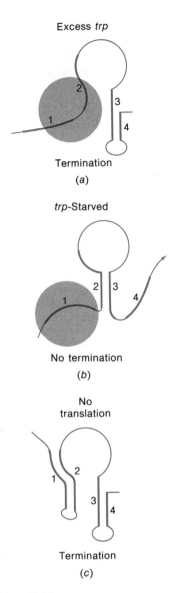

Excess *trp*

Termination

(a)

*trp*-Starved

No termination

(b)

No
translation

Termination

(c)

**Figure 26-18**
Model for attenuation in the *trp*
operon showing ribosome and leader
RNA. (*a*) All amino acids including excess tryptophan are present so that
stem-loop 3·4 is present. (*b*) Cells are
selectively starved for tryptophan so
that the ribosome stops prematurely
at the tandem TRP codons. Under
these conditions, stem-loop 2·3 can
form, and this is believed to lead to the
disruption of stem-loop 3·4, permitting
attenuation. (*c*) Where no translation
occurs, as when the leader AUG
codon is replaced by an AUA codon,
stem-loop 3·4 is intact and termination
in the leader is favored.

mispairing in the 3·4 stem, and all of these lower transcription termination
in the leader to some extent. The absence of translation of the leader
should do nothing to perturb this structure (see Figure 26-18), and consistent with this, it is found that changing the initiator codon of the leader
peptide by a single base prevents read-through, as discussed earlier. It is
known that selective starvation for either tryptophan or arginine stimulates
transcription read-through. Selective starvation for these amino acids
should result in ribosome stalling in the region of bases 54 to 62. The
resulting rupture of the base-pairing 1·2 structure would make region 2
available for pairing with region 3. This would encourage disruption of the
stem and loop 3·4 structure, as suggested in Figure 26-18, resulting in transcription read-through. In the presence of an adequate supply of all amino
acids, translation would proceed beyond this critical region, so that region
2 would not be available for base pairing (Figure 26-18), a situation again
favoring transcription termination at the attenuator.

This attenuator mechanism of control is amazingly simple, since it involves no proteins other than those normally used for transcription and
translation. One might expect such a simple and effective mechanism to be
used repeatedly for other operons involved in amino acid biosynthesis. An
indication that the attenuator mechanism might be functioning would be
the finding that the biosynthetic enzymes for a particular amino acid were
increased when charging of the cognate tRNA was reduced. Indeed, *for the
several other amino acid biosynthetic pathways in E. coli for which
tRNA charging is involved in regulation, attenuator mechanisms have
been found.* Thus analysis of the leader region preceding the structural
genes for the biosynthesis of three such amino acids in *E. coli*, phenylalanine, histidine, and leucine, reveals structures with the features of this type
of control mechanism. These three leader sequences contain several consecutive codons for phenylalanine, histidine, or leucine, respectively, followed by sequences that can be organized into stem and loop structures
that could signal termination. Even the multivalently controlled *thr* (biosynthesis of threonine) and *ilv* (biosynthesis of isoleucine and valine)
operons of *E. coli* exhibit this mechanism of operon expression. Thus the
leader sequence of the *thr* operon specifies a 21-amino acid peptide containing 8 threonine codons and 4 isoleucine codons, and that of the *ilv*
operon specifies a 32-amino acid peptide, of which 14 are either leucine,
isoleucine, or valine. Comparable studies of one of the genes involved with
arginine biosynthesis (*argF*) failed to show attenuation. This gene, which
is required for arginine biosynthesis, appears to be exclusively regulated by
a repressor mechanism. By contrast, the *his* operon has no repressor and
appears to be regulated exclusively with an attenuator. These contrasting
findings provoke the question as to why, in instances such as the *trp* operon, both types of transcription controls are used. To some extent this may
be an evolutionary accident. In this connection, it is appropriate to mention that other investigations indicate that the trpR protein is also a repressor for the *trpR* gene and the *aroH* gene, which encodes one of three isozymes that catalyze the first reaction in aromatic biosynthesis. Two
conceivable advantages have been suggested for the tryptophan operon having preserved two control switches in series. First, for a system with two
such metabolic switches, the amplification factor between fully on and

fully off should be the product of what one could achieve with either switch alone. Second, under some conditions, for example, during very rapid protein synthesis it is possible that a higher level of amino acid is required to keep the tRNA at an adequately charged level.

Attenuation was first observed for the early right and early left transcripts of the bacteriophage λ, where relief from termination requires a phage-encoded protein (see Chapter 27). A good deal is known about the details of the mechanism for relieving attenuation in λ, but the degree of involvement of the translation machinery is still uncertain.

## Regulation of Genes for RNA Polymerase Subunits and Ribosome Proteins

Over 100 genes in *E. coli* are involved in the synthesis of the RNAs and proteins that comprise the enzymatic machinery for transcription and translation. In terms of bulk alone, the relevant gene products make up between 20 and 40 percent of the dry cell mass. In rapidly growing cells, about 85 percent of the RNA is ribosomal, 10 percent is tRNA, and most of the remainder is mRNA. The various RNAs and proteins are produced according to their need. For ribosomes and tRNA, this results in a synthesis rate that is roughly proportional to the growth rate; the relative amounts of the 3 rRNAs (16 S, 23 S, and 5 S), the 40 or so tRNAs, and the 50 ribosomal proteins is consistent with the stoichiometric needs for making ribosomes. For RNA polymerase, this results in equimolar amounts of $\beta$ and $\beta'$ approximately equal in sum to the number of $\alpha$ subunits. The dissociable $\sigma$ subunit of RNA polymerase is produced in somewhat less than stoichiometric amounts. The synthesis of RNA polymerase subunits is loosely coupled to the synthesis of ribosomal protein subunits. Demands for large amounts of all these RNAs and proteins and precise regulation of their relative amounts have led to complex gene arrangements and controls at both the transcription and translation levels. These controls are only partly understood at the present time.

**Control of Stable rRNA Synthesis by the *rel* Gene.** *E. coli* cells under conditions of rapid growth contain about $10^4$ ribosomes. The maximum rate of reinitiation at the ribosomal gene promoter is about 1 per second. In rapid growth, *E. coli* can duplicate once every 20 minutes, which would allow for the synthesis of only about 1200 molecules of rRNA if there were only one gene for ribosomal RNA. In fact, there are seven copies for ribosomal RNA in the bacterial chromosome, which makes it possible for rRNA synthesis to maintain the necessary pace under conditions of rapid growth. The rRNA operons are dispersed at seven locations around the circular *E. coli* chromosome. Each operon is transcribed into one long transcript that is processed into a single 16 S, 23 S, 5 S, and usually at least one transfer RNA (see Figure 26-19 and Chapter 21). The order of the genes on the operon starting from the initiation site for transcription is 16 S, 4 S, 23 S, and 5 S (see Chapter 21 for further discussion). In some of the operons, additional 4 S genes for tRNA are located downstream from the 5 S gene. All the known ribosomal RNA operons appear to have two promoters located in tandem.

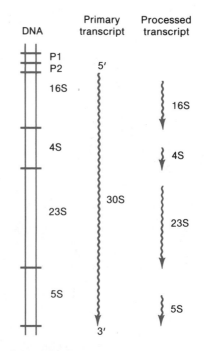

**Figure 26-19**
A typical *rrn* operon contains two promoters and genes for 16 S, 23 S, and 5 S rRNA and a single 4 S tRNA gene. The four fully processed RNAs are derived from a single intact 30 S primary transcript.

As stated earlier, rRNA synthesis is usually maintained at a rate that is proportional to the gross rate of protein synthesis. In a normal wild-type cell, when protein synthesis is limited, e.g., by amino acid availability, there is a rapid rise in ppGpp concentration from about 50 to 500 $\mu M$ (see Figure 26-20). Concomitantly, there is an abrupt cessation of rRNA synthesis. This is part of the syndrome known as the _stringent response_. If amino acids are reintroduced into the growth medium, the ppGpp concentration falls rapidly (half-life about 20s), and the rate of rRNA synthesis rises. In a _relA_ mutant cell, neither the rapid rise in ppGpp concentration nor the cessation of rRNA synthesis are seen when amino acids are removed. The _relA_ gene encodes a protein that is involved in ppGpp synthesis. Another type of mutant called _spoT_ shows the normal rise in ppGpp level on amino acid starvation, but the level of ppGpp falls much more slowly on readdition of amino acids to the growth medium. Correlated with this, the rate of rRNA synthesis also increases very slowly on readdition of amino acids. Apparently, the concentration of ppGpp is regulated by a careful balance between its rate of synthesis (controlled by the _rel_ gene) and rate of breakdown (controlled by the _spoT_ gene). Strong support that _ppGpp directly inhibits rRNA synthesis_ comes from cell-free synthesis studies in which the DNA-directed synthesis of rRNA has been shown to be strongly and selectively inhibited by 100- to 200-$\mu M$ ppGpp.

The synthesis of ppGpp has been studied both in crude cell-free extracts of _E. coli_ and in a partially purified system to determine what factors influence its rate of synthesis. It was found that _ppGpp is synthesized on the ribosome from GTP in the presence of the protein encoded by the wild-type relA gene_. Maximum synthesis occurs in the presence of ribosomes associated with mRNA and uncharged tRNA with anticodons specified by the mRNA. If the uncharged tRNA bound to the ribosome acceptor site (A site) is replaced by charged tRNA, the rate of ppGpp is greatly lowered. If uncharged tRNA anticodons are not complementary to mRNA codons exposed on the ribosome for protein synthesis, ppGpp synthesis does not occur. This set of observations suggests the mechanism whereby amino acid charging of tRNA controls the rate of rRNA synthesis (see Figure 26-21). First, uncharged tRNA that is codon-specific for the exposed codons on the mRNA becomes bound to the ribosome acceptor site, creating a situation unfavorable for protein synthesis but favorable for ppGpp formation. Second, the ppGpp diffuses and binds to RNA polymerase or other factors involved in rRNA transcription. It has been hypothesized that ppGpp acts by binding to RNA polymerase, altering its structure, and thereby affecting its ability to recognize certain promoters. Final proof that this mechanism is correct will probably require the isolation of RNA polymerase mutants that are not affected by ppGpp.

The observations on ppGpp involvement in rRNA synthesis show that this small molecule is an important control factor regulating rRNA synthesis, but it does not eliminate the possibility that other factors exist that also affect the ribosome level.

In vivo and in vitro evidence exists that the inhibitory effect of ppGpp on transcription extends to most tRNA and ribosomal protein genes. The effect of ppGpp on mRNA synthesis is mixed; some syntheses are stimulated and some are inhibited at elevated levels of ppGpp. The DNA-directed synthesis of the RNA polymerase subunits $\beta$ and $\beta'$ is inhibited in

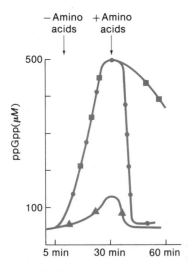

**Figure 26-20**
ppGpp concentration under normal conditions, after amino acid starvation, and after readdition of amino acids (● = wild-type cells; ▲ = _relA⁻_ cells; and ■ = _spoT⁻_ cells.

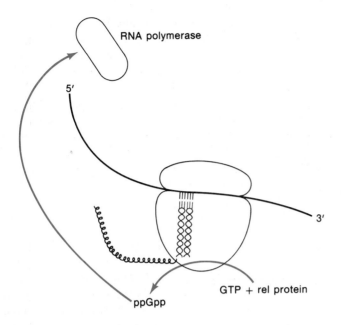

RNA polymerase

5'

3'

GTP + rel protein

ppGpp

**Figure 26-21**
Schematic diagram of ppGpp synthesis and the hypothesized mechanism
for its action. ppGpp is synthesized on
the ribosome when there is a peptidyl-
tRNA on the P site of the ribosome
and an uncharged tRNA on the A site.
The ppGpp probably inhibits rRNA
synthesis by complexing with the RNA
polymerase. Poly(U) in the presence of
uncharged phenylalanine tRNA will
stimulate ppGpp synthesis even in the
absence of a growing chain of polyphenylalanine.

vitro by ppGpp, although conditions leading to elevated ppGpp levels in
vivo do not appear to immediately inhibit $\beta$ and $\beta'$ synthesis. Attempts to
explain this seemingly contradictory result are made below.

**Translational Control of Ribosomal Protein Synthesis.** In exponentially
growing *E. coli* cells, *synthesis rates of all ribosomal proteins (except L7,
L12, and S6) are identical and coordinately regulated in response to environmental conditions.* M. Nomura and others have suggested a second
mode for the regulation of r-protein synthesis, namely, that *free r-proteins
inhibit the translation of their own mRNA* and that *as long as the assembly of ribosomes removes r-proteins, the corresponding mRNA escapes this
feedback inhibition.* This hypothesis has been tested in vitro using a protein-synthesizing system with various template DNA molecules carrying
r-protein genes and in vivo by examining the effect of overproduction of
certain r-proteins on the synthesis of other r-proteins using various recombinant plasmids. By these means it was found that certain r-proteins (L1,
L4, L10, S4, S7, and S8) selectively inhibit the synthesis of other r-proteins
whose genes are part of the same operon as their own and that this autogenous inhibition occurs at the level of the translation of mRNA rather than
at the level of transcription (see Figures 26-22 through 26-24). In vitro translation experiments have confirmed this scheme most directly for the regu-

| ← $P_\alpha$ | | ← $P_{SPC}$ | | ← $P_{S10}$ |
|---|---|---|---|---|

L17　$\alpha$　|S4|　S11　S13　L15　L30　S5　L18　L6　　|S8|　S14　L5　L24　L14　S17　L29　L16　S3　( S19　L22　)L2　L23　|L4|　L3　S10

**Figure 26-22**
A giant cluster containing 25 r-protein genes and the $\alpha$ gene for RNA polymerase. This cluster is divided into three operons with promoters $P_\alpha$, $P_{SPC}$, and $P_{S10}$. The corresponding messages are subject to autogenous inhibition at the translation level by S4, S8, and L4 protein, respectively. S8 only regulates genes downstream from L5. L14 and L24 translation is
also autogenously regulated by one or both of these proteins.

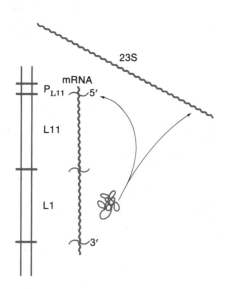

**Figure 26-23**
Schematic diagram explaining autogenous inhibition by L1. L1 can form a complex with either the 5' end of its own mRNA or with 23 S rRNA. If there is an excess of L1 over the available 23 S rRNA, then L1 will bind to the 5' end of the mRNA and inhibit both L1 and L11 synthesis.

latory protein (L1), which acts at the 5' end of the mRNA; the inhibition by L1 can be abolished by addition of 23 S rRNA (Figure 26-23). The 23 S rRNA complexes with the L1 protein, thereby eliminating it as an inhibitor. These results support and extend the proposed model of feedback regulation.

Most of the r-protein genes exist in clusters or operons with promoters at one end, although the possibility of some significant internal promoters cannot be excluded at this time. A summary of information on the structure and regulation of five of these operons is presented in Figures 26-22 through 26-24. In some cases, individual operons are regulated by one of the encoded r-proteins. In other cases, operons appear to be subdivided into "units of regulation," and individual units of regulation are regulated by one of the gene products. A meaningful overall pattern of control appears to be emerging that is consistent with the requirement to balance the relative amounts of rRNA and r-proteins synthesized.

**Regulation of RNA Polymerase Synthesis.** The genes for the RNA polymerase subunits β and β' (*rpoB* and *rpoC*, respectively) and L7/12 and L10 (*rplJ* and *rplL*, respectively) belong to a single transcription unit (see Figure 26-24). The gene for the α subunit (*rpoA*) appears to be cotranscribed with four ribosomal protein genes coding for the proteins S13, S11, S4, and L17 (see Figure 26-22). This suggests that *the regulatory system for RNA polymerase synthesis and that for ribosomal protein synthesis are shared.*

*Despite the use of a common promoter, the synthesis of the RNA polymerase proteins and certain nearby r-proteins is not always coordinately regulated.* Only the β operon will be discussed because it is better understood. For brief times, synthesis of L7/12 and L10 is inhibited in vivo

**Figure 26-24**
DNA and transcripts in the β operon. Only a small number of transcripts read through the attenuator (ATT) located 69 nucleotides beyond the 3' end of the *rplL* gene. The readthrough transcripts are cleaved by RNAse III about 200 nucleotides beyond the end of the *rplL* gene. L10 r protein and RNA polymerase are believed to inhibit translation of the promoter proximal and promoter distal messages, respectively. L10 may specifically inhibit its own synthesis. Normally about 5 times as much L7/L12 is synthesized as L10. Ribosomes can initiate independently at the initiation site for L7/L12, and it is possible that L7/L12 may regulate this initiation. Recently, Bruckner and Matzura have shown that another promoter two genes upstream from the $P_{L10}$ promoter may be the true initiation point for this operon in vivo.

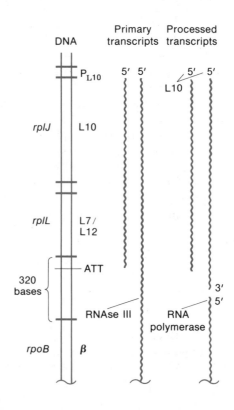

under conditions of amino acid deprivation, while the synthesis of $\beta$ and $\beta'$ is not. Furthermore, the RNA polymerase inhibitor rifampicin causes a transient stimulation of $\beta$ and $\beta'$ synthesis, but not of L10 and L7/12. This evidence for lack of coordinate regulation implies the existence of a separate regulatory device for controlling $\beta$ and $\beta'$ subunit synthesis. The genetic region containing L10, L7/12, and the start of the gene for $\beta$ have been sequenced. There are 320 untranslated bases between the *rplL* gene and the downstream *rpoB* gene; this is a suspiciously large segment for an intercistronic region, suggesting that it functions as more than just a spacer between cistrons. Under normal growth conditions, only 20 percent of the transcripts that initiate at the L10 promoter $P_{L10}$ read through the entire operon. The remainder terminate 69 nucleotides beyond the 3' end of the *rplL* gene (see Figure 26-24). The nonattenuated transcript is normally cleaved by RNAse III in the intercistronic region at a point about 200 nucleotides beyond the end of the *rplL* gene. This divides the mRNA for the operon into two segments, one encoding the r-proteins and the other the two RNA polymerase proteins. It has already been mentioned that L10 is an autogenous regulator inhibiting translation of its own synthesis. Clearly this translational control would have no direct effect on the translation of the message for $\beta$ and $\beta'$. Recently it has been discovered that RNA polymerase holoenzyme inhibits the translation of $\beta$ and $\beta'$. Thus both segments of the operon appear to be regulated at the translation level by different specific autogenous regulatory proteins. The finding that RNA polymerase inhibits the synthesis of its own subunits immediately suggests an explanation for the differential in vivo effect of amino acid deprivation and rifampicin on expression of the r-protein genes and the polymerase protein genes. Whereas both ppGpp or rifampicin should inhibit initiation of transcription that occurs at the $P_{L10}$ promoter, the binding of either rifampicin or ppGpp to RNA polymerase could interfere with its ability to act as an autogenous inhibitor. This should have a stimulatory effect on $\beta$ and $\beta'$ synthesis and would involve all the messages for $\beta$ and $\beta'$ that have already been synthesized, as well as any that are in the process of being made.

Another possible mode of regulation of this complex operon that has to be considered is the provisional transcription stop occurring after L7/12. Varying growth conditions may modulate this stop signal, although no mechanism for this has been proposed.

The discussion on prokaryotes has focused almost exclusively on *E. coli* because most of what is known on this subject has been discovered from studies of this particular enterobacteria. It seems likely that a thorough understanding of the regulatory phenomena of *E. coli* will help in an understanding of other cells, whether they are prokaryotic or eukaryotic. In the next section, the focus is shifted to a simple eukaryote, *Saccharomyces cerevisiae*.

## SACCHAROMYCES CEREVISIAE, A EUKARYOTIC PROTIST

The yeast known as *Saccharomyces* is found in soil and on the surface of many leaves and fruits with a high sugar content. *S. cerevisiae*, or baker's yeast, is a species used commercially in the leavening of bread. *S. cerevisiae*, like *E. coli*, is a facultative aerobe capable of growing either aerobically or anaerobically. On simple, defined medium, doubling times can be

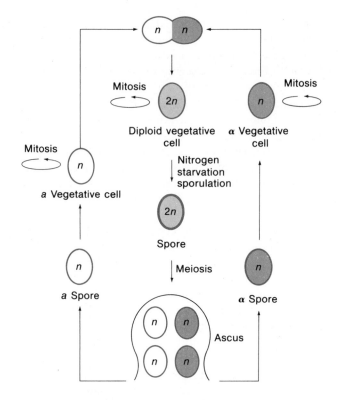

**Figure 26-25**
Life cycle of the yeast *Saccharomyces cerevisiae*.

as short as 90 minutes. Like many fungi, *S. cerevisiae* can exist either in a haploid (which has an *a* or *α* mating type) or a diploid state. In the haploid state, it contains 17 chromosomes with a gross nuclear DNA content about 5 times that of *E. coli* ($\sim 1 \times 10^{10}$ daltons). In either the haploid or diploid state, vegetative multiplication occurs by an asymmetric process known as *budding*. Haploid cells of opposite types, when mixed, can fuse to form diploid cells (see Figure 26-25). Diploid cells that are *a/α* can undergo sporulation. Limited nutrients and potassium acetate in the medium trigger this differentiation process. During sporulation, the chromosomes duplicate and then go through a meiotic segregation. After meiosis, each diploid cell yields an ascus that contains four haploid spores, two of type *a* and two of type *α*. The four haploid spores can be separated by micromanipulation and normally multiply true to type. Certain strains, known as homothallic, undergo changes in mating type much more frequently than abnormal heterothallic strains.

The ability of both haploid and diploid cells to undergo mitosis and the rapid doubling time in chemically defined media have made *S. cerevisiae* an extremely convenient microorganism for both genetic and biochemical investigations. Presently, it is the most intensively studied eukaryotic protist, and consequently, it promises to be the best understood in the near future. Mutants can be readily isolated from haploid cultures and can be readily assayed for their dominant or recessive character by diploidization with wild-type haploid cells of the opposite mating type. Complementation studies between mutant alleles are easily carried out. Linkage maps between different genetic markers are determined by analyzing the distribution of meiotic products (i.e., haploid spores) after sporulation of the appropriate heterozygous diploids.

Yeast cells have more than 10 times the volume of *E. coli* cells. They are near spherical in shape and bounded by a cytoplasmic membrane that is surrounded by a thick polysaccharide cell wall. Structures visible in the electron microscope include a nucleus, 50 or so mitochondria, microsomes as well as ribosomes, a Golgi apparatus with secretory vesicles, and several types of granular and vesicular inclusions. The number of mitochondria, microsomes, and ribosomes fluctuates widely with growth conditions, reflecting the presence of regulatory devices that control their numbers.

As in all eukaryotes, the nuclear membrane leads to a separation of the biochemical activities of transcription and translation. Attenuator-type control mechanisms, which necessitate the close coupling of transcription and translation, clearly could not exist in eukaryotes. However, this separation of transcription from translation permits more elaborate schemes for the processing of nascent transcripts prior to their translation. Nascent transcripts destined to become messenger RNA are modified by capping the 5' end and are tailed by 3'-terminal polyadenylic acid just as they are in higher eukaryotes (see Chapter 21). Splicing of nascent transcripts occurs in some cases during message synthesis, as well as in tRNA synthesis. The mechanics of splicing, which include a removal of certain internal segments of the nascent transcript followed by a mending process, are discussed elsewhere in this text (see Chapters 21 and 28). From the little that is known about processing of message in yeast, it seems likely that processing will play a highly significant role in regulating yeast gene expression. Once a processed transcript reaches the cytoplasm, its capacity for translation is restricted to reading the first AUG and adjoining codons closest to the 5' end of the RNA. Only the 5' proximal initiator codon is recognized as a translation start. This limits the value of polycistronic mRNAs (not known to exist in eukaryotes) and probably accounts for the greater dispersal of genes serving common pathways that are found in yeast.

A comparison of the gene arrangements and messages for histidine biosynthesis in *E. coli* and *S. cerevisiae* typifies the contrast in gene arrangement and use. In *E. coli*, all 10 histidine biosynthetic enzymes are encoded by a single polycistronic mRNA that is used for the synthesis of the corresponding individual polypeptide chains. In yeast, genes for three of these activities (steps 2, 3, and 10 in the biosynthesis) are clustered at the *HIS4* locus; the remainder are dispersed in other regions of the genome. The *HIS4* locus yields a single transcript that specifies a single polypeptide chain of molecular weight $9.5 \times 10^4$. This polypeptide chain folds into a multifunctional enzyme that carries out three enzymatic functions (Figure 26-26) that in *E. coli* are executed by separate proteins.

## Galactose Metabolism and Some Other Aspects of Catabolism

The genes for galactose utilization present another example where some clustering of common functions takes place. Four enzyme activities are required for galactose utilization. Three of these activities are specified by three tightly linked genes (*GAL7*, *GAL10*, and *GAL1*) on chromosome II (Figure 26-27), whereas the fourth, galactose transport, is specified by a gene (*GAL2*) located on chromosome XII. Classic genetic analysis has shown that both positive and negative regulatory genes coordinately affect the appearance of all four enzyme activities. A single polycistronic mRNA for

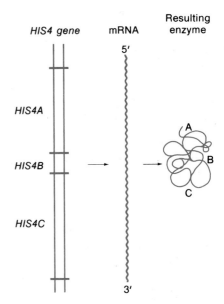

**Figure 26-26**
The *HIS4* gene encodes three enzyme activities. It produces a single transcript that results in a single enzyme. The enzyme is multifunctional, carrying out steps 2 (*B*), 3 (*A*), and 10 (*C*) in histidine biosynthesis.

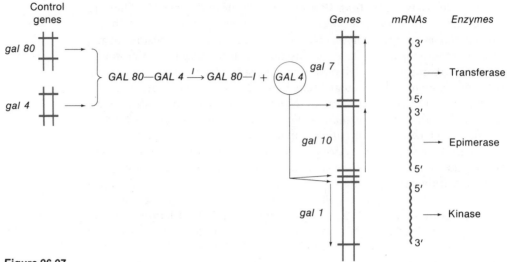

**Figure 26-27**
Hypothetical scheme for the regulation of enzymes involved in galactose metabolism. Three structural genes synthesize distinct mRNAs and enzymes (transferase, epimerase, and kinase). Synthesis requires the *GAL4* transcriptional activator. GAL4 protein is synthesized constitutively from the *GAL4* gene. However, it is inactive in the absence of inducer (I) because of complex formation with the GAL80 protein. The GAL4 protein can be released from this complex in the presence of inducer, which forms a complex with the GAL80 protein. The inducer is either galactose or, more likely, a metabolic derivative of galactose. Arrows next to the genes indicate directions of transcription.

all three genes of the *GAL7-GAL10-GAL1* cluster has not been found, nor has a single polypeptide chain representing all three enzymatic activities been found as in the case of the *HIS4* gene. The *GAL7* and *GAL10* genes are transcribed from the same DNA strand; whereas the *GAL1* gene is transcribed from the complementary DNA strand. Transcription of the *GAL* cluster originates from three separate promoters, one for each gene. Discrete transcripts are made for the synthesis of three distinct proteins that contain the different enzyme activities. In view of the independent transcription of these three genes, it is not clear at this point what biochemical advantages result from their clustering. The arrangement may have subtle advantages or may be an evolutionary vestige.

As in *E. coli*, the expression of the *GAL* genes in yeast is coordinated during both induction and steady-state synthesis. Induction requires the presence of galactose and the absence of glucose in the growth medium. The requirement for galactose reflects a complex-specific regulatory mechanism for the *GAL* genes. The requirement for the absence of glucose reflects a general carbon-catabolite repression mechanism also encountered in *E. coli*, but whose mechanism is likely to be quite different. In yeast, carbon-catabolite repression can result from enzyme inactivation (e.g., fructose diphosphatase) and/or inhibition of transcription or translation.

The immediate specific regulator of the *GAL7-GAL10-GAL1* cluster is the gene product of the unlinked *GAL4* locus, which appears to function as a transcriptional activator. Haploid cells with a mutant gal4* gene cannot be induced to synthesize the encoded *GAL7-GAL10-GAL1* RNAs. In the diploid heterozygous state, a mutant gal4 gene is recessive to the wild-type

---

*In yeast, wild type is represented by capital letters and mutant type by lowercase letters.

allele. Expression of the *GAL4* gene is under negative control by a second, unlinked gene *GAL80*. At first it was thought that the *GAL80* gene produced a repressor that inhibited transcription of the *GAL4* gene. However, recent evidence suggests that the GAL80 protein inhibits function of the GAL4 protein at a posttranslational level. The evidence for this comes from a study of galactokinase expression in certain regulatory mutants. Some *gal80* mutants have been isolated that allow for constitutive expression of galactokinase, i.e., high-level expression either in the presence or absence of galactose. Such mutants are usually recessive to wild-type *GAL80* alleles in the diploid state, as might be expected. Other *gal80* mutants (*gal80$^s$*) have been isolated that lead to permanent repression of galactokinase expression. These mutants are believed to produce an altered gal80 protein that is insensitive to galactose. As would be expected, *gal80$^s$* mutants are dominant to wild-type. Through a series of cleverly conceived synchronous mating experiments, it has been shown that the gal80$^s$ protein inactivates preexisting GAL4 protein even in the presence of galactose. By contrast, wild-type GAL80 protein only inactivates preexisting GAL4 protein in the absence of galactose. These results suggest that GAL80 protein inactivates the GAL4 protein by direct complex formation, with galactose or a derivative of galactose serving to block this complex formation and thereby activate the GAL4 protein. Other supportive experiments for this model indicate that GAL4 protein synthesis is not regulated by GAL80 protein.

A third locus, *GAL3*, may be involved in inducer synthesis, the true inducer being an intermediate in galactose metabolism rather than galactose itself. This conclusion stems from several observations. First, induction of galactose metabolism is greatly retarded in mutant *gal3* strains, 2 to 4 days being required to attain full expression after addition of galactose, in contrast to 1 hour in a *GAL3* strain. Second, in *GAL3* strains, mutations within the galactose gene cluster have no effect on the coordinate expression of other genes in the cluster. However, in mutant *gal3* strains, mutations in any of the genes of the cluster completely eliminate expression of the other two genes.

Addition of glucose to yeast cells growing on galactose not only represses synthesis of the galactose-metabolizing enzymes, but also rapidly and irreversibly inactivates these enzymes. The repression effect may be similar to the phenomenon known as the glucose effect or catabolite repression first described in bacteria (and discussed earlier). However, the inactivation effect occurs in yeast, but not in *E. coli*; it is termed *catabolite inactivation* to distinguish it from catabolite repression. Both effects are general, involving a wide variety of enzymes that are capable of converting their substrates to metabolites which the cell can obtain independently and more readily by the metabolism of glucose. Surprisingly little is known about the precise mechanisms leading to catabolite repression and inactivation. It is not known if cAMP plays a key role in yeast, as it does in catabolite repression in bacteria. Nor is it clear why catabolite-induced enzyme inactivation is found in yeast cells that are already capable of catabolite repression of enzyme synthesis. Inactivation leads to a more rapid disappearance of enzyme activity in growing populations than would be achieved by the gradual dilution of an enzyme after repression of its synthesis. Nitrogen-catabolite repression also occurs in yeast and *E. coli*, affecting nitrogen metabolism by enzyme repression and/or enzyme inactivation.

## Regulation of Ribosome Synthesis

About 7 percent of the chromosomal DNA of yeast codes for rRNA species. There are about 110 copies of rRNA cistrons per haploid yeast cell, each containing the genetic information for the synthesis of a single 5 S, 5.8 S, 18 S, and 26 S ribosomal RNA. Unlike the situation in bacteria, tRNA genes have not yet been found within the chromosomal rRNA cistrons, and interestingly enough, regulation of gross tRNA synthesis in yeast does not follow the same pattern as regulation of rRNA synthesis. In *E. coli*, the concentration of ribosomes is proportional to growth rate over a wide range, so that any change in growth rate has an exaggerated effect on ribosome synthesis. By contrast, in yeast the ribosome content appears to depend relatively little on growth rate, except under drastic conditions. Furthermore, the compound ppGpp, which has been found in yeast, appears to have little to do with regulating synthesis. Nevertheless, both ribosomal RNA and ribosomal protein synthesis decline sharply in less than 2 hours in cells completely deprived of an essential amino acid, a situation resembling the stringent response in *E. coli*. Under the same conditions, gross tRNA and mRNA synthesis do not show a significant decline, indicating a degree of specificity. Under certain conditions, there is no coordination between ribosomal RNA synthesis and ribosomal protein synthesis. For example, when a growing culture is shifted from 23° to 36°C, there is an immediate and significant inhibition of mRNA synthesis for ribosomal proteins, but only a slight inhibition of nascent rRNA synthesis. Normally, the inhibition of mRNA-encoding ribosomal proteins is a transient effect that disappears 60 to 90 minutes after the temperature upshift. However, there are several temperature-sensitive mutants in which the synthesis of mRNA for ribosomal proteins is permanently turned off at the nonpermissive temperature. In such instances, the synthesis of ribosomal proteins declines to 10 percent after 1 hour, while the transcription of ribosomal precursor RNA continues at 95 percent of its normal rate. Under conditions such as those described earlier, where synthesis of the r-protein mRNA is inhibited but synthesis of precursor rRNA is not, one finds a lack of processing and a rapid degradation of the rRNA. This suggests the existence of one or two additional control mechanisms that keep the ratios of mature rRNA and r-proteins in balance.

## Determination of the Mating Type in Yeast

The location of a gene or genetic element in the chromosome may be critical in determining its potential for expression. Interest in this type of control has been greatly stimulated by the discovery of _insertion elements_ that mediate the integration and excision process of relatively large genetic regions. In its simplest form, the insertion sequence (IS) contains no known genes related to insertion function and is usually shorter than two kilobases (at least in bacteria). Insertion sequences are _transposable_ segments of DNA that can insert into several sites in a genome by a poorly understood mechanism that frequently does not involve the use of the cell's general recombination system (*rec*). More complex transposable elements known as _transposons_ (Tn) behave formally like IS elements, but contain additional genes unrelated to insertion function. Most known transposons contain repetitious sequences at their extremities. These are arranged in inverted or parallel order, but the former seems to be more common. Insertion sequences and transposons of many different types are probably universal, since they have been found in all living cells that have been carefully examined. They are most useful in permitting systematic inversions or translocations of genetic material. The degree of target specificity observed for different insertion sequences and transposons varies enormously. Bacteria and yeast contain a multitude of transposable elements; only yeast uses such elements as a means of determining mating type.

Mating type in yeast involves two transposable genetic segments. It is not clear if these are typical transposable elements in terms of their structure or mechanism of transposition, but they are of major importance in terms of the profound effects they have on cellular metabolism. As stated earlier, haploid yeast has two mating types, *a* and *α*, and only cells of the opposite mating type can fuse to produce an *a/α* zygote. The *a* and *α* mating types are determined by a pair of alleles, *MATa* and *MATα*, situated at a locus on the right arm of chromosome III. Each allele appears to control the expression of a different set of functions. To explain how *MATa* and *MATα* control different but apparently complementary functions and the fact that recombination between certain mutant alleles is very infrequent, it was hypothesized that *MATa* and *MATα* contain different DNA sequences, and further, that the switching from one mating type to the other must be effected by the transposition of *a*- or *α*-specific sequences from another genetic locus elsewhere on chromosome III to the mating locus. Genetic analysis has led to the identification of two loci that may serve as the source of these transposable sequences: *HMLα*, situated near the end of the left arm of chromosome III, is responsible for the *a* to *α* interconversion, and *HMRa*, situated on the right arm of chromosome III, is necessary for the reverse process.

The nature of the homology between the mating-type genes was determined by electron microscopic analysis of heteroduplex structures formed between cloned DNA fragments corresponding to each of these loci. A schematic representation of these results is presented in Figure 26-28. Regions with common lettering are identical in sequence. The transposable segments *Yα* and *Ya* show no homology. Whereas it is clear that the *Yα* and *Ya* regions are transposed from the *HMLα* and *HMRa* regions, respectively, it is not known whether some of the *W*, *X*, or *Z* regions are also transposed. The mechanism of transposition is not known, but it must involve a partial duplication, since the transposed region is not lost from the *HML* and *HMR* regions. A mechanism similar to gene conversion has recently been proposed. The transposed regions appear to contain gene segments that are expressed only when moved to the *MAT* locus.

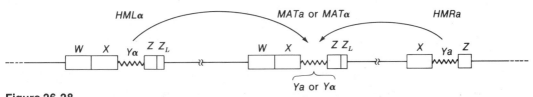

**Figure 26-28**
Yeast chromosome III showing the *HML*, *MAT*, and *HMR* loci. The three loci show the same orientation. W is a region common to *MAT* and *HML*, but not found at *HMR* (~750 bp). X is a region found at *MAT*, *HML*, and *HMR* (~700 bp). Ya is a specific substitution found at *MATa* and *HMRa* (~600 bp). Yα is a specific substitution found at *HMLα* and *MATα* (~750 bp). Z is a region found at *MAT*, *HML*, and *HMR* (~250 bp). $Z_L$ is a region common to *MAT* and *HML* and not found at *HMR* (~70 bp).

## SELECTED READINGS

Beckwith, J. R., and Zipser, D. (Eds.). *The Lactose Operon*. New York: Cold Spring Harbor Laboratory, 1970.

Jacob, F., and Monod, J. Genetic regulatory mechanism in the synthesis of proteins. *J. Mol. Biol.* 3:318, 1961. This is the classic work that has had more influence on the field of regulation than any paper before or since.

Leighton, T. J., and Loomis, W. F. (Eds.). *The Molecular Genetics of Development: An Introduction to Recent Research on Experimental Systems*. New York: Academic Press, 1981. Contains up-to-date comprehensive reviews on yeast and other eukaryotic systems.

Miller, J. W., and Reznikoff, W. S. (Eds.). *The Operon*. New York: Cold Spring Harbor Laboratory, 1978.

The *lac* operon and other bacterial and viral operons revisited.

Monod, J., Changeux, J.-P., and Jacob, F. Allosteric proteins and cellular control systems. *J. Mol. Biol.* 6:306, 1963. The classic work.

Oxender, D. L., Zurawski, G., and Yanofsky, C. Attenuation in the *E. coli* tryptophan operon: Role of RNA secondary structure involving the tryptophan codon region. *Proc. Natl. Acad. Sci. USA* 76:5524, 1979. This paper describes the way in which the structure of the *trp* leader RNA was deduced.

Zubay, G. Regulation of transcription in prokaryotes, their plasmids, and viruses. *Cell Biol.* 3:154, 1980. Recent and referenced.

## PROBLEMS

1. What facts originally led Jacob and Monod to suggest the existence of repressor in *lac* operon regulation?

   a. Cis dominance of the *i* gene
   b. Trans dominance of the *i* gene
   c. Cis dominance of the *o* locus
   d. Isolation of *i* repressor

   Explain.

2. Consider a negatively controlled operon with two structural genes (*A* and *B*, for enzymes A and B), an operator gene (*O*), and a regulatory gene (*R*). The first line of data below gives the enzyme levels in the wild-type strain after growth in the absence of, or in the presence of, the inducer. Complete the table at the bottom of the page for the other cultures.

3. Most mutations in the *lac* operator that give the $o^c$ phenotype have more than a single base change.

4. In a diploid situation, would you expect a $crp^+$ to be dominant to a $crp^-$. Referring to *lac* expression, describe the phenotypes of $crp^-$, $crp^+$, and $crp^+/crp^-$.

5. Effect of cAMP and CAP on DNA-directed in vitro synthesis of $\beta$-galactosidase. Dialyzed cell-free extracts (all $i^-$) were used for DNA-directed synthesis of $\beta$-galactosidase. Fill in the last column where values are not given.

| Source of Bacterial Extract (All Strains Used Were $i^-$) | cAMP | CAP | $\beta$-Galactosidase (Relative Values) |
|---|---|---|---|
| $crp^-$ | − | − | 5 |
| $crp^-$ | + | − | |
| $crp^-$ | − | + | |
| $crp^-$ | + | + | |
| $crp^+$ | − | − | |
| $crp^+$ | + | − | |
| $crp^+$ | − | + | |
| $crp^+$ | + | + | 100 |

| Strains | Uninduced | | Induced | |
|---|---|---|---|---|
| | Enz. A | Enz. B | Enz. A | Enz. B |
| **Haploid** | | | | |
| 1. $R^+O^+A^+B^+$ | 1 | 1 | 100 | 100 |
| 2. $R^+O^cA^+B^+$ | | | | |
| 3. $R^-O^+A^+B^+$ | | | | |
| **Diploid** | | | | |
| 4. $R^+O^+A^+B^+/R^+O^+A^+B^+$ | | | | |
| 5. $R^+O^cA^+B^+/R^+O^+A^+B^+$ | | | | |
| 6. $R^+O^+A^-B^+/R^+O^+A^+B^+$ | | | | |
| 7. $R^-O^+A^+B^+/R^+O^+A^+B^+$ | | | | |

6. Allolactose is the natural inducer of the *lac* operon. It results from the limited action of $\beta$-galactosidase on lactose. In a cell that is $Z^-$, what would be the relative transacetylase concentration
   a. After no special treatment?
   b. After addition of lactose?
   c. After addition of IPTG?
   Compare with wild type.

7. When lactose is added to a culture of *E. coli* growing on glucose, the enzymes involved in lactose metabolism
   a. Will be synthesized because lactose is the natural inducer for the *lac* operon.
   b. Will not be synthesized because of catabolite repression in the presence of glucose.
   c. Will be synthesized partially and then turned off at the translational level.
   d. Will not be affected whether there is glucose present or not.
   Give the best answer.

8. Which of the following mutant genes or loci should give rise to amber-suppressible mutations in the *lac* operon: *i, o, p, z, y, a*.

9. Three mutants altered in the machinery of catabolite repression are isolated and found to have the characteristics listed in the table (below).
   What is the most likely alteration in each mutant (one sentence or less)?

10. The products of *partial* RNAse T1 digestion of *E. coli* terminated *trp* leader RNA were analyzed on a *native* polyacrylamide gel. Three bands, designated A, B, and C, were resolved. Each band was eluted and run separately on *denaturing* polyacrylamide gels. Native band A was resolved to 3 bands of 141 (A1), 71 (A2), and to residues (A3), band B to a single band of 34 residues, and band C to 3 bands of 44 (C1), 25 (C2) and 19 (C3) residues. How do these results bear on the proposed secondary structure of *trp* operon leader RNA.

11. The isolation of a cis dominant $o^c$ mutation for the histidine operon was reported several years ago. Explain why this claim must have been wrong. What is the most likely explanation for the mutation?

12. Certain antibiotics inhibit protein synthesis by binding to the ribosome but do not produce a rise in ppGpp concentration. Suggest an explanation.

13. Certain genes for structural proteins in *E. coli* do not have extended mRNA lifetimes, but they do have elaborate translational control mechanisms. Why is this advantageous? Give an example.

14. Defend the statement that for most types of genes, the extended messenger lifetimes favor translational control mechanisms.

15. A mutant in *gal4* of yeast leads to constitutivity for galactose fermentation in haploid cells. When crossed with a wild-type haploid cell, the resulting diploid is still constitutive. Describe the nature of this mutation at the molecular level.

16. Why is clustering of functionally related genes more common in bacteria than in yeast?

17. Why are multifunctional enzymes more common in yeast than in bacteria?

18. Bacteria, like yeast, carry transposons that control the expression of certain host genes. One of the best understood is the transposon that controls expression of the *H2* gene for a flagellar antigen in *Salmonella*. This transposon is 995 base pairs in length and is bounded by a 14-bp inverted repeat sequence. A homologous recombination event between the 14-bp inverted repeat sequences would result in the inversion of the DNA segment between them. About two-thirds of the inverted region (572 bps) contains the *hin* gene, which encodes a protein required for the inversion. The *hin* gene is active in either orientation but the *H2* gene is active only in one orientation of the transposon. (See the display at the bottom of the page.)

   Of the three genetic regions, *hin* gene, *H2* gene, and IR region, which would you expect to be cis dominant and which would you expect to be trans dominant to defective mutant alleles? In the haploid state, what would be the phenotype of the following:
   a. *hin*⁻ in the A configuration.
   b. *hin*⁻ in the I configuration.
   c. *H2*⁻ in either configuration.

| Mutant | *lac* Operon | Arabinose Operon | Effect of Added cAMP | Dominance |
|--------|-------------|------------------|----------------------|-----------|
| A | Not inducible | Not inducible | Both inducible | Recessive |
| B | Not inducible | Not inducible | No effect | Recessive |
| C | Not inducible | Inducible | No effect | Cis dominant |

A configuration: Active orientation

I configuration: Inactive orientation

# PART V

# SPECIAL ASPECTS OF BIOCHEMISTRY

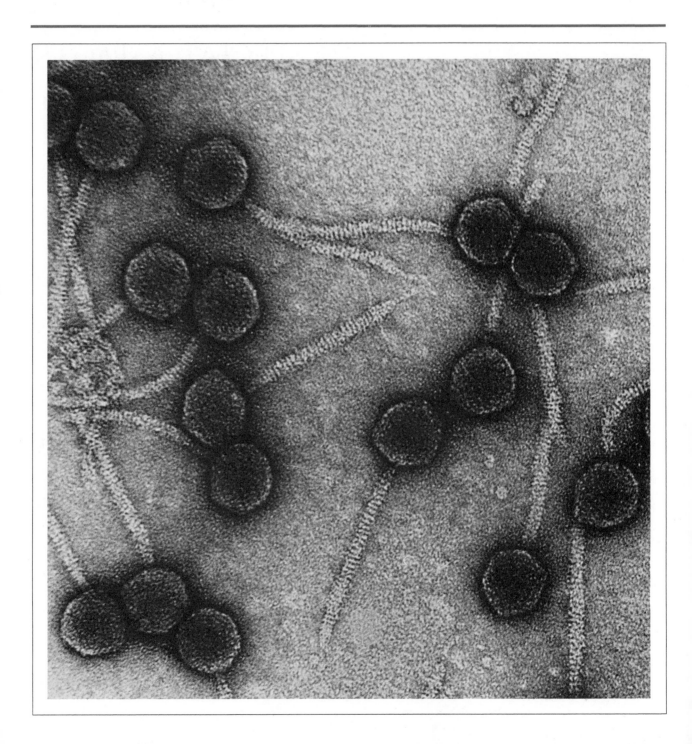

# 27

# THE MOLECULAR BIOLOGY OF BACTERIOPHAGE LAMBDA

Soon after the discovery by Avery and colleagues that the chemical basis of heredity was DNA (1944), it became fashionable for physicists and chemists entering biology to work on relatively simple replicative organisms, in particular, the bacteriophages. Here the marriage of genetics and biochemistry, known as molecular biology, has produced remarkable progress in furthering our understanding of the fundamental phenomena of living systems. Neither parent alone would have been capable of bringing forth such a splendid offspring. In our discussion of λ bacteriophage in this chapter we shall stress how genetic analysis suggested what biochemical experiments would be fruitful and biologically relevant. Biochemistry, in turn, served to solidify and prove genetic hypotheses, although it, too, has indicated which genetic approaches should be taken and which mutants should be searched for. This general methodology of attacking a biochemical problem with the help of genetics is currently being applied in most areas of biology, and it is therefore important for students of biochemistry to understand it.

All bacteriophages contain a single chromosome composed of RNA or DNA, never both. RNA phages are small, usually containing single-stranded nucleic acid (about 4 kb). Some small single-stranded DNA phages have also been found, but the majority of known phages contain double-helix DNA that varies in size from about 40 to 200 kb/strand. The λ bacteriophage is on the small end of this spectrum. When it infects a cell, only the DNA enters the cell. Infection can lead to an active _vegetative state_, in which progeny phage are produced, or it can lead to a dormant so-called _prophage state_, in which the viral chromosome becomes integrated as a linear part of the host chromosome. The dual lifestyle of λ, the

_Electron micrograph of bacteriophage λ. (Courtesy Roger Hendrix.)_

**Table 27-1**

Symbols Used to Refer to Host Functions Important to λ and
λ Functions and Sites (further symbols are referred to in text)

**Phage Functions and Sites**

**Functions**

| | |
|---|---|
| *cI:* | Necessary for establishment and maintenance of immunity; lambda repressor |
| *cII, cIII:* | Necessary for establishment of immunity |
| *cro:* | Negative control of immunity and early gene expression |
| *O, P:* | DNA replication genes |
| *N:* | Positive regulator of early gene transcription |
| *Q:* | Positive regulator of late gene transcription |
| *int:* | Prophage integration and excision |
| *xis:* | Prophage excision |

**Sites**

| | |
|---|---|
| *att:* | Site of prophage integration and excision |
| *cos:* | Left and right cohesive ends of the λ DNA molecule |
| $o_L$, $o_R$: | Operator sites controlling transcription of genes to left and right of immunity |
| $p_L$, $p_R$: | Early leftward and rightward promoters |
| $p_{RE}$, $p_{RM}$: | Promoters for establishment and maintenance of cI synthesis and immunity |
| *nutL, nutR:* | N utilization sites in the left and right arms |
| $t_{R1}$: | Termination site for early right transcription |
| $t_{L1}$: | Termination site for early left transcription |
| *ori:* | Origin of DNA replication |

**Host functions:**

| | |
|---|---|
| *rho:* | Transcription termination factor |
| *recA:* | Necessary for general recombination and induction of SOS functions |
| *lexA:* | Necessary for expression of SOS functions |
| *nusA, nusB, nusE:* | Necessary for N function |
| *gyrA, gyrB:* | DNA gyrase necessary for supercoiling |
| *lig:* | DNA ligase necessary for making circular DNA |

Note: Further symbols are referred to in text.

ability of the virus to serve as a carrier of host genes, and the relatively small size of the viral genome are all important reasons why the λ bacteriophage has become the most popular of all phages for scientific investigation.

In this chapter we will discuss the biochemical and genetic approaches that have been used in the investigations of λ bacteriophage. It will be necessary to refer to a large number of phage and host functions and sites; for convenience, a number of these are defined in Table 27-1 and can be referred to during the course of reading this chapter.

## EARLY STEPS IN λ INFECTION

The bacteriophage λ is a DNA virus whose natural host is *Escherichia coli*. The virion contains linear double-stranded DNA of molecular weight 30 megadaltons, the 5′ ends of which are extended by 12 bases and are complementary to each other. The DNA is injected through the phage tail, with the right 5′ extension (called the cohesive end, *cosR*) first entering the bacterium. By a mechanism that is still elusive, the left cohesive end (*cosL*), once it enters the cell, immediately finds *cosR* and anneals to it, forming a broken circular molecule, or Hershey circle. This molecule is acted upon sequentially by two host enzymes. First, DNA ligase seals the breaks, using energy provided by NAD. In this reaction, the free 5′-phosphate becomes covalently linked to the 3′-hydroxyl of the other end of the λ DNA molecule (Figure 27-1). This reaction involves formation of an enzyme-adenylate complex in which NAD acts as the adenylate donor (Chapter 20).

The second enzyme to act upon λ DNA is an *E. coli* topoisomerase II, DNA gyrase, which introduces negative twists into the covalently closed DNA circles. In reacting with λ DNA, gyrase cuts the two DNA strands, rotates them about each other, and reseals them. ATP is consumed in this process. Ordinarily, the broken DNA ends are constrained and do not move freely around each other. DNA gyrase rotates the DNA strands in one direction until about 300 supertwists (see Chapter 20) have been introduced into the phage chromosome. The enzyme is composed of two subunits encoded by the *gyrA* and the *gyrB* genes of *E. coli*. GyrA is a 105-kilodalton protein that is sensitive to the drug nalidixic acid, and gyrB is a 95-kilodalton protein that is inhibited by coumermycin.

Both DNA ligase and gyrase play roles in the physiology of *E. coli*. Mutations in the *lig* gene have been isolated that result in a temperature-sensitive enzyme. Mutant cells fail to replicate their DNA correctly at

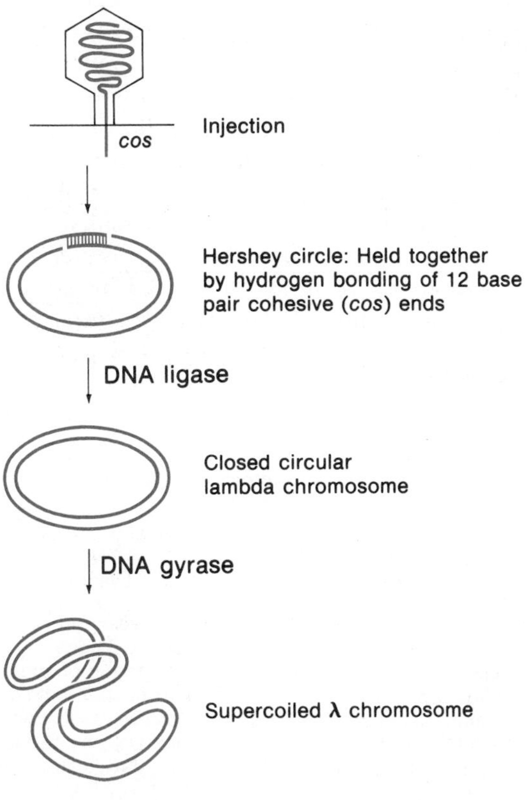

Injection

*cos*

Hershey circle: Held together
by hydrogen bonding of 12 base
pair cohesive (*cos*) ends

DNA ligase

Closed circular
lambda chromosome

DNA gyrase

Supercoiled λ chromosome

**Figure 27-1**
Early steps in lambda infection.

elevated temperatures. The accumulation of low-molecular-weight DNA molecules at the nonpermissive (elevated) temperature indicates that the conversion of these replicative intermediates to full-length DNA requires functional ligase. The role of gyrase in bacteria is less clear. *E. coli* is sensitive to the drugs coumermycin and nalidixic acid, and drug-resistant mutants of *E. coli* have altered gyrase subunits that are resistant to the drug in vitro. There is some evidence that DNA replication is affected by these gyrase inhibitors. In addition, the sensitivity of the *lac* operon promoter to coumermycin suggests that DNA superhelicity greatly augments the activity of at least some bacterial promoters (see Chapter 26).

## THE DEVELOPMENT OF λ

*After infection, λ development may proceed by means of one of two pathways, the lytic pathway or the lysogenic pathway* (Figure 27-2). The former is characterized by extensive replication of the λ chromosome and expression of most of its genes. Phage heads and tails are produced, the λ DNA is packaged, and the cells are lysed with the release of about a hundred infectious phage particles per cell. In the lysogenic pathway, replication is not extensive and expression of the viral genome is limited to a small region. The phage chromosome inserts into the bacterial chromosome and remains as an inert prophage that is repressed by phage-encoded repressor until alterations in the bacterial physiology result in its activation. At this point, the prophage excises from the host chromosome and enters the lytic pathway, replicates and packages its DNA, and destroys its host. The λ life cycles will be discussed in detail below.

The choice between the lytic or lysogenic pathway depends on a number of factors, not all of which are well understood. The mechanism by which λ regulatory circuits react to small changes in the state of the cell has fascinated and continues to fascinate researchers in this field. The mechanisms of regulation that λ has employed have served as models for understanding the regulatory circuits of *E. coli* and as a paradigm for the possibilities of a developmentally controlled system.

We will discuss below the general organization of the λ genome and then examine in detail some of the regulatory circuits. *The genetic organization of the λ map (and the maps of most other bacterial viruses) reflects the sequential expression of functions during phage development*. In addition, as we will see in more detail later, structural genes for products and their site of action frequently are found close together on the genetic map.

**Figure 27-2**
Lambda development: lytic or lysogenic growth.

# THE GENETIC SYSTEM OF λ

## Viral Mutants

Part of the power of the genetic system developed for λ study stems from the early realization of a variety of techniques for selection and screening of mutants. _Mutants define the genes of λ; study of infections with mutant phages leads to an understanding of the pathway of development_, just as studies of the accumulation of metabolic precursors in appropriate bacterial mutants lead to an understanding of many metabolic pathways (see, for example, Chapter 22).

Plaque-morphology mutants were the first to be isolated for λ; these, of course, can only be in nonessential functions. To look at essential phage functions, one must utilize "_conditionally lethal_" mutations, ones in which both conditions to grow the mutant phage stock and conditions to study the defect (the lethal condition) can be found. For λ, these have been of two general classes: (1) _temperature-sensitive mutations_, which are defective for growth at high temperatures but are able to grow at lower temperatures (Table 27-2), and (2) _nonsense mutations_, which lead to premature termination of an essential polypeptide chain (Table 27-3). Nonsense mutants can be isolated by their ability to make phage plaques on one host ($sup^+$) but not on another ($sup^0$).

## Complementation

Complementation between phage mutants, to determine if they carry mutations in the same gene, is a very simple experiment. Nonpermissive cells are infected simultaneously with the two mutant phages, and the burst of phages (i.e., the average yield of phages per infected cell) produced is assayed on a permissive host; if the mutations are in different genes, they should _complement_, so that mixed infection should give close to the normal wild-type yield of both phages. The emerging phages resulting from a mixed infection of phages with defects in different genes should retain the same properties as the parent phage, and they should still carry the conditionally lethal mutations (Figure 27-3). Occasionally, phages will emerge, however, that are able to grow on a nonpermissive host; these result from recombination rather than complementation.

# STEPS INVOLVED IN LYTIC GROWTH

A brief summary of the important functions for the lysogenic and lytic pathways is given in Figure 27-4. For lytic growth, the regulatory genes $N$ and _cro_ lead to DNA replication and expression of late functions, which in turn promote packaging of DNA in new phage heads and cell lysis. During the establishment of a lysogen, repressor acts to shut off synthesis of the lytic functions, and Int promotes integration of the viral genome into the host chromosome.

## N and cro Genes Are the First to Be Expressed

_In the lytic pathway, the phage genes are expressed sequentially, beginning with the immediate early functions N and cro_ (Figure 27-5). These genes are transcribed divergently from two phage promoters: $p_L$, situated to

**Table 27-2**
Detection of λ Temperature-Sensitive Mutations

| | Burst Size (Average Number of Phages per Cell on Lysis) | |
| --- | --- | --- |
| | 30°C | 42°C |
| λ+ | 100 | 100 |
| λOts | 50 | 0.01 |
| λEts | 70 | 0.05 |

Note: While wild-type λ grows reasonably well at both 30 and 42°C, mutations can be isolated that are unable to grow at one temperature extreme but can be propagated at a different temperature.

**Table 27-3**
Growth of λ Carrying Nonsense Mutations in Essential Genes

| | Relative Efficiency of Plaque Formation | |
| --- | --- | --- |
| | _sup_$^0$ | _supF_ |
| λ+ | 1.0 | 1.0 |
| λNam | $10^{-8}$ | 1.0 |
| λSam | $10^{-5}$ | 1.0 |

Note: By comparing the titer of λ plated on a $sup^0$ (carrying no nonsense suppressor) and _supF_ (carrying an _amber_ suppressor) strain, one can identify and study nonsense mutations in essential functions. N is an essential positive regulatory function; S is necessary for cell lysis.

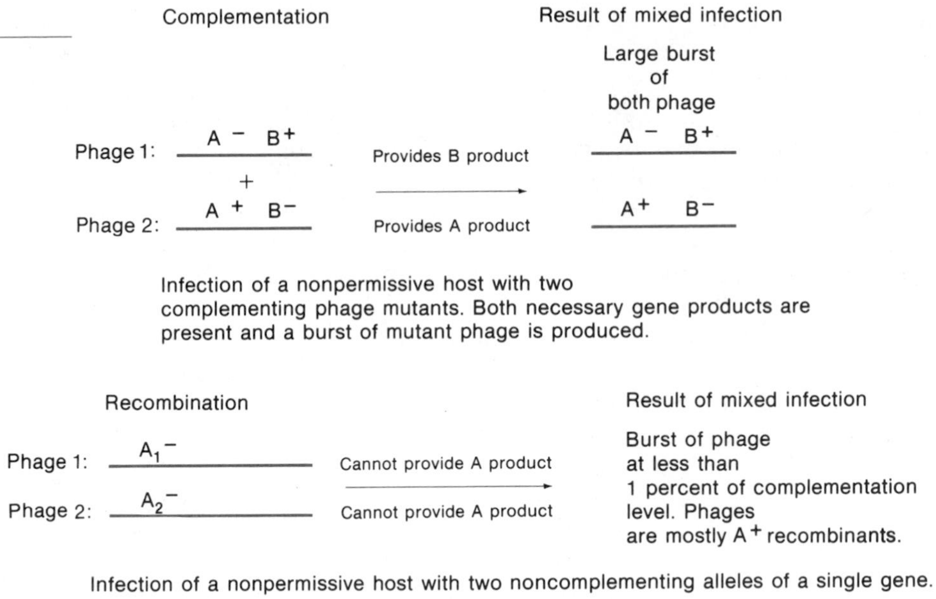

Complementation | Result of mixed infection

Large burst
of
both phage

Phage 1: $\dfrac{A^- \quad B^+}{}$    Provides B product    $\dfrac{A^- \quad B^+}{}$

$+$

Phage 2: $\dfrac{A^+ \quad B^-}{}$    Provides A product    $A^+ \quad B^-$

Infection of a nonpermissive host with two
complementing phage mutants. Both necessary gene products are
present and a burst of mutant phage is produced.

Recombination | Result of mixed infection

Phage 1: $\dfrac{A_1^-}{}$    Cannot provide A product

Phage 2: $\dfrac{A_2^-}{}$    Cannot provide A product

Burst of phage
at less than
1 percent of complementation
level. Phages
are mostly $A^+$ recombinants.

Infection of a nonpermissive host with two noncomplementing alleles of a single gene.

(a)

|   | 1 | 2 | 3 | 4 | 5 | 6 | 7 |
|---|---|---|---|---|---|---|---|
| 1 | − | + | − | + | + | + | + |
| 2 | + | − | + | + | − | + | + |
| 3 | − | + | − | + | + | + | + |
| 4 | + | + | + | − | + | − | − |
| 5 | + | − | + | + | − | + | + |
| 6 | + | + | + | − | + | − | − |
| 7 | + | + | + | − | + | − | − |

Seven uncharacterized λ mutants have been grown under nonpermissive conditions either alone
or with another λ mutant. Good growth of the phage is indicated by a plus sign and reflects com−
plementation between the two mutant phages.
From these data, the mutations can be divided into three complementation groups, which may
represent three different λ genes. *Mutations—A: 1,3; B: 2,5; C: 4,6,7.*

(b)

**Figure 27-3**
Complementation among lambda mutants. (a) Different outcomes when cells
are mixedly infected by two phage mutants with defects in different genes or in
the same genes. (b) Sample complementation assay used to determine the
number of complementation groups.

the left of the immunity region (containing genes *rex* and *cI*), and $p_R$,
situated just to its right. Cro acts to dampen the activity of these promoters,
beginning a few minutes after infection, when the cro protein accumulates.
This modulation must be important for λ developments, since *cro* mutations block phage growth. DNA sequences that lie promoter distal to *N*
($t_{L1}$) and *cro* ($t_{R1}$) signal the termination of transcription. N protein acts as
a positive regulator of λ gene expression by suppressing transcription termination at these and other terminators located elsewhere on the phage
chromosome (Figure 27-6). Before we examine the mechanism of action of
N protein, a brief review of transcription termination is in order (also see
Chapters 21 and 26).

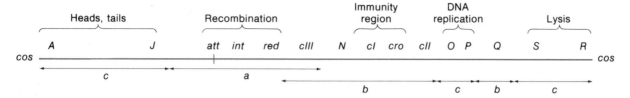

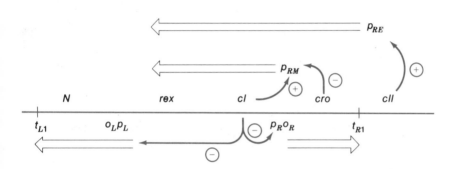

**Figure 27-5**
Regulation of lambda early gene expression. Heavy arrows indicate regions and directions of transcription. Light arrows indicate (in color) points of activation + or repression − by *cro, cl,* or *cll.* (1) Repressor synthesis from *cl* is regulated at (a) $p_{RE}$ (the establishment promoter) by *cll,* which acts as a *positive* factor in the initiation of transcription (initiation from $p_{RE}$ leads to a large burst of repressor synthesis on infection); and (b) $p_{RM}$ (the maintenance promoter), regulated in a positive fashion by repressor itself and in a negative fashion by *cro.* (2) Transcription from $p_L$ and $p_R$ to synthesize the genes necessary for lytic growth are negatively regulated at the closely linked operator sites $o_L$ and $o_R$ by cl and cro proteins.

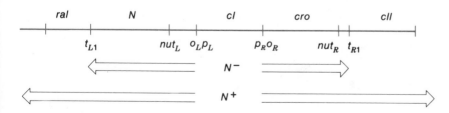

**Figure 27-6**
Antitermination by lambda N function. Transcription beginning at $p_L$ and $p_R$ stops at the first terminator encountered unless both N and the site for N action, *nut,* are available.

## Transcription Termination Is Frequently Sensitive to the Host Rho Protein

*Terminators fall into at least two general classes: those which act with RNA polymerase itself and those which require an additional E. coli protein, Rho, for their activity.* The difference can be demonstrated and, in fact, was first detected in vitro. The distinction has biological significance, since terminators that appear to be Rho-independent in vitro prove to be active in *rho*-defective mutants. The Rho-dependent terminators are inactive when Rho is absent in vitro as well as in vivo in *rho* mutants. The two classes of terminators are sequence-related; both involve regions of DNA with considerable dyad symmetry (see Chapter 26). The DNA symmetry is thought to result in the formation of stem-loop structures in the RNA transcript that are essential to the terminator process. Transcripts made in vitro and composed of base analogs that pair weakly terminate poorly (e.g., the substitution of IMP for GMP produces this result) presumably because the formation of the RNA stem is impeded. Similarly, mutations in the DNA dyad symmetry that reduce the stability of the potential RNA stem inhibit termination (Figure 27-7).

The mechanism of action of Rho is as yet not completely defined. After RNA chain initiation and elongation, RNA polymerase may encounter a Rho-dependent terminator, such as $t_{R1}$, at which point there is a pause in RNA synthesis (Figure 27-8). In the absence of Rho, transcription eventually resumes; in its presence termination is more likely to occur. In vitro, the termination reaction requires the hydrolysis of ATP. Rho also catalyzes the hydrolysis of ATP in the presence of a variety of RNAs, including the homopolymer poly(C). The most reasonable model for Rho action is that

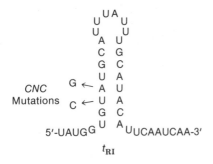

**Figure 27-7**
Structure of the lambda $t_{R1}$ terminator. Only RNA corresponding to the transcribed strand is shown. *cnc* mutations, which disrupt the formation of the hairpin of the terminator, reduce termination at $t_{R1}$.

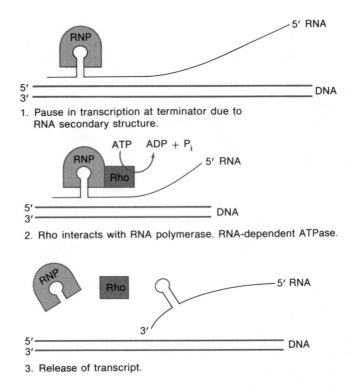

**Figure 27-8**
Three steps involved in transcription termination.

1. Pause in transcription at terminator due to RNA secondary structure.

2. Rho interacts with RNA polymerase. RNA-dependent ATPase.

3. Release of transcript.

Rho interacts with the messenger RNA. When a transcription complex (RNA, DNA, and RNA polymerase) pauses at a Rho-dependent terminator, it becomes susceptible to Rho action. Rho may compete successfully for the 3'-OH terminal of a transcript and use the energy provided by ATP (or other ribonucleoside triphosphates) to dissociate the complex. *E. coli* temperature-sensitive *rho* mutants have been isolated, and these are conditionally lethal for growth. However, the mutants display a bizarre phenotype. They are defective not only for transcription termination, but also for oxidative phosphorylation, recombination, UV-damage repair, and chemotaxis. It is not known whether the entire phenotype can be attributed to a failure in transcription regulation or whether Rho participates directly in these other reactions. Rho is a hexamer composed of identical 50,000-dalton polypeptides, and in vitro it is known to carry out the reactions just described, namely, transcription termination and RNA-dependent ATP hydrolysis. It cannot be excluded that Rho might catalyze other reactions if provided with the appropriate cofactors or substrates.

## The Activity of a Transcription Terminator May also Be Affected by Translation of the Involved Transcript

The attachment of ribosomes to RNA may influence its secondary structure and prevent or enhance the formation of the stem and loop at the terminator. This type of regulation of transcription termination or attenuation is seen in amino acid biosynthetic operons, such as the *trp* operon (see Chapter 26). A second mechanism by which translation regulates transcription is revealed by polar mutations. Polar mutations are translation-termination sequences (i.e., nonsense codons) that not only block the expression of the mutant gene, but also of promoter distal genes. It appears that the termination of translation causes termination of transcription in the vicinity of the nonsense codon. Since the polarity of nonsense triplets is suppressed in *rho* mutants, it must be the case that polypeptide chain termination with concomitant release of ribosomes provokes transcription termination at Rho-dependent sites. The presence of ribosomes may directly affect the RNA secondary structure in instances of polarity as it does in attenuation (see Chapter 26). Alternatively, since the action of Rho requires free RNA, ribosomes may block the access of Rho to certain transcripts. While some potential transcription terminators within an operon may normally never be active (only being revealed by premature polypeptide chain release), others may be regulated by translation frequency. The expression of genes in an operon is often not uniform; a sort of "natural polarity" sometimes (but not always) exists, where the promoter proximal genes are expressed more frequently than promoter distal ones. Natural polarity also seems to be affected by *rho* mutations.

## N-Gene Product Is an Antitermination Factor

*The λ N-gene product suppresses transcription termination at both phage and bacterial terminators, acting at terminators that are Rho-dependent as well as at terminators that are Rho-independent.* Although N is a freely diffusible protein, its action in the cell is mainly, if not exclusively, to cause the elongation of the $p_L$ and $p_R$ transcripts. Thus, after λ infection and the synthesis of N, other bacterial transcripts terminate normally, while, at the same time, the $p_L$ and $p_R$ transcripts become resistant to termination. *The specificity of N is due to the nut (N utilization) sequences that lie in the phage DNA,* one, *nutL,* some 50 base pairs to the left of $p_L$, the other, *nutR,* approximately 250 base pairs to the right of $p_R$, between *cro* and $t_{R1}$ (Figure 27-9). The sequences are similar and include a region of dyad symmetry with a 17 of 18 base pair identity. Without the *nut* sequences, N cannot suppress transcription termination, and the absence of *nut* regions in bacterial operons is why bacterial transcripts are not acted upon by N product. Point mutations in *nutL* are known. They are characterized by normal transcription initiation at $p_L$ and the synthesis of wild-type levels of N product, but the failure of N to suppress termination at $t_{L1}$. Since the $p_L$ operon does not carry functions vital to phage growth, and since the N product can still act to elongate the $p_R$ transcript, *nutL* mutants are viable. The localization of *nutR,* aside from sequence homology with *nutL,* rests on two experiments. First, deletions of the region between *cro* and $t_{R1}$ block the action of N. Second, this region has been cloned and inserted next to the bacterial promoter for the *gal* operon. Transcripts originating at *gal* now will no longer terminate when N is provided by λ. *The nut sequences are thus both necessary and sufficient for N action.*

Earlier, we discussed the various steps in transcription termination and the factors that influence its efficiency. We can conceive of N protein acting at any one or more parts of the termination reaction. Although the mechanism of N protein action is not known, it is enlightening to review how the problem is being approached at both the biochemical and genetic levels.

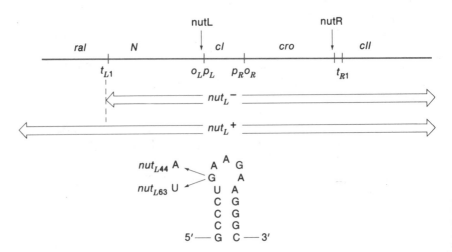

**Figure 27-9**
The N utilization site, *nut.* When *nutL* is mutant, N cannot suppress termination at $t_{L1}$. The $nut_{L44}$ and $nut_{L63}$ are two mutant alleles of the *nutL* site that block N action to antiterminate at $t_{L1}$. The primary sequence of normal and mutant *nutL* RNA is shown in the form of a possible secondary structure.

**Biochemical Analysis of N Protein Action.** The product of the *N* gene was isolated by means of an in vitro coupled transcription-translation system. The system is composed of ribosomes and crude supernatant fractions of *E. coli*, but it has been freed of mRNA by preincubation at 37°C. When supplemented with λ DNA, the system produces λ mRNA and polypeptides. The synthesis of some of these polypeptides requires the addition of N product, since they are encoded by genes whose transcription is blocked by terminators. By assaying for the synthesis of endolysin, the λ gene R product, N was purified to homogeneity and determined to be a small (14-kilodalton) basic protein. The next step in the analysis of N action was to fix the N protein to agarose and to pass an *E. coli* extract through this N-agarose complex. By this means, proteins with affinity for N were separated from other bacterial proteins. Two *E. coli* proteins bound to N, one of 69 kilodaltons and another of 25 kilodaltons. The identity of the latter is unknown, but the 69-kilodalton protein was previously isolated as a factor, called factor L, that stimulated the synthesis of a number of polypeptides in a coupled transcription-translation system (mentioned in Chapter 26). To further dissect the components of the N-mediated reaction, L protein was fixed to agarose, and bacterial factors with an affinity for L were isolated. Interestingly, RNA polymerase core enzyme (see Chapter 21) was isolated. The affinity of L for core is quite specific; L that binds to the core competes with the sigma factor protein (see Chapter 21). One interpretation of these data is that a complex between L and core polymerase is formed after transcription is initiated and the sigma subunit falls off the elongating RNA polymerase molecule. At some point, N must interact with this complex. How the complex becomes specific for phage operons is not revealed by the biochemical analysis, since the *nut* DNA sequence that confers this specificity does not appear to be an obligatory component of these reactions.

Preliminary experiments suggest that N will function in a crude transcription system that is free of ribosomes to antiterminate transcription of phage operons. This observation raises some questions when we examine the genetic approach to understanding N action.

**Genetic Analysis of N Gene Function.** <u>To identify the factors with which N interacts to suppress termination, a genetic analysis of mutations that affect the N reaction were isolated</u>. Since it had been observed that lambda N mutants do not kill their hosts, a series of bacterial mutants was selected as survivors of λ lytic infection or induction. Among these were strains in which the ability of N to act was blocked (see Table 27-4). Some of these mutations, called *nus*, for *N* undersupplied, were located in genes whose products would be expected to participate in termination. Thus the mutations *mar* and *nusC* lie in *rpoB*, which encodes the RNA polymerase β subunit; *nusD* is in the *rho* gene; and *nusA* is a mutation, of which there is but a single example, in L protein. The *nusB* mutation affects a 14-kilodalton polypeptide whose biochemical role is not known, but which may be required for bacterial growth.

One bacterial mutation that blocks N activity is difficult to reconcile with the in vitro analysis. The *nusE* mutation resides in ribosomal cistron *rpsJ* and alters protein S10 of the small ribosomal subunit. The existence of this mutation supports the notion that the ability of ribosomes to block transcription termination by distorting the structure of mRNA might be

**Table 27-4**

*nus* Mutations Block N Action and Reveal Functions Necessary
for N Action

| *nus* Gene | Product | Known Interactions and Function |
|---|---|---|
| *A* | L factor (69 kilodaltons) | Transcription-translation stimulated in vitro<br>Affinity for RNA polymerase core enzyme<br>Transcription termination |
| *B* | 14-kilodalton polypeptide | Unknown |
| *C, Mar*\* | RNA polymerase subunit | Transcription<br>Interacts with the product of *NusA* |
| *D* | Rho | Transcription termination |
| *E* | Ribosomal protein S10 | Translation |

\*Custom forces us to use these symbols. Actually, the gene concerned in both cases is *rpoB*, as explained in the text.

involved in the mechanism of action of N. However, the in vitro experiments suggesting that N is functional in a ribosome-free transcription system are not easy to reconcile with this idea.

Starting with the *nus* mutants, it is possible to then select bacterial or phage suppressor mutations, i.e., bacterial mutations that permit λ growth in *nus* mutants or phage mutants able to grow in *nus* hosts. The bacterial mutation *nusB101* is a *nusB* mutation which, instead of preventing N action, appears to stimulate it. *nusB101* allows λ to grow on *nusA* mutants. *RA4* is a mutation in *rplK* which, like *nusB101*, permits λ growth in a *nusA* mutant; *rplK* encodes L11, a component of the 50 S ribosomal subunit. L11 is known to play a role in the stringent response to amino acid starvation (see Chapter 26) by interacting with the stringent factor. Like *nusE*, *RA4* suggests a role for ribosomes in N-mediated antitermination.

λ*punA* is a mutation in the lambda *N* gene that allows the phage to propagate in *nusA* hosts. A suppressor mutation can indicate the presence of a protein-protein interaction between the products of the mutant and suppressor genes. In the case of the *punA* mutation, the genetic evidence for interaction between N and NusA fits well with the biochemical data that indicates affinity between the L and N proteins (Figure 27-10).

### Early Genes under the Control of the $p_L$ Operon

*The expression of the immediate early genes N and cro is followed by the expression of two sets of early genes, one under the control of the $p_L$ promoter and the other within the $p_R$ operon.* Aside from N, no gene under $p_L$ control is vital for the lytic growth of λ under normal conditions, since the entire region can be deleted without preventing the phage lytic cycle. The function of many of these genes is not known, which, of course, increases their fascination. Some functions are identified only by a polypeptide product rather than by an enzyme activity or even a mutation. The order of known cistrons in the $p_L$ operon is shown in Figure 27-11; it is possible that other undetected cistrons also reside in the operon. The assignment of a polypeptide product to a phage region has been very useful in

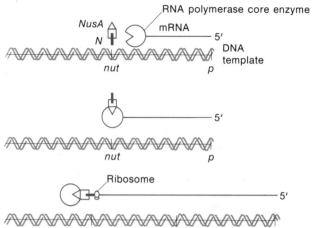

**Figure 27-10**
Hypothesis explaining the antitermination by N protein. (1) N-protein binds host function *NusA*. *NusA* binds RNA polymerase core enzyme. (2) Formation of termination-resistant complex occurs at *nut* site. (3) Transcription passes through terminator. Antitermination may involve *E. coli* ribosomes.

counting the number of cistrons in a particular region. This is accomplished by UV irradiating *E. coli* sufficiently to block transcription of bacterial DNA and then infecting with λ in the presence of radioactive amino acids. Under ideal conditions, only phage-encoded polypeptides become labeled; these may be analyzed after electrophoresis on polyacrylamide gels (Figure 27-12). Comparison of two different phages, one of which carries a DNA deletion, may reveal a difference in the polypeptide pattern, indicating which polypeptide is encoded by the deleted DNA. In those instances where *amber* mutations exist, the loss of a polypeptide and the recovery of an *amber* fragment gives an unequivocal assignment of a polypeptide product to a particular gene. Otherwise, the situation may be more complex. A deletion may result in the loss of several polypeptides, either because the deleted region contains several cistrons or, alternatively, because a control gene has been deleted. For example, deletion of gene *N* would result in the loss of all or a reduction in phage-encoded polypeptides except the product of the *cro* gene. More recently it has become possible to sequence large regions of DNA, and soon we may simply read the DNA sequence of λ and deduce the polypeptides it encodes from the open reading frames.

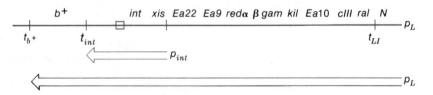

**Figure 27-11**
The lambda $p_L$ operon.

*Gene or Protein: Known Interactions and Function*

| | |
|---|---|
| *int, xis:* | Integration and excision functions (described in text) |
| *redα, redβ:* | Used for generalized recombination of λ; role in maturation |
| *ral:* | Restriction alleviation |
| Ea9 Ea10 Ea22 | General effects on host physiology; identified as polypeptides from this operon by deletion analysis |
| *gam:* | Inhibits *E. coli* recombination function *recBC*, as described in the text |
| *cIII:* | Stimulates lysogenic response |
| *kil:* | Host killing |

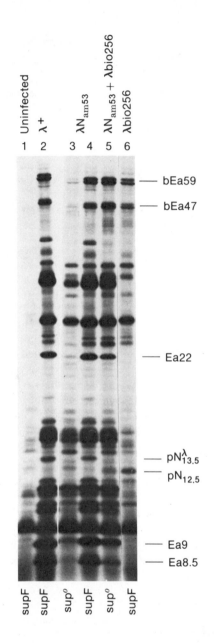

**Figure 27-12**
Proteins made in ultraviolet-irradiated *E. coli* after infection with lambda. Cells were irradiated, infected with phage, and labeled with $^{35}$S-methionine during the period 5 to 10 min after infection. The cells were lysed and the proteins were denatured and separated by electrophoresis on polyacrylamide gels. (*Lane 1*) Uninfected: Inactivation of DNA by UV irradiation blocks gene expression. (*Lane 2*) Infection with undamaged lambda DNA results in expression of phage genes. (*Lane 3*) Infection with a lambda *N* amber mutant, *sup⁰* host (i.e., containing no suppressor). Most phage proteins require N product to be expressed. (*Lane 4*) Infection of *sup⁺* (containing a suppressor) host with lambda *N* amber. *N* mutation is suppressed and phage proteins are synthesized. (*Lane 5*) Infection with lambda *N* and lambda *bio256* mutants. Lambda *bio256* is *N⁺* and complements lambda *Nam*. Note appearance of band *Ea22*. (*Lane 6*) Infection with lambda *bio256*. Absence of *Ea22* indicates that the *Ea22* gene is removed by the *bio256* deletion-substitution. (From Shaw et al., *Proc. Natl. Acad. Sci. USA 75:* 2225, 1978. Used with permission.)

## Early Genes under the Control of the $p_R$ Operon

In the $p_R$ operon, the immediate early *cro* gene is separated from a block of early genes including *cII* by the rho-dependent terminator $t_{R1}$ (Figure 27-13). The $t_{R1}$ terminator is suppressed by N function, but it is relatively inefficient; approximately 50 percent of the $p_R$-initiated transcripts pass into the early gene region in lambda *N* mutants. The early gene block is followed by a much stronger terminator, $t_{R2}$, which is also suppressed by N product. Because of the different terminator strengths, an *N* mutant, upon infecting a cell, replicates but does not kill the host cell; the killing functions lie distal to $t_{L1}$ and $t_{R2}$. This formation of an autonomously replicating plasmid is confined to *N* mutants and does not play a role in the normal physiology of the virus.

## DNA Replication

The products of the phage *O* and *P* genes specifically promote the replication of the phage chromosome, which replicates bidirectionally from a single origin. *Lambda replicates in two modes: (1) theta replication, by which monomeric circles replicate to yield monomeric circular products, and (2) rolling-circle replication, in which a monomer spins out a concatameric linear product* (see Figure 27-14 and Chapter 20). Rolling-circle replication is known to be strongly inhibited by exoV, a DNA exonuclease encoded by host genes *recB* and *recC*. For the phage to replicate by this mode, it must block the action of exoV. This is the role of the phage *gam* gene. Although normally, theta replication alone can sustain phage growth, there are conditions under which rolling-circle replication is required. Monomeric circles cannot be packaged into phage heads; they must recombine into dimers in order to appear in progeny phage. *In the absence of recombination, the inactivation of exoV by gam becomes essential for phage development.*

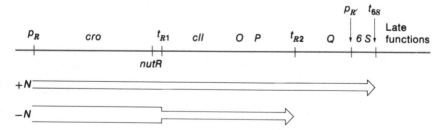

**Figure 27-13**
The $p_R$ operon.

| Gene | Known Interactions and Function |
|---|---|
| *cro* | Negative regulator of $p_R$ and $p_L$ functions (see text) |
| $t_{R1}$ | Transcription terminator; the termination site is leaky and allows some transcription to continue, even in the absence of N |
| $t_{R2}$ | Transcription terminator |
| *nutR* | N utilization site |
| *O, P* | DNA replication functions (see text) |
| *Q* | Positive regulator of late functions |
| *cII* | Positive regulator of repressor synthesis (see text) |

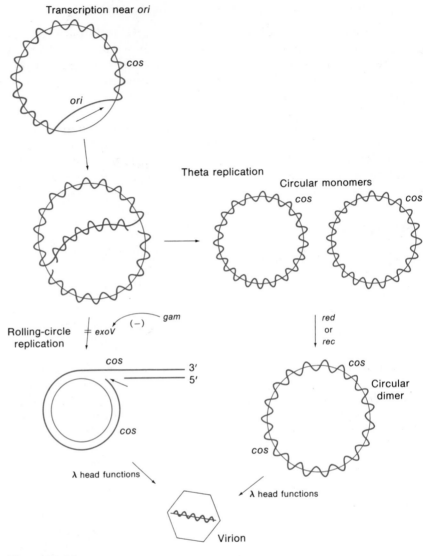

**Figure 27-14**
Two modes of λDNA replication: *ori* is the site for initiation of replication and *red* or *rec* refers to phage or host recombination functions, respectively; exoV is an exonuclease that attacks DNA at free ends; *gam* inactivates exoV; *cos* is the site of fusion of the free ends on mature DNA.

The $O$ and $P$ gene products are absolutely required for phage replication, in both the theta and in the rolling-circle modes. In addition, transcription in the region of the origin is required for replication; $p_R$-defective mutants fail to replicate even when supplied in trans by O and P proteins from a normal phage. Lambda DNA replication has not yet been duplicated in vitro, although the P protein has been purified to homogeneity. The isolation of P protein utilized an assay in which the single-stranded DNA virus φX174 served as template for DNA synthesis with purified *E. coli* replication proteins. P protein inhibits φX174 DNA synthesis by binding to the dnaB protein. The dnaB-P product complex is presumably active for λ replication but not for φX174 duplication. Analysis of phage replication in a set of lambdoid phages with different replication systems allows a dissection of the specificity of lambda O protein for the origin and for P product. Lambda can utilize $\phi80$ P function but not $\phi80$ O function to replicate. $\phi80$ can utilize neither lambda O function nor P function. Recombinants were made between the two phages having a portion of gene O from each. A hybrid O protein with the amino terminus of lambda O and the carboxyl terminus of $\phi80$ O will replicate $\lambda$, but only in the presence of $\phi80$ P function. The reciprocal hybrid, which carries the amino terminus of $\phi80$ O and the carboxyl terminus of lambda O, replicates $\phi80$ and can use either $\lambda$ or $\phi80$ P function. This analysis suggests that the product of the O gene binds to the phage origin by means of its amino-terminal portion; the carboxyl end presumably recognizes P product.

Lambda DNA replication requires the host genes *dnaE*, *dnaG*, and *dnaA* in addition to *dnaB* (see Chapter 20). Other *E. coli* genes also necessary for viral replication have been detected as bacterial mutants unable to sustain $\lambda$ growth. The products of these genes (*dnaJ* and *dnaK*) have not yet been identified.

## Late Gene Expression Is Positively Regulated by the Product of the Q Gene of λ

*Beyond tR$_2$ is gene Q, whose product is required for transcription of the λ late genes.* Q protein acts as an antagonist of transcription termination analogous to N protein (Figure 27-15). Q appears to act specifically to antiterminate the small constitutive (<u>constitutive</u> means produced at all times) 6 S transcript, allowing transcription to pass beyond the Rho-independent 6 S terminator and into phage regions encoding the head and tail genes (see Figure 27-18).

The transcription of $\lambda$ late genes is essentially uniform, initiating at or near the 6 S promoter ($p_R'$) and continuing without modulation rightward into the *b* region, where a terminator resistant to Q protein is located. It is not known if this site is also the locus of the terminator of leftward transcription, and there may be, in fact, a cluster of such sites. Possibly the function of the sites is to prevent transcription of the nonsense strands of DNA of the $p_L$ operon and of the late gene region of the $\lambda$ chromosome.

Although late gene transcription is uniform, the late gene products are synthesized at dramatically different rates. There are about 1000 times more E monomers (the major $\lambda$ head protein) than D monomers. This posttranscriptional regulation might operate at the level of mRNA stability, with transcripts encoding E protein being more resistant to degradation than D transcripts. Alternatively, the attachment of ribosomes to E transcripts might be more efficient than attachment to D transcripts. The detailed sequence of this region has not been reported, and even if the sequences at the translation initiation regions of the transcripts were known, it is not clear that one could relate them unambiguously to translation efficiency (see Chapter 25).

## THE LYSOGENIC PATHWAY

<u>*The decision to enter the lysogenic pathway rather than the lytic pathway is favored by conditions preventing good phage growth.*</u> Thus when the bacterial host is growing in a poor carbon source (meaning a carbon source

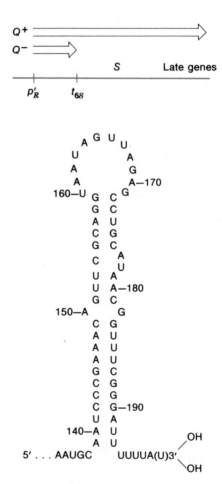

**Figure 27-15**
Positive regulation of late genes by Q. The structure of 6 S RNA at the termination site for 6 S RNA; $t_{6S}$ is shown. $p_{R'}$ is the constitutive promoter for 6 S RNA.

that is difficult to metabolize), or when the multiplicity of phage infection (refers to the ratio of phages to bacteria when the two are mixed) is high ($\geq 10$), the phage initiates the process of converting the infected cells into lysogens. The influence of the bacterial medium on λ development appears to be mediated by the level of intracellular cyclic AMP, which is highest when the carbon source is poorly metabolizable. Cyclic AMP regulates the transcription of many bacterial operons (see Chapter 26), either negatively or positively, but it has not been shown to be involved directly in the activity of any phage promoter. Consequently, it is believed that the expression of a bacterial function that antagonizes lysogenization is negatively regulated by cyclic AMP. A candidate for such a function is the product of the host *hfl* gene, which appears to inhibit lambda cII protein (the product of gene *cII* is essential for the lysogenic response).

The second factor favoring lysogenization, high multiplicity of infection, also works by influencing cII activity. The active form of cII is an oligomer. Because of this, and because cII is an unstable protein, the cII monomer levels produced on infection with one or a few phages are insufficient to generate active enzyme. Teleologically, this aspect of λ regulation makes excellent sense. The high ratio of phages to bacteria indicates the exhaustion of the viral food supply and the advantage of entering the quiescent prophage state until the restoration of optimal growth conditions or until the death of the lysogen appears imminent.

When the concentration of cII protein is sufficient, activation of two phage promoters occurs (Figure 27-16). These are $p_{RE}$ and $p_{int}$, which are responsible for the expression of *cI* and *int*, respectively. cII promotes transcription from these promoters in a fashion analogous to that of CAP protein (see Chapter 26) by binding to specific DNA sequences at or near the promoter and facilitating the binding of RNA polymerase or the initiation of transcription. Unlike CAP protein, however, no small molecule is required to activate cII. In addition to its positive regulatory role, cII, through its action at $p_{RE}$, inhibits the transcription of the $p_R$ operon. This inhibition is manifested as premature lysis by cII phage mutants. It might be caused by RNA polymerase binding at $p_{RE}$ and sterically hindering the rightward movement of RNA polymerases from $p_R$. Alternatively, attempts to transcribe DNA convergently lead to suppression of the least powerful promoter, which, in this case, is $p_R$. Evidence for both mechanisms exists.

The requirement for an additional factor to initiate transcription at $p_{RE}$ and $p_{int}$ is suggested by a glance at their sequences (Figure 27-16). Although they closely resemble each other, they share little homology with other known promoters, either in the $-10$ or in the $-35$ region (see Chapter 21). An analysis of mutations in the two promoters provides some insight into their properties. The $p_{int^c}$ mutation creates a more reasonable facsimile to a $-10$ consensus sequence and renders the $p_{int}$ promoter cII-independent. Evidently the absence of a consensus sequence in the $-35$ region is not fatal. The *cy* mutations, which eliminate $p_{RE}$ activity, were isolated as clear-plaque mutants (clear-plaque mutants indicate an inability to lysogenize) that were cis dominant. That is, they failed to transcribe the *cI* cistron and to synthesize λ repressor even when provided with *cII* and *cIII* functions by a coinfecting helper phage. From this, it was deduced that *cy* mutants represent alterations in the site at which cII and cIII proteins act, the $p_{RE}$ promoter. The sequences of many *cy* mutations are now

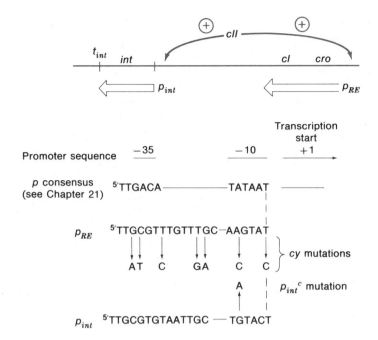

**Figure 27-16**
*cII* stimulates transcription initiation at $p_{RE}$ and $p_{int}$ promoters. (*Above*) The sites of stimulation. (*Below*) The promoter sequences compared with the most favored consensus sequence.

known and may be seen to fall into two categories. One class, *cyL*, appear to lie in the −10 region of $p_{RE}$. The *cyR* mutations are in the −35 region. The two classes are distinguishable physiologically. The *cyL* mutants are slightly responsive to cII, whereas the defect in the *cyR* mutants is not at all alleviated by cII protein. From this, and from the location of the mutations, it is likely that *cyR* defines the site at which cII protein interacts with DNA at $p_{RE}$, although other experiments will have to be performed to prove this hypothesis.

The high rate of synthesis of λ repressor after infection is readily explained by the efficiency of the $p_{RE}$ promoter in the presence of cII. The synthesis of int protein during the cII-dominated lysogenic development requires a closer analysis, since the *int* cistron is transcribed in its entirety during both the lytic and lysogenic cycles. Nevertheless, the *int* sequences in the $p_L$ transcript are not translated efficiently, while those in the $p_{int}$ transcript produce large quantities of int product. The expression of *int* when it is required, i.e., during the lysogenic response or after prophage induction, and not when its presence might prove harmful to phage reproduction, i.e., during the autonomously replicating lytic cycle, is accomplished as follows.

The $p_L$-initiated transcript is elongated beyond $t_{L1}$ under the influence of N product and is resistant to terminators (Figure 27-17). Because of this, it includes both the *int* gene sequences and sequences in the *b* region that are distal to a terminator, $t_{int}$. The cII-directed $p_{int}$-initiated transcript cannot extend beyond $t_{int}$ and does not enter *b* farther than about 170 bases beyond the attachment site. It is the inclusion of these sequences in the $p_L$ transcript that prevents the translation of the *int* mRNA. First, mutations lying in or beyond $t_{int}$ ("*sib*") permit int synthesis from the $p_L$ transcript. Second, after λ integration by means of recombination at the λ attachment site, a circular permutation of the phage genes occurs so that the *b* region is no longer included in the $p_L$ transcript. After prophage induction, int expression from $p_L$ does occur; such expression is required for prophage excision and the production of viable phages. *This novel mechanism of gene control by sequences promoter-distal to a gene has been termed retroregulation* (Figure 27-17). In the case of the *int* gene, retroregulation is a cis effect of the *b* sequences on the translation of the *int* sequences of the same transcript. In a ribonuclease III-deficient host, *int* retroregulation does not occur, suggesting that the *b* sequences may contain a recognition site for this endoribonuclease. An initial scission by RNAse III leads somehow to degradation of the *int* sequences (possibly by RNAse II, which is a 3' exonuclease); pulse-chase experiments indicate that int mRNA is some threefold less stable when associated with *b* sequences (Figure 27-17).

Possibly influencing the translation of *int* mRNA is the overlap between the *xis* and *int* cistrons. The 7-carboxy terminal amino acids of the Xis protein are translated within *int* mRNA in the −1 translation reading frame. The translation of *xis* mRNA could hinder the attachment of ribosomes to the Shine-Dalgarno sequence of *int* mRNA, thus modulating the expression of *int*.

In summary, integration is counterproductive for an infecting phage undergoing the lytic cycle. The *int* gene is transcribed from $p_L$, but not translated. In the lysogenic pathway, where integration is a necessity for phage replication, the presence of cII product activates $p_{int}$ and thus the expression of *int*. The $p_{int}$ transcript does not include the necessary sequences for the translation of *xis*, and so maximal expression of *int* is obtained. Prophage excision requires large amounts of Xis and lesser amounts of Int. The full *xis* cistron is included in the $p_L$ transcript and is efficiently translated. Sufficient expression of the *int* gene is ensured by the removal of the *b* region from the $p_L$ operon in an integrated lysogen.

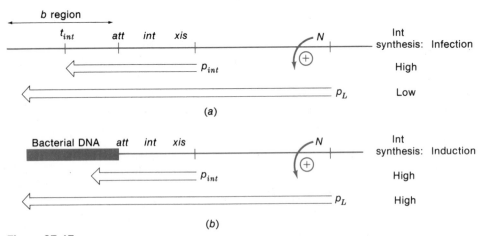

**Figure 27-17**
Retroregulation of *int* synthesis in (a) infecting phage or (b) prophage (after induction). In infecting phage, the *b* region is contiguous with the $p_L$ operon; $p_{int}$ transcripts terminate at $t_{int}$, $p_L$ transcripts extend further into *b*; *int* expression is high from $p_{int}$, low from $p_L$. In a prophage, the *b*-region sequences are not included in either the $p_L$ or $p_{int}$ transcripts, and the expression of *int* from both promoters is high.

# MAINTENANCE OF THE PROPHAGE STATE AND CONTROL OF *cI* GENE EXPRESSION

*During the lysogenic response, the $p_R$ operon comes eventually to be inhibited by two negative regulators, the cro and cI repressors.* With the shutoff of $p_R$ activity, the synthesis of cII product comes to a halt, and the levels of the unstable cII protein rapidly fall. The $p_{int}$ and $p_{RE}$ promoters return to quiescence. The high rate of *cI* expression characteristic of $p_{RE}$ is replaced by a much slower rate of cI product synthesis that is maintained in the integrated prophage state. The expression of the *cI* gene now derives from the only active promoter in the prophage, $p_{RM}$. The lower rates of cI product synthesis from $p_{RM}$ are not due to a difference in the strengths of the $p_{RM}$ and $p_{RE}$ promoters, but rather to the fact that the $p_{RM}$ transcript initiates much closer to *cI* than the $p_{RE}$ transcript and, in fact, lacks the Shine-Dalgarno sequence (ribosome binding site; Chapter 25) necessary for efficient translation of the *cI* cistron. This explains why the *cI* sequences of the $p_{RM}$ transcript are translated only about 10 percent as efficiently as when these sequences are present in the $p_{RE}$ transcript.

*The repression of prophage* λ *by the cI repressor results from a finely controlled series of DNA-protein interactions designed with two purposes: (1) to keep the prophage tightly repressed, and (2) to allow a rapid and irreversible response to inducing agents.* The cI repressor is a dimer composed of identical subunits. It binds by means of the amino-terminal end to the λ operators that regulate the activity of the $p_L$ and $p_R$ promoters. *The two operators, $o_L$ and $o_R$ are each composed of three binding sites with strong sequence homology.* The binding sites are 17 base pairs in length and display dyad symmetry (see Chapter 26 for the significance of dyad symmetry in repressor binding); they are separated by spacers of 6 to 7 base pairs. cI repressor binds to λ DNA in the major groove (see Chapter 18) and largely on one side of the helix.

*The affinity of the different sites for cI repressor is not identical, nor is the effect of cI repressor binding to the sites.* The formation constant $K_f$ for sites $o_{R1}$ and $o_{R2}$ is about $3 \times 10^8$ $M$; that for $o_{R3}$ is some 25 times less (see Chapter 7 for a definition of $K_f$). This difference means that under physiological conditions, $o_{R3}$ is probably not occupied. The apparent high affinity of $o_{R2}$ is the consequence of cooperative binding interactions between the cI repressor dimers at their carboxyl terminals. Thus, if $o_{RI}$ is eliminated by a mutation (called *vir*), the affinity of $o_{R2}$ for cI repressor falls to that of $o_{R3}$. Similarly, if the binding of cI repressor amino-terminal fragments to the operators is quantitated, $o_{R2}$ and $o_{R3}$ have about equal efficiencies of binding (Figure 27-18).

The binding of cI repressor to $o_{R2}$ and $o_{R1}$ inhibits the activity of $p_R$ by providing steric hindrance to the access of RNA polymerase to the promoter. The cI repressor bound at $o_{R2}$, however, stimulates the activity of $p_{RM}$ located within the operator cluster, presumably by an interaction between the bound cI repressor and the polymerase. Repressor is therefore its own inducer. Binding of cI repressor at $o_{R3}$ can be achieved at very high repressor levels; this blocks the activity of $p_{RM}$. This autogenous regulation ensures that the concentration of repressor is low enough (100 cI repressor monomers per cell) to allow for rapid induction under the appropriate circumstances but high enough to repress the prophage completely.

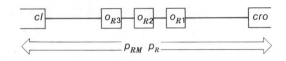

| Repressor Protein | Concentration of Repressor | Operator Sites | | | Effects on Promoters | |
|---|---|---|---|---|---|---|
| | | $O_{R1}$ | $O_{R2}$ | $O_{R3}$ | $p_R$ | $p_{RM}$ |
| cI | Low | + | + | − | ↓ | ↑ |
| | High | + | + | + | ↓ | ↓ |
| cro | Low | − | − | + | ↑ | ↓ |
| | High | + | + | + | ↓ | ↓ |

+ = site occupied; − = site unoccupied; ↑ = promoter stimulation; ↓ = promoter repression.

**Figure 27-18**
Control of the lambda immunity region by cI and cro proteins.

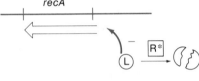

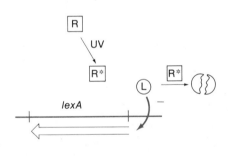

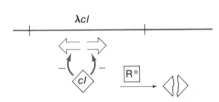

**Figure 27-19**
SOS response leading to prophage induction. ( R = RecA protein; R* = active protease form of RecA protein; L = lexA protein; cleaved inactive lexA protein; cI = λcI repressor; and = cleaved inactive cI repressor.) Colored arrows indicate transcription processes that are inhibited by either L or cI.

Also under the control of $p_{RM}$ is the expression of the *rex* gene, whose product blocks the growth of phage T4 *rII* mutants (T4 *rII* mutants can grow on ordinary bacteria, but not on lysogenic strains; T4 wild-type can grow on either type of bacteria). Little is known about the product of the *rex* gene; it appears to be dispensable for λ propagation either as a phage or in the prophage form.

## PROPHAGE INDUCTION

### Induction Is Initiated by Destruction of cI Repressor

The induction of prophage λ occurs in response to agents that damage the DNA of the lysogen. *Prophage induction is one of a set of responses to DNA damage known collectively as the SOS response.*

The *SOS pathway* involves the increased synthesis of at least a dozen different proteins. Induction of the SOS pathway occurs after exposure of bacteria to UV irradiation, mitomycin C, or a variety of carcinogens that produce chemical changes in the DNA. Direct interference with DNA replication has the same effect; thymine starvation or inactivation of the bacterial DNA replication functions, such as *dnaB* or *lig*, also induces the SOS response. Finally, the introduction of damaged DNA into an undamaged bacterium (UV-irradiated F factor or phage P1) induces the SOS system.

The SOS response falls into two general categories: (1) the induction of bacteriophage λ and other temperate bacteriophages, and (2) the enhanced production of RecA protein, an increase in the rate of mutagenesis and in the ability to repair UV-damaged DNA, and extensive filamentation.

*The key element in the SOS response is the induction of recA protease activity brought about by some as yet undefined product of damaged DNA.* The "activated" RecA proceeds to cleave a limited number of specific repressor proteins (Figure 27-19). Lambda repressor is a target of activated RecA, and its proteolysis leads to prophage induction. The product of the *lexA* gene is also cleaved by activated RecA. LexA is a repressor of many bacterial genes: *recA*, *uvrA*, *uvrB*, *umuC*, *sfiA*, and a number of genes whose products are not yet defined, known collectively as the *din* (for <u>d</u>amage <u>in</u>ducible) genes. In addition, LexA protein is autoregulated; i.e., <u>it</u> inhibits its own synthesis. It is the interplay of these gene products that gives the SOS response. The product of the *sfiA* gene is apparently an inhibitor of septation; in the absence of septation, cells form long filaments. *UvrA* and *uvrB* are genes encoding DNA repair enzymes. The product of *umuC* is likely to be what is termed an "error-prone polymerase," i.e., some replication function that can utilize damaged DNA as template, but which frequently introduces a mismatched nucleotide into the product. The *umuC* function would then be mutagenic, but also capable of rescuing genetically UV-irradiated chromosomes. This function, as well as the other components of the SOS system, are under enthusiastic investigation at the moment, and it does not take great foresight to imagine how an understanding of SOS will be relevant to a great many other important biological processes.

*Prophage induction results from cleavage of the cI repressor by the E. coli recA product, which becomes an active protease after DNA damage.* The site of cleavage is between the amino-terminal and carboxyl-terminal domains of the cI repressor in a region of the protein that is rather extended. Mutations in this portion of the *cI* molecule can readily be isolated; these *ind* mutations render the prophage noninducible. The substrate for the RecA protease is the cI monomer, but because the repressor concentration is low, the monomer-dimer equilibrium ratio is about 1:2, and the destruction of a monomer leads to the further dissociation, and inactivation, of dimers. This has the effect of steepening the dose-response curve for prophage induction (see Figure 27-20).

## Cro Repressor Action Is Required in the Early Stages of Induction

*The two immediate phage responses to loss of cI repressor are the cessation of cI-repressor synthesis and the appearance of cro repressor under the control of the $p_R$ promoter.* The Cro repressor binds to the λ operators at a subset of the contact points utilized by the cI repressor (see Figure 27-18). *In contrast to cI repressor, it fills $o_{R3}$ first and then $o_{R2}$ and $o_{R1}$ at higher concentrations. The effect of either repressor at $o_{R3}$ is the same; they block the activity of $p_{RM}$.* In addition, Cro bound at $o_{R3}$ stimulates, albeit slightly, the activity of $p_R$. When the $o_{R2}$ and $o_{R1}$ sites are filled by Cro, the activity of $p_R$ is inhibited, and thus the expression of the *cro* gene is autoregulated. However, the concentration of Cro during induction is sufficient to inactivate $p_{RM}$ and to prevent cI-repressor synthesis; this effect makes the induction process irreversible.

*Repression of both the $p_L$ and $p_R$ promoters by Cro occurs during the lytic cycle and is required for phage production.* Unmodulated activity of the $p_L$ operon leads to the hyperproduction of the phage Ea10 polypeptide, and this interferes with both phage and host metabolism. Modulation of the $p_R$ operon is also essential for phage growth, since *cro* mutants deleted for Ea10 are nevertheless inviable. Very little is known about this aspect of λ regulation.

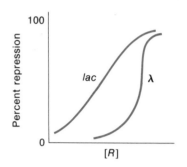

**Figure 27-20**
Comparison of *lac* operon and prophage lambda induction. Small changes in λ repressor concentration cause full prophage induction. *lac* induction requires much larger changes in *lac* repressor concentrations.

# PHAGE INTEGRATION AND EXCISION INVOLVE A PROCESS OF SITE-SPECIFIC RECOMBINATION

Since the autonomous replication of λ phage requires the products of the phage *O* and *P* genes, whose expression is repressed in a lysogen, *prophage λ must be propagated by the host replicational machinery. To do this, λ integrates into the host chromosome*, forming a covalent bond between the viral and host DNAs. Formation of infectious virions after induction requires detachment of the prophage DNA from the *E. coli* chromosome. Failure to excise does not block viral DNA replication or the expression of the viral genome, but because the prophage is circularly permuted, with the *cos* site located centrally, packaging splits the viral chromosome into two inviable particles.

Phage integration and excision are understood in considerable detail. They involve site-specific recombinations between the phage and host attachment sites, *attPOP'* and *attBOB'*, or hybrids of the two (see Figure 27-21). They are both reciprocal recombinational events; i.e., the participating DNA helices undergoing the exchange reaction appear as products without net loss of nucleotide bonds. No DNA synthesis is required. The enzymatic components are limited. DNA gyrase is needed to convert the λ DNA into a supertwisted circle, although this may be obviated in vitro by running the reaction at low ionic strength. This suggests that the role of superhelicity is to force the melting of the reacting portions of the phage chromosome, a situation that is duplicated in vitro by reducing salt con-

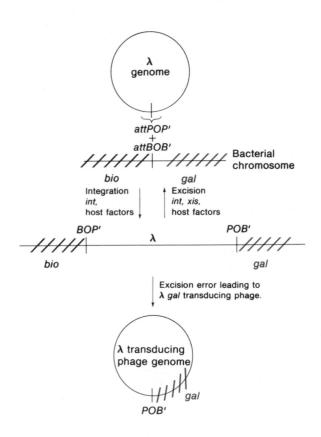

**Figure 27-21**
Lambda site-specific recombination. Region in color represents *gal* region of the bacteria.

centrations. A host protein called IHF, composed of two low-molecular-weight subunits, binds to the phage DNA at several sites within *attPOP'*. The product of the phage *int* gene ($M_r = 40,000$) binds both to the phage and to bacterial attachment sites, interspersing with IHF at the former. Finally, the product of the *xis* gene is required for prophage excision, but not for integration. In fact, it inhibits the integration reaction both in vitro and in vivo.

The analysis of site-specific recombination has been made possible by the isolation of transducing phages from lysates of induced lysogens. Rarely, an excision error is made and the phage genome is excised along with bacterial DNA to one side and/or the other of the integrated prophage (see Figure 27-21). The resultant λ*gal*- or λ*bio*-transducing phages are detected by their ability to convert *gal⁻* or *bio⁻* bacteria to wild-type. The two phage types can be recombined at their attachment sites to form phages that transduce both *gal* and *bio*.

The recombinational properties of attachment sites carried by the transducing phages are like neither the bacterial nor phage attachment sites. Recombination between λ*gal* or λ*bio*, which carry the left prophage attachment site *attBOP'* and right prophage attachment site *attPOB'*, requires Xis, as well as Int and IHF. This is not surprising, since it involves the same attachment-site pair as prophage excision. Some recombination events occur with very poor efficiency, i.e., *attBOP'* or *attPOB'* by *attBOB'*. The four sites are thus clearly different. However, all the recombinations have several features in common: all require IHF and Int, intermediates are not normally seen in the reactions, and the crossover points are the same.

The attachment sites have all been sequenced, and their structure confirms the genetic properties of the attachment sites and indicates the molecular basis for them. All attachment sites have a common core (O) of 15 base pairs within which the crossover takes place (see Figure 27-22). Breakage and exchange occur between nucleotides $-2$ and $-3$ on the upper strand and between nucleotides $+4$ and $+5$ on the lower strand, creating a 7-base-pair overlap (see Figure 27-22). The experimental evidence for placing the crossover points as indicated comes from some beautifully subtle genetic and biochemical studies. The former entailed sequencing attachment sites formed by the rare integration of λ at "abnormal" integration sites in the bacterial chromosome. These events yield attachment sites whose cores often diverge from wild-type. The divergence never occurs to the left of nucleotide $-3$ or to the right of nucleotide $+5$. The in vitro analysis rests on $^{32}$P "suicide" experiments between one highly radioactive parental molecule and a nonlabeled partner. (Note that the term suicide experiment was originally coined by G. Stent. After $^{32}$P has been incorporated into a DNA molecule, the molecule will self-destruct as a result of $^{32}$P decay. Stent used this to show that a single-stranded DNA virus such as bacteriophage ϕX174 is much more susceptible to suicide than a double-stranded DNA virus such as λ.) After attachment-site recombination was performed and the $^{32}$P atoms were permitted to decay ($^{32}$P has a half-life of about 15 days), the fragmented DNA was separated on sequencing gels. Nucleotides contributed by the labeled partner disappear from the pattern on the sequencing gels, defining the contribution of this parent. The results of this experiment were entirely consistent with the genetic analysis; the same crossover points were demonstrated by the two techniques.

Outside the cores, the sequences of the attachment sites are quite different. The phage *att* is rich with repeats, inverted repeats, a passable Pribnow box, and a potential terminator (see Chapter 21). The importance of the flanking *attPOP'* sequences is probably as recognition sites for Int and IHF. DNA protection studies indicate that Int binds at two sites to the left of the core, a site at the core-arm junction, and a site to the right of the core. IHF binds at sites in *attPOP'* that contain the sequence 5'-ATTGATA-3'. Deletion of the outermost binding sites alters the recombination properties of *attPOP'*, supporting the relevance of the binding studies to the in vivo reaction. Remarkably, the total number of nucleotides protected for DNAse digestion (a rough measure of the DNA region binding protein) by Int and IHF totals some 250 base pairs from −150 to +100. The importance of the core in the recombination reaction is in providing homology between the attachment sites. That the nucleotides in the 7-base-pair overlap region must be identical in the reacting attachment-site pairs comes from the studies of attachment-site core mutants (Table 27-5). One such mutant, isolated from a secondary attachment site, bears three nucleotide changes in the overlap region; it recombines very poorly

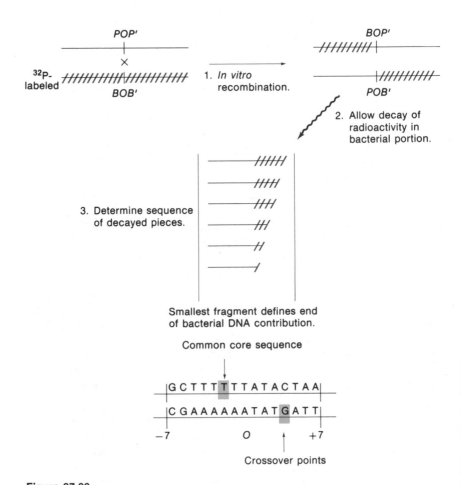

**Figure 27-22**
Determination of cross-over points in lambda *att*. Region in color represents the radioactively labeled DNA.

**Table 27-5**
Site-Specific Recombination: The Effects of Changes in
the Recombination Site

| | Frequency of Recombinant Phage | |
| --- | --- | --- |
| | *attBOB'* | *attBOB'saf* |
| *attPOP'* | 11 | 0.1 |
| *attPOP'saf* | 0.09 | 8.1 |

Note: The *saf* mutation changes three nucleotides in the overlap region of the λ attachment site. Thus *attPOP'* and *attBOB'* contain identical sequences in the overlap region of the attachment site; *attPOP'* and *attBOB'saf* differ by three base pairs, while *attPOP'saf* and *attBOB'saf* have identical overlap sequences.

with a wild-type attachment site. Two core mutants of this type recombine well, however, indicating that homology rather than sequence recognition is vital. More mutants need to be analyzed before the role of each nucleotide in the reaction can be assigned, but that prospect does not seem too far off.

Although all the enzymatic components of the reaction have been purified to homogeneity, the specific role that each plays in recombination is not yet known. Int will carry out a nonspecific topoisomerase reaction under incubation conditions that do not favor recombination; a low level of specific nicking-closing activity at attachment sites also can be demonstrated. The function of IHF is, if anything, even less evident. Its role is not confined to integration and excision. *E. coli* mutants defective for IHF (*himA* and *hip*) are viable. They show phage-related phenotypes (in addition to failing to undergo site-specific recombination); λ forms a clear plaque on *himA* and *hip* mutants, indicating an absence of cI-repressor synthesis (note that normally λ gives turbid plaques, which reflects the fact that some infected cells lysogenize, whereas others lyse. Clear plaques are formed when no lysogeny takes place). The effects of the *himA* and *hip* mutations on host physiology are hidden. The mutants are not recombination-deficient for homologous recombination of the phage or host; perhaps some other aspects of genetic recombination will prove to depend on IHF.

## SELECTED READINGS

Craig, N. L., and Roberts, J. W. *E. coli* recA protein-directed cleavage of λ repressor requires polynucleotide. *Nature* 283:26, 1980.

Gellert, M. DNA topoisomerases. *Ann. Rev. Biochem.* 50:879, 1981.

Hershey, A. D. (Ed.). *The Bacteriophage Lambda*. Cold Spring Harbor Laboratory, Cold Spring Harbor, N.Y., 1971.

Little, J. W., and Mount, D. W. The SOS regulatory system of *Escherichia coli*. *Cell* 29:11, 1982.

*Lambda Two*. Cold Spring Harbor Laboratory, Cold Spring Harbor, N.Y., in press.

Nash, H. Integration and excision of bacteriophage λ: The mechanism of conservative site-specific recombination. *Ann. Rev. Genet.* 15:143, 1981.

Ptashne, M., Jeffrey, A. Johnson, A. D., Maurer, R., Meyer, B. J., Pabo, C. O., Roberts, T. M., and Sauer, R. T. How the λ repressor and *cro* work. *Cell* 19:1, 1980.

Ward, D. F., and Gottesman, M. E. Suppression of transcription termination by phage lambda. *Science* 216:946, 1982.

## PROBLEMS

1. A set of mutations has been isolated in a region of a phage chromosome by different types of mutagenesis. Two different mutants were coinfected into cells, and the phages released were titered for total phage and wild-type recombinants. The results of such an experiment are given at the bottom of the page (the top number = total phages per milliliter; the bottom number = wild-type recombinants per milliliter).
   a. Which pairs of mutations show complementation? How many genes are defined by this set of mutations?
   b. Which mutations show no recombination? What might one conclude from such a result?
   c. Construct a possible genetic map for this region of the phage chromosome.

2. A set of strains has been constructed in which the *lac* operon can be transcribed from a prophage λ promoter; the cognate *lac* promoter has been removed. Different amounts of prophage DNA are present in each strain. No prophage is $N^+$, although N function can be supplied *in trans*. Indicate what prophage components remain, based on the ability of the strain to make β-galactosidase.
   a. Strain 1 is *lac*$^+$ only after UV treatment. N product is not required.
   b. Strain 2 is *lac*$^+$ without UV treatment. N product is required.

   c. A *rho*$^-$ derivative of strain 2 is *lac*$^+$ without UV treatment and no longer requires N product.
   d. Strain 3 when *rho*$^-$ is *lac*$^+$ without UV treatment and N product is not required. When *rho*$^+$, it is *lac*$^-$ even when N product is present.

3. Structures of O polypeptides are shown schematically to indicate segments derived from λ (*line*) and from φ80 (*box*). The amino-terminal segment is to the left. Fill in the missing spaces.

|  | O Protein | *ori* Specificity | *P* Specificity |
|---|---|---|---|
| λ | ——— | | |
| φ80 | ▭ | | |
| λ:φ80 | ——▭ | | |
| φ80:λ | ▭— | | |

4. Four different types of λ clear mutations are known. While each forms a clear plaque and very few, if any, lysogens by itself, some pairs of clear mutations can complement to produce lysogens. For the following pairs of phage infections, predict from the roles of the loci as described in the text whether complementation will occur (i.e., will lysogens form?). If so, which input phage will be found in the resulting lysogen?

| | | | | **Mutants** | | | | | |
|---|---|---|---|---|---|---|---|---|---|
| | **A** | **B** | **C** | **D** | **E** | **F** | **G** | | **Wild-Type Phage** |
| **A** | $\dfrac{<10^2}{<10^2}$ | | | | | | | | |
| **B** | $\dfrac{<10^2}{<10^2}$ | $\dfrac{<10^2}{<10^2}$ | | | | | | | |
| **C** | $\dfrac{<10^2}{<10^2}$ | $\dfrac{10^{10}}{10^7}$ | $\dfrac{<10^2}{<10^2}$ | | | | | | |
| **D** | $\dfrac{10^6}{10^6}$ | $\dfrac{10^9}{10^7}$ | $\dfrac{10^6}{10^6}$ | $\dfrac{<10^2}{<10^2}$ | | | | | |
| **E** | $\dfrac{10^{10}}{10^7}$ | $\dfrac{10^{10}}{10^7}$ | $\dfrac{10^{10}}{10^8}$ | $\dfrac{10^{10}}{10^7}$ | $\dfrac{10^3}{10^3}$ | | | | |
| **F** | $\dfrac{<10^2}{<10^2}$ | $\dfrac{10^{10}}{10^6}$ | $\dfrac{<10^2}{<10^2}$ | $\dfrac{<10^2}{<10^2}$ | $\dfrac{10^6}{10^6}$ | $\dfrac{<10^2}{<10^2}$ | | | |
| **G** | $\dfrac{10^6}{10^6}$ | $\dfrac{10^6}{10^6}$ | $\dfrac{10^{10}}{10^7}$ | $\dfrac{10^{10}}{10^7}$ | $\dfrac{10^{10}}{10^8}$ | $\dfrac{10^{10}}{10^6}$ | $\dfrac{<10^2}{<10^2}$ | | |
| **Wild-type** | $\dfrac{10^{10}}{10^{10}}$ | | | | | | | | $\dfrac{10^{10}}{10^{10}}$ |

a. λ*cI* and λ*cII*
b. λ*cII* and λ*cIII*
c. λ*cy* and λ*cI*
d. λ*cy* and λ*cII*

5. It is known that Int and Xis are specific for a particular set of attachment sites. *Att* mutants wipe out integration and excision activity; sequencing of these mutants has indicated that they represent a single base-pair deletion in the run of A–T base pairs in the common core. Their lack of activity might be due to destruction of homology or to lack of the appropriate *int-xis* recognition sequence. What experiments might one do to distinguish between these possibilities?

6. Indicate whether int and xis are synthesized by the following mutant and wild-type phages, following prophage induction or infection.

|  | Infection | | Induction | |
|---|---|---|---|---|
|  | Int | Xis | Int | Xis |
| a. λ*cII*$^+$*sib*$^+$ | | | | |
| b. λ*cII*$^-$*sib*$^+$ | | | | |
| c. λ*cII*$^+$*sib*$^-$ | | | | |
| d. λ*cII*$^-$*sib*$^-$ | | | | |
| e. λ*cI*$^-$*sib*$^+$ | | | | |
| f. λ*cII*$^-$$p_{int}$$^c$ | | | | |

7. Do lambda *int* mutants make clear or turbid plaques? Explain.

8. Explain the following:
a. Lambda lysogens are not induced by UV irradiation in *recA*$^-$ strains.
b. *clind*$^-$, a mutation in the lambda *cI* gene, blocks UV induction of prophage.
c. *tif* is a *recA* mutation that causes prophage induction at elevated temperatures without UV irradiation.
d. Would a *tif* strain lysogenic for λ*clind*$^-$ be induced at 42°C? At 42°C after UV irradiation?
e. Certain *lexA* mutations result in the constitutive expression of most of the SOS response system. Prophage induction, however, is unaffected.

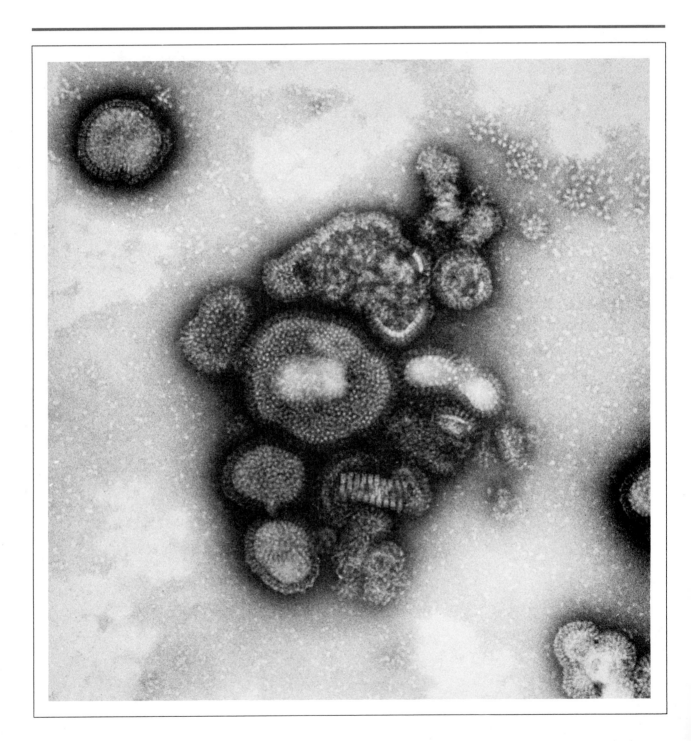

# 28

# BIOCHEMISTRY OF ANIMAL VIRUSES

For thousands of years, viruses were a scourge upon the animals that populated this planet—causing diseases in epidemic form as well as illnesses that took their steady toll in both suffering and life. So it is not surprising that for 60 years (following their discovery in the closing years of the last century), work with animal viruses was virtually synonymous with studies of the diseases they cause. The turning point was the development of methods for culture in vitro of susceptible animal cells. This ability to regulate infection in time and space opened the way to quantitative studies of the interactions of viruses with genetically homogeneous populations of cells free of interference from the host's immune system, and it allowed the use of the techniques of molecular biology to investigate aspects of virus-cell interaction that are inaccessible in whole animals.

## THE DIVERSITY OF ANIMAL VIRUSES

_Animal viruses display an extravagance of sizes and shapes_ (Figure 28-1). The simplest consist of a genome a few thousand nucleotides long usually associated with basic proteins and surrounded by a single protein shell, or capsid. The structural units forming the capsid are composed of one or a small number of polypeptides that are invariably virus-coded. When viewed in the electron microscope or by x-ray diffraction, the individual structural units (called _capsomers_) are often found to be arranged with icosahedral symmetry, i.e., in a shape that has 20 equilateral triangular faces, 12 vertices where the corners of the triangles meet, and 30 edges where the sides of the triangles touch. By definition, each triangular face must contain three asymmetric units. Thus a minimum of 60 individual

_Electron micrograph of particles of influenza virus, showing their varied shapes and fringe of surface spikes. (Courtesy Dr. N. Wrigley of the National Institute for Medical Research, London.)_

asymmetric units are needed to construct an icosahedron. In practice, the number of asymmetric units is often greater than 60. The rules by which these units can be packed within a triangular face are simple (Caspar, 1965) and provide elegant descriptions of the capsid organizations of many viruses. It seems that an icosahedral arrangement of capsomeres provides a highly economical and efficient way to transport viral genomes from one cell to another. Not only are the final structures highly stable, but they also have the potential for self-assembly, thereby eliminating or reducing the number of viral genes necessary for the mere mechanical packaging of particles. It is presumably because of these virtues that icosahedral coats are found in viruses as distantly related as polyoma virus (which contains a genome of double-stranded DNA), reovirus (segmented double-stranded RNA), and poliovirus (single-stranded RNA).

Popular though the icosahedral arrangement may be, it has not been universally adopted. Instead, many RNA viruses carry their genomes in the form of a flexible tube (*nucleocapsid*) consisting of structural polypeptide

**Figure 28-1a**
Representative DNA Viruses

| Name | Configuration | Size (nm) | Genome | Size of Genome (KB) |
|---|---|---|---|---|
| Poxvirus | 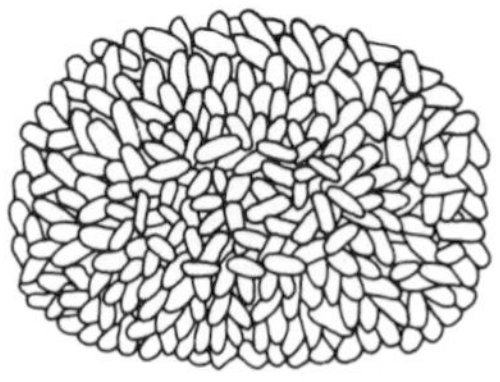 | 300 × 200 × 100 | Double-stranded DNA | 275 |
| Iridovirus | 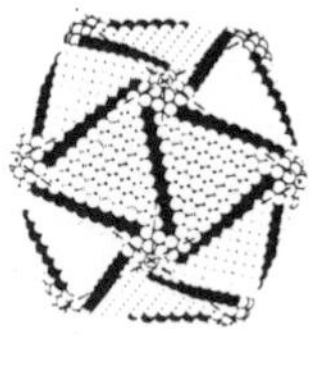 | Capsid 190 | Double-stranded DNA | 200 |
| Herpesvirus |  | Envelope 150; capsid 100 | Double-stranded DNA | 180 |
| Adenovirus | 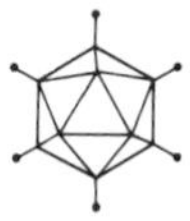 | 70–80 | Double-stranded DNA | 36 |
| Papovavirus |  | 55 | Double-stranded DNA | 5.2 |
| Parvovirus | | 20 | Single-stranded DNA | 1.6 |

subunits arranged in a helical fashion around the RNA. Invariably the nucleocapsids are themselves packed within a lipoprotein envelope—a modified form of plasma membrane whose lipid components are derived from the host cell and whose polypeptide components are usually virus-coded. This type of arrangement is common to viruses such as influenza and retroviruses that mature by budding through cellular membranes. Finally, there exist several kinds of viruses with large genomes (e.g., herpesviruses, poxviruses) that have correspondingly more complex structures, containing many different polypeptides arranged in several coats surrounding the genome. Sometimes an envelope also is present.

*It is on the basis of their morphology, and on the size and nature of their genomes, that animal viruses are classified into families and genera.* Each genus has a rubric that describes its members. Adenoviruses for example are characterized as nonenveloped, icosahedral viruses with a capsid 80 nm in diameter consisting of 252 capsomers with fibers projecting from the vertices and enclosing a double-stranded DNA molecule about 36,500

**Figure 28-1b**
Representative RNA Viruses

| Name | Configuration | Virion Size (nm) | Genome Size (KB) | Number of RNA segments | Virion-Associated Transcriptase |
|------|---------------|------------------|------------------|------------------------|--------------------------------|
| Paramyxovirus (e.g., measles) |  | 100–300 | 2 | 1 | + |
| Orthomyxovirus (influenza virus) |  | 80–120 | 22 | 8 | + |
| Retrovirus |  | 100–200 | 9 | 2 identical copies per particle | + |
| Rhabdovirus | 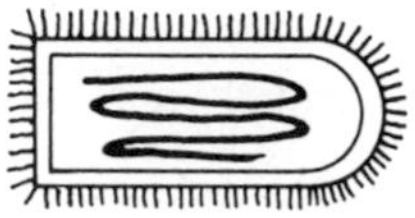 | 70–200 | 12 | 1 | + |
| Reovirus |  | 70–80 | 21 | 10 | + |
| Picornavirus (e.g., poliovirus) | | 20–30 | 7 | 1 | − |

nucleotide pairs in length. However, orthomyxoviruses, which include influenza virus, can be described as a genome consisting of eight separate pieces of RNA with a total length of 22,000 nucleotides, each piece wrapped as a helical nucleocapsid. The surrounding lipoprotein envelope is 80 to 120 nm in diameter and contains a virus-coded hemagglutinin and a neuraminidase. It is remarkable that such simple descriptions should provide an accurate system of classification. Nevertheless, it is a fact that viruses with similar morphologic features display similar physiologic and pathogenic properties. For example, adenoviruses have been isolated from species of animals as diverse as frogs, turkeys, and humans and from parts of the world as distant as Auckland and Anchorage. Yet, irrespective of their origins, they display the same morphologic features, interact with cultured cells in similar ways, possess genomes whose organization is correspondent even down to small details, and induce much the same sorts of disease in their diverse animal hosts. The individual viruses within the genus are distinguished by such secondary characteristics as host-range and serologic properties. Similar statements can be made for each of the 30 or so genera of animal viruses so far recognized.

The depth of knowledge of the molecular biology of animal viruses varies widely from genus to genus. Some, particularly those that grow poorly in tissue culture or are highly pathogenic, remain almost completely unexplored. Others, particularly those which are possibly implicated in or provide credible model systems for tumorigenesis, have become the objects of intense investigation. Thus any comprehensive account of the biochemical behavior of animal viruses would be both highly uneven and extremely lengthy. Fortunately, however, *it is possible to group together various genera of animal viruses according to the strategy they use to express their genomes in infected cells*. This simplifying concept, first described by David Baltimore in 1971, provides the framework on which the rest of this chapter is based.

## STRATEGIES USED BY ANIMAL VIRUSES

The process of infection of animal cells by animal viruses consists of many individual events that can be clustered as follows:

1. Absorption of virus to cells, uncoating of the viral genome, and its subsequent transport to the appropriate cellular location.
2. The synthesis and translation of viral mRNAs.
3. The replication of viral genomes.
4. The assembly of progeny particles and their release from the cell.

The first and last of these are formally equivalent to the disassembly and assembly of viral structural units. Consequently, the details of the processes reflect the morphology and complexity of the particular virus and are almost idiosyncratic for the genus involved. The central events of infection from a molecular biologist's point of view are the replication and expression of viral genomes, and it is on these two functions that Baltimore has built his classification scheme.

No matter to what genus they belong, all animal viruses share one common property: they use cellular ribosomes and presumably also cellular initiation factors and tRNAs to translate their mRNAs. However, different viruses use very different mechanisms to synthesize mRNAs. It is these differences that allow the division of the animal virus kingdom into several classes.

## DNA-Containing Viruses that Replicate in the Nucleus

Conventional DNA viruses, such as papovaviruses and adenoviruses, have a double-stranded DNA genome and replicate in the nucleus of the host cell. The cellular RNA polymerase II is used to synthesize viral RNAs, which are subsequently capped, polyadenylated, spliced, and transported to the cytoplasm. The situation is only slightly more complicated with parvoviruses, which contain a single-stranded DNA genome. Following introduction into the nucleus of the host cell, the single-stranded DNA molecule is converted to a double-stranded form, after which parvovirus mRNA is transcribed in much the same way as described for papovaviruses and adenoviruses.

## DNA-Containing Viruses that Replicate in the Cytoplasm

The best-studied example is vaccinia, a member of the poxvirus genus. Because the cellular enzymes involved in transcription, capping, and so forth are wholly confined to the nucleus, DNA viruses that replicate in the cytoplasm are obliged to carry into the cell their own set of enzymes for the synthesis and processing of mRNA. Almost a dozen enzymes—including DNA-dependent RNA polymerase—have been demonstrated within purified vaccinia virions. The enormous size and complexity of poxviruses (Figure 28-1) is easily understood in this light.

## Conventional RNA Viruses with a Single-Stranded RNA Genome of Positive Polarity

The term _positive polarity_ is used to describe mRNA or any DNA or RNA molecule that has the same sequence as mRNA; the DNA or RNA strand complementary to messenger RNA is said to have _negative polarity_. The best-known example of a positive-strand RNA virus is poliovirus, a member of the picornavirus family. Poliovirus is not obligated to synthesize mRNA in order to initiate an infection; rather, the genomic RNA released from virions is itself a functional message. As described in a later section, the molecule of RNA that is liberated from the virus particle becomes attached to cellular ribosomes and is translated into a single, large precursor protein that is subsequently cleaved to generate the mature viral proteins.

## Double-Stranded RNA Genomes and Negative-Stranded RNA Viruses

This category consists of viruses with *double-stranded RNA genomes* (reovirus) *and* viruses with a single-stranded RNA genome that is complementary to mRNA, i.e., *negative-stranded RNA viruses* (paramyxo-, orthomyxo-, and rhabdoviruses). In order to initiate synthesis of viral proteins upon infection, such viruses first must synthesize mRNAs. Cells are not in the habit of using RNA as a template for synthesis of more RNA; thus the host cell lacks the enzymatic machinery needed to transcribe the infecting viral RNA. Viruses in this category are obliged, therefore, to carry within the virion all the enzymes needed for synthesis of viral mRNA.

### Retroviruses

Finally, there are the *retroviruses*, which have a single-stranded RNA genome that after infection is converted by the virion-associated RNA-dependent DNA polymerase into a double-stranded DNA copy. The viral DNA integrates into the cellular genome and subsequently is transcribed much like any cellular gene.

From what has been said, it becomes clear that only a few classes of animal viruses use entirely conventional ways to synthesize mRNA. All the others, either because they replicate in places in the cell where the appropriate enzymes are not to be found (e.g., poxviruses) or because they use novel mechanisms to transpose information from one nucleic acid molecule to another (e.g., negative-stranded virion RNA to positive-stranded mRNA; virion RNA to double-stranded DNA), need to carry into the cell the necessary specialized enzymes. The discovery of virion-associated polymerases (made initially with poxviruses by Kates and McAuslan in 1967 and 3 years later with retroviruses), was far more than a laboratory curiosity. It marked an important step in rationalizing the classification of animal viruses as a whole and provided a firm intellectual framework for the study of their molecular biology. In the rest of this chapter we will consider in more detail the biochemical events that occur in cells infected with representative members (adenoviruses and SV40, poliovirus, influenza virus, and retroviruses) of four of the major classes of viruses just defined.

## ADENOVIRUSES

More than 80 different adenoviruses have been isolated—all with remarkably similar physical properties (see Tooze, 1980). The host-range of each of these viruses is restricted to one or, at most, a few closely related animal species. It is therefore possible to divide the genus into groups consisting of human, simian, bovine, canine, murine, avian, and frog strains according to the natural host species. These host-specific groups are then subdivided by serologic tests. By this means, 36 different serologic strains of human adenovirus have been identified; these serotypes have in turn been organized into affinity groups that differ from each other in length of DNA, G + C content, size of virion proteins, oncogenicity in experimental animals, and epidemiologic behavior.

No matter what their serotype or natural host, all adenoviruses contain a genome of about 36,500 nucleotide pairs that has two unusual features: the 5' ends of both strands of DNA are covalently attached to a protein, and the terminal hundred or so nucleotides at each end of the genome are inverted duplications of one another. As we shall see, both these rather bizarre arrangements reflect the mechanism the virus uses to replicate its DNA.

## The Permissive Cycle

Irrespective of their classification or serotype, the genomes of adenoviruses are organized in a similar fashion. However, most biochemical studies have been performed with human adenovirus serotype 2 growing in cultured human cells, and this virus has come to be regarded as the prototype of all adenoviruses. The time course of infection is shown in Figure 28-2. The productive cycle may be thought to consist of two major phases: early and late. The early-to-late transition occurs at about 10 h after infection and coincides with the onset of viral DNA synthesis.

*During the early phase of infection, cellular metabolic processes (DNA, RNA, and protein synthesis) gradually decline as the quantity of viral gene products slowly rises.* Hybridization of RNA extracted from infected cells to isolated segments of the adenovirus genome (obtained by digestion of viral DNA with restriction endonucleases) shows that only a small portion of the viral genome is transcribed at early times after infection. The early viral mRNA is complementary to seven far-flung portions distributed between both strands of viral DNA (see Figure 28-3). With the exception of the newly discovered early region 2B, which is transcribed from the remote promoter preceding region 2A, transcription of each of the early gene blocks begins at a separate promoter. RNA polymerase (probably

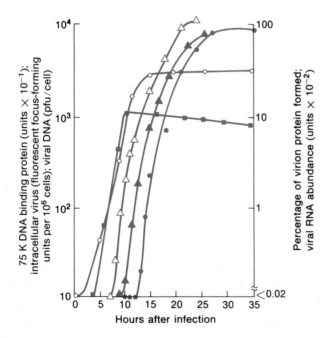

**Figure 28-2**
Time course of adenovirus–2 productive infection in suspension culture of KB cells; ( ● ) intracellular virus measured as fluorescent focus-forming units per $10^6$ cells; ( ○ ) total virus-specific RNA measured by hybridization of labeled RNA to adenovirus-2 DNA; ( △ ) synthesis of viral DNA (data from Green et al., 1971); ( ▲ ) virion protein (hexon antigen) measured by complement fixation; ( ■ ) 72K DNA binding protein. (Redrawn from Philipson and Lindberg, 1974.)

host RNA polymerase II) presumably binds at or near each of these promoters, transcribes the entire length of the region, and then terminates. *The primary transcripts from each of the early regions can be spliced in a variety of ways to yield alternate species of mRNA that code for different polypeptides.* During splicing, the capped 5′ end segment and the polyadenylated 3′ end segment are always conserved and one or more internal tracts of RNA are deleted. As in all eukaryotic RNAs, the first two nucleotides of the intervening sequence are GU and the last two are AG. In the case of all early regions so far examined in detail, the internal tracts that are deleted always contain protein termination codons. Thus the several viral proteins coded by an early region share peptide sequences with one another. The catalog of early mRNAs and proteins has continued to expand as more and more minor species are discovered. With one exception, however, the functions of the early proteins are completely unknown. The interesting exception is a 72,000 dalton (72 K) DNA-binding protein coded by early region 2A.

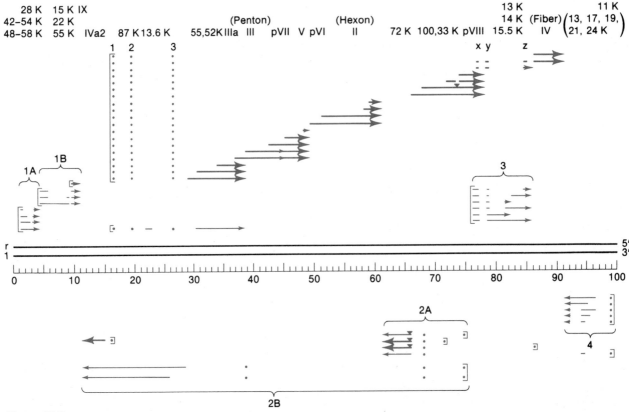

**Figure 28-3**
A functional map of the human adenovirus-2 genome. The 36,500 nucleotide-pair chromosome is divided into 100 map units. Arrows indicate the direction of transcription. Vertical brackets mark the location of promoters. Thin arrows represent early mRNAs; thick arrows represent late mRNAs. Gaps in the arrows represent intervening sequences removed from the mRNAs by splicing. Early regions 1A, 1B, 2A, 2B, 3, and 4 are labeled. The tripartite late leader segments are labeled 1, 2, and 3. The correlation between mRNAs and proteins is derived from cell-free translation of purified mRNAs; the proteins are listed at the top of the figure. The DNA strands are labeled *r* or *l* to indicate the direction of transcription.

For many years it was believed that each of the early regions worked not only autonomously but independently of the others. Recent evidence makes this view untenable. First, there is a definite temporal sequence in which the early transcription units become active. Thus the very first transcripts detected are those whose 5′ ends map at position 16.5; these are closely followed by transcripts of early region 1A and then by those from the early regions 1B, 3, and 4, with early region 2 always the last to be copied. Furthermore, mutations in early region 1A completely abolish the expression of early regions 1B, 2A, 2B, 3, and 4. It therefore seems that adenovirus gene products coded between 1.3 and 4.4 map units (i.e., early region 1A) are required for the expression of other early genes. The molecular mechanisms by which this control is achieved are unknown.

Toward the end of the early phase of infection, additional regions of the viral genome begin to be expressed, particularly the genes lying on the l strand between 16.2 and 11.1. *However, it is not until viral DNA synthesis gets underway that the full weight of late viral transcription is felt* by the cell. The most dominant late transcript by far is that which originates from the major late promoter at 16.5. Sluggishly active early in infection, this promoter becomes intensely active at late times. It is recognized specifically by RNA polymerase II and controls a large transcription unit that encompasses all the major late genes coding for structural proteins. Transcripts initiating at position 16.5 stretch over 28,000 nucleotides to a termination site near the right end of the viral genome. During passage of the polymerase down this long tract, nicks are introduced at any one of five major and several minor sites; poly(A) is then added at these sites (represented by arrowheads in Figure 28-3). Splicing subsequently occurs in a way that conserves the capped 5′ end and the gene or genes proximal to the poly(A) addition site. Thus late mRNAs typically contain a tripartite leader about 200 nucleotides long composed of discrete shorter segments derived from map positions 16.6, 19.6, and 26.6. A small proportion of the late mRNAs are more complex in structure and contain more than three leaders. Whatever their number, however, the late leaders are always attached to the 5′ side of any of about 20 genes that comprise the main body of the late transcript. Many of these genes code for structural proteins, such as hexon, penton, and fiber, which form the icosahedral shell of the virus and carry its major antigenic determinants (see Figure 28-3).

*By comparison with early mRNAs, splicing of late mRNAs involves removal of vast tracts of intervening sequences.* Despite their great size, the boundaries of these intervening sequences are marked by the dinucleotides GU and AG, just like many other eukaryotic splice points, and it therefore seems unlikely that the general rules for adenovirus late splicing will turn out to be radically different from those used by conventional cellular mRNAs. The most plausible mechanism involves (1) looping out of the sequences between the leaders themselves and between the rightmost leader and the coding sequences of the nascent mRNAs, and (2) intramolecular recombination (see Figure 28-4a). This model requires specific parts of the primary transcript located several thousand nucleotides apart to be brought into close proximity. While this could be achieved in theory by intramolecular base pairing, the necessary complementary sequences do not seem to exist at the splice sites. Several people have therefore proposed that the upstream and downstream sites may be brought together by pairing

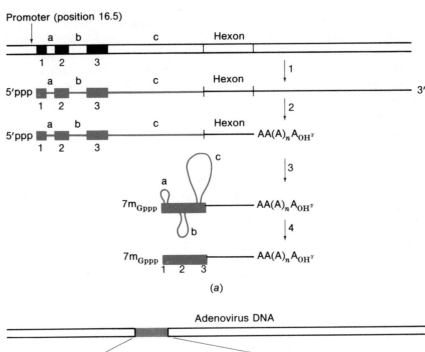

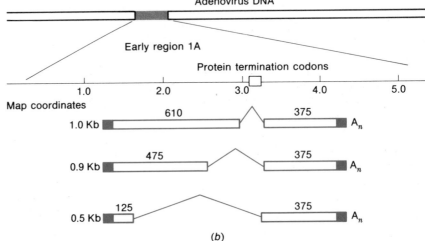

**Figure 28-4**

(a) Synthesis of late segmented adenoviral mRNA. The adenovirus-2 genome is represented by the pair of horizontal lines at the top of the figure, on which are indicated the locations of the major late promoter, the three sets of sequences present at the 5' terminals of all late mRNAs (1, 2, and 3), and the hexon gene. In step 1, transcription yields a large RNA molecule initiated at this promoter and containing the 5' terminal sequence. This is first polyadenylated at specific sites, in this example, at the 3' end of the hexon sequences. Polyadenylation may precede termination of transcription; i.e., after nicking of the transcript and addition of poly(A) (step 2), the polymerase may continue transcribing down to the end of the template. The sequences 1, 2, and 3 are then imagined to be brought together in some way such that the sequences not found in the mRNA (a, b, and c) loop out (step 3). Specific cleavage and ligation of this intermediate would then generate the mature mRNA (step 4). At some time, probably after polyadenylation and before splicing, the 5' terminus becomes capped. (From J. Tooze, *The Molecular Biology of DNA Tumor Viruses,* Cold Spring Harbor Press, New York, 1980. Reprinted with permission.) (b) Alternate splicing patterns of early region 1A: ■ = nontranslated regions; □ = translated regions; ∧ = splices. The figures over the RNAs represent the number of bases in each of the various segments of mRNA.

not with themselves, but with a separate effector molecule. The most likely candidates for such effectors are the small nuclear RNAs (snRNAs) that are ubiquitous in mammalian cells. The sequence of at least one of these small cellular RNAs is complementary to the consensus sequences at splice points, and it is possible that this complementarity is used to pull the two sides of the splice junction together. Such possible effectors may be coded not only by the cell, but also by adenoviruses. Two small genes that map around position 30 on the viral genome code for small RNAs called $VA_I$ and $VA_{II}$. By contrast to all other viral genes, these are transcribed by RNA polymerase III. Both species of RNA are made throughout infection, and by late times they are present in high concentration. The function of the RNAs is unknown, but it seems a reasonable guess that they may be involved in splicing.

## Control of Gene Expression by Splicing

Virtually all primary transcripts of the adenovirus genome are polycistronic in one of two respects: either they contain overlapping coding sequences that are sorted out by splicing, or they consist of a nonoverlapping series of potential coding regions arranged one after the other along the RNA (any one of these coding regions then can be brought into a position to be translated near the 5′ end of the final mRNA by removal of the intervening sequences).

In general, the early genes give rise to transcripts of the first class. Early region 1A provides a good example of the versatility that this arrangement can bring. The region encodes a set of closely related proteins that are translated from three mRNAs generated by alternative patterns of splicing of a common primary transcript (see Figure 28-4b). This transcript contains two long coding regions in different frames that are separated from one another by a cluster of protein synthesis termination codons near position 3.1. Splicing removes these codons, connects the two coding regions, and brings them into the same reading frame. The primary transcript contains a single splice acceptor site (at position 3.3) that can be joined to any one of three donor sites. In this way, three different mRNAs are fashioned, and they code for proteins identical in sequence at their N and C terminals but different in size (and presumably in function). Finally, the fact that the relative abundances of the different region 1A mRNAs change during the course of infection suggests that control of gene expression may be mediated in part at the level of splicing. The mechanism by which different coding sequences are preferentially selected into mRNA is unknown.

There is evidence that regulation of late gene expression also may be mediated at the level of splicing. When human adenoviruses infect simian cells, the infection is aborted at a late stage because of a lack of fiber protein. Very few transcripts are detectable that contain fiber sequences, and even those which are made have an aberrant structure. Long tracts of nucleotides between the tripartite leader and the sequence coding for the fiber protein are not removed. Because these extra coding sequences are adjacent to the 5′ leader, the aberrant transcripts probably would not function as fiber mRNA during translation. Therefore, the infection aborts because of a specific inability of simian cells to splice a particular viral mRNA. Interestingly, adenoviral mutants have been isolated that can grow in simian cells. The mutations restore the full production of fiber mRNA, allow the normal elimination of intervening sequences, and map in the genetic region encoding the DNA-binding protein. This is the first indication that a viral protein may be directly involved in splicing.

In vitro studies using adenovirus DNA and cell-free extracts competent to carry out RNA synthesis indicate that initiation of RNA synthesis from the major late promoter does not require the presence of other viral RNA or protein products. Initially this result was perplexing since in vivo results show that expression of late proteins does not occur in early infection. In fact, the major late promoter is active soon after infection. However, the resulting transcripts abort near map position 60–70 in contrast to the situation late in infection, when elongation continues to map position 99 (see Figure 28-3). Furthermore, the transcript that is made early is processed differently, with respect to both polyA site selection and splicing in such a way that functional messengers for late proteins are not formed.

Thus it appears that late gene expression is regulated by the mode of elongation and processing of the main late transcript and has little if anything to do with the rate of initiation of the transcript.

### DNA Synthesis

Synthesis of adenovirus DNA begins about 8 h after infection and reaches a maximum approximately 8 h later. Initiation of DNA synthesis begins asynchronously at both 5'-ends of the linear DNA molecule and proceeds in both directions until it reaches the other end (see Figure 28-5). It should be remembered that the 5' ends of the two DNA strands are mirror images of one another; each contains a protein linked to the 5'-end and an identical sequence of 100 or so bases.

In Chapter 20 it was indicated that DNA synthesis requires both a primer and a template. In the case of adenovirus DNA synthesis, the primer is not another nucleic acid chain but unusually a serine $\beta$-hydroxyl group of a viral-encoded protein known as adenovirus terminal protein (pTP). The initiation process begins with the formation of a phosphodiester linkage between the serine hydroxyl group of the pTP and dCTP. The pTP protein active during initiation has an $M_r$ of 80,000. During maturation this is cleaved to a protein with an $M_r$ of 50,000, and this latter protein is the one found attached to the 5' ends of the mature viral DNA.

In vitro replicating complexes of adenovirus DNA contain both the $\alpha$ and the $\gamma$ DNA polymerases (see Chapter 20) and both enzymes appear to be required for replication. The presence of dideoxythymidine triphosphate ($d_2TTP$), which inhibits $\beta$ and $\gamma$ polymerases but not $\alpha$ polymerase, severely inhibits replication in vitro of the viral DNA. This implicates the $\gamma$ enzyme in viral replication, since the $\beta$ enzyme is not present in the in vitro replicating complexes. The $\gamma$ enzyme is also involved in mitochondrial DNA replication, but not in nuclear DNA synthesis or SV40 viral DNA synthesis where the $\alpha$ enzyme is used for replication. Recent in vivo and in vitro studies show that the $\alpha$ DNA polymerase is also required for adenovirus DNA synthesis. The presence of aphidicolin, a highly selective inhibitor of $\alpha$ DNA polymerase, blocks adenovirus DNA synthesis, but only at substantially higher concentrations than are needed

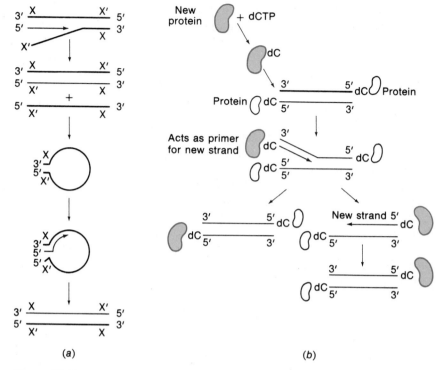

**Figure 28-5**
(a) A model of synthesis of adenoviral DNA in which synthesis of a new strand displaces the parental strand. If the sequences of the inverted terminal repetition base-pair form a panhandle intermediate, then the double-stranded panhandle has the same terminal structure as parental viral DNA and will presumably be recognized by the enzyme complex responsible for initiation of viral DNA synthesis. Although not shown here, *r* strands displaced during the first step in synthesis could obviously form an analogous panhandle intermediate, with the same terminal double-strand sequence. (b) A similar displacement mechanism in which the possible role of the 5′ terminal protein in initiation of adenoviral DNA synthesis is illustrated. (From J. Tooze, *The Molecular Biology of DNA Tumor Viruses*, Cold Spring Harbor Press, New York, 1980. Reprinted with permission.)

to inhibit SV40 or host DNA synthesis. Apparently the $\alpha$ enzyme is involved in a modified or protected form. One of the challenging problems of the future is to determine why both the $\alpha$ and $\gamma$ polymerases are required. As far as is known this is a unique situation with adenovirus.

## Assembly

By late times after infection, cellular macromolecular metabolism has been completely replaced by viral DNA, RNA, and protein synthesis. The cell's nucleus becomes stuffed with viral polypeptides, capsomers, and virus particles, both completed and partially formed. Unfortunately, little is known of the mechanism by which the major structural subunits of the virus shell are formed or how they interact with the viral genome to yield completed particles. However, it is known (1) that viral DNA is packed into particles starting from the left end, with the DNA sequence located 280 to 380 nucleotides from the end playing an essential role, and (2) that several structural polypeptides are cleaved proteolytically as the viral particles mature. Eventually, about 36 h after infection, the cells begin to disintegrate and the newly formed virus is liberated into the medium.

### Conclusion

The present state of affairs with adenoviruses is as follows. We understand a good deal about the way that adenovirus particles are organized and the way that an adenovirus genome is arranged. We also have a rough idea of the temporal order in which its genes are expressed. However, there are many important areas where our knowledge is scanty or nonexistent. For example, we have no idea why DNA replication appears to have a requirement for both the α and the γ DNA polymerases. Second, we do not know how viral gene expression is controlled. In the early phase of infection, it is clear that proteins coded by early region 1A are required for expression of early regions 2, 3, and 4. The details of this process are unknown. Even the most obvious change in expression of viral genes—the early/late shift that allows production of vast amounts of late proteins—is not understood in molecular terms. Finally, we do not know why the arrangement of genes in the adenovirus genome is so complex. Why, for example, are the early regions dispersed and placed in opposite orientations rather than being sequestered in one location, as is often the case with prokaryotic viruses? And why should almost all the late genes be placed under the control of a single, powerful promoter? These and similar questions bear on the origins of adenoviruses and their subsequent evolution—a topic that is too speculative to be discussed here.

## THE PAPOVAVIRUSES: STRUCTURE OF SV40 AND POLYOMA VIRUS GENOMES

Perhaps the most intensively studied of all animal viruses are two small viruses belonging to the papovavirus genus: polyoma and SV40. Extremely similar to one another in genomic size, arrangement, and expression, the two viruses also share many biological properties. Neither seems to cause any overt disease in its natural hosts (mice for polyoma virus; rhesus monkeys for SV40), but both induce tumors in experimental animals and, as we shall see, transform cells in vitro with high efficiency. Both viruses have small genomes, grow readily in cell culture, and are easy to purify. This combination of properties has proved to be highly attractive for molecular biologists, and for the last 10 years, polyoma virus and SV40 have been objects of intense interest.

The genomes of both viruses consist of a single molecule of double-stranded, covalently closed circular DNA, just over 5,000 nucleotide pairs in length. It will become obvious from what follows that polyoma virus and SV40 must be derived from a common ancestor. However, although some remnants of base sequence remain in common, the two viral genomes have diverged to such an extent that they are by and large nonhomologous.

The outer icosahedral shell of the virus particles, built in each case from three virus-coded proteins, encloses the viral chromosome, which consists of DNA complexed with host-derived histones in a familiar nucleosome arrangement. Removal of the histones causes the naked viral DNA to adopt a negative superhelical form, whose properties allow rapid purification by any of several standard techniques. Comprehensive restriction endonuclease cleavage maps of both viral genomes were obtained in the early 1970s, and this led rapidly to the deciphering of the complete nucleotide sequences and to the mapping on the genomes of the major functional domains.

## The Early Region

*The early region of SV40 occupies about half the genome and codes for two proteins; that of polyoma virus codes for three proteins* (see Figure 28-6a). The mRNAs encoding the early proteins seem to be identical at their 5' and 3' ends, but differ in their pattern of splicing. Of these various RNAs, that coding for SV40 small-t protein is the simplest (see Figure 28-6b). It is colinear with the viral genome from its 5' end (near map position 67) to approximately map position 54, where a chain-terminating codon occurs. Translation of this open reading frame yields a protein approximately 20 K in size. Downstream from the coding sequence is the splice junction formed by the removal of 66 nucleotides of intervening sequence, and there then follows a long untranslated region that is entirely colinear with the viral genome to the poly (A) addition site, which maps at position 17 on the viral genome. Thus the total length of the mRNA is some 2,700 nucleotides, of which only 456 are translated. Paradoxically, the second SV40 early mRNA is smaller in size but codes for a much larger protein—large-T antigen, about 90 K in size. Like small-t mRNA, its 5' end maps near position 67 on the viral genome. However large-T mRNA is colinear with the genome only as far as position 60 before a splice-junction occurs, which is formed by removal of an intervening sequence between map positions 60 and 54. From position 54, both small-t and large-T mRNAs are again colinear with the genome as far as the end of the early region. The intervening sequence, which is absent from large-T mRNA, contains terminators of protein synthesis in all three phases. The removal of this sequence leaves an mRNA which, unlike small-t mRNA, is translated essentially all the way to its 3' end at map position 17. The small-t and large-T proteins therefore share N-terminal sequences but differ at their carboxy terminals.

The early region of polyoma virus is slightly more complicated. In addition to small-t and large-T proteins, there is a third virus-coded protein called middle-T. It shares amino-terminal sequences with large-T and small-t antigens but differs at its carboxy terminus. The mRNA for middle-T polyoma virus is formed by splicing the common amino-terminal coding sequences to a second open reading frame that occurs immediately downstream from the small-t gene and out of phase from the coding sequences for large-T antigens (see Figure 28-6b). This arrangement, in which a segment of DNA contains coding sequences in two out of the three possible reading frames, is found in prokaryotic viruses such as φX174 and appears to reflect the virus' efforts to obtain the maximum use from its genome.

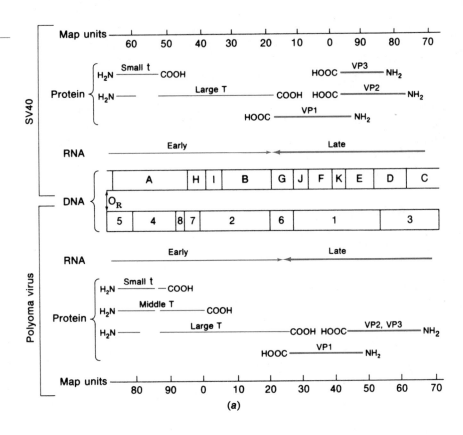

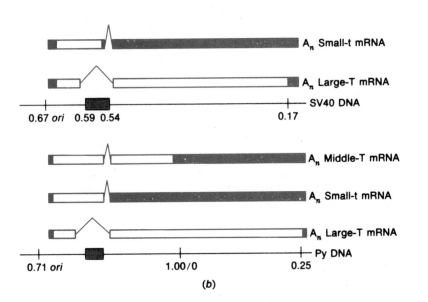

**Figure 28-6**
(a) Comparison of the maps of the SV40 and polyoma virus genomes in relation to the Hin dII / III cleavage maps. The genome is shown as a linear map starting at the origins of DNA replication ($O_R$). Also shown are the early and late transcription regions, the coding regions for viral proteins, and the map positions of mutations. (Adapted from M. Fried and B. E. Griffin, Organization of the genomes of polyoma virus and SV40, *Adv. Cancer Res.* 24:67, 1977. Used with permission.) (b) mRNAs found in cells infected by SV40 or polyoma virus: ■ = nontranslated regions; □ = translated regions; ∧ = splices; ▦ = regions of DNA deleted in viable early deletion mutants. (From P. Q. Rigby, The transforming genes of SV40 and polyoma, *Nature* 282:781, 1979. Reprinted with permission.)

## The Late Region

*Both SV40 and polyoma virus code for three proteins, VP1, VP2, and VP3,
which are structural components of the icosahedral capsid* and which are
synthesized in appreciable amounts only at late times during infection.
Together the three genes coding for these proteins occupy most of the
remaining half of the viral genomes and are arranged in the opposite orientation to the early genes (see Figure 28-6a). The late regions of polyoma
virus and SV40 both code for mRNAs that are heterogeneous at their 5′
ends and are therefore best regarded as families than as discrete species.
The reasons for this heterogeneity are two. First, there seem to be several
alternate sites at the beginning of the late region at which the first nucleotide of the late mRNAs can be laid down. Second, polyoma virus in particular seems to synthesize giant nuclear precursors to late mRNA consisting of
several complete tandem transcripts of the entire late strand of the viral
DNA. It is thought that such giant transcripts may be spliced into mRNA
in a way that causes repeats of the 5′ end sequences to be brought together
to form a reiterated leader (see Figure 28-7). Heterogeneity arises because
different mRNAs have different numbers of copies of the leader. Despite
these complexities, the arrangement of late genes is not difficult to understand. (1) The major structural protein, VP1, is coded by an uninterrupted
sequence of 1,092 nucleotides located in SV40 DNA between 97 and 17
map units. VP1 mRNA consists of these sequences spliced to an upstream
leader about 200 nucleotides in length derived from around position 67.
(2) The coding sequences for VP2 and VP3 roughly correspond to the intervening sequences mapping between the late leader and the coding sequence
for VP1. However, the common termination signal for VP2 and VP3 lies 110
nucleotides beyond the initiation signal for VP1. Hence the three genes

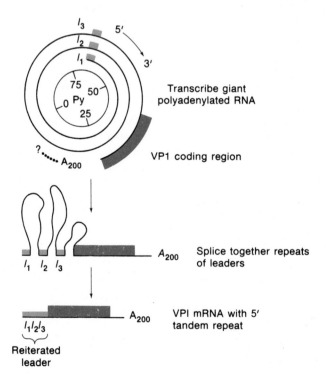

**Figure 28-7**
Generation of polyoma virus VP1
mRNA. Late in infection, polyoma
virus DNA is transcribed into large
nuclear RNA molecules that represent
several complete circuits of the viral
genome. The mechanism by which
this RNA is thought to be spliced into
mRNA is shown in the figure. The
mature mRNA contains a 5′ leader
with tandem repeats of a sequence
that occurs only once in the genome.

overlap, and an examination of the sequence shows that this small region of the SV40 genome codes for parts of three proteins in two different reading frames. (3) The entire sequence of VP3 is a subset of VP2. The exact structure at the 5' end of VP3 mRNA is not yet known.

## The Control Region

*The stretch of DNA between the 5' ends of the early and late mRNAs contains the controlling elements for both DNA replication and transcription in polyoma virus and SV40.* How these controlling elements work is not understood in detail, but the following general points are well established.

First, in viral chromosomes, this region (around position 67 in SV40, and position 71 in polyoma) is less well protected than the rest of the genome from digestion by nucleases. Perhaps this openness facilitates binding of polymerases and the proteins that control them.

Second, the control regions of polyoma virus and SV40 DNA are very similar in sequence. Such a high degree of local homology, which extends also to other members of the papovavirus group, cannot be the result of chance. It reflects not only the common origins of the viruses, but also suggests that the control regions in different viruses may interact with similar host proteins.

Third, the arrangement of the control region is not simple. There is an abundance of palindromic and symmetric sequences as well as tandem repeats. It is apparent that the region is capable of forming a variety of structures, including double-stranded loops (cruciform structures), and it seems likely that these may serve as recognition sites for proteins.

Fourth, analysis of the genome of deletion mutants of SV40 shows that the origin of DNA replication lies within a tract of less than 100 nucleotides. Small deletions or even single base changes within this region give rise to cis-acting mutants that cannot replicate in mammalian cells. Such mutant DNAs are usually propagated in *E. coli* attached to plasmid sequences.

Fifth, analysis of similar deletion mutants indicates that the promoter for early RNA synthesis maps about 30 nucleotides upstream from the 5' end of early mRNA. Like many other eukaryotic genes, the early region of SV40 contains at this position an AT-rich sequence whose presence is required for efficient and accurate initiation of transcription. RNA polymerase II appears to begin transcription at a fixed distance downstream from this sequence. The late promoter appears to obey different rules. No AT-rich region is immediately apparent on the *l* strand, and the 5' ends of late mRNAs show a degree of heterogeneity that is difficult to reconcile with a fixed site of promotion.

Sixth, one of the products of the early region—large-T antigen—binds specifically and with high affinity to three sites in the control region. T antigen is known to be involved both in initiation of DNA replication and in repression of early mRNA synthesis. Both these effects are likely to be mediated through binding to the control region.

Finally, there is evidence for an additional site in the SV40 genome important for early transcription. About 200 bps upstream from the initiation site for early transcription there is a tandem repeated sequence 72 base pairs long. A mutant in which one of the 72 bp repeated units has been deleted replicates normally. Extension of this deletion into the second repeated unit, however, impairs replication as well as early transcription. It is surprising to find a site so far removed from the promoter to be essential to the initiation of transcription. Most intriguing is the recent finding that the 72 bp sequence can stimulate transcription from other genes artificially translocated into the same piece of DNA. In particular it has been shown that rabbit β-globin gene expression is greatly enhanced if it is present in the same segment of DNA as the 72 bp segment. The "enhancer" action on initiation of β-globin gene transcription can occur over enormous distances, as far as 1400 bps upstream or 3300 bps downstream from the 72 bp segment.

## The Lytic Cycle

As in adenovirus infection, _gene expression of polyoma virus and SV40 changes as the lytic infection proceeds_. During the first few hours, the early promoter efficiently directs the synthesis of the various early mRNA species. Synthesis of late mRNAs occurs sluggishly, if at all, at this stage and only becomes dominant as viral DNA replication gets underway, some 12 to 18 h after infection. The mechanism of this early/late shift is not understood, but the following scheme is plausible. During the first few hours of infection, early transcription is dominant (perhaps because of the presence of a well-defined promoter) and mRNA is produced that codes for large-T antigen. As the level of large-T antigen rises, early transcription is throttled down by T antigen–mediated autoregulation, and viral DNA synthesis begins. As the number of progeny molecules increases, the late promoter, which is not subject to repression by T antigen, becomes more and more dominant.

Initiation of each round of viral DNA synthesis requires a functional T antigen and an intact set of binding sites near the origin, but beyond that the mechanism of initiation is unknown. As we shall see, viral DNA replicates bidirectionally with forks growing from the origin in opposite directions at approximately equal rates. Thus, when purified molecules of replicating viral DNA are examined in the electron microscope, they can be seen to contain two branch points, two open circles of equal length (the replicated portions), and a superhelical loop (the unreplicated region). The fact that the unreplicated portion is superhelical while the replicated portions are not can be explained as follows. The superhelical turns in purified SV40 and polyoma virus DNAs are, in a sense, artifacts. In the cells, the DNA molecules, complexed with histones, are probably flat; only after removal of the histones during purification of the DNA do superhelices form. Thus one possible explanation for the failure to find superhelical turns in replicated regions of replicative intermediates of viral DNA is that these segments are not bound to histones in vivo. However, it is known that newly synthesized histones become associated with replicating viral DNA. Furthermore, if new histone synthesis is blocked by cycloheximide, parental histones distribute themselves asymmetrically, with 80 to 90 percent preferentially segregated with the "leading" sides of both replicating forks. Together these results strongly indicate that newly replicated DNA rapidly forms a chromatin-like configuration, with "old" histones lining up on one side of the fork and newly synthesized histones on the other. When the histones are removed from the unreplicated segments of the molecules, superhelices form. However, the replicated portions of the molecules contain discontinuities in the daughter DNA strands near the replicating forks. These nicks, or gaps, could act as "swivels" through which the superhelical twists can be dissipated after the histones have been removed.

Because of their structural features, it is possible to fractionate replicating molecules on the basis of their buoyancy in cesium chloride gradients containing an intercalating dye such as ethidium bromide. Because superhelical turns are present only in the unreplicating portion of the molecules, "young" replicative molecules contain a great proportion of their sequences in the superhelical form. They therefore band in cesium chloride–ethidium bromide equilibrium gradients at a position close to that of superhelical DNA. As replication proceeds, more and more of the sequences of the molecules are transferred to the nonsuperhelical region. There, topologically unconstrained, they are free to bind greater quantities of ethidium bromide per nucleotide pair. As replication proceeds, the number of superhelical turns in the molecules becomes smaller and smaller and the replicative intermediates band in cesium chloride–ethidium bromide gradients at positions that are progressively closer to that of nicked circular DNA.

Two kinds of experiments prove that DNA replication begins at a specific site on the genomes of polyoma virus and SV40. First, when "young" replicating molecules are treated with a restricting endonuclease, such as Eco-R1, that cleaves the viral DNA at only 1 site, linear structures containing a small "bubble" are seen by electron microscopy. The midpoint of the bubble, which consists of the DNA sequences that have been replicated, is always located at the same position in the viral genome. Molecules in a more advanced state of replication contain larger bubbles, and the lengths of both arms of the flanking unreplicated DNA are concomitantly and equally reduced (see Figure 28-8). These results mean that the genomes of polyoma virus and SV40 each contain a single site at which DNA synthesis is initiated, that replication is bidirectional, and that the replication forks move in opposite directions at about the same rate. The origin has been located on the SV40 physical map at a distance 0.33 fractional lengths from the Eco-R1 site (i.e., map position 33 or 67). On polyoma virus DNA, the origin is located at position 71, near the junction of HpaII DNA restriction fragments 3 and 5. In 1972, Danna and Nathans devised a second technique to pinpoint the replication origin in SV40 DNA. They used restriction enzymes to analyze pulse-labeled replicating forms of viral DNA. In molecules that complete replication during the pulse, regions of the DNA near the origin will be least radioactive and those near the terminus most radioactive. Conversely, molecules that begin replication during the pulse will be more radioactive near the terminus. In analyses of SV40 DNA replication, Danna and Nathans concluded that replication begins near position 67.

The replicating forks move apart at about the same rate, so termination occurs 0.5 fractional lengths away from the origin (i.e., at position 17 in SV40 DNA). It seems that replication terminates merely when the replication forks meet opposite the initiation site, rather than at any specific DNA sequence, since some deleted viral DNAs that replicate will lack that portion of the viral genome where termination normally occurs.

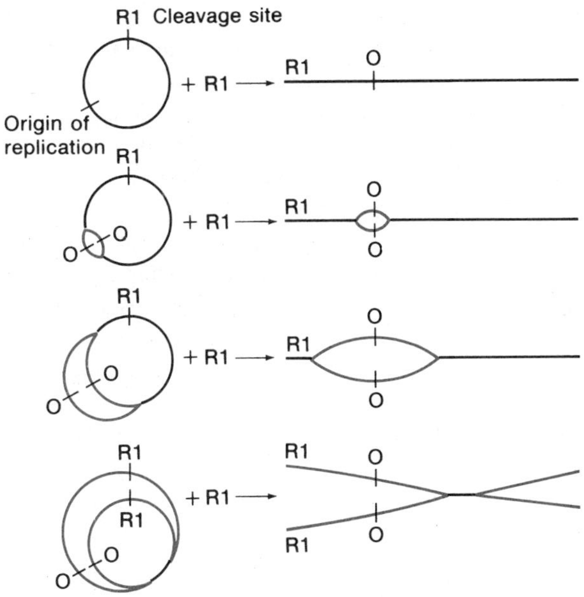

**Figure 28-8**
Diagrammatic representation of the forms of DNA seen when viral DNA that has replicated to varying extents is cleaved by a restriction endonuclease. This experiment shows that there is a unique origin of replication and that the daughter chains grow at approximately the same rate in opposite directions. — = unreplicated DNA; —— = replicated DNA; O = origin; R1 = R1 cleavage site.

By contrast to adenoviruses, where infection leads to dramatic inhibition of cellular macromolecular synthesis, SV40 and polyoma virus stimulate the host cells into making increased quantities of DNA, RNA, and protein. There is an especially marked enhancement of the activity of a variety of cellular enzymes involved in synthesis of DNA, and cells resting in the $G_1$ phase of the cell cycle are induced to resume growth and to enter the S phase. Many lines of evidence point to large-T antigen as the stimulating agent. Most direct are experiments in which purified T antigen was introduced into resting cells by microinjection; within 12 h, all the cells had begun a round of DNA synthesis.

So far, virtually *all the interesting biological properties of early viral-coded proteins have been attributed to the large-T antigens* of polyoma virus and SV40. No specific functions in the lytic cycle have been assigned to the small-t protein of SV40 or to the small-t or middle-t proteins of polyoma virus. However, it is important to remember that the analysis of events occurring during the permissive cycle is carried out under artificial conditions in vitro. The cells used for infection may in no way resemble in their physiology or state of differentiation the cells normally encountered by the viruses in an animal. All these proteins may therefore play an essential role in allowing the viruses to survive in their natural hosts.

Late during infection with polyoma virus or SV40, the cytoplasm of permissive cells contains large quantities of RNAs that code for VP1, VP2, and VP3—the structural components of the virions. The manufacture of these proteins accounts for 10 to 20 percent of the total protein synthesis in cells in the terminal stages of lytic infection. The capsid proteins are synthesized in the cytoplasm but are transported rapidly to the nucleus for assembly into progeny virus particles. The mechanism by which mature virions are formed is largely unknown, but because SV40 and polyoma are viruses of simple structure, it has often been suggested that self-assembly processes may play a large role in the production of progeny particles. Empty shells consisting of virus capsid proteins appear to form readily, and although the relationship between empty capsids and infectious virus has not been firmly established, the available data indicate clearly that empty capsids are not breakdown products of infectious virus particles but are precursors of them. Mature DNA, presumably associated with host histones, is withdrawn at random from the pool of viral genomes and becomes encapsidated within preassembled shells of capsid proteins. Eventually large crystalline arrays of progeny virions form, arranged on various cellular membranes as monolayers of tightly packed particles. At about this stage in the virus growth cycle, the nucleus of the cell stains very brightly with fluorescent antiserum raised against intact virus particles, and in the case of SV40, the cytoplasm begins to show signs of vacuolization 48 h after infection. The cells round up and detach from the substrate and the lytic cycle is complete.

## TRANSFORMATION BY ADENOVIRUSES, POLYOMA VIRUS, AND SV40

Adenoviruses, polyoma virus, and SV40 each have their preferred host cells for lytic infection: human, mouse, and monkey, respectively. Infection leads to a productive response that yields viral progeny and concomitantly causes the death of the cell. If, however, the viruses are used to infect cultured cells of a different animal species (e.g., rat), there occurs an abortive or nonproductive response during which very little or no progeny virus is produced and the cells survive. In a fraction of the cells undergoing nonproductive infection, segments of viral DNA become integrated into the cellular genome. *The expression of this integrated DNA causes the cells to change their pattern of growth and to acquire properties similar to those of tumor cells.* The change is essentially permanent. The cells can be cloned and grown indefinitely; they are "immortal" and continue to express the malignant phenotype. This phenomenon is known as *transformation*.

### The Transformed Phenotype

Typically, untransformed cells will grow in culture only on a solid support (the surface of a petri dish, for example) and in the presence of relatively high concentrations of nutrients. Even then, they will divide only as long as the culture is sparse. When the cell density increases beyond a critical point, the growth rate decreases sharply; in fact, the cells of some untransformed lines stop dividing altogether once they have formed a confluent monolayer.

Under the same conditions, transformed cells continue to multiply and may reach cell densities up to twenty times higher than those of untransformed cells. It is this sort of differential growth that provided the basis of one of the first assays used for transformation. By picking and subculturing transformed cells, it is possible to establish clonal lines of transformants and to ask in what ways such cells differ from their untransformed parents.

**Table 28-1**
Properties of Cells Transformed by SV40 or Polyoma Virus

**Growth**

High or indefinite saturation density*

Different, usually reduced, serum requirement*

Growth in agar or Methocel suspension-anchorage independence*

Tumor formation upon injection into susceptible animals

Not susceptible to contact inhibition of movement

Growth in a less oriented manner*

Growth on monolayers of normal cells*

**Surface**

Increased agglutinability of plant lectins*

Changes in composition of glycoproteins and glycolipids

Tight junctions missing

Fetal antigens revealed

Virus-specific transplantation antigen

Different staining properties

Increased rate of transport of nutrients

Increased secretion of proteases or activators*

**Intracellular**

Disruption of the cytoskeleton

Changed amounts of cyclic nucleotides

**Evidence of virus**

Virus-specific antigenic proteins detectable

Viral DNA sequences detected

Viral mRNA present

Virus can be rescued in some cases

Note: Transformed cells show many, if not all, of these properties, which are not shared by untransformed parental cells.

*Several of these properties have formed the basis of selection procedures for isolating transformants.

*The differences between transformed and untransformed cells can be classified in three ways: changes in cell growth, changes in cell-surface properties, and genetic alterations* (see Table 28-1). These changes are too great in number and too diverse in quality to be caused directly by viral gene products; in the case of polyoma virus and SV40, for example, there are many more changes reported than there are virus-coded proteins to cause them. Clearly, then, most of the observed alterations must be pleiotropic or must be secondary and tertiary effects that have occurred as a consequence of some primary event. It is the aim of much current work with the DNA tumor viruses to discover the molecular nature of this primary event.

## Events During Transformation

Stable transformation of nonpermissive cells by DNA tumor viruses is an inefficient process; at best, only a few percent of the infected cells show a permanent change in phenotype. However, in the case of SV40 and polyoma virus, a much greater proportion of the population is stimulated to divide and to behave for a few generations like transformed cells. Two lines of evidence indicate that this phenomenon, which is called abortive transformation, depends on the presence of functional large-T antigen. First, it can be elicited in part by microinjection of the purified antigen into resting cultures of untransformed cells. Second, mutations in the gene for large-T antigen abolish the effect. Gradually and over a number of generations, most of the abortively transformed cells return to their original state; i.e., they lose T antigen and slowly cease dividing. Out of the cells that have been abortively transformed, there emerges a subpopulation that is permanently transformed. These stable transformants retain forever the properties that were so fleetingly displayed by the abortive transformants.

Why some cells become stable transformants and some do not is not understood. However, the phenomenon of abortive transformations severely complicates experiments aimed at a molecular analysis of the events that lead to stable transformation. Any results obtained by biochemical analysis of the infected cell population are likely to reflect events occurring in the large population, which will not become transformed, rather than in the smaller population, which will.

Exactly what happens during the transformation process is therefore a matter for speculation. However, the following scenario seems reasonable. Soon after infection of nonpermissive cells with adenovirus, SV40, or polyoma virus, the early viral genes begin to be expressed. There is often (even in the case of adenoviruses) stimulation of host DNA synthesis, and a few rounds of viral DNA replication may occur. However, the late phase of viral infection never becomes firmly established, and in most cells, the infective process is aborted. *In a small proportion of cells, however, viral DNA becomes integrated into the host cell chromosome. In a sense, integration is the cornerstone to transformation.* Once established in the cell's genome, the viral DNA is passed on to the cell's descendants like any other cellular gene. The integrated DNA continues to express early viral functions, and it is the continued presence of early viral proteins that causes the cell to display a transformed phenotype.

### Integration

Although it has not been possible to follow directly the molecular events that occur during integration, we have a good idea what those events must be from comparative studies of the arrangement of viral DNA sequences integrated in different cell lines. To analyze the structure of integrated viral genomes, high-molecular-weight DNA is extracted from a number of independently transformed cell lines and cleaved at specific sites with one or more restriction endonucleases. The resulting fragments of DNA are fractionated by electrophoresis through agarose gels, denatured in situ, transferred to nitrocellulose sheets, and hybridized to highly purified viral DNA that has been radiolabeled to extremely high specific activity. By this technique, which was developed by Southern in 1975, the presence of as little as

$10^{-13}$ g of viral DNA can be detected in the transformed cell DNA. By analyzing 5 $\mu$g of cellular DNA, it is therefore possible to recognize as few as 0.01 copies of SV40 DNA per diploid equivalent of cellular DNA.

The major value of the method, however, is that it yields restriction maps of the integrated viral DNAs. For example, if restriction endonucleases are chosen that have no cleavage site within the viral genome, the number of radioactive bands detected in the digest of cellular DNA will be a minimum estimate of the number of viral DNAs inserted into the transformed cell's genome. However, when restriction endonucleases are used that cleave the viral DNA at known positions, it is possible to deduce the site on the viral genome at which integration occurred. Enzymes of this type also yield information about the location of cleavage sites in the flanking cellular sequences (see Figure 28-9).

The major conclusions drawn from experiments of this type are:

1. Viral DNA sequences are covalently joined to those of the host.

2. Different cell lines contain different numbers of insertions of viral DNA. In some, only one insertion of viral DNA is present; in others, as many as 20 are present. There is no clear relationship between the number of insertions and the degree to which the full transformed phenotype is expressed.

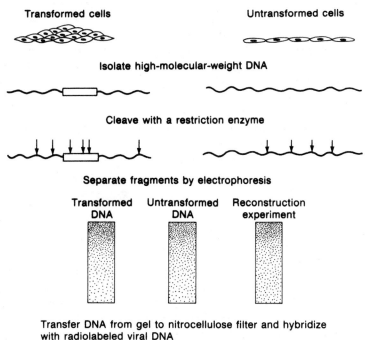

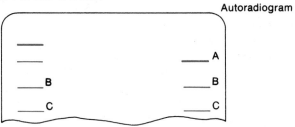

**Figure 28-9**
Scheme for determining the arrangement of viral DNA integrated into the genome of transformed cells. High-molecular-weight DNA is extracted from isogenic cultures of transformed and untransformed cells and cleaved (in this case an enzyme that is known to cleave the viral DNA at three places) with a restriction enzyme. The resulting fragments are separated by electrophoresis through an agarose gel, transferred to a nitrocellulose filter, and hybridized to radiolabeled viral DNA. The positions of the viral DNA fragments are then detected by autoradiography. Always included in the experiment is a reconstruction in which a small quantity of viral DNA is mixed with untransformed DNA. The position of the viral bands in the transformed DNA is compared with the position of authentic viral bands in the reconstruction experiment. In the example shown, the transformed cell genome yields four bands that contain viral DNA, instead of the three formed in the reconstruction experiment. This result would be consistent with a simple pattern of integration in which the integration event had occurred somewhere within the sequences of fragment A. As expected, no viral sequences are detected in the untransformed cell DNA.

3. There is no specificity apparent in the sites of integration within either the host or the viral DNA. Thus the integrated sequences are carried on different chromosomes in different host cells, and the junctions between viral and host DNA map at different sites in the viral genome.

4. The integrated viral sequences sometimes are not colinear with those of intact viral DNA. Examples of inversions, duplications, insertions, and deletions all have been found.

5. At least in some cases, insertion of viral DNA sequences leads to a rearrangement (probably a deletion) of cellular sequences at the integration site.

6. There is no evidence of homology between viral and cellular DNAs at the integration site. Sequence analysis of integrated viral sequences cloned and propagated in prokaryotic vectors has not revealed any features or sequences common to different transformed cell lines. In general, the transition between cellular and viral sequences is abrupt and clean, with no detectable homology between viral and cellular DNAs and no local duplications at the integration sites.

7. A large proportion of SV40 and polyoma virus transformed cell lines contain tandem repetitions of viral DNAs organized in a direct head-to-tail fashion. A similar arrangement is found in some adenovirus transformed cell lines except that the viral sequences are amplified together with their neighboring cellular DNA to form long tandem arrays consisting of alternating tracts of cellular and viral sequences. The mechanism by which such tandem arrays form is not known. However, two possibilities have been discussed. First, direct head-to-tail concatenates of viral DNA may be generated in abortively infected cells by rolling-circle replication (see Chapter 20) of the SV40 or polyoma viral genomes. This mode of replication has been observed at very late times during permissive infection; if it also were to occur after infection of nonpermissive cells, it could create the direct precursors to integrated concatenates. Second, as we shall see in the next section, tandem arrays might be generated by recombination either before or after integration.

## Expansion and Contraction of Integrated Viral Sequences

All cells transformed by adenovirus 2 and most cells transformed by SV40 and polyoma virus carry viral DNA sequences only in the integrated form. However, a small proportion of cells in certain lines transformed by the latter two viruses also contain "free" viral DNA molecules. Apparently, the sequences integrated in these lines are in some way unstable, so they release copies of the viral genome. Once liberated from the integrated state, the free copies of viral DNA begin to replicate, and their number per cell can reach as high as $10^4$.

The proportion of cells in the population in which these events happen is always small—typically between 0.01 and 1 percent. However, the numbers can be increased by treatment of the cells with physical agents, such as x-rays, drugs, such as mitomycin C or 5-bromodeoxyuridine, that produce derangements in DNA metabolism, or in the case of SV40, by fusing nonpermissive transformed cells with permissive monkey cells, a procedure that apparently causes destabilization of integrated genomes to occur more efficiently. Release of viral DNA is believed to occur by the mechanism shown in Figure 28-10. The viral origin is activated in the presence of large-T antigen and permissive cell factors to form a polytene-like or onionskin structure. Replication of this sort not only increases the viral copy number in a confined space, but also creates gapped templates ideally suited for recombination. If, as is commonly the case, the integrated viral DNA is arranged as a head-to-tail tandem, recombination between homologous sequences can occur, and this may have one of two possible results. If recombination occurs between the points marked B′ and A in Figure 28-10, the integrated set will expand by one unit length of viral DNA. If recombination occurs between A and B or A′ and B′, then a free copy of the viral DNA will be generated. This model accounts for all the facts known about the release of viral DNA from the integrated state, and it provides an alternative mechanism to rolling-circle replication for the generation of long, tandem repeats of viral DNA.

## Viral Gene Expression in Transformed Cells

*Not all viral genes are required for transformation.* Several lines of evidence (reviewed in Tooze, 1980) show that only the early region of SV40 and polyoma virus and only early regions 1A and 1B that map at the left end of the adenovirus genome are required. For example, when purified or cloned restriction fragments of the viral DNAs are used to transfect cells, those fragments which span the regions just mentioned are sufficient to cause transformation. Even when the entire viral genome is present in transformed cells, as a general rule only early viral genes are expressed. In the case of adenoviruses and SV40, the expression of these genes remains

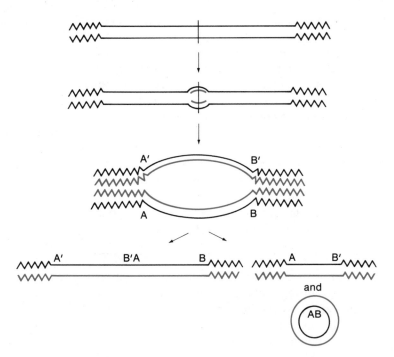

**Figure 28-10**
A model of the generation of tandem repeats and for the excision of SV40 DNA from the host chromosome. Activation of the viral origin of DNA replication generates a polytene or onionskin region. The resulting structure could provide an appropriate substrate for recombination. If recombination occurs between points A and B or A′ and B′, a free copy of viral DNA will be released; if it occurs between B′ and A, a tandem repeat will be generated.

under viral control. Thus the viral RNAs that are found in the cytoplasm are identical in structure to those found early during lytic infection. Similarly, the viral-coded polypeptides are identical to those found early during lytic infection. Lines of SV40 transformed cells, therefore, contain both large-T and small-t proteins; adenovirus transformed cells always contain the proteins coded by early regions 1A and 1B.

The situation in polyoma virus transformed cells is rather different. Only two (small-t and middle-T) of the three early gene products are regularly detected. Large-T antigen is often found only in a truncated form that lacks the C-terminal amino sequences. This situation appears to occur for the following reason: by and large, cells transformed by polyoma virus are more permissive to the virus than their SV40 transformed counterparts are to SV40. The simultaneous presence in the cells of a functional T antigen and permissive factors may lead to excision and loss of viral sequences by the mechanism discussed previously. There is therefore a selection for cells containing a large-T antigen that is nonfunctional for viral DNA replication but retains those properties important for the maintenance of the transformed phenotype. One way that such a situation can arise is for integration of polyoma virus DNA to occur within the gene for T antigen, thereby inactivating its C-terminal segment. Several examples of such transformed cells have been found. They contain forms of T antigen mRNA different from those present in lytically infected cells. Instead of the conventional viral sequences at the 3′ end, they contain sequences that are transcribed from the host DNA immediately flanking the integrated viral genome. In such cells, therefore, viral signals for termination of transcription and for polyadenylation have been replaced by those of the host.

### The Role of Early Viral Proteins in Transformation of SV40 Transformed Cells

*Several lines of evidence strongly indicate that functional large-T antigen is required to maintain cells in a transformed state* (see Martin, 1980). First, all cells transformed by SV40 express the protein. Second, many revertants selected for loss of their transformed phenotype no longer express functional T antigen. Third and most convincingly, many cell lines transformed by temperature-sensitive mutants of T antigen (*tsA* mutants) display a transformed phenotype only at permissive temperature. Unfortunately, however, this statement is not universal. Not all cells transformed by *tsA* mutants behave in this fashion—a finding that has caused an extended debate and much confusion. Although all the necessary information is not yet available, it now seems likely that cells transformed by *tsA* mutants that are not temperature-sensitive for transformation either produce a large amount of the protein or have acquired a cellular mutation that renders them independent of T antigen.

*The role in transformation of the second early protein coded by SV40—small-t antigen—is also controversial.* Mutants deficient in this protein seem to be able to transform growing but not resting cells. However, the resulting transformants tend not to express the full spectrum of transformed properties unless they are grown in high concentrations of serum. Small-t antigen may therefore provide a maintenance function that can be provided alternatively by serum.

## Polyoma Virus Transformed Cells

*By contrast to SV40 transformed cells, a large-T antigen of polyoma virus is not required for maintenance of the transformed state.* Thus transformants can be induced by transfecting cells with polyoma virus DNA fragments that lack the C-terminal portion of the coding sequences for large-T antigen. Furthermore, as discussed earlier, several transformed cell lines have been isolated in which the gene for large-T antigen has been inactivated by cleavage during integration. While these data do not exclude the possibility that the N-terminal portion of large-T antigen plays some role, *there is good evidence that the phenotype of polyoma virus transformed cells is caused by middle-T antigen.* Deletions in the proximal portion of the early region of the viral DNA that affect small-t and middle-T, but not large-T antigen abolish transformation. Deletions in the middle of the early region that affect middle-T and large-T antigens, but leave small-t antigen intact reduce drastically the efficiency of transformation. Deletions in the distal portion of the early region affect large-T antigen, but produce no alteration in transforming efficiency. By subtraction, these results indicate that the protein responsible for establishment, and presumably for maintenance, of transformation is middle-T antigen. The data do not reveal whether small-t antigen is also needed, but small-t antigen certainly is not sufficient to cause transformation.

## Adenovirus Transformed Cells

The left-hand portion of the adenovirus genome codes for a number of proteins (present data indicate at least eight). Which of these is responsible for transformation is unknown.

## The Molecular Basis for Transformation

If transformation by polyoma virus and SV40 is caused by middle-T and large-T antigens, respectively, the obvious question is how do these proteins work? Three hypotheses have been advanced; each is intuitively persuasive, but for none is the evidence entirely compelling. In brief, these hypotheses are that the papovavirus transforming proteins act:

1. On the plasma membrane
2. To stimulate cellular DNA synthesis
3. To reduce the cell's requirement for growth factors in serum

The first hypothesis draws its strength from the many reported observations of differences in membrane structure and function between normal and transformed cells (see Tooze, 1980). However, there are no convincing data to show that these changes are causative rather than consequential.

The second hypothesis stands on firmer ground. There is good evidence that SV40 large-T antigen can cause initiation of DNA synthesis at the viral origin; there are indications that the antigen can bind to specific sequences of cellular DNA; and there are present in the cellular genome tracts of nucleotides whose sequence closely resembles that of the SV40 origin. All this has led to the idea that T antigen is a transforming protein by virtue of its ability to activate cellular origins of DNA synthesis.

Another provocative observation which may provide a clue as to how T antigen is involved in transformation is the finding that large T is strongly associated with a 53 K protein that is of cellular origin. This latter protein is also presented in normal cells but in greatly reduced amounts. It is possible that large-T antigen stimulates a synthesis of the 53 K protein and that the latter is instrumental in the transformation process. A wide variety of transformed cells and tumor cells appear to have elevated levels of this 53 K protein. Despite their similarity, current evidence favors the idea that polyoma virus transforms cells by a very different mechanism which involves the middle-T protein described above.

The final hypothesis stems from the observation that when normal cells are grown in the presence of substances such as epidermal growth factor, their behavior resembles that of transformed cells. It is entirely possible, therefore, that the transforming proteins of papovaviruses either carry an activity that mimics growth factors or somehow renders the cells more sensitive to their action. The evolutionary value to the virus of such an arrangement is obscure, especially in view of the fact that neither virus causes tumors in its natural host.

## PICORNAVIRUSES

The picornavirus family (pico = small; rna = ribonucleic acid) consists of a very large number of viruses arranged in five genera: enteroviruses, rhinoviruses, caliciviruses, cardioviruses, and equine rhinoviruses. Although there are differences between these genera in such properties as pH stability and buoyant density, all the picornaviruses are fundamentally extremely similar both in structure and in the way they express and replicate their genomes.

Poliovirus, the type species of _enteroviruses_, which constitute the type genus of the family, has served for many years as the model for biochemical analysis of the picornaviruses.

### The Structure of the Poliovirus Particle

Poliovirus particles consist of a molecule of single-stranded RNA enclosed in a simple, icosahedral capsid 27 to 28 nm in diameter and consisting of three major polypeptides VP1, VP2, and VP3. There are 60 structural subunits in the particle, each containing a single copy of these three polypeptides. There are two other types of polypeptides in the particle: VP4, which is present in several copies per particle and seems to be located on the inner surface of the capsid, and a single copy of the small, basic protein VPg, which is covalently attached to the 5′ end of the virus RNA through a phosphotyrosine linkage.

_The virus RNA is a continuous single-stranded molecule of the same 5′-3′ polarity as mRNA synthesized during infection._ In fact, as we shall see, the infecting RNA itself serves as a mRNA during the early stages of infection. The 3′ end of the RNA carries a poly(A) tract 100 to 150 nucleotides long that is not added posttranscriptionally but is an integral part of the viral genome. The RNA is 7,700 to 7,800 nucleotides long, and the sequence of approximately half of these is known.

## The Infectious Cycle

Poliovirus multiplies rapidly and to high titer in many primate cell lines. Cells of most other species are nonpermissive because they lack the specific receptors to which the virus absorbs. However, such nonsusceptible cells may be productively infected with naked virus RNA.

**Translation.** All essential events during the replicative cycle seem to occur in the cytoplasm, since cells treated with actinomycin D (which inhibits DNA-directed transcription) and enucleated cells both support growth of the virus. After release of the viral genome from infecting particles, it is likely that the linkage between VPg and RNA is cleaved by host-coded enzymes to yield a functional mRNA. Within a short time after infection, viral RNAs are detected in large polysomes (up to 35 ribosomes per mRNA) and viral proteins are synthesized. By 3 to 4 h after infection, the only protein synthesis detected in the cell is directed by viral mRNA. Analysis of proteins synthesized at this time reveals the presence of a complex set of polypeptides whose summed molecular weights exceed the theoretical coding capacity of the viral genome by a factor of at least four. The key to solving this paradox was the observation that only a few high-molecular-weight polypeptides are synthesized during a short labeling period; these proteins rapidly disappear during a chase and several smaller polypeptides appear. These results strongly suggested that the smaller products were generated by proteolytic cleavage of precursors. Confirmation of this idea came from tryptic peptide maps that showed that the amino acid sequences of the final products were subsets of those present in transient precursors. Furthermore, infected cells treated with an amino acid analog or inhibitors of proteases were unable to process the high-molecular-weight precursors at the normal rate. Under such circumstances, a polypeptide was detected with a molecular weight of approximately 210,000 daltons. This is the size expected if *translation of poliovirus RNA is initiated at a unique site near the 5' end of the genome, continues the entire length of the coding region,* and generates a single polycistronic protein that subsequently undergoes cleavage. In the absence of inhibitors, very little full-length polypeptide is detectable. It is therefore postulated that the first cleavages occur before synthesis of the polyprotein is complete. The protease(s) that accomplish these early cleavages appear to be cell-encoded. The products of this initial round of synthesis/cleavage are three proteins with molecular weights of 115,000 (NCVP-la), 37,000 (NCVP-X), and 85,000 (NCVP-lb) (see Figure 28-11).

It is possible to map the poliovirus genome by taking advantage of the fact that there is only one site for initiation of polypeptide synthesis. When a drug such as pactamycin, which inhibits initiation of polypeptide synthesis, is added to infected cells, those proteins which have been already initiated will be completed, but no new polypeptides will be started. If such runoff synthesis is allowed to occur in the presence of radioactive amino acids, then the amino terminus of the polyprotein (coded by the 5' region of the poliovirus mRNA) will be labeled to a lower extent than the carboxy end. Thus the proteins that are derived by cleavage from the amino terminal region will carry less radioactivity than those from the carboxy end of the polyprotein. By this means it was shown that the order of genes in

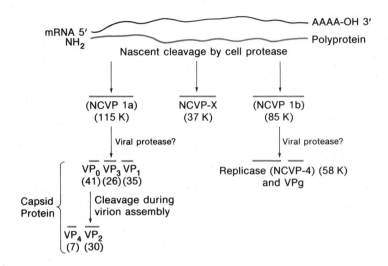

**Figure 28-11**
The processing of poliovirus polyprotein by proteolytic enzymes. The numbers in brackets represent molecular weights in thousands. The designation NCVP and VP stand for noncapsid viral protein and virion protein, respectively.

poliovirus RNA in increasing distance from the initiation site is NCVP-1a, NCVP-X, NCVP-1b.

The additional proteolytic reactions required to produce the stable final products of translation (capsid proteins, VPg, and replicase; see Figure 28-11) take several minutes to occur. The enzymes involved in this process have not been identified, but at least one protease appears to be encoded by the virus. Although the complete sequence of the precursor proteins is not available, from analyses of the amino and carboxy terminals of capsid polypeptides, it seems likely that cleavage occurs between $\alpha$-helix maker-breaker pairs at exposed sites in the precursor. Thus proline, glycine, serine, and arginine residues—all strong $\alpha$-helix breakers—are commonly found at new amino terminals. In addition to the proteins shown in Figure 28-11, numerous other short-lived proteins have been found in infected cells. Some of these might be nonfunctional products generated by aberrant cleavages, although it is possible that alternative cleavages are used to produce related proteins of slightly different function. In addition, there is evidence from in vitro translation experiments that poliovirus mRNA may contain a second, rarely used initiation site for protein synthesis. If this site were used in vivo, its products would contribute further confusion to an already complex picture.

## Inhibition of Cellular RNA and Protein Synthesis

*Soon after infection with poliovirus, there is a profound and rapid inhibition of cellular RNA and protein synthesis.* Later, when assembly of viral progeny begins, the synthesis of cellular DNA is affected. Later still, the cell loses all capacity to synthesize any macromolecules, viral or host. The phenomenon of host shutoff has been under investigation for many years, but still there is no satisfying explanation of how it occurs. The known facts are (1) translation of the viral genome is necessary to cause inhibition of cellular protein synthesis. Thus if cells are infected with UV-irradiated virus, no shutoff occurs. If drugs such as cycloheximide or puromycin, which inhibit protein synthesis, are present during the early stages of infection and then are removed, cellular protein synthesis resumes for a short time at the normal rate before inhibition sets in. (2) A viral structure protein may be involved in shutoff. Temperature-sensitive mutants of poliovirus defective in synthesis of the structural proteins of the particle are unable to inhibit cellular protein synthesis at the nonpermissive temperature. (3) Inhibition of protein synthesis does not result from degradation of host cell mRNAs, since these can be extracted from infected cells in undiminished quantities in a translatable form. (4) At least in the case of poliovirus, inhibition is not due to viral mRNA outcompeting that of the host, since host shutoff occurs even when viral RNA synthesis is prevented. Competition between viral and host mRNA may occur during infection by other picornaviruses, however, such as EMC.

An interesting hypothesis, with some experimental support, is that poliovirus might inactivate an initiation factor required for host protein synthesis. Extracts of infected cells contain an activity that causes the slow inactivation of an initiation factor that seems to be required for translation of cellular mRNAs but not for poliovirus mRNA. The factor that is inactivated interacts with the cap structure of eukaryotic mRNAs. Poliovirus mRNA, lacking a 5' cap, would not be expected to need the factor for its translation.

Less is known of the mechanism by which poliovirus inhibits cellular RNA synthesis. The primary effect seems to be a depression of synthesis of the 45 $S$ nucleolar ribosomal precursor RNA, which, like the inhibition of protein synthesis, is dependent on the expression of the viral genome.

## Replication of Viral RNA

*Replication of poliovirus RNA begins about an hour after infection and occurs in two steps: (1) the synthesis of a "minus strand," i.e., an RNA strand complementary to the infecting genome, and (2) the subsequent synthesis of new positive strands using the minus strand as a template.* Very little is known about how replication begins—in particular, how the infecting RNA is released from ribosomes so that synthesis of the minus strand can begin. Transcription occurs in the conventional direction; i.e., synthesis of the minus strand must initiate at the 3' end of the plus-strand template, on the poly(A) tract. The primer for the synthesis could be oligo(U). Since minus strands are known to be linked to VPg, however, a more appealing hypothesis is that a UTP residue attached to VPg serves as primer. The polymerase responsible is a virus-coded protein with a molecular weight of about 60 K; little else is known of its properties because it has proven extraordinarily difficult to purify the enzyme in an active form that is dependent on exogenously added template. However, it has been suggested that during strand elongation, the polymerase remains attached to VPg, causing the nascent strand to peel off, thereby preventing it from annealing to the plus-strand template (see Figure 28-12). Synthesis of plus strands would occur from the minus strands by the same mechanism.

All these replicative events take place in complexes (replication intermediates, or RIs) attached to smooth membranes in the cytoplasm. The RI consists of an RNAse-resistant double-stranded core plus several single-stranded nascent chains attached to the template by hydrogen bonds (see Figure 28-12). The exact amount of this hydrogen bonding is an unsolved matter.

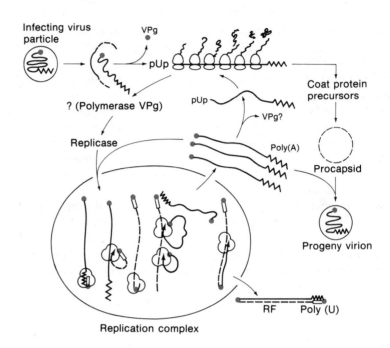

**Figure 28-12**
Replication scheme of poliovirus.
VpG is shown as ●; the solid and
dotted lines represent plus- and
minus-strand RNAs, respectively.
(Adapted from E. Wimmer, The Ge-
nome-Linked Protein of Picorna-
viruses: Discovery Properties and
Possible Functions, in R. Perez-
Bercoff (Ed.), *The Molecular Biology
of Picornaviruses*, Plenum, New
York, 1978. Used with permission.)

Mature RIs contain about five to six strands growing simultaneously in the same direction, and most RIs synthesize exclusively plus strands. Only a small number of minus strands are ever found in infected cells. The mechanism by which this transcriptional regulation is achieved is unknown; however, a second type of membrane-bound complex has been found in poliovirus-infected cells that is smaller than the predominantly plus-strand RI and, more pertinently, is rich in newly synthesized minus-strand RNA. Thus two qualitatively different forms of RI (and presumably two different forms of replicase) may exist in the cell.

Also found late in infection is an entirely double-stranded form of viral RNA consisting of one plus and one minus strand held together by hydrogen bonds. This double-stranded RNA (previously called RF, replicative form) is probably a by-product of replication rather than an essential intermediate. If VPg were to dissociate from the nascent strand after initiation of RNA synthesis, strand elongation might occur normally, but the nascent strand may remain annealed to the template, generating double-stranded RNA.

### Assembly

The progeny plus-strand RNA may either serve as template for further RNA synthesis or as mRNA, if the VPg is removed by cleavage, or it may be encapsidated. The viral capsid is essentially composed of 60 cleaved NCVP-1a molecules. It seems likely that uncleaved molecules of NCVP-1a aggregate initially into groups of five. Cleavage then occurs to give five copies each of VP0, VP1, and VP3, which remain clustered into a 14 S particle. Twelve such pentamers aggregate to form empty capsids that associate with plus-strand RNA by an unknown mechanism. Then follows a series of secondary proteolytic cleavages (notably conversion of VP0 to VP2+VP4), and these result in the formation of mature virus particles. By comparison with other animal viruses, the growth cycle of poliovirus is extremely brief and extraordinarily efficient; the entire process is complete within 8 h, and yields in excess of 100,000 particles per cell are not uncommon.

## MYXOVIRUSES

It may seem odd to choose to discuss influenza virus as the example of a negative-strand virus. After all, it was only recently that the nature of its genome became firmly established, and other negative-strand viruses exist that are inherently simpler or are biochemically better understood. Nevertheless, *no virus currently circulating has had a larger impact on the human population than influenza*. It has caused disease for hundreds of years, frequently in the form of epidemics of varying severity that have spread from country to country and continent to continent affecting humans of all age groups. Such epidemics cannot be controlled or prevented by the vaccines currently available, chiefly because the virus has the ability to vary the structure of two surface antigens—hemagglutinin and neuraminidase—in such a way that antibodies effective against a particular strain of influenza virus may not confer immunity against viruses that arise subsequently. Despite considerable effort, the molecular nature of this antigenic variation has remained, until recently, elusive. However, with the advent of recombinant DNA technology and of monoclonal antibodies, work on influenza viruses has burgeoned to the point where it now seems possible to decipher the pathways by which antigenic variation occurs.

### The Influenza Virus Particle

Particles of influenza virus are not uniform in size or shape; many are roughly spherical with diameters of 100 to 120 nm; others are "blobbier" and slightly bigger in one dimension than another; a few are greatly elongated (see Figure 28-13). Common to all, irrespective of their shape, is an outer, lipid bilayer envelope that is pierced by some 700 to 900 glycoprotein spikes (see Figure 28-14). The spikes are of two different shapes—the more common rods, 16 nm in length, which carry hemagglutinating activity, and the more rare mushroom-shaped projections, which carry neuraminidase activity. Both forms of spike can be released from the envelope by treatment with proteolytic enzymes, although in each case a hydrophobic stump is left anchored in the lipid bilayer.

### Hemagglutinin, Antigenic Drift, and Antigenic Shift

Hemagglutinin is the major surface antigen of influenza and plays the most prominent role in attaching the virus to specific cellular receptors. It is coded by an mRNA specified by one of the eight segments of RNA present in the particle. Although hemagglutinin is the product of a single gene, it consists of two polypeptide chains HA1 and HA2 of molecular weights 50,000 and 25,000, respectively, which are held together by bridging cys-

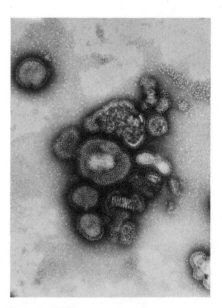

**Figure 28-13**
Electron micrograph of particles of influenza virus showing their varied shape and fringe of surface spikes. (Courtesy of Dr. N. Wrigley of the National Institute for Medical Research, London.)

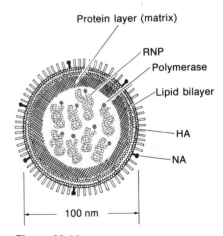

**Figure 28-14**
Diagrammatic representation of influenza virus. The helical ribonucleoprotein (RNP) consists of HP protein tightly bound to viral RNA.

teine residues. The two subunits are created by proteolytic cleavage of the full-length polypeptide, and this cleavage is necessary for infectivity.

*Pandemics of influenza occur in the human population when a major shift in the antigenic structure of HA allows the virus to infect people who are immune to previously circulating strains.* To understand this process of antigenic shift, one obviously would like to compare the complete amino acid sequences of several different pandemic strains. However, such information has proven difficult to obtain by conventional protein chemistry. Recently, however, several groups have cloned in prokaryotic vectors full-length complementary DNA (cDNA) copies of the HA genes of several influenza virus strains, and complete DNA sequences have been established. Despite the fact that the virus strains analyzed differ radically in their antigenic behavior, the general organization of the HA genes is in all cases remarkably similar. Thus the coding portion of the gene exists as a continuous block of nucleotides, so that splicing out of intervening sequences is not involved in HA expression. The sequence of amino acids, deduced from that of the DNA, shows that HA invariably contains three distinct hydrophobic regions. One of these forms the N-terminus of the primary translation product. It is cleaved from the mature product and is believed to be a "signal" sequence responsible for the insertion of the nascent polypeptide into the membrane. A second hydrophobic region, located at the C-terminus, is involved in anchoring the mature HA in the viral envelope. The third region lies in the center of the molecule, and after proteolytic cleavage, it forms the new N-terminus of the small subunit (HA2) that is thought to interact during infection with the membrane of the target cell. The major antigenic sites are located in HA1 (on the exposed portions of the spike), and it is variation in these sites that allows the virus to escape from neutralizing antibody. However, the sequences of HA1 proteins of the different pandemic strains so far examined have diverged to such an extent that it is impossible to produce a geneologic tree that would define the evolutionary relationship between them.

Major shifts in antigenic structure occur approximately at 10-year intervals and define the beginning and end of each pandemic era. The mechanism by which shift occurs is not understood. However, myxoviruses are known to be able to exchange segments of their genome with one another, a process known as *reassortment*. Antigenic shift may therefore occur when a human influenza virus swops hemagglutinin genes with an influenza virus endemic to another species.

In between these major shifts, the virus undergoes a series of smaller antigenic changes. As the human population becomes immune to infection by extant strains of influenza virus, so the pressure rises to select variants which, by displaying small but significant changes in antigenicity, can evade the immune response. This process is known as *antigenic drift*.

The changes responsible for antigenic drift accumulate with time, and field strains isolated several years apart from within a single pandemic era show a surprising number of amino acid substitutions—so many in fact that it has not been possible so far to correlate the locations of antigenic sites with the positions of amino acid substitutions.

To simplify the system, attempts have been made to mimic antigenic drift in vitro. The virus is grown in the presence of either high-titer neutralizing antibody or monoclonal antibodies directed against single antigenic sites in HA. Variants that are resistant to inactivation by the antibodies arise at a frequency of $10^{-5}$ to $10^{-6}$. Peptide mapping and nucleotide sequencing show that such variants arise as a consequence of changes limited both in number and location within HA1. In particular, the region around amino acid 143 seems to be capable of changing in a way that leads to alterations in antigenic behavior. However, additional regions of HA1 also show alterations, although with lower frequency. Thus the total number of antigenic sites may be greater, their topologic relationship more complex, and the selection pressures more subtle than we appreciate at present. Nevertheless, the capacity to select a sequential series of isogenic mutants and analyze them at the molecular level is a major advance that represents the best hope for dissecting the process of antigenic variation in influenza virus.

## Neuraminidase

The second glycosylated protein projecting from influenza virus particles is a neuraminidase (N). Like HA, it is a virus-coded protein ($M_r = 60,000$) that can undergo antigenic shift and drift. However, the contribution of neuraminidase to the overall antigenicity of the virus is quite low, and antibody directed against neuraminidase generally does not affect virus infectivity. Consequently, the protein has not received quite as much attention as HA, and far less is known of its chemistry. The function of neuraminidase during the virus growth cycle is not completely clear. It hydrolyzes sialic acid residues from HA receptors on the cell surface—a process that may allow release of the virus from budding points on the infected cell membrane.

## Inside the Envelope

Lining the inner surface of the lipid bilayer is a shell 6 nm thick of nonglycosylated protein known as matrix. About 3,000 to 4,000 molecules of matrix protein ($M_r = 25,000$) are present per virion; it is therefore the most abundant polypeptide, accounting for some 40 percent by weight of the protein of the virus particle. Details of its structure and function are unknown.

Occupying the center of the virus particle is the viral core, which has five components: RNA (see below); small quantities of three nonglycosylated proteins, P1, P2, and P3, with molecular weights of 80,000 to 100,000 (one of these is an RNA-dependent RNA polymerase; the functions of the others are unknown); and a nonglycosylated protein (NP), with a molecular weight of 60,000, that is present in large quantities (1000 molecules per particle) and binds to RNA in a regular manner to form what appear in the electron microscope as cyclic coiled filaments. It is on the basis of the antigenicity of this protein that strains of influenza virus isolated from humans are divided into types A, B, and C. Unlike hemagglutinin and neuraminidase, the ribonucleoprotein appears to be antigenically identical in different members of each of the three influenza types.

## The Viral Genome

*The genome of influenza virus consists of eight differently sized pieces of RNA which together code for at least nine different proteins.* Seven of the segments each appear to code for one protein (see Table 28-2). The eighth and smallest segment of RNA codes for two nonstructural polypeptides: NS1 ($M_r = 25,000$), which accumulates in the nucleus, and NS2 ($M_r = 11,000$), which is found in the cytoplasm. The two genes are arranged as an overlapping out-of-phase tandem, with the sequences coding for the N-terminal region of NS2 overlapping the C-terminal region of NS1 by about 150 nucleotides. The single piece of RNA codes for two separate mRNAs (one of which is derived by splicing), which are translated in different reading frames.

The complete nucleotide sequences of several segments of RNA derived from several strains of influenza virus have been determined by analysis of cloned copies of cDNA. While the coding sequences show considerable variation from strain to strain, the untranslated sequences at the 5′ and 3′ ends of the genes are highly conserved both between strains and between different segments derived from a single strain. Presumably this conservation reflects the involvement of the ends of genomic RNAs in transcription and replication.

## Transcription

*The mRNAs that code for the nine influenza virus–coded proteins are of the opposite polarity to the RNAs of the viral genome.* Thus almost as the first step during infection, the viral genome must be transcribed into mRNA. As far as is known, cells are unable to use RNA as a template for synthesis of further RNA. Thus *influenza virus, like all other negative-strand viruses, carries into the cell as part of the infecting virion an RNA-dependent RNA polymerase* (probably protein P1). It is this enzyme which is responsible for the synthesis of the first copies of viral mRNA. By contrast to other negative-strand viruses, however, synthesis of influenza virus mRNA depends on the simultaneous transcription of host DNA sequences by cellular RNA polymerase II. Thus if infected cells are treated with actinomycin D or α-amanitin, which inhibit host polymerase II, no transcription of the influenza genome occurs and the infection aborts.

The following explanation of this phenomenon is consistent with the properties displayed by influenza virus RNA-dependent RNA polymerase in vitro and the events that occur in the infected cells. The 5′ ends of influenza virus mRNAs are not virus-coded. Instead, the 5′-terminal methylated caps and the adjacent 12 to 14 nucleotides are donated to the viral mRNAs by newly synthesized capped cellular RNAs. The host-derived primer becomes covalently attached to the 5′ end of viral mRNA, which is therefore a chimera (see Figure 28-15). In all probability, the synthesis of these "primers" constitutes the α-amanitin–sensitive step for viral mRNA production. The fact that viral mRNAs will accept only newly synthesized 5′ ends rather than those already existing in the nucleus would seem to indicate that the transcription of viral and cellular RNAs is physically linked in some way.

**Table 28-2**
Properties of influenza A
virus RNAs

| Gene Number | Chain Length in Nucleotides | Gene Product |
|---|---|---|
| 1 | 5,100 | P1 |
| 2 | 3,670 | P2 |
| 3 | 4,230 | P3 |
| 4 | 2,490 | HA |
| 5 | 2,370 | NP |
| 6 | 2,400 | N |
| 7 | 930 | M |
| 8 | 820 | NS1/NS2 |

Adapted from R. D. Barry and B. W. J. Mahy, The influenza virus genome and its replication, *Br. Med. Bull.* 35:39, 1979.

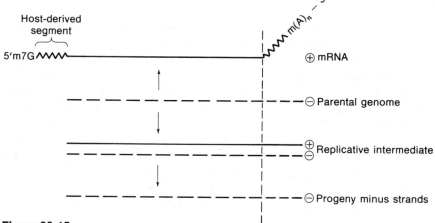

**Figure 28-15**
Transcription of a segment of influenza viral RNA. The vertical dashed line indicates the site near the 5' end of the genome where mRNA transcription terminates. This site is not recognized during synthesis of the plus strand component of replicative intermediate.

Not all the sequences of the genomic viral RNAs are transcribed into mRNAs. At least 17 nucleotides at the 5' terminus of each of the eight segments of viral RNA are not represented in mRNA. Thus polyadenylation appears to occur at a site that does not correspond with the 5' end of the genomic RNAs, but at a position that maps a short distance from it.

## Replication

Polyadenylated mRNAs are not the only positive-strand viral RNAs present in infected cells. A second class exists, consisting of full-length nonpolyadenylated transcripts that are the presumed templates for replication. Unlike viral mRNAs, synthesis of full-length positive-strand RNA is sensitive to inhibitors of protein synthesis and is presumably dependent on the continued production of virus-coded proteins. This type of RNA first appears in the infected cell within 30 min, and then accumulates rapidly in the cytoplasm, so that approximately 2000 copies are present after 2 h. At about this time, copies of mature genomic RNA begin to appear, an observation that is consistent with the idea that nonpolyadenylated full-length positive-strand RNA serves as a template. The detailed enzymology of none of these steps is understood.

## Regulation

Three observations indicate that transcription of influenza virus RNA is regulated. First, different viral mRNAs are produced at different rates during infection; mRNAs for P1, P2, and P3 are synthesized throughout infection at a slow but constant rate; mRNAs for NP and NSI are synthesized preferentially early; at late times during infection, the three mRNAs for HA, neuraminidase, and matrix proteins are the major products of transcription. In general, the amounts of viral proteins synthesized at any given time reflect the rates at which the various viral mRNAs are synthesized. It therefore seems that control of gene expression is modulated at the level of transcription. How this occurs is unknown.

Similarly, there is no explanation for the shift from mRNA production to synthesis of full-length positive strands that occurs early during infection. The most likely hypothesis is that one of the proteins made early during infection acts as an antiterminator, allowing the RNA-dependent RNA polymerase to transcribe beyond the poly(A) addition site. Finally, in contrast to mRNAs, full-length positive strands all begin to be produced at maximum rate in approximately equal amounts at the same time after infection. Thus the synthesis of two forms of RNA derived from the same template is controlled in different ways. Presumably this could occur at the level of initiation (it is not known, for example, whether full-length positive-strand RNAs carry a host-derived cap). However, the possibility remains open that different virus-coded polymerases are involved.

## Assembly

*Assembly of virus particles takes place at the cell membrane and coincides with their release from the cell.* The first stage appears to be the replacement at specific areas of the membrane of host proteins by viral antigens. Visible projections of HA and neuraminidase develop as a layer of matrix protein is laid down on the inner surface of the lipid bilayer. Ribonucleoprotein accumulates in these areas of morphologic alteration and virus particles then bud off.

How these events are coordinated is not known. Particularly obscure is the mechanism used by the virus to ensure that it has a complete set of genes. The chances that the correct eight ribonucleoprotein segments would be selected at random from the intracellular pool are unrealistically low. However, there is no evidence of physical linkage between the segments, and a wealth of genetic evidence exists showing that reassortment of genomic segments occurs with high frequency during mixed infection. It may therefore be that to ensure a complete set of genes, each virion would contain more than eight segments of RNA. Thus the packing of equimolar amounts of each segment, while true at the population level, would not necessarily apply to individual particles.

## RETROVIRUSES

The major features of retroviruses are:

1. *The viral genome is carried in virus particles in the form of single-stranded RNA, which is converted during infection to double-stranded DNA.*

2. *This DNA must integrate into cellular DNA for virus production to occur.*

3. Infection in culture generally does not lead to death of the cell; instead, the cell continues to grow and divide with each daughter carrying integrated viral DNA.

4. Such integrated viral DNA sequences may be transmitted in animals from parent to progeny by means of germ-line cells (vertical transmission).

5. Such vertically transmitted viruses when "induced" (i.e., released from the cell in an infectious form) often will replicate only in the cells of a heterologous animal species.

6. Some highly oncogenic viruses carry transforming genes (*onc* genes) derived from the host cell.

7. There are three viral genes essential for replication: *gag*, which codes for internal structural protein of the virus, collectively known as group-specific antigens; *pol*, which codes for an RNA-dependent DNA polymerase; and *env*, more pleiomorphic than the two others, which codes for envelope glycoproteins. Viruses that carry all three of these genes are replication-competent and are termed nondefective.

8. Many retroviruses exist which lack some or all of these essential genes. All such defective viruses depend on nondefective helper viruses for replication.

Retroviruses of many different sorts have been isolated from birds, mice, rats, reptiles, cats, hamsters, baboons, gibbons, pigs, cows, guinea pigs, and humans. However, the best-studied (see Bishop, 1978) by far are those isolated from birds (the avian retroviruses) and mice (murine retroviruses). Each of these groups consists of several sorts of viruses which can be distinguished from one another on the basis of their oncogenic properties and genomic structures.

The retroviruses with the largest genomes (10 KB) are the nondefective sarcoma viruses of chickens, which contain a complete set of viral genes and therefore are competent for replication. However, they carry two additional genetic elements: *src*, a gene that codes for a protein kinase, and the *C region*, about 450 nucleotides in length which is common to all avian retroviruses and carries an efficient promoter that causes viral products to be made in large quantities. The order of genes in the RNA of nondefective Rous sarcoma viruses (RSV) of chickens is 5′ *gag-pol-env-src*-C$^{3'}$ (see Figure 28-16). The genomic RNA is a capped single-stranded RNA of positive polarity, with a 3′-terminal poly(A) tail.

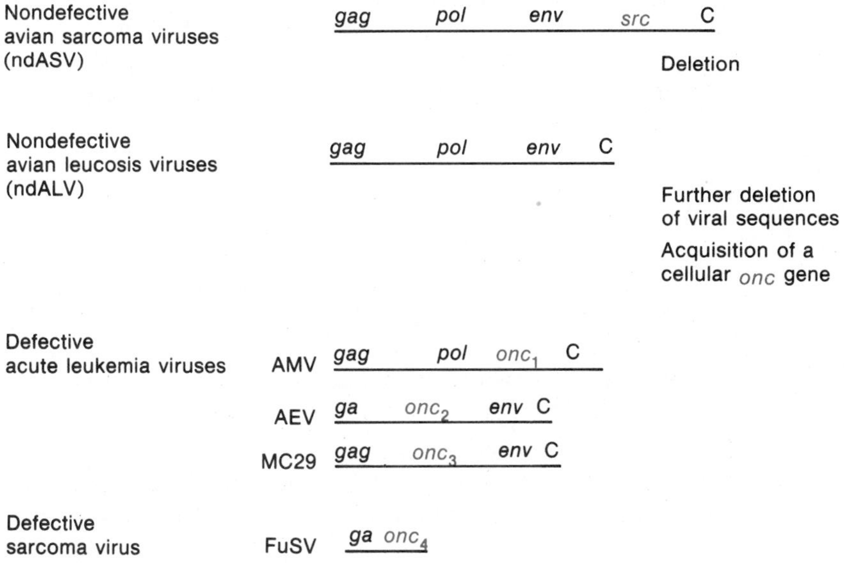

**Figure 28-16**
Genome structure of various defective and nondefective avian retroviruses. The cellular *onc* gene that becomes integrated (at various positions) into the genome of acute leukemia viruses differs from one virus to another. Acquisition of different *onc* genes accounts for the target-cell specificity of each virus:
AMV = avian myeloblastosis virus;
AEV = avian erythroblastosis virus;
FuSV = Fujinami sarcoma virus; and
MC29 is a myelocytomatosis virus (MCV).

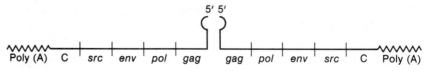

**Figure 28-17**
The genome of avian sarcoma virus. Two identical molecules of single-stranded RNA are joined by hydrogen bonding at or near their 5′ terminals. However, the exact pairing between the two strands has not been elucidated.

Retroviruses stand apart from all other animal viruses in that their genome is diploid, consisting of two identical molecules of single-stranded RNA held together by hydrogen bonds at or near their capped 5′ terminals (see Figure 28-17). The significance of this structure is not presently understood.

In addition to the genomic RNAs, retrovirus particles also contain several species of low-molecular-weight RNA, including 7 s and 5 s RNA derived from host ribosomes, and a variety of host tRNAs. In the case of avian sarcoma viruses, the most important of these is a molecule of tRNA$^{trp}$ that is bound to the viral genome at a site about 100 nucleotides from the 5′ end. This species of tRNA serves as a primer for the initiation of viral DNA synthesis. While it is peculiar for tRNA to serve as a primer for DNA synthesis, there seems to be nothing significant about the particular kind of tRNA used, since tRNA$^{pro}$ carries out the analogous reaction in a mouse leukemia virus. Each haploid unit of the viral genome is terminally redundant, with about 20 nucleotides repeated at the 5′ and 3′ ends. All other retroviruses are genetically less complex than the nondefective avian sarcoma viruses (ndRSV). In a strictly formal sense, they can be regarded as a set of deletion/substitution mutants of a retrovirus whose genome closely resembles that of ndRSV. This is not to imply that the other viruses shown in Figure 28-16 are all related to RSV by homology. Such is not the case; their genomes are merely organized along basically similar lines and the sequences of nucleotides is often very different.

Each taxonomic group of retroviruses contains several classes of viruses that are distinguished from one another by their oncogenic properties.

The *sarcoma viruses* produce sarcomas in vivo and transform fibroblasts in vitro. They are divided into four subdivisions:

1. ndRSV, whose genome, as already discussed, consists of the genes *gag-pol-env-src*-C.

2. Replication-defective ASVs that contain *onc* or *src*. These viruses are identical to ndRSV in their genome structure except that part of the *env* gene has been deleted. They therefore need a helper virus to replicate.

3. Sarcoma viruses of birds, which are defective in all three essential genes and contain transforming genes that are nonhomologous to *src*.

4. The mammalian sarcoma viruses, which also are defective for all three essential genes and contain any one of a variety of transforming genes that generally are linked to the remnants of the *gag* gene.

The *acute leukemia viruses* cause a variety of rapidly developing tumors (erythroblastosis, myeloblastosis, and other leukemias) in animals and may transform fibroblasts or hemopoietic cells in culture. Most of these viruses are defective in all three essential viral genes and contain an oncogene (*onc*) that differs from one type of virus to another. *The nature of the onc gene determines what type of cell that virus will transform.* All acute leukemia viruses are defective and rely for replication on functions supplied by helper viruses.

The *lymphatic leukemia viruses* (avian leukosis viruses, or ALV), by contrast, are nondefective and often serve as helpers for defective viruses of the other groups. They do not transform cells in culture and cause tumors (lymphatic or erythroid leukemias) in animals only after extremely long latent periods. Recent evidence indicates that *these viruses cause tumors, not by carrying an onc gene into the cell, but by integrating next to an endogenous cellular onc gene and thereby activating it.*

All retroviruses, whatever their oncogenic potential or transforming capacity, share a similar architecture. They all mature at the cell surface and bud with the central nucleoid becoming wrapped in an envelope derived from the host cell's plasma membrane.

A diagram of an idealized retrovirus is shown in Figure 28-18. The envelope is derived from the cell membrane. However, by contrast with influenza viruses, retroviruses carry host as well as viral antigens. The spikes protruding through the membrane are composed of two virus-coded glycoproteins, gp85 and gp37, that are linked by disulfide bridges and are probably derived from the same polypeptide precursor. Together these glycoproteins are required for attachment of the virus to the host cells, and they are responsible for the host range of the virus. Lining the envelope is a core shell consisting of several virus-coded proteins. It encloses the central nucleoid, which contains the viral genome and the RNAs and proteins bound to it. Included among these are highly basic charge-shielding proteins as well as enzymes such as RNA-dependent DNA polymerase.

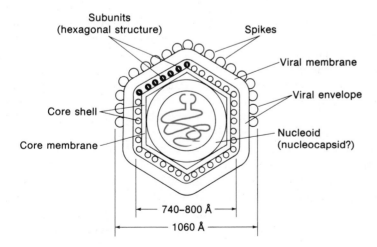

**Figure 28-18**
A diagrammatic representation of a retrovirus particle.

The Replication Cycle

Rous sarcoma virus (a nondefective avian sarcoma virus) is the best stud-ied retrovirus and will serve to exemplify the steps involved in replication.

The replication cycle may be divided into three parts: (1) the pre-integrative stage, when the virus is engaged in synthesizing a complete double-stranded DNA copy of its genome, the provirus; (2) the integration event itself; and (3) the postintegrative stage, when the viral genome is transcribed, viral proteins are synthesized, and progeny virus particles are assembled.

**Preintegrative Events.** *The key to the events that precede integration lies in the activity of the virus-coded enzyme RNA-dependent DNA polymer-ase (reverse transcriptase)*, which was first detected in 1970 in virions of avian and murine retroviruses (Baltimore, 1980; Muzutani and Temin, 1970). It is now known to be a universal constituent of retroviruses, whether or not they are oncogenic. Its discovery fulfilled an important prediction made in 1964 on the basis of experiments with actinomycin by Temin, but never really taken seriously, that RNA tumor viruses multiply through a DNA intermediate. As we shall see, in the years since the discov-ery of RNA-dependent DNA polymerase, an overwhelming mass of experi-mental evidence has proven beyond a shadow of doubt that the provirus hypothesis is correct.

Reverse transcriptase is encoded by the *pol* gene and carries two major enzymatic activities: a DNA polymerase capable of copying RNA or DNA templates and a ribonuclease (RNAse H) that specifically removes the RNA from DNA:RNA hybrids. Synthesis of the first (minus) strand of DNA begins near the 5′ end of the viral genome, at the place where the tRNA primer is bound (Figure 28-19). The growing DNA strand, still covalently attached to the primer, is extended through the redundant sequence to the 5′ end of the template. Part or all of the RNA in the resulting DNA:RNA hybrid is then degraded by RNAse H. The redundant sequence in the newly synthesized DNA is thus exposed and is able to form a bridge to the complementary sequence at the same or another RNA molecule. Synthesis of minus-strand DNA is now free to proceed along the RNA molecule, displacing the primer tRNA. Long before synthesis of the minus strand is completed, however, the plus strand of DNA begins to be made. Much less is known about its synthesis, although it is generally assumed that the plus strand is copied from minus-strand template, perhaps using as a primer oligonucleotides generated by the action of RNAse H on the minus-strand RNA hybrid. One of the first plus-strand segments to be synthesized is a 300-nucleotide sequence complementary to the part of the minus strand where initiation of DNA synthesis took place. The 300-nucleotide DNA therefore is complementary to both the 5′ and 3′ ends of the RNA, to the primer binding site, and perhaps to the tRNA itself. When the minus strand reaches the end of the viral RNA, a second "bridge" can be formed to the primer binding site within the 300-nucleotide plus-strand segment. The minus strand would then be extended further using the 300-nucleo-tide segment as template. In turn, the plus strand is extended; it ultimately consists of a full-length copy of the minus strand. Reverse transcription of

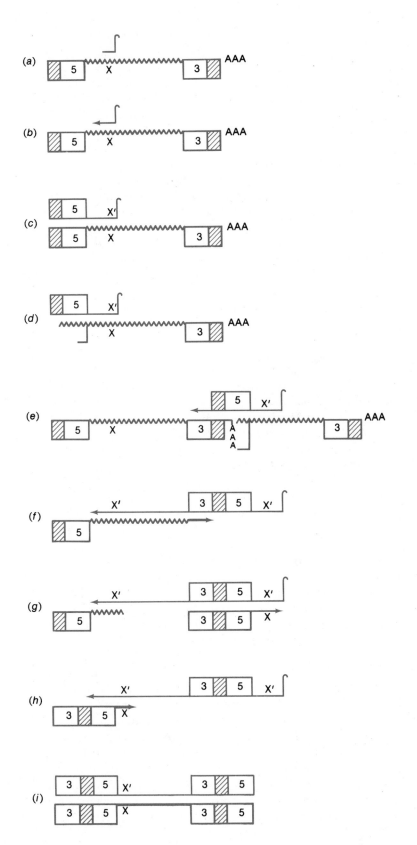

**Figure 28-19**
A model for the generation of double-stranded DNA carrying two copies of LTR. The sequence X is a marker for the plus strand; the complementary sequence X' occurs on the minus strand. (*a–c*) Synthesis of minus-strand DNA from the genomic RNA template using a tRNA primer (represented by inverted J in step *a*). (*d*) Degradation of the RNA portion of the resulting DNA:RNA hybrid by RNAse H. (*e*) Bridge between the newly synthesized segment of minus-strand and repeated sequences at the 3' end of genomic RNA. (*f*) Extension of minus-strand DNA (*leftward*) and initiation of plus-strand DNA (*heavy arrow moving rightward*). (*g*) Completion of 300-nucleotide fragment of plus-strand DNA. (*h*) Bridge between sequences repeated in minus-strand DNA and in the 300-nucleotide fragment. (*i*) Completion of synthesis and tidying up: ▭▨▭ = LTR, with sequences from the 5' and 3' ends of the viral genome separated by the terminal repeat.

retroviral RNA therefore produces linear duplex DNA, several hundred nucleotides longer than the RNA itself, in which both ends contain an identical sequence (called the long-terminal repeat, LTR). Restriction endonuclease analysis and direct DNA sequencing show that the LTR contains 250 to 1,200 nucleotide pairs (the exact size varies from retrovirus to retrovirus) of the region unique to the 3' terminus of viral RNA, a single copy of the short sequence present at both ends of the RNA, and about 100 to 150 nucleotides from a region unique to the 5' terminus.

The synthesis of linear duplex DNA occurs in the cytoplasm of the infected cell, presumably in partially degraded virions. A portion of the linear duplex DNA enters the nucleus and a portion of the molecules are then converted into circular forms. This process seems chiefly to occur by either of two mechanisms. Direct joining of the ends would produce circles carrying two copies of the LTR; homologous recombination between the redundant ends would produce circles carrying one copy of the LTR. Both sorts of molecules have been isolated, cloned, and sequenced. Although direct evidence is lacking, it seems likely that these circular forms of retrovirus DNA are the substrates for integration.

**Integration.** Integration of proviral DNA occurs at multiple sites within the cellular genome with respect to host sequences; *integration therefore appears to be random or quasi-random.* By contrast, the junction points map in the viral genome at a specific set of sequences within the LTR. The resulting integrated copy of proviral DNA is colinear with viral RNA and carries LTR sequences at both ends.

Determination of the sequences at the ends of integrated proviral DNAs cloned in bacteria has shown that during integration: (1) a few base pairs (usually 2) are eliminated from the ends of both LTRs, and (2) different proviruses are flanked by different cellular sequences. However, for any given provirus, the four to six nucleotides of host DNA at both ends of the insertion are identical; (3) this duplication of host sequences occurs during integration.

These observations have led to the model of retrovirus integration shown in Figure 28-20. Here the substrate is shown as a circular molecule containing one copy of LTR. It can, however, easily be adapted for viral templates in other topologic forms. The salient steps (see Figure 28-20) are (1) the introduction of single-stranded nicks on opposite strands of the circular viral DNA, two bases from each end of the LTR; (2) the introduction of single-strand nicks four bases apart in cell DNA; (3) the invasion of the nicks in the viral DNA by the ends of host DNA, to form a recombination intermediate; and (4) the formal resolution of the recombinational structure by denaturation of the LTR sequence and filling in of the resulting single-strand gaps by extension from the exposed 3' ends.

The resulting integrated proviral DNA lacks two bases from the LTR and is flanked by direct repeats of host DNA, four base pairs in length. This structure closely resembles that of transposable elements inserted into the genomes of bacteria, yeast, and *Drosophila*—a similarity that raises the possibility, as yet unproven, that proviruses can be moved to new genomic sites by direct translocation of proviral DNA sequences.

**Expression of the Viral Genome.** *Once installed in the cellular genome, the proviral DNA is transcribed by host RNA polymerase II using a single powerful promoter located in LTR.* Some of the primary transcripts appear to be transported into the cytoplasm in a form that is capped and poly-adenylated but not spliced; these RNA molecules will form the genome of progeny virions. The remainder follows the more traditional pathway of eukaryotic mRNA processing and is exported from the nucleus in several spliced forms; these are the mRNAs that bind to ribosomes and direct

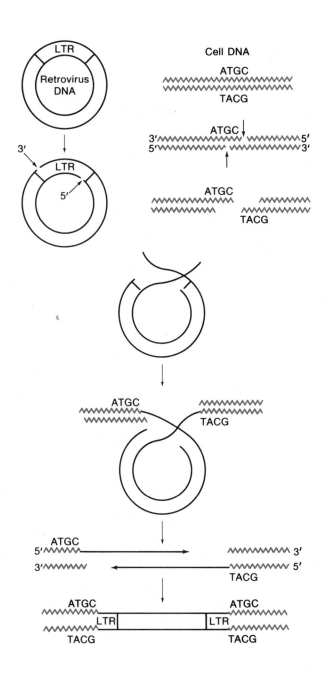

**Figure 28-20**
A model for integration of viral DNA. A circular proviral DNA with one LTR is nicked on opposite strands two nucleotides from the end of the LTR. Host DNA is nicked at two sites on opposite strands four nucleotides apart. The free 5′ ends of cell DNA are joined to the 3′ ends of viral DNA. The resulting recombinational intermediate is resolved by denaturation of the 600-nucleotide LTR and the gapped linear DNA is repaired. Both copies of LTR are identical, and each presumably contains an active promoter. Thus not only are viral genes expressed at high levels, but also perhaps, "downstream" cellular genes. (Adapted from C. Shoemaker et al., Structure of cloned circular Maloney murine leukemia virus DNA molecule containing inverted segment: implications for retrovirus integration, *Proc. Natl. Acad. Sci. USA* 77:3932, 1980. Used with permission.)

translation of viral proteins. The three or four primary translation products of these mRNAs are cleaved and modified posttranslationally to yield multiple proteins.

***gag* Gene.** gag mRNA is indistinguishable from virion RNA in size and structure. The primary product of translation (in the case of ASV) is pr76$^{gag}$, which is modified, phosphorylated, and cleaved into a set of small-core proteins found in virions.

***pol* Gene.** pol mRNA appears to differ from gag mRNA only in that the termination codon for pr76$^{gag}$ and termination codons further downstream are removed by splicing. The resulting mRNA is translated into a protein called pr180$^{pol}$, which is a precursor of RNA-dependent DNA polymerase.

***env* Gene.** The env protein is expressed from a spliced subgenomic mRNA 5.4 KB in length that consists of approximately 350 nucleotides derived from the 5′ end of the primary transcript joined to sequences near the beginning of the segment coding for *env*. The nascent env protein is inserted into the plasma membrane, glycosylated, and then cleaved into two subunits that remain connected by S–S bonds.

***src* Gene.** src mRNA is multiply spliced and contains two leader regions, one consisting of the 350 nucleotides from the 5′ end of the genome and the other derived from the *env* region. The mRNA is translated into a phosphorylated protein, src, which is not required for viral replication, but which can cause cells to assume a malignant phenotype.

## Maturation of the Virus Particles

The viral core is assembled in the cytoplasm from a full-length unspliced transcript and core proteins. The outer envelope is applied to the core as it buds through the cell membrane. The final stages of cleavage and morphogenesis take place outside the cell. By and large, cells infected with retroviruses are not killed. Instead, they leak virus at a steady rate and continue to grow and divide.

## Malignant Transformation by Retroviruses

*Most transforming retroviruses tested code for protein kinases that have the novel property of adding phosphate groups to tyrosine residues in their substrate proteins* rather than the conventional serine or threonine. The best-studied of these kinases is coded by the *src* gene of RSV, but analogous enzymes are coded by the *onc* genes of various retroviruses.

Extensive studies with both deletion and temperature-sensitive mutants of RSV have shown that the presence of an active *src* gene product is necessary for transformation of cells to a malignant state. The mechanism by which the enzyme brings about this dramatic change in cell behavior is not known. One possibility is that the phosphorylation of tyrosine in cellular proteins is the event that triggers the malignant phenotype. The problem is to pinpoint which of the several cellular proteins that become phosphorylated are important in transformation.

*The src gene of ASV is closely related to nucleotide sequences (usually called c-src) present at the level of one or a few copies per cell in normal cells* of birds, fish, humans, and other vertebrates. The conservation of the sequences across large spans of evolutionary time suggests that the sequences code for a protein that serves a necessary function in normal cellular metabolism. Small quantities of *c-src* transcripts are found on the polysomes of normal cells, which also contain a protein kinase ($M_r = 60,000$) that is structurally related to the viral gene product. *The amount of the protein in normal cells is at least 50-fold lower than that of the viral product in transformed cells.* This raises the possibility that transformation by retroviruses is caused not by the introduction of a novel viral protein into a cell, but by elevated levels of a normal cellular protein. It is thought that all the highly oncogenic retroviruses may have acquired, by recombination, cellular genes that cause transformation when they are expressed at high levels.

This idea has been tested by coupling a cloned *c-src* gene (in this case isolated from normal mouse cells) to the strong promoter contained in the LTR sequence of a provirus. The resulting recombinant was found to transform cells with the same efficiency as cloned authentic proviral DNA. This result suggests (1) that malignant transformation merely is a consequence of an increase in the quantity of intracellular *c-src*, (2) that *c-src* may be involved in some cellular function such as maintenance of cell division, and (3) that genes placed under the control of LTR are expressed efficiently in cells. The experiment predicts that *naturally occurring sarcoma retroviruses might be created by recombination between a nontransforming proviral DNA and the cellular c-src sequence.* This prediction has been verified. A similar hypothesis probably explains the origin of other transforming retroviruses. Thus the rapidly acting acute leukemia viruses have been shown to carry their own sets of transforming cellular sequences (*onc* genes) that are specific for hematopoietic cells. The molecular mechanism that enables a given *onc* gene to transform a particular type of cell is not yet understood.

By contrast, the weakly oncogenic viruses (ALV) may not normally carry transforming genes. The occasional tumors induced by such viruses are thought to arise either as a consequence of rare recombination with other viruses or cellular genes or as a consequence of integrating viral DNA in a way that causes a cellular oncogene to be expressed at high levels. Current hypotheses revolve around the idea that the second copy of the LTR (see Figure 28-19) acts as a promoter to increase the expression of a downstream cellular oncogene. It has been shown that the cells of nearly all tumors induced in birds by the weakly oncogenic leukemia viruses carry a copy of the viral genome integrated in a specific location—just upstream from a known cellular oncogene. (This same oncogene had previously been identified as part of the genetic machinery of another avian tumor virus, the acute leukemia virus known as MC29.) In the tumor cells, the cellular oncogene has been harnessed to the efficient leukemia virus promoter, so that it is efficiently transcribed into mRNA. It is the resulting high levels of the onc protein that presumably cause the cells to assume a malignant phenotype.

### Endogenous Viruses

It is perhaps not surprising that viruses which, in order to replicate, must integrate into the genome of their animal hosts should occasionally find their way into the DNA of cells of the germ line. Such events appear to be rare, but once established, such endogenous viral genomes are transmitted from parent to offspring like other cellular genes and evolve in contact with them. In fact, it is possible to use the viral sequences as genetic markers and as indicators of evolutionary relationships between different species.

The number of endogenous viruses varies widely from animal species to species. In humans, for example, none have so far been detected, while in chickens, sequences of at least 10 distinguishable endogenous viruses have been recognized. Many of these viral genes are silent under normal circumstances. In many cases they appear to be under the control of cellular regulatory genes that may, at certain sites, separate from the endogenous viral DNA. Often, however, the viruses can be activated by a variety of chemical and physical agents, as well as by hormonal and immunologic stimuli. The mechanism of this induction is not known.

Finally, for reasons that also are unknown, many endogenous retroviruses will not replicate in cells of the animal species in which they are endogenous. They are said to be _xenotropic_. Consequently, the isolation of novel endogenous viruses often demands the use of cells of exotic species that are permissive for the virus. The finding of such species is more a matter of serendipity than of logic.

## SELECTED READINGS

Bishop, M. M.  Retroviruses. _Ann. Rev. Biochem._ 47:35, 1978.

Capsar, D. L. D.  Design Principles in Virus Particle Construction. In F. L. Horsfan and I. Tamm (Eds.), _Viral and Rickettsial Infections of Man_, 4th Ed. Philadelphia: Lippincott, 1965. P. 51.

Karess, R. E., and Hanafusa, H.  Viral and cellular _src_ genes contribute to the structure of recovered avian sarcoma virus transforming protein. _Cell_ 24:155, 1981.

Temin, H. M.  Structure, variation and synthesis of retrovirus long terminal repeat. _Cell_ 27:1, 1981.

Tjian, R.  T Antigen binding and the control of SV40 gene expression. _Cell_ 26:1, 1981.

Tooze, J.  _The Molecular Biology of DNA Tumor Viruses_. New York: Cold Spring Harbor Press, 1980. P. 383.

Wimmer, E.  The Genome-Linked Protein of Picornaviruses: Discovery Properties and Possible Functions. In R. Perez-Bercoff (Ed.), _The Molecular Biology of Picornaviruses_. New York: Plenum, 1978.

## PROBLEMS

1. What strategies do different types of animal viruses use to synthesize mRNA?

2. Why do certain animal viruses carry as essential parts of their particles, enzymes whose function it is to transpose information from one sort of nucleic acid molecule to another?

3. The naked nucleic acids isolated from some animal viruses are infectious. Would you expect the following viruses to yield infectious nucleic acids?
   a. Retroviruses
   b. SV40
   c. Influenza

d. Poliovirus

e. Poxvirus

4. What ways do viruses use to maximize the utilization of their genomes?

5. Would you expect to find in uninfected cells an enzymatic activity that can remove the terminal protein attached to the 5′ end of the poliovirus genome. If so, why?

6. Compare the patterns of splicing shown by the early and late genes of papovaviruses and adenoviruses.

7. How do human influenza viruses evolve to elude the host's immune system?

8. It has been said that "retroviruses have committed acts of piracy on the cellular genome." Explain.

9. What are the current theories concerning the action of the transforming proteins of SV40 and polyoma virus?

10. How do the proteins coded by the oncogenes of retroviruses appear to differ from the transforming proteins of SV40 and polyoma viruses?

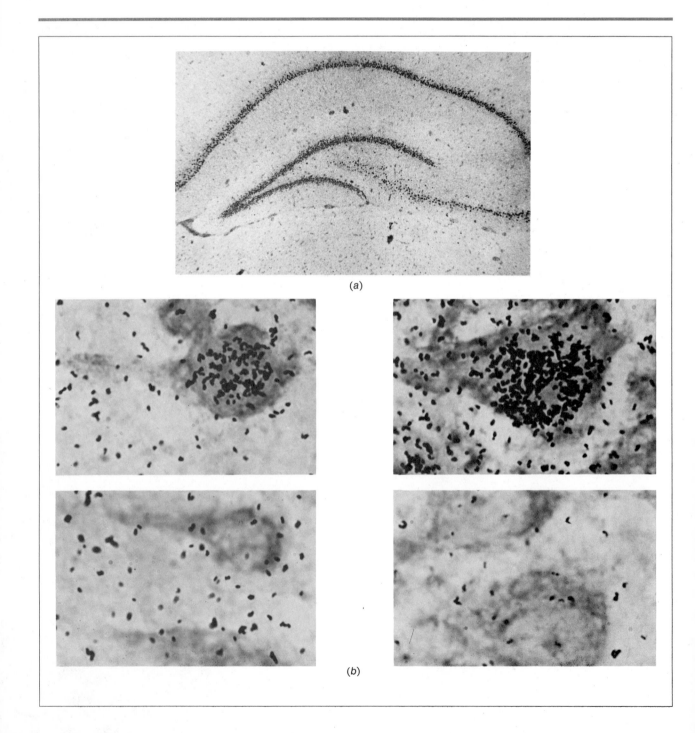

(a)

(b)

# 29

# HORMONE ACTION

Communication among cells of multicellular organisms is achieved by chemical messengers that pass between cells. These chemicals serve to coordinate the metabolic activity of the various tissues, allow the organism to adapt to a changing environment, and prepare it for reproduction. One major class of chemical messengers is liberated by neurons in response to stimulation. If these compounds diffuse to neighboring cells to activate specific responses, they are called _neurotransmitters;_ examples include acetylcholine, epinephrine, norepinephrine, dopamine, and γ-aminobutyrate. If, however, they are liberated into the circulatory system, they are referred to as _neurohormones;_ examples include hypothalamic regulatory proteins and hormones of the neurohypophysis. Another major class of chemical messengers is known as _hormones_, which were originally defined as molecules synthesized in specialized ductless glands and carried by the circulatory system to other parts of the organism where they evoke specific responses. In many cases, hormone synthesis and secretion are regulated either directly or indirectly by neurohormones or neurotransmitters. Likewise, hormones have profound effects on neural activity. These interactions between the nervous system and endocrine glands form feedback loops that serve to integrate the metabolic activities of the organism.

The goal of molecular endocrinology is to understand the biochemical processes leading from hormone synthesis to hormone action. In this chapter we will concentrate on the principles involved, using examples from many different endocrine systems, without describing the vast array of physiological processes that are regulated by hormones; this latter aspect is the subject of endocrinology texts. Three major topics will be covered here: (1) hormone synthesis and its control, (2) hormone-receptor interactions and their sequelae, and (3) regulation of hormone response.

_Localization of steroid receptors by autoradiography. Rats were injected with radioactive estradiol. Then frozen sections of the brain were prepared, mounted on slides, and dipped in photographic emulsion. After several weeks, the slides were developed. The location of the silver grains reveals which cells took up the hormone. (a) A low magnification of the hippocampul region. (b) A higher-power magnification showing the location of the receptors in the nuclei of specific cells. (Reprinted with permission of B. McEwen of the Rockefeller Institute.)_

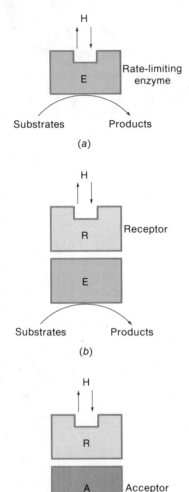

**Figure 29-1**
Possible mechanisms of hormone action. *(a)* The hormone (H) could theoretically activate an enzyme (E) directly as an allosteric effector. *(b)* Alternatively, there might be a separate binding protein for the hormone, called a receptor (R), which then activates an enzyme. *(c)* Another possibility interposes an acceptor protein (A) between the receptor and enzyme.

The molecular aspects of hormone action are similar to many other processes that involve the relay of extracellular chemical signals into cellular responses. For example, the biochemical distinctions between the action of *pheromones* (that travel between organisms), hormones (that travel between tissues), and neurotransmitters (that travel between cells) are slight; the differences between these classes of chemical stimuli relate largely to the distance they travel. In fact, several compounds that were initially discovered as hormones are now being shown to act as neurotransmitters as well. In addition to the classical hormones, many *polypeptide growth factors* circulate in the blood and act on target cells in a manner analogous to hormones.

It is likely that many of the processes involved in cell differentiation are mediated by hormone-like substances that are secreted by one cell and modify the developmental potential of neighboring or distant cells. That such substances exist is well-established, but they have been difficult to isolate and study because they are present in minute amounts. Nevertheless, analysis of the action of these compounds will likely fit into the intellectual framework of molecular endocrinology. Virus-infected cells synthesize a hormone-like macromolecule called *interferon* that diffuses to neighboring cells and stimulates them to synthesize proteins that will protect them against viral infection. The action of interferon shares many features with hormone action.

## HORMONE ACTION IS MEDIATED BY RECEPTORS

Initially it was thought that hormones might bind directly to specific rate-limiting enzymes, thereby either activating or inactivating them (Figure 29-1a). However, when this idea was not confirmed experimentally, the concept of a receptor was introduced. *Receptors* are proteins that bind a hormone with high specificity and affinity. *Each hormone binds with high affinity to biochemically distinct receptors. Binding of the hormone induces a conformational change in the receptor, and this change perturbs other macromolecules.* In the simplest scheme, the receptor might be coupled to a rate-limiting enzyme (Figure 29-1b). This idea also proved to be too simple, at least for the case that we know most about—namely, the activation of adenylate cyclase. In this case, another protein is interspersed between receptor and rate-limiting enzyme. The protein that is activated by the receptor is called an *acceptor*, and it is responsible for directly mediating enzyme activation (Figure 29-1c). All hormones that activate (or inactivate) adenylate cyclase are thought to function by means of receptors and acceptors. However, the molecular mechanism of action of many hormones remains unknown; thus the simpler schemes shown in Figure 29-1 may turn out to be applicable in some hormonal systems. Likewise, more complicated trains of events also may be discovered. Consider the potential advantages and disadvantages of each of the schemes shown in Figure 29-1. Also note that the complexes shown are depicted as permanently coupled. In fact, each of the components is thought to be in equilibrium, perhaps as depicted in Figure 29-2. Now you can see the potential for multiple receptors to impinge on the same acceptors or for different acceptors to have different effects on the same enzyme. Receptors may act catalytically such

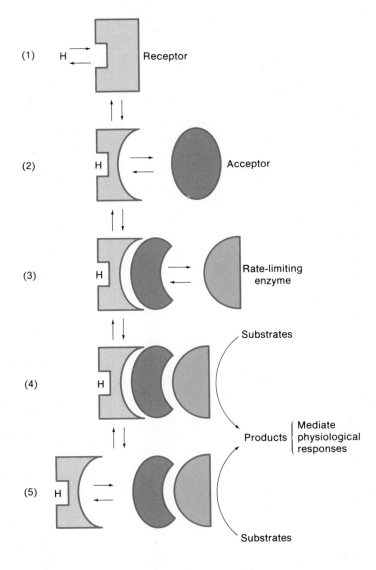

**Figure 29-2**
Generalized scheme of hormone-receptor action. A hormone (H) plus receptor interact reversibly (1) to produce a hormone-receptor complex (2) with an altered conformation that allows the receptor to interact with an acceptor molecule to form a hormone-receptor-acceptor complex (3). This process activates the acceptor (indicated by another conformational change) so that it can modulate the activity of a rate-limiting step (4) in metabolism or transcription. The activated acceptor may be able to stimulate the enzyme in the absence of the receptor (5) if the receptor acts catalytically.

that one receptor may activate many acceptor molecules. The kinetics of interaction of the various components also may be important for a dynamic control of enzyme activity.

The important point is that *all hormones act by binding to macromolecular receptors that are located either on the cell membrane or inside of responsive cells.* Binding of the hormone induces a conformational change in the receptor; this conformational change perturbs other macromolecules (acceptors) and elicits a chain of events that leads to a vast array of cellular changes ranging from alterations in enzyme activities to changes in gene expression. With time, these effects may lead to profound alterations in cell growth, morphology, and function. A detailed description of the molecular events leading from hormone-receptor binding to activation of rate-limiting enzymes constitutes one of the major goals of molecular endocrinology.

Table 29-1 lists many of the common hormones of vertebrates. Structurally, they range from amino acid derivatives, such as epinephrine and melatonin, to cholesterol-derived steroid hormones, to large polypeptides.

**Table 29-1**
Vertebrate Hormones*

| Endocrine Organ | Hormone | Structure | Function |
| --- | --- | --- | --- |
| Pineal | Melatonin | N-Acetyl-5-methoxytryptamine | Regulates circadian rhythms |
| Hypothalamus† | Corticoliberin (Corticotropin-releasing factor, CRF; or CRH) | Polypeptide (41 residues) | Stimulates TSH, GH, and prolactin secretion |
| | Gonadoliberin (Gonadotropin-releasing factor, GnRF; or LH-RH) | Polypeptide (10 residues) | Stimulates LH and FSH secretion |
| | Prolactoliberin (Prolactin-releasing factor, PRF) | Unknown | Stimulates prolactin secretion |
| | Prolatostatin (Prolactin-release inhibiting factor, PIF) | Unknown | Inhibits prolactin secretion |
| | Somatoliberin (Somatotropin-releasing factor, SRF; or GH-RF or GH-RH) | Unknown | Stimulates somatotropin secretion |
| | Somatostatin (Somatotropin-release inhibiting factor, SIF) | Polypeptide (14 residues) | Inhibits somatotropin secretion |
| | Thyroliberin (Thyrotropin-releasing factor, TRF; or TRH) | Polypeptide (3 residues) | Stimulates thyrotropin, somatotropin, and prolactin secretion |
| Pituitary | | | |
| Neurohypophysis | Oxytocin (ocytocin) | Polypeptide (9 residues) | Uterine contraction, milk ejection |
| | Vasopressin (antidiuretic hormone, ADH) | Polypeptide (9 residues) | Blood pressure, water balance |
| | Melanocyte-stimulating hormones (MSH) | α Polypeptide (13 residues) β Polypeptide (18 residues) γ Polypeptide (12 residues) | Pigmentation |

| Gland | Hormone | Chemical nature | Function |
|---|---|---|---|
| Adenohypophysis | Lipotropin (LPH) | $\beta$ Polypeptide (93 residues) $\gamma$ Polypeptide (60 residues) | Fatty acid release from adipocytes |
| | Corticotropin (adrenocorticotropic hormone, ACTH) | Polypeptide (39 residues) | Stimulates glucocorticoid synthesis |
| | Thyrotropin (thyroid-stimulating hormone, TSH) | 2 Polypeptides ($\alpha$, 13,600 daltons; $\beta$, 14,700 daltons) | Stimulates thyroid hormone synthesis |
| | Somatotropin (growth hormone, GH) | Polypeptide (191 residues) | General anabolic effects; stimulates release of growth factors |
| | Prolactin | Polypeptide (197 residues) | Stimulates milk synthesis |
| | Lutropin (luteinizing hormone, LH) | 2 Polypeptides ($\alpha$, 13,600 daltons; $\beta$, 14,900 daltons) | Ovary: luteinization, progesterone synthesis; testis: interstitial cell development, androgen synthesis |
| | Follitropin (follicle-stimulating hormone, FSH) | 2 Polypeptides ($\alpha$, 13,600 daltons; $\beta$, 23,000 daltons) | Ovary: follicle development, ovulation, estrogen synthesis; testis: spermatogenesis |
| Thyroid | Thyroxine and triiodothyronine | Iodinated dityrosine derivatives (see Figure 29-13) | General stimulation of many cellular reactions |
| Parathyroid | Calcitonin | Polypeptide (32 residues) | $Ca^{2+}$ and $P_i$ metabolism |
| | Parathyroid hormone (parathyrin, PTH) | Polypeptide (84 residues) | $Ca^{2+}$ and $P_i$ metabolism |
| Alimentary tract‡ | Gastrin | Polypeptide (17 residues) | Stimulates acid secretion from stomach and pancreatic secretion |
| | Secretin | Polypeptide (27 residues) | Regulates pancreas secretion |
| | Pancreozymin (cholecystokinin) | Polypeptide (33 residues) | Secretion of digestive enzymes |
| | Motilin | Polypeptide (22 residues) | Controls gastrointestinal muscles |
| | Vasoactive intestinal peptide (VIP) | Polypeptide (28 residues) | Gastrointestinal relaxation; inhibits acid and pepsin secretion |
| | Gastric inhibitory peptide (GIP) | Polypeptide (43 residues) | Inhibits gastrin secretion |
| | Somatostatin | Polypeptide (14 residues) | Inhibits gastrin secretion; inhibits glucagon secretion |

(*Continued*)

**Table 29-1** (*Continued*)

| Endocrine Organ | Hormone | Structure | Function |
|---|---|---|---|
| Pancreas | Insulin | 2 Polypeptides (21 and 30 residues) | Glucose uptake, lipogenesis, general anabolic effects |
| | Glucagon | Polypeptide (29 residues) | Glycogenolysis, release of lipid |
| | Pancreatic polypeptide | Polypeptide (36 residues) | Glycogenolysis, gastrointestinal regulation |
| | Somatostatin | Polypeptide (14 residues) | Inhibition of somatotropin and glucagon release |
| Adrenal cortex | Glucocorticoids | Steroids (cortisol, corticosterone) | Many diverse effects on protein synthesis and inflammation |
| | Mineralocorticoids | Steroids (aldosterone) | Maintains salt balance |
| Adrenal medulla | Epinephrine | Tyrosine derivative (see Figure 29-9) | Smooth-muscle contraction, heart function, glycogenolysis, lipid release |
| | Norepinephrine | Tyrosine derivative (see Figure 29-9) | Arteriole contraction, lipid release |
| Ovary | Estrogens | Steroids (estradiol, estrone) | Maturation and function of secondary sex organs |
| | Progestins | Steroids (progesterone) | Ovum implantation, maintenance of pregnancy |
| Testis | Androgens | Steroids (testosterone) | Maturation and function of secondary sex organs |
| Placenta | Estrogens / Progestins | Steroids | Maintenance of pregnancy |
| | Choriogonadotropin | 2 Polypeptides (α, 13,500 daltons; β, 26,500 daltons) | Similar actions to LH and FSH |
| | Choriomammotropin (placental lactogen) | Polypeptide (191 residues) | Similar to prolactin |
| | Relaxin | 2 Polypeptides (22 and 32 residues) | Muscle tone |
| Liver | Angiotensin§ | Polypeptide (8 residues) | Responsible for essential hypertension |
| Kidney | 1,25-dihydroxyvitamin D$_3$ | Steroid | Calcium uptake, bone formation |

*Only the more common hormones of known structure are listed.

†Most of the hypothalamic releasing factors are also called hypothalamic regulatory hormones.

‡Many of these peptides are also bound in the brain, where they may modulate neural activity.

§The liver secretes α$_2$globulin, which is cleaved by renin, a kidney enzyme, to give a decapeptide, proangiotensin, from which the carboxy terminal dipeptide is removed to give angiotensin.

Many polypeptide hormones have several different names or abbreviations; these are indicated in Table 29-1. In the case of steroid hormones, generic names, which refer to an entire class of compounds that interact with a specific receptor, will frequently be used. For example, estrogen is a generic name that includes the specific compounds 17$\beta$-estradiol, estrone, diethylstilbestrol, etc. Likewise, progestin, androgen, glucocorticoid, and mineralocorticoid are generic names.

*The receptors for all polypeptide hormones and most amino acid derivatives are located on the cell membrane, whereas the steroid hormones and thyroid hormones pass through the membrane to interact with cytoplasmic or nuclear receptors.* Interaction of hormones with membrane receptors usually elicits the formation of a diffusible second messenger (e.g., cyclic AMP) within the cell. As noted in Chapter 8, cAMP activates a protein kinase, which in turn phosphorylates specific proteins, thereby activating or inactivating them. The substrates of protein kinase are frequently rate-limiting enzymes in metabolic pathways; however, some of them also may be involved in the regulation of gene expression. The receptors for steroid hormones and thyroid hormones are thought to regulate gene expression by binding to specific DNA sequences adjacent to certain genes, thus activating or inactivating their transcription.

*Different cell types may respond quite differently to the same hormone-receptor interaction.* The response depends on which enzymes are available for phosphorylation (in the case of cAMP-mediated hormonal responses) and which genes are capable of being expressed (in the case of steroid hormones). For example, glucagon stimulates glycogenolysis in hepatocytes and lipid mobilization in adipocytes. Likewise, estrogens stimulate genes coding for egg white proteins in the avian oviduct and genes coding for egg yolk proteins in the avian liver. These differences in response are ultimately determined by which genes are activated (or inactivated) during development; the mechanisms involved in this process of gene commitment are not well understood.

Because of the diversity of hormones and their effects, it would not be surprising for a single vertebrate cell to have distinct receptors for a specific growth factor, insulin, thyroid hormone, a steroid hormone, and one or more hormones that regulate cAMP levels.

## HORMONE SYNTHESIS

*The synthesis of many hormones is ultimately regulated by the brain through neurohormones.* For example, the synthesis of aldosterone, cortisol, and corticosterone, the major hormones of the adrenal cortex, is stimulated by corticotropin (ACTH), which is liberated from the pituitary gland in response to the neurohormone corticolibrin (or corticotropin regulatory hormone, CRH), and this is secreted from hypothalamic nerves in response to inputs from the central nervous system (Figure 29-3). The elevation in serum glucocorticoids (e.g., cortisol) is detected by glucocorticoid receptors in the hypothalamic region of the brain, and this inhibits the release of corticolibrin, thus completing a feedback loop (Figure 29-4). The figure also shows several other feedback loops that serve to maintain blood glucose and sodium at appropriate levels. Similar pathways are operative for

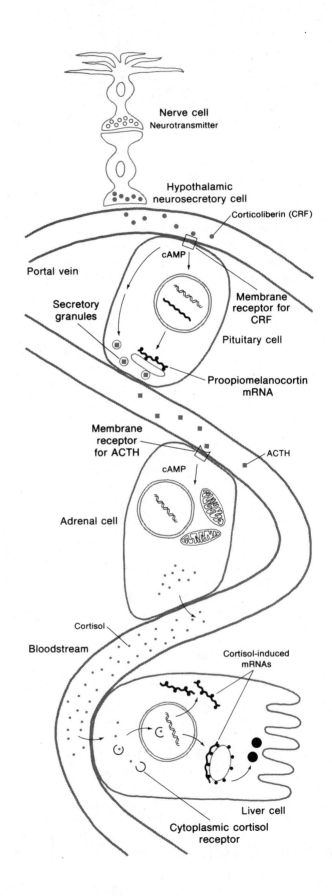

**Figure 29-3**
A hormonal cascade. A neurotransmitter released from specialized brain cells stimulates a hypothalamic neurosecretory cell to secrete corticoliberin, which travels through portal veins to the pituitary, where it activates receptors of target cells leading to the synthesis and secretion of corticotropin (ACTH). This pituitary hormone travels through the bloodstream to the adrenals, where it binds to membrane receptors that activate adenylate cyclase. Cyclic AMP production in these cells stimulates steroid hormone (e.g., cortisol) biosynthesis. The steroids that are released into the blood bind to cytoplasmic receptors in other cells (e.g., liver). These receptors subsequently migrate into the nucleus and activate transcription of specific genes. These new gene products code for proteins that stimulate a variety of metabolic activities.

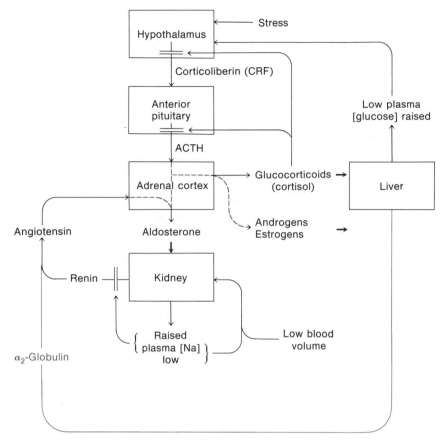

**Figure 29-4**
Feedback loops involved in
metabolic regulation.

regulation of gonadal steroid synthesis and thyroid hormone synthesis. The synthesis and secretion of pancreatic, alimentary, and parathyroid hormones are regulated predominantly by nutrients (glucose and fatty acids) and minerals (calcium and phosphorus).

## Steroid Hormones Are Derived from Cholesterol by Stepwise Removal of Carbon Atoms and Hydroxylation

Hormonal activation of steroid hormone biosynthesis involves the transport of cholesterol into the mitochondria followed by cleavage of the side chain. Because most of the cholesterol exists as cholesterol esters in the cytoplasm, hydrolysis of the fatty acid esters by _cholesterol ester hydrolase_ is the first step in providing substrates for steroid biosynthesis. Cleavage of the side chain in the mitochondria is performed by an enzyme complex called _cholesterol desmolase_, which hydoxylates the side chain in two positions and then cleaves it to yield pregnenolone, as discussed in Chapter 15 and shown schematically in Figure 29-5. These are the rate-limiting steps in steroid biosynthesis in both the adrenals and gonads. Thus it is not surprising that they are regulated by corticotropin in the adrenals and follicle-stimulating hormone and luteinizing hormone in the gonads. Each of these peptide hormones activates adenylate cyclase. Cholesterol ester hydrolase is known to be activated by cAMP-dependent protein kinase. The regulation of desmolase is less well understood. Steroid hormones, unlike all other hormones, are not stored for subsequent secretion. Consequently, _the circulating level of steroids is controlled largely by the rate of steroid biosynthesis_. Nevertheless, if an animal is stressed, for example, the serum concentration of corticosterone can rise several fold within a few minutes. This rapid response includes the time necessary for signals to pass from hypothalamus to pituitary (CRH) and pituitary to adrenal cortex (ACTH), as well as activation of steroid biosynthesis.

The step-by-step biosynthesis of steroid hormones from cholesterol ($C_{27}$) was presented in Chapter 15 and is summarized in Figure 29-5 with emphasis on the number of carbon atoms. Note that pregnenolone ($C_{21}$) and progesterone ($C_{21}$) are intermediates in the synthesis of all the major adrenal steroids, including cortisol ($C_{21}$), corticosterone ($C_{21}$), and aldosterone ($C_{21}$), as well as intermediates in the synthesis of the gonadal steroid hormones: testosterone ($C_{19}$) and estradiol ($C_{18}$). Because the synthesis of these hormones follows a common pathway, a defect in the activity or amount of an enzyme along the pathway can lead to both a deficiency in the hormones beyond the affected step and an excess of the hormones or metabolites prior to that step. Deficiencies in each of the six enzymes involved in the conversion of cholesterol to aldosterone have been observed in humans. Each deficiency gives rise to a characteristic steroid hormone imbalance with telling clinical consequences. For example, a deficiency in 17-hydroxylase gives rise to inadequate levels of cortisol as well as inadequate levels of androgens and estrogens, with severe effects on sexual maturation. A deficiency in the next enzyme along the pathway, 21-hydroxylase, blocks the synthesis of adrenal glucocorticoid and mineralocorticoid hormones and leads to an overproduction of testosterone by the adrenals. The overproduction of androgens is due to metabolic shunting of progesterone into the sex steroid pathway as a result of the absence of the enzyme required for the glucocorticoid and mineralocorticoid pathways. The synthesis of androgens is exacerbated by the lack of feedback inhibition of cortisol on the hypothalamus, resulting in the chronic production of CRH and corticotropin and perpetual activation of steroid biosynthesis. A consequence of excessive androgen production in female offspring is masculinization of the genitals (a condition referred to as _female pseudohermaphrodism_, i.e., a genetic female with male appearance). This reversal of sexual phenotype is explained by the fact that during embryonic development of mammals, the external genitalia develop from common primordia. In the absence of hormonal stimulation, they will develop into female structures, but androgens will direct their development into male structures.

### Testosterone Is Both a Hormone and a Prohormone

The high levels of _testosterone normally produced by the testes have a major role in the growth and function of many tissues in addition to the reproductive organs_. Essentially all the sexual differences of nonreproductive tissues, such as muscle, liver, and brain, are a consequence of androgen action. Although testosterone is the major circulating androgen, many tar-

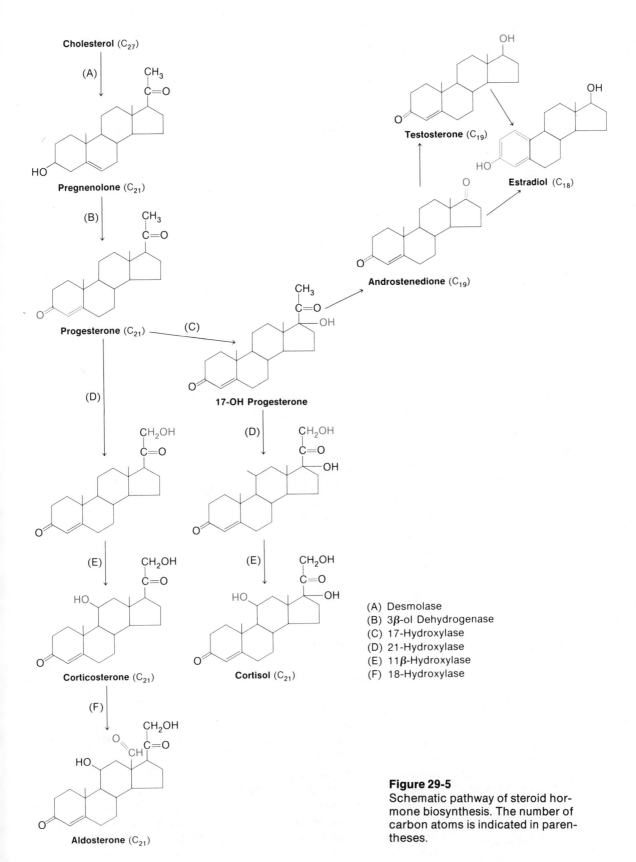

**Figure 29-5**
Schematic pathway of steroid hormone biosynthesis. The number of carbon atoms is indicated in parentheses.

(A) Desmolase
(B) 3β-ol Dehydrogenase
(C) 17-Hydroxylase
(D) 21-Hydroxylase
(E) 11β-Hydroxylase
(F) 18-Hydroxylase

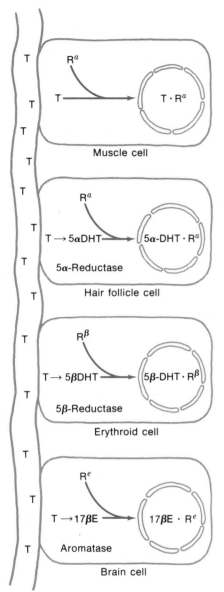

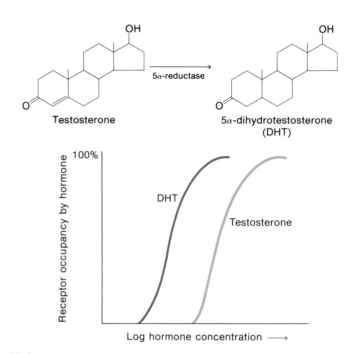

**Figure 29-6**
Conversion of testosterone to 5 α-dihydrotestosterone. 5 α-DHT has a higher affinity for the androgen receptor than testosterone, as revealed by receptor-binding studies.

**Figure 29-7**
Metabolic conversion of testosterone by target cells. Testosterone (T) is the predominant androgen in the bloodstream. When testosterone enters the cytoplasm of target cells, it can be metabolized in a variety of different ways. It can either bind directly to androgen receptors ($R^a$) or be reduced to 5 α-dihydrotestosterone (5α-DHT), which then binds to $R^a$ with higher affinity. Other target cells reduce testosterone to 5 β-dihydrotestosterone (and other 5 β metabolites), which bind to a distinct receptor ($R^\beta$). Yet other cells convert testosterone into an estrogen, 17 β-estradiol, which binds to estrogen receptors ($R^e$).

get cells reduce this steroid to _5α-dihydrotestosterone_, a steroid that binds to the androgen receptor with a higher affinity than testosterone (Figure 29-6). When _5α-reductase_ is defective, the androgen receptors are only partially activated and a full androgen response is not obtained. The consequence of a deficiency of this enzyme is a reduced level of this more potent androgen; hence those tissues which require high levels of androgens for normal development, such as primordia of external genitalia, do not develop normally (they resemble the female phenotype, a clinical condition called _male pseudohermaphrodism, type 2_).

In the example given earlier, testosterone can be considered a prohormone of a more active androgen that binds to the same receptor as testosterone. However, testosterone also can be a prohormone of metabolites that bind to different receptors. For example, some of the effects of androgens on erythropoiesis are due to reduction of testosterone to 5β-dihydrotestosterone within certain blood precursor cells. _5β-Dihydrotestosterone_ binds to a receptor that is distinct from the androgen receptor that binds testosterone and 5α-dihydrotestosterone (Figure 29-7). Testosterone also influences erythropoiesis by stimulating the kidney to synthesize and secrete a hormone-like growth factor called _erythropoietin_. Another striking example of testosterone as a prohormone occurs in the brain, where testosterone influences neural development and activity (e.g., male-specific mating behavior, bird songs, and territorial displays). In these cases the receptors for testosterone are actually estrogen receptors. Thus in the brain, testosterone is metabolized to 17β-estradiol by _aromatase_, the same enzyme involved in 17β-estradiol synthesis in the female ovary (Figure 29-7).

## Vitamin D Is a Precursor of the Hormone 1,25-dihydroxyvitamin $D_3$

Vitamin D is essential for normal calcium and phosphorus metabolism. Vitamin $D_3$ is formed from 7-dehydrocholesterol by ultraviolet photolysis in the skin. Insufficient exposure to sunlight gives rise to a deficiency in vitamin $D_3$ and produces *rickets*, a condition characterized by weak, malformed bones. Vitamin $D_3$ is inactive, but it is converted into an active hormone by two hydroxylations. The first hydroxylation occurs in the liver and gives rise to 25-hydroxyvitamin $D_3$, abbreviated $25(OH)D_3$; the second hydroxylation occurs in the kidney and gives rise to the active product 1,25-dihydroxyvitamin $D_3$, or $1,25(OH)_2D_3$ (Figure 29-8). The hydroxylation in the kidney is stimulated by parathyroid hormone (PTH), which is secreted from the parathyroid gland in response to low calcium. In the presence of adequate calcium, $25(OH)D_3$ is converted to an inactive metabolite, $24,25(OH)_2D_3$. The active derivative of vitamin $D_3$ is considered a hormone because it is transported from the kidneys to target cells, where it binds to receptors that migrate into the nucleus and activate transcription. $1,25(OH)_2D_3$ stimulates calcium transport by intestinal cells and increases calcium uptake by osteoblasts (bone cells).

## Catecholamines Are Synthesized from Tyrosine and Are Stored in Granules

The catecholamines epinephrine and norepinephrine are synthesized from tyrosine in the medullary portion of the adrenal glands. The biosynthetic pathway is shown in Figure 29-9. The first step, which involves oxidation of

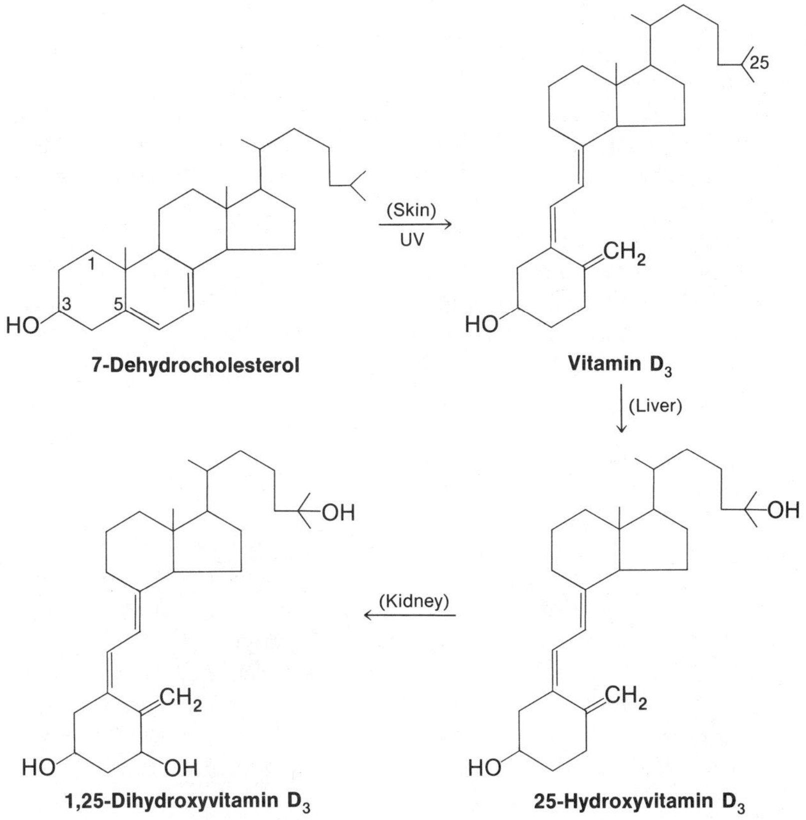

**Figure 29-8**
Hydroxylation of vitamin $D_3$ by the liver and kidney converts it into an active hormone, 1,25-dihydroxyvitamin $D_3$.

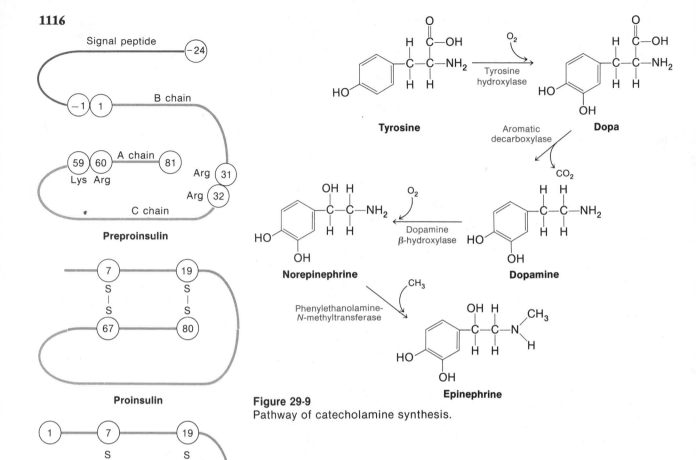

**Figure 29-9**
Pathway of catecholamine synthesis.

**Figure 29-10**
Biosynthesis of insulin. Insulin is synthesized by membrane-bound polysomes in the $\beta$ cells of the pancreas. The primary translation product is preproinsulin, which contains a 24-residue signal peptide preceding the 81-residue proinsulin molecule. The signal peptide is removed by signal peptidase, cutting between Ala ($-1$) and Phe ($+1$), as the nascent chain is transported into the endoplasmic reticulum. Proinsulin folds and two disulfide bonds cross-link the ends of the molecule as shown. Before secretion, a trypsin-like enzyme cleaves after a pair of basic residues 31, 32 and 59, 60; then a carboxypeptidase B-like enzyme removes these basic residues to generate the mature form of insulin.

tyrosine to 3,4-dihydroxyphenylalanine (dopa), is catalyzed by tyrosine hydroxylase. This is the rate-limiting enzyme in the pathway, and its activity is controlled by a cAMP-dependent protein kinase. Epinephrine and norepinephrine are stored in chromaffin granules. These granules contain catecholamine and ATP in a molar ratio of 4:1 complexed with protein. Neural stimulation of the medulla is mediated by acetylcholine, which binds to receptors on the membranes of chromaffin cells and leads to a local depolarization and an influx of $Ca^{2+}$, and this results in the fusion of some chromaffin granules with the cell membrane and, consequently, extrusion of a packet of catecholamines and ATP into the circulation.

## Polypeptide Hormones Are Synthesized as Precursors

*All polypeptide hormones for which the mechanism of synthesis is known are synthesized as precursors.* The primary translation products of the mRNAs coding for insulin, glucagon, prolactin, growth hormone, gastrin, parathyroid hormone, vasopressin, and corticotropin have been characterized by cell-free translation, and in many cases, the mRNAs that code for these hormones have been cloned into bacterial plasmids and sequenced. In all these cases, *the primary translation product contains 20 to 30 predominantly hydrophobic amino acid residues at the amino terminus that function as a signal to direct the nascent polypeptide chain through the endoplasmic reticulum.* During the process of being transported across the membrane, the signal peptides are removed by signal peptidase, and asparagine-linked glycosylation may also occur. *In most cases, the polypeptide hormone still exists as a precursor (sometimes called a prohormone) after removal of the signal peptide.* The precursors travel from the endoplasmic reticulum to the Golgi area, where they are condensed into secretory granules. *Further cleavage of the prohormones occurs prior to secretion into the serum.* For example, the two polypeptides of insulin are synthesized as part of a single polypeptide chain (proinsulin) that folds back upon itself and is stabilized by disulfide bonds; then trypsin-like and carboxypeptidase-like enzymes remove a connecting peptide (C-peptide) to convert proinsulin into active insulin (Figure 29-10).

The search for the precursor to corticotropin (ACTH) led to the discovery that this hormone, as well as β-lipotropin, melanocyte-stimulating hormone (MSH), endorphin, and enkephalin, is derived from a common precursor called *proopiomelanocortin* (Figure 29-11). This precursor is synthesized in several different cell types of the pituitary. Each of the hormones and neuropeptides is separated by a pair of basic amino acids. Cleavage at these sites by a trypsin-like enzyme generates the active hormones; however, the cleavage pattern appears to be different in different cell types, so some cells may produce more of one hormone than another.

The peptides synthesized by neurosecretory cells are often much larger than the final active products. For example, oxytoxin and vasopressin (each a nonapeptide) are derived from proteins called *proneurophysins* that are about 160 and 215 amino acids long, respectively. After cleavage of the hormone from proneurophysin, the active hormone remains associated with the neurophysin until secretion. The neurophysin may serve in these instances as a carrier protein within the neurosecretory cells. Somatostatin, one of the hypothalamic releasing hormones, has recently been shown to be the carboxyl-terminal 14 amino acids of the precursors having 121 or 125 amino acids. Many of the other small polypeptide hormones are probably synthesized in a similar manner. It has been suggested that the smallest polypeptide hormone, thyrotropin-releasing factor (TRF), which is com-

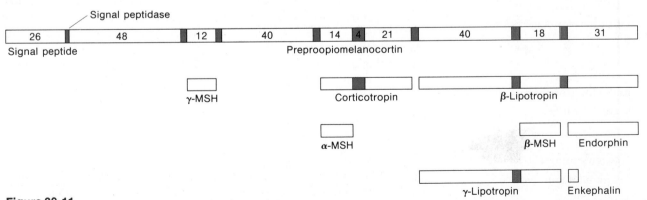

**Figure 29-11**
Processing pathway of preproopiomelanocortin. This precursor is cleaved into a variety of active peptides. The cleavage sites are generally pairs of basic amino acids, although one site contains four (indicated in color). The number of amino acids in each polypeptide is indicated. The processing pathway varies in different cell types.

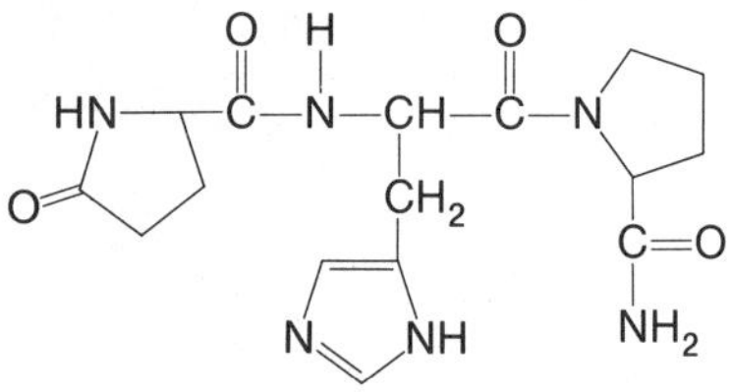

**Pyroglutamylhistidylprolinamide
thyroliberin** (TRF)

**Figure 29-12**
Thyroliberin. Also called thyrotropin
releasing factor. The mechanism of
synthesis of this hypothalamic hormone remains uncertain.

**Table 29-2**
Homologous Polypeptide Hormones

Prolactin

Placental lactogen

Growth hormone

Thyrotropin

Luteinizing hormone

Follicle-stimulating hormone

Choriogonadotropin

Insulin

Relaxin

Insulin-like growth factors (IGF-I, IGF-II)

Nerve growth factor

Glucagon

Vasoactive intestinal peptide

Gastric inhibitory peptide

Secretin

Gastrin

Pancreozymin

posed of only three amino acids (Figure 29-12), may be synthesized enzymatically without the direct involvement of mRNA and ribosomes; however, direct evidence for or against this hypothesis is not yet available.

Some of the polypeptide hormones contain two polypeptide chains that
are derived from separate mRNAs, e.g., thyroid-stimulating hormone
(TSH), luteinizing hormone (LH), follicle-stimulating hormone (FSH), and
choriogonadotropin (CG). The assembly of the two chains by disulfide
bond formation occurs shortly after synthesis, probably in the endoplasmic
reticulum or in the Golgi. It is noteworthy that all these hormones share a
nearly identical $\alpha$ subunit but have different $\beta$ subunits that provide receptor specificity. Each of these hormones activates adenylate cyclase.

Many of the polypeptide hormones fall into homologous families
(Table 29-2). The similar chemical structures of different members of a
family suggest that they probably evolved from a common gene by gene
duplication and subsequent divergence. This process, along with a parallel
divergence of receptors, could lead to increasingly fine regulation of metabolic processes during evolution. Because of their homology, polypeptides
within a hormone family often show some affinity for all the receptors
within that family. For example, most of the actions of prolactin are mimicked by placental lactogen. Likewise, many of the actions attributed to
pharmacologic doses of insulin are probably due to insulin binding to homologous growth factor receptors.

## Thyroid Hormones Are Derived from
## Iodinated Thyroglobulin

Thyroxine ($T_4$) and the more potent triiodothyronine ($T_3$) are cleaved
from a large precursor protein called _thyroglobulin_. It exists as a dimer of
660,000 daltons. Thyroglobulin is secreted into the lumen of the gland,
where several of the tyrosine residues are iodinated in one or two positions
by a peroxidase; then two iodinated residues condense as shown in Figure
29-13. Secretion of thyroid hormones is initiated by endocytosis of the
thyroglobulin, followed by fusion of the endocytotic vesicles with lysosomes. Lysosomal enzymes degrade the thyroglobulin, liberating triiodothyronine and thyroxine into the circulation. Less than five molecules of
$T_3$ and $T_4$ are generated from each molecule of thyroglobulin. Thyroglobulin is a storage protein for iodine and can be considered a prohormone of
active circulating thyroid hormones. Thyroid hormone secretion is stimulated by thyrotropin (thyroid-stimulating hormone, TSH), a pituitary hormone that activates adenylate cyclase. TSH release is stimulated by thyrotropin-releasing factor (TRF), and release of both these hormones is
inhibited by $T_3$, $T_4$, and somatostatin.

## REGULATION OF HORMONE
## CONCENTRATION IN THE BLOOD

_The fraction of receptors occupied by hormone can fluctuate greatly and is
ultimately determined by the concentration of "free" hormone in the
blood._ Some of the major determinants of hormone concentration are the
rate of hormone secretion from endocrine cells, the concentration of serum
carrier proteins, and the rate of hormone clearance, degradation, or metabolic inactivation.

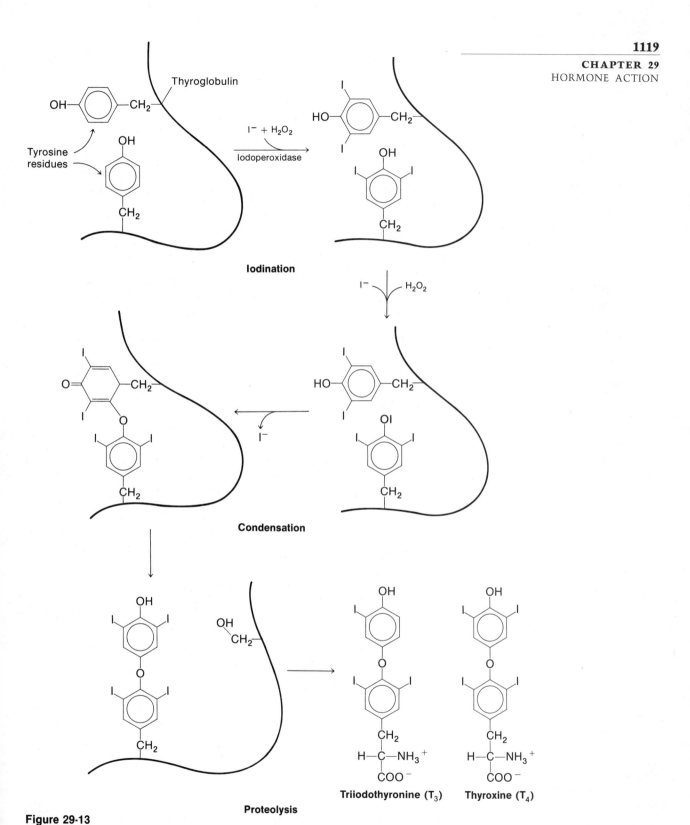

**Figure 29-13**

Pathway of thyroxine and triiodothyronine synthesis. Thyroid cells actively transport iodine ($I^-$), which is incorporated into the tyrosine residues of thyroglobulin by the enzyme iodoperoxidase. After condensation of iodinated tyrosine residues, the thyroglobulin is proteolytically degraded, liberating thyroxine and triiodothyronine.

Most vertebrate hormones, except for the steroid hormones, are stored in membrane-bound granules. Upon appropriate stimulation, usually by another hormone (e.g., TRF acting upon TSH-synthesizing cells) or as a result of neural excitation (e.g., sympathetic innervation of epinephrine-synthesizing cells of adrenal medulla), the membranes of these granules fuse with the cellular membrane that borders the capillaries and the contents are liberated into the circulation. The mechanisms involved in secretory granule movement and fusion with the membrane are poorly understood. However, secretion appears to be coupled with a breakdown of phosphatidylinositol, elevation of "free" intracellular calcium, and an increase in cyclic GMP. *Because of these hormonal stores, it is possible for the serum concentration of hormone to increase very rapidly upon appropriate stimulation.* Stimulation of hormone secretion is usually coupled with an increase in hormone synthesis, so that the hormonal stores are replenished.

*Most hormones have a half-life in the blood of only a few minutes because they are cleared or degraded very rapidly.* Polypeptide hormones are removed from circulation by serum and cell-surface proteases, endocytosis followed by lysosomal degradation, and glomerular filtration in the kidney. Steroid hormones are taken up by the liver and metabolized by hydroxylation, sulfation, or conjugation with glucuronic acid to inactive forms that are then excreted into the bile duct or back into the blood for removal by the kidneys. Catecholamines are metabolically inactivated by O-methylation, deamination, and conjugation with sulfate or glucuronic acid.

*Thyroid hormones and most steroid hormones exist in the serum in association with carrier proteins,* called appropriately thyroxine-binding globulin, transcortin (for cortisol), and sex-steroid-binding protein. These proteins have high affinity ($K_d \sim 10^{-8}$ to $10^{-9}$ $M$) for their respective hormones. They presumably function to buffer the concentration of "free" hormone and to retard hormone degradation and excretion. These carrier proteins are distinguishable from the intracellular receptors.

The serum concentrations of many hormones fall in the range of $10^{-12}$ to $10^{-7}$ $M$. The biological assays that were originally used to measure these low concentrations of hormones have now been supplemented or completely replaced by extremely sensitive *radioimmunoassays* (RIA). The essential ingredients of an RIA are an antibody that specifically binds the hormone and a radioactively labeled hormone. The principal of an RIA is competition of the unlabeled hormone (test sample or standards) with radioactively labeled hormone for a limiting amount of antibody. The sensitivity of these assays is limited mainly by the specific radioactivity of hormone. It is not unusual to be able to measure $10^{-12}$ to $10^{-15}$ mol of hormone with these assays.

## HORMONE RECEPTOR INTERACTIONS

### Agonists and Antagonists

*Receptors transduce chemical signals; hence, they must have at least two binding sites: one for the hormone and one for the macromolecular acceptor.* The interaction of receptors with hormones is analogous to the interaction of allosteric enzymes with effectors in that binding of the hormone to one site on the receptor affects the functional activity (conformation) of another site on the same molecule. Since receptors are essential for hormone action, it follows that target cells are those cells which have receptors and that the magnitude of the response in target cells should be related to the number of functional receptors.

The initial binding of hormones to their receptors, be they polypeptide hormones interacting with membrane receptors or steroid hormones interacting with soluble receptors, can be thought of in terms of Michaelis-Menten enzyme kinetics described in Chapter 4. That is, one can measure and assign an affinity ($K_d$) to the formation of a hormone-receptor complex ($H \cdot R$) that is equivalent to the formation of an enzyme-substrate complex ($E \cdot S$). The formation of $H \cdot R$ is reversible, as is $E \cdot S$, but unlike $E \cdot S$, there is no alternative degradation pathway leading to products. Because the affinity of receptors for hormones is generally very high ($K_d$ values range from $10^{-7}$ to $10^{-12}$ M), it is usually possible to measure $H \cdot R$ directly by using radioactive hormones and rapidly separating $H \cdot R$ from unbound hormone.

*Hormones that bind to receptors in a productive manner are called agonists.* Chemical modification of various constituents of natural hormones usually leads to a family of agonists with either increased or decreased affinity relative to the natural hormone. These modifications not only help define the stereochemistry of the active site on the receptor, but they also have practical implications for effective drug design. *Compounds that bind to receptors in a nonproductive manner are called antagonists* (or antihormones). These compounds compete with agonists for binding and are generally structurally related to agonists.

Although most members of an agonist/antagonist family are usually chemically related, there are numerous examples of what appear to be totally unrelated compounds that have similar biological activity. In many cases these compounds are stereochemically similar, if not chemically related, but in other cases even the stereochemical relationship is not obvious. The three most abundant natural estrogens are 17$\beta$-estradiol, estrone, and estriol; these compounds are obviously related, but they differ by a factor of 10 in their affinity for estrogen receptors (Figure 29-14). Diethylstilbestrol (DES), however, is a synthetic estrogen chemically unrelated to steroids, yet it has an affinity for estrogen receptors that is comparable with that of 17$\beta$-estradiol, and it promotes all the biological effects of natural estrogens. During the 1950s, millions of pregnant women were treated with DES to prevent miscarriage. Now it is becoming apparent that this exposure increases the incidence of genital malformation and cancer in their chil-

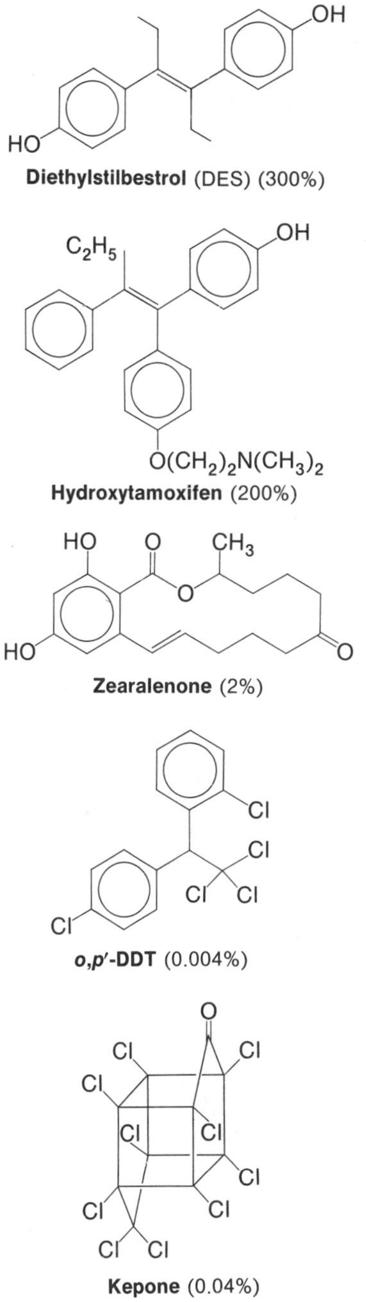

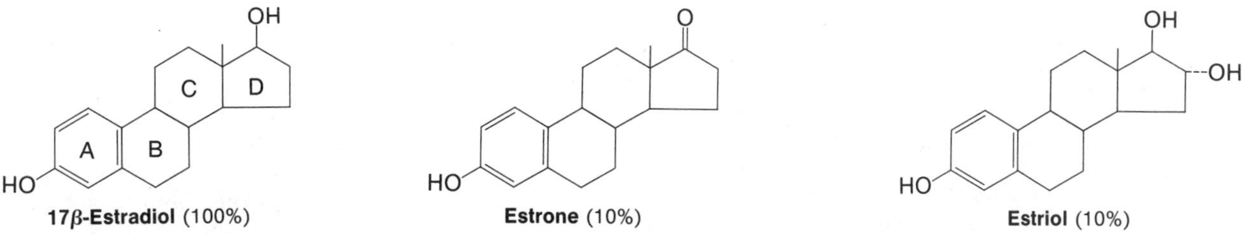

**Diethylstilbestrol** (DES) (300%)

**Hydroxytamoxifen** (200%)

**Zearalenone** (2%)

*o,p'-*DDT (0.004%)

**Kepone** (0.04%)

**17$\beta$-Estradiol** (100%)

**Estrone** (10%)

**Estriol** (10%)

**Figure 29-14**
Structures of some natural and some synthetic estrogens. The relative receptor-binding activity is shown in parentheses, with 17 $\beta$-estradiol being 100 percent.

dren. It is not clear, however, whether the harmful effects of DES are associated with its estrogenic activity or whether perhaps nonestrogenic metabolites of DES are the culprit. Hydroxytamoxifen is related to DES, but it is an estrogen antagonist used for the treatment of steroid-dependent tumors. Although DES and hydroxytamoxifen are not chemically related to steroids, three-dimensional models reveal that the two phenolic rings of these compounds lie in approximately the same position as the A and D rings of $17\beta$-estradiol. Likewise, the binding of zearalenone, an estrogen produced by the fungus *Fusarium* can be envisaged with stereochemical models. The estrogenic activity of zearalenone can be a serious problem when this fungus infests corn or other livestock foods. Several insecticides, including *o,p'*-DDT and Kepone, have been shown to have weak estrogenic activity; the structural relationship of these compounds to steroids is even less apparent (Figure 29-14).

The interaction of catecholamine receptors with agonists and antagonists illustrates some common principles. These receptors, which are found on smooth muscle, heart, liver, and brain cells, normally respond to epinephrine or norepinephrine. It is now clear that there are at least four distinct catecholamine receptors, designated $\alpha_1$, $\alpha_2$, $\beta_1$, and $\beta_2$ adrenergic receptors, that can be distinguished on the basis of their affinities for different agonists and antagonists. Many agonists have some activity with each receptor; in contrast, there are antagonists that are quite specific in blocking one receptor but not another. The greater specificity of the antagonists can be thought of in terms of an active-site domain. An agonist must fit precisely into the active site, whereas an antagonist may bind to a peripheral domain that overlaps the active site (see Figure 29-15). These peripheral domains may be very different on $\alpha$ and $\beta$ receptors and hence allow discrimination by antagonists. Thus, by using the appropriate combination of agonists and antagonists, a particular catecholamine receptor can be activated.

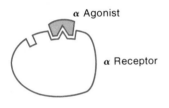

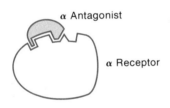

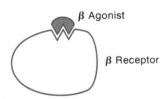

**Figure 29-15**
Binding of agonists and antagonists to specific receptors. $\alpha$ Agonists bind to $\alpha$ receptors with high affinity, and $\beta$ agonists bind to $\beta$ receptors with high affinity; they bind to the heterologous receptors as well, but with lower affinity. The $\alpha$ antagonist, however, is specific for the $\alpha$ receptor because it recognizes a domain that overlaps the active site of the agonists in addition to a peripheral site that is specific for one of the receptors.

## Hormone Receptors Are Identified and Purified by Affinity Techniques

Hormone receptors can be identified by *affinity labeling*. The principle is to covalently cross-link a radioactively labeled hormone with its receptor, either by using a bifunctional reagent that reacts with both the hormone and the receptor or by modifying the hormone so that it is chemically reactive and will form a covalent bond when it binds in the active site of the receptor. A particularly effective approach is to incorporate a latent chemical group into the hormone that can be activated by light only after it has bound to its receptor; photoactivation minimizes spurious chemical reactivity with other proteins.

Receptors for insulin, luteinizing hormone, thyroid hormones, and $\beta$-adrenergic hormones have been identified utilizing these methods. The insulin receptor is composed of four subunits with the composition $\alpha_2\beta_2$, where $\alpha$ is 130,000 daltons and $\beta$ is 90,000 daltons; the subunits are held together by disulfide bonds, as illustrated schematically in Figure 29-16. The receptors for thyroid, $\beta$-adrenergic, and luteinizing hormones appear to be single polypeptide chains in the 50,000- to 100,000-dalton range.

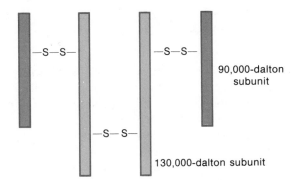

**Figure 29-16**
Hypothetical structure of the insulin receptor. Each receptor is composed of four chains: two $\alpha$ and two $\beta$. The number and positions of disulfide bonds are unknown.

Steroid hormone receptors are generally purified by *affinity chromatography*, a procedure in which an agonist or antagonist is covalently attached to an insoluble support, frequently agarose, and then the solubilized receptor preparation is passed through a column containing this material. Because of the high affinity of the receptor for the immobilized hormone, it binds tightly while the contaminating proteins pass through the column. The receptor is then eluted with an agonist or antagonist of higher affinity. Receptors for estrogens, progestins, and glucocorticoids have been purified by this technique. These steroid receptors have molecular weights in the 100,000-dalton range and they have an ellipsoidal shape. Purified receptors not only allow detailed biochemical characterization, but they also provide a means for generating specific antibodies. These antibodies then allow one to localize and quantitate receptors in the absence of hormone.

## MEMBRANE RECEPTORS

### Many Hormones Stimulate Adenylate Cyclase

*Epinephrine (in muscle) and glucagon (in liver) activate glycogenolysis by a cascade mechanism that commences with the stimulation of adenylate cyclase, which converts ATP to cAMP plus inorganic pyrophosphate.* The cAMP then promotes the dissociation of the catalytic and regulatory subunits of a protein kinase. The now-active catalytic subunit phosphorylates phosphorylase kinase, which then activates glycogen phosphorylase (see Chapter 8). The elucidation of this pathway provided the first molecular explanation of enzyme activation by covalent modification as well as insight into the molecular mechanism of hormone action. Remarkably, this entire pathway can be reproduced in preparations of lysed cells; this has greatly simplified elucidation of the individual steps. It was soon learned that the first step in this cascade required a membrane fraction. It was surmised, and later confirmed, that adenylate cyclase was a membrane protein. Since hormone receptors had not yet been identified when this pathway was first elucidated, it seemed possible that these hormones might directly activate adenylate cyclase. However, it soon became apparent that many hormones and neurotransmitters activate adenylate cyclase (Table 29-3), making a direct interaction unlikely. Then it was assumed that distinct receptors for each of the hormones somehow interact with adenylate cyclase. However, this notion also had to be abandoned when genetic and biochemical experiments implicated yet another protein.

**Table 29-3**
Hormones that Activate or Inhibit Adenylate Cyclase

**Activators:**

Corticotropin (ACTH)

Calcitonin

Catecholamines (acting on $\beta_1$ and $\beta_2$ receptors)

Choriogonadotropin

Follicle-stimulating hormone (FSH)

Glucagon

Gonadotropin-regulating hormone (GnRH)

Luteinizing hormone (LH)

Lipoprotein (LPH)

Melanocyte-stimulating hormones (MSH)

Parathormone (PTH)

Secretin

Thyrotropin regulatory hormone (TRH)

Thyrotropin (TSH)

Vasoactive intestinal peptide (VIP)

Vasopressin

**Inhibitors:**

Angiotensin

Catecholamines (acting on $\alpha_2$ receptors)

## Three Membrane Proteins Transduce Extracellular Hormone Binding to Intracellular Cyclic AMP Production

One approach to defining a biochemical pathway is to isolate mutants in each step of the pathway. The number of complementation groups frequently defines the number of different gene products (proteins) involved in the pathway. [Recall that mutants within a complementation group will not complement each other when combined in the same cell, whereas mutants of different complementation groups will restore function when combined within the same cell (e.g., see Chapter 22).] This approach is more difficult in mammalian cells than in microorganisms, but nevertheless, it has been instrumental in defining the pathway linking receptors to adenylate cyclase. Mutants defective in various steps in this pathway can be readily selected in cells that cannot tolerate prolonged exposure to cAMP. Three of the complementation groups correspond to proteins already discussed: $R^-$ mutants lack functional receptors, $C^-$ mutants lack functional adenylate cyclase, and $PK^-$ mutants lack functional protein kinase. $G^-$ mutants lack a functional GTP-binding protein that will be discussed later. *UNC* refers to mutants in which receptor binding and activation of adenylate cyclase are uncoupled, but all the proteins mentioned above are present. A specific protein corresponding to *UNC* has not yet been described. The fact that cells carrying any of these mutations can grow normally is consistent with the notion that hormones stimulate specialized functions rather than maintain basal metabolic activities.

Biochemical experiments with these mutants and with normal cells provide a picture of the components involved and a preliminary description of their function (Figure 29-17). The hormone receptor is embedded in the lipid bilayer of the membrane, and the hormone binding site faces the outside of the cell. The receptor can diffuse laterally within the membrane. When the receptor binds an agonist, it undergoes a conformation change that allows it to activate the GTP-binding protein called G. Since GTP cannot penetrate the membrane, the GTP-binding site must face the inside of the cell. When GTP is bound, G stimulates adenylate cyclase. G also has a GTPase activity that prevents it from being permanently activated. To obtain maximal G activity, the nonhydrolyzable GTP analog guanosine-5'-($\beta,\gamma$-imido)triphosphate (GppNHp) is frequently used. When the hormone receptor complex activates G to bind GTP, there is a reciprocal inhibitory effect of G on the receptor that lowers the affinity of the receptor for the agonist. Both the negative effect of the GTP-binding protein on hormone binding and the hydrolysis of GTP serve to limit the extent of adenylate cyclase activation when the hormone is not at saturating concentrations.

Adenylate cyclase requires a divalent cation for activity. In the presence of $Mg^{2+}$, the physiological cation, catalytic activity depends on G; however, with $Mn^{2+}$, the enzyme activity can be measured without G. Assays with $Mn^{2+}$ thus allow one to measure activity in mutants and during purification without relying on other protein factors. It is noteworthy that a mutation in G, adenylate cyclase, or cAMP-dependent protein kinase eliminates all responses of this class of hormone receptors, indicating that there is a single, unbranched pathway leading from the receptors to protein kinase.

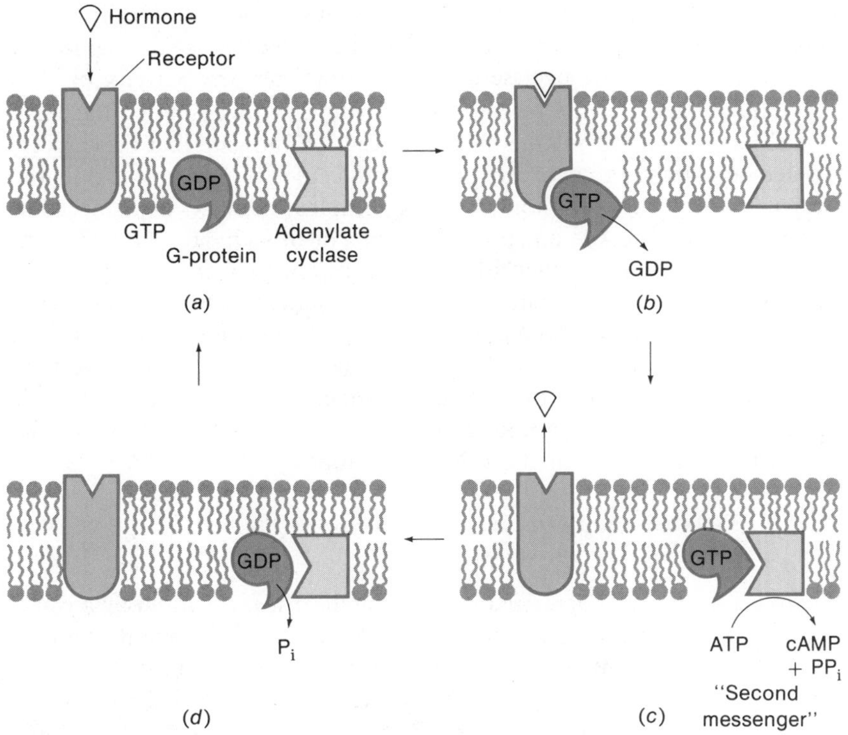

**Figure 29-17**
Possible orientation of a membrane receptor, G protein, and adenylate cyclase.
A cycle of adenylate cyclase activation and inactivation by a membrane receptor. The hormone-receptor complex stimulates G protein to bind GTP, which
then allows it to activate adenylate cyclase. Binding GTP also may dissociate
the receptor from the G protein. Hydrolysis of GTP inactivates G and leads to
dissociation of the G from adenylate cyclase.

The receptors, G, and adenylate cyclase are free to diffuse within the
membrane, and all three may rarely be coupled. Evidence for this notion
comes from experiments in which cells lacking adenylate cyclase but carrying functional receptors are fused with cells lacking receptors but carrying
adenylate cyclase. Shortly after fusion, the hybrid cells become capable of
responding to hormone by producing cAMP, indicating that the receptors from one membrane can diffuse into the other to activate adenylate
cyclase.

### Activation of a GTP-Binding Protein Is the Primary Site of Action of Many Membrane Receptors

Techniques for isolating membrane proteins and inserting them into other
membranes are just beginning to be developed. One exciting application of
this technique is the rescue of mutant cells by supplying the missing membrane protein. For example, detergent extracts made from membranes containing a functional G protein can be mixed with cells lacking this protein
(G⁻ cells) to restore hormonal activity. This technique has been extended
to study the interaction of receptors with G by mixing two detergent extracts—one containing only functional $\beta$-adrenergic receptors and the other
only functional G and then assaying for activated G by inserting these

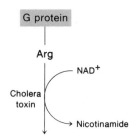

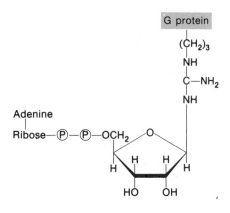

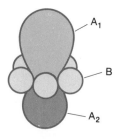

**Cholera toxin**

**Figure 29-18**
Cholera toxin catalyzes the ADP-ribo-sylation of G protein. The ADP-ribose moiety of $NAD^+$ is transferred to an arginine residue of G protein. Cholera toxin is composed of seven subunits; only the $A_1$ subunit enters the cell and is catalytically active.

components into $G^-$ cells and assaying for adenylate cyclase activity. Only when both membrane extracts and a potent agonist are used is significant activation of adenylate cyclase achieved. Moreover, the activity of G is maintained even after the receptor is inactivated with a potent antagonist. This observation and the kinetics of the reaction indicate that these receptors act catalytically to activate many G molecules. This event is thus the primary action of the hormone-receptor complex; hence G is the acceptor protein described in the opening paragraphs of this chapter. G is the first acceptor molecule to be identified. There are many intriguing questions that can be asked now that membrane components can be dissociated and recombined in a functional manner. For example, how long does G remain active after activation by the hormone receptor? Is the GTPase activity dependent on the interaction of G with adenylate cyclase? Are specific phospholipids required for productive interaction of receptors and G? What is the molecular defect in the *UNC* mutation?

## The GTP-Binding Protein Is Activated by Cholera Toxin

*Cholera*, a disease characterized by extreme diarrhea, is caused by a bacterial toxin consisting of an $A_1$ subunit, an $A_2$ subunit, and five B subunits. The $A_1$ subunit of cholera toxin gains entry into cells when the B subunits bind to membrane gangliosides, type $G_{M1}$. Once inside the cell, the $A_1$ subunit catalyzes the cleavage of NAD and attachment of the ADP-ribose moiety to an arginine residue of G protein (Figure 29-18). This process, called *ADP-ribosylation*, leads to permanent activation of G by inactivating the GTPase activity. Excretion of sodium and water by intestinal cells is regulated by hormones that activate adenylate cyclase. Prolonged activation of adenylate cyclase by cholera toxin leads to excessive water loss and dehydration. The cholera toxin-stimulated binding of radioactive NAD to membrane proteins allowed the identification of G as a 42,000-dalton protein. Diphtheria toxin acts similarly by penetrating cells and ADP-ribosylating one of the elongation factors (translocase, an enzyme that also binds and hydrolyzes GTP) involved in protein synthesis.

## Other Mechanisms by Which Membrane Receptors Act

Although many membrane receptors activate adenylate cyclase (Table 29-3), other receptors inhibit adenylate cyclase, some receptors raise intracellular calcium levels, and for a few the mechanism of action is completely unknown (Table 29-4). Identification of the acceptor proteins, the enzymes that are modulated by the acceptors, and the second messengers for these hormones comprise a challenging area of research.

Angiotensin and epinephrine (acting on $\alpha_2$ receptors) inhibit adenylate cyclase by a mechanism that requires GTP and is potentiated by sodium ions. These hormones can be grouped with a much larger class of compounds, including opiates, some neurotransmitters, and adenosine, that inhibit adenylate cyclase in a GTP- and sodium-dependent manner. The requirement for GTP implies the existence of a G-like protein; thus there may be a G protein that activates as well as a G protein that inactivates adenylate cyclase. This dual system of regulating cAMP synthesis provides another means of integrating all the different environmental stimuli that impinge on cells. Cyclic AMP levels also can be regulated at the level of degradation. For example, insulin is known to lower cAMP levels in some situations by activating *cAMP phosphodiesterase*, the enzyme that degrades cAMP. Activation involves cAMP-independent phosphorylation of the enzyme, but the kinase involved and the intermediate events are unknown.

A frequent and rapid response of hormone-receptor interaction involves mobilization of calcium. There are many cellular enzymes and processes that are activated by calcium. Normally the intracellular calcium concentration is very low, but it can increase rapidly in response to a wide variety of stimuli impinging on membrane receptors, including epinephrine ($\alpha_1$ receptors), vasopressin, and thyrotropin. The acceptor in these cases is unknown, but an early consequence of receptor-hormone interaction is breakdown and resynthesis of *phosphatidylinositol*, a relatively minor membrane phospholipid. This cycle, sometimes called a *PI cycle* (illustrated in Figure 29–19), may be coupled with calcium mobilization. A critical question is whether calcium activates the PI cycle or vice versa. The increase in calcium activates the calcium-binding protein *calmodulin*, which in turn stimulates a variety of enzymes, including a calcium-dependent protein kinase. The unsaturated diglyceride formed in this cycle, along with calcium and phosphatidylserine, activates yet another kinase, called *protein kinase C*. The diglyceride formed in the PI cycle may also be degraded to monoglycerides plus arachidonic acid; the latter is a substrate for prostaglandin synthesis.

Another membrane event that is stimulated by catecholamine receptors ($\beta$ receptors) is methylation of phosphatidylethanolamine to form phosphatidylcholine. This conversion is carried out by two membrane-bound *phospholipid methyltransferases* using $S$-adenosylmethionine as a methyl donor. The result is reorientation of the lipids in the membrane, which leads to a local increase in membrane fluidity. The phosphatidylcholine is then rapidly degraded by phospholipase $A_2$, giving rise to lysophosphotidylcholine and fatty acid. These metabolites may mediate some of the actions of catecholamines.

## Regulation of Hormone Response

The steady-state concentration of membrane receptors is determined, like all proteins, by the rates of receptor synthesis and degradation. In most cases, the factors that determine the rate of receptor synthesis are un-

**Table 29-4**
Hormones that Do *Not* Affect Adenylate Cyclase

Catecholamines (acting on $\alpha_1$ receptors)
Placental lactogen
Growth hormone
Insulin
Oxytocin
Prolactin
Somatostatin

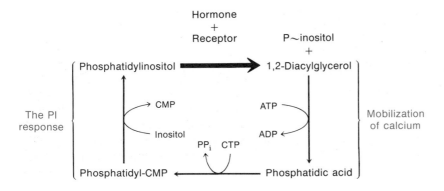

**Figure 29-19**
The phosphatidylinositol cycle. Receptor activation stimulates the degradation of phosphatidylinositol; the degradation products stimulate calcium mobilization in the cytoplasm by mechanisms that are not yet understood. The increase in available $Ca^{2+}$ stimulates a wide variety of enzymes. The phosphatidylinositol is rapidly resynthesized and is often referred to as the PI response.

known; however, there are a few good examples of receptor synthesis being stimulated by heterologous hormones. For example, FSH not only stimulates estradiol synthesis, but also promotes a 100-fold induction of LH receptors in granulosa cells. These LH receptors then respond to the ensuing surge of LH that triggers ovulation. There are also less common examples of autostimulation of receptor levels. Prolactin has a dramatic stimulatory effect on the level of its own receptors in the liver. The more common effect of hormones impinging on their own receptors is to promote receptor degradation, a process that usually involves internalization of receptors. Another means of reducing receptor activity without physically degrading them is to uncouple them from G-protein activation. This mechanism has been studied most thoroughly with $\beta$-adrenergic receptors. The decrease in surface receptors that occurs after stimulation allows a large immediate hormonal effect followed by a lower sustained effect.

### Maximum Cellular Response to Epinephrine Is Short-Lived Because of Desensitization

Exposure of cells to $\beta$-adrenergic agonists, such as epinephrine, leads to an initial sharp rise in cAMP levels; however, cAMP levels are not maintained; they fall nearly to basal levels within an hour or so. Furthermore, if the agonist is removed and the cells are challenged again within a few hours, the secondary response is submaximal. *This phenomenon, whereby prior exposure to an agonist leads to an acute decrease in responsiveness, is called desensitization.* Desensitization is associated with both an uncoupling of receptors from adenylate cyclase activation and a decrease in the number of receptors accessible to hormone binding. It occurs only after productive receptor function; hence it is not achieved with antagonists. The consequence of desensitization is that exposure to a hormone can result in a transient activation of cellular events rather than chronic activation.

### Chronic Activation of Polypeptide Hormone Receptors Leads to Their Internalization and Degradation

Binding of insulin, LH, calcitonin, EGF, GH, TRH, and glucagon to their respective receptors promotes the clearance of these receptors from the surface as well as the characteristic hormonal response. If cells are stimulated again before membrane receptors are replenished, the response is reduced. *This hormone-mediated loss of receptors is frequently called down-regulation.* It is distinct from desensitization in that it represents a loss of receptors rather than a reorientation or uncoupling of receptors. In addition, the loss and recovery occur over a much longer time scale, sometimes days, and receptor synthesis is essential for recovery. The net effect of down-regulation is, however, similar to desensitization in that chronic exposure to a high level of hormone leads to a decreased cellular response.

Binding of peptide hormones and growth factors to receptors stimulates their internalization. First, the hormone-receptor complexes begin to cluster together, as if they were being cross-linked; then these clusters tend to aggregate in membrane structures called *coated pits*, so-called because the

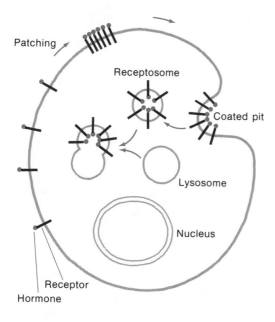

Patching

Receptosome

Coated pit

Lysosome

Nucleus

Receptor

Hormone

**Figure 29-20**
Down-regulation of receptors by
endocytosis. Continuous activation
of receptors leads to patching and
endocytosis of the receptor-hormone
complexes. The endocytotic vesicles,
sometimes called receptosomes,
fuse with lysosomes, where the contents are degraded. Receptosomes
also may allow entry of receptors into
other cell compartments, such as
the nucleus.

intracellular side of the membrane depressions, or pits, are coated with a
scaffolding protein called _clathrin_. Small vesicles bud off from the coated
pits that trap receptors and ligands (Figure 29-20). These vesicular structures, sometimes called _receptosomes_, migrate within the cell and associate
with other membrane structures known collectively as GERL (Golgi-endo-
plasmic reticulum-lysosomes). After a few hours they fuse with lysosomes,
at which point the lysosomal enzymes degrade both hormone and receptor.
Replacement of hormone receptors requires protein synthesis. Hormone
action is generally assumed to occur while the receptors are on the cell
surface, but it is not clear whether receptor function ceases during endocytosis or whether endocytosis may be essential for the action of some hormones.

**Down-Regulation of Insulin Receptors
Is Associated with Obesity**

Most membrane receptors are in excess over the enzymes they regulate. As
a consequence, the biochemical response is not proportional to the number
of receptors occupied by hormone. Instead, the dose-response curve for
activity is displaced (toward lower hormone concentration) relative to receptor occupancy, as shown diagramatically in Figure 29-21. These relationships give rise to the concept of _spare receptors_. With insulin, it is normal
to achieve maximum physiological response (e.g., activation of glucose uptake) with only approximately 10 percent of the surface receptors occupied.

Obesity is correlated with high circulating insulin levels (even after a
fast), and this leads to down-regulation of receptor levels. Because of the
spare receptors, this does not limit the maximum response possible, but
rather shifts the dose-response curve toward higher hormone concentrations (see Figure 29-21). Half-maximal response occurs when a certain
number of receptors are occupied, but it takes more hormone to achieve

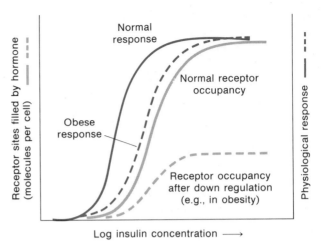

**Figure 29-21**
Binding and dose-response curves for insulin in normal and down-regulated cells. Note that the response curves are shifted to the left of the binding curves; i.e., maximal response occurs when a fraction of the available receptors are filled. With obesity there is a prolonged exposure of cells to insulin, leading to a decrease in the number of receptors per cell (down-regulation). The consequence is that the insulin response curve shifts to the right; i.e., for a normal response, a higher concentration of insulin is required.

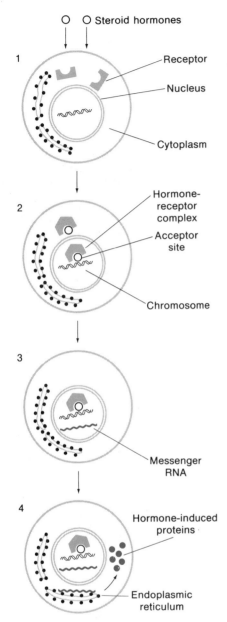

**Figure 29-22**
Receptors for steroid hormones activate transcription. Steroid hormones diffuse through the membrane and bind to cytoplasmic receptors, which become activated and then migrate into the nucleus, where they bind to specific acceptor sites. This interaction leads to the activation (or sometimes inhibition) of the rate of transcription of specific genes. The messenger RNA that is synthesized is translated into proteins.

this occupancy when there are fewer receptors. The net effect of down-regulation, in the case of obesity, is that the biochemical consequences of high insulin levels are blunted. Thus down-regulation is a homeostatic mechanism.

## STEROID AND THYROID HORMONE RECEPTORS

Receptors for all classes of steroid hormones, including 1,25-dihydroxyvitamin $D_3$, are soluble cytoplasmic proteins. Steroid receptors are not very abundant; even in target cells there are only $10^3$ to $10^5$ receptors per cell, which means that purification is an arduous task. Formation of a hormone-receptor complex leads to activation of the receptor that most likely corresponds to a conformational change in the protein. Activation proceeds readily at 37°C, but may take hours at 0°C. The activated receptor has an increased affinity for chromatin and redistributes from the cytoplasm to the nucleus of the cell (Figure 29-22). This process is frequently called *translocation*, but it does not require energy or RNA or protein synthesis. In some cases, e.g., with estrogen receptors, two activated receptors form a dimer. As a consequence, the nuclear form of the estrogen receptor sediments as a 5 S particle compared with 4 S for the unactivated cytoplasmic form. Cells that contain steroid receptors can be visualized by autoradiography, a technique in which cells are exposed to radioactive steroids, and then fixed onto slides and coated with photographic emulsion. After developing the slides the presence of silver grains over the nucleus identifies those cells with receptors.

The nature of the chromatin binding sites is poorly understood. Activated receptors have an increased affinity for double-stranded DNA; in fact, this property can be used effectively for receptor purification. However, the binding is of low affinity ($K_d \simeq 10^{-4}\ M$) and appears to be sequence-independent. It is generally assumed that there is a small number of DNA sequences within the genome that bind receptors with high affinity ($K_d < 10^{-10}\ M$), but they are not observed in DNA-binding assays because of the vast number of low affinity sites. Another view is that steroid receptors recognize specific chromatin acceptor proteins and that these proteins recognize specific DNA sequences. Evidence that steroid receptors bind to specific sites on chromatin comes from the visualization of nuclear ecdysone receptors. Ecdysone leads to visible puffs in polytene chromosomes of dipteran insects, and these puffs have long been associated with transcriptional activation of specific genes. When sites of ecdysone receptor binding are visualized (Figure 29-23), there is a clear association of ecdysone with only those puffs which are induced by ecdysone.

Thyroid hormone receptors differ from steroid receptors in that they appear to be permanently associated with chromatin regardless of whether they are filled with hormone.

## Steroid and Thyroid Hormone Receptors Regulate Transcription of Specific Genes

Binding of steroid receptors to chromatin or binding of thyroid hormones to nuclear receptors modulates the rate of transcription of specific genes. The effect of these receptors on transcription is usually stimulatory, but several examples of inhibitory actions also have been reported. A list of some of the specific mRNAs known to be regulated by steroid and thyroid hormones is presented in Table 29-5. This list includes many specialized proteins as well as housekeeping enzymes.

The first assays of hormonal effects on mRNA concentration relied on cell-free translation systems (typically prepared from reticulocytes or wheat germ) in which the synthesis of a specific protein is proportional to the amount of mRNA added; then techniques were devised to measure specific mRNA sequences by hybridization with radioactive complementary DNA (cDNA). The translation assays measure only functional mRNA, whereas the hybridization assays measure total mRNA sequences. In most cases, both assays give the same answer, indicating that most of the mRNA is functional. Steroid hormones can have dramatic effects on the concentration of mRNAs within a cell; e.g., ovalbumin mRNA increases from about 10 to about 50,000 molecules per cell in response to estrogens. Such an increase in mRNA concentration could be due to an increase in mRNA synthesis, a decrease in mRNA degradation, or both.

Direct demonstration that a hormone affects mRNA synthesis requires a transcriptional assay. One recent approach has been to isolate nuclei from hormone-stimulated cells and allow the endogenous RNA polymerases to complete RNA chains in the presence of radioactive nucleotides. The labeled RNA is then hybridized to plasmids that contain the gene sequences of interest. The percent of the RNA that hybridizes is considered to be proportional to the number of functional RNA polymerases on the

**Figure 29-23**
Association of ecdysone receptors with specific puffs. *Drosophila* salivary glands were treated with ecdysone. A few minutes later, the ecdysone was covalently linked to the receptor by ultraviolet light. Then the nuclear hormone-receptor complexes were visualized with fluorescent antibodies prepared against ecdysone. The salivary chromosomes of *Drosophila* are amplified about 1000-fold, which makes them easily visible. When specific genes are activated, the chromosomal region corresponding to that gene expands or puffs. The ecdysone receptors are localized over those puffs which normally appear in response to this hormone. (Photographs kindly supplied by O. Pongs.)

gene at the time the nuclei were isolated. In cases where they have been studied, steroid hormones have been shown to increase the number of polymerases actively transcribing the gene in question. However, in some cases, the activation of transcription does not completely account for the accumulation of mRNA, suggesting that the hormones may increase mRNA stability as well. Figure 29-24 shows an example of the correlation between nuclear receptor levels, the rate of transcription, and the accumulation of mRNA. In this example, the rate of conalbumin gene transcription is directly proportional to nuclear estrogen receptor levels, whereas ovalbumin transcription is not.

How steroid hormone receptors regulate transcription is not known. They presumably influence the rate of RNA polymerase initiation, but direct evidence is lacking. Proteins in addition to receptors may also be essential because inhibiting protein synthesis specifically blocks hormone-activated transcription of certain genes. In the example shown in Figure 29-24, inhibition of protein synthesis with cycloheximide blocks conalbumin and ovalbumin gene transcription even though receptors migrate into the nucleus and total transcription rates are nearly normal.

**Table 29-5**
Regulation of Specific Genes by Steroid and Thyroid Hormones

| | | | |
|---|---|---|---|
| **Glucocorticoids** | | **Progestins** | |
| Tyrosine aminotransferase | Liver | Avidin | Oviduct |
| Tryptophan oxygenase | Liver | Ovalbumin | Oviduct |
| Glutamine synthase | Liver, retina | Conalbumin | Oviduct |
| Phosphoenolpyruvate carboxykinase | Kidney | Uteroglobin | Uterus |
| Ovalbumin | Oviduct | **Androgens** | |
| Conalbumin | Oviduct (liver)* | $\beta$-Glucuronidase | Kidney |
| $\alpha$-Fetoprotein ($\downarrow$) | Liver | Aldolase | Prostate |
| $\alpha_{2\mu}$-Globulin | Liver | Prostate-binding proteins | Prostate |
| Metallothionein | Liver | Ovomucoid | Oviduct |
| Proopiomelanocortin ($\downarrow$) | Pituitary | Ovalbumin | Oviduct |
| Mammary tumor virus | Mammary gland | **1,25-Dihydroxyvitamin D$_3$** | |
| **Estrogens** | | Calcium-binding protein | Intestine |
| Ovalbumin | Oviduct | **Ecdysone** | |
| Conalbumin | Oviduct (liver) | Dopa-decarboxylase | Epidermis |
| Ovomucoid | Oviduct | Vitellogenin | Fat body |
| Lysozyme | Oviduct | Larval serum protein I | Fat body |
| Vitellogenin | Liver | **Thyroid hormones** | |
| apo-VLDL | Liver | Carbamyl phosphate synthase | Liver |
| Glucose-6-P-dehydrogenase | Uterus | Growth hormone | Pituitary |
| | | Prolactin ($\downarrow$) | Pituitary |
| | | $\alpha$-Glycerophosphate dehydrogenase | Liver (mitochondria) |
| | | Malic enzyme | Liver |

*In the liver, the product of the conalbumin gene is called *transferrin*.
Note: ($\downarrow$) means that mRNA levels are decreased by hormone.

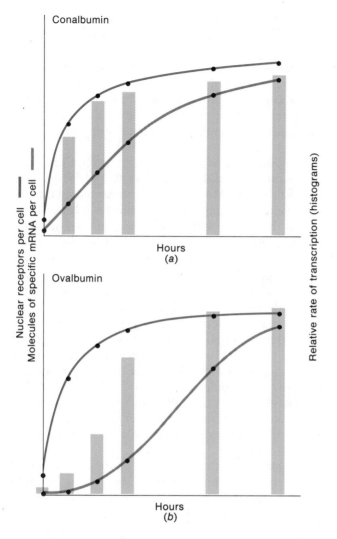

**Figure 29-24**
Kinetics of hormone response in chick oviduct. Chickens were treated with estrogen, and the accumulation of nuclear estrogen receptors as well as two specific gene products, conalbumin mRNA (*a*) and ovalbumin mRNA (*b*), were measured. In addition, the relative number of RNA polymerases transcribing these genes was determined. Both these genes are transcribed in the same cells. The graphs show that with the conalbumin gene, the rate of transcription is proportional to the number of nuclear estrogen receptors; however, with the ovalbumin gene, the relationship between receptors and transcription is more complex. It has been postulated that multiple receptor binding sites must be filled to fully activate ovalbumin gene transcription, whereas only a single site need be filled to activate conalbumin gene transcription. The accumulation of the respective mRNAs is a function of the rate of transcription and the half-lives of the mRNAs.

Different receptors can regulate the same genes. In the chick oviduct, estrogen, progestin, androgen, and glucocorticoid receptors can activate transcription of the conalbumin gene. The simplest explanation of this observation is that each of these classes of receptors recognizes a common chromatin acceptor site near the conalbumin gene.

Several steroid-responsive genes have been cloned and introduced into foreign cells by calcium phosphate-mediated cell transfection. In some cases, these introduced genes respond normally to the appropriate hormonal stimulation. In vitro mutagenesis of these genes followed by cell transfection should facilitate the identification of DNA sequences required for receptor binding and activity. Direct analysis of receptor-DNA binding is also becoming feasible.

## Regulation of Responsiveness to Steroid Hormones

The fertilized egg is not known to respond to many hormones; yet subsequent development produces cells that are capable of responding to a wide variety of hormones. *It is likely that the acquisition of hormone respon-*

*siveness is correlated with the activation of genes coding for specific receptors,* but in most cases the signals (hormones) responsible for receptor synthesis are unknown.

The metamorphosis of an amphibian tadpole to an adult frog is triggered by thyroid hormones. Tadpole liver cells do not respond to estrogen, whereas shortly after exposure to thyroid hormones they become capable of synthesizing the egg yolk protein vitellogenin in response to estrogen. Thyroid hormones may induce functional estrogen receptors in this system.

The avian oviduct is first recognizable at about 9 days of embryonic development in the chicken. Between days 10 and 14, estrogen receptors accumulate to adult levels, but the oviduct shows only limited response to estrogen. After exposure of the oviduct to estrogen, it becomes fully responsive to estrogen; this is an example of positive or up-regulation of responsiveness. Estrogen also induces a qualitative change in responsiveness to other classes of steroid hormones; after about 2 days of exposure to estrogen, the oviduct becomes responsive to progestins and glucocorticoids as well. The mechanisms involved in these changes in hormonal responsiveness are not understood; they may involve changes in receptors, accessory proteins, or chromatin structure of the responsive genes. An important point about all these examples is that the changes in responsiveness are stable in the absence of the primary estrogen stimulus. Thus exposure of the oviduct to estrogen for 2 days leaves the oviduct permanently committed to respond to progesterone. This is quite different from the hormonal induction of mRNA, which ceases after estrogen levels fall. Similar irreversible steroid effects are the essence of phenotypic sex determination by androgens. Only if androgens are present at a critical time during development are they effective.

It has been postulated that changes in cell (or gene) commitment may require DNA replication. During replication, changes in chromatin proteins or DNA (e.g., methylation of cytosine residues) may be imposed that are then stably inherited by daughter cells. Hormones may influence these processes, and hence they may be effective only when these critical cell divisions occur.

## GROWTH FACTORS

*A number of growth factors have been discovered in recent years (Table 29-6) that have the general property of stimulating cell division of specialized cells.* All the factors listed in Table 29-6 are proteins that bind to membrane receptors and are internalized by endocytosis. Some factors, e.g., EGF, NGF, and PDGF, elicit a PI response and stimulate calcium mobilization. Many stimulate protein phosphorylation, although none are known to activate adenylate cyclase. However, the link between the membrane events and stimulation of DNA synthesis, which occurs many hours after exposure, is unknown. The discovery that EGF stimulates phosphorylation of tyrosine residues of several membrane proteins, including its receptor, raises the possibility that this factor may impinge on the same growth-control mechanisms as the virally encoded kinases that phosphorylate tyrosine residues and lead to cell transformation. Growth hormone exerts its major physiological effects by stimulating secretion of growth factors such

**Table 29-6**

**1135**

**Table 29-6**
Some Common Vertebrate Growth Factors

| Growth Factor | Molecular Weight | Source | Effects |
|---|---|---|---|
| Epidermal growth factor (EGF) | 6400 | Male salivary gland | Epithelial cell growth, tooth eruption, eyelid opening |
| Nerve growth factor (NGF) | 13,000 | Male salivary gland | Development of sensory and sympathetic neurons |
| Erythropoietin | 23,000 | Kidney | Stimulates hematopoiesis |
| Colony-stimulating factors | 1500–150,000 | Many | |
| Thymosin | 12,200 | Thymus | Proliferation of T-lymphocytes |
| Platelet-derived growth factor (PDGF) | 13,000 | Platelets | Proliferation of mesenchymal cells (wound healing) |
| Fibroblast growth factor (FGF) | 13,400 | Pituitary, brain | Mesenchymal cell proliferation |
| Insulin-like growth factor I (IGF-I), somatomedin C | 7500 (basic) | Liver | Sulfate uptake by cartilage |
| Insulin-like growth factor II (IGF-II), somatomedin A, multiplication stimulating activity (MSA) | 7500 (neutral) | Liver | Proliferation of mesenchymal cells |

as somatomedins. The receptors for many of the somatomedin-related factors also bind insulin, although at lower affinity; thus some of the activities originally attributed to insulin are now being reassigned to more potent growth factors. Future research will undoubtedly reveal many similarities between the mechanism of action of hormones and growth factors. Interactions between them are likely to play an important role in the control of development.

## ENDOCRINOPATHIES

*Disease states in humans related to endocrine dysfunction can be broadly grouped into (1) overproduction of a particular hormone, (2) underproduction of a hormone, and (3) decreased target-tissue sensitivity to a hormone.*

### Hormone Overproduction

Most cases of hormonal overproduction are associated with hypertrophy of the endocrine organ, frequently owing to a tumor. Pituitary neoplasms usually affect the production of only one pituitary hormone as a result of the cancerous proliferation of a cell specialized in the synthesis of that hormone. Examples include giantism (acromegaly), which is associated with excessive production of growth hormone, and *Cushing's syndrome*, which is usually due to an overproduction of corticotropin (ACTH). Adrenal and parathyroid tumors leading to overproduction of various adrenal steroids and parathyroid hormones also have been described. Occasionally, tumors of other organs (*ectopic tumors*) synthesize and secrete peptide hormones, as in corticotropin synthesis by certain lung tumors. The excessive production of thyroid hormones in *Grave's disease* is associated with an enlarged thyroid gland, but in this case a circulating immunoglobulin (IgG) that mimics the activity of thyroid-stimulating hormone is implicated. Finally, we have already discussed the case of excessive secretion of steroid intermediates when specific enzymes of adrenal steroid biosynthesis are at inadequate levels.

## Hormone Underproduction

A wide variety of defects could potentially lead to underproduction of a hormone, ranging from a decrease in glandular tissue (e.g., *juvenile-onset diabetes* is associated with a decrease in pancreatic islet cells synthesizing insulin) to an inactive or defective gene coding for a hormone or enzyme necessary to synthesize that hormone. In many cases, a decrease in hormone production is established; e.g., in diabetes insipidus, circulating vasopressin levels are low, leading to excessive water loss by the kidneys, but the molecular mechanisms leading to this defect are unknown. In these diseases, the gene may be nonfunctional, or the precursor may not be processed properly, or there may be an amino acid substitution that renders the hormone inactive or leads to rapid degradation. The techniques of molecular biology should reveal the bases of these disorders, just as they do in analysis of hemoglobinopathies. Other defects in hormone synthesis are associated with inadequate supply of precursors; e.g., iodine is essential for thyroid hormone synthesis and its absence leads to goiter (hypertrophy of the thyroid owing to high concentrations of thyrotropin resulting from the lack of normal feedback inhibition of $T_3$ and $T_4$ on the hypothalamus). Likewise, synthesis of vitamin D requires ultraviolet irradiation of 7-dehydrocholesterol, and without it, the formation of $1,25(OH)_2D_3$ cannot occur and rickets ensues. A rare cause of rickets involves a defect in the kidney enzyme that converts $25(OH)D_3$ to $1,25(OH)_2D_3$, the active form of vitamin D. We have already mentioned deficiencies in adrenal enzymes necessary for glucocorticoid and mineralocorticoid synthesis (*Addison's disease*).

## Target-Cell Insensitivity

The most dramatic examples of target-cell insensitivity are due to lack of receptors. *Complete testicular feminization* is due to absence of androgen receptors in all target tissues. The consequence is phenotypic expression of female characteristics by genotypic males. As pointed out earlier, lack of the target-cell enzyme 5α-reductase leads to a less severe form of the same syndrome. A rare form of dwarfism (Laron dwarfs) is associated with high plasma levels of growth hormone, but somatomedin levels are low; consequently, a defect in growth-hormone receptors is suspected. Occasionally, antibodies form against membrane receptors and compete with hormones for binding. In some forms of adult-onset diabetes mellitus, there are circulating antibodies that bind to insulin receptors; these people show little or no response to added insulin. A remarkable finding, however, is that the antibody itself promotes insulin effects.

*Interferons are proteins liberated by cells in response to viral infection that circulate to neighboring cells and stimulate them to make antiviral proteins*, as shown schematically in Figure 29-25. Three types of interferons have been described: leukocyte ($\alpha$), fibroblast ($\beta$), and immune ($\gamma$). The names refer to the types of cells that synthesize them. The actions of $\alpha$ and $\beta$ type interferons have been studied most and will be described below; $\gamma$ appears to have quite different properties.

Upon entering an animal cell, viruses lose their protective coat, commence to replicate their nucleic acids, and synthesize more coat proteins for eventual packaging into mature viruses. In the process, they generally monopolize cellular enzymes and usually kill their host. While replicating,

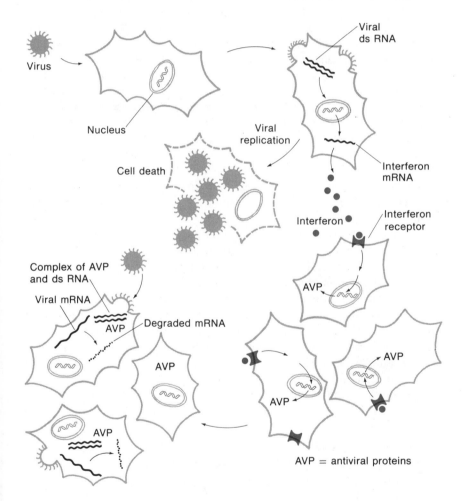

**Figure 29-25**
Induction and action of interferon. Interferon mRNA is induced by double-stranded RNA (dsRNA) that is generated by viruses during replication. The mRNA is translated into interferon, which binds to membrane receptors of other cells, stimulating them to synthesize antiviral proteins (AVP). These AVPs are activated by dsRNA; they inactivate viral mRNA and prevent its translation (see Figure 29-26).

viruses also induce the host cell to synthesize and secrete interferon. The inducer of interferon synthesis is thought to be *double-stranded RNA* because these molecules are the most potent synthetic inducers. A polymer of inosine and cytosine, poly(I·C), is one of the most frequently used synthetic inducers. The idea is that all viruses, be they single-stranded or double-stranded RNA or DNA viruses, generate small amounts of double-stranded RNA during their replication, and these molecules then activate interferon mRNA synthesis. Interferons are secreted; hence they are synthesized as precursors by membrane-bound polysomes.

Fibroblast and leukocyte interferons are proteins of about 20,000 daltons. Analysis of cDNA and genomic interferon clones has revealed that there is 1 gene for fibroblast interferon but 10 or so for leukocyte interferon. The sequences of the different interferons are homologous, but they have remarkably different potencies, the significance of which is only beginning to be appreciated. These proteins have high affinity for their corresponding receptors; they are effective at concentrations of $\sim 10^{-11}$ M. Although interferon receptors have not been characterized, they are generally assumed to be membrane proteins. The interaction of interferon with their receptors leads to the induction of mRNAs for several antiviral proteins.

*Two of the best-characterized antiviral proteins are oligoadenylate synthase and an eIF$_2$ protein kinase* (Figure 29-26). Both these enzymes are activated by double-stranded RNA; thus they are dormant until virus infection. Oligoadenylate synthase catalyzes the synthesis of short oligomers of adenosine linked by $2' \rightarrow 5'$ phosphodiester bonds instead of the usual $3' \rightarrow 5'$ bonds found in all other RNA molecules. These oligomers, often referred to as 2-5A, activate an endonuclease that degrades RNA. The eIF$_2$ protein kinase phosphorylates the smallest ($\alpha$) subunit of initiation factor 2, thereby inhibiting protein synthesis. These activities presumably protect the cell by degrading and preventing the translation of viral mRNAs. However, neither enzyme is specific for viral mRNA; thus it remains to be determined how the specificity is generated or whether in some cases the infected cell may kill itself along with the unprotected virus.

Much of the excitement about interferons stems from the observation that they protect cells against essentially all viruses. Although the medical implications are obvious, very few clinical trials have been performed to date because of the difficulty in isolating sufficient quantities of interferon. Now that many interferon genes have been isolated and expressed in microorganisms, the biological mechanism of action and medical value of these molecules can be evaluated.

## INVERTEBRATES PROVIDE MODEL SYSTEMS FOR STUDYING HORMONAL CONTROL OF DEVELOPMENT

Considering the wide diversity of invertebrates, one could reasonably expect a profusion of hormones and receptors. Although there are many intriguing systems, the biochemical mechanisms involved are generally unknown. Insect hormones are an exception; they have been studied extensively because of the role they play in molting and metamorphosis and, more recently, because of their economic importance in the regulation of insect populations.

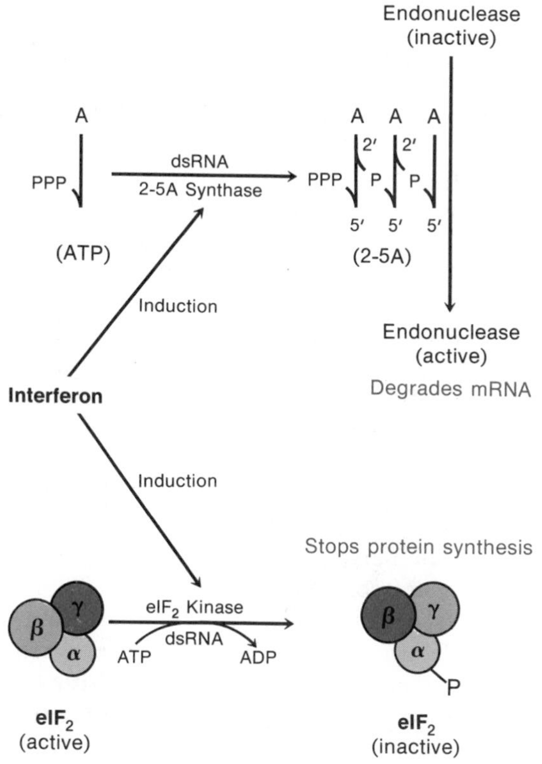

**Figure 29-26**
Activation of two antiviral proteins.

Arthropods are characterized by their hard, cuticular exoskeleton. In order to grow, insects must shed their old cuticle (molting) and synthesize a new, larger one. They typically repeat this process several times before adulthood. In some classes, e.g., flies and moths, the adults are physically unrelated to the larvae; in these species, a pupal phase is interspersed between the last larval molt and adulthood. During the pupal phase, the adult organs develop from small patches of predetermined cells called *imaginal disks*, while the larval tissues are largely destroyed, a process called *metamorphosis*. Molting and metamorphosis are controlled by at least five hormones: three polypeptide hormones, a steroid hormone, and a novel class of terpene derivatives (Figure 29-27). The molting process is initiated by an increase in the secretion of the steroid *ecdysone* from the prothoracic gland in response to a polypeptide brain hormone called *prothoracicotropic hormone* (PTTH). The structure of this hormone has not been determined, but it activates adenylate cyclase and its action can be mimicked by cAMP. Ecdysone then acts on the epidermis to stimulate the synthesis of a new cuticle by binding to receptors that migrate to the nucleus and activate specific genes. The newly synthesized cuticle detaches from the old cuticle, and digestive enzymes are secreted to partially degrade the old cuticle. The new cuticle is soft, to allow for rapid expansion, but it is eventually hardened by a process in which the cuticular proteins become covalently cross-linked. This hardening is stimulated by *bursicon*, a polypeptide hormone liberated from the neurohemal organs associated with abdominal ganglia. Bursicon is thought to activate adenylate cyclase. At the end of the molt, another polypeptide, *eclosion hormone*, triggers behavioral patterns essential for shedding of the old cuticle. Eclosion hormone raises cyclic GMP levels dramatically, and its effects can be mimicked by exogenous cGMP, but not by cAMP. It is not yet clear, however, whether eclosion hormone receptors directly activate guanylate cyclase. The type of cuticle formed at each molt depends on the titer of _juvenile hormone_ liberated from the brain. It is likely that juvenile hormone acts in the nucleus in a manner analogous to that of ecdysone and other steroid hormones. Juvenile hormone or ecdysone also regulates the synthesis of egg yolk proteins in many insect species. This sketch reveals several features found in many hormonally regulated systems. A polypeptide hormone of neural origin controls the synthesis and secretion of a steroid hormone, which then acts on a target tissue. In addition, hormones with antagonistic effects (ecdysone and juvenile hormone) impinge on the same target tissues.

Because of their simplicity, invertebrate systems provide excellent model systems for studying hormonal effects on neural cell development and function. During metamorphosis of a caterpillar into a moth, much of the body musculature is revamped by a process involving destruction of larval muscles and differentiation of new muscles from the imaginal disks. During this process, the nerves, which are conserved from larva to adult, must develop new motor connections and must respond to new sensory inputs. Changing titers of ecdysone (Figure 29-28) are primarily responsible for triggering these changes in nerve growth and synapse formation. A dramatic example of the hormonal effects on the dendritic branching pattern of a single neuron is shown in Figure 29-28. The effects of ecdysone in these relatively simple systems provide valuable clues for understanding how steroid hormones may affect the development and function of vertebrate neurons.

β-Ecdysone (steroid)

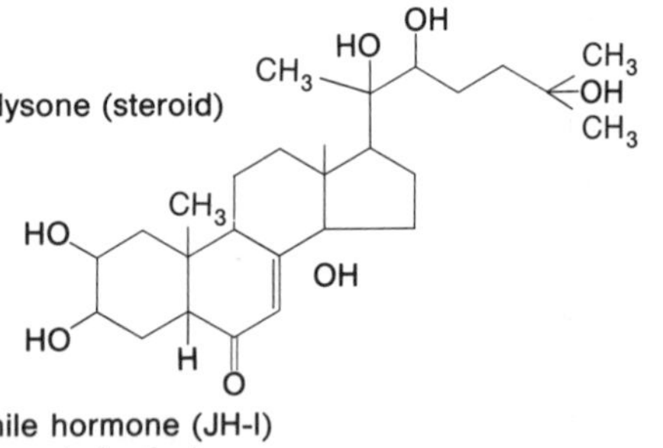

Juvenile hormone (JH-I)
(terpene derivative)

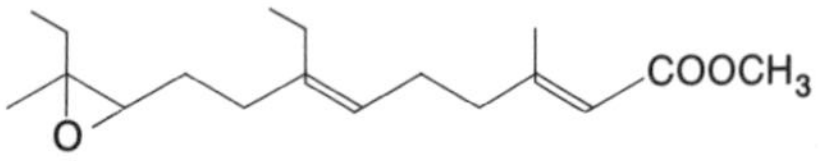

Prothoracicotropic hormone (PTTH)
(polypeptide ~4500 daltons)

Bursicon
(polypeptide: 20,000–40,000 daltons)

Eclosion hormone
(polypeptide: ~8000 daltons)

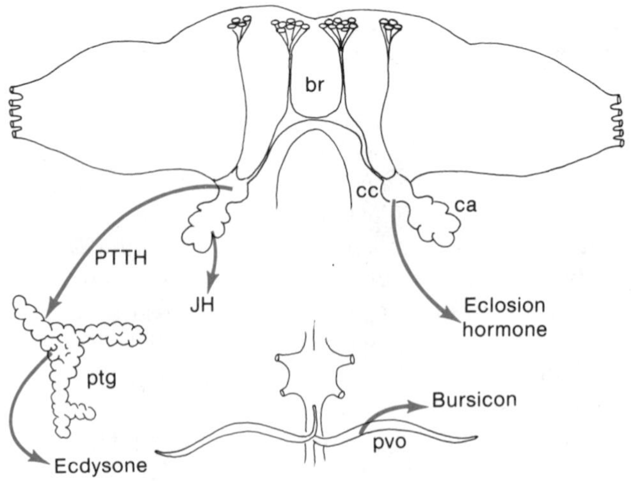

Insect Hormones

**Figure 29-27**
Some common insect hormones and their sites of synthesis and action. br, brain; cc, corpus cardiacum; ptg, prothoracic gland; pvo, perivisceral organ; PTTH, prothoracicotropic hormone; JH, juvenile hormone.

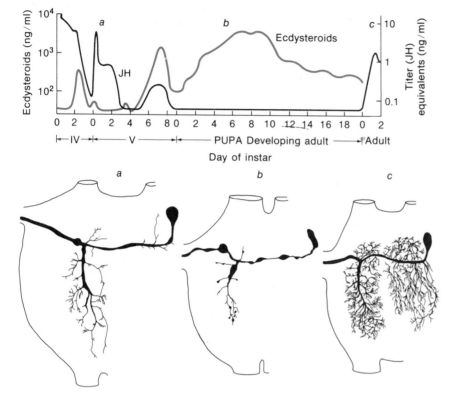

**Figure 29-28**
Hormonal effect on insect neuron branching pattern. During metamorphosis, larval muscles degenerate and adult muscles develop. The neurons, however, are preserved. Thus larval neurons must make new connections to adult muscles, and they must respond to new sensory inputs. This figure shows the branching pattern of a single neuron from the abdominal ganglion of the tobacco hornworm *Manduca sexta*. The three stages were taken at the developmental times indicated: (*a*) larval fifth instar, (*b*) pupa, and (*c*) adult. The titers of ecdysone and juvenile hormone throughout this period are also shown (on a log scale). [Reprinted with permission from James Truman and Lynn Riddiford. (*Upper part*) Interaction of Ecdysteroids and Juvenile Hormone in the Regulation of Larval Growth and Metamorphosis of the Tobacco Hornworm. In J. A. Hoffman (Ed.), *Progress in Ecdysone Research*, Elsevier/North Holland Biomedical Press. P. 410. (*Lower part*) *Science* 192:477, 1980.]

# PLANT HORMONES CONTROL CELL GROWTH AND DIFFERENTIATION

Plant hormones (also frequently referred to as growth factors) coordinate the growth and development of plants. The major regulatory factors discovered to date fall into five classes: auxins, cytokinins, gibberellins, abscisic acid, and ethylene. However, unlike animal hormones, there are many different natural compounds within some classes that have hormonal activity, e.g., over 60 gibberellins have been discovered.

The structures of some representative plant hormones are depicted in Figure 29-29. All the well-characterized plant hormones are low-molecular-weight compounds; indeed, one hormone, ethylene, is a gas. Polypeptide hormones have not been described for higher plants, although some of the mating factors of lower plants and fungi are polypeptides. Each class of compounds elicits many diverse responses, and there is considerable interaction among different hormones in the control of physiological processes. Furthermore, it is not surprising to find that the same process is controlled by different hormones (or hormonal combinations) in different species. These considerations make it difficult to analyze, and to generalize about, the mechanism of plant hormone action.

Studies on the mechanism of <u>auxin</u> action are the most advanced. These hormones are synthesized from tryptophan in the apical buds of growing shoots. They stimulate growth of the main shoot, but inhibit the development of lateral shoots; this has led to the practice of pinching off the apical buds to stimulate the formation of bushy plants. Auxins bind to membrane proteins of about 40,000 daltons. The affinity of auxins for these proteins, their location, and their abundance on target cells support the

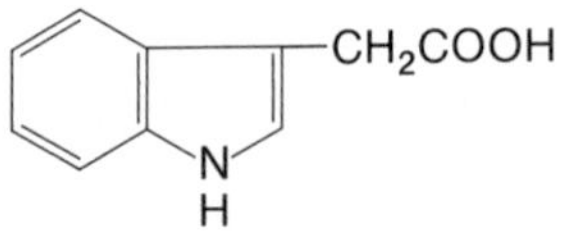

**Auxin**
(indole acetic acid)

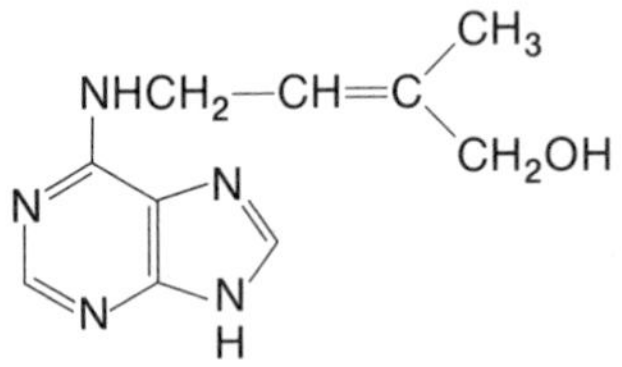

**Cytokinin**
(zeatin)

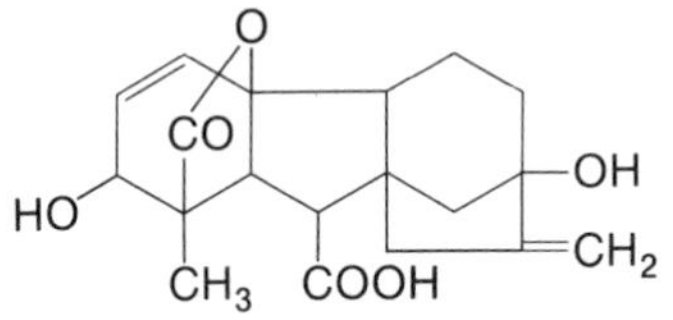

**Gibberellin**
(gibberellic acid)

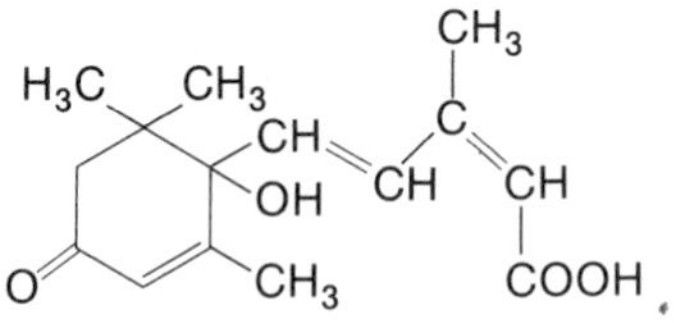

**Abscisic acid**

$$H_2C{=}CH_2$$
**Ethylene**

**Figure 29-29**
Structures of some common plant
hormones.

view that they may be the receptors that mediate auxin action. Auxin stimulation of cell growth can be divided into two phases. The earliest response is an increase in proton transport out of the cell, which occurs after a lag of a few minutes. This proton pump is thought to be coupled with a membrane ATPase. It is not clear whether the auxin receptor interacts directly with the ATPase or whether other intermediates are involved. It is possible that lowering the extracellular pH stimulates enzymes that partially degrade the cell wall, thereby loosening it and allowing for cellular expansion. A subsequent effect of auxins is to increase the synthesis of proteins and nucleic acids, resulting in sustained growth; e.g., auxin increases the amount of cellulose mRNA during pea cell expansion. All the auxin effects are blocked by inhibitors of protein synthesis.

Auxins are transported from cell to cell in a unidirectional manner rather than through the plant circulatory system. Because of their lipid solubility and weakly acidic character, auxins lend themselves to this type of transport in response to chemiosmotic gradients. Asymmetrical growth is a consequence of these gradients in auxin concentration. For example, the curvature of plants toward the light (*phototropism*) is due to transport of auxins away from the light, thereby stimulating more rapid growth of the cells on the darker side of the shoot. Auxins also can be stored; e.g., esters and amides of indole acetic acid exist in seeds. These stores are subsequently used during early stages of seed germination.

*Cytokinins* are adenine derivatives that are produced in the roots and promote growth and differentiation of numerous tissues. Cytokinins can overcome the auxin-mediated inhibition of lateral shoots; in other tissues these two hormones can act synergistically. In plant-cell cultures, the ratio of auxins to cytokinins in the medium is important. A high ratio promotes root differentiation and growth, a low ratio promotes shoot differentiation and growth, while an intermediate concentration of both hormones promotes growth of undifferentiated cells with neither roots nor shoots. Plant cells are much more plastic in their developmental potential than are animal cells. Indeed, normal plants can be grown readily from single tissue-culture cells. The preceding example indicates that hormones can have a profound effect on the developmental direction of uncommitted cells. Thus, depending on their hormonal milieu, plant cells can display a wide range of responses to a given hormone. While this diversity of responses is fascinating, it has also made it very difficult to define the primary site of action of plant hormones.

*Gibberellins* also promote shoot elongation, and frequently they act synergistically with auxins. Gibberellins stimulate the accumulation of amylase mRNA in germinating seeds, suggesting that their receptors or a second messenger may act at the genetic level.

*Abscisic acid* is antagonistic to many of the other plant hormones. It inhibits germination, growth, bud formation, and leaf senescence. Wilting stimulates abscisic acid synthesis in the chloroplasts of mesophyll cells in the leaves; the abscisic acid in this case is a stress signal that stimulates the guard cells to close and thus minimize water loss through the stomata. Abscisic acid has a rapid inhibitory effect on $K^+$ uptake by guard cells; thus it may act by means of membrane receptors to modulate ion pumps.

For many years, plant physiologists were reluctant to consider the concept of a gaseous hormone. Although a receptor for *ethylene* has not yet been identified, its effects on plants are comparable to those of many of the established hormones. Ethylene plays an important role in transverse rather than longitudinal growth of plants by redirecting auxin transport. It also stimulates fruit ripening and flower senescence, and it inhibits seedling growth. Ethylene is synthesized from *S*-adenosylmethionine (SAM), as illustrated in Figure 29-30. The conversion of SAM to 1-aminocyclopropane-1-carboxylic acid, the immediate precursor to ethylene, is stimulated by auxins, cytokinins, wounding, and anaerobiosis. Thus once again we see the interplay of hormones in the regulation of cell activity.

Some plant tumors may result from uncontrolled hormone production. Infection of plant wounds by some strains of *Agrobacterium* leads to tumorous growth at the site of wounding. Recent experiments reveal that these bacteria carry a large plasmid, the Ti plasmid, that is transferred to plant cells; part of the plasmid (the transforming region, or T region) becomes integrated into the plant genome. This T DNA encodes several genes that promote rapid cell proliferation. Interestingly, inactivation of one of these genes promotes the growth of shoots at the site of infection, while inactivation of another gene promotes the growth of roots at the site of

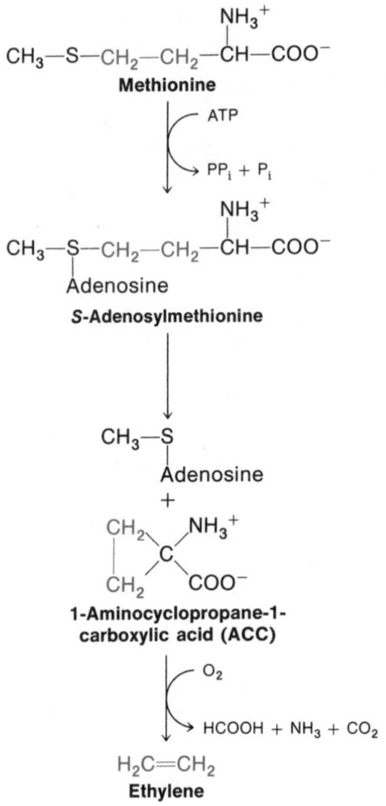

**Figure 29-30**
Pathway of ethylene synthesis in plants.

infection. This situation is reminiscent of the effects of auxins and cytokinins on callus differentiation. Thus it is likely that these gene products control auxin and cytokinin accumulation. When both hormones are overproduced, an undifferentiated tumor results, but when the ratio of hormones is unbalanced, either root or shoot differentiation predominates.

Many of the plant hormones and chemical analogs are of commercial interest because of their potential in synchronizing the time of harvest as well as increasing the yield and quality of food crops.

## SELECTED READINGS

*Annual Reviews of Biochemistry* and *Annual Reviews of Physiology*. Over 40 volumes in each series with many relevant reviews in the area of hormone receptors and hormone action. Provides good access to primary literature.

Baxter, J. D., and MacLoed, K. M. Molecular Basis for Hormone Action. In P. K. Bondy and L. E. Rosenberg (Eds.), *Metabolic Control and Disease*. Philadelphia: Saunders, 1980. Pp. 104–160. A useful summary.

deGroot, L. J., et al. (Eds.). *Endocrinology,* 3 Vols. New York: Grune and Stratton, 1979. Over 2000 pages of comprehensive treatment; primarily from a medical point of view.

Guillemin, R. Peptides in the brain: The new endocrinology of the neuron. *Science* 202:390, 1978. Nobel laureate speech.

O'Malley, B. W., and Birnbaumer, L. *Receptors and Hormone Action*, 3 Vols. New York: Academic Press, 1977. Review articles by many authors cover most aspects of hormone action.

*Recent Progress in Hormone Research*. New York: Academic Press. An annual publication with nearly 40 volumes. A good place to find a recent summary.

Riddiford, L. M., and Truman, J. W. Biochemistry of Insect Hormones and Insect Growth Regulators. In *Biochemistry of Insects*. New York: Academic Press, 1978. Pp. 307–357.

Sutherland, E. W. Studies on the mechanism of hormone action. *Science* 177:401, 1972. Nobel laureate speech related to discovery of cAMP as second messenger.

## PROBLEMS

1. Why are all known hormone receptors proteins? Could other macromolecules serve as receptors? Or as acceptors?

2. What advantages or disadvantages accrue to each of the hypothetical receptor-mediated pathways depicted in Figure 29-1?

3. Draw the feedback loops that are involved in the regulation of androgen synthesis in the testes or estrogen synthesis in the ovaries. What other inputs are required for cyclic production of estrogen?

4. Sex determination in mammals can be divided into two stages: (1) differentiation of the uncommitted gonad (ovotestis) into either testis or ovary, and (2) differentiation of identical internal structures (Wolffian and Müllerian ducts) and identical external structures (urogenital sinus, tubercle, etc.) into appropriate male or female organs. Normally there is no confusion between genetic and phenotypic sex because the Y chromosome of males (XY) codes for hormone-like proteins (called Y antigens because they have only been identified immunologically). These antigens are postulated to direct the differentiation of the ovotestis into a testis which then synthesizes androgens that direct the differentiation of the remaining sex organ primordia in the male direction. Thus, in the absence of the Y antigen or androgens, the embryo will develop into a female. With this highly simplified background and your knowledge of steroid biosynthesis and action, describe the various defects that could lead to an ambiguity between genetic and phenotypic sex, i.e., pseudohermaphrodites.

5. List several reasons why polypeptide hormones are synthesized as precursors.

6. What sort of experiments would you propose to ascertain the mechanism of thyroliberin (TRH) synthesis?

7. Imagine that you want to establish a radioimmunoassay for a hormone that you have discovered. How would you proceed? Draw your standard curve. Describe the sensitivity of your assay and how you calculated it. What artifacts might you encounter?

8. Using the scientific literature, describe the purification of a hormone receptor. What steps provide the most purification? What criteria are used to document purity? Are the purified receptors still active?

9. Make a table listing each of the mutants involved in the pathway leading from receptor to cyclic AMP-dependent protein kinase. Describe biochemical assays that would allow you to distinguish each complementation group.

10. Describe an approach that might lead to the identification of an acceptor protein for steroid hormone receptors.

11. Draw hypothetical dose-response and receptor-binding curves for steroid hormones. What would you conclude if the response curve were displaced to the left or right of the binding curve? Suppose it were superimposed on the binding curve?

12. Differential hormone response depends ultimately on which genes are expressible in a particular cell. The process by which a gene becomes expressible is sometimes referred to as *gene commitment*. What mechanisms may be involved in gene commitment?

13. The steady-state concentration of a protein or an mRNA is determined by its rate of synthesis ($S$) and its rate of degradation, usually expressed as a half-life ($t_{1/2}$). If $S$ and $t_{1/2}$ are constant, then [protein] or [MRNA] $= (S \times t_{1/2})/\ln 2$. Assuming that the half-life of insulin mRNA is 24 h, what rate of synthesis is required to accumulate 1000 molecules per cell. How often does RNA polymerase II initiate transcription on the insulin genes to accomplish this rate of synthesis? What assumptions did you have to make? If each insulin mRNA is translated 10 times a minute and insulin has a half-life of 1 h, what would be the steady-state concentration of insulin produced by a single $\beta$ cell? How many $\beta$ cells would it take to maintain a serum concentration of $10^{-10}$ $M$, assuming a serum volume of 5 liters?

14. What sort of observations would lead you to believe that there were several different receptors that recognize the same hormone? How would you prove it? What advantages are there for having multiple receptors for the same hormone? What kind of evidence would prove that there is a single receptor that mediates all responses of a particular hormone?

15. Interferon is not usually considered a hormone, yet it shares some properties with peptide hormones. What are they?

16. How might interferon-induced antiviral proteins selectively prevent viral replication without killing the cell?

17. Inhibitors of protein synthesis have been shown to block both the rapid and the slow auxin-mediated growth responses. How would you explain these observations?

18. Imagine that you have discovered a new plant growth factor. How would you decide whether it was a nutrient or a hormone? How would you decide whether it was a member of a preexisting class of hormones or represented a new class?

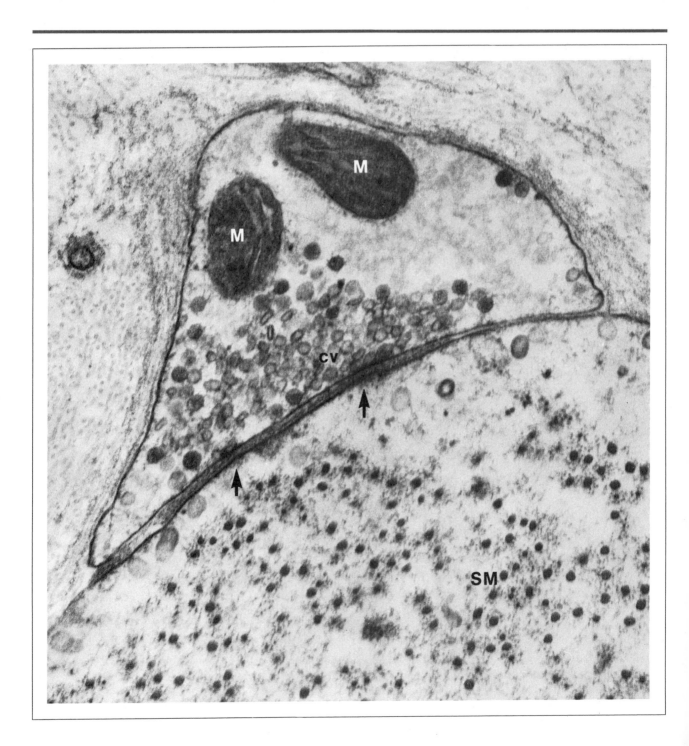

# 30

# NEUROTRANSMISSION

The structures of biological membranes and the mechanisms by which ions and metabolites are transported across them were considered in Chapters 16 and 17. Clearly, however, cells must do much more than simply assimilate nutrients and expel waste materials. Unicellular, free-living organisms such as bacteria and protozoa must be able to navigate and successfully compete in environments that often contain many perils to their livelihood. Multicellular organisms, however, have the additional problem of coordination and signaling among many diverse, differentiated cell types to ensure the efficient functioning of the individual as a whole. Interactions of a cell with its environment and other cells that are necessary for these types of communication must obviously involve processes associated with the cytoplasmic membrane. In many cases, transmembrane transport is integrally involved in these signaling processes between cells and their environment.

One type of communication mechanism in animals was considered in Chapter 29. Hormones released in one part of the body interact with receptors, which are often membrane-bound, to affect processes occurring in other cells that are often quite distant in space from the cells releasing the signal. This mechanism allows for efficient coordination of the wide variety of metabolic events continually taking place in higher organisms. Another, quite different sensory phenomenon takes place in the retina cells of the eye of vertebrate animals. Light energy is converted by these cells into nerve impulses which are translated by the brain into an image of our surroundings. This interconversion involves transmembrane movements of ions and will be considered in Chapter 31.

*Neuromuscular junction on vascular muscle in the mollusc Aplysia (a sea slug). This electron micrograph shows details of the specialized membranes and organelles associated with a nerve terminal and its target smooth muscle fiber. The small clear vesicles (CV) are involved in the release of neurotransmitter serotonin (5-hydroxy-tryptamine). Serotonin is released at the active zones (arrows), diffuses across the intercellular cleft, binds to receptors on the muscle membrane, and causes contraction of the fiber. Thick contractile myofilaments are distributed throughout the muscle cytoplasm, each surrounded by a number of much thinner filaments (M = mitochondrion and SM = smooth muscle fiber cut in cross section). (Magnification approximately 50,000×.) (Micrograph Courtesy C. Price)*

In this chapter, we shall examine the mechanism by which nerve impulses are propagated to illustrate how excitable tissues function and interact in higher organisms. As will be seen, the fundamental mechanisms revealed by studies of membrane transport in both prokaryotic and eukaryotic cells (Chapter 17) can account in large part for the seemingly more complex phenomena of nerve-impulse propagation and transmission.

## NERVE-IMPULSE PROPAGATION

As early as the late eighteenth century, from experiments by Galvani and Volta, it was suspected that the transmission of nerve impulses and muscular contraction involved electric signals. In 1902, J. Bernstein first proposed that the unequal distribution of $K^+$ across the nerve-cell membrane and the selective permeability of this membrane for this ion were responsible for a _resting potential_ known to exist in nerve and muscle fibers. He further believed that excitation of a nerve cell involved a transient collapse in this selective permeability such that other ions were able to penetrate the nerve-cell membrane and abolish the resting potential. If these changes in ion permeability could move down the axon of a nerve cell and be transmitted to other cells, they could provide the basis for the propagation of nerve impulses.

In the 1930s, the isolated _squid giant nerve axon_ became available for experimentation, and its size was especially amenable to electrophysiologic measurements. Experiments pioneered by A. L. Hodgkin and A. F. Huxley soon established the essential ionic movements associated with impulse propagation, and the resultant local changes in the membrane potential could be measured. It became clear that the transmembrane transport of ions was important in nerve cells for their signaling function and for the actual signal conductance itself. Relatively little is still known at the molecular level about how these changes in ion permeability in nerve cells are accomplished. Somewhat more is known about how these signals are chemically transmitted between nerve cells and from nerves to muscle; these mechanisms will be considered as well. An examination of such mechanisms is likely to provide insights into ways in which all different types of cells can communicate with each other in response to changes in their environments.

### The Resting Membrane Potential

To understand how nerve impulses are generated along the axon of a nerve cell (Figure 30-1), the basis for electric potentials that exist across the neuronal membrane must first be considered. An unequal distribution of ionic species across a biological membrane that is permeable to these molecules can result in a transmembrane electric potential, $\Delta\psi$ (Chapter 17). For a membrane system permeable to several ionic species, the numerical value of $\Delta\psi$ can be approximated by the _Goldman equation_, derived by D. E. Goldman in 1943:

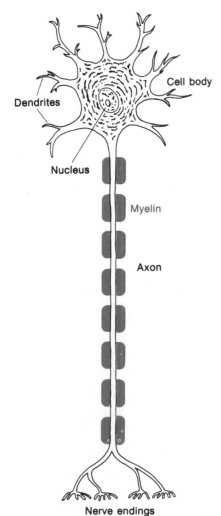

**Figure 30-1**
Schematic diagram of a typical motor neuron (a nerve cell conducting impulses to muscle cells).

$$\Delta\psi \text{ ("in" relative to "out")} = \frac{2.3RT}{F} \log_{10}\left(\frac{\Sigma P_c[\text{C}]_{\text{out}} + \Sigma P_a[\text{A}]_{\text{in}}}{\Sigma P_c[\text{C}]_{\text{in}} + \Sigma P_a[\text{A}]_{\text{out}}}\right) \quad (1)$$

where C and A are univalent cations and anions, respectively, and $P_c$ and

$P_a$ refer to their _permeability coefficients_* across the membrane of interest. Since multivalent ions are generally not quantitatively significant in contributing to $\Delta\psi$ in resting neuronal membranes, they are usually ignored in calculating $\Delta\psi$. If the membrane is selectively permeable to one ion only, for example C, Equation (1) reduces to the familiar _Nernst equation_:

$$\Delta\psi = E_c = \frac{2.3RT}{F}\log_{10}\frac{[C]_{out}}{[C]_{in}} \tag{2}$$

where $E_c$ refers to the equilibrium electric potential of C.

It is the unequal distribution of protons and other cations that gives rise to transmembrane potentials that can drive ATP synthesis and secondary active transport in many cells (Chapter 17). In nerve cells, a similar situation exists as summarized in Table 30-1. Thus the external environment of the nerve cell contains a high concentration of sodium ions and a low concentration of $K^+$, while the reverse is true for the cytoplasm (_axoplasm_). Furthermore, the extracellular concentration of $Cl^-$ is 5 to 10 times that of the axoplasm. A number of other impermeant anions, largely organic molecules, proteins, and nucleic acids, maintain an approximate charge neutrality in the axoplasm. _Resting nerve cells are highly permeable to $K^+$ but not $Na^+$, and it is this selective permeability that allows an electric potential to develop across the membrane in the presence of a $K^+$ concentration gradient_. The unequal distributions of $K^+$, $Na^+$, and $Cl^-$ are partially determined by passive processes leading to what is called a _Donnan equilibrium_ and partly by the activity of the $Na^+$-$K^+$ ATPase (Chapter 17).

The passive distribution of ions across a membrane was predicted by F. Donnan. He theorized that if a hypothetical cell having membrane-impermeable anions inside (such as proteins and nucleic acids) were placed in a KCl solution, $Cl^-$ would diffuse into the cell, down its concentration gradient, accompanied by an equivalent amount of $K^+$ to maintain electroneutrality, until a state of equilibrium was achieved. If $K^+$ were the counterion present initially inside the cell, Donnan predicted that the equilibrium concentration of $K^+$ inside would be much higher than that of $Cl^-$ (and higher than that of $K^+$ outside the cell if a sufficient concentration of nondiffusible anions were present inside). More specifically, Donnan showed that the final equilibrium concentrations of $K^+$ and $Cl^-$ inside and outside the cell would be related by the following equation:

$$\frac{[K^+]_{in}}{[K^+]_{out}} = \frac{[Cl^-]_{out}}{[Cl^-]_{in}} \tag{3}$$

This relationship describes a Donnan equilibrium (Figure 30-2). Because both $K^+$ and $Cl^-$ are permeable to the resting axonal membrane, their unequal distributions (Table 30-1) can be partly explained by this passive process. _The $Na^+$ gradient, however, is due both to the relative impermeability of the membrane to this ion and to the $Na^+$-$K^+$ ATPase that actively pumps $Na^+$ out of the cell_. The activity of this enzyme also alters somewhat the value of the $K^+$ gradient from that predicted by a Donnan equilibrium.

---

*The permeability coefficient is equal to the diffusion coefficient $D$ divided by the width of the membrane $l$.

**Table 30-1**
Ionic Concentrations Inside (Axoplasm) and Outside (Blood) the Squid Giant Axon

| Ion | Inside (mM) | Outside (mM) |
|---|---|---|
| $Na^+$ | 50 | 440 |
| $K^+$ | 400 | 20 |
| $Cl^-$ | 40–150 | 560 |

Adapted from S. W. Kuffler and J. G. Nicholls, _From Neuron to Brain_. Sunderland, Mass.: Sinauer Associates, 1976.

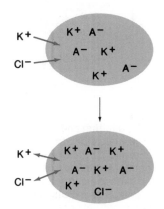

**Figure 30-2**
If a cell containing the potassium salt of a nondiffusible anion ($A^-$) is placed in a KCl solution, $K^+$ and $Cl^-$ will diffuse into the cell (_above_) until a Donnan equilibrium is reached [_below_; also see Equation (3)].

The predicted resting membrane potential $\Delta\psi$ across the axonal membrane can be calculated from Equation (1) using the values in Table 30-1, the permeabilities of $Na^+$ and $Cl^-$ relative to $K^+$ (0.04 and 0.45, respectively), and assuming an intracellular $Cl^-$ concentration of 50 m$M$:

$$\Delta\psi = 60 \log\left(\frac{20 + 0.04(440) + 0.45(50)}{400 + 0.04(50) + 0.45(560)}\right) = -62 \text{ mV}$$

This value is close to the experimentally measured value of the resting membrane potential across a squid axonal membrane.

## The Action Potential and Its Propagation

The use of giant axons from squid nerve cells in electrophysiologic experiments has greatly aided our understanding of the electrical events that take place during nerve stimulation. An experimental apparatus for measuring changes in the potential across the membrane of such an axon, which has a diameter of approximately 0.5 mm, is schematically illustrated in Figure 30-3. It consists of a pair of stimulating electrodes connected to a current source and a second pair of recording electrodes located slightly farther down the axonal segment. The latter electrodes are connected to a sensitive recording device, such as an oscilloscope, and can be used to measure time-dependent changes in the membrane potential.

Initially, the potential measured in the system shown in Figure 30-3 is about $-60$ mV; i.e., the resting membrane potential (Figure 30-4). If a brief current is applied at the stimulating electrodes, a time-dependent change in the membrane potential may be recorded on the oscilloscope, as shown in Figure 30-4a. *This so-called action potential only occurs if the stimulus is sufficient to depolarize the membrane by about 20 mV* (i.e., to about $-40$ mV). Weaker stimuli give small local potential changes, while current pulses greater than this _threshold_ value give a curve similar in shape and height to that shown in Figure 30-4a, independent of the magnitude of the stimulus. During the development of the action potential, the value of $\Delta\psi$ across the axonal membrane rises in about 1 ms to nearly $+40$ mV. This is followed by a somewhat slower return to the resting potential, during which time the membrane potential drops transiently below the resting value, to about $-75$ mV. This is close to the value predicted by the Nernst equation if the membrane were permeable only to $K^+$ (the potassium equilibrium potential) and is referred to as _hyperpolarization_.

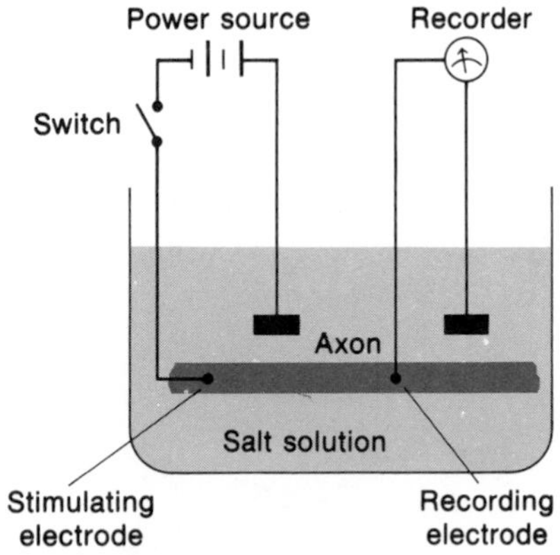

Classic experiments by A. L. Hodgkin and A. F. Huxley have established that the *changes in membrane potential occurring during nerve stimulation are due to transient changes in the permeability of the membrane to $Na^+$ and $K^+$ ions*. As illustrated in Figure 30-4b, the rapid rise in $\Delta\psi$ to a positive value is accompanied by a large increase in the relative permeability of $Na^+$, while the return of the membrane to the resting potential is correlated with inactivation of $Na^+$ permeability and a transient increase in $K^+$ permeability. An important conclusion from these observations is that *the permeabilities of the axonal membrane for $Na^+$ and $K^+$ depend on the membrane potential*. Thus depolarization above the threshold, leading to a more positive $\Delta\psi$, first leads to an increased permeability of $Na^+$ followed by inactivation of this phenomenon and an increase in the

**Figure 30-3**
A device for eliciting and recording action potentials along the squid giant nerve axon. Brief closure of the switch connected to the stimulating electrode causes a current pulse into the axoplasm. If an impulse is generated, resultant potential changes can be detected by the recording electrode, which is connected to an oscilloscope or other recording device. (Adapted from S. L. Wolfe, *Biology of the Cell*, 2d ed., Wadsworth, Belmont, Calif., 1981.)

membrane permeability of $K^+$. The latter events tend to hyperpolarize the membrane, increasing the negative $\Delta\psi$ and decreasing the $Na^+$ permeability.

As mentioned in the preceding section, the membrane potential depends on the relative permeabilities and concentration gradients of electrolytes across the membrane [Equation (1)]. The resting potential is largely dependent on the $K^+$ gradient, since unstimulated nerve membranes have a high permeability only for this cation. At the height of depolarization, however, the membrane is much more permeable to $Na^+$ than to $K^+$. Because the $Na^+$ gradient is opposite to that of $K^+$, $Na^+$ influx causes the membrane potential to become positive rapidly, approaching but never reaching the value it would have if the membrane were permeable only to $Na^+$. This can be seen if one substitutes the values of $[Na^+]_{in}$ and $[Na^+]_{out}$ into Equation (2). A value of $+57$ mV is thereby obtained for $\Delta\psi$ if the membrane were permeable only to $Na^+$. This value is never attained because an increased permeability of the membrane to $K^+$ follows the change in $Na^+$ permeability, and the opening of the $Na^+$ "channel" is only transient (Figure 30-4b).

The ionic movements leading to an action potential can therefore be summarized as follows:

1. Stimulation leads to an influx of $Na^+$ into the axoplasm, down its electrochemical gradient, owing to an increased permeability of the membrane to this cation.

2. The change in $\Delta\psi$ resulting from this flow increases the membrane permeability of $K^+$, which flows out of the cell, down its electrochemical gradient; this reestablishes a negative $\Delta\psi$ and is accompanied by inactivation of the influx of $Na^+$.

3. The membrane potential eventually becomes sufficiently negative to return both $K^+$ and $Na^+$ permeabilities to their normal values, and $\Delta\psi$ resumes its resting level.

It should be pointed out that _the ion fluxes that accompany these events and lead to changes in $\Delta\psi$ are actually very small compared with the concentrations of $Na^+$ and $K^+$ inside and outside the cell._ From the capacitance of a squid giant axon membrane, it can be calculated that for a 100-mV change in $\Delta\psi$ (from $-60$ to $+40$ mV), only about $10^{-12}$ mol $Na^+$ per $cm^2$ of cell surface need enter the cell, while an equal amount of $K^+$ must leave the axoplasm to return the potential to the resting value. This corresponds, for example, to about one in $10^6$ molecules of $K^+$ leaving the cell per action-potential spike, or to a change in the intracellular $K^+$ concentration of only 0.0001 percent.

The events shown in Figure 30-4 record fluctuations in $\Delta\psi$ and ionic currents at one point on the axonal membrane as a function of time during passage of an action-potential wave through this point. _The action potential, however, is conducted down the axon as a wave of depolarization-repolarization events_ through the following mechanism: depolarization of a given area of the membrane causes current to flow in the axoplasm from the more positive (depolarized) region to neighboring regions. This, in turn, triggers an action potential across the neighboring section of mem-

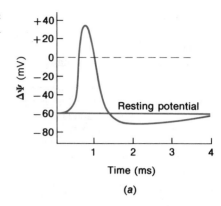

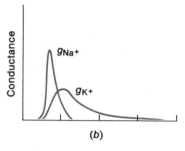

**Figure 30-4**
(a) A typical action potential that might be recorded by the instrument in Figure 30-3 if the stimulating current is sufficient to depolarize the membrane by at least 20 mV. The membrane potential eventually returns to its resting level of about $-60$ mV within about 5 ms.
(b) Changes in relative $Na^+$ permeability ($gNa^+$) and $K^+$ permeability ($gK^+$) as a function of time during passage of the action potential in (a) along a point on the axonal membrane. Depolarization is correlated with an increase in $gNa^+$, while repolarization is accompanied by a decrease in $gNa^+$ and a transient increase in $gK^+$. Note that the membrane potential transiently becomes more negative than the resting value (hyperpolarizes) until $gK^+$ returns to its normal value. The time of appearance of the action potential after stimulation at $T = 0$ depends on the distance between the recording and stimulating electrodes. (Adapted from A. L. Hodgkin and A. F. Huxley, A quantitative description of membrane current and its application to conduction and excitation in nerve, _J. Physiol._ 117:500, 1952.)

**Figure 30-5**
Nerve impulses in unmyelinated nerves are conducted through local current movements which propagate the action potential. A portion of the resting axonal membrane is shown at the top. Arrival of an action potential causes local depolarization of the membrane (*middle*) which is propagated from left to right by local currents shown by the arrows. (Adapted from A. L. Hodgkin, *Proc. R. Soc. Lond.* [*Biol.*] 148:1, 1957.)

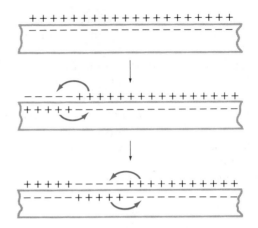

brane, and so forth (Figure 30-5). Thus the action potential provides a mechanism whereby the transmitted signal is constantly amplified to maintain a constant amplitude. In a regular cable without amplification, the propagated pulse decreases with distance due to resistance and leakage. Without the action potential, a current pulse would therefore be reduced to an insignificant level after traveling a very short length along the axon. In the nerve cells of invertebrate animals, as well as in many cells in vertebrates, nerve impulses are therefore conducted along axons and dendrites by these local currents and action potentials.

The axons of many nerve cells of higher animals, however, are also surrounded by a multilayered *myelin sheath*, each layer consisting of a typical lipid bilayer membrane (Figure 30-6). This myelin insulation is interrupted at intervals, the spacing of which depends on the fiber diameter, by the so-called *nodes of Ranvier* (Figure 30-7). Nerve impulses in these types of nerve fibers are conducted in a *saltatory* manner, with action potentials "jumping" from node to node where the axonal membranes are in direct contact with the extracellular fluid (Figure 30-7). The insulation provided by the myelin sheath allows for efficient current conduction within the axoplasm by preventing signal loss, and as a result, this type of impulse propagation can be much more rapid than that observed in unmyelinated fibers of similar diameter. Indeed, impulse propagation velocities of *over 100 m/s* have been recorded in myelinated nerve fibers.

### The Existence of Separate Channels for K⁺ and Na⁺ in Nerve Cell Membranes

By a slight modification of the experimental setup shown in Figure 30-4 it is possible to hold the membrane potential across an axonal membrane constant at any predetermined value. This is accomplished by connecting the recording electrodes to a device called a *feedback amplifier*, which compensates for any potential change sensed by these electrodes by applying a current to keep the voltage constant. In such a *voltage-clamped* situation, ionic movements can be inferred from the amount of current necessary to hold $\Delta\psi$ at its predetermined value.

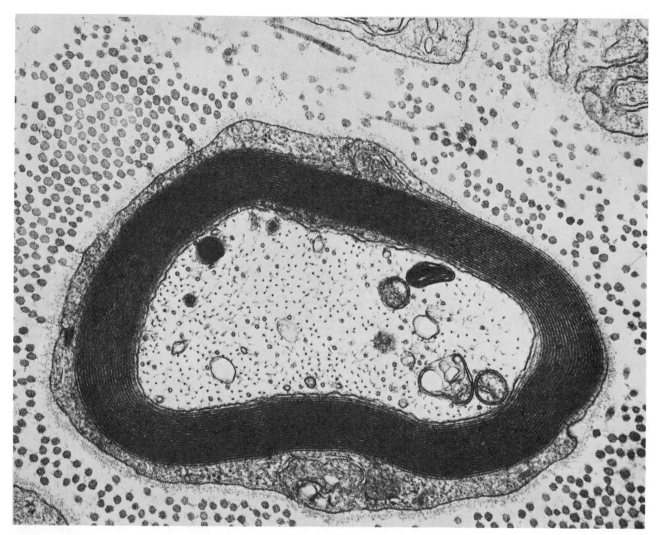

**Figure 30-6**
Cross section of a myelinated nerve axon from the superior cervical ganglion of the rabbit. Note the multilayered membrane of the myelin sheath that serves as an electric insulator. (Micrograph courtesy Dr. T. Lentz.)

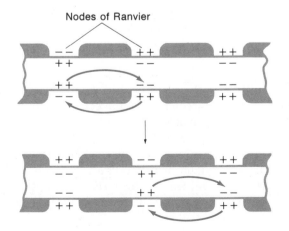

Nodes of Ranvier

**Figure 30-7**
Schematic illustration of a longitudinal cross section of a myelinated axon. The sheath is interrupted at various intervals by the nodes of Ranvier, where the cytoplasmic membrane is exposed to the surrounding fluid. Nerve impulses are conducted in a saltatory manner in myelinated axons, as illustrated in this schematic diagram. Impulse propagation is from left to right and the arrows show the accompanying local current movements.

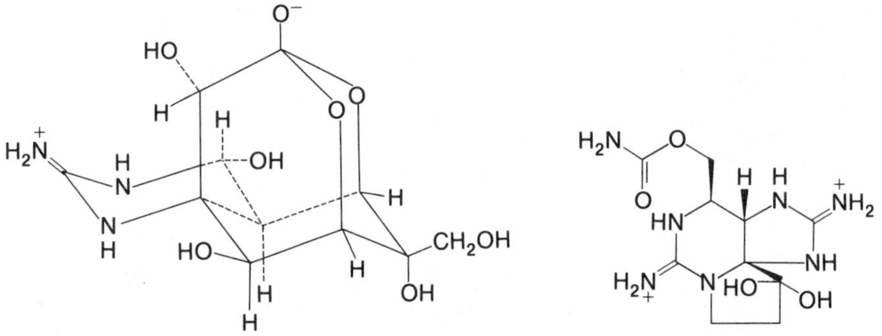

(a)

(b)

**Figure 30-8**
(a) The structure of tetrodotoxin, a compound that specifically blocks $Na^+$ channels in nerve cell membranes. This extremely toxic compound is found in the liver and ovaries of the Japanese puffer fish (*Spheroides rubripes*). (b) The structure of saxitoxin, a compound found in certain marine dinoflagellates ("plankton"), which are constituents of the so-called red tide. Mussels and clams that have fed upon these organisms are therefore extremely poisonous, and commercial shell fishing is banned in areas where these dinoflagellates appear. Saxitoxin is also a $Na^+$-channel blocking agent, and both compounds most likely interact with the channel through their positively charged guanidino group.

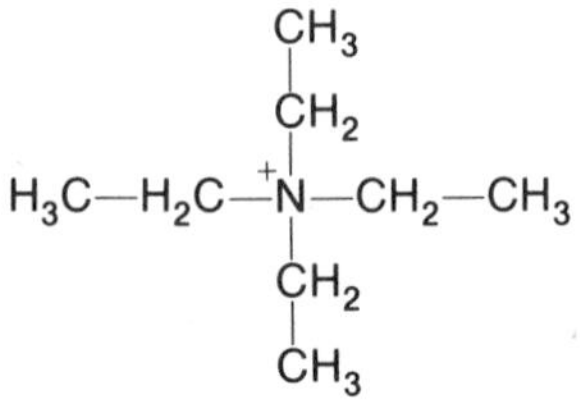

**Figure 30-9**
Tetraethylammonium ion, a compound that specifically inhibits $K^+$ channels in nerve cell membranes. The ethyl groups presumably sterically hinder passage of this cation through the channel, "plugging" the channel and therefore blocking the transport of $K^+$.

It was the use of such a voltage clamp by Hodgkin and Huxley that allowed them to deduce the movements of $Na^+$ and $K^+$ that accompany the appearance of the action potential. Because the changes in membrane permeability to $Na^+$ and $K^+$ were not superimposable in time, it was tentatively concluded that two different "channels" were involved in the transmembrane movements of these ions. This conclusion has since been confirmed by the discovery of compounds that specifically block the conductance of the nerve membrane to either $Na^+$ or $K^+$. *Tetrodotoxin* and *saxitoxin* (Figure 30-8) are both nerve poisons that specifically inhibit the transmembrane movement of $Na^+$ by binding to the outside of nerve membranes without affecting $K^+$ permeability, as measured in voltage-clamp experiments. Similarly, *tetraethylammonium ions* (Figure 30-9) have been shown to bind to the axoplasmic membrane surface and to specifically block the outward flow of $K^+$.

*A large body of evidence has now accumulated that supports the suggestion that $K^+$ and $Na^+$ movements through axonal membranes are mediated by separate channels that act as gated pores* (Chapter 17). During a typical action potential lasting about 5 ms, it has been calculated that *at least 60 $Na^+$ ions pass through a single $Na^+$-specific channel* (i.e., about 12,000 per second). Since the neuronal $Na^+$-$K^+$ ATPase normally pumps less than 200 $Na^+$ ions per second out of the cell, this pump does not seem to be directly involved in the development of the action potential. Instead, a process involving passive $Na^+$ diffusion through a transmembrane pore appears to best explain these rapid $Na^+$ fluxes. Furthermore, voltage-clamp studies by B. Hille have established that *molecules such as the $K^+$-monohydrate complex with dimensions larger than about 0.3 to 0.5 $nm^2$ in cross section* (the size of a monohydrated $Na^+$ ion) *do not pass readily* *through the Na+ channel.* However, smaller molecules, such as monohydrated Li+ ions, and cations approximately the same size as a hydrated Na+ ion, such as hydroxylamine and hydrazine, do. This observation, that is, exclusion based on size rather than on chemical structure, is more characteristic of a pore than a specific membrane permease (Chapter 17).

Additional properties of the Na+ channel have been inferred by extensions of the specificity studies just outlined. Thus the pH dependence of Na+ permeation as well as other experimental observations suggest that a *negatively charged carboxylate group* ($—COO^-$) *is present in the pore* and is involved in interacting with the cationic molecules that can traverse the Na+ channel. It is presumably this interaction that allows the positively charged guanidinium groups of tetrodotoxin and saxitoxin (Figure 30-8) to bind to the channel, but the size of these poisons prevents their passage through the membrane and results in blockage of Na+ conduction. An additional observation is that methylamine, a cation with dimensions that should allow its more or less free passage through the pore, is nevertheless much less permeable than Na+. This observation has been interpreted as evidence that the sodium channel has one or more oxygen atoms at its narrowest point, including the carboxylate group. Monohydrated Na+, hydroxylamine, and hydrazine are thought to form hydrogen bonds with these oxygen atoms through their $H_2O$, $—OH$, and $—NH_2$ groups, respectively, thereby facilitating passage of these cations through the channel. The relative impermeability of methylamine can thus be explained by the inability of its $—CH_3$ moiety to participate in this interaction (Figure 30-10).

It has recently been possible to estimate the number of Na+ channels in a variety of nerve types by the use of radioactively labeled tetrodotoxin or saxitoxin molecules. From these studies it has become apparent that in unmyelinated nerve fibers, the density of these channels in the membrane is exceedingly low. For example, the olfactory nerve of the garfish has only about 30 to 40 Na+ channels per square micrometer of membrane, which corresponds to only about 0.2 percent of the total surface area of the phospholipid bilayer. In the squid giant axon, which has a fiber diameter some 2500 times that of the garfish olfactory nerve, this value is only increased to several hundred Na+ channels per square micrometer, or about 2 percent of the surface area. The situation is quite different, however, in myelinated nerves of vertebrate animals. In this case, Na+ channels are found in significant numbers only at the nodes of Ranvier, as might be anticipated from the mechanism of impulse propagation in these types of nerve cells (Figure 30-7). Here, the channel density is on the order of $10^4$ channels per square micrometer corresponding to approximately 60 percent of the total membrane surface area. Indeed, Na+ fluxes in these regions of myelinated nerves have been estimated to be 10 to 100 times larger than those in their unmyelinated counterparts during propagation of the action potential.

In contrast to the Na+ channel, much less is known about the properties of the channel responsible for the outward flow of K+ from the axon during nerve excitation. This is due partly to the fact that there are even fewer K+ channels than Na+ channels in the axonal membrane (about one-tenth the number in the squid giant axon). Furthermore, tightly binding specific inhibitors of the K+ channel analogous to the poisons of the Na+ channel have not been found. Nevertheless, it has been demonstrated

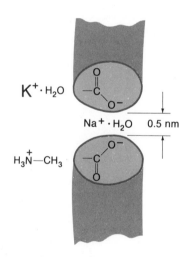

**Figure 30-10**
Schematic representation of a sodium channel in a nerve cell membrane. The protein(s) comprising the channel are believed to form a pore with dimensions such that only molecules with a diameter $\leq 0.5$ nm will pass through (excluding ions such as monohydrated K+). The impermeability of methylamine, which is about the size of monohydrated Na+, suggests that other interactions, such as hydrogen bonding to one or more oxygen atoms within the channel, are also necessary for penetration of an ion through the pore to occur.

that the $K^+$ pore is quite specific, barring the passage of cations both smaller ($Na^+$) and larger ($Cs^+$; tetramethylammonium ions) than potassium. It seems likely, therefore, that both the diameter of the channel and specific interactions of channel components with the $K^+ \cdot H_2O$ complex confer cation selectivity on this channel as well.

### The Mechanism of $Na^+$ Channel Function

How do the channels selective for $Na^+$ and $K^+$ ions "sense" the membrane potential and react by opening or closing during various stages of action-potential propagation? Although this question is still far from being answered in molecular detail, a number of clues are emerging from recent work on the gating properties of these channels and their purification and reconstitution into artificial membrane systems. The results of these investigations suggest that *ion channels in nerve and muscle cell membranes can assume more than one conformation, one of which is more permeable to the ion in question than the others*.

A fairly recent observation that bears on this question is the discovery of *gating currents*, first predicted by Hodgkin and Huxley. Under appropriate conditions, a small current opposite in direction to that carried by $Na^+$ during the opening of the $Na^+$ channels can be detected during the development of an action potential. This current is short in duration (0.1 ms) and precedes the opening of the sodium channels (as detected by the inward movement of $Na^+$) (see Figure 30-11). One attractive model is that the $Na^+$ channel has a "built in" or closely associated "voltage sensor" that has a large dipole moment. Depolarization of the membrane, which elicits the action potential, would then result in the displacement or rearrangement of this dipole in response to the electric field. This process would be detected as the gating current. The change in orientation of this dipole could then be the "trigger" by which the $Na^+$ channel is opened, presumably by means of a conformational change in this protein.

Additional evidence for voltage-dependent conformational changes that allow gating of the $Na^+$ channel has been obtained using nerve poisons isolated from North African scorpions and the sea anemone. These toxins, which are basic polypeptides, bind to sites on the $Na^+$ channel that are distinct from those which bind the channel-blocking agents tetrodotoxin and saxitoxin. They appear to exert their physiological effects by *slowing inactivation* of the $Na^+$ channel after its initial activation (opening) during the action potential. Binding of scorpion toxin to $Na^+$ channels in nerve and muscle-cell membranes has been shown to be *voltage-dependent,* suggesting that this toxin recognizes a conformation of the channel that also depends on the membrane potential. This observation has led W. A. Catterall to recently hypothesize that toxins of the scorpion type may interact with the voltage-sensor component of the channel, perhaps the same moiety responsible for the gating current described earlier. Toxin binding would thus reflect the conformational state of the channel, which, in turn, depends on $\Delta\psi$. These and other experiments have led to the proposal that the $Na^+$ channel can exist in at least three conformations, depending on $\Delta\psi$: *closed* (resting $\Delta\psi$), *open,* and *inactive*. The latter two states are responsible for the transient increase and then decrease in $Na^+$

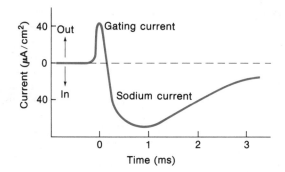

**Figure 30-11**
Illustration of the gating current that precedes the inward-directed Na$^+$ current associated with depolarization during the action potential. The gating current is believed to be related to the voltage-dependent opening of the Na$^+$ channels and perhaps may reflect rearrangement of a dipolar "voltage sensor" associated with the channel gate. (Adapted from C. M. Armstrong and F. Bezanilla, Charge movement associated with the opening and closing of the activation gates of the Na channels, *J. Gen. Physiol.* 63: 533, 1974.)

permeability observed during the action potential. Hyperpolarization as a result of increased K$^+$ permeability then recycles the channel to its closed conformation.

Perhaps the most promising approach to understanding structure-function relationships in the ion channels of nerve-cell membranes will be to purify these proteins and to study their properties in artificial membrane systems, as we described for transport permeases in Chapters 16 and 17. Significant progress has already been made in this direction. For example, M. A. Raftery and coworkers have purified a tetrodotoxin-binding protein from the _electric organ_ of the electric eel (*Electrophorus electricus*) that is rich in Na$^+$ channels. A major component of this preparation was a protein with a polypeptide-chain molecular weight of about 260,000. Similarly, W. A. Catterall and collaborators have purified a saxitoxin receptor from rat brain that consists of two polypeptides, $\alpha$ and $\beta$, with molecular weights of about 270,000 and 38,000. They also have demonstrated that similar-sized polypeptides can be labeled in cultured neuroblastoma cells with a photoactivatable derivative of scorpion toxin. Thus Na$^+$ channels from quite different sources appear to be proteins of similar structure.

It is to be anticipated that reconstitution of such purified preparations into phospholipid bilayers will reveal whether these toxin-binding proteins are sufficient to form a functional Na$^+$ channel and, if so, whether one or more subunits are necessary for this activity. Reconstitution into liposomes of detergent-solubilized nerve membrane preparations that have not been completely purified has already been achieved. Proteoliposomes so obtained are selectively permeable to Na$^+$ and show the expected responses to the various toxins affecting the Na$^+$ channel in nerve cell membranes. The use of planar bilayer systems that can be electrically manipulated (Chapter 17) along with purified Na$^+$-channel proteins should provide systems in which to probe the detailed molecular events associated with Na$^+$ movements during nerve excitation.

## Synaptic Transmission: Acetylcholine

Nerve impulses, propagated along the axon by the mechanisms outlined in the preceding sections, must be transmitted between nerve cells or from nerve cells to muscle or glandular tissues in order for their effects (e.g., contraction or secretion) to take place. This type of *intercellular communication can occur either by an electrical mechanism of coupling or by chemical transmission across specific connections between nerve cells and target cells called synapses.* An example of electrical transmission was given in Chapter 17. Direct cell-cell contact by means of *gap junctions* can allow action potentials to be transmitted by ionic mechanisms between cells connected in this manner. The more common mechanism, however, involves the release of specific chemicals, called *neurotransmitters*, at the *presynaptic membrane* of one cell in response to membrane depolarization and their diffusion to *receptors* in the *postsynaptic membrane* of the recipient cell, where a new action potential may be generated.

The process of synaptic transmission is depicted schematically in Figure 30-12 for *acetylcholine*, the excitatory neurotransmitter at vertebrate neuromuscular junctions (motor end plates). Acetylcholine is synthesized in nerve cells by the enzyme *cholineacetyltransferase* (Figure 30-13) and is packaged in units ("quanta") of $10^3$ to $10^4$ molecules within *synaptic vesicles*, which are abundant near the cytoplasmic membrane of the presynaptic axon. The arrival of an action potential triggers a large increase in the permeability of the presynaptic membrane to $Ca^{2+}$, and this ion flows into the axoplasm down its chemical gradient. Fusion of the synaptic vesicles with the plasma membrane and the concomitant release of acetylcholine into the *synaptic cleft* are promoted by this increase in intracellular $Ca^{2+}$. Several hundred synaptic vesicles empty their contents into the synaptic cleft in a typical neuromuscular synaptic junction by this mechanism in response to a single action potential. The resultant large increase in the local concentration of acetylcholine is "sensed" by a protein, the *acetylcho-*

**Figure 30-12**
Schematic diagram of a synaptic junction in which acetylcholine is the chemical transmitter. Arrival of an action potential at the terminus of the presynaptic cell (*top*) stimulates $Ca^{2+}$ uptake, which triggers release of acetylcholine (ACh) from vesicles near the terminus of the presynaptic cell. Release is accomplished by vesicle fusion with the plasma membrane, and the interaction of acetylcholine with its receptors in the postsynaptic membrane triggers depolarization, thus propagating the action potential in the postsynaptic cell (*bottom*).

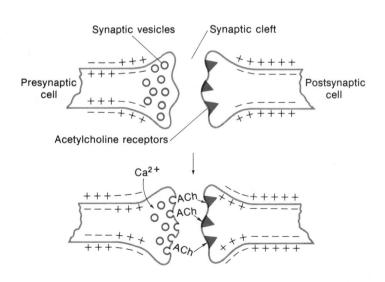

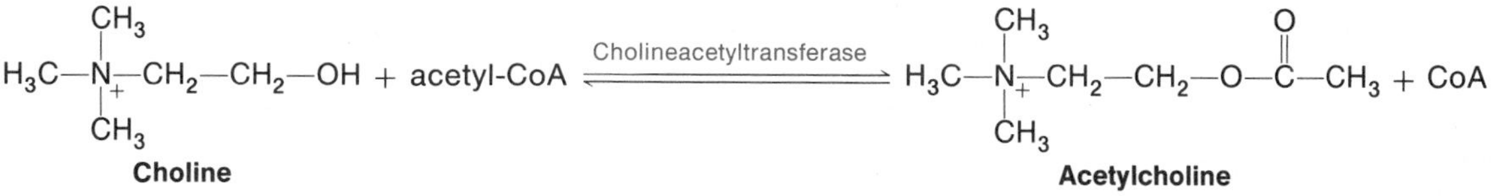

Choline + acetyl-CoA ⇌ (Cholineacetyltransferase) Acetylcholine + CoA

line receptor, located in the cytoplasmic membrane of the postsynaptic cell. Binding of the neurotransmitter to many receptor molecules triggers an action potential in the recipient cell. The acetylcholine is subsequently rapidly degraded into acetate and choline by an enzyme in the synaptic cleft called *acetylcholinesterase*, and the resting potential of the postsynaptic membrane is rapidly restored.

How does the binding of a neurotransmitter to its receptor promote depolarization of the postsynaptic membrane? One of the first clues came from studies by B. Katz and collaborators who showed that *acetylcholine increases the ionic permeability of the postsynaptic membrane*. Subsequent studies with radioactive tracers and by the voltage-clamp technique established that *the membrane permeabilities to both Na$^+$ and K$^+$ were increased simultaneously by the action of acetylcholine*. Because the electrochemical Na$^+$ gradient is somewhat larger than that of K$^+$ across postsynaptic membranes, depolarization results from the inward flow of Na$^+$, tending to collapse $\Delta\psi$. This local perturbation in $\Delta\psi$ is enough to initiate a new action potential in the recipient neural or muscular membrane *as long as a sufficient number of receptor molecules bind the neurotransmitter*. In fact, as we shall see in a later section, *the receptor itself contains the ion channel* through which both K$^+$ and Na$^+$ can flow. The number of occupied receptors at a given moment thus dictates the magnitude of the inward flow of Na$^+$ and the resulting magnitude of the change in the membrane potential. Rapid hydrolysis of acetylcholine by acetylcholinesterase bound to the postsynaptic membrane then quickly reduces the number of transmitter-receptor complexes and repolarizes the membrane until a new action potential triggers the release of more acetylcholine quanta from the presynaptic membrane.

The elucidation of the steps just outlined has been aided greatly by the use of specific inhibitors of both acetylcholinesterase and the acetylcholine receptor. Compounds such as *eserine* (Figure 30-14a) block the hydrolytic enzyme and thus can be used to study the effects of acetylcholine in *cholinergic* systems (those using acetylcholine as a transmitter) under conditions where this neurotransmitter cannot be hydrolyzed. Likewise, certain organophosphorus compounds efficiently inhibit acetylcholinesterase by forming stable covalent intermediates with an active-site serine in the enzyme. The widely used insecticides *parathion* and *malathion* are examples of this class of nerve poisons (Figure 30-14b). Neuromuscular junctions exposed to acetylcholinesterase inhibitors are paralyzed because the persistent presence of acetylcholine prevents repolarization of the postsynaptic membrane to restore its excitability. In fact, such a situation eventually results in the acetylcholine receptor becoming *desensitized*, i.e., remaining closed to ion flow for long intervals even in the presence of the neurotransmitter.

**Figure 30-13**
Acetylcholine is synthesized by cholineacetyltransferase, which esterifies choline with an acetyl group from acetyl-CoA.

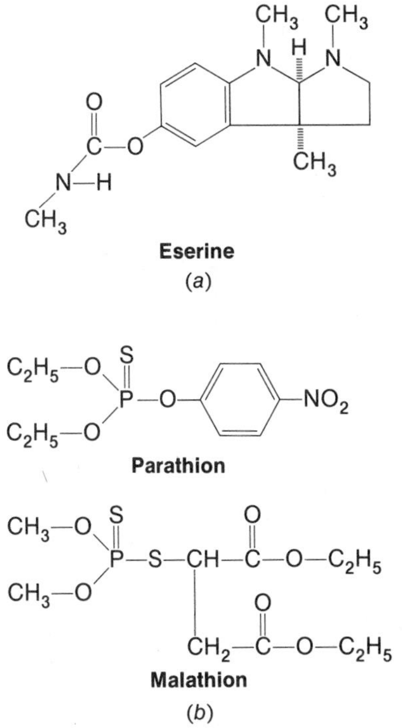

**Eserine**
(a)

**Parathion**

**Malathion**
(b)

Acetylcholinesterase inhibitors

**Figure 30-14**
(a) Eserine, or *physostigmine,* is an alkaloid that forms a relatively stable covalent carbamoyl intermediate with an active-site serine on the enzyme that is hydrolyzed only very slowly. (b) Parathion and malathion are organophosphorus compounds that also form stable covalent complexes with the active-site serine of acetylcholinesterase. They are widely used agriculturally as insecticides.

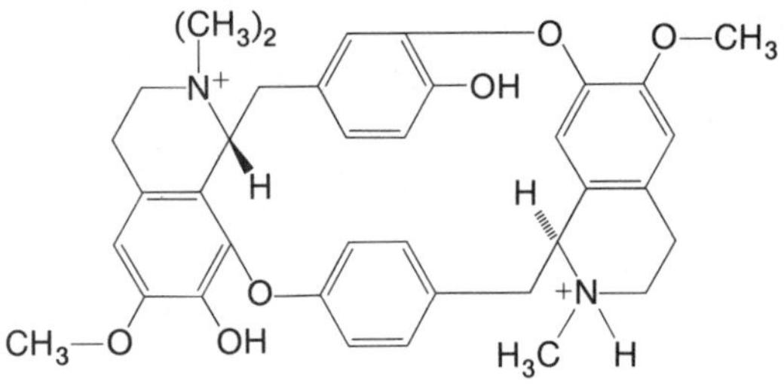

**Figure 30-15**
The structure of $d(+)$-tubocurarine, an active component of the neurotoxin curare. This compound binds to the acetylcholine-binding site on its receptor, preventing synaptic transmission and subsequent depolarization of the postsynaptic cell membrane.

Specific blocking agents of the acetylcholine receptor include *d-tubocurarine*, an active component of the neurotoxin curare (Figure 30-15), and the snake venom poisons *α-bungarotoxin* (from snakes of the genus *Bungarus*) and *cobratoxin*. The latter are small basic proteins with masses around 7000 daltons. All three of these substances interact noncovalently with the receptor and interfere with acetylcholine binding, thus blocking depolarization of the postsynaptic membrane. They are referred to as *antagonists* of the cholinergic systems. Another type of acetylcholine receptor-inhibitor is exemplified by the divalent cation *decamethonium* (Figure 30-16), which "locks" the ion channel of the receptor in the open state and thus leads to a constant depolarization of the recipient cell membrane. Such compounds, referred to as *agonists*, mimic the effect of acetylcholine but cannot be rapidly inactivated, thus blocking resensitization of the postsynaptic membrane. These substances have allowed workers to investigate properties of the acetylcholine receptor in the "open" and "closed" states and their effects on the permeability of the postsynaptic membrane. They also have been useful in the purification of this protein, as we shall describe shortly.

## Other Neurotransmitters

A number of other compounds have been implicated as neurotransmitters in addition to acetylcholine. The best-documented examples of these are the *catecholamines*, certain *amino acids* (and derivatives), and a variety of *peptides*. The catecholamines are all derived biosynthetically from L-tyrosine (Figure 30-17) and include the hormones *norepinephrine* and *epinephrine* (adrenaline). Because these compounds are also synthesized in the adrenal gland (see Table 29-1), neurons that use these substances as chemical transmitters are said to be *adrenergic*. Sympathetic nerve fibers that innervate smooth-muscle cells in internal organs such as the heart, spleen, and gut have been shown to release norepinephrine at their terminals by mechanisms similar to those used by cholinergic neurons. Norepinephrine and related amines also have been shown to serve as neurotransmitters in a number of nerve pathways in the brain. Like acetylcholine, catecholamines may be inactivated by chemical modifications. Inactivation may be effected by a methylation reaction or by an oxidation reaction catalyzed by the enzyme *monoamine oxidase* (Figure 30-17). In some cases, they also can be resorbed through the presynaptic membrane after their release, providing an additional mechanism for removal from the synaptic cleft.

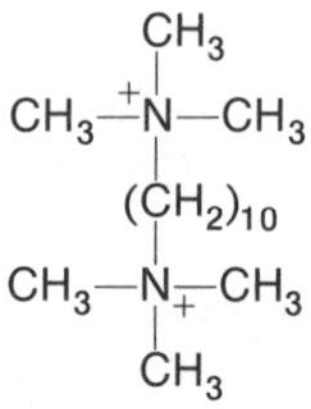

**Figure 30-16**
Decamethonium ion, an agonist of cholinergic systems. This compound binds to the acetylcholine receptor, but because it cannot be degraded, it causes persistent depolarization of the postsynaptic membrane.

The hydroxylation of tyrosine, which is the first unique step in the biosynthesis of catecholamines, is catalyzed by *tyrosine hydroxylase* and yields the compound *3,4-dihydroxyphenylalanine* (L-DOPA). The neurological disorder *Parkinson's disease* is associated with an underproduction in the human brain of the catecholamine transmitter *dopamine*, which is derived from L-DOPA by a decarboxylation reaction (Figure 30-17). L-DOPA has therefore been found to be an effective drug in many instances in the treatment of Parkinson's disease. Interestingly, overproduction of dopamine in the brain also occasionally occurs and appears to be associated with psychological disorders such as schizophrenia. In this case, *dopamine-receptor blocking drugs*, such as *chlorpromazine* (Figure 30-18), have been found to be useful therapeutic agents.

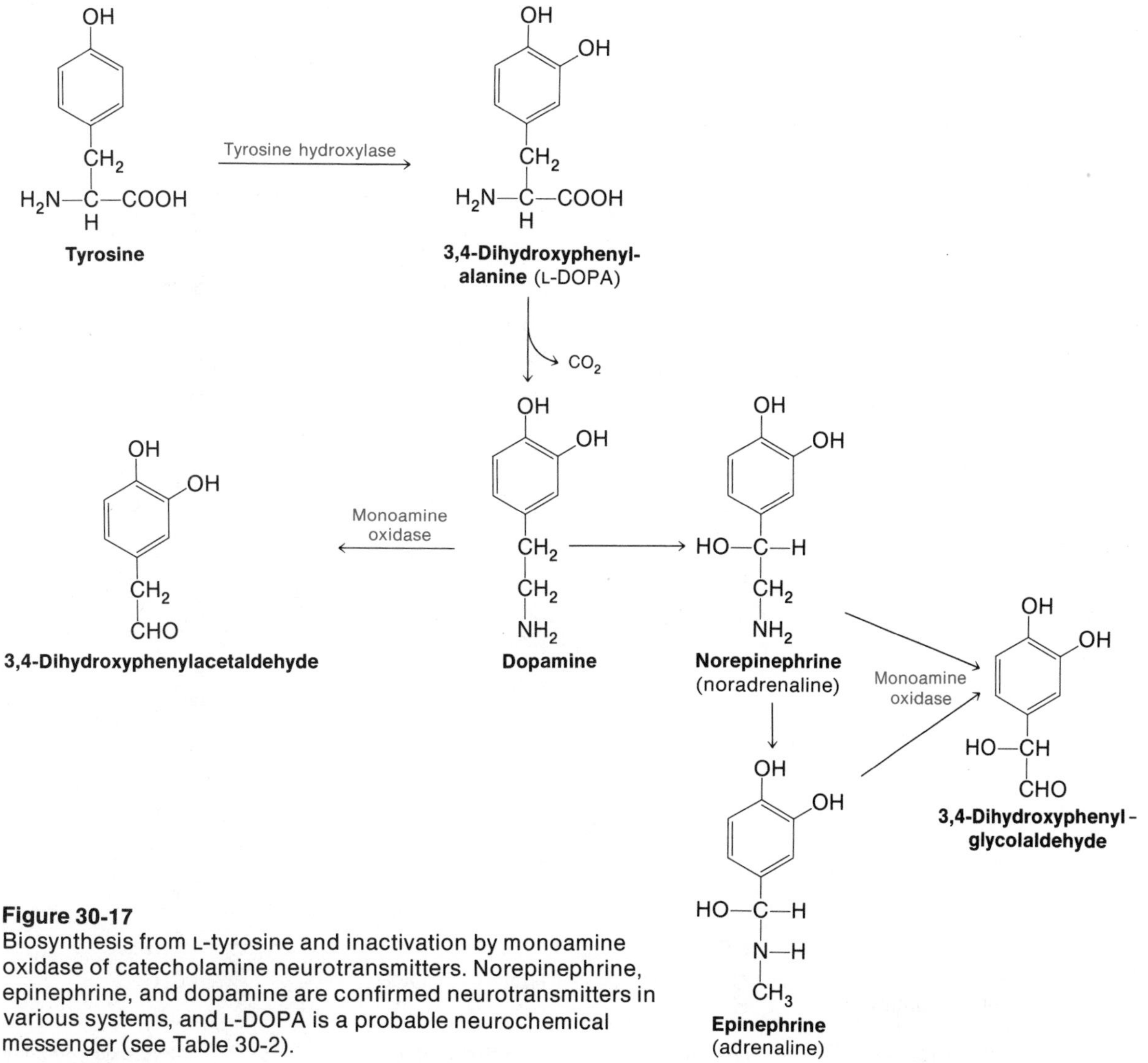

**Figure 30-17**
Biosynthesis from L-tyrosine and inactivation by monoamine oxidase of catecholamine neurotransmitters. Norepinephrine, epinephrine, and dopamine are confirmed neurotransmitters in various systems, and L-DOPA is a probable neurochemical messenger (see Table 30-2).

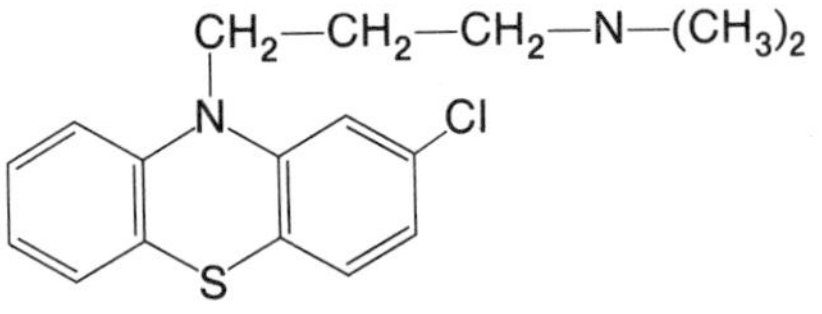

**Figure 30-18**
The structure of chlorpromazine, a drug that blocks dopamine receptors and which has been used in the treatment of psychological disorders such as schizophrenia.

Amino acids that are believed to have roles as neurotransmitters include *glutamic acid* and *glycine*. The amino acid derivatives *histamine* (synthesized by the decarboxylation of histidine), *5-hydroxytryptamine* (or *serotonin;* derived from tryptophan), and *gamma-aminobutyric acid* (or *GABA;* a decarboxylation product of glutamic acid) have all been shown to be transmitters in various systems as well. The reactions involved in the biosynthesis of these compounds are diagramed in Figure 30-19. Gamma-aminobutyrate is used most often as an *inhibitory transmitter.* Interaction of this compound with its receptor on many postsynaptic membranes results in a large *increase in the membrane permeability to Cl⁻ and/or K⁺ ions.* This inhibits the postsynaptic cell, often by hyperpolarizing its membrane (recall, for example, that the equilibrium potential of K⁺ is more negative than the resting potential). *Most target tissues, in fact, are innervated by more than one type of nerve fiber, each using a different neurotransmitter. This allows for different signals to be relayed to the recipient cells, some stimulatory and others inhibitory.* Given the com-

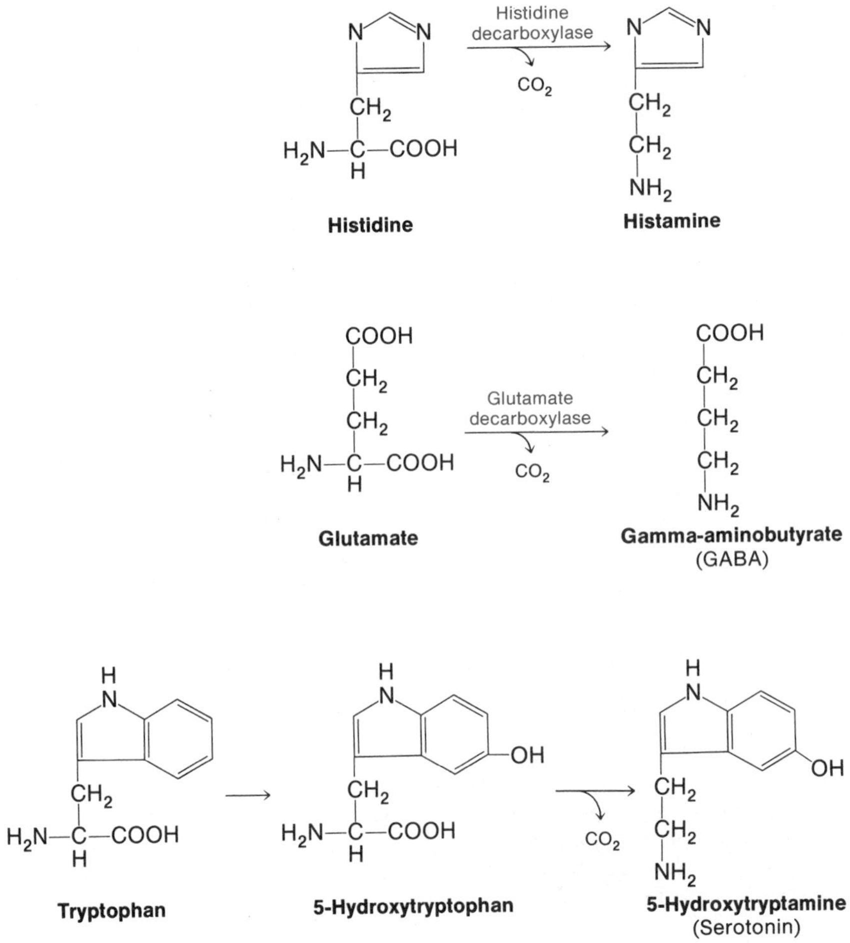

**Figure 30-19**
Biosynthesis of the neurotransmitters histamine, gamma-aminobutyrate (GABA), and serotonin (5-hydroxytryptamine).

**Table 30-2**
Neurochemical Messengers

| Compounds | Status* |
|---|---|
| 1. Acetylcholine | C |
| 2. Catecholamines | |
| Norepinephrine (noradrenaline) | C |
| Epinephrine (adrenaline) | C |
| L-DOPA | P |
| Dopamine | C |
| Octopamine | P |
| 3. Amino acids (and derivatives) | |
| Glutamate | C |
| Aspartate | Pos |
| Glycine | C |
| Proline | Pos |
| Gamma-aminobutyrate (GABA) | C |
| Tyrosine | Pos |
| Taurine | Pos |
| Alanine | Pos |
| Cystathione | Pos |
| Histamine | C |
| Serotonin (5-hydroxytryptamine) | C |
| 4. Peptides | |
| Substance P | P |
| Cholecystokinin | P |
| Neurotensin | P |
| Enkephalins | P |
| Somatostatin | P |

*C = confirmed neurotransmitter;
P = probable; Pos = possible.

plexity of the nervous systems of higher animals, additional neurotransmitters are likely to be discovered. Some compounds currently thought to be chemical transmitters are listed in Table 30-2.

## The Acetylcholine Receptor

The structure and properties of the acetylcholine receptor are by far the best understood among all neurotransmitter receptors. One reason for this is their abundance in postsynaptic membranes found in the electric organs of the electric eel (*Electrophorus*) and the electric ray (*Torpedo*). These organs contain stacks of cells (*electroplaxes*), each cell of which receives nerve endings on one side, but not on the other. Release of acetylcholine by the presynaptic nerve endings thus depolarizes the postsynaptic membrane by virtue of the binding of neurotransmitter to the membrane-bound receptors, while the other face of the cell remains at its resting potential. In this way, each cell, when stimulated, can attain a potential difference of over 100 mV between its two faces, and thousands of such stacked cells, present in the electric organ, can consequently emit an electric discharge of several hundred volts.

Acetylcholine receptor-rich membranes can easily be isolated from the electric organs of *Electrophorus* and *Torpedo*. These membranes contain from 10 to $50 \times 10^4$ receptor sites per square micrometer, as measured by the binding of radioactively labeled antagonists such as $\alpha$-bungarotoxin. Electron microscopy and x-ray diffraction have further revealed that such membrane fragments have a similar density of particles, each about 8 nm in diameter, arranged in a regular hexagonal lattice (Figure 30-20). Each particle is actually a rosette, apparently consisting of four to six subunits arranged around a central axis, as shown in Figure 30-20. The density of these oligomeric protein molecules on the membrane, their size (see below), and their interaction with antibodies leave little doubt that they are the acetylcholine receptors themselves. Furthermore, closed membrane vesicles derived from such membrane fragments ("microsacs") can be made permeable to $Na^+$ by adding agonists of the acetylcholine receptor, an effect that is blocked by antagonists such as the snake venom toxins.

Purification of the acetylcholine receptor to apparent homogeneity has been achieved by solubilization of electroplax membranes of *Torpedo* by nonionic detergents and affinity chromatography on columns containing covalently bound cobratoxin. Alternatively, preparations of similar purity can be obtained by fractionation of membrane fragments using sucrose density-gradient centrifugation and subsequent removal of peripherally associated proteins by treatment at high pH values. Both these procedures yield a glycoprotein consisting of four subunits (40, 50, 60, and $65 \times 10^3$ daltons named $\alpha$, $\beta$, $\gamma$, and $\delta$) in a molar ratio of $2:1:1:1$. The simplest structure for the acetylcholine receptor is therefore a pentamer ($\alpha_2\beta\gamma\delta$) with a mass of about 255,000 daltons. This value agrees with hydrodynamic measurements of the molecular weight of the purified receptor in detergent, as well as with the dimensions and apparent subunit composition of the particles seen by electron microscopy of acetylcholine receptor-rich membranes (Figure 30-20). Interestingly, the four different subunits of this protein exhibit significant amino acid sequence homologies near their $NH_2$ terminals, indicating that they all arose by duplication and divergence of a single ancestral gene.

Reactive affinity labels that covalently bind to the acetylcholine receptor have been prepared by chemically modifying known agonists and antagonists of cholinergic systems. In most cases examined, the $\alpha$-subunit was labeled by these treatments. These results suggest that this subunit contains at least part of the binding site (or sites) for acetylcholine in the receptor complex. The purified receptor from *Torpedo* has been functionally reconstituted into vesicles consisting of lipids isolated from electroplax membranes. *These proteoliposomes became permeable to $Na^+$ in the presence of an acetylcholine analog, while $\alpha$-bungarotoxin blocked $Na^+$ permeability.* The availability of this experimental system should allow detailed studies of structure-function relationships in the receptor protein, e.g., the determination of which subunits are necessary to form the ion channel and whether agonist/antagonist binding sites can be separated from the channel-forming parts of the molecule. Because all five subunits appear to span the postsynaptic membrane, as recently demonstrated by their susceptibility to tryptic hydrolysis at both membrane surfaces, it is possible that the channel corresponds to the central "hole" in the rosette seen by electron microscopy of electroplax postsynaptic membranes (Figure 30-20).

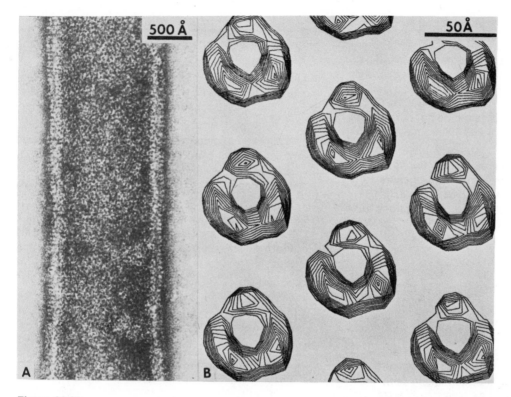

**Figure 30-20**
A view of the arrangement of acetylcholine receptors on electroplax membranes from
*Torpedo californica*. (a) Membrane tubes formed spontaneously from membrane vesicles
showing a crystalline surface lattice of acetylcholine receptors. Stained with uranyl acetate. (b) Computer filtration of micrographs similar to that in (a) showing the arrangement
of protein density in a single receptor around the central depression (presumed to be the
ion channel). (From J. Kistler, R. M. Stroud, M. W. Klymkowsky, R. A. Lalancette, and
R. H. Fairclough, *Biophys. J.* 37:371, 1982.) (Micrograph courtesy Dr. R. Stroud. Used
with permission.)

Considerable evidence points to a gated-pore-type mechanism for ion
permeability conferred by the acetylcholine receptor, as is the case for the
$Na^+$ channel. At least two conformations of the protein, corresponding to
"open" and "closed" states, have been recognized in the presence of
agonists and antagonists, respectively. Furthermore, voltage-clamp studies
of muscle fibers containing cholinergic receptors in the presence of agonists
have revealed current pulses having a square shape and a constant amplitude similar to those seen with bacterial porins and certain ionophore antibiotics in artificial membrane systems (Chapter 17). If these events correspond to the opening and closing of individual channels, as seems likely,
then it can be calculated that *about $10^4$ molecules of $Na^+$ flow through the
receptor in vivo in the millisecond or so that it is open.* A pore model
allowing more or less free diffusion of ions through the channel in its open
state rather than a slower, carrier-mediated mechanism is therefore favored
for the acetylcholine receptor.

## SELECTED READINGS

Agnew, W. S., Moore, A. C., Levinson, S. R., and Raftery, M. A.  Identification of a large molecular weight peptide associated with a tetrodotoxin binding protein from the electroplax of *Electrophorus electricus*. *Biochem. Biophys. Res. Commun.* 92:860, 1980. The purification of a tetrodotoxin-binding polypeptide with a molecular weight of 260,000 from electroplax membranes.

Hartshorne, R. P., and Catterall, W. A.  Purification of the saxitoxin receptor of the sodium channel from rat brain. *Proc. Natl. Acad. Sci. USA* 78:4620, 1981. This preparation consists of two polypeptides with molecular weights of 270,000 and 38,000, both of which are also labeled by a photoreactive derivative of scorpion toxin.

Katz, B.  *Nerve, Muscle and Synapse.* New York: McGraw-Hill, 1966. Although over 15 years old, this classical book offers an excellent overview of the subject.

Keynes, R. D.  Ion channels in the nerve-cell membrane. *Sci. Am* 240:126, 1979. Reviews properties of the $Na^+$ and $K^+$ channels in nerve cell membranes.

Kistler, J., Stroud, R. M., Klymkowsky, M. W., Lalancette, R. A., and Fairclough, R. H.  Structure and function of an acetylcholine receptor. *Biophys. J.* 37:371, 1982. Recent review of the structure and function of this receptor in *Torpedo californica* electroplax membranes.

Kuffler, S. W., and Nicholls, J. G.  *From Neuron to Brain.* Sunderland, Mass.: Sinauer Associates, 1976. This book covers aspects of neural transmission and neurophysiology in an easily readable form.

## PROBLEMS

**1.** a. Calculate the membrane potential $\Delta\psi$ across a resting nerve cell membrane in the presence of tetrodotoxin. (Assume that the resting permeability to $Na^+$ is the result of a small, steady-state level of "open" $Na^+$ channels.)

  b. What would the value of $\Delta\psi$ be if the resting axonal membrane were permeable only to $Cl^-$? (Assume an axoplasmic $Cl^-$ concentration of 50 m$M$.)

**2.** Explain why a nerve cell membrane exhibits an "all or none" response (i.e., action potential) independent of the magnitude of an electric or chemical stimulus (above a threshold value).

**3.** List the criteria for demonstrating that a particular compound acts as a neurotransmitter in a given system, assuming that the mechanism of synaptic transmission in most instances is analogous to that found in cholinergic systems.

**4.** Solubilization of $Na^+$ channels from nerve membranes can be achieved by treatment with sodium cholate. These channels, as well as other proteins solubilized by this procedure, can be incorporated into proteoliposomes such that each artificial vesicle receives only one, or at most a few, solubilized proteins. "Open" $Na^+$ channels are known to allow the diffusion of $Cs^+$, a dense monovalent cation, through the membrane, although the rate of this diffusion is much slower than that of $Na^+$. *Veratridine,* an alkaloid that binds to the $Na^+$ channel (see structure below), causes persistent activation of these channels in the reconstituted state. On the basis of these observations, devise a procedure for purifying proteoliposomes containing the $Na^+$ channel from a heterogeneous population of reconstituted vesicles.

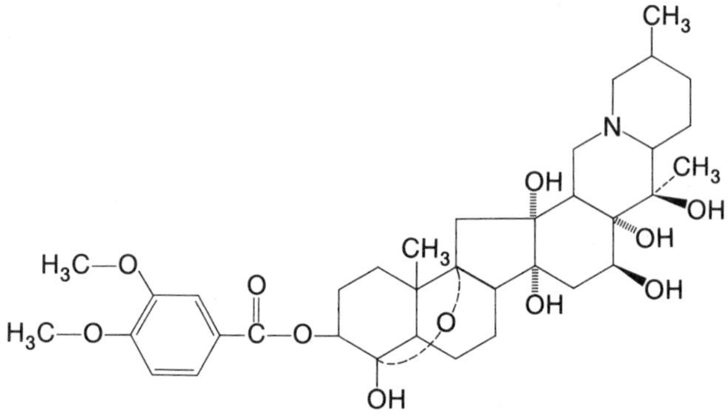

**Veratridine**

**5.** In reconstituted transport systems, it is often important to demonstrate that the rate of transport is similar to that observed in vivo; i.e., that the transport protein is fully functional in the reconstituted state. For systems in which in vivo fluxes are very rapid (e.g., cation flux through the acetylcholine receptor), it is often difficult to measure these rates directly in reconstituted vesicles. Thallous ion ($Tl^+$) is known to pass readily through the "open" state of the acetylcholine receptor. It also very efficiently quenches the fluorescence emission of the fluorophore 8-amino-naphthalene-1,3,6-trisulfonate (ANTS), which is relatively impermeable to phospholipid bilayers.

a. Outline a series of experiments to measure $Tl^+$ fluxes in reconstituted proteoliposomes containing purified acetylcholine receptors using this information.

b. Actual $Tl^+$ fluxes into proteoliposomes containing an average of 2 acetylcholine receptor channels per vesicle in the presence of agonists have been measured to be 200 moles $\cdot$ liter$^{-1}$ $\cdot$ sec$^{-1}$. If the average inner diameter of such vesicles is 400 Å, calculate the number of $Tl^+$ ions transported per second by each activated acetylcholine receptor channel. How does this value compare with the rate of $Na^+$ flux measured in vivo?

6. Excitable cells, such as neurons, are not limited to multicellular higher organisms. For example, ciliated unicellular protozoa such as *Paramecium* are known to undergo a behavioral response called the "avoidance reaction." When such cells collide with an obstacle, the direction of beating of their cilia (which normally propel the cell in a forward direction) reverses, and the protozoa "back up" to avoid repeated collisions with the obstacle. After a short time, the cells resume their normal forward motion in a new direction. *Paramecium* is known to have a "resting potential" across its cytoplasmic membrane of about $-30$ mV when swimming normally. The avoidance reaction and the direction of ciliary beating in this organism are critically dependent on the extracellular concentration of $Ca^{2+}$ ions (e.g., no $Ca^{2+}$, no response). By analogy with the mechanism of action potential production in nerve cells, postulate a reasonable mechanism for the avoidance reaction in *Paramecium* using this information.

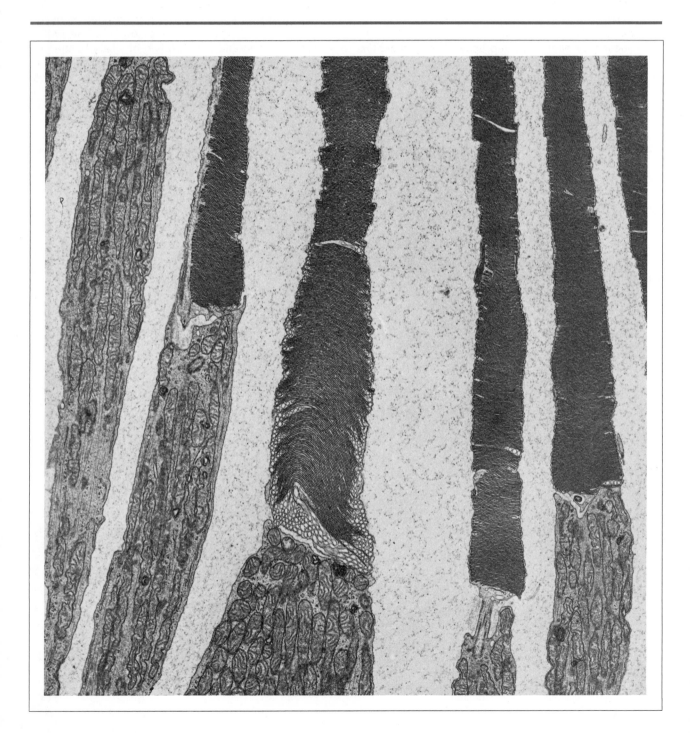

# 31

# VISION AND OTHER REACTIONS INVOLVING LIGHT

The reactions that photosynthetic organisms use to capture the energy of sunlight were discussed in Chapter 11. Here we turn to some of the other ways that living things use light and are affected by it. The topic that probably comes to mind first is vision. Because of its importance to humans, the biochemistry of vision has been studied actively, but it still is poorly understood. Some of the other topics that we shall discuss (bacteriorhodopsin, circadian rhythms and phytochrome, bioluminescence, and photodamage and repair of nucleic acids) are just beginning to be explored, and information on them is still very limited.

## VISION

### The Visual Pigments Are Found in Membranes in Rod and Cone Cells

The photochemistry of vision is different from that of photosynthesis. Perhaps this is not surprising, for the biological role of vision is quite different. Photosynthetic organisms use light as a source of energy; animals that can see use light to obtain information. However, *vision and photosynthesis both start with the excitation of an electron from one molecular orbital to another orbital of higher energy*. The excited molecule must then undergo a transformation to a metastable product. Since chlorophyll can undergo such a transformation with a quantum yield near 1.0, it is not hard to imagine an eye that uses electron-transfer reactions like those that work so well in photosynthesis. In fact, one can make a synthetic eye that works very much that way: a silicon diode array television camera tube. However,

*A low-magnification electron micrograph (7400×) of the photoreceptor cells in a human retina. (Courtesy Toichi Kuwabara.)*

**Figure 31-1**
Thin-section electron micrograph of a rod cell in a rabbit retina. Part of the outer segment is shown at the top and part of the inner segment at the bottom. (D = disks; M = mitochondrion; C = cilium; 59,200×) (Courtesy Dr. Ann H. Bunt.)

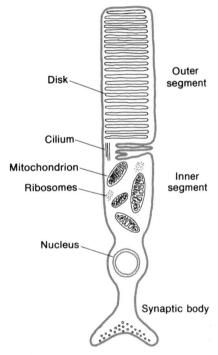

**Figure 31-2**
Schematic diagram of a rod cell. The orientation of the cell is the same as that in Figure 31-1. In the retina, many rod cells are stacked side-by-side, with the outer segments all pointing out to the periphery of the eye. Light enters the cells end on through the inner segment after passing through several layers of other neural cells. In cone cells, the outer segments are shorter and are conical instead of cylindrical.

the eyes of multicellular animals on earth are different. Instead of a chlorophyll, their light-sensitive pigment is a complex called _rhodopsin_, which consists of a protein, _opsin_, and a linear polyene, _11-cis-retinal_. Instead of undergoing oxidation when it is excited, the retinal isomerizes.

The eyes of vertebrates are marvelously complex organs, with many different types of specialized cells. Light rays entering the eye are refracted by the _cornea_, the clear tissue at the front of the eye. The light traverses an aqueous chamber and reaches the lens, which is densely packed with proteins called _crystallins_. Adjustments in the shape of the lens focus a sharp optical image onto the _retina_, a thin layer of tissue that lines the back of the eye. The retina is a neural tissue with several different layers of cells. Some of these cells, _the rod cells and the cone cells, contain the visual pigments. Other cells make synaptic connections to the rods or cones and to additional neural cells that carry impulses to the brain_.

In both rods and cones, the light-sensitive molecules are collected in a layered system of membranes at one end of the cell (Figures 31-1 and 31-2). The membranes form by invaginations of the cytoplasmic (plasma) membrane near the middle of the cell. In cone cells, the membranes remain contiguous with the cytoplasmic membrane. In rods, they pinch off to form a stack of autonomous flattened vesicles, or _disks_, in the _outer segment_ of the cell. The outer segment of each rod contains 500 to 2000 of these disks. This region of the cell is connected by a thin cilium to the _inner segment_, which is packed with mitochondria and ribosomes. The basal part of the cell ends in a synaptic junction with another neural cell called a _bipolar cell_. In a living rod cell, disks are constantly forming at the base of the outer segment and moving in a file to the tip of the outer segment, where they are sloughed off and phagocytosed by the underlying epithelial cells. It takes about 10 days for a disk to make this journey.

*Rod cells are specially adapted for vision in dim light. Cones serve for visual acuity, as well as for perception of color.* Animals such as owls, which have very high visual sensitivity in dim light but cannot distinguish colors, have only rod cells. Some animals, such as pigeons, have only cones and are skilled at distinguishing colors in bright light but inept at night vision. Primates have both rods and cones, with rods considerably outnumbering cones in all but the central portion of the retina.

## Rhodopsin Consists of 11-*cis*-Retinal on a Protein, Opsin

Figure 31-3 shows the structures of 11-*cis*-retinal and its more stable isomer all-*trans*-retinal. The retinals are related to the alcohol *retinol*, or *vitamin A₁*. These compounds cannot be synthesized de novo by mammals, but they can be formed from carotenoids, such as β-carotene, which are abundant in carrots and some other vegetables. A deficiency of vitamin A causes night blindness, along with a serious deterioration of the eyes and a number of other tissues.

There are animals with eyes in several different phyla, including Mollusca, Arthropoda, and Annelida, in addition to Chordata. The eyes of arthropods differ from those of mollusks and chordates anatomically, and they apparently originated independently, after the phyla had separated in

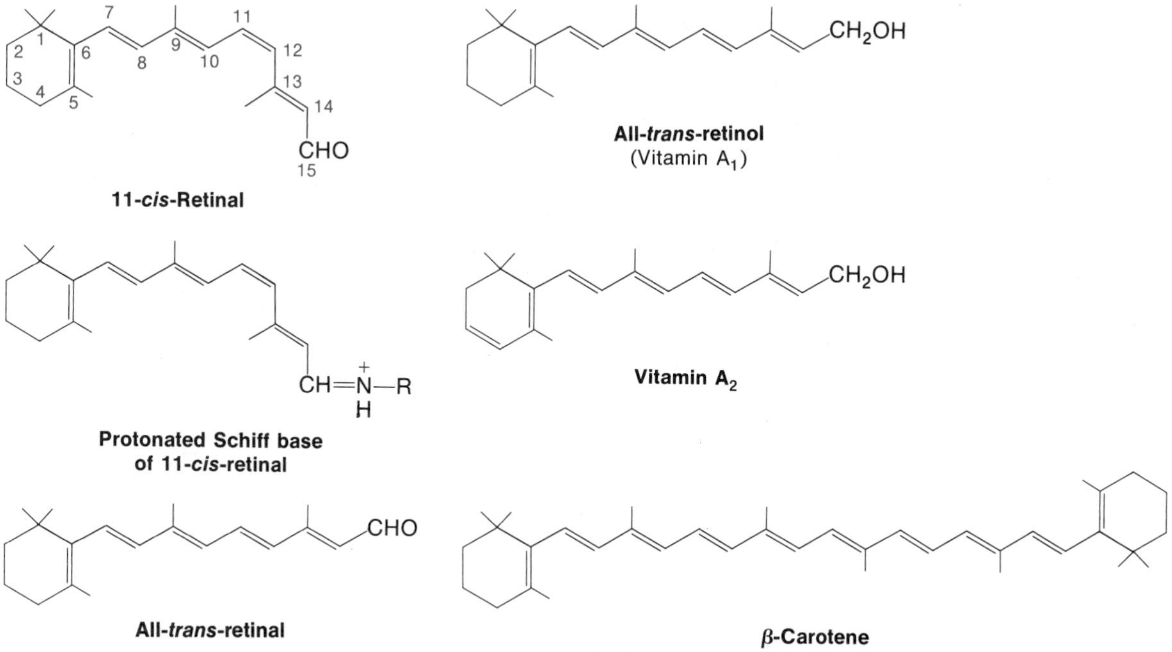

**Figure 31-3**
Structures of retinals, retinols, and β-carotene. The structure of 11-*cis*-retinal *(top)* indicates the numbering system used for the carbons. The molecules are shown here as being planar for simplicity, but the most stable conformer of 11-*cis*-retinal is twisted about the C(6)-C(7) and C(12)-C(13) single bonds to reduce steric crowding of the methyl groups. In rhodopsin, 11-*cis*-retinal is bound by a protonated Schiff's base linkage.

evolution. Some animals, such as sea turtles or amphibians in certain stages of development, use a retinal that has an additional double bond in the ring. The parent molecule in this case is vitamin $A_2$, or 3,4-dehydroretinol (Figure 31-3). *In all cases, however, the photochemically active protein complex is made from the 11-cis isomer of the aldehyde.* 11-*cis*-Retinals must somehow be particularly fit for the task. (Some unicellular organisms have light-sensitive organelles that contain a different type of pigment; little is known of the biochemistry of these primitive receptors, however.)

Opsin is a hydrophobic intrinsic membrane protein with a molecular weight of about 38,000. The retinal is bound to a lysyl-$\varepsilon$-$NH_2$ group of the protein by a *Schiff's base*, or aldimine linkage (Figure 31-3). Rhodopsin also contains several carbohydrate groups bound near the amino terminal end of the protein. Labeling studies of disks isolated from rod outer segments indicate that rhodopsin is a transmembrane protein. The carboxyl-terminal end of the protein is exposed to the cytoplasmic space outside the disk, and the amino end is exposed to the space inside the disk, with hydrophobic stretches of the protein probably making several excursions back and forth across the lipid bilayer. The lysine that holds the retinal is in the carboxyl-terminal part of the protein (probably 53 residues from the end) in a region that is not exposed to the solvent. An individual disk from a rat rod cell contains on the order of 30,000 molecules of rhodopsin, and these account for about 95 percent of the protein in the membrane.

In solution, 11-*cis*-retinal absorbs maximally near 380 nm, but in rhodopsin, the peak is at 500 nm (Figure 31-4). *The absorption spectrum of rhodopsin is essentially identical to the spectrum of the sensitivity of the rod cells* (after correction for absorption in the cornea and lens). *The perception of color by cones depends on the fact that there are three different types of cones with different absorption spectra.* One of these absorbs blue light (440 nm) maximally, the second absorbs green (530 nm), and the third absorbs yellow (570 nm). In some species of animals, the third component absorbs maximally at 630 nm. However, all cones contain 11-*cis*-retinal bound to proteins that are similar to opsin. Some animals use another strategy for color vision. Their eyes have carotenoid-containing oil droplets of three colors, and these act as filters in front of the receptor cells.

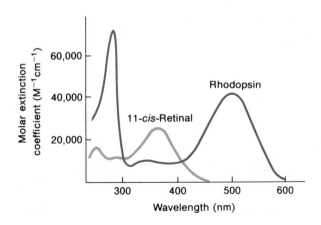

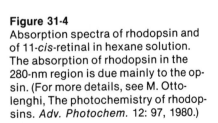

**Figure 31-4**
Absorption spectra of rhodopsin and of 11-*cis*-retinal in hexane solution. The absorption of rhodopsin in the 280-nm region is due mainly to the opsin. (For more details, see M. Ottolenghi, The photochemistry of rhodopsins. *Adv. Photochem.* 12: 97, 1980.)

## Light Isomerizes the Retinal of Rhodopsin to All-*trans*

The discovery that rhodopsin contains the 11-cis isomer of retinal came as a surprise. In the early 1950s, R. Hubbard, G. Wald, and their coworkers found that rhodopsin decomposes into retinal and the apoprotein opsin when the retina is exposed to light. The retinal that was released turned out to be the all-trans isomer. When opsin was mixed with a crude preparation of retinal, rhodopsin was regenerated. Crystalline all-*trans*-retinal, however, did not bind to opsin. The regeneration achieved with the crude preparation proved to be due to a small amount of the 11-cis isomer in the mixture. Since opsin binds 11-*cis*-retinal in the dark but releases all-*trans*-retinal after illumination, Hubbard and Kropf concluded that *the action of light in vision is to isomerize the chromophore about its C(11)-C(12) double bond*. This change in structure must somehow be translated into an electrophysiological signal that can be transmitted to the brain.

## Transformations of Rhodopsin Can Be Detected by Changes in Its Absorption Spectrum

The release of all-*trans*-retinal from opsin takes several minutes and is too slow to be an obligatory step in visual perception. It appears to be a step in the regeneration of the active form of rhodopsin. The all-*trans*-retinal probably leaves the rod cell and is isomerized back to the 11-cis form before it returns. The enzyme that catalyzes the isomerization from all-trans to 11-cis has still not been identified, although several proteins that bind retinals or retinols have been found in the retina and other tissues. In the eyes of invertebrates, all-*trans*-retinal does not come off the opsin at all but is isomerized back to the 11-cis form on the protein photochemically and probably also enzymatically.

To explore the changes in rhodopsin that precede the release of all-*trans*-retinal from the protein, T. Yoshizawa and G. Wald measured the optical absorbance changes that occurred when they illuminated rhodopsin at low temperatures. Illumination at liquid $N_2$ temperature (77 K) caused the absorption band of the rhodopsin to shift from 500 to 543 nm. The product of this transformation is now called *bathorhodopsin*. Bathorhodopsin is stable indefinitely at 77 K, but if it is warmed above about 130 K in the dark, it decays spontaneously to a species that absorbs maximally at 497 nm. This is called *lumirhodopsin*. If the sample is warmed further to about 230 K, lumirhodopsin decays to *metarhodopsin I*, which absorbs at 478 nm. Above about 255 K, metarhodopsin I decays to *metarhodopsin II*, which absorbs at 380 nm. These transformations are outlined in Figure 31-5.

Subsequent kinetic measurements by other investigators have shown that rhodopsin passes through the same series of states if it is excited with a short flash at physiological temperatures. The numbers on the right side of Figure 31-5 give approximate half-times for the transformations at 37°C. The agreement between the kinetic studies and the low-temperature experiments may seem routine, but it might not have turned out this way at all. The trapping of a metastable state at low temperature can depend on the

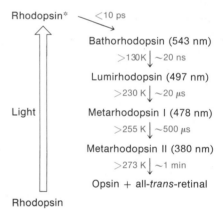

**Figure 31-5**
Photochemical transformations of rhodopsin. The numbers in parentheses indicate the optical absorption maxima of the intermediates that have been characterized. The numbers on the right are approximate half-times for the conversions in rod outer segment membranes near 37°C. With isolated rhodopsin, some of the reactions are slower and are multiphasic. The steps following the formation of bathorhodopsin have progressively higher thermal activation energies. They can be blocked by lowering the temperature below the points indicated on the left.

thermodynamic properties of the state, as well as on its position in the kinetic sequence. For example, side products that normally are unimportant could accumulate when the normal pathway is blocked by lowering the temperature. In fact, another state (*hypsorhodopsin*) can be obtained if rhodopsin is illuminated at extremely low temperatures (4 K), and its role in the kinetic sequence is still controversial.

*How can the absorption spectrum of rhodopsin go through such wild changes from 500 to 543 nm, and eventually to 380 nm?* In thinking about this, one is reminded that rhodopsin's initial absorption spectrum is already shifted 120 nm compared with the spectrum of free 11-*cis*-retinal (Figure 31-4). Further, the absorption maxima of the cone pigments vary from 450 to 630 nm, in spite of the fact that the pigments all contain 11-*cis*-retinal. The explanation for these spectral differences depends partly on the fact that the N of the Schiff's base linkage in rhodopsin is protonated and is therefore positively charged (Figure 31-6). The protonation state of the N has been demonstrated clearly by resonance Raman spectroscopy. When the retinal absorbs light and is raised to an excited state, the electron density on the nitrogen increases, and positive charge moves to the opposite end of the molecule, as represented roughly in Figure 31-6. Because of the redistribution of charge, the relative energies of the excited and ground states are extremely sensitive to the positions of other charged, dipolar, or polarizable groups nearby. An arrangement of charged groups that stabilizes the excited state relative to the ground state will shift the absorption spectrum to longer wavelengths.

There must be at least one charged group near the Schiff's base in rhodopsin, the anionic counterion that is needed to balance the positive charge on the nitrogen. The anion is likely to be an amino acid residue of the protein rather than a free ion, because the Schiff's base linkage is buried in the protein and is not accessible to the solution. *The structure of the protein thus can determine the distance between the counterion and the nitrogen, and this will affect the absorption spectrum of the pigment profoundly.* Moving the counterion closer to the N will shift the absorption spectrum to shorter wavelengths. However, this seems not to be the whole story. Studies of the absorption spectra of the complexes of opsin with a variety of synthetic derivatives of retinal suggest that the binding site contains an additional negatively charged or dipolar group, probably near C-12 and C-14 of the retinal. A specific model incorporating this idea is described in the following discussion.

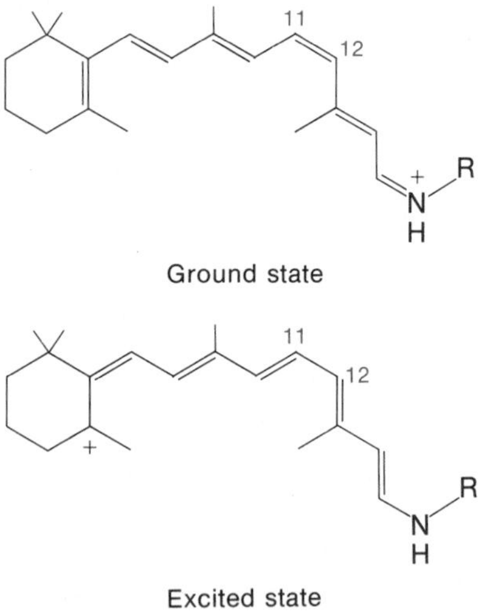

## Isomerization of the Retinal Causes Other Groups in the Protein to Move

Let us now consider bathorhodopsin, which appears to be the first metastable product of the photochemical reaction. If bathorhodopsin is excited with long-wavelength light at 77 K, it can be converted back to rhodopsin. Resonance Raman measurements support the view that the retinal in bathorhodopsin has isomerized to the all-trans form, but that it continues to be held as a protonated Schiff's base. Additional evidence that bathorhodopsin contains all-*trans*-retinal has been obtained by studying a modified form of rhodopsin, *isorhodopsin*, which contains the 9-cis isomer of retinal. When isorhodopsin is illuminated, it gives rise to bathorhodopsin within 10 ps,

Ground state

Excited state

**Figure 31-6**
Excitation of the protonated Schiff's base causes a movement of positive charge from the N toward the ring. The C(11)-C(12) bond loses much of its double-bond character. The valence bond diagrams shown here should not be taken too literally, because they neglect other formal states that also make significant contributions to the molecule, particularly in the excited state. Even so, the diagrams give a good qualitative picture of the redistribution of electrons that occurs when the molecule is excited.

just as rhodopsin (11-cis) does. The only common product that could form directly from both the 9-cis and 11-cis isomers is the all-trans isomer.

Isomerization of the retinal Schiff's base can occur when the molecule is excited with light, because the C(11)-C(12) bond loses much of its double-bond character in the excited state. The valence-bond diagrams of Figure 31-6 illustrate this point qualitatively. In the ground state of rhodopsin, the potential energy barrier to rotation about the C(11)-C(12) bond is probably on the order of 30 kcal/mol. This barrier essentially vanishes in the excited state. In fact, molecular orbital calculations suggest that the energy of the excited state may be minimal when the C(11)-C(12) bond is twisted by about 90 degrees. The 11-cis and all-trans molecules thus have a common excited state. When the molecule decays from the excited state to a ground state, it can end up in either of these isomeric forms. It turns out that the excited state decays to bathorhodopsin (all-trans) about 67 percent of the time and to rhodopsin (11-cis) about 33 percent of the time. Isorhodopsin (9-cis) and other isomers are formed only in small amounts ($\sim$1 percent). Similar ratios of products are obtained no matter whether one starts by exciting rhodopsin or bathorhodopsin.

The preference for the formation of bathorhodopsin may be due partly to the fact that steric factors make this the most stable isomer intrinsically, but it probably also reflects the guidance of the protein. Remember, however, that opsin does not bind all-*trans*-retinal in the dark. This suggests that the isomerization of the retinal causes changes in the structure of the protein surrounding the binding site. Figure 31-7 shows a model that B. Honig, T. Ebrey, and their colleagues have suggested based on this idea. In this model, the movement of the protonated N of the Schiff's base causes changes in the electric charge near the counterion ($A_1^-$) and near another charged group ($A_2^-$). The protein then relaxes by adjusting the positions of protons and other nuclei nearby.

The pronounced shift of the spectrum to longer wavelengths in bathorhodopsin can be explained if the isomerization results in a movement of the Schiff's base N, increasing the distance between the positively charged N and its counterion (Figure 31-7). Such a movement would destabilize the ground state of bathorhodopsin relative to the excited state, as explained earlier for rhodopsin.

The high energy barrier to the isomerization of rhodopsin in the dark is physiologically important because it limits the noise in our perception of light. If the barrier were low enough to be overcome thermally, the discrimination of light from dark would be more difficult.

The reorganization of rhodopsin set in motion by the isomerization of the retinal continues as the system relaxes through the states lumirhodopsin, metarhodopsin I, and metarhodopsin II. However, very little is known about the details of this reorganization. Physical measurements indicate that isolated rhodopsin undergoes substantial changes in protein conformation during the transition from metarhodopsin I to metarhodopsin II. The extent of these changes and the kinetics of the formation of metarhodopsin II are sensitive to the type and number of phospholipids attached to the protein. Conformation changes also undoubtedly occur in this step when the rhodopsin is in place in the disk membrane, but they are subtler and may be confined to small regions of the protein. The nitrogen of the retinal Schiff's base loses its proton and positive charge, and a different

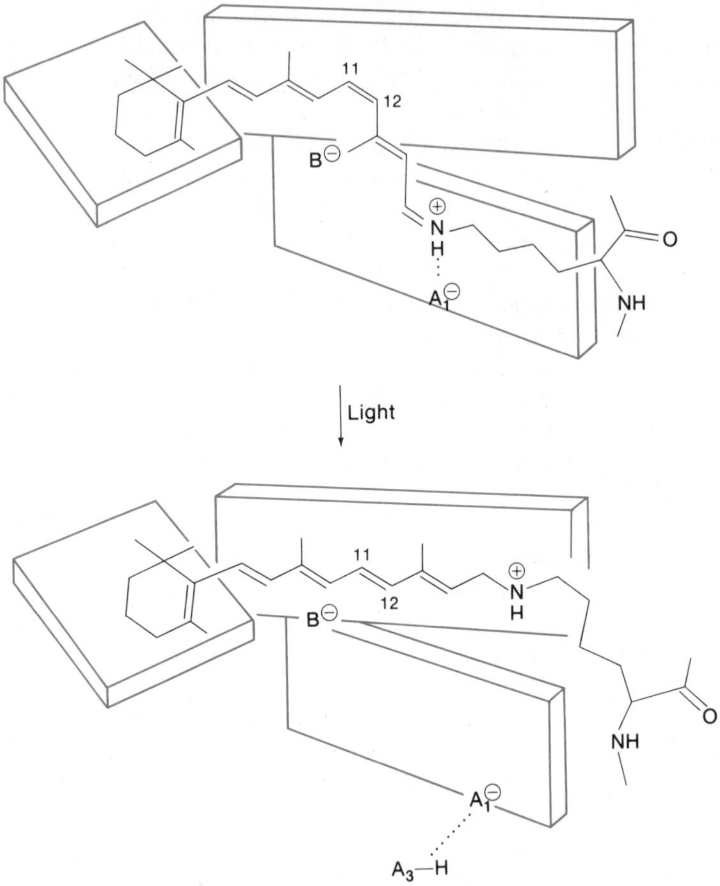

**Figure 31-7**
*(Top)* Model for the region around the retinal in rhodopsin. The proton of the
Schiff's base N is hydrogen-bonded to a counterion, $A_2^-$. Another anionic or dipolar group $B^-$ is postulated to be located near C-12 and C-14, to explain the absorption spectra of rhodopsin and its analogs. The retinal is twisted about the
C(12)-C(13) and C(6)-C(7) single bonds. (Free retinal is known to be twisted in this
way, but the amount of twisting in rhodopsin has not been established.) *(Bottom)* Model for bathorhodopsin. The ring end of the retinal is presumed to be
locked in position. Isomerization about the C(11)-C(12) double bond flips the protonated N and its positive charge away from the counterion $A_1^-$. The separation
of charge accounts for the shift of the absorption spectrum to longer wavelengths. The isomerization leads to proton movements. $A_1^-$ could obtain a proton from an acidic group ($A_3H$); and the proton of the Schiff's base could be
pushed to a basic group ($A_2^-$). The lysine attached to the retinal also must adjust to the new position of the N. (For more details, see B. Honig et al., An external point-charge model for wavelength regulation in visual pigments, *J. Am.
Chem. Soc.* 101: 7084, 1979, and Photoisomerization, energy storage, and
charge separation: A model for light energy transduction in visual pigments and
bacteriorhodopsin, *Proc. Natl. Acad. Sci. USA* 76: 2503, 1979.)

group takes up a proton from the solution. The loss of the charge on the N accounts for the shift of the absorption maximum to 380 nm, where unprotonated Schiff's bases of all-*trans*-retinal absorb. The Schiff's base linkage also becomes accessible to reagents in the aqueous solution during or shortly after the formation of metarhodopsin II.

The formation of metarhodopsin II is fast enough to be an obligatory step in visual transduction. It clearly is associated with changes in the interactions between rhodopsin and its surroundings. A reasonable hypothesis, therefore, is that the changes in protein structure allow metarhodopsin II to initiate an interaction with some other component of the disk membrane.

## Absorption of a Photon Causes a Change in the Na⁺ Conductivity of the Cytoplasmic Membrane

The human eye is amazingly sensitive. After being in the dark for a time, one can perceive continuous light that is so weak that an individual rod absorbs a photon, on the average, only once every 38 minutes. *A weak flash of light is detectable if approximately six rods each absorb one photon.* This means that *the absorption of a single photon by any one of the approximately $3 \times 10^7$ molecules of rhodopsin in a rod must be sufficient to excite the cell* and trigger a neuronal response. The combined responses of about six rods can elicit the sensation of seeing.

*The electrophysiological response of a rod or cone to light involves a change in the permeability of the cytoplasmic membrane to Na⁺ ions.* In the inner segment and basal parts of the cell, the cytoplasmic membrane contains a Na⁺/K⁺ pump, which moves Na⁺ out of the cell and K⁺ in using ATP as an energy source (ion pumps have been discussed in detail in Chapter 17). K⁺ can diffuse back out of the cell relatively freely, but the reentry of Na⁺ is restricted. The unequal pumping and diffusion of the two ions causes the membrane to become negatively charged by about 20 mV on the inside relative to the outside. Electrophysiological measurements by W. Hagins have shown that Na⁺ flows outside the cell from the inner segment to the outer, where it reenters through channels that are selectively permeable to Na⁺ (Figure 31-8). Once back inside, the Na⁺ flows to the inner segment to be pumped out again. The round trip takes on the order of a minute. When a rod cell of a vertebrate's retina is excited with light, some of the Na⁺ channels in the outer segment suddenly close, decreasing the inward movement of Na⁺. The accumulation of Na⁺ outside causes an increase in the electrical potential across the cytoplasmic membrane. This *hyperpolarization* causes a change in the movement of an unidentified chemical transmitted to the *bipolar cell* that joins synaptically with the rod. The bipolar cell then sends a signal to a *ganglion* cell, the third in the heirarchy of retinal neurons.

*The hyperpolarization of the rod and the response of the bipolar cell are graded effects.* Absorption of a single photon causes the membrane potential to increase by about 1 mV, with the effect peaking about 1 s after the excitation. Up to about 100 photons per rod, the more light the cell absorbs, the larger the amplitude of the hyperpolarization and the faster the hyperpolarization occurs. The reaction of the ganglion cell, however, is

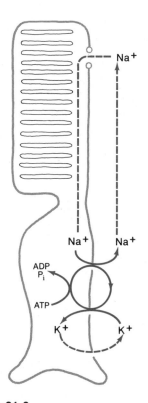

**Figure 31-8**
Na⁺ is pumped out of the inner segment (*solid arrows*) and diffuses back in through channels in the outer segment (*dashed arrows*). In rods of vertebrates, the channels are held open in the dark, and they close in the light. In invertebrates, the channels open in the light.

all-or-none. When the ganglion cell is triggered, it responds with an action potential that proceeds to the brain. Sensitivity to weak light is enhanced by having multiple rods connected to single bipolar cells and multiple bipolar cells connected to single ganglion cells. This enhancement sacrifices visual acuity. The output of cone cells is not summed in this way.

*Analogous events occur in the eyes of invertebrates, except that light causes an increase in the Na⁺ permeability of the receptor cell rather than a decrease.* In the rods of vertebrates, the absorption of a single photon decreases the current of $Na^+$ into the outer segment by about 3 percent. The inflow of approximately $10^6$ $Na^+$ ions is transiently prevented. Exactly how many $Na^+$ channels have to close in order to achieve this effect is uncertain, but estimates have ranged from 25 to 1000. Since only one molecule of rhodopsin is excited, *the cell must have a mechanism for amplifying the effect of light* by a substantial factor. Note also that the cytoplasmic membrane of the rod outer segment is not contiguous with the disk membranes that hold the rhodopsin (Figure 31-1). This suggests that the excitation of rhodopsin causes a change in the concentration or activity of a *diffusible transmitter*, which moves from the disk to the cytoplasmic membrane. The identity of the transmitter is not yet known, but there are two major candidates: $Ca^{2+}$ ions and 3',5'-cyclic-GMP.

### The Effect of Light May Be Mediated by Guanine Nucleotides

The rod outer segment disk membrane contains a *phosphodiesterase*, which hydrolyzes 3',5'-cyclic-GMP (cGMP) to 5'-GMP. Unlike rhodopsin, this is a peripheral membrane protein that can be readily solubilized. When the disk membrane is illuminated, the activity of the phosphodiesterase increases. The outer segment also contains a *guanylate cyclase*, which forms cGMP from GTP, but this enzyme appears not to be greatly affected by light. Illumination thus results in decrease in the cGMP content of the cell. Each rhodopsin that is excited appears to be able to activate on the order of 500 molecules of the phosphodiesterase.

The activation of the phosphodiesterase by illumination requires the presence of GTP and is associated with the binding of GTP to a second protein, which has been called the *G protein*, or *transducin*. The GTP displaces a molecule of tightly bound GDP from the G protein. Like the phosphodiesterase, the G protein is a peripheral membrane protein. It may be the complex of GTP and the G protein that is responsible for the activation of the phosphodiesterase (Figure 31-9). First, the binding of GTP is stimulated many times when the disk membrane is excited with light. Some 500 GTP-protein complexes are formed for each rhodopsin that is excited. This is comparable with the number of phosphodiesterase molecules that are activated. The action of light in stimulating GTP binding and in activating the phosphodiesterase is undoubtedly mediated by rhodopsin, because the effectiveness of illumination at various wavelengths parallels the absorption spectrum of rhodopsin. More significantly, the same effects can be obtained by exposing the soluble proteins to purified, illuminated rhodopsin that has been incorporated into phospholipid vesicles. The purified complex of GTP and the G protein is capable of activating the phosphodiesterase in the absence of rhodopsin. Finally, the bound GTP is gradually hydrolyzed to form bound GDP and free inorganic phosphate, and the phosphodiesterase returns to its resting, inactive state as this occurs. (The hydrolysis of the bound GTP may require the presence of still another soluble protein.) Activation of the phosphodiesterase can be prolonged indefinitely if, instead of adding GTP, one adds a nonhydrolyzable analog of GTP.

The role of cGMP in visual transduction is still far from clear. One speculation is that the conductivity of the $Na^+$ channels is regulated by a phosphorylation that is catalyzed by a kinase and reversed by a phosphatase. Suppose that phosphorylation makes the channels conductive. If the kinase is activated by cGMP, a drop in the cGMP level would lead to a decrease in the $Na^+$ conductivity. In agreement with this model, the injection of extra cGMP into rod outer segments inhibits the hyperpolarization caused by light and even causes a depolarization of the membrane. An attractive feature of the model is its analogy to the hormonal control mechanisms that are mediated by cyclic AMP (Chapter 29). An objection to the model is that a continuous expenditure of ATP would be needed, because the kinase and phosphatase would both have to be active when the eye is in the dark. An alternative hypothesis that avoids this expenditure is that the phosphatase is inactive in the dark and is switched on by a decrease in the cGMP concentration.

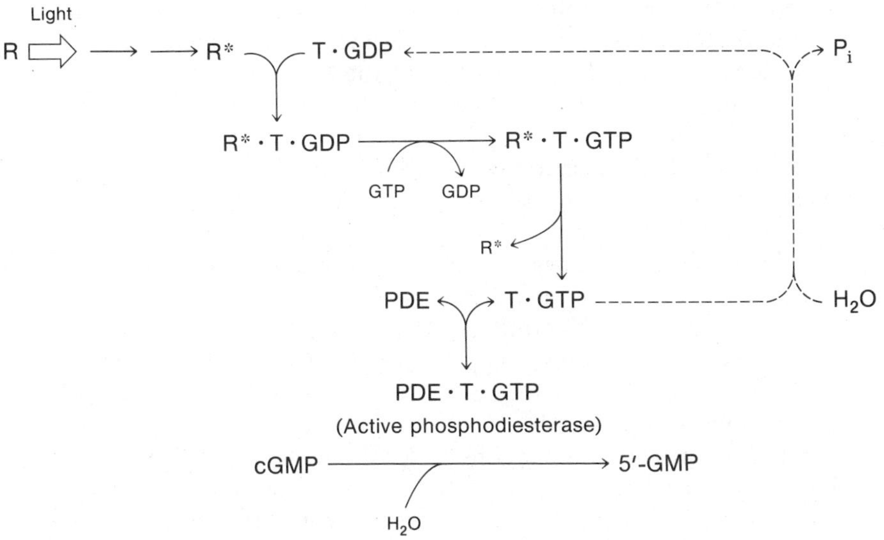

**Figure 31-9**
Working hypothesis on the activation of the phosphodiesterase by light. Light converts rhodopsin (R) to a metastable form (R*). R* reacts catalytically with the G protein (T), which has bound GDP. The reaction causes T to release the GDP and to bind GTP instead. The T·GTP complex activates the phosphodiesterase (PDE), which hydrolyzes cGMP to GMP. R* can go on to react with additional T·GDP complexes. The activation switches off when T·GTP decays to T·GDP by the action of a GTPase *(dashed arrow)*. R* also decays to an inactive form within about 10s (not shown). (For more details, see B. Fung et al., Flow of information in the light-triggered cyclic nucleotide cascade of vision, *Proc. Natl. Acad. Sci. USA* 78: 152, 1981.)

## Another Possibility Is that the Diffusible Transmitter Is Ca$^{2+}$

Ca$^{2+}$ ions appear to participate in some way in the control of the Na$^+$ channels. If Ca$^{2+}$ is injected into rod outer segments, the Na$^+$ conductivity is decreased. Conversely, the introduction of a Ca$^{2+}$ buffer inhibits the hyperpolarization caused by light. Illumination of intact retinas results in the release of a large amount of Ca$^{2+}$ from the rod outer segments into the extracellular space. It is likely, although not proven, that the Ca$^{2+}$ comes from inside the disks, which accumulate Ca$^{2+}$ in substantial concentrations in the dark. These observations suggest that Ca$^{2+}$ could be the diffusible transmitter that moves from the disk to the cytoplasmic membrane and causes the Na$^+$ channels to close. In this hypothesis, cGMP and GTP could play secondary regulatory roles in modulating the response. Isolated disk membrane vesicles release little or no Ca$^{2+}$ when they are illuminated, but the significance of this is not clear; perhaps an essential component is lost during the isolation of the disks.

## Rhodopsin Can Move Around in the Disk Membrane

Each molecule of rhodopsin that is converted to metarhodopsin II is capable of activating some 500 molecules of the cGMP phosphodiesterase within about 0.5 s, probably by promoting the formation of a comparable number of complexes between GTP and the G protein. This must mean that either rhodopsin or the G protein, or both, can move about rapidly enough to encounter hundreds of reaction partners per second. Rhodopsin is firmly embedded in the disk membrane and the G protein is attached to the surface of the membrane, so they need to diffuse only in two dimensions in the plane of the membrane. Is it likely, however, that they could move this rapidly? Actually, evidence that rhodopsin can diffuse rapidly within the membrane was obtained prior to the discovery of the phosphodiesterase and the GTP-binding protein.

The initial evidence came from studies of _linear dichroism_ of the rod outer segment disks. A material is said to have linear dichroism if the strength of its optical absorption, when measured with polarized light, depends on the orientation of the polarizer. (See Chapter 11 for a brief introduction to polarization.) Molecules that are in solution usually do not exhibit linear dichroism because they are free to tumble about. At any given time, the solution contains molecules with all possible orientations relative to the polarizer, so the absorbance of the solution will not change if one rotates the polarizer. Suppose, however, that all the molecules are fixed in position with their molecular axes aligned; such a system generally does exhibit linear dichroism. An explanation for this was suggested in Chapter 11. In the case of retinal and its Schiff's base, light is absorbed best when the polarization makes the electric vector of the light parallel to the long axis of the molecule. One can get a feeling for this by examining the valence bond diagrams of Figure 31-6. In order to convert the molecule from the ground state to the excited state, the electric field of the light must move electron density away from the ring end of the molecule and in the direction of the nitrogen.

*The retinas of vertebrates are ideally suited for measurements of linear dichroism, because the rod cells are neatly aligned with their long axes perpendicular to the plane of the tissue.* Suppose that one sends a weak beam of polarized light through the rods from the side (Figures 31-10a and b). The polarization then could be made either parallel or perpendicular to the long axis of the cells. The planes of the disk membranes in the rod outer segments are aligned perpendicular to the cell axis, so the polarization of the light will be either perpendicular or parallel to the membrane surface. It turns out that the *rhodopsin absorbs much more strongly when the light is polarized parallel to the membranes than it does when the polarization is perpendicular.* This means that the long axes of the retinal groups must be held more or less parallel to the plane of the membrane. Now suppose that one sends the light beam through the cells end on (Figures 31-10c and d). The polarization must be parallel to the disk membranes, no matter how the polarizer is turned. With light coming through end on, the absorbance of the rhodopsin turns out to be independent of the direction of the polarization. This means that the retinal groups are not aligned in any particular direction within the plane of the membrane, but instead can take on all possible orientations in this plane.

It is interesting to note in passing that the orientation of the retinal groups in the retinas of vertebrates optimizes our visual sensitivity for unpolarized light. Most of the light reaching the retina passes through the rods end on and is absorbed well, whatever its polarization (Figures 31-10c and d). *Insects, however, can distinguish between horizontally and vertically polarized light.* This can be advantageous, because light that is reflected by smooth surfaces is partially polarized. *Insects may make this distinction by having separate sets of cells with differently aligned retinals.*

How can one use measurements of linear dichroism to see whether the rhodopsin molecules move about within the membrane? Suppose that the retina is exposed to a flash of polarized light that enters the cells end on. Some of the retinal groups in the disk membranes will be oriented parallel to the polarization, and some will not. The light will selectively excite those which are parallel to the polarization, because these have the highest absorbance. Most of the molecules that are excited will be converted to bathorhodopsin and the series of states that descend from it. Their absorption spectrum will change. If one measures the absorbance changes at a short time after the excitation flash using polarized measuring light that passes through the cells end on, the absorbance changes will depend on how the polarization of the measuring light is oriented with respect to the excitation polarization. The polarized measuring light selectively measures

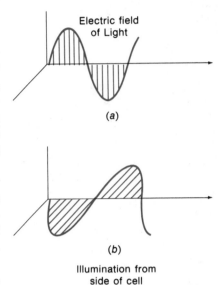

Electric field
of Light

(a)

(b)

Illumination from
side of cell

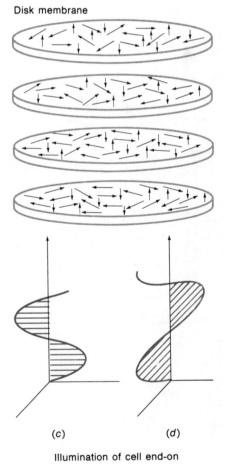

Disk membrane

(c)    (d)

Illumination of cell end-on

**Figure 31-10**
When rod cells are illuminated from the side, the light can be polarized either perpendicular *(a)* or parallel *(b)* to the planes of the disk membranes. The absorbance of the rhodopsin is much greater for parallel polarization *(b)* than it is for perpendicular *(a)*. When the cells are illuminated end on, the electric field of the light has to be parallel to the membranes, no matter how the light is polarized (c and d). With end-on illumination, the absorbance of the rhodopsin is independent of the polarization. These observations show that the 11-*cis*-retinal groups (*small arrows in disks*) are held parallel to the plane of the membrane but can point in any direction in this plane.

molecules that have a particular orientation. Before the excitation flash, there is no linear dichroism for measurements made end on; after the flash, there is. However, if the rhodopsin molecules can rotate in the plane of the membrane, the orientations of the molecules that were excited will in time become randomized, and the linear dichroism that is induced by the excitation will disappear. By measuring the kinetics of this disappearance, it is possible to determine how rapidly the rhodopsin rotates.

Experiments of this sort were done by R. Cone. Rhodopsin proved to rotate surprisingly rapidly. The time required for a rotation was found to be about 20 $\mu$s. This means that the environment of rhodopsin in the membrane must be highly fluid.

To determine how rapidly rhodopsin can diffuse laterally in the membrane, Cone and P. Leibman independently used a microspectrophotometric technique. They excited individual rod outer segments end on near one edge of the disks through a microscope. Initially, the excitation caused absorbance changes only in the region that was illuminated, but within a few seconds after the excitation, the absorbance changes became uniformly distributed throughout the disk. The only satisfactory explanation of the results is that rhodopsin can diffuse rapidly in the plane of the membrane. The diffusion coefficient can be calculated to be about $5 \times 10^{-9}$ cm$^2$/s. At this speed, rhodopsin molecules would collide with each other $10^5$ to $10^6$ times a second, and it would take only about a second for a single rhodopsin to collide with most of the molecules of the G protein in the disk.

## BACTERIORHODOPSIN

*Halobacterium halobium* is a red colored halophilic (salt-loving) bacterium that thrives in tidal salt flats. Its red color comes from two components of the cytoplasmic membrane: carotenoids and a purple pigment-protein complex that bears a striking resemblance to rhodopsin. The purple complex, *bacteriorhodopsin*, has been studied extensively by W. Stoeckenius and colleagues. Bacteriorhodopsin is not distributed randomly throughout the cell membrane. It is collected in patches that are held together by strong interactions among the bacteriorhodopsins. The patches, or *purple membranes*, can be isolated relatively simply, if one disrupts the cells by putting them in distilled water. Bacteriorhodopsin is the only protein in the purple membrane. It forms a highly ordered two-dimensional array that is almost crystalline in nature. Electron diffraction studies have shown that the individual proteins are folded into seven $\alpha$-helical stretches extending from one side of the membrane to the other (see Chapter 2).

Like rhodopsin, bacteriorhodopsin contains retinal bound to a lysine as a protonated Schiff's base. The retinal isomerizes when it is excited with light. However, bacteriorhodopsin differs from rhodopsin in that the retinal starts out in the all-trans form rather than the 11-cis form. The product of the isomerization appears to be the 13-cis isomer. Like rhodopsin, the excited bacteriorhodopsin progresses through a series of metastable states with different absorption spectra. The state that is formed first resembles bathorhodopsin in having an absorption spectrum that is shifted markedly to longer wavelengths. A later state resembles metarhodopsin II in that the retinal Schiff's base has lost its proton. A major difference, however, is that the transformations of bacteriorhodopsin are cyclic. The retinal returns to the all-trans form, and the complex relaxes back to its original state and is ready to operate again.

The role of bacteriorhodopsin in *H. halobium* appears to be to pump protons across the membrane (this has also been discussed in Chapter 17). In the course of relaxation of the excited molecule, protons are taken up from the solution on the inside of the cell and released on the outside. Exactly how many protons are translocated during each cycle of bacteriorhodopsin is not yet entirely clear, but the best current estimate is 2. One of these could be the proton that is initially on the N of the Schiff's base, because this proton is definitely lost and then replaced again in the course of the cycle. The isomerization of the retinal could move this proton with the nitrogen from a position where it equilibrates with the internal solution to a position where it equilibrates with the outside. Figure 31-7 illustrates this idea. The figure is drawn for the isomerization of 11-*cis*-retinal to all-*trans*-retinal that occurs in rhodopsin, but the principles are the same. Note that the proton of the Schiff's base is pushed to $A_2^-$ after the isomerization. From here it could be conducted to the solution on one side of the membrane. If the retinal then returns to its original conformation, the proton could be replaced by one that came from $A_3H$ on the other side of the membrane. This scheme does not explain how a second proton could be pumped simultaneously; no compelling explanation for this has been obtained.

As in other photosynthetic bacteria, the protons that are pumped across the membrane by bacteriorhodopsin in the light can reenter the cell by way of a proton-linked ATPase, generating ATP. *H. halobium* also contains a transport system that moves $Na^+$ ions out of the cell in exchange for protons coming in. Removal of $Na^+$ is a major chore for an organism that lives in water containing 4 $M$ NaCl.

## CIRCADIAN RHYTHMS AND PHYTOCHROME

*Most eukaryotic organisms have endogenous rhythms of metabolic activity with periods of about 24 hours.* These *circadian*, or *diurnal*, rhythms are triggered by light, and under natural conditions they are locked in phase with the 24-h cycle of day and night. Many organisms also have seasonal cycles that are set by the length of the day. In mammals, the synchronization of circadian and seasonal rhythms depends mainly on visual responses of the retina, but organisms that lack nervous systems depend on other types of photoreceptors. Judging from the effectiveness of light of various wavelengths, the photoreceptive pigment in some of these organisms appears to be a flavin or flavoprotein. In higher plants, the receptor is a pigment-protein complex called *phytochrome*. In addition to the synchronization of diurnal and seasonal rhythms, phytochrome has been implicated in literally hundreds of other responses to light in plants. These include the germination of seeds, the development of flowers, movements of leaves, changes in ion transport, and synthesis of numerous enzymes. Some of its effects involve changes in gene expression, including changes in the rate of synthesis of ribulose-diphosphate carboxylase, but others seem too rapid for this and probably reflect a more immediate site of action.

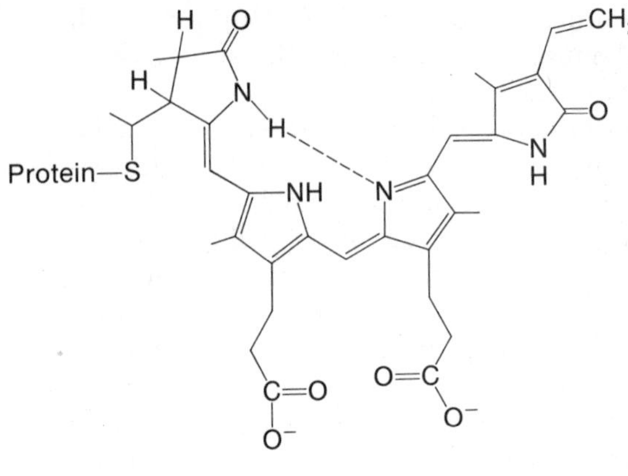

**Figure 31-11**
Structure of phytochrome in the $P_R$ form. The pigment is bound to the protein as a cysteine thioether. The covalent structure of the molecule is well-established, but the conformation is still uncertain. The conformation shown here is based on spectroscopic measurements. (For more details, and for proposals concerning the structure of $P_{FR}$, see J. C. Lagarias and H. Rapoport, Chromopeptides from phytochrome. The structure and linkage of the $P_r$ form of the phytochrome chromophore, *J. Am. Chem. Soc.* 102:4821, 1980, and P.-S. Song, Q. Chae, and J. D. Gardner, Spectroscopic properties and chromophore conformations of the photomorphogenic phytochrome. *Biochim. Biophys. Acta* 567:479, 1979.)

The protein in phytochrome is a dimer of subunits with molecular weights of 120,000. The pigment, an open-chain tetrapyrrole, has two interconvertible forms. When the molecule is in the form shown in Figure 31-11 (the $P_R$ form), phytochrome absorbs maximally near 670 nm. Excitation with light converts phytochrome from $P_R$ to the other form ($P_{FR}$), which absorbs maximally at longer wavelengths (720 nm). (The subscripts R and FR stand for "red" and "far red".) $P_{FR}$ is evidently the biologically active form of phytochrome, but its structure is still unknown, and exactly what it does is not clear.

Phytochrome decays slowly from $P_{FR}$ back to $P_R$ in the dark, and it also can be returned to $P_R$ immediately by exciting it with light. Because $P_R$ absorbs maximally at 670 nm and $P_{FR}$ at 720 nm, phytochrome can be cycled back and forth between the inactive and active forms by excitation flashes that alternate between the two wavelengths. Continuous illumination sets up a steady-state mixture, in which the amount of $P_{FR}$ depends on the color of the light. Since chlorophyll absorbs red light better than it does far-red light (Chapter 11), the color of the light reaching a leaf depends on whether the leaf is shaded by other leaves. The information collected by phytochrome could therefore be useful in controlling the direction and rate of growth of the plant.

## BIOLUMINESCENCE

Conceptually, *bioluminescence is just the reverse of photosynthesis. A reduced substrate reacts with $O_2$ or another oxidant and is converted to an oxidized product in an excited electronic state. The excited molecule then decays to the ground state, emitting light.* The process occurs in several groups of bacteria and fungi; in marine invertebrates such as sponges, shrimp, jellyfish, and sea pansies; and in a variety of terrestrial creatures, including earthworms, centipedes, and insects. The bacteria that emit light generally live symbiotically with fish in special luminous organs. In some cases, the evolutionary benefits of bioluminescence seem clear: fireflies use it for communication; fish, to attract or locate prey, or to confuse predators. In other cases, such as in fungi, the benefits are not obvious, and one is struck by the expense of the process. The energy that is released in the oxidation-reduction reaction could, in principle, be directed into ATP production instead of being emitted as light.

The substrates that undergo oxidation with the emission of light are all called, as a class, _luciferins_, although their structures vary among different bioluminescent species. The oxidized products are termed _oxyluciferins_, and the enzymes that catalyze the oxidations are called _luciferases_. Figures 31-12 and 31-13 show the structures of the luciferins and oxyluciferins used by two species, the sea pansy (_Renilla reniformis_) and the firefly (_Photinus_). The figures also indicate likely intermediates in the oxidation reactions. Although the luciferins have rather different structures in the two species, the chemistry appears to be basically much the same. $O_2$ reacts with a heterocyclic ring, forming a peroxide at a carbon that is bound to an N and to a carboxylic amide or ester. The peroxide cyclizes to give a cyclic peroxide, or _dioxetanone_, with the release of the amine or alcoholic group that was bound to the carboxyl. The cyclic peroxide then decomposes with the release of $CO_2$, forming the oxyluciferin in an excited electronic state.

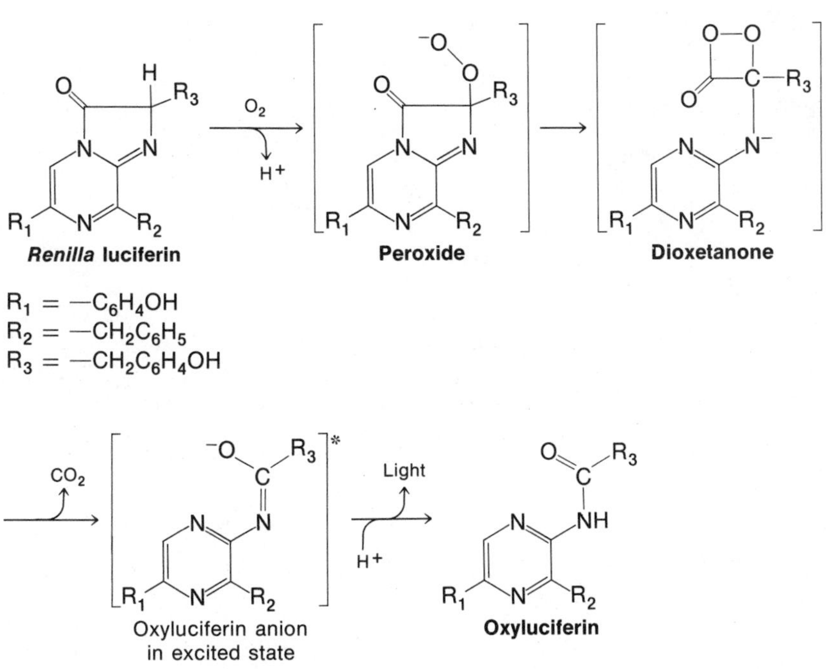

$R_1 = -C_6H_4OH$
$R_2 = -CH_2C_6H_5$
$R_3 = -CH_2C_6H_4OH$

**Figure 31-12**
Bioluminescent reactions in the sea pansy _Renilla reniformis_. The structures in brackets are plausible enzyme-bound intermediates, but they have not been identified conclusively. They are based partly on analogous nonenzymatic reactions of dioxetanones and on the observation that one of the O atoms of the $CO_2$ that is released comes from the $O_2$.

One difference between fireflies and *Renilla* is that the luciferase of fireflies activates the carboxyl group of the luciferin by forming an acyl-adenylate (Figure 31-13). This step is similar to the activation of amino acids for protein synthesis (Chapter 25), but it is unique to fireflies among the bioluminescent systems that have been studied. In *Renilla* luciferin, the homologous carboxyl group is prepared in an activated form in advance (Figure 31-12).

Luminescent bacteria use a long-chain aliphatic aldehyde such as decanal as a luciferin, oxidizing it to the carboxylic acid. $FMNH_2$ is oxidized simultaneously to FMN. Again, a cyclic peroxide appears likely to be an intermediate in the oxidation. The excited product is probably a derivative of FMN, but it has not been identified conclusively.

In vivo, the molecule that actually emits the light is sometimes not the oxyluciferin itself, but a pigment on another protein to which the oxyluciferin transfers its energy. The emitter generally luminesces with a higher quantum yield than the oxyluciferin does, as well as at a different wavelength. This is conceptually the reverse of the light-harvesting antenna systems of photosynthesis. The emitter of the luminescent bacterium *Photobacterium phosphorium* has been identified as a pterin derivative.

## PHOTODAMAGE AND REPAIR OF NUCLEIC ACIDS

Ultraviolet (UV) light is strongly absorbed by nucleic acids, with serious consequences. UV irradiation causes mutations and death in prokaryotes, and it can cause skin cancer in humans. Most of the damage that it does

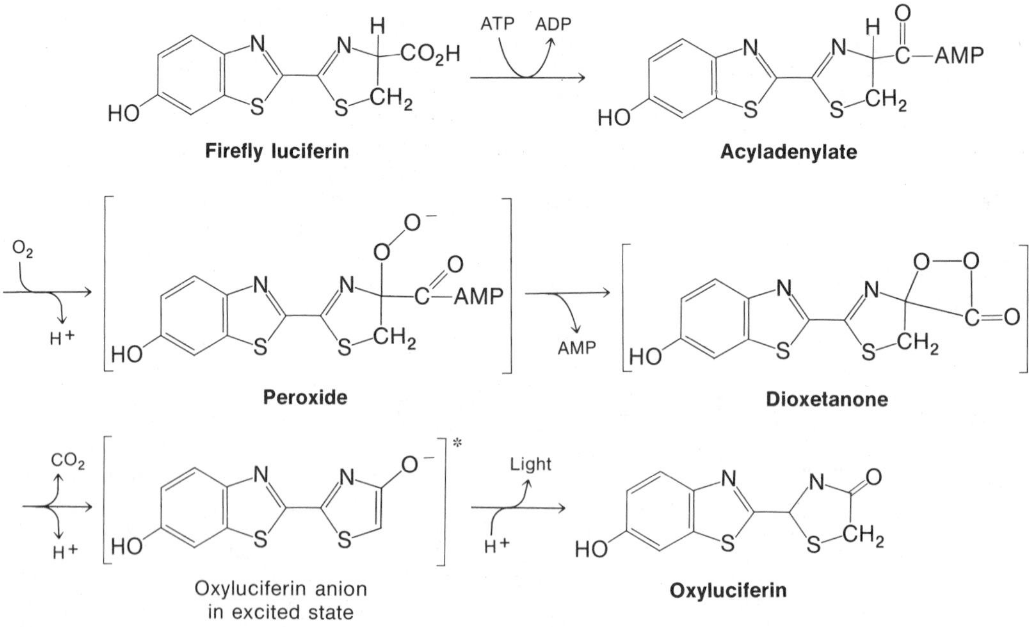

**Figure 31-13**
Bioluminescent reactions of fireflies. The excited oxyluciferin that emits light is believed to be a dianion in which the phenolic and enolic oxygens are both ionized. The ionization state can be deduced by comparing the emission spectrum of the luminescence with those of model compounds. The phenol is shown protonated here for simplicity.

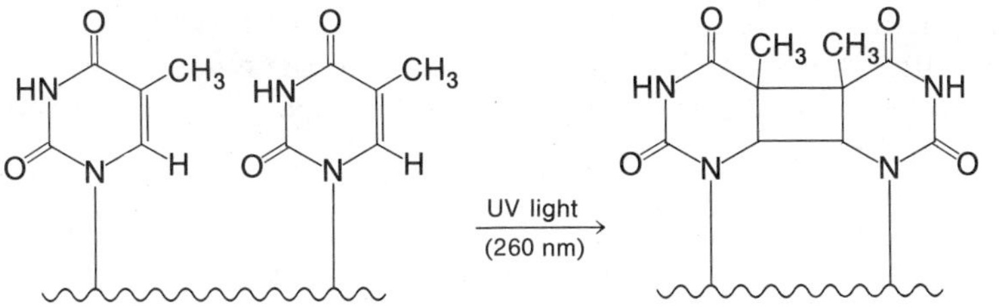

**Figure 31-14**
UV light causes adjacent thymines (or cytosines) in DNA to dimerize. The rippled lines at the bottoms of the structures represent the deoxyribose-phosphate backbone of the DNA strand. The dimer is folded so that the two heterocyclic rings are almost parallel.

results from the formation of *cyclobutane-type dimers* between adjacent thymines in DNA (Figure 31-14). Similar dimers also form in lower amounts between cytosines and between thymine-cytosine pairs. The dimers are extremely stable, and they block the normal replication and transcription of the DNA.

The harmful effects of UV light are not as much of a problem today as they probably were during the early development of life on earth. Today ozone in the atmosphere screens out essentially all the sunlight at wavelengths below 300 nm. This short-wavelength light is the most damaging to nucleic acids, because the absorption spectra of the pyrimidines peak near 260 nm. Ozone is formed by the action of UV light itself on $O_2$. Before the development of plants, the atmosphere contained little or no $O_2$; the surface of the earth was exposed to intense UV radiation, and living things must have had to remain under shelter in order to survive. However, even sunlight in the 300- to 350-nm region can be extremely damaging. It is not surprising, therefore, that cells have developed ways of repairing the lesions caused by UV light.

One repair mechanism is a nuclease that cuts out the pyrimidine dimers and patches in a normal piece of DNA, using the undamaged complementary strand as a template (discussed in Chapter 20). Humans with the disease *xeroderma pigmentosum* have a deficiency of this repair system and are prone to developing skin cancers. Many types of cells also contain an enzyme called the *photoreactivating enzyme* that uses the energy of light to decompose pyrimidine dimers in DNA. Photoreactivating enzyme has been purified to homogeneity from several microorganisms. The enzyme from *Escherichia coli* contains bound RNA and carbohydrate, but no other recognized cofactor; that from *Streptomyces griseus* contains another unidentified chromophore. The enzyme from *E. coli* binds specifically to DNA that contains a cyclobutane-type pyrimidine dimer, forming a complex that absorbs maximally near 350 nm. When the complex is irradiated in this wavelength region, the pyrimidine dimer separates into the two monomers, repairing the DNA. Model studies of the photochemical reactions of thymine dimers in solution suggest that the reaction could involve electron transfer either to or from the dimer in the excited complex. *E. coli* contains only 10 to 20 molecules of the photoreactivating enzyme per cell, but this suffices to increase by several orders of magnitude the intensity of UV radiation that the cell can survive. A cell evidently can be killed if only a few pyrimidine dimers are not repaired.

Photodamage to nucleic acids also can occur as a result of the absorption of light by other molecules. In some cases, the excited molecule, or *sensitizer*, reacts directly with the nucleic acid. More commonly, the excited sensitizer undergoes intersystem crossing to a triplet state and then reacts with $O_2$ to form singlet $O_2$, which attacks the nucleic acid (Chapter 11). Potential sensitizers include many biological molecules, such as flavins, and a variety of nonphysiological dyes.

## SELECTED READINGS

Cone, R. Rotational diffusion of rhodopsin in visual photoreceptor membrane. *Nature* 236:39, 1972. Excitation of the retina with polarized light creates a transient dichroism. The decay of the dichroism is explained by rotation of rhodopsin.

Fung, B. K., Hurley, J. B., and Stryer, L. Flow of information in the light-triggered cyclic nucleotide cascade of vision. *Proc. Natl. Acad. Sci. USA* 78:152, 1981. Discussion of the roles of cGMP, GTP, phosphodiesterase, and the G protein ("transducin") in the response of rod outer segments to light.

Honig, B., Ebrey, T., Callender, R., Dinur, V., and Ottolenghi, M. Photoisomerization, energy storage, and charge separation: A model for light energy transduction in visual pigments and bacteriorhodopsin. *Proc. Natl. Acad. Sci. USA* 76:2503, 1979. Isomerization of the retinal could cause movement of the protonated N of the Schiff's base linkage. This can explain the capture of energy and the movement of protons in rhodopsin and bacteriorhodopsin.

Pratt, L. H. Phytochrome: Function and Properties. In K. C. Smith (Ed.), *Photochemical and Photobiological Reviews*, Vol. 4. New York: Plenum Press, 1979.

Pp. 59–124. A review of the diverse biological phenomena controlled by phytochrome.

Stoeckenius, W. Purple membrane of halobacteria: A new light-energy converter. *Acc. Chem. Res.* 13:337, 1980. A review of the structure and function of bacteriorhodopsin and the purple membrane.

Sutherland, J. C. Photophysics and photochemistry of reactivation. *Photochem. Photobiol.* 25:435, 1977. Discussion of possible mechanisms of the photoreactivating enzyme, which breaks thymine dimers photochemically.

Wald, G. The molecular basis of visual excitation. *Nature* 219:800, 1968. Wald's Nobel Prize address, describing early work on the isomerization of retinal in rhodopsin.

Ward, W. W. Energy Transfer Processes in Bioluminescence. In K. C. Smith (Ed.), *Photochemical and Photobiological Reviews*, Vol. 4. New York: Plenum Press, 1979. Pp. 1–57. A review of bioluminescence, covering the luciferins and luciferases of numerous species and discussing examples of energy transfer from the excited oxyluciferin to other light-emitting species.

## PROBLEMS

1. Draw traces showing how the optical absorbance of a fresh suspension of rod outer segment disks might change as a function of time when the suspension is excited with a short flash of light at 37°C. Show the absorbance at (a) 545 nm and (b) 480 nm. Select the time scale for each trace judiciously, so that the traces illustrate the kinetics of the major absorbance changes that occur at the two wavelengths. (You may need to use two traces with different time scales at each wavelength to show both the initial absorbance change and its decay.)

2. If rhodopsin is illuminated at 500 nm at .77 K, the absorbance of the sample at 500 nm decreases. If the sample is then illuminated at 550 nm (still at 77 K), the absorbance at 500 nm increases again. Explain.

3. Why is the absorption spectrum of metarhodopsin II so different from that of metarhodopsin I?

4. Draw traces showing how the membrane potential of a rod cell changes with time when the cell is excited with (a) one photon, and (b) two photons. Assume that each photon converts one molecule of rhodopsin to bathorhodopsin and on to metarhodopsin II. (The probability of conversion actually is only about 0.67.)

5. a. Draw a scheme illustrating how a change in cGMP concentration could result in an increase in the $Na^+$ conductivity of the rod outer segment cytoplasmic membrane. Use your imagination, but base your scheme on what is known of metabolic control by cyclic nucleotides in other systems.

b. Select a feature of your scheme that is not discussed extensively in the text and suggest ways of testing this feature experimentally.

6. A frog retina is excited with a flash of polarized light that passes through the rods end on. The flash causes optical absorbance changes at 500 nm, which are measured with polarized light that also passes through the rods end on.

   a. Draw traces showing the kinetics of the absorbance changes measured with light polarized parallel to the excitation polarization and with light polarized perpendicular to the excitation polarization. The traces should cover the time period up to 100 $\mu$s after the excitation.

   b. Similar experiments are done with a retina that has been treated with glutaraldehyde, which causes cross-linking of the rhodopsins in the membrane. Draw traces showing the kinetics of the absorbance changes expected in this case. Assume that the cross-linking immobilizes the rhodopsin molecules but does not otherwise affect their photochemical transformations.

7. The action of phytochrome probably does not involve cGMP or cAMP, because cyclic nucleotides have not been found to play regulatory roles in plant metabolism. It could involve changes in membrane permeability. How would you study this possibility?

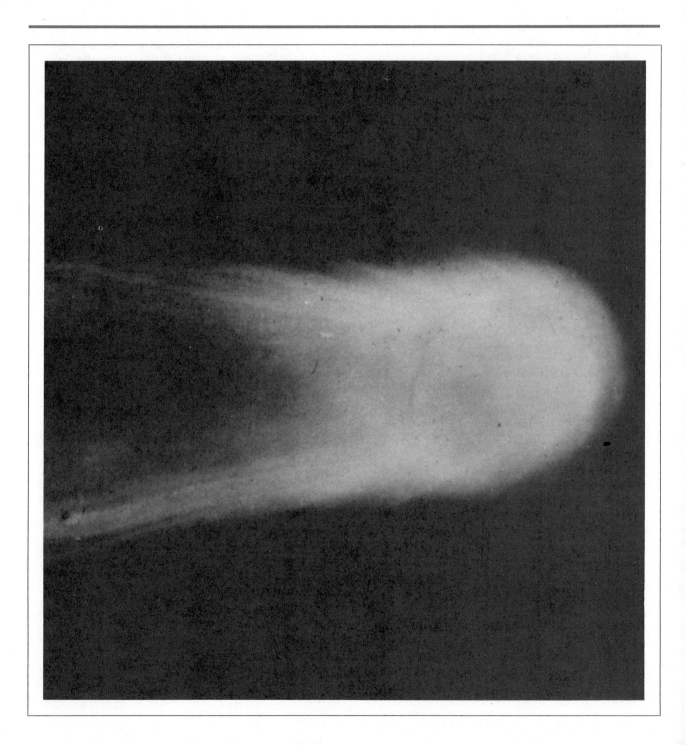

# ORIGINS OF LIFE

*David Usher*
*Cornell*

The previous chapters have emphasized that biochemistry is largely a study of complex organic molecules that are self-organizing and self-replicating. These molecules are synthesized and interconverted by enzyme-catalyzed processes in which the enzyme interacts at specific locations of these structures. The ultimate driving force for these reactions is the solar energy absorbed by specific chromophores in photosynthetic organisms. What has not been emphasized is how complex biomolecules initially formed and how life processes spontaneously resulted from their interactions.

*Life on Earth today is the product of nearly 4 billion years of evolution and, consequently, may have only a superficial resemblance to the earliest living systems.* The original forms of life no longer are present because they were unable to compete with the higher organisms that evolved from them. In addition, it appears unlikely that life as we know it will be found elsewhere in our solar system. No life was discovered in the vicinity of the two Viking landers on Mars, and there is only a remote possibility that life or processes leading to the origins of life will be found on the satellites of Jupiter and Saturn. Mercury and Venus are too hot for life to exist, while Uranus, Neptune, and Pluto are too cold. There are levels of the atmospheres of Jupiter and Saturn warm enough for life, but it is an unlikely possibility because the turbulent winds and convection currents would carry these living systems to upper or lower levels, where they would be destroyed. It appears likely that some of the chemical reactions that took place prior to the origins of life have taken place on Titan, the largest moon of Saturn. Compounds generally believed to be essential for the origins of life, including HCN, HC≡CCN, $(CN)_2$, HC≡CH, and HC≡CC≡CH, were detected in its atmosphere by the Voyager I probe in 1980. The low

*Halley's Comet. Comets and meteorites may have been important sources of carbon, hydrogen, and nitrogen compounds of the primitive Earth.*

surface temperature of Titan (95 K) probably precludes the origin of life on this planet.

Even though it is unlikely that it will be possible to test theories of the origin of life by studying life on other planets in our solar system, it is an increasingly active research area. *The goal of these experimental studies is to establish the reaction conditions on the primitive Earth and to determine if biological molecules will form under these conditions.* Since there is little hope of exactly duplicating the process of the origin of life on Earth, the objective of these experiments is to determine whether there exist reasonable pathways by which life processes may have originated. If convincing processes can be demonstrated for the origin of life on Earth, then it would be reasonable to conclude that life is a natural outcome of the chemistry of the universe. This finding would suggest (1) that there must be abundant life in the universe and (2) that humans may not be one of the more highly evolved forms of life in the universe. The reasoning and experiments that form the basis for the study of the origins of life are outlined in this chapter.

## WHAT IS LIFE?

The object of the search must be defined before the search can begin. "What is life?" is not a trivial question because if the first living forms evolved spontaneously from their environment, they must have differed very little from it. Consequently, sophisticated tests would be needed to distinguish the living from the nonliving. One definition of life is that it is a process capable of *self-duplication* and *mutation*. Self-duplication provides for the formation of new generations of the new life form, while mutation permits it to evolve, under the pressure of natural selection, to forms that can compete more readily for energy and nutrients. This definition of life was derived directly from observation of the life processes on Earth.

*A more general definition of life is that living systems have the property of cycling bioelements to achieve complexity in a process driven by solar energy.* This is a more general definition that allows one to envision life in many forms. One form might be minicomputers that replicate their individual parts, yet constantly reorganize themselves into more efficient structures. Here the semiconductor chip would be the fundamental building block or bioelement. This model for life has been pictured in science fiction to be the next stage of evolution on Earth. Once humans design an intelligent computer, it may eventually be able to operate more efficiently than humans and may possibly control them.

Crystalline inorganic polymers have been proposed as the starting point for the origins of life. It is suggested that *clays*, polyaluminum silicates, had defects in their structures and that these defects were the source of information (genetic material) for primitive life. These clays replicated under the pressure of natural selection to form structures with a more complex pattern of defects. The clays also catalyzed the reactions of organic molecules, including the synthesis of proteins and nucleic acids. At some point in this process, the nucleic acid–protein combination became more efficient at storing genetic information and catalyzing reactions and it took over the business of life. The latter process is comparable to that noted earlier, where computers devised by humans become more efficient than humans.

The controversial point in this theory is the evolution of the clays to structures with a more complex pattern of defects. It has not been established that clays evolve, and the limited number of clay structures makes it difficult to imagine what these more complex clays might be. There is no question that clays on the primitive Earth may have catalyzed organic processes that eventually led to the origin of life. In fact, some clay mineral-organic complexes may have had a specific catalytic or electron-transport function in primitive life processes. _Ferredoxins_, enzymes that contain iron sulfur clusters, are a contemporary example of a class of proteins that are a combination of organic and inorganic segments. However, the rich and varied chemistry of organic structures makes them much more attractive than minerals as the basis for primitive life. Therefore, _in the remainder of this chapter it will be assumed that reactions of organic compounds in the presence of water and minerals led to the origin and proliferation of life._

## UNDERLYING ASSUMPTIONS

Since little is known about the chemistry of the primitive Earth when life originated and there is essentially no knowledge of the biochemistry of the first living systems, a model must be devised before experiments can be performed. A few of the assumptions that provide the basis for this model include the following.

1. Life originated on Earth. Models based on the supposition that living systems were transported here, either by other intelligent life or by comets, are almost impossible to test in the laboratory and so will not be considered.

2. The biochemistry of the initial living system was based on some of the biochemical subunits that are of importance to contemporary biochemistry. It is assumed that amino acids and nucleotides had a central role in the initial life forms, but it is recognized that the biochemistry was much simpler and also less efficient than that of contemporary life.

3. Life originated as a consequence of the spontaneous formation of complex molecules on the primitive Earth. These complex structures were formed from simple molecules present in the Earth's atmosphere or were transported to the Earth by comets and meteorites. The chemical processes leading to life's origins took place in the presence of water and minerals, where the pH of the aqueous phase was approximately 7. The pH of the ocean today is controlled at 8 by the buffering action of clay minerals. It appears likely that the same buffers maintained the nearly neutral pH of the oceans 4 billion years ago when life originated.

4. Chemical evolution was driven by solar energy in the form of direct irradiation by ultraviolet and visible light or indirectly in the form of lightning (electric discharges and shock waves). Some thermal energy was supplied by volcanic action and the heat resulting from radioactive decay. An estimate of all the available energy sources on the primitive Earth is given in Table 32-1.

**Table 32-1**
Energy Sources on the Primitive Earth

| Source | Energy (cal/cm$^2$/yr) |
|---|---|
| Total solar radiation: | |
| Ultraviolet light | 260,000 |
| <300 nm | 3,400 |
| <250 nm | 563 |
| <200 nm | 41 |
| <150 nm | 2 |
| Electric discharges | 4 |
| Shock waves | 1 |
| Radioactivity | 1 |
| Volcanoes | 0.1 |

Adapted from S. L. Miller and L. E. Orgel, _The Origins of Life on the Earth_, Prentice-Hall, Englewood Cliffs, N.J., 1974. P. 55. Used with permission.

# POSSIBLE STAGES IN CHEMICAL EVOLUTION

The synthesis of the biological molecules essential for the origin of life proceeded in the 0.5 billion year period after the formation of the Earth. Since it is difficult to model this process in a laboratory experiment, various stages in the process are investigated separately.

*Stage 1*   The conversion of simple inorganic and organic compounds into the building blocks for biopolymers. _Electric discharges, ultraviolet light_, and other energy sources are used to initiate the formation of amino acids, purine and pyrimidine bases, and sugars or their precursors.

*Stage 2*   The formation of ordered biopolymers from the monomer units produced in stage 1. The search for _templates_ to direct the formation of the ordered polymers is an important aspect of this research. The reactions and catalysts used ideally should preferentially select one structural type (e.g., α-amino acids) from closely related structures (e.g., β-amino acids) which are also formed in the nonspecific stage 1 processes.

*Stage 3*   The self-replication of the biopolymers formed in stage 2 in a process that may have taken place in the first living system. At the present time, the only experiments in which stage 3 has been investigated have made use of an intact biological system (e.g., a virus) that is separated into its molecular components. The regeneration of the intact biological system is then investigated using only those biopolymers which direct the synthesis of other biopolymers. For example, a viable RNA virus has been regenerated from its RNA, the RNA polymerase, nucleotides, and helper enzymes of the host bacterium of the virus.

It should be emphasized that the experimental studies in this area are fragmentary, and often there is no well-defined connection between each stage. For example, the formation of nucleotides from nucleosides will be discussed, even though there has been limited success in devising a prebiotic synthesis of nucleosides. Consequently, it will not be possible for us to provide a neat, well-defined picture of the origins of life in this chapter.

# FORMATION OF OUR GALAXY AND SOLAR SYSTEM

Insight into the origin of life on Earth resulted from studies of the constituents of the galaxy to which our Earth belongs. In addition to billions of stars, the galaxy is also made up of _planets_, _planetesimals_ (solid objects much smaller than planets), _dust_, _simple molecules_, and _atoms_. These constituents are arrayed in the shape of a disk, with the bulk of the mass present in a central bulge in the form of massive stars and interstellar dust. Tenuous spiral arms extend out from the disk, and our solar system is located in the periphery of one of the arms.

The galaxy was formed by the collapse of a giant rotating dust cloud. As the mass collapsed to the center of the cloud, the angular momentum of the system was conserved by the cloud forming into the shape of a disk with appreciable mass distributed at great distance from the massive center. Protostars first formed in the center of the disk, and these ignited when the temperature and pressure became high enough to initiate the fusion of hydrogen.

Our sun formed in the more tenuous outer regions of the dust cloud billions of years after the formation of the initial stars near the galactic center. The condensation of this portion of the cosmic dust cloud may have been triggered by the explosion of a nearby star to a supernova. The local dust cloud collapsed to a large spinning disk of dust and molecules from which the sun and planets eventually condensed. The inner planets, i.e., those near the sun, are more dense because only the heavier elements condense at the higher temperatures in the vicinity of the sun. The much lower temperatures ($< -150°C$) at the orbit of Jupiter favor the condensation of the light gases, including hydrogen, helium, methane, and ammonia.

The Earth continued to accrete significant amounts of dust and planetesimals for about 0.1 billion years after it initially condensed. The impact craters on the moon are direct evidence for this continued bombardment of the Earth and moon with this material (Figure 32-1). *There was a preferential accretion of the heavier elements in the early stages of the formation of the Earth. Consequently, the lighter elements, including those essential for the origins of life (C, H, N, C) were swept up by the orbiting Earth during the later stages of the accretion process. The partial segregation of these essential ingredients in the crust of the Earth set the stage for the origins of life.*

**Figure 32-1**
Impact craters on the moon.

## COSMOCHEMISTRY

### Interstellar Molecules

The chemistry of our solar nebula can be deduced from a spectroscopic observation of the molecular species present in interstellar gases in our galaxy and in the other galaxies in the universe. These include an abundance of carbon compounds, such as CO, $CH_4$, $CH_2O$, HCN, and $HC{\equiv}CCN$, together with $H_2O$ and $NH_3$ (Table 32-2). Many of these same compounds are considered to have initiated the abiotic chemical processes leading to the origins of life. Most of these carbon and nitrogen compounds were probably not transferred intact to the atmosphere of the primitive Earth, but their presence in the solar nebula provides strong support for the hypothesis that compounds of this type can be readily formed and they were likely starting materials for the formation of biological molecules on the primitive Earth. *The widespread occurrence of these organic compounds in our galaxy and other galaxies suggests strongly that carbon chemistry is not only the chemistry of the cosmos, but probably the basis for the origin of life elsewhere in the universe as well.*

### Meteorites

Meteorites, relics of the condensation of the solar dust cloud to the solar system, also provide clues concerning the chemistry of the primitive Earth. There was a large influx of meteorites during the first 0.5 billion years after the Earth formed. These 4.6-billion-year-old meteorites are still striking the

**Table 32-2**
Some Interstellar Molecules

| | |
|---|---|
| HCN | $CH_3OH$ |
| $H_2O$ | $HCONH_2$ |
| $H_2S$ | $CH_3CN$ |
| $SO_2$ | $CH_3CHO$ |
| $H_2C{=}O$ | $CH_2{=}CHCN$ |
| $HN{=}C{=}O$ | $CH_3NH_2$ |
| $H_2C{=}S$ | $CH_3C{\equiv}CH$ |
| $NH_3$ | $HC{\equiv}CC{\equiv}CCN$ |
| $CH_2{=}NH$ | $HCO_2CH_3$ |
| $HC{\equiv}CCN$ | $CH_3CH_2OH$ |
| $HCO_2H$ | $CH_3OCH_3$ |
| $NH_2CN$ | $CH_3CH_2CN$ |
| $CH_4$ | $HC{\equiv}CC{\equiv}CC{\equiv}CCN$ |
| | $HC{\equiv}CC{\equiv}CC{\equiv}CC{\equiv}CCN$ |

Earth today, and *those few which contain carbon possess organic molecules that have some of the same monomeric units present in biological molecules*. Some meteorites also contain clays, which are formed only in the presence of water, suggesting that these specimens may be a fragment of much larger bodies, such as asteroids, that had sufficient mass to attract and maintain water on their surfaces. The $^{13}C/^{12}C$, $^2H/^1H$, and $^{15}N/^{14}N$ ratios of the organic material in meteorites vary with the organic components present, indicating a number of different pathways for the formation of these bodies in the solar nebula. The presence of $Fe^{2+}$ indicates that the meteorites were formed in a reduced environment and that they brought this reducing capability to the crust of the primitive Earth. Meteorites are depleted in carbon, hydrogen, and nitrogen relative to the composition of the universe, which is also indicative of their being fragments of a larger body that accreted from the more dense elements in the solar system.

The carbon compounds present in the carbonaceous meteorites include alcohols, aldehydes, amines, amino acids, hydrocarbons, ketones, purines, pyrimidines, and other heterocyclic compounds. These relatively complex organic molecules were probably formed from the simpler compounds prevalent in the interstellar medium (Table 32-2). Some of the reactions leading to the formation of the meteoritic carbon compounds may have taken place on the surfaces of dust particles at the high temperatures and pressures of the contracting solar nebula. Additional chemical reactions may have occurred on the colder planetesimals from which the meteorites were derived.

It is not clear whether the organic compounds in meteorites were transferred intact to the primitive Earth. The Earth's crust was probably very hot after its initial formation owing to high levels of radioactivity. The carbon compounds present in meteorites that impacted on the hot crust were pyrolyzed to $CO$, $CO_2$, $COS$, $CH_4$, $CH_3CH{=}CH_2$, $CH_3COCH_3$, $CH_3CN$, $C_6H_6$, and higher-molecular-weight hydrocarbons and heterocycles. As will be seen later, many of these compounds have been postulated as the starting materials for the synthesis of biomolecules on the primitive Earth. The bulk of the organic compounds arriving after the Earth's crust cooled were probably not changed on impact. Since many of the compounds present in carbonaceous meteorites form in prebiotic simulation experiments, it probably did not matter if the carbon compounds present in meteorites were pyrolyzed to simpler organics on impact with the Earth. Compounds similar to those present in the meteorites would have been formed rapidly from the gases released by pyrolysis.

## Comets

*Comets also served as a source of carbon, hydrogen, and nitrogen compounds on the primitive Earth*. They probably formed in the colder regions of the solar nebula at the time that the solar system formed. The orbits of known comets are eccentric and not necessarily in the plane of the solar system. It is believed that many comets have orbits that extend beyond Jupiter and Saturn and therefore have not been detected. Comets visible from the Earth were probably projected into eccentric orbits as a result of gravitational interaction with Jupiter or Saturn so that they pass near the sun. These comets cross the Earth's orbit and may collide with it.

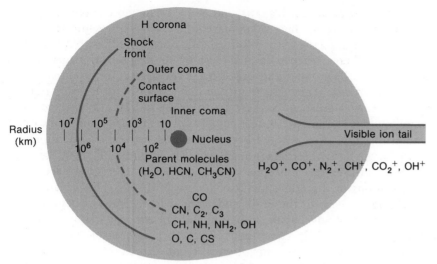

**Figure 32-2**
A model for the chemical processes in a comet.

The *"dirty ice"* model is the one most widely accepted for comets. The comet nucleus consists mainly of ice with dust, rocks, and organic molecules embedded in the ice matrix. In addition, there may be a core consisting of more dense stoney material. When a comet approaches the sun, the solar radiation vaporizes the icy coating to release dust particles and molecular species, which are photolyzed to ionic species and radicals (Figure 32-2). The dust, neutral molecules, ions, and radicals constitute the cometary tail, which often extends some 10 million km from the head. Periodic meteorite showers occur when the Earth crosses the trail of dust and larger particles sloughed off a comet. Some of the stoney meteorites are believed to have been a comet's core from which all the volatiles have been vaporized as a result of its orbiting close to the sun.

The known molecular constituents of comets (Figure 32-2) have all been detected in the interstellar medium. This observation is consistent with the postulate that comets are composed mainly of the less dense molecular species and that they condensed in the colder regions of the solar nebula distant from the sun. Comets may have brought significant amounts of water and cyano compounds when they collided with the primitive Earth. The icy portion of the comet vaporized completely as it passed through the Earth's atmosphere and only the stoney core would have impacted with the Earth's crust.

## THE ATMOSPHERE OF THE PRIMITIVE EARTH

Comets and meteorites may have been important sources of the constituent compounds of the atmosphere of the primitive Earth. Comets brought with them large amounts of _water, carbon dioxide, hydrogen cyanide, and cyano compounds_. Pyrolysis of the carbonaceous material present in meteorites resulted in the formation of compounds similar to those present in comets, along with aliphatic and aromatic hydrocarbons. The principal atmospheric constituents from these two sources were probably carbon dioxide and water vapor.

*Degassing* of the Earth's crust was probably the most important source of atmospheric gases. There would have been extensive volcanic activity on the young Earth owing to the heating of the crust by radioactivity. In addition, holes were punched in the crust by the impacting planetesimals, which caused lava flows and degassing of the crust. *Carbon dioxide and water vapor were probably the predominant gases resulting from outgassing the Earth's crust,* since these are the main compounds emitted from contemporary volcanoes. Smaller amounts of the reduced forms of carbon and nitrogen also were formed, because the crust had some reducing equivalents in the form of ferrous iron and sulfides. Information about the composition of the Earth's atmosphere was obtained from our knowledge of the compositions of the atmospheres of Venus and Mars. These planets are comparable in size to the Earth and flank it in the solar system. Consequently, compounds that are present in both the atmospheres of Venus and Mars were very likely constituents of the atmosphere of the primitive Earth. Carbon dioxide is a major constituent of the atmospheres of both these planets, and it was probably a major component of the primitive Earth's atmosphere. *Carbon dioxide is now a minor constituent of the present-day atmosphere because it was deposited as limestone as water weathered the Earth's crust.* Both Mars and Venus have small amounts of water in their atmospheres. Mars probably had much larger amounts in an earlier era, as evidenced by the dried up river channels observed on the planet (Figure 32-3). The water on Venus was lost over the past 4.6 billion years by its photolysis to hydrogen atoms, which escaped from the upper atmosphere of the planet. *The cold trap in the upper atmosphere of the Earth prevents extensive amounts of water from reaching levels where it can be photolyzed.* The cold trap on Venus does not prevent the loss of water because of the extremely high temperature of its atmosphere owing to its close proximity to the sun and because the carbon dioxide and water vapor in its atmosphere decrease the radiative loss of heat (greenhouse effect).

From these data one can derive a model for the primitive Earth at the time when the prebiological formation of biological molecules took place. *There was an atmosphere consisting of water vapor and carbon dioxide that contained lesser amounts of carbon monoxide, methane, and nitrogen. Large bodies of water started to accumulate and these had a pH of 7.5 ± 1.5* depending on how effectively the clay minerals buffered and calcium ions precipitated the dissolved carbon dioxide. *Smaller amounts of ammonia, hydrogen sulfide, hydrogen cyanide, and simple organic compounds also were dissolved in the primitive oceans and lakes.* Most of the radioactivity of the heavy elements was probably dissipated, so that the temperature of the Earth was probably only slightly warmer than it is today. The presence of water vapor would have allowed for charge separation between the clouds and the Earth's surface, so thunderstorms, with the resulting lightning and shock waves, would have been prevalent. In addition, ultraviolet light of wavelengths between 220 and 400 nm would have reached the surface of the Earth, and even shorter-wavelength light would have been absorbed by the molecules of the atmosphere.

*The challenge to the chemist is to determine what chemical processes took place under these conditions and how these processes led to the formation of biopolymers and the eventual origin of life.*

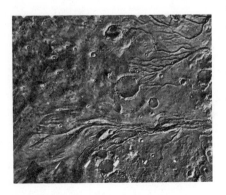

**Figure 32-3**
Channels on Mars believed to be caused by ancient rivers that have since evaporated.

Plausible reaction conditions on the primitive Earth were derived previously from our understanding of the origin of our galaxy and solar system. The conditions place some constraints on the laboratory experiments that can be used to simulate the chemical events on the primitive Earth leading to the origins of life. It would be possible to constrain the chemical system further if we knew the biochemistry of the first life on Earth. This information is not available; however, there is strong evidence for the presence of life on Earth 3.5 billion years ago. It is helpful to our understanding of the processes leading to the origin of life to know that _life was present on the Earth only 1 billion years after it formed._

_Evidence for this ancient life, together with the chemical processes on the primitive Earth, has been found in studies of ancient sedimentary rocks._ The oldest known sedimentary rocks are located at Isua in Greenland. These rocks are 3.8 billion years old and in many respects resemble recent sedimentary rocks. The presence of sedimentary rock 3.8 billion years old indicates that there was weathering of the Earth's crust at that time. This suggests that a hydrosphere similar to the contemporary hydrosphere, minus molecular oxygen, was present on the Earth over 3.8 billion years ago.

Graphitized carbon is present in the Isua sediments, and its distribution and isotopic constitution suggests that it may have been originally present in microorganisms. The Isua sediments have been heated to such high temperatures (500°C) since they were deposited that it appears unlikely that it will be possible to prove that the carbon is biological in origin. In fact, it appears unlikely that it will be possible to find intact organic molecules that are greater than 3 billion years old in any sedimentary deposits because the rocks in which these compounds are embedded have been subjected to extreme conditions of heat and pressure. Consequently, these organic molecules would have undergone nonbiological transformations.

Fossilized remains of microorganisms would be expected to survive longer than the organic constituents of these microorganisms. A 3.5 million year old _stromatolite_, the fossilized remains of microbial colonies, is the oldest evidence of life. Some stromatolites are still forming in the brackish water of Shark Bay, Australia (Figure 32-4). These dome-shaped structures form because limestone, sand, and bits of rock stick to the slimy coating of each successive generation of bacterial growth. A new layer of the microorganisms grows on top of the sand and debris until the stromatolite grows so far out of the water that the microorganisms do not have sufficient moisture to survive. The fossilized stromatolite, which was found by radioactive dating techniques to be 3.5 billion years old, was located in North Pole, Australia. Morphologic studies on this stromatolite revealed its similarity to contemporary stromatolites. Since the algae prevalent in contemporary stromatolites are photosynthetic, it is assumed that the microorganisms responsible for the formation of this ancient stromatolite also were photosynthetic. This indicates that these microorganisms were quite highly evolved 1 billion years after the formation of the Earth. Therefore, the origins of life probably took place much earlier, about $4 \pm 0.2$ billion years ago. Thus the chemical processes resulting in the origins of life may have

**Figure 32-4**
Contemporary stromatolites in Shark Bay, Australia. Stromatolites grow in this brackish bay because microbial predators cannot survive in the salty environment.

been completed in less than 0.5 billion years. This is a short time if one considers it took another 2 billion years for the evolution of eukaryotes and the first dinosaurs appeared 4 billion years after the Earth formed.

*Microfossils* also have been identified in other ancient rocks. Filamentous bacteria have been detected in thin sections of stromatolites that have been shown to be 3 billion years old. Filamentous structures, which resemble contemporary cyanobacteria, were encapsulated in sedimentary rock that is 2.6 billion to 2.8 billion years old. These and other data provide a picture of the evolution of these primitive life forms to the more complex microorganisms prevalent on the Earth today.

Some tentative conclusions can be made concerning the biochemistry of the first forms of life. Since *ancient microorganisms appear to be similar to contemporary microorganisms*, it is unlikely that their biochemistry is substantially different from that prevalent today; e.g., it was not based on silicon instead of carbon. Since it appears very likely that the 3.5-billion-year-old stromatolite from North Pole, Australia is the remains of a colony of photosynthetic bacteria, it is conceivable that either chlorophyll or carotene derivatives were utilized to absorb the sunlight needed to form carbohydrates. There was concern that these pigments would not survive the short-wavelength radiation prevalent at that time on the primitive Earth. Recent studies have shown that contemporary microbial communities that form a mat of cells are protected from ultraviolet light by the top layer of cells. In addition, radiation damage is reversed in the dark and by visible light. Since colonies of matting microorganisms and algae effect the formation of stromatolites, the formation of these structures in the presence of ultraviolet light may not be surprising.

## SYNTHESIS OF BIOMONOMERS

The simple biomonomers essential for the origins of life may have been formed from the components of the primitive atmosphere in reactions driven by the energy sources listed in Table 32-1. The precursors of these biomonomers also may have been brought to Earth by comets (e.g., HCN and nitriles) or by the pyrolysis of carbonaceous meteorites (HCN, nitriles, aliphatic and aromatic hydrocarbons) when they impacted on the Earth's hot crust. Alternatively, some of the biomolecules present in carbonaceous meteorites may have reached the Earth intact if they accreted after the crust cooled.

*Lightning and the shock waves resulting from lightning were probably the most effective energy sources for the conversion of the components of the primitive atmosphere into the precursors for biomonomers*. The energy is released close to the surface of the ocean and the products formed can be easily dissolved and shielded from ultraviolet light. Ultraviolet light caused the syntheses of some biomonomers; however, it also effected the photodecomposition of organics to simpler carbon and nitrogen derivatives. Thermal energy was important in the initial stages of chemical evolution when the Earth's crust was strongly heated by the rapid decay of radioactive elements, but most of the radioactive isotopes were short-lived and the crust cooled rapidly.

No attempt will be made to discuss the formation of the biomonomers found in carbonaceous meteorites. The source of these meteorites is still a subject of considerable debate, so any discussion of the formation of these organic compounds in these yet-to-be-determined environments would be too speculative to be meaningful. Consequently, we will limit our discussion to the chemical processes that may have been taking place on the primitive Earth.

## Amino Acids

**Formation in Electric Discharges.** The first experiment that dramatically demonstrated that it is possible to perform laboratory investigations of the origins of life was performed by Stanley Miller while working as a graduate student with Harold Urey. Miller arced an electric discharge for 1 week through a mixture of methane, ammonia, and water and discovered that amino acids formed. Hydroxy acids, aldehydes, and hydrogen cyanide are produced efficiently by the discharge, which suggests that the observed products are formed by their condensation in the aqueous reaction mixture (Figure 32-5). The relative yields of hydroxy acids and amino acids depended on the concentration of ammonia in the system. The levels of ammonia present in the primitive lakes and oceans are a subject of considerable disagreement. Ammonia may have been photolyzed to nitrogen and hydrogen by solar ultraviolet light; however, there may have been a low steady-state concentration of ammonia resulting from the hydrolysis of hydrogen cyanide and other nitriles.

About half the 20 primary amino acids are produced directly in electric-discharge experiments with $CH_4$, $N_2$, $NH_3$, and $H_2O$ as the starting materials. Methionine is formed when hydrogen sulfide is added to the reaction mixture. Some of the aromatic amino acids may be produced when aromatic hydrocarbons are added to the reaction mixture. Aromatic hydrocarbons are formed by the pyrolysis of methane and carbonaceous meteorites.

Primitive Earth simulation experiments in which methane and ammonia are the major constituents have been criticized on the grounds that most geochemical models predict that the amounts of these highly reduced forms of carbon and nitrogen will be much lower than the amounts of carbon dioxide, carbon monoxide, and nitrogen. The primitive atmosphere must have contained reducing equivalents in some form to yield amino acids, since no biomolecules or their precursors are formed when a mixture of carbon dioxide, water, and nitrogen is sparked. Amino acids are formed

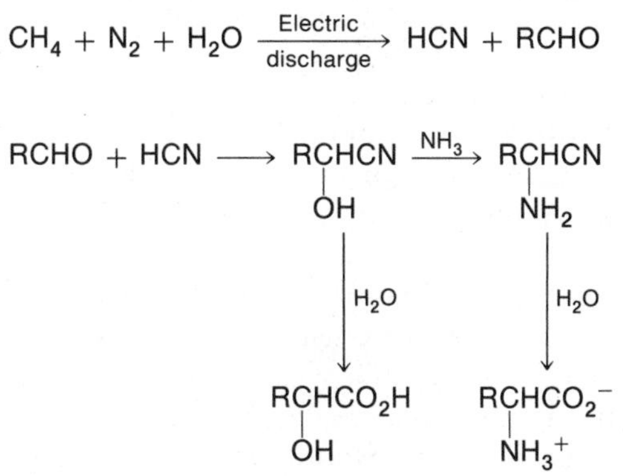

$$CH_4 + N_2 + H_2O \xrightarrow[\text{discharge}]{\text{Electric}} HCN + RCHO$$

$$RCHO + HCN \longrightarrow \underset{\underset{OH}{|}}{RCHCN} \xrightarrow{NH_3} \underset{\underset{NH_2}{|}}{RCHCN}$$

**Figure 32-5**
Synthesis of amino acids in an electric discharge.

when hydrogen is added to the mixture of carbon dioxide, water, and nitrogen, or when a mixture of carbon monoxide, water, nitrogen, and hydrogen is sparked. The amino acid yields are lower using these starting mixtures, while formaldehyde and HCN are major reaction products.

Amino acids also have been produced by the action of heat (900°C), gamma rays, shock waves, and short-wavelength ultraviolet light on postulated primitive atmospheres. From these findings it may be concluded that *amino acids were produced as a consequence of any form of energy interacting with the constituents of the primitive atmosphere.* Consequently, it appears very likely that amino acids were present in significant amounts on the primitive Earth.

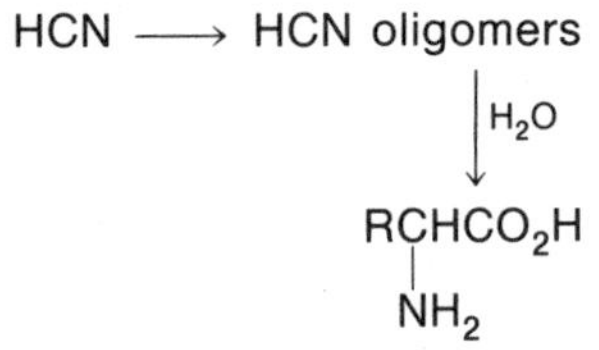

**Amino Acids from HCN.** *Amino acids are released by hydrolysis of the oligomers that form by the self-condensation of hydrogen cyanide in aqueous solution* (Figure 32-6). Hydrogen cyanide is produced in appreciable amounts by the action of electric discharges or shock waves on simulated primitive atmospheres. Pyrolysis of organic nitrogen compounds is another potential source of hydrogen cyanide on the primitive Earth. Glycine, alanine, and aspartic acid, together with five other amino acids that are not among the 20 proteinaceous amino acids, have been identified. Numerous other ninhydrin-positive compounds have been detected using an amino acid analyzer. These findings indicate the HCN alone may have been a direct source of some of the amino acids essential for the origins of life.

**Amino Acids in Carbonaceous Meteorites.** Many of the principal amino acids extracted from carbonaceous meteorites are identical with those obtained in primitive Earth simulation experiments. The meteoritic amino acids include glycine, glutamic acid, alanine, aspartic acid, valine, $\beta$-alanine, and $\alpha$-aminoisobutyric acid. These amino acids are present as racemic mixtures in meteorites, a result that proves that the amino acids were not present as a consequence of terrestrial contamination. This finding does not prove that the amino acids were formed abiotically, since the half-life for the racemization of amino acids is about $10^6$ years at 0°C and the meteorites are about $10^9$ years old. However, the presence of $\beta$-alanine and $\alpha$-aminoisobutyric acid in the meteorites, amino acids that are not utilized in contemporary life, provides support for the contention that these organic compounds were formed in abiological processes. The *abiotic formation of a similar ensemble of amino acids both on a planetesimal that fragmented into meteorites and in primitive Earth simulation experiments provides strong support for the validity of the simulation experiments.* Unfortunately, no conclusions can be drawn concerning the atmosphere of the primitive Earth because amino acids are formed from such a diverse array of starting materials.

The Earth's crust was very hot during the period of intense meteoritic bombardment because of the heat generated by the large numbers of impacting meteorites and the radioactive decay of short-lived isotopes present. Most of the amino acids reaching the Earth by means of meteorites were probably pyrolyzed to hydrogen cyanide, hydrocarbons, and carbon dioxide if they impacted on the crust while it was still hot. Those meteorites which arrived during the latter stages of accretion would not have been pyrolyzed, and the amino acids present in them would have been released by weathering. Thus meteorites would have served as a source of the starting materials for the abiotic synthesis of amino acids themselves.

HCN $\longrightarrow$ HCN oligomers

$\downarrow$ H$_2$O

RCHCO$_2$H
|
NH$_2$

**Figure 32-6**
Amino acids from HCN.

**Selection of the α-Amino Acids.** The facile abiotic synthesis of amino acids does present one problem. These syntheses are not restricted to the formation of α-amino acids; β-amino acids, hydroxy acids, and simpler carboxylic acids also are formed. If the simulated abiotic syntheses correctly reflect events on the primitive Earth, then there must have been some mechanism for the selection of the α-amino acids from all the other acids present. Since α-amino acids form *metal chelates* more readily than any of the other acids, this process may have served to concentrate the α-amino acids. The chelates may have been selectively precipitated or bound to a mineral surface, and their higher-level concentrations resulted in the selective synthesis of polymers that contained mainly α-amino acids. The possibility of the selection of α-amino acids in a template-directed reaction is discussed below.

## Nucleic Acid Bases

*Hydrogen cyanide and cyanoacetylene were the most likely starting materials for the synthesis of the nucleic acid bases* on the primitive Earth. These bases also may have been brought intact to the primitive Earth by meteorites.

**Purines.** When hydrogen cyanide condenses under the conditions described previously for the synthesis of amino acids, it also yields purines and pyrimidines. Hydrolysis of HCN oligomers yields *adenine* and *5-aminoimidazole-4-carboxamide*. There is tentative evidence for the formation of xanthine and hypoxanthine as well (Figure 32-7). Purines also may have been produced from *5-aminoimidazole-4-carbonitrile* (Figure 32-7), a compound produced by the photolysis of diaminomaleonitrile (Figure 32-7).

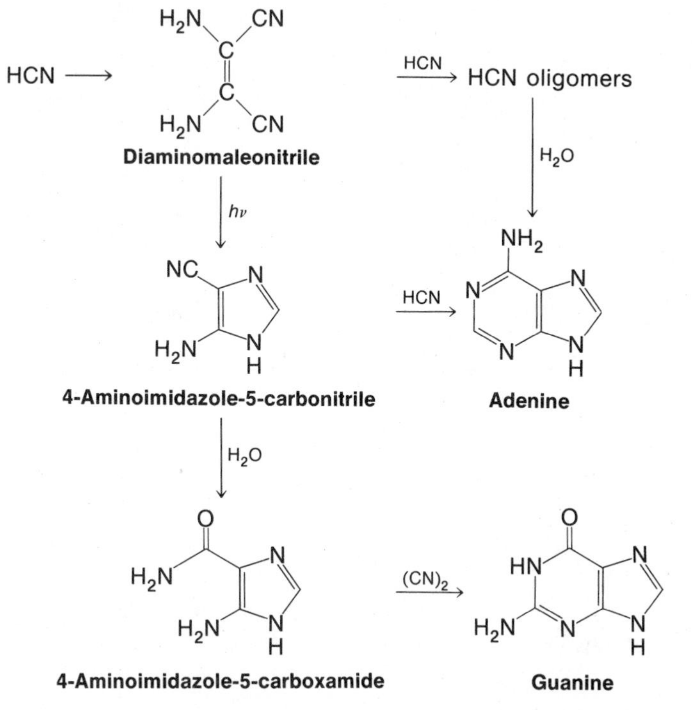

**Figure 32-7**
Purines from hydrogen cyanide.

These syntheses proceed if the concentration of hydrogen cyanide exceeds $10^{-2}$ $M$. This is the minimum concentration of cyanide required for the formation of diaminomaleonitrile, the key intermediate in both the reaction pathways outlined in Figure 32-7. The hydrolysis of hydrogen cyanide to formamide and formic acid proceeds at a faster rate than its condensation to diaminomaleonitrile if the cyanide concentration is less than $10^{-2}$ $M$.

Hydrogen cyanide may have been produced in a variety of ways on the primitive Earth, but it seems unlikely that a $10^{-2}$ $M$ solution of cyanide was present if the oceans had the same volume as they do today. This concentration of cyanide may have existed shortly after the oceans first formed, when their volume was low and the amount of hydrogen cyanide was high owing to a high rate of impact of comets and meteorites. The hydrogen cyanide may have been concentrated in certain areas as a result of a high frequency of electrical storms. Alternatively, since the eutectic concentration of hydrogen cyanide is 74.5 mol% at $-23.4°C$, the cyanide would have been concentrated sufficiently for diaminomaleonitrile formation if the water droplets and lakes containing it were cooled to $-20°C$. One can imagine the oligomerization of dilute cyanide solutions each winter as water droplets or lakes that contain hydrogen cyanide freeze. In the summer, the photolysis of diaminomaleonitrile proceeds along with the continued synthesis of hydrogen cyanide. Since hydrogen cyanide is more volatile than water, a concentration mechanism based on the evaporation to near dryness of a lake that contains small amounts of cyanide would not give concentrated cyanide solutions.

**Pyrimidines.** Pyrimidines also are released by the hydrolysis of HCN oligomers (Figure 32-8). The most significant of the three pyrimidines identified at this time is *orotic acid*, because it is photodecarboxylated to *uracil*, one of the pyrimidines present in RNA. The nucleoside and nucleotide derivatives of orotic acid are also photochemically converted to the corresponding uracil derivatives.

*Cyanoacetylene*, a major product formed by the action of an electric discharge on methane-nitrogen mixtures, is the starting material for an

**Figure 32-8**
Pyrimidines from hydrogen cyanide.

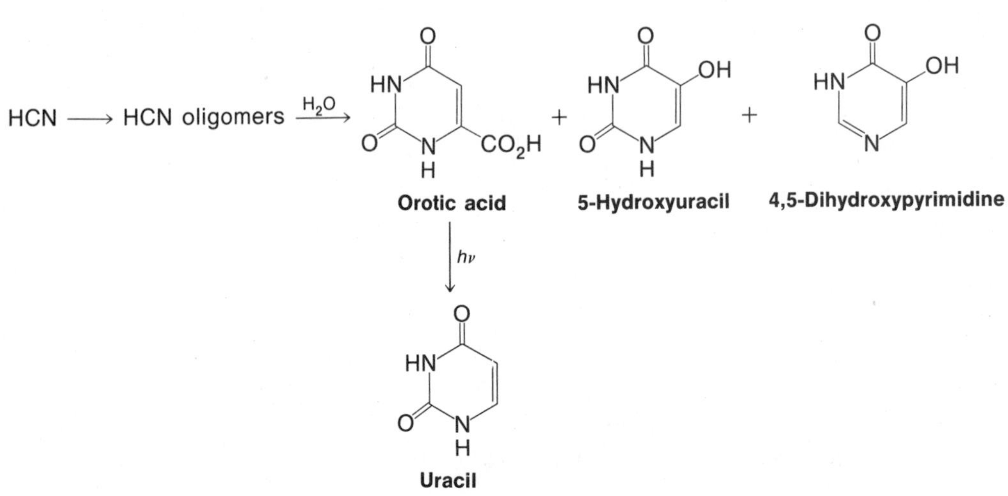

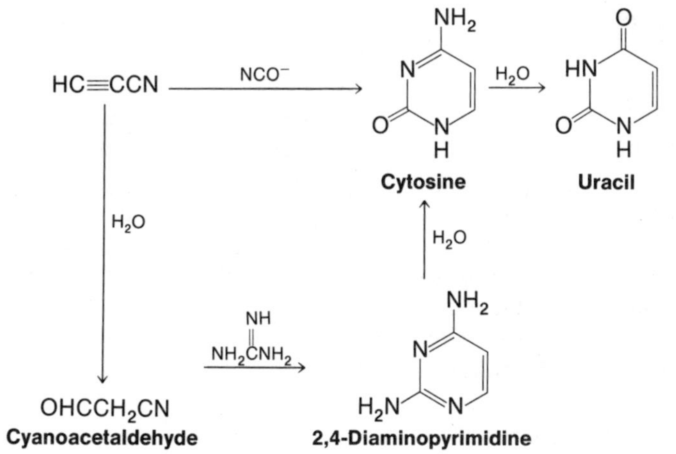

**Figure 32-9**
Pyrimidines from cyanoacetylene.

alternative pyrimidine synthesis (Figure 32-9). *Cytosine*, a pyrimidine base present in both RNA and DNA, is formed directly by the reaction of cyanoacetylene with *cyanate*. *Uracil* is formed by the hydrolysis of cytosine. The validity of this reaction sequence as prebiotic synthesis of pyrimidines has been questioned because both cyanoacetylene and cyanate are rapidly hydrolyzed in mildly basic solutions. These hydrolytic reactions would limit the possibility of the bimolecular reaction of cyanate and cyanoacetylene on the primitive Earth.

An alternative pyrimidine synthesis includes the hydrolysis of cyanoacetylene to *cyanoacetaldehyde* as the first step (Figure 32-9). The cyanoacetaldehyde reacts with guanidine in aqueous solution to give 2,4-diaminopyrimidine, a compound that is hydrolyzed to cytosine and uracil.

The efficiency of the synthesis of pyrimidines from cyanoacetylene is much greater than from hydrogen cyanide. However, hydrogen cyanide is formed in greater yield in primitive Earth simulation experiments. As a consequence, it is not possible to assess whether hydrogen cyanide or cyanoacetylene was the more important starting material for the synthesis of pyrimidines.

**Purines and Pyrimidines in Carbonaceous Meteorites.** The presence of both purines and pyrimidines in carbonaceous meteorites has been reported by several groups. However, there is some controversy concerning the identity of the specific bases present. Further investigations will be needed to assess whether meteorites were an important source of the pyrimidine and purines needed for the origins of life.

## Sugars

Ribose is an essential structural element in the backbone of RNA, and it probably had a similar role in the primitive RNA used in the first life forms. *Formaldehyde was probably the starting material for the synthesis of ribose and other sugars* on the primitive Earth. Formaldehyde is formed readily in primitive Earth simulation experiments by the photolysis of $CH_4$-$H_2O$, $CO$-$H_2O$, or $CO_2$-$H_2O$-$Fe^{2+}$ mixtures and by the passage of an electric discharge through $CH_4$-$H_2O$ or $CO$-$H_2O$-$N_2$ mixtures.

The condensation of formaldehyde to sugars is catalyzed by divalent cations, alumina, and clays under mildly basic conditions. The reaction proceeds by the stepwise condensation of formaldehyde to a dimer (glycolaldehyde), a trimer, and a tetramer. The tetramer then condenses with the smaller units to generate pentoses, hexoses, heptoses, and so forth.

There are several questions that must be resolved before this route can be accepted as a prebiotic synthesis of ribose. Since the sugars formed in this synthesis decompose under the reaction conditions, there must have been some mechanism for their stabilization. This may have been the formation of a more stable adduct such as a nucleoside. Alternatively, this problem may have been circumvented if there was a constant synthesis of formaldehyde that resulted in the formation of constant levels of ribose for incorporation into nucleic acids.

A second problem is the facile reaction of formaldehyde with hydrogen cyanide to give the cyanohydrin of formaldehyde, a reaction that could have inhibited the synthesis of nitrogen-containing biomolecules as well as sugars. This would not have been as great a problem if the hydrogen cyanide and formaldehyde were synthesized in large amounts at different stages of the process of chemical evolution. The concentration of hydrogen cyanide may have been greater in the early stages of chemical evolution, just after the period of extensive meteoritic and cometary impact on the primitive Earth. Larger amounts of formaldehyde may have been produced later, after meteoritic infall decreased and the Earth's crust cooled, so that the pyrolysis of carbonaceous meteorites to HCN also decreased.

A third difficulty with this synthesis is the observation that many different sugars are formed and the yield of ribose is only in the 3 to 4 percent range. If ribose was the only sugar present initially in the nucleoside, its rate of formation must have been accelerated by some catalytic process or else there was a mechanism by which ribose was selectively removed from the mixtures of sugars that were formed. So far, neither possibility has been demonstrated in the laboratory.

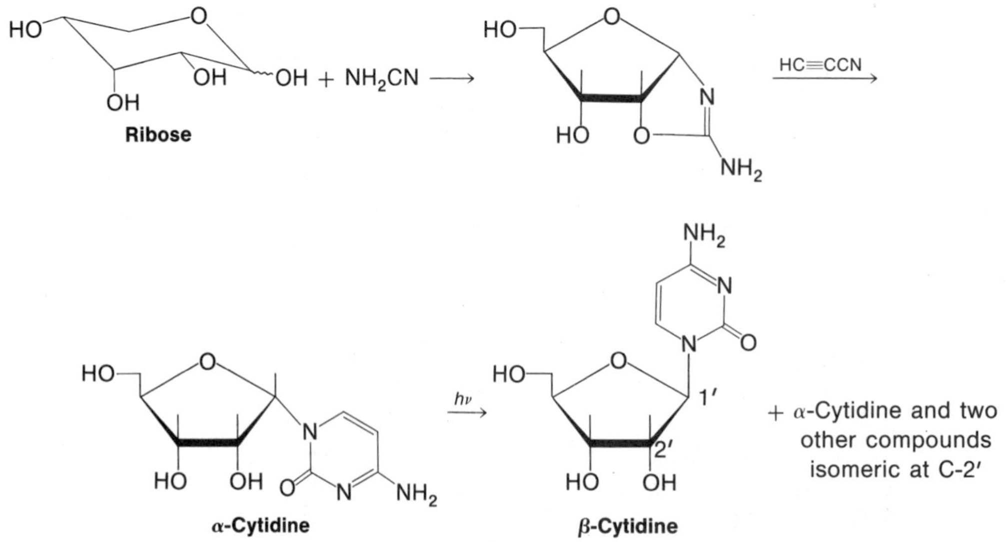

**Figure 32-10**
A synthesis of β-cytidine.

## Nucleosides and Nucleotides

*The first life forms probably contained nucleic acids for the storage of genetic information.* This information-storage function may have been performed by RNA, with the DNA in contemporary cells evolving from RNA after life originated. Consequently, there must have been a pathway by which the nucleotide building blocks of RNA were synthesized.

**Nucleosides.** There has been some success in elucidating pathways for the prebiotic synthesis of nucleosides. Purine nucleosides have been prepared by *dry-phase heating* of the heterocyclic bases with ribose in the presence of divalent ions such as magnesium or calcium. In this synthesis, an aqueous solution of the reactants is evaporated to dryness and heated in the dry state in a synthetic procedure that models the evaporation to dryness of a prebiotic lake or pond. Yields range from 2 to 15 percent of a mixture of $\alpha$- and $\beta$-anomers depending on the purine base used in the reaction. Pyrimidine nucleosides are not formed in this process.

The $\alpha$-anomer of cytidine is formed in good yield by the reaction of cyanoacetylene with the adduct formed between ribose and cyanamide (Figure 32-10). Photolysis of the $\alpha$-anomer gives a mixture of four products that are isomeric at C-2', including the desired $\beta$-anomer. The lack of specificity in the final stage of the synthesis detracts from an otherwise elegant prebiotic synthesis.

The possibility of the direct synthesis of pyrimidine nucleotides by the reaction of pyrimidines with phosphate esters of sugars was suggested by the observation of a reaction between barbituric acid and ribose-5-phosphate (Figure 32-11). In the initial adduct, which is formed at 37°C and pH 5.5, the barbituric acid is bound to the phosphate grouping. Incubation of this adduct for 1 week under the same reaction conditions resulted in the formation of a low yield of the *N*-nucleotide of barbituric acid by the intramolecular migration of the barbituric acid moiety from the phosphate group to the 1' position of ribose. This would constitute a plausible prebiotic route to pyrimidine nucleotides if uracil and cytosine also react with ribose-5-phosphate to give the corresponding nucleotide.

The phosphorylation of nucleosides to nucleotides has been studied both in solution and in the dry state. Condensing agents such as *cyanamide* have been used in solution-phase studies in which an activated adduct of the phosphate is formed which then phosphorylates the nucleoside (Figure 32-12). The phosphorylation often proceeds nonspecifically and in low yield, thus decreasing its attractiveness as a potential prebiotic synthesis. The desired selectivity can be observed when other condensing agents are used, especially if metal ion or mineral surface catalysis is part of the phosphorylation process.

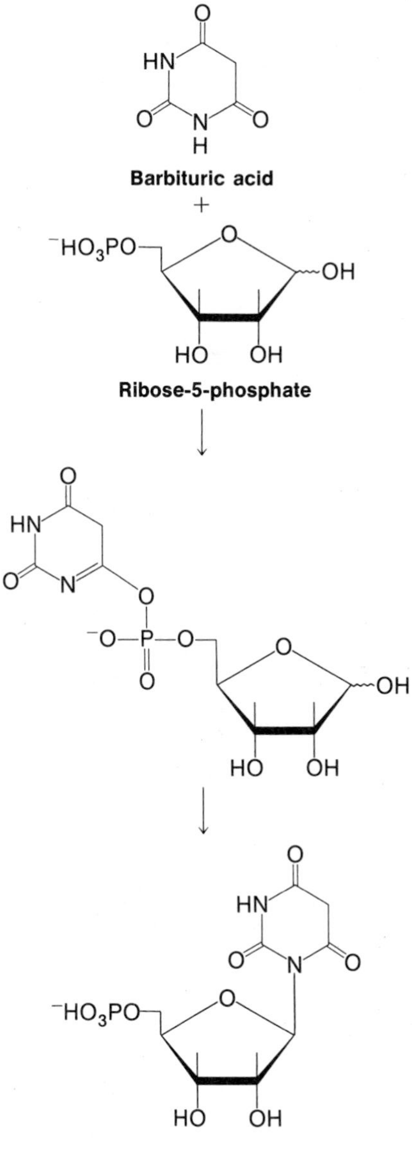

**Barbituric acid**
+
**Ribose-5-phosphate**

**Figure 32-11**
Reaction of barbituric acid with ribose-5-phosphate.

**Figure 32-12**
Cyanamide-mediated phosphorylation.

Solid-state reactions, which model the evaporation of a lake or pond to dryness and the subsequent heating of the precipitated salts in the dry state, provide an efficient prebiotic nucleotide synthesis. Dry heating of the nucleoside at 60°C to 100°C with inorganic phosphate and urea gives initially the _5'-phosphate_ as the major reaction product along with traces of the 2'- and 3'-phosphates. The _2',3'-cyclic phosphate_ is the major product on further heating of the reaction mixture (Figure 32-13). Addition of $Mg^{2+}$ to the mixture accelerates the rate of formation of pyrophosphates so that the major product is the nucleoside 5'-diphosphate, while low yields of the 5'-triphosphate are observed (Figure 32-13). The total yield of phosphorylated derivatives exceeds 90 percent, a result that suggests that this is a plausible prebiotic pathway for nucleotide synthesis. Since both the 3'- and 5'-phosphates may be used directly in polynucleotide synthesis and the 2'-phosphate can be converted to the 3'-phosphate by means of the 2',3'-cyclic diester, all the simple phosphorylated products may be utilized for the synthesis of oligonucleotides. The polyphosphates and 2',3'-cyclic phosphate may have served as a prebiotic energy source for the formation of polynucleotides on the primitive Earth.

### Lipids

_Membranes are the key components of all cells; they allow the cell to maintain an internal environment different from the external environment._ This segregation of organic materials and nutrients also was essential in the cell of primitive life, so that _membranes must have had an important role in the origin of life._ The main structural component of the contemporary membrane is the phospholipid moiety.

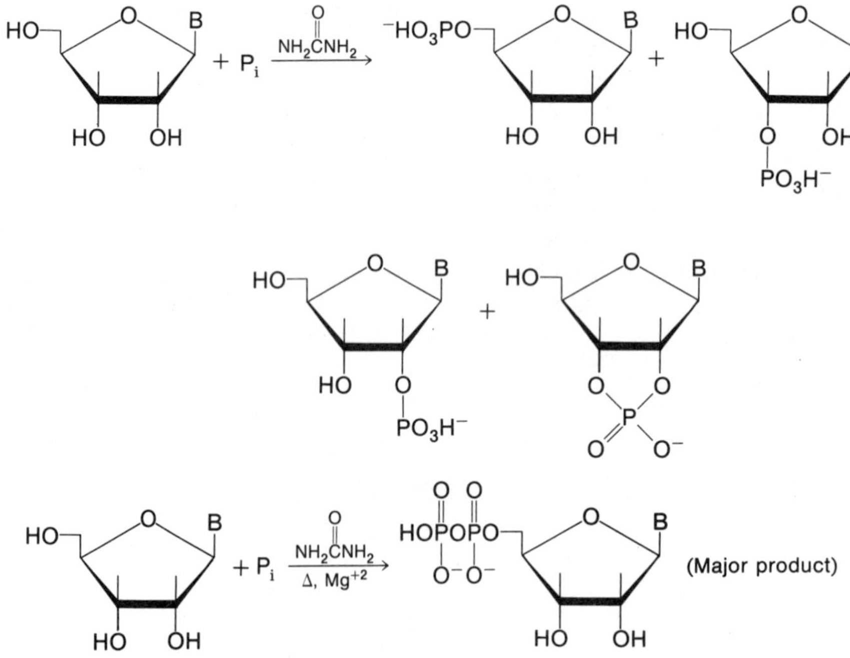

**Figure 32-13**
Thermal synthesis of nucleotides under anhydrous conditions.

The prebiotic formation of lipids has not been extensively investigated, but the limited experiments that have been performed suggest that lipid-like materials probably formed spontaneously on the primitive Earth. For example, when an electric discharge is passed through a mixture of methane and water, a mixture of carboxylic acids is obtained that contains 2 to 12 carbon atoms. However, all the acids with more than 6 carbon atoms have branched chains instead of the linear chains found in the lipids present in the membranes of contemporary cells. Linear fatty acids and hydrocarbons with up to 18 carbon atoms are produced when carbon monoxide and hydrogen are heated at high pressures to 200°C to 400°C in the presence of an iron catalyst (*Fischer-Tropsch synthesis*). The Fischer-Tropsch reaction has been suggested as a possible source of the complex carbon compounds in the solar nebula, but it is unlikely that the high temperatures and pressures required for this synthesis were prevalent on the primitive Earth. Since carboxylic acids have been detected in carbonaceous meteorites, they may have been synthesized by a Fischer-Tropsch process in the solar nebula and then brought to the Earth in the latter stages of meteoritic infall.

The condensation of fatty acids with glycerol to form glycerides and the condensation of fatty acids, phosphate, and glycerol to form *phospholipids* have been performed using prebiotic reaction conditions. *The only obstacle remaining to a convincing prebiotic phospholipid synthesis is the lack of a plausible abiotic synthesis of linear $C_{12}$ to $C_{18}$ fatty acids.*

## Coenzymes

The diversity of structure and function observed for the coenzymes is unique among all the classes of biomolecules. These compounds facilitate a wide array of group transfers and redox reactions in contemporary biochemical systems. The prevalent structural features are the heterocyclic ring system (often adenine) and, less frequently, the nucleotide grouping. One feels that the coenzymes were assembled from an assortment of organic molecules formed during the prebiotic era. It appears as if a molecular tinkerer fabricated a new useful structure out of these "spare parts" every time a different catalytic agent was needed.

There are two theories for the incorporation of coenzymes into the first living systems. In the conventional view, the coenzymes, or their structural elements, were formed abiotically at the same time as the purines and pyrimidines. This accounts for the presence of adenine, a purine that forms readily, as a frequently observed structural unit. The cofactors facilitated the inefficient biochemical processes in the earliest forms of life. The efficiency of early life was greatly enhanced as enzymes evolved that accelerated the reactions between the coenzymes and substrate molecules.

An alternative theory is based on the hypothesis that both the catalytic and information-storage processes of the early life forms were performed by nucleic acids. Reaction catalysis by nucleic acids was, according to this view, facilitated by the coenzymes incorporated into the nucleic acids. At a later evolutionary stage, when proteins took over the catalytic function of the cell, vestiges of these primitive RNA molecules were utilized in concert with the proteins because the proteins did not possess the functionality necessary to catalyze all the cellular processes. In fact, it has been suggested that histidine evolved from a nucleic acid precursor. This amino acid, which is present at the catalytic site of many enzymes, reflects this proposed nucleic acid ancestry in its contemporary biosynthesis, which proceeds from ATP.

Both these theories postulate that *the synthesis of coenzymes took place at the same time as the abiotic synthesis of the purine and pyrimidine bases.* Very little progress has been reported on the abiotic synthesis of coenzymes. This is probably due more to a lack of effort than to a lack of success.

Limited success has been obtained in the synthesis of the *nicotinamide* ring system. Nicotinamide isomers are obtained in low yield when an electric discharge is passed through a mixture of ethylene and ammonia. The reaction probably proceeds by means of hydrogen cyanide and pyridine, since pyridine is also a reaction product. Nicotinamide isomers are formed by the action of an electric discharge on a pyrimidine–hydrogen cyanide mixture (Figure 32-14, Sequence 1). An alternative synthesis is the addition of cyanoacetaldehyde to propioaldehyde in the presence of ammonia. Propioaldehyde and cyanoacetylene, the precursor to cyanoacetaldehyde, are both formed in discharge reactions. The nicotinonitrile that is formed is readily hydrolyzed to nicotinamide (Figure 32-14, Sequence 2).

The direct synthesis of nicotinonitrile in an electric discharge (Sequence 1, Figure 32-14) is simpler than the multistep process outlined in

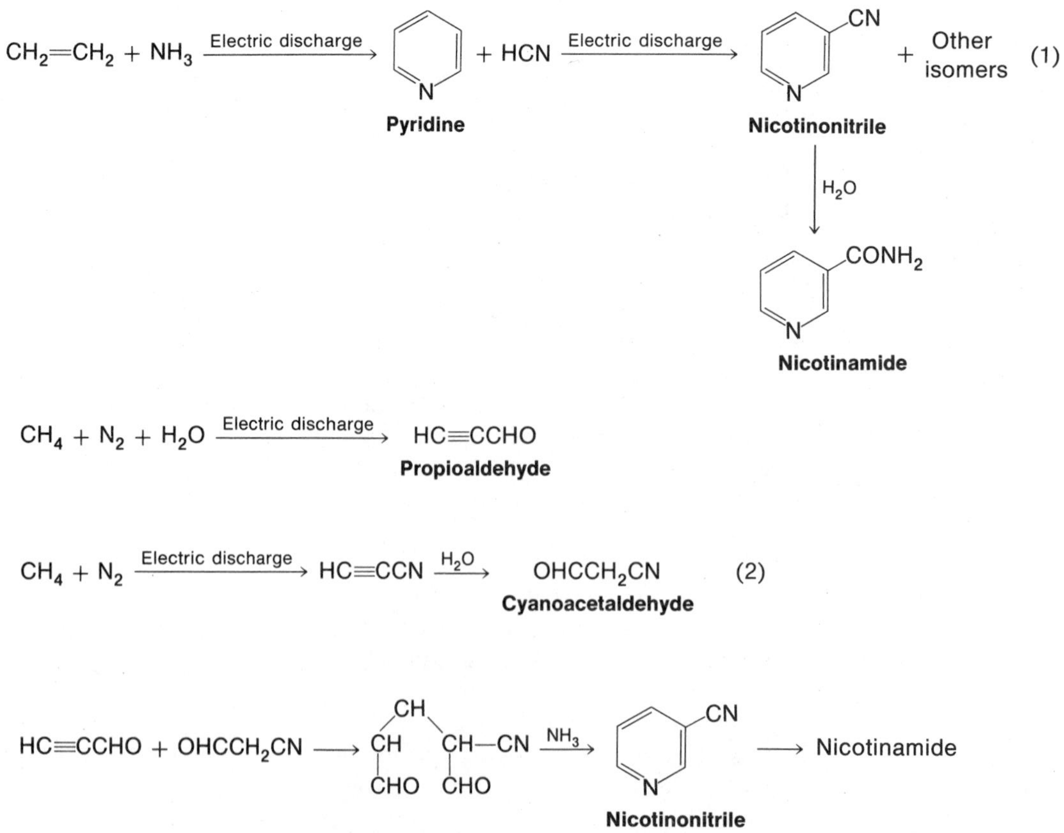

**Figure 32-14**
Abiotic nicotinamide synthesis.

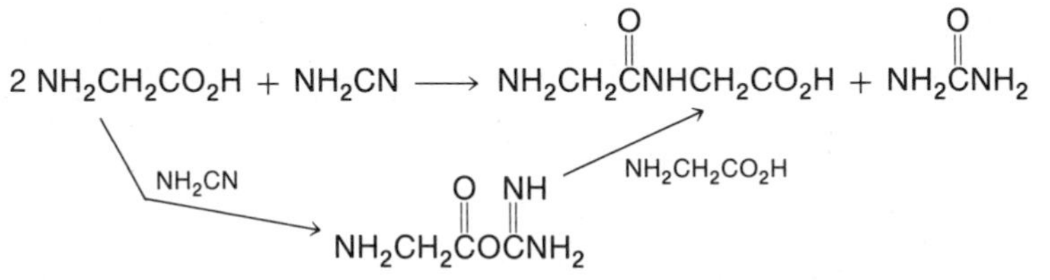

$$2 \ NH_2CH_2CO_2H \ + \ NH_2CN \ \longrightarrow \ NH_2CH_2\overset{\overset{\displaystyle O}{\|}}{C}NHCH_2CO_2H \ + \ NH_2\overset{\overset{\displaystyle O}{\|}}{C}NH_2$$

with a branch from $NH_2CN$ leading to $NH_2CH_2\overset{\overset{\displaystyle O}{\|}}{C}O\overset{\overset{\displaystyle NH}{\|}}{C}NH_2$ and from there via $NH_2CH_2CO_2H$

**Figure 32-15**
Formation of diglycine with cyanamide.

Sequence (2) of Figure 32-14, but the yields in Sequence (2) are greater than those of Sequence (1). It is not possible to ascertain which is the more likely primitive Earth synthesis.

*Nicotinamide-adenine dinucleotide* has been prepared in up to 15 percent yields from the solid-phase reaction of ATP and nicotinamide mononucleotide in the presence of magnesium and imidazole. The direct synthesis of nicotinamide mononucleotide has not been demonstrated, but presumably it would be readily prepared by the phosphorylation of the corresponding nucleoside. A low-yield (0.8 percent) synthesis of the nucleoside has been described.

These reactions suggest that it is possible to synthesize NAD under prebiotic conditions. The low yields in this multistep sequence indicate that the reaction conditions used probably differ from those prevalent on the primitive Earth when NAD was synthesized.

## Condensing Agents

*Chemical condensing agents may have linked together biomonomers into biopolymers on the primitive Earth.* Most effect this condensation by removing a molecule of water from two monomeric units to bring about the formation of a dimer, with the free energy of hydration of the condensing agent driving the reaction. This process is illustrated in Figure 32-15 with *cyanamide*, a postulated prebiotic condensing agent. The mechanism given is consistent with the chemistry of carbodiimides, a related class of condensing agents.

**Cyanamide and Related Structures.** Cyanamide has been evaluated extensively as a condensing agent for the formation of peptides and nucleotides. It has been detected in the interstellar medium (Table 32-2) and is formed under prebiotic conditions by the irradiation of *ferrocyanide* in the presence of ammonia and halide ions. This facile synthesis suggests that cyanamide may have been present on the primitive Earth.

Cyanamide effects the condensation of amino acids to dipeptides at pH 5 or less. This is a much more acidic environment than is assumed to have been likely for the primitive ocean (pH 7.5 ± 1.5). The absence of polypeptide formation and the high acidity required for reaction detracts from its plausibility as a peptide-forming reagent on the primitive Earth.

The oligomerization of deoxythymidine-5′-phosphate is brought about in low yield by aqueous solutions of cyanamide at pH 7. Cyanamide is more effective when it is used in the presence of a nucleotide triphosphate under the dehydrating conditions that might result from a pond drying up on the primitive Earth. The synthesis of pyrophosphates, deoxyoligonucleotides, polypeptides, and glycerides has been demonstrated using solid-phase synthesis conditions. The solid-phase synthesis of polypeptides is not a plausible prebiotic reaction because it requires an initial pH that is less than or equal to 4. Both the solution-phase synthesis at pH 7 and the other solid-phase reactions are plausible prebiotic syntheses.

**Cyanogen, Cyanoformamide, and Cyanate.** Cyanogen is formed readily by the photolysis of (or passing an electric discharge through) hydrogen cyanide. Hydrolysis of cyanogen yields cyanoformamide, which then decomposes to cyanide and cyanate (Figure 32-16). These three chemically related condensing agents exhibit comparable efficiency as condensing agents.

*Cyanogen* effects the phosphorylation of uridine, cytidine, and adenosine in low yields. Cyanoformamide, cyanate, and cyanogen all bring about the cyclization of uridine-3'-phosphate to the corresponding cyclic phosphate, a simple phosphorylation reaction, in 10 percent yields. None of these reagents has been effective in condensing simple nucleotides to oligonucleotides.

Pyrophosphate is formed by the action of cyanate on the mineral *hydroxyapatite*. Presumably this reaction proceeds by the formation of carbamoyl phosphate, a phosphorylating agent that reacts with a second phosphate anion to give pyrophosphate (Figure 32-16). Cyanate also will effect the condensation of glycine to diglycine in about 10 percent yields when heated with apatite in the dry state.

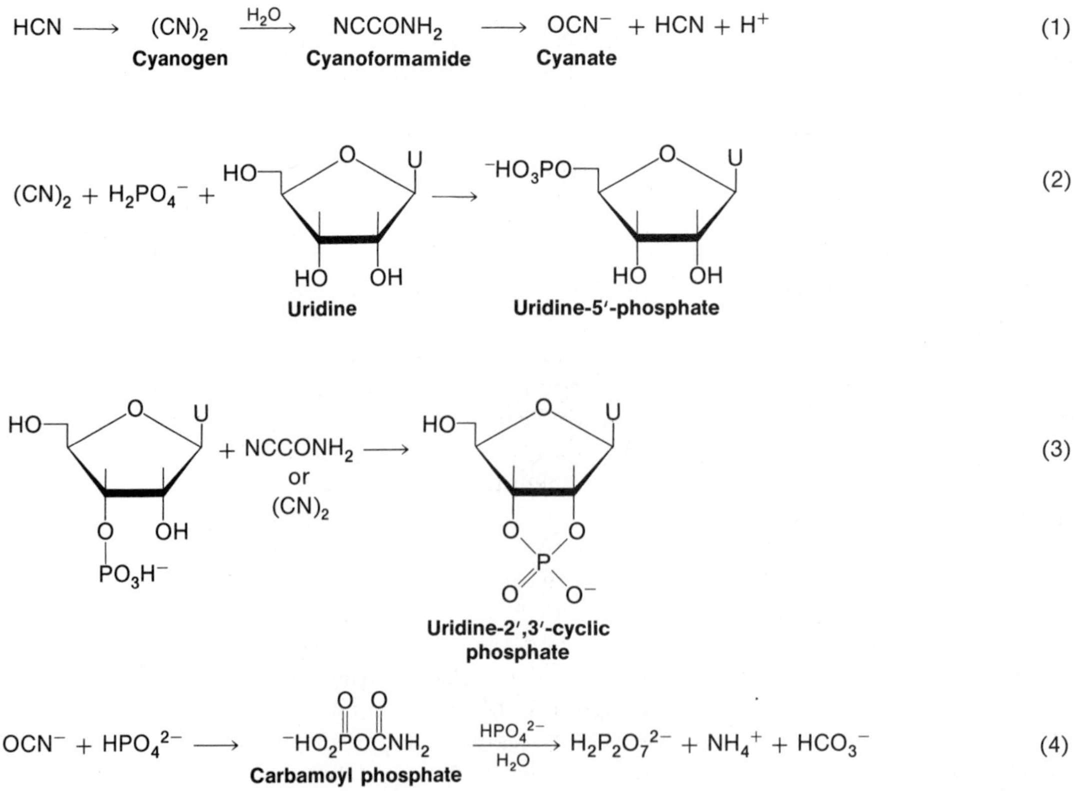

**Figure 32-16**
Cyanogen, cyanoformamide, and cyanate as prebiotic condensing agents.

$$HCN \longrightarrow (CN)_2 \xrightarrow{H_2O} NCCONH_2 \longrightarrow OCN^- + HCN + H^+ \qquad (1)$$

Cyanogen    Cyanoformamide    Cyanate

$$(CN)_2 + H_2PO_4^- + \quad \text{Uridine} \longrightarrow \text{Uridine-5'-phosphate} \qquad (2)$$

$$\text{(uridine with } PO_3H^- ) + NCCONH_2 \text{ or } (CN)_2 \longrightarrow \text{Uridine-2',3'-cyclic phosphate} \qquad (3)$$

$$OCN^- + HPO_4^{2-} \longrightarrow {}^-HO_2POCNH_2 \xrightarrow[H_2O]{HPO_4^{2-}} H_2P_2O_7^{2-} + NH_4^+ + HCO_3^- \qquad (4)$$

Carbamoyl phosphate

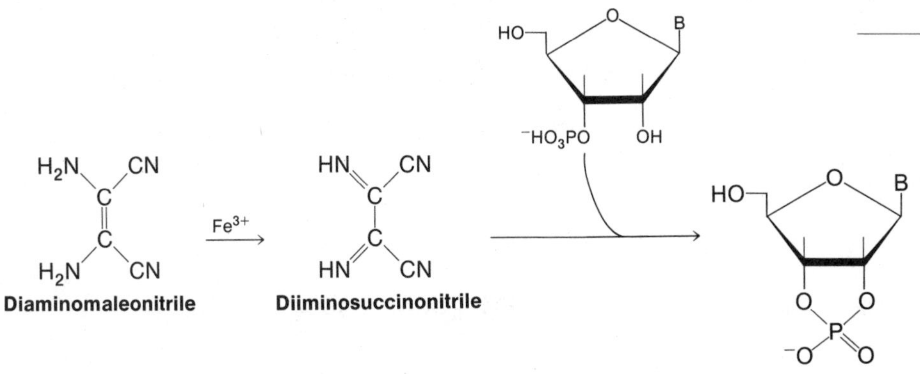

Since cyanogen, cyanoformamide, and cyanate do not effect the direct formation of oligonucleotides, they could be considered as important prebiotic condensing agents only to the extent that the polymerization of the nucleoside cyclic 2′,3′-phosphate derivatives were a major route for the formation of nucleic acids.

**Figure 32-17**
Phosphorylation with diiminosuccinonitrile. Unpublished results from the laboratory of J. P. Ferris.

**Diaminomaleonitrile.** Diaminomaleonitrile has been shown to be a plausible prebiotic condensing agent. Since diaminomaleonitrile is an intermediate in the formation of purines, pyrimidines, and amino acids from hydrogen cyanide, it is reasonable to assume that it also may have served as a condensing agent on the primitive Earth. It brings about the condensation of glycine to diglycine in 5 percent yields, but it does not effect the phosphorylation of uridine. _Diiminosuccinonitrile_, an oxidation product of diaminomaleonitrile, is a more likely condensing agent, since it would be expected to give the nitrogen analog of a mixed anhydride on reaction with carboxylate or phosphate. These mixed anhydride analogs react to form peptide or phosphoric acid ester linkages, respectively. An example of the formation of the phosphoric ester linkage is the cyclization of uridine-3′-phosphate to the corresponding 2′,3′-cyclic phosphate (Figure 32-17).

**Trimetaphosphate and Polyphosphates.** _Linear inorganic polyphosphates_ and the cyclic phosphate trimer _trimetaphosphate_ are effective prebiotic condensing agents. Their prebiotic syntheses proceed by the thermolysis of phosphate to yield linear polyphosphates, which in turn cyclize to trimetaphosphate in aqueous medium.

Glycine reacts with trimetaphosphate at pH 9 to 11 to give a 20 percent yield of diglycine. The yield decreases markedly at pH 7. Imidazole enhances the yield of diglycine in the higher pH range, and small yields of triglycine also are obtained. The reaction is initiated by the attack of the glycine amino group on the trimetaphosphate, thus explaining the pH dependency of the diglycine formation. It is apparent from the reaction mechanism (Figure 32-18, Sequence 1) that one would expect to obtain oligopeptides. The requisite cyclic acylphosphoramide would not form readily with the longer peptide chains. In addition, the amino group of the initial product dimer is phosphorylated, so it cannot participate in peptide bond formation.

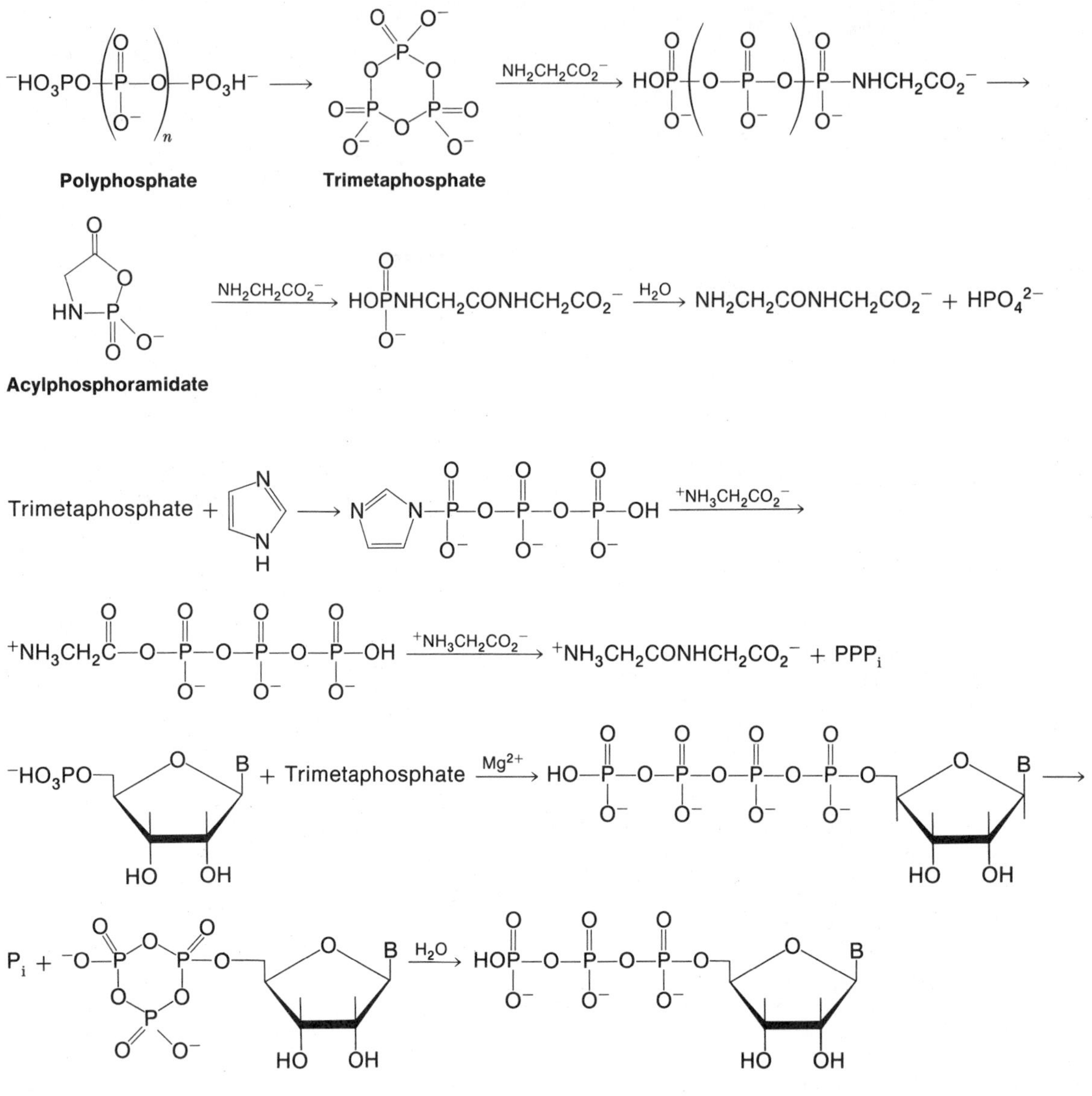

**Figure 32-18**
Trimetaphosphate as a condensing agent.

Oligopeptides are formed when glycine reacts in the solid state with trimetaphosphate or linear polyphosphates in the presence of imidazole or magnesium ions. A series of oligomers up to the decamer are obtained. The reaction may proceed by the attack of imidazole on the phosphoric anhydride linkage to give a phosphoramidate. This reacts with the carboxylate group of glycine to give the mixed anhydride that acylates the free amino group of glycine or glycine oligomers. This or related pathways in which the glycine amino group is _not_ phosphorylated account for the formation of oligoglycines in this synthesis (Figure 32-18, Sequence 2).

Nucleoside-5′-polyphosphates are formed by the dry-phase reaction of nucleoside-5′-phosphates with trimetaphosphate in the presence of $Mg^{2+}$

(Figure 32-18, Sequence 3). These *nucleoside polyphosphates may have served as an energy source for primitive life forms, just as ATP does in contemporary biochemical systems*. These polyphosphates degrade rapidly to the corresponding triphosphates, which in turn decompose more slowly to diphosphates and monophosphates.

## BIOSYNTHETIC PATHWAYS AS A GUIDE TO PREBIOTIC CHEMISTRY

*The earliest forms of life used preformed biomolecules in their replicative and biosynthetic processes. As living systems proliferated*, the supply of these preformed molecules diminished. *Only those life forms which evolved catalysts (enzymes) to accelerate the synthesis of these essential molecules from other precursors survived*. As early as 1945, Norman Horowitz outlined a mechanism by which natural selection could have extended biosynthetic sequences a step at a time, starting from the final step in the synthesis. If the primitive metabolic pathways evolved from prebiotic syntheses, some steps in the current metabolic pathways may be *"chemical fossils"* of reactions that occurred on the primitive Earth. Consequently, the plausibility of a proposed prebiotic pathway is greater if it can be correlated with a contemporary biosynthetic process.

There is a good correlation between the contemporary biosyntheses of purines and pyrimidines and some steps in their proposed prebiotic syntheses. Central intermediates in the prebiotic syntheses of purines are 5-aminoimidazole-4-carboxamide and the corresponding nitrile (Figure 32-7). The same imidazole, in the form of ribotide, is the precursor to purines in contemporary living systems. Purine nucleotides also are formed by the direct reaction of the purine with a ribose nucleotide in the "salvage" pathway for nucleotide synthesis. This biosynthetic route is very similar to the prebiotic synthesis of nucleotides, which proceeds by the direct condensation of ribose with the purine base.

The decarboxylation of the nucleotide of orotic acid is one of the steps in the biosynthesis of the nucleotides of uracil and cytosine. The same decarboxylation is a key reaction in the prebiotic synthesis of uracil from HCN (Figure 32-8). In addition, aspartic acid, the starting material for the biosynthesis of orotic acid, is produced in a variety of prebiotic experiments. It would not have been a major change for early life to evolve a system for the biosynthesis of orotic acid from the readily available aspartic acid once the low supply of preformed orotic acid limited the growth of primitive life forms.

## ELEMENTAL ABUNDANCES

*Metal ions* have a central role in contemporary biochemical systems as *cofactors, components of enzymes, and enzyme regulators*. The same metal ions probably had similar functions in primitive life, but with much diminished efficiency and specificity because their chemical properties were not modulated by associated proteins. Specific examples of the catalytic role of metal ions in the polymerization of nucleotides will be discussed later.

**Table 32-3**
Elemental Abundances in Seawater

| $>10^6$ n$M$ | $10^6$–$10^2$ n$M$ | $10^2$–1 n$M$ |
|---|---|---|
| H, O, Na, Cl, Mg, S, K, Ca, C, N | Br, B, Si, Sr, F, Li, P, Rb, I, Ba | Mo, Zn, Al, V, Fe, Ni,* Ti, U, Cu, Cr,* Mn, Cs, Se, Sb, Cd, Co, W |

Note: There is no evidence that the underlined elements are utilized in contemporary living systems.

*Only used in higher life forms.

Data from F. Egami, Minor elements and evolution. *J. Mol. Evol.* 4:113, 1974.

The elements abundant in living systems today generally reflect their relative amounts in the ocean (Table 32-3) and, to a lesser extent, in the Earth's crust. Since it is very likely that life originated in the presence of water, the elements required for life were probably established from their concentrations in the primitive oceans 4 billion years ago. This does not mean that there is a direct correlation between the elemental abundances in the oceans and in contemporary life. The following factors also must be considered.

1. The elemental abundances in the oceans today do not reflect an earlier era when the Earth's atmosphere was not strongly oxidizing. For example, at that time, a significant portion of the iron would have been present in the more soluble $Fe^{2+}$ oxidation state. This would account for the observation that iron is prevalent in contemporary biochemical systems, yet it has a very low concentration in seawater. Conversely, chromium, which is relatively abundant in seawater, would have been expected to be present as the $Cr^{3+}$ form in the primitive seas. The $Cr^{3+}$ oxidation state forms a very insoluble hydroxide, so chromium would not have been abundant in the early oceans. Chromium is an essential element for higher life forms only, a dependence that probably evolved after life originated.

2. Some elements had similar functions in early living systems, but natural selection resulted in the choice of a few that had superior reactivity. For example, divalent manganese, nickel, cobalt, and zinc may have had similar functions, as illustrated by the observation that these metals can replace the zinc ions in zinc metalloenzymes. The reconstituted enzyme is usually less efficient when a metal other than zinc is bound at the active site, indicating that zinc possesses the optimal properties for that system.

3. The carbon, sulfur, nitrogen, and phosphorus content of life is strongly enriched relative to the ocean. Since it is unlikely that the concentrations of these elements in the primitive seas were as great as they are in living systems, life may have originated at the *edge of the sea* or in an *environment of sediments* and phosphate minerals where the concentration of these substances was high.

When a _chiral_ compound, such as an amino acid, is formed in an achiral environment by the use of achiral physical or chemical agents (as in an atmospheric sparking experiment), equal amounts of the two _enantiomers_ must be formed. If the spontaneous _resolution_ of a _racemic_ mixture into its separate enantiomers is an uncommon event, then why is it that proteins contain only L-amino acids and nucleic acids only D-ribose? Workers in this field have tried to answer this question in two different ways. Perhaps there was some slight asymmetric influence present in the prebiotic environment that gave rise to a small excess of one enantiomer over the other. It is not hard to identify slight asymmetries even today: the sense of the earth's rotation and magnetic field, the spin of the neutrino, differential circular polarization of light between the morning (when it is cold and synthesis may be slow) and the evening (when it is warmer owing to the lag of time in the earth's cooling and synthesis may be more rapid). These and a variety of other phenomena may indeed lead to small excesses of one enantiomer over the other in derived organic materials, but in order to account for the high enantiomeric purity of most organic compounds in nature today, this small difference must have been subsequently amplified. Alternatively, it is easy to see why it may be advantageous for an organism to employ only one type of enantiomer in its nucleic acids and proteins, so perhaps what we see today is a type of "frozen accident." According to this view, at some early stage, a slight selective advantage was gained by a set of molecules that happened to contain L-amino acids and D-ribonucleotides, and this set eventually triumphed over all other sets, some of which may have contained other combinations of enantiomeric types. It can be seen that both these explanations invoke a mechanism for the amplification of optical purity by natural selection. They differ as to whether a slight excess of one enantiomer over the other is a necessary starting point.

At first glance it may seem that if an organism used D-amino acids in its proteins, it would not be in competition with those which used L-amino acids.* So why should there not still be D-amino acid–containing life today? There are two factors that affect this argument: (1) amino acids can be racemized (either enzymatically or nonenzymatically), and competition for limited amounts of preformed amino acids therefore is possible; and (2) an organism that uses (say) L-amino acids can use raw materials salvaged from the bodies of earlier organisms that also used L-amino acids. This would provide an initial advantage for a completely L-amino acid community of organisms.

Samples of _d_ and _l_ crystalline quartz show some enantiomeric selectivity in binding amino acids, but this effect is unlikely to have played much part in the generation of a sizable enantiomeric excess. In any given bed of quartz, there will be approximately equal amounts of the two forms, and on average there can be little or no residual differential effect on the two enantiomers of any given amino acid.† In any event, these experiments

---

*Bacterial cell walls and some polypeptide antibiotics do contain D-amino acids, but L-ribonucleotides are not found in nature.

†Miller and Orgel reported that in a count of 6,404 crystals from one bed of quartz, 50.05% were _l_ and 49.95% were _d_. The difference was statistically insignificant.

lose some of their relevance when it is realized that the half-life for racemization of an amino acid is quite short on the geologic time scale. For alanine at pH 7, it is about 11,000 years at 25°C. Isoleucine goes a bit more slowly, and phenylalanine somewhat faster.

Spontaneous resolution is possible for a material that undergoes equilibration between the enantiomers in solution, but that crystallizes in a pure *d* or *l* form. A good example of this type of compound is tri-*o*-thymotide (see Figure 32-19). The molecule is propeller-shaped and rapidly equilibrates in solution between *d* and *l* forms:

$$d\text{-crystals} \longleftarrow d \xrightleftharpoons{\text{Solution}} l \longrightarrow l\text{-crystals}$$

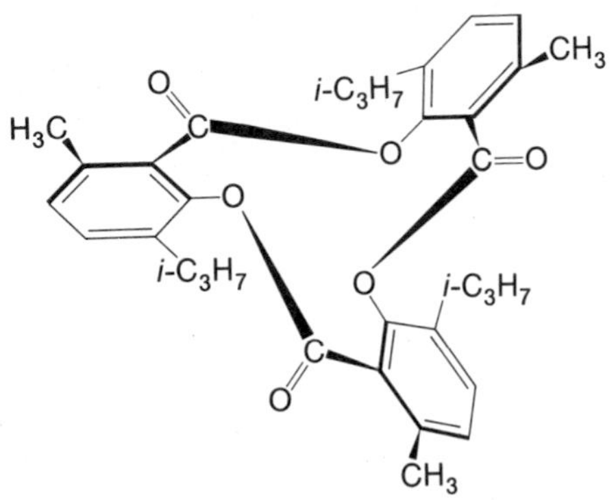

Any given crystal is chiral and consists of molecules that are either all *d* or all *l*. If a saturated solution is seeded with a crystal of one form (or the solution is allowed to evaporate slowly until the first small crystal appears), the remainder of the material can be induced by the action of the seed to crystallize in the same form. As one form crystallizes out, the equilibrium state of 50 percent of each enantiomer is rapidly reestablished in the solution. Of course, if this experiment is repeated many times (without the artificially added seed), the results will come out favoring neither enantiomer on the average. Tri-*o*-thymotide is interesting in that the crystal can trap other organic molecules, and if they are chiral, a partial resolution can be achieved in this way. This is merely a model for resolution; tri-*o*-thymotide is not a likely prebiotic molecule, but similar arguments can be applied to crystals of urea. The molecule itself is achiral, but urea forms chiral crystals and can incorporate organic molecules in the chiral crystal lattice.

**Figure 32-19**
The structure of tri-*o*-thymotide. Crystals of this substance contain either all *d* or all *l* forms, but in solution the two forms rapidly interconvert. Under the right conditions, this compound can be made to undergo spontaneous resolution.

However, these experiments miss the most significant point. If there is a clear selective advantage for an organism to use only one enantiomer in its biochemical processes, then sooner or later one enantiomer will be all that is used. Such an advantage could come from a necessary economy in the generation of specific catalysts or from avoiding the perturbation of the orderly structure of a double helix of nucleic acid or the α helix of proteins. If there already existed a strong selection for a single enantiomer of ribose in ribonucleotides and RNA, it is possible that the derived proteins would be forced to limit themselves to a single enantiomer. Perhaps the chirality of the nucleic acid and the required geometry of the translation reaction would prevent one set of enantiomers from being incorporated into the growing peptide chain.

Experiments testing some of these ideas have already been carried out. For instance, it has been shown that the coupling of D-adenosine with an activated form of D-adenylic acid occurs readily on a D-poly(U) template, but when L-adenosine is substituted for the D-enantiomer, the reaction fails. Alternatively, selection for a single enantiomer in nucleic acids could have been on the basis of resistance to hydrolysis. A right-handed helix of all D-nucleotides may hydrolyze less rapidly than one that contains a mixture of D- and L-ribonucleotides. There is some experimental evidence to support this view. An analogous selection during nonenzymatic synthesis or degradation of proteins has been studied in some detail. Synthesis of polypeptide from an amino acid that readily forms an α helix tends to give runs of several amino acids that have the same chirality. The structure of the preexisting $\alpha$ helix appears to select more molecules with the same chirality. Likewise, hydrolytic degradation of peptides is inhibited in regions of $\alpha$ helix, so that if the regular helical form is interrupted by the presence of a run of randomly arranged D- and L-amino acid residues, hydrolysis is more likely to occur in this region. The resulting peptides will be shorter, but perhaps enriched in one enantiomer. This is more a mechanism for the amplification of a small difference in the concentration of two enantiomers than for the spontaneous resolution of a racemic mixture.

We conclude that *it is not difficult to see why an organism should have made use of one enantiomer* of ribose and one set of enantiomeric amino acids. What we still do not know is whether there is some reason, other than a chance event that became frozen, why the enantiomers that we happen to use are D and L, respectively.

## POLYMERIZATION OF BIOMONOMERS

### The First Genetic Material

We have seen in earlier chapters that in contemporary life the flow of information is generally from nucleic acids to proteins. The nucleic acids represent the repository of the genetic message, and they are not expected to interact in a major way with the environment. Instead, they specify the production of different proteins, which actually do most of the work—catalytic, structural, or mechanical. However, there is not complete agreement that the earliest living forms were based on molecules that at all resembled today's nucleic acids and proteins. As mentioned earlier, it has even been suggested that something like a clay may have been the earliest genetic material, where "information" was passed from generation to generation in the form of *lattice defects* in the clay structure. An additional feature of this system is that the genetic material could be its own phenotype and thus could itself interact directly with the environment—a *direct-acting gene*. At some point in time, of course, the present system must have taken over. A genetic role for clay is an ingenious idea, but there is an understandable tendency among experimentalists to simplify the overall problem by reducing the number of possible genetic systems under consideration. Since some workers feel that a reasonable pathway for the origin of life can already be seen, paved with nucleic acids and protein, we will emphasize these classes of molecules in the following discussion. Certainly a great deal of information has been amassed about nucleic acids and proteins, and this can be drawn on for the design of new experiments.

We can now consider the joining together of monomer units to make polymeric strands. We will assume that there is a sufficient supply of condensing agents and of monomeric amino acids and nucleotides. *It is important to distinguish between template-directed polymerization and the nontemplate variety*. Note that the latter term is not necessarily synonymous with "random" polymerization, since there may be some self-ordering property, however slight, in the monomers themselves. However, this latter type of ordered synthesis is unlikely to give rise to a self-reproducing system of molecules, where a daughter is made that preserves information originally present in a parent strand and where, over several generations, that information gradually changes by the processes of *mutation* and *selection.* A distinction also can be made between "dry-state" synthesis and synthesis that occurs in aqueous solution, but since many dry-state reactions appear to proceed best under slightly moist or alternately moist and dry conditions, this distinction is not always very meaningful.

We will first discuss the polymerization of nucleotides, both non-template and template-directed, and then turn to the formation of peptides from amino acids in the absence of nucleotides. (This order of presentation is not intended to endorse either a "nucleic acid first" or a "protein first" school of chemical evolution. Indeed, a complete separation of the chemical reactions and functions of amino acids and of nucleotides in the early days of this planet seems unlikely.) Finally, the coupling of these two classes of biomolecules will be discussed in light of recent suggestions for the origin of the genetic code.

## Polynucleotides: Nontemplate Reactions

The polymerization of monomeric nucleotides requires the removal of water. This can be accomplished either by using conditions of low humidity (usually combined with elevated temperatures) or, as we have already seen, by adding a condensing agent, such as a polyphosphate or cyanamide. In some experiments, an activated nucleotide has been prepared ahead of time, and this has then been added to the polymerization mixture. A *5'-phosphorimidazolide* (Figure 32-20, Reaction 1) is a typical example of such an activated nucleotide, and these imidazolides can be formed under simulated prebiotic conditions (see Figure 32-24). Other 5' activating groups include carboxylates or polyphosphates. The activating group is merely a good leaving group and can be displaced by a ribose hydroxyl group of another nucleoside. This will result in the formation of a phospho-diester bond. If a similar attempt is made to activate a ribonucleoside-2'-phosphate or a ribonucleoside-3'-phosphate, the adjacent hydroxyl group on the ribose will displace the imidazole in a rapid intramolecular reaction and a 2',3'-cyclic phosphate will result (Figure 32-20, Reaction 2). The same product will be obtained whatever the means of activation employed. This is of some importance, for nonenzymatic hydrolytic degradation of RNA eventually yields a mixture of nucleoside-2'-monophosphate and nucleoside-3'-monophosphate (Chapter 20). It therefore seems probable that if RNA-like molecules were undergoing alternate degradation and repolymerization (in the absence of enzymes), nucleoside-2'(3')-phosphates and the related 2',3'-cyclic phosphates would have been the most abundant of all monomer nucleotides. This is not to say that the 5'-phosphate monoesters would have been totally absent—they could have been made, for example, from the 2'(3')-esters by hydrolysis and rephosphorylation—but they would probably have been present in relatively low concentration compared with the cyclic phosphates. These nucleoside-2',3'-cyclic phosphates are still "activated" (they have a large negative standard free energy of hydrolysis at pH 7) compared with a phosphate monoester and can be polymerized, but they are not as active as, for example, a 5'-phosphorimidazolide. *The situation presents an interesting choice to the experimentalist: 5' activated esters can give longer oligomers in polymerization experiments, but they may not have been present in great abundance on the primitive Earth.*

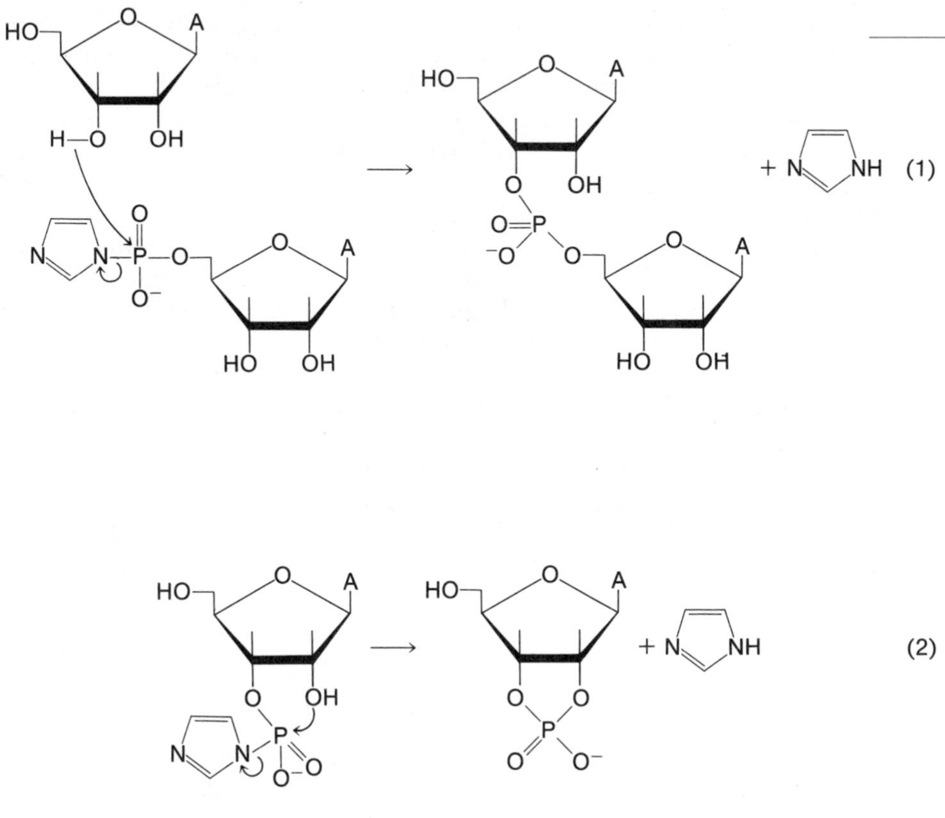

*The 2′,3′-cyclic phosphates may have been abundant, but they tend to give shorter oligomers.*

It has been found that polymerization of 2′,3′-cyclic phosphates can be brought about in many ways, e.g., by heating the triethylammonium salt at 138°C for 48 h, or by evaporating a solution of a 2′,3′-cyclic phosphate in ethylenediamine or propylenediamine buffer and holding the residue at around room temperature for a few days. If imidazole buffer is used, then mild heating of the residue is needed to obtain a reasonable yield of oligomer. This dry-state synthesis has been investigated in some detail. Adenosine-2′,3′-cyclic phosphate gives oligomers up to at least the 13-mer (0.7 percent), with about 5 percent being heptamer and higher, although polymerization appears to be inhibited if sodium salts are present. The ratio of 3′,5′ to 2′,5′ internucleotide bonds formed is about 1.8:1. The possibility of this type of reaction occurring on the primitive Earth was underscored by an experiment performed in the Anza-Borrego Desert in California in July of 1972. A solution of adenosine-2′,3′-cyclic phosphate in aqueous ethylenediamine buffer at pH 9.5 was poured (at night) over a number of different surfaces, such as rock, sand, and glass, and left out for 41 h. About 4.5 percent of material longer than the dimer was found among the products. In what may seem at first to be an unrelated reaction, oligouridylates (27 percent dimer, 7 percent trimer) can be made by holding a mixture of uridine, urea, and ammonium dihydrogen phosphate at 100°C for 11 days. However, it is likely that this reaction goes by way of the

**Figure 32-20**
The formation of an internucleotide bond from a nucleoside and a nucleoside-5′-phosphorimidazolide (Reaction 1). A 3′-phosphorimidazolide would rapidly give a 2′,3′-cyclic phosphate (Reaction 2).

intermediate formation of uridine-2′,3′-cyclic phosphate, which has been isolated from the reaction products in 45 percent yield. Cyanamide has been used as a condensing agent to give oligothymidylates (70 to 90 percent dimer and higher oligomers) from thymidine-5′-phosphate. This reaction proceeds best under drying conditions at elevated temperatures in the presence of an acid salt such as ammonium chloride. Circular oligomers that have no 3′ or 5′ ends appear to be the major products of this reaction.

The polymerization of nucleotides in aqueous solution in the absence of a template has proved to be difficult. This is understandable when we remember that in order to make an internucleotide bond, a ribose hydroxyl at a concentration that may be only a few hundredths molar must compete successfully with 55 $M$ water for an activated phosphate group. Nevertheless addition of $Pb^{2+}$ or $Zn^{2+}$ ions to an aqueous solution of a nucleoside-5′-phosphorimidazolide does allow the formation of a modest yield of small oligomers (Figure 32-21). The bonds formed are 80 to 90 percent the 2′,5′ isomer. Presumably the metal ions complex two or more nucleotides in such a way that the ribose hydroxyls of one molecule are held close to the phosphorus of another so that reaction can occur more readily. It is interesting that $Mg^{2+}$ speeds up hydrolysis of the phosphorimidazolide but does not catalyze diester bond formation.

## Polynucleotides: Template Reactions

Another way of trying to bring the potential partners of the reaction into proximity with each other is to use a *template* that consists of a complementary polynucleotide (Figure 32-22). If this could be done with precision and in such a way that the Watson-Crick rules of base pairing were always preserved (A with U, C with G), then we would have a nonenzymatic mechanism for copying the information present in a polynucleotide strand. This important goal has not yet been reached, but significant progress has been made.

The first template-directed joining reaction of nucleotides was recorded by Naylor and Gilham in 1966. They showed that *hexathymidylate* molecules [(pdT)$_6$] could be lined up on a poly(A) template, and when a water-soluble carbodiimide was added, 5 percent of the dodecamer [(pdT)$_{12}$] was produced. This reaction was not in itself considered prebiotic, largely on account of the nature of the condensing agent that was used, but it did serve to show that reactions of this type could succeed. It was soon found that whereas poly(U) was able to enhance the coupling of monomer derivatives of adenosine, poly(A) was quite without effect on the coupling of monomeric derivatives of uridine. A similar situation was

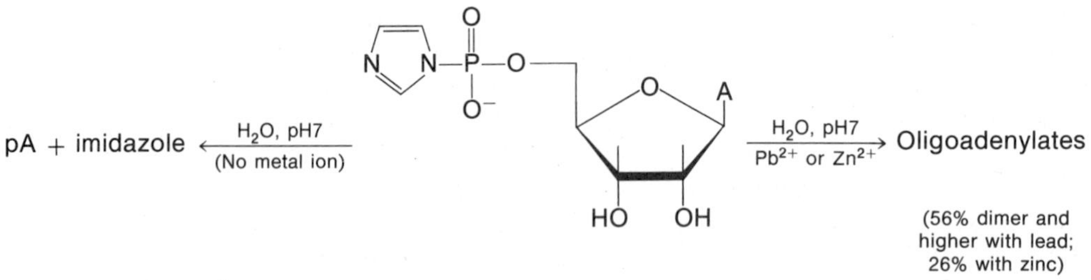

**Figure 32-21**
The effect of a metal ion on the polymerization of adenosine-5′-phosphorimidazolide.

pA + imidazole $\xleftarrow[\text{(No metal ion)}]{\text{H}_2\text{O, pH7}}$ $\xrightarrow[\text{Pb}^{2+} \text{ or Zn}^{2+}]{\text{H}_2\text{O, pH7}}$ Oligoadenylates

(56% dimer and higher with lead; 26% with zinc)

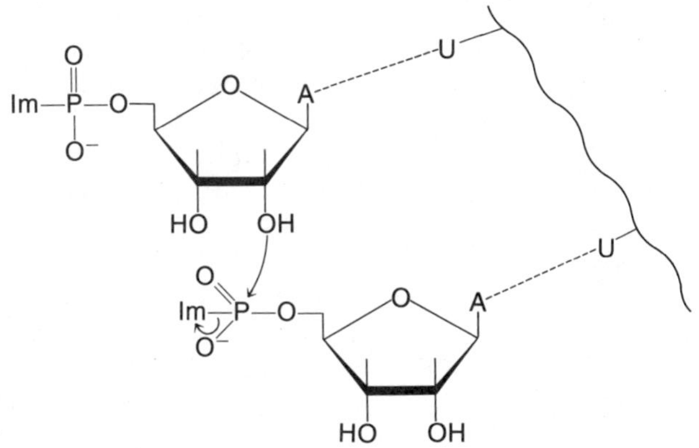

**Figure 32-22**
The use of a poly (U) strand as a template to line up adenosine-5′-phosphorimidazolide monomers prior to reaction (ImH = imidazole).

found for G and C; that is, poly(C) enhanced monomer G coupling, but poly(G) had no effect on monomer C coupling. These observations can be understood on the basis of the greater tendency of purines compared with pyrimidines to self-stack in aqueous solution. Even when they are not joined by a ribose phosphate diester backbone, the flat purine bases tend to stack together, just as they do in the double helix. By contrast, the stacking energy for pyrimidines is much less, and it has not been possible to demonstrate stable complex formation in any poly(purine):monomer(pyrimidine) system, even at 0°C. There is another complication in that the complex formed from poly(U) and a monomeric or oligomeric adenosine derivative [and sometimes from poly(C) and monomer G] is actually a *triple* rather than a double helix. The stoichiometry is always A:2U, and each triplet of bases is held together by hydrogen bonds, as shown in Figure 32-23. The poly(U) strand (denoted U1) is antiparallel to both the A strand and the other poly(U) strand. Thus the U2 strand is parallel to the A strand. When monomer adenosine units are used, it would perhaps be more accurate to refer to a "stack" of A units, but even here some polarity (5′ end to 3′ end) is evident from the direction in which the separate ribose groups all line up.

The monomer adenosine-2′,3′-cyclic phosphate or oligoadenylates that end in a 2′,3′-cyclic phosphate will join together when they are allowed to complex with poly(U) in an ethylenediamine buffer at pH 8. Coupling between two adjacent units occurs with a typical yield of about 25 percent after 5 days at 2°C. In contrast to the related nontemplate "solid-state" reaction that gave a preponderance of the 3′,5′ internucleotide bond, here the new bond formed is almost entirely the "unnatural" 2′,5′-isomer (96

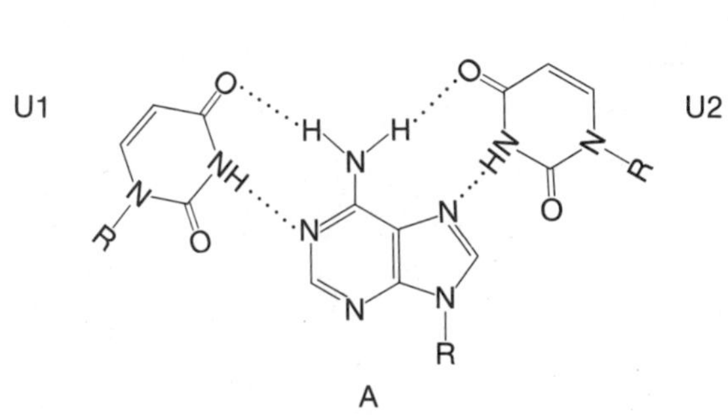

**Figure 32-23**
The geometry of the bases in the triple helix of stoichiometry A:2U.

percent). An explanation for this is given below. Although monomeric uridine derivatives will not couple on a poly(A) template, _oligomers_ of uridylate from the hexamer on up do form a complex with poly(A) [just as hexathymidylate will complex with poly(A)], and coupling will therefore occur if there is an activated phosphate group at one end of the template-bound oligomer. If this activated group is a 2′,3′-cyclic phosphate, the new internucleotide bond formed is again about 95 percent the 2′,5′-isomer.

Orgel and co-workers have examined the base specificity of a number of related reactions. In a coupling reaction that was driven by the addition of carbodiimide, poly(U) was found to enhance the coupling of pA with A, but not with U, C, or G. Similarly, poly(C) enhanced the coupling of pG with G, but not with U, C, or A. A remarkable discrimination has been found for the reaction of guanosine-5′-phosphorimidazolide (ImpG) on a poly(C) template when $Zn^{2+}$ ions are present. When the coupling of ImpG with G, A, U, or C was tested, the correct base (G) was incorporated with an error rate of only 0.5 percent. With $Pb^{2+}$ in place of $Zn^{2+}$, the error rate rose to about 10 percent. These metal ion–catalyzed reactions are significant in several other respects. First, the lead and zinc ion–catalyzed reactions of ImpG on poly(C) both yield oligomers of G up to 30 to 40 units in length, with the lead reaction giving 85 percent of material that has a mean chain length of 17 units. When the poly(C) template is still present but the metal ions are absent, the total yield of oligomer is about 50 percent, and the product has an average chain length of only about 5 units. Second, whereas the uncatalyzed and lead ion–catalyzed reactions both give a preponderance of 2′,5′-linked oligomers, the zinc ion–catalyzed reaction gives very largely 3′,5′-linked material. The exact function of the metal ions is not known. As if to emphasize our lack of understanding of these reactions, lead ions also catalyze the coupling of adenosine-5′-phosphorimidazolide (ImpA) on poly(U), but in this case they give 75 percent 3′,5′ linkages. In this system, zinc ions do not appear to have any further catalytic effect over that attributable to the poly(U) alone. _The zinc ion–catalyzed ImpG/poly(C) system is significant in its efficiency and in its being the first template-directed synthesis that gives almost entirely a 3′,5′-linked oligomer as product._

Additional interest in zinc has come from a study of the binding of nucleoside-5′-phosphates to a bentonite clay. Bernal suggested in 1951 that clays may have been important in chemical evolution because of their ability to adsorb and thus concentrate organic materials from dilute solution. Reactions may then take place in the relatively high local concentration that would exist on the face of the clay particle. It has been found that the metal ions associated with clay minerals have a marked effect on the binding of nucleotides to the clay. Adsorption of adenosine-5′-phosphate to a series of homoionic bentonites (clays that have been treated so that they contain a single type of exchangeable metal ion) increased in the series:

$$\text{Alkali metals} < Mg^{2+} < Co^{2+} < Ni^{2+} < Cu^{2+} < Zn^{2+}$$

It also was observed that adenosine-5′-phosphate bound to the zinc bentonite 10 to 20 times more strongly than did the 2′- or 3′-phosphate isomers.

These binding and polymerization results are two pieces of a plausible yet incomplete picture for the prebiotic synthesis of polynucleotides in which zinc serves both to bind and to catalyze the oligomerization of the nucleotides. *The observation that all contemporary DNA and RNA polymerases are zinc enzymes is consistent with $Zn^{2+}$ as an early catalytic agent.* In a modern enzyme, not only can the protein directly participate in the catalytic process, but also the efficiency and selectivity of the $Zn^{2+}$ as a catalyst is presumably increased as a consequence of its coordination with the protein ligands.

Most of the template reactions that have been described so far occur in a *triple helix*. This can result when an oligomer or stack of monomers complexes with a homopolymer. However, a homopolymer is not a very likely candidate for the prebiotic Earth, so attention is now turning to more realistic systems, such as double helices and copolymers. Unfortunately, the synthetic problems here are more complex, and progress therefore has been slow. Mixed dimers that are activated at the 5′ position can be prepared under mild conditions (Figure 32-24) and have been shown to couple on an appropriate alternating-sequence polymer template.

However, there are some unexpected variations in coupling efficiency that suggest subtle differences in the structures of the complexes. The dimers ImpGpU and ImpUpG both self-couple quite efficiently (20 to 25 percent of trimer and higher oligomers) on the alternating-sequence template poly(A-C), but ImpCpA couples with only moderate efficiency (5 percent trimer and higher) and ImpApC with poor efficiency (0.6 percent trimer and higher) on the alternating-sequence template poly(U-G). The ratio of 3′,5′ to 2′,5′ bonds produced is also highly variable and does not correlate in any obvious way with the overall yield. In spite of these present difficulties, this type of approach is worth pursuing, for results obtained with an alternating-sequence copolymer in a double helix should eventually prove to be more applicable to a "real" random-sequence polymer than results that come from a homopolymer in a triple helix. These experiments also allow pyrimidines to be carried into synthesis by the attached purine, whereas they would never have made it alone. There still remains the question of the origin of the template strand, but this does not present much difficulty. It could have been formed in one of the nontemplate reactions discussed earlier, and there is even a demonstrated mechanism for concentrating longer oligonucleotides from a mixture of chain lengths by selective adsorption onto hydroxyapatite. The longer molecules stick more effectively and therefore better resist being washed away. A similar process is often used for the separation of a mixture of oligonucleotides by column chromatography.

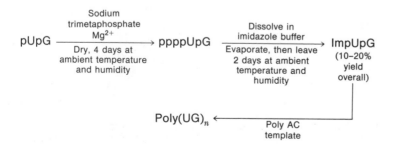

**Figure 32-24**
Formation of the 5′-imidazolide of a dinucleotide under simulated prebiotic conditions and its subsequent coupling on a complementary template.

## Early Polynucleotides and the 2′,5′-Linkage

We have seen that in the products of many template-directed reactions there is a preponderance of the "unnatural" 2′,5′-internucleotide linkage. This is particularly true for template-directed polymerization of 2′,3′-cyclic phosphates, and in this case it is clearly the geometry of the reaction that is responsible for the result. When a 5′-hydroxyl group attacks a 2′,3′-cyclic phosphate of an adjacent nucleotide unit, the geometry of the helix ensures that the incoming 5′-oxygen is in a direct line with the phosphorus atom and the 3′-oxygen of the cyclic phosphate. Under mildly basic conditions, only the 3′-oxygen can be displaced (an "in-line" attack), and the result is a 2′,5′-linked product (Figure 32-25).

However, the reverse of this process also can occur, i.e., attack on phosphorus by the 3′-hydroxyl to displace the 5′-oxygen, and this happens much more rapidly when the 2′,5′-linked nucleotides are bound in a helix. In contrast, in a helical 3′,5′-linked oligomer, the internucleotide bonds are strongly stabilized against cleavage. These experimental observations suggest how the daily cycles in temperature and humidity could have given rise to the gradual formation of complementary sequences of largely 3′,5′-linked oligonucleotides. We start with a mixture of nucleoside-2′,3′-cyclic phosphates. During the day, this undergoes (nontemplate) solid-state polymerization to give random copolymers that contain 3′,5′- and 2′,5′-internucleotide bonds in the ratio of about 2:1. This mixture dissolves in evening dew or rain, and at night the temperature may drop sufficiently so that short helices can form between oligonucleotides that happen to have complementary sequences. The next day the temperature increases slowly, and helical 2′,5′ bonds will degrade most readily, nonhelical 2′,5′ and 3′,5′ bonds less so, while the helical 3′,5′ bonds will be most stable. As the temperature rises further, most of the short helices will melt into separate strands, and eventually the solution will dry out. At this stage, the oligomers that are left have a higher percentage of 3′,5′ bonds, but are shorter in length than previously. Another round of dry-state synthesis can now take place, followed that night by another pairing of complementary sequences. Next day

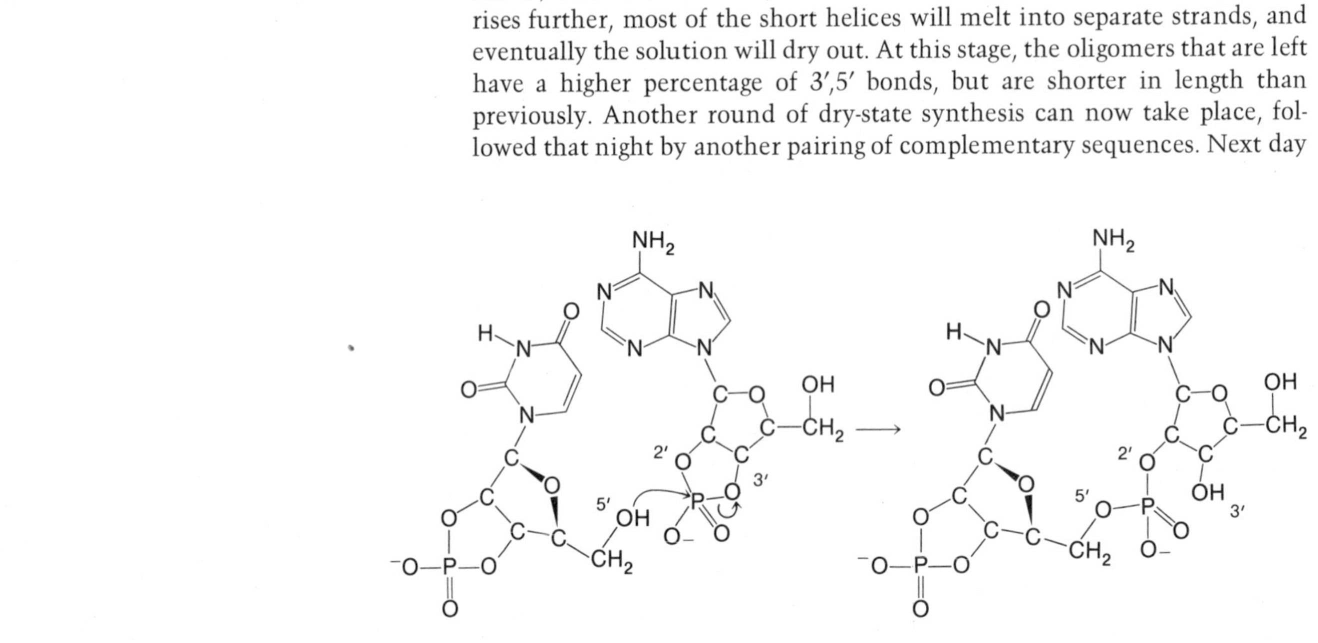

**Figure 32-25**
The geometry of attack on a 2′,3′-cyclic phosphate by the 5′-hydroxyl group of an adjacent nucleotide unit. Both nucleotides are bound in a helix, and the in-line attack results in the formation of the 2′-5′ internucleotide bond.

there will be another round of degradation and further synthesis. It must be emphasized that some aspects of this cyclical scheme have not yet been demonstrated experimentally, but in theory it could provide longish double-stranded oligomers that are mostly 3′,5′-linked.

We have seen that in the template-directed coupling of 2′,3′-cyclic phosphates, the formation of the 2′,5′ bond is inevitable, but that it is not necessarily a problem. We also have seen that this geometric restriction does not apply to template-directed reactions of activated nucleoside-5′-phosphates, and that under some circumstances the 3′,5′ bond can be formed directly. It is interesting to remember that contemporary RNA polymerases all give 3′,5′-linked products and all require 5′-activated nucleotides as substrates.

## RNA Before DNA

*Circumstantial evidence suggests that RNA-like polymers probably preceded DNA.* First, ribose, but not 2′-deoxyribose, has been made in prebiotic simulation experiments. Second, the polymerization of ribonucleotides is easier than that of deoxyribonucleotides. Third, in modern biochemistry, deoxyribonucleotides are made from ribonucleotides.‡ Fourth, nonenzymatic hydrolytic degradation of unprotected RNA to monomer units occurs relatively easily. The monomer units can then be easily reassembled into a new and potentially more useful (and stable) sequence. Chemical degradation of DNA into monomer nucleotide units is difficult (although cleavage of the purine and pyrimidine bases is readily achieved). Last, there is a reasonable evolutionary sequence of events that starts with the use of ribonucleoside-2′,3′-cyclic phosphates (which give much 2′,5′-linked product); use of activated ribonucleoside-5′-phosphates (which can more easily give a high yield of the potentially more stable 3′,5′-linked oligomers). Finally activated 2′-deoxyribonucleoside-5′-phosphates were used [which give as product a polymer (DNA) that is still more stable toward hydrolysis than is double-helical 3′,5′-linked RNA] (Figure 32-26). Any crude polymerases that aided the second process would have required only minor changes in structure to be able to aid the third. By this time there may well have evolved crude DNAses that could degrade the DNA for possible recycling of the monomer units.

Although many sugars can be formed under the conditions that give rise to ribose, there is experimental evidence to suggest that the incorporation of β-ribonucleosides into a polynucleotide strand is a favored process. It was mentioned earlier that ImpA on a poly(U) template will react with adenosine, but it will not couple nearly as well with 2′-deoxyadenosine, α-adenosine, 3′-deoxyadenosine, or arabinosyl adenine. Some workers have pointed out that ribose (and deoxyribose) are not indefinitely stable in the presence of basic amines. This is true, but conversion to the nucleosides would have eliminated the immediate cause of this instability by masking the aldehyde carbonyl group.

---

‡The observation that ribonucleotides are precursors of deoxyribonucleotides in contemporary biosynthesis could be taken to support the opposite argument: i.e., that DNA preceded RNA. This would follow from a strict application of Horowitz's argument that biochemical pathways have evolved "backwards."

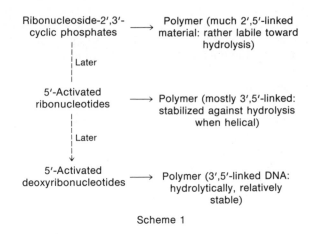

Figure 32-26
A hypothetical scheme for the evolution of the use of different types of monomers in polynucleotide formation.

## Polypeptides

Most suggestions for nontemplate polypeptide synthesis envision the polymerization of amino acids, either through the application of heat under drying conditions or by the addition of condensing agents. Fox and co-workers have synthesized thermal copolymers of amino acids from mixtures of the monomers. In the early work, rather high temperatures were used (150°C to 180°C), but in some cases temperatures below 100°C were found to be sufficient if the reactions were allowed to proceed for extended periods of time. These "thermal proteinoids" appear to contain linkages that are not found in proteins, as well as the normal peptide bond, and since several polyfunctional amino acids were present in the original mixture of monomers, this is probably not very surprising. The use of high temperatures in the early experiments was criticized on the grounds that such temperatures probably would not have existed over extended regions of the earth's surface, and in addition, prolonged exposure to temperatures in this range would tend to cause decomposition of any peptides that may have been produced. The mixtures of pure amino acids that were employed are not very realistic, but this is a type of criticism that can be leveled against many prebiotic simulation experiments. This point is taken up in more detail later. A high percentage of lysine has been used in some recent work, but at present there is no good "prebiotic" synthesis of this amino acid.

The presence of clay can aid peptide bond formation in these dry-state reactions, and workers at NASA's Ames Research Center have found that when cycles of temperature and humidity are used, the yield is further enhanced. *Homoionic bentonite clays* were investigated, and a $Cu^{2+}$-substituted clay was found to be somewhat better than $Ni^{2+}$ or $Zn^{2+}$ at promoting the synthesis of oligoglycine and oligoalanine from the monomeric amino acids. The maximum yields and chain lengths were not very high, however, reaching 6 percent of total oligoglycines with a maximum chain length of 5 units, and 2 percent of alanylalanine. The highest temperature employed during each cycle was 94°C. When a polyribonucleotide or a small amount of histidylhistidine was added to a similar kaolinite clay system, the yield of oligoglycine was found to increase about two to four times. In contrast to these successful dry-state reactions, it has been shown, at least for the formation of lysyllysine from the monomer, that clays alone cannot help at all if one attempts to carry out peptide bond formation in aqueous solution at temperatures up to 90°C. This result further supports two independent groups of workers who tried and failed to repeat a reported polymerization of aspartic acid on kaolinite in aqueous solution at 90°C. Not only had it been claimed that this reaction proceeded in high yield, but also (and even more surprisingly) that the L-enantiomer was incorporated into polymer more rapidly than the D.

Various condensing agents have been used to make peptides from amino acids. As one example, cyanamide together with ATP and aminoimidazolecarboxamide in a dry-state reaction at 90°C for 24 h gave 5 percent $Gly_n$ (up to the tetramer), 17 percent $Ile_n$ (up to the dimer), or 66 percent $Phe_n$ (up to the tetramer). Somewhat better results came from the use of polyphosphates as a condensing agent. In a solid-state reaction that required cyclic trimetaphosphate, $Mg^{2+}$, and imidazole, glycine gave 25 percent of total oligomers, with 1.1 percent higher than the heptamer, after 10 days at 65°C. After only 4 h at 100°C, oligomers up to the decamer were produced. By contrast, an earlier reaction between glycine and trimetaphosphate that was run in aqueous solution at mildly alkaline pH gave only low yields of the dimer and trimer (see Figure 32-18 for the mechanisms of these reactions). A reaction of a very different type occurs when the hexadecyl ester of glycine spreads in a monolayer on a water–air interface. Hydrophobic interactions between the hydrocarbon tails cause the ester molecules to come into close enough proximity that a fairly efficient synthesis of oligoglycine can occur. This reaction has some resemblance to the template reactions discussed in ensuing paragraphs.

In the peptide syntheses that we have considered so far, there is no obvious copying or translation of information. Except that there may be some slight self-ordering property in the growth of a polypeptide that affects the sequence of amino acids, the products of these reactions will tend to be random copolymers. There have been suggestions that a polypeptide could act as a template for the production of a complementary polypeptide, and thus by a repetition of this process, a more or less faithful copy of the original could be produced. Such schemes are usually based on interactions between amino acid side chains. Thus a positively charged side chain (e.g., lysine) may select a carboxylate anion (aspartate or glutamate) in the side chain of its "complement." A hydrophobic group or neutral group may select another hydrophobic or neutral group directly. It could be argued that a highly accurate copying mechanism is unnecessary: the function of the resulting polypeptide will perhaps not be much changed if aspartate is substituted for glutamate, or valine for leucine. Some parts of these suggestions seem plausible, and the idea as a whole would have obvious importance to the "protein first" school of chemical evolution, but there is as yet little experimental evidence that really suggests that such a scheme should work. The enzyme-catalyzed nonribosomal synthesis of some polypeptide antibiotics that has been described (see Chapter 23) cannot be used to champion this cause. The product polypeptides are very much simpler molecules than the proteins that are used to form them. For a synthesis of this type to be accepted as evidence in support of protein self-replication, it would be necessary to demonstrate either that assembly of the polypeptide antibiotics could be carried out by equivalently simple polypeptides or that the polypeptide products could somehow specify the formation of proteins as complex as those which made them. Both possibilities seem highly unlikely.

Such is the abundance of amino acids that are formed in spark-discharge and related experiments that some production of polypeptides on the primitive Earth seems unavoidable, granted present assumptions concerning the primitive atmosphere. If an orderly evolution of these polypeptides seems out of the question, then what part did they play in the origin of life, and what is the relevance of these polymerization reactions? We are now 4 billion years removed in time and cannot really know, but perhaps some of them acted as weak nonspecific catalysts that helped to establish the operations of replication and translation. These operations soon would have evolved the ability to make their own polypeptides that were better catalysts for the same reactions, and the original makeshift peptides would no longer have been of any use, except perhaps as a reservoir of amino acids.

## Coevolution of Polynucleotides and Polypeptides: The Origin of Translation

*Proteins can be marvelous catalysts, but they cannot self-reproduce; nucleic acids are well-suited to self-replication, but at least in their present form, there is not much else that they can do.*\* For instance, it does not seem likely that they would be able to catalyze the formation of their own building blocks from simpler precursors if the reservoir of endogenous nucleotides became depleted. *If the ability of a polynucleotide to replicate could be coupled in the right way with the ability of a polypeptide to act as a catalyst, both parties to this cooperation would benefit.* The polynucleotide could specify the formation of the polypeptide by a process of translation, and the polypeptide in turn could catalyze replication of the polynucleotide and possibly also the translation process. According to a theoretical analysis by Eigen and Schuster, for a system to be capable of evolving to still greater complexity, this particular set of mutually helpful interactions is not enough. However, the process of translation, as envisioned here, *is* an essential part of the more complex schemes that can evolve. *The importance of a working translation system is so great that for an increasing number of people, the origin of translation has become almost synonymous with the origin of life.*

Not all the experimental work that links amino acids with nucleotides has been concerned with translation. Aminoacyl adenylates (Figure 32-27) are part amino acid and part nucleotide and are highly reactive by virtue of the mixed carboxylic-phosphoric anhydride linkage. They also are intermediates in contemporary protein biosynthesis (Chapter 25) and are therefore of intrinsic interest. When such an aminoacyl adenylate is dissolved in aqueous buffer, it hydrolyzes rapidly to give the amino acid and pA, together with about 30 percent of di- and tripeptide. However, in the presence of the clay mineral montmorillonite, or when absorbed into *micelles* or reversed micelles, these aminoacyl adenylates can give polypeptides up to 40 to 50 amino acid units long. The total yield of polypeptide in some cases is as high as 95 percent. Although this is an impressive result, it has not yet been shown to be directly relevant to the process of translation. It is true that the adenylate group is being made use of as a handle, with which the

---

\*Recently T. Cech has shown that certain rRNA precursors in Tetrahymena are capable of self-splicing. This observation may require us to revise our views on the ability of polynucleotides to act as catalysts.

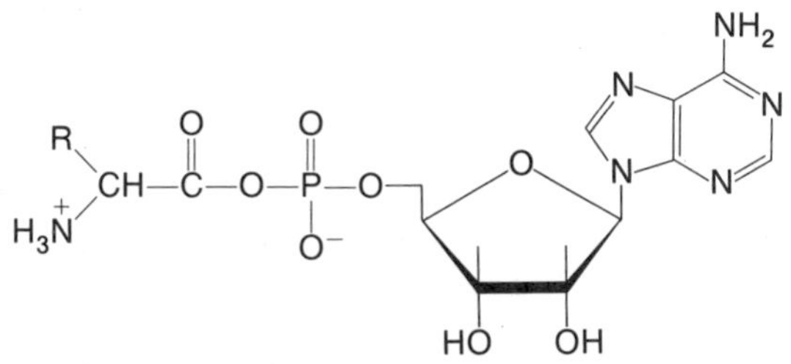

**Figure 32-27**
An aminoacyl adenylate.

amino acids can be made to line up so that they can react, but it is not really being made use of as a specific handle. Indeed, in one related case where an attempt was made to utilize the specific base-pairing capability of the adenylate group, the results were disappointing: phenylalanyl adenylate gave no more peptide in the presence of poly(U) than in its absence.

**Origin of the Genetic Code.** Many theories for the origin of translation and the genetic code have been proposed. Some are more philosophical and some more chemical, but by and large they can be put in one or the other of two categories according to their author's view of how particular codons came to be assigned to particular amino acids. There is the *"frozen-accident"* school, and there is the *"specific-interaction"* school. According to the former, the assignment is purely a random event, but once this event occurred, the system that used it and was able to carry out translation would have had a large selective advantage over any nontranslating system; the assignment would have become fixed. As an example of how this could occur, suppose the sequence of nucleotides in a small RNA molecule caused it to fold up in such a way that a pocket was generated that exactly matched a particular amino acid, say, alanine, which therefore bound to this site. In a different place on that RNA molecule there may have been an exposed triplet of nucleotides, say, AGC, that allowed the RNA to bind to a message strand at the "codon" GCU and thus incorporate alanine at the corresponding position on a growing peptide chain. One codon for alanine could thus have become fixed as GCU, even though there was no special interaction between alanine and its codon or anticodon. This frozen-accident theory was dominant until quite recently, but lost some ground when it was demonstrated that *there is a small but significant correlation between the polarity and hydrophobicity of an amino acid and the same properties of its anticodon nucleotides.*

The effect is most clearly seen with the homocodonic amino acids—those amino acids which have a codon that is a set of three identical bases, e.g., CCC or UUU. The comparison of properties between amino acids and their anticodons has been explored in several ways. In one set of experiments, Weber and Lacey measured the $R_F$ values of amino acids and dinucleotides on paper chromatography in a solvent that contained a high salt concentration (Figure 32-28). The salt was present to mask the extreme polarity of the charged phosphate residue in each dinucleotide. A relationship was found between the $R_F$ values for a particular amino acid, and the dinucleotide that was *complementary* to the first two bases of its codon (see Chapter 24 for a discussion of the wobble base in the third position). Thus one of the most hydrophobic amino acids is phenylalanine, and the most hydrophobic dinucleotide is ApA; both the codons for Phe start with UU. The most polar amino acid is lysine, and the most polar dinucleotide is UpU; both the codons for Lys start with AA. There are some anomalies

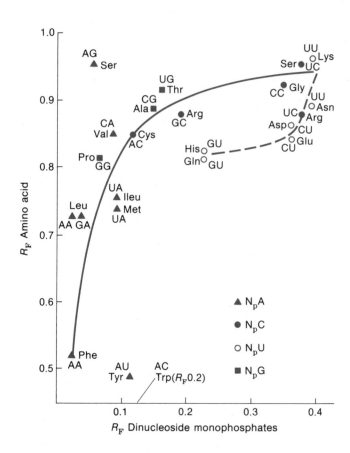

**Figure 32-28**

$R_F$ values (the solvent was 10 vol % saturated ammonium sulfate, pH 7.0) of amino acids plotted versus the $R_F$ values of the dinucleoside monophosphates representing the first two letters (3′ → 5′ direction) of their anticodons. (From A. L. Weber and J. C. Lacey, Jr., Genetic code correlations: Amino acids and their anticodon nucleotides, *J. Mol. Evol.* 11:199, 1978. Reprinted with permission.)

in the results, but plausible explanations can be seen for most of them. There have been many previous attempts to demonstrate specific interactions or relationships between amino acids and their codons or anticodons, but it is only with these more recent results that the weight of the evidence seems to be turning in favor of the specific-interaction theory.

*It is one thing to show that an amino acid side chain and its modern anticodon have some properties in common, but it is quite another to incorporate this information into a working translation model of protein synthesis.* The ideal sort of reaction in which to examine the effects of specific interactions between nucleotides and amino acids would be an efficient oligonucleotide-directed peptide synthesis reaction that conceivably could have occurred around the time of the origin of life. Unfortunately, we have not yet discovered any such reaction.

In 1972, Nakashima and Fox claimed to have identified conditions under which a codonic interaction affected the outcome of a peptide formation reaction. They prepared thermal copolymers of amino acids that contained a preponderance of lysine, added one of the four homopolymers [poly (A), (C), (U), or (G)], and tested the ability of the resultant complex to catalyze the formation of peptide from the aminoacyl adenylates of glycine, proline, lysine, or phenylalanine (the four "homocodonic" amino acids). The products of this reaction were not actually shown to be peptides, but this is a reasonable assumption. Under some conditions, the results appeared to favor codonic interactions [e.g., the thermal polypeptide complex that contained poly(A) seemed to enhance the formation of peptides from Lys-pA a little more than did the complex that contained poly(U), poly(C), or poly(G)], but under other conditions, there was either no obvious preference or else one that seemed to be anticodonic. Some of this variability appears to have come from the different conditions, especially of concentration, employed in the different experiments, but even in repeat runs, the results were quite variable. In the series of experiments in which a general codonic preference was claimed, the individual aminoacyl adenylates were not used at the same concentrations. So this type of experiment has not been very convincing and anyway probably should be carried out with an amino acid derivative that is less reactive than an aminoacyl adenylate. In general, the less reactive a compound, the greater selectivity it tends to show in reaction.

**Translation.** *There are two basic requirements for a working translation system: (1) the selection of a particular amino acid, and (2) the efficient formation of the peptide bond.* A possible mechanism for the first of these has been suggested by Hopfield. Fortunately, it is testable by experiment. Consideration of the nucleotide sequences of different tRNA molecules from *E. coli* led him to the tentative conclusion that earlier versions of these molecules may have been able to fold up in a different way. The important feature of this alternative conformation is that the nucleotides of the anticodon would sit quite close to the amino acid that is esterified at the 3′ terminus (Figure 32-29a). In the usual structure of a modern tRNA, the anticodon is as much as 70 Å away from the amino acid (Figure 32-29b). We have seen that a hydrophobic amino acid has a hydrophobic anticodon, and thus it appears possible that a mutual interaction of the esterified amino acid with its anticodon ("like attracts like") could, for example, stabilize the amino acid ester bond against hydrolysis. This would result in a small preference for a hydrophobic amino acid to remain in the site adjacent to a hydrophobic anticodon. There are ways in which this effect could be amplified, and some additional specificity could come from more subtle interactions between the shape or polarity of the amino acid side chain and the anticodon nucleotides. It is already known that the hydrolysis rate of an amino acid ester can be affected by the presence of a polynucleotide, e.g., the hydrolysis rate of 2′(3′)-glycyl adenosine is de-

**Figure 32-29**
(*a*) A possible alternative conformation for an early tRNA. The normal conformation is shown in (*b*). (From J. J. Hopfield, Origin of the genetic code: A testable hypothesis based on tRNA structure, sequence, and kinetic proofreading, *Proc. Natl. Acad. Sci. USA* 75:4334, 1979. Reprinted with permission.)

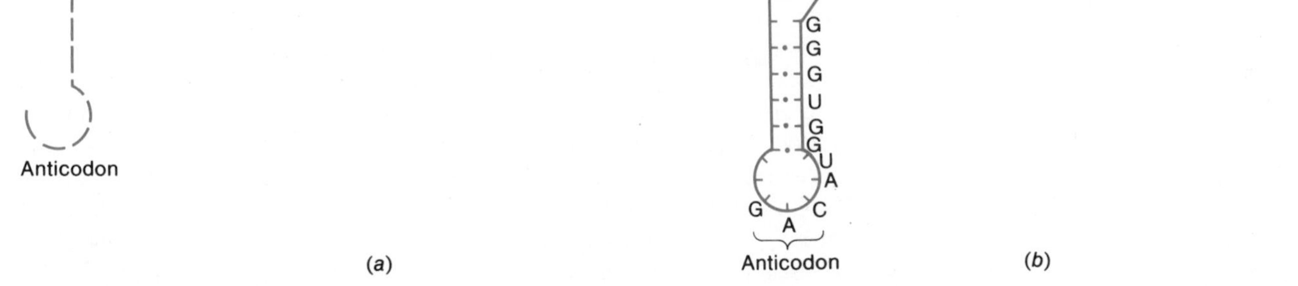

(*a*)

(*b*)

creased by a factor of four when the nucleoside ester is bound in a triple helix with 2 poly(U). However, UUU is not an anticodon for glycine, and in any case, the most likely explanation for the effect is rather mundane. Electrostatic repulsion of hydroxide ions by the phosphate anions on the poly(U) strands also will decrease the concentration of hydroxide in the vicinity of the amino acid ester bonds. This is unlikely to be an effect that is specific for a particular amino acid or a particular polynucleotide.

**Peptide Bond Formation.** Turning now to the peptide-bond-formation step, we find that although many theories have been published that are concerned with the origin of the genetic code, remarkably few of them indicate precisely how this key reaction is supposed to take place. It is therefore hard to put them to an experimental test. What should we look for in such a theory? Actually the requirements can be spelled out simply enough. We should remember that whatever mechanism we choose, it should be possible for it to have evolved by way of a series of reasonable steps to the present system of protein biosynthesis. For example, it has been pointed out by Crick that a primordial doublet code is unacceptable, since all the information present in a message would be rendered useless when the transition to a triplet code was made. Ways around this type of objection can be found, but they tend to be cumbersome.

As far as peptide bond formation itself is concerned, we need to consider both the thermodynamic and kinetic factors. Is there enough driving force to make it go, and if so, will it go fast enough to be useful? An amino acid ester is sufficiently activated for our purposes. The equilibrium constant for the formation of a dipeptidyl ester from two template-bound aminoacyl esters is probably about 3,000 at 25°C and pH 7. This is perfectly adequate to drive a peptide-synthesis reaction. Aminoacyl adenylates could be even better, but one often pays for higher reactivity in that the material is hydrolyzed more rapidly when it is dissolved in water.

In modern protein biosynthesis, peptide bond formation occurs when the amino group of an aminoacyl-tRNA attacks the ester group of a peptidyl-tRNA. In this reaction, one molecule of GTP is hydrolyzed (another molecule of GTP is hydrolyzed during translocation; see Chapter 25), but as we have just seen, this cannot be considered necessary actually to drive the peptide-bond-formation step itself. Indeed it has been reported that the translation process does work in the absence of GTP, but very slowly. So esters have enough driving force, but can they be made to react at a reasonable rate in a prebiotic reaction without the aid of modern enzymes? All we can say for certain is that 2'(3')-glycyl adenosine does show a modest increase in the rate of peptide bond formation (Figure 32-30) when it is complexed to poly(U). The templating effect of the poly(U) is more evident at low concentrations of ester, where the ordinary bimolecular reaction of one amino acid ester with another is very slow. The template should serve to concentrate the adenosine units out of dilute solution, as well as to align them for reaction. Perhaps this reaction is not very efficient because the reactants are too "floppy"; i.e., they are not held rigidly enough in the right position for reaction to occur. The study of intramolecular reactions as models of enzyme action has shown that what may appear to be a small increase in the rigidity of a system can give a large increase in rate.

Consideration of the stereochemical requirements of the attack of an amino group on an ester bond leads to the conclusion that for reaction to be possible, the distance apart of the two esterified oxygens (Figure 32-31)

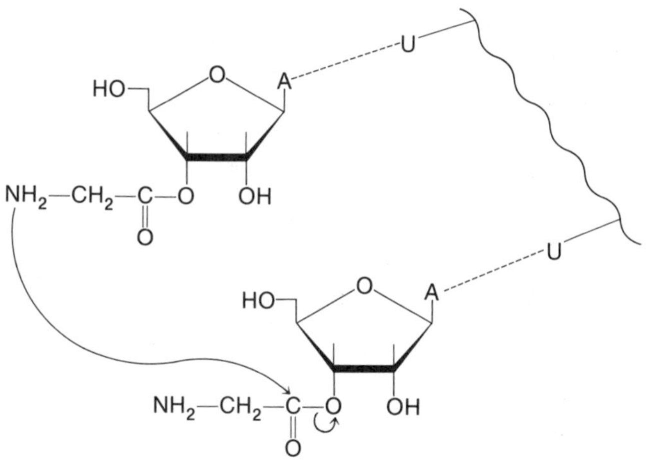

**Figure 32-30**
The formation of glycylglycine from two adjacent molecules of 2′ (3′)-glycyl adenosine that are complexed with poly(U). The complex is a triple helix, but in the diagram the second poly(U) strand has been omitted.

should not be greater than about 6 Å. For example, this rules out efficient peptide bond formation between amino acids that are esterified to 2′-hydroxyls on opposite strands of a double-helical RNA in the A form. The closest approach of any two 2′-OH groups in this structure is about 10 Å. An interesting possibility, however, is reaction between two amino acids, one of which is esterified to a 3′ terminus of an oligonucleotide and the other to the 5′ terminus of an adjacent oligomer. The ester oxygens in this case are only 2.5 Å apart.

The reaction of 2′(3′)-glycyl adenosine mentioned earlier actually gives more *diketopiperazine* (DKP) than diglycine. A diketopiperazine is a cyclic dimer that forms when the free amino group of a dipeptidyl ester attacks its own ester bond in an intramolecular reaction (Figure 32-32). This can happen so rapidly that the dipeptidyl ester may not survive long enough to be converted to tripeptide. If the dipeptidyl ester stage is safely passed, the N terminus will then tend to be sufficiently distant from the C-terminal ester that this type of cyclization reaction becomes unimportant. It has been pointed out that this unwanted reaction to make DKP is blocked if the first amino acid to enter peptide synthesis (i.e., the N-terminal amino acid) carries a group such as acetate or formate. It is possible that the use of *N*-formylmethionine to initiate protein synthesis in prokaryotes is a remnant from a time when there had to be a relatively simple chemical method for preventing DKP formation.

**Selection of α-Amino Acids in a Template Reaction.** The process of bringing amino acids into reaction by means of oligonucleotide carriers that bind to a template has other advantages in a prebiotic setting than just the formation of peptides by translation. The formation of α-amino acids in prebiotic simulation experiments is accompanied by the production of a variety of other compounds, such as β-amino acids, simple carboxylic acids, and amines. Why should these compounds not interfere with the formation of a peptide chain? After all, the primitive Earth environment was unlikely to have provided pools that contained pure amino acids and pure nucleotides just waiting to get acquainted. Fortunately, we can invoke the special characteristics of a template synthesis to provide a partial answer to this problem. First, a simple amine that cannot become esterified to an oligonucleotide could attack the peptidyl ester bond of a growing chain and so interrupt the synthesis, but it would have to do so in an ordinary

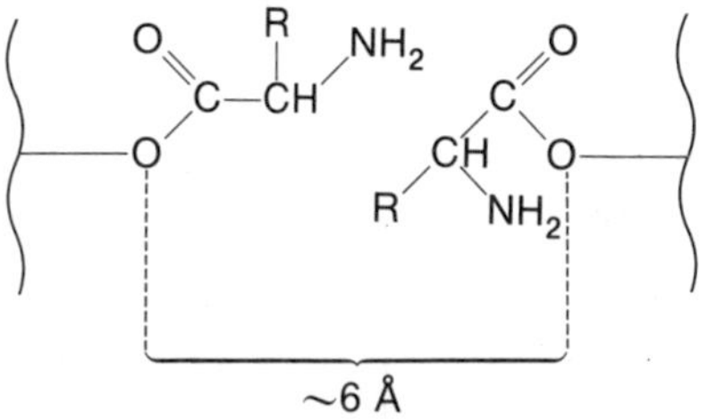

**Figure 32-31**
In order for two amino acid esters to be able to react and form a peptide bond, the ester oxygens should be no farther than about 6 Å apart.

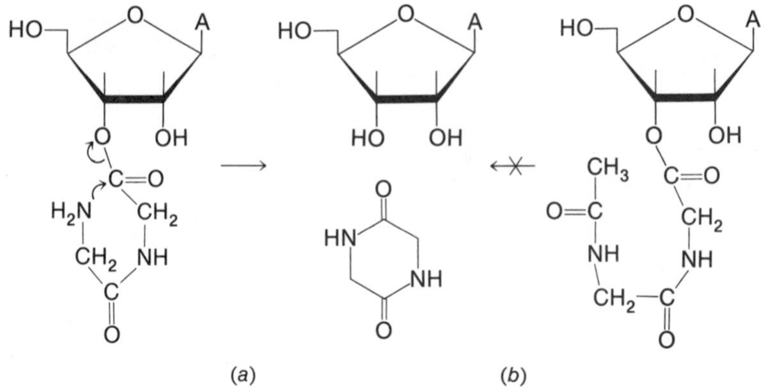

**Figure 32-32**
(a) The formation of diketopiperazine
by an intramolecular cyclization. (b)
The reaction is prevented by the presence of the *N*-acetyl group.

bimolecular reaction with no advantage from the templating effect. The
effective concentration of the amino group of an amino acid in an efficient
template-directed reaction could be several hundred molar, and this would
easily swamp any unwanted interference from a simple amine at a concentration of a few hundredths molar. A simple carboxylic acid could become
esterified to an oligonucleotide carrier, but it would have no amino group,
and so as long as the direction of translation is from N terminus to C
terminus, it could not participate or interfere with the formation of peptide, except at the N-terminal end. As we have just seen, an N-terminal
formate or acetate could be a distinct advantage. Discrimination against β-
or γ-amino acids could occur because of the stereochemical requirements of
the template-directed reaction itself, but until this reaction has been identified, we cannot test this point. Indeed, *it is hard to avoid the feeling that
when an efficient template-directed peptide-bond-formation reaction is
discovered, it will reveal a great deal more about related phenomena (such
as the origin of the genetic code) than we had previously expected to find
out from this source alone*.

## MEMBRANES AND COMPARTMENTS

If we could go back in time about 4 billion years and look around for some
evidence of life, we would probably start by trying to find objects that
resembled cells. We would be unlikely to award the title of "living" to a
puddle or small pond, even though it may at that moment possess a system
of self-reproducing molecules and be the site of synthesis of any number of
interesting compounds. This is only partly semantic prejudice, for such a
puddle or small pond has some features that make it ill-adapted to any
further development. A sudden downpour of rain could easily disperse the
contents of a special puddle to an extent that the component molecules
would no longer be able to interact. A pond is not a unit that can compete
for limited resources, nor can it evolve by the processes of replication,
mutation, and selection. The formation in such a place of a crude replicase
enzyme (by translation of the sequence of bases in an oligonucleotide)
would result in an increased rate of replication of *all* the oligonucleotides
in that pond, unless the replicase were able to recognize and therefore
duplicate only its own gene. This latter feat would be asking quite a lot of
a primitive enzyme. Similarly, the formation of a superior "enzyme" for
catalyzing translation could result in an enhanced translation of all the
oligonucleotides in the pool and the accumulation of useless polypeptides.

If a self-reproducing system were enclosed instead in a cell-like structure, some of these problems would disappear. Dispersion of the contents would be unlikely to result from a shower of rain, and although recognition of its own gene by a replicase would still be advantageous, it would no longer be absolutely essential—particularly if the cell were able to grow and divide, so that there would be a finite chance of one of the progeny *not* containing any nonsense RNA that was being unnecessarily copied and translated. *A cell is a compartment, and even an early cell must therefore have had a membrane that separated it from everything "outside."* § What material could have served as a membrane to encapsulate the first "living" set of molecules? Contenders include some form of lipid, the thermal polypeptides of Fox, or coacervate droplets. Analogy with the modern cell would favor the lipid vesicle as a container for this purpose, but each of these materials can be induced to show some morphologic similarity to a cell. For example, when thermally produced polymers of amino acids are boiled in water and then cooled, microspheres of about 2 $\mu$m diameter are formed in abundance. However, it does not seem very likely that these could have served as the first cells, for they are too leaky: small molecules can pass in or out far too readily. Coacervate droplets were investigated by Oparin, but they were usually formed from materials such as gum arabic and gelatin, and these would have required the prior existence of living organisms for their formation. However, it is possible that similar structures could be formed from abiotic polymers.

## Lipid Vesicles as the First Cells?

*At present it seems most likely that the first membrane was made largely of lipid.* Lipids can form bilayer vesicles relatively easily, and these vesicles not only are self-repairing, but they can trap small polar molecules. Recent experiments by Deamer and Hargreaves have given considerable support to this idea. Mixtures of glycerol, fatty acids, and ortho-phosphate (sometimes with silica or clay) were subjected to cycles of temperature and humidity in a similar way to that described for the formation of oligonucleotides and oligopeptides. Among the products were mono- and diglycerides and a small amount of phosphatidic acid. Under the same conditions, these lipid products appeared to form vesicles, and when the external water was allowed to evaporate, the vesicles opened and were able to exchange material with the outside world. When the system was rehydrated, the vesicles again formed their separate units. If the pH of the outside solution became different from what it was when the vesicles had reclosed, the resulting pH gradient could be used to drive the uptake of various compounds through the membrane and into the vesicle. For example, if the interior were at a lower pH than the exterior, a weak-base amine in its neutral unprotonated form could cross the membrane, become protonated, and thus be trapped inside as the lipid-insoluble cation. Other schemes for facilitating the transport of ionic molecules have been suggested by Stillwell. Amino acids could be rendered lipid-soluble by formation of a Schiff base with a long-chain aldehyde; nucleotides could be introduced as complexes with metal ions, and a sugar as a Schiff base with a long-chain amine (Figure 32-33). Once inside, the amino acids could be liberated by hydrolysis of the Schiff base

---

§An alternative way of preventing the dispersal of a set of interacting molecules could be to have them adsorbed onto the outside of a suitable object, such as a mineral grain.

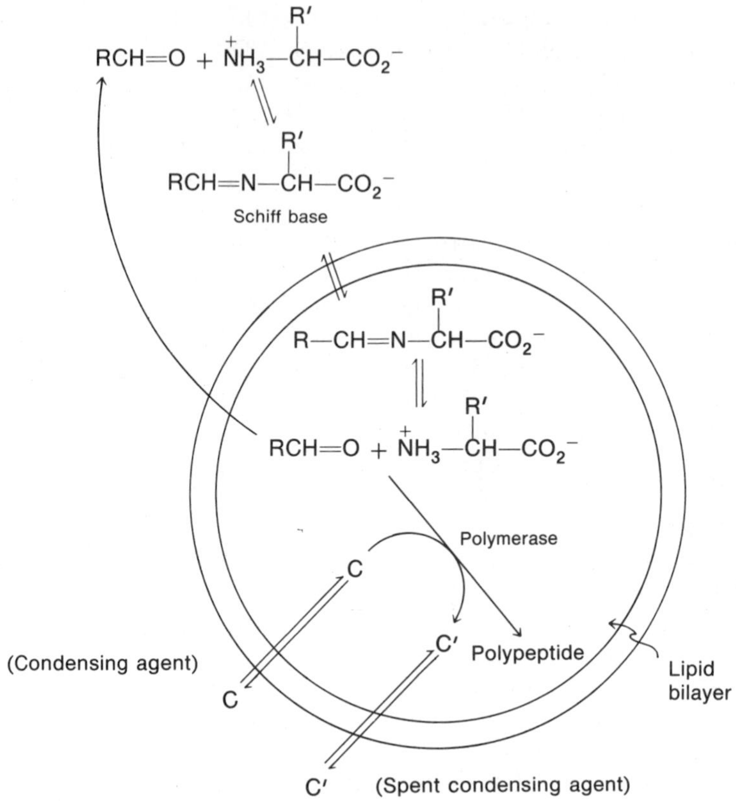

**Figure 32-33**
The facilitated diffusion of amino acids into a lipid vesicle. The amino acids form a Schiff base with a long-chain fatty aldehyde and after passage through the membrane are liberated by hydrolysis. The amino acids can then be polymerized through the action of a polymerase; energy for this reaction is supplied by a condensing agent C. (Adapted from W. Stillwell, *Facilitated diffusion as a method for selective accumulation of materials from the primordial oceans by a lipid-vesicle protocell, Origins of Life* 10: 277, 1980. Used with permission.)

and retained inside by polymerization. This latter step would require the introduction of a condensing agent and the removal of the products of its hydrolysis, as well as the presence of a polymerase catalyst.

In spite of what has been accomplished so far, we seem to be still a long way away from showing in detail how a working cell could have arisen. We will need to demonstrate not only that a functioning genetic system can be constructed, but that it can be incorporated into a compartment such as a lipid vesicle.

## CATALYSIS ON THE PRIMITIVE EARTH

We have mentioned the possibility that *polypeptides that were produced in nontemplate reactions may have acted as early nonspecific catalysts.* They could, for instance, have acted as acids and bases in much the same way as urea is believed to aid the phosphorylation of uridine by ammonium phosphate. They also could have participated in more complex schemes, such as tightening a double helix of complementary oligonucleotides by binding in the major or minor groove. However, what may we reasonably expect in the way of help from other components, chemical or physical, present at that time? *By far the most important potential catalysts were metal ions.* The remarkable catalytic effect of $Zn^{2+}$ in an oligonucleotide-forming reaction was mentioned earlier, but even this is not as impressive as the effect of metal ions in some nonbiotic examples. Hydration of phenanthroline cyanide is about $10^9$ times faster in the presence of $Cu^{2+}$ than in its absence. The reaction is believed to proceed by attack of a copper-bound hydroxyl group onto the adjacent cyanide group (Figure 32-

34). This factor approaches the theoretical limit for a rate increase that comes solely from the enforced juxtaposition of two reactants. It is possible, then, that similar factors for metal ion catalysis may be found to apply to some plausibly prebiotic reactions. Another important way of speeding up a sluggish reaction is the application of heat. We do not know for sure what the temperature was on the primitive Earth, but it is likely that it was not much different from today. There would, of course, have been daily and probably longer-term cycles in temperature, but the daily cycles would have come at much shorter intervals than 24 h. There is little agreement about the exact length of the day, say, 4 billion years ago, but there is a general feeling that it was very much shorter than today, perhaps only 5 or 10 h long. We have seen that day-to-night variations of temperatures can be important in some reactions of lipids, nucleotides, and amino acids, and parts of the Earth today experience quite large fluctuations in temperature on a daily basis. In the Namib Desert, the surface temperature was found to range from a high of 54°C to a low of 6°C in one period of 24 h. Rhode Island beach sand can reach 66°C around noon on a summer's day, and surface temperatures well over 75°C have been measured in Death Valley. These temperatures are perfectly adequate for much of the prebiotic chemistry that we have been considering.

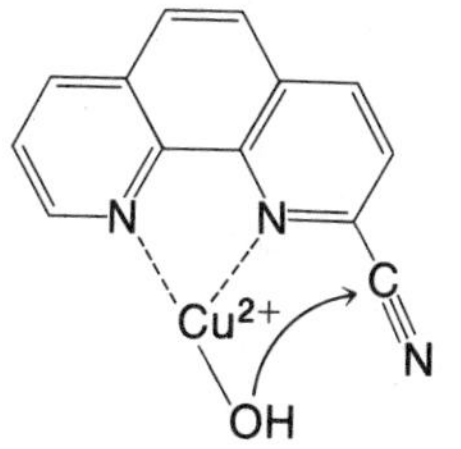

**Figure 32-34**
Intermediate in the copper ion-catalyzed hydrolysis of phenanthroline cyanide.

## BIOCHEMISTRY OF EARLY LIFE

From the origin of the Earth until the appearance of the first eukaryotes was a period of 3.1 billion years. During this time, our planet saw the formation of the first living cells, the development of aerobic photosynthesis, and the consequent transformation of the atmosphere. The exact timing of these and other significant events in the history of life is still debated—even the composition of the atmosphere of the early Earth is still not known with any certainty—but nevertheless it is possible to sketch some of the main developments and indicate approximate times for their occurrence.

*The first living organisms appeared around 4 billion years ago, give or take a few hundred million years. They most likely were heterotrophs, using preformed organic compounds from the outside world* and assembling them into necessary proteins, nucleic acids, and other molecules by means of available condensing agents such as cyanamide or polyphosphates. Ammonia, too, would have been taken up from the outside. It is not clear to what extent these organisms could have originated in the open oceans; Miller and Orgel have calculated that even if all the carbon that is presently on the surface of the earth were converted to $C_3$ organic compounds and dissolved in the ocean, the resulting solution would be only about 0.01 $M$ in organic material. This is an upper limit, so a tidal pool or a pond would appear to offer a better habitat for the first heterotrophs, since materials present could be concentrated by evaporation. Eventually some important ingredient in the local environment would become depleted, and those organisms which evolved the ability to make it from a previously unused precursor would now have a selective advantage. Eventually the precursor, too, would be in short supply, and again a selective advantage would accrue to those organisms which were able to make this precursor from yet another compound present in the environment. As discussed earlier in this chapter, repetition of this process could result in the generation of reasonably complex metabolic pathways. These pathways would require energy to drive them; one possible source of energy is the breakdown of glucose to lactate or ethanol. This pathway of anaerobic glycolysis is thought to have occurred rather early in the history of life. It is inefficient; it yields only two molecules of ATP for every molecule of glucose consumed compared with another 34 molecules of ATP for those organisms which complete the oxidation by means of the tricarboxylic acid cycle, and yet these same aerobic organisms merely append the more efficient process onto the end of the glycolytic sequence.

We can accept without serious question that *glycolysis evolved when the atmosphere was still deficient in oxygen*, but recent phylogenetic work has suggested that it may be no more ancient than one or two other pathways for producing energy. The nucleotide sequences of parts of 16 S rRNA molecules from a large number of prokaryotes were compared, and a phylogenetic tree was drawn up based on the extent of nucleotide-sequence homology for the different species. It was found that clostridial bacteria—strict anaerobes that use the glycolytic sequence—may have evolved from an ancestral line no earlier than did the green or purple photosynthetic bacteria. These bacteria do not evolve oxygen during photosynthesis; instead they use hydrogen sulfide as a reducing agent and excrete elemental sulfur or sulfate, respectively. The discovery in 1980 of a number of 3.4-billion to 3.5-billion-year-old stromatolites in Western Australia was mentioned earlier in this chapter. They reasonably confirm that photosynthetic organisms were in existence by that time; again there is no need to postulate that these stromatolites were formed by oxygen-releasing algae (cyanobacteria). The green and purple photosynthetic sulfur bacteria, together with clostridia and four or five other groups (classified together as the eubacteria), are believed to have diverged from a separate group, the archaebacteria, very early in time. Representatives of this latter group are the halophilic bacteria and the *methanogens*. Whereas the halophiles are photosynthetic and appear to have evolved rather recently, the *methanogens still employ a method for energy production that could well have been used by some of the earliest organisms: the exergonic reduction of carbon dioxide by hydrogen to give methane (Figure 32-35). The development of nitrogen-fixing ability also must have occurred before there was significant oxygen in the atmosphere*. Nitrogenase enzymes are highly sensitive to traces of oxygen and can be completely inactivated by exposure to a 5% oxygen atmosphere for a few minutes.

Significant quantities of dissolved sulfate must have been produced by the action of the photosynthetic purple sulfur bacteria. These autotrophic organisms reduce carbon dioxide to sugars in a light-driven reaction that results in hydrogen sulfide being oxidized to sulfate. About 2 to 3 billion years ago there appeared sulfate-respiring bacteria, heterotrophs that obtained energy by reversing this process, i.e., oxidizing organic foodstuffs while reducing sulfate back to sulfide. The dating comes from careful measurements of $^{34}S/^{32}S$ isotope ratios in pyrite that was deposited at different times. The action of sulfate-respiring bacteria yields sulfide (and hence pyrite) that is depleted in $^{34}S$ compared with the original sulfate.

*The combined effects of the methanogens, nitrogen fixation, hydrogen escape to space, and hydrogen-utilizing photosynthetic bacteria would have caused a gradual depletion in the amount of free hydrogen available in the atmosphere*. Some replenishment occurred from volcanic gases, but this source would have been insufficient to support large populations of

**Figure 32-35**
A dividing methanogen.

hydrogen-requiring organisms. *There was still an enormous reservoir of potential reductant: water, and any organism that could use it in a photosynthetic reaction that converted carbon dioxide to sugars and other organic compounds would have had an enormous selective advantage.* The problem was solved about 2.4 billion years ago by the *cyanobacteria*. They evolved a second photosystem that could use water as a reductant, and *with aerobic photosynthesis now under way, significant amounts of oxygen began to be released into the Earth's atmosphere.* However, for the next few hundred million years, the oxygen content of the atmosphere remained considerably below today's level. Only when the soluble ferrous ion that was dissolved in the oceans had been oxidized to ferric—which then precipitated as complex ferric oxides—did the partial pressure of oxygen increase rapidly. At this point, virtually all atmospheric "prebiotic" syntheses must have ceased, but heterotrophs could now evolve the ability to use atmospheric oxygen for the complete oxidation of organic foodstuffs to carbon dioxide and water with an unprecedented increase in the amount of energy stored as ATP. Anaerobic species either would have had to adapt or else would have had to retire to special habitats out of contact with oxygen.

Carlin has described an interesting linear relationship between the natural logarithm of the number of adenylate units at the 3' end of a mRNA from a given species and the time since the origin of that species (Figure 32-36). What biochemical significance the polyadenylate tail has is still unclear, but extrapolation of the line back to a poly(A) tail of one unit gives a date for the origin of mRNA (origin of life?) as $3.85 \pm 0.2$ billion years ago. Application of this (admittedly speculative) method also suggests that the eukaryotic mitochondrion originated about 2.1 billion years ago. Fossil evidence suggests that eukaryotes themselves did not appear until about 1.45 billion years ago, long after significant amounts of oxygen had appeared in the atmosphere (at about 2.0 billion years ago). The correspondence of these dates (2.1 billion years ago for mitochondria and 2.0 billion years ago for oxygen) may be fortuitous, but it is at least interesting to remember that the mitochondrion is the site of the citric acid cycle and oxidative phosphorylation, both quintessential aerobic processes, and that there is reasonably good evidence that eukaryotic cells came about as a result of *endosymbiosis*—the mutually beneficial capture of one organism into another.

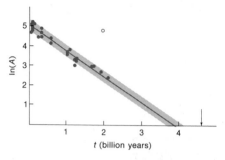

**Figure 32-36**
The relationship between the natural logarithm of short-term labeled poly(A) size and the approximate time of origin for different organisms. The arrow indicates the approximate time for the origin of the Earth. (From R. K. Carlin, Poly(A): A new evolutionary principle, *J. Theor. Biol.* 82:353, 1980. Reprinted with permission.)

## SELECTED READINGS

Day, W. *Genesis on Planet Earth*. East Lansing, Mich.: Talos, 1979. Good coverage of a broad range of the relevant literature, new and old.

*Evolution: A Scientific American Book*. San Francisco: Freeman, 1978. A compilation of articles published in *Scientific American*.

Goldsmith, D., and Owen, T. *The Search for Life in the Universe*. Menlo Park, Calif.: Benjamin/Cummings, 1980. One of the best of the new books on the subject; authoritative and highly readable.

Horowitz, N. H. On the evolution of biochemical syntheses. *Proc. Natl. Acad. Sci. USA* 31:153, 1945. A discussion of biochemical pathways as evidence for prebiological syntheses. This theory was suggested before Stanley Miller demonstrated the formation of amino acids in electric discharges.

Miller, S. L., and Orgel, L. E. *The Origins of Life on the Earth*. Englewood Cliffs, N.J.: Prentice-Hall, 1974. An excellent critical discussion of chemical evolution.

Walker, J. C. G. *Evolution of the Atmosphere*. New York: Macmillan, 1977. A discussion of the formation of the primitive Earth and atmospheres.

# INDEX

A band of sarcomere, 118
Abietic acid, 567, 568
Abscisic acid, 1142, 1143
Absorbance, 415
Acceptor protein, and hormone action, 1104–1105
Acetals, 438
2-Acetamidofluorene, affecting DNA bases, 768
Acetate, entry into TCA cycle, 329
Acetoacetate, 486
  formation from tyrosine, 891
Acetoacetyl-CoA, 217–218
Acetoacetyl-CoA thiolase, 486
Acetogenins, 545, 569–570
Acetohydroxyacid synthase, 848, 851
Acetyohydroxybutyrate, 851
Acetolactate, 851
Acetone, 486
Acetylcholine, 878
  as neurotransmitter, 1158–1160
  receptor for, 1158, 1163–1165
    blocking agents of, 1160
Acetylcholinesterase, 1159
Acetyl-CoA, 481, 486, 487, 498
  as activator of pyruvate carboxylase, 344
  fate in glyoxylate cycle, 352–355
  formation from α-ketoadipate, 897
  in TCA cycle, 326–330
    alternative sources for, 345–352
Acetyl-CoA acetyltransferase, 349
Acetyl-CoA-ACP transacylase, 493
Acetyl-CoA : 1-alkyl-2-
    lysoglycerophosphocholine
    transferase, 521
Acetyl-CoA carboxylase, 225, 490–491, 499–500
  concentrations of, 500
S-Acetyldihydrolipoyl groups, interaction of, 223

N-Acetylgalactosamine, 528
N-Acetylglucosamine, 528
  as substrate of lysozyme, 171
N-Acetylglutamine-γ-semialdehyde, 835
Acetylhydrolase, 521
N-Acetylmuramic acid, 445
  as substrate of lysozyme, 171
N-Acetylneuraminic acid, 445, 528, 584
N-Acetylornithine, 835
Acetyl phosphate, hydrolysis of, free energy change in, 272
O-Acetylserine sulfhydrylase, 838
Acetylthiamine pyrophosphate, 199
Acid, dissociation of, 270
Acid hydrolysis of proteins, 40
Acid mucopolysaccharides, 443–444
Acidity constant, 270
Aconitase, 330, 337
  in eukaryotic cells, 354
Aconitate hydratase, 330, 337
ACP. See Acyl carrier protein
Acridine, interaction with DNA, 762
ACTH. See Corticotropin
Actin, 20, 119–120
  cross-bridges with myosin, 120
  in microfilaments, 602
Actinomycin D
  inhibiting RNA metabolism, 820, 821
  interaction with DNA, 761
  D-valine in, 868
Action potential, 1150–1152
Activation-energy barrier, 136
Active site of enzymes, 154
Active transport across membranes, 262
Acylation of enzymes, 148
Acyl carrier protein (ACP), 216, 219–220
  of Escherichia coli, 491
Acyl carrier protein thioester, 488–490
Acyl carrier protein transacylase, 493
Acyl-CoA : cholesterol acyltransferase
    (ACAT), 548, 557

Acyl-CoA dehydrogenases, 214
Acyl-CoA synthase, 348, 479
1-Acylglycerol-3-phosphate, 513
1-Acylglycerol-3-phosphate
    acyltransferase, 511, 513
Adaptive changes in enzyme levels, 500
Adaptor hypothesis of translation, 909
Addison's disease, 1136
Adenine, 287, 663
  formation from hydrogen cyanide, 1203
Adenine phosphoribosyltransferase, 710
Adenohypophysis, hormones of, 1107
Adenosine, 287, 791
Adenosine 3',5'-cyclic monophosphate
    (cAMP), 311–313
  biosynthesis of, 719–720
    stimulation of, 476
  interaction with catabolite activator
    protein, 992–993
  intracellular production of, hormone
    action in, 1124–1125
  and prostaglandin activity, 542
Adenosine diphosphate (ADP), 271, 287
  ATP/ADP exchanger, 640–641, 653
  oxidative phosphorylation to ATP, 325
  ratio to ATP, and TCA cycle flux, 339
Adenosine diphosphate glucose, 452
Adenosine kinase, 712
Adenosine monophosphate (AMP), 271, 287, 662
  biosynthesis of, 703–704
  cyclic. See Adenosine 3',5'-cyclic
    monophosphate
Adenosine 5'-phosphate. See Adenosine
    monophosphate
Adenosine 5'-phosphosulfate (APS), 839
Adenosine triphosphatase. See ATPase
Adenosine triphosphate (ATP), 286–289
  aerobic production of, 323–361. See
    also Tricarboxylic acid cycle